ORGANIC CHEMISTRY

An Acid–Base Approach

ORGANIC CHEMISTRY

An Acid–Base Approach

MICHAEL B. SMITH

CRC Press
Taylor & Francis Group
Boca Raton   London   New York

CRC Press is an imprint of the
Taylor & Francis Group, an **informa** business

CRC Press
Taylor & Francis Group
6000 Broken Sound Parkway NW, Suite 300
Boca Raton, FL 33487-2742

© 2011 by Taylor and Francis Group, LLC
CRC Press is an imprint of Taylor & Francis Group, an Informa business

No claim to original U.S. Government works

Printed in India by Replika Press Pvt. Ltd.
10 9 8 7 6 5 4 3 2

International Standard Book Number: 978-1-4200-7920-3 (Hardback)

Library of Congress Cataloging-in-Publication Data

Smith, Michael B., 1946 Oct. 17-
   Organic chemistry : an acid-base approach / Michael B. Smith.
      p. cm.
   Includes bibliographical references and index.
   ISBN 978-1-4200-7920-3 (hardcover : all. paper)
   1. Chemistry, Organic--Textbooks. 2. Organic acids--Textbooks. 3. Chemical reactions--Textbooks.
I. Title.

QD251.3.S654 2011
547--dc22
                                 2010026467

Visit the Taylor & Francis Web site at
http://www.taylorandfrancis.com

and the CRC Press Web site at
http://www.crcpress.com

Contents

Preface

Perhaps the most common fable about organic chemistry is the one often heard in the first few days of the course: Organic chemistry is a memorization course. Memorize what you need and regurgitate it on the test and you will pass. This assessment is wrong on so many levels that it serves as a starting point to describe this book. One reason for the fable is that it may be difficult to ascertain a fundamental theme that ties organic chemistry together. The other issue is the fact that organic chemistry builds from the first day to the last and is not compartmentalized. For this reason, a theme is even more important.

Is there a common, underlying theme for organic chemistry that allows one to understand it rather than simply memorize? In this book, acids and bases provide the theme. It is important to understand the relationships among functional groups. How are they similar and what structural motifs make them react differently? What is the relationship between structure and chemical reactivity? These questions point to understanding not only the chemical reactions that populate an organic chemistry book, but also the mechanisms used to describe those reactions.

The theme of this book is the premise that many if not most reactions in organic chemistry can be explained by variations of fundamental acid–base chemistry concepts. Moreover, the individual steps in many important mechanisms rely on acid–base reactions. The ability to see these relationships makes understanding organic chemistry easier and, perhaps more importantly, allows one to make predictions about reactivity when memory fails.

Apart from the acid–base theme, this book uses several techniques to assist students in their studies. A list of concepts that the student should know before reading the chapter is listed at the beginning of virtually every chapter. A list of concepts that should be known after the chapter is completed is also provided at the beginning of most chapters. Each chapter concludes with this same list of the concepts that should now have been learned, and a correlation of each concept with the homework problems is provided. The intent is to help identify concepts that need work apart from those that are readily understood.

Each chapter has a number of embedded problems. They will help the student think about the concept and possibly review something from a previous chapter that is pertinent, with the goal of better understanding that section. The answers to these problems are provided at the end of each chapter. Regular homework problems are given with the intention of offering help in understanding the concepts in the chapter. For additional material and a solutions manual, please see the book Web site (http://www.crcpress.com/product/isbn9781420079203).

Experimental details are provided for key reactions to introduce the reaction. The point is to show that the experiment shows a result and then we try to understand that result in the context of the structural features of the reactive species. Mechanisms are given as part of the discussion of key reactions. The mechanism is discussed in most cases first as a walk-through of the reaction to understand how the transformation occurred, and then structures for the walk-through are provided that constitute the mechanism. Many of these mechanistic steps will involve acid–base reactions—both Brønsted–Lowry and Lewis.

The book begins with an introduction to organic chemistry that describes what it is as well as individuals who have contributed to building organic chemistry as a scientific discipline. Chapter 2 provides a bridge between the acid–base discussions in a typical general chemistry book and those in this organic chemistry book. An understanding of bonding is critical to understanding the structure of organic molecules; this is provided in Chapter 3, followed by Chapter 4 with an introduction to alkanes, the basic structure of organic molecules, and the fundamental rules of nomenclature. Chapter 5 introduces the concept of functional groups as well as several important functional groups that appear in organic molecules. This chapter includes an extension of the basic nomenclature rules to include functional groups.

Chapter 6 returns to the acid–base theme to give a direct correlation between acid–base reactions and organic chemical reactions that involve the functional groups just introduced in Chapter 5. Chapter 7 extends the acid–base concept of equilibrium reactions to discuss bond energetics and kinetics. Rotation about single covalent bonds leads to different orientations of atoms and groups within a molecule, known as conformations; this is presented in Chapter 8. Chapter 9 introduces the concept of chirality and the relationship between chirality and structure.

Chapter 10 introduces the acid–base chemistry of molecules that contain the C=C and C≡C functional groups. Related reactions that do not fall under the acid–base category are also presented. Chapter 11 uses nucleophiles, which are loosely categorized as specialized Lewis bases, in reactions with alkyl halides. These are substitution reactions. Chapter 12 shows the acid–base reaction of alkyl halides that leads to alkenes (an elimination reaction), and Chapter 13 ties Chapters 11 and 12 together with a series of simplifying assumptions that allows one to make predictions concerning the major product.

Chapter 14 introduces methods used to identify the structure of organic molecules: mass spectrometry, infrared spectroscopy, and nuclear magnetic resonance spectroscopy. This chapter is in the middle of the book, but it may be presented in either semester, and homework problems associated with this chapter are incorporated throughout. All homework problems requiring the information in Chapter 14 are clearly marked and segregated from the other problems. Therefore, this chapter may be introduced in either semester and the associated homework problems can be easily found.

Chapter 15 provides a brief introduction to commonly used organic molecules that also have a carbon–metal bond—organometallics. Chapter 16 introduces the fundamental characteristics of molecules that contain the carbon functional group, along with a review of the nomenclature of carbonyl-containing molecules. Carbonyl compounds are often prepared by oxidation reactions, and several key oxidation reactions are discussed in Chapter 17. Oxidation reactions of a few other functional groups are included. Chapter 18 elaborates the chemical reactions of the carbonyl-containing molecules known as aldehydes and ketones. This chemistry is dominated by the acyl addition reaction introduced in Chapter 16.

Chapter 19 in effect continues the chemistry of carbonyls by introducing reduction reactions. Carbonyl-containing molecules, as well as molecules that contain other functional groups, may be reduced to different functional groups, and such reactions are discussed in this chapter. Chapter 20 continues carbonyl chemistry by discussing chemical reactions of the carboxylic acid derivatives introduced in Chapter 16, with a focus on acyl substitution reactions.

Chapter 21 discusses the concept of aromaticity as well as the nomenclature and the specialized chemical reactions of aromatic compounds such as benzene and its derivatives. This chapter comes late in the book, with the notion that the chemistry of aliphatic compounds is simply used more often. The acid–base theme is continued with the recognition that the fundamental substitution chemistry associated with benzene derivatives may be explained by the reaction of aromatic rings as Lewis bases or nucleophiles with electrophilic reagents. The reactions of benzene derivatives with strong bases and good nucleophiles are also presented.

Chapter 22 returns to carbonyl chemistry and a discussion of the acid–base properties of carbonyl compounds. The proton on the α-carbon (directly attached to the carbonyl) is slightly acidic and removal with a suitable base leads to an enolate anion. Enolate anions react as nucleophiles in aliphatic substitution, acyl addition, and acyl substitution reactions.

Chapter 23 begins a discussion of multifunctional molecules that will conclude the book and, in many courses, may be considered as special topic material, beginning with simple conjugated dienes and conjugated carbonyl compounds. Ultraviolet spectroscopy is discussed in the context of identifying conjugated compounds. Chapter 24 continues this discussion with pericyclic reactions that involve dienes and other multifunctional compounds. Chapter 25 is a stand-alone chapter to introduce the retrosynthetic approach to synthesis. It is presented as part of the discussion on multifunctional molecules, but it may be taught at any point in the course. As with spectroscopy, synthesis problems are clearly marked homework problems in several chapters and are segregated from the other problems so that they are easily found.

The book concludes with three chapters that involve multifunctional molecules. Heterocycles are introduced in Chapter 26, with nomenclature and structure as well some elementary chemical reactions of heterocyclic compounds. Amino acids and peptides are discussed in Chapter 27 and carbohydrates and nucleic acids are discussed in Chapter 28. These last three chapters are, arguably, the most important to biological chemistry.

This book constitutes an important introduction to organic chemistry. The reactions and mechanisms contained herein describe its most fundamental concepts, which are used in industry, biological chemistry and biochemistry, molecular biology, and pharmacy. These concepts constitute the fundamental basis of life processes, which means that they are pertinent to the study of medicine. For that reason, most chapters end with a brief section that describes biological applications for each concept. The last two chapters (27 and 28) have more than one section that discusses biological applications because these molecules make up the proteins, enzymes, cellular structure, and DNA and RNA found in living systems. It is hoped that this course will provide the skills to go to the next level of study using this fundamental understanding of acids and bases applied to organic transformations and organic molecules.

Acknowledgments

There are many people to thank for a book such as this one. First, let me thank my students, whose enthusiasm and interest have pushed me to develop better ways to communicate and better ways to describe organic chemistry. I thank my colleagues in the organic chemistry community, particularly the synthetic organic community, for allowing me to keep my knowledge and skills honed over the years.

In addition, myriad discussions with my colleagues have helped me to become a better organic chemist and be better able to communicate that knowledge. I particularly want to thank Spencer Knapp (Rutgers University) and John D'Angelo (Alfred University) for providing most helpful reviews of this manuscript. I thank George Majetich (University of Georgia), Fred Luzzio (University of Louisville), and Tyson Miller (University of Connecticut) for discussions over the years that helped me to crystallize certain aspects of this book, particularly the acid–base theme and how to present it.

All structures and reactions were drawn using ChemBioDraw Ultra, v. 11.0.1 I thank CambridgeSoft, Inc. for a gift of this software. All 3-D drawings and molecular models were prepared using Spartan06 software, v.1.0.1. I thank Warren Hehre, Sean Ohlinger, and Wavefunction, Inc. for the gift of this software.

I thank my editor, Lance Wobus of Taylor & Francis, for giving me the opportunity to write this book and working with me over these many months. His support, vision, and ability to solve the inevitable issues that arise in such a project have been essential for getting this book into print. I also thank Dr. Fiona Macdonald for her support in getting this book off the ground and David Fausel and Judith Simon for their excellent work in converting the manuscript to the finished book. I also want to thank Jim Smith, who was instrumental to this project in its early days. I thank Warren Hehre (Wavefunction, Inc.), who introduced me to Lance Wobus and who has provided many discussions over the years that helped me in writing the book.

Finally, I thank my wife, Sarah, and my son, Steven, for their support and encouragement over the years. That support has helped more than can be properly expressed in words.

Michael B. Smith is professor of chemistry in the Department of Chemistry at the University of Connecticut at Storrs. His research interests focus on the identification of bioactive lipids from the dental pathogen *Porphyromonas gingivalis,* exploration of the use of conducting polymers as a reaction medium for chemical transformations, development of fluorescent probes for the detection of cancerous tumors, and the synthesis of phenanthridone alkaloids. He is the author of volumes 6–12 of the *Compendium of Organic Synthetic Methods* and author of the fifth and sixth editions and upcoming seventh edition of *March's Advanced Organic Chemistry.* He is also the author of the first and second editions and upcoming third edition of the graduate level textbook *Organic Synthesis,* as well as several monographs.

Dr. Smith received his PhD in organic chemistry from Purdue University in 1977, a BS in chemistry from Virginia Polytechnic Institute in 1969, and an AA from Ferrum College in 1967. Postdoctoral work at the Cancer Research Institute at Arizona State University and at Massachusetts Institute of Technology preceded his appointment at the University of Connecticut.

The
Author

Introduction

Congratulations, you are taking organic chemistry! It is likely that you are a science major, in a class of students with a wide range of interests and career choices. **Why is organic chemistry important?** The answer lies in the fact that every aspect of life—mammalian and nonmammalian as well as plant and microscopic life—involves organic chemistry. In addition, many of the products used every day (pharmaceuticals, plastics, clothing, etc.) involve organic molecules. Organic chemistry holds a central place in chemical studies because its fundamental principles and its applications touch virtually all other disciplines. Several years ago, a T-shirt at an American Chemical Society meeting in Dallas sported the logo "Chemistry: The Science of Everything." Organic chemistry is certainly an important player in that science.

Most organic chemistry textbooks have a brief section to describe how organic chemistry developed as a science. I was a graduate student when I first read an organic chemistry book that presented some historical facts as part of the normal presentation of facts. The book was Louis Fieser's (United States; 1899–1977) *Advanced Organic Chemistry*.[1] This book gave a perspective to my studies and helped me to understand many of the concepts better. I believe that putting a subject into its proper context makes it easier to understand, so I am introducing an abbreviated history of organic chemistry as the beginning to this book. I will include material from Fieser's book and also from a book on the history of chemistry by Leicester.[2] It is important to remember the great organic chemists of the past, not only because their work is used today but also because it influences the way that we understand chemistry.

1.1 A Brief History of Organic Chemistry

Humans have used practical applications of chemistry for thousands of years. The discovery and use of folk medicines, the development of metallurgical techniques, and the use of natural dyes are simple examples. For most of history, humans were able to use simple chemicals or a complex mixture of chemicals without actually understanding the science behind them. Organic chemistry became a defined science (the chemistry of carbon compounds) in the nineteenth century, but organic compounds have been known and used for millennia. Plants have been "milked," cut, boiled, and eaten for thousands of years as folk medicine remedies, particularly in Africa, China, India, and South America. Modern science has determined that many of these plants contain organic chemicals with effective medical uses, and indeed many of our modern medicines are simply purified components of these plants or derivatives of them made by chemists.

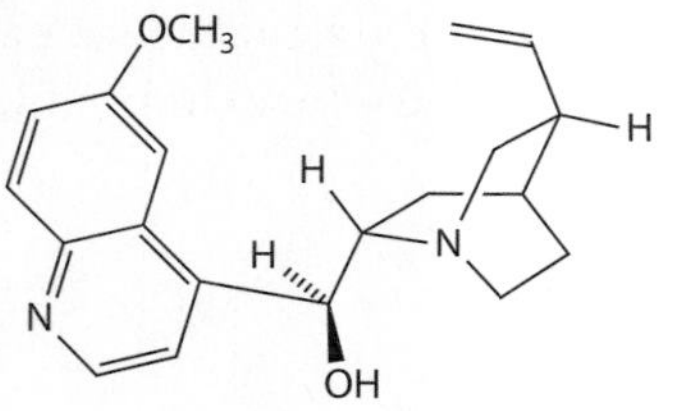

The bark of the *Cinchona* tree has been chewed for years to treat symptoms of malaria, and it was later discovered that this bark contains quinine (**1**), which is a modern medicine. Ancient Egyptians ate roasted ox liver in the belief that it improved night vision. Later it was discovered that ox liver is rich in vitamin A (**2**), a chemical important for maintaining healthy eyesight. An ancient antipyretic treatment (this means that it lowers a fever) involved chewing willow bark, and it was later discovered that this bark contained the glycoside salicin (**3**), a derivative of salicylic acid (**4a**). Eventually, chemists learned how to make new organic molecules rather than simply isolating and using those that were found in nature, although isolation from nature remains an important source of new compounds.

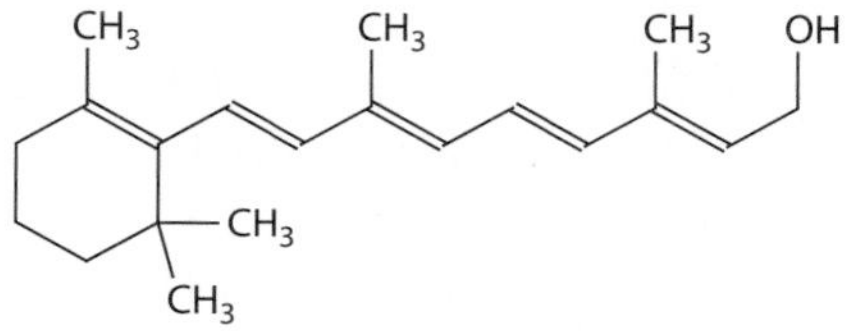

In the mid-nineteenth century a new compound was synthesized (chemically prepared from other chemicals) called acetylsalicylic acid (**4b**; better known as aspirin), and it was found to be well tolerated by patients as an effective analgesic (this means that it reduces some types of pain). These few examples are meant to represent the thousands of folk medicine remedies that have led to important medical discoveries. All of these involve organic compounds.

3

4a R = H   **4b** R = [structure with C=O and CH₃]

Plants have provided humans with many organic chemicals or mixtures of chemicals that are useful for purposes other than medicine. Ethyl alcohol (**5**) has been produced by fermentation of grains and fruits and consumed for thousands of years in various forms, including a beer consumed by ancient Egyptians. The symbols (**1–5**) used to represent the chemicals require some explanation, and ethyl alcohol is a simple example. Each "line" is a chemical bond. Therefore, C–C is a carbon–carbon bond and "–" is used as a shorthand notation to represent that bond. The symbol C–N is a carbon–nitrogen bond, C–O is a carbon–oxygen bond, and O–H is an oxygen–hydrogen bond. In **5**, –OH represents –O–H. Each intersection of bonds, such as [ᘐ], is a carbon atom. Various groups may be attached to these carbon atoms (OH, NH₂, CH₃, etc.). This notation, called **_line notation_**, is used to draw most of the chemical structures in this book.

5

Line notation uses one line for each bond, so C=C indicates two bonds between the carbon atoms (a carbon–carbon double bond), and C=O indicates two bonds between carbon and oxygen (a carbon–oxygen double bond). Similarly, C≡C represents three bonds between the two carbon atoms, a carbon–carbon triple bond. Many molecules have double and triple bonds. Examples are compounds **1–4**, as well as compounds **6** and **7**.

Line notation is used for simple compounds such as **5**; however, as shown in structures **1–3**, it is applicable to very complicated structures. In ancient India, in Java, and in Guatemala certain plants provided a deep blue substance used to color clothing. In recent times, the main constituent was found to be indigo (**6**). The ancient Phoenicians used an extract from a snail (*Murex brandaris*) found off the coast of Tyre (a city in modern-day Lebanon) to color cloth. It was a beautiful and very expensive dye called Tyrian purple. This dye was so prized that Roman emperors used it to color their clothing, and for many years no one else was permitted to wear this color (hence the term "born to the purple"). The actual structure of the organic chemical Tyrian purple is **7**. Note that the only difference between indigo and Tyrian purple is the presence of two bromine atoms in the latter. Structural differences that on the surface appear to be minor can lead to significant changes in the physical properties of organic compounds, such as color.

Organic chemicals have also been used in an unethical manner. The plant *Atropa belladonna* (deadly nightshade) has been used for centuries as a poison. The principal "poison" in this plant was found to be atropine (**8**), which is also found in the *stems* of tomato plants (but not the tomato itself).

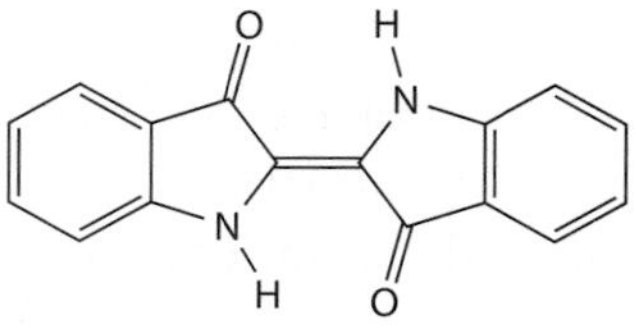

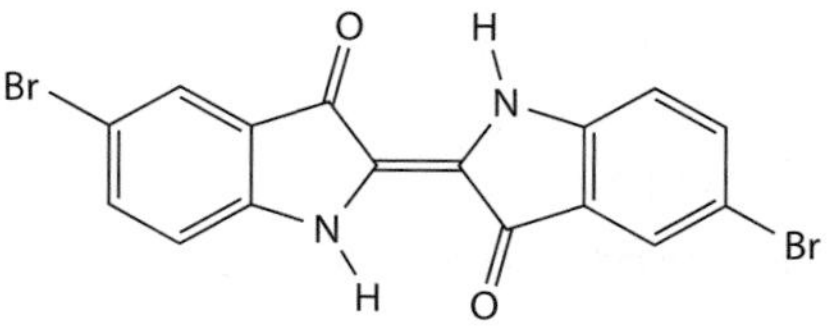

These few examples show that organic chemicals are important to humans and have been for a very long time. For most of this time, however, the actual chemical structure of these compounds was unknown. Indeed, the fact that the chemicals were discrete molecules was unknown. It was known that a multitude of materials could be obtained from natural sources, primarily from living organisms. This knowledge sparked a curiosity that eventually led to modern organic chemistry as a science. In the following paragraphs, a few of the chemists who advanced organic chemistry as a science are introduced.

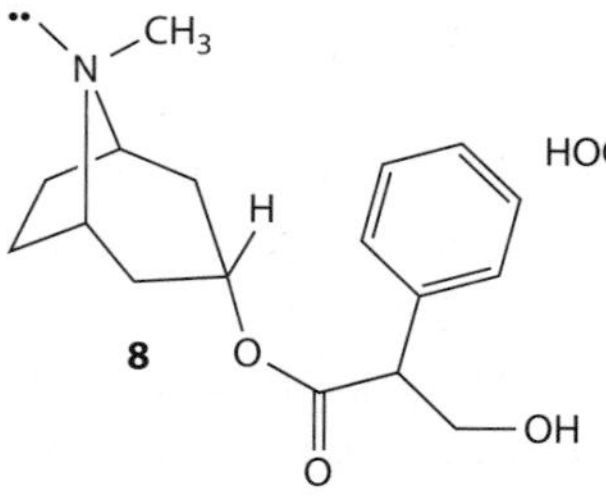

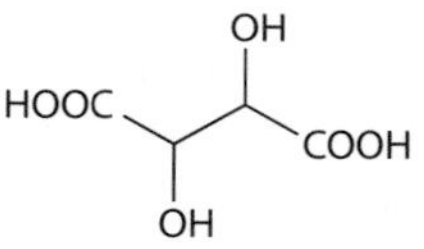

As pointed out earlier, natural materials have been used for many years. It was not until the eighteenth century that people began to look for specific chemicals in these natural materials. One of the first to report discrete chemicals was **Carl Wilhelm Scheele** (Sweden; 1742–1786), who isolated acidic components from grapes and lemons by forming precipitates with calcium or lead salts, and then added mineral acids to obtain the actual compounds. The compound from grapes is now known to be tartaric acid (**9**) and that from lemon is now known to be citric acid (**10**). Scheele also isolated uric acid (**11**) from urine.

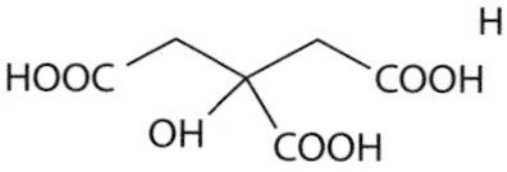

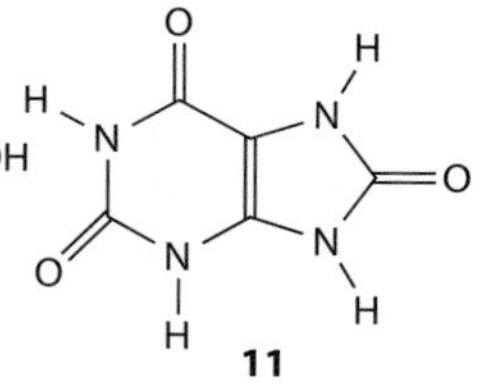

Friedrich W. Sertürner (Germany; 1783–1841) isolated a compound from opium extracts in 1805 that is now known to be morphine (**12**). In 1815, **Michel E. Chevreul** (France; 1786–1889) isolated a material from skeletal muscle now known to be creatine (**13**), which has been used as a dietary supplement despite the observation that it can cause kidney damage and muscle cramping. He also elucidated the structure of simple soaps, which are metal salts of fatty acids (**14**) and gave the name butyric acid (**15**) to the carboxylic acid found in rancid butter.

In 1818–1820, **Pierre J. Pelletier** (France; 1788–1842) and **Joseph Caventou** (France; 1795–1877) isolated a poisonous compound from Saint

Ignatius' beans (*S. ignatii*) that they called an alkaloid (an alkali-like base) now known to be strychnine (**16**; found in the seeds of the nux vomica tree [*S. nux-vomica*] and also from related plants of the genus *Strychnost*). Alkaloids are a large group of diverse compounds that contain nitrogen and are primarily found in plants. Although difficult to define because of their structural diversity, alkaloids are commonly assumed to be basic nitrogenous compounds of plant origin that are physiologically active. The practice of isolating specific compounds (now known to be organic compounds) from natural sources continues today. Such compounds are called "natural products."

Clearly, an early task in organic chemistry was to isolate pure compounds from natural sources and then attempt to identify them. Initially, the compounds were purified (usually by crystallization) and characterized as to their physical properties (melting point, boiling point, solubility in water, etc.). It was not until much later (mid- to late nineteenth century and even into the early twentieth century) that the structures of most of these compounds were known absolutely. **Justus Liebig** (Germany; 1803–1873) perfected the science of analysis of organic compounds, based on the early work of **Antoine Lavoisier** (France; 1743–1794).

Late in the eighteenth century, Lavoisier made a monumental contribution to the science of chemistry that was important to understanding organic chemistry. He used the discovery that air was composed mainly of oxygen (O_2) and nitrogen (N_2) and burned natural materials in air. He discovered that carbon in the burned material was converted to carbon dioxide (CO_2) and that hydrogen in the material was converted to water (HOH). By trapping and weighing the carbon dioxide and the water, he was able to calculate the percentage of carbon and hydrogen in molecules, which led to a determination of the empirical formula. Because we now know that organic molecules are composed mainly of carbon and hydrogen, this ***elemental analysis*** procedure was and is an invaluable tool for determining the structure of organic molecules.

In 1807, a Swedish chemist named **Jöns J. von Berzelius** (Sweden; 1779–1848) described the substances obtained from living organisms as **organic compounds**, and he proposed that they were composed of only a few selected elements, including carbon and hydrogen. Because all organic compounds known at that time had been isolated as products of "life processes" from living organisms (hence the term organic), Berzelius and **Charles F. Gerhardt** (France; 1816–1856) described what was known as the *vital force theory*. This theory subscribed to the notion that "all organic compounds can arise with the operation of vital force inherent to living cells." The vital force theory was widely believed at the time.

In 1828 **Friedrich Wöhler** (Germany; 1800–1882) synthesized (prepared from other chemicals) the organic molecule urea (**18**) from chemicals that had not been obtained from living sources. When he heated ammonium cyanate (**17**), the product was urea (**18**). Urea is a component of urine and also a component of bird droppings that have been used for centuries as fertilizer. This work, along with that of others, was contrary to the vital force theory because it showed that an organic compound could be produced from a "nonliving" system. However, it was not until **Pierre-Eugene-Marcellin Berthelot** (France; 1827–1907) showed that all classes of organic compounds could be synthesized that the vital force theory finally disappeared.

Synthesis of organic molecules (preparation of a more complex organic compound from more structurally simple compounds in several chemical steps) began in the mid-nineteenth century, and many compounds were prepared. **Hermann Kolbe** (Germany; 1818–1884) prepared ethane (CH_3CH_3) by electrolysis of potassium acetate ($CH_3CO_2^-K^+$) and Sir **Edward Frankland** (England; 1825–1899) prepared butane ($CH_3CH_2CH_2CH_3$) from iodoethane (CH_3CH_2I) and zinc (Zn). **Charles A. Wurtz** (France; 1817–1884) discovered *amines* in 1849, and **August W. von Hofmann** (Germany/England; 1818–1892) prepared many amines as well as ammonium salts by an acid–base reaction of the amine with a mineral acid. Amines are compounds containing nitrogen. Conceptually, they are carbon derivatives of ammonia. This means that a thought experiment can replace a hydrogen atom in NH_3 with a carbon group to give $C–NH_2$, which is an amine.

At about the same time, **Alexander W. Williamson** (England; 1824–1904) showed how ethers (*ethers* contain the C–O–C linkage) could be prepared from the potassium salt of an alcohol (an alcohol contains a C–O–H unit). The potassium salt is the conjugate base of that alcohol, $C–O^-K^+$, and the oxygen atom of that species reacts with an alkyl halide, which contains a C–X bond. In this representation, X is a halogen and *alkyl* is the term used for a unit containing carbons and hydrogen atoms. The nomenclature for all of these compounds will be described in Chapters 3 and 4.

In 1863, **William H. Perkin** (England; 1838–1907) prepared the first commercially useful synthetic dye (a dye made by humans), mauveine (**19**), which showed a purple color that had *not* been previously isolated from nature. In 1869, the synthesis of a natural dye was reported by **Carle Graebe** (Germany; 1841–1927) and **Carl Liebermann** (Germany; 1882–1914). They prepared the natural dye alizarin (**20**) from anthracene (**21**), which is obtained from petroleum distillates. **Adolf von Baeyer** (Germany; 1835–1917) was the first to synthesize the dye indigo (**6**). Aspirin (**4**) was first prepared in the mid-nineteenth century and commercialized later in that century. The synthesis of the various dyes and of aspirin was important to the economies of both England and Germany in the late nineteenth and early twentieth centuries. Clearly, a major step in organic chemistry involved the chemical synthesis of the compounds that could be isolated from nature, and this knowledge was expanded to prepare compounds that were not known in nature. Once accomplished, these products were commercialized and this led to the development of chemical industries.

$$
\begin{array}{cc}
\begin{array}{c}
\text{H} \\
| \\
\text{H}-\text{C}-\text{H} \\
| \\
\text{H}
\end{array}
&
\begin{array}{c}
\text{H} \\
\!\cdot\!\cdot \\
\text{H}\!:\!\text{C}\!:\!\text{H} \\
\!\cdot\!\cdot \\
\text{H}
\end{array}
\\
\mathbf{22} & \mathbf{23}
\end{array}
$$

By the mid-nineteenth century, chemists were beginning to understand that organic molecules were discrete entities and they were able to prepare them. Determining the structures of these compounds (how the atoms are connected together), however, posed many problems. The idea of *valence* was introduced by **C. W. Wichelhaus** (1842–1927) in 1868. Aleksandr M. Butlerov (Russia; 1828–1886) introduced the term *chemical structure* in 1861. There was no accepted method to determine how atoms in a molecule were arranged in a molecular structure. In 1859, **August Kekulé** (Germany; 1829–1896) suggested the idea of discrete valence bonds. It was actually **Jacobus H. van't Hoff** (the Netherlands; 1852–1911) and **Joseph A. Le Bel** (France; 1847–1930) who deduced that *when carbon appeared in organic compounds, it was connected to four other atoms and the shape of the atoms around carbon was tetrahedral.* In other words, carbon is joined to other elements by *four chemicals bonds,* as in **22** (methane), where each line connecting the atoms represents a chemical bond as mentioned earlier (see compound **5**).

In 1859, however, precisely how these four other atoms were attached remained unknown. The concept of a bond was vague and largely undefined. It was not until 1916 that **Gilbert N. Lewis** (United States; 1875–1946) introduced the modern concept of a *bond formed by sharing two electrons.* He called a bond composed of shared electrons pairs a *covalent bond.* **Erich Hückel** (Germany; 1896–1980) developed theories of bonding and orbitals and also speculated on the nature of the C=C unit. However, **Alexander Crum Brown** (England; 1838–1922) first wrote a "double bond": C=C, for ethylene ($H_2C=CH_2$), in 1864. **Emil Erlenmeyer** (Germany; 1825–1909) wrote acetylene ($HC \equiv CH$) with a triple bond in 1862.

Understanding covalent bonds allows us to understand how organic molecules are put together and that each covalent bond consists of two electrons shared by the two atoms in the covalent bond. Returning to methane (**22**), the four covalent bonds to carbon could now be represented as **23,** where each ":" represents two shared electrons. A structure such as **23** is commonly known as a *Lewis electron dot structure,* after G. N. Lewis. In 1923, Lewis suggested that *a molecule that accepts an electron pair should be called an acid* and *a molecule that donates an electron pair should be called a base*. These are called *Lewis acids* and *Lewis bases* to this day.

Understanding where electrons are in an organic molecule and how they are transferred is important for an understanding of both the structure of molecules and also their chemical reactions. In 1925 two physicists, **W. Karl Heisenberg** (Germany; 1901–1976) and **Erwin Schrödinger** (Austria; 1887–1961) described the orbital concept of molecular structure. In other words, they introduced the idea of *orbitals in chemistry and bonding* (see Chapter 3). Today we combine these ideas by saying that orbitals contain electrons and that orbital interactions control chemical reactions and explain chemical bonding. Clearly, this area of organic chemistry involved identifying the chemical structure of organic molecules and relating that structure to organizations of atoms held together by shared electrons.

In structures **1** and **16**, the lines used for chemical bonds have been replaced with **solid wedges** or with **dashed lines**, which are used to indicate the spatial relationship of atoms and groups within these compounds. The solid wedge indicates that the group is projected *in front of the plane* of the page, and the dashed line indicates that the group is projected *behind the plane* of the page. This three-dimensional representation corresponds to the spatial relationship of the atoms or groups and is known as the *stereochemistry* of that atom or group. Stereochemistry is also a part of the structure of many organic compounds. It was not until the mid- to late twentieth century that the stereochemistry of organic compounds could be accurately determined, although its discovery dates to the mid-nineteenth century.

Indeed, the concept of stereochemistry is almost as old as organic chemistry itself. In 1848, **Louis Pasteur** (France; 1822–1895) found that tartaric acid (**9**) existed in two forms that we now know differ only in their ability to rotate plane-polarized light in different directions (they are examples of stereoisomers). Pasteur observed that the crystals had a different morphology, defined here as their external structure, and he was able to separate these two forms by peering through a microscope and using a pair of tweezers to separate them physically. Because of this difference, the two forms of tartaric acid are considered to be different compounds, now called *enantiomers* (see Chapter 9).

Van't Hoff found that alkenes existed as a different type of stereoisomer that is not an enantiomer, but is also discussed in Chapter 9. Pasteur, Van't Hoff, and Le Bel are widely considered to be the founders of stereochemistry. **Emil**

Fischer (Germany; 1852–1919) studied carbohydrates in the late nineteenth and early twentieth centuries and made many major contributions to understanding not only their chemistry, but also their structures and stereochemistry. Many scientists have helped develop this concept into the powerful tool it is today, including **John Cornforth** (Australia/England; 1917–), **Vladimir Prelog** (Yugoslavia/Switzerland; 1906–1998), and **Donald J. Cram** (United States; 1919–2001).

As in the beginning of organic chemistry, isolation of organic compounds from natural resources continues to be important. New organic molecules are isolated from natural sources such as terrestrial and marine plants, fungi, and bacteria, as well as some animals. **G. Robert Pettit** (United States; 1929–) and **S. Morris Kupchan** (United States; 1922–1976) are two organic chemists who have discovered many new and interesting organic compounds—many with potent biological activity against cancer and other human diseases. The synthesis of organic compounds has also continued unabated since the nineteenth century. Over the years, increasingly more complex molecules have been synthesized.

It is useful to examine a handful of syntheses of organic molecules to show the structural challenges and also how more sophisticated methods and reagents have become an important part of synthesis. The choice of the compounds presented here is largely due to the book by **Elias J. Corey** (United States; 1928–),[3] who described the theory and practice of modern organic synthesis. In 1904, **William H. Perkin, Jr.** (England; 1860–1929) synthesized α-terpineol (**24**) and, in 1917, Sir **Robert Robinson** (England; 1885–1975) synthesized tropinone (**25**). Terpineol is isolated from pine oil and used as an ingredient in perfumes, flavors, and cosmetics. The synthesis of tropinone was important because that synthesis led to the synthesis of atropine (**8**), which was used as an immediate antidote after exposure to certain chemical nerve agents used in gas attacks during World War I and was in short supply. Atropine is still used today.

In 1929, **Hans Fischer** (Germany; 1881–1945) synthesized protoporphyrin (hemin, **26**), and quinine (**1**) was synthesized in 1944 by **Robert B. Woodward** (United States; 1917–1979) and **William von E. Doering** (United States; 1917–). Hemin contains the unit found in hemoglobin, the oxygen-carrying component of blood, and quinine is an effective antimalarial drug. In 1951, Sir **Robert**

Robinson and **Robert B. Woodward** synthesized strychnine (**16**), and ever more complex organic molecules are being synthesized today.

A huge number of syntheses have been reported in the last 50 years that have contributed enormously to organic chemistry. Apart from simply synthesizing the molecules, many chemists develop new chemical reactions as well as new chemical reagents (molecules that induce a chemical transformation in another molecule). There is no question that another area of organic chemistry involves developing new or modified chemical reactions to the point that virtually any molecule can be prepared. Understanding chemical reactions and the reagents used in those reactions and developing new reagents and reactions are a critical part of the synthesis of organic molecules (including those shown here) and have profound influences in all areas of organic chemistry.

Prior to the late 1940s and 1950s, chemists did not really understand *how* chemical reactions occurred. In other words, what happened during the bond making and bond breaking processes remained a mystery. Understanding these processes, now called **reaction mechanisms**, required an enormous amount of work in the period of the late 1940s throughout the 1960s, and it continues today. Pioneers in this area include **Frans Sondheimer** (Germany; 1926–), **Saul Winstein** (Canada/United States; 1912–1969), Sir **Christopher K. Ingold** (England; 1893–1970), **John D. Roberts** (United States; 1918–2001), **Donald J. Cram** (United States; 1919–2001), **Herbert C. Brown** (England/United States; 1912–2004), **George A. Olah** (Hungary/United States; 1927–), and many others. They studied reactions that have reactive ionic intermediates such as **carbocations** (a carbon having three covalent bonds and a positive charge on carbon; a positively charged carbon atom) or **carbanions** (a carbon having three covalent bonds and two extra electrons on carbon; a negatively charged carbon atom). Another type of intermediate called a carbon radical (a carbon having three covalent bonds and one extra electron) has been identified.

An **intermediate** is a transient and usually high-energy molecule that is formed initially and then reacts further to give a final, more stable product. The nature and structure of these intermediates were determined, and methods were developed to ascertain the presence of these intermediates and how long they were present in the reaction (in other words, how reactive they were). The idea of reaction **kinetics** was developed for organic chemistry, describing the rate at which products were formed and reactants disappeared. This information gives clues as to how the reaction proceeded and what, if any, intermediates may be involved.

Roald Hoffman (Poland/United States; 1937–) and **Robert Woodward** (United States; 1919–1979) and **Kenichi Fukui** (Japan; 1918–1998) described the concept of frontier molecular orbitals and the use of orbital symmetry to explain many reactions that did not appear to proceed by ionic intermediates. The concept of reaction mechanism allows a fundamental understanding of how organic reactions work, and it is a relative latecomer to the study of organic chemistry. It is perhaps the most important aspect, however, because understanding the mechanism of chemical reactions allows chemists to predict products and reaction conditions without having to memorize everything.

Finally, **how does a chemist know the structure of any organic chemical? How are organic compounds isolated and identified?** In early work, inorganic materials such as metal salts and acids or bases were added to force precipitation of organic compounds. In other cases, liquids were distilled out of "organic material" or solids were crystallized out. In the 1950s, **Archer J. P. Martin** (United States; 1910–2002) and **Richard Synge** (England; 1914–1994) developed the concept of *chromatography;* this allowed chemists to separate mixtures of organic compounds conveniently into individual components.

Light has always been an important player in chemistry. In the early to mid-twentieth century, ultraviolet light was shown to interact with organic molecules at certain wavelengths. In the 1940s and 1950s, molecules were exposed to infrared light and found to absorb only certain wavelengths. Identification of which wavelengths of light were absorbed and correlation of this information with structure comprised a major step in the identification of organic molecules. Even today, *ultraviolet spectroscopy* and *infrared spectroscopy* are major tools for the identification of organic compounds.

In the 1950s and especially in the 1960s, practical instruments were developed that were based on the discovery that organic molecules interact with electromagnetic radiation with wavelengths in the radio signal range if the molecules are suspended in strong magnetic fields. Initially, it was discovered that hydrogen atoms interacted in this manner and chemical differences could often be discerned. If the different hydrogen atoms in an organic molecule could be identified, the chemical structure could be puzzled together, giving a major boost to the identification of organic compounds. This technique is now known as *nuclear magnetic resonance* (NMR) and it is one of the most essential tools for an organic chemist. With the power of modern computers, we can now use NMR to determine the numbers and types of carbon atoms, nitrogen atoms, fluorine atoms, lithium atoms, and many more in an organic molecule. To do this, we use stable natural isotopes of these atoms: ^{13}C, ^{15}N, ^{19}F, ^{6}Li, etc. It is noteworthy that the important modern tool of medicine (MRI or magnetic resonance imaging) is in reality an NMR technique applied to medicine, and it was developed in the 1970s.

The 1950s and especially the 1960s and 1970s (although the origins of this method date to the 1890s) saw the development of instruments that could exploit the concept of bombarding an organic molecule with a high-energy electron beam to induce fragmentation of that molecule and then identifying these fragments to give important structural formation. This technique is known as *mass spectrometry* and this methodology has been greatly expanded and modified in recent years to become a very powerful tool for structural identification of organic molecules. Other tools are constantly being developed and each of the techniques mentioned has "cutting edge" methodology that allows a chemist to probe very complex structures.

Other methods include the use of ***X-ray crystallography*** (known for many years) to identify crystalline molecules. When X-rays interact with a molecule with a distinct crystal structure, the resulting X-ray scattering patterns can often be analyzed to provide clues to its chemical structure. With modern computer technology, a picture of the structural features of a molecule can be produced and the methodology has been expanded to include structures of proteins and even biologically active small molecules docked to a protein. With modern electron tunneling microscopes, pictures of atoms have been made.

Using these techniques to give more information and develop new techniques is another major area of organic chemistry. The work continues as new areas of organic chemistry are discovered, including studies in biochemistry and the exploration of the interface between molecular biology and organic chemistry.

1.2 The Variety and Beauty of Organic Molecules

Section 1.1 described how organic chemistry came to be a science. **Why is it important?** You are alive because of chemical reactions involving organic molecules, and the science we know as biology results from organic chemical reactions of simple and complex organic chemicals. The DNA (deoxyribonucleic acid) and the proteins in your body are organic molecules. They are made and they work by a series of organic chemical reactions. Proteins are large structures composed of many small organic chemical units known as amino acids, such as serine (**27**), and DNA is made up of many units called nucleotides, such as cytosine (**28**).

27   28   29   30

If you are blinking an eye while reading or moving your arm to turn the page, that nerve impulse from your brain was induced, in part, by one of several important organic molecules called neurotransmitters. One important neurotransmitter is acetylcholine (**29**). If you see this page, the light is interacting with a photopigment in your eye called rhodopsin, which releases retinal (**30**) upon exposure to the light. Retinal reacts with a lysine fragment (another amino acid; see Chapter 27) of a protein as part of the process we call vision. Note the similarity of retinal to vitamin A (**2**), which is simply the reduced form of **30.** Oxidation and reduction are discussed in Chapters 16 and 18, respectively.

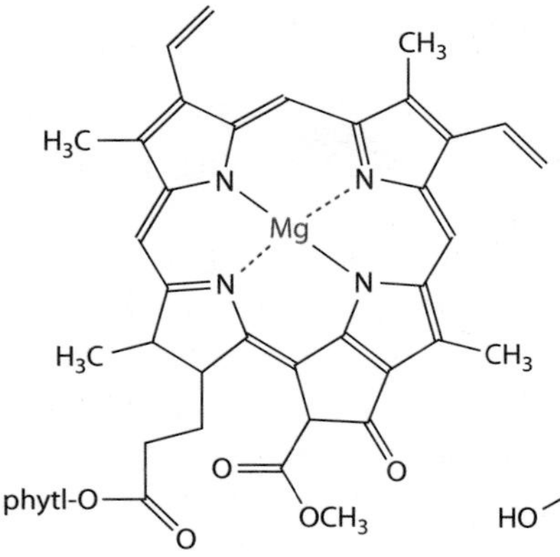

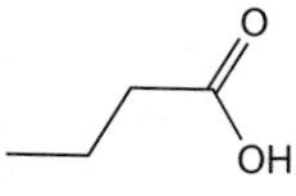

31　　　　　　　　　　**32**　　　　　　　　　　**33**

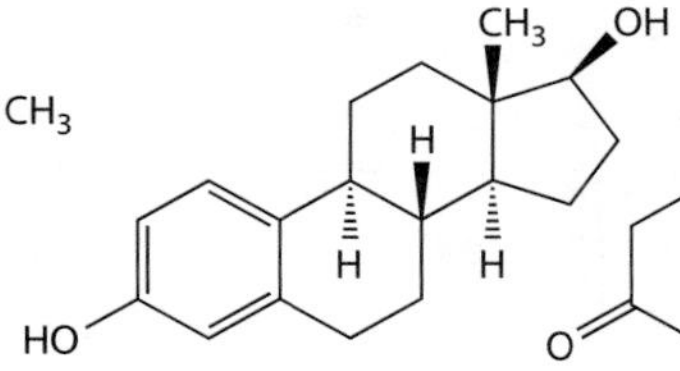

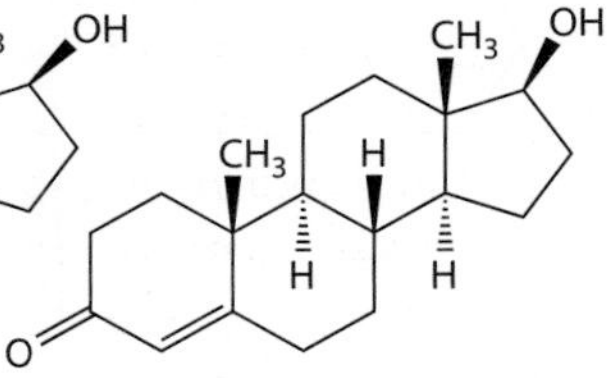

15

What you see, at least the color associated with what you see, is usually due to one or more organic molecules. If the trees and grass in your yard appear green, one of the chemicals responsible is called chlorophyll A (**31**). The "------" in **31** means there is an interaction between N and Mg (a coordinate bond) rather than a formal covalent N–Mg bond. Many other things about human physiology involve organic chemistry. There are more fundamental physiological influences of organic chemicals. If you are female, one of the principal sex hormones for your gender is β-estradiol (**32**), but if you are male, your principal sex hormone is testosterone (**33**). It should be noted that each gender has both of these hormones (and others), but in quite different proportions. Note also that the chemical structures of estradiol and testosterone are related.

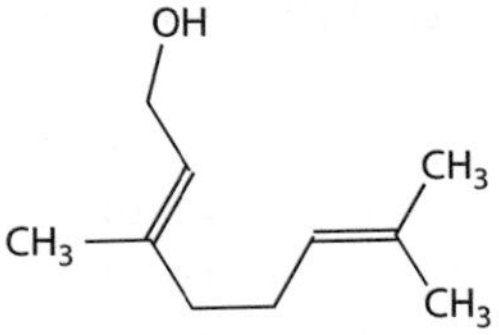

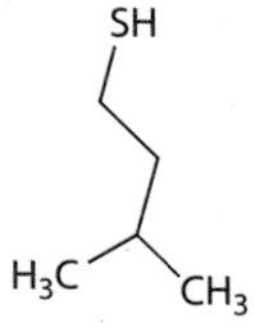

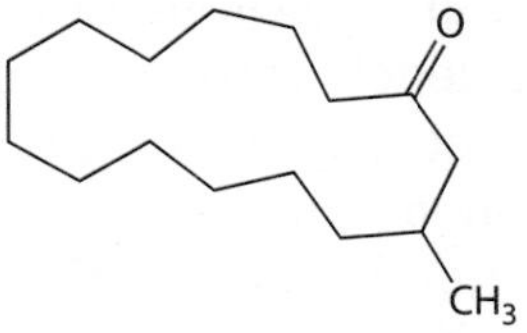

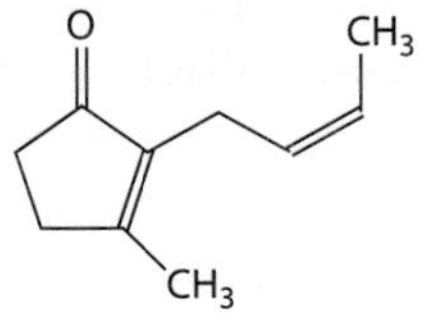

34　　　　　　　**35**　　　　　　　**36**　　　　　　　**37**

Smells are a very important part of life. **But what are smells?** They are the interaction of organic chemicals with olfactory receptors in your nose. If your feet have not been washed recently, you probably detect a pungent odor. In such a situation, your feet exude a chemical called butyric acid (**15**), among other things, with obvious deleterious social effects. If you walk into a garden and smell a rose, one of the chemicals in that aroma is called geraniol (**34**), which interacts with those olfactory receptors. If a skunk has ever sprayed your dog or cat, many organic chemicals are part of the spray, including mercaptan (also called a thiol) **35**. Clearly, this is an unpleasant smelling organic chemical and your pet was very unhappy. If you are wearing musk cologne, it may contain muscone (**36**) if it is natural musk (scraped from the hindquarters of a male musk deer). If you are wearing a jasmine perfume, it probably contains jasmone (**37**), which is part of the essential oil of jasmine flowers. These are clearly pleasant-smelling and foul-smelling organic molecules.

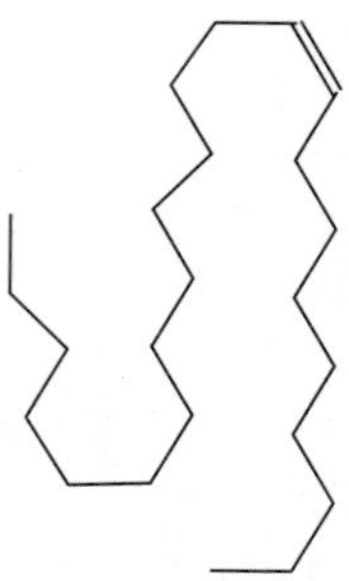

38 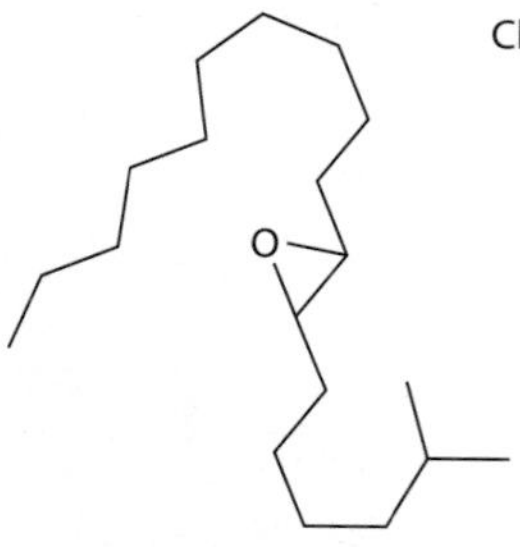   39 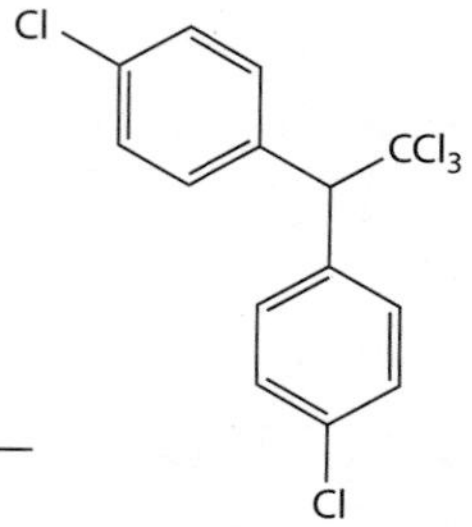   40 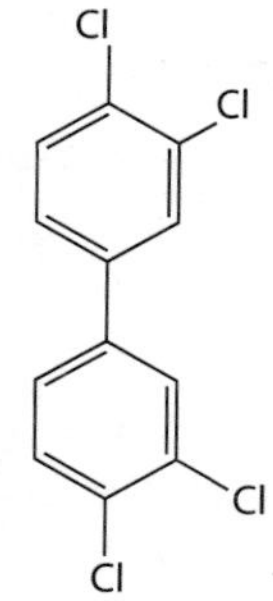   41

Many things around you involve subtle uses of organic molecules. If you see a housefly, know that it uses a chemical called a pheromone (in this case, muscalure, **38**) in order to attract a mate and reproduce. The American cockroach (hopefully there are none in your dorm) similarly attracts a mate by exuding **39**. To control insect pests, we sometimes use the pheromone of that pest to attract it to a trap. Alternatively, we can spray insecticides such as DDT (**40**; the chemical name is 1,1,1-trichloro-2,2-*bis*(4-chlorophenyl)ethane, but DDT comes from the trade name, p,p′-**d**ichloro**d**iphenyl**t**richloroethane) directly on plants, sometimes with devastating environmental consequences. Nonetheless, without some form of pest control and plant growth promoters, we may not be able to feed our enormous and growing population.

The term PCB (**41**; short for **p**oly**c**hlorinated **b**iphenyl) is in the news as an environmental pollutant. PCBs have been used as pesticide extenders, but also in transformers and as stabilizers in poly(vinyl chloride) coatings (PVC coatings). When PCBs leach into soil and water, they have serious environmental consequences. Yet, chemicals such as these have played a positive role in the industrialization of humans, as well as the negative role described. Understanding these chemicals, how they work, and when to use them is obviously important and a thorough understanding of organic chemistry is important. This understanding and the research that flows from it are important if we are to develop new and environmentally safer compounds.

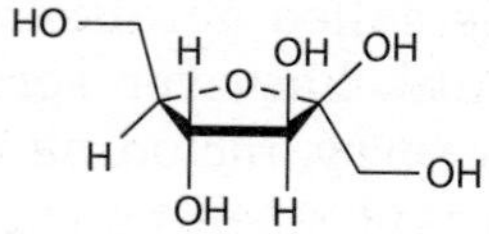

42 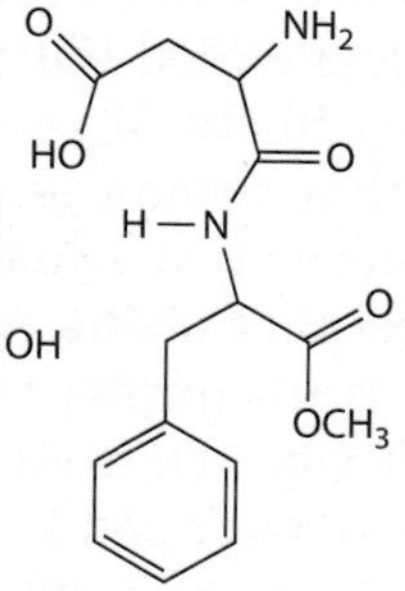   43 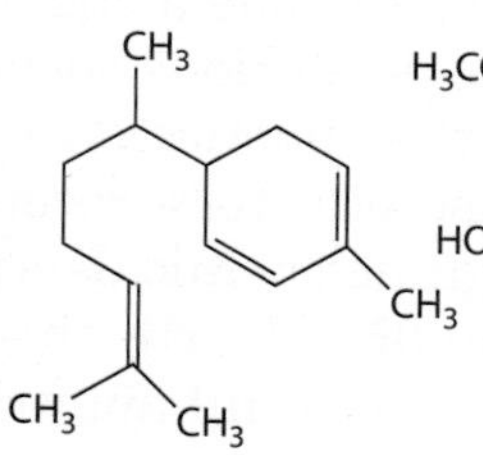   44 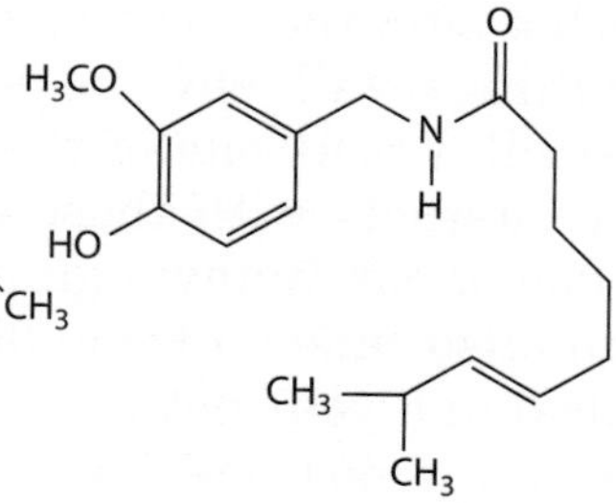   45

Eating is obviously an important part of life, and the taste of the food is important. **What are tastes?** They are the interaction of organic chemicals (and other chemicals as well) with receptors on your tongue, although *smell is also associated with taste.* **Are you drinking a soda? Does it taste sweet?** If it is not a diet soda, it probably contains a sugar called fructose (**42**), but if it is a diet drink it may contain one of the sugar substitutes such as aspartame (**43**). Different chemicals in different foods have their own unique tastes. **Do you like the taste of ginger?** The active ingredient that gives ginger its spicy taste is an organic compound called zingiberene (**44**). **Do you like the taste of red chili peppers?** If so, the hot taste is due to an organic chemical called capsaicin (**45**). In both cases, these chemicals interact with your taste buds to produce that characteristic taste. Capsaicin is also found in some topical creams used to alleviate symptoms of arthritis and muscular aches.

46   47   48   49

Most medicines used today are organic chemicals. Clearly, this is of vital importance to the health and well-being of humans. **Do you have a headache after reading all of this stuff?** If so, you are probably looking for a bottle of aspirin (**4**) or Tylenol (which contains acetaminophen, **46**). **Have you been to the dentist recently?** If so, you might have had a shot of Novocain (**47**; the actual name of this chemical is procaine hydrochloride) so that you would not feel the pain (it is a local anesthetic). If you have recently been ill, your physician may have given you a prescription for an antibiotic. Commonly prescribed antibiotics could include amoxicillin (**48**), penicillin G (**49**), or even a tetracycline antibiotic such as aureomycin (**50**).

Clearly, much more serious and devastating diseases can afflict humans. **Has a friend or relative been treated for cancer?** The physician might have used vinblastine (**51**) to treat the cancer. **Do you smoke?** If so, you are breathing nicotine (**52**) as well as many other organic compounds into your lungs, which then make their way into your bloodstream. **Have you heard of the use of AZT for the treatment of AIDS?** The structure of AZT is shown in **53**.

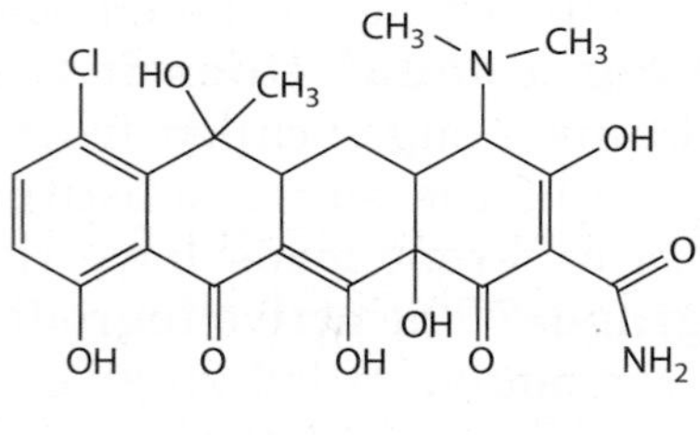

50 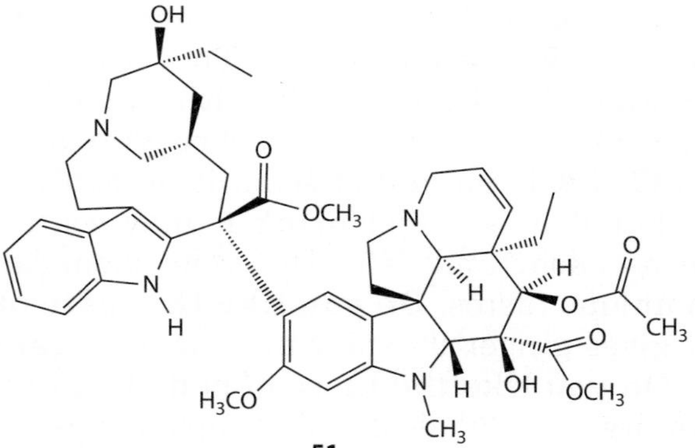   51

Finally, organic molecules can touch vast areas of your life, often in subtle ways. When I say they touch you, I mean that quite literally. **Are you wearing clothes?** If so, you might be wearing a synthetic blend of cloth made from rayon (cellulose acetate, **54**). This is a polymer (a large molecule made by bonding many individual units together). The "n" beside the bracket represents the number of repeating units (this is common nomenclature for all polymers). You might be wearing nylon or, specifically, something made from nylon 66, which has the chemical structure **55**.

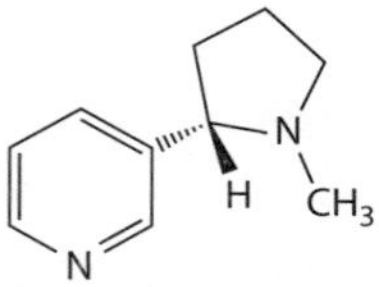

52 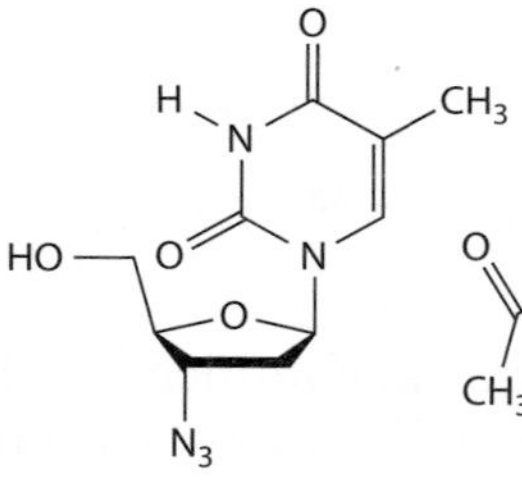   53 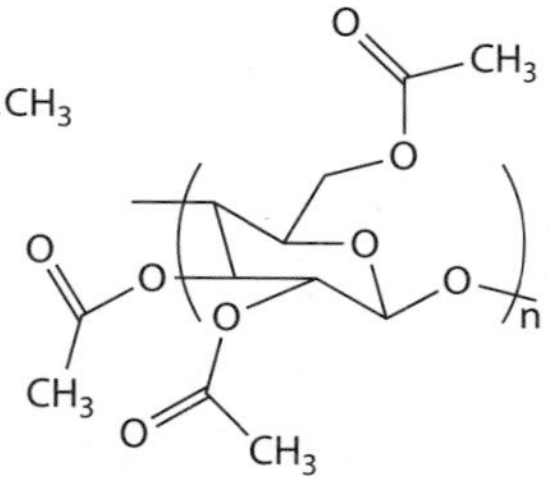   54 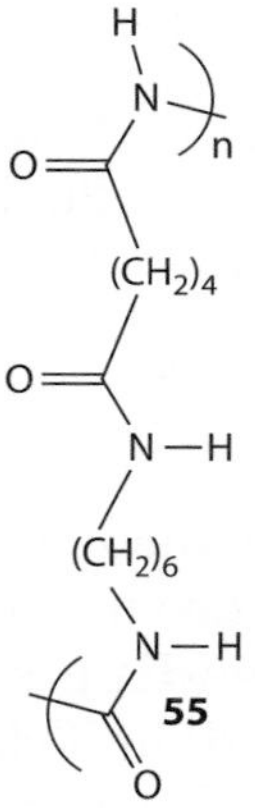   55

Have you ever heard of Teflon? It finds uses in many machines and devices that you use every day. It has the structure shown in **56**. Natural rubber is obtained from the sap of certain trees, and it has been used for automobile tires and other things. Although tires now have a more complex composition, natural rubber is poly(isoprene) (**57**). You might be using a piece of paper to describe your thoughts about organic chemistry at this moment. If so, you are probably writing on something with cellulose (**58**) in it. Notice that rayon is essentially a derivative of cellulose, the main constituent of wood fiber. When you crumple up the paper and throw it into a "plastic" waste container, that container might be made of poly(ethylene) (**59**).

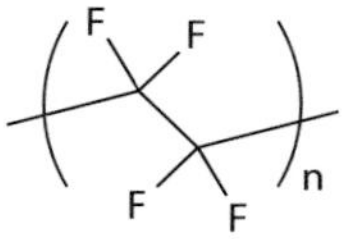

| 56 | 57 | 58 | 59 |

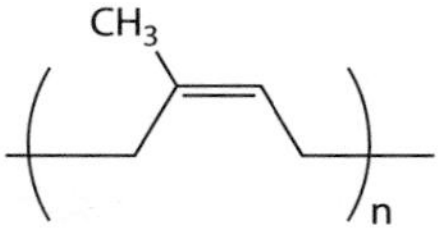

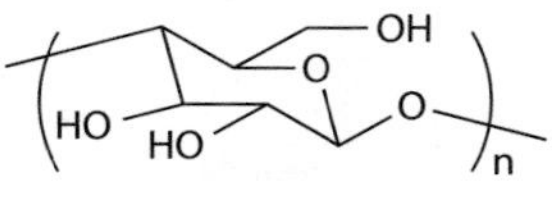

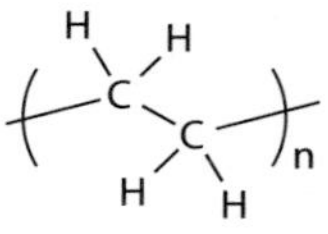

I have thrown a lot of structures at you. **Why?** Organic chemistry is all around you and it is an integral part of your life. Understanding these things is not just to get you into the program of your dreams; it can help you make informed choices. Every day of our lives, an organic molecule touches virtually every aspect of our lives. This is why you are sitting there reading this book. I hope that understanding the concepts in its pages will not only help you in your career, but also will help you understand the beauty that surrounds you. It might also help you understand the dangers that surround you in the form of pollution, illicit drugs, or chemical weapons. I hope that understanding organic chemistry will help you understand some of the debate that swirls around these subjects and possibly help you contribute to new solutions for these problems. Perhaps you will find a solution to one of these problems or make an important contribution. The journey begins here. Good luck!

References

1. Fieser, L. F., and Fieser, M. 1961. *Advanced organic chemistry,* 1–31. New York: Reinhold Publishing Corp.
2. Leicester, H. M. 1956. *The historical background of chemistry,* 172–188. New York: John Wiley & Sons.
3. Corey, E. J., and Cheng, X-M. 1989. *The logic of chemical synthesis.* New York: John Wiley & Sons.

Why Is an Acid–Base Theme Important?

The operational theme of this book is that understanding acid and base chemistry will lead to a better understanding of organic chemistry. Applying acid–base principles to many reactions allows one to understand and predict the reaction rather than simply memorizing it. To examine this premise, the acid–base reactions found in general chemistry must be reexamined to determine how they can be applied directly or when they must be modified. *In most general chemistry books, an acid–base reaction is defined as one in which a proton is transferred from an acid to a base, and a Brønsted–Lowry acid is a proton donor (HB), whereas a Brønsted–Lowry base is a proton acceptor (A–).* The species formed when a proton is removed from an acid is the conjugate base (B⁻), and the species formed when a proton is added to a base is the conjugate acid, HA:

$$HB(aq) \ + \ A^-_{(aq)} \ \rightleftarrows \ B^-(aq) \ + \ HA_{(aq)}$$

To begin this chapter, you should know the following from a general chemistry course:

- **Define and recognize the structures of simple Brønsted–Lowry acids and bases as well as simple Lewis acids and bases.**
- **Understand the definitions of conjugate acid and conjugate base.**
- **Understand the fundamentals of acid–base strength in aqueous media.**

- **Understand the role of water in acid–base equilibria.**
- **Understand K_a and pK_a and how to manipulate them.**
- **Recognize classical mineral acids.**
- **Recognize classical Brønsted–Lowry bases.**

This chapter will review the principles of acid–base chemistry from general chemistry in order to make a link with modern organic chemistry. The chapter will introduce the theme of acid–base chemistry as a basis for understanding each chemical transformation where it is appropriate.

When you have completed this chapter, you should understand the following points:

- **There is an inverse relationship between K_a and pK_a, and a large K_a or a small pK_a is associated with a stronger acid.**
- **It is important to put water back into the acid–base reaction for aqueous media.**
- **The fundamental difference between a Brønsted–Lowry acid and a Lewis acid is that the former is a proton that accepts electrons and the latter is any other atom that accepts electrons. The fundamental difference between a Brønsted–Lowry base and a Lewis base is that the former donates electrons to a proton and the latter donates electrons to another atom.**
- **Curved arrows are used to indicate electron flow from a source of high electron density to a point of low electron density.**
- **Acid strength in an acid X–H is largely determined by the stability of the conjugate acid and base and the strength of the X–H bond. For example, HI is a stronger acid than HF due to a weaker X–H bond and charge dispersal in the larger anionic conjugate base. An acid such as perchloric acid is stronger than sulfuric acid due to charge dispersal in the conjugate base that results from resonance.**
- **Base strength is largely determined by the stability of the conjugate acid and base and the electron-donating ability of the basic atom. Water is a stronger acid than ammonia because hydroxide ion is more stable than the amide anion and the OH bond is weaker than the NH bond. Ammonia is a stronger base than water because oxygen is more electronegative than nitrogen, and the ammonium ion is more stable than the hydronium ion.**
- **Reaction of a Lewis acid and a Lewis base leads to an ate complex as the product.**
- **Acid–base reactions have biological relevance.**

2.1 Acids and Bases in General Chemistry

In 1884, Santa Arrhenius (Sweden; 1859–1927) defined an acid as a material that can release a proton or hydrogen ion (H^+). When hydrogen chloride (HCl) is dissolved in water, a solution is formed via ionization that leads to hydrated hydrogen ions and hydrated chloride ions. Using the original definition, a base (then called an alkali) is a material that can donate a hydroxide ion (^-OH). Sodium hydroxide in water solution ionizes to hydrated sodium ions and hydrated hydroxide ions. This concept led to a definition of acids and bases as reported independently in 1923 by Thomas M. Lowry (England; 1874–1936) and Johannes Nicolas Brønsted (Denmark; 1879–1947). *According to this definition, an acid is a material that donates a proton and a base is a material that can accept a proton*[1,2]—*the Brønsted–Lowry definitions mentioned before.* If a high concentration of hydrogen ions is observed when an acid is added to water (an aqueous solution of the acid), that acid must have an *ionizable hydrogen* (known as a **proton**). Such acids typically take the form H–X, where X is an atom other than carbon or hydrogen (O, S, N, etc.).

2.1 Write out the structures of hydrochloric acid, hydrobromic acid, sulfuric acid, and nitric acid.

According to Brønsted and Lowry, an acid–base reaction is defined in terms of a proton transfer. By this definition, the reaction of HCl in water is the following:

$$HCl(aq) \quad + \quad H_2O \quad \rightleftharpoons \quad Cl^-(aq) \quad + \quad H_3O^+_{(aq)}$$

Water is the base in this reaction. Acid and base reactions in general chemistry are always done in water. When a Brønsted–Lowry acid such as HCl is placed in water, a proton is transferred to a water molecule (water is the base); a conjugate acid is formed (the hydronium ion H_3O^+) as well as a conjugate base (the chloride ion). In neutral pure water (no acid is present), the hydrogen ion concentration is about 1.0×10^{-7} M (pH of 7). An increase in the concentration of hydrogen ions above 1.0×10^{-7} M gives an acidic solution, with a pH less than 7. If the pH is greater than 7, it is considered a basic solution.

2.2 What is the conjugate base formed when HCl reacts with ammonia?

If the criterion for acid strength is the extent of ionization in water, the mineral acids HCl, HBr, HI, H_2SO_4, HNO_3, and $HClO_4$ are all strong acids. Likewise, strong bases are extensively ionized in water; common strong bases include NaOH (soda lye), KOH (potash lye), LiOH, CsOH, $Mg(OH)_2$, $Ca(OH)_2$, and $Ba(OH)_2$. Using the definitions given previously, weak acids and weak bases will ionize to a lesser extent in water relative to the strong acids or bases. Indeed, weak acids are solutes that react reversibly with water to form H_3O^+ ions.

There are two categories: (1) molecules containing an ionizable hydrogen atom such as nitrous acid (HNO_2), and (2) cations such as the ammonium ion (NH_4^+). Weak bases are defined as solutes that react with water molecules to acquire a H^+ ion and leave an hydroxide ion (HO^-) behind. Once again, all definitions relate to ionization in water.

An example of a weak base is the reaction of ammonia plus water to give the ammonium cation and hydroxide anion. (Amines will be discussed in Chapter 5, Section 5.6.2.) It is known that they are weak bases. Examples of weak acids that will be discussed in this book are acetic acid (CH_3COOH), boric acid (H_3BO_3), butanoic acid ($CH_3CH_2CH_2COOH$), formic acid ($HCOOH$), hydrocyanic acid (HCN), uric acid ($HC_5H_3N_4O_3$), and the fatty acid known as stearic acid ($[CH_2(CH_2)_{15}CH_2COOH$, an 18-carbon acid]). See Chapter 5 (Section 5.9.3) for a definition of carboxylic acids such as these and the structures of these acids. Weak bases include ammonia (NH_3) and amines such as methylamine (CH_3NH_2), triethylamine [$(CH_3)_3N$], and pyridine (C_5H_5N) (discussed in Section 5.6.2 of Chapter 5). Once again, the relative strength of these acids and bases is defined by their reaction with water.

2.3 **In the reaction of nitric acid and KOH, which atom in nitric acid accepts the electron pair from the base, and which atom in KOH donates the electrons to that proton?**

2.4 **If the ammonium ion is an acid that reacts with a suitable base, what is the conjugate base of this reaction?**

Several of the weak acids and the weak bases are carbon based (carbon atoms in the structure), although the structures may not yet be familiar. It is reasonable to assume that such acids and bases will play a prominent role in the chemistry to be discussed throughout this book. Two main points are to be made. *First, many organic acids and bases are insoluble in water, so the classical Brønsted–Lowry definition does not formally apply. Second, a very weak acid can react with a particularly strong base, and a very weak base may react with a particularly strong acid.* If an organic molecule is a very weak base, for example, an acid–base reaction occurs only if that base is treated with a very strong acid. The acid–base pair must be defined in order to evaluate the reaction. An acid–base equilibrium in general chemistry usually deals with strong acid–strong base, weak acid–strong base, or strong acid–weak base interactions.

2.2 Acids and Bases in Organic Chemistry

General chemistry defines acids and bases in terms of their ionization in water. *The chemistry discussed in this book will show that many organic compounds are insoluble in water.* Although the acetic acid and ethanol discussed in the previous section are soluble in water, many other weak acids or very weak acids are completely insoluble, as are many bases or very weak bases. A solvent other than water is required to put such acids or bases into solution, and this "other" solvent is typically another organic compound.

Diethyl ether ($CH_3CH_2OCH_2CH_3$), hexane ($CH_3CH_2CH_2CH_2CH_2CH_3$), and dichloromethane (CH_2Cl_2) are common solvents. A study of an acid–base reaction in organic solvents must begin with understanding how the organic solvent interacts with the acid, the base, the conjugate acid, and the conjugate base. Because many of the strong mineral acids are insoluble in organic solvents, a mixture of water and an organic solvent is used. In other words, a water–organic solvent mixture is required to make the acid–base components soluble. The issue of solubility may force one to choose a different type of acid or base. The examples that follow illustrate this problem.

2.5 What are the common mineral acids?

$$HCl(aq) + H_2O \longrightarrow H_3O^+_{(aq)} + Cl^-(aq)$$

$$CH_3CH_2OH(aq) + H_2O \rightleftharpoons CH_3CH_2O^-(aq) + H_3O^+_{(aq)}$$

Water reacts with HCl, and the organic acid known as formic acid (HCOOH, **1**) also reacts with water as a weak acid, as shown. Formic acid is a much weaker acid than HCl. When **1** reacts with water, the conjugate base is the formate anion, **2**, and the conjugate acid is the hydronium ion. If **1** is a weaker acid than HCl, the equilibrium for **1** + H_2O lies to the left in the reaction shown when compared to the reaction of HCl + H_2O in Section 2.1. Note the (aq) term indicates solvation by the solvent water. Note also that *the term "reaction" is used for the acid–base equilibrium.* The acid–base equilibria shown for HCl and HCOOH are chemical reactions that generate two products: the formate anion (HCOO⁻, **2**) and the hydronium ion (from **1**) or the hydronium ion and the chloride ion (from HCl).

Just as water reacts with HCOOH or with HCl, water reacts with ethanol (CH_3CH_2OH). As with HCl and with HCOOH, water is the base and ethanol is the acid in this reaction. Ethanol is known to be much weaker than formic acid, so the equilibrium for this reaction lies almost entirely to the left. By comparing three different acids for ionization in water, an order of acid strength is established: HCl > HCOOH > ethanol. This analysis is more or less identical to the concepts used in general chemistry, but with different acids. If an acid and/or base is insoluble in water, how is this analysis applicable? Simply change the solvent from water to an organic solvent!

Consider the reaction of formic acid using diethyl ether (**3**) as a solvent rather than water. Note the (*ether*) term in the reaction, indicating solvation by ether rather than by water. The organic solvent diethyl ether (**3**) is also the base, just as water is the base in reactions with HCl or HCOOH (**1**). The basic atom in **3** is the oxygen, and it reacts with the weak acid (formic acid) to give the formate anion in **4** (the conjugated base) and the protonated form of ether (**5**, an oxonium ion), which is the conjugate acid. The equilibrium lies to the left in the reaction because that formic acid is a weak acid in ether; however, *the definition of strong or weak acid relating to dissociation in water no longer applies*. It is the same reaction, in principle, but the solvent is different and that influences the position of the equilibrium. In fact, *the equilibrium for formic acid lies further to the left in ether than in water, so formic acid is a weaker acid in ether than in water.* This is an important statement because it suggests that *the strength of the acid depends on the base with which it reacts,* which is critical for establishing whether the equilibrium lies to the left or to the right.

If HCl is placed in anhydrous (no water) methanol (**6**), where methanol is the solvent, the oxygen of the methanol is the only available base for reaction with the acid HCl. The oxygen of methanol reacts with HCl to form an oxonium ion **7** (the conjugate acid), with chloride ion as the counter-ion (the conjugate base). The term (*ether*) is used for solvation in the formic acid reaction to show the analogy to the water reaction, but the (*methanol*) designation is omitted for **6**. Although *it is understood that all components are solvated by the solvent methanol, the solvation term is usually omitted in organic chemistry and it will be omitted throughout this book.* The important point of this reaction is the same as that for the formic acid reaction in ether. **The definition for ionization in water does not apply, but ionization is still the key to the acid–base reaction regardless of solvent.** This leads to the idea that *an acid reacts with a base, regardless of solvent, but the extent of ionization changes with the solvent and with the base.* In fact, the degree of ionization changes as the acid and the base are changed and also as the solvent is changed.

2.6 In the reaction of 5 with water, indicate which H in 5 is the acid and which atom in water is the base.

2.7 Which is the stronger base, water or hydroxide ion? Justify your answer.

Why are these observations important? Remember from general chemistry that the equilibrium constant, K_a (acidity constant), is given by the following equation when the reaction is done in water:

$$K_a = \frac{[H^+]\ [B^-]}{[HB]} \quad \text{and for HCl in water} \quad K_a = \frac{[H_3O^+]\ [Cl^-]}{[HCl]}$$

However, the actual equation for the equilibrium constant is

$$K_a = \frac{[\text{conjugate acid}]\ [\text{conjugate base}]}{[\text{acid}]\ [\text{base}]} \qquad \text{so,} \qquad K_a = \frac{[H_3O^+]\ [Cl^-]}{[HCl]\ [\textbf{base}]}$$

where the base in the HCl reaction is actually the water from the solvent.

Where is the term for water in the first set of equations? In general chemistry, the water is removed from the equation to simplify calculations. ***However, when water is removed from the equilibrium constant equation, this act also removes the base from the acid–base equation.*** If water is not the solvent and there is a different base in the reaction, the base certainly must be put back in the equation. In other words, the equilibrium constants for the formic acid reaction with ether as well as the methanol and HCl reaction must use the K_a equations:

$$K_a = \frac{[HCOO^-]\ [\text{ether·H}^+]}{[HCOOH]\ [\text{ether}]} \qquad \text{and} \qquad K_a = \frac{[Cl^-]\ [CH_3OH_2]}{[CH_3OH]\ [HCl]}$$

This analysis relies on the Brønsted–Lowry definitions of an acid and a base as used in general chemistry, as well as an extension of the fundamental ionization equations to solvents other than water. ***What is the point?*** Water is not used for many reactions found in organic chemistry, and it is important to *change the focus from acids that ionize in water to the concept that acids react with bases in order to ionize.* This concept is the basis for many of the reactions that will be introduced in succeeding chapters. For the moment, it is important to remember that both the acid and the base must be identified in a reaction based on their reactivity.

2.8 **What is the base in the reaction that generates hydronium ion and chloride ion from HCl?**

2.9 **In the reaction of methanol with HCl, which is the Brønsted–Lowry acid?**

2.3 How Are the Two Acid–Base Definitions Related?

A Brønsted–Lowry acid is defined as a proton donor, and a Brønsted–Lowry base is a proton acceptor. A Lewis acid accepts an electron pair from a Lewis base, which is an electron pair donor. There may be confusion about how these definitions are related rather than about just how they differ. Part of the problem may lie with removing the base (water) from acid–base reactions done in water. When HCl reacts with water, the Brønsted–Lowry definition states that the proton (H^+)

is "donated" to water, forming the hydronium ion. In this reaction, the oxygen atom of the water "accepts" the proton. How does an oxygen accept a proton?

2.10 If ammonia were to react with HCl rather than water, which atom accepts the proton?

What happens when water accepts a proton to form the hydronium ion? The answer is that a new σ-covalent bond (see Chapter 3) is formed between the hydrogen atom and the oxygen atom, represented as O–H. This reaction is shown using a Lewis dot formula for water, and one of the unshared electron pairs on oxygen reacts with H+ to form the new bond to the proton. This reaction uses a **blue curved arrow,** which indicates that oxygen donates two electrons to H+ to form a new O–H bond. The color blue will be used to indicate an electron-rich species or with curved arrows to indicate electron donation. *The curved arrow formalism is used in chemical reactions to indicate transfer of electron density in order to form a new chemical bond.* A proton has no electron density associated with it, so it accepts the electron density from the electron-rich oxygen atom. In other words, oxygen is the base and the proton is the acid. The resulting hydronium ion has three bonds to oxygen, making it electron deficient; it takes on a positive charge as shown (see formal change in Chapter 5, Section 5.5).

2.11 Draw the Lewis dot formula for ammonia and then for HCl.

Further in this reaction, one electron pair on oxygen is donated to the electron-deficient proton to form a new O–H bond. Therefore, the oxygen atom "accepts" a proton by donating two electrons to form the new bond. Remember that the Lewis definition of a base is an electron donor and a Lewis base donates electrons to an atom other than hydrogen. However, *in this case, the base donated two electrons to a proton. At this point, the definitions merge.* A Brønsted–Lowry base accepts a proton by donating two electrons to that proton. Why two electrons? Chapter 3 will discuss the fact that a covalent bond requires two electrons. Therefore, bases may be defined as two-electron donors to an electron-deficient center. If the base donates electrons to a proton, it is a Brønsted–Lowry base. If the base donates electrons to an atom other than hydrogen, it is formally a Lewis base. The key concept is that *a base is a two-electron donor.*

To donate two electrons, a base must have an excess of electron density. Therefore, it makes sense that molecules containing oxygen or nitrogen react as bases. The oxygen atom of hydroxide and the nitrogen atom of ammonia are examples. The more easily an atom is able to donate electrons, the more basic that atom should be, ignoring all other factors. This statement will be discussed in more detail in Chapter 6.

2.12 Can the phosphorus atom function as a Lewis base? The sulfur atom? The sodium atom in Na⁺?

A Brønsted–Lowry acid is a proton donor, whereas a Lewis acid is an electron pair acceptor. If all bases donate two electrons, *the electron flow is from the base to the acid, rather than from the acid to the base.* Therefore, an acid does **not** donate the proton, but rather the proton is "attacked" by the base to form a new bond to the proton, as shown previously for water and H⁺ giving the hydronium ion. As noted, a blue curved arrow is used to indicate the flow of electrons, and *the electron flow is always from a source of high electron density to a point of low electron density.* In this case, the direction of the arrow is **always** *from* the base *to* the proton of the acid, as illustrated. Although the color **blue** is used for the electron-rich base as well as the electron flow represented by the arrow, the color **red** is used for the electron-deficient acid—here, a proton. This convention is used throughout this book.

2.13 Why is it incorrect to draw the arrow from H to oxygen in this reaction?

What has been learned from this discussion? A base is electron rich and will donate two electrons to an electron-deficient atom, such as a proton, to form the conjugate acid. The reaction with water used H⁺, a free proton. There are no free protons. The protons in common acids such as HCl or HCOOH have the hydrogen atom attached to another atom by a covalent bond (see Chapter 3, Section 3.3). Chapter 3 discusses bond polarization; the hydrogen atom in HCl is polarized such that H is electron deficient and positive (a proton-like atom), whereas the Cl is electron rich and negative (see Section 3.7). Therefore, *a base does not react with a free proton but rather with an electron-deficient hydrogen atom that is attached to another atom,* as in HCl. When the base donates two electrons to the hydrogen atom, it literally pulls the proton away.

What happens to the two electrons in the bond between H and Cl? That bond will break as the two electrons from the base are used to form a new bond (such as O–H), and the two electrons in the former H–Cl bond will migrate toward the chlorine atom, forming chloride ion, which is the conjugate base. A curved arrow represents the electron flow in this process, and the entire reaction is written as the following:

The electron-rich atom in water is the oxygen atom (the basic atom), and it donates two electrons to the electron-deficient hydrogen of HCl (the acid), forming a new O–H bond in the hydronium ion, which is the conjugate acid. As the O–H bond is formed by using the two electrons from the oxygen, the H–Cl bond breaks and the two electrons in that bond are transferred to chlorine, forming the chloride ion, which has a negative charge. The δ+ hydrogen atom in H–Cl is an acid consistent with identification of H as an electron-deficient atom.

2.14 In a reaction of HCl and ammonia, identify the electron-donating atom of the base and draw the reaction using the curved arrow formalism, giving the products of the reaction.

If the acid–base reaction is written with the electron pairs and the arrows, as shown for water and HCl, the Lewis base definition is quite useful. The electron-rich molecule is the base, and the electron-rich atom donates two electrons. The molecule bearing the electron-deficient atom (hydrogen) is the acid. For reactions of organic molecules, it is essential to identify electron-rich and electron-poor components of molecules, to understand the electron flow, and to understand how to predict the products. That process begins with making the transition to thinking in terms of Lewis bases/Lewis acids rather than Brønsted–Lowry acids and bases.

2.4 Acid and Base Strength

It is important to identify the acid and the base in the reactions that we have discussed. Sometimes an acid–base reaction does not work very well because the K_a favors the acid and the base rather than the conjugate acid and the conjugate base. If the equilibrium constant K_a for a given acid–base reaction is unfavorable, changing the base may make the K_a more favorable. In other words, changing the base to one that is stronger for a given acid should shift K_a toward the conjugate acid and the conjugate base. To do this, however, it is necessary to differentiate one base from another by comparing the electron-donating capability (the base strength). In other words, one base may be stronger than another, and one must be able to predict relative basicity. Likewise, two acids can be differentiated based on their relative strength.

This analysis can be probed further. Why is one base stronger than another? Why is one acid stronger or weaker than another? In previous sections, pK_a is used to differentiate acid strength where $pK_a = -\log K_a$. What information does that number convey? If one acid has a $pK_a = 1$ and a second acid has a $pK_a = 8$, the numbers indicate that the first acid is stronger. The more important question is why the first acid is stronger. The structure of the acid is important, but an acid–base reaction is an equilibrium process.

To probe acid or base strength, the structure of the acid must be compared with the structure of its conjugate acid, and the structure of the base compared with that of its conjugate base. This comparison will provide information about K_a, the relative position of the equilibrium. Determining the K_a for two different acids that react with the same base will make it possible to identify the stronger acid. Similarly, determining the K_a for two reactions in which a common acid reacts with two different bases will make it possible to identify the stronger base. Comparisons of this type will reveal structural differences of the acid/conjugate acid and conjugate/conjugate base pair *for both reactions*. This information will give clues for structural features that explain why one acid is made stronger or weaker than the other.

2.15 For a reaction of sulfuric acid and NaOH, write out the acid conjugate acid pair and also the base conjugate base pair.

Most of the concepts mentioned in the following narratives will be discussed in detail in later chapters. It is assumed that some knowledge of the fundamental concepts is carried over from general chemistry. Although it is a bit irritating to talk about concepts that will be discussed later, this is an introduction to bridge knowledge of acid–base chemistry with reactions to be discussed in the remainder of the book. Key concepts will be repeated and reinforced in the context of organic chemistry reactions. Practical aspects of the key concepts will be discussed in terms of specific questions. Why is one compound more acidic than another or why is one base weaker than another? The answers to such questions will lay the foundation for making the transition from acid–base reactions in general chemistry to acid–base reactions in organic chemistry.

$$\text{H-F} \quad \text{BASE} \quad \rightleftharpoons \quad \text{H-BASE} \quad \text{F}^- \quad K_a = \frac{[\text{H-Base}]\,[\text{F}^-]}{[\text{HF}]\,[\text{BASE}]}$$

$$\text{H-I} \quad \text{BASE} \quad \rightleftharpoons \quad \text{H-BASE} \quad \text{I}^- \quad K_a = \frac{[\text{H-Base}]\,[\text{I}^-]}{\text{HI}[\text{BASE}]}$$

2.4.1 Why Is HI More Acidic Than HF?

This question is answered by examining the equilibria for both reactions, using a common but unspecified base to keep the focus on the two acids. The statement that HI is a stronger acid than HF indicates that the K_a for the HI reaction is larger than the K_a for the HF reaction. Saying that HI is more acidic because K_a is larger simply repeats the question. If K_a is larger for HI, pK_a is smaller. The proper question is why the K_a for HI is larger than the K_a for HF. The reaction of HF with the base gives H-BASE as the conjugate acid and fluoride ion as the conjugate base. Similarly, reaction with HI gives H-BASE, but the conjugate base is chloride ion. Both sides of each equation must be examined. In HF and HI, hydrogen is the electron-deficient atom, so it makes sense to examine the strength of the HF bond and the HI bond (Chapter 3, Section 3.6). The bond dissociation energy of the H–I bond is 298.3 kJ mol^{-1}, making it weaker than the H–F bond at 569.7 kJ mol^{-1}.[3]

2.16 Is it possible for the BASE in these reactions to be diatomic hydrogen?

2.17 Based on the data for HF and HI, estimate which has the weaker bond: HCl or HBr.

Why does a weaker X–H bond make K_a larger? The covalent radii of iodine and fluorine can be compared to make an initial comparison. Covalent radii are estimated from homonuclear bond lengths, which are pertinent to covalent molecules such as HF and HI. The covalent radius of I (135 pm) is much greater than that of F (71 pm),[4] so the bond distance between H and I will be longer than the bond distance between H–F. In other words, if there is a bond between H–F and H–I, the distance between the H and F nuclei along that bond is less than the distance between the H and I nuclei, due to the much larger radius of the iodine. This is the bonding distance and, generally, *a longer bond is a weaker bond*. If the H–I bond is weaker, it should break more easily when attacked by the base, pushing the reaction to the right, toward the conjugate acid and base. This means that K_a is larger for the reaction with HI than the K_a for HF. Therefore, a weaker bond is consistent with a more acidic compound.

2.18 What does the unit "pm" represent?

It is not enough to examine only the strength of HI and HF in the acids. The conjugate bases, fluoride ion and iodide ion, must also be examined. If one is more stable than the other, the more stable ion will be less reactive. What does this mean? In an equilibrium reaction, the conjugate acid and conjugate base will react to give the acid and the base. In this example, the fluoride ion reacts with the conjugate acid or iodide ion reacts with the conjugate base. In other words, the reactivity refers to the reverse reaction of the acid–base equilibrium. It is known that the iodide ion is less reactive because it is larger and less able to donate electrons (it is a weaker base) when compared with the fluoride ion.

Another way to state this fact is to say that the iodide ion is more stable than the fluoride ion. It is more stable, and it is less reactive. If the iodide ion is less reactive, there is a higher concentration of iodide relative to fluoride, which means that the equilibrium in the HI reaction is pushed more to the right, and the equilibrium for HF is pushed further to the left. Remember from Section 2.2 that the equilibrium constant K_a is calculated taking the ratio of the molar concentrations of the conjugate acid and conjugate base divided by molar concentrations of the acid and the base. In other words, if there is a higher concentration of the conjugate base (iodide ion), K_a is larger, and HI will be the stronger acid.

2.19 From general chemistry, can fluorine have more than eight electrons?

Why does a more stable conjugate base make K_a larger? Why is the iodide ion more stable and less reactive? First, compare the size of the ions using the ionic radii rather than the covalent radii used before. The ionic radius of the iodide ion (I^{-1}) is reported to be 215 pm and that of the fluoride ion (F^{-1}) is 136 pm.[4] In a different experiment based on thermochemical data, the ionic radii of the iodide ion are reported to be 211 pm and that for fluoride ion is 126 pm.[5] It is clear that the iodide ion is significantly larger. Imagine that the fluoride ion and iodide ion have the same charge (an excess of two electrons and a total of eight electrons). The charge is dispersed over a much larger area in the larger iodide ion, making it more difficult to donate those electrons to the acid. Therefore, the larger size of the ion and dispersal of charge over a larger area make the ion less reactive and less basic.

If the conjugate base is less reactive (more stable), there will be a larger concentration of conjugate base and less of the original acid–base pair. In other words, K_a is larger. If a conjugate base is more reactive (less stable), there will be a lower concentration of the conjugate base/conjugate acid and a higher concentration of the acid/base, so K_a is smaller (a weaker acid).

Based on this analysis, HI is a stronger acid because the equilibrium constant (K_a) is pushed to the right (K_a is larger) because of the weaker HI bond. In addition, the larger size of the iodide ion disperses the charge and makes that conjugate base less reactive.

2.4.2 Why Is Water More Acidic Than Ammonia?

The reported pK_a of water is 15.74[6] and ammonia has a reported a pK_a of 38.[7] Why is water so much more acidic? In both cases, the electrophilic atom is a hydrogen atom, O–H in water versus N–H in ammonia. As shown in the equilibrium where water and ammonia react with an unspecified but generic BASE that give the same conjugate acid (H-BASE), water leads to hydroxide ion as the conjugate base and ammonia leads to the amide anion as the conjugate base. The bond strength of the O–H and N–H bonds must be determined, as well as the relative stability (reactivity) of the conjugate bases, hydroxide versus amide.

$$\text{H-OH} \quad \text{BASE} \quad \xrightleftharpoons{\qquad} \quad \text{H-BASE} \quad \text{HO}^-$$

$$\text{H-NH}_2 \quad \text{BASE} \quad \xrightleftharpoons{\qquad} \quad \text{H-BASE} \quad \text{H}_2\text{N}^-$$

The reported bond dissociation energy of H–O is 429.9 kJ mol^{-1}, and the bond dissociation energy of H–N is <338.9 kJ mol^{-1}.[3] These numbers indicate that the O–H bond is stronger and should be more difficult to break; however, this is inconsistent with water as the stronger acid. Perhaps O–H versus N–H bond strength of the acid–base pair is not the only criterion for the reaction! In fact, *this is a misuse of bond dissociation energies.* The bonds made and broken for the base and conjugate base are ignored, so the two bond dissociation energies do not take into account all bond making and bond breaking (see Chapter 7, Section 7.2). If the relative strength of the O–H versus the N–H bonds does not predict amide as the stronger base, it is prudent to examine the conjugate bases because it is known that NH_2^- is more basic than HO^-. If the amide anion is more basic, it is more reactive.

Why is hydroxide ion less basic than the amide anion? Oxygen is more electronegative than nitrogen, which indicates that oxygen tends to retain electrons more than nitrogen (see Chapter 6, Section 6.4). This statement means that if oxygen retains electron density, it is less likely to donate electrons (it is a weaker base). In addition, the ionic radius of the hydroxide ion is 252 pm and that of the amide ion is 168 pm. In the hydroxide ion, the charge is dispersed over a larger area than it is in the amide anion, so hydroxide is more stable and less reactive.

2.20 Can ammonia be considered as a Brønsted–Lowry acid?

2.21 From general chemistry, define electronegativity.

If the hydroxide anion is a weaker base, the reaction with H-BASE is poor, so there is a greater concentration of H-BASE and hydroxide and a lower concentration of BASE + water. If the amide anion is a stronger base, the reaction with H-BASE is good, so there is a lower concentration of H-BASE and amide and a greater concentration of BASE and ammonia. This means that the K_a for that reaction of water + BASE is larger than the K_a for the reaction of ammonia + BASE.

In this example, the relative basicity (stability–reactivity) of the conjugate base plays a major role in determining relative base strength. The base strength obviously influences the relative stability (reactivity) of the conjugate acid. The point of this discussion is to reinforce the idea that simply looking at the left side of the reaction (the acid–base pair) and asking which is the stronger bond is *not* sufficient. To compare the strength of two acids or of two bases, both acid–base reactions must be examined, and the natures of the acid, the base, the conjugate acid, and the conjugate base must be examined.

2.22 How many unshared electrons are on (a) the nitrogen of ammonia, and (b) the oxygen of water?

2.4.3 Why Is Ammonia a Stronger Base Than Water?

The nitrogen atom of ammonia is a better electron donor (a stronger base) than the oxygen atom of water. Once again, the acid–base equilibrium will be examined, using HCl as the acid in both cases. The reaction of ammonia and HCl gives the ammonium ion as the conjugate acid and the chloride ion as the conjugate base. For the reaction with water, the conjugate acid is the hydronium ion, and the conjugate base is the chloride ion. The pK_a of the hydronium ion is estimated to be –2, the pK_a of HCl is –7, and the pK_a of NH_4^+ is 9.24.[8]

$$H_3N: \quad H\text{-}Cl \quad \rightleftharpoons \quad H_3\overset{+}{N}:H \quad Cl^-$$

$$H_2O: \quad H\text{-}Cl \quad \rightleftharpoons \quad H_2\overset{+}{O}:H \quad Cl^-$$

As will be seen in Chapter 3 (Section 3.7), oxygen is more electronegative than nitrogen, so oxygen will retain electron density more efficiently than nitrogen. This observation is consistent with the nitrogen atom of ammonia being a better electron donor (it does not retain electrons as well) and more basic. This means that if an atom retains electrons more efficiently, it cannot donate electrons as efficiently. In other words, it is a weaker base.

If the hydronium ion is a stronger acid than ammonia, then the equilibrium for that reaction will be pushed to the left (smaller K_a), whereas the equilibrium for the ammonium ion will be more to the right (larger K_a). The reaction of ammonia and HCl has a larger K_a, with formation of more ammonium chloride, so ammonia reacts with HCl better than water, consistent with ammonia as a stronger base. Therefore, ammonia is a stronger base because it is better able to donate electrons, but also because the conjugate acid is more stable (less reactive) when compared to the conjugate acid formed from water.

2.4.4 Why Is $^-NH_2$ More Basic Than NH_3?

The amide anion has a charge of –1, which means that there is an excess of electron density on the nitrogen when compared to ammonia. In ammonia, there are three covalent bonds, and although there is one unshared pair of electrons, the molecule is neutral (no charge). In the amide anion, there are two pairs of unshared electrons. The amide anion is more basic simply because there is a higher concentration of electron density that can be donated to an acid. In general, a base that has a charge of –1 should be a stronger base than the uncharged base. Note that the amide anion is the conjugate base of ammonia, when ammonia undergoes reaction as an acid. A proton must be removed from ammonia to form the amide anion. The amide anion must be a stronger base than ammonia.

2.4.5 Why Is Sulfuric Acid Less Acidic Than Perchloric Acid?

The full acid–base equilibrium for both reactions is shown, again using a generic base so that the conjugate acid is the same in both equations. The reported pK_a of perchloric acid is –10 and that of sulfuric acid is about –1.9.[8]

Both perchloric acid and sulfuric acid have an O–H group, and the acidic proton is part of an O–H unit for both acids. Any differences in bond strength may be due to differences in the sulfur versus the chlorine. The covalent radii of S is 102 pm and that of Cl is 99 pm,[4] so there is minimal difference. If there is little difference in bond strength in the acids, there may be more subtle factors; a simple expedient is to examine the conjugate bases. The hydrogen sulfate anion has an ionic radius of 221 ppm, whereas the perchlorate anion has an atomic radius of 225 ppm. There is little difference in the size of the anions.

There is one difference in the HSO_4^- anion and the ClO_4^- anion relative to the fluoride ion, iodide ion, hydroxide ion, or amide anion discussed previously. There are π-bonds. The π-bonds are discussed in Chapter 5 (Section 5.1), so this concept is premature for a proper understanding of this concept. However, π-bonds should have been discussed in general chemistry: S=O indicates the presence of two bonds between S and O, a so-called σ-bond and a π-bond. The σ-bond is stronger than the π-bond because the σ-bond is formed by sharing electron density on a line between the two nuclei. The π-bond is formed sharing electron density between parallel adjacent p-orbitals on S and O (see the diagram) and, because less electron density is shared, the bond is weaker. This is shown in **8**, which represents the bonding in HSO_4^-; **9** represents the bonding in the perchlorate anion. A full discussion of this concept will wait until Chapter 5, but **8** is meant to show that the π-orbitals can overlap and the excess electron density (the negative charge) is dispersed over four atoms (S and the three O atoms). This type of charge dispersal is also called charge delocalization and is given the name **_resonance._**

Examination of **9** shows that there is another oxygen and an extra p-orbital. Therefore, the charge is dispersed over five atoms (the Cl and four O atoms). In other words, the charge is more delocalized in perchlorate than in the hydrogen sulfate anion, which makes the perchlorate anion more stable. Greater charge dispersal due to resonance leads to greater stability (less reactive), just as charge dispersal due to the larger size of an anion leads to greater stability. If the perchlorate anion is more stable, the equilibrium is pushed to the right (larger K_a), consistent with perchloric acid as the stronger acid. Therefore, sulfuric acid is a weaker acid, in part, because the perchlorate anion has greater resonance stability than the hydrogen sulfate anion, so it is less reactive as a base.

2.23 **Make an educated guess as to whether nitric acid is more or less acidic than perchloric acid.**

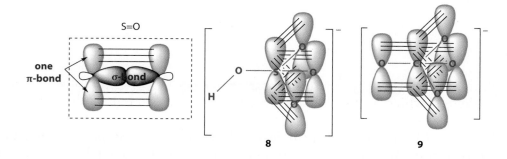

2.5 Lewis Acids and Lewis Bases

A typical general chemistry course introduces Lewis acids (electron pair acceptors) and Lewis bases (electron pair donors) as well as Brønsted–Lowry acids and bases. Note that defining acids as electron pair acceptors and bases as electron pair donors bypasses the need for ionization in water and simply focuses on transfer of electrons from one species to another. Apart from acids and bases, electron transfer is the fundamental requirement for a chemical reaction. A Lewis base must be electron rich in order to donate electrons. Likewise, a Lewis acid must be electron deficient. In this book, Lewis bases will be confined to molecules that contain oxygen, nitrogen, a halogen atom, and sometimes sulfur or phosphorus. The structures of each Lewis base will be discussed in the context of each chapter. There are, however, some simple guidelines for identifying Lewis acids.[9,10]

Group 13 atoms such as boron and aluminum can only form three covalent bonds and remain neutral, so they are electron deficient. When boron or aluminum atoms that have three bonds accept an electron pair from a Lewis base, the resulting fourth bond will provide the two electrons needed to satisfy the octet rule. This reaction forms the so-called Lewis acid–Lewis base complex, which is a charged complex called an **ate complex** (**10**). A typical reaction is shown between BF_3 and ammonia, in which the Lewis base (ammonia) donates two electrons to the electron-deficient boron. This forms the ate complex, in which the boron and the nitrogen share two electrons. Note that the nitrogen takes on a positive charge in the complex and that the boron takes on a negative charge. The nitrogen donates electrons and becomes electron deficient, whereas the boron accepts electrons to become more electron rich. The arrow in **10** represents what is called a **dative bond,** which is indicative of the electron-donating nature of the bond in a Lewis acid–Lewis base ate complex.

2.24 Draw the reaction, with curved arrow, and the ate complex formed (using Lewis electron dot structures) for the reaction of aluminum chloride ($AlCl_3$) and PH_3.

Transition metal salts have a metal atom that is capable of assuming multiple valences, which means that the salt may form compounds with two, three, four, five, or six bonds to other atoms. In many cases, these compounds function as a Lewis acid. Although transition metals will not be used extensively, a brief overview in the context of some things learned in general chemistry is appropriate. The relative order of Lewis acidity is not always straightforward because the electrophilic atom (here a metal) may have different valences and several groups or atoms (known as **ligands**) may be attached to the metal. A number of general statements can be made, however.

Lewis acids usually take the form MX_n, where X may be a halogen atom or X may be a molecule that contains oxygen, nitrogen, or phosphorus (other X units will be introduced later in the book). The metal is M and n in X_n is the number of units attached to M. Reaction with a Lewis base leads to formation of a conjugate acid–base adduct (MX^- B^+, the ate *complex*). Lewis acids and Lewis bases are discussed in more detail in Chapter 6, Section 6.5.

1. In MX_n ($n < 4$) acidity arises from the central atom's requirement for completion of an outer electron octet by accepting one or more pairs of electrons from the base. Acidity is diminished when two electron pairs are required. Group 13 acids are more acidic than transition metal acids: $BF_3 > AlCl_3 > FeCl_3$.
2. The acidity of M will decrease within any group with increasing atomic volume (effectively, with increasing atomic number) owing to the weaker

alkalosis. The normal range for blood pH is 7.36–7.44. The body can generate acid that can change this pH, making blood or tissue more acidic as a result of the natural breakdown of fats (Chapter 20, Section 20.2). This is particularly true when fats are used for energy rather than carbohydrates (Chapter 28, Section 28.1), although other processes can generate acids.

If there is insufficient bicarbonate to compensate for the extra acid, acidosis can occur. Formally, acidosis is a significant decrease in pH of extracellular fluid.[10] This condition can occur due to both respiratory and metabolic abnormalities. Respiratory acidosis occurs when breathing abnormalities result in CO_2 retention and an elevation in P_{CO2} in alveoli and arterial blood (known as hypercapnia).[11] The term P_{CO2} refers to the partial pressure of CO_2 in the pulmonary alveoli during respiration.[11] Retention of CO_2 can result from inadequate ventilation during anesthesia, certain conditions that result from central nervous system disease, or from drug use, and it is observed with emphysema.[11] Metabolic acidosis occurs with starvation, uncontrolled diabetes mellitus with ketosis, and with electrolyte and water loss due to diarrhea.[11]

Formally, alkalosis is a significant increase in the pH of extracellular fluid.[11] Both respiratory alkalosis and metabolic alkalosis are known. Respiratory alkalosis results from hyperventilation, which produces lowered P_{CO2} and higher pH of extracellular fluids.[11] Anxiety is the usual cause of hyperventilation. Metabolic alkalosis can result from the loss of gastric juices that are rich in HCl from excessive sodium bicarbonate ingestion, and it is associated with potassium ion deficiency.[11]

Dependence on acid–base reactions occurs in other aspects of human biology. Local anesthetics are commonly used to block pain in procedures such as filling a tooth by a dentist. Many local anesthetics will block nerve conduction by reducing membrane permeability of sodium ions.[12] A common local anesthetic is lignocaine (also called lidocaine, **15**). The nitrogen atom, marked in **blue** in this amine (see Chapter 6, Section 6.4.1), is a weak base, with a pK_a of 7.86.[12] Lidocaine is usually distributed as an aqueous solution of the HCl salt (**16**), which is the conjugate acid of the reaction of **15** with HCl.[12] At physiological pH, which is usually between 7.36 and 7.44, most of the drug will exist in its ionized form (**16**), which is believed to combine with the excitable membrane to inhibit sodium permeability.[12] Some of the drug will exist in the un-ionized form (**15**) because, in the acid–base equilibrium, both the base (**15**) and its conjugate acid (**16**) will be present. It is believed that **15** more easily penetrates the lipid barrier around and within the nerve tissues.[12]

2.25 Write out the acid–base reaction between 15 and HCl.

15   16

References

1. Lowry, T. M. 1923. *Chemistry and Industry* 42:43.
2. Brønsted, J. N. 1923. *Recueil des Travaux Chimiques* 42:718.
3. Lide, D. R., ed. 2006. *CRC handbook of chemistry and physics,* 87th ed., 9-56–9-57. Boca Raton, FL: Taylor & Francis.
4. Huheey, J. E. 1972. *Inorganic chemistry, principles of structure and reactivity,* 184–185, Table 5.1. New York: Harper & Row.
5. Lide, D. R., ed. 2006. *CRC handbook of chemistry and physics,* 87th ed., 12–27. Boca Raton, FL: Taylor & Francis.
6. Harned, H. S., and Robinson, R. A. 1940. *Transactions of the Faraday Society* 36:973.
7. Buncel, E., and Menon, B. C. 1977. *Journal of Organometallic Chemistry* 141:1.
8. Smith, M. B., and March, J. 2007. *March's advanced organic chemistry,* 6th ed., 359–362, Table 8-12. Hoboken, NJ: Wiley-Interscience.
9. Satchell, D. P. N., and Satchell, R. S. 1969. *Chemical Reviews* 69:251.
10. Satchell, D. P. N., and Satchell, R. S. 1971. *Quarterly Reviews of the Chemical Society* 25:171.
11. Meyers, F. H., Jawetz, E., and Goldfien, A. 1972. *Review of medical pharmacology,* 431–432. Los Altos, CA: Lange Medical Pub.
12. Grahame-Smith, D. G., and Aronson, J. K. 1984. *The Oxford textbook of clinical pharmacology and drug therapy,* 551–552. Oxford, England: Oxford University Press.

Answers to Problems

2.1

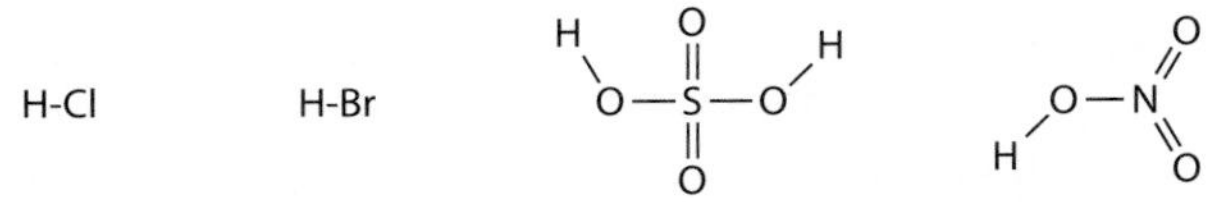

2.2   The chloride ion, Cl^-.

2.3   The hydrogen on the OH unit of nitrile acid accepts the electron pair from the base, and the oxygen atom in KOH donates the electrons to that proton.

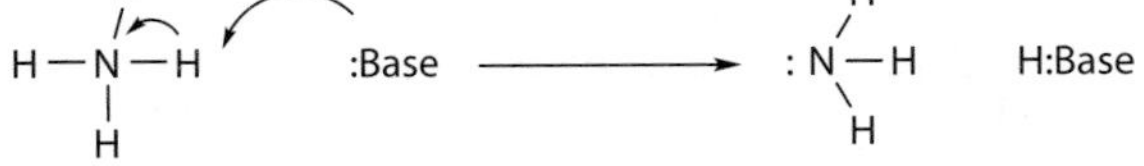

2.4   Ammonia is the conjugate base.

2.5   HCl, HBr, HI, H_2SO_4, and HNO_3 are probably the most common.

2.6   The indicated hydrogen atom in **1** is the acidic proton and the oxygen of water is the basic atom.

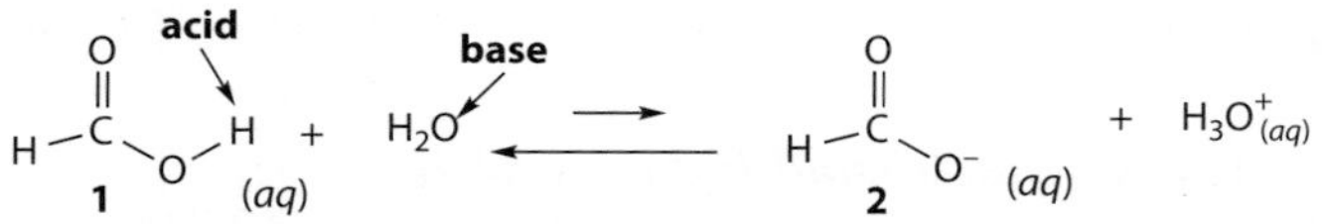

2.7 The oxygen atom in hydroxide has a charge of −1 and will be electron rich relative to the oxygen atom in neutral water. In other words, there is a higher concentration of electron density on the charge oxygen atom when compared to the neutral oxygen atom. Hydroxide is the more basic. In addition, hydroxide is the conjugate base of water, so we expect hydroxide to be more basic.

2.8 Water is the base.

2.9 HCl is the Brønsted–Lowry acid.

2.10 The nitrogen atom of ammonia accepts the proton, to form ammonium chloride.

2.11

2.12 Both P and S have unshared electrons that can be donated. In compounds that contain trivalent P (three 2-electron bonds) or divalent S (two 2-electron bonds), both are considered to be Lewis bases. The sodium atom in Na$^+$ is electron deficient and cannot function as a Lewis base.

2.13 Hydrogen is the electron-deficient atom and oxygen is the electron-rich atom. The arrow is always from the electon-rich atom to the electron-poor atom; therefore, it must be oxygen to hydrogen.

2.14

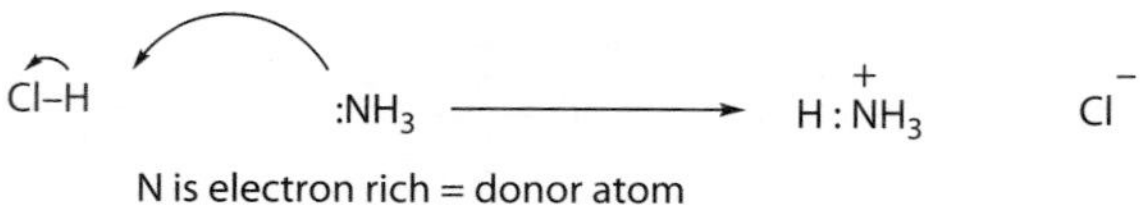

N is electron rich = donor atom

2.15

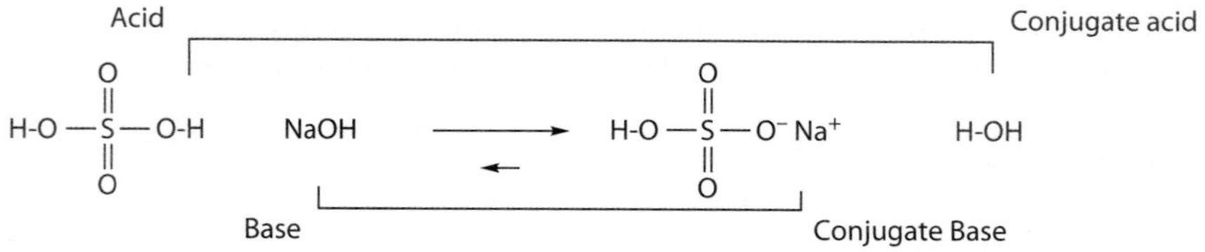

2.16 No! Diatomic hydrogen has no electrons that can be donated. The electrons are tied up in a covalent bond between the two hydrogen atoms (see Chapter 3, Section 3.3, for a discussion of covalent bonds).

2.17 Because Br is larger than Cl, the HBr bond is expected to be weaker and the bromide ion is expected to disperse charge more efficiently; thus, bromide should be more stable than the chloride ion. Therefore, HBr should be a stronger acid.

2.18 The unit pm is a picometer, which is 1×10^{-12} m, or 0.01 Å.

2.19 No! The outer shell of the second row can only hold eight electrons, so fluorine can never have more than eight.

2.20 Yes! Because ammonia has a hydrogen atom attached to the nitrogen, that hydrogen is slightly electron deficient and, with a strong enough base, ammonia is a Brønsted–Lowry acid.

2.21 The generally accepted definition is that electronegativity is the ability of an atom in a molecule to attract electrons to itself.

2.22 There is one unshared pair of electrons on the nitrogen of ammonia, so there are two unshared electrons. There are two unshared pairs of electrons on the oxygen of water, so there are four unshared electrons.

2.23 Because nitric acid forms the nitrate anion, there are only three atoms for charge dispersal via the π-bonds. Nitrogen is also much smaller than chlorine, so the nitrate anion is expected to be smaller than the perchlorate anion. Both of these observations suggest that nitric acid is less acidic than perchloric acid, which is correct.

2.24

2.25

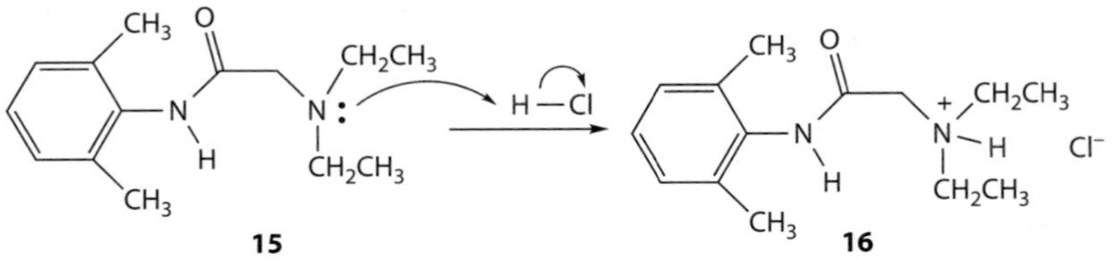

Correlation of Homework with Concepts

- There is an inverse relationship between K_a and pK_a, and a large K_a or a small pK_a is associated with a stronger acid: 40, 41, 42, 44, 45, 47.
- It is important to put water back into the acid–base reaction for aqueous media: 8, 28.

- The fundamental difference between a Brønsted–Lowry acid and a Lewis acid is that the former has a proton that accepts electrons and the latter has any other atom that accepts electrons. The fundamental difference between a Brønsted–Lowry base and a Lewis base is that the former donates electrons to a proton and the latter donates electrons to another atom: 1, 2, 4, 5, 9, 11, 12, 15, 20, 25, 26, 29, 30.
- Curved arrows are used to indicate electron flow from a source of high electron density to a point of low electron density: 3, 10, 13, 14, 24, 25, 26, 30.
- Acid strength in an acid X–H is largely determined by the stability of the conjugate acid and base and the strength of the X–H bond. For example, HI is a stronger acid than HF due to a weaker X–H bond and charge dispersal in the larger anionic conjugate base. An acid such as perchloric acid is stronger than sulfuric acid due to charge dispersal in the conjugate base that results from resonance: 6, 17, 18, 21, 26, 27, 35, 36, 42, 44, 45, 46.
- Base strength is largely determined by the stability of the conjugate acid and base and the electron-donating ability of the basic atom. Water is a stronger acid than ammonia because the hydroxide ion is more stable than the amide anion and the OH bond is weaker than the NH bond. Ammonia is a stronger base than water because oxygen is more electronegative than nitrogen, and the ammonium ion is more stable than the hydronium ion: 7, 16, 19, 22, 23, 31, 32, 37, 38, 39, 43, 46.
- Reaction of a Lewis acid and a Lewis base leads to an ate complex as the product: 21, 22, 24, 29, 30, 31, 33, 34.

Homework

2.26 For each of the following, write the complete acid–base equilibrium and then write out the equation for K_a. For each reaction, draw a curved arrow for the acid–base pair to indicate the flow of electrons during the reaction:
(a) $HNO_3 + H_2O$
(b) $H_2O + NH_3$
(c) $HBr + NH_3$
(d) $HCl + H_2O$
(e) $Cl_3C–H + NaNH_2$

2.27 Briefly explain why HBr (pK_a of −9) is more acidic than HCl (pK_a of −7).

2.28 When HCl is dissolved in water, an equilibrium expression can be written. Write out the K_a expression excluding water, as you would do in a calculation.

Now write out the expression but include the concentration term for water. Why is excluding the water term a problem?

2.29 Draw the structure of the conjugate base formed if each of the following reacts as an acid:
(a) NH_4+
(b) H_3COH
(c) HNO_3
(d) HBr
(e) NH_3
(f) H_3CH

2.30 Boron trifluoride is considered to be a Lewis acid. What product is formed when BF_3 reacts with ammonia? Which atom of ammonia is the base? Draw out a reaction with a curved arrow to show the flow of electrons for this reaction.

2.31 Explain why ammonia is a stronger Lewis base than water.

2.32 Phosphine (PH_3) is the phosphorus analog of ammonia. Which is the stronger Lewis base, ammonia or phosphine? Explain your answer.

2.33 What is an ate complex?

2.34 Draw out the structure of the complex expected to form when $AlCl_3$ reacts with dimethyl ether, which has the structure $H_3C–O–CH_3$ (see Chapter 5, Section 5.6.4).

2.35 Briefly explain why methane (CH_4, which is a carbon with four 2-electron covalent bonds; pK_a of 48) is a much weaker acid than ammonia (pK_a of 38).

2.36 For each series, indicate which is likely to be the strongest acid. Justify your choice:
(a) CH_4 CH_3NH_2 CH_3OH NaF
(b) HF   HCl   HBr   HI

2.37 Explain why the fluoride ion is a stronger base than the iodide ion.

2.38 Explain why the nitrate anion (NO_3^-) is a weaker base than the hydroxide ion. You must draw out the structure of the nitrate anion to answer this question.

2.39 Which of the following is the more basic: H_3C^- or F^-?

2.40 Determine the pK_a for each of the following:
(a) $K_a = 1.45 \times 10^5$
(b) $K_a = 3.6 \times 10^{-12}$
(c) $K_a = 6.7 \times 10^{-31}$
(d) $K_a = 18$
(e) $K_a = 3.8 \times 10^{14}$

2.41 Which of the following is likely to have the smallest pK_a in a reaction with H_2O?
(a) HCl
(b) HF
(c) H_2O
(d) NH_3

2.42 Which of the following is likely to have the largest K_a in a reaction with H_2O?
(a) HCl
(b) HF
(c) H_2O
(d) NH_3

2.43 Identify the basic atom in NaF. Justify your choice.

2.44 Explain the following data: The pK_a of HF is 3.17 and the pK_a of HI is −10. *Just saying that one is more acidic is **not** correct.* Why is one more acidic?

2.45 Consider the two compounds **A** and **B**. One has a pK_a of about 4.8, whereas the other has a pK_a of about 46. Which is more acidic? Why?

2.46 When comparing the reaction of formic acid with NaOH and methanol with NaOH (both structures are shown), the equilibrium concentration of the conjugate base of formic acid is greater than the equilibrium concentration of the conjugate base of methanol.
(a) Explain why the concentration of the conjugate base from methanol is lower than from formic acid.
(b) When comparing the two reactions, how does the greater concentration of conjugate base from formic acid influence K_a?
(c) Which is likely to be more acidic in a reaction with NaOH, formic acid or methanol?

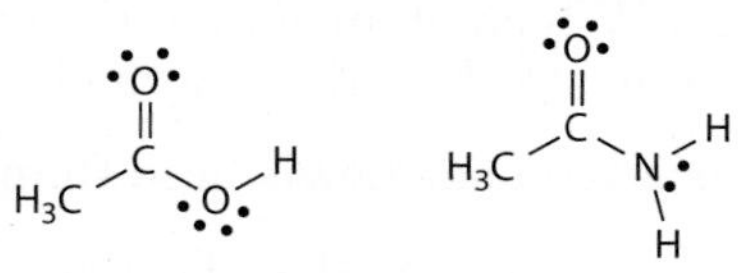

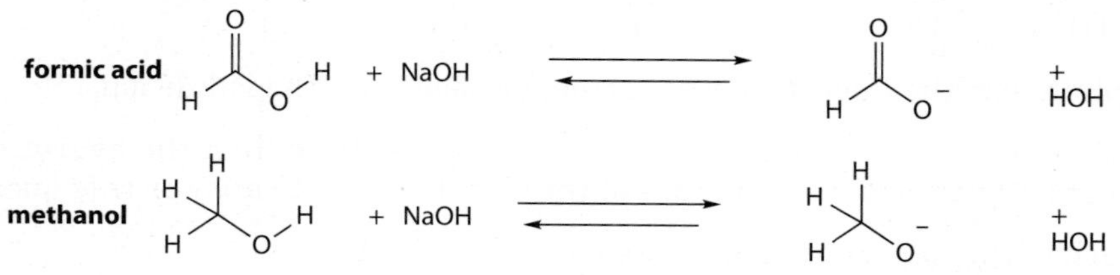

Bonding

The most fundamental concept in organic chemistry is the nature of the bond between two carbon atoms or between carbon and another atom. For the most part, these are covalent bonds to carbon, although ionic bonds will be seen. However, most common organic molecules are characterized by the presence of covalent bonds. The place to begin a study of organic chemistry is with chemical bonds of carbon to carbon, carbon to hydrogen, or carbon to other atoms.

To begin this chapter, you should know the following from a general chemistry course:

- **the electronic configuration of elements in the first two rows of the periodic table**
- **the shape of s- and p- atomic orbitals and how they relate to electronic configuration**
- **the differences among s-, p-, and d-orbitals**
- **the difference between an ionic bond and a covalent bond**
- **a sense of difference in the size of the elements and their respective ions**
- **that bonds are made of electrons**
- **the concept of electronegativity**

This chapter will build on the principles just listed, in order to define covalent bonds and then extend this definition to discuss polarized covalent bonds. The hybridization model of covalent bonding will be introduced, as well as the concept of functional groups and the shape of molecules of the first two rows of the periodic table.

When you have completed this chapter, you should understand the following points:

- **s-Orbitals are spherically symmetrical and p-orbitals are "dumbbell" shaped.**
- **Electrons are found in orbitals at discrete distances from the nucleus in an atom.**
- **Electrons in the bond of a molecule are located between two nuclei and are at energy levels different from that in an unbonded atom.**
- **Ionic bonds are formed by electrostatic attraction of two atoms or groups that have opposite charges.**
- **Covalent bonds are made of two electrons that are mutually shared between two atoms.**
- **Mixing atomic orbitals forms hybrid molecular orbitals, and the number of s- and p-orbitals used to form the hybrid determines the hybridization (sp^3, for example).**
- **Organic molecules generally have a backbone of carbon–carbon covalent bonds.**
- **Electronegativity for an atom increases to the right and up the periodic table.**
- **Polarized bonds are formed when two atoms are bonded together and one is more electronegative.**
- **Polarized covalent bonds are generally weaker than nonpolarized covalent bonds.**
- **Reactions are driven by making and breaking bonds, which release or require energy.**
- **The VSEPR model is used to predict the three-dimensional shape around an individual atom.**
- **There are biological applications of covalent bonds.**

3.1 The Elements

Organic molecules are made of carbon and hydrogen atoms, for the most part. Although many elements in the periodic table form bonds to carbon, most compounds discussed in this book have bonds between carbon and elements found in the second period of the periodic table (Li → F). However, there will be many examples of bonds formed by carbon to other atoms. Bonding is the key issue in this chapter, but to understand a chemical bond, information must be collected about the elements that make up that bond.

3.1.1 What Is Electronic Configuration?

Atoms are discrete entities that differ from one another by the number of protons, neutrons, and electrons that make up each atom. Protons and neutrons are found in the nucleus, of course, but reactions involving organic molecules do not involve transfer, gain, or loss of protons or neutrons. Chemical reactions involve the transfer of electrons, which are the important non-nuclear constituents of an atom.

It is known that the motion of electrons has some characteristics of wave motion. The motion of an electron is expressed by a wave equation, which has a series of solutions; each solution is called a *wave function*. Each electron may be described by a wave function whose magnitude varies from point to point in space.[1] The particular solution to the so-called **Schrödinger wave equation**, for a given type of electron, is determined by the following equation[2]:

$$H\psi = E\psi \tag{3.1}$$

where H is a mathematical operator called the Hamiltonian operator, E is the numerical value for the energy, and ψ is a particular wave function.

The Hamiltonian operator is the general form of the kinetic and potential energies of the system. If certain assumptions are made, it is possible to use the Schrödinger equation to generate different wave functions. The relationship between orbitals and the Schrödinger equation is apparent when its solutions are represented as the waves shown in Figure 3.1a[3] for various values of ψ that correspond to different energies. The first five solutions to the wave function are shown in Figure 3.1a, and each wave is correlated with the space volume pictorial representations in Figure 3.1b. The amplitude of the wave is the wave function (ψ) and it has a ***maximum (represented by +)*** and a ***minimum (represented by –)***, and each point in space can be represented by spatial coordinates (x,y,z).[4] The point at which the wave changes its phase is referred to as a ***node***. Using Cartesian coordinates for a point, the wave function is described by $\psi(x,y,z)$, the position of the electron in space.

The Heisenberg uncertainty principle states that the position and momentum of an electron cannot be simultaneously specified. *It is only possible to determine the probability that an electron will be found at a particular point relative to the nucleus.* Because the exact position of the electron is unknown (there is uncertainty as to its position), the probability of finding the electron in a unit volume of three-dimensional space is given by $/\psi(x,y,z)/^2$. The position is expressed as $/\psi/^2 d\tau$, which is the probability of an electron being in a small element of the volume $d\tau$.[1] This small volume can be viewed as a charge cloud if it contains an electron. *The charge cloud represents the region of space where we are most likely to find the electron in terms of the (x,y,z) coordinates.* These charge clouds are known as orbitals. Plotting ψ versus distance from the nucleus in Figure 3.1 leads to the familiar s-, p-, and d-orbitals. As the distance from the nucleus increases, the value of the wave function becomes smaller.

Nodes are generated as the amplitude of the wave changes from + to –. The wave functions for electrons of different energies can be described in terms of

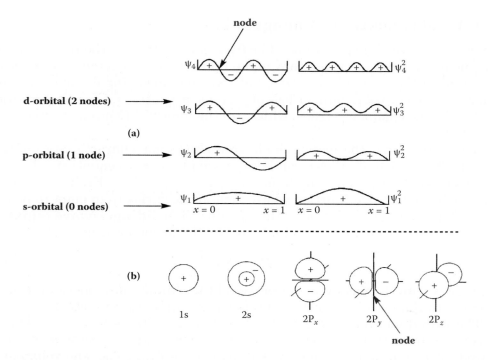

(a)

(b)

Figure 3.1 Wave function solutions and orbital picture for those wave functions. (Tedder, J. J., Nechvatal, A., and Pitman, M. A. 1985. *Pictorial Orbital Theory,* 2. Essex, England: Longman Science & Technology. Pearson Limited. 1985. Reproduced with Permission.)

the number of nodes. If there are zero nodes, there is no change in sign and the wave function remains positive. The electron cloud is drawn as a spherically symmetrical orbital—an s-orbital in Figures 3.1 and 3.2. When there is one node, a p-orbital is described with a "dumbbell" shape (see Figures 3.1 and 3.2). In the (x,y,z) coordinate system, the single node could be in the x plane, the y plane, or the z plane. All three are equally likely.

Therefore, three identical p-orbitals must be described: p_x, p_y, and p_z. In other words, there are three p-wave functions that describe three different places where an electron has the same energy (the p-orbitals are said to be triply *degenerate*). When there are two nodes, d-orbitals are generated. There are five ways to generate wave functions that have two nodes in the (x,y,z)

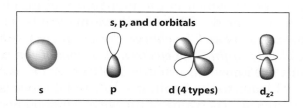

Figure 3.2 Common representations of the s- and p-orbitals from Figure 3.1, along with d-orbitals.

coordinate system, leading to d_{xy^-}, d_{xz^-}, d_{yz^-}, $d_{x^2-y^2}$, and d_{z^2}-orbitals. The shapes of four of the d-orbitals along with the d_{z^2}-orbital are shown in Figure 3.2. For the most part, f-orbitals will not be used in this book.

A description of electrons in atoms and molecules usually involves the type of orbital in which those electrons are found (s-electrons, p-electrons, d-electrons, and f-electrons). Orbitals with lower principal quantum numbers are more stable (lower in energy). This means that an electron in a 1s-orbital is lower in energy than an electron in a 2s-orbital, which is lower than an electron in a 3s, etc. Referring to Figure 3.1b, it is clear that the "size" of the 2s-orbital is greater than that of the 1s-orbital. This difference represents the fact that the 2s-electrons are found further from the nucleus than a 1s-orbital. In other words, it takes more energy to "hold" a 2s-electron to the nucleus than to "hold" a 1s-electron.

3.1 Based on Figures 3.1 and 3.3, describe the shape of the s-orbital.

The energy levels in Figure 3.1 represent the region in space where electrons are found relative to the nucleus. There are several quantum levels in atoms, particularly in high atomic mass elements. This means that there are different energy levels associated with each type of electron shell, so there are different types of s-orbitals: 1s, 2s, 3s, 4s, etc.; they are similar in shape but differ in energy (their relative distance from the nucleus). Likewise, there are 2p-, 3p-, and 4p-orbitals, and 3d-, 4d-, 5d-, and 4f-, 5f-, 6f-orbitals. The description of the way electrons are distributed in the atomic orbitals of an element is known as its *electronic configuration*. The electronic configuration of each atom will show the type, quantum level, and number of electrons.

Chemical reactions involve the gain, loss, or transfer of electrons—particularly *valence electrons,* which are those found in the outermost orbitals (those furthest away from the nucleus); they are more weakly bound than electrons in orbitals closer to the nucleus. In order to describe reactions, it is convenient to describe atoms in terms of their electronic configuration, which describes the total number of electrons, the orbital where the electrons can be found, and the number of electrons in each orbital. Clearly, a convenient organization of the elements is required in order to find the information we need. This organization is known as the periodic table of the elements.

The periodic table is arranged more or less by chemical reactivity, using the number of electrons in the outermost shell of the element and the energy of those outermost (valence) electrons. In effect, elements are arranged according to their valence orbitals. The periodic table currently lists 109 elements. The first attempt to categorize elements in this manner was by Dmitri Ivanovich Mendeleev (Russia; 1834–1907), in the nineteenth century. The first row of elements (H, He) have only the spherical s-orbitals, but the second row (Li, Be, B, C, N, O, F, He) has the 1s-orbital and the 2s- and 2p-orbitals are in the outermost shell. The third row introduces 3s- and 3p-orbitals, and d-orbitals appear in the fourth row. Each shell will have one s-, three p-, five d-, and seven f-orbitals (1, 2, 3, 4), and the d- and f-orbitals accept more electrons or give up more electrons than a p-orbital. Indeed, elements with d- and f-orbitals are characterized by multiple valences. This stands in sharp contrast to the second-row elements, which have only 2s- and 2p-orbitals and a single valence.

Organic chemistry is the study of carbon compounds and most of the molecules studied in this course will contain C–H, C–C, C–O, C–N or C–halogen bonds. Therefore, the focus will be on s- and p-orbitals, with some discussion of d-orbitals; f-orbitals are not relevant to most discussions. The d-orbitals are relevant in organic molecules that contain C–P and C–S bonds, for a few C–metal bonds, and for bonds to halogen (Cl, Br, I). When carbon bonds to elements on the left side of the periodic table (Li, Mg, Na, K), highly reactive organometallic compounds are formed. The term **organometallic** means *molecules composed of a metal with one or more attached carbon atoms*. Other organometallic compounds that use transition metals, including Cu, Pb, Sn, and Zn, will be discussed for some specialized reactions, and Pt, Pd, Cr, Ti, Mn, Ag, Mo, Os, Ir, and Hg are important metals that will be used in organic chemistry.

3.2 What is the total number of electrons for an atom in a fully filled second row? What is that atom?

The electronic configuration of each element is described by the order in which the various orbitals fill as the number of electrons increases for each element of the periodic table. *Each individual orbital can hold no more than two electrons.* Electrons have the property of spin, which is associated with a magnetic dipole. The spin quantum number was introduced in general chemistry, and it is one of four quantum numbers where self-rotation of the electron gives rise to an angular momentum vector. In other words, each electron will have spin and the symbol ↑ is used to indicate an electron with a certain spin quantum number. A single orbital containing two electrons, also known as a ***filled orbital***, is represented by two opposed arrows (↑↓). This symbol indicates that when an orbital contains two electrons, those two electrons are spin paired (as electrons in Figure 3.3). Note that if two electrons occupy one orbital, spin

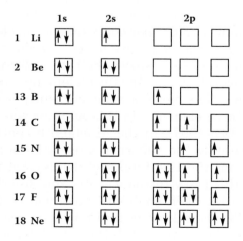

Figure 3.3 Filled orbitals for elements in the second row.

pairing is lower in energy than if two electrons of the same spin are forced to occupy the same orbital.

One of the concepts learned in general chemistry is the **Pauli exclusion principle**, which states that *if there are several orbitals of equal energy (such as the three 2p-orbitals), each orbital will fill with one electron before any orbital contains two.* This principle is shown in tabular form in Figure 3.3 for lithium, beryllium, boron, carbon, nitrogen, oxygen, fluorine, and neon as the 2s- and three 2p-orbitals are filled. In Figure 3.3, a "box" is used as a mnemonic representation of the order in which orbitals fill, and arrows are used to represent the order in which the electrons are added. Attempting to fill one orbital with two spin-paired electrons before filling the next orbital would require more energy. Electrons have like charges, and two electrons will repel if they are in the same orbital. In other words, adding two electrons to two 2p-orbitals is lower in energy than adding two electrons to a single 2p-orbital, so electrons "fill" orbitals to conserve energy. This concept of filling orbitals with electrons in ascending order of orbital energy until all available electrons have been used is known as the **aufbau procedure**. Note that *aufbau* is a German word that translates to "building up." The order in which electrons fill is generalized by the mnemonic:

$$1s \rightarrow 2s \rightarrow 2p \rightarrow 3s \rightarrow 3p \rightarrow 4s \rightarrow 3d \rightarrow 4p \rightarrow 5s \rightarrow 4d \rightarrow 5p \rightarrow 6s \rightarrow 4f \rightarrow 5d \rightarrow 6p$$

Note that orbitals generally fill s $\rightarrow$ p, but in the fourth row, the 4s-orbital fills before the 3d-orbitals (there are five 3d-orbitals). Likewise, the 5s-orbital fills before the 4d-orbitals and the 4f-orbitals do not fill until after the 5p- and 6s-orbitals, but before the 5d-orbital.

These tools are used to determine the electronic configuration of any element in the periodic table. In the first row of the table, hydrogen has an electronic configuration $1s^1$ and helium is $1s^2$. The "1" represents the row of the periodic table, the "s" represents the orbital, and the superscript "1" represents the number of electrons in that orbital. With two electrons in the 1s-orbital of helium, the first row is filled. The next element (lithium) begins the second row and the 2s-orbitals begin to fill. The electron configuration of lithium is $1s^2 2s^1$. This is followed by beryllium ($1s^2 2s^2$), which fills the 2s-orbital. The next element begins to fill the 2p-orbitals and boron has the configuration $1s^2 2s^2 2p^1$, carbon is $1s^2 2s^2 2p^2$, and continues to the noble gas neon with a configuration $1s^2 2s^2 2p^6$. This fills the second row with a total of two electrons in the 1s-orbital and eight electrons in the 2s-/2p-orbitals. In the third row, sodium begins to fill the 3s-orbital ($1s^2 2s^2 2p^6 3s^1$) and continues to argon, with a configuration of $1s^2 2s^2 2p^6 3s^2 3p^6$. The electronic configuration of elements through at least the first three rows should be known, because this will help to understand bonding in the elements that will be encountered in organic molecules.

3.3 Write the electronic configurations of N, O, P, and S.

3.4 Explain why neon has a configuration of $1s^2 2s^2 2p^6$, where six electrons are in the 2p shell.

3.5 Name the elements in the third row of the periodic table.

3.1.2 The Nature of Atomic Orbitals

The term "chemical reaction" typically refers to the transfer of electrons from one atom to another. It is therefore important to understand the electronic character of each atom or bond. Electrons in an element or in a bond between two atoms are assumed to reside in orbitals. As electrons are added to s- or p-orbitals in Figure 3.3, the simple pictorial images of atomic orbitals shown in Figure 3.2 represent atomic orbitals. In other words, the s-, p-, d-, and f-orbitals are used for electrons in atoms such as elemental carbon, nitrogen, sulfur, etc. A bond occurs between two atoms, and the electrons in that bond are associated with a molecule, not a single atom. Because bonds are found in molecules, the orbitals used to form a bond in a molecule are *different* from the orbitals found in elemental atoms. Therefore, it is important first to understand atomic orbitals before extending those principles to the bonds in a molecule.

To "construct" an atom, there is a 1s-orbital and then a 2s-orbital. As mentioned before, there are three 2p-orbitals in the second row of equal energy and they have the same shape. It has also been established that the space volume for each electron is described in a Cartesian (three-coordinate) system, x,y,z. If the 1s-, 2s-, and all three 2p-orbitals are superimposed onto the tricoordinate system, the result is Figure 3.4, which shows the spatial relationships of the orbitals in an elemental atom, one to the other.

The 1s-orbital is represented by the spherical green dot at the center. The 2s-orbital is represented by the black circle (meant to represent a sphere) that

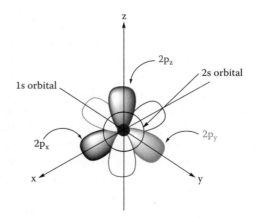

Figure 3.4   Orbital diagram with 1s-, 2s-, and 2p-orbitals.

is larger than the 1s-sphere, showing its greater distance from the nucleus. The nucleus is the convergence point of the tricoordinate system. The three 2p-orbitals in the figure are labeled as **red** for the $2p_z$-, **blue** for the $2p_x$-, and **yellow** for the $2p_y$-orbitals. As the distance from the nucleus increases, less energy is required to remove an electron that would reside in that orbital. This means that it is easier to lose an electron from a 2s-orbital or a 2p-orbital than from a 1s-orbital. The purpose of this simplistic picture is to give a mental image that a 2p-electron is more easily removed than the 1s- or 2s-electrons because it is further from the nucleus. It is also intended to show that the electrons are closely associated with the atomic nucleus.

3.6 Is it easier to lose an electron from a 3p-orbital or from a 4s-orbital?

3.2 What Is a Chemical Bond? Ionic versus Covalent

A chemical reaction between atoms or groups of atoms will usually produce new combinations of atoms held together by what is called a chemical bond. When one element reacts with another it does so via its electrons, not by the protons and neutrons in the nucleus. The resulting bond between the atoms is composed of two electrons. Two major types of bonds will be considered. A **covalent bond** is formed by the mutual sharing of valence electrons between two atoms. In other words, sharing electron density holds the atoms together. An **ionic bond** is formed by transfer of electrons from one atom to another, resulting in ions (+ and –) that are held together by electrostatic attraction.

A typical ionic bond is found in LiF, where the positively charged Li is electrostatically bound to the negatively charged F. To participate in an ionic bond, an atom or group of atoms must be electron rich or electron poor and have a charge of – or +. The key question in this example is whether Li gains or loses electrons as LiF is formed, and the same must be asked of F. In one sense, this is a bogus question because fluorine exists as F–F not as F, and the formation of Li–F is more complicated than simply mixing a Li atom with a F atom. However, it is reasonable to ask in such a reaction if Li is more likely to donate electrons or accept electrons, and a similar question can be asked for F.

To assist in visualizing electron transfer, it is useful to refer to the **octet rule**, which states that in the second row, a maximum of eight electrons can occupy the valence shell (the Ne configuration). In the first row, a maximum of two electrons can occupy the valence shell (the He configuration), and the second row can accommodate a maximum of eight electrons to give the Ne configuration (a total of 10 electrons). This is an important consideration because the noble gas configurations (having a filled outer electronic shell) are particularly stable (low in energy, where more stable is taken as an indication of low reactivity). It is therefore reasonable to assume that *there is an energetic preference for transferring electrons to attain the noble gas configurations.*

If one 2s valence electron is lost from lithium during electron transfer processes, the result is Li^+ with a $1s^2$ configuration (the He configuration). This is energetically more favorable than lithium gaining an extra electron to give Li^-, with a $1s^2 2s^2$ configuration. *During electron transfer processes, transfer of the negatively charged electron from elemental Li does not lead to helium, but to a positively charged lithium ion (Li^+) with a $1s^2$ configuration.* In other words, Li^+ is more stable than Li^-.

A similar electron transfer process can be applied to the fluorine atom in LiF. If electron transfer removes an electron from F to give F^+, the electronic configuration is $1s^2 2s^2 2p^4$. If electron transfer adds one electron to a fluorine atom, however, the result is F^-, which has the $1s^2 2s^2 2p^6$ configuration (the Ne configuration). *Once again, addition of an electron to fluorine does not give Ne, but rather the fluoride ion F^- with the configuration $1s^2 2s^2 2p^6$.* From an electronic standpoint, F^- is more stable than F^+. Based on these analyses, it is reasonable to assume that an atom with a valence electronic configuration such as Li (group 1) will lose an electron during an electron transfer process, and an atom with a configuration such as fluorine (group 17) will gain one electron. *Note that the energy required for the loss of one electron from an atom is called its **ionization potential**. The energy required for the gain of one electron into an atom is called its **electron affinity**.* This means that the energy gained or lost for an atom is a measurable quantity. Electron transfer to form ions is the basis for the known ionic bonding in many molecules composed of alkali metals (groups 1 and 2) and halogens (in group 17): LiF, NaCl, KBr, NaI, etc.

3.7 Make an educated guess concerning the bonding in NaCl: ionic or covalent?

3.8 Draw the electronic configuration of neon.

3.9 Is it easier for rubidium to lose one electron or gain one electron?

As mentioned earlier, carbon has an electronic configuration of $1s^2 2s^2 2p^2$ with four valence electrons. If a bond is formed between a carbon atom and another, is the electron transfer model used for LiF applicable? There are many examples of molecules that contain carbon atoms, and the bonds to those carbon atoms are usually not ionic. The bonds in molecules that contain carbon are usually formed by *sharing* electrons with another atom in what is known as a ***covalent bond***, rather than by complete transfer of electrons to form an ionic bond.

A covalent bond is defined as the mutual sharing of electrons between two atoms. Why? Carbon has the electronic configuration $1s^2 2s^2 2p^2$, so there are four electrons in the outmost shell. To achieve the helium configuration by electron transfer ($1s^2$), four electrons must be lost (*fourth* ionization potential). To achieve the neon configuration ($1s^2 2s^2 2p^6$), four electrons must be gained (*fourth* electron affinity). In both cases, the energy requirements are absurd. If one electron is lost, carbon would generate C^+ ($1s^2 2s^2 2p^1$), whereas if one electron is gained, C^- would be formed ($1s^2 2s^2 2p^3$). In both cases, these are high-energy species, and such electron transfer reactions do not easily occur with elemental carbon. Therefore, it is unlikely that an elemental carbon atom will form bonds by the same type of electron transfer as found with LiF.

The preceding discussion means that for reactions in this book, elemental carbon such as that found in diamond or graphite will *not* be converted into carbon-containing molecules. Instead, atoms in a molecule that contains carbon will react with another molecule to change the groups attached to carbon. Therefore, electron transfer is not with elemental carbon but rather with a carbon-containing molecule. This distinction will be elaborated on in later sections, but to say that carbon "transfers electrons and shares electrons" does **not** mean that atomic carbon will react with other atoms or other compounds or form a bond to become covalent. If electrons are found in orbitals, *the orbitals for atomic carbon are different from the orbitals for covalent carbon in a molecule; carbon in the element is different from carbon in a molecule.* The distinction is that the bonding in organic molecules is covalent and it involves carbon, but elemental carbon is not readily converted to such compounds. This means the direct conversion of elemental carbon (graphite or diamond, for example) into an organic molecule such as methane is unlikely. Both graphite and diamond are known for their stability, not for their ability to react with things. The fundamental conclusion is that the placement of electrons in a bond to carbon in a molecule (molecular carbon) is different from that in elemental carbon. This bonding is described in the next section.

It is important to point out that there are many examples of carbon atoms that participate in ionic bonds. This means that there are molecules that contain C^+ and molecules that contain C^-. An ionic carbon (+ or −) is generated by breaking bonds in a molecule rather than by electron transfer to atoms of elemental carbon. Such reactions will be described in Chapter 10 (Section 10.2) and Chapter 11 (Section 11.4).

3.3 The Covalent Carbon–Carbon Bond

Elemental carbon in the form of diamond or graphite is well known. Carbon is also found in carbon dioxide (CO_2), which is a molecule. Carbon is found in all living things, and millions of molecules contain carbon. As pointed out earlier, the orbitals for carbon in a molecule are different from the orbitals for carbon in diamond or graphite. It is known that the fundamental skeleton of the molecules containing carbon (organic molecules) is made up of carbon atoms held together by covalent bonds between the carbon atoms. Organic chemistry is described as the chemistry of carbon, literally and figuratively. This section will discuss factors that are important for understanding bonding to carbon.

When electrons are concentrated in orbitals on a single atom, the electrons are said to be "localized" on that atom (see Figure 3.4). When an atom is part

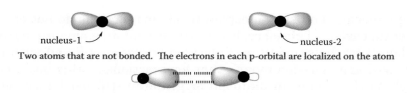

Two atoms that are not bonded. The electrons in each p-orbital are localized on the atom

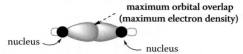

As the atoms are brought within bonding distance, the orbitals of one atom must be directed towards the orbital of the other in order to share the electrons to form a covalent bond

maximum orbital overlap
(maximum electron density)

nucleus   nucleus

When the covalent bond is formed, the orbital on each atom has been distorted - the orbital is directed towards the other atom. The electrons are now shared between the atoms and the greatest concentration of electron density is between the nuclei (**black** atoms)

Figure 3.5   A mnemonic to describe overlapping orbitals used to form a covalent bond.

of a covalent bond, however, it **shares** electrons with the other atom; that is, *the electrons in the bond are not localized on one atom, but rather are shared by both atoms and concentrated between the atoms.* Figure 3.5 illustrates this concept, where the electrons that constitute the bond are concentrated *between both nuclei* rather than localized on one or the other.

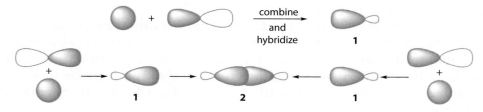

For an isolated atom of a given element, the electrons are in atomic orbitals (Figures 3.1, 3.2, and 3.4), but in a molecule, the electrons reside in different orbitals known as **molecular orbitals**. One way to visualize a molecular orbital is to "mix" atomic orbitals to indicate the directionality of the new orbital toward the atom to which the bond is formed. Mixing a spherical s-orbital and a dumbbell-shaped p-orbital, for example, leads to a hybrid orbital, **1**. This new molecular orbital is more or less a hybrid of the s- and p-orbitals from which it was derived, with characteristics of both.

Note the directionality of **1**. Assume that a bond is formed between carbon and a hydrogen atom. Remember that the outer shell electrons of carbon are in the 2p-orbitals of the valence shell, and the electron associated with hydrogen is in an s-orbital, which is the valence shell. *In the C–H bond of a molecule, the electron density must be concentrated between the C and H nuclei if there is a covalent bond.* In other words, in a C–H bond, the electrons shared by carbon are directed toward the hydrogen atom. Likewise, the electrons shared by hydrogen will be directed toward the carbon atom. The carbon in this bond mixes an s-orbital and a p-orbital to give what is known as a hybrid orbital, with directionality, as in **1**. The directionality simply reflects the fact that **1** is used to form a covalent bond, and the shared electron density will be directed toward the other atom in that bond.

A covalent bond is described as the overlap of two hybrid orbitals. If **1** represents a hybrid orbital of a molecular carbon atom, then overlap of two carbon atoms to give a C–C bond is represented by orbital picture **2**. This picture represents a covalent bond known as a *sigma bond, which is defined as a bond between two atoms where the electron density is concentrated symmetrically on a line between the two nuclei.*

3.10 Is a covalent C–C sigma bond relatively strong or relatively weak? Why or why not?

When two identical atoms share electrons in a covalent bond, most (but not all) of the electron density is equally distributed between the two nuclei (in the "space" between the two atoms). The electron density that constitutes the bond is located in orbitals between the nuclei. This means that the charge density due to the electrons is greatest between the two nuclei. There is a mathematical probability of finding electron density in locations other than between the nuclei, as reflected in the orbital pictures in Figure 3.5. This orbital picture describes a **sigma bond (σ-bond)**, as defined before. The orbitals used to form covalent bonds in a molecule are called **molecular orbitals** (and are different from atomic orbitals) and they will be discussed in Section 3.4.

The strength of a sigma bond is related to the amount of electron density concentrated between the nuclei. The strongest bonds should occur between two identical atomic nuclei (C–C, for example), where the two electrons are shared equally by both atoms. This means that there is a **symmetrical distribution of electron density** between the two nuclei. There are two electrons in the sigma bond, which is commonly called a **single bond** between the two atoms, as in a carbon–carbon single bond. This statement is repeated because it is extremely important: *Each sigma bond is composed of two shared electrons.* Sigma bonds also occur between carbon and many different atoms, including not only C–H and C–C, but also C–O, C–N, O–H, N–H, C–X (X = halogen), C–S, etc.

3.11 Briefly describe why it is important that electron density lie between the nuclei.

A key parameter in a covalent bond is the **bond length**. Bond length is the measured distance between the nuclei of each atom participating in the bond (the internuclear distance) and it has been measured in angstroms (Å), where $1 \text{ Å} = 1 \times 10^{-10}$ m. Modern measurements are reported in picometers (pm; 1×10^{-12} m, or 0.01 Å). Therefore, 1 Å = 100 pm. A longer bond length usually indicates a weaker bond, and a shorter bond should be stronger. This is reasonable if a stronger bond has more electron density between the nuclei, and a weaker bond has less electron density between the two nuclei (see Figure 3.5). A longer bond has less electron density per unit length than if the bond is short, meaning it is easier to disrupt the bond (easier to break); it is weaker. Conversely, a shorter bond will have more electron density per unit length, and it is stronger.

Bond length is the result of a balance between the repulsion of the two positively charged nuclei (one from each atom forming the bond) and the attraction of negatively charged valence electrons for the positively charged nucleus. It is also a function of the size of the atoms. *The internuclear distance between two atoms of larger atomic radius will be greater than that between two atoms of smaller atomic radius.* The size of each nucleus (number of protons and neutrons), the number of valence electrons, and the nature of the orbitals in which they reside determine the bond length. A simple model illustrates bond length. If two nuclei are widely separated, the energy of repulsion and attraction between them is zero. Now imagine a model where two nuclei approach each other. As they approach, the electrons of one atom attract the nuclei of the second atom once they are close enough. This force will bring the two nuclei closer together. At some critical point, however, the two nuclei are too close and the two like charges will repel, forcing the atoms apart. *The energetic point at which these two forces balance is said to be the bonding distance or bond length.*

3.12 Based on the relative size of the atoms involved, guess which is the shorter bond: C–O or C–C.

All covalent bonds are not the same. Bonds between different nuclei will have different strengths and the distance between the nuclei will be different. The bond length and bond energy of a C–C bond are different from those of an O–O bond or a C–O bond. All carbon–carbon bonds are not the same. A C–C bond such as $R_3C–CH_3$ (the R represents any carbon group) will be different from the C–C unit in $R_2CHCH_2C–CH_2R$, for example. This fact is clearly shown by examining the bond lengths of the units shown in Table 3.1.[5]

The molecule types listed (alkane, ether, alcohol, etc.) are introduced in Chapter 4 and Chapter 5, Section 5.7.). Note that the shortest bonds in Table 3.1 that involve carbon are the C–C and C–H bonds. When an atom other than carbon or hydrogen is attached to carbon, the bond is generally longer and often weaker (this will be explained in Section 3.7). Both the OH and NH bonds are rather short, relative to bonds of various atoms to carbon. The O–O bond (a peroxide) is not particularly long, but it is a very weak bond.

An organic molecule is usually composed of many atoms, and the atoms are connected by covalent bonds. For an atom of a pure element that is not involved in bonding, the electrons reside in atomic orbitals closely associated only with that atom at discrete distances from the nucleus. These atomic orbitals are the s-, p-, and d-orbitals discussed in Section 3.1.2. In a molecule, the atoms of each element share electron density with another atom, and the electrons reside in orbitals that are different from the familiar atomic s-, p-, or d-orbitals. In other words, ***molecular orbitals*** are different in energy when compared to the corresponding atomic orbitals. Note that elements that are categorized as diatomic, including H_2 or F_2, and such entities are molecules, so the H–H and F–F bonds are covalent bonds. The shared electron density associated with one atom in the bond is directed toward the other atom in the bond.

3.13 Is the C–C bond more likely to be covalent or ionic?

Table 3.1
Typical Bond Lengths for Single Covalent Bonds

Bond	Molecule Type	Bond Length (Å)	Bond Length (pm)
C–I	Alkyl halide	2.139	213.9
C–Br	Alkyl halide	1.938	193.8
C–Cl	Alkyl halide	1.767	176.7
C–C	Alkane	1.541	154.1
O–O	Hydrogen peroxide (HOOH)	1.48	148
C–N	Amine	1.472	147.2
C–O	Ether	1.426	142.6
C–F	Alkyl halide	1.379	137.9
C–H	Alkane	1.094	109.4
N–H	Amine	1.012	101.2
O–H	Alcohol	0.97	97
O–H	Water	0.958	95.8

Source: Lide, D. R., ed. 2006. *CRC Handbook of Chemistry and Physics,* 87th ed., 9-1–9-15. Boca Raton, FL: Taylor & Francis.

3.4 Molecular Orbitals

Hydrogen normally exists as a diatomic molecule, H_2, with a covalent bond connecting the two hydrogen atoms. Section 3.3 established that the orbitals of a molecule are different from those of an individual atom. Is it possible to use the hydrogen atom and its atomic orbitals to make predictions about the orbitals in a molecule? In one model, the atomic orbitals of the atom are mixed and combining two such atoms may lead to a molecule. This idea of mixing atomic orbitals to form a molecular orbital is called the ***linear combination of atomic orbital (LCAO) model,*** and in some cases it helps predict the relative energy of these molecular orbitals.

Beginning with two individual hydrogen atoms; each has one electron in a 1s-orbital (see Figure 3.6). If the 1s-orbitals from each hydrogen atom (there are two) are mathematically "mixed," two new orbitals are formed. The new orbitals are associated with the dihydrogen molecule (two molecular orbitals for H–H), *but they must be of a different energy* than the orbitals from the atoms. This statement means that the electrons shared between the two atoms are no longer associated with an individual atom, so they must be at a different energy relative to each hydrogen nucleus. This difference is represented by two

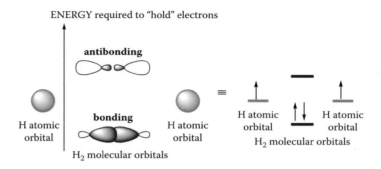

Figure 3.6   Molecular orbital diagram for diatomic hydrogen.

different orbitals at different energy levels. Note that *mixing two atomic orbitals generates two molecular orbitals, so orbitals are not gained or lost by this process.* Each orbital represents an energy level, and because the orbitals must be of a different energy, one orbital must be lower in energy and one is higher when compared to the atomic orbitals.

Each orbital is shown as a "line" that corresponds to a particular energy in Figure 3.6. Each orbital of hydrogen has a single electron, represented by the *vertical arrow*. The linear combination of two atomic orbitals generates two molecular orbitals, one of higher energy and one of lower energy. The two electrons must be added to the molecular orbitals, beginning with the lowest energy orbital and working toward the higher energy orbitals (the valence orbitals). It only makes sense that the electrons in the molecule will distribute themselves in the lowest energy arrangement, which leads to an electron pair in the lower orbital, as shown.

In other words, after adding the first electron, spin pairing in the lower energy orbital is lower in energy than adding the second electron to the higher energy orbital. This orbital is said to be a **bonding molecular orbital** and implies that the molecule uses the two electrons to form a bond; the electrons are *shared* between the two atoms. The higher energy orbital is an empty molecular orbital (no electrons), but it is often referred to as an **antibonding orbital**. An antibonding orbital is an energy level that normally has no electrons in it and is populated only when sufficient energy is provided to move an electron or if an electron is added from an external source.

3.5 Tetrahedral Carbons and sp³ Hybridization

Mixing the atomic s-orbitals of two hydrogen atoms to form hybrid molecular orbitals seems to predict the bonding in diatomic hydrogen. There are limitations, however, and if the two atoms to be mixed have both s- and p-orbitals, problems arise. When the atomic orbitals of two carbon or two oxygen atoms are mixed using the LCAO method, the results are disappointing, and simply mixing the 1s-, 2s-, and 2p- atomic orbitals from each carbon atom does ***not*** adequately describe the molecular orbitals in a carbon–carbon bond. This section will describe models used that give a more accurate picture of bonding in organic compounds.

Examining the electron configuration of atomic carbon reveals that there are four valence electrons (two electrons in the 2s-orbital and two electrons in one each of two 2p-orbitals. In order to achieve the stable noble gas configuration, carbon forms four covalent bonds. Sharing electrons with four other atoms gives carbon the eight electrons it requires. Because carbon requires four electrons, it shares one electron from four other atoms and provides one electron to each of those four atoms. This gives a total of eight electrons around carbon, satisfies the octet rule, and leads to a stable organic molecule.

3.14 Using the principle that each atom donates one shared electron to a covalent bond, how many bonds do you think N will form and remain electrically neutral?

3.5.1 Mixing Carbon Atomic Orbitals Does *Not* Predict Real C–C Bonding!

Figure 3.6 shows the LCAO method for generating molecular orbitals of diatomic molecules such as H_2. In real molecules, the atomic orbitals of elemental carbon are not really transformed into the molecular orbitals found in methane (CH_4). Figure 3.6 represents a mathematical model that mixes atomic orbitals to predict molecular orbitals. Molecular orbitals exist in real molecules and the LCAO model attempts to use known atomic orbitals for atoms to predict the orbitals in the molecule. Molecular orbitals and atomic orbitals are very different in shape and energy, so it is not surprising that the model used for diatomic hydrogen fails for molecules containing other than s-orbitals.

When the orbitals for two carbon atoms are mixed to form a C–C bond, the core electrons in the 1s-orbital are omitted because they are not involved in bonding. Two electrons in the 2s-orbital remain, but they are found in only two of the three 2p-orbitals. When these atomic orbitals are "mixed" using the LCAO model, the valence orbitals are derived from the 2s- and 2p-orbitals, generating two "2s-type" molecular orbitals and six "2p-type" molecular orbitals. This model predicts *unshared electrons* in the valence shell (atoms with unshared electrons of this type are called radicals; see Chapter 7, Section 7.4.3).

This solution is *not* correct, as is apparent when the chemical bonds in many real organic molecules are examined. The four C–H bonds in methane (CH_4) are identical, for example, and the four bond lengths and bond strengths are identical. The LCAO model did *not* make a correct prediction; to obtain a correct solution, the mathematical model must be modified. To form a covalent bond, an atom must share electrons with another atom. This means that the position of the electrons relative to the nucleus is different from their positions in the elemental atom; they are located in a molecular orbital.

3.5.2 sp³ Hybridization

Atomic carbon is not changed to molecular carbon in real life, and the LCAO model fails to predict bonding in molecules accurately. **How is the atomic orbital– molecular orbital model modified to accommodate the covalent bonding in carbon?** In reality, a molecule that has covalent bonds is usually prepared from another molecule that also has covalent bonds. The bonding in molecules is understood, so the correct answer is known. Methane forms four identical bonds directed to the corners of a regular tetrahedron! A model has been developed that will transform the 2s- and three 2p atomic orbitals of carbon into the four identical molecular orbitals used for covalent bonding to carbon in a molecule. The 2p-orbitals are higher in energy than the 2s-orbital, and this model conceptually "promotes" an electron in the 2s-orbital to the empty 2p-orbital. This means that the energy of the 2s-orbital and the 2p-orbital is mathematically modified, and the result is four identical molecular orbitals (**3**) that are hybrids of the atomic orbitals. This promotion model is a variation of forming hybrid orbitals discussed earlier.

It is *not* possible to "promote" an electron in a real atom of carbon to give a real organic molecule. A diamond does not promote an electron from one orbital to another to transform itself into something else. It is possible, however, to start with a mathematical description of the atomic orbitals in carbon and have the program generate a solution based on the electrons being moved into four identical orbitals in molecular carbon.

In Figure 3.7, a 2s-orbital and the three degenerate 2p-orbitals of carbon are mixed together and transformed into four identical "hybrid" molecular orbitals. Mixing one s- and three 2p-orbitals results in a hybrid called an *sp³ hybrid molecular orbital*, **3**. The alternative view shows the filled 2s-orbital and the two electrons in the 2p-orbitals. *Promoting* a 2s-electron and making all four new orbitals of the same energy leads to the four sp³ hybrids. The hybridization

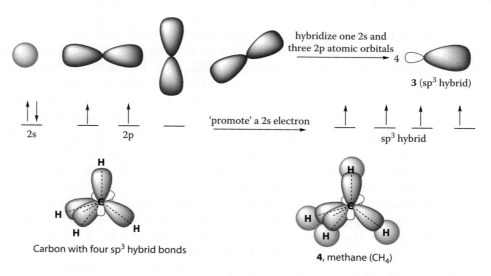

Carbon with four sp³ hybrid bonds

4, methane (CH_4)

Figure 3.7   sp³ Hybrid bonding in methane where carbon is in violet and hydrogen in red.

process is illustrated by the "blue lobes" for electrons in the 2p-orbitals "donated" by carbon and the "red lobe" the electrons in the 2s-orbital. When the electron is "promoted," the result is the violet sp³ hybrid molecular orbital, **3**.

Figure 3.7 also shows a carbon atom with four sp³ hybrid orbitals directed to the corners of a tetrahedron, reflecting the known geometry of the four bonds in the molecular carbon atom. This tetrahedral shape is derived from measurement of bond angles in the actual molecule methane, where there are four identical H–C–H bond angles = 109°28′, the angles of a regular tetrahedron. The bonding picture of methane is **4**, using the four sp³ hybrid bonds (sigma bonds) between carbon and the s-orbitals of four hydrogen atoms. Note that it is not possible to distinguish the "donation" of any individual atom.

3.5.3 Carbon in a Molecule of Methane. Four Bonds Attached to a Central Locus

The hybridization model is useful, but atomic carbon does not really hybridize and change into something else. In methane and in other organic molecules, the molecule forms the best bonds possible with the atoms, given the three-dimensional requirements of the molecule and the orbitals that are available. When four atoms are attached to a central carbon, internuclear repulsion of electrons in the bonds and the nuclei of the various bonded atoms pushes the atoms apart. The repulsive energy is minimized when the atoms are at the corners of a regular tetrahedron. In other words, in methane the tetrahedral shape represents an energy minimum for positioning the four hydrogen atoms about the central carbon, when all the hydrogen atoms are attached to the carbon. The overlapping orbitals used to form these bonds do not include the 1s-orbitals of carbon. Only the outermost (valence) orbitals (2s and 2p for carbon) can overlap with the hydrogen to form the covalent bonds. Based on overlap of orbitals using atomic carbon as a model, an alternative model is shown in Figure 3.8.

Each of the 1s-orbitals of the four hydrogen atoms will overlap with all three of the 2p-orbitals as well as the 2s-orbital to form the four identical covalent bonds. The model in Figure 3.8 illustrates the bonding by using the tetrahedral geometry of the bonds around the central carbon atom and overlap of all available orbitals. Imagine that the central carbon atom is the central point that has four hydrogen atoms approaching from the corners of a regular tetrahedron to form the four identical bonds. Each hydrogen atom (represented by a spherically symmetrical s-orbital) is close enough to the carbon so that its s-orbital will overlap with all orbitals of carbon that are available. Remember that if the carbon and hydrogen atoms are too close, the nuclei will repel, so the bonding distance is determined by overlap of orbitals balanced by internuclear repulsion.

The geometry of the hydrogen atoms in methane is tetrahedral, so a tetrahedral model is used. Each hydrogen atom will overlap all three p-orbitals and also the 2s-orbital, which occupies space at some distance from the carbon nucleus, to form the bond. A hydrogen atom will ***not*** approach so close to carbon

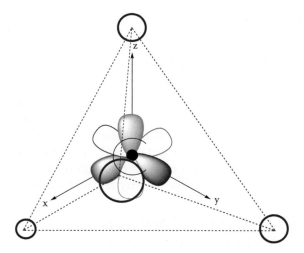

Figure 3.8.   Three-dimensional diagram for bonding in methane.

that it can overlap with the 1s-orbital, however. Each covalent bond between carbon and hydrogen is formed by overlap of the hydrogen s-orbital with *three* 2p-orbitals and *one* 2s-orbital; sp³ is a reasonable description of the bonding. *The model in Figure 3.8 is not real in the sense that four hydrogen atoms do not fly into a carbon atom to form methane. This model simply takes a molecule of methane, arranges the hydrogen atoms to the corners of a tetrahedron, and asks which orbitals of carbon can be used to form the bonds.* In order to describe the bonding, the term **sp³** indicates overlap with one s-orbital and all three p-orbitals. This is an alternative way to visualize the same sp³ hybrid bond that was formed via the previous electron promotion model in Section 3.5.2.

3.15 Predict the hybridization of carbon in the molecule CCl₄ (carbon tetrachloride).

Is this a reasonable description of the bonding in methane? Before introducing a model based on this idea, an experimental technique called **photoelectron spectroscopy** (PES) must be introduced. PES directs an electron beam at the surface of molecules and "dislodges" electrons in the valence shell. Using this technique with methane leads to a spectrum (Figure 3.9) that shows that there are **two** types of bonds as indicated by the presence of one peak at 13.6 eV and another at 23.1 eV.[6–8] An eV is an electron volt, which is converted into joules (1.602×10^{-19} J/eV) or into kilocalories per mole (3.883×10^{-20} kcal/mol/eV, and 1 kcal = 4.187 KJ). If higher energy is used to displace electrons, a third band is found at 290 eV. This latter band is correlated with the 1s-electrons and the band at 23.1 eV is due to the 2s-electrons. The band at 13.6 eV is due to the three degenerate 2p-orbitals.[6–8] This experiment suggests that electrons in the 1s- and 2s-orbitals and the 2p-orbitals have different energies and that **both** are used in bonding to the hydrogen atoms. This picture of bonding in methane is the basis of the model in Figure 3.8.

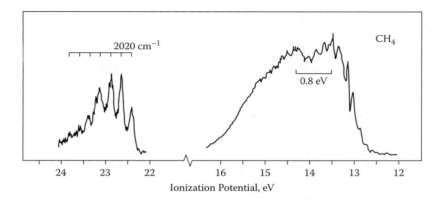

Figure 3.9 Photoelectron spectroscopy scan of methane. (Reprinted with permission from Brundle, C. R., and Robin, M. B. 1970. *Journal of Chemical Physics* 53:2196. Copyright 1970, American Institute of Physics.)

Methane (**4**, CH_4) is the prototypical model used to understand a covalent carbon. As noted several times, the four hydrogen atoms are distributed around carbon in the ***shape of a regular tetrahedron*** (see Figure 3.10). The bond angles have been measured to be 109°20′ for all H–C–H bond angles. All four C–H bond lengths are the same (1.094 Å; 109.4 pm).[9] *Remember that these facts are determined experimentally and may not be intuitively obvious.* The structure of methane is shown as a **Lewis dot formula** (**4a**) in Figure 3.10. In this representation, each covalent bond is represented by two dots for the two electrons. *This particular representation does not show the proper spatial relationship of*

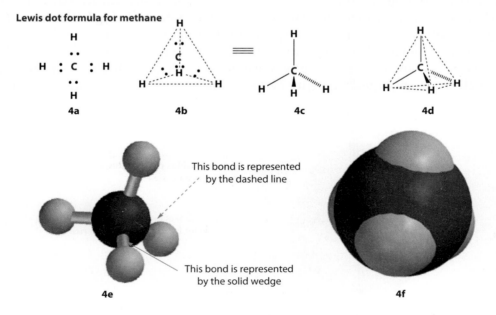

Figure 3.10 Representations of the structure of methane.

the hydrogen atoms relative to carbon. If the hydrogen atoms are placed in a tetrahedral array around the central carbon, as in **4b**, there is a better perspective of the shape of methane.

An alternative drawing is presented where the edges of the tetrahedron have been removed and only the atoms and bonds are shown, **4c**. *Rather than using dots for the electrons in a covalent bond, each bond is represented as a line.* Note that the closest hydrogen atom appears to be projected out of the page and is represented by a **solid wedge** in the tetrahedron structure to the right in Figure 3.10. The hydrogen atom that is the most distant in the tetrahedron appears to be projected behind the page, and this is represented by a **dashed line** in that structure. The other two hydrogen atoms are connected by solid lines and are in the plane of the page. *This system of solid lines, wedges, and dashed lines is the universal model used for representing the three-dimensional bonding pattern of molecules in two-dimensional space (on a page) in organic chemistry.*

Note that the four hydrogen atoms are distributed to the corners of an imaginary tetrahedron in three-dimensional space around the carbon of **4d**, which is "buried" in the center of the tetrahedron. In addition, a three-dimensional ***molecular model*** of methane is shown where "cylinders" represent the bonds and spheres represent the atoms (**4e**), which is the so-called "ball-and-stick" model. The bonds represented by wedges and dashed lines in the two-dimensional model are marked in **4e**. A so-called "space-filling model," **4f**, is also shown to illustrate the relative size of the atoms. Both types of molecular models (ball and stick or space filling) will be used for other molecules in various places in the book.

The model of methane shown in Figure 3.11 is a molecular model known as an ***electron density potential map***. The **blue** areas indicate regions where there is a large negative value of the potential (***low electron density***) and the **red** areas represent large positive values of the potential (***high electron density***). The **blue** correlates with the position of the hydrogen atoms in Figure 3.11, and the **red** more or less correlates with the position of the covalent bonds. The other colors show intermediate potentials. Note that the higher concentration of electron density (**red**) appears *between* the hydrogen atoms and the central

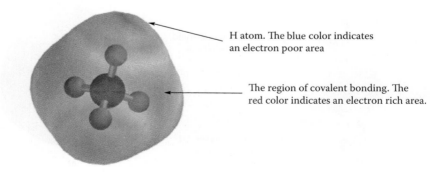

H atom. The blue color indicates an electron poor area

The region of covalent bonding. The red color indicates an electron rich area.

Figure 3.11   An electron density potential map of methane.

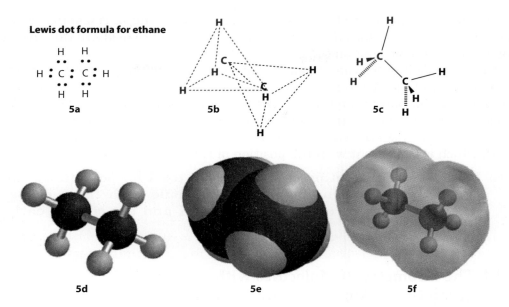

Figure 3.12 Structural representations showing the covalent bonding in ethane.

carbon atom. This model is shown to reinforce the idea that the electron density is concentrated in the bonds, which are between the atoms. Also note the tetrahedral shape of the electron density map in Figure 3.11.

The principles used to describe the bonding for methane can be extended to larger molecules that have additional carbon atoms and carbon–carbon bonds rather than only the C–H bonds found in methane. Ethane (**5**) has a covalent bond between two carbon atoms, for example, and *both* carbons have four bonds directed to the corners of a tetrahedron. The structure of ethane is shown in Figure 3.12, first with the Lewis dot formula for ethane (**5a**) and then as a structure made by overlapping two tetrahedrons of carbon (**5b**), each carbon with three hydrogen atoms and the fourth bond between carbon and carbon. The carbon–carbon bond is represented by overlap of the tetrahedrons in **5b**.

Each carbon atom and the three attached hydrogen atoms in ethane are distributed with a tetrahedral geometry around the carbon. The H–C–H bond angle for each carbon atom is 109.3°, the C–H bond length is measured to be 1.107 Å (110.7 pm), and the C–C bond length is 1.536 Å (153.6 pm).[9] Slight differences in bond angle and bond length are noted when compared to methane. Figure 3.12 also contains the chemical structure where the bond is shown without the contrivance of the tetrahedron edges, but the three-dimensional wedges and dashed lines are used to show the three-dimensional shape (**5c**). It is very important to remember that this three-dimensional array of atoms around the carbons in ethane is similar to that for ensembles of carbon and hydrogen found in most organic molecules. Both the bond length and the bond angle are determined by the nature and size of the atoms or groups attached to carbon.

From the standpoint of bonding, the important thing to remember is that there are four bonds for each carbon, composed of sp³ hybrid orbitals. All the bonds are covalent, and there are no unshared electrons on carbon. The two molecular models of ethane are shown for comparison with the two-dimensional structures, and the electron density potential map is also shown. The ball-and-stick model (**5d**) shows the relative position of the atoms, and the space-filling model (**5e**) shows the relative size of the atoms. Note the concentration of electron density between the two carbon atoms and the carbon and hydrogen atoms in **5f**, which is consistent with the position of the covalent bonds.

3.16 **Using a representation similar to that used for 5c with two carbons, draw the structure of the molecule with three carbons where each of the end carbons has 3H and the central carbon has 2H.**

3.5.4 The Geometry of Molecules with a Second-Row Atom as Central Locus

The discussions of methane and ethane make it clear that, for certain aspects of organic molecules, the two-dimensional representation of a simple line drawing does not provide sufficient information. A three-dimensional model must be made of molecules to understand them truly because the three-dimensional character of a molecule is important to many aspects of chemical reactivity. The tetrahedral shape in methane can be used to predict the three-dimensional shape of other molecules made from second-row elements. It is called the **valence shell electron pair repulsion model (VSEPR).** *Although the VSEPR model is useful for simple molecules, it does not give a completely accurate picture of the structure and is useful only for molecules made from second-row elements and hydrogen.* It is, however, a good place to begin. If a molecular modeling program is not available, the model kits that are usually suggested for an organic chemistry course will give more information.

Organic molecules are known that have other atoms in the second row of the periodic table in covalent bonds to carbon—specifically, C, O, N, and F. Other atoms in the second row also form covalent compounds. The previous section showed that carbon forms covalent bonds to four groups or atoms. Nitrogen is in group 15 and forms three covalent bonds, just as oxygen is in group 16 and forms two covalent bonds. The shape of a molecule is tetrahedral with respect to carbon. Using methane as a model, the shape of other molecules formed with atoms in the second row may be predicted, in accord with the number of atoms or groups attached to carbon, nitrogen, or oxygen, as well as the number of unshared electrons.

As noted, this "tetrahedral" model is known as the **VSEPR model.** In this model, carbon, nitrogen, and oxygen covalent compounds are assumed to have a tetrahedron shape, with C, N, or O at the center and the groups or atoms surrounding the central atom. In this model, all unshared electron pairs must be included. The electrons cannot be seen—only the atoms; however, the electrons exert an influence on the shape. *The shape of each molecule is determined by the relative positions of the atoms.*

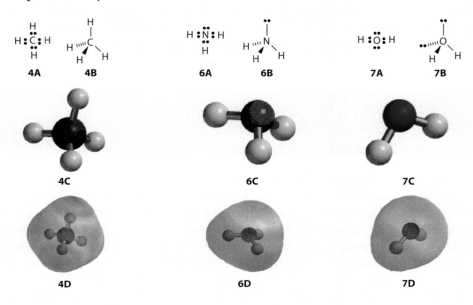

4A   4B   6A   6B   7A   7B

4C   6C   7C

4D   6D   7D

The three molecules of interest are methane (**4A**), ammonia (**6A**), and water (**7A**), shown first in the Lewis electron dot representations. Using the VSEPR model, these three molecules are drawn again using the wedge-dashed line notation. Methane (CH_4, **4B**) has no unshared electrons on carbon but there are electrons in the C–H covalent bonds. *Assume* that repulsion of the electrons in the bonds leads to a tetrahedral arrangement to minimize electronic repulsion. Ammonia (H_3N, **6B**) has a tetrahedral array around nitrogen if the electron pair is taken into account. If only the atoms are viewed, however, **6B** has the *pyramidal* shape shown. Water (HOH, **7B**) has two electron pairs that occupy the corners of a tetrahedral shape, as shown.

If only the atoms are viewed, **7B** is a *bent (angular)* molecule. Molecular models of each molecule are shown to emphasize the shape predicted by this model. The tetrahedral shape of methane is clear in **4C**, where the central carbon is surrounded by the four hydrogen atoms. In ammonia, the nitrogen atom is **blue** in **6C** and in water (**7C**) the oxygen atom is **red**. The electron density maps of each molecule are shown as **4D, 6D,** and **7D**. In methane, the highest concentration of electron density is in the bonds rather than on the atoms, as expected for a molecule with covalent bonds and no unshared electrons. Ammonia has an unshared electron pair on nitrogen, and **6D** shows a high concentration of electron density on that atom. Similarly, the two unshared electron pairs on oxygen of water are clearly reflected in the electron density map (**7D**). The two-dimensional models **4B, 6B,** and **7B** are intended to illustrate the position and directionality of the electron pairs, which contributes to the overall shape of the molecule.

3.17 Estimate the shape of a molecule if it has the structure OF_2.

The VSEPR model is a useful place to begin thinking about the three-dimensional nature of organic molecules. *Remember, however, that this model does a poor job of accurately predicting bond lengths and bond angles because it underestimates the importance of electron pairs and does not take into account the size of the atoms or groups attached to the central atom.*

3.6 How Strong Is a Covalent Bond? Bond Dissociation Energy

The strength of a covalent bond is directly related to the electron density between the atoms, and that strength is usually reported as a bond energy. Breaking a bond liberates the amount of energy that is required to keep the atom together. This amount of energy is considered to be stored in a bond and it is released via homolytic bond cleavage. For a covalent bond X–Y, there are two ways to break that bond. In one, both electrons in the bond are transferred to one atom. In this case, the two electrons are transferred to Y as the bond breaks, generating a cation (X^+) and an anion (X^-). This is called **heterolytic bond cleavage**. Note the use of the curve double-headed arrow to indicate transfer of two electrons as the bond breaks. The alternative breaks the covalent bond and transfers one electron to each of the atoms, generating a radical (a species with one unpaired electron). This is known as **homolytic bond cleavage**, and in this case, homolytic cleavage of X–Y leads to two radicals, X• + Y•. Note the use of the curved single-headed arrows to indicate transfer of the electrons as the bond breaks.

When a bond is broken or formed, the energy required is known as the **bond dissociation enthalpy** ($D°$ or, more commonly, $H°$ for a bond broken or formed in a reaction). It is also called **bond dissociation energy**; for convenience, the values listed will be used for both heterolytic and homolytic bond cleavages.

Heterolytic bond cleavage   X —Y $\longrightarrow$ X^+ (cation) Y^- (anion)

Homolytic bond cleavage   X —Y $\longrightarrow$ X• (radicals)   Y•

A useful place to begin a discussion of bond strength compares ionic bonds and covalent bonds. A great deal of energy is required to disrupt the ionic bond in the ionic material sodium chloride, NaCl. Sodium chloride melts at 801°C and it boils at 1413°C. Melting point is the disruption of the crystalline structure that changes its physical state from a solid to a liquid, and boiling point is the disruption of the intramolecular forces in the liquid that change its physical state from a liquid to a gas. No bonds are broken in either the melting or boiling processes. In other words, the ionic bond is not broken at these temperatures.

It is clear that a great deal of energy is required to break the NaCl bond to convert it to individual atoms or to ions. The bond dissociation energy of the

Cl–Na bond is formally reported to be 98 kcal mol^{-1} and represents the amount of energy released when that bond is broken.[10] In an organic compound, the bond dissociation energy of the C–C bond is reported to be 145 kcal mol^{-1}.[11] Based on this analogy, one expects a very high temperature to be required for cleavage of a C–C bond, and the numbers suggest that the C–C bond is more difficult to break than the NaCl bond.

This is not true! Most C–C bonds are completely disrupted at temperatures between 400 and 600°C. The covalent bond is inherently much weaker than an ionic bond *in terms of its ability to undergo chemical reactions.* **Disruption of the bond is not just "ripping" the two atoms apart, but rather a chemical process where electrons are transferred, leading to the bond being broken.** This process requires much less energy (heat) than simple cleavage of the bond into its two individual components (the process described before for NaCl). The stability gained by sharing two electrons between nuclei is inherently weaker than the powerful ionic attraction between two ions of opposite charge, and such bonds are easier to break *when they are involved in chemical reactions* (electron transfer process with other atoms or groups of atoms).

3.18 Is LiCl likely to dissociate completely at a temperature of 600°C?

If two atoms could be brought together in a direct manner to form a bond, the energy required to form that bond is the same as that required to break the bond. Unfortunately, in organic chemistry, two elements rarely come together in a direct manner to form a bond. In most cases, a derivative of the element is involved in the bond-making or bond-breaking process. Such derivatives have an atom that is part of a structure containing several atoms (i.e., a molecule).

Covalent bonds are found in millions of organic molecules that participate in a multitude of organic reactions. In principle, each one will require a slightly different energy for cleavage or formation of a particular bond. If a bond is broken in one molecule, a bond is usually formed in a new molecule. This is a chemical reaction and the molecules in which the bonds are broken are called **reactants**, whereas the molecules in which bonds are formed are called **products**. A general formula can be given for a reaction where reactants (A–B and C) are converted to products (A and B–C), as represented in Figure 3.13. A certain amount of energy is required to break the A–B bond (called H°$_{reactants}$) and a certain amount of energy is required to form the B–C bond (called H°$_{products}$). In this process, there is a *change* in bond dissociation energy, represented by ΔH°, where the symbol Δ represents *change in*. This value is determined by subtracting the bond

Figure 3.13 Fundamental components of a chemical reaction.

Table 3.2
Bond Dissociation Energies of Common Bonds

Bond	Bond Strength ($H°_{298}$), kcal mol^{-1}	Bond	Bond Strength ($H°_{298}$), kcal mol^{-1}
I–I	36.5	Br–H	87.4
F–F	37.5	C–Cl	95
Br–Br	46.3	Cl–H	103.2
C–I	50	H–O	104.2
Cl–Cl	58	O–O	119.1
C–Br	67	C–C	145
H–I	71.4	C–S	167
N–H	75	C–N	184
C–H	80.6	N–N	225.9
S–H	82.3	C–O	257.3

dissociation energy for the reactants ($H°_{AB}$) from the bond dissociation energy for the products ($H°_{BC}$). This equation leads to a general formula for $\Delta H°$:

$$\Delta H° = H°_{products} - H°_{reactants} \text{ and for this reaction} = H°_{BC} - H°_{AB}$$

The value $H°$ for a bond is the amount of energy released when that bond is broken or the amount of energy that is required to form that bond. Typical values are listed in Table 3.2 and are reported in kilocalories per mole at 298°C.[12]

A stylized example is a reaction between I$^-$ and C–Br to give Br$^-$ and C–I, represented by the following:

In this reaction, the product is C–I and the reactant is C–Br, so the formula is $\Delta H° = H°_{C-I} - H°_{C-Br} = 50 - 67 = -17$ kcal mol^{-1}.

If there are no other factors, the negative sign of $\Delta H°$ indicates that this reaction will *release* 17 kcal mol^{-1} of energy and is exothermic. In other words, this is expected to be an energetically favorable process based solely on bond dissociation energies.

3.19 Calculate $\Delta H°$ for a process with I$^-$ and R_3C–Cl that breaks a C–Cl bond and makes a C–I bond, generating Cl$^-$. Ignore the chloride ion and the iodide ion.

There are many different types of organic molecules, and the C–C bond in each type of compound will have a slightly different environment. In addition

Table 3.3
Correlation of Bond Strength for Bonds in Typical Organic Molecules

Bond	Bond Strength ($H°_{298}$), in kcal mol^{-1}	Bond	Bond Strength ($H°_{298}$), kcal mol^{-1}
I–CH$_3$	56	H–C(CH$_3$)$_3$	92
Br–CH$_3$	70	H–CH(CH$_3$)$_2$	95
Cl–CH$_3$	84	H–CH$_2$CH$_3$	98
H$_2$N–CH$_3$	87	H–CH$_3$	104
HO–CH$_3$	91	F–CH$_3$	109

Source: Lide, D. R., ed. 2006. *CRC Handbook of Chemistry and Physics,* 87th ed., 9-1–9-15. Boca Raton, FL: Taylor & Francis.

to C–C bonds, C–O, C–Cl, and C–N bonds, as well as many others, occur in organic molecules. The bond strengths of some simple organic molecules are shown in Table 3.3.[5] The structural formulas used in Table 3.3 will be described in Chapters 4 and 5. For the moment, concentrate on the bond and the number of carbons in each structure and the point of attachment of the hydrogen or other atom.

Note that when carbon is attached to atoms other than carbon or hydrogen, the bond is weaker, except for F–CH$_3$ and HO–CH$_3$. As the carbon atom becomes more substituted (more carbon atoms attached to it), the strength of the H–C bond is diminished. The reasons for these trends will be discussed shortly.

3.7 Polarized Covalent σ-Bonds

When certain atoms are collected into discrete units, they often have special physical and/or chemical properties. Such units are called *functional groups*. The C=C unit of alkenes and the C≡C unit of alkynes are introduced in Chapter 5 (Sections 5.1 and 5.2, respectively), and they will be the first examples of functional groups. *The C–C unit of an alkane is not considered to be a functional group because it is the "backbone" of virtually all organic molecules.* These special arrays of atoms can include atoms other than carbon or hydrogen, and the presence of these other atoms (call them **heteroatoms**) can lead to new functional groups. This section will describe many of the common functional groups that *contain single covalent bonds* and are found in organic molecules. Functional groups that contain multiple bonds to carbon will be introduced in Chapter 5. Before discussing functional groups containing heteroatoms, it is important to introduce an important effect that the heteroatom will impart to the bond and the molecule, bond polarization, and polarity.

H 2.2												
Li 1.0	Be 1.5							B 2.0	C 2.6	N 3.0	O 3.4	F 4.0
Na 0.9	Mg 1.3							Al 1.6	Si 1.9	P 2.2	S 2.6	Cl 3.2
K 0.8	Ca 1.0	Sc 1.3	Ti-V 1.5–1.6	Cr-Mn 1.6–1.7	Fe-Ni 1.8–1.9	Cu 1.9	Zn 1.7	Ga 1.8	Ge 2.0	As 2.2	Se 2.6	Br 3.0
												I 2.7

Figure 3.14 Electronegativities of selected atoms in the periodic table. (Lide, D. R., ed. 2006. *CRC Handbook of Chemistry and Physics,* 87th ed., 9–83. Boca Raton, FL: Taylor & Francis.)

3.7.1 Electronegativity

The property of an atom to attract electrons is called **electronegativity**. Atoms have different abilities to attract and "hold" electrons. In general, electronegativity of an element increases across the periodic table as one goes to the *right* (left to right toward fluorine) and *up* (bottom to top) the table. This means that oxygen is more electronegative than carbon and chlorine is more electronegative than bromine. Figure 3.14 shows the electronegativity of many elements that appear in organic molecules, arranged according to the periodic table.[13] Interestingly, carbon is more electronegative than hydrogen by a small amount. *For most organic molecules, however, the electronegativities of carbon and hydrogen are assumed to be the same. This is only an assumption, but it is very useful in many cases.* This statement will be discussed in detail in connection with spectroscopy in Chapter 14.

Inspection of Figure 3.14 reveals that trends in electronegativity to the right or left and up or down are very clear. The trend in electronegativity going across the periodic table is O > N > C > B; going down the periodic table, the trend is O > S and N > P, for example. Diagonally (C → P → Se, for example), however, trends are not clear. It is easy to recognize that fluorine is more electronegative than carbon, but bromine is also slightly more electronegative than carbon. Inspection across the diagonal C→P→Se→I shows that the electronegativities change from 2.6→2.2→2.6→2.7. Clearly, there is no trend, and it is not obvious how comparisons can be made without specific numbers. Therefore, when citing trends in chemical properties such as bond strength—or even acid strength in Chapter 2—the trend is restricted to across or up and down the periodic table.

3.20 **Rank the following atoms by their electronegativity: Li, B, C, and O.**

3.7.2 A Covalent Bond between Carbon and a Heteroatom. Bond Polarization

When two identical atoms are connected by a covalent bond and each atom has the same atoms or groups attached (such as $H_3C–CH_3$ in ethane), those atoms have identical electronegativities. In ethane, the covalent bond connects two atoms of the same electronegativity, which means the electron density is equally and symmetrically distributed between both nuclei. In other words, one atom does not have more electron density than the other. This type of covalent bond is represented by **8**, where the black dots represent the atoms of the covalent bond.

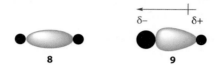

If a bond is formed between two atoms that have different electronegativities, however, the picture is quite different. The electron density is *not* equally distributed between the nuclei but is distorted *toward* the more electronegative atom, as in **9**. The net result of this electron distortion is that one atom has more electron density than the other, making one atom more negative and one atom more positive. Such a bond is said to be **polarized—*a polarized covalent bond***. This bond can be represented as (+)-------(–) or by a specialized arrow (+---->) where the + part of the arrow is on the more positive atom and the arrow (→) points to the more negative atom (see **9**). Because these are covalent and not ionic species, the (+) and (–) do not indicate charges. The atom is said to be *partially negative* or *partially positive,* so δ+ and δ– are used (see **9**). A molecule containing a polarized bond is said to be **polar** if certain conditions are met (see below). A molecule that does not have a polarized bond is categorized as **nonpolar**.

3.7.3 Bond Polarity, Bond Moments, and Bond Strength

A polarized covalent bond has some interesting characteristics that affect both the physical and chemical properties of a molecule. A polarized covalent bond has the property of *bond polarity,* which leads to a dipole moment for the bond. The electronegativity of both carbon atoms in a C–C bond is identical, so there is no dipole and the bond is *nonpolar*. If one compares C–N and C–F, however, the F is more electronegative than the C, and N is also more electronegative than C. Both C–F and C–N are polarized covalent bonds. Fluorine is more electronegative than nitrogen, so fluorine will attract more electron density from the bond than will nitrogen. If one compares the bonds, the electron density in C–F (**10**) is distorted to a greater extent than in C–N (**11**). This means that the C–F bond is more polarized than the C–N bond. The *magnitude* of this bond polarization is measurable and it is called ***bond moment*** measured as the ***dipole moment***.

Dipole moment has a magnitude and a direction, and it is measured in Debye units (1 Debye = 3.33564×10^{-30} coulomb-meters). These properties are best seen in actual molecules **12, 13, 6,** and **14**. H–F (**12**) shows a clear dipole moment from the less electronegative H atom toward the more elec-tronegative F atom. Note that the dipole moment lies along the C–F bond. In the molecule $F_3C–F$, the C–F bond is polarized because fluorine (**13**) is more electronegative than carbon. Note the dipole moment in **13**, which lies along the C–F bond, toward the fluorine.

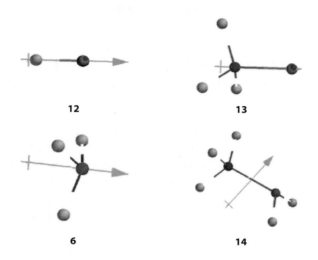

Although the direction is obvious for an individual bond, it is not obvious for an entire molecule. In ammonia (**6**), each N–H bond is polarized with a dipole moment toward the N. For the molecule, however, the dipole is the sum of all three N–H bond moments, which have both a magnitude and a direction. In ammonia, the direction of the dipole for each N–H bond is toward N, and the dipole for the molecule is toward nitrogen. The dipole for the molecule does not lie along an individual bond, but rather bisects the base of the pyramidally shaped molecule, as shown in **6**. A final example is the molecule $H_3C–NH_2$ (**14**) with polarized C–N and H–N bonds. Note that the dipole moment does not lie along one bond, but rather seems to be tilled by the C–N bond, suggesting that C–N is more polarized than N–H. Polarity is discussed in Chapter 5 (Section 5.8) in connection with polarization in molecules and how that polarity influ-ences both reactivity and physical properties. Polarity in solvents is discussed in Chapters 11 and 13.

Table 3.4 shows the dipole moment for a number of bonds commonly found in organic molecules.[14] In the simple cases **10** and **11** (C–F and C–N), the direc-tion of the dipole moment is toward F and toward N, which means that F and N are $\delta-$. The magnitude for a typical C–F bond is 1.79 Debye units and for a typical C–N bond it is 0.45 Debye units; the C–F bond is more polar, as pre-dicted.[14] The C–I bond is interesting. The electronegativity of iodide does not suggest a large dipole, but Table 3.4 clearly shows a significant dipole moment for C–I. Factors other than electronegativity differences contribute to dipole

Table 3.4
Bond Dipole Moments for Common Bonds in Organic Chemistry

Bond	Dipole Moment (Debye)	Bond	Dipole Moment (Debye)
C–C	0.0	H–N	1.31
C–H	0.3	C=N	1.4
H–I	0.38	H–O	1.51
C–N	0.45	C–I	1.65
C–O	0.74	C–F	1.79
H–Br	0.78	C–Br	1.82
C–S	0.9	C–Cl	1.87
H–Cl	1.08	H–F	1.94
		C=O	2.4

Source: Dean, J. A. 1987. *Handbook of Organic Chemistry,* 3-28–3-29, Tables 3-6 and 3-7. New York: McGraw-Hill.

moment. In all cases, the direction of the dipole of a C–X bond is toward the heteroatom, X. In other words, the direction of the dipole is toward the atom of higher electron density.

3.21 Determine the more polarized covalent bond: C–F or C–B.

In general, a more polarized bond will have less electron density concentrated ***between*** the nuclei. If the electron density is distorted more toward one atom than the other, a cross-section of the bond will show that there is less electron density between the nuclei. This does not necessarily mean that it will be a weaker bond because other factors contribute to bond strength, but it does contribute to bond strength. In Table 3.5, it is clear that more energy is associated with a C–F bond than with a C–N bond, even though the C–F bond is more polarized. In a chemical reaction that involves transfer of electrons from one atom to another, it is usually easier to break a polarized bond than a nonpolarized bond, *if one atom of the bond can readily accept electrons.* This statement is illustrated by Figure 3.15, where the polarized C–Cl bond breaks and the two electrons in that bond are transferred to Cl, forming the chloride ion.

The bond strengths of several typical polarized bonds are listed in Table 3.5.[15] Comparing the bond strength of C–C (88 kcal mol^{-1}) with C–I (55.5 kcal mol^{-1}) shows that the highly polarized C–I bond is much weaker. See the discussion in Chapter 7 (Section 7.2) for more information on using bond strength. Comparing the bond strength of C–N (79 kcal mol^{-1}) and C–F (108 kcal mol^{-1}), however, shows that the C–F bond is stronger.

Why is the C–F bond stronger when it is more polarized? The length of the C–F bond is listed as 1.379 Å (137.9 m) in Table 3.1, and the length of the

Table 3.5
Bond Strength of Various Polarized Covalent Bonds

Bond	Bond Strength (kcal mol⁻¹)	Bond	Bond Strength (kcal mol⁻¹)
C–I	55.5	N–H	85
C–Br	68	C–C	88
C–N	79	H–O	102.3
C–O	80	C–F	108
C–Cl	80.8	C=O	175
C–H	81		

Source: Lide, D. R., ed. 2006. *CRC Handbook of Chemistry and Physics,* 87th ed., 9-54–9-74. Boca Raton, FL: Taylor & Francis.

C–N bond is 1.472 Å (147.2 pm). The fluorine atom is quite small. In covalently bound atoms, the covalent atomic radius of fluorine is 0.64 Å (64 pm) and that of nitrogen is 0.70 Å (70 pm).[16] Because the C–F bond is shorter in a covalent molecule, the electron density is dispersed over a smaller area and the "electron density per unit length" is greater. This means that the C–F bond is stronger despite the fact that the F is more electronegative. *The take-home lesson here is that bond strength is a function of several parameters. One cannot assume that the more polarized bond is weaker.*

A comparison of a nonpolarized bond such as C–C with a polarized bond such as C–Cl is a useful exercise for completing this section. There is no polarization in the C–C bond and both atoms are *neutral,* which means they do not have δ+ or δ–. Compare this with C–Cl, *which also has two neutral atoms,* but the carbon is δ+ and the chlorine is δ–. Imagine that the positive carbon in the C–Cl bond comes into contact with a negatively charged species (X⁻). In Figure 3.15, the negative species (X⁻) donates electrons to the positive carbon, breaking the C–Cl bond (note the use of the curved electron transfer arrows). This process transfers the electrons in that bond to the more electronegative chlorine (which is δ– in the original bond) and forms a new bond between carbon and the "incoming" atom. This is the basis of a chemical reaction, which would ***not*** occur if X⁻ collided with a molecule that contained only a C–C bond. With no bond polarization, the bond is much stronger and, if the bond breaks in a C–C bond, the carbon cannot readily accommodate the negative charge (excess

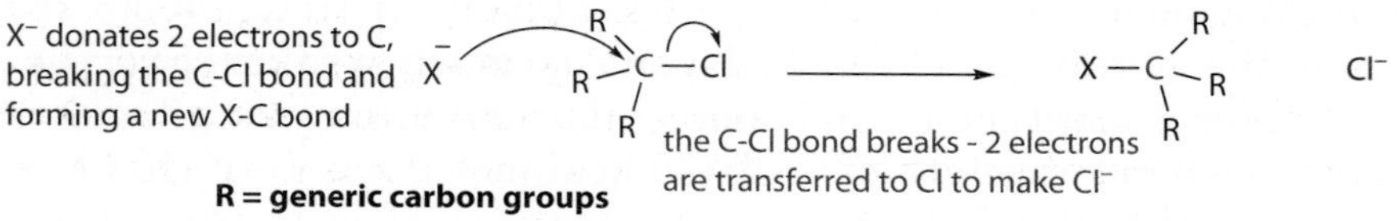

X⁻ donates 2 electrons to C, breaking the C-Cl bond and forming a new X-C bond

the C-Cl bond breaks - 2 electrons are transferred to Cl to make Cl⁻

R = generic carbon groups

Figure 3.15 Donation of electrons to an electropositive carbon, with bond making and bond breaking.

electrons). This concept will be used many times in chapters to come. The real purpose of this section is to introduce the idea that chemical reactions involve making and breaking bonds. The chemistry of molecules containing polarized bonds is richer than the chemistry of molecules containing nonpolarized bonds.

3.22 Based *only* on bond strength, which bond is easier to break: C–Br or C–O? Is this valid?

3.8 Biological Relevance

It is difficult to show biological applications with such a general topic. Enzymes are obviously important to biological processes, however, and it is known that some enzyme reactions accelerate the biological process by forming a covalent bond between the enzyme and the substrate. A substrate such as **15** reacts with an enzyme (labeled E), for example, that has an attached reactive nucleophile (X). The nucleophile reacts with the carbonyl unit to give what is known as a tetrahedral intermediate, **16**. This type of structure is a key intermediate in acyl substitution reactions, which will be discussed in Chapter 15 (Section 15.8). The "Y" group is expelled as the corresponding anion, Y⁻. In this reaction, "Y" is known as a *leaving group* (see Chapters 11 and 15), and the new product is the enzyme-bound acyl derivative **17**. This covalently bound enzyme (C–X–E) is now available for use in various enzymatic reactions.

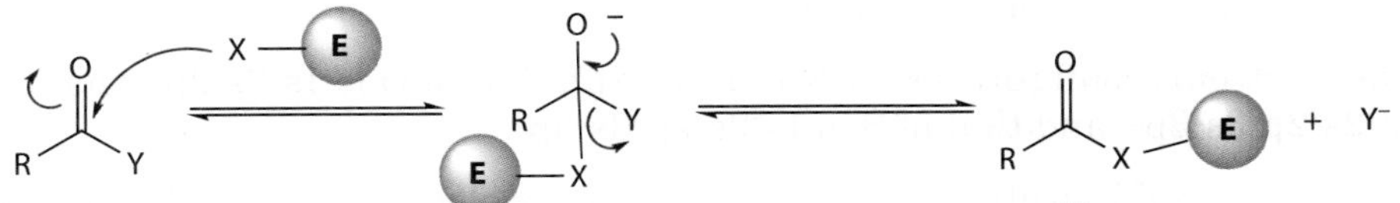

References

1. Coulson, C. A. 1982. *The shape and structure of molecules,* 2nd ed., revised by McWeeny, R., 3. Oxford, England: Clarendon Press.
2. Puddephatt, R. J. 1972. *The periodic table of the elements,* 6. Oxford, England: Clarendon Press.
3. Tedder, J. J., Nechvatal, A., and Pitman, M. A. 1985. *Pictorial orbital theory,* 2. Essex, England: Longman Science & Technology.
4. Coulson, C. A. 1982. *The shape and structure of molecules,* 2nd ed., revised by McWeeny, R., 5, Figure 3. Oxford, England: Clarendon Press.
5. Lide, D. R., ed. 2006. *CRC handbook of chemistry and physics,* 87th ed., 9-1–9-15. Boca Raton, FL: Taylor & Francis.
6. Brundle, C. R., and Robin, M. B. 1970. *Journal of Chemical Physics* 53:2196, American Institute of Physics.
7. Baker, A. D., Betteridge, D., Kemp, N. R., and Kirby, R. E. 19671. *Journal of Molecular Structure* 8:75.

3.20 Because electronegativity increases to the right in the periodic table, the relative order of electronegativity should be Li < B < C < O.

3.21 Because F is more electronegative than B, one predicts that C–F is the more polarized covalent bond.

3.22 Based *only* on bond strength, the C–Br bond (68 kcal/mol) is weaker than the C–O bond (80 kcal/mol) and C–Br should be easier to break. This is not necessarily a valid method because it ignores relative bond polarity, stability of final products, etc.

Correlation of Concepts with Homework

- **s-Orbitals are spherically symmetrical and p-orbitals are "dumbbell" shaped: 1, 24, 39, 42.**
- **Electrons are found in orbitals at discrete distances from the nucleus in an atom: 2, 3, 4, 5, 6, 8, 23, 38, 41, 44.**
- **Electrons in the bond of a molecule are located between two nuclei and are at energy levels different from those in an unbonded atom: 26, 27.**
- **Ionic bonds are formed by electrostatic attraction of two atoms or groups that have opposite charges: 7, 9, 18, 25, 30, 37.**
- **Covalent bonds are made of two electrons that are mutually shared between two atoms: 10, 11, 12, 13, 14, 29.**
- **Mixing atomic orbitals forms hybrid molecular orbitals, and the number of s- and p-orbitals used to form the hybrid determines the hybridization (sp³, for example): 15, 28, 45, 46, 47.**
- **Organic molecules generally have a backbone of carbon–carbon covalent bonds: 16, 34.**
- **Electronegativity for an atom increase to the right and up the periodic table: 20, 21.**
- **Polarized bonds are formed when two atoms are bonded together but one is more electronegative: 35, 36, 43.**
- **Polarized covalent bonds are generally weaker than nonpolarized covalent bonds: 40.**
- **Reactions are driven by making and breaking bonds, which releases or requires energy: 19, 22, 32, 33.**
- **The VSEPR model is used to predict the three-dimensional shape around an individual atom: 17, 31.**

Homework

3.23 Give the electronic configuration for the following atoms: Al, He, Be, Mg, Cl, Br, Ti, and Cu.

3.24 Draw a comparative diagram for a 1s-orbital and a 2s-orbital.

3.25 Based on electron availability in the valence orbital, suggest a simple reason why potassium metal tends to be more reactive than sodium metal.

3.26 Explain why the Pauli exclusion principle is important for filling orbitals as we go across the second row of the periodic table (Li→Be→B→C→N→O→F).

3.27 Briefly explain why a 4s-orbital will fill before a 3d-orbital.

3.28 For each of the following, indicate which carbon atoms are sp³ hybridized, with the hint that an sp³ hybridized carbon cannot have multiple bonds (no X=X or X≡X).

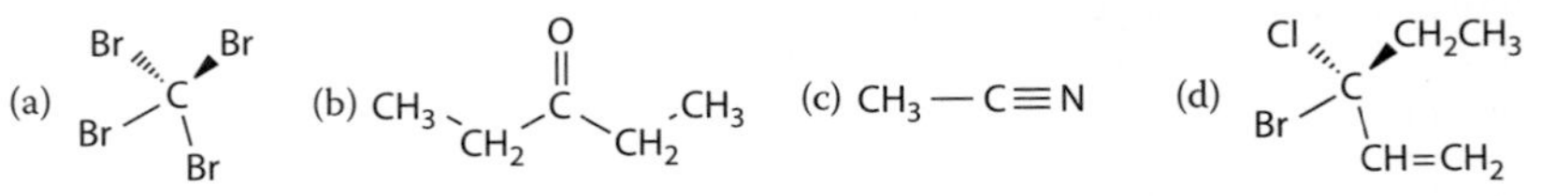

(a) Br—C(Br)(Br)—C—Br   (b) CH₃—C(=O)—CH₂—C—CH₂—CH₃   (c) CH₃—C≡N   (d) Cl—C(CH₂CH₃)—Br—CH=CH₂

3.29 Indicate whether the highlighted bond (in **red**) is expected to be covalent or ionic.

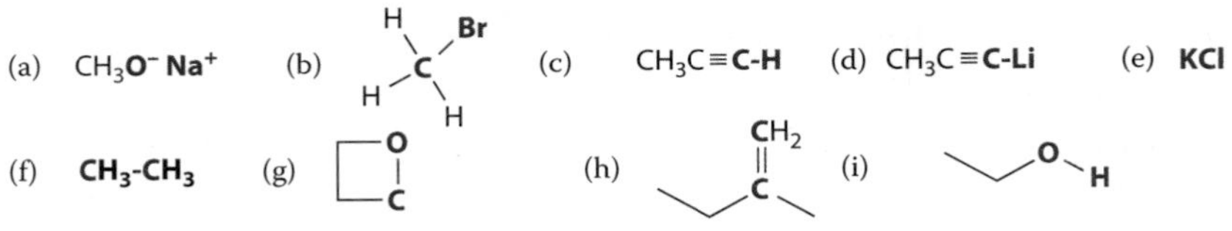

(a) CH₃O⁻ Na⁺   (b) H,H—C—Br,H   (c) CH₃C≡C-H   (d) CH₃C≡C-Li   (e) **KCl**

(f) **CH₃-CH₃**   (g) [O—C ring]   (h) [CH₂ structure C]   (i) [O—H ether structure]

3.30 The first ionization potential of Li is 5.392 eV, Na is 5.139 eV, and K is 4.341 eV—all in the gas phase. What do these numbers tell you about the relative reactivity of each element?

3.31 Use the VSEPR model to predict the shape for each of the following. Draw each one using line notation:
(a) CCl_4
(b) CH_3OH
(c) CH_3OCH_3
(d) $(CH_3)_4N^+$
(e) $CH(CH_3)_3$
(f) $ClCH_2Cl$

3.32 Calculate $\Delta H°$ for each of the following hypothetical reactions. Determine if they are exothermic or endothermic.

(a) H-Br + C-O ⟶ C-Br + H-O
(b) C-C + I₂ ⟶ 2 C-I
(c) O-H + C-C ⟶ C-O + C-H
(d) C-N + H-I ⟶ C-I + N-H

3.33 Explain the trend in the following bond dissociation enthalpies: I–CH₃ (C–I = 56 kcal/mol); Br–CH₃ (C–Br = 70 kcal/mol); Cl–CH₃ (C–Cl = 84 kcal/mol).

3.34 Based on the structure given for ethane (CH₃–CH₃) in Figure 3.12, suggest structures for propane (CH₃CH₂CH₃) and butane (CH₃CH₂CH₂CH₃).

3.35 Indicate the more electronegative atom in each of the following:
(a) C–N
(b) N–O
(c) C–H
(d) Cl–Br
(e) B–C
(f) Li–C
(g) C–F
(h) N–H
(i) H–Cl

3.36 Indicate if each bond is polarized covalent or nonpolarized covalent:
(a) C–N
(b) N–O
(c) C–H
(d) C–F
(e) C–C
(f) Li–C

3.37 Draw the structure of each molecule using the VSEPR model and then indicate the general direction of the dipole moment. Indicate whether the dipole moment is zero:
(a) CH_3Cl
(b) CH_3OH
(c) CCl_4
(d) $C(CH_3)_4$
(e) $ClCH_2Br$
(f) Cl_3CH

3.38 What is the electronic configuration for F^+.

3.39 Which of the following best describes the shape of (a) a p-orbital and (b) a d-orbital?

3.40 Is the C–I bond stronger or weaker than C–F? Briefly explain your answer.

3.41 Briefly explain why lithium tends to react with halogens to form the ionic species Li^+.

3.42 Which of the following best describes the shape of a 2p-orbital? (a) tetrahedral; (b) spherical; (c) pyramidal; (d) dumbbell shaped; (e) trigonal.

3.43 Which of the following is the **least** polarized bond?
(a) C–C
(b) C–N
(c) C–O
(d) C–F

3.44 What is the electronic configuration of Na^+?
    (a) $1s^2 2s^2 2p^2$
    (b) $1s^2 2s^2 2^3$
    (c) $1s^2 2s^2 2p^4$
    (d) $1s^2 2s^2 2p^5$
    (e) $1s^2 2s^2 2p^6$

3.45 Which hybrid orbital is likely to be higher in energy: mixing a 1s-orbital and a 2p-orbital or a 2s-orbital with a 2p-orbital?

3.46 Using the electron promotion model for carbon as a guide, suggest hybrid orbitals formed for a C–O covalent bond.

3.47 What is the hybridization of the carbon atom in Cl_3C^+, with three covalent C–Cl bonds and a positive charge on carbon?

Alkanes, Isomers, and an Introduction to Nomenclature

4

Carbon is only one atom in the periodic table. Why is carbon special? From a chemical perspective, carbon forms covalent single, double, and triple bonds to another carbon and it forms covalent bonds to a variety of other atoms. This ability to bond with itself leads to an almost limitless number of organic molecules (molecules that contain carbon) and is one of the key reasons why the chemistry of carbon comprises an entire branch of chemistry. Millions of different molecules have covalent bonds to carbon. This chapter will discuss the structural features of a class of organic molecules that contain only carbon and hydrogen (hydrocarbons) and have no double or triple bonds. The chapter will also introduce a method for giving each different molecule an individual name.

To begin this chapter, you should know the following:

- **the fundamental nature of atoms (Chapter 3, Sections 3.1 and 3.2)**
- **covalent bonding between carbon and carbon or carbon and hydrogen (Chapter 3, Sections 3.3, 3.5 and 3.6)**
- **sp^3 hybridization (Chapter 3, Section 3.5)**
- **how to draw simple structures and the connectivity of atoms (Chapter 3, Section 3.5)**

- that molecules with only C and H are considered to be nonpolar (Chapter 3, Section 3.7)
- that carbon forms four covalent bonds in neutral molecules (Chapter 3, Section 3.5)
- the VSEPR model for carbon (Chapter 3, Section 3.5)

This chapter will introduce the first class of organic molecules, alkanes, which are hydrocarbons. The concept of isomers, different connectivity within organic molecules, and rules for naming organic molecules will also be presented.

When you have completed this chapter, you should understand the following points:

- Hydrocarbons are molecules that contain only carbon and hydrogen, and alkanes are hydrocarbons that have only carbon–carbon single bonds and carbon–hydrogen bonds with the generic formula C_nH_{2n+2}.
- The structure of hydrocarbons is effectively tetrahedral about a central carbon atom.
- Molecules with different connectivity are considered to be different molecules. Isomers are molecules with the same empirical formula and the same number of atoms, but the atoms are attached in different ways.
- Combustion analysis is used for determining the percent of C and percent of H in an alkane, which is then used to calculate the empirical formula.
- The IUPAC rules of nomenclature involve identification of the longest continuous carbon chain as assigning a prefix that correlates with the number of carbon atoms. The class name for a molecule is designated by a suffix, which is "ane" for alkanes. Rules are established based on number of carbons, assignment of the lowest number(s), and position on the longest carbon chain for naming substituents. Groups attached to the longest continuous chain are known as substituents, and give the suffix "yl" for alkane base substituents. Complex substituents are treated as a substituent-on-a-substituent.
- Cyclic alkanes use the prefix "cyclo."
- There is a biological relevance for some alkanes.

4.1 The Fundamental Structure of Alkanes Based on the sp³ Hybrid Model

Molecules in which all carbon atoms are sp^3 hybridized (see Chapter 3, Section 3.5) and in which there are only single covalent bonds are perhaps the simplest organic molecules. Such molecules are hydrocarbons—molecules that contain *only* carbon and hydrogen. It is known that carbon forms bonds to other carbon atoms and molecules can be formed that contain chains, branches, rings, etc.

Organic molecules may contain virtually any number of carbon–carbon bonds, even hundreds. There are virtually no limits. In the simplest cases, each carbon–carbon bond is a covalent σ-bond as described in Chapter 3. Each sp^3 hybridized carbon in a hydrocarbon has a valence of four, so a total of four covalent bonds must be attached to each carbon for that atom to remain neutral (no + or – charge). A hydrocarbon that has covalent bonds only between sp^3 carbons or between sp^3 carbon and hydrogen is known as an **alkane**. The suffix **"ane"** in this term defines the compound as an alkane. *A bond to hydrogen occupies every covalent site to carbon that is not occupied by a bond to carbon*. This latter fact leads to a **general formula for alkanes**:

$$C_nH_{2n+2}$$

where n is an integer: 1, 2, 3, 4,...

When $n = 1$, the alkane formula is CH_4; when $n = 3$, it is C_3H_8; when $n = 100$, it is $C_{100}H_{202}$, and so on. This formula defines the ***maximum number of hydrogen atoms*** found in an alkane having a particular number of carbons. In an alkane, there can never be more hydrogen atoms than the number obtained with this formula; there may be fewer, but never more. The three structures shown have the formula C_6H_{14}, C_9H_{20}, and $C_{32}H_{66}$. All three structures fit the C_nH_{2n+2} formula ($n = 6$, $n = 9$, and $n = 32$), and they are categorized as alkanes.

4.1 Determine whether a molecule with the formula $C_{10}H_{20}$ is an alkane.

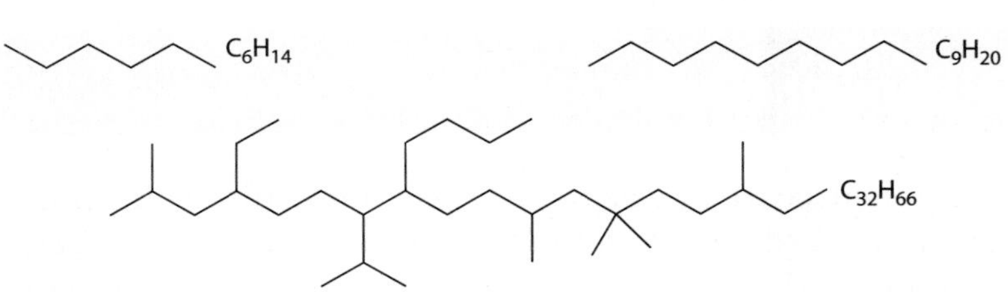

It is important to know the various methods used for drawing the structures of alkanes. Initially, focus on alkanes that have linear chains. This means that all of the carbon atoms are connected in a chain, with no branches. The alkane with a linear chain of four carbon atoms has the butane (see Section 4.3), and it is used as a fuel in butane lighters. The actual structure is represented in several different ways as **1**. The symbol "**:**" is used to represent the two electrons in each covalent bond, giving the Lewis electron dot formula, **1A**. Note that the four carbon atoms are bonded together in a chain, the valence of carbon is four, and all remaining valences on the carbon atoms are filled with hydrogen atoms. Structure **1B** is a simpler way to represent the structure, where a line represents the two electrons, but **1B** is equivalent to the Lewis dot structure **1A** without the need to draw all those dots.

An alternative way to draw this molecule, called line drawings, was introduced in Chapter 1 (Section 1.1), and this method is used for the three alkanes shown previously. Each carbon is represented as an intersection of two lines, and each carbon–carbon bond becomes a "line." Each line is a two-electron covalent (sigma) bond. In other words, C–C is a carbon–carbon bond and "–" is used as a shorthand notation to represent that bond. Each intersection of bonds such as [$\diagup\diagdown$] is taken to be a carbon atom. The line representing the C–C bond to each carbon is counted as one contributor to the valence of 4, and the remaining valences are assumed to be hydrogen atoms. Structure **1C** shows the structure of butane using this **line notation**. In **1C**, C1 and C4 are connected to only one other carbon by a single line. There are three remaining valences, so both C1 and C4 are $-CH_3$ units. Similarly, C2 and C3 are connected to two other carbon atoms, as indicated by the two lines. For those carbon atoms, the two remaining valences are filled by hydrogen atoms, so both C2 and C3 are $-CH_2-$ units. Line drawings of organic compounds are used throughout the book.

4.2 Draw the linear alkane with six carbons, using line notation.

1A	**1B**	**1C**

4.2 Millions of Hydrocarbons: Alkanes

In Section 4.1, a hydrocarbon was defined as a molecule containing only carbon and hydrogen and *a hydrocarbon with the general formula C_nH_{2n+2} is called an alkane*. Alkanes that have more than one carbon are composed of carbon atoms bonded to other carbon atoms by sigma covalent bonds, with all remaining valences for each carbon taken up by covalent bonds to hydrogen atoms. Carbon can be bonded to one other carbon (**2**) or form chains of carbons atoms: two, three, four, and so on (see **3, 4, 5**, respectively). If other carbon atoms are attached to carbons in the chain, molecules with "branches" are formed (as in **6**). Long linear chains containing one or many branches of different lengths and composition allow the possibility of an enormous number of alkanes as the number of carbon atoms increases. Note that the more simplistic "line drawings," **4A, 5A,** and **6A** (known as line notation) are used to represent structures **4–6**.

Many different alkanes may have the same formula. For the formula C_6H_{14}, for example, one molecule has six carbon atoms in a linear chain, but two others have five carbons in a linear chain with a one-carbon branch at either the second or the third carbon of the chain. All have the same formula, but different structures with different connectivity. They are different molecules. Molecules with the same empirical formula but different connectivity are called isomers, and they will be discussed in Section 4.5

All molecules discussed in this chapter will contain only carbon and hydrogen. There are linear chains of carbon atoms, and carbons can branch from the linear chain. The carbon branches are known as **_alkyl groups_**. Line notation is used to draw linear alkanes as well as branched alkanes. For example, **4A** is identical to **4, 5A** is identical to **5**, and **6A** is identical to **6**. Structure **4A** is a three-carbon alkane, and **5A** is a four-carbon alkane where the carbons are connected in a linear chain. For **6A**, there is a five-carbon linear chain, with two one-carbon groups attached to that chain.

4.3 Use line notation to draw a nine-carbon straight-chain (unbranched) alkane.

2   3   4   5   6

4A   5A   6A

As noted before, many different structures are possible for a given empirical formula such as $C_{10}H_{22}$ because of the many isomers that are possible. Each of the possible structures will be a different molecule. How can the empirical formula be correlated with a specific molecule (a specific isomer)? A good method for correlating an empirical formula and a molecule relies on physical properties (see Chapter 5, Section 5.8.2). If the boiling points of all possible structures for a molecule within a given family of compounds are known, for example, a match may be found.

For an alkane that has no branching and has an empirical formula of C_3H_7, the C3 alkane has a boiling point of $-42.1°C$, the C6 alkane has a boiling point of $68.7°C$, the C9 alkane has a boiling point of $150.8°C$, and the C12 alkane has a boiling point of $216.28°C$. If an unknown is known to be a C3, C6, C9, or C12 alkane and the boiling point of that unknown is found to be $69°C$, comparison with the listed possibilities allows its identification as the C6 alkane with a

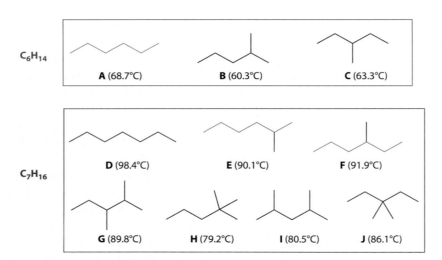

Figure 4.1 Comparison of boiling points for selected branched and nonbranched alkanes.

formula of C_6H_{14}. Even if the measured boiling point is off by several degrees (±5°C, for example), the large difference in boiling point for the given choices makes it obvious which formula correlates with the unknown alkane.

There is a fundamental problem with this approach, however. If the alkane is branched (carbon groups are attached to the long chain of carbon atoms), there are other structural possibilities. Several isomers of the formula C_6H_{14} (**A–C**) and of the formula C_7H_{16} (**D–J**) are shown in Figure 4.1 using line notation. It is easy to distinguish **A** from **D** in Figure 4.1, using only the boiling point, because there is a large difference. It is not easy to distinguish **B** from **C**, however, because there is only a 3°C difference. For the C_7H_{16} series, it is even more difficult to distinguish **E**, **F**, and **G** by boiling points because they are close. In general, nonbranched alkanes can usually be distinguished by boiling points, and alkanes that differ in the number of carbons can also be distinguished. Distinguishing between branched alkanes with the same number of carbons may be problematic.

4.4 Draw all isomers for the C7 compound in Figure 4.1 as the CH structures drawn for 3, 4, and 5 previously.

4.3 Combustion Analysis and Empirical Formulas

The alkane formula is used to identify hydrocarbons according to their structure. **How is the formula for an unknown determined? Can it be proven that a hydrocarbon has a formula that fits the C_nH_{2n+2} rule?** A common experiment burns a compound in the presence of oxygen, making use of Antoine Lavoisier's (France; 1743–1794) discovery from the eighteenth century (see Chapter 1, Section 1.2). The carbon in an organic molecule is efficiently converted to carbon dioxide (CO_2) when burned in oxygen, and this gas can be trapped and weighed. Similarly, the hydrogen atoms in the molecule are converted to water vapor, which can also be trapped and weighed. If the unknown organic compound is accurately weighed and the trapped water and CO_2 are also weighed, it is possible to calculate the percentage of carbon and hydrogen in the original molecule that was burned, and these data are used to calculate the number of carbon and hydrogen atoms.

Knowing the number of carbon and hydrogen atoms allows one to calculate the **empirical formula**. An example of this process takes 0.36 g of an unknown organic compound. When burned, trapping and weighing the products give 0.594 g of water and 1.078 g of CO_2. First examine the carbon dioxide product, with knowledge that the molecular weight of CO_2 is 44. Because the atomic weight of carbon is 12, the ratio of carbon to carbon dioxide (C to CO_2) is 12/44 = 0.2727. This number means that 27.27% of the weight of CO_2 is due to carbon. If the weight of CO_2 trapped in this experiment is 1.078 g, then the weight of carbon in the CO_2 is 1.078 × 0.2727 = 0.294 g.

Similarly, the molecular weight of water is 18, and the atomic weight of hydrogen is 1. Because there are two hydrogen atoms in water, the ratio of hydrogen to water (H to H_2O) is 2/18 = 0.1111 (11.11% of the weight of water is due to hydrogen). Because the weight of water trapped in this experiment is 0.594 g, the weight of hydrogen in the water is 0.594 × 0.1111 = 0.066 g. Both the CO_2 and the water came from the carbon and the hydrogen in the original organic sample and, based on the amount of CO_2 and water trapped, the 0.36 g sample contained 0.294 g of carbon and 0.066 g of hydrogen. The original weight of the unknown was 0.36 g, so the percentage of each element in the unknown can be calculated. The percent of carbon is 0.294/0.36 = **0.817 = 81.7%** and the percent of hydrogen is 0.066/0.36 = **0.183 = 18.3%**.

How can this percentage be translated to a formula? Assume that there are 100 g of the unknown just to make it easy (this is an arbitrary choice). The calculations showed that the sample contained 81.7% of carbon and 18.3% of hydrogen. Therefore, in a 100 g sample, there would be 81.7 g of carbon and 18.3 g of hydrogen in the unknown. The atomic mass of carbon is 12 and the atomic mass of hydrogen is 1. With the weight of each element and the atomic mass, the number of moles of carbon and hydrogen in the sample can be calculated. The moles of carbon in 100 g of unknown is 81.7/12 = 6.81. The moles of hydrogen in 100 g of unknown is 18.3/1 = 18.3.

Dividing the smaller number into both number will give the molar ratio of each element. This calculation gives 6.81/6.81 = 1 and 18.3/6.81 = 2.69 or 2.69H/1C. This ratio can also be expressed as 2.69 H:1 C or $C_1H_{2.69}$. Clearly, there are no fractional atoms, so simply multiply this ratio by a whole number to obtain a whole number value for all elements. In this case, a multiplication factor of 2 gives fractional carbons, so the multiplication factor is 3, which leads to 2.69 H × 3 and 1 C × 3 or 8.07 H (rounded to 8) and 3 C. The empirical formula is therefore C_3H_8.

This is the **empirical formula** and many isomers are possible for this formula. If the molecule is known to be an alkane, it *must* have a formula of C_nH_{2n+2}. The C_3H_8 formula fits the alkane formula. This overall process of burning a sample to trap the CO_2 and water, determining empirical formula and thereby molecular formula, is known as **combustion analysis** or often simply as C, H, N analysis. In order to correlate the empirical formula to the structure of the unknown, more information must be known, such as the boiling point information shown in Figure 4.1.

4.5 Calculate the percent of C and percent of H for the formula $C_{16}H_{30}$.

4.6 Determine the grams of CO_2 and H_2O from combustion of 0.348 g of a sample that has 94.34% C and 5.66% H.

4.4 The Acid or Base Properties of Alkanes

The theme of this book is acids and bases, so it is reasonable to ask if an alkane is an acid or a base. There are no electrons to donate, so an alkane is not a base. There are certainly hydrogen atoms in an alkane, but no base has been discussed that is strong enough to remove a proton from an alkane. To be more precise, *alkanes are not bases and they are remarkably weak acids*.

  If an alkane loses a proton to a powerful base, the only available hydrogen atoms are connected to a carbon, and the C–H bond is a rather strong bond. If the C–H unit in **2** loses a proton as an acid, the conjugate base would be **7**. It can be stated categorically that the equilibrium for this reaction lies far to the left (K_a is *very* small), which means there is an extremely low concentration of **7**. The equilibrium constant (acidity constant, K_a) is less than 10^{-40}, for a pK_a of >40. The actual value of K_a will depend on the base that is used, but this generic value indicates that **2** is an *extremely weak* acid. If **7** were to form, it would be a remarkably strong base (very reactive and rather unstable). As noted, **2** does not contain a lone electron pair, and all electrons are tied up in covalent bonds, so it is not a base.

  Why are acid–base properties presented for alkanes? The answer is because it is important to define relative acidity as it relates to other compounds. Establishing that alkanes are considered to be extremely weak acids or bases gives a benchmark for categorizing other weak protonic acids and bases. In Chapter 15, a reaction will be discussed that generates an alkane *as a conjugate acid*. Categorizing alkanes as very weak acids rather than nonacids will help explain that reaction (see Sections 15.3 and 15.5.2).

4.5 Isomers

The concept of isomers was briefly introduced in Section 4.2. A chain of carbon atoms in an alkane can be represented by the structural formula $CH_3(CH_2)_nCH_3$, where n is an integer of the series $n = 0$ to ∞. Methane (CH_4) does not fit this structural formula, although it is, of course, an alkane. For all other alkanes, the structural formula shown generates a series of linear alkanes having only σ-bonds involving C and H, and all compounds will fit the formula C_nH_{2n+2}. When $n = 0$, for example, the alkane is $CH_3–CH_3$; n may be very large, so the linear chain can be an enormous number. This book will not deal with molecules that have such enormous carbon chains, but carbon chains up to 100 carbons are well known. This is an important point because *linear alkanes are used as the basis for a nomenclature system of organic molecules described in Section 4.6*.

Any linear chain for a given alkane may have one carbon or several carbon atoms branched from the linear chain of carbon atoms. The idea of generating carbon chains of different lengths with different numbers of attached carbon branches generates many different molecules. It is also possible to keep the total number of carbon atoms constant and vary the connectivity of those carbon atoms. A simple illustration is Figure 4.2, which shows molecules that have a total of seven carbons and the empirical formula C_7H_{16}. A total of eight *different* structural arrangements, drawn in line notation, are eight different organic molecules that differ only in the *attachment* of the atoms.

Another way to recognize the differences is to say that the molecules have different constitutions. These different compounds have the same empirical formula, but they differ in their attachment of atoms. They are called **isomers** (in this case, *structural or constitutional isomers*) of the formula C_7H_{16}. An analysis of the isomers in Figure 4.2 shows that compound **A** has a linear chain of seven carbons, compounds **B** and **C** have a linear chain of six carbons, compounds **D–G** have a linear chain of five carbons, and **H** has a linear chain of four carbons. Compounds **B–H** are *branched* because carbon groups are attached to the linear chain. Note that compounds **A** (**blue**), **B** (**red**), **C** (**green**), and **D** (**purple**) are redundant structures marked with the same color in Figure 4.2.

A systematic approach may be used to generate the isomers. Beginning with six linear carbons and moving one carbon "down the chain," different isomers

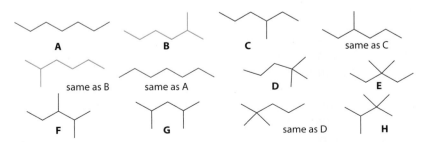

Figure 4.2   Structural variation in alkanes with a total of seven carbon atoms.

are generated. Putting the one-carbon unit at the second carbon and the fifth carbon gives the same structure (**B**). Similarly, putting a one-carbon unit at the third carbon and the fourth carbon gives the same structure (**C**). There is no "left and right" in these structures because you can pick them up and try to put one on the other. ***Use a model kit to make both forms of B shown in Figure 4.2. Try to "superimpose" the models to show that the two forms of B are the same. Do the same experiment with D and E to show they are different.*** Any alkane of a given formula can give rise to many different isomers if there are a sufficient number of carbon atoms. Each different isomer will have different (although closely related) physical properties (see Chapter 5, Section 5.8.2).

Short linear chains of carbon atoms can be attached to a longer linear chain to form branched molecules. Therefore CH_3-, CH_3CH_2-, $CH_3CH_3CH_2-$, etc. units may be the branches from a longer carbon chain. It is also possible to attach branched carbon chains to a longer linear chain, which will generate even more complex molecules. The ability of carbon to form linear and branched chains is a property that leads to millions of possible isomers and hundreds of millions of possible alkanes.

The number of possible isomers is directly related to the total number of carbon atoms in the molecule. An alkane with four carbons (C_4H_{10}) will have only three structural isomers. An alkane with seven carbons (C_7H_{16}) will have eight isomers (see Figure 4.2). Alkanes with the formula $C_{12}H_{26}$ will have 355 isomers. $C_{15}H_{32}$ alkanes will have 4,347 isomers; $C_{25}H_{52}$ alkanes will have about 36.8×10^6 structural isomers, and $C_{40}H_{82}$ alkanes can have about 62.5×10^{12} different structural isomers. The numbers of isomers for a given formula can be calculated by a finite recursive formula,[1–3] but that will not be used in this book. More importantly at this moment is the fact that the number of different alkanes is a huge number, and each one must have a unique name.

4.7 Draw eight *different* isomers of an alkane with the formula $C_{20}H_{24}$.

4.8 Draw all possible isomers for an alkane having six carbons.

4.6 Naming Millions of Isomers: Rules of Nomenclature. The IUPAC Rules of Nomenclature

If there are a vast number of alkanes—certainly several millions—each unique structure requires a unique name. The nomenclature system used today is based on the number of carbon atoms in straight-chain alkanes. It relies on a "code" to designate each type of molecule or functional group, if one is present. (Functional groups will be described later, beginning in Chapter 5.) The focus in this chapter is exclusively on alkanes. Each carbon atom in the linear chain of an alkane will receive a number if a group or atom is attached to that particular carbon, and its position is identified by that number as part of the name. Groups attached to the straight chain of carbon atoms are called *substituents*.

A hydrocarbon substituent that has sp³ hybridized carbon atoms (Chapter 3, Section 3.5) is known as an ***alkyl group*** or an ***alkyl substituent***.

To accommodate the variations in structure, a set of rules has been devised that are universally used to name common organic molecules. The organization that supervises these rules is the *International Union of Pure and Applied Chemistry* (I.U.P.A.C., or just IUPAC).

4.6.1 Prefixes and the Longest Unbranched Chain

The system for naming organic molecules begins with the first 20 straight-chain alkanes (C_1 to C_{20}), which are listed in Table 4.1. A "code" is required to indicate that the molecule is an alkane. The code used is a unique *suffix* and for alkanes that suffix is **"ane."** A *prefix* is added to indicate how many carbons are in the alkane chain. These prefixes are loosely based on the

Table 4.1
Nomenclature for Alkanes with Linear Chains of Carbon Atoms[4]

Number of Carbons	Structure	IUPAC Name
1	CH_4	methane
2	CH_3CH_3	ethane
3	CH_3CH_2CH3	propane
4	$CH_3(CH_2)_2CH_3$	butane
5	$CH_3(CH_2)_3CH_3$	pentane
6	$CH_3(CH_2)_4CH_3$	hexane
7	$CH_3(CH_2)_5CH_3$	heptane
8	$CH_3(CH_2)_6CH_3$	octane
9	$CH_3(CH_2)_7CH_3$	nonane
10	$CH_3(CH_2)_8CH_3$	decane
11	$CH_3(CH_2)_9CH_3$	undecane
12	$CH_3(CH_2)_{10}CH_3$	dodecane
13	$CH_3(CH_2)_{11}CH_3$	tridecane
14	$CH_3(CH_2)_{12}CH_3$	tetradecane
15	$CH_3(CH_2)_{13}CH_3$	pentadecane
16	$CH_3(CH_2)_{14}CH_3$	hexadecane
17	$CH_3(CH_2)_{15}CH_3$	heptadecane
18	$CH_3(CH_2)_{16}CH_3$	octadecane
19	$CH_3(CH_2)_{17}CH_3$	nondaecane
20	$CH_3(CH_2)_{18}CH_3$	icosane (eicosane)
21	$CH_3(CH_2)_{19}CH_3$	hentriacontane
22	$CH_3(CH_2)_{20}CH_3$	docosane
30	$CH_3(CH_2)_{28}CH_3$	triacontane
40	$CH_3(CH_2)_{38}CH_3$	tetracontane
50	$CH_3(CH_2)_{48}CH_3$	pentacontane
100	$CH_3(CH_2)_{98}CH_3$	hectane
132	$CH_3(CH_2)_{130}CH_3$	dotricontahectane

The prefix that indicates the number of carbon atoms in the longest straight chain is indicated in purple.
The suffix -ane indicates that the class name is alkane is shown in green.

Latin word for a given number, but there are exceptions. A one-carbon unit has the prefix **"meth,"** two carbons are **"eth,"** three carbons are **"prop,"** four carbons are **"but,"** and five, six, seven, eight, nine, and ten carbons are derived from the Latin terms **"pent," "hex," "hept," "oct," "non,"** and **"dec,"** respectively.

These linear alkanes are followed by the equivalent of $1 + 10$, $2 + 10$, $3 + 10$, etc. The prefixes are **"undec"** (11), **"dodec"** (12), **"tridec"** (13), **"tetradec"** (14), **"pentadec"** (15), **"hexadec"** (16), **"heptadec"** (17), **"octadec"** (18), **"non-adec"** (19), and **"icos"** (20). By this system, a 12-carbon straight-chain alkane is called **dodecane**. The seven-carbon straight-chain alkane is **heptane** and the 20-carbon straight-chain alkane is called **icosane**. The structures and names of the first 20 alkanes are shown in Table 4.1; the structures are drawn using the $CH_3(CH_2)_nCH_3$ formulas—known as **condensed notation**.

Given this fundamental system of using a **prefix to give the number of carbons** and a **suffix to describe the class of molecules** (and/or functional group), a set of rules allows virtually any organic molecule to be named. This nomenclature system uses the so-called the IUPAC selection rules[4]:

1. **Determine the longest continuous chain of carbon atoms that contains the functional group of highest priority and assign the proper prefix to indicate the number of carbon atoms.**
2. **Determine the class of compounds to which the molecule belongs and assign the proper suffix. For straight-chain, saturated hydrocarbons, the class name is alkane and the suffix is "ane."**
3. **For alkanes, determine the longest straight chain present, using a prefix for the number of carbons as in Table 4.1 and the suffix "ane."**

For a molecule with a linear chain of five carbon atoms, using Table 4.1, the name is pentane, which is the five-carbon straight-chain alkane. ***The first step in nomenclature is to take a given structure and assign a base name. Certainly, it is just as important to take a name and draw the appropriate structure***. Pentane is drawn as the condensed structure, and the same structure is drawn using the line notation described before. The protocol for line notation is worth repeating (see Section 4.1). In line notation, each carbon in the chain is a point and a line connects the dots to indicate a carbon–carbon covalent bond. Each carbon has a valance of 4 (i.e., each carbon forms four bonds), so carbon 1 in the line notation has one bond to a carbon and the remaining three valences are understood to be hydrogen atoms. In other words, as drawn, C1 is CH_3-. Likewise, C2 shows two bonds and the remaining two valences are hydrogen atoms, so C2 is $-CH_2-$, and C2 is also $-CH_2-$. The line notation is a convenient method of drawing structures and will be used throughout the text.

$CH_3(CH_2)_3CH_3$ same as $CH_3CH_2CH_2CH_2CH_3$ same as 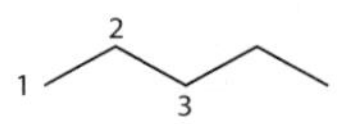   **pentane**

4.6.2 Substituents

When the longest unbranched chain has a branch (another group attached), that group or atom is called a **substituent**. If the substituent has only sp^3 hybridized carbon and hydrogen atoms, it is called an **alkyl group** or an **alkyl substituent**. Another rule covers the name of such compounds:

> 4. **Alkanes that have carbon groups attached to the longest unbranched chain (called substituents) are known as branched chain alkanes. When that branch is an alkane fragment, it is known as an alkyl group or an alkyl substituent.**

Alkyl substituents are the carbon branches attached to the longest carbon chain. When the substituents are hydrocarbon fragments, they require the same prefix used in Table 4.1 to indicate the number of carbon atoms, but a different suffix is required. Branched chain alkanes are named in two parts: The terminal portion of the name of the longest straight chain present in the compound (known as the parent chain) and the name of the group constituting the branch (known as a substituent for a substituting group) or the side chain *precede* the name of the parent chain. In the parent alkane, the suffix "ane" indicates an alkane, but the "ane" is dropped and replaced with "yl" to indicate an alkyl substituent. The nomenclature system for alkyl substituents is shown in Table 4.2.

In this table, the carbon atom that serves as the point of attachment for the substituent to the main chain is shown in **orange**. The prefix for a C1–C20 substituent is meth→icos as in Table 4.1. Combining the "yl" suffix and the alkane prefixes indicates that a one-carbon substituent is methyl, a two-carbon substituent is ethyl, a three-carbon substituent is propyl, a four-carbon substituent is butyl, etc.

Table 4.2
Common Alkyl Substituents

Structure	IUPAC Name	Common Name
CH_3-	methyl	methyl
CH_3CH_2-	ethyl	ethyl
$CH_3CH_2CH_2-$	1-propyl	*n*-propyl
$CH_3CH_2CH_2CH_2-$	1-butyl	*n*-butyl
$(CH_3)_2CH-$	2-methylethyl	isopropyl
$(CH_3)_2CHCH_2-$	2-methylpropyl	isobutyl
CH_3CH_2-CH- $\quad\quad\quad\mid$ $\quad\quad\;\;CH_3$	1-methylpropyl	*secondary*-butyl (*sec*-butyl)
$CH_3CH_2CH_2-CH-$ $\quad\quad\quad\quad\;\mid$ $\quad\quad\quad\quad CH_3$	1-methylbutyl	*secondary*-pentyl
$(CH_3)_3C-$	1,1-dimethylethyl	*tertiary*-butyl (*tert*-butyl)
$(CH_3)_3CCH_2-$	2,2-dimethylpropyl	isoamyl (isopentyl)

In each structure, the highlighted carbon (C) is the point of attachment to the longest continuous chain.

The suffix "yl" comes from work by J. Liebig and F. Wöhler (see Chapter 1, Section 1.2), who adopted it from the Greek word *hyle,* which translates to *material.* Note that the "yl" suffix is used in the generic term alkyl. Apart from the IUPAC names shown in Tables 4.1 and 4.2, a few common names for simple alkyl fragments have been used for many years and must be learned. Table 4.2 shows common alk**yl** substituents with both their IUPAC and common names. The common name for straight-chain fragments such as propyl or butyl adds "*n*," which means "normal." Therefore, propyl becomes *n*-propyl and butyl is *n*-butyl. The fragments *sec*-isopropyl, butyl, *tert*-butyl, and isoamyl are specialized terms.

The term "tertiary" indicates that two methyl groups are on the next-to-last carbon, and "iso" indicates one methyl group on the next-to-last carbon. The term secondary (*sec*) indicates one methyl on the second carbon away from the point of attachment, except that $(CH_3)_2CH-$ is known as isopropyl. These terms become unwieldy if there are more than five carbons, so they are used only for relatively small fragments. The common names are used only for a parent hydrocarbon such as isobutane or *tert*-butylhexane. In the IUPAC system, the systematic names must be used exclusively (labeled IUPAC name), and ***common names are not mixed with IUPAC names***. The IUPAC system for naming the common alkyl substituents will be discussed in Section 4.6.3.

4.9 Draw isobutane and *tert*-butylhexane using line notation.

In principle, an alkyl substituent can be attached to one of several different carbon atoms of the longest linear chain. For example, if a compound is named methylhexane, it is reasonable to ask, "Which methylhexane?" The methyl group can be attached at several positions of the six-carbon linear chain, and different points of attachment will lead to different isomers. Different isomers require unique names. How can the position of the substituent on the linear chain be identified?

The hexane chain (marked in **purple** in **8**) is numbered to give the **methyl** group (the one-carbon alkyl substituent is marked in **orange** in **8**) the lower number. The nomenclature rule states that the numbering sequence is based on the nearest locant, which is the carbon bearing the branch. In other words, the locant is the carbon bearing the substituent. ***Number the longest linear chain from one end to the other by Arabic numerals, and assign the lowest number to the substituent (the nearest locant).***

Two possible numbering sequences are shown for **8** from each end of the molecule. One sequence places the **methyl** group at C3, the other at C4. In the former case, C3 is the nearest locant; in the latter case, C4 is the nearest locant. The nearest locant at C3 takes precedent because the lower numbering scheme places the **methyl** at C3. Therefore, the name of **8** is 3-**methyl**hexane. This point is emphasized by compound **9**, which is a structure that *could* be numbered in such a way that it would be called 8-methyldecane. **This is incorrect!** Compound **9** is named 3-methyldecane due to the requirement that numbering the chain gives the substituent the smaller number (C3 in **9** is the nearest locant). Note that if one could "pick up" an "8-methyldecane" and place it on a model of 3-methyldecane so that the methyl groups overlapped, *both methyl groups are attached to C3. Make a model to test this!*

4.10 Briefly explain why 9 is not named 2-ethylpentane.

4.11 Draw the structure of 4-methylheptane.

The rule stating that the numbering sequence is based on the nearest locant (lowest number is assigned to the substituent) is reinforced with examples **10** and **11**. In **10**, there is an eight-carbon chain (**oct**ane) with a one-carbon substituent (methyl). There are two ways to number the eight-carbon chain, and the preferred one gives the methyl group a number of 4. The name of **10** is 4-methyloctane—*not* 5-methyloctane. Alkane **11** is drawn on the page in such a way that the eye is drawn to an eight-carbon linear chain, but the *longest* linear chain is 10, so **11** is a decane. If **11** is a decane, then there is a three-carbon substituent, a propyl group (see Table 4.2). Numbering the 10-carbon chain to give propyl the lowest number leads to 5-propyldecane rather than 6-propyldecane.

12A   **12B**

The molecule marked **12A** is a seven-carbon straight-chain alkane named heptane. However, **12A** is color coded in a peculiar manner such that a methyl group is attached to the end of a six-carbon linear chain. Can an alkyl substituent be added to the end of a linear chain? Adding a methyl group in this fashion gives a seven-carbon unbranched chain, which is heptane, of course. Adding a methyl group to the end of a chain simply extends the linear chain.

Do not be fooled by this! Because the methyl group is on the last carbon, the *longest continuous chain* is now seven carbons, and this is *not* 1-methylhexane, despite the way in which it is drawn. Compound **12A** is heptane. Examine **12B**, which appears to have a six-carbon linear chain with a methyl group drawn at a different angle. The longest linear chain is seven, so **12B** is heptane. Both line drawings represent the *same molecule. The angle at which a substituent is attached is not important in the line drawings. Look at the points of attachment to determine the structure of the molecule*. If the atoms are the same and the points of attachment are the same, the structures must be identical (like **12A** and **12B**).

4.6.3 Multiple Substituents

When the longest unbranched chain has two or more substituents, new problems arise. There is more than one locant. How is the substituent number determined? In what order are the substituents arranged in the name, and how is their position indicated on the longest chain? There is a rule for these structural motifs:

> **5. In branched hydrocarbons having more than one identical substituent, as in 13 with three methyl groups, name each substituent using rule 4 and Table 4.2, and assign an Arabic number for the position of each substituent on the longest unbranched chain.**

This rule means that the rule *assigns the lowest sequence of numbers such as the locant closest to the end of a chain receives the lowest number*. However, it is possible to have two or more substituents that are identical. In other words, there could be two methyl substituents or three ethyl substituents. The rules state that another prefix is used with the substituent: *di* for two, *tri* for three, *tetra* for four, *penta* for five, and *hexa* for six identical substituents. In this book, molecules with more than six identical substituents are very rare and will be ignored for now. This rule means that two methyl groups are indicated as dimethyl, three ethyl groups are triethyl, and five methyl groups will be pentamethyl. When a compound has more than one substituent, but the substituents are the same, first determine the name of the substituent and then insert the multiplying prefix. If there are two or more identical alkyl substituents, the one to be assigned the lower number is the one cited first in the name.

Examples of compounds with multiple, identical substituents include **13, 14**, and **15**. All three have one-carbon substituents (methyl). In **13**, the "first point of difference" is identified when the numbering in purple shows 2,4,5-trimethyl sequences, whereas the numbering in green shows a 4,5,7-trimethyl sequence. The numbering leads to a difference, and the lowest number sequence is 2,4,5 rather than 4,5,7. This is just another way of saying that the preferred numbering sequence begins with the nearest locant for the first methyl group encountered. In **13**, the nearest locant is C2 rather than C4. For **13**, the longest linear chain is eight, so this is an octane. The closest locant is the methyl group at C2, so 2,4,5-trimethyl is used. Can the molecule be named 2,4,5-methyloctane? **No!** *Every substituent must receive a number, even if there are more than one of the same substituent*. Because **13** has three methyl groups, it is a trimethyloctane, and the name is 2,4,5-trimethyloctane; where *commas separate the numbers and a hyphen separates the number from the rest of the name*.

13   **14**   **15**

Compound **14** has two methyl groups attached to a decane chain: one at C2 and one at C8. The nearest locant is at C2, so **14** is named 2,8-dimethyldecane. For compound **15**, the two nomenclature possibilities are 2,3-dimethyldecane and 8,9-dimethyldecane. The nearest locant is at C2, so the smaller set of numbers is 2,3 (rather than 8,9) and 2,3-dimethyldecane is the correct name for **15**.

4.12 **Draw the structure for 3,3,5,6-tetramethyltetradecane using line notation.**

Alkane **16** poses a problem. The longest unbranched chain is 17, so it is a heptadecane, with a methyl and an ethyl group. In terms of sequencing the substituents in the name, ethyl will come before methyl, but the nearest locant is 5 for either group. Therefore, it could be 5-ethyl-13-methylheptadecane or 13-ethyl-5-methylheptadecane. The rule states that if the same substituent numbers are obtained in either direction, numbering should be in the direction giving the lowest number to the first named substituent. Therefore, **20** is 5-ethyl-13-methylheptadecane.

16

Halogens are not functional groups. Halogen atoms can be attached to the carbon of virtually any alkane structure. When a fluorine, chlorine, bromine, or iodine is attached to a carbon chain, the molecule is named to show that a halogen is present; however, ***there is no unique suffix for the halogen***. Halogen atoms are considered to be substituents and **a halogen substituent is named by dropping the "ine" ending of each halogen and replacing it with "o"** (i.e., fluoro, chloro, bromo, iodo). Given this rule, the other aspects of naming follow the same protocol used for any substituent. In the examples shown, the nearest locants are C1 in **17** and C2 in **18**. Therefore, **17** is 1-iodo-3-methylheptane and **18** is 5-bromo-2-chlorohexadecane.

These names raise a new problem, however. Why is iodo listed before methyl in **17** and bromo before chloro in **18**? Numbering the substituents is established by determining the nearest locant that gives the lowest set of numbers, as described previously. However, ***when there are different substituents, they are arranged alphabetically in the name with their appropriate number***. Using this rule, *i* before *m* leads to 1-iodo-3-methyl in **17**—*not* 3-methy-1-iodo. Similarly, in **18**, *b* before *c* gives 5-bromo-2-chloro, despite the fact that the chlorine atom has the smaller number.

Two more examples are shown to reinforce the rules in this section. In **19**, the longest chain is 12, so it is a dodecane. There are two 2-carbon (ethyl) substituents. The nearest locant is C3, so numbering leads to 3,7-diethyl rather than 6,10-diethyl, and **19** is named 3,7-diethyldodecane. Compound **20** has a chain of

11, so it is an undecane, and there are three different substituents: methyl, propyl, and chloro. There are two chloro groups, so it is dichloro. The two numbering
possibilities lead to 3,6,7,11 or 1,5,6,9; however, the closest locant is at carbon 1,
so 1,5,6,9 is preferred. The name of **20** is 1,9-dichloro-6-methy-5-propylundecane.
Note that the alphabetical arrangement is c→m→p, ignoring the "di" of dichloro.
Also note that each chlorine atom receives a number (3 and 9).

4.13 Draw the structure of 3,4-dibromo-3-methylheptane.

4.14 Draw 6,6-dichloro-9-ethyl-2,4,4-trimethyltetradecane.

        Some alkyl groups are so common that they have *common names* pre-dating
development of the IUPAC rules. These common names are listed in Table 4.2
along with the IUPAC name. **One should always use the IUPAC nomenclature**, but some common names appear so often that one must be able to recognize
them. ***The IUPAC rules indicate that common names should only be used
for the parent alkane, and not for substituents. Substituents are named
according to the IUPAC rules listed above.*** As shown in Table 4.2, the four-
carbon substituent named 1,1-dimethylethyl is known as a *tertiary*-butyl group
(or *tert*-butyl). The term *tertiary* refers to the fact that the carbon attached to
the main chain is tertiary (i.e., it bears three carbon entities). The 1,1-dimethyl-
propyl substituent is called *tertiary*-amyl, where amyl is an old term for pentyl.
The three-carbon fragment 1-methylethyl group is called isopropyl; the term
"iso" refers to a carbon at the end of the branch chain that has one hydrogen and
two methyl groups. The four-carbon 2-methylpropyl is called isobutyl and the
3-methylbutyl fragment is called isopentyl (sometimes isoamyl).
        When a methyl group is attached to the C1 carbon of the branch chain and four
or more carbons are in the substituent, the common term used is *secondary* or *sec*.
The 1-methylpropyl fragment is therefore *sec*-butyl and the 1-methylbutyl fragment
is *sec*-pentyl or *sec*-amyl. A special fragment, the 2,2-dimethylpropyl fragment, is
known as neopentyl (see Table 4.2). It is important to emphasize that these common names should *not* be used when a molecule is given an IUPAC name. From
time to time, however, common names will be mentioned in this text.

**4.15 Draw the structure of "2-*tert*-butyloctane" (not the actual name)
        and then give it the proper IUPAC name.**

        Why are the IUPAC names for substituents listed with common names?
In order to name such substituents, the longest linear chain is numbered
from one end to the other by Arabic numerals, *from the point of attachment
to the longest linear chain.* In other words, identify the substituent and begin

numbering the substituent chain from the point of its attachment. The direction of numbering is chosen to *give the lowest numbers possible to the side chains on the substituent.* Using this rule, *tert*-butyl is 1,1-dimethylethyl and *sec*-butyl is 2-methylpropyl.

4.16 Give the IUPAC prefix for the longest chain in the isopropyl group, the *sec*-butyl group, and the *tert*-butyl group.

4.6.4 Complex Substituents

In some structures, the longest unbranched chain has a substituent, and this substituent has one or more substituents. In other words, the branch has branches. The IUPAC names *tert*-butyl (1,1-dimethylethyl) and *sec*-butyl (2-methylpropyl) from the previous section are examples, and those names require further explanation. Another rule is required to name such compounds:

> 6. **If a complex substituent is present on the longest continuous chain, count the number of carbon atoms in the longest continuous part of that side chain and use the proper prefix. The name of a complex substituent is considered to begin with the first letter of its complete name (take the longest chain of the substituent from the point of attachment to the longest unbranched chain and ignore "di," "tri," etc.).**

In **21**, the longest unbranched chain is 17, and there is a methyl substituent and an ethyl substituent. Using the rule that substituents are arranged alphabetically, **21** is a 13-ethyl-5-methylheptadecane. However the group at C8, in the box, is a complex substituent. The point of attachment on the 17-carbon linear chain is at C8, and there is a four-carbon linear chain attached to C8. The butyl substituent is numbered, beginning from the point of attachment, and that carbon is labeled C1. Therefore, **21** is a 8-butyl-13-ethyl-5-methylheptadecane. However, the butyl substituent has methyl substituents. The carbon attached to C8 is numbered 1. The butyl side chain, in the box, has three methyl groups attached at C1 and C3 as the side chain is numbered. The complex side chain is therefore a 1,1,3-trimethylbutyl, and the entire unit is attached to C8 of the heptadecane. The final name of **21** is 8-(1,1,3-trimethylbutyl)-13-ethyl-5-methylheptadecane.

Complex side chains are named by choosing the longest chain from the point of attachment to the longest linear chain that gives rise to the base name (here, heptadecane), numbering the side chain, and inserting the secondary substituents based on that numbering. In order to set the complex substituent apart from the other substituents, parentheses are used, as in **21**. Note that the 1,1,3-trimethylbutyl group in **21** is alphabetized by the "b" because it is a butyl group (ignoring the "tri" and the "methyl").

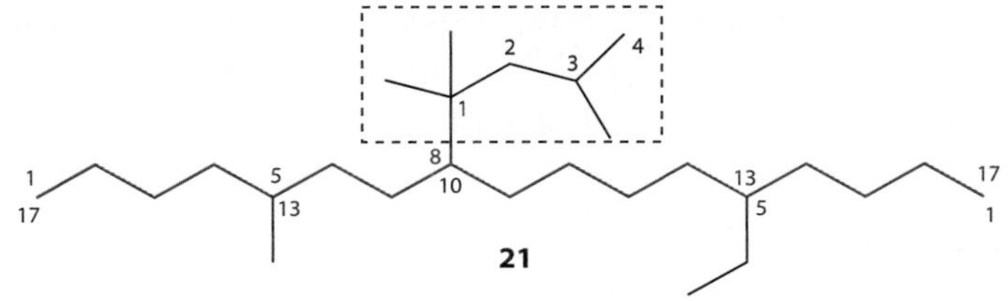

21

Several corollaries are used if chains of equal length are competing for selection as the main chain in a saturated branched acyclic hydrocarbon. The main chain is that which has the greatest number of side chains; the chain whose side chains have the lowest numbered locants; the chain having the greatest number of carbon atoms that have smaller side chains (methyl or ethyl rather than a complex substituent, as discussed in **6**); or the chain having the least branched side chains.

In some cases, the molecule contains two or more complex substituents that have the same number of carbons in the longest chain, but they are different. An example is the three pentyl groups in **22**. This compound is a tripentylicosane, but the secondary substituents are different. At C7, there is a 1-methylpentyl, there is a 4-methylpentyl at C8, and a 3-methylpentyl group is at C8. In such cases, the rule states that priority for citation is given to the substituent that contains the lowest locant at the first cited point of difference—in this case, the 1-methylpentyl, followed by 3-methylpentyl, followed by 4-methylpentyl. The name of **22** is 7-(1-methylpentyl)-9-(3-methylpentyl)-8-(4-methylpentyl)icosane.

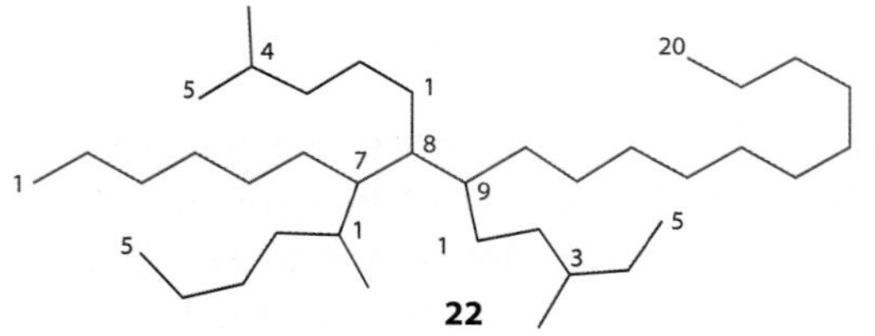

22

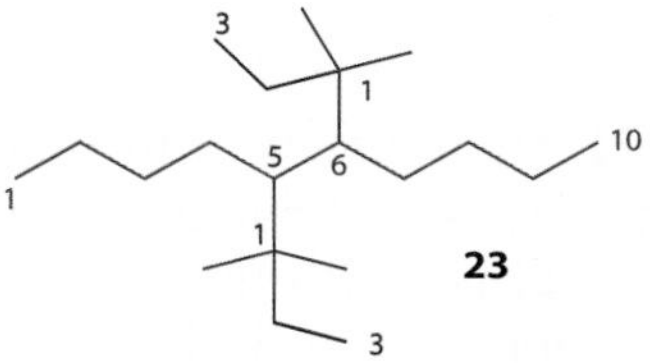

23

A related case occurs when there is more than one complex substituent, but they are identical. When identical substituents are substituted in the same way, they may be indicated by the appropriate multiplying prefixes *bis, tris, tetrakis, pentakis,* etc. The complete expression denoting such a side chain may be enclosed in parentheses or the carbon atoms in side chains may be indicated by prime numbers. For **23**, a decane, there are two 1,1-dimethylpropyl groups at C5 and C6, so the term *bis* is used. Compound **23** is therefore 5,6-*bis*-(1,1-dimethylpropyl)decane.

4.17 Give the proper IUPAC name for each of the following:

(a)

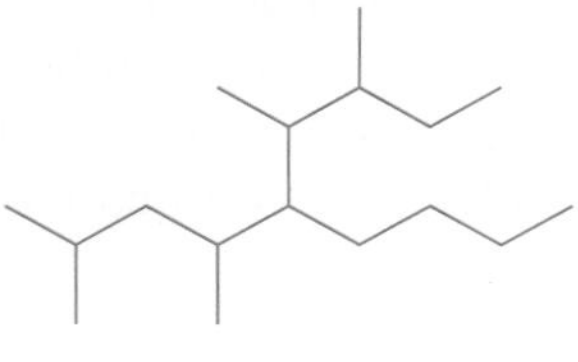

(b)

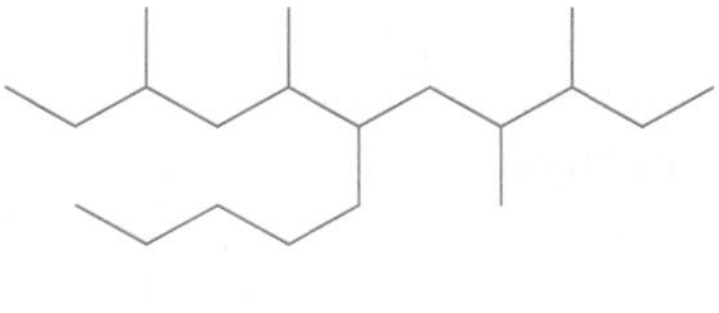

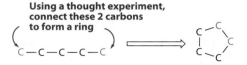

Figure 4.3   Cyclic alkanes are possible.

4.7 Rings Made of Carbon. Cyclic Compounds

In all examples discussed so far, the molecules were formed from linear chains of carbon atoms with or without substituent branches attached to the linear chain. Alkanes are known to have rings of carbon atoms. Using a "thought experiment," imagine that the "terminal" carbons of the linear alkane pentane are connected to form a carbon ring (as illustrated in Figure 4.3). This thought experiment generates a ring, and such compounds are known as cyclic compounds. If all the carbon atoms are sp^3 hybridized, they are *cyclic alkanes*.

In an acyclic alkane (acyclic means there is no ring), every carbon has the maximum number of hydrogen atoms attached to it (determined by the fact that each carbon must have four bonds). In a cyclic alkane, two carbon atoms must be joined to form a ring, and there are *two fewer hydrogen atoms* when compared to an acyclic alkane. Because of this, the **general formula for cyclic alkanes** is

$$C_nH_{2n}$$

where n is an integer: 2, 3, 4,…..

The smallest ring alkane that is possible has three carbon atoms, so for a cyclic alkane the integer n *must be* ≥3. The formula is C_3H_6 when $n = 3$ and $C_{100}H_{200}$ for $n = 100$, and so on. In the absence of substituent, each carbon in a cyclic alkane will have two hydrogens attached (a CH_2 unit, known as a **methylene** unit) and that carbon will be connected to two other carbons in the ring. Figure 4.4 shows the basic structure and the names of 12 cycloalkane rings (cyclic alkanes) that have 3–12 carbon atoms. Note that **general formula for cyclic alkanes, C_nH_{2n}, is the same as the general formula for alkenes** (see Chapter 5, Section 5.1), which are characterized by a C=C unit.

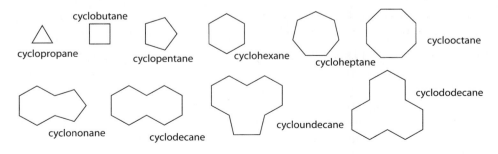

Figure 4.4   Cyclic hydrocarbons of 3–12 carbons.

Alkenes have not yet been discussed, but as a preview, imagine an unknown compound with the formula C_5H_{10}. Structural possibilities include a linear five-carbon alkene or a five-carbon cyclic alkane, which means that *they are isomers*. Cyclopentane and pentane are not isomers because they have different empirical formulas.

4.18 Draw the structure of the cyclic alkane that has 15 carbons in the ring.

4.19 Write a structural isomer of cycloheptane that does not have a ring (see Chapter 5 for a hint).

There is a nomenclature dilemma for cyclic alkanes. The suffix for the name must be *ane* because they are alkanes. The normal prefix will indicate the presence of 3–12 carbons for the alkanes in Figure 4.4, so the names should be propane to dodecane. Those names have already been taken for the acyclic alkanes! **To distinguish between the linear 12-carbon molecule (dodecane) and the 12-membered cyclic (ring) alkane, use the term "cyclo."** If the ring is viewed as a **cycle**, then the prefix **"cyclo"** is used in front of the alkane name (a prefix in front of the carbon number prefix). The three-membered ring alkane therefore becomes cyclopropane and the 12-membered ring alkane becomes cyclododecane (see Figure 4.4).

4.20 Draw the structure of cyclohexadecane.

24   25

Cyclic alkanes can have substituents attached, and the nomenclature must identify the nature and the position of each substituent. The position of a substituent when it is attached to a cyclic alkane is assigned the lowest possible number, as with acyclic alkanes. However, in **24**, this six-membered cyclic alkane (cyclohexane; marked in **green**) has one methyl substituent (marked in **orange**). The name is methylcyclohexane. Technically, it is 1-methylcyclohexane, but there is only one group attached to the ring, so the "1" is obvious and it is omitted. With 1,3-dimethylcyclohexane (**25**), however, both numbers must be included to specify the position of the methyl groups (in **orange**) on the ring (in **green**) and the relationship of the groups one to the other. The ring is numbered clockwise from the "top" methyl group to give the substituents the lowest possible combination of numbers. Numbering in the opposite direction (counterclockwise) from the top methyl in **25** would give 1,5-dimethylcyclohexane, *but this is an incorrect name*. If different substituents are attached to the ring, assign the lowest combination of numbers and list the group names on the ring in alphabetical order.

4.21 Write a structural isomer of methylcyclooctane that has a nine-membered ring, a six-membered ring, and a seven-membered ring; then, name all three compounds.

4.22 Draw the structure of 1-chloro-3-ethyl-4,5,6-trimethylcyclononane.

Cyclic alkanes that have multiple substituents require that the groups are prioritized in order to identify C1. The rule states that substituents should be given the lowest possible number. Examination of **26** shows that it is a six-membered ring, so the base name is cyclohexane; there are a chloro substituent, an ethyl substituent, and two methyl substituents (dimethyl), so it is a chloro-ethyl-dimethylcyclohexane, alphabetizing the groups. Four numbering schemes are shown, where **26A** is numbered beginning at the more substituted position and numbers toward the methyl group to give 5-chloro-1-ethyl-1,2-dimethylcyclohexane. Structure **26B** also begins numbering at the more substituted position, but numbers toward the chlorine to give 3-chloro-1-ethyl-1,6-dimethylcyclohexane. Comparing these two schemes, **26A** gives the lower set of numbers, consistent with numbering substituents to give the lower set of numbers. In other words, if ethyl and one methyl are at C1, number to give the second methyl and also the chloro group the lower possible number.

Structure **26C** is numbered to give the substituent the lower numbers based on the methyl being at C1 and would give 4-chloro-2-ethyl-1,2-dimethylcyclohexane. This scheme gives the same absolute value of numbers relative to 5-chloro-1-ethyl-1,2-dimethylcyclohexane (**26A**). Structure **26D** is numbered such that chloro is at C1, and the substituents receive the lower numbers to give 1-chloro-3-ethyl-3,4-dimethylcyclohexane, which is a higher set of numbers relative to **26A** and **26C**. To distinguish them, compare the second and third numbers based on ethyl at C1 or chloro at C1: 1:1:2:5 for **26A** versus 1:2:2:4 for **26C**. It seems clear that **26C** gives the lower set of secondary and tertiary numbers, so **26** is named 4-chloro-2-ethyl-1,2-dimethylcyclohexane. This analysis is clearly more convoluted and relies on subtle differences to establish the name. The names of most examples in this book will be more straightforward.

Another convoluted example is similar to the analysis of **26**. Cyclic alkane **27** could be named 1,1-dibromo-3-ethyl-4-methylcyclopentane, 1,1-dibromo-4-ethyl-3-methylcyclopentane, 3,3-dibromo-1-ethyl-5-methylcyclopentane, 4,4-dibromo-1-ethyl-2-methylcyclopentane, 3,3-dibromo-5-ethyl-1-methylcyclopentane, or 4,4-dibromo-2-ethyl-1-methylcyclopentane. The lowest combination of numbers is either 1,1-dibromo-3-ethyl-4-methylcyclopentane or 1,1-dibromo-4-ethyl-3-methylcyclopentane, but this choice leads to another dilemma because it could be 3-ethyl-4-methyl or 4-ethyl-3-methyl. To distinguish the latter, rely on the alphabetizing scheme of b→e→m. In this case, the alphabetized rank ordering places ethyl before methyl and that numbering scheme leads to 1,1-dibromo-3-ethyl-4-methylcyclopentane for **27**.

In other words, when two different alkyl groups can receive the same numbers, number them by alphabetical priority. Compound **28** is another example, and possible names are 1-(1-methyethyl)-2-propylcycloheptane or 2-(1-methylethyl)-1-propylcycloheptane. Note that the (1-methylethyl) substituent is treated as a methylethyl group for alphabetizing. Because "m" is before "p," in the alphabet, the proper name of **28** is named 1-(1-methylethyl)-2-propylcyclo-heptane. The name for **28** uses 1-methylethyl rather than isopropyl (Table 4.2). Remember that common names and IUPAC names should not be mixed.

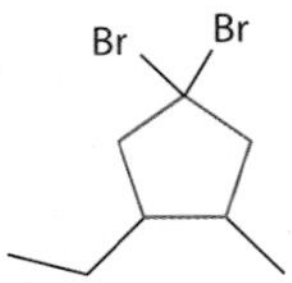

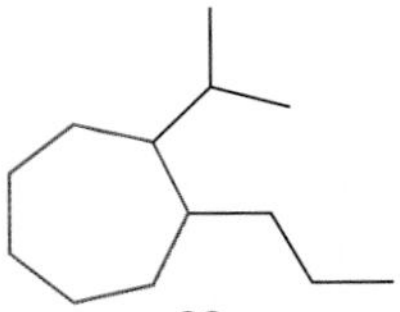

A different subtle naming problem arises when comparing **29** and **30**. Alkane **29** has a cyclopentane unit and **30** has a cyclooctane unit. However, **29** has a seven-carbon acyclic unit attached to the five-membered ring, and **30** has a six-carbon acyclic unit attached to the eight-membered ring. What is the parent unit in these structures? The rule states that one must count the carbons in the ring and also in each noncyclic unit. *If the number of carbon atoms in the ring is equal to, or greater than, the number in the non-cyclic unit, the compound is named as cyclic unit, a cycloalkane. If the number of carbon atoms in the largest noncyclic unit is greater than the number in the ring, the compound is named as an acyclic alkane, and the ring is treated as a substituent.* Note that a cyclopropane unit treated as a substituent is cyclopropyl, a four-carbon ring substituent is cyclo-butyl, a five-carbon ring substituent is cyclopentyl, etc. Therefore, **29** is 1-cyclo-pentylhexane, and **30** is (1-ethylbutyl)cyclooctane.

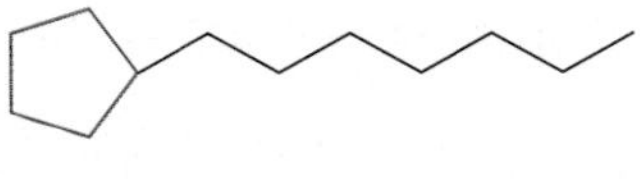

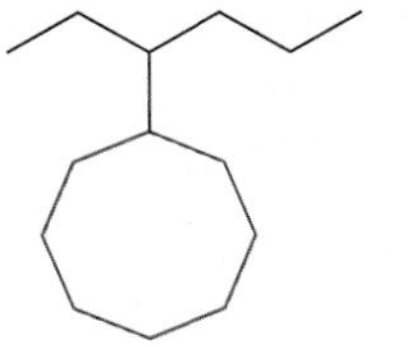

4.23 Give the IUPAC name for the following compound:

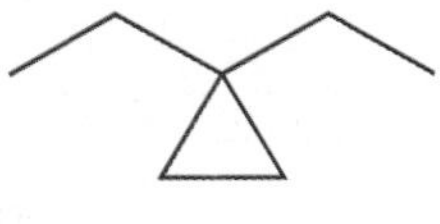

4.8 Biological Relevance

Due to their remarkably poor chemical reactivity, it may be surprising that alkanes have a place in biological systems. Two examples are the formation of alkanes during lipid peroxidation and the use of microorganisms to digest and remove hydrocarbons, as in cleaning up oil spills.

It has been reported that alkanes are formed by peroxidation (see Chapter 17 for oxidation reactions) of unsaturated fats, which leads to fatty acid hydroperoxides that usually decompose.[5,6] In one study, the levels of ethane and pentane in human breath were measured as markers of lipid peroxidation in patients who smoked, suffered from HIV infection, or suffered from inflammatory bowel disease.[7] Breath alkane output and other lipid peroxidation parameters are significantly reduced with antioxidant vitamin supplementation.[7]

From the standpoint of chemical reactions, alkanes are remarkably unreactive. It is possible to burn them, and this is done every day in the form of gasoline, which is mainly a complex mixture of hydrocarbons. Long straight-chain and branched-chain alkanes are major components of crude oil, accounting for their presence in gasoline and fuel oil. The process of "cracking" crude oil breaks down and separates long complex hydrocarbons into mixtures of hydrocarbons that have simpler structures and lower boiling points and vapor pressures. The cracking process simply heats the crude oil to high temperatures to break carbon–carbon bonds, producing many smaller branched chain alkanes, among other products.

All of us are familiar with the unfortunate oil spills that occur from time to time. Cleaning up these hydrocarbons is a problem, but patents have been issued for highly absorbent hollow spheres of beeswax that bind hydrocarbons, providing nutrients to naturally occurring microbes that create enzymes to biodegrade contaminants.[8]

Patents have been recorded for enzymatic degradation of fuel oils. Organisms that utilize alkanes are used to degrade petroleum pollutants cometabolically (cometabolism is a process in which microorganisms transform compounds not utilized for growth or energy into other products) or by direct metabolism.[9] The metabolism of petroleum compounds is enhanced through oxygenase-catalyzed reactions, where an oxygenase is an enzyme that catalyzes the incorporation of molecular oxygen into its substrate.

The alkane methane has an unusual relationship with biology. It is a greenhouse gas and is known to be 10 times more effective than carbon dioxide in contributing to climate change. Methane is produced by decaying vegetation and is referred to as swamp gas in some regions. It is a by-product of the digestion of mammals, such as cows. Methane is known to be trapped in marine sediments as methane hydrate, which is a crystalline solid consisting of methane molecules surrounded by a cage of water molecules. Methane hydrate is stable in ocean floor sediments at depths of greater than about 1,000 feet, where the methane hydrate is kept very cold and under high pressure. There is a theory that warming of the Earth 250 million years ago, caused by excessive volcanic

4.11

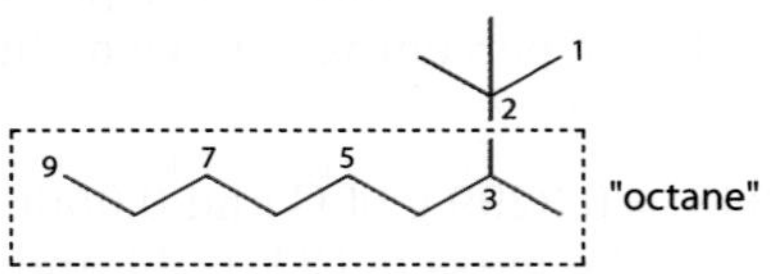
1
2
9
7
5
3
"octane"

This molecule is named 2,2,3-trimethylnonane

4.12

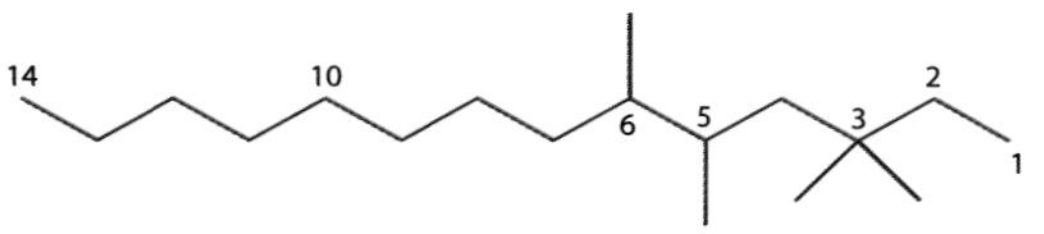
14
10
6
5
3
2
1

4.13

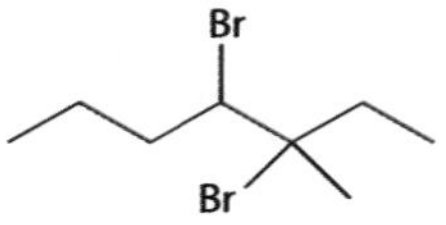
Br
Br

3,4-dibromo-3-methylheptane.

4.14

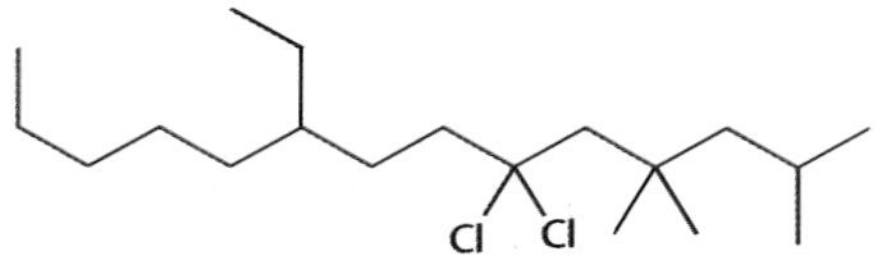
Cl
Cl

6,6-dichloro-9-ethyl-2,4,4-trimethyltetradecane

4.15

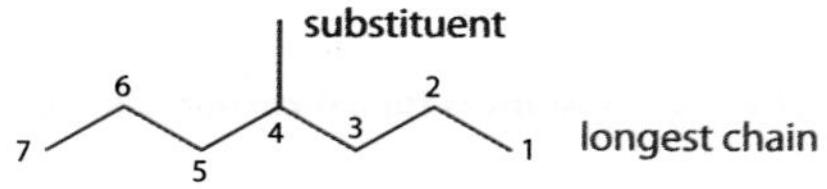
substituent
6
2
7
4
3
1
5
longest chain

4.16

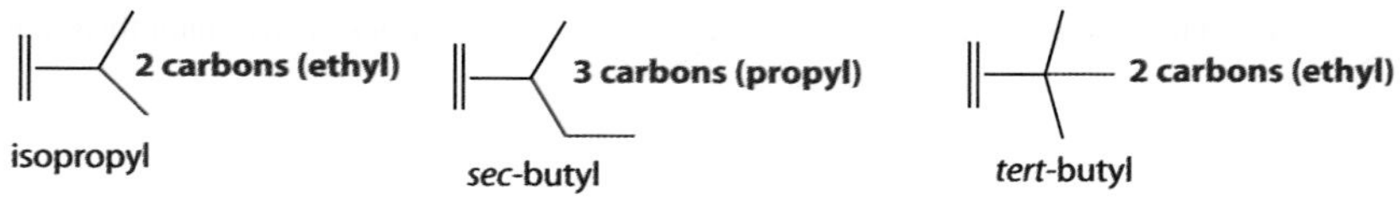
2 carbons (ethyl)
isopropyl
3 carbons (propyl)
sec-butyl
2 carbons (ethyl)
tert-butyl

4.17 (a) 5-butyl 2,4,6,7-tetramethylnonane; (b) 3,4,7,9-tetramethyl-6-pentylundecane.

4.18

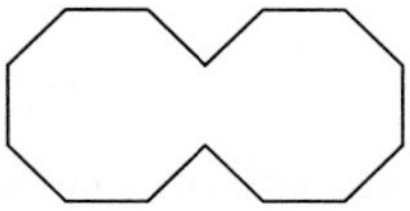

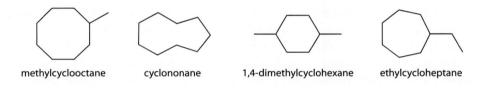

cyclopentadecane

4.19

1-heptene

4.20

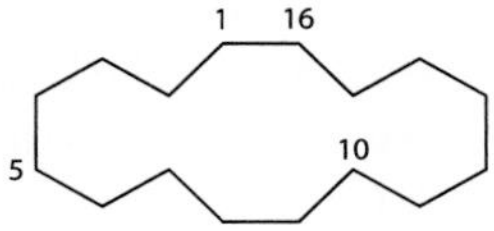

cyclohexadecane

4.21

methylcyclooctane   cyclononane   1,4-dimethylcyclohexane   ethylcycloheptane

4.22

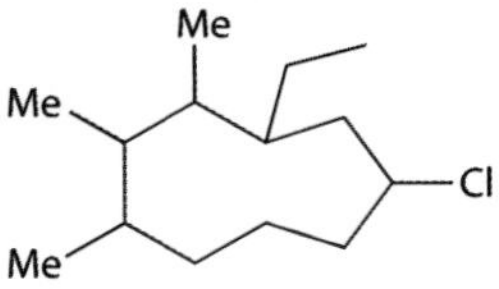

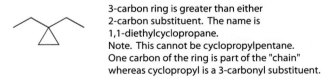

1-chloro-3-ethyl-4,5,6-trimethylcyclononane.

4.23

3-carbon ring is greater than either
2-carbon substituent. The name is
1,1-diethylcyclopropane.
Note. This cannot be cyclopropylpentane.
One carbon of the ring is part of the "chain"
whereas cyclopropyl is a 3-carbonyl substituent.

Correlation of Concepts with Homework

- **Hydrocarbons are molecules that contain only carbon and hydrogen, and alkanes are hydrocarbons that have only carbon–carbon single bonds and carbon–hydrogen bonds with the generic formula C_nH_{2n+2}: 1, 2, 3, 24, 27, 32, 39.**
- **The structure of hydrocarbons is effectively tetrahedral about a central carbon atom: 37, 45.**
- **Molecules with different connectivity are considered to be a different molecule. Isomers are molecules with the same empirical formula and the same number of atoms, but the atoms are attached in different ways: 4, 7, 8, 26, 30, 31, 42, 43.**
- **Combustion analysis is used for determining the percent of C and percent of H in an alkane, which is then used to calculate the empirical formula: 5, 6, 33, 40.**
- **The IUPAC rules of nomenclature involve identification of the longest continuous carbon chain as assigning a prefix that correlates with the number of carbon atoms. The class name for a molecule is designated by a suffix, which is "ane" for alkanes. Rules are established based on the number of carbons, assignment of the lowest number(s), and position on the longest carbon chain for naming substituents. Groups attached to the longest continuous chain are known as substituents and given the suffix "yl" for alkane base substituents. Complex substituents are treated as a substituent-on-a-substituent: 9, 10, 11, 12, 13, 14, 15, 16, 17, 24, 25, 26, 27, 28, 29, 41, 44.**
- **Cyclic alkanes use the prefix "cyclo": 18, 19, 20, 21, 22, 23, 29, 34, 37, 38.**

Homework

4.24 Draw the structure of butane, decane, and tridecane using line notation for making chemical structures.

4.25 (a) Which of the following molecules have an IUPAC name that uses the prefix "prop"?

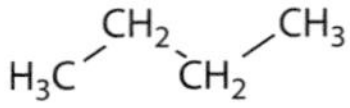

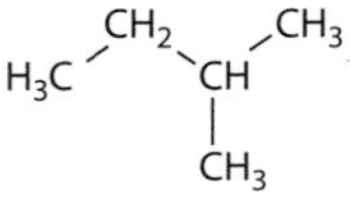

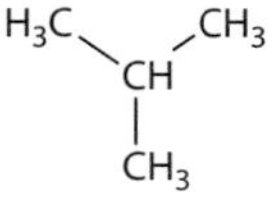

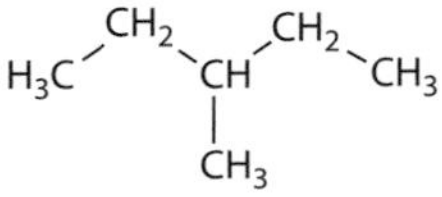

(b) Indicate which of the following substituents have the term propyl as part of the name:

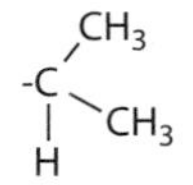

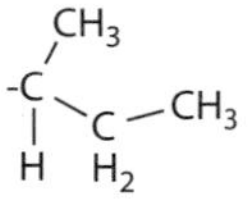

(c) Indicate which of the following molecules have a prefix of "pent":

C_5H_{10} $\qquad$ C_6H_{14} $\qquad$ C_5H_8 $\qquad$ $C_{100}H_{202}$ $\qquad$ $C_{60}H_{120}$

4.26 (a) Indicate, by letter, which of the following molecules are isomers with the empirical formula C_8H_{18}:

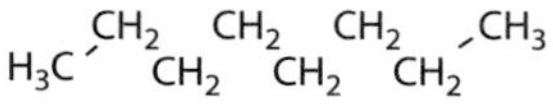

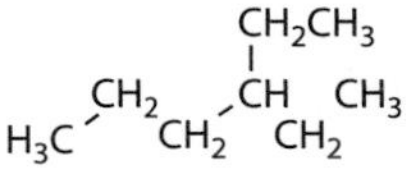

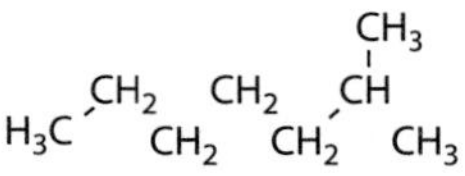

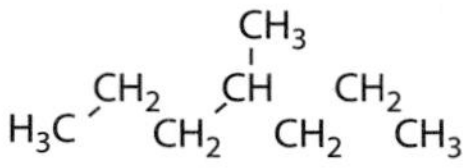

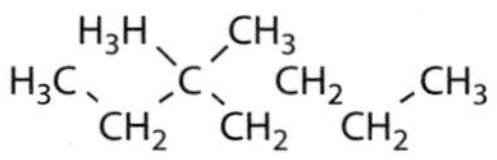

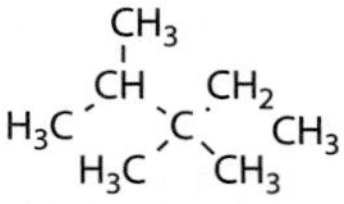

(b) Indicate which of the following molecules are isomers with the formula C_6H_{14}:

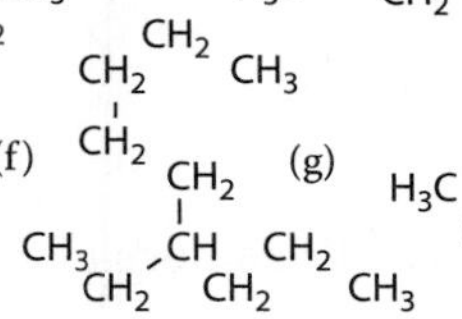

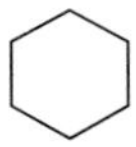

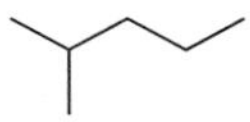

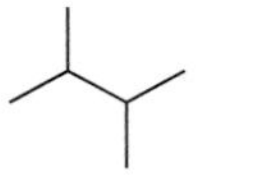

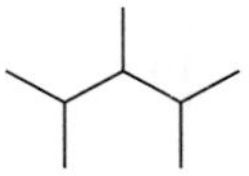

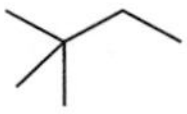

4.27 Identify all of the following formulas that are alkanes:

C_5H_{10} $\qquad$ C_5H_{12} $\qquad$ C_5H_8 $\qquad$ $C_{100}H_{202}$ $\qquad$ $C_{60}H_{120}$

4.28 Give the correct IUPAC name (no common names) for each of the following:

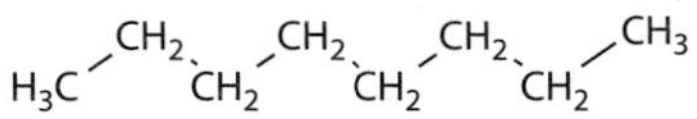

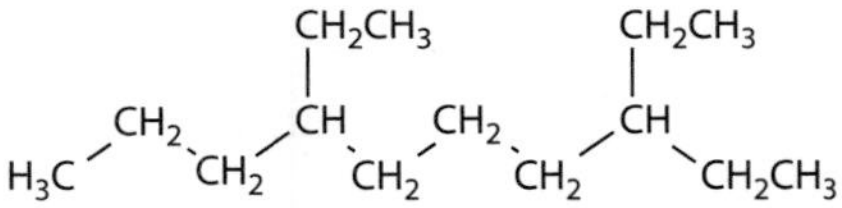

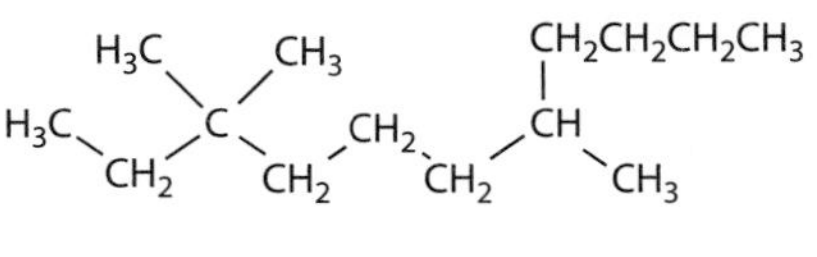

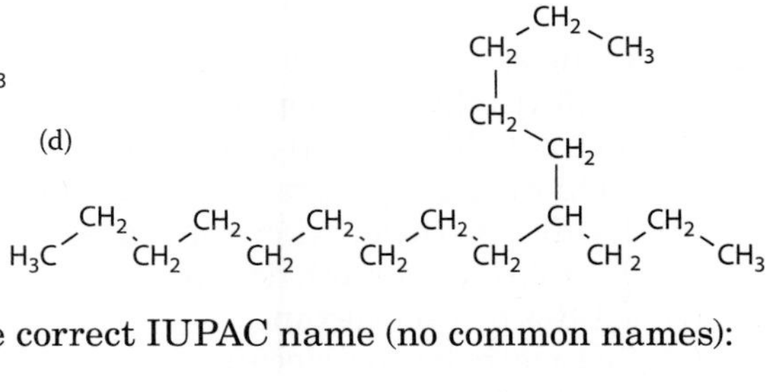

4.29 For each of the following, give the correct IUPAC name (no common names):

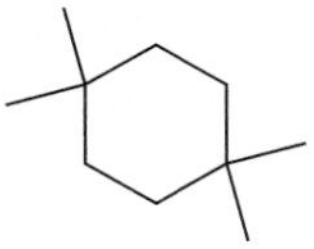

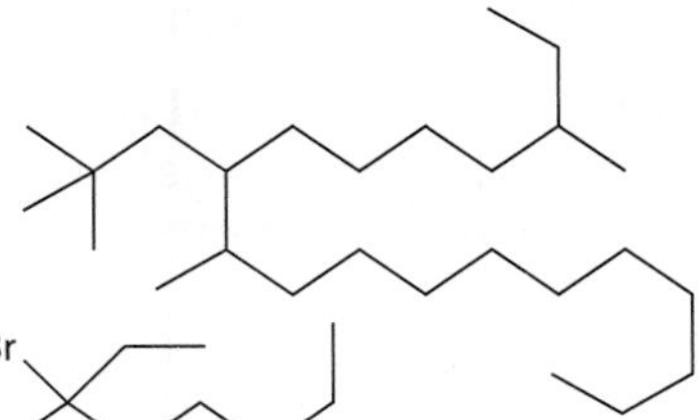

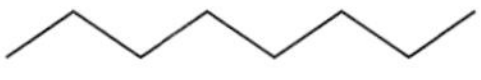

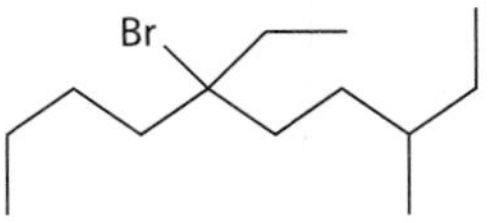

4.30 (a) Draw **six** *different* isomers of 3,3-dimethylheptane. Do NOT include 3,3-dimethylheptane. Give the IUPAC name for each isomer you draw.

   (b) Draw **six** *different* isomers of 1-bromooctane. Give the IUPAC name for each isomer you draw.

   (c) Draw the structure, using line notation, for **five** *different* isomers of the formula C_6H_{14}. Name each structure and do **not** use common names.

4.31 (a) Identify all *isomers* of 3,5-dimethylheptane:

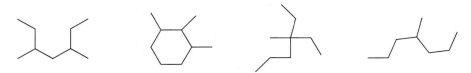

   (b) Identify all isomers of 2-methylhexane:

   (c) Identify all isomers of cycloheptane:

   (d) Identify all molecules that are isomers of cyclohexane:

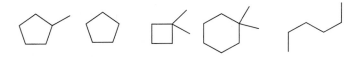

4.32 Draw the structure for each of the following using line notation:

   (a) 1,2,3-triethylcycloheptane
   (b) 5-(1,2-dichlorobutyl)-2-methylhexadecane
   (c) 1-chloro-2,2,4,4-tetramethylhexane
   (d) 1,4-di(1-methylethyl)cyclohexane
   (e) 2,2-dibromo-3-methyloctane
   (f) 1,1-diethylcyclohexane
   (g) 5-(1-methylpropyl)decane
   (h) 2,2,3,3,4,4-hexamethylheptane

4.33 Calculate the **empirical formula** for a molecule having only C and H with the following combustion analysis. A sample weighing 0.6000 g was burned in the presence of oxygen to give 0.7692 g of water and 1.8827 g of CO_2.

4.34 Indicate which of the following *substituents* has ethyl for the basis of the name from the point of attachment to the main chain:
(a) $-CH_3$
(b) $-CH_2CH_3$
(c) $-CH_2CH_2CH_2CH_3$
(d) $-CH(CH_3)_2$
(e) $-CH_2CH_2CH_3$
(f) $-C(CH_3)_3$

4.35 Indicate the **prefix** that is used for the molecule $(CH_3)_2CH(CH_2)_2CH_2OH$: meth, eth, prop, but, pent, or hex.

meth-   eth-   prop-   but-   pent-   hex-

4.36 Briefly explain why a cyclohexane ring with an attached ethyl group is named ethylcyclohexane but a cyclohexane ring with an attached dodecane chain is named cyclohexyldodecane. Draw both of these molecules.

4.37 Draw each of the following in line notation:
(a) $CH_3CH_2CH_2CH_2CH_2CH_3$
(b) $(CH_3)_2CHCH_2CH_2CH_3$
(c) $CH_3CH_2CH_2CH_2CH_2CH=CH_2$
(d) $CH_3CH_2C\equiv C(CH_2)_8CH_3$

4.38 Calculate the percent of C and H as well as an empirical formula for each of the following using the combustion analysis provided:
(a) Combustion of 0.52 g of an unknown organic compound gives 0.6688 g of water and 1.6344 g of CO_2.
(b) Combustion of 0.81 g of an unknown organic compound gives 0.85789 g of water and 2.6208 g of CO_2.
(c) Combustion of 1.04 g of an unknown organic compound gives 1.4779 g of water and 3.2116 g of CO_2.
(d) Combustion of 0.12 g of an unknown organic compound gives 0.1656 g of water and 0.4114 g of CO_2.

4.39 Give the proper systematic name for each of the following; do **not** use common names:

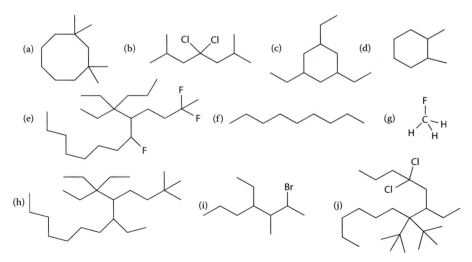

4.40 Give the proper IUPAC name for each of the following:

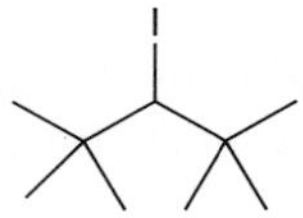

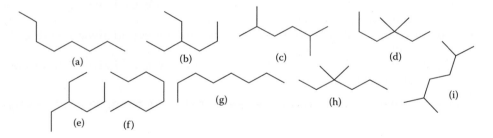

4.41 Draw each of the following fragments and assign it an IUPAC label rather than the common name shown:
(a) *tert*-butyl
(b) isopropyl
(c) isobutyl
(d) *sec*-butyl
(e) *tert*-amyl
(f) amyl

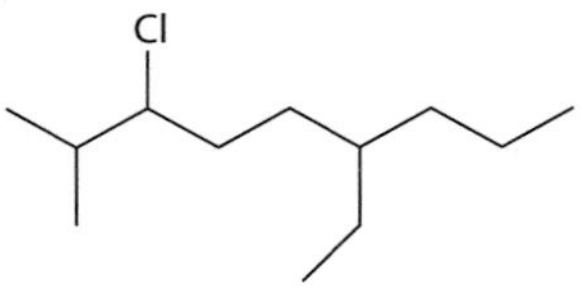

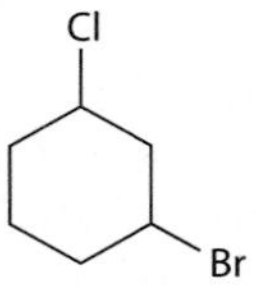

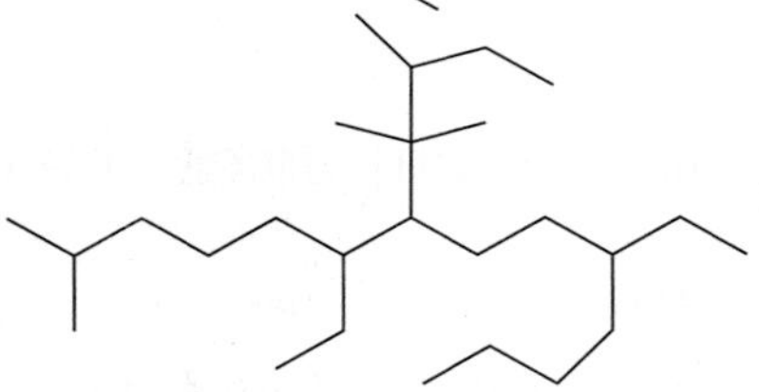

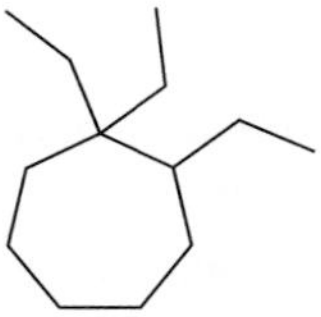

4.42 Determine which of the following structures are isomers and which are identical:

4.43 Draw 10 different structural isomers that have the formula C_9H_{20} and give each its IUPAC name.

4.44 Draw the correct structure for each of the following alkyl halides:
(a) 3,3,5-trichlorodecane
(b) 2,6-dimethyl-3-fluoroheptane
(c) 2,2-dichloro-4,4-dibromooctane

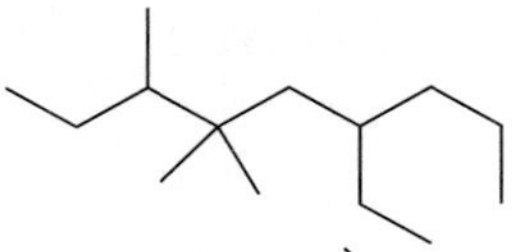

4.45 Briefly explain why the Cl–C–Cl bond angle in dichloromethane is larger than the H–C–H bond angle in methane.

Functional Groups

5

When certain atoms are collected into discrete units, they have special physical and/or chemical properties. Such units are known as **functional groups**. The C=C unit of alkenes and the C≡C unit of alkynes are examples of hydrocarbon functional groups. The C–C unit of an alkane is *not* considered to be a functional group because it is the "backbone" of virtually all organic molecules. Functional groups can include atoms other than carbon or hydrogen and the presence of these other atoms (call them **heteroatoms**) leads to new functional groups.

To begin this chapter, you should know the following:

- **covalent bonding between carbon and carbon and carbon and hydrogen (Chapter 3, Sections 3.3 and 3.5)**
- **the definition of a heteroatom (Chapter 3, Section 3.7)**
- **polarized covalent bonds between carbon and heteroatoms (Chapter 3, Section 3.7)**
- **the identity and characteristics of s-orbitals, p-orbitals, and σ-bonds (Chapter 3, Sections 3.1 and 3.3–3.5)**
- **how to name any alkane with alkyl or halogen substituents (Chapter 4, Sections 4.6 and 4.7)**
- **the valency of atoms in the first and second rows of the periodic table to form covalent bonds in neutral molecules (Chapter 3, Sections 3.1–3.3)**
- **the VSEPR model for all atoms in the first and second rows of the periodic table (Chapter 3, Section 3.5.4)**

This chapter will introduce hydrocarbon organic molecules that involve π-bonding. Compounds composed of carbon bonded to other atoms (heteroatoms) such as oxygen and nitrogen have both σ- and π-bonds. The rules of nomenclature will be extended to accommodate each new functional group. The physical properties that result from polarized covalent bonds and π-bonds will be discussed, using simple hydrocarbons as a starting point.

When you have completed this chapter, you should understand the following points:

- **A π-bond is formed by "sideways" overlap of p-orbitals on adjacent sp^2 hybridized atoms, and it is composed of a strong σ-bond and a weaker π-bond.**
- **Alkenes are hydrocarbons that have a C=C unit, and the generic formula for an alkene is C_nH_{2n}.**
- **The C≡C unit is composed of one strong σ-bond and two π-bonds that are orthogonal to each other. Alkynes are hydrocarbons that have a C≡C unit, with the generic formula C_nH_{2n-2}.**
- **Functional groups are discrete collections of atoms that usually contain π-bonds and/or polarized bonds.**
- **Hydrocarbons with multiple π-bonds include dienes, diynes, and allenes.**
- **Alkenes, alkynes, alcohols, amines, ethers, ketones, aldehydes, and carboxylic acids have various structures.**
- **Amines contain a covalent bond to nitrogen and undergo fluxional inversion.**
- **If the functional group contains a carbon, that carbon receives the lowest possible number. Each functional group has a unique suffix. Alkenes end in "ene," alkynes in "yne," and alcohols in "ol." Amines are named as amines and ethers are named as ethers. Ketones end in "one," aldehydes in "al," and carboxylic acids in "oic acid."**
- **Many physical properties are influenced by the presence of polarized bonds within molecules.**
- **The electron-donating ability of a functional group can be used to predict the reactivity with positively charged atoms.**
- **The unshared electrons of heteroatoms and the electrons in a π-bond can function as Lewis bases or Brønsted–Lowry bases.**
- **Formal charge can be used to determine the credibility of a given structure and whether it is neutral or bears an ionic charge.**
- **Dispersal of charge over several atoms via aligned p-orbitals is called resonance and leads to greater stability.**
- **Benzene is a unique cyclic hydrocarbon that is more stable than expected due to resonance delocalization.**
- **When benzene is a substituent in a molecule, it is called phenyl.**
- **Alkenes and functional groups are important in biological systems.**

5.1 Introducing a Functional Group: Alkenes

5.1.1 sp² Hybridized Carbon—Three Attached Atoms and a New Type of Bond

5.1.1.1 The π-Bond

In Chapter 3, Section 3.5, molecular orbitals described the bonding in alkanes using the hybridization model. Specifically, sp³ hybrid orbitals overlap to form a sigma-covalent bond (a σ-bond). It is possible to have two covalent bonds between adjacent carbon atoms, a carbon–carbon double bond. One of the two bonds is the usual σ-bond, but the other is called a π-bond. Hydrocarbons that contain one π-bond are called **alkenes**. In other words, an alkene will have a C=C unit. Each carbon atom of the C=C unit will have four bonds, but only three of the bonds are σ-bonds, and the fourth bond is a π-bond.

The σ-bonds are formed using what are called **sp² hybrid molecular orbitals**. Using a hybridization model, one of the 2s-electrons of carbon is promoted to form the hybrid, but only three hybrids of the sp² hybrid orbitals are used to form three σ-bonds to other atoms. Using this model, formation of a σ-bond between two sp² hybridized carbon atoms leaves an unused orbital on each carbon, and each orbital has an extra electron. When the two carbon atoms are connected by the σ-bond, the unused electrons are shared between the two atoms to form another covalent bond. This type of covalent bond is different from the σ-bond, and it is known as a π-bond. The σ-bonds are due to overlap of the sp² hybrid orbitals, as shown in Figure 5.1, and the overall geometry is trigonal planar about each carbon. The "extra" molecular orbital on each carbon, after formation of the σ-bond using the sp² hybrid orbitals, is perpendicular to the plane of the atoms. *After formation of the σ-bonds, the only way adjacent sp²-hybridized carbon atoms can share electron density is by sideways overlap, as shown in Figure 5.1, to form what is called a π-bond.*

From Figure 5.1, it is clear that less electron density is shared by the adjacent carbon atoms in the π-bond than in the σ-bond and the π-bond is weaker. In a σ-bond, the shared electron density is on a line between the two nuclei, whereas in a π-bond, the shared electron density is between two parallel orbitals. The net result of the sp² hybridization model is two bonds between the carbon atoms—a weak π-bond and a strong σ-bond: a carbon–carbon double bond.

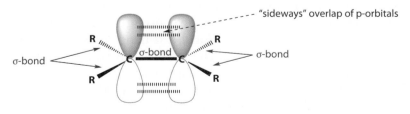

Figure 5.1   Interaction of two sp² hybridized carbons to form a π-bond.

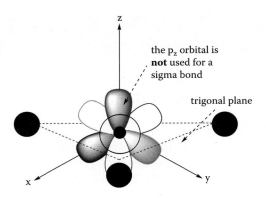

Figure 5.2 Covalent bonding for an sp^2 carbon atom.

As with sp^3 hybridization, the sp^2 hybridized carbon can be viewed in a different way, based on the number of atoms available to form σ-covalent bonds and the total number of orbitals available for bonding on that carbon. When four atoms are attached to carbon, the most efficient (lowest energy) way to arrange them is in the shape of a regular tetrahedron, with the carbon atom as the central locus. When three atoms are around carbon and carbon is the central locus, the lowest energy arrangement has the three atoms in a planar triangle with carbon at the center (this is called *trigonal planar geometry*). All four atoms are in the same plane.

In Figure 5.2, three atoms surround carbon as they approach to form three σ-bonds. All three atoms are in one plane (say the x,y plane); they can overlap *only* with the 2p$_x$- and 2p$_y$-orbitals (**blue** and **yellow**) but *not* with the 2p$_z$-orbital (**red**). These atoms can also overlap with the 2s-orbital, but cannot approach closely enough to overlap the 1s-orbital of carbon. In Figure 5.2, this is illustrated as three "incoming" atoms overlapping with two of the 2p-orbitals and the 2s-orbital. This would mean that each σ-covalent bond is formed using two 2p-orbitals and one 2s-orbital; this is described as an **sp^2 hybrid molecular orbital**. If three atoms are attached to carbon by covalent (sigma) bonds, these three bonds are formed by using three identical sp^2 hybrid orbitals. In this picture, the third 2p-orbital (the p$_z$-orbital in **red** in Figure 5.2) is perpendicular to the plane formed by the three atoms connected to carbon, and it is *not* used for bonding.

5.1 Describe the hybridization of O in the C=O unit.

The goal of this exercise is to form a carbon–carbon bond. If two of the sp^2 hybridized carbon atoms just described are brought together, one sp^2 hybrid orbital on each carbon will overlap to form a σ-bond (most of the electron density is concentrated on a line between the nuclei) and this is represented by the usual lines, wedges, and dashed lines in Figures 5.1 and 5.3. The other two sp^2 hybrid orbitals on each carbon are used to form σ-bonds to other atoms, represented by "R." The term "R" refers to any alkyl substituent, so it is any carbon

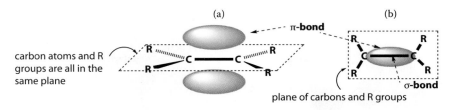

Figure 5.3 A π-bond in an alkene where (a) is a "side" view and (b) is a "top" view.

group. When the σ-bonds are formed, the unused $2p_z$-orbitals of each carbon (see Figure 5.2) are parallel to each other as shown in Figure 5.1. These two orbitals are close enough that they can share electron density, *but not along a line between the nuclei.* The only way the two p-orbitals can overlap (share electron density) is by "sideways" overlap, as in Figure 5.1; the resulting bond is illustrated in Figure 5.3.

This sideways overlap between two adjacent and parallel p-orbitals is a new type of bond. It is called a π-**bond** and is very different when compared to a σ-bond. Just as with a p-orbital, the probability of finding electron density is equal for **both** lobes of the π-bond, not part on top and part on the bottom. However, when two p-orbitals share electrons this way, *less* electron density is concentrated between the two nuclei than with a σ-bond. The π-bond is (a) different from and (b) weaker than the σ-bond. Two electrons are located in this π-bond. It is important to reiterate *that the overlap of these two electrons is not as efficient as if the electron density were concentrated on a line between the nuclei, so the π-bond is weaker than the sigma bond.*

The molecular model of $H_2C=CH_2$ shown as **1A and 1B** is an electron density map, where the red color indicates a higher concentration of electron density and blue indicates a lower concentration. It is clear from **1A** that electron density is concentrated between the carbon atoms, which is consistent with the idea of a π-bond. In **1B**, the model is turned to show both sides of the alkene, and it shows electron density on both sides, which is consistent with a π-bond as shown in Figure 5.3. From **1B**, it is also clear that the π-bond will be orthogonal (perpendicular) to the σ-bonds that attach the three atoms mentioned before. If the molecule is "turned" so that one is looking down on the π-bond, as in Figure 5.3b, the two carbon atoms and four R groups are shown to be coplanar and one lobe of the π-bond can be seen. **The π-bond constitutes a second bond to carbon**, and the two carbon atoms are joined by a **double bond** represented as C::C in a Lewis electron dot structure, or as C=C. As stated previously, one bond is a strong σ-bond and the second is the weaker π-bond.

5.2 **If an alkene π-bond were to undergo a chemical reaction, which bond would donate electrons: the σ- or the π-bond? Briefly explain your choice.**

5.3 **Describe the geometry of a molecule $R_2C=O$.**

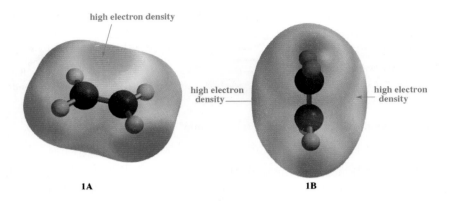

1A   1B

5.1.1.2 Alkene Nomenclature: "ene" for Alkenes

An alkene is a hydrocarbon that contains at least one C=C unit, which is the functional group. A functional group is a collection of atoms with certain chemical properties. Different compounds that contain the same functional group will have specific different physical properties, but they will have many chemical properties that are identical. Compounds with one functional group often have different chemical properties when compared to compounds that have a different functional group. A method is required to identify the different functional groups, which leads back to nomenclature.

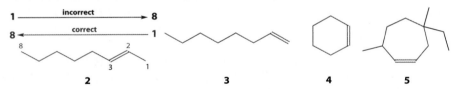

2   3   4   5

The nomenclature system developed in Chapter 4 allows one to name any alkane. The same fundamental system must be used for naming alkenes, but a modification is required to indicate the presence of the functional group. *A suffix is assigned to each functional group and a number is used to designate the position of the functional group in the chain or ring*. For those hydrocarbon molecules containing a double bond (an alkene), the suffix is taken from the **class name for an alkene ("ene")**. In other words, *the suffix for an alkene is -ene, and the longest linear chain must contain the C=C unit*. Alkene **2** has an eight-carbon chain ("oct") and the C=C unit with the π-bond is part of the longest continuous chain. In other words, there is an eight-carbon unbranched chain that contains the C=C unit, so this molecule is named octene. **More than one molecule can be called an octene** because the C=C unit can be anywhere along the eight-carbon chain.

This fact leads to the possibility of different isomers, and each requires a different name. **The name must specifically designate the position of the π-bond by *numbering the chain so that the first carbon of the π-bond receives the lowest possible number*.** In this example, numbering from the "left" leads to 6-octene, but numbering from the "right" leads to 2-octene, as

shown. The proper name of **2** is 2-octene. The number can also be placed immediately in front of the "ene" term, so **2** is also named oct-2-ene. Alkene **3** is an
octene but the position of the C=C unit is different than in **2**. Using the same
protocol, **3** is named 1-octene. Note that there is another isomer of **2**, but this
will be discussed in Chapter 9, Section 9.4. For the moment, simply focus on the
position of the C=C unit and the fundamental nomenclature of alkenes.

5.4 Draw the structures of both 3-octene and 4-octene.

Cyclic alkenes are known compounds. As with cyclic alkanes, the parent
name for cyclic alkenes is based on the number of carbon atoms in the ring
derived from the analogous linear alkene, but the prefix "cyclo" is added to the
name. The six-carbon cyclic molecule that contains a C=C unit is called cyclohexene (**4**). Substituents are named in the usual manner, and one carbon of the
C=C unit must be C1. The choice of which C=C carbon becomes C1 is dictated
by giving the substituents the lowest number. Compound **5** is a cycloheptene,
but the substituents can be numbered 6-ethyl-3,6-dimethyl or 4-ethyl-4,6-di-
methyl. The smaller numbers dictate the priority, which also gives the ethyl
group the lower number, and **5** is 4-ethyl-4,6-dimethylcycloheptene. Note that
the 1-heptene is unnecessary in the name.

**5.5 Draw the structures of cyclobutene, cycloheptene, and cyclopen-
tadecene.**

5.1.2 Alkenes: The C=C Functional Group

In an alkene, three atoms surround each individual carbon; one of the three
atoms is the other carbon connected by a σ-bond and a π-bond. Assume that
the other atoms are hydrogen and the structure is $H_2C=CH_2$, which is named
ethene (the common name is ethylene). As noted previously, the three atoms
connected to each sp² hybridized carbon of the C=C unit will have a trigonal
planar geometry. If ethene has this geometry for both carbons of the C=C unit,
all four hydrogen atoms and both carbon atoms are in one plane, with the
π-bond perpendicular to that plane. The H–C–H and the H–C=C bond angles
are predicted to be close to those of a right triangle, 120°. This geometry is
apparent when ethene is shown from the "top" as in **1C** and from the "side" as
in **1D**. Note that molecular models **1A** and **1B** were ethene. All four hydrogen
atoms and the two carbon atoms are clearly coplanar in these models.

1C   1D

The bond lengths for the C=C bond and the C–C bond can be compared to
determine which is longer. The two carbon atoms in a C=C unit should be held
more closely together by two covalent bonds than within a C–C unit with only

one covalent bond. A typical C=C bond distance is 133.7 pm (1.337 Å) and a typical singly bonded C–C bond distance is 154.1 pm (1.541 Å).[1] *In general, a double bond is shorter than a single bond, which means that the internuclear distance between the two carbon atoms is shorter.*

In an alkane, every carbon has the maximum number of hydrogen atoms attached to it (determined by the fact that each carbon must have four bonds). In an alkene, each carbon of the C=C unit has four bonds, but two carbon atoms are connected by two bonds, a σ- and a π-bond. *The presence of a C=C unit means that there are two fewer hydrogen atoms when compared to an alkane.* Because of this, the **general formula** for alkenes is C_nH_{2n}, where n is an integer: 2,3,4. The integer n *cannot* be 1, but when $n = 2$, the formula becomes C_2H_4; when $n = 3$, it is C_3H_6; when $n = 100$, it is $C_{100}H_{200}$, and so on. *Note that the generic formula for an alkene is the same as that for a cyclic alkane* (**Chapter 4, Section 4.7**). Indeed, 1-hexene (**6**) is an *isomer* of cyclohexane (**7**) because the formula is the same: C_6H_{12}. In both cases, there are two fewer hydrogen atoms relative to the six-carbon alkane **8**, which has the alkane formula C_6H_{14}.

5.6 Briefly explain why alkanes are called "saturated" hydrocarbons, but alkenes are considered to be unsaturated.

5.7 Draw eight different isomers with the formula C_7H_{14}, where half of the structures are alkenes and the remainder are cyclic alkanes.

6 (C_6H_{12})   alkene   **7** (C_6H_{12})   cyclic alkane   **8** (C_6H_{14})   alkane

5.2 Another Hydrocarbon Functional Group: Alkynes

5.2.1 sp Hybridization and the Triple Bond

Just as a carbon atom is attached to four other atoms via sp³ hybrid orbitals in alkanes and to three other atoms via sp² hybrid orbitals with one π-bond in alkenes, it is possible to generate molecules where carbon is attached to only two other atoms where at least one is another carbon atom. There is a σ-bond to one atom and a σ-bond to another carbon, but there are two bonds—additional bonds between the carbon atoms. Focusing only on the two carbons, there are one σ-bond and two π-bonds that constitute a carbon–carbon triple bond. In this triple bond, each carbon atom is said to be *sp hybridized*.

When only two other atoms are attached to a carbon atom, the hybridization on each of the two carbon atoms must change. In alkenes, there is one π-bond

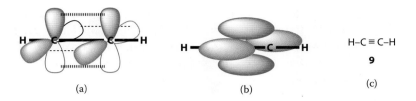

(a)   (b)   (c)

Figure 5.4   Two π-bonds in an sp hybridized carbon–carbon bond.

between two sp² carbons and three σ-bonds to each carbon. The one π-bond is formed by sideways overlap of the unused p-orbital on each sp² carbon. When two carbon atoms are involved in a triple bond, each sp hybridized carbon will use one hybrid orbital to form a σ-bond to another atom and one hybrid to form a σ-bond to the other carbon of the triple bond, represented as C:::C (Lewis dot formula) or as C≡C. Formation of the σ-bonds leaves ***two*** unused p-orbitals on each carbon.

When an σ-bond is formed between two sp hybridized carbon atoms, the unused p-orbitals overlap in a sideways manner to form two π-bonds, as shown in Figure 5.4 for the simplest alkyne, ethyne (acetylene, **9**), with the formula C_2H_2. The sp hybridization can be predicted using the electron promotion model. In this model, each carbon will promote the 2s-electrons to form the sp hybrid orbital, but only two of the resulting hybrids are used to form σ-bonds. The remaining two electrons of the p-orbitals are used to form the two π-bonds, as shown in Figure 5.4. The p-orbitals on the atom are mutually perpendicular (see Figure 5.2), so the two π-bonds formed by overlap of those orbitals must be perpendicular to each other when two π-bonds are between adjacent carbon atoms. The overlap of the p-orbitals is shown in Figure 5.4a, and the resulting π-bonds are shown in Figure 5.4b. The line notation structure is shown as **9** in Figure 5.4c. The carbon atoms and all the σ-bonds form a linear array that is apparent in **9**.

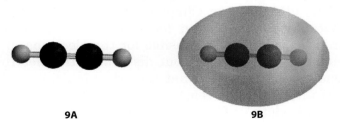

9A   9B

The molecular model of ethyne (**9A**) clearly shows the linear relationship of the atoms, which is due to the sp hybridization and the presence of two π-bonds. An electron density map of acetylene is also shown in **9B**. The higher concentration of red corresponding to higher electron density is apparent in the region between the carbon atoms *surrounding* the molecule, consistent with two mutually perpendicular π-bonds. Compare this model with the alkenes **1A** or **1B** (ethene; Section 5.1.1), where the electron density is concentrated above and below the plane of the atoms. The position of the electron density in **9B** is consistent with Figure 5.4. The carbon–carbon σ-bond is illustrated in Figure 5.4 as a solid dark line and the two π-bonds are generated by overlap

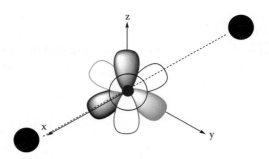

Figure 5.5   Covalent bonds to an sp hybridized carbon.

of the p-orbitals. From this figure and from **9B** it is clear that each π-bond is orthogonal to the σ-bond and also orthogonal to the other π-bond.

As with sp³ hybridization and sp² hybridization, sp hybridization can be examined from the standpoint of the atoms required to form two σ-bonds and the orbitals that are available on carbon to form those bonds. The lowest energy arrangement of two atoms as they approach carbon to form σ-bonds, when carbon is the central locus, has two atoms on either side of carbon, in a straight line (linear geometry), as seen in Figure 5.5. As with sp³ and sp² models, this picture is superimposed on the 1s/2s/2p-orbital diagram for carbon; the two "incoming" atoms can overlap with only one of the 2p-orbitals (say $2p_x$) and also the 2s-orbital. This means that a single covalent bond is formed using one 2p-orbital and one 2s-orbital and that two p-orbitals are not used. The orbital used to form this bond is a new type of hybrid orbital called an **sp orbital** because it is formed from one p-orbital and one s-orbital. Using two sp hybrid orbitals would form the two covalent σ-bonds.

There is an important point about triple bonds that is pertinent to their chemical reactions (discussed in Chapter 10, Section 10.6). *Because the two π-bonds are mutually perpendicular, they essentially react independently of one another.* In other words, it is possible for the electrons in one π-bond to interact with another molecule without disturbing the electrons in the second π-bond. One π-bond "knows" what the other is doing and the electrons in one π-bond can influence the reactivity of the other; however, each π-bond reacts largely independently of the other. This point is clarified in Chapter 10, Section 10.6.

5.8 **Is it possible for one π-bond to react without disrupting the second π-bond? Why or why not?**

5.2.2 Alkynes: The C≡C Functional Group

A hydrocarbon (only carbon and hydrogen atoms) that contains two π-bonds in a C≡C unit is called an alkyne. The C≡C unit is a second type of functional group (C=C was the first one). It has characteristic reactions that lead to different products relative to alkanes and alkenes, although some of the chemical reactions are quite similar to those of alkenes. As with alkanes and alkenes, alkynes form carbon chains and branched carbon chains; however, stable cyclic compounds of carbon atoms with a triple bond must be larger than those with a C=C unit in order to accommodate the linear nature of C≡C group. A molecule can have more than one triple bond.

The so-called triple bond of an alkyne has one σ-bond and two π-bonds that are mutually perpendicular. Compared to the alkane formula, there are four fewer hydrogens in an alkyne because of those two π-bonds and two fewer than in an alkene with a C=C unit. The **general formula** for an alkyne reflects this fact:

$$C_nH_{2n-2}$$

where n is an integer: 2, 3, 4,…. The integer n **cannot** be 1, but when $n = 2$, the formula is C_2H_2; when $n = 3$, the formula is C_3H_4; when $n = 100$, the formula is $C_{100}H_{198}$, and so on.

5.9 A molecule has two π-bonds and a formula $C_{12}H_{22}$, but it is not an alkyne. Offer an explanation.

5.2.3 Nomenclature for Alkynes: "yne" for Alkynes

The generic name for the class of hydrocarbons containing a triple bond (two π-bonds between adjacent carbon atoms) is alkyne. For those hydrocarbon molecules containing a carbon–carbon triple bond (an alkyne), the suffix is taken from **the class name for an alkyne ("yne")**. Alkyne **10** has a linear eight-carbon chain ("oct") containing a triple bond, so it is an alkyne, and the suffix is "yne," so it is an octyne.

As with alkenes, there are several possible isomers, and the position of the C≡C unit must be identified. The name specifically designates the position of the triple bond by *numbering the chain so that the first carbon of the triple bond receives the lowest possible number*. In this example, numbering from the left would make it 6-octyne, but numbering from the right makes it 2-octyne. The proper name of **10** is 2-octyne. The number can also be placed immediately in front of the "yne" term, so **10** could also be called oct-2-yne. An isomer (**11**) positions the triple bond at a different place in the eight-carbon chain, requiring a different number, and **11** is named 3-octyne. Note that, in both cases, the carbon of the triple bond that is closest to the terminus of the chain gives the lower number and is preferred.

There is a difference between alkyne **12** and **10** or **11** in that the first two carbons of the chain constitute the triple bond, and there is a hydrogen atom attached to one carbon of the triple bond. In **10** and **11**, there is a carbon on both carbons of the triple bond. Alkyne **12** is an example of a **terminal alkyne** (the triple bond occupies the terminal position and there is at least one H atom attached), whereas **10** and **11** are known as **internal alkynes**. Alkyne chains can have substituents, and in **12** there is a methyl substituent and two bromine substituents. The numbering of the triple bond dictates the numbering of the substituents, so the methyl group is at C3 and both bromine atoms are attached to C7. The name of **12** is 7,7-dibromo-3-methyl-1-heptyne.

5.10 Draw the structure of 3-decyne; draw an isomer that has a linear chain of only eight carbons and name it.

5.11 Draw the structure of 5-chloro-3,3,6-trimethyl-oct-1-ene and of 6,6-diethyl-9-iododec-3-yne.

    The now familiar system is used for cyclic molecules that contain a triple bond, but the linear nature of the triple bond unit makes it difficult to put a triple bond into a small ring. A ring that contains a triple bond has severe distortion of the sp^3 hybridized carbon atoms if the ring is too small (see Chapter 8, Section 8.5, for a discussion of strain in cyclic molecules). For this reason, *it is not possible to isolate cyclobutyne, cyclopentyne, or cyclohexyne although they may exist as transient intermediates in certain reactions.* This limitation is apparent when the 5-membered ring alkyne cyclopentyne (**13**) is compared with a 12-membered ring alkyne, cyclododecyne (**14**). The small size of the ring in **13** leads to some distortion of the triple bond away from the expected linear array (see **13A**). Such destabilizing strain makes it difficult for **13** to exist. In **14**, however, the larger cavity of the 12-membered ring leads to a nearly linear structure for the triple bond (see **14A**); this is indicative of much greater stability. When the ring has eight or more carbons (generally), it is possible to isolate cyclic alkynes and naming is required.

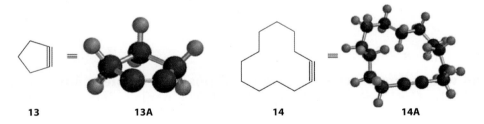

     13   **13A**   **14**   **14A**

5.12 Draw the structure of 4-ethylcyclododecyne.

5.3 Hydrocarbons with Several Multiple Bonds

Compounds are known that have more than one multiple bond. Multiple double bonds can be incorporated, as well as multiple triple bonds, and it is possible to mix alkene and alkyne units within a molecule. When two methyl groups are substituents, the term "dimethyl" is used (see Chapter 4, Section 4.6). Similarly, a molecule with two bromine atoms uses the term "dibromo." It is therefore reasonable that the prefix *di* should be used when two alkene units or two alkyne units are in one molecule. A molecule containing two C=C units is called a diene and a molecule containing two C≡C units is called a diyne.

The terms "di," "tri," "tetra," "penta," etc. can be used, just as in compounds that contain multiple substituents. A molecule with one C=C unit and one C≡C is called an en-yne. Compound 15 is an example of a diene, and nomenclature rules require that *both* alkene units must be included in the longest eight-carbon chain, so it is an octadiene. A number must be provided for each C=C unit, giving the lowest combination. The position of the C=C units in 15 dictates the position of any substituents on those chains.

Therefore, **15** is 4-methyl-1,4-octadiene and ***not*** 5-methyl-4,7-octadiene. Diynes are similarly named, as in **16**, where both C≡C units are contained in the longest chain, so it is a hexadiyne. The chain is numbered to give both C≡C units the lowest combination of numbers. The position of the C≡C units in **16** dictates the position of any substituents on those chains. Diyne **16** is named 3,3-diethyl-1,5-hexadiyne. Cyclic molecules such as 1,3-cycloheptadiene (**17**) are named using these protocols. In these examples, the primary carbon skeleton and/or ring system is marked in **purple**, with substituents in **green**.

When an alkene and an alkyne are in the same molecule, the priority is not necessarily straightforward. In general, numbers as low as possible are given to carbon atoms bearing double and triple bonds. If there is a further choice, the double-bond atoms are assigned the lowest numbers. En-yne **18** can be named either 3-propylhept-1-en-5-yne or 5-propylhept-6-en-2-yne. The former is the lower number for both groups, so **18** is named 3-propylhept-1-en-5-yne. For **19**, however, the possible names are 3-propylhex-1-en-5-yne or 5-propylhex-5-en-1-yne. *There is no clear choice and, in such a case, the C=C unit receives the lower number;* thus, **19** is named 3-propylhex-1-en-5-yne.

5.13 Draw the structure of 1,3-cyclopentadiene, of 3-ethyl-1,5-hexadiene, and of 6-bromo-1,3-hexadiyne.

5.14 Draw the structures of 1,3,6-octatriene, 1,4,7-octatriyne, and hex-4-en-1-yne.

5.15 Give the name of a molecule with a linear carbon chain of nine carbons with a C≡C unit beginning at C2, a methyl group at C6, and a –$CH_2CH_2CH=CHCH_3$ unit at C4.

Another class of hydrocarbons has multiple π-bonds: the **allenes**. The parent compound is **20** and is named allene. Formally, **20** is a diene and the IUPAC name is 1,2-propadiene or propan-1,2-diene. Allene is an example of a **cumulene**—*a molecule that has cumulative π-bonds*. The term "cumulative" indicates that there are three or more adjacent sp^2 or sp hybridized carbon atoms. Note that the central carbon has two π-bonds and is sp hybridized, whereas the two flanking carbons are sp^2 hybridized. This arrangement requires that the two π-bonds are perpendicular—one to the other—as shown in **20B**, which positions the methylene units (the –CH$_2$– units) in different planes.

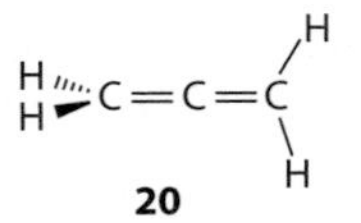

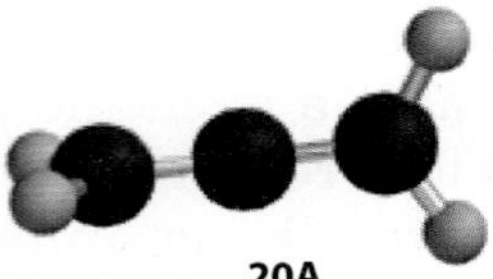

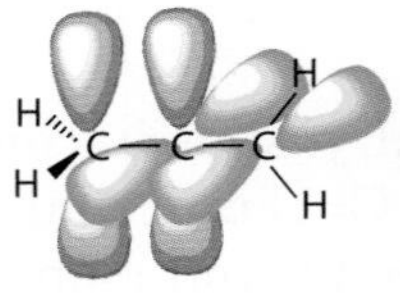

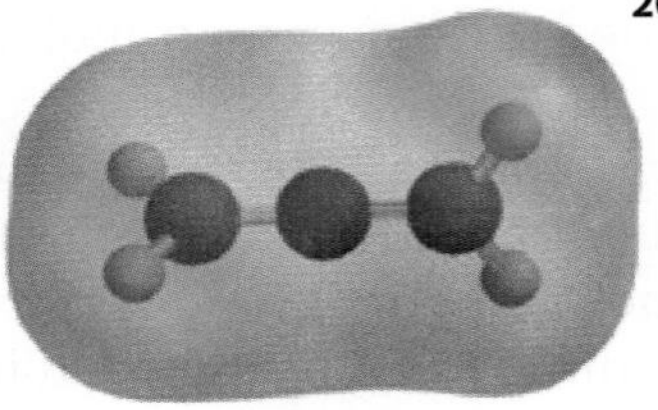

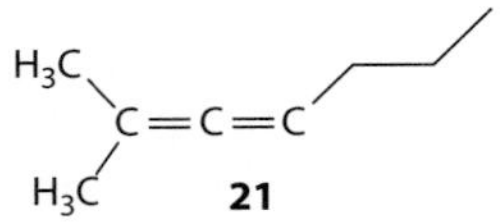

Molecular model **20A** shows that the two hydrogen atoms on one terminal carbon (the CH$_2$ unit) are perpendicular to the hydrogen atoms on the other terminal carbon (the other CH$_2$ unit). The electron density map of allene (see **20C**) shows the higher concentration of electron density (red color) is between the carbon atoms, but in different planes. This model confirms that the two π-bonds are perpendicular to one another. Substituted allenes are also named as dienes. For example, compound **21** has the C=C=C unit as part of a seven-carbon chain, so it is a heptadiene. Giving the cumulative diene unit the lower numbers leads to 2-methyl-2,3-heptadiene as the name for **21**.

5.4 Reactivity of Polarized Covalent σ-Bonds

Many functional groups include atoms other than carbon and hydrogen. Any atom other than carbon or hydrogen, such as oxygen, nitrogen, sulfur, chlorine, etc., is called a heteroatom. Before discussing functional groups that contain heteroatoms, it is important to introduce an important effect that the heteroatom will impart to any single or multiple bond in the molecule: bond polarization and the resultant polarity of the molecule.

To begin this discussion, compare a nonpolarized bond such as C–C with a polarized bond such as C–Cl in Figure 5.6. Polarized covalent bonds were introduced in Chapter 3 (Section 3.7). First, C–C is not polarized, which means that

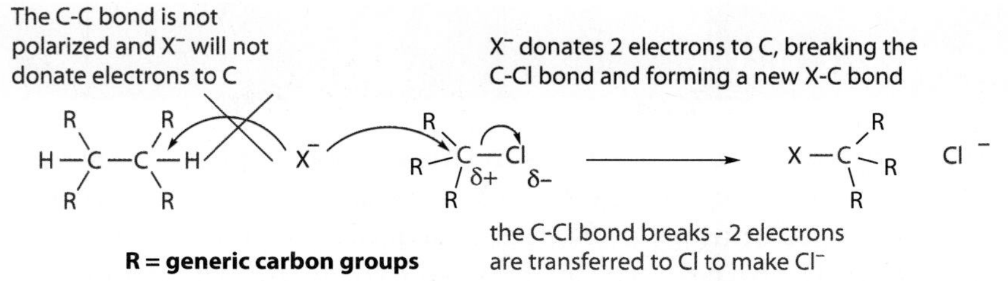

The C-C bond is not polarized and X⁻ will not donate electrons to C

X⁻ donates 2 electrons to C, breaking the C-Cl bond and forming a new X-C bond

R = generic carbon groups

the C-Cl bond breaks - 2 electrons are transferred to Cl to make Cl⁻

Figure 5.6 Donation of electrons to an electropositive carbon, with concomitant bond making and bond breaking.

both carbons are *neutral* (they have no formal charge) and there is no partial charge. Compare C–C with C–Cl, *which is also neutral* but has a $\delta+$ on the carbon and $\delta-$ on the chlorine. Imagine that a negatively charged species (X^-) collides with carbon of the nonpolar carbon (in **green** in the figure) to form a new C–X bond. The C–C bond will not break because it is rather strong and there is no bond polarization. In other words, if the C–C bond breaks upon collision with X^-, two products would be formed in the reaction: $X–CHR_2$ and $R_2CH:^-$, an unstable species known as a carbanion (see Chapter 7, Section 7.4).

Now imagine that a negatively charged species (X^-) collides with the $\delta+$ carbon of the C–Cl bond. In Figure 5.6 (note the use of the electron transfer arrows), the negative species (X^-) donates two electrons to the $\delta+$ carbon, breaking the C–Cl bond. In effect, the X^- unit is a Lewis base and the $\delta+$ carbon atom is a Lewis acid, at least from the standpoint of electron donation and electron accepting abilities. As first defined in Chapter 2 (Section 2.6), *when a species donates two electrons to carbon, it is known as a nucleophile*. (Nucleophiles are discussed in more detail in Chapter 6, Section 6.7, and Chapter 11, Section 11.3.)

In Figure 5.6, the species X^- reacts as a nucleophile. When the X–C bond is formed, the two electrons in the C–Cl bond are transferred to the more electronegative chlorine (which is $\delta-$). In other words, a new bond is formed between carbon and the "incoming" atom X, breaking the C–Cl bond. This process is the basis of a chemical reaction, and it is possible because of the polarized C–Cl bond. This reaction does **not** happen if X^- collides with a molecule containing only the C–C bond; there is no bond polarization, so there is not as much attraction between the atoms, and this bond is much stronger. If the C–C bond were to break, carbon cannot readily accommodate the negative charge (excess electrons). This is another way of saying that *alkanes do not easily undergo chemical reactions*. This concept will be used many times in chapters to come. The real purpose of this section is to introduce the idea that chemical reactions involve making and breaking bonds. Based on the discussion in this section, the chemistry of molecules containing polarized bonds is expected to be richer than the chemistry of molecules containing nonpolarized bonds.

5.16 Based *only* on bond strength, which bond is easier to break: C–Br or C–O? Is this valid?

5.5 Formal Charge

The structures drawn for all molecules in this section imply that they are neutral (they do not have a positive or negative charge). Is it possible to prove this statement? It is possible to analyze a chemical structure for charge by doing a calculation called **formal charge**, which involves examining the difference between the last number of the group number, which is the number of valence electrons, and eight. Remember that eight is the maximum number of valence electrons for the second row (Chapter 3, Section 3.1). Because carbon is in group 14 of the periodic table, it has four valence electrons and requires four electrons to give it a total of eight, which would give it the same electronic configuration as neon (this completes the second row octet). If carbon forms four bonds by sharing four electrons with four atoms, it will remain electrically neutral (no charge). Similarly, nitrogen, in group 15, has five valence electrons and requires three to complete the octet (attain the neon configuration). Therefore, nitrogen forms three bonds by sharing electrons with three atoms and it remains electrically neutral.

Using this criterion, oxygen is in group 16 and will form two bonds just as fluorine is in group 17 and will form one bond. To determine if the number of covalent bonds to a given atom will generate a charge, calculate **formal charge (FC)**. A specific formula is used:

FC = ending number of the group number − 1/2 (number of shared electrons)

− (number of unshared electrons)

Note that counting half the shared electrons is the same as counting the number of bonds. Begin with C2 in **22**, recognizing that carbon is a group 14 atom and there are four covalent bonds. Because each covalent bond is composed of two shared electrons, C2 has eight shared electrons (four bonds) and zero unshared electrons. The formal charge for C2 is calculated by $FC^{C2} = 4 - 1/2(8) - 0 = 4 - 4 = 0$. The calculated formal charge for C2 is zero (it is neutral). Similar calculations show that none of the carbons in **22** have a charge. If some atoms have a charge and other atoms do not, the charge is calculated for each atom in a molecule and ***the charge for the molecule is calculated by summing all charges for the atoms***.

Figure 5.7 shows the calculations for **22** and **23** and the formal charge for the individual atoms. Oxygen has a formal charge of −1 in **22**, which means that a negative charge resides on the oxygen; all other atoms in **22** have a formal charge of zero. The charge on the *molecule* is the sum of all formal charges on the atoms, which is −1 for **22**. Molecule **23** in Figure 5.7 shows that O^2 has a formal charge of −1, the nitrogen has a formal charge of +1, and all other atoms have a formal charge of zero. When these are added together, the formal charge for *molecule* **23** is **zero**. This is an important example because it shows that individual atoms can have charge, but if they neutralize one another, the formal charge for the molecule can be zero (a neutral molecule). Formal charge

$$O = 6 - (0.5)(2) - 6 = -1$$
$$C^1 = 4 - (0.5)(8) - 0 = 0$$
$$C^2 = 4 - (0.5)(8) - 0 = 0$$
$$\text{all } H = 1 - (0.5)(2) - 0 = 0$$

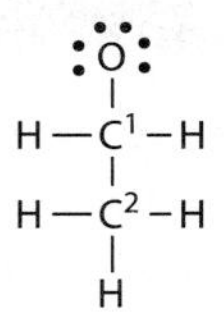

$$H-C^1-H$$
$$H-C^2-H$$
$$H$$

$$H-C^2-H$$
$$H-N-H$$
$$H$$

$$O^1 = 6 - (0.5)(4) - 4 = 0$$
$$O^2 = 6 - (0.5)(2) - 6 = -1$$
$$C^1 = 4 - (0.5)(8) - 0 = 0$$
$$C^2 = 4 - (0.5)(8) - 0 = 0$$
$$N = 5 - (0.5)(8) - 0 = +1$$
$$\text{all } H = 1 - (0.5)(2) - 0 = 0$$

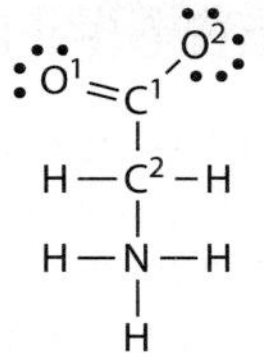

Formal charge for the molecule =   **22**
$$-1 + 0 + 0 + 0 + 5(0) = -1$$

23

Formal charge for the molecule =
$$0 - 1 + 0 + 0 + 1 + 5(0) = 0$$

Figure 5.7 Calculation of formal charge for two molecules.

is a very useful concept because it allows inspection of any structure to determine whether it is a neutral molecule or a charged molecule and whether there is charge on any atom in that molecule.

5.17 Calculate the formal charge for $CH_3\ddot{O}{:}$.

5.18 Determine the formal charge of bromine and carbon in CH_3Br (bromomethane; also called bromoform).

5.19 Determine the formal charge of all atoms in methylamine.

5.20 Determine the formal charge of the oxygen in ethanol.

5.6 Heteroatom Functional Groups

Carbon forms single covalent bonds to many different heteroatoms. The valence of the heteroatom will determine how many atoms may be attached to that heteroatom. For example, oxygen has a valence of two and must form an X–O–X species. One possibility is that a C–O–H unit will be formed, where ***the OH unit is called a hydroxyl group***. When OH is incorporated into a hydrocarbon molecule in place of one of the hydrogen atoms, the new molecule is called an alcohol. Oxygen also forms *a **C–O–C unit and molecules containing this unit are called ethers***. Nitrogen has a valence of three and can form three types of species containing at least one C–N bond: $R-NH_2$, R_2NH, and R_3N, where "R" represents a carbon group. ***These nitrogen-containing units are known as amino groups, and a molecule containing an amino group is called an amine***.

The OH unit, the C–O–C unit, and the amine units are examples of ***functional groups: collections of atoms that impart physical and chemical characteristics to a molecule***. These particular **functional groups** contain polarized covalent bonds and their chemical reactions will differ from the C=C and C≡C units discussed previously. This section will show several functional groups based on single bonds to nitrogen and oxygen; bonds to halogen atoms will also be included.

5.6.1 Molecules with a Carbon–Halogen Bond (Alkyl Halides)

The introduction to functional groups noticeably skipped halogens. The reason is quite simple: **halogen substituents are not considered to be functional groups,** and therefore no suffix is required for the name. Halogens are considered to be substituents, as noted in Chapter 4 (Section 4.6.2). Organic molecules that contain a carbon–halogen bond (C–F, C–Cl, C–Br, or C–I) are known collectively as **alkyl halides**. These molecules have a halogen atom joined to carbon by a polarized and relatively weak covalent σ-bond. Because the valence of a halogen atom is one, F, Cl, Br, and I form one covalent bond to carbon and the halogen has three unshared electron pairs. The chemistry of a C–X unit usually involves a nucleophile attacking the carbon to form a new bond and breaking the C–X bond (see Figure 5.6 and Chapter 11). Typical alkyl halides are trichloromethane (**24**, also known as chloroform), dichloromethane (**25**, also known as methylene chloride), and 6-fluoro-3-iododecane, **26**.

5.6.2 Amines

Nitrogen is in group 15 of the periodic table, has five electrons in its valence shell, and requires only three electrons to complete the octet. It obtains these three electrons by sharing with three other atoms via covalent sigma bonds. Most organic compounds containing nitrogen have three covalent bonds to nitrogen in the neutral molecule. This arrangement leaves an unshared electron pair on nitrogen. When the molecule is NH_3 (ammonia), there are three N–H σ-bonds. When the molecule is an amine, NR_3, there may be N–H σ-bonds and N–C σ-bonds. Indeed, amines are compounds that are characterized by one or more C–N bonds.

One way to think about the structure of amines is to use a thought experiment in which the hydrogen atoms in ammonia are replaced with alkyl groups to form N–C σ-bonds. Amines are organic molecules with three variations: RNH_2, R_2NH, or R_3N, where R represents virtually any alkyl group, although other groups may be attached to nitrogen. Amines have an unshared electron pair. Ammonia is known to be a base because of its ability to donate an electron pair to a positive species such as a proton (H+), as described in Chapter 2 (Section 2.2). Likewise, amines are known to react as bases (see Chapter 6, Section 6.4).

The $-NH_2$, NHR, or NR_2 units can be treated as substituents when they are attached to another unit, and they are known as *amino* groups. However, an organic molecule containing nitrogen groups such as these is an **amine**. To describe the structural variations in amines **27–29**, the terms *primary, secondary,* and *tertiary* must be introduced:

- A primary amine has one carbon and two hydrogen atoms on nitrogen (RNH_2).
- A secondary amine has two carbons and one hydrogen atom on nitrogen (R_2NH).
- A tertiary amine has three carbon and no hydrogen atom on nitrogen (R_3N).

The structure of a **primary amine** is shown as **27**, a **secondary amine** as **28**, and a **tertiary amine** as **29**. Using the VSEPR model (Chapter 3, Section 3.5.4), the three-dimensional shape of each amine with respect to nitrogen is predicted to be similar to that of ammonia. Assume that only the C, H, or N atoms can be observed, but not the electron pair. When viewed, the amine takes on a **pyramidal** shape with nitrogen at the apex. Indeed, amines are considered to be pyramidal, with the unshared electron pair projected from the apex of the pyramid (see **30**). Note that the C–N–C or C–N–H bond angles will vary with the size of the alkyl group.

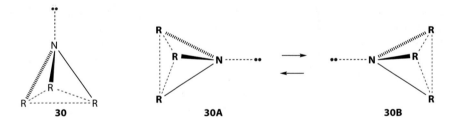

Amines such as **30** do not have a rigid pyramidal structure, but rather flip like an umbrella in the wind with nitrogen at the apex of a flipped umbrella and the electron pair at the handle. This phenomenon is known as **fluxional inversion** and is represented by the interconversion of **30A** and **30B**. This rapid inversion has important implications for the chemistry of amines (see Chapter 9, Section 9.7.3, for a discussion of this property of amines with respect to chirality).

5.21 Draw a four-carbon primary amine, a five-carbon secondary amine, and a six-carbon tertiary amine using line notation.

5.22 Draw a diagram to illustrate the fluxional inversion found in ammonia.

There are many different amines, all containing at least one C–N bond. The IUPAC nomenclature system treats the amine unit as a substituent attached

to the longest hydrocarbon chain. The name drops the "e" ending of the hydrocarbon chain and replaces it with the term **"amine."** The one-carbon primary
amine is methanamine, the two-carbon primary amine is ethanamine, the five-
carbon primary amine is pentanamine, and so on. Any group attached to nitrogen other than the longest hydrocarbon chain is indicated by the terms *N*-alkyl,
N,N-dialkyl, or *N*-alkyl-*N*-alkyl. In other words, groups could be *N*-methyl,
N,N-diethyl, or *N*-ethyl-*N*-methyl. By this system, dimethylamine is called
N-methylmethanamine. Amine **32** is named (*N,N*-diethyl)-1-pentanamine,
where the "1" indicates the position of the nitrogen on the longest chain. The
longest carbon chain is pentane, and the two ethyl substituents on nitrogen are
indicated by *N,N*- as shown. Similarly, **33** is *N*-ethyl-2-methyl-1-ethanamine,
and **34** is 1-butanamine. Note that the methyl substituents on the carbon chain
in **33** are treated the same as any substituent on the longest continuous chain
that also includes the functional group (here, the amine unit).

    Amines can also be named using common names, and they appear so often
that the system must be noted. The system is simple, in that the alkyl groups
are identified and that term is followed by the word "amine." Using this system,
32 is diethyl pentylamine, **33** is ethyl isobutylamine, and **34** is butylamine.
When several alkyl groups are present, they are named alphabetically and
then followed by the separate word "amine."

**5.23 Draw the structures of methanamine, ethanamine, and
pentanamine.**

5.24 Draw the structure of *N*-methyl-3,4-diethyl-2-decanamine.

5.6.3 Alcohols

Oxygen is in group 16 of the periodic table and has six electrons in its valence
shell, so it requires only two electrons to complete the octet. The valence of
oxygen is two, so it forms two covalent bonds, with two unshared electron pairs
on oxygen. Forming two oxygen–hydrogen σ-bonds generates water, H–O–H.
In a thought experiment, one or both hydrogen atoms are replaced with carbon
groups (forming C–O σ-bonds). Replacing one hydrogen generates an alcohol
and replacing both hydrogen atoms generates an ether.

    Water has the structure H–O–H and the VSEPR model assumes that the
geometry around oxygen is roughly tetrahedral, so placing two hydrogens and
two electron pairs on the oxygen leads to structure **35** for water. The VSEPR
model assumes that the electron pairs are invisible, so the atoms will determine the shape. This leads to the conclusion that the H–O–H molecule is "bent"
or **angular**.

If one hydrogen atom of water is replaced with a carbon group—CH_3 (called a methyl group; see Chapter 4, Section 4.6.2), for example—the result is CH_3OH (**36**, called methanol as described later). If the focus is on the atoms in H–O–C, **36** is angular just like water. The OH unit attached to carbon in **36** is treated as a functional group and is called a **hydroxyl group**. A carbon molecule containing an OH group (hydroxyl functional group) is called an **alcohol**. *An alcohol is an organic molecule containing an OH group*.

The O–H bond is polarized such that oxygen is partially negative and the hydrogen is partially positive. This bond polarization is consistent with the known fact that the hydrogen atom attached to oxygen in an alcohol is slightly acidic (see Chapter 6, Section 6.2); indeed, alcohols will react with a suitable base. *Alcohols are Brønsted-Lowry acids*, which accounts for many of the chemical reactions of alcohols that will be discussed later. There are countless alcohols. Typical examples are methanol (**36**; common name of methyl alcohol, also called wood alcohol); ethanol (CH_3CH_3OH; common name of ethyl alcohol) from grain and found in liquors, wines, etc.; and isopropanol, $(CH_3)_2CHOH$, which is also called rubbing alcohol.

In the IUPAC system, alcohols are named using the carbon prefix and the **suffix "ol"** (taken from the generic name alcoh**ol**). The alcohol is identified by attachment of the oxygen of an OH functional group to the longest linear carbon chain. Essentially, one identifies the hydrocarbon chain, drops the "e" if it is an alkane backbone, and adds the "ol." *The carbon chain is numbered such that the carbon bearing the oxygen of the OH unit has the lowest possible number. Giving the oxygen-bearing carbon the lowest number supercedes the number and placement of alkyl or halogen substituents.*

5.25 Based on the IUPAC name, draw the structure of ethanol.

Substituents on the longest carbon chain are numbered after assigning the lowest possible number to the position of the carbon bearing the OH unit. Alcohol **37** has the OH unit connected to a six-carbon chain, so it is a hexanol. Numbering to give the OH the lowest number results in 2-hexanol, regardless of the positions of the two substituents. Because it is a 2-hexanol, **37** is named 5-bromo-3-methyl-2-hexanol. Note that the numbering is different from simple alkanes because the position of the OH group dictates the numbering. In **38**, the OH is connected to a six-carbon chain, but the OH is at C1 to receive the lowest number, so it is a 1-hexanol. There is a 1-methylethyl group at C3 (see Chapter 4, Section 4.6.2) and two methyl groups at C2, so **38** is 3-(1-methylethyl)-2,2-dimethyl-1-hexanol. In **39**, there is a 12-carbon hydrocarbon chain that does not contain the OH unit. The longest unbranched carbon chain that is connected to the OH oxygen is 11, so this is an undecanol. The OH is at C3 in the lowest numbering scheme, so **39** is 4-butyl-3-ethyl-3-undecanol.

Cyclic alcohols are also possible and **40** has the OH group attached to a cyclohexene ring, so it is a cyclohexanol. The carbon of the ring that bears the OH is always C1, and the ring is then numbered to give the substituents the lowest numbers. Therefore, **40** is 2,4-dimethylcylcohexanol. Because the oxygen-bearing carbon is always C1, the "-1-" is omitted.

When two hydroxyl units are incorporated into the same molecule, it is a dihydroxy compound (two alcohol units); this type of molecule is called a diol (two OH: diol). When there are three hydroxyl units, it is a triol; tetraols, pentaols, etc. are known. The nomenclature for a diol identifies the longest chain that bears **both** OH units, and gives the two carbon atoms that bear the OH units the lowest possible number. Typically, the name of the hydrocarbon chain precedes the term "diol," with numbers to identify the positions of the hydroxyl groups. For example, $HOCH_2CH_2CH_2CHOH$ is 1,4-butanediol and $CH_3CH(OH)$ $CH_2CH(CH_3)CH_2CH_2CH_2OH$ is 4-methyl-2,7-heptanediol.

5.26 Draw the structure of 3,3,5-trichloro-1,7-decanediol.

The sulfur analog of alcohols (RSH rather than ROH), thiols are occasionally used in this book, but not often. Mercaptan is the common name for a thiol. Thiols are named using the hydrocarbon portion of the alkyl unit with the suffix thiol. Therefore, CH_3SH is methanethiol and $CH_3CH_2CH_2CH_2SH$ is butanethiol. Methanethiol is also known as methyl mercaptan. Thiols are foul-smelling compounds, and the smell of natural gas is usually 1 part per million (ppm) of methanethiol in the gaseous hydrocarbon propane. Thiols react similarly to alcohols, but there are differences. Sulfur is a larger atom when compared to oxygen, and it has d-orbitals, whereas oxygen does not. Both of these factors lead to differences in chemical reactivity, but these differences are not important for most of the reactions observed in this book. The hydrogen atom attached to oxygen in an alcohol is acidic (see Chapter 6, Section 6.2), and the proton on sulfur in a thiol is also acidic. Thiols are generally more acidic than an alcohol. The pK_a of a typical thiol is about 10, whereas the pK_a of a typical alcohol is about 15–18. Just as there are diols, there are dithiols. Ethanedithiol is $HSCH_2CH_2SH$ and butanedithiol is $HS-(CH_2)_4-SH$. Dithiols are used in reactions with aldehydes and ketones in Chapter 18, Section 18.6.7.

5.6.4 Ethers

Organic molecules that contain a C–O–C unit, an oxygen atom with two alkyl groups attached and no hydrogen atoms on the oxygen, are called ethers (see **41**). The divalent oxygen has two covalent σ-bonds to carbon, and two unshared electron pairs. Using the VSEPR model, there is a tetrahedral geometry around oxygen and the C–O–C unit is angular, with two electron pairs at the other corners of the tetrahedron (see **41**). **While ethers are structurally related to alcohols and water,** *they are not chemically related*. Alcohols have an O–H unit and ethers do not, so *ethers are* **not** *Brønsted–Lowry acids.* As will be seen later, this critical difference makes ethers relatively unreactive under a variety of conditions. Indeed, ethers are commonly used as solvents in which organic chemical reactions take place. Ethers are, however, Lewis bases (see Chapter 6, Section 6.4.3).

The recommended method for naming identifies a long chain and a shorter linear chain attached to the oxygen; the longer chain is the parent. The oxygen bearing the shorter chain is treated as a substituent. If the shorter alkyl chain is a butyl, for example, the "yl" used thus far for carbon substituents is replaced with "oxy," so the shorter chain is identified as butoxy. In other words, the alkyl group becomes alkoxy. Similarly, OCH_3 is methoxy, OCH_2CH_3 is ethoxy, etc. Using these protocols, **42** is 4-ethoxydecane. In **43**, the longest chain is 5, with a branching methyl group. Therefore, **43** is named 1-butoxy-3-methylpentane. Note that **44** is a symmetrical ether, with two propyl groups flanking the oxygen.

For symmetrical ethers, each alkyl group can be identified, followed by the word *ether*. The IUPAC name for **44** is 1-propoxypropane, but because it is symmetrical, it can also be named dipropyl ether. Similarly, the IUPAC name for **45** is 1-cyclopentoxycyclopentane, but it is symmetrical and can be called dicyclopentyl ether. Dimethyl ether is the name for **41** (methoxymethane is the IUPAC name). Diethyl ether (ethoxyethane) is another simple example; this ether was used as a general anesthetic many years ago but often induced convulsions in patients. Diethyl ether was abandoned when better and safer anesthetics were discovered.

5.27 Draw the structure of diethyl ether.

5.7 Acid–Base Properties of Functional Groups

When an electron-rich functional group comes into close proximity to an electron-deficient species (an acid), *donation* of two electrons to the electron-deficient center leads to a new sigma bond. This process is illustrated in Figure 5.8, where it is compared to a typical Lewis base–Lewis acid reaction between a generic base, B:, and BF_3 to give the complex shown (see Chapter 2, Section 2.5 and Chapter 6, Section 6.5). *The concept of electron flow involves the transfer of two electrons from the electron rich base to the electron poor boron to give the Lewis-acid-Lewis base complex formed as a new bond*.

The **blue** arrow indicates transfer of two electrons from the base to the acid. Protonic acids (H^+) react with a base to give a conjugate acid, as shown.

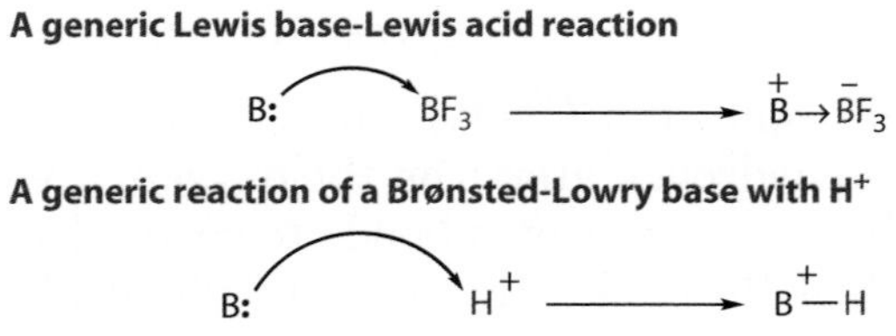

A generic Lewis base-Lewis acid reaction

$$B: \quad\curvearrow\quad BF_3 \quad\longrightarrow\quad \overset{+}{B}\rightarrow\overset{-}{B}F_3$$

A generic reaction of a Brønsted-Lowry base with H⁺

$$B: \quad\curvearrow\quad H^+ \quad\longrightarrow\quad \overset{+}{B}{-}H$$

Figure 5.8 Electron donating capability of a base.

In a Brønsted–Lowry system, the base donates two electrons to the proton to form a new B–H covalent bond (see Chapter 2, Section 2.2). Note that both a Brønsted–Lowry base and a Lewis base donate electrons—the former to a proton and the latter to a different electron-deficient atom. ***In both cases and universally, a base is a two-electron donor***.

This chapter has discussed alkenes, alkynes, alkyl halides, amines, alcohols, and ethers. In the context of an acid–base theme, it is reasonable to ask if these functional groups have acid or base properties that are noteworthy. The π-bonds of an alkene and an alkyne are electron rich. In addition, heteroatoms such as the chlorine of the alkyl halide, the nitrogen of the amine, and the oxygen atom of both the alcohol and the ether are electron rich. It is reasonable to assume that all of these functional groups may function as electron donors. In other words, the π-bonds of alkenes or alkynes, as well as heteroatoms, may all react as bases. ***The halogen, nitrogen, and oxygen all have electron pairs and are expected to be better electron donors than a π-bond. In other words, an alkene or an alkyne is expected to be a weaker base than functional groups that have a heteroatom***.

Oxygen has an electronegativity value of 3.5 (see Chapter 3, Section 3.7.1) and is more electronegative than chlorine or nitrogen, which both have an electronegativity value of 3.0. In part, alcohols and ethers are weaker electron donors than an amine because they "hold" their electrons more efficiently. Indeed, an amine is a good electron donor and a good base. Halides are much weaker bases because they are much larger in size, and electronegativity also plays a role in the case of fluorine. In terms of base strength, amines > ethers > alcohols > halides > alkenes > alkynes. This order of basicity can be presented in terms of the structure of each functional group.

Order of Basicity

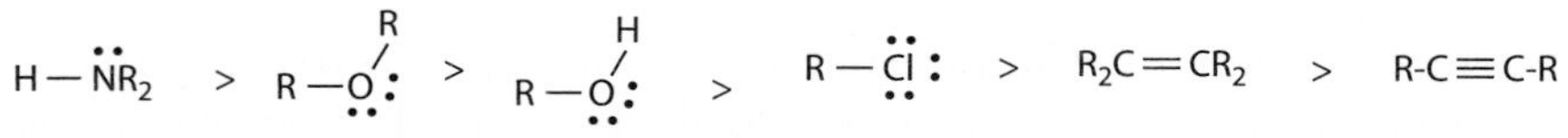

$$H{-}\overset{..}{N}R_2 \quad > \quad R{-}\overset{R}{\underset{..}{\overset{/}{O}}}\!: \quad > \quad R{-}\overset{H}{\underset{..}{\overset{/}{O}}}\!: \quad > \quad R{-}\overset{..}{\underset{..}{Cl}}\!: \quad > \quad R_2C{=}CR_2 \quad > \quad R{-}C{\equiv}C{-}R$$

If an amine reacts as a base with HCl, the two electrons are donated from N to H to form a new covalent bond in the conjugated acid, **46**. The conjugate base results from breaking the H–Cl bond with transfer of both electrons to chloride, to form the conjugate base, the chloride ion. Ammonium salt **46** is rather stable and it is a weak acid. The HCl is a much stronger acid than **46**, and the K_a for this reaction is expected to be large as it is written, so the amine is a good base. The ether and the alcohol are weaker electron donors than the amine. If the ether reacts with HCl, the conjugate acid is oxonium ion **47**, whereas the alcohol will give oxonium ion **48**. In both cases, the oxonium ion is a very strong acid, probably stronger than HCl, which suggests a smaller K_a and weaker acids.

Recall that the conjugate acid of H_2O is H_3O^+, the very acidic hydronium ion. The carbon groups on oxygen are electron releasing because the oxygen is more electronegative. With two carbon groups on oxygen, the ether is expected to have a more electron-rich oxygen relative to oxygen of the alcohol, with only one electron-releasing carbon group. Therefore, it is anticipated that ethers are stronger bases than alcohols. Both alcohols and ethers react as weak bases, but generate relative strong conjugate acids. If an alkyl halide reacts with HCl, the product is ion **49**, which would be a quite strong acid, and, as the reaction is written the equilibrium lies far to the left. In other words, alkyl halides are very weak bases with HCl.

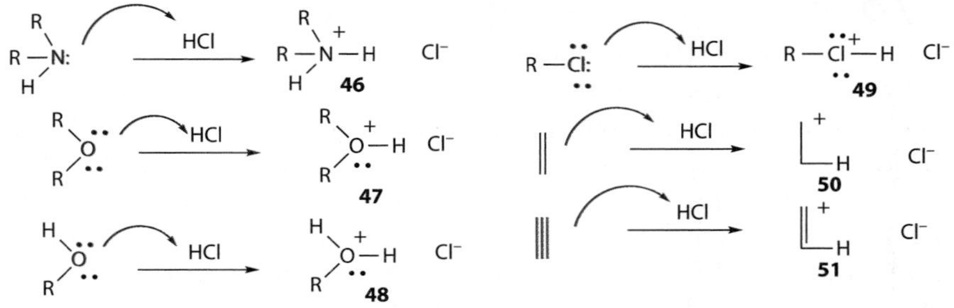

When a π-bond of an alkene or an alkyne reacts with HCl, the two-electron donor is the π-bond; however, to donate electrons to HCl, the π-bond must break. The two electrons of the π-bond are donated to the acidic H of HCl, forming a new covalent C–H bond and generating chloride ion. The interesting part of this reaction is the fact that such a reaction would form a bond to only one of the two carbon atoms, leaving a positive charge on the other, as shown in **50**. A species with a positive charge on carbon is called a carbocation (also called a carbenium ion) and is very reactive, as shall be discussed in Chapters 7 and 10. An alkene is a relatively weak base, and it will react only with strong mineral acids such as HCl to form species such as **50**. This means that the reaction requires a strong acid to compensate for the weak basicity of the alkene π-bond. An alkyne reacts similarly with HCl, to give carbocation **51**, but there is a positive charge on the sp^2 carbon in **51**. This cation is less stable than **50** and very reactive (see Chapter 10, Section 10.6.1). The two π-bonds of the alkyne make the alkyne a weaker electron donor relative to an alkene. In other words, an alkene is a stronger base than an alkyne.

5.28 Calculate the formal charge on the chlorine atom in 49 and on both carbon atoms in 50.

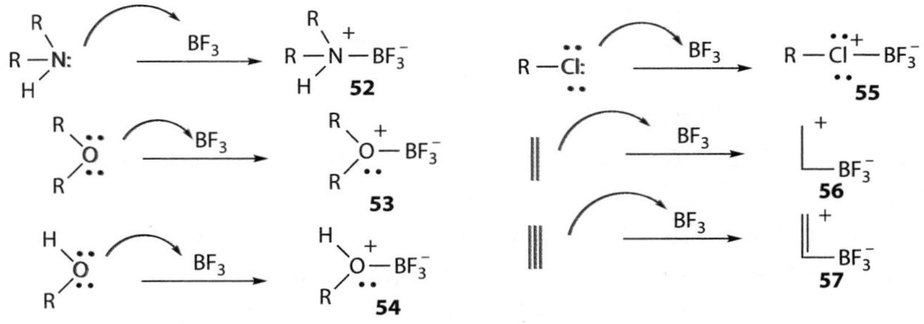

The goal of this discussion is to suggest that either a heteroatom or a functional group that is electron rich may react as a base by donating electrons to a Brønsted–Lowry acid. However, if the electron-donating power is lower (a weak base), it will only react with strong acids. If the base is a good electron donor (strong base), it will react with strong acids and many weak acids. It is also clear from these reactions that the stability (reactivity) of the conjugate acid will play a major role in determining the relative base strength of an electron-donating species. This is exactly analogous to the analysis in Chapter 2, where both the acid–conjugate acid and base–conjugate base structure/stability had to be analyzed to determine how well the reaction occurred. For now, it is known that amines are good Brønsted–Lowry bases, ethers and alcohols are moderate bases, alkyl halides are quite weak Brønsted–Lowry bases, and alkenes and alkynes are weak bases that react only with strong acids.

The previous discussion indicated that functional groups and particularly heteroatoms are Brønsted–Lowry bases. Can they also function as Lewis bases? This question is easily answered by examining the reaction of the electron-donating functional groups with Lewis acids such as boron trifluoride (BF_3). The amine is a good electron donor and reacts readily with BF_3 to form the corresponding ate complex **52** (recall Section 2.5, Chapter 2). Both ethers and alcohols react readily with BF_3 to form **53** and **54**, respectively. Both ate complexes are relatively stable, and amines and ethers are considered to be good Lewis bases, but an amine is usually a stronger Lewis base than an ether.

Similar electronic reasons discussed for Bronsted–Lowry bases predict that an ether is a stronger Lewis base than an alcohol. Interestingly, the alkyl halide is a reasonably good Lewis base with BF_3, forming **55**. Although **55** is much more reactive (less stable) than **54** or **53**, a species such as **55** can be generated and used in chemical reactions. In the case of an alkene and an alkyne, the carbocations **56** and **57** are quite unstable, and if formed at all will be very reactive (see Chapter 10, Section 10.6). Compared with the heteroatom functional groups, alkenes and alkynes are poor Lewis bases. Nonetheless, there are reactions in which they do react as Lewis bases (see Chapter 10, Section 10.6).

In all of the cases examined, the electron-donating power of the functional group results from the fact that they are electron rich. However, an alcohol is characterized by the OH functional group, which has a proton attached to the oxygen, and in the presence of a good base, alcohols react as acids. Alcohols are **amphoteric**, reacting as an acid in the presence of a strong base and a base in the presence of a strong acid. When alcohol **58** is mixed with the amide anion

(NaNH$_2$), a strong base, the hydrogen is removed from the alcohol to form the conjugate base **59** (known as an **alkoxide**). Why?

When $^-$NH$_2$ reacts as a base, the conjugate acid is ammonia, NH$_3$. When the alcohol is compared with the conjugate acid ammonia, the alcohol is a stronger acid, making **59** a weaker base than -NH$_2$, using the axiom that a stronger acid generates a weaker conjugate base. When **58** is mixed with sulfuric acid, however, the oxygen of **58** reacts as a base to form the conjugate acid **60** and the resonance stabilized conjugate base hydrogen sulfate. When sulfuric acid is compared with the alcohol **58**, sulfuric acid is the stronger acid; therefore, when the alcohol reacts, it is classified as a base. Sulfuric acid can also be compared with **60**, but this comparison requires knowledge that the oxonium ion is a weaker acid.

When does an alcohol react as an acid? It reacts when it is mixed with a base that generates a conjugate acid that is much weaker than the alcohol. Likewise, the alcohol reacts as a base when mixed with an acid that generates a conjugate base that is weaker than the alcohol.

5.29 Draw all resonance forms of the hydrogen sulfate anion.

5.8 Polarity and Intermolecular Forces

5.8.1 Noncovalent Bonding Interactions

There are forces that lead to attraction between molecules, even when there are no polarized bonds. Imagine that two alkane molecules come close together and the electron density associated with the covalent bonds is attracted to the positive nuclei in the atoms of the molecules, as illustrated by Figure 5.9. In effect, subnuclear particles change polarity when they come into close proximity with other particles. When two molecules come close and there is no dipole, the electrons reorganize to some extent due to polarization so that there is a higher concentration of electron density on one side than another, leading to a very small *induced dipole* (illustrated in Figure 5.9a). This small induced dipole allows a small attraction between molecules, as indicated in Figure 5.9b.

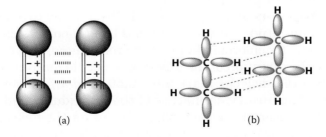

(a)   (b)

Figure 5.9   Attractive London forces in ethane.

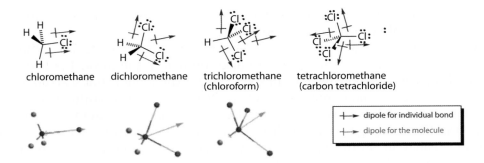

Figure 5.10   Dipole moments for chlorinated methane compounds.

The attractive forces between the particles of one molecule and the particles of a second molecule are extremely weak and are known as **London forces**, after Fritz Wolfgang London (German-born American; 1900–1954) and are also known as **van der Waals forces**, after Johannes Diderik van der Waals (Holland; 1837–1923). Because *the induced dipole results from close contact, the larger the surface area of the molecule is, the greater is the van der Waals interaction*. This is a weak force, and it can be disrupted by application of only small amounts of energy. This statement means that two molecules held closely together by these forces are easily separated at low energy (low temperatures).

The concept of dipole moment is introduced in Chapter 3, Section 3.7. The δ+ and δ– atoms determine the direction of the dipole moment for an individual bond. Because molecules are three dimensional, however, the direction of the dipole moment for a molecule depends on the shape of that molecule. This is illustrated in Figure 5.10 for four chlorinated methane compounds.

In chloromethane, the tetrahedral shape is clear, but there is only one polarized bond and the dipole for the molecule is easily predicted. In dichloromethane, however, there are two bond moments, and the dipole for the molecule is the vector sum of these two bond moments (magnitude and direction). The dipole is shown. For trichloromethane (chloroform), the magnitude and direction of the three polarized C–Cl bonds lead to the molecular dipole moment shown. Carbon tetrachloride is interesting. There are four C–Cl bonds with equal bond polarization and dipole moments. Summing all four dipole moments for the bonds, which are directed to the corners of a regular tetrahedron, leads to a dipole moment of zero because the magnitudes of the individual bond moments cancel.

Molecular models for chloromethane, dichloromethane, and trichloromethane are given to show the direction of the dipole of molecule more clearly. Calculated dipoles for these three molecules are 2.87, 2.50, and 1.72 Debye, respectively, and it is clear that the directional nature of the individual bond dipoles plays a role in the overall magnitude of the dipole moment for the molecule. With three chlorine atoms directed to different regions of space, chloroform is the least polar of the three molecules, despite the presence of three polarized bonds.

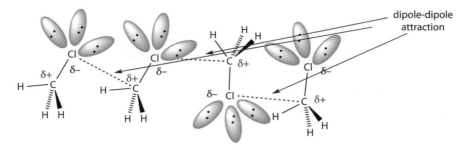

Figure 5.11 Attractive dipole–dipole interactions in chloromethane.

5.30 Estimate the direction of the dipole moment in BrCCl$_3$.

When a molecule has a dipole moment, it is considered to be polar. The larger the magnitude of the dipole moment *for the molecule* can be associated with the polarity of the molecule. Polarity in molecules leads to certain consequences that relate both to reactivity and its physical properties. The differences that arise between polarized and nonpolarized molecules give important clues as to the properties and identity of those molecules.

When the polarized atoms in an alkyl halide such as the molecule chloromethane come into close proximity to the dipole of another molecule of chloromethane, as shown in Figure 5.11, the δ+ carbon of one molecule is attracted to the δ– chlorine of the second molecule. This electrostatic attraction is called a **dipole–dipole interaction** and it is *stronger* than London forces, which are also present. As a consequence of this attraction, more energy (heat) is required to separate two molecules held together by dipole–dipole interactions than is required to separate two molecules held in close proximity by simple London forces. In other words, the boiling point required for molecules with dipole–dipole interactions is higher. **The dipole–dipole interaction does not have to be between identical molecules; it can be between two entirely different molecules**. Dipole–dipole interactions can be envisioned between two alkyl halides, two ethers, or even two amines.

A special type of dipole–dipole interaction occurs when one atom of the dipole is a hydrogen, as in the O–H bond of an alcohol. When the OH of one alcohol (such as methanol, **36**) comes into close proximity with the OH unit of a second molecule of methanol (illustrated in Figure 5.12), there is a strong dipole–dipole attraction between the positively polarized hydrogen and the negatively polarized oxygen. Figure 5.12 is meant to represent an array of *many* methanol molecules attracted to each other by this type of hydrogen bond. Because an O–H bond is generally more polarized than a C–O bond, the hydrogen atom is rather small compared to carbon, and a positively polarized hydrogen is acidic, the O------H interaction shown in Figure 5.12 is particularly strong. It is much stronger than a common dipole–dipole interaction and is given a special term: a **hydrogen bond**.

Because the hydrogen bond is rather strong, it takes much more energy to disrupt it. In the cases mentioned here, more energy (heat) must be applied to

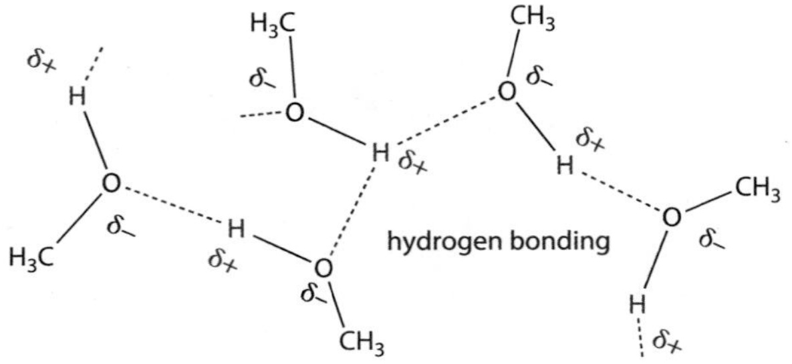

Figure 5.12   Hydrogen bonding in methanol.

separate associated methanol molecules than is required to separate associated chloromethane molecules, and the least amount of energy will be required to separate associated alkane molecules (such as ethane, shown in Figure 5.9). All of this means that the boiling point of methanol is higher than that of chloromethane, which is higher than that of ethane.

Water (H–O–H) is a common solvent and is frequently used in reactions with organic molecules; therefore, it is important to mention that water can form strong hydrogen bonds to itself as well as to molecules that contain a polarized X–H bond (X is usually O, S, N, etc.). Examples are water with alcohols, water with amines, and water with carboxylic acids—but *not* with alkane or alkenes because there is no polarized bond in those molecules. A reasonably strong hydrogen bond can also be formed between water and compounds that do not contain an X–H unit but do have a highly polarized functional group. These include water and ketones or aldehydes (see the following section) and, to a lesser extent, ethers.

5.31 Which molecule will form the stronger hydrogen bond to water–hydrogen bond: methylamine (CH_3NH_2) or methanol (CH_3OH)?

5.8.2 Physical Properties Derived from Noncovalent Interactions

A **physical property** of an organic compound can be measured experimentally and is characteristic of that molecule (or a collection of like molecules). The temperature at which a compound boils or melts, for example, is a physical property of the molecule. Physical properties can be used to characterize and identify a compound because the complete set of physical properties of an individual molecule is usually unique. The functional group is also very important in determining the physical properties of a molecule. An alcohol has physical properties different from those of an amine, for example, or of a carboxylic acid. This section will tie together the concept of bond polarity introduced in Chapter 3, Section 3.7, with the physical properties that result from the presence of these polar bonds.

5.8.2.1 Solubility

"Like dissolves like" is an old axiom in chemistry. The term "dissolve" is formally defined as "to cause to pass into solution" or "to break up." If there are two compounds and at least one of them is a liquid, one dissolves in the second compound to form a solution, or one compound is dispersed into the second. The term "like dissolves like," in the context of this section, means that polar compounds are likely to dissolve in other polar compounds, but not very well in nonpolar compounds. In terms of solubility, one functional group in a molecule of less than five carbons is considered polar, whereas one functional group in a molecule of greater than eight carbons is much less polar and often is nonpolar.

Compounds of five, six, or seven carbons are difficult to categorize. An alcohol with only two carbons and water are both polar, and such an alcohol will likely dissolve in water. Water is polar and an alkane is nonpolar, so they are not expected to be mutually soluble. A nonpolar compound will dissolve in a nonpolar liquid, but it is usually not very soluble in a polar liquid. In other words, one alkane should dissolve in another alkane, or in most hydrocarbons, but not in the polar molecule water.

The term "solubility" refers to one molecule **dissolving** in another. This means that the molecules mix together such that one phase (one layer) is formed. If two things are not mutually soluble (like oil and water), two phases (two layers) are formed. Note that oil is a complex mixture of hydrocarbons. To some extent, the term soluble can be quantified. Two grams of compound A might dissolve in 100 g of compound B, but not in 50 g. Fifty grams of molecule C might dissolve in 25 g of compound B, but 1 g of C is observed to dissolve in no less than 200 g of D. Solubility differs with the compound and with the amount of material, so a formal definition must be agreed upon. *The formal definition of solubility is the "mass of a substance contained in a solution, which is in equilibrium with an excess of the substance."* A more practical but arbitrary definition defines the number of grams of one molecule (the **solute**) that can be dissolved in 100 g of the second compound (the **solvent**).

This is dependent upon the temperature because more solute can usually be dissolved in a hot solvent than in a cold solvent. The temperature of the solution is therefore provided with the solubility information. One molecule can be *partially soluble* in another so that some of it dissolves in 100 g, but two phases can still be formed. The word **miscible** is often used interchangeably with **complete solubility**. Imagine that 1, 10, 50, or 100 g of acetone will mix with 100 mL of water to form a single layer. Acetone is said to be infinitely soluble in water and it is also said to be miscible. Imagine that a different molecule was partially soluble in 100 mL of water. This might mean that if 5 g were added, only 4.5 g would dissolve and the remaining 0.5 g would not go into solution. That concentration may be sufficient to allow a chemical reaction or to extract one compound from another. In some ways, solubility is a relative term, despite the formal definition.

In general, a nonpolar molecule is one that has no polarized bonds or a dipole moment and only noncovalent van der Waal's interactions. Alkanes are a typical example. A polar molecule will have dipole–dipole interactions or hydrogen bonding interactions. If a polar molecule is attracted to another polar molecule, then this aggregation of polar molecules will allow them to mix with each other. A molecule capable of hydrogen bonding or generating a dipole–dipole interaction (methanol in water, for example) should be soluble in a molecule that is also capable of forming a hydrogen bond or a dipole interaction. The reason why nonpolar molecules are soluble in each other is more subtle. Only London forces are at work between two alkanes (see Figure 5.8), leading to the association of one alkane with another alkane.

61

In general, the number of carbon and hydrogen atoms and the number of functional groups will determine solubility, as hinted at before. Imagine that each carbon and hydrogen adds a nonpolar component to the molecule and that each functional group (OH, C=O, NH, etc.) adds a polar component to the molecule. If the number of carbon and hydrogen atoms is very large (say, 100 carbons) and there is one C=O unit, the C/H units will "win"; the molecule will be largely nonpolar and will be soluble in nonpolar solvents. If there are three OH units and three carbon atoms (as in glycerol; see **61**), the large proportion of polar groups makes the molecule very polar, and it is very soluble in polar solvents such as water. If a molecule contained 100 carbon atoms and 90 OH units, it would be very polar and very likely soluble in water.

A very rough guideline for solubility in water is that one heteroatom *func-tional group* (not halogen) in a molecule of less than five carbons will be soluble in polar solvents. The presence of one functional group in a molecule of four to seven carbons *may* impart solubility, but with greater than eight carbons water solubility will be minimal or absent. Table 5.1[2] contains data about the solubility (measured in grams of solute per milliliter of water) of various molecules in solvents. If there are no heteroatom functional groups, the molecule is nonpolar and probably insoluble in water. Every molecule will have a slightly different solubility in various solvents (acetone, methanol, ethanol, diethyl ether) (see Table 5.1).

5.32 **Should decanol (nine CH_2 units, one CH_3 unit, and one OH unit) be miscible with hexane (two CH_3 units and four CH_2 units)?**

5.8.2.2 *Adsorption*

The property of **adsorption** is formally defined as the binding or adherence of gases, liquids, or dissolved substances to the surface of a solid. In effect, the atoms in a molecule are attracted to the atoms of another molecule (a solid) and "stick" there. The interactions required for a compound to adhere to a solid are generally the same as seen in liquid–liquid interactions: London forces,

Table 5.1
Solubility of Organic Compounds in Various Solvents

Molecule	Solvent	Solubility (g/100 g solvent)
1,4-Pentanediol	Water	Completely miscible
1-Pentanol	Diethyl ether	Completely miscible
1-Pentanol	Ethanol	Completely miscible
1-Pentanol	Water	2.7
1-Pentene	Diethyl ether	Completely miscible
1-Pentene	Ethanol	Completely miscible
1-Pentene	Water	0.014
1-Pentene	Water	Insoluble
3-Pentanone	Water	3.4
Pentane	Diethyl ether	Completely miscible
Pentane	Water	Completely miscible
Pentane	Water	0.036
Pentanoic acid	Ethanol	Completely miscible
Pentanoic acid	Water	2.4
Pentanoic acid	Diethyl ether	Completely miscible
Methane	Diethyl ether	104 mL gas/mL
Methane	Ethanol	47.1 mL gas/mL
Methane	Water	3.3 mL gas/mL
Methanol	Ethanol	Completely miscible
Methanol	Water	Completely miscible
Methylamine	Benzene[a]	10.5
Methylamine	Water	959 mL gas/mL water

Source: Lide, D. R., ed. 2006. *CRC Handbook of Chemistry and Physics,*
       87th ed., 3-1–3-523. Boca Raton, FL: Taylor & Francis.
[a] Benzene (see Section 5.10 and Chapter 20).

dipole–dipole interactions, and hydrogen bonding. In general, the compound does not permanently adhere to the solid, but binds reversibly. If it binds in an equilibrium, some is adsorbed to the solid and some is not.

Just as with the attractive forces described before, adsorption is very strong if a hydrogen bond is possible and very weak if only London forces are possible.

Molecules of a liquid can stick to the surface of the solid and the stronger the intermolecular interaction is, the more "sticky" the liquid is on the solid. A polar compound would be expected to stick on a polar solid more tightly than a nonpolar compound would stick. If the polar solid is silica gel (SiO_2), a polar alcohol would stick very well but a nonpolar alkane would not. This property is important for separating molecules and also in many chemical reactions. These principles will be discussed in more detail for specific applications (liquid–solid chromatography, catalytic hydrogenation, certain metal catalyzed reactions) as they arise.

5.8.2.3 *Boiling Point*

Boiling point is defined as the temperature at which a liquid and the vapor (gas) above it are in equilibrium. At normal atmospheric pressure, the boiling point of a liquid will be the temperature at which the liquid is at equilibrium with the atmosphere above the liquid (atmospheric pressure). Another way to say this is that boiling point is the temperature at which molecules leave the liquid phase and are moved into the gas phase. Several factors are important in determining the boiling point of a liquid. The number of atoms in a molecule and the number and type of heteroatom functional groups will play an important role.

Comparing the alkanes in Table 5.2,[2] the boiling point increases steadily as more carbon and hydrogen atoms are added. Mass is not the only thing to influence boiling point. Clearly, boiling point varies with the nature of the functional group. When the molecular weights of the molecules are close, as in methane–methanol–formic acid–methylamine, the functional groups lead to a vast difference in boiling point. The key reason for these differences is the intermolecular forces discussed previously.

At the molecular level, boiling point is the temperature where intermolecular forces (London forces, dipole–dipole interactions, and hydrogen bonding) are disrupted to produce single molecules that are not associated with other molecules. It is important to note that energy is required to break these associations. Because a hydrogen bond is stronger than a simple dipole–dipole interaction, it is reasonable to assume that a higher temperature (more heat) is required to boil a molecule containing hydrogen bonds (such as an alcohol) than is required to boil a molecule containing only simple dipole–dipole interactions (such as a ketone).

In Table 5.2, 1-pentanol, 2-pentanol, and 3-pentanol all boil at higher boiling points than chloropentane or the ethers. Because the chlorine and bromine atoms add mass, there is a natural increase in the boiling point. When the alcohols are compared with the ethers, the molecular weights are close, but there is a large difference in the boiling point. This difference is a reasonable measure of the influence of hydrogen bonding. It is also clear that branched alcohols and branched ethers have a lower boiling point, in part due to diminished London forces because those molecules have a smaller surface area.

If only London forces are at work (as in pentane and other molecules without functional groups) only a small amount of energy is required to overcome

Table 5.2
Boiling Points of Common Organic Molecules

Molecule	bp (°C)	Molecular Weight
methane	−161.5	16.04
ethane	−88.6	30.07
propane	−42.1	44.10
butane	−0.5	58.12
pentane	36.1	72.15
hexane	68.7	86.18
heptane	98.4	100.21
octane	125.7	114.2
water	100.0	18.01
methane	−161.5	16.04
methanamine	−6.3	31.06
methanol	64.7	32.04
methanoic acid (formic acid)	100.8	46.03
pentane	36.1	72.15
1-pentene	30.0	70.14
1-pentyne	30.2	68.11
1-chloropentane	108.4	106.59
3-chloropentane	97.5	106.59
1-bromopentane	129.8	151.05
3-bromopentane	118.6	151.05
1-methoxybutane	63.2	88.14
2-methyl-1-methoxyethane	55.0	88.14
1-ethoxypropane	63.21	88.14
1-pentanamine	104	87.17
1-pentanol	137.8	88.15
2-pentanol	119.0	88.15
3-pentanol	115.6	88.15
pentanoic acid (chapter 15, section 15.4)	185.5	102.13
1,5-pentanediol	242.5	104.15

Molecules capable of only London forces are in **green**
Molecules capable of hydrogen bonding are in **blue**
Molecules capable of dipole–dipole are in **purple**

Source: Lide, D. R., ed. 2006. *CRC Handbook of Chemistry and Physics,*
87th ed., 3-1–3-523. Boca Raton, FL: Taylor & Francis.

their association in the liquid. Because the associations between molecules are weak in this case, the boiling point is very low and only a small amount of heat is required to disrupt the associative forces in the liquid to generate molecules in the gas phase. If more than one hydrogen-bonding group is available (see 1,4-pentanediol), more energy is required to disrupt the increased hydrogen bonding, and a much higher boiling point is the result.

The influence of hydrogen bonding is perhaps best seen when water (molecular weight: 18), methane (molecular weight: 16), and methanol (molecular weight: 32) are compared. Methane boils at −161.5°C, water at +100°C, and methanol at +64.7°C. Water and methane are about the same molecular weight, but the hydrogen-bonded water boils 261° higher. Methanol has a molecular weight almost twice that of water, but it boils almost 35° lower. This is an indication of greater hydrogen bonding in water than in methanol.

5.33 Which should have the higher boiling point: an eight-carbon alcohol or an eight-carbon ether? Briefly explain.

It is important to mention that as the pressure is lowered, the temperature required for the liquid to come into equilibrium is also lowered. This means that the boiling point is lowered as the pressure is lowered (liquids boil at a lower temperature under vacuum).

5.8.2.4 Melting Point

The melting point is defined as the temperature at which the solid phase of a molecule and the liquid phase of that molecule are in equilibrium. It is the temperature at which the strong forces that hold molecules together in the solid phase are disrupted, but the forces that keep molecules together in the liquid stated are not disrupted. In general, the melting point of a compound increases as the molecular weight of that compound increases. A molecule with a molecular weight of 300 is expected to melt at a higher temperature than a molecule with a molecular weight of 100. **It is not that simple, however, because a molecule in the solid phase often exists in a regular array of atoms: a crystal lattice**.

The shape, geometry, and packing of molecules within this lattice have a great effect on the melting point. A compact molecule such as (a) in Figure 5.13 may pack into a rectangular array such as (b). If a "dot" is used for each molecule, this shape is (c) and, as shown in (d), several of these units can pack together to form a rigid and strong structure (a crystal structure). A significant amount of energy is required to disrupt a structure like (d), which is expected to have a relatively high melting point. If a molecule has an extended structure, such as (e), it will not pack into a regular structure. If molecules are only loosely associated in the solid phase, the attractive forces are easily disrupted. Such a compound will have a low melting point. The accompanying Table 5.3[2] gives the structures and melting points for some common organic molecules.

Table 5.3 shows that the melting point increases as the molecular weight of the alkanes increases. There is only loose packing in the solid state of these

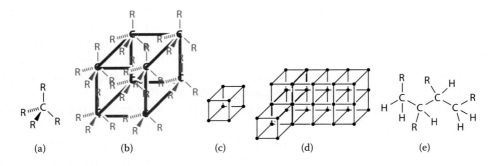

(a)   (b)   (c)   (d)   (e)

Figure 5.13   Packing organic molecules.

Table 5.3
Melting Points of Common Organic Molecules

Molecule	Melting Point (°C)	Molecular Weight
pentane	−129.7	72.15
hexane	−95.4	86.18
heptane	−90.6	100.21
octane	−56.8	114.23
nonane	−53.5	128.26
decane	−29.7	142.29
undecane	−25.6	156.31
dodecane	−9.6	170.41
tridecane	−5.4	184.37
tetradecane	5.9	198.40
pentadecane	9.9	212.42
nonane	−53.5	128.26
3,3-diethylpentane (A)	−33.11	128.26
2,3-dimethylheptane (B)	−116	128.26
2,2-dimethylheptane (C)	−113	128.26
2,6-dimethylheptane (D)	−102.9	128.26
2,2,5-trimethylhexane (E)	−105.78	128.26

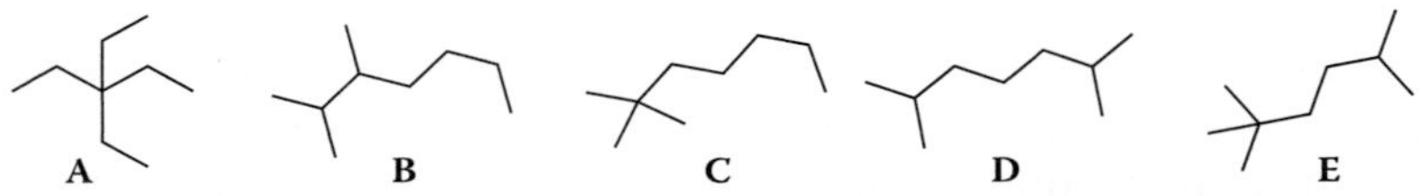

A   B   C   D   E

Source: Lide, D. R., ed. 2006. *CRC Handbook of Chemistry and Physics,* 87th ed., 3-1–3-523. Boca Raton, FL: Taylor & Francis.

alkanes, and the melting point of alkanes is >0°C only at the 14-carbon alkane (tetradecane). The ability for a molecule to pack is more subtle and depends on the overall shape of the molecule. If a molecule is shaped more like a ball or has a regular structure, as with 3,3-diethylpentane (see Chapter 4, Section 4.6, for alkane nomenclature), the melting point is higher than if it has an irregular shape, as in the nine-carbon alkane (nonane). As carbon atoms are branched from the linear chain of carbon atoms, the shape becomes more irregular and the melting point is lower.

5.34 **Which compound should have the higher melting point: CCl_4 or CH_3Cl? Briefly explain.**

5.9 Functional Groups with Polarized π-Bonds

Formation of a π-bond between two atoms other than carbon or between carbon and a heteroatom is well known. In principle, a π-bond may form between any atom that has available p-orbitals and a valence greater than one.

Carbon forms a π-bond to another carbon (C=C), to an oxygen (C=O), to sulfur (C=S), to nitrogen (C=N), and to other atoms. It does *not* form one to hydrogen or a halogen and remain neutral. Molecules that contain N=N bonds, N=O bonds, and O=O bonds will be encountered from time to time. The discussion of alkenes and alkynes in Sections 5.1 and 5.2 showed that the C=C bond distance is shorter than the C–C bond distance and that the C≡C bond is shorter than the C=C bond. These facts are the basis of an analogy that is extended to other types of π-bonds. The C=O bond is shorter than the C–O bond, the C=N bond is shorter than the C–N bond, and the N=N bond is shorter than the N–N bond.

Nitrogen and oxygen atoms are attached to carbon in C–N and C–O by covalent σ-bonds (single bonds) in the functional groups amines, alcohols, and ethers. It is also possible to form multiple bonds between a carbon atom and some heteroatoms, including N and O. Double bonds between carbon and oxygen, nitrogen, or sulfur are possible, with one strong σ-bond and one weaker π-bond. It is also possible to form triple bonds between carbon and nitrogen with one strong σ-bond and two π-bonds.

5.9.1 A Double Bond to Oxygen—The Carbonyl Group

Just as there are carbon–carbon double bonds, there are carbon–oxygen double bonds, with one π-bond and one σ-bond between a carbon and oxygen. The carbon–oxygen double bond, represented as C=O, is called a **carbonyl**. Both the carbon and the oxygen are sp² hybridized and because oxygen forms two bonds, two unshared electron pairs reside on oxygen. As seen in Figure 5.14, the unshared electrons are orthogonal to the π-bond and coplanar with the atoms. The R–C–O bond angle is about 120°—consistent with the sp² hybridization, making the carbon, oxygen, and the hydrogen atoms all coplanar.

A molecular model of formaldehyde shows the typical trigonal planar geometry of a carbonyl compound. An important feature of a carbonyl is that oxygen makes the bond polarized (carbon is δ+ and oxygen is δ–), so the electron density of both the σ-bond and the π-bond is distorted toward oxygen. Combined with the fact that oxygen can accommodate excess electron density, the π-bond of a carbonyl is relatively easy to break.

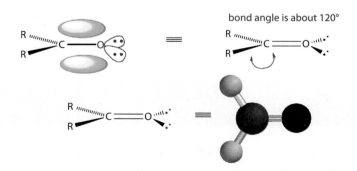

Figure 5.14   The carbonyl group.

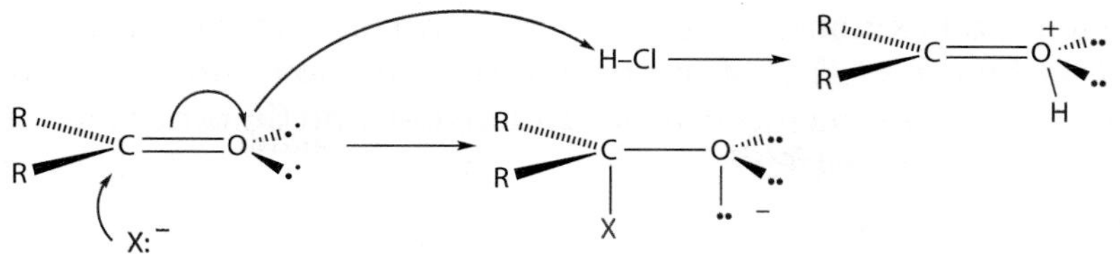

Figure 5.15 Donation of electrons to a carbonyl with cleavage of the π-bond.

The C=C unit of an alkene and the C=O unit of a carbonyl are similar in that they react as Brønsted–Lowry bases in the presence of a strong acid. In the case of the carbonyl, the oxygen is the base and it is protonated to form the oxonium ion shown in Figure 5.15. Both a C=C unit and a C=O unit have a π-bond. However, the C=C unit of an alkene is *not* polarized and is not easily attacked by a negative (anionic) species. As pointed out previously, a *carbonyl is polarized.* Because of the bond polarization, a negatively charged species (X^-) reacts by donating two electrons to the carbon of the C=O unit.

Reaction at the carbonyl carbon forms a new σ-bond to that carbon, the π-bond breaks, and the two electrons in the π-bond are transferred to the electronegative oxygen, thus making an anionic species (see Figure 5.15 and note the use of the electron flow arrows). In this reaction, X^- is a nucleophile because it donates two electrons to carbon. The curved arrows show the flow of *pairs* of electrons—***from X^- to*** carbon to form a new bond and ***from*** the bond between carbon and oxygen ***to*** an unshared pair on oxygen alone. Note that this process (a reaction) reacts a neutral species with an anion to form a new species with a negative charge.

5.35 Draw the product obtained when $H_2C=O$ reacts with H^+.

5.9.2 Ketones and Aldehydes

There must be three covalent bonds to carbon in a carbonyl, so two additional atoms other than oxygen are required (see Figure 5.14). If at least one of the groups is a hydrogen atom, there is a H–C=O unit (see the box in **62**; also written as–CHO) and the molecule to which it is attached is called an ***aldehyde***. Structurally, an aldehyde has one hydrogen and one carbon group attached to the carbonyl carbon, although formaldehyde has two hydrogen atoms ($H_2C=O$). If two carbon groups (two alkyl groups) are attached to the carbonyl carbon, the generic formula is $R_2C=O$ and the molecule is called a ***ketone*** (see **63**). *In both aldehydes and ketones, **the functional group is the carbonyl, C=O**.*

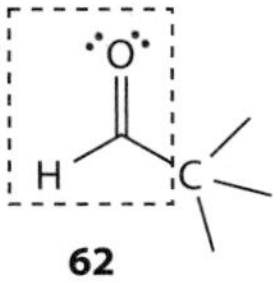

62 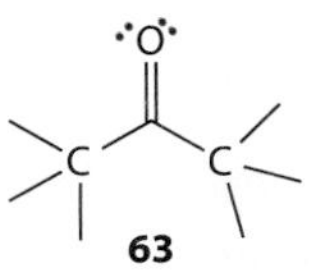   **63**

Aldehydes and ketones show a wide variety of structural variation. Compound **64** has the name of methanal (the common name is formaldehyde), **65** is ethanal (the common name is acetaldehyde), and **66** is named 2-propanone (the common name is acetone).

5.36 Determine the formal charge of oxygen and the carbonyl carbon in 3-pentanone (see 64, where R = ethyl).

formal (formaldehyde)   ethanal (acetaldehyde)   2-propanone (acetone)
(an aldehyde)   (an aldehyde)   (a ketone)
64   **65**   **66**

A ketone contains a carbonyl group attached to two alkyl groups (as in **66**). *The functional group is the **carbonyl** unit* C=O and the suffix for ketones derives from the last three letters of ket**one**: **"one."** When the longest chain is numbered, the *carbonyl carbon* receives the lowest possible number. The six-carbon straight-chain ketone **67** is 3-hexanone. Substituents are handled in the usual manner, and compound **68** is 1-chloro-3-ethyl-8-methyl-4-decanone (the decanone chain is marked in **purple** and the substituents are marked in **green**). Note that for some simple ketones, the common name is used virtually all the time and one almost never observes the IUPAC name (**acetone**, used as a commercial solvent, is an example). Cyclic ketones are possible and the carbonyl carbon is part of the ring. In **69**, the carbonyl unit is part of a seven-membered ring, and the usual protocols for rings apply. Ketone **69** is a cyclo-heptanone, the carbonyl carbon is C1, and the substituents are given the lowest possible number. Therefore, the name is 7-bromo-3,3-dimethylcycloheptanone.

67   **68**   **69**

When the carbonyl is compared to C=C or C≡C, the C=O unit has a higher priority. Therefore, a molecule containing a carbonyl and an alkene is named an "ene-one" and an alkyne-ketone is named an "yne-one." The carbonyl is also higher in priority than an alcohol. The reason for this priority is that the alcohol has one bond to O, whereas the carbonyl has two bonds to O (see the Cahn–Ingold–Prelog rules in Chapter 9, Section 9.3.1).

5.37 Draw 3,6-dibromo-2-ethyl-4-decanone.

5.38 Draw the structure of 5-ethyl-4-hydroxy-2-octanone.

An aldehyde is a molecule with one alkyl group and one hydrogen attached to a carbonyl, but *the functional group is still the **carbonyl** unit* C=O. The suffix for aldehydes derives from the first two letters of *al*dehyde: **"al."** As with ketones, the carbonyl carbon takes priority and is given the lowest possible number, which *in all cases* will be 1 because all aldehydes have a hydrogen attached to the carbonyl unit. For this reason, the number is omitted. Substituents attached to the aldehyde chain are named in the usual manner; the carbonyl carbon is always numbered 1 and **70** is 5-chloro-3-ethyldecanal (the decanal chain is marked in **purple** and the substituents in **green**).

5.39 Draw the structure of 2,2,4,4-tetramethyl-5-(2-chloro-1-propyl)-decanal.

As with ketones, the carbonyl of an aldehyde has a higher priority than the C=C of an alkene, the C≡C of an alkyne, or the OH of an alcohol. Note that the aldehyde unit is drawn using a shorthand version, –CHO. *When the aldehyde unit (-CHO) is attached to a ring, a major modification in the name is required.* Aldehyde **71**, for example, is *not* named cyclohexanal. This name makes no sense because an aldehyde must have the CHO carbon labeled as C1. Instead, the term **carboxaldehyde** is used and **71** is named cyclohexanecarboxaldehyde. This nomenclature rule should be used for all molecules where the CHO unit is attached to a ring.

5.40 Draw the structure for 3,4-dimethylhex-4-ynal.

5.41 Draw the structure of 3-chlorocyclopentane-1-carboxaldehyde.

5.9.3 Carboxylic Acids and Resonance

An important functional group has a carbon atom (alkyl group) attached to a carbonyl (C=O) functional group, but a hydroxyl (OH) group is also attached via the oxygen. This structural arrangement is shown in the box on structure **72**. It is also written as –COOH or CO_2H. This unit is the **carboxyl** functional group and it is the major structural feature of the class of organic molecules known as **carboxylic acids**. The carboxyl group has an O–H unit (see **72A**) attached to a carbonyl, which has a weaker and polarized π-bond and a stronger σ-bond. The two oxygen atoms and the two carbon atoms are coplanar due to the presence of the π-bond.

The interesting feature of the carboxyl functional group is the presence of the highly polarized O–H unit where the hydrogen is δ+ (see **72A**). This molecule is known as ethanoic acid, but the common name is acetic acid. This molecule is also drawn as the molecular model (**72B**) to illustrate the spatial arrangement of the groups. The polarization induced by the carboxyl oxygen makes the carboxyl carbon atom very positive, which leads to the oxygen of the OH unit being negatively polarized and the hydrogen positively polarized, as shown. In other words, the proton is acidic (see Chapter 2, Section 2.2). The O–H unit of a carboxylic acid is more polarized than the OH unit of an alcohol (C^{acid}–O–H >>

C^{alcohol}–O–H). In other words, the proton of the CO_2H unit is a much stronger acid (pK_a = 1–5) than the proton of an alcohol (pK_a = 16–18). As introduced in Chapter 2 (Section 2.4), ***the greater acidity of the carboxylic acid is largely due to the stability of the conjugate base that is formed from 72.***

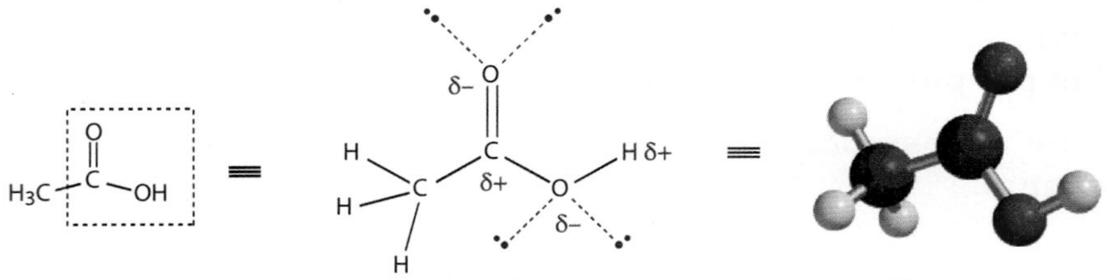

72   **72A**   **72B**

There are many carboxylic acids, and they are found in such common products as vinegar and lemon juice (lemon juice contains citric acid; see structure **10** in Chapter 1, Section 1.1). Two simple examples of carboxylic acids are **73**, which has the common name of formic acid (found in some ant venoms), and **72**, which has the common name acetic acid (a dilute solution in water is called vinegar). Lactic acid (**74**; the IUPAC name is 2-hydroxypropanoic acid) is found in sour milk. An urban myth claims that lactic acid is responsible for muscle soreness after strenuous exercise; however, this has been disproved, although it is involved in metabolism.

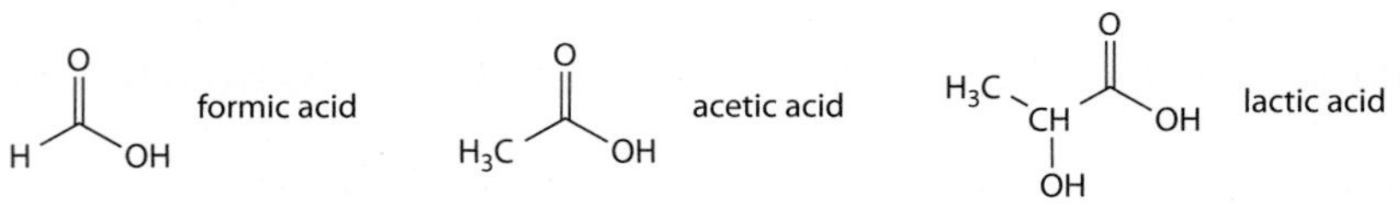

73   **72**   **74**

A carboxylic acid contains a carbonyl unit (C=O), but the functional group is the carboxyl unit, –COOH. The nomenclature rule is the same as that for aldehydes in that the longest continuous chain *must* contain the COOH unit, with the carboxyl carbon as C1. The carbonyl carbon of the COOH group must receive the lowest possible number, "1," so it is omitted from the name. The suffix for carboxylic acids is **"oic acid"** and the word "acid" is separated from the first part of the name. The eight-carbon acid (**75**) is octanoic acid. When the carboxyl unit is attached to a ring, the rules are again similar to those given for an aldehyde because the carbonyl carbon cannot be part of the ring, but must be attached to the ring. Compound **76** is named cyclohexane carboxylic acid. Substituents are numbered relative to the carbonyl carbon of the COOH unit, so **77** is 4-ethyl-3-(1,1-dimethylbutyl)heptadecanoic acid (the heptadecanoic acid chain is marked in **purple** and the substituents are marked in **green**; secondary substituents are in **yellow**).

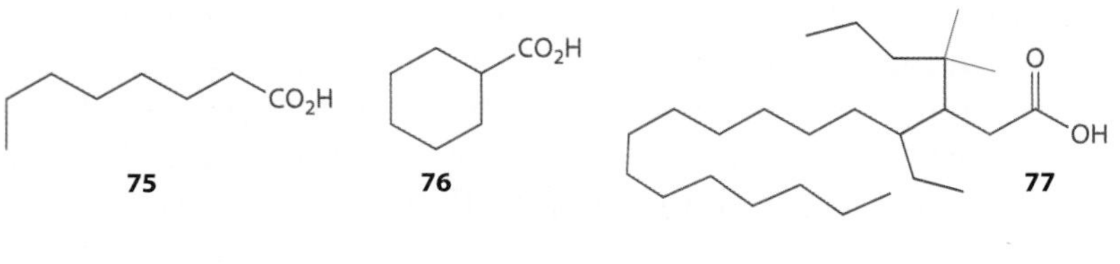

75   **76**   **77**

A carboxylic acid with two COOH units is known as a dicarboxylic acid, or dioic acid. The molecule $HO_2C-(CH_2)_3-CO_2H$, for example, is named 1,5-pentanedioic acid. These compounds and more about the nomenclature of carboxylic acids are presented in Chapter 16 (Section 16.4).

5.42 Draw the structure of 1-methylcyclohexane-1-carboxylic acid.

5.43 Draw the structure of 3-chloro-4-methylheptanoic acid.

5.44 Draw the structure of 3,4-diethyl-1,6-hexanedioic acid.

When the word *acid* is read or heard, thoughts usually turn to a proton (H^+), which is a hydrogen atom with no electron. If H^+ is considered to be the symbol for an acid, then a hydrogen atom that is positively polarized (it is $\delta+$) is "close to an acid." The more positively polarized the hydrogen is, the closer it is to H^+ and the more acidic. This is a crude and overly simplistic analogy, but it is meant to convey the idea that the larger the $\delta+$ on a hydrogen atom is, the more acidic that proton is likely to be. *It is the acidity of the proton in the O–H unit of the COOH functional group that gives the class of molecules its name of* ***carboxylic acid***.

The acidic proton of the O–H unit dominates the chemistry of the carboxyl, despite the many chemical reactions that involve the π-bond. The acidic hydrogen is removed by a base in a classical acid–base reaction shown for ethanoic acid (**74**) in Figure 5.16. The base donates two electrons to hydrogen, forming a new base–H bond, and the two electrons in the O–H σ-bond (which is broken) are transferred to oxygen (more electronegative) to generate an anion. This anion is called a **carboxylate anion, 78**, and it is the conjugate base of the carboxylic acid. The common name of ethanoic acid is acetic acid, and the carboxylate anion is often called the **acetate anion** or just acetate. *The process shown in Figure 5.16 represents the main reaction of carboxylic acids (an acid–base reaction).*

5.45 Determine the formal charge on both oxygen atoms in the acetate anion (acetate; derived by removing the acidic hydrogen from acetic acid; see the structures in Figure 5.15).

A carboxylic acid can be called an O–H acid, as can an alcohol (see Chapter 6, Section 6.2.1); however, acetic acid is more acidic than methanol. Formic acid (HCOOH) has a pK_a of about 3.2, whereas methanol has a pK_a of about 16. The acidic proton in both cases is part of an O–H unit, so why is **72** more acidic? The

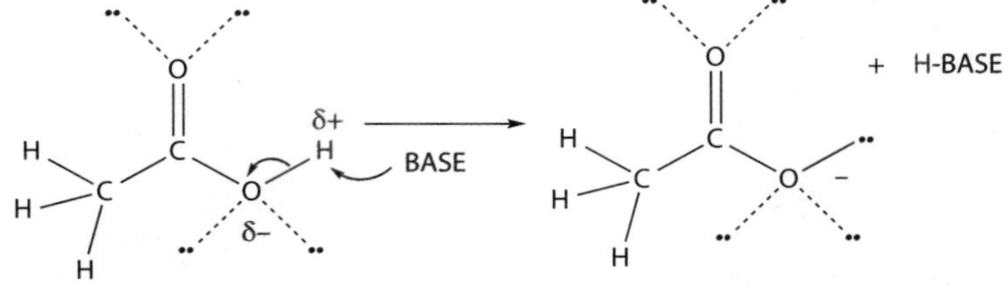

Figure 5.16 Reaction of a carboxylic acid with a base.

polarizing C=O unit withdraws electrons from the OH of the carboxylic acid, making that O–H bond a bit weaker than the O–H unit in methanol, but it is unlikely that this sufficiently accounts for the approximately 10^{13} difference in acidity. Remember that this is an equilibrium reaction and that pK_a is determined by the position of the equilibrium, so the conjugate base formed in each reaction should be examined.

This analysis leads to a discussion of the important concept called **resonance**, which will be used in many places in this book. Reaction of formic acid (acetic acid) and a base gives a carboxylate anion (the formate anion, **79A**) as the conjugate base. Note that the three atoms (O=C–O) are connected. One atom (O) has a negative charge, which can be viewed as a "full" p-orbital (a p-orbital containing two electrons). In addition, there is an adjacent π-bond (C=O). In Figures 5.1–5.3, it is clear that two adjacent and parallel orbitals overlap and share electron density to form a π-bond. The concept that adjacent orbitals share electron density can be extended from two orbitals to three, four, or even more. All that is required is that the orbitals are on adjacent atoms and are parallel. *Therefore, when three orbitals are on adjacent atoms and are parallel, electron density is shared between all three atoms.*

This sharing of electron density over three or more orbitals can be called *delocalization of electron density* (as represented in **79B**). In other words, the two electrons are not localized on the negatively charged oxygen as in **79A**, but rather are delocalized over three atoms by overlaps of two orbitals of the π-bond with the orbital on oxygen. There are a total of four electrons distributed over three atoms in **79B**. *Delocalization of electron density over several atoms in this manner is called resonance.*

The oxygen atoms are more electronegative, so more electron density is concentrated on the oxygen atoms rather than on the carbon. This is represented by the δ+ and δ– in **79B**. *The delocalized structure is lower in energy than a structure that has the charge localized on a single atom.* Therefore, **79B** is a more stable structure than localized structure **79A**. The charge dispersal is also shown in the electron density map for the formate anion **79C**, which indicates the higher electron density (**red**) on the two oxygen atoms. Note that the electron density is concentrated on both oxygen atoms, rather than on one of the oxygen atoms.

79A　　79B　　80

79C　　81A　　81B

Structure **79B** is a more accurate representation of the formate anion, where the charge is ***delocalized*** over three atoms. *It is difficult to draw **79B**, so the resonance delocalization is represented by two structures, as shown in **80** with a **double-headed arrow to indicate resonance**. The two structures labeled **80** represent **one** resonance-stabilized anion, rather than two different molecules.* Resonance will be seen many times in connection with intermediates of various reactions and, in all cases, resonance leads to a more stable species.

5.46 The species $H_2C=CH–CH_2^{\oplus}$ is also resonance stabilized. Draw a picture similar to structure 79B to show the resonance and indicate on which carbon atom or atoms the positive charge is higher.

Dispersal of a charge over a larger area (over three atoms in **79A and 79B**) makes the anion less reactive (more stable). Carboxylic acids undergo acid–base reactions, and **80** is the conjugate base formed when **73** reacts with a base. If the conjugate base of an acid–base reaction is more stable, there is a greater concentration of products and K_a is larger (the species is more acidic, as noted in Chapter 2, Section 2.4, and Chapter 6, Section 6.3). This is apparent when the acid–base reaction of methanol is examined.

When the weak acid methanol reacts with a base, the conjugate base is the methoxide anion (**81**), where the charge is localized on a single oxygen, as represented by **81A**. The electron density map of methoxide (see **81B**) clearly shows that the highest concentration of **red** (electron density) is localized on the oxygen. The resonance delocalization that is apparent in **79C** leaves each of the two oxygen atoms with less electron density than is observed in **81B**. In other words, **81** is a stronger base. This fact indicates that methoxide is more reactive as an electron donor, so it is a stronger base and in the acid–base reaction of methanol there will be a lower concentration of products (K_a is smaller and pK_a is larger, so methanol is less acidic). The reaction of a carboxylic acid with a base generates a *resonance-stabilized* carboxylate anion. Because the conjugate base (the formate anion) is more stable, K_a is shifted to the side of products and is larger (pK_a is smaller) and formic acid is a stronger acid than methanol.

5.47 Is the neutral acid HCOOH resonance stabilized?

Remember from Chapter 2 that the relative acidity of two acids is measured by the position of the equilibrium constant, K_a. The value of K_a is determined by the relative concentrations of the conjugate base–conjugate acid and the acid–base concentrations. Therefore, a discussion of the stability/reactivity/structure of an acid must include the structure/reactivity/structure of its conjugate base. The analysis must also include the base and its conjugate acid. In the case of the salt of carboxylic acids, the resonance stability of the carboxylate anion (the conjugate base) leads to increased stability, which increases K_a. Therefore, resonance is associated with the word stability, as in resonance stability. Resonance stability of the carboxylate anion is usually cited as a key factor in the acidity of carboxylic acids when compared with other compounds.

the net charge on each individual atom, making the entire array of atoms more stable and less reactive. The energy saved by resonance, relative to a non-resonance-stabilized molecule, is called **resonance energy** or **delocalization energy**. In the case of benzene, the difference between the actual bond dissociation energy of 49.8 kcal mol^{-1} and the *estimated* bond dissociation of cyclohexatriene (85.8 kcal mol^{-1}) is taken to be the resonance energy of benzene (36 kcal mol^{-1}). Benzene is said to be more stable than cyclohexatriene by this amount of energy.

5.49 Which do you think is more stable: benzene or cyclohexane? Briefly explain your answer.

5.10.2 Benzene as a Substituent: The Phenyl Group

Benzene is a discrete molecule that is the parent of an entire class of molecules called aromatic hydrocarbons (see Chapter 21). Naming benzene derivatives is different from naming other types of molecules, and the nomenclature system is discussed in Chapter 21. A benzene ring can be attached to a carbon chain, however, and in such cases the benzene unit is a substituent. An example is **95**, where the benzene ring is attached to C5 of 2-octanone. Note that this benzene substituent has the formula C_6H_5 and it is attached to the carbon chain by a carbon–carbon bond. The C_6H_5 unit is called *phenyl* and the name of **95** is 5-phenyl-2-octanone. In the nineteenth century, Auguste Laurent (France; 1808–1853) suggested the name *phene* for benzene (from the Greek, meaning "to illuminate").[3] Benzene had been found in the illuminating gas used at the time. Although benzene replaced phene as a name, phene became phenyl when benzene was a substituent.

A final note about nomenclature is in order for benzene. Drawing a benzene ring occupies a lot of space, so a shorthand representation is used in many structures. In **96**, the shorthand symbol "Ph" is used to represent a phenyl substituent. In the older chemical literature, the Greek symbol ϕ (phi) was sometimes used. The Ph representation will be used often in this book, and compound **96** is named 5-chloro-2,6-diphenyloct-2-ene.

5.50 Draw the structure of 3,5-diphenyl-2-octanone; tetraphenylmethane.

5.51 Draw the structure of 3-phenyl-1-pentanol using the symbol Ph for the phenyl substituent.

5.11 Biological Relevance

Ethene (common name ethylene) was introduced in Section 5.1. It has been used as an inhalation anesthetic to induce general anesthesia. What is the biological relevance of this alkene? In fact, there are several important aspects. The first is historical. What remains of the Oracle of Delphi is found in ruins of the temple of Apollo on Mount Parnassus, near the ancient city of Delphi in Greece. Many in the ancient world consulted the oracle about important matters, from farmers planting their crops to Alexander the Great. The oracle was a priestess, known as a Pythia, who communicated with the gods while in a trance-like state. The Pythia sat in a small room as vapors from cracks in the floor washed over her, causing her to fall into a trance. The vapors reportedly had a sweet smell. A priest would take money from a visitor, who would ask a question of the Pythia. The answer was usually very obscure, but the priest would translate the Pythia's words.[4] The painting, "The Priestess of Delphi," was painted by the Hon. John Collier in 1891.

This painting is "The Priestess of Delphi" by the Hon. John Collier and was painted in 1891. Reproduced from Wikipedia. This image is in the public domain and its copyright has expired. This applies to the United States, Australia, the European Union and those countries with a copyright term of life of the author plus 70 years.

Geologists have discovered that water in a spring near the ancient site of the oracle contains the hydrocarbons methane, ethane, and ethylene (ethene).[4] Ethane and methane were found in pieces of travertine, a limestone stalactite deposited by an ancient spring.[4] In the days of the Pythia, colliding tectonic plates near the Temple of Apollo are believed to have generated sufficient heat to vaporize the hydrocarbons, which were extruded as vapors in the chamber of the oracle. If those vapors included ethylene, the Pythia may have been in a state of ethylene narcosis.[5] Ethylene may produce states of euphoria and memory disturbances. Overexposure can lead to loss of consciousness and even death due to hypoxia.[6]

Polarity and solubility are critical to the design and delivery of the drugs used as medicines, an area of chemistry known as medicinal chemistry. The so-called ***partition coefficient*** (distribution coefficient) is used as a measure of the ability of the drug to pass through relatively nonpolar lipid membranes from the highly polar environment of blood serum (mostly water). This correlation has been used to predict the activity of potential drugs, but is valid only when solubility and transport by diffusion though a membrane are important. The partition coefficient, P, is defined as

$$P = \text{drug in an organic phase/drug in an aqueous phase}^7$$

The values of the partition coefficient are usually measured using water or a phosphate buffer at pH 7.4 (the pH of blood) against 1-octanol.[7] A large value of P is taken as an indication that the compound will diffuse into lipid membranes and fatty tissue, whereas a low value indicates that it will not easily diffuse. Also, a large value of P is associated with more water-insoluble compounds, usually caused by a higher percentage of nonpolar organic fragments.

Pharmacokinetics is the science that concerns itself with what the body does to a drug. The distribution coefficient is an important factor because a drug must first pass through lipid bilayers in the intestinal epithelium in order to be absorbed after oral ingestion. Pharmacodynamics is the science of what a drug does to the body; it is known that hydrophobic drugs tend to be more toxic because they tend to be retained longer, have a wider distribution within the body, are somewhat less selective in their binding to proteins, and, finally, are often extensively metabolized.

Obviously, DNA is important to modern biology, and it is known that DNA exists as a double-stranded helix. DNA contains many polarized functional groups, and hydrogen bonding is an important property of these organic molecules. In the helix, a nucleotide (bearing the heterocyclic base adenine as described in Chapter 28) can hydrogen bond with a thymine of the other strand (an A–T base pair) and a guanine base in one strand can hydrogen bond with a cytosine base in the other strand (a G–C base pair).

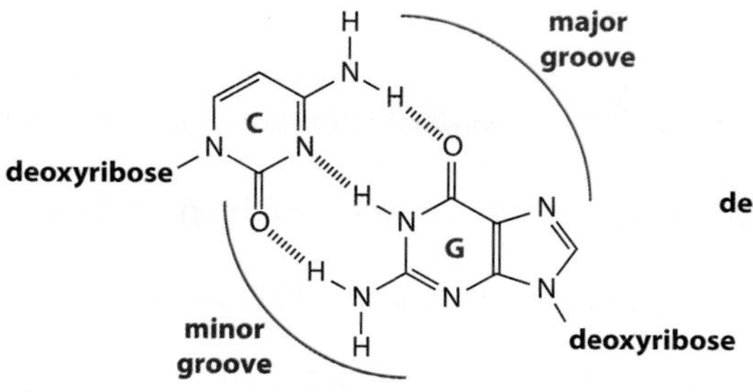

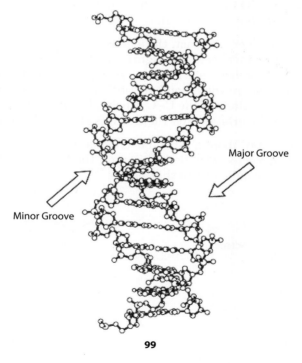

97 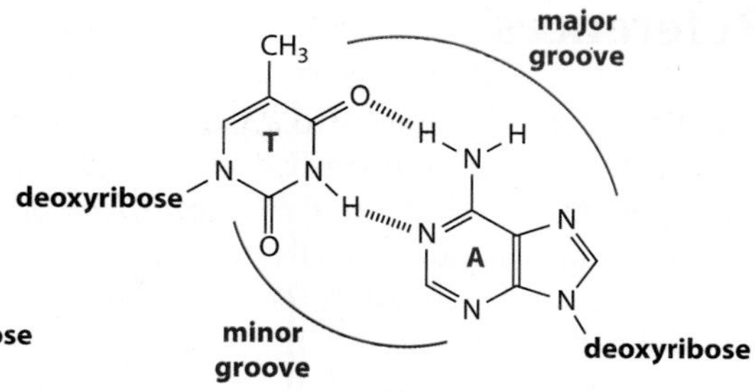   98

(Hecht, S. M., ed. 1996. *Bioorganic chemistry: Nucleic acids*, 6–7, Figure 1-4. New York: Oxford University Press. By permission of Oxford University Press.)

The strands must be antiparallel (one strand is 3'→5' while the other is 5'→3', as described in Chapter 28) to maximize hydrogen bonding because the stereogenic nature of the ribofuranose and deoxyribofuranose units leads to a "twist" in the polynucleotide backbone. The term "stereogenic" is described in Chapter 9, Section 9.1. These hydrogen bonding base pairs are called ***Watson–Crick base pairs***; the C–G pair is shown in **97** and the A–T base pair is shown in **98**.[8] The inherent chirality (see Chapter 9) of the D-ribofuranose and the D-deoxyribofuranose leads the β-form of DNA to adopt a right-handed helix (see **99**).

99

(Hecht, S. M., ed. 1996. *Bioorganic chemistry: Nucleic acids*, 9, Figure 1-8. New York: Oxford University Press. By permission of Oxford University Press.)

References

1. Weast, R. C., and Astle, M. J., eds. 1978. *CRC handbook of chemistry and physics,* 59th ed., F-215. Boca Raton, FL: CRC Press, Inc.
2. Lide, D. R., ed. 2006. *CRC handbook of chemistry and physics,* 87th ed., 3-1–3-523. Boca Raton, FL: Taylor & Francis.
3. Leicester, H. M. 1956. *The historical background of chemistry,* 177. New York: John Wiley & Sons.
4. de Boer, J. Z., Hale, J. R., and Chanton, J. 2001. *Geology* 29:707–710.
5. Roach, J. 2001. *National Geographic* August 14, 2001.
6. Clayton, G. D., and Clayton, F. E., eds. 1981–1982. *Patty's industrial hygiene and toxicology: Volumes 2A, 2B, 2C: Toxicology,* 3rd ed., 3199. New York: John Wiley & Sons.
7. Thomas, G. 2000. *Medicinal chemistry, an introduction,* 123. New York: John Wiley & Sons, Ltd. Equation reproduced with permission. Copyright Wiley-VCH Verlag GmbH & Co. KGaA.
8. Hecht, S. M., ed. 1996. *Bioorganic chemistry: Nucleic acids,* 6. New York: Oxford University Press.

Answers to Problems

5.1 The hybridization of O in the C=O unit is sp² because it is one part of an array of two atoms connected by a π-bond.

5.2 If an alkene π-bond were to react, the electrons in the π-bond would react because they are weaker (held less tightly) than the electrons in the σ-bond. The electrons in the σ-bond are localized along a line between the C nuclei, whereas the electrons in the π-bond are shared between the two carbon atoms above and below the plane of the C–C σ-bond. Because of this, there is only partial overlap of the orbitals and there is less shared electron density.

5.3 Because the geometry about a π-bond is trigonal planar, we expect that R₂C=O will be a planar molecule with both R groups, the C, and the O all in one plane and the π-bond perpendicular to that plane.

5.4

3-octene    4-octene

5.5

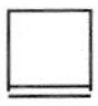

cyclobutene    cycloheptene 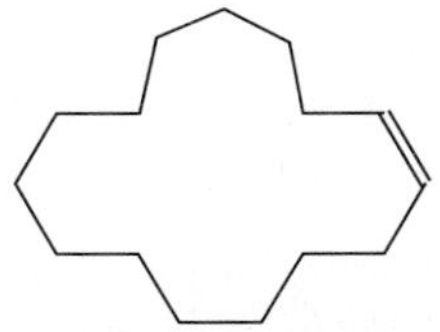   cyclopentadecene

5.6   Alkanes are called "saturated" hydrocarbons because they have the generic formula C_nH_{2n+2} and, with the valence of each carbon = 4, one cannot attach any other atoms to the carbons. For an alkene, the generic formula is C_nH_{2n}, which means that one could add two more atoms before it becomes saturated, so an alkene is not a saturated molecule because of the number of atoms attached to the carbons that are present.

5.7

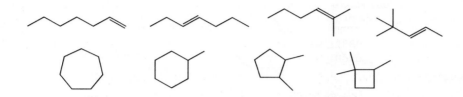

5.8   Because one π-bond in an alkyne is perpendicular to the second π-bond, it is possible to react one without touching the other. Although they are perpendicular, reacting one π-bond will likely create a charged intermediate (usually a cation) and the cation will certainly "feel" the presence of the π-bond perpendicular to it. This will influence the stability of the cation and possibly its reactivity even if the second π-bond does not formally become involved in the reaction.

5.9   A molecule with two π-bonds and a formula $C_{12}H_{22}$ that is not an alkyne probably contain two alkene units (C=C). All that is necessary is that two π-bonds be present. Such a molecule is called a diene (see Chapter 3).

5.10

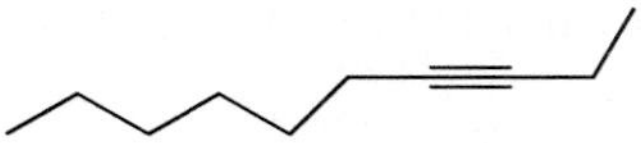

3-decyne 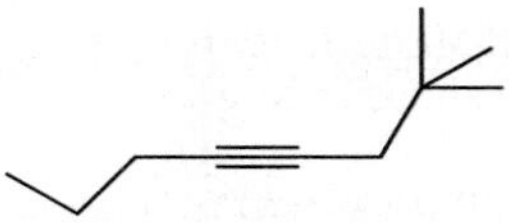   **2,2-dimethyl-4-octyne**

5.11

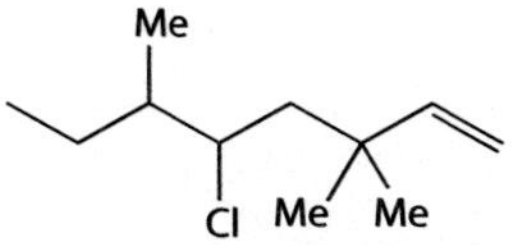

Me

Cl   Me   Me

5-chloro-3,3,6-trimethyl-oct-1-ene 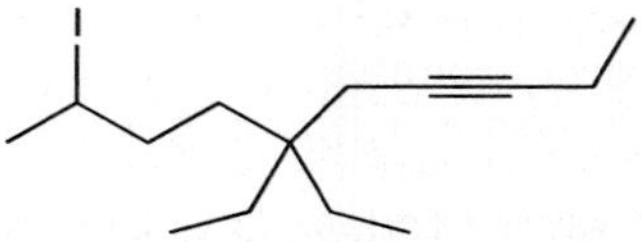   **6,6-diethyl-9-iododec-3-yne**

5.12

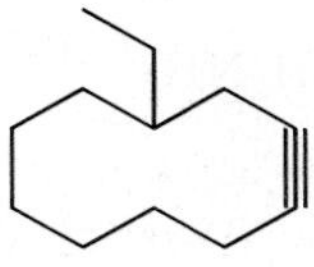

4-ethylcyclododecyne.

5.13

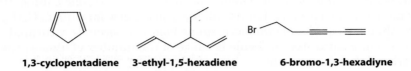

1,3-cyclopentadiene   **3-ethyl-1,5-hexadiene**   **6-bromo-1,3-hexadiyne**

5.14

5.15 4-(2-Methylpentyl)-non-7-en-2-yne.

5.16 Based *only* on bond strength, the C–Br bond (68 kcal/mol) is weaker than the C–O bond (80 kcal/mol) and C–Br should be easier to break. This is not necessarily a valid method because it ignores relative bond polarity, stability of final products, etc.

5.17 Oxygen is $6 - \frac{1}{2}(2) - 6 = -1$. Carbon and all $H = O_1$ so the format change is -1 for the molecule.

5.18 For bromomethane, CH_3Br, the formal charge for C is $C = 4 - 1/2(8) - 0 = 0$; for Br, $Br = 7 - 1/2(2) - 6 = 0$. Therefore, the formal charge for this molecule is 0.

5.19 For CH_3NH_2, the formal charge is $C = 4 - 1/2(8) - 0 = 0$; for all five H, $H = 1 - 1/2(2) - 0 = 0$; and for N, $N = 5 - 1/2(6) - 2 = 0$. Therefore, the formal charge for this molecule is 0.

5.20 The formal charge for O in ethanol is $O = 6 - 1/2(4) - 4 = 0$.

5.21

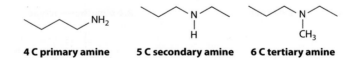

4 C primary amine   **5 C secondary amine**   **6 C tertiary amine**

5.22

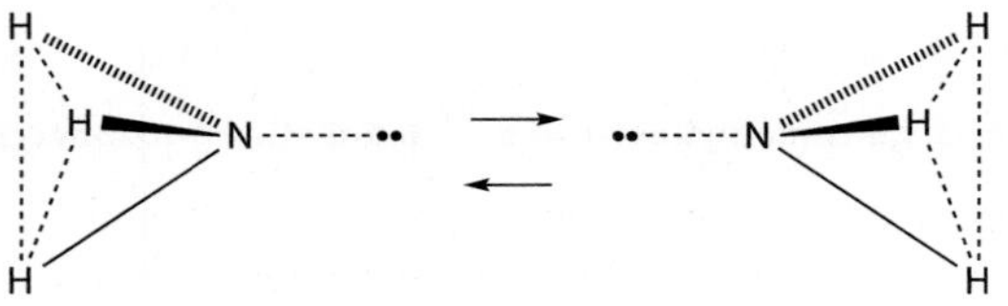

5.23 CH_3NH_2, $CH_3CH_2NH_2$, $CH_3CH_2CH_2CH_2CH_2NH_2$

5.24

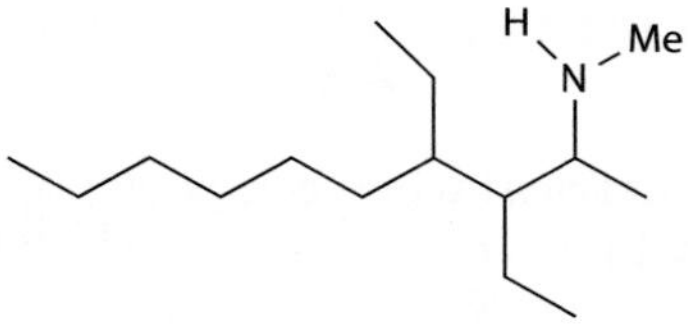

2-(*N*-methylamino)-3,4-diethyldecane

5.25 CH_3CH_2OH

5.26

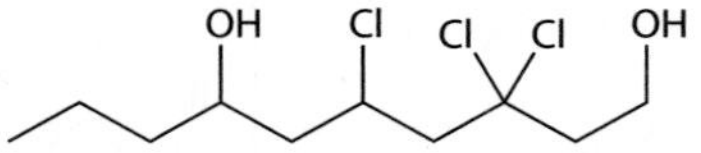

5.27 $CH_3CH_2OCH_2CH_3$

5.28

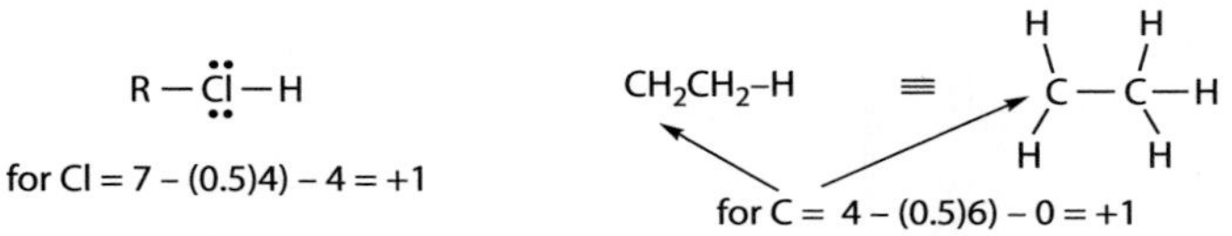

$$R - \ddot{C}l - H$$

for Cl = 7 – (0.5)4 – 4 = +1

$$CH_2CH_2\text{-}H \quad \equiv \quad \overset{H}{\underset{H}{C}} - \overset{H}{\underset{H}{C}} - H$$

for C = 4 – (0.5)6 – 0 = +1

5.29

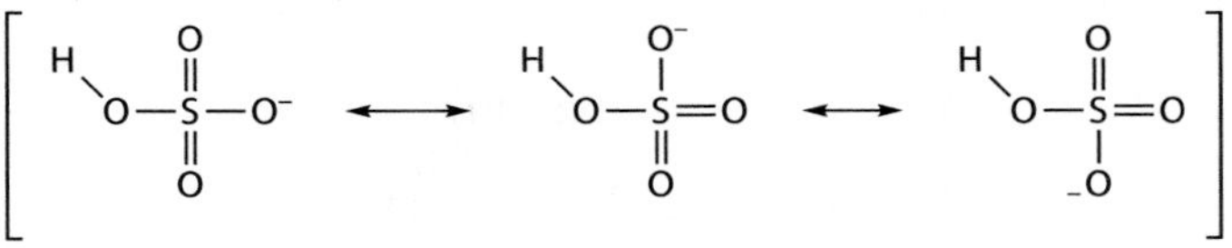

5.30

bromotrichloromethane

There are 3 C-Cl dipoles in the "direction" of the tetrahedral "base." The C-Br dipole is in the opposite direction and diminishes the net dipole due to the chlorine atoms, but the direction remains that shown.

5.31 Because O is more electronegative than N, we assume that the O–H bond is more polarized than the N–H bond. If this is true, then the $\delta+$ dipole on H in an O–H unit is greater than that in a N–H unit and the H of OH is more acidic. Therefore, we anticipate greater hydrogen bonding with the OH unit, which means that methanol will form stronger hydrogen bonds.

5.32 The 10 C atoms found in decanol will more than balance the one OH unit, so this molecule will be relatively polar–polar. Therefore, it should be miscible with hexane and relatively soluble in water.

5.33 The eight-carbon alcohol with a H-bonding O–H unit will have a higher boiling point than an eight-carbon ether, which has only dipole–dipole interactions. For example, 1-octanol has a boiling point of 194.45°C and dibutyl ether has a boiling point of 52°C.

5.34 Carbon tetrachloride, CCl_4, is more symmetrical when compared to CH_3Cl, and those molecules should pack into a crystal lattice more efficiently. Therefore, we predict that CCl_4 has the higher melting point. Note also that CCl_4 has a higher formula weight. At –23°C, CCl_4 melts at a much higher temperature than chloromethane at –97°C.

5.35

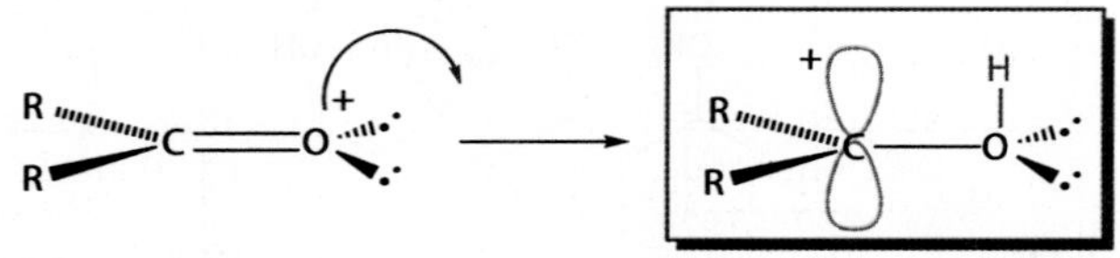

5.36 For 3-pentanone, the formal charge for the carbonyl C is C = 4 – 1/2(8) – 0 = 0; for O, O = 6 – 1/2(4) – 4 = 0. Therefore, the formal charge for this molecule is 0.

5.37

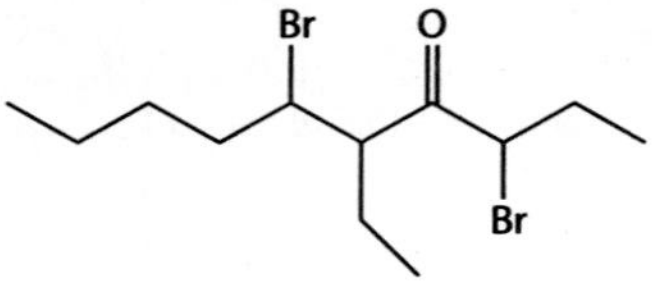

3,6-dibromo-5-ethyl-4-decanone

5.38

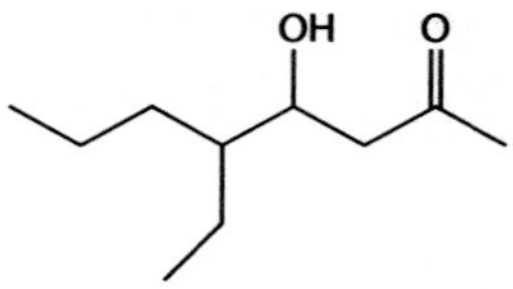

5.39

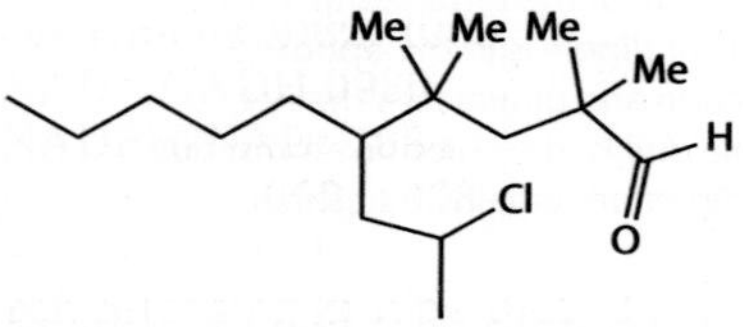

2,2,4,4-tetramethyl-5-(2-chloro-1-propyl)-decanal

5.40

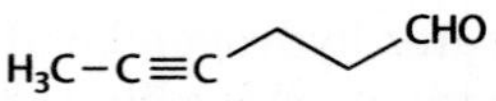

5.41

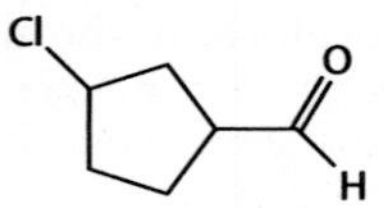

3-chlorocyclopentane-1-carboxaldehyde

5.42 The positive charge is greatest on the two terminal carbon atoms of this cation due to resonance. This is called an allyl cation.

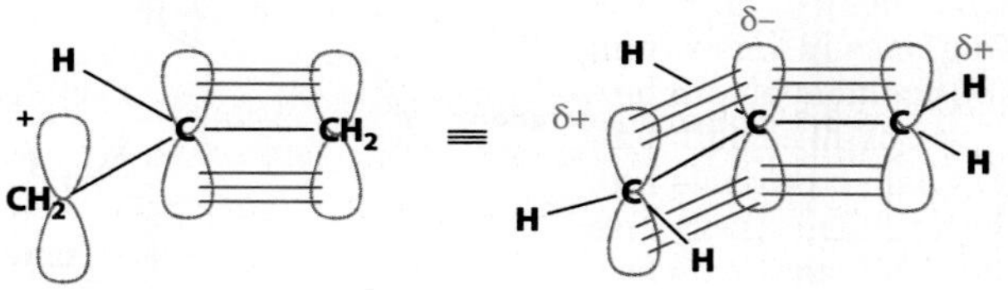

5.43 For the acetate anion, the formal charge for the O doubly bonded to C (the carbonyl O) is O = 6 − 1/2(4) − 4 = 0; for the singly bonded O, O = 6 − 1/2(2) − 6 = −1. Therefore, the formal charge for this molecule is −1, with the charge residing on an oxygen atom.

5.44 No! The formal charge in HCOOH is 0. There is no opportunity for resonance because there is no vacant p-orbital, as there is in the carboxylate anion.

5.45

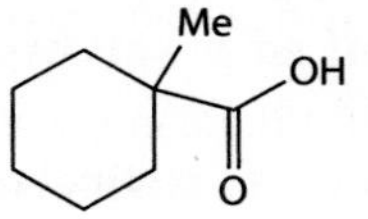

1-methylcyclohexane-1-carboxylic acid

5.46

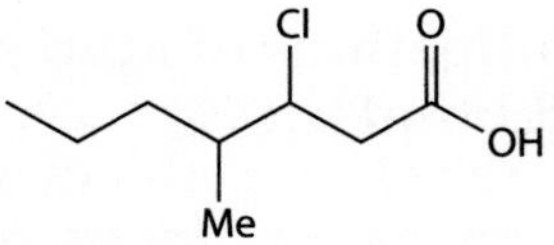

3-chloro-4-methylheptanoic acid

5.47

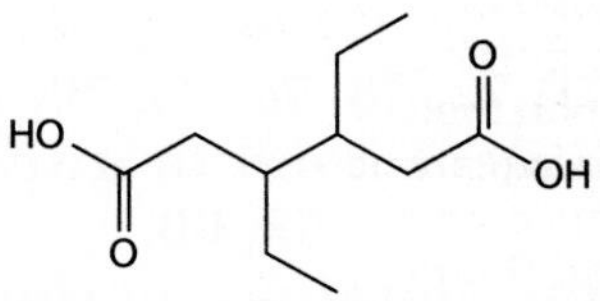

5.48 Cyclohexadiene is not resonance stabilized. There are π-electrons, but there is not a continuous array and it is not possible for one "end" of the π-system to interact (delocalize electrons) to the other "end."

5.49 Benzene is more stable than cyclohexane. Benzene is stabilized by electron delocalization of the π-electrons. Cyclohexane has no π-electrons and there is no possibility for electron delocalization or extra stability.

5.50

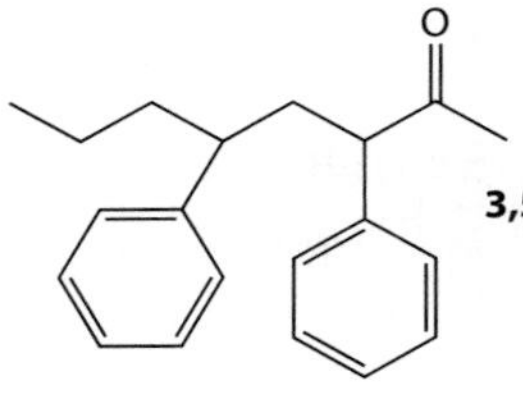

3,5-diphenyl-2-octanone

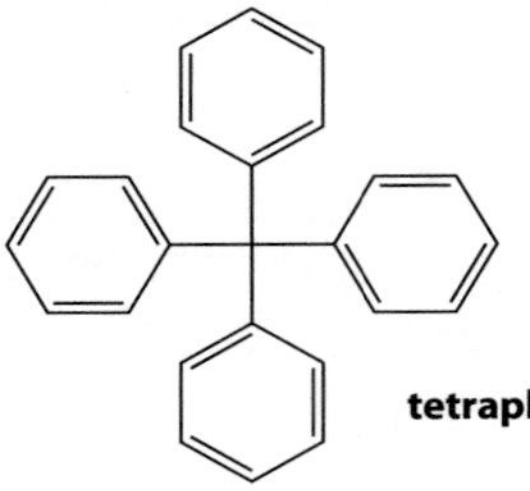

tetraphenylmethane

5.51

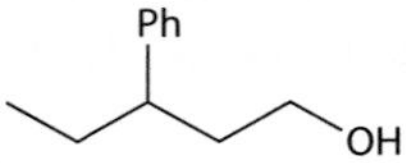

3-phenylpentan-1-ol

Correlation of Concepts with Homework

- A π-bond, formed by "sideways" overlap of p-orbitals on adjacent sp^2 hybridized atoms, is composed of a strong σ-bond and a weaker π-bond: 1, 2, 3, 52, 79.
- Alkenes are hydrocarbons that have a C=C unit, and the generic formula for an alkene is C_nH_{2n}: 4, 5, 6, 7, 53, 54.
- The C≡C unit is composed of one strong σ-bond and two π-bonds that are orthogonal to each other; alkynes are hydrocarbons that have a C≡C unit, with the generic formula C_nH_{2n-2}: 8, 9, 10, 11, 12, 68, 69, 70, 72.
- Functional groups are discrete collections of atoms that usually contain π-bonds and/or polarized bonds: 3, 4, 9, 10, 11, 12, 13, 14, 15, 16, 21, 24, 25, 26, 27, 35, 37, 38, 39, 40, 41, 42, 45, 46, 47, 52, 59, 61, 62, 63, 66, 67, 68, 69, 70, 71, 72, 73, 74, 75, 76, 77, 79, 82, 84.
- Hydrocarbons with multiple bonds include dienes, diynes, and allenes: 14, 15, 68, 72, 73, 84.
- The structures of alkenes, alkynes, alcohols, amines, ethers, ketones, aldehydes, carboxylic acids: 3, 4, 9, 10, 11, 12, 13, 14, 15, 16, 21, 24, 25, 26, 34, 37, 38, 39, 40, 41, 45, 46, 47, 50, 51, 53, 59, 62, 63, 66, 67, 68, 69, 70, 71, 72, 73, 74, 75, 76, 77, 84.
- Amines contain a covalent bond to nitrogen and undergo fluxional inversion: 21, 22, 23, 24, 59, 60.
- If the functional group contains a carbon, that carbon receives the lowest possible number. Each functional group has a unique suffix. Alkenes end in "ene," alkynes in "yne," and alcohols in "ol." Amines are named as amines and ethers are named as ethers. Ketones end in "one," aldehydes in "al," and carboxylic acids in "oic acid": 4, 5, 10, 11, 12, 13, 14, 15, 23, 24, 25, 26, 27, 37, 38, 39, 40, 41, 45, 46, 47, 50, 51, 60, 68, 69, 70, 72, 73, 74, 75, 76, 77, 78.

- **The electron-donating ability of a functional group can be used to predict the reactivity with positively charged atoms: 17, 31, 56, 57, 80, 83.**
- **Formal charge can be used to determine the credibility of a given structure and whether it is neutral or bears an ionic charge: 18, 19, 20, 28, 36, 43, 44, 86.**
- **Many physical properties result from the presence of polarized bonds within molecules: 33, 34, 58, 62, 63, 66, 81, 82, 85.**
- **The unshared electrons of heteroatoms and the electrons in a π-bond can function as Lewis bases or Brønsted–Lowry bases: 35, 65, 80, 83.**
- **Dispersal of charge over several atoms via aligned p-orbitals is called resonance and leads to greater stability: 29, 42, 48, 49, 55, 64, 65, 80, 83.**
- **Benzene is a unique cyclic hydrocarbon that is more stable than expected due to resonance delocalization: 48, 49, 78.**
- **When benzene is a substituent in a molecule, it is called phenyl: 50, 51, 70, 72, 73, 74, 76.**

Homework

5.52 Categorize all atoms highlighted in **red** for each of the following molecules as to its hybridization:

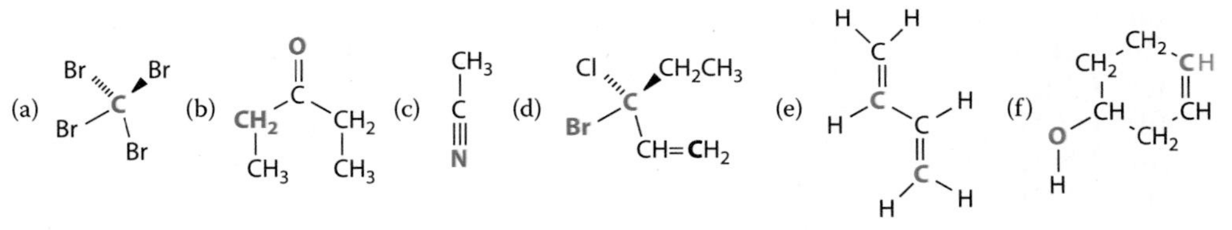

5.53 Categorize each of the following formulas as consistent with an alkane, alkene, or alkyne:
- (a) $C_{10}H_{20}$
- (b) $C_{20}H_{42}$
- (c) C_8H_{14}
- (d) $C_{16}H_{34}$
- (e) C_9H_{16}
- (f) $C_{100}H_{200}$

5.54 Draw each of the following in line notation:
- (a) $Ch_3CH_2CH_2CH_2CH_2CH_3$
- (b) $(CH_3)_2CHCH_2CH_2CH_3$
- (c) $CH_3CH_2CH_2CH_2CH_2CH=CH_2$
- (d) $CH_3CH_2C\equiv C(CH_2)_8CH_3$

5.55 Examine each of the following cations and draw a second structure with an X=X bond (resonance stabilized) where it is appropriate. If no X=X structure is possible, briefly explain why not:

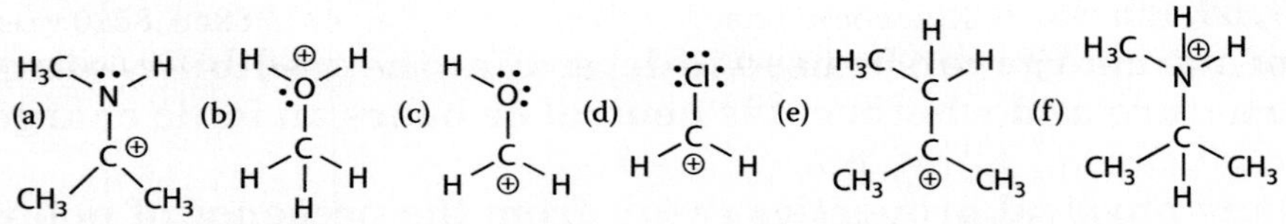

5.56 Indicate the more electronegative atom in each of the following:
   (a) C–N
   (b) N–O
   (c) C–H
   (d) Cl–Br
   (e) B–C
   (f) Li–C
   (g) C–F
   (h) N–H
   (i) H–Cl

5.57 Indicate if each bond is polarized covalent or nonpolarized covalent:
   (a) C–N
   (b) N–O
   (c) C–H
   (d) C–F
   (e) C–C
   (f) Li–C

5.58 Draw the structure of each molecule using the VSEPR model and then indicate the general direction of the dipole moment. Indicate if the dipole moment is zero:
   (a) CH_3Cl
   (b) CH_3OH
   (c) CCl_4
   (d) $C(CH_3)_4$
   (e) $ClCH_2Br$
   (f) Cl_3CH

5.59 Draw both fluxional isomers for the amine shown. Are they the same? Explain your answer.

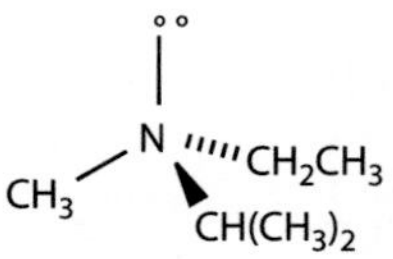

5.60 (a) Draw a primary amine that has four carbon atoms.
    (b) Draw a secondary amine that has six carbon atoms.
    (c) Draw a tertiary amine that has three carbon atoms.

5.61 (a) Draw three different five-carbon alcohols.
    (b) Draw three different six-carbon ethers.

(c) Draw a four-carbon secondary alcohol, a three-carbon primary alcohol, and a six-carbon tertiary alcohol.

(d) Is it possible to draw a two-carbon secondary alcohol? A three-carbon tertiary alcohol?

5.62 Rank order each of the following lists, a–e, according to their boiling point, lowest to highest:

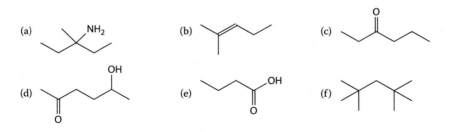

(a)

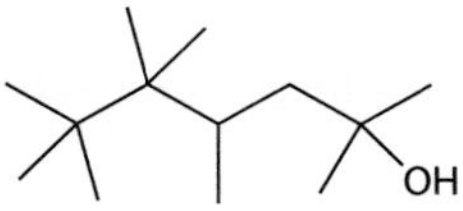

(i) 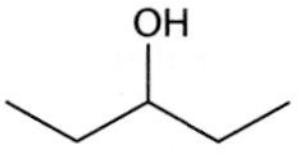   (ii) 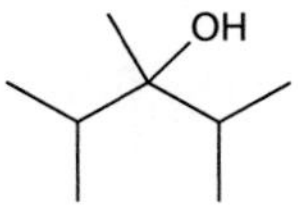   (iii)

(b)   NH$_2$ CO$_2$H   OH

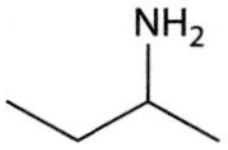

(i) 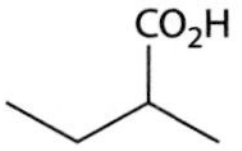   (ii) 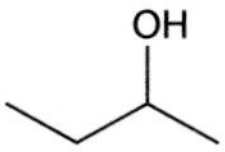   (iii)

(c)   CH$_3$Cl   CH$_3$NH$_2$ CH$_3$COOH

(i)   (ii)   (iii)

(d)   OH   CH$_3$ CH$_3$

OH   CH$_3$ OH

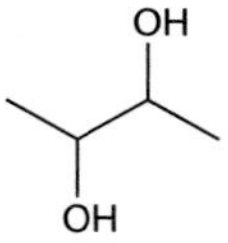

(i) 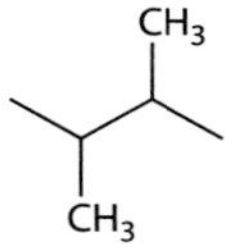   (ii) 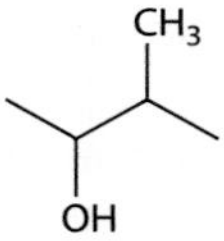   (iii)

(e)   OH   OCH$_2$CH$_3$ CH$_2$CH$_2$CH$_2$OH

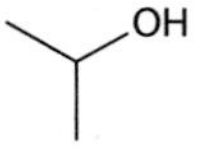

(i) 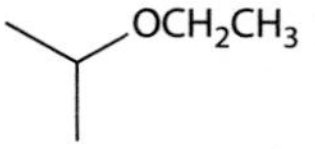   (ii) 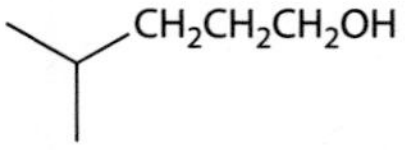   (iii)

5.63 Indicate whether the boiling point of the following molecules is most influenced by London forces, dipole–dipole interactions, or hydrogen bonding:

(a)   NH$_2$ (b)   (c)

(d)   OH   (e)   OH   (f)

O   O

5.64 Indicate which of the following anions might be stabilized by resonance and offer a brief explanation of your choices:

(a) CH$_3$O$^-$ (b)   (c)   O   (d)   O$^-$

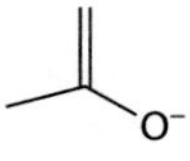

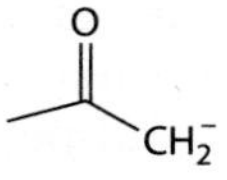

O$^-$ CH$_2^-$ O

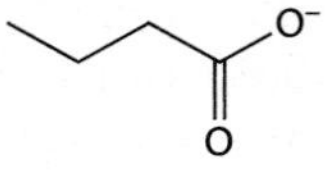

5.65 The following molecule is called tropolone and it reacts very similarly to a carboxylic acid when exposed to base. Offer an explanation.

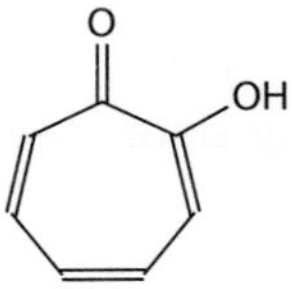

5.66 Indicate whether the following molecules are likely to be soluble or insoluble in water:

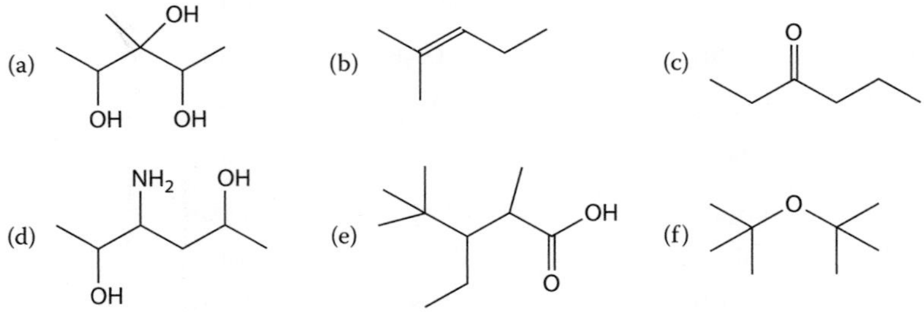

5.67 Determine which of the following structures are isomers:

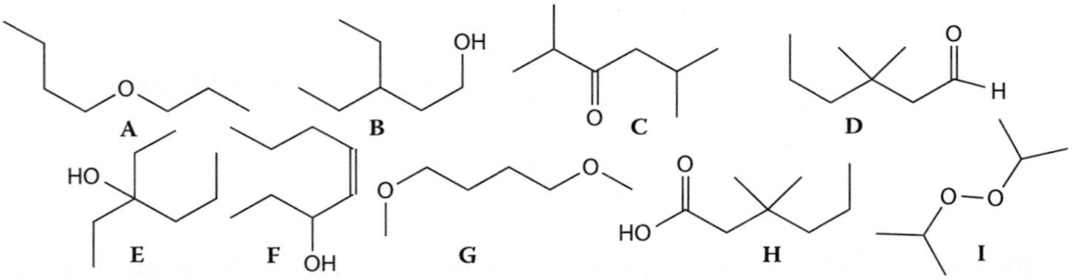

5.68 Give the IUPAC name for each of the following:

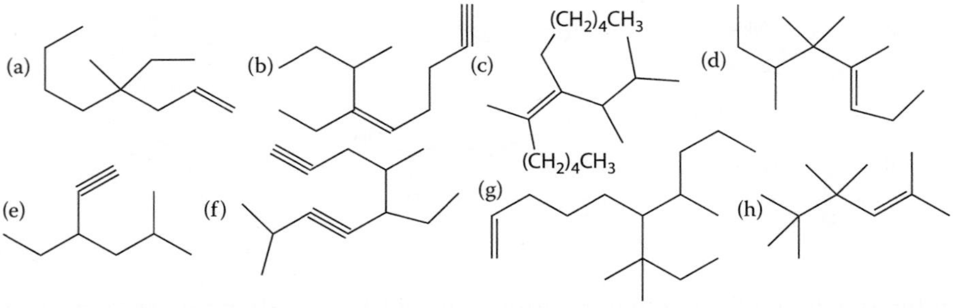

5.69 Draw the correct structure for each of the following:
   (a) 5-(2,2-dimethylbutyl)-2-hexadecene
   (b) 4,5,6,7-tetraethyl-2-dodecyne
   (c) 7,8-di(1,1-dimethylethyl)-1,3-pentadecadiene

(d) 1,3,3,5,5,6-hexamethylcyclohexene
(e) 1-cyclopropyl-2-ethylcycloheptene
(f) 5,5-diethyl-3-nonyne

5.70 Give the correct IUPAC name for each of the following cyclic alkenes:

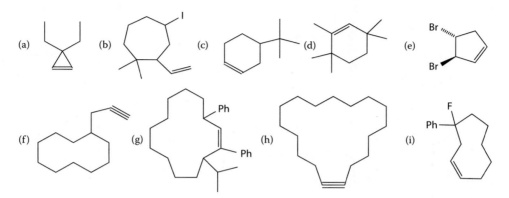

5.71 Draw the structure and provide the IUPAC name for 12 different molecules that have the empirical formula $C_{10}H_{20}$. Only *five* of these structures can contain a C=C unit.

5.72 Provide the unique IUPAC name for each of the following molecules:

5.73 Give the name for each of the following molecules:

(a) 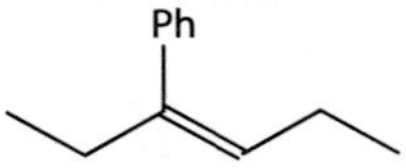

(b) 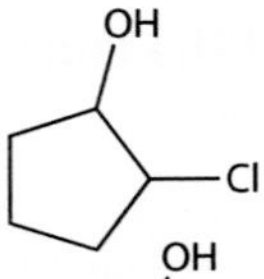

(c) 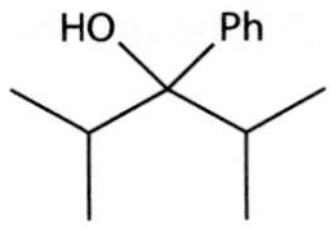

(d) 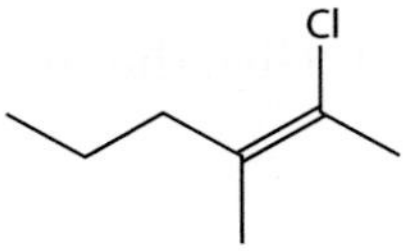

(e) 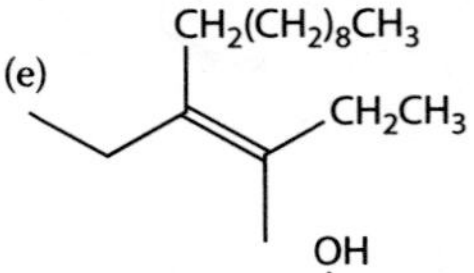

(f) 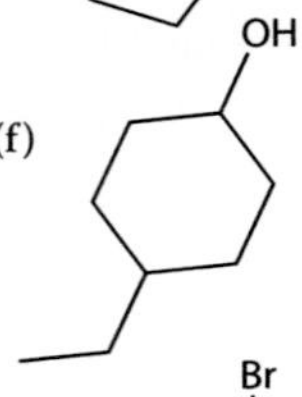

(g) 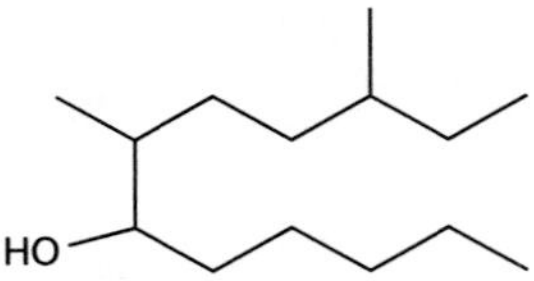

(h) 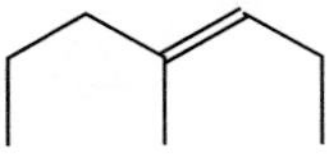

(i) 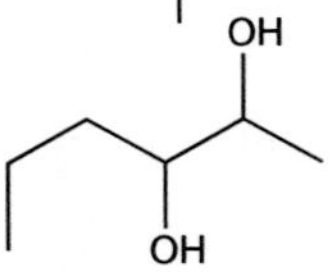

(j) 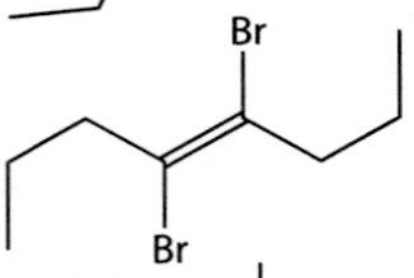

(k) 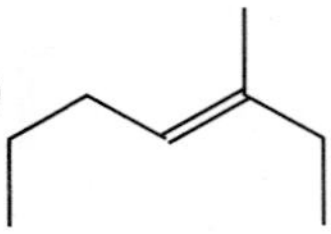
(l) 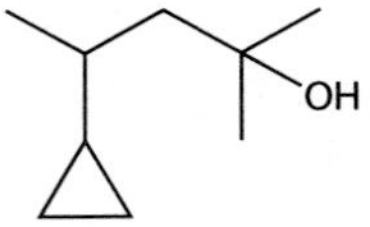

(m) 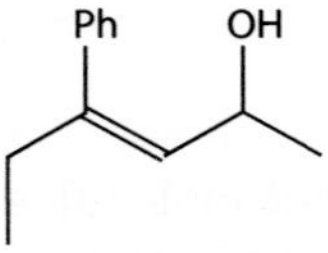

(n) 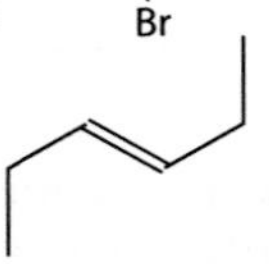
(o) 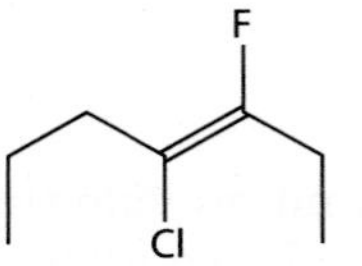

(p) 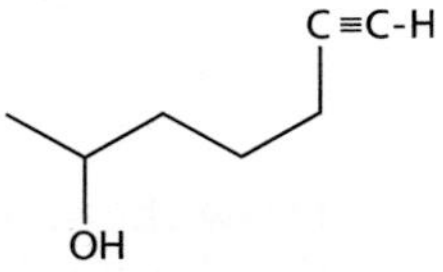

5.74 Draw the structure for each of the following:
(a) 2-methyl-1-cycloheptanol
(b) 5,6-diphenyl-2-heptanol
(c) hex-2-en-1-ol
(d) 5-(3-ethylhexyl)-8-chloro-1-pentadecanol
(e) 3,4,5-heptanetriol
(f) 1,2,3,4,5,6-hexamethylcyclohexanol
(g) 4-phenyl-1,8-octanediol
(h) 3-chloronon-8-en-1-ol

5.75 Give the correct IUPAC names for the following amines:

(a) 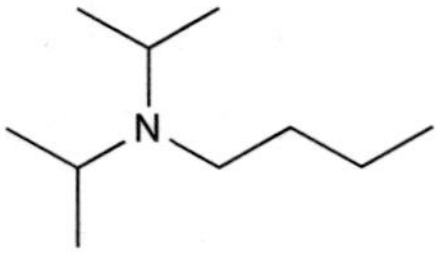

(b) 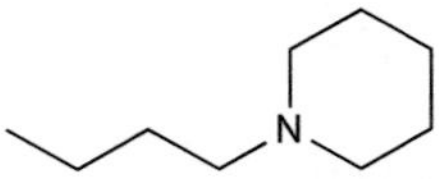

(c) 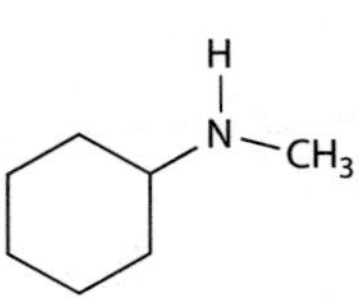

(d) 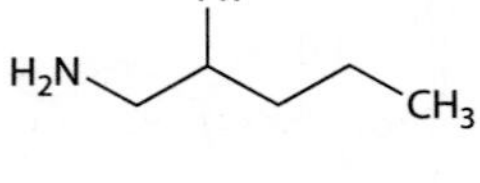

(e) 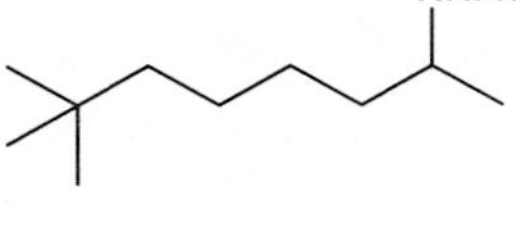

(f) 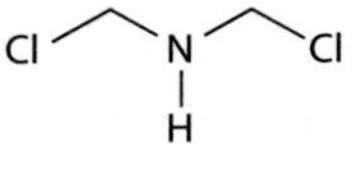

5.76 Name the following ketones and aldehydes:

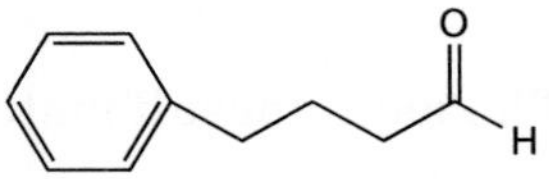

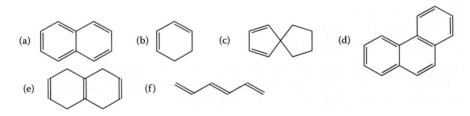

5.77 Draw the structures of the following molecules:
(a) 8-phenyloctanoic acid
(b) 3,3,6,6-tetrabromohexadecanoic acid
(c) 2,5-dimethylhexanedioic acid
(d) 3-chlorocyclohexane-1-carboxylic acid
(e) 2,5-dimethylcyclopentane-1-carboxaldehyde

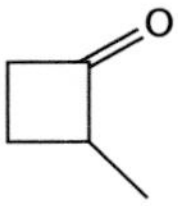

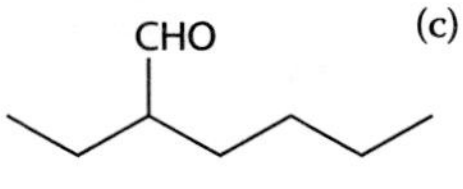

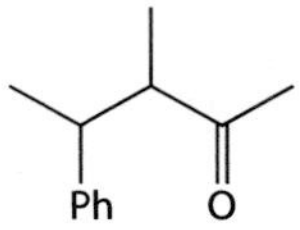

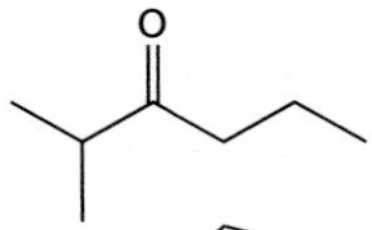

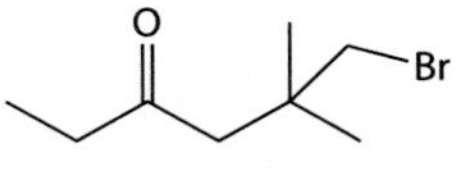

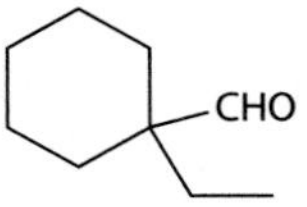

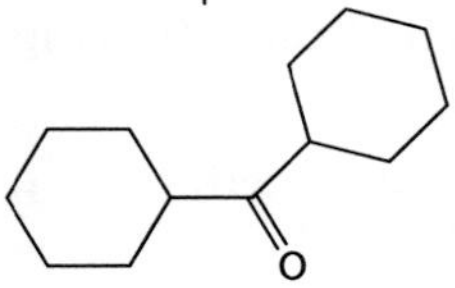

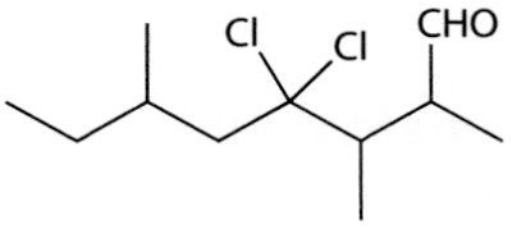

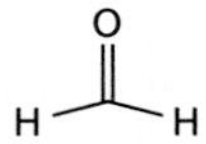

5.78 Based on the special stability shown by benzene, indicate which of the following molecules is stabilized in the same way. Why or why not?

(a) (b) (c) (d)

(e) (f)

5.79 Briefly explain why the π-bond in the C=C unit of ethene ($H_2C=CH_2$) is weaker than the σ-bond in the C=C unit.

5.80 Methanesulfonic acid has a much lower pK$_a$ when compared to acetic acid:

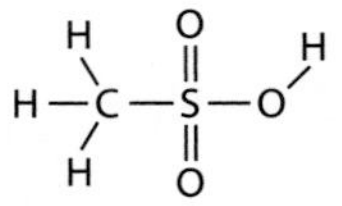

methanesulfonic acid 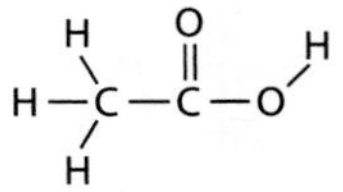   acetic acid

(a) Draw the products expected acid–base reaction between methanesulfonic acid and NaOH, and also for the reaction of acetic acid and NaOH.
(b) Discuss why methanesulfonic acid has a lower pK$_a$ relative to acetic acid. There is no need to do a calculation here because the answer should be a discussion-type answer.

5.81 Which of the following molecules is the most polar molecule?
  (a) CH_4
  (b) CH_3F
  (c) CF_4
  (d) $CH_2=CH_2$

5.82 Which of the following molecules are capable of hydrogen bonding?

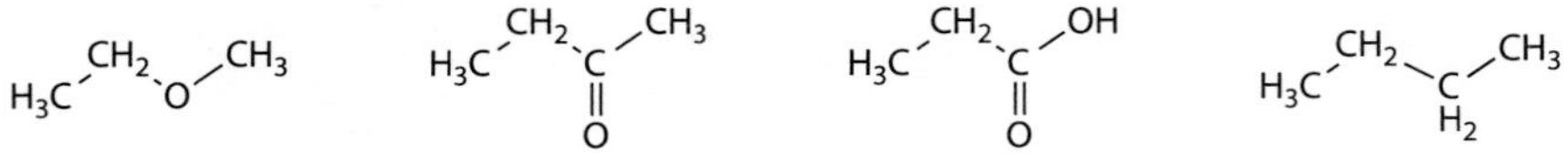

5.83 The methanesulfonate anion $(H_3C-\overset{\overset{O}{\|}}{\underset{\underset{O}{\|}}{S}}-O^-)$ is *less basic* than the methoxide anion H_3C-O^-. Draw *all* resonance structures for *all* molecules that exhibit resonance. Use the structures you have drawn to explain why the methanesulfonate anion is less basic.

5.84 Ketenimine is an interesting molecule with the structure $H_2C=C=NH$.
  (a) Give the hybridization for all three carbon atoms.
  (b) Draw a diagram that indicates the relative positions of any and all σ-bonds and π-bonds.

5.85 Which of the following alkanes is likely to have the *highest* boiling point? Justify your answer.
  (a) CH_3CH_3
  (b) $CH_3CH_2CH_3$
  (c) $CH_3CH_2CH_2CH_2CH_3$
  (d) $CH_3CH_2CH_2CH_2CH_2CH_3$

5.86 Calculate the formal charge for all atoms marked in **bold letters**, and then calculate the formal charge for each molecule. Do **not** assume unshared electrons are present unless they are shown or indicated.

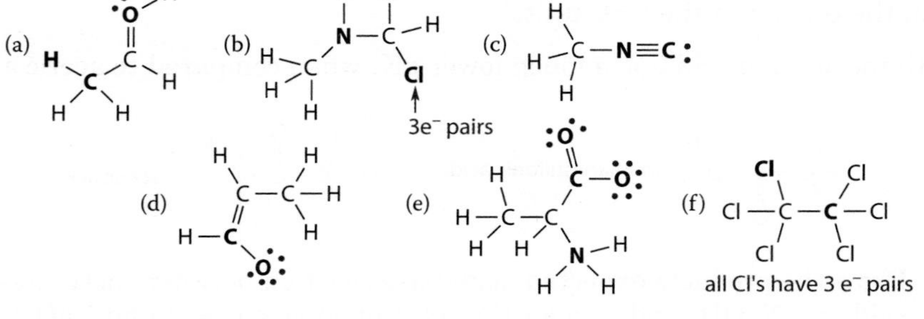

Spectroscopy Problem

This problem is to be done only after Chapter 14 has been read and understood.

5.87 List the important absorption peaks found in the infrared for each of the following. Also list the chemical shift in the proton NMR for the protons on the carbon connected to each functional group and give the multiplicity.

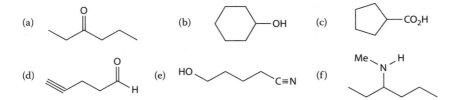

Acids, Bases, Nucleophiles, and Electrophiles

6

Perhaps the most common reaction type in all of organic chemistry is the acid–base reaction. The introduction of acid–base chemistry in Chapter 2 is an effort to bridge what has been learned in general chemistry with what will come in organic chemistry. Some of the discussion from Chapter 2 will be repeated, but in the context of the functional groups introduced in Chapter 5.

One can reasonably argue that most of the reactions encountered in this book involve acid–base chemistry in one form or another. Indeed, generic acid–base reactions and the factors that drive the acid–base equilibrium have been referred to several times. In Chapter 5, the carboxyl group is the key functional group in carboxylic acids, which are taken as the prototypical organic acids. For the most part, carboxylic acids (RCO_2H) are weaker than the mineral acids encountered in general chemistry (HCl, H_2SO_4, HNO_3, etc.).

It is important to modify views of acids to include very weak acids, as introduced in Chapter 2. This means that the functional groups introduced in Chapter 5 must be considered, including alcohols, ketones, terminal alkynes, and even primary and secondary amines. If these functional groups are weak acids, then an acid–base reaction requires the use of a powerful base to generate the corresponding conjugate base. It is also possible that a weak base will react with a very strong acid, generating the corresponding conjugate acid.

ion is a quite potent acid (much stronger than methanol) and the equilibrium is shifted to the left, which means that K_a is much smaller. If **1** is a very weak acid relative to **3**, then water is a rather weak base in this reaction—much weaker than sodium amide. If water is a weak base relative to sodium amide, then K_a is large for the reaction with sodium amide and small for the reaction with water.

An important point of this discussion is to show that *the strength of the acid depends on the strength of the base.* Methanol is a very weak acid when water is the base, but it is a stronger acid when sodium amide is the base. When methanol is dissolved in water, virtually no conjugate base (methoxide) is present in this reaction. However, when methanol is mixed with sodium amide, there is a significant concentration of **2** in the reaction.

Note the important implication of these two reactions. The acid has been kept the same (methanol), and the base is varied. It is clear that methanol is a stronger acid in a reaction with sodium amide than it is in a reaction with water. The important lesson of this discussion is repeated: *The strength of the acid depends on the strength of the base.*

A cautionary statement must be added: *Other factors influence acidity* of a given acid. These factors include solvent, stability of the ionic products formed, solubility, and the nature of the carbon group attached to the heteroatom. For the time being, however, the simple analysis involving bond polarization of the acid and relative stability of the conjugate base is useful and allows a reasonable prediction of relative acidity.

6.3 Draw the acid–base reaction of 1 (a) with $NaNH_2$ as the base, and (b) with $CH_3C{\equiv}C{:}^-Na^+$ as the base.

6.2.2 Transition States and Intermediates: Reaction Curves

For any reaction, including acid–base reactions, the progress of that reaction can be correlated with changes in energy, particularly at the point where the reaction begins. One parameter that determines the course of the reaction is the change in free energy, ΔG°, which is a function of the change in enthalpy (ΔH°) as well as the change in entropy (ΔS°) for the reaction. Enthalpy (H°) is the bond energy for the acid A–H, but the bond energy of all bonds that are made or broken must be examined to determine the change in enthalpy. It is important to point out that enthalpy is not the key parameter in this process. The energy (ΔG°, called the free energy) of the entire system must be determined, which includes enthalpy as well as entropy and temperature. The relationship of these parameters is well defined.

An important measure of whether or not a chemical reaction will proceed is the change in **standard free energy ($\Delta G°$)**, which *assumes that the reaction is done under standard conditions:* in solution at a concentration of 1 M and at 1 atm of pressure for gases. The standard free energy is calculated from enthalpy (H°) and entropy (S°) by the following equation:

$$\Delta G° = \Delta H° + T\Delta S°$$

For the acid–base reaction of **1** with sodium amide to give **2**, enthalpy changes are estimated from the bond dissociation energies. Calculation of $\Delta H°$ for making and breaking bonds was discussed in Chapter 3, Section 3.6 and that chapter listed bond dissociation energies in Tables 3.2 and 3.3. The value for $\Delta H°$ is determined by taking the bond dissociation energy for the products and subtracting the energy for the starting materials. This leads to a general formula for $\Delta H°$: $\Delta H° = H°_{products} - H°_{reactants}$ and:

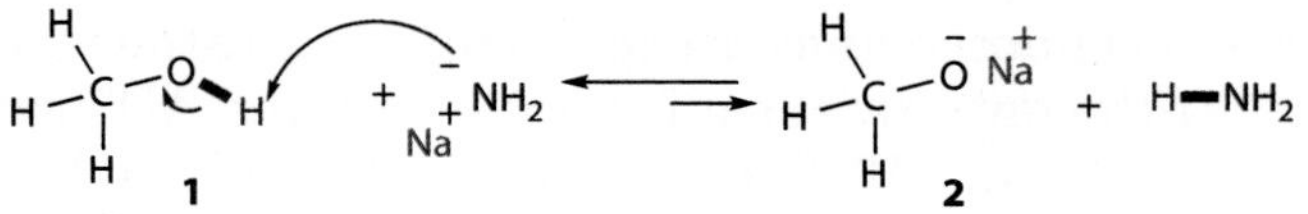

Inspection of the reaction of **1** with sodium amide makes it clear that the O–H bond is broken and an H–N bond is formed. If the ionic bonds are ignored, $\Delta H° = H°_{NH} - H°_{OH}$. The value H° for OH is 104.2 kcal mol^{-1} and the H° for NH is 75 kcal mol^{-1}, so $\Delta H°$ is $75 - 104.2 = -29.2$ kcal mol^{-1}.

The entropy term ($\Delta S°$) measures the "disorder" of a given system. In practical terms, if the number of particles for a reaction remains the same or decreases, the change in entropy is small. If the number of particles greatly increases during the course of a reaction, then $\Delta S°$ increases. The term "T" is temperature. In many cases, the "entropy term" is *assumed* to be very small relative to the enthalpy term, and it is often ignored. If the entropy term is assumed to be zero, then $\Delta G° = \Delta H°$ and estimating the energy of a reaction is quite simple. The entropy term is obviously important, but in Tables 3.1 and 3.2 in Chapter 3, values of H° are measured in kilocalories per mole, so $\Delta H°$ will have units of kilocalories per mole; however, the majority of chemical reactions in this book have an entropy term measured in *calories* (cal), rather than in kilocalories.

An example is a process with a $\Delta H°$ of 23 kcal (23,000 calories) and the accompanying entropy ($\Delta S°$) is 30 calories (typical for reactions done at normal temperatures between 25 and 100°C). Note that 25°C is 298.15 K, because the T term is measured in Kelvin, not centigrade. The $T\Delta S°$ term is small relative to the $\Delta H°$ term (typically, less than 5%), but with the temperature term this may lead to a significant error in the actual value of $\Delta G°$. However, conclusions about the spontaneity of the overall reaction are usually unaffected if the entropy term is ignored. It is important to emphasize that *ignoring the entropy term is an assumption* and in many cases leaving out the entropy term can lead to erroneous conclusions. For most of the reactions encountered in this book, however, assuming that TΔS° is zero (so that ΔG° = ΔH°) gives a good estimation of how a reaction will behave. It is certainly a good *first approximation*. The calculated value of ΔH° for the reaction of **1** with sodium amide assumes that ΔS° is small, so at room temperature, assume ΔG° ~ ΔH° ~ –29.2 kcal mol^{-1}.

6.4 If ΔH° = 21.5 kcal, ΔG° = 22.4 kcal, and T = 298K, what is the value of ΔS° in kilocalories? How serious is the error introduced into a ΔG° calculation if we ignore the ΔS° term?

A negative value of ΔH° indicates a **spontaneous exothermic reaction (also, an exergonic reaction)**. In other words, the bond-making and bond-breaking processes of the reaction generate more energy than is required to initiate the reaction. A positive value of ΔH° indicates a **nonspontaneous endothermic reaction (an endergonic reaction)**. In other words, the bond-making and bond-breaking processes of the reaction generate less energy than is required to initiate the reaction. The importance of this statement can be overestimated. In fact, it is *not* a positive or negative ΔH° that determines whether a reaction is endothermic or exothermic, but rather a positive or negative value of ΔG°. If the entropy term is very small and ΔH° is relatively large, then using only the value of ΔH° to estimate ΔG° is reasonable and gives a good correlation with the reaction. If the entropy term is large and/or the ΔH° term is rather small, then entropy *cannot be ignored*. If the difference between an endothermic prediction and an exothermic prediction is less than 2–4 kcal, then it is wise to determine the entropy term (TΔS°) and use it in the ΔG° estimate.

It is important to know when a reaction is exothermic or endothermic, but that information does not necessarily indicate *if* the reaction will proceed and under what conditions. When a stick of dynamite explodes, that chemical reaction is clearly exothermic. However, it is possible to manufacture the chemicals in a stick of dynamite without its blowing up. What makes the dynamite blow up? More pertinent to this discussion is the question: **How is it possible to handle a starting material that gives a violently exothermic reaction?** To make dynamite blow up, *energy* must be applied (a match with a long fuse, a spark—detonator cap and electricity source, or some other source of energy). In other words, energy must be applied to the dynamite to *initiate* the exothermic chemical reaction, and *this is independent of whether the reaction is exothermic or endothermic*. The same is true of virtually all chemical reactions. Energy must be applied to the reaction (heat, light, etc.) to start it, at least in most cases. Clearly, factors other than ΔG° and exothermic/endothermic behavior of a reaction are important. This section will introduce some of these factors, particularly those posed by the "dynamite" problem.

If a reaction has a negative ΔG° and is exothermic, a graphic method can display this information and compare it to an endothermic reaction. **It is possible to follow the energetic changes for a chemical reaction and relate them to ΔG° and exothermic/endothermic behavior.** Figure 6.1 shows a typical reaction and an *energy curve* that correlates with that reaction. At the left of the curve (the bottom coordinate is labeled "reaction coordinate"), the starting materials have not yet begun to react, and this energy represents

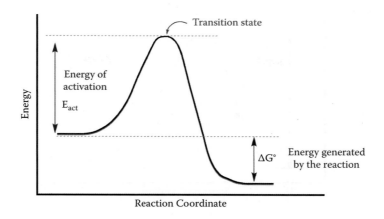

Figure 6.1   Typical energy curve for an exothermic reaction.

the energy inherent to those compounds. At the far right of the reaction coordinate, the reaction has been completed and the energy inherent to the products is shown. In this case it is lower in energy than that of the starting materials. Note that the difference in these two energy points corresponds to $\Delta G°$ and, because it is negative, this reaction is exothermic. Figure 6.1 is a typical reaction curve for an exothermic reaction.

Following the curve from left to right, the reaction proceeds from reactant to product. To begin the reaction, an energy increase is required to reach a certain point (labeled transition state) where the reaction begins, and then the energy curve falls away toward the energy of the final product. As the starting materials (reactants) mix together, a certain amount of energy (labeled **activation energy**, E_{act}) is required to initiate the bond-breaking/bond-making process. In other words, a certain amount of energy is required to begin breaking bonds. Until that energy requirement (equal to E_{act}) is met, no chemical reaction takes place.

One way to define energy of activation is as the amount of energy supplied to a system so that the chemical reaction can begin. Where does this energy come from? In most cases, energy added to the system by heating causes the kinetic energy of the molecules, including solvent molecules, to increase. Molecular collisions transfer energy from one molecule to another. If the energy released is equal to or greater than E_{act}, the reaction begins (i.e., bond making and bond breaking are initiated). When E_{act} is attained, this energy correlates with the transition state energy, the reaction begins, and the curve levels off as shown. Once the bond-breaking/bond-making process begins (at the apex of the curve; the highest energy point) this particular reaction is exothermic (it generates energy equal to $\Delta G°$), so it will continue spontaneously. In other words, once the bond-breaking/making process begins, products begin to form spontaneously if they have a lower incipient energy.

Figure 6.2 is a typical endothermic reaction curve, where the energy of the product is higher than that of the reactant. Once energy equal to E_{act} is applied to the starting materials, the transition state is reached; the reaction begins and

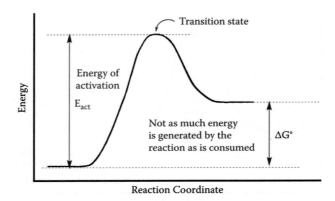

Figure 6.2   Typical energy curve for an endothermic reaction.

the energy curve falls as the reaction proceeds to the lower energy products. The bond making/breaking process in this case generates a product whose energy is *not* equal to E_{act}, however, because the energy gained by forming the product is less than E_{act}. In other words, the inherent energy of the product is higher than that of the starting material. Energy must be continually added to keep the reaction going or else the reaction will stop. Because the amount of energy required to start the reaction (E_{act}) is larger than the amount of energy required to sustain the reaction (ΔG°), the reaction will go to products only as long as an amount equal to E_{act} is available. A reaction may proceed to products even when it is endothermic, but energy must continually be added or it will stop.

6.5 Is the product of an endothermic reaction higher or lower in energy than that of the starting materials?

A last, very important point must be made. When the relative energy of products is discussed, products with a lower energy are said to be more stable than those with higher energy. All discussions in this book will assume that lower energy correlates with greater stability. In terms of ΔG°, a lower energy product formed via an exothermic reaction will have a larger ΔG° than reaction that generates a product with a higher energy. If ΔG° is larger, the lower energy product is more stable in terms of reactivity.

Energy of activation is a very important concept. It is the "match" referred to in the dynamite example. Once the reaction starts, the match (here E_{act}) can be removed if the reaction is exothermic. *The energy of activation E_{act} is formally defined as the difference between the highest point of the energy curve that defines the conversion of starting materials to products.* It is the amount of energy required to initiate (activate) a chemical reaction. A reaction may be exothermic, but nothing happens until an amount of energy equal to the activation energy is added to the system. It is possible that a reaction is exothermic, but the E_{act} is so large that the reaction simply does not proceed under normal conditions. It is also possible that the E_{act} is so low that the reaction will begin under very mild conditions.

6.6 If $\Delta H°$ for one reaction is –21 kcal and that for a second is –10 kcal, determine which reaction is likely to occur at a lower temperature if E_{act} for the first reaction is 35 kcal mol^{-1} and E_{act} for the second reaction is 44 kcal mol^{-1}.

6.2.3 Reaction Curve for an Acid–Base Reaction

It may not be clear from the preceding section how the energy of the reaction correlates with the acid–base reaction that converts **1** to **2**. The calculated value of $\Delta H°$ for the reaction of **1** with sodium amide assumes that $\Delta S°$ is small; that term is ignored, so at room temperature assume that $\Delta G° \sim \Delta H° \sim -29.2$ kcal mol^{-1}. What does this value mean? If the reaction is exothermic, the reaction of **1** with sodium amide is more favorable than the reaction of **2** with ammonia.

In Figure 6.3, it is apparent that the activation energy for the reaction of **1** + NaNH$_2$ is much smaller than the activation energy for the reverse reaction (**2** + NH$_3$). This is entirely consistent with a large value of K_a, and **1** is much more acidic than ammonia. If the reaction were endothermic, the activation energy for the reverse reaction (**2** + NH$_3$) would be less than the activation energy for the reaction of **1** + NaNH$_2$, and the reverse reaction might be very favorable. Curves such as the one in Figure 6.3 may be generated when the stability and reactivity of the acid–base and conjugate acid–conjugate base are correlated with energy. The values of K_a and pK_a are used to estimate the position of the curve so that one acid–base reaction can be compared with another. To

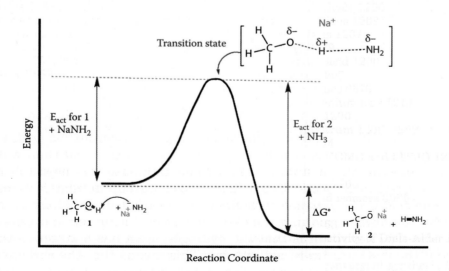

Figure 6.3   Energy curve for the reaction of methanol with sodium amide.

understand the dynamics of each individual reaction, the change in energy for the reaction must be examined.

The transition state structure is shown in Figure 6.3. As the nitrogen atom of $^-NH_2$ donates electrons to the acidic proton, a new H–N bond begins to form. This event is indicated by the dashed line, H------N. At the same moment, the O–H bond of the acid (**1**) begins to break, again indicated by the dashed line O------H. This point on the energy curve is the transition state, and it represents the energy at which bond breaking and bond making are initiated. Once started, the products alkoxide **2** and ammonia are formed. Analysis of the transition states of reactions will be important for many reactions in this book.

6.2.4 Other OH Acids

Water and alcohols have an acidic proton attached to oxygen—hence the term "OH" acid. Alcohols are rather weak acids, slightly weaker than water for the most part. However, two classes of relatively strong OH organic acids—carboxylic acids (**4**) and sulfonic acids (**5**)—have pK_a values generally in the range of 1–5, depending on the nature of the R group. Why are water and alcohols weak acids, but **4** and **5** considered to be strong acids? Carboxylic acids have a carbonyl group (C=O) attached to the O–H unit, and sulfonic acids have an O_2S (sulfonyl) group attached to the O–H unit. In **4**, the polarized C=O unit leads to a larger δ– oxygen in the OH unit. In turn, this inductive effect (through-bond and through-space inductive effects; see below) makes the attached hydrogen more positive, and the acid is more acidic. The carbon atom in methanol (**1**) is also δ+. Remember that the positive carbon made only a small difference in the pK_a of alcohols when compared to water, and methanol has a pK_a of about 15.2 (the pK_a is 16–18 for other alcohols). This explanation relies on simple bond polarization of the OH unit in the alcohol and electron-releasing effects of the carbon group in the alkoxide product.

Formic acid (HCOOH, **6**) is a specific example of a carboxylic acid, with a pK_a of 3.75. Formic acid is clearly a significantly stronger acid than water or methanol, and resonance stability of the conjugate base was used to explain this fact in Chapter 5, Section 5.9.3. The carboxyl group in **6** has a δ– oxygen atom connected to a δ+ carbonyl carbon, and that carbon draws electron density away from the oxygen, making the O–H bond more polarized and the proton more acidic. The net result is that the hydrogen atom has a larger δ+ than is possible if the C=O unit were not present. This means that the carbonyl unit (C=O) is electron withdrawing relative to the oxygen atom, making the OH bond more polarized and the proton more acidic. In methanol (**1**), the carbon group attached to the oxygen is electron releasing, making the O–H bond less polarized and the proton less acidic.

Another important factor that influences acidity is hydrogen bonding. In **6**, the δ+ proton forms hydrogen bonds more effectively than the proton in methanol. Hydrogen bonding effectively elongates the O–H bond, making hydrogen more δ+ and therefore more acidic. However, the more δ+ the proton is, the more extensive is the hydrogen bonding. Although hydrogen bonding exacerbates the acidity, it does not account for the fundamental reason why **6** is so much more acidic than **1**. To explain this difference fully, the nature of the conjugate acid and conjugate base must be examined.

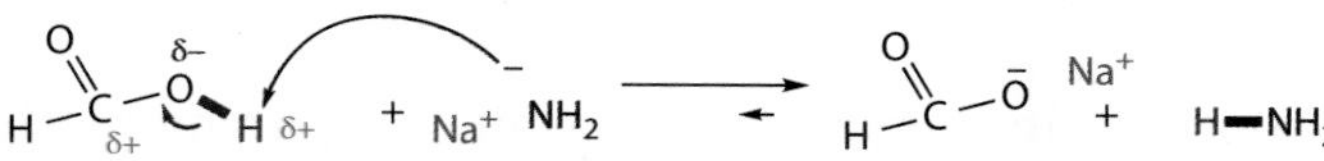

The stability of the product (the conjugate base) is a major contributing factor to the position of the equilibrium (K_a) in the reactions of formic acid and methanol. The difference in K_a for these reactions is the true measure of the relative acidity of carboxylic acids relative to alcohols. When formic acid (**6**) reacts with sodium amide (the base), removal of the proton leads to the formate anion (**7**) as the conjugate base; similar reaction of methanol (**1**) gives methoxide (**2**). In methoxide, the two electrons associated with the negative charge are localized in an orbital on the oxygen (see **8**). In formate **7**, however, the electrons are not localized on one oxygen atom, but rather delocalized over three atoms by resonance. Inspection of **7** shows that the electrons in the orbital on oxygen are adjacent to the π-bond of the carbonyl. If all three contiguous orbitals are aligned parallel (as in **9**), the two electrons in the π-bond and the two electrons associated with the negative charge are shared (*delocalized*) over those three adjacent orbitals on the three atoms. Delocalization of electron density in this manner is called ***resonance*** or resonance delocalization (introduced in Chapter 5, Section 5.9.3).

Dispersal of electron density over several atoms leads to a lower net energy than concentrating charge on a single atom. This electron delocalization (resonance) is represented by the two structures shown (**9A** and **9B**) and the "double-headed" arrow between the two structures. *The double-headed arrow indicates resonance and is different from the two opposite arrows used to indicate a reversible reaction.* Structure **9C** represents both resonance contributors, where the central carbon atom is sp^2 hybridized and the π-cloud extends over three atoms. Four electrons are dispersed in that π-cloud. The electron potential map for the formate anion (**9D**) shows the red color (higher electron density) dispersed over the three atoms, with a higher concentration of electron density on both oxygen atoms. The diagram is consistent with the electrons of the negative charge being delocalized by resonance. This resonance delocalized formate anion **9** is lower in energy and more stable than the localized methoxide anion **2**. *If 9 is more stable than 2, the equilibrium for 6 shifts to favor products (larger K_a), consistent with the fact that formic acid is more acidic than methanol in a reaction with sodium amide.*

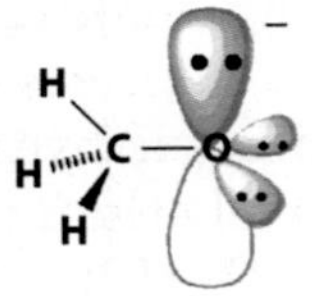

Three main concepts influence acidity when comparing methanol with formic acid: (1) the carbonyl group polarizes the O–H bond, (2) the OH of carboxylic acids hydrogen bonds to a greater extent, and (3) the conjugate base (formate, **9**) is more stable than methoxide (**2**), driving the equilibrium to the right. These concepts are the same as those introduced in Chapter 2 to explain acid–base reactions and the importance of the acidity constant, K_a.

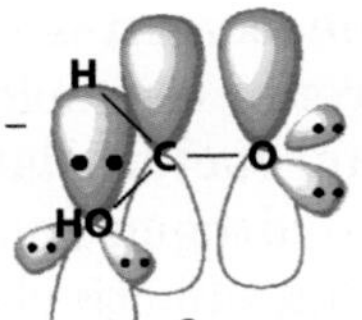

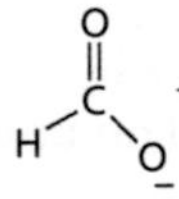

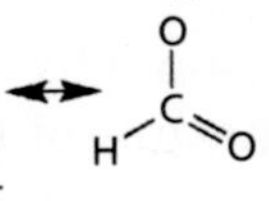

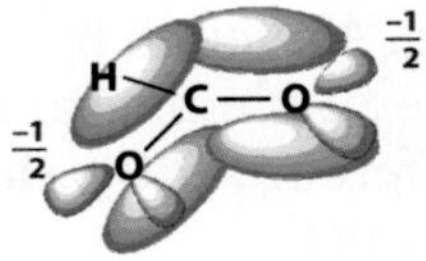

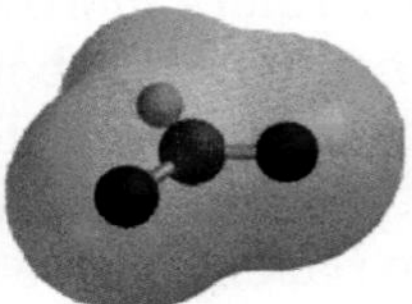

6.7 **Draw the two resonance forms for the anion derived from propanoic acid.**

6.8 **Briefly discuss why the pK of HCl in water is –6.1, but that of sulfuric acid (for loss of the first proton) is about –3.**

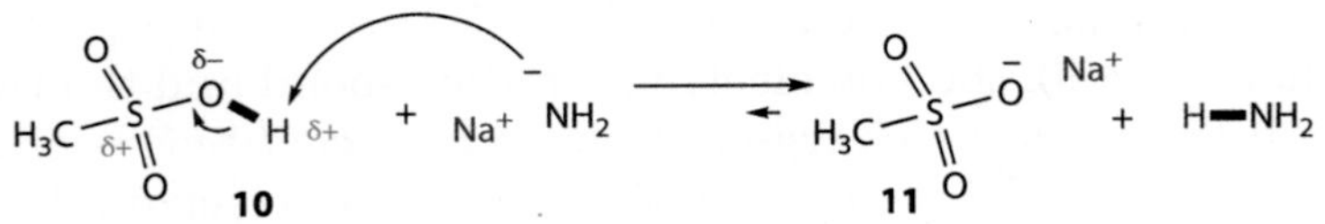

It is known that sulfonic acids are more acidic than the corresponding carboxylic acid. If the C=O group of a carboxylic acid polarizes the adjacent O–H bond, the S=O bond of a sulfonic acid is also polarized. Indeed, simply substituting O=S=O for C=O in the previous discussion concerning the OH group gives a reasonable explanation for the acidity of a sulfonic acid. Using a simple analogy, the sulfur acid analog of formic acid (HCOOH, pK_a of 3.75) is $HOSO_2OH$, which is sulfuric acid and clearly much more acidic (pK_a of about –3). This greater acidity is due to the fact that the sulfonyl group polarizes an adjacent OH bond to a greater extent than an adjacent carbonyl group. In addition, the sulfate anion (the conjugate base) is stabilized by resonance to a greater extent than a carboxylate anion (**9**).

Sulfuric acid is not really an organic acid, but it is useful for comparative purposes because it represents the simplest analogy (like comparing water and methanol). A comparison of sulfonic acids with carboxylic acids—all organic molecules—leads to a more realistic idea of relative acid strength. The pK_a for acetic acid (ethanoic acid, MeCOOH) is 4.76 (see Table 6.1) and the pK_a for methanesulfonic acid ($MeSO_2OH$) is –1.9.[1] The reaction of methanesulfonic acid

Table 6.1
pK_a Values of Common Carboxylic Acids

Acid	pK_a	Acid	pK_a
Formic acid	3.75	Chloroacetic acid	2.87
Acetic acid	4.76	Bromoacetic acid	2.90
Propanoic acid	4.87	Iodoacetic acid	3.18
Butanoic acid	4.82	Dichloroacetic acid	1.26
2,2-Dimethylpropanoic acid (pivalic acid)	5.03	Dibromoacetic acid	1.39
4-Methylpentanoic acid	4.79	2-Methylpropanoic acid	4.85
Phenylacetic acid	4.31	2-Methytlbutanoic acid	4.76
Benzoic acid[a]	4.20	2-Chloropropanoic acid	2.84
4-Methylbenzoic acid[a]	4.36	3-Chloropropanoic acid	3.99
3-Chlorobenzoic acid[a]	3.99	2-Chlorobutanoic acid	2.88
4-Methoxybenzoic acid[a]	4.49	4-Chlorobutanoic acid	4.50
4-Nitrobenzoic acid[a]	3.44	3-Chlorobutanoic acid	3.83

Source: Lide, D. R., ed. 2006. *CRC Handbook of Chemistry and Physics,* 87th ed., 8-43–8-52. Boca Raton, FL: Taylor & Francis.
[a] See Chapter 15, Section 15.4.

(**10**) with sodium amide forms conjugate base **11** (a sulfonate anion; in this case, sodium methanesulfonate).

Three items are important. First, the sulfonate anion (the conjugate base **11**) is resonance stabilized to a greater extent than the formation anion (**9**); thus, all things being equal, **11** should be more stable (less reactive) than **9** and the K_a for **10** is larger than the K_a for **6** (**10** is a stronger acid). Second, sulfur is larger than carbon, which contributes to greater dispersal of charge in the anion **11**. This leads to increased stability and poorer reactivity for the conjugate base. Third, the O–S bond is more polarized than the O–C bond, and the increased size of sulfur relative to carbon suggests that the bond is weaker. In other words, the OH bond in **10** is likely to be weaker than the OH bond in **6**, and the methanesulfonate anion is likely to be more stable (less reactive) than the formate anion. The presence of the SO_2 unit leads to a more acidic compound. *In general, sulfonic acids are more acidic than carboxylic acids.*

6.9 Draw all resonance contributors to the methanesulfonate anion, 11.

6.10 Which is more acidic: butanesulfonic acid or butanoic acid?

6.2.5 Amphoterism

The definition of a Lewis acid is an electron pair acceptor, and the acceptor atom in the Brønsted–Lowry acids just described is a hydrogen atom. If the reactions of **1**, **6**, and **10** with a base are viewed as simple chemical reactions, the Brønsted–Lowry and Lewis definitions converge in the sense that a hydrogen atom or another atom accepts an electron pair from a base. The definition of a Lewis base is an electron pair donor. In the Brønsted–Lowry chemical reactions, the base donates electrons to the proton, to form a new covalent bond. If the focus is shifted from the proton of the O–H unit to the oxygen atom, the oxygen atom of water or alcohols has unshared electron pairs. In the presence of a base that is much stronger than the conjugate base of the alcohol, these molecules may react as bases.

In other words, *if the base that reacts with the alcohol (ROH) gives a conjugate acid with a pK$_a$ that is much smaller than the pK$_a$ of the alcohol, then that base will be a stronger base than the alkoxide conjugate base of the alcohol, RO$^-$.* This fact means that if an alcohol reacts with a base that is much stronger than the alkoxide, the alcohol will react as an acid. *If an alcohol reacts with a base that is much weaker than an alkoxide, however, the oxygen of the alcohol may react as a base.* Think about this statement! If the base is much weaker than the alkoxide, the conjugate acid of that base must be much stronger than the alcohol. This observation is the key to understanding the difference in reactivity.

With the exception of methanol (pK$_a$ of 15.2) and ethanol (pK$_a$ of 15.7), the pK$_a$ of an alcohol is in the range of 16–18. Therefore, alcohols will behave as acids in the presence of relatively strong bases whose conjugate acids have pK$_a$ values that are approximately >20 (a very general cutoff point). *In the presence of very strong acids (acids that have a pK$_a$ that is typically <5–10), alcohols will definitely react as bases.* If an alcohol such as ethanol (**12**; pK$_a$ of 15.7) is mixed with a strong acid such as HCl (pK$_a$ of −6.1), the oxygen of ethanol will donate electrons to the acidic proton of HCl, and the products are a conjugate acid (oxonium salt **13**) and a conjugate base (Cl$^-$, the counter-ion of the oxonium cation). If ethanol is mixed with sodium amide, however, the nitrogen of amide anion is much more basic than the oxygen of ethanol, and the conjugate acid is ammonia, with a pK$_a$ of about 25. In this reaction, the conjugate base is the ethoxide anion **14**. In the presence of this stronger base, ethanol reacts as an acid.

6.11 Draw the products of a reaction between 1-butanol and HCl.

The property of a compound to react as either an acid or a base is called ***amphoterism***. Alcohols, water, and other compounds that function in a similar manner are referred to as ***amphoteric compounds***. *It is possible to generalize and say that an amphoteric compound such as ethanol will function as a base in the presence of an acid with a* pK_a *significantly lower than itself. It will function as an acid when a molecule has an electron-donating atom that is a stronger base than its own conjugate base (EtO−).* Carboxylic acids are much more acidic than alcohols ($pK_a < 5$), and a very strong acid indeed (perhaps, $pK_a < 8$–10) is required before they will react as bases. A few examples of such a reaction will be used in this book (see Chapter 20, Section 20.5.2).

6.2.6 Is a Hydrogen Attached to a Carbon Acidic?

The answer is yes, but the hydrogen atom on the carbon must be polarized δ+. In general, the hydrogen of a C–H bond in an alkane is a very weak acid indeed, with a $pK_a > 40$. Compare this number with the pK_a of the weak acid ammonia (pK_a of about 25), which has the N–H unit. For a C–H unit, an electron-withdrawing group such as a carbonyl (C=O), a sulfonyl (S=O), a carboxyl (CO$_2$−), or a cyano (C≡N) must be attached. The hydrogen atoms attached to the carbon α to the functional group (see the **red** hydrogen atoms in **15** and **16**) are weakly acidic.

The bond polarization in both **15** and **16** shows that the hydrogen atom attached to the carbon adjacent to the carbonyl (the so-called α-carbon) is polarized δ+. The pK_a values for a ketone such as acetophenone (**15**) is 18.6, but a different electron-withdrawing groups (other than C=O) will lead to a pK_a of these α-hydrogens in the range of 16 or 18–25 (see Chapter 22, Section 22.2, for nomenclature and a discussion of these compounds). Such compounds are weak acids (weaker than alcohols), and the base required for an acid–base reaction must be very strong. *The base used must generate a conjugate acid that is weaker than the acid with which it reacts; in this case,* $pK_a > 25$–27.

If two electron-withdrawing groups are present in a molecule, such as the two carbonyl groups in **16**, the indicated hydrogen is more polarized and more acidic (the pK_a of **16** is about 9). In later chapters, carbon acids will be discussed, and they are particularly useful because their conjugate bases are carbanions (a carbon atom bearing a negative charge). The reaction of **15** and a strong base generates carbanion **17** (an enolate anion, which will be discussed in Chapter 22, Section 22.2). This enolate anion is rather stable because it is resonance stabilized, with the charge delocalized on carbon (**17A**) and on oxygen (**17B**). Enolate anions will be very important players in the formation of carbon–carbon bonds by chemical reactions yet to come. This chemistry requires a rather detailed explanation and understanding of chemical reactions not yet discussed, so further examination of these carbon acids will be delayed until Chapter 22.

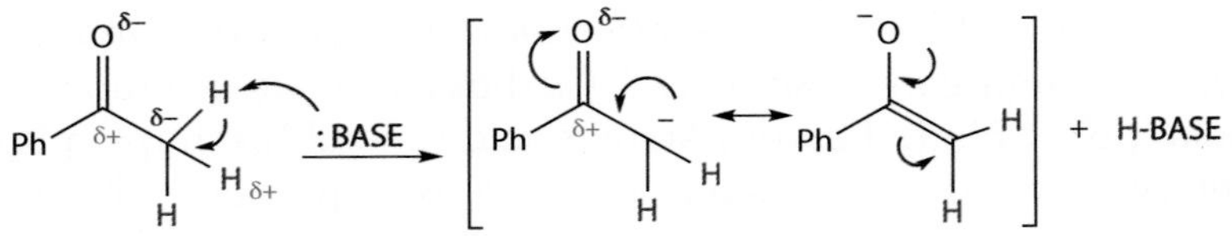

Another carbon compound has an acidic proton, and the acid–base reaction is quite common. Alkynes (Chapter 5, Section 5.2) are categorized as internal alkynes (R–C≡C–R) or terminal alkynes (R–C≡C–H). The hydrogen atom of a terminal alkyne is a weak acid (pK_a of about 25), and strong bases such as sodium amide ($NaNH_2$) remove that proton to give an alkyne anion (R–C≡C:⁻). In this reaction, the alkyne anion is the conjugate base of the alkyne, which is a weak acid. Alkyne anions are useful nucleophiles in many reactions (see Section 6.7).

6.12 Draw the product formed when 1-pentyne reacts with $NaNH_2$, and draw the corresponding conjugate acid and conjugate base.

6.3 Factors That Influence the Strength of a Brønsted–Lowry Acid

There are many carboxylic acids, many sulfonic acids, and many alcohols. Several factors influence the relative strength of an acid, and most of these factors were introduced in Chapter 2 in the context of general chemistry definitions of acids and bases. This section will focus on those parameters that can be identified as making an organic acid stronger or weaker.

6.3.1 Stability of the Conjugate Base

The reaction of an acid and a base will generate a conjugate base and a conjugate acid. Some of the discussion of acid–base strength in the previous sections focused attention on the structure, stability, and reactivity of the conjugate base. If these reaction products are stable, there is an increased concentration of products relative to the reactants, described by saying that the equilibrium is shifted to the right. This is a long way of saying that K_a is larger and the acid is stronger. For a general discussion, it is reasonable to look for factors that stabilize conjugate acids and conjugate bases.

As seen in Chapter 2, Section 2.4, the size of the conjugate base and whether or not it is stabilized by resonance are two important factors that influence charge dispersal. If the charge of a conjugate base is dispersed over a greater area, it is more stable (lower reactivity), as discussed in Section 6.3. The resonance stability of an ion is determined by the extent of delocalization, as with the formate ion (**9**) or the methanesulfonate anion (**11**). The size of the anion is important, as when two different size conjugate bases are compared: iodide from HI versus fluoride from HF. It is known that HI is more acidic, in large part because the much larger iodide anion disperses the charge over a larger area relative to fluoride ion, making iodide more stable (less reactive). Bond strength can be compared, as with H–O in water versus H–N in ammonia versus H–O in the hydronium ion, using the examples discussed before. The order of relative bond strength (and relative acidity) is that O–H for hydronium is greater than O–H for water, which is greater than N–H for ammonia. In general, anything that leads to a more stable product will shift the equilibrium to products and a larger K_a.

6.13 **Compare the conjugate bases of nitromethane (CH_3NO_2) and methanol (CH_3OH). Which is the stronger base and why? Compare the relative acidity of nitromethane and methanol.**

6.3.2 Structural Variations. Inductive Effects

When the relative acid strength of carboxylic acids and alcohols was compared in Section 6.2, the influence of the group attached to the oxygen (C–O–H) was mentioned in connection with the strength of the O–H bond. In a carboxylic acid, that C is the sp^2 carbon of a carbonyl group, but that C is the sp^3 carbon of an alkyl group in alcohols. Clearly, the nature of these groups has an influence on the pK_a of the proton in the O–H unit of the acid. This effect is called an *inductive effect* (see Chapter 3, Section 3.7), where electron density is donated toward the acidic proton or pulled away from the acidic proton. In general, "pushing" electron density toward the proton (electron releasing) makes the O–H bond stronger and the proton is less acidic. Conversely, "pulling" electron density away from the proton (electron withdrawing) makes the O–H bond weaker and the proton is more acidic. This flow of electrons also influences the stability of the anion after the proton is removed, which influences the position of the acid/base-conjugate–acid/conjugate base equilibrium, which is important to acid strength.

Formic acid (**6**) can be compared with ethanoic acid (acetic acid, **18**) because both have a carbonyl group attached to the O–H unit. Formic acid has a pK_a of 3.75 and acetic acid has a pK_a of 4.76. The only structural difference is the presence of a hydrogen or a methyl group attached to the COOH unit. Relative to hydrogen, the carbon group is electron releasing because carbon is slightly more electronegative than hydrogen (see Chapter 3, Section 3.7). This difference in electronegativity makes the carbon slightly electron releasing (see **18A**) in the direction of the arrow. If the methyl group in **18A** releases electrons to the $\delta+$ carbon of the carbonyl, less electron density is pulled from the oxygen and, in turn, less electron density is pulled from the hydrogen. In other words, the hydrogen atom is less $\delta+$ (less polarized) and less acidic. As described, this effect occurs through the bonds by pushing and pulling electron density and it is called a ***through-bond inductive effect*** (graphically illustrated in **18A**).

Another view of inductive effects is via the electron density maps for formic acid (see **6A**) and acetic acid (see **18B**). Remember that the **red** area indicates higher electron density and the **blue** area lower electron density. Although it is a little difficult to see, the blue area over the acidic hydrogen atoms in **6A** and **18B** are clearly visible, and the blue area for **6A** is somewhat larger and slightly **bluer**. This difference is an indication that formic acidic is more acidic, which is consistent with the concept that acidity is influenced by the presence of the methyl group. Using **18A** as a model, the electron-releasing characteristics of the carbon group of the methyl unit are consistent with a less polarized OH and a less acidic species.

What is the result if an electron-withdrawing group or atom is attached to a carbonyl carbon? The structural difference between acetic acid (**18**, pK_a of 4.76) and chloroacetic acid (**19**, pK_a of 2.87) is the presence of the chlorine on the sp^3 carbon that is attached to the sp^2 carbon of the carbonyl (the so-called α-carbon). The pK_a values indicate that **19** is more acidic. A bond polarization model of **19** (see **19A**) shows a different electron flow and a different induced dipole when compared to **18A**. Chlorine is more electronegative than carbon and this makes the C–Cl bond polarized with a $\delta-$ Cl and a $\delta+$ C. The result of this polarized bond is that the $\delta+$ carbon of the C–Cl unit is adjacent to the carbonyl carbon, which is also polarized $\delta+$.

To compensate, the carbonyl carbon will draw more electron density from oxygen, making the O–H bond more polar (H is more $\delta+$) and more acidic. This observed effect in **19** is described via a ***through-bond effect***. If chloroacetic acid is drawn differently (see **19B**), however, the acidic hydrogen and the chlorine are seen to be close together in space. The $\delta-$ chlorine is attracted to the $\delta+$ hydrogen in what constitutes an intramolecular hydrogen bond. The hydrogen bonding will pull the proton closer to chlorine, such that the O–H bond is elongated. Formation of a hydrogen bond between H and Cl in **19B** leads to a relationship that resembles a five-membered ring. The H and Cl ***are not*** connected, but rather simply close together.

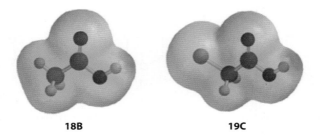

18   19   19A   19B

This bond elongation makes the O–H bond more polar (H is more δ+) and more acidic. This effect is a ***through-space*** effect due to the attraction of the chlorine for the acidic hydrogen. This effect is possible *only* if the two polarized atoms (with opposite charges) are in close proximity. There is rotation about single covalent bonds that allows these atoms to approach, as will be discussed in Chapter 8, Section 8.1. ***This through-space effect is much more important than the through-bond effect described previously***. The electron density maps of acetic acid (**18B**) and chloroacetic acid (**19C**) are compared. Note that the area of blue color in **19C** is greater and more intense, which is indicative of a more positive proton and a more acidic compound. Also note that the red color is shifted to the chlorine atom in **19C**, which is indicative of electron density shifted toward the chlorine and away from the OH unit. These models are consistent with an electron-withdrawing chlorine group, making **19** more acidic than **18**.

6.14 Do you predict that 2-nitroethanoic acid (HOOCCHNO$_2$) is stronger or weaker than ethanoic acid (acetic acid)? Briefly explain.

6.15 Draw the structures of 2-chlorobutanoic acid, 3-chlorobutanoic acid, and 4-chlorobutanoic acid.

18B   19C

Table 6.1[2] shows the correlation of several carboxylic acids with pK$_a$ to show how structural changes and the presence of heteroatom groups influence acidity. There is a reasonable correlation of inductive effects with the distance of the acidic proton from the atom or group responsible for the electron-releasing or -withdrawing effect. Inductive electron-withdrawing effects are important when the δ– atom is attached to the α- or the β-carbon. This trend is clear when 2-chlorobutanoic acid, 3-chlorobutanoic acid, and 4-chlorobutanoic acid are compared. When the electron-withdrawing chlorine is on C2 (2-chlorobutanoic acid), the pK$_a$ is 2.84, making it significantly more acidic relative to butanoic acid (pK$_a$ of 4.82). If it is assumed that the through-space effect dominates, then a simple comparison of **20–22** gives a reasonable explanation of the difference in acidity.

For **20**, the acidic proton and the δ– chlorine atom are close in space, leading to a significant intramolecular through-space interaction as well as a through-bond effect. This arrangement mimics a five-membered ring through internal hydrogen bonding. When the chlorine atom is on C3 (3-chlorobutanoic acid), the through-space hydrogen-bonding effect requires a conformation such as **21**, but the chorine atom is further away and the arrangement mimics a six-membered ring. In this arrangement, the chlorine atom is further away from the proton and the effect is diminished relative to **20**. The pK$_a$ is 3.99—more acidic than butanoic acid but less acidic than 2-chlorobutanoic acid. In 4-chlorobutanoic acid, the chlorine atom is so far away that a through-space interaction demands an arrangement that approximates a seven-membered ring (see **22**). The inductive effect is diminished, as confirmed by the pK$_a$ of 4.50, which is only slightly more acidic than butanoic acid.

As the electron-withdrawing group is further removed from the carboxyl group, the effect diminishes. The chlorinated butanoic acid derivatives suggest that the practical limit of this effect is reached when the electron-withdrawing group is on the third carbon atom (C4) with respect to the carbonyl. The effect is noticeable when the halogen is attached to C2, C3, or C4 relative to the carbonyl carbon of the carboxylic acid. Table 6.1 also shows that chloroacetic acid is stronger than bromoacetic acid and that the electron-withdrawing ability of the halogens seems to be Cl > Br > I. This trend correlates very well with the electronegativity of the halogens (see Chapter 3, Section 3.7).

20   **21**   **22**

The halogens are electron withdrawing, and their presence leads to stronger acids. The effects of electron-releasing alkyl groups can also be examined, and they make the acid weaker (larger pK$_a$). This trend is apparent when formic acid (**6**) and acetic acid (**18**) are compared in Section 6.2. Acetic acid (CH$_3$COOH) has a pK$_a$ of 4.76, and the presence of a methyl group on the α-carbon (propanoic acid) makes the acid weaker (pK$_a$ of 4.87). The presence of two methyl groups on the α-carbon (dimethylacetic acid) has little effect (pK$_a$ of 4.85), but the presence of three methyl groups on the α-carbon (trimethylacetic acid, the common name of which is pivalic acid) leads to diminished acidity (pK$_a$ of 5.03).

6.16 Draw acetic acid, methylacetic acid, dimethylacetic acid, and trimethylacetic acid. Are these the IUPAC names?

6.17 Discuss the relative acidity of 2-methoxybutanoic acid and 4-methoxybutanoic acid.

6.3.3 Solvent Effects

The pK$_a$ values given for all organic molecules in this chapter are based on their reaction in water, which means that water is the base in those reactions. If a different solvent is used, the pK$_a$ is different. If another base is added to the water solution, the pK$_a$ may be different. Remember that acid strength depends on the strength of the base, which influences the position of the equilibrium and K$_a$. In the reaction of acetic acid (ethanoic acid, **18**) and water, the conjugate acid is H$_3$O$^+$ and the conjugate base is the acetate anion (**24**). In this reaction, water reacts as the base, but water is also the *solvent* and it has a profound effect on the course of the reaction.

Water is a very polar molecule, and it is capable of solvating and separating ions (see Chapter 13, Section 13.2). The transition state for the acetic acid–water reaction is shown as **23**, and it is a highly polarized and ionic transition state. Once the reaction begins, water will solvate the acetate ion (**24**) and the hydronium ion (H$_3$O$^+$), which serves to separate the developing ions. Solvation means that the solvent surrounds each ion. This concept is explained in more detail in Section 13.2 (Chapter 13). Separation of the ions pushes the reaction to the right because it facilitates formation of the products (higher concentration of products and larger K$_a$). The net result is that acetic acid is a stronger acid in water than in a solvent that cannot generate a polarized transition state to assist the ionization.

If one molar equivalent of water is used as a base in a reaction with **18**, but diethyl ether is used as a solvent, acetic acid is a weaker acid. As a solvent, ether does *not* separate ions, so ionization of **23** to form **24** and the hydronium ion will be slower than in water. Ether does not separate charge because the δ− oxygen can coordinate with the acidic proton, but the δ+ carbon of ether is a tetrahedral atom and sterically inaccessible for efficient coordination with the oxygen atom of **18**. *Because the solvent does not facilitate charge separation or stabilize the ionic products, acetic acid is weaker (smaller K$_a$, larger pK$_a$) in ether than in the water solvent.* The ability to separate ions and "drive" the reaction to the right is an important effect in acid–base reactions. This means the solvent may participate in the reaction, although it does not appear in the final product. Water facilitates ionization and charge separation more efficiently than other solvents and acetic acid is stronger in water than in diethyl ether. Throughout this book, examples of solvent effects will be pointed out in the context of each different type of reaction in the appropriate chapter.

6.18 Both water and methanol are solvents that contain the acidic OH unit. With this in mind, briefly explain why acetic acid is stronger in water than in methanol.

6.4 Organic Bases

Just as there are organic acids, there are also organic bases. The fundamental definition of a base is "a species that contains an atom capable of donating two electrons to an electron-deficient center." Such a reaction will form a new covalent bond to a hydrogen atom or a new dative bond to another atom (see Chapter 2, Section 2.5). For the most part, the stronger organic bases contain a heteroatom. Any atom that has unshared electron pairs in a neutral molecule can potentially function as a base. Likewise, a formal anion, which has a negative charge localized on an atom, can function as a base. The nature of various organic bases will be explored.

6.4.1 The Nitrogen Atom as a Base: Amines

In this book, the most common organic bases are amines, which are molecules that contain a nitrogen with one unshared electron pair (see **25**; Chapter 5, Section 5.6.3, and Chapter 27, Section 27.1). If an amine such as **25** reacts with a Brønsted–Lowry acid such as HCl, nitrogen donates two electrons to hydrogen to form a new N–H bond, and the product is ammonium chloride (**26**). There are four covalent bonds to nitrogen in **26**, so it has a formal charge of +1 (see Chapter 5, Section 5.5). The ammonium unit, R_3NH^+, is the conjugate acid of this reaction.

There are three structural types of amines, as discussed in Chapters 5 and 27: primary amines (RNH_2), secondary amines (R_2NH), and tertiary amines (R_3N). Base strength differs for these amines and also for two different secondary amines or two different tertiary amines. A comparison of ammonia (NH_3) with methanamine (**27**; methyamine) shows that **27** is more basic. In the reaction of **27** with water, the conjugate acid is ammonium salt (**28**) and the conjugate base is the hydroxide. The K_a for this reaction is small, as indicated by the equilibrium arrows; this means that the equilibrium lies to the left. A consequence of this observation is that the hydroxide ion is more basic than amine

27, and ammonium salt **28** is more acidic than water. As written, the reaction represents the K_a for a reaction in water, and K_a is small. In other words, water is a weak acid in the presence of the amine.

Remember from Chapter 2 (Section 2.2) that the reverse process for **27** or **28** is the reaction of the conjugate acid with the conjugate base in an acid–base reaction. Therefore, it is possible to look at the reverse reaction of ammonium salt **28** with hydroxide, where the stronger acid–base pair is on the right, so that K_a for the reaction of **28** and hydroxide is larger than the K_a for the reaction of **27** with water. In the equation for **28**, K_a can be determined for the reaction of **28** to give the amine **27**, and this reaction of the conjugate acid–conjugate base uses the term K_{BH}:

$$K = [BH^+]/B, \text{ but } K_{BH} = B/BH^+ \text{ for the reaction } B + H^+ \rightarrow BH^+$$

and $pK_{BH} = -\log K_{BH}$. The K_{BH} can be determined from pK_{BH} by the formula $K_{BH} = 10^{-pK_{BH}}$.

If K_{BH} is large, then the equilibrium favors a higher concentration of **27** relative to **28**, which means that the amine clearly does **not** react with water. *In other words, 27 is a weak base.* Alternatively, if K_{BH} is small, then there is a higher concentration of **28** relative to **27**, which means that the amine reacts with water, and the amine is a relatively strong base.

The term K_{BH} is used to evaluate the basicity of a base such as an amine; *a larger value of K_{BH} (small pK_{BH}) indicates a weak base, and a smaller value of K_{BH} (large pK_{BH}) indicates a stronger base. It is therefore possible to use the pK of the acid–base reaction as a measure of which species is more basic.* The pK_{BH} of ammonia is 9.2, whereas the pK_{BH} for methylamine is 10.64. A base with pK_{BH} of 10.64 is a stronger base than one with a pK_{BH} of 9.2. To determine the pK_{BH} of methylamine (**27**), examine the reaction of methylammonium ion **28** and water but **not** the free amine. The pK_{BH} for this reaction is known to be 10.64[3], which means the K_{BH} is $10^{-10.64} = 2.29 \times 10^{11}$. These data indicate that the equilibrium lies far to the left and that **27** is a relatively weak base when compared to hydroxide.

It is possible to compare **27** to dimethylamine (**30**), where the pK_{BH} of dimethylamine is 10.73[6] for the ammonium salt **29**. The K_{BH} for **30** is $10^{-10.73} = 1.86 \times 10^{11}$, so it is also a weak base where the equilibrium with the ammonium salt (**29**) is more to the left than is the equilibrium for **28**. The interpretation of this comparison is that **30** is a stronger base than **27**.

6.19 **Calculate the pK_{BH} for an amine with a $K_{BH} = 3.92 \times 10^6$.**

In the discussion of inductive effects for acids (Section 6.3.2), alkyl groups such as methyl are classified as electron-releasing groups. Inductive effects are important in bases as well. Ammonia is a base, and a thought experiment replaces one hydrogen of ammonia with a methyl group to give methylamine, **27**. The methyl group in **27** releases electron density to nitrogen, so the electron density on nitrogen is greater in methylamine than in ammonia, as represented in Figure 6.4.

The carbon groups release electron density to nitrogen. What does this statement really mean? If nitrogen is more electronegative than carbon, the bond polarization is such that there is more electron density on nitrogen. Ammonia also has a polarized bond, but there is less electron density on a hydrogen atom than on a carbon atom N; thus, relative to H, carbon is said to be electron releasing. If there is more electron density on nitrogen, it should be a stronger base. In principle, the presence of one electron-releasing alkyl group makes a primary amine a stronger base than ammonia. If one alkyl group makes a primary amine a stronger base by an inductive effect, then two alkyl groups should make a secondary amine an even stronger base. The pK_a of dimethylamine[6] (**30**) is 10.73, as shown.

This fact is completely in line with the arguments made concerning inductive effects, and Figure 6.4 shows two alkyl groups releasing electrons to

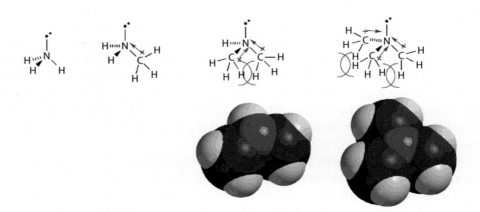

Figure 6.4 Inductive effects in amines.

nitrogen. If two alkyl groups are better for increasing electron release to nitrogen, then three alkyl groups must be the best—**Is this correct?** The pK_a for trimethylamine (**31**), which gives ammonium salt **32** as the conjugate acid, is 9.75.[6] In other words, trimethylamine is a *weaker* base than dimethylamine *and* methylamine. Tertiary amine **31** is only a slightly stronger base than ammonia despite the presence of three electron-releasing alkyl groups. Note that the three methyl groups take up quite a bit of space, which means that these groups effectively repel one another as they compete for space. This repulsion due to competition for the same space is shown in Figure 6.4 as crossed curved lines, and it is known as a *steric interaction* between the adjacent alkyl groups or as *steric hindrance*. Molecular models are provided in Figure 6.4 for dimethylamine and for trimethylamine to show the steric interaction of the methyl groups. The steric hindrance at the nitrogen makes it more difficult for nitrogen to approach another atom in a chemical reaction.

Steric hindrance is not the only effect that influences basicity. An important property of amines is **fluxional inversion** (see Chapter 9, Section 9.7.3), which means that trimethylamine exists as an equilibrating mixture of **31A** and **31B**. In other words, the pyramidal structure of an amine flips back and forth, much like an umbrella inverting in a windstorm. This fluxional inversion occurs rapidly at normal temperatures, so the actual structure of the amine is a mixture of *both* structures. Remember that the nitrogen must collide with the proton in order to react as a base. With the methyl groups moving around the nitrogen in this manner, there is significant steric inhibition as the lone electron pair approaches any positive center, and reaction as a base is more difficult. Apart from these effects, when the reaction occurs to give **32**, the hydrogen atom takes up more space than a pair of electrons. This forces the methyl groups in **32** more closely together, increasing the steric hindrance and shifting the equilibrium back toward **31**.

In relatively polar solvents, hydrogen bonding also increases the effective "size" of the groups around nitrogen, further inhibiting its reactivity. The net result is that **31** is a weaker base than a primary or secondary amine *in solution*. *Because of steric hindrance and fluxional inversion, trimethylamine (and tertiary amines in general) are less basic than predicted.* It is important to repeat that this effect is important for amines that are in solution rather than in the gas phase.

Based on inductive effects and steric effects, it is possible to categorize the base strength of amines in a general sense. The trend is $NH_3 < R_3N < RNH_2 < R_2NH$, where secondary amines are the strongest bases and ammonia is the weakest base. Steric hindrance can play a subtler role in many amines, so this trend must be viewed as general, not absolute.

Me   Me Me   Me Me

N—H   N—H   N—H

Me   Me   Me Me

33   **34**   **35**

For example, compare the pK_a of piperidine (**33**; pK_{BH} = 11.12)[4] and 2,6-dimethylpiperidine (**34**; pK_{BH} = 11.07), where **34** is a slightly weaker base than piperidine. Despite the fact that both are secondary amines, the two methyl groups provide some steric hindrance as the amine approaches the proton of an acid. 2,2,6,6-Tetramethylpiperidine (**35**) is also a secondary amine, but the four methyl groups severely inhibit the reactivity of the electron pair on nitrogen and a pK_{BH} of 1.24.[4] This small pK_{BH} means that **35** is an extremely weak base when compared to piperidine or dimethylpiperidine. When steric hindrance is severe, as in **35**, the effect on base strength can be dramatic, even when the groups are not attached directly to nitrogen.

6.20 **Briefly discuss whether 2-(N,N-di-tert-butylamino)butane is a stronger or weaker base than 2-(N-methylamino)butane and justify your choice.**

If an electron-withdrawing group is attached to nitrogen, the electron-withdrawing inductive effects should make the molecule a weaker base because electron density is removed from the nitrogen. The most common electron-withdrawing group attached nitrogen is probably a carbonyl group (see Chapter 5, Section 5.9.1), which leads to a functional group called an amide (see Chapter 16, Section 16.7). An example is **36**, which has the common name of acetamide. The δ+ carbonyl carbon pulls electron density away from nitrogen, as shown, so the nitrogen cannot donate electrons to an acid as well as in an amine. In other words, a primary amide (NH_2 attached to the carbonyl) is a weaker base than a primary amine (see **27**).

Examination of **36A**, the electron potential map for acetamide, shows that the higher concentration of electron density (**red**) is on the oxygen rather than on the nitrogen. This model illustrates withdrawal of electron density from nitrogen toward the oxygen by the carbonyl, making the nitrogen less basic. If an amide reacts as a base, a stronger acid is required to protonate the amide than is required to protonate an amine. If acetamide reacts with HCl, for example, the product (the conjugate acid) is an acyl ammonium salt, **37**. Compare this reaction with that of methylamine (**27**) and HCl in. The pK_a of ammonium salt **37** is 10.62,[4] so the equilibrium for the amide reaction lies to the left and the amide is a much weaker base than the amine. This is due primarily to the electron-withdrawing effects of the carbonyl group.

6.21 **Draw the conjugate acid of a reaction between N-methylbutanamide (36 is ethanamide) and HCl.**

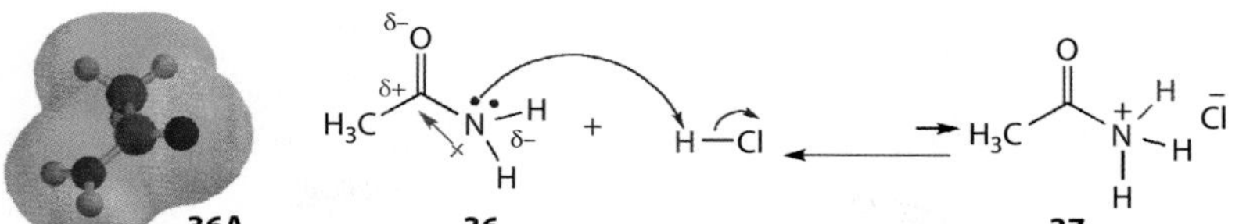

As noted before, the solvent has a profound effect on the strength of a base. In the reaction of methylamine (**27**) and HCl, the transition state for this reaction is **38** and the final product is ammonium salt **39**. The pK$_a$ for **39** is 10.64 when the solvent is water.[4] In transition state **38**, the neutral amine begins to develop charge on the nitrogen, and water will solvate this ion and separate it from the counter-ion (chloride ion). Because water separates the developing charges, the reaction is driven to the right, making dimethylamine a *stronger* base in water than in a solvent that does not separate charge (such as diethyl ether).

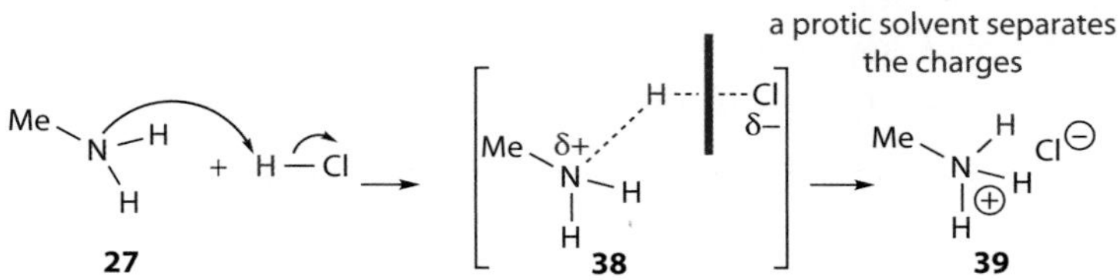

If the equilibrium shifts to the right (large K$_a$, small pK$_a$)—meaning that the amine readily reacts with the acid—the amine is stronger. Water also solvates the ammonium salt product. The lesson from this example is simple, but it has important implications for many organic chemical reactions. *The relative basicity of an amine is effectively decreased by doing the reaction in a solvent that does not solvate and separate ions.* In order to increase the basicity of the amine, water or another solvent that solvates and separates ions may be used.

6.22 The nitrogen atom of amines can attack the carbon atom in bromomethane. Briefly discuss whether the reaction of diethylamine (Et$_2$NH) and bromomethane proceeds faster or slower in water when compared to ether.

6.23 Using only simple arguments, predict whether the conjugate acid of ethylamine (EtNH$_2$) is more or less stable than that of triethylamine (Et$_3$N). Draw both products.

Amines are good Lewis bases, donating two electrons to electron-deficient atoms to form a dative bond in the expected ate complex (see Chapter 2, Section 2.5). Diethylamine (*N*-ethylethanamine, **40**) readily reacts with the strong Lewis acid to form ate complex **41**. Note the dative bond between nitrogen and aluminum and the direction of the arrow *from* the electron-rich nitrogen atom *to* the electron-deficient aluminum atom. Similarly, tertiary amine **42** (*N*-ethyl-*N*-methylethanamine) reacts with boron trifluoride (BF$_3$) to form ate complex **43**. Amines will be used as Lewis bases several times in this book.

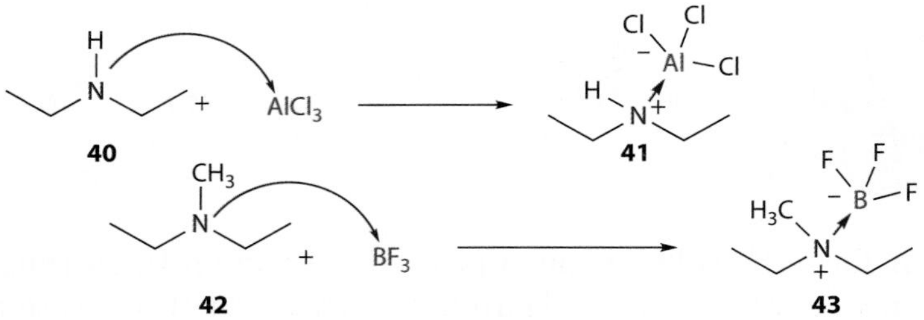

40 **41**
42 **43**

6.4.2 Alcohols Are Bases

In Section 6.2.5, alcohols were seen to be amphoteric. This section is simply a reminder that alcohols react as bases. In the presence of the strong acid HCl, alcohols such as ethanol (**12**), methanol, or butanol will react to form an oxonium ion, **13**, from ethanol. Oxonium ions such as **13** are rather strong acids, and the K_a for this reaction generally lies to the left. In an aqueous solvent, water may stabilize the ions and separate them, which will influence the position of the equilibrium. In general, oxonium ions are transient and highly reactive species.

6.24 Draw the product formed when cyclopentanol reacts with HBr.

base acid
12 **13**

6.4.3 Ethers Are Bases

As discussed in Section 6.2.5, water and alcohols are amphoteric compounds that function as either an acid or a base under the proper conditions. The reactions of ethanol (**12**) as both an acid and a base were discussed in that section, in the context of the amphoteric nature of alcohols. Ethers are molecules with two alkyl groups flanking oxygen and they have no acidic protons. Ethers are weak acids because all hydrogen atoms are attached to carbon rather than oxygen. However, the oxygen atoms of ethers react as bases. Diethyl ether (**44**) is a typical acyclic ether. The nomenclature for ethers is presented in Chapter 5, Section 5.6.4.

6.25 Draw the structure of dipentyl ether.

6.26 Give the structure for (butyl)(1-methylethyl) ether.

6.27 Draw the structure of (2,2-dimethylpropyl)(2-methylbutyl) ether.

If **44** reacts as a Brønsted–Lowry base with a strong acid such as HCl, the conjugate acid product is an oxonium salt (**46**). Note the similarity of this reaction with that of water and HCl or alcohols with HCl. The conjugate base is formally the chloride ion. Oxonium salt **45** is related to the hydronium ion (H_3O^+) in that two hydrogen atoms have been replaced with alkyl groups, and it is an obvious analog of the oxonium salt derived from ethanol (see **13**). Oxonium salts such as **45** are very strong acids and loss of H^+ will regenerate the neutral (uncharged) ether, so the equilibrium for this reaction lies far to the left (K_a is very small), unless **45** reacts with something else to give a different product. In comparison to amines, diethyl ether is a rather weak base. *As with water and alcohols, identifying diethyl ether as a weak base means that it reacts as a base only with very strong acids.* Diethyl ether will *not* react with water or with an alcohol (both weak acids) to give the oxonium salt because they are rather weak acids. With strong acids such as HCl, HBr, HI, H_2SO_4, etc., diethyl ether (as well as other acyclic ethers, R–O–R) reacts to give small concentrations of **45** at equilibrium. The reaction of ethers with HI or HBr will be discussed in Chapter 11, Section 11.8.

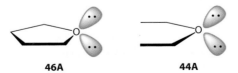

A cyclic ether such as tetrahydrofuran (THF, **46**; see Chapter 26, Section 26.4.2) will react in a similar manner with HCl to give oxonium salt (**47**). Once again, the equilibrium lies to the left (K_a is small) and THF is considered to be a weak base. However, THF is a stronger base than diethyl ether because the carbon groups attached to oxygen in THF are "tied back" (see **46A**), whereas the carbon groups in diethyl ether have more freedom to move around (see **44A**). The result is that the electrons on oxygen in THF are more available for donation (due to less steric hindrance) and this makes THF a stronger base.

Another key point must be stated. Oxonium ions such as **45** and **47** are much higher in energy relative to the ether starting materials and are highly reactive, so there will be a low concentration of **45** or **47** and K_a is small. If the unstable **45** or **47** reacts with some other species, the equilibrium will shift to accommodate loss of the oxonium ion product. In this way, all of the ether starting material may be used despite the low equilibrium concentration of the conjugate acid products. This will be discussed in Chapters 10, 11, and 19.

Ethers are better Lewis bases than alcohols, although they are weaker Lewis bases than amines. Ethers are stronger Lewis bases than alcohols because there are two electron-releasing carbon groups attached to oxygen, making oxygen more electron rich and a better electron donor. The reaction of 2-propanol (**48**) and aluminum chloride ($AlCl_3$), for example, gives the expected ate complex, **49**. Tetrahydrofuran (THF, **46**) reacts with boron trifluoride (BF_3) to give ate complex **50**.

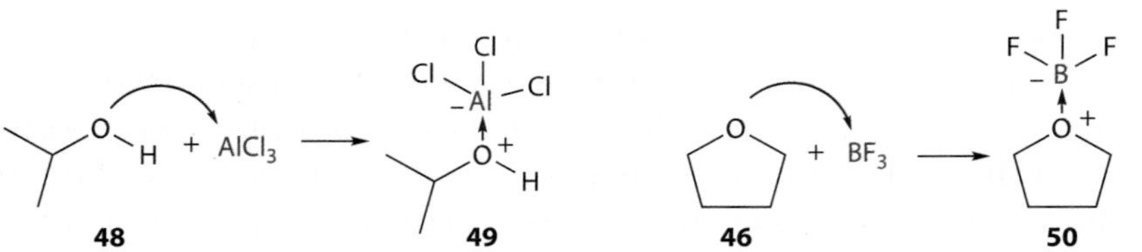

6.4.4 Carbonyl Compounds Are Bases

The oxygen of carbonyl compounds such as ketones and aldehydes (Chapter 5, Section 5.9.2) has unshared electrons that react as a Brønsted–Lowry base or a Lewis base. When an aldehyde or ketone reacts with a Brønsted–Lowry acid (H^+) the conjugate acid that is formed is a new kind of oxonium ion: a protonated carbonyl (see **52** and Chapter 18, Section 18.2). The carbonyl oxygen is a relatively weak base in this reaction, requiring a rather strong acid before reaction occurs. The oxygen atom in butanal (**51**) reacts with HCl, for example, to form the conjugate acid (**52**). Ion **52** is known as an **oxonium ion** or an **oxocarbenium ion** and is resonance stabilized, as shown.

When the oxocarbenium is drawn as resonance contributor **52A**, it is an oxonium salt, clearly related to oxonium salts derived from ethers or alcohols in the preceding sections. Oxonium ion **52** is actually more stable than **13** or **47** (see preceding discussion) because it is resonance stabilized, and the positive charge can be delocalized on carbon (see **52A** and **52B**). The resonance form that has a positive charge on carbon (**52B**) is considered to be an oxygen-stabilized carbocation (hence the term "oxocarbenium ion"; see Chapter 16, Section 16.3) and it can react with nucleophiles. Oxocarbenium ions will be generated in several chemical reactions to be presented in later chapters, and their reactions with suitable nucleophiles will be examined at that time. Ketones behave similarly.

6.28 Draw the product of a reaction between cyclopentanone and HCl.

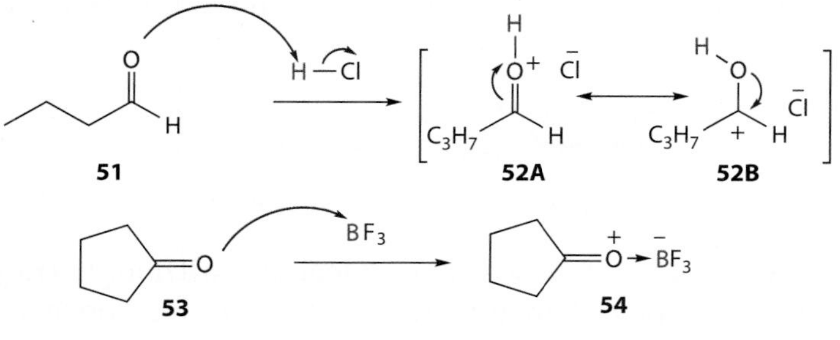

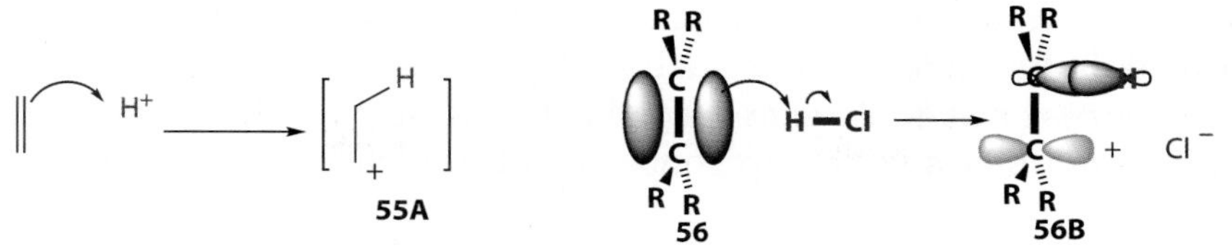

Figure 6.5   Reaction of a π-bond with a strong Brønsted–Lowry acid.

When an aldehyde or ketone reacts with a Lewis acid such as boron tri-fluoride (BF_3), an ate complex is formed. The reaction of cyclopentanone (**53**) and BF_3 leads to **54**, which may react with another species, as will be seen in Chapter 16 (Section 16.3).

6.4.5 The π-Bonds of Alkenes Are Bases

An alkene has a C=C unit with a π-bond that is not polarized. The π-bond extends in space above and below the plane of the molecule, as in Figure 6.5. The π-bond is an electron-rich species, and those electrons can be *donated* to a positive center, initiating a chemical reaction. In other words, *an alkene is a weak Brønsted–Lowry base*. The π-bond is a weak Brønsted–Lowry base and reaction occurs only with a strong acid. This acid–base reaction is illustrated by ethene reacting with a proton H^+ to give **55A**. Donation of the electrons *from* the π-bond (**blue** arrow, see **56**) *to* the **acid (H of HCl)** leads to forma-tion of a new σ-C–H bond (see **56**), and a positive carbon C^+ (a carbocation; see Chapter 10, Section 10.2) on the other carbon atom of the C=C unit. *Note that the positive carbon atom is represented as a p-orbital that has no electrons. Note also that the double-headed arrow indicates transfer of two electrons from C=C to H^+ to form a new σ-covalent bond.*

Because two electrons are required to form a σ-bond, both electrons of the π-bond are used to form the C–H bond, making the other carbon atom of the C=C unit electron deficient, with a formal charge of +1 (see Chapter 5, Section 5.5). This reaction therefore generates a carbocation as the product (**55**). Note that carbocations are highly reactive species that are not isolated but react with another species to give a more stable product. This chemistry is discussed in Chapter 10 (see Sections 10.1–10.3).

6.29 Draw the charged product initially formed when cyclopentene reacts with HCl.

What is the relative strength of an alkene as a base? In other words, what acids will react with an alkene? If 1-pentene (**57**) is used as a typical example, experiments show that this alkene does not react with water under neutral con-ditions, so water ($pK_a = 15.8$) is not a strong enough acid. Ethanol ($pK_a = 17$) does not react with 1-pentene under neutral conditions. The acid HCN ($pK_a = 9.31$) is stronger than an alcohol or water, but experiments have shown that simple alkenes give no reaction with HCN. Acetic acid (pK_a = 4.76) is a significantly stronger acid, but simple alkenes such as 1-pentene either do not react at ambient temperatures or at best react very slowly. These experiments establish that alkenes are weak bases in the presence of water, alcohols, HCN, and carboxylic acids.

A strong mineral acid such as HCl, with a pK_a of –7, reacts rapidly with 1-pentene to give an excellent yield of **59** as the isolated product, however, with small amounts of **60**. These alkyl halides are not the initial product of the acid–base reaction. It has been proven that formation of **59** proceeds by initial formation of the positively charged and highly unstable carbocation **58**, followed by reaction with chloride ion (see Chapter 10, Section 10.2, for a discussion of carbocation intermediates). This reaction of alkenes and mineral acids will be discussed in great detail in Chapter 10. From the standpoint of this chapter, this discussion supports the hypothesis that alkenes react as Brønsted–Lowry bases to generate positively charged species called carbocations, but only with strong acids, typically with pK_a values of 1 or less.

alkenes react with Lewis acids, but in general those reactions are not as useful. For that reason, the reactions of an alkene with BF_3, $AlCl_3$, or other Lewis acids will be ignored. Alkynes react with Brønsted–Lowry acids and also Lewis acids to form carbocations, but the resulting carbocation is attached to an sp^2 hybridized carbon (a so-called vinyl carbocation). A proper discussion requires more than a cursory treatment and vinyl carbocations will be shown in Chapter 10 (Section 10.6).

6.5 Lewis Acids and Lewis Bases

In the cases discussed so far in this chapter, the main focus of the discussion is on a proton (H^+) or a "proton precursor" ($H^{\delta+}$) as the electron-deficient partner in the reaction. A base is electron rich and can donate electrons to any suitable electron-deficient center, not just a proton. When a base donates electrons to a proton, it is called a Brønsted–Lowry base, but when it donates those electrons to another electron-deficient atom, it is called a Lewis base. *The formal definition of a Lewis base is that it is an electron pair donor. The atom that accepts electrons from the Lewis base is called a Lewis acid, and its definition is that it is an electron pair acceptor.* The reaction between Lewis acids and Lewis bases is common in organic chemistry, and many of the reactions to be discussed fall into this category. For that reason, this section will discuss Lewis acid–Lewis base reactions in more detail, although some of the discussion will be a review of concepts presented earlier.

6.5.1 Why Are Brønsted–Lowry and Lewis Acids and Bases Separated?

The answer to the question posed in the section title was introduced in Chapter 2 (Section 2.3). There are two working definitions of acids and bases. The Brønsted–Lowry definition focuses on protons because the acid is H^+. A Brønsted–Lowry acid is a proton donor and a Brønsted–Lowry base is a proton acceptor. A more general definition is the Lewis acid, which is an electron pair acceptor, whereas a Lewis base is an electron pair donor. If the base donates electrons to a proton, it is called a Brønsted–Lowry base, but it is still an electron donor. In water, transfer of a proton is related to pH, so the Brønsted–Lowry definition is quite useful. In organic chemistry, the focus must be on the electron transfer process and making or breaking bonds. This latter focus is more consistent with the Lewis definitions.

There is one important distinction. When a base such as $^-NH_2$ donates electrons to the electron-deficient proton of HCl, for example, a conjugate acid is formed: $H–NH_2$, with the new covalent N–H bond. This reaction breaks the H–Cl bond, generating Cl^-. There are *two products* because the acid–base reaction breaks the bond between the proton and the other atom and forms the conjugate acid. When a Lewis base such as ammonia reacts with a Lewis acid such as BF_3, however, a dative bond is formed to give the Lewis acid–Lewis base complexes (an ate complex), so there is one product rather than two. This statement is an overgeneralization, but it offers a useful starting point to allow the two definitions to be distinguished from one another. These points will be emphasized in this section.

6.5.2 Lewis Acids

18   **61**

When a Brønsted–Lowry acid such as acetic acid (**18**) is compared with a classical Lewis acid such as boron trifluoride (BF_3, **61**), it is important to look for differences as well as similarities. In acetic acid, the atom that accepts electron density from the base is hydrogen (marked in **red**), so reaction with a base will generate two products: the acetate anion and the conjugate acid. In boron trifluoride, boron is the electron-deficient atom (also marked in **red**) and reaction with a Lewis base will give one product: the ate complex. In general, any atom other than hydrogen or carbon that accepts an electron pair and forms a new bond may be categorized as a Lewis acid, but if a product other than an ate complex is formed, one should be careful about the categorization.

This section will focus on Lewis acids. The relative strength of the Lewis acid depends on the nature of the atom and the groups attached to that atom, as introduced in Chapter 2 (Section 2.5). Group 13 atoms such as boron and aluminum have three valence electrons, so only three covalent bonds can be formed for the resulting molecule to remain neutral. Lewis acids such as boron or aluminum complete the octet around those atoms by reaction with a Lewis base that can donate two electrons. Therefore, BF_3 is an acid and so are BBr_3, BCl_3, and $B(CH_3)_3$. In Section 6.4.1, secondary amine **40** and tertiary amine **42** reacted to give ate complexes **41** or **43**. When trimethylamine (**31**) reacts with boron trifluoride, the ate complex is **62A** (also drawn as **62B** with the dative bond). This reaction is a classical example of a Lewis acid–Lewis base reaction. Differences among fluorine, chlorine, and methyl will have an effect on the strength of the Lewis acid, on the ability to form the ate complex, and on the relative stability of that complex. Aluminum compounds such as $AlCl_3$ behave similarly, and transition metal compounds such as $FeCl_3$, $SnCl_4$, or $TiCl_4$ are generally categorized as Lewis acids.

6.30 Draw the product expected to form when BH_3 reacts (a) with ammonia, and (b) with diethyl ether.

In general, the inductive effects that played a role in Brønsted–Lowry acids also influence the relative acidity of Lewis acids. Comparing BF_3 and $B(CH_3)_3$, for example, fluorine is electron withdrawing whereas methyl is electron releasing. The δ+ boron atom is more electron deficient when a fluorine is attached than when a carbon atom is attached, making BF_3 a stronger acid. This is not the only consideration, however, especially when transition metal compounds are involved. Transition metals have d- and f-orbitals, which have a great influence on reactivity. In addition, multiple valences are possible for most transition metals, leading to further complications of Lewis acid strength.

The focus here is on two major factors that determine the strength of a Lewis acid where all Lewis acids have an atom that can accept one or more electron pairs. *(1) In general, an atom that can accept only one electron pair (such as B) is more acidic than an atom that can accept more than one electron pair (such as Fe or Sn). (2) If two Lewis acids are in the same group of the periodic table (say, group 13), then acidity decreases as the size of the atom increases (increasing atomic volume down the periodic table).* This means that BCl_3 is a stronger acid than $AlCl_3$, which is stronger than $GaCl_3$. This order is also linked to better availability of d-orbitals in the heavier atoms. Compounds derived from group 13 atoms (B, Al, Ga, In) are strong and well-known Lewis acids. As seen in the preceding paragraph, the order of acidity is $BX_3 > AlX_3 > GaX_3 > InX_3$. In most of the reactions in this book that require a Lewis acid, a boron or aluminum compound will be used.

Many transition metal salts are Lewis acids. In general, they are weaker Lewis acids than group 13 acids such as BX_3 and AlX_3. Lewis acids derived from compounds in groups 11 (Cu, Ag, Au) and 12 (Zn, Cd, Hg) usually form MX_2 species that are moderately strong. Using the atomic volume rule, relative acid strength (from strongest to weakest) is $CuX_2 > AgX_2 > AuX_2$ and $ZnX_2 > CdX_2 > HgX_2$. In general, "X" is a halogen. Group 3 atoms (Sc, Y, La, Ac) are relatively strong Lewis acids that form MX_3 compounds. Using the atomic volume rule, $ScX_3 > YX_3 > LaX_3$, allows predictions of acid strength in group 3.

Group 13 compounds are generally more acidic than group 3 compounds. Group 14 (C, Si) forms MX_4 compounds (like methane, CH_4) that have no Lewis acid properties. Group 4 compounds (Ti, Zr, Hf) also form MX_4 compounds and they are weak Lewis acids. Atoms in group 15 (N, P) are usually Lewis bases, but MX_5 and MX_3 compounds derived from the group 15 element Sb are strong Lewis acids, with SbX_5 stronger than SbX_3 (expanded octet in MX_5 and d-orbitals are used). In group 5 (V, Nb, Ta), MX_5 compounds are Lewis acids, and NbX_5 compounds are very strong Lewis acids. The Lewis acids in group 6 (Cr, Mo, W) and group 7 (Mn, Tc, Re) tend to be relatively weak.

6.31 Predict whether Et_3B is a stronger or weaker Lewis acid than Et_3Al, where Et = ethyl.

6.32 Predict whether $CuBr_2$ is a stronger or weaker Lewis acid than $CdBr_2$ and whether $NbCl_5$ is stronger or weaker than $TiBr_4$.

6.5.3 Lewis Bases

Formally, a Lewis base is any compound that has an atom capable of donating electrons to an electron-deficient center, other than a proton. The most common Lewis bases in organic chemistry are amines such as NMe_3 (N is the donating atom; Me = methyl) and phosphines such as PMe_3 (P is the donating atom), which are rather good Lewis bases. Many oxygen compounds behave as weak Lewis bases, including water, alcohols, ethers, ketones, aldehydes, and even carboxylic acid derivatives in some cases. Some of these reactions have been introduced in previous sections.

In general, basicity increases going *up* the periodic table and to the *left*. This means that an amine (R_3N) is a stronger Lewis base than a phosphine (R_3P) and an ether (ROR) is a stronger Lewis base than the sulfur analog of an ether (a sulfide, RSR). Similarly, an amine (R_3N) is a stronger Lewis base than an ether (ROR). The usual inductive and steric effects noted in previous sections determine the relative strength of base when comparing several amines or several ethers, but the simple periodic table rules just stated help determine relative basicity of different types of compounds.

6.33 Predict whether diethyl ether (Et–O–Et) or dimethyl sulfide (Me–S–Me) is the stronger Lewis base.

Traditional bases such as hydroxide ($^-$OH), or amide ($^-$NH$_2$) are classified as Lewis bases if they donate electrons to an electron-deficient center other than H$^+$. The extent of reaction of a Lewis base with a Lewis acid is usually determined by the stability of the product formed for the reaction, the Lewis acid–Lewis base complex. The nature of the ate complex will be discussed in the next section and will help determine which of the Lewis bases mentioned here will be most useful for a given reaction.

6.5.4 The Lewis Acid–Lewis Base Complex

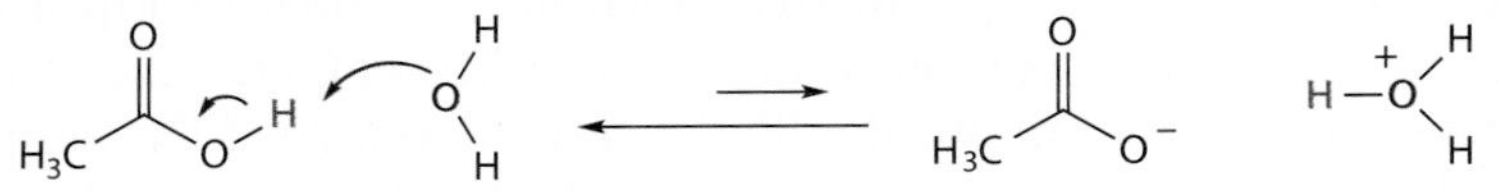

When a Brønsted–Lowry acid such as acetic acid (**18**) reacts with a base such as water, the products are the hydronium ion (H$_3$O$^+$; **3**, the conjugate acid) and the acetate anion (**24**, the conjugate base). The acetate anion (**24**) is resonance stabilized, just at the formate anion (**9**) is resonance stabilized, as discussed in Section 6.2.4. The stability of the conjugate base is an important consideration in the overall acidity of carboxylic acids such as **18**. When a Lewis base such as trimethylamine reacts with an acid that does not have a proton, such as boron trifluoride (BF$_3$) or triethylborane (Et$_3$B), a dative bond is formed between the atom donating the electrons and the atom accepting the electrons, as shown for the reaction of **31** with **61** to give **62**. The ate complex product can also be described as a **zwitterion** (a dipolar ion), a molecule that contains both a positive and a negative charge.

The reaction shown for trimethylamine and BF$_3$ is a general reaction in that the ate complex is the end-product of the reaction of a Lewis acid and a Lewis base. As with other chemical reactions, the ate-complex product can be long lived or transient due to instability or proximity to another reactive species. The two electrons donated by nitrogen (in **blue** in **31**) and used to form the bond between B and N (**blue** in **62A**) are represented by an arrow (see **62B**). The arrow indicates a charge-transfer complex, as noted several times, where the arrow in **62B** formally represents a dative bond.

6.34 Draw out the resonance contributions to the acetate anion, 24.

6.35 Draw the product when diethylamine reacts with triethylborane (BEt$_3$), using the dative bond formalism.

Lewis acid–Lewis base reactions may be reversible, if the ate complex is not very stable. Solvents play an important role in the relative stability of the ate complex. If the dative bond is relatively weak, a polar protic solvent such as water can separate the charges. This would regenerate the original Lewis acid and Lewis base, making the reaction reversible with the equilibrium lying to the left. For this reason, Lewis acid–Lewis base reactions are usually done in nonaqueous solvents such as diethyl ether. Note that water can function as a Lewis base and that this may interfere with the planned reaction.

6.6 A Positive Carbon Atom Can Accept Electrons

It is clear that atoms other than hydrogen can be electron deficient and function as electron pair acceptors. Can a carbon atom function as a Lewis acid? The answer is yes, if the definition is modified somewhat. Various reactions generate carbocation intermediates (see **55** and **58**) and a Lewis base can certainly donate electrons to that positive carbon. A species that donates electrons to carbon is called a **nucleophile** (see Section 6.7), so an electron donor that reacts with **55** or with **58** is a nucleophile. In addition to carbocations, which are charged species, the carbon atom in a polarized bond is electron deficient, and a nucleophile could donate electrons to the δ+ carbon. This is the basis of many organic reactions to be discussed, particularly in Chapter 11. The fundamental concept of a species donating electrons to a δ+ carbon is introduced in this section, with the goal of relating this chemical reactivity to the Lewis acid–Lewis base definitions used in previous sections.

In this chapter, a molecule with an electron-deficient hydrogen atom is a Brønsted–Lowry acid and a molecule bearing a nonhydrogen electron-deficient center is a Lewis acid. Organic chemistry deals with the chemistry of carbon. *When the electron-deficient atom is carbon, it may be categorized as a Lewis acid just as any other electron-deficient atom, but the product of a reaction is not an ate complex.* A carbon that accepts two electrons is defined as ***electrophilic,*** which formally means "electron loving" and that carbon is considered to be an ***electrophile*** in a reaction with a nucleophile. Anything that "loves" electrons must be electron deficient and will have electrons donated *to* it. A wide range of compounds have electrophilic carbon atoms, and identifying such atoms is very useful in predicting chemical reactions.

One very important electrophilic center is the positive carbon that resides in a carbocation (see **63**). This highly reactive species (see Chapter 10, Section 10.2) reacts with an atom that donates two electrons (:B) to form a new bond to that carbon (**64**). In this reaction, :B is a nucleophile, and the **red** carbon atom in **63** is an electrophile. Methods that lead to formation of carbocations are discussed in Chapters 10 and 11. *Note that product 64 is not an ate complex, but rather a neutral covalent compound.* Formally, **63** is not a Lewis acid because it does not give the usual Lewis acid–Lewis base product. However, **63** has an

than water (HOH), an alkoxide (RO⁻) is a stronger nucleophile than an alcohol (ROH), and an amide anion (R_2N^-) is a stronger nucleophile than an amine (R_2NH). The general trend is for nucleophilic strength to increase going *down* the periodic table and to the *left*. This means that iodide is a stronger nucleophile than chloride, and the phosphorous in trimethylphosphine (**67**) is a stronger nucleophile than the nitrogen in trimethylamine (**31**). It also means that R_2N^- is a stronger nucleophile than RO⁻.

6.38 **Decide whether MeO⁻ or HO⁻ is the stronger nucleophile and whether acetate or ethoxide is stronger.**

The relative strength of a nucleophile also depends on the nature of the δ+ center (the electrophilic center) with which it reacts. If Me_2N^- is the nucleophile, it will have a different nucleophilic strength with a carbonyl compound such as acetone than it will with an alkyl halide such as iodomethane. This is exactly analogous to acids and bases because acetic acid is a much stronger acid in the presence of a strong base like hydroxide than in the presence of a weak base like water. The role of changing solvent will be discussed in the context of specific reactions, and the reaction of a nucleophile with the carbonyl of an aldehyde or ketone will be discussed in Chapter 18. The role of steric hindrance around the electron-donating atom will also be discussed in the context of specific reactions, but the influence on reactivity is similar to the discussion of basicity of primary, secondary, and tertiary amines in Section 6.4.1. Both of these properties influence nucleophilic strength, but a broader discussion is required than is possible at this point.

6.8 Biological Relevance

Heavy-metal poisoning is a serious problem. Arsenic or thallium (used in rat poisons), mercury (from some fish and some industrial products), or lead (lead-based paints) can cause severe illness or death if ingested. One method used to diminish the amount of metal in the human body is chelation therapy, which takes advantage of a Lewis acid–Lewis base reaction by utilizing compounds that form ate complexes with the metal via electron-donating groups such as nitrogen or sulfur. The goal is to form a complex that can be excreted, thus removing or diminishing the amount of metal in the body. Note that chelating agents may be toxic and there are many adverse effects. The most common chelating agents are dimercaprol (dimercapto succinate, DMSA) and 2,3-dimercapto-1-propanol (Succimer, Chemet), which is also approved for pediatric lead poisoning and for mercury poisoning.

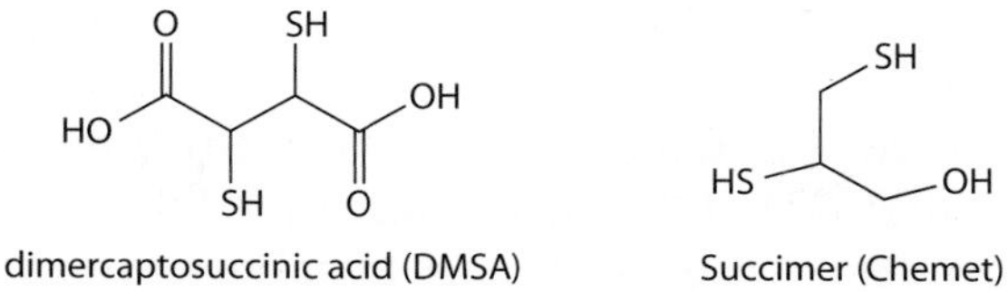

dimercaptosuccinic acid (DMSA)

Succimer (Chemet)

The mechanism of arsenic poisoning is illustrated by the following cartoon. Arsenic forms a coordination complex with thiol units of a protein. Thiols (R–SH) are the sulfur analogs of alcohols, and much of the chemistry is similar. When DMSA is administered, the therapy relies on DMSA forming a better coordination (ate) complex with the metal. Formation of the new complex will clear the metal from the protein, and the DMSA complex should be soluble in water, thus allowing the body to excrete the complex in the urine. The Lewis base–Lewis acid properties of the metal (Lewis acid) and the chelating agent (DMSA, the Lewis base) are the basis of this therapy.

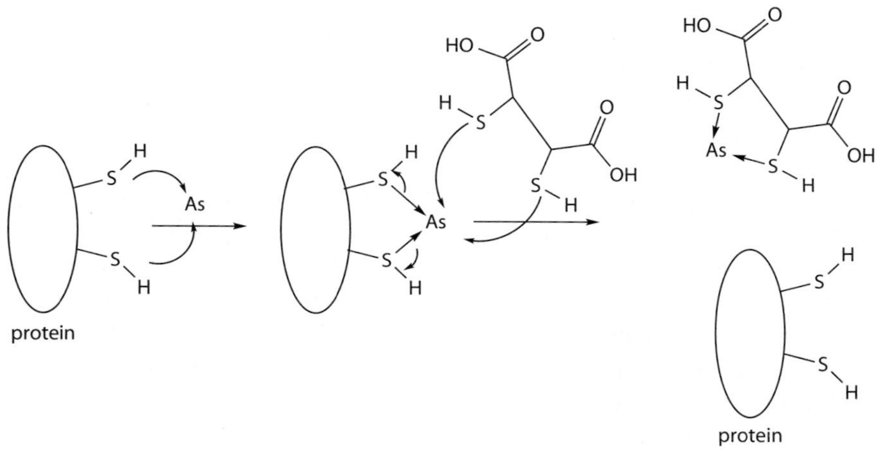

protein

protein

Amino acids will be discussed in Chapter 27, but there are acidic amino acids that play an important role in certain biological functions. There are also neutral amino acids, including two that have an amide unit as part of the side chain. Two acidic amino acids are **aspartic acid** (**68**, abbreviated Asp) and **glutamic acid** (**69**, abbreviated Glu).

One example that illustrates the importance of amino acids involves insulin-like growth factor binding protein (IGFBP)-5 (a secreted protein that binds to IGFs and modulates IGF actions) interacting with the nuclear histone–DNA complex. Mutations were made that changed acidic amino acid residues to neutral residues or a polar to a basic residue and it was found that transactivation activity was greatly diminished. **Transactivation** is an increased rate of gene expression triggered either by endogenous cellular or by viral proteins called **transactivators**. Transactivation is also a technique used in molecular biology to control gene expression by stimulating transcription. A study by Zhao et al.[5] showed that IGFBP-5 is present in a complex containing histone and DNA in the nucleus and has a functional transactivation domain in its *N*-terminal region.

Several amino acids (known as residues) in this region, including Glu8, Asp11, Glu12, Glu30/Pro31, Glu43, Glu52, and Gln56, are critical for its transactivation activity. Note that the numbers after each amino acid abbreviation indicate the position of that residue in the protein. Gln is the abbreviation for glutamine (**70**) and Pro is the abbreviation for proline (**71**). The amino acid abbreviations are presented in Chapter 27, Section 27.3.2 (see Table 27.1). Presumably, the acidic nature of the carboxylic acid side chains plays a key role because transactivation domains are often rich in acidic amino acids.[6]

Important chemical reactions mediated by enzymes, which are proteins, are often controlled by the interaction of a nucleophile with an electrophile. Proteins are polymeric structures made of amino acids units (called amino acid residues). Important nucleophilic groups include the hydroxyl (OH) group of the amino acid serine (**72**), the thiol unit (SH) of cysteine (**73**; see Chapter 27) and the nitrogen of the imidazole group of histidine (**74**). Note that the term nucleophilic in this biochemical context does not necessarily indicate reaction at carbon, but rather indicates a two-electron donor. The electron-deficient center (called an electrophilic site or electrophilic group) may be a metal such as Mg^{+2}, Mn^{+2}, Fe^{+3}, or an ammonium unit ($-NH_3^+$). In some cases, the electron-deficient center is a carbonyl group (C=O) in which the electron donor is indeed a nucleophile because the reaction occurs at the carbonyl carbon.

Acid–base reactions are ubiquitous in biological transformations. One example is the hydrolysis of a cholesterol ester (**76**) (see Chapter 20, Sections 20.2 and 20.5) with an enzyme known as *cholesterol esterase* (**75**).[7] At the active site, the enzyme contains a carboxylate anion residue, an imidazole residue from the amino acid histidine (see Chapter 27, Section 27.3.2, and Chapter 26, Section 26.1.1), and an alcohol residue. The reaction with **76** is an acyl addition substitution of the alcohol unit with the ester unit to give **77**. Acyl substitution is discussed in Chapter 20, Section 20.2, and in Chapter 16, Section 16.8.

This chapter discusses acid–base reactions, however. The reaction of **75** with **76** is initiated by an acid–base reaction of the carboxylate unit with the imidazole to generate a carboxylic acid unit in **77**. The alkoxide unit in **77** regenerates the carbonyl unit to complete the acyl substitution reaction to give cholesterol (**78**) and the ester unit in **79**. Note that the alcohol unit in **76** is converted to an ester in **79** and the ester unit in **76** is converted to an alcohol in **78**. This is called a transesterification reaction (Chapter 20, Section 20.5.3).

Another acid–base reaction of **79** with water, which is in the cellular medium, generates **80**, which regenerates **75** by loss of a carboxylate unit, RCO_2^-. Note that both **77** and **80** are known as tetrahedral intermediates, which will be important in Chapter 20. While the focus of the reaction is conversion of a cholesterol ester to cholesterol and regeneration of the carboxylate unit derived from **76**, the reaction is driven by acid–base reactions that occur on the enzyme, cholesterol esterase. Cholesterol is a part of mammalian cell membranes and it is very important for membrane permeability and fluidity. Cholesterol is an important precursor for the biosynthesis of bile acids and some fat-soluble vitamins.

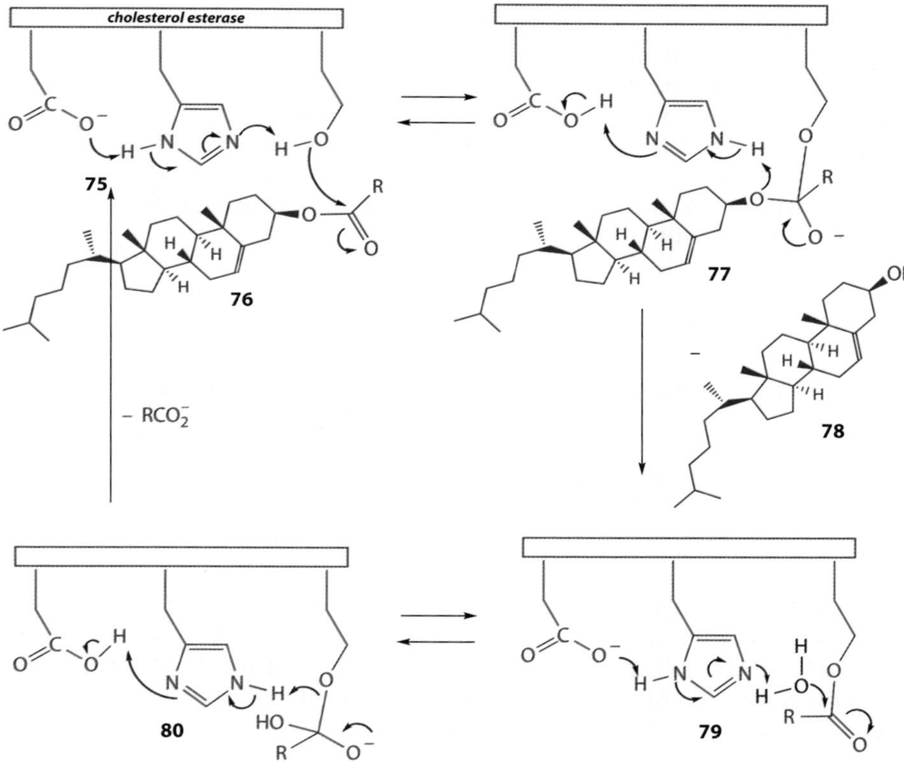

(Reprinted in part with permission from Sutton, L. D. et al. 1991. *Biochemistry* 30:5888. Copyright 1991 American Chemical Society.)

References

1. Stewart, R. 1985. *The proton: Applications to organic chemistry,* 17. Orlando, FL: Academic Press.
2. Lide, D. R., ed. 2006. *CRC handbook of chemistry and physics,* 87th ed., 8-43–8-52. Boca Raton, FL: Taylor & Francis.

3. Stewart, R. 1985. *The proton: Applications to organic chemistry,* 102. Orlando, FL: Academic Press.
4. Dean, J. A. 1987. *Handbook of organic chemistry,* 8-2. New York: McGraw-Hill.
5. Zhao, Y., Yin, P., Bach, L. A., and Duan, C. 2006. *Journal of Biological Chemistry* 281:1418–1419.
6. Ma, J., and Ptashne, M. 1987. *Cell* 51:113–119.
7. Sutton, L. D., Froelich, S., Hendrickson, H. S., and Quinn, D. M. 1991. *Biochemistry* 30:5888, the American Chemical Society.

Answers to Problems

6.1   Because $pK_a = -\log K_a$, then $pK_a = -\log 6.34 \times 10^{-8} = -(-7.2) = 7.2$. Because $K_a = 10^{-pKa}$, then $K_a = 10^{-11.78} = 1.66 \times 10^{-12}$.

6.2   The OH bond in water is more polarized than the NH bond of ammonia because oxygen is more electronegative than nitrogen. This means that the H in water will have a greater $\delta+$ dipole than H in ammonia and is more reactive with a base. Reaction with water generates hydroxide (HO^-) and reaction with ammonia generates (H_2N^-). The fact that oxygen is more electronegative than nitrogen suggests that having the charge on oxygen in hydroxide will be more stable than the charge on nitrogen in H_2N^-. These two factors suggest that water should be more acidic than ammonia.

6.3

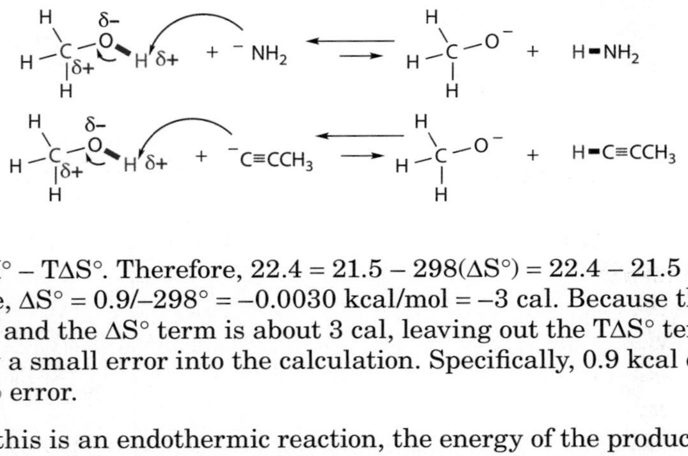

6.4 $\Delta G^\circ = \Delta H^\circ - T\Delta S^\circ$. Therefore, $22.4 = 21.5 - 298(\Delta S^\circ) = 22.4 - 21.5 = -298^\circ\Delta S^\circ$ Therefore, $\Delta S^\circ = 0.9/-298^\circ = -0.0030$ kcal/mol $= -3$ cal. Because the ΔG° term is 22.4 kcal and the ΔS° term is about 3 cal, leaving out the $T\Delta S^\circ$ term will introduce only a small error into the calculation. Specifically, 0.9 kcal out of 22.4 kcal $= 4\%$ error.

6.5   Because this is an endothermic reaction, the energy of the products must, by definition, be higher in energy than the energy of the starting materials.

6.6   The reaction with the lower E_{act} (35 kcal) will occur at a lower temperature. The value of ΔH° does not play a role in initiating the reaction, which is measured by E_{act}.

6.7

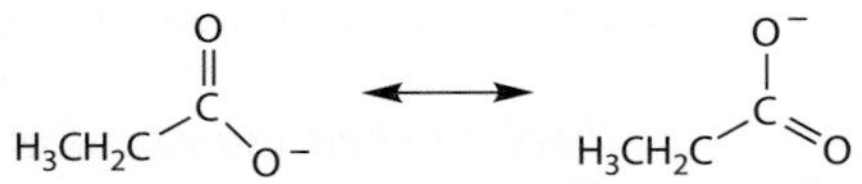

6.8   This question states that HCl is more acidic. If we look at the two reactions,
      we find that the hydrogen sulfate anion is resonance stabilized and should
      be more stable than the chloride ion. The HCl bond is more polarized than
      the OH bond, however, and the chloride ion can be solvated more efficiently
      by the solvent (water). For these two reasons, HCl in water tends to be the
      stronger acid.

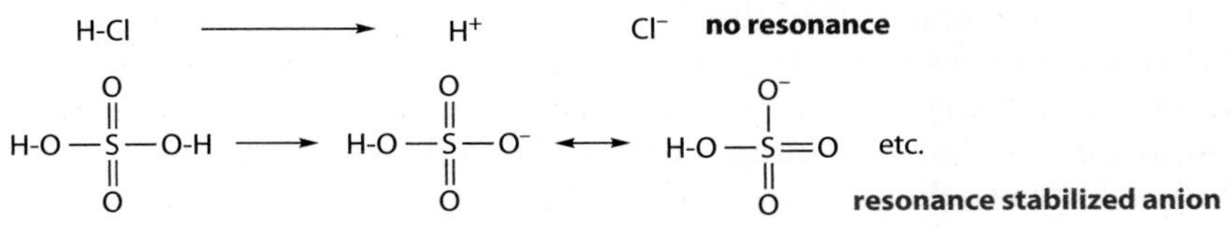

6.9

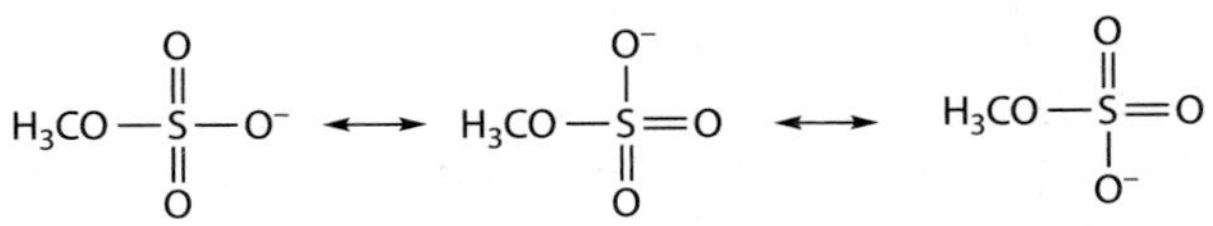

6.10 In general, sulfonic acids are more acidic than carboxylic acids. The S–O–H
      unit is more polarized than the C–O–H unit, and the sulfonate anion product
      has more resonance stability than the carboxylate anion. Therefore, butanesul-
      fonic acid is the more acidic.

6.11

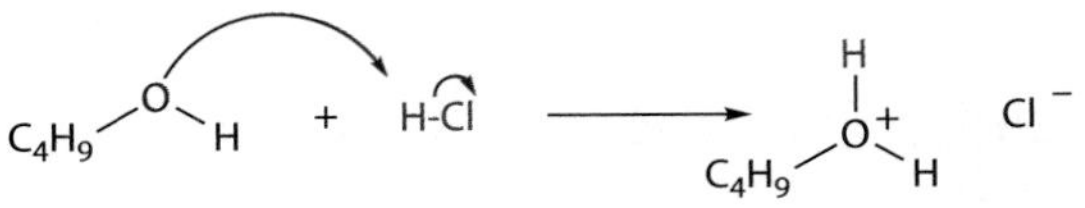

6.12

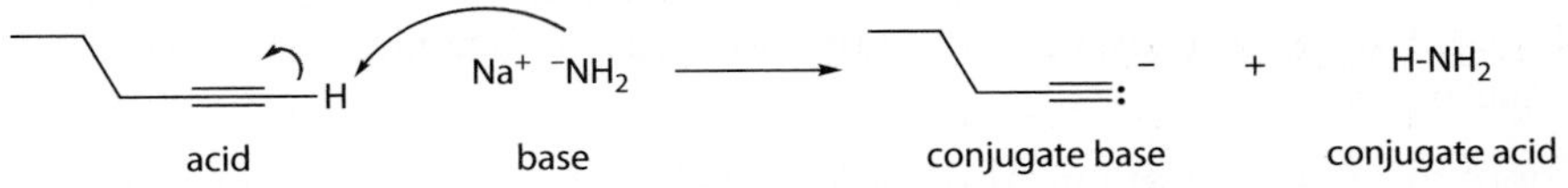

6.13 The pK_a of nitromethane is known to be about 10, whereas that of methanol
      is about 16. The two conjugate bases are shown. The carbanion formed from
      nitromethane is a resonance stabilized anion, where the charge is delocalized
      on both oxygen atoms. When the proton is removed from oxygen in methanol,
      the charge is localized on oxygen. However, notice that in one resonance con-
      tributor of the nitromethane anion, the negative charge is localized on carbon,
      whereas it is on oxygen in methoxide (MeO−). Oxygen is more electron nega-
      tive and clearly able to accommodate the charge better than carbon, even with
      delocalization. The carbanion is less stable than the oxygen anion, so it is
      better able to donate its electrons; it is a stronger base. Because a strong acid
      gives a weak conjugate base and vice versa, we expect that the stronger base
      ($^-CH_2NO_2$) will be derived from the stronger acid, nitromethane.

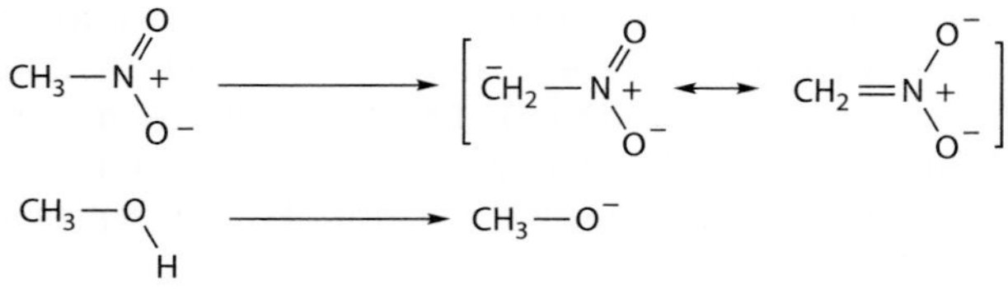

It is important to note that this is a trick question at this stage because it involves subtleties in acidity that we have not fully examined. It is included to show that examining only one feature, such as stability of the conjugate base, and ignoring other factors can be misleading. This will be fully explored in later chapters.

6.14 2-Nitroethanoic acid is stronger than ethanoic acid. The presence of the electron-withdrawing nitro group makes that molecule more acidic. Because the nitro group is close to the carboxyl unit, there is maximum effect for the electron-withdrawing effects.

6.15

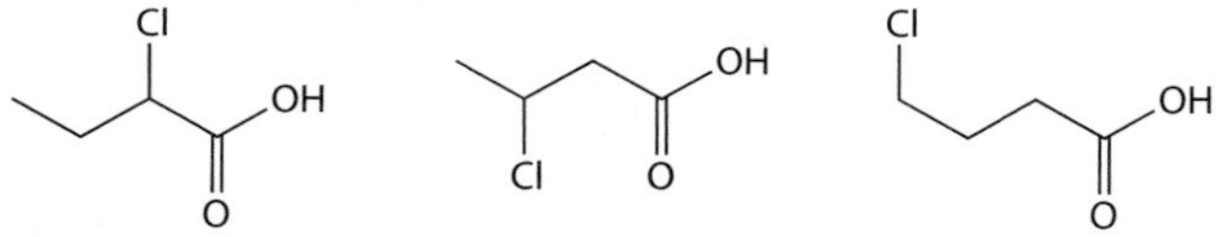

6.16 No, the names given are the common names. The IUPAC names are ethanoic acid for acetic acid, propanoic acid for methylacetic acid, 2-methylpropanoic acid for dimethylacetic acid, and 2,2-dimethylpropanoic acid for trimethylacetic acid.

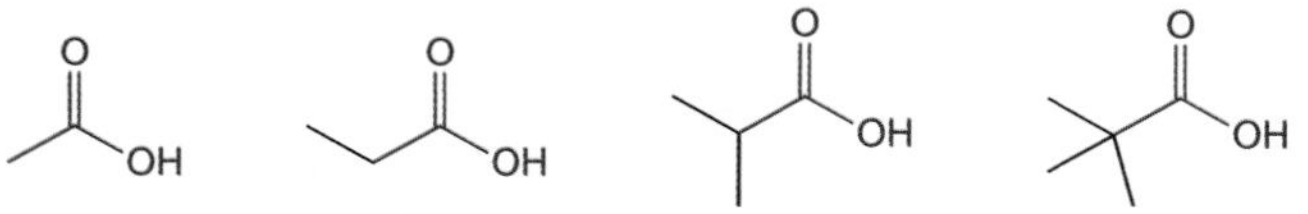

6.17 2-Methoxybutanoic acid is a stronger acid because the electron-withdrawing OMe unit is closer to the carboxyl unit. The through-space effect can occur via a five-membered cycle as shown by structure **26**. In 4-methoxybutanoic acid, the electron-withdrawing unit is too far away to provide a significant effect (see structure **32**).

6.18 There are two fundamental reasons. First, water can hydrogen bond more effectively with the acidic proton in acetic acid, making it easier to remove. Second, water is much better at solvating anions, making it easier to form the carboxylate anion product.

6.19 Because pK_{BH} = –log K_{BH}, then pK_{BH} = –log 3.92×10^6 = –(6.59) = –6.59.

6.20 2-(*N,N*-di-*tert*-butylamino)butane is a weaker base than 2-(*N*-methylamino)butane. The tertiary amine is more sterically hindered, making it more difficult for the nitrogen to approach a potential acidic site. In addition, a tertiary amine reacts with an acid to form an ammonium salt such as R_3NH^+, whereas the secondary amine forms a salt such as $R_2NH_2^+$. The salt derived from the secondary amine is better solvated because there are two N–H units, making formation of the salt easier (it is a better base).

6.21

6.22 This reaction occurs to produce a charged species, the ammonium bromide shown. Water is known to solvate and separate charged species, but ether cannot do this. Separation of charge in this reaction (+ and –) separates the ammonium unit from the bromide, which helps in the formation of the product, pushing the reaction to the right. In ether, charge separation is inhibited, which makes it more difficult to form the charged products. Therefore, we expect that this reaction is faster in water than in ether.

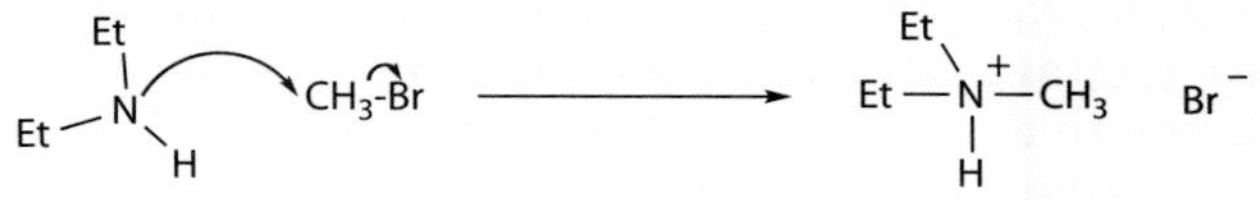

6.23 Using very simple inductive arguments, the conjugate acid from triethylamine has three electron-releasing ethyl groups that will diminish the net charge on nitrogen and stabilize it. Because the conjugate acid derived from diethylamine has only two electron-releasing ethyl groups, it is expected to be stabilized to a lesser degree.

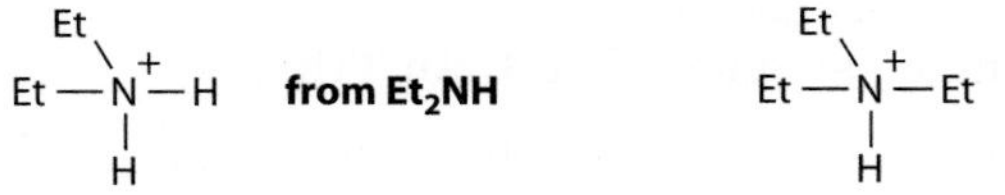

6.24

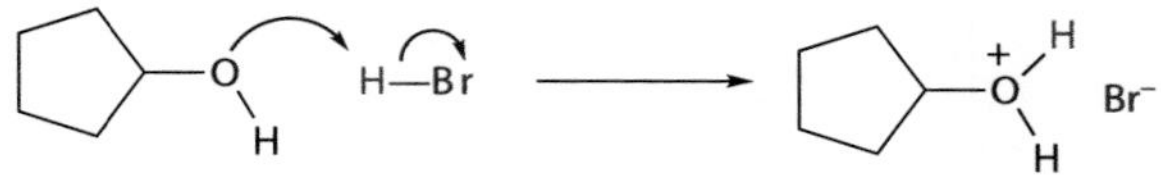

6.25

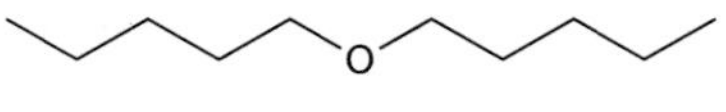

dipentyl ether

6.26

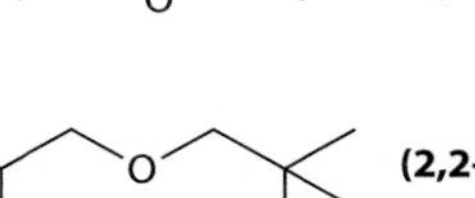

(butyl)(1-methylethyl) ether

6.27

(2,2-dimethylpropyl)(2-methylbutyl) ether

6.28

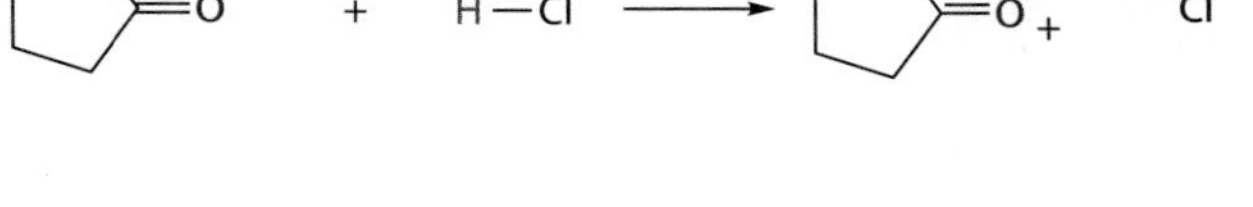

6.29

6.30

6.31 As the atomic volume of an atom increases (effectively, the atomic number increases), there is weaker attraction between the nuclear charge and the incoming electron pairs and the acidity decreases. Because the atomic number of Al > B, AlEt$_3$ is expected to be a weaker acid. In addition as the atomic number increases, orbital contractions lead to different atomic orbitals being closer together, leading to more effective overlap of hybridized orbitals and, with Al, d-orbitals are available. All of these things lead to a decrease in acid strength: AlEt$_3$ is a weaker acid than BEt$_3$.

6.32 More electron density is required to complete the octet in Cd than in Cu, making CuBr$_2$ more acidic, and the atomic volume of Cd is greater than that of Cu (see problem 6.21). Titanium, in group 14, tends to form more covalent species and tends not to be a strong acid. The group 15 element Nb forms the strong Lewis acid NbCl$_5$ because of an expanded octet and the use of d-orbitals. Therefore, NbCl$_5$ is a much stronger acid than TiBr$_4$.

6.33 Diethyl ether is the stronger base. The sulfur atom is larger, making the net charge per unit centimeter for S less than for O. This means that less electron density is available for donation: a weaker base.

6.34

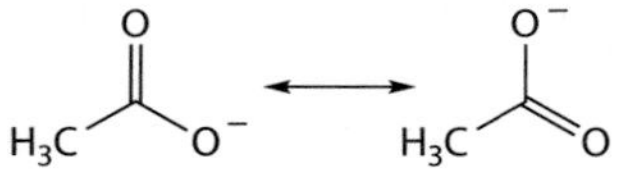

6.35

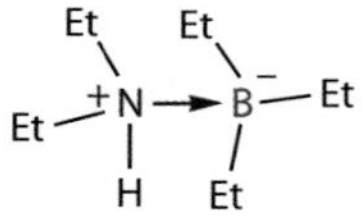

6.36

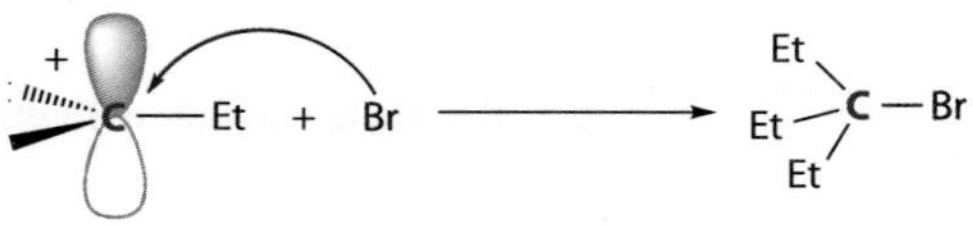

6.37

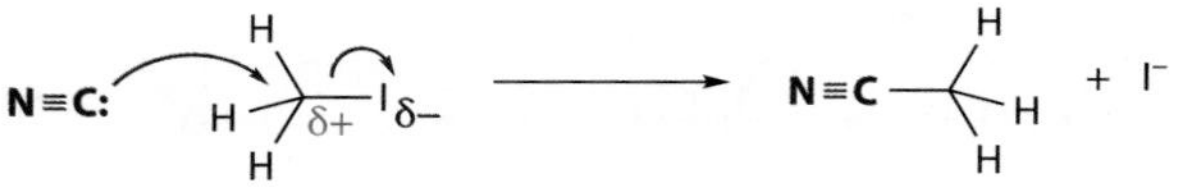

6.38 The answer really depends on what each nucleophile is reacting with. However, because the electron-releasing methyl group "pushes" electron density toward oxygen, we expect that more electron density on MeO⁻ will make it more reactive (more nucleophilic) than hydroxide (HO⁻). Acetate is a resonance stabilized anion, so there is less electron density available for donation (weaker nucleophile). In ethoxide, the electron density is effectively concentrated on oxygen and more available for donation: a stronger nucleophile.

Correlation of Concepts with Homework

- **Common organic acids are carboxylic acids, sulfonic acids, and alcohols. Carboxylic acids are more acidic than alcohols, but sulfonic acids are more acidic than carboxylic acids: 10, 42, 68.**
- **The equilibrium constant for an acid–base reaction that involves organic acids is the same as any other acid–base reaction, K_a, where K_a = products/reactants, and pK_a = –log K_a: 1, 39, 40.**
- **The relative strength of an acid depends on the strength of the base with which it reacts: 3, 11, 41, 51, 64.**
- **The free energy of reaction is indicative of the spontaneity of that reaction (endothermic or exothermic), and the activation energy is a measure of the transition state energy. $\Delta G° = \Delta H° - T\Delta S$: 4, 5, 6, 65, 66, 67.**
- **Stability of a conjugate base is linked to its reactivity, and greater stability can result from charge dispersal due to increased size of that species or resonance stability. In both cases, K_a is larger: 7, 8, 9, 13, 34, 45, 47, 55, 56.**
- **Inductive effects are very important for determining the variations in pK_a with structural changes in carboxylic acids. As electron-withdrawing substituents are positioned further away from the carboxyl group, the effect diminishes. An electron-withdrawing group attached to oxygen in an OH unit makes the hydrogen more acidic, and an electron-releasing group attached to oxygen in an OH unit makes the hydrogen less acidic. Terminal alkynes are weak acids: 2, 12, 14, 15, 16, 17, 21, 43, 44, 48, 49, 52, 68.**
- **The solvent can play a significant role in determining K_a for an acid, particularly when comparing polar ionizing solvents with nonpolar solvents: 18, 50, 59, 63, 64.**
- **Compounds that behave as an acid in the presence of a strong base and a base in the presence of a strong acid are called amphoteric compounds: 24, 46.**
- **Alcohols, ethers, aldehydes and ketones, and alkenes are Brønsted–Lowry and Lewis bases: 25, 26, 27, 28, 29, 33, 34, 57, 58.**

- **Common organic bases include amines, phosphines, and alkoxide anions. The basicity of an amine is measured using pK$_{BH}$:** 19, 53, 54.
- **The basicity of amines is influenced by the electron-releasing effects of alkyl groups attached to nitrogen as well as by the steric hindrance imposed by those alkyl groups to anything approaching the nitrogen. An electron-withdrawing group on the nitrogen of an amine will diminish its basicity:** 20, 23, 69, 70, 71.
- **Lewis acids are electron pair acceptors and Lewis bases are electron pair donors:** 31, 32, 35, **72, 73, 74.**
- **Electrophiles are carbon molecules that react with electron-donating compounds:** 36, 37, 38, **62.**
- **Nucleophiles are molecules that donate electrons to carbon:** 22, 36, 37, 38, **60, 61, 62, 63.**

Homework

6.39 Calculate the pK$_a$ given the following values for K$_a$:
   (a) 6.35×10^{-6}
   (b) 12.1×10^7
   (c) 18.5×10^{-12}
   (d) 9.2×10^{-3}
   (e) 10.33×10^8
   (f) 0.08×10^{-3}

6.40 Calculate the K$_a$ given the following values for pK$_a$:
   (a)   6.78
   (b)   −3.2
   (c)   23.5
   (d)   10.3
   (e)   35.8
   (f)   −11.1

6.41 Draw the complete reaction of each acid listed with (1) NaOH and then (2) NaNH$_2$, showing all starting materials and all products:
   (a)   propanoic acid
   (b)   2-methyl-2-propanol
   (c) $HC(CN)_3$
   (d)   methanesulfonic acid

6.42 Draw the structure of the following acids:
   (a)   3,3-diphenylbutanoic acid
   (b)   4-chloro-2-methylpentanoic acid
   (c)   5,5-diethyloctanesulfonic acid
   (d)   hex-4Z-enoic acid

6.43 Carboxylic acid **A** has a smaller pK_a than carboxylic acid **B**. Suggest a reason for this observation.

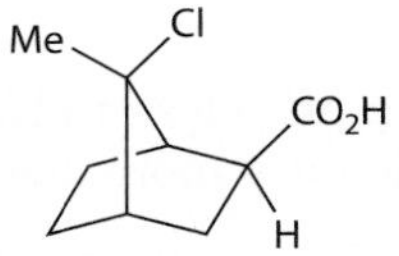

A 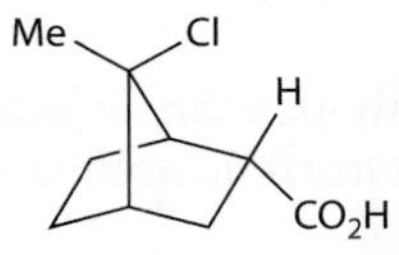   **B**

6.44 Suggest a reason why *cis*-2-chlorocyclohexanecarboxylic acid (**A**) is slightly more acidic than *trans*-2-chlorocyclohexanecarboxylic acid (**B**).

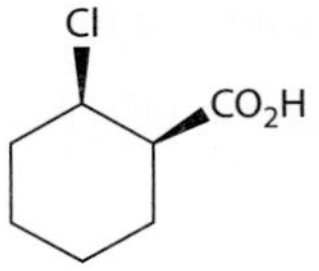

A 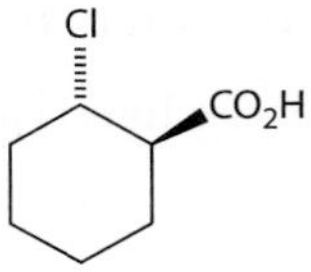   **B**

6.45 Draw all resonance structures for those anions that are resonance stabilized and indicate which are not resonance stabilized.

(a)   (b) 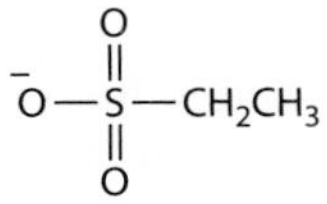   (c) 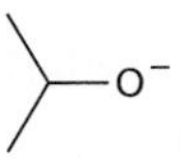   (d) 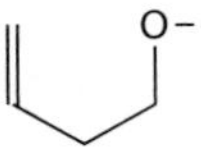   (e) 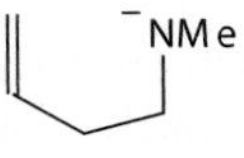   (f)

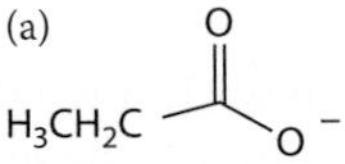

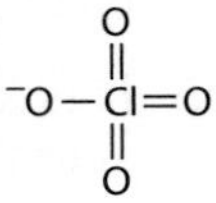

6.46 Alcohols are known to be amphoteric. Predict whether 1-propanol will be an acid or a base or will be neutral in the presence of each of the following:
(a) NaOH
(b) HCl
(c) water
(d) ethanol
(e) NaNH$_2$
(f) 2-butanone
(g) methane
(h) H$_2$SO$_4$

6.47 Draw the conjugate base formed by the reaction of LiNH$_2$ with each of the following. If that product is resonance stabilized, draw all resonance contributors:
(a) 2-pentanone
(b) hexanal
(c) 2,4-pentanedione
(d) 1-propanol
(e) propanoic acid

6.48 Maleic acid (A) is known to have a pK_a of about 1.8, whereas fumaric acid (B) has a pK_a of about 3. Explain.

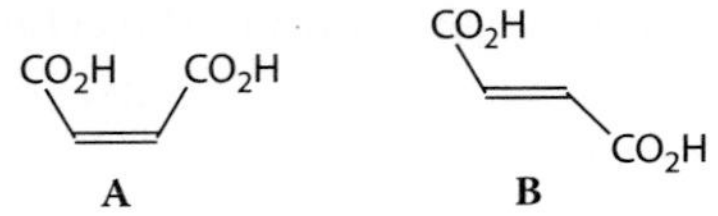

6.49 Explain why 5-bromopentanoic acid has a pK_a close to that of pentanoic acid, whereas 2-bromopentanoic acid is significantly more acidic.

6.50 Discuss whether propanoic acid is more acidic in diethyl ether or in diethylamine.

6.51 If an absolutely pure sample of HCl is sealed in an ampoule whose inner surface is neutral, is HCl considered to be an acid under these conditions?

6.52 In Table 6.1, butanoic acid is listed as a weaker acid than acetic acid. Why? In that table, formic acid is seen to be a stronger acid than acetic acid. Why?

6.53 Draw the product formed when triethylamine (Et_3N) reacts with the following:
(a) HCl
(b) BF_3
(c) propanoic acid
(d) $AlCl_3$
(e) methanesulfonic acid
(f) $FeCl_3$

6.54 Briefly explain how and why we can use pK_a with an amine to describe how basic it is.

6.55 It is known that the conjugate base of diethylamine is less reactive with a weak acid than is the conjugate base of diisopropylamine. Draw both conjugate bases and briefly explain the difference in reactivity.

6.56 Draw the structure of:
(a) triphenylphosphine
(b) butylphosphine
(c) ethylphenylphosphine
(d) 1,2-(diphenylphosphino)ethane

6.57 Draw the conjugate acid formed when each of the following reacts with HCl:
(a) 3-pentanone
(b) dimethyl ether
(c) diethylamine
(d) water
(e) cyclopentanecarboxaldehyde
(f) CH_3–S–CH_3
(g) 2-propanol
(h) oxetane
(i) piperidine
(j) cyclopentanone
(k) acetic acid

6.58 Draw the product for each of the following reactions:
(a) 2-butanone and $AlCl_3$
(b) diethyl ether and BF_3
(c) triethylamine and BEt_3

    (d)   prop-2-enal (called acrolein) and BF_3
    (e)   chloroethane and $FeBr_3$
    (f)   tetrahydrofuran and ZnI_2

6.59 Briefly discuss whether diethylamine will be more basic when it reacts with HCl in diethyl ether or in ethanol.

6.60 Briefly discuss the relative order of nucleophilic strength for the following when they react with iodomethane: $^-NH_2$, ^-OH, and ^-F. Justify your choice.

6.61 It is known that ammonia (NH_3) is a much weaker nucleophile in its reaction with acetone than is the amide anion ($^-NH_2$). Suggest reasons why this is so.

6.62 Draw the product expected when each of the following nucleophiles reacts with bromoethane: NaOMe, $CH_3C{\equiv}C{:}Na$, NaCN, and NaI. Draw the product expected when iodomethane reacts with each of these four nucleophiles.

6.63 When we react $CH_3C{\equiv}C{:}Na$ with iodomethane, we use THF as a solvent and not water. Why?

6.64 When sodium methoxide (NaOMe) is mixed with methanol, methoxide reacts as a base and not a nucleophile. Offer an explanation why this is the case.

6.65 Using the following bond dissociation energy data, calculate the free energy for each reaction and determine if it is spontaneous or nonspontaneous, assuming that the reaction will have a reasonable activation energy to proceed (which may or may not be correct): O–H (104.2), N–H (75), C–H (104), C–I (56), and C–O (91). Ignore ONa and NaI and ignore all ionic bonds. All bond dissociation energies are in kilocalories per mole.

    (a) CH_3OH + NH_3 $\longrightarrow$ CH_3O^- + $^+NH_4$
    (b) Cl_3CH + H_2O $\longrightarrow$ Cl_2C^- $^+OH_3$
    (c) CH_3ONa + ICH_3 $\longrightarrow$ CH_3OCH_3 (ignore NaI)

6.66 Based only on the bond dissociation energy data, use a $\Delta G°$ calculation to estimate if the following reaction may be reversible: C–I (56), C–O (91). Ignore all ionic bonds. All bond dissociation energies are in kilocalories per mole.

$$CH_3OH \;+\; CH_3I \;\longrightarrow\; H_3C{-}\overset{\displaystyle H}{\underset{\displaystyle CH_3}{O}}^{+} \;\; I^-$$

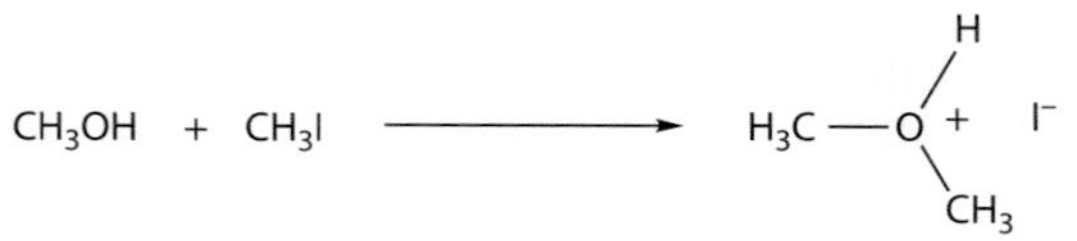

6.67 Based only on the bond dissociation energy data and a $\Delta G°$ calculation, can you decide which of the following acids is more acidic in a reaction with hydroxide ion? Why or why not? O–H (104.2). Ignore all ionic bonds. All bond dissociation energies are in kilocalories per mole:
    (a)   HCOOH (formic acid)
    (b) CH_3OH (methanol)
    (c) CH_3SO_3H (methanesulfonic acid)

6.68 Draw the structure of hexanesulfonic acid, 3-methylbutanesulfonic acid, and 2-chlorobutanesulfonic acid. Of these three, indicate which may be the stronger acid.

6.69 Briefly explain why methanamine is more basic than acetamide.

6.70 Is N-chloromethanamine ($Cl-NHCH_3$) more or less basic than methanamine? Explain!

6.71 Which amine is likely to react faster with formic acid: $(CH_3CH_2)_3N$ or $(Me_3C)_3N$? Draw both reactions with the conjugate acid–base that is formed and explain your answer.

6.72 Which of the following should react faster with BF_3: trimethylamine or trimethylarsine (Me_3As)? Explain!

6.73 Which of the following should react faster with $AlCl_3$: diethyl ether or fluoromethane? Draw the products expected from both reactions and explain your answer.

6.74 Draw the product expected from a reaction of mercuric chloride ($HgCl_2$) and dimethylamine.

Spectroscopy Problems

These problems are to be done only after Chapter 14 has been read and understood.

6.75 Briefly describe how one can use infrared (IR) and proton nuclear magnetic resonance (NMR) to distinguish between propanoic acid and 1-propanol.

6.76 When triethylamine reacts with HCl to form triethylammonium chloride, the IR changes dramatically. Briefly describe the differences in these two molecules that would allow them to be distinguished by IR.

6.77 The proton of the OH unit in 2-propanol can resonate between 1 and 5 ppm in the proton NMR, in deuterochloroform, as the concentration changes. Why?

6.78 As the structures of carboxylic acids change, the acidic proton can resonate between 10 and 20 ppm, depending on the concentration. Briefly explain this phenomenon.

6.79 Briefly describe the differences in the IR and proton NMR that will allow us to distinguish among trimethylamine, *N*-methylaminoethane, and 1-aminopropane.

6.80 Predict the proton NMR for formic acid versus formal (formaldehyde). Contrast and compare them.

Chemical Reactions, Bond Energy, and Kinetics

7

What are your thoughts when you hear the word chemistry? It is a very broad term, but it will be used in a narrow sense for the purposes of this chapter. When one does "chemistry" on a molecule, a chemical reaction occurs. **A chemical reaction is the transformation of one chemical or collection of chemicals into another chemical or collection of chemicals.** The acid–base reactions discussed in previous chapters are all chemical reactions. The emphasis has been on the electron transfer definition (two electron donors and two electron acceptors) of acids and bases. The chemical reactions presented in subsequent chapters involve electron donation to make or break covalent bonds. The fundamental principles for bond making and bond breaking in acid–base reactions will be extended to many other reactions.

Understanding key characteristics of chemical reactions will help understand how reactions work, why they work, what energy characteristics drive a reaction, whether there is an intermediate, whether the reaction proceeds in one step or several steps, and what the final isolated product is. Is there more than one isolated product? If so, which is major and which is minor? The concepts that will begin to answer the questions were introduced in Section 6.2.2 of Chapter 6, but this chapter will provide more detailed explanations.

To begin this chapter, you should know the following:

- **the structure and nomenclature of functional groups (Chapter 5, Sections 5.1–5.3, 5.6, and 5.9)**
- **the chemical bond in general, and the polarized covalent bond in particular (Chapter 3, Sections 3.3, and 3.7; Chapter 5, Section 5.4)**
- **the factors that influence making and breaking bonds (Chapter 3, Section 3.6; Chapter 5, Section 5.4)**
- **the fundamentals of acid–base equilibria and pK_a (Chapter 2, Sections 2.1–2.2; Chapter 6, Section 6.1)**
- **an introductory knowledge of bond dissociation energy (Chapter 3, Section 3.6)**
- **the free energy ($\Delta G°$) equation: $\Delta G° = \Delta H° – T\Delta S°$ (Chapter 3, Section 3.6; Chapter 6, Section 6.2)**
- **the terms exothermic (exergonic) and endothermic (endergonic) (Chapter 6, Section 6.2)**
- **an introductory recognition of an energy-reaction curve (Chapter 6, Section 6.2.3)**

This chapter will introduce the concepts associated with the energy of a given reaction. The correlation between free energy and spontaneity will be continued with reactions other than acid–base reactions. Transition states will be discussed as well as intermediates. Rate of reaction will be discussed, along with the concept known as reaction order.

When you have completed this chapter, you should understand the following points:

- **A chemical reaction involves making and breaking bonds, with the gain, loss, or transformation of a functional group.**
- **Enthalpy is the term used to measure bond dissociation energy, and $\Delta H°$ for a reaction is the difference in bond dissociation energy for bonds made (products) minus that for bonds broken (starting materials). This is products minus reactants.**
- **The free energy of a reaction determines if the reaction is spontaneous (exothermic) or nonspontaneous (endothermic). The free energy ($\Delta G°$) is calculated: $\Delta G° = \Delta H° – T\Delta S°$.**
- **The energy of activation (E_{act}) is the energy required to initiate the bond-making or bond-breaking process of a chemical reaction. The transition state for a reaction is the highest energy portion of a reaction curve and corresponds to the point in a reaction where bonds in the starting materials are "partly broken" and bonds in the products are "partly made."**

- An intermediate is a transient, high-energy product that is not isolated but is converted to the final isolated products. The most common intermediates are carbocations, carbanions, and radicals.
- A mechanism is a step-by-step map that shows the presence or absence of intermediates and the total number of steps required to convert the string materials to the products.
- Reversible chemical reactions have an E_{act} (reverse) that is equal to or less than E_{act} (forward), for the forward and reverse reactions. The position of the equilibrium is measured by the equilibrium constant K, defined as products divided by reactants.
- In ionic reactions, an atom with an excess of electrons (– or δ–) donates two electrons to an atom that is electron deficient (+ or δ+) to form a new bond. When this occurs, a bond in the original molecule must break, with transfer of the two electrons in that bond to one of the atoms in the starting material.
- A first-order reaction rate is obtained by plotting the natural logarithm (ln) of the concentration of a starting material versus time. The rate of a second-order reaction is obtained by plotting the ln of a concentration term that includes both reactants versus time. The rate equation for a first-order reaction is rate = k [A], where k is the rate constant. The rate equation for a second-order reaction is rate = k [A] [B], where k is the rate constant.
- The half-life of a reaction is the time required for half of the starting materials to react.
- The fundamental parameters of organic chemical reactions can be applied to biological transformations.
- Simple chemical energy parameters apply to biological systems.

7.1 A Chemical Reaction

Previous chapters that discuss acid–base reactions are full of the term "chemical reaction," or "reaction." A chemical reaction involves making new chemical bonds and breaking old chemical bonds. This discussion of chemical reactions will begin with the covalent bond formed by sharing two electrons (Chapter 3, Section 3.3). Each bond has an inherent energy (bond dissociation enthalpy; see Chapter 3, Section 3.6), and that energy is released when the bond is broken. Likewise, it will cost that amount of energy to make the bond. These facts can be used to make predictions about the efficacy of a chemical reaction.

In general, a chemical reaction is followed by monitoring changes in the structure of a molecule, and this usually includes modification or replacement of the functional group. A few examples illustrate such modifications, including the acid–base reaction discussed in Chapter 6, in which sodium hydroxide (the base) reacts with the acidic proton of butanoic acid (**1**, the acid) to give sodium butanoate (**2**, the conjugate base). A different type of reaction is the conversion of the aldehyde hexanal to the alcohol 1-cyano-1-hexanol (**3** to **4**), in which the cyanide ion of sodium cyanide is a nucleophile that forms a covalent bond to the carbonyl carbon of **3** to form an alkoxide product.

Hydrolysis (an acid–base reaction between the alkoxide base and the hydronium ion as the acid) converts the alkoxide to alcohol **4** in a second chemical reaction. Two reactions introduced in Chapter 6 include nucleophilic reactions with alkyl halides and the reaction of an alkene as a Brønsted–Lowry base. The former reaction is illustrated by the reaction of potassium iodide with 1-bromohexane (**5**) to give 1-iodohexane (**6**), and the latter by the reaction of 2-ethyl-1-butene (**7**) with hydrogen bromide to give an alkyl halide, 3-bromo-3-methylpentane, **8**. In the reaction of the alkene, the π-bond of **7** is broken as a hydrogen atom is transferred to the less substituted carbon of the C=C unit, and a bromine atom is transferred to the more substituted carbon of the C=C unit. These four reactions are different, but all involve transfer of two electrons with bond breaking and bond making to give the products shown. All involve modification of the original molecule (**the starting material**) and formation of a new molecule (**the product**) with the gain of one or more atoms or groups, loss of one or more atoms or groups, or transformation of one functional group into another.

7.1 Draw the same reactions shown for 1, 3, 5, and 7 but use phenylacetic acid (2-phenylethanoic acid), 4,4-diphenylbutanal, 2-iodohexane, and 2-phenyl-1-hexene, respectively. In each case give the new product.

Many chemical reactions can be categorized by their correlation with a given functional group. Do all alkenes react similarly? Do ketones react similarly to or differently from alkenes? Are there similarities in the way alkenes or alkynes react? *The goal is to find ways in which functional groups react, and then look for similarities and differences in the fundamental reactivity with other functional groups.* The similarities and differences in the way functional groups react will define much of what is called organic chemistry.

It is convenient to categorize reactions. The reaction of butanoic acid (**1**) with sodium hydroxide mentioned earlier is also an example of an **acid–base reaction** to give a conjugate base, **2**. When one group or atom replaces another group or atom, as in the conversion of bromopentane (**5**) to iodopentane (**6**), the first group **substitutes** for the second and this is categorized as an **aliphatic nucleophilic substitution reaction**. When the nucleophile attacks a carbonyl (of **3**)—the so-called **acyl carbon**—the π-bond is broken and a new covalent bond is formed between the nucleophile and the acyl carbon, as in **4**. This type of reaction is known as **nucleophilic acyl substitution**: Cyanide adds to the acyl unit and the product is **4**, formed by breaking the π-bond and forming a new bond to CN.

When hydrogen bromide (HBr) reacts with (**7**), the H–Br bond is broken and new bonds are formed between carbon and bromine as well as carbon to hydrogen. In effect, H and Br **add** to the π-bond, and as the π-bond is broken, two single covalent bonds are formed in **8**. The sp^2 carbons of the C=C unit are transformed to sp^3 carbons. This is called an **addition reaction**, although it is actually an acid–base reaction (see Chapter 10, Section 10.2). Another type of reaction can be included, in which 2-bromo-2-methylpentane (**9**) reacts with a strong base such as sodium hydroxide (ethanol is the solvent here). In this reaction, the bromine atom and a hydrogen on the adjacent carbon are lost (loss of hydrogen to the hydroxide base is an **acid–base reaction**) and the starting material is transformed to an alkene, **10**. The elements of H and Br are **eliminated** from **9** and the overall process (loss of H via an acid–base reaction and subsequent loss of Br) is called an **elimination reaction**. The sp^3 bonds of C–Br and C–H are converted to sp^2 bonds in a new π-bond.

There is no pretense that all of the reactions presented or the details of each one are understood at this point. Presentation of these reactions, as well as giving them a category name, is meant to introduce chemical reactions and to show the variety of reaction types. There is also a need to introduce the terms and fundamental transformations that will be used in the remaining chapters of this book.

In each reaction, look for the groups involved in the reaction, where electrons go when a bond breaks, and from where electrons come when a new bond is formed. It is also important to note one additional aspect of each reaction. If a carbon–carbon bond is being formed, as in the reaction of cyanide and hexanal to give **4**, this is called a **carbon–carbon bond-forming reaction**. If no carbon–carbon bond is formed, but the functional group is simply changed or modified, it is called a **functional group transformation** or a **functional**

group interchange reaction. Within the context of each reaction, keep these two broader categories in mind. Many functional group interchange reactions will be found, but only a few reactions that form carbon–carbon bonds. This is important because if a molecule containing carbon–carbon bonds (an organic molecule) is to be prepared, carbon–carbon bonds must be made and it is important to make a note of those reactions.

7.2 Bond Dissociation Enthalpy and Reactions

As discussed in Chapter 3, Section 3.6, the energy released when a σ-covalent bond is cleaved (bond breaking) is called the **bond dissociation enthalpy** (referred to as bond dissociation energy). That amount of energy is released when the bond is broken. That amount of energy is also required to **make** the bond. The bond dissociation energy for a given bond is given the label H°, or **enthalpy**. The value of H° for a typical C–C bond is 145 kcal mol^{-1} and that for the C–Cl bond is 95 kcal mol^{-1}. Less electron density is concentrated between the nuclei in the polarized C–Cl bond (see Chapter 3, Section 3.7), and it is weaker than the nonpolarized C–C bond.

Bond polarization is not the only parameter, and the exact energy of a bond depends on the bond length (a function of the size of the atoms), electronegativity considerations, solvent in some cases, and other bonds that are connected to the atom of interest. These subtleties are apparent when comparing C–H bonds in different molecules. The bond dissociation energy for a generic C–H bond is 80.6 kcal mol^{-1}, but the C–H bond in methane (H–CH$_3$) has a bond dissociation energy of 104 kcal mol^{-1}, the C–H in ethane (H–CH$_2$CH$_3$) is 98 kcal mol^{-1}, the C–H for the secondary carbon in propane (H–CHMe$_2$) is 95 kcal mol^{-1}, and the C–H for the tertiary carbon in 2-methylpropane (H–CMe$_3$) is 92 kcal mol^{-1}. If bond dissociation energy is used to help understand a specific chemical reaction, accuracy demands using the bond dissociation energy for a specific molecule, not a generic value for a given bond. However, the generic bond dissociation energy values may be used for comparisons in many cases without a significant error.

Bond dissociation energy is one parameter used to predict whether or not a chemical reaction is likely to proceed to give a particular product. One example is the reaction of the iodide ion (from the *ionic* compound NaI) and chloromethane (**11**) to give chloride ion and iodomethane (**12**). Note that THF (tetrahydrofuran; see Chapter 5, Section 5.6.4), shown under the reaction arrow, is the solvent and it does not participate in the bond-making and bond-breaking process. In this reaction, the C–Cl bond is broken in **11** (the **starting material**, always drawn

Table 7.1
Selected Bond Dissociation Energy Values for Bonds in Specific Molecules

Bond	Bond Strength ($H°_{298}$), kcal mol^{-1}	Bond	Bond Strength ($H°_{98}$), kcal mol^{-1}
I–CH$_3$	56	H–C(CH$_3$)$_3$	92
Br–CH$_3$	70	H–CH(CH$_3$)$_2$	95
Cl–CH$_3$	84	H–CH$_2$CH$_3$	98
H$_2$N–CH$_3$	87	H–CH$_3$	104
H$_3$C–CH$_3$	88	F–CH$_3$	109
HO–CH$_3$	91		

on the left side of the equation) and a new C–I bond is formed in **12** (the **product**, always drawn on the right side of the equation). Table 7.1 (identical to Table 3.3 in Chapter 3) shows that the bond strength for the C–Cl bond in chloromethane is 84 kcal mol^{-1} and the bond strength of C–I in iodomethane is 56 kcal mol^{-1}.

The ***change in energy during a reaction*** is of interest. Therefore, the *difference* in energy between the products and the reactants (starting material) is the important calculation. The difference in two values of H° (the change in energy) will be ΔH°, where ΔH° is $H°_{(product)} - H°_{(starting\ material)}$. For the given reaction, ΔH° is

$$\Delta H° = H°_{(product)} - H°_{(starting\ material)}$$

$$\Delta H° = H°_{(I-C)} - H°_{(C-Cl)}$$

$$\Delta H° = 56\ \text{kcal mol}^{-1} - 84\ \text{kcal mol}^{-1} = -28\ \text{kcal mol}^{-1}$$

A negative sign for ΔH° means that energy is *released* during the reaction; this is an **exothermic process (exergonic)**. Such a reaction is said to be spontaneous because it produces enough energy during the course of the reaction to be self-sustaining, once it has started. If ΔH° has a positive value, the process is **endothermic (endergonic)** and energy must be put into the system for it to continue. An endothermic process is usually not spontaneous because less energy is produced during the reaction than is required to keep it going. Therefore, *a spontaneous reaction should continue to produce a product once it has started. A nonspontaneous reaction requires more energy than is supplied by breaking and making bonds in order to continue, and will likely require more vigorous reaction conditions.*

As noted, the terms exergonic and endergonic are used for these reactions. An exergonic process is one that releases energy from the system to the surroundings; an exergonic reaction is a chemical reaction that releases energy in the form of work (exothermic). An endergonic process is one in which the system absorbs energy from the surroundings, and an endergonic reaction is a chemical reaction that absorbs energy in the form of work (endothermic). *This book will use exothermic and endothermic to describe chemical reactions.*

7.2 Calculate $\Delta H°$ for a reaction if the energy required to form a product is 100 kcal mol^{-1} and that required to break bonds in the starting materials is 68 kcal mol^{-1}.

The value of $\Delta H°$ is often correlated with predictions about whether a given reaction will generate heat and be spontaneous (exothermic) or require added heat in order to keep going (endothermic; nonspontaneous). *This simple calculation is misleading, because it is the change in free energy ($\Delta G°$) that determines if a reaction is spontaneous, and not $\Delta H°$.* This point will be elaborated on in Section 7.5. In addition to $\Delta H°$, the other important factors are the change in entropy ($\Delta S°$), and the reaction temperature in degrees Kelvin. These factors will be discussed in the sections immediately following this one.

7.3 Transition States

Imagine a reaction where the starting materials are A and B, and the final product is A–B, with a new bond joining atom A to atom B. This is a one-step reaction, meaning that A and B react to give A–B directly. The normal representation of this reaction is A + B → A–B. At some point in the reaction, the bond between A and B must begin to form. It is not yet a formal bond, but the bond-making process has begun and is represented by [A-----B]; this is known as the *transition state.* **A transition state is *not* a product.** It cannot be isolated, or even observed. It is the logical midpoint of a reaction and it is an energy associated with making and breaking bonds as the reaction proceeds. *One way to define transition state is at the point in the reaction where the starting materials begin to react, and begin the transition to the product.*

$$\text{A} \;+\; \text{B} \longrightarrow \left[\text{A} \text{-----} \text{B} \right] \longrightarrow \text{A} - \text{B}$$

Returning to the conversion of **11** to **12** in Section 7.2, this is a one-step chemical reaction. The transition state for the reaction of chloromethane and iodide to give iodomethane must be **13**. Note the use of dashed lines to indicate that the C–Cl bond is being broken (Cl----C) and that the C–I bond is being formed (C----I). The transition state is shown in a bracket, to set it apart from the actual molecules in the reaction. The concept of transition states helps to describe reactions that do not have a formal intermediate (see Section 7.4).

7.3 Draw the transition state for the reaction of 2-bromopropane and NaI, using 13 as a guide.

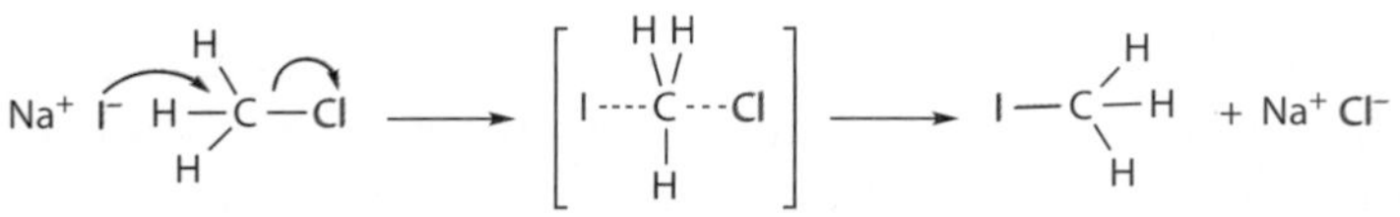

                11   **13**   **12**

7.4 Reactive Intermediates

The conversion of **11** to **12** illustrates a reaction that involves direct conversion of starting material to product. In other words, there is only one chemical reaction, or one step, and the reaction is said to be **synchronous**. There are reactions in which an unstable product is formed, and that product undergoes a second chemical reaction to give a more stable product. In other words, *the starting materials do not give the final product directly, but rather give a transient product prior to the formation of the final product. Such transient products are known as intermediates*.

A generic reaction illustrates such a process. Starting material A reacts with B to give a product C. However, C is not isolated from this reaction. When the reaction is complete, it is clear that C reacts with additional B to give an isolated product, D. In this overly generalized example, C is a transient species that is not isolated, but it is so high in energy (unstable) that it reacts with B before it can be isolated. A transient and relatively high-energy product such as [C] is an **intermediate**, where the brackets in this case indicate a transient species, again set apart from the molecules participating in the reaction. *An intermediate is therefore a reaction product that is not isolated, but reacts to give another more stable product*. Another way to say this is that *an intermediate is a transient product*. Note that there are two reactions, not one: [C] is the product of the first reaction and D is the product of the second reaction. If the focus is on starting materials and isolated product, then A + B → D via a two-step process.

$$A \;+\; B \;\longrightarrow\; \left[\, C \,\right] \;\xrightarrow{\;\;B\;\;}\; D$$

(starting materials)   intermediate   (final product)

Before discussing any reaction, it is important to understand (1) whether a reaction proceeds in a synchronous manner (one step with no intermediate) or by an intermediate in two or more steps, and (2) if an intermediate is involved, what the nature of the intermediate, C, is. This information is determined experimentally, and it is not intuitively obvious. For the reactions presented in this book, this information will be provided. In general, three types of intermediates will be discussed in this book. *Intermediate C may be a carbocation, a radical, or a carbanion.*

Intermediates are common in organic chemistry, but they cannot be isolated in the vast majority of reactions. They may be detected by various experimental techniques. *For reactions in this book, it will be assumed that intermediates cannot be isolated because they are unstable and extremely reactive.* Three fundamental types of intermediates will appear many times in this book: **carbocations**, **carbanions**, and **radicals**. Cations are **electron-deficient** species that bear a positive charge and quickly react with a species that can donate electrons. Carbocations are species where a positive charge resides on carbon (carbon has a formal charge of +1) and they react with an electron-donating species. *Carboctions are also called carbenium ions*.

An anion is electron rich and easily donates electrons to an electron-deficient species. **A carbanion** is a species in which a negative charge resides on carbon (carbon has a formal charge of –1), and it is electron donating. When a carbanion reacts with another carbon atom, it is classified as a nucleophile.

A radical is a species with **one unshared electron**, which makes it very reactive, and it will react to form a bond in a manner different from that of an anion or a cation. An anion will donate two electrons, and a cation will accept two electrons to form a covalent bond. A radical only has one electron, and it will react with another radical to form a covalent bond, where each species donates one electron. Radicals also react with a neutral species A–B to form a new s-covalent bond and a new radical: $X\bullet + A–B \rightarrow X–A + B\bullet$. This description is intended to show that radicals react differently. Because the focus of this book is the chemistry of carbon, the following discussions will focus on carbon intermediates.

7.4.1 Carbocations (Carbenium Ions)

A cation is an electron-deficient species that has a formal charge of +1. A cation is attracted to and reacts with a species that can donate two electrons (such as an anion). A carbon atom that bears a positive charge is called a **carbocation** (a **carbenium ion**). A carbocation is formed when a covalent bond to carbon is broken in such a way that two electrons are transferred to one atom and the carbon receives no electrons during the transfer. This process is called a **heterolytic cleavage**. A carbocation has the structure **14** and the central carbon atom is clearly electron deficient.

14A   14B

The three σ-covalent bonds in **14** provide only six electrons, but carbon requires eight to be electrically neutral because it is in group 14 of the periodic table, so carbocation **14** will have a formal charge of +1 (see Chapter 5, Section 5.5). This charged species has only three covalent bonds, is high in energy, unstable, highly reactive, and difficult to isolate. In other words, it is an intermediate. The positive charge is **localized** on carbon, and the charge is associated with the empty p-orbital on that carbon as shown in **14B**. If this is a correct representation, *the carbon of a carbocation is sp² hybridized (see Chapter 5, Section 5.1) and must have trigonal planar geometry* (see **14A**). This p-orbital in **14B** is considered to be the region in space above and below the plane of the carbon and hydrogen atoms where electron density can be accepted to form a new bond. *A carbocation will react with another species that can donate the two missing electrons to give it eight, satisfy the valence requirements of carbon, and form the fourth bond to make carbon tetravalent.* Note the similarity of this description to that for a Lewis acid in Chapter 2.

A carbocation is observed in the reaction sequence in which 2-methyl-2-chloropropane (**15**) reacts with a Lewis acid (AlCl$_3$), in the presence of potassium iodide (KI), to give 2-methyl-2-iodopropane, **16**. This transformation is explained by initial reaction of the electron-rich chlorine atom of **15** as a Lewis base with the potent Lewis acid AlCl$_3$ (see Chapter 2, Section 2.5). When the C–Cl bond breaks, both electrons are transferred to aluminum (a heterolytic bond cleavage), leaving behind a positive charge on the carbon atom, which is carbocation **17**. Note the arrow going *from* **the electron-rich** bond to the chlorine atom *to the* **electron-deficient** aluminum in AlCl$_3$ to form cation **17** and the counter-ion, AlCl$_4^-$. Carbocation **17** is a reactive intermediate that accepts two electrons from the iodide ion (added to the reaction in the form of potassium iodide, KI) to give 2-methyl-2-iodopropane, **16**, which is the final observed product.

Alkyl halide **15** does *not* react with KI directly, but forms an initial transient product (**17**) that reacts with the iodide ion in a second chemical step. The carbocation **17** is an intermediate in this reaction and the iodide ion is a nucleophile. The sequence of events in the reaction of **15** and AlCl$_3$ to give **17**, which reacts with iodide to give **16**, is the **mechanism** of this reaction. A mechanism is nothing more than a reaction sequence that includes each reactive intermediate produced on the way to the final product. The concept of mechanism will be elaborated in Section 7.8.

Carbocations can be generated in many different ways, as will be seen in succeeding chapters, but they all have certain characteristics. In summary, *carbocations are high-energy intermediates, and relatively high activation energies are required for their formation (see Section 7.6). Once formed, carbocations have a low activation energy for reaction with the nucleophile (this is what the term highly reactive really means).*

7.4 Draw the structure of a carbocation where all the "R" groups in 16 are ethyl groups.

7.4.2 Carbanions

An anion is a species that has an excess of electrons and bears a formal charge of –1. When the negative charge resides on carbon, it is called a **carbanion**. In general, carbanions are formed by breaking a covalent bond in such a way that two electrons are transferred to the carbon involved in that bond, and the second atom receives no electrons during the transfer (a **heterolytic cleavage**). A generic carbanion is shown as **18**, with three covalent bonds between C and R, and *a pair of electrons in a p-orbital*. When the R group in **18** is R (an alkyl group), a formal charge calculation places a charge of −1 on the carbon atom shown ($4 - 0.5 \times 6 - 2 = -1$). The two unshared electrons reside in a p-orbital. Therefore, *an anion can be viewed as a "filled" p-orbital that is localized on an atom, in this case carbon. This contrasts with a carbocation, which is effectively an "empty" p-orbital localized on a carbon atom.*

Because electrons in covalent bonds are repelled by the lone electron pair, the three-dimensional structure shown for **18** resembles a "squashed" tetrahedron (see Chapter 3, Section 3.5.4, for a discussion of the VSEPR model). Carbanion **18** is high in energy (has a relatively high E_{act} for its formation), unstable, and very reactive; it is an intermediate. Because a carbanion has an excess of electrons, it behaves as an electron donor and reacts with a species that is electron deficient. *If 18 reacts with an electron-deficient carbon atom, it is classified as a nucleophile. If 18 reacts with the acidic proton of a Brønsted-Lowry acid, it is classified as a base.* Carbanions are not as prevalent in this book as carbocations, and the structure of a carbanion is not as simple as that presented for **18**. For these reasons, a discussion of carbanions in a chemical reaction will be delayed until they are encountered (see alkyne anions in Chapter 11, Section 11.3.7, and Chapter 18, Section 18.3; Grignard reagents in Chapter 18, Section 18.4; and enolate anions in Chapter 22, Sections 22.2 and 22.4).

7.5 Draw the structure of a carbanion where two of the "R" groups in 20 are methyl groups and the third is a phenyl.

7.4.3 Radicals

In preceding sections, a carbocation has been described as a carbon atom bearing an "empty" p-orbital (no electrons) and a carbanion as a carbon atom with a "filled" p-orbital (two electrons). When a p-orbital on any given atom has only **one unshared electron**, it is called a *radical,* and a carbon radical is represented as $R_3C\bullet$ (see structure **19**). With three covalent bonds and one extra electron, $R_3C\bullet$ is a high-energy species and a very reactive intermediate. The single electron in the orbital (see **19** and **20**) will slightly repel the electrons in the covalent bonds, so one might expect a squashed tetrahedron (pyramidal) shape (**19**). *There is evidence, however, that a planar structure (20) is probably the low-energy structure rather than the pyramidal structure, at least for the methyl radical ($H_3C\bullet$; 20 where R = H).*

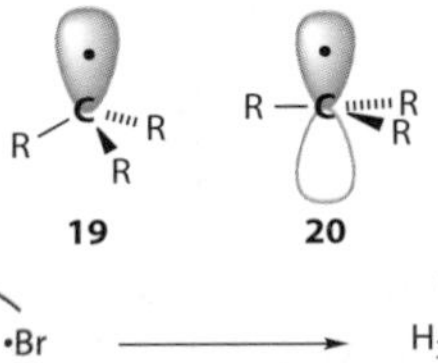

19   20

21   22

One way to form a carbon radical is by a chemical reaction between a neutral species such as methane and a preformed radical such as Br• (**21**). Methods that generate Br• will be discussed in Chapter 11 (Section 11.9). In this reaction, the bromine radical donates a single electron (note the **single-headed arrow**; much like a fishhook) to one hydrogen atom of methane, which donates one electron from the covalent C–H bond (see the **arrow**). When this occurs, the other electron in the C–H bond is transferred to the carbon and the resulting products are a new carbon radical (methyl radical **22**) and H–Br. Note that there are two electrons in the H–Br covalent bond: *one derived from the bromine radical and one from the C-H bond on methane*.

Radicals can be formed by breaking a covalent bond in such a way that one electron is transferred to each of the two atoms involved in that bond (called **homolytic cleavage**). The bonds in some molecules easily undergo homolytic cleavage. One example is the O–O bond of a peroxide, which has the structure R–O–O–R. Most bonds are very resistant to homolytic cleavage, preferring **heterolytic cleavage** (breaking a bond in such a manner that both electrons are transferred to one atom and no electrons are transferred to the other). Radicals will appear in several reactions to be discussed in this book; in particular, see Chapter 10, Section 10.8, and Chapter 11, Section 11.9. Radicals are not used as frequently as carbocations and carbanions, so a formal discussion of radical reactions will be delayed until they are encountered.

7.6 Draw the structure of a radical where two of the "R" groups in 22 are hydrogen atoms and the third group is phenyl.

7.5 Free Energy. Influence of Enthalpy and Entropy

Chemical reactions are driven by changes in energy. Heating imparts energy to the molecules in a reaction, which will increase the energy of the system. Molecules absorb energy from their environment by collision with the sides of the reaction vessel (a flask) or with another atom or molecule, and this energy is utilized in bond making and bond breaking. Monitoring changes in energy is important for following the progress of a chemical reaction. The energy of the entire system ($\Delta G°$; called the free energy) is the important parameter, and it is a function of enthalpy as well as entropy and temperature. The relationship of these parameters is well defined and will be presented in this section.

7.5.1 The Free Energy Equation

Whether or not a chemical reaction will proceed spontaneously is determined by the change in **standard free energy ($\Delta G°$)**. The standard free energy is calculated from the change in enthalpy ($H°$) and the change in entropy ($S°$) by the *Gibbs free energy equation:* $\Delta G° = \Delta H° + T\Delta S°$. This calculation *assumes that the reaction is done under standard conditions:* in solution with a concentration of 1 M and at 298.15 K.

As first presented in Section 6.2.3 of Chapter 6, the entropy term ($\Delta S°$) measures the disorder of a given system. In practical terms, if the number of particles for a reaction remains the same or decreases, the magnitude of the entropy term is small. If the number of particles greatly increases during the course of a reaction, then entropy increases. In the preceding reaction that converted **11** (chloromethane) to **12** (iodomethane), changing chlorine for iodine does not change the overall number of particles, and the size and shape are not greatly changed. The change in entropy should be small. For most reactions in this book, small changes in the skeleton of the molecule lead to small changes in entropy, so an assumption that the change in entropy is small for most reactions in organic chemistry is usually correct. However, this is an assumption, and there are reactions where the change in entropy can have a large influence on the reaction. Such cases will be pointed out at the appropriate time. The term "T" is temperature, measured in degrees Kelvin.

If the entropy term is ***assumed*** to be zero, then $\Delta G° \approx \Delta H°$ and estimating the progress of a reaction is quite simple. How valid is this assumption? In Table 7.1, values of H° were measured in kilocalories per mole, so $\Delta H°$ will have units of kilocalories per mole, whereas the majority of chemical reactions in this book have an entropy term measured in *calories* (cal) rather than in kilocalories. When the entropy is only a few calories, the T$\Delta S°$ term is smaller than the $\Delta H°$ term (typically less than 5%, ***but it is not zero***) and ignoring it will not introduce a large error. ***Ignoring the entropy term is an assumption*** and, in some cases, leaving out the entropy term will lead to a significant error. For the most part, assuming that T$\Delta S°$ is zero so that $\Delta G° \approx \Delta H°$ is a good *first approximation*. This information should indicate if a reaction is exothermic (spontaneous) or endothermic (nonspontaneous), allowing one to begin the planning of reaction conditions, the time required for the reaction to be completed, and if special safety precautions might be necessary. If the difference between an endothermic prediction and an exothermic prediction is less than 2–4 kcal mol^{-1}, then it is wise to calculate the entropy term (TΔS).

7.7 If $\Delta H° = 21.5$ kcal mol^{-1}, $\Delta G° = 22.4$ kcal mol^{-1}, and T = 298 K, what is the value of $\Delta S°$ in kilocalories per mole? How serious is the error introduced into a $\Delta G°$ calculation if the $\Delta S°$ term is ignored?

7.5.2 Spontaneity of a Reaction

In Section 7.2, a negative value of $\Delta H°$ was associated with a spontaneous or exothermic reaction (heat is generated during the reaction by the bond-making and bond-breaking processes). If it is assumed that $\Delta H° \approx \Delta G°$, then a positive value of $\Delta H°$ is associated with a nonspontaneous or endothermic reaction. From the standpoint of planning a reaction, this calculation suggests that heat must be added continually for the reaction to proceed because the bond-making and bond-breaking processes will generate insufficient heat. As pointed out in Section 7.2, it is *not* a positive or negative $\Delta H°$ that determines whether a reaction is endothermic or exothermic, but rather a positive or negative value of $\Delta G°$.

Estimating $\Delta G°$ is useful for a comparison of two different reactions. If one reaction has a negative $\Delta G°$ and a second reaction has a positive $\Delta G°$, the first reaction is assumed to be faster and more facile (easier). This is not always true because the amount of energy required to get things started (E_{act}) must be known (see the following section), but it is a good place to begin. The free energy calculation is particularly important if a starting material can react by two different reaction pathways that compete with one another, and in reversible reactions. If one reaction is more exothermic than the other, it is assumed that the more exothermic reaction will lead to the major product. This is correct only for reactions that have a transition state that is closer in energy to the products than the starting material, and the E_{act} must be known or estimated. This idea will be clarified in later sections dealing with specific reactions.

7.8 Determine if a reaction is spontaneous or not when $\Delta H°$ is known to be -30 kcal mol^{-1}.

7.6 Energetics. Starting Materials, Transition States, Intermediates, and Products on a Reaction Curve

Categorizing a reaction as exothermic or endothermic is important, but that information does not necessarily tell us **whether** the reaction will proceed or under what conditions it will proceed. Chapter 6 (Section 6.2.1) indicated that a certain amount of energy is required to initiate the reaction (heat, light, etc.) that is independent of $\Delta G°$ and the exothermic/endothermic behavior of a reaction.

The energy requirement for a given reaction can be graphically presented by what is known as an **energy curve**. Figure 7.1 shows an **energy curve** that

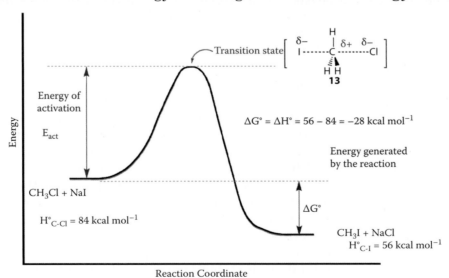

Figure 7.1 Typical energy curve for an exothermic reaction.

correlates with the reaction of sodium iodide with chloromethane (**11**) to give iodomethane (**12**) and sodium chloride. This is a nucleophilic substitution reaction (see Chapter 11, Section 11.2). The abscissa in this energy curve is labeled "reaction coordinate" and the ordinate is labeled "energy." At the left of the curve, the starting materials (NaI and CH_3Cl, **11**) have not yet begun to react, and the "line" is an energy level that represents the energy inherent to the starting materials, CH_3Cl and NaI. At the far right of the reaction coordinate, the reaction has been completed and the energy inherent to the products (CH_3I, **12**, and NaCl) is shown. In this reaction, the bond energies of the ionic compounds NaI and NaCl are ignored, and only the bonds involving the organic compounds are considered. A direct comparison of these two energies gives $\Delta G°$.

The difference in these two energy points corresponds to $\Delta G°$ ($-28\,kcal\,mol^{-1}$), which means that the energy of the product is lower than that of the starting materials. Because $\Delta G°$ is negative, this reaction is exothermic. Figure 7.1 is a typical reaction curve for an exothermic reaction.

As the curve is followed from left to right, the reaction proceeds from starting material to product. As the starting materials mix together, it takes a certain amount of energy (labeled **activation energy**, E_{act}) to initiate the bond-breaking or bond-making process. Energy added to the system (heating) initiates molecular collisions and that energy is transferred from one molecule to another. The energy of the system must equal or exceed E_{act} or no chemical reaction takes place. When the energy of the system equals or exceeds E_{act}, bond breaking and bond making begin and this point is labeled the *transition state* (see **13**). This is the highest point of the curve in Figure 7.1. ***The energy of activation is, therefore, the amount of energy supplied to a system so the chemical reaction can begin.***

The **transition state** is represented by **13**, which shows the C–Cl bond beginning to break as C----Cl and the C–I bond beginning to form as C---I. A transition state cannot be isolated or even observed. At the transition state, the reaction begins, and then the energy curve falls away toward the energy of the final product and the reaction will be endothermic or exothermic. The reaction curve shown in Figure 7.1 is an exothermic reaction, so it will continue once it begins (it is spontaneous).

Figure 7.2 is a typical endothermic reaction curve, where the energy of the products is higher than that of the starting materials. The reaction between NaCl and iodomethane is the reverse of the reaction presented in Figure 7.1 to give chloromethane and sodium iodide. Once energy equal to E_{act} is applied to the starting materials, the transition state is reached (again represented by **13**), in which the C–Cl bond begins to form (C----Cl) and the C–I bond begins to break (C----I). The products are lower in energy than the transition state, but in this case higher in energy relative to the starting materials because $\Delta G°$ is +28 (a positive number), indicating an endothermic reaction. Note that $\Delta G°$ is *not* equal to E_{act}, so the reaction does not produce enough energy to keep going. In other words, it is a nonspontaneous reaction and it may continue to give the final product, but energy must be continually added or it will stop.

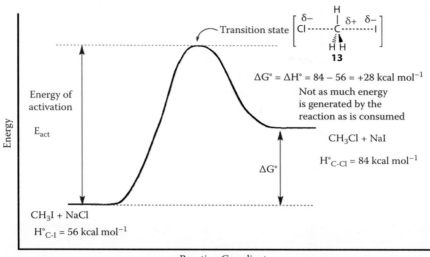

$\Delta G° = \Delta H° = 84 - 56 = +28$ kcal mol^{-1}

Not as much energy is generated by the reaction as is consumed

CH$_3$Cl + NaI

H°$_{C-Cl}$ = 84 kcal mol^{-1}

$\Delta G°$

CH$_3$I + NaCl

H°$_{C-I}$ = 56 kcal mol^{-1}

Reaction Coordinate

Figure 7.2 Typical energy curve for an endothermic reaction.

7.9 Draw the reaction curve for an endothermic reaction and label the transition state.

7.10 Is the product of an endothermic reaction higher or lower in energy than that of the starting materials?

A comparison of the energetics of the reaction in Figure 7.1 and that in Figure 7.2 makes it is clear that ***iodide will displace chlorine from chloromethane, but chloride ion will not displace iodide from iodomethane***. The first reaction is exothermic and spontaneous, whereas the second is endothermic and nonspontaneous. The reaction of iodide ion and chloromethane is not expected to be reversible because the chloride ion + iodomethane reaction does not generate enough energy to overcome the activation barrier. This simple analysis has provided a great deal of information about these reactions, which is the point of this entire section.

- **When referring to the relative energy of products, products with a lower energy are said to be more stable than those with higher energy. It will always be assumed that lower energy correlates with greater stability and poorer reactivity.**
- **In terms of $\Delta G°$, a lower energy product formed via an exothermic reaction will have a larger $\Delta G°$ than a product with a higher energy. Because $\Delta G°$ is larger, the lower energy product is more stable in that more energy is required to make it do something.**
- **Transition states cannot be isolated or observed. *Transition states are assumed to be the logical mid-point of a reaction, but they cannot be observed.***

It is possible that a reaction is exothermic, but the E_{act} is so large that the reaction simply does not proceed under normal conditions. It is also possible that a reaction is endothermic, but the E_{act} is so low that the reaction can be made to proceed under very mild conditions. Remember that $\Delta G°$ should be calculated to determine if the reaction is endothermic or exothermic, but the activation energy will determine the conditions required to initiate the reaction.

7.11 If $\Delta H°$ for one reaction is –21 kcal mol^{-1} and that for a second is –10 kcal mol^{-1}, determine which reaction is likely to occur at a lower temperature if E_{act} for the first reaction is 35 kcal mol^{-1} and E_{act} for the second reaction is 44 kcal mol^{-1}.

7.7 Competing Reactions

Suppose there are two competing reactions, such as when a reactant, A, can go either to product, B, *or* to product, C, under the same conditions:

C ⟵————— A —————⟶ B

With some experimental work, the E_{act} for the reaction that converts A to C is shown to be 5 kcal mol^{-1} and the E_{act} for the conversion of A to B is 10 kcal mol^{-1}. Note that these energy values are not intuitively obvious, but are determined in the laboratory by experiments. These competing reactions are represented in Figure 7.3, where the **red** reaction is the conversion of A to B and the **blue** reaction is the conversion of A to C. These data suggest that one of these reactions is more likely to occur than the other. The reaction with the lowest E_{act} will begin first.

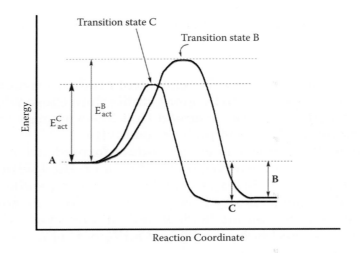

Figure 7.3 Reaction curves for competing reactions A to B and A to C.

The energy of the final products B or C is also very important. In this case, product B and product C are close in energy, but the reaction to produce B is more exothermic than the one to produce C. If the applied energy is only 6 kcal mol^{-1}, the product will be C and there will be no B because that reaction does not have enough energy to begin. Therefore, under some reaction conditions, the reaction of A to give C is fast, whereas the reaction of A to give B is slow or difficult. **Whether or not this statement is true depends entirely on how much energy is provided to the reaction (the reaction conditions).** In other words, by controlling the amount of energy applied to the reaction, it is possible to produce C as the exclusive product, whereas heating the reaction to a high temperature may give a mixture of B and C as products. Throughout this book, one reaction or process may be preferred to another. The activation energy of each competing reaction is one of the key pieces of information used to make such a determination.

A cautionary note is that some reaction products are determined by the stability of the products rather than the activation energies. When the energy of one transition state relative to another leads to one product over another, the reaction is under kinetic cntrol. When the relative energy of one product relative to another determines the major product, the reaction is under thermodynamic control. These terms will be explained in greater detail later. *The way this paragraph is phrased should suggest that it is possible to control reaction conditions to drive some reactions to one product or another.* Hopefully, this will make more sense as real chemical reactions are discussed.

When a reaction has an intermediate, there are two or more chemical steps, each with a product and an E_{act}. The E_{act} for each of these steps is reflected in the reaction curve. For a reaction $A + B \rightarrow [C] \rightarrow D$, where [C] is a reactive intermediate, the energy curve in Figure 7.4 is a combination of two reaction curves, one to form C and the second to form D. The activation energy (marked E_{act}^1) represents conversion of the starting materials A and B to the point marked TS_1 (first

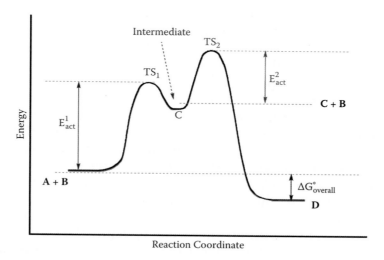

Figure 7.4 Energy curve for a reaction that has one intermediate.

transition state). The bond-making or bond-breaking process begins at TS_1 and leads to an energy minimum that corresponds to product C in Figure 7.4. Reaction product C is an **intermediate**. This intermediate is lower in energy than the transition state but much higher in energy than the starting material.

Because intermediate C is unstable and highly reactive, it begins to react with B and the energy rises again to a new maximum (a second transition state, marked TS_2) with a second energy of activation (E_{act}^2). This is the energy required for C to begin reaction with B, and it is clear that the necessary energy is rather low, consistent with the instability (high reactivity) of C. Once C begins to react, the energy curve falls to the lower energy of the product, D. Figure 7.4 therefore represents *two reactions*—each with an energy of activation and each with a separate transition state. Note that the energy of activation for the second reaction is much smaller than the first, and that the second reaction is exothermic. This accounts for the spontaneity of the second reaction, and the high reactivity of the intermediate.

When the reaction is done, all that is observed is that the starting materials are transformed into the isolated products. In other words, A and B are heated and the product is D. The intermediate (C) may be detected using certain experimental techniques, as mentioned before, but it is rarely, if ever, isolated.

7.12 **Draw the reaction curve for an exothermic reaction with two intermediates (the energy of intermediate 2 is lower in energy than intermediate 1) where the energy of transition state 2 is lower in energy than transition state 1, but the energy of transition state 3 is higher than transition state 2 and lower than 1.**

7.8 Mechanisms

When cyclopentene reacts with HCl, the isolated product is chlorocyclopentane. Does cyclopentene give chlorocyclopentane directly after reaction with HCl, with no intermediate? With no knowledge of the reaction, the answer is a simple, "I don't know." To find the answer, an experiment must be done in which it is discovered that cyclopentene does not go directly to chlorocyclopentane, but rather a high-energy product is formed that disappears over time to give chlorocyclopentane. ***Perhaps the most important part of this question is to understand that the first experiment observes the isolated product from an unfamiliar reaction. Another experiment is usually required to confirm or refute any assumption about the presence of an intermediate.***

The initial question leads to the experimental observation that the reaction proceeds by a transient product, an intermediate. This approach leads to certain generalizations that can be applied to many chemical reactions. Identification of all intermediates in a reaction allows one to define the step-by-step process by which starting materials are converted to products, and ***the step-by-step listing of intermediates is called the mechanism of that reaction.***

For all reactions in this book, the following protocol will be used to discuss a mechanism:

1. **Examine the products and compare them with the starting material.**
2. **If no intermediates can be found experimentally, it will be stated that there are none.**
3. **Make an educated guess on where the product came from, one chemical step at a time.**
4. **When an experiment shows that there is an intermediate, try to deduce its structure based on the electron transfer properties of the starting materials and the nature of the product.**
5. **Use experimental data to be certain that the analysis is correct.**

The step-by-step diagram of how a reaction works is known as its *mechanism*.

If one is asked to write down a mechanism, what is required? *For the reaction of cyclopentene and HCl, the mechanism must show on paper all of the chemical reactions and intermediate products that transform cyclopentene and HCl into chlorocyclopentane.* First, the product of the reaction must be known. In this case, the product is chlorocyclopentane. Ask the question: Does the π-bond in cyclopentene break and the new C–H and C–Cl bonds in chlorocyclopentane form simultaneously, directly from cyclopentene? As noted previously, a series of experiments discovered that chlorocyclopentane is *not* formed directly, but that an unknown product appears during the course of the reaction and then disappears as chlorocyclopentane forms. This transient product is the intermediate. The experiment suggests that cyclopentene must react with HCl to give an intermediate product, which is then converted to chlorocyclopentane in a second step. How?

Inspection of chlorocyclopentane indicates that a new C–Cl bond is formed, and the Cl must come from HCl. The reaction leads to cleavage of the H–Cl bond. Similarly, the proton in chlorocyclopentane must come from HCl, and it is clear that the π-bond is broken. Indeed, the H and Cl are incorporated on the two carbons of the C=C unit. What kind of reaction is this? One instantly recognizes that HCl is an acid, and if it reacts with cyclopentene, then the C=C unit probably reacts as a Brønsted–Lowry base. Electron donation by the acid–base reaction will break the π-bond (the source of the two electrons) and form a new σ-covalent C–H bond to electron-deficient H of HCl (the acid). As the C–H bond is formed, the H–Cl bond breaks, and both electrons in that covalent bond are transferred to Cl to give chloride ion. *The analysis of this reaction is based on the knowledge that there is an intermediate, and it leads to the conclusion that carbocation 23 must be the intermediate. The structural difference between 23 and chlorocyclopentane is formation of a C-Cl bond, so the second step must involve the chloride ion donating two electrons to the positive carbon.*

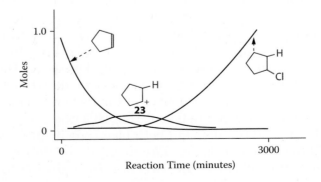

What type of experiment is possible for a reaction that will suggest the presence of an intermediate? One can monitor the disappearance of the starting material cyclopentene, as the concentration diminishes during the reaction. In other words, the number of moles per liter of solution added to the reaction is known, and the number of moles per liter of starting material at different times can be determined. Thus, once the reaction has started, one can monitor loss of starting material as a function of time. It is also possible to monitor the appearance of the product, chlorocyclopentane. When an experiment is carried out, assume that one can observe a transient product as it begins to form, before the appearance of chlorocyclopentane. The concentration of this product increases to a certain level but eventually disappears as chlorocyclopentane grows in concentration.

This statement is represented by the graph in Figure 7.5, where the molar concentration of starting material decreases from an initial 1.0 mol/L to near 0 toward the end of the reaction. The molar concentration of the product is zero at the beginning, and remains zero through a significant part of the reaction. In the "middle," chlorocyclopentane begins to form and its molar concentration rises to eventually reach about 1.0 mol/L, the theoretical yield based on 1.0 mol of cyclopentene. In this graph, an unknown product appears during the course of the reaction and then disappears; this corresponds to the intermediate. Production of chlorocyclopentane follows formation of this transient product, which is the intermediate carbocation **23**. When this experiment is considered in the context of the conversion of cyclopentene to chlorocyclopentane, it is a clue to explain how the reaction works.

Carbocation **23** is an intermediate in a process that involves two chemical reactions. These reactions are represented by the energy diagram in Figure 7.6, where the intermediate carbocation **23** is shown as a high-energy species (higher in

Figure 7.5   Disappearance of starting material cyclopentene and formation of chlorocyclopentane.

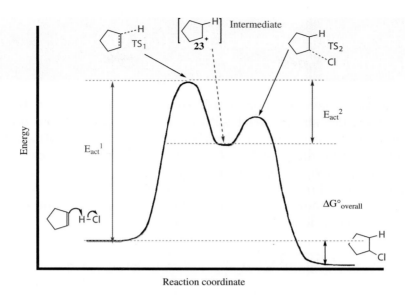

Figure 7.6 Energy diagram for the reaction of cyclopentene and HCl.

energy than the products, but lower in energy than either transition state). The transition states for both reactions are shown as the high-energy points of the reaction curve. In transition state TS_1, the π-bond is breaking and the new C–H bond is beginning to form. For transition state TS_2, the chloride ion is beginning to attack the positive carbon, so the C–Cl bond is beginning to form. As drawn, this overall reaction is exothermic. The real point of Figure 7.6 is the association of actual chemical structures with key energy points or energy terms on the reaction curve. Note that E_{act}^2 is much lower than E_{act}^1, accounting for the instability (higher reactivity) of the intermediate (**23**) as it proceeds exothermically to product.

7.13 Write out the mechanism for the reaction of 2,3-dimethyl-2-butene with HBr.

Identification of **23** allows construction of a *step-by-step road map of exactly how the two reactions work to give the final product*. This road map is called a **mechanism. *A mechanism is simply a method of identifying how reactions work by charting the presence or absence of intermediates and the total number of reaction steps required to convert a starting material into the final isolated product***. Many mechanisms will be presented throughout the book for different reactions. For the reaction of cyclopentene with HCl, analysis began with the product and worked back to the starting materials, with a knowledge of whether or not there is an intermediate. This analysis suggested the mechanism of the reaction. Specifically, the analysis suggests that cyclopentene *reacts with HCl to form a carbocation, which then reacts with chloride ion to give chlorocyclopentane. Drawing the carbocation, and the arrows to transfer electrons from the π-bond to H, breaking HCl, and then **23** reacting to form the final product (see above the mechanism of this reaction).*

7.9 Why Does a Chemical Reaction Occur? Defining a "Reactive" Center

Chemical reactions have been presented in a very general sense, but the issue of why two molecules should react with each other has not been addressed. A reactive center is the atom in a molecule that reacts with another atom. Most of the reactions to be discussed (*not all*) involve ***ionic reactions***, where one or both reactants have a formal charge (positive or negative; carbocations or carbanions) or are polarized ($\delta-$ or $\delta+$). If the simplifying assumption is made that most reactions occur when polarized bonds or ionic reactants are present, certain trends in chemical reactivity are apparent.

One of the most basic precepts in science is that a negative charge will attract a positive charge and that two like charges (positive to positive or negative to negative) will repel. If most organic reactions are assumed to involve ionic or polarized atoms, then this basic precept can be used to predict the course of reactions. In the reaction of chloromethane (**11**) and sodium iodide presented in Section 7.3, the C–Cl bond is polarized with the carbon atom $\delta+$ and the chlorine $\delta-$. If chloromethane is mixed with iodide ion (I^-), which is clearly a negatively charged anion, negative attracts positive and iodide is most strongly attracted to $C^{\delta+}$, which is the carbon attached to the chlorine. Iodide has a negative charge, so it has an excess of electron density. If iodide collides with the electron-deficient ($\delta+$) carbon atom of chloromethane, two electrons from iodide will be transferred to carbon, forming a new C–I bond in the product, **12**.

*Forming a new C–I bond **will not** give 10 electrons around carbon (five bonds) because the C–Cl bond breaks as the C–I bond is formed.* If one new C–I bond is formed, then one bond in the original molecule must break. The bond that breaks is the weakest bond (C–C). Chorine easily accommodates an excess of electrons, and when the C–Cl bond breaks, both electrons in that bond are transferred to chlorine, forming the chloride ion, Cl^-. In other words, Cl will be lost completely from the molecule. Initiation of the bond-making and bond-breaking process is represented by transition state **13**. Remember that the C–Cl bond breaks rather than C–H because C–Cl is weaker (see Table 7.1) and also because chlorine can accommodate the two electrons better than hydrogen.

In contrast with the reaction just described, iodide ion will ***not*** react with methane (CH_4) to form **12**. There are no polarized bonds in methane (only C–H bonds), so there is no great attraction between carbon and iodide. Even if iodide collides with carbon (which will certainly occur), there is no weak bond to break that will allow transfer of the electrons to another atom. A colloquial way to say this is that there is no place to put the electrons if the bond is broken. Formally, the energy required to form a C–I bond and break a C–H bond is too high for this reaction to occur, but a simplistic question—**Where do the electrons go?**—allows a prediction that this reaction should be very difficult, especially when compared to the reaction of iodide and **11**. If the electrons of the C–H bond cannot be transferred to hydrogen when iodide collides with carbon, the reaction does **not** occur.

7.10 Reversible Chemical Reactions

A reversible chemical reaction is one where the compounds normally defined as products react to regenerate the compounds normally defined as starting materials, and the two reactions are competitive. All of the Brønsted–Lowry acid–base reactions discussed in previous chapters are equilibrium reactions, defined by an equilibrium constant K. They are acid–base reactions, so K_a is used, but the principle is fundamentally the same. The fundamental principles of equilibria and reversible reactions can be applied to other systems.

7.10.1 Energy of Activation for the Reverse Process

$$A \;+\; B \;\rightleftharpoons\; C \;+\; D$$

In the reaction of A + B to give C + D, there are two competing reactions. The first is the chemical reaction of A and B, and the products are C and D. The second chemical reaction is the reverse of the first one, where C reacts with D to give products A and B. Going from left to right, the energy curve in Figure 7.7 defines the reaction (A + B → C + D) and E_{act} with the starting materials on the left and the products on the right. Looking from right to left, however, there is

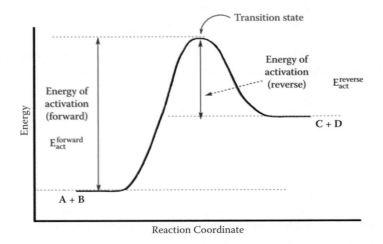

Figure 7.7 Forward and reverse energy of activation.

another E_{act} that is smaller than the E_{act} for the forward reaction of C and D to give A and B. The reverse reaction is formally endothermic, so it takes less energy to go back to A and B than it does to form C and D, initially. The first criterion of a reversible reaction, therefore, is the presence of a low E_{act} for the reverse process relative to the forward process. Note that the reverse E_{act} need not be smaller than the forward E_{act}—only close in energy.

If a reaction is reversible, both products and reactants will be present during that reaction. If the reaction of A + B to give C + D is reversible, all four components—A, B, C, and D—will be present. In principle, the reaction will reach an equilibrium condition and at that point the amounts of A–D will be relatively constant. If the reaction comes to equilibrium (designated by the "stacked" forward and reverse reaction arrows), the position of the equilibrium is defined by the **equilibrium constant, K** (K_{eq}). The value of K is determined by the concentrations of all species, **K = products/reactants**, where ***products are on the *right* side of the reaction and *reactants* are on the *left* side of the equation***. In this particular reaction, K = [C][D]/[A][B].

If the value of K is very small, the bottom term (reactants) is larger than the top term (products). As a practical matter, this is interpreted to mean that A does not react with B to give products. The reaction does not work well as written, and the reverse reaction is more favorable. If the value of K is large, the top term (products) is greater than the bottom term (reactants) and then the reaction proceeds to products as written. In other words, A reacts with B to give C + D as the products. If the value of K is about 1, the reaction has an equal mixture of products and reactants at equilibrium.

In order to simplify the numbers, pK is converted to the negative log of K (pK = –log K). If K = 2,560, pK = –log 2,560 = –3.408. If the pK is 5.4, then K = 10^{-pK} = $10^{-5.4}$ = 4×10^{-6}). This is exactly analogous to the definitions of K_a and pK_a used for acids in Chapter 2, Sections 2.1 and 2.2, and in Chapter 6, Section 6.1. The most important thing to remember is that K and pK are *inversely* proportional, so a large value of K correlates with a small pK, and a small value of K correlates with a large pK. If the pK of a reaction is 15, the equilibrium is to the left (small value of K), but if the pK is 1.2, the equilibrium is to the right (large value of K).

If a reaction is identified as reversible and there is an equilibrium constant (K), K is related to the free energy ($\Delta G°$). It is known that

$$K = e^{-\dfrac{\Delta G°}{RT}}$$

or

$$\Delta G° = -RT\,(\ln K) = -2.303\,RT\,(\log K)$$

where R = 1.986 cal/deg mol, T = temperature in Kelvin, and e = 2.718 (base of natural logarithms). With these equations, the values of K for two reactions can be compared and the values of $\Delta G°$ used to predict which one might be more favorable.

7.14 Calculate K for a reaction where ΔH° is 68.2 kcal mol^{-1} and the reaction temperature is 91°C.

Another factor to be considered is whether the compounds of the right side of the equation (by definition, the products) are more reactive (less stable) than those on the left side of the equation (normally, the reactants). If the reaction of chloromethane (**11**) and sodium iodide occurred, it would generate iodomethane (**12**) and sodium iodide. For reasons that will be apparent in Chapter 11 (Section 11.3), the iodide ion is a better nucleophile than the chloride ion, as illustrated by Figures 7.1 and 7.2. This means that the value of K for the conversion of **11** to **12** is >>1 for the reaction, and the reverse reaction has a K of <<1. In other words, iodine reacts with bromomethane to give product, but bromide does not react with iodomethane to give bromomethane.

This fact is an experimental observation that allows the value of K to be correlated with nucleophilic strength, and iodide is a better nucleophile than bromide in this reaction. In other words, the reaction of iodide with chloromethane is more favorable than the "reverse" reaction (chloride reacting with iodomethane) because the two molecules on the left of the equation are more reactive than those on the right side of this equation. Several examples of reactions that are potentially reversible will be seen, but the reaction parameters change with the reaction and the reactants, and the position of the equilibrium (the value of K) changes.

7.10.2 Not All Reactions Are Reversible

Several points can be made that concern equilibrium reactions:

- When the equilibrium constant for a reaction (K) is extremely large, the reaction is effectively irreversible.
- It is possible that a reaction will generate products that cannot react with each other regardless of the reaction conditions.
- It is possible that one product may escape from the reaction medium (a low molecular weight gas, for example).
- If the two reactants are less stable than the products, the equilibrium constant will be large (>>1).
- If the bonds formed in the product are much stronger than those broken in the reactant, the equilibrium usually shifts to the right.
- If the E_{act} for the reaction going to the right (forward) is much smaller than that going to the left (reverse), the reaction is effectively irreversible.
- A large K is usually correlated with products (right side of equation) being less reactive than the reactants (left side of equation).

As noted previously, some chemical reactions produce one or more products that are gases. The reaction of 2-chloro-2-methylbutane (**24**) and KOH, for example, gives 2-methyl-2-butene (**25**) and HCl gas, and a gas may escape from the reaction medium. However, the HCl is formed in the presence of the basic species hydroxide, so HCl will react with HCl to form KCl and water. The reaction of HCl with KOH effectively consumes one product, which inhibits the reverse reaction, but also consumes one of the reactants. Therefore, a minimum of two equivalents of hydroxide is required in this reaction. One equivalent will react with **24**, and the second will react with the HCl product as it is produced. HCl is removed from the reaction, which "drives" the reaction to the right, making it effectively irreversible.

7.11 Kinetics

Once a chemical reaction has been identified, it is reasonable to question how fast it will go. How long does it take until the reaction is complete: a minute? an hour? a day? a week? Apart from the obvious information about how long one will have to monitor the reaction, other information can be obtained from knowledge of how fast a reaction goes and how long it takes. In many cases, such knowledge can give insight into the mechanism of a reaction and help to identify intermediates in the reaction. The number of intermediates (which defines the number of chemical steps in the transformation) is clearly important because this leads to the mechanism of the reaction and a better understanding of the process.

7.11.1 Reaction Rate and First-Order Reactions

The **rate of reaction** is a parameter that measures how quickly a reaction proceeds. In other words, how long does it take to consume a starting material and convert it to product? An analogy is the rate of speed for an automobile, which is defined in terms of a unit of distance traveled per unit of time. For a chemical reaction, a different parameter per unit of time is required: the change in molar concentration of either the starting materials (reactants) or the products. A simple reaction in which A is transformed into B (A → B), with no intermediate, will be used to illustrate the concept.

   The concentration of the starting material (how many moles of reactant per liter of solvent) is known, and that initial number of moles must *decrease* during the reaction as that starting material A is consumed to form the product B. Because there is no product at the beginning of a reaction, measuring the number of moles of product being formed during the reaction as a function of time also allows one to assess the rate of a reaction. However, if multiple steps occur in the mechanism of the reaction, there may be problems with this

Table 7.2
Loss of Starting Material
A as a Function of Time

[A]	Time (min)
1.0	0
0.9	100
0.54	300
0.4	400
0.21	680
0.12	900
0.09	1080
0.05	1240
0.02	1400
0.01	1700

measurement. In a simple case, a rate is determined by measuring the initial concentration (in moles or moles/liter) of a reactant (A) and then measuring the concentration of A at certain time intervals as it is consumed over a known time period. At the beginning of this particular reaction (time = 0 seconds), the concentration of A is determined to be 1.0 *M*. After 100 minutes, the concentration of A is measured to be 0.9 *M,* after 300 minutes it is measured to be 0.54 *M,* etc. The measurements are continued until A has disappeared and a table of data is generated: concentration of A and time (in minutes), as in Table 7.2. The term [A] is used, where the brackets indicate molar concentration, so [A] is the molar concentration of reactant A for each time point in the reaction.

The rate of a reaction is a dynamic property because the concentration of A changes during the course of the reaction. Because the concentration changes as a function of time, which also changes, Table 7.2 represents data for a differential equation, **rate = –d[A]/dt**, where the rate is proportional to the concentration of A and also to time. The differential equation is

$$\text{rate} = -\int_{[A]_0}^{[A]_t} \frac{d[A]}{A} = k \int_0^t dt$$

where $[A]_0$ is the concentration of A at time = 0, and $[A]_t$ is the concentration of A at any specified time. When this differential equation is integrated such that [A] is $[A]_0$ at t = 0 and it is $[A]_t$ at t = "end time," the expression obtained is

$\ln [A]_0/[A]_t = k (t_{\text{end time}} - t_0)$ and if $t_0 = 0$ (as defined), then $\ln [A]_0/[A]_t = k\, t$

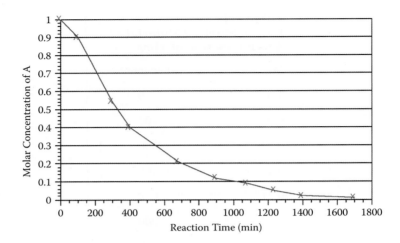

Figure 7.8   Plot of the data in Table 7.3.

Because concentration is proportional to time (conc. $\propto$ t), setting the rate of change in concentration equal to the rate of change of time requires a proportionality constant, k (for time changing from t = 0 to t = some measured time). In this case, conc. = k t. *The proportionality constant (k) is defined as the **rate constant**.* ***Note that the small k is used to represent the rate constant, whereas a capital K is used to indicate an equilibrium constant.***

The value of k is determined by plotting some function of A versus time, and the result is the graph shown in Figure 7.8 (the data come from Table 7.2). It is clear that the plot is a curve, not a straight line, as the concentration of A diminishes as the reaction proceeds. To obtain a value for k, there must be a straight-line plot. One solution plots ln $[A]_0/[A]_t$ versus time to give a straight line (see Figure 7.9), and the slope of this line gives k, the rate constant. Note that Figure 7.9 shows the "best" straight line, which is common in such measurements due to measurement errors in the experiment. In this case, the slope is 2.8×10^{-3} mol/L/min. Note the units of the rate constant, which show a change in concentration per unit of time. This rate constant and particularly how it was calculated may give valuable information about how the reaction works.

7.15 **Briefly explain why knowing the rate constants for two competing reactions occurring at the same time in the same reaction flask might be important.**

Figure 7.8, the original plot of [A] versus time, and Figure 7.9, the plot of ln [A] versus time, are characteristic of what is called a **first-order reaction**. A first-order reaction is one is which the concentration of A is the only concentration term in the equation. This is an important piece of information because the preceding equations lead to a **definition of the rate of the first-order reaction as *rate* = *k [A]*.** This expression is called the **rate equation** for the reaction. How fast the reaction proceeds depends only on the concentration of reactant A, rather than on the interaction with another molecule in the rate-determining (the slowest) step. This rate equation

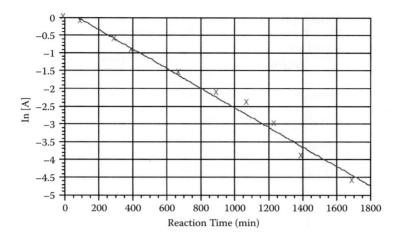

Figure 7.9 Plot of ln [A] versus time from data in Table 7.3.

suggests that reactant A essentially falls apart to give the product in the key step. This statement may not be true in terms of how the product is formed, but it does give useful information that may rule out many possible mechanisms. It places the focus on those processes that initiate decomposition of A. This may be the single greatest value in determining the rate equation, beyond the knowledge of how long one must wait for the reaction to be completed.

7.16 **Examine the first-order rate constant and determine whether the rate would be increased, decreased, or remain unchanged if [A] were to be increased by a factor of 100.**

7.11.2 Second-Order Reactions

Another type of reaction is quite different from the one just discussed, and it is illustrated by the following, where reactant A must react with B to form a new bond and a new product, A–B.

$$A + B \longrightarrow A{-}B$$

Assume that the rate data in Table 7.3 are obtained for this reaction.[1] When [A] is plotted against time, the curve is shown in Figure 7.10, which shows that the rate of the reaction depends on the concentration of both A and B. In other words, simply plotting the concentration of A versus time or the concentration of B versus time does not give a curve that describes the reaction. The concentrations of both A and B must be plotted against time to obtain the correct result for formation of A–B. To calculate the rate constant, *both* reactants are important for the reaction to occur, and the rate expression is found to be $- d[A]/dt = k\,[A]\,[B]$, or rate $= k\,[A]\,[B]$.

Table 7.3
Rate Data for a Typical Second-Order
Reaction

[A]	[B]	Time (min)
0.00980	0.00486	0
0.00892	0.00398	178
0.00864	0.00370	273
0.00792	0.00297	531
0.00724	0.00230	866
0.00645	0.00151	1510
0.00603	0.00109	1918
0.00574	0.00080	2401

The differential equation for this reaction is solved to give a complex expression:

$$1/[A]_0 - [B]_0 \ln [B]_0[A]_t/[A]_0[B]_t = k\, t$$

where the $[\ ]_0$ terms denote the initial concentrations of A or B (at time = 0), and the $[\ ]_t$ terms indicate the concentrations of A and B at a specified time. To obtain a straight-line plot, time is plotted against the integrated rate expression $\ln [A][B]_0/[A]_0[B]$. This second-order plot is shown in Figure 7.11 and the slope is calculated to be $5.93 \times 10^{-4}\ M^{-1}\mathrm{sec}^{-1}$, which is the second-order rate constant.

What do these calculations mean in terms of a mechanism? If the rate expression is rate = k [A] [B], it means that A must react with B for the reaction

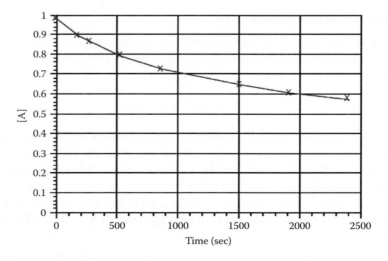

Figure 7.10   Plot of [A] versus time in a second-order reaction.

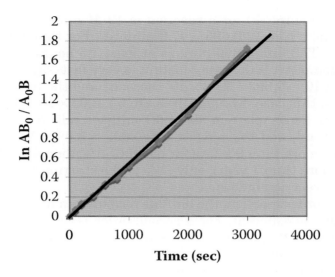

Figure 7.11   Second-order plot for the reaction of A and B.

to occur. Any mechanism and any intermediates proposed must take this into account.

7.17 Examine the second-order rate constant and determine whether the rate would be increased, decreased, or remain unchanged if [A] were to be increased by a factor of 100.

7.11.3 Half-life

The half-life of a reaction is the amount of time required for half of the starting material to react. This is a useful device for determining how long it takes for a reaction to proceed to completion. Assume that it takes 100 minutes for 50% of a starting material to be converted to product. If the initial concentration of a reactant X is 1.0 *M,* then the concentration of X after 100 minutes will be 0.5 *M.* This is the concentration of starting material that remains after one half-life. After another 100 minutes, 50% of the remaining X will react, and the concentration of X is 0.25 *M* (after the second half-life). Another 100 minutes are required to bring the concentration to 0.125 *M,* and after a total reaction time of 500 minutes (five half-lives) the concentration of X is 0.0313 *M.* In other words, 0.9787 mol of X have reacted (97.87%) and, for a reaction to be "complete," it must be allowed to proceed for more than five half-lives.

7.18 Determine how many moles of A have reacted, after five half-lives, if the initial concentration of A is 1.8 *M.*

The half-life for a first-order reaction (see Section 7.11.1) is given the symbol $t_{1/2}$ and is calculated from the simple formula:

$$\text{Half-life (first order)} = t_{\frac{1}{2}} = k/\ln 2 = k/0.693$$

If the rate constant (k) for a reaction is 12 M min^{-1} (moles per liter per minute), then the $t^{1}/_{2}$ is 12/0.693 = 17.3 minutes. For this reaction, five half-lives will be 5 × 17.3 min = 865 min, so at least 1.5 hours is required for the reaction to be complete. If the rate constant is very fast, say, 1.4 × 10^{-5} M sec^{-1}, then $t^{1}/_{2}$ = (1.4 × 10^{-5})/0.693 = 2.02 × 10^{-5} seconds and the reaction will rapidly be completed. In this case, five half-lives are 5 × 2.02 × 10^{-5} sec = 1 × 10^{-4} sec. The reaction will be complete in about 0.1 milliseconds. A slow rate constant is 5.6 × 10^{6} M sec^{-1}, and $t^{1}/_{2}$ = (5.6 × 10^{6})/0.693 = 8.08 × 10^{6} seconds (= 93.5 days). If it takes more than 93 days for one half-life, then after four half-lives, or 374.1 days (more than 1 year), only 93% of the starting materials will have reacted—a slow reaction indeed.

For a second-order reaction, the half-life formula is a little different because the rate equation is different. For a second-order reaction (A + B) where the initial concentration of A is the same as B (i.e., $[A]_0 = [B]_0$), the time required for half the starting materials to react is *inversely proportional to the initial concentration*. The half-life for a second-order reaction is

$$\text{Half-life (second order)} = t_{\frac{1}{2}} = 1/k[A]_0$$

The second-order half-life is used the same way as the first-order half-life. The units for a second-order half-life (M^{-1} sec^{-1}) are different, but five half-lives or more are required for a reaction to be "complete." If k = 2 M^{-1} sec^{-1} and [A] = 1, then $t^{1}/_{2}$ = 0.5. For this reaction, five half-lives is 2.5 sec. Similarly, if k = 3 × 10^{-3} M^{-1} sec^{-1} and [A] = 1, then $t^{1}/_{2}$ = 333.3 and if k = 3 × 10^{5} M^{-1} sec^{-1} and [A] = 1, then $t^{1}/_{2}$ = 3.3 × 10^{-6}.

7.12 No Reaction

How long does it take for a reaction to be completed? This is an important question because it defines the term *no reaction*. To say that mixing two reactants does not give a product indicates that there is no reaction, which is drawn as the following:

A + B —————————▶ No Reaction or N.R.

Suppose the rate of this reaction is such that the half-life is 1.5 × 10^{8} minutes. That means it will take 2.5 million hours to get a 50% yield, which is more than 104,166 days or about 285 years. Five half-lives are required to obtain >97% yield, which is 1,427 years. Obviously, this is absurd and this example is defined as no reaction.

Are you willing to wait 10 years for a reaction to be complete? How about 1 year? These are silly questions. Allowing minutes, hours, or even several days for a reaction to be complete is probably reasonable, but waiting longer than a few weeks except under special circumstances is not. The point is that, for all practical purposes, reactions are defined as working or not working (reaction or no reaction) by the time required for their completion (essentially measured by the half-life). In many cases, an excellent argument can be made that the energy of a process is such that it cannot happen, and there is no reaction. However, in effect this means that the reaction is so slow that nobody will wait long enough to find out. Another point to consider is that a reaction may be very slow, but a catalyst may be found that will increase the rate of the reaction to the point that it is practical and useful. Recognizing that a reaction is slow may be a clue that research should be done to find a suitable catalyst.

When there are two competing reactions, the tendency is to think that one process will occur and that the other is impossible. This is usually incorrect. More commonly, one process is simply slower than another and, given enough time, a product may be observed from the slower reaction. *It is very important to differentiate reaction rates that are so slow that they are unlikely to compete with another reaction in a given time from those that are so slow that no product will be formed given the length of time the reaction is monitored*. For example, in the example where A could be transformed to either B or C (see Section 7.4.3), the high activation energy of the A to C transformation led to the conclusion that only product B is formed. If the temperature of the reaction is increased, both B and C may be formed.

Suppose the rate of the A to B transformation is 20,000 M min^{-1} and the rate of the A to C transformation is 10 M min^{-1}. If the reaction proceeds for 1,500 minutes (slightly more than 24 hours), it is likely that no C will be formed. If the reaction is allowed to proceed for 30,000 minutes, however, some product C maybe formed. If the temperature of the reaction is low and it is allowed to proceed for only a short time, it is likely that only B and no C will be formed. What does this discussion have to do with this section? If the reaction is stopped after 1,500 minutes, it can be said that the reaction A → C gives no reaction, but if it is stopped after 30,000 minutes, the reaction A → C is said to be slow.

How fast is fast? How slow is slow? *The previous discussion may be a bit confusing at this point in the book because specific examples of competing reactions will not be discussed until later chapters.* However, it is important to know that when there are two competing reactions and one is slow and the other is fast, reaction time is an important parameter. Also, when there are two competing reactions, categorizing one as "no reaction" simply means that no product is expected from that reaction in the time frame specified. The term "no reaction" should provoke the question: Is the reaction really impossible or just too slow to be observed?

7.19 **Do you think that a reaction that produced a molecule with a half-life of 6.2×10^{15} seconds would be considered commercially viable? Why or why not?**

7.13 Biological Relevance

$$E + S \underset{k_2}{\overset{k_1}{\rightleftarrows}} ES \underset{k_4}{\overset{k_3}{\rightleftarrows}} E + P$$

In chemical reactions, a catalyst participates in a reaction either to initiate the reaction or influence the rate of the reaction, and it is regenerated during the course of the reaction. Therefore, only a small amount of the catalyst must be used. Enzyme catalyzed reactions are ubiquitous in biological systems. Figure 7.12 shows an energy profile of a catalyzed and an uncatalyzed reaction.[2] Clearly, the catalyzed reaction is more complex in that it has more steps, but complexation with the enzyme catalysts leads to a lower activation energy, and there is a similarity to Figure 7.1. Indeed, catalysis by an enzyme can

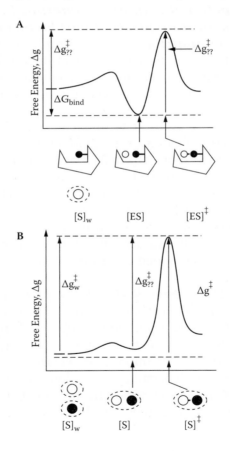

Figure 7.12 (A) Free energy profile for an enzymatic reaction (a) and that for the corresponding solution reaction. The figure describes the free energies associated with kcat/KM and kcat. (B) Description of the energetics of a reference solution reaction that is not catalyzed. (Reprinted in part with permission from Warshel, A. et al. 2006. *Chemical Reviews* 106:3210. Copyright 1991 American Chemical Society.)

be correlated with the principles that govern the chemical reactions discussed in this book. It is just organic chemistry! It is more complex, but it is organic chemistry. For enzymatic reactions, the rate of a reaction has been called the velocity of the reaction, but rate constants can be determined for each process.

A reversible chemical reaction in this context is between an enzyme and a substrate (a molecule of biological importance). It has been observed that at low substrate concentration, the reaction velocity is proportional to the substrate concentration and the reaction is first order with respect to substrate.[3] Increasing the concentration of the substrate causes the reaction rate to diminish, and there is a point where it is no longer proportional to the substrate concentration and, indeed, the rate of the enzyme-catalyzed reaction becomes constant and independent of substrate concentration. Here, the reaction is zero-order (*a zero-order reaction is independent of the concentration of the reactant, so a higher concentration of reactants will not increase the rate of reaction*) with respect to the substrate and the enzyme is said to be saturated with substrate (saturation). This effect is described by a process in which the enzyme (E) reacts with substrate (S) to form a complex ES, which then breaks down to regenerate the enzyme and products (P).[2] Both reactions are reversible, with the rate constants that are indicated k_1–k_4. This reaction has been analyzed to give the following:

$$\frac{[S]\,([E] - [ES])}{[ES]} = \frac{k_2 + k_3}{k_1} = K_M \quad \text{and,} \quad [ES] = \frac{[E]\,[S]}{K_M + [S]}$$

where the second equation has a constant KM, which replaces the rate constant term as shown, and is called the *Michaelis–Menten constant*. Further manipulation of the equation leads to the *Michaelis–Menten equation*, which defines the quantitative relationship between the enzyme reaction rate and the substrate concentration [S], if both V_{max} and K_M are known, where $V_{max} = k\,[E]$ (V_{max} is the maximum velocity for formation of the complex ES):

$$\text{Velocity} = v = \frac{V_{max}\,[S]}{K_M + [S]}$$

It is clear that the fundamental rate expressions presented in Section 7.11, as well as the equilibrium expressions in Section 7.10 and Chapter 2, Section 2.2, are applied to enzyme reactions. In effect, treating an enzyme reaction as an organic chemical reaction allows analysis of a given transformation, which leads to a better understanding of the relevant biology.

This chapter discussed key intermediates in organic chemistry. Similar or identical intermediates arise in biological transformations, particularly radicals and carbocations:

There are cellular oxidants that are derivatives of oxygen (called reactive oxygen species, ROS), and are constantly produced in our cells. Superoxides and hydroxyl radicals are particularly important ROS, generated by mitochondria via the release of electrons from the electron transport chain and the reduction of oxygen molecules to superoxides ($O_2\bullet$). The enzyme *superoxide dismutase* can transform a superoxide into hydrogen peroxide (H_2O_2), but reaction of hydrogen peroxide transition metal ions such as iron or copper produces the most reactive

ROS, hydroxyl radicals (•OH) (the so-called **Fenton reaction**). Cytochrome P450 complexes in the endoplasmic reticulum generate superoxides to metabolize toxic hydrophobic compounds and phagocytes produce superoxides, hydrogen peroxide and hydroxyl radicals to kill infectious microorganisms and cancer cells.[4]

Phagocytes kill harmful microorganisms by releasing ROS. According to Salganik, "NADPH supplies electrons, required for the reduction of oxygen and the formation of ROS. In turn, NADP+ receives electrons from the pentose cycle pathway by NADPH oxidase through cytochrome b245."[4]

An excessive amount of ROS can interfere with the beneficial effects. Detoxification by cytochrome P450 occurs by release of ROS to oxidize toxic hydrophobic substances (lipid soluble; see Chapter 20, Section 20.2) to hydrophilic (water soluble) compounds that can be removed from the body. An excess of ROS may also interfere with this important function. Note that these radical intermediates are formed by enzyme mediate chemical reactions and are essential to life, and that they may be related to "increased formation of ROS may play an important role in carcinogenesis, atherosclerosis, diabetes, emphysema, cataracts and neurodegenerative diseases."[4]

References

1. Daniels, F., and Alberty, R. A. 1967. *Physical chemistry,* 3rd ed., 331. New York: John Wiley & Sons.
2. Warshel, A., Sharma, P. K., Kato, M., Xiang, Y., Liu, H., and Olsson, M. H. M. 2006. *Chemical Reviews* 106:3210.
3. Lehninger, A. L. 1970. *Biochemistry,* 153–154. New York: Worth Publishing Inc.
4. Salganik, R. I. 2001. *Journal of the American College of Nutrition* 20:464S–472S.

Answers to Problems

7.1

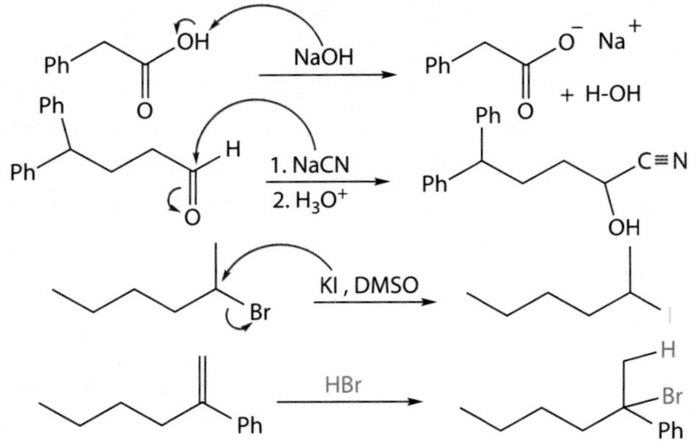

7.2 $\Delta H° = H°_{prod} - H°_{react} = 100 - 68 = 32$ kcal mol^{-1}.

7.3

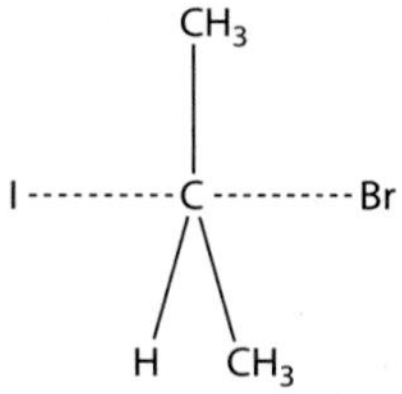

$I \cdots\cdots\cdots C \cdots\cdots\cdots Br$

with CH_3 up, H and CH_3 down

7.4

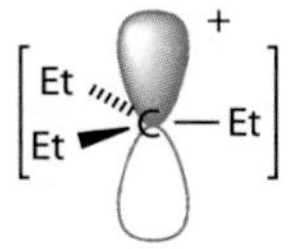

$\begin{bmatrix} Et_{\text{\tiny////}} \\ Et \blacktriangleright C-Et \end{bmatrix}^+$

7.5

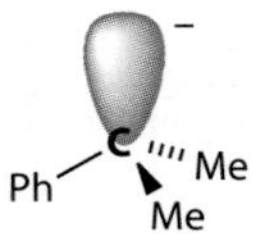

$Ph \diagup C_{\text{\tiny////}} Me$, Me with $-$ charge

7.6

$Ph - C_{\text{\tiny////}} H$, H with radical dot

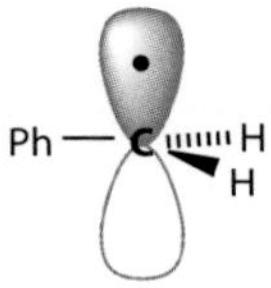

7.7 $\Delta G^\circ = \Delta H^\circ - T\Delta S^\circ$. Therefore, $22.4 = 21.5 - 298(\Delta S^\circ) = 22.4 - 21.5 = -298^\circ \Delta S^\circ$. Therefore, $\Delta S^\circ = 0.9/-298^\circ = -0.0030$ kcal mol^{-1} = -3 cal. Because the ΔG° term is 22.4 kcal mol^{-1} and the ΔS° term is about 3 cal, leaving out the $T\Delta H^\circ$ term will introduce only a small error into the calculation. Specifically, 0.9 kcal mol^{-1} out of 22.4 kcal mol^{-1} = 4% error.

7.8   Because the ΔH° term is negative, this indicates an exothermic (spontaneous) reaction.

7.9   The curve shown in Figure 7.4 illustrates this process.

7.10 Because this is an endothermic reaction, the energy of the products must, by definition, be higher in energy than the energy of the starting materials.

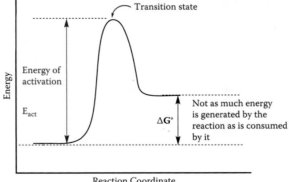

7.11 The reaction with the lower E_{act} (35 kcal/mol) will occur at a lower temperature. The value of $\Delta H°$ does not play a role in initiating the reaction, which is measured by E_{act}.

7.12

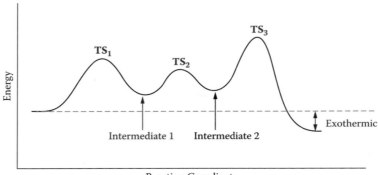

7.13

7.14 $\Delta G° = -RT (\ln K) = -2.303\ RT (\log K)$; $°K = °C + 273.1$. Assume that $\Delta S°$ is zero, so $\Delta G° = \Delta H°$. Therefore, $68.2 = -2.303(1.986)(364.1) \log K$, and $\log K = 68.2/-1665.3$, and $\log K = -0.041$. Then, $K = 10^{-\Delta G°/RT} = 10^{-(-0.041)} = 1.099$. $K = 1.099$.

7.15 If one reaction proceeds faster than another, then the faster reaction should lead to the major product of the reaction. Knowing this allows us to predict major versus minor products. If the rate difference is quite large, we may obtain one product in very good yield.

7.16 Increasing or decreasing the concentration of A in a first-order reaction has no effect on the rate constant; therefore, it will remain unchanged.

7.17 For a second-order reaction, the rate is dependent on reaction partners A and B. Increasing the concentration of one increases the overall rate, so increasing the concentration of A by 100 should increase the rate by a factor of 100.

7.18 For an initial concentration of 1.8 M, half will be consumed after one half-life = 0.9 M. Therefore, we expect that 0.9 mol has reacted after one half-life; 0.9 + 0.45 = 1.35 after two; 0.9 + 0.45 + 0.225 = 1.575 after three; 0.9 + 0.45 + 0.225 + 0.113 = 1.688 after four; and 0.9 + 0.45 + 0.225 + 0.113 + 0.0561.744 after five. After five half-lives, 1.744 mol of the initial 1.8 have reacted; this means that 96.9% of A has reacted.

7.19 No! If we convert this half-life to years, we find that it is 1.97×10^8 years. Clearly, 0.2 billion years is a long time to wait for a commercial product.

Correlation of Concepts with Homework

- A chemical reaction involves making and breaking bonds, with the gain, loss, or transformation of a functional group: 1, 20, 21.
- Enthalpy is the term used to measure bond dissociation energy, and $\Delta H°$ for a reaction is the difference in bond dissociation energy for bonds made (products) minus that for bonds broken (starting materials). This is products minus reactants: 2, 22.
- The free energy of a reaction determines if the reaction is spontaneous (exothermic) or nonspontaneous (endothermic). The free energy $(\Delta G°)$ is calculated: $\Delta G° = \Delta H° - T\Delta S°$: 7, 8, 10, 11, 24, 26, 32, 35.
- The energy of activation (E_{act}) is the energy required to initiate the bond-making/bond-breaking process of a chemical reaction. The transition state for a reaction is the highest energy portion of a reaction curve and corresponds to the point in a reaction where bonds in the starting materials are "partly broken" and bonds in the products are "partly made": 3, 9, 12, 24, 25, 27, 28.
- An intermediate is a transient, high-energy product that is not isolated but is converted to the final isolated products. The most common intermediates are carbocations, carbanions, and radicals: 4, 5, 6, 27.
- A mechanism is a step-by-step map that shows the presence or absence of intermediates and the total number of steps required to convert the string materials to the products: 13, 36, 37, 38.
- Reversible chemical reactions have an E_{act} (reverse) that is equal to or less than E_{act} (forward), for the forward and reverse reactions. The position of the equilibrium is measured by the equilibrium constant K, defined as products divided by reactants: 14, 33.
- In ionic reactions, an atom with an excess of electrons (– or δ–) donates two electrons to an atom that is electron deficient (+ or δ+) to form a new bond. When this occurs, a bond in the original molecule must break, with transfer of the two electrons in that bond to one of the atoms in the starting material: 40, 41.
- A first-order reaction rate is obtained by plotting the natural logarithm (ln) of the concentration of a starting material versus time. The rate of a second-order reaction is obtained by plotting the ln of a concentration term that includes both reactants versus time. The rate equation for a first-order reaction is rate = k [A], where k is the rate constant. The rate equation for a second-order reaction is rate = k [A] [B], where k is the rate constant: 15, 16, 17, 29, 30, 31.
- The half-life of a reaction is the time required for half of the starting materials to react: 18, 19, 34, 39.

Homework

7.20 Draw the reaction and final product or products for the reaction of sodium cyanide, sodium amide, and sodium iodide with
  (a) 3-pentanone
  (b) 2-iodo-4-phenylpentane
  (c) cyclopentanecarboxaldehyde

7.21 Taking the transformation of **7** to **8** as a guide, draw the product expected when
  (a) 1-pentene reacts with HI
  (b) 1,2-dimethylcyclobutene reacts with HBr
  (c) 3,4-dimethyl-3-hexene reacts with HCl

7.22 Use the energy values in Table 7.1 to determine the value of $\Delta H°$ for the following **hypothetical** reactions (none of these transformations actually occur). Determine if each is endothermic or exothermic:

(a) CH_3I + CH_3OH $\longrightarrow$ $H_3C-O{\overset{H}{\underset{CH_3}{}}}^+$

(b) CH_3CH_3 + CH_3NH_2 $\longrightarrow$ $\overset{H}{\underset{+}{N}}{\overset{H}{\underset{H}{}}}$

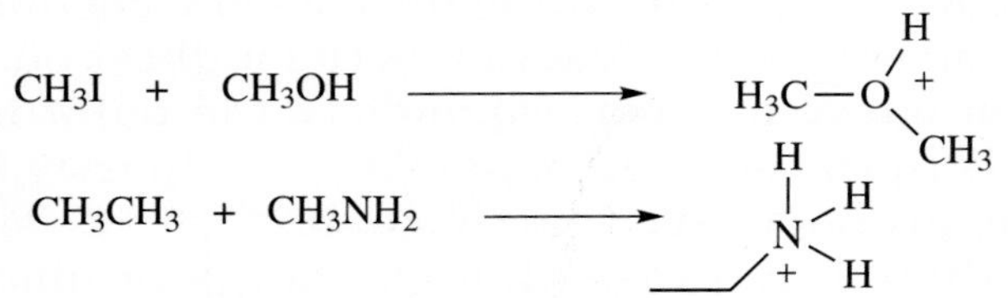

(c)   F-F + $(CH_3)_3C$-CH_3 $\longrightarrow$ CH_3F + $(CH_3)_3C$-F

(d) Me_3CH + I^- $\longrightarrow$ Me_3C-F

7.23 Calculate $\Delta G°$ for a process where $\Delta H° = 56$ kcal mol^{-1}, T = 100°C, and $\Delta S° = 3.2$ cal^{-1}. The $\Delta S°$ term is insignificant and the $T\Delta S°$ term is relatively small. Calculate the percentage of the $T\Delta S°$ term relative to the $\Delta H°$ term. Calculate how much the temperature must be changed in order to make the $\Delta G°$ term equal to -10 kcal mol^{-1}. Is this realistic and achievable goal? Why or why not?

7.24 For a given reaction, explain the relationship between the energy of activation and the transition state.

7.25 Comment on whether or not it is possible to "see" a transition state. Why or why not?

7.26 When a reaction has a negative value for $\Delta H°$, it is said to be spontaneous. Does this mean that simply mixing the reactants together will automatically produce the product in a reaction that generates heat? Why or why not?

7.27 A reaction with 2-bromo-2-methylpropane generates a cation intermediate by direct loss of the bromine. Draw the transition state for this reaction as well as the cation intermediate.

7.28 Acetonitrile (CH_3CN) can react with a base (B:) to generate a carbanion. Draw a reasonable structure for this carbanion as well as the transition state that leads to this intermediate. Comment on the relative stability of this intermediate.

7.29 Given the following experimental data, determine the first-order rate constant and the half-life:

[A]	Time (sec)
2.0	0
1.5	25
0.95	50
0.7	100
0.52	150
0.40	200
0.29	250
0.23	300
0.19	350
0.12	400

7.30 Plot this data to obtain the second order rate constant.

Plot This Data to Obtain the Second Order Rate Constant

[A]	[B]	Time (sec)
0.00980	0.0486	0
0.00892	0.0398	17.8
0.00864	0.0370	27.3
0.00792	0.0297	53.1
0.00724	0.0230	86.6
0.00645	0.0151	151.0
0.00603	0.0109	191.8
0.00574	0.0080	240.1

7.31 Calculate the half life, given the following data. For second order reactions assume $[A]_0 = 0.5$.
   (a) a first-order reaction where $k = 1.2 \times 10^{-6}$
   (b) a second-order reaction where $k = 4.5$
   (c) a first-order reaction where $k = 5.8 \times 10^3$
   (d) a second-order reaction where $k = 9.25 \times 10^{-4}$
   (e) a first-order reaction where $k = 0.6 \times 10^{-9}$
   (f) a second-order reaction where $k = 3.44 \times 10^{12}$

7.32 Calculate $\Delta G°$ given the following equilibrium constants for reactions at 25°C:
   (a) 2.5
   (b) 1.55×10^{-6}

(c) 8.77×10^{-9}
(d) 4.4×10^{5}
(e) 1.23×10^{18}
(f) 10.45×10^{-3}

7.33 Calculate the equilibrium constants given the following values of $\Delta H°$ for reactions at 25°C, assuming that $\Delta S°$ is less than 1 cal in all cases:
(a)   −1.5 kcal mol⁻¹
(b)   100.3 kcal mol⁻¹
(c) -4.5×10^{4} cal mol⁻¹
(d)   18.5 kcal mol⁻¹
(e)   −33 kcal mol⁻¹
(f) -12.5×10^{6} kcal

7.34 (10 pts.) The half-life for a certain reaction is 8 hours. Estimate how many half-lives and how many hours would be required for the reaction to go to at least 98% completion.

7.35 (11 pts.) For reaction A, $\Delta G°$ is −200 kcal mol⁻¹; for reaction B, $\Delta G°$ is −20 kcal mol⁻¹. For reaction A, $\Delta G^{\ddagger} = +100$ kcal mol⁻¹; for reaction B, $\Delta G^{\ddagger} = +10$ kcal mol⁻¹. Both reactions are run at room temperature without any other source of heat or energy, and the available energy at room temperature is about 25 kcal mol⁻¹. Briefly discuss which, if either, of these reactions will go to products.

7.36 Based on the conversion of cyclopentene to chlorocyclopentene in Section 7.7, give the mechanism for the following reaction:

7.37 Work backward from the given product to the starting material to provide a mechanism for the following reaction, with the experimental knowledge that there are two intermediates and three chemical steps in the process:

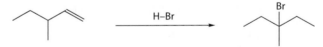

7.38 Given the concentration curve that is provided for loss of starting material and appearance of product, speculate on whether or not this reaction is likely to have an intermediate:

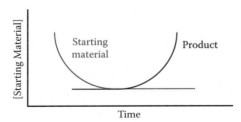

7.39 Two competing reactions lead to either C or D from the starting materials A and B:

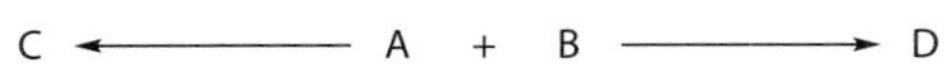

$$C \longleftarrow \quad A \; + \; B \quad \longrightarrow D$$

If k_1 is the rate constant for the formation of C, k_2 is the rate constant for formation of D, and $k_1 = 3.6 \times 10^{-2}$ mol L^{-1} sec^{-1}, but $k_2 = 3.5$ mol L^{-1} sec^{-1}, calculate the half-life for each reaction and briefly discuss which product (C or D) is likely to be the major product. Assume that both are second order reactions, and that the initial concentration is 1 mol L^{-1}.

7.40   Which of the following is likely to be the best two-electron donor in a reaction with CH_3Cl? Explain.
  (a) CH_3O^-
  (b) HO^-
  (c) Cl^-

7.41   For each of the following, draw reaction arrows to indicate a two-electron (two-headed arrow) or a one-electron (single-headed arrow, like a fishhook) transfer that corresponds to all bonds broken and all bonds formed. Be sure to indicate the correct direction for each arrow:

  (a) I^- CH_3-Cl $\longrightarrow$ CH_3-I Cl^-

  (b) $I\cdot$ $\cdot CH_3$ $\longrightarrow$ $I-CH_3$

  (c) $(CH_3)_3C +$ Cl^- $\longrightarrow$ $(CH_3)_3C-Cl$

There are no pertinent spectroscopic problems for this chapter.

Rotamers and Conformation

8

The incipient kinetic energy increases as molecules absorb energy from their surroundings and they move around. Each molecule will undoubtedly collide with another molecule of the same type, a molecule of solvent, or the walls of the container. Each collision will dissipate excess energy to another molecule, to the solvent or to the sides of the container. Such molecular motion is one way in which molecules in solution transfer heat. Molecules also dissipate energy by molecular vibration. One vibration mode changes the shape of the molecule by stretching covalent bonds, bending bonds, or internal rotation around single covalent σ-bonds. Rotation about carbon–carbon bonds positions atoms and groups in the molecule at different relative positions in space. Some of these positions may lead to interactions of the electrons in the bonds or interactions between atoms or groups induced by those units being too close together (known as **steric hindrance**).

Many different rotations are possible for each, and it is possible to describe an overall "shape" for the molecule, called a **conformation**. There may be many conformations for a given molecule, some higher in energy and some lower. Of all the possibilities, one or at best a few conformations will be lower in energy than the others, and the molecule will spend most of its time in these low-energy conformations. Understanding the low-energy conformation of a molecule is critical to understanding both its properties and its chemical reactivity.

To begin this chapter, you should know the following:

- **the nature of a covalent σ-bond (Chapter 3, Sections 3.2 and 3.3)**
- **the nature of a covalent π-bond (Chapter 5, Sections 5.1 and 5.2)**
- **differentiation of a cyclic compound from an acyclic compound (Chapter 4, Section 4.7)**
- **how to name cyclic and acyclic compounds, particularly hydrocarbons (Chapter 4, Sections 4.6 and 4.7)**
- **the nature and nomenclature of functional groups (Chapter 5, Sections 5.1, 5.2, 5.6, and 5.9)**
- **the VSEPR model for predicting the shape of simple molecules (Chapter 3, Section 3.6)**
- **the concept of electronic repulsion between bonds (Chapter 3, Sections 3.3, 3.5, and 3.7)**
- **the concept of equilibrium constant K, other than K_a (Chapter 7, Section 7.10)**

This chapter will introduce the concepts associated with the shape (conformation) of a molecule due to rotation about carbon–carbon single bonds. The influence of steric hindrance for determining conformation will be discussed. The concept of pseudorotation for cyclic compounds will be introduced, as well as the resulting conformations. The low-energy conformation for cyclic and acyclic compounds will be identified and also the relationship of energy considerations and conformation. Learning how to draw the various conformations is an important component of this study.

When you have completed this chapter, you should understand the following points:

- **A rotamer is one position of bonds, atoms, and groups in a molecule, generated by rotation around a single bond. A conformation for a molecule represents one position of bonds, atoms, and groups in the molecule. Bonds in acyclic molecules rotate to generate many rotamers; this generates many conformations for the molecule. The rotamer where two atoms or groups are 180° apart is called an *anti* rotamer and the rotamer where two atoms or groups eclipse each other is called an eclipsed rotamer.**
- **When rotation around a bond brings two atoms or groups on adjacent atoms, as well as the electrons in those bonds close together in space, competition for that space and electronic repulsion of the bonds lead to repulsion. This increases the energy of that rotamer. The energy of repulsion in eclipsed rotamers is called torsion strain. When substituents are attached to the atoms involved in the bond that rotates and they eclipse, the energy of that rotamer increases, effectively giving an energy barrier to rotation. When heteroatom substituents are attached to the atoms**

- involved in the bond that rotates, internal dipole interactions and hydrogen bonding can bias the rotation to favor one rotamer.
- Introduction of a π-bond leads to flattening of the molecule in the region of the π-bond. No rotation occurs about multiple bonds such as C=C, C≡C, C=O, C=N–, etc.
- Cyclic alkanes undergo pseudorotation because rotation by 360° is not possible. Pseudorotation in cyclic alkanes leads to many conformations. Cyclopropane is planar, with relatively weak "banana" bonds. The lowest energy conformation of cyclobutane is a "puckered" conformation. The lowest energy conformation of cyclopentane is an "envelope" conformation. The lowest energy conformation of cyclohexane is an equilibrating mixture of two "chair" conformations.
- In cyclohexane, substituents prefer the equatorial position in the chair conformation, where they are in the equatorial position, because that conformation is lower in energy. Steric interactions of axial substituents in substituted cyclohexanes leads to A-strain in the chair conformation of substituted cyclohexanes, making the conformation with the most A-strain higher in energy. A-strain occurs in chair cyclohexane derivatives between an axial group and axial hydrogen atoms at C1, C3, and C5.
- Medium-sized rings (8–13) are less stable (higher in energy) because of transannular interactions within the cavity of the ring. The lowest energy conformation for eight-membered rings and rings with an even number of carbons is the "crown" conformation. Rings with an odd number of carbon atoms tend to be higher in energy than those with an even number due to a "twist" in the ring. Large rings (>14 carbons) are lower in energy than medium-sized rings because the cavity of the ring is so large that transannular strain is diminished.
- Heterocyclic ring systems of three to six members assume conformations that are similar to the cycloalkanes, but the conformation is influenced by the nature of the heteroatom and the number of lone electron pairs.
- Conformation is important in biological systems.

8.1 Rotamers

Organic molecules constantly absorb energy from their environment, and excess energy is constantly dissipated by intermolecular collision. Dissipation of energy also occurs by molecular vibrations, and some vibrations involve rotation about carbon–carbon single bonds.

examining the relative energies of such key rotamers, a map of the rotation around a given bond can be constructed. Such rotamer maps provide valuable information about the shape of a given molecule. Remember that the term ***conformation*** is used to describe the *shape of a molecule*. For a molecule with several bonds, many conformations are possible. In other words, there are *many* rotamers for a molecule with several bonds and *many* conformations. Some conformations may be lower energy than others and more abundant. The shape of a molecule may be difficult to describe because each conformation may have a somewhat different shape.

Remember that rotamers result from rotation around bonds when a molecule dissipates energy absorbed from its environment. By lowering the temperature of the environment, the energy available to the molecule is diminished and rotation becomes increasingly difficult. If the temperature is lowered so that the available energy is less than the steric interaction mentioned earlier, rotation about 360° will stop, and the rotamer will be frozen out. **At normal temperatures, however, there is plenty of energy for rotation about carbon–carbon bonds, and organic molecules should be considered as dynamic species that have a large population of different rotamers.** For the purposes of this book, always assume that there is free rotation about a single covalent bond, unless otherwise noted.

8.1 Draw the rotamer of 2 that occurs when R^1 is rotated 90° counterclockwise.

8.2 Estimate the number of rotamers possible for compound 2.

8.1.1 Rotation around Carbon–Carbon Bonds in Alkanes: Ethane

The simplest example of an organic molecule with one carbon–carbon bond is ethane (**3**), which is drawn as its sawhorse projection in Figure 8.2 (also see the molecular models for **1** and **2**). Focus on the C–C bond, where each carbon atom is attached to three hydrogen atoms. Imagine holding the "left" or "back" carbon atom and rotating around the bond by twisting the "right" or "front" carbon atom clockwise by 360° (see Figure 8.2), in increments of 60°. This process generates several rotamers (**3–8**) and the rotation can be followed by focusing on the **red** hydrogen atom.

The most notable feature in **3, 5,** and **7** is the observation that the hydrogen atoms on the front carbon are in between the hydrogen atoms on the back carbon atom. These three are referred to as ***staggered rotamers***. In **4, 6**, and **8**, the hydrogen atoms on the front carbon overlap (eclipse) those on the back carbon. These are referred to as ***eclipsed rotamers***. It may be difficult to see the overlap in **4, 6**, and **8**, and drawing each rotamer from a different perspective may make this clearer.

8.3 Draw conformation 5 as a Newman projection.

In the sawhorse diagram used in Figure 8.2, a focus on **1** and **3** shows the molecule is tilted at an angle to show both carbon atoms as well as the groups

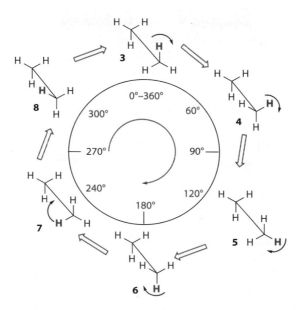

Figure 8.2   Rotation about the carbon–carbon bond in ethane using sawhorse diagrams.

and atoms attached to each carbon. In the corresponding Newman projection (**2**), the bond of interest is viewed head-on so that one carbon atom is in front and the second is in the rear, as shown. In order to see both carbons, the atom in front is represented as a dot and the one in the rear is represented as a circle, as described previously. Because each carbon is tetrahedral, the bonds radiate from these two carbons—three to the front and three to the rear. This view is particularly useful for identifying different interactions of groups attached to the carbon atoms in different rotamers.

    This reiteration of sawhorse and Newman diagrams is presented because the rotamers for ethane in Figure 8.2 are repeated as Newman projections in Figure 8.3 (see **9–14**). Clockwise rotation around the carbon–carbon bond for ethane is shown in Figure 8.3, generating rotamers **10–14** with Newman projection **9** as the starting point. For convenience, the **red** hydrogen is again used as a marker to follow this rotation. Correlation of Figures 8.2 and 8.3 shows that **3** is analogous to **9, 4** to **10, 5** to **11, 6** to **12, 7** to **13**, and **8** to **14**. Molecular models for each Newman projection are also provided in Figure 8.3 to show the spatial relationships of the hydrogen atoms.

8.4　Draw all rotamers for 1,2-dichloroethane generated by rotation of 60° and focusing attention on the chlorine atoms.

    The rotamer shown in sawhorse form (**3**) and also in Newman projection (**9**) is the ***staggered* rotamer**. When the hydrogen atom on the front and top (in **red**) is compared with the hydrogen in the back and on the bottom, it is apparent that they are 180° from each other. The term "staggered" is used to denote the relationship of two atoms or groups in a rotamer that are not eclipsed. Both

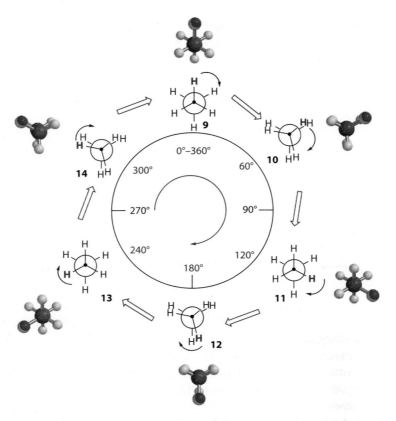

Figure 8.3 Rotation about the carbon–carbon bond in ethane using Newman projections.

the front and back carbons of ethane have three hydrogen attached atoms (i.e., three identical atoms). Because of this, the same staggered rotamer is generated three times during the rotation.

Following the **red** hydrogen allows the staggered rotamers (**9, 11**, and **13**) to be distinguished, *but this is completely artificial*. In fact, such structures cannot be distinguished in a real molecule of ethane. If the **red** hydrogen were converted to a deuterium atom (^{2}H), because deuterium is different from a proton, this would constitute a real label that might allow the rotation to be tracked, assuming one could distinguish ^{2}H from ^{1}H. In fact, rotation occurs so rapidly that even with the deuterium label, tracking rotamers as in Figure 8.3 is possible only at extremely low temperatures, and even then it is difficult for a molecule such as ethane.

8.5 Draw the staggered rotamer of 1-deuterioethane (one hydrogen atom of ethane is replaced with a deuterium) using D for the deuterium atom.

In contrast to **3** or **9**, sawhorse diagram **4** and Newman projection **10** show a rotamer where the hydrogen atoms (and the bonds attaching them to the carbons) *eclipse* each other. The hydrogen atom on the front and top (**red** H) and the hydrogen atom on the rear and top in **10** line up one behind the other (eclipse) in the Newman projection. These atoms are close together in space, and this is called an ***eclipsed* rotamer**. The *bonds* connecting the C–H units in **10** are closer together than they are in **9**. Because each bond is made up of two electrons, it is reasonable to assume that when these electrons are close in space there will be electronic repulsion (like charges repel). ***The electronic repulsion due to overlap of bonds is called torsional strain***.

Because the hydrogen atoms and the bonds in **10** repel, they are pushed away from each other, and energy must be expended to keep them together. This repulsive energy is sometimes called ***torsional energy*** (see Section 8.1.2), but more commonly the term torsional strain is used to indicate this energy in combination with the steric strain discussed previously. When the hydrogens are staggered, as in **9**, torsional strain is greatly diminished. This means that the *eclipsed rotamers are higher in energy than the staggered rotamers*. Rotamer **10** is higher in energy than rotamer **9**, where the hydrogen atoms are further apart and do not repel. As with the staggered rotamer, an eclipsed rotamer is encountered more than once during the rotation and rotamers **10, 12**, and **14** are identical.

Representation **3A** is a molecular model of the staggered rotamer as a sawhorse diagram and **4A** shows the eclipsed rotamer in which the hydrogen atoms eclipse, also as a sawhorse diagram. In this view, the hydrogen atoms in **4A** appear rather close together. This fact is clearer in **3B** and **4B**—the Newman projection molecular models of **3** and **4**, respectively. Space-filling molecular models show the staggered rotamer as **3C** and eclipsed rotamer as **4C**. A comparison of these structures with Figure 8.3 clearly shows that the hydrogen atoms eclipse (are very close together) in **4C**, whereas they are opposite each other in **3C**.

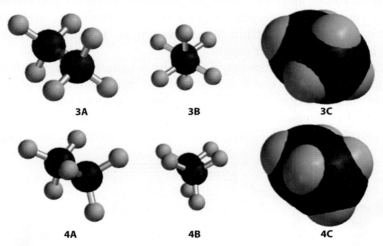

3A   3B   3C

4A   4B   4C

8.6 Draw the eclipsed rotamer for 1,2-dichloroethane in Newman projection.

8.7 Briefly discuss why 4B is higher in energy than 3B.

8.8 Using Spartan model, draw *anti*-1,2-dichloroethane in Newman projection, but as a space-filling model.

8.1.2 Torsional Strain: Steric Hindrance and Energy Barriers

If a snapshot of ethane could be produced while rotation occurs about the C–C bond, the atoms would be frozen in space. Imagine such a "frozen" rotamer by making a model using a modeling kit. In effect, the rotamers drawn in Figures 8.2 and 8.3 represent such snapshots. If 10,000 snapshots of the molecule were taken and compared, a large number of them would show the molecule as a staggered rotamer and far fewer would show an eclipsed rotamer. This thought experiment leads to the question of why the eclipsed rotamers are less abundant. The answer is that they are higher in energy, which makes it more difficult for them to exist when compared to the lower energy *anti* rotamers. This preference for lower energy rotamers leads to the observation that the molecule will spend most of its time in the lowest energy forms, the *anti* rotamers. In other words, the majority of the "thought snapshots" will be of staggered rotamers of ethane.

If sufficient energy is present to cause the molecule to rotate freely around the carbon–carbon bond, *both* eclipsed and staggered rotamers will be present, but the eclipsed rotamers are higher in energy. As the C–C bond in ethane is rotated clockwise (arbitrary) through 360° (60° at a time), six rotamers are generated, as in Figure 8.3 and **9–14**. To go from one staggered rotamer to the next during the rotation, the molecule *must pass through an eclipsed rotamer*. Because the eclipsed rotamer is higher in energy, **there is an energy barrier to rotation** that is measured to be 2.9 kcal mol⁻¹ (13.95 kJ mol⁻¹).[1–4] If the staggered rotamer **9** is taken as the standard, the energy of the eclipsed rotamer (**10**) will be 2.9 kcal mol⁻¹ (13.95 kJ mol⁻¹) higher in energy than **9**. *Rotamers 9 and 10 define the upper and lower energy limits for rotation around the carbon-carbon bond*. Many rotamers are possible between these extremes, and each one will have an energy value higher than **9** but lower than **10**, as the atoms and bonds move closer together or further away.

Identifying the energy associated with the rotamers in Figure 8.3 allows one to plot angular rotation of the bond versus the relative energy of each rotamer generated by that rotation. The resulting energy curve (note the relative positions of the **red** hydrogen atom) is essentially a map that defines the energy barriers to rotation and the magnitude of those barriers (see Figure 8.4).[1–4]

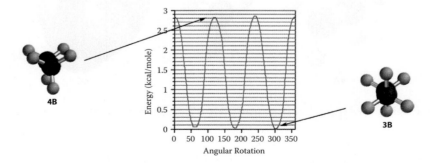

Figure 8.4 Energy barriers to rotation in ethane (plot of energy versus angular rotation). Reprinted by permission from Macmillan Publishers Ltd: Weinhold F. *Nature*. 2001, 411, 539. Figure 1.

It is clear that complete rotation around the carbon–carbon bond in ethane must generate three eclipsed rotamers, each with an energy barrier of 2.9 kcal mol^{-1} (13.95 kJ mol^{-1}). Rotation slows down when these eclipsed rotamers are encountered because each presents a barrier to rotation. Rotation in ethane is therefore said to be "hindered" rather than "free."

8.9 **If we compare X–CH$_2$CH$_2$–X, where X = Me and X = CMe$_3$, briefly discuss which will have the higher energy barrier to rotation around the C–C bond.**

8.2 Longer Chain Alkanes: Increased Torsional Strain

The energy barriers that are associated with certain rotamers of ethane occur for all carbon–carbon bonds in all molecules. Once other atoms or groups are introduced, there are new rotamers to consider and different energy barriers to worry about. As a practical matter, rotamers of substituted alkanes can be categorized into two fundamental types—X–C–C–H (**15**) and X–C–C–X (**16**)—where rotation occurs about a single C–C bond, and X can be any substituent. With this simple classification, the interactions X and H or X and X are important, and it is assumed that the X–H interactions in the second case are much less. If this analysis is extended slightly to include Y–C–C–X (**17**), then the Y–X steric interactions must be examined, assuming the X–H and Y–H interactions are much less. In other words, ethane can be used as a standard for alkanes where the X–X interactions using **16** are H–H. *Other alkanes or compounds with substituents are treated as substituted ethane derivatives and the important interactions are X-X in **16** and X-Y in **17**.*

In propane (**18**), the Me–H interaction (X–Y) as well as the H–H interaction are important. In butane (**19**), the X–X interactions are Me–Me and H–H and the X–Y interaction is Me–H. Extending this model to other systems, the X–Y interactions in chloroethane (**20**) are Cl–H; in 1,2-dichloroethane (**21**), there are two X–X interactions (Cl–Cl and H–H) as well as an X–Y interaction (Cl–H). In chloropropane (**22**), there are three X–Y interactions (Cl–Me, Cl–H, and Me–H), as well as the X–X interaction (H–H). Clearly, as the number of atoms increases, the interactions increase. For practical reasons, attention will be focused on the steric interactions of the larger groups: Me–Me rather than H–H or Cl–Me rather than Me–H or H–H, etc.

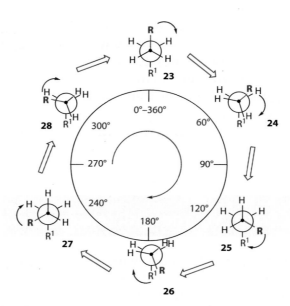

Figure 8.5 Rotation about carbon–carbon bonds in long-chain alkanes.

There is rotation about all carbon–carbon bonds in alkanes, which can be examined using a generic alkane RCH_2CH_2R', which is the equivalent of **16** when R = R' or **17** when R and R' are different. This alkane is shown as its staggered rotamer in Newman projection **23**. As with ethane, rotation of the indicated bond in **23** clockwise through 360° (60° at a time) leads to a series of rotamers (follow the motion of the **red** R group relative to R^1 in **green**) **23–28**. In **23**, the R and R' groups are as far apart as possible, but rotation through 60° gives an eclipsed rotamer (**24**) where R and R' *do not eclipse each other,* but R eclipses a H and R' eclipses a different hydrogen. The eclipsing bonds and atoms in **24** make it higher in energy than **23** (see Figure 8.5).

Rotation by another 60° generates a staggered rotamer (**25**) that is different from **23** in that R and R' are closer together. This means that some steric and electronic repulsion will make **25** higher in energy than **23**. There are no eclipsed bonds, so both **23** and **25** are *lower* in energy than **24**. The next 60° rotation generates an eclipsed rotamer (**26**) where R and R' completely eclipse each other; this is clearly the highest energy rotamer—higher than **25** and certainly higher than either staggered rotamer (see Figure 8.5). Note the similarity of Figure 8.5 to Figure 8.1 (some R groups are replaced with H in Figure 8.5).

Rotamer **26** is an eclipsed rotamer, but it is higher in energy than eclipsed rotamers **24** and **28** because the R/R' groups are eclipsed, whereas in **24** and **28**, R eclipses H and R' eclipses H. This higher energy eclipsed rotamer is known as the *syn* **rotamer**, and it poses the highest energy barrier to rotation in this molecule. When **26** is rotated by 60°, another staggered rotamer is obtained (**27**), which is the same energy as **25** when R and R' are the same. Another rotation by 60° generates another eclipsed rotamer (**28**), which also has a R–H and a R'–H interaction. Whether it is of the same energy as **24**, higher, or lower depends on whether R and R' are the same, close in structure, or very different, as with **25** and **27**. A last 60° rotation completes the cycle to give **23** again. Note that **23** is a staggered rotamer, but the R/R' groups are as far apart as possible. In staggered rotamers **25** and **27**, the R/R' groups are closer together in space than they are in **23**. Therefore, **23** is the lowest energy staggered rotamer, and the groups are 180° apart. For this reason, **23** is identified as the ***anti* rotamer**. To distinguish staggered rotamers **25** and **27** from **23, 25** and **27** are known as **gauche rotamers**.

Examining **26** suggests that as the size of the carbon groups R and R^1 increases, the atoms in those groups begin to compete for the same space. *Repulsion of atoms due to their proximity in space is called **steric repulsion***, as noted in previous examples. This **steric repulsion** or **steric hindrance** is one of the more important concepts in organic chemistry, and it will be encountered many times. The existence of steric repulsion means that more energy is required to force the atoms in R and R^1 into close proximity during the rotation than is required when those atoms are further apart (as in **23**). The energy due to steric hindrance is often called **steric energy**.

8.10 **Draw a conformation of pentane in Newman projection, sighting down the C2–C3 bond with the methyl and ethyl groups staggered but not in the *anti* conformation.**

8.11 **Briefly explain why the energy barrier for 16 or 20 is lower in energy than that for 18.**

Structures **15, 16**, or **17** and specifically Figure 8.5 can be used to examine key rotamers for simple alkanes (see Chapter 4 for nomenclature and a discussion of alkanes). Propane is symmetrical (a central CH_2 unit with a methyl group on either end), and rotation about either the C1–C2 or the C2–C3 bond will generate an identical energy-rotation map. An *anti* rotamer of propane (**15**, X = Me or **23**, where R = H and R' = Me) has the methyl group on C2 *anti* to one hydrogen on C1. The methyl–hydrogen eclipsing interaction leads to the highest energy eclipsed rotamers (see **24**, R = Me, R' = H). Because there is only one methyl group, the energy requirements are similar to ethane in that there are only two important rotamers (the methyl and hydrogen are eclipsed or staggered); these are repeated as rotation occurs around the carbon–carbon bond, staggered rotamer **29A**, and an eclipsed rotamer **29C**. The space-filing molecular models (**29B** and **29D**, respectively) show the steric interactions in propane. The energy map shows an energy barrier for the methyl–hydrogen interaction (about 3.5 kcal mol^{-1}) that is higher than the hydrogen–hydrogen interaction in ethane (about 2.9 kcal mol^{-1}). This result is consistent with the idea that the Me–H steric interaction is greater than that of a H–H interaction.

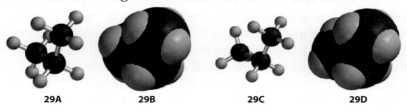

29A   29B   29C   29D

In butane (RCH_2CH_2R'; **16**, X = Me or **23**, R = R′ = methyl), three bonds that must be considered for rotation are C1–C2, C2–C3, and C3–C4. **Hereafter, the symbol Me will be used to represent methyl (–CH_3).** Examination of the C2–C3 bond shows butane to be symmetrical with respect to these two carbons (Me–CH_2CH_2–Me), making the C1–C2 and C3–C4 bonds identical. The goal is to find the highest energy barrier to rotation in butane and, to do this, C1–C2 is compared with C2 –C3 (C2–C3 versus C3–C4 can also be used).

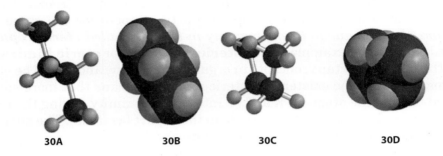

30A   30B   30C   30D

An ethyl group is larger than a methyl group, but *the steric demands of an ethyl group are not significantly greater than those of a methyl group*. The methyl group is a single carbon with three hydrogen atoms attached. An ethyl group is a carbon bearing two hydrogen atoms and one methyl group. The ethyl group is attached to a carbon (C2) via the CH_2 unit and the CH_3–CH_2–C2 unit will be in an *anti* orientation because that is the lowest energy rotamer for that bond. *The ethyl-H interaction is not much greater than a methyl-H interaction because the Me-CH_2-unit of ethyl will have the methyl group projected "away" from the interactions involved for C1-C2*. Because the ethyl group does not give a significantly greater steric effect than methyl, the steric energy for the Et–H interaction is about the same as that for the Me–H interaction. This contrasts with the Me–Me interaction in **30C** (**26**, where R = R^1 = Me; note this interaction in the space-filling molecular model **30D**), where the methyl groups have their maximum interaction. In terms of energy, the Et–H interaction will be close to the Me–H interaction observed with propane (about 3.5 kcal mol^{-1}), whereas the Me–Me interaction is about 5.9 kcal mol^{-1}.[5] The *anti* rotamer of butane is **30A** (**23**, where R = R^1 = Me) and it is taken as the standard because it is the lowest energy rotamer of butane. Rotamer **30A** is lower in energy because the steric interaction of the methyl groups is minimized, as seen by a comparison of space-filing molecular models **30B** and **30D**.

Figure 8.6 shows the proper picture for the key rotamers, where *anti* rotamer **30A** is drawn as **23** (R = R′ = Me) and the *syn* rotamer **30C** is drawn as **26** (R = R′ = Me). If the front carbon of the *anti* rotamer is rotated 60° clockwise, eclipsed rotamer **24** is generated (R = R′ = Me) with two Me–H interactions (each is measured to be about 3.5 kcal mol^{-1}). This is higher in energy than the *anti* rotamer, but lower in energy than the *syn* rotamer. Rotation by another 60° generates a new staggered rotamer, **25** (R = R′ = Me), shown as model **30E**. This rotamer is different from that obtained with propane or ethane in that the two methyl groups are close together in space. The spatial proximity of the methyl groups leads to a steric interaction, despite the fact that there are no eclipsing bonds or atoms. Although this staggered rotamer is lower in energy than any eclipsed rotamer, it is higher than the *anti* rotamer **30A**.

Staggered conformation **30E** is a ***gauche* rotamer**. The measured energy for this rotamer is about 0.9 kcal mol^{-1}. The spatial relationship of the methyl groups is perhaps best seen in molecular model **30F**, which can be compared with **30B** and **30D**. Continued rotation gives *syn* rotamer **30C** and then a second gauche rotamer (**27**, R = R′ = Me) and then eclipsed rotamer **28** (R = R′ = Me). Note that there are two gauche rotamers that are equal in energy and two eclipsed rotamers that are equal in energy, but the latter are lower in energy than the *syn* rotamer **30C**.

8.12 Draw both gauche rotamers for 1,2-dichloroethane in Newman projection.

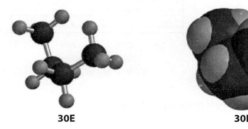

30E   **30D**

There is one last point to be made that will be important for an effect observed with cyclic alkanes (see Sections 8.5 and 8.6). The gauche steric interaction is a "through-space" interaction. In other words, the methyl groups are not directly connected to each other, but rather attached to adjacent carbon atoms. The steric interaction arises because they are close together in space during the rotation, which leads to some steric repulsion.

In Figure 8.6, the highest barrier to rotation is 5.86 kcal mol^{-1}, reflecting the Me–Me interaction in *syn* rotamer **30C**. There are smaller energy barriers for eclipsed rotamers **24** and **28**, and the "valleys" for gauche rotamers **25** and **27** are higher in energy than the energy minimum for the *anti* rotamer, **23**. Clearly this is a complex rotation map, but the *anti* rotamer **23** is lowest in energy and should constitute a large percentage of the rotamer population of butane.

As the length of the alkane chain increases in length from pentane to hexane (six carbons), to heptane (seven carbons), and so on, energy-angular rotation maps can be constructed for each bond in the molecule. In all cases, the rotamer that leads to the largest energy barrier will define the rotational barrier for that molecule. In the alkanes examined, ***the anti-rotamer is assumed to be the lowest energy rotamer for every carbon-carbon bond in the alkane***. Taking one representation of decane as an example (see **31A**), assume that every C–C bond exists as the *anti* rotamer. This assumption leads to the zigzag structure shown, which is one ***conformation*** of decane. In effect, it is the collection of rotamers for all bonds. This conformation is drawn again as **31B** using line notation. In previous sections, a major point was the free rotation about carbon–carbon bonds. If all the bonds in an alkane such as **31A** and **31B** rotate,

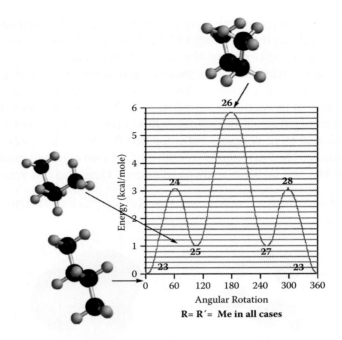

Figure 8.6   Angular rotation-energy map for butane.

there are potentially an infinite number of rotamers for each bond and an infinite number of conformations for the molecule. This should make a long-chain alkane "floppy" because the bond rotations should be almost random.

there are potentially an infinite number of rotamers for each bond and an infinite number of conformations for the molecule. This should make a long-chain alkane "floppy" because the bond rotations should be almost random.

31A   **31B**

It is now clear that some rotamers are lower in energy than others, leading to one or a least a few conformations that will be energetically preferred. The zigzag structure shown for decane is taken to be the low-energy conformation for the molecule (the "all *anti*" conformation). This is not necessarily correct, but it is a reasonable assumption to treat decane as structure **31B** rather than as a conglomeration of amorphous conformations. This zigzag structure is assumed to be prevalent in all long-chain alkanes and long-chain alkyl substituents used in this book.

8.13 Draw the zigzag structure of heptane using simple line notation, and then use a molecular modeling program or a model kit to show that same structure as a space-filling model.

8.3 Conformations of Alkenes and Alkynes: Introducing π-Bonds

Section 8.2 describes conformations for alkanes that arise from generating low-energy *anti* rotamers at each carbon–carbon bond for a given set of substituents. In all alkanes, the carbon structure contains nothing but sp³ carbon atoms, which are tetrahedral. The presence of sp² carbon atoms from an alkene unit will lead to flattening about each carbon.

In Chapter 5, Section 5.1, the C=C unit of an alkene was shown to be planar because each sp² carbon atom of the double bond has a trigonal planar geometry, and all four atoms connected to those carbons lie in the same plane. If the C=C unit is incorporated into a chain of sp³ carbons, as in **32A** (*cis*-3-hexene), the C=C unit "flattens" a portion of the carbon chain, as shown. This flattening of this region is even more apparent in the space-filling molecular model (**32B**), where the C=C unit leads to a "bend" in the carbon chain. In **32A**, the sidedness of the two ethyl groups attached to the C=C unit is apparent—another result of the planar alkene unit—as well as the inability of the groups to rotate about the C=C unit.

8.14 **Draw two isomers of 3-hexene in line notation, one with two ethyl groups on the same side of the C=C unit and one with the two ethyl groups on opposite sides of the C=C unit. Use a molecular modeling program or a model kit to show both structures as space-filling models.**

8.15 **Briefly discuss whether there is a significant interaction between the methyl groups attached to the C=C unit in 2-methyl-2-pentene.**

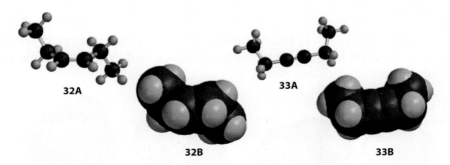

32A   33A

32B   33B

As discussed in Chapter 5, Section 5.2, the alkyne unit (C–C≡C–C) is linear, with all four carbon atoms in a line. The effect on the conformation of a hydrocarbon is to flatten the molecule in the region of the triple bond, to an even greater extent than seen with alkenes. The conformation of 4-hexyne is apparent in **33A**, but the flattening effect is quite clear in the space-filling molecular model, **33B**. The zigzag structure of the alkyl portion of the molecule changes to a linear structure at C2–C4. This is the normal effect on the conformation of a molecule that contains a triple bond. When the triple bond unit of an alkyne is treated as a substituent, its linear nature provides very little steric interaction when brought close to other alkyl groups. This is general, and it is expected that an alkyne unit will lead to minimal steric interaction for a given rotamer of that molecule.

8.16 Briefly discuss whether there is a steric interaction between the methyl groups of 2-butyne.

8.4 Influence of Heteroatoms on the Rotamer Population

In Sections 8.2 and 8.3, the main interactions of alkyl substituents and hydrogen atoms were steric interactions because dipole–dipole interactions and hydrogen bonding are not possible. If heteroatoms or functional groups containing heteroatoms are introduced into a carbon chain, however, the polarizing influence of these atoms and groups may have a profound influence on the rotamer populations of individual bonds and the conformation of molecules.

8.4.1 Introduction of Halogens

Halogens (F, Cl, Br, I) are interesting substituents because they have the same valence as hydrogen. This means that a halogen can replace a hydrogen in a carbon chain without changing the hybridization of any carbon in that chain and the tetrahedral geometry around each sp³ carbon is retained. Halogens are larger than hydrogen atoms, however, so the bond angles around the tetrahedral carbon bearing the halogen will be different. Compare the three-dimensional model of methane (**34**) with that of chloromethane (**35**). In **34**, all of the H–C–H bond angles are the same: 109.47°. In **35**, however, the Cl–C–H bond angle is 106.90°, one H–C–H bond angle is 109.00°, and the other H–C–H bond angle is 117.62°. The larger chlorine atom has caused one bond angle to open up, forcing the others to close down.

34   35   36A   36B

This is the typical effect when a large atom is attached to carbon and it influences the rotamer populations in alkyl halides (organic compounds bearing halogen atoms). In addition to the effect on the bond angles, halogens have three lone pairs of electrons that influence the rotamer population of adjacent bonds. Those electrons occupy space, making the halogen atom even larger than is apparent by simply drawing Cl on a bond.

1-Chloroethane is a simple example of a molecule that contains one halogen atom. The Newman projection shown is for an eclipsed rotamer where the chlorine atom eclipses a hydrogen atom (**36A**). Rotation leads to the lower energy *anti* rotamer, **36B**. The radius of the chlorine atom is rather large, and it has three electron lone pairs, which contributed to making the chlorine larger. The increase in steric hindrance in **36A** for chlorine relative to hydrogen is apparent. Note the similarity of **36** with eclipsed and staggered propane. Figure 8.5 can be used to predict the key rotamers, if R = Cl and R′ = H.

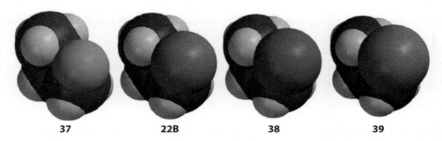

37   22B   38   39

When other halogen atoms are incorporated into an organic molecule, the size of the halogen will influence the steric hindrance in various rotamers. The covalent radius of fluorine is 0.64 Å, that of chlorine is 0.99 Å, that of bromine is 1.14 Å, and that of iodine is 1.33 Å.[6] Clearly, the presence of bromine or iodine will lead to even larger energy differences for rotamer populations than those noted for chlorine. The smaller fluorine will have a smaller effect. Indeed, the fluorine atom shows a steric interaction similar to that of hydrogen, although the covalent radius of hydrogen is only 0.37 Å.[6] The steric differences in the halogens are illustrated by the space-filling molecular models of the higher energy eclipsed rotamer of fluoropropane (**37**), chloropropane (**22B**), bromopropane (**38**), and iodopropane (**39**). Note the similarity of these four models to that of butane (**30D**), although the sizes of the halogens vary relative to methyl.

When two atoms or groups are attached to the same carbon, those groups or atoms are said to be *geminal* or to have a geminal relationship. The term comes from the Latin word *gemini* (twin). This relationship is seen in **40** (2,2-dichlorobutane).

8.17 Do the chlorine atoms in 2,3-dichlorohexane have a geminal relationship?

H_3C Cl Cl CH$_3$ **40** *geminal* chlorine atoms   Cl Cl **21** *vicinal* chlorine atoms

When two chlorine atoms (or any other atom or group) are attached to two adjacent carbons (a consecutive arrangement, as in 1,2-dichloroethane, **21**), the chlorine atoms are said to have a ***vicinal*** (neighboring or consecutive) attachment. Vicinal is derived from the Latin word *vicinus,* which means neighbor. This molecule has vicinal chlorine atoms (more precisely, chlorine atoms on vicinal carbons). There is an *anti* rotamer for 1,2-dichloroethane, a *syn* rotamer, and gauche rotamers (**16**, X = Cl). Experiments have determined that a Cl–Cl interaction in the *syn* rotamer of 1,2-dichloroethane (see **23**, R = R′ = Cl) is worth about 4.5 kcal mol^{-1}, whereas a Me–Me interaction in the *syn* rotamer of butane (**23**, R = R′ = Me) is worth about 5.8 kcal mol^{-1}. Therefore, in terms of steric interactions, a chlorine group (even with its electrons) is smaller (occupies less space) than a methyl group. Two halogen atoms on adjacent carbons is simply the effect of having two relatively large substituents, with a significant steric interaction between those atoms.

8.18 Draw the high- and low-energy rotamers for 1,1,2-dichloroethane.

8.19 Briefly discuss whether the energy diagram for rotation about the C–C bond of 1,1,1,2-tetrachloroethane should look more like that of ethane or that of Figure 8.6.

8.4.2 Effects of OH or NH Groups in Alcohols or Amines

The OH and NH units of alcohols or amines are large enough to have significant steric effects that may influence rotamer populations. Both the O–H and N–H units also contain polarized hydrogen atoms that are capable of hydrogen bonding to the oxygen or nitrogen atom of another OH or NH unit or with another heteroatom in a different molecule. Rotamers are formed in such molecules, as with all other molecules. The presence of hydrogen bonding in a rotamer may bias the rotamer population to favor that rotamer. In effect, a rotamer that is destabilized by a steric interaction between two groups may be stabilized by a hydrogen bond elsewhere in that molecule. The two effects are working at cross purposes, and the energy gained or lost by each interaction will determine which rotamer predominates.

41   **42A**

The introduction of one hydroxyl group (OH) into a molecule, as in 2-butanol (**41**), has a relatively simple steric effect on rotamer populations. There is a *syn* rotamer with an eclipsing methyl–methyl (**41B**), another rotamer with an eclipsing methyl–OH (**41D**), and a third eclipsing rotamer, **41F**. In addition, there are the three staggered rotamers: one with a methyl–methyl gauche interaction (**41A**), one with a methyl–hydroxyl gauche interaction (**41C**), and **41E**, where the three groups are relatively close to each other. These interactions will produce a complex energy rotamer diagram, but **41** appears to be normal in that the lower energy rotamers (**41A** and **41C**) are derived from diminished steric interactions.

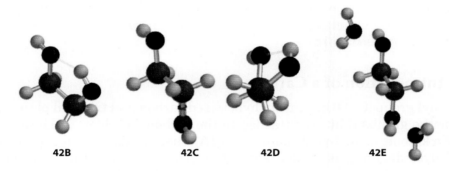

41A **41B** **41C** **41D** **41E** **41F**

When two OH units are introduced into the same molecule, such as with ethylene glycol **42**, there is a profound difference in the rotamer populations when compared with **41**. As mentioned in Chapter 5 (Section 5.6.3), a molecule containing two OH groups is called a diol; 1,2-ethanediol (**42**; known as ethylene glycol, a common ingredient in automobile antifreeze) is a simple example. Note that **42** can be correlated with **16**, X = OH. Diol **42** has two OH units on adjacent carbon atoms, so it is a vicinal diol.

42B **42C** **42D** **42E**

Molecular models are shown for three rotamers of ethylene glycol: *syn* rotamer, **42B**; *anti* rotamer, **42C**; and one gauche rotamer, **42D**. The two OH groups are as far apart as possible in **42C**, and based solely on steric interactions, this should constitute the lowest energy rotamer. As seen in the *syn* rotamer (**42B**), however, an intramolecular O----H–O hydrogen bond can partially stabilize the *syn* conformation, offsetting some of the steric interaction. The hydrogen bond cannot entirely overcome this eclipsing interaction, however. Examination of a staggered rotamer that is close to a gauche rotamer (**42D**) shows that this slight rotation allows retention of the hydrogen bond while alleviating most of the steric hindrance of the eclipsing OH units. Indeed, **42D** is the more stable rotamer for ethylene glycol **in the absence of a hydrogen-bonding solvent**. In other words, the extra stability of the hydrogen bond in **42D** will bias the rotamer population such that it accounts for the highest percentage of rotamers found for ethylene glycol. For all practical purposes, ethylene glycol exists in the gauche conformation.

The preceding bold-print statement about solvent concerning **42D** is rather interesting. If ethylene glycol is dissolved in water, *intermolecular* hydrogen bonds are possible between the OH unit of the alcohol and the OH unit of water. This means that the *anti* rotamer (**42C**) will be the lowest energy because extensive intermolecular hydrogen bonding effectively makes the OH groups much larger than normal. Structure **42E** illustrates this point, showing each OH unit hydrogen bonded to water molecules.

Introduction of an NHR or NH$_2$ unit (from a secondary or primary amine) will have a similar hydrogen bonding effect, but to a lesser degree. The amine group cannot hydrogen bond as effectively as the OH group. When a single amine unit is present, there can be no intramolecular hydrogen bonding, only steric interactions, as in alcohol **41**. The NH$_2$ group appears to be only slightly less than a methyl in terms of its steric interactions.

When two NH units are present, as in 1,2-diaminoethane (NH$_2$CH$_2$CH$_2$ NH$_2$; **16**, X = NH$_2$; ethylenediamine, a vicinal diamine), intramolecular hydrogen bonding is possible just as in 1,2-diols. Because the N–H bond is less polarized than the O–H bond, there is weaker hydrogen bonding in diamines than in diols. The effect on the rotamer population is essentially the same, however.

8.20 **Briefly discuss whether gauche ethylene glycol will have a greater or lesser percentage of the rotamer population when compared to ethylenediamine.**

8.4.3 Introduction of a Carbonyl

A carbonyl group (C=O) has an sp^2 hybridized carbon and trigonal planar geometry for the groups that are attached to the carbonyl carbon. If the carbonyl is part of a ketone, as in 3-pentanone (see **43A**), there is localized flattening due to the sp^2 hybridization, as is apparent in the space-filling molecular model (**43B**). The C2–C*–C4– bond angle is approximately 120°, where C* is the carbonyl carbon. This clearly influences the overall shape of the molecule, bringing the hydrogen atoms on C2 and C4 closer together than in pentane, for example. As with other heteroatoms, electron pairs on oxygen contribute to the interaction of the oxygen with adjacent atoms or groups. An aldehyde unit (–CHO) will have a similar effect when it acts as a substituent in a given rotamer, but a hydrogen is always attached to the carbonyl, so an interaction such as that noted for C4–C6 in **43B** is not possible.

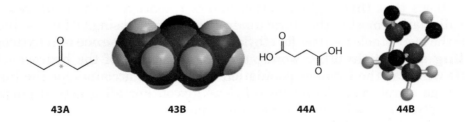

| 43A | 43B | 44A | 44B |

Carboxylic acids contain the carboxyl group (COOH), which contains a carbonyl. The COOH unit is rather large and it is also capable of hydrogen bonding. When a molecule contains two carboxyl units such as 2,4-butanedioic acid (**44A**; see **16**, X = CO$_2$H), intramolecular hydrogen bonding can play a role, just as it does with diols. Internal hydrogen bonding in a gauche rotamer (see **44B**) leads to greater stability for that rotamer in the absence of hydrogen-bonding solvents.

8.21 Speculate as to whether hexanal will have greater or less steric interactions than 2-hexanone.

8.22 Speculate as to whether the gauche rotamer of 2-hydroxybutanoic acid will be stabilized by hydrogen bonding.

8.5 Cyclic Alkanes

8.5.1 Are Cyclic Alkanes Flat?

Preceding sections discussed the conformation of acyclic molecules (no rings), where the low-energy conformations result from rotation about carbon–carbon bonds and it was assumed that each rotamer was the low-energy rotamer for that bond. In all cases for the acyclic molecules presented, complete rotation around carbon–carbon bonds is possible. This "free" rotation is hindered only by steric and electronic interactions of the substituents on the carbon chain, leading to a preference for some rotamers over other, higher energy rotamers. *In a cyclic compound, the carbon atoms are confined to a ring and a carbon-carbon bond in a ring cannot rotate around 360° without breaking those bonds. It is possible, however, to "partially rotate" the bond (in other words rotate by some angle less than 360°, clockwise or counterclockwise).* Such a partial rotation is called **pseudorotation**.

The simplest drawings of cyclic alkanes are elementary geometric figures. Using line-notation drawings, the three-membered ring cyclopropane (**45**) is shaped like a triangle, the four-membered ring cyclobutane (**46**) is a square, the five-membered ring cyclopentane (**47**) is a pentagon, the six-membered ring cyclohexane (**48**) is a hexagon, the seven-membered ring cycloheptane (**49**) is a heptagon, and so on.

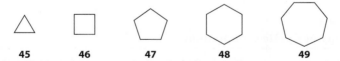

| 45 | 46 | 47 | 48 | 49 |

When these simple geometric shapes are used, all the rings appear to be flat. With the exception of 45, they are not! It is well known that every carbon atom in these cyclic structures *should* retain its tetrahedral three-dimensional shape. If this is true (and it is), then the "flat" structures **45–49** do ***not*** accurately represent the structure of these cyclic alkanes. If the planar structures were correct, the C–C–C bond angle for cyclopropane would be 60°; however, that angle is 90° for cyclobutane, 108° for cyclopentane, 120° for cyclohexane, and 128.6° for cycloheptane. These angles are calculated from the angles of the regular polyhedral structures used to represent the structures (triangle, square, etc.), which come from the formula for a planar polygon with n sides; angle = 180° × [$(n - 2)/n$]. Note that drawing the cyclic molecules as planar rings implies a shape or conformation for each molecule. Cyclic molecules do have conformations, but it must be determined if the planar conformation is highest or lowest in energy for a given ring.

8.23 Draw "flat" structures of cyclooctane, cyclodecane, and cyclotridecane.

8.24 Calculate the bond angles in cyclodecane (10 carbons).

The bond angle for a tetrahedral carbon is 109°28′ in methane and if this is a typical value for sp^3 hybridized carbon in a ring, many of the planar structures **45–49** must have significant distortion in the geometry of that ring. The *difference in the bond angle* expected for a tetrahedral carbon in methane and the bond angles for the carbon atoms in the *planar* structures is *about* 49, 19, 1, 11, and 19°, respectively. The greater the distortion from the ideal tetrahedral bond angles is, the greater is the energy required for the molecule to assume a conformation with planar geometry and the less stable that structure should be. Adolph von Baeyer (Germany; 1835–1917) proposed this idea in the late nineteenth century, and to this day the energy induced in a molecule due to distortion of the bond angles is called **Baeyer strain**. *The formal definition of Baeyer strain is the conformational energy arising in a molecule by distortion of the bond angles away from the tetrahedral ideal.*

By this analysis, planar cyclopentane should be most stable, then cyclohexane, cyclobutane (which is about the same as cycloheptane), and finally cyclopropane. However, *this is not the correct order for the inherent stability of cyclic alkanes*. The energy inherent to each ring is shown in Table 8.1.[7] The data in this table clearly show that cyclopropane is the highest in energy, but they also show that cyclohexane is lower in energy than cyclopentane. Indeed, cyclopentane and cyclohexane are the more stable (lowest energy) cyclic alkanes in the series **45–49** *because cyclic alkanes are not planar*. The pseudorotation mentioned before leads to conformations that are lower in energy than

Table 8.1
Enthalpy of Cyclic Compounds

Compound	$\Delta H°$ (KJ mol⁻¹): Liquid	$\Delta H°$(KJ mol⁻¹): Gas
Cyclopropane	35.2	53.3
Cyclobutane	3.7	27.7
Cyclopentane	−105.1	−76.4
Cyclohexane	−156.4	−123.4
Cycloheptane	−156.6	−118.1
Cyclooctane	−167.7	−124.4

Source: Lide, D. R., ed. 2006. *CRC Handbook of Chemistry and Physics,* 87th
ed., 5-24–5-37. Boca Raton, FL: Taylor & Francis.

the planar structure. The three-dimensional structures of cyclic alkanes (their low-energy conformation) result from pseudorotation within each ring system. Before looking at these structures, however, there is one more source of strain in planar cyclic molecules that must be examined.

8.5.2 Torsion (Pitzer) Strain

A cyclic molecule such as cyclopentane does ***not*** exist in a planar conformation (i.e., as a flat ring, **47**). A molecular model of this flat ring is shown in **47A**, and it is clear that all of the C–H bonds eclipse. Just as in eclipsed ethane or butane, these interactions raise the energy of the conformation. Going around the ring, each bond eclipses its neighbor and all hydrogen atoms on adjacent carbon atoms eclipse. The space-filling molecular model of planar cyclopentane, **47C**, more clearly shows the proximity of the hydrogen atoms. The strain energy arising from *eclipsing* interactions in cyclic molecules of this type is called **Pitzer strain**, although it may also be called **torsional strain**.

Pitzer strain in any planar cyclic alkane makes that conformation higher in energy and relatively unstable, and the molecule will not exist in that form if pseudorotation leads to a lower energy conformation. In the case of cyclopentane, pseudorotation about the carbon–carbon bonds alleviates the Pitzer strain and leads to a twisted structure such as **47B**. This structure is discussed in more detail later, but for the moment note that the eclipsing interactions are largely removed, and the conformation is much lower in energy when compared to **47A**. The space-filling molecular model of this pseudorotated structure is **47D**, where the relief of Pitzer strain is apparent when compared to **47C**.

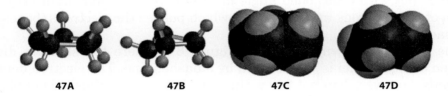

47A	**47B**	**47C**	**47D**

Pseudorotation in cyclic molecules may lead to many possible conformations, depending on the size of the ring (the number of carbons forming the ring) and the number and nature of substituents attached to the ring. *Of all the conformations possible for a cyclic alkane, the planar structure is one of the highest energy and least likely to form because of the high energy demands just described.*

8.5.3 Conformations of C3–C5 Cycloalkanes

The lowest energy shape or conformation is different for each cyclic alkane. Beginning with cyclopropane, line drawing **45** suggests there is considerable strain in this molecule because the bond angles are distorted to 60° (Baeyer strain) and all of the C–H bonds eclipse (torsion strain). Trying to "twist" this three-membered ring about any given C–C bond by pseudorotation is very

Figure 8.7 Bond distortion in the strained cyclopropane ring.

difficult, however; in fact, virtually no pseudorotation is possible. A molecular model of cyclopropane is provided as **45A**. Note that the model is flat and that all the C–H bonds overlap and the hydrogen atoms eclipse. Figure 8.7 also shows the space-filling molecular model **45B** for comparison. If a model kit is available, build a model of cyclopropane. The "bonds" will bend as the third bond is connected. The model is under a lot of strain, as is the real molecule. Be careful: It is easy to break the model because the atoms must be forced together.

This observation suggests that it *costs* energy to make this strained ring. Try to twist any two carbons of the model; there is little or no movement of the third carbon. If cyclopropane cannot rotate, the molecule is more or less locked into a planar structure that is rather high in energy. Although high in energy, cyclopropane is a known compound that can be prepared and isolated. It is very reactive in several chemical reactions, however, because of the strain. The C–C–C bond angle is locked at 60°, so the electrons in each bond must distort to "get away" from each other. In other words, forcing the bonds in cyclopropane close together with a 60° bond angle means there is significant electronic distortion of the electrons in those bonds, which pushes the electrons (bonds) away from each other.

The result is shown as **45C** in Figure 8.7, where the shortest distance between carbon nuclei is represented by a dashed line. The electron density in the bonds, however, does not follow this line, but rather is pushed out of linearity. To describe these bonds, cyclopropane is said to have "banana" bonds, but they look a little bit like π-bonds. ***They are not π-bonds!*** There is less electron density between the carbon nuclei than in a normal σ-bond due to the distortion, however, and ***the bonds are weaker than a normal alkane carbon–carbon bond***. If the bond is weaker, it is easier to break.

8.25 Based on its bonding, is cyclopropane classified as an alkane or an alkene?

It is clear from this discussion that cyclopropane is a special alkane in that it possesses high energy, weak bonds, and an unusual geometry. If one carbon is added to the ring to make cyclobutane, the ring is larger and it has more flexibility relative to cyclopropane. The bond angles are about 90° in planar cyclobutane (**46**), so there is less Baeyer strain than in cyclopropane. There is a significant amount of torsion strain due to eclipsing bonds, however, because there are more bonds in cyclobutane than in cyclopropane.

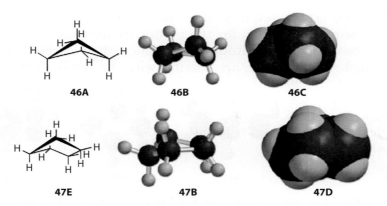

Greater flexibility due to the larger size of the ring allows some pseudorotation of the C–C bonds, which distorts the ring into a "puckered" conformation, **46A** (drawn in line notation). This conformation is also called the **butterfly conformation**. The molecular model for this puckered cyclobutane is **46B**, which shows that slight pseudorotation diminishes Baeyer strain as well as Pitzer strain, making that conformation more stable than the planar form. Relief of Pitzer strain is better seen in the space-filing molecular model **46C**. The lowest energy form of cyclobutane is taken to be the puckered or butterfly conformation.

Cyclopentane (**47**) has one more carbon than cyclobutane and this gives the molecule even more flexibility for pseudorotation. The simple line drawing of cyclopentene is **47**, but pseudorotation causes the molecule to distort, bringing some atoms closer together while taking other atoms further apart—resulting in **47E**, the so-called **"envelope" conformation**. The molecular model of planar cyclopentane (**47C**) shows the eclipsed hydrogen atoms on each of the five carbon atoms that give the torsion strain. Interestingly, the bond angles for planar cyclopentane are 108°—very close to the ideal tetrahedral angle of 109°28′—so there is little Baeyer strain. Molecular model **47B** shows that pseudorotation diminishes the torsion strain by pushing the eclipsing hydrogen atoms further apart. Although torsion strain is greatly diminished in **47B**, the bond angles are decreased to 105°, which is a small increase in Baeyer strain. It appears that relief of torsion strain competes with an increase in Baeyer strain. Which is the dominant factor? *The observation that cyclopentane exists primarily in the envelope conformation (47A) is an indication that relief of torsion strain is more important than the increase Baeyer strain is for cyclopentane. Cyclopentane exists primarily in the envelope conformation.*

8.26 **Briefly explain why changing the bond angles to 105° increases Baeyer strain in 47B.**

8.5.4 Conformationally Mobile Cyclohexane

Cyclohexane has one more carbon atom than cyclopentane and the greater flexibility allows for significant pseudorotation. Just as with smaller rings, the planar form of cyclohexane (**48A**) is very high in energy, and it has both Baeyer strain and torsion strain (see molecular model **48C**). Note that **48C** is drawn to mimic a Newman projection for the planar conformation. To alleviate the strain in cyclohexane, pseudorotation about the carbon–carbon bonds leads to many more conformations than are possible with the smaller rings.

48A   48B   48C   48D   48E

One conformation appears to be lowest in energy. This conformation is shown as molecular model **48D** in the same Newman-type projection as **48C**. This conformation is thought to look a little like an easy chair, and it is called a ***chair conformation***. The chair shape can be seen in **48B** or in molecular model **48E**, which views the conformation from the side. ***Note that 48D is the molecular model of 48B, and that 48D and 48E are identical, but simply viewed from a different perspective***.

The experimentally measured H–C–H bond angles of cyclohexane are tetrahedral, or about $109.5°$. There is virtually no Baeyer strain in the chair conformation of cyclohexane (**48D**). There is very little torsion strain in cyclohexane. The lack of torsion strain and Baeyer strain makes chair conformation (**48B**) the lowest energy conformation of cyclohexane.

Two drawings are provided that appear to be different chair conformations: **48B** and **48F**. The C1 and C4 carbon atoms are marked in both structures; C1 in **48B** is up whereas C1 is down in **48F**. Likewise, C4 is down in **48B** but up in **48F**. Conformations **48B** and **48F** are identical in structure and shape, and they are identical in energy. Twisting the bonds (pseudorotation) in cyclohexane will interconvert **48C** into **48F** and back again. In other words, chair conformations **48B** and **48F** are in equilibrium and because they are of the same energy, the equilibrium constant (K_{eq}; see Chapter 7, Section 7.10.1) is unity ($K_{eq} = 1$). This means that there is a 50:50 mixture of **48C** and **48F**. The equilibrium constant (K_{eq}) for this molecule is defined as $K_{eq} = [\textbf{48F}]/[\textbf{48B}]$, where [**48C**] and [**48F**] are the molar concentrations of each conformation.

Make a model of chair cyclohexane 48B, take hold of the up carbon (C1), and twist it down. Then, take hold of the down carbon (C4) and twist it up. This exercise will generate chair conformation **48F**. If the model is held by two adjacent carbon atoms and twisted back and forth in a rotational motion, this action mimics pseudorotation about that bond. This exercise generates both chair conformations, along with many others.

48B   48F

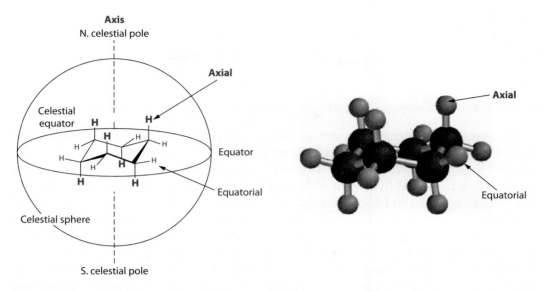

Figure 8.8 Axial and equatorial hydrogen atoms in chair cyclohexane.

Chair conformation **48F** is shown again in Figure 8.8. Twelve bonds connect hydrogen atoms; six are in the vertical plane (**red** hydrogen atoms), and the other six bonds are in the horizontal plane (**blue** hydrogen atoms). If cyclohexane is viewed as a planet, the bonds in the vertical plane are aligned in the direction of the axis and are called *axial* **bonds**. Those bonds in an approximate horizontal plane circling the equator of the molecule are called *equatorial* **bonds**. The hydrogen atoms attached to the axial or equatorial bonds are referred to as *axial hydrogen atoms* or *equatorial hydrogen atoms*. A molecular model is included in Figure 8.8 to show the spatial relationship of axial and equatorial bonds in the actual molecule

Structure **48A** is a line drawing of planar (flat) cyclohexane; it represents a conformation that is very high in energy, but it can be used to make a point. There are two sides: top and bottom. The six hydrogen atoms on top of the molecule are marked in red, and six more are on the bottom of the molecule (in **blue**). When **48A** is converted to the chair conformation (**48F**), three of the top hydrogens are axial and three are equatorial. This means that the hydrogen atoms on the top in **48F** (in **red**) show an *alternating pattern: axial–equatorial–axial–equatorial–axial–equatorial*.

48A   **48F**

Likewise, the bottom of **48A** has six hydrogen atoms (in **blue**) and in **48F** three of those hydrogen atoms are axial and three are equatorial. The same alternating

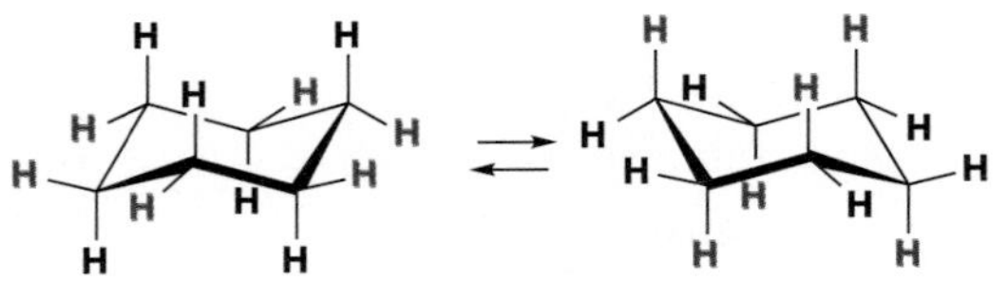

Figure 8.9 Interconversion of axial-equatorial hydrogen atoms in chair cyclohexane.

axial–equatorial pattern appears on the bottom as on the top of cyclohexane. It is useful to identify the hydrogen atoms on the same side and those on the opposite side for Figure 8.9, which uses a different color scheme to illustrate an interesting point. All of the axial hydrogen atoms on both the top and bottom of the chair on the left are marked in **black**, whereas all equatorial hydrogen atoms are marked in **violet**. Pseudorotation twists the ring to convert the chair on the left to the chair on the right. A comparison of the two structures shows that all the **black** axial hydrogen atoms in one chair are **violet** equatorial hydrogen atoms in the other. In addition, all **violet** equatorial hydrogen atoms become axial. *The pseudorotation that converts one chair to another also converts all axial to equatorial and all equatorial to axial.* This interconversion is sometimes called a "ring flip," but there is no flip. One chair is converted into the other by pseudorotation (twisting). This phenomenon will play a significant role in much of the chemistry of six-membered rings to be discussed later.

8.27 Identify all the axial hydrogen atoms in 48F.

8.28 Draw both chair conformations of chlorocyclohexane: one with an axial chlorine and one with an equatorial chlorine.

The chair forms of cyclohexane are the lowest energy conformations and that planar cyclohexane is probably the highest in energy. There are other conformations of cyclohexane and the most important are those that occur during the pseudorotation of one chair to another.

Beginning with chair **48F** in Figure 8.10, twisting one end (**red**) of the ring upward will flatten that region of the molecule to give what is known as a *half-chair* conformation, **50**. The five coplanar carbon atoms increase both

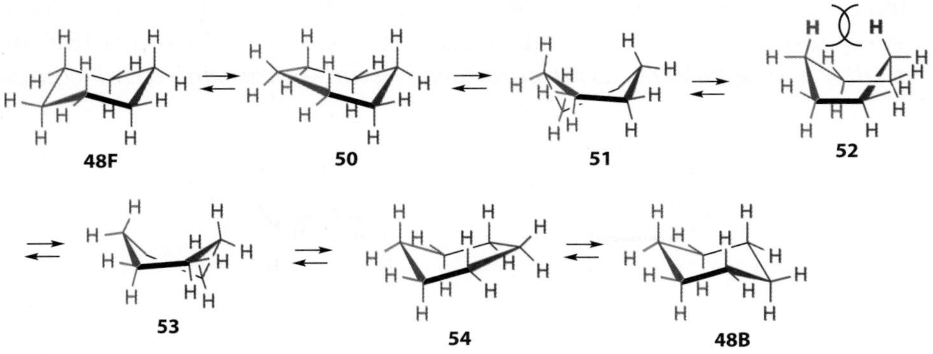

Figure 8.10 Important conformations of cyclohexane.

torsion strain and Baeyer strain and make **50** high in energy. If the twisting motion (follow the **red** atoms and bonds) that converted **48F** to **50** is continued, a twist is put into the molecule to give what is called a ***twist-boat*** conformation, **51** (also known as a ***twist*** conformation). This twist-boat is higher in energy than the chair, but much lower in energy than the half-chair. If the twisting motion is continued, conformation **52** results. This is called a ***boat*** conformation because it looks a little like the paper boat that a child might make. The boat conformation is discussed in more detail later.

Once the **red** end of the chair has twisted up (follow the motion from **48F** to **50** to **51** to **52**), the carbons on the other side of the molecule (**blue**) begin to twist down, as shown for the conversion of boat **52** to a new twist-boat, **53**. Twist-boat **53** should be of the same energy as the other one (**51**). Further twisting generates another half-chair (**54**, which is analogous to **50**), and completing the twisting motion generates the second chair conformation, **48B**.

The energy curve for this pseudorotation is shown in Figure 8.11, and the energy of each conformation is shown, taking the chair conformation as the reference point. The half-chair is measured to be about 10.8 kcal mol^{-1} (45.2 kJ mol^{-1}). The next highest energy conformation is the boat, which is about 6.7 kcal mol^{-1} (28.0 kJ mol^{-1}). The twist-boat is lower in energy (about 5.4 kcal mol^{-1}; 2.6 kJ mol^{-1}) and the planar conformations are so high in energy that they are omitted from Figure 8.11.

It is known that the twisting motion to convert chair **48F** to twist-boat **52** can proceed *without* forming a half-chair, which is another reason to omit it from Figure 8.11. Of the many relevant conformations in Figure 8.11, chair conformations **48F** and **48B** are lowest in energy, and cyclohexane will spend most of its time as equilibrating chair conformers.

The boat conformation (**52**) is rather interesting. Torsion strain in the four coplanar carbon atoms makes **52** higher in energy than the chair conformations. The two ends of the boat (C1 and C4) are sometimes labeled the bow and

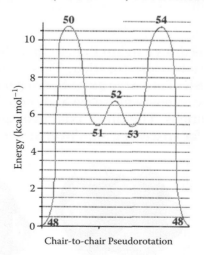

Chair-to-chair Pseudorotation

Figure 8.11 Chair-to-chair interconversion by pseudorotation.

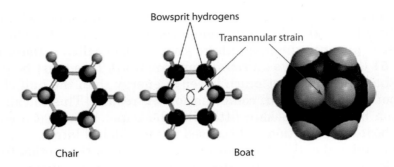

Figure 8.12   Transannular interactions in boat cyclohexane.

the stern, and it is apparent that two of the hydrogen atoms attached to these carbons are relatively close to each other. These "bowsprit" hydrogen atoms compete for the same space over the cavity of the cyclohexane ring. This is seen more clearly by comparing a view of chair cyclohexane in Figure 8.12, looking directly down on the cavity of the ring for the chair, with the same view of the boat conformation. The two hydrogen atoms are partially inside the ring cavity, which is a steric interaction called **transannular strain**. This interaction is clearer in the space-filling molecular model of the boat conformation in Figure 8.12. This new type of steric interaction is important because the boat conformation is *higher* in energy than the chair or the twist boat due to the increase in Baeyer strain, torsion strain, and now transannular strain.

8.29   Briefly comment on why the middle of Figure 8.11 has only one peak when there are six C–C bonds in cyclohexane.

8.6   Substituted Cyclohexanes

8.6.1 $A^{1,3}$-Strain

Chair conformation **48F** is drawn again to note that the axial hydrogen atoms (**red**) on the top appear to be close together. This is more apparent in space-filling molecular model **48G**. The three axial hydrogen atoms on top are *not* on adjacent carbons, but they compete for space and there is some steric repulsion between these hydrogen atoms. The three axial hydrogen atoms on the bottom of **48F** (in **blue**) compete for space in a similar manner.

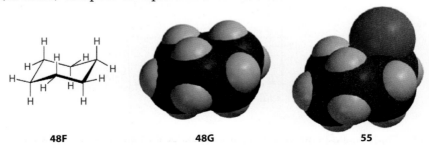

48F   **48G**   **55**

This interaction is not very large. If one of the axial hydrogen atoms is replaced with the larger chlorine atom, however, the interaction is much greater. The space-filling molecular model of chlorocyclohexane (**55**) is shown in a direct comparison with **48G**. The axial hydrogen atoms that sterically repel the large chlorine atom are on C3 and C5 with the chorine on C1, so they have a 1–3 relationship. This steric interaction is sometimes called a 1,3-diaxial interaction, but more commonly this repulsion is called ***A-strain*** or ***$A^{1,3}$-strain***. *The axial hydrogen atoms and the axial chlorine atom are not on adjacent atoms, so the steric interaction is literally across the six-membered ring. This type of steric strain is generically called **transannular strain**, so $A^{1,3}$-strain is one type of transannular strain.*

Chlorocyclohexane **55** is an example of a monosubstituted cyclohexane, and a substituent will have a major influence on the conformational mobility of the six-membered ring. Monosubstituted cyclohexane (**56**) has an equilibrating chair conformation **57**. In **56**, the R group is axial and there is **$A^{1,3}$-strain** due to the interaction of the R group with the two axial hydrogen atoms (see chlorocyclohexane, **55**). To be clear, the **$A^{1,3}$-strain** (or just A-strain) in a monosubstituted cyclohexane is due to the interaction of the R group with the axial hydrogens on that "side" of the molecule (see **48F**) in Section 8.5.4. In **57**, the R group is equatorial, and there is no $A^{1,3}$-interaction because any interaction of the three axial hydrogen atoms on the top of the molecule is taken to be the standard interaction (see molecular model **48G**). The R group occupies more space than the hydrogen atoms, so the R–H steric interaction must be greater than that of H–H.

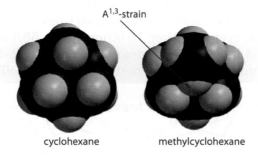

In other words, the axial H–H–H interactions are the standard and energy increases when the hydrogen atoms are replaced with substituents due to the transannular interaction. The increase in A-strain is apparent when the space-filling molecular model of chair methylcyclohexane (**56**, R = Me) in Figure 8.13 is

$A^{1,3}$-strain

cyclohexane   methylcyclohexane

Figure 8.13   Comparison of A-strain in cyclohexane and methylcyclohexane.

a relatively small number of low-energy conformations is the most important. The conformations are similar to those observed with cyclohexane, but there are some interesting differences.

8.7.1 Some Torsion Strain Is Unavoidable

The two-dimensional representation of cycloheptane as a planar molecule was **49** in Section 8.5.1. A chair conformation is possible that relieves the Baeyer strain and torsion strain, and two conformations are possible (**49A** and **49B**). The top hydrogen atoms are marked **blue** and the bottom hydrogen atoms are marked **red**. There is some strain for the axial-like hydrogen atoms, but the larger size of the ring allows those hydrogen atoms to be further apart. Closer examination revels that two of the carbons of cycloheptane are nearly coplanar (the flat part of **49A** and **49B**).

This flattening is due to the presence of an odd number of carbons in the ring and it means that there will be some torsion strain due to eclipsing bonds and atoms in this form of cycloheptane. Some twisting of the ring can occur to relieve this strain, but such pseudorotation may increase strain elsewhere in the molecule. This increase in strain makes conformations **49A** and **49B** for cycloheptane higher in energy than the chair conformations of cyclohexane.

There is also a boat-like conformation, **49C**, but the flat part of the seven-membered ring has diminished transannular strain because the hydrogen atoms are a little further apart. The strain energies for chair cycloheptane and boat cycloheptane are close, and one does not greatly predominate over the other. There are several other conformations for cycloheptane, as there are for cyclohexane, because the size and flexibility of the ring has increased; however, at this point, no other conformations for cycloheptene will be discussed.

8.33 Speculate as to whether boat cycloheptane is higher or lower in energy than boat cyclohexane.

8.7.2 Medium-Sized Rings. Another Type of Transannular Strain

For cycloalkanes that contain eight carbons or more, the size of the cavity is larger, leading to a greater number of conformations that are available by pseudorotation. This conformational mobility brings the hydrogen atoms *within the cavity* of the ring into closer proximity when the total number of carbon atoms

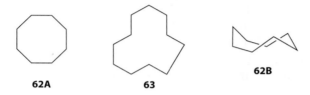

64

attempt to join the
ends to make an
eight-membered ring

Figure 8.14   Attempt to form an eight-membered ring by cyclization.

in the ring is 8–13. This is another type of **transannular strain,** as introduced in boat cyclohexane (see Figure 8.13) and $A^{1,3}$-strain (Figure 8.14). This
new type of transannular strain in medium-sized rings (8–13 carbon atoms) is
a significant contributing factor to making them much higher in energy when
compared to cyclic compounds that have 5–7 carbon atoms.

62A   63

62B

Two medium-sized cyclic alkanes are cyclooctane (**62A**) and cycloundecane
(**63**), shown as two-dimensional (planar) conformations that do not represent
their three-dimensional structure. With larger rings, there are several relatively
low-energy conformations. Cyclooctane, for example, is known to have 10 symmetrical conformations and **62B** (called the **"boat-chair"**) is identified as an
important low-energy conformation. Careful analysis is required to determine
which conformation or conformations are lower in energy for medium-sized and
large ring compounds. For this reason, this book will focus on conformations of
the smaller rings.

**8.34 Reproduce 62A as a space-filling model (computer model or a
model kit). Rotate the model so that you can sight down the cavity
of the ring and comment on transannular interactions.**

Before moving on, a greater discussion of the transannular strain in medium-sized rings is in order. The molecular model of a conformation called crown-
cyclooctane (**62C**) shows that some of the hydrogen atoms are actually pointing
inside the cavity of the ring. Looking at cyclooctane from a different perspective
allows one to sight down the cavity of the ring, as in **62D**. From this perspective,
the hydrogen atoms appear to be tilted into the ring. The space-filling model
62E shows that these hydrogen atoms are relatively close together. **The steric
interaction of these hydrogen atoms *inside the ring* is the source of the
transannular strain.**

This type of "across-the-ring" (*transannular*) steric interaction occurs in 8–13
carbons, making them higher in energy relative to rings of 5–7 carbons, where
the hydrogen atoms cannot pseudorotate inside the ring cavity. An important consequence of this strain is that any reaction that attempts to join together the two
ends of an eight-carbon chain (see Figure 8.14) *must* look like transition state **64**

at the instant the new bond is being formed and the ring is made. This assumption is based on the *Hammond postulate,* which assumes a late transition state (see Chapter 10, Section 10.2.1).

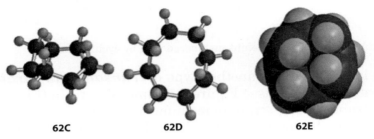

| 62C | 62D | 62E |

This assumption means that a transition state such as **64** assumes the conformation of an eight-membered ring, and the transannular strain will inhibit approach of the two ends and make the ring-closing reaction much more difficult. Indeed, transannular strain makes formation of 8- to 13-membered rings by cyclization reactions very difficult. In principle, transannular strain should not inhibit formation of large rings (>14) because the cavity is large. Bringing two reactive ends together is difficult, however, if they are separated by a large number of atoms. Therefore, formation of large rings by this technique remains problematic and such reactions will not be discussed in detail in this book.

8.35 Determine whether there is transannular strain in cyclononane, even when the odd carbon makes one end a little flat.

8.36 Look at cyclooctadecane and speculate on the possibility of transannular strain.

8.8 Cyclic Alkenes

There are interesting conformational differences between cyclic alkenes and cyclic alkanes. The C=C unit of an alkene is composed of two sp² hybridized carbon atoms and each atom has a trigonal planar geometry (see Chapter 5, Section 5.1). Therefore, the C=C unit *and the bonds attached to that unit* are all coplanar (i.e., the C=C unit is flat).

| 66A | 66B | 42B |

The flat representation of cyclohexene (**66A**) does not show this effect, but a molecular model (**66B**) clearly shows flattening in the region of the C=C unit.

When **66B** is compared to the molecular model of cyclohexane (**42B**), the flattening of the ring is quite apparent.

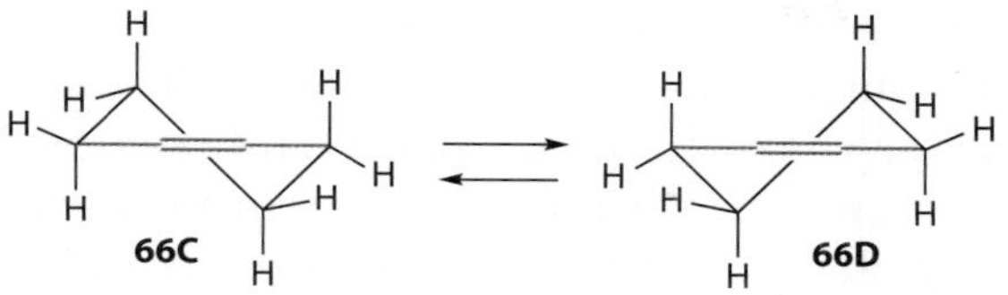

Cyclohexene can be drawn in what appears to be a half-chair conformation, **66C**, which is the conformational consequence of the ring flattening. Four carbons of cyclohexene are coplanar (see **66C**), with one of the other carbons up and one down relative to the C=C. The coplanar carbon atoms are marked in **red**. Which carbon is up and which is down is arbitrary because **66D** exists in equilibrium with another "flattened-chair" conformation, **66C**. The presence of the double bond inhibits the conformational mobility of the ring, and the two carbon atoms toward the back effectively flip back and forth between **66C** and **66D**.

8.9 Introducing Heteroatoms into a Ring

Previous sections showed various cyclic hydrocarbons and the conformations that are possible with these rings. It is also possible to place heteroatoms *within the ring,* which generates several new classes of compounds. Several of these compounds were introduced in Chapter 5 (Section 5.6) in connection with recognition of the groups and their nomenclature, but there is a more formal discussion in Chapter 26, Section 26.4. This section will briefly introduce the conformations available to the more common cyclic molecules that contain oxygen or nitrogen.

8.9.1 Oxygen

Cyclic molecules that have an oxygen *in* the ring rather than **attached to** the ring are ethers (see Chapter 5, Section 5.6.2). Cyclic ethers in a three-membered ring, a four-membered ring, a five-membered ring, and a six-membered ring are very common.

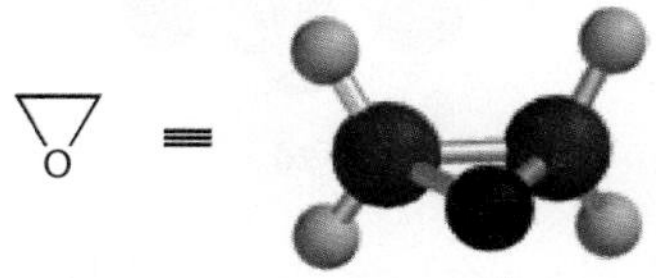

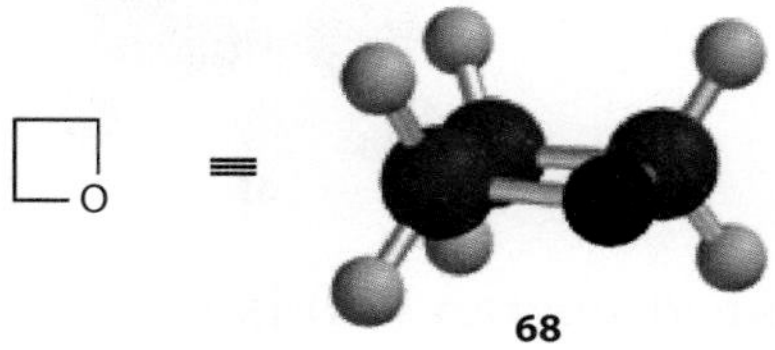

A three-membered ring that contains one oxygen is called an oxirane (see Chapter 26, Section 26.4.2). The simplest example is oxirane itself (**67**), and derivatives with substituents at the carbon atoms are known as oxiranes or epoxides. Epoxide is the common name for this class of cyclic ethers. This three-membered ring ether is essentially flat, but the bonds are "bowed" as in cyclopropane, **45A** (see Section 8.5.3). The molecular model for oxirane (**67**) shows this planarity, but the model does not show the bowed bonds. The C–O bond is slightly shorter than the C–C bond (1.436 versus 1.467 Å; 143.6 and 146.7 pm, respectively), making it a bit more strained than cyclopropane; it is very reactive in many chemical reactions. Oxetane is the four-membered ring ether, **68**. Derivatives of **68** that have substituents attached to the carbon atoms but contain the four-membered ring ether unit are called oxetanes. The ring is somewhat larger than the ring in oxirane, so there is less Baeyer strain, and oxetanes are less reactive than oxiranes. The ring is slightly puckered (as seen in **68**), similarly to cyclobutane (**46**); however, the C–O bonds (1.409 Å; 140.9 pm) are shorter than the C–C bonds (1.507 Å; 150.7 pm), thus causing the ring to be slightly distorted. Oxetanes do undergo some chemical reactions, but few will be discussed in this book.

Tetrahydrofuran (**69**, abbreviated as THF) is the five-membered ring ether (see Chapter 26, Section 26.4.2). The name is derived from the **heterocycle** parent, furan (see Chapter 26, Section 26.1.1). Tetrahydrofuran is somewhat flatter than cyclopentane (see **47** in Section 8.5.2), but the lowest energy conformation of THF is a slightly distorted envelope conformation (see **69**). The distortion comes from the different bond lengths for the C–O bonds (1.407 Å; 140.7 pm) when compared to the C–C bond lengths (1.513 Å; 151.3 pm). It is important to note that the lone electron pairs on oxygen occupy space and will influence the overall conformation of the ring. Tetrahydrofuran is a cyclic ether, but there is virtually no Baeyer strain and, like acyclic ethers, it is not very reactive. This fact and its somewhat polar nature make THF an ideal solvent in many chemical reactions to be discussed in later chapters. The fundamental THF ring skeleton will also be seen in the ring of carbohydrate derivatives, such as ribose (a furanose), discussed in Chapter 28, Section 28.1.

8.37 Draw the structures of 2-ethyloxirane and 3-methyloxetane.

8.38 Draw the structure of 2,3-dimethyltetrahydrofuran.

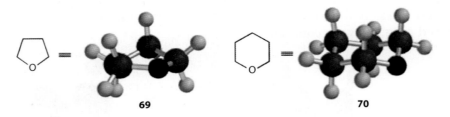

69   70

Tetrahydropyran (**70**) is the six-membered ring ether derived from pyran (see Chapter 26, Section 26.4.2). The lowest energy conformation of tetrahydropyran is essentially a chair, similar to cyclohexane (**48**), and **70** shows the chair shape with the oxygen in the ring. A major difference is the *absence* of one axial hydrogen (the CH_2 unit of cyclohexane is replaced with O) that is present in cyclohexane. Tetrahydropyran is also relatively unreactive (like THF), but it is more expensive, has a higher boiling point, and is used less often. The pyran ring will be encountered again in Chapter 28 with the pyran derivatives known as pyranose carbohydrates.

8.9.2 Nitrogen

Molecules that have a nitrogen atom in the ring are amines (see Chapter 5, Section 5.6.2, and Chapter 27, Section 27.1). Cyclic amines behave essentially the same as acyclic amines in terms of their chemistry, although there are subtle differences. Cyclic amines are known with a three-membered ring, a four-membered ring, a five-membered ring, and a six-membered ring. These cyclic amines are quite common.

Aziridine (see Chapter 26, Section 26.4.1) is a three-membered ring amine. As with cyclopropane and oxirane, the three-membered ring is relatively rigid and exits in a planar conformation (see **71**). The nitrogen atom has a lone electron pair projected away from the ring, and the hydrogen atom attached to nitrogen is projected away from the ring and also away from the electron pair (see **71**). There is the usual fluxional inversion about nitrogen (see Chapter 9, Section 9.7.3), as with other amines, but it is somewhat slower because the three-membered ring is not very flexible.

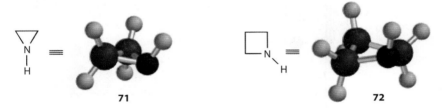

71   **72**

Azetidine (see Chapter 26, Section 26.4.1) is a four-membered ring amine, and its conformation is slightly puckered (see **72**). As with aziridine, the lone electron pair on nitrogen and the hydrogen atom attached to nitrogen project away from the ring, and there is fluxional inversion. The ring is slightly distorted due to the difference in the C–N bond length (1.458 Å; 145.8 pm) when compared to the C–C bond length (1.518 Å; 151.8 pm). As with aziridines, azetidines will not appear very often in this book.

8.39 Draw the structures of *N*-phenylaziridine and of *N*,2-diethylazetidine.

Pyrrolidine (see Chapter 26, Section 26.4.1) is a five-membered ring amine and has a conformation that is similar to the expected envelope (see **73**). The position of the electron pair and the hydrogen attached to nitrogen are projected away from the ring, and fluxional inversion occurs. The five-membered ring is more conformationally mobile than smaller cyclic amines. The pyrrolidine ring is a very important component of many natural products and this ring system appears in several places in this book.

Piperidine (see Chapter 26, Section 26.4.1) is a six-membered ring amine, and exists in a chair-type conformation (see **74**) as its low-energy form. The position of the electron pair and the hydrogen attached to nitrogen depends on the chair conformation. In **74A**, the lone electron pair is equatorial and the hydrogen atom attached to nitrogen is axial. If this chair conformation undergoes pseudorotation to the other chair, the hydrogen atom is changed to equatorial and the lone pair will be axial. Because a hydrogen atom is larger than an electron pair, one expects that the chair conformation with the hydrogen equatorial (**74B**) will be more stable. Indeed, calculations suggest that **74B** is more stable than **74A** by about 1.3 kJ mol^{-1}—a small but significant difference.

Because fluxional inversion occurs at nitrogen, **74A** and **74B** are considered to be fluxional isomers. The issue of where the electron pair is to be found in a given chair conformation is not very important in simple cases. Fluxional inversion will tend to favor the more stable form, which is usually taken to be **74B**. The six-membered ring is more conformationally mobile than pyrrolidine derivatives and there are other conformations, as seen with cyclohexane in Section 8.5.4. The piperidine ring is also important in many natural products and in the preparation of other molecules.

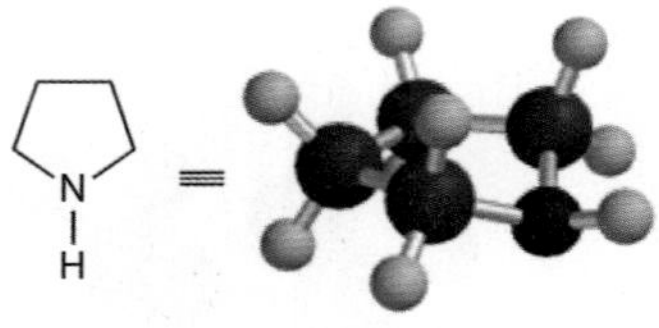

73

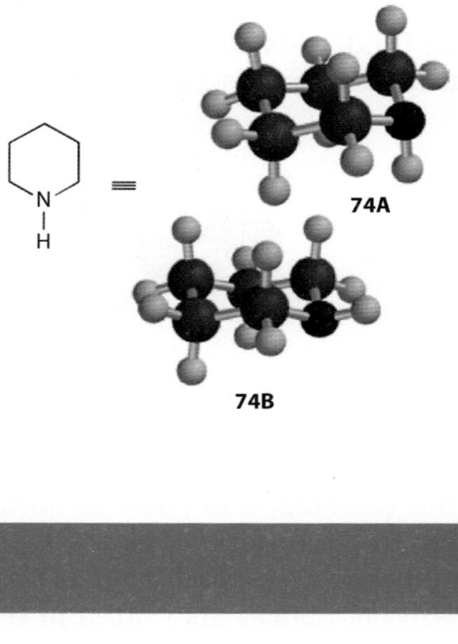

74A

74B

8.10 Biological Relevance

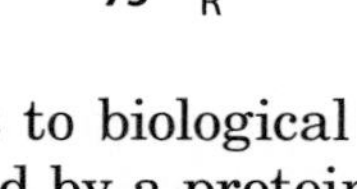

75

Conformation is very important to biological systems such as DNA and proteins. Indeed, the shape assumed by a protein is critical to its biological function. The same can be said for DNA, as well as other biomolecules. Chapter 27 will discuss amino acids (see **75**). When amino acids react to form long chains of repeating amino acid units, such molecules are called peptides, and each amino acid unit is connected by an amide bond (O=C–N; see Chapter 27, Section 27.3), which is called a peptide bond in a protein. An example is tripeptide **76**, which is formed from three amino acids (alanine–valine–serine, or ala–val–ser; see Chapter 27, Section 27.4). Rotation occurs around both the C–N bond and the C–C bond (see **76A**). Each $C_{carbonyl}$–N–C_α unit defines a plane due to the planar nature of the amide bond linkage (Chapter 27, Section 27.4.1) and rotation is somewhat restricted. A rotational angle for the atoms is defined. In **76B**, the tripeptide is drawn as a wire-frame molecular model. The rotational angle for the C–N bond is labeled by φ, which is the angle defined by rotation about that bond. The angle ψ is defined by rotation about C–C=O bond.

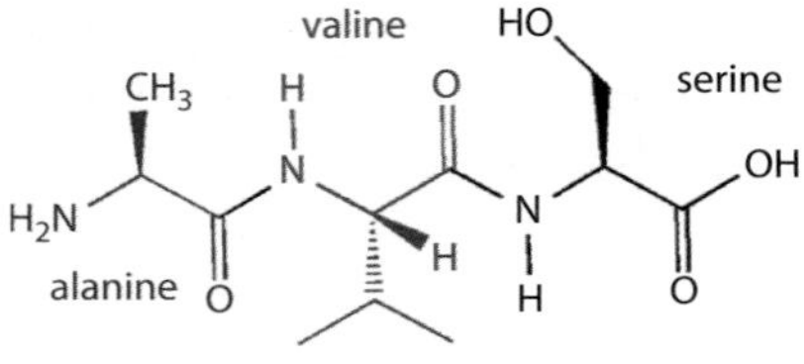

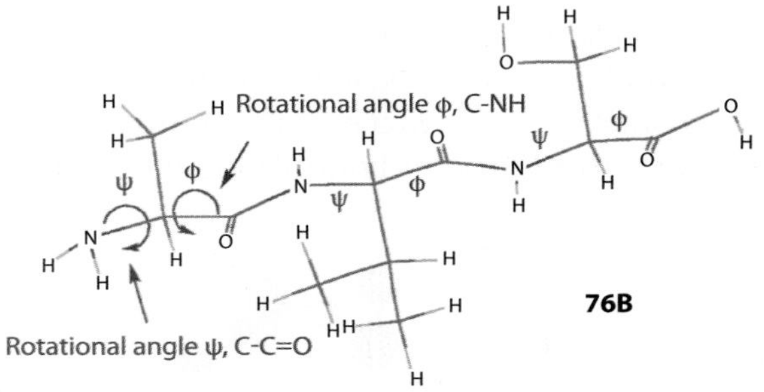

Each peptide bond in a peptide tends to be planar, which is a consequence of the planar nature of the amide unit. The R groups in **75** have a great influence on the magnitude of angles ψ and φ, and these angles of rotation define the conformation for that portion of the peptide. Structure **76B** shows that the amide unit of one amino acid residue is *anti* to the amide unit of the adjacent amino acid residue. Peptide **76B** also shows the carbonyl of the valine residue is *anti* to the carbonyl of the valine residue, which is *anti* to the carbonyl of the serine residue. The consequences of this observation are that a peptide chain assumes this alternating or *anti* pattern.

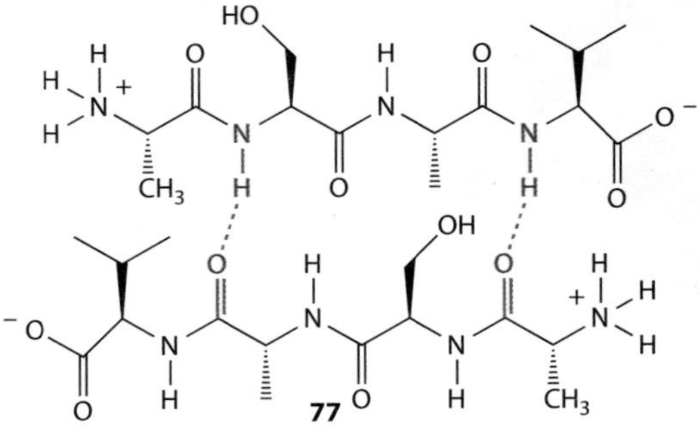

If the two dipeptide units are brought into close proximity, as shown in **77**, the oxygen of one carbonyl can form a hydrogen bond to the proton on the amide nitrogen of the second dipeptide. For this to occur, however, one dipeptide must have an *anti* orientation. This type of hydrogen bonding occurs in long-chain peptides, and when combined with the planar nature of the amide units, the *anti* orientation of adjacent amide carbonyls and the magnitude of the angles ψ and φ lead to a rather unique structure for the peptide. It forms an α-helical structure as shown in the poly(alanine) peptide **78**.

The α-helix structure was proposed by Linus Pauling (United States; 1901–1995) and Robert Brainard Corey (United States; 1897–1971) in 1951. The hydrogen atoms have been omitted from **78** so that the helical structure is easier to see. Helix **78A** is shown in a linear manner to show the coiling of the structure, and **78B** is viewed by sighting down the helical chain. In **78B**, it is clear that the structure coils. In a long-chain peptide composed of l-amino acid residues, a right-handed helix (called an α-helix) is formed, where the hydrogen atom on the amide nitrogen is hydrogen bonded to the oxygen of the carbonyl on the fourth amino acid residue. The hydrogen bonds stabilize the α-helix structure, which is an example of the **secondary structure** of a peptide. Formally, the secondary structure is the amount of structural regularity in a peptide that results from hydrogen bonding between the peptide bonds.

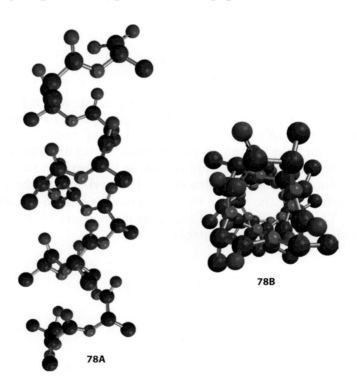

78A

78B

References

1. Aston, J. G., Isserow, S., Szasz, G. J., and Kennedy, R. M. 1944. *Journal of Chemical Physics* 12:336.
2. Mason, E.A., and Kreevoy, M. M. 1955. *Journal of the American Chemical Society* 77:5808.

3. Mason, E. A., and Kreevoy, M. M. 1957. *Journal of the American Chemical Society* 79:4851.

4. Weinhold, F. 2001. *Nature* 411: 539.

5. Kagan, H. 1979. *Organic stereochemistry,* 50. New York: Halsted Press.

6. Dean, J. A. 1987. *Handbook of organic chemistry,* 3-122–3-126, Table 3-10. New York: McGraw-Hill.

7. Lide, D. R., ed. 2006. *CRC handbook of chemistry and physics,* 87th ed., 5-24–5-37. Boca Raton, FL: Taylor & Francis.

Answers to Problems

8.1

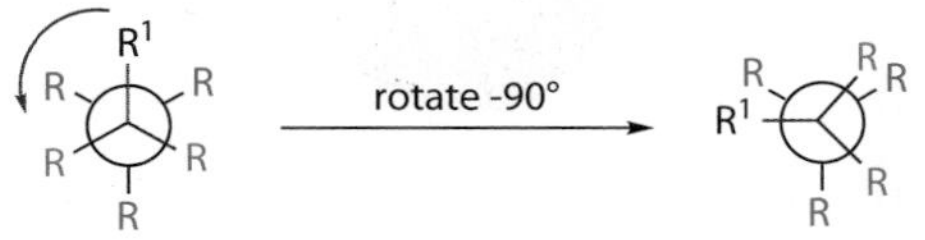

8.2   There are literally an infinite number of rotamers for **2** because one can imagine an infinite number of rotation angles.

8.3

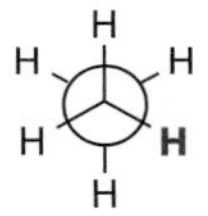

8.4

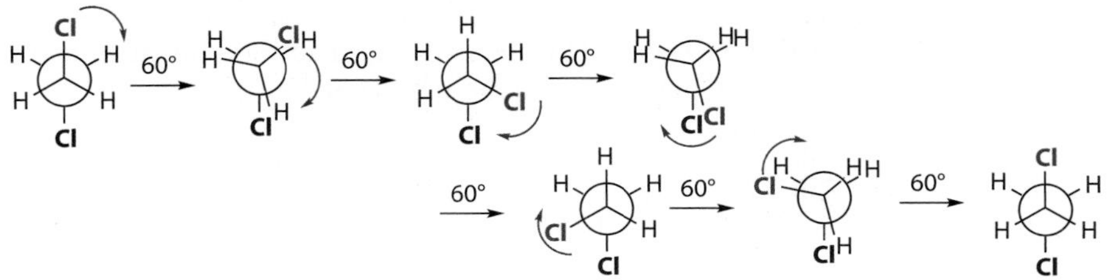

8.5

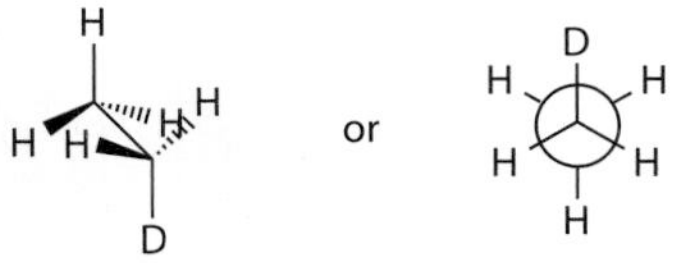

8.6

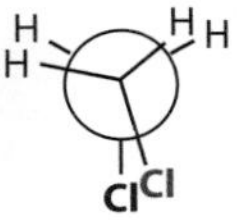

8.7   Rotamer **4B** has all the hydrogen atoms eclipsing so they are close together and repel. When looking at **4A**, where the atoms are staggered and further apart, there is little repulsion.

8.8

8.9   The CMe₃ group (*tert*-butyl) is much larger than methyl, so the energy barrier to rotation around the indicated bond will be much greater, making rotation more difficult. The steric interaction of two *t*-butyl groups will be much greater than that of two methyl groups, so the energy barrier is greater.

8.10

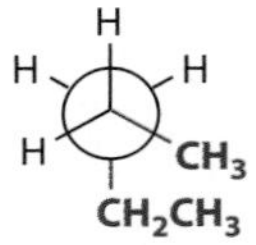

$$\text{H} \quad \overset{\text{H}}{\underset{\underset{\text{CH}_2\text{CH}_3}{\text{CH}_3}}{\diagup\!\!\!\!\bigcirc\!\!\!\diagdown}} \quad \text{H}$$

8.11 The R and R¹ groups completely eclipse in **18**, maximizing the steric repulsion. Those groups are further apart in **16** and **20** and the interaction is between R and H as well as R¹ and H. The R–R¹ interaction is much larger, as seen in Figure 8.6, than the sum of the R–H and R¹–H interactions because the R and R¹ groups have no place to "get away from each other."

8.12

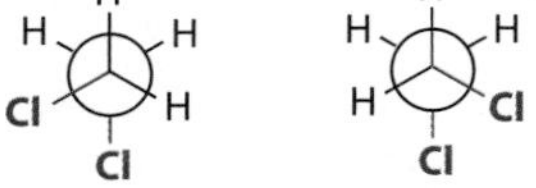

8.13

8.14

cis-3-hexene   *trans*-3-hexene

8.15 In 2-methyl-2-pentene, the Me–C–Me bond angle is about 120°, and this is sufficiently large that there is minimal interaction between the methyl groups. It is probably not zero, but small enough that we usually ignore it.

8.16 No! The two methyl groups are 180° apart because they are connected to different carbons of the linear C≡C unit. Therefore, there is zero interaction between them.

8.17 No! The chlorines are on adjacent carbon atoms, not on the same carbon atom.

8.18 The high-energy rotamer has two of the chlorine atoms eclipsing, but there are two staggered rotamers that constitute the low-energy rotamers.

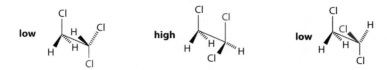

8.19 There are Cl–Cl interactions and Cl–H interactions, but they are degenerate. Therefore, one conformation will have a Cl–Cl eclipsing interaction and the other will have the two chlorine atoms staggered so that the energy diagram will look more like ethane, except for the magnitude of the energy barriers.

8.20 Because the hydrogen bonding is stronger in the diol than it is in the diamine, one expects a larger percentage of the more stable gauche diol.

8.21 Because the carbonyl of an aldehyde such as hexanal has a H attached, one anticipates that the steric crowding will be somewhat less than in a ketone, where two sp³ atoms are attached to the carbonyl carbon.

8.22 As shown in the accompanying figure, the OH unit can easily hydrogen bond, helping to stabilize the gauche rotamer of 2-hydroxybutanoic acid:

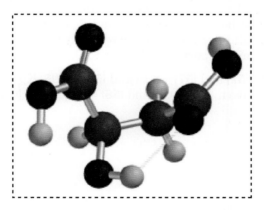

8.23

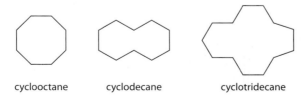

cyclooctane   cyclodecane   cyclotridecane

8.24 For cyclodecane, there are 10 sides to the polygon, so $n = 10$; therefore, the angle = $180° \ [(10 - 2)/10] = 180° \ [8/10] = 180° \ [0.8] = 144°$.

8.25 Based on its bonding with hybridization, which is less than sp^3 but more than sp^2, it appears to be closer to an alkene, but it is classified as a cyclic alkane because the definition of an alkene requires at least one π-bond.

8.26 As the bond angles in the envelope form of cyclopentane distort from planar cyclopentane, the hydrogen atoms are further apart, which decreases torsion strain. Make a model and use a ruler to measure the distances between hydrogen atoms in both conformations to convince yourself of this. To form this conformation, however, requires changing the bond angles to something other than the 108° found in planar cyclopentane. This increases Baeyer strain, but the energy savings in torsion strain more than compensates. Use a molecular model or a model kit to determine that the bond angles change when going from planar to envelope cyclopentane.

8.27

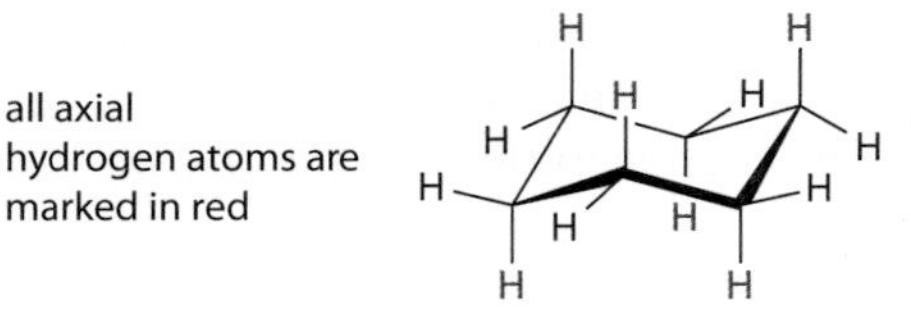

all axial
hydrogen atoms are
marked in red

8.28

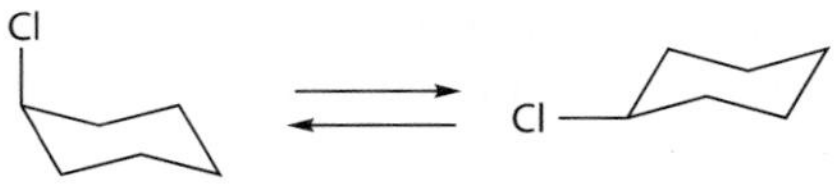

8.29 The boat and twist-boat conformations of cyclohexane are in equilibrium. There are six bonds in cyclohexane, so pseudorotation around each of these bonds will induce a boat–twist-boat equilibration, leading to six boat and six twist conformations. Because all have the same energy, only the maximum and minimum are shown in the middle portion of Figure 8.10.

8.30 Because $K_{eq} = [\mathbf{56}]/[\mathbf{57}]$ and $[\mathbf{57}] = 0.15$ and $[\mathbf{56}] = 0.95$, $K_{eq} = [0.95]/[0.15] = K_{eq} = 6.33$. In the second part, $[\mathbf{57}] = 0.77$ and $[\mathbf{56}] = 0.22$; $K_{eq} = [0.22]/[0.77] = K_{eq} = 0.286$.

8.31

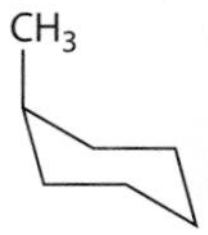

8.32 The diequatorial conformation is much lower in energy because the diaxial conformation will have significantly greater A-strain.

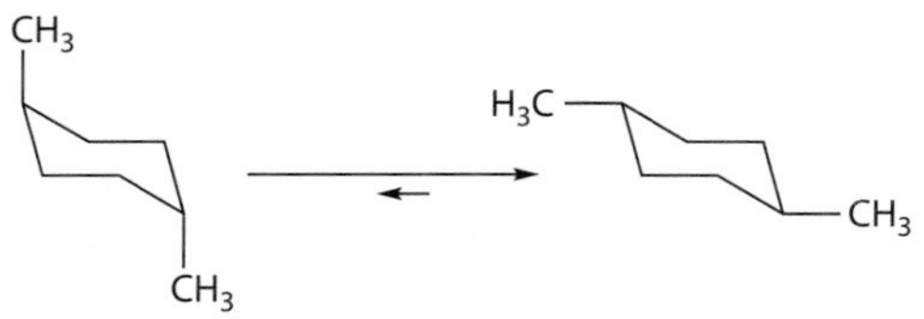

8.33 Two carbons in boat cycloheptane are essentially coplanar and eclipsing, but the cross-ring interaction with the CH_2 group is minimal because of the "flattening." For that reason, the flagpole interaction in boat cyclohexane is probably greater, and the two atoms are closer together in the smaller ring. Therefore, we expect that boat cycloheptane is lower in energy than boat cyclohexane.

8.34

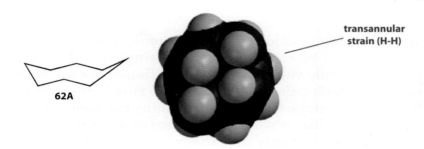

62A

transannular strain (H-H)

8.35 If cyclononane exists in the "crown" structure shown, then the transannular strain indicated is expected to be present and help destabilize the structure. It is also possible that the molecule can partially alleviate some strain by pseudorotation into a different conformation.

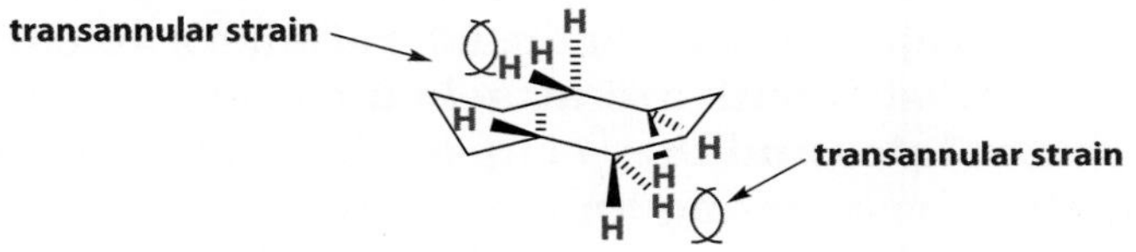

transannular strain

transannular strain

8.36 As drawn, it appears that the molecule might have transannular strain. However, the ring is so large that the hydrogen atoms rarely come into close contact, so there is minimal transannular strain.

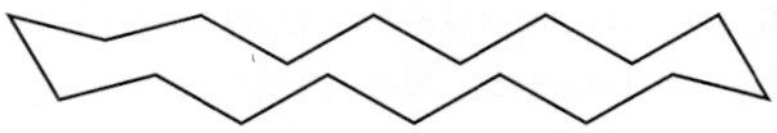

8.37

2-ethyloxirane

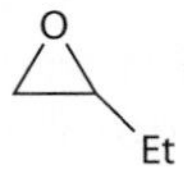

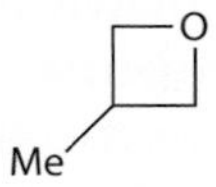

Et

3-methyloxetane

Me

8.38

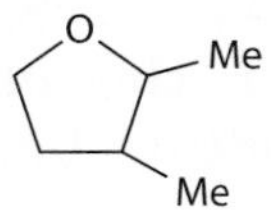

2,3-dimethyltetrahydrofuran

Me

Me

8.39

N-phenylaziridine   Ph–N (aziridine ring)   Et–N (azetidine ring with Et)   **N,2-diethylazetidine.**

Correlation of Concepts with Homework

- A rotamer is one position of bonds, atoms, and groups in a molecule, generated by rotation around a single bond. A conformation for a molecule represents one position of bonds, atoms, and groups in the molecule. Bonds in acyclic molecules rotate to generate many rotamers; this generates many conformations for the molecule. The rotamer where two atoms or groups are 180° apart is called an *anti* rotamer and the rotamer where two atoms or groups eclipse each other is called an eclipsed rotamer: 1, 3, 4, 5, 8, 10, 11, 12, 13, 17, 63, 66, 67, 68, 69, 70, 71, 76, 77, 86, 87, 88, 89.

- When rotation around a bond brings two atoms or groups on adjacent atoms, as well as the electrons in those bonds close together in space, competition for that space and electronic repulsion of the bonds lead to repulsion. This increases the energy of that rotamer. The energy of repulsion in eclipsed rotamers is called torsion strain. When substituents are attached to the atoms involved in the bond that rotates and they eclipse, the energy of that rotamer increases, effectively giving an energy barrier to rotation. When heteroatom substituents are attached to the atoms involved in the bond that rotates, internal dipole interactions and hydrogen bonding can bias the rotation to favor one rotamer: 2, 6, 7, 9, 11, 12, 15, 16, 18, 19, 20, 22, 26, 41, 45, 46, 47, 49, 55, 56, 58, 59, 60, 61, 62, 64, 65, 69, 72, 74, 75, 76, 77.

- Introduction of a π-bond leads to flattening of the conformation in the region of the π-bond. No rotation occurs about multiple bonds such as C=C, C≡C, C=O, C=N–, etc.: 14, 15, 16, 21, 22, 63, 66, 67, 68, 69, 70, 71, 76, 77, 86, 87, 88, 89.

- Cyclic alkanes undergo pseudorotation because rotation by 360° is not possible. Pseudorotation in cyclic alkanes leads to many conformations. Cyclopropane is planar, with relatively weak "banana bonds." The lowest energy conformation of cyclobutane is a "puckered" conformation. The lowest energy conformation of cyclopentane is an "envelope" conformation. The lowest energy conformation of cyclohexane is an equilibrating mixture of two "chair" conformations: 23, 24, 25, 26, 44, 51, 78, 79.

- In cyclohexane, substituents prefer the chair conformation where they are in the equatorial position because it is lower in energy. **Steric interactions of axial substituents in substituted cyclohexanes leads to A-strain in the chair conformation of substituted cyclohexanes, making the conformation with the most A-strain higher in energy. A-strain occurs in chair cyclohexane derivatives between an axial group and axial hydrogen atoms at C1, C3, and C5: 27, 28, 29, 30, 31, 32, 42, 43, 48, 50, 52, 53, 54, 81, 82, 83, 84, 89.**

- **Medium-sized rings (8–13) are less stable (higher in energy) because of transannular interactions within the cavity of the ring. The lowest energy conformation for eight-membered rings and rings with an even number of carbons is the "crown" conformation. Rings with an odd number of carbon atoms tend to be higher in energy than those with an even number due to a "twist" in the ring. Large rings (>14 carbons) are lower in energy than medium-sized rings because the cavity of the ring is so large that transannular strain is diminished: 33, 34, 35, 36, 79, 85, 86.**
- **Heterocyclic ring systems of three to six members assume conformations that are similar to the cycloalkanes, but the conformation is influenced by the nature of the heteroatom and the number of lone electron pairs: 37, 38, 39, 40, 90.**

Homework

8.40 Draw the structure for each of the following:
   (a) *cis*-2,3-dimethyloxirane
   (b) *N*,3-diethylpyrrolidine
   (c) 4-hydroxypiperidine
   (d) 3-chlorooxetane
   (e) *N*-phenyl-2-methylaziridine
   (f) *trans*-3,4-dimethyltetrahydrofuran

8.41 Which of the following represents a **gauche** conformation?

8.42 Which of the following conformations has the *least* **transannular** steric interactions, assuming each molecule is locked in the conformation shown?

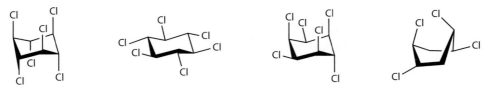

8.43 Take each structure as it is drawn and assume that it is frozen into that conformation. Examine each frozen structure and determine the one with the **most axial** bromine atoms.

8.44 Which of the following molecules is likely to have the **greatest** amount of **Baeyer strain** (also called angle strain)?

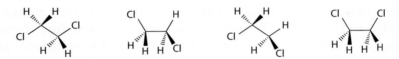

8.45 Which of the following are **anti** conformations?

8.46 On the diagrams provided, fill in the appropriate bonds and all atoms to show gauche 1,2-dichloroethane. In other words, all three diagrams will show the same gauche conformation of 1,2-dichloroethane.

8.47 Briefly explain why ethylene diamine ($H_2N–CH_2CH_2–NH_2$) may exist primarily as the *anti* rotamer rather than the gauche rotamer in methanol. **Draw both conformations as part of your answer.**

8.48 Look at the molecule drawn in the box as a chair. Note the numbering scheme for the carbons. Draw in the bonds on the flat ring, using wedges and dashed lines, to indicate the proper arrangement of atoms for this molecule. **For the flat ring, the top is the one facing you as you look at the sheet of paper.** The top for the chair form is as indicated.

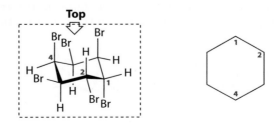

8.49 Which of the following are gauche conformations?

8.50 Which of the following is the *lowest* energy conformation?

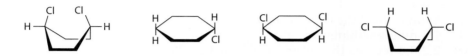

8.51 The space-filling models for cyclopropane and hexachlorocyclopropane are shown. Which has the weaker C–C bond? **Justify your answer.**

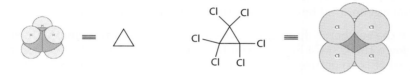

8.52 Which of the following molecules have *two* axial bromines in at least one of the chair conformations.

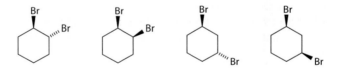

8.53 Which boat conformation has the *most* **transannular** steric interactions, assuming each molecule is locked in the conformation shown?

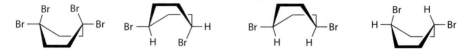

8.54 On the diagrams provided, fill in the appropriate atoms on the proper carbon and in the proper axial or equatorial position. These are two chair conformations for the same molecule, not isomers. Note the numbering scheme and assign the proper atoms to the properly numbered carbons in each chair. Which is lower in energy: A or B?

8.55 Draw the *syn* and *anti* rotamers for each of the following in Newman projection:
   (a)   butane along the C2–C3 bond
   (b)   1,2-difluoroethane
   (c)   2,2,3,3-tetramethylbutane along the C2–C3 bond
   (d)   1,2-dimethoxyethane
   (e)   pentane along the C1–C2 bond
   (f)   1-chloropropane along the C2–C3 bond

8.56 Draw the *syn* and *anti* rotamers for each of the following in a sawhorse projection:
   (a)   butane along the C2–C3 bond
   (b)   1,2-difluoroethane
   (c)   2,2,3,3-tetramethylbutane along the C2–C3 bond
   (d)   1,2-dimethoxyethane
   (e)   pentane along the C1–C2 bond
   (f)   1-chloropropane along the C2–C3 bond

8.57 Draw the *syn* rotamer of 1-bromopropane along the C2–C3 bond. Draw each rotamer generated by rotation of the *syn* rotamer by 60° clockwise through 360°.

8.58 Draw the sawhorse diagram of *syn*- and *anti*-2,2,5,5-tetramethylhexane sighting down the C3–C4 bond. If you have access to a computer modeling program, convert this structure to its 3D structure (both a ball and a stick drawing and a space-filling model). Comment on the steric hindrence found in each rotamer.

8.59 The energy barrier for rotation about the C2–C3 bond in butane is 5.86 kcal mol⁻¹. Make a model of the eclipsed-syn rotamer of butane and hold it by the C2–C3 bond. Rotation about the C1–C2 bond and the C3–C4 bond is possible, and the model indicates that the methyl groups do not touch. The model is misleading, so you must offer an explanation for why the C2–C3 barrier is so high when the model suggests it is not too bad. Use a space-filling model of the *syn* rotamer of butane to help you.

8.60 The gauche rotamer of 1,2-dichloroethane (1.2 kcal mol⁻¹) is slightly higher in energy than the gauche rotamer of butane (1.0 kcal mol⁻¹). Offer an explanation.

8.61 Draw a sawhorse diagram of what you believe to be the low-energy and the high-energy conformation for each of the following, using the **red** bond as a focus:

8.62 Using line notation, draw heptane in its all-*anti* (zigzag) conformation. Draw it again, but make every C–C bond into the highest energy syn rotamer. Compare and contrast the two conformations.

8.63 Draw the lowest energy conformation for octane, 3*Z*-octene, 3*E*-octene, and 1,3,5,7-octatetraene. Compare and contrast these four structures. A three-dimensional model may help.

8.64 Although we stated that 1,2-ethanediol exists primarily as a gauche rotamer, neat (no solvent) diol has a significant concentration of the *anti* rotamer. In the gas phase, 1,2-diol exists almost exclusively as the gauche rotamer. Explain.

8.65 If we sight down the C1–C2 bond of 1-propanol, we can draw the usual *syn* and *anti* rotamers (draw them). When 1-propanol is dissolved in water, there is a higher percentage of the *anti* rotamer than when 1-propanol is dissolved in hexane. Explain.

8.66 Offer an explanation why *trans*-cyclohexene is not expected to be a very stable molecule. Explain why cyclohexyne is not a molecule that is stable under normal conditions.

8.67 Generating a C=C unit such as that found in **A** is very difficult, whereas forming the C=C unit in **B** is rather easy. This observation is often called Bredt's rule. Explain.

8.68 Chapter 5, Section 5.10, introduced benzene, where every carbon in a six-membered ring is sp^2 hybridized. We do not see a molecule where every carbon is sp hybridized. Explain.

8.69 Draw a Newman projection for what you consider to be the lowest energy rotamer of 3,4-dimethyl-1,5-hexadiene and briefly explain your choice.

8.70 Briefly describe the structure expected for 1,3,5,7-octatetrayne.

8.71 Briefly describe the lowest energy conformation for 3,4-dichlorooctane.

8.72 There are several bonds in 2,3-dibromo-4-chlorohexane. Choose the bond that will show the greatest steric hindrance to rotation, draw it in Newman projection, and briefly discuss your choice.

8.73 Draw all isomers of the formula $C_6H_{12}Cl_2$ that have six-linear carbon atoms. Indicate which of these isomers have geminal dichloride and which have vicinal dichloride structures.

8.74 The major rotamer of 2,2,5,5-tetramethyl-3,4-hexanediol has the OH units anti, focusing on the C3–C4 bond, even in solvents containing water. This is not what is observed with 1,2-ethanediol. Discuss why one observes this difference.

8.75 Explain why butylhexyl ether should assume a zigzag conformation even with the presence of the oxygen in the chain.

8.76 Describe the structure of the prevalent conformation of 3-hydroxybutanoic acid in hexane solvent and give reasons for your choice.

8.77 Discuss the conformation of the following molecule, paying particular attention to the overall conformation as well as the position of the aldehyde (CHO) units:

8.78 Rank order the following molecules from highest to lowest in terms of Baeyer strain (use "≈" if appropriate). Rank order the molecules in terms of torsion strain. Base your answer on the conformation given for the question and assume it does not change. Briefly explain your order for each type of strain.

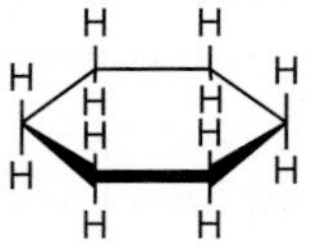

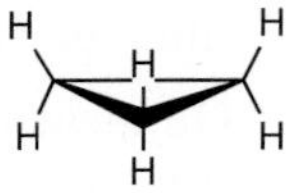

8.79 Are there transannular interactions in the following molecules? If yes, designate the interaction or interactions

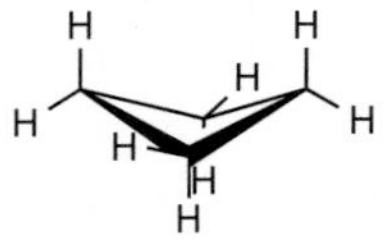

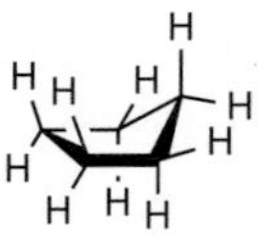

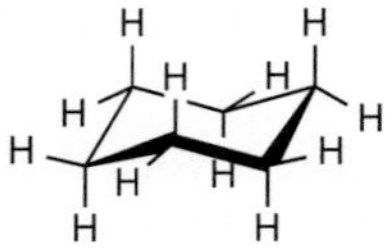

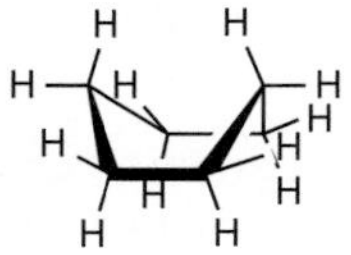

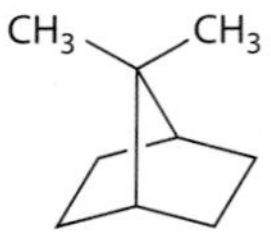

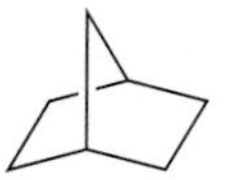

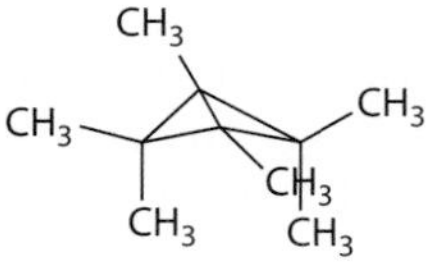

8.80 Draw one isomer of hexamethylcyclohexane in a chair conformation where
  (a) all methyl groups are axial
  (b) all methyl groups are equatorial
  (c) three are axial and three are equatorial
  (d) two are axial and four are equatorial
  Draw the flat representation for (a)–(d).

8.81 It is known that there are two boat conformations for 1,4-dimethylcyclohexane. Briefly explain why *trans*-1,4-dimethylcyclohexane has approximately equal amounts of the two boat conformations (draw them), but *cis*-1,4-dimethylcyclohexane has a large percentage of one boat with only small amounts of the other (draw both).

8.82 It is known that 1,2-di-*tert*-butylcyclohexane exists in a boat conformation to a large extent. Draw a chair conformation for this molecule and suggest a reason why this should be so.

8.83 Draw *cis*-1,3-dimethylcyclohexane, *cis*-1,3-dibromocyclohexane, and *cis*-1,3-diisopropylcyclohexane as ball-and-stick molecular models. Turn them around so that you are sighting down the cavity of each ring. Comment on the relative transannular steric interaction for these three molecules.

8.84 Of *cis*- and *trans*-1,2-diisopropylcyclohexane, which molecule will exist largely in one chair conformation and which will exist as a roughly equal mixture? Draw both molecules.

8.85 In cycloheptane, the energies of the chair and boat conformations are about equal. This is not true with 1,1,4,4,5,5-hexamethylcycloheptane, however.

Briefly discuss why not and suggest in which conformation this molecule will spend most of its time.

8.86 There is a reaction that converts a ketone carbonyl (C=O) to an alcohol unit (CH–OH) called a reduction. The conversion of the alcohol back to the ketone is called an oxidation. It is known that reduction of cyclooctanone to cyclooctanol (draw both structures) is relatively difficult, whereas the oxidation of the alcohol to the ketone is relatively easy. Explain why.

8.87 Draw both chair conformations for *cis*-1,4-dimethylcyclohex-2-ene. Indicate which you believe is the lower in energy.

8.88 The conversion of methylenecyclohexane to methylcyclohexane (draw both structures) requires energy because the product is higher in energy than the starting alkene. Briefly explain why this is the case.

8.89 Briefly explain why *cis*-1,4-cyclohexanedicarboxylic acid has a relatively high percentage of a boat conformation.

8.90 There is a well-known conformational effect observed in pyran derivatives called the anomeric effect. Whereas methoxycyclohexane (**A**) exists primarily in the chair conformation with the OMe unit equatorial, 2-methoxypyran (**B**) exists primarily in the chair conformation where the OMe is axial. Although the real explanation for this is nontrivial, suggest a simple reason that is consistent with why **B** might exist with an axial OMe unit.

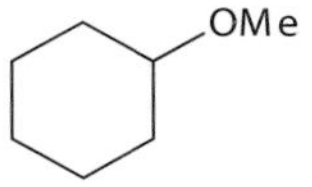

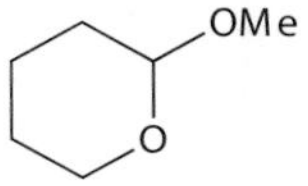

Spectroscopy Problems

No subject will be presented in sufficient detail concerning spectroscopy that will allow problems that relate to conformations.

Stereoisomers

Chirality, Enantiomers, and Diastereomers

9

This chapter focuses on a special class of isomers that differ only in the arrangement of various atoms and groups in space. When two molecules have the same atoms and groups and the same empirical formula, they are isomers. When they have the same points of attachment (the same connectivity), but differ in the spatial arrangement of those groups, they are different molecules and members of a new class of isomers known as *stereoisomers*. For this chapter, the use of models is strongly recommended, and that may include molecular modeling with a computer. The ability to "see" molecules move in three dimensions is a critical skill that must be developed.

To begin this chapter, you should know the following:

- **how to name organic molecules, based on the nomenclature rules introduced in Chapters 4 and 5**
- **the VSEPR model for drawing structures (Chapter 3, Section 3.5.4)**
- **the nature of a covalent σ-bond (Chapter 3, Sections 3.3, 3.5–3.7)**
- **the nature of a π-bond (Chapter 5, Sections 5.1 and 5.2)**
- **how to assemble and identify constitutional isomers (Chapter 4, Sections 4.5 and 4.7)**
- **how to draw molecules so that they are clearly identified (Chapters 4 and 5)**

- **the concept of rotation about bonds (Chapter 8, Sections 8.1)**
- **conformations of acyclic molecules (Chapter 8, Sections 8.1–8.4)**
- **conformations of cyclic molecules (Chapter 8, Sections 8.5–8.7)**
- **physical properties associated with organic compounds (Chapter 5, Section 5.8)**
- **how to manipulate models to "see" the shape of molecules and compare one with another (talk with your instructor)**

This chapter will introduce the concepts of asymmetry and chirality as they apply to stereoisomers. Enantiomers are nonsuperimposable mirror images that are different compounds, identifiable only by differences in the physical property known as specific rotation. Enantiomers arise when a molecule has one or more atoms (including carbon) with different substituents (from different substituents for carbon). Such atoms are known as stereogenic (chiral) atoms (most of the examples in this book will deal with a stereogenic carbon atom). With more than one stereogenic center, another type of stereoisomer results known as a diastereomer. All of these are types are stereoisomers, and a nomenclature system is in place to correlate the structure with what is known as absolute configuration.

When you have completed this chapter, you should understand the following points:

- **Two or more molecules with the same empirical formula, same points of attachment for the atoms or groups, but different spatial arrangement of those atoms or groups are called stereoisomers. A stereogenic carbon has four different atoms or groups attached, and it has a nonsuperimposable mirror image known as an enantiomer.**
- **A polarimeter measures the angle that an enantiomer can rotate plane polarized light. Observed rotation (α) is the value recorded directly from a polarimeter. Specific rotation ($[\alpha]$) is the standardized value for optical rotation, where $[\alpha] = \alpha/l \cdot c$.**
- **The absolute configuration (R or S) of an enantiomer is determined by applying the Cahn–Ingold–Prelog selection rules and the "steering wheel" model.**
- **The percentage of enantiomeric excess (% ee) is a measure of the excess of one enantiomer over another, where 0% ee indicates a racemic mixture and 100% ee indicates a single pure enantiomer.**
- **When a molecule has two or more stereogenic (chiral) centers, there are a maximum of 2^n stereoisomers, where n = the number of chiral centers. When a molecule has two or more chiral centers, diastereomers are possible. Diastereomer is the term for two or more stereoisomers that are not superimposable and not mirror images. A diastereomer that has symmetry such that**

its mirror image is superimposable is called a *meso* compound.
If there is no symmetry, cyclic molecules can have enantiomers
and diastereomers. If there is symmetry in one diastereomer,
cyclic compounds can have *meso* compounds.
- Alkenes can exist as stereoisomers and are named using the *E/Z*
  or *cis/trans* nomenclature.
- In substituted cyclic molecules, one side of the ring (relative to
  the substituent) is considered to be one group and the other side
  of the ring is considered to be a second group. If the groups
  (sides) are identical, the cyclic molecule is achiral. The terms *cis*
  and *trans* can be used to indicate the relative stereochemistry
  of diastereomers.
- Bicyclic compounds are named using a "pinwheel" model and
  the number of atoms in each chain is indicated in brackets, with
  numbers such as [x.y.z].
- Because amines undergo fluxional inversion, nitrogen is not
  considered to be chiral and amines are treated as equilibrating
  mirror images (racemates). In some polycyclic amines, fluxional
  inversion is impossible, and nitrogen can be chiral, with the
  lone electron pair considered as the fourth group.
- There are biological aspects of stereochemistry.

9.1 Stereogenic Carbons and Stereoisomers

In Chapter 5 (Section 5.1.1.2), 2-octene was named, but it was noted, without
explanation, that this molecule has another isomer. Remember that the C=C
unit is rigid and there is no rotation about that bond as there is around C–C
single bonds in alkanes (see Chapter 8, Section 8.1). Once attached to the C=C,
a group is "locked" on one side of the double bond or the other, leading to different isomers. These particular isomers are discussed in Section 9.4.

In molecules that contain only sp^3 atoms, free rotation occurs as described
in Section 8.1. However, once atoms are attached in a specific manner to an
sp^3 atom, as in **1**, the only way to change the relative and specific positions
of groups A–D is to make or break bonds. In that sense, once attached, the
absolute positions of the groups are fixed. If the A–D groups attached to the
sp^3 center are different from each other, a new type of stereoisomer is possible,

based on generating the mirror image of the original compound. The sp³ atom of interest is called a chiral atom or a stereogenic atom. The sp³ atom is usually carbon in this book, but it does not have to be. The main criterion is that the sp³ atom must have *four different atoms or groups attached to it, and its mirror image must not be superimposable.*

9.1.1 One Stereogenic Carbon

A stereogenic atom is one that has no two identical atoms or groups attached. For an sp³ hybridized atom with a tetrahedral structure such as **1**, there are four different atoms or groups (A–D) attached to a central sp³ hybridized atom (X). The central atom (X) is called a chiral center or a stereogenic center and the molecule will have the property of being chiral.[1] Molecule **1** has no symmetry, so it is asymmetric. Indeed, the atom X in **1** is often called an asymmetric atom. The simplest method of determining the presence or absence of symmetry may be to make the mirror image of the molecule. If the two structures are completely superimposable, they are the same. If they cannot be superimposed, they are different.

The photo in Figure 9.1 shows a familiar view: the reflection of a tree is a still pond. The tree image in the pond is the mirror image of the tree. In order to view the mirror image of a molecule such as **1**, simply reflect each atom as shown for **1** and **2**. A mirror is shown for effect, but **2** is the reflection of **1** and the two structures are mirror images. Both **1** and **2** are drawn again, side by side (see the boxed structures) to give a more familiar perspective. Examination of both structures reveals that no two points of the molecule or its mirror image (**1** and **2**) superimpose, one on the other. If there is symmetry, at least two points will superimpose. Molecules **1** and **2** do not superimpose, so they are considered to be nonsuperimposable mirror images, *and they are different molecules.*

Mirror image

Figure 9.1 Mirror images and a stereogenic center.

Structures **1** and **2** are isomers because they have the same empirical formula, but they have the same constitution (same point of attachment). According to the simple definition of an isomer in Chapter 4, this fact poses a problem. They do have the same points of attachment, but they differ in the spatial arrangement of the atoms, so they are ***stereoisomers***. Therefore, nonsuperimposable mirror images are different stereoisomers and they are recognized as enantiomers. ***In other words, an enantiomer is a stereoisomer that has a nonsuperimposable mirror image.***

The mirror image of a molecule or an object that has symmetry will be superimposable, and the two structures can be merged. In other words, the molecule and its mirror image constitute only one thing; they are identical. The mirror image of a thing that is unsymmetrical, as with **1** and **2**, cannot superimpose and it is different from the original. A potential shortcut to making a mirror image of a molecule is to test for the presence of symmetry. Your two hands are asymmetric. It appears that the right hand is the mirror image of your left hand, and they cannot be made to match up point for point. The palms can be touched, but it is not possible to make one hand superimpose so that both palms are facing in the same direction and the backs of both hands face in the same direction. There is no symmetry; your right hand is effectively the mirror image of your left hand and they are different. The fact that they are different leads to the property called ***handedness***. The same logic applies to a molecule that is asymmetric.

9.1 Indicate the stereogenic carbon atom or atoms, if any, in 3-chloro-4,5-dimethylnonane, 3,3-diethylpentane, and 3-methyl-5,6-diphenyl-1-hexanol.

How can symmetry be detected in a molecule? It is detected in much the way it is found in the things around us. Figure 9.2 shows the letter "W." The line down the center of the letter illustrates the presence of a plane of symmetry. In other words, the left side of the "W" can be folded over to superimpose on the right side of the "W," so the left side is the mirror image of the right side. This property indicates the presence of a plane of symmetry in the letter "W."

Figure 9.2 Planes of symmetry.

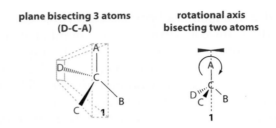

plane bisecting 3 atoms   rotational axis
(D-C-A)   bisecting two atoms

Figure 9.3 The search for a plane of symmetry and a symmetrical rotational axis.

Another example is the table. Once again, there is a plane of symmetry that bisects the table and the left side is identical to the right side.

How are these observations applied to a molecule and its mirror image? A "W" and its mirror image are shown. It is easy to see that they are completely superimposable. The two images represent one "W"—not two. Similarly, the table and its mirror image are shown, and it is clear that the image on the left is superimposable on the image on the right. They are identical. The images in Figure 9.2 show that the presence of a plane of symmetry will lead to the mirror image of that object being superimposable on the original.

If the same logic is applied to molecule **1**, it is not possible to define a plane through any three of the atoms (including the stereogenic center) such that the atoms or groups on either side of the plane are the same. In other words, if a plane is drawn through A, C, B in **1** (Figure 9.3), C does not reflect into D. In addition, it is not possible to place a line through any two atoms (including the stereogenic center) and rotate by $360°/n$ (where n is an integer less than and not equal to 360) such that superimposable rotamers are formed. When $n = 360°$, the same structure is generated, so the only values of n used are $<360°$. Rotation of **1** along the C–A bond axis by 60, 120, 180°, etc. will not generate a superimposable molecule, but something different. The carbon atom in **1** is a stereogenic carbon.[2] In older literature, such carbon atoms were called **asymmetric** or **chiral** carbons. The term stereogenic was introduced to make the concepts more general.

The definition given for a stereogenic center recognizes the presence of a stereogenic carbon or a chiral center by the absence of symmetry. Methane and dichloropropane (**3**) do not have a stereogenic carbon because they are symmetrical molecules. When a carbon atom is attached to four identical groups or atoms (the hydrogen atoms in methane), the tetrahedral arrangement of these ligands (marked in dotted lines in **red**) about carbon leads to a plane of symmetry (in **yellow**) that bisects the central carbon and two of the hydrogen atoms. Of the two remaining hydrogen atoms, one projects to the front (in **green**) and the other to the back of this plane (in **violet**). If the **green** H is reflected through the plane, it will superimpose on the **violet** H, so the atoms on either side of the plane (marked in **yellow**) are identical (both H), and there is a **plane of symmetry**.

methane   plane of symmetry   1,1-dichloropropane

To test this idea using a model kit, construct methane and then construct the mirror image of methane; then, try to superimpose the two models. It is easy to do this and the fact that methane and its mirror image superimpose indicates that there is symmetry in the molecule and they are the same molecule. The same is true for 1,1-dichloropropane (**3**), where a plane bisects the H–C–C unit and puts one chlorine atom in front of the plane (in **green**) and one chlorine atom behind the plane (in **violet**). It is symmetrical so the mirror image of 1,1-dichloropropane is completely superimposable on the original. Dichloropropane **3** and its mirror image are the same compound. *If a molecule has symmetry, its mirror image will be superimposable.*

If symmetry is not easily spotted, the reliable test for a molecule such as 1,1-dichloropropane (**3A**) is the first one presented: Examine its mirror image. For **3A**, the mirror image is **3B**, where the mirror is shown for convenience. This experiment is easily done with a model kit by making **3A** and then carefully constructing the mirror image (**3B**). The test is to try to put one model on the other to see if they superimpose. If **3A** is "lifted" by the hydrogen atom (**red**) and turned by 180°, it sits down on **3B** for a perfect match. In other words, there is symmetry and they are identical; there is no stereogenic carbon (**the two structures drawn represent one molecule, not two**).

9.2 Indicate which of the following molecules has or have a plane of symmetry: 3,3-dichloropentane, 3-chloroheptane, and 1-chlorobutane.

9.3 Draw each of the following molecules and its mirror image and determine whether it is an enantiomer or not: 2-bromo-2-chloropentane, 2,2-dimethylhexane, and 4-(1-methylethyl)octane.

3A   3B   4   5

reflect Cl into Cl both   reflect Cl into Cl in back
in front and in back   and CH$_3$ into CH$_3$ in front

2-Chlorobutane (**4**) is an example of an asymmetric molecule. The highlighted carbon is attached to four different groups or atoms: a chlorine atom, an ethyl group, a methyl group, and a hydrogen atom. These four different groups and atoms are color coded for convenience. The mirror image of **4** is **5**, and the two are easily compared to determine if the two structures are superimposable.

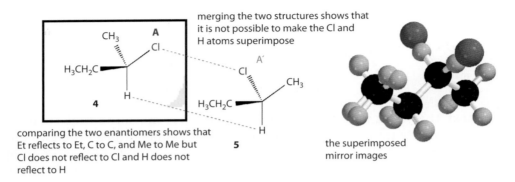

Figure 9.4 Attempt to superimpose enantiomers of 2-chlorobutane (**4** and **5**).

In mirror image (**5**), each point is reflected from **4** Cl to Cl, H to H, C to C, ethyl to ethyl, etc. Make a model of **4** and then of **5**, pick up the model of **5** and try to set it upon **4**, turning it any way that is possible as long as no bonds are broken. The methyl and ethyl groups will match, but this will place the chlorine on one side in **4** and on the other side in **5**. *It is not possible to make all of the atoms and groups match up,* so the two molecules are not superimposable. Figure 9.4 shows an attempt to make the models superimpose, using both line drawings and molecular models. It is clear that **4** and **5** are **not** superimposable. **If they cannot be made to match, they are different molecules**.

These two molecules have the same atoms and the same attachment of atoms, but **4** and **5** are two different isomers of 2-chlorobutane. They differ in the spatial orientation of the atoms, so they are stereoisomers and cannot be interconverted except by breaking bonds. As defined before, a molecule and its nonsuperimposable mirror image are called ***enantiomers***. 2-Chlorobutane **4** and its mirror image **5** are enantiomers of each other. *In general, when an sp³ carbon is stereogenic (a chiral center), interchanging any two groups or atoms of the four will generate a nonsuperimposable stereoisomer. The formal definition of an enantiomer is a molecule that has a nonsuperimposable mirror image*. A molecule that has a stereogenic carbon and generates at least one pair of enantiomers is said to be **chiral**.[1]

9.4 Draw both enantiomers of 3-bromo-3-methylheptane.

9.1.2 Fischer Projections

The ability to determine if a molecule and its mirror image are superimposable is key for recognition of enantiomers. Because a model is not always available, the way the molecule is drawn on paper is important. This book uses line notation with wedges and dashes to indicate stereochemistry (see Chapter 1, Section 1.1, and Chapter 4, Section 4.1), as with **4A**. Occasionally, other representations are useful. Emil Fischer (Germany; 1825–1919; Nobel laureate, 1902) was a remarkable chemist whose work influences chemistry even today. His name is used for a particular method for drawing stereoisomers.

4A   4B   4C   4D

Enantiomer **4A** of 2-chlorobutane shows the stereochemistry of the stereo-genic carbon using line notation, and it is easy to determine that the four groups or atoms attached to the stereogenic carbon are ethyl, methyl, chlorine, and hydrogen. The stereogenic carbon can be viewed from a different perspective: as one edge of a tetrahedron. The molecule is twisted to project the methyl and the hydrogen atom out from the front of the page (solid wedges in **4B**), with the chlo-rine atom and the ethyl group projected behind the page (dashed lines in **4B**). If the wedges and dashes in **4B** are replaced with simple lines, **4C** is obtained where two groups are attached to the vertical line and two are attached to the horizontal line. Note that the four groups appear to be attached to a cross. Drawing a mol-ecule this way constitutes a **definition** in which the horizontal line is projected in front of the page (as drawn in **4B**) and the vertical line is projected behind the page. *Representation **4C** is called a Fischer projection.* A molecular model (**4D**) is shown to give a three-dimensional view of **4C**.

Fischer projections are not used as extensively as they once were. However, Fischer projections of a stereogenic center may be convenient because simply switching any two groups on the horizontal line **or** the vertical line generates the enantiomer of **4C**. Switching both the vertical *and* the horizontal (i.e., both lines) does *not* generate the enantiomer. Interchanging groups on the hori-zontal line *or* the vertical line (here the groups on the horizontal line in **4C**) gives its mirror image, **5A**, which is the enantiomer. This may be a convenient method to generate and evaluate two structures for superimposability. Fischer projections will be used from time to time in this book, but line notation will be used most of the time, as in **4A**.

4C   5A

9.5 Draw the structure of 3-bromohexane and its enantiomer in Fischer projection.

9.2 Specific Rotation: A Physical Property

Previous sections made it clear that two enantiomers (**4** and **5** for 2-chlorobutane) are different molecules. It is one thing to draw pictures, but it is quite another experimentally to verify the validity of the premise that the two structures are different molecules. In the case of enantiomers, there is a method for distinguishing the two enantiomers based on a difference in one physical property of the two molecules. The method is derived from the interaction of the chiral molecules with plane-polarized light.

9.2.1 Physical Properties of Enantiomers

Two enantiomers differ in their spatial arrangement of atoms. They have the same boiling point, same melting point, same solubility in various solvents, same refractive index, same flash point, same adsorptivity, etc. Virtually all methods for separating two different compounds rely on differences in physical properties, but all the physical properties listed are identical for each enantiomer. Therefore, separation techniques based on physical properties, such as distillation, crystallization, simple chromatography, etc., cannot be used to separate a mixture of two enantiomers. There is one physical property in which enantiomers differ: their interaction with polarized light, which is filtered so that all the light is in a single plane (plane-polarized light).

When light is passed through a polarizing filter (see Figure 9.5), all the light that leaves the filter is in one plane. When that light interacts with an enantiomer that is placed in its path, the plane of the light is changed and the change in plane can be detected, measured in degrees. The plane of light is rotated either to the right (clockwise) or to the left (counterclockwise). In Figure 9.5, it is rotated clockwise from the viewpoint of the observer. *One enantiomer will rotate the plane of light counterclockwise, but the opposite enantiomer will rotate that plane of the light clockwise*. If the plane of light can be detected before and after it interacts with the chiral molecule, the angle of rotation can be determined for each enantiomer. *This rotation of plane polarized light (clockwise or counterclockwise) is the only physical property in which two enantiomers are different*. The instrument used to detect this rotation is called a *polarimeter* and the degree of rotation is called the *observed rotation*.

9.2.2 The Polarimeter and Observed Rotation

A polarimeter is a device for measuring the angle of rotation of plane polarized light for solutions containing chiral molecules. Figure 9.6 shows a diagram for a polarimeter. Note that modern instruments obtain readings in digital form, but

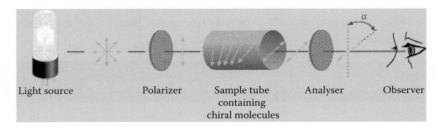

Light source   Polarizer   Sample tube containing chiral molecules   Analyser   Observer

Figure 9.5 Plane polarized light and its interaction with an enantiomer. (Provided by JASCO, Inc. With permission.)

the old style polarimeter is shown to explain the components. There is a light source and a polarizing filter. A *solution* of the enantiomer is placed between the polarizing light source and an eyepiece. As the observer looks into the eyepiece, the light passes directly *through* the solution containing the enantiomer (sometimes the neat liquid is used). If the enantiomer is a solid, it must be dissolved in a solvent before it can be analyzed. Even if it is a liquid, the enantiomer is usually dissolved in a solvent. *The solvent **cannot** be a chiral molecule.* In other words, the solvent cannot be a molecule that contains a stereogenic atom because its rotation would "swamp out" that of the molecule under examination.

The concentration of the enantiomer in the solvent is determined in grams per milliliter. This solution is added to a sample tube and placed into the polarimeter at the appropriate time. As the plane polarized light passes through the instrument *without the sample,* the plane of light is adjusted to 0° on an appropriate scale (see Figure 9.6). The sample is placed into the instrument. Sighting down the tube (through the solution), the plane of light is adjusted to determine the angle of rotation. Not only is the angle measured, but also the *direction* (clockwise or counterclockwise). The magnitude of this angle is measured and its direction is called the ***observed rotation*** and given the symbol α. In modern instruments, this number is read directly from the instrument rather than using the cumbersome array shown in Figure 9.6. Normally, a (+) is used for a clockwise rotation and a (−) is used for a counterclockwise rotation. A typical number read from the

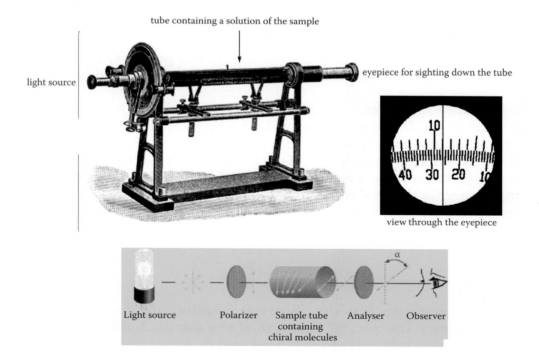

Figure 9.6 A traditional polarimeter. (The polarimeter picture and eyepiece diagram are reprinted with permission from Bellingham + Stanley Ltd. The polarimeter diagram is reprinted with permission from JASCO, Inc.)

polarimeter will therefore be recorded as (+)-23° or (–)-56°. A molecule that rotates plane polarized light in this manner is said to be *optically active*.

An important reminder about observed rotation is that one enantiomer will have a (+) rotation and its enantiomer will have a (–) rotation *of exactly the same magnitude*. Unfortunately, the observed rotation will change with the solvent used, with the concentration of the enantiomer, with the length of the container used to hold the solution, and even with temperature. This means that a person measuring the observed rotation of a compound with one instrument is likely to record a *different* rotation than someone using a different instrument with different parameters. The only way to be certain that two different observations were obtained for the same enantiomer is to standardize the method.

9.2.3 Specific Rotation

The magnitude of rotation measured in degrees (α), obtained directly from the polarimeter, is influenced (changed) by the solvent, the concentration, the temperature, the length of the polarimeter tube holding the sample, and the wavelength of the plane polarized light. Therefore, a reading of optical activity on one instrument may be different on another instrument for the same molecule. A person in a different country may use different conditions at a different temperature to measure the optical activity and observe a different value for degree of rotation. A standardized method is required that takes into account the differences in measurement conditions.

The rotation obtained by this method is called *specific rotation* (given the symbol $[\alpha]_D^{20}$), and it is considered to be **a physical property reported for optically active (chiral) molecules**. Specific rotation should be the same for a given compound regardless of the instrument, size of the sample cell, or the concentration. Several parameters must be recorded. The D in this symbol refers to the D-line of sodium, when a sodium light source is used. This is the yellow line that appears in the visible spectrum with a wavelength of 589 nm. If a different wavelength of light is used from a different light source, the wavelength of light is recorded in place of D. The 20 on the bracket is the temperature (in degrees centigrade) at which the measurement was made. Specific rotation is calculated from the observed rotation α (taken directly from the polarimeter). The formula used for specific rotation is

$$[\alpha]_D^{20} = \frac{\alpha}{lc}$$

In this calculation, α **is the observed rotation** (the angle measured by the polarimeter). The term **"*l*" is length of the sample holder** (the cell that holds the sample solution) and it is measured in decimeters (dm). Most polarimeters have sample tubes that are 0.5, 1.0, 5.0, or 10.0 dm in length. The **"*c*" term is concentration** of the enantiomer in solution and is measured in grams of enantiomer per milliliter of solvent). If the observed rotation for a given compound is +102° at a concentration of 2.1 g/mL in ethanol, in a 5.0-dm cell, the specific rotation is

$$[\alpha]_D^{20} = \frac{+102}{5.0 \cdot 2.1} = \frac{+102}{10.5} = +9.71.$$

The specific rotation for this example is reported as

$$[\alpha]_D^{20}, +9.71 \ (c \ 2.1, \text{ethanol})$$

Reporting the specific rotation this way gives the number, the concentration, and the solvent (c indicates concentration in ethanol). With this information, anyone will be able to compare the observed rotation for the enantiomer with that reported by someone else.

9.6 Determine the specific rotation of an enantiomer when α = –58.1°, c = 0.52, and l = 5 dm.

As mentioned, the specific rotation is a physical property of the enantiomer. One enantiomer will have a specific rotation with a clockwise rotation (+) and the other enantiomer will have a specific rotation with counterclockwise rotation (–). 2-Butanol has two enantiomers. One has a $[\alpha]_D^{20}$ of +13° (neat) and is named (+)-2-butanol. The term ***neat*** refers to the fact that *no solvent is used* and pure 2-butanol is placed in the polarimeter cell to measure α. The other enantiomer is (–)-2-butanol with $[\alpha]_D^{20}$, –13° (neat) and it may be called *ent*-2-butanol, where *ent* means enantiomer. The *ent* nomenclature is usually applied to more complicated chiral molecules, and it will rarely be used in this book. As with this example, enantiomers will have the *same magnitude* for specific rotation, but they will have the opposite sign [(+) or (–)].

For 2-butanol, one enantiomer will have a specific rotation of –13° and the other enantiomer (the mirror image of the first one) will have a specific rotation of +13°. The (+) and (–) terms refer to the sign of the specific rotation for a pure enantiomer. If 2-butanol is a 1:1 mixture of (+) and (–), it is named (±)-2-butanol or *racemic*-2-butanol (often abbreviated *rac*). The term racemic will be described in the next section.

9.7 Draw both enantiomers of 2-butanol.

9.8 A sample of 3-phenyl-1-pentanol is labeled as having a specific rotation of 26.8°, but the sign is missing. Is there a way to tell whether it is + or – without putting the sample into a polarimeter?

When there is a sample of a pure compound uncontaminated by its enantiomer, this compound is said to be ***enantiopure***. The compounds labeled (–)-2-butanol and (+)-2-butanol are each considered to be enantiopure.

9.2.4 Mixtures of Enantiomers and Racemates

If a mixture of enantiomers is prepared by adding known amounts of each enantiomer, the specific rotation of the mixture can be determined because *specific rotation of each enantiomer is additive,* using the sign of the rotation. In other words, if (+)-2-butanol has a specific rotation of +13° and (–)-2-butanol has a specific rotation of –13°, a 50:50 mixture of (+)- and (–)-2-butanol [labeled (±)-2-butanol] will have a specific rotation of zero:

$$[\alpha]_D^{20} \text{ (mixture)} = 0.5 \,(+13°) + 0.5 \,(–13°) = +6.5° + –6.5° = 0.$$

This 50:50 mixture of two enantiomers of a single molecule is called a **racemic mixture** *or a* **racemate**. When this specific *mixture* of enantiomers occurs, the compound is said to be **chiral, racemic**, or simply **racemic**. When 2-butanol is labeled as (±)-2-butanol, it is a chiral, racemic mixture, or it can simply be said that 2-butanol is racemic. A sample of only one enantiomer is said to be **enantiopure** [100% of (–) or 100% of (+)]. If a mixture is not a 50:50 mixture of enantiomers, as when a chemical reaction makes 2-butanol, one enantiomer may be present to a greater extent than the other.

By methods that will not be discussed, it is determined that a reaction mixture contains 72% of (+)-2-butanol and 28% of (–)-2-butanol. Because the values of specific rotation in the mixture are additive, the value of the specific rotation can be predicted:

$$[\alpha]_D^{20} \text{ (mixture)} = 0.72 \,(+13°) + 0.28 \,(–13°) = +9.36° + –3.64° = +2.62$$

There is a slight excess of (+)-2-butanol in the mixture, and the sign of the specific rotation for the mixture is positive.

Imagine a chemical reaction that produces a product; the percentage of each enantiomer formed in that reaction is unknown, but the specific rotation of the product is measured to be –10.5°. To determine the percentage of each enantiomer, begin with the knowledge that specific rotation is additive. If the product is 2-butanol, the specific rotation of one pure enantiomer of 2-butanol must be obtained. Someone must prepare or isolate the pure enantiomer, measure its specific rotation, and then report it in the literature. **Should you intuitively know the specific rotation of each enantiomer?** No! This is not intuitively obvious; it is an *experimentally determined* property, so someone has to do the experiment.

If there is no report of the specific rotation in the chemical literature, enantiopure 2-butanol must be made in an unambiguous manner and the specific rotation determined. The specific rotation for (+)-2-butanol is reported to be +13°, but data for (–)-2-butanol cannot be found in the literature. With the specific rotation of one enantiomer in hand, one of the fundamental precepts of stereoisomers is that the (–) enantiomer will have a specific rotation of equal magnitude but opposite sign to the (+) enantiomer. In other words, the specific rotation for (–)-2-butanol must be –13°. It is now possible to set up the equation:

$$[\alpha]_D^{20} \text{ (mixture)} = x \,(+13°) + y \,(–13°) = –10.5.$$

The term "x" is the percent of (+)-2-butanol, and "y" is the percent of (–)-2-butanol. Because specific rotation is additive, x + y = 1. Therefore,

$$= x \,(+13°) + y \,(–13°) = –10.5$$

$$= (1 – y)(+13°) –13°y = –10.5°$$

$$= +13° -13°y -13°y = -10.5°$$

$$= -26°y = -10.5° - 13°$$

$$= -26°y = -23.5° = y = \frac{+23.5°}{-26°} = 0.904$$

This calculation means that if the mixture obtained from a reaction has a specific rotation of $-10.5°$, it contains 90.4% of $(-)$-2-butanol [remember that the specific rotation of the mixture is $(-)$, so the $(-)$ enantiomer predominates] and 9.6% of $(+)$-2-butanol. Remember, *there must be a reported specific rotation for the molecule of interest before the relative percentages can be calculated.*

9.9 Calculate the percent of each enantiomer of an unknown (X) if the specific rotation of (+)-X is +109.3° and the specific rotation of the mixture is +27.7°.

If there is a 90.4:9.6 mixture of enantiomers of 2-butanol, the molecule is not enantiopure and it is not a racemic mixture since it is not a 50:50 mixture. This type of enantiomeric mixture is referred to as ***chiral, nonracemic***. In this context, there is another term that must be mentioned: ***enantiomeric excess***. If there is a 50:50 mixture, no single enantiomer is in excess of the other, so there is zero (0) enantiomeric excess. This is expressed as percent of enantiomeric excess, or % ee, so if there is zero enantiomeric excess, there is 0% ee. If there is only one enantiomer (it is enantiopure), there is a 100% ee of that enantiomer over the other one. The scale for % ee therefore ranges from 0% ee (50:50 mixture) to 100% ee (100% of one enantiomer). This means that the simple scale in Table 9.1 can be constructed.

Table 9.1 focuses on an excess of enantiomer A, but enantiomer B is calculated in an identical manner. A linear plot (graph) of Table 9.1 is used to determine % ee for a given mixture (see Figure 9.7). If the mixture has a purity of 90% ee, there is 95% of A and 5% of B. A mixture of 98% ee A is about 99% of A and 1% of B. The use of % ee is now widespread for reporting the enantiomeric purity of chiral, nonracemic mixtures. If the actual ratio of enantiomers is reported, it is called the enantiomeric ratio (% er).

9.10 For the 90.4:9.6 mixture of 2-butanol, Figure 9.7 indicates a 95% ee for (–)-2-butanol. Determine the percent of each enantiomer in a mixture if the % ee is measured to be 69.7% ee for the (+) enantiomer.

9.11 Calculate the % ee of a molecule that has (a) 93% of one enantiomer and 7% of the other, (b) 99.4% of one enantiomer and 0.6% of the other, and (c) 55% of one enantiomer and 45% of the other.

Table 9.1
Correlation of % ee with Relative Amounts of
Each Enantiomer in a Mixture

% ee A	% Enantiomer A	% Enantiomer B
0	50	50
10	55	45
20	60	40
30	65	35
40	70	30
50	75	25
60	80	20
70	85	15
80	90	10
90	95	5
100	100	0

9.3 Absolute Configuration (*R* and *S* Nomenclature)

So far, the enantiomers have been named (+)-2-butanol and (–)-2-butanol, where the (+) and (–) refer to specific rotation. To describe the specific structure of the enantiomer, however, more information is required: in part, a name. The designations (+) and (–) provide no structural information to reveal the absolute configuration of each enantiomer. One enantiomer of 2-butanol is drawn as **6**, both in line notation and as a Fischer projection. The other enantiomer of

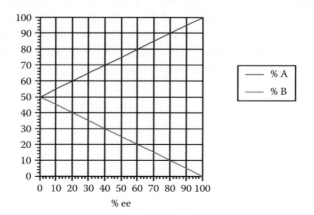

Figure 9.7   The percent composition of enantiomers as a function of % ee.

2-butanol is **7**. Which enantiomer correlates with (+)-2-butanol: **6** or **7**? In the Fischer projection of **6**, the OH is on the right side, so does that mean **6** is (+)? Not necessarily! In fact, ***there is no correlation between specific rotation and the absolute configuration (the specific location of groups attached to a stereogenic center)***.

The specific rotation has a (+) sign, which indicates a clockwise rotation of plane polarized light, but specific rotation is a physical property that is measured with a polarimeter. The (+) in (+)-2-butanol only means that this enantiomer rotates plane polarized light clockwise. *This measurement indicates nothing about how the atoms are connected*. In other words, the only way to determine the actual structure of a given enantiomer is to devise a set of standardized rules to name each enantiomer that allows us to draw that enantiomer properly. Once the enantiomer has been named, the name is correlated with the structure. Remember: the name is chosen by an arbitrary set of rules, but the physical property is not arbitrary.

Any name that allows two enantiomers to be distinguished must be very specific. Taking 2-butanol as an example, each enantiomer must have a name that can be translated to a structure. Because the only difference is stereochemistry, the method chosen to name such compounds indicates the position of atoms or groups one to the other (absolute configuration). The terms R (*rectus;* Latin for "right") and S (*sinister;* Latin for "left") are used, and the two enantiomers of 2-butanol will be named (R)-2-butanol (see **6**) and (S)-2-butanol (see **7**). Assignment of R and S can be done only with a set of rules that allows analysis of each structure: ***Assign a priority of importance to each of the atoms attached to the stereogenic carbon. The spatial arrangement of these atoms will determine the absolute configuration (R or S).*** To repeat, these rules have nothing to do with the physical properties.

9.3.1 Cahn–Ingold–Prelog Priority Rules and the Steering Wheel Model

There is a set of rules for naming organic molecules (see Chapter 4, Section 4.6), and there is a set of rules for assigning the ***absolute configuration*** of a given stereogenic center, which becomes part of the name for that enantiomer. All *atoms* attached to a stereogenic carbon atom are inspected with a simple goal: The so-called Cahn–Ingold–Prelog selection rules are used to assign the priority of each atom.

A simple example is 1-bromo-1-chloroethane, where **8** and **9** are the structures of the two enantiomers (nonsuperimposable mirror images). The stereogenic carbon is the focus of the examination, and the atoms or groups attached to that stereogenic carbon are bromine (**red**), chlorine (**green**), hydrogen (**cyan**), and methyl (**blue**). **The *atoms* attached to the stereogenic carbon are examined—not the group—so Br, Cl, H, and C are compared.** *The model assigns one atom as the highest priority (is most important) and another as the lowest priority (is least important).* A logical way to assign a priority is on the basis of atomic mass for each **atom**. This definition poses a problem for methyl. *Do not use the mass of the group (-CH₃) but rather use the mass of the atom attached to the stereogenic carbon (in this case the carbon atom). If the priority is based on atomic mass, it is essential that the assignment is based only the atom attached to the stereogenic center. Therefore, compare the atomic masses of Br, Cl, H, and C.* The order, according to descending atomic mass, is $Br > Cl > C > H$. The priority letters **a, b, c**, and **d** are assigned for each atom, with **"a" the highest priority and "d" the lowest priority**; $Br =$ **a**, $Cl =$ **b**, $C =$ **c**, and $H =$ **d**. These letters are assigned to a tetrahedral representation of **8** (see **10A**).

| **8** | **10A** | **10B** | **10C** | **10D** |

The positions of (**a**)–(**d**) are fixed in **10A** because they represent the fixed positions (absolute configuration) of the four groups in space. If model **10A** is "picked up" (make a model with the four colors shown and twist it as described), it can be twisted around to generate **10B, 10C**, or **10D**. It does not make any difference where (**a**) resides, the *absolute arrangement* of the other groups relative to (**a**) is the same. It must be! No bonds have been broken, so all orientations represent the identical spatial arrangement of atoms in **8**.

To assign the absolute configuration, an *assumption* must be made. The tetrahedral model **10A** has the four different colored atoms to represent (**a**)–(**d**) and it is turned so that the lowest priority atom (**d**) is pointed *180° from the viewer (i.e., behind the plane of the page,* to the rear of the tetrahedron). Imagine holding **10A** by the (**d**) atom and tilting it back, behind the plane of the paper. This action will give **10E**. When (**d**) is projected behind the plane of the paper, (a)–(c) are projected toward the front of the plane of the paper. In effect, one point of the tetrahedron is pointed behind the plane of the paper, and the base of the tetrahedron, which is made up of three atoms, is projected to the front. Two tetrahedral models are shown to illustrate this point.

Imagine that a curved arrow is drawn from the highest priority atom (**a**), pointing to the next highest priority atom (**b**) and finally to (**c**). This imaginary arrow generates the arc of a circle, with the center being the stereogenic atom, which is also connected to **d** (projected behind the plane of the paper). This view looks like the steering wheel of an old-time car, and this representation is called the ***steering wheel model*** (represented by **10F**). In the case of **10F**, the imaginary arrow from **a** → **b** → **c** goes in a counterclockwise direction, and it is labeled the (*S*) configuration for **8**. The enantiomer of **8** is **9** and, using the same priority scheme, the arrow from **a** → **b** → **c** proceeds in a clockwise rotation (see **11**), and this is labeled the (*R*) configuration. ***The absolute configuration for an enantiomer is defined as: Assign priorities a-d for atoms connected to a stereogenic center, and if the direction a → b → c is clockwise, the absolute configuration is (R) but if the direction is counterclockwise, the absolute configuration is (S).*** The steering wheel model is used to assign absolute configuration and the (*R*) or (*S*) configuration becomes part of the name. The correct name for enantiomer **8** is therefore (*S*)-1-bromo-1-chloromethane and **9** is (*R*)-1-bromo-1-chloroethane.

The term (*R*) represents the absolute configuration seen in **8** and the term (*S*) represents the absolute configuration seen in **9**. ***Which structure is (+)-1-bromo-1-chloroethane, 8 or 9?*** *Given only the name, it is impossible to know!* The names (*R*) and (*S*) are determined by an arbitrary (but universally accepted) set of rules. The (+) or (–) label is the specific rotation, a physical property. Therefore, if only the names (*R* or *S*) are given, that information does not allow one to correlate the structure with the sign of the specific rotation. It is necessary to identify **8**, isolate it, put it into a polarimeter, and then determine its specific rotation. Alternatively, the same can be done for **9**. *Only after the physical measurement of each named compound can specific rotation of an enantiomer be correlated with the absolute configuration of that enantiomer.* **This is a very important lesson for this chapter.**

It is time to return to the initial problem posed for **6** and **7**: the enantiomers of 2-butanol. Using the steering wheel model, the absolute configuration of each enantiomer of 2-butanol can be determined. Obviously, one is *R* and the other is *S*. The two enantiomers in Fischer projection are **6A** and **7A** and these are also drawn in tetrahedral forms **6B** and **7B**. Note the use of the shorthand terms Et (for ethyl, –CH$_2$CH$_3$) and Me (for methyl, –CH$_3$). Using the atomic mass priority scheme, ethyl, methyl, hydrogen, and OH are attached to the stereogenic carbon in **6**. As noted, *only the atoms attached to the stereogenic carbon* are used: C, C, H, and O.

(4) **If the atom being considered is part of a π-bond, each bond is counted as being attached to a substituent (two atoms for a double bond, and three atoms for a triple bond).**

There is one more instance where the given rules break down. When one of the groups attached to the stereogenic atom is an electron pair rather than another atom, the electron pair is always given the lowest priority (d). This will be discussed in Section 9.7.3, where it will make more sense.

9.3.2 Relationship of Specific Rotation and Absolute Configuration

In Section 9.2, specific rotation and absolute configuration were identified as two completely different things. Specific rotation is a physical property and the absolute configuration is a name determined by a set of rules. It is mentioned again because it is a potential point of confusion and an easy mistake to make. Simply put, an enantiomer with a (+) specific rotation could have either an (*R*) or an (*S*) absolute configuration. Looking back at 2-butanol, the specific rotation of (2*R*)-butanol (**6**) is +13° and that of (2*S*)-butanol (**7**) is –13°. This means that the clockwise absolute configuration (*R*) correlates with a counterclockwise (–) specific rotation; this is (–)-(2*R*)-butanol, and the opposite is true for the enantiomer, (+)-(2*S*)-butanol.

Do not be fooled by the fact that a clockwise/counterclockwise rotation is used to measure both parameters. Most importantly, **the name (2R)-butanol should not lead to any assumption about specific rotation. The name (–)-2-butanol should not invoke any assumption about absolute configuration.** To find out the specific rotation of (2*R*)-butanol, take a bottle of pure (2*R*)-butanol, put it into a polarimeter, measure the observed rotation, and then calculate the specific rotation. There is no other way to know unless you do it or someone has reported it in the chemical literature.

9.4 Alkenes

Alkenes were introduced in Chapter 5 (Section 5.1). They are characterized by a C=C unit containing four groups or atoms. Because rotation about the C=C unit is not possible, atoms or groups attached to the C=C unit are effectively locked in space. Therefore, *alkene stereoisomers are possible, but they are not enantiomers*.

9.4.1 Alkene Stereoisomers

2-Butene, $CH_3CH=CHCH_3$, is a simple example of an alkene that exists as a stereoisomer. Structurally, 2-butene contains a planar C=C unit with two methyl groups attached to it. Rotation around the rigid C=C unit is *impossible* (the π-bond ensures that it is locked into position), so the methyl groups can be attached on the same side (as in **26A**) or on opposite sides (as in **27A**). The two alkenes are *different molecules* (isomers), with different physical properties. They have the same empirical formula and the same constitution with respect to atom attachment. They differ in their structure only by the relative position of the methyl groups in space, which are on opposite sides or the same side of the molecule. They are stereoisomers. **Remember that a stereoisomer is a molecule that has an identical empirical formula with another molecule (an isomer) and differs from the other molecule (or several molecules) only in the spatial arrangement of its atoms.**

It is **impossible** to interconvert **26** and **27** by rotating bonds or twisting atoms. Use a model kit to make a model such as the molecular models **26B** and **27B** and try to superimpose them! It is clear that they are different. *Only a chemical reaction can change one into the other, and that requires making and breaking bonds.* If **26** and **27** are really different compounds, then their physical properties should be different. Indeed, they differ slightly in boiling point (37°C for **26** and 37–38°C for **27**), melting point (–140°C for **26** and –180°C for **27**), and so on. Clearly these alkenes are closely related, but their stereochemical difference leads to differences in their physical properties. The idea of sidedness in an alkene may be difficult to correlate with **26** or **27**.

In **26C**, the alkene is "tipped on its side" such that both hydrogen atoms are projected toward the front (marked in **green**) and both methyl groups are projected to the rear (marked in **yellow**). In **27C**, one H and one methyl are projected to the front (in **green**) and the other methyl group and other hydrogen atom are projected to the rear (in **yellow**). These drawings indicate that the **green** atoms/groups are on the same side and the **yellow atoms**/groups are on the same side. Therefore, the two methyl groups are on the same side in **26C** but the methyl groups are on opposite sides in **27C**. The term "side" is replaced with the term "face" in many cases, such that the two methyl groups are on the same face in **26C**, but on opposite faces in **27C**. It is important to recognize the sidedness of groups with respect to a C=C unit.

Because **26** and **27** are different molecules, each requires a unique name. Both molecules are 2-butene, however, so the stereoisomer name must simply reflect the stereochemical differences of the methyl groups. There are no stereogenic centers, so the *R/S* nomenclature cannot be used. There are two methods used in nomenclature that distinguish **26** from **27**: the *cis/trans* nomenclature and the *E/Z* nomenclature.

9.4.2 *cis* and *trans* Nomenclature

In **26** and in **27**, each carbon of the C=C unit (C1 and C2) has a methyl group and a hydrogen. The relative positions of those two groups cannot be changed in either **26** or **27**. In other words, **26** cannot be transformed into **27** by bond rotation. In **26A**, the methyl groups are on the same side of the C=C unit, but in **27A** the methyl groups are on opposite sides of the C=C unit. **This leads to a definition for naming the stereoisomers:** *If two like groups are on the same side of an alkene, the molecule is a cis-alkene. If two like groups are on opposite sides of an alkene, the molecule is a trans-alkene.* By this definition, **26** is *cis*-2-butene and **27** is *trans*-2-butene. A key word in these definitions is "like." The *cis/trans* **nomenclature applies** *only* **when identical groups are on each carbon of the C=C unit, such as XYC=CXZ, where X is on both carbon atoms**. When the same group is on a single carbon, as in $X_2C=CYZ$, there are no stereoisomers because the two possible structures are superimposable. An example of this latter occurrence is 2-methyl-2-pentene, **28**. One carbon of the C=C unit has two identical groups, both methyl. One methyl group is *cis* to the ethyl group, but a methyl is also *trans* to the ethyl, so **28** has no stereoisomers. Indeed, when two identical groups are on the same carbon of the C=C unit, there is no possibility for *cis/trans* isomers or for the *E/Z* isomers described in the next section.

            28   **29**   **30**

If the groups to be compared on each carbon of the C=C unit are not the same, the *cis/trans* nomenclature does not apply. In 2-pentene, for example, both a methyl group and an ethyl group are attached to the C=C unit. Two stereoisomers are possible, **29** and **30**. Because the groups are not identical in **29** or **30**, *cis* or *trans* cannot be used.

9.16 Draw the structures of *cis*-3-heptene and *trans*-3-heptene.

9.17 Does 1,1,3-trichloro-2-methyl-1-pentene have *cis/trans* stereoisomers?

9.4.3 *E* and *Z* Nomenclature

The *cis–trans* nomenclature is quite useful, but in many cases determining *cis* versus *trans* is impossible because no two groups are identical, as shown earlier for **29** and **30**. An alternative method has been developed that determines the relative priority of groups attached to the C=C unit and compares those priorities. For the 1-chloro-1-pentenes **31** and **32**, one chlorine, one propyl, and two hydrogen atoms are attached to the C=C units of these stereoisomers. No two atoms or groups are the same, so *cis* or *trans* cannot be used. The new system for naming stereoisomers is called the *E/Z* system. The term *E* comes from the German word *entgegen,* which means opposite or apart. The term *Z* comes from the German word *zusammen,* which means together. What constitutes "apart" and what constitutes "together" are determined by the priority rules introduced in Section 9.3, the Cahn–Ingold–Prelog (CIP) priority rules.

To determine the stereochemistry, compare sidedness with groups or atoms on one side or the other of the C=C unit. However, there must be something to compare. This system compares the higher priority group on each carbon of the C=C unit independently from one another. **Each carbon atom of the C=C unit is examined independently** by asking which of the two attached atoms or groups is higher in priority, using an atom-by-atom comparison. The goal is to find the highest priority ***atom*** on each of the two sp^2 carbons. For **31**, carbon C1 of the C=C unit has a chlorine atom and a hydrogen atom, whereas the other carbon (C2) has a propyl group and a hydrogen atom. For C1, when you compare chlorine with H, chlorine is clearly the higher priority. For carbon C2 of the C=C unit, a H is attached to C2 as well as a carbon (of the propyl group). Using the Cahn–Ingold–Prelog rules discussed in Section 9.3 for stereogenic atoms, it can be determined that carbon has a higher atomic number than hydrogen, so the propyl group has the higher priority.

The chlorine atom on C1 has the higher priority and the carbon on C2 has the higher priority, and those priority groups are on opposite sides. To use the *E/Z* nomenclature, **compare the sidedness of the two high-priority groups on the C=C unit**. (See the sidedness analysis used for **26** and **27** in Section 9.4.2.) In **31**, the chlorine and the propyl group are on *opposite* sides of the C=C unit. This stereochemistry is designated *E,* and **31** is named 1-chloro-1*E*-pentene. Using a similar analysis, the chlorine and propyl groups are on the *same* side in **32**, so *Z* is used, and this is named 1-chloro-1*Z*-pentene.

Another example compares the isomeric 4-chloro-3-ethyl-2-butenes **33** and **34**. To determine the *E/Z* name, the methyl on one carbon (in **green**) is higher in priority that the hydrogen atom on that carbon. On the other carbon of the C=C unit, the chloromethyl (in **red**) is higher in priority than ethyl (C^{ClHH} > C^{CCH}), so **33** is 4-chloro-3-ethyl-2*Z*-butene and **34** is *E*-4-chloro-3-ethyl-2-butene.

It is important to emphasize that although *Z* and *cis* are both derived from groups on the same side of the double bond, they arise from completely different definitions. ***Never*** **mix** *E/Z* **and** *cis/trans* **nomenclature.** ***Never*** **assume that** *Z* **is** *cis* **or that** *E* **is** *trans.*

9.18 Using the *E/Z* nomenclature, assign the proper name to 30 and to 31.

9.19 Draw the structure of 5-ethyl-3-(4-chloro-1-butyl)-oct-2*E*-ene.

9.20 Draw the structures of hex-2*E*-ene, 3,5-dimethyloct-3*Z*-ene, and 1-chloro-1*Z*-heptene.

1-Bromo-1,2-dichloro-1-butene illustrates the point about *cis/trans* versus *E/Z*. The two stereoisomers for this compound are **35** and **36**. There are two identical atoms—the chlorine atoms on either carbon of the C=C unit—so *cis* or *trans* is appropriate. In **35**, the two chlorine atoms are on opposite sides of the molecule, so it is named *trans*-1-bromo-1,2-dichloro-1-butene; the chlorine atoms are on the same side in **36**, so it is named *cis*-1-bromo-1,2-dichloro-1-butene. Using *E/Z* nomenclature, the bromine and the chlorine are the two priority groups and in **35** the priority groups are on the same side, so it is 1-bromo-1,2-dichloro-1*Z*-butene. In **36**, the two priority groups are on opposite sides, so the name is 1-bromo-1,2-dichloro-1*E*-butene. Clearly, the *trans* alkene is the *Z* alkene and the *cis* alkene is the *E* alkene. Choose one name or the other, but do not mix them. Formal IUPAC name uses the *E/Z* nomenclature.

9.4.4 *cis* and *trans* Substituents Attached to Rings

The *cis/trans* nomenclature can be used for substituents that are attached to rings. ***The E/Z nomenclature cannot be used for cyclic compounds.*** A carbon ring is flexible, but 360° rotation is not possible (see Chapter 8, Section 8.5, for a discussion of pseudorotation in cyclic compounds). Because complete rotation about the C–C bond is impossible, substituents on that ring are fixed (locked) onto one side of the ring or another. Therefore, two substituents can be on the same side of a ring (as in 1,2-dimethylcyclopentane, **37**) or on opposite sides of that ring (as in 1,2-dimethylcyclopentane, **38**). Alternatively, the methyl groups in **37** are on the same face, but the methyl groups in **38** are on opposite faces.

The solid wedges in **37** and **38** show the group projected out of the paper and the dashed lines show groups projected behind the paper. **If two like groups are on the same side of a ring, it is a *cis*-cycloalkane and if the like groups are on opposite sides of the ring, it is a *trans*-cyclic alkane.** Therefore, cyclic alkane **37** is named *cis*-1,2-dimethylcyclopentane and **38** is *trans*-1,2-dimethylcyclopentane.

An important note that concerns substituents on cyclic compounds must be reiterated: The *E/Z* nomenclature just discussed is ***never*** used with cyclic compounds, only with alkenes.

9.21 Draw the structure of *cis*-1,3-dibromocyclopentane.

9.22 Give the correct name for a cyclic alkane of nine carbons with an ethyl group at C1 and a methyl group at C4 when the methyl and ethyl are on opposite sides of the ring.

9.5 Diastereomers

A new category of stereoisomers is possible when a molecule contains more than one stereogenic center. In terms of absolute configuration, every stereogenic center is treated as an independent unit and assigned the appropriate *R* or *S* absolute configuration. However, the presence of more than one stereogenic center poses new problems.

9.5.1 Two Stereogenic Centers

Molecules containing more than one stereogenic center, such as 2,3-dichloropentane, have more stereoisomers than may be expected. Structures **39** and **40** are enantiomers (2,3-dichloropentane), shown as sawhorse diagrams **39A** and **40A**, and also as Fischer projections **39B** and **40B** (the view seen for **39A** and **40A** from the top of each molecule). The name of **39A** is (2*S*,3*R*)-dichloropentane and **40A** is (2*R*,3*S*)-dichloropentane. Note the use of Me for methyl and Et for ethyl. Representation **39B** and its mirror image (**40B**) are stereoisomers and nonsuperimposable mirror images, so by definition they are enantiomers. Using the Fischer projections, "pick up" **40B** and rotate it by 180°. The chlorine atoms will match, as will the hydrogen atoms, but the ethyl and methyl groups will not match. This thought experiment is reproduced in Figure 9.8, which attempts to superimpose the three-dimensional molecular models to ascertain whether or not **39A** and **40A** are superimposable (see Section 9.1.1 and **4** and **5**).

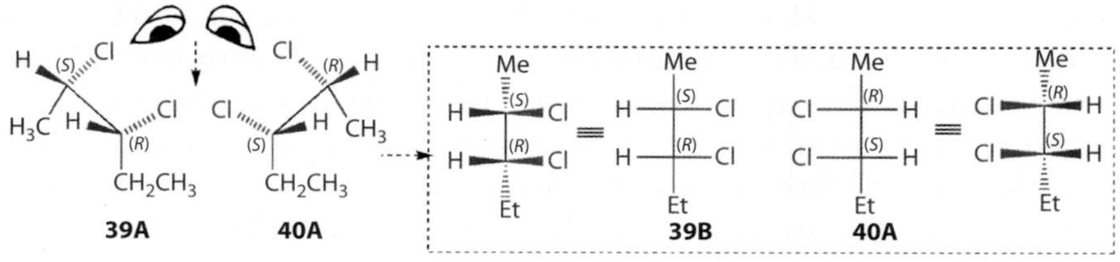

The molecular models of the two enantiomers are shown individually and then they are superimposed so that all atoms that are identical can merge.

Figure 9.8 Attempt to superimpose enantiomers **39** and **40**.

The resulting model is quite cluttered, but although the methyl group and the ethyl group overlap in this view, the chlorine and hydrogen atoms clearly do not.

Enantiomers **39** and **40** are two different stereoisomers. The presence of the second stereogenic center allows a different arrangement of atoms for 2,3-dichloropentane, shown in the new stereoisomers **41** and **42**. Changing the stereochemistry at C2 in **39A** from (S) to (R) leads to **41A** [(2R,3R)-dichloropentane], which is *different* from either **39A** or **40A**. Making a model of all three compounds and attempting to superimpose them can prove this.

When **39A** and **41A** are drawn side by side, all attempts to move the molecules together so that all identical atoms that can merge will fail (one chlorine atom and one hydrogen do not match). Therefore, **39** and **41** are ***different molecules, but they are not enantiomers because they are not mirror images of one another.*** They are stereoisomers, however! The mirror image of **41A** is **42A**, and these two molecules are enantiomers. Note that the new molecule (**42A**) differs from **41A** and also from **39A** and **40B**.

Molecules **41** and **42** are enantiomers, as are molecules **39** and **40**, but **41** is different from **39** or **40**, as is **42**. It is clear these molecules are isomers with the same connectivity; however, they differ in the spatial arrangement of groups and atoms, so they are stereoisomers. ***When two stereoisomers are not mirror images of each other (not enantiomers) and are not superimposable (not the same molecule), they are called diastereomers.*** Molecule **39** is a diastereomer of **41** and **42**, and **40** is a diastereomer of **41** and **42**. ***The formal definition of a diastereomer is, therefore, a nonsuperimposable, nonmirror image.*** This appears to be a strange definition because if an apple is compared with an orange, they are nonsuperimposable, nonmirror images. ***However, this definition of diastereomer applies only to two stereoisomers that are not the same molecule and are not mirror images of each other.***

9.23 Draw all four stereoisomers for 3,4-dimethylheptane in Fischer projection and label enantiomers and diastereomers.

Of course, many molecules with two or more stereogenic centers have diastereomers. Remember that every diastereomer can have a mirror image, which constitutes two stereoisomers. A molecule with two stereogenic centers will have up to two diastereomers and each will have a mirror image. This means there is a maximum total of four stereoisomers for a molecule with two stereogenic centers. *This observation can be extended to all molecules that have stereogenic centers. For a given number of stereogenic centers (say n) there will be a **maximum** of 2^n stereoisomers.*

A molecule with two stereogenic centers has 2^2 or 4 stereoisomers. A molecule with 4 stereogenic centers will have a maximum of 2^4 or 16 stereoisomers. If a molecule has 9 stereogenic centers, the maximum number of stereoisomers will be 2^9 or 512 stereoisomers. If 512 stereoisomers does not seem like a large enough number, look at a molecule with 28 stereogenic centers; 2^{28} means 2.684×10^8 stereoisomers (that is, more than 268 million stereoisomers) *for one constitutional isomer of a single empirical formula.* For that same empirical formula, there may be other molecules with different connectivity, and additional isomers are possible. Clearly, the presence of multiple stereogenic centers and the many stereoisomers that are possible is a concept that cannot be ignored.

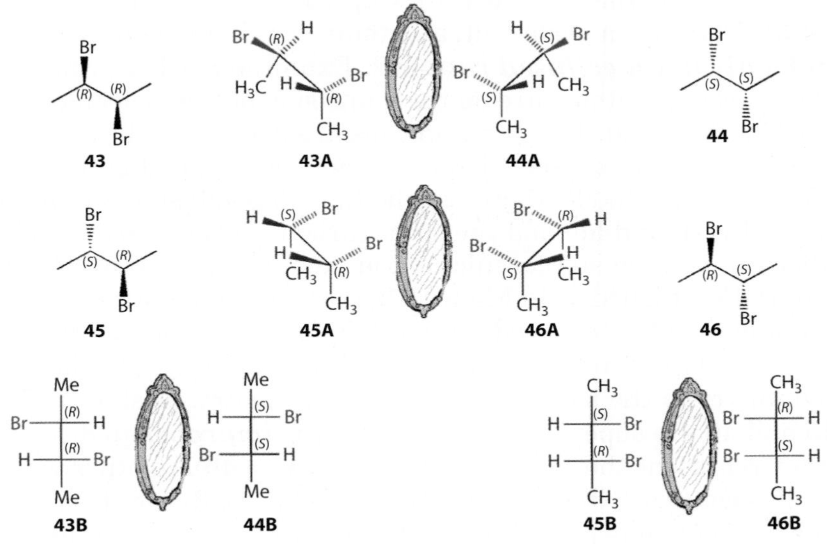

9.5.2 *Meso* Compounds. Symmetry in Molecules with Stereogenic Centers

The 2^n rule from the previous section gives the *maximum* number of stereoisomers for a molecule with more than one stereogenic center. There are never more than 2^n stereoisomers, *but it is possible to have fewer stereoisomers if a molecule with two or more stereogenic centers has symmetry.* When there is more than one stereogenic carbon, similar groups may be attached that make one part of the molecule identical to another, although each individual stereogenic carbon is asymmetric with respect to that center. One such case is 2,3-dibromobutane. The line drawing (**43**) of one enantiomer of this molecule is drawn as a sawhorse diagram (**43A**) and in Fischer projection (**43B**). The mirror image **44** is also shown in Fischer projection (**44B**) and as a sawhorse diagram (**44A**).

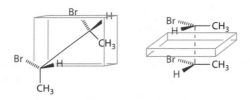

Figure 9.9 Plane of symmetry in dibromide **45** (= **46**).

Similar drawings are provided for diastereomer **45** (**45A** and **45B**), and another line drawing is provided for **46** (**46A** and **46B**), which is the mirror image of **45**. Careful inspection of **45** and **46** reveals something different from previous stereoisomers: Structure **45** is ***superimposable*** on **46**, which means that these two structures are the same (they represent one molecule, **45** = **46**). Make a model of both **45A** and **46A**. Pick up **45A**, rotate it by 180°, and "lay" it on top of **46A**. They are a perfect fit; all atoms match up. ***Note that the atoms will match only in an eclipsed rotamer.*** Examine Fischer projection **45B** to see that both bromine atoms are on the same side of the molecule, as are both hydrogens. The "up" and the "down" groups are both methyl.

If the eclipsed rotamer (see Chapter 8, Section 8.1) in Figure 9.9 is examined, a slice down the middle (between the C2–C3 bond) shows that one carbon is attached to Br, H, and Me and the other carbon is also attached to Br, H, and Me. In other words, each stereogenic carbon atom has the same attached atoms and groups. If **45** is turned as in Figure 9.9, the "top" and "bottom" are seen to be identical (but only in this eclipsed rotamer). Because the top and the bottom are identical in the eclipsed rotamer, the top reflects perfectly into the bottom and there is symmetry in the molecule (a plane of symmetry, as shown). The mirror images **45** and **46** are superimposable so that it is ***improper*** to draw **45** *and* **46** because they represent the same structure (i.e., **45** = **46**). In other words, this is one molecule—not two. Structure **46** will be used hereafter for this molecule.

When a molecule has a structure with identical atoms or groups on the top and bottom (one side can be perfectly reflected into the other side), that molecule has a plane of symmetry (see Figure 9.9). When such symmetry occurs, the mirror image of one stereoisomer is superimposable on itself. Such a stereoisomer is called a ***meso compound***. Because 2,3-dibromobutane has two stereogenic centers, the 2^n rule predicts a maximum of four stereoisomers. Symmetry in one stereoisomer means that **46** is a *meso* compound, so there are only three stereoisomers (the two enantiomers **43** and **44** and the *meso* compound **46**). *It is important to point out that 46 and 43 are diastereomers. Likewise, 46 and 44 are diastereomers, but 43 and 44 are enantiomers.*

9.24 Determine the maximum number of stereoisomers for 3,4-diphenylhexane and indicate whether this is the actual number.

9.25 Decide whether 3,4-hexanediol has a *meso* compound. Draw it as well as the structure of *meso*-2,3-butanediol.

9.26 Name the three stereoisomers 43, 44, and 46, using *R/S* to assign absolute configuration.

It is important to reiterate that when a molecule has symmetry, it will have *fewer* stereoisomers than predicted by the 2^n rule. Look for a plane of symmetry when two adjacent chiral carbons have exactly the same groups or atoms attached to them. Such a compound will have a *meso* structure. There is no symmetry in 4-methyl-3-hexanol, and there are two diastereomers, **47** and **48**. Each diastereomer will have an enantiomer, so the 2^n rule predicts the four different stereoisomers. In the context of this section, one stereogenic carbon has an ethyl group, an OH, a hydrogen atom, and a 2-butyl group, whereas the other stereogenic carbon has a methyl group, an ethyl group, a hydrogen atom, and a CH(OH)Et group. Clearly, they are different, so there is no chance for a *meso* compound.

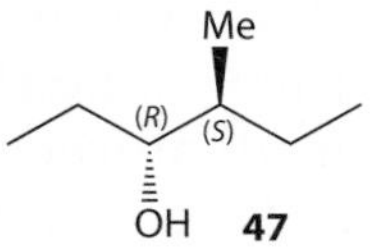

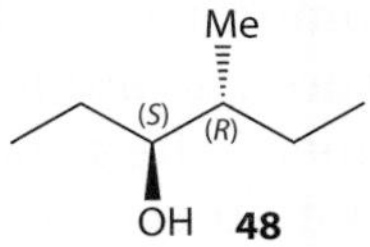

9.5.3 Molecules with Three or More Stereocenters

Going from one to two stereogenic centers in a molecule led to greater complexity with respect to the number of stereoisomers. Molecules with three or more chiral centers are even more complex, in accord with the 2^n rule, where three stereogenic centers lead to a maximum of eight stereoisomers.

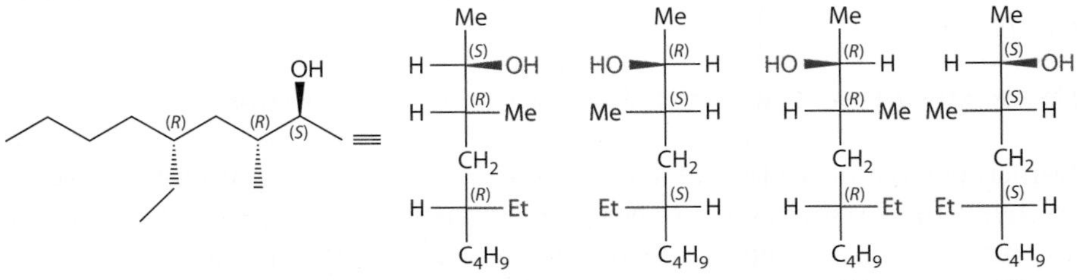

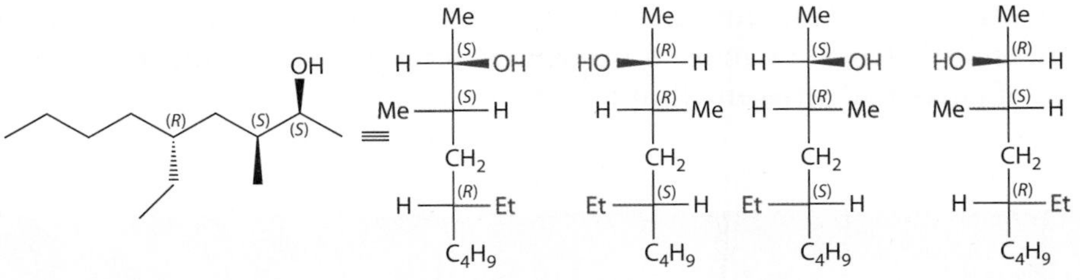

An example is 5-ethyl-3-methyl-2-nonanol. There are three stereogenic carbons, so a maximum of $2^3 = 8$ stereoisomers is expected, which includes all diastereomers and all pairs of enantiomers. **How many diastereomers are there?** One diastereomer is **49**, drawn in the extended conformation **49A** and also in Fischer projection (**49B**), and **50** is its enantiomer. Changing the position of the OH group (inverting its stereochemistry) generates diastereomer **51** and **52** is its enantiomer. Now invert the stereochemistry of the methyl group in **49A** to give **53A** (also drawn in Fischer projection **58B**) and this diastereomer **56** has its enantiomer **59**. Finally, invert the position of the ethyl group in **50** to give diastereomer **51**, and its enantiomer is **52**.

Structures **49–56** are the eight stereoisomers expected from four stereogenic centers by the 2^n rule ($2^4 = 8$): four diastereomers and each diastereomer has an enantiomer. Clearly, 5-ethyl-3-methyl-2-nonanol generates a more complex set of stereoisomers than compounds with only two stereogenic centers. The principles used to distinguish diastereomers and enantiomers are the same, but more patience is required to find all the stereoisomers. A systematic approach, such as that used before, is usually a good idea: Take the first stereogenic center as it is drawn, invert it, and then make the enantiomer. Go to the next stereogenic center and invert it, keeping the other centers the same, and then make the mirror image. Work down the chain and repeat the process for all stereogenic centers. This will usually give all possible stereoisomers *as long a check is made for redundant structures.*

9.27 **Draw both 50 and 54 in the extended conformation that shows them to be mirror images of 49A and 53A, respectively.**

9.28 **Draw 51, 52, 55, and 56 in their extended conformations.**

9.29 **Draw all stereoisomers of 3-bromo-2,5-dichloroheptane in Fischer projection.**

9.30 **Draw the *meso* compounds of 3,4,5-trichloroheptane.**

When there are more than two stereogenic centers, great care must be exercised in drawing various structures and testing them for enantiomers, diastereomers, and the presence of *meso* compounds. As the number of stereogenic centers increases, the complexity of this problem and the difficulty in sorting out structures that are the same or different become more challenging. Molecules with three or more stereogenic centers are encountered in the discussion of carbohydrates in Chapter 28.

9.6 Stereogenic Centers in Cyclic Molecules

Cyclic molecules may have stereogenic centers and it is possible to generate enantiomers and/or diastereomers. Problems of identifying the stereogenic center and the number of stereoisomers arise with some cyclic molecules that do not as acute in acyclic molecules. This section will focus on these problems.

Methylcyclohexane (**57A**) has one carbon atom in the ring that is connected to a methyl group, a hydrogen atom, and two carbons that are part of the six-membered ring. ***The methyl-bearing carbon in methylcyclohexane is not stereogenic.*** The two carbons in the ring (marked with a **red** dot in **57**) are considered to be identical; cyclohexane rings exist as equilibrating chair conformations, and both chair conformations must be examined. In one conformation, the methyl is axial (**57B**) and in the other the methyl is equatorial (**57C**). A plane of symmetry can be drawn along a line between C1 and C4. In effect, the carbons on one-half of the six-membered ring constitute one group (see **57A**) and the carbons on the other half of the ring constitute a second group.

Using the Cahn–Ingold–Prelog selection rules is perhaps a better way to compare the two **red** carbon atoms. The "top" carbon atom in **57A** is C^{CHH} and the "bottom" red carbon is also C^{CHH}, so there is no point of difference. Going to the next carbon atom on each side, the assignments remain C^{CHH} and C^{CHH}. Attempts to go to the next atom in each chain lead to the same carbon. Therefore, comparing each "side" of the ring does not lead to a point of difference and each side is identical. In this case, each side is considered to be a group and both groups are the same, so C1 is not stereogenic. ***The symmetry associated with the six-membered ring makes methylcyclohexane achiral***. Such symmetry is observed with many cyclic compounds. The difference between cyclic and acyclic is that symmetry can occur in a cyclic structure due to its rigidity (see Chapter 8, Section 8.5), whereas free rotation is possible in an acyclic structure.

Methylcyclopentane (**58**) does not contain a stereogenic center due to the inherent plane of symmetry. 1,2-Dimethylcyclopentane (**59**) is a different matter, however. Two stereogenic carbons lead to two diastereomers: the *cis*-diastereomer **60A** and *trans*-diastereomer **61**. The proper name of **60A** is (1*R*,2*S*)-dimethylcyclopentane and **61** is (1*R*,2*R*)-dimethylcyclopentane. The enantiomer of **60A** is (1*S*,2*R*)-dimethylcyclopentane, **60B**, and the enantiomer of **61** is (1*S*,2*S*)-dimethylcyclopentane, **62**. The *trans*-diastereomer **61** is drawn as the planar conformation in Figure 9.10, and the mirror image is **62** similarly drawn to emphasize the "sidedness" of the methyl groups.

There is no symmetry in these structures and *trans*-diastereomer **61** and its mirror image **62** are *not* superimposable, so they are enantiomers (two

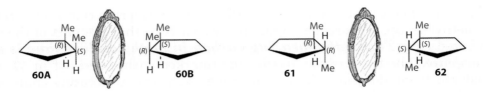

Figure 9.10   Dimethylcyclohexane stereoisomers.

stereoisomers). *cis*-Dimethylcyclohexane is drawn in a similar manner in Figure 9.10, with the planar view of **60A**, which is compared with **60B**. If **60B** is simply rotated counterclockwise by 180°, it will superimpose with **60A**. *cis*-1,2-Dimethylcyclopentane is a *meso* compound, **60**, and constitutes only one stereoisomer. 1,2-Dimethylcyclopentane has two stereogenic carbons, but instead of the four possible stereoisomers predicted by 2^2, there are only three stereoisomers because of the presence of symmetry in one stereoisomer. Stereoisomers **61** and **62** are enantiomers, but these two stereoisomers are diastereomers of **60**.

9.31 Draw both the *cis*- and *trans*-diastereomers for 1-bromo-2-chloro-cyclopentane; label and name them.

Disubstituted cyclohexane derivatives present a more complex system because there are two chair conformations as well as other conformations. *For all comparisons of enantiomers and diastereomers of cyclohexane derivatives, **assume** that the chair conformation is the only one of interest (see Chapter 8, Section 8.5.4).* Even with this simplifying assumption, *both* chair conformations must be examined for *each* diastereomer. An example is 1,2-dimethylcyclohexane, which has a *cis*-diastereomer (**63**) and a *trans*-diastereomer (**64A–64E**). It appears that **63** has a plane of symmetry by simply looking at the planar structure, and this is correct. In other words, **63** is a *meso* compound. However, **64A** is labeled as having no symmetry, which indicates it should be a mixture of enantiomers. Is this correct? One enantiomer of *trans*-diastereomer **64A** [(1*R*,2*S*)-dimethylcyclohexane] has two chair conformations: **64B** and **64C**. There is one axial and one equatorial methyl group in each chair conformation, and in the flat representation (**64A**) a line is drawn between the methyl groups that *suggests* that the structure should be examined for the presence of a plane of symmetry.

There is no symmetry, despite the appearance of the flat structure **64A**. Looking at the chair conformation **64B** and its mirror image **64D** shows there is no symmetry. For convenience, C1 in these structures is labeled **red** and C2 is labeled **blue**. It is possible to "pick up" **64B** and rotate it clockwise by 180° so that the methyl groups on the **red** carbons superimpose, but one **blue** carbon bearing a methyl projects to the front in **64B** and to the back in **64D** if this thought experiment is performed. The conclusion is that **64B** and **64D** are enantiomers.

The examination must continue, however, because **64B** is in equilibrium with **64C**. The *trans*-diastereomer cannot be labeled as having no symmetry until **64C** and its mirror image **64E** are inspected. Note that **64E** is in equilibrium with **64D**. Once again, there is no plane of symmetry because **64C** does not superimpose on **64E**. For this experiment, **64B** and **64E** and also **64C** and **64D** are compared before deciding if the molecule has symmetry. The preceding protocol is necessary to be certain that none of these structures are superimposable and *there is no plane of symmetry in **64***. Technically, the same analysis should be done for **63** to determine conclusively that the chair conformations are superimposable and that it is indeed a *meso* compound. The plane of symmetry is clear in **63**, however. If it is not clear and obvious, do the analysis.

9.32 Is *trans*-1,4-dimethylcyclohexane a *meso* compound?

9.7 Stereogenic Centers in Complex Molecules

Many organic compounds have two, three, or more rings joined together in different ways. It is common to find several substituents, different functional groups, and many stereogenic centers. The fundamental ideas discussed in preceding sections apply to complex molecules. This section is meant only as a brief introduction to some of these structures and how to approach them.

65   66   67   68

9.7.1 Bicyclic Molecules: Nomenclature

Many organic molecules have two rings joined together. Some examples are **65, 66**, and **67**. Both **65** and **66** have rings joined together by a common "edge," whereas **67** has two rings joined together by sharing a common "face." The base name of these compounds requires an explanation. Molecule **65** is a bicyclo[4.4.0] decane, **66** is a bicyclo[4.3.0]nonane, and compound **67** is a bicyclo[2.2.1]heptane. The substituents are ignored so that only the fundamental name of each compound is given.

In each case, imagine a pinwheel with three arms (**red, blue, green**) such as **68**, where the total number of atoms in all rings defines the base name (decane, nonane, heptane). Molecules **65, 66**, and **67** are color-coded to **68** to see the correlation with each arm. Each arm represents a chain or "bridge" of atoms (2, 3, 4, 5, etc.) that is connected to bridgehead atoms. The number of atoms in each bridge of the molecule is included in the brackets.

For **65**, only two of the three loops shown in **66** are present. The "bridge-head carbons" (those holding the rings together) are not counted, so there are four carbons in one loop, four carbons in the second loop, and no third loop (zero carbons). There are a total of 10 carbons in both rings so it is a decane, and the term **bicyclo** is used to indicate the presence of *two* rings fused together; it is therefore a bicyclodecane. The number of atoms in each loop is put into a bracket, making the base name bicyclo[4.4.0]decane and if the substituents are included, **65** is 4-chloro-1-methylbicyclo[4.4.0]decane. Note that the bridgehead carbon bearing the methyl is C1 and the ring is numbered to give chlorine the lower number.

Using similar nomenclature, compound **66** has four carbons in one loop, three carbons in the second loop and zero carbons in the third loop, with a total of nine carbon atoms. Therefore, it a bicyclo[4.3.0]nonane and, including the substituents, **66** is 1-(1-methylethyl)-7,7-dimethylbicyclo[4.3.0]nonan-3-ol. In this case, the isopropyl group (1-methylethyl) at the bridgehead is connected to C1 and the numbering goes to the smaller ring first, making the carbon bearing the OH C3 and that bearing the *geminal*-dimethyl unit C7. In compound **67**, there are three loops: two carbons in both the first and second loops and one carbon in the third, and there are a total of seven carbons in both rings. Therefore, it is a bicyclo[2.2.1]heptane and **67** is named 7-chloro-7-ethyl-1-methylbicyclo[2.2.1]heptan-2-ol. In this case and others where two rings share a face, the larger six-membered ring is numbered first (C1–C6) and the smaller loop is then counted (C7). Nomenclature for these systems can be very complex, and the focus will remain on relatively simple cases for the foreseeable future. Note that the absolute stereochemistry is given for **65–67**. As a separate exercise, determine *R/S* configuration for these compounds using the rules in Section 9.3 to confirm the assignments.

9.33 Draw the structure of bicyclo[3.3.2]decane; of bicyclo[5.3.0] decane; of bicyclo[1.1.1]pentane.

9.7.2 Bicyclic Molecules: *R/S* Configuration

A relatively simple molecule such as **69** is an example of a molecule that is somewhat more complex than those presented so far. Each stereogenic carbon (there are three: **red, blue**, and **green** in Figure 9.11) is treated independently. Because rings are involved, each "side" of the ring attached to a stereogenic center is treated as an independent group (see **57A** in Section 9.6). For the red carbon, there is a methyl group and three carbon "arms," so the four carbon atoms are C^{HHH}, C^{CHH}, C^{CHH}, and C^{CCH}.

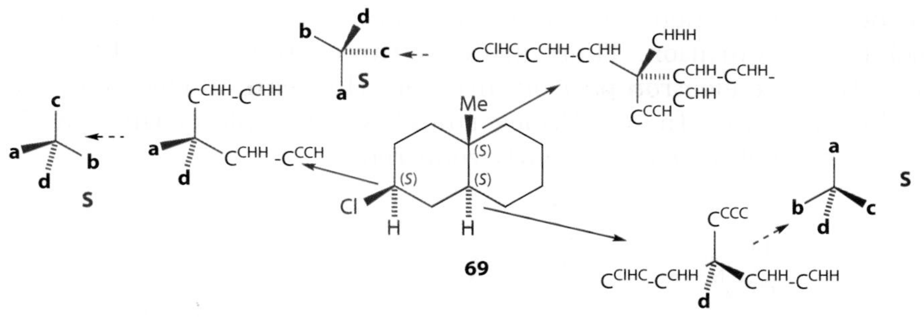

Figure 9.11 Absolute configuration of stereogenic centers in **69**.

Clearly, the methyl group is the lowest priority (d) and because atomic mass cannot distinguish the others at the point of difference, the number of carbons is considered, making the C^{CCH} carbon the highest priority. To distinguish C^{CHH} and C^{CHH}, go down the chain to the next carbon and C^{CHH} is found for both arms. Continue along those arms and a point of difference is found at C^{ClCH} and C^{CHH}. Because Cl > C, this is an (*S*) stereogenic center. The **blue** carbon is also a stereogenic center (see **69** in Figure 9.11) and analysis shows it to be an (*S*) stereogenic center, as is the **green** carbon. The lesson of this example is that each stereogenic center is treated as a unique entity with everything else in the molecule being part of the attached substituents for that particular analysis. Absolute configuration in all molecules is determined in a similar manner, despite the complexity.

9.34 **Draw the structure of a molecule with a five- and six-membered rings fused together with a *trans* ring juncture, and then add a chlorine at one bridgehead position, a methyl on the carbon closest to Cl in the five-membered ring, and an ethyl group on the six-membered ring two carbons away from the chlorine. The carbon bearing chlorine should be (*S*), that bearing methyl should be (*R*), and that bearing ethyl should be (*S*). Then determine whether the bridgehead carbon bearing H is (*R*) or (*S*).**

9.7.3 Polycyclic Amines and Electron Lone Pairs

When a bicyclic molecule has a nitrogen in one of the rings, it is an amine, and the nitrogen atom may be a stereogenic center. As pointed out in Chapter 5 (Section 5.6.2), amines undergo fluxional inversion (see **70A** and **70B**). Examination of these structures shows that **70A** is the mirror image of **70B**, which is essentially a racemic mixture with a specific rotation of zero. The absolute configuration at nitrogen is labeled for **70A** and for **70B**, but the *R/S* label is irrelevant due to the fluxional inversion. In bicyclic amine **71**, however, fluxional inversion at the nitrogen is impossible because it is part of a rigid structure. If the lone electron pair is counted as a "group," there are four different attached groups and this nitrogen is a stereogenic center.

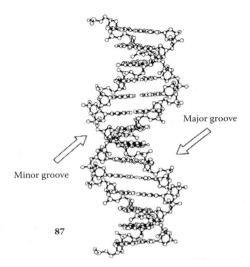

87

The stereochemistry of alkenes (*E* or *Z*) is also important in many biological processes. Retinol (vitamin A$_1$, **88**) and dehydroretinol (vitamin A$_2$, **89**) are found in animal products such as eggs, dairy products, and animal livers and kidneys, as well as fish liver oils. These retinoids are known to act as signaling molecules that regulate aspects of cell differentiation, embryonic development, growth, and vision.[4] In mammals, retinol is obtained by cleavage of the tetraterpenoid β-carotene (**90**) at the highlighted bond, usually in the intestine, to give the aldehyde retinal (**91**), which is reduced to **88**.

(Dewick, P. M. 2002. *Medicinal Natural Products*, 2nd ed., 231, Figures 5.72 and 5.73. West Sussex, England: Wiley. Copyright Wiley-VCH Verlag GmbH & Co. KGaA. Reproduced with permission.)

The biological process known as vision involves conversion of retinol (**88**) to retinal (**91**) via an oxidation reaction (see Chapter 17) that involves the cofactor NADP+ (also see Chapter 19, Section 19.6).[4] Note the *E* geometry of the C11–C12 double bond in **88** (the numbering is based on numbering that begins in the cyclohexene unit, as shown). The focus of this chapter has been on stereochemistry, and the key process here has been the change of an *E*-alkene to a *Z*-alkene using an enzyme. Coordination to an enzyme and the interaction with light will change the alkene stereochemistry back to *E*. Enzymatic conversion of the C11–*E* double bond to the C11–*Z* double bond (11-*cis*-retinal, **92**) allows reaction with an amine unit on a protein called opsin to form an imine (a Schiff base, **93**; see Chapter 18, Section 18.7.1 for a discussion of imines). This is the red visual pigment, and exposure to light (hν) converts the C11–*Z* double bond back to the C11–*E* double bond in **94**; this triggers a nerve impulse to the brain.[4] Conversion of the Schiff base **94** back to retinal (**91**) allows the process to cycle again.

References

1. Eliel, E. L., Wilen, S. H., and Mander, L. N. 1994. *Stereochemistry of organic compounds,* 1194. New York: John Wiley & Sons.
2. Eliel, E. L., Wilen, S. H., and Mander, L. N. 1994. *Stereochemistry of organic compounds,* 1208. New York: John Wiley & Sons.
3. Hecht, S. M., ed. 1996. *Bioorganic chemistry: Nucleic acids,* 6. New York: Oxford University Press.
4. Dewick, P. M. 2002. *Medicinal natural products,* 2nd ed., 230–231, Figures 5.72 and 5.73. West Sussex, England: Wiley.

Answers to Problems

9.1

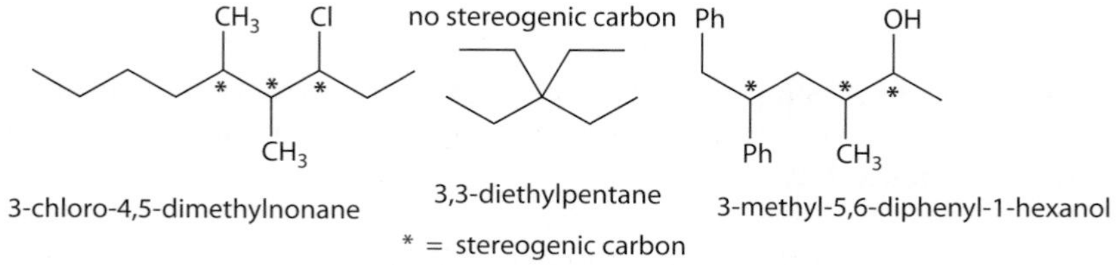

CH3
Cl
CH3
no stereogenic carbon
Ph
OH
Ph
CH3
3-chloro-4,5-dimethylnonane
3,3-diethylpentane
3-methyl-5,6-diphenyl-1-hexanol
* = stereogenic carbon

9.2

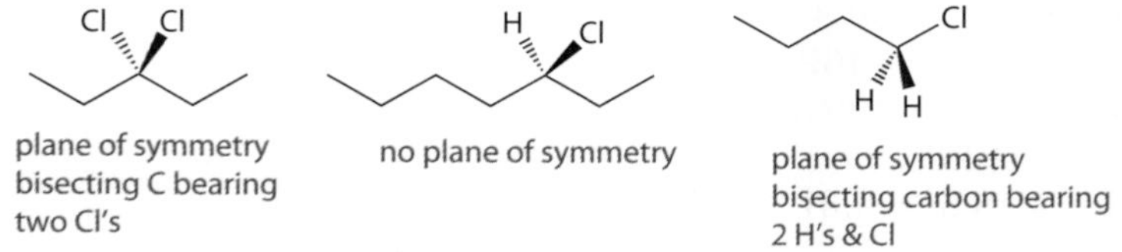

Cl
Cl
H
Cl
Cl
H
H
plane of symmetry bisecting C bearing two Cl's
no plane of symmetry
plane of symmetry bisecting carbon bearing 2 H's & Cl

9.3

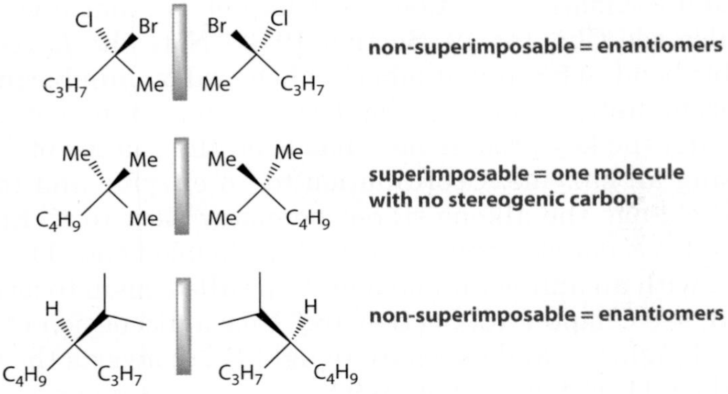
Cl
Br
C3H7
Me
Br
Cl
Me
C3H7
non-superimposable = enantiomers
Me
Me
C4H9
Me
Me
Me
Me
C4H9
superimposable = one molecule
with no stereogenic carbon
H
C4H9
C3H7
H
C3H7
C4H9
non-superimposable = enantiomers

9.4

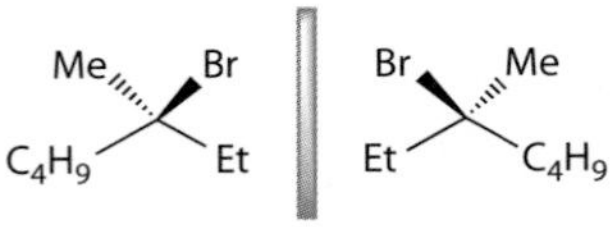
Me
Br
C4H9
Et
Br
Me
Et
C4H9

9.5

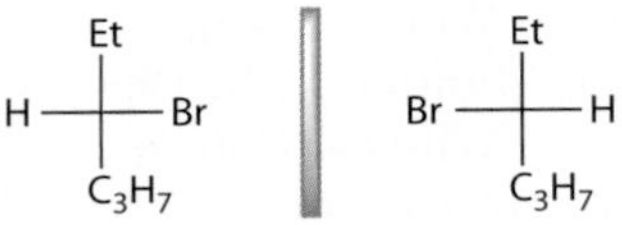
Et
H
Br
C3H7
Et
Br
H
C3H7

9.6 $[\alpha] = \frac{\alpha}{ld} = \frac{-58.1}{50.52} = \frac{-58.1}{2.6} = -22.3_5$

9.7

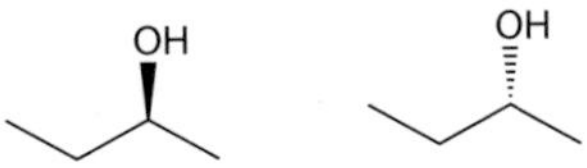
OH
OH

9.8 No! Specific rotation is a physical property and the direction of rotation (sign) is part of that property. It must be measured in a polarimeter and cannot be estimated or guessed at.

9.9 $[\alpha]_D^{20}$(mixture) = x (+109.3°) + y (–109.3°) = 27.7 and specific rotation is additive, x + y = 1. Therefore,
= x (+109.3°) + y (–109.3°) = 27.7°
= (1 – y)(+109.3°) – 109.3°y = 27.7°
= +109.3° – 109.3°y – 109.3°y = 27.7°
= –218.6°y = 27.7° – 109.3°
= –218.6°y = –81.6°
= y = –81.6°/218.6° = 0.37.3 = 37.3% of (–)X
Therefore, 37.3% of (–)X and 62.7% of (+)X.

9.10 Figure 9.7 indicates that 69.7% ee = 84% of the (+) enantiomer and 16% of the (−) enantiomer.

9.11 For a 93%:7% mixture, 88% ee; for a 99.4%:0.6% mixture, 99% ee; for a 55%:45% mixture, 11% ee.

9.12

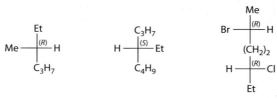

3R-methylhexane   4S-ethyloctane   5R-chloro-2R-bromoheptane

9.13

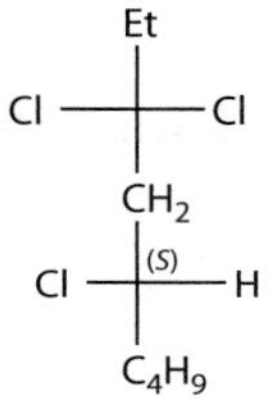

3,3,5S-trichlorononane

9.14

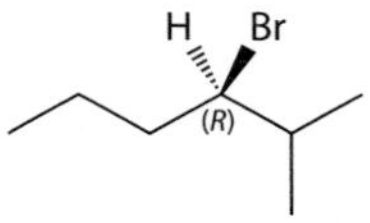

3R-bromo-2-methylhexane

9.15

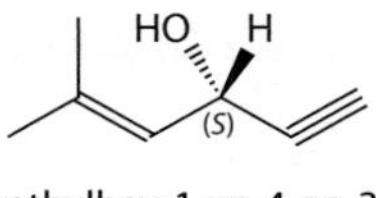

5-methylhex-1-yn-4-en-3S-ol

9.16

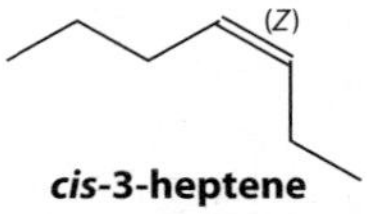

cis-3-heptene 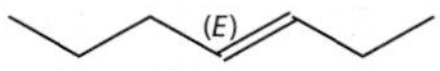   **trans-3-heptene**

9.17

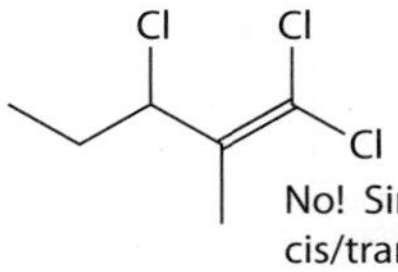

1,1,3-trichloro-2-methyl-1-pentene

No! Since the terminal carbon of the C=C unit has 2 Cl groups, cis/trans isomers are not possible for this molecule

9.18 Compound **30** = pent-2*E*-ene and compound **31** = pent-2*Z*-ene.

9.19

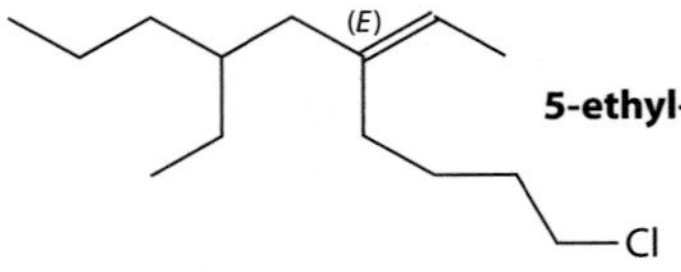

5-ethyl-3-(4-chloro-1-butyl)-oct-2*E*-ene

9.20

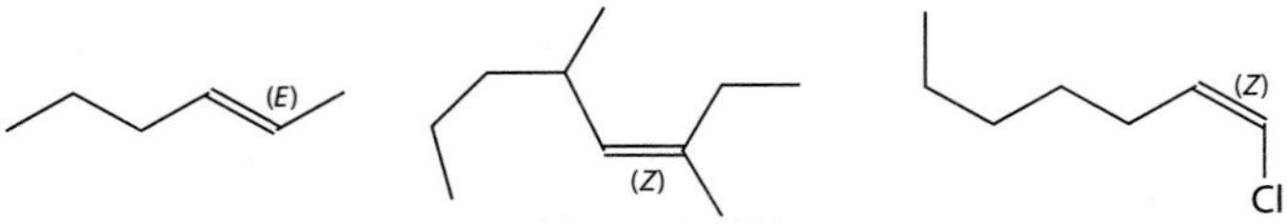

hex-2*E*-ene   3,5-dimethyloct-3*Z*-ene   1-chloro-1*Z*-heptene

9.21

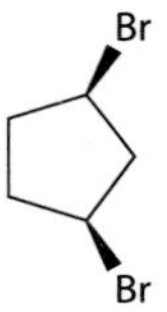

cis-1,3-dibromocyclopentane

9.22

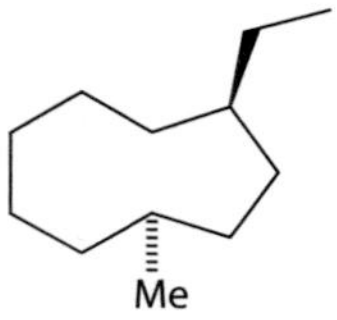

trans-1-ethyl-4-methylcyclononane

9.23

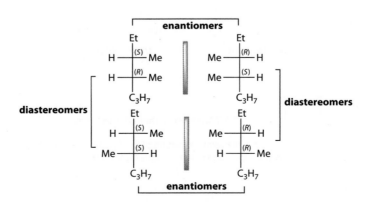

9.24

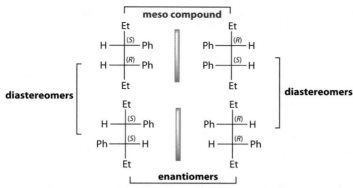

meso compound

diastereomers

diastereomers

enantiomers

$2^2 = 4$ **stereoisomers maximum, but the presence of the meso compound means that there are only three stereoisomers**

9.25

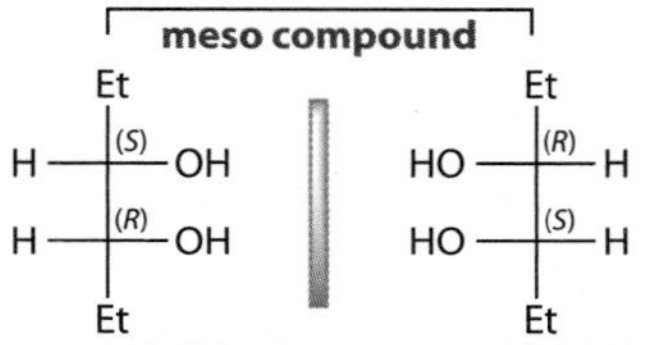

meso compound

This diastereomer of 3,4-hexanediol has a superimosable mirror image. Therefore, it is a meso compound.

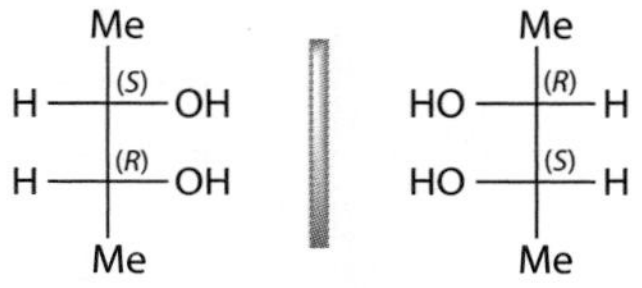

This is *meso*-2,3-butanediol

9.26

2*R*,3*R*-dibromobutane   2*S*,3*S*-dibromobutane 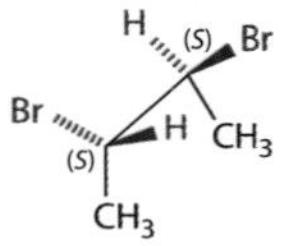   2*R*,3*S*-dibromobutane

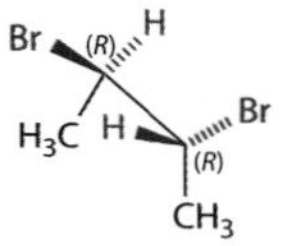

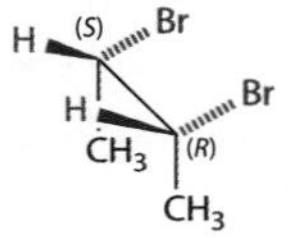

9.27

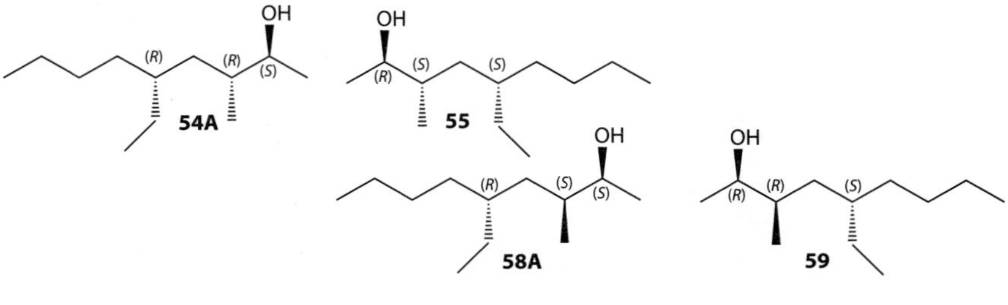

54A   **55**

58A   **59**

9.28

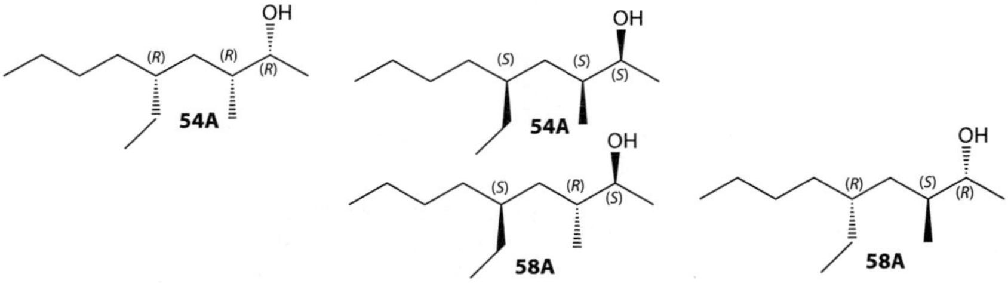

54A   **54A**

58A   **58A**

9.29

There are three stereogenic carbons, $2^3 = 8$ stereoisomers marked in red. Take **A** and interchange C_2, C_3, and then C_5 stepwise to give 12 structures, but 4 are redundant, as shown.

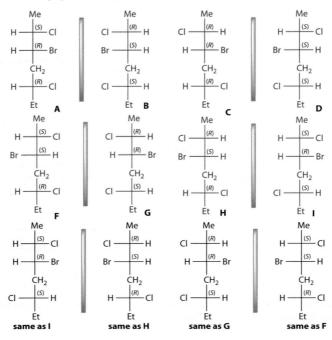

A   **B**   **C**   **D**

F   **G**   **H**   **I**

same as I   same as H   same as G   same as F

9.30

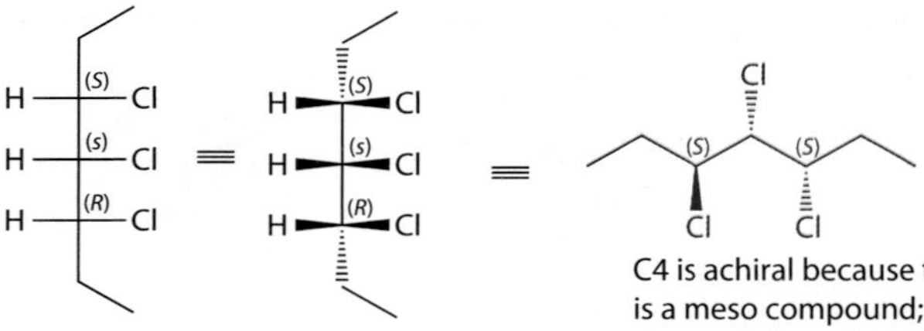

C4 is achiral because this is a meso compound; ignore the designation in the Fischer projection

9.31

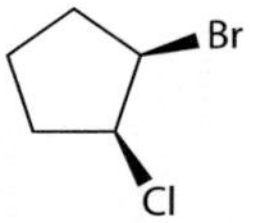

cis-1-bromo-2-chlorocyclopentane 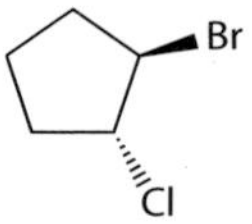   trans-1-bromo-2-chlorocyclopentane

9.32

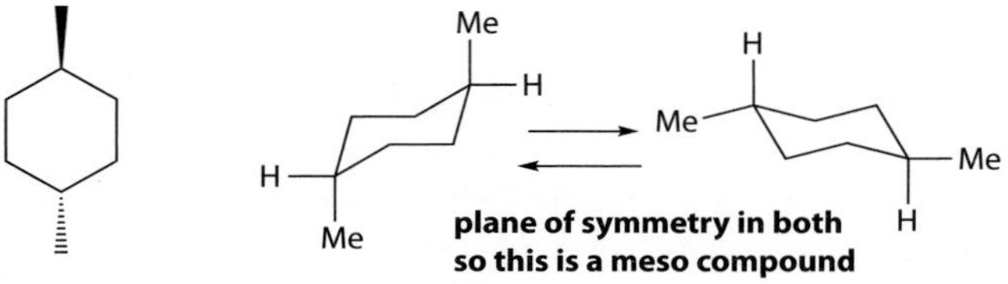

plane of symmetry in both so this is a meso compound

9.33

bicyclo[3.3.2]decane   bicyclo[5.3.0]decane   bicyclo[1.1.1]pentane

9.34

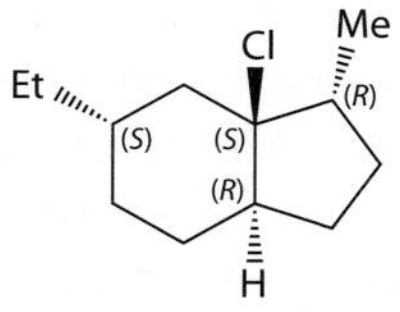

9.35

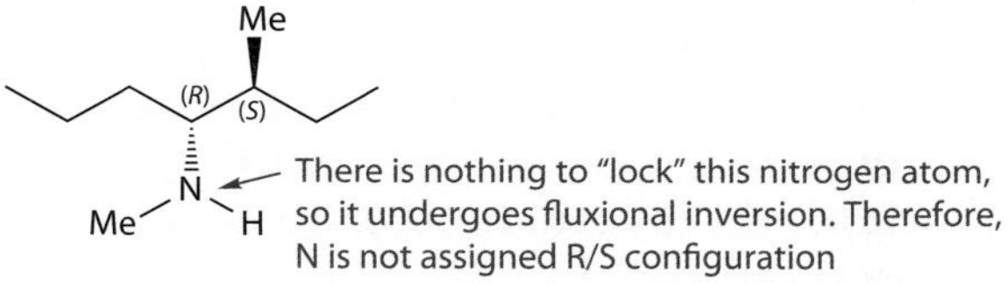

There is nothing to "lock" this nitrogen atom, so it undergoes fluxional inversion. Therefore, N is not assigned R/S configuration

Correlation of Concepts with Homework

- Two or more molecules with the same empirical formula, same points of attachment for the atoms or groups, but different spatial arrangement of those atoms or groups are called stereoisomers. A stereogenic carbon has four different atoms or groups attached, and it has a nonsuperimposable mirror image known as an enantiomer: 1, 2, 3, 4, 5, 7, 36, 44, 47, 61.

- A polarimeter measures the angle that an enantiomer can rotate plane polarized light. Observed rotation (α) is the value recorded directly from a polarimeter. Specific rotation ($[\alpha]$) is the standardized value for optical rotation, where $[\alpha] = \alpha/l \bullet c$: 38, 6, 8, 9, 39, 40, 41, 42, 43, 45, 51, 52, 53, 54, 76, 81.

- The absolute configuration (R or S) of an enantiomer is determined by applying the Cahn–Ingold–Prelog selection rules and the "steering wheel" model: 12, 13, 14, 15, 37, 46, 49, 50, 55, 61, 64, 65, 68, 69, 70, 71, 73, 74.

- The percentage of enantiomeric excess (% ee) is a measure of the excess of one enantiomer over another, where 0% ee indicates a racemic mixture and 100% ee indicates a single pure enantiomer: 10, 11, 42, 52, 80, 81, 82.

racemic mixture and 100% ee indicates a single pure enantiomer: 10, 11, 42, 52, 80, 81, 82.
- When a molecule has two or more stereogenic (chiral) centers, there are a maximum of 2^n stereoisomers, where n = the number of chiral centers. When a molecule has two or more chiral centers, diastereomers are possible. Diastereomer is the term for two or more stereoisomers that are not superimposable and not mirror images. A diastereomer that has symmetry such that its mirror image is superimposable is called a *meso* compound. If there is no symmetry, cyclic molecules can have enantiomers and diastereomers. If there is symmetry in one diastereomer, cyclic compounds can have *meso* compounds: 23, 24, 25, 26, 27, 28, 29, 30, 48, 49, 59, 60, 62, 63, 67, 68, 69, 70, 71, 75, 78, 79.
- Alkenes can exist as stereoisomers and are named using the *E/Z* or *cis/trans* nomenclature: 16, 17, 18, 19, 20, 56, 57, 58.
- In substituted cyclic molecules, one side of the ring (relative to the substituent) is considered to be one group and the other side of the ring is considered to be a second group. If the groups (sides) are identical, the cyclic molecule is achiral. The terms *cis* and *trans* can be used to indicate the relative stereochemistry of diastereomers: 16, 17, 18, 19, 20, 63, 64, 84.
- Bicyclic compounds are named using a "pinwheel" model and the number of atoms in each chain is indicated in brackets, with numbers such as [x.y.z]: 21, 22, 31, 32, 64, 66, 72.
- Because amines undergo fluxional inversion, nitrogen is not considered to be chiral and amines are treated as equilibrating mirror images (racemates). In some polycyclic amines, fluxional inversion is impossible, and nitrogen can be chiral, with the lone electron pair considered as the fourth group: 35, 83, 84.

Homework

9.36 Which of the following molecules are chiral?
  (a) FCH_2CH_2Br
  (b) CCl_4
  (c) $CHBrF$
  (d) $BrCHClOH$

9.37 Each of the four groups shown is connected to one stereogenic carbon atom; indicate the highest priority group using the Cahn–Ingold–Prelog selection rules:
  (a) $-CH_2CH_2Br$ $-CH_2CHBrCH_3$ $-CH_2CH_2CH_2CH_2OH$ $-CH_2F$
  (b) $-CH_2CH_2OH$ $-CH_2CHBrCH_3$ $-CH_2CH_2CH_2CH_2I$ $-CH_2CH_3$

9.38 There are two enantiomers, A and A'. Enantiomer of A has a specific rotation of +70°. A mixture of A and A' has a measured specific rotation of −35°. Calculate the percent of both A and A'.

9.39 Indicate which of the following parameters do/do *not* contribute to specific rotation:
(a) path length
(b) group priority
(c) concentration
(d) α

9.40 If $[\alpha]$ for the (R) enantiomer of a molecule is $-50°$, what is $[\alpha]$ for the (S) enantiomer?
(a) $+50°$
(b) $-50°$
(c) $+5°$
(d) $-5°$
(e) $0°$

9.41 What is the $[\alpha]$ for a mixture of enantiomers when the mix is 60% R and 40% S and $[\alpha]$ for R is $-20°$?
(a) $+10°$
(b) $-10°$
(c) $+20°$
(d) $-20°$
(e) $0°$
(f) $+4°$
(g) $-4°$

9.42 What is the $[\alpha]$ for a racemic mixture?
(a) $+100°$
(b) $-100°$
(c) $+50°$
(d) $-50°$
(e) $0°$

9.43 If the specific rotation of a pure enantiomer is $+100°$ and the specific rotation of a mixture of both enantiomers is $-20°$, what is the ratio of the two enantiomers in the mixture?
(a) 50:50
(b) 20:80
(c) 30:70
(d) 40:60
(e) 10:90

9.44 Which of the following is the enantiomer of ($2R$)-bromohexane?

9.45 Which of the following solvents cannot be used for determining $[\alpha]$ of a compound in a polarimeter?

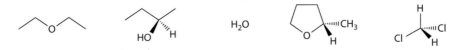

9.46 For each of the following molecules, assign all stereogenic carbons with the correct R or S configuration:

(a)

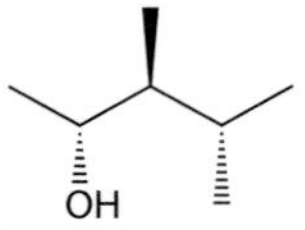

(b)

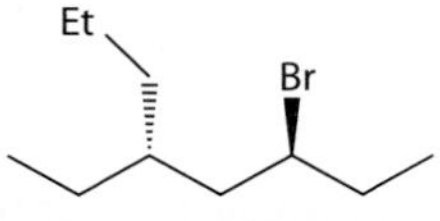

(c)

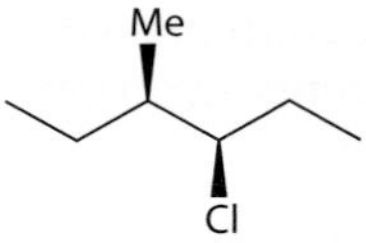

(d)

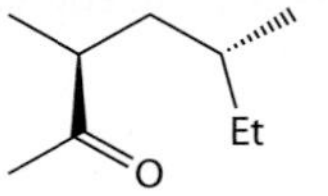

(e)

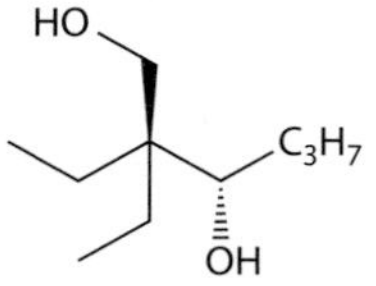

(f)

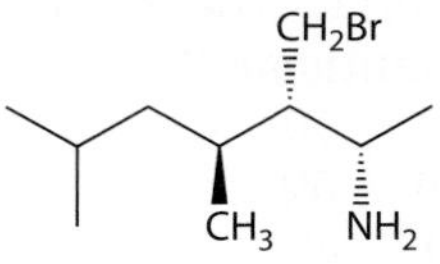

9.47 Determine which of the following has a superimposable mirror image and which has an enantiomer:

(a)

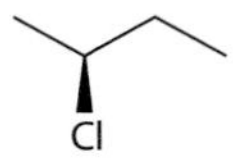

(b)

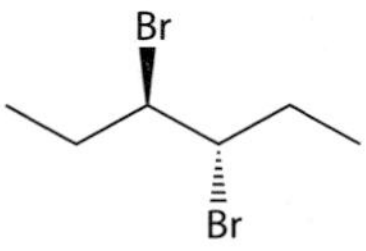

(c)

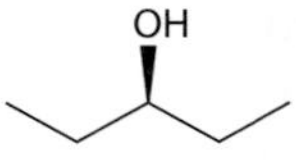

(d)

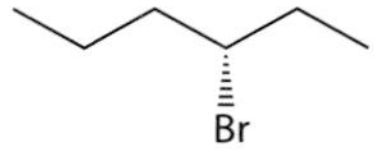

(e)

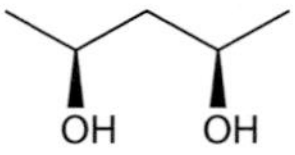

(f)

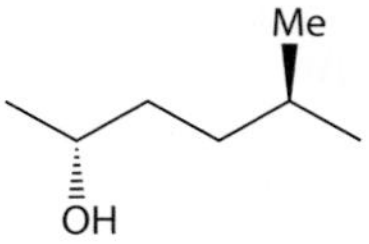

9.48 What is the maximum number of stereoisomers possible for the compound shown? Indicate each stereogenic center with an asterisk:

(a)

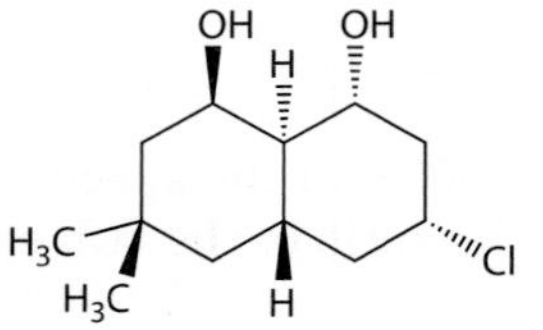

$2 \qquad 2^2 \qquad 2^3 \qquad 2^4 \qquad 2^5 \qquad 2^6$

(b)

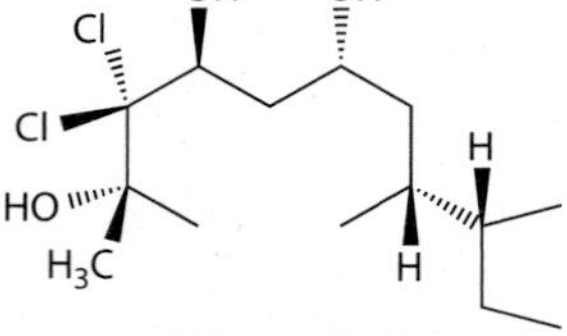

$2^3 \qquad 2^4 \qquad 2^5 \qquad 2^6 \qquad 2^7 \qquad 2^8$

9.49 Draw each of the following in Fisher projection:
(a) 3R-bromo-2S-hexanol
(b) 4R-methyldodecane
(c) 2R,3S,4R-trichlorooctane

(d) 3S-heptanol
(e) 3R-ethyl-3-methyloctane
(f) hept-1-en-3R-ol

9.50 Determine the absolute configuration of each stereogenic carbon in the following molecules:

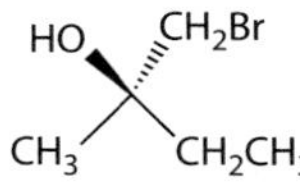

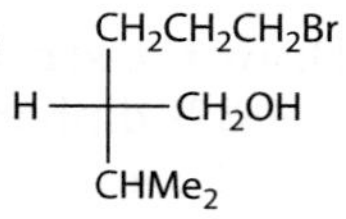

9.51 Determine the specific rotation for each of the following:
(a) $\alpha = +18°$, c = 1.1 g/mL in a 25 cm tube
(b) $\alpha = -176°$, c = 0.3 g/mL in a 50 cm tube
(c) $\alpha = -1.4°$, c = 5.4 g/mL in a 25 cm tube
(d) $\alpha = +94°$, c = 2.3 g/mL in a 30 cm tube

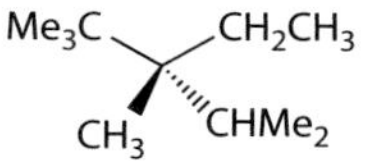

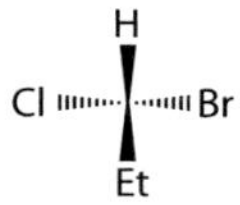

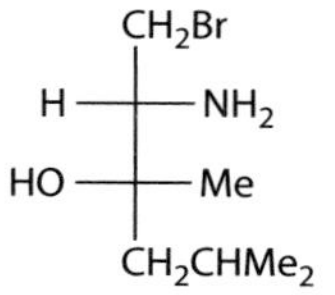

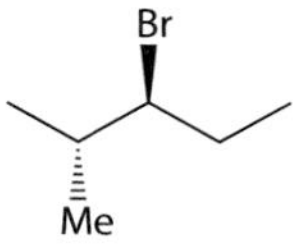

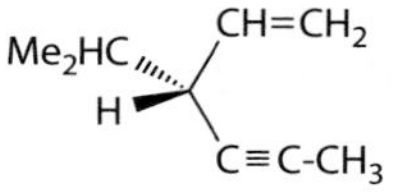

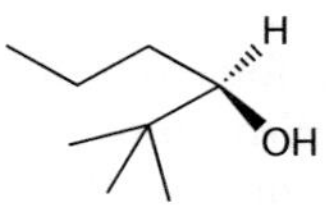

9.52 Calculate the percentage of each enantiomer and the % ee for the mixture given the following information:
(a) $[\alpha]_D^{20} = +18.6°$ for the S enantiomer and $[\alpha]_D^{20} = -2.5°$ for the mixture.
(b) $[\alpha]_D^{20} = -166°$ for the R enantiomer and $[\alpha]_D^{20} = -154°$ for the mixture.
(c) $[\alpha]_D^{20} = -45°$ for the S enantiomer and $[\alpha]_D^{20} = +27°$ for the mixture.
(d) $[\alpha]_D^{20} = +208°$ for the R enantiomer and $[\alpha]_D^{20} = -118°$ for the mixture.

9.53 A reaction is known to proceed with complete racemization of the product. If $[\alpha]_D^{20} = +30°$ for the R enantiomer and it reacts to give a racemate, what is the specific rotation of the product? Briefly explain.

9.54 A reaction is known to proceed with 100% inversion of the absolute configuration. If $[\alpha]_D^{20} = -77°$ for the S enantiomer and it reacts to give a product with complete inversion, can you make a prediction about the sign of the specific rotation? About the absolute configuration of the product? Briefly explain.

9.55 Determine the absolute configuration of each stereogenic carbon in the following molecules:

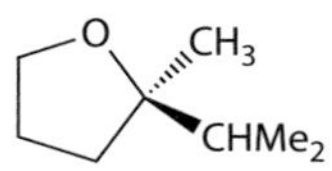

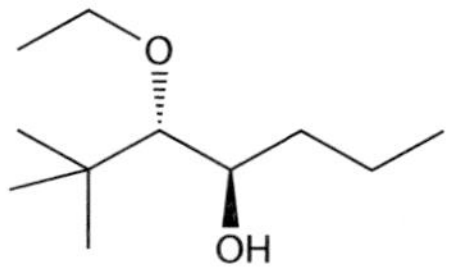

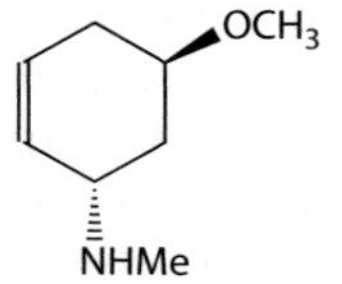

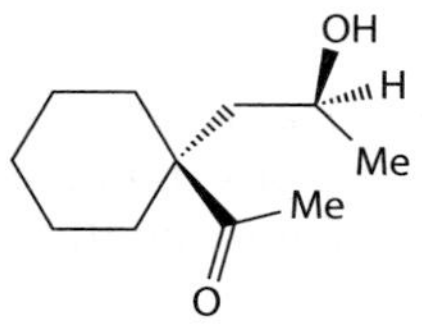

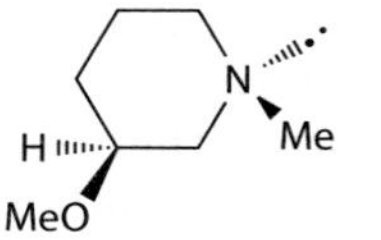

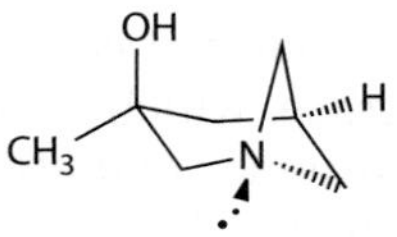

9.65 Give an unambiguous IUPAC name to each of the following:

(a)

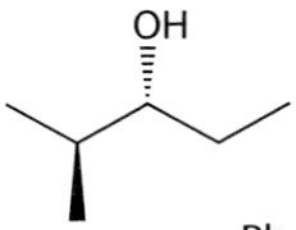

(b)

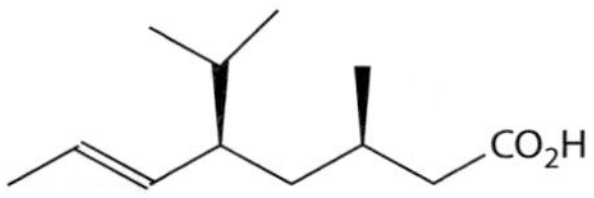

(c)

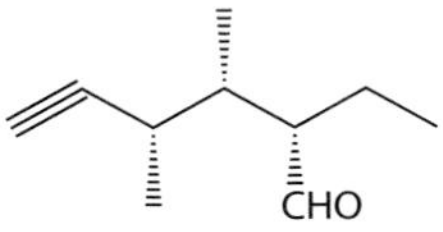

(d)

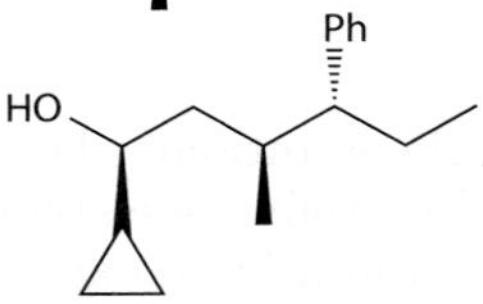

(e)

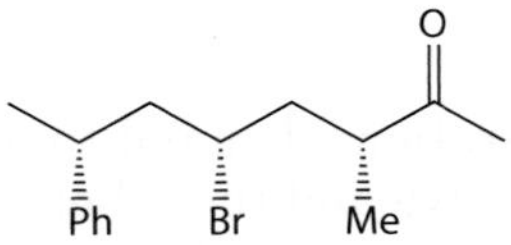

(f)

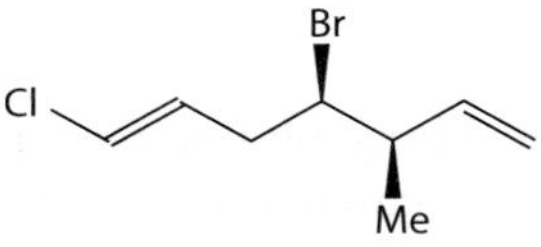

9.66 Name each of the following:

(a)

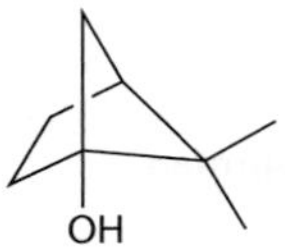

(b)

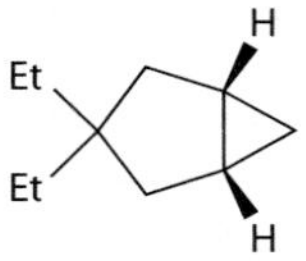

(c)

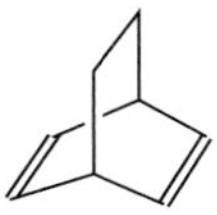

(d)

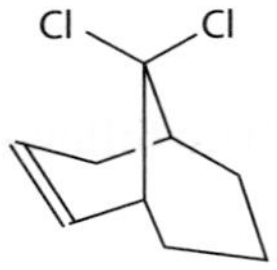

(e)

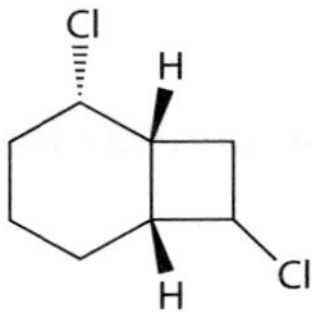

(f)

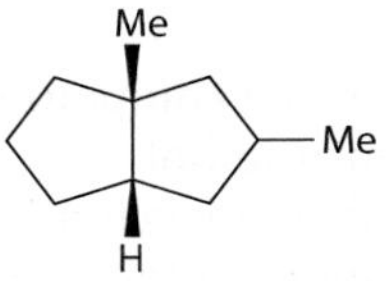

9.67 Which of the following are *meso* compounds?

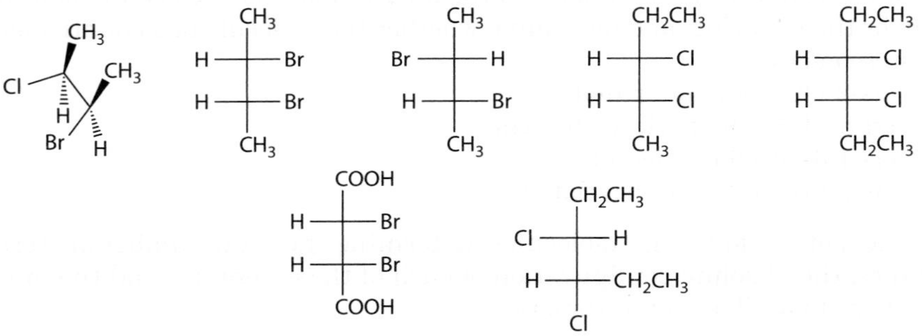

9.68 Which of the following correspond to 2R,3R-dibromobutane and its mirror image?

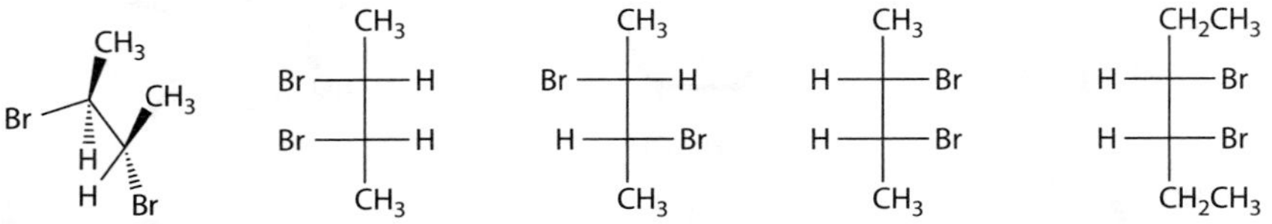

9.69 Determine the absolute configuration of all stereogenic centers in the following molecules:

9.70 Draw all **different** stereoisomers of 3,4-dichlorohexane and give each one the correct unique **name**.

9.71 Draw all stereoisomers of 2-bromo-3-methylhexane in extended conformation. Give the correct IUPAC name for each of the stereoisomers you have drawn.

9.72 Which of the following is bicyclo[3.3.1]nonane? Draw bicyclo[3.2.2]nonane.

9.73 Determine the absolute configuration (*R* or *S*) for all stereogenic centers in the following molecules.

9.74 Determine the absolute configuration (R or S) for all stereogenic centers in the following molecules:

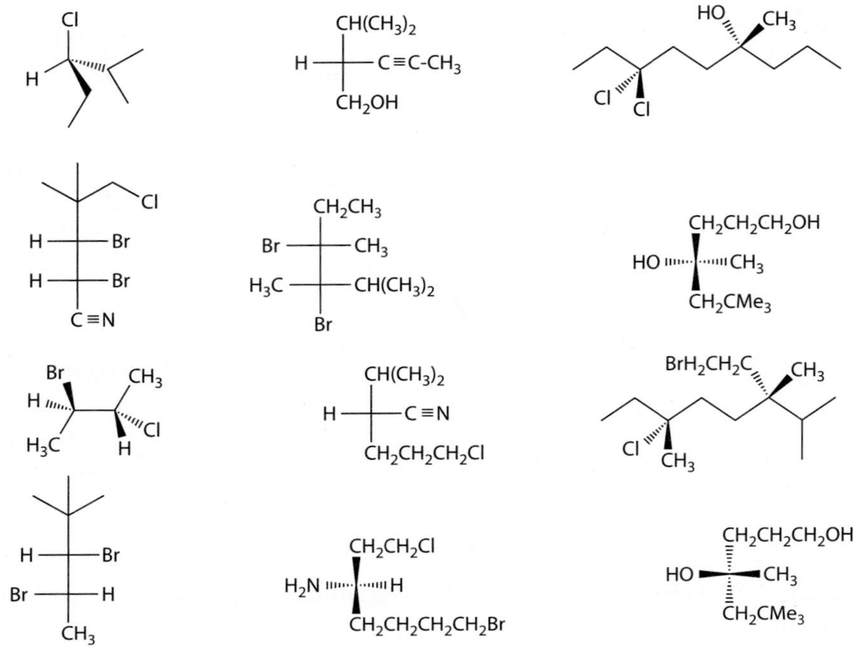

9.75 Which of the following are *meso* compounds?

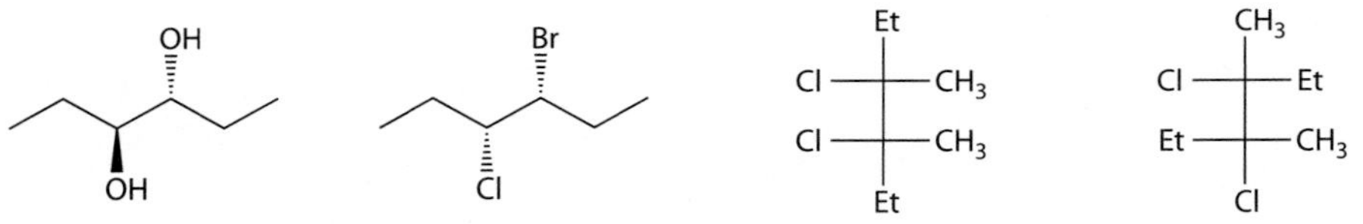

9.76 A molecule with one stereogenic center was observed to have a specific rotation of zero. What is the term used to describe such an observation?
(a) enantiomer
(b) diastereomer
(c) racemic
(d) enantipure

9.77 The reaction of KOH in ethanol with 2R-bromo-3S-methylpentane (specific rotation = +30°) gives only 2Z-3-methyl-2-pentene. What is the specific rotation of pure 2Z-3-methyl-2-pentene? Explain!
(a) +15°
(b) −15°
(c) 0°
(d) +30°
(e) −30°

9.78 Which of the following corresponds to *meso*-dibromobutane?

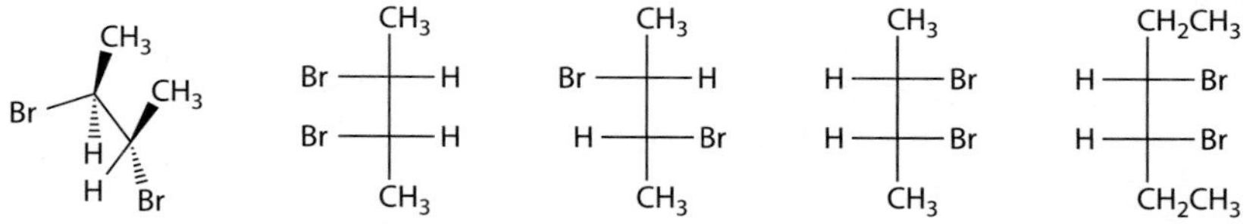

9.79 Draw the specific stereoisomer for each molecule, in Fischer projection.
  (a) 3S,4S-dichloroheptane
  (b) 2R-bromo-3S-methylhexane
  (c) 4R-phenyl-3R-heptanol
  (d) 3R,4R-dibromohexane
  (e) 2R,5R-hexanediol
  (f) 3S,4S,5R-heptanetriol
  (g) 2S-bromo-5S-methylhexane
  (h) 2,3R,4R,5-tetramethylhexane

9.80 Convert each of the following ratios to % ee for the (R) enantiomer:
  (a) 82:18 *R:S*
  (b) 55:45 *R:S*
  (c) 99:1 *R:S*
  (d) 75:25 *R:S*

9.81 A chemical reaction produced 2-phenyl-2-hexanol. If the pure (+)-(R)-alcohol has a specific rotation of −100° (this is not the correct value) and the reaction product has a specific rotation of +91°, calculate the relative percentage of (R) and (S) enantiomers. Now calculate the % ee for the reaction.

9.82 Give a brief explanation of why reaction that produces a racemic mixture of a product is said to proceed with 0% ee.

9.83 It is known that amines undergo fluxional inversion that leads to loss of optical activity. Phosphines (PR_3) are analogs of amines in that phosphorus is immediately beneath nitrogen in the periodic table. Yet the phosphine shown has a specific rotation that can be measured if the temperature is kept below 135°C. Above 135°C, the phosphine loses optical activity (specific rotation goes to zero). What does this indicate about fluxional inversion in this phosphine?

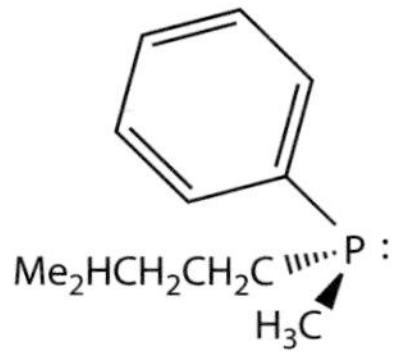

9.84 Determine the absolute configuration for each nitrogen atom in the following
       molecules:

(a)   (b)   (c)   (d)

Spectroscopy Problems

There are no pertinent spectroscopic problems for this chapter.

Acid–Base Reactions of π-Bonds

10

Previous chapters defined and discussed several functional groups. The fundamental issues of equilibria associated with acid–base reactions are presented, showing that the strength of the acid is dependent upon the base and vice versa. The difference between a Brønsted–Lowry base that reacts with molecules such as HCl bearing an electropositive proton and a Lewis base that reacts with electropositive atoms other than hydrogen (B, Al, etc.) is explained. Electron-donating species that react with carbon are separated into a unique category labeled as nucleophiles. It is time to use this information to explore how functional groups react and what products are formed. In other words, it is time to explore the chemical reactivity of organic molecules, which is how one molecule is changed into another. The discussion will begin with alkenes (see Chapter 5, Section 5.1) and alkynes (Chapter 5, Section 5.2)

To begin this chapter, you should know the following:

- **IUPAC nomenclature for alkanes, alkenes, alkynes, and alkyl halides (Chapter 4, Sections 4.6 and 4.7; Chapter 5, Sections 5.1, 5.2, 5.6, and 5.9)**
- **how to recognize and name alkenes and alkynes (Chapter 5, Section 5.1 and 5.2)**
- **the nature of σ-bonds (Chapter 3, Sections 3.3, 3.5–3.7)**
- **the nature of π-bonds (Chapter 5, Sections 5.1 and 5.2)**
- **definition and recognition of the structures of Brønsted–Lowry acids and bases (Chapter 6, Sections 6.1 and 6.2)**

- **definition and recognition of the structures of Lewis acids and bases (Chapter 6, Section 6.5)**
- **definition of a nucleophile (Chapter 6, Section 6.7)**
- **the role of bond strength in acid–base equilibria (Chapter 6, Section 6.3)**
- **the role of conjugate acid and conjugate base stability in acid–base equilibria (Chapter 6, Sections, 6.1 and 6.3)**
- **how to recognize the presence of resonance (Chapter 5, Section 5.9.3)**
- **the role of resonance in the stability of molecules (Chapter 5, Sections 5.9.3 and 5.10.1)**
- **the structure of reactive intermediates such as anions, carbocations, and radicals (Chapter 7, Section 7.4)**
- **understanding and identification of intermediates in a reaction-energy curve (Chapter 7, Section 7.4 and 7.6)**
- **the concept of mechanism (Chapter 7, Section 7.8)**
- **the acid–base nature of alcohols and amines (Chapter 6, Sections 6.2 and 6.4)**
- **the fundamentals of conformation applied to acyclic molecules (Chapter 8, Sections 8.1–8.4)**
- **the fundamentals of conformation applied to cyclic molecules of ring sizes of three to six (Chapter 8, Section 8.5)**
- **stereogenic centers and chirality (Chapter 9, Section 9.1)**
- **absolute configuration (Chapter 9, Section 9.3)**
- **diastereomers (Chapter 9, Section 9.5)**

This chapter will build on principles introduced in previous chapters and show applications to common chemical reactions of two important hydrocarbon functional groups: alkenes and alkynes. The chapter will also introduce several new chemical reagents (compounds that react with an alkene or alkyne to give a new molecule), as well as several new types of reactions. The theme of acid–base chemistry will be used as a basis for understanding each chemical transformation where it is appropriate. Mechanisms that are the step-by-step processes by which one molecule is transformed into another by tracking the intermediates will also be discussed. The concept of mechanism was introduced in Chapter 7 (Section 7.8).

When you have completed this chapter, you should understand the following points:

- **Alkenes react as Brønsted–Lowry bases in the presence of strong mineral acids, HX. The reaction of alkenes and mineral acids (HX) generates the more stable carbocation, leading to substitution of X at the more substituted carbon. This constitutes the mechanism of the reaction, but it is given the name Markovnikov addition.**

- More highly substituted carbocations are generally more stable. An increase in the number of resonance contributors will increase the stability of a carbocation.
- If a carbocation is formed adjacent to a carbon that can potentially be a more stable carbocation, assume that rearrangement occurs faster than the reaction in which the carbocation traps the nucleophile. In other words, if rearrangement can occur to give a more stable carbocation, that carbocation is assumed to rearrange.
- Weak acids such as alcohols and water can add to an alkene in the presence of a strong acid catalyst. Acid-catalyzed reaction with alkenes and water gives the Markovnikov alcohol and similar reaction with alcohols gives the Markovnikov ether.
- Chlorine, bromine, and iodine react with alkenes to give the three-membered ring halonium ion, which reacts with the halide nucleophile to give *trans*-dichlorides, -dibromides, or -diiodides. Alkynes react to give vinylhalonium ions that lead to vinyl dihalides.
- HOCl and HOBr react with alkenes to give halohydrins.
- In ether solvents, borane adds to alkenes via a four-centered transition state to give an alkylborane. Borane adds by a *cis* addition that places the boron on the less substituted carbon as the major product. Alkylboranes react with $NaOH/H_2O_2$ to give an anti-Markovnikov alcohol.
- Mercuric acetate and water react with alkenes via a mercury-stabilized carbocation to give a hydroxy alkyl-mercury compound. Reduction of the C–Hg bond with $NaBH_4$ leads to the Markovnikov alcohol.
- The reaction of an alkene and a peroxyacid gives an oxirane (an epoxide) and a carboxylic acid.
- Alkynes react with mineral acids HX to give vinyl halides, and with nucleophiles in the presence of an acid catalyst to give vinyl compounds.
- Alkynes react with an acid catalyst and water or with mercuric salts and water to give an enol, which tautomerizes to a ketone.
- Hydroboration of alkynes leads to an enol after treatment with $NaOH/H_2O_2$, and tautomerization gives an aldehyde from terminal alkynes, or a ketone from internal alkynes.
- Dilute potassium permanganate and osmium tetroxide react with alkenes to give a manganate ester or an osmate ester, respectively. Both of these products are decomposed under the reaction conditions to give a vicinal (1,2)-*cis*-diol.
- Reaction of an alkene with ozone leads to a 1,2,3-trioxolane, which rearranges to a 1,2,4-trioxolane (an ozonide). Subsequent cleavage gives an aldehyde, ketone, or carboxylic acid product. When an

ozonide contains a C–H unit, oxidation with hydrogen peroxide leads to a carboxylic acid, but reaction with dimethyl sulfide leads to an aldehyde.

- Radicals react with alkenes to form a new radical that can react further to give addition reaction products. The initial reaction will generate the more stable radical, and rearrangement is not observed. In the presence of peroxides, alkenes react with HBr to give the alkyl bromide having Br on the less substituted carbon. This is called anti-Markovnikov addition.
- Alkenes can be polymerized under both radical and acid-catalyzed conditions. Polymerization of functionalized and substituted alkenes leads to a variety of commercially important polymers.
- Many functional group transformations are possible via addition reactions, leading to many new reactions.
- A molecule with a particular functional group can be prepared from molecules containing different functional groups by a series of chemical steps (reactions). This process is called synthesis: The new molecule is synthesized from the old one (see Chapter 25).
- Spectroscopy can be used to determine the structure of a particular molecule and can distinguish the structure and functionality of one molecule when compared with another. See Chapter 14.
- Alkene chemistry is observed in biological systems.

10.1 Alkenes and Acid–Base Chemistry

Acids and bases are defined in Chapters 2 and 6, as are the factors that influence acid–base equilibria (K_a). The definition of a base in Chapter 2 is extended to organic molecules that are capable of donating electrons in Chapter 6 (Section 6.4) and it includes alkenes (Section 6.4.5). As seen in Chapter 5, an alkene has a C=C unit with a π-bond that is not polarized. The π-bond extends in space above and below the plane of the molecule, as in Figure 10.1. The fundamental basis of this chapter is that π-electrons can be *donated* to a positive center, initiating a chemical reaction. In other words, the π-bond reacts as is a base in the presence of a suitable acid, so an alkene reacts as a base.

This acid–base reaction is illustrated by the reaction of ethene with the Brønsted–Lowry acid HCl. Donation of the electrons from the π-bond (**blue arrow**), which is electron rich, to the **acidic proton** of HCl (**H**) leads to formation of a new C–H σ-bond, in **1**. *Note that the double-headed arrow in blue indicates transfer of two electrons from C=C to H⁺ to form a new σ-covalent bond.* The **base** is given a **blue** color to indicate that it is **electron rich** and reacts as

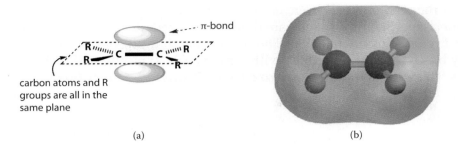

Figure 10.1 A π-bond in an alkene where (a) is a stylized cartoon view and (b) is a molecular model view of ethylene, $CH_2=CH_2$, where the red color indicates the higher concentration of electron density associated with the π-bond.

the **electron donor**. The **acid** (**H** of HCl) is given a **red** color to indicate that the acid is **electron deficient** and **accepts an electron pair** from the **base**. Two electrons are required to form a σ-bond, and both electrons of the π-bond are used to form the C–H bond, leaving an electron-deficient carbon with a formal charge of +1 (Chapter 5, Section 5.5).

In other words, this reaction generates a carbocation intermediate, **1** (see Chapter 7, Section 7.4, to review intermediates). An alkene reacts with a suitable acid to form a new σ-bond in a carbocation intermediate, but as the C–H bond is formed, the H–Cl bond must break. Note that the **chlorine atom** of HCl is marked in **green**, as is the **bond** connecting H and Cl. The curved arrow indicates that both electrons in the H–Cl bond are transferred to chlorine to form the **chloride ion** (**Cl⁻**) as the H–Cl bond breaks.

$$\left\| \quad \underset{H \frown Cl}{} \quad \longrightarrow \quad \left[\quad \overset{H}{\underset{+}{\Big|}} \quad \right] \; Cl^- \right. \quad \mathbf{1}$$

If an alkene reacts as a base, what acids will react with an alkene? In other words, what is the relative base strength of an alkene? This issue was addressed in Chapter 6, Section 6.4.5, by identifying the pK_a of acids that react with an alkene. This strategy will be repeated here. If cyclopentene is used as a typical example, experiments show that this alkene does not react with water under neutral conditions, so water (pK_a = 15.8) is not a strong enough acid. In other words, mixing cyclopentene and water gives no reaction (see Chapter 7, Section 7.12, to define "no reaction"). Likewise, methanol (pK_a = 15.54) does not react with cyclopentene under neutral conditions. Reaction with an acid such as HCN (pK_a = 9.31) fails because experiments show that alkenes such as cyclopentene give no reaction or a very slow reaction with HCN at low temperatures. Acetic acid (pK_a = 4.76) is a significantly stronger acid, but simple alkenes such as cyclopentene do not react at ambient temperatures or at best react very slowly.

These experiments indicate that *alkenes are weak bases in the presence of water, alcohols, HCN, and carboxylic acids*. On the other hand, experiments show that a strong mineral acid such as HCl, with a pK_a of -7, reacts rapidly with cyclopentene. Experiments also indicate that an intermediate carbocation (**2**) is formed in the first step of the reaction. The hypothesis that alkenes react as Brønsted–Lowry bases is proven, but only with strong acids, typically with pK_a values of 1 or less. This means that **alkenes are weak Brønsted–Lowry bases**.

The reaction of cyclopentene with the strong mineral acid HCl gives **3** as the final product. For a moment, imagine that the presence of carbocation **2** as an intermediate is *not* known. It is known, however, that the product **3** has a new H and Cl incorporated into the structure and that the π-bond has been broken. There are two possible choices to explain this observation. One possibility is that the π-bond and the H–Cl bond break simultaneously to generate **3** directly, with no intermediate. The alternative is a stepwise reaction with an intermediate that is formed by initial reaction of the alkene with HCl, and then the second atom is incorporated in a separate chemical reaction. It is known that HCl is a strong acid, so it is reasonable to assume that if the alkene reacts, it must react as a base. The experimental evidence indicates that an intermediate is formed, so the alkene must react with the acidic H of the HCl to form that intermediate. A second step is then required where the chlorine reacts with that intermediate to form **3**. *Once it is known that an intermediate is formed, this product-driven analysis leads to a description that is an essentially correct mechanism.*

Experiments have established that the reaction of cyclopentene and HCl produces intermediate carbocation **2**, written in a bracket to show that it is a reactive (transient) product. The intermediate is the product of the initial reaction, but it is unstable and reacts further to give the product that is isolated in a second chemical step. What is the fate of the carbocation intermediate? The answer is that a second chemical reaction between the carbocation and the counter-ion occurs to form a new σ-bond. In the case of **2**, the chloride counter-ion donates two electrons to the positive carbon to form a new σ-bond (C–Cl) in chlorocyclopentane: **3**, as shown. The *chloride ion* is a **nucleophile** (see Chapter 6, Section 6.7). If the overall sequence is examined, the π-bond breaks, the H–Cl bond breaks, and new C–H and C–Cl bonds are formed in two separate chemical steps.

One way to look at this process is to say that the elements H and Cl *added* to the π-bond, so it can be called an ***addition reaction***. In reality, the alkene reacts as a Brønsted–Lowry base with the strong acid to form the C–H bond and a carbocation intermediate. In a second chemical step, the chloride counter-ion is a nucleophile that attacks the electropositive carbon of the carbocation to form a C–Cl bond. *The conversion of cyclopentene to 2 and then to 3 constitutes the mechanism of this reaction, which is the step-by-step chemical sequence of reactions that transforms cyclopentene to chlorocyclopentene.* The mechanism of this reaction, as described in Chapter 7 (Section 7.8), is a two-step process in which cyclopentene reacts with HCl to give carbocation **2**, which reacts with chloride ion in a second chemical step to give **3**.

10.1 If cyclohexene is mixed with ethanol, CH_3CH_2OH, will there be an acid–base reaction? If cyclohexene is mixed with HBr, will there be an acid–base reaction? If the answer to either question is yes, draw the reactive intermediate and the final product.

In this book, the experimental data for the presence or absence of a reactive intermediate will be identified. In other words, you will be told that the reaction of an alkene and HCl generates a carbocation intermediate. With that experimental knowledge, an analysis of products, reagents, and starting material should lead to a reasonable mechanism. This *product-driven analysis* will be used hereafter for many different reactions.

10.2 Carbocation Intermediates

The reaction of cyclopentene and HCl illustrates the fundamental reaction of virtually any alkene and a strong mineral acid. In a different experiment, cyclopentene reacted with HBr to give bromocyclopentane as the isolated product, in 83% yield.[1] The remainder of the 100% is unreacted starting materials or minor products that are not identified here. For the purpose of this chapter, ignore them. Note that the percentage yield is obtained by experiment. *Should you know this number before doing the experiment or reading this book? No!* Although algorithms are studied with the goal of predicting percentage yield by computer simulation, the percentage yield is an experimental number. *For reactions in this book, do not worry about the actual number, but rather, identify the product. If there are several products, determine which product is the major product and which is minor, based on a mechanistic understanding of each reaction.*

Cyclopentene is a symmetrical alkene, which takes on meaning when it is compared to the reaction of an unsymmetrical alkene such as 2-methyl-2-butene. An unsymmetrical alkene will have different atoms or groups attached to the carbons of the C=C unit. In 2-methyl-2-butene, one carbon of the C=C unit has a methyl and a hydrogen, and the other carbon has two methyl groups. If 2-methyl-2-butene reacts with HBr in the same way as cyclopentene, the intermediate is a carbocation, and subsequent reaction with the nucleophilic bromide ion will form an alkyl bromide. However, the reaction may generate two different products—**4** and **5**—via two different carbocation intermediates. When the percentage yield of products from this reaction is determined experimentally, it is clear that **5** is the major product. Why? An analysis of the mechanism for formation of **4** and also for formation of **5** will provide a prediction of the major product based on understanding each carbocation intermediate.

10.2.1 Regioselectivity and Carbocation Stability

2-Methyl-2-butene reacts with HBr to give **5** as the major product by an acid–base reaction in which the π-bond reacts as a base to form a new σ-C–H bond as the π-bond is broken. This alkene is unsymmetrical, so the σ-C–H bond can form to either carbon of the C=C unit, which will generate different carbocation intermediates (carbocation **6** or carbocation **7**). Using the mechanism presented for cyclopentene as a guide, carbocations **6** and **7** react with the nucleophilic chloride ion, and the isolated products are **4** and **5**, respectively. *Experiments show that 5 is the major product, indicating a clear preference for the formation of carbocation 7 over formation of carbocation 6. If the reaction of 2-methyl-2-butene and HBr gives 5 as the major product and 7 is the intermediate that leads to 5, then 7 must be a more stable carbocation than 6.* Note that **7** has three carbon groups attached to C+ [R_3C+], whereas **6** has only two carbon groups attached to C+ [R_2HC+].

It is time to define an important structural feature of carbocations. A carbocation with three attached carbon groups is a **tertiary carbocation**, [R_3C+]. A **secondary carbocation** has two attached carbon groups, [R_2HC+]. A **primary carbocation** has one attached carbon group, [RH_2C+], and a **methyl carbocation** has no attached carbon groups, [H_3C+]. The observed fact that carbocation **7** is formed in preference to carbocation **6** suggests that the tertiary carbocation must be more stable than the secondary. Indeed, *a tertiary* *carocation is known to be more stable than a secondary carbocation. Similarly, a secondary carbocation is more stable than a primary carbocation, which is more stable than a methyl carbocation.*

How does this information lead to an understanding of the reaction of the alkene and HBr? Remember that an acid–base reaction is an equilibrium, and an equilibrium tends to favor the more stable product (see Chapter 2, Section 2.2, and Chapter 7, Section 7.10). Because the reaction of 2-methyl-2-butene with HBr is categorized as an acid–base reaction, the first reaction that forms the carbocation will be reversible. If this reaction is reversible, it is an equilibrium reaction, and the more stable carbocation **7** will be favored over the less stable carbocation **6** based on formation of the thermodynamically more stable product. In other words, the reaction is under *thermodynamic control* and it generates the thermodynamically more stable product. Formation of the more stable tertiary carbocation can also be rationalized using the **Hammond postulate**, which states that a reaction under thermodynamic control will have a late transition state and that the energy of the transition state will be influenced by the structure and nature of the product. *In this case, the product is the carbocation, so the Hammond postulate suggests that relative stability of each carbocation product should play a major role is determining which carbocation is formed preferentially.*

Assuming that the reaction will follow the lowest energy pathway to the more stable tertiary carbocation, then **7** will be formed in greater abundance relative to **6**. This means that product **5** will be the major product and product **4** the minor. *From this analysis, a generalized assumption can be made that when an alkene reacts with an acid (HX), the reaction will generate the more stable carbocation intermediate; the more stable carbocation is the one with the most substituents attached to the positive carbon.* To summarize: the reactions of an alkene and a protonic acid HX will react with selectivity for the formation of the more highly substituted alkyl halide.

10.2 Draw the intermediate and final products when methylcyclopentene reacts with HCl.

The experiment just described and the conclusion drawn from it are consistent with a significant energy difference between a tertiary and a secondary carbocation. Can this energy difference be validated experimentally? Yes! Several experiments show an energy difference of about 12–15 kcal mol^{-1} (50.2–62.8 kJ mol^{-1}) between the less stable **6** (which is a secondary cation) and the more stable **7** (which is a tertiary cation). Using these experimental data, the structures of a methyl carbocation and generic primary (1°), secondary (2°), and tertiary (3°) carbocations (**8–11**, respectively) may be compared. In all four carbocations, C+ is an sp^2 hybridized carbon (see Chapter 5, Section 5.1), and that carbon has three σ-covalent bonds and a trigonal planar geometry. The overall structure has a formal charge of +1, but that charge resides on the sp^2 hybridized carbon atom. An sp^2 hybridized carbon is expected to have a p-orbital that is orthogonal to the plane of the atoms, so in one sense the positive charge of the carbocation can be viewed as an "empty p-orbital," as shown in structures **8–11**.

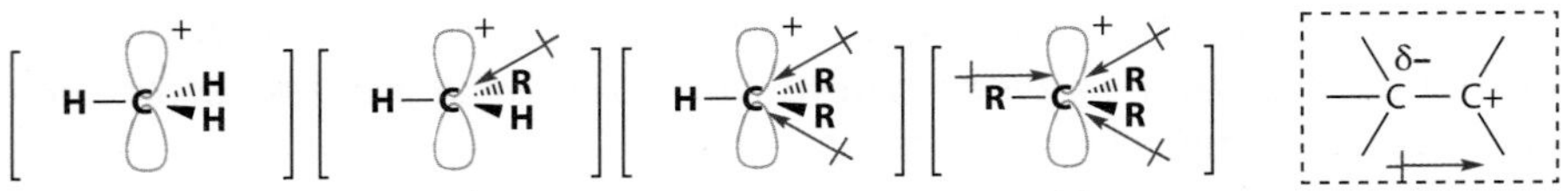

The carbon substituents (alkyl groups) attached to C+ are electron releasing relative to that electron-deficient carbon. Why? The positive charge on carbon leads to an induced dipole on the adjacent atoms, so if there are carbon groups attached to C+, each attached carbon will be δ– (see **12**). This dipole leads to a shift in electron density *from* the δ– carbon *to* C+, which is consistent with the statement that *an alkyl substituent is electron releasing, and stabilizing.* Therefore, the more alkyl groups that are attached to the sp² carbon, the more electron density is shifted toward the positive center (an inductive effect), which *diminishes the net charge on that carbon.* Lower charge is associated with greater stability, so a carbocation with a smaller net charge on the carbon should be more stable and easier to form. This means that a tertiary carbocation with three carbon substituents is more stable than a secondary carbocation with only two carbon substituents. Indeed, the general order of stability using this protocol is the following:

more stable $3° > 2° > 1° > CH_3$ **less stable**

The methyl carbocation and simple alkyl primary carbocations are remarkably unstable. Indeed, they may not form at all in the reactions discussed in this chapter. *A working assumption will be made that, given a choice in reactions with alkenes, methyl carbocations and primary carbocations will not be formed.*

Another useful fact is associated with stability of intermediates. *A more stable intermediate is less reactive, and a less stable intermediate is more reactive. Note also that a less stable intermediate is associated with a higher energy and the more stable intermediate with a lower energy.* In practical terms, this means that a secondary carbocation is more reactive than a tertiary carbocation. This also means that the transition state to a tertiary carbocation is lower in energy than the transition state to a secondary carbocation. Because a methyl carbocation is extremely unstable and difficult to form, it is expected to have a very high energy transition state, but will react very rapidly if it does form. The high energy associated with the transition state for this carbocation is basis for the assumption that methyl carbocations and primary carbocations will not form.

If a 2° carbocation is more reactive than a 3° carbocation, why is **5** the major product rather than **4**? The answer is that, under thermodynamic conditions, carbocation **7** is formed preferentially to **6**, and it reacts quickly enough that **5** is formed to a great extent relative to **4**. In other words, **7** is a very reactive intermediate and, although **6** may be more reactive than **7**, there is a lot more of **7** and it reacts fast enough that it gives the major product.

The reaction of alkenes with mineral acids was discovered in the later part of the nineteenth century. Scientists working during that time did not understand how reactions worked at the molecular level. The mechanisms presented in this book were unknown then, and reactions had to be described in terms of structure and products. In 1869, Vladimir Vassilyevich Markovnikov (Russia; 1838–1904) observed that many different alkenes reacted with HCl or HBr and always gave products in which the halogen atom was on the most substituted carbon and the hydrogen atom was on the less substituted carbon. This observation is now known as ***Markovnikov's rule***. Today, it is understood that this observation is correct because the alkene first reacts with the HX reagent to give the most stable carbocation via an acid–base reaction, which then reacts with the halide counter-ion (the nucleophile) in a second chemical step. A generalization is usually made that alkenes react with acids HX via Markovnikov addition or in a Markovnikov manner.

10.3 **Comment on the stability of the two carbocations formed by reaction of HCl with 2,3,3-trimethyl-1-pentene.**

10.4 **Draw the structure of the Markovnikov product formed when 3-ethyl-2Z-hexene reacts with HCl.**

Returning to alkyl bromides **4** and **5**, the bromine atom is attached to different carbon atoms. The two products formed by this reaction are isomers, but have a different attachment of the atoms. Such isomers are called **regioisomers**. *A regioisomer is the product formed by a reaction when two or more sites for positioning a group or substituent are possible as a consequence of that reaction. When 2-methyl-2-butene reacts with HBr, there are two possible products—4 and 5—so these two bromides are regioisomers.* If the addition of HX to an alkene generates one regioisomer in preference to the other, the reaction is said to be highly **regioselective** (of two possible regioisomers, one predominates as the major product but both are formed). To say that an alkene reacts with acids HX via Markovnikov addition is to say that it proceeds with regioselectivity for the isomer with halogen on the more substituted carbon of the C=C unit.

If *E*-2-hexene (**13**) reacts with HCl, there are two possible alkyl bromide products because there are two carbocation intermediates: regioisomers **14** and **15**. Both **14** and **15** are secondary carbocations, so there is no energy difference to drive the reaction to favor one intermediate over the other. It is therefore anticipated that *both* will be formed, in approximately equal amounts, in the absence of any other information. Carbocation intermediate **14** reacts with the nucleophilic bromide ion to give 2-bromohexane as the product, and intermediate **15** reacts to give 3-bromohexane. *The most important point is that when the reaction generates two carbocation intermediates of equal stability, the reaction should give two products.* There is no obvious reason to choose one product as major and the other as minor, so the prediction is that both are formed in roughly equal amounts. Indeed, *the laboratory experiment must be done in order to predict the actual percentages.*

In one experiment, the reaction of 2-hexene and aqueous HBr gave about 55% of 2-bromohexane and 27% of 3-bromohexane.[2] The remainder of the 100% is starting material and some minor products that are not discussed. Contrast this result with the reaction of 1-hexene and HBr, which gives an 88% yield of 2-bromohexane via the clearly more stable secondary carbocation.[2] The reaction with 1-hexene allows a prediction of the major product based on the mechanism, but a similar analysis does not easily predict a preference for 2-bromohexane from 2-hexene. The point of these experimental data is to show that when there is no obvious difference in energy for the intermediates, a simple mechanistic analysis does not necessarily predict the correct outcome. Other factors must be at work to bias the reaction toward one product over the other, but with the information given for the reaction, it is not obvious what those factors may be. Therefore, in all such cases, *assume that both products are formed, in roughly equal amounts.*

10.5 Are regioisomers possible when 3-hexene reacts with HBr?

In the examples discussed to this point, only HCl and HBr have been used. The acid HF (hydrofluoric acid) can react with alkenes, but much more slowly than the other acids mentioned before (it is a weaker acid; see Chapter 2, Section 2.4). Using published values for pK_a, the mineral acids can be rank ordered as to their acid strength: HF ($pK_a = 3.20$)[3] is less acidic than HCl ($pK_a = -2.2$), which is less acidic than HBr ($pK_a = -4.7$), and HI ($pK_a = -5.2$) is the most acidic in this series. Note the large difference in pK_a between HF and HCl, and remember that acetic acid ($pK_a = 4.78$; see earlier discussion) did not react with a simple alkene. HI is expected to react with alkenes rapidly, as will HBr and HCl, but the diminished acidity of HF may explain its poor reactivity in this reaction. This is not the only factor contributing to relative reactivity with alkenes, but it is a useful one for making predictions.

10.6 What is the major product formed when 1-ethylcyclohexene reacts with HI?

Apart from HCl, HBr, and HI, there are certainly other strong mineral acids, including nitric acid (HNO_3; $pK_a = -1.3$), sulfuric acid (H_2SO_4; $pK_a = -5.2$), and perchloric acid ($HClO_4$; $pK_a = -4.8$). If 1-pentene reacts with any of these acids, the product will be the secondary cation, **16**, but the nucleophilic counter-ion changes. When HBr reacts with 1-pentene, the counter-ion is bromide (Br^-), which is rather nucleophilic, and the product is 2-bromopentane. However, if nitric acid reacts, the nucleophile is the nitrate anion (NO_3^-), which is resonance stabilized and is less likely to donate electrons (less nucleophilic). The product of a reaction with nitric acid in which the nitrate anion reacts as a nucleophile with **16** is **17**. When 1-pentene reacts with sulfuric acid, the nucleophile is the hydrogen sulfate anion (HSO_3^-) and reaction with perchloric acid gives perchlorate (ClO_4^-); both are extensively stabilized by resonance and not very nucleophilic. It is anticipated that the reaction of 1-pentene and sulfuric acid will give **18** and that with perchloric acid will give **19**.

10.7 Write all resonance contributors for NO_3^-, HSO_3^-, and ClO_4^-.

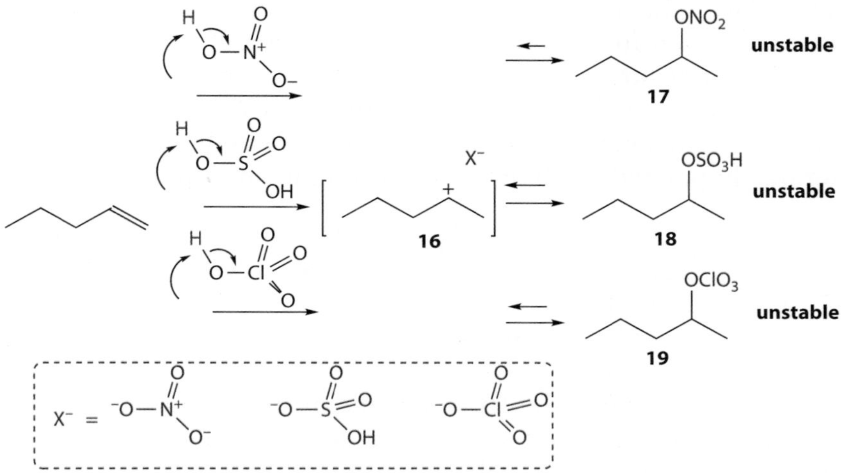

In fact, it is difficult to isolate **17, 18,** or **19** because they are unstable in this medium and highly reactive. This fact is not necessarily obvious upon first glance and without some knowledge of the experimental results. It is known from experiments that nitrates such as $R–ONO_2$, sulfates such as $R–OSO_3H$, and perchlorates such as $R–OClO_3$ (**17, 18,** and **19,** respectively) decompose in such a way that the carbocation is formed again. Another way to state this fact is to say that the C–O bond in the nitrate, the hydrogen sulfate, and the perchlorate makes them excellent leaving groups. Thus, it is easy to break the bond to such groups because they generate stable anions after "leaving" the molecule. Regeneration of the resonance stabilized anion (**17–19**) and the carbocation is facile in solvents that have an OH unit, such as water or an alcohol. This is particularly true when water is present (see Chapter 13).

Another reaction is possible in these systems: an elimination process that generates an alkene called an E1 reaction, but this will be discussed in Chapter 12, Section 12.4. Reaction of an alkene with these acids to give the carbocation does not give the type of products that are obtained from HCl, HBr, or HI. As a practical matter, the reaction of alkenes with strong acids HX to give stable isolable products that contain the nucleophile is limited to the mineral acids HX = HCl, HBr, and HI. *In other words, the reaction of an alkene with HBr, HCl, or HI will give an alkyl halide as the final isolated product. The reaction of an alkene with nitric acid, sulfuric acid, or perchloric acid will not give the corresponding nitrate, hydrogen sulfate, or perchlorate. Rather, it will give a product that results from secondary reactions of the intermediate carbocation. This statement will be discussed in more detail in Section 10.3.*

10.8 Draw the final product that is expected if 3-methyl-2-pentene reacted with sulfuric acid to give a stable sulfate product.

10.2.2 Carbocation Rearrangements

In accord with the principles just discussed, 3-methyl-1-pentene should react with HBr to form the more stable secondary carbocation **20**. Based on the mechanism presented in Section 10.2.1, the reaction of carbocation **20** and the nucleophilic bromide ion should give 2-bromo-3-methylpentane. It does not! When the experiment is done, the major product isolated from this reaction is **21**, 3-bromo-3-methylpentane. If this reaction proceeds via a carbocation **20**, something has happened to **20** *before* the bromide ion could react.

In order to understand why **21** is the major product, examine the mechanism. Look at the final product and use the knowledge that the reaction of an alkene and HBr proceeds by a carbocation intermediate to look for changes in the reaction pathway. First, ask what carbocation is expected when HBr reacts with 3-methyl-1-pentene. The answer is carbocation **20**, which is the secondary carbocation that is formed preferentially to the regioisomeric primary carbocation. The bromine atom in **21** clearly comes from the **bromide ion**, which must react with a carbocation, so the carbocation precursor to **21** must be **22**. In other words, bromide **21** cannot arise from **20**, but must result from tertiary carbocation **22**. Note that **22** is a tertiary carbocation, whereas **20** is a secondary carbocation.

In terms of structure, **22** differs from **20** in that the positive charge has been shifted from C2 to C3. The only way to transform **20** into **22** is to shift the **hydrogen atom** on the tertiary carbon at C3 in **20** to the adjacent carbon atom (C2) in **22**. The initial reaction of the alkene and **HBr** must give carbocation **20** (note the position of the **hydrogen atom derived from HBr** in **20**). If **22** is the correct intermediate to **21**, then the initially formed **20** is converted to **22** by simply moving a **hydrogen atom** from one carbon to the adjacent carbon. By so doing, a less stable secondary carbocation is converted to a more stable tertiary carbocation, an *exothermic* process by 12–15 kcal mol^{-1} (50–63 kJ mol^{-1}).

The conclusions drawn from this analysis indicate that the *mechanism* of this reaction is initial donation of electrons from the C=C unit of 3-methyl-1-

pentene to **HBr** to give **20**, followed by migration of the **hydrogen atom** in **20** to give **22**, and, finally, nucleophilic reaction of **bromide ion** with the carbocation to give **23**. *This mechanism is shown, and this reaction follows a three-step (three-reaction) process, with two different intermediates. The conversion of 3-methyl-1-pentene to 21 is said to proceed by a rearrangement: specifically, a 1,2-shift of the hydrogen atom.*

The **hydrogen atom** does not migrate by itself, but rather it is the bonding electrons to the **hydrogen atom** that move in a carbocation intermediate; this process is said to be a *1,2-hydride shift*. *The clear message from this reaction is that carbocations can rearrange to a more stable carbocation by movement of an atom or group from the adjacent atom.*

Section 10.2.1 shows that tertiary cations are more stable than secondary cations, which are more stable than primary cations, all by about 12–15 kcal mol^{-1} (50–63 kJ mol^{-1}). *Based on difference in stability, assume that if a primary or secondary carbocation is generated on a carbon adjacent to a carbon that would be a more stable carbocation if one atom or group was moved, that group will move.* This is precisely what was observed in the rearrangement of **20** to **22**. The rearrangement can be tracked by examining the movement of the hydrogen atom. Generation of a more stable carbocation explains why the hydrogen atom migrates, but how does it move?

Note the single bond between the sp^2 hybridized carbon atom (C+) and the adjacent sp^3 carbon in **20**. Rotation about this bond is possible (see Chapter 8, Section 8.1.1) and, in one rotamer, the p-orbital of the positive center is parallel to the adjacent sp^3 hybrid orbitals of the C–H bond (**20B**). The hydrogen atom does not move by itself, but rather the bonding electrons in the adjacent and parallel sp^3 orbitals begin to migrate toward the electron-deficient carbon, eventually giving **22B**, leaving an empty p-orbital on the tertiary carbocation with an adjacent sp^3 C–H bond.

The midpoint of this migration is the *transition state* (see Chapter 7, Section 7.3) represented by **23**. In **23**, the hydrogen atom is carried along as the bonding electrons migrate, and the initial sp^3 bond begins to rehybridize to a p-orbital and the initial orbital begins to rehybridize to the new sp^3 orbital.

In other words, it is the migration of the bonding electrons in the adjacent bond that "moves" a hydrogen from the carbon adjacent to the positive carbon to gener-ate a more stable carbocation. This is a chemical reaction, and it is *exothermic by 12–15 kcal mol⁻¹ (50–63 kJ mol⁻¹) when the migration generates a tertiary carbocation from a secondary cation.* The "curly arrow" shown between **20** and **22** is used to highlight the rearrangement.

If **21** is the major product observed when 3-methyl-1-pentene reacts with HBr, but **20** is the initially formed carbocation, *the rearrangement to **22** must be faster than the reaction of bromide ion and **20*** (see Chapter 7, Section 7.11, for an introduction to reaction kinetics). Based on this observation, which is known to be the normal reactivity for such alkenes, *a simple assumption can be made. If a carbocation is generated next to a carbon atom that can potentially bear a more stable cation, assume that the rearrangement occurs faster than any other process.* This is not always true, but it is true most of the time and is a good working assumption for problems encountered in this book.

10.9 Draw the final product formed when 3-phenyl-1-pentene reacts with HCl.

10.10 Draw the final product of the reaction between HI and 3,5-dimethylcyclopentene.

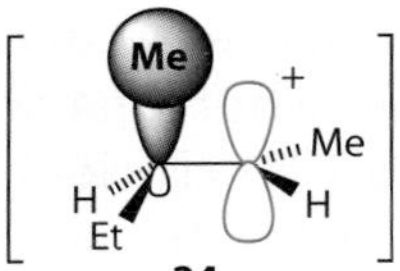

In the rearrangement of 20 to 22, migration of the bonding electrons moved a hydrogen atom, but it is also possible to draw a rotamer that has a methyl group parallel to the p-orbital (see **24**)? Why do the electrons of the C–H bond move in preference to the electrons of the C–methyl bond? A simple answer is that moving the smaller hydrogen atom requires less energy and is generally faster. *This is an overstatement* because the energy differences involved are usu-ally quite small, and other factors are involved; however, it is essentially cor-rect for most simple systems. *Therefore, assume that, given a choice, the smaller group or atom will migrate in preference to a larger group or atom.* Exceptions involve substituent groups that can stabilize charge by delocalization, but they will not be discussed at this point. When such groups are introduced, this sub-ject will be revisited, but for the moment assume that the smaller group will migrate.

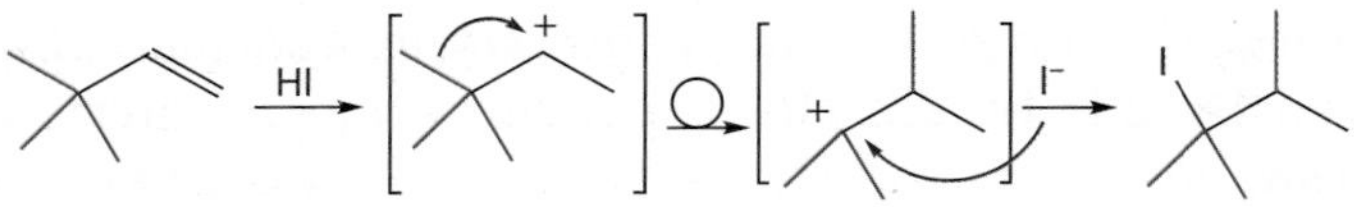

It is clear from the conversion of **20B** to **22B** that migration of electrons in the sp³ hybrid orbitals of an adjacent bond initiates the rearrangement. Therefore, the hydrogen atom is essentially "along for the ride." Based on the discussion in the preceding paragraph, rearrangement of atoms other than hydrogen may occur, including methyl, ethyl, etc., if a more stable carbocation is formed and a "smaller" group or atom is unavailable. The preceding discussion made a case for why the methyl group did not migrate because the smaller hydrogen atom was more facile. *If there are no hydrogen atoms on carbon atoms adjacent to C+, then alkyl groups can migrate. If the carbocation formed by migration of an alkyl group is more stable than the one obtained by migration of a hydrogen atom, then the alkyl group will migrate. Few of the latter will be used in this book.*

When 3,3-dimethyl-1-butene (**25**) is treated with HI, the initially formed carbocation is **26**, as expected. No hydrogen atoms are on the adjacent carbon bearing the methyl groups, but the bonding electrons to a methyl group can move to form the more stable tertiary carbocation **27**. Remember that the bonding electrons in the σ-C–C bond that is parallel to the empty p-orbital of C+ will migrate toward the positive carbon (note the **arrow**). Therefore, the methyl group attached to the σ-bond "is carried by" the bonding electrons to give the more stable tertiary carbocation **27**. The rearrangement is tracked by monitoring the position of the methyl group. Carbocation **27** reacts with nucleophilic iodide to give the final product, **28** (2-iodo-2,3-dimethylbutane). This particular rearrangement is called a *1,2-methyl shift* because the methyl group moves from one carbon to the adjacent carbon. In carbocation rearrangements, hydrogen atoms and simple alkyl groups can move by a 1,2-shift of the corresponding bonding electrons if a more stable carbocation results from the rearrangement. In this book, *the migration must occur on adjacent atoms, and 1,3- or 1,4-migrations will NOT be observed.*

10.11 Draw the final product of the reaction between 3-ethyl-3-methyl-1-pentene and HCl, and show the complete mechanism.

10.3 Alkenes React with Weak Acids in the Presence of an Acid Catalyst

Section 10.1 discussed strong acids such as HCl or HBr reacting with alkenes via an acid–base reaction to form carbocation intermediates that lead to an addition product (see **5**). It was noted at that time that water and alcohols are weak acids and do *not* react directly with alkenes. Is it possible to design an experiment that allows weak acids to react? The answer is yes.

It is known that alkenes react with an acid to give a carbocation, but water is not a sufficiently strong acid for the reaction to occur. On the other hand, if a highly reactive carbocation is generated independently, it is known that the oxygen atom of water is a sufficiently strong nucleophile to form a new C–O covalent bond. *This statement means that water is a good enough nucleophile to react with the carbocation, but too weak an acid to react with the neutral alkene.* Why not mix the alkene with water and a mineral acid in order to take advantage of the acid–base reaction to add the nucleophile of our choice? The acid must be chosen so that its counter-ion is a weaker nucleophile than water, and the discussion in Section 10.2.1 concerning sulfuric acid, nitric acid, and perchloric acid suggested those acids as possibilities. Indeed, reaction of an alkene with any of these three acids will generate a carbocation, along with a counter-ion that is resonance stabilized and a poor nucleophile.

This hypothesis is easily tested. In one experiment, mixing 2-methyl-2-butene with water and sulfuric acid gave 2-methyl-2-butanol (**29**) in 74% isolated yield.[4] The remaining 26% was probably starting material or secondary minor products. Is it possible to construct a mechanism from this experiment based on formation of **29**? Clearly, the OH group in **29** is from a molecule of water that reacts with an intermediate, and water must add to carbon via a carbocation. No hydroxide (HO⁻) exists in this reaction medium, so *water is the only source of oxygen and the OH group.* If water reacts as a nucleophile with a carbocation intermediate, a new C–O bond is formed because oxygen is the nucleophilic atom, and that product is an oxonium in (C–OH₂⁺). This means that the C–OH unit in **29** must arise from a $C-OH_2^+$ intermediate, and the $C-OH_2^+$ intermediate must arise from reaction of water with a carbocation.

The oxonium ion constitutes another intermediate that is different from the carbocation, and its presence requires another step in the mechanism in order to lose the acidic proton to give the neutral alcohol (an acid–base reaction). To obtain the tertiary alcohol structure, the intermediate must be a tertiary carbocation, and that arises from the reaction of the 2-methyl-2-butene starting material with H⁺ of the acid. Therefore, the alkene will react with acid to form a carbocation, which reacts with water and then loses a proton in an acid–base reaction to give **29**.

This analysis translates into a mechanism where the C=C unit in 2-methyl-2-butene reacts as a base with the acidic proton of sulfuric acid (H_2SO_4) *as the first step,* generating the more stable tertiary carbocation **30** rather than the less stable regioisomeric secondary carbocation. Only two nucleophiles are present in the medium: water and the hydrogen sulfate anion; water is (a) present in a greater amount and (b) a better nucleophile. Therefore, the electron-rich oxygen atom of water donates two electrons to the positive carbon in **30** to form a C–O σ-bond in oxonium ion **31**. Oxonium ions were introduced in Chapter 6 (Section 6.4) in connection with an introduction to acids and bases. *The oxonium ion is quite acidic (compared with the hydronium ion H_3O^+) because it is the conjugate acid of a weakly acidic alcohol, ROH.* The reaction of **31** with either unreacted alkene or unreacted water, in an acid–base reaction, will give the alcohol **29**, *which is the conjugate base of 31.* This statement means that the alkene or water is the base and **31** is the acid. Water is shown as the base in the reaction.

The overall process converts the alkene to an alcohol and essentially adds the elements of water to the C=C unit. This reaction is known as hydration. It is important to observe that *in the conversion of 31 to 29, H^+ is regenerated.* Note that H^+ is used in the mechanism rather than H_2SO_4. This is an expedient used to focus on the reaction of the alkene and water and suggests that the source of the acid is unimportant. This latter comment is not completely true, of course, but for the purpose of a simple mechanism, it is sufficient. Therefore, the reaction is **catalytic in acid**, and only a small amount (less than one molar equivalent) of sulfuric acid must be added.

The mechanism that is shown is general for the *acid-catalyzed hydration of alkenes.* A number of different acids can be used. Rather than the strongly oxidizing sulfuric acid, *p*-toluenesulfonic acid (**32**, TsOH; see Chapter 6, Section 6.2.4, and Chapter 16, Section 16.9) is more commonly used because it is not oxidizing and generally soluble in organic solvents. In effect, tosic acid is an "organic" acid. This variation is illustrated by the reaction of cyclohexene with water in the presence of a catalytic amount of tosic acid, generating cyclohexanol, **33**. Note also that alkenes are often insoluble in water, so a cosolvent such as acetone or ether is normally used, as shown:

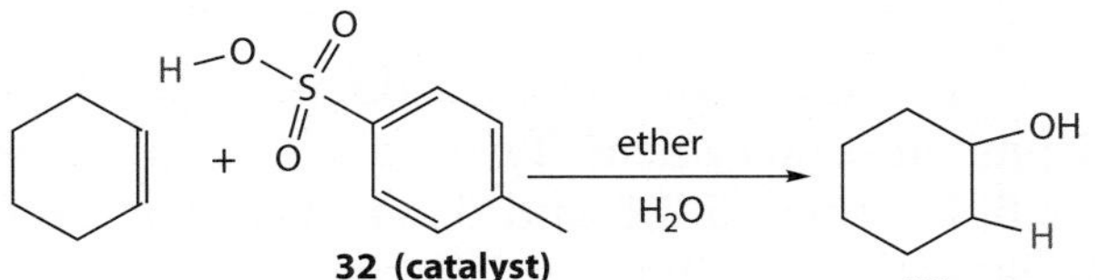

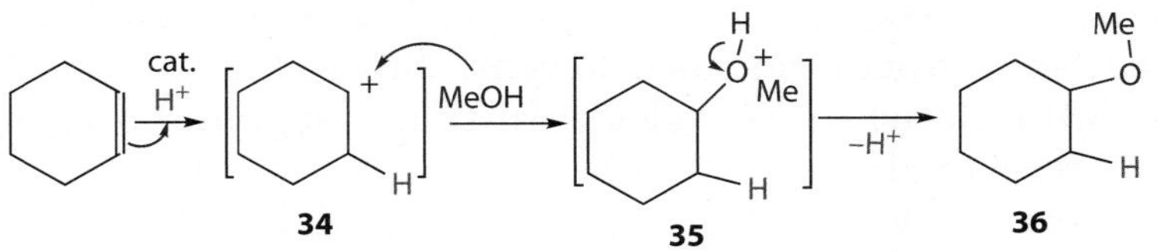

It is clear from the conversion of 2-methyl-2-butene to 2-methyl-2-butanol (**29**) that addition of a strong acid produces the expected carbocation intermediate, such as **30**. Once the carbocation is formed, it will react with the most readily available nucleophile; in the case of water, this leads to the oxonium ion and then to the alcohol. If the same reaction is done in methanol rather than water, the nucleophile will be the oxygen atom of methanol. The reaction of a carbocation with an alcohol will also generate an oxonium ion, but loss of a proton leads to an ether. The reaction of cyclohexene and methanol in the presence of an acid catalyst, for example, initially gives the expected carbocation (**34**), but in the methanol solvent, the nucleophilic is the oxygen atom of methanol. The reaction of methanol with **34** generates oxonium ion **35**. Note the use of H⁺ in the mechanism rather than TsOH. When the proton of the acidic oxonium ion is lost to methanol or cyclohexene in an acid–base reaction, the product is an ether, **36** (methoxycyclohexane; also known as cyclohexyl methyl ether). Note also that **35** is the conjugate acid of **36** if the ether reacts as a base (see Chapter 6, Section 6.4.3).

10.12 **Draw the product formed when 2-phenyl-1-butene is treated with a catalytic amount of tosic acid in aqueous ether.**

10.13 **Draw the product formed when 3,3-dimethyl-1-hexene is treated with a catalytic amount of tosic acid in methanol.**

The important point to be made in this section is that alkenes react with weak nucleophiles, in the presence of an acid catalyst. This approach can be extended to other systems, so an alkene will react with a variety of specific nucleophiles, even weak nucleophiles.

10.4 Alkenes React as Lewis Bases

In the previous section, the C=C unit of an alkene reacted as a Brønsted–Lowry base with a Brønsted–Lowry acid. These are, of course, acids that have a proton (see Chapter 6, Sections 6.2.1 and 6.2.4). Can the alkene react as a Lewis base? The answer is yes (see Chapter 6, Section 6.4.5). With this in mind, consider the following experiment. When cyclohexene is mixed with elemental bromine in carbon tetrachloride as a solvent, the product is *trans*-1,2-dibromo-cyclohexane (**38**) isolated in 57% yield.[5] In this case, quite a bit of unreacted cyclohexene is recovered.

Bromine is certainly not a Brønsted–Lowry acid, but can diatomic bromine be categorized as a Lewis acid? Bromine and indeed all of the diatomic halogens are polarizable. This means that Br–Br covalent bond will become polarized with a δ+ Br and a δ– Br when brought into proximity of an electron-rich species such as the π-bond of an alkene (see **37**). In the proximity of a δ+ bromine, the electron-rich π-bond of the alkene will donate electrons to this electrophilic bromine atom. This is formally analogous to the C=C unit reacting as a Lewis base with respect to the bromine. *A cautionary note is that the final* *product is not an ate complex, as with traditional Lewis bases. However, from the standpoint of bromine, the alkene π-bond certainly qualifies as an electron donor.* Therefore, the Lewis base analogy will be used for reactions of this type, where the alkene π-bond donates two electrons to an atom other than hydrogen.

The reaction of an alkene with diatomic bromine is quite facile, often occurring at room temperature in a few minutes. Based on previous discussions, breaking the π-bond and forming a new C–Br bond is expected to generate a carbocation intermediate. It does not! In fact, detailed experiments into the mechanism of this reaction show that there is indeed an intermediate cation, but it is not a carbocation in the solvent CCl_4. Any mechanism must also address the stereochemical issue that the bromine atoms in **38** are *trans*. These issues are discussed in the following section.

10.4.1 Polarizable Diatomic Halogens React with Alkenes

The reactivity of bromine with cyclohexene is explained in terms of the polarizability of the bromine. Indeed, diatomic halogens such as chlorine (Cl–Cl), bromine (Br–Br), and iodine (I–I) are highly polarizable. In the absence of another molecule, there is no difference in electronegativity between the atoms in Br–Br and Cl–Cl, or I–I. When the halogen is in close proximity to an electron-rich atom or group, however, the halogen atom closest to the electron source becomes δ+ and the other becomes δ–. This means that the atoms are polarized via an induced dipole. In the case of cyclohexene, the electron source is the π-bond and if it donates electrons to the positive halogen, the C=C unit reacts as a Lewis base.

What about diatomic fluorine, F–F? For many years, elemental fluorine was thought to be too reactive and too dangerous for reaction with alkenes.[6,7] To void such problems, fluorine is typically mixed with an inert gas such as nitrogen or argon.[8] Diluted in this manner, fluorine does react with alkenes, but the yields are often poor and, in some cases, solvents for the alkene, such as methanol, participate in the reaction.[9] 1-Phenylpropene ($PhCH=CH_2$), for example, reacted with fluorine in methanol to give 51% of the corresponding difluoride, along with 49% of 2-fluoro-1-methoxy-1-phenylpropane. The problems associated with fluorine lead to a simplification. *In this chapter, alkene reactions are reported only for chlorine, bromine, or iodine but not fluorine.*

10.14 **Draw the structure of the difluoride obtained from 1-phenylpropene and also draw 2-fluoro-1-methoxy-1-phenylpropane.**

It was stated that experimental evidence confirmed the presence of an intermediate, and the *trans* relationship of the bromine atoms in the dibromide **38** suggests that the bromine atoms are incorporated in a stepwise manner. ***This statement means that an intermediate is formed that contains one bromine, and the intermediate must react with the second bromine to form the final product in a different chemical reaction.*** Further, the *trans* relationship of the bromine atoms suggests that one face in the intermediate is blocked such that, for a second reaction to occur, the **second bromine** must approach from the face opposite the **first bromine** atom. The conclusion is that the alkene reacts with the polarized **bromine** to form a C–Br bond in an intermediate, and the **second bromine** attacks the second carbon atom from the face opposite the **first bromine** to give a *trans*-dibromide. This analysis is not consistent with formation of a carbocation, which is planar due to the sp^2 hybridization of the positively charged carbon. Using these observations, it is possible to speculate on the nature of the transition state for the reaction with bromine with the alkene.

When an alkene reacts with Br–Br, the π-bond breaks as both of the electrons of the π-bond are transferred to **Br$^{\delta+}$** to form a new C–Br bond, which should leave behind a positive charge on the carbon. As the new C–Br σ-bond is made, the Br–Br σ-bond must break, and the two electrons of the Br–Br bond are transferred to the **negatively polarized bromine** to form bromide ion (**Br$^-$**). In the *transition state* for this reaction (**39**; see Chapter 7, Section 7.3, for transition states), the second carbon of the alkene π-bond develops positive character as the π-electrons are transferred to **bromine**. In principle, this process would lead to the usual secondary carbocation.

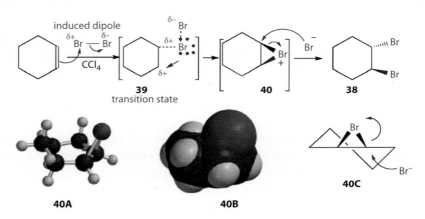

39
transition state

40

38

40A

40B

40C

However, this positive charge develops on carbon in the presence of the **bromine** atom, with has unshared electron pairs. The **bromine** donates two electrons to that positive carbon (called **back-donation**) to form a second C–Br bond in **40** and a three-membered ring with a formal charge of +1, with a **bromide ion** as the counter-ion. *Cation **40** is the experimentally detected cation intermediate mentioned before.* The product is **not** the carbocation (with the positive charge on carbon), but rather the three-membered ring cation **40**, called a **bromonium ion** (one type of halonium ion), with the positive charge on bromine. A similar reaction occurs with all of the halogens. In reactions with alkenes, bromine gives a bromonium ion, chlorine gives a chloronium ion, iodine gives an iodonium ion, and, generically, a halogen gives a **halonium ion**. *It is important to understand that **40** is generated in this nonpolar solvent because it is more stable than the carbocation intermediate.*

The formal charge on ion **40** is +1, but the charge resides on the bromine atom and the individual C–**Br** bond remains polarized C$^{\delta+}$–**Br**$^{\delta-}$. Bomine is more electronegative than carbon, so *both carbon atoms of the three-membered ring in **40** are electrophilic because the C–Br dipole leads to two δ+ carbon atoms.* In other words, *two* carbons are attached to **bromine** in **40,** *both* are electrophilic, and both may react with a nucleophile such as the **bromide ion (Br⁻)**, which is also formed in the initial reaction with C=C. In **40**, the large **bromine** atom must reside on one side of the cyclohexane ring or the other, due to the nature of the three-membered ring and the inability to undergo C–C bond rotation (see Chapter 8, Section 8.1). As drawn, the bromine atom is on the "top" as the molecule in **40**. The sidedness is clearer in the molecular model **40A** and particularly obvious in the space-filling model **40B**. When **bromide ion** attacks an electrophilic carbon atom of the three-membered ring, it must do so from the *sterically less hindered side opposite the first bromine atom* (as in **40C**), leading to formation of a new C–Br bond, with cleavage of the three-membered ring to form the *trans* dibromide. Note that none of the diastereomer (the *cis*-dibromide) is formed in this reaction.

Although this solvent contains halogen, CCl₄ does **not** react. This means that Cl does not appear in the product when bromine reacts with cyclohexene in the solvent CCl₄.

10.15 Draw the secondary carbocation you expect from the reaction of cyclohexene and diatomic bromine.

10.16 Draw the intermediate and final products formed when 2-methyl-1-butene reacts with I₂.

When cyclohexene reacts with bromine, the product is *trans*-1,2-dibromocyclohexane. There are two possible diastereomers of 1,2-dibromocyclohexane: the *cis*- and *trans*-isomers, but only *trans*-1,2-dibromocyclopentane (**38**) is formed in the reaction. (See Chapter 9, Section 9.5, to review diastereomers.) Only the *trans*-diastereomer formed, so this reaction is *diastereospecific*. This stereochemical preference has been confirmed in the reactions of many alkenes, over many years. If cyclohexene is viewed from the side, as in **41**, it is clear that the initial reaction with bromine must deliver the **Br** of the bromonium ion to one side of the ring or the other. It can be either on the top or the bottom because there is no facial bias in the C=C unit of cyclohexene.

If bromine is arbitrarily drawn on the bottom, as in **42** (compare this structure with **40C**, with the bromine on the top), nucleophilic attack by the **bromide ion** will occur from the opposite top face because that is less sterically hindered. This back-side attack leads to the *trans* stereochemistry in **43**. Note that attack from the top or bottom can occur with equal facility, but a *trans*-dibromide is produced in both cases. Attack from the top face leads to one enantiomer and attack from the bottom face leads to the other enantiomer; attack from either face occurs with equal facility. *The reaction must produce a racemic mixture.*

10.17 Draw *cis*- and *trans*-1,2-dibromocyclohexane, along with each mirror image.

10.18 **Draw the bromonium resulting from reaction with 41 that places the bromine on the "top" relative to 42. Draw the resulting dibromide and compare it with 43 to see that they are enantiomers.**

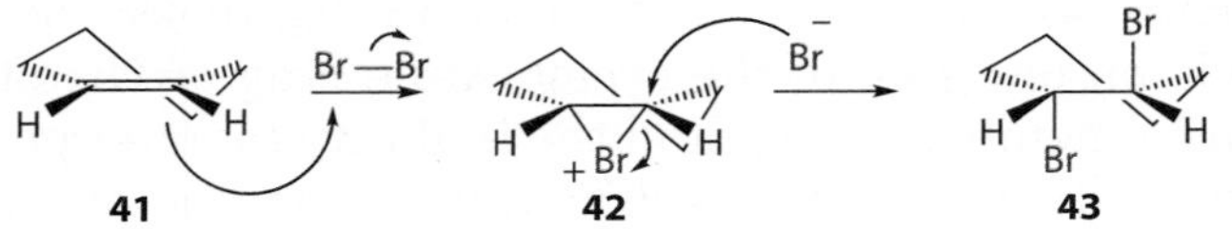

The stereochemistry of **43** (**38**) is easy to see because the cyclohexane ring cannot rotate about individual C–C bonds. Does this same stereochemical bias occur with acyclic alkenes? When an acyclic alkene reacts with a halogen, the product is an acyclic dihalide, and free rotation is possible about those bonds. However, the answer is "yes": *Acyclic alkenes react with the same stereochemical bias because the mechanism of reaction of an alkene and diatomic halogen is the same for both acyclic and cyclic alkenes.* This selectivity is demonstrated with the simple acyclic alkene, 2-butene; however, an analysis requires an examination of each stereoisomer, *cis*-2-butene and *trans*-2-butene (see Chapter 9, Section 9.4). When *cis*-2-butene reacts with bromine, the product is a racemic mixture, (2*S*,3*S*)-dibromobutane along with (2*R*,3*R*)-dibromobutane (see **44**). ***Two new stereogenic centers are created by this reaction.*** (See Chapter 9, Section 9.3, to review absolute configuration.) When *trans*-2-butene reacts with bromine, however, the product is a racemic mixture of (2*S*,3*R*)-dibromobutane and (2*R*,3*S*)-dibromobutane (see **45**), which are drawn a second time as the eclipsed rotamer (Chapter 8, Section 8.1) to show their relationship to 44. ***Dibromides 44 and 45 are diastereomers*** (Chapter 9, Section 9.5).

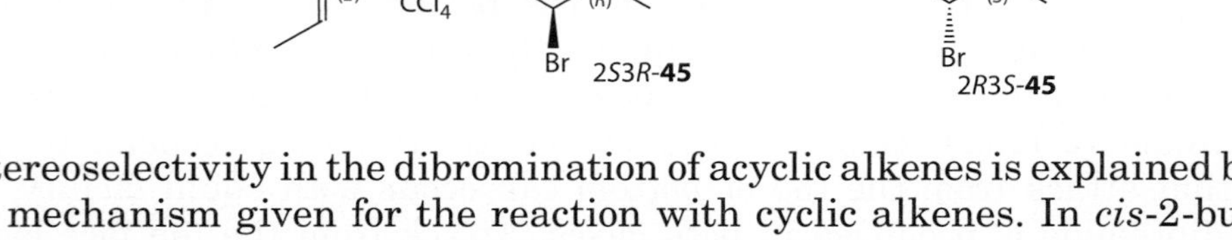

Stereoselectivity in the dibromination of acyclic alkenes is explained by the same mechanism given for the reaction with cyclic alkenes. In *cis*-2-butene, the two methyl groups are "locked" on one side of the C=C unit because there is no rotation around those carbon atoms. *The key to the stereoselectivity is the fact that the stereochemical relationship of the groups on the C=C unit is retained in the transition state that leads to the bromonium ion—in the bromonium ion and in the final product.* When *cis*-2-butene reacts with bromine, bromonium ion **46** is formed, which is arbitrarily drawn with the bromine

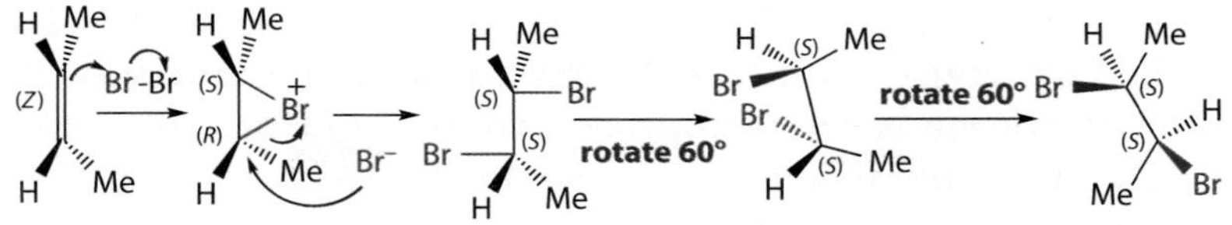

Figure 10.2 Stereospecificity in the bromination of *cis*-2-butene.

on the right, although there is nothing to distinguish one side from another (to be discussed later). If the two methyl groups are locked on one side in the alkene, they will also be locked in position in the transition state and in the **bromonium ion** product, **46** (see Figure 10.2) because the three-membered ring prevents rotation.

This is very important. The counter-ion in **46** is the **bromide ion (Br⁻)**. As seen previously, **Br⁻** reacts with **46** via back-side (*anti*) attack at carbon on the face opposite the bromine atom in **46**, which fixes the stereochemistry of the two bromine atoms as *anti*. If the stereochemistry of the methyl groups is fixed in the transition state leading to the bromonium ion and *anti* attack fixes the stereochemical relationship of the two bromine atoms, the stereochemistry at each new stereogenic center must be fixed in the dibromide product as shown. In this case, it is (2S,3S).

Of two possible diastereomeric dibromides, only one is formed. *As noted before, bromination of alkenes is said to be a diastereospecific **reaction** and dihalogenation is generally diastereospecific. Once the dibromide product is formed, there is free rotation around the carbon–carbon bonds, and **44A** is drawn to show other rotamers in Figure 10.2. What is the point of redrawing this molecule? It is essential to recognize the diastereomer as any of its possible rotamers. In all representations of **44A**, the absolute configuration remains (2S,3S), which *is the only reliable way to identify the product.*

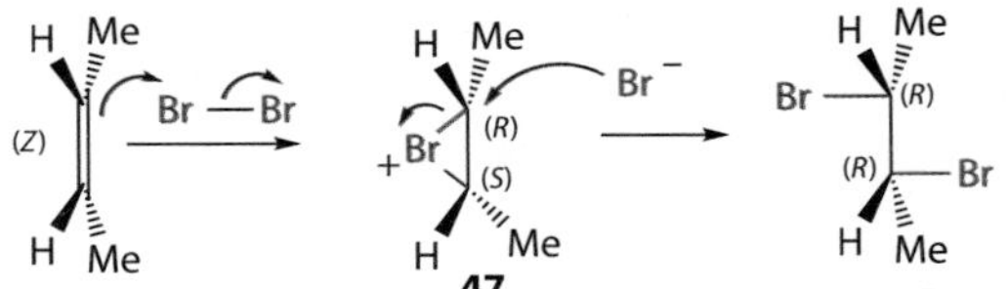

Imagine that diatomic bromine reacts with *cis*-2-butene such that the bromonium ion has a bromine on the "left" of the C=C unit, as in **47**. Subsequent attack of the bromonium ion from the face opposite the bromine atom will generate the (2R,3R) diastereomer (i.e., **44B**), *which is the enantiomer of **44A** generated from bromonium ion **46**, in which bromine was on the "right."* This analysis could easily be done with the alkene drawn in a way that bromine attacks from top and bottom (see **41**). In other words, the bromine may add to either face of the alkene with equal facility. Because the reaction of bromine and an alkene may occur from either side of the C=C with equal facility, a mixture of enantiomers is formed; a racemic mixture (Chapter 9, Section 9.2.4). This reaction is described by saying that there is **no facial selectivity** in the addition of bromine to the alkene because approach from either face is equally likely. *A racemic mixture is formed when there is no facial selectivity, but the reaction is still diastereospecific because only one diastereomer is formed.*

10.19 Draw the bromonium ion intermediate and final product for "bromine on the bottom" from *E*-3-hexene.

10.20 Draw both enantiomers of the product formed when chlorine reacts with *E*-3-methyl-3-heptene and then draw the diastereomer that is *not* formed.

10.21 Draw both enantiomers generated by the reaction of *E*-2-butene and chlorine.

10.22 Determine which of the stereoisomers of 3-hexene (*cis* or *trans*) reacts with chlorine to generate a *meso* compound (see Chapter 9, Section 9.5.2).

10.4.2 Reaction with Aqueous Solutions of Halogens (Hypohalous Acids)

In the preceding section, diatomic bromine, chlorine, or iodine reacted with an alkene in a nonaqueous solvent such as carbon tetrachloride. Exploring a reaction usually requires experimenting with reaction conditions, including changing the solvent. A reasonable experiment with an alkene that involves chlorine might use an aqueous solution saturated with chlorine gas.

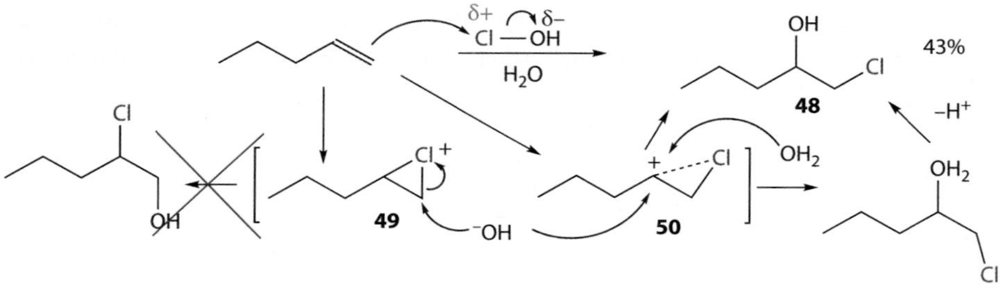

It is known that dissolving chlorine in water leads to a solution that contains hypochlorous acid (HOCl) and bromine dissolved in water contains hypobromous acid (HOBr). In one experiment, 1-pentene is mixed with chlorine and water (HOCl in aqueous media) and the major product is 1-chloro-2-pentanol (**48**), in 43% isolated yield.[10] In the previous section, chlorine reacted with 1-pentene in a nonaqueous solvent such as carbon tetrachloride to give a dichloride. To ascertain why this reaction is different, the first useful observation is that HOCl is in solution rather than Cl–Cl. The polarization of HOCl is HO$^{\delta-}$–Cl$^{\delta+}$, where chlorine is the electrophilic atom. The π-bond of an alkene should react with the positive chlorine atom, and cleavage of the Cl–H bond will give hydroxide ion, which is a nucleophile in this reaction. If the reaction is done in water, the oxygen of atom is also a nucleophile.

As noted in Section 10.4.1, a nucleophile attacks a halonium ion at the less substituted carbon. If chloronium ion **49** forms in this particular reaction, attack by the nucleophilic hydroxide ion at the less hindered carbon of **49** gives 2-chloro-1-pentanol. However, the isolated product is **48** (1-chloro-2-pentanol), and formation of **48** must result from attack at the *more* substituted carbon atom of **49**. Such a product is not consistent with nucleophilic attack with chloronium ion **49**.

In the reaction with 1-pentene, formation of **48** is only consistent with formation of a secondary carbocation such as **50** rather than **49**. Carbocation **50** may react with hydroxide to give **48** directly, but it is more likely that it will react with water to give an intermediate that leads to **48**. If **49** forms when bromine reacts with an alkene in CCl_4, why does **50** form when the reaction is done in water? Carbocation **50** forms in water, with some back donation by chlorine. Water not only generates HOCl, but it also separates charge and stabilizes charge by solvation (also see Chapter 13, Section 13.2, for similar solvent effects). Because HOCl is generated in the presence of water, it is anticipated that **50** is more stable in the aqueous medium than **49**, and attack by hydroxide at the positive carbon gives **48**. Note that in a large excess of water (water is the solvent), **50** can also be attacked by water, and loss of a proton from the oxonium ion (hydroxide or water can function as the base in this reaction) also gives **48**.

For reactions of chlorine or bromine in water, *assume* that the reaction proceeds in aqueous media via the secondary ion, and attack by the nucleophile leads to the major product. *In other words, for reactions of bromine and chlorine in aqueous media, assume the OH unit will be attached to the more substituted carbon atom.*

10.23 Write out the mechanism that describes the proposed reaction of water with **50**.

10.24 Draw both regioisomeric products formed in the reaction between chlorine and 4-ethyl-3,6,6-trimethyl-3Z-heptene when it is done in aqueous THF. Determine which will be major, based on our assumptions.

10.4.3 Borane as a Lewis Acid (Hydroboration)

Boron trifluoride (BF_3) is a classic Lewis acid. Borane, BH_3, is a highly reactive boron compound that also functions as a Lewis acid in the presence of a suitable electron-donating species.

51

Borane is usually prepared by a simple chemical reaction. Sodium boro-hydride ($NaBH_4$) is a reducing agent, and it will be discussed in Chapter 19 (Section 19.2.2). It reacts with boron trifluoride (BF_3) to give a volatile product, borane. Borane is a reactive species that is usually written as BH_3, but it is actually a dimeric species called diborane (**51**) that has *hydrido bridges* (bridging hydrogen atoms). There is an equilibrium between borane and **51**, but for reactions presented in this section, the monomeric species (BH_3) and the dimeric species B_2H_6 are used interchangeably. Boron is in group 13 (three valence electrons); with only three covalent bonds in the neutral species, it is electron deficient and a Lewis acid (see Chapter 2, Section 2.5).

Diethyl ether is a Lewis base, and it reacts with borane to form an ate complex ($H_3B·OEt_2$) known as borane etherate, which is more stable and easier to handle than the gaseous borane. It is commonly used rather than generating borane with $NaBH_4$ and BF_3•etherate. Borane is an interesting molecule because the B–H bond is polarized such that the boron is δ+ and the hydrogen is δ− because hydrogen is more electronegative than boron (see Chapter 3, Section 3.7). With this knowledge, consider the following experiment:

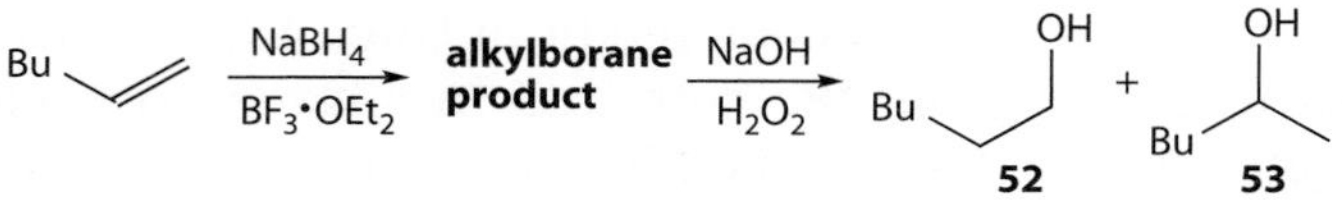

1-Hexene reacts with borane to give a new product known as an **alkylborane**, which is listed as unknown for the moment. The alkylborane product is treated with NaOH and H_2O_2 in a second chemical reaction and the major product obtained after this two-reaction process (two-step process) is 1-hexanol (**52**), isolated in 81% yield.[11] The second reaction with hydroxide and peroxide will be discussed later in this section, but for now focus attention on the new borane product of the first reaction. The remainder of the 100% is alcohol **53** as a minor product, and this will be explained later.

In the reaction with 1-hexene, borane reacts as a Lewis acid and the alkene is a Lewis base. Donation of two electrons from the π-bond to the boron is accompanied by cleavage of the π-bond and formation of a new C–B σ-covalent bond. In addition, a hydrogen atom from boron must be transferred to carbon, forming a new C–H σ-covalent bond as the H–B bond is broken. So far, this is similar to the reaction of an alkene with HBr. *However, experiments have shown that there is no intermediate, so there is no carbocation.* If there is no intermediate, the C–H σ-bond must be formed almost simultaneously with the C–B σ-bond. *Based on the experimental evidence that there is no intermediate, the reaction of borane with an alkene is said to be a* **concerted reaction. Formally, it classified as concerted asynchronous rather than concerted synchronous.** *A reaction that has no intermediate is called a concerted reaction. If the bond-making and bond-breaking events do not occur simultaneously, the reaction is said to be asynchronous.*

To understand this reaction, the unknown alkylborane product must be identified. If there is no intermediate, the reaction of borane with the alkene gives the unknown alkylborane product in a single step with only a transition state to describe the process. An unknown alkylborane product is listed, but there are two alcohol products after the second reaction, which means there must be two alkylborane products. Note that alcohols **52** and **53** are regioisomers, with the OH unit attached to different carbon atoms. The position of the OH units corresponds to the position of carbon atoms in the C=C unit of the alkene. The alkylborane products contain C–B bonds, and it is reasonable to assume that the OH unit in the alcohol product replaces boron in the alkylborane. Therefore, the boron may be attached to either of the two carbon atoms associated with the C=C unit and the two alkylborane products are **56** and **57**.

Borane product **56** must be the precursor to alcohol **52** and borane product **57** is the precursor to alcohol **53**. Before alcohol **52** is formed as the major product, the reaction shows a preference for the formation of **56**. There is no intermediate, so the transition states for each alkylborane product (**54** and **55**, respectively) must be examined to explain the preference for **56** over **57**. This preference for the less substituted product is sometimes referred to as an **anti-Markovnikov orientation** (see Section 10.2.1).

In the transition state for a concerted reaction, the p-bond and the B–H bond break as the new C–B and new C–H bonds are formed. This is called a **_four-center transition state_**, represented by **54** or **55** for the two alkylborane products. Remember that the B is δ+ and the H of borane is δ–, so the alkene attacks the boron and the hydrogen atom is transferred to carbon. The only way to rationalize **56** as the major product is for transition state **54** to be lower in energy than transition state **55**. Why should this be the case?

A comparison of **54** and **55** shows that steric hindrance builds between the butyl group of the alkene and the BH$_2$ unit of borane in **55**, whereas this steric interaction is minimized in **54**. In other words, there is more steric hindrance in **55** than in **54**, so **55** is higher in energy than **54**. Because transition state **54** is lower in energy, it will lead to alkylborane **56**, the observed major product, whereas transition state **55** leads to organoborane **57**. Note that both products have a single alkyl group attached to the boron and are known as monoalkylboranes (RBH$_2$). Boranes that have two alkyl groups attached to boron are dialkylboranes (R$_2$BH) and those with three alkyl groups are called trialkylboranes (R$_3$B). The products of this reaction are monoalkylboranes.

In general, the greater the steric bulk of substituents on the alkene C=C unit is, the greater is the steric hindrance in the transition state and the greater the preference for attaching boron to the *less substituted carbon*. Therefore, *in reactions with alkenes, borane will become attached to the less substituted carbon atom via a four-center transition state.*

In the reaction of 1-hexene to give **56** as the major product, an essential component is easy to overlook. The ether solvent is important, and the solvent *glyme* is commonly used. The formal name of glyme is 1,2-dimethoxyethane and it has the structure $CH_3OCH_2CH_2OCH_3$. If borane is simply mixed with 1-hexene with no solvent, no reaction occurs unless the mixture is heated to about 180–200°C. In ether solvents, however, the reaction occurs rapidly at ambient temperatures (i.e., room temperature). In fact, ethers *catalyze* the reaction of borane with an alkene and, although the ether structure does not appear in the product, the ether is not simply a solvent. This is probably because the oxygen atom of an ether is a Lewis base (remember the ate complex formed from borane and diethyl ether) and reacts with diborane to form a highly reactive complex that reacts with the alkene. All reactions of alkenes with borane in this chapter will use an ether solvent.

10.25 Show the monoalkylborane product formed when methylcyclo-pentene reacts with borane in an ether solvent.

10.26 Draw the ate complex that results when borane **56** reacts with diethyl ether.

A second important feature of this reaction has to do with the structure of monoalkylborane **56** and the number of molar equivalents of the alkene that react with BH_3. Initial reaction of 1-hexene and borane gives monoalkylborane **56**. One of the B–H units of **56** can react with a second molar equivalent of 1-hexene. In other words, the monoalkylborane RBH_2 also reacts as a Lewis acid to give a dialkylborane R_2BH, **58**. Because **58** also has a B–H unit, reaction with a third molar equivalent of 1-hexene leads to the trialkylborane **59**. Therefore, one molar equivalent of borane can react with three molar equivalents of an alkene such as 1-hexene to form a trialkylborane, **58**. Three equivalents of the alkene are used because the alkylborane products react with the alkene and it is possible to obtain a mixture of alkylborane products. For terminal alkenes, the reaction tends to favor formation of di- or trialkylboranes. With very hindered alkenes, it is possible to stop this reaction at the dialkylborane or even the monoalkylborane stage, but in general the reaction proceeds to the trialkylborane. *For convenience, assume that the reaction of borane (BH_3) and any alkene gives the trialkylborane.* Using an excess of the alkene is a convenient way to ensure that the reactions proceed to the trialkylborane.

When the alkene precursor is hindered (highly substituted), it may be possible to isolate monoalkylboranes or dialkylboranes. Three alkylboranes should be discussed because they are also used in reactions with alkenes. When 2-methyl-2-butene reacts with borane, the product is the dialkylborane **60**, and its common name is disiamylborane (from the common name di-*secondary*-isoamylborane). Similar reaction of borane with 2,3-dimethyl-2-butene leads to a monoalkylborane, **61**. The common name of this product is thexylborane (from tertiary-hexylborane).

Finally, reaction of borane with 1,5-cyclooctadiene gives a product where two of the B–H units have reacted with the two C=C units across the ring to give a bicyclic product (see Chapter 9, Section 9.7), **62**. The name of this dialkylborane is 9-borabicyclo[3.3.1]nonane, which is often abbreviated 9-BBN or drawn as shown later. Nomenclature rules for bicyclic systems such as this are described in Chapter 9, Section 9.7.A. There are a total of nine atoms (nonane) but C9 has been replaced with boron (9-bora). All three of these boranes react as Lewis acids with an added alkene.

An important reason for using **60–62** is the stereochemical consequence of the borane–alkene relative to the same reaction with BH$_3$. Reaction of methylcyclopentene with 9-BBN (**62**), for example, leads to two regioisomers via transition states **62** and **64**, and the diagrams suggest that **63** is less sterically hindered. Because 9-BBN is so bulky, **63** is much less hindered and therefore lower in energy. The lower energy transition state will form faster and preferentially, and the second chemical step will convert it to **65** as the major product with <0.1% of the regioisomer. The same reaction with BH$_3$ gives about 85–90% of **65** and about 10% of the regioisomer. The larger and bulkier 9-BBN provides greater selectivity in the transition state due to increased steric hindrance, which means greater regioselectivity in formation of the product.

10.27 Draw the structure of 1,5-cyclooctadiene.

10.28 Draw the products formed when 60, 61, and 62 react with cyclopentene, assuming one equivalent of borane reacts with one equivalent of alkene.

Note that the B and the H of the borane are on the same side of the ring in **65**, which is a consequence of the four-center transition state and concerted asynchronous delivery of B and H to the C=C unit. This reaction constitutes a *cis* addition of borane to the alkene, where the B and the H add *cis* (on the same side). In the case of methylcyclopentene, the *cis* addition of B and H leads to a *trans* relationship between the BH$_2$ unit and the methyl group in **65**, but the *hydroboration reaction is a* cis *addition.*

10.29 Draw the major product formed when 9-BBN reacts with 2,3-dimethyl-3-hexene.

10.4.4 Oxidation of Alkylboranes to Alcohols

Organoboranes are formed from the reaction of alkenes and BH_3 or, in some cases, with other organoboranes. Are organoboranes useful in other reactions? The experimental example in Section 10.4.3 reports that monoalkylboranes **56** and **57** are converted to alcohols **52** and **53** by treatment with hydrogen peroxide and sodium hydroxide. This transformation is a formal oxidation (see Chapter 17). Several other reactions can transform an alkylborane into different functional groups, but this chapter will focus only on oxidation to an alcohol.

After many experiments, Herbert C. Brown (United States; 1912–2004; Nobel Prize, 1979) and co-workers discovered that reaction of **59** (the trialkylborane derived from 1-hexene) with a mixture of hydrogen peroxide (H_2O_2) and aqueous sodium hydroxide (NaOH) gave an alcohol (in this case 1-hexanol, **52**) and boric acid [$B(OH)_3$]. Comparing the borane with alcohol product shows that the OH has "replaced" the boron, but experiments have shown that the mechanism for this transformation of **59** to **52** involves several steps. Initial reaction is at boron rather than carbon, and the transformation involves a *rearrangement from boron to oxygen*. The sequence begins with the reaction of HOOH and NaOH to give sodium hydroperoxide ($Na^{+-}OOH$), which attacks the boron (a Lewis acid) to form "ate" complex **66** (the Lewis acid–Lewis base complex). This product is not isolated because it undergoes a boron-to-carbon alkyl shift (a rearrangement), with loss of hydroxide (HO^-) to give **67**.

In other words, *the alkyl group on boron migrates to oxygen*. If HOO^- attacks boron twice more, two additional B→O alkyl shifts transfer the remaining alkyl groups from boron to oxygen. The final product is **68**. If hydroxide attacks boron in **68**, the RO^- group (an alkoxide) is displaced (three times in three successive steps) and final neutralization with aqueous acid in a final step liberates boric acid and three equivalents of the alcohol (ROH, in this case **52**) from the alkoxide. The reaction sequence just described converts the organoborane to the corresponding alcohol.

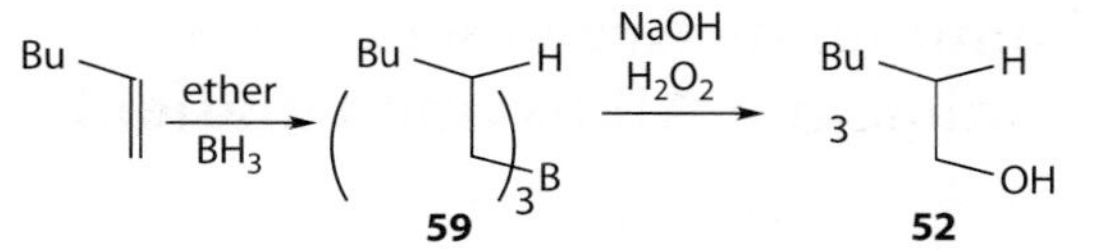

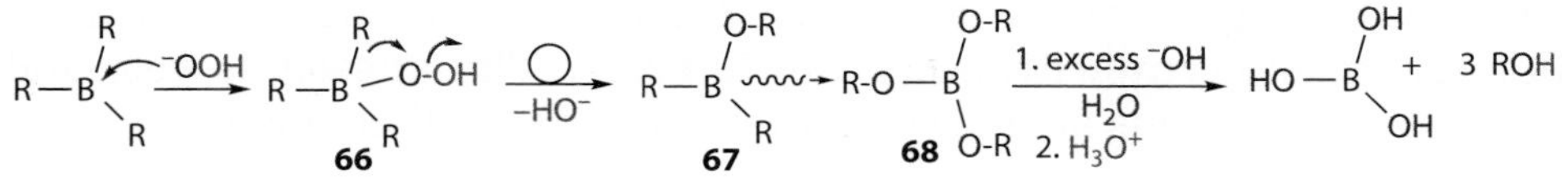

This mechanism is interesting because attack of HOO⁻ and of HO⁻ are at boron, not at carbon. This fact is known by experiments with alkylboranes such as **69**, derived from methylcyclopentene, and subsequent reaction with NaOH/H_2O_2. The final product is the *trans*-2-methylcyclopentanol, **70**, where the OH unit and the methyl group are *anti* to each other (B and H are *cis* or *syn* in **69**, just as the boron and the methyl were *anti* in **65** from the preceding). If attack of HOO⁻ occurs at carbon, replacing boron with OH will proceed with inversion of configuration at carbon (see the S_N2 reaction in Chapter 11, Section 11.2). In order to explain the experimental results, initial attack must occur at boron rather than at carbon, followed by a B → O rearrangement that occurs with *retention of configuration*. In other words, the stereochemistry relationship of the boron unit with other substituents is retained in the final alcohol product.

10.30 Draw *all* major products formed when 3-ethyl-2-pentene is treated with (a) 9-BBN; (b) NaOH, H_2O_2.

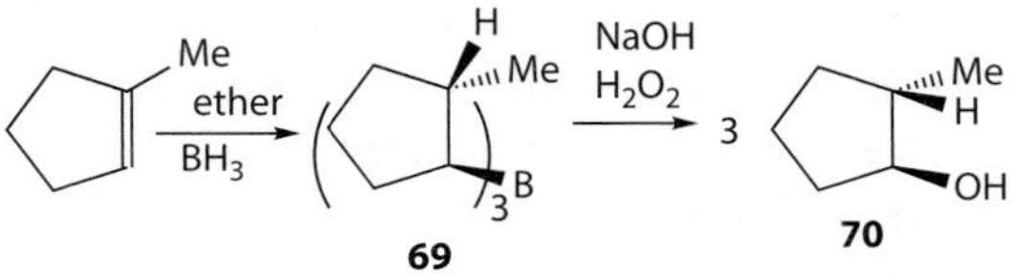

There is another way to present this multistep reaction. The reactions are "stacked" on an arrow, and a number is provided to indicate each separate reaction. For methylcyclopentene, the first reaction with borane (reaction 1) is followed by reaction with NaOH and H_2O_2 (reaction 2). When presented this way, the alkene starting material reacts with borane to give **69**, which is not shown, and in a second reaction **69** is converted to **70**. Only the starting material and the final product are shown, however, so the focus is on the isolated product rather than the mechanism. This method of "stacking" reactions is used throughout this book.

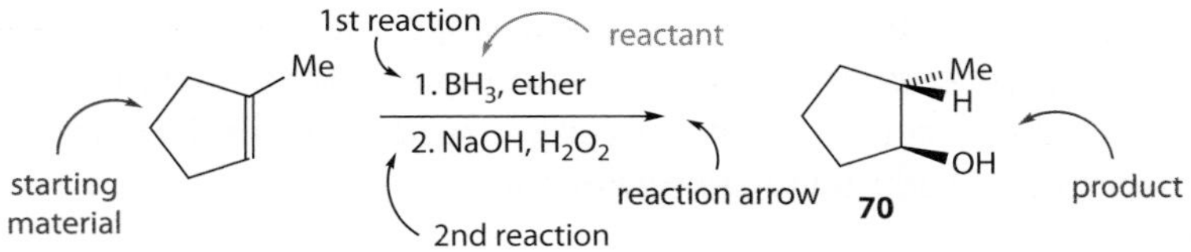

10.4.5 Mercuric Compounds Are Lewis Acids (Oxymercuration and Alkoxymercuration)

It is clear from the preceding section that alkenes react as Lewis bases with polarizable dihalogens and with a classical Lewis acid such as borane. Can similar reactions occur with other Lewis acids? The answer is yes, and an important class of Lewis acid is composed of mercury(II) compounds.

Mercuric salts (HgX_2) are Lewis acids that react readily with alkenes, and experiments show that the reaction proceeds by formation of a carbocation intermediate. When mercuric acetate—$Hg(OAc)_2$; see the structure and Chapter 20, Section 20.5—reacts with 1-hexene in a THF–water mixture, the isolated product is a mercuric compound identified as **71** (with a C–Hg bond) and it is formed by reaction of water with a carbocation intermediate. The initial acid–base reaction is the donation of the π-electrons to form a new C–Hg covalent bond, leaving behind a carbocation at the other carbon of the C=C unit.

In the cited experiment, the mercury is removed in a second chemical step in which **71** is treated with NaOH and then $NaBH_4$, and 2-hexanol (**53**) is isolated after this step in 68% yield.[12] Sodium borohydride is a reducing agent, and its chemistry is explained in more detail in Chapter 19 (Section 19.2.2). In this reaction, $NaBH_4$ reduces the C–Hg bond to give a C–H bond and the Hg is transferred to the boron atom. The transformation of the $NaBH_4$ is not shown so that the focus remains on the organic fragments. *Cleavage of the C–metal bond and conversion to a C–H bond is known as hydrogenolysis.*

The structure of **71** indicates that the Hg is at the less hindered carbon atom, and experiments show that there is a carbocation intermediate. The reaction with mercuric acetate can be related to the reaction of HBr in that the more stable carbocation is formed as a new C–Hg or C–H bond is formed. It is known that alkenes react with an acid catalyst and water to produce an alcohol (see Section 10.3). Using this as an analogy, the OH unit in **71** is derived from the reaction of the nucleophilic water with an intermediate carbocation.

Formation of the secondary alcohol **71** clearly indicates that reaction of the alkene and the Lewis acid gives the more stable secondary carbocation. Indeed, this is the mechanism for this transformation, as discussed in further detail later. The overall transformation is a hydration process that adds water to the more substituted carbon of the alkene (sometimes called a Markovnikov addition), followed by removal of mercury, and is called **oxymercuration–demercuration** or simply **oxymercuration**.

Based on the previous analysis, 3-methyl-1-hexene reacts with mercuric acetate to give secondary carbocation intermediate **72**. Is this carbocation unusual relative to previously discussed carbocations, such as **30** in Section 10.3? Note that mercury is a transition metal with d-orbitals that can donate electron density to the carbocation, stabilizing that positive center via what is known as *back-donation* (similar to what is observed with the halogens in the formation of halonium ions). Despite the manner in which is it drawn, this carbocation is not a three-membered ring, but rather a secondary carbocation that is stabilized by back-donation from the mercury atom. *The dashed line (----) indicates significant coordination between the carbon and mercury.*

The positive carbon in **72** is a secondary carbocation that is proximal to a potential tertiary carbocation. If **72** were formed by reaction of the alkene with water and a catalytic amount of an acid (see Section 10.3), it should rearrange to the more stable tertiary cation (see Section 10.2). *However, it does not. Experimentally, the observation that rearrangement does not occur leads to the conclusion that there is significant stabilization (back-donation by the d-orbitals) by mercury.* The solvent in this reaction contains water, and water reacts as a nucleophile to form oxonium salt **73**, which loses a proton in a simple acid–base reaction to give alcohol **71**. In this latter reaction, the base is water or the alkene. The reaction of the alkene with mercuric acetate is an attractive sequence because *it generates a carbocation that does not rearrange,* but sodium borohydride (NaBH$_4$) is required to remove the mercury (see preceding discussion). When **71** is treated with sodium borohydride in aqueous NaOH, the product is the alcohol 3-methyl-2-hexanol—the alcohol formed from the alkene with no rearrangement.

3-Methyl-1-hexene is a terminal alkene, and formation of the more stable secondary cation leads to a single major product. However, oxymercuration of an unsymmetrical internal alkene leads to a mixture of products. Oxymercuration of 3,3-dimethylcyclopentene, for example, gives a mixture of **74** and **75**. Although rearrangement did not occur, two regioisomers are formed because both of the possible carbocation intermediates are secondary. The two mercury-stabilized carbocations are essentially equal in stability, so both are formed and subsequent reaction with water leads to the mixture of alcohols shown after reduction with sodium borohydride.

Attack at C2 is more sterically hindered due to the *geminal*-dimethyl group, so **75** *may* be formed in greater amount. This observation is difficult to predict without experimental data, so *in the oxymercuration of unsymmetrical alkenes, anticipate a mixture of both possible alcohols.* In other words, *assume that oxymercuration of unsymmetrical internal alkenes will give a 1:1 mixture of two alcohols unless there is a compelling reason for one to predominate, such as electronic stabilization, severe steric hindrance, or resonance effects.* Exceptions to this assumption will rarely be encountered in this book.

10.31 Write out the acid–base reaction between **73** and 3-methyl-1-hexene. Write out the acid–base reaction between **73** and water.

10.32 Draw both carbocation intermediates derived from oxymercuration of 4,4-dimethyl-2*E*-heptene in water.

10.33 Draw the final product or products formed when 4,4-dimethyl-5-phenyl-2*Z*-pentene is treated with 1. $Hg(OAc)_2/H_2O$; 2. $NaBH_4$.

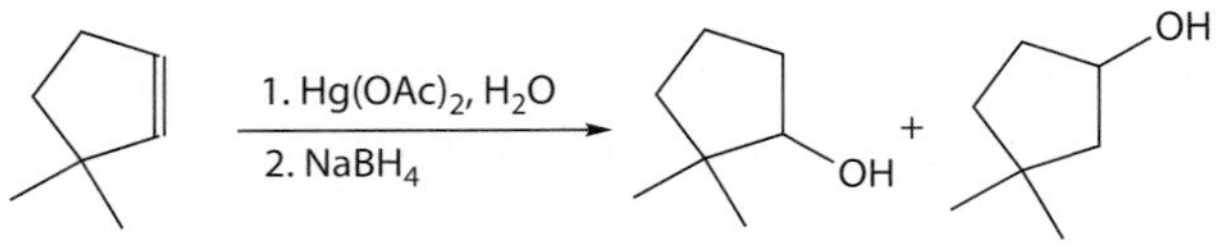

A variation of this reaction is known as ***alkoxymercuration***. It simply involves using an alcohol as the solvent rather than water. The mechanism is identical to that given before except that ROH replaces water. For example, the reaction of 3-methyl-1-pentene with mercuric acetate and ethanol, followed by reaction with sodium borohydride, leads to 2-methyl-1-ethoxycyclopentane (**76**). Many alcohols can be used in this reaction, but alcohols such as methanol, ethanol, 2-propanol, etc. that are commonly used as solvents are used most often.

10.34 Suggest a mechanism for the formation of the mercuric-alcohol precursor to **76** via alkoxymercuration.

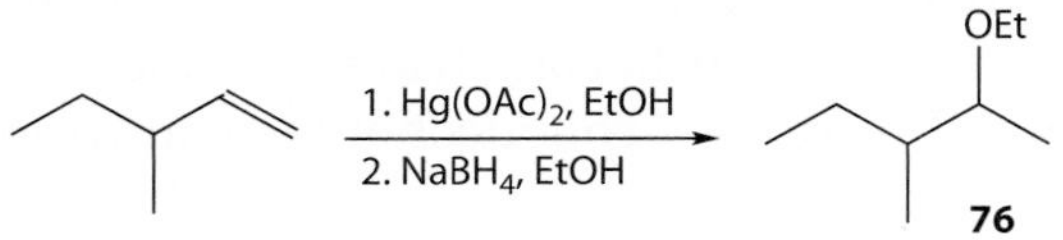

10.5 Alkenes React as Lewis Bases with Electrophilic Oxygen. Oxidation of Alkenes to Oxiranes

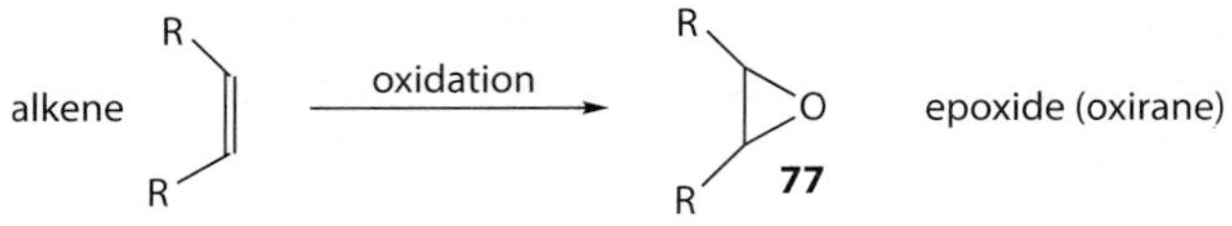

An oxirane, also known as an epoxide (**77**), is a three-membered ring ether. The highly strained three-membered ring, analogous to a cyclopropane ring (Chapter 8, Section 8.5.3), makes an epoxide highly reactive. Contrary to most other ethers, they are easily opened by nucleophiles. How are epoxides prepared? An important method for the preparation of epoxides is the oxidation of an alkene.

Much of this chapter focused on the acid–base reactions of alkenes, so it is reasonable to ask if a similar approach can be used in reactions that convert alkenes to epoxides. Formally, the conversion of an alkene to **77** is an oxidation that requires an oxidizing agent (see Chapter 17). Hydrogen peroxide (**78**) is an important oxidizing agent, and it is probably the most common peroxide. A peroxide is characterized by a weak oxygen–oxygen bond: −O−O− (see Chapter 3, Section 3.6). In a useful thought experiment, hydrogen peroxide is *conceptually* viewed as the "peroxy" derivative of water (see the **blue**-colored atoms), where the peroxide unit is the O−O bond. If water is the *imaginary* parent of **78**, then an alcohol (ROH) is the *parent* of an alkyl hydroperoxide (ROOH), **79**.

With this same analogy, the peroxide derivative of an ether (ROR), **80**, is a dialkyl peroxide (ROOR). There is also a peroxide derivative of a carboxylic acid (RCOOH), peroxycarboxylic acid (RCOOOH) **81**. Examination of the bond polarization in **78–81** reveals that one oxygen of the O−O unit is δ− and the other oxygen is δ+. Radical reactions are usually associated with the O−O bond of peroxides **79** and **80**. (See Chapter 7, Section 7.4.3, for an introduction to radicals.) In the presence of a Lewis base, bond polarization leads to ionic reactions for **81**. In this section, the Lewis base is the π-bond of an alkene.

As a rule, alkenes do not react with **78–80** unless there is another reagent present—specifically, a transition metal. This reaction will not be discussed further. In sharp contrast, peroxycarboxylic acids such as **81** react directly with alkenes. Peroxycarboxylic acids **81** are named by adding the term *peroxy* to the name of the carboxylic acid (see Chapter 5, Section 5.9.3 and Chapter 16, Section 16.4). Using the common names, the peroxy analog of formic acid is peroxyformic acid (**82**), and others include peroxyacetic acid (**83**), peroxytrifluoroacetic acid (**84**), peroxybenzoic acid (**85**), and *meta*-chloroperoxybenzoic acid (abbreviated mCPBA, **86**). Peroxycarboxylic acid **85** is a derivative of the aromatic carboxylic acid benzoic acid (PhCOOH), and the carboxylic acid precursor to **86** is clearly another aromatic carboxylic acid. (The nomenclature and structural features of benzoic acid and other aromatic carboxylic acid derivatives will be discussed in detail in Chapter 21, Section 21.2.) The salient feature of peroxyacids **82–86** is the presence of the electrophilic oxygen atom mentioned previously, which will react with an alkene.

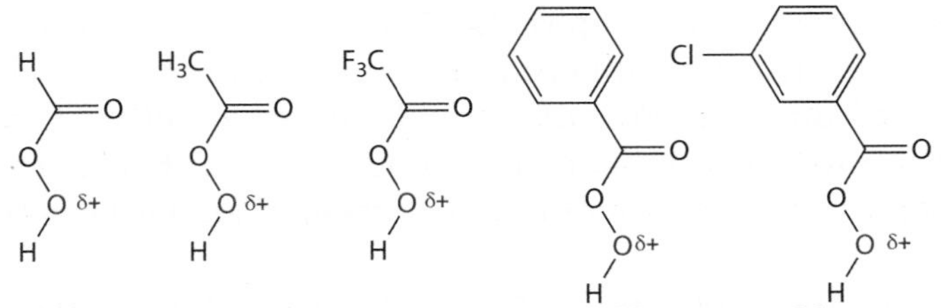

82   83   84   85   86

The reaction of peroxyacids with alkenes is illustrated by the experiment that reacts cyclopentene with peroxyacetic acid (**83**). There are *two* isolated products: cyclopentene oxide (**88**) in 57% yield and acetic acid (**89**).[13] The yield is only 57% because the remainder of the 100% is unreacted alkene. Labeling experiments (replacing the natural isotope ^{16}O with the less abundant isotope ^{18}O) show that the oxygen atom of the epoxide in **88** comes from the **OH oxygen** in the peroxyacid (labeled in **red** in **83** and **88**).

Experiments have failed to identify an intermediate for this reaction. To explain the observed products, a complex transition state is required. The oxygen atom labeled in **red** in **83** is δ+ because bond polarization extends from the carbonyl unit (C=O). That oxygen reacts as a Lewis acid in the presence of an alkene. If there is no intermediate in this reaction, it is concerted and the transition state must involve the formation of both bonds between C and **O** in **88**, as well as a transfer of electrons to form the carboxylic acid from the peroxy acid. This transition state is represented as **87**.

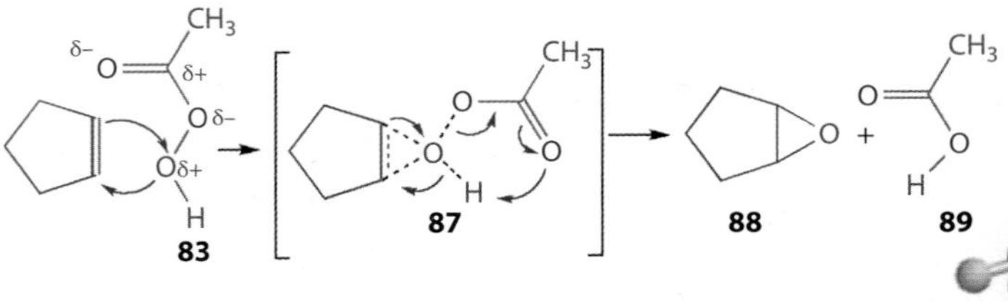

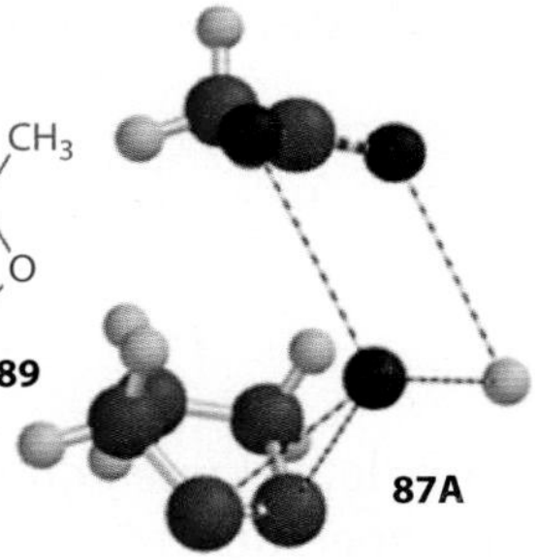

It is clear from **87** that the electron transfer process is complicated. The electronic walk-through begins with donation of two electrons from the π-bond to the electrophilic oxygen, which is accompanied by breaking the weak **O–O** bond (Chapter 3, Sections 3.6 and 3.7). The electrons from the O–O bond are transferred to the carbonyl. The carbonyl oxygen donates two electrons to the hydrogen atom (the carbonyl oxygen is a base), which forms a new O–H bond and breaks the original O–H bond, as shown in **87**. The molecular model of this transition state is shown as **87A**, in which the pertinent bonds are being made and broken. The net result of this electron transfer process is formation of a **C–O** bond to the remaining carbon of the C=C unit. In other words, the first C–O bond is formed; the other carbon of the C=C unit becomes more positive and is "attacked" by the **oxygen** as the O–H bond of the peroxyacid is broken.

This process generates the epoxide unit (the three-membered ring ether) in **88**. Formation of the epoxide and transfer of the hydrogen atom to oxygen also generate a carboxylic acid: in this case, acetic acid (**89**). *The reaction of a peroxyacid and an alkene generates an epoxide and a carboxylic acid by what is believed to be a concerted synchronous mechanism for simple alkenes that involves a "butterfly" transition state such as 87.* The carboxylic acid product is the acid "parent" of the peroxyacid (acetic acid from peroxyacetic, formic acid from peroxyformic, etc.). In some cases, the carboxylic acid by-product can react with the epoxides and generate secondary products that are unwanted. Epoxides undergo ring-opening reactions in the presence of acids, just as other ethers react (see Chapter 6, Section 6.4.3). To circumvent this problem, buffers are added to react with the carboxylic acid by-product and suppress secondary reactions. Common buffers include the salt of the carboxylic acid. Therefore, if acetic acid is the product, sodium acetate $CH_3C_3O^-Na^+$ will be added to buffer the reaction.

The epoxidation reaction works with many alkenes, but it is slow with monosubstituted (terminal) alkenes. Why do terminal alkenes react more slowly with peroxycarboxylic acids? The reaction rate (Chapter 7, Section 7.11) of two alkenes can be examined in order to determine if one is a stronger Lewis base. As a rule, the alkene best able to donate electrons should undergo epoxidation faster if the Lewis acid–base analogy is correct. Alkyl groups are electron releasing, and more alkyl substituents on a C=C unit lead to greater electron density in the π-bond, which makes that alkene a stronger Lewis base.

Therefore, a tetrasubstituted alkene ($R_2C=CR_2$) should react faster with peroxyacids than a trisubstituted alkene ($R_2C=CHR$), which reacts faster than a disubstituted alkene. Monosubstituted alkenes such as $RHC=CH_2$ react slowest in this reaction. In fact, simple alkenes such as 1-pentene can be very difficult to epoxidize under normal conditions because the reaction is very slow. In order to simplify this reaction, *assume that all alkenes given as problems in this book can be epoxidized.* However, in reality, there are vast differences in the reaction rate among various alkenes.

10.35 Draw the transition state and all products of the reaction between 2-methyl-1-pentene and peroxytrifluoroacetic acid.

10.36 Draw the products formed when 3-ethyl-2-hexene reacts with *meta*-chloroperoxybenzoic acid (**84**).

In recent years, peroxyacids have been less available than they once were. In addition, high-purity hydrogen peroxide is rather dangerous to handle and difficult to obtain. This fact has forced manufacturers to use dilute aqueous solutions of hydrogen peroxide to prepare peroxides; this gives lower purity peroxide products and lower yields of epoxides when they react with alkenes. The by-product of such reactions is a carboxylic acid, which can sometimes react with the epoxides product unless a buffer is used. Remember also that terminal alkenes are difficult to epoxidize using peroxyacids.

For all of these reasons, alternative epoxidation reagents have been developed, including a class of compounds known as **dioxiranes**. A specific example is dimethyl dioxirane (**90**). This reagent requires a co-reagent (an oxidizing agent) such as potassium peroxomonosulfate ($KHSO_5$). Commercially available Oxone® ($2KHSO_5 \cdot KHSO_4 \cdot K_2SO_4$) is a common source of $KHSO_5$. Oxone reacts with ketones and sodium bicarbonate to generate **90**, and, in the presence of an alkene, an alkene is converted to an epoxide. A typical example is the reaction of cyclohexene with Oxone and acetone (see Chapter 5, Section 5.9.2) to give cyclohexene oxide, **91**.

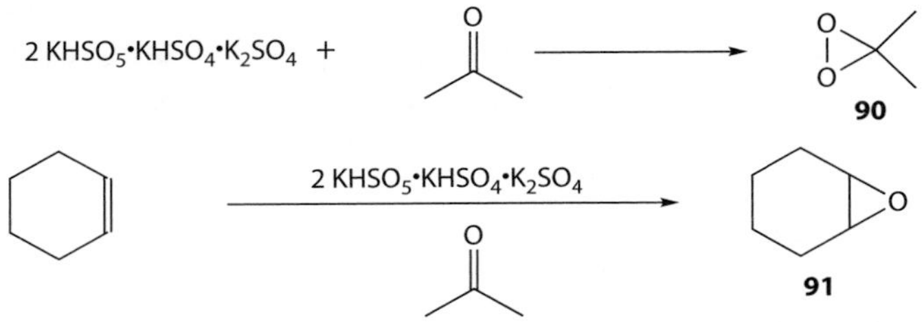

10.6 Alkynes React as Brønsted–Lowry Bases or Lewis Bases

Preceding sections focused on acid–base reactions of the C=C unit of alkenes with various acids. Because the π-bond of the alkene reacts, it is reasonable to assume that the π-bond of an alkyne may also react with these reagents, and there should be great similarities in the chemical reactivity. Indeed, this is the case. In the experiment where 3,3-dimethyl-1-butyne is treated with HCl in acetic acid, the major product is 2-chloro-3,3-dimethyl-1-butene (**92**), in up to 75% yield, although the acetic acid reacted with the product to give other products (ketones).[14] This secondary reaction will not be discussed so that the focus will be on the C=C unit in the product (**92**) and the position of the chlorine in that alkene.

When the halogen is directly attached to an sp^2 carbon of an alkene, the compound is known as a vinyl halide, so **92** is classified as a vinyl chloride. Formation of **92** is clearly similar to the alkene reaction, but it is clear from the structure of **92** that only one of the π-bonds in the alkyne reacts and the second π-bond remains in the product. Despite an obvious similarity to reactions of alkenes, the second π-bond in an alkyne leads to some interesting differences in product formation.

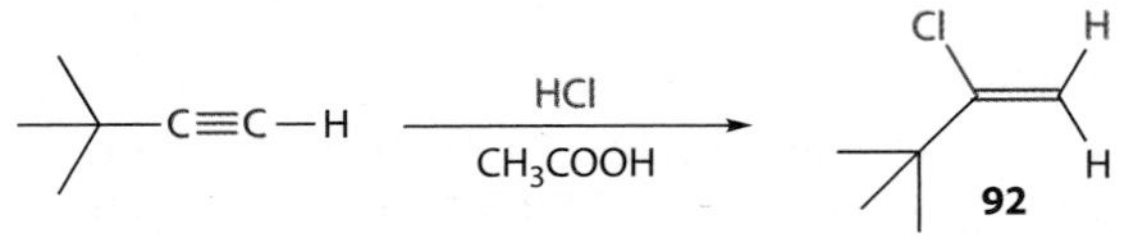

If one π-bond of an alkyne reacts as a base, it is quite reasonable to ask if it is a stronger or weaker base relative to the π-bond of an alkene. *An alkyne is a weaker base than an alkene.* It is known that one π-bond of a carbon–carbon triple bond is about 8.8 kcal mol⁻¹ (36.8 kJ mol⁻¹) weaker than the π-bond in an alkene. In other words, the alkyne π-bond has less electron density and it is less able to donate electrons to an acid (i.e., it is a weaker base). Some differences in chemical reactivity between alkynes and alkenes can be predicted by understanding this difference in base strength.

10.6.1 Alkynes as Brønsted–Lowry Bases (Reaction with Acids, HX)

Inspection of **92** shows that the chlorine is attached to the more substituted sp² carbon, so the reaction with HCl may be termed a Markovnikov addition to the triple bond. If the alkyne reacts as a Brønsted–Lowry base with HCl in a manner similar to alkenes from Section 10.2, the intermediate will be a carbocation. There are two π-bonds in an alkyne, and if only one π-bond reacts with HCl, the second π-bond of the C≡C unit should remain, meaning that a C=C⁺ intermediate must be formed. When the cationic center is on an sp² carbon, it is called a **vinyl carbocation**. In a vinyl carbocation, the positive charge resides on a carbon atom that is part of a C=C unit: a vinyl unit. The two possible vinyl carbocations are secondary vinyl carbocation **93** and primary vinyl carbocation **94**. As with any other carbocation, the secondary vinyl carbocation **93** is more stable than the primary vinyl carbocation **94**, and the more stable carbocation is formed preferentially.

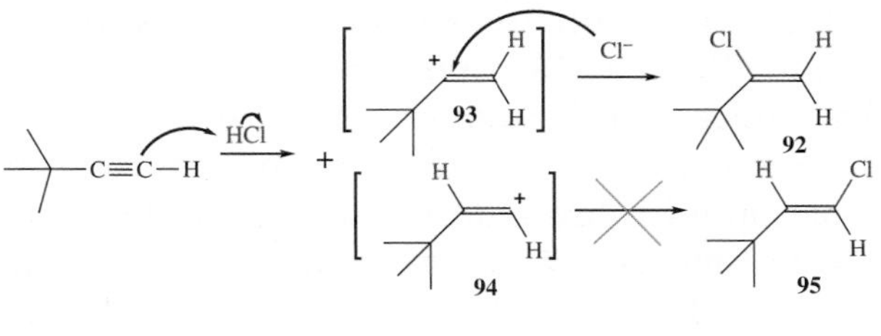

Once formed, a vinyl carbocation will react with the most available nucleophile, which in this case is the chloride ion. The product from reaction of chloride ion with **93** is 2-chloro-3,3-dimethyl-1-butene. Because experiments indicate that the major product is **92**, it is clear that the secondary vinyl carbocation **93** is a more stable intermediate when compared to the primary vinyl carbocation **94** that would give **95**. This is consistent with assumptions concerning carbocation stability.

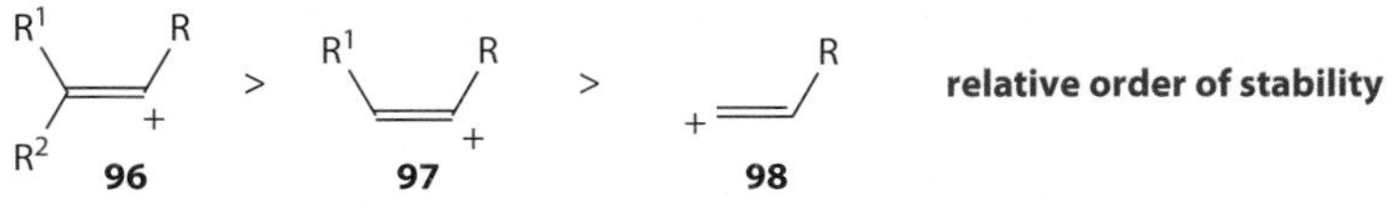

relative order of stability

The stability of vinyl carbocations generated from alkynes parallels that of the carbocations generated from alkenes; secondary vinyl carbocation **96** is more stable than secondary carbocation **97** because there are more groups attached to the C=C unit, and **97** is more stable than primary vinyl carbocation **98**. As with alkenes, the reaction of unsymmetrical alkynes with an acid will preferentially give the more stable carbocation, which reacts with the nucleophile to give the alkene product.

If carbocation **93** leads to the major product, this fact raises an interesting question. There are methyl groups on the carbon atom attached to the vinyl carbocation, and a 1,2-methyl shift would generate a tertiary carbocation. The fact that **92** is the major product indicates that rearrangement does not occur. Does this observation mean that the secondary vinyl carbocation **93** is more stable than the tertiary carbocation? No! In fact, the vinyl carbocation is significantly **less** stable than the tertiary carbocation, but that also means it is more reactive. Vinyl carbocation **93** is so reactive that trapping the nucleophilic chloride ion is much faster than any rearrangement. The take-home lesson of this example is that *vinyl carbocations do not rearrange as aliphatic carbocations do.* Once a vinyl carbocation intermediate is formed, it quickly reacts with the available nucleophile.

10.37 Draw both carbocation intermediates formed when HCl reacts with 3,3-dimethyl-1-pentyne, and draw the final major product.

10.38 Draw the product expected if 93 undergoes a 1,2-methyl shift to form a tertiary carbocation (*this does not actually occur*).

The reaction of the unsymmetrical alkyne 2-hexyne with HCl poses a regiochemical problem. Two carbocations are possible—**99** and **100**—and both are secondary vinyl carbocations and are expected to be of essentially equal stability. It is anticipated that both will be formed in roughly equal amounts, and subsequent reaction with bromide ion will give two products: **101** (3-chloro-2-hexene) and **102** (2-chloro-2-hexene). Experimentally, the reaction of HCl with 2-hexyne generates **101** and **102** in roughly a 50:50 ratio.[15] However, the reaction of the alkyne unit with HCl will generate carbocations **99** and **100** as a mixture of *E*- and *Z*-isomers in both cases. (Nomenclature for *E*- and *Z*-alkenes was described in Chapter 9, Section 9.4.3.)

When chloride ion reacts with **99**, it actually reacts with two compounds: *E*-**99** and *Z*-**99**, so two products will form: *E*-**101** and *Z*-**101**. Therefore, **101** will be a mixture of *E*-3-chloro-2-hexene and *Z*-3-chloro-2-hexene; similarly, **102** will be a mixture of *E*-2-chloro-2-hexene and *Z*-2-chloro-2-hexene. Although the isomeric *E*-alkenes are generally more stable and will account for a higher percentage of product than the isomeric *Z*-alkenes, it is difficult to predict product ratios and *assume* that the unsymmetrical alkyne will give four products.

10.39 Draw the *Z*-isomer of 99 and the *Z*-isomer of 101.

10.40 Draw all carbocation intermediates formed when HCl reacts with 2,7-dimethyl-3-octyne, and draw the final major product or products.

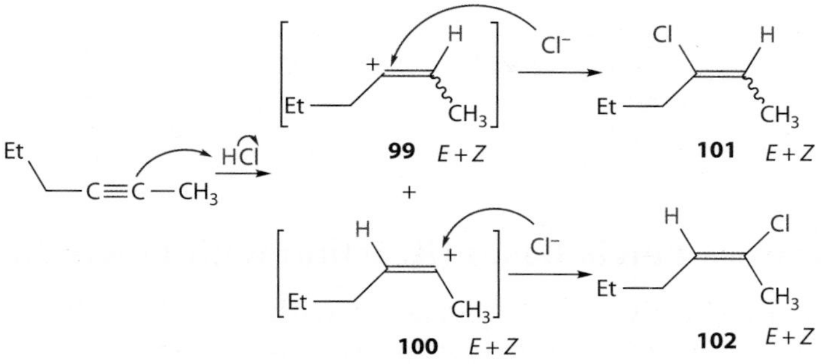

10.6.2 Alkynes as Brønsted–Lowry Bases (Reaction with Weak Acids)

Because alkynes are weaker bases than alkenes and given the discussion in Section 10.1 concerning the reactivity of alkenes, it is not surprising that alkynes do not react directly with weak acids such as water, alcohols, etc. As with alkenes in Section 10.3, alkynes can be made to react with a weak acid such as water if a strong acid catalyst is added to generate the reactive vinyl carbocation intermediate *in situ*. Therefore, *hydration of alkynes* with aqueous acid is a viable reaction.

When 1-hexyne is treated with a catalytic amount of **sulfuric acid** in an aqueous solvent, initial reaction with the **acid** gives the expected secondary vinyl carbocation **103**, and the most readily available nucleophile in this reaction is water (from the aqueous solvent). **Nucleophilic addition of water** to **103** leads to the vinyl oxonium ion **104**. Loss of a proton in an acid–base reaction (the water solvent is the base) generates a product (**105**) where the **OH unit** is attached to the C=C unit, an *enol*. Enols are unstable and an internal proton transfer converts enols to a carbonyl derivative, an aldehyde, or a ketone. This process is called *keto-enol tautomerization* and, in this case, the keto form of **105** is the ketone 2-hexanone (**106**). (Enols are discussed in more detail in Chapter 18, Section 18.5.) Note that the **oxygen of the OH** *resides on the secondary carbon due to preferential formation of the more stable secondary carbocation followed by reaction with water, and tautomerization places the* **carbonyl oxygen** *on that same carbon, so the product is a ketone.* When a disubstituted alkyne reacts with water and an acid catalyst, the intermediate secondary vinyl cations are of equal stability and a mixture of isomeric enols is expected; each will tautomerize, so a mixture of isomeric ketones will form.

10.41 Write out the acid–base reaction between water and 105.

10.42 Draw both enol products expected when **5-methyl-2-hexyne** is treated with an acid catalyst in aqueous THF. Also draw the carbonyl products expected from both enols.

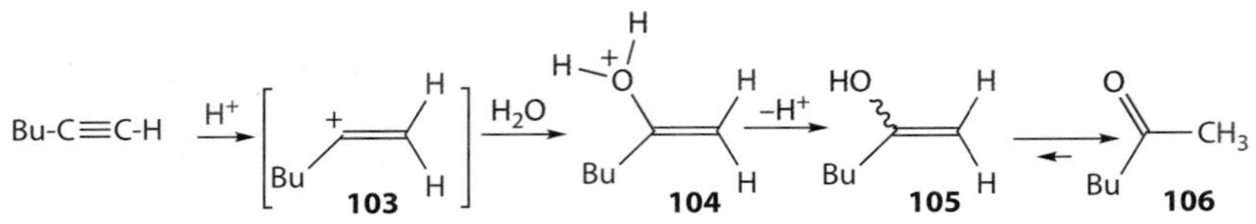

10.6.3 Alkynes as Lewis Bases (Reaction with Lewis Acids)

Oxymercuration occurs with an alkyne as with an alkene, but differences in reactivity lead to a modification in the procedure. For reasons that will not be discussed, a mixture of mercuric sulfate ($HgSO_4$) and mercuric acetate [$Hg(OAc)_2$] is used. When 1-heptyne is treated with this mixture in aqueous solvent, the initially formed enol (**107**) tautomerizes to 2-heptanone (**108**), which is isolated in 80% yield.[15] Note that the ketone product mentioned in connection with vinyl chloride **92** in Section 10.4.5 results from formation of an enol. There is an important difference in the oxymercuration of alkynes and alkenes that is notable in this transformation. The mercury reacts with the alkyne, but the mercury is lost when the enol is formed and *there is no need to add $NaBH_4$ in a second step. This observation is general for oxymercuration of alkynes under these conditions.* The more stable secondary vinyl carbocation is an intermediate, but the vinyl–mercury compound formed by reaction with the carbocation is unstable in the presence of water, so the enol is the product.

10.43 Draw the enol and the ketone expected in the reaction of **3,3-dimethyl-1-butyne** with an aqueous solution of HCl.

10.44 Draw the enol and ketone product formed when **cyclopentyl-ethyne** is treated with mercuric sulfate and mercuric acetate in water containing phosphoric acid.

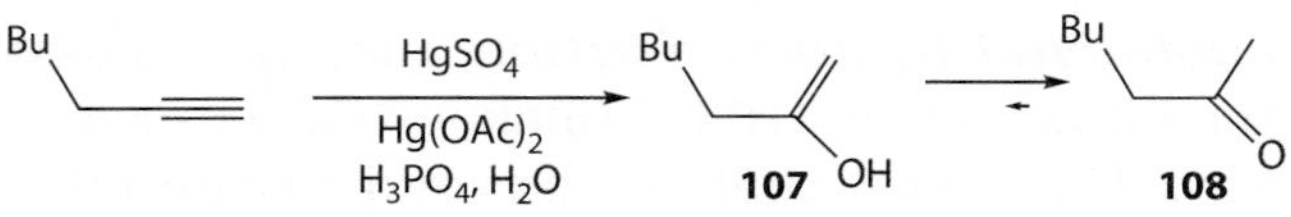

The analogy with alkene chemistry can be continued in that alkynes react with bromine, chlorine, or iodine, but only one of the two π-bonds is used. The reaction is known as ***dihalogenation of alkynes***. When 2-hexyne reacts with chlorine (Cl_2), the alkyne reacts as a Lewis base and the isolated product is the vinyl dichloride **110** (*E*-1,2-dichloro-1-pentene). Formation of this product is explained by an intermediate "vinyl-chloronium ion," **109**, which is analogous to the halonium ions formed from alkenes in Section 10.4.1. As with alkenes, the chloronium ion reacts with the nucleophilic **chloride ion** (**Cl⁻**) via *anti* attack on the face opposite the **first chlorine atom,** so the final product is *trans.* The fundamental difference between alkenes and alkynes is that the second π-bond remains after the alkyne reacts, so the chloronium ion formed by this reaction is a vinyl chloronium ion, and the final product is a vinyl dichloride.

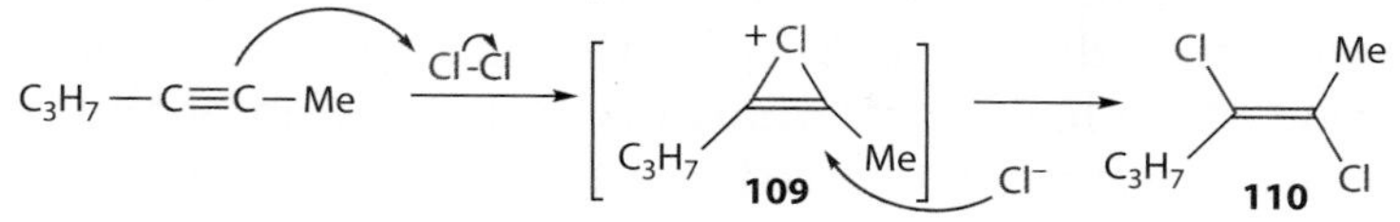

Dihalogenation of alkynes gives a dihalogenated alkene, which is also susceptible to reaction with bromine, chlorine, or iodine. Tetrahalo derivatives are available from dihalogenated alkenes (vinyl dihalides). When 1-pentyne reacts with one molar equivalent of diatomic bromine, **111** is the product. Because alkenes are also subject to reaction with halogens, **111** can react with a *second molar equivalent of bromine* to give 1,1,2,2-tetrabromopentane, **112**.

Alternatively, 1-pentyne can be treated directly with *two molar equivalents of diatomic bromine* to give **112**. In this latter case, the process is stepwise, where **111** is formed *in situ* (within the reaction medium) and then reacts with the excess bromine. Dibromide **111** may also react with a different halogen, such as diatomic chlorine, to give 1,2-dibromo-1,2-dichloropentane, **113**. A number of highly halogenated alkanes may be prepared in this manner, beginning with alkyne starting materials and using two equivalents of one halogen or using one halogen followed by a different one.

10.45 Draw the intermediate and final products formed when 1-cyclohexyl-1-butyne reacts with iodine.

10.46 Draw the product or products formed when 1-phenyl-1-butyne reacts with one molar equivalent of iodine and then with one molar equivalent of bromine.

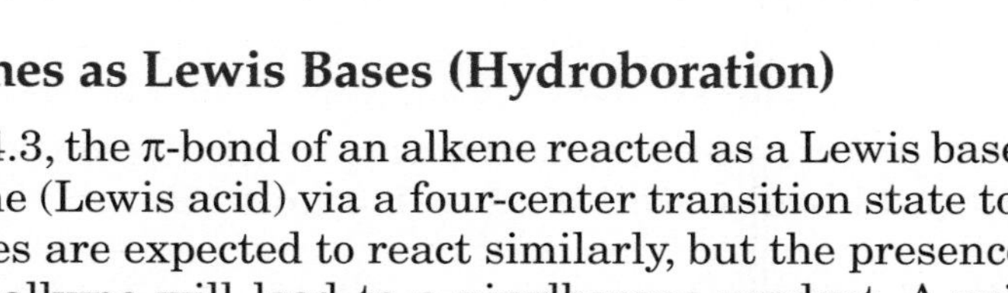

10.6.4 Alkynes as Lewis Bases (Hydroboration)

In Section 10.4.3, the π-bond of an alkene reacted as a Lewis base with the H–B unit of a borane (Lewis acid) via a four-center transition state to give an alkylborane. Alkynes are expected to react similarly, but the presence of the second π-bond of the alkyne will lead to a vinylborane product. A primary vinylborane can be ($RCH=CHBH_2$), but this can react with additional alkyne to give a secondary vinylborane, $(RCH=CH)_2BH$. In principle, a tertiary vinylborane may form. For the sake of simplicity, this discussion will assume formation of a primary vinylborane, but it is likely that the product is actually a secondary vinylborane. An example is the reaction of 1-pentyne (a terminal alkyne) with BH_3, where a mixture of two vinylborane regioisomers can be formed **114** and **115**, but the major product is **115**. As with the reaction of alkenes, the four-center transition state leads preferentially to formation of the less sterically hindered product in which boron attaches to the less substituted carbon to give **113**.

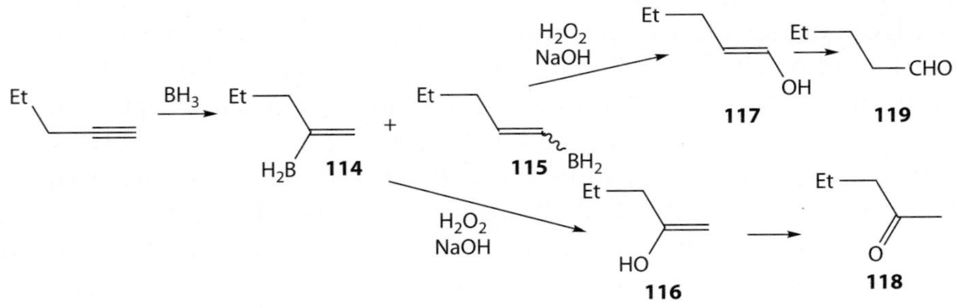

When the alkylboranes formed by reaction with an alkene are oxidized with $NaOH/H_2O_2$, the product is an alcohol. When the vinylborane unit is similarly oxidized in **115** (or **114**), replacement of boron with the OH gives an enol product, **117** or **116**, which tautomerizes to the corresponding carbonyl derivative (see Section 10.4.5 for the similar reaction in oxymercuration). Enol **116** tautomerizes to 2-pentanone (**118**) and **117** to butanal (**118**). Because vinylborane **115** is the most prevalent vinylborane, aldehyde **119** is the major isolated product after oxidation. Note the anti-Markovnikov orientation of the boron and the oxygen, analogous to the regioselectivity shown by hydroboration of alkenes.

The hydroboration of disubstituted (internal) alkynes leads to a mixture of two ketones. When 2-pentyne reacts with borane, two vinylboranes are formed in virtually equal amounts: **120** and **121**. There is no significant difference in steric hindrance to make one transition state favored over the other, so both vinylboranes are formed. Subsequent oxidation leads to the isomeric ketone products (from their respective enols): 2-pentanone (**122**) and 3-pentanone (**123**). To conclude, *hydroboration of a terminal alkyne leads to an aldehyde as the major product, whereas hydroboration of an internal alkyne gives a mixture of two isomeric ketones.*

10.47 Draw the 2 four-center transition states from reaction of 1-pentyne and borane that lead to 114 and 115.

10.48 Draw the final product or products formed when 4-phenyl-1-butyne is treated with (a) 9-BBN; (b) H_2O_2, NaOH.

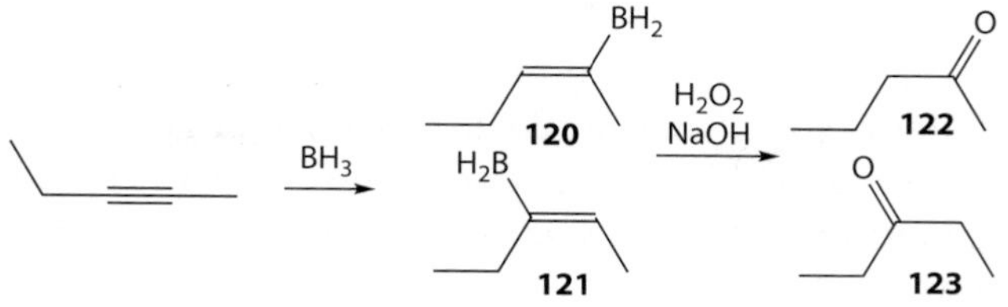

10.7 Reactions That Are Not Formally Acid–Base Reactions

In several of the reactions in previous sections, one heteroatom (halogen, oxygen) was added to one side of a π-bond and a hydrogen or another heteroatom to the other side. There is another type of reaction in which heteroatoms are incorporated on both carbons of a π-bond. A specific example incorporates two hydroxyl (OH) units. In one experiment, cyclohexene is treated with OsO_4 (osmium tetroxide) in anhydrous *tert*-butyl alcohol (2-methyl-2-propanol) at 0°C. After standing overnight, a 45% yield of *cis*-1,2-cyclohexanediol (**124**) was obtained.[16] Analysis of this reaction shows that two OH units added and that both oxygen atoms were derived from osmium tetroxide. This reaction is termed a ***dihydroxylation***. Furthermore, the two OH units have a *cis* relationship.

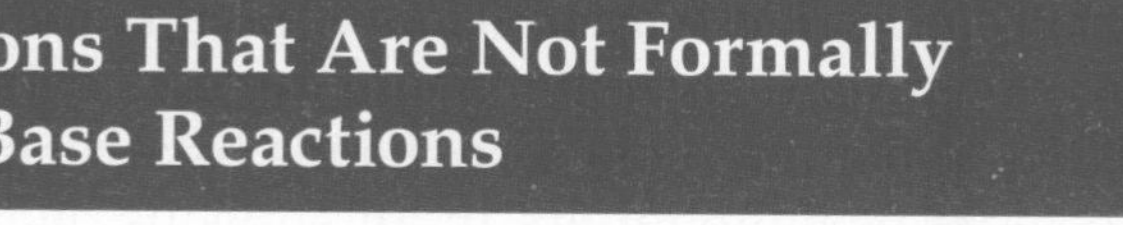

10.7.1 Dihydroxylation of Alkenes

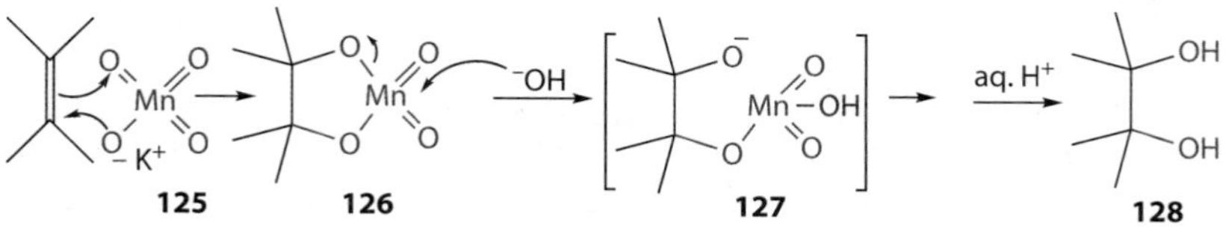

Osmium tetroxide is not the only reagent that converts alkenes to 1,2-diols (the reaction is called *dihydroxylation*). The reaction of an alkene with dilute aqueous potassium permanganate ($KMnO_4$) does the same thing. This discussion will begin with potassium permanganate and then move on to osmium tetroxide. The actual mechanism of these reactions is different from anything seen so far. The permanganate ion is drawn out as **125**, and it reacts with the

C=C unit of an alkene by what is known as 1,3-dipolar addition (a *cycloaddition reaction*—a pericyclic reaction governed by *molecular orbital theory,* which is introduced in Chapter 24, Section 24.1). A discussion of this mechanism will not be presented, but to explain dihydroxylation, it is necessary to understand that the initial reaction converts the alkene to the cyclic product, **126**, by a *concerted process*.

An example is the reaction of 2,3-dimethyl-2-butene with potassium permanganate to make two C–O bonds. This product (**126**) is called a *manganate ester*. The concerted nature of the reaction of the alkene is important because **125** reacts to form **126** in such a way that *the rate of bond formation for each C–O bond is close, so they have a cis relationship.* Indeed, cyclic product **126** has a *cis* relationship of the two atoms that form bonds to carbon. An alkene is normally added to potassium permanganate in an aqueous hydroxide solution.

To convert **126** to the diol product, the hydroxide ion attacks the manganese, donating two electrons to Mn to form an O–Mn σ-bond and opening the ring as shown to give **127**. In the presence of water and hydroxide, further attack at manganese followed by hydrolysis of the reaction (a second reaction with aqueous acid) gives a 1,2-diol, in this case 2,3,-dimethyl-2,3-butanediol **128**. The overall transformation is conversion of the C=C unit of an alkene to a 1,2-diol (a vicinal diol).

It is important to understand that the reaction of potassium permanganate to give a diol must be done in a relatively dilute solution (typically, 0.1–0.5 *M*) and the temperature must be kept relatively cold (usually room temperature or lower). If the concentration is too high and the temperature too great, oxidative cleavage (Chapter 17, Section 17.4) can occur and a variety of products can be formed, so there is much less diol and more unwanted by-products. The terms "hot" and "cold" "high" or "low" concentration are relative, and the actual values of these parameters depend on the particular alkene in the reaction. In general, however, cold and dilute conditions will lead to the diol, whereas hot and concentrated conditions lead to oxidative cleavage (much like the final products observed with the ozonolysis discussed in the next section).

10.49 **Draw the intermediate and final products formed when cycloheptene is treated with KMnO$_4$ in the presence of aqueous NaOH followed by hydrolysis.**

129   130   124

Osmium tetroxide (OsO$_4$, **129**) also reacts with an alkene via 1,3-dipolar addition to give what is known as an osmate ester, **130**. If the reaction is done in an aqueous sodium thiosulfite (NaHSO$_3$) or *tert*-butanol, the final product is a vicinal diol analogous to the reaction of permanganate and hydroxide. The reaction of cyclohexene and osmium tetroxide, for example, gives osmate ester **130** with a *cis* relationship of the two C–O bonds in the five-membered ring. In the presence of NaHSO$_3$, the osmate ester is converted to the diol, **124**. The overall transformation of C=C to diol is identical to the reaction of permanganate, although the intermediates and reaction conditions are different. Osmium tetroxide is more expensive than permanganate, but the yields tend to be higher and there are fewer by-products.

If the reaction of an alkene with OsO$_4$ is done in the presence of *tert*-butylhydroperoxide (Me$_3$C–OOH) or *N*-methylmorpholine-*N*-oxide (**131**, NMO), the reaction is catalytic in osmium. Both reagents react with the osmate ester to generate the diol, and they also regenerate the OsO$_4$ reagent. Therefore, if 1-pentene reacts with a 5% aqueous solution of OsO$_4$ that contains NMO, the final product is pentane-1,2-diol, **132**. This variation of the reaction is superior to that described for cyclohexene and, when **131** is used, as little as 1% of osmium tetroxide can be used.

For the reaction of either potassium permanganate or osmium tetroxide with an alkene, the manganate ester or the osmate ester is formed in what is effectively a *cis* addition. Cyclohexene gives the *cis* diol, **124**, and there is none of the *trans* diol, which means that **dihydroxylation is a diastereospecific reaction**; of the two diastereomers (*cis* and *trans*), only the *cis* product is formed. Note that dihydroxylation occurs from both sides of the C=C unit, so the reaction gives both enantiomers. Only one diastereomer is formed (diastereospecific), but that diastereomer is racemic.

The overall *cis* addition of two OH units to the C=C unit of the alkene is easy to see when a cyclic alkene is used, but the reaction with an acyclic alkene is also diastereospecific. If hex-2*E*-ene (**133**) reacts with OsO$_4$, osmate ester **134** is formed where the stereochemistry of the two alkyl groups attached to the C=C unit has been preserved in the *cis* addition. There is no facial bias, so both enantiomers of the osmate ester are formed: **134** and **135**. In other words, formation of the osmate ester can occur from either the top or the bottom, as shown, so the final product will be racemic. Subsequent reaction with NMO leads to diols **136** and **137**, which have the (2*S*,3*S*) and (2*R*,3*R*) absolute configurations. Similarly, the reaction with permanganate via the manganate ester gives the *cis* diol.

10.50 Draw the product formed by treatment of 3-phenyl-2Z-pentene with (a) OsO$_4$; (b) aq. NaHSO$_3$.

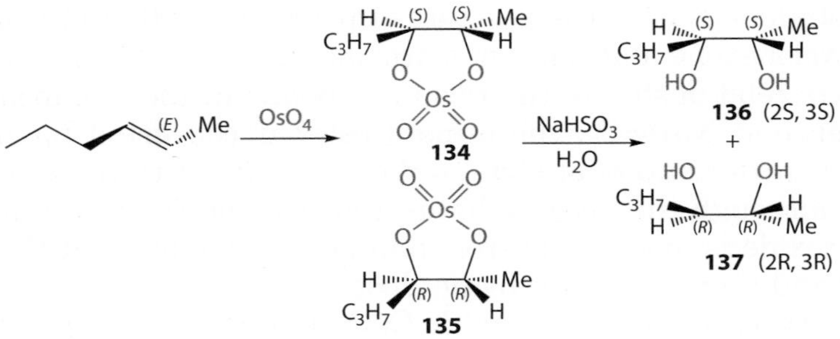

10.7.2 Ozonolysis

Ozone (O_3) is another common molecule that reacts as a 1,3-dipole. Ozone is formed by electrical discharge (lightning) in the atmosphere (which contains O_2). This process can be mimicked in the laboratory by discharging a spark in an oxygen stream under controlled conditions. Ozone is a resonance-stabilized species that has the structure **138**. Two of the resonance forms have the positive and negative charge at the terminal oxygen atoms, making them **1,3-dipoles**, and they can be compared with the permanganate ion and osmium tetroxide as a 1,3-dipole. It should be noted, however, that the single bond resonance structures of **138** probably contribute very little to the structure and are used here to keep the analogy with permanganate and osmium tetroxide.

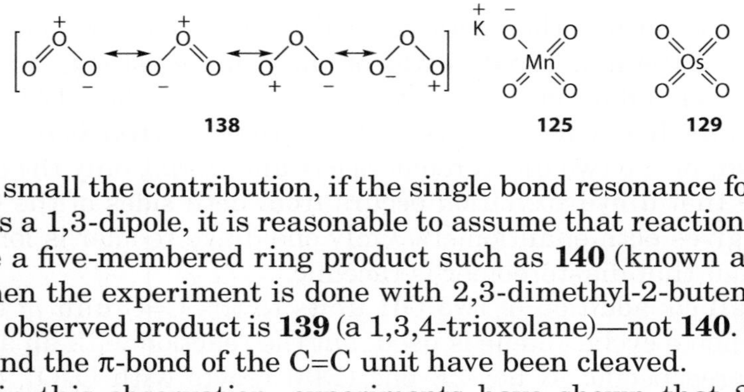

However small the contribution, if the single bond resonance forms of ozone are viewed as a 1,3-dipole, it is reasonable to assume that reaction with an alkene will give a five-membered ring product such as **140** (known as a 1,2,3-trioxolane). When the experiment is done with 2,3-dimethyl-2-butene and ozone, however, the observed product is **139** (a 1,3,4-trioxolane)—not **140**. Clearly, both the σ-bond and the π-bond of the C=C unit have been cleaved.

To explain this observation, experiments have shown that 2,3-dimethylbutene reacts with ozone to give **140** by initial dipolar addition; however, a rearrangement occurs to give **139**, which is commonly called an **ozonide**. When **139** is treated with aqueous hydrogen peroxide in a second chemical step, two molar equivalents of acetone are formed. The overall transformation cleaved both bond of the C=C unit, converting each carbon to a carbonyl. The overall process is called **oxidative cleavage** and will be elaborated in Chapter 17, Section 17.4.

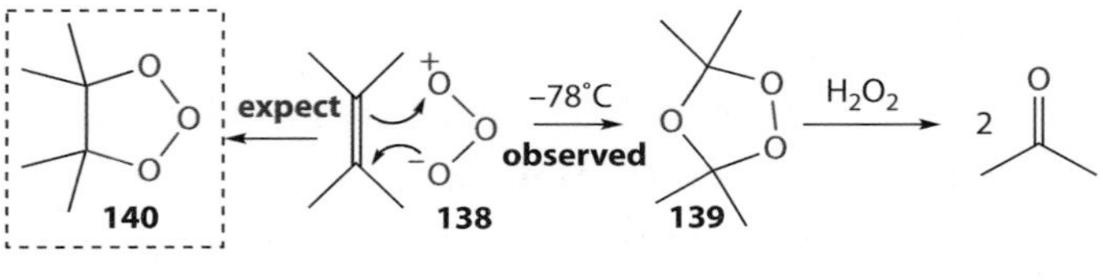

   The experimentally determined mechanism for this transformation is called the *Criegee mechanism,* after Rudolf Criegee (Germany; 1902–1975), who proposed it. It begins with a 1,3-dipolar addition of ozone (**138**) to the alkene to give the 1,2,3-trioxolane, **140**. Experiments have shown that this product is very unstable, even at temperatures as low as –78°C, and one of the weak O–O bonds breaks to give an ionic species, **141**. The alkoxide unit in **141** initiates cleavage of the C–C bond, leading to two intermediates represented by **142**. (This type of reaction will be seen again in Chapter 18, Sections 18.1 and 18.2.)

   These two molecules could drift apart, but dipole–dipole interactions keep them close, particularly at low temperatures, and the alkoxide unit of one fragment subsequently attacks the acyl carbonyl of the other to give **143**. Once formed, the alkoxide in **143** rapidly reacts with the C=O carbon to complete the reorganization (rearrangement) to the ozonide, **139**. *The reaction works because reaction of the alkoxide oxygen in **142** with the carbonyl carbon (acyl addition; see Chapter 18) occurs faster than the separation of these two molecules.* This fact is not intuitively obvious, but it explains the experimental results and it has been confirmed with other experiments.

   If the focus is on the C=C unit of 2,3-dimethyl-2-butene, the dipolar addition reaction cleaves the first bond (the π-bond), and the rearrangement of the 1,2,3-trioxolane to the 1,2,4-trioxolane cleaves the second bond (the σ-bond). The conversion of an alkene to an ozonide is known as ***ozonolysis***, and it is an example of an ***oxidative cleavage reaction***. The ozonide is usually not isolated, but a second chemical step is performed in the same flask. When treated with hydrogen peroxide, the products of this reaction are 2-propanone (acetone) and a second molecule of acetone. This statement is phrased this way because an unsymmetrical alkene will give two different ketones. In effect, the C=C unit is cleaved and each carbon is oxidized to a C=O unit. The mechanism described for this reaction is consistent with known chemistry of ketones and aldehydes and other carbonyl-bearing functional groups. A full discussion of carbonyl chemistry will be presented in Chapters 17 and 19.

10.51 **Draw the 1,2,3-trioxolane and ozonide expected when 2-methyl-2-hexene reacts with ozone.**

   Many alkenes have a hydrogen atom attached to the C=C unit, and this structural feature causes a complication in the ozonolysis sequence. When

2-methyl-2-pentene reacts with ozone, the initial 1,2,3-trioxolane product is **144**, but this rearranges to ozonide **145**. If **145** is treated with hydrogen peroxide as previously, one of the cleavage products is acetone, as expected; however, the other product is propanoic acid. This result appears to be unusual and, in the ozonolysis of 2,3-dimethyl-2-butene, one might expect that treatment of ozonide **145** with hydrogen peroxide should give acetone and the aldehyde. *However, aldehydes are susceptible to oxidation by many reagents, including oxygen in the air (see Chapter 17 for oxidations).*

If a simple aldehyde such as propanal is exposed to air, it is oxidized to a carboxylic acid, propanoic acid. If air can oxidize an aldehyde, then formation of propanal in the presence of a stronger oxidizing agent such as hydrogen peroxide will also oxidize the aldehyde to a carboxylic acid. Therefore, *if an ozonide is formed from an alkene that has a hydrogen on a C=C carbon and then treated with an oxidizing agent such as hydrogen peroxide, the final product is a carboxylic acid—not an aldehyde.*

Is it possible to isolate the aldehyde product from ozonide **145**? It was discovered that if **145** is treated with dimethyl sulfide, the products are acetone and propanal, rather than acetone and propanoic acid. Hydrogen peroxide is an oxidizing agent and dimethyl sulfide is a reducing agent when used with an ozonide. *When there is a hydrogen atom on the C=C unit, ozonolysis followed by oxidation leads to a carboxylic acid, but ozonolysis and then reduction lead to an aldehyde. The presence of a hydrogen atom on the C=C unit, and thereby in the ozonide, is the key to predicting the difference between oxidation or reduction of an ozonide.* Zinc and acetic acids have also been used to reduce the ozonide to an aldehyde.

10.52 **Draw the product or products formed when 2-methyl-3-phenyl-2-pentene is treated with (a) O$_3$; (b) Me$_2$S.**

10.53 **Draw the ozonolysis sequence that converts 3-methyl-3-heptene to 2-butanone and butanal.**

Ozonolysis of a cyclic alkene leads to oxidative cleavage, but the two carbonyl fragments are connected, so there is only one product rather than the two observed from ozonolysis of 2,3-dimethyl-2-butene or 2-methyl-2-pentene. When cycloheptene, for example, is treated with ozone and then with zinc and acetic acid, oxidative cleavage leads to the α,ω-dialdehyde **146** (1,7-heptanedial or 1,7-heptanedicarboxaldehyde).

10.54 **Draw the product formed when 1-methyl-1-cyclohexene is treated with (a) O$_3$; (b) H$_2$O$_2$.**

Most of the reactions prior to ozonolysis have built molecules, whereas ozonolysis breaks molecules apart. Why is this useful? In the past, identification of complex molecules relied on chemical reactions rather than on spectroscopic methods (see Chapter 14), and ozonolysis is a convenient method to simplify complex molecules. If there is an alkene unit in the molecule, cleavage will generate two fragments that may be easier to identify. In addition, formation of **146** from cycloheptene is an important process. If one requires an α,ω-disubstituted molecule such as **146**, how can it be made? For the most part, chemical reactions occur within two to three bonds of a functional group, but in **146** the two aldehyde units are separated by five carbon atoms. Cleavage of a cyclic alkene, however, easily generates the α,ω-disubstituted derivative, and ozonolysis is an important method for the preparation of such compounds.

10.8 Non-ionic Reactions: Radical Intermediates and Alkene Polymerization

In the early sections of this chapter, alkenes reacted with acids to form a carbocation. Once the carbocation is formed, it reacts with other electron-donating species, including the alkene itself. Continuous reaction of an alkene to form a new carbocation allows an alkene to continue reaction until a large molecular-weight material known as a polymer is generated. Alkenes also react with radicals to give new compounds. (See Chapter 7, Section 7.4.3, for the structure of radicals.) When an alkene reacts with a radical, the product is another radical. This process can be controlled in many cases to produce polymers. This section will serve as a brief introduction to the chemistry of radicals.

10.8.1 Formation of Radicals

A carbon radical can be viewed as a trivalent species containing a single electron in a p-orbital (described in Chapter 7, Section 7.4.3). A radical contains one electron in an orbital, and it can theoretically be tetrahedral, planar, or "in between." It is generally conceded that carbon radicals without significant steric encumbrance are planar, as represented by **147**. In terms of its reactivity, radical **147** may be considered electron rich or electron poor. In most of its reactions, the electron-deficient characterization is useful for predicting products. In other words, *radicals such as this are not nucleophilic.*

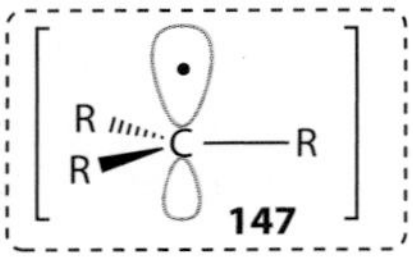

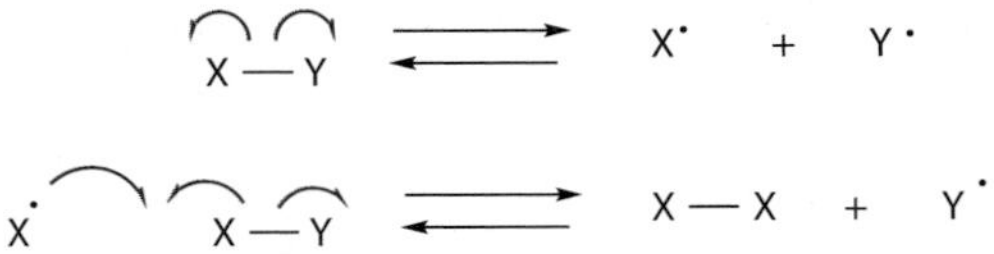

Radicals can be formed in several ways. Many involve dissociative **homolytic cleavage** (one electron is transferred to each adjacent atom from the bond) as a key step, as depicted by X–Y, leaving two radical products. (Homolytic cleavage is introduced in Chapter 7, Section 7.4.3.) Another major route to radical intermediates involves the equilibrium reaction of a radical (X•) and a neutral molecule (X–Y), producing a new radical (Y•) and a new neutral molecule (X–X). The equilibrium constant (Chapter 7, Section 7.10) will determine if there is a significant concentration of the radicals, and the position of the equilibrium depends on both the relative bond strength of X–Y and the relative stabilities of X• and Y•. Raising the temperature of the reaction usually shifts the equilibrium toward a higher concentration of radicals in homolytic cleavage reactions.

A number of reagents generate radicals when heated or exposed to light (the symbol for a photon of light is hv). (Peroxides were introduced in connection with epoxidation reactions in Section 10.5.) When heated, many peroxides such as alkyl hydroperoxide **79** and dialkylperoxide **80** undergo homolytic cleavage to generate radicals RO• + •OH or two molar equivalents of RO•, respectively. One example is the heating of *tert*-butylhydroperoxide (**148**) to give the *tert*-butoxy radical and the hydroxyl radical. Another example is dibenzoyl peroxide (**149**), which gives two molar equivalents of the acyl radical **150** when heated. Peroxide **149** is classified as a diacyl peroxide, but acyl compounds of this type will be discussed in Chapter 16 (Section 16.7) in the context of carboxylic acid derivatives. A useful reagent that will generate radicals is an **azo compound** called *azobis*-isobutyronitrile (AIBN, **151**). When heated, AIBN easily decomposes to produce two molar equivalents of the radical **152**, along with nitrogen gas, which escapes from the reaction. Once a radical is formed, it can react with another molecule that is in the reaction medium.

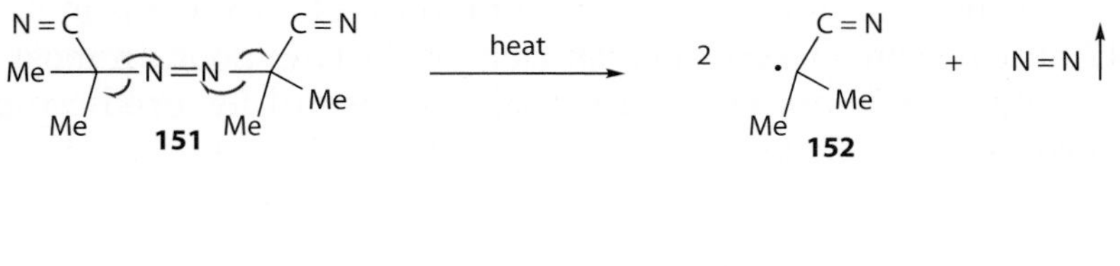

10.8.2 Anti-Markovnikov Addition to Alkenes

HBr reacts with an alkene to give the more substituted (and more stable) carbocation intermediate, and the nucleophile is incorporated at that position (see Section 10.2). In experiments designed to further probe this reaction, hydrogen bromide (HBr) is added to undec-10-enoic acid (**153**) in a hydrocarbon solvent, but benzoyl peroxide (**149**) is added to the reaction. When the product is isolated, 11-bromoundecanoic acid (**154**) is obtained in 70% yield.[17] The bromine is attached to the *less substituted carbon*. Because Markovnikov's rule places the hydrogen atom on the less substituted carbon atom of the C=C unit and the bromine on the more substituted, formation of **154** is exactly the opposite result—an ***anti-Markovnikov addition.***

In order to understand this reaction, first probe the nature of the mechanism. Does the reaction proceed by a carbocation? The answer is certainly no! If a carbocation were formed, the reaction must follow Markovnikov addition, consistent with formation of the more stable secondary carbocation. Therefore, the reaction must follow a different mechanism, presumably with a different intermediate, but what? *Note that a peroxide is added, and this appears to be responsible for changing the normal reaction of HBr and an alkene.*

If the mechanism is different (suggested by formation of a different product), then some chemical process must occur *before* HBr can react with the alkene. Logically, this event involves the new additive, the peroxide, which is known to undergo homolytic bond fragmentation to produce radicals. If radicals are involved, does the bromine go in first or second? If a secondary radical is assumed to be more stable than a primary radical (as with carbocations, which are electron deficient), the most reasonable mechanism generates a bromine radical (Br•), which reacts with the C=C unit of the alkene. The bromine must add to the C=C unit before the H in order to generate a secondary radical, and this will place bromine on the less substituted carbon.

The mechanism is shown and it begins with heating **149** to give the radical, which reacts with HBr to form benzoic acid and the **bromine radical, Br•**. The **bromine radical** reacts with the alkene unit in **153** to give a secondary carbon radical rather than a less stable primary radical. The secondary radical is expected to be more stable than a primary radical, for the same inductive-effect reasons used to discuss carbocation stability in Section 10.2. In the radical process, a covalent bond is formed, but there are two electrons in the new σ-bond. One electron is donated by **Br•** and the other by the C=C π-bond, leaving behind an unpaired electron on the carbon, a carbon radical.

All that remains is for radical **155** to react with the hydrogen atom from another molecule of **HBr** to give **154** and another equivalent of the **bromine**

radical. Note that if two **bromine radicals** react, they will generate the neutral molecule bromine; the radical process is stopped and must be initiated again before any additional product is produced. This mechanistic sequence is described as a ***chain radical reaction***. It is important to note that when a reaction generates a radical product, the process continues; when the reaction generates a neutral product, the radical reaction stops.

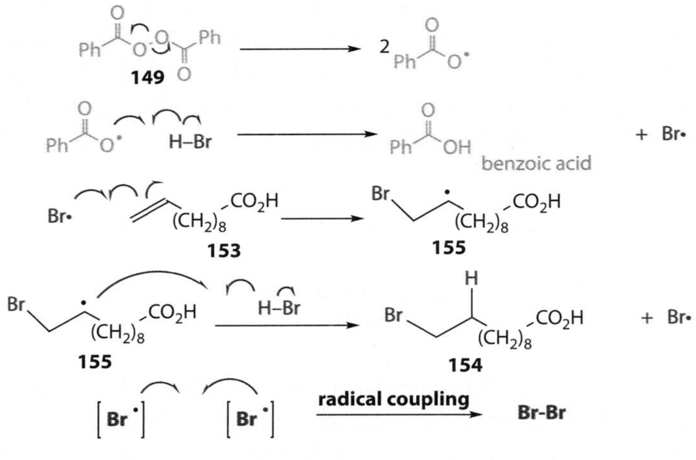

Initial cleavage of the peroxide first generates the radical. The reaction does not proceed to products until a radical is formed, so this is called the chain ***initiation step***. The entire process is called a radical chain reaction because once the radical is formed, reactions continue to form a product until there are no longer reactive radicals. The bromine radical reacts with the alkene and the product is another radical, **155**, that is not isolated. Radical **155** reacts with HBr to remove the hydrogen atom and give the neutral product **154**, along with another Br•. Each step in the sequence that produces a reactive radical that goes on to react with either HBr or alkene is called a chain ***propagation step*** because production of product is propagated. When there are no more radicals, the reaction sequence stops. Therefore, the step where two radicals react to produce a neutral molecule (Br_2) is called a chain ***termination step***.

10.55 **Draw the final product of the following reaction sequence: cyclopentylethene + benzoyl peroxide in the presence of HBr.**

The reaction of an alkene with HBr and peroxide gives the bromide product with the bromine on the less substituted carbon in an anti-Markovnikov addition reaction. *This reaction work wells with HBr, but not with HCl or HI because the bromine radical reacts in a selective manner. Differences in the reactivity of halogen radicals are addressed in Chapter 11, Section 11.9.*

One additional example of this reaction can be used to illustrate an important difference between addition reactions that proceed by radicals and those that proceed by carbocation intermediates. The reaction of 3-methyl-1-pentene and HBr in the presence of di-*tert*-butyl peroxide ($Me_3COOCMe_3$, another peroxide radical precursor) leads to **157** via a radical chain mechanism and radical intermediate

156. Note that, unlike carbocation intermediates, *radical intermediates such as 156 do not rearrange.* By contrast, when 3-methyl-1-pentene reacts with HBr without the peroxide, reaction gives the expected secondary carbocation **158**, which rearranges to the more stable tertiary carbocation **159**. Final coupling with bromide ion gives **160**. Therefore, reaction with HBr via the carbocation leads to **160** and reaction with HBr and peroxide via the radical leads to **157**.

Although radicals such as **156** do not rearrange, the carbon radical can extract a hydrogen atom from another atom in the molecule, leading to a different radical. Such internal atom-exchange reactions of radicals may appear to be rearrangements because the radical is at a different position, but they are not. For the examples discussed in this book, *it is assumed that all carbon radicals do not undergo rearrangement or atom transfer reactions.*

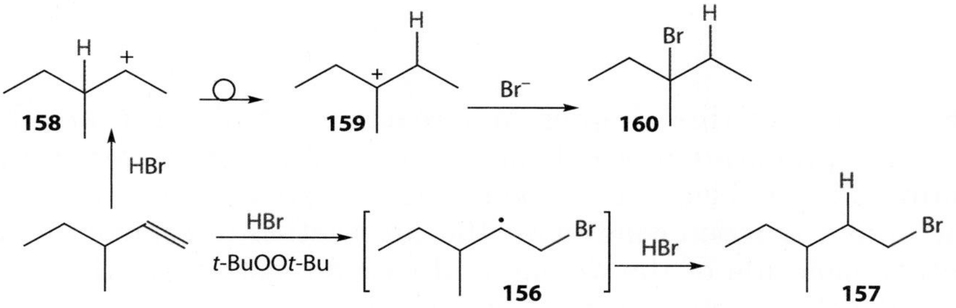

10.8.3 Alkene Polymerization

Radicals react with an alkene to form a carbon racial (see **156**). When the C=C unit of an alkene reacts with a radical X• to form a radical intermediate (X–C–C•), this radical can react with another molecule of the alkene to form another radical (X–C–C–C–C•). In this chain propagation reaction, another reaction with more alkene leads to X–C–C–C–C–C–C•. If this process continues, the final product will be a large molecule known as a ***polymer,*** $-(C–C)_n-$, where n is a large number, such as 500 or 2,000, that represents the number of times that unit is repeated.

In other words, there are 500 repeating C–C units or 2,000 repeating C–C units. ***A polymer is simply a high molecular weight substance***. A *low polymer* will have a molecular weight below about 10,000–20,000, whereas a *high polymer* will have a molecular weight between 20,000 and several million. A **monomer** is defined as the single molecule precursor to a polymer. A monomer is polymerized to form a polymer. An alkene such as **161** is considered to be a monomer if polymerization converts it to polymer **162**.

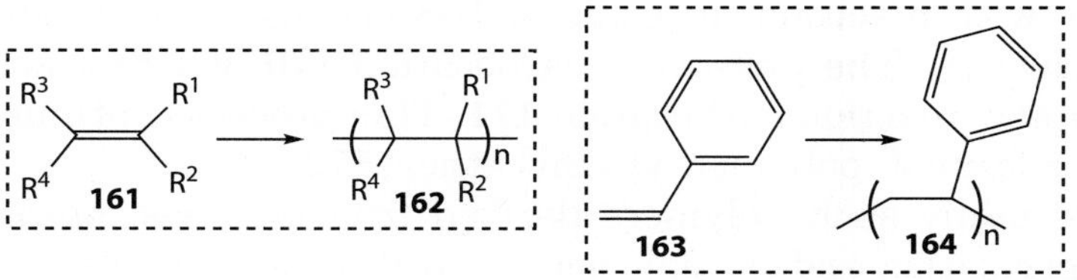

A specific example uses purified styrene (**163**) heated in a special receptacle at 125°C for several days. The product is polystyrene, **164**. The n in **164** indicates the number of times the styrene monomer unit is repeated, which varies with the way the polymer is formed. Under the reaction conditions just described, the polymer has a molecular weight of about 150,000. If the monomeric C_8H_8 unit (mass = 104) repeats, a polymer weight of 150,000 would correspond to $n \approx (150{,}000/104)$, which is $\approx 1{,}440$. When benzoyl peroxide (**149**) is heated in the presence of ethene (ethylene), radical addition will continue until the ethene is consumed, and the final product is polyethylene, **165**, which is an example of a linear polymer.

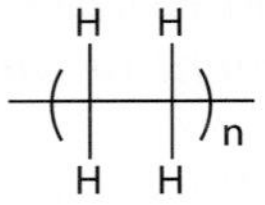

165

In the context of this chapter and reactions of alkenes, an alternative method for the preparation of polymers is possible. In the presence of strong acids, many alkenes form carbocation intermediates, as in Section 10.2, but under the proper reaction conditions, the carbocation intermediate may react with another molecule of the alkene and cationic polymerization is possible. Alkenes that polymerize under cationic conditions usually contain electron-releasing groups attached to the C=C unit. Examples are substituted alkenes such as isobutylene (**166**; 2-methyl-1-propene) or methyl vinyl ether (**168**). The reaction of isobutylene (**166**) and sulfuric acid leads to poly(isobutylene), **167**, which is also known as butyl rubber.

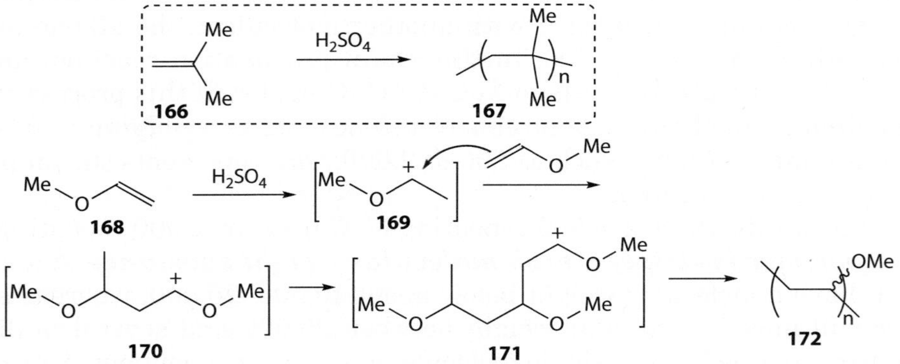

The polymerization process can be examined in more detail using **168**, which reacts with sulfuric acid to form carbocation **169**. Carbocation reacts preferentially with a second molecule of **168** because that alkene is the best nucleophilic species. The product is carbocation **170**, which reacts with more **168** to give a new reaction carbocation **171**. This process continues until a linear polymer is formed: poly(methyl vinyl ether) **172**.

There are many such polymers derived from alkenes, prepared by both radical polymerization and cationic polymerization techniques. Some examples are **173** [poly(vinyl chloride), or PVC]; **174** [poly(acrylonitrile), trade names are Orlon® or Creslan®]; **175** [poly(tetrafluoroethylene), DuPont's trade name for this material is Teflon®]; and **176**, poly(vinyl acetate).

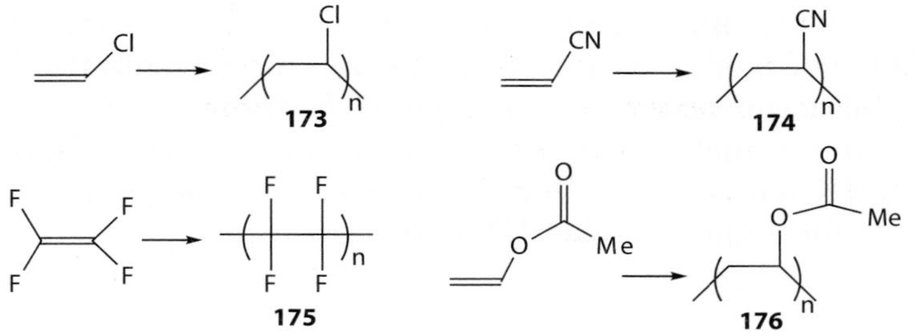

There are many different types of polymers. In addition to the linear polymers such as polyethylene, there are *branched polymers* and *cross-linked polymers*. *Copolymers* can be prepared by polymerizing one alkene in the presence of another. When styrene and acrylonitrile are polymerized in the same reaction vessel, for example, a copolymer such as **177** is formed. These copolymers can be random copolymers, where there is no definite sequence of monomer units, but they can also be regular copolymers, where there is a regular alternating sequence of each monomeric unit.

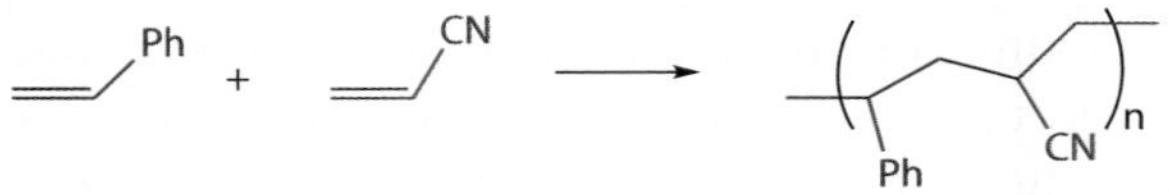

Block copolymers can also be formed. In these polymers, there is a sequence (or block) of one monomeric unit followed by a sequence (block) of the second monomeric unit, and these blocks repeat. In one example, a block copolymer of styrene (S) and acrylonitrile (A) could look like –A–A–S–S–S–S–S–S–S–S–S–A–A–A–A–A–A–A–S–S–S–S–S–S–S–S–S–S–S–, etc.

There are further definitions for polymers. A *graft copolymer* links together two different polymers. Two individual polymers are exposed to gamma- or x-radiation to produce the grafted copolymer. A *thermoplastic* is a polymer that softens when it is heated, but it usually describes a polymer that passes through a specified sequence of property changes as it is heated. An *elastomer* is a polymer that gives a physical state between its glass transition temperature and its liquefaction temperature. Finally, there are *polymer blends,* which are obtained when two polymers are mixed together by mechanical means.

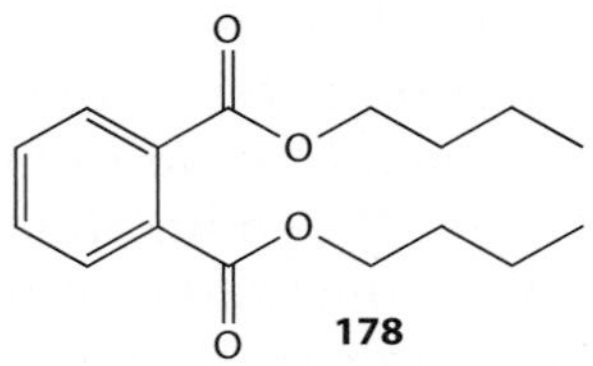

Polymers are often rigid and difficult to use. Compounds known as plasticizers can be added during the polymerization process. When a liquid such as dibutyl phthalate (**178**; see Chapter 16, Section 16.7, for esters) is added to poly(vinyl chloride), for example, a flexible polymer is produced. This material is known as Tygon® and is commonly seen in the laboratory as clear "plastic" tubing. When Tygon is used to transfer organic solvents from one place to another, the plasticizer may be leached out of the tubing and will subsequently be observed as a "product." Most organic chemists who carry out research have obtained an NMR spectrum of dibutyl phthalate at some point in their careers (see Chapter 14 for a discussion of NMR spectroscopy).

10.9 Synthetic Transformations

This chapter has introduced several new chemical reactions. These reactions can be used to modify existing molecules by changing the functional group. In later chapters, molecules can be built by making carbon–carbon bonds. Making a carbon–carbon bond usually requires modification of the functional group to accommodate the reaction. All of this will be the subject of later chapters. The process of modifying molecules via several chemical steps, or building up large molecules from smaller ones, is called chemical synthesis, or just synthesis. The process of using specific chemical reactions for synthesis in organic chemistry will begin here, and it will expand as each new type of reaction is introduced. *Chapter 25 introduces an approach to synthesis called retrosynthetic analysis. The reader should read that chapter to understand the principles of synthesis before reading this section or attempting to do problems associated with synthesis.*

In this particular chapter, carbon–carbon bonds are not made, but the reactions exchanged or modified one functional group for another (or others)—a **functional group exchange**. Some reactions cleave carbon–carbon bonds, but such transformations also fall into the category of a functional group exchange reaction. To simplify identification of each type of transformation, a retrosynthetic disconnection is shown: X ⇒ Y. Note that the reverse arrow indicates a disconnection (see Chapter 25, Section 25.1). In this notation, X is the target molecule being made, and Y represents the precursor molecule from which it is made. In the case of alkenes and HBr, the disconnection indicates that alkyl halides can be made from alkenes. In a synthesis, recognizing this relationship is essential.

The addition of an acid such as HX to an alkene leads to formation of an alkyl halide. If the acid catalyzed addition of water or an alcohol is examined, the product is an alcohol or an ether. Oxymercuration also leads to addition of water to an alkene. The functional group exchange can be generalized as shown, where X = Cl, Br, I, OH, OR. When an alkene reacts with a halogen, the product is a vicinal dihalide. The functional group exchange is shown where X = Cl, Br, I. When bromine in water or chlorine in water is reacted with an alkene, the product is a halohydrin. The functional group exchange is that shown where X = Cl or Br. Hydroboration of an alkene followed by oxidation leads to the less substituted alcohol, which leads to a regiochemically different functional group exchange, as shown in the illustration.

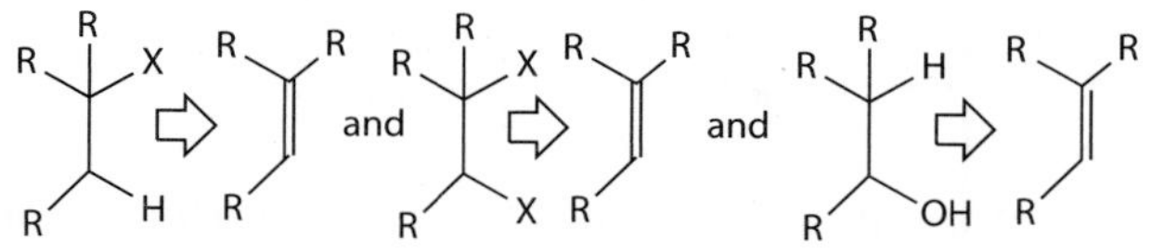

Reactions that probe these disconnections are presented in the homework problems. The disconnection diagrams should be read as follows. Alkyl halides are prepared from alkenes, with regioselectivity. Similarly, alcohols are prepared from alkenes with regioselectivity. A chemical reaction involves treatment of an alkene with an acid HX to give the alkyl halide. Another chemical reaction is treatment of an alkene with aqueous acid to give the more substituted alcohol, or with borane and then basic hydrogen peroxide to give the less substituted alcohol.

The oxidation reactions of either permanganate or osmium tetroxide with an alkene lead to a vicinal diol. The functional group exchange for this process is that shown. In other words, a diol is obtained by dihydroxylation of an alkene. Oxidative cleavage reactions such as ozonolysis eventually lead to an aldehyde, ketone, or carboxylic acid, depending on the substituents attached to the C=C unit of the alkene. The appropriate functional group exchange is shown, where R = H, alkyl, or OH. The disconnection implies that two carbonyl compounds (ketones or aldehydes or carboxylic acids) are prepared by cleavage of an alkene.

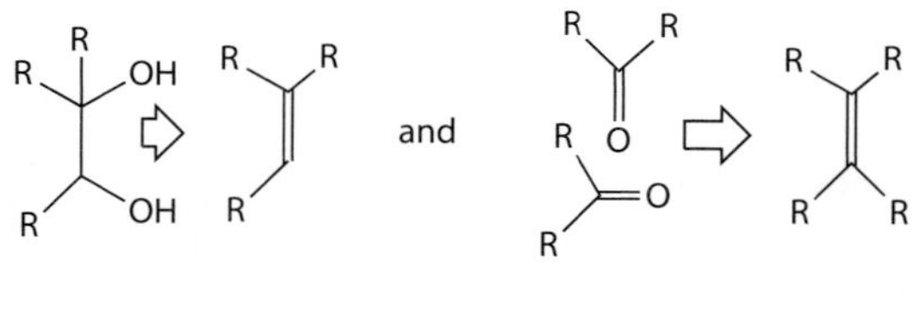

10.10 Biological Relevance

The biogenetic pathway for production of the steroid lanosterol (**181**) from acetyl coenzyme A (acetyl CoA) is shown in Figure 10.3.[18–21] One alkene unit in the polyene squalene (**179**) is epoxidized by the enzyme *squalene monooxygenase* to give squalene-2,3-epoxide (**180**), which is the precursor to lanosterol (**181**). The enzyme *lanosterol synthetase* accomplishes this biotransformation, and lanosterol is ultimately converted to cholesterol (**183**) via the intermediate compound known as desmosterol, **182**.[22] From the standpoint of this chapter, epoxidation of the alkene unit in **179** to give **180** is an obvious analogy to the epoxidation reactions discussed in Section 10.5.

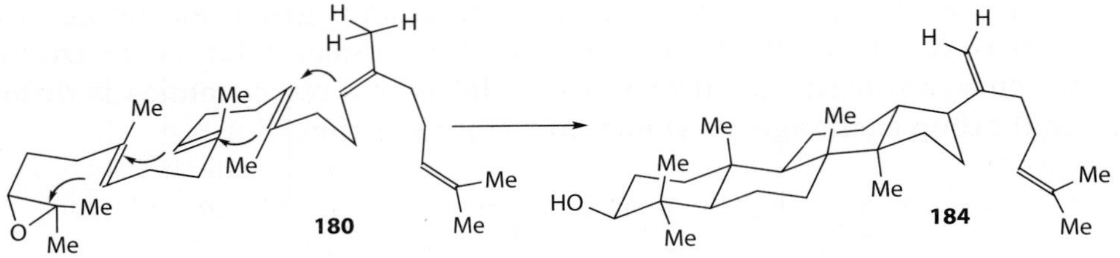

Further, it is easy to imagine that the enzyme *lanosterol synthetase* generates an intermediate from the epoxide or an alkene unit that reacts with another alkene unit. This new intermediate reacts with the next alkene unit, and so on until all four rings are formed. This is formally analogous to an alkene being converted to a carbocation and then reacting with another alkene, which functions as a nucleophile is an addition-type reaction. This process is suggested by the conversion of **180** to **184**, showing the conformational bias of each six-membered ring as it is formed. This scheme suggests that the transition state for each reaction assumes the chair conformation of the new ring formed in each step.

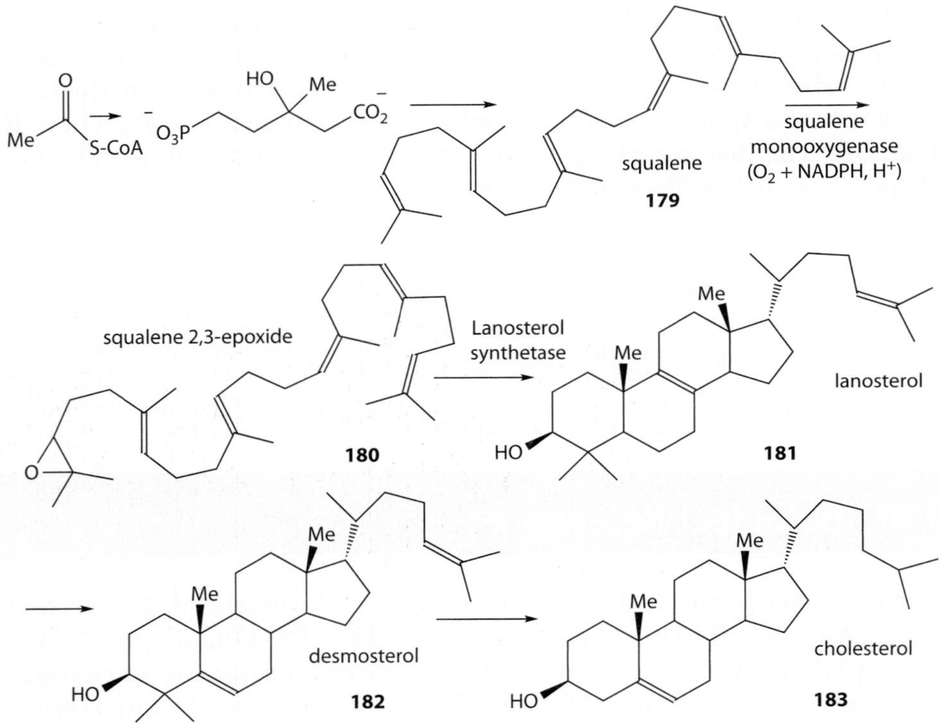

Figure 10.3   The biogenesis of cholesterol from acetyl coenzyme-A. (Reprinted in part with permission from van Tamelen, E. E. et al. 1966. *Journal of the American Chemical Society* 88:4752. Copyright 1966 American Chemical Society; Wendt, K. et al. 2000. *Angewandte Chemie International Edition* 39:2812. Copyright Wiley-VCH Verlag GmbH & Co. KGaA. Reproduced with permission.)

Another biochemical example is the hydration reaction of 2,3-dienoyl thioesters to 3-ketoacyl thioesters by an enzyme.[23] Similarly, Alipui, Zhang, and Schulz[24] observed that 3-ketooctanoyl-CoA **186** was formed from 2,3-octadienoyl-CoA (**188**) when incubated with the enzyme *crotonase*. As seen in the following diagram, possible mechanisms for this enzymatic transformation include reaction of the alkyne unit in **185** with water to give enol **186**, which tautomerizes to the ketone product (**188**). Alternatively, the alkyne can be converted to an allene (**187**), and enzymatic hydration leads to **186**, which tautomerizes to the observed ketone unit. The discussion of polymers in this chapter is pertinent to the biopolymers known as DNA, RNA, and it is pertinent to various enzymes. Enzymes are poly(amino acids) and DNA or RNA are poly(nucleic acids). These polymers are discussed in Chapters 27 and 28, respectively.

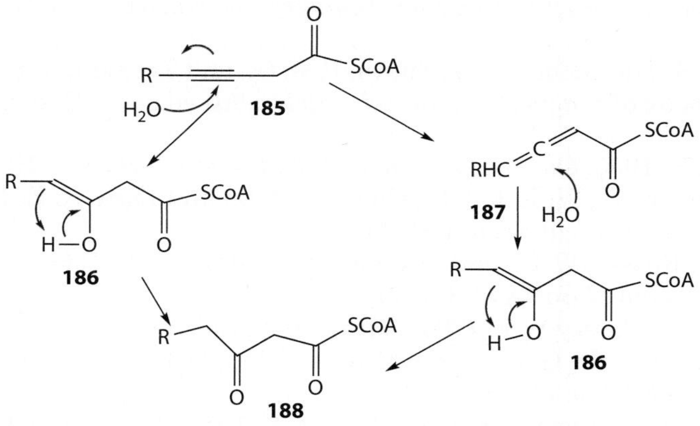

(Reprinted from Alipui, O. D., Zhang, D., and Schulz, H. *Biochemical and Biophysical Research Communications* 292:1171–1174. Copyright 2002, with permission from Elsevier.)

References

1. Michael, A., and Zeidler, F. 1911. *Justus Liebigs Annalen der Chemie* 385:227.
2. Landini, D., and Rolla, F. 1980. *Journal of Organic Chemistry* 45:3527–3529.
3. Lide, D. R., ed. 1995. *CRC handbook of chemistry and physics,* 76th ed., 8–43. Boca Raton, FL: CRC Press, Inc.
4. Adams, R., Kamm, O., and Marvel, C. S. 1918. *Journal of the American Chemical Society* 40:1950–1955.
5. Barluenga, J., Yus, M., Concellón, J. M., and Bernad, P. 1981. *Journal of Organic Chemistry* 46:2721.
6. Purrington, S. T., Kagen, B. S., and Patrick, T. B. 1986. *Chemical Reviews* 86:997–1018.
7. Grakauskas, V. 1971. *Intra-Science Chemistry Reports* 5:85.
8. Humiston, B. 1919. *Journal of Physical Chemistry* 23:572.
9. Merritt, R. F. 1967. *Journal of the American Chemical Society* 89:609.

10. Glavis, F. J., Ryden, L. L., and Marvel, C. S. 1937. *Journal of the American Chemical Society* 59:707.

11. Furniss, B. S., Hannaford, A. J., Smith, P. W. G., and Tatchell, A. R., eds. 1994. *Vogel's textbook of practical organic chemistry,* 5th ed., 543–544. Essex, England: Longman.

12. Furniss, B. S., Hannaford, A. J., Smith, P. W. G., and Tatchell, A. R., eds. 1994. *Vogel's textbook of practical organic chemistry,* 5th ed., 546. Essex, England: Longman.

13. Rickborn, B., and Gerkin, R. M. 1971. *Journal of the American Chemical Society* 93:1693.

14. Fahey, R. C., Payne, M. T., and Lee, D.-J. 1974. *Journal of Organic Chemistry* 39:1124–1130.

15. Thomas, R. J., Campbell, K. N., and Hennon, G. F. 1938. *Journal of the American Chemical Society* 60:718.

16. Furniss, B. S., Hannaford, A. J., Smith, P. W. G., and Tatchell, A. R., eds. 1994. *Vogel's textbook of practical organic chemistry,* 5th ed., 548–549. Essex, England: Longman.

17. Furniss, B. S., Hannaford, A. J., Smith, P. W. G., and Tatchell, A. R., eds. 1994. *Vogel's textbook of practical organic chemistry,* 5th ed., 576. Essex, England: Longman.

18. Clayton, R. B. 1965. *Quarterly Review of the Chemical Society* 19:168.

19. van Tamelen, E. E., Willett, J. D., Clayton, R. B., and Lord, K. E. 1966. *Journal of the American Chemical Society* 88:4752.

20. Corey, E. J., Russey, W. E., and Ortiz de Montellano, P. R. 1966. *Journal of the American Chemical Society* 88:4750.

21. Corey, E. J., and Russey, W. E. 1966. *Journal of the American Chemical Society* 88:4751.

22. Wendt, K. U., Schulz, G. E., Corey, E. J., and Liu, D. R. 2000. *Angewandte Chemie International Edition* 39:2812.

23. Branchini, B. R., Miesowicz, F. M., and Bloch, K. 1977. *Bioorganic Chemistry* 6:49–52.

24. Alipui, O. D., Zhang, D., and Schulz, H. 2002. *Biochemical and Biophysical Research Communications* 292:1171–1174.

Answers to Problems

10.1

There is no acid–base reaction when ethanol is mixed with cyclohexene.
There is an acid–base reaction between cyclohexene and HBr, to give a carbocation.

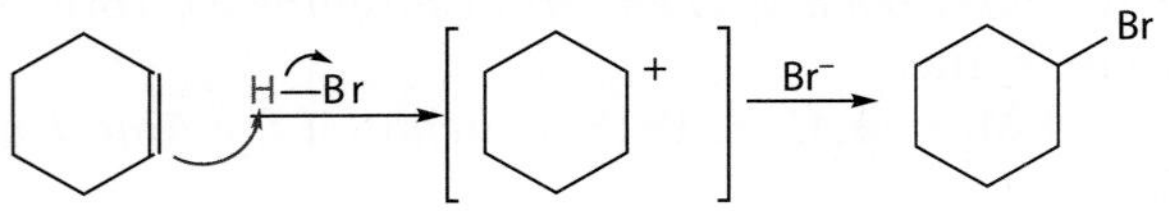

10.2

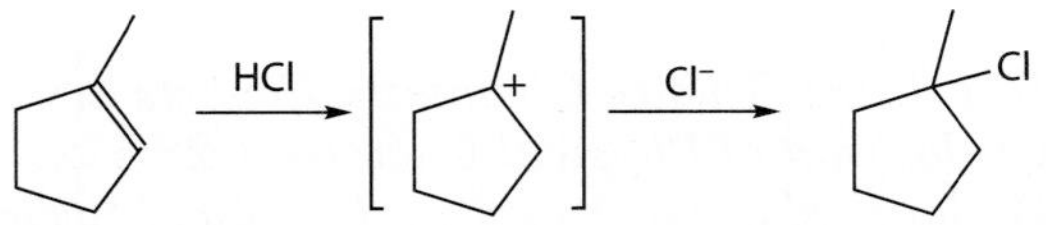

10.3

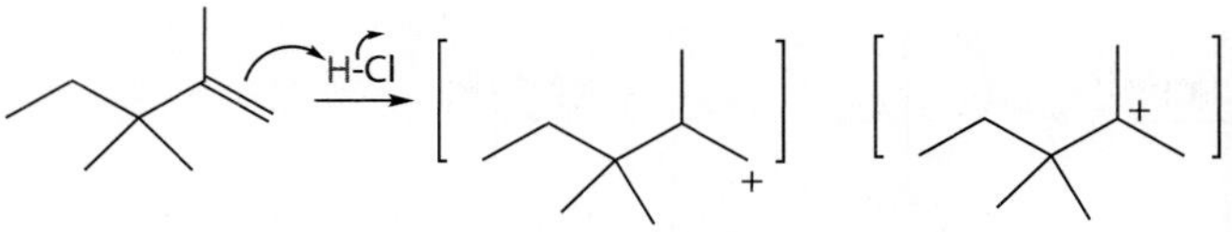

Less stable primary carbocation

More stable tertiary carbocation leads to the major product. There is no rearrangement

10.4

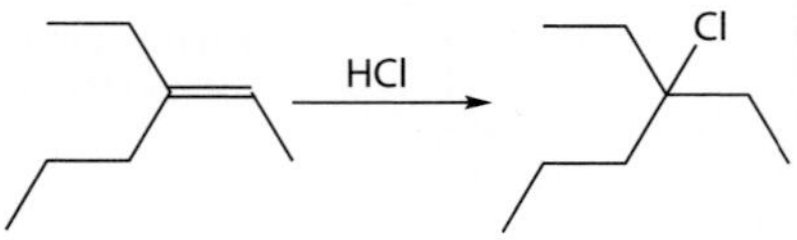

$$\xrightarrow{\text{HCl}}$$

10.5

$$\xrightarrow{\text{HBr}} \quad [\quad + \quad] \quad \xrightarrow{\text{Br}^-}$$

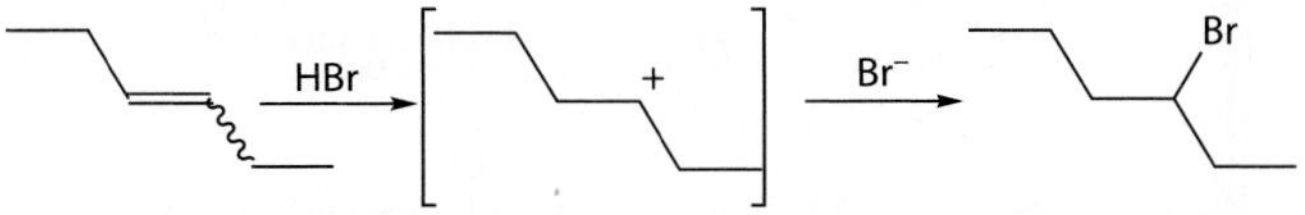

3-Hexene is symmetrical so a single secondary carbocation is formed, leading to one bromide. Therefore, there are no regioisomers.

10.6

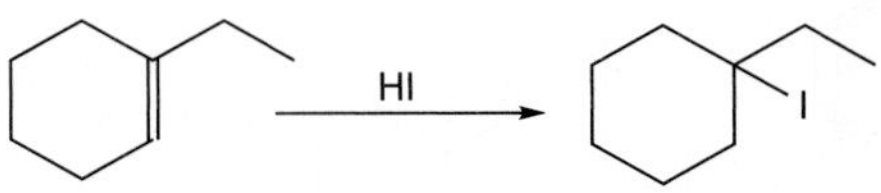

$$\xrightarrow{\text{HI}}$$

10.7

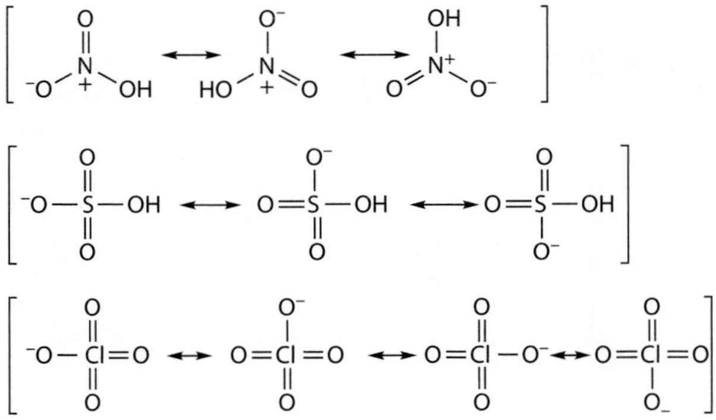

10.8

$$\xrightarrow{H_2SO_4}$$

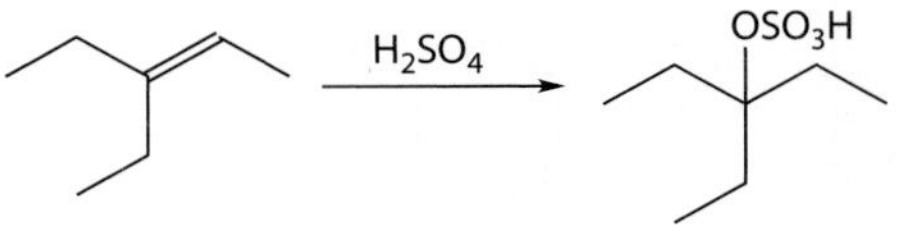

10.9

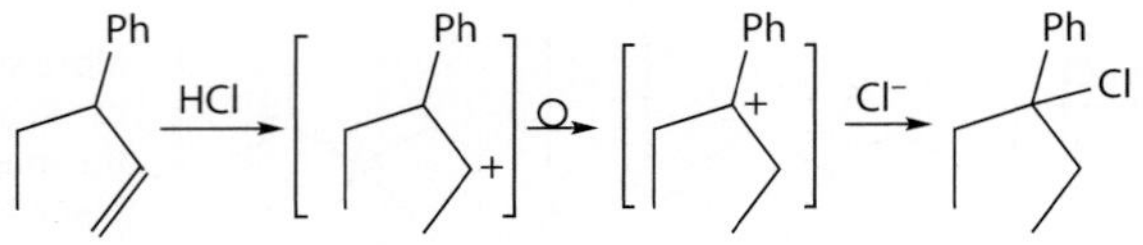

10.10

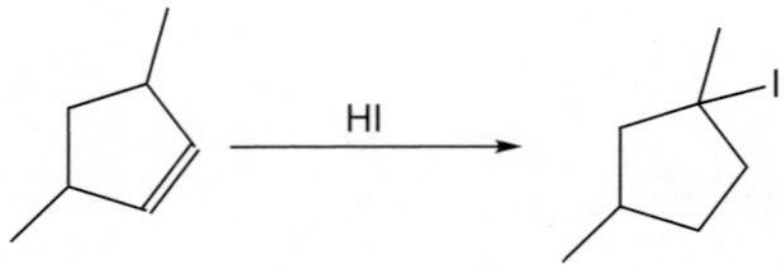

Rearrangement of the initially formed tertiary carbocation

10.11

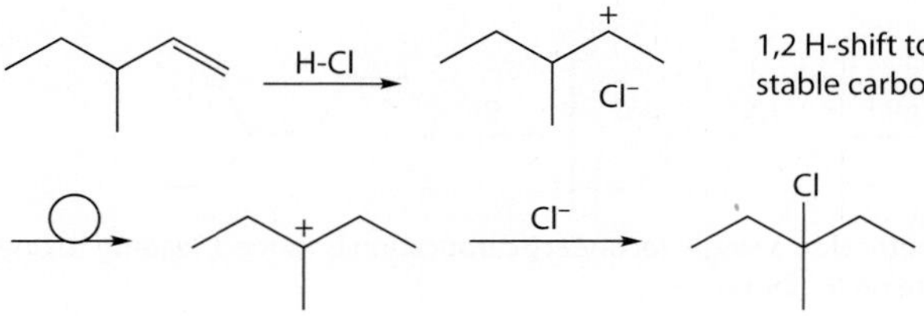

1,2 H-shift to a more stable carbocation

10.12

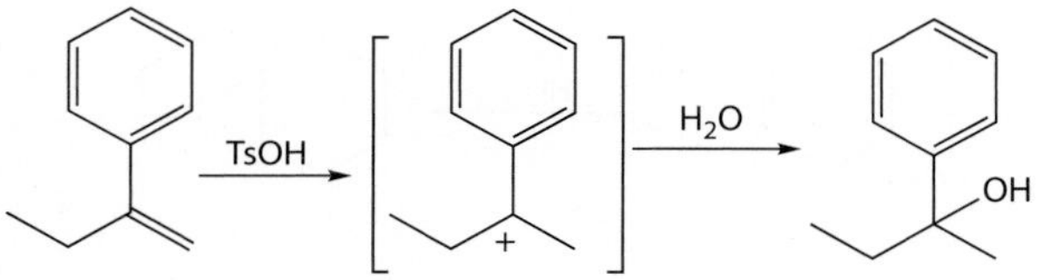

10.13

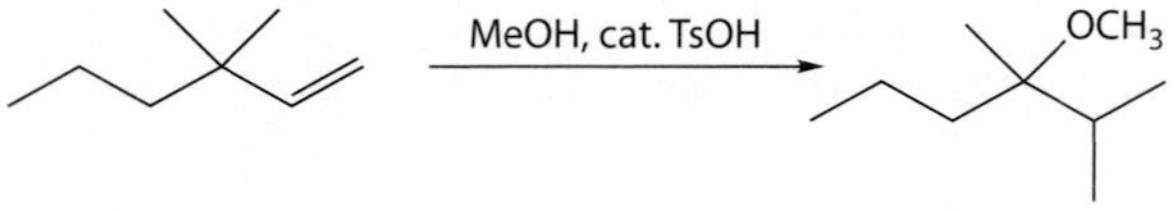

via 1,2-methyl shift from a secondary carbocation to a tertiary carbocation

10.14

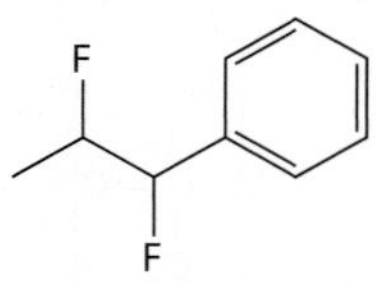

1,2-difluoro-1-phenylpropane

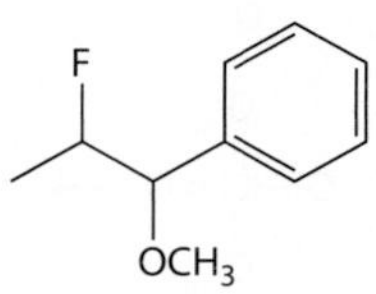

2-fluoro-1-methoxy-1-phenylpropane

10.15

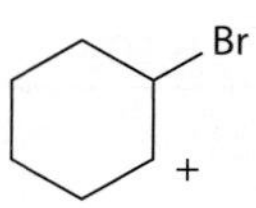

10.16

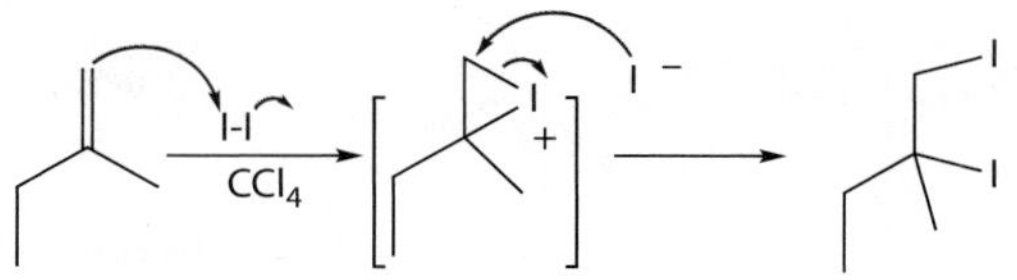

10.17

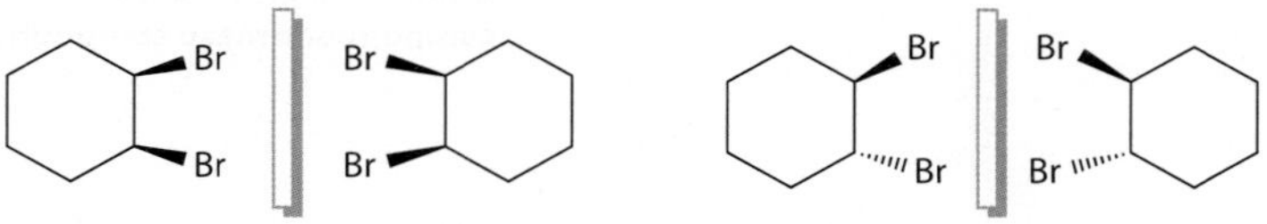

10.18

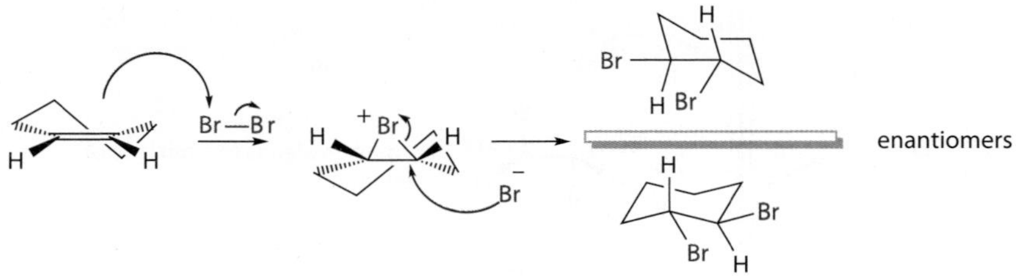

enantiomers

10.19

50

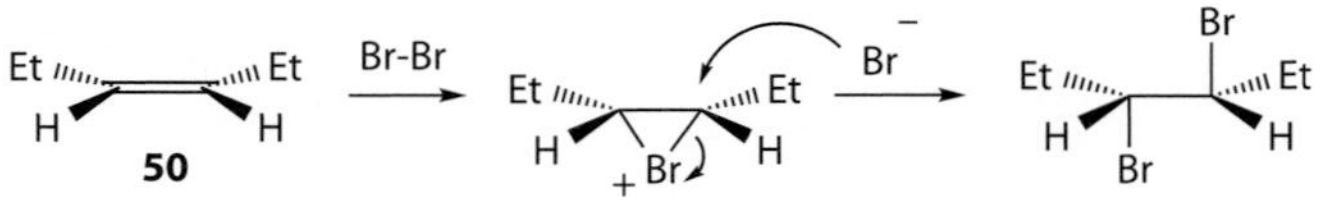

10.20

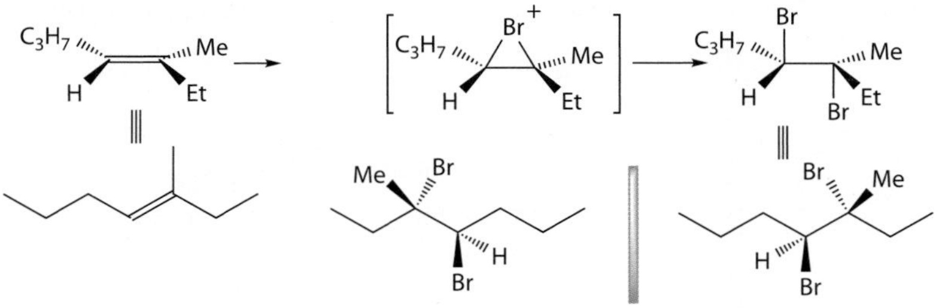

These are the two enantiomers - *RS* and *SR*

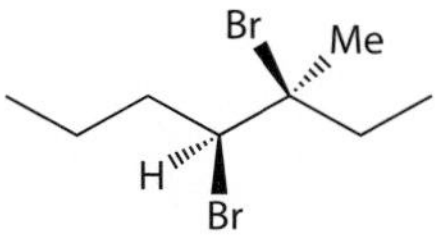

This is one enantiomer of the diastereomer not formed -
the *SS* and *RR* stereoisomers

10.21

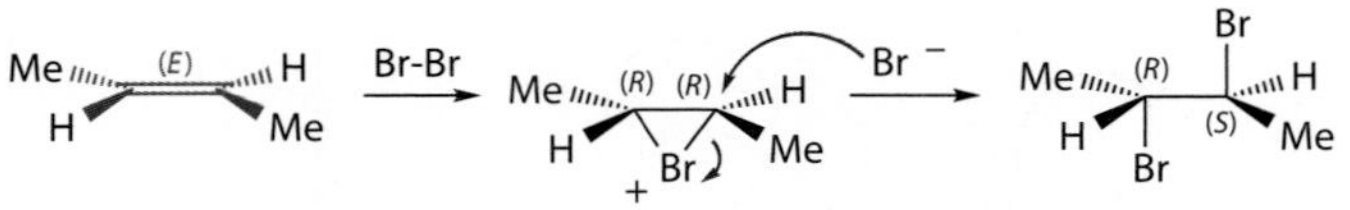

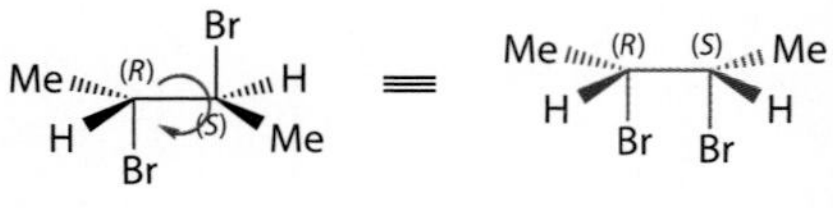

As seen by rotation of the C-C bond, the dibromide is a meso compound. Therefore, this is a trick question since *trans*-2-butene reacts with bromine to give a one compound, a meso compound.

10.22

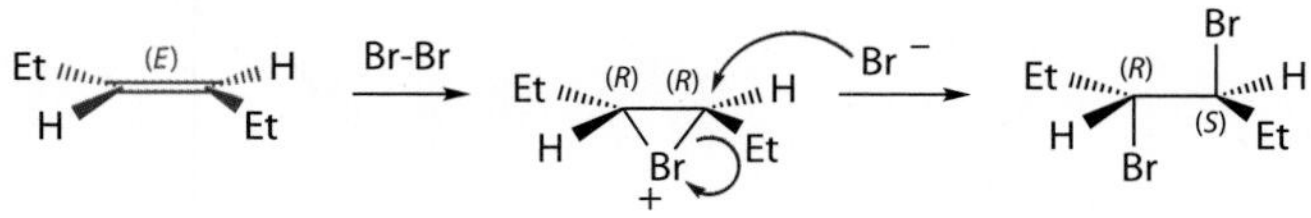

meso from the *trans*-isomer

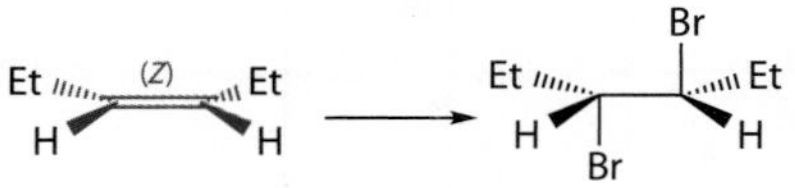

cis-alkene leads to a mixture of enantiomers

10.23

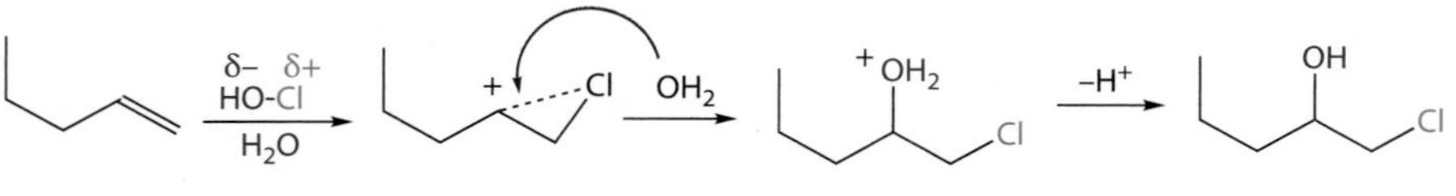

10.24

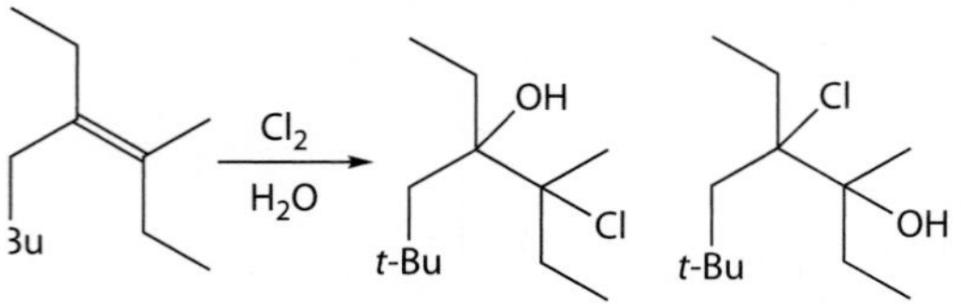

Since an intermediate tertiary carbocation is formed from both carbons of the C=C unit, both regioisomers are expected as products in about equal amounts

10.25

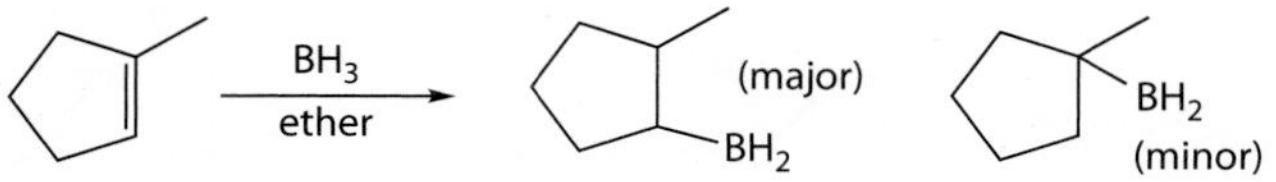

(major)

(minor)

10.26

$H_3CH_2CH_2CH_2C$

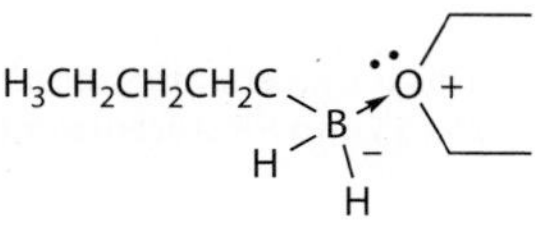

10.27

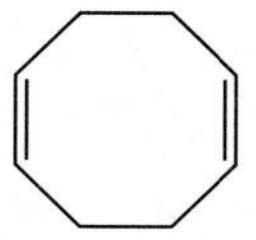

10.28

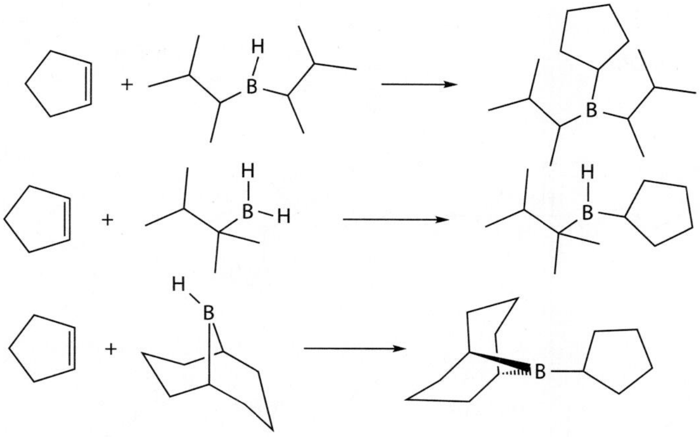

10.29

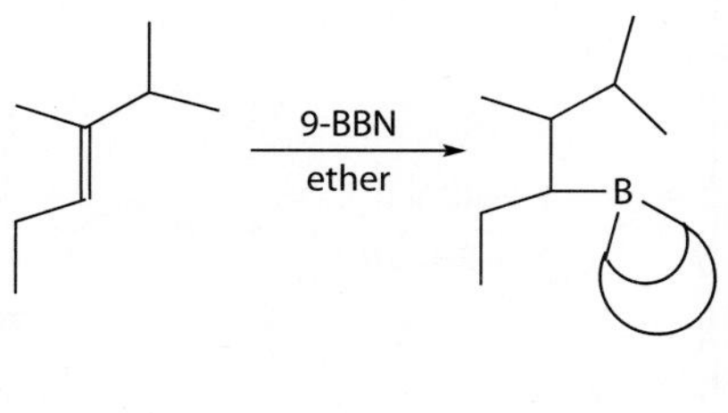

10.30

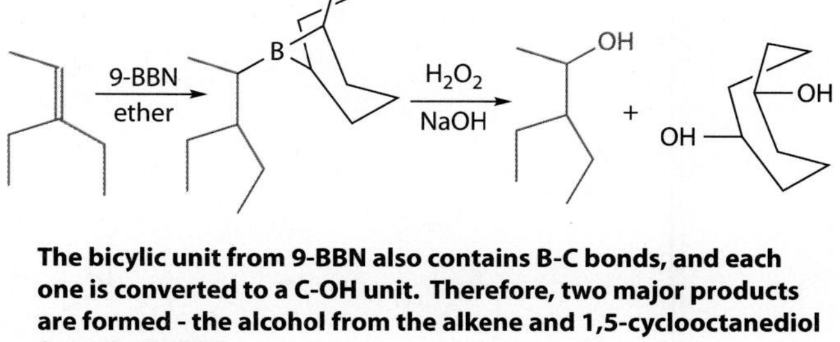

**The bicylic unit from 9-BBN also contains B-C bonds, and each
one is converted to a C-OH unit. Therefore, two major products
are formed - the alcohol from the alkene and 1,5-cyclooctanediol
from the 9-BBN**

10.31

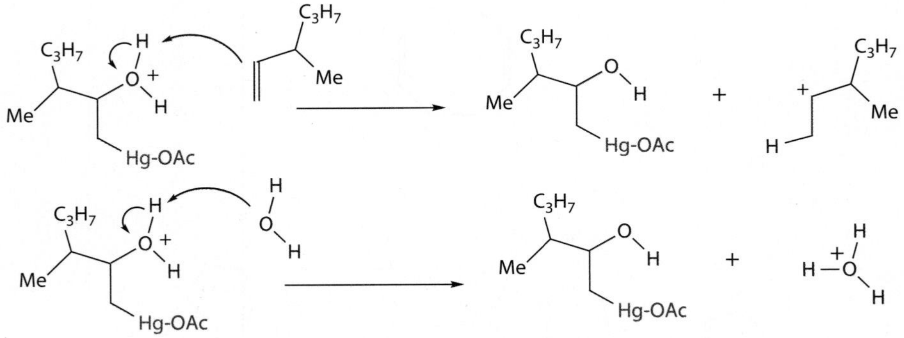

10.32

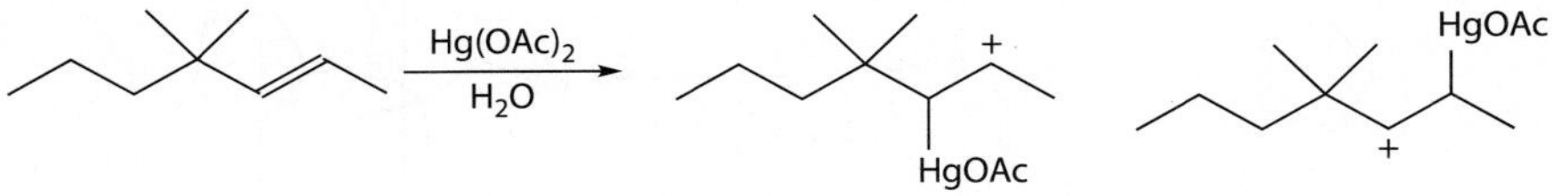

10.33

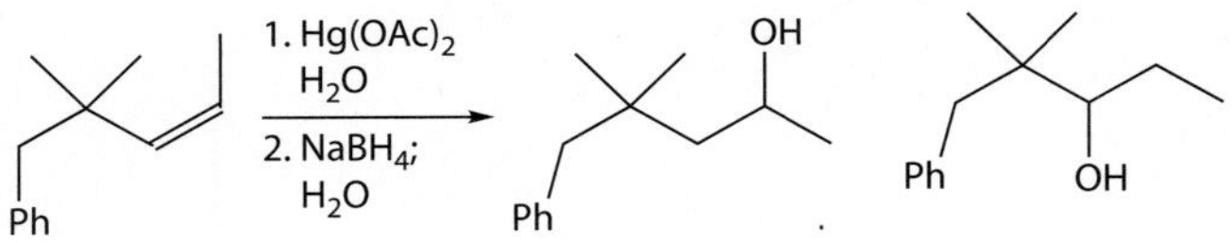

10.34

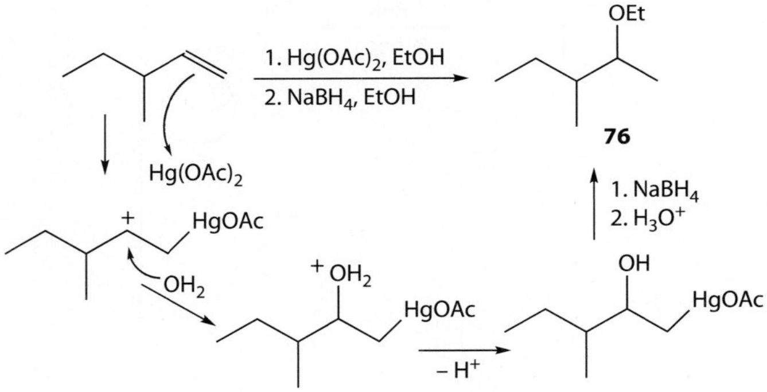

10.35

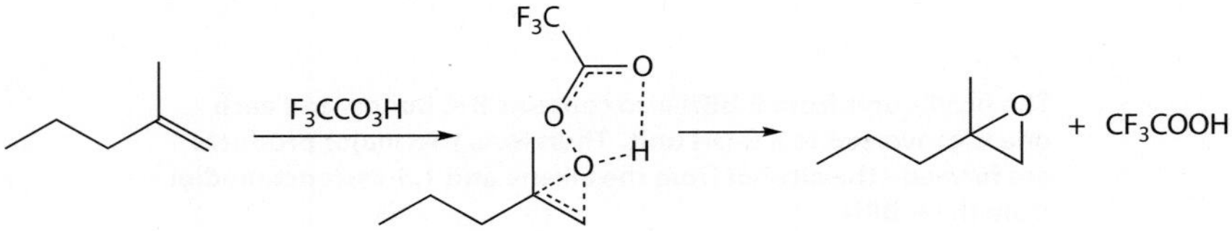

10.36

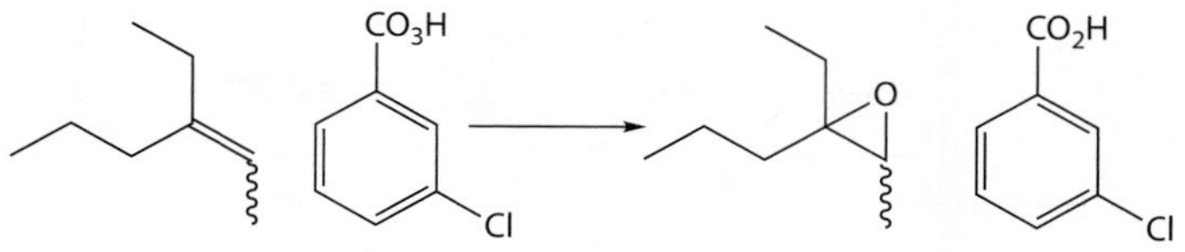

10.37

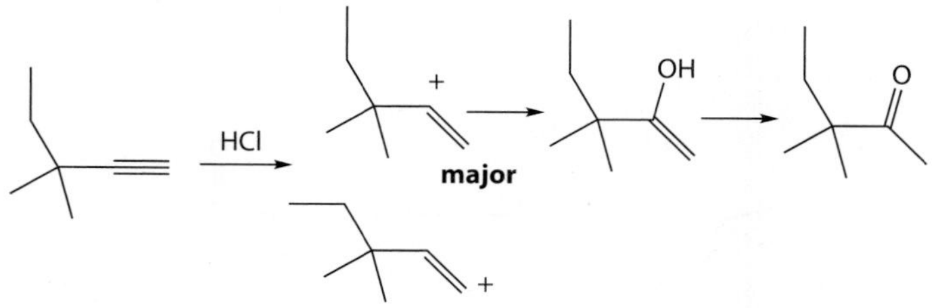

10.38

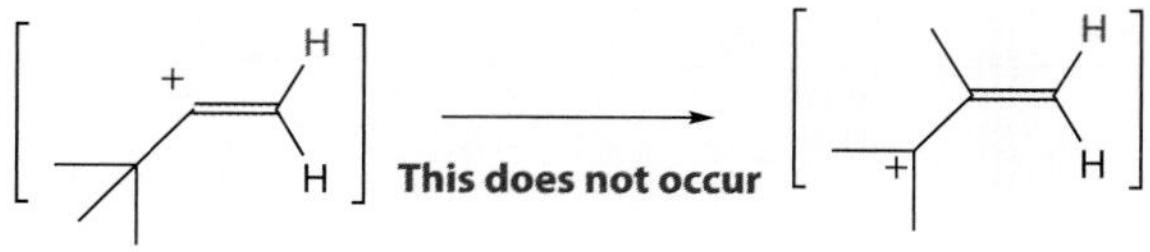

10.39

10.40

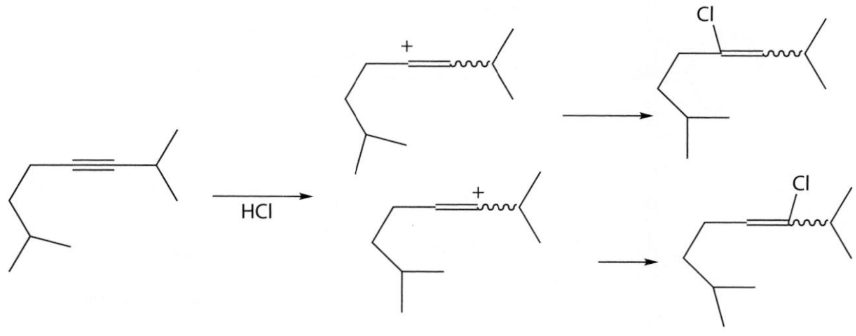

10.41

10.42

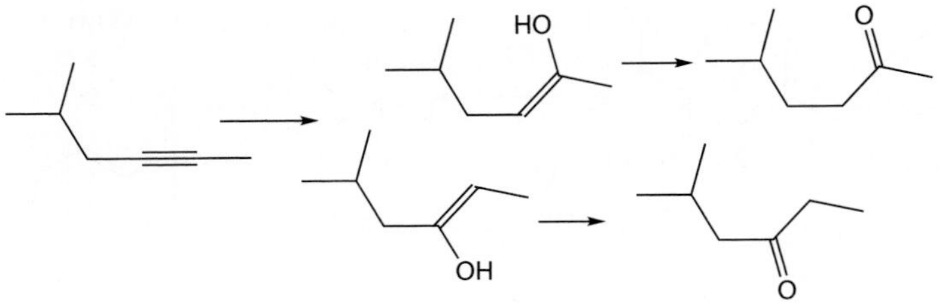

10.43

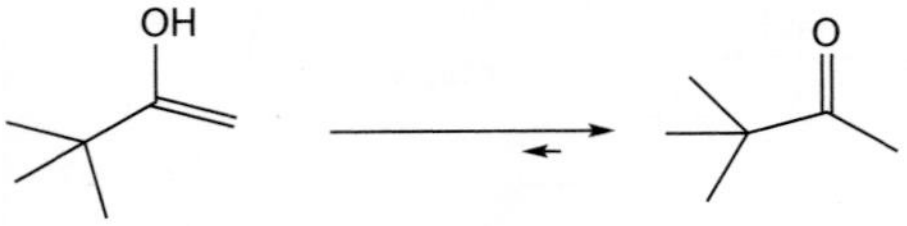

10.44

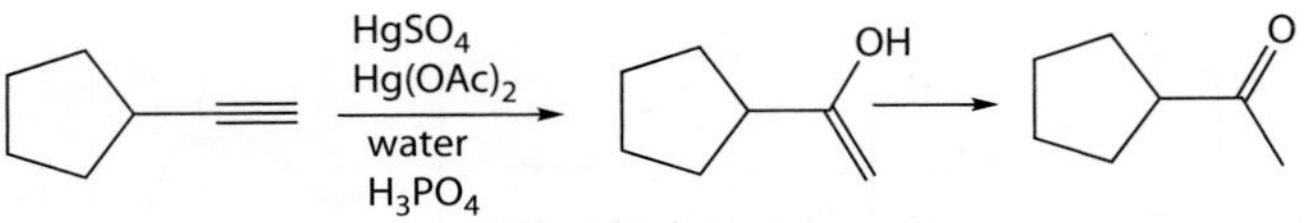

10.45

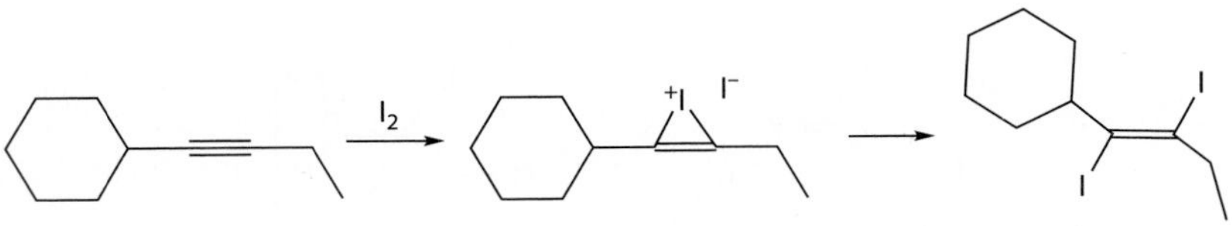

10.46

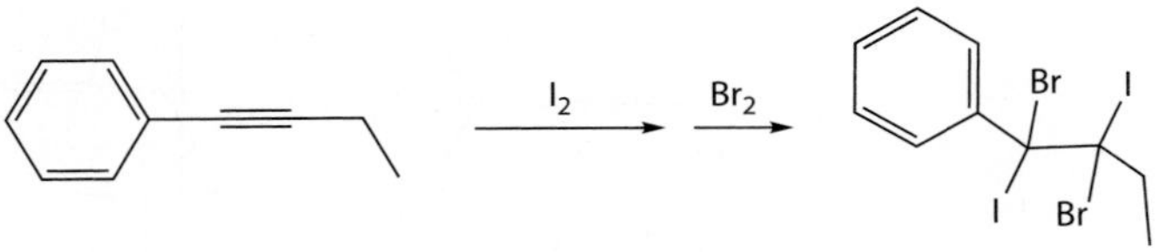

10.47

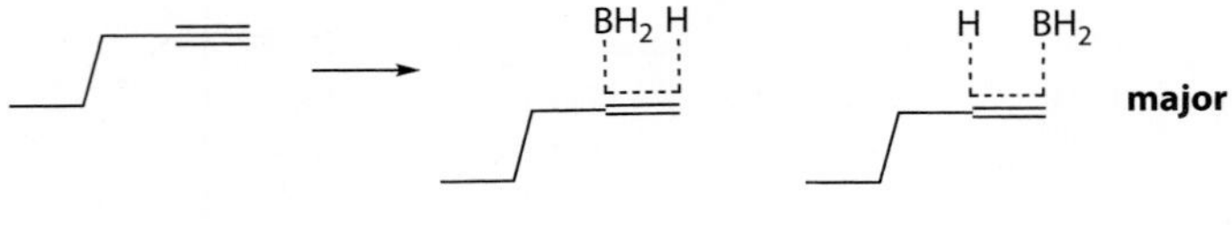

10.48

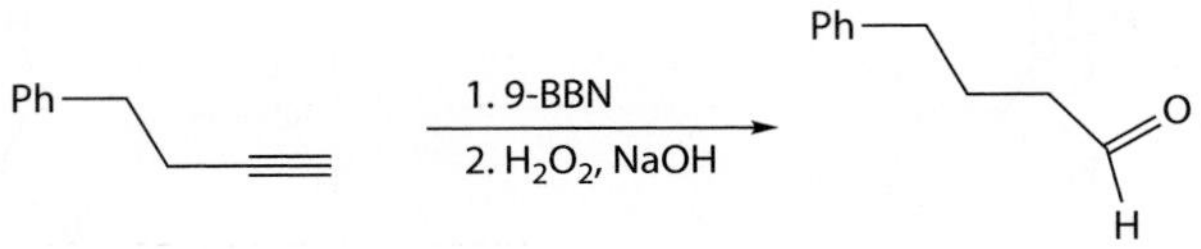

10.49

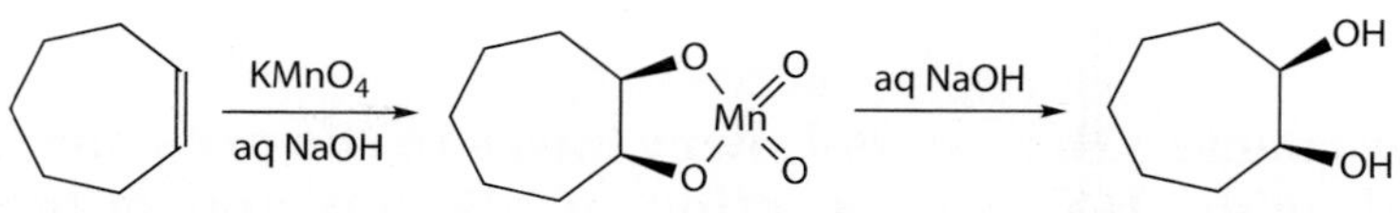

10.50

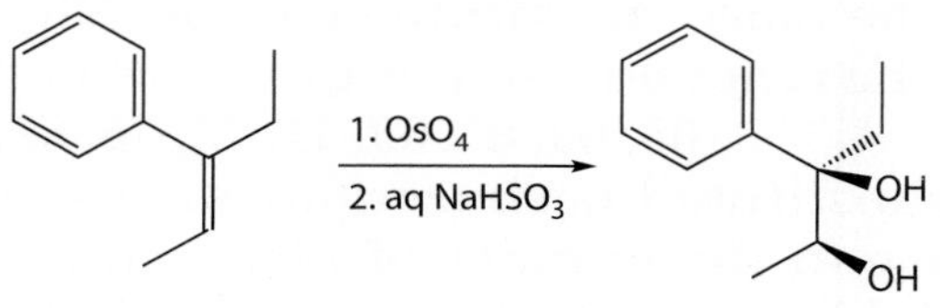

10.51

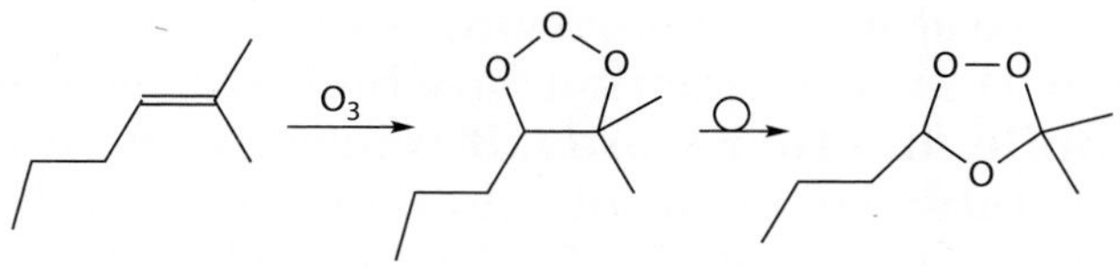

10.52

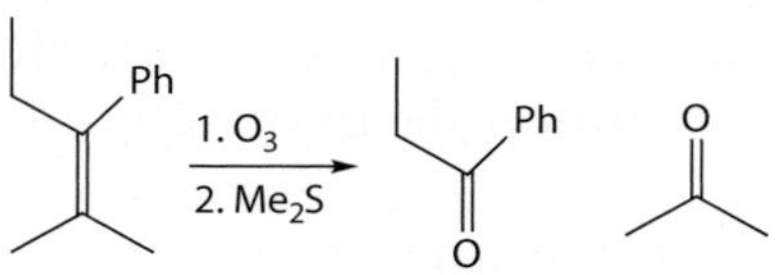

10.53

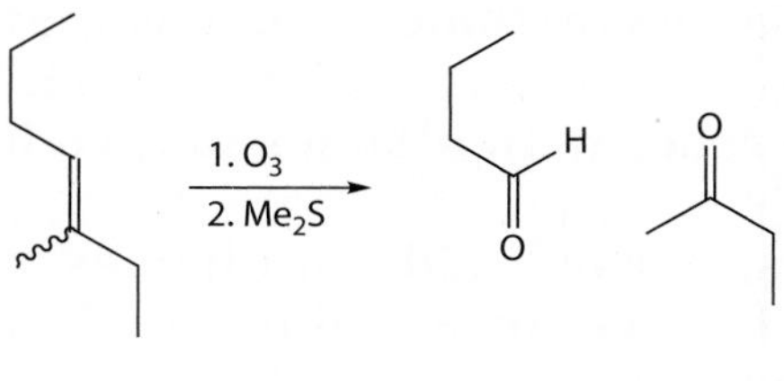

10.54

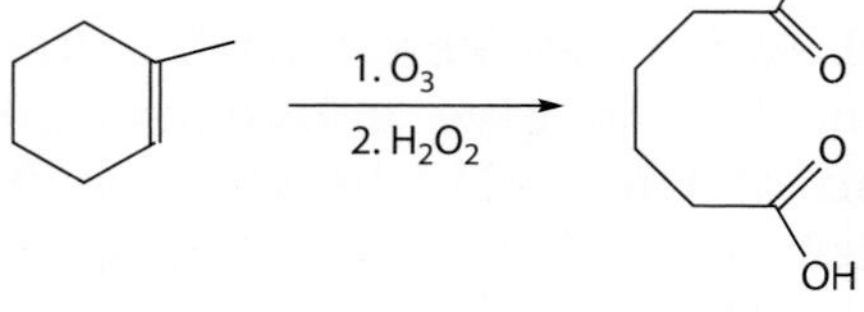

10.55

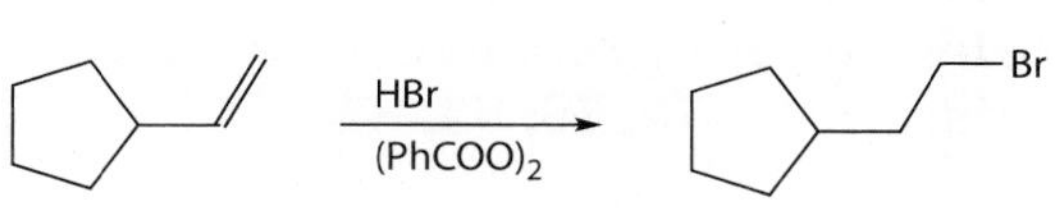

Correlation of Concepts with Homework

- Alkenes react as Brønsted–Lowry bases in the presence of strong mineral acids, HX. The reaction of alkenes and mineral acids (HX) generates the more stable carbocation, leading to substitution of X at the more substituted carbon. This constitutes the mechanism of the reaction, but it is given the name Markovnikov addition: 1, 2, 3, 4, 5, 6, 65, 66, 67, 69, 71, 72, 75, 82, 83, 99, 100, 116.
- More highly substituted carbocations are generally more stable. An increase in the number of resonance contributors will increase the stability of a carbocation: 7, 61, 68.
- If a carbocation is formed adjacent to a carbon that can potentially be a more stable carbocation, assume that rearrangement occurs faster than the reaction in which the carbocation traps the nucleophile. In other words, if rearrangement can occur to give a more stable carbocation, that carbocation is assumed to rearrange: 9, 10, 11, 13, 60, 62, 70, 71, 80, 103, 106.
- Weak acids such as alcohols and water can add to an alkene in the presence of a strong acid catalyst. Acid-catalyzed reaction with alkenes and water gives the Markovnikov alcohol and similar reaction with alcohols gives the Markovnikov ether: 8, 12, 81, 82, 83.
- Chlorine, bromine, and iodine react with alkenes to give the three-membered ring halonium ion, which reacts with the halide nucleophile to give *trans* dichlorides, dibromides, or diiodides. Alkynes react to give vinylhalonium ions that lead to vinyl dihalides: 14, 15, 16, 17, 18, 19, 20, 21, 22, 64, 73, 74, 75, 82, 83, 98, 108.
- HOCl and HOBr react with alkenes to give halohydrins: 23, 24, 75, 82, 83.
- In ether solvents, borane adds to alkenes via a four-centered transition state to give an alkylborane. Borane adds by a *cis* addition that places the boron on the less substituted carbon as the major product. Alkylboranes react with $NaOH/H_2O_2$ to give an anti-Markovnikov alcohol: 25, 26, 27, 28, 29, 30, 47, 48, 78, 79, 81, 82, 83, 101.
- Mercuric acetate and water react with alkenes via a mercury-stabilized carbocation to give a hydroxy alkyl-mercury compound. Reduction of the C–Hg bond with $NaBH_4$ leads to the Markovnikov alcohol: 31, 32, 33, 34, 82, 83, 84.
- The reaction of an alkene and a peroxyacid gives an oxirane (an epoxide) and a carboxylic acid: 35, 36, 59, 82, 83.
- Alkynes react with mineral acids HX to give vinyl halides, and with nucleophiles in the presence of an acid catalyst to give vinyl compounds: 39, 40, 75, 76, 83, 84.

- Alkynes react with an acid catalyst and water or with mercuric salts and water to give an enol, which tautomerizes to a ketone: 37, 38, 42, 43, 78, 83, 97, 102.
- Hydroboration of alkynes leads to an enol after treatment with NaOH/H$_2$O$_2$, and tautomerization gives an aldehyde from terminal alkynes, or a ketone from internal alkynes: 44, 78, 83.
- Dilute potassium permanganate and osmium tetroxide react with alkenes to give a manganate ester or an osmate ester, respectively. Both of these products are decomposed under the reaction conditions to give a vicinal (1,2)-*cis*-diol: 49, 50, 83, 96.
- Reaction of an alkene with ozone leads to a 1,2,3-trioxolane, which rearranges to a 1,2,4-trioxolane (an ozonide). Subsequent treatment with hydrogen peroxide or with dimethyl sulfide leads to an aldehyde, ketone, or carboxylic acid product. When an ozonide contains a C–H unit, oxidation with hydrogen peroxide leads to a carboxylic acid, but reaction with dimethyl sulfide leads to an aldehyde: 51, 52, 53, 54, 78, 82, 83, 117.
- Radicals react with alkenes to form a new radical that can react further to give addition reaction products. The initial reaction will generate the more stable radical, and rearrangement is not observed. In the presence of peroxides, alkenes react with HBr to give the alkyl bromide having Br on the less substituted carbon. This is called anti-Markovnikov addition: 55, 88, 105.
- Alkenes can be polymerized under both radical and acid-catalyzed conditions. Polymerization of functionalized and substituted alkenes leads to a variety of commercially important polymers: 77.
- Many functional group transformations are possible via addition reactions, leading to many new reactions: 57, 58, 59, 63, 64, 75, 78, 82, 83, 93, 96, 104.
- A molecule with a particular functional group can be prepared from molecules containing different functional groups by a series of chemical steps (reactions). This process is called synthesis: The new molecule is synthesized from the old one (see Chapter 25): 85, 86, 89, 90, 91, 92, 93, 94, 95, 96, 107, 108, 109, 110.
- Spectroscopy can be used to determine the structure of a particular molecule and can distinguish the structure and functionality of one molecule when compared with another. See Chapter 14: 111, 112, 113, 114, 115, 116, 117, 118, 119, 120, 121.

Homework

10.56 Which of the following alkenes has the strongest π-bond? Explain your choice.

10.57 Which of the following reagents give *cis* addition in a reaction with 1-methylcyclopentene?
(a) Br_2
(b) HCl
(c) BH_3
(d) CH_3CO_3H
(e) H_2O

10.58 Which of the following molecules react with HBr to give a vinyl bromide?

10.59 Which of the following reagents convert an alkene to the corresponding epoxide?
(a) CH_3OH
(b) CH_3CO_2H
(c) CH_3CO_3H
(d) NaOH

10.60 Which of the following is the most stable carbocation? *Assume* that there is *no* rearrangement and explain your choice.

10.61 Which of the following is the more stable carbocation? *Assume* that there is *no* rearrangement and explain your choice.

10.62   Which of the following alkenes do NOT give a rearrangement product upon
         reaction with water and H⁺?

10.63   Which of the following reagents give anti-Markovnikov addition to alkenes
         or alkynes?
         (a) HCl
         (b) OsO_4
         (c) Br_2
         (d) BH_3
         (e) $H_2O/HgCl_2$

10.64   What is the product formed when 2-butyne reacts with diatomic chlorine
         (Cl_2)? Explain your choice.
         (a) a *cis* dichloride
         (b) a *trans* dichloride
         (c) an *E* mono-chloride
         (d) a *Z* mono-chloride

10.65   Inadvertently, cyclohexene was treated with a mixture of HBr and H_2SO_4
         rather than pure HBr. Nonetheless, the major isolated product was the
         expected bromocyclohexane. Briefly explain.

10.66   Briefly discuss why 2,3-dimethylbutene might react faster with HCl than
         does 2-butene. Draw the mechanism for both reactions as part of your
         answer.

10.67   Under strictly controlled conditions where exactly one equivalent of HCl is
         added to 1,1-diphenyl-1,5-hexadiene, the major product results from addition
         of HCl to only one of the C=C units. Draw the product that is formed and
         briefly discuss why that C=C unit reacts with such a high preference.

10.68   A carbocation is formed when 2-methyl-2-propene is reacted with a catalytic
         amount of acid, and a carbocation is also formed when acetone is treated
         with a catalytic amount of acid. Draw both carbocations and comment on
         their relative stability.

10.69   When methylcyclopentene is treated with HCl, the isolated product is
         1-chloro-1-methylcyclopentane. When methylcyclopentene is treated with
         concentrated sulfuric acid, however, methylcyclopentene is usually isolated
         rather than a different product. (This will be discussed in Chapter 12.)
         Briefly discuss why there is such a difference when sulfuric acid is a stron-
         ger acid than HCl.

10.70   Draw the product expected when 1-(2-phenylcyclobutyl)-1-cyclohexene is
         treated with HBr and give a mechanism to support your answer.

10.71   Draw the final products obtained when 1-methoxycyclopentene is treated
         with aqueous HCl and give a mechanism for the formation of this product.

10.72 A reaction called the Johnson polyene cyclization (based on the Stork–Eschenmoser hypothesis) converts triene **A** into the polycyclic molecule **B**. When first discovered, an initially formed carbocation at one end of the polyene reacted with a nearby alkene to form a ring containing a new cation. This reacted with another nearby alkene, etc. The reaction was plagued by low yields and formation of polymeric material and decomposition products. This transformation required many years to perfect and two improvements were the use of a cyclopentenol unit on the "left" to initiate the sequence, which became a cyclopentene unit, and an alkyne unit on the "right" to end the sequence by generating a methyl ketone. Briefly discuss why these two improvements helped the problems inherent to this reaction.

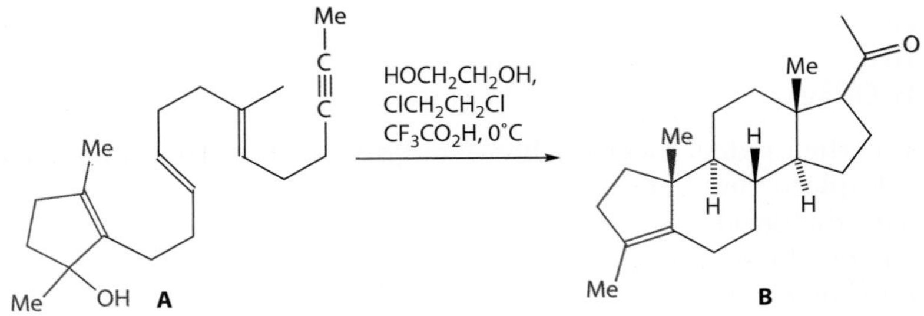

10.73 Bromine is a diatomic molecule that reacts with cyclopentene to form 1,2-dibromocyclopentane. Nitrogen is also diatomic but does not give a similar reaction. Why not?

10.74 Molecule **A** gives a remarkably poor reaction with bromine and very little of the 1,2-dibromide is formed. Suggest a reason why **A** is so unreactive that includes the mechanism of this reaction.

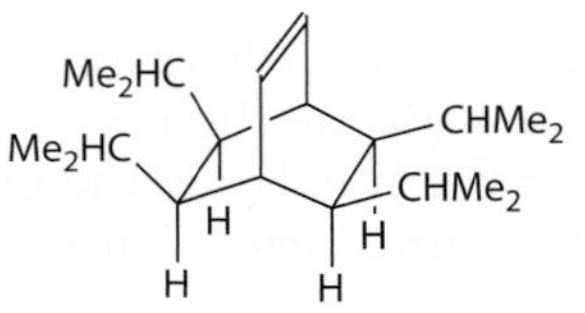

10.75 Give the major product of the following reactions:

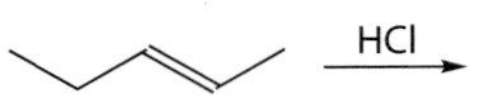

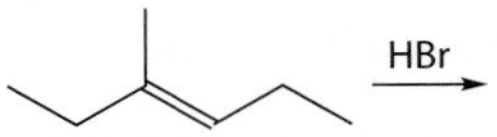

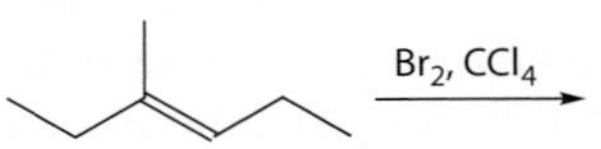

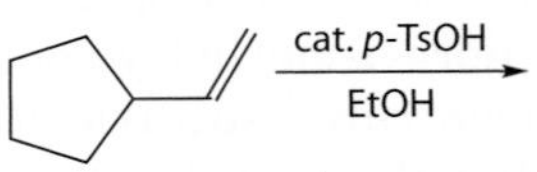

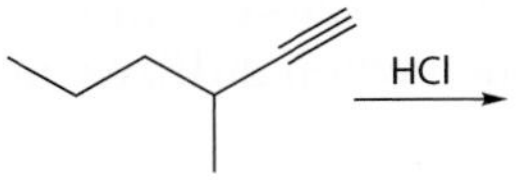

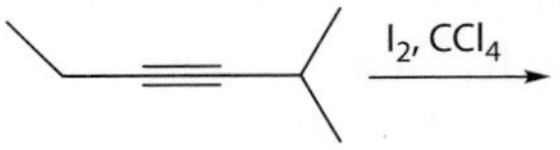

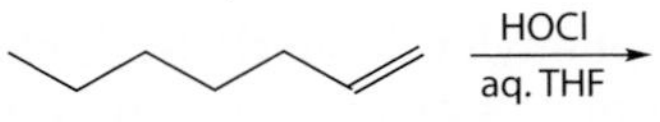

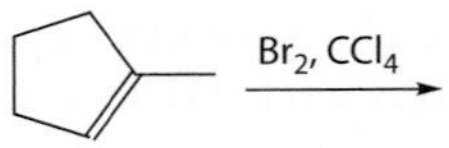

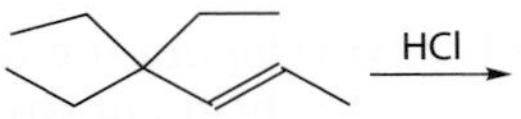

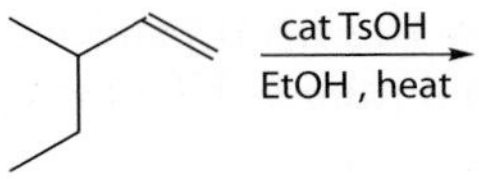

10.76   Draw the product expected when 1-pentyne is treated with a catalytic amount of HCl in ethanol solvent. Draw the mechanism for formation of your product.

10.77   Assume that you can obtain cyclooctyne. Draw it using line notation. Draw the product formed when cyclooctyne reacts with Br_2 in CCl_4. Predict whether this bromination reaction should be exothermic or endothermic and offer a reason for your choice.

10.78   Give the major product of the following reactions:

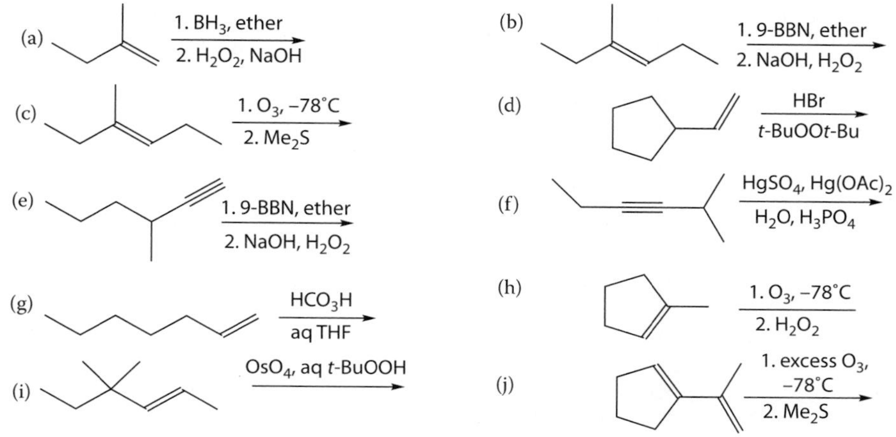

(a)   1. BH_3, ether
        2. H_2O_2, NaOH

(b)   1. 9-BBN, ether
        2. NaOH, H_2O_2

(c)   1. O_3, −78°C
        2. Me_2S

(d)   HBr
        t-BuOOt-Bu

(e)   1. 9-BBN, ether
        2. NaOH, H_2O_2

(f) $HgSO_4$, $Hg(OAc)_2$
       H_2O, H_3PO_4

(g) HCO_3H
       aq THF

(h)   1. O_3, −78°C
        2. H_2O_2

(i) OsO_4, aq t-BuOOH

(j)   1. excess O_3,
        −78°C
        2. Me_2S

10.79   Hydroboration of molecule **A** with 9-BBN and then oxidation leads almost exclusively to a single diastereomeric alcohol. Draw the product and the transition state leading to this product. Justify your answers.

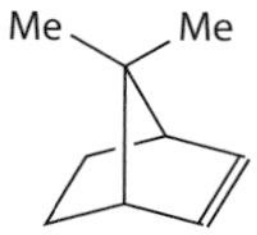

Me   Me

A

10.80   The reaction of the alkene shown with aqueous acid leads to an anomalous product. Contrary to what you have learned about cationic rearrangements, secondary-to-secondary and tertiary-to-tertiary rearrangements are possible under some circumstances. With this in mind, predict the structure of the product using a mechanism to do so.

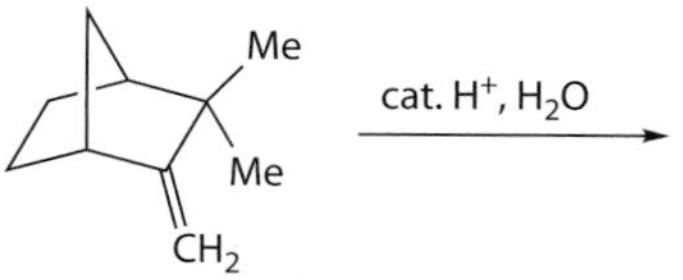

Me
cat. H^+, H_2O
Me
CH_2

10.81   Give the major product formed from each molecule after treatment with diborane and then oxidation with NaOH and hydrogen peroxide:
(a) 1-ethylcycloheptene
(b) 2-phenyl-1-butene
(c) 3,4-diethyl-3-hexene
(d) 3,3-dimethyl-1-hexyne

10.82 Give the major product or products for each of the following reactions. If there is no reaction, indicate by "N.R." Remember stereochemistry where it is appropriate.

(a) I_2, CCl_4

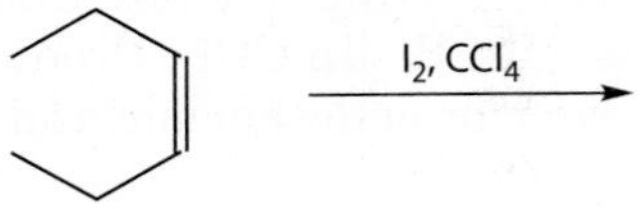

(b)   1. O_3, −78°C   2. H_2O_2

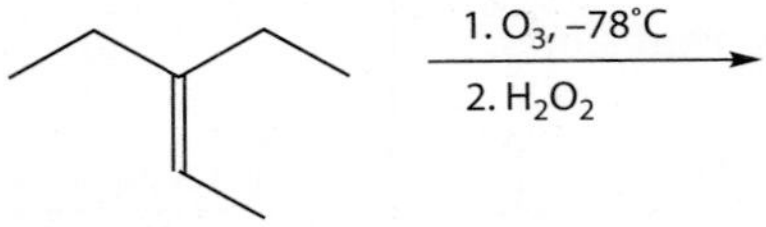

(c) OsO_4, Me_3COOH

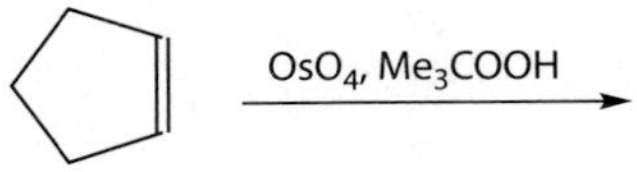

(d) Br_2, CCl_4

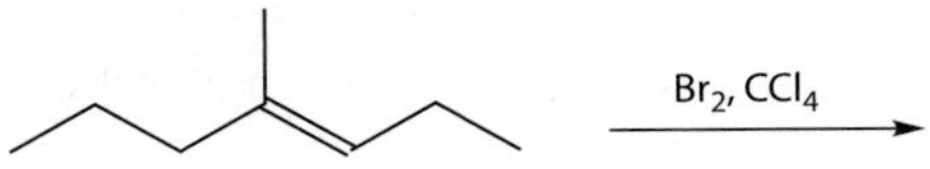

(e)   ether   NaOH H_2O_2

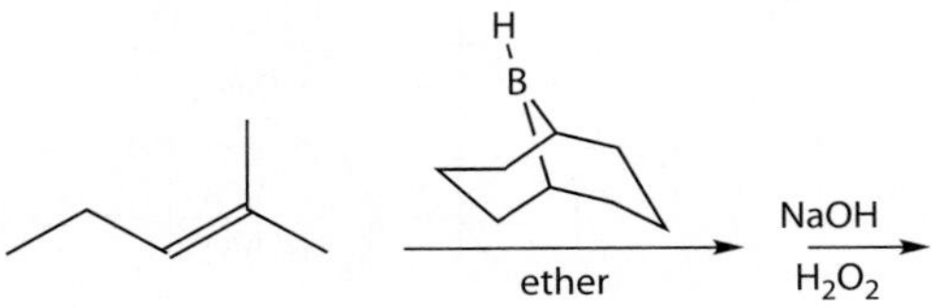

(f)   HOCl

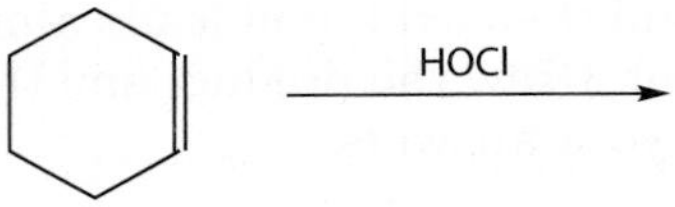

(g)

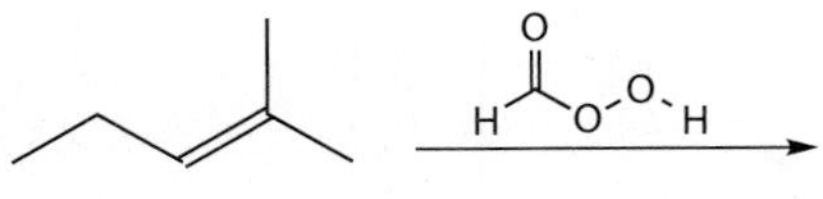

(h) Br_2, CCl_4

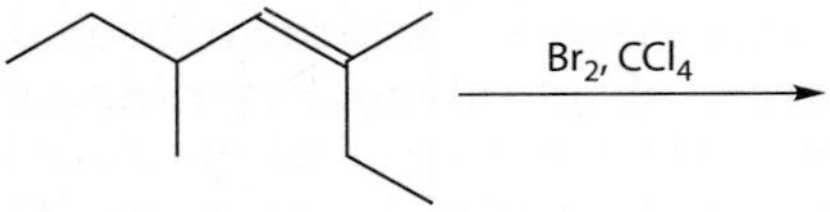

(i)   HBr

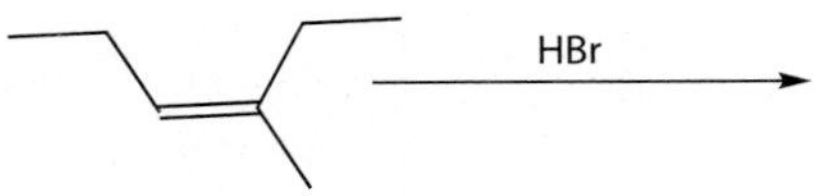

(j)

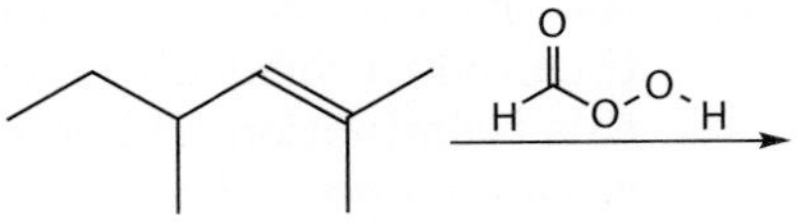

(k) I_2, CCl_4

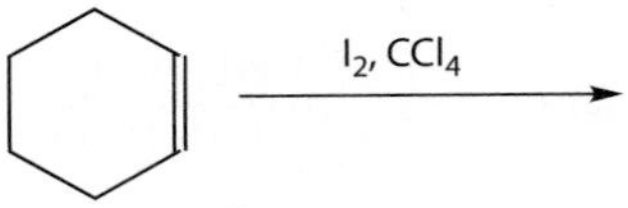

(l)   catalytic H_2SO_4 H_2O

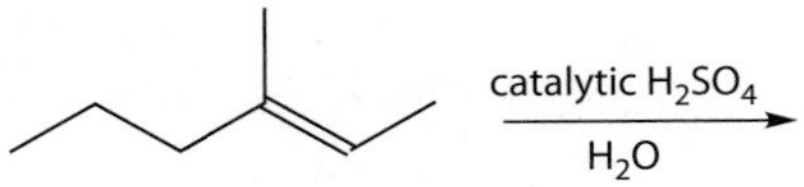

(m)   HBr

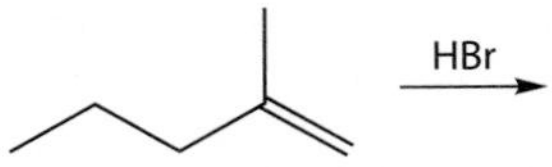

(n)   HCl

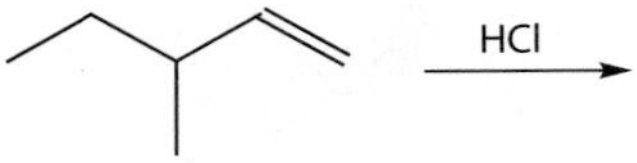

(o) Br_2, CCl_4

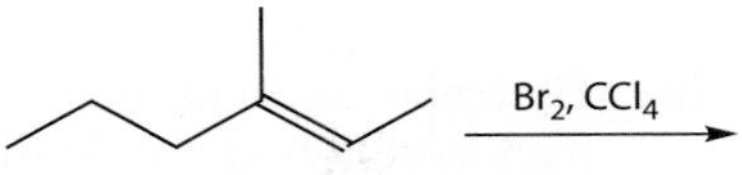

(p)   HBr

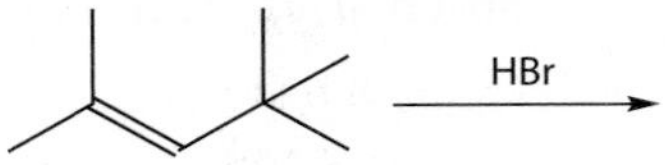

(q) Cl_2, CCl_4

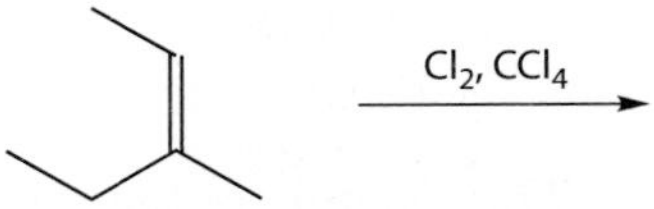

(r)   1. O_3, −78°C   2. CH_3SCH_3

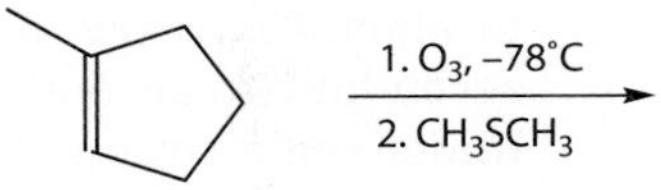

(s) CH_3CO_3H

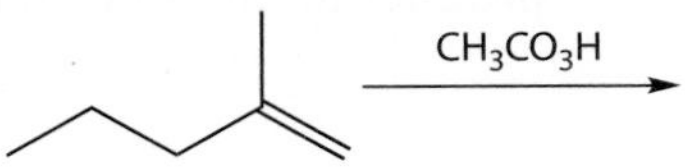

10.83   Give the major product for each reaction:

(a) 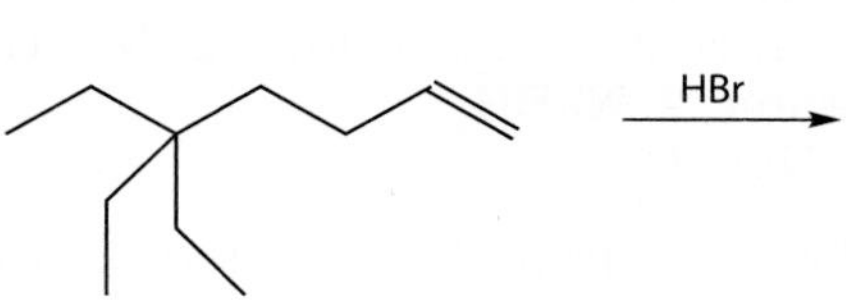 $\xrightarrow{\text{HBr}}$ (l) 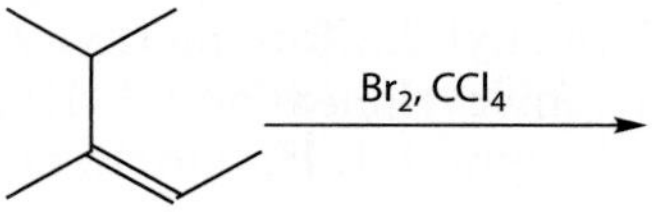 $\xrightarrow{\text{Br}_2,\text{CCl}_4}$

(b) $\xrightarrow[\text{CCl}_4]{\text{Br}_2}$ (m) 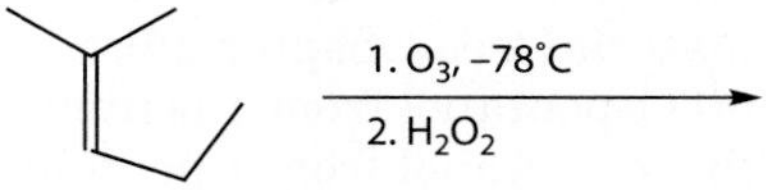 $\xrightarrow[\text{2. H}_2\text{O}_2]{\text{1. O}_3,\,-78°\text{C}}$

(c) $\xrightarrow{\text{I}_2,\,\text{CCl}_4}$ (n) $\xrightarrow{\text{H}_2\text{O, HgSO}_4}$

(d) $\xrightarrow{\text{HOCl}}$ (o) $\xrightarrow[\text{NaHSO}_3,\text{H}_2\text{O}]{\text{OsO}_4}$

(e) $\xrightarrow{\text{H}_2\text{O, cat. H}^+}$ (p) 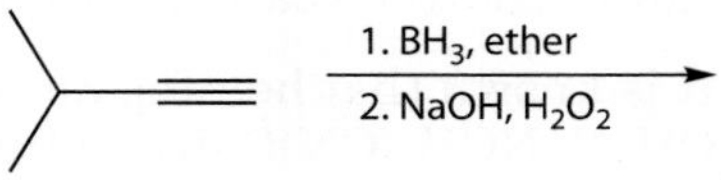 $\xrightarrow[\text{2. NaOH, H}_2\text{O}_2]{\text{1. BH}_3, \text{ether}}$

(f) $\xrightarrow{\text{H}_2\text{O}}$ (q) 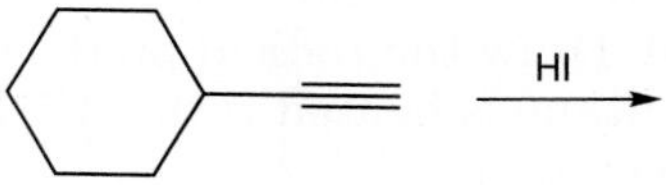 $\xrightarrow{\text{HI}}$

(g) $\xrightarrow{\text{2 HBr, CCl}_4}$ (r) 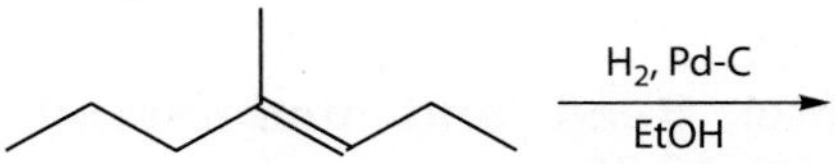 $\xrightarrow[\text{EtOH}]{\text{H}_2,\text{ Pd-C}}$

(h) $\xrightarrow{\text{HCl}}$ (s) 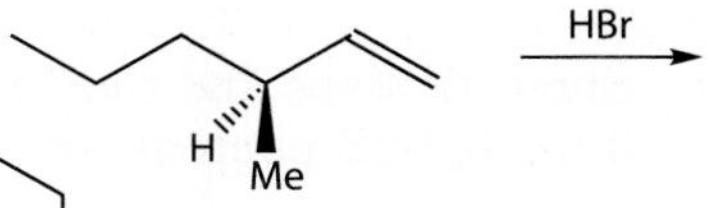 $\xrightarrow{\text{HBr}}$

(i) $\xrightarrow[\text{2. NaBH}_4]{\text{1. HgCl}_2, \text{H}_2\text{O}}$ (t) 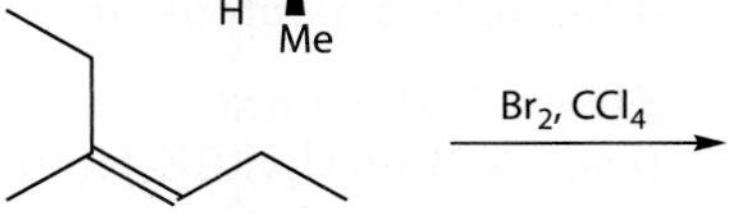 $\xrightarrow{\text{Br}_2, \text{CCl}_4}$

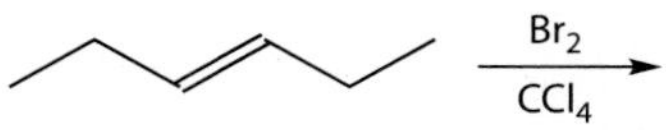

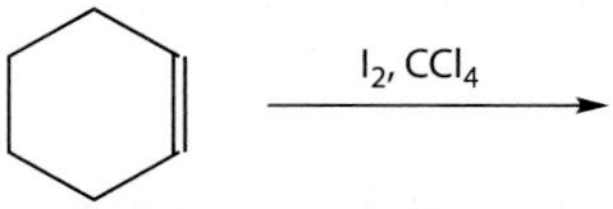

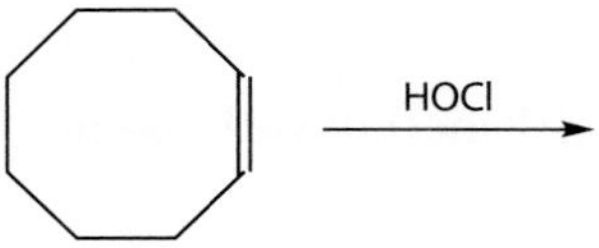

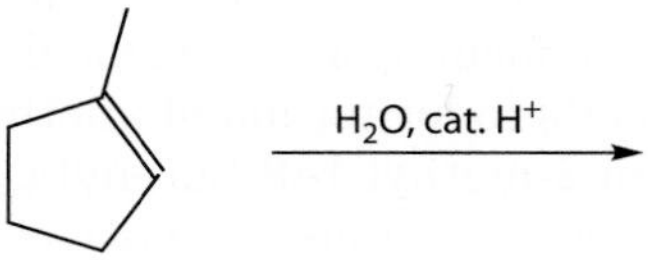

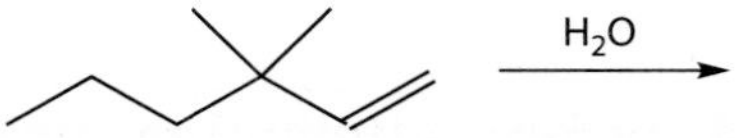

(j) $\xrightarrow[\text{2. H}_2\text{O}_2]{\text{1. O}_3,\,-78°\text{C}}$ (u) 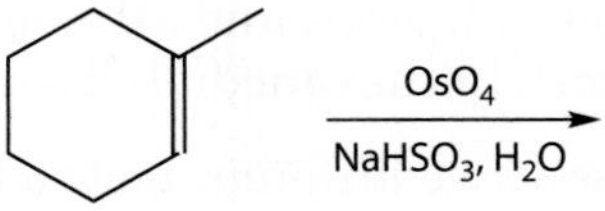 $\xrightarrow[\text{NaHSO}_3, \text{H}_2\text{O}]{\text{OsO}_4}$

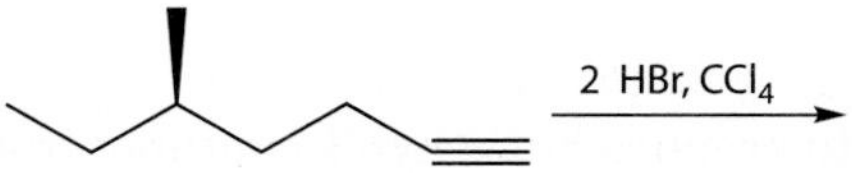

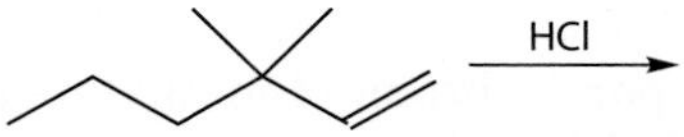

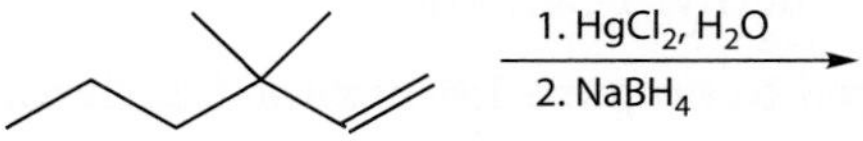

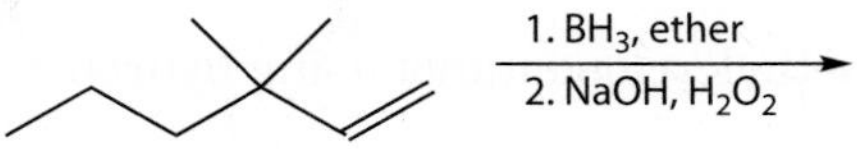

(k) $\xrightarrow[\text{2. NaOH, H}_2\text{O}_2]{\text{1. BH}_3, \text{ether}}$ (v) 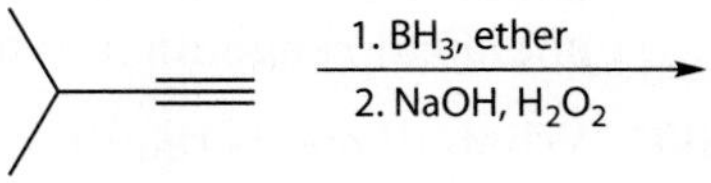 $\xrightarrow[\text{2. NaOH, H}_2\text{O}_2]{\text{1. BH}_3, \text{ether}}$

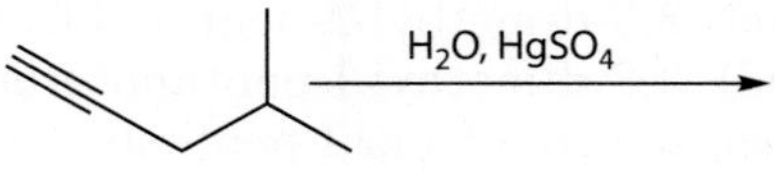

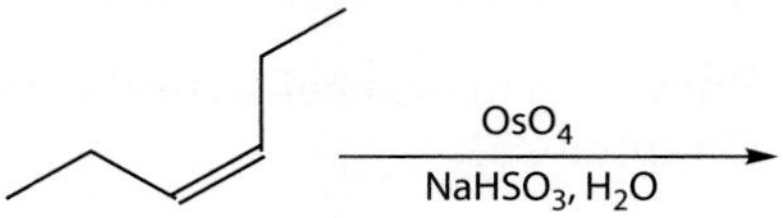

10.84   Give the major product formed from each of the following reactions:
        (a) cyclohexene + 1. Hg(OAc)$_2$, water; 2. NaBH$_4$
        (b) 6-phenyl-2,3,3-trimethyl-2-hexene + 1. Hg(OAc)$_2$, methanol; 2. NaBH$_4$
        (c) 1-ethylcyclohexene + 1. Hg(OAc)$_2$, water; 2. NaBH$_4$
        (d) 1-hexyne + 1. Hg(OAc)$_2$, ethanol; 2. NaBH$_4$

10.85   Show a brief synthesis for each of the following from the indicated starting
        material (see Chapter 25):
        (a) 1-pentanol from 1-pentene
        (b) 2-pentanol from 1-pentene
        (c) 3,3-dimethyl-2-pentanol from 3,3-dimethyl-1-pentene
        (d) 3,3-dimethyl-1-pentanol from 3,3-dimethyl-1-pentene
        (e) butanal from 1-pentene
        (f) 2-hexanone from 4-methyl-3-octene

10.86   Show a synthesis of 2,3-dibromo-2,3-diiodohexane acid from 2-hexyne (see
        Chapter 25).

10.87   Briefly discuss why the cationic polymerization of 1-methoxyethene might be
        slower than the cationic polymerization of ethene.

10.88   It is known that heating an alkene with *azo-bis*-isobutyronitrile [Me$_2$C(CN)
        CH$_2$N=NCH$_2$(CN)CMe$_2$; called AIBN] leads to formation of a radical
        [(CN)Me$_2$C•] and reaction with an alkene leads to formation of another radi-
        cal. Draw the radical product expected when 2-methyl-1-(3-butenyl)cyclo-
        pentane is heated with AIBN and comment on the regioselectivity of radical
        formation.

**Note: Read   and   understand   Chapter   25   before   attempting   problems
10.89–10.96.**

10.89   Show all steps and reagents required to prepare 3-methyl-3-pentanol from (a)
        3-methyl-2Z-pentene and (b) 3-methyl-1-pentene.

10.90   Show all steps and reagents required to prepare 3-bromo-3-methyl-pentane
        from (a) 3-methyl-2Z-pentene and (b) 3-methyl-1-pentene.

10.91   Show all steps and all reagents required to prepare 1,6-hexanedicarboxalde-
        hyde (1,6-hexanedial) from cyclohexene.

10.92   Show two different methods, with all steps and reagents, to prepare 1-hexene
        oxide (1-butyloxirane) from 1-hexene. Different methods mean different
        chemical reagents and also different types or classes of chemical reagents.

10.93   What alkene is required to prepare 2-methylcyclopentanol using hydroboration?

10.94   Describe all organic starting materials and all reagents required to prepare
        3-pentanone and butanoic acid from a single alkene.

10.95   Describe all organic starting materials and all reagents required to prepare
        1-cyclobutyl-1-butanone and 3-methylbutanal from a single alkene.

10.96   Describe two different reagents with appropriate reaction conditions that can
        be used to convert 2E-hexene to the corresponding diol. Name it.

10.97 Give a complete mechanism for the following reaction:

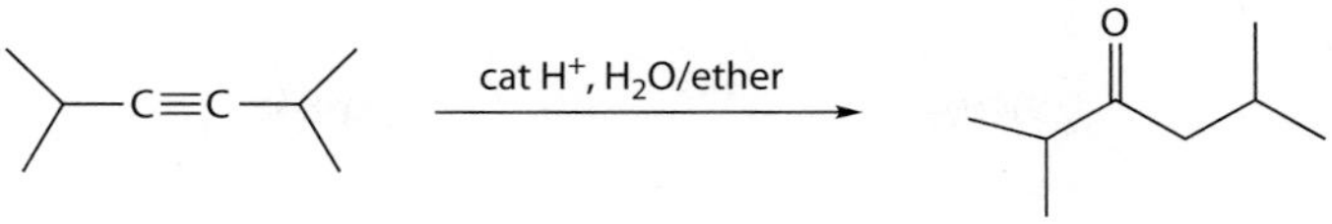
C≡C
cat H+, H2O/ether
O

10.98 Explain why the reaction of *trans*-3-hexene and bromine is diastereospecific. Draw the product of this reaction.

10.99 Give the structure for each of the following:
(a) *E*-2,9-dibromo-3-methyl-4-(2,2-dimethylpropyl)-3-nonene
(b) 1,2-difluorocyclohexene
(c) *cis*-3,4-dimethyl-3-octene
(d) 1,5-heptadiene
(e) 4-phenyl-1-hexyne

10.100 Give the complete mechanism for the following transformation:

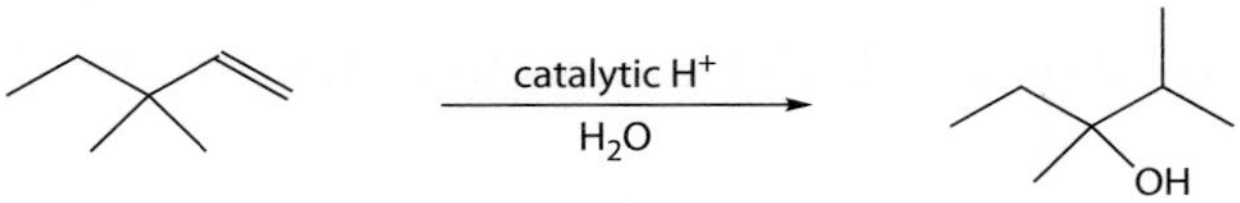
catalytic H+
H2O
OH

10.101 Briefly explain why the reaction of 3,3-dimethyl-1-pentene and BH_3 in ether leads to a borane with the boron atom at C1 rather than C2. Draw this product.

10.102 Give the complete mechanism for the following reaction:

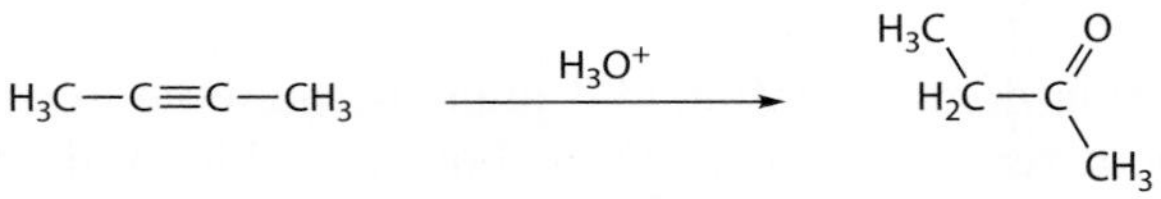
H3C—C≡C—CH3
H3O+
H3C
O
H2C—C
CH3

10.103 Give the complete mechanism for the following reaction:

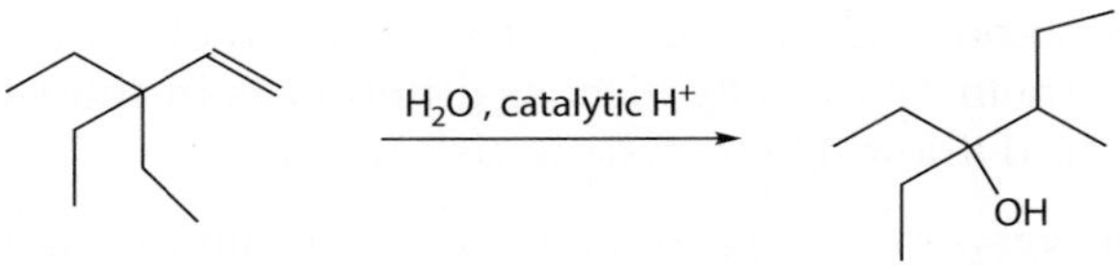
H2O, catalytic H+
OH

10.104 Which of the following reagents react with cyclopentene to give only the *trans* product? Explain.
(a) HBr
(b) Br_2
(c) BH_3
(d) H_2/Pd
(e) $SOCl_2$

10.105 Give a detailed mechanism for the following reaction:

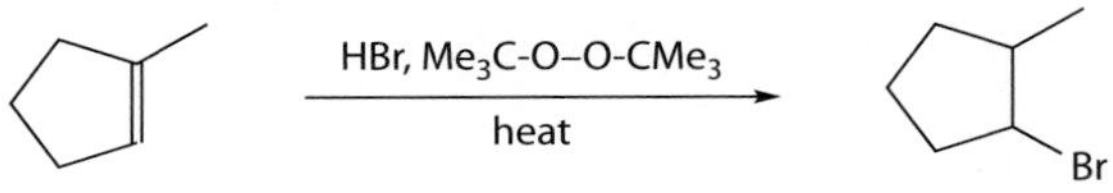
HBr, Me3C-O-O-CMe3
heat
Br

10.106 Give a complete mechanism for the following reaction:

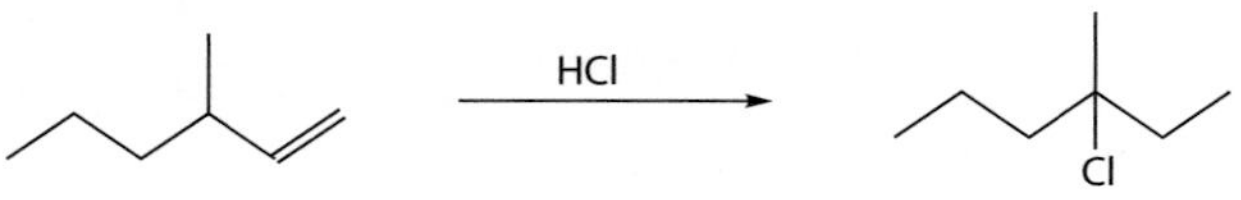

Synthesis Problems

Note: Chapter 25 must be read and understood before attempting the following problems.

10.107 Show a synthesis of 1,6-hexanedioic acid from cyclohexene.

10.108 Show a synthesis of 3-ethoxy-3-methylhexane from 3-methyl-2-hexene.

10.109 Show a synthesis of 2,3-dipropyloxirane from *cis*-3-hexene.

10.110 Show a synthesis of 3-methyl-3-hexanol from 3-methyl-1-hexene.

Spectroscopy Problems

Note: Chapter 24 must be read and understood before attempting these problems.

10.111 Oxymercuration-demercuration of 3-phenyl-1-pentene gives an alcohol that is formed without rearrangement. Draw both possible products and use differences in the IR and proton NMR to compare the unrearranged and the rearranged products in order to distinguish them.

10.112 Hydroboration and oxidation of 2-pentyne can lead to two possible products. Draw both of them and briefly discuss differences in their IR and proton NMR that would allow one to distinguish them.

10.113 When 3,4-diethyl-3-hexene is treated with potassium permanganate and aqueous hydroxide, the expected product is a diol. It is possible for oxidative cleavage of the C=C unit to occur if the solution is too concentrated or heated too much. Draw the expected diol and the oxidative cleavage product or products and use IR and proton NMR spectroscopic differences to distinguish them.

10.114 When methylenecyclopentene is reacted with HBr and a peroxide, the bromide product is not the product formed when HBr is added without a peroxide. Spectroscopy can be used to verify this result. Draw both the regioisomeric products of this reaction and use IR and proton NMR to distinguish between them.

10.115 Using the provided spectral data, identify the structure of this compound.
Mass spectral data: M = 74 (100%), M + 1 = 75 (4.44%), M + 2 = 76 (0.30%).
The molecular ion is weak relative to the base peak.

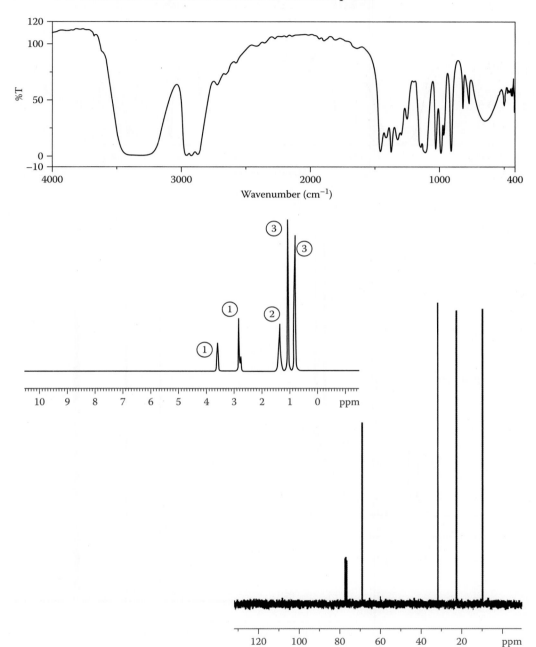

10.116 Using the provided spectral data, identify the structure of this compound with the formula $C_6H_{14}O$:

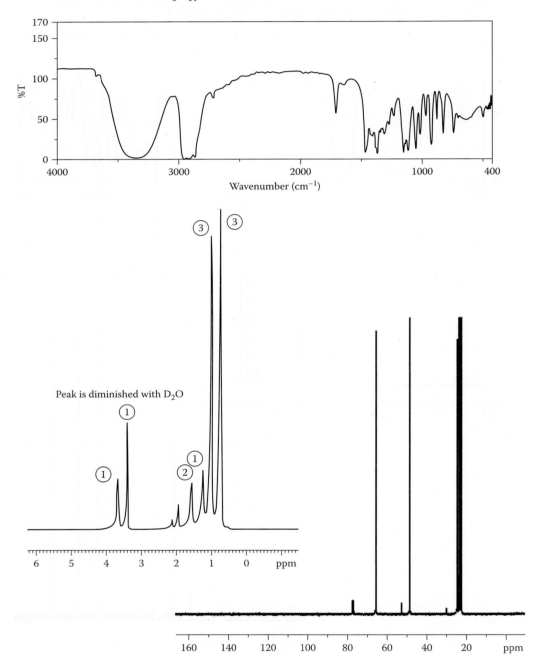

10.117 Identify the following molecule based on the MS, IR, and proton NMR data:
102 (P+2, 0.42%), 101 (P+1, 6.66%), 100 (P, 100%)
MS: 29 (31), 42 (68), 43 (22), 55 (50), 57 (21), 58 (45), 71 (100), 85 (8), 100 (very small)

IR: 3045–2863, 1480–1432, 1262–1214 cm^{-1}
^{1}H NMR: 2.9 (m, 1H), 2.73 (t, 1H), 2.45 (d, 1H), 1.53 (m, 2H), 1.45–1.39 (m, 4H), 0.92 (t, 3H) ppm

10.118 Identify the following molecule based on the MS, IR, and proton NMR data:
166 (P+2, 98%), 165 (P+1, 6.66%), 164 (P, 100%)
MS: 41 (25), 43 (93), 55 (59), 56 (15), 57 (11), 71 (32), 85 (100), 135 (8), 137 (7), 164 (very small)
IR: 2964–2864, 1515–1435, 1382–1344, 1302, 1271–1263, 1166–1168, 1048–1014, 615–610 cm^{-1}
^{1}H NMR: 3.46 (d, 2H), 1.45 (m, 1H), 1.43 (m, 4H), 0.9 (broad t, 6H) ppm

10.119 Identify the following molecule based on the MS, IR, and proton NMR data:
144 (P+4, 35%), 142 (P+2, 68%), 141 (P+1, 6.66%), 140 (P, 100%)
MS: 41 (37), 42 (44), 55 (100), 68 (48), 69 (23), 104 (9), 106 (3), 140.0 (very small)
IR: 2994–2845, 1459–1436, 1348–1248, 1053–1000, 981–836, 1048–1014, 663, 469 cm^{-1}
^{1}H NMR: 3.55 (broad t, 4H), 2.0–1.35 (broad m, 6H) ppm

10.120 A molecule **A** has the following data:
MS: 152 (1.1%, P+2), 150 (1.1%, P), 71 (100%, B), 70 (18%), 55 (59%), 43 (96%), 42 (16%), 41 (42%), 29 (28%), 27 (21%)
IR: 2930–2970, 1460, 1138
HNMR: 1.80 (2H, q), 1.74 (6H, s), 1.06 (3H, t)
Identify the molecule **A**. Suggest the structures for two reasonable starting materials that can be used to form **A** by reaction with HBr. Give the structure of the actual starting material given the following data:
MS: 71 (2.1%, P+1), 70 (70%, P), 55 (100%, B), 53 (8%), 42 (30%), 41 (24%), 39 (25%), 29 (23%), 27 (15%)
IR: 2885–2970, 1651, 1466, 1377, 887, 869 cm^{-1}
HNMR: 4.67 (2H, m), 2.02 (2H, q), 1.73 (3H, s), 1.03(3H, t) ppm
Note that the proton NMR has been simplified in terms of multiplicity to accommodate the fact that the text has not described NMR beyond first-order spectra.

10.121 The product of an ozonolysis reaction is a single product, **A**, that has the following spectra. Identify its structure:
MS: 107 (7.8%, P+1), 106 (100%, P), 105 (94%), 78 (16%), 77 (93%), 51 (37%), 50 (18%)
IR: 3031, 2860, 2820, 2738, 2696, 1703, 1664, 1697, 1584, 1456, 1391, 1311, 1204, 1168, 828, 46, 686, 650 cm^{-1}
HNMR: 10.02 (1H, s), 7.87–7.56 (5H, m) ppm
The precursor to **A** was treated with ozone at –78°C and the resulting ozonide was treated with dimethyl sulfide to give **A**. Given the following spectral data, draw the structure of the precursor as well as the ozonide intermediate:
MS: 181 (14.5%, P+1), 180 (100%, P), 179 (81%), 178 (44%), 165 (31%), 89 (15%), 76 (10%)
IR: 3079, 3060, 3034, 3021, 1698, 1578, 1496, 1452, 1030, 985, 973, 966, 910, 767, 693, 541, 526 cm^{-1}
HNMR: 7.48–7.21 (10H, m), 7.15 (2H, s) ppm

Nucleophiles

Lewis Base-Like Reactions at sp³ Carbon

11

Chapter 10 discusses the reaction of a nucleophile with a carbocation in the second step of a two-step mechanism, where the carbocation is generated by the reaction of a mineral acid with an alkene (Section 10.2). In Chapter 6 (Section 6.7), a nucleophile is defined as a species that donates two electrons to carbon, to form a new covalent σ-bond to carbon. A carbocation has a positively charged carbon atom, and a nucleophile such as chloride ion, bromide ion, or iodide ion donates two electrons to that electron-deficient atom. The Lewis base analogy for a nucleophile is appropriate in these reactions, in which a carbocation **1** reacts with a nucleophile, iodide ion, to form a new C–I bond in **2**.

The reaction of a nucleophile with an electrophilic carbon atom is not limited to carbocations. A nucleophile can also donate electrons to a polarized carbon atom ($C^{\delta+}$) such as the one in **3**, where the presence of an electronegative atom X (such as Cl or Br) generates an induced dipole at carbon. The nucleophilic iodide ion donates two electrons to the positive carbon in **3** to form a new C–I σ-bond in **4**. However, if a new bond is formed to a carbon that has four covalent bonds, one of those bonds must break. The relatively weak C–X bond breaks as the C–I bond is formed, and the products are alkyl iodide **4** and the X⁻ ion. The conversion of **3** to **4** constitutes a new type of reaction, a substitution at an sp³ hybridized carbon.

removing hydrogen atoms via a radical process that leads to substitution and formation of alkyl chlorides and alkyl bromides. Both NBS and NCS can be used for controlled radical bromination or chlorination of allylic and benzylic systems.

- A molecule with a particular functional group can be prepared from molecules containing different functional groups by a series of chemical steps (reactions). This process is called synthesis: The new molecule is synthesized from the old one (see Chapter 25).
- Spectroscopy can be used to determine the structure of a particular molecule and can distinguish the structure and functionality of one molecule when compared with another. See Chapter 14.
- Substitution reactions are important in biological transformations.

11.1 Alkyl Halides, Sulfonate Esters, and the Electrophilic C–X Bond

An old and interesting experiment mixes **5** with sodium iodide (NaI) using acetone as a solvent. This mixture is heated at the boiling point of acetone (refluxed) and the isolated product is 1-iodo-3-methylbutane (**6**), in 66% yield.[1] Sodium bromide (NaBr) is formed during the reaction, as the sodium iodide is consumed. In terms of the structural changes, iodide substitutes for the bromine, producing bromide ion (Br⁻). Iodide is a nucleophile because it reacts at $C^{\delta+}$, breaking the C–Br bond and transferring the electrons in the C–Br bond to bromine. This transformation constitutes a new type of reaction that is called **nucleophilic aliphatic substitution**. In other words, a nucleophile (iodide) attacks an sp³ carbon atom (aliphatic) and replaces bromine with iodine (substitution). This reaction occurs with alkyl halides. New methods of the preparation of alkyl halides will be introduced in later sections, but Chapter 10 showed that an alkyl halide is prepared from an alkene.

11.1 **Draw the products formed when (a) 1-hexene reacts with HBr, (b) 1-methylcyclopentene reacts with HCl, and (c) 3-methyl-1-hexene reacts with HI.**

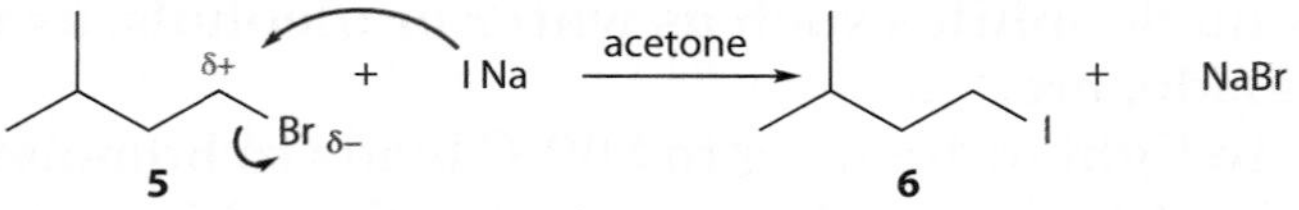

An alkyl halide (R–X) has a halogen atom (X) attached to a carbon atom (C–F, C–Cl, C–Br, and C–I) and the halogen is treated as a substituent in the

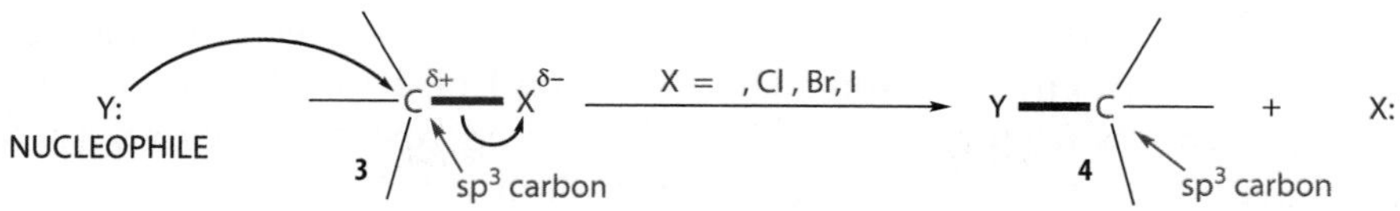

Figure 11.1 Nucleophilic attack at an sp³ carbon bearing a halide leaving group.

nomenclature system (see Chapter 4, Section 4.6). Polarized covalent bonds were introduced in Chapter 3 (Section 3.7); in all alkyl halides, the halogen is more electronegative than the attached carbon, leading to a dipole that makes the carbon atom electrophilic (it is δ+), as shown in Figure 11.1. Chlorine, bromine, and iodine are rather large atoms (see Chapter 5, Section 5.6.1, for molecular models of simple alkyl halides); this makes the C–X bond elongated and rather weak and thus easy to break.

The bond polarization and inherent weakness of the C–X bond combine to make this species susceptible to reaction with a nucleophile, Y⁻. If the nucleophile donates two electrons to the electropositive carbon atom, a new C–Y bond will form and the C–X bond will break, generating X⁻, as shown in Figure 11.1 for **3→4**. This is a substitution reaction in which a nucleophile displaces another group at an aliphatic (sp³) carbon atom: **nucleophilic aliphatic substitution**. *Alkyl halides react with a suitable nucleophile by nucleophilic aliphatic substitution.*

Another class of compounds, known as sulfonate esters (see Figure 11.2), is derived from sulfonic acids RSO_3H (see Chapter 16, Section 16.9). The C–O bond in Figure 11.2 is polarized such that the carbon is electrophilic, analogous to alkyl halides. Sulfonate esters are relatively easy to form, and they react with nucleophiles because the sulfonate anion (RSO_3^-) is displaced as readily as a halide in this reaction. As with alkyl halides, the nucleophile attacks the δ+ carbon to form a new C–Y bond, while breaking the C–O bond to form a sulfonate anion, RSO_3^-, which is resonance stabilized.

11.2 **Draw the structure of 2R-bromohexane, the butanesulfonate ester of cyclopentanol, and *trans*-2-iodomethylcyclopentane.**

11.3 **Draw all resonance forms for the hydrogen sulfate anion, the conjugate base of sulfuric acid. Now replace the OH of the hydrogen sulfate anion with OR, where R is methyl, and draw all resonance forms associated with the sulfonate anion in Figure 11.2.**

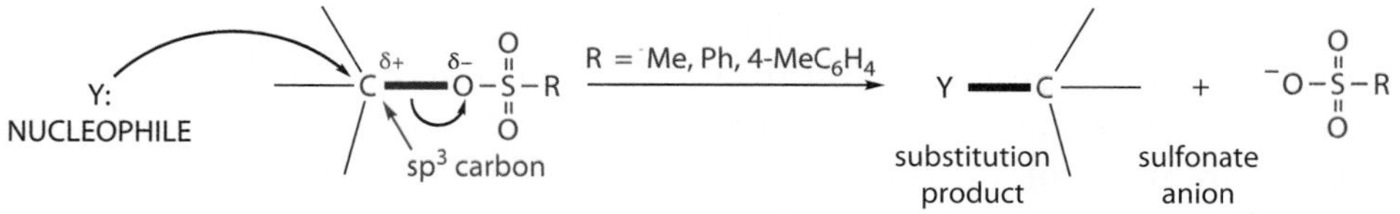

Figure 11.2 Nucleophilic attack at an sp³ carbon bearing a sulfonate ester leaving group.

In Figures 11.1 and 11.2, the nucleophile attacks (donates two electrons like a Lewis base) the sp³ carbon atom; the C–X bond of **1** (or the C–O bond in **2**) is broken, and the two electrons in the C–X bond are transferred to X. The group X departs (leaves) to become an independent ion, X⁻ (chloride ion, bromide ion, iodide ion, or a sulfonate anion), and X is referred to as a ***leaving group***. *Although the X group is actually displaced by the nucleophile, it is referred to as a leaving group because it "leaves" the sp³ carbon to become a separate molecule or ion.* The X leaving group does not spontaneously "fly off" or leave; it is expelled by the nucleophile at the time of reaction, so the incoming nucleophile effectively "kicks out" the leaving group, but only after collision with the electropositive carbon. The conversion of **5** to **6** is a nucleophilic aliphatic substitution reaction, where bromine is the leaving group and iodide is the nucleophile. The iodide ion attacks the δ+ carbon atom in **5**, displacing bromide ion to form alkyl iodide **6**. The positive sodium ion is there to counterbalance the charge; note that the sodium ion in NaI is transferred to the bromide ion product (NaBr).

11.2 Nucleophiles and Bimolecular Substitution (the S$_N$2 Reaction)

Figures 11.1 and 11.2 show that a nucleophile donates electrons to an sp³ carbon that is polarized δ+. The introductory reaction that converts **5** to **6** shows that the carbon bearing the bromine in 5 is δ+, the bromine atom is a good leaving group, and iodide of NaI is a nucleophile. Clearly, this is a nucleophilic aliphatic substitution reaction. Certain characteristics of this process that are important to understanding the overall reaction are elaborated upon in this section.

11.2.1 Nucleophilic Approach to the Electrophilic Carbon

An early experiment by Hughes (Edward D. Hughes, England (Wales); 1906–1962), Ingold (Christopher K. Ingold, England; 1893–1970), and a student named Masterman reacted (+)-2S-bromooctane (**7**) with sodium ethoxide (NaOEt) in ethanol and obtained the corresponding ether, 2R-ethoxyoctane, **8**.[2] This experiment introduces alkoxide nucleophile (RO⁻), which is the conjugate base of an alcohol (see Chapter 6, Section 6.2). The electron-rich oxygen of the ethoxide ion (⁻OCH₂CH₃) attacks the carbon bearing the bromine in **7**, displacing the bromine. Ethanol is a solvent and does not appear in the final product.

The interesting observation is that the product, ether **8**, is also optically active, but the absolute configuration is *R*. The actual experiment reacted **7**, with a measured specific rotation of −24.54°, with sodium ethoxide to give **8**; the specific rotation was measured to be +17.1°. The inversion of configuration is observed by monitoring the specific rotation (see Chapter 9, Section 9.2) of both the reactant (**7**) and the final product (**7**). The change in sign indicated that inversion of configuration occurred during the conversion of the *S*-bromide to the *R*-ether. The experimental data show that **complete inversion of configuration is observed**. Apart from the substitution of a leaving group with a nucleophile, a mechanism must explain this result.

11.4 Draw the acid–base reaction of ethanol with the base sodium amide, NaNH$_2$.

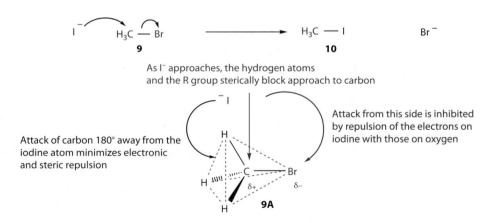

To develop a mechanism that explains the reactions of **5** and of **7**, perhaps it is best to use a very simple alkyl halide such as bromomethane, **9**. In the reaction of bromomethane with iodide, the product is iodomethane (**10**), which clearly shows that reaction occurs at the bromine-bearing carbon. *Note: Iodomethane is toxic. Overexposure may result in pulmonary edema, depression of the CNS, irritation of the lungs and skin, and effects on the kidneys. In addition, effects include nausea, vomiting, vertigo, ataxia, slurred speech, drowsiness, skin blistering, and eye irritation.* Why does the nucleophile attack the bromine-bearing carbon atom? How can inversion of configuration be explained? First, ask if this reaction has an intermediate. Many years of experimentation have shown that *nucleophilic aliphatic substitution does not proceed by an intermediate.* Therefore, any explanation must focus on the transition state of the reaction.

For iodide to displace bromide, it must collide with the polarized carbon atom bearing the bromine atom. The δ+ polarization of that carbon leads to some electrostatic attraction between the electron-rich nucleophile and the electron-deficient carbon. If the iodide ion must **collide** with the sp^3 carbon atom marked in **red** (see **9** in Figure 11.3), the angle of approach is important because that carbon is part of a three-dimensional molecule with a tetrahedral geometry. If the electron-rich iodide approaches the δ+ carbon atom from the

Figure 11.3   Nucleophilic attack on an aliphatic carbon.

direction of the polarized δ– bromine (the **green** arrow), there is electrostatic repulsion because both atoms are electron rich. There is also steric repulsion of the large iodide ion with the large bromine atom, which makes the reaction more difficult. Approach of iodide to carbon over one of the hydrogen atoms (see the **violet** arrow) is also difficult due to a steric interaction (steric hindrance).

The most reasonable approach that minimizes both electronic effects and steric effects is from the bottom of the tetrahedron, or 180° (*anti*) relative to the iodine atom (see the **blue** arrow). Such an approach angle constitutes what is usually called ***back-side attack***. *When a nucleophile approaches an electrophilic sp^3 carbon atom, it will do so by back-side attack, 180° away from the leaving group.*

Back-side attack imposes an important stereochemical consequence: complete inversion of configuration. Remember that the conversion of **7** to **8** proceeds with complete inversion of configuration. This inversion of configuration via back-side attack is the model used to explain experiments by Paul Walden (Latvia; 1863–1957) and it is called **Walden inversion** in his honor. **Is inversion of configuration intuitively obvious?** No! *Inversion of stereochemistry is determined by the observation that the sign of the specific rotation in the product changed relative to a chiral starting material, and this key experiential observation led to the back-side-attack model now used for aliphatic substitution reactions.*

It was previously stated that there is no intermediate in the conversion of **7** to **8** or **9** to **10,** so any explanation of the reaction must involve the transition state. The reaction involves back-side attack of the nucleophile to form a new C–I bond and break the C–Br bond; the transition state is the pentacoordinate species **11** (a ***pentacoordinate transition state***). A practical view of this transition state defines it as the midpoint of the bond-making and bond-breaking processes. As the C–I bond is being formed, the hydrogen atoms are pushed away from the incoming iodide due to steric repulsion. At the midpoint of the reaction, those hydrogen atoms are coplanar with the carbon so that **11** is pentacoordinate (five atoms around carbon). In **11**, five atoms surround the central carbon, so it is sterically crowded and relatively high in energy. It is important to point out that there are ***not five covalent bonds*** to carbon, but rather ***three covalent bonds with one bond made as another is broken.*** *Remember that representation **11** is not an intermediate but rather a transition state.* It is not isolated or even observed. As first discussed in Chapter 7 (Section 7.3), the transition state is the high-energy point of an energy-reaction coordinate curve (see Figure 11.4).

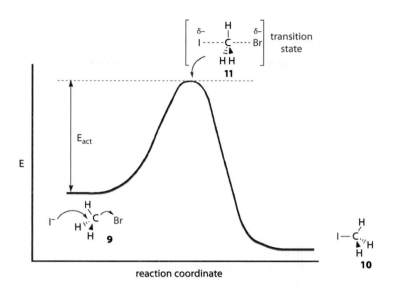

Figure 11.4 Energy of activation for the reaction of iodide with bromomethane.

The energy required to generate **11** from the starting materials is the energy of activation (E_{act}) for the reaction (see Chapter 7, Section 7.6). The higher the E_{act} is, the more difficult it will be to reach the transition state and for the overall substitution reaction to occur. As the reaction continues, the C–I bond is fully formed and the bromide departs to give a new product (iodomethane) and bromide ion. Because NaI is the source of iodide, the counter-ion for bromide is also sodium (NaBr).

Observe what has happened at carbon. Because the motion of the hydrogen atoms is away from the incoming nucleophile, the configuration of the carbon atom they are attached to is **inverted** in the product. Because iodomethane does not have a stereogenic center, the inversion cannot be detected, but *this transition state model of inversion nicely explains the earlier inversion of configuration of 7 to 8. Indeed, a pentacoordinate transition state such as 11 is taken as the transition state for all nucleophilic aliphatic substitution reactions, and they all proceed with 100% inversion of configuration (consistent with back-side attack of the nucleophile).*

11.5 Draw the transition state and final product when 3S-chloro-2-methylhexane reacts with KI.

The reactions presented thus far are examples of **nucleophilic aliphatic substitution** reactions, in which a nucleophile substitutes for a leaving group at an aliphatic carbon. Nucleophilic aliphatic substitution proceeds with back-side attack and inversion of configuration in a bimolecular process. In other words, the nucleophile collides with the electrophilic carbon atom to initiate the reaction. A shorthand symbol is used to describe this reaction: S_N2, *where S means substitution, N means nucleophile, and 2 means bimolecular, or nucleophilic bimolecular substitution.* Once a reaction is identified as S_N2, back-side attack and inversion of configuration are understood to be part of that process. *Transition state **11** is now referred to as the S$_N$2 transition state* and sometimes as the Walden inversion transition state. In this book, **11** will be called the S$_N$2 transition state.

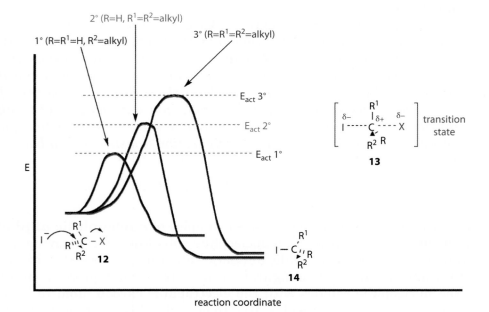

The alkyl halide in an S$_N$2 reaction may be primary, secondary, or tertiary, and differences in reactivity are illustrated with a generic halide (**12**) in a reaction with the nucleophile iodide, where the pentacoordinate transition state is **13** and the final product is **14**. For bromomethane (X = Br, R^1, R^2, R^3 = H), there is some steric hindrance in **13**, so it is relatively high in energy. This steric energy is taken as the standard, and replacing hydrogen atoms in **12** with alkyl groups will increase the energy of transition state **13**. If bromoethane is used (X = Br, R^1 = Me, R^2 and R^3 = H), the methyl group in **13** provides an increased steric hindrance and the reaction is expected to be more difficult because the transition state energy is higher, which means that E$_{act}$ is higher. This fact is shown graphically as an increase in the activation energy for the reaction in Figure 11.5.

Figure 11.5 Energy of activation for the S$_N$2 transition state as a function of the reacting halide.

For the secondary halide 2-bromopropane (X = Br, R^1 and R^2 = Me, R^3 = H), the two methyl groups provide even more steric crowding in **13**, making the energy of that transition state higher (higher E_{act}). In the real world, it is known that the reaction can still proceed, but it is slower than with bromomethane or a primary halide. For the tertiary halide 2-bromo-2-methylpropane (X = Br, R^1, R^2, R^3 = Me), the steric crowding in the pentacoordinate transition state is so high, E_{act} so high, and ***the reaction so slow that it is assumed that it cannot occur***. These structural variations are graphically illustrated in Figure 11.5, where the energy required to attain transition state **13** (E_{act}) increases with the steric bulk around the electrophilic center.

Remember that increased steric hindrance in the transition state, which correlates with a higher E_{act}, makes the reaction more difficult (slower). The steric hindrance is in the transition state—not in the alkyl halide. In other words, the nucleophile may collide with a 3° carbon atom, but the activation energy for the reaction is so high that the S_N2 reaction does not occur. *The energy of activation is so high for a tertiary halide that even when the nucleophile collides with the electrophilic carbon atom, there is not enough energy to begin the bond-making or bond-breaking process.* The S_N2 reaction occurs with primary and secondary halides; however, a primary alkyl halide will react faster than a secondary alkyl halide. A relative order of reactivity for alkyl halides **12** in the S_N2 reaction is therefore: methyl- > 1° > 2° >>>> 3°. This order is supported by the data in Table 11.1,[3] which uses the rate of reaction for bromomethane as a standard and compares the rate of reaction for the other halides to this standard.

If the relative rate for bromomethane is 1, a *smaller* number for the relative rate means the reaction is *slower,* and it is assumed that it is slower because the activation barrier for the reaction is higher. A higher number indicates that the reaction is faster, suggesting that the transition state for that reaction

Table 11.1
Reaction of Alkyl Halides with Potassium Iodide

$$R-Br \ + \ KI \ \longrightarrow \ R-I \ + \ KBr$$

R–X	Relative Rate
CH_3Br	1
CH_3CH_2Br	3.3×10^{-2}
$(CH_3)_2CHBr$	8.3×10^{-4}
$(CH_3)_3Br$	5.5×10^{-5}
$(CH_3)_3C\text{-}CH_2Br$	3.3×10^{-7}
$CH_2=CHCH_2Br$	2.3
benzyl $-CH_2Br$	4.0

Source: Reprinted in part with permission from Streitwieser, A., Jr. 1956. *Chemical Reviews* 56:571. Copyright 1956 American Chemical Society.

is lower in energy. The molecule neopentyl bromide (1-bromo-2,2-dimethylpro-pane, fifth entry in Table 11.1; **12**, where R^1=R^2=H but R^3=CMe$_3$) is interesting because it is a primary halide, but it reacts even more slowly than the tertiary halide. Although it is a primary alkyl halide, CMe$_3$ is a very large group that leads to enormous steric hindrance in the transition state **13**, making the reaction very difficult.

In other words, neopentyl bromide reacts even more slowly than the tertiary bromide because the transition state is more sterically crowded and higher in energy. The result with neopentyl bromide stands in sharp contrast to both allyl bromide (3-bromo-1-propene, CH$_2$=CHCH$_2$Br) and benzyl bromide (PhCH$_2$Br; note the use of "Ph" to abbreviate the benzene ring), which show *increased* rates of the reaction. If it is assumed that the activation energy is lower for these reactions, why is it lower? The π-bond in both of these molecules helps to expel the bromide, as illustrated by **15**. In other words, the π-bond participates in the reaction, in the transition state. This "π-bond-assisted" increase in the rate of the reaction makes the S$_N$2 reaction for both the allylic halide and the benzyl halide faster than a simple primary alkyl halide.

11.6 Draw the transition state 13 for 1-bromo-2E-pentene in a reaction with iodide ion.

11.7 Draw the transition state for the reaction of neopentyl bromide and KI.

15

11.2.2 Reaction Rate

Table 11.1 shows the relative rate for S$_N$2 reactions in which the nucleophile is the same, but the structure of the alkyl halide partner is changed. (Rate of reaction was discussed in Chapter 7, Section 7.11.) The rate of the S$_N$2 reaction of bromomethane with KI is used as the standard for comparison to all other alkyl halides. Analysis of the S$_N$2 reaction indicates that for transition states such as 11 or 13 to be formed, the nucleophile must *collide* with the electrophilic carbon atom. A collision requires two components (nucleophile and halide), and the rate of the reaction is proportional to the concentration of both components. The rate at which this reaction occurs is described by the following expression:

$$\text{Rate (S}_N\text{2)} \propto [\text{nucleophile}] \, [\text{halide}]$$

A reaction with this type of rate expression is **second order** (see Chapter 7, Section 7.11.2). To change the proportionality to an equality, the expression must contain a proportionality constant, k:

$$\text{Rate (S}_N2) = k \text{ [nucleophile] [halide]}$$

This proportionality constant is called the ***rate constant*** (see Chapter 7, Section 7.9). A large rate constant means that the starting material is rapidly converted to product. *For a second-order reaction, the rate constant is determined from a plot of time versus a concentration term that includes both starting materials.* Formally, the "2" in an S_N2 reaction indicates that it follows second-order kinetics, but a second-order reaction is usually associated with the collision process of the nucleophile with the electrophilic carbon atom. *As a practical matter, the fact that the S_N2 reaction is second order means that if the reaction is slow, increasing the concentration of the nucleophile will increase the rate of the reaction (the rate constant remains the same). Similarly, increasing the concentration of the halide will increase the rate.*

Therefore, if one molar equivalent of KI reacts with one molar equivalent of 2-bromopentane at a given rate, the identical reaction of one molar equivalent of 2-bromopentane with 10 molar equivalents of KI should have a rate that is 10 times greater. Table 11.1 reports relative rates rather than absolute values of k (the actual rate constant). Note that the relative rate is calculated by taking the rate of the bromomethane reaction as the standard (relative rate = 1) and all other rates are reported as their ratio relative to bromomethane.

11.8 Determine how much faster the rate will be if 1 mol of bromomethane is treated with 2.5 mol of KI in THF.

11.2.3 The Role of the Solvent

Ethanol is used as a solvent in the reaction that gives **7** in Section 11.2.1. The solvent is an essential component to most reactions. Not only does the solvent modulate energy gain and loss, but it also solubilizes the reactants to keep them in one phase, at least in most cases. The solvent plays a subtler and yet profound role in the S_N2 reaction. To understand that role, it is convenient to organize solvents into two categories: *polar* or *nonpolar* and then *protic* or *aprotic*.

Normally, a ***polar solvent*** is one with a substantial dipole; a nonpolar solvent tends to have a small dipole or none at all. Another measure of polarity is the **dielectric constant**, which is the ability of a solvent to conduct charge and, in effect, to solvate ions. For substitution reactions, a high dielectric constant is associated with a polar solvent and a low dielectric is associated with a less polar solvent. The dielectric constants for several common solvents are shown in Table 11.2.[4] A ***protic solvent*** is one that contains an acidic hydrogen (O–H, N–H, S–H, essentially a weak Brønsted–Lowry acid), whereas an aprotic solvent does not contain an acidic hydrogen. If a solvent is protic, it will have a polarized bond, but there are degrees of polarity that are important to the progress of different reactions.

Table 11.2
Dielectric Constants for Common Solvents

Protic Solvent	Dielectric Constant	Aprotic Solvent	Dielectric Constant
Polar			
HOH	78.5	$H(C=O)NH_2$	109.0
DOD	78.25	DMSO	46.68
EtOH (80% aq)	67	DMF	36.71
HCO_2H	58.5	MeCN	37.5
MeOH	32.7	$MeNO_2$	35.87
EtOH	24.55	HMPA	30
n-BuOH	17.8	Acetone	20.7
NH_3	16.9	Ac_2O	20.7
t-BuOH	12.47	Pyridine	12.4
Phenol	9.78	o-Dichlorobenzene	9.93
$MeNH_2$	9.5	CH_2Cl_2	9.08
$MeCO_2H$	6.15	THF	7.58
		Ethyl acrylate	6.02
		Chlorobenzene	5.71
		Ether	4.34
		CS_2	2.64
		NEt_3	2.42
		Benzene	2.28
		CCl_4	2.24
Nonpolar			
		Dioxane	2.21
		Hexane	1.88

Source: Lowry, T. H., and Richardson, K. S. 1987. *Mechanism and Theory in Organic Chemistry,* 3rd ed., 297–298. New York: Harper and Row. Reprinted by permission of Pearson Education, Inc., Upper Saddle River, NJ.

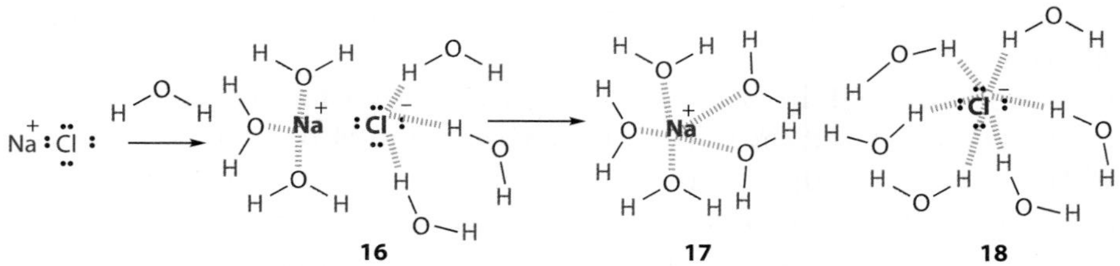

Figure 11.6   Solvation of NaCl in water.

11.9 Draw the chemical structures of all solvents listed in Table 11.2.

To understand the role of a solvent in these reactions, begin with a common nucleophile such as chloride ion, which is readily available in common table salt, sodium chloride. When NaCl is added to water, the electropositive hydrogen atom is attracted to the negative chloride ion and the electronegative oxygen atom is attracted to the positive sodium. As the polarized hydrogen atom in water hydrogen bonds to the chloride, it "pulls" on the chloride as the polarized oxygen atom of water associates with and thereby "pulls" on the sodium, represented by **16**. Therefore, in the presence of water the charges begin to separate. Eventually, water molecules encroach between the two atoms and the sodium ion is completely surrounded by water (**17**), as is the chloride ion (see **18**). This phenomenon is called **solvation** (see Figure 11.6) and it occurs because the hydrogen atoms (δ+) of water are attracted to the chlorine ion, and the oxygen atoms (δ−) of water are attracted to the sodium ion. A polar, protic solvent like water has two "poles," so it is possible to solvate *both cations and anions* (**17** and **18**).

A polar and protic solvent such as water can be contrasted with an aprotic solvent such as ether. The oxygen atom is δ−, and it can certainly coordinate to cations. The δ+ atom is a tetrahedral carbon, however, and, due to steric repulsion between the atoms, it is difficult for a negative anion or a negatively polarized atom to approach the carbon. Therefore, an aprotic solvent such as diethyl ether may solvate cations but not anions (see Figure 11.7) and *there can be no separation of charges (ions)*. A simple experiment to examine the difference between these two solvents is to observe that sodium chloride will dissolve in water, but not in diethyl ether. The more polar the solvent is in

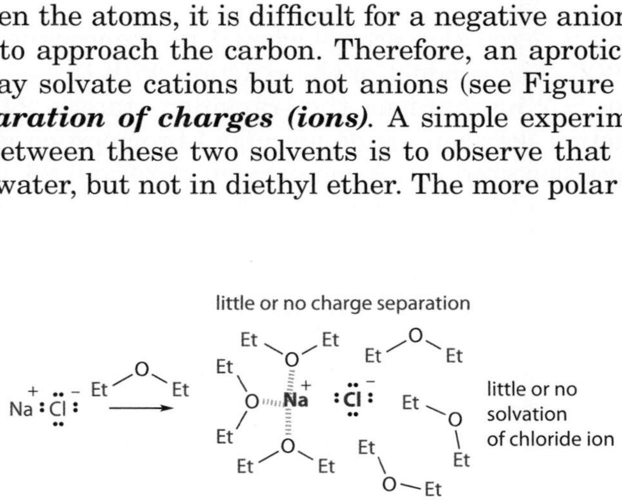

Figure 11.7   NaCl in diethyl ether.

terms of its ability to solvate both anions and cations, the more efficient will be the solvation effect.

Transition state **11** in Section 11.2.1 was shown for the conversion of **9** to **10**. The charge distribution is such that the iodide is δ−, the central carbon is δ+, and the leaving group X is δ−. If the solvent is water, the iodide ion will be solvated (surrounded by water molecules) **before** it collides with the sp³ carbon atom. This means that solvation will impede the collision of iodide with the carbon atom, and the S$_N$2 reaction will be slower. Water solvates both the anion and cation and the net result is that the S$_N$2 reaction is slower.

If the solvent is changed to diethyl ether, the negatively charged nucleophile is **not** well solvated, and it is easier to collide with the carbon atom. This suggests that *an aprotic solvent is preferred in S$_N$2 reactions* because the initial reaction with the nucleophile and the carbon atom occurs more readily. The most common polar aprotic solvents used in S$_N$2 reactions are diethyl ether, tetrahydrofuran (THF), dimethyl sulfoxide (DMSO; Me₂S=O), and dimethylformamide (DMF). An example that uses DMF as a solvent is the reaction of 1-chlorophenylmethane (benzyl chloride, **19**) with potassium iodide to give benzyl iodide, **20**.

11.10 Draw the structures of diethyl ether, tetrahydrofuran (THF), and dimethylformamide (DMF).

11.11 Draw the solvated transition state for the reaction of 19 with KI if water is the solvent.

An interesting S$_N$2 reaction occurs when 1-bromopropane (**21**) reacts with dimethylamine (**22**). The product of the reaction with the amine is **23** and if it proceeds by an S$_N$2 mechanism, the transition state is **24**. The transition state for the reaction between iodide ion and 1-bromopropane is **25**, which is analogous to transition state **11**. A comparison of **24** and **25** shows a different charge distribution. In **25**, a polar protic solvent such as water will surround the incoming negatively charged nucleophile and inhibit attack at carbon, as just described, and slow the reaction. In the reaction with dimethylamine, however, *the incoming nucleophile is neutral*. In transition state **24**, the nitrogen takes on a positive charge and the bromide takes on a negative charge. A solvent that separates charge will therefore increase the rate of reaction because it leads to products. If the reaction of **21** and **22** to give ammonium salt **23** is done in water, the rate of the reaction is *faster* than when it is done in an aprotic solvent such as THF. This stands in sharp contrast to the reaction of potassium iodide and **21**, which is faster in the aprotic solvent THF than it is in a protic solvent such as water. The reason for this difference is the different charge distribution in the transition state and the ability of the solvent to solvate and separate charge.

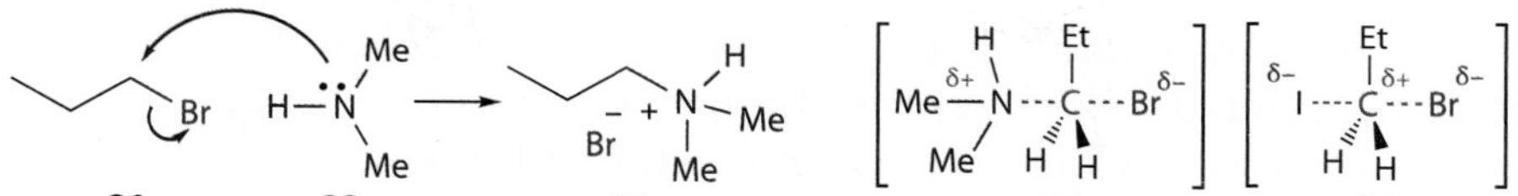

11.12 In which solvent is the reaction of 1-iodobutane and diethylamine faster: aqueous THF or anhydrous THF?

11.2.4 Leaving Groups in the S_N2 Reaction

Why are halides and sulfonates considered to be good leaving groups? The C–X bond (X = halogen) is polarized and it is relatively weak. Moreover, when groups such as Cl, Br, or I leave, they form a very stable halide ion. Because halogens are electronegative, they can easily accommodate the negative charge, and the larger ions disperse the charge more effectively than smaller ions over a larger surface area, again leading to stabilization (see Chapter 6, Section 6.3). Therefore, when a nucleophile attacks the sp^3 carbon of the halide, breaking the C–X bond generates a very stable ionic product, which facilitates the substitution. Most halides are good leaving groups, with iodide the best and fluoride the poorest leaving groups in this series.

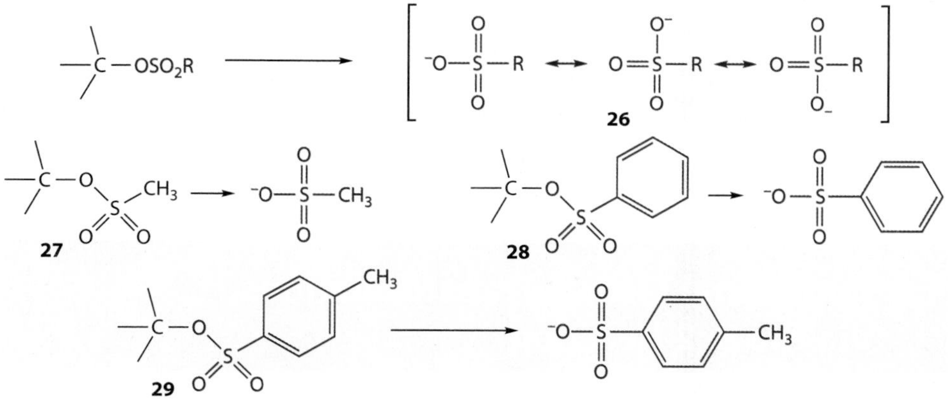

In Figure 11.2, a sulfonate ester is the leaving group, and reaction with a nucleophile breaks the weaker C–O bond to generate a sulfonate anion (**26**). This anion is resonance stabilized and quite stable, which means it is relatively unreactive as a nucleophile. The combination of a weak C–O bond and the stability and poor reactivity of the sulfonate anion makes sulfonate esters good leaving groups. The most common sulfonate esters (see Chapter 20, Section 20.11) are those derived from methanesulfonic acid, benzenesulfonic acid, or 4-methylbenzenesulfonic acid. Formation of sulfonate esters from sulfonic acids or sulfonyl chlorides will be discussed in Section 11.7.3 and in Chapter 20,

Section 20.11.2. However, the use of sulfonate esters as leaving groups is important and some knowledge of their structure is required now. For the moment, it is sufficient to observe that the leaving group of a methanesulfonate ester (**27**; known as a mesylate) is the methanesulfonate anion. The leaving group of a benzenesulfonate ester (**28**) is the benzenesulfonate anion, and the leaving group of a 4-methylbenzenesulfonic ester (**29**; known as a tosylate) is the 4-methylbenzenesulfonate anion.

11.13 Briefly comment on the relative order of leaving-group ability for these fragments: Cl, CH$_3$, OCH$_3$, and NMe$_2$.

The $-OSO_2R$ unit of a sulfonate ester is a leaving group, as explained in Section 11.1.2. Any statement concerning alkyl halides with respect to leaving group ability in preceding discussions largely applies to sulfonate esters. When the mesylate (methanesulfonate ester) of 1-butanol (see **30**) reacts with potassium iodide in THF, the products are 1-iodobutane (**32**) and the potassium salt of the methanesulfonate anion. The transition state is the expected pentacoordinate structure, **31**. In general, sulfonate esters are produced from the corresponding alcohol by reaction with a sulfonyl chloride RSO$_2$Cl (see Chapter 20, Section 20.11.2). Therefore, if a primary or secondary alcohol can be prepared, the OH unit can be converted to a leaving group suitable for S$_N$2 reactions. Two preparations of alcohols were introduced in Chapter 10: hydration reactions (Sections 10.3 and 10.4.5) and the hydroboration reaction of alkenes (Section 10.4.4).

11.14 Draw the transition state and final product when the benzene-sulfonate ester of 4-phenyl-2S-butanol reacts with KI in THF.

11.3 Functional Group Transformations via the S$_N$2 Reaction

There are many examples of the S$_N$2 reaction; it is considered to be one of the more important processes in organic chemistry. This is true because there are many different types of nucleophiles, and the S$_N$2 reaction can incorporate many different functional groups into a molecule. In addition, one functional group can be transformed into a different functional group with inversion of configuration.

11.3.1 Halide Nucleophiles

An S$_N$2 reaction requires that a nucleophile react with an alkyl halide or a sulfonate ester. There are many examples of atoms or groups that react as nucleophiles. The conversion of **5** to **6** in Section 11.1 involved the S_N2 displacement of bromide by the nucleophilic iodide ion. In general, the order of nucleophilic strength for halides in S_N2 reactions is $I^- > Br^- > Cl^- > F^-$. In fact, an iodide ion will react with most primary and secondary alkyl bromides or alkyl chlorides to give an alkyl iodide.

When the reagent is NaI in the solvent acetone (2-propanone; see Chapter 16, Section 16.2), this transformation is known as the Finkelstein reaction, named after Hans Finkelstein (Germany; 1885–1938). Because iodide is a better nucleophile than the bromide ion or the chloride ion, it is unlikely that bromide or chloride will displace iodide ion from an alkyl iodide to give the alkyl bromide or the alkyl chloride. (This point was made in Chapter 7, Section 7.6.) In other words, iodide can displace bromide or chloride, but bromide or chloride will *not* displace iodide. In general, fluoride is a poor leaving group in the S_N2 reaction and it will not be used.

11.15 Offer an explanation of why fluoride ion is a poor nucleophile.

It is interesting to observe that iodide (a good nucleophile) can displace another iodide that is a good leaving group. An early experiment by Edward Hughes and co-workers reacted radiolabeled NaI with 2*S*-iodooctane (**33**), and the initial product was 2*R*-iodooctane, **34**, where the iodine atom has the radiolabel. The ability to distinguish the two iodine atoms allowed them to follow how racemization occurred, because the final product was racemic 2-iodooctane (a 1:1 mixture of **33** and **34**).[5] This result is explained by the fact that the product accompanying 2-iodooctane is the iodide ion. While iodine is a leaving group, the iodide ion is also a nucleophile, so once **34** is formed, it reacts with the nucleophilic I^- to give **33**. Because **33** reacts to give **34** and **34** subsequently reacts to give **33**, the 2-iodooctane that is isolated from the reaction is racemic. Each individual S_N2 reaction proceeds with inversion, but because both forward and reverse reactions are equally possible, the net result is a mixture of **33** and **34**. This is somewhat unusual and most reactions of nucleophiles with electrophilic substrates lead to inversion of configuration in the final isolated product. *When the leaving group is also a good nucleophile, however, remember the configurational problem just described for iodide as a possible complicating factor.*

11.3.2 Alkoxide Nucleophiles (the Williamson Ether Synthesis)

When an alcohol (ROH) reacts with a base, the product is an alkoxide (RO^-). The alkoxide is formally the conjugate base of the alcohol and is, of course, a base. In the presence of electrophilic carbon compounds, alkoxides are also nucleophiles, as pointed out in Section 11.2.1 for the reaction of (2*S*)-bromooctane (**7**) with sodium ethoxide to give **8**. Alkoxides react with primary and secondary halides to form ethers in what is known as the **Williamson ether synthesis**, named after Alexander W. Williamson (England; 1824–1904).

Another example places the focus on the alcohol 1-hexanol (**35**), which is treated with sodium metal to give the conjugate base, alkoxide **36**, which is then treated with iodoethane to give 1-ethoxyhexane (**37**) in 46% yield.[6] Note the use of Et as an abbreviation for ethyl. Alkoxide **36** is the nucleophile in this reaction and the S_N2 reaction with the electrophilic carbon of iodoethane leads to the substitution product, **37**. Sodium metal is an unusual reagent to convert the alcohol to the alkoxide, based on discussions presented in this book. Nowadays, different bases can be used with an alcohol to form the alkoxide, but the pK_a of the conjugate acid of that base must be significantly weaker than the pK_a of an alcohol (see Chapter 6, Sections 6.1–6.3). Bases such as sodium hydride (NaH) or sodium amide (NaNH$_2$) are commonly used, as are organometallic reagents that will be presented in Chapter 15 (see Sections 15.3 and 15.5.2).

Because this is an S_N2 reaction, the alkoxide must react with a primary or secondary halide. There are not very many restrictions on the alcohol precursor to the nucleophile, and alkoxides from 1°, 2°, and 3° alcohols can be used.

11.16 **Explain why the reaction of sodium ethoxide with 2-bromo-2-methylpentane does not give an ether by an S_N2 reaction.**

11.17 **Draw the final product when 3-methyl-3-pentanol is treated with (a) NaH in THF and then (b) benzyl bromide.**

11.3.3 Amine Nucleophiles

Amines react as nucleophiles with alkyl halides in S_N2 reactions, as seen in Section 11.2.1 for the conversion of **21** to **23**. This example is repeated, but the isolated product is tertiary amine **38** rather than ammonium salt **23**. Displacement of bromide ion by the nucleophilic nitrogen in dimethylamine (**22**) leads to formation of *N,N*-dimethylammonium-1-aminopropane, **23**, as described previously. However, this salt is a weak acid, and it is formed in the presence of amine **22**, which is a base (Chapter 6, Section 6.4.1) as well as a nucleophile. A simple acid–base reaction occurs between **23** and **22** to generate the neutral amine (*N,N*-dimethylaminopropane, **38**) along with dimethylammonium bromide. This reaction may not be quite as simple as shown, however.

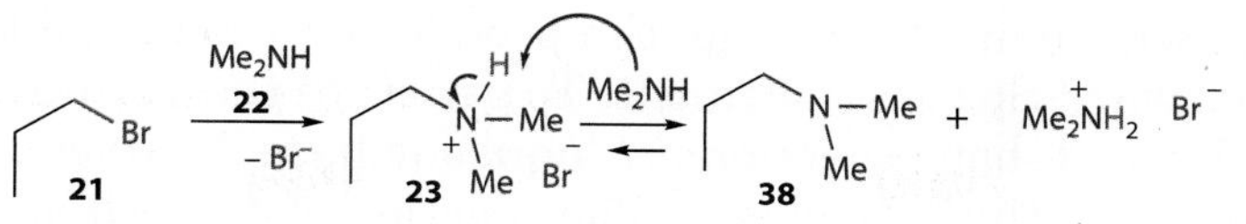

There are problems when primary amines react with structurally simple alkyl halides. If a primary amine such as butanamine (**39**) reacts with iodomethane, the initial product is *N*-methylbutanamine, **40**, via the corresponding ammonium salt. This secondary amine is *more reactive* than primary amine **39** (it is a stronger nucleophile and a stronger base), so it can competitively react with iodomethane to form a tertiary amine (*N,N*-dimethylbutanamine, **41**), also via an ammonium salt. It is also possible for **41** to react with iodomethane to give *N,N,N*-trimethylbutanammonium iodide, **42**. When this reaction occurs with an amine and iodomethane (or any MeX substrate), it is called **exhaustive methylation**. A more generic term is **polyalkylation** of the amine. Polyalkylation occurs because the amine product formed in the initial S_N2 reaction with the alkyl halide is more nucleophilic than the amine starting material and it competes for the alkyl halide. Polyalkylation should always be considered as a problem in reactions of amines with alkyl halides.

11.18 **Draw the final product when benzylamine (PhCH$_2$NH$_2$) is treated with a large excess of iodoethane.**

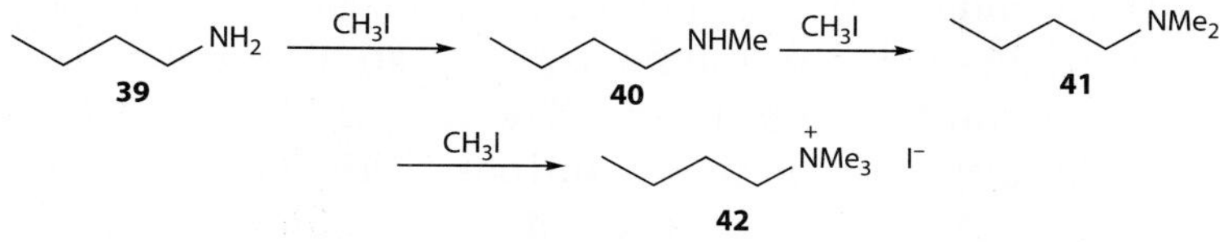

Polyalkylation makes incorporation of a nitrogen into a molecule rather unattractive using the S_N2 reaction with an amine. Using a different nitrogen nucleophile in the reaction with the alkyl halide is an idea that may circumvent the problem of polyalkylation, *but it must be possible to convert that product to an amine. Such nitrogen-containing nucleophiles that are not amines are known as amine surrogates.* One such modification uses the molecule phthalimide (**43**), which is the imide (see Chapter 20, Section 20.6.5) derived from phthalic acid (**47**). If **43** is treated with a strong base such as sodium amide, the phthalimide anion (**44**) is formed as the conjugate base (ammonia is the conjugate acid of this reaction), and it is a good nucleophile in reactions with alkyl halides.

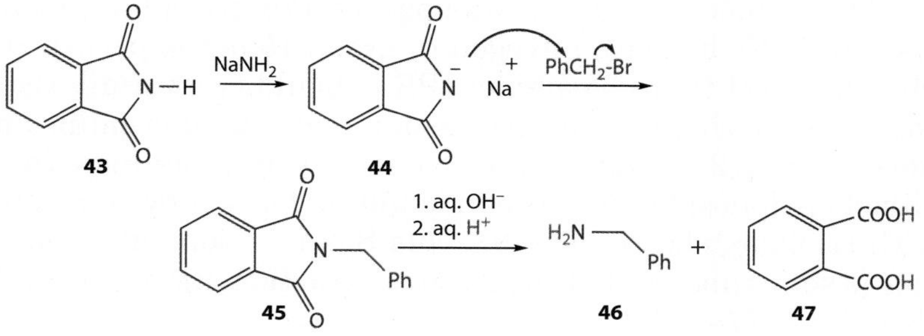

If **44** reacts with benzyl bromide, the product is *N*-benzylphthalimide (**45**) via a straightforward S_N2 displacement. To generate the amine, the imide can be hydrolyzed by acid–base reactions (*1*. aqueous base *2*. aqueous acid) to give the amine (**46**) and phthalic acid (**47**). The reaction of **45** with aqueous hydroxide is an example of an acyl substitution reaction that will be introduced in Chapter 16 (Section 16.8). For several reasons, it will not be discussed at this time. A ***better*** procedure has been developed that treats **45** with hydrazine (NH_2NH_2) to generate amine **46** and a molecule known as a hydrazide, **48**. Hydrazide **48** is easily separated from amine **46**. A hydrazide is one of those specialized chemicals encountered from time to time, but not discussed in detail.

This is clearly a much longer sequence to make a primary amine than simple reaction with ammonia. It does not suffer from problems of polyalkylation, however, and the yield of the amine for the two-step sequence is often higher than the direct reaction with an amine. The fact that this procedure is used more often to make aliphatic primary amines than the reaction of ammonia with an alkyl halide suggests the severity of the problems associated with polyalkylation. *The preparation of secondary or tertiary amines often requires the use of the S_N2 reaction with amines, however.* Other amine surrogates are known, including the azide ion (Section 11.3.4) and the cyanide ion (Section 11.3.5), but both give primary amines as the end product.

11.19 Draw a reaction sequence that prepares (2*S*)-aminopentane from (2*R*)-bromopentane.

11.3.4 Phosphine Nucleophiles

Alkylphosphines (PR_3) are analogs of the parent phosphine (PH_3) in the same ways that alkylamines (NR_3) are analogs of the parent ammonia (NH_3). Phosphorus is directly beneath nitrogen in group 15 of the periodic table, so it is reasonable to expect that phosphines (PR_3) should react similarly to amines (NR_3). In most cases, this is a correct assumption, and phosphines react with alkyl halides by an S_N2 mechanism in the same way as amines. In one example, benzyltriphenylphosphonium bromide (**50**) is formed by reaction of benzyl bromide with triphenylphosphine (**49**). This S_N2 reaction with phosphine and alkyl halides is sometimes called the Arbuzov reaction, named after Alexander Arbuzov (Russia; 1877–1968).

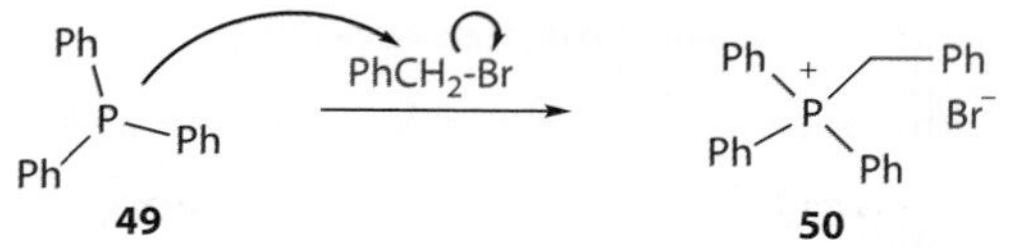

This reaction is used to convert tertiary phosphines such as **49** to phosphonium salts, which will be discussed in the Wittig reaction in Chapter 22, Section 22.10. Primary phosphines (RPH_2) and secondary phosphines (R_2PH) also react to give the corresponding phosphonium salt, but they are less common and few, if any, examples of these compounds will be discussed.

11.20 Draw the product of a reaction between triethylphosphine and (2*R*)-iodopentane.

11.3.5 Azide

In Section 11.3.3, phthalimide (**43**) reacted with alkyl halides and was a surrogate for an amine. Other nitrogen surrogates are available to prepare primary amines. One is a class of compounds known as azides, and the nucleophile is the azide ion, **51**. Azide ion reacts with alkyl halides via an S_N2 reaction to give an alkyl azide, which can subsequently be converted to an amine. The azide anion (**57**) is commercially available as sodium azide (NaN_3). An example reacts 1-bromopropane (**21**) with sodium azide in THF, and the product is alkyl azide **52** and sodium bromide (NaBr). For the azide group to function as an amine surrogate, the N_3 group must be **reduced** with sodium borohydride, $NaBH_4$ (see Chapter 19 for reduction reactions) to give amine **53**. More commonly, the azide is treated with hydrogen gas and a metal catalyst such as palladium (see Chapter 19, Section 19.3, for catalytic hydrogenation), as shown:

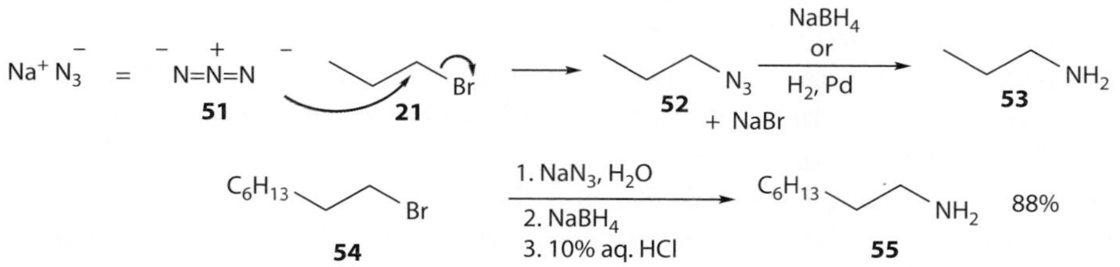

The key feature of this surrogate is that it is easy to incorporate azide into a molecule and reduction gives the amine. A common way to represent this multistep sequence is to "stack" the reagents for each reaction above and below the reaction arrow as shown for **54**. The first reaction (step 1) gives a product that is isolated and then reacts with the reactants in step 2, and that product reacts in step 3. There are three different reactions, and it is important to add the numbers 1–3. If the numbers are omitted and all reagents are simply placed on the arrow, it implies that all reagents are in the flask at the same time. Note that it makes no difference if the reagents are placed above or below the arrow, so no significance should be placed on their top or bottom position. Using this

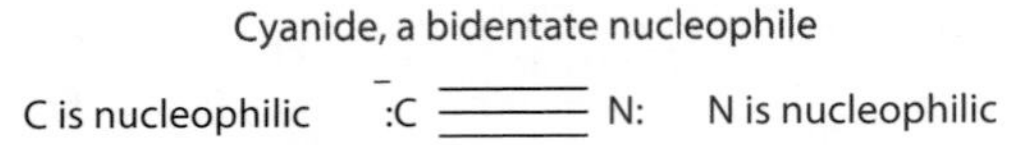

Cyanide, a bidentate nucleophile

C is nucleophilic   :C $\equiv\!\!\equiv$ N: N is nucleophilic

Figure 11.8   Bidentate cyanide.

representation, 1-bromooctane (**54**) is converted to octylamine (**55**) in 88% yield for the three steps shown.[6] The third step is a hydrolysis step for the sodium borohydride, which will be explained in Chapter 19 (Section 19.2.2).

    It is important to understand that alkyl azides may be thermally or photochemically unstable, and sometimes explosive, so great care is always exercised when this chemistry is used. If one exercises the proper caution, however, azide substitution reactions are a reasonable method for making primary amines from primary and secondary alkyl halides.

11.21   Show three different reaction sequences that will convert 1-bromobutane to 1-aminobutane.

11.3.6   Cyanide and Acetylide Nucleophiles (Making a Carbon–Carbon Bond)

The cyanide ion, $N\equiv C^-$, is the conjugate base of hydrocyanic acid (HCN; pK_a = 9.31). Cyanide is a relatively weak base, but it is a good nucleophile and it gives S_N2 reactions with alkyl halides. Indeed, reaction of 1-bromohexane (**56**) with NaCN in aqueous methanol gives a 73% yield of heptanenitrile, **57**.[7] This is a straightforward S_N2 reaction, but close inspection of the cyanide ion (see Figure 11.8) suggests a complicating factor: There is an electron pair on *both* carbon and nitrogen, so the C≡N unit is a *bidentate nucleophile*. This means that both the carbon and the nitrogen atoms may react as nucleophiles. *In general, the nitrogen is more electronegative and is less able to donate electrons when compared to carbon.*

    If cyanide forms a new σ-bond to carbon, the newly formed C–C bond is stronger than the C–N bond, which breaks, so bond energetics (see Chapter 7, Section 7.5) usually favor the reaction of carbon as the nucleophile. *A formal charge calculation (see Chapter 5, Section 5.5) shows the negative charge to be on carbon.* In general, a charged carbon should be more nucleophilic than a neutral nitrogen, *as long as the positive counter-ion is sodium or potassium.* With more covalent counter-ions such as Li or Ag, nitrogen can react preferentially as a nucleophile to give molecules known as isonitriles (this chemistry will not be discussed). *In all reactions given in this book, assume that cyanide reacts as a carbon nucleophile.* Note also that water and methanol are used as the solvents, despite the admonition in Section 11.2.3 that an aprotic solvent should be used. Water is an acceptable solvent for primary halides (this will be explained in Section 11.4.2 and in Chapter 13, Section 13.4).

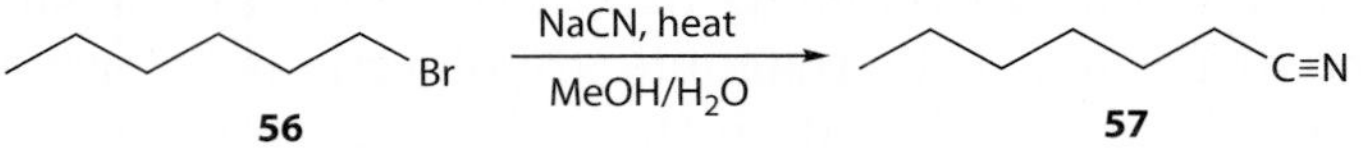

11.22 Draw the final product when *trans*-2-bromomethylcyclohexane and NaCN are heated in THF.

Using cyanide ion as a nucleophile in an S_N2 reaction is important because *a new carbon–carbon bond is formed that extends the carbon chain by one carbon.* Because it is an S_N2 reaction, the reaction proceeds with inversion of configuration. This means that the stereochemistry of the product is predictable. The reaction of the *p*-toluenesulfonate (tosyl) ester of 4-phenyl-2*S*-butanol (**58**) with KCN, for example, leads to nitrile **59** (the potassium salt of the tosylate anion is the other product). Note that the reaction proceeds with 100% inversion of configuration, and the product also has the (*S*) absolute configuration. Replacing the C–O bond in **58** with the C–C bond in **65** does not change the priorities of the groups attached to the stereogenic carbon (see Chapter 9, Section 9.1), although inversion does occur. The point here is that the structure of starting material and product should be examined, rather than just the name of each.

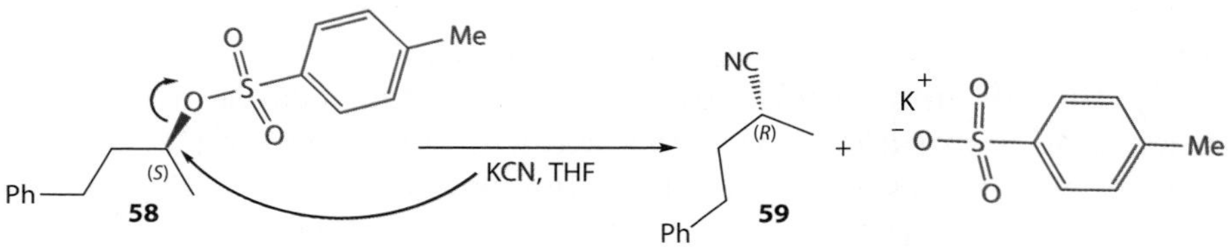

In Chapter 6 (Section 6.2.6), the hydrogen atom of a terminal alkyne was identified as a weak acid that reacts with strong bases such as $NaNH_2$ to give the corresponding conjugate base, an alkyne anion. When propyne (**60**) reacts with sodium amide, for example, the products are the alkyne anion **61** (the conjugate base) and ammonia (the conjugate acid). Alkyne anions are good nucleophiles in S_N2 reactions. Displacement of the tosylate group in **62** by the propyne anion leads to **63** with the expected inversion of configuration. Alkyne anions are also important nucleophiles because *a new carbon–carbon bond is formed.* Note that a terminal alkyne (**60**) is converted to an internal alkyne (**63**) using this reaction.

11.23 Draw the final product when 2*S*-iodo-3*R*-methylheptane reacts with the anion generated from 3,3-dimethyl-1-pentyne.

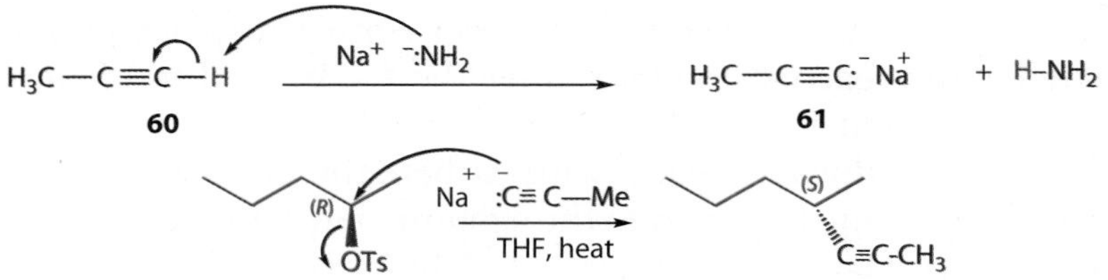

11.4 A Tertiary Halide Reacts with a Nucleophile When the Solvent Is Water

In Section 11.2.2, the rate of reaction for a tertiary halide in an S_N2 reaction is shown to be prohibitively slow, due to steric hindrance in the requisite pentacoordinate transition state. When 2-bromo-2-methylpropane (**64**) is heated in anhydrous THF (no water) with KI, for example, only unreacted **65** and KI are isolated. *There is no reaction* (see Chapter 7, Section 7.12, for a definition of "no reaction"). Many different experiments have been done with this reaction, including using different nucleophiles, different solvents, and different reaction conditions. Interestingly, when **64** is heated in water, 2-methyl-2-propanol (**65**) is isolated in low yield. This is clearly a substitution reaction (Br for OH), but it cannot be an S_N2 reaction. The reaction occurred with a tertiary halide, which is contrary to the requirement of a high-energy pentacoordinate transition state such as **13** (see Figure 11.5). *Many experiments have demonstrated that this reaction follows first-order rather than second-order kinetics and there is an intermediate in an overall two-step reaction.*

Clearly, the OH unit in **65** is derived from water because hydroxide ion is **not** present and is **not** added. Because water is the source of the OH unit, the oxygen atom of water must replace bromine, which is a substitution. However the kinetics and the presence of an intermediate indicate that this is not an S_N2 reaction. If water does not directly displace the leaving group in an S_N2 reaction, the alternative is that the leaving group must depart before water reacts with carbon. If this mechanistic evaluation is correct, the reaction follows a different mechanism. Experiments have shown that the reaction has a carbocation intermediate. The following sections will describe this new type of substitution reaction. Note that *the reaction solvent includes water,* which plays a key role in this reaction.

11.4.1 Ionization to a Carbocation

Consider the conversion of **64** to **65**. The C–Br bond must be broken and water must attack the bromine-bearing carbon to form a C–O bond. The experimental data show that this is not an S_N2 reaction, so water cannot displace bromine directly. The only way for the reaction to occur is for the C–Br bond to break *first,* which means that bromine must leave *before* water reacts. If true, then the OH in **65** arises from water, but if water reacts with carbon, the product is an oxonium ion, $C-OH_2^+$. Therefore, **65** must arise from the oxonium ion, which arises from the reaction of water with a carbocation intermediate. The carbocation must arise by loss of bromine from **64**. This walk-through suggests the mechanism for this reaction.

Does the bromine atom in **64** spontaneously fly away from the carbon? No! This is a silly question because **64** is a quite stable molecule. *It is, however, possible for the bromine to be pulled off, but what can exert a pull on the bromine?* The only difference between the reaction that converted **64** to **65** and the reaction of NaI in ether that gives no reaction is the presence of water. What is special or different about water? Refer to Table 11.2 to see that water is polar and protic. As seen in Section 11.2.3 (Figure 11.6), the δ+ hydrogen of water can coordinate with the δ− Br atom of **64**, allowing water to "pull" on bromine. If water is pulling on bromine, eventually the C−Br bond breaks, with transfer of both electrons in that bond to form the bromide ion. This process will generate a carbocation intermediate with a formal charge of +1 on the reactive carbon.

The reaction in water just described involves *ionization* of the C−Br bond to form the bromide ion and carbocation **66**. Once formed, **66** is highly reactive with any species that donates two electrons to carbon (a nucleophile), as shown in Chapter 10, Section 10.2. The most abundant nucleophile in this reaction is the oxygen atom of water. If that oxygen reacts as a nucleophile with the positive carbon in **66**, oxonium ion **67** is formed. The protons in **67** are acidic, so it is a strong Brønsted–Lowry acid (see Chapter 6, Section 6.2.1). Water is amphoteric and it will react with **67** as a Brønsted–Lowry base. Transfer of the proton (marked as −H$^+$) as shown gives the observed alcohol product, **65** (the conjugate base of **67**) and the hydronium ion (the conjugate acid of water). *This is the mechanism for the overall transformation of halide to alcohol, which is clearly a nucleophilic substitution, but mechanistically very different from the S$_N$2 reaction.* The mechanism involves an *ionization* reaction to give a key reactive intermediate, the carbocation.

Special circumstances are required to ionize the carbon–bromine bond in **64** to generate **66**. *Water must be present if the reaction is to proceed at a reasonable rate.* Although ionization is possible in other protic solvents such as ethanol and acetic acid, *and it does occur,* ionization is much slower because these other solvents are less polar and less able to solvate both cations and anions and thereby separate them. *Water or solvents mixed with water are the best media for ionization reactions.*

In an aqueous medium, ***water molecules surround 64***. The δ+ hydrogen coordinates with the δ− bromine of **64** and starts to "pull" on it, as in **68**. This action elongates the C−Br bond, making the bromine more δ+, which increases coordination with the hydrogen atom of water. This, of course, increases the pull. As the bromine is pulled away from the carbon, more water molecules begin to surround the bromine, as well as the carbon (see **69**). This process

continues until the C–Br bond is completely broken and water molecules completely surround the carbocation (**70**) as well as the bromide ion (see **71**). This
process of pulling on the leaving group and then surrounding the resulting ions
is called **solvation**, as noted for the solvation of NaCl in Figure 11.6. Solvation
is very efficient in water (see Table 11.2), but not in other protic solvents such
as ethanol. This observation is apparent when NaCl is seen to dissolve easily in
water but not in pure ethanol or in the aprotic solvent diethyl ether.

Aprotic solvents such as diethyl ether can only solvate cations and thus
cannot solvate and separate ions. Ethanol is protic, but the experiment with
NaCl strongly suggests that it is much poorer at solvating and separating ions
when compared to water. *For the purpose of this book, assume that water is
the only solvent that efficiently solvates and separates both cations and anions.*
Methanol and ethanol do solvate and separate ions, but the process is usually
so slow that another reaction occurs first. This statement will be elaborated
upon in Section 11.6. Therefore, if **64** is placed in an aqueous medium and
heated, solvation to the carbon cation may occur. If water is not present, ionization to the cation is, at best, slow; at worst, it does not occur at all. *To reiterate, assume that if water is present in the reaction medium, ionization is a facile
and competitive process.*

There is a problem with this ionization model. Alkyl halides are essentially
insoluble in pure water (see Chapter 5, Section 5.8.2). To get around this problem, *aqueous solvents* (a mixture of water and an organic solvent) are used
to provide sufficient solubility for the halide and also provide a source of water.
Solutions used for ionization of halides include aqueous acetone, aqueous ethanol, aqueous methanol, and aqueous THF.

11.4.2 Relative Stability and Reactivity of Carbocations

In Section 11.4.1, ionization of the C–Br bond in tertiary halide **64** gives carbocation **66**. Carbocations were discussed in Chapter 10, Section 10.2, in connection with the acid–base reaction of an alkene with acids such as H–X (HCl,
HBr, etc.). To understand formation of a carbocation in a substitution reaction,
remember that the stability of a carbocation is related to the number of substituents attached to the positive carbon. The formation of carbocations from
alkenes was described in Chapter 10, Section 10.2, as was the relative stability
of carbocations.

66A

A carbocation such as **66** has only three σ-bonds, which makes it sp^2 hybridized. These atoms attached to these bonds assume a trigonal planar geometry (see **66A**) and the positive charge resides in a p-orbital that is orthogonal to the plane of the atoms, which is effectively a carbon atom with an **empty p-orbital**. Because the cation is planar, there is nothing to bias approach of another molecule to the positive carbon. In other words, an incoming molecule can approach from the "top" of **66A** or the "bottom" of **66A** with equal ease.

In Chapter 10, Section 10.2, carbon substituents (alkyl groups) were described as electron releasing, so the more alkyl groups that are attached to the sp^2 carbon, the lower the net charge on that carbon is. Lower charge is associated with greater stability, so a tertiary cation with three carbon substituents should be more stable that a secondary cation with only two carbon substituents. The general order of stability using this protocol is tertiary > secondary > primary > methyl. This order of stability correlates with the relative rate of ionization of an alkyl halide to form a carbocation. In other words, ionization of a tertiary alkyl halide is faster than a secondary alkyl halide, and for all practical purposes ionization of a primary alkyl halide is so slow that it is classified as no reaction in most aqueous media.

The more stable a cation is, the less reactive it is and, conversely, the less stable a cation is, the more reactive it is. This means that, although a methyl cation is very reactive if it forms, this higher energy carbocation has a higher activation barrier to formation, relative to a more stable ion such as a tertiary carbocation. As noted, this information leads to the conclusion that it is easier to form a tertiary cation by ionization of a tertiary halide than it is to form a secondary carbocation by ionization of a secondary halide. Continuing the trend, it should be very difficult to form a primary cation from a primary halide such as 1-bromopentane (**72**) and very difficult indeed to form a methyl carbocation from iodomethane or bromomethane.

Assume that ionization to a primary carbocation or a methyl carbocation is simply too slow to be relevant. This is a very important concept. If **72** is dissolved in aqueous acetone, generation of the unstable cation **73** will be very slow because the energy required to form it is quite high. This does **not** mean that it is impossible to form **73**. It simply means that the rate is very slow. This is indicated with an "X" for the conversion of **72** to **73**, which signifies that the conversion does not occur *at a significant and competitive rate. This point is worth repeating. In this book, assume that primary alkyl halides do not form primary carbocations by ionization in aqueous solvents.*

11.24 Comment on the stability of the cation formed by ionization of 1-bromo-1-phenylethane.

approach equally well from the "top" or from the "bottom" because there is nothing to make one "face" of the cation different from the other. If iodide approaches **76** from the bottom, **77** is formed, but if iodide approaches from the top, **78** is formed. *The loss of enantiopurity and formation of a racemic mixture results from forming a planar carbocation with no facial differentiation to an incoming nucleophile.*

11.25 **Draw the intermediate and the final product formed when (2S)-bromo-2-phenylpentane is treated with KI in aqueous THF.**

This is a general phenomenon. If a chiral, nonracemic halide is converted to a carbocation, the product will be a racemic mixture. *The S_N1 reaction proceeds with no stereoselectivity and produces racemic products.* This sharply contrasts with the S_N2 reaction that proceeded with 100% inversion of configuration due to back-side attack.

11.5 Carbocation Rearrangements

Carbocation rearrangements were discussed in Chapter 10 (Section 10.2.2) in connection with the reaction of an alkene with an acid. Rearrangements certainly occur in the carbocation intermediates generated in an S_N1 reaction. In an old experiment performed by Friedrich Auguste Kekulé (Germany/Belgium; 1829–1896) in 1879, 1-bromopropane (**21**) was treated with the powerful Lewis acid $AlCl_3$. The final isolated product was 2-bromopropane, **81**.[8] This was an unexpected result at the time. The rearrangement was explained in Chapter 10, but it will be repeated in part here.

Aluminum chloride is a powerful Lewis acid that reacts with the bromine atom in **21** (the Lewis base), and the initial product is the primary cation **79**. In previous sections, it was assumed that a primary carbocation cannot form. This reaction is obviously an exception to that assumption. The reason for the difference is the powerful Lewis acid aluminum chloride, which does indeed react with the bromine to give a carbocation. This is a Lewis acid–Lewis base reaction, and does not involve ionization. However, carbocation **79** does not lead to the final product. The final product is **81**, and this product must arise from secondary carbocation **80**.

The only way to explain the transformation of the initially formed **79** to **80** is by a rearrangement. Specifically, a 1,2-hydride shift (see Chapter 10,

Section 10.2.2) transforms the primary carbocation to the more stable second-ary cation **80**, which then reacts with bromide ion to give **81**. With the excep-tion of the skeletal rearrangement, this is an S_N1 reaction. The mechanism of carbocation rearrangements is discussed in Chapter 10, but the main principle is worth repeating: *If a primary or secondary cation is generated on a carbon adjacent to a carbon that would be a more stable cation if one atom or group was moved, that group will move.* Remember that the "curly arrow" between **79** and **80** indicates the rearrangement.

As in Chapter 10, *if a cation is generated next to a carbon atom that is poten-tially a more stable carbocation, assume that the rearrangement occurs faster than any other process.* This is not always true, but it is true most of the time and it is a good working assumption for problems encountered in this book. This type of rearrangement is not limited to the alkyl halide–Lewis acid reac-tion done by Kekulé, and it occurs with ionization reactions of alkyl halides. Ionization of halides in aqueous media, although somewhat slow, occurs and once a carbocation is formed, rearrangement is possible. An example is the dis-solution of **82** in aqueous THF with KI that gives the rearranged iodide **83** as the major product of the reaction. Because **82** is not very soluble in water, the THF is used as a cosolvent.

Alkyl groups migrate if a more stable cation is formed, but this is true pri-marily when a hydrogen atom is not available for migration. When 3-bromo-2, 2-dimethylpentane (**84**) is heated with KI in aqueous THF, the initially formed cation is **85**. No hydrogen atoms are on the adjacent carbon bearing the methyl groups, but if a methyl group moves, tertiary cation **86** is formed. Remember that it is the electrons in the σ-bond that migrate toward the positive carbon, so the atom or group attached to that bond is, in many ways, "along for the ride" (there are many cases where the migrating group influences the facility of the migration, but they will not be discussed here). Therefore, the 1,2-methyl shift leads to **86**, which reacts with iodide to give the final product, **87** (2-iodo-2,3-dimethylpentane). Again, as discussed in Chapter 10, assume that *when there is more than one group or atom that can migrate, it is always the smaller atom or group that migrates.*

11.26 Draw the final product of the reaction between 2-bromo-1-methylcyclopentane and KI in aqueous THF.

11.27 Draw the final product of and mechanism for the reaction when 3-ethyl-4-bromo-3-methylhexane is heated with KI in aqueous THF.

11.6 Solvolysis Reactions of Alkyl Halides

If 1-bromopentane is heated with anhydrous ethanol, there is no perceptible reaction after several days. If 2-bromo-2-methylpentane (**88**) is similarly heated for a day or two, there is little perceptible reaction. However, if this latter reaction is heated for several days, or even several weeks, there is a reaction, and the product is 2-ethoxy-2-methylpentane, **90**. This tertiary halide *cannot* react by an S_N2 mechanism because the activation energy barrier for that transition state is too high, and the primary halide did not react at all. Formation of the ether indicates that an alcohol unit replaces the bromide (a substitution), and the only rational mechanism that fits these data is an S_N1 reaction. However, there is no water.

Remember from Table 11.2 that ethanol is a protic solvent. Although ionization and stabilization of charge are not as facile as in water (i.e., they are *slow*), this can occur. Prolonged heating leads to slow ionization of the halide to carbocation **89**, which reacts quickly with ethanol to give **90**, and loss of the proton to ethanol in an acid–base reaction gives the ether product, **91**. Replacement of alkyl halides with solvent in this way is called ***solvolysis*** *and occurs most often with protic solvents such as alcohols*. This reaction is a reminder that it is ***assumed*** that water is the only solvent that will facilitate ionization, but the assumption ignores other protic solvents and it ignores the rate of various reactions. Solvolysis is usually slower if water is not present, but it can occur. Indeed, *hydrolysis is simply solvolysis where the solvent is water*.

11.28 Write out the acid–base reaction of ethanol with 89 and then the acid–base reaction of the resulting oxonium salt with ethanol to produce 91.

11.29 What product is formed when 3-iodo-3-ethylheptane is heated for several days in (a) refluxing methanol, and (b) refluxing water? Write the complete mechanism for both reactions.

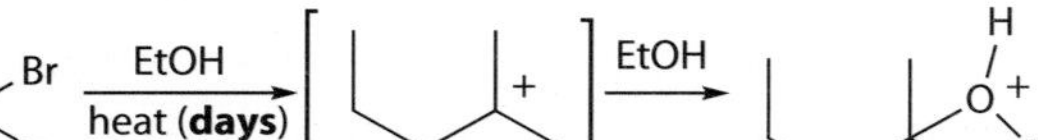

11.7 Preparation of Halides and Sulfonate Esters by Substitution Reactions

In both S_N1 and S_N2 reactions, the leaving group is the halogen of an alkyl halide or the sulfonate group of a sulfonate ester. Both alkyl halides and sulfonate esters are prepared from alcohols. In Chapter 10, alcohols were prepared by the hydration reaction of alkenes, by oxymercuration–demercuration of alkenes, or by hydroboration of alkenes. Other methods can be used to prepare alcohols, and they will be discussed at a later time. This section will describe several of the reactions used to convert alcohols to halides or sulfonate esters.

11.7.1 Alcohols React as Bases with Mineral Acids

Both S_N2 and S_N1 substitution reactions are presented in this chapter. Alcohols are converted to alkyl halides by either of these substitution reactions, but an extra step is involved in the mechanism because OH is a very poor leaving group. The extra step converts OH into a leaving group—water—by reaction with a Brønsted–Lowry acid.

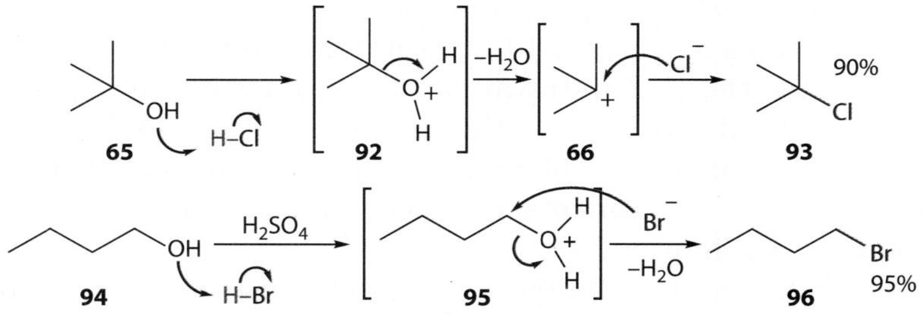

When 2-methyl-2-propanol (*tert*-butyl alcohol, **65**) is treated with concentrated HCl, 2-chloro-2-methylpropane (2-chloro-2-methylpropane; *tert*-butyl chloride, **93**) is isolated in 90% yield.[9] Similarly, when 1-butanol (**94**) is treated with 48% HBr in the presence of sulfuric acid, a 95% yield of 1-bromobutane (**96**) is obtained.[10] In both reactions, the oxygen of the alcohol reacts as a Brønsted–Lowry base in the presence of the protonic acids, HCl, or sulfuric acid. The fact that alkyl halides are produced clearly indicates that these are substitution reactions. In previous sections, tertiary halides gave substitution reactions when a nucleophilic halide ion reacted by an S_N1 mechanism that involved ionization to a carbocation prior to reaction with the halide. Primary halides react with a nucleophilic halide ion by an S_N2 mechanism. It is reasonable to assume that tertiary alcohols and primary alcohols will react similarly, *if* the OH unit is converted to a leaving group.

The OH unit does not leave as hydroxide ion (HO⁻) either by ionization or by direct displacement. *Hydroxide is a very poor leaving group.* In other words, *a nucleophile cannot displace OH from an alcohol by direct substitution.*

In the reaction of both **65** and **94**, an acid is present and, because alcohols are amphoteric (Chapter 6, Section 6.2.5), the oxygen atom of an alcohol can react as a Brønsted–Lowry base. This acid–base reaction of the alcohol and a strong acid gives an oxonium ion (the conjugate acid), along with the counterion of the acid (the conjugate base). Therefore, reaction of **65** with the acidic proton of HCl will give **92** and the chloride ion; **94** reacts with HBr to give **95** and the bromide ion. Note that sulfuric acid is added to the second reaction to facilitate protonation of the primary alcohol, but this additive is not always required.

The $-OH_2^+$ unit in each oxonium ion intermediate is effectively a water molecule bound to an alkyl group. Water, of course, is a stable and neutral molecule, so both **92** and **95** have the good leaving group water. The water is displaced by a nucleophilic halide ion. It is reasonable to assume that **102** reacts by an S_N1 mechanism and **105** reacts by an S_N2 mechanism. Because **92** is a tertiary system, it is more likely to undergo ionization (in the protic medium) to give a carbocation, **66**, via loss of water. Once **66** is formed, rapid reaction with the nucleophilic chloride ion gives **93**. Oxonium ion **95** is a primary system, and ionization to a primary cation is very slow. Therefore, nucleophilic displacement of water by the bromide ion is much faster, and the products are the bromide ion and **96** in an S_N2 reaction. *Tertiary alcohols react with HCl, HBr, or HI to give a tertiary alkyl halide by an S_N1 process and alcohols primarily react with HCl, HBr, or HI to give an alky halide by an S_N2 process.*

Secondary alcohols are ignored in these two examples. Secondary alcohols may react by *either* the S_N2 pathway or the S_N1 pathway, depending on the solvent. In many if not most cases, the halide is formed by both S_N1 and S_N2 mechanisms, depending on the solvent. An approach for estimating the product formed when different mechanisms are in competition will be discussed for S_N1 and S_N2 reactions in Chapter 13.

Because cationic intermediates are possible for secondary alcohols, rearrangement must be considered. For tertiary carbocations generated from a tertiary alcohol, rearrangement is usually not a problem. For primary alcohols, the mechanism does not involve a carbocation, so once again rearrangement is not a problem. Secondary systems are another matter. *If an S_N1 mechanism operates in the conversion of a secondary alcohol to the corresponding halide, rearrangement is a distinct possibility and must be considered.*

11.30 Write the final product when 2,2-dimethylbutanol is treated with conc. HCl.

11.7.2 Sulfur and Phosphorous Halide Reagents

Although the reaction of alcohols and mineral acids is usually quite efficient, there are times when the use of mineral acids in chemical reactions must be avoided. Therefore, the availability of alternative methods for the preparation of halides from alcohols is essential. To prepare an alkyl chloride from an alcohol, any reagent must contain chlorine atoms, just as a brominating agent must contain bromine. The most commonly used reagents are inorganic molecules because the by-products are easily washed away with water, easily separated in other ways, and will usually not contaminate the organic solvent containing the products. Several common inorganic reagents are used as halogenating agents with alcohols. *Fluorides will not be prepared because they tend to be less reactive than the other halides in nucleophilic aliphatic substitution reactions.* The focus will be on chlorides and bromides and then will be expanded to include iodides.

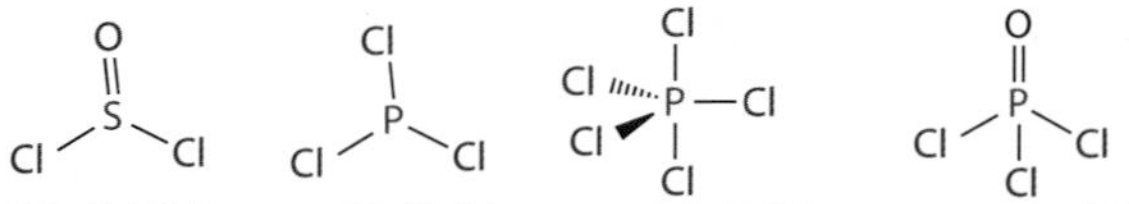

97 (SOCl$_2$)   **98** (PCl$_3$)   **99** (PCl$_5$)   **100** (POCl$_3$)

The most common chlorinating reagents are sulfur and phosphorous halides, and the sulfur reagent thionyl chloride (**97**) is used quite often. Three phosphorus-based reagents are also commonly used: phosphorus trichloride (**98**), phosphorus pentachloride (**99**), and phosphorus oxychloride (**100**). The condensed notation for these reagents is shown below each structure (SOCl$_2$, PCl$_3$, PCl$_5$, and POCl$_3$). In all cases, an alcohol is converted to an alkyl chloride, ROH → RCl. The sulfur or the phosphorus atom is electrophilic in all cases, and it is attacked by the electron-donating oxygen atom of an alcohol. *In effect, the alcohol oxygen functions as a Lewis base with sulfur or phosphorus as the Lewis acid.*

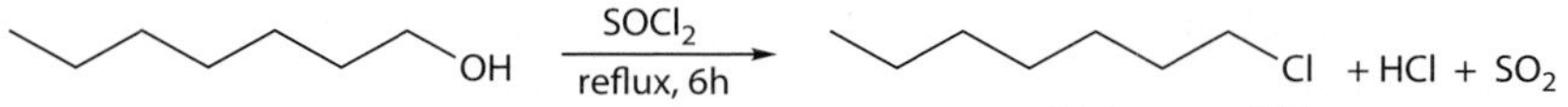

101   **102**   77%

A typical use of these reagents is the reaction of 1-heptanol (**101**) with thionyl chloride. After refluxing for 4 hours, 1-chloroheptane (**102**) is isolated in 77% yield,[11] and the by-products of the reaction are HCl and sulfur dioxide (SO$_2$). This is clearly a substitution reaction in which the OH unit is replaced by Cl. The term *reflux* indicates that the reaction is heated at the boiling point of the mixture, and the vapors are condensed and returned to the flask rather than condensed and removed as in a normal distillation. Clearly, the Cl comes from SOCl$_2$, but the substitution requires loss of the OH unit. Presumably, OH is converted to a leaving group.

When thionyl chloride reacts with an alcohol such as 1-heptanol, the oxygen atom of the alcohol donates two electrons to the electrophilic sulfur atom in **97** to give an oxonium ion, **103**. The initial reaction with the alcohol induces an electron flow that breaks the π-bond of the S=O unit to form an S–O$^-$ species. In a second reaction, the O$^-$ donates electrons back to sulfur to regenerate the S=O bond and HCl is lost. If a chlorine donates electrons to the acidic proton in **103**, this intramolecular process leads formation of HCl and formation of chlorosulfite **104** (heptyl chlorosulfite in this case).

Alternative mechanisms are possible for loss of HCl, but this one appears to be the most logical. This is a different type of intermediate than discussed so far, but the SO_2Cl unit is a leaving group. Remember that one chlorine atom in **97** is converted to HCl, and the alcohol is converted to **104**. Therefore, the only nucleophilic species available in **104** is the chlorine atom, which donates electrons to the carbon atom bearing the oxygen (which is electrophilic); this intramolecular process generates **102**, as a molecule of sulfur dioxide (O=S=O) is lost. Therefore, the leaving group SO_2Cl falls apart, and the OH unit in **101** leads to **102** (1-chloroheptane), HCl, and SO_2 as a result of the reaction with thionyl chloride. This is not really an S_N2 or an S_N1 reaction.

This conversion of **101** to **102** is a substitution, but the intermediacy of **104** leads to an intramolecular transfer of chlorine to displace sulfur dioxide. Reactions such as this are classified as ***internal nucleophilic substitution***, which is abbreviated by the symbol S_Ni.

This halogenation reaction can be done with many different alcohols, including 1°, 2°, or 3° alcohols. Chlorination of alcohols is not limited to using thionyl chloride because PCl_3, PCl_5, and $POCl_3$ also convert alcohols into alkyl chlorides. The mechanism is slightly different for the phosphorus reagents and will not be presented here. However, the overall transformation is essentially the same in that the oxygen atom of the alcohol donates electrons to phosphorus, and there is formation of a P=O bond with transfer of the halogen to carbon.

11.31 Draw the final product formed when (a) cyclopentanol reacts with PCl_3, (b) 3-ethyl-3-pentanol reacts with PCl_5, and (c) cyclopentanemethanol reacts with $POCl_3$.

An interesting phenomenon is noted when the alcohol contains a stereogenic carbon (see Chapter 9, Section 9.1). When 2R-pentanol (**105**) is treated with thionyl chloride, chlorosulfite **106** is formed, and transfer of a chlorine to carbon generates the observed product, (2R)-chloropentane (**107**). Examination of the absolute configuration of these products indicates that when chloride attacks the carbon to displace SO_2, the reaction proceeds with ***retention of configuration***. This result is not necessarily obvious, but it is the data obtained from the experiment. *A general assumption can be made that the reaction of a chiral alcohol with thionyl chloride will give the alkyl chloride with retention of configuration, if nothing else is added to the reaction.* In other words, the absolute configuration of the alcohol is retained in the chloride product ***when thionyl chloride is used and no other reagents are added***.

11.32 Draw the final product formed when 5,5-dimethyl-2S-hexanol is heated with thionyl chloride.

Another interesting experiment repeats the reaction of a chiral alcohol with thionyl chloride, but an amine such as pyridine or triethylamine is added (see Chapter 26 for a discussion of amines). When (2R)-pentanol (**105**) reacts with thionyl chloride, but now in the presence of triethylamine, the 2-chloropentane product is isolated and has the 2S configuration, **108**. The reaction occurs with inversion of configuration rather than retention of configuration, and this result suggests an S$_N$2 reaction.

What is the difference? The second reaction has an amine in the reaction. Assume that chlorosulfite **106** is the intermediate in both reactions. In the S$_N$i mechanism given before, the chlorine atom reacts with a proton on the oxygen to generate HCl. Therefore, in the reaction of **105** and thionyl chloride alone, HCl is a product and *no chloride ion is present*. In other words, there is no external nucleophile to react with **106**. When triethylamine, which is a base, is present, the HCl will react with the base to form the conjugate acid of the amine, triethylamine hydrochloride, Et$_3$NH$^+$Cl$^-$. In other words, when the amine is present in the reaction, chloride ion is also in the reaction medium in the form of the ammonium chloride, and chloride ion is a nucleophile. An S$_N$2 displacement of the chlorosulfite unit by chloride, as shown by the arrows in **106**, proceeds with the expected inversion of configuration, as observed in the final product (2S)-chloropentane, **108**.

In reactions with thionyl chloride, addition of an amine leads to inversion of configuration. With no amine, the reaction proceeds with retention of configuration, so we can control the stereochemistry of the hydroxyl group by adding an amine base or keeping the reaction free of the amine. Therefore, **105** is converted to **107** upon reaction with thionyl chloride alone, but it is converted to **108** upon reaction with thionyl chloride in the presence of an amine.

Thionyl bromide (SOBr$_2$) is the bromine analog of thionyl chloride, and it reacts essentially the same way to convert alcohols to bromides. The reagents PBr$_3$ and PBr$_5$ both convert alcohols to the corresponding bromide. *Although thionyl chloride reacts with chiral, nonracemic halides to give chlorides with either retention or inversion of configuration, thionyl bromide gives racemic products in both cases.* This is important! Thionyl bromide reacts with alcohols to give the bromide, but the reaction produces a mixture of enantiomers and does *not* give one enantiomer selectivity. If **105** reacts with thionyl bromide, the product is therefore assumed to be *racemic* 2-bromopentane. *Only thionyl chloride gives the stereoselectivity noted. Assume that reactions with thionyl bromide, PCl_3, PCl_5, $POCl_3$, PBr_3, or PBr_5 do not give clean inversion or retention, but rather racemic halides.*

11.33 **What is the product when cyclopentanol reacts with (a) $SOBr_2$, and (b) PBr_3?**

11.34 **What is the product formed when (4*R*)-methyl-(2*S*)-heptanol reacts with (a) $SOCl_2$ + NEt_3, (b) PBr_3 + NEt_3, (c) $POCl_3$, and (d) PBr_3?**

In the previous discussions, no reagents are listed that contained iodine. Sulfur and phosphorus iodides are not very stable. In order to convert an alcohol to an alkyl iodide, a mixture of reagents must be used. The most common method reacts an alcohol such as cyclopentanol (**109**) with elemental iodine and red phosphorus. Note that white phosphorus is pyrophoric, which means that it spontaneously ignites in air. Under these conditions, iodocyclopentane (**110**) is the product. The reagent responsible for making the iodide is probably PI_3, which is unstable but can be used when generated *in situ* (produced during the reaction without isolation) by this method. This is the preferred method for making alkyl iodides.

11.35 **What is the product formed when 3-hexanol reacts with iodine and red phosphorus?**

11.7.3 Preparation of Sulfonate Esters from Alcohols

The structure and nomenclature of sulfonate esters (see **113, 114**, or **115**) are described in Chapter 20 (Section 20.11.2). It is also true that sulfonate esters are good leaving groups in the substitution reactions described in this chapter (see Section 11.2.4). Sulfonate esters are prepared by the reaction of sulfonic acids with alcohols—much the way that carboxylic acid esters are prepared from carboxylic acids and alcohols (described in Chapter 20, Section 20.11.2). More commonly, sulfonate esters are prepared by the reaction of a sulfonyl chloride (see **112**) with an alcohol. This reaction is also described in Chapter 20. This section presents only a simple preview of that chemistry, with the goal of showing that it is easy to convert alcohols into sulfonate esters, which are then useful as leaving groups in substitution reactions. The formal mechanism of these reactions will be discussed in Chapter 20.

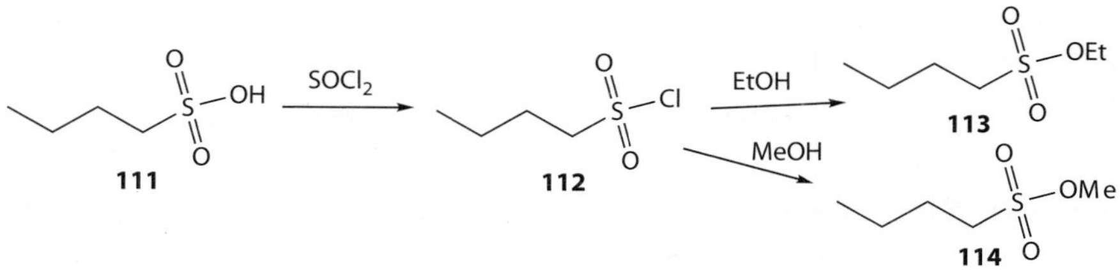

When a sulfonic acid such as butanesulfonic acid (**111**; see Chapter 16, Section 16.9) is treated with thionyl chloride or phosphorus trichloride, the product is a sulfonyl chloride. In this case, **111** is converted to what is known as butanesulfonyl chloride, **112**. When **112** reacts with an alcohol, a sulfonate ester is produced. Treatment of **112** with ethanol leads to **113** (ethyl butanesulfonate) and treatment with methanol leads to **114** (methyl butanesulfonate). A variety of sulfonate esters can be prepared in this manner. It is also possible to react a sulfonic acid with an alcohol under acid catalysis conditions to give sulfonate esters, but that reaction is not discussed until Chapter 20.

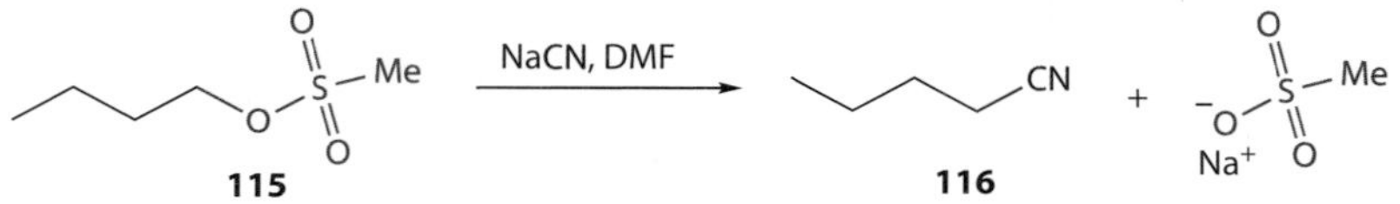

If methanesulfonyl chloride, or another simple derivative, reacts with a more complex alcohol, the OH is converted into a leaving group, which can then be reacted with a nucleophile in S_N2 or S_N1 reactions. An example is the reaction of methanesulfonate ester **115** (prepared from 1-butanol and methanesulfonyl chloride) with NaCN in DMF (classical S_N2 conditions). Displacement of the sulfonate group leads to formation of pentanenitrile (**116**). The methanesulfonate anion is the leaving group. Sulfonate groups are, in general, excellent leaving groups in these reactions—often better than the analogous bromide or chloride. Due to the availability of methanesulfonic acid and *p*-toluenesulfonic acid, mesylates (such as **115**) and tosylates (esters of 4-methylbenzenesulfonic acid) are the most commonly used groups.

11.36 **Write the reaction that forms the *p*-toluenesulfonate ester from cyclopentanol and the appropriate sulfonic acid.**

11.37 **Show the reaction sequence, including all reagents and products, that converts 2-butanol to 2-cyanopentane (2-methylbutanenitrile) and involves the formation of a mesylate.**

11.8 Reactions of Ethers

In general, ethers are unreactive to most chemical reagents. An exception is the reaction of ethers with strong mineral acids such as HI or HBr. The three-membered ring ethers (epoxides or oxiranes), however, are notable exceptions to this trend of reactivity. Epoxides are discussed in Chapter 26, and the strain inherent to the three-membered ring leads to a variety of chemical reactions.

11.8.1 Ethers React as Brønsted–Lowry Bases

Ethers such as diethyl ether and tetrahydrofuran (THF) are commonly used as solvents because they do not react with alkoxide bases, hydroxide, or many other reagents that will be encountered in later chapters. Ethers do not react directly with common nucleophiles such as cyanide, azide, or halides in the S_N2 reactions discussed in this chapter.

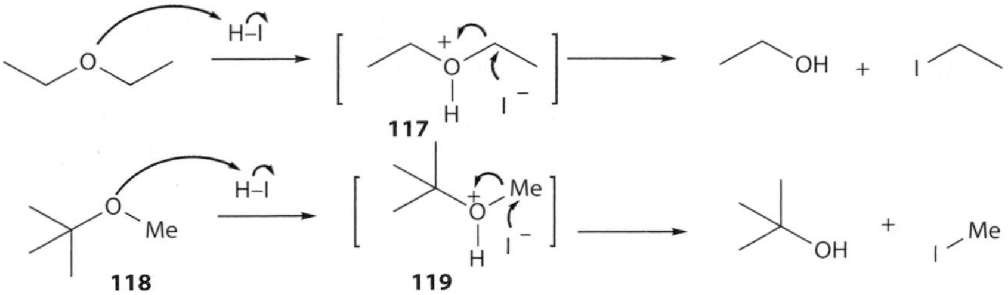

However, ethers do react with very strong acids such as HI and HBr. The reaction with HCl is quite slow and is usually not observed. When diethyl ether reacts with HI, however, an acid–base reaction generates the oxonium ion **117**. The counter-ion is the nucleophile iodide, and an S_N2 reaction at the carbon connected to oxygen leads to iodoethane and ethanol. In this reaction, protonation of the ether oxygen generates the ethanol leaving group in **117**. When *tert*-butyl methyl ether (2-methyl-2-methoxypropane, **118**) reacts with HI, oxonium ion **119** is formed by the acid–base reaction. This oxonium ion is unsymmetrical, and the iodide may react with the primary carbon atom of the methyl group or the tertiary carbon of the *tert*-butyl group. An S_N2 reaction is possible only if iodide attacks the less sterically hindered site, and the products are iodoethane and *tert*-butyl alcohol. Anisole (methylphenyl ether, **120**) will be discussed in Chapter 21, but it is used here to illustrate this type of reaction. When **120** reacts with concentrated HBr, the products are bromomethane and phenol, **121**.

11.38 Write out a mechanism and the final product for the reaction of 2-methyltetrahydrofuran and HI.

11.39 Draw the oxonium ion formed by reaction of anisole and HBr and give a brief explanation for why the bromide attacks the methyl group in that oxonium ion.

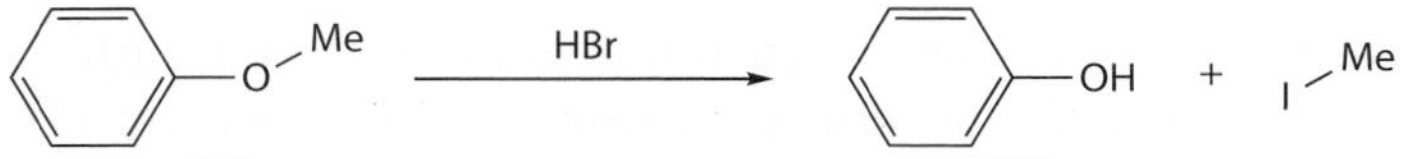

11.8.2 Epoxides React as Brønsted–Lowry Bases

Epoxides are three-membered ring ethers, as described in Chapter 10, Section 10.5. The strain inherent to that small ring makes it relatively easy to open the ring and relieve that strain. In other words, epoxides react with a variety of reagents when an acyclic ether will give no reaction with the same reagent. An epoxide such as **122**, for example, will react with HI in the same way as a simple ether, but it also reacts with much weaker acids such as HCl. The initially formed oxonium ion (**123**) can react with iodide, but there are two carbon atoms of the three-membered ring. Reaction at C^a will lead to **124**, whereas reaction at C^b will lead to **125**.

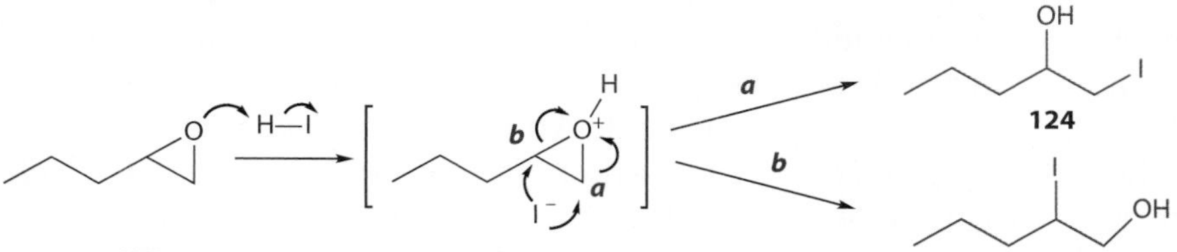

The reaction with iodide ion could be S_N2, but HI is a protic medium, so the three-membered ring in **123** opens and iodide attacks the more substituted carbon to give **125**. Whether **124** or **125** is the major product depends on the substituents and on the solvent, but assume that a terminal epoxide will give an oxonium ion that leads to the nucleophile at the more substituted carbon. Therefore, the reaction of **122** and HI will give **125** as the major product. Contrary to the ethers in Section 11.8.1, epoxides do react with HCl or weaker acids.

11.40 Draw the expected major product formed when 122 reacts with HCl.

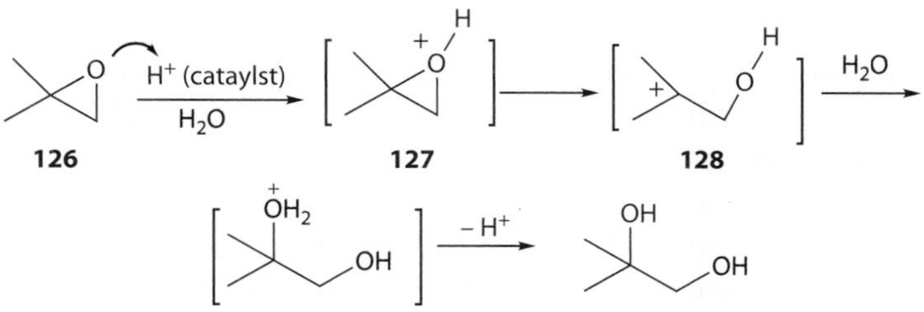

Epoxides will not react directly with water or with alcohols because they are weak acids. However, in the presence of a strong acid catalyst, an epoxide will react first to form an oxonium ion, and then with the weak nucleophile to open the three-membered ring. In the case of **126**, reaction with the acid catalyst will generate oxonium ion **127**. In the presence of the water solvent, the oxonium ion tends to open to give tertiary carbocation **128**, and the oxygen atom of water attacks the positive carbon of **128** to give **129**. Loss of the proton from **129** gives diol **130** and regenerates the acid catalyst. If **126** reacts with ethanol and an acid catalyst, an identical mechanism will generate an ether product.

11.41 Draw the mechanism and final product formed when 126 reacts with ethanol and an acid catalyst.

11.8.3 Nucleophilic Substitution Reactions of Epoxides

Section 11.8.1 clearly stated that ethers do not react directly with nucleophiles. Nucleophiles do react with epoxides, however, due to the strain inherent to the three-membered ring. Such reactions can be viewed as S_N2-type reactions in which the nucleophile will attack the less sterically hindered (less substituted) carbon of an epoxide. When **126** is heated with sodium azide (NaN_3), for example, attack at the less substituted carbon atom opens the ring and generates alkoxide **131** as the major product. An acid–base reaction is required as a second chemical step to convert the alkoxide to its conjugate acid, alcohol **132**. Interestingly, tetrahydrofuran (THF) is used as the solvent, graphically illustrating the difference in reactivity between the epoxide and this five-membered ring cyclic ether.

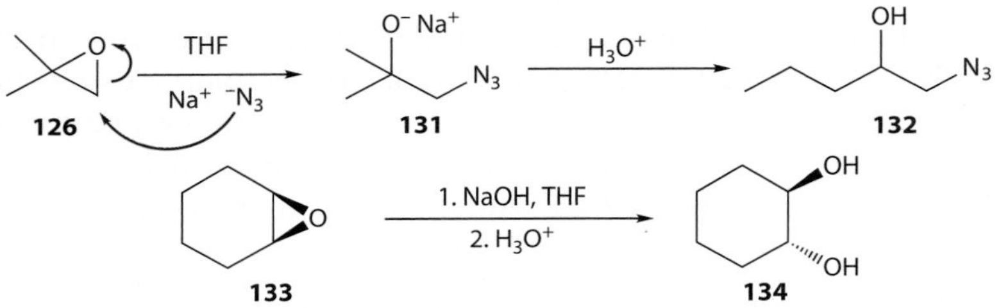

Many nucleophiles react with epoxides. The reaction of an epoxide and NaOH will lead to a diol, and the reaction with sodium cyanide will give a product containing an OH unit and a CN unit (a cyanohydrin). When **133** reacts with NaOH, for example, the product after hydrolysis is cyclohexanediol (**134**). Note the *trans* relationship of the two hydroxyl units, which result from backside attack of hydroxide at the epoxide. When 2(*S*)-2-ethyloxirane (**135**) reacts with NaCN in DMF (the solvent), the resultant alkoxide (**136**) shows retention of the stereogenic center at C2 because attack occurs at the less substituted C1 atom. An aqueous acid workup gives the cyanohydrin product, **137**.

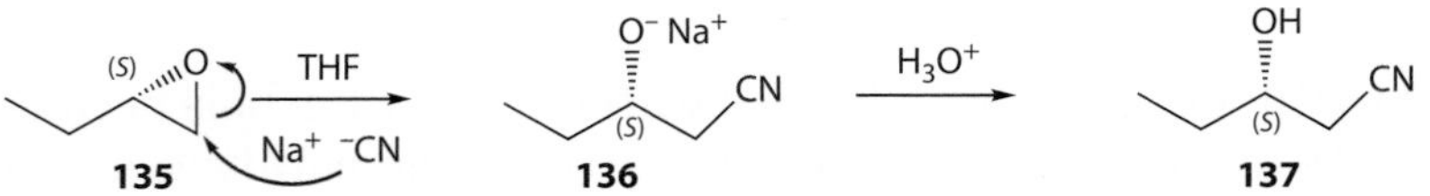

When an epoxide does not have a clearly less substituted carbon atom, reaction with a nucleophile will lead to regioisomers. When **138** reacts with NaCN in DMF, for example, attack at either carbon atom of the three-membered ring is equally facile. In other words, attack will occur at both carbon atoms to give a mixture of two products—**139** and **140**—after the aqueous acid workup. In the absence of other information, the reaction is expected to give a 1:1 mixture of **139** and **140**. *Assume that the reaction of a nucleophile with an unsymmetrical epoxide will give a mixture of regioisomeric products.*

11.42 Draw all products expected when sodium azide reacts with the epoxide of 4-methylcyclooctene followed by an aqueous acid workup. The methyl group is *trans* to the epoxide group in the starting epoxide.

11.9 Free Radical Halogenation of Alkanes

Another method for preparing alkyl halides involves a reaction without carbocation or anion intermediates. The reaction involves radical intermediates, so the mechanism of this process is quite different from those seen previously. The reaction of an alkene with HBr in the presence of a peroxide gives an anti-Markovnikov bromide, as discussed in Chapter 10, Section 10.2. That reaction involves a radical intermediate.

In the new reaction for preparing alkyl halides, formation of an alkyl halide proceeds by cleavage of C–H bonds in such a manner that one electron of the bond goes to one atom, and the other electron goes to the second atom. As noted in Chapter 7 (Section 7.4.3), this process is termed homolytic cleavage, and the product of homolytic cleavage is an intermediate called a radical. A radical is defined as a species that has an atom bearing a single unshared electron. Radicals are highly reactive and their reactions are sometimes difficult to control, from the standpoint of giving only one product. In the following sections, a few examples of radical reactions that can be controlled to give alkyl chlorides and alkyl bromides are discussed.

11.9.1 Homolytic Cleavage of Halogens

The reactions discussed so far commonly involve breaking polarized bonds when mixed with a suitable reagent. As discovered in Chapter 10, Section 10.4.1, *the dihalogens are polarizable in the presence of electron-rich or electron-poor species, and they also have relatively weak bonds.* Many experiments confirm that dihalogen bonds (X–X) break in the presence of heat or light, via homolytic cleavage, to generate two radicals, X•. Once the radical is formed, it can react with many different compounds. A radical is so reactive that it can even react with the C–H bond of an alkane.

the alkene. Both syntheses are presented, based on the two retrosyntheses. Which is better? The answer depends largely on which route gives the best yield of product, with the fewest side products. Without experimental data, both look viable; the HBr reaction sequence is shorter, so that would likely be chosen.

The third problem involves the synthesis of ether **165**, and examination of the designated alcohol starting material **166** suggests two disconnections. The first places the oxygen on the primary carbon and the C_a synthetic equivalent would be tertiary bromide **170**. The second disconnection places the oxygen on the tertiary carbon, and C_a on a primary carbon, leading to bromide **21**. Both disconnections suggest a Williamson ether synthesis (Section 11.3.2), but because the required S_N2 reaction cannot occur at tertiary carbon, the use of **170** is not possible for the S_N2 reaction. On the other hand, tertiary alcohol **166** is the designated starting material. Therefore, the synthesis becomes quite simple, reacting the tertiary alcohol with a base to generate the alkoxide, which then reacts with 1-bromopropane.

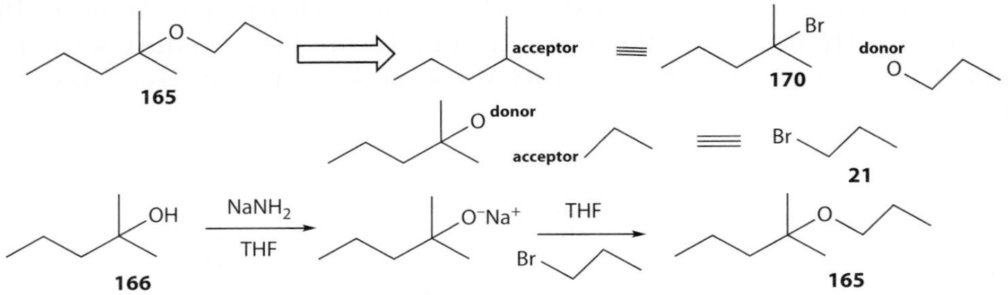

11.11 Biological Relevance

Substitution reactions occur in many biological processes. *bis*(chloroethyl) sulfide, otherwise known as mustard gas, is **171**. It is a sulfide (containing a C–S–C unit and the sulfur analog of an ether), but the presence of the primary alkyl chloride units led to significant reactivity and its use as a poison gas in World War I. Nucleophilic species react with the chlorides, leading to the toxic effects. It is cytotoxic, mutagenic, and a vesicant; on exposure, it will cause large blisters on exposed skin. The cytotoxicity and mutagenic properties arise by reaction with heterocyclic bases in DNA. These properties led to chemical modification of **171** and development of the amine derivative **172**, which is *N*-methyl *bis*(chloroethyl)amine. This is a so-called nitrogen mustard, and it is one of the first anticancer drugs known to be a DNA intercalating agent. The mechanism of this anticancer activity is known.

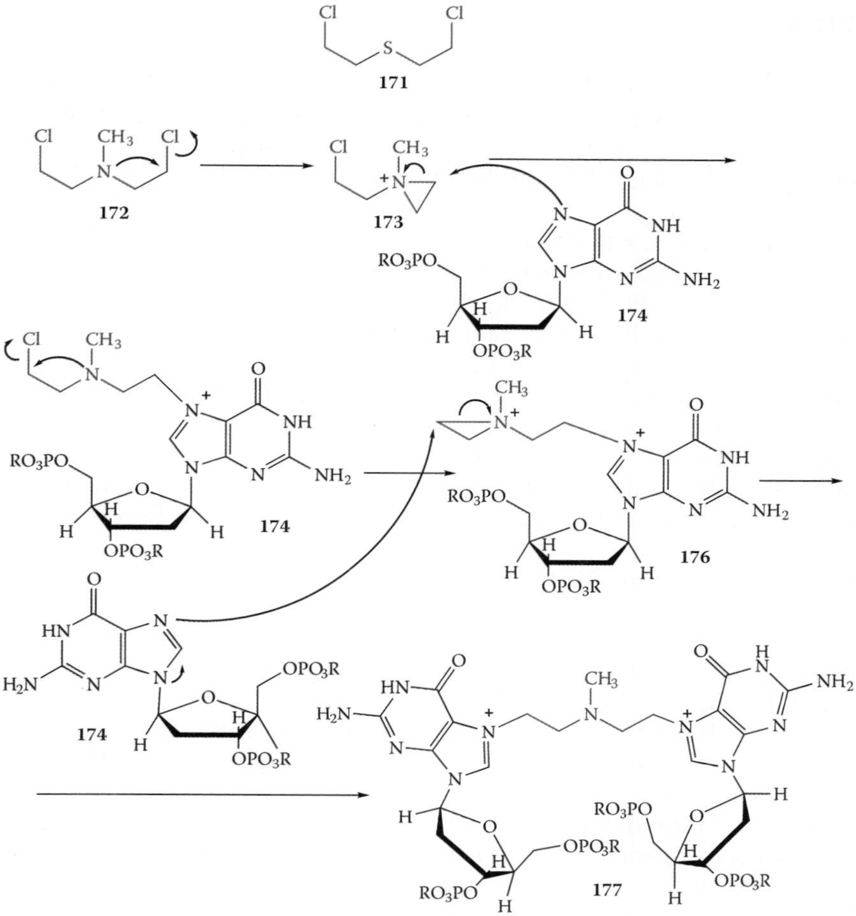

Nitrogen mustard **172** first reacts internally by an S_N2 reaction of the amine group at the primary chloride unit to give aziridinium salt **173**.[14] (Aziridines are discussed in Chapter 26, Section 26.4.1.) The three-membered ring aziridinium unit is susceptible to attack by nucleophiles, and the nitrogen of a nucleobase of DNA easily reacts. The example shows the N^9-nitrogen of a guanine that is part of a DNA strand, and S_N2-like reaction at the aziridinium salt leads to **174**. Hence, **172** is an alkylating agent. The cross-linking ability arises when the tertiary amine unit in **174** reacts with the other primary alkyl chloride to give aziridinium salt **175**. A second molecule of DNA (**174**) reacts to form **176**. If two different strands of DNA react, the nitrogen mustard links the two strands as indicated by **177**. If the DNA is double-stranded, this reaction leads to inter-calation of the nitrogen mustard. The nitrogen mustards remain an active area of research today.

References

1. Furniss, B. S., Hannaford, A. J., Smith, P. W. G., and Tatchell, A. R., eds. 1994. *Vogel's textbook of practical organic chemistry,* 5th ed., 572, Exp. 5.62. Essex, England: Longman.
2. Hughes, E. D., Ingold, C. K., and Masterman, S. 1937. *Journal of the Chemical Society* 1196–1201.
3. Streitwieser, A., Jr. 1956. *Chemical Reviews* 56:571.
4. Lowry, T. H., and Richardson, K. S. 1987. *Mechanism and theory in organic chemistry,* 3rd ed., 297–298. New York: Harper and Row.
5. Hughes, E. D., Juliusburger, F., Masterman, S., Topley, B., and Weiss, J. 1935. *Journal of the Chemical Society* 1525–1529.
6. Furniss, B. S., Hannaford, A. J., Smith, P. W. G., and Tatchell, A. R., eds. 1994. *Vogel's textbook of practical organic chemistry,* 5th ed., 584, Exp. 5.73. Essex, England: Longman.
7. Furniss, B. S., Hannaford, A. J., Smith, P. W. G., and Tatchell, A. R., eds. 1994. *Vogel's textbook of practical organic chemistry,* 5th ed., 772, Exp. 5.193. Essex, England: Longman.
8. Fieser, L. F., and Fieser, M. 1961. *Advanced organic chemistry,* 653. New York: Reinhold Pub.
9. Furniss, B. S., Hannaford, A. J., Smith, P. W. G., and Tatchell, A. R., eds. 1994. *Vogel's textbook of practical organic chemistry,* 5th ed., 556, Exp. 5.49. Essex, England: Longman.
10. Furniss, B. S., Hannaford, A. J., Smith, P. W. G., and Tatchell, A. R., eds. 1994. *Vogel's textbook of practical organic chemistry,* 5th ed., 561–562, Exp. 5.54. Essex, England: Longman.
11. Furniss, B. S., Hannaford, A. J., Smith, P. W. G., and Tatchell, A. R., eds. 1994. *Vogel's textbook of practical organic chemistry,* 5th ed., 558, Exp. 5.52. Essex, England: Longman.
12. Fieser, L.F., and Fieser, M. 1961. *Advanced organic chemistry,* 120–121. New York: Reinhold Pub.
13. Furniss, B. S., Hannaford, A. J., Smith, P. W. G., and Tatchell, A. R., eds. 1994. *Vogel's textbook of practical organic chemistry,* 5th ed., 578–579, Exp. 5.68. Essex, England: Longman.
14. Rajski, S. R., and Williams, R. M. 1998. *Chemical Reviews* 98:2723.

Answers to Problems

11.1

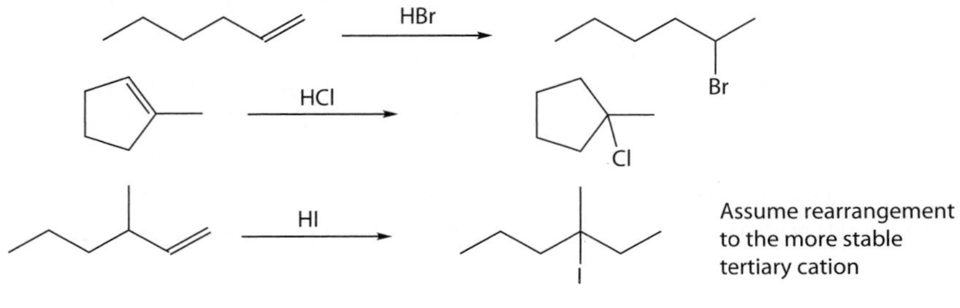

Assume rearrangement to the more stable tertiary cation

11.2

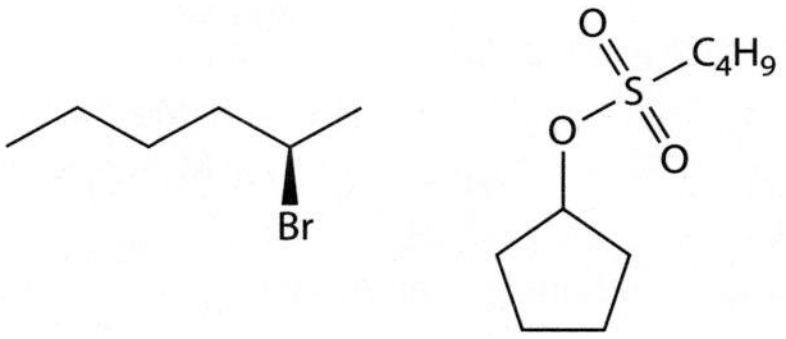

11.3

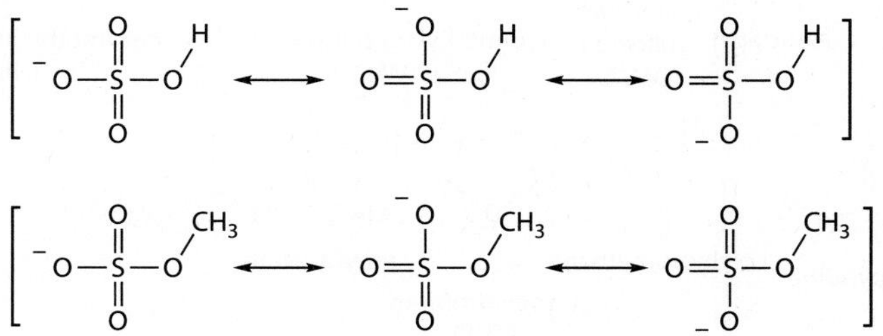

11.4

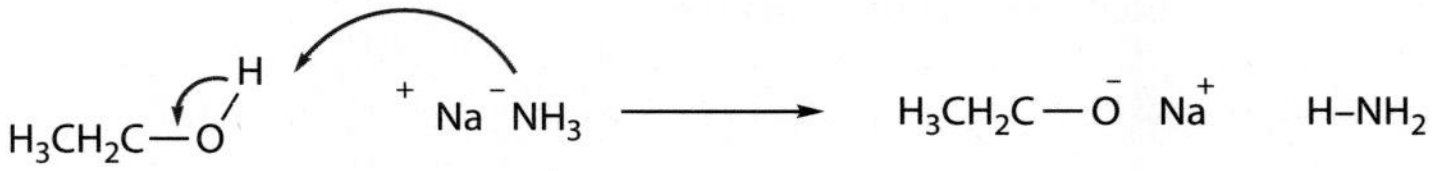

11.5

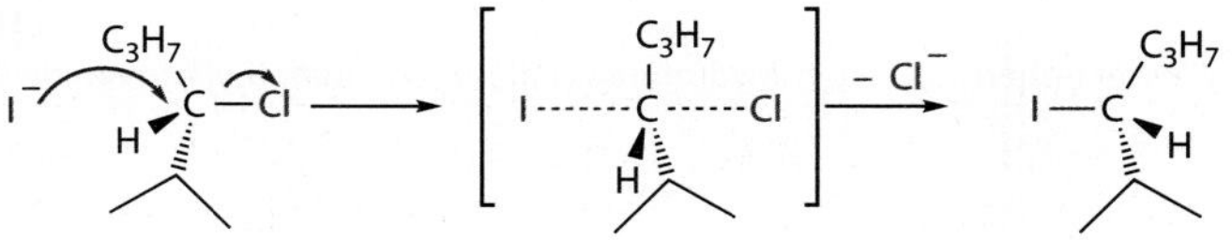

11.6

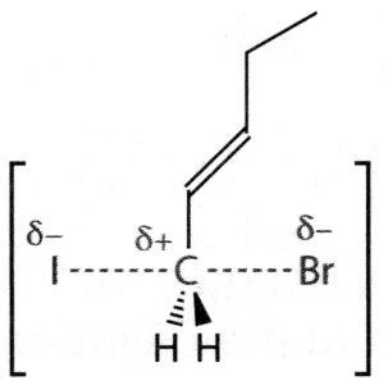

11.7

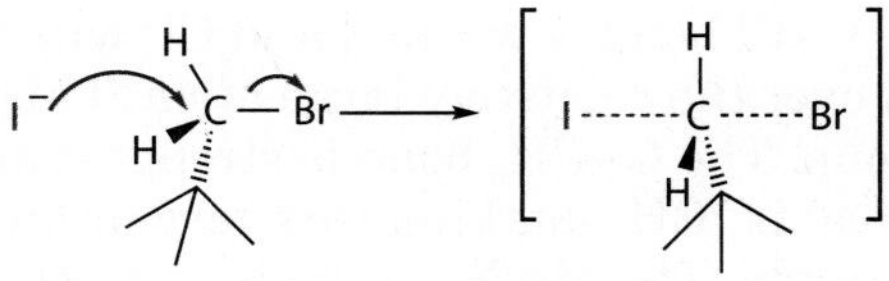

The bulky *tert*-butyl group makes this transition state very high in energy.

11.8   Two and a half times faster.

11.9

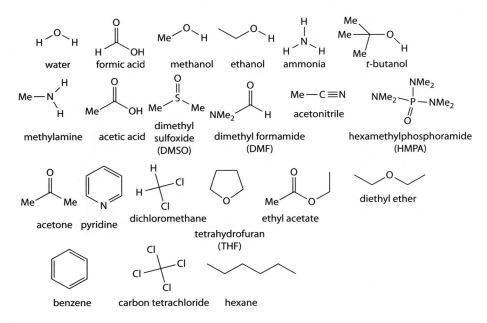

water   formic acid   methanol   ethanol   ammonia   *t*-butanol

methylamine   acetic acid   dimethyl sulfoxide (DMSO)   dimethyl formamide (DMF)   acetonitrile   hexamethylphosphoramide (HMPA)

acetone   pyridine   dichloromethane   tetrahydrofuran (THF)   ethyl acetate   diethyl ether

benzene   carbon tetrachloride   hexane

11.10

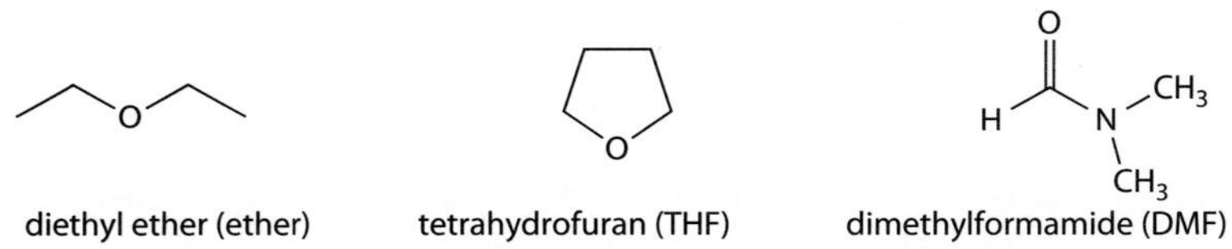

diethyl ether (ether)   tetrahydrofuran (THF)   dimethylformamide (DMF)

11.11

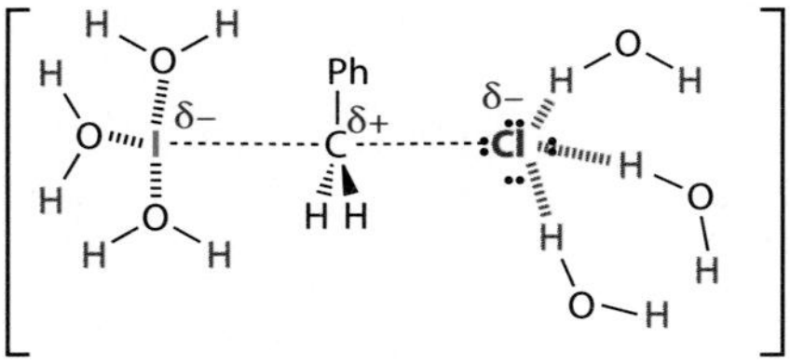

11.12   The reaction should be faster in aqueous THF. The transition state develops a positive charge on nitrogen and a negative change on the leaving group, so water will accelerate charge separation, which leads to the product. See transition state **13**.

11.13   Of these four fragments, Cl is the most electronegative and best able to accommodate charge, the C–Cl bond is weaker, and Cl⁻ will be very stable and the charge dispersed over the relatively large atom. This leads to Cl being the best leaving group. The C–CH₃ bond is strongest and the anion formed after methyl "leaves" is ⁻CH₃, making this very unfavorable so methyl will be the worst leaving group. The Me₂N⁻ group is somewhat less stable and more reactive than MeO⁻, but the C—O bond is weaker than the C—N bond. Because OR is a better leaving group than NR$_2$, the final order of leaving group ability should be Cl ≫ OMe > NMe$_2$ ⋙ Me.

11.14

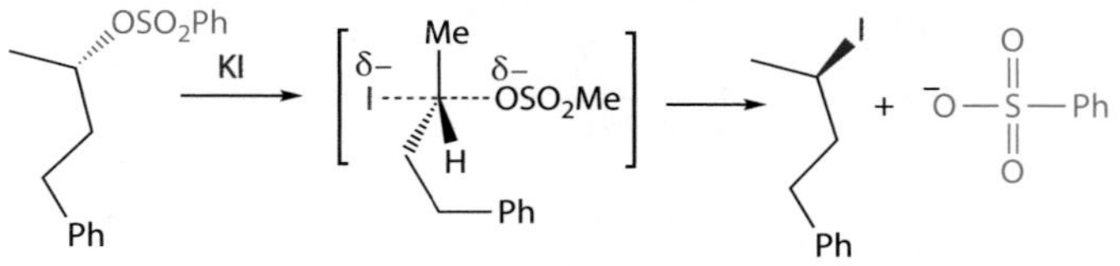

11.15 Fluorine is a very electronegative atom. The fluoride ion effectively holds electrons because of this electronegativity. If fluoride "holds" electrons, they are not available for donation to a carbon atom—the definition of a nucleophile. Therefore, the high electronegativity of fluoride makes it unable to donate electrons effectively, which makes it a poor nucleophile.

11.16 The reaction of sodium ethoxide with 2-bromo-2-methylpentane requires that ethoxide collide with a tertiary carbon for an S$_N$2 reaction. The steric hindrance for a tertiary halide in an S$_N$2 transition state is so high that the reaction does not proceed at any reasonable rate. Therefore, we conclude that there is no S$_N$2 product.

11.17

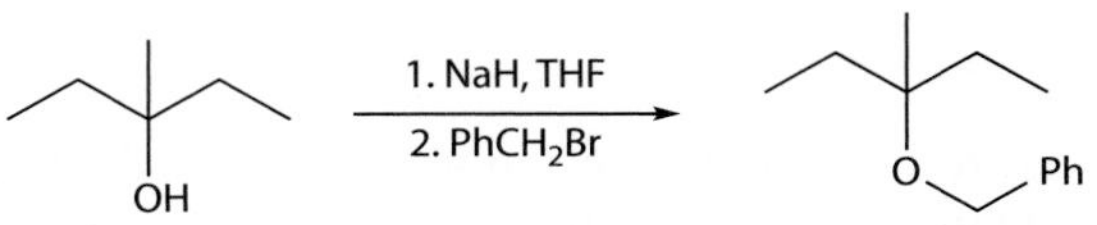

11.18

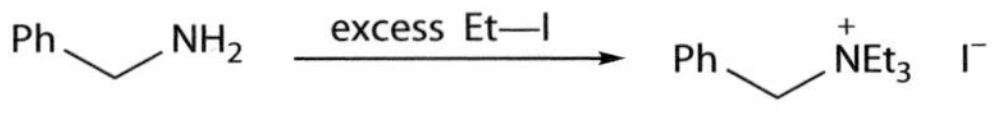

11.19 Initial S$_N$2 inversion of the bromide by phthalimide creates the C—N bond with the correct absolute configuration. Reduction of the imide unit with hydrazine liberates the amine.

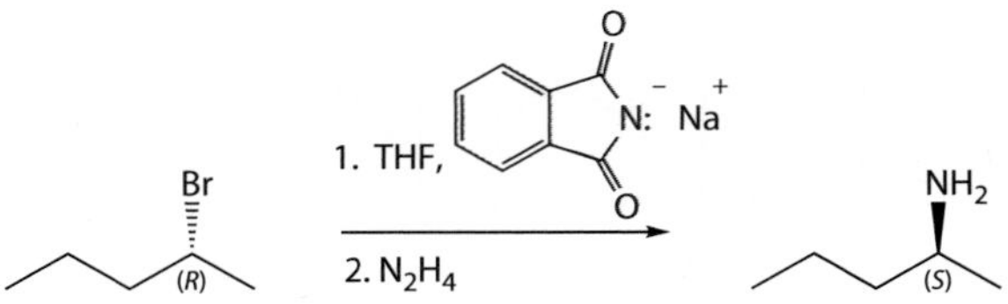

11.20

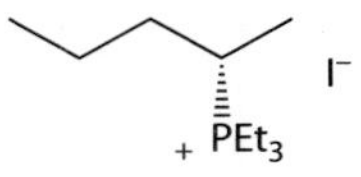

11.21

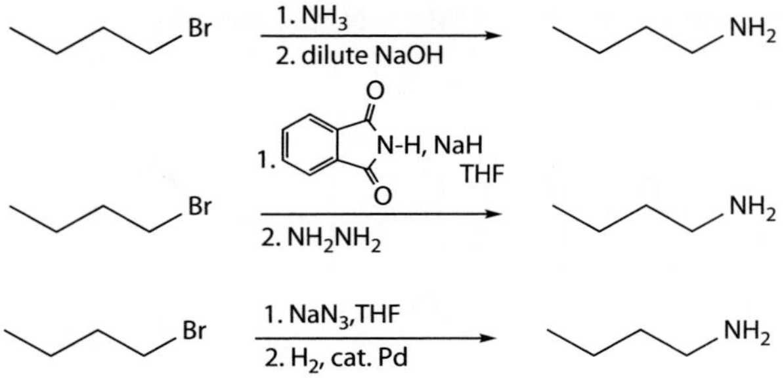

11.22

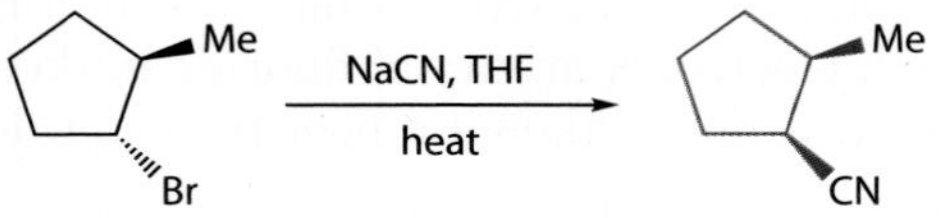

11.23

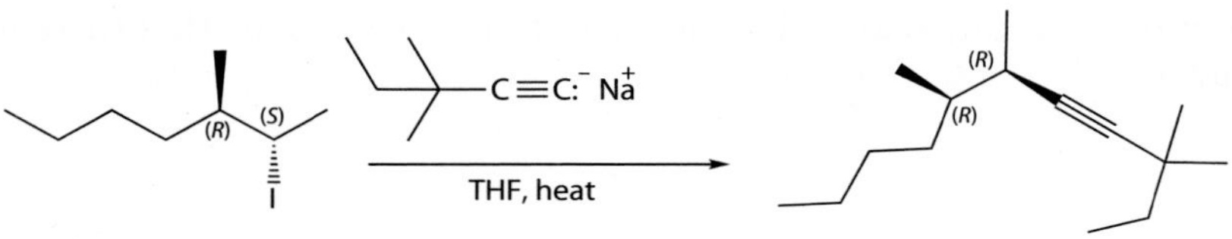

11.24 Ionization leads to the benzylic cation shown. It is resonance stabilized because the charge can be delocalized into the benzene ring. This imparts extra stability, making the cation easy to form and quite stable.

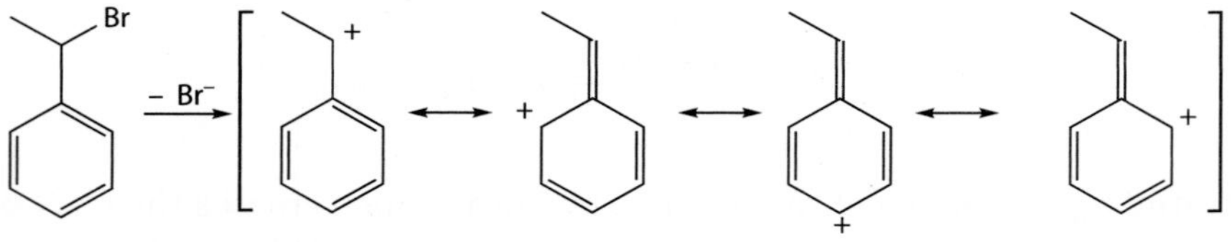

11.25

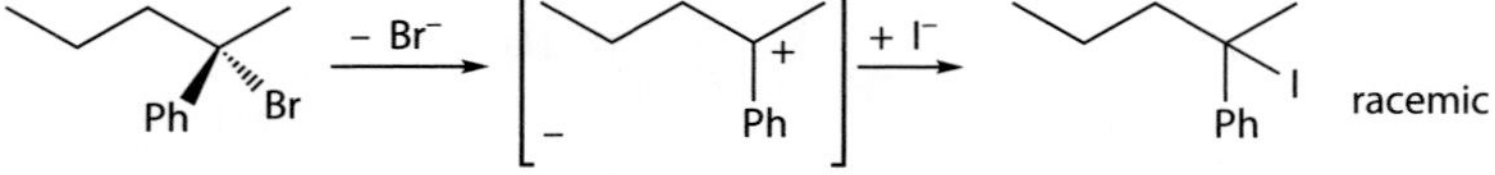

11.26

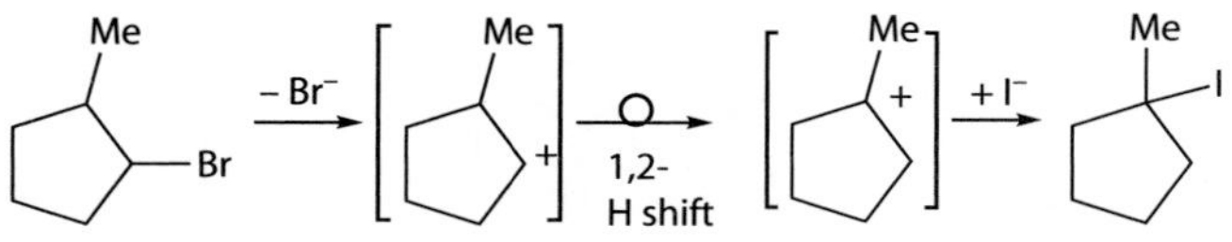

11.27

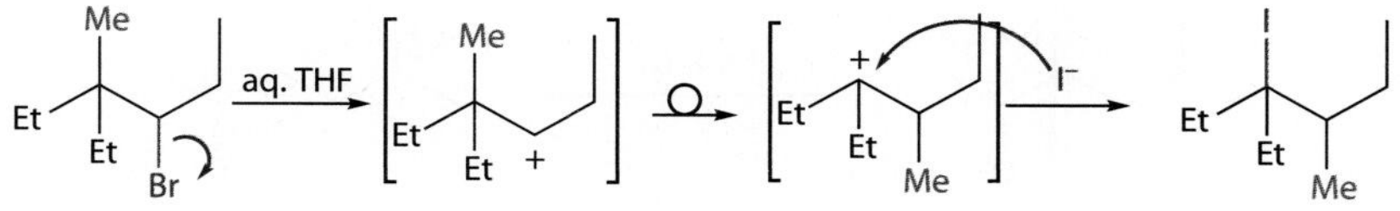

11.28

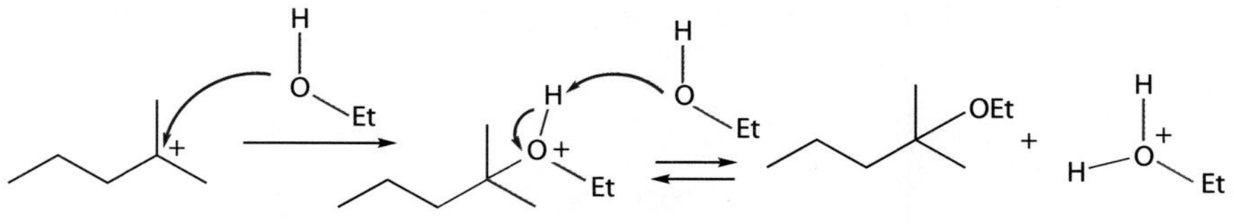

11.29

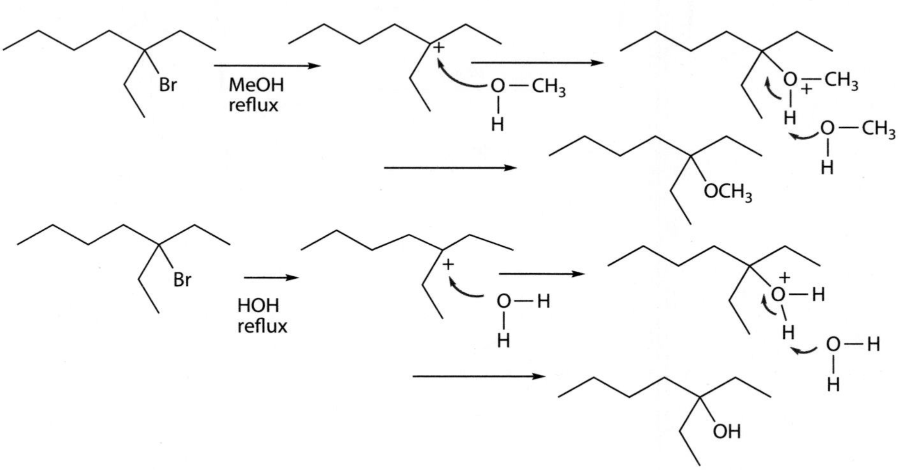

11.30

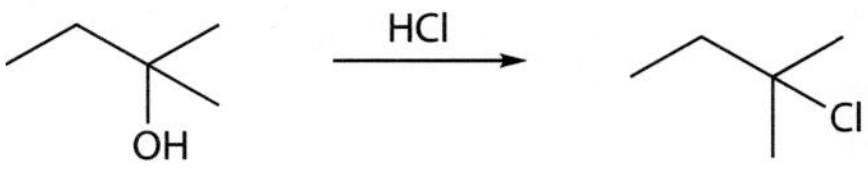

11.31

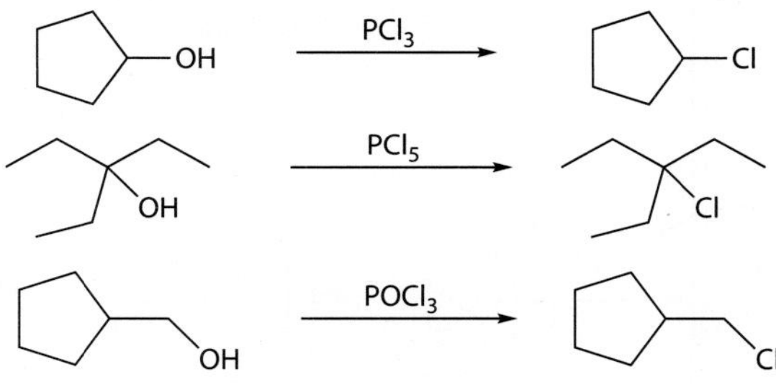

11.32

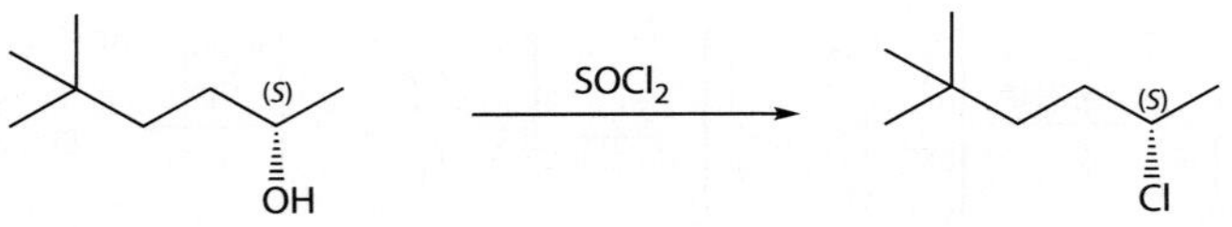

11.33

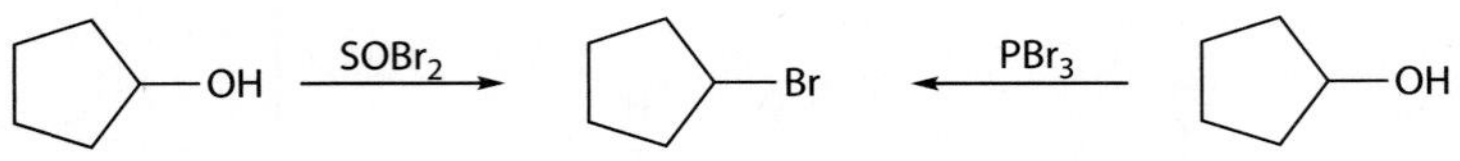

11.34

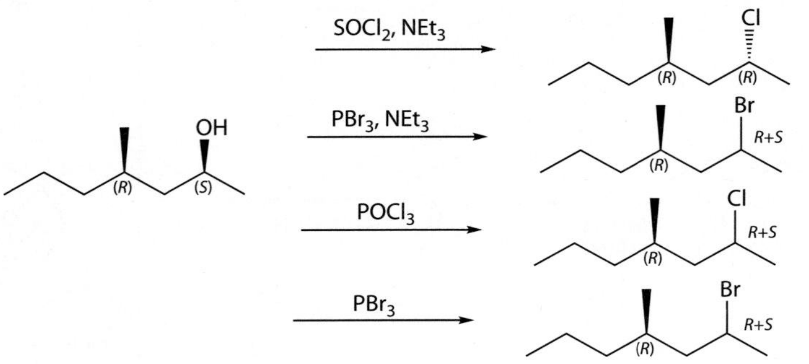

11.35

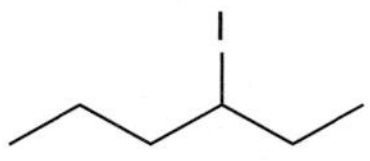

11.36

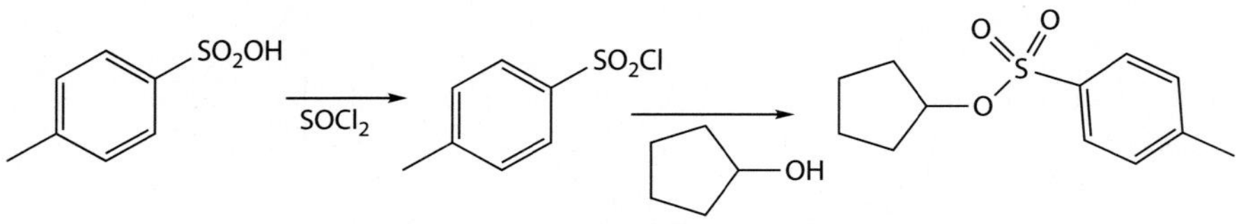

11.37

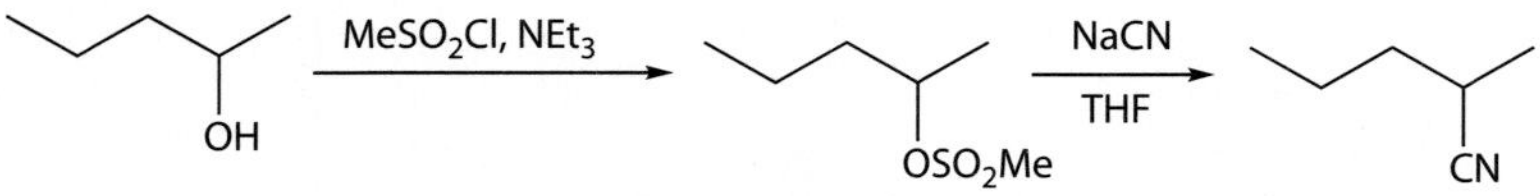

11.38

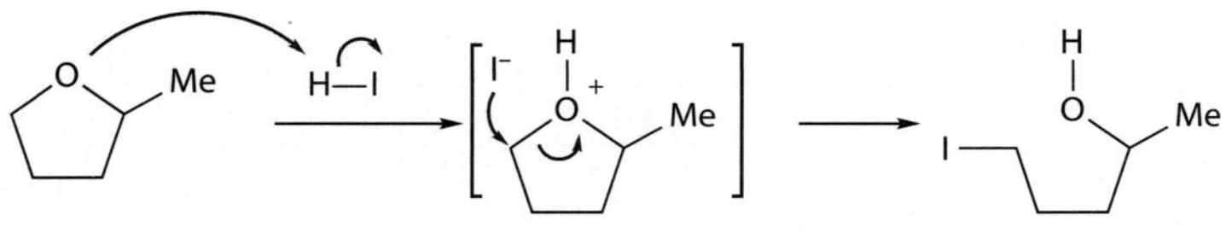

11.39   Formation of the oxonium ion is accompanied by formation of the nucleophilic bromide ion. Bromide ion attacks the sp³ hybridized and less hindered methyl group in an S_N2 reaction rather than the sp² carbon of the benzene ring, which requires significantly higher energy to achieve the requisite transition state. Therefore, the product is bromomethane and phenol.

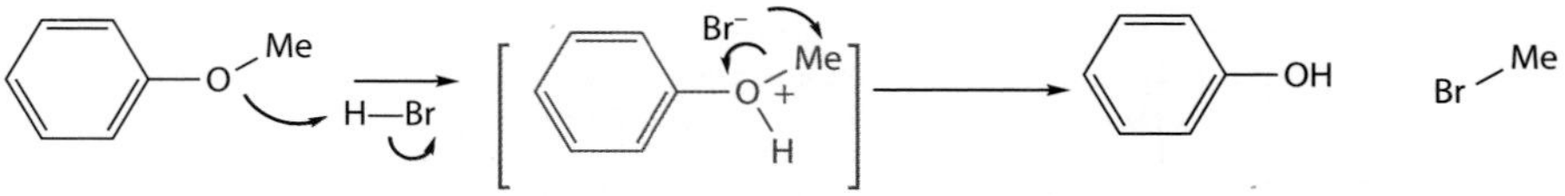

11.40

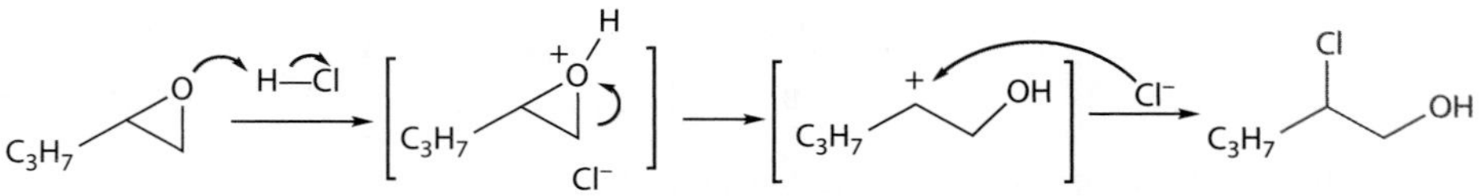

11.41

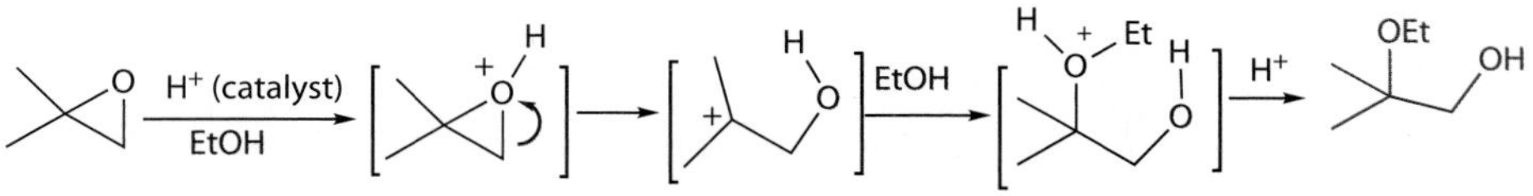

11.42   The azide ion attacks both carbons of the epoxide because they are equally substituted. The result is two regioisomeric products. Further, the azide ion attacks the face opposite the epoxide oxygen and on the same face as the methyl group, giving the stereochemistry shown.

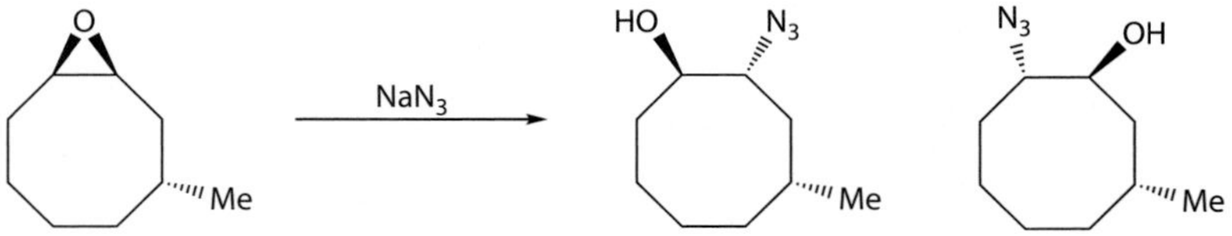

11.43

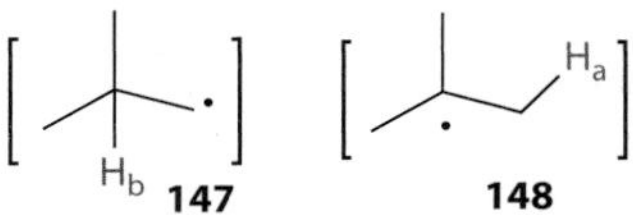

The larger number of carbon groups on the tertiary radical **147** should lead to greater stability. Therefore **147** should be more stable than **148**. If **147** is more stable, it should be easier to form, which suggests that **147** should have a lower E_{act} relative to **148**.

11.44

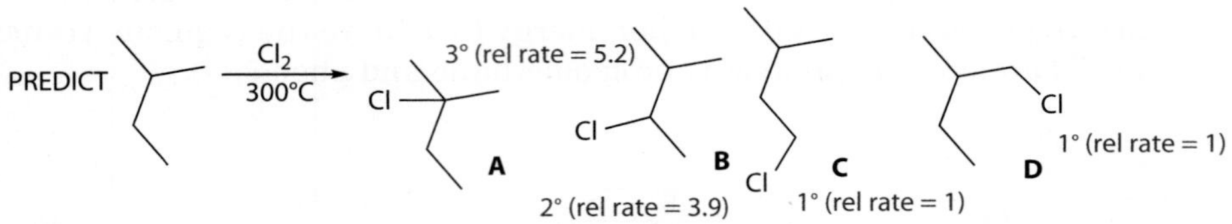

A, 1 3° H = 1×5.2 = 5.2   % **A** (5.2/22)×100 = 23.6%
B, 2 2° H = 2×3.9 = 7.8   % **B** (7.8/22)×100 = 35.5%
C, 3 1° H = 3×1 = 3.0   % **C** (3.0/22)×100 = 13.6%
D, 6 1° H = 6×1 = 6.0   % **D** (6.0/22)×100 = 27.4%
                     ─────
                     22.0

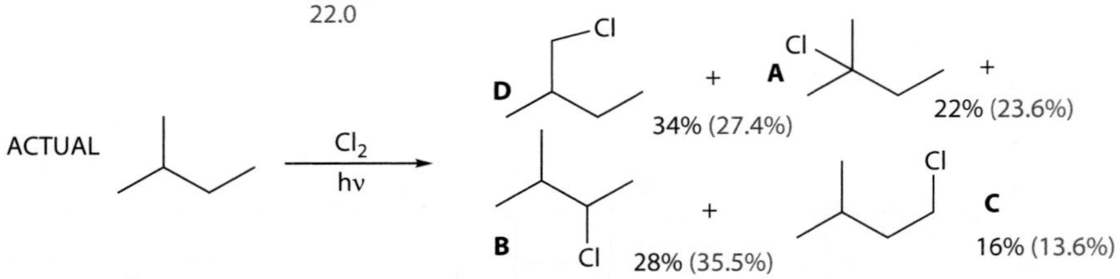

The predicted values are not accurate, and the actual major product is **D** rather than the predicted **B**. Nonetheless, the prediction indicates that there will be a mixture, **and** the predicted numbers show a higher percentage of **A** from the tertiary hydrogen atom, which is generally correct.

11.45

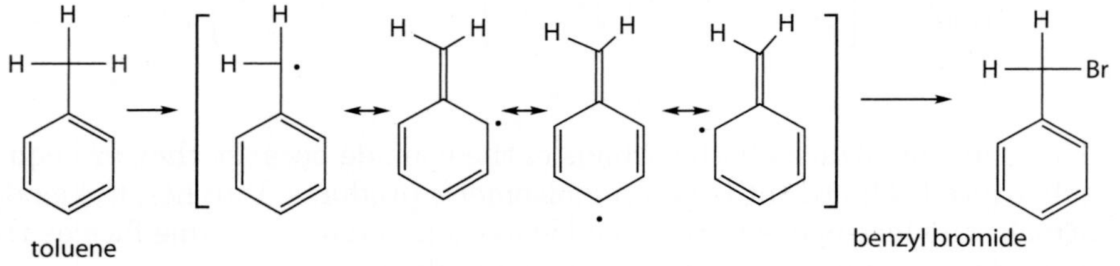

toluene   benzyl bromide

11.46

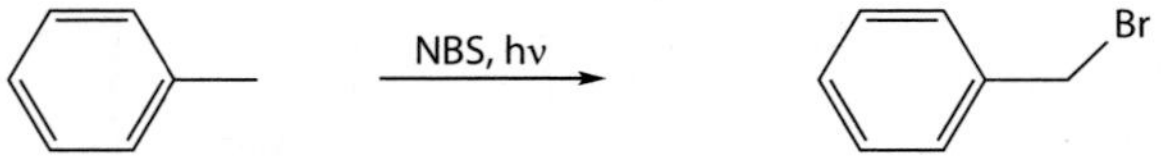

NBS, hv

11.47

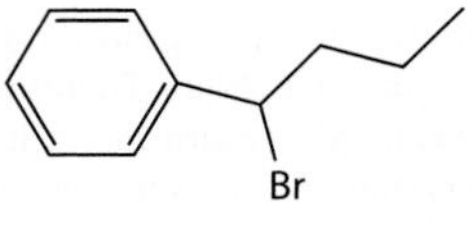

Reaction occurs at the benzylic position

Correlation of Concepts with Homework

- The C–X bond of alkyl halides and sulfonate esters is polarized such that the carbon has a positive dipole. Halides and sulfonate anions are good leaving groups. Nucleophiles attack primary and secondary alkyl halides, displacing the leaving group in what is known as aliphatic, bimolecular nucleophilic substitution, the S_N2 reaction. The S_N2 reaction follows second-order kinetics, has a transition state rather than an intermediate, and proceeds via back-side attack of the nucleophile on the halide and inversion of configuration: 2, 3, 4, 14, 36, 62, 63, 88, 89.

- Due to steric hindrance in the pentacoordinate transition state, tertiary halides do not undergo the S_N2 reaction and primary halides undergo the reaction fastest, with secondary halides reactive but less so than primary halides: 5, 6, 7, 8, 11, 12, 14, 16, 17, 18, 19, 20, 22, 23, 51, 52, 53, 54, 55, 56, 57, 63, 64, 65, 66, 68, 69, 72, 73, 74, 75, 86.

- Most S_N2 reactions are faster in aprotic solvents and slower in protic solvents. Water tends to promote ionization. In protic media, particularly aqueous media, ionization of tertiary halides occurs to give a carbocation intermediate (more slowly with secondary halides): 9, 10, 11, 12, 49, 55, 69, 70, 77, 86.

- Carbocation intermediates can be trapped by nucleophiles in what is known at an S_N1 reaction. An S_N1 reaction proceeds by ionization to a planar carbocation containing an sp^2 hybridized carbon, follows first-order kinetics, and proceeds with racemization of a chiral center. Carbocations are subject to rearrangement to a more stable cation via 1,2 hydrogen or alkyl shifts: 24, 25, 26, 27, 29, 50, 65, 67, 76, 77, 78, 79, 80, 81, 83, 84, 85, 87, 88, 89.

- A variety of nucleophiles can be used in the substitution reactions, including halides, alkoxides, amines, phosphines, azides, cyanide, acetylides, and enolate anions: 2, 3, 13, 14, 36, 62, 63, 88, 89, 105.

- Alkyl halides are prepared by the reaction of alcohols with mineral acids (HCl and HBr) or with reagents such as thionyl chloride, thionyl bromide, PX_3, and PX_5. Alcohols are also prepared by the reaction of alkenes with HCl, HBr, or HI; however, alkyl sulfates, alkyl nitrates, and alkyl perchlorates from those mineral acids tend to be unstable: 1, 28, 30, 31, 32, 33, 34, 35, 61, 82, 88, 89.

- Ethers are generally unreactive except with strong acids such as HI and HBr, which leads to cleavage of the ether to an alcohol and an alkyl halide. Epoxides are particularly reactive with nucleophiles, which open the three-membered ring at the less substituted carbon. Epoxides also react with an acid catalyst and weak nucleophiles such as water or alcohols, as well as with cyanide, azide, etc.: 38, 39, 40, 41, 42, 58, 59, 60, 65, 82, 88, 89, 95, 105.

- Exposure to light or heating to 300°C leads to homolytic cleavage of diatomic chlorine and bromine to give chlorine or bromine radicals. Chlorine and bromine radicals react with alkanes, removing hydrogen atoms via a radical process that leads to substitution and formation of alkyl chlorides and alkyl bromides. Both NBS and NCS can be used for controlled radical bromination or chlorination of allylic and benzylic systems: 43, 44, 45, 46, 47, 48, 90, 91, 92.

- A molecule with a particular functional group can be prepared from molecules containing different functional groups by a series of chemical steps (reactions). This process is called synthesis: The new molecule is synthesized from the old one (see Chapter 25): 19, 21, 22, 23, 27, 93, 94, 95.

- Spectroscopy can be used to determine the structure of a particular molecule and can distinguish the structure and functionality of one molecule when compared with another. See Chapter 14: 96, 97, 98, 99, 100, 101, 102, 103, 104, 105,

Homework

11.48   Which of the following alkanes give only one alkyl chloride upon reaction with chlorine and light?

11.49   Which of the following solvents will allow second-order reactions to proceed at a faster rate than first-order reactions?

11.50   Which of the following is the least stable carbocation? Justify your choice.

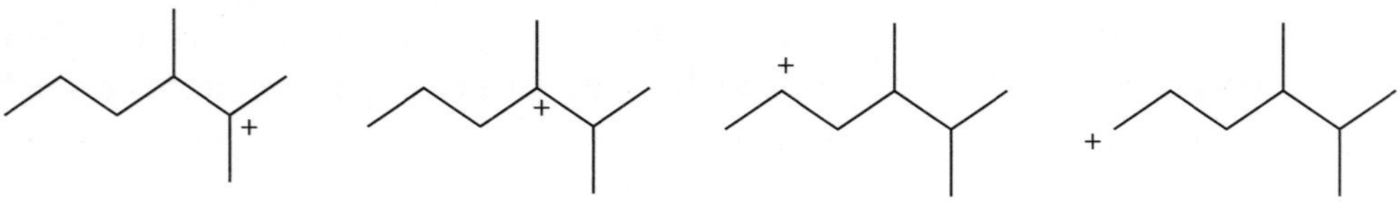

11.51   Which of the following is the most reactive halide when treated with NaI in dry ether? Explain.

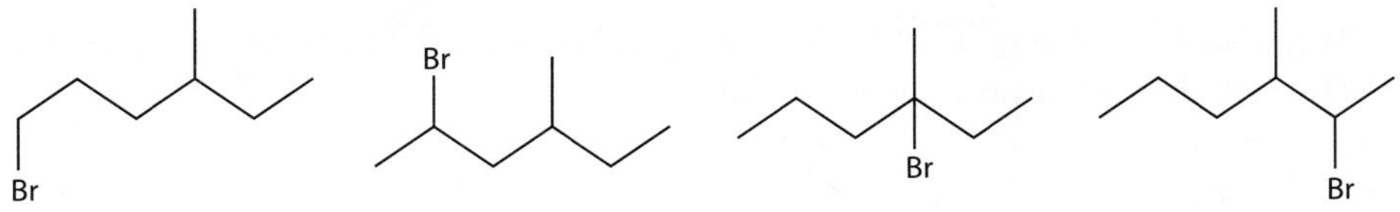

11.52   Which of the following reactions will proceed with 100% inversion of configuration? Explain.

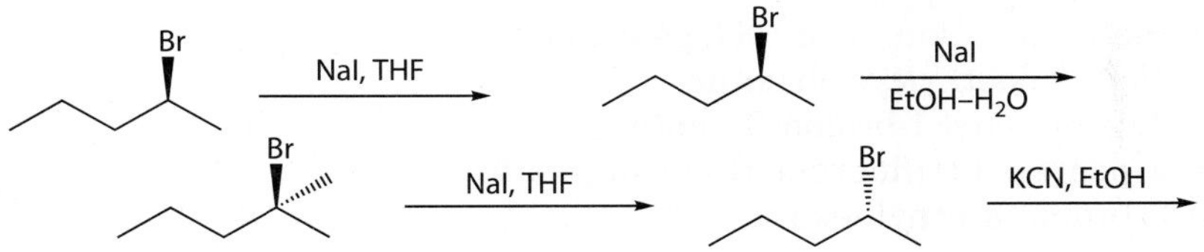

11.53   Which of the following nucleophiles will react fastest with CH_3I in THF? Explain your choice.
(a) CH_3O^-
(b) H_2O
(c) CH_4
(d) CH_3OH
(e) HSO_4^-

11.54   If the specific rotation of a pure alkyl halide is +100° and a chemical reaction with NaI generates a new alkyl halide with a specific rotation of 0°, which reaction best describes this result?
(a) S_N1
(b) S_N2
(c) acid–base
(d) radical
(e) addition

11.55   Briefly explain why 1-bromobutane might undergo a very slow S_N2 reaction in water to give an alcohol rather than an S_N1 reaction.

11.56   Briefly explain why the reaction of 2S-bromobutane + NaCN in ether gives a product that is *not* racemic.

11.57   Draw the transition state that would be required for the S_N2 reaction of 2-bromo-2-methylpropane and KI in diethyl ether. Use this diagram to explain **why** this combination of reactants gives **no reaction under these conditions**.

11.58   2-Phenyloxirane reacts with methanol and an acid catalyst to give 2-ethoxy-2-phenylethanol rather than 2-ethoxy-1-phenylethanol. Draw both products and suggest a reason why this product is formed.

11.59   The reaction of 2,2,5,5-tetramethyltetrahydrofuran with an acid catalyst in aqueous media leads to a diol. Suggest a mechanism for this reaction.

11.60   The reaction of oxirane and HI is significantly faster than the reaction of oxetane and HI. Draw the product of each reaction and suggest a reason for this difference in rate.

11.61   What is the structure of the halide formed when 2S-butanol is treated with thionyl chloride and triethylamine?

11.62   Draw the structure of the following molecules:
(a) 2R-bromo-4-phenylhexane
(b) methanesulfonate of cyclopentanol
(c) *cis*-2-iodoethylcyclohexane
(d) 4,4-diphenyl-1-bromo-3-heptanone
(e) 2-pentanol trifluoromethanesulfonate
(f) 3-bromo-3-ethylhexane

11.63   Briefly explain why methoxide (MeO⁻) is a relatively poor leaving group but methanesulfonate (MeSO₃⁻) is a good leaving group in nucleophilic substitution reactions.

11.64   Briefly explain why the S_N2 reaction of 1-bromo-2-cyclohexylethane with KI has a relatively short half-life in THF, but the identical reaction with 1-bromo-2-cyclohexyl-2-methylpropane has a particularly long half-life.

11.65   Give the major product for each of the following reactions:

(a) 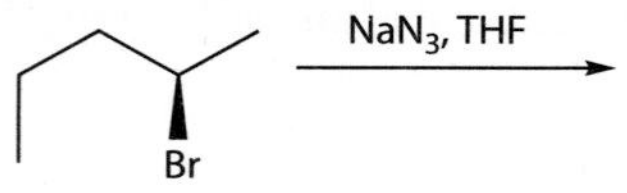   NaN₃, THF

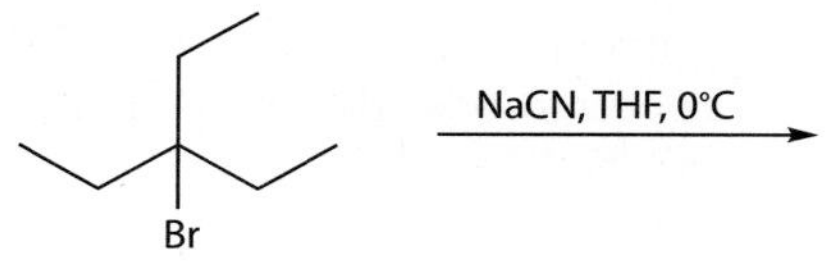

(b) 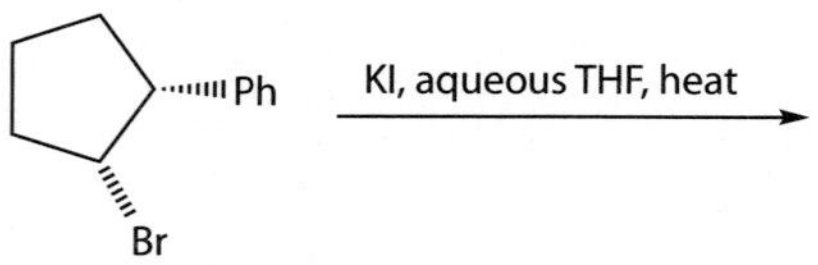   ⅢPh   KI, aqueous THF, heat
      Br

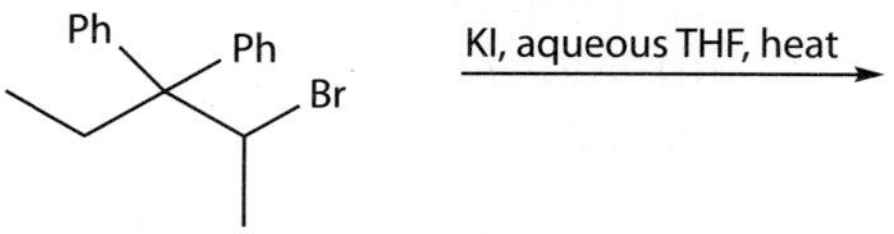

(c) 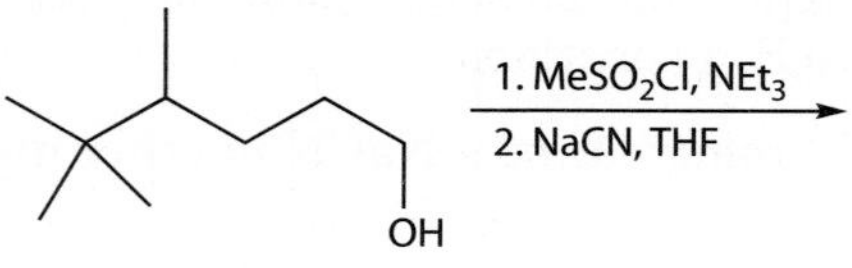   1. MeSO₂Cl, NEt₃
      2. NaCN, THF
      OH

(d)   NaCN, THF, 0°C
      Br

(e)   Ph   Ph   KI, aqueous THF, heat
            Br

(f) 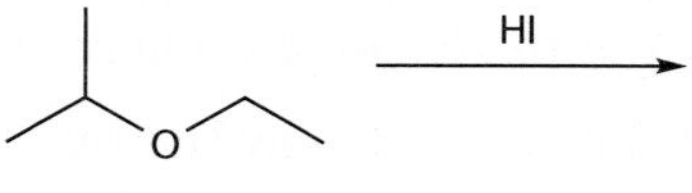   O   HI

11.66   When 1,3-dibromo-4,4-dimethylpentane is treated with KI in THF, the product formed results from reaction with only one of the two bromides. Draw the product and discuss why that bromine reacts preferentially.

11.67   Experimental data suggest that 2-bromopentane reacts with KI in aqueous THF by both S_N2 and S_N1 mechanisms. However, when the reaction is done in ethanol, the product is formed almost excessively by an S_N2 mechanism. Because ethanol is a polar protic solvent, why is there such a difference in the mechanism of this reaction in two different solvents? Only water can ionize the halide to the cation, so an S_N1 reaction can occur. Ionization is very slow in ethanol, so the S_N2 process wins.

11.68   An interesting S_N2 reaction between an alkyl halide (RX) and a nucleophile (NaX) in anhydrous ethanol has a half-life of 20 hours. Assume that RX costs \$55/gram and NaX costs \$0.02/pound. Assuming we need five half-lives for completion of the reaction, we must wait 100 hours (4.1 days) for the reaction to be completed. Discuss modifications to this procedure used for this reaction that might allow us to diminish the time required for this reaction to be completed.

11.69   We want to prepare 1-iodopentane by an S_N2 reaction between 1-bromopentane and KI. Look up the boiling points of 1-iodopentane, and potential solvents DMF, THF, acetone, diethyl ether. With knowledge that the solvent with the highest dielectric constant is likely to be the best solvent for the S_N2 reaction, choose a solvent for this reaction and briefly discuss the pros and cons of each solvent that led you to your choice.

11.70   Briefly explain why diethylamine is a stronger nucleophile with 1-bromopentane when water is used as a solvent when compared to THF.

11.71   The nitrate anion (NO_3^-) is not a very good nucleophile in the S_N2 reaction. Briefly explain why this is the case.

11.72   A common "trick" used by organic chemists when an S_N2 reaction has a slow rate of reaction is to add NaI. When 2-bromo-3-methylhexane reacts with NaN_3, for example, the reaction rate can be significantly increased by adding NaI to the reaction. Why and how does this help the reaction?

11.73   Cyanide is a bidentate nucleophile. Draw the product that results from a reaction with 1-bromopentane where carbon is the nucleophile. Draw the product that results from the reaction where nitrogen is the nucleophile. Calculate the formal charge, if any, on both products. Assuming there is a mixture of these two products, suggest a simple way that you might separate them.

11.74   Draw the major product for the reaction of 1-pentyne + 1. $NaNH_2$, THF; 2. 2R-bromopentane.

11.75   Exhaustive methylation (polyalkylation) is a common problem when an amine reacts with iodomethane. Polyalkylation is less of a problem when 2-bromobutane reacts with amines. Why?

11.76   Fluorine is the most electronegative element we have discussed, and the C–F bond is certainly polarized. Nonetheless, in aqueous solvents, bromine is a much better leaving group in the S_N1 reaction than fluorine. Why?

11.77   It is clear that we need water for efficient ionization of alkyl halides to initiate an S_N1 process. In all the reactions we have discussed, we used two solvents: Water was mixed with THF, ethanol, acetone, etc. Why is it necessary to use the cosolvent in these reactions?

11.78   When 7-bromo-1,3,5-cycloheptatriene is mixed with KCN in aqueous THF, one observes an S_N1 reaction. Draw the intermediate for this reaction and the final product. Comment on the stability of the intermediate.

11.79   Treatment of 4-phenyl-3-buten-2-ol with KI and an acid catalyst in aqueous THF leads to two iodide products in a rapid reaction, whereas treatment of 4-phenylbutan-2-ol under the same conditions leads to one product in a significantly slower reaction. Draw all products and suggest a reason why the first one is faster.

11.80   When 2R-butanol is treated with an acid catalyst in aqueous THF in the presence of KI, the resulting iodide is racemic. When 4,4-diethyl-2,2-dimethyl-3-(1-methylethyl)-3S-hexanol is treated with an acid catalyst and KI in aqueous THF, the resulting iodide is a mixture of R + S, but it is not racemic. In fact, the S-iodide is favored over the R-iodide. Suggest a reason for these observations.

11.81   Explain the following observation and give the complete mechanism as part of your answer:

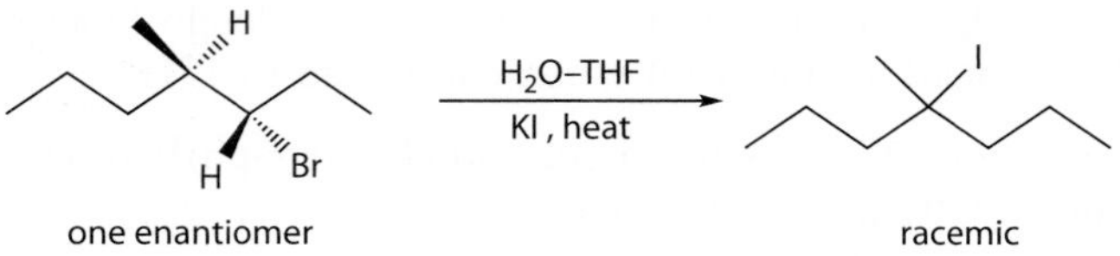

one enantiomer   racemic

11.82   Give a complete mechanism for the following reactions:

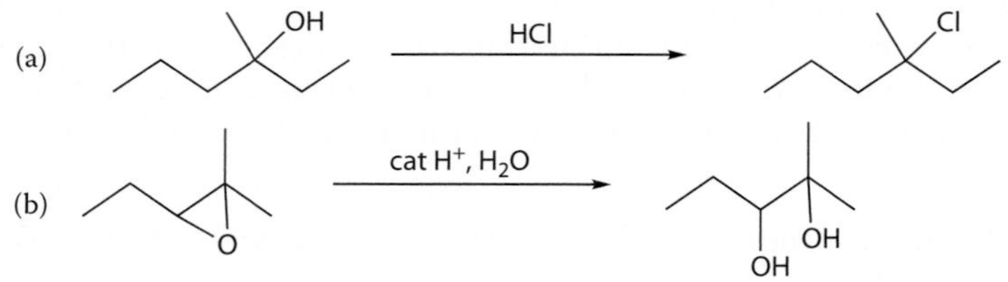

11.83   In most S$_N$1 reactions, adding a large excess of the nucleophile does not have a significant influence on the overall rate of the reaction. In one S$_N$1 reaction that you run, however, adding a 10-fold excess of the nucleophile leads to a threefold increase in the rate of the reaction. Briefly discuss the implications of this observation to the generic rate law given for an S$_N$1 reaction.

11.84   A well-known reaction in organic chemistry is the pinacol rearrangement of diols to ketones. Another old reaction is the so-called Demjanov rearrangement that converts alcohols such a 1-cyclopropyl-1-methanol to cyclobutanol. Draw both of these molecules. Both are acid-catalyzed reactions. Based on a knowledge of the S_N1 reaction, suggest a mechanism for the following transformations:

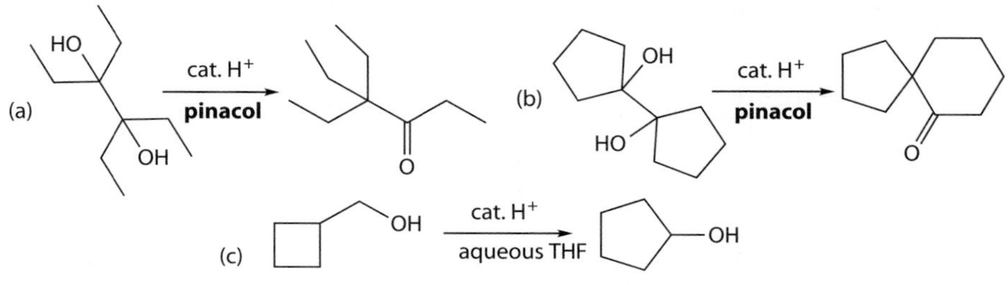

11.85 The reaction of a primary amine with nitrous acid (HO–N=O, which is formed by mixing sodium nitrite [NaNO$_2$] and HCl) leads formation of a diazo compound (R–N$_2^+$ = R–N≡N$^+$). This unstable molecule fragments to form a carbocation. What is the other product and why is cation formation so facile? Alkyl diazo compounds are often explosive and great care must be exercised during this reaction. Why should molecules with the formula R–N$_2^+$ be explosive? If we treat 1-aminocyclopentane with HONO in an aqueous medium, the product is cyclopentanol. Draw the product when each of the following amines is treated with NaNO$_2$ and HCl in aqueous THF:
(a) 2-aminopentane
(b) 1-amino-2-methylcyclohexane
(c) 1-amino-2,2-dimethylpropane

11.86 When 3R-(N,N,N-triethylammonium)hexane was treated with KI in aqueous THF, the final product (3-iodohexane) was isolated as a mixture of 2R and 2S enantiomers. There was a slight excess of the 2S enantiomer. We normally expect that an S_N2 reaction is faster than an S_N1 reaction and that ammonium salts undergo S_N2 reactions faster in aqueous media. It is known that S_N2 reactions proceed with clean inversion. This reaction gives anomalous results. Discuss the mechanism of this reaction in light of the experimental observations and focus on why we observe partial racemization.

11.87 The Wagner–Meerwein rearrangement occurs when **A** is treated with acid. The product is **B**. This is a classical reaction in organic chemistry. Offer a mechanism that will convert **A** to B.

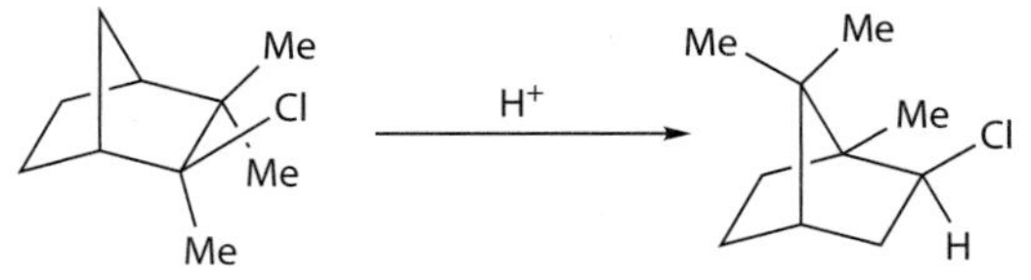

11.88 Give the major product expected from each of the following reactions:

(a)
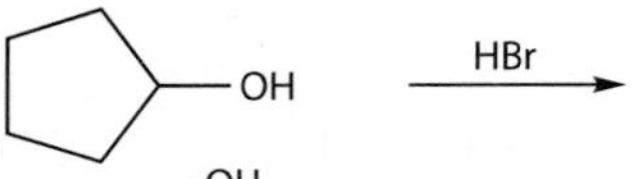
OH
HBr

(b)
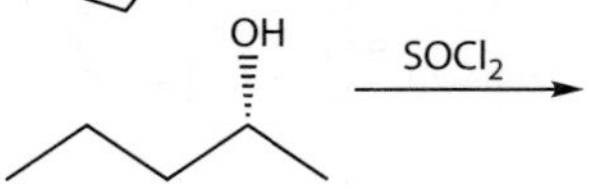
OH
SOCl2

(c)
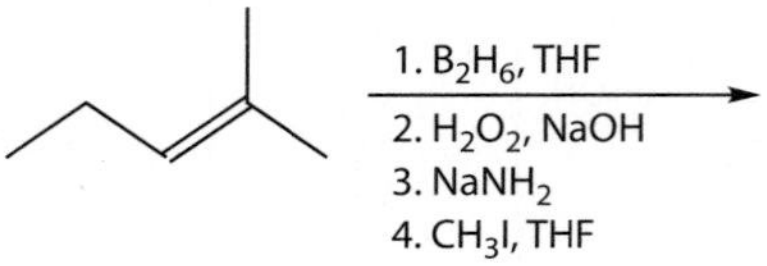
1. B2H6, THF
2. H2O2, NaOH
3. NaNH2
4. CH3I, THF

(d)
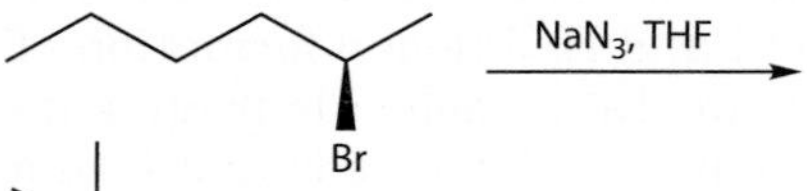
Br
NaN3, THF

(e)
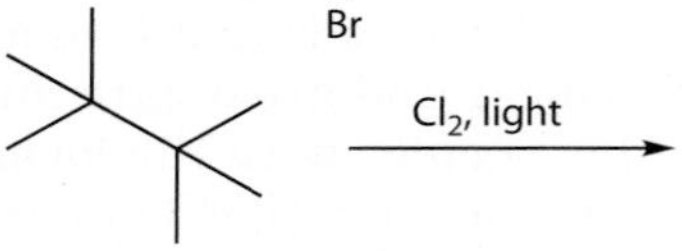
Cl2, light

(f)
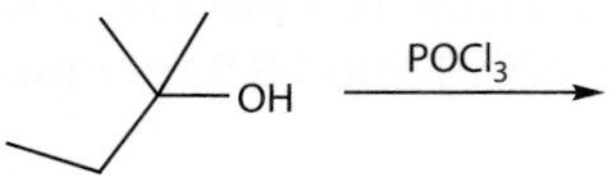
OH
POCl3

(g)
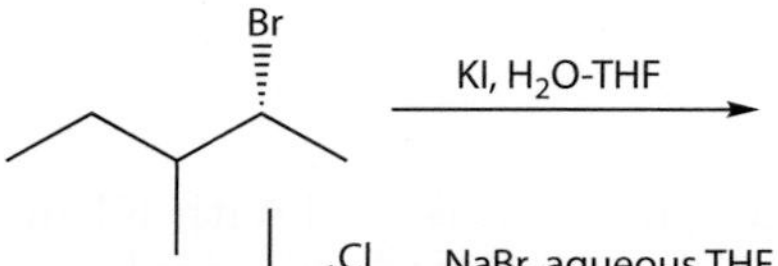
Br
KI, H2O-THF

(h)
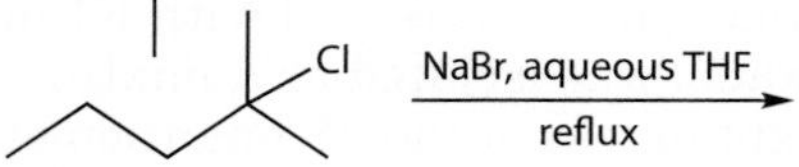
Cl
NaBr, aqueous THF
reflux

(i)
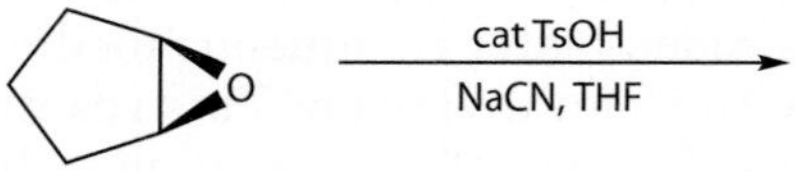
O
cat TsOH
NaCN, THF

(j)
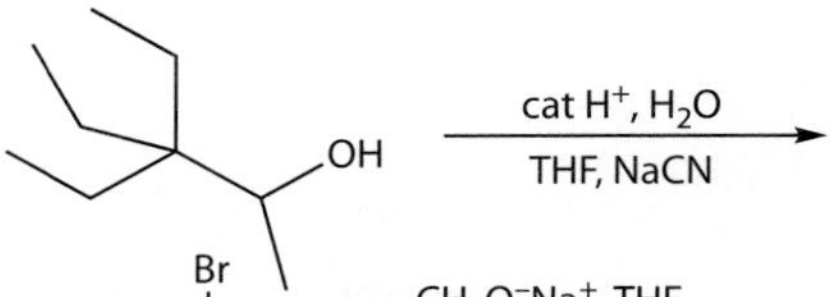
OH
cat H+, H2O
THF, NaCN

(k)
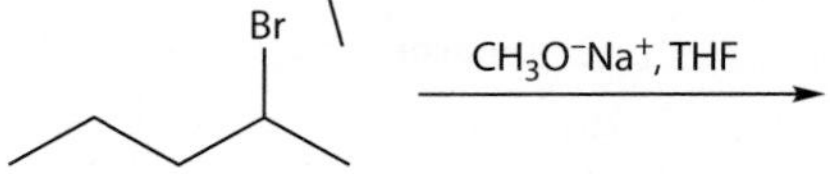
Br
CH3O−Na+, THF

(l)
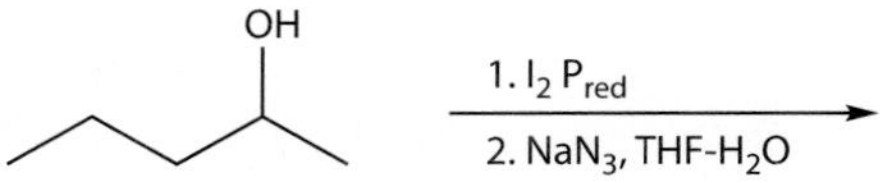
OH
1. I2 Pred
2. NaN3, THF-H2O

(m)
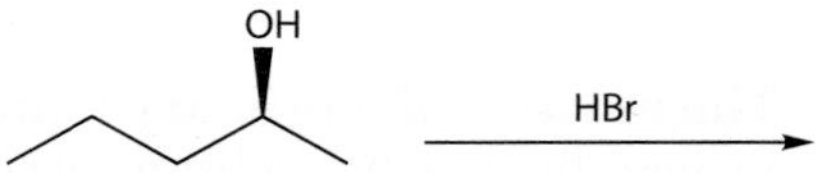
OH
HBr

(n)
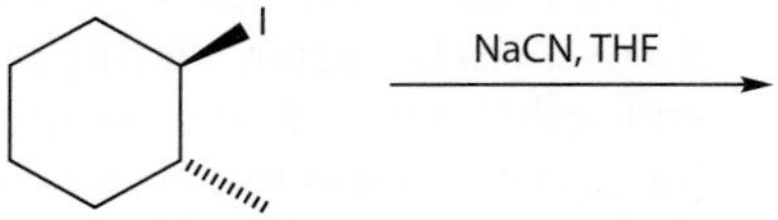
I
NaCN, THF

(o)
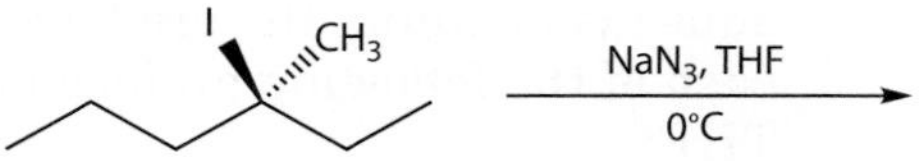
I
CH3
NaN3, THF
0°C

(p)
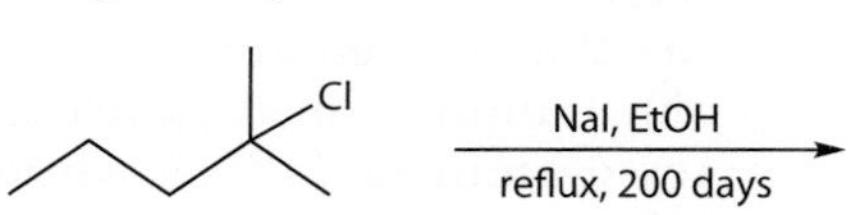
Cl
NaI, EtOH
reflux, 200 days

(q)
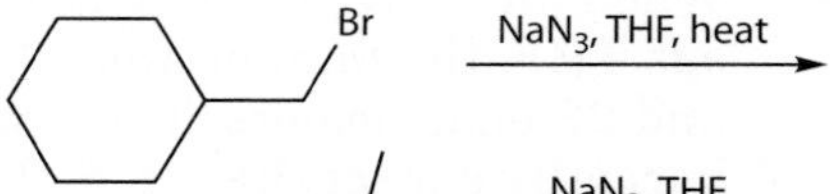
Br
NaN3, THF, heat

(r)
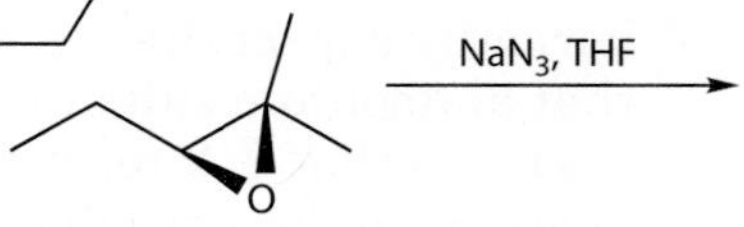
O
NaN3, THF

11.89 Give the major product for each of the following reactions:

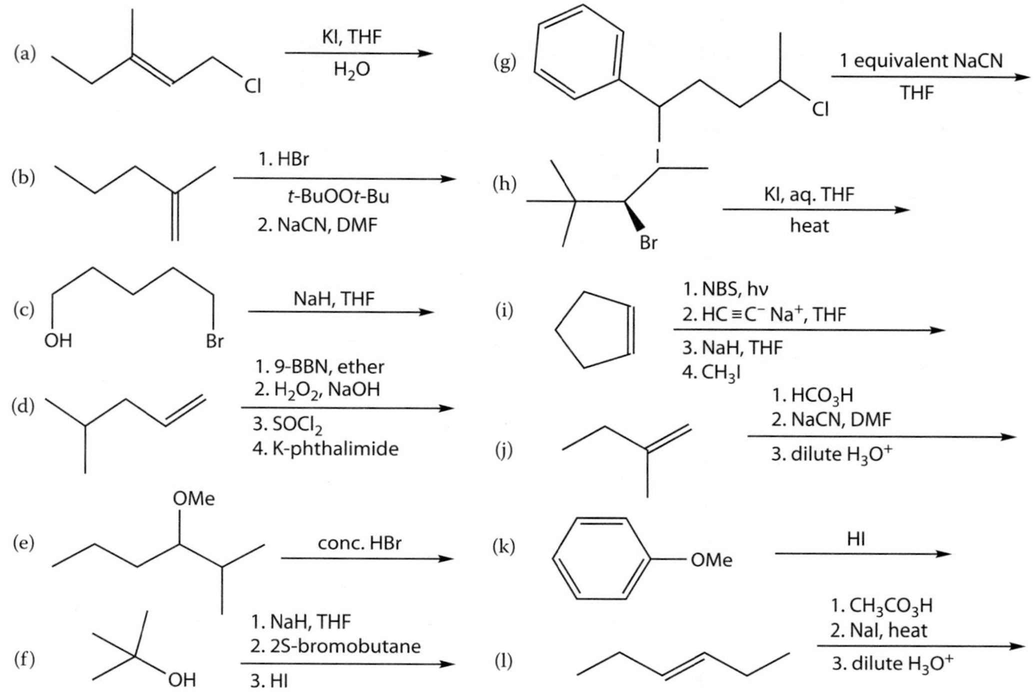

(a) KI, THF / H₂O

(b) 1. HBr / t-BuOOt-Bu / 2. NaCN, DMF

(c) NaH, THF

(d) 1. 9-BBN, ether / 2. H₂O₂, NaOH / 3. SOCl₂ / 4. K-phthalimide

(e) conc. HBr

(f) 1. NaH, THF / 2. 2S-bromobutane / 3. HI

(g) 1 equivalent NaCN / THF

(h) KI, aq. THF / heat

(i) 1. NBS, hv / 2. HC≡C⁻ Na⁺, THF / 3. NaH, THF / 4. CH₃I

(j) 1. HCO₃H / 2. NaCN, DMF / 3. dilute H₃O⁺

(k) HI

(l) 1. CH₃CO₃H / 2. NaI, heat / 3. dilute H₃O⁺

11.90 The reaction of cyclohexane and chlorine upon exposure to light gives essentially one product, whereas similar reaction with hexane gives three products. Explain.

11.91 Treatment of 2-pentene with NBS and light can give three different products. Draw them and explain this lack of selectivity.

11.92 Treatment of 3-methylpentane with chlorine and light gives four different products (draw them), but treatment of 3-phenylpentane gives almost exclusively one product. Draw this last product and explain the difference in reactivity for these two molecules.

Synthesis Problems

Do not attempt the following until you have read and understood Chapter 25.

11.93 Suggest a synthetic route to each of the following molecules. Provide all reagents and show all products:
  (a) ethyl 4-aminobutanoate from ethyl 4-hydroxybutanoate
  (b) 2S-dimethylaminopentane from 2R-pentanol
  (c) *cis*-1-amino-3-ethylcyclohexane from *trans*-3-ethylcyclohexanol
  (d) 2-(*N*-benzylamino)cyclopentane from bromocyclopentane

11.94   Show a synthetic sequence for each of the following molecules:
        (a) dibutyl ether from 1-butanol
        (b) 2-methylhexanenitrile from 2-hexanol
        (c) 1-ethoxy-2, 3-dibromopropane from allyl bromide
        (d) tert-butyl benzyl ether from tert-butyl methyl ether

11.95   In each of the following cases, give a reasonable synthetic route showing all
        reagents, reactants, and products:

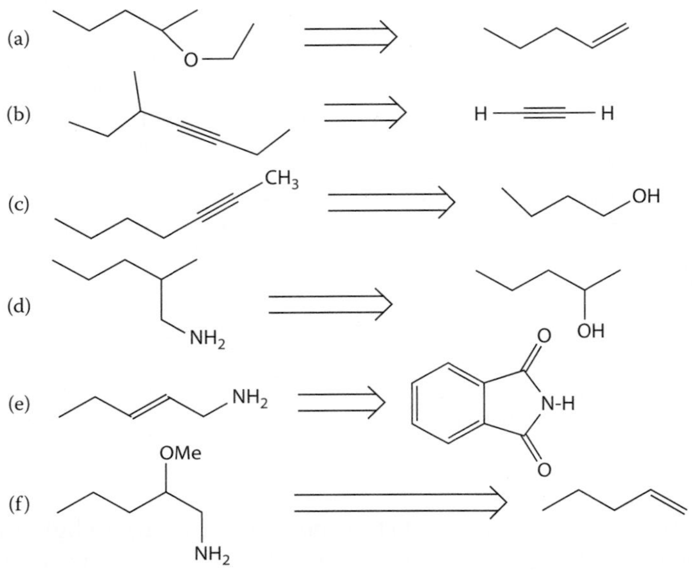

Spectroscopy Problems

**Note: Do not attempt these problems until you have read and understood
Chapter 14.**

11.96   Briefly describe differences in the IR and proton NMR that will allow you to
        distinguish between 2-pentene and 1-pentene.

11.97   The reaction of 2S-bromohexane and potassium cyanide leads to a nitrile.
        Draw it and then discuss characteristics of the IR and proton NMR that will
        allow you to identify it.

11.98   The reaction of 3-methyl-2-pentanol with an acid catalyst in aqueous THF
        can give two possible products: the starting alcohol and a rearranged alcohol.
        Draw both structures and discuss differences in their IR and proton NMR
        that would allow you to distinguish them.

11.99 Given the theoretical spectral data shown, provide a structure. The exact mass is 73.0892.

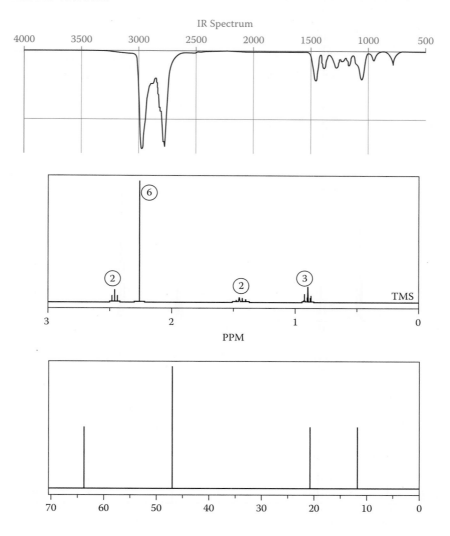

11.100 Given the theoretical spectral data shown, provide a structure. The mass
spectral data are M = 102, 100%; M+1 = 103, 6.66%; M+2 = 104, 0.42%.

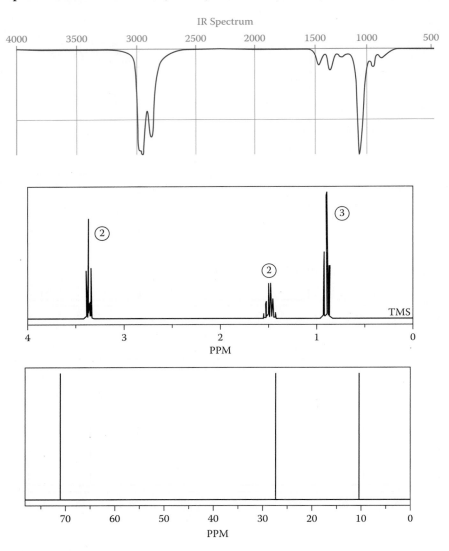

11.101 Given the spectral data, provide a structure. The exact mass is 96.0939.

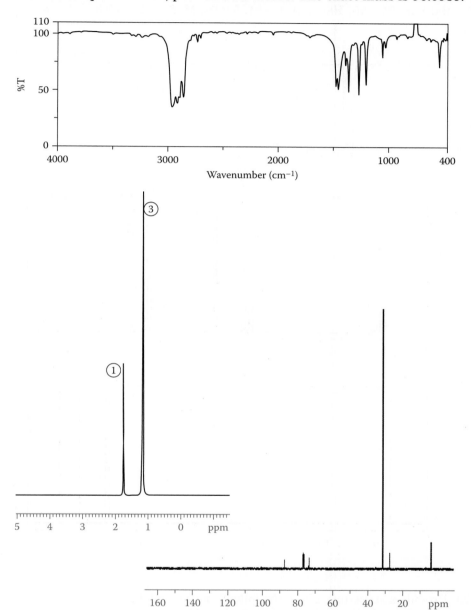

11.102 Given the theoretical spectral data, provide a structure. The mass spectral data are M = 83, 100%; M+1 = 84, 5.92%; M+2 = 85, 0.15%.

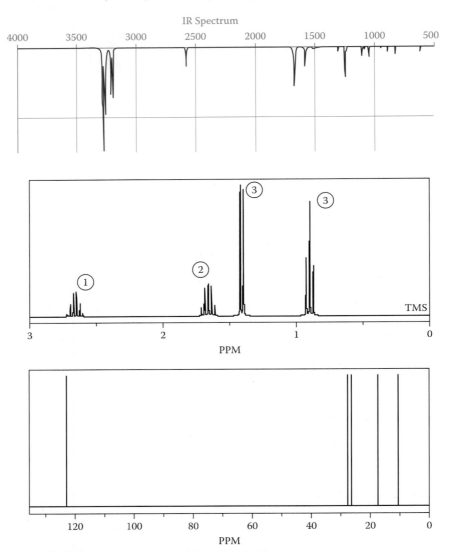

11.103 Provide a structure, given the following spectral data:
MS: M = 118 (100%), M+1 = 119 (6.66%), M+2 = 120 (0.62%)
43 (36), 45 (100), 55 (29), 69 (29), 70 (89), 71 (20), 87 (22), 118 (very small)
IR: 3270, 2870–2854, 1472–1438, 1410, 1068–1028 cm⁻¹
1H NMR: 4.13 (d, 1H), 1.83–1.77 (m, 2H), 1.70 (d, 3H), 1.49–1.39 (m, 2H), 1.33 (m, 2H), 0.9 (broad t, 3H) ppm

11.104 Provide a structure given the following spectral data:
MS: M = 164 (100%), M+1 = 165 (6.66%), M+2 = 165 (98%)
43 (100), 55 (12), 57 (14), 85 (68), 164 (very small)

IR: 2961–2931, 2874–2754, 1467–1458, 1378–1303, 1290–1194, 539–432, 465 cm−1

1H NMR: 3.54 (broad s, 2H; this peak is greatly diminished when the sample is washed with D_2O), 3.49 (broad s, 4H), 1.34 (q, 2H), 0.86 (7, 3H), 0.8 (s, 3H) ppm

11.105 A molecule **A** has the formula C_5H_{10} and shows a broad peak at 2900 cm^{-1} and a moderate peak at 1650 cm^{-1}:

MS: M = 164 (100%), M+1 = 165 (6.66%), M+2 = 165 (98%)

proton NMR: 5.81 (m, 1H), 4.97 (m, 1H), 4.93 (m, 1H), 2.0 (m, 2H), 1.43 (m, 2H), 0.91 (t, 3H) ppm

The reaction of this molecule with a catalytic amount of sulfuric acid in aqueous THF gives a modest yield of a new molecule **B** that has the formula $C_5H_{12}O$. This new molecule shows a very strong peak in the IR at 3683–2933 cm^{-1}. It has a proton NMR of 3.8 (m, 1H), 1.92 (broad s; this peak is diminished when treated with D_2O, 1H), 1.52–1.29 (m, 2H), 1.28 (m, 2H), 0.93 (broad t, 3H) ppm. When **B** reacts with NaH in THF, a slightly exothermic reaction occurs. This solution is then treated with iodomethane and the final product is **C**, which has a formula of $C_6H_{14}O$, no prominent peaks in the IR, and a proton NMR of 3.30 (s, 3H), 3.01 (m, 1H), 1.42 (m, 2H), 1.33 (m, 2H), 1.18 (broad d, 3H), 0.90 (broad t, 3H) ppm.

Base-Induced Elimination Reactions

12

At first glance, alkyl halides do not appear to be acids or bases. In fact, the hydrogen atom three bonds removed from the halogen is δ+ and slightly acidic because of the bond polarization induced by the halogen. The proton, referred to as the β-hydrogen (on the β-carbon, where the halogen is attached to the α-carbon), is a weak acid and it reacts only with a strong Brønsted–Lowry base.

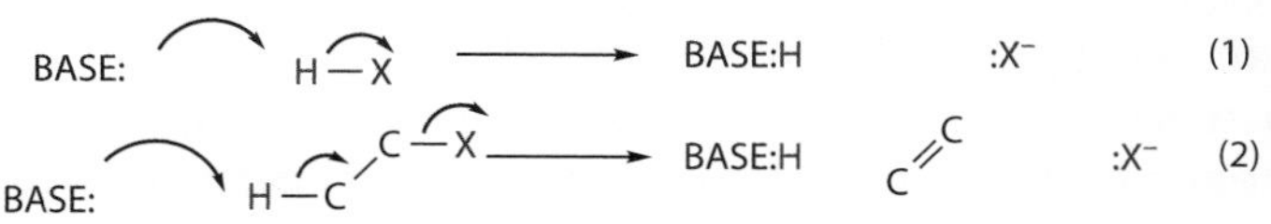

BASE: H—X ⟶ BASE:H :X⁻ (1)

BASE: H—C C—X ⟶ BASE:H C=C :X⁻ (2)

A typical Brønsted–Lowry acid–base reaction is shown in reaction (1), where the base donates electrons to the proton, which is directly connected to the leaving group X. After the H–X bond is broken, X⁻ is the conjugate base. The reaction of HCl with NaOH is typical of reaction type (1). If the acidic proton and the leaving group are separated by two carbon atoms, as shown in (2), a related acid–base reaction is possible that gives the conjugate acid (BASE:H) along with the conjugate base, X⁻. However, removal of the acidic proton leads to transfer of electron density through the intervening carbon groups to form an alkene (C=C) unit as the X⁻ group leaves.

An alkyl halide contains a C–X bond, where X is Br, Cl, or I, and this bond polarization extends to the β-hydrogen, which takes on a small δ+. In other words, the X–C–C–H unit in alkyl halides has a δ– halogen attached to a carbon (the α-carbon). This is δ+, making the next carbon (the β-carbon) δ–. Extending this bond polarization to the β-hydrogen makes that proton δ+, so it is slightly acidic. In the presence of a **strong** base, the acid–base reaction suggested in (2) removes the β-hydrogen and leads to formation of a π-bond with loss of the halogen. Therefore, the acid–base theme of chemical reactions is extended to very weak acids such as alkyl halides, leading to a functional group transformation of an alkyl halide to an alkene.

To begin this chapter, you should know the following:

- **the structure and nomenclature of alkyl halides (Chapter 4, Section 4.6; Chapter 5, Section 5.6.1)**
- **the structure and nomenclature of alkenes and alkynes (Chapter 5, Sections 5.1 and 5.2)**
- **the fundamental Brønsted–Lowry acid–base reaction of a base with an acid HX (Chapter 2, Sections 2.1 and 2.2; Chapter 6, Section 6.1)**
- **definition and recognition of a base (Chapter 6, Section 6.4)**
- **differences in base strength (Chapter 6, Sections 6.3 and 6.4)**
- **the role of bond polarization and dipole to generate a $^{\delta+}C–X^{\delta-}$ species (Chapter 3, Section 3.7)**
- **that an alkyl halide has a polarized C–X bond, where X is negative and C is positive (Chapter 3, Section 3.7)**
- **that bimolecular reactions are favored when there is no water present in the reaction (Chapter 11, Section 11.2)**
- **that ionization of an alkyl halide to a carbocation is competitive when water is present in the reaction (Chapter 11, Sections 11.4 and 11.6)**
- **that tertiary alkyl halides do not undergo S_N2 reactions due to steric hindrance in the transition state (Chapter 11, Sections 11.1 and 11.2)**
- **that a nucleophile can donate two electrons to a positive carbon or to a carbon with a positive dipole to form a new bond to that carbon (Chapter 6, Section 6.7; Chapter 11, Section 11.3)**
- **how to recognize intermediates (Chapter 7, Section 7.4)**
- **understanding and identifying intermediates and transition states in a reaction-energy curve (Chapter 7, Sections 7.3 and 7.4)**
- **how to recognize a transition state (Chapter 7, Section 7.3)**
- **the fundamentals of kinetics (Chapter 7, Section 7.11)**
- **how to determine the relative stability of carbocation intermediates (Chapter 10, Section 10.2; Chapter 11, Section 11.4.2)**
- **the basics of conformation applied to both acyclic molecules and cyclic molecules of ring sizes of three to six (Chapter 8, Sections 8.1 and 8.4–8.6)**
- **how to recognize a stereogenic center and assign the absolute configuration (Chapter 9, Sections 9.1 and 9.3)**
- **how to recognize and name diastereomers (Chapter 9, Sections 9.5 and 9.6)**

This chapter will introduce the reaction of a base with an alkyl halide, with a hydrogen atom that is on the β-carbon relative to a halide leaving group; the β-carbon is sp³ hybridized. Bond polarization makes the β-hydrogen atom slightly acidic, and reaction with the base expels the leaving group with formation of a new π-bond. In other words, the alkyl halide is converted to an alkene. The reaction just described is a classical acid–base reaction, with the β-hydrogen atom of an alkyl halide reacting as a weak acid. With tertiary halides and an aqueous medium, ionization to a carbocation may occur before loss of the leaving group, but the product is also an alkene. These types of reactions are generically known as elimination reactions.

When you have completed this chapter, you should understand the following concepts:

- **When alkyl halides are heated with a base in protic solvents, an E2 reaction occurs to give an alkene. The E2 reaction follows second-order kinetics; requires a base and an acidic β-hydrogen on the halide; does not have an intermediate, but rather a transition state; and gives the more substituted alkene as the major product.**
- **The E2 reaction is diastereospecific and if an enantiopure halide is used, a single stereoisomeric alkene (*E* or *Z*) will be formed.**
- **Prediction of the product in an elimination reaction for a cyclic halide requires a detailed analysis of the conformations for that ring system.**
- **Ionization to a carbocation, in the presence of a base, can give unimolecular elimination to the more substituted alkenes, the E1 reaction.**
- **When a base is tethered to a leaving group, removal of a β-hydrogen occurs via an intramolecular process and a *syn* rotamer to give the less substituted alkene as the major product. Thermal elimination of tetraalkylammonium hydroxides, without solvent, usually proceeds by *syn* elimination.**
- **1,3 Elimination is known, and a typical example is decarboxylation of 1,3 diacids or β-keto acids.**
- **When vinyl halides are heated with a strong base, an E2 reaction occurs to give an alkyne:**
- **A molecule with a particular functional group can be prepared from molecules containing different functional groups by a series of chemical steps (reactions). This process is called synthesis: The new molecule is synthesized from the old one (see Chapter 25).**
- **Spectroscopy can be used to determine the structure of a particular molecule and can distinguish the structure and functionality of one molecule when compared with another (see Chapter 14).**
- **Elimination reactions are important in biological systems.**

12.1 Bimolecular Elimination

In Chapter 11, several different nucleophiles reacted with alkyl halides or sulfonate esters via both S_N1 and S_N2 conditions. Alkoxides (RO⁻) are important nucleophiles, and the S_N2 reaction of an alkoxide with an alkyl halide was the basis of the Williamson ether synthesis in Chapter 11, Section 11.3.2. Based on the poor reactivity of tertiary halides with nucleophiles in the S_N2 reaction, reading Chapter 11 may lead to the conclusion that an alkoxide such as sodium ethoxide will give no reaction when mixed with a tertiary halide. ***This is incorrect.*** A reaction that forms the basis of this chapter occurs because alkoxides such as sodium ethoxide are strong bases as well as nucleophiles.

When 2-bromo-2-methylpropane (**1A**) is refluxed with sodium ethoxide (the conjugate base of ethanol) in ethanol, the isolated product is an alkene, 2-methyl-2-propene (**2**) along with sodium bromide and ethanol.[1] The Williamson ether synthesis in Chapter 11, Section 11.3.2, is the reaction of an alkoxide with an alkyl halide to give an ether. In Chapter 11, it is also clear that the transition state of an S_N2 reaction is so high for a tertiary halide that the reaction does not occur. Therefore, it is not a surprise that **1A** does not give an ether, but formation of an alkene indicates that this is a different reaction. The key to understanding this reaction is the observation that ethoxide is both a nucleophile and a base and that an alkene and ethanol (the conjugate acid of ethoxide) are the products.

A comparison of the starting material and the product shows that both Br and a hydrogen atom are lost from **1A**. Structurally, C–Br and C–H bonds are broken, and a π-bond is formed. The bromine atom is converted to the bromide ion in NaBr, and ethoxide reacts with the acidic hydrogen atom (on the β-carbon relative to bromine, so it is called a β-hydrogen) to form ethanol. In this reaction, the ethoxide ion reacts as a base to form a new O–H bond in the conjugate acid (ethanol) and bromine is a leaving group that gives bromide ion as the conjugate base. This reaction is a bit different in that the acid–base reaction also produces a neutral molecule, the alkene. *The conversion of the tertiary bromide 1A to alkene 2 is an acid–base reaction.* In practical terms, the alkyl halide loses a hydrogen atom as well as a bromine to form the alkene product, so the overall transformation is called an **elimination reaction** (the elements of H and Br are eliminated to form a π-bond).

Ethoxide is a nucleophile as well as a base, but why did the reaction with **1A** give an alkene product? The activation barrier for an S_N2 reaction is too high due to steric hindrance in the transition state (Chapter 11, Section 11.2). The polarization of the bromine atom on the α-carbon is δ−, and the induced dipole on the second carbon away from the bromine is also δ−, which leads to a δ+ hydrogen (see **1B**). This small positive dipole makes the hydrogen on the β-carbon slightly acidic. Remember that the β-hydrogen atom is the hydrogen atom attached to the β-carbon.

12.1 Show all positively polarized hydrogen atoms, relative to the iodine atom, in 3-iodo-2,3-dimethylpentane.

In Chapter 11, hydroxide was not used as a nucleophile in reactions with alkyl halides. The reason is now known. Hydroxide is a rather strong base, and it can react to form alkenes. This is illustrated by the reaction of hydroxide ion (NaOH) with 2-bromo-2-methylpentane (**3**) to give alkene **5** as a product. The negatively polarized oxygen in hydroxide is most strongly attracted to the δ+ α-carbon. A collision between these two species does not lead to an S_N2 reaction because the tertiary carbon is sterically hindered, making the energy for the pentacoordinate transition state too high (see Chapter 11, Section 11.2). In other words, collision at the tertiary carbon does not produce enough energy to initiate a substitution reaction, and the S_N2 reaction does not occur.

Although substitution cannot occur with hydroxide reacting as a nucleophile, it is also a base. As with **1A**, the bond polarization in **3** leads to a δ+ β-hydrogen, so the hydroxide is also attracted to the β-hydrogen. The collision of hydroxide with the β-hydrogen initiates an acid–base reaction that removes the β-hydrogen with concomitant formation of **5**.

Is there an intermediate in the conversion of **3** to **5**? Experimentally, *no intermediate has been detected under the conditions at which the reaction is normally done.* If there is no intermediate, then the characteristics of the reaction are described by the transition state for the reaction. The transition state is defined by donation of electrons from the hydroxide to the β-hydrogen to form a new O–H covalent bond, and the electrons in the C–H bond are transferred toward the δ+ β-carbon. In other words, as electron density increases on the β-carbon, electron density migrates toward the δ+ carbon (the α-carbon in **3**) to form a second bond to carbon, a π-bond.

Formation of this new bond demands that the C–Br bond break, with bromide ion as the leaving group. If there is no intermediate, all of the bond making and bond breaking occurs simultaneously (this is a *synchronous* reaction) and the transition state for the elimination reaction is represented as **4**. Note that *the new π-bond forms between the α- and β-carbons* (see 2-methyl-1-propene; isobutylene, **5**). The initial reaction of hydroxide and the β-hydrogen is a collision process, and experimental kinetic data show it to be a *second-order reaction* (see Chapter 7, Section 7.11.2).

Because the overall transformation is an elimination reaction, with loss of H and Br, the symbol "E," for elimination, is used. It is a bimolecular elimination reaction, so the symbol 2 is used. *The elimination reaction of an alkyl halide with a base is known as an E2 reaction.* The conversion of **3** to **5** is an example of an E2 reaction, and **4** is a typical E2 transition state. *A generalized E2 transition state is drawn as 6, where a base reacts with the β-hydrogen and a leaving group X is lost to form the new π-bond.*

12.2 Draw the elimination product expected when 2-bromo-1,1-dimethylcyclopentane is treated with hydroxide ion and heated.

A close inspection of 2-bromo-2-methylpentane (**3**) reveals that something has been ignored. The hydrogen atoms on the methyl groups in **3** are also β-hydrogen atoms. The final product **5** indicates that those hydrogen atoms are not removed to give an alkene product. Why? Another example is used to answer this question. When 1-bromo-1-methylcyclopentane (**7**) is heated with sodium ethoxide in ethanol, 1-methylcyclopentane (**9**) is the major product—not methylenecyclopentane (**8**). There are two β-hydrogen atoms (H$_a$ and H$_b$) in **7**, but the major product **9** results from removal of H$_b$.

Note that H$_a$ is a slightly stronger acid because it is attached to a secondary carbon, whereas H$_b$ is attached to a tertiary carbon. Alkyl groups on carbon release more electron density, so H$_a$ has a slightly larger δ+ than H$_b$, and this fact makes H$_a$ a slightly stronger acid. The E2 reaction is an acid–base reaction, so the more acidic proton H$_a$ should react fastest. If H$_a$ is removed faster, then **8** should be the major product. It is not! Therefore, *a factor other than the acidity of the β-hydrogen atom must be important in leading to the formation of 9.*

If a comparison of the starting materials does not predict the major product, what about a comparison of the two alkene products? Alkene **9** has three carbon substituents attached to the C=C unit, whereas **8** has only two carbon substituents. A carbon substituent (an alkyl group) is electron releasing, and the C=C unit with more alkyl groups has more electron density released to the π-bond, making the bond stronger and the alkene more stable. *In other words, the more substituents that are attached to the C=C unit of an alkene, the more stable it will be,* as discussed in Chapter 5 (Section 5.1.2). Therefore, a trisubstituted alkene is more stable than a disubstituted alkene, and a tetrasubstituted alkene is more stable than a trisubstituted alkene.

In short, alkene **9** is more stable than alkene **8**, and this observation is consistent with the E2 reaction of **7**. The transition states for these two products are **10** for alkene **9** and transition state **11** for alkene **8**. Consider the Hammond postulate (see Chapter 10, Section 10.2.1), which can be stated as *the transition state for a given step resembles the side of the reaction to which it is closer in energy.*[2,3]

If the reaction is controlled by factors influencing the starting material (the stronger acid), the reaction should have an early transition state (closer in energy to the starting materials and largely by the relative acidity of the different protons) and give **8** as the major product. Because alkene product **9** is the major product, the reaction does not have an early transition state. If stability of the products is more important in the reaction, the energy of the transition state is closer to that of the product and factors that influence stability of the product should dominate, consistent with a late transition state and formation of **9**. The experimental fact that **9** is the major product is interpreted to mean *that the E2 reaction proceeds by a late transition state, so factors that influence the stability of the products are more important than differences in acidity in the starting material.* Note that the E2 reaction favors formation of the more substituted product.

12.3 **Draw the major product expected when KOH is heated with 3-bromo-2,3-dimethylhexane in a suitable solvent.**

The bottom line of this discussion is summed up in the statement that *an E2 reaction of an alkyl halide and a base gives the more highly substituted alkene as the major product.* Returning to **3**, removal of a β-hydrogen atom from a methyl group will give a disubstituted C=C unit, whereas **5** is trisubstituted and more stable. Based on this analysis, **5** is the predicted major product. In the nineteenth century, chemists did not understand the mechanism of the E2 reaction just presented and could only describe different reactions in terms of the products that were formed and the starting materials that were used. Alexander Mikhaylovich Zaitsev (also spelled Saytzeff or Saytzev; Russia; 1841–1910) was a Russian chemist from Kazan who recognized the selectivity of elimination reactions for the more substituted product and formulated what is known as Zaitsev's rule (also known as Saytzeff's rule or Saytsev's rule).

Zaitsev's rule is "If more than one alkene can be formed by an elimination reaction, the more stable alkene is the major product." This means that the E2 reaction is regioselective because of two alkene products; one is formed as the major product. Note that this says *major* product—not *the only* product. Although small amounts of **8** may be formed in the E2 reaction of **7**, the major product will be **9**. The percentage of each remains unknown without doing the experiment, so it is only possible to make predictions about major and minor products.

Examination of transition states **10** and **11** reveals another key feature of the E2 reaction. The leaving group for the reaction (bromide) is 180° opposite the hydrogen being removed (the β-hydrogen, H_b). In other words, those two atoms have an *anti* relationship. Why? As the base removes H_b, negative charge develops on the β-carbon. Remember that *there is no intermediate, so there is no carbanion. In other words, the hydrogen atom is not removed to form a carbanion.* As the C–H bond begins to break, the electron density in that bond migrates toward the α-carbon, which forms a new π-bond, and expels the bromide ion. In effect, the electrons in the C–H bond displace the leaving group via back-side attack (as far away as possible for the negatively polarized bromine). Therefore, *a requirement of the E2 reaction is that the β-hydrogen removed by the base is anti to the leaving group.* Implicit in this statement is that it is possible to form the requisite *anti* rotamer (for a discussion of rotamers, see Chapter 8, Section 8.1).

12.4 Draw the two possible transition states that lead to an alkene product when KOH is heated with 2-bromo-3-ethylpentane in a suitable solvent. Indicate which one leads to the major product.

12.5 Draw the final product formed when 1-bromo-2-ethylheptane is treated with KOH in ethanol.

An E2 reaction with 3-bromo-4-ethyl-2-methylhexane (**12**) and potassium *tert*-butoxide ($Me_3CO^-K^+$) in 2-methyl-2-propanol (Me_3COH; *tert*-butanol) reveals another issue. Two alkene products are formed: **13** and **14**. Both are trisubstituted alkenes; based on stability of the two products, **13** and **14** should be roughly equal in energy. In other words, the transition state energies should be about the same, which means that both alkenes are formed in approximately equal amounts. Indeed, for two products of equal substitution and presumed equal stability, **assume** they are formed as a 1:1 mixture, unless there are extenuating circumstances.

In the reactions of **1, 3, 8,** and **12**, three different bases are used. In general, the bases used for an E2 reaction are NaOH or KOH in water or an alcohol solvent, or the alkoxide RO^- derived from the alcohol solvent. In other words, use $NaOCH_3$ in CH_3OH, $NaOCH_2CH_3$ in CH_3CH_2OH, or $(CH_3)_3CONa$ in $(CH_3)_3CHOH$.

Primary halides, such as 1-bromopentane (**15**) usually give very poor yields of an alkene when treated with KOH in ethanol. Based on experimental results with primary alkyl halides, *assume that primary alkyl halides do not give an alkene under normal E2 conditions.* The activation energy required to generate a transition state such as **16** (for an E2 reaction of **15**) is so high that the E2 reaction is quite slow relative to another reaction such as substitution. It is not impossible, just slow. Remember that hydroxide and ethoxide are nucleophiles as well as bases, and one can imagine a facile S_N2 reaction at the primary carbon, which is faster than the E2 reaction. If any reaction occurs when **15** reacts with hydroxide in ethanol, it will likely be the S_N2 product, 1-pentanol via reaction with hydroxide, or 1-ethoxypentane via reaction with ethoxide. If any alkene is formed, it is usually a very minor product.

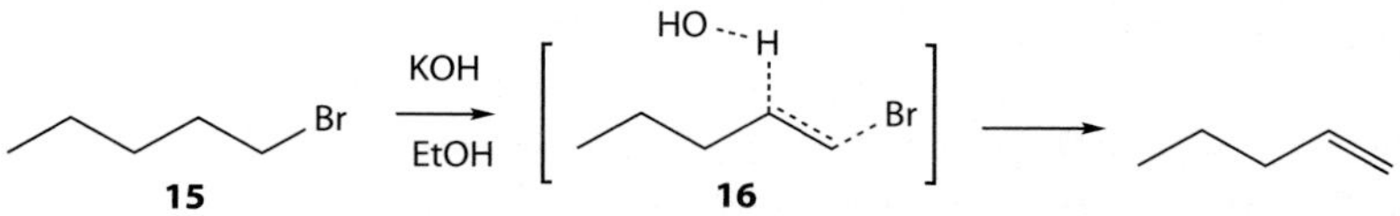

An E2 reaction can be summarized by saying that tertiary halides react with strong bases to give alkenes. Secondary halides sometimes give a mixture of E2 and S_N2 products, but the product ratio depends on the solvent. In protic solvents such as ethanol, the E2 product is usually the major product. In general, it is very difficult for a primary halide to undergo an E2 reaction except under special conditions that use specialized bases, and those conditions will not be discussed in this book.

12.6 **Draw the reaction and the product that is probably formed when sodium ethoxide reacts with 1-bromopentane, 15.**

12.7 **Comment on why the reaction of 3-bromo-2-methyl-4-phenylpentane with KOH in ethanol leads to more 4-methyl-2-phenyl-2-pentene than 2-methyl-4-phenyl-2-pentene.**

12.2 Stereochemical Consequences of the E2 Reaction

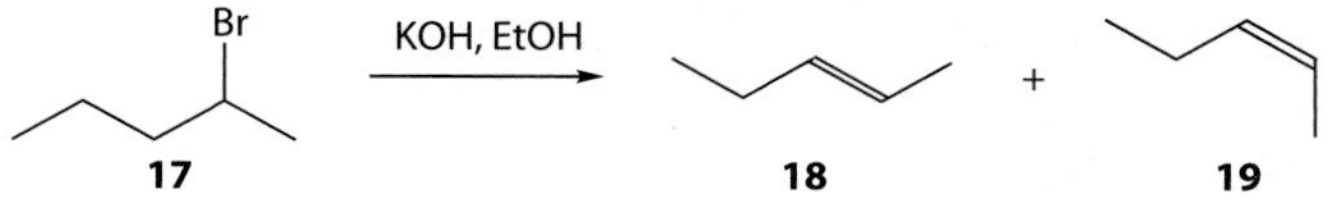

The alkenes formed by an E2 reaction in the previous section cannot exist as *E*- or *Z*-isomers. The reaction of 2-bromopentane (**17**) to give 2-pentene, however, requires a closer look. The more highly substituted product predicted by

Zaitsev's rule is 2-pentene rather than 1-pentene, but there are two 2-pentenes. Under normal E2 reaction conditions, a secondary halide will give both possible stereoisomers: *E*-2-pentene (**18**) and *Z*-2-pentene (**19**). Unless there are extenuating circumstances, the transition state that leads to **18** is about equal in energy to the one that leads to **19**. Therefore, it is anticipated that both will be formed in roughly equal amounts.

However, there is a different instability between these two disubstituted alkenes. An *E*-alkene tends to be lower in energy than a *Z*-alkene due to diminished steric interactions. In other words, the methyl and ethyl groups in **19** are closer together and, because there is no rotation about a C=C bond, that steric interaction is greater than the methyl–H or ethyl–H interactions found in **18**. Because alkene **18** is lower in energy than **19**, that transition state should be more favorable and **18** should form faster via the E2 mechanism. It is therefore anticipated that both **18** and **19** are formed, but **18** will be the major product.

12.8 Draw the E2 transition states that lead to 18 and also to 19, remembering the stereochemistry required to generate each alkene.

The reaction is more complicated if a chiral, nonracemic halide such as 2*R*-bromo-3*R*-methylpentane (**20**) is treated with KOH in ethanol because the E2 reaction is known to be ***diastereospecific***. In other words, one diastereomeric alkyl halide will generate the *E* or the *Z* alkene (*one, not both*), and the other diastereomeric alkyl halide will give only the other alkene stereoisomer. The more stable alkene is the one generated between C2 and C3. Therefore, the transition state for an E2 reaction of **20** is **21**, where the leaving group and the β-hydrogen are *anti,* as discussed in Section 12.1. This model demands one rotamer of **20** in which the β-hydrogen atom and the bromine leaving group are *anti,* and this leads to the requisite transition state. In other words, the E2 can only occur via that particular rotamer.

The (2*R*,3*R*) absolute configuration indicates the stereochemical placement of the groups in the *anti* rotamer are fixed, as shown. Once in the appropriate *anti* rotamer, the bond-making or bond-breaking process begins, and the stereochemical positions of all groups are essentially "locked" in transition state

21. According to the Hammond postulate, the transition state is assumed to be closer in energy to the alkene product. Therefore, *the stereochemical orientation of the groups in 21 will be the same as will be found for those in the alkene product*, and transition state **21** must give 3-methyl-2*E*-pentene (**22**).

The two methyl groups in **20** (marked in **violet**) are "on the same side" in transition state **21** and they will be on the same side in the alkene product, **22**. The requisite *anti* transition state dictates that an enantiopure halide such as **20** will give one and only one stereoisomeric alkene. *The E2 reaction is diastereospecific*. The E2 reaction of stereoisomer **20** is diastereospecific, which also means that the identical reaction of 2*S*-bromo-3*R*-methylpentane (**23**, the diastereomer of **20**) with KOH will lead to transition state **24**, which gives 3-methyl-2*Z*-pentene (**25**), as shown.

12.9 Draw the E2 transition state and final major product formed when 3*S*-iodo-4*R*-(1-methylethyl)heptane reacts with sodium ethoxide in refluxing ethanol.

Returning to 2-bromopentane **17**, which is drawn without stereochemistry (no wedges or dashes), it must be assumed that it is a racemic mixture (see Chapter 9, Section 9.2.4) because there is only one stereogenic center. The two pertinent β-hydrogen atoms in **17** are drawn with stereochemistry (**red** and **violet**) to show that either one may be removed by the KOH. The E2 reaction is diastereospecific, but there are two β-hydrogen atoms on β-carbon (marked in **red** and in **violet**), and removal of either will lead to a 2-pentene product. A rotamer is possible where each β-hydrogen atom is *anti* to the bromine (see **26** and **27**).

Transition state **26** leads to the *Z* alkene (**19**), whereas **27** leads to the *E* alkene (**18**). If **17** is racemic, removal of the two β-hydrogen atoms will give a mixture of *E*- and *Z*-isomers. Even if the halide is enantiopure, the carbon bearing the β-hydrogen is not stereogenic, so both alkenes are produced. To obtain a stereochemically pure alkene, the starting halide must have substituents so that both α- and β-carbons are stereogenic (diastereomeric), and the molecule must be enantiopure. In other words, *an enantiopure diastereomer leads to a stereochemically pure alkene. A mixture of diastereomers leads to a mixture of E- and Z-alkenes.*

both β-hydrogens are equatorial

No E2 reaction is possible

12.10 Label the two hydrogen atoms on C3 of 2*R*-bromopentane H$_a$ and
H$_b$; show both E2 transition states and the product expected for
removal of H$_a$ and then H$_b$.

If this principle is taken to its ultimate conclusion, there should be halo-
cyclohexanes for which an E2 reaction is impossible. 2,6-Dimethyl-1-bromocy-
clohexane (**34**) is such a case. To satisfy the relative stereochemistry of the
two methyl groups (*cis* to each other) and the bromine (*anti* to the two methyl
groups), the bromine atom must be axial in one chair conformation with two
axial methyl groups (**34A**), but equatorial in the other chair conformation that
has two equatorial methyl groups (**34B**). Only conformation **34A** has an axial
bromine atom required for an E2 reaction, but both β-hydrogen atoms (H$_a$ and
H$_b$) are equatorial. No β-hydrogen atoms are *trans,* diaxial to an axial bromine,
so there is no E2 reaction. When **34** is heated with ethanol KOH, there is no E2
reaction. The carbon bearing the bromide in **34** is very sterically hindered, so
an S$_N$2 reaction is very unlikely; certainly the substitution will be very slow.

12.13 Draw the major product, if any, formed when 1*R*-bromo-2*R*-
ethyl-5*R*-methylcyclohexane is heated with KOH in ethanol.

12.4 Unimolecular Elimination

Both the S$_N$2 reaction in Chapter 11 (Section 11.2) and the E2 reaction discussed before follow second-order kinetics (see Chapter 7, Section 7.11.2) and
are bimolecular. The E2 reaction involves a collision of the base with the
β-hydrogen atom to initiate the elimination. In effect, collision of the base with
the β-hydrogen will initiate the elimination sequence that expels the leaving
group. Similarly, collision of a nucleophile with the alkyl halide will "kick out"
the leaving group for an S$_N$2.
  Chapter 11, Section 11.4, discussed the S$_N$1 reaction—a unimolecular reaction in which ionization of the halide to a carbocation occurs with an assist by
water (in an aqueous solvent)—and that carbocation reacts with a nucleophile
in a second step. The slow step is ionization of the halide to a carbocation; the
reaction follows first-order kinetics and the overall reaction is considered to be
a unimolecular process. This section will examine a unimolecular reaction that
generates a carbocation in the presence of a base. If elimination occurs via an
initial ionization to a carbocation, followed by removal of the β-hydrogen by the
base, the reaction is termed a unimolecular elimination, E1.
  If 2-bromo-2-methylbutane (**35**) reacts with KOH in aqueous THF, water
facilitates ionization to the bromide ion and carbocation, **36**. Remember that
water will solvate both ions (see Chapter 11, Section 11.6). *The hydrogen on the*
β-carbon to the positively charged carbon of carbocation 36 (marked H$_a$) is more
acidic that the β-hydrogen in the neutral molecule, 35. Polarization of the C–C
bond in the carbocation is greater than the C–Br bond of a neutral alkyl halide,
so the β-hydrogen is more polarized. It is therefore more easily removed by a
base (it is more acidic).
  In other words, the magnitude of the positive charge in **36** leads to a larger
induced dipole on H$_a$ that makes it more acidic. If hydroxide collides with H$_a$,

the resulting acid–base reaction leads to formation of a conjugate acid (H–OH) and transfer of electrons from the C–H bond to the positive carbon to give an alkene, 2-methyl-1-propene (**37**). *This reaction proceeds by formation of a carbocation, so the slow step is ionization with loss of bromide to give the carbocation, and removal of the β-hydrogen is a much faster step.* Because the acid–base reaction that removes H$_a$ to give **37** is fast compared to ionization, **the reaction follows first-order kinetics** (see Chapter 7, Section 7.11.1). The overall reaction is a unimolecular process and it is termed an **E1 reaction (unimolecular elimination).**

In this reaction, **36** is an intermediate carbocation and the reaction medium contains both water and hydroxide, which are nucleophiles. If a nucleophile attacked carbocation **36**, this would constitute an S$_N$1 reaction. Therefore, it is reasonable to ask if an S$_N$1 reaction using water or hydroxide as a nucleophile can compete with the E1 reaction. This important question is resolved by considering whether the hydroxide ion is more attracted to H$_a$ or to C+ in **36**. If hydroxide collides with the positive carbon in **36**, this nucleophilic reaction is formally analogous to S$_N$1 and the product will be 2-methyl-2-butanol. In fact, the attraction of hydroxide to C+ is largely a function of the solvent, and in a protic solvent such as water, elimination is usually faster. If a base is rather nucleophilic and the reaction is done in an aprotic solvent, the S$_N$1 reaction competes with E1 if there is a carbocation intermediate. If the reaction is done in a protic solvent, elimination is usually faster than in the S$_N$1 reaction, but this obviously depends on the nucleophile.

Because reaction conditions used for an E1 reaction often favor S$_N$1 rather than E1, it is difficult to find a "pure" E1 reaction. In general, this is true, but there are exceptions when the base used in the reaction is a poor nucleophile or if the S$_N$1 product is unstable and leads to a reversible reaction. If cyclohexanol (**38**) is treated with concentrated sulfuric acid, the observed product is cyclohexene in a very fast reaction. This mechanism involves an acid–base reaction of the oxygen from the OH unit (the base) with the sulfuric acid to form an oxonium ion, the conjugate acid. Loss of water from the oxonium ion gives carbocation **39** and the hydrogen sulfate anion, which is highly stabilized due to resonance and is not very nucleophilic. If the S_N1 reaction is faster, the product should be alkyl sulfate **41**; however, **41** is very unstable, so this reaction is not very favorable (see Chapter 20, Section 20.12). In other words, the instability of **41** leads to an equilibrium that favors **39**.

On the other hand, the hydrogen sulfate anion is sufficiently basic that it can remove a β-hydrogen from **39**, leading to the cyclohexene product. The major product of this reaction is cyclohexene (**40**), the E1 product. In most E1 reactions, an aqueous medium is used to generate the carbocation, so water is available as a nucleophile. The E1 product can be formed when secondary and tertiary alcohols react with concentrated sulfuric acid or concentrated perchloric acid. When a good nucleophile is present, as in a dilute aqueous solution of sulfuric acid, S_N1 products usually dominate (such as the alcohol product when a carbocation reacts with water). In general, the E1 reaction is a "nuisance" in that E1 products are formed as minor products when a carbocation is generated, even when the S_N1 product is the major one.

12.14 Draw 2-methyl-2-butanol.

12.15 Draw the oxonium ion formed when cyclohexanol (38) reacts with sulfuric acid.

12.16 Draw the product formed when cyclopentanol is treated with perchloric acid.

12.5 Intramolecular Elimination

There is another type of elimination, and the contrast with an E2 reaction is clear in the experiment that heats ammonium salt **42** [2-(trimethylammonium) butane hydroxide] to about 200°C, neat (no solvent). The product is 1-butene (**43**) isolated in 95% yield.[4] The remainder of the product (5%) was a mixture of *cis*- and *trans*-2-butene (**44**). Water and triethylamine are also products of this reaction. This is clearly an elimination reaction, but the major product is the less stable alkene, which is exactly opposite from the results obtained in an E2 reaction. There are two β-hydrogen atoms in **42**—H_a and H_b—but the only way to obtain water, triethylamine and **43** is by removal of H_a from the methyl group of **42**.

The base is hydroxide, which must remove the β-hydrogen H_a. The reaction is heated without a solvent and, with no solvation effects, the hydroxide ion is coordinated to the positive nitrogen atom of the ammonium salt by an ionic bond. In other words, with no solvent, the hydroxide ion is tethered to the nitrogen of the positively charged NMe_3 unit. One of the products is trimethylamine, which suggests that the NMe_3 unit must also be the leaving group. If hydroxide is tethered to the nitrogen, an E2 *anti* transition state is not possible, and the only way hydroxide can react with a β-hydrogen is by an intramolecular process where the C–H bond of the methyl–proton and the ammonium unit eclipse each other (they have a *syn* relationship), as shown in **45**. This type of reaction is known as an *intramolecular elimination (Ei reaction), which is defined as a reaction in which two groups leave at about the same time and bond to each other as they do so.* This is a reaction that follows first-order kinetics (see Chapter 7, Section 7.11.1).

12.17 **Using 45 as a model, draw the *syn* rotamer that is required to remove H_a in 42.**

45

In the E2 reaction, the β-hydrogen and the leaving group have an *anti* orientation in the transition state that leads to the alkene product. This orientation requires an *anti* rotamer that places the leaving group and the β-hydrogen atom 180° apart. The base and the halide are separate molecules (they are not tethered together), so the E2 reaction is an intermolecular process. In the reaction of **42**, the base is tethered, so H_a must be removed by an intramolecular process via an eclipsed rotamer.

The generic model **45** illustrates the reaction of **42** (R = ethyl and hydrogen). The hydroxide base (Y) must attack the β-hydrogen H_a if an elimination is to occur. In **45**, the base (Y) is tethered to the leaving group X, whereas in an E2 reaction the base is a molecule separate from the alkyl halide. If the reaction proceeds via **45**, the base must attack the β-hydrogen atom intramolecularly, and this requires an eclipsed rotamer. When **45** achieves the eclipsed rotamer, Y reacts with the β-hydrogen in an acid–base reaction, inducing loss of the leaving group (X) to form an alkene. *This type of intramolecular elimination requires an eclipsed rotamer for the acid–base reaction to occur.*

In Chapter 8 (Section 8.1), it was clear that an eclipsed rotamer is higher in energy than a staggered rotamer, so intramolecular elimination will require higher temperatures than *anti* elimination (the E2 reaction). In general, this is true. The following sections will describe two intramolecular elimination processes and the characteristics of these reactions. Before discussing why this type of elimination is favored or why the less substituted alkene is formed, it is important to understand how a base may be tethered to a leaving group. There are several ways to do this. One approach is illustrated by the reaction of 2-(*N,N*-dimethylamino)butane (**46**), in which the nucleophilic nitrogen displaces iodide in an S_N2 reaction with iodomethane to give ammonium iodide salt **47**. This reaction was described in Chapter 11, Section 11.3.3. If there is no solvent, the positive and negative charges attract to form a *tight ion pair,* which is characterized by little or no charge separation.

In other words, the charged ions stay together by electrostatic attraction and they are not separated by solvation. Iodide is not basic enough to remove the proton, so there is no opportunity for an acid–base reaction and subsequent elimination. The iodide ion must be changed to a basic ion. To accomplish this, treatment of **47** with silver oxide with a trace of water leads to tetraalkylammonium hydroxide **42**. This reaction exchanges the iodide ion for hydroxide ion, and when there is no solvent, the hydroxide remains with the ammonium ion via electrostatic attraction.

The hydroxide ion is a base, and if an aqueous solvent is used in this reaction, ionization leads to an E2 reaction via the solvent separated hydroxide ion. However, *if there is no solvent, ion pairing effectively tethers hydroxide to the positive nitrogen and a β-hydrogen is available only in an eclipsed rotamer.* Examination of **42** shows that there are two β-hydrogen atoms: H_a and H_a. An eclipsed rotamer is required for the reaction that removes H_a and a different eclipsed rotamer is required for the removal of H_b.

Structure **48A** is the eclipsed-rotamer required for removal of H_a, and **48B** is the *syn* rotamer required for removal of H_b. As noted before, it is clear that an eclipsed rotamer is higher in energy than a staggered rotamer due to greater steric interactions of the groups (Chapter 8, Section 8.1). For that reason, reasonable concentrations of the **48A** and **48B** are attained only when the reaction medium is heated. In general, an alkyl ammonium hydroxide must be heated to about 200°C for the elimination reaction to occur. The eclipsed rotamer with the least amount of steric hindrance is preferred.

The methyl–ethyl interaction in **48B** makes the steric hindrance higher, and that rotamer is higher in energy when compared to the interactions in **48A**. If **48A** is the lower energy eclipsed rotamer for removal of β-hydrogen H_a, then **48A** will lead to the major product, 1-butene (**43**). Alkene **43** is the less stable alkene. *An E2 reaction always gives the more substituted (more stable) alkene product in an acyclic system via an anti rotamer (which implies an anti transition state), but this intramolecular elimination generates the less substituted (less* *stable) alkene because it must go through an eclipsed rotamer (which implies an eclipsed transition state).*

The intramolecular elimination just illustrated is under kinetic control, rather than thermodynamic control, and the transition state leading to removal of a β-proton from the less substituted β-carbon is lower in energy, so reaction occurs faster. This is the fundamental difference between E2 reactions and intramolecular elimination reactions. The transformation of amine **42** to alkene **43** is called ***Hofmann elimination***, named after August Hofmann (Germany; 1818–1892). Other elimination reactions proceed by an intramolecular elimination mechanism, but Hofmann elimination is the only one that will be discussed.

Before leaving this subject, one more question should be asked. If **42** is heated in aqueous ethanol, what is the product? In an aqueous medium, the positively charged ammonium ion and the negatively charged hydroxide ion are solvent separated. Heating leads to an intermolecular reaction in which the hydroxide removes H_b via an E2 reaction to give the more substituted alkene, **44**. It is also possible to make an argument that H_b is removed via an E2 process, if one argues that H_b is sterically hindered by the ammonium unit. This will force the reaction to occur via removal of H_a. This particular E2 reaction is shown. When the reaction is heated without solvent (neat), the eclipsed elimination mechanism is assumed to dominate the reaction. When the ammonium hydroxide is in a solvent, there can be a competition between E2 reactions for removal of H_b or removal of H_a. Steric hindrance will determine which leads to the major product. For the purpose of this book, *assume that the Hofmann elimination and related reactions proceed via an eclipsed rotamer and intramolecular elimination.*

12.18 Write out the reaction of 42 with hydroxide ion as solvent separated ions, giving the expected E2 transition state and E2 product.

12.19 Draw the product formed when 2-iodo-3-methylpentane is reacted with 1. NMe$_3$; 2. AgO/H$_2$O; 3. heat to 200°C.

12.6 1,3 Elimination: Decarboxylation

Another important elimination reaction involves an eclipsed rotamer. It proceeds by what is effectively an intramolecular acid–base reaction. Carboxylic acids with a carbonyl at C3 (the carboxyl carbonyl is C1) lose CO_2 to form an enol

12.25   Which of the following reaction conditions are consistent with an E2 reaction?
(a) KOH, EtOH
(b) NaCl, ether
(c) NaNH$_2$, NH$_3$
(d) EtOH, NaOEt
(e) aqueous H$_2$SO$_4$
(f) H$_2$O, 0°C

12.26   Which of the following are *cis*-alkenes?

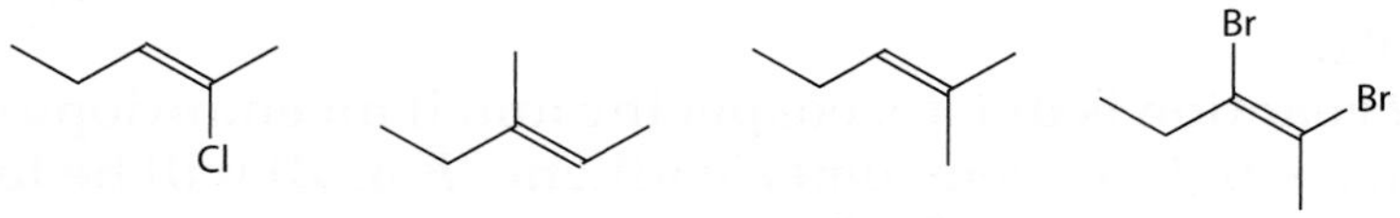

12.27   Which of the following is the most stable carbocation?

12.28   Provide an IUPAC name for each of the following:

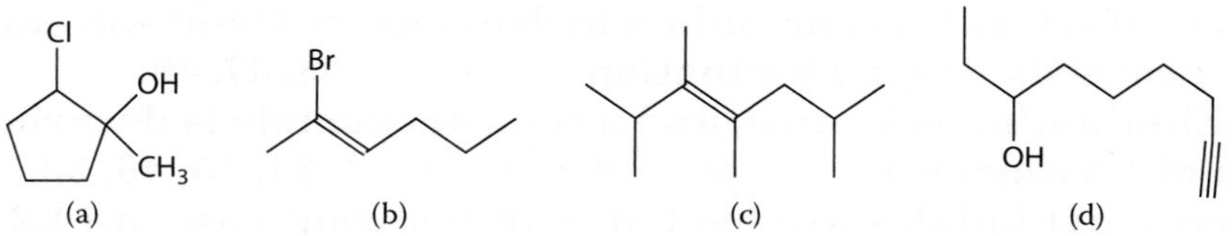

(a)   (b)   (c)   (d)

12.29   Which of the following alkyl halides give 2,3,4-trimethyl-2-pentene upon reaction with KOH in ethanol?

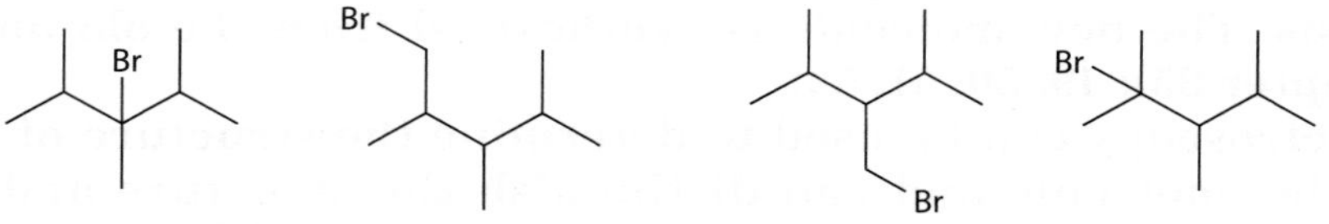

12.30   Which of the following are basic enough to induce an E2 reaction?
(a) Cl⁻
(b) HO⁻
(c) CH$_3$O⁻
(d) I⁻
(e) CH$_4$

12.31   When 2S-bromopentane is treated with KOH and ethanol, a mixture of *E*- and *Z*-2-pentene is formed as the major product. Explain why a mixture of stereoisomers is formed from this enantiopure halide.

12.32   Suggest a reason why treating 2-bromo-bicyclo[2.1.1]hexane (draw it) with KOH and ethanol leads to the less substituted alkene rather than the more substituted alkene (draw both alkenes)—despite the fact that these are E2 conditions. This observation led to what is known as Bredt's rule for reactions that produce an alkene unit in small bicyclic molecules.

12.33   When the alcohol shown is treated with NaH, two products are possible (**A** and **B**). Draw both of them and comment on the mechanism for the formation of each one. Given the reaction conditions, speculate on whether we expect more **A** or more **B** to be formed.

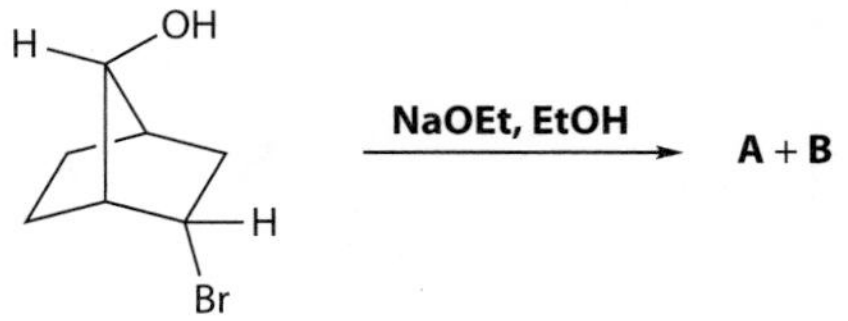

12.34   Briefly explain why the decarboxylation of 2-ethylmalonic acid requires a lower reaction temperature than the decarboxylation of 2-ethylbut-3-enoic acid.

12.35   For the following reaction, draw all products that result from an E2 reaction, and draw the transition state for removal of the hydrogen that leads to the major product. Ignore stereoisomers as being different products.

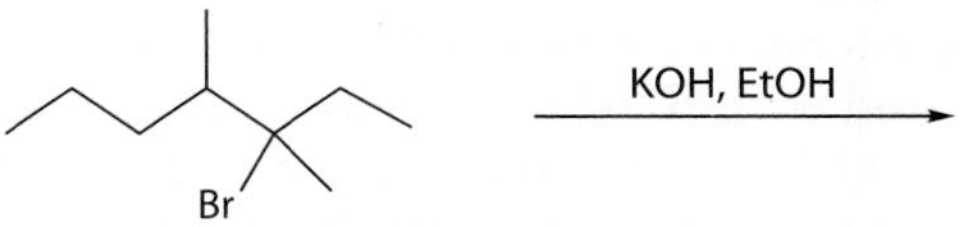

12.36   The following gives no reaction after 24 hours. Use mechanistic arguments for the two possible reaction pathways to explain why neither one works to give a product.

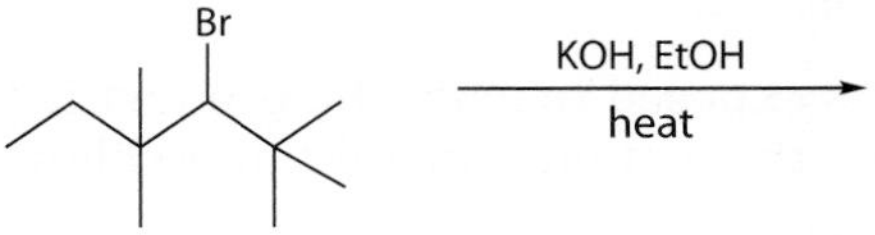

12.37   Explain why the reaction of cyclohexanol + concentrated HBr gives 1-bromocyclohexane but the reaction of cyclohexanol + concentrated sulfuric acid gives cyclohexene.

12.38   Briefly explain why heating 2-bromo-3methylheptane leads to 3-methyl-1-heptene rather than 3-methyl-2-heptene via the Hoffman elimination sequence.

12.39   Draw the major product expected from each of the following reactions:

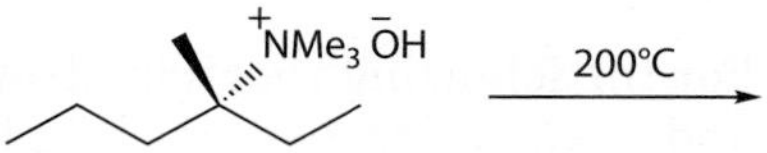

12.40   When 2,3-dibromopentane was treated with KOH in ethanol, the product was 2-bromo-2-pentene, not 2-bromo-3-pentene. Explain. When 2-bromo-2-pentene was treated with KOH in ethanol at 25°C, there was little reaction, but when it was heated to the reflux temperature of ethanol for several hours, 2-pentyne was formed. Explain why the higher temperature was required.

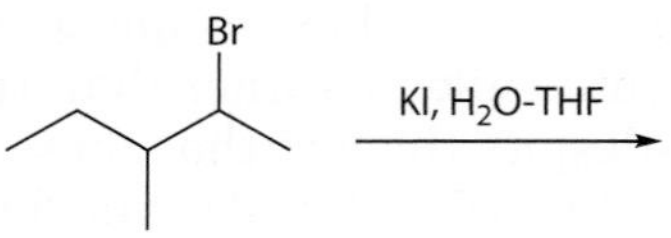

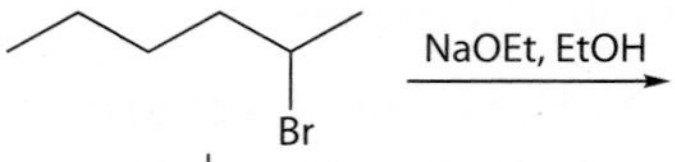

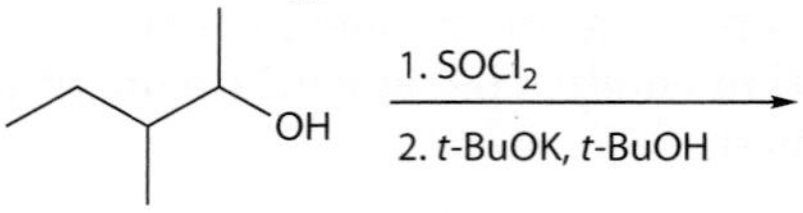

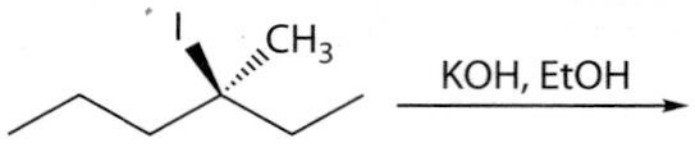

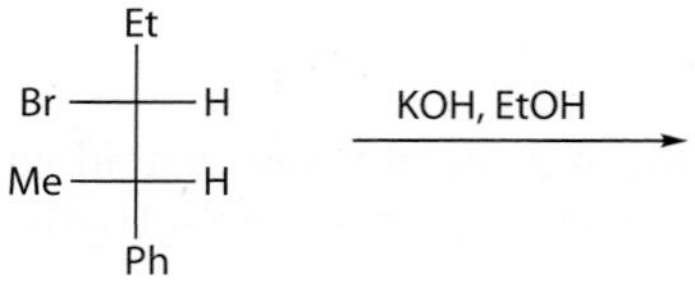

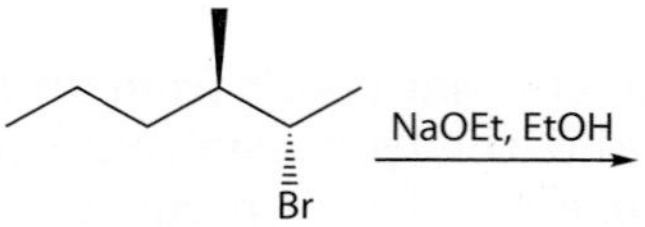

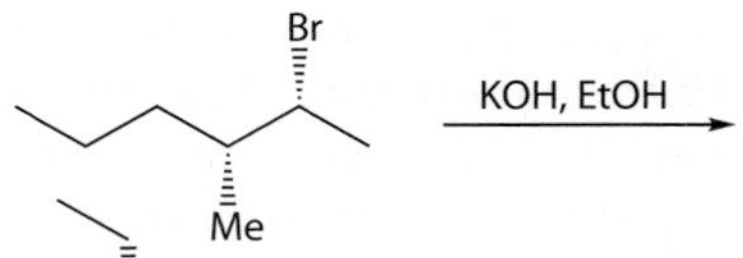

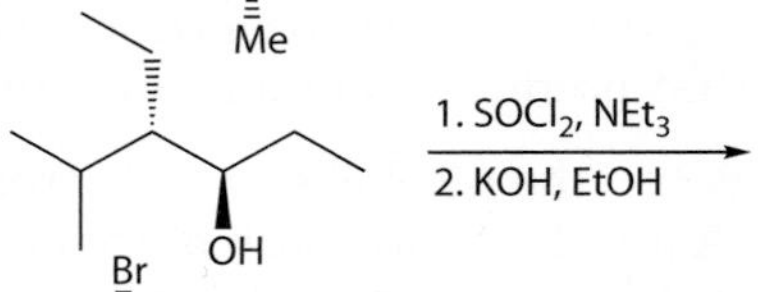

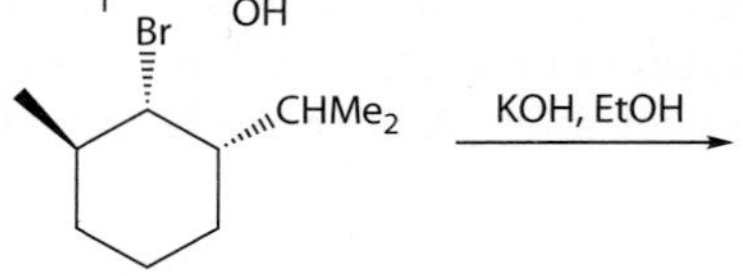

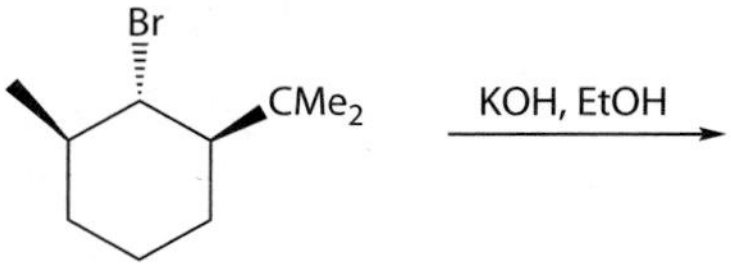

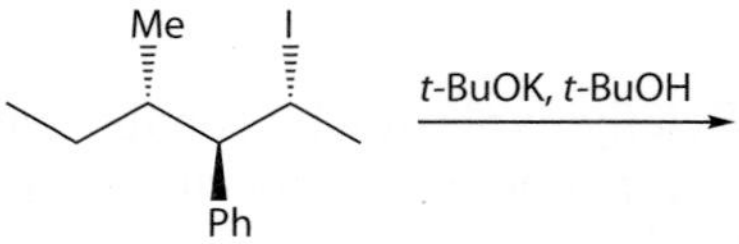

12.41   2,3-Dibromo-2,4,4-trimethylpentane does not generate an alkyne when heated with an excess of KOH in ethanol. Explain.

12.42   3-Bromo-2,4-dimethyl-2-pentene does not form an alkyne when heated with KOH in ethanol, but rather forms an allene. Draw the allene product and explain why it is formed.

12.43   When Ph_3C-OH is exposed to acid, the product is the extremely stable trityl cation. Draw this cation and offer an explanation for the stability.

12.44   When cyclohexanol is treated with concentrated sulfuric acid, the product is cyclohexene, and the reaction has a short half-life. When **A** is treated with concentrated sulfuric acid, however, no alkene is formed. Draw the expected alkene product and then speculate on why **A** is so unreactive in an E1 reaction.

12.45 Explain why heating 2-methyl-1,4-butanedioic acid does not lead to decarboxylation.

12.46 Suggest a reason that will explain why decarboxylation of **A** requires significantly higher temperatures than decarboxylation of **B**.

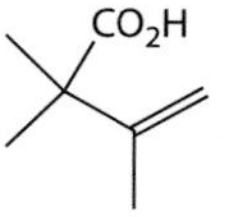

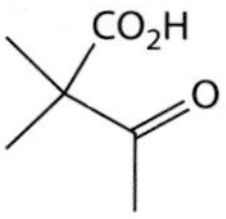

12.47 Heating the molecule shown (an amine *N*-oxide) leads to an alkene in what is known as Cope elimination. Suggest a mechanism that will explain this elimination reaction and suggest a major product based on your mechanism.

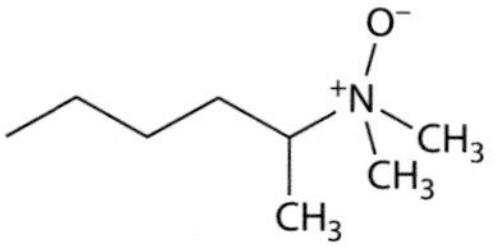

12.48 Heating **A** (a carboxylic acid ester; see Chapter 20, Section 20.5) to about 450°C leads to elimination to give 2-methyl-1-butene. Provide a mechanism for the reaction and speculate on why the elimination requires such high temperatures when compared to the Hofmann elimination.

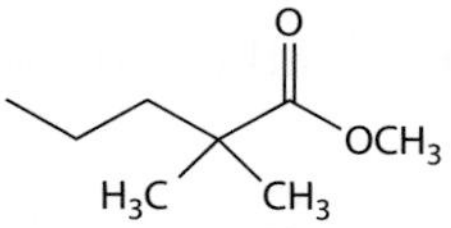

12.49 Provide a reaction sequence that will convert **A** into **B**:

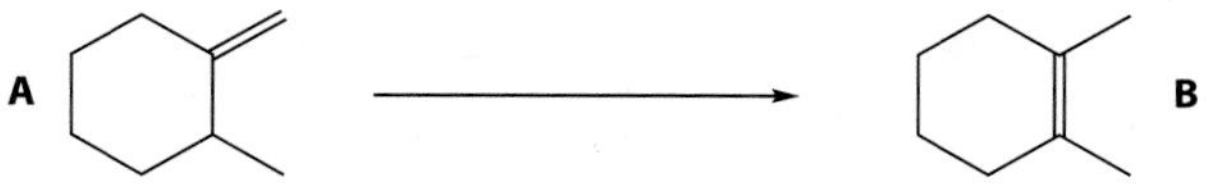

12.50 Provide a reaction sequence that will convert **A** into **B**:

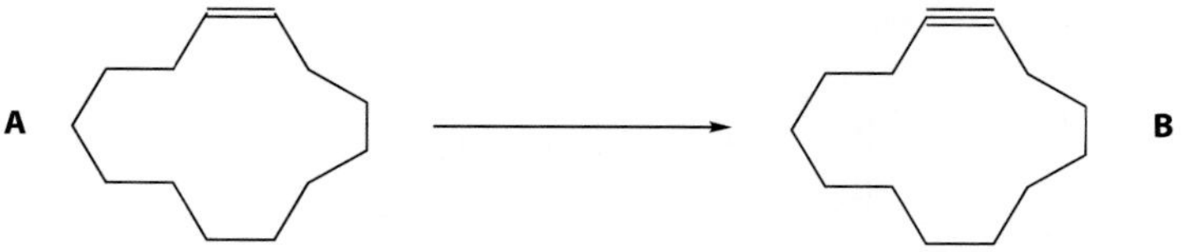

12.51 When the alkyl bromide shown is treated with KOH in ethanol, little or no elimination occurs. Speculate on a reason that explains why the halide shows such poor reactivity in an E2 reaction:

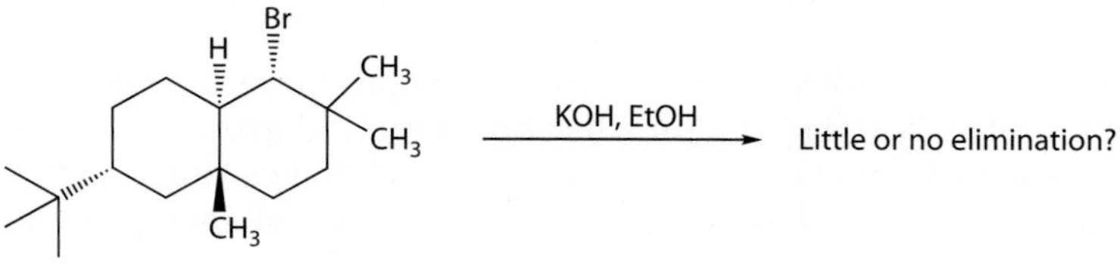

12.52 When bromo-alcohol **A** is reacted with the base sodium hydride (NaH), the reaction produces an alkoxide. Draw the structure of this alkoxide. However, the alkoxide is not the isolated product. Instead, another reaction occurs that generates a 10-membered ring that has a C=O unit and a C=C unit, in what is known as Grob fragmentation. Suggest a mechanism that will lead to the ketone–alkene cyclic product.

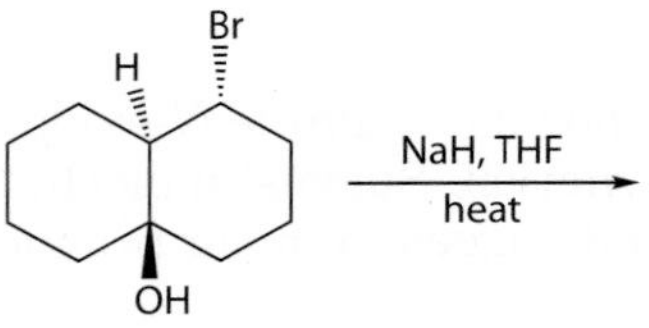

Synthesis Problems

Do not attempt these problems until you have read and understood Chapter 25.

12.53 Describe a synthesis of 2-methyl-2-hexene from 2-methyl-2-aminohexane that uses the Hofmann elimination sequence. Show all reactions, reagents, and products.

12.54 In each case, give a reasonable synthetic route showing all reagents, reactants, and products.

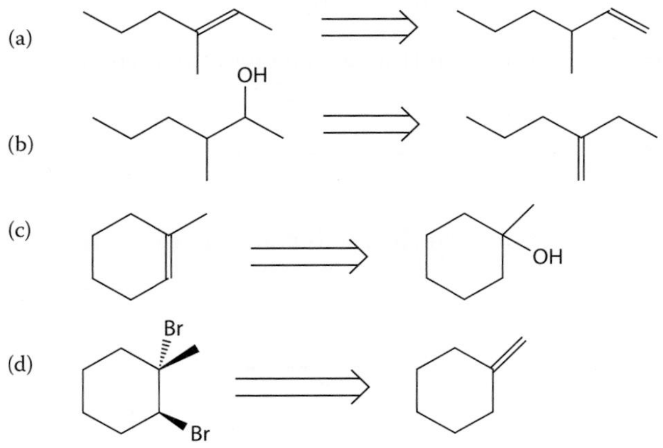

Spectroscopy Problems

Do not attempt these problems until you have read and understood Chapter 14.

12.55 An E2 reaction of 2-bromobutane gives two products. The major product shows a proton NMR spectrum that has a methyl group as a doublet, with a relatively small value of *J*. The minor product shows that the proton NMR spectrum has an ethyl group. Draw the structure of each product.

12.56   The reaction of 2-bromobutane and sodium methoxide in THF gives a mix-
        ture of an ether product and an alkene product. Draw the structure of these
        products and describe differences in the IR sepctrum and the proton NMR
        spectrum that will allow you to identify each one.

12.57   Identify the following molecule, given the following theoretical spectral data.
        Mass spectra data are M = 96, 100%; M + 1 = 97, 7.77%; M + 2 = 98, 0.30%.

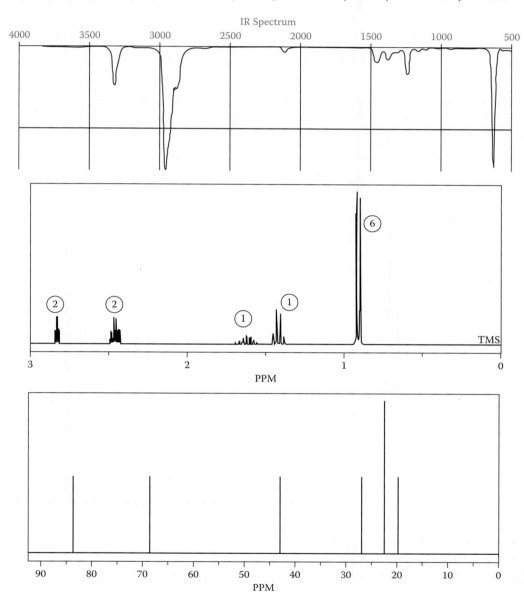

12.58 Identify the following molecule, given the following spectral data. Mass spectral data are M = 96, 100%; M + 1 = 97, 7.77%; M + 2 = 98, 0.30%.

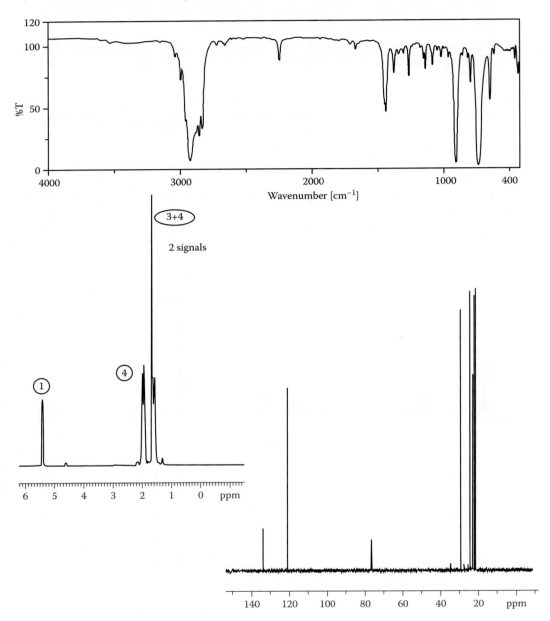

12.59 Identify the following molecule, given the following spectral data:
MS: M = 112 (100%), M + 1 = 113 (8.88%), M + 2 = 114 (0.39%)
MS: 43 (100), 55 (59), 70 (15), 71 (95), 83 (12), 97 (7), 112 (very small)
IR: 3077, 2964–2882, 1820, 1640, 1464–1438 cm^{-1}
^{1}H NMR: 5.81 (m, 1H), 5.00 (m, 1H), 4.99 (m, 1H), 1.93 (m, 2H), 1.23 (q, 2H), 0.83 (s, 6H), 0.82 (t, 3H) ppm

12.60   Identify the following molecule, given the following spectral data:
MS: M = 82 (100%), M + 1 = 83 (6.66%), M + 2 = 84 (0.22%)
MS: 41 (16), 67 (100), 79 (8), 81 (18), 82 (P, 31.3), 83 (P + 2, 2.4)
IR: 3044, 2953–2849, 1658, 1446 cm^{-1}
^{1}H NMR: 5.30 (narrow t, 1H), 2.36–2.11 (broad m, 4H), 1.90 (m, 2H), 1.72 (s, 3H) ppm

12.61   A molecule **A** has a formula $C_6H_{14}O$, with the following spectral data:
IR: 3373 cm^{-1}
^{1}H NMR: 2.04 (broad s, 1H), 1.44 (t, 2H), 1.38 (m, 2H), 1.20 (s, 6H), 0.93 (t, 3H) ppm
Compound **A** reacts with PBr_3 to give **B**, and then **B** reacts with KOH in hot ethanol to give **C**, which has the following spectra:
MS: parent ion at m/z 84
IR: 1676–1667 cm^{-1}
^{1}H NMR: 5.11 (m, 1H), 1.97 (m, 2H), 1.68 (d, 3H), 1.60 (s, 3H), 0.93 (t, 3H) ppm
Identify the structures of **A–C**.

Substitution and Elimination Reactions Can Compete

13

Chapters 11 and 12 discuss reactions of alkyl halides to give either substitution or elimination products. It is clear from Chapter 12 that elimination occurs when the nucleophile is also a strong base and when substitution is inhibited due to steric hindrance. There are many cases in which substitution and elimination compete, particularly when the substrate is a secondary alkyl halide. The solvent plays an important role in these reactions, and solvent identification is a key parameter for distinguishing bimolecular versus unimolecular (ionization) processes. The nature of the alkyl halide (1°, 2°, or 3°) is important, as is the strength of the nucleophile and whether or not that nucleophile can also react a strong base. This chapter will discuss those factors that influence both substitution and elimination, as well as introduce several assumptions that will help make predictions as to the major product.

To begin this chapter, you should know the following:

- **structural differences and nomenclature for alkyl halides (Chapter 4, Section 4.6; Chapter 5, Section 5.6.1)**
- **structural differences and nomenclature for alkenes (Chapter 5, Section 5.1)**
- **how to identify stereoisomers and to assign *R/S* configuration (Chapter 9, Sections 9.1 and 9.3)**

- the basics of conformation applied to both acyclic and cyclic molecules (Chapter 8, Sections 8.1, 8.4–8.6)
- understanding and identification of intermediates and transition states and a reaction-energy curve (Chapter 7, Sections 7.2, 7.3, 7.5 and 7.6)
- the fundamentals of kinetics (Chapter 7, Section 7.11)
- definition and recognition of a nucleophile and understanding that a nucleophile can donate two electrons to a positive carbon or to a carbon with a positive dipole to form a new bond to that carbon (Chapter 6, Section 6.7; Chapter 11, Sections 11.2 and 11.3)
- how to determine the relative stability of carbocation intermediates (Chapter 10, Section 10.2; Chapter 11, Sections 11.4 and 11.5).
- the fundamentals of Brønsted–Lowry acid–base reactions (Chapter 2, Section 2.2; Chapter 6, Sections 6.1 and 6.2).
- definition and recognition of a base and understanding differences in base strength (Chapter 6, Sections 6.3 and 6.4)
- differences in nucleophilic strength (Chapter 6, Section 6.7; Chapter 11, Section 11.3)
- the S_N2 transition state and the fundamental mechanistic details of the S_N2 reaction (Chapter 11, Section 11.2)
- the S_N1 mechanism and the formation of carbocations (Chapter 11, Section 11.4)
- the E2 transition state and the fundamental mechanistic details of the E2 reaction (Chapter 12, Section 12.2)
- the E1 mechanism (Chapter 12, Section 12.4)

This chapter will show that the potent ionizing ability of water is a key to predicting bimolecular versus unimolecular reactions. Aprotic solvents favor bimolecular reactions. It is assumed that water is required for ionization. Primary alkyl halides undergo S_N2 reactions under most conditions, whereas tertiary alkyl halides never undergo S_N2. If a nucleophile is classified as the conjugate base of a strong acid, it is a weak base and will probably not induce an E2 reaction.

When you have completed this chapter, you should understand the following points:

- Solvent effects are important for substitution and elimination reactions. Solvents can be categorized as polar or nonpolar and also as protic or aprotic. Protic solvents have an acidic proton.
- Water is assumed to favor ionization and promote $S_N1/E1$ reactions over $S_N2/E2$ reactions. Other protic solvents tend to favor bimolecular reactions, but slow ionization is possible with long reaction times. Aprotic solvents favor bimolecular reactions, particularly S_N2 reactions.

- Both E2 and E1 reactions are acid–base reactions. The E2 reaction requires a strong base, but E1 reactions occur even with weak bases.
- The structure of the alkyl halide plays an important role. Tertiary halides cannot undergo S_N2 reactions. Primary halides almost always undergo S_N2 reactions because ionization and elimination reactions are very unfavorable.
- Secondary halides can undergo both unimolecular and bimolecular reactions, as well as substitution and elimination. Elimination is favored in protic solvents and substitution is favored in aprotic solvents.
- The assumptions may fail for primary halides in aqueous media and for solvolysis reactions of tertiary halides.

13.1 A Few Simplifying Assumptions

Four distinct reactions are presented in Chapters 11 and 12. The S_N2 and E2 reactions are bimolecular, and the S_N1 and E1 reactions are unimolecular. Remember that bimolecular reactions follow second-order kinetics and unimolecular reactions follow first-order kinetics (see Chapter 7, Section 7.11). In addition to these four reaction types, three types of halides (primary, secondary, and tertiary) may be substrates for reaction with a nucleophile and/or a base. For a given halide and set of conditions, it is possible to predict which of these four reactions will predominate *if* a few simplifying assumptions are made. These assumptions take the form of questions, and answering these questions allows one to make a reasonable prediction of the major product in a given reaction.

To what extent does the competition between substitution and elimination influence the reactions? In reactions of tertiary halides, unimolecular processes dominate in protic solvents (especially water and aqueous solvents), and substitution is usually faster than elimination. In aprotic solvents, bimolecular substitution is not observed for tertiary halides due to the high energy required to form the pentacoordinate transition state (see Chapter 11, Section 11.2). Under conditions that favor bimolecular reactions and in the presence of a suitable base, elimination is the dominant process.

Table 13.1 shows the competition between bimolecular substitution (S_N2) and bimolecular elimination (E2) for a series of alkyl bromides that react with a nucleophile that is also a base. There is a clear preference for S_N2 over E2 with primary halides, and the preference for E2 over S_N2 for tertiary halides is also apparent. In this case, secondary halides give predominantly E2, but it will be seen that this preference is changed as the solvent as well as other reaction conditions is changed.

Table 13.1
Relative Rates of Substitution versus Elimination for Alkyl Bromides

Halide	K (S_N2)	k (E2)	% Alkene
EtBr	172	2.6	0.9
n-PrBr	54.7	5.3	8.9
i-PrBr	0.058	0.237	80.3
t-BuBr	0.1	4.17	97

Source: Reprinted with permission from Hine, J. 1962. *Physical Organic Chemistry,* 2nd ed., 203. New York: McGraw-Hill. Copyright 1962 McGraw-Hill.

13.2 Protic versus Aprotic and Water

The first important question deals with the solvent used in the reaction. **Is the solvent water or does it contain water?** If the answer is no, solvation and separation of *both* cations and anions is very slow and generally very unfavorable. Water facilitates ionization, but ionization is very slow in a solvent that does not contain water; in such solvents, bimolecular processes may dominate. *If there is no water present, assume that the reaction proceeds by a bimolecular mechanism: S_N1 or E2. If water is present in the solvent, ionization is possible. When water is the solvent or a cosolvent, assume that the reaction proceeds by a unimolecular mechanism: S_N1 or E1.*

What is special about water? A solvent should solubilize all reactants and also absorb excess heat that may be liberated by the reaction. A very important property of a solvent is polarity, which largely determines its ability to solvate and separate ions (solvation). A good measure of the ability to separate ions is the dielectric constant.[2–4] Table 13.2 shows the dielectric constants for several common solvents, where a high dielectric constant indicates that the solution can conduct a current, which is associated with an increase in ion formation (i.e., the solvent facilitates ionization).[5] This table also appears as Table 11.2 in Chapter 11.

Table 13.2 is further divided into two classes of solvents: those that contain an acidic proton (X–H, **protic**) and those that do not possess an acidic proton (**aprotic**). Within each class, the higher dielectric constant is associated with the more polar solvent. *The essential difference between protic and aprotic solvents is the ability of protic solvents to solvate both cations and anions, whereas aprotic solvents efficiently solvate only cations.* **Ionization is favored only when both ions are solvated, allowing them to be separated.** The O and the H atoms of the polarized OH unit in protic solvents such as water or an alcohol are not sterically encumbered, which allows the δ-O to solvate cations and the δ+H to solvate anions via hydrogen bonding. The δ- atom of aprotic

Table 13.2
Dielectric Constants and Relative Polarity of Common Organic Solvents

Protic Solvent	Dielectric Constant	Aprotic Solvent	Dielectric Constant
Polar			
HOH	78.5	$H(C{=}O)NH_2$	109.0
DOD	78.25	DMSO	46.68
EtOH (80% aq)	67	DMF	36.71
HCO_2H	58.5	MeCN	37.5
MeOH	32.7	$MeNO_2$	35.87
EtOH	24.55	HMPA	30
n-BuOH	17.8	Acetone	20.7
NH_3	16.9	Ac_2O	20.7
t-BuOH	12.47	Pyridine	12.4
Phenol	9.78	o-Dichlorobenzene	9.93
$MeNH_2$	9.5	CH_2Cl_2	9.08
$MeCO_2H$	6.15	THF	7.58
		Ethyl acrylate	6.02
		Chlorobenzene	5.71
		Ether	4.34
		CS_2	2.64
		NEt_3	2.42
		Benzene	2.28
		CCl_4	2.24
Nonpolar			
		Dioxane	2.21
		Hexane	1.88

Source: Lowry, T. H., and Richardson, K. S. 1987. *Mechanism and Theory in Organic Chemistry,* 3rd ed., 178–179. New York: Harper and Row. Reprinted by permission of Pearson Education, Inc., Upper Saddle River, NJ.

solvents can participate in the solvation of cations, but the positive dipole is sterically hindered and anion solvation is poor. Alcohols have a carbon group attached to the oxygen, which is tetrahedral and more sterically hindered than an O–H unit. For this reason, alcohols are less polar and less able to solvate cations and anions than water, which has two O–H units. Ethers are even poorer at solvation than alcohols.

If the solvent is water or if it contains water, the bimolecular (collision) processes between a neutral substrate and a charged nucleophile (such as nucleophilic acyl addition reactions and nucleophilic displacement with alkyl halides) are slower due to solvation effects. On the other hand, water is an excellent solvent for the solvation and separation of ions, so unimolecular processes (which involve ionization to carbocations; see Chapter 11, Section 11.6) may be competitive. If the solvent is protic (ethanol, acetic acid, methanol), ionization is possible, but much slower than in water. However, ionization can occur if the reaction is given sufficient time to react. In other words, ***ionization is slow, but not impossible***. An example of this statement is the solvolysis of alcohols presented in Chapter 6 (Section 6.4.2). *Based on this observation, assume that ionization (unimolecular reactions) will be competitive in water, but not in other solvents, leading to the assumption that bimolecular reactions should be dominant in solvents other than water.* ***This statement is clearly an assumption, and it is not entirely correct because ionization can occur in ethanol, acetic acid, and so on; however, the assumption is remarkably accurate in many simple reactions and it allows one to begin making predictions about nucleophilic reactions.***

13.1 Which is more polar according to Table 13.2: the protic solvent methanol or the aprotic solvent DMF? Which is more polar: the protic solvent acetic acid or the aprotic solvent THF?

13.3 Nucleophilic Strength versus Base Strength

The next question deals with the nature of the nucleophile. **Is the nucleophile a strong base?** The strength of various nucleophiles in the S_N2 reaction is discussed in Chapter 11, Sections 11.2 and 11.3. Base strength differs somewhat when compared to nucleophilic strength, as described in Chapter 6, Sections 6.3 and 6.4. Both the E2 and the E1 reactions are acid–base reactions, with respect to removal of the β-hydrogen from an alkyl halide or carbocation, respectively (see Chapter 12). If the nucleophile is a weak base (i.e., it is the conjugate base of a strong acid), elimination is very slow, if not impossible, and substitution will be the major process. However, if the nucleophile is also a strong base, *both* substitution *and* elimination are possible.

For the purposes of this book, a nucleophile is classified as a strong base using the old axiom that "a strong acid gives a weak conjugate base." The nucleophiles chloride or bromide (Cl^- or Br^-), for example, are conjugate bases of HCl and HBr, respectively. Both HCl and HBr are strong acids, so *chloride and bromide are considered to be weak bases and it is assumed that they will not induce elimination.* The nucleophiles hydroxide (HO⁻) or methoxide (CH₃O⁻) are conjugate bases of water and methanol. Both are relatively weak acids (pK_a = 15.8 and 15.3, respectively), so it is *assumed* that hydroxide or methoxide are strong bases that can induce an elimination reaction.

There are clearly "in-between" bases such as cyanide (⁻CN) from HCN, which is a moderately strong acid (pK_a = 9.31), making cyanide a moderately strong base. Experimental evidence shows that cyanide ion does not induce an elimination unless the molecule contains a very acidic proton, and simple alkyl halides do not undergo E2 reactions in the presence of cyanide. *Assume that a strong base, as determined by the weak acid–strong base rule, will be strong enough to use in an E2 reaction and a weak base will not.* These simplistic rules will fail from time to time, but they work well with most of the nucleophiles and bases used in this book.

13.2 **Draw the acid–base reaction of water and of methanol in a reaction with NaNH₂.**

13.3 **Draw the conjugate base of nitric acid. Is this base strong enough to induce an E2 reaction with an alkyl halide? What is the conjugate base of methylamine? Is it strong enough to induce an E2 reaction?**

13.4 The Nature of the Halide

The third question deals with the nature of the halide. **Is the halide primary, secondary, or tertiary?** In reality, *this question asks whether the structure of the halide is compatible with the predicted reaction type.* In other words, an S$_N$2 reaction is predicted, but is it possible for the halide to undergo an S$_N$2 reaction? Table 13.1 shows a distinct difference between primary and tertiary alkyl halides in the E2 reaction. This statement means that the S$_N$2 transition state for a primary alkyl halide with a nucleophile such as methoxide ion that is also a strong base is significantly lower in energy than the E2 transition state. More bond-breaking and bond-making processes occur in the E2 transition state relative to the S$_N$2 transition state, which is consistent with a slower reaction.

Substitution is faster than elimination for primary halides, a premise supported by the data in Table 13.1. Based on this analysis, the conditions suggest S$_N$2 and/or E2. Can a primary halide undergo an S$_N$2 reaction? The answer is yes. Can a primary halide undergo an E2 reaction? For the conditions described here, the answer is no. In reality, this means that the E2 reaction is so slow relative to the S$_N$2 that the S$_N$2 reaction is faster and leads to the major product.

Primary cations are very unstable relative to a tertiary carbocation. More to the point, the transition state energy required to form a primary carbocation is much higher in energy than the transition state energy required to form a tertiary carbocation. This means that it is very difficult to form a primary carbo-cation (see Chapter 10, Section 10.2.1). In practical terms, substitution reactions of a primary alkyl halide occur faster than ionization to the primary carboca-tion. This difference is so great that ***primary carbocations effectively do not form under solvolysis conditions***. Specifically, the S_N2 reaction is much faster than the S_N1, and the previous discussion demonstrates that the S_N2 process is faster than the E2 reaction. The S_N2 reaction is also much faster than the E1 for a primary halide because the E1 reaction also requires ionization to a primary carbocation. *This discussion leads to an assumption that, even if water is present, assume that substitution is much faster than elimination and that bimolecular reactions are much faster than unimolecular reactions.* In other words, *assume that primary halides undergo S_N2 reactions faster than any other reaction.*

13.4 **Draw the S_N2 transition state for the reaction of sodium methox-ide with 1-bromopropane. Draw the E2 transition state for these reactants.**

Chapter 11 (Section 11.2) shows that the S_N2 transition state for tertiary alkyl halides is so high in energy that tertiary halides do not undergo S_N2 reactions. Therefore, it is possible to state categorically that *if a tertiary halide reacts with nucleophile in a solvent that does not facilitate ionization (such as ether), ionization is so slow that one can assume there is no reaction.* If a tertiary halide is treated with a base, the S_N2 reaction cannot occur, but an E2 reaction is facile if there is a β-hydrogen (see Table 13.1). If the ionizing solvent water is used, assume that ionization to the tertiary carbocation is a competitive pro-cess, so S_N1 or E1 reactions may occur. Note that the E2 reaction is competitive even in water because water is a protic solvent (see Section 13.5). If the solvent is protic, but does not contain water (such as ethanol or acetic acid), ionization is very slow. Solvolysis can occur if the reaction is allowed a long reaction time (see Section 7, Section 7.12), but for a reaction time of a few hours or a day or so, there is usually no reaction with simple alkyl halides.

The analyses in this section indicate that *primary alkyl halides tend to undergo S_N2 reactions, whereas tertiary alkyl halides can undergo all reactions except S_N2, and that the final product depends on the nature of the nucleophile/base and the solvent.* Therefore, assumptions are made based on structure and reactivity. *Assume that a primary halide will always undergo S_N2, but not S_N1, E1, or E2. Assume that a tertiary halide will never undergo an S_N2 reaction, but may undergo S_N1, E1, or E2.* Conspicuously absent from this discussion are sec-ondary halides, which can undergo both substitution and elimination.

13.5 What about Secondary Halides?

In the previous section, the reactivity for primary and tertiary halides is well defined, but not so for secondary halides. In fact, secondary halides can undergo both bimolecular and unimolecular reactions and both substitution and elimination. Making an educated guess as to the product distribution demands analysis of another experimental parameter (i.e., another question).

For a secondary halide in a reaction with a base, with water as the solvent, ionization is a competitive process. Most of the time, the S_N2 is faster than the S_N1 reaction because direct attack at the α-carbon is more facile than ionization, but the extent of direct substitution versus ionization and then trapping with a nucleophile depend on the strength and nature of the nucleophile. If the nucleophile is a weak base, the S_N2 reaction will dominate in an aprotic solvent. If the nucleophile is a strong base, elimination competes with substitution, and a mixture of S_N2 and E2 products is predicted. In water, it is not obvious whether ionization will lead to the major product, although it is assumed that in aqueous media the S_N1 reaction will dominate.

In an aprotic solvent, both S_N2 and E2 reactions are favorable for a secondary halide. Are there reaction conditions that will favor substitution over elimination or elimination over substitution? **Yes! The nature of the solvent will favor one over the other, so a last question may be asked:** *Is the solvent for the reaction of a secondary halide protic or aprotic?* Aprotic solvents favor bimolecular reactions and, because substitution is a faster process than elimination, *assume that aprotic solvents favor substitution.* If a secondary halide is treated with sodium methoxide in THF, the solvent is aprotic, and the S_N2 reaction is assumed to give the major product. This is the Williamson ether synthesis in Chapter 11, Section 11.3.2.

On the other hand, *if the solvent is protic (such as methanol or ethanol), elimination is known to be faster than substitution* because solvation of the nucleophile/base will slow down the S_N2 rate of reaction. Therefore, *assume that protic solvents favor elimination.* If the solvent for the reaction of sodium methoxide and a secondary halide is changed to methanol (a protic solvent), an E2 reaction will predominate to give the alkene.

It has been noted that even in an aqueous solution of hydroxide, an E2 reaction may be faster than an E1 reaction. *Why does an E2 reaction occur for a secondary halide in aqueous hydroxide? Water is a protic solvent, so it is anticipated that the E2 reaction is faster than the reaction.*

13.5 Draw the product of a reaction between sodium ethoxide and 2-bromobutane in (a) THF and (b) ethanol.

13.6 Strength and Limitations of the Simplifying Assumptions

From the previous discussions, four questions are used to predict the major product of a reaction, based on an analysis of reaction conditions, the substrate, and reactants. *Remember that this protocol is based on working assumptions and can give only an educated guess as to the major product and that the protocol fails in some cases.* Nevertheless, it is a reasonable approach to begin analyzing reaction with respect to reaction conditions and structure:

1. *Does the solvent contain water? (yes or no)*

   If yes, assume that unimolecular processes (S_N1, E1) are faster than bimolecular processes (S_N2, E2).

   If no, and particularly in aprotic solvents, assume that bimolecular processes (S_N2, E2) are faster than unimolecular processes (S_N1, E1).

2. *Is the nucleophile categorized as a strong base? (yes or no)*

   If yes, assume that elimination reactions are possible (E2, E1), but other factors are required to determine whether elimination is faster or slower than substitution.

   If no, then assume that elimination (E2, E1) is not possible.

3. *Is the alkyl halide 1°, 2°, or 3°?*

   If 1°, assume that S_N2 predominates, regardless of solvent, and that S_N1, E1, and E2 do not occur.

   If 3°, assume S_N2 is impossible, but that S_N1, E1, or E2 is possible, depending on the answers to questions 1 and 2.

   If 2°, all reactions are possible, but if the answer to question 2 is yes, see question 4.

4. *Is the solvent protic or aprotic?*

   If protic, assume that elimination is faster than substitution.

   If aprotic, assume that substitution is faster than elimination.

The validity of these assumptions can be tested with a few examples. If **1** is treated with KI in THF, ask the following questions: Does the solvent contain water? No. Therefore, assume a bimolecular reaction (S_N2 or E2). Is the nucleophile a strong base? No. Iodide is not very basic and E2 is not competitive. The halide is primary, so the major product of this reaction should result from an S_N2 reaction.

13.6 Write the structure of the product.

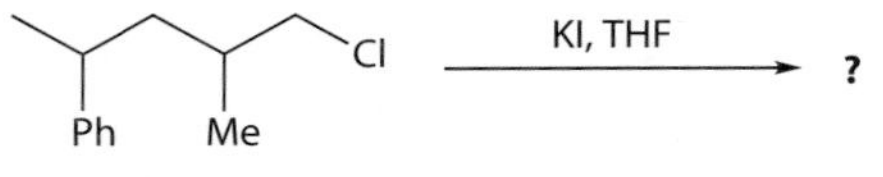

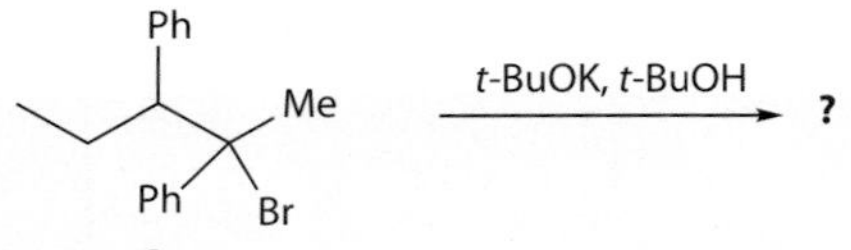

When **2** reacts with potassium *tert*-butoxide in *tert*-butanol, the same questions are asked. Does the solvent contain water? No. The solvent is protic but does not contain water, so assume a bimolecular reaction (S_N2 or E2). Is the nucleophile a strong base? Yes. Alkoxides are rather good bases and an E2 reaction is competitive, so *both* E2 and S_N2 are possible. Halide **2** is a tertiary halide, which means that E2 is possible but not S_N2. In addition, the solvent is protic, which favors elimination over substitution. An alkene is predicted to be the major product.

13.7 Write the structure of the product formed from 2.

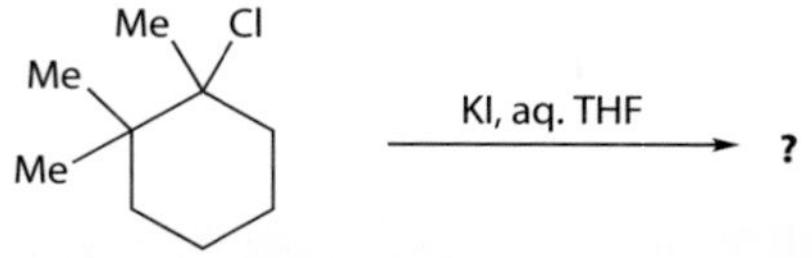

In the reaction of **3**, first ask if water is present. The term "aqueous THF" indicates a mixture of water and the solvent tetrahydrofuran (an ether), so the answer to question 1 is yes. Therefore, *assume* that S_N1 or E1 is faster than S_N2 or E2. Is iodide a strong base? No. Therefore, assume that E1 is not competitive and reaction conditions point to an S_N1 reaction. The halide is tertiary and an S_N1 reaction is feasible. Therefore, it is predicted that iodide will replace chloride in an S_N1 reaction. Because the intermediate carbocation is tertiary, no rearrangement is anticipated (see Chapter 10, Section 10.2.2).

13.8 Write the structure of the product formed in the reaction of 3.

13.9 Briefly discuss what changes would occur in the product if 2-chloro-1,1-dimethylcyclohexane were heated with KI in aqueous THF. Draw the product.

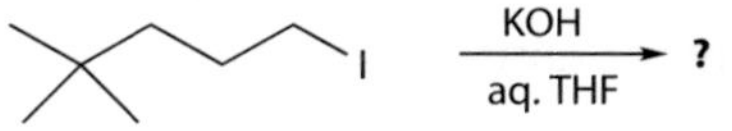

It is important to remember that all of these "questions" are really assumptions and are not universally correct. Understanding each possible mechanism and trying to apply those principles to the given reaction is essential. These four questions are useful tools and they work most of the time, at least in this book, ***but not always***. When **4** is reacted with KOH in aqueous THF, for example, water is present in the solvent, so assume that S_N1/E1 is faster than S_N2/E2. The hydroxide ion is a strong base, so elimination is possible. The fact that **4** is a primary halide, however, means that elimination is not viable, so substitution is preferred to elimination. It is also true that S_N1 reaction requires ionization to a carbocation, which is not a competitive process for primary halides. The analysis points to S_N1 or E1, but *both* of these reactions are unlikely with a primary halide.

In Section 13.4, it is assumed that primary halides will undergo S_N2 reactions even in aqueous media and even with a base present. **Why?** Water slows down the bimolecular process, *but it does **not** stop it.* Ionization of a primary halide is very slow because the energy required to form a primary carbocation is high and it is relatively unstable. To put this in colloquial terms, ionization is so slow that the slowed down S_N2 still wins. For this reaction of **4**, therefore, the final product is the S_N2 product. The working assumptions lead to an incorrect conclusion in this case. ***Always check to make sure that any assumption is reasonable for the individual molecule under consideration.***

13.10 Write the structure of the product formed from 4 under the given reaction conditions.

13.7 When Do the Assumptions Fail?

The reaction of a primary alkyl halide in aqueous media led to an incorrect conclusion when using the four questions. Are there other failures? Yes! ***The assumptions based on the questions are overly general.*** When the structural features of a substrate do not favor one product over another, the assumptions have poor predictive power. When there are subtle electronic or steric features of a substrate, the assumptions do not take these features into account. If a reactant is a moderately strong base, elimination may occur; if reactions are slow due to solvent effects, electronic effects, or steric effects, the assumptions do not give a good correlation with the actual reaction.

The working assumption for tertiary halides is that ionization does not occur at a significant rate in any solvent but water. In one sense, this is correct. However, when a tertiary halide such as **5** is heated in a protic solvent such as ethanol, a reaction occurs very slowly. For the reaction of **5** with KI in heated ethanol, first ask if the solvent contains water. No! This suggests a bimolecular process such as S_N2 or E2. Because iodide is not a strong base, assume that E2 is not possible, but **5** is a tertiary alkyl halide, so S_N2 is not possible. The working assumptions lead to the conclusion that there is no reaction.

Is this correct? It depends on how long the reaction is allowed to proceed. If the reaction is stopped after 1 or 2 hours, the answer is, indeed, no reaction. If the reaction is heated for several days or several weeks, two different products may be observed: 2-iodo-2-methylbutane and 2-ethoxy-2-methylbutane. Both products arise by an S_N1 process. Given sufficient time, slow ionization to a tertiary carbocation is followed by trapping with iodide ion or with ethanol. This process is known as solvolysis (see Chapter 11, Section 11.6). This contradicts the assumption that ionization can only occur in water because it is an *assumption,* as pointed out in the earlier section and thus not always correct. Indeed, ionization occurs in protic solvents such as ethanol, but it is slow relative to ionization in water. If 2-bromobutane is heated with KI in ethanol, the S_N2 product (Chapter 11, Section 11.2) is formed, indicating that the S_N2 reaction is much faster than the very slow ionization reaction in ethanol.

On the other hand, the S_N2 reaction cannot occur with **5**, but given a sufficiently long reaction time, ionization will occur. Once formed, the carbocation will react with the available nucleophiles. Iodide is a better nucleophile in this case, but ethanol is used as a solvent, so it is not entirely clear which product will predominate. Because the allyl iodide product is also subject to ionization and is a carbocation, it is anticipated that the ether product will be major.

13.11 Draw 2-iodo-2-methylbutane and 2-ethoxy-2-methylbutane.

13.12 Draw the formal mechanism for ionization of 5 and reaction with ethanol to give the ether.

13.13 Draw the S$_N$2 product formed when 2-bromobutane is heated with KI.

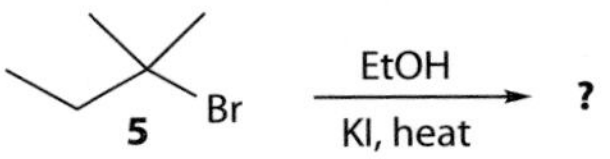

To summarize, the working assumptions in this chapter have value, but are somewhat limited, and *they apply only to reactions of alkyl halides and alkyl sulfonate esters.* They work quite often, but may fail for primary halides in aqueous media and for solvolysis reactions of tertiary halides in particular. The take-home lesson is to use the assumptions to begin an analysis, check to be certain that they are reasonable, and then make an educated guess. Check each guess against the literature or do the experiment to determine the actual reaction products. However, even if the prediction is incorrect, the exercise forces a review of mechanism, structure, and reactivity, as well as reaction conditions.

References

1. Hine, J. 1962. *Physical organic chemistry,* 2nd ed., 203. New York: McGraw-Hill.
2. Parker, A. J. 1962. *Quarterly Review of the Chemical Society* 16:163.
3. Parker, A. J. 1969. *Chemical Reviews* 69:1.
4. Cowdrey, W. A., Hughes, E. D., Ingold, C. K., Masterman, S., and Scott, A. D. 1937. *Journal of the Chemical Society* 1252.
5. Lowry, T. H., and Richardson, K. S. 1987. *Mechanism and Theory in Organic Chemistry,* 3rd ed., 177–181. New York: Harper and Row.

Answers to Problems

13.1   DMF is more polar than methanol and THF is more polar than acetic acid.

13.2

$$H_2O + NaNH_2 \longrightarrow NaOH + H\text{-}NH_2$$

$$CH_3OH + NaNH_2 \longrightarrow NaOCH_3 + H\text{-}NH_2$$

13.3   The nitrate anion is a resonance-stabilized conjugate base of a strong acid. It is a weak base and should not induce an E2 reaction under normal conditions. On the other hand, the amide base derived from methylamine is a strong base (methylamine has a pK$_a$ of about 25), and this powerful base should easily induce an E2 reaction.

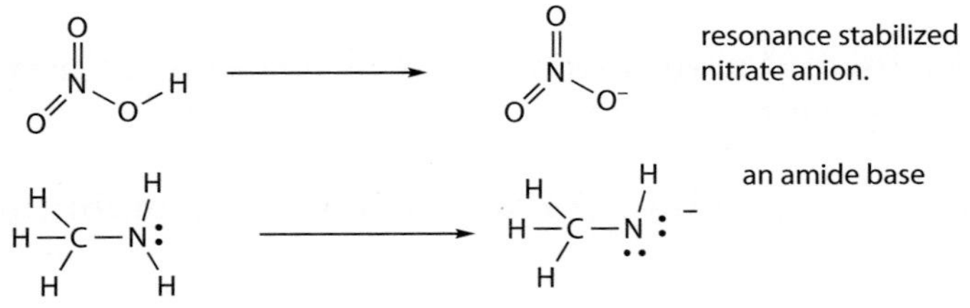

resonance stabilized nitrate anion.

an amide base

13.4

S$_N$2 transition state

E2 transition state

13.5

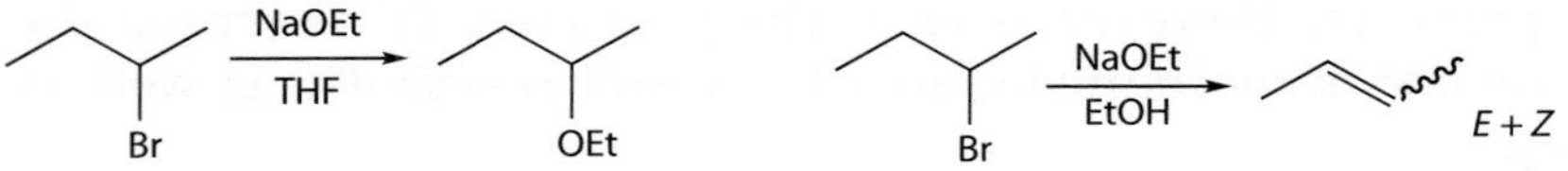

NaOEt

THF

OEt

NaOEt

EtOH

E + Z

13.6

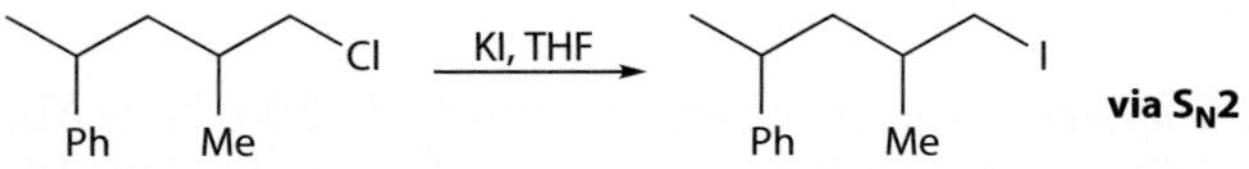

KI, THF

via S$_N$2

Ph   Me

Ph   Me

13.7

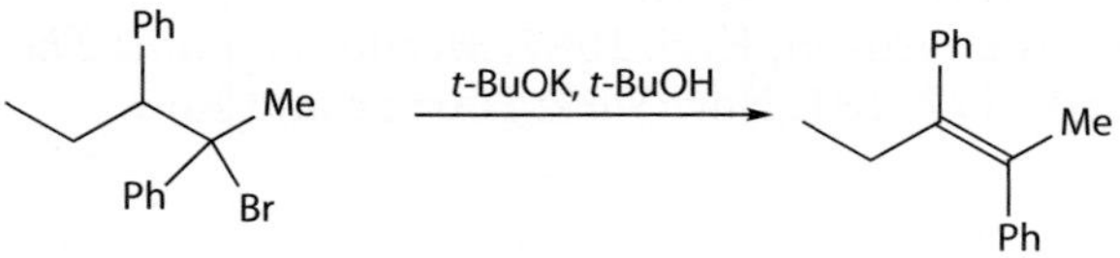

Ph

t-BuOK, t-BuOH

Me

Ph   Br

Ph

Me

Ph

13.8

Me   Cl

Me

Me

KI, aq. THF

Me   I

Me

Me

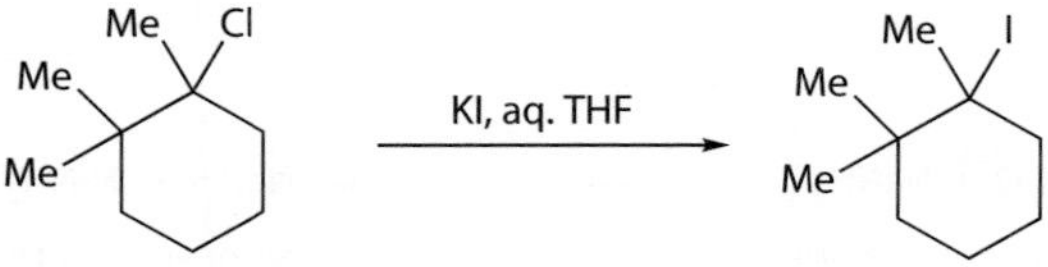

13.9 Ionization of the chloride to give a secondary carbocation under the S_N1 conditions is followed by a rearrangement (a 1,2 methyl shift) to give the more stable tertiary carbocation. Trapping the nucleophilic iodide gives the indicated product. The major difference between this alkyl halide and **3** is the fact that ionization leads to a carbocation that can rearrange, where ionization of **3** gives a tertiary carbocation that does not.

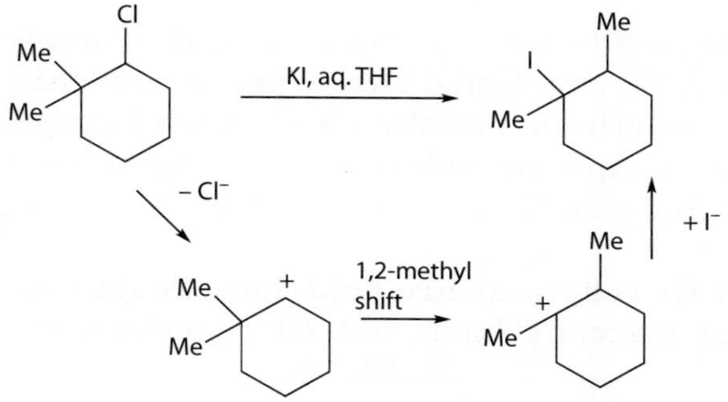

13.10

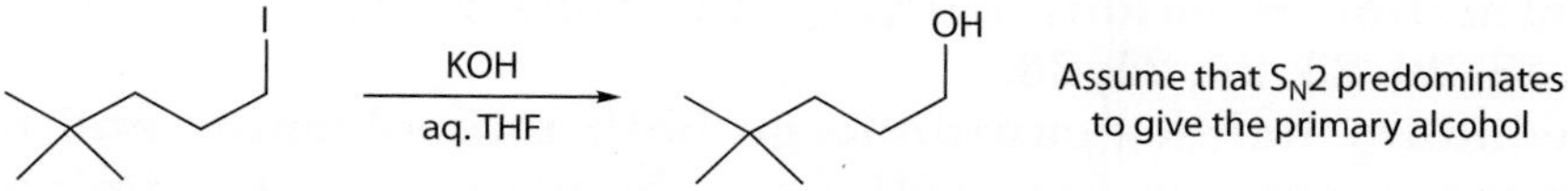

Assume that S_N2 predominates to give the primary alcohol

13.11

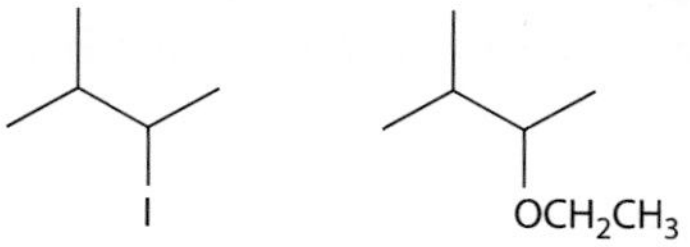

13.12

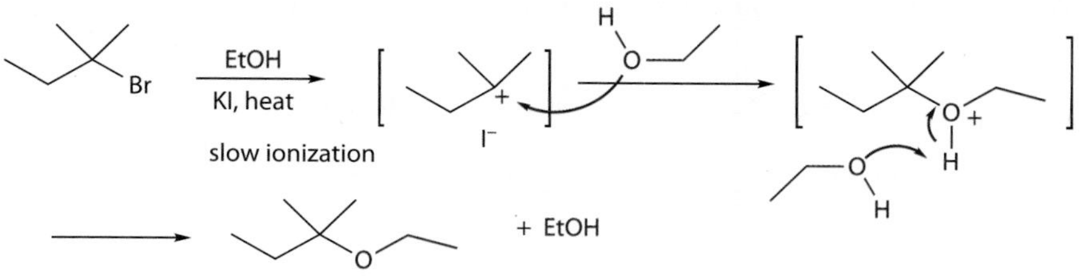

13.13

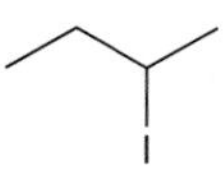

Correlation of Concepts with Homework

- **Solvent effects are important for substitution and elimination reactions. Solvents can be categorized as polar or nonpolar and also as protic or aprotic. Protic solvents have an acidic proton: 1, 14, 15, 18.**
- **Water is assumed to favor ionization and promote S_N1/E1 reactions over S_N2/E2 reactions. Other protic solvents tend to favor bimolecular reactions, but slow ionization is possible with long reaction times. Aprotic solvents favor bimolecular reactions, particularly S_N2 reactions: 4, 5, 7, 8, 9, 10, 11, 12, 16, 17, 18, 22, 23, 24, 26.**
- **Both E2 and E1 reactions are acid–base reactions. The E2 reaction requires a strong base, but E1 reactions occur even with weak bases: 2, 3, 4, 6, 7, 9, 21, 25, 26.**
- **The structure of the alkyl halide plays an important role. Tertiary halides cannot undergo S_N2 reactions. Primary halides almost always undergo S_N2 reactions because ionization and elimination reactions are very unfavorable: 6, 7, 8, 9, 10, 11, 12, 16, 18, 20, 23, 24, 25, 26.**
- **Secondary halides can undergo both unimolecular and bimolecular reactions, as well as substitution and elimination. Elimination is favored in protic solvents and substitution is favored in aprotic solvents: 11, 16, 18, 23, 25, 26.**
- **The assumptions may fail for primary halides in aqueous media and for solvolysis reactions of tertiary halides: 10, 11, 12, 26.**

Homework

13.14   Which of the following are considered to be protic solvents?

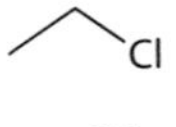

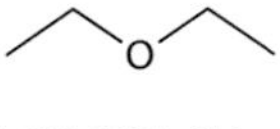

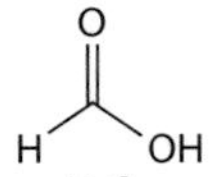

CCl_4 $CH_3CH_2OCH_2CH_3$ CH_3CH_2OH NH_3 Me_3COH

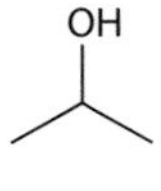

13.15   Which of the following are aprotic solvents?

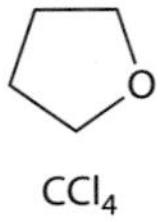

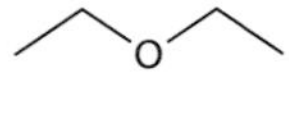

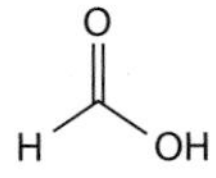

CCl_4 $CH_3CH_2OCH_2CH_3$ CH_3CH_2OH NH_3 DMF

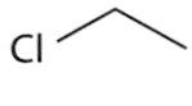

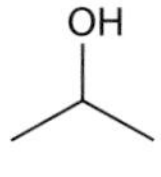

13.16 What product is formed when 2*R*-bromobutane reacts with NaI in THF?

13.17 Which nucleophile reacts fastest with CH_3I in THF?
(a) CH_3O^-
(b) H_2O
(c) CH_4
(d) CH_3OH
(e) HSO_4^-

13.18 Offer an explanation for why 2-bromobutane undergoes an S_N2 reaction, but 2-bromo-2-methylbutane does not when reacted with KI in ether.

13.19 Briefly explain why the presence of water favors first-order reactions.

13.20 Rank order each of the following from the more stable to the less stable. Assume no rearrangement occurs. Judge each structure as drawn.

13.21 Which of the following bases are suitable for use in an E2 reaction with a tertiary halide?
(a) NaOEt
(b) H_2O
(c) CH_4
(d) $NaNH_2$
(e) HNO_3

13.22 If 10 times the equivalents of KI is added to the reaction of KI and bromomethane in THF, what is the effect on the rate?
(a) increased by 2
(b) decreased by 2
(c) no effect
(d) increased by 10
(e) decreased by 10

13.23 Give the complete mechanism for the following reaction:

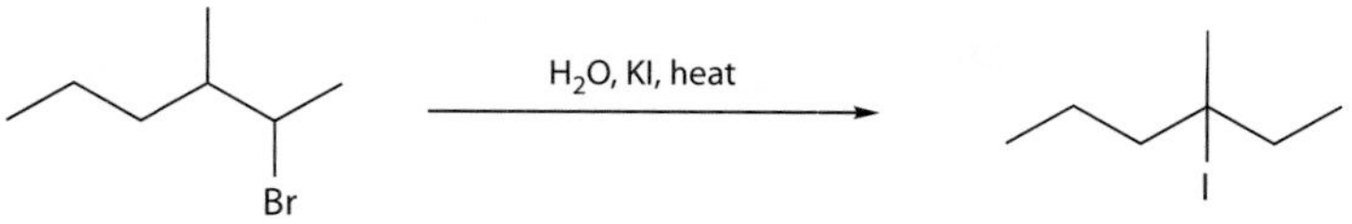

13.24 Give a complete mechanism for the following reaction:

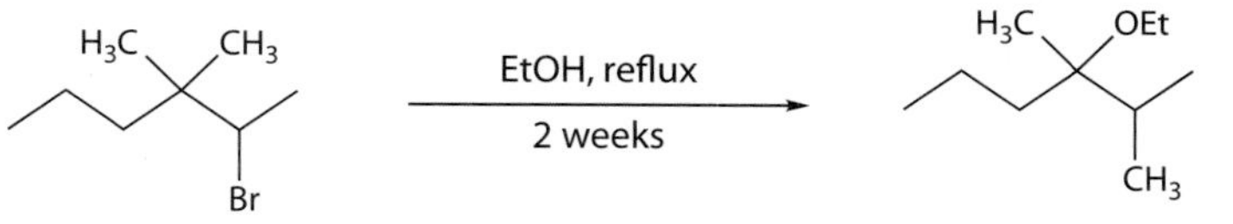

13.25 Give the major product for each of the following transformations, with the correct stereochemistry where appropriate. Give the structure of **A**, then react **A** with the reagent shown to give **B**. React the bromide with KI to give **C** and then react **C** with the reagent shown to give **D**. Finally, react **D** with the reagent shown to give **E**. In addition, determine if product **B** is racemic or if it is one enantiomer. Determine whether product **E** is racemic or is one enantiomer.

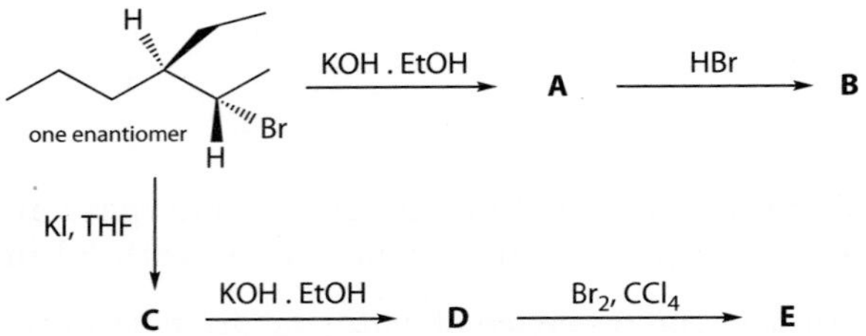

13.26 Using the simplifying assumptions presented in this chapter, predict the major product for each of the following reactions. Indicate no reaction by "N.R." if that is appropriate.

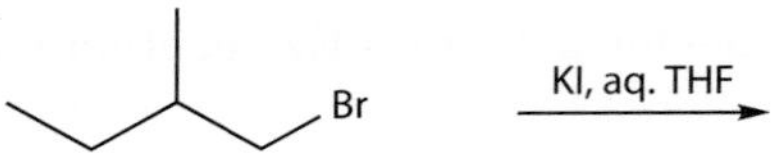

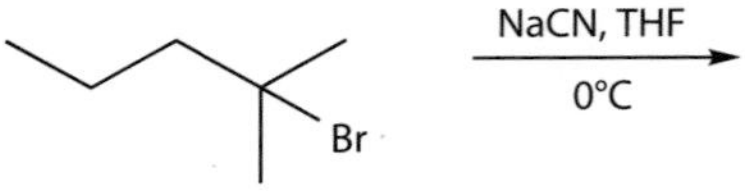

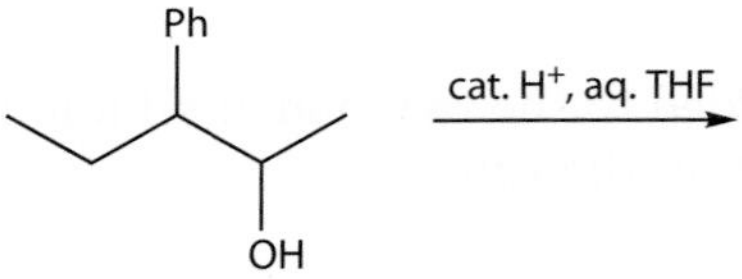

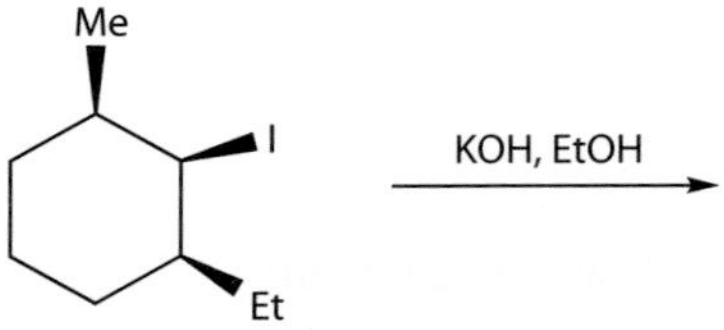

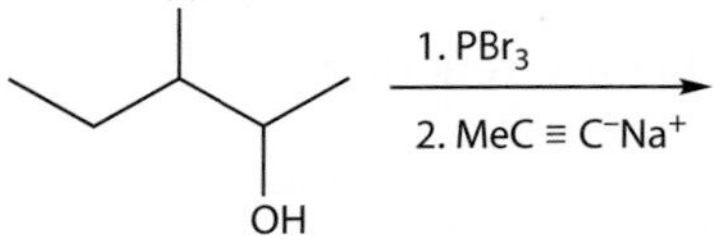

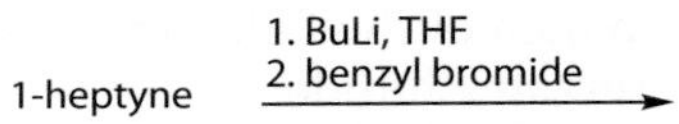

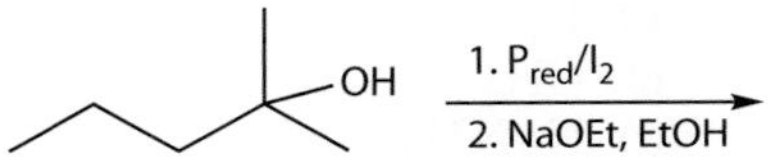

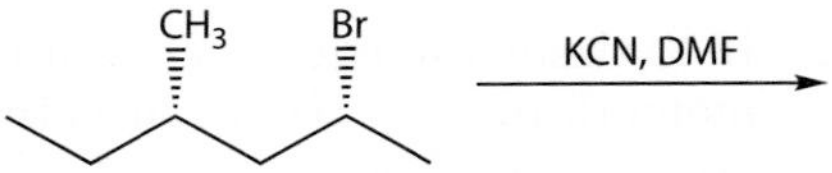

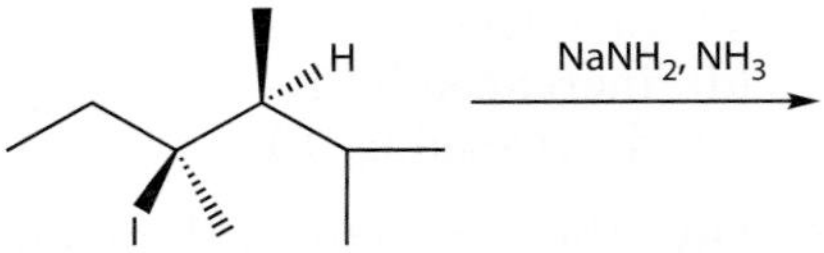

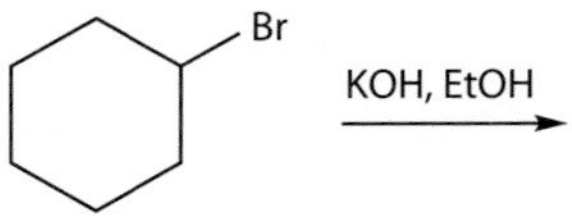

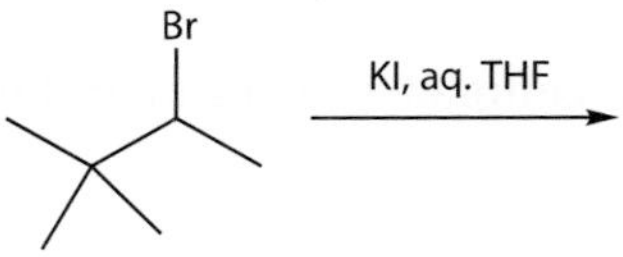

(m)
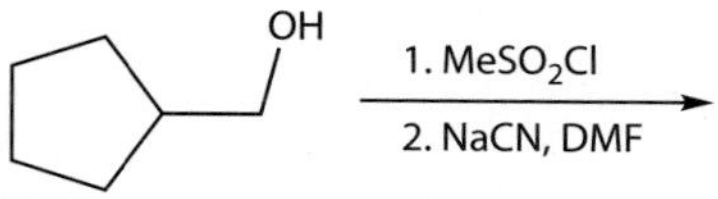
OH
1. $MeSO_2Cl$
2. NaCN, DMF

(n)
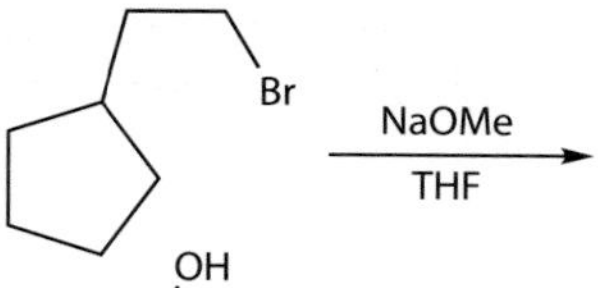
Br
NaOMe
THF

(o)
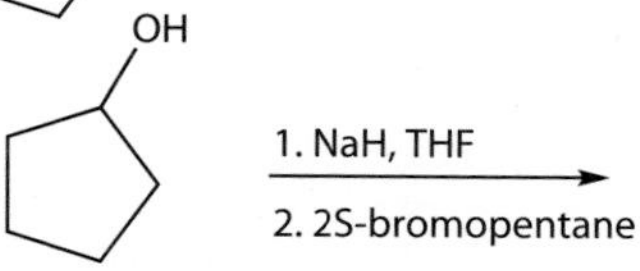
OH
1. NaH, THF
2. 2S-bromopentane

(p)
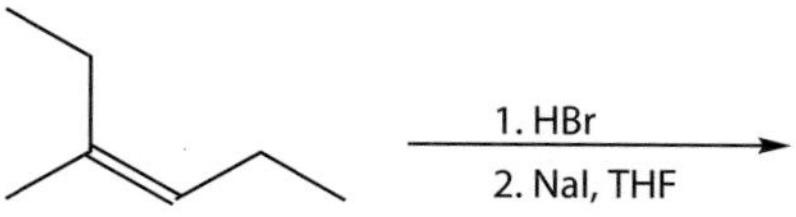
1. HBr
2. NaI, THF

(q)
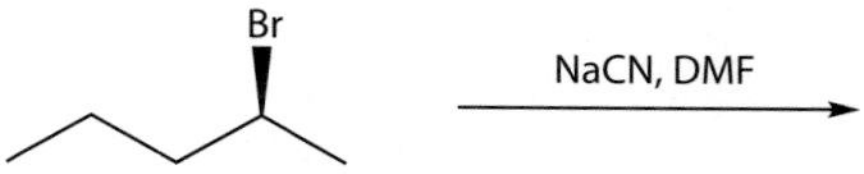
Br
NaCN, DMF

(r)
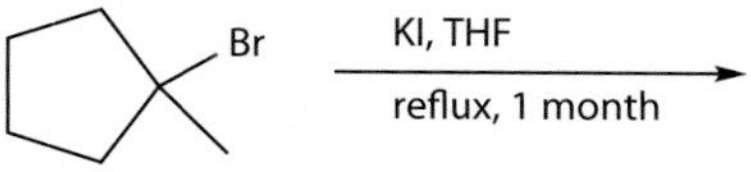
Br
KI, THF
reflux, 1 month

(s)
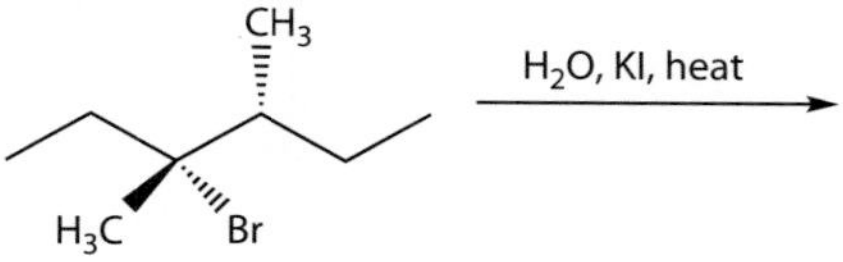
CH_3
H_3C
Br
H_2O, KI, heat

Note: Synthesis problems for these reactions are found in Chapters 11 and 12. No spectroscopy problems are provided in this chapter.

Spectroscopic Methods of Identification

14

Many molecules with a variety of chemical structures are discussed in this book. **Is it possible to verify that these structures are correct?** The answer is yes, and several methods are available to an organic chemist. The most useful ones involve irradiating a molecule with an electron beam or with electromagnetic radiation of various wavelengths and then observing the energy that emerges from the molecule. In Chapter 23, the interaction of a molecule with ultraviolet light gives information about the structure of conjugated molecules (Section 23.3).

The interaction of a molecule with other sources of energy will also give structural information. If a molecule is subjected to so much energy that bonds break, the molecule fragments into smaller pieces. Determining which bonds break and the composition of the fragments gives structural information. Exposure to infrared (IR) light causes the bonds in a molecule to vibrate but not break. This gives structural information concerning the types of bonds within the molecule. The number and types of hydrogen atoms or the number and types of carbon atoms within a molecule are determined by yet another technique. It is important to emphasize that when a compound is irradiated with energy, the energy that emerges from the compound is analyzed in order to gain information about the atoms and bonds. When all of the information from all of the techniques is combined, the result is a very accurate picture of the structure of a given molecule.

To begin this chapter, you should know the following:

- **the structure and bonding of organic molecules and all functional groups (Chapters 4, 5, 15, 19 and 20)**
- **the nomenclature of all functional groups in this book (Chapters 4, 5, 15, 19 and 20)**
- **the strength of covalent bonds (Chapter 3, Sections 3.3 and 3.6)**
- **polarized covalent bonds (Chapter 3, Section 3.7)**
- **the concept of isomers (Chapter 4, Section 4.5)**
- **rotamers and conformations (Chapter 8, Sections 8.1 and 8.3–8.7)**
- **_E_ and _Z_ nomenclature for alkenes (Chapter 9, Section 9.3)**
- **the fundamentals of conjugated and nonconjugated systems (Chapter 23, Sections 23.1 and 23.2)**
- **the concept of keto-enol tautomerism (Chapter 10, Sections 10.4.4 and 10.4.5; Chapter 22, Sections 22.1 and 22.2)**
- **the fundamentals of the electromagnetic spectrum (general chemistry)**
- **the structure of a radical and radical ions (Chapter 7, Section 7.4.3; Chapter 18, Section 18.4)**
- **the most fundamental concepts of magnetism (from general chemistry)**
- **the characteristics of a π-bond in alkenes and alkynes (Chapter 5, Sections 5.1 and 5.3).**
- **the characteristics of a π-bond in carbonyl compounds Chapter 5, Section 5.9; Chapter 15, Sections 15.1–15.3 and 15.5–15.7)**
- **aromaticity and aromatic compounds (Chapter 21, Sections 21.1, 21.7, and 21.8)**

This chapter will discuss three methods for structure determination that involve the use of different types of energy. When a molecule is subjected to a high-energy electron beam, fragmentation occurs and the fragments are observed in a technique called mass spectrometry. When a molecule is irradiated with infrared light, the molecule absorbs that energy and it is dissipated via vibrations rather than bond cleavage. That energy is measured as it emerges from the compound being irradiated and recorded. The energy of vibration for a bond depends on the atoms connected to the bond and the strength of that bond. Therefore, infrared spectroscopy is a good method for determining the functional groups within a molecule.

When a molecule is irradiated with a radio signal in the presence of a strong magnetic field, the energy is absorbed by hydrogen atoms or carbon atoms, leading to a change in their spin state. The energy that emerges from that molecule is detected as a signal that gives information about the connectivity of that hydrogen atom within the molecule. This technique is called nuclear magnetic resonance spectroscopy (NMR spectroscopy for short) and it allows one to probe the natures and types of hydrogen atoms or carbon atoms within a molecule.

When you have completed this chapter, you should understand the following points:

- **Chemists make use of the electromagnetic spectrum to probe the structure of organic molecules.**
- **Bombarding a molecule with 70 eV of energy leads to formation of a molecular ion that allows determination of the empirical formula based on the presence of isotopes of common elements found in organic molecules. This is called mass spectrometry. Fragmentation of the parent ion leads to daughter ions, and determining the mass of lost fragments can assist in structure determination.**
- **High-resolution mass spectrometry is valuable for determining the empirical formula of a molecule based on the exact mass of the molecular ion.**
- **The M+1 and M+2 isotopic peaks can be used to determine an empirical formula.**
- **Absorption of infrared light leads to molecular vibrations, particularly in bonds that are polarized and/or exhibit a change in dipole upon absorption of the light. Observation of the effects of infrared radiation of the molecular vibrations induced by infrared light is called infrared spectroscopy.**
- **The absorbances of most common functional groups appear in the functional group region, whereas all molecules exhibit vibrations characteristic of their structure that appear in the fingerprint region. Matching peaks in the infrared spectrum with functional group correlation tables allows one to determine the presence or absence of functional groups in an organic molecule.**
- **When an organic molecule is placed in a powerful magnetic field, the nucleus of each hydrogen atom behaves as a tiny magnet and sets up a small magnetic field that can be aligned with or against the large magnetic field. When this proton is irradiated with energy in the radio signal range, that energy, for a given magnetic field strength, flips its spin state. The energy absorbed when the precessional frequency of the hydrogen nucleus and the external field are in resonance is measured and displayed as absorption peaks. This is called nuclear magnetic resonance spectroscopy, and it is used to determine the structure of organic molecules.**
- **Substituents attached to carbon atoms bearing protons alter the local magnetic environment and lead to different magnetic fields for each different type of proton. Each different proton will lead to a different signal in the NMR spectrum. The ppm scale is established for NMR spectroscopy and tetramethylsilane (TMS) is used as a zero point standard.**

644 Organic Chemistry: An Acid–Base Approach

- **Protons attached to carbons bearing electron withdrawing groups appear downfield. Integrating the area under each different proton allows one to establish the ratio of hydrogen peaks and, with the empirical formula, the number of different protons may be obtained.**
- **π-Bonds react to an applied magnetic field by developing a small opposing local field that orients a molecule in a large magnetic field. This leads to the development of an opposing magnetic field that leads to significant downfield or upfield shifts in the chemical shift and is called magnetic anisotropy.**
- **Neighboring protons split a proton signal into n+1 peaks. This multiplicity allows one to determine how many neighboring protons are present for a given signal. The coupling constant J defines the distance between the peaks.**
- **Carbon-13 NMR spectroscopy can be used to count the number of different carbon atoms in an organic molecule.**
- **There are biological applications of spectroscopy.**

14.1 Light and Energy

At its core, spectroscopy is a technique by which it is observed how radiation of a particular energy is altered by interaction with a molecule. Although the term "light" in this chapter is less appropriate than "energy," examining the electromagnetic spectrum is the logical place to begin a discussion.

14.1.1 Energy as a Wave

Electromagnetic radiation is energy transmitted through space in the form of waves. If the energy of this electromagnetic radiation is determined and the various components labeled, the result is the electromagnetic spectrum. Visible light is a relatively small range of energy in the electromagnetic spectrum. If energy is a wave, certain parameters of that wave are equated with common energy parameters.

Two waves are drawn in Figures 14.1A and B. **Wavelength** is defined as the distance between the crests of each wave. A longer wavelength, as in A, is associated with lower energy. As the wavelength is decreased (as in B), the energy of the wave increases. Therefore, A is considered to be low-energy radiation and B is high-energy radiation. Radiation is also defined by the amount of time it takes to complete one cycle of the wave, or the distance required to complete one wave. This is the **frequency** of the wave (shown in Figure 14.1). When the cycle or frequency is displayed in seconds, one cycle/second is defined as 1 hertz (Hz). Note that if frequency is 1 Hz in A and 1 Hz in B, the wave moves a greater

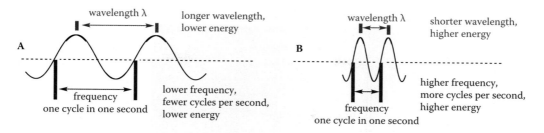

Figure 14.1 Energy as a wave.

distance in 1 second for A than for B. If frequency is defined by distance using centimeters (cm), then one cycle/cm = 1 cm^{-1}. If both A and B show a frequency of 1 cm^{-1}, then it takes more time for one cycle of A than for one cycle of B. If radiation is of higher frequency, it will have more waves per second, so the wavelength is shorter. This observation is equated with higher energy.

Wavelength multiplied times frequency equals the speed of light ($\nu \lambda = c$), so they are inversely proportional and related by the speed of light, c ($c = 3 \times 10^{10}$ cm sec^{-1}), $\nu = c/\lambda$. If frequency is given the symbol ν (in hertz) and wavelength is given the symbol λ (in centimeters), then it is known that energy (E) is proportional to frequency (ν) and it is related by Planck's constant, h (where $h = 6.626 \times 10^{-34}$ J Hz^{-1}), so that $E = h\nu$. If the frequency and/or wavelength is known, the energy of the wave may be calculated and thereby the energy of the electromagnetic radiation being used.

14.1 Determine the frequency in hertz for a wave of 60×10^6 cycles/ second.

14.2 Calculate the wavelength that corresponds to a frequency of 8,500 Hz. Calculate the energy of this wave.

14.1.2 The Electromagnetic Spectrum

An energy continuum stretches from very low energy radiation such as radio signals to very high energy radiation such as x-rays and gamma rays. Radio waves are low in energy, so they have a long wavelength. Conversely, gamma rays are extremely high in energy with a very short wavelength. The light that is seen by humans comprises only a tiny portion of the electromagnetic spectrum. The visible spectrum is easily seen when visible light is reflected through a prism, for example, or in a rainbow with red light on one end and violet on the other (see Figure 14.2).[1]

The energy of red light in this spectrum is measured to be about 35.75 kcal mol^{-1}, whereas the energy of the violet light is about 71.5 kcal mol^{-1}. The red light is lower in energy than the violet light. If this energy for wavelength is converted from kcal mol^{-1} to angstroms (Å; 1×10^{-8} cm), red light appears at 8000 Å and the violet light appears at 4000 Å. Therefore, the larger number in wavelength corresponds to lower energy. Finally, convert

Visible Spectrum

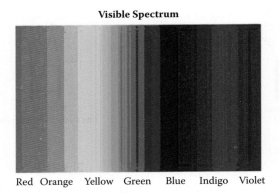

Red Orange Yellow Green Blue Indigo Violet

Figure 14.2 The visible spectrum of light.

wavelength to nanometers (nm or v; 1×10^{-7} cm) so that red light appears at 800 nm and violet light at 400 nm. Once again, the larger number correlates with the lower energy. Although irradiation with visible light gives information about structure in some cases, it will not be used because it is of marginal value for the identification of most of the organic molecules discussed in this book.

14.3 Determine the color of the light that absorbs at 600 nm.

If the electromagnetic spectrum is expanded to include other types of energy in addition to visible light, the result is Figure 14.3. Infrared radiation and ultraviolet radiation are important tools for probing the structure of organic molecules, and infrared radiation (infrared means below red) is lower in energy than visible light or ultraviolet light (ultraviolet means beyond or above violet). In Chapter 24 (Section 24.3), ultraviolet light is used to induce photochemical transitions in the presence of some organic compounds, such as dienes, and other chemical reactions are possible. Infrared light is much lower in energy, and absorption by an organic molecule will induce molecular

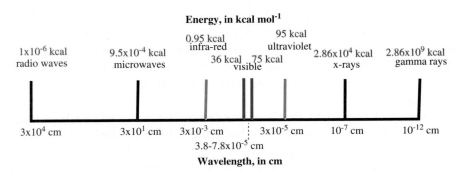

Figure 14.3 Electromagnetic spectrum measured in wavelength (centimeters) and energy (kilocalories/mole).

vibrational transitions (see later discussion) rather than chemical reactions (making and breaking bonds).

14.2 Mass Spectrometry

The previous section discussed the electromagnetic spectrum, with a focus on visible light, infrared light, and ultraviolet light. Structure determination is not possible unless the empirical formula is known. One method to obtain this information employs much higher energy than even ultraviolet light. The goal is to supply enough energy to fragment the molecule such that the mass of the initial molecule is obtained, as well as the mass of various fragments generated from the original molecule.

Therefore, the first probe of molecular structure uses a high-energy electron beam. Impact with an organic molecule leads to ejection of an electron to form a detectable, high-energy species called a radical cation. This high-energy radical cation fragments into smaller radical cation fragments that also may be detected. Knowledge of the fragmentations gives structural information. In addition, detection of the molecular radical cation, missing only an electron, allows one to determine the mass of that fragment and thereby the mass of the original molecule. This technique is called mass spectrometry.

14.2.1 The Mass Spectrometer

The instrument used to bombard molecules with an electron beam is called a mass spectrometer. Figure 14.4 shows a diagram of a very simple instrument where a sample is introduced into an entry port known as the source and then bombarded with a high-energy electron beam. Electron bombardment converts the molecule to one or more radical cations, which are charged species.

In Figure 14.5, three radical cations are formed—m_1, m_2, and m_3—from an original molecule. These fragments are swept into a chamber that is positively charged. Passing each radical cation through a positively charged sector accelerates the ion via repulsion of like charges, and each ion will then pass into a curved tube that sits inside a powerful magnet. Assume that each ion has a different mass but that all three are accelerated equally.

The heavier mass ion m_1 has insufficient kinetic energy to pass through the magnetic sector, so it will not reach the detector. It essentially "crashes" into the side of the tube and is lost (not detected). This is analogous to a person on a bicycle in a velodrome trying to negotiate a highly banked race track at too low a speed. The person may fall off a steeply banked track if he or she is not going fast enough. The other extreme is represented by a very light ion, m_3, that possesses too much kinetic energy, also crashes into the side, and is also not detected. This is analogous to a car trying to take a curve at too high a speed, causing it to push into the wall. Ion m_2 possesses a mass such that acceleration

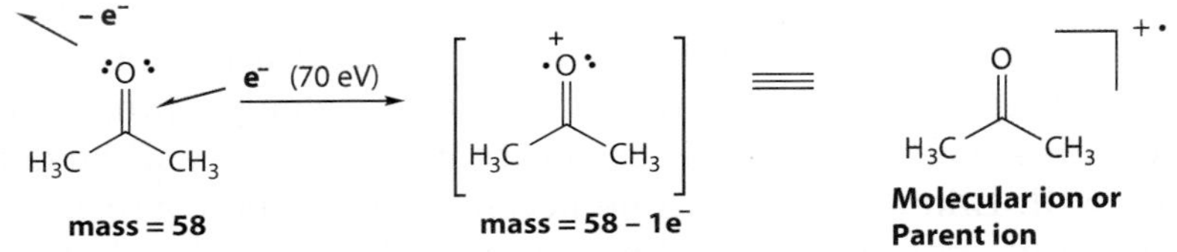

Figure 14.6   Formation of the molecular ion of acetone.

Removal of a single electron will have minimal effects on the overall mass of the molecule, so if the mass of acetone is 58, the mass of the molecular ion will be 58 minus the mass of a single electron. If the charge on the ion is +1, then the mass of the parent ion is taken to be that of the original molecule. It is not possible to measure the mass to a level that allows the loss of a single electron to be observed, so the mass of the molecular ion is taken to be that of the original molecule. Therefore, *as long as the charge is +1, the mass of the molecular ion will correspond to the mass of the original molecule; i.e., its molecular weight*.

14.5 What is the m/z for the parent ion for hexane, 3-pentanone, and diethylether?

The molecular ion is the first ion formed in the mass spectrometer, derived from the molecule that is introduced into the instrument. To be detectable, the ion must not fragment into smaller ions before it reaches the detector. The molecular ion is a very high energy species, however, and that energy may be dissipated by fragmentation into smaller pieces by breaking bonds, as depicted in Figure 23.7, *before it reaches the detector*.

The initially formed parent ion (**1**) is a radical cation, so it usually fragments via a homolytic process (see the **green** and **blue** single-headed arrows) of the carbon–carbon bond. In this case, homolytic fragmentation may lead to two new ions—**2** and **3**—but not at the same time. Either ion **2** will form and be detected or ion **3** will form and be detected. Ion **2** has an m/z of 43 and methyl ion **3** has an m/z of 15. If **1** is the parent ion, ions **2** and **3** are referred to as *daughter ions* because they must result from fragmentation of the parent or molecular ion. If all three ions are detected, the result is the mass spectrum of acetone shown in Figure 14.8.

When the m/z for each ion is observed and the charge for each is *assumed* to be +1, the mass spectrum reveals the mass of the original molecule to be 58.

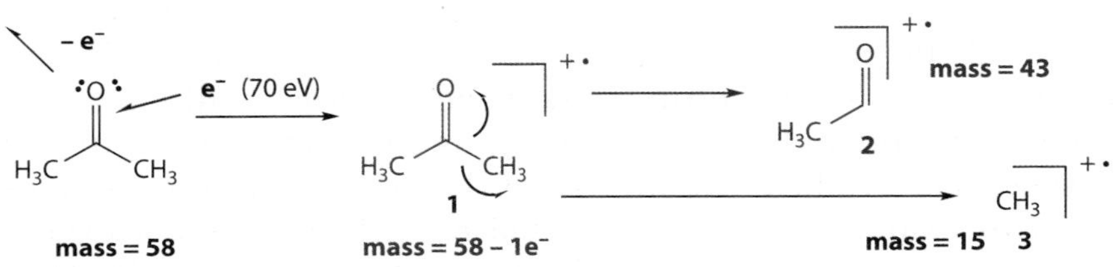

Figure 14.7   Fragmentation of the parent ion of acetone.

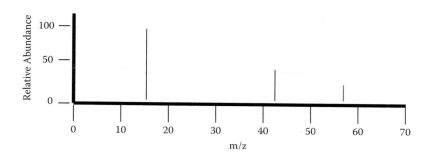

Figure 14.8 Mass spectrum of acetone.

Indeed, observation of the parent ion with an m/z of 58, with a charge of +1, gives the molecular weight of the starting compound—in this case, acetone. Fragmentation gives two ions of m/z 43 and m/z 15. If the structure of each daughter ion is known, this will contribute to an identification for the structure of the parent. Alternatively, knowledge of the structure of each fragment that is lost from the parent will aid the identification.

14.6 Using Figures 14.7 and 14.8 as a model, predict the fragmentation and mass spectrum of 2-butanone.

The relative abundance of each of the three ions in Figure 14.8 is different. The parent ion is not the most abundant ion (the base ion), but rather the base ion appears at m/z 15. Determining the most abundant ion and taking the ratio of all other ions to it leads to the mass spectrum in Figure 14.8. This is an important point because ions are rarely reported if their relative abundance is less than 5% of *B,* but there are exceptions. The relative percentage of ions, including which ion is the base peak, will vary with different molecules, and the mass spectrum of different ions with different intensities should be unique for a given molecule. Molecules that are similar in structure are expected to have similar fragmentation patterns in their mass spectra, whereas molecules that have different functional groups are expected to exhibit different fragmentation patterns.

14.2.3 High-Resolution Mass Spectrometry and Determining a Molecular Formula

Detection of radical cation fragments in modern mass spectrometers has been refined to the point where it is possible to determine the mass of a fragment to four or five decimal places. In other words, a molecular ion may be determined as 100.1251, or better. With this level of accuracy, a molecular formula may be obtained directly from the molecular ion. However, the mass used for this determination requires clarification. In the periodic table, the atomic mass of carbon is shown to be 12.0107 amu, hydrogen is 1.00794 amu, oxygen is 15.9994 amu, and nitrogen is 14.00674 amu. Interestingly, chlorine is 35.4527 amu and bromine is 79.904 amu. These are the masses of the elements as found in nature. However, carbon in nature is not just carbon-12 (^{12}C), but rather a mixture of three isotopes: ^{12}C, ^{13}C, and ^{14}C.

Table 14.1
Isotopic Abundance of Common Isotopes and
Isotopic Masses

Isotope	Relative Abundance (%)	Isotope Mass
^{12}C	100	12.00000
^{13}C	1.1	13.00336
^{1}H	100	1.00782
^{2}H	0.106	2.01410
^{14}N	100	14.00307
^{15}N	0.38	15.00010
^{16}O	100	15.99491
^{17}O	0.14	16.99913
^{18}O	0.20	17.99916
^{33}S	100	31.97207
^{32}S	0.78	32.9716
^{34}S	4.40	33.96786
^{35}Cl	100	34.96885
^{37}Cl	32.5	36.96590
^{79}Br	100	78.91839
^{81}Br	98.0	80.91642

As shown in Table 14.1, ^{12}C is the most abundant isotope for carbon. For the chemical reactions discussed in this book, using 12.0107 for the atomic mass is appropriate. In mass spectrometry, molecules are separated based on mass, which means that isotopes are differentiated. For that reason, the exact mass or isotopic mass of each isotope must be used. The exact masses of the various isotopes are shown in Table 14.1. If the mass of a molecule is determined to be 100.1251, that ion will be composed of the most abundance, lowest mass isotopes for each atom: ^{12}C, ^{1}H, ^{16}O, ^{14}N, etc.

High-resolution mass spectrometry provides a high level of accuracy for the molecular ion, but how is this translated to a formula? If the exact mass is divided by 12.00000 (Table 14.1), this number will represent the maximum number of carbon atoms permissible for that formula. For a molecular ion mass of 100.1251, 100.1251/12 = 8.3437. This means that eight carbon atoms will have a mass of 96.0000 and 4.1251 mass units remain. This calculation suggests H_4. The alkane formula represents the maximum number of hydrogen atoms permissible for a given number of carbons, and it is C_nH_{2n+2}. Each carbon atom must have four covalent bonds, however, so if there are eight carbon atoms, a formula of C_8H_4 makes no sense. Therefore, try C_7, which gives 7×12.0000 mass units = 84.0000, for a difference of 16.1251. From Table 14.1, the mass of H is 1.00782, and 16.1251/1.00782 = 16. Therefore, a formula of C_7H_{16} is reasonable and fits the alkane formula.

Another molecule has a molecular ion of m/z 100, but the exact mass is 100.0887. Using the same approach, C_8 is ruled out, and if C_7 is used, $C_7 + H_{16} = 100.1251$. At this level of measurement, the mass is off by 0.0364. A difference of more than four or five units (0.0004 or 0.0005) suggests that the formula is incorrect. In other words, the molecule does not have seven carbons. If there are six carbon atoms, the $C_6 = 72.0000$, and $100.0887 - 72 = 28.0887$—but, clearly, there are not 28 H atoms.

Of the common atoms, N and O are the most logical. If there are 2N (28.0074), there are no H atoms, which is silly. If there is one N (14.0037), then $C_6N = 86.0037$, and $100.0087 - 86.0037 = 13.9963$, for H_{14}. A formula of $C_6H_{14}N$ cannot be correct because the alkane formula is C_6H_{14}. If there is one O (15.9949), then $C_6O = 87.9949$, and $100.0087 - 87.9949 = 12.0138$. This correlates with H_{12}, so the formula $C_6H_{12}O$ is reasonable.

These examples suggest a drawn out process to arrive at a reasonable formula. This correlation is easily accomplished with a computer, as long as there is a library of possible structures with the exact mass of each. For example, for a molecule with m/z 100, the exact mass possibilities are C_5H_5Cl (100.0080), $C_5H_{10}NO$ (100.0762), $C_6H_{12}O$ (100.0888), $C_5H_8O_2$ (100.0524), $C_6H_{14}N$ (100.1126), $C_6H_{14}N$ (100.1126), and CCl_2F (100.9361).[3] The computer attached to the mass spectrometer will display the experimental exact mass of 100.0890 along with the seven possibilities shown. Within the range of 0.0005, the only possibility is $C_6H_{12}O$. For problems used in this book, this latter method will be used when the exact mass is provided, along with several possibilities for that formula.

14.7 **The exact mass of a molecule is determined to be 73.0891. Suggest a molecular formula.**

14.8 **The exact mass for a molecular ion is measured to be 100.0520. Identify the formula.**

Another use for the exact mass is that it is possible to track the loss of certain key fragments. In organic molecules, cleavage of the molecular ion will generate daughter ions, as described in Section 14.2.2. The mass of each fragment lost from the parent ion can be monitored using mass. For example, loss of a CH_3 (methyl) corresponds to formation of a daughter ion at m/z = 15. However, loss of a methyl group also corresponds to loss to 15 mass units from the molecular ion (M-15). The molecular ion of **4** (M = 148), for example, shows an M-15 peak at 133 for ion **5**. If the exact mass of 5 is determined, as well as the exact mass of the molecule ion **4**, **5** is measured to be 133.04097. With a value of 148.0654 for **4**, it is clear that the 15.02346 fragment lost is a methyl group.

Table 14.2
Exact Mass of Common Fragments in Organic Molecules

Mass	Fragment	Mass	Fragment
15.0235	CH_3	29.0391	CH_3CH_2
16.0187	NH_2	31.0184	CH_3O
17.0265	NH_3	41.0391	$CH_2=CHCH_2$
18.0106	H_2O	43.0547	$(CH_3)_2CH$
27.0235	$CH_2=CH$	43.9898	CO_2
28.0061	N_2	57.0704	C_4H_9
28.0313	$CH_2=CH_2$	77.391	C_6H_5
29.0027	CHO	91.0547	C_7H_7

Source: McLafferty, F. W. 1963. *Mass Spectral Correlations, Advances in Chemistry Series,* ed. Gould, R. F. Washington, D.C.: American Chemical Society.

This analysis is taken further if the exact mass is determined for all significant daughter ions. In the case of **4**, loss of Cl leads to **6**. An interesting fragmentation is the cleavage of **4** to give **7** and **8**. Analysis of **8** shows that a C_4H_9 fragment (**7**) is lost from **4**. The exact mass determination allows a reasonably precise analysis of the fragment as C_4H_9. There is nothing in the C_4H_9 fragment or loss of 57.0702 mass units to identify that fragment as *tert*-butyl. It may be *n*-butyl, *sec*-butyl, or *tert*-butyl. Other evidence is required to make that determination. Table 14.2 shows the exact mass of several common fragments that are found in organic molecules and may be lost as fragments in the mass spectrum.

14.9   **5-Methyl-3-hexanone is observed to lose a fragment with the mass 29.0391. What is the exact mass expected for 5-methyl-3-hexanone? Identify the fragment that is lost.**

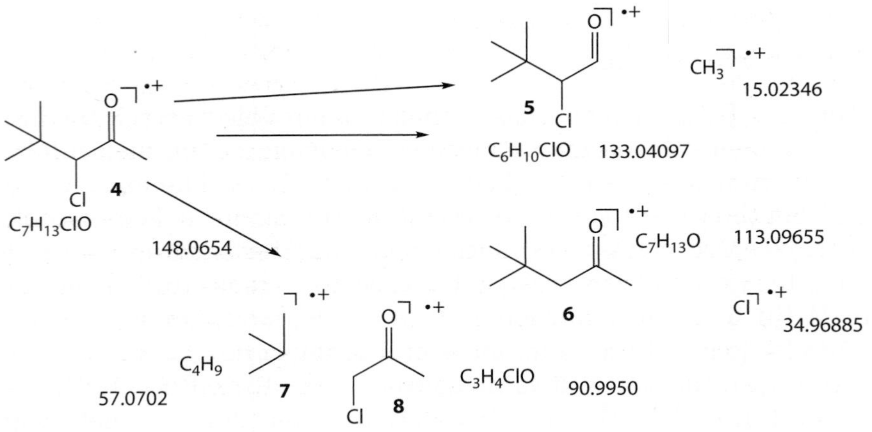

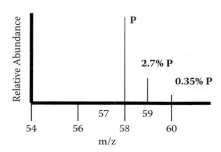

Figure 14.9   Explanation of the parent ion region for acetone.

14.2.4 Isotopic Peaks and Determining a Molecular Formula

In Section 14.2.2, ion **1** was shown to have the same mass as the starting compound (acetone), so the parent should be the highest mass ion in the spectrum. A closer look at the mass spectrum for the region close to the m/z 58 ion for acetone, however, shows an m/z 59 peak and an m/z 60 peak (see Figure 14.9). Such peaks arise because isotopes exist for many of the common elements, as shown in Table 14.1. For example, carbon exists as ^{12}C, the naturally most abundant isotope, as well as ^{13}C and ^{14}C. Similarly, the most abundant form of hydrogen is ^{1}H, but deuterium (^{2}H) and tritium (^{3}H) exist as well. Table 14.1 shows several elements that often appear in organic molecules, along with their common isotopes and relative abundance. Although also common in organic molecules, fluorine (^{19}F), phosphorus (^{31}P), and iodine (^{127}I) do not have natural abundance isotopes that are important.

The mass spectrometer can separate ions according to their mass, so it differentiates an ion containing ^{12}C from one containing ^{13}C. If acetone has a mass of 58, the molecular ion is composed of ^{12}C, ^{1}H, and ^{16}O. If an acetone molecule is formed from ^{13}C, ^{2}H, and ^{17}O, the relative abundance of ^{13}C is about 1%, with three carbon atoms, and this ion is detectable. Based on this analysis, the mass of parent ion for acetone composed of *the lowest mass and most abundant isotopes will appear in the molecule ion* at m/z 58.

Isotopes other than ^{12}C, ^{1}H, and ^{16}O may be present in a molecule of acetone. The relative abundance of ^{2}H is only 0.016, and even if there are six hydrogen atoms, the resulting m/z peak that arises from deuterium is small. There is only 0.04% of ^{17}O and if a molecule has only one oxygen atom in its structure, the m/z peak resulting from that isotope is too small for detection. If a molecule of acetone has a ^{13}C or a ^{15}N atom, with relative abundance of 1.11 and 0.28, respectively, presence of ions comprising minor isotopes is easily detected. In the actual spectrum of acetone, another peak appears at one mass unit greater than the parent (P or M), at m/z 59. It is called the **P+1 peak** or the **M+1 peak**, which is usually due to ^{13}C and ^{2}H. If a molecule has several carbon atoms, there is 1.11% of ^{13}C for each ^{12}C, which means that the ^{13}C contribution is detectable. For a molecule of acetone that includes a ^{13}C, the m/z is 59, or M+1. Nitrogen has a ^{15}N isotope that contributes to the M+1 peak when a compound contains one or more nitrogen atoms.

Other isotopes have higher mass isotopes. If a molecule of acetone has an ^{18}O atom, the relative abundance of that oxygen isotope is 0.20. This is at the threshold of detectability for small molecules, but this ion will appear at m/z 60 for acetone. This ion is two mass units higher than the parent and is called the **M+2 peak** (also known as the **P+2 peak**). *Both the M+1 and M+2 peaks are due to formation of compounds using isotopes of lesser abundance.* For acetone, the M+1 peak at 59 has an abundance of 2.7% relative to the molecular ion, M, for acetone in Figure 14.9. Similarly, the M+2 peak at 60 has an abundance of 0.35% relative to M. *The presence of isotopes leads to peaks of higher mass than the molecular ion, M, whose composition is based on the most abundant isotopes.*

The isotopic ratio of the elements is fixed and predictable, as seen in Table 14.1. If the number of carbon atoms in a molecule is known, the intensity of the M+1 peak that results from the presence of ^{13}C is predictable. The ratio of ^{13}C to ^{12}C is 1.11, so the intensity of the M+1 peak is 1.11 × #C. Using isotopic ratios, similar predictions are possible for other elements. This observation is important because the ability to predict the number of carbon atoms in an unknown molecule allows determination of the formula. The isotopic ratio is a known quantity, and measurement of the M+1 peak relative to M (i.e., M/M+1) will calculate the number of carbon atoms found in the molecular ion. The molecular ion also provides the molecular weight if $z = +1$, so the molecular formula is available for a given sample. For molecules that contain only C, H, N, and O, the two formulas used to calculate a formula are the following:

$$M+1 = (\#C)(1.11) + 0.38\ (\#N)$$

and

$$M+2 = [(\#C)(1.11)]^2/200 + 0.20(\#O)$$

In the M+1 formula, the first term uses the isotopic ratio for ^{13}C to ^{12}C, and the second term uses that for ^{15}N to ^{14}N (Table 14.1). In the M+2 formula, the first term determines the amount of ^{14}C relative to ^{12}C, and the second gives the amount of ^{18}O relative to ^{16}O. These isotopes are used because they usually give detectable peaks in the mass spectrum.

14.10 Predict the M, M+1, and M+2 patterns for 2-butanone and for triethylamine.

Obviously, some molecules have both carbon and nitrogen and some contain no nitrogen atoms at all. Therefore, the formula for M+1 has two unknowns. To solve this problem, working ***assumptions*** must be made. *Assume that if a molecule contains an even number of nitrogen atoms (0, 2, 4, etc.), the molecule will have an even mass. Assume that if a molecule contains an odd number of nitrogen atoms (1, 3, 5, etc.), the molecule will have an odd mass.* For the molecules encountered in this book, it is somewhat unusual for a molecule to have three or more nitrogen atoms, except for some heterocycles that are briefly described in Chapter 26. Therefore, the most obvious assumption is that a molecule will have zero, one, or two nitrogen atoms.

For the molecules presented in this book, the majority of simple molecules have zero nitrogen atoms or one nitrogen atom. Therefore, *begin with the assumption that an unknown molecule with an odd mass has one nitrogen atom and an unknown molecule with an even mass has zero nitrogen atoms*. This allows a substitution of 0 or 1 into the M+1 formula, so the equation may be solved for the number of carbon atoms. This assumption fails fairly often in real life, but it is a good place to start; for the purposes of doing homework, it will work almost all the time.

14.11 Determine the formula of an unknown molecule that shows the following information in the mass spectrum: M (m/z = 126) 100%, M+1 (m/z = 127) 8.88%, M+2 (m/z = 128) 0.59%.

Isotopic ratios are useful when sulfur, chlorine, or bromine is present in an organic molecule. The M+1 and M+2 formulas are not used, in a formal sense. Because bromine is composed of two isotopes—^{79}Br and ^{81}Br—in almost equal amounts, the M+2 peak for a molecule containing one bromine is about 100% of M (see Table 14.1). This is clear and obvious in the mass spectrum. Similarly, the ^{37}Cl isotope is about one-third of the ^{35}Cl isotope. Therefore, the presence of one chlorine atom in a molecule makes the M+2 peak in the mass spectrum about 30–35% of M. Finally, ^{34}S is about 4.4% of ^{32}S, so the M+2 peak of a molecule containing one sulfur atom will be about 4 or 5% of M. *These atoms are instantly recognizable when there is only one heteroatom in the molecule*. The M+2 formula is not used because the larger abundance of the Cl, Br, and S isotopes makes it impossible to detect the small abundance the ^{18}O isotope.

14.12 Briefly discuss the significance of the following: M (m/z = 92) 100%, M+1 (m/z = 93) 4.44%, M+2 (m/z = 94) 33%.

14.2.5 Mass Spectral Fragmentation as a Structural Probe (Common Fragmentations)

When the parent ion fragments into smaller daughter ions, determining the structure of each new ion can be problematic. There are techniques for doing this; however, for the purposes of this discussion, only a simple "thought" mass spectrometer is available, so the options are limited. If relatively small fragments are lost from the molecular ion, it is sometimes possible to equate that small mass with a specific molecular fragment, as pointed out in Section 14.2.3. This means that small fragments have such small masses that a very limited number of structures is possible, and the possibilities are easily correlated with known organic fragments. In the case of acetone from before, for example, a fragment is lost with a mass of 15 (see Figure 14.8). The only common fragment with m/z 15 that is logically obtained from acetone in the mass spectrum is a methyl group, CH_3. This fragment is also easily lost from many organic compounds.

If the mass spectrum of an unknown shows a daughter ion that is 15 mass units smaller than the molecular ion M (**the term M-15 is used**), *assume* that a methyl group has been lost. As pointed out in Section 14.2.3, exact mass determination of the daughter ions will make the identification more precise. Figure 14.10

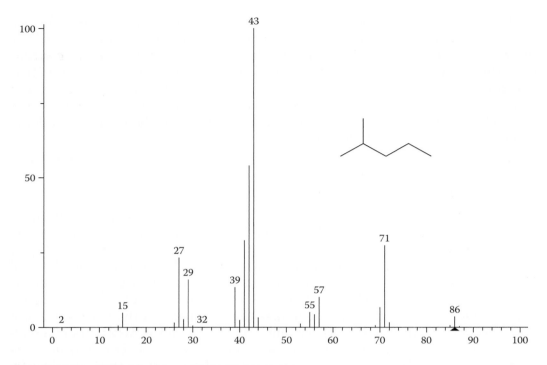

Figure 14.10 Mass spectrum of 2-methylpentane.

shows a mass spectrum of 2-methylpentane, where the molecular ion appears at m/z = 86. Another peak at m/z = 71 corresponds to M-15 and is associated with loss of a methyl group. Note that major fragmentations usually occur at branch points rather than within the straight-chain sections of the alkane backbone.

A daughter ion that is 18 mass units less than the molecular ion is associated with loss of a water molecule (H_2O; m/z = 18). This observation, of course, raises the question of which organic molecules can lose a molecule of water. One answer is that alcohols contain an OH unit and lose water in the mass spectrum in most cases. The daughter ion is the **M-18 peak**. Loss of water often means that the molecular ion of alcohols may be very weak and, with tertiary alcohols, is often missing altogether. The mass spectrum of 3-pentanol (see Figure 14.11) shows essentially no molecular ion at m/z 89, but there is a weak M-18 peak at m/z 70. The base peak at m/z 59 corresponds to M-29 (loss of an ethyl group). A detectable M-18 peak is often indicative of an alcohol.

There are many functional groups, of course, and a proper discussion of each one in terms of mass spectral fragmentation is beyond the scope of this section and this chapter. However, it is important to know that molecules lose discrete fragments that can be identified in many cases. Recognition of these fragmentation patterns gives structural clues. A last example is the mass spectrum of 3-pentanone (see Figure 14.12), where the molecular ion at m/z 86 is prominent. The M-15 peak is weak, but α-cleavage leads to loss of an ethyl group, M-29. Ketones undergo an interesting fragmentation called α-cleavage. The bond on either side of the carbonyl in the molecular ion of 2-pentanone (see **9**)

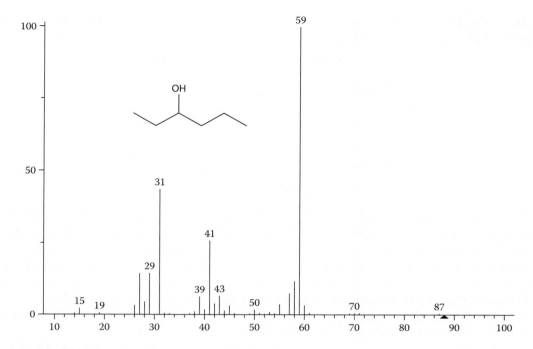

Figure 14.11   Mass spectrum of 3-pentanol.

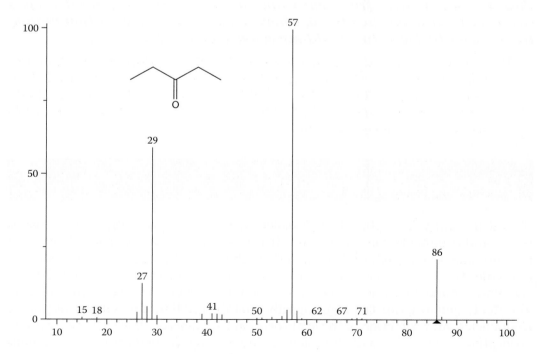

Figure 14.12   Mass spectrum of 3-pentanone.

cleaves to form an oxonium ion. When the bond to methyl (in **red**) is broken, a methyl group is lost (M-15) and oxonium radical cation **12** is generated.

Similarly, cleavage of the bond on the other side of the carbonyl (in **blue**) leads to loss of a propyl group (M-43, where m/z 43 corresponds to propyl, C_3H_7) and formation of oxonium radical cation **11**. Identification of these fragmentations provides a great deal of structural information. When an alkyl fragment of at least three carbons is present, as in **9**, a special fragmentation called a **McLafferty rearrangement** leads to one new ion and *one neutral fragment*. In other words, this cleavage gives an even mass m/z, whereas most cleavages lead to odd mass fragments from parents with an even mass.

This cleavage is named after Fred W. McLafferty (United States; 1923–). Shown is the McLafferty rearrangement for **9**, where ethene (ethylene, with a mass of 28) is lost and ion **10** is formed. This ion is resonance stabilized, as shown, which accounts for the facility of the McLafferty rearrangement. *Note that loss of a neutral fragment from the molecular ion generally leads to an even-mass ion such as 10, whereas loss of a radical cation daughter ion usually leads to an odd-mass ion such as 11 or 12*.

14.13 A molecule known to be a ketone has the formula $C_9H_{18}O$, shows a parent ion at m/z = 142, and has only two major fragmentation peaks at m/z = 127 and m/z = 85. The m/z = 127 is particularly large and the m/z = 85 is the base peak. Briefly discuss possible structures for this ketone.

14.3 Infrared Spectroscopy

So much energy is applied to the molecule in mass spectrometry that bonds break and fragmentation occurs. In infrared spectroscopy, the molecule is given the equivalent of a good, swift kick and, as the applied energy emerges from the molecule, the consequences are observed. Not enough energy is applied to do chemistry (make and break bonds), but enough is provided to cause the bonds to vibrate. In general, functional groups absorb infrared light and vibrate at unique frequencies, and detection of these absorption frequencies allows one to correlate absorption frequencies with functional groups. Knowledge of functional groups is a crucial structural determination component.

14.3.1 Absorbing Infrared Light (Molecular Vibrations)

The interaction of infrared light with organic molecules occurs most often at frequencies between 4000 and 400 cm^{-1}. For a simple diatomic molecule, the bond of interest must be identified as symmetrical or asymmetric. A symmetrical bond has two identical atoms, as in H_2, O_2, or N_2, whereas an asymmetric bond will have two different atoms. Diatomic molecules with an asymmetric bond will usually have a dipole, as long as there is a difference in the electronegativity of the bonded atoms. A greater electronegativity difference between the two atoms usually leads to a larger dipole moment for that bond (see Chapter 3, Section 3.7). A symmetric bond should have little or no dipole and, hence, little or no dipole moment. *Symmetrical bonds with no dipole do not absorb infrared light very well and usually give a weak absorption peak because there is no change in the dipole moment. An asymmetrical bond will strongly absorb infrared light and that absorption leads to a change in the dipole moment and a strong absorption peak*. *The greater the change in the dipole moment, the stronger will be the absorption.*

Infrared radiation is essentially an alternating electrical field that interacts best with fluctuations (vibrational frequency) in the dipole moment of the molecule. The dipole moment changes in a repetitive manner and the molecule vibrates. Fluctuation of vibrational frequency is greater when there is a change in dipole moment for a given bond. Absorption of the infrared radiation leads to a change in the amplitude of molecular vibration. Note that molecules without a permanent dipole may absorb infrared light if there is a bending vibration or another vibration that produces a change in dipole moment.

What happens when a diatomic molecule absorbs infrared light? The molecule will absorb the energy and the atoms and bonds connecting those atoms will vibrate to release the excess energy. Such vibrations will have a characteristic frequency (ν) that depends upon the mass of the two atoms in the bond and the strength of that bond. As a simple model, consider this diatomic molecule to have two masses (m_1 and m_2), connected by a bond (approximated by a spring) separated by distance r_0. The two masses connected by a spring will behave as a harmonic oscillator when infrared light is absorbed, so **Hooke's law** will apply: $F = -f\,\Delta r$, where F = force, f = a proportionality factor (the force constant), and Δr is the change in distance between the two atoms of the molecule. Hooke's law is named after Robert Hooke (England; 1635–1703).

In a diatomic molecule, Δr will be small and this will result in a simple harmonic motion; this type of system is referred to as a harmonic oscillator. Assuming that this vibrational motion is equivalent to a harmonic oscillator allows the frequency of the oscillation to be calculated by

$$\nu = \frac{1}{2\pi}\frac{f}{\mu}$$

where μ is the reduced mass [$m_1 m_2/m_1+m_2$].

In this equation, f = force constant and μ = reduced mass. The mass (m) used in this equation is the mass of the atom, given by m = atomic weight/Avogadro's number, where Avogadro's number = 6.02252×10^{23} per gram mole. The force constant (f) is proportional to the strength of the covalent bond linking m_1 and m_2. This model indicates that as the force constant increases (the spring or bond becomes stronger), the frequency of the vibration will increase. As the reduced mass (μ) increases, the vibrational frequency will decrease. If diatomic hydrogen (H_2) is compared with diatomic deuterium ($^2H_2$), for example, H_2 has a smaller μ and is calculated to have a vibrational frequency of $\nu = 4160$ cm^{-1}. The heavier deuterium has a larger μ and has a calculated $\nu = 2993$ cm^{-1}. The force constants (f) for each of these two molecules are similar but not identical.

Later in this section, a C–C bond is shown to absorb at 1300–1800 cm^{-1}, whereas a C=C bond absorbs at 1900–1500 cm^{-1}. The stronger C=C bond has a larger force constant and therefore a greater frequency of absorption for two atoms of the same mass. This indicates that atoms of different masses, with different bond strengths, will have different absorption frequencies in the infrared. A C–O bond will absorb with a different frequency than a C–N bond, and a C≡C bond will absorb differently than a C=O bond or a C=C bond. The bonds noted here either constitute or are part of functional groups, and vibrational differences are used to identify functional groups in an organic molecule.

14.14 **Calculate the reduced mass of the C–O bond where $m_o = 2.65\ 7 \times 10^{-23}$ and $m_c = 1.994 \times 10^{-23}$.**

14.15 **Comment on which bond will have the greater force constant: C≡C or C=C. Which will absorb at lower energy?**

When a diatomic molecule absorbs infrared light, it becomes a diatomic oscillator, leading to several energy levels, as shown in Figure 14.13.[4] When a

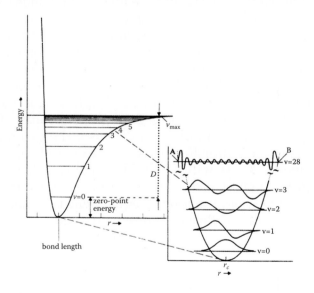

Figure 14.13 Energy levels for an anharmonic diatomic oscillator. (Isaacs, N. S. 1995. *Physical Organic Chemistry,* 2nd ed., p. 31, Figure 1.6. Pearson Education Limited. 1995. Longman, Essex, UK. With permission.)

molecule is in the ground state (it has not absorbed infrared light), the energy is at a minimum (see the dotted line in Figure 14.13), but when it absorbs a photon of light, energy reaches the first quantum level. Several such transitions are possible. A molecule tends to absorb light and show a single transition from $n = 0$ to $n = 1$ in most cases, and in the infrared spectrum this occurs between 4000 and 400 cm^{-1}, as mentioned earlier. If a diatomic molecule functions as an anharmonic oscillator, the energy absorptions are more complicated, leading to "fine structure" and line broadening in the infrared absorption.

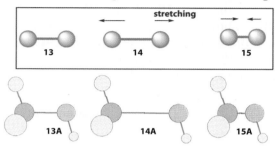

Most organic molecules have bonds that bend and vibrate under normal conditions at ambient temperatures. The infrared spectroscopy experiment irradiates the molecule with infrared light, and energy is absorbed when the stretching or bending vibration matches the applied energy. In other words, when the applied IR light is the same as the stretching or bending vibration of a bond, that energy is absorbed, and the absorption is detected and displayed as a peak in the spectrum.

Remember that for strong absorption of infrared light, the bond in a diatomic molecule must show a strong change in dipole moment, which is correlated with several different types of vibrational motions. The discussion will be restricted to a two atom–one bond system such as **13** and three atom–two bond systems such as **16**. Imagine that the bond in **13** is stretched out (elongated) and is compressed (shortened), as shown in **14** and **15**. This vibration constitutes a stretching vibration of a particular frequency, depending on the mass of the two atoms. The stronger the bond is, the more difficult it will be to stretch that bond in this manner. This type of bond will be part of a larger molecule, such as the C–O bond in methanol provided as a real example. The normal C–O bond is shown in **13A**, and the vibration will stretch the atoms away from one another in **14A** and toward one another in **15A**. These cartoons illustrate the vibrational stretching absorption of the C–O bond after that bond absorbs infrared light of the correct energy (i.e., the correct wavelength).

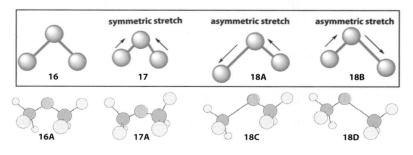

When three atoms and two bonds are present, as in **16**, each bond will have a "diatomic" interaction, but there is also the possibility of interactions among the three atoms, where vibrations of both bonds must be considered. A different type of stretch occurs with **16** in that there may be a symmetrical stretch (**17**) or an asymmetric stretch (**18**). In principle, there are two asymmetric stretches: **18A** and **18B**. These will occur at different frequencies. Once again, this model is shown for a real molecule—dimethyl ether. It is difficult to represent these vibrations properly because the molecular center of mass tends to remain the same, forcing the central atom to move. The models used are **16, 17,** and **18,** but the atomic motion has been exaggerated. Model **16A** shows the normal bonds; **17A** shows the stretch, where both oxygen atoms compress toward the carbon. In **18C,** one oxygen compresses toward carbon while the other expands away from carbon; **18D** shows the same stretch in which the oxygen atoms show the opposite motion.

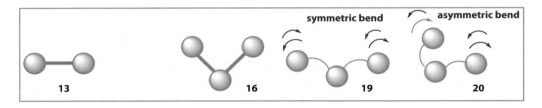

Another variation has three different atoms connected by bonds, such as the H–O–C bond in methanol (**8**). Bending a diatomic molecule such as **13** is rather difficult. With **16,** there are two bonds and different modes for bending. There are a symmetrical (**19**) and an asymmetric (**20**) bend, as shown. Once again, each of these occurs at a different frequency, although these vibrations are usually lower in energy than the stretching vibrations.

Absorption of infrared light of a given frequency leads to a vibration that will generate a "peak" in the infrared spectrum. There is usually more than one absorption in an infrared spectrum for a given functional group, so several peaks are common in the infrared spectrum for an individual bond. Because there may be several different atoms and several types of bonds, there are many stretching or bending vibrations for a given molecule. Other types of vibrations can occur as well, and a typical infrared spectrum will contain many peaks. The real point of this discussion is to show that for a given diatomic or tri-atomic unit (a functional group), there are several different vibrational modes (bending, stretching, etc.) and that each will occur at a unique frequency. These vibrational frequencies emerge from the molecule after interaction with the applied infrared light, and detection of the emerging vibrational frequencies led to the infrared spectrum. In general, stretching vibrations are stronger than bending vibrations, leading to a stronger signal (larger peak).

14.3.2 The Infrared Spectrophotometer

Figure 14.14 shows a schematic of a very simple infrared spectrophotometer. The infrared light beam is split by a prism into two beams of equal intensity.

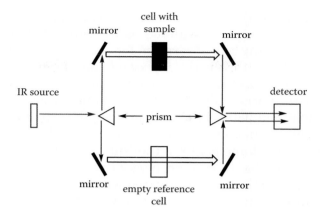

Figure 14.14 Schematic for a simple infrared spectrophotometer.

One beam passes through a cell containing a reference (air, solvent, etc.) so that none of the light is absorbed. The other beam passes through a cell containing the sample, which will absorb some of the infrared light. When these two beams are recombined, the signal that emerges from the reference cell is electronically subtracted by the detector from the signal that emerges from the sample cell. The amount of infrared light absorbed at various wavelengths is displayed as absorption peaks in the infrared spectrum (see Figure 14.15 for hexane). (Remember that spectrum is singular and the plural is spectra.)

The normal operating procedure for taking an infrared spectrum places the molecule of interest on or in a sample cell. The sample cell is often two clear plates made of pressed NaCl or KBr, and a liquid sample is sandwiched between the plates. These plates are sensitive to water, which should be avoided during handling the plates and during sample preparation. If the compound of interest is a liquid, a drop is placed on the salt plates (the sample is said to be

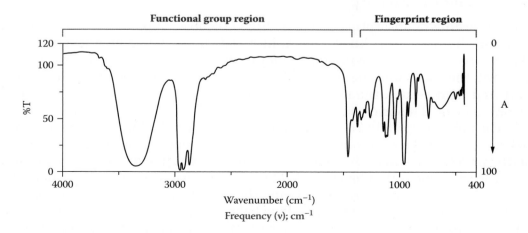

Figure 14.15   Infrared spectrum of 2-butanol.

analyzed ***neat***—no solvent) and placed in the spectrophotometer. The reference is usually air, and the sample is scanned as the infrared light is varied from 4000 to 400 cm^{-1}, recording any absorption for each wavelength. The combined absorption-wavelength pattern is the ***infrared spectrum***.

Occasionally, the compound is suspended in Nujol (a heavy mineral or paraffin oil containing primarily long-chain alkanes, C20 to C40, with major peaks between 2950 and 2800, 1465 and 1450, and 1380 and 1370 cm^{-1}). This suspension is called a "Nujol mull," and it is placed on the salt plates and then scanned. Although the alkane peaks of the Nujol may obscure part of the spectrum, it is usually possible to discern peaks due to the functional groups as well as a significant portion of the infrared spectrum. To obtain an infrared spectrum of a solution of a compound in a solvent, the solvent should not contain water, which absorbs in the region of 3300 cm^{-1}, and it should *not* have a functional group that has a strong absorption in the infrared, such as a ketone, alcohol, or amine (see Section 14.3.4). When the compound of interest is a solid, it is dissolved in a solvent that will have minimal infrared absorption in the functional group region of an infrared spectrum (see Section 14.3.3).

Chloroform or carbon tetrachloride is commonly used for this purpose. A solution of the sample is placed in a special sample holder that is essentially two salt plates (NaCl or KBr) separated by a suitable spacer such that a cavity of known volume is created. Two cells similarly prepared (same volume; known as matched cells) are required. The reference cell is loaded with the solvent and no sample, and the sample cell is loaded with a solution of the molecule of interest. These are placed in the infrared spectrophotometer and analyzed in the normal way. In this technique, either the instrument must electronically subtract out absorption peaks due to the solvent, or the user must mentally subtract out those peaks.

14.16 Comment on the possibility of using acetone to wash KBr or NaCl infrared plates.

14.3.3 The Infrared Spectrum

The recording of absorption peaks obtained by absorption of a range of infrared light frequencies is known as an infrared spectrum, as mentioned earlier. There are a series of "peaks" or "valleys," depending on the point of view. In general, the spectrum is analyzed by starting at the "top" and, as absorption of infrared light occurs, the line approaches the "bottom" as a stronger signal—hence, a peak. This is explained next.

The infrared spectrum of hexane is shown in Figure 14.15. There are two horizontal scales: one in reciprocal centimeters (the frequency scale, ν) and the other in μ (the wavelength scale in microns or micrometers, which is 10^{-6} m). There are also two vertical scales: absorbance (A) and percent transmittance (% T). If all of the infrared light is absorbed by the molecule, the value of A = 100%, which means that no light is transmitted, and % T = 0. If no light is absorbed by the sample, there is 100% transmittance (% T = 100) and 0% A. ***A larger value of A is associated with a stronger peak and a smaller value of A with a weaker peak.***

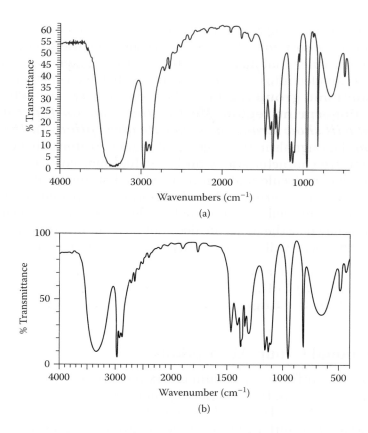

Figure 14.16 Infrared spectrum of 2-propanol: (a) linear in wavelength; (b) linear in frequency.

An infrared spectrum can be displayed in two ways: linear in wavelength or linear in frequency. Two infrared spectra for 2-propanol are shown in Figure 14.16. In spectrum (a), which is linear in wavelength, note that the functional group region is expanded, making it easier to see the OH stretch at about 2500 cm^{-1}, but more difficult to see the C–O stretch at about 1000 cm^{-1}. When compared with spectrum (b), which is linear in frequency, the right side of the spectrum is expanded, making it easier to see the C–O stretch that absorbs in the 900–1200 cm^{-1} region, but the region between 4000 and 1500 cm^{-1} is condensed. As will be discussed later, it is a bit easier to see the functional groups in spectrum (a), whereas it is easier to see individual vibrational variations for a molecule in spectrum (b). This latter distinction is useful for identifying one specific compound when compared to another or to a library of possibilities. Note that the micron scale (linear in wavelength) is not used in modern spectroscopy because it is not proportional to energy. It will be seen in older literature, however.

Returning to Figure 14.15, which is 2-butanol (a typical aliphatic alcohol), there is a cluster of strong peaks that absorb at about 3000–2800 cm^{-1}, a moderately strong set of peaks at 1470–1380 cm^{-1}, and another weak peak at about 710 cm^{-1}. There is also a very strong peak at about 3300 cm^{-1}. The absorptions at 3300 cm^{-1} are due to the OH group, but the other absorptions mentioned are due to

of the absorption band are also important, and several spectra are shown to illustrate the major functional groups that will be encountered in this book.

The hexane spectrum (Figure 14.15) shows that the C–H stretch at 2850–2960 cm^{-1} and the signals at 1350–1470 cm^{-1} correlate with the absorptions in Table 14.3. Benzene derivatives and other aromatic compounds usually show absorption for the C–H units at 3000–3100 cm^{-1}, but also at 675–870 cm^{-1} (in the fingerprint region). Note the subtle shift of the C–H absorption to lower energy for the aromatic compounds. Other compounds that have a C–H absorption are those for alkenes and alkynes. The C≡C–H absorption is at lower energy than the C=C–H absorption, which is lower in energy than the C–C–H absorption for alkanes.

14.18 Briefly explain why the C=O absorption is at higher energy than the C–O absorption in Table 14.3.

If the focus remains on hydrogen atoms, there is the OH unit of an alcohol or a carboxylic acid as well as the N–H unit of an amine. The hydrogen absorption for an OH is at 3610–3640 cm^{-1}, which is higher in energy than the C–H absorptions because of the mass of the atoms and the relative strength of the bonds (see Section 14.3.1). Hydrogen bonding for the O–H unit changes the effective O–H bond distance, which leads to a range of absorption frequencies. The result of this phenomenon is that a hydrogen-bonded OH signal is rather broad and very strong (see Section 14.3.5). The hydrogen absorption for NH is in generally the same position as that for OH, but is weaker and not as broad because the N–H unit cannot hydrogen bond as effectively as the OH unit. Finally, aldehydes have a hydrogen atom attached to the carbonyl and this absorption appears at about 2817 cm^{-1}, which is much lower in energy than what is seen for other C–H absorptions.

Most organic compounds have carbon–carbon bonds. The signals for C–C single bonds appear at 1200–800 cm^{-1} (they are not listed in Table 14.3), but they are very weak and of little value for compounds discussed in this book. Note that these signals appear in the fingerprint region and, because they are weak, identifying them may be difficult. The C$_6$ unit of benzene rings and other aromatic compounds appears in the fingerprint region; however, benzene does not really have a C–C unit, but rather a resonance delocalized system that influences the hybridization of those bonds. In Table 14.3, these signals appear at 1600 and 1500 cm^{-1}, and there are also several characteristic signals that appear in the fingerprint region. These signals can usually give a clue that an aromatic ring is present.

Although the C–C bond of an alkane or an alkyl unit is usually weak, the C=C unit gives a moderate to strong signal in the infrared, at 1640–1680 cm^{-1}. Sometimes, the signals are so weak that there is ambiguity, although in most cases their presence is clear and diagnostic. The alkyne unit (C≡C) also gives rise to a moderate to weak signal, but it appears at 2100–2260 (note that it is higher in energy than the C=C signal because the alkyne unit is stronger), which is in a region of an infrared spectrum that does not have many signals. This point is noted in Section 14.3.5, but it is important to state it here.

Internal alkynes often have a weak C≡C absorption because there is only a small change in dipole moment for the vibration. *A **symmetrical internal**

alkyne such as 2-butyne may show no signal at all. The alkyne C≡C for terminal alkynes is usually apparent. Because the peak is often weak, one must be careful not to confuse it with a nitrile (C≡N) or a C–D signal, although the CN signal is usually moderate to strong. Deuterium-labeled compounds are not often seen in this book; nevertheless, the point is important because $CDCl_3$ is often used as a solvent for NMR spectroscopy (see Section 14.4) and it is common practice to take an NMR spectrum and then use that sample to obtain the infrared spectrum. The C–D stretch appears at about 2240 cm^{-1}.

14.19 **Discuss the possibility that another absorption in an infrared spectrum is due to the C–O–H unit of the alcohol.**

14.20 **Briefly describe differences in an infrared spectrum that will allow one to distinguish 1-pentyne from 1-pentene.**

The next important type of bond involves carbon and oxygen, and both the C–O unit and the C=O unit (the carbonyl) are known. The C–O unit is found in alcohols, in ethers, and in esters. From Table 14.3, the C–O stretching modes are usually found between 1080 and 1300 cm^{-1} as strong or moderate peaks. Unfortunately, the C–O absorption appears in the fingerprint region, which means it may be obscured by other peaks. In most simple alcohols and ethers, the C–O absorption is strong and easy to find. In many cases, however, there is sufficient ambiguity that its identification can be problematic. Both the OH of an alcohol and the carbonyl of an ester (see later discussion) show signals other than the C–O. Therefore, it is relatively easy to identify these functional groups. An ether has only the C–O bond as the functional group. *The ether unit must be suggested from the presence of a C-O, but more commonly it is inferred by the absence of an OH or a C=O absorption, and with the aid of additional information such as NMR data* (see Section 14.4).

The carbonyl unit is usually very easy to identify. The C=O unit of aldehydes, ketones, carboxylic acids, and esters absorbs at 1690–1760 cm^{-1}, and it is a strong absorption peak. For most aliphatic derivatives, the C=O absorption is centered about 1725 cm^{-1}. When the carbonyl is conjugated to a C=C unit or a benzene ring (conjugated carbonyl derivatives), the absorption is shifted to lower energy and the C=O stretch is found at about 1695 cm^{-1}. In these cases, ***the functional group is the carbonyl, but this absorption peak does not distinguish between aldehydes, ketones, esters, etc. if only this signal is used***.

The CH signal for an aldehyde absorbs at 2817 cm^{-1} (see earlier discussion), and when the carbonyl absorption is also present, this is usually a good indicator of the presence of an aldehyde. The carbonyl of an acid chloride absorbs at higher energy, at about 1802 cm^{-1}. Because few other absorptions appear in this region, a single absorption signal is usually taken as a diagnostic for the presence of an acid chloride. An anhydride will show *two* absorption peaks in this region—for the two carbonyl units—usually centered on 1818 and 1750 cm^{-1}.

As mentioned before, the carbonyl unit of a carboxylic acid absorbs around 1725 cm^{-1}. The carboxyl unit (COOH) also contains an OH unit, however. The OH unit of an alcohol absorbs at about 3610–3640 cm^{-1} as a strong, relatively broad peak. The OH of a carboxylic acid absorbs between 2500 and 3000 cm^{-1} as a *very* broad and strong absorption because the OH of the acid unit is more extensively hydrogen bonded than that of an alcohol. This absorption peak, along with the carbonyl absorption, is very diagnostic of a carboxylic acid.

14.21 **Is infrared spectroscopy the best tool for distinguishing between 2-hexanone and methyl pentanoate?**

14.22 **Oxidation of a primary alcohol with Jones reagent usually gives the carboxylic acid rather than the aldehyde. Describe how one can use infrared spectroscopy to distinguish these two possible products.**

Amines are another important class of compounds that are characterized by C–N bonds and N–H bonds for the primary and secondary amines. The N–H signal absorbs essentially in the same place as the OH signal, at 3300–3500 cm^{-1}, but is usually weak or moderate in strength because amines do not hydrogen bond as extensively as an alcohol. The OH signal by comparison is very strong. An interesting phenomenon is observed with amines. A primary amine has the NH$_2$ unit, and asymmetric vibrations of the two N–H units lead to *two* N–H absorptions. Therefore, a primary amine should show a doublet (two peaks) in this region.

This is not always the case because the peaks can merge into one broad peak due to hydrogen bonding, but two peaks are often observed. Because a secondary amine has only one N–H unit, it absorbs as a singlet (one peak) in this region. Tertiary amines do not have an N–H, and there is no absorption in this region. For the data presented in this and other chapters, a tertiary amine unit will not have N–H absorption in the infrared spectrum and its presence must be inferred from mass spectral data or from the empirical formula (see Sections 14.2.3 and 14.2.4).

Amines also have a C–N unit, but it gives a relatively weak signal that appears in the fingerprint region (1180–1360 cm^{-1}). In general, it is not used as a diagnostic tool because it is difficult to identify. It is also noted that imines (see Chapter 5, Section 5.9.4) have a C=N unit, which absorbs at 1690–1640 cm^{-1} and is usually strong enough to be observed. This signal appears at lower energy than most carbonyl signals, although it can sometimes be confused with a conjugated carbonyl. Imines are not included in Table 14.3.

The C≡N unit of a nitrile is an important functional group, and the triple bond of the cyano group is expected to be in the same general region of an infrared spectrum as the C≡C unit of an alkyne. Indeed, the C≡N unit absorbs at 2210–2260 cm^{-1}, usually as a moderate to strong, sharp peak. In general, however, a typical alkyne C≡C unit absorbs at about 2268 cm^{-1} and a typical C≡N unit at 2252 cm^{-1}. Conjugation will shift the cyano signal to 2240–2220 cm^{-1}.

14.3.5 Correlation of Functional Groups with Infrared Spectroscopy Absorption Bands

This section will correlate the information from Section 14.3.4 with actual infrared spectra. It is important to know what the absorption looks like as well as the frequency of the absorption. When is a peak strong enough to be considered, and what constitutes a strong, weak, or medium absorption peak? Of the many peaks in a typical infrared spectrum, there may be a very weak peak at 1725 cm^{-1}. It is not considered to be a carbonyl, however, unless it is a "strong" signal. Similarly, a weak peak at 3525 cm^{-1} is not diagnostic of an alcohol OH unless it is strong and relatively broad. Therefore, it is important to know what types of absorption to look for vis-à-vis a given functional group.

Figure 14.17 shows the infrared spectra for an alkane (A, 3-methylpentane), an alkene (B, 1-methylcyclohexene), and an alkyne (C, 16-methyl-1-hexadecyne). Note that the CH region is essentially the same in all three spectra, but there is a C≡CH absorption in Figure 14.17C. The C=C signal in Figure 14.17B is clearly visible, as is the absorption for the C≡C unit in Figure 14.17C. As noted,

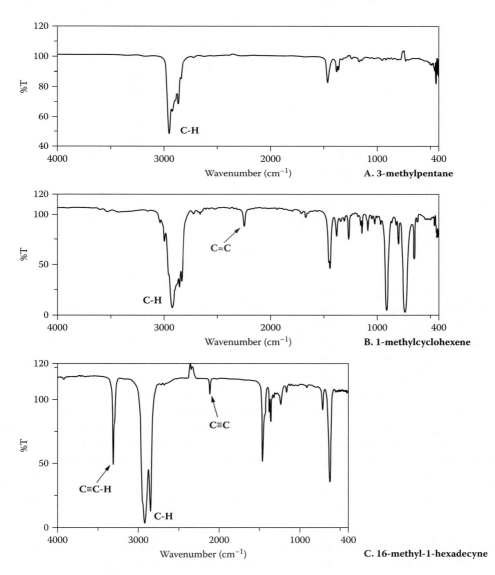

Figure 14.17   Infrared spectra of typical hydrocarbons.

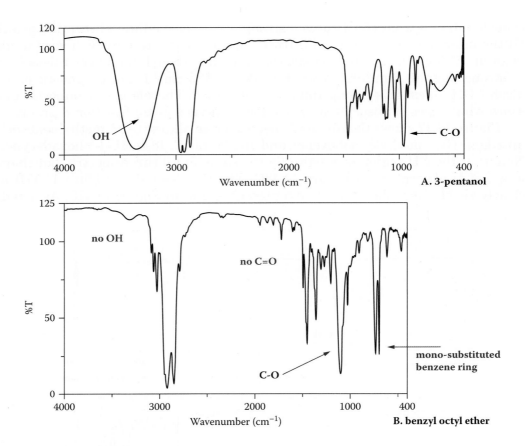

Figure 14.18   Infrared spectra of an alcohol and an ether.

the alkyne spectrum also contains the C≡CH signal at about 3300 cm^{-1}. Note the line shape and intensity of several of these important, but weak-to-moderate intensity diagnostic peaks.

Figure 14.18 shows the infrared spectrum of a common alcohol (A, 3-pentanol) with its prominent OH absorption as well as the strong C–O absorption in the fingerprint region. As a comparison, the infrared spectrum of benzyl octyl ether is also shown (B), with the more prominent C–O absorption. The out-of-plane bending vibrations for a monosubstituted benzene ring are also noted. For a molecule that has a molecular formula with a single oxygen atom, the presence of an ether unit must be inferred by the absence of an OH signal and the absence of a C=O signal. The C–O unit may not be obvious because it appears in the fingerprint region, so the ***absence*** of the OH and C=O peaks suggests an ether in this particular case. Note that both compounds have several peaks in the functional group region.

There are several types of molecules with carbonyl-containing functional groups. Both ketones and aldehydes have carbonyl absorption at about 1725 cm^{-1}, and it is not possible to distinguish them based on this peak. The infrared spectra of 4-methyl-2-pentanone (A) and butanal (B) in Figure 14.19 clearly

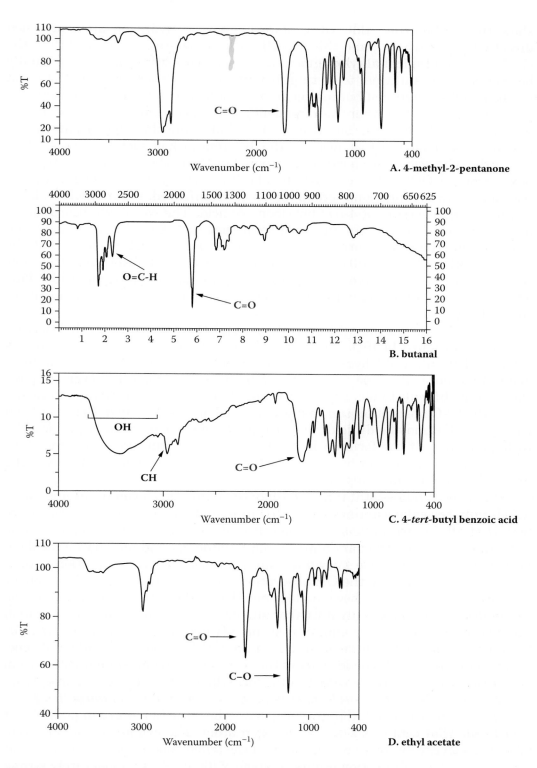

Figure 14.19 Infrared spectra of carbonyl-containing compounds.

show this to be true. Aldehydes, however, have a hydrogen atom attached directly to the carbonyl, and this absorbs at lower energy (2817 cm^{-1}) than the C–H units of a typical alkyl fragment as shown. This weak signal can usually be seen and it is sufficient to distinguish an aldehyde from a ketone. Note that the infrared spectrum of butanal is an older style infrared spectrum, which is linear in wavelength rather than frequency. This spectrum is shown not only to highlight the C–O stretch but also to show that many older spectra are in the literature, and one must be aware of the different look of such spectra.

Carboxylic acids have a carbonyl group, but the carbonyl is part of the carboxyl group (COOH). The carboxyl has both a carbonyl and an OH unit, and both absorb in the functional group region of the infrared. Figure 14.19C shows the infrared spectrum of 4-*tert*-butylbenzoic acid. The carbonyl absorption at about 1730 cm^{-1} is indistinguishable from that of 4-methyl-2-pentanone and butanal, but the OH absorption is very broad and distinctive. Note that the OH absorption (2500–3300 cm^{-1}) partly obscures the CH absorption. It is very broad due to the extensive hydrogen bonding found in carboxylic acids. This absorption, in conjunction with the carbonyl peak, is absolutely diagnostic for the carboxyl unit.

Esters are carboxylic acid derivatives, and the spectrum of ethyl acetate is shown in Figure 14.19D. The carbonyl absorption does not distinguish this compound from an aldehyde or a ketone, but there is the C–O absorption at about 1200 cm^{-1}. Because this is in the fingerprint region, however, its position can be difficult to identify. This is clearly the case for methyl pentanoate, where the C–O absorption can easily be missed or misidentified. Based only on the infrared, it may be difficult to distinguish an ester from an aliphatic aldehyde or ketone. If the formula is known, however (from mass spectrometry), the identification is easier because an ester has two oxygen atoms, whereas the aldehyde or ketone has only one.

The three fundamental types of aliphatic amines are primary amines with a NH$_2$ unit, secondary amines with a NH unit, and tertiary amines, which have no NH at all. Figure 14.20 shows the infrared spectrum of the primary amine butylamine (A), the secondary amine dibutylamine (B), and tertiary amine diisopropylethylamine (C), also known as Hünig's base. The NH absorption is relatively weak compared to the OH absorption seen in Figure 14.18A for alcohols. The primary amine (A) shows a clear doublet in the NH region, whereas the secondary amine shows only a weaker singlet. This is sufficient to distinguish the NH$_2$ unit of a primary amine from the NH unit of a secondary amine. For the tertiary amine (C), there is no NH unit and that region of an infrared spectrum shows no signal. The identity of the tertiary amine must be inferred from the presence of one nitrogen atom in the formula and the lack of NH absorption or a C≡N unit. ***In a case such as this, negative evidence is compelling***.

The cyano group, found in nitriles, is another functional group possible for a molecule that contains a single nitrogen atom. The cyano functional group (C≡N) has a very distinctive absorption in the triple bond region similar to that for alkynes. This absorption occurs at about 2260 cm^{-1} as a moderately strong and sharp peak, as seen in Figure 14.20D (hexanenitrile; an old-style infrared

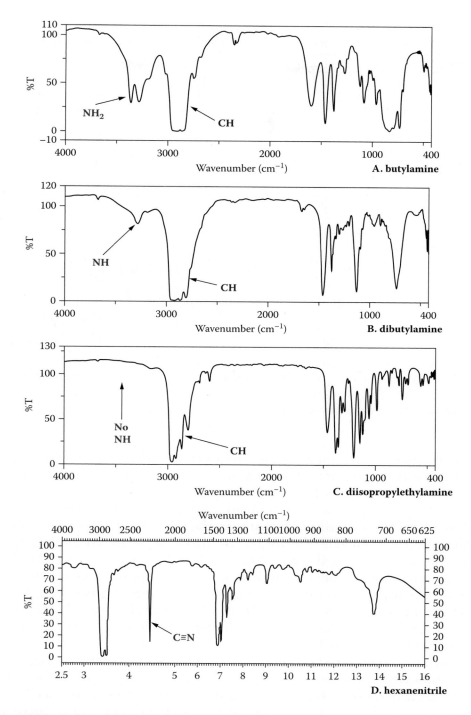

Figure 14.20 Infrared spectra of amines and a nitrile.

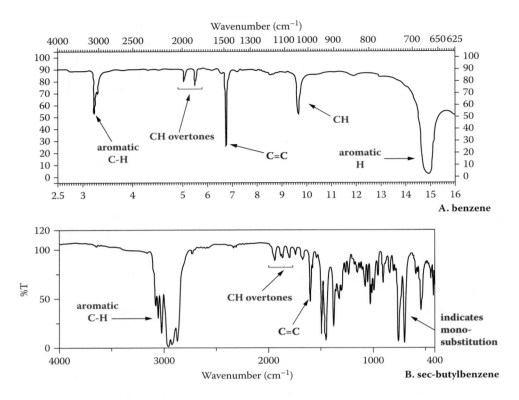

Figure 14.21 Infrared spectra of simple aromatic derivatives.

spectrum). There is a slight difference between a C≡C unit and a C≡N unit, as seen in Table 14.3, but a molecule containing one nitrogen atom in the formula showing this absorption is very likely a nitrile.

Aromatic hydrocarbons such as benzene and substituted benzene derivatives have a few distinguishing features in the infrared. This section concludes with the spectra of benzene (Figure 14.21A, an old-style infrared spectrum) and *sec*-butylbenzene (2-phenylbutane, Figure 14.21B). Benzene has C–H absorptions that are clearly evident, as well as a C=C absorption at about 1600–1630 cm^{-1}. There is little to indicate that this is an aromatic hydrocarbon. Note the very weak signals at about 1750–2000 cm^{-1}. These are known as C–H overtone absorptions, and they usually appear only with benzene derivatives. They are very weak and may easily be missed.

In this case, there are two small peaks. There is a strong absorption peak at about 680 cm^{-1}. This peak is also characteristic of benzene, but it appears in the fingerprint region. If the spectrum of benzene is compared with that of *sec*-butylbenzene (Figure 14.21), the spectra are very similar, but there are a few differences. The normal C–H region is more complex because there is a methyl group in addition to the aromatic C–H absorptions. The overtone region is slightly more complex, but it is difficult to pick out a specific pattern without "blowing up" that region of the spectrum. The most notable difference is the two large signals at about 680 and 740 cm^{-1}. These are characteristic of a monosubstituted benzene ring; in fact, signals in this region are used to distinguish monosubstituted, disubstituted, and trisubstituted benzene derivatives. Those absorption peaks are not included in this discussion.

14.4 Nuclear Magnetic Resonance Spectroscopy

Modern organic chemists rely on nuclear magnetic resonance spectroscopy for day-to-day identification of organic molecules. It is a powerful technique that allows one to "count" protons, identify the chemical environment (actually, the magnetic environment) of different protons, and predict how many neighbors each proton has. With this information, as well as the empirical formula and functional group information obtained from mass spectrometry and infrared spectroscopy, determination of the structure of organic molecules is usually feasible.

14.4.1 The Hydrogen Atom Is a Magnet

For all practical purposes, a hydrogen nucleus is a charged particle (a proton) possessed of a property called "spin." As with any spinning charge, an intrinsic magnetic field is generated, as shown in the Figure 14.22. The charge

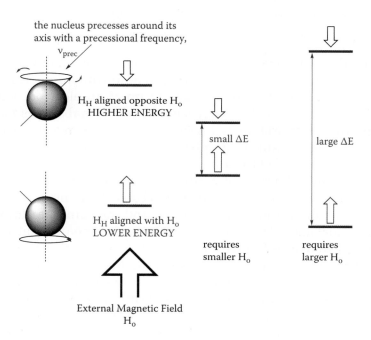

Figure 14.22 The energy gap generated when a spinning proton changes its spin state relative to an external magnetic field, H_o.

is small, so the magnitude of the magnetic field generated by the proton is small. The proton spins off-axis, and it precesses around the axis at a certain frequency—the *precessional frequency*. This information becomes important if a proton—a weak magnet—is placed into a large external magnetic field represented by H_o.

As with any two magnets, the large magnet will influence the smaller magnet, and the field of the smaller magnet will orient itself relative to the larger magnet (H_o). Two nonrandom orientations are possible: The proton magnetic field H_H may be aligned with H_o or opposed to H_o (see Figure 14.22). If H_H is aligned with H_o, this will be a lower energy arrangement than when H_H is opposed to H_o. This constitutes an energy gap (ΔE) between the higher energy state and the lower energy state. If energy equal to ΔE is applied to the proton, which normally resides in the lower energy state, it "flips spin states." This means that *the proton magnetic field changes from aligned to opposed (low energy to high energy)*. Absorption of energy ΔE occurs only at a particular magnetic field and when this occurs they are said to be in *resonance*. If the magnetic field increases or decreases, ΔE changes, as shown in Figure 14.22. The orientation of molecules in a magnetic field is statistical, and only a few molecules must change their spin to give rise to an absorbance signal.

If there is a second hydrogen atom, as in a molecule, that proton is also a weak magnet, and it will generate a slightly different magnetic field from the first proton (H_H and $H_{H'}$). If each proton generates a different magnetic field, the fields will have different strengths. Therefore, a different ΔE is required for resonance with H_H than with $H_{H'}$. This is the fundamental basis for distinguishing different types of hydrogens in proton NMR.

14.23 Determine all different types of hydrogen atoms in 2-butanone.

In the opening paragraph a proton is said to have the property of spin. General chemistry usually discusses the spin quantum number, I, which is the orientation of the intrinsic angular momentum of an electron, with values of $+\frac{1}{2}$ or $-\frac{1}{2}$. The hydrogen nucleus may assume additional values for I. An atom with an even number of protons or an even number of neutrons will not exhibit spin. That is, the spin quantum number will be zero. If an element has an odd number of protons or an odd number of neutrons, then I will have a positive value. If there is one odd proton or neutron, I = 1/2; with two odd particles, I = 1; with three odd particles, I = 3/2, and so on. Table 14.4 shows several common nuclei, the number of protons and neutrons, and the spin quantum number.

The proton nucleus (^1H) has a spin of 1/2, which means that there are two energy levels, as shown in Figure 14.23. Absorption of the proper energy leads to transition from the lower energy level to the higher, so there is *one signal per absorption*. When I = 1, however, there are three possible energy levels, or orientations, as shown. Indeed, the **number of orientations = 2I + 1**, where I is the spin quantum number. Three orientations allow three possible transitions (Figure 14.23) and three signals per absorption. In proton NMR, each distinct type of hydrogen will give rise to one signal, and the spectrum will be relatively easy to interpret. Deuterium (^2H) shows three signals per

magnetic field is required to make the proton come into resonance for a given v_{rf}. One way to describe this effect is to say that the electrons surrounding the nucleus **shield** it and require a larger value for H. Such a nucleus (here a proton) is said to be shielded. ***The greater the electron density around the proton, the more shielded it will be, requiring a larger value of H***. If the proton has less electron density around it, it is less shielded (***deshielded***) and a smaller value of H will be required. ***Shielding and deshielding are easily correlated with electron releasing and electron withdrawing groups***.

An electron-releasing group will shield the proton, making it absorb at a higher magnetic field. Conversely, an electron-withdrawing group will deshield the proton, making it absorb at a lower magnetic field. High field and low field are marked on the NMR spectrum in Figure 14.25. High field indicates that a larger magnetic field is required for the proton to come into resonance, and low field indicates that a small magnetic field is required. The region near the zero point (TMS) is at high field (more shielded; greater magnetic field strength), whereas the further the signal appears to the left of TMS, the less shielded it will be (it is deshielded) and it will absorb at a lesser magnetic field strength. The statements are summarized as follows:

1. **Sigma electrons shield the nucleus and the proton absorbs at higher field.**

   Alkyl groups are electron donating. Methyl groups usually absorb at 0.9 ppm; $-CH_2-$ groups (methylene) are found at about 1.1–1.3 ppm and methine protons ($-CH-$) are found at about 1.5 ppm. Virtually all protons connected to a carbon bearing a heteroatom or functional group will resonate *downfield* of these signals.

2. **Electronegative atoms deshield the nucleus and the proton absorbs at low field (downfield).**

   When the proton is connected to a carbon connected to oxygen, nitrogen, halogen, or sulfur, the bond polarization is such that the proton has less electron density and it is deshielded. That proton will appear downfield relative to methyl, methylene, and methine of an alkane. The more polarized the C–X bond is, the further downfield the proton will appear.

3. **Functional groups that are classified as electron withdrawing will deshield the nucleus and the proton will absorb at low field.**

   Cyano groups, carbonyl-bearing functional groups, and nitro groups are all electron withdrawing because they have a δ+ atom connected to the carbon bearing the proton of interest. This pulls electron density away from the proton and deshields it, and that signal will appear at lower field.

Figure 14.27 shows the position of protons connected to various functional groups in the proton NMR. The carbonyl group shifts the signal downfield to about 2.1 ppm relative to methyl, and oxygen shifts the signal to about 3.3 ppm relative to methyl. Oxygen and the halogens also shift the signal downfield by similar amounts, and it can be difficult to distinguish these signals using only

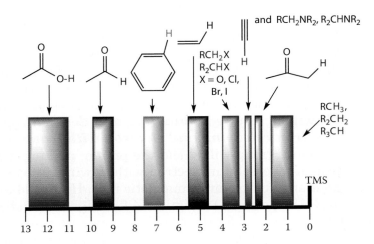

Figure 14.27 General absorption patterns for protons attached to functional groups in the proton NMR.

proton NMR. Figure 14.27 also shows the signals from protons connected to alkene units (C=C), benzene rings, the aldehyde proton, and the acidic proton of a carboxylic acid.

Note that these latter signals are far downfield. The C=C unit is clearly not as polarized as the carbonyl (C=O), so **why are the alkene protons further downfield relative to the proton on a H–C–C=O unit?** These are the α-protons on a ketone or an aldehyde—not the aldehyde hydrogen, O=C–H. Note that the alkyne proton is upfield of the alkene. **Why should this be so?** To answer these questions, something other than simple electron release and withdrawal is at work. The functional groups in question all contain π-electrons, and the answers to the questions are found in what constitutes a fourth rule:

4. **Pi electrons have spin and therefore generate a magnetic field that opposes H$_o$.**
   Just as σ-electrons generate a secondary field (have a magnetic moment), π-electrons also generate a secondary magnetic field that influences H$_o$. The greater the concentration of π-electrons is, the greater will be the secondary magnetic field.

If ethylene (ethene) is examined, the π-electrons generate a secondary magnetic field, as shown in Figure 14.28, which opposes H$_o$ in the region of the π-bond. Although it may not be obvious, it is true that the field generated by the more polarizable π-electrons is greater than that of the sigma electrons, so the π-secondary field dominates the interaction with H$_o$. In other words, the dominance of the field of the π-electrons leads to a nonrandom orientation of the planar ethylene molecule such that the carbon and hydrogen atoms are perpendicular to H$_o$. Interestingly, the protons connected to the π-bond are not in the region of the secondary field in opposition to H$_o$, but rather in that

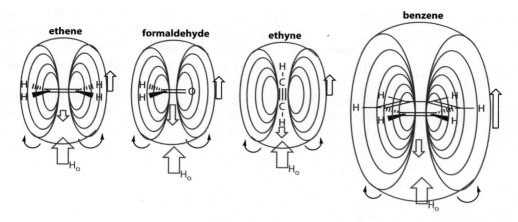

Figure 14.28 Magnetic anisotropy for ethene, ethyne, formaldehyde, and benzene.

region aligned with H_o. *This means that H_π augments H_o and H for the nucleus requires less of a magnetic field [H = H_o − H_π], so that signal will appear downfield (lower field).* In fact, protons connected to a C=C unit appear between 5 and 6 ppm in most cases and the effect of the secondary field generated by the π-electrons is responsible for this downfield shift. The protons are deshielded by the π-electrons. This phenomenon is called ***magnetic anisotropy***.

A proton connected to a benzene ring appears at about 7.1 ppm, according to Figure 14.27. Six π-electrons set up the secondary field shown in Figure 14.28 and, again, the interaction of the π-electrons orients the benzene molecule such that all atoms are perpendicular to H_o. The protons on the benzene ring are deshielded, but the secondary field is larger due to the presence of six electrons rather than two. The proton resonates even further downfield when compared with the proton attached to the C=C unit of an alkene.

14.27 Comment on whether or not the protons attached to a pyrrole ring are deshielded due to magnetic anisotropy.

A proton connected directly to a carbonyl constitutes an aldehyde functional group. This proton resonates at about 9 or 10 ppm, far downfield from the alkene proton. The magnetic anisotropy for the carbonyl in formaldehyde is shown in Figure 14.28, but the two π-electrons in a carbonyl will not exhibit a significantly larger secondary field than the two π-electrons of a C=C unit. The aldehyde proton is far downfield because ***the carbonyl is also polarized, whereas the C=C unit is not***. In other words, the anisotropy effect of the π-electrons and the electron-withdrawing properties of the carbonyl operate independently, and the effects are additive to deshield the proton even more.

The signal for the carboxyl proton is about 11–15 ppm according to Figure 14.27. That proton is not connected directly to the carbonyl, and the effect of magnetic anisotropy should be minimal. Certainly the OH bond is polarized, but so is the OH of an alcohol and that appears between 1 and 5 ppm. The carboxyl proton appears so far downfield because it is part of a carboxylic acid and the acidic proton has a large δ+ hydrogen atom. In addition, it is extensively hydrogen bonded, which increases the bond polarization and makes the proton even more deshielded. These effects combine to create a very deshielded proton that resonates far downfield. The more hydrogen bonded the carboxyl proton is, the further downfield it will appear. In general, the range is 11–13 ppm.

A terminal alkyne proton resonates at about 2.3 ppm. It is upfield of the alkene protons, although there are two π-bonds rather than one. The two π-bonds are orthogonal, one to the other; all four electrons (both π-bonds) must interact with the external magnetic field H_o to generate the secondary magnetic field. Both π-bonds are involved only when the molecule is oriented as shown in Figure 14.28, and the secondary magnetic field opposes H_o along the line of the H–C–C–H bonds. The second π-bond in an alkyne therefore changes the orientation of the molecule to about 90° relative to the way an alkene orients with only one π-bond. The proton of the terminal alkyne is in the shielding portion of the secondary field and *upfield relative to the alkene.*

This section shows that chemical shift is used to estimate the proximity of a proton to a heteroatom or a functional group. For purposes of identifying an unknown, the ability to correlate a proton signal with its attachment or proximity to a functional group is an important piece of information. This concept is elaborated upon in Section 14.5. With the chemical shift explanations in hand, Table 14.5 gives the approximate chemical shifts for hydrogen atoms attached to various functional groups. This chemical shift information will be applied to structure proof in later sections.

14.4.4 Peak Areas and Integration: Counting Hydrogen Ratios

Figure 14.29 shows a proton NMR spectrum of a compound that has four signals representing four different types of hydrogens. If the number of hydrogen atoms represented by each absorption peak is known, this effectively *counts* the number of different kinds of protons. It is possible to integrate the area under each peak, and this number should be proportional to the number of hydrogen atoms that contribute to that absorption peak. The larger the peak area is, the greater the number of hydrogen atoms will be. An exact proton count is impossible unless the total number of hydrogens in the molecule is known; if there is only one peak, there is no clue as to how many hydrogen atoms that peak represents. If there is more than one peak, however, the ratio of the peak areas may be taken and used in conjunction with the empirical formula (obtained via mass spectrometry; see Section 14.2.4) to obtain a reasonable count of the number of different types of hydrogen atoms.

There are horizontal line traces along with the proton peaks in Figure 14.29. At each peak, the line trace rises to a new plateau. If the height of these traces is measured, beginning where the peak starts to rise and ending where the peak levels out again, a number is obtained that is not the peak area, but rather correlates with the peak area. This is the peak integration. Modern instruments report the integration in digital form. In Figure 14.29, peak A

Table 14.5
Proton NMR Spectroscopy Correlation

Proton NMR Chemical Shifts			
cyclopropane	0.2	primary, RCH_3	0.9
secondary, R_2CH_2	1.3	tertiary, R_3CH	1.5
vinylic, C=C-H	3.5–5.9	alkynyl, C≡C-H	2.0–3.0
aromatic, Ar-H	6.0–8.5	benzylic, Ar-C-H	2.2–3.0
allylic, C=C-C-H	1.7	fluorides, F-C-H	4.0–4.5
chlorides, Cl-C-H	3.0–4.0	bromides, Br-C-H	2.5–4.0
iodides, I-C-H	2.0–4.0	alcohols, HO-C-H	3.4–4.0
		α-H of alcohols	
ethers, C-O-C-H	3.3–4.0	esters, RCOO-C-H	3.7–4.1
		H on carbon of alcohol unit	
carboxylic acids, HOOC-C-H	2.0–2.2	esters, ROOC-C-H	3.7–4.1
α-H of acids		α-H of esters	
carbonyls, O=C-C-H	2.0–2.7	aldehydes, O=C-H	9.0–10.3
α-H of aldehydes & ketones		aldehyde proton	
hydroxyl, O-H	1.0–5.5	phenols, ArO-H	4.0–12.0
enols, C=C-O-H	15.0–17.0	carboxylic acids, RCOO-H	10.5–15.0
		acidic proton of the OH	
amino, R-N-H	1.0–5.0	amines, N-C-H	2.5
proton on nitrogen		α-proton of amines	

Methyl Signals for Common Fragments

To determine signals for functional groups to a methylene, add 0.4 to these numbers
To determine signals for functional groups to a methine, add 0.6 to these numbers

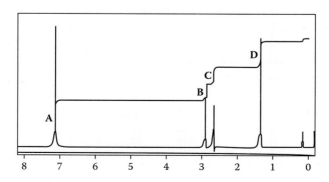

Figure 14.29 A typical proton NMR spectrum with four different signals representing four different kinds of protons.

integrates for 26 mm, peak B for 9.2 mm, peak C for 9.1 mm, and peak D for 9.7 mm. If all numbers are divided by the smallest integration (9.1 units), the ratio of A:B:C:D = 2.86:1:1:1.6.

There are no fractional hydrogen atoms, so the numbers are multiplied by an integer that will give close to whole numbers. Multiplication by 2 leads to a ratio of 5.7:2:2:3.2. The ratio may be 5:2:2:3 or 6:2:2:3. The mass spectral data will provide a formula of $C_{10}H_{12}O$, so the ratio must be 5:2:2:3. This ratio suggests 5+2+2+3 hydrogen atoms, which is H_{12} and corresponds to the exact number of hydrogen atoms in this particular molecule. This is a "real life" example, so the ***integration does not always give us the exact integer, making comparison with the mass spectrum empirical formula essential***.

This example also points out that the ratio may be any whole number multiple of 5:2:2:3 (any multiple of 12 hydrogens). Therefore, 10:4:4:6 (24 H), 15:6:6:9 (36 H), etc. must also be considered, but this information is tempered by knowledge of the empirical formula from mass spectrometry. If the empirical formula is $C_{20}H_{24}O_2$, for example, then the integration for this molecule is 10:4:4:6. The first goal is to establish the integration ratio and then compare it with the empirical formula to be certain that the number of hydrogen atoms match.

14.28 For a formula of $C_9H_{10}O_2$, the proton NMR spectrum gives three signals with the following integration values: 71.5:28.6:42.9. Determine the relative ratio of these three signals.

14.4.5 Neighboring Protons Influence the Magnetic Field: Multiplicity

Returning to the example in Figure 14.29, four signals correspond to four different kinds of hydrogen atoms. In this figure, these four signals are singlets (single peaks). Three different signals in Figure 14.30 correspond to three different kinds of hydrogen atoms (see structure **21**). Each signal is a cluster of peaks rather than a singlet. Why? Signal A is three peaks rather than one, signal B is four peaks rather than one, and signal C is one broadened peak. Figure 14.30 shows the ratio of peaks obtained by integrating the peak areas,

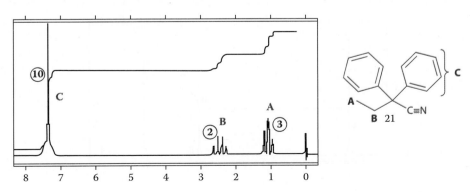

Figure 14.30 Proton NMR spectrum of 2,2-diphenylbutanenitrile.

and the ratio of A:B:C is 1.5:1:5; if the empirical formula is $C_{16}H_{15}N$, then the ratio of A:B:C must be 3:2:10. Signal A integrates for three protons, signal B for two protons, and signal C for ten protons.

Why are several peaks associated with each signal? Signal A is called a *triplet* that integrates to three protons, signal B is called a *quartet* that integrates to two protons, and signal C is one peak (a *singlet*) that integrates to 10 protons. ***Each of the peak clusters corresponds to one signal (one type of hydrogen), but more than one peak is present for each different hydrogen.*** To understand why there are multiple peaks per signal, examine structure **21** (2,2-diphenylbutanenitrile). Hydrogen atoms A are attached to a carbon that is also connected to the carbon-bearing hydrogen atoms B. Protons H_A and H_B are separated from each other by three covalent bonds. In other words, *protons A and B are on adjacent carbon atoms* and, ***when two protons are on adjacent carbon atoms separated by three bonds they are called neighbors.*** When hydrogen atoms are neighbors, the magnetic field of one will influence the magnetic field of the other.

14.29 Identify those protons that are neighbors in 2,5-dimethyl-3-hexanone.

Why do protons on neighboring carbon atoms lead to multiple peaks? Remember that H_A functions as a small magnet in the presence of the external field (H_o), but H_B is also a small magnet. If H_A and H_B are close enough, it is reasonable that the magnetic field exerted by H_B may influence that of H_A. Adjacent spin systems either oppose or reinforce the applied field. This means that the H_A field may be aligned or opposed to the field of H_B, in addition to its interaction with H_o. If H_A leads to a signal in an NMR spectrum due to H_o, then that signal will be "split" into two signals by its interaction with H_B. Indeed, *every "neighbor" will split each signal into two signals (a doublet).*

If there are two identical neighbors, then H_A is split into three signals (a *triplet*), and if there are three neighbors, H_A is split into four signals (a *quartet*). This is observed in Figure 14.30, where H_A has two neighbors and appears as a triplet. Its neighbor is H_B (there are two identical H_B protons). Similarly, H_A is the neighbor to H_B and there are three identical H_A protons. In other words, H_B has three identical neighbors and H_B appears as a quartet.

In simple cases where all the neighbors are identical, a proton with "n" neighbors will appear as "n+1" peaks. **Molecules that exhibit this behavior are said to have first-order spectra.** Therefore, a proton with one neighbor resonates as a *doublet,* two neighbors lead to a *triplet,* three neighbors lead to a *quartet,* and four neighbors lead to a *pentet.* If a proton has no neighbors, it will appear as a *singlet* (one single peak). This splitting is illustrated in Figure 14.31. Pascal's triangle is used to estimate how the peaks should appear. As the signals split, they split *symmetrically* so that the "inner" peaks overlap. This leads to a 1:2:1 signal for a triplet, a 1:3:3:1 signal for a quartet, a 1:4:6:4:1 signal for a pentet, etc. These are very characteristic and give important structural clues, as will be seen later. When there are several peaks in a signal and the number is unknown or cannot be determined, such peaks are called *muliplets.*

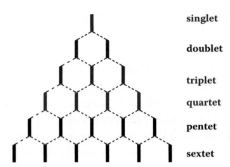

Figure 14.31   Pascal's triangle, used in conjunction with the n + 1 rule.

14.30 What is the multiplicity of the –2– group in ethyl methyl ether, of the CH$_2$ group in propane, and of the CH$_3$ group in ethanol?

Returning to Figure 14.30, there is an equidistant separation of the three peaks of the triplet for H$_A$ and the four peaks for H$_B$ are also separated by the same distance. The separation between peaks is measured in hertz, and this distance is called the ***coupling constant (given the symbol J)***. For Figure 14.30, there are two neighboring protons (H$_A$ and H$_B$) and the coupling constant for these two neighbors is J$_{AB}$. Typically, the value of J ranges from close to 0 to 8–10 Hz. The value of J depends on the dihedral angle of separation between the two neighbors' protons. In Figure 14.32, curve (a) shows the H–H vicinal coupling constants as a function of dihedral angle of the C–H bond, and curve (b) is the coupling constant calculated with the cos^2 function shown, which is known as the Karplus–Conroy equation. This chart is used to estimate the value of J for a given system, where the experimental values usually lie between the two curves a and b[6,7]:

$$J = a \cos^2 \Phi - 0.28$$

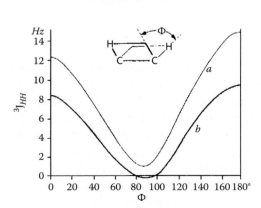

Figure 14.32   Vicinal coupling constants as a function of the dihedral angle (Φ) of the CH bonds. Breitmaier, E.; Wade, J. *Structure Elucidation by NMR in Organic Chemistry: A Practical Guide*, Wiley, Chichester, 1993, p. 43, Figure 2.16. Copyright Wiley-VCH Verlag GmbH & Co. KGaA. Reproduced with permission.

The two curves in Figure 14.32 are obtained from $a = 8.5$ for Φ up to 90°, whereas $a = 9.5$ for Φ above 90°.[7] The equation is named in honor of Martin Karplus (Austria/United States; 1930–) and Harold Conroy (United States).

In the cases cited, a dihedral angle of 10° corresponds to a J of about 7–8 Hz, whereas a dihedral angle of 60° corresponds to J of about 1.5–2 Hz. Note that **when the *dihedral angle is zero, there is a coupling constant of about 8 Hz* whereas *a dihedral angle of 90° shows a coupling constant close to zero*.** Returning to Figure 14.30, the fact that J_{AB} is the same for the H_A signal and also the H_B signal indicates that they are neighbors (their **spins are coupled**). This information is for structure determination.

14.31 Estimate the coupling constant for a H–C–C–H unit when the angle ϕ is 43°.

The neighbor effect described is actually called **spin–spin coupling**. *Assume that if protons are separated by four or more bonds, there is no neighbor effect (no coupling, and thus very little splitting).* This is *incorrect because some protons separated by four or more bonds may couple due to their spatial position in the molecule. Such interactions are known as long range coupling, but will observed only occasionally in this book.* **In addition, if protons are separated by a heteroatom such as O, N, or S, then assume there is no neighbor effect.** The coupling constant and knowing that two signals are coupled is one of the most important pieces of information obtained from the proton NMR.

With a knowledge of the chemical shift for a given signal and the integration for that signal, knowing the number of neighbors suggests a structure. For example, in Figure 14.33, the triplet–quartet signal for two types of protons

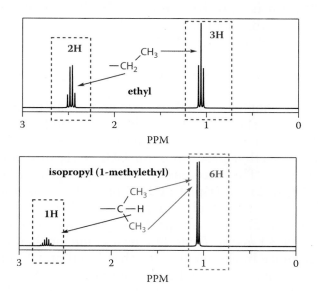

Figure 14.33 Ethyl and isopropyl patterns in the NMR.

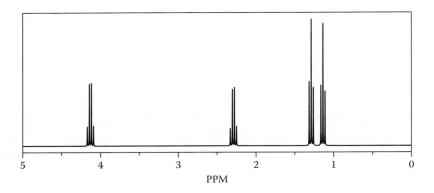

Figure 14.34 Proton NMR spectrum of ethyl propanoate.

integrates to 2:3. This is a typical pattern for **an ethyl fragment (–CH$_2$CH$_3$).** If this pattern appears in an unknown NMR, immediately think "ethyl." This and another common pattern (isopropyl) are shown in Figure 14.33. **A singlet that integrates to three hydrogens is usually a methyl group that has no neighbors. A singlet that integrates to six hydrogens is usually a *gem*-dimethyl unit such as CH$_3$–C–CH$_3$,** but a **doublet that integrates to six hydrogens is usually an isopropyl unit (CH$_3$–CH–CH$_3$).**

Indeed, a typical isopropyl unit will show a doublet that integrates to six hydrogens at about 1–1.2 ppm and a signal downfield split into at least seven peaks (a septet). If there are two signals that integrate to two hydrogens each and there are two triplets, this is characteristic of a –CH$_2$–CH$_2$– unit. **A *tert*-butyl group will appear as a singlet that integrates to nine hydrogens at about 0.9–1.0 ppm. A broad singlet that integrates to five hydrogens at about 7.0–7.2 ppm is usually a monosubstituted phenyl group.** With these few examples, a great deal of structural information is available, and this information will be used in Section 14.5.

14.32 What is the multiplicity for the CH$_2$ units in 1,2-diphenylethane?

Before predicting chemical structure, two points can be made; one concerns chemical shift and the other concerns coupling constants. Figure 14.34 shows the proton NMR spectrum of ethyl propanoate. There are two ethyl groups in this molecule. One ethyl group has an OCH$_2$CH$_3$ unit and the other has an O=C–CH$_2$CH$_3$ unit. They are easily distinguished by ***chemical shift***. A quartet is at 4.13 ppm and a triplet at 1.26 ppm. There is another quartet at 2.32 ppm and a second triplet at 1.14 ppm. The downfield shift for protons proximal to an oxygen places them past 3.5 ppm, whereas protons next to a carbonyl appear past 2.1 (using Table 14.5 as a guide for chemical shift). Simply based on this difference in chemical shift, the signal at 4.13 ppm is associated with the OCH$_2$ unit, and the methyl group of the OCH$_2$CH$_3$ unit is assigned to 1.26 ppm. Obviously, the methylene attached to the carbonyl appears at 2.32 and the methyl of the O=C–CH$_2$CH$_3$ group resonates at 1.14 ppm. Using the

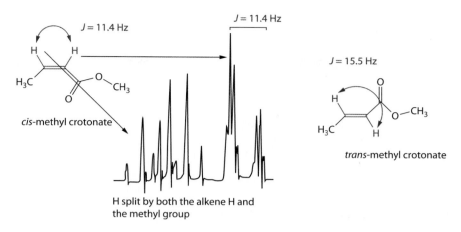

Figure 14.35 Coupling constants for *cis*- and *trans*-methyl crotonate. (Fraser, R. R., and McGreer, D. E. 1961. *Canadian Journal of Chemistry* 39:505–509. Copyright 1961 NRC Canada or its licensors. Reproduced with permission.)

chemical shift information in Table 14.5, there is a clear difference in chemical shift that allows easy identification.

Another structural example is *cis*- versus *trans*-methyl crotonate, shown in Figure 14.35.[8] Distinguishing between *cis*- and *trans*-methyl crotonate is easily done by examining the coupling constants for the two protons on the C=C unit. The coupling constant for *cis*-methyl crotonate is 11.4 Hz for the C=C protons, and the coupling constant for the C=C protons in *trans*-methyl crotonate is 15.5 Hz. If the coupling constants are measured, those values can be used to distinguish these isomeric compounds.

Figure 14.35 shows the alkene protons for *cis*-methyl crotonate. There is also coupling between the terminal methyl and the proton on C3 (measured to be 7.0 Hz), although the protons are four bonds apart. This type of coupling (mentioned before) is called **long-range coupling** and may be used for structural identification. There is also a very small coupling constant between the protons on the methyl group and the proton on C3. Note that each of the two peaks is split into a quartet with a tiny coupling constant. Long-range coupling is shown here to illustrate that more information is often available in the NMR spectrum than is discussed in this introduction. Coupling constants are an invaluable tool for structure determination; however, a full discussion for all types of molecules found in various chapters is beyond the scope of this book, so *J* values will be used in homework problems only occasionally.

A few coupling constants are useful for structure determination in relatively simple molecules. For the C=C unit of an alkene, the *trans* protons have a relatively large coupling constant of $J = 12$–18 Hz, whereas the *cis* protons show a coupling constant of $J = 6$–12 Hz, as shown in Figure 14.36.[9] This means that the *cis* or *trans* relationship of the protons on a C=C unit is relatively easy to ascertain from the coupling constants, as in Figure 14.35. For an unknown alkene with J = 17 Hz for the C=C protons, that molecule will have a *trans* relationship for those protons. Note that the *geminal* protons have a small value for J.

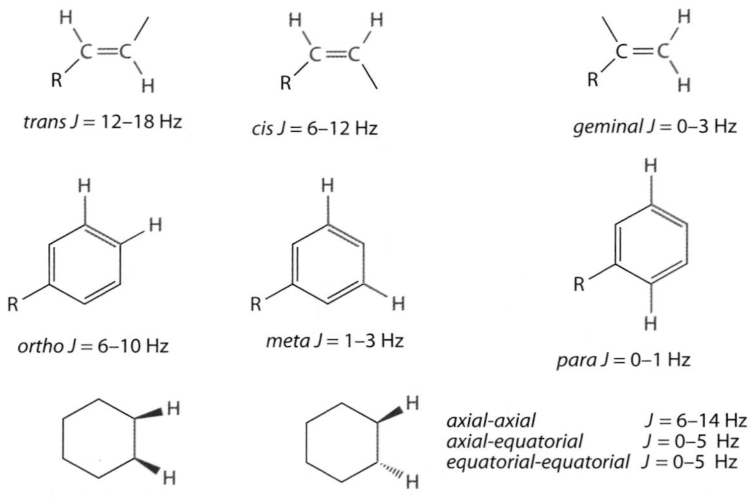

Figure 14.36 Typical coupling constants for simple proton systems.

The aromatic region of a molecule may be quite complicated due to the coupling of the *ortho, meta,* and *para* protons. Modern instruments can usually distinguish these coupling constants, and the *ortho* protons have a relatively large value. Knowledge of the relationship of the protons on an aromatic ring helps identify the position of substituents as to an *ortho, meta,* or *para* relationship, for example. A molecule with *para* substituents will have only *ortho* coupled protons, whereas a *meta* relationship will show one uncoupled proton as well as two *ortho* couplings and one *meta* coupling. The analysis may be rather elaborate when the substituents are different, and it is not always easy to identify the relationship of these protons.

Finally, cyclohexane derivatives show different coupling constants for protons that are axial or equatorial, as indicated in Figure 14.36. Analysis of structure using coupling constants is a powerful tool, when such data are available. As indicated previously, coupling constant data will be provided occasionally with homework problems, but only for relatively simple cases (*cis-, trans-*alkenes, *cis-, trans-*cyclohexanes, and some substituent patterns on benzene rings).

14.4.6 Predicting Chemical Structure by Proton NMR Spectroscopy

Once the fundamentals of proton NMR spectroscopy are known, the next step is to use this information to predict what the NMR spectrum should look like for a given structure. It is difficult to translate chemical shift, integration, and multiplicity for an unknown spectrum into a structure if there is no notion of what the spectrum for a real molecule should look like. Therefore, in order to predict the proton NMR spectrum, a few examples are presented of known structures, beginning with 2-butanone.

The presence of a single C≡C unit indicates an alkyne. Molecules that have only one oxygen and no other heteroatoms may have only one of four functional groups. Alcohols, ethers, ketones, and aldehydes are the only possibilities, and this simplifies the structural analysis quite a bit. Similarly, a simple molecule with one nitrogen and no other heteroatoms is probably a nitrile or an amine, but if it is an amine it can be primary, secondary, or tertiary.

Structural information is available from the "index of hydrogen deficiency." This is essentially an analysis based on the alkane generic formula; there are two types: one for molecules without nitrogen and those for molecules containing nitrogen. The formulas are

$$\Omega_{noN} = \frac{2\#C + 2 - \#H - \#X}{2} \qquad \Omega_N = \frac{2\#C + 3 - \#H - \#X}{2}$$

If these formulas are renamed the "number of rings and/or π-bonds" in a molecule, a value of one means one ring *or* one π-bond. If there is no oxygen or nitrogen, the molecule may be a cyclic alkane or it may be the π-bond of an acyclic alkene. If there is one oxygen, then a value of one could indicate the π-bond of a C=O unit (a carbonyl). A value of two could mean two C=C units, two C=O units, a C=C and a C=O unit, two rings, or a ring with a C=C or a C=O. A value of two could also mean the presence of a triple bond, either C≡C or C≡N. A value of four usually indicates a benzene ring, for example, and a value of eight is consistent with two benzene rings. Knowledge of the number of rings or π-bonds may narrow the choices for a structure when used in conjunction with the IR and NMR data. As the structure becomes more complex, the value of this parameter is consistent with many more possibilities.

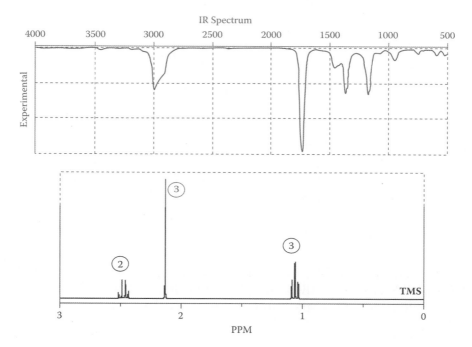

Compound 14.5.1

This problem is analyzed by calculating the empirical formula from the mass spectral data. The M+1 peak indicates the presence of four carbons, and M+2 indicates one oxygen. The difference of $72 - C_4O = 72 - 64 = 8$, so the formula is C_4H_8O. The rings and π-bonds calculation generates a value of one. The infrared spectrum clearly shows the presence of a carbonyl at about 1740 cm^{-1}, which accounts for the double bond, and there is no sign of the aldehyde CHO. The proton NMR spectrum does not show an aldehyde proton at 9 or 10 ppm, so the compound is a ketone. The NMR spectrum shows a singlet that integrates to three protons (a methyl group) and the chemical shift indicates that it is connected to the carbonyl. There is another methyl group that appears as a triplet (two neighbors) and the quartet at 2.4 ppm is the other part of an ethyl signal, with the $-CH_2-$ connected to the carbonyl. These fragments lead to the structure of this compound, 2-butanone.

14.33 **Show the calculations that convert the M, M+1, and M+2 data into the formula for compound 14.5.1 and then show the calculation for the number of rings or π-bonds.**

Compound 14.5.2

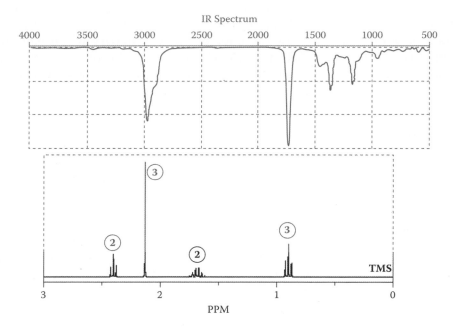

The exact mass is 86.0727, and the possibilities include $C_3H_2O_3$ (86.0003), $C_3H_4NO_2$ (86.0242), $C_4H_6O_2$ (86.0368), C_4H_8NO (86.0606), $C_5H_{10}O$ (86.0732), and C_6H_{14} (86.1096). The mass spectrum of compound 14.5.2 has a formula C_5H_8O and there is one ring or π-bond. The infrared spectrum shows the presence of a carbonyl but no aldehyde CH. The NMR spectrum does not contain the aldehyde proton, so this is a ketone. As with the previous example, there is a singlet that integrates to three protons at about 2.1 ppm, indicating a methyl group attached to the carbonyl. There is a methyl group that appears as a triplet, so it has two neighbors. There is no quartet, however—only a triplet that integrates to two protons at about 2.4, indicating a $-CH_2-C=O$ unit; this methylene is attached to the methyl group. A multiplet at about 1.6 ppm integrates to two protons, and this is a $-CH_2-$ unit that connects the methylene and the methyl. This constitutes a propyl group attached to the carbonyl and the compound is 2-pentanone.

Compound 14.5.3

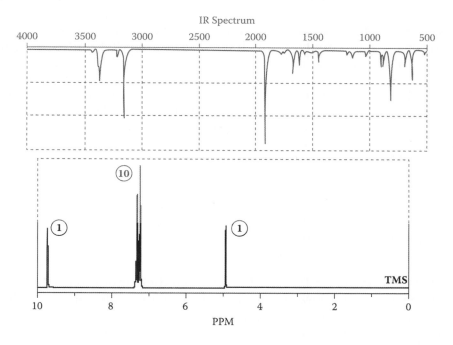

The formula for compound 14.5.3 is $C_{14}H_{12}O$ and it has nine rings or π-bonds. It is likely there are two benzene rings to give such a high number. The infrared spectrum shows a carbonyl signal at about 1710 cm^{-1} and a C–H signal at 2815 cm^{-1}, which indicates an aldehyde. This is confirmed by the signal at 9.9 ppm in the NMR. The infrared spectrum shows peaks at about 760 and 700 cm^{-1} that indicate benzene rings, and the peak 7.3 ppm in the NMR spectrum confirms this identification. The integration of 10 protons shows that there are two monosubstituted benzene rings. The remaining signal in the NMR spectrum is a singlet that integrates to one proton at about 4.8 ppm, indicating that it has n neighbors and is close to more than one functional group. The fragments are two Ph units—a CH unit and a CHO. The only way to combine these fragments leads to diphenylacetaldehyde as the structure for this compound.

Compound 14.5.4

MS: M (73) 100%, M+1 (74) 4.81%, M+2 (75) 0.10%
IR: 3225, 3279, 2941–2817, 1639, 1460, 1058, 952, 840, 763 cm^{-1}
^{1}H NMR: 2.55 (d, 2H), 1.6 (m, 1H), 1.1 (s, 2H; this peak is diminished when treated with D_2O), 0.9 (d, 6H)

The formula for compound 14.5.4 is $C_4H_{11}N$; note the odd mass for the parent that suggests the presence of an odd number of nitrogen atoms. There are no rings or π-bonds for this formula. The infrared spectrum shows a doublet at around 3250 cm^{-1} in the infrared, which is indicative of a primary amine ($-NH_2$). The NMR spectrum shows a signal at 0.9 ppm that integrates to six protons and is a doublet. It is likely a *gem*-dimethyl unit. The doublet at 2.55 ppm is a doublet (one neighbor) and its chemical shift indicates that it is attached to the nitrogen ($-CH_2-NH_2$). The signal at about 1.1 ppm is diminished upon addition of D_2O; it is likely due to the protons on the nitrogen and integrates to two protons. The *gem*-dimethyl and the multiplet at 1.6 ppm suggest an isopropyl unit, and combining $CHMe_2$ and $-CH_2-NH_2$ gives compound, 2-methyl-1-aminopropane.

Compound 14.5.5

MS: M (102) 100%, M+1 (103) 5.55%, M+2 (104) 0.55
IR: 2980–2850, 1724, 1205, 1143, 1010 cm^{-1}
^{1}H: 3.9 (s, 3H), 2.6 (m, 1H) 1.1 (d, 6H) ppm

The formula for compound 14.5.5 is $C_5H_{10}O_2$ and there is only one ring or π-bond. The infrared spectrum shows a carbonyl group but no aldehyde peaks. There is no COOH proton and there is no indication of an acid chloride, an anhydride, or an amide. This is either a ketone or an ester, although it may also be an ether–ketone. The doublet at 1.1 ppm that integrates to six protons (a *gem*-dimethyl unit) combined with the multiplet at 2.6 ppm that integrates to one proton indicates the presence of an isopropyl unit ($CHMe_2$). The chemical shift of the signal at 2.6 suggests a CH unit attached to a carbonyl. There is a singlet that integrates to three protons at 3.9 ppm. This is a methyl group and its chemical shift is consistent with attachment to an oxygen, so it appears this is a methyl ester ($-COOCH_3$). Combining the isopropyl unit attached to a carbonyl via the CH unit and the COOMe unit gives methyl 2-methylpropanoate as the structure for this compound.

Compound 14.5.6

M = 116, 100%; M+1 = 117, 6.66%; M+2 = 118, 0.62%

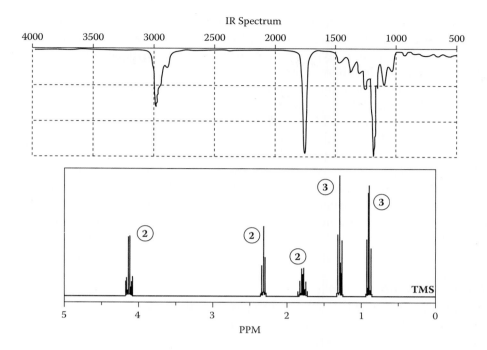

Compound 14.5.6 has a formula of $C_6H_{12}O_2$ and it has only one ring or π-bond. The infrared spectrum indicates a carbonyl, but it is not an aldehyde, an acid, an acid chloride, an anhydride, or an amide. Therefore, it is likely a ketone or an ester. The NMR spectrum has more peaks than the previous example, but the quartet that integrates to two protons at 4.1 ppm suggests a methylene attached to an oxygen (–CH_2–O), so this probably is an ester. There are two triplets that integrate to three protons and one of them is coupled to the quartet at 4.1, so there is a CH_3CH_2O unit (an ethyl ester). A triplet integrates to two protons at 2.35 ppm, suggesting a methylene attached to a carbonyl. The methyl group is probably attached to a methylene, which should give a quartet, but the signal at 1.8 ppm is a multiplet. These three signals are consistent with a propyl unit. If the propyl unit is attached to the carbonyl via the –CH_2– and combined with the OEt unit, ethyl butanoate is the structure.

Compound 14.5.7

MS: exact mass of the molecular ion = 68.0626
IR: 3220, 2985–2778, 1635, 1449, 1031, 990, 694 cm^{-1}
^{1}H NMR: 5.8 (broad singlet, 2H), 2.45–2.1 (broad m, 4H), 2.05–1.7 (broad m, 2H) ppm

14.34 What is the empirical formula of compound 14.5.7 based on the mass spectral data?

Compound 14.5.7 has two rings or π-bonds. The infrared signal at about 1635 cm⁻¹ suggests an alkene unit and the peak at about 3220 cm⁻¹ also suggests a C=CH. This is a hydrocarbon and the C=C unit accounts for only one ring or π-bond. The NMR spectrum shows alkene signals at about 5.8 ppm that integrate to two protons and there is a multiplet that integrates to two protons centered at 1.9 ppm. There is another multiplet centered at about 2.3 ppm that integrates to four protons. There is no methyl group. This type of pattern, which lacks the triplet of a terminal methyl group, is common in cyclic compounds. The single ring of a monocyclic compound accounts for the other ring or double bond. There are two –CH$_2$– units that are very similar and another –CH$_2$– that is different. The slight downfield shift of the four protons suggests that they are close to the C=C unit. This pattern suggests a structure –CH$_2$–CH=CH–CH$_2$–. Adding the other methylene and making it into a ring leads to cyclopentene as the structure.

Compound 14.5.8

MS: M = 78, 100%; M+1 = 79, 3.33%; M+2 = 80, 33%
IR: 2941, 1449, 1282, 893, 784, 722, 649 cm⁻¹
¹H NMR: 3.5 (t, 1H), 1.9 (m, 1H), 1.0 (t, 1.5H) ppm

14.35 After obtaining a formula, determine whether there are any rings or π-bonds.

Compound 14.5.8 has an M+2 peak that is indicative of a chlorine atom. There are three carbon atoms, leading to a formula of C$_3$H$_7$Cl. The infrared spectrum shows no functional groups. The NMR spectrum shows an integration that sums to 3.5, but there are seven protons in the formula. Therefore, the triplet at 1.0 ppm must integrate to 3H, the multiplet at 1.8 ppm must integrate to 2H, and the triplet at 3.4 ppm must integrate to 2H. The methyl is not attached to the methylene at 3.4 ppm, which must be attached to the Cl to be so far downfield. This analysis leads to –CH$_2$Cl and –CH$_3$, which is attached to a –CH$_2$. Therefore, there is a propyl unit and the structure is 1-chloropropane.

14.5.2 Identifying Bifunctional Molecules

When a molecule contains more than one functional group, the spectra may be more complex. The influences on chemical shift may be more complicated, and the signals in the proton NMR spectrum may be further downfield than observed with monofunctional molecules. This section examines compounds of this type.

Compound 14.5.9

Exact mass of the molecular ion is 72.0580. Possibilities are $C_2H_2NO_2$ (72.0085), $C_3H_4O_2$ (72.0211), $C_3H_8N_2$ (76.0688), C_4H_8O (72.0575), $C_4H_{10}N$ (72.0814), and C_5H_{12} (72.0939). In addition, the mass spectrum shows a very weak peak at m/z 72, but a prominent daughter ion at m/z 54.

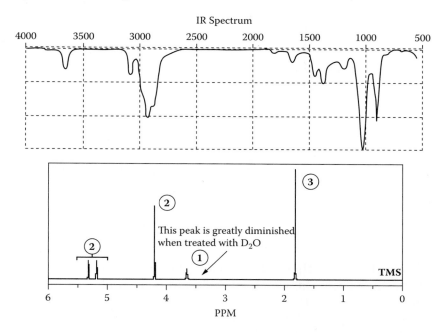

Compound 14.5.9 has the formula of C_4H_8O and there is one ring or π-bond. The infrared spectrum shows that there is no carbonyl, but there is a strong OH signal and the M–18 signal in the mass spectrum is consistent with an alcohol. The infrared spectrum does not show an obvious alkene signal, so structures that contain a ring should be considered. However, the proton NMR spectrum shows a two-proton signal at 5.2–5.4 ppm that is clearly due to alkene protons. Assume that a C=C unit accounts for the ring or π-bond and that this molecule is an alkene–alcohol. The signal at 3.7 ppm is greatly diminished with D_2O and this bit of information identifies that signal as the OH proton. There is a methyl group at 1.8 ppm, which is not consistent with a methyl attached to an alkyl fragment but rather with a C=C unit. The signal at 4.20 is a methylene and it must be attached to the oxygen of the OH unit to be so far downfield. The fragments are $C=CCH_2OH$, a $C=C$–Me, but only two protons on an alkene. This is consistent with 2-methyl-3-propen-1-ol as the structure.

Compound 14.5.10

M = 75, 100%; M+1 = 76, 3.71%; M+2 = 77, 0.25%

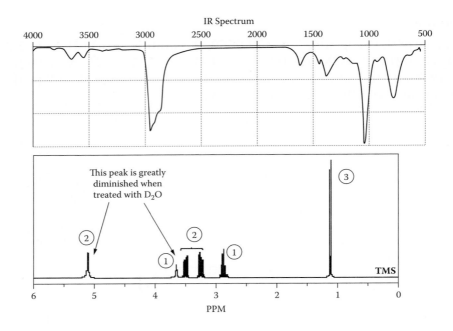

14.36 Are there rings or π-bonds in compound 14.5.10?

Compound 14.5.10 is another bifunctional molecule and it has a formula of C_3H_9NO. Note the odd mass, which suggests an odd number of nitrogen atoms. There is no nitrile peak in the infrared spectrum and no carbonyl, but there is a very broad peak in the OH/NH region that stretches into the C–H region. This peak does not appear to be a COOH, so the nitrogen unit is likely to be an amine, and the oxygen unit is either an alcohol or an ether. The infrared spectrum suggests an alcohol. The NMR spectrum shows only two broad signals at about 3.6 and 5.2 ppm, and both signals are diminished when treated with D_2O, so they must be NH and/or OH. Because there is only one oxygen, there can be only one OH, so *assume* that the other two protons are part of a primary amine, $-NH_2$. The doublet at 1.0 ppm integrates to three protons, so it is a methyl attached to a CH unit. The multiplet centered at about 3.4 ppm integrates to two protons (CH_2), and there is a multiplet at about 2.9 ppm that integrates to 1H, which suggests a CH unit. Summing all fragments gives C_3H_9NO—NH_2—OH—C_2H_4 (the Me–CH) = C_3H_9NO—C_2H_7NO—CH_2. The CH_2 must be attached to the OH via oxygen. Combining these fragments leads to 2-amino-1-propanol as the structure.

Compound 14.5.11

The exact mass of the molecular ion is 138.0790. Possibilities are $C_4H_2N_4O_2$ (137.9940), $C_6H_2O_4$ (137.9953), $C_7H_6O_3$ (138.0317), $C_7H_{10}N_2O$ (138.0794), $C_8H_{10}O_2$ (138.0681), $C_8H_{12}NO$ (138.0919), $C_9H_{14}O$ (138.1045), $C_9H_{16}N$ (138.1284), and $C_{10}H_{18}$ (138.1409).

IR: 3030–2899, 2260, 1667, 1439, 1408 cm^{-1}
^{1}H NMR: 3.4 (broad m, 6H), 2.0 (broad m, 4H) ppm

This is a difficult problem. Compound 14.5.11 has an even mass, but the exact mass indicates that the formula is $C_7H_{10}N_2O$. There are four rings or π-bonds. The infrared spectrum shows a peak indicative of a nitrile at about 2260 cm^{-1} that accounts for two π-bonds. The peak at 1667 cm^{-1} could be a conjugated carbonyl, but in fact it is an amide peak. How did the analysis lead to that conclusion? There are no NH signals or OH signals in the infrared, ruling out alcohols, primary and secondary amines, and amides. The structure must therefore be a ketone plus a tertiary amine or a tertiary amide. The carbonyl signal does not point to a ketone, and the structure is assumed to be a tertiary amide where the carbonyl accounts for one more π-bond. The NMR spectrum does not help much. The spectrum has six protons for the signal at 3.4 ppm and another broad signal at 2.0 worth four protons. The broad nature of the peaks suggests methylene units of a ring, and the two-proton signal is probably a CH_2. Assume the CH_2 is connected to both the nitrile and the carbonyl and a C_4H_8 must be accounted for. This is consistent with a pyrrolidine ring, accounting for the last ring or π-bond, and if a pyrrolidine amide is assumed, the structure is 1-cyanoacetylpyrrolidine.

14.37 Draw the structure of 1-cyanoacetylpyrrolidine.

Compound 14.5.12

M = 114, 100%; M+1 = 101, 6.66%; M+2 = 102, 0.62%

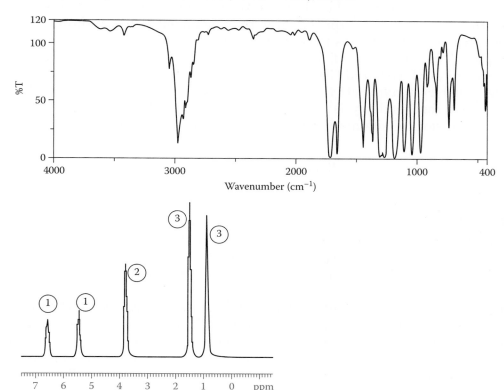

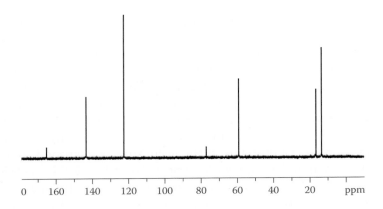

Compound 14.5.12 has a formula of $C_6H_{10}O_2$ and there are two rings or π-bonds. The infrared spectrum shows a conjugated carbonyl at 1695 cm^{-1} and also a medium peak at 1650 cm^{-1} consistent with an alkene. There is no sign of an aldehyde, acid, acid chloride, anhydride, or amide. This molecule is a conjugated ketone or a conjugated ester. With the two oxygen atoms and one ring or π-bond accounted for by the C=C unit, it is reasonable to assume a conjugated ester. The NMR spectrum shows two different alkene protons between 5.5 and 6.8 ppm, each integrating to one proton, consistent with a CH=CH unit. The signal that integrates to two protons at 3.8 ppm must be a methylene group attached to an oxygen, consistent with the assumption that this is an ester. The signal that integrates to three protons at 1.3 ppm is also a methyl group, and it is consistent with attachment to a C=C unit. Another methyl group at 0.9 ppm is part of an ethyl group, coupled with the signal at 3.8 ppm. The fragments of this molecule are O=C–OCH$_2$CH$_3$ (assuming an ethyl ester), CH=CH, and CH$_3$. The only way to put these fragments together is as ethyl crotonate (ethyl but-2-enoate). As presented, determination of *E*- or *Z*-stereochemistry is difficult because no coupling constants are provided, and several signals in the infrared spectrum make the interpretation ambiguous.

14.38 Based on the NMR coupling constants, is this molecule *cis* or *trans*?

Compound 14.5.13

   MS: M = 108, 100%; M+1 = 109, 4.44%; M+2 = 110, 34%
   IR: 2967–2817, 1400, 1294, 1117, 746, 669 cm^{-1}
   ^{1}H NMR: 3.6–3.85 (m, 4H), 3.5 (t, 2H), 1.1 (t, 3H) ppm

Compound 14.5.13 clearly has a chlorine atom in the formula because of the M+2 peak, and the M+1 leads to C_4. The M+2 formula is used to estimate oxygens, but 108 – C_4Cl (83) = 25 H, which is clearly nonsense. In most simple organic compounds, the formula will contain a nitrogen or an oxygen in addition to the C, H, or halogen. Because the mass is even, assume there is one oxygen. The presence of hydrogen atoms is certain and the missing atom probably has a mass of less

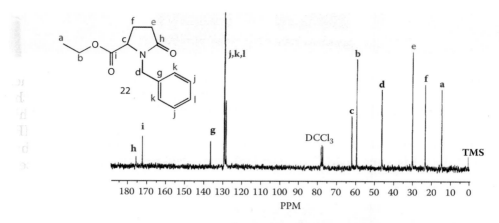

Figure 14.46 ^{13}C NMR spectrum of ethyl *N*-benzyl-2-pyrrolidinone-5-carboxylate.

to nitrogen or the ester oxygen appear between 45 and 65 ppm, as expected. The carbons in the benzene ring appear close together at about 129 ppm, but there is a small peak at 136 ppm. This is the carbon that has no hydrogen atoms attached, as indicated. There are two carbonyl carbons—the lactam carbonyl and the ester carbonyl—and they appear furthest downfield. All carbon atoms that do not bear a hydrogen, such as a carbonyl and the quaternary carbon of the benzene ring, are weaker than the others. This is a typical observation and assists in identifying each carbon.

At about 77 ppm, there is a cluster of three carbon atoms. It turns out that this is the carbon of deuterochloroform (CDCl$_3$), the solvent used to obtain this spectrum. Because deuterium has a spin of 1 (see Section 14.4.1), it will cause the carbon to split into a triplet (see Section 14.4.5). This signal is ignored because it is not part of the sample under analysis.

Carbon-13 NMR spectroscopy is a powerful tool for structural analysis, particularly when it is used in conjunction with proton NMR spectroscopy, infrared spectroscopy, and mass spectrometry. Many other NMR spectroscopy techniques can provide a wealth of information. Because the objective of this section is to give an introduction to spectroscopy, rather than to give a complete spectroscopy course, the discussion will stop here. These are the primary tools for structural analysis. A few problems will include ^{13}C NMR spectroscopy, but this brief introduction will not allow its full potential to be exploited. If proton NMR spectroscopy is understood, a brief and specialized course will give a greater knowledge of ^{13}C NMR spectroscopy.

14.40 **The empirical formula is C$_4$H$_8$O, and the IR spectrum and proton NMR spectrum are unavailable. The mass spectrum shows an M-15 peak. Fortunately, the ^{13}C spectrum is available, and there are peaks at 57.8, 76.8, 136.0, and 118.2 ppm. Are the spectral data provided consistent with CH$_3$OCH$_2$CH=CH$_2$ or with 2-methyloxetane? Draw both compounds and indicate which molecule is consistent with the ^{13}C NMR.**

14.7 Biological Relevance

A biological application of NMR spectroscopy is known as magnetic resonance imaging (MRI). In effect, an NMR spectrum is taken of a person or a part of the person. The result is a high-quality image of the inside of the human body. This noninvasive technique is a great help to medicine. Figure 14.47 shows an MRI of the spine and cervical and sagittal regions of a human,[11] and it is clear that this level of detail is very good. The cervical vertebrae are those immediately

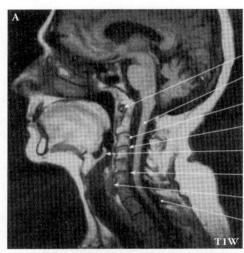

Anterior arch of C1

Cervical cord

CSF, ligament, and cortical bone

Bulging disc

Epiglottis

Osteophyte and disc material

Anterior osteophyte

Spinous process

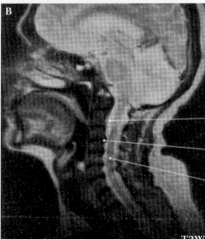

CSF in thecal sac

Osteophyte and disc protrusion

Cervical cord

Figure 14.47 MRI: Spine, cervical, sagittal. These images show the relative change in the CSF from (A) nearly black to (B) very bright. The spinal cord is outlined best with the T1W image. (Bushong, S. C. 1996. *Magnetic Resonance Imaging. Physical and Biological Principles,* 2nd ed., 352, Figure 24-14. St. Louis: Mosby. Reproduced with permission from Stewart C. Bushong ScD, FAAPM, FACR.)

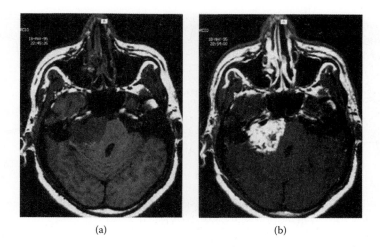

(a)   (b)

Figure 14.48 Signal enhancement of intracranial lesion upon administration of 0.1 mmol/kg of a Gd^{3+} agent: (a) precontrast image and (b) postcontrast image. Note the increase in intensity of the mass to the left in (b). (Reprinted from Rajan, S. S. 1998. *MRI. A Conceptual Overview,* 71, Figure 5.4. New York: Springer. With kind permission of Springer Science and Business Media.)

behind the skull, and the sagittal plane is an imaginary plane that vertically divides the body into left and right portions. It may be difficult to distinguish one tissue type from another, however, or to bring out detail for one organ or tissue type. The difference between a cancerous tumor and normal brain tissue may be so small, for example, that the tumor will not be detected by MRI.

Chemicals injected into a patient that make it easier to distinguish tissue types are known as **contrasting agents**; they typically make one or more tissue types appear brighter and lead to greater contrast. Alternatively, a tissue may appear darker and again there is significant contrast. A typical contrasting agent is an organic complex of a paramagnetic metal ion such as gadolinium (Gd and Gd^{3+} compounds), ferric compounds (Fe^{3+}), or manganese compounds (Mn^{2+}).[12]

Heavy metals may be toxic, but complexation with other molecules or ions often diminishes the toxicity and facilitates clearance of those metals from the body. This technique is used with the contrast agents given in Figure 14.49. If one tissue type complexes the metal more than another, that tissue type (or organ) "stands out" from the background. Most tumors, for example, have a greater Gd uptake than the surrounding tissues, leading to a stronger signal (the tumor stands out from the background).[13] Figure 14.48 shows enhancement of an intracranial lesion by administration of a Gd^{3+} contrast agent.[14]

Some typical contrast agents[15] are Gd-EDTA, Gd-DTPA, and Gd-DOTA, which are useful for imaging the central nervous system, as is gadodiamide. These compounds are shown in Figure 14.49. Ferric ammonium citrate (sold commercially as Geritol) is used to image the stomach and upper small intestine. Mangafodipir trisodium is used for lesions of the liver. Note that the organic fragments of the contrast agents are amino acid derivatives (Chapter 27, Section 27.3), phosphates (Chapter 20, Section 20.12), or amides (Chapter 20, Section 20.6).

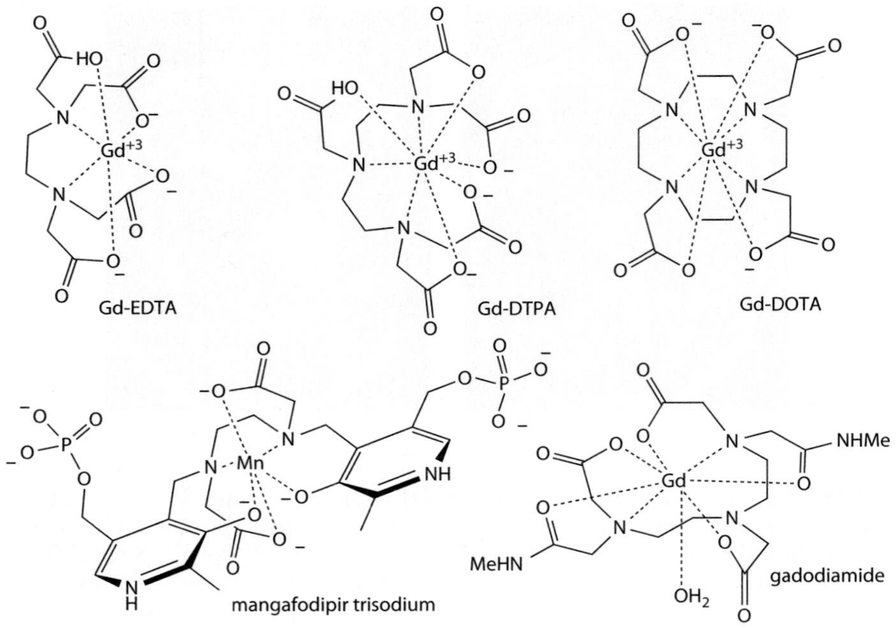

Figure 14.49 MRI contrast agents.

References

1. Taken from http://wfc3.gsfc.nasa.gov/MARCONI/images-basic/spectrum.jpg
2. Silverstein, R. M., Bassler, G. C., and Morrill, T. C. 1991. *Spectrometric identification of organic compounds,* 5th ed., 4, Figure 2.2. New York: John Wiley & Sons, Inc.
3. McLaffety, F. W. 1963. *Mass spectral correlations, Advances in chemistry series,* ed. Gould, R. F. Washington, D.C.: American Chemical Society.
4. Isaacs, N. S. 1995. *Physical organic chemistry,* 2nd ed., 31, Figure 1.6. Essex, England: Longman.
5. Friebolin, H. 1991. *Basic one- and two-dimensional NMR spectroscopy,* 8, Figures 1–6. Weinham, Germany: VCH Publishers.
6. Breitmaier, E., and Wade, J. 1993. *Structure elucidation by NMR in organic chemistry, a practical guide,* 43, Figure 2.16. Chichester, England: Wiley.
7. Jackman, L. M., and Sternhell, S. 1969. *Applications of NMR spectroscopy in organic chemistry,* 2nd ed. New York: Pergamon Press.
8. Fraser, R. R., and McGreer, D. E. 1961. *Canadian Journal of Chemistry* 39:505–509.
9. Silverstein, R. M., Bassler, G. C., and Morrill, T. C. 1991. *Spectrometric identification of organic compounds,* 5th ed., 221, Appendix F. New York: John Wiley & Sons, Inc.
10. This compound is 5R-methyl-1-[(N-(2R)-carboethoxyl-3-methylpropylamino) methyl]-2-pyrrolidinone. See Chen, P., Tao, S., and Smith, M. B. 1998. *Synthetic Communications* 28:1641.
11. Bushong, S. C. 1996. *Magnetic resonance imaging. Physical and biological principles,* 2nd ed., 352, Figure 24-14. St. Louis: Mosby.

12. Rajan, S. S. 1998. *MRI. A conceptual overview,* 66. New York: Springer.
13. Hornak, J. P. 2008. *The basics of MRI,* chap. 1. Online book. http://www.cis.rit.edu/htbooks/mri/
14. Rajan, S. S. 1998. *MRI. A conceptual overview,* 71, Figure 5.4. New York: Springer.
15. Hornak, J. P. 2008. *The basics of MRI,* chap. 12. Online book. http://www.cis.rit.edu/htbooks/mri/

Answers to Problems

14.1 A frequency of 60×10^6 cycles/second corresponds to 60 million Hz or 60 MHz.

14.2 The wavelength is $= \gamma = (3 \times 10^{10})/8500 = 3.53 \times 10^6$ cm. The energy is $= 6.626 \times 10^{-34}$ J Hz^{-1} (8500 Hz) $= 5.63 \times 10^{-30}$ J.

14.3 Based on Figure 14.2, 600 nm lies in the middle region of the visible spectrum, so we anticipate that the color will be green.

14.4 The mass for the $C_3H_4^{+2}$ fragment is 40, but the charge is +2, so the m/z value = $40/2 = 20$.

14.5 The formula for hexane is C_6H_{14} and the parent ion will show m/z = 86. The formula for 3-pentanone is $C_5H_{10}O$ and the parent ion will show m/z = 86. Diethyl ether has a formula of $C_4H_{10}O$ and the parent ion will show m/z = 74.

14.6 The parent ion of 2-butanone has m/z = 72 and it can fragment on both sides of the carbonyl. One cleavage leads to daughter ions m/z = 43 and m/z = 29. Cleavage on the other side of the carbonyl leads to daughter ions m/z = 15 and m/z = 57. All five of these ions are drawn on the simulated mass spectrum.

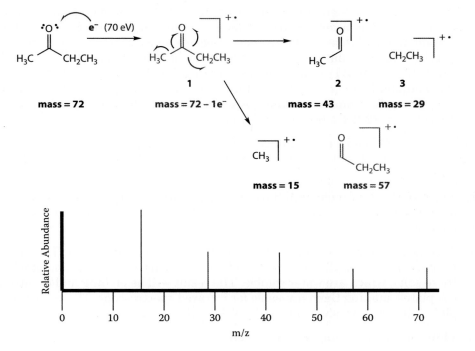

14.7 $C_4H_{11}N$.

14.8 $C_5H_8O_2$.

14.9   114.1044 and loss of $-CH_2CH_3$.

14.10 For 2-butanone, with a formula of C_4H_8O, the parent ion will appear at m/z = 72 and that will be taken as 100%. Because there are four carbons and zero N, M+1 = 1.11 (4) = 4.44. The M+2 is $4.44^2/200 + 0.2$ (for 1 O) = 0.30. For triethylamine, with a formula of $C_6H_{15}N$, the parent ion will appear at m/z = 101 and will be taken as 100%. There are six carbons and one N, so M+1 = 6 (1.11) + 0.38 = 7.04. There is zero O, so M+2 = $6.66^2/200 = 0.22$.

14.11 Because the mass of the parent is 126, it is even and we assume zero N. Taking the M+1 signal, #C = 8.88/1.11 = 8. Using this in the M+2, we find that $8.88^2/200 = 0.39$. Because M+2 = 0.59, we assume there is an oxygen, so #O = (0.59 − 0.39)/0.2 = 1. We have $C_8O = 112$. Because the mass is 126, we solve for the number of H atoms by difference = 126 − 112 = 14. Therefore, the formula is $C_8H_{14}O$.

14.12 Because 92 is an even mass, we assume zero N and M+1 = 4.44 shows the presence of four carbon atoms. Most striking is the M+2 peak, which is 33% of the P. This is a clear indication that Cl is present. We cannot find an O if it is present because the 33% swamps all other contributions. However, $C_4Cl = 83$. Note that we use the mass of the chlorine isotope (35) and not the molecular weight of 35.45. This means that the M+2 peak is due to ^{37}Cl. Using this information we solve for H and find that 92 − 83 = 9. If this number had been very high, showing a ridiculously large number of H atoms, then we would know there is another atom present. As it is, the amount of nine hydrogen atoms is consistent with four carbons and the formula is C_4H_9Cl.

14.13 Because it is a ketone, α-cleavage is prominent. There are no peaks that correspond to a McLafferty rearrangement, suggesting that there are no chains of carbon atoms of three atoms or more. Because m/z = 85 corresponds to P-57 (a butyl group) and that group cannot be three or four carbon atoms in length, we think of a *tert*-butyl group. The m/z = 127 peak is P-15 for loss of methyl and it is abundant, suggesting "a lot of methyls." This is consistent with *tert*-butyl and, given the simplicity of the spectrum, a reasonable structure is di-*tert*-butyl ketone (2,2,4,4-tetramethyl-3-pentanone).

14.14 For the C–O bond, reduced mass = $m_o m_c/m_o + m_c$ = (2.657 × 10⁻²³) × (1.994 × 10⁻²³)/(2.657 × 10⁻²³) + (1.994 × 10⁻²³) = 5.298 × 10⁻⁴⁶/(4.651 × 10⁻²³) = 1.139 × 10⁻²³.

14.15 The stronger C≡C bond will have the larger force constant and it will require more energy for vibration. Therefore, the C=C bond will absorb at lower energy.

14.16 Acetone is infinitely soluble in water and it rapidly absorbs water. Therefore, washing KBr or NaCl plates with acetone will bring water into contact with these salts, which are water soluble. The result will be etching and/or pitting of the plates, making them unusable for infrared spectroscopy. These pressed salt plates should be washed only with solvents that do not contain water.

14.17 1100 cm⁻¹ is in the fingerprint region.

14.18 The C=O bond is stronger than the C–O bond, requiring greater energy for it to vibrate. Therefore, it absorbs at higher energy than the C–O bond.

14.19 The C–O–H unit of an alcohol will have the OH absorption in the functional group region, but it will also have a strong C–O absorption in the fingerprint region.

14.20 The C=C unit of 1-pentene gives a moderate peak at 1640–1680 cm⁻¹ and the C≡C unit of 1-pentyne gives a signal at 2100–2260 cm⁻¹. These differences allow us to distinguish the two molecules easily.

14.21 No! The carbonyl peak appears in virtually the same place for both molecules. Although the ester contains the C–O unit and this will not be present in the ketone, the C–O absorption appears in the fingerprint region and can be difficult to identify. Therefore, it is not a reliable diagnostic tool.

14.22 For both the aldehyde and the acid, the carbonyl unit absorbs at about 1725 cm⁻¹. The OH of a carboxylic acid absorbs between 2500 and 3000 cm⁻¹ as a very broad and strong absorption and this is missing from the aldehyde spectrum. This can be used to distinguish them absolutely.

14.23 There are three different kinds of hydrogens in 2-butanone: two different methyl groups and the methylene group. These protons are marked in the figure.

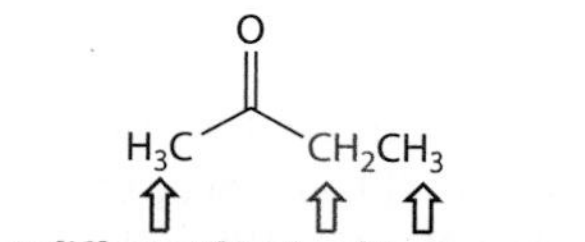

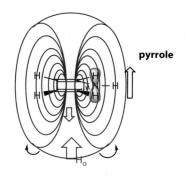

3 different kinds of hydrogens

14.24 The normal isotope of boron is ¹¹B, which means that the atom will have five protons and five neutrons, with five electrons. To make ¹¹B requires addition of an extra neutron, so there are five protons and six neutrons. Because there is one extra particle, I = 1/2.

14.25 Because the spin = 1/2, there are two orientations and one signal per ¹³C nucleus.

14.26 The ppm for this signal is determined by ppm = 438/(500 × 10⁶) = 0.876 × 10⁻⁶ = 0.876 ppm.

14.27 Because pyrrole is aromatic, with six π-electrons, it will set up the anisotropy field shown and this will make the protons on the pyrrole ring appear downfield.

14.28 The ratio is taken to be $(71.5/28.6) + (28.6/28.6) + (42.9/28.6) = 2.5:1:1.5$. This sum adds up to five protons. Because the formula indicates that 10 protons are present, we multiply this ratio by 2 to obtain 5:2:3. Therefore, there are three different kinds of hydrogen atoms in this molecule: five of one kind, two of a second kind, and three of a third kind.

14.29 In 2,5-dimethyl-3-hexanone, there are five different kinds of protons, labeled H_a–H_e in the figure. The six methyl protons, H_a, and the single proton, H_b, are neighbors. The methylene protons H_c and proton H_d are neighbors, but H_d is also neighbors with the six methyl protons H_e.

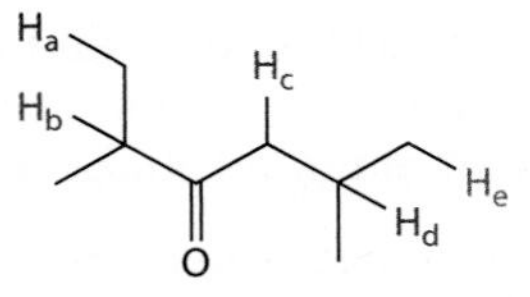

14.30 Quartet; quartet; triplet.

14.31 J is about 4 Hz.

14.32 Singlet. All of the protons are identical, so there are no neighbors.

14.33 For an even mass formula, M+1 = 4.44 = 1.11 × #C, so #C = 4.44/1.11 = 4. M+2 = 0.30 = (4.44 × 4.44)/200 + 0.2 #O, so #O = (0.3 – 0.1)/0.2 = 1 O. C4 = 12 × 4 = 48 and O = 16. Because the parent has a mass of 72, the number of hydrogen atoms must be 72 – 48 – 16 = 8. Therefore, the formula is C_4H_8O. The number of rings or π-bonds is (2 × 4) + 2 – 8 – 0/2 = 1.

14.34 The empirical formula is C_5H_8.

14.35 The rings or π-bonds are (2 × 3) + 2 – 7 – 1/2 = 0. Therefore, there are no rings or π-bonds.

14.36 The rings or π-bonds are (2 × 3) + 3 – 9 – 0/2 = 0. Therefore, there are no rings or π-bonds.

14.37 It is an amide.

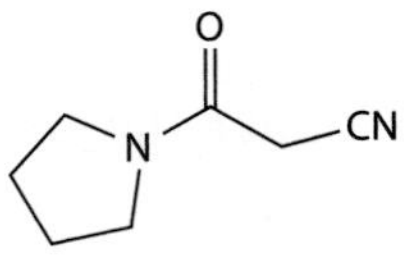

14.38 Based on the NMR coupling constants, this compound is *trans*.

14.39 According to Figure 14.42, a signal at 155 ppm is in the range where aromatic carbon is observed, so it is consistent with an unknown compound that has a benzene ring.

14.40

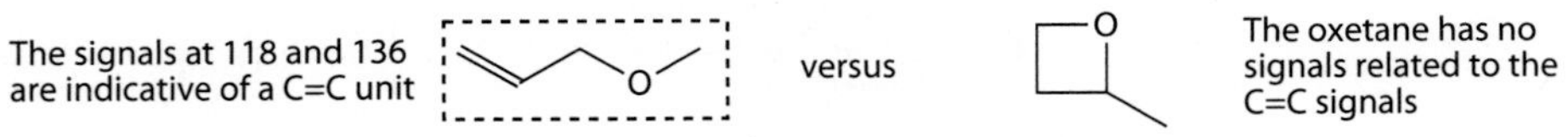

The signals at 118 and 136 are indicative of a C=C unit

versus

The oxetane has no signals related to the C=C signals

Correlation of Concepts with Homework

- Chemists make use of the electromagnetic spectrum to probe the structure of organic molecules: 1, 2, 3, 35, 36, 37, 87, 90, 91, 92, 93, 94, 95, 96, 97, 98, 99.
- Bombarding a molecule with 70 eV of energy leads to formation of a molecular ion that allows determination of the empirical formula based on the presence of isotopes of common elements found in organic molecules. This is called mass spectrometry. Fragmentation of the parent ion leads to daughter ions, and determining the mass of lost fragments can assist in structure determination: 4, 5, 6, 13, 43, 46, 48, 51, 52, 53, 54, 55, 56, 71, 86.
- High-resolution mass spectrometry is valuable for determining the empirical formula of a molecule based on the exact mass of the molecular ion: 7, 8, 9, 46, 49, 50, 54, 92, 93, 99.
- The M+1 and M+2 isotopic peaks can be used to determine an empirical formula: 10, 11, 12, 13, 33, 34, 41, 42, 43, 44, 45, 46, 47, 48, 51, 70, 82, 85, 86.
- Absorption of infrared light leads to molecular vibrations, particularly in bonds that are polarized and/or exhibit a change in dipole upon absorption of the light. Observation of the effects of infrared radiation of the molecular vibrations induced by infrared light is called infrared spectroscopy: 14, 15, 16, 57, 58, 78, 84, 90, 91, 92, 93, 94, 95, 96, 97, 98, 99.
- The absorbances of most common functional groups appear in the functional group region, whereas all molecules exhibit vibrations characteristic of their structure that appear in the fingerprint region. Matching peaks in the infrared spectrum with functional group correlation tables allows one to determine the presence or absence of functional groups in an organic molecule: 17, 18, 19, 20, 21, 22, 59, 60, 61, 62, 63, 64, 73, 75, 83.
- When an organic molecule is placed in a powerful magnetic field, the nucleus of each hydrogen atom behaves as a tiny magnet and sets up a small magnetic field that can be aligned with or against the large magnetic field. When this proton is irradiated with energy in the radio signal range, that energy, for a given magnetic field strength, flips its spin state. The energy absorbed when the precessional frequency of the hydrogen nucleus and the external field are in resonance is measured and displayed as absorption peaks. This is called nuclear magnetic resonance spectroscopy, and it is used to determine the structure of organic molecules: 24, 25, 66, 67, 68, 90, 91, 92, 93, 94, 95, 96, 97, 98, 99.
- Substituents attached to carbon atoms bearing protons alter the local magnetic environment and lead to different magnetic

fields for each different type of proton. Each different proton will lead to a different signal in the NMR spectrum. The ppm scale is established for NMR spectroscopy and tetramethylsilane (TMS) is used as a zero point standard: 23, 24, 25, 67, 68, 74, 76, 77.

- Protons attached to carbons bearing electron withdrawing groups appear downfield. Integrating the area under each different proton allows one to establish the ratio of hydrogen peaks and, with the empirical formula, the number of different protons may be obtained: 23, 28, 65, 69, 80, 81, 88.

- π-Bonds react to an applied magnetic field by developing a small opposing local field that orients a molecule in a large magnetic field. This leads to the development of an opposing magnetic field that leads to significant downfield or upfield shifts in the chemical shift and is called magnetic anisotropy: 27, 72, 77, 93.

- Neighboring protons split a proton signal into n+1 peaks. This multiplicity allows one to determine how many neighboring protons are present for a given signal. The coupling constant J defines the distance between the peaks: 29, 30, 31, 32, 38, 69, 74, 76, 77, 89.

- Carbon-13 NMR spectroscopy can be used to count the number of different carbon atoms in an organic molecule: 39, 40, 90, 91, 92, 93, 94, 95, 96, 97, 98, 99.

Homework

14.41 Given the following molecular ion region data, determine a formula for each structure. For each formula, calculate the number of rings and/or double bonds:
   (a) M (100) 100%, M+1 (101) 6.66%, M+2 (102) 0.42%
   (b) M (149) 100%, M+1 (150) 11.46%, M+2 (151) 0.62%
   (c) M (96) 100%, M+1 (97) 7.77%, M+2 (98) 0.30%
   (d) M (96) 100%, M+1 (97) 6.66%, M+2 (98) 0.42%
   (e) M (110) 100%, M+1 (111) 8.88%, M+2 (112) 0.39%
   (f) M (83) 100%, M+1 (84) 5.91%, M+2 (85) 0.15

14.42 The measured mass of an ion is m/e 66 for an organic compound that may or may not contain O and/or N. First deduce possible formulas that fit m/e 66. If the high-resolution mass is measured to be 66.0459, which formula best corresponds to this mass? Show your work.

14.43 Draw the molecular ion region expected for bromomethane, including m/e values and percent relative to P.

14.44 Draw the molecular ion region expected for 1,2-dibromo-3-chlorobutane.

14.45 What is the ratio of M to M+1 for butane, acetone, 2-hexanol, toluene, triethylamine, and butanenitrile?

14.46 What is the relative intensity of M+1 to M in $C_{100}H_{202}$?

14.47 An M:M+1 ratio in a mass spectrum is 100:24. How many carbon atoms are present?

14.48 Draw the molecular ion region for $^2HCCl_3$, for $^1HCCl_3$, for $^1H^{13}CCl_3$, and for $^2H^{13}CCl_3$.

14.49 Determine a reasonable empirical formula based on the following exact mass determinations for the molecular ion:
(a) 72.0939
(b) 72.0211
(c) 72.0575
(d) 58.054
(e) 58.0419
(f) 58.0657

14.50 Based on the exact mass of the daughter ion, identify the fragment lost from the original molecule. Suggest a cleavage that would lead to that daughter ion:
(a) M = 86.0732 and the daughter ion has a mass of 57.0341
(b) M = 86.0732 and the daughter ion has a mass of 68.0626
(c) M = 114.1045 and the daughter ion has a mass of 58.0419
(d) M = 122.0368 and the daughter ion has a mass of 77.0392

14.51 Draw the molecular orbital diagram for ethylene and then decide which electron is most likely to be lost during ionization in the mass spectrometer. Draw a possible structure for the resulting radical cation.

14.52 Draw the structure for the radical cation formed from 3-pentanone in the mass spectrum. (Put the radical and positive charge on appropriate atoms.) Draw the daughter ion resulting from cleavage of the X–C^α bond and also from the C^α–C^α bond, where X = C=O.

14.53 The molecular ion formed from naphthalene in the mass spectrum is very stable. Draw this ion, with the radical and charge, and explain why this ion is so stable.

14.54 An ester gave the mass spectrum shown. The high-resolution spectrum gave m/e 116.0833.
(a) Suggest a formula. How many ring and/or double bonds does the ester have? (In other words, is the ester saturated or unsaturated?)
(b) Suggest formulas for the ions of m/e 29, 43, and 71.

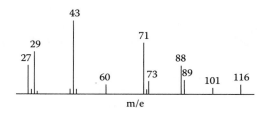

14.55 A carboxylic acid gave the following mass spectrum. Suggest a structure and justify your choice.

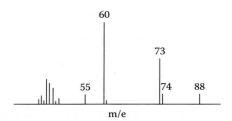

14.56 The following ketone has a formula of $C_9H_{10}O$. Based on the mass spectrum shown, assign a structure and justify your choice based on key daughter ions.

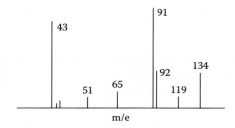

14.57 As more energy is added to a vibrating spring system, will the frequency of the vibrations change? Why or why not?

14.58 Assuming a simple diatomic molecule, obtain the frequencies of the absorption band from the force constants given:
(a) $k = 5.1 \times 10^5$ dynes cm^{-1} for C–H in ethane
(b) $k = 5.9 \times 10^5$ dynes cm^{-1} for C–H in ethyne
(c) $k = 7.6 \times 10^5$ dynes cm^{-1} for C–C in benzene
(d) $k = 17.5 \times 10^5$ dynes cm^{-1} for C≡N in CH$_3$CN
(e) $k = 12.3 \times 10^5$ dynes cm^{-1} for C=O in formaldehyde

14.59 Following is an infrared spectrum of a compound that is known to have either structure A or B. Which structure is *not* consistent with the spectrum? Why?

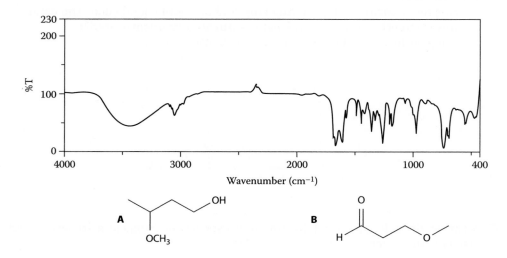

14.60 Following is a carbon NMR spectrum of a compound that is known to have structure A, B, or C. Which structure is consistent with the spectrum? Why?

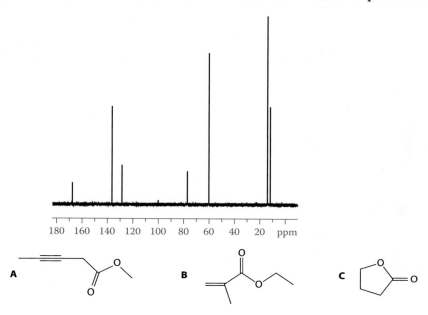

14.61 A very dilute solution of *cis*-cyclopentane-1,2-diol in CCl_4 shows bands at 3620 and 3455 cm^{-1}. Explain.

14.62 What infrared absorptions would be expected for each of the following?
(a) 1-butyne
(b) 2-butyn-1-ol
(c) 1,2-dichloroethyne
(d) 3-cyanobutanoic acid

14.63 How would the spectra of the following compounds differ?
(a) methylenecyclopentane and methylcyclopentene
(b) cyclopentanone and pent-3-enal

14.64 Provide a logical structure for compounds (a)–(e) based on their IR spectrum:
(a) C_4H_9Br

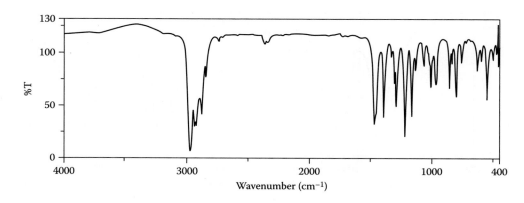

(b) $C_{12}H_{11}N$

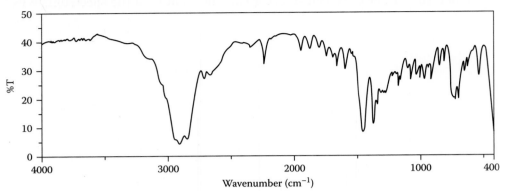

(c) $C_4H_8O_2$

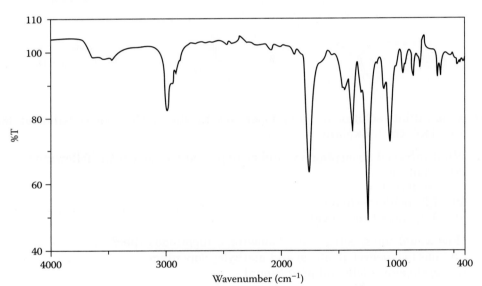

(d) C_3H_7NO

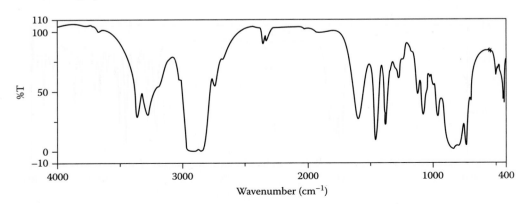

(e) $C_5H_{10}O$

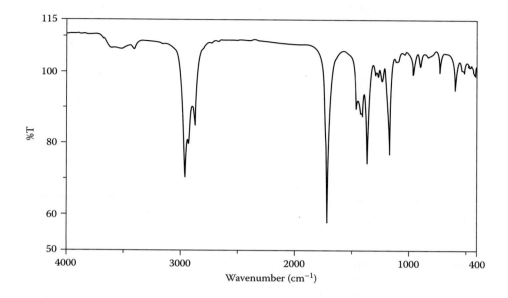

14.65 Two signals in the proton NMR spectrum appear at 345 and 350 Hz in a 60 MHz field. Calculate the position of both signals in hertz in a 270-MHz instrument and then in a 600-MHz instrument. Calculate the position of these signals in parts per million.

14.66 The proton NMR spectrum of a compound with the formula $C_4H_7ClO_2$ is shown, and it has obvious overlapping peaks. There is a strong signal at about 1720 cm^{-1} in the IR. Deduce the structure.

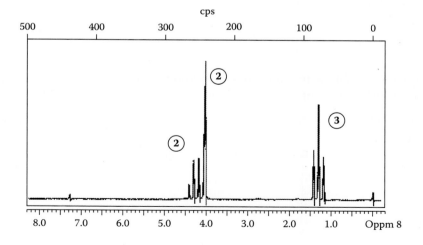

14.67 The carbon NMR spectra for two isomeric compounds with the formula $C_5H_{10}O$ are shown in the following. The IR spectrum for both molecules shows a band at 1725 cm^{-1}. Deduce a structure for each compound.

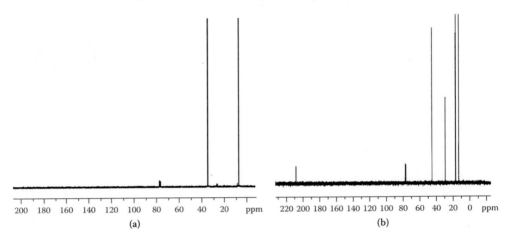

(a)   (b)

14.68 The proton NMR spectrum for a compound shows a singlet at 1.1 ppm (integrates to 3 H) and a singlet at 3.8 pm (integrates to 1 H). There are two possible structures: methyl 2,2-dimethyl ethanoate or 1,1-dimethylethyl ethanoate. What is the correct structure? Justify your answer.

14.69 Careful analysis of the proton NMR spectrum of the molecule shown in the following, obtained at –50°C, indicates that there are two different doublets for the circled methyl group. Offer an explanation.

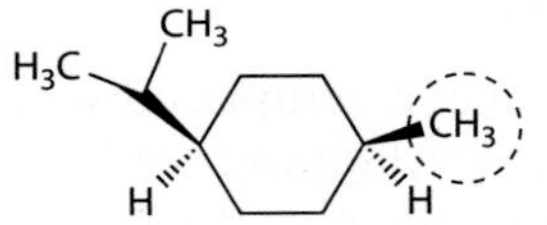

14.70 Which of the following structures are consistent with a mass spectrum having an M+2 = 98% of M? Explain.

14.71 Which of the following molecules are expected to give an m/z = 179 peak in the mass spectrum, where the parent ion is m/z = 208?

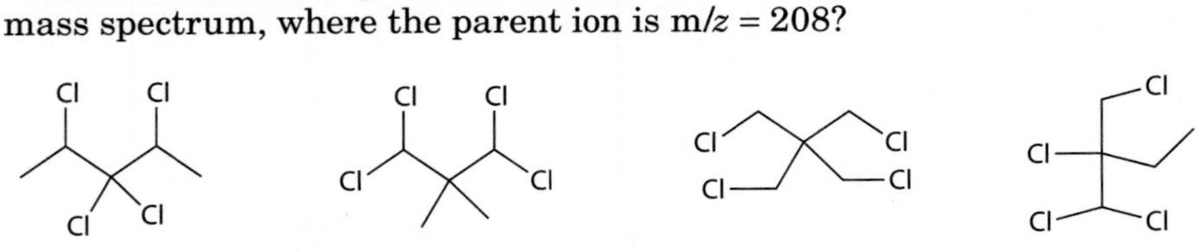

14.72 Which of the following molecules exhibit magnetic anisotropy in the proton NMR spectrum? Indicate the structural fragment that is responsible for each positive answer.

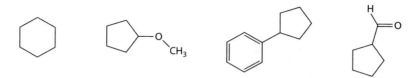

14.73 A molecule with a formula $C_6H_{12}O$ shows a strong peak at 1725 cm^{-1} and another weaker peak at 2815 cm^{-1}. Which of the following are structures consistent with the data? Briefly explain each choice.

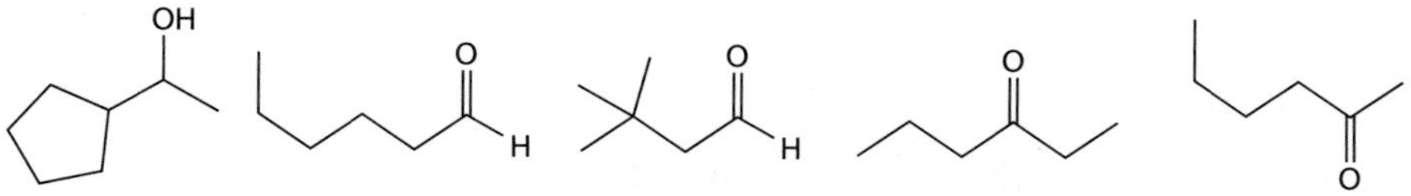

14.74 One of the four following structures exhibits a singlet at 5.9 ppm as well as a triplet at 1.0 ppm and a quartet at 3.5 ppm. Which molecule is consistent with this proton NMR? Explain your choice.

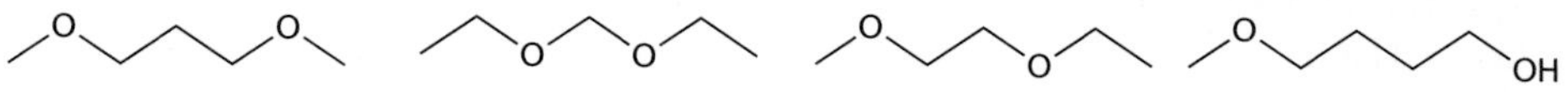

14.75 Which of the following molecules give no peaks at 2850–2960 cm^{-1} in the infrared?

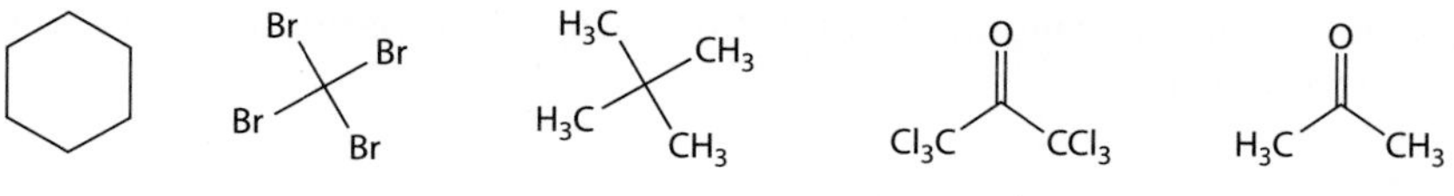

14.76 Examine the molecules shown.
(a) How many different signals will appear in the proton NMR spectrum of **A**?
(b) How many different signals will appear in the proton NMR spectrum of **B**?
(c) Give *specific differences* in the proton NMR spectrum that will allow you to distinguish between **A** and **B**.

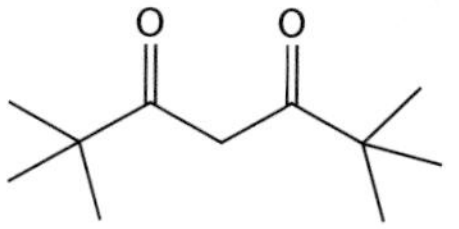

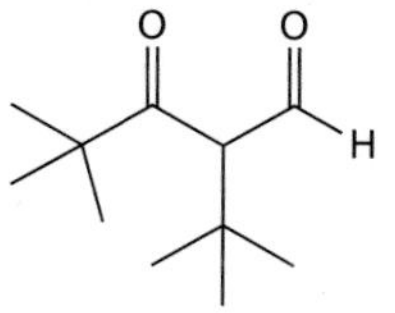

Your answer should include number of signals, multiplicity, and chemical shift.

14.77 Which of the following molecules are expected to show *no peaks* in the 7–8 ppm region of a proton NMR spectrum?

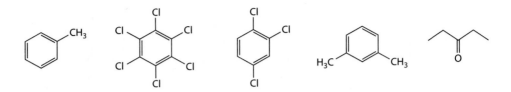

14.78 Which of the following solvents *cannot* be used on an infrared salt plate? Explain.
(a) CCl_4
(b) H_2O
(c) aq. acetone
(d) THF
(e) hexane

14.79 (a) Briefly explain why the proton for ethene absorbs at about 5.4 ppm in the proton NMR spectrum, whereas the proton for ethyne absorbs about 2.3.
(b) Briefly explain why the H of the aldehyde H–C=O unit appears at about 9.4 ppm, whereas the H of the alcohol H–C–O appears at about 3.5 ppm in the proton NMR spectrum.

14.80 Which of the following solvents cannot be used as a solvent in proton NMR spectrum? Explain.
(a) CH_3OH
(b) D_2O
(c) $CDCl_3$
(d) CH_2Cl_2
(e) CCl_4

14.81 How many *different* signals will 2,2-dimethyl-1-propanol have in the proton NMR spectrum?
(a) 0
(b) 1
(c) 2
(d) 3
(e) 4
(f) 5
(g) 6

14.82 Which of the following atoms contribute to the M+1 signal in the mass spectrum?
(a) ^1H
(b) ^{13}C
(c) ^{34}S
(d) ^{35}Cl
(e) ^{81}Br
(f) ^3H

14.83 For a molecule with a molecular formula of C_5H_8, which functional groups does the infrared spectrum shown suggest are consistent with these data and are possible?

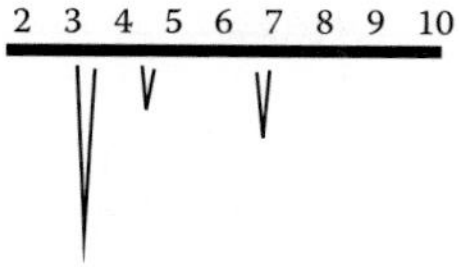

ketone	aldehyde	alcohol	ether
1° amine	2° amine	3° amine	nitrile
acid	alkyne	alkene	alkane

14.84 Which IR absorption is higher in energy?
   (a) 700 cm^{-1}
   (b) 1000 cm^{-1}
   (c) 1800 cm^{-1}
   (d) 3200 cm^{-1}

14.85 At what m/z does the molecular ion for C_5H_9Cl appear? Explain.
   (a) 104
   (b) 105
   (c) 106
   (d) 107

14.86 In alcohols, the mass spectrum often shows that the parent is missing, but that there is an M-18 peak corresponding to loss of water. With this in mind, briefly explain why $CH_3(CH_2)_5NH_2$ might show a peak at m/z 84.

14.87 You have two bottles (A and B), but the labels have fallen off and blown away. You know that one is 3-pentanone and the other is methyl butanoate. You take a mass spectrum and an infrared spectrum of each to help determine their identity. Describe how you would distinguish A and B.

14.88 Which of the following molecules give a signal in the proton NMR spectrum that is *upfield* of an −OCH_3 signal?

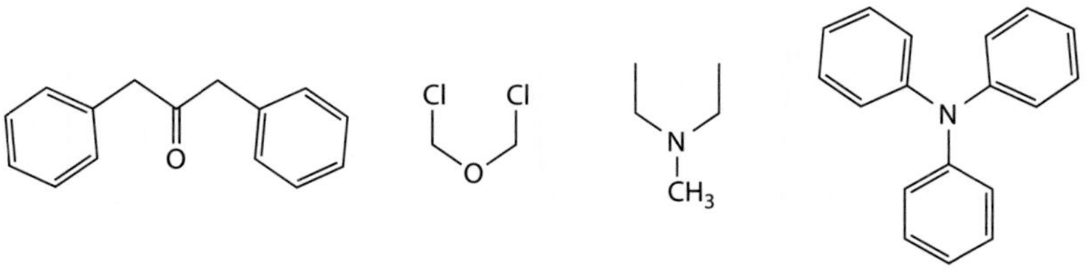

14.89 Which molecules give no peaks at all in the proton NMR spectrum?

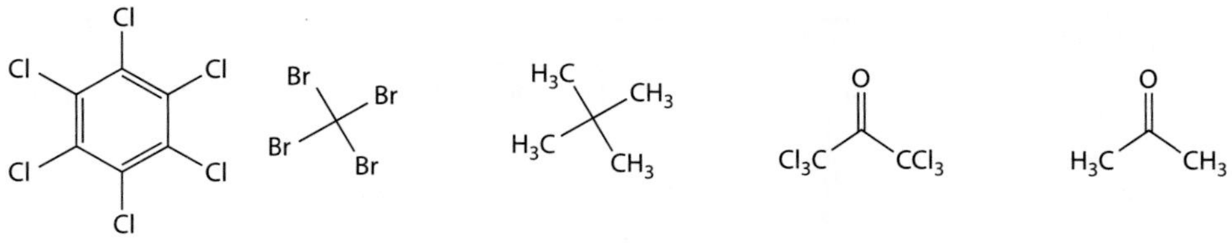

14.90 Give the structure for the molecule with a formula of $C_7H_{14}O$ and the following spectra:

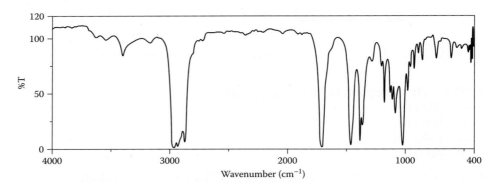

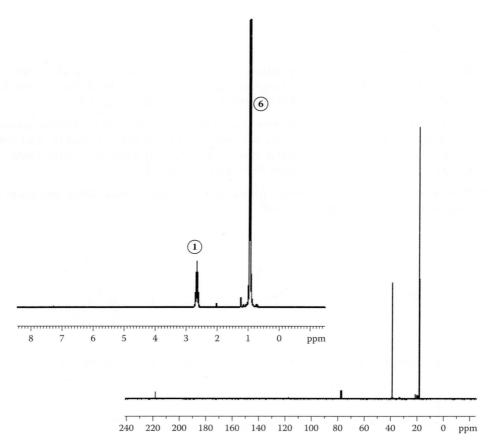

14.91 Give the structure for the molecule with the following theoretical spectra:
M = 146, 100%; M+1 = 147, 8.88%; M+2 = 148, 0.794%.

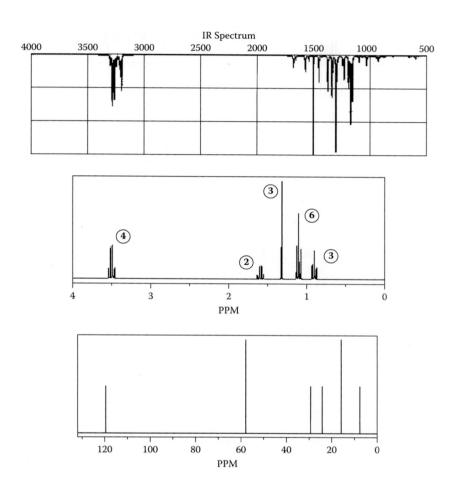

14.92 Give the structure for the molecule with the following theoretical spectra. The exact mass for the molecular ion is 115.1362:

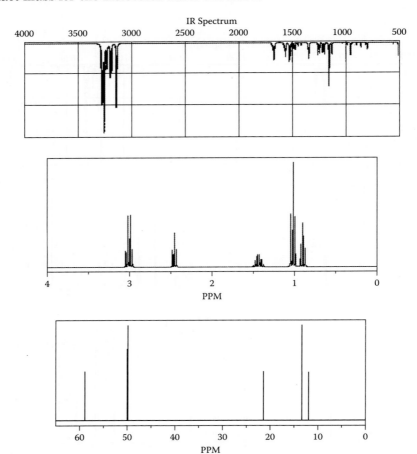

14.93 Give the structure for the molecule with the following spectra:
Exact mass for the molecular ion = 82.0419
IR: 2980, 1695, 1250, 740 cm^{-1}
^{1}H NMR: 6.57 (m, 1H), 6.07 (d, 1H), 2.98 (t, 2H), 2.07 (m, 2H) ppm
^{13}C NMR: 209.0, 164.9, 134.2, 34.0, 29.1 ppm

14.94 Give the structure for the molecule with the following spectra:
MS: M (164) 100%, M+1 (165) 12.21%, M+2 (166) 0.945%. The molecular ion is rather weak, but there is a prominent daughter ion at m/z 146 and another at m/z 77.
IR: broad peak at 3050, 3000–3860, 1460 cm^{-1}
^{1}H NMR: 7.54–7.38 (broad s, 5H), 3.65 (s, 1H; the peak is diminished when treated with D$_2$O), 1.77 (q, 4H), 0.90 (t, 6H) ppm
^{13}C NMR: 128.8, 128.1, 126.0, 144.4, 78.6, 34.2, 7.8 ppm

14.95 Give the structure for the molecule with the formula C$_9$H$_{12}$O$_2$ and the following spectra:
IR: weak peaks at 3015–2860, 1740, 1620, 1280, 1140, 780 cm^{-1}
^{1}H NMR: 7.93–7.34 (m, 4H), 4.30 (q, 2H), 2.34 (s, 3H), 1.29 (t, 3H) ppm
^{13}C NMR: 165.9, 142.7, 129.8, 128.9, 127.1, 60.9, 21.3, 14.3 ppm

14.96 Give the structure for the molecule with the following spectra:
MS: M (178) 100%, M+1 (179) 15.54%, M+2 (180) 1.21%
IR: 3118–3020, 1632, 1498, 765, 691 cm⁻¹
¹H NMR: there is only one broad peak at 7.38–7.59 (m, 1H) ppm
¹³C NMR: 132.3, 128.4, 128.3, 17.4, 89.7 ppm

14.97 Give the structure for the molecule with the formula $C_8H_{17}NO$ and the following spectra:
IR: 3265, 3235, 3175, 1956, 1638, 1426 cm⁻¹
¹H NMR: 8.03 (broad s, 1H; somewhat diminished when treated with D_2O), 3.81 (m, 1H), 2.13 (t, 2H), 1.53–1.51 (m, 4H), 1.00 (d, 6H), 0.90 (t, 3H) ppm
¹³C NMR: 173.8, 44.7, 36.5, 27.8, 23.2, 21.7, 13.4 ppm

14.98 Give the structure for the molecule with the formula C_9H_9ClO and the following spectra:

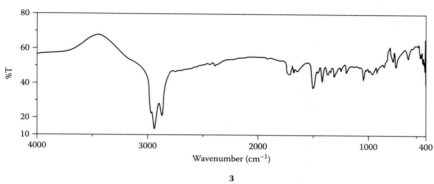

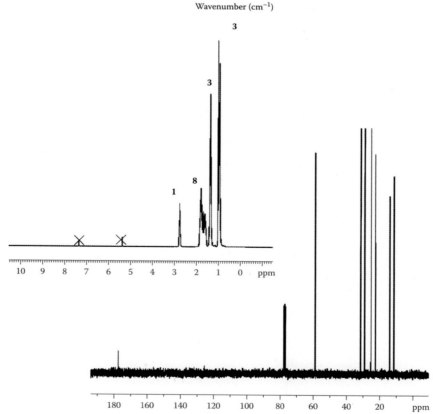

14.99 Give the structure for the molecule with the following theoretical spectra. The exact mass of the molecular ion is 96.0939.

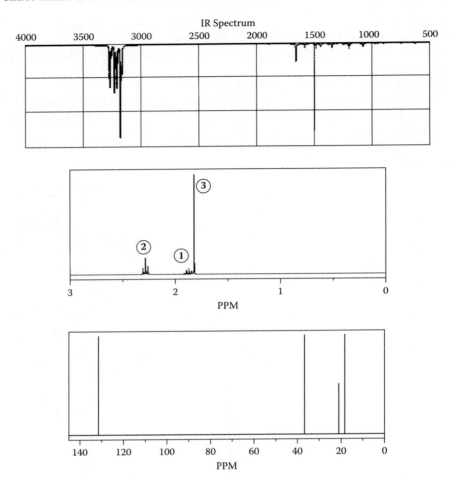

Organometallic Reagents

15

Thus far, organic molecules have been discussed that contain halogen, nitrogen oxygen, sulfur, and, of course, carbon and hydrogen. A carbon–boron bond is important in hydroboration (Chapter 10, Section 10.4.3) as is a carbon–mercury bond (Section 10.4.5). The carbon–mercury bond is an example of a class of compounds known as organometallics (organic molecules that incorporate one or more metal atoms). Indeed, carbon can also form bonds to many metals to give a species C–M, where M is the metal. The more common metals include lithium, magnesium, and copper; organometallic compounds based on these metals are introduced in this chapter. Other organometallic compounds will be introduced as they are required.

To begin this chapter, you should know the following:

- **the fundamental properties of group 1 and 2 metals, as well as group 11 and 12 metals (general chemistry; Chapter 3, Section 3.1)**
- **functional groups with polarized covalent bonds (Chapter 5, Sections 5.4, 5.6, and 5.9)**
- **structural differences and nomenclature for alkyl halides (Chapter 4, Section 4.6; Chapter 5, Section 5.6.1).**
- **the structure of metal halides (general chemistry; Chapter 3, Section 3.2)**
- **bond polarization based on electronegativity and how to identify it (Chapter 3, Section 3.7)**

- the structural features that constitute a weak acid and the reaction of a strong base with a weak acid (Chapter 5, Section 5.7; Chapter 6, Section 6.3)
- how to define and recognize a base and base strength (Chapter 2, Section 2.4; Chapter 6, Sections 6.3 and 6.4)
- how to define and recognize Lewis acids and Lewis bases and how they react (Chapter 2, Section 2.5; Chapter 6, Section 6.5)
- how to define and recognize a nucleophile and understand that a nucleophile can donate two electrons to a positive carbon or to a carbon with a positive dipole to form a new bond to that carbon (Chapter 6, Section 6.7; Chapter 11, Sections 11.2 and 11.3)
- the nature of radicals and carbanions (Chapter 7, Sections 7.4.2 and 7.4.3)
- simple reactions of radicals (Chapter 10, Section 10.8; Chapter 11, Section 11.9)
- the S_N2 transition state and the fundamental mechanistic details of the S_N2 reaction (Chapter 11, Sections 11.1 and 11.2)
- the fundamentals of equilibrium versus nonequilibrium reactions (Chapter 6, Section 6.1; Chapter 7, Section 7.10)
- chirality and stereogenic centers (Chapter 9, Sections 9.1, 9.3, and 9.5)

This chapter will show that alkyl halides react with lithium or magnesium to generate a new C–metal bond in an organomagnesium or an organolithium compound. The bond polarization is such that the carbon atom in the C–M bond of these organometallic compounds is $\delta-$. This property leads to reactions of organolithium and organomagnesium compounds as nucleophiles as well as bases. Organolithium reagents react with cuprous salts, Cu(I), to give a different class of organometallic known as organocuprates, which react as nucleophiles with alkyl halides to give alkanes.

When you have completed this chapter, you should understand the following points:

- **A Grignard reagent is an organomagnesium compound formed by the reaction of an alkyl halide with magnesium metal.**
- **Grignard reagents are strong bases, but they do not react with alkyl halides very well unless a transition metal catalyst is present.**
- **Organolithium reagents are formed by the reaction of alkyl, aryl, or vinyl halides and lithium metal. Organolithium reagents react with alkyl halides to give a new organolithium reagent via metal–halogen exchange.**
- **Organolithium reagents are strong bases and good nucleophiles.**
- **Organocuprates can be prepared from organolithium reagents and cuprous halides. Organocuprates react with alkyl halides to form a new C–C bond in the product.**

- **A molecule with a particular functional group can be prepared from molecules containing different functional groups by a series of chemical steps (reactions). This process is called synthesis: The new molecule is synthesized from the old one (see Chapter 25).**
- **Spectroscopy can be used to determine the structure of a particular molecule and can distinguish the structure and functionality of one molecule when compared with another (see Chapter 14).**
- **There are biological applications of organometallic compounds.**

15.1 Introducing Magnesium into a Molecule

In previous chapters, cyanide was used as a readily available and excellent nucleophile in S_N2 reactions (Chapter 11, Section 11.2). Similarly, terminal alkynes are weak acids that react with bases such as sodium amide to give the corresponding alkyne anion (Section 11.3.6). Alkyne anions are easy to form and they also react as nucleophiles in S_N2 reactions. Both the cyanide anion and an alkyne anion are important because they are *carbon nucleophiles and react to form carbon–carbon bonds*. Another class of carbon nucleophiles is formed by the reaction of alkyl halides R–X with alkali metals (where X = halogen; see Chapter 5, Section 5.6.1). The alkali metals (groups 1 and 2) include sodium, lithium, and magnesium, and the product of this reaction contains a carbon–metal bond (C–Na, C–Li, C–Mg). Such compounds are generically called *organometallics*. An early, important class of organometallic reagents is the organomagnesium reagents, commonly called Grignard reagents.

How can a metal be incorporated into an organic molecule? If the metal is magnesium, it is surprisingly easy. Magnesium reacts directly with alkyl halide $R_3C–X$ (X = Cl, Br, I) to give an organometallic product, **1**, if the proper solvent is used. Organometallic **1** is known as a ***Grignard reagent***, named after Françoise Auguste Victor Grignard (France; 1871–1935; Nobel Prize, 1912), who prepared and characterized these compounds; it is clear that magnesium has inserted into the C–Br bond of the alkyl halide. The work of Grignard is derived from earlier work by Philippe Antoine Barbier (France; 1848–1922), who studied similar reactions, but did not characterize the organomagnesium compound that was formed.

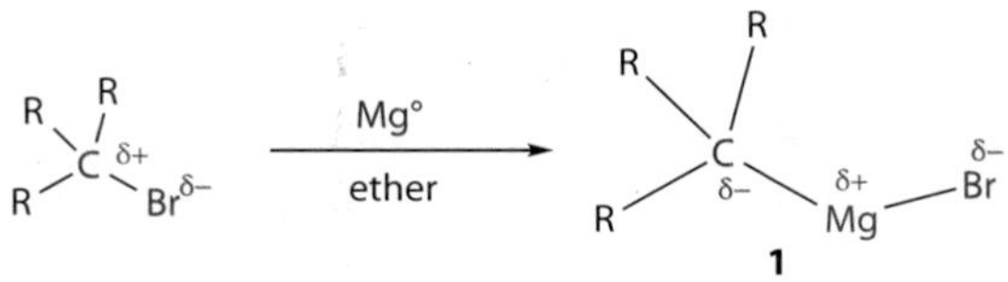

Grignard found that premixing magnesium metal and a halide generated a reagent that reacted with aldehydes and ketones. These Grignard reagents react with ketones or aldehydes in what is called a ***Grignard reaction,*** which will be discussed in Chapter 18 (Section 18.4) in the context of the chemistry of aldehydes and ketones. This discovery was critically important for two reasons. First, it constituted a new method to make carbon–carbon bonds and, second, it showed that an organomagnesium reagent is formed as the reactive species.

$$\boxed{^{\delta+}\text{C}\!-\!\text{Br}^{\,\delta-}}\ \text{CH}_3\text{CH}_2\text{Br}\ \ +\ \ \text{Mg}^\circ\ \xrightarrow{\ \text{ether}\ }\ \text{CH}_3\text{CH}_2\text{MgBr}\ \boxed{^{\delta-}\text{C}\!-\!\text{Mg}^{\,\delta+}\!-\!\text{Br}}$$
$$\qquad\qquad\qquad\qquad \textbf{2}\qquad\qquad\qquad\qquad\qquad\qquad\qquad \textbf{3}$$

Grignard won a Noble Prize in 1912, only 12 years after the initial discovery of compounds such as **1**. The importance of the discovery was immediately apparent because reagents such as **1** constituted a new class of organic compounds where the carbon is polarized $\delta-$. In bromoethane, and in all common alkyl halides, the dipole induced by bromine leads to $^{\delta+}\text{C}\!-\!\text{Br}^{\delta-}$, as shown in **2**. This bond polarization is the result of the heteroatom being more electronegative than carbon. The S_N2 reaction in Chapter 11 (Section 11.2) and the E2 reaction in Chapter 12 (Section 12.1) exploited the fact that alkyl halides have a $\delta+$ carbon. The periodic table shows that ***carbon is more electronegative than magnesium*** (see Chapter 3, Section 3.7.1, for a discussion of electronegativity). This means that the electron density in a C–Mg bond is distorted toward carbon, making carbon $\delta-$ and magnesium $\delta+$ (see **3**). No one had characterized such a reagent prior to Grignard's work and, in modern terminology, it means that carbon can function as a base or a nucleophile. In effect, Grignard reagents such as **1** or **3** are ***reactive carbanion equivalents***.

The reaction between magnesium metal and an alkyl halide works best in the presence of an ether solvent such as diethyl ether or THF. The oxygen of the ether is as a Lewis base in this system, and it forms a coordination complex with the magnesium to generate a species that looks like **4**. The **arrow bond** (formally known as a **dative bond**; see Chapter 2, Section 2.5) is *less than a formal covalent bond* and results from donation of electrons from the electronegative oxygen atom the electron-deficient magnesium. Remember from Chapter 3 (Section 3.2) that group 2 elements usually react by electron transfer (their first ionization potential is low). If **4** donates one electron to the halogen atom (X) in the alkyl halide (R–X), the magnesium is deficient by one electron and becomes a **radical cation** (+ •), as shown in **5**. The halogen atom (X) accepts the one electron to become electron rich, and the resulting species is a **radical anion** (– •), **6**.

The halide radical anion (**6**) dissociates to give a carbon radical (R•, where a radical is defined as a reactive intermediate with one electron in a p-orbital; see Chapter 7, Section 7.4.3) and a halide anion, X⁻. If X⁻ and **5** react with each other, a Mg–X bond is formed and the magnesium complex becomes a radical (**6**) rather than a radical cation. Note that the radical cation can accept electron density from the electron-rich X⁻. If **6** subsequently reacts with the carbon radical (R•), Grignard reagent (**7**) is generated. ***This sequence is taken to be the mechanism of Grignard reagent formation from alkyl halides*** **1**. The ether stabilizes the C–Mg–X bond as shown in **7** and thereby stabilizes the Grignard reagent. The fact that an ether can stabilize a Grignard reagent by reacting as a Lewis base is particularly important for stabilization of radical cation and radical intermediates. This overall reaction that converts **2** to **3** is called a ***metal insertion reaction*** and it occurs with most alkyl halides. Benzyl chloride (**8**) reacts with magnesium to form **9**, for example, and 2-iodoheptane (**10**) reacts to form **11**. Primary, secondary, and tertiary halides react with magnesium metal to form the corresponding Grignard reagent.

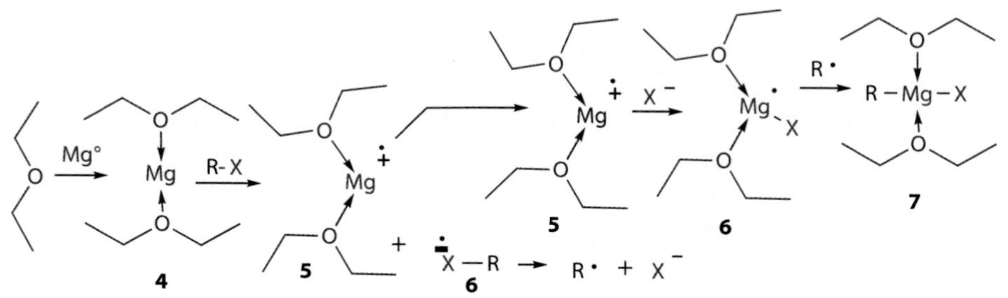

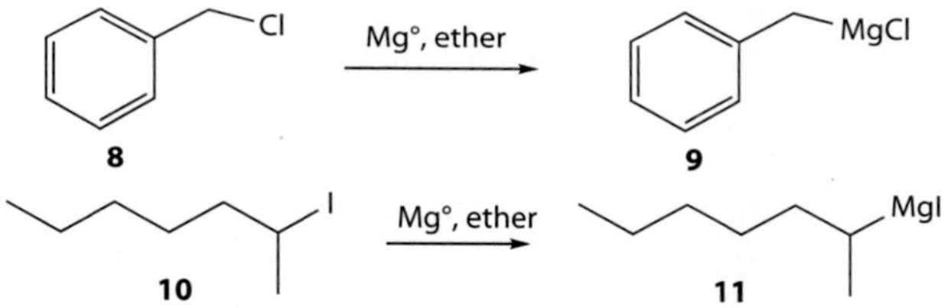

15.1 Write out the product formed when 2-bromo-2-methylhexane reacts with magnesium metal in diethyl ether.

15.2 Write out the structures of diethyl ether and THF.

15.3 Draw the Grignard reagent formed when 3-bromo-3-ethylheptane reacts with magnesium metal in diethyl ether.

Unfortunately, the Grignard reagent drawn as **1** is not such a simple structure. Grignard reagents actually exist as an equilibrium mixture of several organometallics, including the starting halide R_3CBr as well as **1**, **12** (a dialkylmagnesium compound), and the dimeric forms **13** and **14**. In addition, some trimeric forms exist in low concentrations. This mixture is called the ***Schlenk equilibrium*** and it constitutes a more complete representation of the Grignard reagent formed when an alkyl halide reacts with magnesium than does **1**. The position of this equilibrium depends on the solvent, but *in ether solvents the "normal" Grignard reagent **1** predominates. For this reason, the RMgX representation is used when discussing Grignard reagents.* This mixture is named after Wilhelm Johann Schlenk (Germany; 1879–1943), who discovered these reactions. Note that Schlenk is also credited as the discoverer of organolithium reagents, which are discussed in Section 15.5.

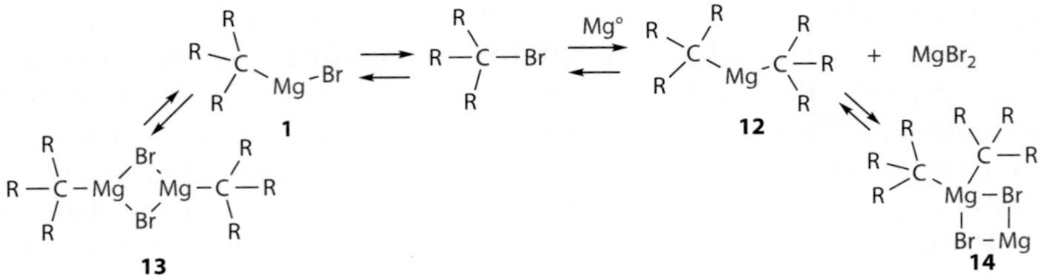

As mentioned, most alkyl halides react with magnesium, and "R" is often used in the structure to show that an alkyl halide (R–X) reacts with magnesium to give the **Grignard reagent (RMgX)**. As shown for **1**, a Lewis acid (MgX$_2$) is also formed as a product. Although the presence of the Lewis acid can influence the Schlenk equilibrium in some cases, those effects will be ignored in all reactions presented in this book.

It is clear from the Schlenk equilibrium that the magnesium is mobile with respect to the C–Mg bond. Consider the Grignard reagent **15** derived from 2R-bromopentane. The magnesium must be inserted into the C–Br bond, but the carbon is a stereogenic center (Chapter 9, Section 9.1). Does formation of a Grignard reagent from a chiral, nonracemic halide lead to a chiral nonracemic Grignard reagent? In most cases, the answer is no. Experimentally, **15** is *not* enantiopure, but it is *racemic*.

In other words, Grignard reagent **15** and all similar reagents exist as an equilibrating mixture of both R and S enantiomers (**15A** and **15B**, a racemic mixture, as indicated by the "squiggle" line in **15** or simply as the straight line). Because this equilibrium consists of both enantiomers of **15**, insertion of magnesium into the C–Br bond of the enantiopure bromide will generate a racemic Grignard reagent. In part, this is due to the mobility of the magnesium in the Schlenk equilibrium where the magnesium "comes and goes" from the carbon of **15**. **The "bottom line" is that a Grignard reagent formed from a chiral, nonracemic halide will not be chiral nonracemic, but rather *chiral racemic*.** (The terms chiral, nonracemic and chiral racemic are defined in Chapter 9.)

15.4 Draw the product formed when each of the following reacts with Mg: 2-iodobutane in ether, 1-bromo-1-butene in THF, and 1- chloropentane in ether.

15.5 Draw the product of a reaction between 2S-iodo-5,5-dimethyl-hexane and magnesium in diethyl ether.

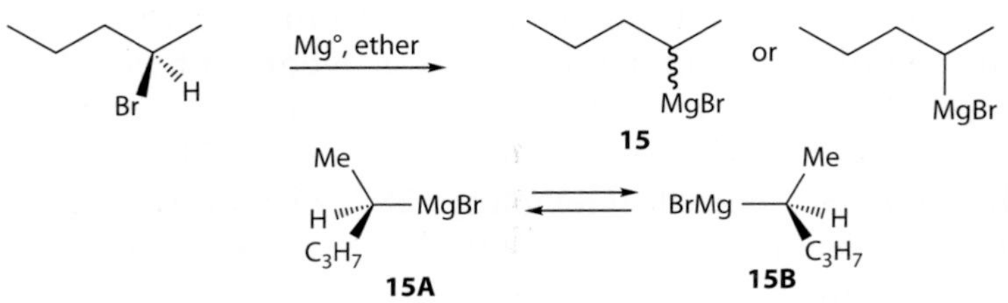

15.2 Reaction of Aryl and Vinyl Halides with Magnesium

Alkyl halides react with magnesium to form Grignard reagents in the presence of an ether solvent. The reaction is not limited to alkyl halides, however; Grignard reagents can be generated from molecules that have a halogen connected to an sp^2 carbon atom, vinyl halides, and aryl halides. Vinyl halides were discussed in Chapter 10 (Section 10.6.1) and aryl halides (halogen connected to a benzene ring) will be discussed in Chapter 21 (Section 21.2). As a rule, diethyl ether is used in reactions with alkyl halides, but ethers that are superior Lewis bases must be used as a solvent for vinyl halides or aryl halides.

The Grignard reagents formed from those halides (C=C–MgX or Ar–MgX) are less stable than those from aliphatic alkyl halides (RMgX). Because they are less stable, more stabilization is required from the solvent. In other words, *Grignard reagents from vinyl halides and aryl halides are readily formed only in solvents that are more basic than diethyl ether*. Tetrahydrofuran (THF; see Chapter 26, Section 26.4.2) is commonly used. A comparison of diethyl ether and THF shows that the alkyl groups in THF are "tied back" into the five-membered ring, and the electrons on oxygen are more available for donation than in diethyl ether. In other words, THF is a stronger Lewis base than diethyl ether.

As mentioned, two classes of halides that attach the halide to a π-bond are alkenyl (vinyl) halides such as **16** and aryl halides such as **17**. When a halogen atom is attached to a π-bond, it is more difficult for the magnesium to insert into the C–X bond. The more basic solvent (THF) provides additional stabilization to the intermediates and to the final Grignard reagent. A simple example of an aryl halide is bromobenzene (**18**), and reaction with magnesium in THF leads to the corresponding Grignard reagent, **19**. Note that the Lewis acid by-product $MgBr_2$ is omitted. *In all reactions that generate a Grignard reagent, only the organometallic product is shown*. Aryl halides are formed by reaction of benzene derivatives with halogens in the presence of Lewis acids (see Chapter 21, Section 21.3), but this chemistry is not discussed until Chapter 21. For the moment, simply use aryl halides to form the appropriate organometallic.

In Chapter 10 (Section 10.6.1), the reaction of HCl or HBr with an alkyne led to vinyl halides (**16**). The term "vinyl" is synonymous with alkene (the C=C unit). As with aryl halides, formation of a Grignard reagent requires a stronger Lewis base, such as THF. Reaction of the vinyl halide 3-chloro-3-hexene (**37**) with magnesium and THF generates the vinyl Grignard reagent 3-hexenyl-magnesium chloride (**21**).

15.6 Draw the structure of the products formed by the reaction of 4-phenyl-1-bromobenzene and magnesium, in THF.

15.7 Draw the alkyne required to generate **20** by reaction with HCl and comment on the stereoselectivity of this reaction.

15.8 Draw the product of a reaction between magnesium and 3-bromo-4-ethyl-3*E*-octene.

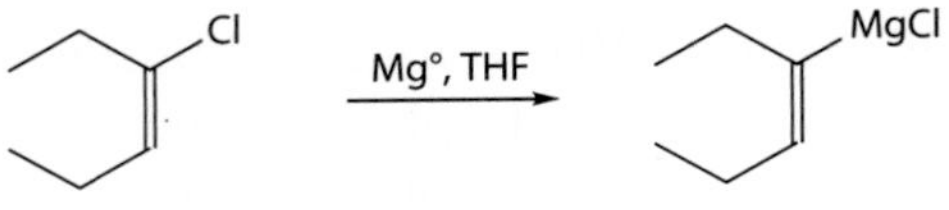

20   **21**

15.3 Grignard Reagents Are Bases

The carbon of the C–Mg–X unit is polarized $\delta-$ and can therefore donate electrons to an electrophilic atom. In other words, Grignard reagents are nucleophiles, and this aspect of reactivity is discussed in Section 15.4 and in Chapter 20. A Grignard reagent also donates electrons to a proton, so it is a Brønsted–Lowry base. How strong is this base? Consider the fact that H_3C^- as a base would react with H^+ to form the conjugate acid, methane (H_4C). Clearly, methane is an extremely weak acid, making H_3C^- an extremely powerful base. If it is assumed that a Grignard reagent reacts in the same manner as C^-, then a Grignard reagent is a powerful base that reacts with very weak acids. Indeed, water, alcohols, terminal alkynes, and even amines react with a Grignard reagent to form the corresponding alkane (the conjugate acid) and the conjugate base of each of the weak acids mentioned.

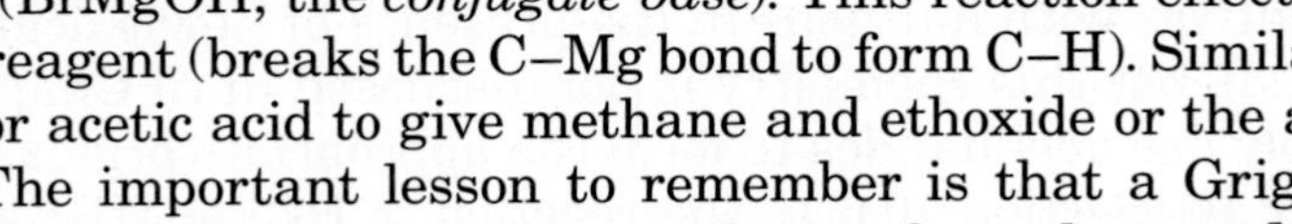

Methylmagnesium bromide (**22**) is formed by reaction of iodomethane and magnesium and will react with water to form methane (the *conjugate acid*) and hydroxide ion (BrMgOH, the *conjugate base*). This reaction effectively destroys the Grignard reagent (breaks the C–Mg bond to form C–H). Similarly, **22** reacts with ethanol or acetic acid to give methane and ethoxide or the acetate anion, respectively. The important lesson to remember is that a Grignard reagent must be prepared in an aprotic solvent and, once formed, **must be kept away from water, alcohols, or any acid**.

15.9 Draw the conjugate base formed when a Grignard reagent $CH_3CH_2CH_2MgX$ reacts with (a) water, (b) methanol, (c) 1-propyne, and (d) dimethylamine.

15.10 Write out the reaction products formed when the Grignard reagent of 1-bromopropane is mixed with acetylene.

15.11 Is a Grignard reagent formed when benzyl bromide is heated with magnesium metal in a mixture of diethyl ethyl and ethanol as the solvent? Explain!

15.4 Grignard Reagents Are Poor Nucleophiles with Alkyl Halides

If a Grignard reagent donates electrons to a carbon atom, it is classified as a nucleophile. In Chapter 11 (Section 11.3.6), the S_N2 reaction of cyanide or of an alkyne anion with alkyl halides were examples of carbon nucleophiles. A Grignard reagent is characterized by a $\delta-$ carbon atom, so it is quite reasonable to ask if a Grignard reagent will react with an alkyl halide. Grignard reagent **24** is derived from 1-bromobutane (**23**). An S_N2 coupling reaction of **24** with 1-bromopropane (**25**) would form the alkane heptane, *if it occurs*. However, the product is *not* heptane. There is either no reaction or **24** decomposes or reacts in other ways. In general, *a simple Grignard reagent derived from an alkyl halide does not react with an alkyl halide to give a coupling product*. If true, then *a Grignard reagent from simple aliphatic alkyl halides is a poor nucleophile in an S_N2 reaction with a simple aliphatic halide*.

An exception to this statement occurs when the Grignard reagent is unusually reactive and/or the alkyl halide is particularly reactive. Benzyl bromide and allyl bromide are examples of such reactive alkyl halides. In addition, the Grignard reagents from benzylic halides or allylic halides are more reactive than those derived from simple aliphatic halides. If the Grignard reagent derived from 1-bromo-2-hexene (see **26**) reacts with benzyl bromide, reaction does occur to give the coupling product, 1-phenyl-4-octene (**27**).

How does one decide whether or not a Grignard will react with an alkyl halide? A *working assumption is required*. **Assume that Grignard reagents from benzylic or allylic halides will react with a benzylic or allylic halide to give a coupling reaction. Also assume that all other Grignard reagents do not give coupling**. Note that aryl Grignard reagents and vinyl Grignard reagents (see **19** and **21**) are *less* reactive than Grignard reagents derived from aliphatic halides. Therefore, **assume that aryl and vinyl Grignard reagents do not react with alkyl halides by coupling**.

15.12 Draw the organometallic formed when benzyl bromide reacts with Mg in ether.

15.13 Draw the product formed when 1-bromo-3-methyl-2-butene reacts with Mg in ether and then reacts with 1-bromo-3-methyl-2-pentene.

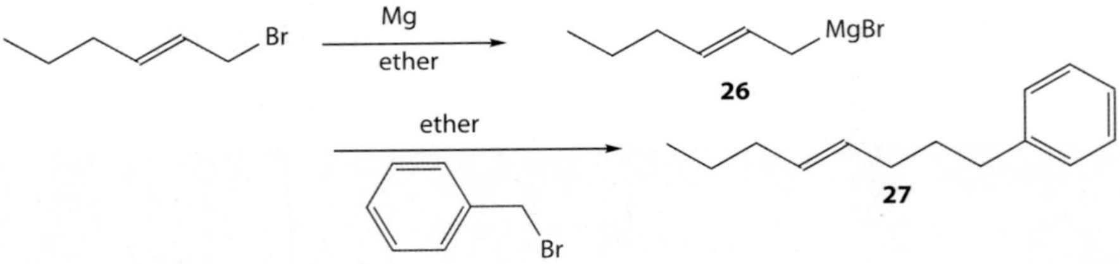

An experimental method allows one to circumvent the generally poor reactivity of Grignard reagents with alkyl halides. *A Grignard reagent reacts with an alkyl halide if a transition metal salt is added to the reaction.* If Grignard reagent **24** is first mixed with cuprous bromide (CuBr) or ferric bromide (FeBr$_3$), a reaction with the metal produces a new organometallic intermediate that is reactive enough to react with alkyl halides. In this specific case, formation of **24** in the presence of either CuBr or FeBr$_3$, followed by addition of **25**, leads to heptane as the major product, as shown. Note that the reaction generates a new carbon–carbon bond, shown in **green** in the illustration. This variation is called the **Kharasch reaction** and the important thing to remember is that *a transition metal is needed in order for common Grignard reagents to react with common alkyl halides*. The reaction is named for Morris S. Kharasch (Ukraine/United States; 1895–1957). With no transition metal additive, *assume* that simple aliphatic Grignard reagents do not react with alkyl halides, but that aliphatic Grignard reagents with CuBr or FeBr$_3$ and alkyl halides do give the coupling product.

15.14 Draw the product formed when the Grignard reagent derived from 2-bromopentane + Mg in ether reacts with 1-iodo-3-methylpentane in the presence of CuBr.

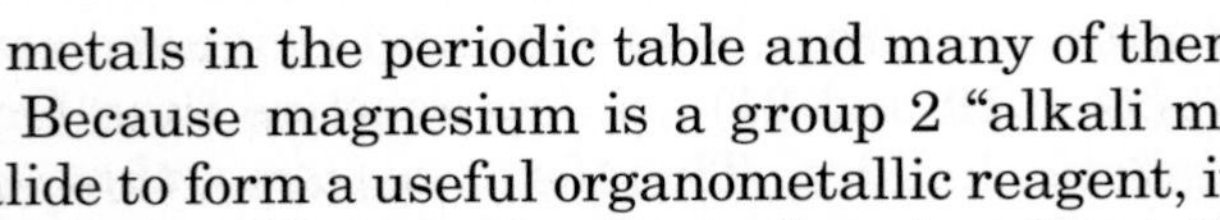

15.5 Organolithium Reagents

There are other metals in the periodic table and many of them react with carbon derivatives. Because magnesium is a group 2 "alkali metal" that reacts with an alkyl halide to form a useful organometallic reagent, it is reasonable to look for other alkali metals that may do similar chemistry. Reaction with the group 1 metal lithium (Li) gives an organometallic called an **organolithium reagent** (RLi).[2] The reaction of an alkyl halide such as 1-bromobutane (**23**) and lithium metal (which exits primarily as dilithium, Li$_2$) forms **28**, where lithium has replaced bromine. This organolithium reagent is known as *n*-butyllithium or 1-lithiobutane, and it is characterized by a C–Li bond. As with Grignard reagents with a C–Mg bond, the C–Li bond is polarized with a δ– carbon and a δ+ lithium (see **28**) because carbon is more electronegative than lithium (carbon is to the right of lithium in the periodic table; see Chapter 3, Section 3.7.1).

15.5.1 The Reaction of Alkyl Halides with Lithium

As just described, 1-bromobutane reacts with lithium metal (Li°) to form the organolithium reagent 1-lithiobutane, **28**. The other product of this reaction is lithium bromide (LiBr). This reaction can be done in solvents that range from diethyl ether to hexane, but ethers are used more often. The reaction can be generalized so that an alkyl halide R–X (**29**) reacts with lithium metal to form an organolithium reagent (**30**) and LiX. Although this reaction is related to the insertion reaction of magnesium with alkyl halides, there are important differences. Magnesium is in group 2 and the C–Mg–X bond of a Grignard reagent is the result of the bivalent (forms two bonds) magnesium. Lithium is in group 1 and is monovalent (forms one bond), so an organolithium reagent forms the C–Li bond found in **28**. The reaction that forms **30** proceeds by a somewhat different mechanism when compared to the reaction with magnesium.

$$R-X \ + \ Li_2 \ \xrightarrow{\text{ether or hexane}} \ R-Li \ + \ LiX$$
$$\mathbf{29} \qquad\qquad\qquad\qquad\qquad \mathbf{30}$$

15.15 **Draw and name the products that result when lithium metal reacts with 2-bromopropane, 2-bromo-1-butene, and 1-iodopentane (all in ether).**

   For a more detailed look at this reaction, iodomethane reacts with lithium metal, which is assumed to exist as a simple dimer (Li–Li). The products are lithium iodide and CH_3Li (methyllithium, **33**). When the lithium dimer comes close to the C–I bond of iodomethane, the polarized C–I bond induces a polarized Li–Li structure (**an induced dipole**) and the transition state of the reaction is taken to be **31**. Rather than transferring two electrons, the Li–Li bond breaks with transfer of only one electron (*homolytic cleavage*; remember that Li is in group 1), which leads to formation of a methyl radical (•CH_3) and a lithium radical (•Li), as well as a lithium cation and an iodine anion (see **32**). Transition state **31** represents the transfer of single electrons to generate radicals. When the methyl radical and the lithium radical combine, each donates its single electron (*note the use in the illustration of the single-headed arrow to indicate transfer of one electron*) to form a new bond. When these two radicals combine, the product is the organolithium reagent (**33**; abbreviated MeLi) and lithium iodide (LiI). This ***single electron transfer mechanism*** effectively interchanges the halogen (iodide) and one lithium atom (Li) and is therefore referred to as a ***metal–halogen exchange*** reaction.

Because an organolithium reagent (RLi) is thought to proceed via a radical intermediate (see •CH₃ in **32**), those radicals may react to form other products (see Chapter 11, Section 11.9, for reactions of radicals). If two radicals come together, they react to form a new molecule. Each radical donates one electron to form a new σ-bond (composed of two electrons) in what is known as ***radical coupling***. The coupling of two bromine radicals in Chapter 11 (Section 11.9.2) was a radical coupling. The coupling of •CH₃ and •Li to form CH₃Li (**33**) is a radical coupling.

However, once •CH₃ is formed it may couple to another methyl radical to generate ethane (CH₃CH₃), which is shown in the illustration in **violet** using single-headed arrows. Radical coupling of two alkyl radicals to form an alkane in the presence of the metal is called ***Wurtz coupling***. This is a competing reaction in many direct reactions of alkyl halides and lithium metal that produce organolithium reagents. Note that the reaction of an organometallic such as a Grignard reagent or an organolithium reagent and an alkyl halide is not Wurtz coupling. Wurtz coupling is the coupling of alkyl radicals to form alkanes that accompanies formation of the organometallic compound. In the reaction of bromobutane with lithium metal to form *n*-butyllithium (**28**), for instance, coupling of two butyl radicals (**34**) leads to octane (**35**) as a significant by-product.

15.16 Write out the organolithium and the Wurtz coupling products expected during the reaction of each of the following with lithium: 1-bromobutane, 2-iodopentane, and 3,4-dimethyl-1-iodohexane.

15.17 Write out the Wurtz product formed when phenyllithium is formed by the reaction of bromobenzene and lithium metal.

Wurtz coupling is not as much of a problem in ether solvents as it is when hydrocarbon (alkane) solvents are used in the reaction of lithium metal and alkyl halides. Unfortunately, many organolithium reagents slowly *react* with ethers to give decomposition products. This means that if an organolithium reagent is formed in ether and allowed to stand for a long period of time, the organolithium reagent will be converted into something else and lost (the molar concentration of the organolithium reagent will be diminished). This reaction does not occur as rapidly in a solvent such as hexane, but Wurtz coupling is more of a problem in the hydrocarbon solvent than it is in ether. To get around these problems, the organolithium reagent is often formed in ether and then the ether solvent is partially removed and replaced with hexane or another hydrocarbon, or a mixed hexane–ether solvent is used.

The preceding discussion certainly suggests that organolithium reagents are highly reactive and their preparation can be problematic. It is therefore important to know that several organolithium reagents are sold commercially, including methyllithium (**33**), *n*-butyllithium (**28**), *secondary*-butyllithium (also known as *sec*-butyllithium; the correct name is 2-lithiobutane), and *tert*-butyllithium (2-lithio-2-methylpropane, **36**).

15.18 **Draw the structure of 2-lithiobutane.**

15.19 **Draw reactions for the preparation of 2-lithiobutane from 2-bromobutane and of 36 from 2-iodo-2-methylpropane.**

As seen in Section 15.4, Grignard reagents typically give poor yields in a reaction with an alkyl halide. Highly reactive Grignard reagents that react with highly reactive halides (such as allyl halides) are the exception, but with simple aliphatic alkylation halides, the yield of coupling product is usually very poor. That section also discussed the fact that this poor reactivity is converted to good reactivity when certain transition metals are added to the Grignard reagent. How reactive is an organolithium reagent with an alkyl halide?

Many organolithium reagents are available and it is known that a tertiary organolithium is much more reactive than the secondary organolithium, which is more reactive than primary organolithium reagents. If *tert*-butyllithium (**36**) is mixed with iodoethane, a rapid reaction takes place to produce ethyllithium (**37**) and *tert*-butyl iodide (2-iodo-2-methylpropane). Although it is not intuitively obvious from the structure, the primary organolithium reagent (**37**) is more stable than the tertiary organolithium reagent (**36**), and the reaction shown favors formation of **37**.

Because of this difference in reactivity and stability, *when a tertiary organolithium reagent such as tert-butyllithium (36) reacts with a primary alkyl halide, a new organolithium reagent is formed if the product is more stable.* **Indeed, *the reaction of 36 with a primary alkyl halide (mostly the iodide) is a very common method of producing organolithium reagents that are not commercially available*.** As mentioned, this ***metal–halogen exchange*** reaction works best when the tertiary organolithium reagents react with a primary iodide (bromides and chlorides are less reactive), but the exchange reaction with secondary halides is much slower and often results in poor yields. Exchange with tertiary halides is even more difficult from the perspective of obtaining a new organolithium reagent that is uncontaminated with other products.

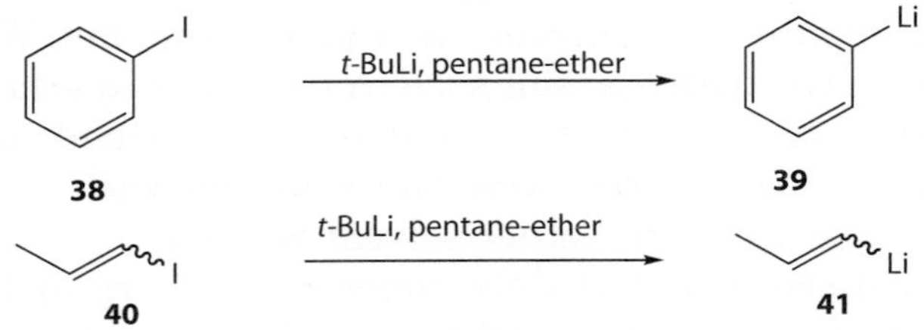

Both aryl halides such as iodobenzene (**38**) and vinyl halides such as 1-bromo-1-propene (**40**) react slowly with lithium metal, but faster with *tert*-butyllithium (**36**), to give the corresponding organolithium reagent (**39**, phenyllithium) or **41** (1-lithio-1-propene), respectively. Note that the vinyllithium reagent (**41**) is formed as a mixture of stereoisomers, as indicated by the "squiggle" line in the illustration. Recall the lack of selectivity in the reaction of magnesium with a chiral alkyl halide (see Section 15.1). The reactions use a shorthand notation for *tert*-butyllithium (**36**): *t*-BuLi. Both aryllithium reagents such as **39** and vinyllithium reagents such as **41** react similarly to other organolithium reagents.

15.20 Draw the starting materials required to prepare 2-lithio-2-pentene.

15.5.2 Organolithium Reagents Are Potent Bases

Organolithium reagents have a carbon connected to a lithium atom: $\delta-$. Therefore, organolithium reagents behave as carbanions, and they are very powerful bases. If methyllithium (**33**) reacts with water, the conjugate base is LiOH (lithium hydroxide) and the conjugate acid is methane, which is a very weak acid indeed. Methyllithium, and all organolithium reagents, are considered to be very strong bases. Section 15.3 discussed the fact that Grignard reagents are strong bases, and it is known that the $\delta-$ carbon of a C–Li bond is greater than the $\delta-$ carbon of a C–Mg bond. Accordingly, organolithium reagents are stronger bases than Grignard reagents. Organolithium reagents are such strong bases that they react with the protons of very weak acids, some with pK_a of >30.

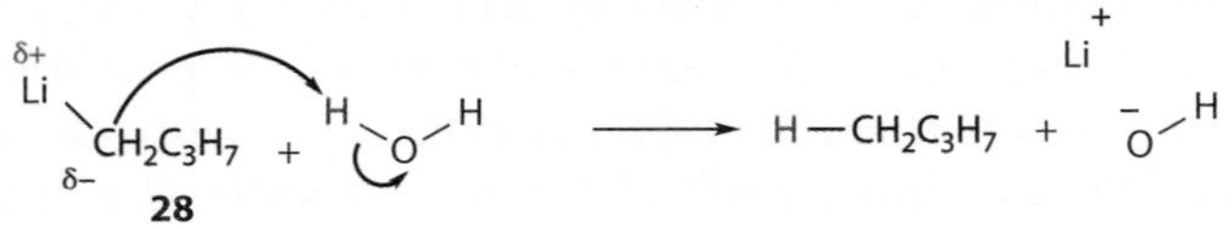

The reaction of methyllithium with the weak acid water (pK_a of 15.8) gives methane, as shown, and *n*-butyllithium (**28**) also reacts with water to give butane and the hydroxide ion (as LiOH). The reaction just described between an acid and an organolithium reagent is sometimes called ***metal–hydrogen***

exchange. It is obviously an acid–base reaction. Alcohols react as acids in the presence of an organolithium reagent. The pK_a of ethanol is about 15.7 and the pK_a of butane is >40, which means that *n*-butyllithium is a remarkably strong base in this system. Therefore, the reaction of **28** and ethanol will give butane and lithium ethoxide as the products. The reaction with ethanol can be violent and exothermic, and the great base strength of *n*-butyllithium makes ethanol a very strong acid in this reaction. Similarly, the reaction with water can be violent and exothermic because water is a slightly stronger acid than ethanol. Obviously, acetic acid is the strongest acid discussed here, and the products are butane and the acetate anion $(CH_3CO_2^-)$.

15.21 **Draw out the reaction of butyllithium and ethanol and show all products.**

15.22 **Draw out the acid–base reaction of butyllithium and acetic acid as described and show all products.**

Organolithium reagents can, in principle, react with any acid with a pK_a *lower* than that of its conjugate acid (an alkane). Because alkanes are generally assumed to have pK_a values of greater than 40, any "acid" with a pK_a of about 35 or lower probably reacts with an organolithium reagent in an acid–base reaction. This statement takes in most organic compounds that have a polarized hydrogen atom.

The basicity of organolithium reagents forces a change in perspective in what constitutes an acid. Many compounds that are known to be neutral and not thought of as acids react with organolithium reagents to give interesting conjugate bases that may be used for other purposes. Alkynes such as 1-propyne (**42**) and amines such as piperidine (**44**) are weak acids with pK_a values close to 25. In both cases, *n*-butyllithium reacts to form the anion and butane as the conjugate acid. Alkyne **42** gives alkyne anion, **43**, and **44** gives lithium piperidide, **45**. Although **45** is initially drawn as the salt to make the connection with the other reactions we have discussed, the actual structure is probably the covalent structure in the box.

Note the use of the shorthand notation for *n*-butyllithium (BuLi) and for butane (BuH). In some cases, a powerful organolithium base such as *tert*-butyllithium (**36**) can deprotonate a compound that is not normally considered to be an acid: benzene, for example. *tert*-Butyllithium reacts with benzene to give the conjugate base phenyllithium (**39**), along with 2-methyl propane (the conjugate acid).

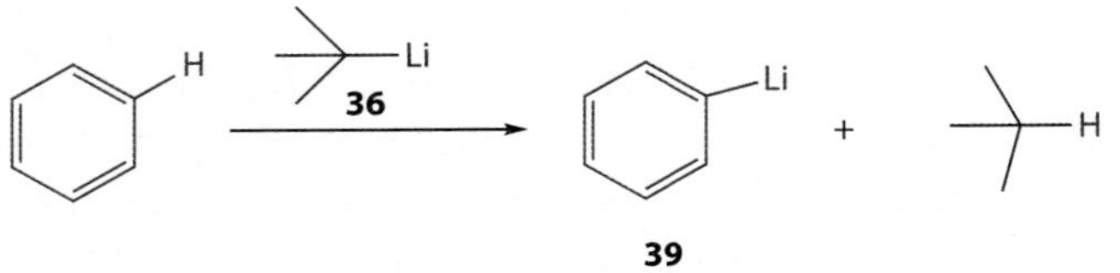

15.23 **Show all products of the reaction between 2-phenylethyne and**
_t_butyllithium; between _n_-butyllithium (1-lithiobutane) and 2,2,6,
6-tetramethylpiperidine and 1-lithiopentane.

The real value of this acid–base reaction is to transform a weak acid into
an anion by using a powerful base: the organolithium reagent. Such anions
behave as nucleophiles in various reactions. In Chapter 11 (Section 11.3.6),
alkyne anions underwent S_N2 reactions with alkyl halides. In Chapter 18
(Section 18.3.2), alkyne anions react with aldehydes and ketones. Both Grignard
reagents and organolithium reagents react as nucleophiles with aldehydes and
ketones (also described in Chapter 18, Section 18.4). Lithium amides such as
45 react as bases with aldehydes or ketones in Chapter 22 (Section 22.3). Many
such examples are discussed in this book.

15.6 Organocuprates

Both Grignard reagents and organolithium reagents have problems in reactions with alkyl halides. This was seen in Section 15.4, where addition of
CuBr or FeBr$_3$ to the Grignard reagent enhanced reactivity with alkyl halides
to give a coupling product. This boost in reactivity occurs because a new organometallic species is formed. In other words, _it is possible to generate a dif-_
ferent organometallic reagent from an initially formed organometallic
reagent, and the new reagent may react directly with an alkyl halide.
When copper is added to an organolithium reagent, the coupling reaction is
more facile than the similar reaction with Grignard reagents. The focus here
is on the reaction of an organolithium reagent with the copper CuBr or CuI.
Both are cuprous salts, or Cu(I).

When an organolithium reagent (RLi) reacts with cuprous iodide (CuI) in
ether at –10°C, a reaction takes place to generate what is called an **organocu-**
prate, R$_2$CuLi (**46**). Organocuprates with this structure are sometimes called
Gilman reagents, after Henry Gilman (United States, 1893–1986). The temperature noted in the reaction is important because many organocuprate reagents
decompose above 0°C. _Note that the reaction of RLi and CuI requires two molar_
equivalents of the organolithium reagent to react with one equivalent of copper.

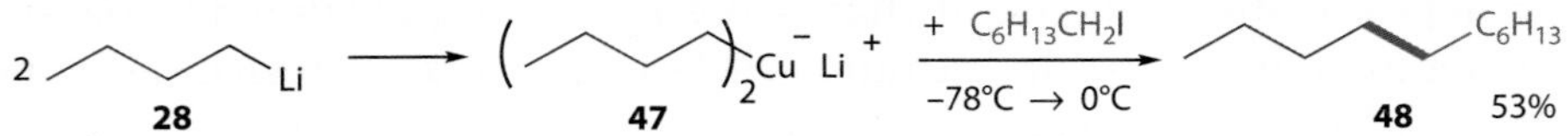

Specifically, if two equivalents of *n*-butyllithium (**28**) react with CuI, the product is lithium dibutylcuprate (**47**). Note the nomenclature that is used, where the alkyl unit of the organolithium reagent is combined with term *cuprate,* which describes the oxidation state of the copper. The carbon attached to copper in **47** is δ– (it is carbanionic), and organocuprates such as **47** are very reactive with most alkyl halides, leading to the coupling product. In one experiment, **47** was formed in THF at –78°C and then treated with 1-iodoheptane as the temperature was allowed to rise to 0°C. The product of this reaction is undecane (**48**), isolated in 53% yield.[3]

This is a very general reaction in that primary, secondary, and tertiary alkyl halides react with primary, secondary, or tertiary organocuprates. The halide can be a chloride, a bromide, or an iodide, but iodides are more reactive than bromides, which are more reactive than chlorides. Another example is the reaction of **49** with lithium diphenylcuprate to give **50**. *The reaction of alkyl halides and organocuprates is the preferred method for coupling an alkyl halide to an organometallic compound.* Many different hydrocarbons can be prepared.

15.24 **Write out the reactions and products formed when benzyl bromide is treated with dipropylcuprate and when 3-bromocyclopentene is treated with dimethyl cuprate.**

15.25 **Write out reactions for the preparation of diphenylcuprate and lithium dimethylcuprate from appropriate precursors.**

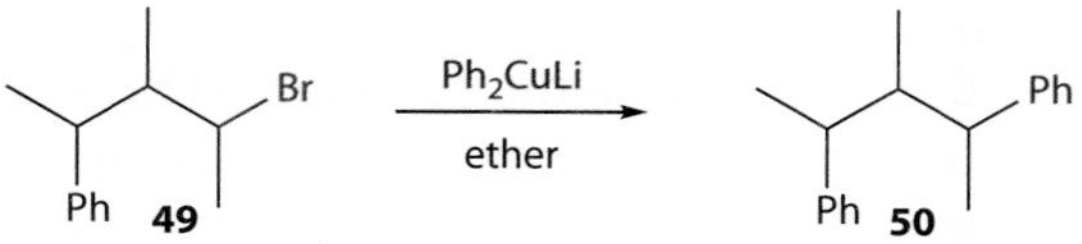

15.7 Organometallic Disconnections

The organocuprate disconnection involves making a carbon–carbon bond by reaction with an alkyl halide. The same disconnection is used for coupling an organolithium reagent, or a Grignard reagent, with an alkyl halide; however, the organocuprate reaction is by far the better one. The example is intended to show that the cuprate is derived from an organolithium reagent. The final coupled product (R–R¹) is therefore derived from another halide (R–X), which is the precursor to the organolithium reagent R–Li. Note that this disconnection represents formation of a carbon–carbon bond, which is important for building molecules (see Chapter 25).

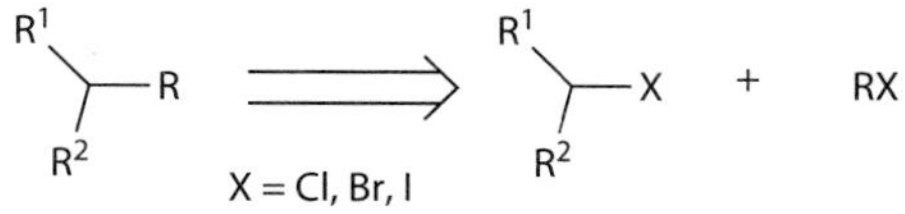

If alkane **51** is the target, a synthesis can be proposed using 2-methylpro-pene, **52**, as the starting material. Clearly, the four carbons of the starting material are involved in the portion of **51** bearing the methyl substituent, so the logical disconnection is at the indicated bond to a straight-chain butyl fragment and a second, four-carbon fragment that structurally corresponds to 2-methylpropene, **52**.

This disconnection gives the two disconnect products, but there is no func-tional group to guide assignment of donor and acceptor sites. Therefore, *assume that one of them must be C^d, a nucleophilic carbon that forces one disconnect product to correspond to a known organometallic compound.* Likewise, if one must be C^a, then it will be an alkyl halide. If the straight-chain butyl frag-ment is assigned as the donor fragment, then lithium dibutylcuprate (**53**) is a reasonable synthetic equivalent. If **53** is chosen, the C^a synthetic equivalent for the other fragment is **54**. Alkyl halide **54** must be derived from the designated starting material, **52**, because it contains the same four carbons with an identi-cal structural motif.

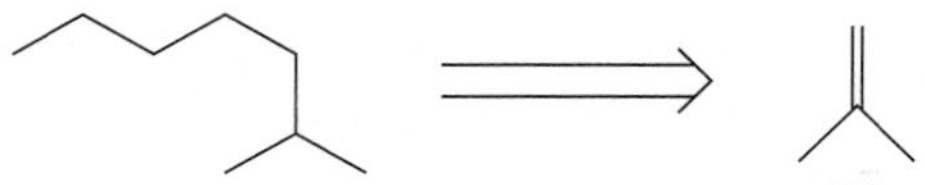

51   52

Functional group transformations are required because no additional car-bon–carbon bonds must be formed. Bromides and alkyl halides in general can be formed from alkenes by reaction with HX (HBr in this case). If **52** reacts with HBr, however, the product is 2-bromo-2-methylpropane rather than **54** (Chapter 10, Section 10.2). It is possible to react HBr with an alkene in the presence of a peroxide (Chapter 10, Section 10.8.2) to give a primary bromide. However, if it is recognized that a bromide is prepared from an alcohol, then **54** may be prepared from **55**, which is directly available from **52** by hydroboration (Chapter 10, Section 10.4.3).

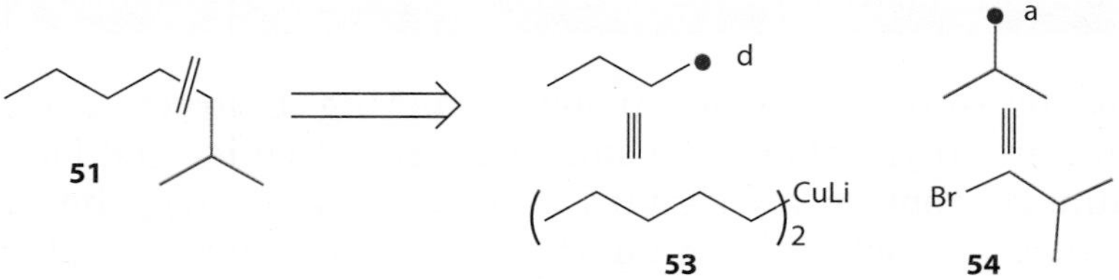

51

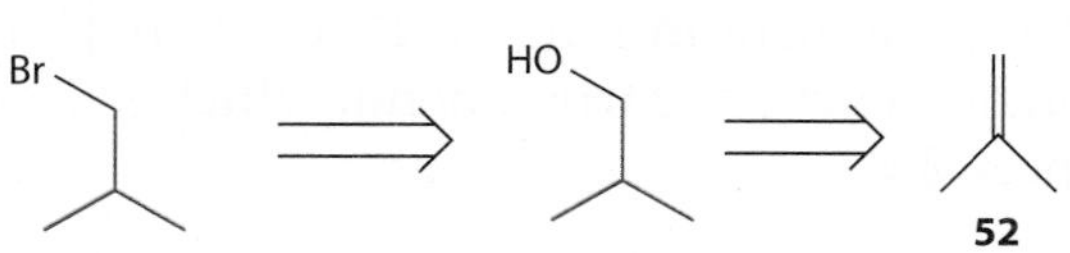

54   55   52

With this disconnection analysis in hand, a synthesis is straightforward. If **52** is treated with borane and then oxidized, the product is alcohol **55**. This alcohol reacts with a brominating reagent such as PBr_3 to give **54**, which then reacts with lithium dibutylcuprate **53** to give the target **51**. Reagent **53** is prepared from *n*-butyllithium and cuprous iodide (see Section 15.6).

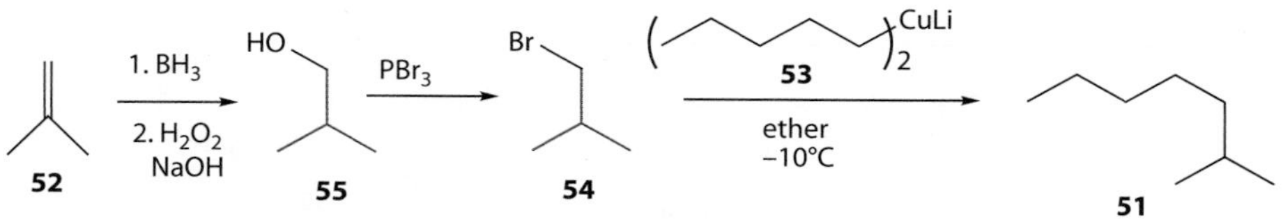

15.8 Biological Relevance

Although Grignard reagents, organolithium reagents, and organocuprates may not be found in living systems, metals are essential. If metal complexes are viewed as organometallics, the biological relevance of these metals can be described. Enzymes are key to life, and many enzymes require metal ions as cofactors. Tyrosinase (an enzyme that catalyzes the production of melanin and other pigments from tyrosine by oxidation, as in the blackening of a peeled or sliced potato exposed to air) and cytochrome oxidase (one of a family of proteins that act as the terminal enzymes of respiratory chains) require Cu^{+2} or Cu^+ as a cofactor.[4] Phosphohydrolases are a class of enzymes that cleave phosphoric acid from phosphate ester linkages to give an orthophosphate. Phosphotransferases are a class of enzymes that include the kinases and that catalyze the transfer of phosphorus-containing groups from one compound to another. Both of these enzymes require Mg^{+2} as a cofactor.[4]

Magnesium is required with some phosphatases (enzymes that remove a phosphate group), carboxylases (enzymes that catalyze decarboxylation or carboxylation), and some proteolytic enzymes (enzymes that cleave peptide bonds). Magnesium is important in several biologically important compounds. One is chlorophyll A (**56**), one of the green pigments found in photosynthetic cells, and it forms a magnesium complex with the porphyrin nitrogen atoms, as shown. Chlorophyll is typically extracted from the leaves of trees and plants. The structure of vitamin B12 is shown in **57**,[5] where the active form is a cobalt complex, and the metal is again coordinated to the nitrogen atoms of a porphyrin system.

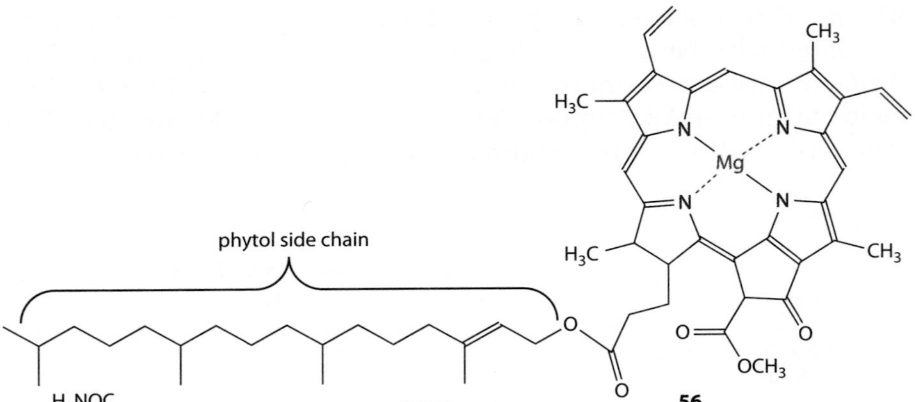

56

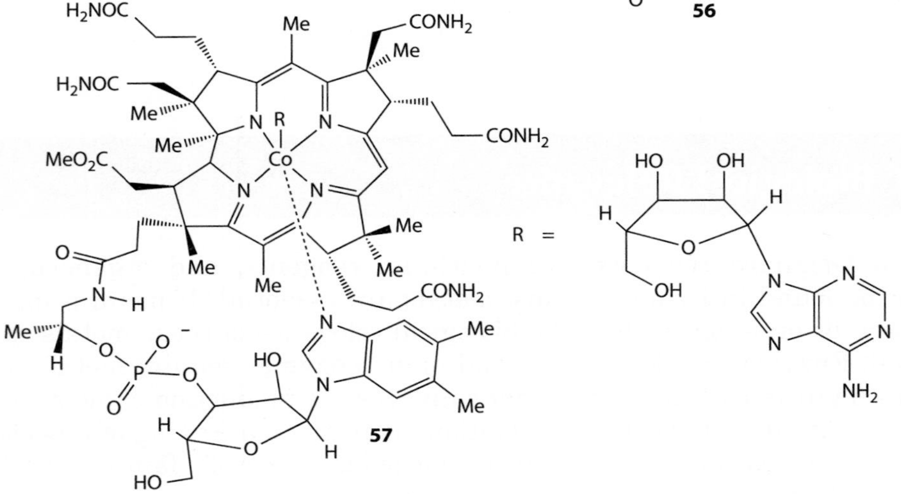

phytol side chain

57

R =

Although it is not an organometallic, lithium metal finds use in medicine. Lithium therapy can "reverse the manic phase of manic depression, and it increases the net re-uptake of certain biogenic amines into nerves, and it is capable of reducing nerve-stimulated release of biogenic amines."[6]

References

1. Smith, M. B., and March, J. 2007. *March's advanced organic chemistry,* 6th ed., 832. New York: Wiley–Interscience.
2. Wakefield, B. J. 1974. *The chemistry of organolithium compounds.* Oxford, England: Pergamon Press.
3. Furniss, B. S., Hannaford, A. J., Smith, P. W. G., and Tatchell, A. R., eds. 1994. *Vogel's textbook of practical organic chemistry,* 5th ed., 483, Exp. 5.10. New York: Longman.
4. Lehninger, A. I. 1970. *Biochemistry,* 149. New York: Worth Publishing.
5. Halpern, J. 2001. *Pure and Applied Chemistry* 73:209–220.
6. Foye, W. O., ed. 1989. *Principles of medicinal chemistry,* 296. Philadelphia: Lea and Febiger.

Answers to Problems

15.1

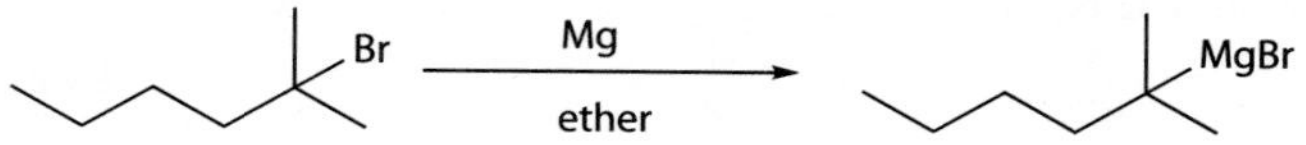

15.2

(diethyl ether)   (tetrahydrofuran, THF)

15.3

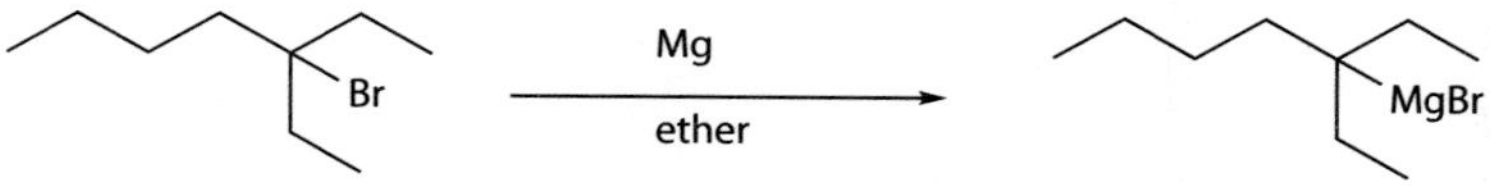

15.4

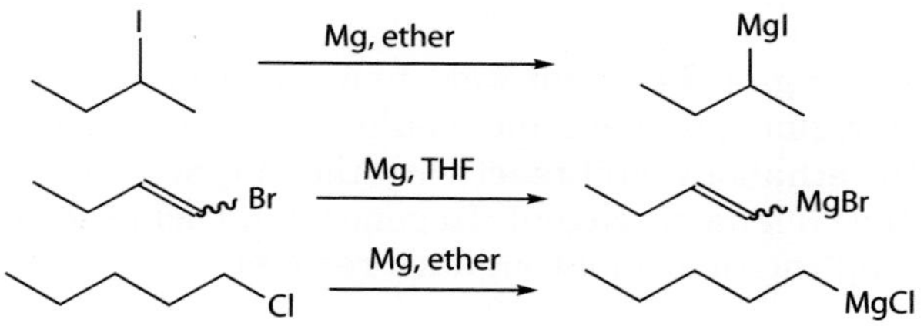

15.5

racemic (*R+S*)

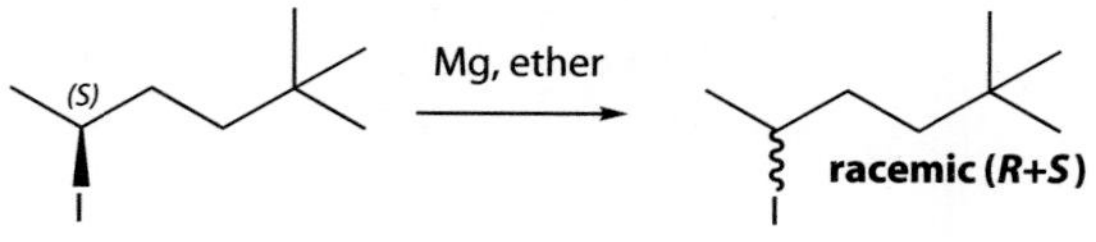

15.6

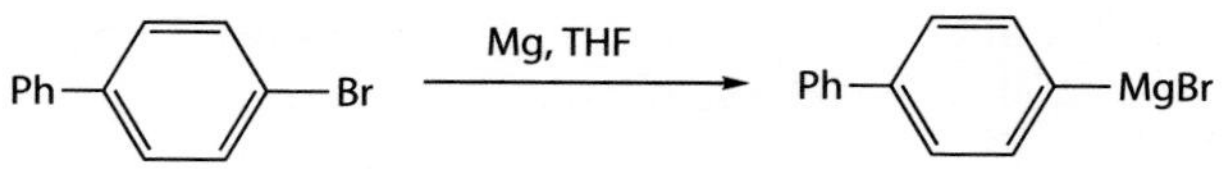

15.7   Because the reaction proceeds by a secondary vinyl carbocation (Chapter 10, Section 10.6.1), we anticipate a mixture of *E*- and *Z*-isomers. The alkyne is symmetrical, so there is only one regioisomer.

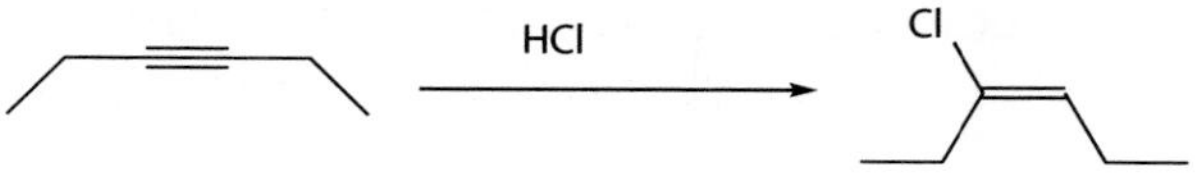

15.8

3-bromo-4-ethyl-3E-octene

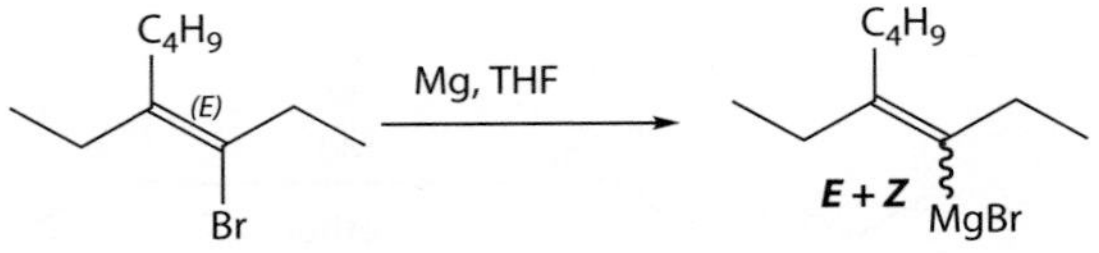

$$\text{Mg, THF}$$

$$E + Z \quad \text{MgBr}$$

15.9

$$CH_3CH_2CH_2MgX + H\text{-}O\text{-}H \longrightarrow CH_3CH_2CH_2\text{-}H + XMg^+ {}^-O\text{-}H$$

$$CH_3CH_2CH_2MgX + H\text{-}O\text{-}CH_3 \longrightarrow CH_3CH_2CH_2\text{-}H + XMg^+ {}^-O\text{-}CH_3$$

$$CH_3CH_2CH_2MgX + H\text{-}C\equiv C\text{-}CH_3 \longrightarrow CH_3CH_2CH_2\text{-}H + XMg^+ {}^-C\equiv C\text{-}CH_3$$

$$CH_3CH_2CH_2MgX + H\text{-}NMe_2 \longrightarrow CH_3CH_2CH_2\text{-}H + XMg^+ {}^-NMe_2$$

15.10

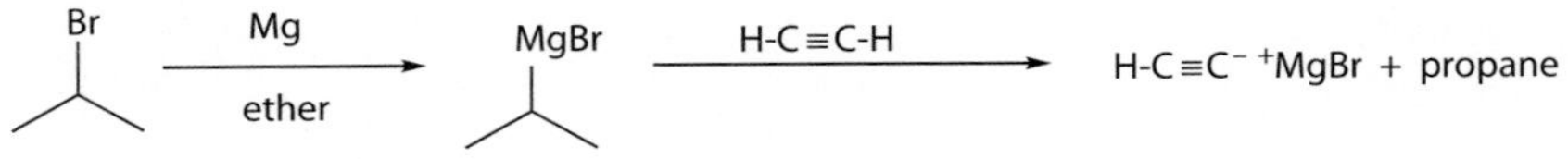

$$H\text{-}C\equiv C^- \, {}^+MgBr + \text{propane}$$

15.11 No! Even if the Grignard reagent were to form, ethanol is a strong acid in the presence of a Grignard reagent and would react via an acid–base reaction. More likely, the ethanol would react with the magnesium or at least suppress formation of the Grignard reagent. In general, no acid solvents should be present when attempting to form a Grignard reagent.

15.12

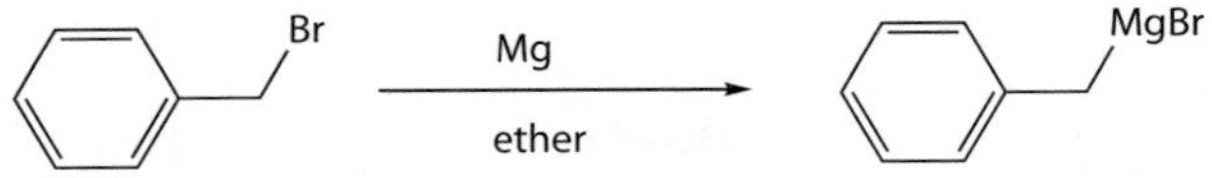

15.13

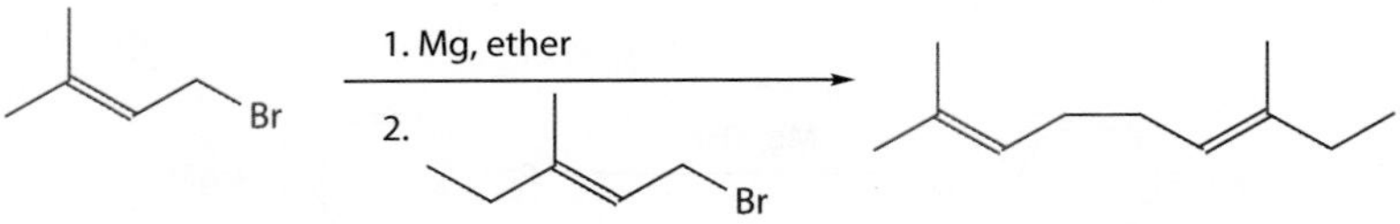

15.14

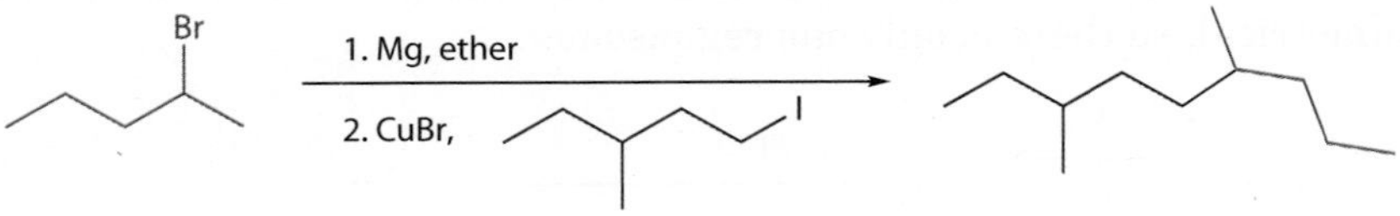

15.15

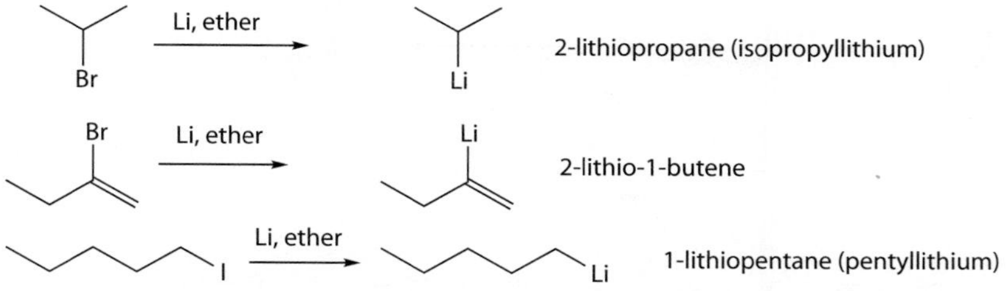
Br
Li, ether
Li
2-lithiopropane (isopropyllithium)
Br
Li, ether
Li
2-lithio-1-butene
I
Li, ether
Li
1-lithiopentane (pentyllithium)

15.16

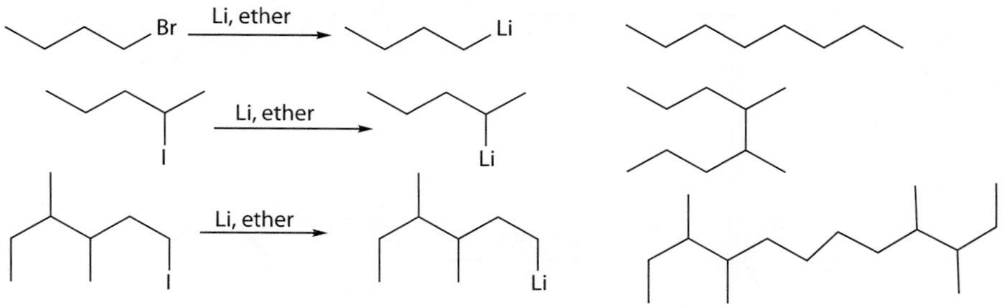
Br
Li, ether
Li
I
Li, ether
Li
I
Li, ether
Li

15.17

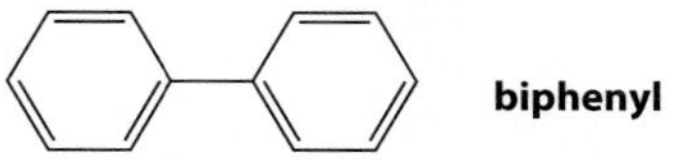
biphenyl

15.18

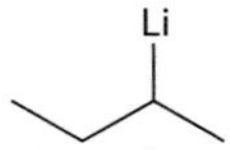
Li

15.19

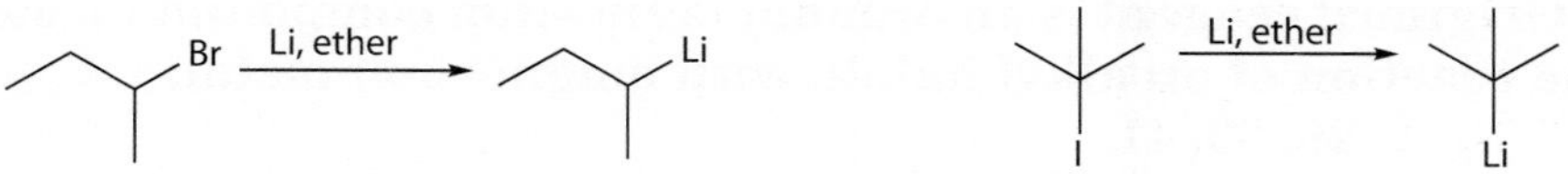
Br
Li, ether
Li
I
Li, ether
Li

15.20

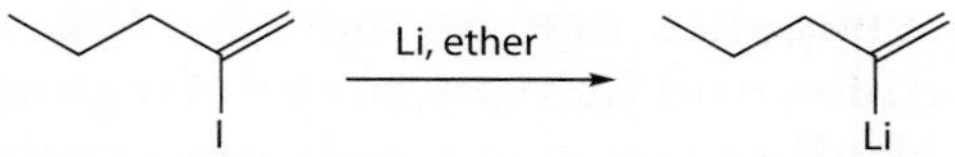
I
Li, ether
Li

15.21

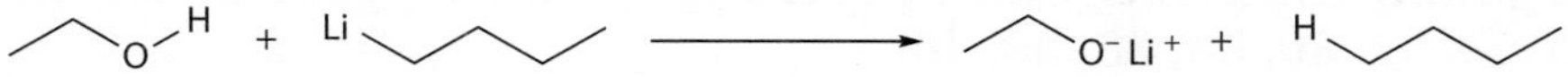
O
H
+
Li
O⁻ Li⁺
+
H

15.22

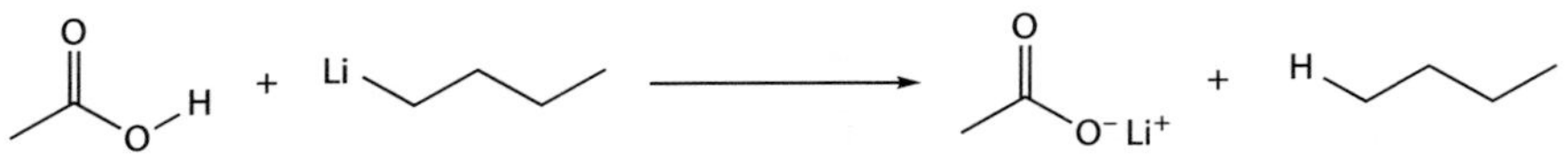

15.23

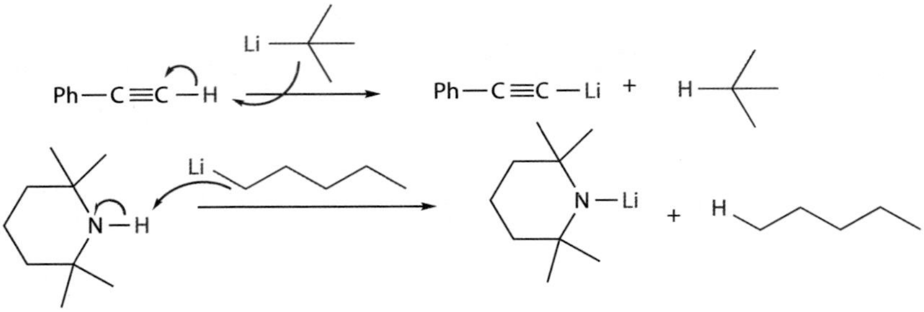

15.24

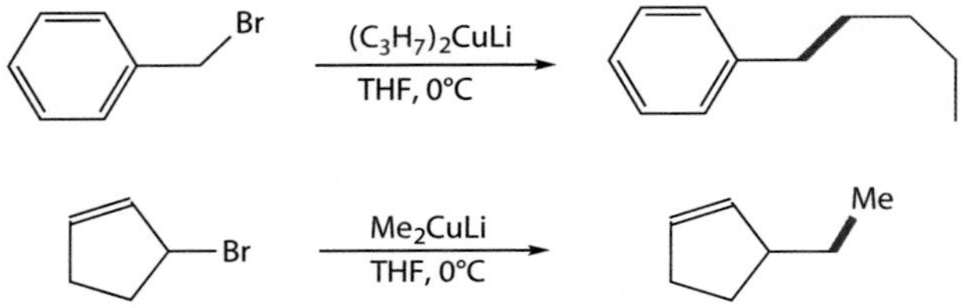

15.25

2 PhLi + CuI $\xrightarrow{\text{ether, }-10°C}$ Ph$_2$CuLi   2 MeLi + CuI $\xrightarrow{\text{ether, }-10°C}$ Me$_2$CuLi

Correlation of Concepts with Homework

- **A Grignard reagent is an organomagnesium compound formed by the reaction of an alkyl halide with magnesium metal:** 1, 2, 3, 4, 5, 6, 7, 8, 12, **26, 33, 41.**
- **Grignard reagents are strong bases, but they do not react with alkyl halides very well unless a transition metal catalyst is present:** 9, 10, 11, 13, 14, 29, 31, **33, 34, 35.**
- **Organolithium reagents are formed by the reaction of alkyl, aryl, or vinyl halides and lithium metal. Organolithium reagents react with alkyl halides to give a new organolithium reagent via metal–halogen exchange:** 15, 16, 17, 18, 19, 20, **27, 28, 33, 35, 36, 38.**
- **Organolithium reagents are strong bases and good nucleophiles:** 21, 22, 23, **30, 33, 38, 39, 40.**

- **Organocuprates can be prepared from organolithium reagents and cuprous halides. Organocuprates react with alkyl halides to form a new C–C bond in the product: 24, 25, 32, 33, 42.**
- **A molecule with a particular functional group can be prepared from molecules containing different functional groups by a series of chemical steps (reactions). This process is called synthesis: The new molecule is synthesized from the old one (see Chapter 25): 37.**
- **Spectroscopy can be used to determine the structure of a particular molecule and can distinguish the structure and functionality of one molecule when compared with another (see Chapter 14): 41, 42, 43.**

Homework

15.26 Draw the product formed when each of the following reacts with magnesium metal in ether or in THF:
   (a) 1-iodohexane
   (b) 2-bromo-2-pentene
   (c) 4-bromo-2-phenylhexane
   (d) bromobenzene

15.27 Draw the product formed when each of the following reacts with lithium metal in ether or in THF:
   (a) 1-iodohexane
   (b) 2-bromo-2-pentene
   (c) 4-bromo-2-phenylhexane
   (d) bromobenzene

15.28 Draw the product formed when each of the following reacts with *tert*-butyllithium in a mixture of ether and hexane:
   (a) 1-iodohexane
   (b) 2-bromo-2-pentene
   (c) 4-bromo-2-phenylhexane
   (d) iodomethane

15.29 Draw the product formed when each of the following reacts with allylmagnesium bromide:
   (a) diethylamine
   (b) 1-aminobutane
   (c) piperidine
   (d) cyclohexanol
   (e) 1-butyne
   (f) dicyclohexylamine
   (g) benzyl bromide
   (h) CuI
   (i) *tert*-butyl iodide

15.30 Draw the product formed when each of the following reacts with *n*-butyllithium:
  (a) diethylamine
  (b) 1-aminobutane
  (c) piperidine
  (d) cyclohexanol
  (e) 1-butyne
  (f) dicyclohexylamine
  (g) benzyl bromide
  (h) CuI
  (i) *tert*-butyl iodide

15.31 Which of the following alkyl halides is expected to give a good yield of coupling product with benzylmagnesium iodide? Explain your answer:

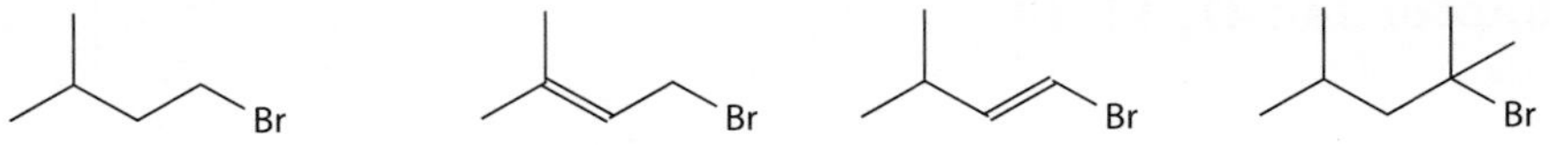

15.32 Draw the product expected when lithium dimethylcuprate reacts with each of the following in ether at –10°C:
  (a) 2-bromo-4-methylhexane
  (b) 3,4-diphenyl-1-iodoheptane
  (c) 2-iodo-2-methylpentane

15.33 Draw the final product, if any, for each of the following reactions:
  (a) 2-iodopentane + 1. Mg/ether; 2. acetylene
  (b) phenylmagnesium bromide + 1. CuBr/THF/–10°C; 2. 2-bromopentane
  (c) 2-bromo-2-butene + 1. Li/THF; 2. CuI/THF/–10°C; 3. iodomethane
  (d) 2-bromo-2-butene + 1. Mg/THF; 2. 1-butyne
  (e) 3-bromocyclopentene + 1. Mg/THF; 2. benzyl bromide
  (f) butylmagnesium chloride + water
  (g) 2-methylhexylmagnesium bromide + 1,2-dimethoxyethane
  (h) 1-iodopentane + 1. Li; 2. 2-iodo-2-methylpentane
  (i) phenyllithium + 1. CuI/THF/–10°C; 2. 2-bromohexane
  (j) *n*-butyllithium + 1. 1-propyne; 2. dilute aqueous acid
  (k) *n*-butyllithium + *N*-methyl-1-aminopentane
  (l) methyllithium + 2,2,4,4-tetramethylhexane

15.34 Briefly explain why one should not use 2-propanol as a solvent to form a Grignard reagent.

15.35 Draw and discuss the product or products expected from a reaction of 2S-bromohexane and 1. Mg/ether; 2. CuBr; 3. iodoethane.

15.36 Briefly explain why we use THF as a solvent to form the Grignard reagent of a vinyl halide when we use ether to form the Grignard reagent of a simple alkyl halide.

15.37 Draw all reaction steps necessary, including all reagents and products, for each of the following syntheses:

(a)

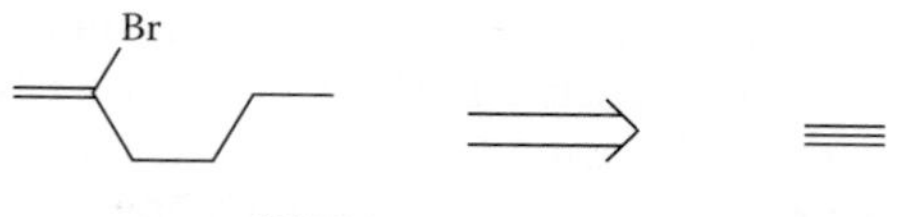

(b)

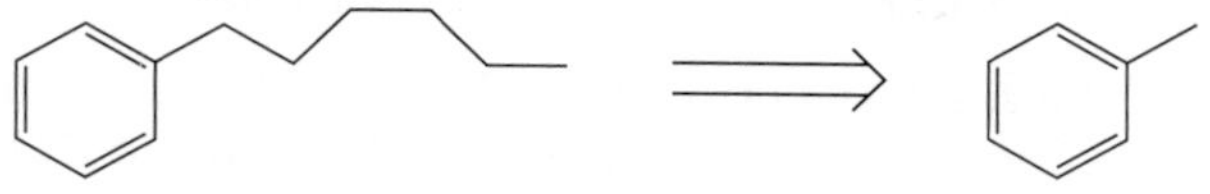

(c)

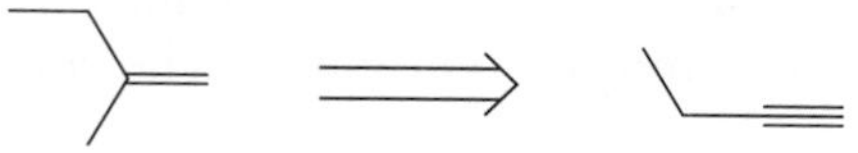

(d)

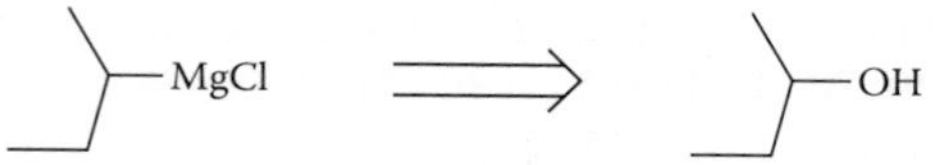

(e)

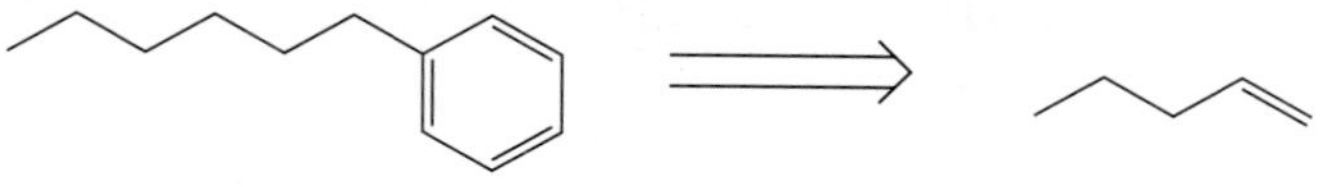

15.38 When *n*-butyllithium (as a hexane solution) is transferred via syringe, it sometimes fumes and smokes but rarely catches fire unless one is very careless. When *tert*-butyllithium (as a pentane solution) is transferred, there is almost always a fire on the tip of the syringe (potentially very dangerous). Briefly explain what this indicates about these two organolithium reagents.

15.39 The reaction of diisopropylamine and *n*-butyllithium in THF generates a product known as lithium diisopropylamide (LDA). The following procedure is used. The diisopropylamine is added to a flask containing dry THF that has been cooled to −78°C and the entire unit is under a positive pressure of nitrogen gas. We then remove the flask from the cooling bath and allow it to warm briefly to about 0°C, where the solution turns pale yellow, and then put the flask back into the bath so that it can cool down to −78°C once again. We are now ready to do chemistry. If we do not warm the flask to 0°C, the solution does not turn yellow. What is the purpose of the positive head pressure of nitrogen? What does the requirement that the reaction temperature be raised from −78°C → 0°C tell you about the acidity of the amine? If we were to add 1-pentyne to the LDA solution and then add allyl bromide, what is the final product?

15.40 A bottle of *n*-butyllithium purchased from a commercial vendor is usually packaged in hexane. After we do a reaction, it is common to find that octane is present. Presumably, this comes from the bottle of *n*-butyllithium. Briefly discuss how octane might be found in this commercial product.

Spectroscopy Problems

Do not attempt these problems until after you have studied Chapter 14.

15.41 A Grignard reagent is prepared by reaction of 1-bromododecane with magnesium in ether. The resulting solution is then treated with ethanol. Can IR be used alone to see if bromododecane was converted to dodecane? Can proton NMR be used alone? Explain the answer in both cases.

15.42 The product of the reaction between the 1-bromo-3-hexyne and lithium diethylcuprate is an alkyne. Draw it. Is there anything in the IR that would allow one to distinguish the starting alkyne and the product alkyne? How about the proton NMR? How about information that can be obtained from the mass spectrum?

15.43 Identify the structure of the following molecule given the following spectral data:
MS: 128 (M, 100%), 129 (M+1, 9.99%), 130 (M+2, 0.36%)
IR: 2966–2870, 1478–1468, 1393–1366 cm^{-1}
^{1}H NMR: 1.44 (m, 1H), 1.14 (m, 2H), 1.14 (t, 2H), 0.88 (d, 6H), 0.86 (s, 9H) ppm

Carbonyl Compounds

Structure, Nomenclature,

Reactivity

16

Several important classes of molecules have the carbonyl unit (–C=O) as a key part of their structure. They are generically known as carbonyl compounds. The chemistry associated with carbonyl compounds is important to many chemical reactions. The carbonyl unit is reactive because it is polarized and also contains a weak and reactive π-bond. This chapter is primarily dedicated to learning the identity, structure, and nomenclature of the key carbonyl-containing functional groups, as well as the most fundamental aspects of their chemical reactivity. These include aldehydes, ketones, carboxylic acids, and carboxylic acid derivatives.

To begin this chapter, you should know the following:

- **the basic rules of nomenclature for alkanes, alkenes, and alkynes (Chapter 4, Section 4.6; Chapter 5, Sections 5.1 and 5.2)**
- **the concept of polarized covalent bonds (Chapter 3, Section 3.7)**
- **the fundamentals of nomenclature for carbonyl compounds (Chapter 5, Section 5.9)**
- **the CIP rules for prioritizing substituents, groups, and atoms (Chapter 9, Section 9.3)**

- the fundamental properties of a π-bond (Chapter 5, Sections 5.1 and 5.2)
- the structure of a carbonyl and that a carbonyl group is polarized (Chapter 5, Section 5.9)
- the carboxyl functional group (Chapter 5, Section 5.9.3)
- the acid–base properties of carbonyl compounds (Chapter 5, Section 5.7; Chapter 6, Section 6.4.4)
- mechanisms (Chapter 7, Section 7.8)
- carbocations (Chapter 10, Section 10.2; Chapter 11, Section 11.4)
- the concept of Lewis acids and Lewis bases (Chapter 6, Section 6.5)
- the concept of a nucleophile (Chapter 6, Section 6.7; Chapter 11, Sections 11.2 and 11.3)
- the concept of a leaving group (Chapter 11, Section 11.2.4)
- the S_N2 transition state and the fundamental mechanistic details of the S_N2 reaction (Chapter 11, Sections 11.1–11.3)

This chapter will revisit the IUPAC nomenclature system for aldehydes, ketones, and carboxylic acids, as well as introduce nomenclature for the four main acid derivatives: acid chlorides, anhydrides, esters, and amides. The chapter will show the similarity of a carbonyl and an alkene in that both react with a Brønsted–Lowry acid or a Lewis acid. The reaction of a carbonyl compound with an acid will generate a resonance stabilized oxocarbenium ion. Ketones and aldehydes react with nucleophiles by what is known as acyl addition to give an alkoxide product, which is converted to an alcohol in a second chemical step. Acid derivatives differ from aldehydes or ketones in that a leaving group is attached to the carbonyl carbon. Acid derivatives react with nucleophiles by what is known as acyl substitution, via a tetrahedral intermediate.

When you have completed this chapter, you should understand the following points:

- **Nomenclature for aldehydes and ketones. Nomenclature of difunctional molecules containing a carbonyl group and an alkene or alkyne, or two carbonyl groups.**
- **The carbonyl of an aldehyde or ketone react with protonic acids as Brønsted–Lowry bases to generate an oxocarbenium ion. The carbonyl of an aldehyde or ketone reacts with Lewis acids as a Lewis base to generate an oxocarbenium ion.**
- **Acid derivatives react with protonic acids as Brønsted–Lowry bases to generate an oxocarbenium ion.**

- **Nucleophiles react with aldehydes or ketones by acyl addition to generate alkoxide in a reaction known as acyl addition. A second chemical step converts the alkoxide product to an alcohol in an acid–base reaction.**
- **Nomenclature for carboxylic acids and dicarboxylic acids.**
- **Structure and nomenclature for acid halides, anhydrides, esters, and amides.**
- **Carboxylic acids react primarily via an acid–base reaction to form carboxylate anions.**
- **Carboxylic acid derivatives react with nucleophiles primarily by acyl substitution via a tetrahedral intermediate.**
- **Spectroscopy can be used to determine the structure of a particular molecule and can distinguish the structure and functionality of one molecule when compared with another (see Chapter 14).**
- **Carbonyl compounds are important in biological systems.**

16.1 The Carbonyl Group

The carbonyl group was introduced in Chapter 5 (Section 5.9), along with a brief introduction to the nomenclature of aldehydes and ketones. This group (the C=O unit) is a polarized group with a $\delta+$ carbon and a $\delta-$ oxygen (see **1**). A bonding description of a carbonyl is shown in **1A**, where there is a strong σ-covalent bond (in **red**) and a weaker π-bond (in **yellow**). As described in Chapter 5, an alkene has a double bond (a C=C unit) with a σ-covalent bond and a π-bond (see **2A**). The carbonyl unit is similar, except that there is a C=O unit rather than a C=C unit. Molecular model **3** is an electron density potential map of the molecule formaldehyde (**1**, R = H), and the **red** color between the C and O indicates the presence of high electron density above and below the plane of the atoms, consistent with the π-bond.

The major difference between an alkene unit (C=C, see **2**) and a carbonyl unit (C=O, see **1**) is the presence of the oxygen and its polarizing influence. Because the carbonyl carbon is $\delta+$, it is said to be **electrophilic** and it reacts with a species able to donate two electrons. In other words, a carbonyl carbon reacts with a **nucleophile** to form a new σ-covalent bond. Nucleophiles were introduced in Chapter 10 (Section 10.2) and discussed again in Chapter 11 (Sections 11.2 and 11.3) in connection with nucleophilic reactions with sp³ carbon atoms of alkyl halides. The carbonyl unit has an sp² hybridized carbon that also reacts with nucleophiles, but in a different manner when compared to a reaction with an alkyl halide.

1

1A

2

2A

3

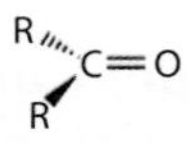

16.2 Aldehydes and Ketones. Nomenclature

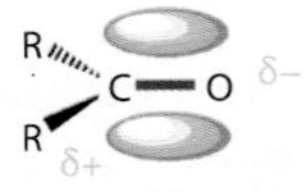

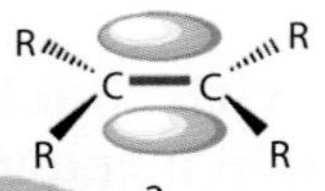

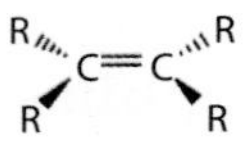

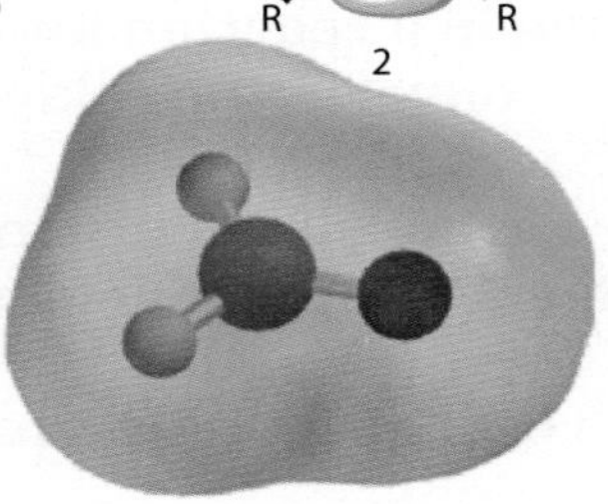

A carbon–oxygen double bond has two additional substituents attached to the carbonyl carbon: either carbon groups or a hydrogen atom. If at least one of these substituents is a hydrogen, the compound is an ***aldehyde*** with a CHO unit (see the boxed unit in **4**). All aldehydes except formaldehyde, which has two hydrogen atoms ($H_2C{=}O$), will have one hydrogen and one carbon group attached to the carbonyl carbon. If two carbon substituents are attached to the carbonyl carbon, the molecule is called a ***ketone*** (see the boxed unit in **5**). *In both aldehydes and ketones,* ***the functional group is the carbonyl,*** **C=O.**

4

5

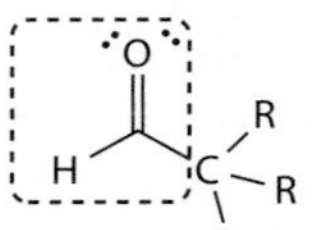

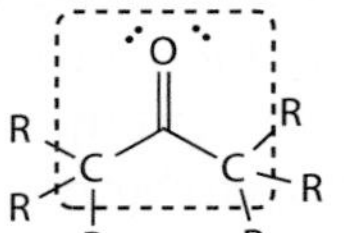

Simple aldehydes and ketones have common names that are so well known they cannot be ignored. Three simple aldehydes are formaldehyde, acetaldehyde, and, when the aldehyde unit is attached to a benzene ring, the molecule is known as benzaldehyde. The simplest ketone is acetone, but other common ketones are benzophenone and pinacolone. The IUPAC system will be described and these names will be given their proper name at that time. However, the common names are used extensively and they must be learned.

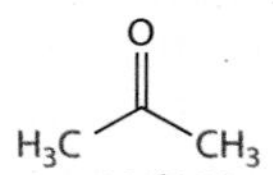

16.1 If reaction occurs at the carbonyl carbon of the C=O unit, which do you anticipate is the more reactive: acetaldehyde or acetone?

formaldehyde

acetaldehyde

acetone

benzaldehyde

benzophenone

pinacolone

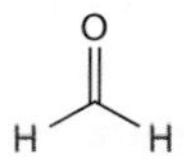

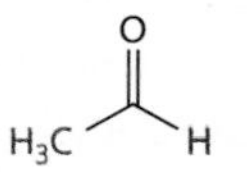

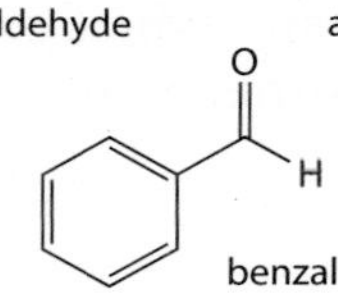

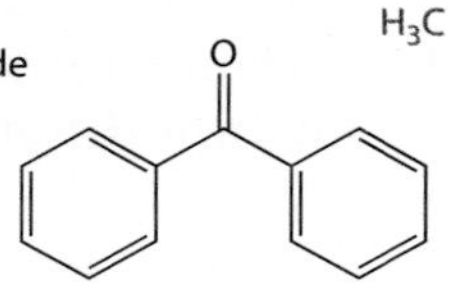

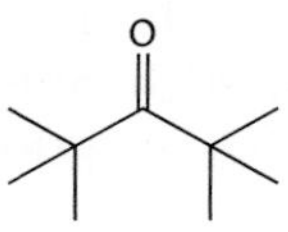

16.2.1 Ketones

The structure of a ketone is characterized by a carbonyl group attached to two alkyl groups (see **1**, R = alkyl). *The functional group is the carbonyl unit,* C=O, and the suffix for ketones is taken from the last three letters of ketone: "one." When the longest chain is numbered, the *carbonyl carbon* must be part of that chain, and it receives the lowest possible number. Based on this rule, the seven-carbon straight-chain ketone **6** is a heptanone. Note the use of "heptan," which is derived from the saturated chain heptane. To generate the name, the final letter "e" is dropped from heptane and replaced by the letters "one." *The position of the carbonyl carbon, which must be included in the longest continuous chain, is given the lowest possible number as part of the name.* Therefore, **6** is 3-heptanone.

Substituents are named in the usual manner. Ketone **7** has a 10-carbon chain that contains the carbonyl carbon, so it is a decanone, and the carbonyl carbon at C4, so it is a 4-decanone. Once the carbonyl carbon is numbered, the carbon atoms bearing the substituents are assigned numbers accordingly. There is a phenyl group at C1, an ethyl group at C3, a bromine at C8, and two methyl groups at C7. Rank ordering the substituents alphabetically, using the assigned numbers, **7** is 8-bromo-3-ethyl-6,6-dimethyl-1-phenyl-4-decanone (the decanone chain is marked in **blue** in the illustration). The IUPAC name for acetone is 2-propanone. The common name is acetone, and it is commonly used in nail polish and nail polish remover, airplane glue, cleaning fluids, paint and varnish, and in some. This commercial product is rarely called 2-propanone, but rather "acetone" is used virtually all the time.

Cyclic ketones have the carbonyl carbon as part of a ring. The nomenclature is straightforward in that "cyclo" is used to denote the ring, the usual prefix is used to indicate the number of carbons in the ring (including the carbonyl carbon), and the suffix is "one." Compound **8** is named cycloheptanone. There is no need to use the number 1, as in 1-cycloheptanone, because the carbonyl carbon must receive the lowest possible number. When there is a substituent, as in **9**, the carbonyl carbon is always C1, so, again, there is no need to use 1. Therefore, **9** is 3-chlorocyclopentanone. *The use of a number to indicate the position of the carbonyl is essential in acyclic ketones such as **6** or **7**, but unnecessary in cyclic ketones such as **8** or **9**.*

16.2 Give the IUPAC names for benzophenone and for pinacolone using the system just described.

16.3 Draw the structure of 3,4,5-trichlorocyclohexanone and of 3*R*-ethyl-4*S*-iodocyclooctanone.

In the previous examples, the base name is derived from a saturated, or alkane, backbone. Therefore, a heptane chain in a ketone becomes heptanone and a tetradecane chain would be tetradecanone. It is also possible to have a heptene or a heptyne backbone, as in 10 and 11. This raises a problem: Which group has the higher priority: the C=O unit or the C=C unit or the C≡C unit? Using the Cahn–Ingold–Prelog selection rules (Chapter 9, Section 9.3.1), the C=O unit is the higher priority, and this means that both 10 and 11 will be named as ketones (suffix is "one"). A C=O unit is higher in priority than a C=C unit, so a molecule is named as an en-one when an alkene and a ketone unit are in the same molecule.

Because 10 has an alkene backbone, it is a heptenone—specifically, a 2-heptenone. Does the "2" in 2-heptenone refer to C=C or C=O? It is supposed to refer to the carbonyl, but there are two functional groups and greater clarity is required. A more precise way to name this compound is "hepten-2-one" so that there is no ambiguity about the C=O unit; however, a number is also required for the C=C unit. Because the carbonyl carbon is C2, the first carbon of the C=C unit is at C3. This number is again placed immediately before the functional group designation, so 10 is hept-3-en-2-one. However, a problem remains: which hept-3-en-2-one? Compound 10 can exist as an E- or Z-isomer (see Chapter 9, Section 9.4.3) and 10 is hept-3E-en-2-one. The stereochemical designation of the alkene unit is a formal part of the name in such compounds.

Compound 11 is an alkyne–ketone and, because the carbonyl has the higher priority, it is named as an yn-one. Compound 11 is a heptyne derivative, so it is named as a heptynone—specifically, heptyn-2-one. If the carbonyl carbon is at C2, the first carbon of the C≡C unit is at C4, so 11 is hept-4-yn-2-one. There are no E- or Z-isomers for the alkyne unit.

16.4 Write out the structure of 1-phenyl-4S-ethyldodec-6Z-en-2-one.

16.5 Draw the structure of dec-1,5-diyn-4-one.

16.2.2 Aldehydes

An aldehyde is the name given to a molecule with one alkyl group and one hydrogen atom attached to a carbonyl (see 12), but *the functional group is still the **carbonyl** unit,* C=O. The suffix for aldehydes derives from the *first* two letters of *al*dehyde: "al." As with ketones, the carbonyl carbon takes priority, must be part of the longest continuous chain, and is given the lowest possible number. *In all cases,* the number of the carbonyl carbon will be 1 because all aldehydes have a hydrogen atom attached to the carbonyl unit. Therefore, *the number is omitted from the name.*

A specific example of an aldehyde is **13**, nonanal. Note the use of "nonan" and "al," where the former is derived from the saturated nonane skeleton. The "e" is dropped and replaced with "al." Substituents attached to the aldehyde chain are named in the usual manner, with the carbonyl carbon always numbered 1. Aldehyde **14** has a 10-carbon alkane chain, so it is a decanal. Using the carbonyl carbon as C1, there is a 1-methylpropyl group at C3 and a methyl group at C5. Therefore, the name is 5-methyl-3-(1-methylpropyl)decanal (the decanal chain is marked in **blue** in the illustration).

16.6 Draw the structure of 2,2,4,4-tetramethyl-5-(2-chloro-1-propyl)-decanal.

As with ketones, an aldehyde may have an alkane, alkene, or alkyne backbone. In all cases, the carbonyl of an aldehyde has a higher priority than the C=C of an alkene or the C≡C of an alkyne. Compound **15** has a 10-carbon chain that includes the C=C and C=O units, but also has a 14-carbon chain that includes only the C=C unit. *The nomenclature rules demand that the longest chain contain both functional groups,* and **15** has an en-al backbone, so the aldehyde suffix is used. Compound **15** is a dec-3-enal (the longest chain is marked in **blue**) with a dimethyl unit at C7 and a hexyl group at C3. Note that the alkene is a *Z*-alkene, so **15** is 7,7-dimethyl-3-hexyldec-3*Z*-enal. In **16**, there is 12-carbon chain that includes only the C≡C unit, but the longest chain that includes *both* C≡C and C=O is marked in **blue**. Because the aldehyde unit is higher in priority, **16** is a nonynal. There is a butyl group at C4 and a methyl group at C3, relative to the aldehyde carbonyl; therefore, **16** is 2-butyl-3-methylnon-8-ynal. Note that the aldehyde unit is drawn using a shorthand version, –CHO, for both **15** and **16**.

When the aldehyde unit (–CHO) is attached to a ring, a major modification in the name is required. An aldehyde has a hydrogen atom attached to the carbonyl carbon, so the carbonyl carbon cannot be part of a ring, but only attached to it. Aldehyde **17**, for example, is ***not*** named cyclohexanal. This name makes no sense because an aldehyde must have the CHO carbon labeled as C1. For compounds that have an aldehyde unit attached to a carbocyclic ring, the suffix **"carboxaldehyde"** is used, and **17** is named cyclohexanecarboxaldehyde.

16.7 Draw the structure for 3,7-diphenylhexadec-10-ynal.

16.8 Draw the structure of 3,7-diethylcyclooctane-1-carboxaldehyde.

16.2.3 Diketones, Dialdehydes, and Keto-aldehydes

Just as there are dienes and diynes, there are compounds that have two ketone units, two aldehyde units, or a ketone and an aldehyde in the same molecule. For two ketone units, "dione" is used and, for two aldehydes, the term is "dial." *The longest chain must contain both carbonyl carbons.* When a ketone and an aldehyde unit are in the longest chain, the aldehyde takes priority and is used as the suffix. The term "oxo" is used to indicate the position of the other carbonyl carbon in that chain. Examples of diketones are **18** and **19**. Diketone (or dione) **18** has a six-carbon chain that contains both carbonyl carbons; numbering to give the carbonyl carbons the lowest numbers leads to 2,4-hexanedione.

   With this numbering scheme, the substituent is at C3, so **18** is named 3-heptyl-2, 4-hexanedione. In **19**, the two ketone units are part of the ring, and the name is 1,3-cyclohexanedione. Dialdehyde (dial) **20** has both carbonyl carbons in a six-carbon chain, so it is a 1,6-hexanedial. Numbering the substituents accordingly leads to 2-ethyl-5-methyl-1,6-hexanedial. For **21**, there is a seven-carbon chain and, because the aldehyde unit takes priority, it is a heptanal. Naming **21** as an aldehyde means that the aldehyde carbonyl is carbon-1, so the ketone carbonyl is at C5. The name of **21** is 5-oxoheptanal.

16.9 Draw the structures of 9,9-dichloro-8-oxo-4-cyclopropyldecanal and 2-methyl-1,5-cyclododecanedione.

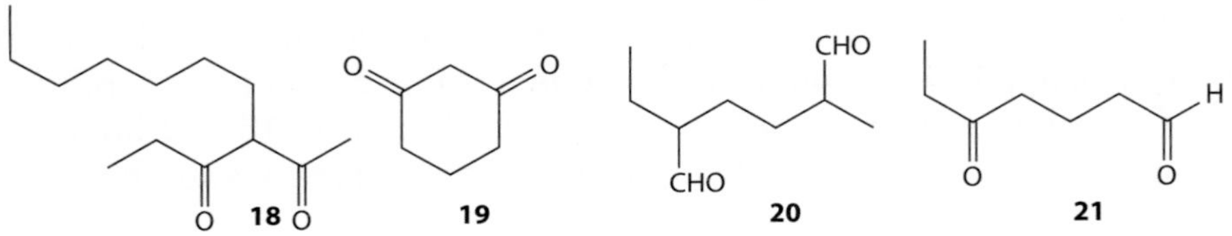

16.3 Chemical Reactivity of Ketones and Aldehydes

The chemical reactivity of a C=O unit is related to the C=C unit discussed in Chapter 10. In that chapter, the C=C unit reacts as a base, and the π-electrons are donated to an appropriate acid. The π-bond of an alkene is a Brønsted–Lowry base in reactions with a protonic acid such as HCl, whereas the alkene reacts as a Lewis base in reactions with a Lewis acid such as boron (in a borane). A π-bond is present in the C=O unit, so similar reactivity is expected. The oxygen atom of a carbonyl unit also has lone electron pairs, so the oxygen atom of the carbonyl is the logical focus of the basicity. In short, the carbonyl of an aldehyde or a ketone reacts with an acid, such as sulfuric acid or *p*-toluenesulfonic acid, and also with a Lewis acid such as boron trifluoride.

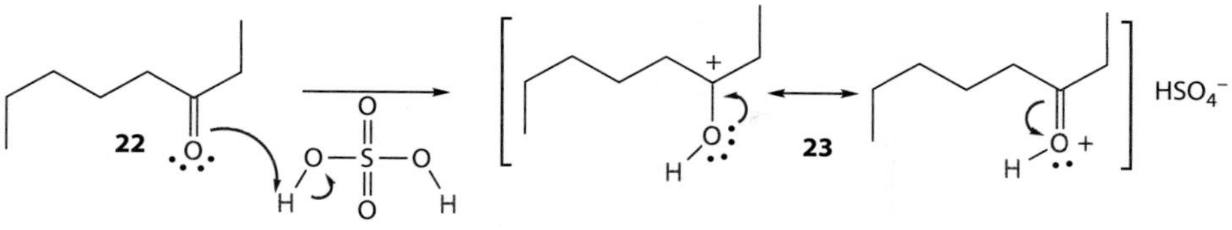

If the π-bond of a ketone such as 3-octanone (**22**) reacts as a base with the Brønsted–Lowry acid sulfuric acid, the conjugate acid is oxocarbenium ion **23**, and the conjugate base is the hydrogen sulfate anion. Ion **23** is formed by cleavage of the π-bond in **22** and donation of the two electrons to form a new O–H bond, leaving a positive charge on the carbonyl carbon. The two electron pairs remaining on oxygen must be taken into account. Those electrons are donated from the oxygen back toward the positive carbon to form a new π-bond, and the oxygen has a formal charge of +1 (see Chapter 5, Section 5.5, for formal charge). This is a resonance-delocalized, oxygen-stabilized carbocation **23** (known as an oxocarbenium ion). Therefore, *an aldehyde or a ketone reacts with a Brønsted–Lowry acid (sulfuric acid, HCl, nitric acid, p-toluenesulfonic acid) as a Brønsted–Lowry base to form a resonance-stabilized oxocarbenium ion.*

Lewis acids react with the carbonyl oxygen of aldehydes or ketones to form the corresponding ate complex (see Chapter 6, Section 6.5). When a 3-octanone **22** reacts with BF$_3$, the boron is a Lewis acid, and the product is charge-transfer complex **24** (an ate complex; see Chapter 6, Section 6.5), which is also stabilized by resonance. The boron has a formal charge of –1 in the ate complex, and the oxygen has a formal charge of +1. The resonance **forms 24** places the positive charge on carbon, so **24** is also classified as an oxocarbenium ion. Alternatively, the π-bond of **22** donates two electrons to the boron, generating the oxocarbenium ion directly, and it will have the oxonium ion as its resonance contributor.

16.10 Draw all resonance forms of the hydrogen sulfate anion.

16.11 Draw the oxocarbenium ion formed when butanal reacts with *p*-toluenesulfonic acid and when 2-pentanone reacts with sulfuric acid.

16.12 Draw the oxocarbenium ion formed when butanal reacts with BF$_3$ and when 2-pentanone reacts with AlCl$_3$.

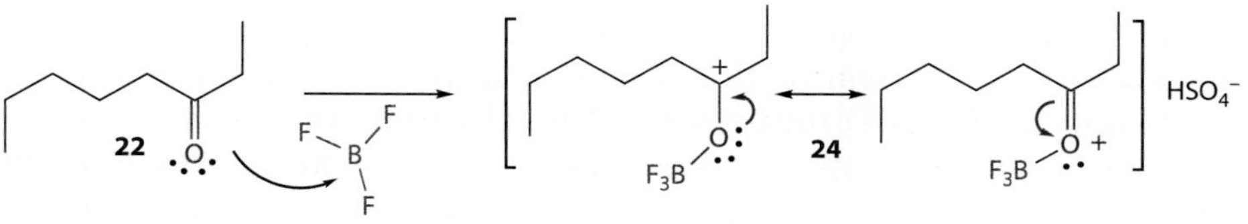

16.13 Draw the carbocation formed when ethene reacts with H^+.

16.14 Write out all resonance forms for the oxocarbenium ion formed when hex-3-en-2-one reacts with *p*-toluenesulfonic acid.

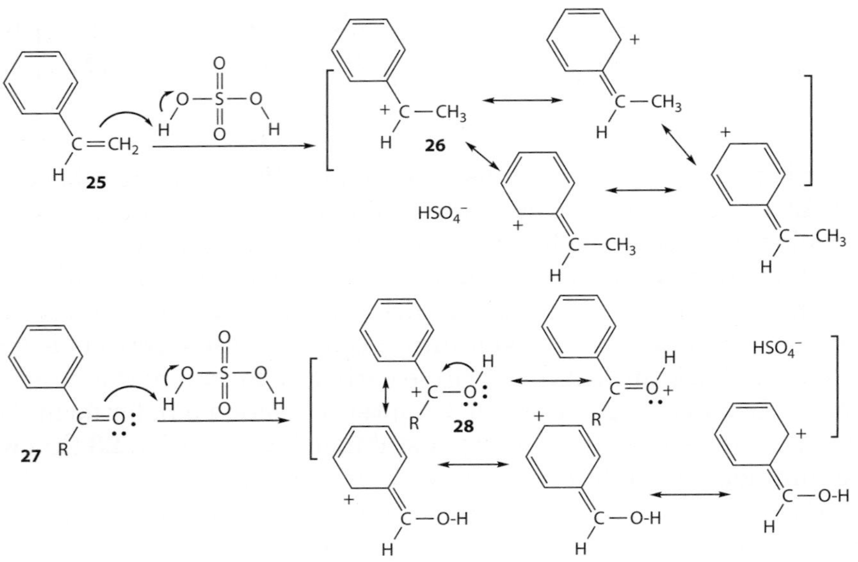

In the reaction with a Brønsted–Lowry acid or a Lewis acid, a carbonyl reacts similarly to alkenes. The major difference is formation of the oxocarbenium ion products, which are resonance stabilized, whereas a simple nonconjugated alkene forms a carbocation (a carbenium ion) that is not resonance stabilized. Some alkenes are exceptions to this statement. When the C=C unit is conjugated to another π-bond, as in styrene (phenylethene, **25**), reaction with an acid generates a resonance-stabilized carbocation, **26**. A conjugated carbonyl compound such as benzaldehyde (**27**) will also form a conjugated oxocarbenium ion (**28**) that is resonance stabilized as shown in the illustration.

Although carbonyls react as a Bronsted–Lowry base or a Lewis base similarly to the chemistry of an alkene, there is a major difference in their reactivity. The C=O group is polarized such that the oxygen is δ– and the carbon is δ+ (see **2**). This polarization, combined with the presence of a weak π-bond, leads to a reaction for carbonyls that is not observed with the C=C unit of alkenes. A nucleophile reacts with the positive carbon of a carbonyl to form a new σ-covalent bond to carbon, with cleavage of the π-bond and transfer of those two electrons to the oxygen.

In the reaction of butanal with a nucleophile (Y), note the *arrow formalism* for transfer of *two electrons* from the electron-rich Y⁻ (the nucleophile) to the electron-deficient carbonyl carbon. In other words, when the nucleophile collides with the δ+ carbon of the carbonyl, a new C–Y bond forms and the weaker π-bond in the C=O unit breaks. The electrons in that bond are transferred to the more electronegative oxygen atom to form an alkoxide, **29** (see the **blue** arrow in the illustration). This overall process is called ***nucleophilic*** *acyl addition* because a nucleophile "adds" to the carbonyl carbon (an **acyl** carbon), and breaks the π-bond. Contrast this reaction with aliphatic substitution at the sp³ carbon of bromomethane (an S_N2 reaction), where formation of the new C–nucleophile bond is accompanied by cleavage of the bond between carbon and the leaving group (see Chapter 11, Section 11.2).

16.15 Draw the transition state and product formed when cyanide ion reacts with iodomethane.

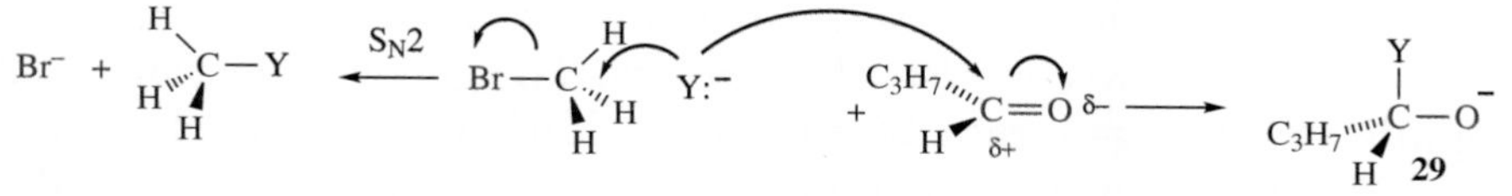

When butanal reacts with a nucleophile, the product is an alkoxide (**29**) that has a formal charge of –1 residing on the more electronegative oxygen (see Chapter 5, Section 5.5, for "formal charge"). *Acyl addition is a new type of reaction and it occurs with a variety of nucleophiles.* Three examples are shown in the illustration, using three different nucleophiles: cyanide ion, the alkyne anion of 1-butyne, and an organometallic compound methylmagnesium bromide (see Chapter 15, Section 15.1). In the first reaction, cyanide ion reacts with benzaldehyde to give the alkoxide product **30**. In the second reaction, the alkyne anion of 1-butyne reacts with 2-pentanone to give the alkoxide product **31**. In the final example, methylmagnesium bromide reacts as a nucleophile with butanal to give alkoxide **32** as the final product.

In Chapter 15, Grignard reagents react as bases, but they are poor nucleophiles with alkyl halides. Grignard reagents are good nucleophiles with ketones or aldehydes, however, and these reactions will be discussed in more detail in Chapter 18. For the moment, the point of this example is to see that *nucleophiles react with ketones and aldehydes by acyl addition.* In each example, the alkoxide products **30–32** are converted to an alcohol via an acid–base reaction where the alkoxide is the base and aqueous acid (H⁺) is used. The alcohol product is the conjugate acid of this reaction and, because the hydronium ion is used as an acid, water is the conjugate base in this second reaction. *This second chemical reaction is necessary in order to isolate a neutral product.*

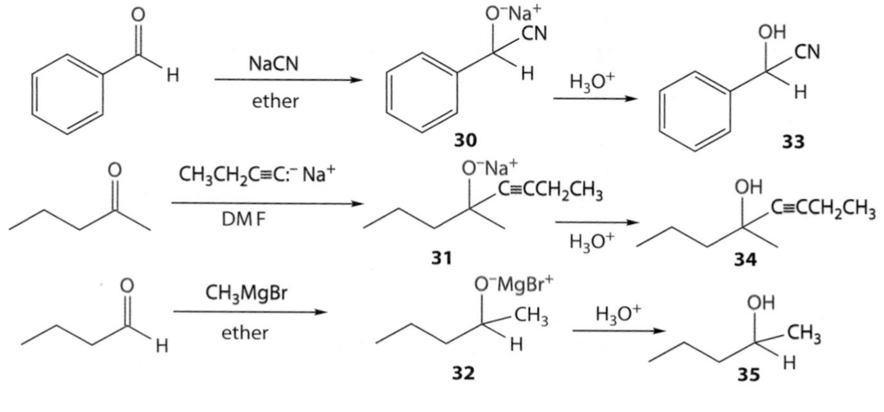

The overall two-step process is illustrated with benzaldehyde and involves the reaction of the aldehyde with NaCN to give **30**, followed by hydrolysis (the term for the acid–base reaction with aqueous acid) to give **33** (known as a cyanohydrin). Similarly, 2-pentanone leads to **31** and hydrolysis gives **34**; butanal reacts with methylmagnesium bromide to give **32**, and hydrolysis gives alcohol **35**.

Are the reactions that give **33–35** three different reactions, or one? *Rather than looking at each individual reaction as a different entity, look at all of them as acyl addition reactions, understand how acyl addition works, and then use any specific nucleophile to complete the reaction.* The generic reaction of **22** with a nucleophile (Y) will generate **29** via acyl addition, and hydrolysis leads to the alcohol **36**. This generic reaction represents reactions of both aldehydes and ketones with a variety of nucleophiles, and formations of **33–35** are simply examples of this general acyl addition. In other words, the three examples shown are one type of reaction, with different nucleophiles. Note that the salient feature of this reaction is formation of a new bond between the nucleophile and the carbonyl carbon.

16.16 **Draw the products that result when each of the following reacts with butanal in THF, with no other added reagents or reaction steps: (a) KCN, (b) NaOEt, (c) HC≡C:⁻, and (d) NEt₃.**

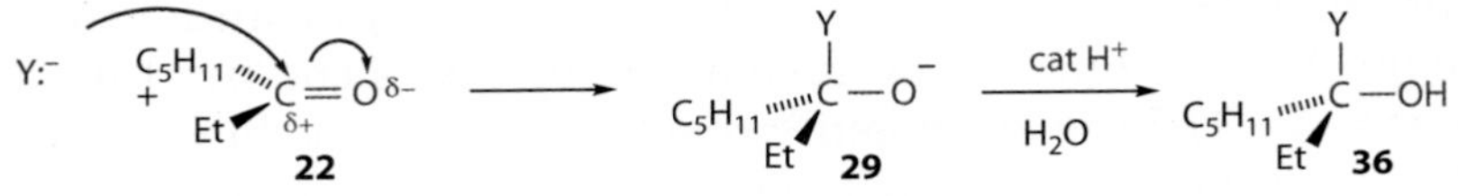

16.4 Carboxylic Acids. Nomenclature and Properties

Carboxylic acids (introduced in Chapter 5, Section 5.9.3) contain a carbonyl unit as well as an OH unit, in what constitutes the carbonyl unit COOH. However, the carbonyl is attached to an alkyl group and also an OH unit. (see **37**). Carboxylic acids react with various Brønsted–Lowry bases, as shown in several places in earlier chapters. Carboxylic acids were also discussed in Chapter 6 (Section 6.2) as strong acids that participate in acid–base reactions.

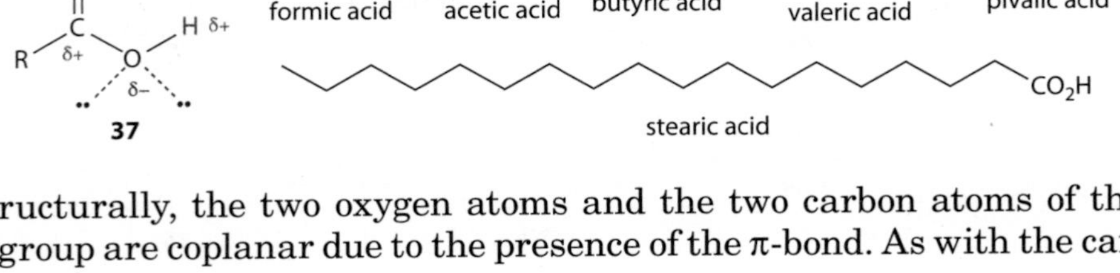

Structurally, the two oxygen atoms and the two carbon atoms of the carboxyl group are coplanar due to the presence of the π-bond. As with the carbonyl unit in aldehydes or ketones, the carbonyl in a carboxylic acid has a weaker and polarized π-bond along with a stronger σ-bond. The highly polarized O–H unit makes that hydrogen atom δ+ (see **37**). The greater bond polarization of the O–H unit leads to an even larger δ+ hydrogen atom than is found in the OH unit of an alcohol (C^{acid}–O–H >> $C^{alcohol}$–O–H). The conjugate base is the carboxylate anion, formed from the COOH unit, and it is resonance stabilized. Formation of the resonance-stabilized anion is a large contributing factor to the acidity of carboxylic acids, as discussed in Chapter 6, Section 6.2. When R = methyl, the molecule is known as ethanoic acid or acetic acid. Nomenclature is discussed in more detail later.

As with aldehydes and ketones, several simple carboxylic acids have common names. Although the emphasis will be on the IUPAC names, some of the common names appear regularly. Four are shown, including the one-carbon acid formic acid and the two-carbon acid acetic acid. Formic acid is a main component of many ant venoms, and acetic acid is a by-product of alcohol oxidation. Formic acid is used in food packaging, flavorings, and hair products. A dilute aqueous solution of acetic acid is called vinegar. It is used in baked goods, catsup, cheese, mayonnaise, and hair-coloring products. The four-carbon acid, butyric acid, is a component of rancid butter. Butyric acid is also found in some perfumes and disinfectants and is used as a flavoring component of candy and caramel. Note the short notation for the carboxyl group: –COOH or –CO_2H. The five-carbon acid is valeric acid, which is found in the human colon. Pivalic acid is a common intermediate in the preparation of pharmaceutical compounds; stearic acid (an 18-carbon acid) is obtained from animal fat and is a common ingredient in candles and soaps.

The IUPAC nomenclature is based on the longest continuous chain (alkane, alkene, or alkyne based) that contains the carboxyl carbon (the –COOH unit). As with aldehydes, the carbon of the COOH group is always at one terminus of the chain and must receive the lowest possible number (1), which is omitted from the name. The suffix for carboxylic acids is **"oic acid,"** with the word "acid" separated from the first part of the name. As with aldehydes and ketones, the carboxyl unit is higher in priority than an alkene or alkyne. Compound **38** is an eight-carbon acid with an alkane backbone, so it is octanoic acid (note the short notation for the carboxyl group). Substituents are numbered relative to the carbonyl carbon of the COOH unit, so **39** is 2,2-dimethyl-5-phenyloctanoic acid. In a similar manner, **40** is named 16-chloro-4-ethylheptadecanoic acid.

The carboxyl group is the highest priority group discussed so far. If a C=C or C≡C group is in a molecule along with the COOH unit, the base name remains "oic" acid. The carbon chain is numbered from the carbonyl carbon, and the C=C or C≡C unit is assigned a number accordingly. Compound **41** is a nonenoic acid and **42** is an octynoic acid. Substituents are named in the usual manner. Acid **41** is 5,6-dibromo-2,2-dimethylnon-5*E*-enoic acid, and **42** is 2-ethyl-7-methyloct-5-ynoic acid. Note that the chain used for the name in **41** includes the carboxyl group and the C=C unit and that, in **42**, the chain used for the name includes both the C≡C and the COOH units.

16.17 Write out the structure of 3-bromo-4-ethylnonanoic acid.

16.18 Write out the structure of 5-phenyl-2,5,6-trimethyldodec-8Z-enoic acid.

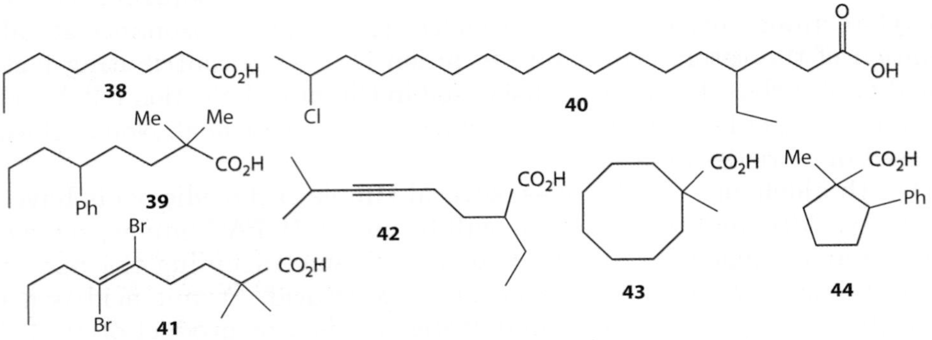

When the carboxyl group is attached to a ring, the name of the compound must be modified the same way as is done with aldehydes. Compound 43, for example, is named 1-methylcyclooctanecarboxylic acid. Similarly, 44 is 1-methyl-2-phenylcyclopentanecarboxylic acid.

16.19 Draw the structure of 1-methyl(cyclohex-2-ene)-1-carboxylic acid.

16.5 Dicarboxylic Acids

Dicarboxylic acids are introduced in this and the next section. Their reactions are discussed in Chapter 20 (Section 20.9), but it is appropriate to give examples of these compounds here. A dicarboxylic acid is named as a *dioic acid*. Many dioic acids of less than 10 carbons have common names, but a few examples will suffice. The C2–C5 dicarboxylic acids are **45–48**. The IUPAC nomenclature requires that the longest chain contain the carbonyl carbon of *both* carbonyl groups, the chain be numbered to give both carbonyl carbons the lowest possible number, and the ending be dioic acid.

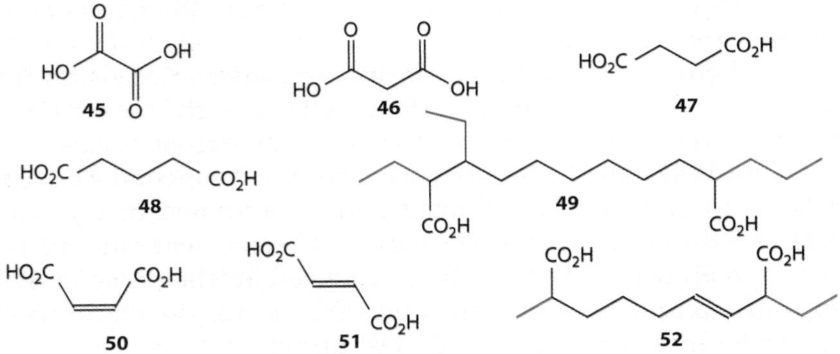

Table 16.1
Common Dicarboxylic Acids

Structure	Common Name	IUPAC Name
$HO_2C–CO_2H$	Oxalic acid	1,2-Ethanedioic acid
$HO_2C–CH_2–CO_2H$	Malonic acid	1,3-Propanedioic acid
$HO_2C–(CH_2)_2–CO_2H$	Succinic acid	1,4-Butanedioic acid
$HO_2C–(CH_2)_3–CO_2H$	Glutaric acid	1,5-Pentanedioic acid
$HO_2C–(CH_2)_4–CO_2H$	Adipic acid	1,6-Hexanedioic acid
$HO_2C–(CH_2)_5–CO_2H$	Pimelic acid	1,7-Heptanedioic acid
$HO_2C–(CH_2)_6–CO_2H$	Suberic acid	1,8-Octanedioic acid
$HO_2C–(CH_2)_7–CO_2H$	Azelaic acid	1,9-Nonanedioic acid
$HO_2C–(CH_2)_8–CO_vH$	Sebacic acid	1,10-Decanedioic acid

Therefore, **45** is 1,2-ethanedioic acid (or simply ethanedioic acid because that is the only possibility). Similarly, **46** is 1,3-propanedioic acid (or just propanedioic acid), **47** is 1,4-butanedioic acid, and **48** is 1,5-pentanedioic acid. Using this nomenclature, **49** has an 11-carbon chain that contains both carboxyl units, so it is a 1,11-undecanedioic acid. The substituents are given the lowest sequence of numbers, so the name of **48** is 2,3-diethyl-10-propyl-1,11-undecanedioic acid.

Because common names are frequently used with simple dicarboxylic acids, a few examples are noted. The first nine dicarboxylic acids have common names that must also be learned. Table 16.1 shows several common dicarboxylic acids, along with their common names and their IUPAC names. Oxalic acid (**45**) is the common name for the dicarboxylic acid with two carboxyl groups directly attached to each other, and malonic acid (**46**) has one $–CH_2–$ (methylene) unit separating the carboxyl units. As the number of methylene "spacers" increases, we see succinic acid (**47**), glutaric acid (**48**), etc. (all common names). The IUPAC names for these compounds are based on the total number of carbon atoms and the use of "di" to indicate the presence of two functional groups.

To indicate the presence of two COOH units, the "oic" acid suffix of a monocarboxylic acid is replaced with "dioic" acid. Oxalic acid therefore becomes 1,2-ethanedioic acid and malonic acid becomes 1,3-propanedioic acid. Similarly, succinic acid is 1,4-butanedioic acid and glutaric acid is 1,5-pentanedioic acid. With two carboxylic units, the dicarboxylic acids with five or fewer carbons have reasonable water solubility, reflecting their highly polar nature and ability to hydrogen bond. As the number of methylene groups increases, the solubility in water decreases as expected.

When the dicarboxylic acid has an alkene backbone, it is an enoic acid. Therefore, **50** is but-2Z-en-1,4-dioic acid and **51** is but-2E-en-1,4-dioic acid. The common name of **50** is maleic acid and **51** is known as fumaric acid. Compound

52 does not have a common name, but using the IUPAC system, it is 2-ethyl-8-methylnon-3E-en-1,9-dioic acid.

16.6 Dicarboxylic Acids Have Two pK$_a$ Values

The single acidic hydrogen atom of carboxylic acids (RCO$_2$H) reacts with Brønsted–Lowry bases in an acid–base reaction, as discussed in Chapter 6 (Section 6.2). The conjugate base in this reaction is the resonance-stabilized carboxylate anion, RCO$_2^-$. The ability to lose this hydrogen (acidity) is indicated by the value of pK$_a$ (a measure of the equilibrium constant, K$_a$, as discussed in Chapter 6, Section 6.1). A smaller value of pK$_a$ reflects a stronger acid and a large pK$_a$ value reflects a weaker acid. An interesting feature of dicarboxylic acids (**53**) is the presence of *two acidic hydrogen atoms;* this means that there are *two pK values:* pK$_1$ and pK$_2$.

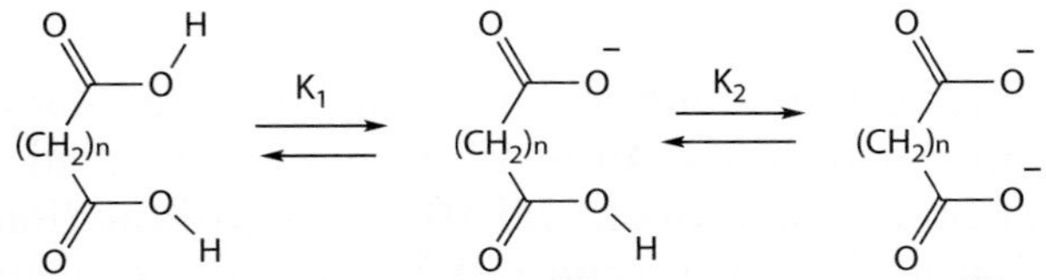

The value of pK$_1$ reflects the equilibrium constant for loss of the more acidic proton to form a monocarboxylate (see **54**, where "n" represents the number of –CH$_2$– units: 0, 1, 2, 3, etc.). The value of pK$_2$ reflects loss of the second proton to give the dianion, known as a dicarboxylate (see **55**). The pK values for several dicarboxylic acids are shown in Table 16.2,[1] along with the analogous monocarboxylic acid. This table clearly shows that the second pK value is rather close to the parent monocarboxylic acid, whereas the first pK is much lower (more acidic). Oxalic acid is much more acidic than formic acid (pK$_1$). This is due to the presence of an electron-withdrawing acyl group proximal to the carboxyl group that influences the acidity of that proton (see Chapter 6, Section 6.3). As the electron-withdrawing carboxyl moves further away from the first one (malonic to succinic to glutamic to adipic, etc.), the pK$_1$ generally increases (weaker acid). Glutaric acid is an exception; however, in general, dicarboxylic acids are more acidic than monocarboxylic acids because of the enhanced inductive effects possible when the second carboxylic is present.

16.20 **Write out the complete acid–base reaction of butanoic acid and sodium methoxide.**

16.21 **Draw the structure of 2-chloro-1-5-pentanedioic acid. Briefly discuss which, if either, of the two COOH units will have the lower pK$_a$.**

Table 16.2
pK Values for Common Dicarboxylic Acids

Structure	pK$_1$	pK$_2$	Common Name	pK of RCO$_2$H
HO$_2$C–CO$_2$H	1.25	3.81	Oxalic acid	3.75 R = H
HO$_2$C–CH$_2$–CO$_2$H	2.85	5.7	Malonic acid	4.76 R = CH$_3$
HO$_2$C–(CH$_2$)$_2$–CO$_2$H	4.21	5.64	Succinic acid	4.87 R = Et
HO$_2$C–(CH$_2$)$_3$–CO$_2$H	4.32	5.42	Glutaric acid	4.83 R = C$_3$H$_7$
HO$_2$C–(CH$_2$)$_4$–CO$_2$H	4.41	5.41	Adipic acid	4.83 R = C$_4$H$_9$
HO$_2$C–(CH$_2$)$_5$–CO$_2$H	4.71	5.58	Pimelic acid	4.85 R = C$_5$H$_{11}$
HO$_2$C–(CH$_2$)$_6$–CO$_2$H	4.52	5.40	Suberic acid	4.89 R = C$_6$H$_{13}$
HO$_2$C–(CH$_2$)$_7$–CO$_2$H	4.53	5.33	Azelaic acid	4.89 R = C$_7$H$_{15}$
HO$_2$C–(CH$_2$)$_8$–CO$_2$H	4.59	5.59	Sebacic acid	4.96 R = C$_8$H$_{17}$

Source: Lide, D. R., ed. 2006. *CRC Handbook of Chemistry and Physics,* 87th ed., 8-43–8-52. Boca Raton, FL: Taylor & Francis.

Loss of the second acidic proton (pK$_2$) is more difficult, as reflected by the larger value in all cases. The opportunity for hydrogen bonding is diminished and the product is a dianion (see **55**), which is more difficult to form and to solvate. As expected, however, the presence of the second carboxyl makes the second proton more acidic in oxalic acid and there are only minor changes in pK$_2$ in longer chain dicarboxylic acids. Once again, the exception is glutaric acid, with a pK$_2$ of 6.08. The ability to form a five-membered ring-like conformation could bring the two carboxylate units closer together (like charges repel), making them higher in energy and more difficult to form.

16.7 Carboxylic Acid Derivatives. Nomenclature and Properties

In four important derivatives of carboxylic acids, the OH unit in **37** is replaced by a halogen, –O$_2$CR, –OR, or –NR$_2$; all have a carbonyl. The first type is an *acid halide*, generated when the OH unit in **37** is replaced with a halogen atom such as chlorine (see **56**, an *acid chloride*). The related molecule where OH is replaced by bromine is called an acid bromide (see **57**). An *acid anhydride* is formed when the OH group is replaced by another acid unit (O$_2$CR), as in **58**. As the name suggests (anhydride = without water), anhydrides are essentially two carboxylic acid units joined together with loss of a water molecule.

If the OH group in **37** is replaced by an OR′ group (from an alcohol), **59** is formed, and it is called an *ester* or a carboxylic ester. An ester is essentially a

Table 16.3
Typical Acid Derivatives

Structure	Parent Acid	Name
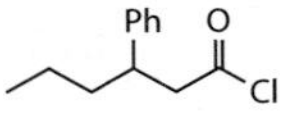 **61**	3-phenylhexanoic	3-phenylhexan**oyl chloride**
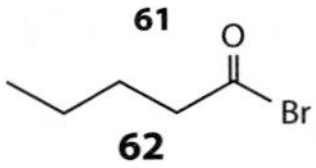 **62**	pentanoic	pentan**oyl bromide**
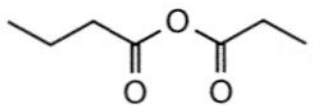 **63**	butanoic	butanic propanoic **anhydride**
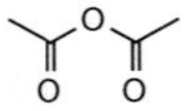 **64**	ethanoic (acetic)	ethanoic **anhydride** (acetic **anhydride**)
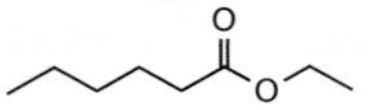 **65**	hexanoic	**ethyl** hexan**oate**
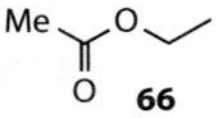 **66**	ethanoic (acetic)	**ethyl** ethan**oate** (**ethyl acetate**)
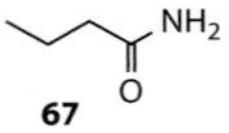 **67**	butanoic	butan**amide**
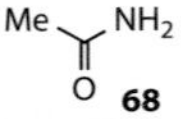 **68**	ethanoic (acetic)	ethan**amide** (acet**amide**)
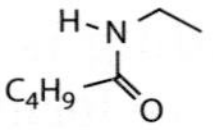 **69**	pentanoic	N-ethyl pentan**amide**
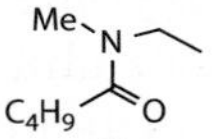 **70**	pentanoic	N-ethyl-N-methylpentan**amide**

combination of a carboxylic acid and an alcohol. Finally, if OH in **37** is replaced with an amine group (NH_2, NHR^1, or NR^1R^2 in **60**), the derivative is called an *amide*. An amide is essentially a combination of a carboxylic acid and an amine. For each of these carboxylic acid derivatives, the unit that has replaced the OH unit is shown in **green** in the illustration. Table 16.3 shows the structure of several acid derivatives, the parent acid, and their names. The suffix for each functional group is highlighted in **blue**. The nomenclature rules for each type of molecule follow.

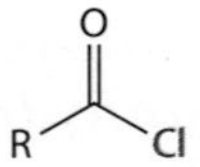

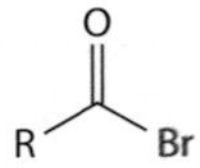

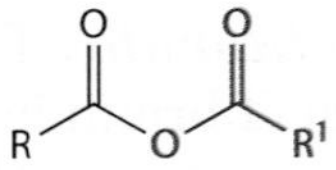

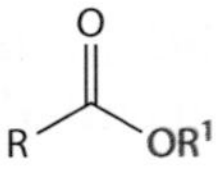

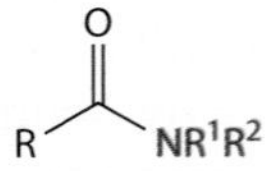

Each of these acid derivatives has a unique system of nomenclature, but all are fundamentally based on the carboxylic acid precursor (the parent acid). Several examples are shown in Table 16.3. Acid chlorides **56** are named by taking the alkyl carboxylic acid name (R in **56**, ending in "oic" acid) and replacing the -oic acid term with **"oyl" chloride**. The first structure in Table 16.3 is a derivative of 3-phenylhexanoic acid. Replacing OH with Cl leads to 3-phenylhexanoyl chloride, **61**. Acid bromides **57** are less common than acid chlorides, but they are named in an identical manner, except that the word bromide is used as with pentanoyl bromide (**62**). The example in Table 16.3 is pentanoyl bromide.

Acid anhydrides have structure **58** and are named by recognizing that the structure consists of two carboxylic acid units. The IUPAC rules demand that the two acids be named sequentially, followed by the word anhydride. For anhydride **63** in Table 16.3, one component is derived from butanoic acid and the other from propanoic acid. The names of the acid are listed alphabetically, so the name of **63** is butanoicpropanoic anhydride. For a general rule of naming, list R and R^1 in **63** *alphabetically,* followed by the word **anhydride**). Anhydride **63** is an *unsymmetrical* anhydride because the two acid components are different. If both acid components are identical, it is a *symmetrical* anhydride such as **64**. Using both acid names for this compound leads to diethanoic anhydride, or simply ethanoic anhydride: When the single name is used, the anhydride is understood to be symmetrical. Anhydride **64** is derived from ethanoic acid, which has the common name acetic acid, so the common name acetic anhydride is often used for **65**.

Esters such as **59** are essentially a combination of a carboxylic acid and an alcohol. The acid part (RCO in **59**) is named after the parent carboxylic acid and the alcohol part (OR^1 in **59**) is named after the alcohol. To make it work, the ***alcohol part is treated as an alkyl substituent***, so an ester with a methanol component is a methyl ester and an ester with a propanol component is a propyl ester. This alcohol part of the name is followed by the acid part, where the "oic" acid part of the name is replaced by **"oate."** **Ethyl** hexan**oate** is the name of **65** (from ethyl alcohol—ethanol—and hexan**oic acid**); ethyl ethan**oate** ($MeCO_2Et$, **66**) is another example. Ethyl ethan**oate** is a special case in the sense that it is derived from ethanoic acid, which has the common name of acetic acid. Therefore, **66** has a common name that is used most often: ethyl acetate. Note the shorthand way of writing ethyl acetate: EtOAc, where Et is ethyl and OAc is the acetate unit O_2CCH_3.

Amides such as **60** are viewed structurally as a combination of an amine and an acid, so there is an amine name and an acid name; however, amides are treated differently than esters. For primary amides, an $-NH_2$ group is attached to the carbonyl (two hydrogens on N; see **60** $R^1=R^2=H$); the name of the acid is changed by replacing the "oic" acid with the word **amide**. Two examples in Table 16.3 are butan**amide** **67** and the ethanoic acid (acetic acid) derivative **68**. Amide **67** is a butanoic acid derivative in which OH has been replaced by NH_2 and the named is butan**amide**. The IUPAC name of **68** is ethan**amide**, but as a derivative of acetic acid, its common name of acet**amide** is used more often.

If the amide has a $-NHR$ group (one hydrogen on N; **60**, $R^1=H$, $R^2=$alkyl), it is called a secondary amide and named similarly to the primary amide except that there is a group on nitrogen whose position must be identified. This is done by using the term *N*-alkyl, where the alkyl group is understood to be attached directly to nitrogen (see the related amine nomenclature in Chapter 5, Section 5.6.2). An example in Table 16.3 is **69**, which is a butanoic acid derivative with an ethyl group attached to nitrogen. Using the IUPAC system, the name of **69** is *N*-ethylpentan**amide**. When the amide has two alkyl groups attached to nitrogen (no hydrogen atoms on N; **56**, R^1=R^2=alkyl), it is known as a tertiary amide and both alkyl groups on nitrogen are designated by using the "*N*" protocol. Example **70** is a pentanoic acid with an ethyl group as well as a methyl group on the nitrogen. The name of **70** is *N*-ethyl-*N*-methyl-pentan**amide**.

The two alkyl groups on nitrogen are different, and the names of those groups are listed alphabetically, using an *N* for each group. The use of *N* is analogous to using a number, where a substituent on carbon is given a number and a group on nitrogen is given an *N*. An example is **71**, which is an octanoic acid derivative, so it is an octan**amide**. There are three methyl groups—on N as well as at C3 and C4—so **71** is an *N*,3,4-trimethyloctan**amide**. To complete the name, we add the chlorine at C8 and the propyl group on nitrogen. Therefore, **71** is named 8-chloro-*N*,3,4-trimethyl-*N*-propyloctan**amide**.

16.22 Draw the structure of the acid bromide of 3-cyclopentylheptanoic acid.

16.23 Draw the structure of 2-phenylheptanoic 3,3,4-trimethylpentanoic anhydride.

16.24 Draw the structure of 2-phenylethyl 3,3-dimethylpentanoate.

16.25 Draw the structure of *N*,3-diethyl-*N*-methyl-5-phenylhexanamide.

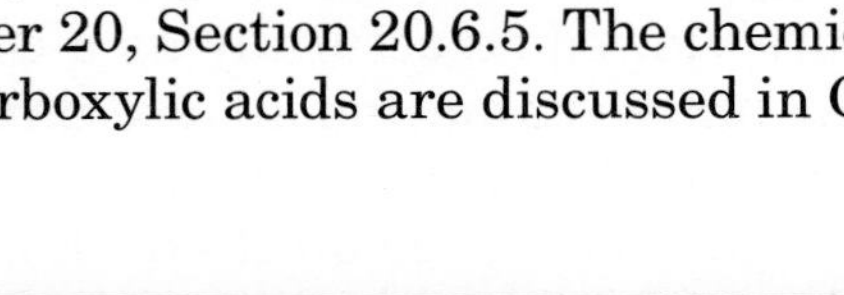

There are cyclic esters (known as lactones), cyclic amides (known as lactams), cyclic anhydrides, and cyclic imides. The structures for these derivatives are presented in Chapter 20, Section 20.6.5. The chemical reactions associated with derivatives of dicarboxylic acids are discussed in Chapter 20 as well.

16.8 Acyl Substitution with Carboxylic Acid Derivatives

In Section 16.3, aldehydes and ketones reacted as a Brønsted–Lowry or Lewis bases with an appropriate acid, but they also react with nucleophiles by acyl addition. It is anticipated that the carbonyl of an acid derivative will behave similarly. Indeed, acid derivatives react with an acid to form an oxocarbenium ion.

When nucleophiles react with acid derivatives, however, there is a significant difference when compared with aldehydes and ketones. Acid derivatives have a *leaving group* attached to the carbonyl, but aldehydes and ketones do not. This difference leads to a new type of reaction that is known as ***acyl substitution***.

When a generic acid derivative, **72**, reacts with sulfuric acid, reaction of the carbonyl mimics that of an aldehyde or ketone, producing oxocarbenium ion **73**, which is resonance stabilized. Remember that the pK_a of sulfuric acid is about -3 and a typical carboxylic acid has a pK_a of about 4.5–4.8 (see Table 16.2). Therefore, in this reaction, the oxygen of carboxylic acid **72** is the ***base*** (the proton acceptor) and sulfuric acid is the ***acid*** (the proton donor).

In principle, this reaction occurs with carboxylic acids, where X = OH in **72**. The reaction occurs for carboxylic acid derivatives **56–60**, where the X group in **72** and in **73** is Cl for an acid chloride, O_2CR for an anhydride, OR for an ester, or NR_2 for an amide. For these derivatives **73**, a heteroatom is attached to the δ+ carbon, and there is additional resonance stabilization due to the lone electron pairs on these atoms. How oxocarbenium ions might react after they are formed has not been discussed, but it is clear that the carbonyl group of an acid derivative can react as a base in the presence of an appropriate acid. Reactions of oxocarbenium ions are discussed in Chapters 18 and 20.

16.26 Draw the resonance forms possible when X = Cl in 73 and when X = OCH_3.

In Section 16.3, a nucleophile "Y" reacted with a ketone or aldehyde (**22**) by acyl addition to give an alkoxide (**29**), and hydrolysis in a second chemical step gave the alcohol product **36**. If the same reaction is done with an acid chloride **72** (X = Cl), the nucleophile reacts with the acyl carbon to form **74**. Remember that "Y" is the nucleophile and "X" is the leaving group. There is a major difference between **29** and **72**. In **29**, two alkyl groups are connected to the carbonyl carbon by C–C σ-covalent bonds, which are rather strong. In **74**, there is one C–C σ-covalent bond to an alkyl group, but there is also a much weaker C–Cl bond. The chlorine atom is a leaving group in **74**, and expulsion of chloride ion leads to formation of **75**.

Returning to the generic structure **72**, as the nature of the X changes, the leaving group ability changes. In **74**, the alkoxide can regenerate the C=O group as shown, with loss of the X group. The presence of a leaving group in acid derivatives such as an acid chloride leads to a new type of substitution reaction in which the chlorine (or another leaving group) is replaced with a nucleophile (Y). The process termed nucleophilic *acyl substitution* generates alkoxide **74** with two potential leaving groups X and Cl, and is known as a *tetrahedral intermediate*. In general, *acid derivatives undergo acyl substitution, whereas aldehydes and ketones undergo acyl addition.*

The acyl substitution reactions associated with acid derivatives are discussed in great detail in Chapter 20. Acid chlorides, anhydrides, esters, and amides all undergo this reaction. Acid chlorides tend to be more reactive than esters because of the greater reactivity of the tetrahedral intermediate. This means that the Cl is a better leaving group when compared to OR. Such differences in leaving group ability are also discussed in Chapter 20.

There is one last reaction to consider. Remember the reaction of a carboxylic acid such as butanoic acid with a base such as NaOH or $NaOCH_3$ described in Chapter 6 (Section 6.2). Sodium methoxide is a good base (Chapter 12, Section 12.1), but as seen in Chapter 11 (Section 11.3.2), methoxide is also a good nucleophile. What happens when butanoic acid reacts with sodium methoxide in ether? The answer is that the acid–base reaction dominates; indeed, *the acid–base reaction is much faster than the acyl substitution reaction.* Therefore, sodium methoxide reacts with butanoic acid to give the sodium salt of butanoic acid (**76**, the conjugate base) and methanol (the conjugate acid). If a potential nucleophile is a potent base, the acid–base reaction will dominate with carboxylic acids. Nucleophilic acyl substitution reactions dominate with acid derivatives, with some exceptions that are discussed in Chapter 22.

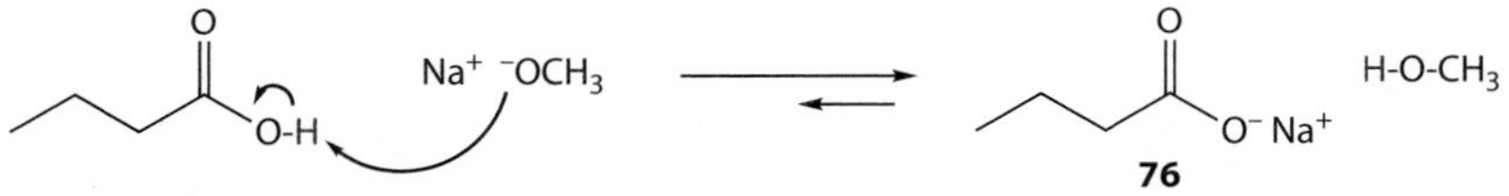

16.9 Sulfonic Acids

Carboxylic acids are an important class of organic acids, but there is another important class known as the sulfonic acids. Sulfonic acids have the general structure shown in **77**, and R may be any alkyl groups (methyl, ethyl, butyl, etc.). (Sulfonic acids were introduced in Chapter 6, Section 6.2.4.) Aliphatic sulfonic acids are named by using the name for the hydrocarbon unit R, with the suffix sulfonic acid. Examples include the one-carbon sulfonic acid **78**. The one-carbon hydrocarbon is methane, and the one-carbon sulfonic acid **78** is named methanesulfonic acid. Similarly, **79** is a three-carbon sulfonic acid named propane sulfonic acid. Substituents are handled much like carboxylic acids, in that the SO_3H unit is part of the longest continued chain. Because **80** is attached to a seven-carbon chain, it is a heptanesulfonic acid. There are two methyl groups and a propyl substituent, and the name is 3,4-dimethyl-1-propylheptanesulfonic acid.

Aromatic derivatives are known, including benzene sulfonic acid (**81**; see Chapter 21). Sulfonic acid **82** is a generic sulfonic acid derived from an aromatic compound (Ar = aryl), where an aryl group is any benzene derivative. The aryl group may also be a polynuclear aromatic compound (see Chapter 21, Section 21.8). The nomenclature for **78** and aryl derivatives is presented in Chapter 21.

However, one aromatic sulfonic acid is rather important. Sulfonic acid **83** has the common name *para*-toluenesulfonic acid (tosic acid), and it is commonly used as an acid catalyst because it is reasonably soluble on organic solvents. Tosic acid has been used several times in previous chapters, including Chapter 10 (Sections 10.3 and 10.6) and Chapter 11 (Section 11.4). Chapter 21 will describe the formal nomenclature of aryl sulfonic acids, but the IUPAC name of **81** is benzenesulfonic acid and **83** is 4-methylbenzenesulfonic acid. However, the common name is used most often for **83**.

Sulfonic acids are more acidic than the analogous carboxylic acid, as briefly introduced in Chapter 6, Section 6.2.4. For example, the one-carbon sulfonic acid (**78**) is compared with ethanoic acid (acetic acid, **84**). The pK_a of **84** is 4.76, whereas the pK_a of **78** is −1.9.[2] Tosic acid has a pK_a of −2.8,[3] compared with a pK_a of 4.2 for benzoic acid (PhCO$_2$H; see Chapter 21). For the most part, sulfonic acids other than **78** and **81** are not used in this book. For that reason, there is no reason to give a table of pK_a values for different sulfonic acids; in virtually all cases, it is assumed that a comparable sulfonic acid is more acidic than the analogous carboxylic acid.

16.10 Biological Relevance

The aldehyde and ketone units appear in some biologically important systems. The ketone unit appears in many carbohydrates (sugars), including ribulose (**85**), fructose (**86**), and sorbose (**87**). d-Ribulose is an intermediate in the fungal pathway for d-arabitol production; as the 1,5-*bis*(phosphate), d-ribulose combines with carbon dioxide at the start of the photosynthetic process in green plants (a carbon dioxide trap). Fructose is found in many fruits and in honey and is used as a preservative for foodstuffs and as an intravenous nutrient. Fructose is also called *fruit sugar* or *levulose*. Sorbose is about as sweet as common table sugar (sucrose) and is used in the manufacture of vitamin C (ascorbic acid, **88**).

Vitamin C is an essential nutrient for humans, as well as many other species of living organisms, and is required for many essential metabolic reactions in all animals and plants. Humans obtain vitamin C from citrus or as a food additive or supplement. A deficiency of vitamin C causes scurvy in humans. Note that vitamin C exists in the form of an enol (Chapter 10, Section 10.4.5), rather than in the keto form.

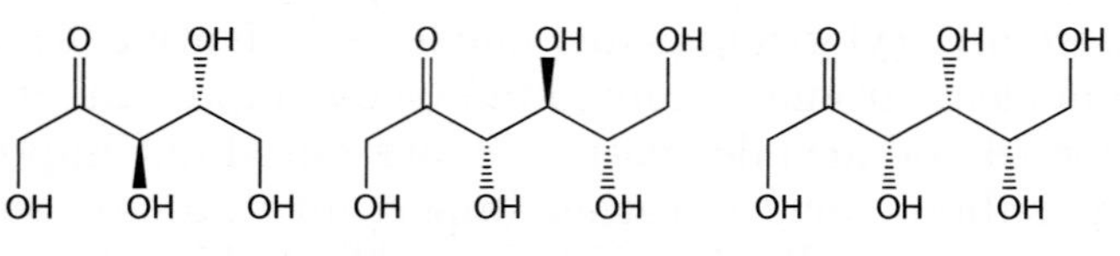

85 (ribulose)   **86** (fructose)   **87** (sorbose) 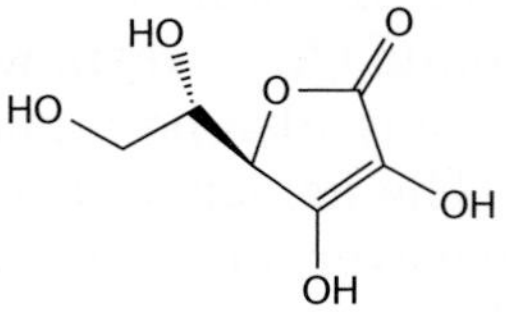   **88**

Carboxylic acid derivatives are ubiquitous. Proteins are characterized by so-called peptide bonds, but they are in reality amide bonds. A simple example is the nonapeptide (nine amino acid residues) **89**. This nonapeptide is composed (reading from left to right) of **alanine–valine–serine–leucine–alanine–phenylalanine–glutamic acid–methionine–histidine**, using only (*S*)-amino acids (see Chapter 27).

Amide bonds are found in many proteins. One is the acyl carrier protein of *Escherichia coli* (see **90**), which contains the peptide backbone, and a 4′-phosphopantetheine unit (in **violet** in the illustration) is attached to a serine residue. Note the amine bonds in the pantothenic acid unit and also the O–P=O unit, which is a phosphate ester (an ester of phosphoric acid). An acyl carrier protein is involved in fatty acid synthesis, linking acetyl and malonyl groups from acetyl coenzyme A and malonyl coenzyme A to form β-keto acid acyl carrier protein (abbreviated as ACP). The widely utilized acetyl CoA is an ester (**91**) attached to coenzyme A. Acetyl CoA is a key intermediate in aerobic intermediary metabolism of carbohydrates, lipids, and some amino acids.

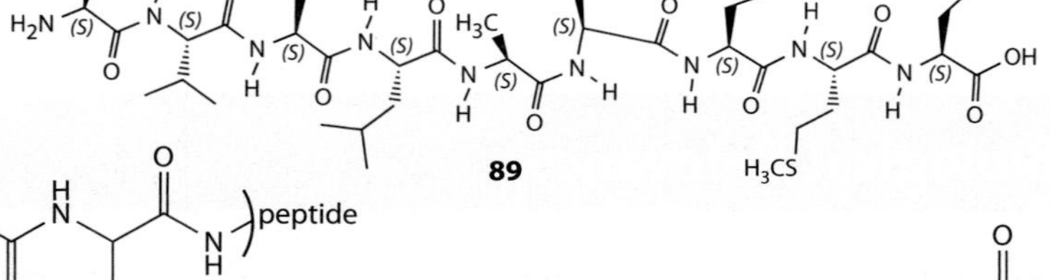

89

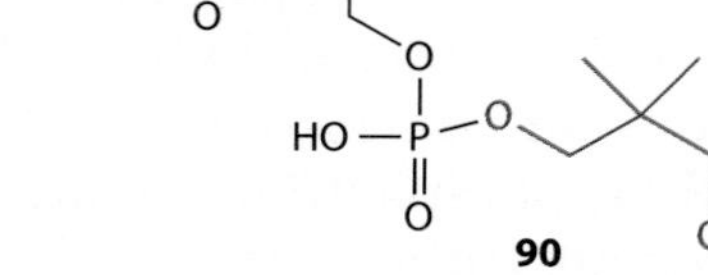

90 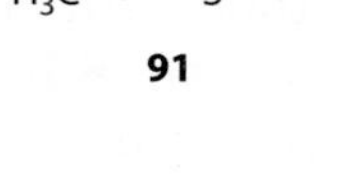   **91**

   Polyketides, with the generic formula $-[CH_2CO_n]-$, are an important class of naturally occurring compounds that include fatty acids and prostaglandins. Fatty acids are long chain carboxylic acids (see Chapter 20, Section 20.2). Common naturally occurring fatty acids include the C14 acid myristic acid (**92**), the C16 acid palmitic acid (**93**), the C18 acid stearic acid (**94**), and the C20 acid arachidic acid (**95**). Myristic acid (tetradecanoic acid) is found in nutmeg. Palmitic acid (hexadecanoic acid) is found in butter, cheese, and meat. It is also a component of palm oil (see illustration).

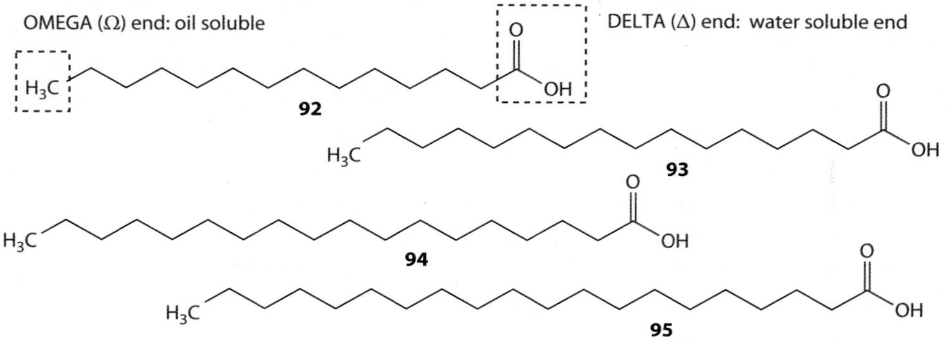

   Stearic acid (octadecanoic acid) is a waxy solid that is a component of animal and vegetable fats (see illustration). Stearic acid is used in cosmetics, as a dietary supplement, and to make candles. Arachidic acid (eicosanoic acid) is found in peanut oil as well as in fish oil. Certain terminology can be used, particularly for fatty acids that are found in nature. The carboxylic acid end is polar, associated with water solubility, and is labeled the delta (Δ) end of the fatty acid. The methyl end is associated with being nonpolar and more soluble in oils and is labeled the omega (Ω) end (see the labels in **92**).

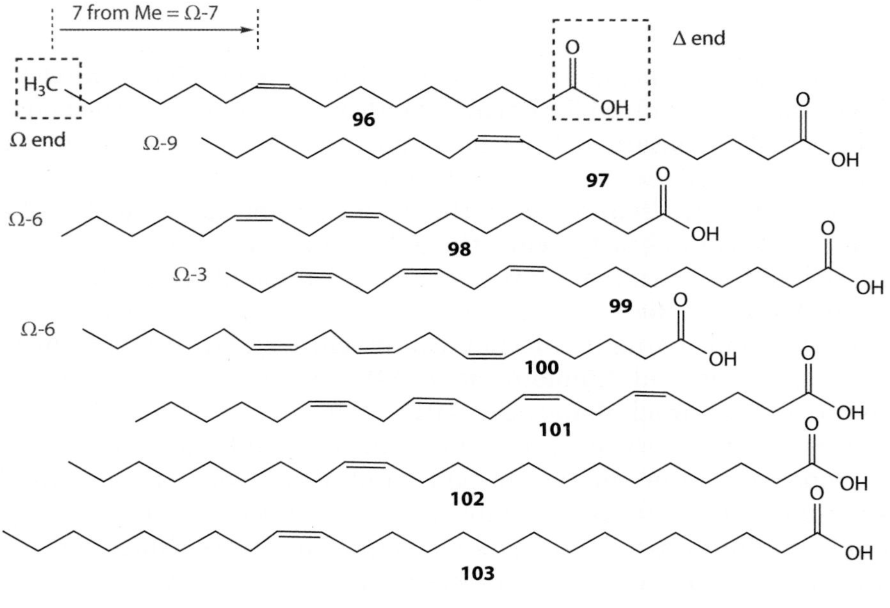

There are many unsaturated fatty acids, characterized by having an alkene unit or diene or polyene units in the long carbon chain rather than the alkane chain found in **92–95**. Common unsaturated fatty acids are palmitoleic acid (**96**; C16), oleic acid (**97**; C18), linolenic acid (**98**; C18), α-linolenic acid (**99**; C18), and γ-linolenic acid (**100**; C18). Other examples include arachidonic acid (**101**; C20), erucic acid (**102**; C22—found in mustard seed), and nervonic acid (**103**; C24—important for the biosynthesis of nerve cell myelin).

Additional nomenclature for the unsaturated fatty acids is used to identify the site of C=C units relative to the terminal methyl group (the omega end). Because the closest double bond to the methyl group in palmitoleic acid (**96**) is 7 carbon atoms away from the methyl, palmitoleic acid is called an **omega–7** (ω–7 or **n–7**) fatty acid. For α-linolenic acid (**99**), the double bond closest to the methyl group is only 3 carbons away, so it is an **omega–3** (ω–3 or **n–3**) fatty acid. Several other acids are suitably labeled.

There are alkyne fatty acids such as crepenynic acid (**104**; C18) and fatty acids that contain branches such as corynomycolic acid (**105**; C32), found in *Corynebactrium diphtheriae,* or tuberculostearic acid (**106**; C19), which is found in *Mycobacterium tuberculosis*. These latter organisms are responsible for diphtheria and tuberculosis, respectively.

The essential fatty acids are unsaturated fatty acids that cannot be made in the body, but are required for normal, healthy functioning of the body. These fatty acids must be obtained from foods such as nuts, sunflower oil or other vegetable oils, and oil-rich fish. A deficiency of these essential fatty acids may result in hyperactivity, reduced growth, or, in extreme cases, death. There are two families of essential fatty acids: the ω-3 and ω-6 fatty acids.

Nutritionally essential ω-3 fatty acids are α-linolenic acid (**99**), eicosapentaenoic acid (**107**), and docosahexaenoic acid **108**). The human body cannot synthesize ω-fatty acids using its biochemical machinery, but it can form **107** and **108** from α-linolenic acid. Linolenic acid (**98**) is an essential fatty acid used by the body to produce arachidonic acid (**101**), which is physiologically significant because it is the precursor for prostaglandins. Prostaglandins activate inflammatory response, the production of pain, and also fever. They are implicated in other biological functions; thromboxane A2 (**109**) is one of the prostaglandins that stimulate constriction and clotting of platelets. Note that ω-9 fatty acids are not classified as essential fatty acids because they can be created by the human body from unsaturated fat (see illustration) and are therefore not essential in the diet.

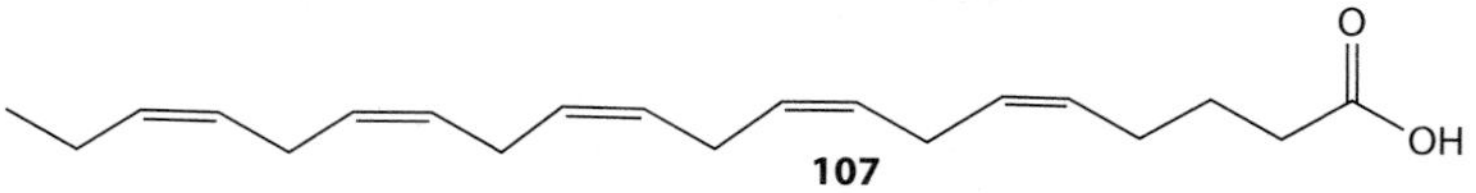

Fatty acids are usually found as an ester of 1,2,3-propanetriol (glycerol). Diesters of glycerol are known as diglycerides (**110**) or triglycerides (**111**). If these compounds (known as lipids) are solids, they are called **fats**; if they are liquids, they are known as **oils**. An important diglyceride is phosphatidic acid (with stearic acid ester units [stearates], marked in **red** in **112**). This is the parent compound for glycerol-based phospholipids. One is phosphatidylcholine (also known as lecithin), which is **112** and contains the trimethylammonium ethanolamine unit. Lecithin is used as an emulsifier, and it keeps cocoa and cocoa butter in a candy bar from separating, for example. It is produced by the liver and is a building block of cell membranes. Lecithin is found in the protective sheaths surrounding brain cells.

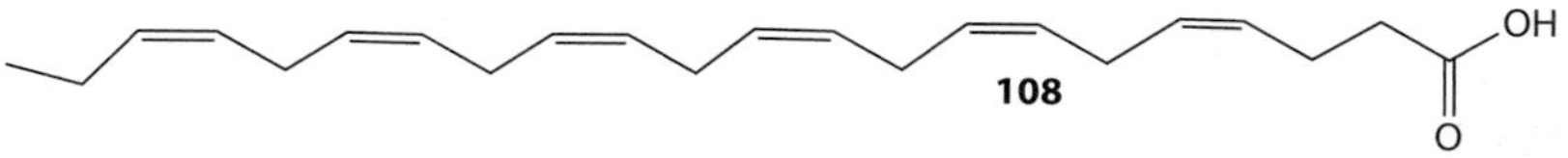

Most natural fats and oils are mixed triglycerides, in which the fatty acid constituents are different (see **114**, where the fatty acids are myristic acid, stearic acid, and palmitoleic acid).[4] A simple triglyceride usually has three identical fatty acids, as in **113**, with three stearic acid units.[2] Naturally occurring triglycerides are usually composed of fatty acids with 16, 18, or 20 carbon atoms. The fats obtained from the food we eat are the usual source of the triglycerides found in our blood plasma, and an excess of triglycerides in blood plasma is called hypertriglyceridemia. An elevated level of triglycerides is linked to the occurrence of coronary artery disease.

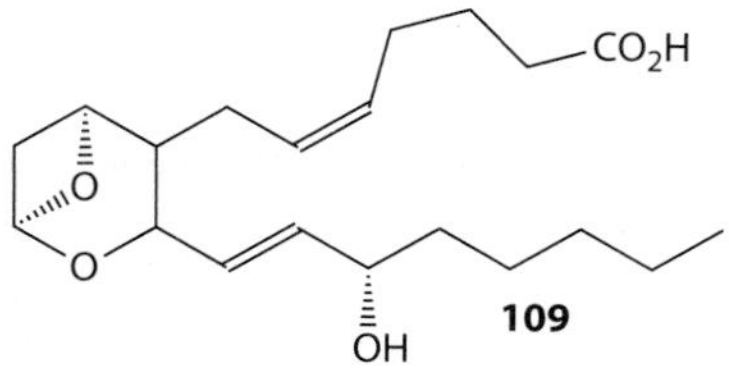

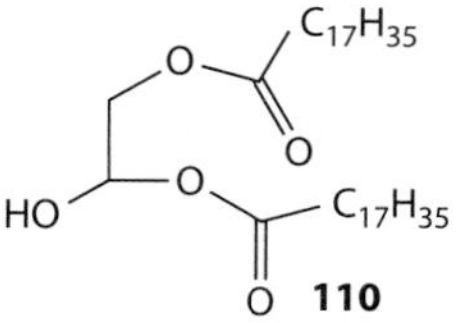

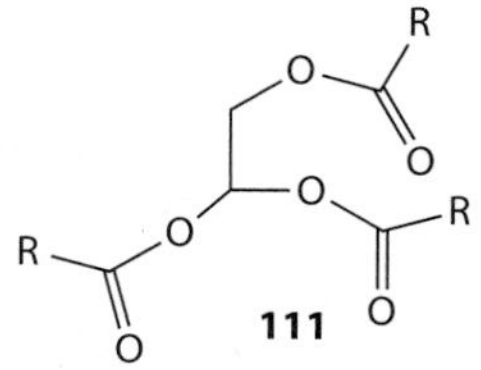

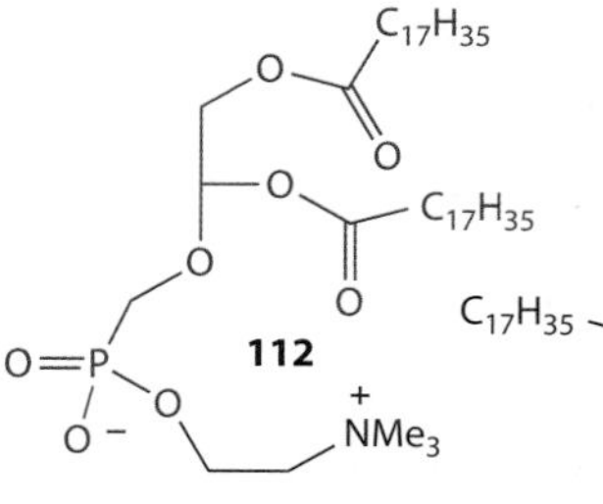

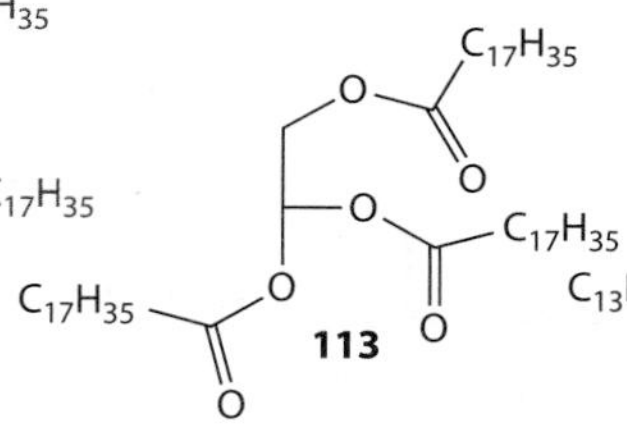

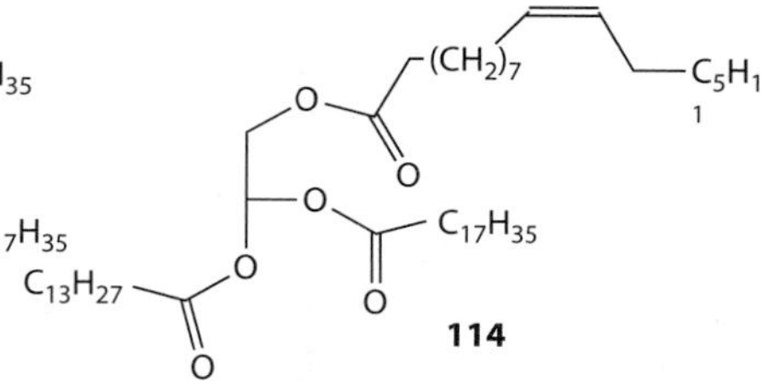

A final example is a sulfonic acid known as taurine, **115**, a common constituent of bile found in the lower intestines of humans. Taurine is implicated in many
key biological functions in humans,[5] and it is found in many "energy" drinks.

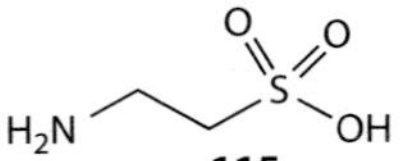

115

References

1. Lide, D. R., ed. 2006. *CRC handbook of chemistry and physics,* 87th ed., 8-43–
   8-52. Boca Raton, FL: Taylor & Francis.
2. Stewart, R. 1985. *The proton: Applications to organic chemistry,* 17. Orlando, FL:
   Academic Press.
3. Guthrie, J. P. 1978. *Canadian Journal of Chemistry* 56:2342–2353.
4. Garrett, R. H., and Grisham, C. M. 1995. *Biochemistry,* 279–284. Ft. Worth, TX:
   Saunders.
5. For example, see Huxtable, R. J., ed. 2008. Taurine in health and disease. In
   Advances in experimental medicine and biology, vol. 359. New York: Kluwer
   Academic Publishers.

Answers to Chapter Problems

16.1   The carbonyl carbon of an aldehyde has a carbon group and a hydrogen atom,
       whereas the carbonyl group of a ketone is attached to two carbon atoms.
       Because the aldehyde carbonyl carbon is less sterically hindered, all things
       being equal, it should be more reactive.

16.2   Benzophenone would be 1,1-diphenylmethanone, but because it is derivative of
       benzene, benzophenone is taken as the IUPAC name (see Chapter 21, Section
       21.2). Pinacolone is 2,2,4,4-tetramethyl-3-pentanone.

16.3

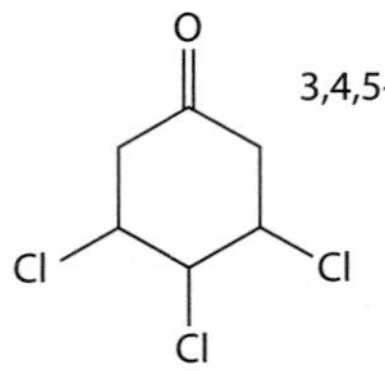

3,4,5-trichlorocyclohexanone   3R-ethyl-4S-iodocyclooctanone

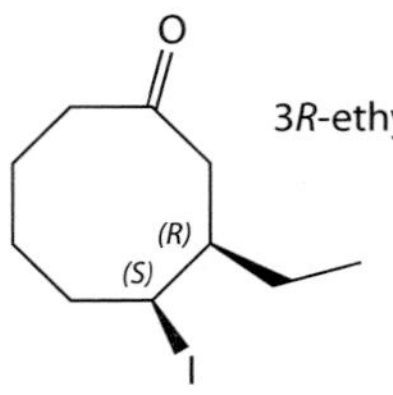

16.4

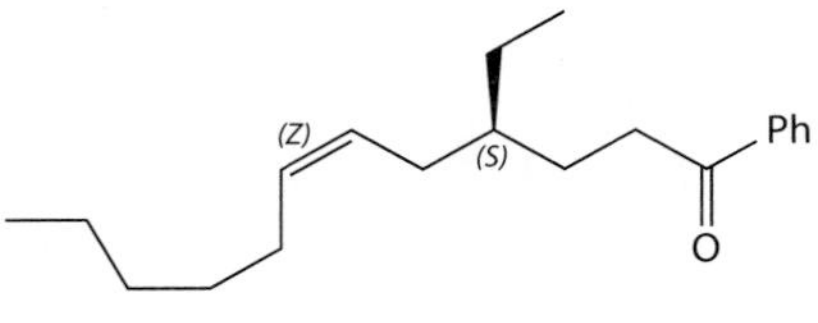

16.5

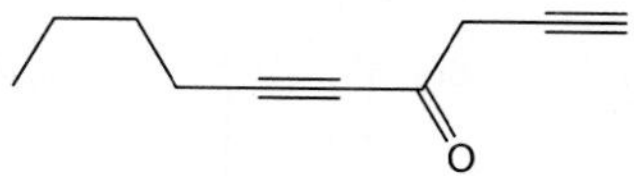

16.6

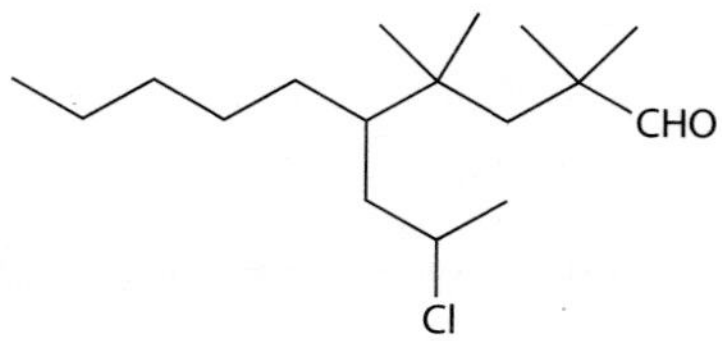

16.7

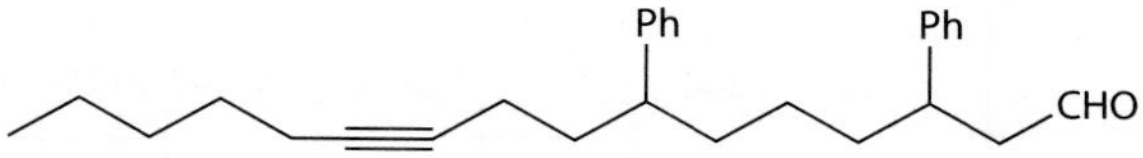

16.8

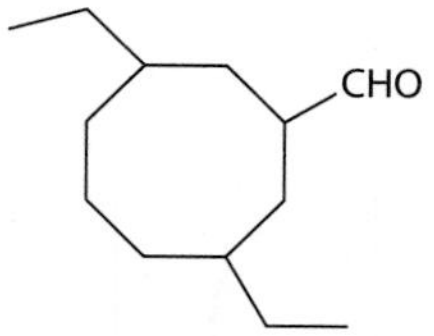

16.9

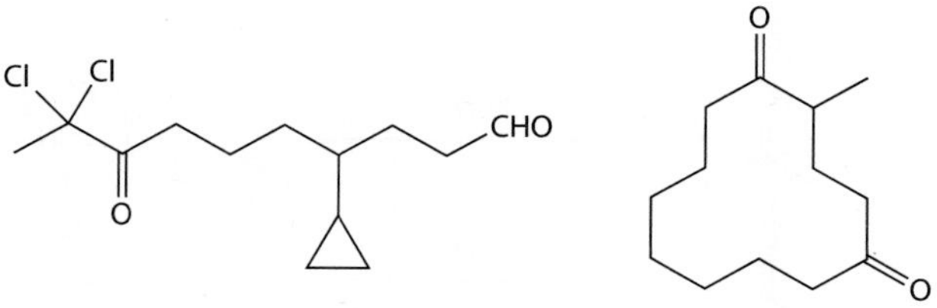

16.10

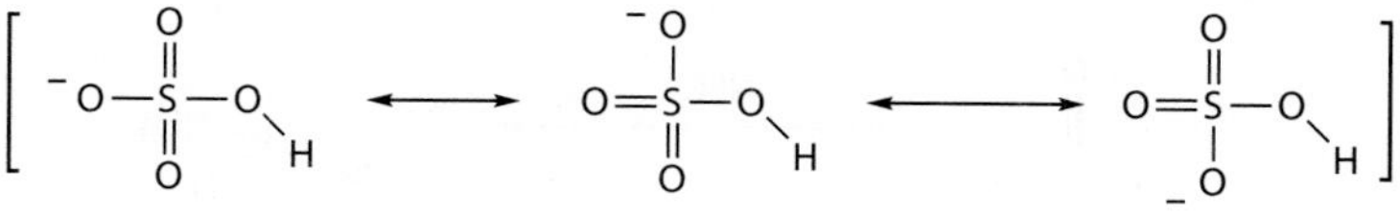

16.11

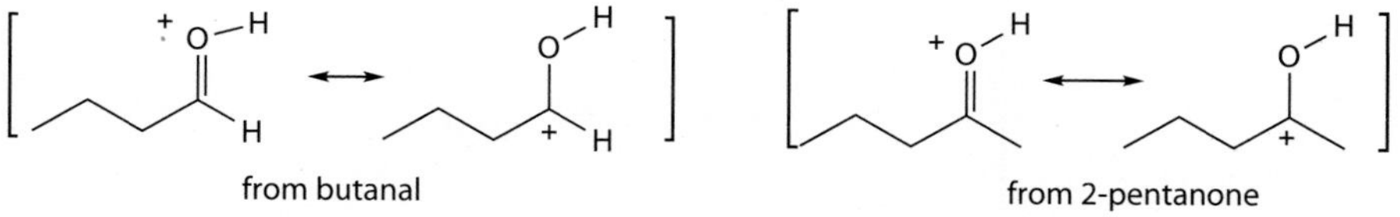

from butanal   from 2-pentanone

16.12

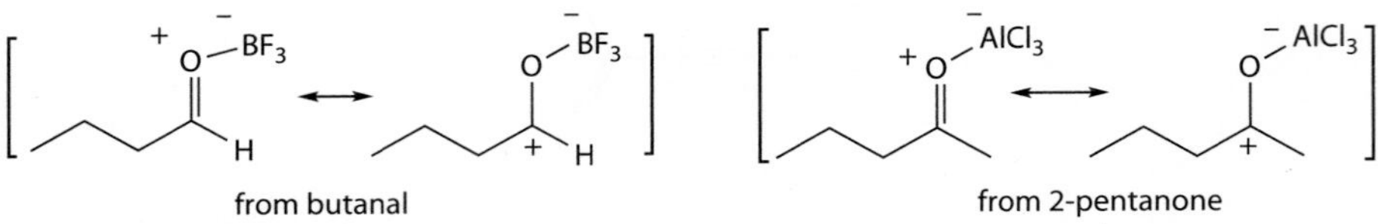

from butanal   from 2-pentanone

16.13

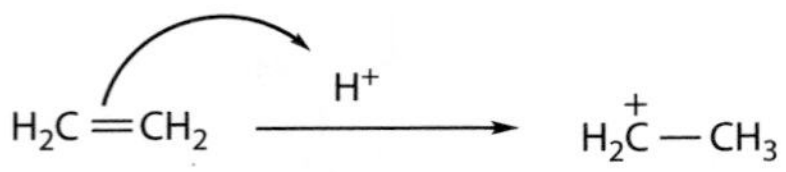

16.14 Note that the 3*E*-isomer was used, but resonance delocalization via the carbo-cation leads to a mixture of *E*- + *Z*-enols.

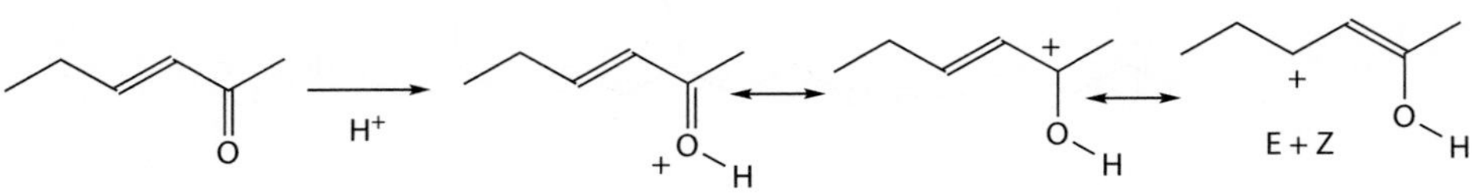

16.15

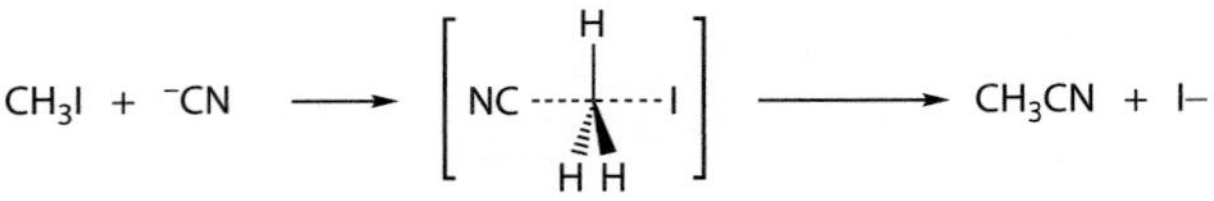

$$CH_3I + {}^-CN \longrightarrow \left[NC \cdots \overset{H}{\underset{H\ H}{|}} \cdots I \right] \longrightarrow CH_3CN + I^-$$

16.16

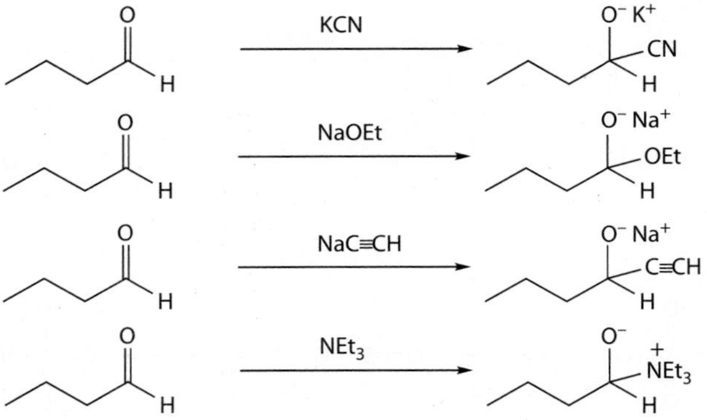

16.17

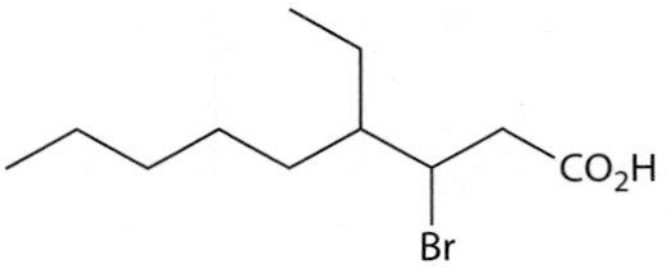

16.18

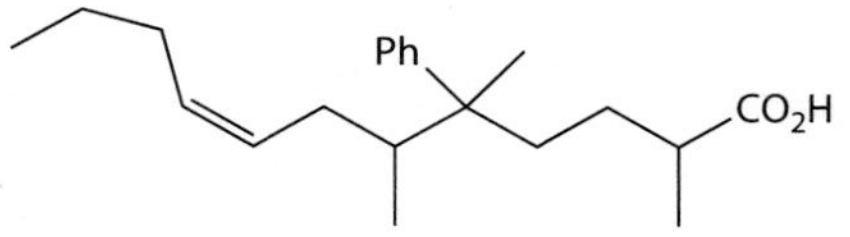

16.19

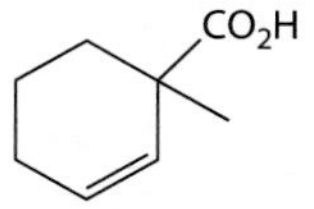

16.20

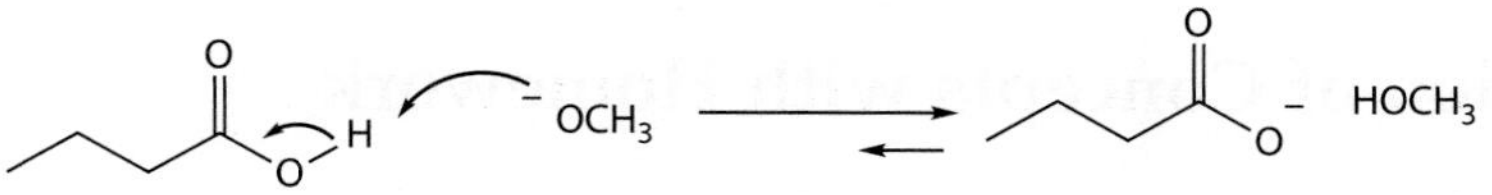

16.21 Due to the usual inductive effects, the COOH unit closest to the Cl will have the lower pK_a. Using IUPAC numbering for this molecule, the C1 carboxyl has the more acidic H.

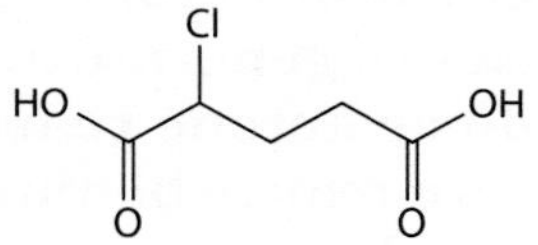

16.22

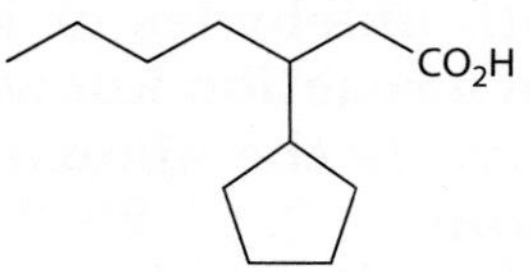

16.23

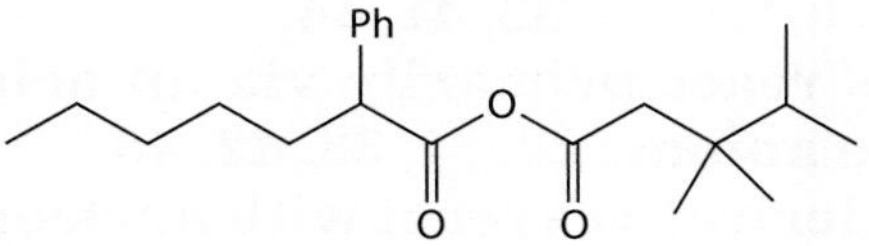

16.24

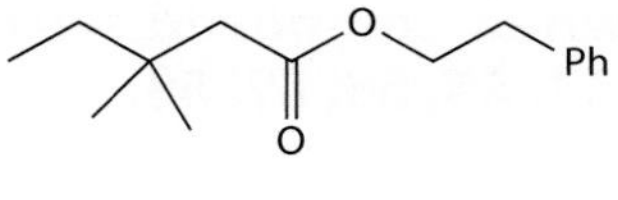

16.25

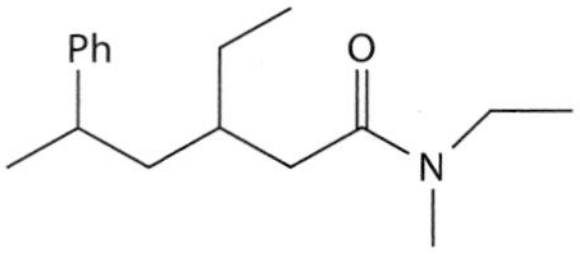

16.26

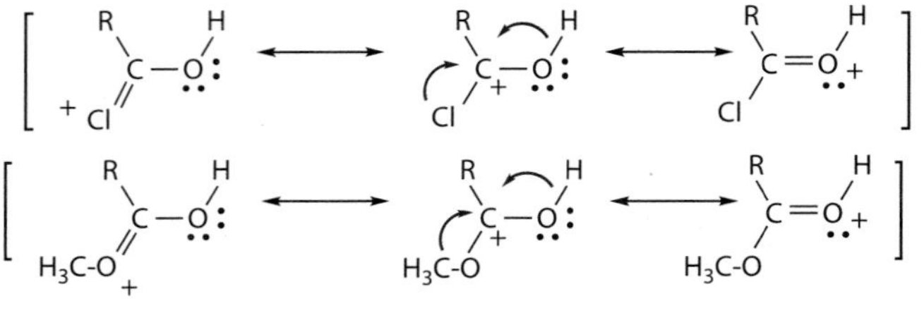

Correlation of Concepts with Homework

- **Nomenclature for aldehydes and ketones. Nomenclature of difunctional molecules containing a carbonyl group and an alkene or alkyne, or two carbonyl groups:** 2, 3, 4, 5, 6, 7, 8, 9, 27, 29, 30, 31, 32, 34.
- **The carbonyl of an aldehyde or ketone reacts with protonic acids as Brønsted–Lowry bases to generate an oxocarbenium ion. The carbonyl of an aldehyde or ketone reacts with Lewis acids as a Lewis base to generate an oxocarbenium ion.**
- **Acid derivatives react with protonic acids as Brønsted–Lowry bases to generate an oxocarbenium ion:** 10, 11, 12, 13, 14, 42, 46, 47, 48.
- **Nucleophiles react with aldehydes or ketones by acyl addition to generate alkoxide in a reaction known as acyl addition. A second chemical step converts the alkoxide product to an alcohol in an acid–base reaction:** 1, 15, 16, 28, 34, 35, 40, 42.
- **Nomenclature for carboxylic acids and dicarboxylic acids:** 17, 18, 19, 31, 37.
- **Structure and nomenclature for acid halides, anhydrides, esters, and amides:** 22, 23, 24, 25, 38, 41, 44.
- **Carboxylic acids react primarily via an acid–base reaction to form carboxylate anions:** 20, 21, 33, 42, 45.
- **Carboxylic acid derivatives react with nucleophiles primarily by acyl substitution via a tetrahedral intermediate:** 26, 36, 39, 42.
- **Spectroscopy can be used to determine the structure of a particular molecule and can distinguish the structure and functionality of one molecule when compared with another (see Chapter 14):** 49, 50, 51, 52, 53, 54, 55, 56, 57, 58.

Homework

16.27 Draw the structure for each of the following:
- (a) 1-cyclopropyl-1-pentanone
- (b) 3-ethenylcyclopentanone
- (c) 8-chloro-5,5-dimethyloct-3Z-enal
- (d) 4,5-dicyclopentylnon-6-yn-2-one
- (e) 3,4,5-trimethylcyclohexanecarboxyaldehyde
- (f) 6-(3,3-dimethylbutyl)octadecane-3-one
- (g) 2-cyclohexylcyclohexanone
- (h) 6-phenylhex-5-ynal

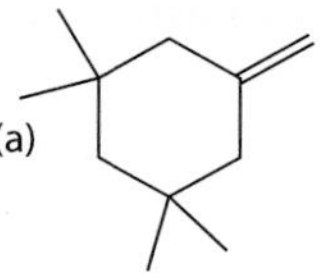

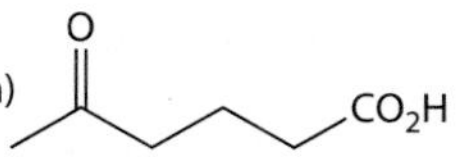

16.28 In an acyl addition reaction with a nucleophile such as sodium cyanide, which is expected to be more reactive: butanal or 2-butanone? Explain.

16.29 Give the IUPAC names for each of the following:

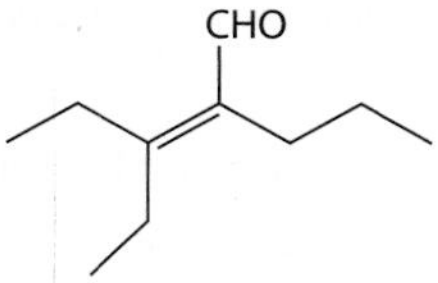

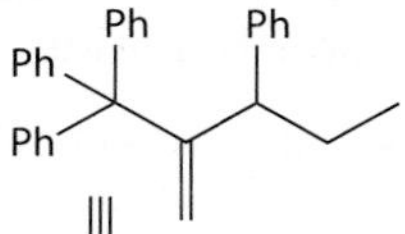

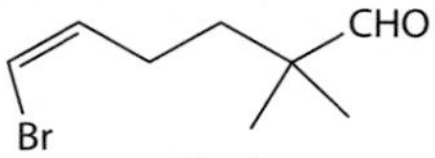

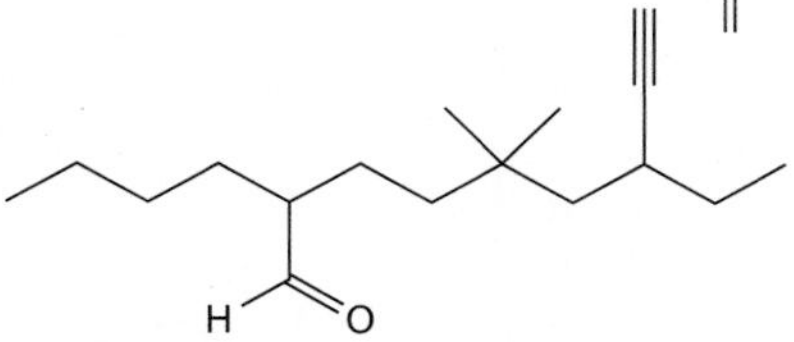

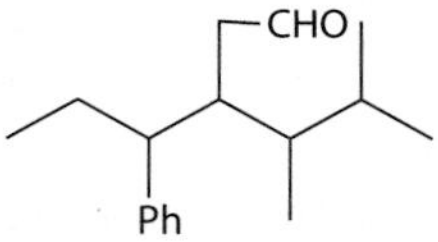

16.30 Explain why the name of the following molecule is **not** 2-cyclopropylpentanal. What is the correct name for this molecule?

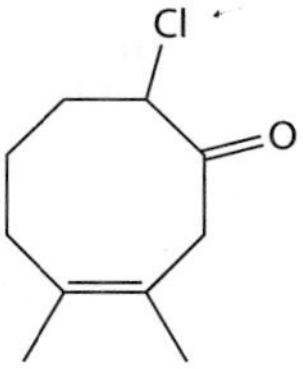

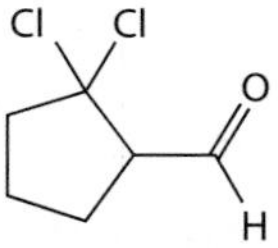

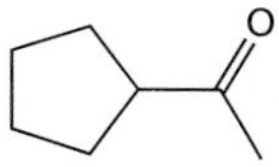

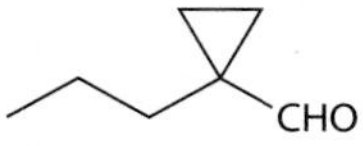

16.31 Based on the concept of higher and lower priority for functional groups, indicate the suffix for each of the following:

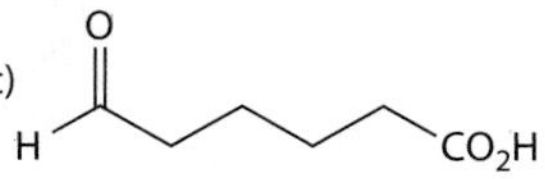

16.32 Common names creep into modern organic chemistry. It is common to name ketones by naming both alkyl groups flanking the carbonyl, followed by the word ketone. Give the correct IUPAC name and draw each structure for the following:
- (a) dibutyl ketone
- (b) methyl ethyl ketone

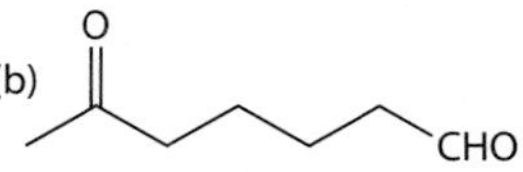

(c) ethyl vinyl ketone
(d) dibenzyl ketone
(e) phenyl propyl ketone

16.33 Suggest a reason why 2-hydroxybutanoic acid may have a lower pK_a when compared to 4-hydroxybutanoic acid.

16.34 Thiocarbonyl compounds are characterized by a C=S unit rather than a C=O unit. Draw the structure of 2-pentanone and then draw the corresponding thioketone by replacing O with S. If this thioketone reacted with a nucleophile such as $CH_3C\equiv C:^-$, what is the expected product?

16.35 Suggest a reason why 3,3,5,5-tetraethyl-4-heptanone is less reactive in acyl addition reactions than 4-heptanone.

16.36 Draw the tetrahedral intermediate expected when ethoxide reacts with N, N-dimethylbutanamide and then draw the tetrahedral intermediate formed when dimethylamide (Me_2N^-) reacts with ethyl butanoate. Compare these structures and suggest which reaction is likely to proceed to amide to ester, or ester to amide.

16.37 Give the correct IUPAC name for each of the following:

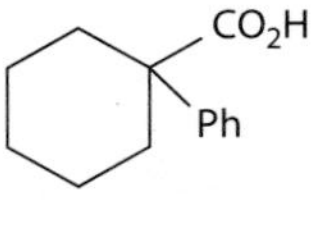

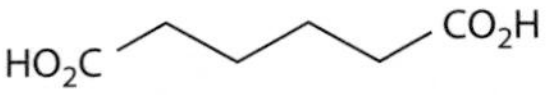

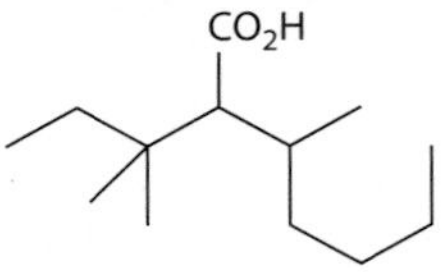

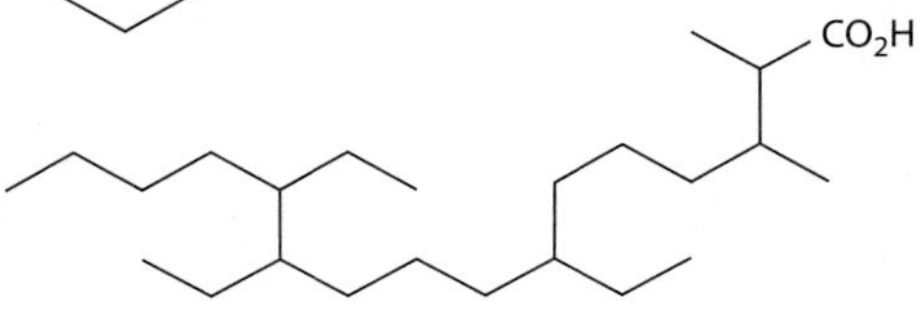

16.38 Give the correct IUPAC name for each of the following:

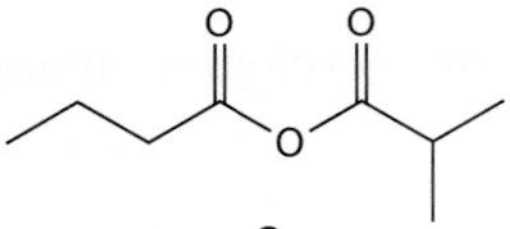

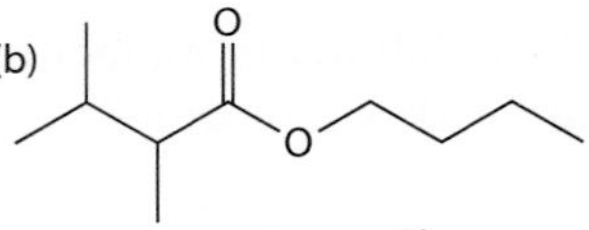

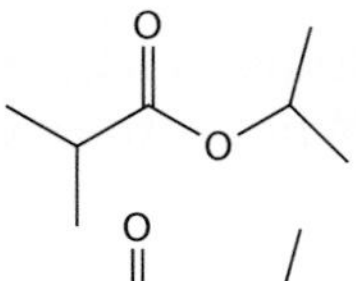

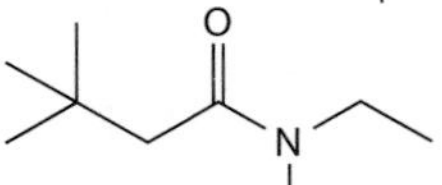

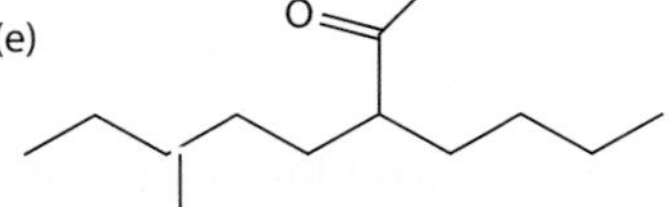

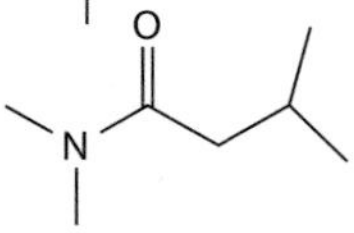

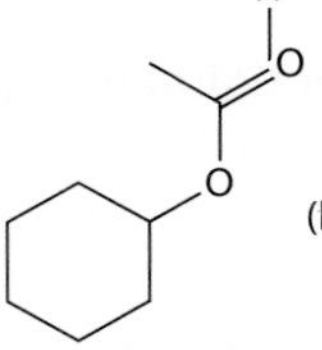

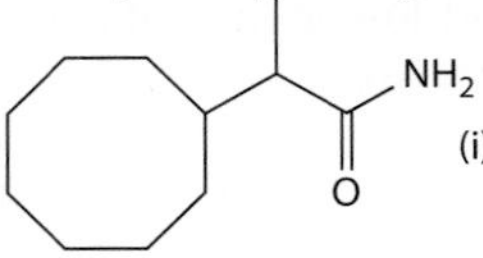

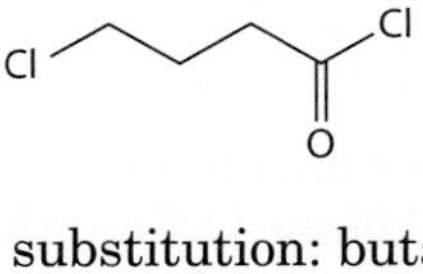

16.39 Which would you anticipate is the more reactive in acyl substitution: butanoyl chloride or butanoyl iodide? Justify your answer.

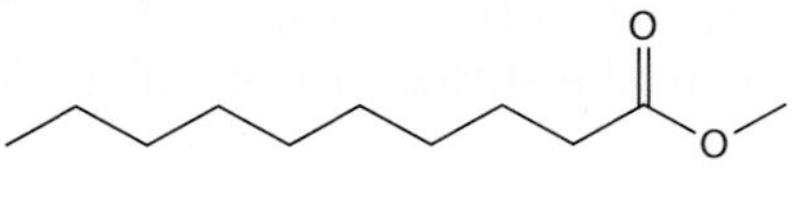

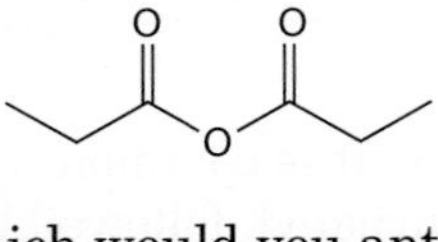

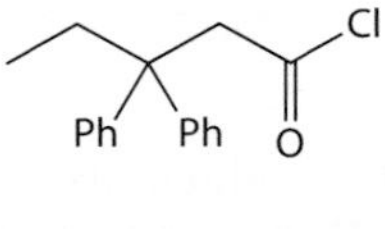

16.40 Which is the better nucleophile in a reaction with 2-butanone, iodide ion, or $CH_3C\equiv C:^-$? Justify your answer.

16.41 Draw the structure for each of the following:
   (a) *N,N*-diphenylhexanamide
   (b) cyclobutyl 3,3-dimethylhexanoate
   (c) dipentanoic anhydride
   (d) hexadec-5*Z*-enoyl chloride
   (e) ethyl oct-4-ynoate
   (f) *N*-chlorobutanamide
   (g) butyl butanoate
   (h) *N*-cyclopropyl-4,4-diphenyldodecanamide
   (i) 4-phenyl-3-cyclohexenyl pentanoate

16.42 Suggest reasonable products for the following reactions:

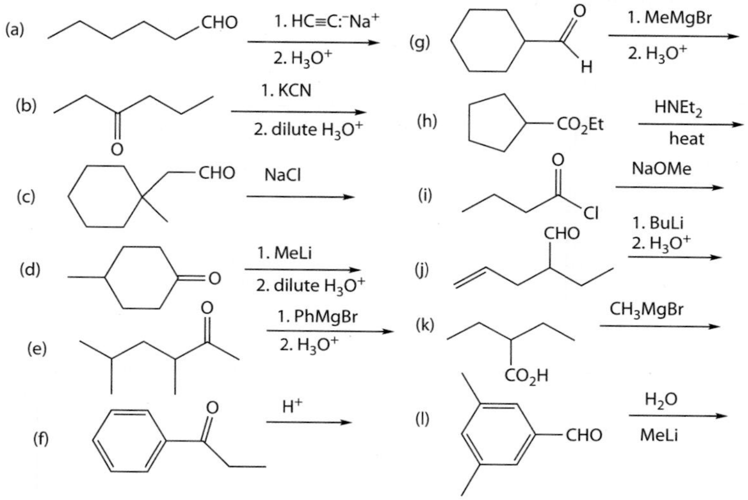

16.43 If *n*-butyllithium were to react with methyl butanoate, what do you think is the most likely reaction? Draw the product and explain.

16.44 A common fat is the octadecanoic acid triester of glycerol (1,2,3-propanetriol). Draw this triester.

16.45 The first pK_a for dicarboxylic acid **A** is lower than that for **B**. Explain.

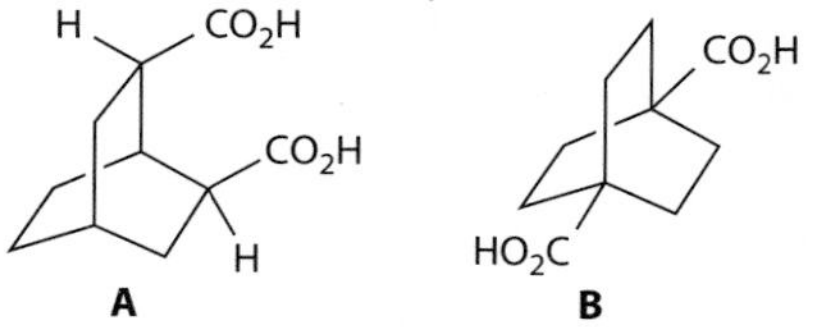

16.46 Based on relative stability of the oxocarbenium ion that is formed, which is more basic: 3-pentanone or methyl pentanoate?

16.47 Draw the acid–base reaction between cyclopentanone and 4-toluenesulfonic acid. Is this a reversible process? Is the K_a for this reaction expected to be much large than 1 or much smaller than 1? Justify your answer.

16.48 Draw all resonance contributors that are possible for the oxonium formed when benzaldehyde reacts with an acid catalyst.

Synthesis Problems

There are no synthesis problems associated with this chapter.

Spectroscopy Problems

You should have read and understood Chapter 14 before attempting these problems.

16.49 A compound has the formula $C_{11}H_{14}O$. Given the following theoretical spectral data, provide a structure:

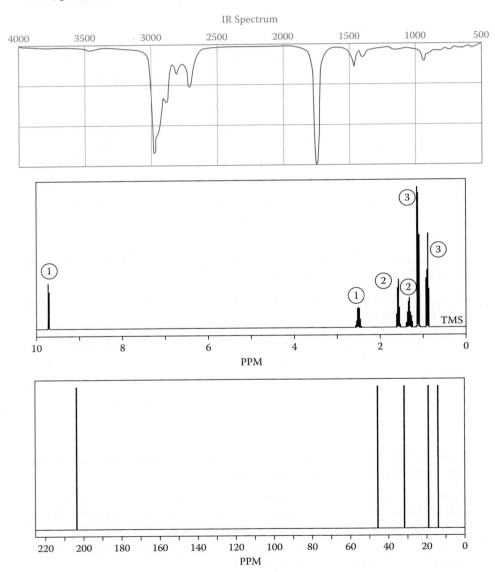

16.50 Given the following theoretical spectral data, give the structure: M = 162,
100%; M+1 = 163, 11.88%; M+2 = 164, 0.91%.

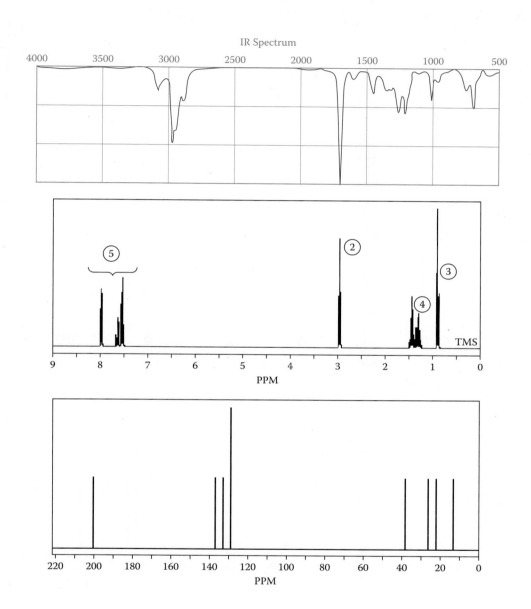

16.51 Given the following theoretical spectral data, give the structure: M = 115,
       100%; M+1 = 116, 7.04%; M+2 = 117, 0.42%.

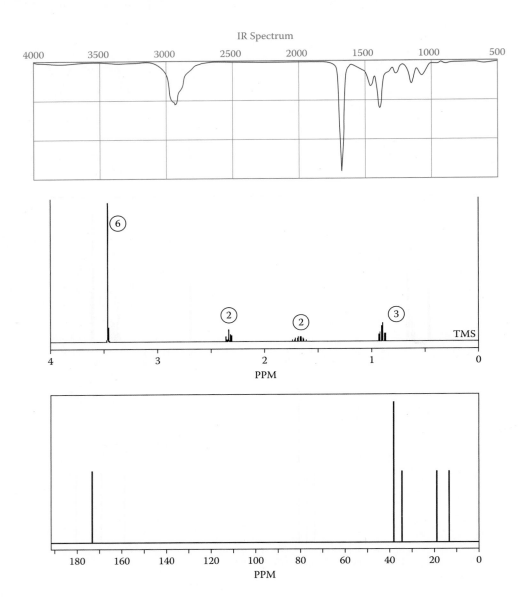

16.52 Given the following spectral data, provide a structure: M = 86, 100%; M+1 = 87, 5.60%; M+2 = 88, 0.33%.

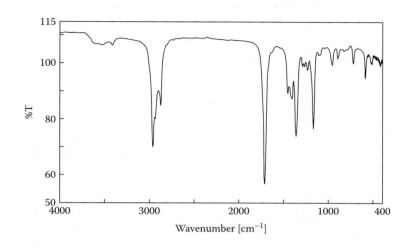

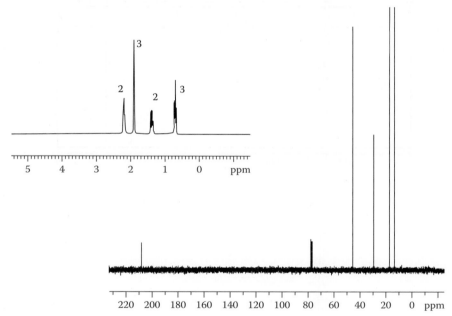

16.53 Given the following spectral data, give a structure:

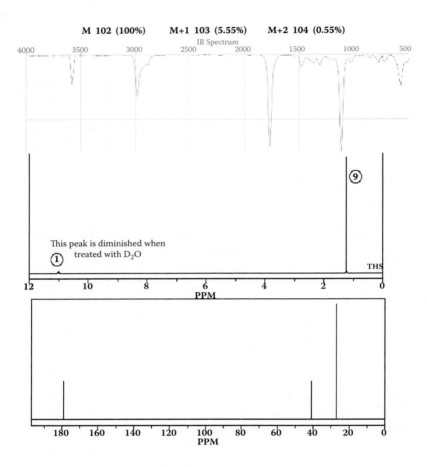

M 102 (100%)   **M+1 103 (5.55%)**   **M+2 104 (0.55%)**

16.54 Given a formula of $C_5H_8O_2$ and the following spectral data, give the structure:

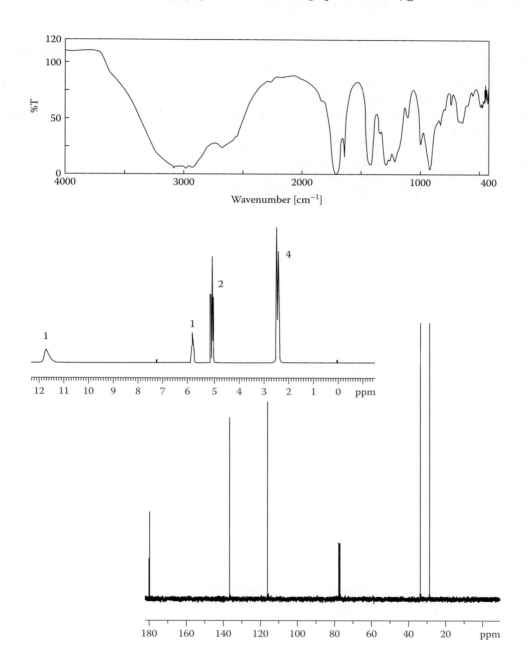

16.55 For a formula of $C_6H_{10}O$, given the following spectral data, provide a structure:
IR: 3082, 3002–2921, 1718, 1642, 1431–1415, 1363 cm^{-1}
1H NMR: 5.79 (m, 1H), 5.02 (m, 1H), 4.97 (m, 1H), 2.52 (t, 2H), 2.34 (m, 2H), 2.15 (s, 3H) ppm

16.56 For a formula of $C_6H_{10}O_2$, given the following spectral data, provide a structure:
IR: 2967–2670, 1712, 1657, 1463, 1418 cm^{-1}
1H NMR: 10.4 (bd s, 1H; disappears when treated with D$_2$O), 5.63 (m, 1H), 5.51 (m, 1H), 3.07 (d, 2H), 2.06 (m, 2H), 0.99 (t, 3H) ppm

16.57 For a formula of $C_6H_{11}ClO$, given the following spectral data, provide a structure:
IR: 2861, 2935, 1800, 1468, 1381 cm^{-1}
1H NMR: 2.87 (t, 2H), 1.72 (m, 2H), 1.59–1.06 (m, 4H), 0.92 (broad t, 3H) ppm

16.58 For a formula of $C_8H_{14}O_3$, given the following spectral data, provide a structure:
IR: 2980–2861, 1813, 1747, 1471, 1389, 1169–965 cm^{-1}
1H NMR: 2.66 (m, 2H), 1.24 (d, 12H) ppm

Oxidation

17

Chapter 16 introduces several carbonyl compounds, including aldehydes and ketones. So far, there has been virtually no mention of how such compounds are prepared. A general class of reactions involves the gain or loss of two electrons, but the structural changes in the product are often measured by whether hydrogen or oxygen is gained or lost. Such reactions are known as oxidation and reduction reactions. Several functional group exchange reactions are classified as oxidations, including the conversion of alcohols to ketones or aldehydes (an oxidation). The conversion of aldehydes or ketones back to alcohols is a reduction. This chapter will introduce a few important oxidation reactions and typical transformations associated with them. (For the oxidation reactions of alkenes, see Chapter 10, Sections 10.5 and 10.7.) Discussion of the reactions that constitute an oxidation of alkenes is essentially a review. A discussion of reduction reactions is presented in Chapter 18.

To begin this chapter, you should know the following:

- **the structure and basic rules of nomenclature for alcohols, aldehydes, ketones, diols, ethers, and carboxylic acids (Chapter 5, Sections 5.6 and 5.9; Chapter 16, Sections 16.2 and 16.5)**
- **the CIP rules for prioritizing substituents, groups, and atoms (Chapter 9, Section 9.3)**
- **polarized bonds (Chapter 3, Section 3.7)**
- **σ-covalent bonds (Chapter 3, Section 3.3)**
- **π-bonds (Chapter 5, Section 5.1 and 5.2)**

- **Brønsted–Lowry acids and bases (Chapter 6, Sections 6.2–6.4)**
- **Lewis bases and Lewis acids (Chapter 6, Section 6.5)**
- **the acid–base properties of alcohols, alkenes, aldehydes, and ketones (Chapter 6, Sections 6.3–6.5)**
- **the fundamental reactions known for alkenes (Chapter 10, Sections 10.2–10.4)**
- **that alkenes are converted to epoxides and to diols (Chapter 10, Sections 10.5 and 10.7.1)**
- **that alkenes undergo oxidative cleavage with ozone (Chapter 10, Section 10.7.2)**
- **the general reactions of carbonyl compounds (Chapter 16, Sections 16.3 and 16.8)**
- **E2 type reactions (Chapter 12, Sections 12.3 and 12.7)**

This chapter will define an oxidation in general terms. Various reagents will be introduced that oxidize an alcohol to an aldehyde or to a ketone. In some cases, alcohols may be oxidized to a carboxylic acid. Oxidation of alkenes will be revisited—specifically, for the oxidation of an alkene to an oxirane or to a 1,2-diol. Oxidative cleavage with ozone will be revisited, and new methods for oxidative cleavage will be introduced.

When you have completed this chapter, you should understand the following points:

- **Oxidation is defined as loss of electrons or gain of a heteroatom such as oxygen or loss of hydrogen atoms. Oxidation number is a convenient method to track the gain or loss of electrons in a reaction.**
- **Chromium(VI) reagents are powerful oxidants. The reaction of a secondary alcohol with chromium trioxide and acid in aqueous acetone is called Jones's oxidation, and the product is a ketone. Chromium oxidation of an alcohol proceeds by formation of a chromate ester, followed by loss of the α-hydrogen to form the C=O unit.**
- **Jones's oxidation of a primary alcohol leads to a carboxylic acid in most cases. A mixture of chromium trioxide and pyridine gives a reagent that can oxidize a primary alcohol to an aldehyde. This is called Collins's oxidation. The reaction of chromium trioxide and pyridine in aqueous HCl leads to pyridinium chlorochromate, called PCC. The reaction of chromium trioxide and pyridine in water leads to pyridinium dichromate (PDC). Both PCC and PDC can oxidize a secondary alcohol to a ketone or a primary alcohol to an aldehyde.**

- **The oxidation of a secondary alcohol to a ketone or a primary alcohol to an aldehyde using dimethyl sulfoxide (DMSO) with oxalyl chloride at low temperature is called the Swern oxidation.**
- **Oxidation of an alkene with osmium tetroxide or potassium permanganate gives a *cis*-1,2-diol.**
- **Oxidation of an alkene with a peroxyacid leads to an epoxide, with a carboxylic acid as the by-product.**
- **Oxidative cleavage of an alkene with ozone leads to an ozonide. Reductive workup with dimethyl sulfoxide or zinc and acetic acid gives ketones and/or aldehydes. Oxidative workup with hydrogen peroxide gives ketones and/or carboxylic acids.**
- **Oxidative cleavage of 1,2-diols with periodic acid or with lead tetraacetate gives aldehydes or ketones.**
- **A molecule with a particular functional group can be prepared from molecules containing different functional groups by a series of chemical steps (reactions). This process is called synthesis: The new molecule is synthesized from the old one (see Chapter 25).**
- **Spectroscopy can be used to determine the structure of a particular molecule and can distinguish the structure and functionality of one molecule when compared with another (see Chapter 14).**
- **Oxidation is important in biological systems.**

17.1 Defining an Oxidation

An oxidation is formally defined as the loss of one or more electrons from an atom or a group. How is it possible to recognize loss of electrons in the course of a reaction? If there are structural changes in the product relative to the starting material, it is possible to associate electron transfer with bond-making and bond-breaking reactions. The structural changes usually include either loss of hydrogen atoms or the replacement of a hydrogen atom bonded to carbon with a more electronegative atom, usually a heteroatom. Such heteroatoms include oxygen, halogen, nitrogen, sulfur, etc.; however, the most common is oxygen.

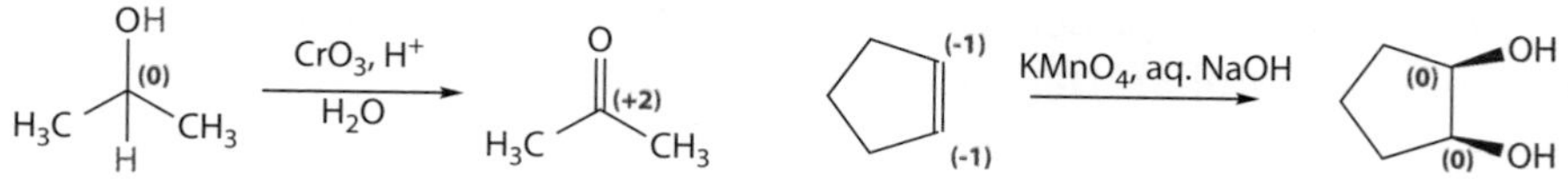

A common oxidation is the conversion of an alcohol such as 2-propanol (**1**) to a ketone (acetone, **2**), and it involves loss of hydrogen atoms from oxygen and from carbon. Similarly, the conversion of an alkene such as cyclopentene (**3**) to a vicinal diol (**4**) (see dihydroxylation in Chapter 10, Section 10.7.1) involves the gain of two oxygen atoms and is also an oxidation. A method that identifies the so-called oxidation state of atoms is useful for identifying whether electrons are gained or lost during the transformation. Oxidation state is a number assigned to the carbon atoms involved in the transformation, and the formal rules for determining oxidation state are the following:

The oxidation state of a carbon is taken to be zero.
Every hydrogen atom attached to a carbon is given a value of –1.
Every heteroatom attached to a carbon is assigned a value of +1.

In alcohol **1**, C2 is attached to two carbon atoms, one hydrogen atom, and one oxygen. Therefore, the oxidation state of C2 = 0 + 0 – 1 + 1 = 0 (0 for each carbon, –1 for the hydrogen, and +1 for the oxygen). In the final product, **2**, C2 is bonded to two carbon atoms, and it has two bonds to oxygen. The oxidation state of C2 in **2** is calculated by 0 + 0 + 1 + 1 = +2. A comparison of the oxidation state of the pertinent carbon atoms in **1** and in **2** reveals that conversion of **1** to **2** involves a transfer of two electrons. Is this a gain or a loss? To go from 0 to +2 requires that two electrons be lost because electrons are negatively charged particles. The conversion of **1** to **2** involves the loss of two electrons, and it is an oxidation. Loss of electrons is an oxidation and a gain of electrons is a reduction (which will be discussed in Chapter 18).

17.1 **Calculate the oxidation numbers given in 3 and 4 and verify that this transformation is an oxidation.**

17.2 **Calculate the oxidation number for the conversion of 1-butene to 2-bromobutane and categorize this transformation as an oxidation or a reduction.**

17.2 Oxidation of Alcohols with Chromium(VI)

The rules of nomenclature for aldehydes and ketones were described in Chapter 16 (Section 16.2). How are aldehydes and ketones prepared? In many cases, the answer is via oxidation of an alcohol. Indeed, one of the most common and most important functional group transformations converts an alcohol to a ketone, as in the oxidation of 2-propanol (**1**) to acetone (**2**). This section will discuss several reagents that accomplish this transformation.

17.2.1 Chromium Trioxide and Related Compounds

Chromium(VI) is a powerful oxidizing agent and several inorganic reagents are characterized by the presence of chromium(VI). The most common is chromium trioxide (CrO_3), which probably exists in a polymeric form $[(CrO_3)_n]$, where the "n" is an integer signifying the number of repeating CrO_3 units.

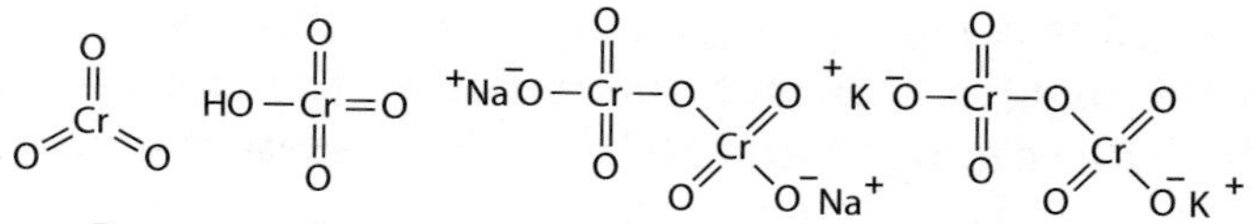

5   **6**   **7**   **8**

Despite this structural feature, chromium trioxide is usually written as the monomer (the single unit) CrO_3, with the structure shown for **5**. Several other reagents also involve chromium(VI), including chromic acid ($HCrO_4$, **6**), sodium dichromate ($Na_2Cr_2O_7$, **7**), and potassium dichromate ($K_2Cr_2O_7$, **8**). When chromium trioxide is dissolved in water, a complex equilibrium is established that includes not only **5**, but also chromic acid (**6**) and the protonated form of the dichromate ion ($Cr_2O_7^{-2}$). When the solution is highly concentrated (only a small amount of water), the equilibrium favors **5**. When the solution is dilute (a large amount of water), the equilibrium favors dichromate. From a practical point of view, an aqueous solution of sodium dichromate will do the same chemistry as a solution of chromium trioxide because a solution of either will contain the same oxidizing agents. From the standpoint of the chemical reactions presented in this book, these reagents will be used interchangeably.

17.2.2 Oxidation of Alcohols

Chromium(VI) reagents are commonly used to oxidize alcohols. Most of these oxidizing agents are soluble in aqueous sulfuric acid, but this poses a problem. Many alcohols are not very soluble in aqueous media, so an organic cosolvent is usually required. A convenient recipe for the oxidation uses a solution of chromium trioxide in aqueous acetone (**2**) in the presence of an acid such as H_2SO_4. This is called the **Jones reagent** and the reaction of this mixture with an alcohol is called **Jones's oxidation**. It is named after Sir Ewart Ray Herbert Jones (E. R. H Jones, England; 1911–2002). The acetone moderates the reaction and helps to solubilize the various reactions found in this oxidation. In a typical experiment using $Na_2Cr_2O_7$ in sulfuric acid, 3-pentanol (**9**) is converted to 3-pentanone (**12**) in 57% yield.[1] *Recognition that in aqueous solution CrO_3 is in equilibrium with $Cr_3O_7^{-2}$ allows the use of CrO_3 as the active oxidizing agent to look at a simplified mechanism.*

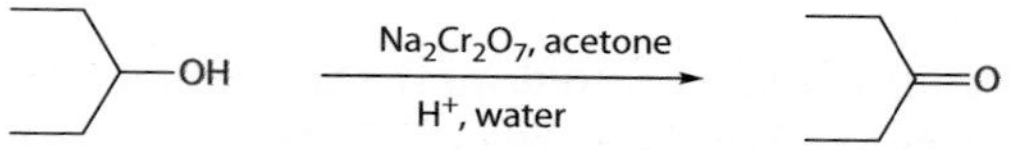

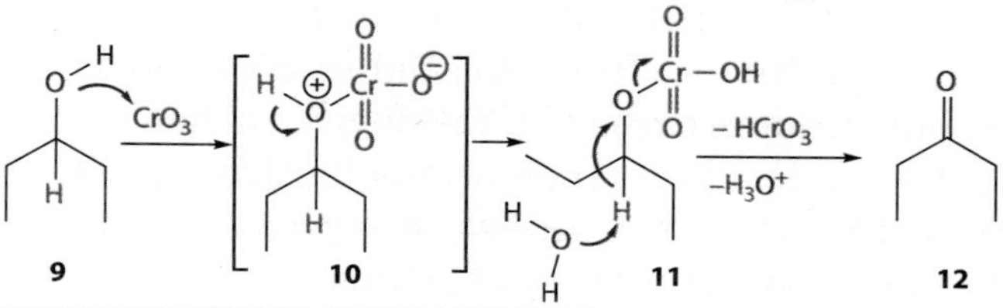

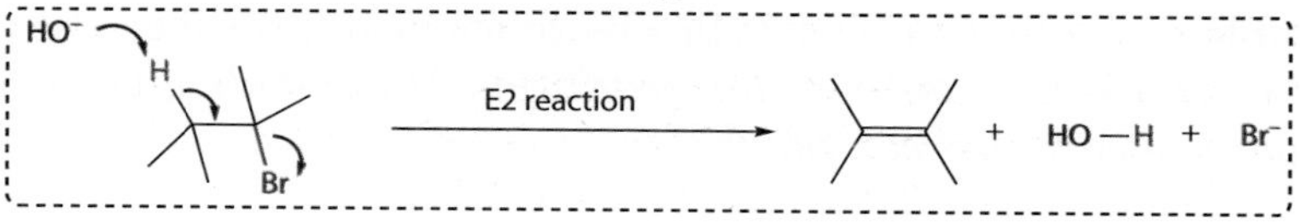

If the product **12** is compared with **9**, it is clear that one hydrogen atom has been lost from the oxygen of **9** and another from the carbon atom. The other products are the hydronium ion and $HCrO_3$, which is unstable and decomposes. To form the hydronium ion, water reacts as a base to remove a hydrogen atom. From Chapter 6 (Section 6.2), it is clear that an alcohol is not a strong enough acid to react with water, so some reaction must occur to convert the alcohol into a compound that has an acidic hydrogen. The key is to recognize that the chromium of CrO_3 is a Lewis acid and that a hydrogen atom is transferred to one oxygen of the chromium by this reaction. If CrO_3 is a Lewis acid, the oxygen of the alcohol must be the Lewis base. The reaction of the alcohol and CrO_3 leads to formation of an ate complex, and a proton transfer to an oxygen of the CrO_3 unit gives an intermediate that is acidic enough to react with water. Indeed, this mechanism is used to explain the oxidation.

The oxygen of the alcohol (a Lewis base) donates two electrons to chromium (Cr is a Lewis acid) to form oxonium salt **10**. Transfer of the acidic proton of the oxonium salt in **10** to the chromate oxygen (the base) leads to a so-called *chromate ester,* **11**. It may not be obvious, but formation of the chromate ester makes the hydrogen atom on the α-carbon acidic (marked in **red** in **11**). Removal of that hydrogen by water is an acid–base reaction that leads to loss of the chromium(III) leaving group, which is an elimination reaction that forms a new π-bond: a carbonyl (C=O). *If the chromium unit is viewed as a leaving group, then the hydrogen α to the oxygen is lost to water and the leaving group is CrO₃H.* This mechanism predicts that the secondary alcohol (**9**) will be converted to 3-pentanone (**12**), which is an oxidation.

Note the similarity of the oxidation of an alcohol to the elimination reaction of **11** via an E2 reaction (shown in the box) for conversion of an alkyl halide to an alkene (discussed in Chapter 12, Section 12.1). Note that the hydrogen atom marked in **red** in **11** is β to the Cr leaving group. This comparison is a useful point for understanding the oxidation of an alcohol to a carbonyl once the chromate ester has been formed. In the case of an E2 reaction, the acidic hydrogen is two atoms removed from the leaving group. In **11**, the hydrogen atom is also two atoms removed from the leaving group, which is the chromium species.

This oxidation can be applied to most primary or secondary alcohols, and any of the chromium(VI) reagents mentioned in the preceding section may be used. When cyclohexanol (**13**) is treated with sodium dichromate (**8**) in aqueous sulfuric acid, cyclohexanone (**9**) is isolated in 85% yield.[1] Yields are often greater than 90%, but in some cases the yields are rather poor when there are groups that sterically hinder the carbon atoms flanking the C–OH unit in the alcohol. Such steric hindrance may interfere with formation of the chromate ester, or removal of the hydrogen atom attached to the C–O unit may be difficult. A simple case of this effect is briefly discussed later (see the oxidation of **24**), but for convenience, *assume that oxidation of alcohols gives good yields of the corresponding carbonyl compound.*

Jones's oxidation is a powerful oxidizing medium for the conversion of alcohols to ketones. Unfortunately, this is such a powerful oxidizing medium that unwanted products are possible due to overoxidation. When a primary alcohol such as 1-pentanol (**15**) reacts with chromium trioxide and aqueous sulfuric acid, it follows the same mechanistic pathway as **9**, with formation of chromate ester **16**. However, experiments show that the yields of aldehyde from primary alcohols can be very low. 1-Propanol is oxidized to propanal, for example, in only 49% yield, and to obtain the product requires a short reaction time.[2] Very often, a carboxylic acid is formed as a second product or even the major oxidation product rather than the aldehyde. It is known that aldehydes are easily oxidized to carboxylic acids, even by oxygen in the air. If a sample of butanal were spilled, for example, it is rapidly oxidized to butanoic acid by air. This oxidation is easily detected as the sharp butanal smell is replaced by the pungent butanoic acid smell. Butanoic acid is found in rancid butter and in "dirty feet," for example.

17.3 Draw the chromate ester and final product of a reaction between 4,4-diethyl-2-hexanol and $K_2Cr_2O_7$ in aqueous acetone.

Formation of an aldehyde such as **17** in the presence of a powerful oxidizing agent such as chromium(VI) is usually followed by rapid oxidation of **17** to the corresponding carboxylic acid, pentanoic acid (**18**). *In other words, Jones's oxidation of simple aldehydes usually gives the carboxylic acid as the major product.* If the reaction mixture is heated, overoxidation to the carboxylic acid is even more rapid. The fact that the Jones reagent rate of oxidation of alcohol to aldehyde is fast can be used to an advantage, and *when acetone is used as a solvent, the rate of oxidation of aldehyde to acid is relatively slow.* Acetic acid (ethanoic acid) serves a similar role in many oxidations.

Therefore, *cold temperatures and short reaction times favor the aldehyde, but long reaction times and heat favor formation of the acid.* This fact is apparent in the reaction of **19** with CrO_3 in sulfuric acid and aqueous acetic acid. When allowed to react several hours and then heated to 100°C, carboxylic acid **20** was isolated in 82% yield.[3] If the number of molar equivalents of oxidizing agent is diminished and the temperature is kept low with a short reaction time, aldehyde **21** is isolated in 59% yield.[3] Therefore, *assume that oxidation* *of a secondary alcohol with chromium(VI) leads to a ketone and oxidation of a primary alcohol leads to an aldehyde, if temperature and time are controlled.*

17.4 Write out the product of a reaction between 3-cyclopentyl-1-propanol and Jones's reagent if the reaction is done at 80°C and a reaction time of 24 hours.

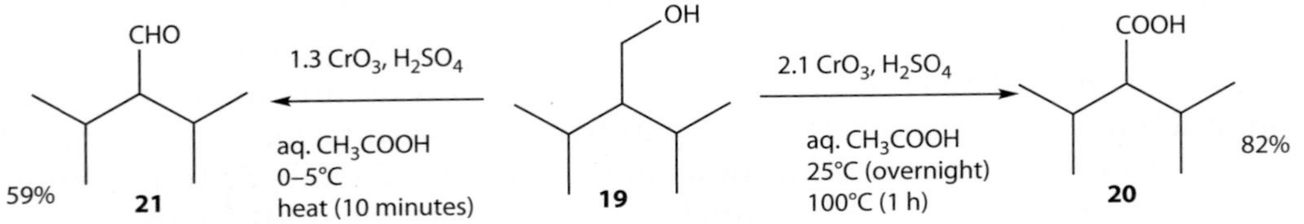

A more subtle aspect of this oxidation involves steric hindrance in the chromate ester, which was mentioned earlier in this section. There is an experimental observation that alcohol **22** (2-methyl-3-pentanol) is oxidized faster than alcohol **24** (2,2,4,4-tetramethyl-3-pentanol). If both alcohols are converted to the corresponding chromate ester (**23** and **25**, respectively), the α-hydrogen (marked in **red**) must be removed in each case to give the ketone. Assume that formation of both chromate esters occurs without difficulty because it is the oxygen of the alcohol that attacks chromium. The surrounding methyl groups in **25** create significant steric hindrance around the α-hydrogen, so it is more difficult for the base (water) to approach that hydrogen.

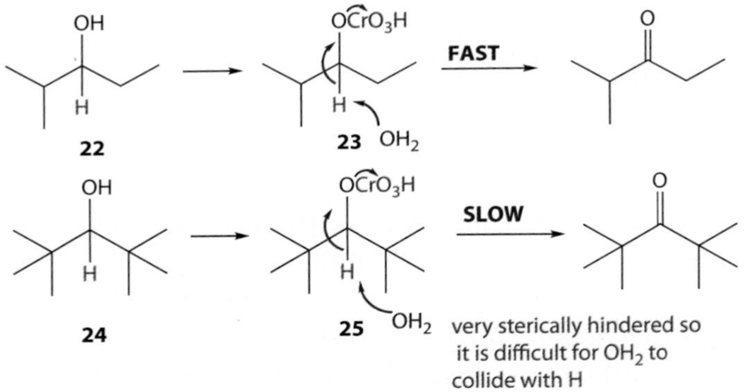

By contrast, the α-hydrogen in **23** is relatively unhindered and is easily removed by the base. This means that the rate of oxidation (see reaction rate in Chapter 7, Section 7.11) is relatively fast for **23** but slower for **25** due to steric hindrance in the chromate ester. In other words, *the steric hindrance in the chromate ester makes it more difficult for the water to react with the α-proton.* Note that the chromate ester is formed in both cases and *the steric hindrance to oxidation occurs in the chromate ester—not in the alcohol.* Alcohols are occasionally found where the steric hindrance is so severe that no oxidation occurs at all. Such cases are unusual, however. It is important to note that tertiary alcohols cannot be oxidized.

17.5 Give the IUPAC name for each ketone product from 22 and from 24.

17.2.3 PCC and PDC

Although primary alcohols are often oxidized to carboxylic acids with Jones' reagent, the aldehyde formed by the initial oxidation may be isolated in many cases, but not all. Isolation of the aldehyde formed by oxidation of a primary alcohol without overoxidation to the carboxylic acid is a desirable goal. One method for isolating the aldehyde is to make the chromium(VI) oxidizing agent less reactive, by reacting chromium trioxide (**5**) with pyridine (**26**) before adding the reagent to the alcohol.

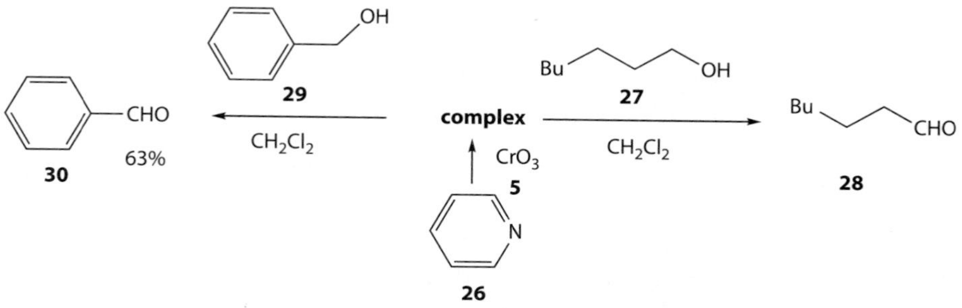

If chromium trioxide is added to pyridine (**26**; see Chapter 26, Section 26.1.2) in dichloromethane (CH_2Cl_2), a dark, viscous complex is formed. Oxidation of unconjugated alcohols with this reagent is variable. A conjugated alcohol (known as an allylic alcohol) has a C=C–C–OH unit; (see Chapter 22). When 1-heptanol (**27**) is added to this CrO_3•pyridine mixture, heptanal (**28**) is isolated, but in only 10% yield.[4] Clearly, this is not a good result. Oxidation of cyclohexanol produces cyclohexanone in 45% yield,[4] however, and 2-octanol is oxidized to 2-octanone in 97% yield.[5]

Analysis of these results led to the conclusion that yields are variable, in part, because of difficulty in isolating the product from the pyridine-chromium complex. This reagent is usually effective for oxidizing allylic alcohols ($RCH=CHCH_2OH$) and benzylic alcohols ($ArCH_2OH$, where Ar is an aromatic ring; see Chapter 20) to aldehydes. An example is the conversion of benzyl alcohol (**29**) to benzaldehyde, **30**, in 63% isolated yield.[4] Oxidation with this CrO_3•pyridine complex in dichloromethane is called ***Collins's oxidation*** (after J. C. Collins, United States). The most striking feature of the Collins oxidation reagent is that it is *less reactive* than the Jones reagent discussed in Section 17.2.2. As mentioned, the yields are sometimes low because it can be difficult to remove the aldehyde product from the viscous CrO_3•pyridine complex. Because of this problem, other chromium-based oxidizing agents have been developed.

17.6 Draw the products expected when 4,5-diphenyl-8-oxo-nonanol reacts with chromium trioxide and pyridine.

17.7 Draw the reactants and products for the oxidation of cyclohexanol and of 2-octanol.

17.8 Write out the product formed when allyl alcohol is oxidized with Collins's reagent and when benzyl alcohol is oxidized with Collins's reagent.

17.9 Write out and name the product of Collins's oxidation of 4-methyl-3-phenylpentanal.

17.10 Write out and name the product of Collins's oxidation of 2-methylcyclopentanone.

Two modified chromium(VI) reagents are widely used. Note that the Collins oxidation reagent mixes chromium trioxide and pyridine, but there is no effort to isolate a specific reagent. The mixture is used for the oxidation. The first of the new reagents is formed by the reaction of chromium trioxide (**5**) with pyridine in aqueous HCl. This reaction generates a specific compound known as pyridinium chlorochromate (PCC, **31**) that is isolated and purified. In this solution, CrO_3 forms $HCrO_4$ (**6**) in dilute aqueous acid, which reacts with HCl to form $HCrClO_3$. Pyridine then reacts as a base with this acidic proton to form PCC.

If the reaction conditions are modified to increase the amount of pyridine in the water solution and the HCl is omitted, the reaction generates pyridinium dichromate (PDC, **32**), presumably by reaction of an excess of pyridine with $H_2Cr_2O_7$. In dilute solution, CrO_3 is in equilibrium with $H_2Cr_2O_7$, and pyridine reacts with both acidic hydrogen atoms to produce PDC. Once again, PDC is a specific compound that is isolated and purified.

Once purified, PCC is added to an alcohol in dichloromethane solvent to do the oxidation. Similarly, purified PDC is added to an alcohol in dichloromethane. Both PCC and PDC are less reactive than chromium trioxide in the Jones reagent, but they are very effective for converting primary alcohols to the aldehyde, in good yield and under mild conditions. Secondary alcohols are readily converted to ketones by both reagents. Dichloromethane is typically used as the solvent with both reagents. An example is the reaction of 2-cyclopentylethanol (**33**) and PCC in dichloromethane, which gave 1-cyclopentylethanal, **34**, in 72% yield.[5]

In a different experiment, the reaction of 4-propylcyclohexanol (**35**) and PDC gave 4-propylcyclohexanone (**36**) in 97% isolated yield.[5] The PCC reagent is significantly more acidic than PDC and this sometimes causes deleterious side reactions. Indeed, PDC was developed, in part, because of the acidity of PCC. This will not be an issue in the examples used in this book. In other words, assume that both reagents are used without a problem in dichloromethane as a solvent.

Note that chromium(VI) is a ***cancer suspect agent***. The handling, use in chemical reactions, and chemical disposal must be carefully monitored any time one of the preceding Cr(VI) reagents is employed. In other words, any time Cr(VI) is around, be very careful. In part, this is a reason for developing alternative oxidizing agents such as the one discussed in the next section.

17.11 Write out the acid–base reaction of pyridine with $HCrClO_3$.

17.12 Write out the product of the reaction of 1-decanol with PDC.

17.13 Write out the product when cyclooctanol reacts with PCC.

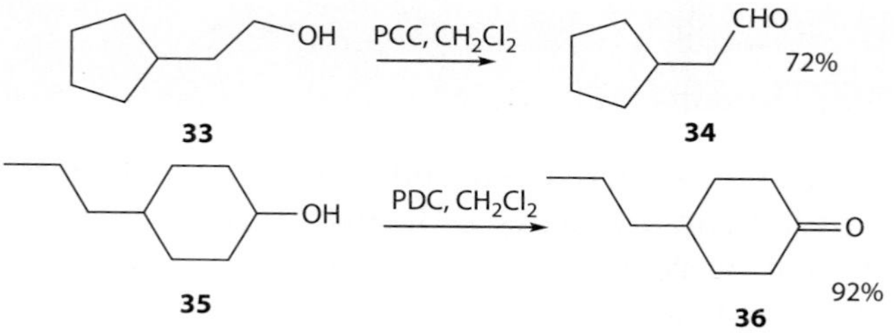

17.2.4 Swern Oxidation

Several "neutral" oxidation reactions do not involve chromium(VI), but readily convert an alcohol to a ketone or an aldehyde. One such reaction uses a reagent derived from the reaction of dimethyl sulfoxide (DMSO, **37**) and oxalyl chloride (**38**; see Chapter 16, Section 16.7). When 2-butanol is mixed with these reagents at −60°C, dimethyl sulfide (MeSMe) is observed as a product and 2-butanone (**43**) is isolated in 78% yield.[6]

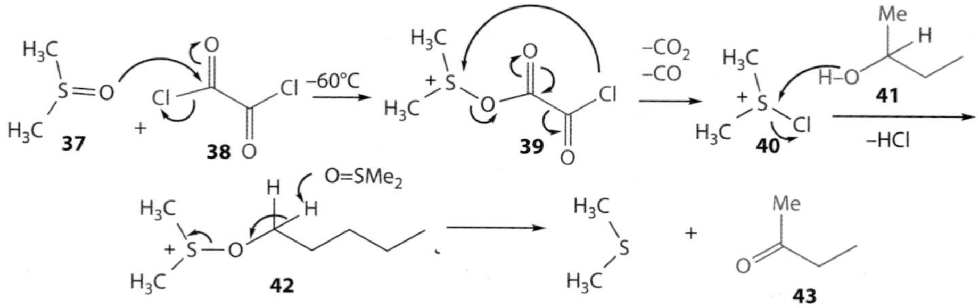

The mechanism proposed to explain these observations shows that the oxygen of DMSO must attack the acyl carbon of oxalyl chloride in an acyl substitution reaction (Chapter 19, Section 19.2) to form an acyl-sulfonium derivative, **39**. This intermediate decomposes with loss of carbon dioxide (CO_2) and carbon monoxide (CO) to give a chlorosulfonium salt, **40**. Sulfonium salt **40** has an electrophilic sulfur that is attacked by the nucleophilic oxygen of an alcohol such as

2-butanol (**41**), at low temperatures (less than −60°C), to give **42**. Intermediate **42** is related to intermediate **16** (the chromate ester of **19** discussed in Section 17.2.2) in that a leaving group (Me$_2$S) is lost and a hydrogen atom (in **red** in **42**) is removed. This particular oxidation of an alcohol to a ketone or aldehyde is called the ***Swern oxidation***, after Daniel Swern (United States; 1916–1982).

17.14 Write out all the products formed by Swern oxidation of 4-methyl-3*E*-penten-2-ol.

17.15 Draw the major product formed by Swern oxidation of 3-phenyl-2-butanol.

17.3 Oxidation of Alkenes

Several reactions of alkenes are discussed in Chapter 10 that involve an acid–base reaction in which the alkene is a Brønsted–Lowry base or a Lewis base. In that chapter, two other reactions transformed an alkene into a vicinal diol or into an epoxide. Both of these reactions are oxidations, although they were discussed in a different context at that time. This section will review the previously discussed chemistry to put it into proper perspective as oxidation reactions.

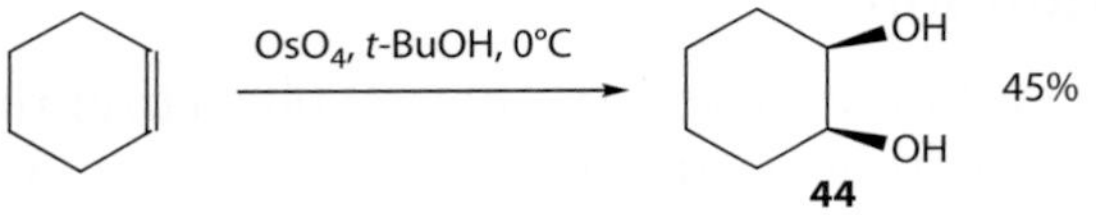

In Chapter 10, an experiment was reported in which cyclohexene was treated with OsO$_4$ (osmium tetroxide) in anhydrous 2-methyl-2-propanol (*tert*-butyl alcohol) at 0°C. After standing overnight, a 45% yield of *cis*-1,2-cyclohexanediol (**44**) is isolated.[7] This reaction is a ***dihydroxylation*** and the two OH units have a *cis* relationship. Alkenes also react with dilute aqueous potassium permanganate to give a *cis* diol.

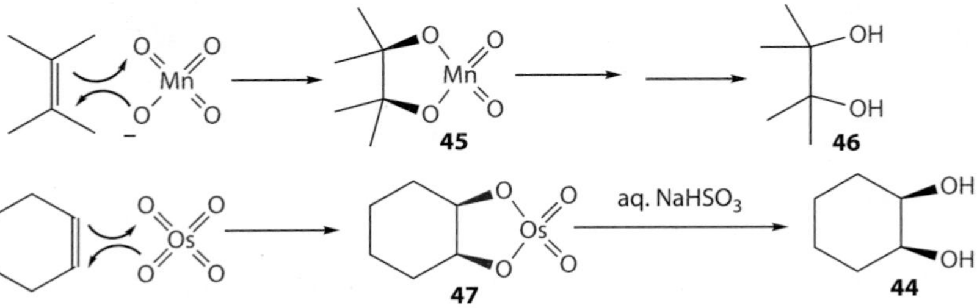

Both potassium permanganate (KMnO$_4$) and osmium tetroxide (OsO$_4$) react with an alkene via 1,3-dipolar addition (Chapter 10, Section 10.7.1) to give what is known as a manganate ester (**45**) or an osmate ester (**47**). In the presence of aqueous hydroxide, **45** reacts to give 1,2-diol **46**. In the osmium tetroxide reaction, aqueous sodium thiosulfite (NaHSO$_3$) or *tert*-butanol is used to convert **47** to diol **44**. The conversion of an alkene to a diol is a formal oxidation, as described for **3** → **4** in Section 17.1.

As described in Chapter 10, if the reaction of an alkene with OsO$_4$ is done in the presence of *tert*-butylhydroperoxide (Me$_3$C–OOH) or *N*-methylmorpholine-*N*-oxide (**49**, NMO), the reaction is catalytic in osmium. Both additives react with the osmate ester to generate the diol, and they are also able to regenerate the OsO$_4$ reagent. Therefore, if 1-pentene (**48**) reacts with a 5% aqueous solution of OsO$_4$ that contains NMO, the final product is pentane-1,2-diol, **50**. As little as 1% of osmium tetroxide can be used.

17.16 Draw the product formed by treatment of 3-phenyl-2Z-pentene with 1. OsO$_4$; 2. aq. NaHSO$_3$.

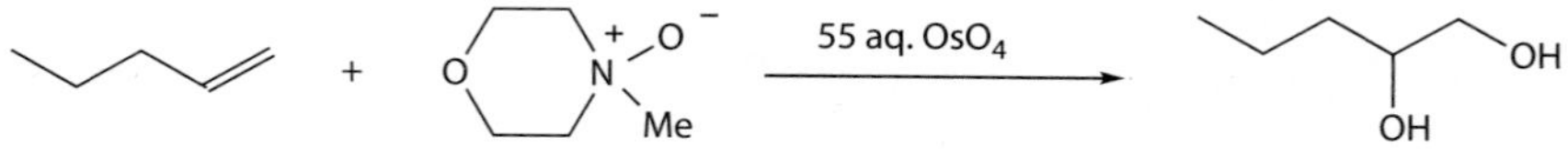

In Chapter 10, Section 10.5, the reaction of an alkene with a peroxyacid generated an epoxide (an oxirane) as well as the carboxylic acid precursor of the peroxyacid. The alkene is oxidized to an epoxide. This oxidation is shown for the preparation of epoxide **52** from a generic alkene and a generic peroxyacid.

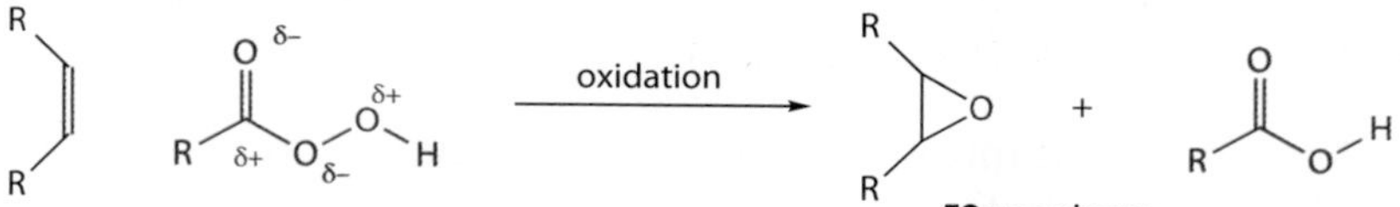

An oxirane (an epoxide, **51**) is a three-membered ring ether (see Chapter 26, Section 26.4.2) and conversion of an alkene to **51** requires an oxidizing agent. The peroxide derivative of a carboxylic acid is a peroxycarboxylic acid (**51**; see Section 10.5), which is an effective oxidizing agent for this reaction. Several peroxyacids can be used as "R" in **51**, including peroxyformic acid (HCO$_3$H, **53**), peroxyacetic acid (CH$_3$CO$_3$H, **54**), and *meta*-chloroperoxybenzoic acid (abbreviated mCPBA, **55**). The carboxylic acid precursor to **55** is an aromatic carboxylic acid. The nomenclature and structural features of benzoic acid and other aromatic carboxylic acid derivatives will be discussed in detail in Chapter 20, Section 20.2.

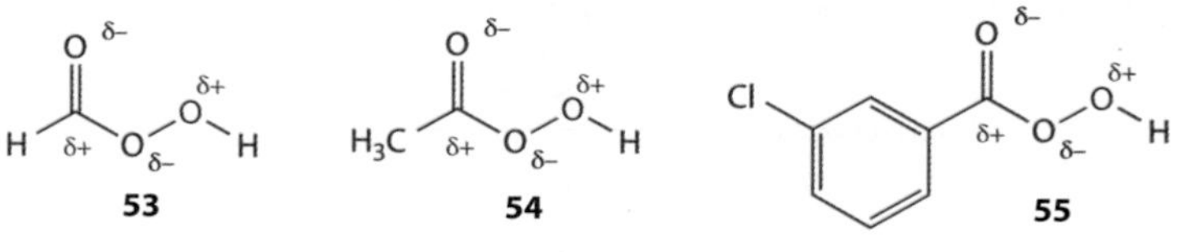

In a typical experiment, cyclopentene reacts with peroxyacetic acid (**54**) to give two isolated products: cyclopentene oxide (**57**) in 57% yield and acetic acid.[8] Experiments have failed to identify an intermediate for this reaction and the transition state must be examined to explain the observed products. For the reaction of cyclopentene, the transition state is **56**. The oxygen atom labeled in **red** in **54** is δ+ because bond polarization extends from the carbonyl unit (C=O). The alkene reacts as a Lewis base, and the oxygen in **54** is the electrophilic atom. If there is no intermediate in this reaction, the transition state must involve the formation of both bonds between C and O in **57**, as well as the transfer of electrons to convert the peroxyacids to the carboxylic acid product. This reaction was described in detail in Chapter 10 (Section 10.5). For the purposes of this chapter, an alkene is oxidized to an epoxide, and the by-product is a carboxylic acid—in this case, acetic acid. The carboxylic acid product is always the acid parent of the peroxyacid (RCO_2H from RCO_3H).

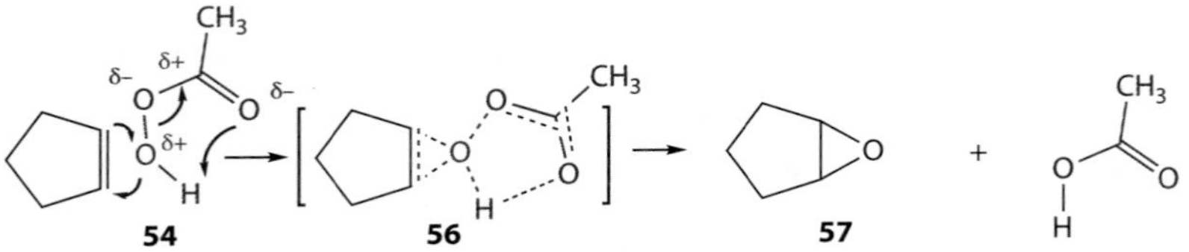

17.17 **Show the oxidation state for the oxygen-bearing carbon atoms in 52 where R = methyl.**

17.18 **Write out the transition state and all products of the reaction between 3,4-dimethyl-3-hexene and mCPBA.**

Remember from Chapter 5 (Section 5.1) that a tetrasubstituted alkene ($R_2C=CR_2$) reacts faster with peroxyacids than a trisubstituted alkene ($R_2C=CHR$), which reacts faster than a disubstituted alkene. The alkenes that react slowest in this reaction are monosubstituted alkenes such as $RHC=CH_2$. However, ***assume that all alkenes given as problems in this book are epoxidized***. In reality, there are vast differences in reaction rate among various alkenes.

Chapter 10 (Section 10.5) described an alternative class of epoxidation reagents known as **dioxiranes**. The cited specific example is dimethyl dioxirane (**58**). This reagent requires a co-reagent (an oxidizing agent) such as potassium peroxomonosulfate ($KHSO_5$). Commercially available Oxone→ ($2KHSO_5 \cdot KHSO_4 \cdot K_2SO_4$) is a common source of $KHSO_5$. Oxone reacts with a ketone such as acetone and sodium bicarbonate to generate **58**; in the presence of an alkene, the alkene is oxidized to an epoxide. An example is the reaction of 2-ethyl-1-pentene (**59**) with Oxone and acetone to give 2-ethyl-2-propyloxirane, **60**.

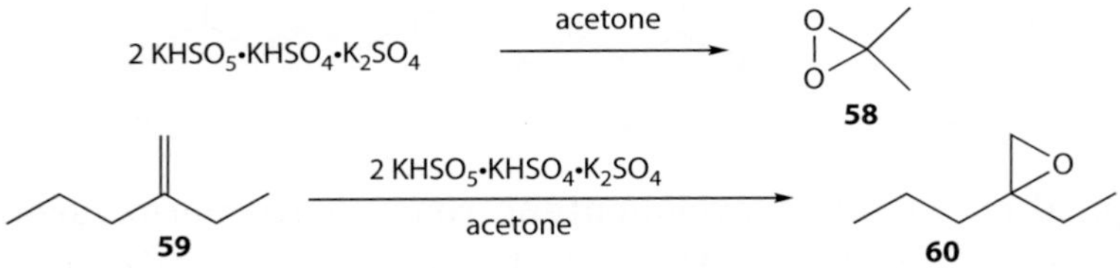

17.4 Oxidative Cleavage

Oxidative cleavage is a process that cleaves covalently bound carbon atoms to give two functionalized carbon atoms. Reactions of this type include oxidative cleavage of a C=C unit, as in the ozonolysis reaction of Chapter 10 (Section 10.7.2) to give carbonyl derivatives. The C–C bond of 1,2-diols may also be oxidatively cleaved to give carbonyl derivatives. This section will revisit ozonolysis and then introduce the oxidative cleavage of diols.

17.4.1 Ozonolysis

In Chapter 10 (Section 10.7.2), alkenes such as 2,3-dimethyl-2-butene reacted with ozone (O_3, **61**) to give a 1,2,3-trioxolane, **62**. This unstable intermediate rearranges to a 1,2,4-trioxolane (an ozonide, **63**), which is the final product. The ozonide is not isolated but rather treated with hydrogen peroxide or another reagent (see following reaction) to generate two carbonyl compounds—in this example, acetone. In other words, ozonolysis of an alkene leads to cleavage of the C=C unit and formation of an aldehyde or a ketone, and sometimes a carboxylic acid. The π-bond is cleaved in this reaction via the dipolar addition, and then the σ-bond is cleaved via the rearrangement step. The net result is cleavage of both bonds of the C=C unit and transformation of each carbon of the C=C unit to a C=O unit, a carbonyl compound (aldehyde, ketone, or carboxylic acid). The conversion of an alkene to an ozonide is known as ozonolysis and is an example of an ***oxidative cleavage reaction***.

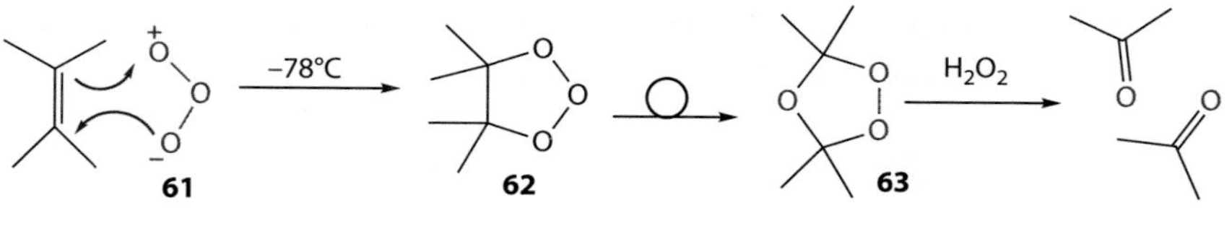

As noted, the ozonide is not isolated, but a second chemical step is performed in the same flask (such as treatment with hydrogen peroxide). Two products result from the ozonolysis of 2,3-dimethyl-2-butene: 2-propanone (acetone, **2**) and a second molecule of acetone. This statement is phrased this way because this is a symmetrical alkene, but an unsymmetrical alkene will give two different ketones. In effect, the C=C unit is cleaved and each carbon is oxidized to a C=O unit. The mechanism of ozonolysis was described in Chapter 10.

17.19 Suggest structures for the two carbonyl products obtained by treatment of 3-ethyl-4-methyl-3-heptene with 1. ozone; 2. hydrogen peroxide.

17.20 Write out the 1,2,3-trioxolane and ozonide expected when cyclohexene reacts with ozone. Write out the reaction and name the final product formed when that ozonide reacts with hydrogen peroxide.

When the alkene has a hydrogen atom attached to the C=C unit, as in 2,4-dimethyl-3-hexene (**64**), reaction with ozone gives the expected 1,2,3-trioxolane product **65** and rearrangement to ozonide **66**. Subsequent reaction of **66** with dimethyl sulfide, a reducing agent in this case, gives 2-butanone and 2-methylpropanal. When **66** is treated with hydrogen peroxide (a good oxidizing agent), the cleavage products are 2-butanone and 2-methylpropanoic acid. In this latter reaction, the initially formed aldehyde is oxidized to the acid. As noted in Chapter 10, when there is a hydrogen atom on the C=C unit, ozonolysis and oxidation lead to a carboxylic acid, but reduction of the ozonide leads to an aldehyde.

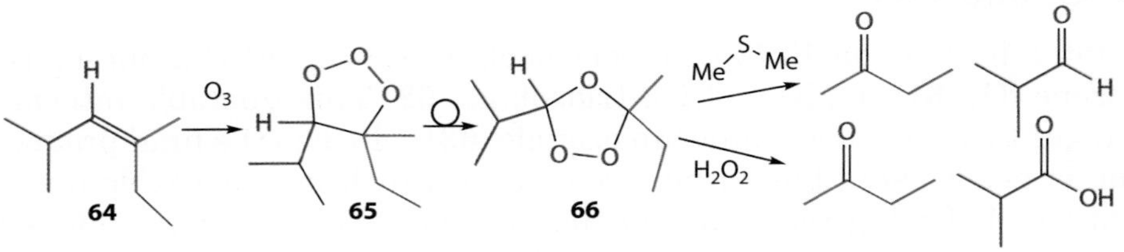

For any alkene, hydrogen peroxide may be used as the oxidizing agent and either dimethyl sulfide or zinc and acetic acid may be used as the reducing agent (see Chapter 18, Section 18.4.3). Ozonolysis of a cyclic alkene leads to oxidative cleavage to a diketone, a dialdehyde, a keto-aldehyde, a keto acid, or a dicarboxylic acid. When 1,3-dimethylcyclopentene (**67**) is treated with ozone and then with zinc and acetic acid, oxidative cleavage leads to keto-aldehyde **68** (2-ethyl-5-oxooctanal).

17.21 Draw the product or products formed when 2-methyl-3-phenyl-2-pentene is treated with 1. O_3; 2. Me_2S.

17.22 Draw the product formed when 1-ethyl-1-cyclononene is treated with 1. O_3; 2. Zn/acetic acid.

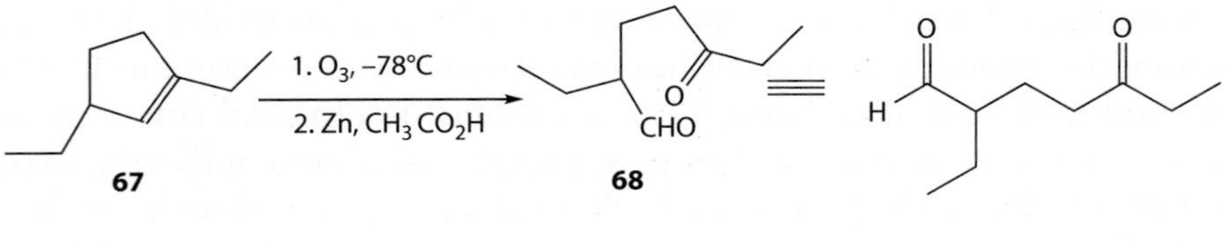

17.4.2 Oxidative Cleavage Reactions of Diols

Section 17.3 reviewed the oxidation of alkenes to give dihydroxylation. Ozonolysis is a good method for the oxidative cleavage of alkenes, but vicinal diols are also subject to oxidative cleavage and the products are aldehydes or ketones. Two common reagents used for this purpose are periodic acid (HIO_4) and lead tetraacetate [$Pb(OAc)_4$].

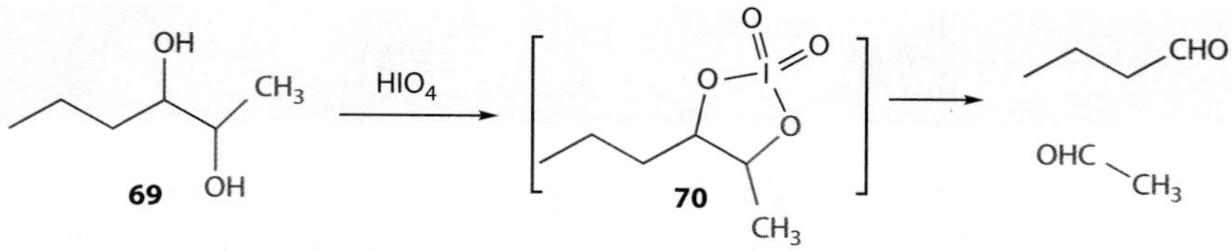

When hexane-2,3-diol (**69**) is treated with HIO$_4$ (periodic acid), a cyclic product is formed (**70**) that decomposes under the reaction conditions to generate butanal and ethanal (acetaldehyde). In other words, oxidative cleavage of diol **69** gives two aldehydes. Experiments have shown that there is little or no over-oxidation to the carboxylic acid, so *periodic acid is a mild and effective method for converting alkenes to two aldehydes via the diol*. A useful variation of this reaction treats a cyclic diol such as 1,2-cyclopentanediol (**71**) with HIO$_4$ to give 1,5-pentanedial, **72**. Suitably substituted diols are cleaved with periodic acid to give ketones or keto-aldehydes. The reaction of 3,4-dimethyl-3,4-hexanediol and HIO$_4$ leads to two ketones. Similarly, 4-ethyl-3,4-octanediol is cleaved to a ketone and an aldehyde.

17.23 Write out the reaction that generates **69** from an appropriate alkene.

17.24 Draw the final product formed when 1,2-dimethylcycloheptene is treated with 1. OsO$_4$, aq. NaHSO$_3$; 2. HIO$_4$.

17.25 Draw the ketone products of the reaction of 3,4-dimethyl-3,4-hexanediol and HIO$_4$.

17.26 Draw the ketone and aldehyde products of reaction of 4-ethyl-3,4-octanediol and HIO$_4$.

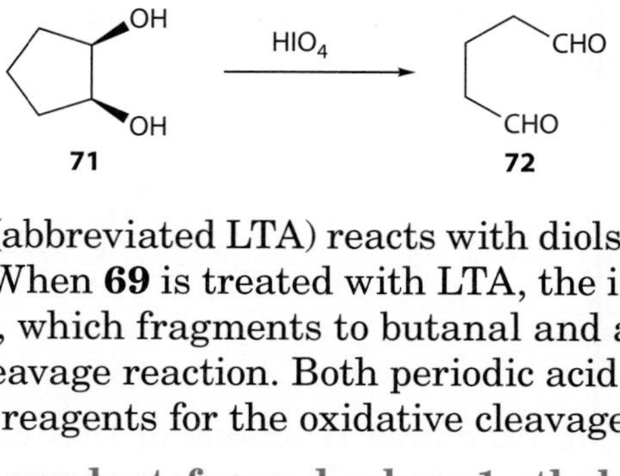

Lead tetraacetate (abbreviated LTA) reacts with diols to give the same kind of oxidative cleavage. When **69** is treated with LTA, the initial cyclic product is cyclic intermediate **73**, which fragments to butanal and acetaldehyde. Periodic acid gives the same cleavage reaction. Both periodic acid and lead tetraacetate are mild and effective reagents for the oxidative cleavage of diols.

17.27 Write out the product formed when 1-ethylcycloheptene reacts with LTA.

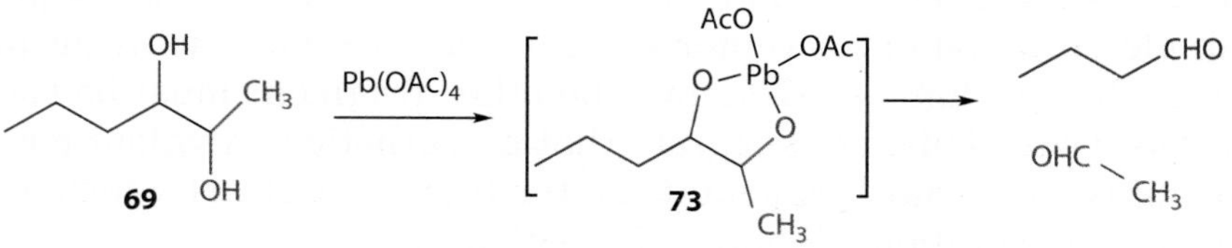

17.5 Summary of Functional Group Exchanges

Several new and interesting functional group transformations are possible via oxidation reactions. To summarize these reactions, the functional group exchange will be shown for each.

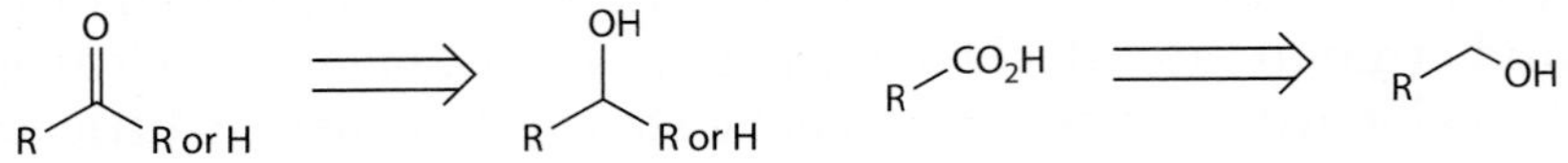

Oxidation of alcohols converts an alcohol to a ketone or an aldehyde. It is also possible to oxidize a primary alcohol to a carboxylic acid. The functional group exchange reactions therefore show a ketone or aldehyde with an alcohol precursor and a carboxylic acid with a primary alcohol precursor.

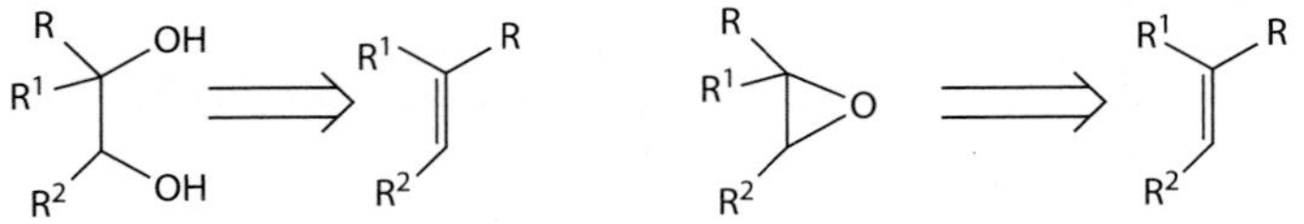

Because alkenes may be oxidized to either a vicinal diol or an epoxide, the functional group exchange reaction shows formation of the diol from the alkene or formation of an epoxide from the alkene. Oxidative cleavage of an alkene or a diol leads to the corresponding aldehyde, ketone, or carboxylic acid. The functional group exchange for the generation of ketones, aldehydes, or acids from alkenes is for ozonolysis. The functional group exchange for generation of aldehydes or ketones from diols complements the ozonolysis reaction, allowing a bit of selectivity in the cleavage.

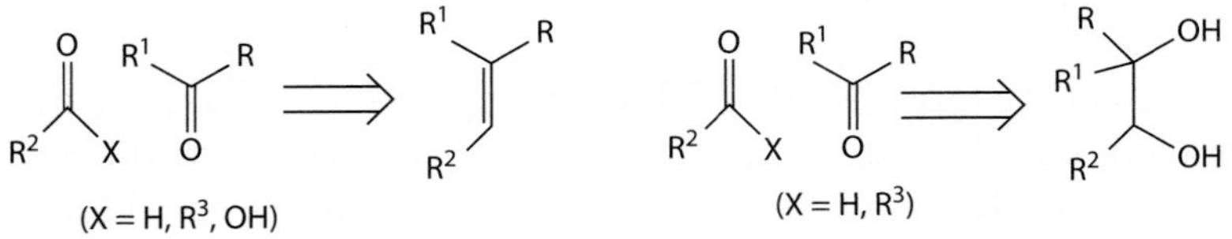

(X = H, R³, OH)

(X = H, R³)

The functional group exchange reactions of this chapter may be applied to synthesis. A synthesis of ketone **74** can be imagined from 1-pentene (**48**). Disconnection of **48** adjacent to the carbonyl carbon leads to the two disconnect products shown. The bond polarization of oxygen makes the adjacent carbon atom electrophilic (an acceptor site), which correlates with an aldehyde. Propanal is the synthetic equivalent. The other fragment must be the nucleophilic or donor site. This fact suggests that a synthetic equivalent for a simple carbon donor is a Grignard reagent (Chapter 15, Section 15.1), which adds to an aldehyde via acyl addition (Chapter 17, Section 17.4).

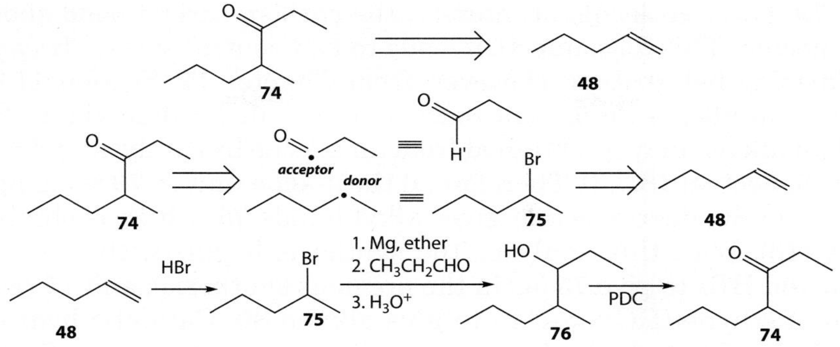

The immediate precursor to the Grignard reagent is alkyl halide **75**, and the bromide is arbitrarily chosen. Bromide **75** is prepared directly from the chosen starting material **48**. Therefore, this analysis leads to the synthesis shown. 1-Pentene (**48**) reacts with HBr to give the more stable carbocation (see Chapter 10, Section 10.2) and then secondary bromide **75**. Reaction with magnesium gives the Grignard reagent, and then reaction with propanal gives alcohol **76** after hydrolysis. Note that the product of acyl addition is the alcohol (Chapter 17, Section 17.4) and an oxidation is required to generate ketone **74**. The mild oxidizing agent PDC is used, but PCC or Jones's oxidation will give the same product.

A second example prepares aldehyde **77** from 1-pentene (**48**). If the carbon–carbon bond is disconnected, it is not obvious from the chemistry studied so far how that bond will be made. On the other hand, from the study of oxidative cleavage, an aldehyde is obtained by oxidative cleavage of a 1,2-diol. Therefore, if **78** is chosen as the precursor, **79** is the precursor to **78**. ***This analysis is different in that it does not require a disconnection, but rather an unusual functional group manipulation that relies on the oxidative cleavage reaction.***

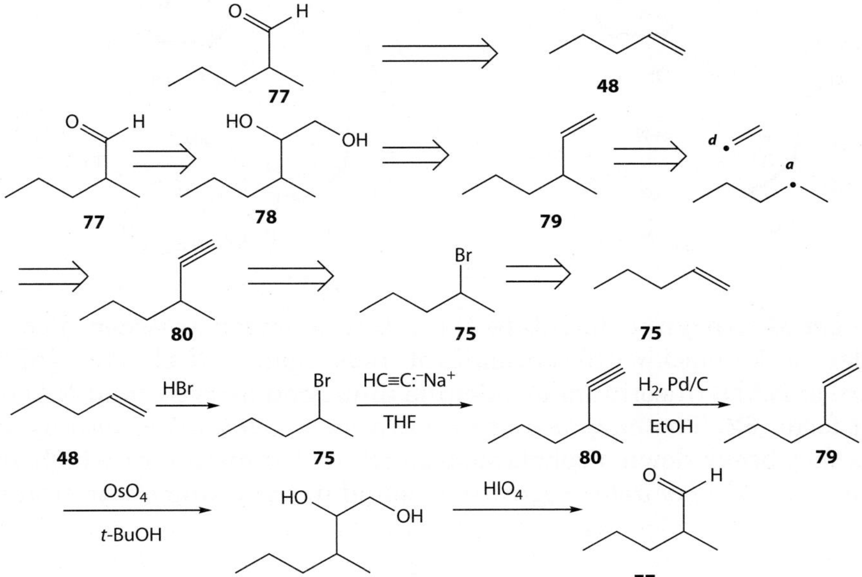

With **79** as the molecule of interest, the carbon–carbon bond shown must be disconnected. This disconnection leads to two neutral species because there is no polarizing heteroatom. However, from Chapter 11 (Section 11.3.6), it is known that an alkyne anion will react with an alkyl halide via an S_N2 reaction, and an alkene may be obtained from an alkyne by catalytic hydrogenation (Chapter 18, Section 18.3.3). Therefore, if the alkene unit in **79** is changed to an alkyne unit in **80**, disconnection gives alkyl halide **75**, which is obtained from 1-pentene (**48**). With this analysis, the synthesis begins with the reaction of 1-pentene and HBr to give **75** (as in the previous synthesis), and **75** reacts with the anion of ethyne ($HC{\equiv}C{:}^-Na^+$) to give alkyne **80**. Catalytic hydrogenation leads to alkene **79**, and dihydroxylation with osmium tetroxide (Section 17.3) gives diol **78**. Oxidative cleavage of this diol with periodic acid gives the target aldehyde, **77**.

17.28 **Show a retrosynthetic analysis and then the synthesis of cyclohexanone from bromocyclohexane.**

17.29 **Show a retrosynthetic analysis and then the synthesis of 7-oxononanal from ethylcycloheptanol.**

17.6 Biological Relevance

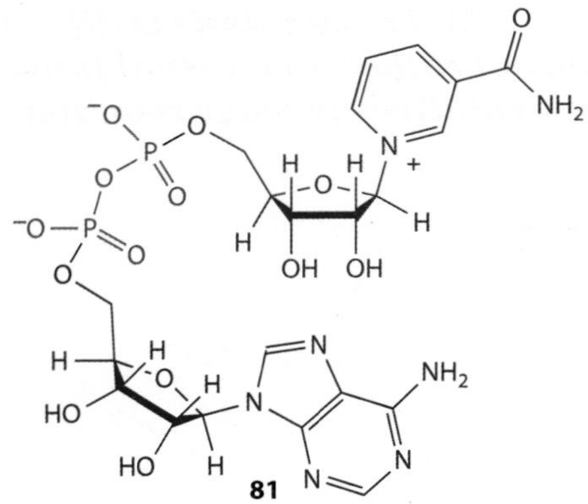

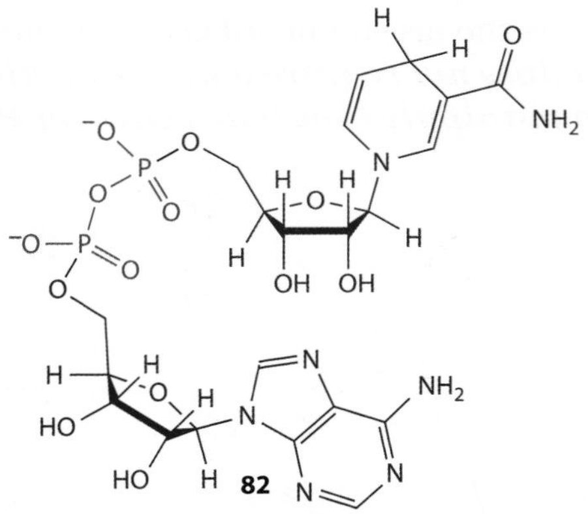

Dehydrogenase enzymes facilitate the interconversion between alcohols and aldehydes or ketones with concomitant reduction of NAD^+ to NADH. The structure of NAD^+ (nicotinamide adenine dinucleotide) is **81**, and NADH is the reduced form, **82**. The enzymes are known as *alcohol dehydrogenases* and in humans they break down alcohols such as ethanol or methanol, which are toxic. In humans, an *alcohol dehydrogenase* is found in the lining of the stomach and in the liver. The dehydrogenation is in reality an oxidation, and *alcohol dehydrogenase* catalyzes the oxidation of ethanol to acetaldehyde:

$$CH_3CH_3OH + NAD^+ \rightarrow CH_3CHO + NADH + H^+$$

The conversion of ethanol to acetaldehyde (ethanal) allows the consumption of alcoholic beverages, but it probably exists to oxidize naturally occurring alcohols in foods or those produced by bacteria in the digestive tract.

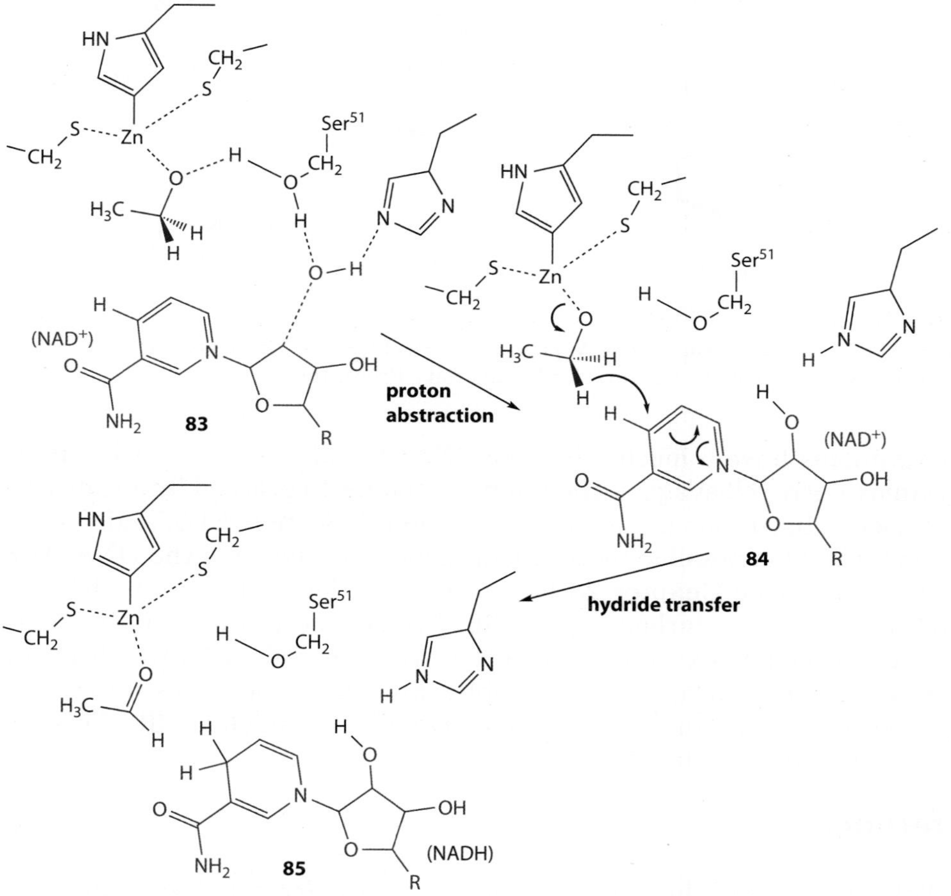

Eklund, H., B.V. Samama, J.P. Bränden, C.I. *Journal of Biological Chemistry* 1982, 257, 14349, © 1982 The American Society for Biochemistry and Molecular Biology.

The mechanism of this oxidation for the enzyme liver alcohol dehydrogenase is shown for the reaction of **83**, where ethanol is bound to the active site of the enzyme to give **84** via proton abstraction and then hydride transfer to generate acetaldehyde (see **85**). NAD⁺ binds to the active site of the enzyme to induce a conformational change (see Chapter 8 for conformation) to close the active site.[9] The oxidation of ethanol to acetaldehyde (ethanal) is accompanied by reduction of NAD⁺ to NADH, as shown in the illustration.

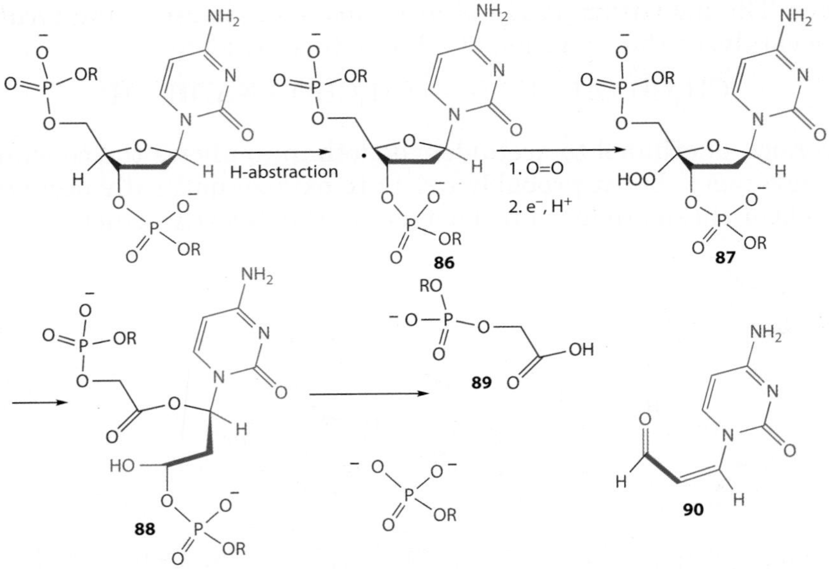

Long, E. C. in *Bioorganic chemistry: Nucleic acids,* Hecht, S. M., (Ed.), Oxford University Press, New York, 1996. By permission of Oxford University Press, Inc.

Organisms have a mechanism for DNA strand scission that is fundamentally an oxidative cleavage, although it is thought to proceed by a radical mechanism (see a discussion of radicals in Chapter 7, Section 7.4.3, and Chapter 10, Section 10.8). One possible mechanism involves cleavage of the DNA strand at the sugar unit (see Chapter 28 for an introduction to DNA) in which a C–H bond breaks to give a carbon radical (**86**).[10] In one mechanism, subsequent reaction with molecular oxygen leads to the hydroperoxide (**87**), which fragments to **88** (several mechanisms are possible), and then undergoes an elimination sequence to give carboxylic acid derivative **89** and aldehyde **90**. This process results in cleavage of the DNA strand.

References

1. Wiberg, K. B. 1965. In *Oxidation in organic chemistry,* part A, ed. Wiberg, K. B., 146–147. New York: John Wiley & Sons.
2. Wiberg, K. B. 1965. In *Oxidation in organic chemistry,* part A, ed. Wiberg, K. B., 151. New York: John Wiley & Sons.
3. Hudlicky, M. *Oxidations in organic chemistry,* 117. Washington D.C.: American Chemical Society.
4. Wiberg, K. B. 1965. In *Oxidation in organic chemistry,* part A, ed. Wiberg, K. B., 154. New York: John Wiley & Sons.
5. Luzzio, F. *Organic Reactions* 53:1
6. Tidwell, T. T. 1990. *Organic Reactions* 39:297. See p. 431.
7. Furniss, B. S., Hannaford, A. J., Smith, P. W. G., and Tatchell, A. R., eds. 1994. *Vogel's textbook of practical organic chemistry,* 5th ed., 548–549, Exp. 5.47. New York: Longman.

8. Rickborn, B., and Gerkin, R. M. 1971. *Journal of the American Chemical Society* 93:1693.
9. Eklund, H., Plapp, B. V., Samama, J. P., and Branden, C. I. 1982. *Journal of Biological Chemistry* 257:14349.
10. Long, E. C. 1996. In *Bioorganic chemistry: Nucleic acids,* ed. Hecht, S. M., 25–26, Figure 1.21. New York: Oxford University Press.

Answers to Problems

17.1   In **3**, the oxidation state of C1 is $0 + 0 + 1 = +1$ and that for C2 is $0 + 0 + 1 + 1$. The two carbon substituents receive a value of zero, and each hydrogen has a value of –1. In **4**, the oxidation state of C1 is $0 + 0 + 1 – 1 = 0$, and this is identical for C2, which also has an oxidation state of 0. Both carbon substituents receive a value of zero because the oxygen contributes +1 and the hydrogen contributes –1. The reaction converts two carbon atoms of the C=C unit in **3** to a HO–C–C–OH unit, so we look at the oxidation level of both carbons. This transformation involves loss of one electron from C1 and loss of one electron from C2, for a net loss of two electrons and this is an oxidation.

17.2   In this reaction, the terminal carbon loses one electron, as does C2. Because there is a net loss of two electrons, this reaction is an oxidation.

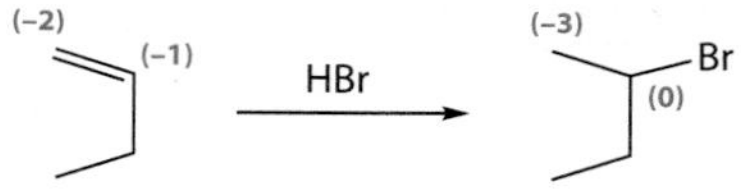

17.3

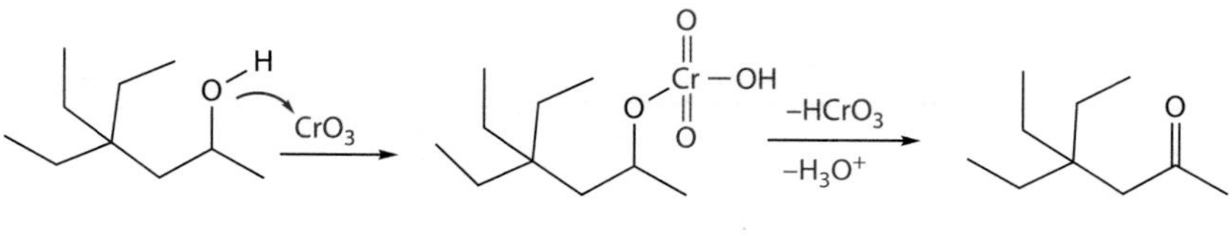

17.4

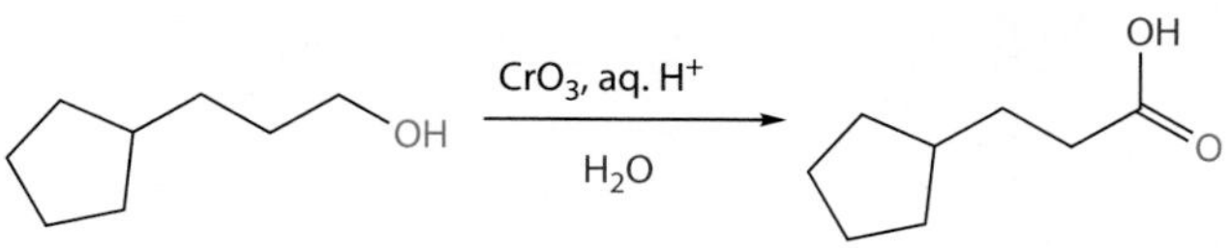

17.5   The name of the first ketone is 2-methyl-3-pentanone. The name of the second ketone is 2,2,4,4-dimethyl-3-pentanone.

17.6

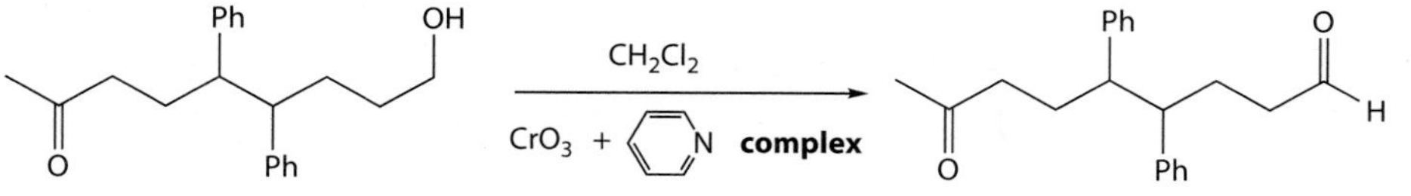

17.7

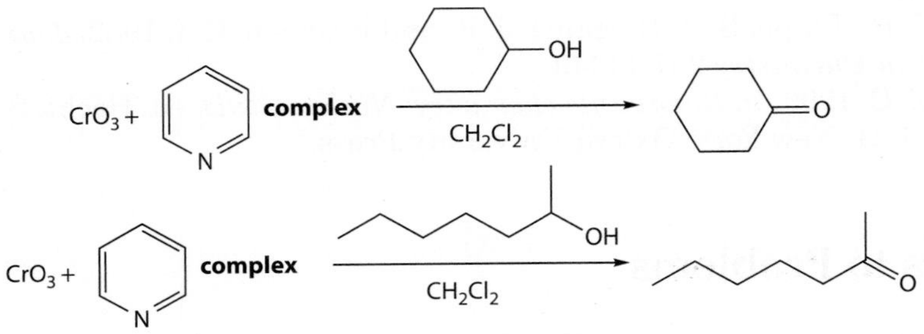

CrO₃ + [pyridine] **complex** → cyclohexanol → cyclohexanone (CH₂Cl₂)

CrO₃ + [pyridine] **complex** → 2-heptanol → 2-heptanone (CH₂Cl₂)

17.8

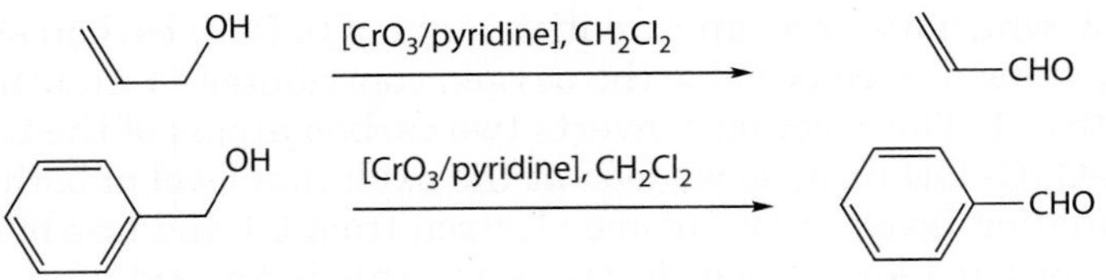

allyl alcohol OH →[CrO₃/pyridine], CH₂Cl₂→ CHO

benzyl alcohol OH →[CrO₃/pyridine], CH₂Cl₂→ CHO

17.9

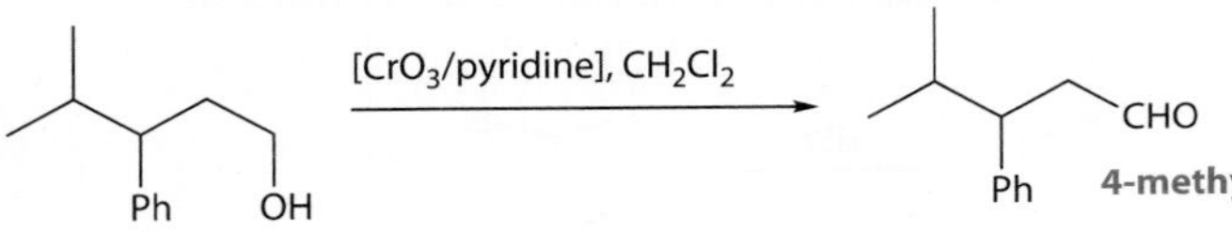

→[CrO₃/pyridine], CH₂Cl₂→

Ph OH   Ph   **4-methyl-3-phenylpentanal** CHO

17.10

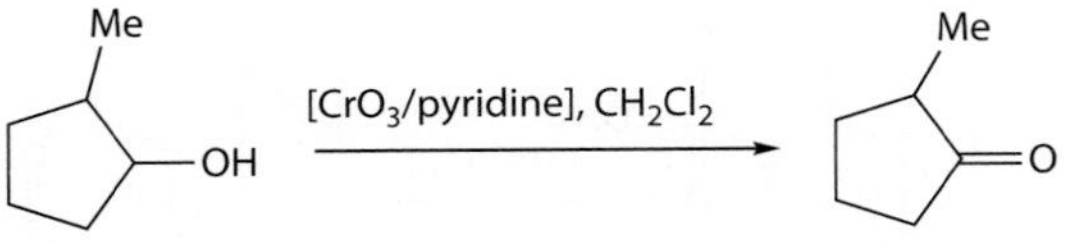

Me   →[CrO₃/pyridine], CH₂Cl₂→   Me   **2-methylcyclopentanone**
   OH   O

17.11

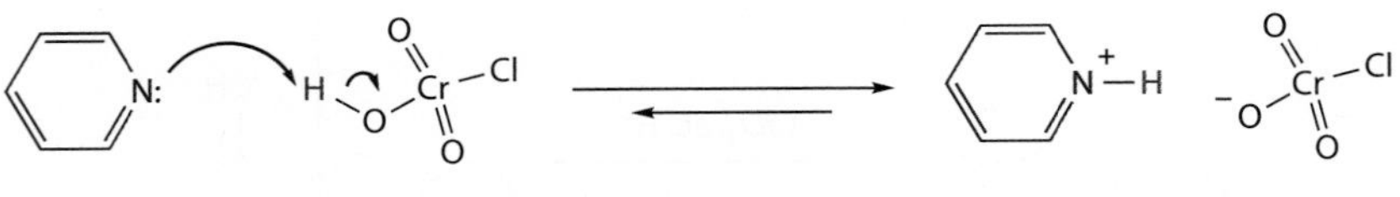

17.12

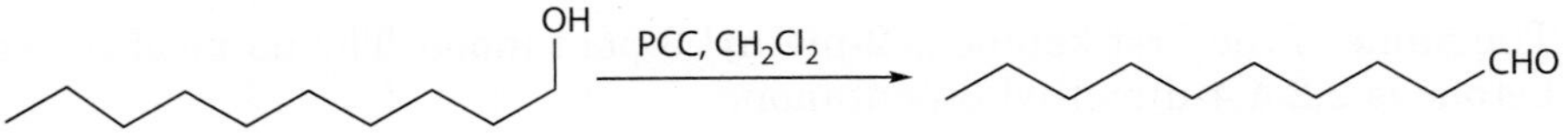

OH →PCC, CH₂Cl₂→ CHO

17.13

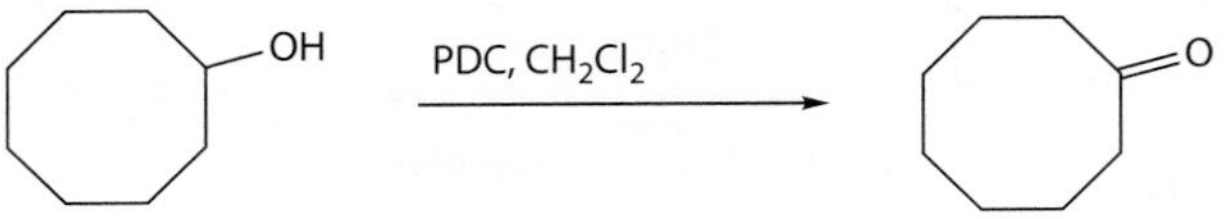

OH →PDC, CH₂Cl₂→ O

17.14

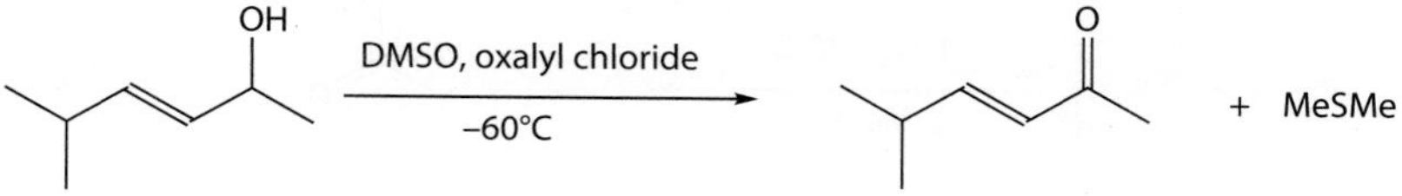

17.15

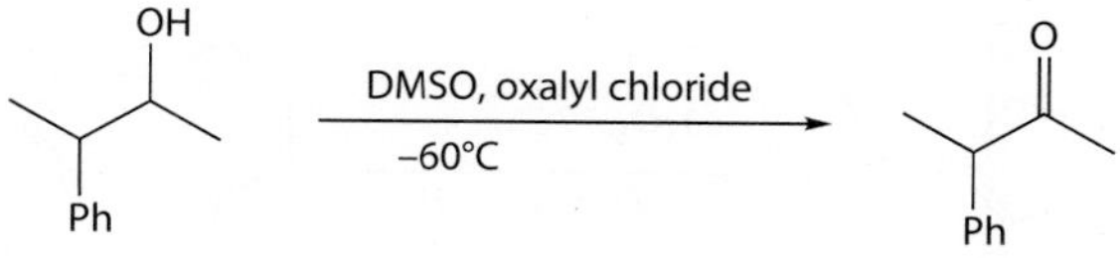

17.16

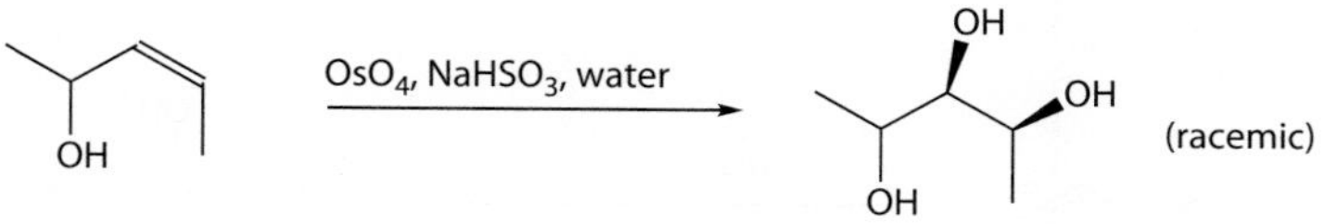

17.17

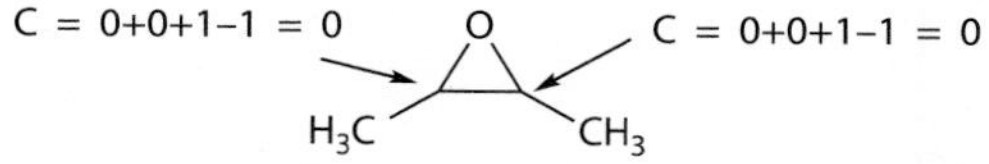

$C = 0+0+1-1 = 0$ $C = 0+0+1-1 = 0$

17.18

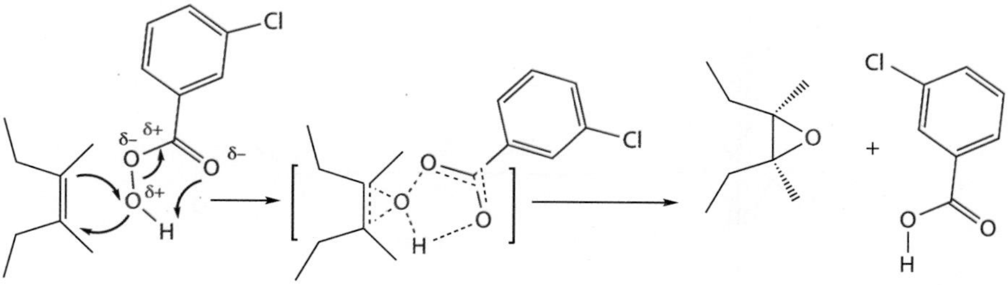

17.19

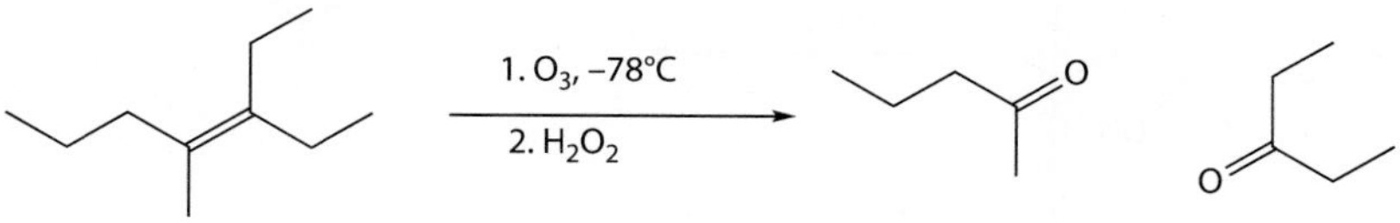

17.20

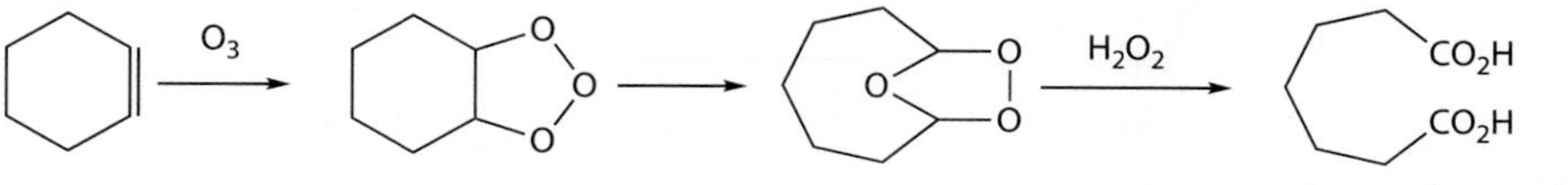

1, 7-hexanedioic acid

17.21

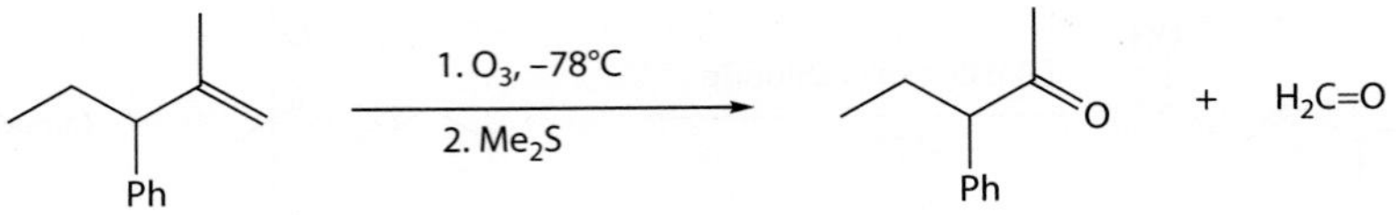

17.22

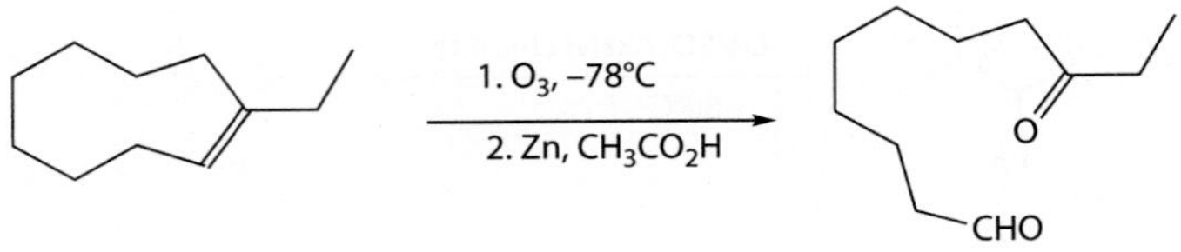

17.23

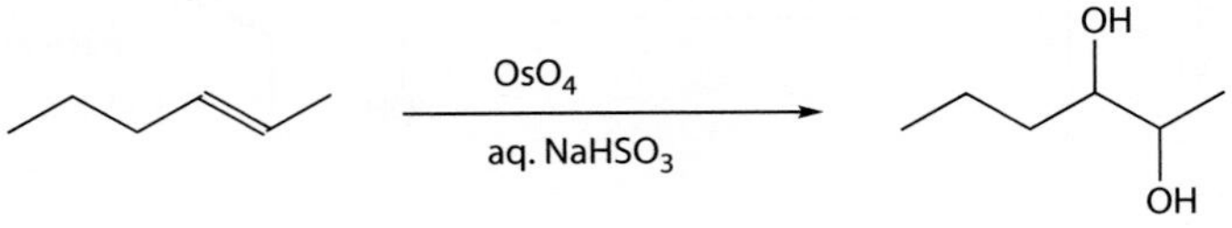

17.24

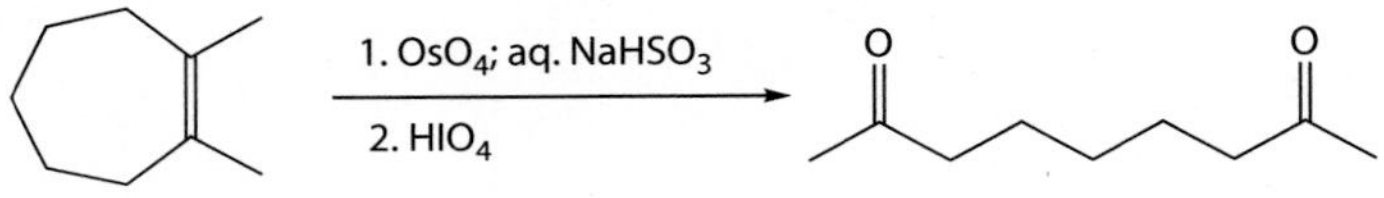

17.25

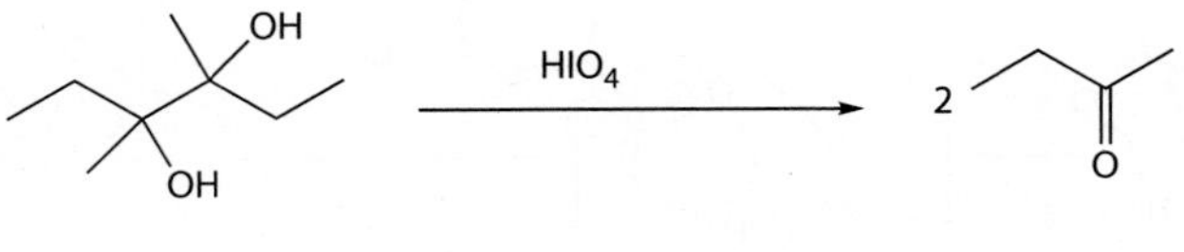

17.26

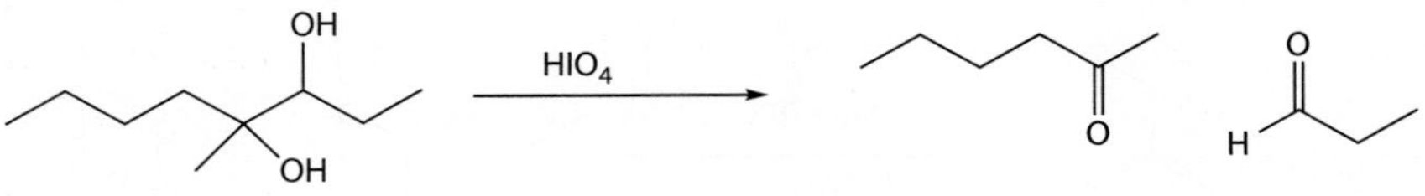

17.27

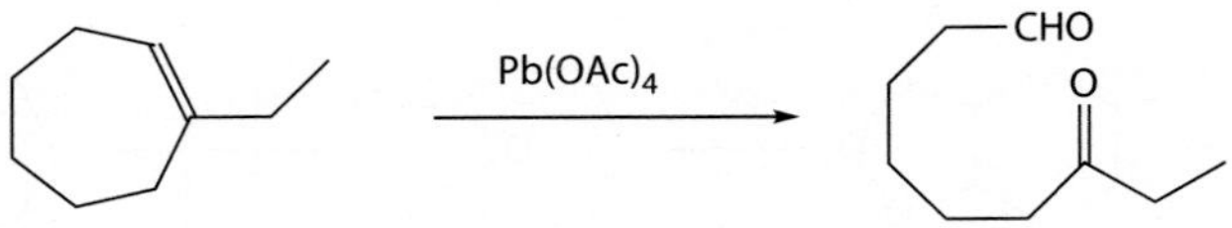

17.28

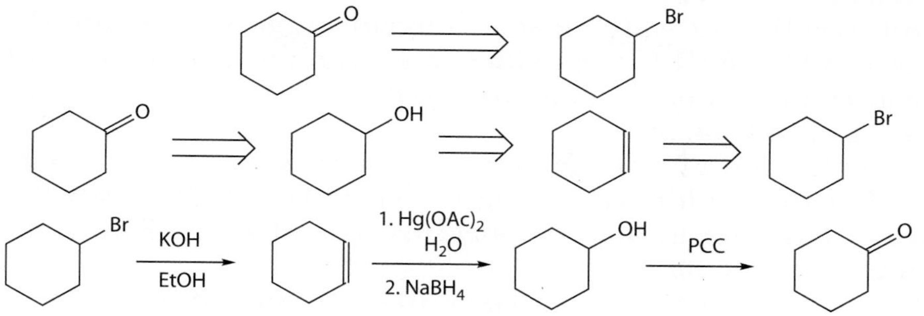

17.29

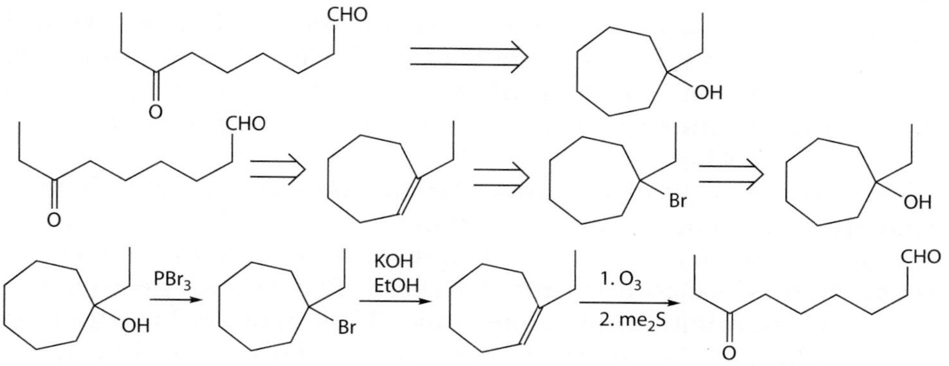

Correlation of Concepts with Homework

- **Oxidation is defined as loss of electrons or gain of a heteroatom such as oxygen or loss of hydrogen atoms. Oxidation number is a convenient method to track the gain or loss of electrons in a reaction: 1, 2, 17, 30, 37.**
- **Chromium(VI) reagents are powerful oxidants. The reaction of a secondary alcohol with chromium trioxide and acid in aqueous acetone is called Jones's oxidation, and the product is a ketone. Chromium oxidation of an alcohol proceeds by formation of a chromate ester, followed by loss of the α-hydrogen to form the C=O unit: 3, 4, 5, 31, 32, 34, 44, 54.**
- **Jones's oxidation of a primary alcohol leads to a carboxylic acid in most cases. A mixture of chromium trioxide and pyridine gives a reagent that can oxidize a primary alcohol to an aldehyde. This is called Collins's oxidation. The reaction of chromium trioxide and pyridine in aqueous HCl leads to pyridinium**

chlorochromate, called **PCC**. The reaction of chromium trioxide and pyridine in water leads to pyridinium dichromate (**PDC**). Both **PCC** and **PDC** can oxidize a secondary alcohol to a ketone or a primary alcohol to an aldehyde: 6, 7, 8, 9, 10, 11, 12, 13, 32, 33, 35, 39, 43, 44, 52.

- **The oxidation of a secondary alcohol to a ketone or a primary alcohol to an aldehyde using dimethyl sulfoxide (DMSO) with oxalyl chloride at low temperature is called the Swern oxidation:** 14, 15, 34, 39.
- **Oxidation of an alkene with osmium tetroxide or potassium permanganate gives a *cis*-1,2-diol:** 16, 39, 44.
- **Oxidation of an alkene with a peroxyacid leads to an epoxide, with a carboxylic acid as the by-product:** 18, 37, 38, 40, 42, 44, 51.
- **Oxidative cleavage of an alkene with ozone leads to an ozonide. Reductive workup with dimethyl sulfoxide or zinc and acetic acid gives ketones and/or aldehydes. Oxidative workup with hydrogen peroxide gives ketones and/or carboxylic acids:** 19, 20, 21, 22, 36, 39, 41, 44.
- **Oxidative cleavage of 1,2-diols with periodic acid or with lead tetraacetate gives aldehydes or ketones:** 23, 24, 25, 26, 27, 39, 44, 46.
- **A molecule with a particular functional group can be prepared from molecules containing different functional groups by a series of chemical steps (reactions). This process is called synthesis: The new molecule is synthesized from the old one (see Chapter 25):** 28, 29, 30, 37, 45, 46, 47, 48.
- **Spectroscopy can be used to determine the structure of a particular molecule and can distinguish the structure and functionality of one molecule when compared with another (see Chapter 14):** 49, 50, 51, 52, 53, 54, 55, 56, 57, 58, 59, 60, 61, 62, 63, 64.

Homework

17.30 Categorize each of the following transformations as an oxidation or a reduction using oxidation numbers:

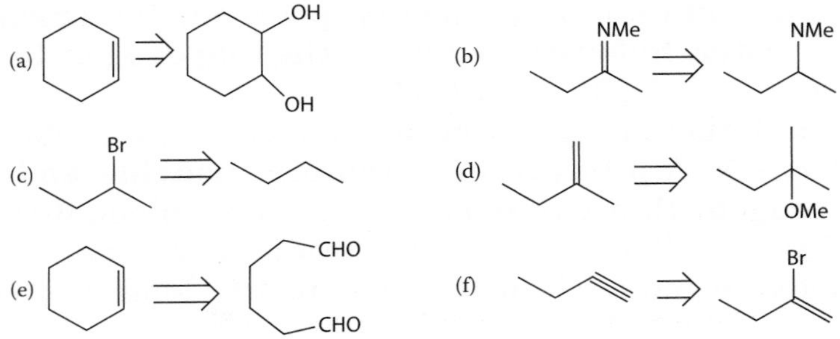

17.31 Briefly explain why Jones's oxidation of 4-methyl-2-pentanol is faster than that of 3,3-dimethyl-2-pentanol.

17.32 Give the major product for each of the following:

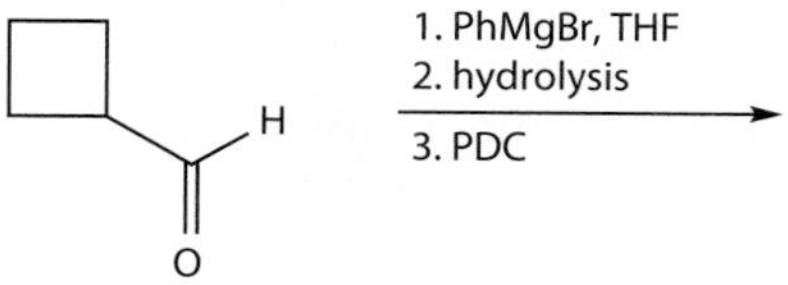

17.33 Briefly discuss whether you believe that the reagent used for the Collins oxidation is the same as PDC.

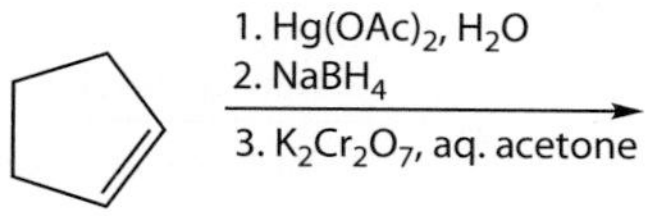

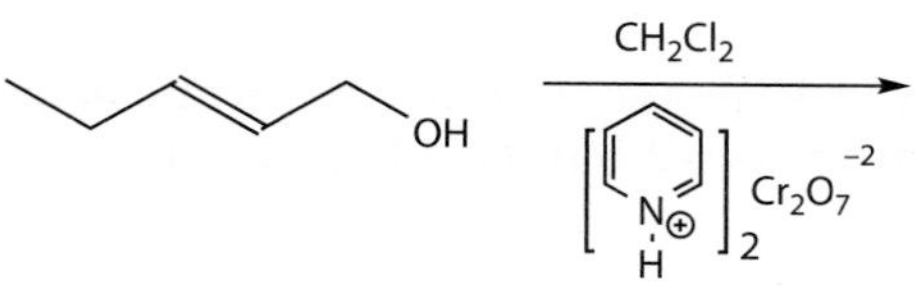

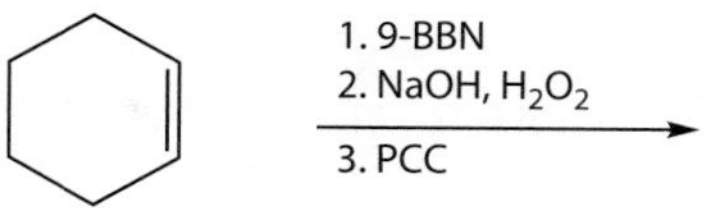

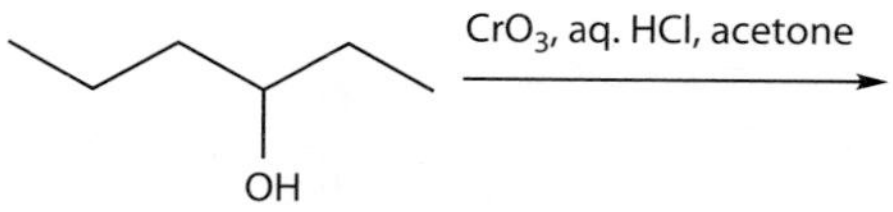

17.34 Give the major product for each of the following:
(a) 4,4-diphenyl-1-hexanol + CrO_3 and aqueous sulfuric acid in acetone
(b) cyclohexanemethanol + oxalyl chloride and DMSO at −60°C
(c) cycloheptanol + $(COCl)_2$/DMSO/−78°C

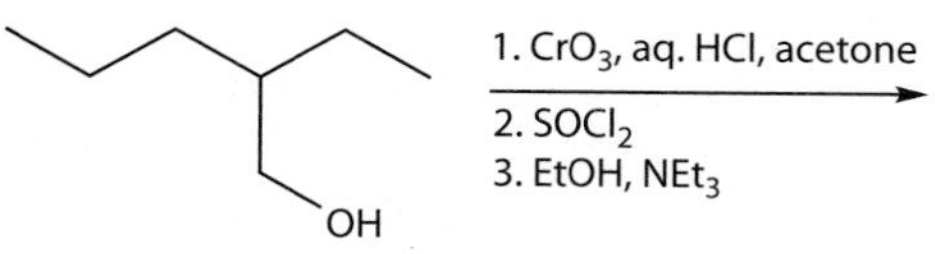

17.35 Lutidine is 2,6-dimethylpyridine. Draw the reagent that should be formed if chromium trioxide is mixed with lutidine in water with HCl, assuming that CrO_3 in water is concentrated and one equivalent of lutidine is used.

17.36 What product or products are expected if 1,5-cyclooctadiene is treated with ozone and then with hydrogen peroxide?

17.37 An alternative method for the preparation of an epoxide reacts a halohydrin (Chapter 10, Section 10.4.2) with a base such as sodium hydride. The resulting alkoxide undergoes an intramolecular Williamson ether synthesis. Draw the product expected when 2-bromo-1-hexanol reacts with NaH in THF. Based on the two carbons of the epoxide, is this transformation a net oxidation or a reduction?

17.38 Explain why 3,4-dimethyl-3-hexene undergoes epoxidation with peroxyacetic acid faster than does 3-hexene.

17.39 Give the major product for each of the following reactions:

(a)
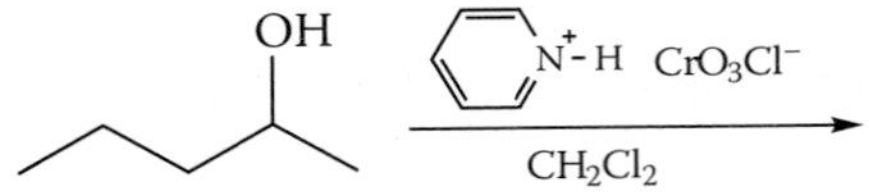
OH
CrO_3Cl^-
CH_2Cl_2

(d)
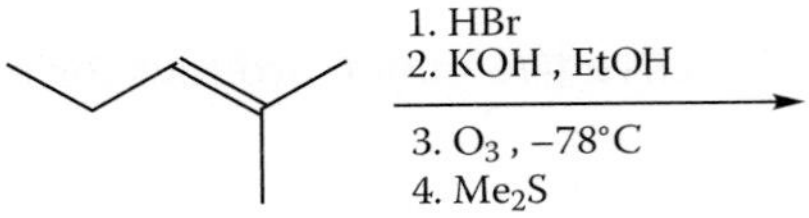
1. HBr
2. KOH , EtOH
3. O_3 , −78°C
4. Me_2S

(b)
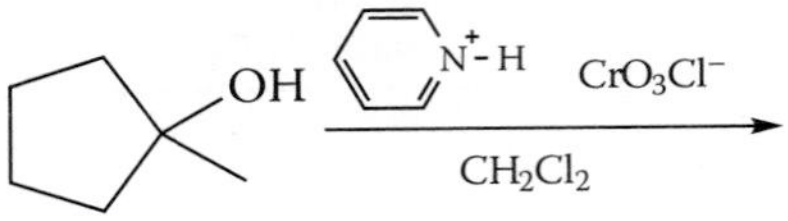
OH
CrO_3Cl^-
CH_2Cl_2

(e)
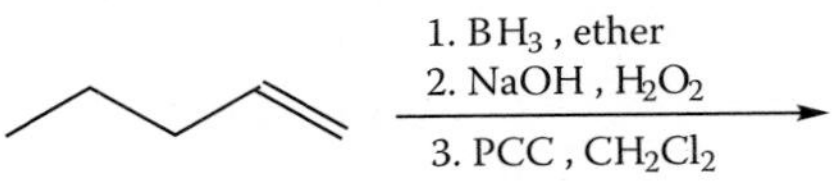
1. BH_3 , ether
2. NaOH , H_2O_2
3. PCC , CH_2Cl_2

(c)
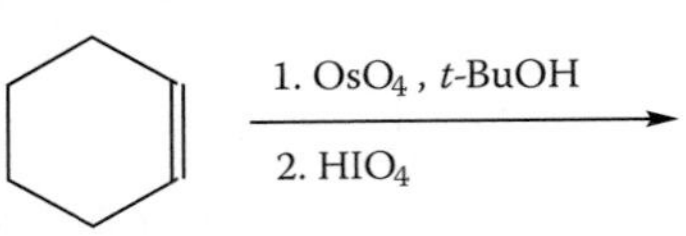
1. OsO_4 , *t*-BuOH
2. HIO_4

(f)
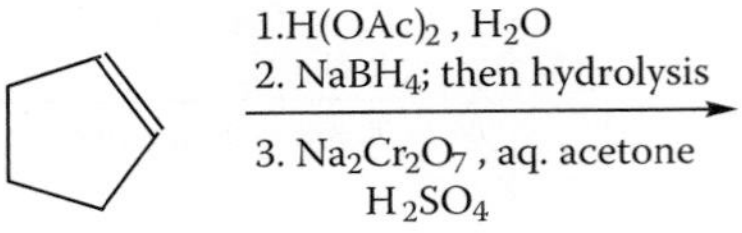
1.$H(OAc)_2$, H_2O
2. $NaBH_4$; then hydrolysis
3. $Na_2Cr_2O_7$, aq. acetone
H_2SO_4

17.40 Peroxytrifluoroacetic acid (CF_3CO_3H) reacts faster than peroxyformic acid (HCO_3H) in an epoxidation reaction with *trans*-2-butene. Offer a reason for why this is so.

17.41 Ozonolysis of a tetrasubstituted alkene such as 1,2-dimethylcyclopentene is faster than ozonolysis of a disubstituted alkene such as cyclopentene. Offer an explanation.

17.42 Epoxidation of 1-methylcyclopentene followed by treatment with aqueous acid (H^+) leads to a 1,2-diol. Suggest a mechanism for conversion of the epoxide to the diol.

17.43 Oxidation of citronellol with PCC can lead to a cyclized alcohol product due to an acid-catalyzed reaction. Once formed, the cyclized alcohol is further oxidized to a ketone, known as pulegone. PCC is rather acidic, which accounts for the reaction. Draw a mechanism that converts the expected oxidation product of citronellol and PCC to cyclized alcohol product and then draw the ketone derived from the cyclized alcohol.

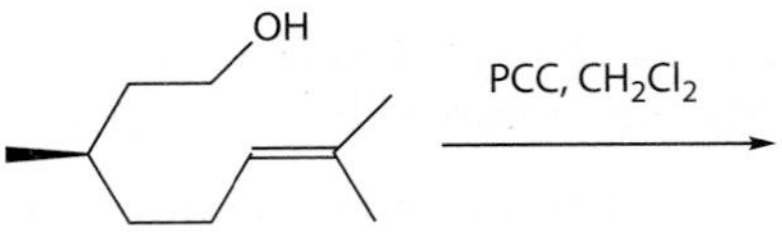
OH
PCC, CH_2Cl_2

17.44 In each case, give the major product of the reaction:

(a)
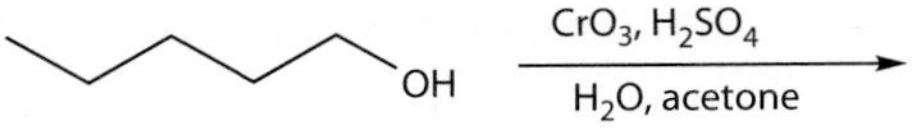
OH
CrO_3, H_2SO_4
H_2O, acetone

(d)
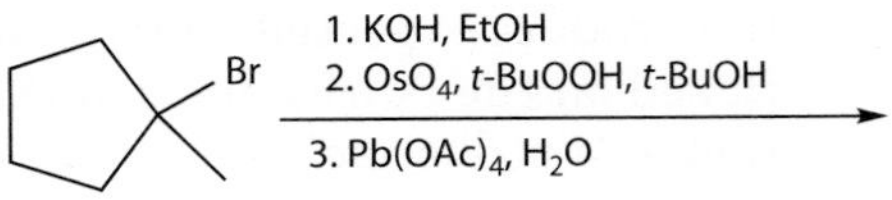
Br
1. KOH, EtOH
2. OsO_4, *t*-BuOOH, *t*-BuOH
3. $Pb(OAc)_4$, H_2O

(b)
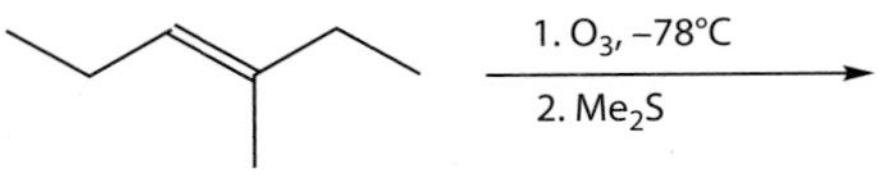
1. O_3, −78°C
2. Me_2S

(e)
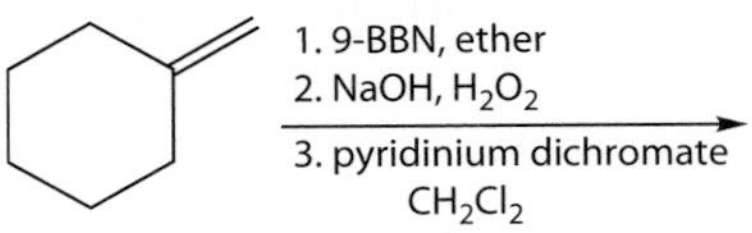
1. 9-BBN, ether
2. NaOH, H_2O_2
3. pyridinium dichromate
CH_2Cl_2

(c)
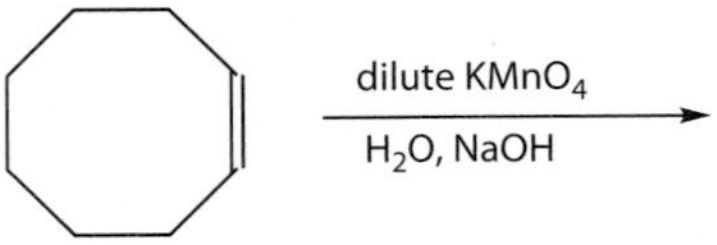
dilute $KMnO_4$
H_2O, NaOH

(f)
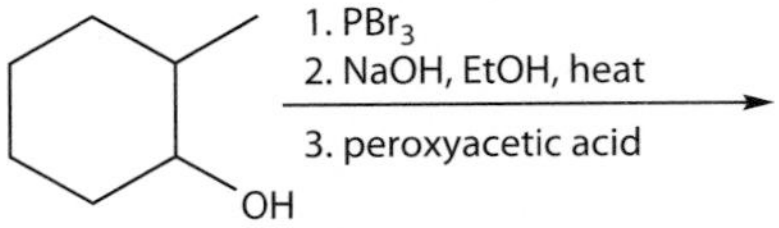
1. PBr_3
2. NaOH, EtOH, heat
3. peroxyacetic acid
OH

Synthesis Problems

Do not attempt these problems until you have read and understood Chapter 25.

17.45 Describe a route to prepare 2,6-heptanedione from 2-methylcyclopentanone.

17.46 Suggest a synthesis that converts (hydroxymethyl)cyclooctane to cyclooctanone.

17.47 Suggest a synthesis of 2-methylhexanal from 2-methyl-2-(trimethylammonium) hexane hydroxide.

17.48 For each of the following, provide a retrosynthetic analysis and also a complete synthesis, with all reagents and intermediate products:

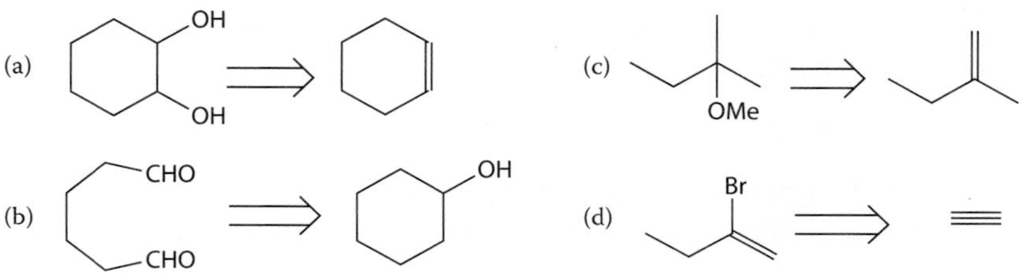

Spectroscopy Problems

Do not attempt these problems until you have read and understood Chapter 14.

17.49 How can one distinguish between cyclohexanone and 2-cyclohexenone using spectroscopy?

17.50 The reaction of 1-pentyne with mercuric sulfate and water, in the presence of mercuric acetate and phosphoric acid, gave a product with a strong signal at 1725 cm^{-1} in the IR and a singlet that integrates to three protons at 2.15 ppm in the proton NMR. What is the product? Offer an explanation for its formation.

17.51 The reaction of 3,4-diethyl-3-hexene and trifluoroperoxyacetic acid did not give the expected epoxide, but rather 4,4-diethyl-3-hexanone. Suggest a mechanism for formation of this product. (Hint: when the reaction was done under identical conditions—except that a large excess of sodium acetate was added—there was a small amount of ketone, but the major product was the epoxide.)

17.52 It is possible to oxidize hex-3-en-2-ol to the corresponding conjugated ketone using PDC. Draw this product and describe what differences exist in the IR and proton NMR between the starting alcohol and the ketone product. In addition, what spectroscopic details in the IR and/or NMR can be used to verify that the product is a conjugated ketone?

17.53   Describe differences in the IR and proton NMR that will allow you to distinguish 3-phenyl-1-pentanal and 3-phenyl-2-pentanone.

17.54   When 4-methyl-1-pentanol is treated with Jones's reagent, it is possible to form an aldehyde and/or a carboxylic acid. Describe differences in the IR and proton NMR that will allow you to distinguish these two products. If both are formed, suggest a simple experimental technique that will allow you to separate them.

17.55   The proton on the carbon bearing the O in dimethyl ether appears at about 3.3–3.5 ppm in the proton NMR. That proton in oxirane appears significantly upfield. Why?

17.56   Briefly explain why the aldehyde H (of the O=C–H unit) appears at 9 or 10 ppm in the proton NMR, but the hydrogen of an alkene (the C=C–H unit) appears at 4.5–5.5 ppm.

17.57   Describe differences between 2,3-butanediol and 1,4-butanediol that will allow one to distinguish them.

17.58   Given the following spectral data, provide a structure:
$C_9H_{18}O$
IR: 2967–2872, 2713, 1728, 1478–1469, 1395 cm^{-1}
^{1}H NMR: 9.74 (s, 1H), 2.63–1.88 (m, 3H), 1.24–1.16 (m, 2H), 1.02 (d, 3H), 0.92 (s, 9H) ppm

17.59   Given the following spectral data, provide a structure:
$C_6H_{14}O_2$
IR: 3368–3347, 2971, 2933–2908, 1660–1655, 1156–1121, 1077–1043 cm^{-1}
^{1}H NMR: 4.28 (broad s, 2H; this peak is diminished when treated with D_2O), 4.2 (m, 1H), 1.64–1.47 (d, 2H), 1.3–1.24 (s, 6H), 1.18 (d, 3H) ppm

17.60   Given the following spectral data, provide a structure:
$C_6H_{12}O$
IR: 3045, 2961–2863, 1483–1468, 1432, 1410, 1132, 1070, 955–917, 847–836 cm^{-1}
^{1}H NMR: 2.9 (m, 1H), 2.73 (m, 1H), 2.45 (m, 1H), 1.53 (m, 2H), 1.45–1.39 (m, 4H), 0.92 (t, 3H) ppm

17.61   Given the following spectral data, provide a structure:
$C_8H_{16}O_2$
IR: 2960–2673, 1706, 1467–1456, 1417, 1063–943 cm^{-1}
^{1}H NMR: 11.67 (broad s, 1H; this signal is diminished when washed with D_2O), 2.38 (m, 1H), 1.52–1.35 (m, 8H), 0.92 (broad t, 6H) ppm

17.62   Given the following spectral data, provide a structure:
C_5H_8O
IR: 2972–2817, 2735, 1692, 1567, 1461–1423 cm^{-1}
^{1}H NMR: 9.52 (s, 1H), 6.94 (m, 1H), 6.12 (m, 1H), 2.38 (m, 2H), 1.13 (t, 3H) ppm

17.63   Given the following spectral data, provide a structure:
$C_6H_{10}O_4$
IR: 3029–2644, 1698, 1464, 1423, 1575, 1317, 1084–1067 cm^{-1}
^{1}H NMR: 12.3 (broad s, 2H; this peak is diminished when treated with D_2O), 2.54 (m, 2H), 1.07 (broad d, 6H) ppm

Reactions of Aldehydes and Ketones

18

This chapter will elaborate on the chemical reactions of the carbonyl functional group introduced in Chapter 16, Section 16.3. As noted in that chapter, the carbonyl group is found in ketones and aldehydes. Section 16.3 introduced acyl addition, but only in the most general items. Reaction of a nucleophile with an aldehyde or ketone gives an alkoxide, and subsequent hydrolysis leads to an alcohol. This chapter will define differences in nucleophiles that lead to different acyl addition products. Other reactions of the carbonyl group, including extensions of the fundamental acyl addition chemistry, will also be discussed. Many new functional group exchange reactions are presented, along with new carbon–carbon bond-forming reactions.

To begin this chapter, you should know the following:

- **the structure and rules of nomenclature for aldehydes and ketones (Chapter 5, Section 5.9.2; Chapter 16, Section 16.2)**
- **the fundamental properties of a π-bond (Chapter 5, Section 5.1)**
- **the structure of a carbonyl and that of a carbonyl group is polarized (Chapter 5, Section 5.9)**
- **the concept of a nucleophile (Chapter 6, Section 6.7; Chapter 11, Sections 11.2 and 11.3)**
- **the concept of a leaving group (Chapter 11, Section 11.2.4)**

- **how to name bifunctional molecules containing a carbonyl group and an alkene or alkyne, or two carbonyl groups (Chapter 16, Section 16.2.3)**
- **the structure and reactivity of carbocations (carbenium ions) (Chapter 10, Section 10.2; Chapter 11, Section 11.4)**
- **the structure and reactivity of an oxocarbenium ion (Chapter 6, Section 6.4.4)**
- **that the carbonyl of an aldehyde or ketone reacts with protonic acids as Brønsted–Lowry bases to generate an oxocarbenium ion (Chapter 6, Section 6.4.4)**
- **rotamers in acyclic compounds and conformations of cyclic and cyclic compounds (Chapter 8, Sections 8.1, 8.4–8.7)**
- **mechanism (Chapter 7, Section 7.8)**
- **competing reactions (Chapter 7, Sections 7.7 and 7.10)**
- **transition states (Chapter 7, Sections 7.3 and 7.6)**
- **the carbonyl of an aldehyde or ketone reacts with Lewis acids as a Lewis base to generate an oxocarbenium ion (Chapter 6, Section 6.5)**
- **that nucleophiles react with aldehydes or ketones by acyl addition to generate an alkoxide product, which is the product of acyl addition (Chapter 16, Section 16.3)**
- **that alkoxides react with aqueous acid to give an alcohol (Chapter 6, Section 6.4.2)**
- **that a Grignard reagent is an organomagnesium compound formed by the reaction of an alkyl halide with magnesium metal (Chapter 15, Sections 15.1 and 15.2)**
- **that organolithium reagents are formed by the reaction of alkyl, aryl, or vinyl halides and lithium metal (Chapter 15, Section 15.5)**
- **that organolithium reagents react with alkyl halides to give a new organolithium reagent via metal–halogen exchange (Chapter 15, Section 15.5.1)**
- **that organolithium reagents are strong bases and good nucleophiles (Chapter 15, Section 15.5)**
- **be able to identify and assign *R/S* configuration to stereogenic centers (Chapter 9, Sections 9.1 and 9.3)**
- **diastereomers (Chapter 9, Section 9.5)**

This chapter will show the similarity of a carbonyl and an alkene in that both react with a protonic acid or a Lewis acid to give a resonance-stabilized oxocarbenium ion that subsequently reacts with a nucleophile. Ketones and aldehydes also react directly with nucleophiles by acyl addition to give an alkoxide product, which can be hydrolyzed to an alcohol.

When you have completed this chapter, you should understand the following points:

- Ketones and aldehydes react with nucleophiles to give substituted alkoxides by acyl addition to the carbonyl to give alcohols in a two-step process: (1) acyl addition and (2) hydrolysis. Nucleophilic acyl addition involves forming a new bond between the nucleophile and the acyl carbon, breaking the π-bond of the carbonyl with transfer of those electrons to the oxygen to give an alkoxide. Addition of an acid catalyst leads to an oxocarbenium ion that facilitates acyl addition.
- Grignard reagents and organolithium reagents react as carbanions with aldehydes and ketones to give alcohols in a two-step process: (1) acyl addition and (2) hydrolysis. The organometallic reagent reacts as a nucleophile with epoxides at the less substituted carbon atom.
- Water adds reversibly to aldehydes and ketones to give an unstable hydrate.
- Alcohols add reversibly to aldehydes and ketones to give a transient hemiacetal or hemiketal, which then reacts with more alcohol to give an acetal or a ketal. Acetal and ketal formation is reversible. An excess of alcohol drives the reaction toward the acetal or ketal, but an excess of water will convert the acetal or ketal back to the aldehyde or ketone. 1,2-Diols react with aldehydes and ketones to give 1,3-dioxolane derivatives and 1,3-diols react to give 1,3-dioxane derivatives.
- Thiols react with aldehydes and ketones to give dithioacetals or dithioketals. 1,2-Dithiols react with aldehydes and ketones to give 1,3-dithiolane derivatives and 1,3-dithiol reactions to give 1,3-dithiane derivatives.
- Primary amines react with aldehydes and ketones to give imines. Secondary amines react with ketones and sometimes with aldehydes to give enamines.
- Hydrazine reacts with aldehydes and ketones to give hydrazones. Hydrazone derivatives react with aldehydes and ketone to give N-substituted hydrazones. Hydroxylamine reacts with aldehydes and ketones to give oximes. Semicarbazone reacts with aldehydes and ketones to give semicarbazides.
- A molecule with a particular functional group can be prepared from molecules containing different functional groups by a series of chemical steps (reactions). This process is called synthesis: The new molecule is synthesized from the old one (see Chapter 25).
- Spectroscopy can be used to determine the structure of a particular molecule and can distinguish the structure and functionality of one molecule when compared with another (see Chapter 14).
- Acyl addition type reactions are very important in biological systems.

18.1 Chemical Reactivity of the Carbonyl Group

Chapter 16 introduced functional groups that contain the carbonyl group and two fundamental reactions associated with aldehydes and ketones: acid–base chemistry of the C=O unit and acyl addition of nucleophiles to the carbonyl carbon. Acyl addition is illustrated by the reaction of a generic carbonyl compound **1** with a Brønsted–Lowry acid (sulfuric acid). The π-bond donates two electrons as a base to the acidic proton to give the conjugate acid, oxocarbenium ion **2**, along with the conjugate base (the hydrogen sulfate anion). The protonated oxygen has a formal charge of +1 (see Chapter 5, Section 5.5, for formal charge), but it is only one resonance contributor to the oxocarbenium ion. In other words, an aldehyde or a ketone reacts as a Brønsted–Lowry base with a Brønsted–Lowry acid (sulfuric acid, HCl, nitric acid, p-toluenesulfonic acid) to form the corresponding oxocarbenium ion.

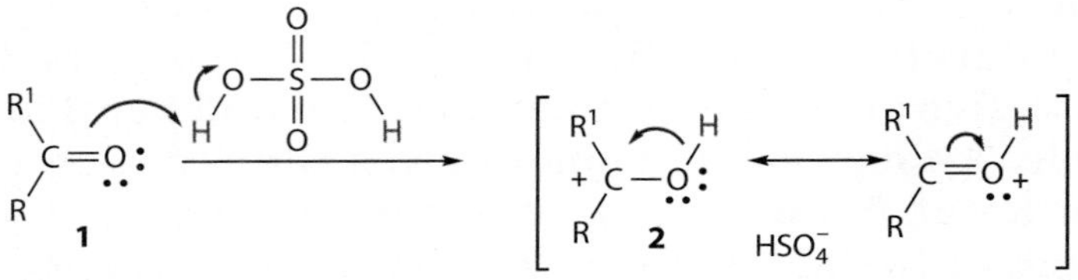

Lewis acids react with the carbonyl oxygen of aldehydes or ketones to form the corresponding ate complex, which is also an oxocarbenium ion. When the generic carbonyl compound **1** reacts with BF_3, charge-transfer complex **3** is the oxocarbenium ion product (see Chapter 16, Section 16.3). This ate complex is also stabilized by resonance, as shown in **3**.

18.1 Draw the oxocarbenium ion formed when cyclopentanone reacts with sulfuric acid.

18.2 Draw the oxocarbenium ion formed when benzaldehyde reacts with BF_3.

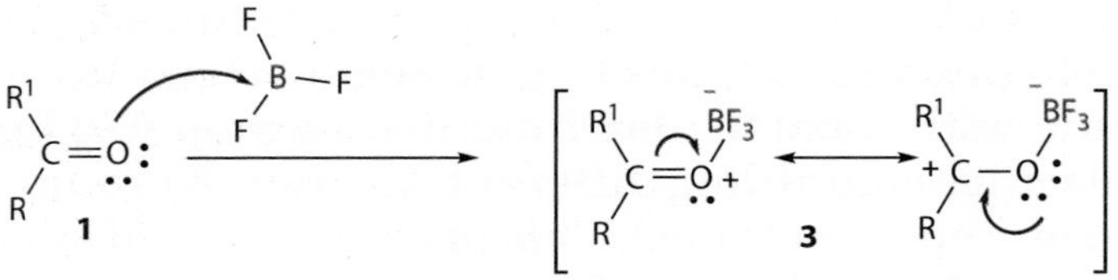

The polarization of the C=O group, where the oxygen is δ− and the carbon in δ+, is responsible for the acid–base reactions, and also for the reaction of aldehydes and ketones with nucleophiles. ***Nucleophilic acyl addition*** occurs when a nucleophile X donates electrons to the carbonyl carbon (the **acyl** carbon), breaking the π-bond. The product of this reaction with an aldehyde or ketones is an alkoxide, **4**. Several different nucleophiles react with aldehydes or ketones at the acyl carbon to generate an alkoxide product. Hydrolysis of the alkoxide leads to the isolated alcohol product, **5**.

18.3 Draw the products that result when 3-pentanone reacts with HC≡C:⁻ in THF, followed by reaction with dilute aqueous acid.

18.2 Reversible versus Irreversible Acyl Addition

The introduction to acyl addition in Chapter 16 avoided or minimized identification of the nucleophiles because nucleophiles of different strength react differently. Acyl addition is simply the reaction of a nucleophile with the carbonyl of an aldehyde or ketone. If nucleophiles of different electron-donating ability (differing nucleophilic strength) react differently, there should be a reasonable evaluation of the strength of different nucleophiles. One measure of nucleophilic strength is certainly whether or not a nucleophile reacts reversibly or irreversibly, at least in terms of product isolation. This discussion begins with the reaction of a weak nucleophile, the chloride ion, which adds to carbonyl compounds reversibly.

Acetone (2-propanone, **6**) is a simple and typical ketone, and mixing acetone with sodium chloride is an easy experiment. If chloride ion reacts with acetone via acyl addition, electron donation to the acyl carbon in the usual manner gives **7**. However, *when the reaction is examined for products, only acetone and chloride ion are detected, but no 7 is detected*. There are two possible reasons: (1) The chloride ion did not add to the carbonyl, or (2) acyl addition occurred but the reaction is reversible and loss of chloride ion from **7** regenerates the starting material. The fact that only starting material is observed indicates that if this equilibrium exists, it lies far to the left as the reaction is written (K is small). If **7** forms at all, the new C–Cl bond is polarized and relatively weak. Chloride is recognized as a leaving group, and the reverse reaction occurs (from right to left as written) by transfer of excess electron density on the oxygen atom in **7** toward the δ+ carbon to regenerate the C=O, with expulsion of chloride ion as a leaving group.

In other words, the alkoxide unit donates electrons to the δ+ carbon to form C=O with loss of chloride ion. This, of course, is the reverse of the initial acyl addition. Since the experimental results of this reaction show that *acetone and NaCl did not give the acyl addition product 7, chloride ion is classified as a weak nucleophile in this reaction*. Indeed, halide ions such as chloride, bromide, or iodide are weak nucleophiles and do not give an acyl addition product with aldehyde or ketones. *Note that the definition of a weak nucleophile is based on the fact that reaction with an aldehyde or a ketone does not give an isolable acyl addition product*.

The results with the weak nucleophile chloride ion and others that are not discussed here suggest that a generalization can be made: *Acyl addition should be reversible* **only** *if the newly formed bond in the product is connected to a good leaving group, and such nucleophiles are classified as weak nucleophiles*.

18.3 Reaction of Aldehydes or Ketones with Strong Nucleophiles

Halide ions are poor nucleophiles and do not give an acyl addition product. Other weak nucleophiles give a low yield of an acyl addition product because the reaction is reversible. If this equilibrium can be controlled, there is a chance the product may be isolated or it may undergo another reaction. An equilibrium is effectively controlled by application of *Le Chatelier's principle,* which states that changes in concentration, temperature, volume, or partial pressure in a chemical system at equilibrium will shift the equilibrium to counteract that change. This principle is named after Henry Louis Le Chatelier (France/Italy; 1850–1936). In acyl addition reactions, changing the concentration and temperature is the most common action.

Based on the discussion in the previous section, a practical definition of a strong nucleophile is one that reacts with an aldehyde or ketone to give a good yield of an acyl addition product, and the reaction should be largely irreversible. In other words, a nucleophile is strong if it reacts to give the acyl addition product in a facile manner. For strong nucleophiles, acyl addition will give **4** as the product. What constitutes a strong nucleophile? Common strong nucleophiles include carbon nucleophiles, such as the cyanide ion and alkyne anions, and organometallic reagents, such as Grignard reagents and organolithium reagents that were introduced in Chapter 15. The following sections discuss reactions that generate **4** irreversibly and are then hydrolyzed to **5**.

18.3.1 Cyanide: A Carbon Nucleophile

The cyanide ion is ubiquitous in the form of sodium cyanide (NaCN) or potassium cyanide (KCN). Acyl addition to an aldehyde or ketone will generate a species **4** (X = CN) in which a new carbon–carbon bond is formed. To understand the reaction of cyanide, it must be classified as a strong or a weak nucleophile. In effect, this requires that cyanide be classified as undergoing a reversible or an irreversible acyl addition.

Sodium cyanide is composed of two ions: the sodium cation and the cyanide anion. As seen in **8**, both carbon and nitrogen have an unshared pair of electrons. Because both atoms have excess electrons, each may react as a nucleophile. A nucleophile with two nucleophilic centers is called a ***bidentate nucleophile***; however, one atom is usually the dominant nucleophilic center. Referring to Chapter 13 (Section 13.3), nucleophilic strength for the reaction with alkyl halides increases to the left in the periodic table, and a charged species is generally more nucleophilic than a neutral species. The formal charge on the cyanide ion is –1 (see Chapter 5, Section 5.5), and it is calculated to be on the carbon. These criteria suggest that carbon is the better nucleophilic center and, in general, most reactions of cyanide occur at the carbon when NaCN or KCN is used.

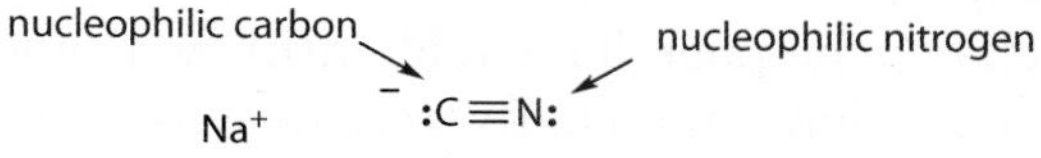

nucleophilic carbon   nucleophilic nitrogen

Na^+ :C≡N:

8

There are exceptions to this rule, however, particularly when the electrons on the carbon of cyanide are "tied up" in a covalent bond. Both silver cyanide (AgCN) and cuprous cyanide (CuCN) have bonds between the metal and carbon (Ag–C or Cu–C) that have significant covalent character. The Ag and Cu ions are not charge dense, and they prefer to coordinate to atoms that are also not dense in charge (the C end of cyanide). If the metal–carbon bond in M–CN is covalent, the electrons on carbon are *shared* and less available for donation, which makes carbon less nucleophilic. In both AgCN and CuCN, the nitrogen atom is more nucleophilic and reaction with an alkyl halide R–X leads to a molecule called an isocyanide (or isonitrile, R–$^+$N≡C$^-$). ***Isocyanides and the reaction of such compounds are not discussed in this book.***
Both sodium and potassium ions form more ionic bonds, and the NaCN and KCN bonds are essentially ionic. Sodium and potassium are charge-dense cations, and they prefer to coordinate to charge-dense anions (the N end of cyanide). The electrons on carbon are more available for donation in these ionic compounds, and carbon is the better nucleophile. ***To be certain that reaction occurs at the acyl carbon, all reactions in this book will use sodium cyanide (NaCN) or potassium cyanide (KCN) as the reagent.***

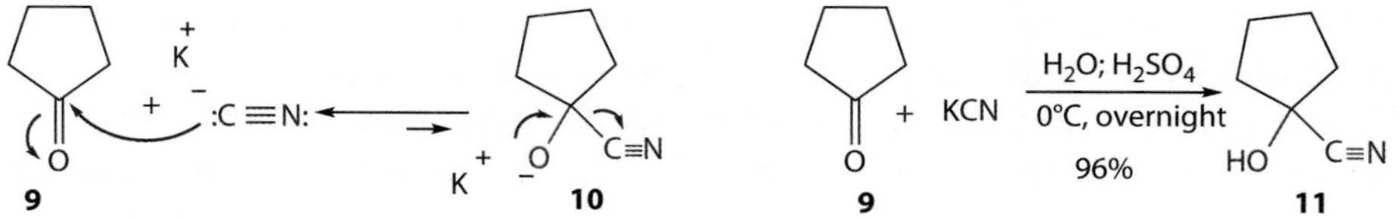

9   **10**   **9**   **11**

Is this analysis of reactivity with NaCN correct? In an experiment that reacts cyclopentanone (**9**) with potassium cyanide, a ***low yield*** of **10** is obtained, consistent with cyanide as a weak nucleophile. Compounds that have both a CN unit and an OH unit are called *cyanohydrins*. The direct acyl addition of cyanide to **9** gives a poor yield of **10**, suggesting that the reaction may be reversible or that cyanide is a relatively poor electron donor, or both. On the other hand,

10 is formed, so cyanide must have some ability to react as a nucleophile, which means that cyanide is a stronger nucleophile than chloride ion.

In a separate experiment, **9** is treated with KCN and sulfuric acid at 0°C and the product is 1-hydroxycyclopentanecarbonitrile, **11**, isolated in 96% yield.[1] The acid-mediated reaction clearly proceeds to product in good yield, so there is a difference in the direct nucleophilic reaction versus the same reaction in an acid medium. From Section 18.1, it is known that a carbonyl reacts with an acid to give an oxocarbenium ion (see **2** and **12**). It is quite reasonable that an oxocarbenium ion, once it is formed, will react with cyanide ion to give **11**. In other words, the positive carbon of oxocarbenium ion **12** reacts with the nucleophile to give **11**.

The explanation used for these two experiments suggests that the reaction of cyanide ion with **9** is reversible with an aldehyde or a ketone, so the yield of cyanohydrin is expected to be poor. The acid-catalyzed reaction, however, leads first to an oxocarbenium ion that easily reacts with cyanide. Therefore, the acid-catalyzed reaction gives a good yield of the cyanohydrin. This difference between direct substitution and acid-catalyzed addition is an important key to understanding and controlling reactions with nucleophiles. Note that when cyanide reacts with the carbonyl unit of **9**, a new carbon–carbon bond is formed as the carbonyl π-bond is broken.

Why is acyl addition of cyanide to an aldehyde or a ketone reversible? Once formed, the C–CN bond is weak enough that the alkoxide unit in **10** can donate electrons back to the $\delta+$ carbon atom to regenerate the C=O unit. As the C=O unit is formed, cyanide is lost as a leaving group, so this reaction is reversible. *Cyanide is a much poorer leaving group when compared to chloride ion, which contributes to making cyanide a better nucleophile*; some **10** is isolated as a product. The reaction of cyclopentanone with KCN and a sulfuric acid catalyst, however, leads to oxocarbenium ion **12**. This reactive intermediate then reacts with the nucleophilic cyanide to give cyanohydrin **11**. Cyanohydrins from both aldehydes and ketones are prepared in this way.

If NaCl is added to **9** in the presence of sulfuric acid, will the acyl addition product be formed? The answer is no! Halide ions are too weak a nucleophile in this particular reaction for the reaction to occur and, in many cases, if the product did form, it would be too unstable to be isolated. In other words, under acid conditions, a species such as CH–C(OH)Cl is too unstable for isolation. In summary, *cyanide ion is a moderately strong nucleophile and acyl addition is facilitated by the presence of an acid catalyst.*

18.4 Draw the product of the reaction between 2-butanone and KCN in the presence of an acid catalyst.

18.3.2 Alkyne Anions

In the acid-catalyzed reaction of cyanide, direct addition of cyanide to a carbonyl is reversible, leading to a poor yield of the acyl addition product in most cases. Addition of an acid catalyst will first form an oxocarbenium ion, shifting the equilibrium so that the weaker nucleophile will react. However, several nucleophiles are categorized as strong nucleophiles that add irreversibly, without the need for a strong acid. One such nucleophile is derived from alkynes. Weak nucleophiles will be discussed after the discussion of the strong nucleophiles.

In Chapter 11 (Section 11.3.6), the hydrogen atom of a terminal alkyne (marked in **red** in **13**) was shown to be weakly acidic, and it reacted with a suitable base to generate a conjugate base (the alkyne anion, **14**). Alkyne anions are classified as strong nucleophiles. A *terminal* alkyne (i.e., $-C \equiv C-H$) is a **carbon acid**. Carbon acids are very important because removal of the acidic proton with a suitably strong base gives a **carbanion** (a carbon nucleophile)—in this case, an alkyne anion: $-C \equiv C-H \rightarrow -C \equiv C:^-$. *Carbon acids give carbanion conjugate bases that are also* **carbon nucleophiles**. Alkyne anions are strong nucleophiles in acyl addition reactions, generating a carbon–carbon bond irreversibly.

$$Me-C\equiv C-H \xrightarrow{\text{Na:NH}_2} Me-C\equiv C:^- \ \overset{+}{Na} \ + \ H:NH_2$$
$$\quad\quad\quad\textbf{13} \quad\quad\quad\quad\quad\quad\quad\quad \textbf{14}$$

The pK_a of a typical alkyne such as 1-propyne (**13**) is about 25, which is approximately as acidic as that of an amine hydrogen. **A contributing factor to the acidity of the alkyne hydrogen is the fact that the carbon is sp hybridized** (see Chapter 5, Section 5.2.1). Because a terminal alkyne is a weak acid, removal of the proton in an acid–base reaction requires a very strong base (a base that generates a conjugate acid with a pK_a that is weaker than the alkyne pK_a). The reaction of sodium amide (Na:NH$_2$) and **13** in an acid–base reaction generates conjugate base **14**, which is known as an **alkyne anion**.

Commonly used bases for this reaction are sodium amide or the so-called amide bases LiNR$_2$ (lithium dialkyl amides; see Chapter 22, Section 22.3). For the time being, Na (or K or Li)$^+$ $^-$NH$_2$ is used, and a discussion of LiNR$_2$ bases is delayed until Chapter 22. A specialized base called **sodium hydride (NaH) is** sometimes used for reaction of alkynes because the conjugate acid is hydrogen gas. An example is the reaction of **15** with sodium hydride, which gives alkyne anion **16** and one-half a mole of hydrogen gas. As shown for the reaction of **15** an aprotic organic solvent like diethyl ether can be used. This is a particularly useful reagent because the by-product (H$_2$) escapes from the medium, driving the reaction to the right.

$$\text{Ph}-C\equiv C-H \xrightarrow[\text{ether}]{\text{NaH}} \text{Ph}-C\equiv C:^-Na^+ + \frac{1}{2}H_2$$
$$\quad\quad\quad \textbf{15} \quad\quad\quad\quad\quad\quad\quad\quad\quad \textbf{16}$$

An alkyne anion is a carbanion, and it is relatively strong base. A typical acid–base reaction is the reaction of alkyne anion **15** (the sodium anion of ethyne) with ethanol to give a conjugate base (ethoxide, **16**) and a conjugate acid (ethyne, also known as acetylene). The equilibrium for this acid–base reaction lies to the right as it is drawn, which is consistent with **17** being a stronger base than **18**. This means that acetylene (pK_a = 25) is a weaker acid than ethanol (pK_a = 17). *From a practical point of view, once the alkyne anion has been formed, contact with alcohol or water should be avoided since the equilibrium (17 ⇌ 18) can diminish the amount of alkyne anion via an acid–base reaction.* In order to use an alkyne anion as a nucleophile with aldehydes or ketones, the best results are obtained in an aprotic solvent such as ether or THF (see Chapter 13, Section 13.2).

18.5 Write out the acid–base reaction that results when acetylene reacts with NaH.

18.6 Draw the products formed when 16 reacts with water, with ethanol, and with propanoic acid.

18.7 Briefly discuss whether ethanethiol (EtSH) can be used as a solvent in reactions involving an alkyne anion.

How nucleophilic is an alkyne anion when compared to the cyanide ion? In one experiment, sodium acetylide (**17**) reacts with 2-butanone (**19**) to give **20** via acyl addition. An 86% yield of **21** is obtained after a second chemical step that treats **20** with aqueous acid (known as an acidic workup).[2] Alkoxide **17** is the initial product of the acyl addition to **19**, and the good yield of **21** suggests that **20** is formed in an essentially irreversible reaction. When **17** is compared with cyanide, the increased ability to undergo irreversible acyl addition to give **20** leads to the conclusion that **17** is a stronger nucleophile. Examination of **20** reveals that a strong C–C bond is formed by acyl addition, and the irreversibility of the reaction suggests that the alkyne unit is a rather poor leaving group. Because **20** is an alkoxide and the conjugate base of an alcohol, treatment with dilute aqueous acid gives the alcohol, **21**.

As first introduced in Chapter 10, Section 10.4.4, this two-step reaction may be condensed so that the starting material (**19**) is treated in one chemical step with **17** and then that product is treated with aqueous acid to give the final product, **21**. This symbolism assumes that the reaction of **17** and **19** in step 1 generates **20**, and the reaction in step 2 with aqueous acid leads to the final isolated product, **21**. Drawing only starting materials and products in this manner focuses the attention on the final product of the overall transformation, *but not the* *mechanism.* **Addition of an alkyne anion proceeds as a normal acyl addition, *and the only difference is the fact that 17 is a carbon nucleophile.***

18.8 Draw the reaction sequence and final product or products resulting from the reaction of 1-phenylethyne with (a) $NaNH_2$, (b) 1-phenyl-1-butanone, and (c) dilute aqueous acid.

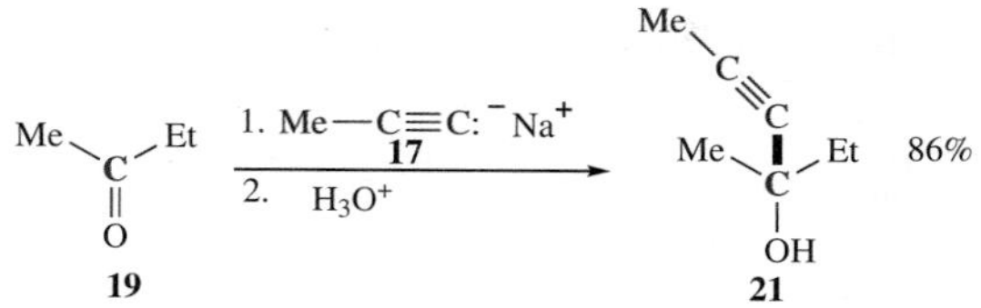

When 2-butanone (**19**) reacts with the alkyne anion of 1-propyne (**17**), acyl addition and the acid workup of the alkoxide product leads to 3-methylhex-4-yn-3-ol (**21**). Formation of **21** adds carbon atoms when compared to starting material **19**, and the complexity of the carbon structure is increased. Another aspect of this reaction must be considered. Alkyne alcohol **21** contains a stereogenic carbon (shown in **green** in the illustration), but the starting materials 2-butanone and propyne do not contain a stereogenic carbon. (Stereogenic centers were discussed in Chapter 9, Section 9.1.)

The acyl addition reaction created a stereogenic center, but the experiment shows that alkyne-alcohol 21 is racemic. Why? If **17** approaches **19** from the "top" (path a), the oxygen is pushed "down" and one enantiomer is formed (**S-21**). If **17** approaches from the "bottom" (path b), however, the oxygen is pushed to the top and the opposite enantiomer is formed (**R-21**). Because there is nothing to bias one side from the other as **17** approaches **19,** both enantiomers are formed in equal amounts. Therefore, a racemic mixture is formed (see Chapter 9, Section 9.2.4). *Nucleophilic acyl addition to aldehydes or ketones is assumed to generate racemic alcohol products in the absence of any additional information.*

18.9 Write the structures, with correct absolute configuration, of the reaction between 3-heptanone and 17.

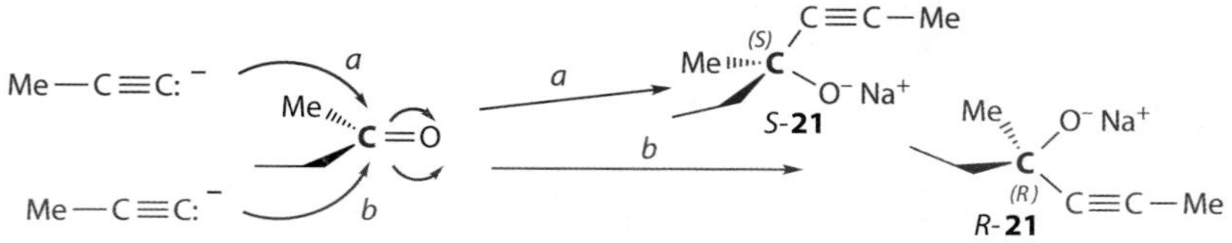

18.4 Organometallic Reagents Are Nucleophiles

In Chapter 15 (Section 15.1), the reaction of an alkyl halide and magnesium gave an alkylmagnesium halide (RMgX), commonly known as a Grignard reagent. At that time, it was pointed out that the carbon of the C–Mg–X unit is δ– and that it reacts as a base. The δ– carbon of a Grignard reagent is also a nucleophile in reactions with aldehydes or ketones. In one experiment, pentanal (**22**) reacted with butylmagnesium bromide in diethyl ether to give an alkoxide product (**23**) via acyl addition. Treatment of **23** with aqueous acid gives an 83% yield of 5-nonanol (**24**).[3] This experiment makes it clear that the **Grignard reagent is a carbanion nucleophile**, and the yield of **24** suggests the acyl addition to form **23** is essentially irreversible.

Grignard reagents are strong nucleophiles in acyl addition reactions. Acyl addition of the negatively polarized carbon to the acyl carbon of the carbonyl forms a new carbon–carbon bond in **24** (shown in **blue** in the illustration). The irreversibility of the acyl addition is explained by formation of a strong carbon–carbon bond in the product that is difficult to break. In other words, *the alkyl group is a very poor leaving group*. As in other acyl addition products, the reaction of the basic **23** with dilute aqueous acid gives the conjugate acid, alcohol **24**. *The reaction of Grignard reagents and aldehydes (or ketones) to give an alcohol product is a two-step process known as a **Grignard reaction**.* This reaction is represented with an arrow and the two steps, as described earlier. There are at least two different ways to represent the reaction, however. The first shows pentanal as the starting material, and it is treated with the Grignard reagent. The second representation starts with bromobutane, forms the Grignard reagent, and then adds the aldehyde. Both are correct. The choice depends on what is identified as the starting material.

18.10 **Draw the structure of the Grignard regent that is abbreviated BuMgBr in the reaction.**

18.11 **Write out all reactions necessary to convert 3-bromohexane to 4-ethyl-3-hexanol.**

Note that aldehyde **22** reacts with the Grignard reagent to produce a secondary alcohol. Ketones react with Grignard reagents to generate tertiary alcohols via the Grignard reaction. An example using a ketone starting material is the reaction of cyclohexanone (**25**) with butylmagnesium bromide to give 1-methyl-1-cyclohexanol (**26**), after an aqueous acid workup, in 90% yield.[4] This acyl addition reaction is general for both aldehydes and ketones with a wide range of Grignard reagents.

18.12 Write out the alkoxide product of the first step in the reaction that converts 25 to 26.

18.13 Write out all reactions necessary to convert 1-iodo-3,3-dimethyl-pentane to 3-(1-methylethyl)-6,6-dimethyl-2-phenyl-3-octanol.

Acyl addition reactions of Grignard reagents with aldehydes or ketones have no intermediate, so the transition state must be examined to understand how the reaction works. The negatively polarized carbon of a Grignard reagent (see methylmagnesium bromide **27**) is attracted to the positive carbonyl of the carbonyl (in acetone, **6**), and the electrophilic magnesium is attracted to the electronegative oxygen of the carbonyl, which leads to a ***four-center transition*** state represented by **28**. This transition state leads to the product that contains a new carbon–carbon bond in **29**. The positively charged MgBr is the counter-ion to the negatively charged alkoxide unit (an ionic bond). *Note that, although this is an overly simplistic mechanistic picture, it predicts the products of Grignard reactions*. The actual mechanism may involve several steps that ultimately lead to the alkoxide.[5] ***In this book, a simplistic four-center transition state model with no intermediate is used in all cases to represent the reaction of Grignard reagents with aldehydes or ketones***.

18.14 Draw the transition state and final product formed when methylmagnesium bromide reacts with cyclobutanone.

Grignard reagents are formed by the reaction of alkyl halides and magnesium metal. Chapter 15, Section 15.5, discussed organolithium reagents RLi, which are formed from an alkyl halide and lithium metal. Organolithium reagents are potent bases, but they are not strong nucleophiles with simple alkyl halides. As with Grignard reagents, ***organolithium reagents are good nucleophiles in reactions with aldehydes and ketones***. The reaction of an organolithium reagent such as *n*-butyllithium (**30**) with a ketone such as of cyclohexanone, for example, gives alcohol **32** in 89% yield after an aqueous acid workup.[6] As with Grignard reagents, acyl addition of an organolithium reagent is irreversible to give **33**. The usual reaction with aqueous acid in a second step is required to generate the alcohol product (1-butylcyclohexanol, **32**).

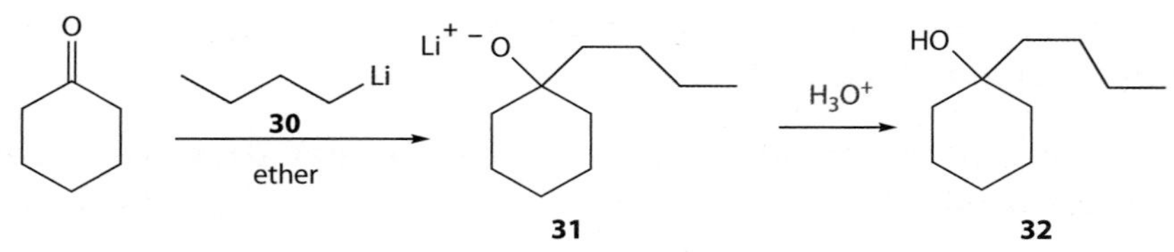

In a different example, cyclopentanone (**9**) reacts with *n*-butyllithium (**30**) to give only 44% of 1-butylcyclopentanol,[7] which is somewhat low. For the most part, acyl addition with organolithium reagents gives a good yield of the alcohol product. Assume that all organolithium reagents are strong nucleophiles and react via acyl addition to give an alcohol product after an aqueous acid workup. This reaction is identical to that described for Grignard reagents and other carbon nucleophiles.

The reaction of epoxides and nucleophiles was introduced in Chapter 11 (Section 11.8.2). Grignard reagents and organolithium reagents react with epoxides at the less sterically hindered carbon atom. The reaction of phenylmagnesium bromide and epoxide **33**, for example, generates alkoxide **34**. An aqueous acid workup converts the alkoxide to the conjugate acid and final product, alcohol **35**.

18.15 Write out the structures of both of these alcohol products.

18.16 Write out the reaction sequence that converts 2-iodobutane to 2-benzyl-3-methyl-2-pentanol.

18.17 Write out the final product formed when the epoxide derived from methylcyclopentene reacts with (a) ethyllithium and (b) aqueous acid workup.

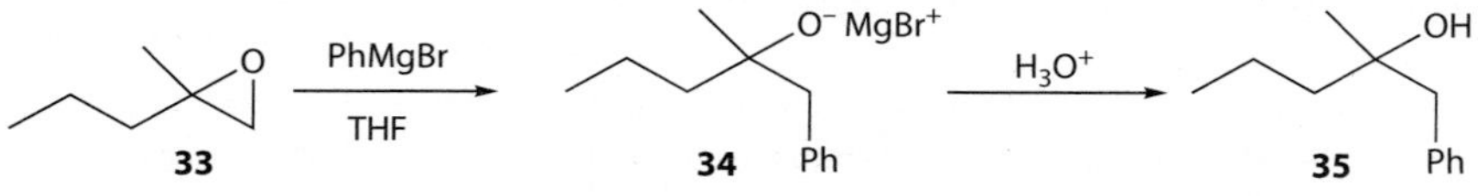

18.5 Water: A Weak Nucleophile That Gives Reversible Acyl Addition

In Section 18.2, the chloride ion was shown to be a weak nucleophile, and there is no acyl addition product at all. In Section 18.3.1, the cyanide anion was classified as a weak nucleophile because the non-acid-catalyzed acyl addition is reversible and the yield of product is low. With an acid catalyst, formation of an oxocarbenium ion allowed the acyl addition product to be isolated in good yield. Other weak nucleophiles give reversible acyl addition, and it is reasonable to ask if those equilibria may be shifted to give an acyl addition product via formation of an oxocarbenium ion. The answer is yes. This statement is demonstrated by the reaction of very weak nucleophiles in acyl addition, water.

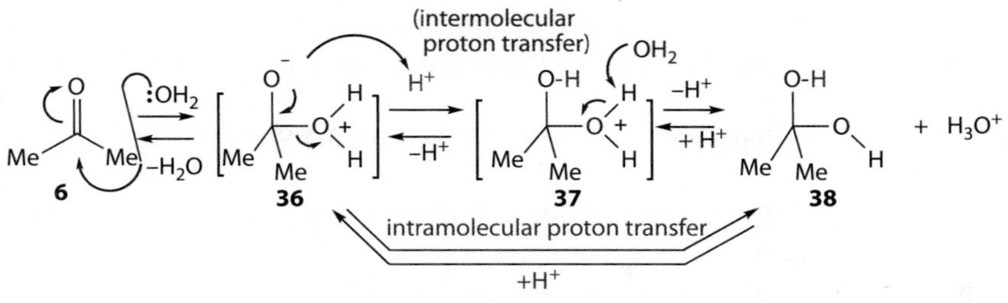

Figure 18.1 Mechanism of carbonyl hydration.

The reaction of water with ketones or aldehydes is a reversible process. Only the ketone or aldehyde starting material is isolated when a ketone or an aldehyde is mixed with water, and no new products are isolated. This result suggests that there is no reaction. Is this correct? The electron-donating oxygen atom of water may react as a nucleophile in reactions with ketones and aldehydes. The product of acyl addition of water to a simple ketone such as acetone (**6**) is oxonium ion **36** (the $^{+}OH_2$ unit). However, the OH_2 unit (water) in **36** is an excellent leaving group and it is easily lost to regenerate acetone and water, as shown in Figure 18.1.

There are several key points. The fact that **6** is isolated from this reaction and there is no product clearly indicates that if addition of water occurs, **36** is unstable and the equilibrium lies far to the left as the reaction is written. Remember the hydronium ion (H_3O^+) that is formed when an acid (H^+) reacts with water (see Chapter 2, Section 2.1). This means that the hydronium ion is the conjugate acid of water when water reacts with an acid, H^+. In **36**, the stable molecule water is released as **6** is regenerated, so *the water unit in **36** is a good leaving group*. In other words, the reaction of water gives a product (**36**) that is unstable and water is considered to be a weak nucleophile. *In general, a neutral nucleophile such as water will add to a carbonyl to give a charged species; the reaction is reversible for most ketones and aldehydes, but not all.*

There is, at best, a small concentration of **36** in the reaction. The hydronium ion (H_3O^+) is a strong acid and, similarly, the protons on the positively charged oxygen in **36** are quite acidic. The alkoxide unit RO^- is a good Brønsted–Lowry base, and *if* it reacts with the acidic proton of a second molecule of oxonium ion **36**, a new product (**37**) would be formed. Because **37** is also an oxonium ion, the proton of that ion is acidic and is removed by even a weak base such as the water present in the medium. Water is amphoteric (see Chapter 6, Section 6.2.5) and reacts as a base with **37** to give **38** and the hydronium ion.

The reaction of **37** with water is a simple acid–base reaction in which water is the base and **37** is the acid, forming **38** as the conjugate base and the hydronium ion as the conjugate acid. Product **38** is called a *hydrate* because it results from the addition of the elements of *water* to the carbonyl. The conversion of **6** to **38** also occurs in a one-step process by simply transferring a proton from the OH_2 unit of **36** to the alkoxide (RO^-) unit in the same molecule (intramolecularly). The pathway described from acetone to the hydrate in Figure 18.1 represents three separate chemical reactions and every step is reversible. *The* *structures in brackets are reactive intermediates* (**36** and **37**) *and the overall sequence shown in Figure 18.1 is the **mechanism of carbonyl hydration**.*

18.18 Draw the mechanism for the hydration reaction of propanal.

The preceding paragraph describes a reaction sequence that converts acetone to a product, **38**. However, this discussion began with the observation that when acetone is mixed with water, **38** is not isolated; rather, acetone is recovered. Indeed, hydrates are not isolated because *hydrates are inherently unstable products*. This fact is not obvious, but it has been verified experimentally. Hydrate **38** is unstable because it loses a molecule of water and regenerates acetone by an interesting mechanism. The α-carbon in **38** is marked in **violet** and it is connected to the carbon (marked in **red**) that bears both OH units. This α-carbon (the carbon atom attached to the carbonyl carbon) is polarized such that the carbon is δ– and the hydrogen is δ+. This means that the hydrogen atom is a *weak* acid.

Attraction between this acidic hydrogen and an oxygen of the hydrate (see **38**) leads to proton transfer to oxygen and loss of water from the hydrate to form a new species (**39**), which is called an **enol**. *An enol is formally defined as a molecule with an OH group attached directly to a C=C unit*. As first observed in Chapter 10 (Section 10.4.5), where enols are formed by oxymercuration of alkynes, enols are very unstable and exist in an equilibrium with the corresponding carbonyl compound. *Experiments show that this equilibrium between the enol form and the carbonyl form lies toward the carbonyl*.

The hydrogen atom of the O–H group in enol **39** (also acidic) is attacked by the π-bond of the C=C unit; the hydrogen atom is transferred to the carbon with cleavage of the O–H bond, as shown in **39**, to form the ketone (**6**). This reaction interconverts an enol and a ketone (the keto form in the equilibrium); however, the equilibrium strongly favors the keto form, and this process is called **keto–enol tautomerism**. The enol is said to tautomerize to the ketone. The carbonyl form is favored over the enol unless there is some special structural feature such as the presence of a second electron-withdrawing group on an α-carbon. **Therefore**, *if an enol is formed in a chemical reaction, assume that it will tautomerize to the carbonyl form, which is the isolated product*.

Hydrates cannot be isolated from the ketones and aldehydes seen most often in this book, but in a few special cases a hydrate is isolated. If the carbon attached to the carbonyl carbon (the α-carbon) has strong electron-withdrawing groups on it—and, particularly, if no hydrogens are on the α-carbon adjacent to the carbon bearing the two OH units—the hydrate may be isolated. Chloral (**40**) is a common name for trichloroethanal (trichloroacetaldehyde is also used as a common name) and it reacts with water to form a stable hydrate, **41**.

Hydrate **41** is stable because there are no α-hydrogen atoms to lose, and the chlorine atoms withdraw electrons from the carbon, which in turn withdraws electrons from the C–O units. Hydrate **41** is known as chloral hydrate—a white, crystalline solid that was first prepared in 1832. It is a depressant used to induce sleep. This compound is more potent when mixed with alcohol and is a very powerful depressant that has been used as a knockout potion or "Mickey Finn." From a structural point of view, it is important to remember that hydrate **41** has no protons on the α-carbon, so enol formation is not possible.

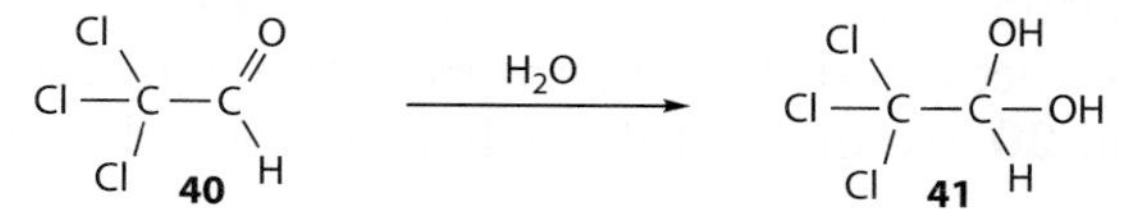

18.19 Draw the structure of the enol form of pentanal.

18.20 Draw the hydrate expected from **2,2,2-triphenylethanal** and speculate on its stability.

18.6 Alcohols: Neutral Nucleophiles That Give Reactive Products

In Section 18.5, water was a weak nucleophile that reacts with aldehydes or ketones to generate hydrates; however, they are unstable and lose water to regenerate the ketone or aldehyde via an enol. Therefore, even if a reaction is devised that will overcome the weak nucleophilic strength of water and force the reaction, the product is unstable. An alcohol is ROH and, from a simple structural point of view, one H of HOH has been "replaced" by an alkyl group. Chemically, this will cause some differences, but there should be many similarities. The oxygen atom of an alcohol is a nucleophile when it reacts with carbonyls, and there is an obvious structural relationship to water. The nucleophilic strength of an alcohol, the reversibility of acyl addition, and the stability of the expected product lead to differences with the water reaction.

18.6.1 Addition of Alcohols to Carbonyls

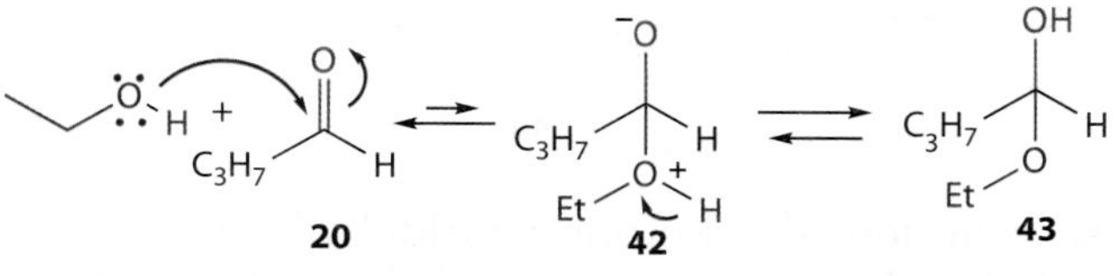

The reactivity of alcohols and carbonyl compounds will be examined using ethanol as the nucleophile. Initial reaction of ethanol with butanal (**20**) leads to the acyl addition product—oxonium ion **42**—analogous to the reaction of water

products, the mechanism requires protonation of the OH unit. Oxonium ion **48** now has the leaving group water, and loss of water gives a new oxocarbenium ion: **49**. A second molecule of ethanol reacts with the positively charged carbon atom in **49** to give oxonium ion **50**, and simple loss of a proton in a final acid–base reaction gives the product acetal **46**.

Neutral intermediate compound **43** has a HO–C–OEt unit and is called a *hemiacetal* (the term "hemi" means "half," so a hemiacetal is half of an acetal). *A hemiacetal is defined as a compound derived from an aldehyde with an OH and an OR group on the same carbon.* The final product **46** has an EtO–C–OEt unit and is called an *acetal. An acetal is defined as a compound derived from an aldehyde, where one carbon is attached to two OR units (a geminal attachment).* Because two molar equivalents of ethanol are required to react with **20** for each mole of **46** formed, this sequence is reasonable as long as there is an excess of ethanol (remember that ethanol is the solvent).

18.23 Draw the hemiacetal and also the acetal formed when phenylethanal reacts with cyclopentanol in the presence of an acid catalyst.

18.24 Draw the hemiacetal and acetal formed when cyclopentane carboxaldehyde reacts with methanol and an acid catalyst.

The sequence shown for the conversion of **20** (butanal) to **46** (the acetal) is the *mechanism of acetal formation and may be applied to any aldehyde or ketone starting material. Note that most of the reactions in the mechanism are acid–base reactions.* The hemiacetal (**43**) is a neutral product, but it reacts with the acid catalyst and with ethanol, so it is unstable to the reaction conditions. This instability makes **43** an intermediate, although it is not charged. There are a total of six intermediates and seven reversible steps. This means that seven steps are required to convert **20** to **46**. Compare that to the acyl addition reactions of strong nucleophiles in Sections 18.3 and 18.4. *The mechanism for this transformation is the logical step-by-step description of a reaction that follows each bond being made or broken and each atom or group being gained or lost. Mechanisms are introduced in Chapter 7 (Section 7.8).* Implicit in the definition of a mechanism is the concept of "arrow pushing," where arrows in a mechanism indicate the flow of electrons from an atom of high electron density (usually indicated in **blue**) to an atom of low electron density.

18.25 Write out the complete mechanism for the reaction of cyclopentane carboxaldehyde with methanol in the presence of an acid catalyst.

In the acid-catalyzed reaction just discussed, how much catalyst is used? The answer depends on the alcohol, the carbonyl compound, the acid, and the concentration of the reactants. Typically, a catalytic amount is 1–10% relative to the starting material used, and only a catalytic amount is required because H^+ is regenerated during the course of the reaction.

What acids are used as catalysts in this reaction? Sulfuric acid may be used, but it is a strong oxidizing acid and this can cause problems. Recall what happens when sulfuric acid is accidentally dripped on a paper towel! The paper chars as it reacts. In other words, the paper decomposes or is transformed into something else. The same thing may happen with the starting materials. Acids such as HCl can be used, but solubility in an organic medium is often a problem. For this reason, an organic acid such as a sulfonic acid (Chapter 16, Section 16.9) is commonly used as a catalyst.

Such an acid is 4-methylbenzenesulfonic acid (**51**; commonly known as *p*-toluenesulfonic acid, **tosic acid**, or *p*-TsOH; this benzene nomenclature system is discussed in Chapter 21, Section 21.2). Another commercially available sulfonic acid is methanesulfonic acid (**52**), commonly abbreviated as MsOH, where Ms = mesyl. Tosic acid is a crystalline solid that is cheap, readily available, and soluble in most organic solvents. Throughout most of this book, the acid catalyst is simply referred to as H^+, but the actual acid is usually sulfuric acid or tosic acid.

18.26 Draw the structure of 4-methylpentanesulfonic acid.

18.6.3 If Acetal Formation Is Reversible, How Is an Acetal Isolated?

In the conversion of **20** to **46**, every step is reversible; this means that all products are in equilibrium. To isolate the acetal product, the equilibrium must be shifted to the right (toward the acetal product) by adjusting certain parameters of the reaction. Remember Le Chatelier's principle (Section 18.3), in which an equilibrium will adjust if one of several parameters is adjusted. These parameters include concentration. Therefore, an increase in the concentration of one reactant will shift the equilibrium toward the products. Ethanol is one of the reactants, and using a large excess will shift the equilibrium toward the acetal. For this reason, the alcohol reactant (in this case, ethanol) is often used as the solvent (if it has a relatively low boiling point). If the alcohol cannot be used as a solvent, it is used in excess relative to the aldehyde.

Another way to shift the equilibrium to the right without adding a large excess of one reactant is to remove a product from the reaction medium. Based on Le Chatelier's principle, removing one product will change the concentration term such that the equilibrium will shift to favor the remaining product. In other words, removing the acetal product (by distillation, for example, if it has a low boiling point) will force the system to equilibrate toward the product and "drive" the reaction toward the acetal. If the boiling point of the acetal is too high to be distilled at reasonable temperatures, other techniques must be used to remove a product.

The mechanism for the conversion of **20** to **46** involves *loss of a water molecule* from **48** to give **49**. If water is removed from the reaction, the equilibrium shifts to the right. It is often impractical to distill water from the reaction, but a drying agent may be added to the reaction to "soak up" the water and remove it from the reaction medium. The drying agent must *not* react with the aldehyde, the alcohol, or the acid catalyst, however. Common drying agents include molecular sieves of 3 or 4 Å (highly porous zeolites that trap water when thermally activated), magnesium sulfate ($MgSO_4$, a mild Lewis acid that is not very useful in this type of reaction), or calcium chloride ($CaCl_2$, a relatively inefficient drying agent that also reacts with many alcohols).

An alternative method for removing water relies on the use of a solvent that forms an *azeotrope* with water. **An azeotrope is defined as a constant boiling *mixture* that often boils at a temperature different from its components**. To understand this definition, a **zeotrope** is defined as a mixture that can be *separated* by distillation.[8] An azeotrope, therefore, is a mixture that *cannot be separated* by distillation. Ethanol, for example, forms an azeotrope with water that boils at 78.2°C, lower than the boiling point of ethanol (78.5°C) or water (100°C).[9] This azeotrope is composed of 95.6% ethanol and 4.4% water.

The problem at hand is how to remove water. Because ethanol and water form an azeotrope, it is possible to remove water from a reaction in this manner, but there are problems. Ethanol and water are miscible (mutually soluble in one another), so the water and ethanol do not separate (form layers) once the mixture has been distilled. If a solvent other than ethanol is used to form an azeotrope water and water is insoluble in that solvent, water would separate after the mixture had been distilled. One such solvent is benzene, which forms a binary (two-component) azeotrope with water (boiling point of benzene is 80.1°C) that boils at 69.4°C and is composed of a 91.1:8.9 benzene:water mixture. (*Caution: benzene is toxic and a suspected carcinogen;* see Chapter 21.) Refluxing the aldehyde or ketone with at least two equivalents of ethanol in benzene will drive the reaction toward the acetal (**46**) and the benzene–water azeotrope can be distilled off.

However, there is a problem. Water and ethanol form an azeotrope, as mentioned before, but benzene also forms a binary azeotrope with ethanol. The benzene–ethanol binary azeotrope boils at 67.8°C and consists of 67.6% benzene and 32.4% ethanol.[10] This means that as the benzene–water azeotrope is distilled, ethanol is also removed. An excess of ethanol is therefore essential.

It turns out that the problem is even more complicated because benzene forms a ternary azeotrope (three components) with ethanol and water that boils at 64.6°C (the lowest temperature yet). This azeotrope consists of 74.1% benzene, 18.5% ethanol, and 7.4% water.[11] All of this means that the reaction is done using benzene as the solvent, with an excess of ethanol; therefore, the ternary azeotrope is distilled and the distillate contains both water and ethanol, and the equilibrium will shift to the right so that the final isolated product is acetal **46**.

18.27 Briefly discuss whether generating the dimethyl acetal of acetaldehyde is feasible using benzene as a solvent.

18.6.4 Ketals

Ketones have been excluded from the discussion so far; although only aldehydes have been shown in the reaction with alcohols, ketones also react with alcohols and an acid catalyst to give similar products that are known as ketals. The term "ketal" was abandoned for some time, and $(RO)_2CR_2$ derivatives from both aldehydes and ketones were known as acetals. **The term *ketal is now recognized as a subclass of acetals*.** For convenience, this book will use "ketal" so that reference to an acetal or a ketal will keep the focus on the aldehyde or ketone precursor.

The reaction mechanism is the same for both aldehydes and ketones. If acetone (**6**) reacts with ethanol in the presence of a catalytic amount of *p*-toluenesulfonic acid, **53** is formed but not isolated. This compound is formally analogous to the hemiacetal derived from an aldehyde, but it is derived from a ketone. Therefore, it is called a ***hemiketal***. The isolated product is **54** (2,2-diethoxypropane)—again, analogous to the acetal; however, the starting material is a ketone rather than an aldehyde, so the final product is called a ***ketal***. ***A ketal is a compound derived from a ketone that contains two OR groups connected to the same carbon.*** The mechanism for ketal formation from ketones is identical to that for acetal formation, and every step in this sequence is reversible. The same methods used to remove water from the reaction that was used to drive the equilibrium toward the acetal in that reaction (see Section 18.6.3) can be used here for conversion of the ketone to the ketal.

Note that the name of **54** is 2,2-dimethoxypropane. Acetals and ketals are named by choosing the longest chain and treating the OR units as alkoxy substituents. Using this system, acetal **46** is named 1,1-diethoxybutane.

18.28 Draw the hemiketal formed when cyclohexanone reacts with methanol and an acid catalyst.

18.29 Draw the structure of the product when 3,5-diphenylcycloheptanone reacts with ethanol and a catalytic amount of tosic acid.

18.30 Write out the complete mechanism for the conversion of **6** to **54** and also for the conversion of cyclohexanone to 1,1-dimethoxycyclohexane, using methanol and an acid catalyst.

18.6.5 An Acetal or a Ketal May Be Converted Back to the Aldehyde or Ketone

Close examination of an acetal or ketal shows that the C=O unit is converted to a C(OR)$_2$ unit. Another way to think about this is to recognize that an acetal or ketal does not undergo acyl addition and indeed is the product of acyl addition. At certain times in a synthesis (see Chapter 25), it is useful to covert a reactive group such as the carbonyl to another unit that does not undergo the reactions of a carbonyl, but may be converted back to a carbonyl later in the synthesis. This step is known as ***protecting the carbonyl*** and an acetal or ketal is a common ***protecting group*** for an aldehyde or a ketone, respectively. Therefore, the ability to convert an acetal or a ketal back to the original aldehyde or ketone is an important transformation.

How is an acetal converted to an aldehyde or a ketal converted to a ketone? The fact that every step in the conversion of an aldehyde to an acetal or a ketone to a ketal, under acid catalysis, is reversible may be exploited by reversing the equilibrium toward the carbonyl derivative. Le Chatelier's principle can be used once again. This time, changing the concentration term requires using an excess of the acetal or ketal product, but this is usually impractical. Remember that water is also a product, so increasing the concentration of water rather than the alcohol should shift the equilibrium back toward the aldehyde or ketone starting material. If removal of water is required to drive the equilibrium toward the acetal or ketal, adding an excess of water should drive the equilibrium in the other direction. ***In order to reverse the reaction that makes the acetal or ketal, use water as a solvent rather than the alcohol***.

In the formation of acetal **20** with an acid catalyst, **48** is a key intermediate that loses a molecule of water to give the resonance-stabilized oxocarbenium ion **49**. If there is an excess of water, as when the alcohol solvent is replaced with water, this oxocarbenium ion will react with the nucleophile that is most available: water. The equilibrium will shift to favor formation of the aldehyde or the ketone rather than the acetal or ketal.

This transformation is illustrated by the reaction of a ketal such as 1,1-dimethoxycyclopentane (**55**) with an acid catalyst and a large excess of water to give cyclopentanone (**9**), along with two equivalents of methanol. An acetal reacts in an identical manner to give the corresponding aldehyde and two equivalents of the alcohol. Removal of methanol from this reaction by distillation drives the reaction to the right with formation of **9**.

18.31 Draw the reaction that prepares **55** from the appropriate reactants.

18.32 Write out the mechanism for the conversion of 3,3-diethoxy-pentane to 3-pentanone.

The key to the reaction of alcohols with aldehydes or ketones is to understand the equilibrium. If it is necessary to generate an equilibrium that favors the acetal or ketal, add a large excess of alcohol and remove the water. If it is necessary to generate an equilibrium that favors the aldehyde or ketone, add a large excess of water. The mechanism is the exact reverse of acetal or ketal formation, except that an excess of water is used and the alcohol is the leaving group.

18.6.6 Cyclic Acetals and Ketals

In preceding sections, aldehydes or ketones reacted with *two* equivalents of alcohol to give an acetal or a ketal. Diols such as 1,2-ethanediol (**56**, ethylene glycol; see Chapter 5, Section 5.6.3) have two alcohol units that are tethered together by a carbon chain. When diols react with aldehydes and ketones, *both* OH units are used to form *cyclic* acetals or ketals. As with acyclic alcohols such as ethanol, the acid-catalyzed reaction is the most efficient. An example is the reaction of butanal (**20**) with ethylene glycol in the presence of *p*-toluenesulfonic acid. Initial reaction with the carbonyl oxygen of **20** gives oxocarbenium ion **47**, as observed in the reaction with ethanol. However, subsequent reaction with one OH of **56** gives oxonium ion **57**, where the other OH unit of the diol unit is "tethered" to the intermediate. In other words, the second OH unit is part of the same molecule and not a separate molecule.

Loss of the acidic proton from oxonium ion **57** gives the hemiacetal **58**. As with the reaction with ethanol, all steps are reversible. In order to generate the hemiacetal from butanal, water is lost as a leaving group, as with the reaction with ethanol. Protonation of the OH oxygen (attached to the carbon derived from the carbonyl group) gives oxonium ion **59**, which loses water to form oxocarbenium ion **60**. At this point, the electrophilic carbon reacts with a second equivalent of OH; however, *the second hydroxyl at the end of the ethylene glycol chain is the OH group and it reacts faster than an OH in a separate molecule of diol 56*. That free OH unit in **60** reacts as shown in the illustration to give **61**, and loss of a proton gives the final product, cyclic acetal **62**.

This five-membered ring acetal is actually named as a **1,3-dioxolane**, and **62** is 2-propyl-1,3-dioxolane. Except for the obvious fact that the product is cyclic, there is only one difference in this reaction and reactions of aldehydes with mono-alcohols: The "second" equivalent of OH comes from the same molecule rather than a different molecule. In other words, the last reaction of an OH unit to form the acetal is an *intramolecular* process rather than an intermolecular process.

Ketones react with diols in a manner identical with aldehydes. An example reacts 2-butanone (**16**) with 1,3-propanediol (**63**) to form cyclic ketal **64**. The cyclic ketal in this case is a six-membered ring containing two oxygen atoms, and it is known as a **1,3-dioxane**. The name of **64** is 2-ethyl-2-methyl-1,3-dioxane.

18.33 Write out the complete mechanism for the reaction of **16** and **63** to give **64**.

18.34 Draw the structure of the product (and name it) when 2-hexanone reacts with 2,3-butanediol in the presence of an acid catalyst.

The cyclic acetals and ketals generated in this section are actually dioxolane and dioxane derivatives when 1,2-diols or 1,3-diols are used. It is often possible to make larger ring acetals or ketals by using 1,4-diols, but 1,5-diols would lead to an eight-membered ring. Formation of eight-membered rings is very difficult (see Chapter 8, Section 8.7). Forming cyclic ketals or acetals from diols where six or more carbons separate the hydroxyl-bearing carbons would lead to very large rings. Formation of such compounds requires specialized techniques that will not be discussed in this book.

18.6.7 Dithioacetals and Dithioketals

In Chapter 5 (Section 5.5.3), it was apparent that thiols are similar to alcohols, at least in terms of nomenclature and structure. Alcohols react with aldehydes and ketones, and it is known that thiols react in a similar manner. If butanal (**20**) reacts with ethanethiol (CH_3CH_2SH) in the presence of an acid catalyst, the reaction mechanism is identical to that for the reaction with ethanol (Section 18.6) except that the sulfur of the thiol is the nucleophile. Reaction with the thiol (rather than an alcohol) leads to **65**, a *thioacetal*. The term "thio" is used to show the presence of a sulfur atom rather than an oxygen atom.

An acetal is named as a dialkoxy derivative, and a thioacetal is named as a dialkylthio derivative. The name of **65** is 1,1-di(ethylthio)butane. Just as ketones react with alcohols to form ketals, they also react with thiols to form the sulfur analog. The mechanism is identical, and the reaction is illustrated by the reaction of cyclohexanone (**23**) with propanethiol to give **66**. This is a **dithioketal** named 1,1-di(propylthio)cyclohexane. The key intermediate in this latter reaction is a hemithioketal.

18.35 **Write out the complete mechanism for the acid-catalyzed conversion of 20 to 65 using ethanethiol.**

18.36 **Write out the complete mechanism and final product for the reaction of 3-pentanone and propanethiol to give the corresponding thioketal.**

1,2-Diols and 1,3-diols react with aldehydes and ketones to form cyclic acetals and ketals (dioxolanes or dioxanes), and 1,2-dithiols and 1,3-dithiols react similarly to form cyclic derivatives containing sulfur atoms. The reaction of 1,2-ethanethiol (HSCH$_2$CH$_2$SH) and benzaldehyde (**67**) leads to **68**. This five-membered ring containing two sulfur atoms is known as a **1,3-dithiolane** and **68** is named 2-phenyl-1,3-dithiolane. Once again, ketones react similarly; when 3-pentanone (**69**) reacts with 1,3-propanedithiol (HSCH$_2$CH$_2$CH$_2$SH), the product is **70**. This product contains a six-membered ring with two sulfur atoms; it is a **1,3-dithiane** named 2,2-diethyl-1,3-dithiane. There is one significant difference in these last reactions: A Lewis acid (BF$_3$) is used as the acid catalyst rather than a protonic acid. Because sulfur has a great affinity for many Lewis acids, this modification is commonly used to generate dithioacetals and dithioketals. Indeed, **65, 66, 68**, and **70** are best prepared using a Lewis acid such as BF$_3$ rather than the Brønsted–Lowry acids shown in those reactions.

18.37 **Write out the complete mechanism for the formation of 68 from 67 and ethanedithiol.**

Dithioacetals and dithioketals can be converted back to the carbonyl under acidic conditions in the presence of an excess of water because every step is reversible (just as with acetals and ketals). Aqueous hydrolysis is more difficult for dithio derivatives when compared to acetals or ketals, and water-compatible Lewis acids such as mercuric (Hg^{+2}) salts are used. Dithioacetal **68**, for example, is converted back to benzaldehyde (**67**) when it is treated with HgCl$_2$ (mercuric chloride) and BF$_3$ (boron trifluoride) in aqueous ether. Similar treatment of **70** gives **69**, and **66** is converted back to **23**.

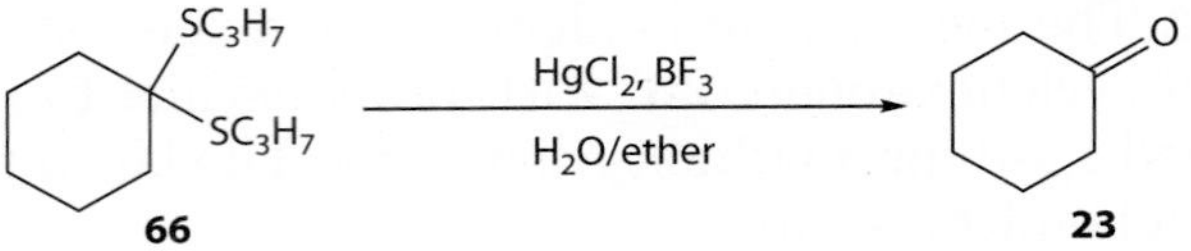

18.7 Amines Are Nucleophiles That React to Give Imines or Enamines

Amines were introduced in Chapter 5 (Section 5.6.2); they have the structure **71**. If **71** has three alkyl groups, it is a tertiary amine. With two alkyl groups and one hydrogen atom, it is a secondary amine, and when it has one alkyl group and two hydrogen atoms, it is a primary amine. There are other nitrogen-containing compounds that are important in organic chemistry; two are an imine (**72**) characterized by a C=N unit and an enamine (**73**) characterized by a C=C–N unit.

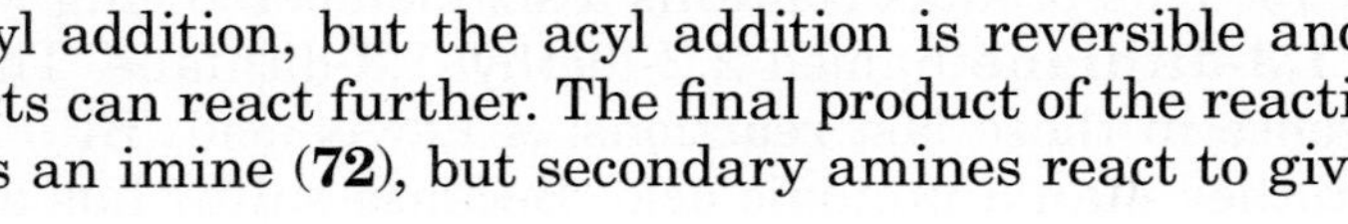

The nitrogen atom in an amine (**71**) is a good electron donor and a better nucleophile than water or alcohols. Indeed, amines react with aldehydes and ketones by acyl addition, but the acyl addition is reversible and the initially formed products can react further. The final product of the reaction with a primary amine is an imine (**72**), but secondary amines react to give an enamine (see **73**).

18.7.1 Primary Amines React to Give Imines

When the nitrogen of an amine reacts with the electrophilic carbon of a carbonyl, a new C–N bond is formed as the π-bond is broken, as in any other nucleophilic acyl addition reaction. If a primary amine such as (1-methylethyl)amine (called isopropylamine, Me$_2$CHNH$_2$) reacts with acetone (**6**), acyl addition leads to an ammonium salt (**74**). This molecule is a *zwitterion* because it contains a negatively charged alkoxide unit as well as the positively charged nitrogen, which is part of the H$_2$NCHMe$_2$ unit. This reaction is reversible because the charged NH$_2$Me$^+$ unit is a good leaving group. As with alcohols, the reversibility of this reaction causes a problem, but such problems are alleviated by the use of an acid catalyst.

18.38 Calculate the formal charge on both oxygen and nitrogen in 74 (from Chapter 5, Section 5.5).

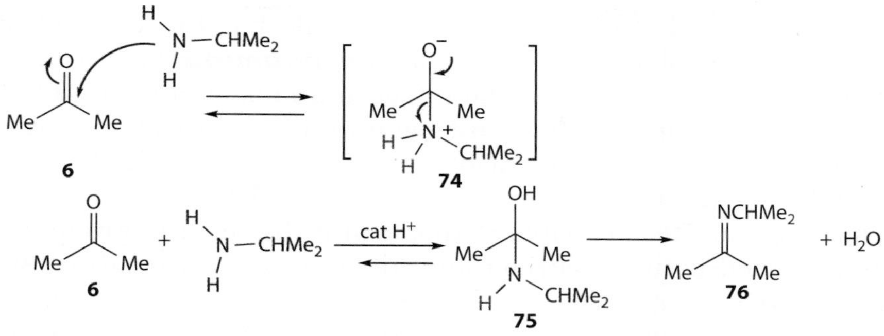

When acetone (**6**) reacts with isopropylamine at 50°C in the presence of HCl, the product is not the amino alcohol **75** derived by protonation of the alkoxide unit in **74** and loss of the proton on nitrogen. Instead, the product is imine **76**, which is isolated in 84% yield after neutralization with NaOH.[12] An analysis of the reaction based on the products shows that the –NCHMe$_2$ unit is derived from the amine, so the nitrogen atom of the amine must have reacted with the acyl carbon of acetone. The water formed during the reaction is derived from the carbonyl oxygen of acetone, requiring the transfer of two protons to that oxygen as with acetal and ketal formation in Section 18.6. There is only one amine unit in **76**, suggesting that loss of water is accompanied by formation of the C=N unit. This walk-through describes the mechanism of the reaction.

To begin the mechanism, first recognize that both the carbonyl oxygen and the amine nitrogen are basic atoms that can react with the acid catalyst. The more basic amine reacts first to form its conjugate acid, the ammonium ion (Me$_2$CHNH$_3$$^+$). This ammonium salt is probably the acid catalyst in the actual imine-forming reaction. In this acid–base equilibrium reaction, the carbonyl oxygen of acetone reacts with H$^+$ in the usual reversible acid–base process to form oxocarbenium ion **76**. The nucleophilic nitrogen atom of the amine reacts with **77** to give ammonium ion **78**.

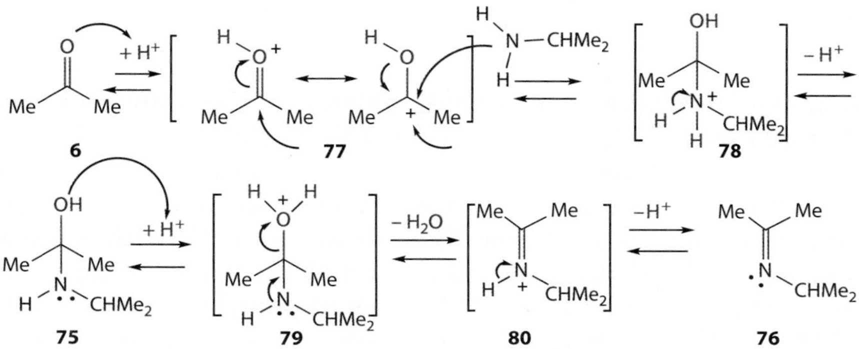

The ammonium salt is a weak acid (compare with ammonium chloride, NH_4Cl; $pK_a = 9.25$) and the rate of reaction is such that unreacted amine is available. The amine reacts as a base with **78**, and loss of the acidic proton from that species leads to the neutral amino–alcohol **75**. Although it is not obvious, amino–alcohol **75** (it is known as a ***hemiaminal***) is unstable to the reaction conditions (it is formally the nitrogen analog of a hemiacetal). If the acid catalyst reacts with the basic nitrogen in **75**, the reaction is reversible, regenerating **78**. If, however, the acid catalyst reacts with the less basic oxygen of the OH unit in an equilibrium reaction, then **79** is formed. The electron pair on nitrogen in **79** is capable of displacing the leaving group (water) to form a new π-bond between carbon and nitrogen (a C=N bond) in **80**, called an ***iminium salt***. The proton on the nitrogen is a weak acid and reaction with a suitable base (such as the amine) gives a neutral product, **imine** (**76**). ***Virtually any aldehyde or ketone will react with a primary amine to give an imine***.

From a conceptual viewpoint, the amine is the nucleophile, leading to incorporation of the nitrogen group in the final imine product (the **blue** atoms in the illustration). **It is usually very difficult to isolate the unstable hemiaminal, so the reaction proceeds to form the more stable imine.** Note that ammonia will react with aldehydes or ketones in a manner that is similar to a primary amine. However, the imine products formed from ammonia are much less stable, as a rule. Imine-forming reactions will focus on reactions of primary amines rather than ammonia.

18.39 Write out the acid–base reaction between 78 and isopropylamine.

18.40 Write out the acid–base reaction between 80 and isopropylamine.

18.41 **Give the final product and write out the complete mechanism for the reaction between benzaldehyde and 1-aminobutane in the presence of an acid catalyst. Do a similar analysis for the reaction with ammonia.**

With unsymmetrical ketones or with aldehydes, the imine can exist as an *E*- or a *Z*-stereoisomer, where *syn* and *anti* indicate the relationship of the alkyl group on nitrogen with respect to the larger alkyl group on the carbon of the C=N unit. (Nomenclature for *E*- and *Z*-isomers with alkenes was discussed in Chapter 9, Section 9.4.3.) Imine **81** is drawn as *E* (*E* with respect to the C=N unit), where the R group on nitrogen and the higher priority isopropyl group are on opposite sides. In **82**, R and the isopropyl group are on the same side, and this is the *Z*-isomer (*Z* with respect to the C=N unit). In general, the *E*-stereoisomer is preferred because it is less sterically hindered, but this depends on the nature of the alkyl groups attached to the C=N unit. Although a mixture of both stereoisomers is generated in virtually all of these reactions, *assume* that the mixture favors the *E* product.

18.42 Briefly explain why the reaction of cyclopentanone and triethylamine does *not* give an imine.

18.7.2 Secondary Amines React to Give Enamines

Aldehydes and ketones react with primary amines to give imines. The iminium salt intermediate **80** immediately precedes the imine product, and it has an acidic proton that is lost from the nitrogen to form the imine. When a secondary amine (HNR_2) reacts with an aldehyde or a ketone, rather than a primary amine, the course of the reaction is identical until the iminium salt intermediate is formed. In a typical experiment, 3-pentanone (**69**) reacts with the secondary amine diethylamine (Et_2NH) and an acid catalyst; however, the isolated product is not an imine with a C=N unit, but rather a mixture of compounds that have an amine group attached to a C=C unit. Such products are enamines, and **83** is formed in 86% yield along with 14% of **84**.[13] Note that enamine **83** is an *E*-isomer and **84** is a *Z*-isomer.

To explain the formation of an enamine, the NEt_2 unit clearly comes from diethylamine, and water has been formed. The oxygen of the water must come from the carbonyl oxygen. An important difference from the reaction with primary amines is that at least one of the protons found in the water product must come from the α-carbon of **69**. However, based on analogy with primary amines, **69** must be protonated to form an oxocarbenium ion, which reacts with the secondary amine. This pathway must lead to formation of an iminium salt. In this case, the iminium salt must lose a proton from the adjacent carbon atom to form the enamine.

This walk-through suggests a mechanism that forms **83** and **84** by a parallel route to the reaction of a carbonyl with a primary amine, in that the amine unit is attached to the acyl carbon and the oxygen atom in **69** is lost as water. Presumably, the loss of water will give an iminium salt that is analogous to **80**, but there is a significant difference. A primary amine has a C=NHR unit in **80** with an acidic proton attached to the nitrogen of the iminium salt. A secondary amine leads to an iminium salt with a $C=CNR_2$ unit and no proton on the nitrogen atom. Mechanistic analysis of the reaction of diethylamine with 3-pentanone (**69**), in the presence of an acid catalyst, leads to oxocarbenium ion **85**. This is followed by reaction with the nucleophilic amine to give **86**, just as with primary amines.

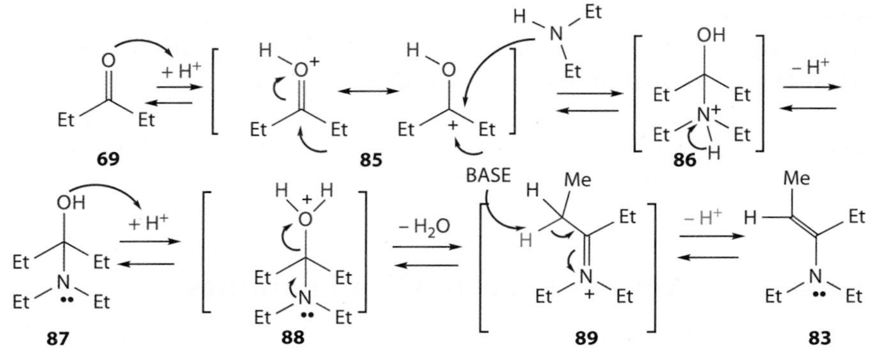

Loss of a proton from the acidic ammonium salt **86** gives hemiaminal **87**. Subsequent loss of water gives iminium salt **89**, but it is clear that **89** does *not* have a proton on nitrogen, but rather two carbon groups (ethyl). *In other words, there is no acidic hydrogen attached to nitrogen in 89*. It is therefore *impossible* for **89** to be converted to a neutral imine such as **76**. However, iminium salt **89** does have a slightly acidic proton that can be removed. Examine the α-carbon in **89** to find that the hydrogen attached to that carbon (shown in **red** in the illustration) is polarized δ+, and it is slightly acidic. That proton in **89** reacts with diethylamine, as shown by the arrows in the illustration, to generate **83**. As noted for **73**, compound **83** is an *enamine* ("en" for alkene and "amine" for the NR$_2$ group). This sequence *is the general mechanism for enamine formation from the reaction of a ketone with a secondary amine*.

The *E*-isomer is generally more stable than the *Z*-isomer due to diminished steric hindrance, so it is assumed that the *E*-isomer is the major product (shown for **83**). Water, a reaction by-product, may be removed to give a better yield of product and azeotropic distillation is used as well as molecular sieves (see Section 18.6.3). Enamines are structurally related to an enol (HO—C=C) in that the heteroatom is directly attached to the alkene unit. Enamines are often isolable compounds, whereas enols tautomerize spontaneously to the carbonyl form. *Note that when imine 76 is formed from iminium salt 80, there is no enamine product*. In fact, the C—H in **80** is much *less* acidic that the N—H unit, so the product is the imine rather than the enamine. It is noted that *there is an equilibrium between an imine and an enamine, known as imine-enamine tautomerism, but it will be ignored in this book*. Many different secondary amines can be used in this reaction, including cyclic amines (see Chapter 26, Section 26.4.1) such as pyrrolidine (**90**), piperidine (**91**), and morpholine (**92**). It is important to note that it is generally easier to form an enamine from a ketone than from an aldehyde.

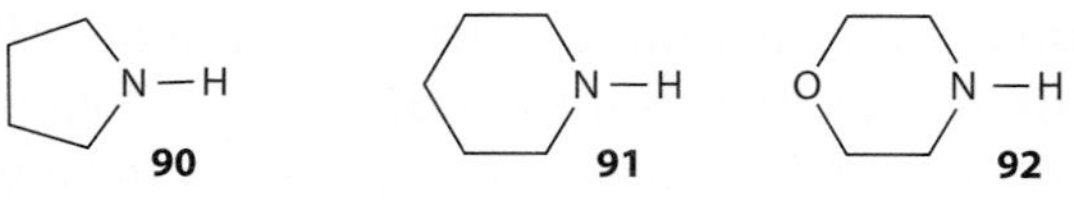

Aldehydes react with secondary amines to give enamines, but by a slightly different pathway that is not necessarily intuitive from the preceding discussion. Aldehydes are so reactive that they normally react with *two equivalents* of the secondary amine to generate a *gem*-diamino compound. In an experiment that illustrates this difference, 3-methylbutanal (**93**) reacts with piperidine to give **94**; however, upon distillation, a molecule of piperidine is lost to give the enamine, **95** (shown for convenience as the *E*-isomer).[14] This reaction is a typical example of the reaction of an aldehyde with a secondary amine. Whereas an aldehyde will react with a primary amine to give an imine, an aldehyde reacts with a secondary amine to give an enamine via the diamine. If care is taken, the diamine may be isolated, but it is assumed that the enamine is the major isolated product.

18.43 **Draw the enamine product that would be formed from intermediate 80.**

18.44 **Draw the structure of the product formed when cyclopentanone reacts with diethylamine in the presence of an acid catalyst.**

18.45 **Draw the structure of the product from the reaction of 3-pentanone and pyrrolidine and also the reaction of 3-pentanone with morpholine in the presence of an acid catalyst.**

18.46 **Draw the *gem*-diamine and final enamine product formed when acetophenone reacts with pyrrolidine.**

18.7.3 Hydroxylamine, Hydrazines, and Related Nucleophiles

Several useful nitrogen-containing molecules react with ketones or aldehydes in a manner that is similar to the reaction of primary amines. Hydroxylamine (**96**) is essentially a functionalized amine, as is hydrazine (**97**)—an important compound that was used as rocket fuel many years ago. Hydrazine is still used as a propellant on board space vehicles and in satellites in order to make course corrections. Hydroxylamine may be explosive when concentrated, and should be handled with care, but it is used for the removal of hair from animal hides and in photographic developing solutions.[15]

All three of these compounds have a primary amine unit ($-NH_2$) and each reacts with an aldehyde or ketone essentially as a "functionalized primary amine" to give a functionalized imine. Hydrazine derivatives are known (**98**; R

= alkyl or aryl) and they react similarly. Note that aryl indicates a substituent that is a derivative of an aromatic compound such as benzene (see Chapter 21). Finally, semicarbazide (**99**) is an "amide-substituted" hydrazine that reacts with aldehydes or ketones to give functionalized imines.

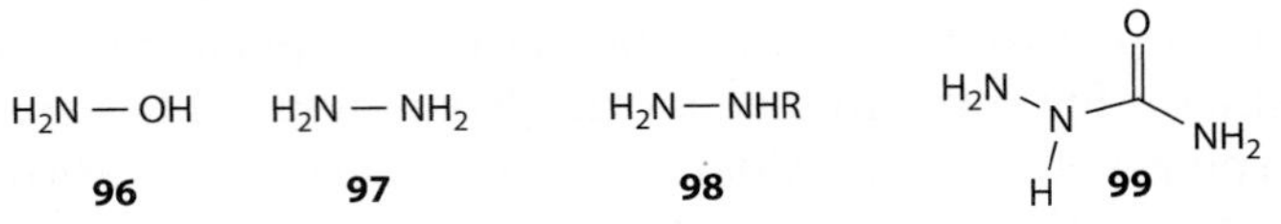

All of these compounds have the structure X–NH$_2$, where X is NH$_2$, NHR, OH, or CONH$_2$, so it is not surprising that they react with ketones and aldehydes by the same mechanism shown for the conversion of acetone (**6**) to imine **76** in Section 18.7.1.

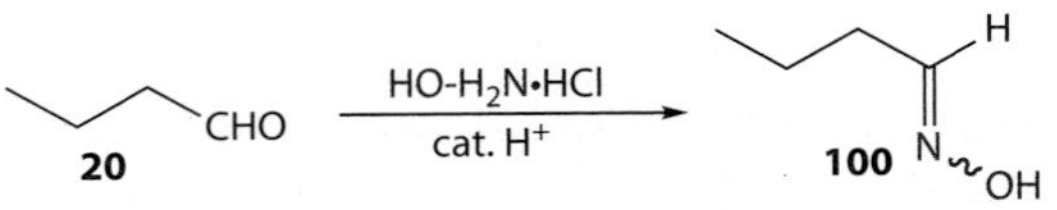

An example is the reaction of hydroxylamine (**96**) with butanal (**20**) to give the imine-like derivative **100** (called an *oxime*) in 81% yield.[16] In this particular experiment, the H$^+$ is an acidic resin called Amberlite IRA-420, but that is not an important issue and here it is treated simply as H$^+$. The hydroxylamine is added as its HCl salt (known as hydroxylamine hydrochloride) because it is easier to handle, and NH$_2$OH is generated during the course of the reaction. Note that the OH group may be on the same side or the opposite side relative to the carbon fragment, generating *Z*- or *E*-isomers, respectively. In **100**, the "squiggle line" connecting N and the OH indicates a *mixture* of *E*- and *Z*-stereoisomers. *Assume that this reaction always gives a mixture*, but in most cases the preferred stereoisomer has the larger group *E* to the OH unit. When a ketone or aldehyde reacts with hydrazine (**97**), the resulting functionalized imine product is called a ***hydrazone***. Cyclohexanone (**23**) reacts with hydrazine, for example, to give hydrazone **101**.

18.47 **Draw both the *E*- and *Z*-isomers of 100.**

18.48 **Draw the structure of the product formed in a reaction between 4,4-diphenylpentanal and hydroxylamine in the presence of an acid catalyst.**

18.49 **Write out the mechanism for the formation of 101, using the conversion of 6 to 76 as a template.**

18.50 **Draw the structure of the product when 2-hexanone reacts with hydrazine.**

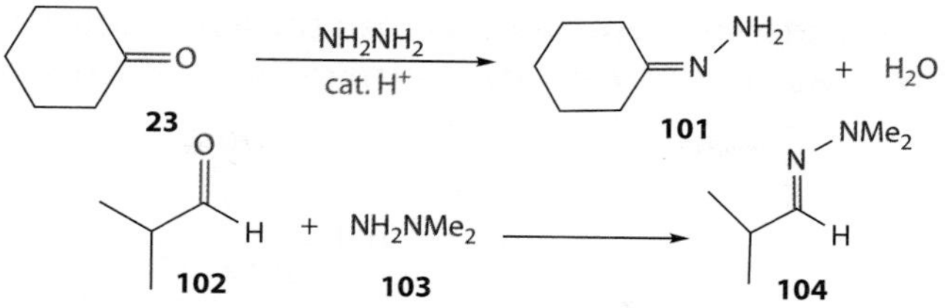

Hydrazine (**97**) is considered to be the parent of substituted hydrazine derivatives, such as **98**, which react with aldehydes and ketones to form substituted hydrazone derivatives. An example is the reaction of *N*,*N*-dimethylhydrazine (**103**) and 2-methylpropanal (isobutyraldehyde, **102**) to give the *N*,*N*-dimethylamine hydrazone **104** in 64% yield.[17,18] Two important *N*,*N*-disubstituted hydrazine reagents are phenylhydrazine (PhNHNH₂) and 2,4-dinitrophenylhydrazine (**106**), which react with aldehydes or ketones to give the corresponding hydrazone.

Both derivatives give hydrazones that tend to be solids and are rather stable. Therefore, a melting point is easily determined and compared with known derivatives for the purposes of identification. Because of their utility as an identification agent, they are usually formed in a test tube; yields are not always important, although these reactions usually proceed in yields >70% and often >90%. A simple example is the reaction of 2,4-dinitrophenylhydrazine (usually abbreviated 2,4-DNP, **106**) and cyclooctanone (**105**), which gives a 2,4-dinitrophenylhydrazone, **107**.

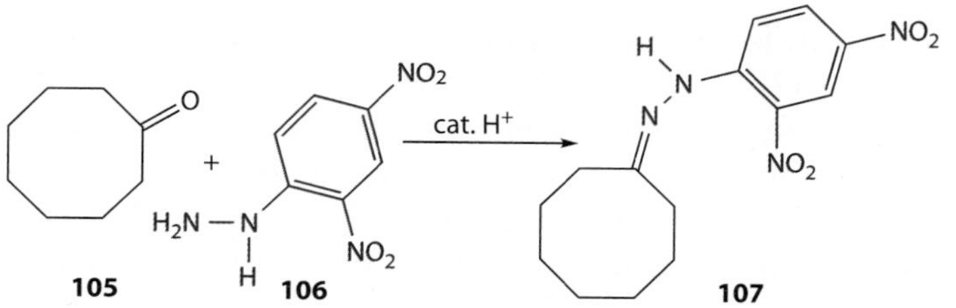

Another highly specialized hydrazine derivative used for derivatization purposes is semicarbazide (**99**), which has a "urea unit" in the molecule. The structure of urea is **108**. The reaction of **99** with acetophenone (**109**; the IUPAC name is 1-phenylethanone) proceeds in an identical manner with all of the other functionalized amines discussed in this section. This reaction generates **110**, which is called a ***semicarbazone***. This is an unusual structure based on compounds discussed previously, but it is essentially a "urea-imine."

18.51 Draw the product and write out the mechanism for the conversion of cyclopentanecarboxaldehyde to its phenylhydrazone.

18.52 Draw the product of the reaction between butanal and *N*-ethylhydrazine in the presence of an acid catalyst.

18.53 Draw the structure of the product formed in a reaction of cyclopentanone and semicarbazide in the presence of an acid catalyst.

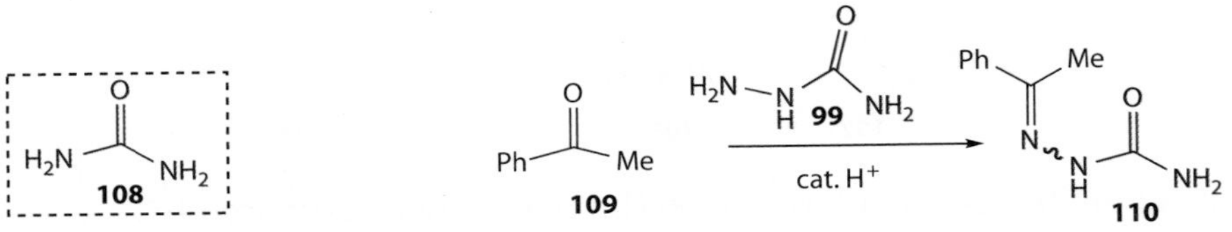

In essence, these specialized reagents are functionalized primary amines
that give functionalized imine products. The reaction is much more favorable
for reactions that form hydrazones and oximes, and these products can be generated in water.

The hydrazone derivatives, oximes, and semicarbazones discussed in this
section are rather specialized compounds. **Why mention them?** Organic chemists rely on spectroscopy (see Chapter 14) to identify the structure of organic
compounds, but prior to the development of spectroscopy, chemical reactions
were used as an identification tool. One such method converts unknown organic
compounds to solid derivatives. The color and melting point of several derivatives are cross-referenced to a library of known derivatives to help identify the
unknown. The conversion of aldehydes and ketones to phenylhydrazones, 2,4-
dinitrophenylhydrazones, oximes, and semicarbazones is one method used for
derivatization.

The four compounds—cyclohexanone (**23**), 3-pentanone (**69**), butanal (**20**),
and benzaldehyde (**67**)—are listed in Table 18.1, which shows the melting
point for each of the four derivatives mentioned for each carbonyl compound.
Comparing the melting point for two or more derivatives allows a single compound to be identified. Note that the oxime of benzaldehyde exists in two forms
(α and β, which are the two diastereomers), and the melting point of each one
is shown in Table 18.1. Those entries labeled "oil" indicate that the derivative

Table 18.1
Hydrazine Derivatives of Ketones and Aldehydes

| Carbonyl Derivative | Melting Point (°C) | | | |
	Phenylhydrazone	2,4-Dinitrophenylhydrazone	Oxime	Semicarbazone
Cyclohexanone (**23**)	81–82	160	91	166–167
3-Pentanone (**69**)	Oil	156	Oil	138–139
Butanal (**20**)	93–95	123	Oil	95.5
Benzaldehyde (**67**)	158	237	35; 130	222

Source: Rappoport, Z., ed. 1981. *CRC Handbook of Tables for Organic Compound Identification*,
Table IX. Boca Raton, FL: CRC Press, Inc.

is not a crystalline solid, but rather a liquid (whose boiling point could be determined and reported).

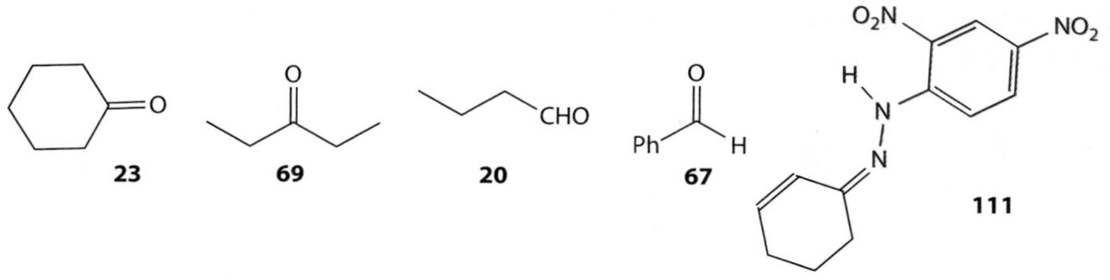

The 2,4-dinitrophenylhydrazone of most ketones and aldehydes will have a
yellow color. A 2,4-DNP product such as **111** that is derived from a conjugated
aldehyde or ketone (in this case, cyclohexenone) tends to have an orange or red
color. The color indicates the presence of the C=C unit in conjugation with the
C=O unit. ***Formation of a red-orange or red 2,4-DNP is a good indication that the aldehyde or ketone precursor is conjugated***. The reasons
for this shift in wavelength of absorbed light in connection with spectroscopy
were discussed in Chapter 14. Chapter 23, which discusses conjugated organic
molecules, will also discuss this phenomenon.

18.54 Identify the unknown that has a 2,4-DNP derivative with a melting point of 122–124°C and a semicarbazone derivative with a
melting point of 94–96°C.

18.8 Carbon–Carbon Bond-Forming Reactions and Functional Group Modification

This chapter provides several important functional group transformations
based on acyl addition of strong nucleophiles or weak nucleophiles to ketones
and aldehydes. Addition of cyanide, alkyne anions, Grignard reagents, or
organolithium reagents to aldehydes or ketones leads to the corresponding alcohol in which a new C–C bond is formed. Acetals and ketals are generated from
the appropriate carbonyl compound by acid-catalyzed reaction with alcohols,
although hydrate formation is not considered a useful transformation except in
specialized cases such as chloral.

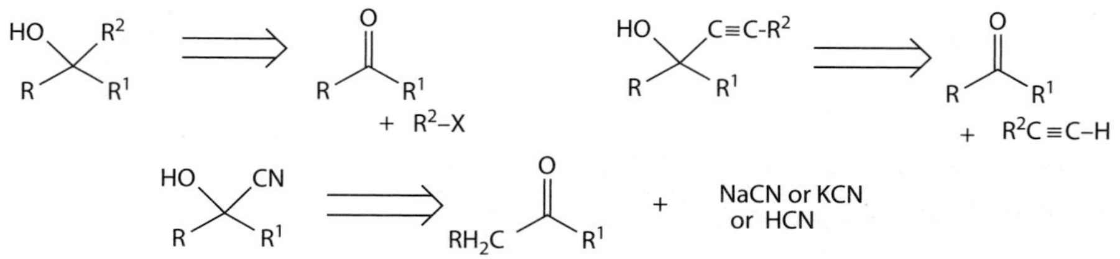

This transformation includes formation of cyclic acetals and ketals and, if R^2SH is used, thioacetals or thioketals. The R and R^1 groups can be alkyl, hydrogen, or aryl. Imines are available from primary amines and enamines from secondary amines. Similarly, R and R^1 can be hydrogen, alkyl, or aryl; R^2 is usually allyl or aryl.

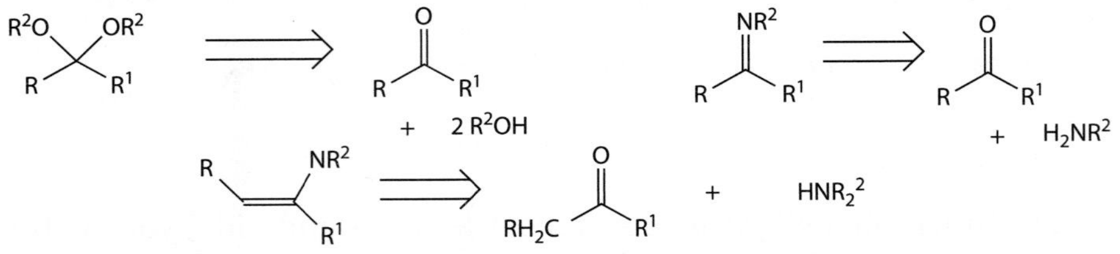

A typical synthesis is the preparation of halide **112** from alkene **113**. It is clear that this synthesis will require both C–C bond-forming reactions as well as functional group exchange reactions.

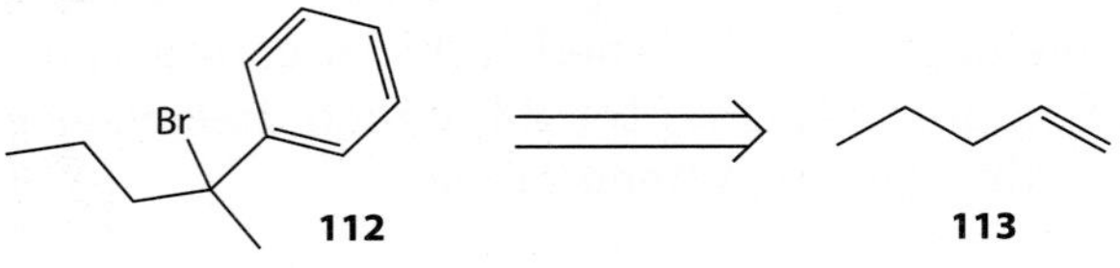

It is clear by comparing **112** and **113** that the benzene ring must be incorporated, and this requires a C–C bond-forming reaction. Because no such reactions have been introduced that generate the requisite bond and a bromide, changing the functional group is necessary. If the bromide is changed to the alcohol unit in **114**, a disconnection of the phenyl group is possible, suggesting an acyl addition reaction. Indeed, the fragment bearing the OH group has a natural bond polarity that leads to C^a, which has the synthetic equivalent 2-pentanone (**115**). With this carbon bearing the C^a, the phenyl carbon will be C^d, making it the equivalent of a carbanion surrogate such as phenylmagnesium bromide, **116**. Addition of the Grignard reagent to 2-pentanone (**115**) will give **114** after hydrolysis. The ketone **115** is derived from **113**, so an OH unit must be incorporated. Therefore, alcohol **117** is the logical precursor to **115**, and it is well known that terminal alkene **113** is converted to **117** by oxymercuration (see Chapter 10, Section 10.4.5).

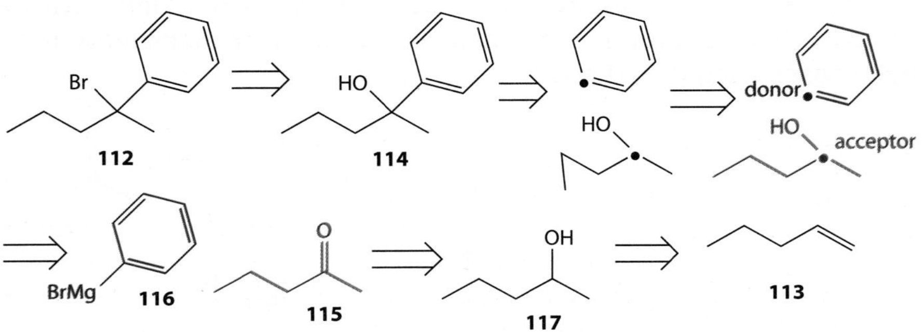

This retrosynthetic analysis leads to the synthesis that begins with oxymercuration of **113** to give **117**, followed by oxidation to the ketone, **115**, with PCC or another oxidizing agent (see Chapter 17, Section 17.2.3). Grignard **116** is used as a nucleophile in an acyl addition reaction (Section 18.4), and an aqueous acid workup gives alcohol **117**. The alcohol is readily converted to bromide **112** with PBr_3 or another brominating agent (see Chapter 11, Section 11.7).

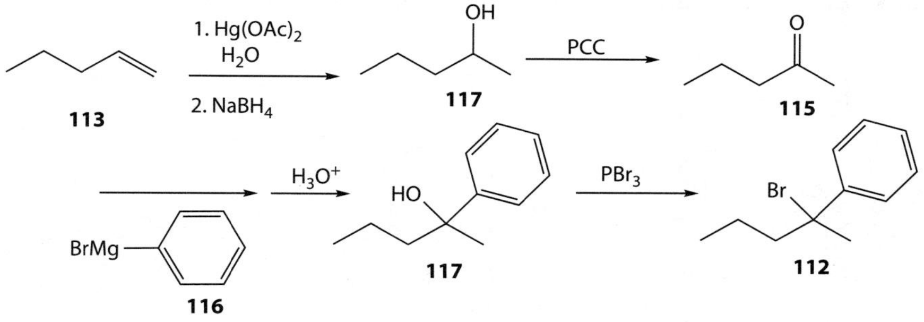

18.9 Biological Relevance

Vitamin B_6 (pyridoxine; also known as pyridoxol, **118**) is an essential growth factor in the diet of many organisms and animals. It forms part of a coenzyme (pyridoxylphosphate) and it is a cofactor for a class of enzymes known as transaminases. A transaminase or an aminotransferase is an enzyme that catalyzes a type of reaction between an amino acid and an α-keto acid. The presence of elevated transaminase levels can be an indicator of liver damage. Vitamin B_6 has both an aldehyde form (pyridoxal, **119**) and an amine form (pyridoxamine, **120**), and it is known that pyridoxal phosphate is a carrier of amino groups and sometimes amino acids.[20]

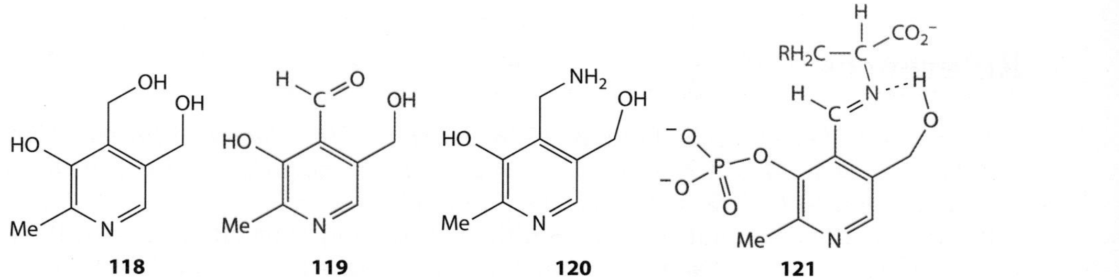

The mechanism involves conversion of the aldehyde group in **119** to an imine, which is called a Schiff base (see **121**) via reaction with ammonia or amine groups. The pyridoxal phosphate reacts with an amino acid via the amine unit to form a Schiff base (imine **122**), and the amino acid is converted to an α-keto acid. The conversion between the aldehyde and amine forms (pyridoxal–pyridoxamine) facilitates the transfer of an amino group from an amino acid to form a keto acid.[20] Enzymatic protonation gives iminium salt **123**, which is deprotonated at the α-carbon to give iminium ion **124** (a new Schiff base that has the carboxyl unit attached to the C=C unit). Hydrolysis of the imine via addition of water gives **125** (see Section 18.7.1) and formation of α-hydroxy salt **126**. Loss of pyridoxamine phosphate **127** gives the protonated α-keto acid **128** and loss of a proton gives the final product, keto-acid **129**.

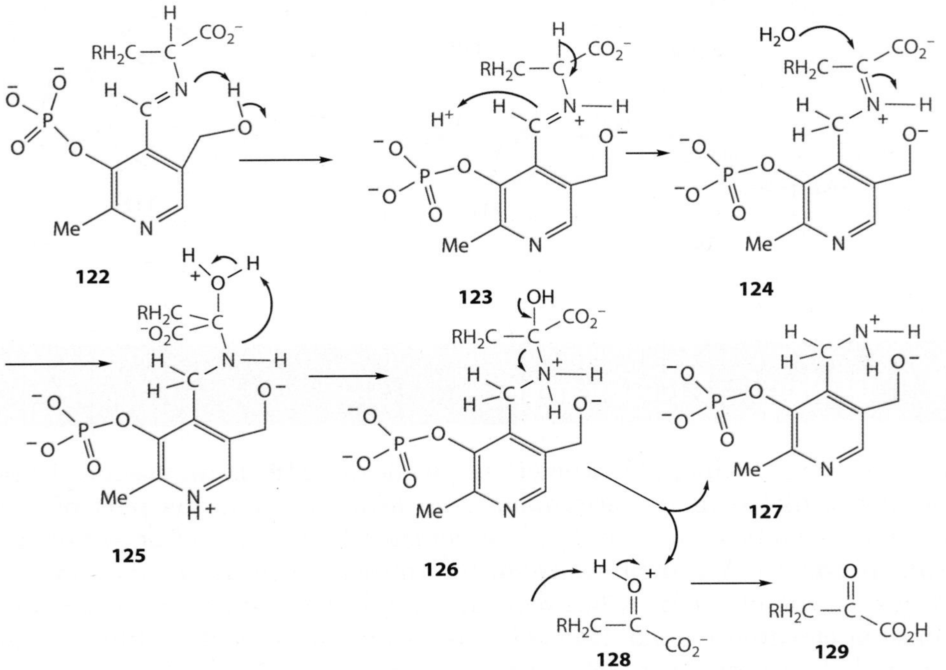

(Reprinted in part with permission from Martell, A. E. 1989. *Accounts of Chemical Research* 22:115. Copyright 1989 American Chemical Society.)

References

1. Ayerst, G. G., and Scholfield, K. 1958. *Journal of the Chemical Society* 4097.
2. Walton, D. R. M. 1969. *Acetylenic compounds: Preparation and substitution reactions,* 148. New York: T. F. Rutledge, Reinhold.
3. Furniss, B. S., Hannaford, A. J., Smith, P. W. G., and Tatchell, A. R., eds. 1994. *Vogel's textbook of practical organic chemistry,* 5th ed., 537, Exp. 5.40. New York: Longman.
4. Furniss, B. S., Hannaford, A. J., Smith, P. W. G., and Tatchell, A. R., eds. 1994. *Vogel's textbook of practical organic chemistry,* 5th ed., 538, Exp. 5.41.
5. Smith, M. B. 2002. *Organic synthesis,* 2nd ed, 586. New York: McGraw-Hill.
6. Stowell, J. C. 1979. *Carbanions in organic synthesis,* 57. New York: John Wiley & Sons.
7. Buhler, J. D. 1973. *Journal of Organic Chemistry* 38:904.

8. Dean, J. A. 1987. *Handbook of organic chemistry,* 4–17. New York: McGraw-Hill.
9. Weast, R. C., and Astle, M. J., eds. 1978. *CRC handbook of chemistry and physics,* 59th ed., D-21. Boca Raton, FL: CRC Press, Inc.
10. Weast, R. C., and Astle, M. J., eds. 1978. *CRC handbook of chemistry and physics,* 59th ed., D-5. Boca Raton, FL: CRC Press, Inc.
11. Weast, R. C., and Astle, M. J., eds. 1978. *CRC handbook of chemistry and physics,* 59th ed., D-37. Boca Raton, FL: CRC Press, Inc.
12. Norton, D. G., Haury, V. E., Davis, F. C., Mitchell, L. J., and Ballard, S. A. 1954. *Journal of Organic Chemistry* 19:1054.
13. Cervinka, O. 1994. In *The chemistry of enamines, part 1,* ed. Rapoport, Z., 223. Chichester, England: Wiley.
14. House, H. O. 1972. *Modern synthetic reactions,* 2nd ed., 571. Menlo Park, CA: Benjamin.
15. Patnaik, P., and Pradyot, P. 2003. *Handbook of inorganic chemicals,* 385–386. New York: McGraw-Hill.
16. Yamada, S., Chibata, I., and Tsurui, R. 1954. *Pharmaceutical Bulletin* 2:59.
17. Wiley, R. H., Slaymaker, S. C., and Kraus, H. 1957. *Journal of Organic Chemistry* 22:204.
    Wiley, R. H., and Irick, G. 1959. *Journal of Organic Chemistry* 24:1925.
18. Rappoport, Z., ed. 1981. *CRC handbook of tables for organic compound identification,* Table IX. Boca Raton, FL: CRC Press, Inc.
19. Martell, A. E. 1989. *Accounts of Chemical Research* 22:115.

Answers to Problems

18.1

18.2

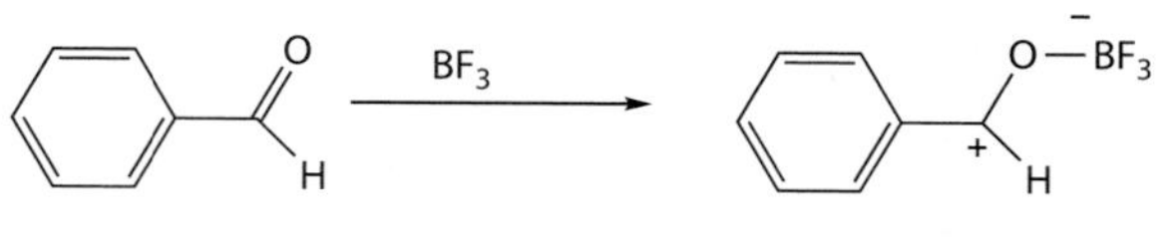

18.3

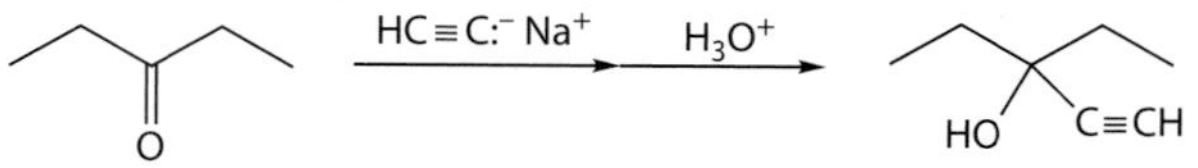

18.4

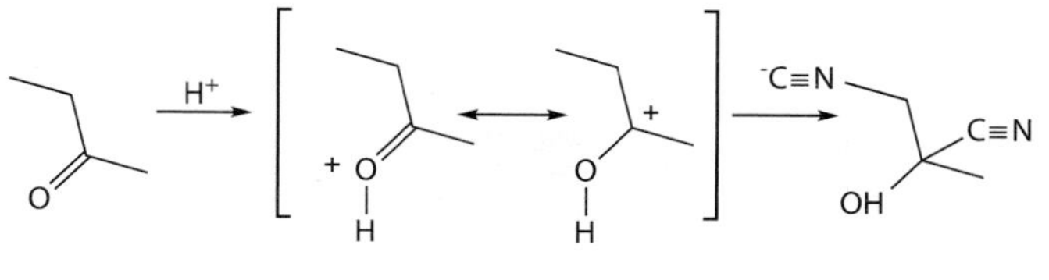

18.5

$$H-C\equiv C-H \xrightarrow{\text{NaH}} H-C\equiv C{:}^{-}\ \overset{+}{Na} + \frac{1}{2}H_2$$

18.6

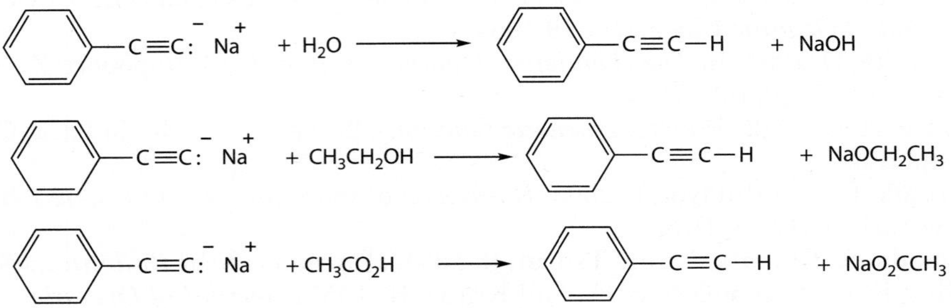

18.7   No! Ethanethiol has an acidic hydrogen attached to sulfur. Once the alkyne anion is formed, it can react with the EtSH. In addition, the base used to generate the alkyne anion will react preferentially with the stronger acid (RSH is a stronger acid than RC≡CH).

18.8

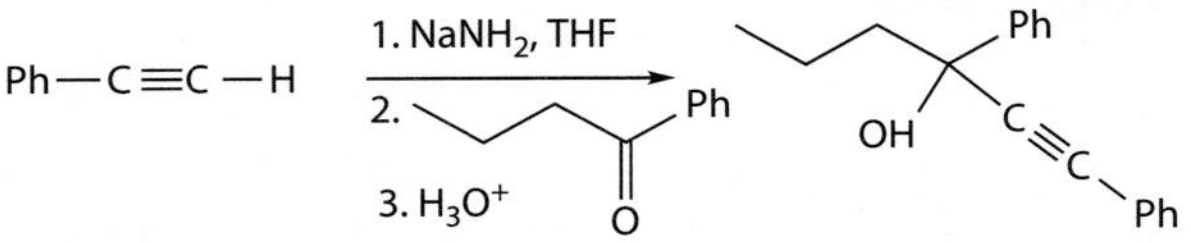

18.9

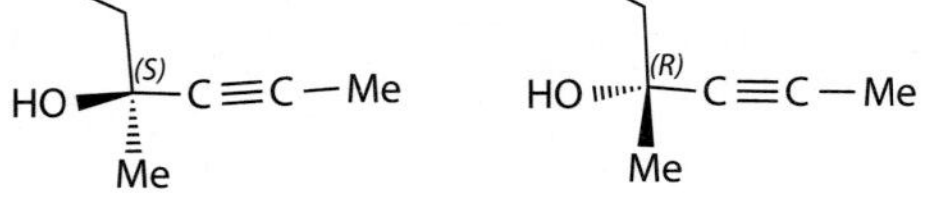

18.10

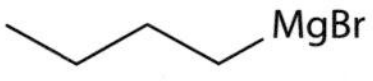

18.11

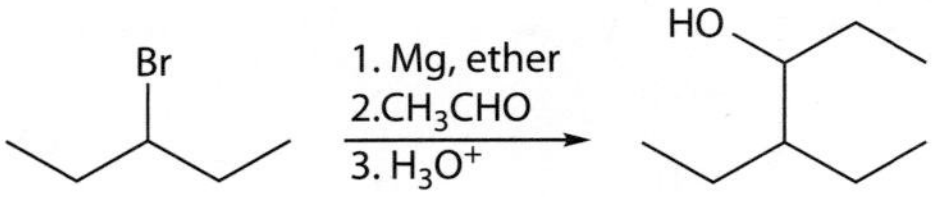

18.12

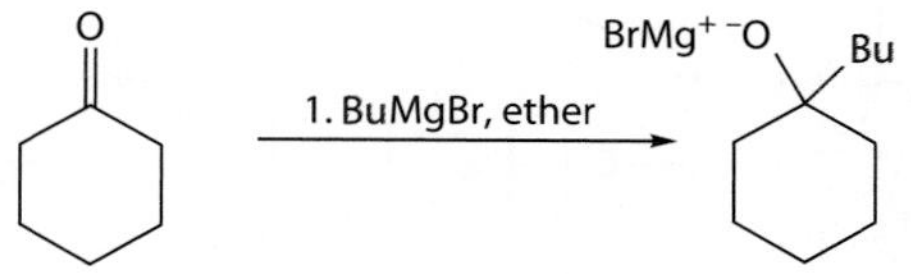
O
1. BuMgBr, ether
BrMg+ -O
Bu

18.13

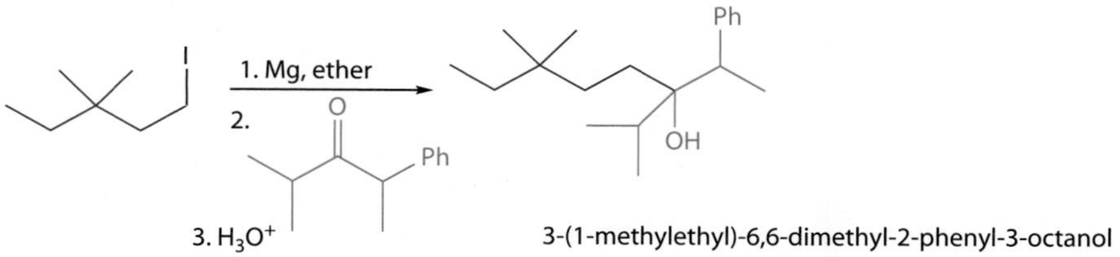
I
1. Mg, ether
2.
O
Ph
3. H3O+
Ph
OH
3-(1-methylethyl)-6,6-dimethyl-2-phenyl-3-octanol

18.14

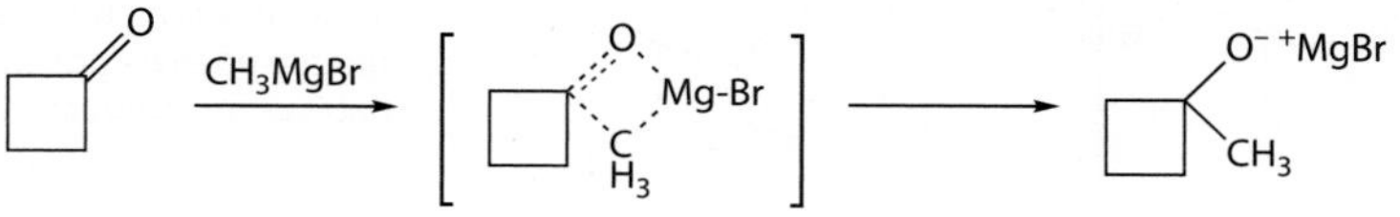
O
CH3MgBr
O
Mg-Br
C
H3
O- +MgBr
CH3

18.15

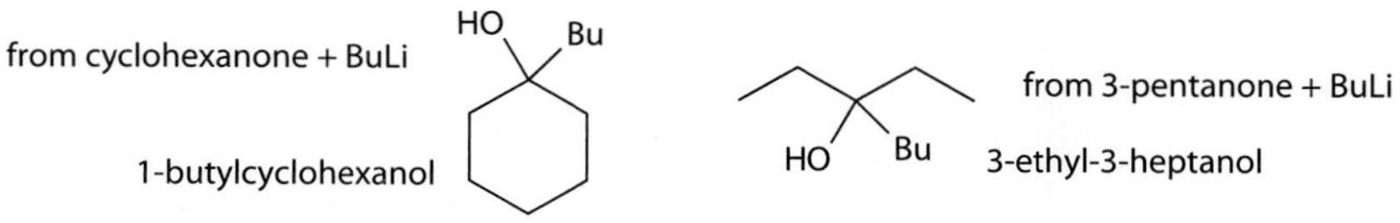
from cyclohexanone + BuLi
HO
Bu
1-butylcyclohexanol
from 3-pentanone + BuLi
HO
Bu
3-ethyl-3-heptanol

18.16

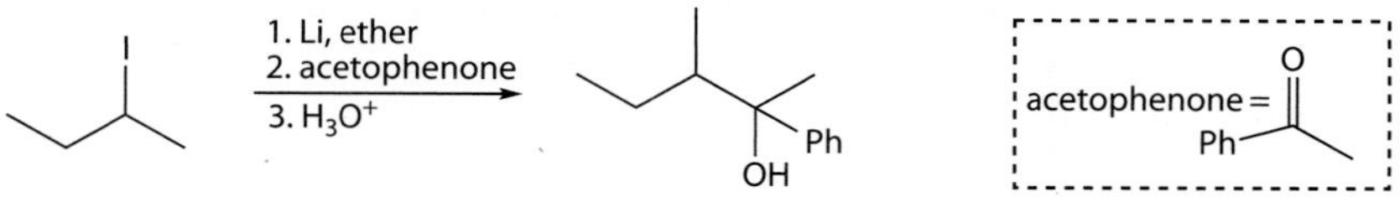
I
1. Li, ether
2. acetophenone
3. H3O+
OH
Ph
acetophenone =
O
Ph

18.17

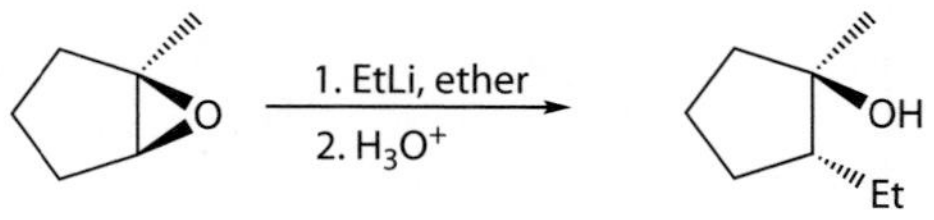
O
1. EtLi, ether
2. H3O+
OH
Et

18.18

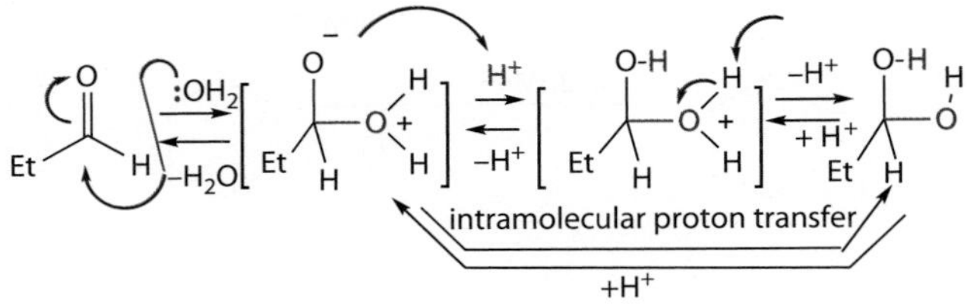

intramolecular proton transfer

$+H^+$

18.19

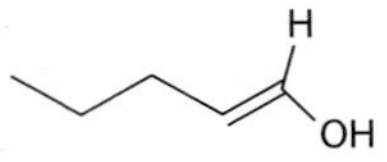

18.20

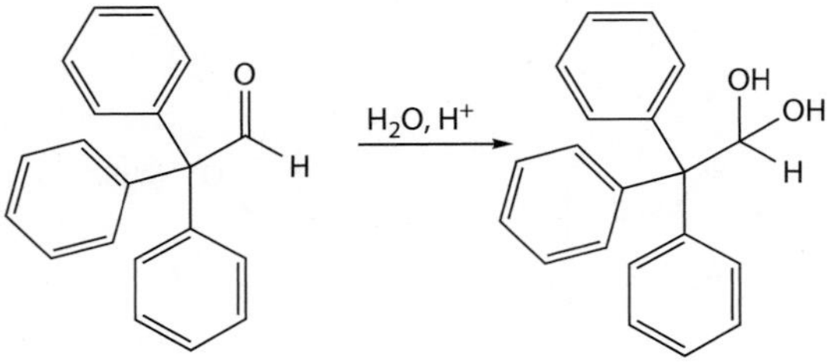

This hydrate should be stable since there is no α-hydrogen, and the phenyl rings should stabilize the product.

18.21

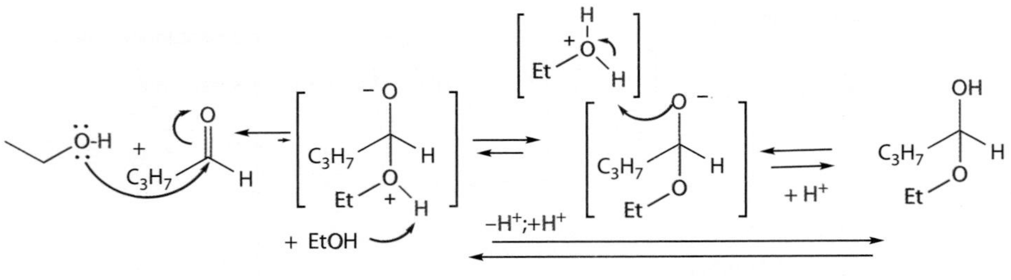

18.22

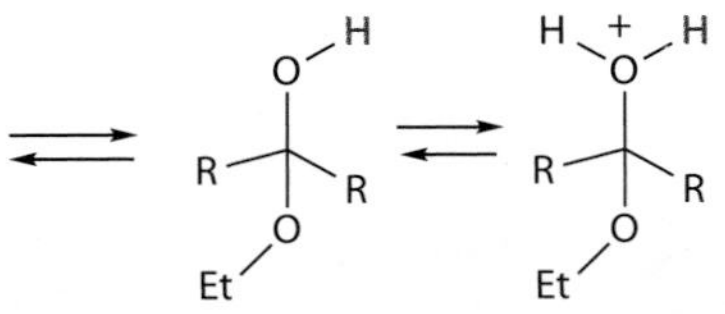

18.23

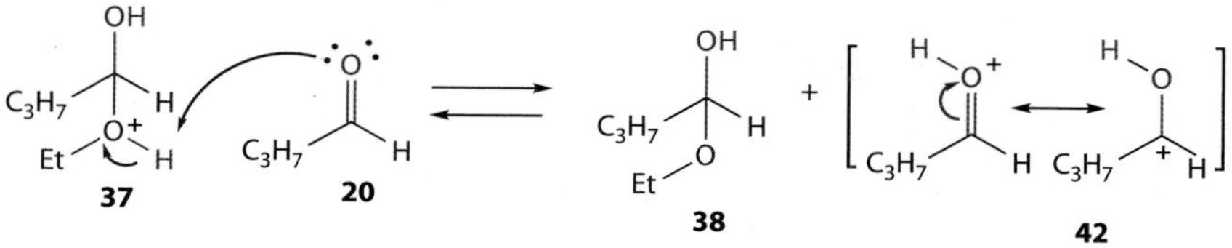

18.24

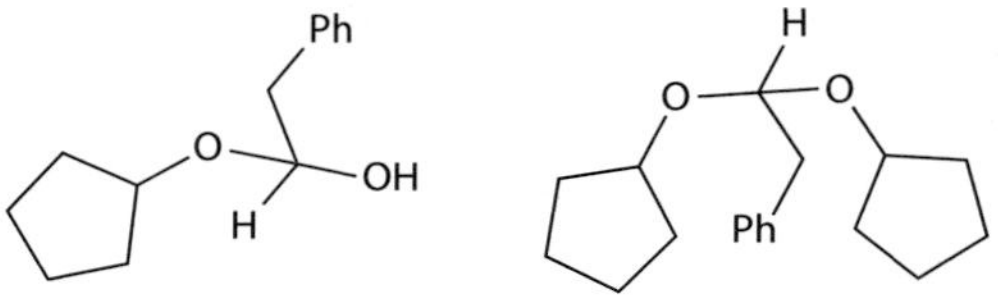

18.25

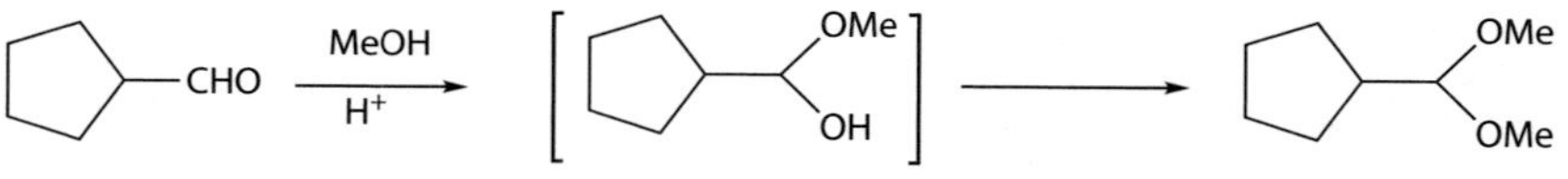

18.26

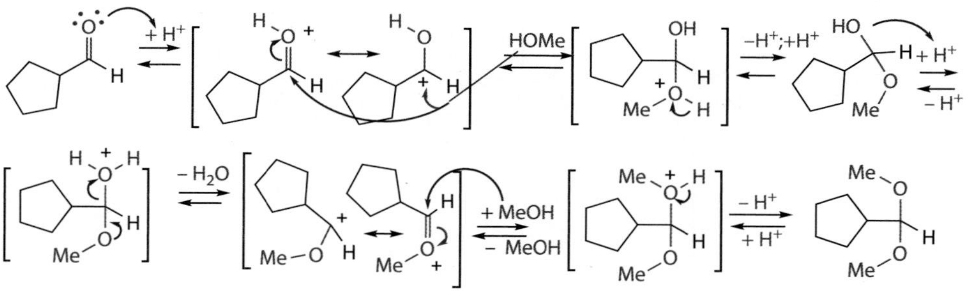

18.27

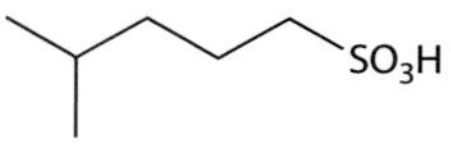

18.28 It is not feasible. The boiling point of acetaldehyde is 20°C, whereas the boiling point of the benzene–water azeotrope is 69.4°C. Clearly, the acetaldehyde would be removed from the reaction mixture and it would be difficult to keep sufficient amounts in solution to allow the reaction to proceed. In addition, the boiling point of the acetal (dimethoxyethane) has a boiling point of 64.5°C, so it is possible that the product would be removed as well. The bottom line is that the starting material and product have boiling points too low for the Dean–Stark procedure to be effective with this solvent.

18.29

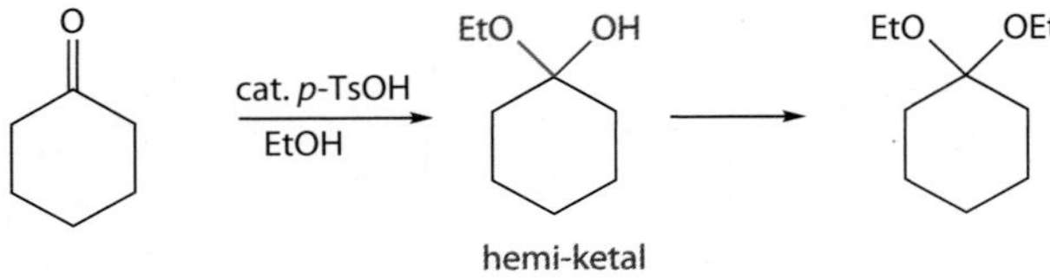

hemi-ketal

18.30

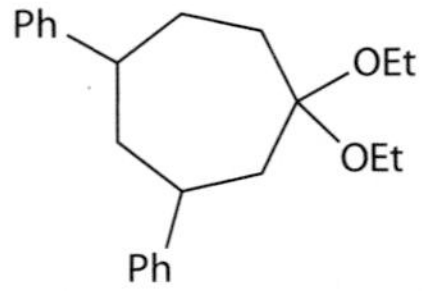

18.31

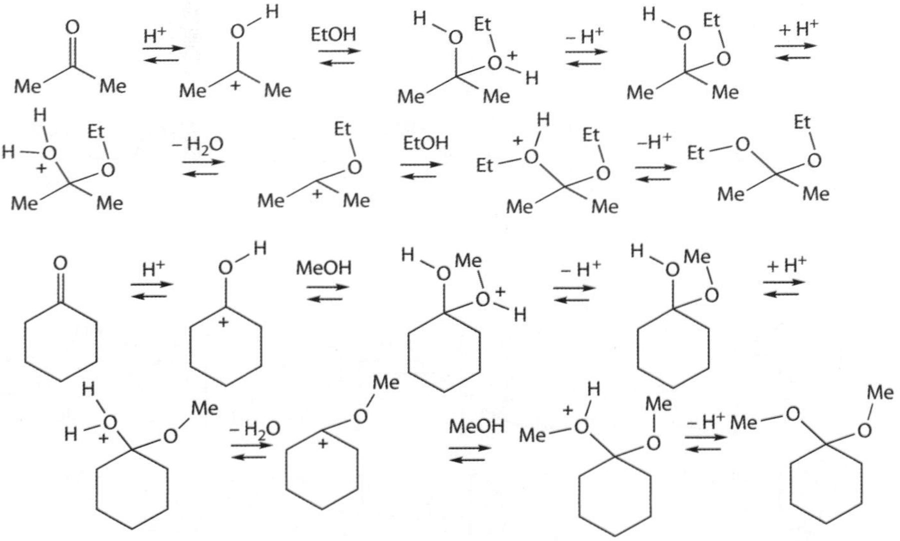

18.32

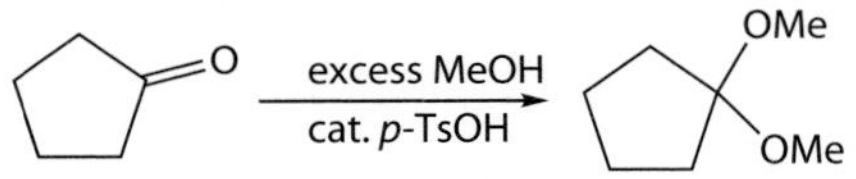

18.33

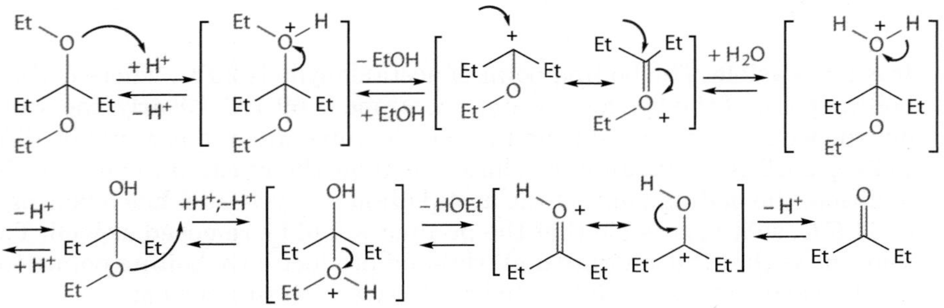

18.34

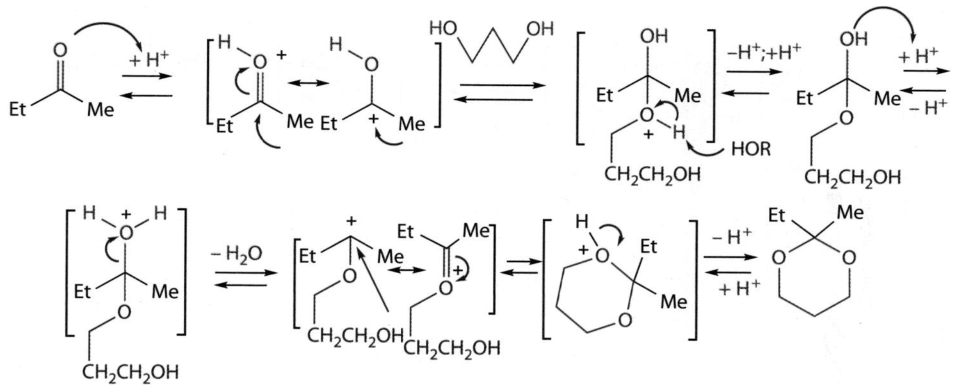

18.35

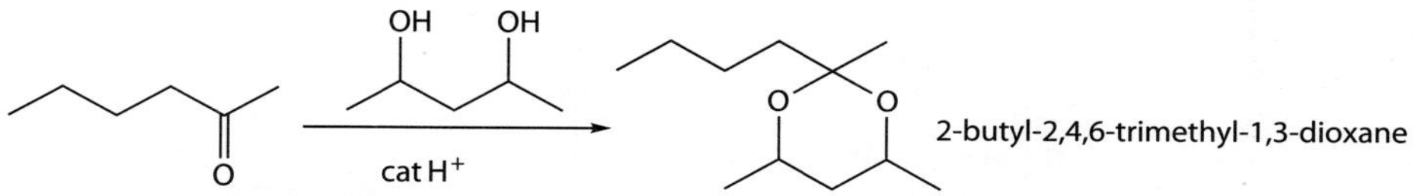

2-butyl-2,4,6-trimethyl-1,3-dioxane

18.36

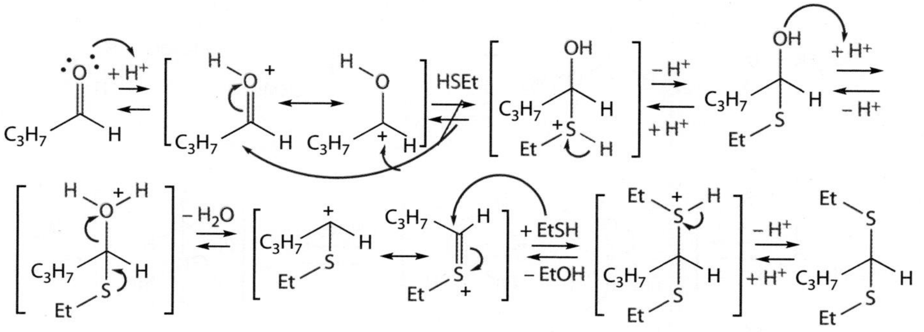

18.37

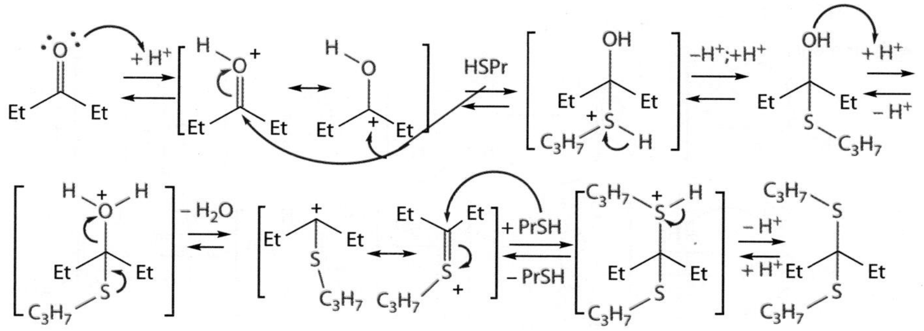

18.38

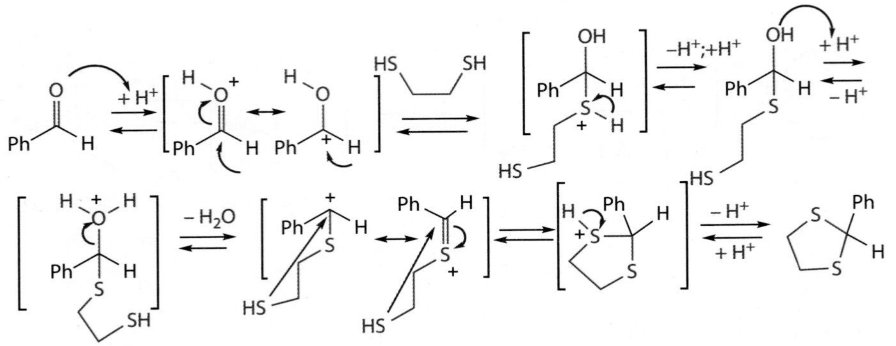

18.39

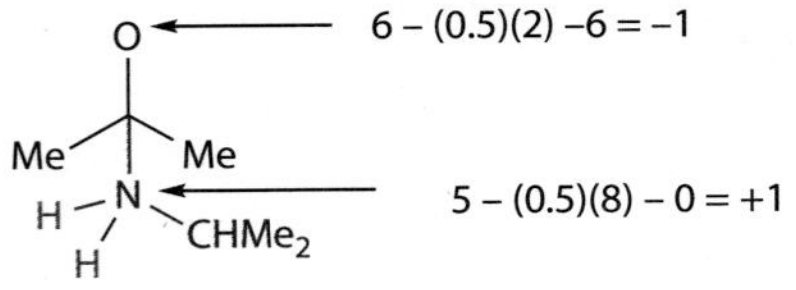

$$6 - (0.5)(2) - 6 = -1$$

$$5 - (0.5)(8) - 0 = +1$$

18.40

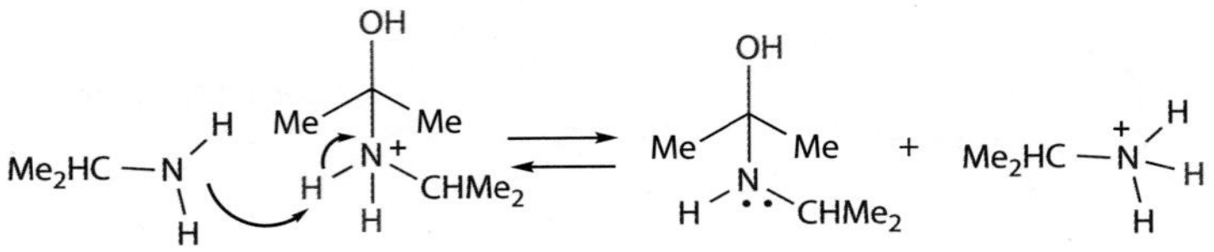

18.41

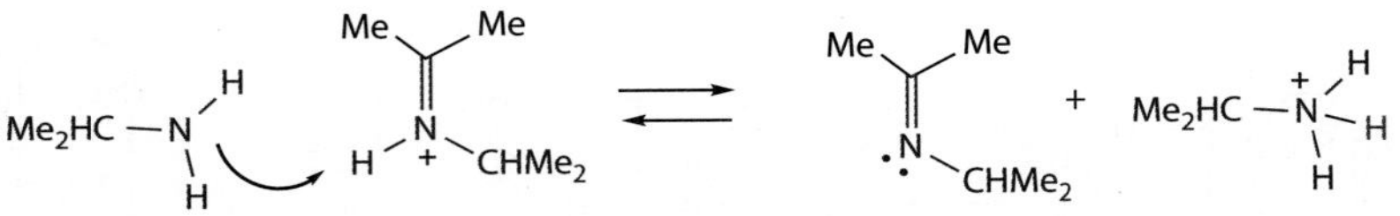

18.42

with butylamine

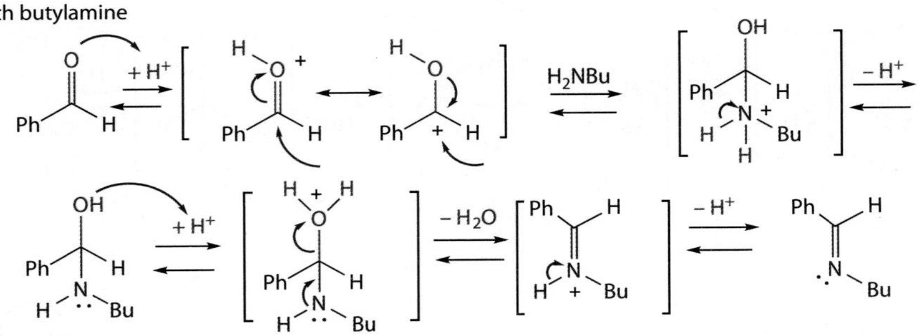

with ammonia

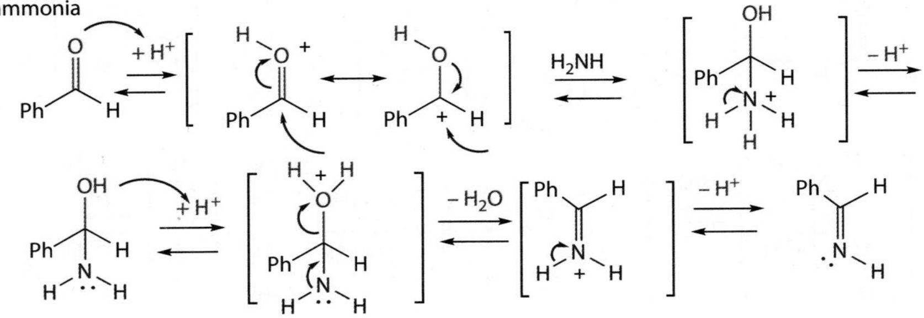

18.43   The key to forming an imine is loss of a proton from the intermediate ammonium and iminium salts. There are no hydrogen atoms attached to nitrogen
in a 3° amine, so it is not possible to complete the mechanistic requirements
for generating an imine.

18.44

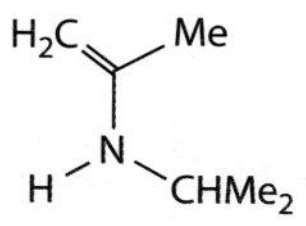

18.45

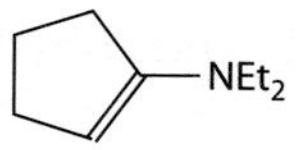

18.46

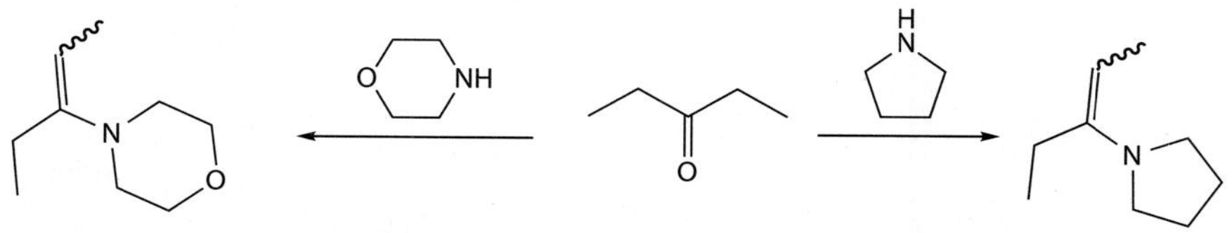

18.47

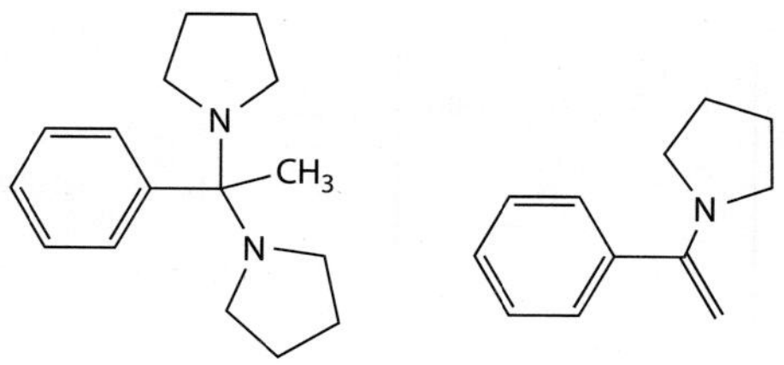

18.48

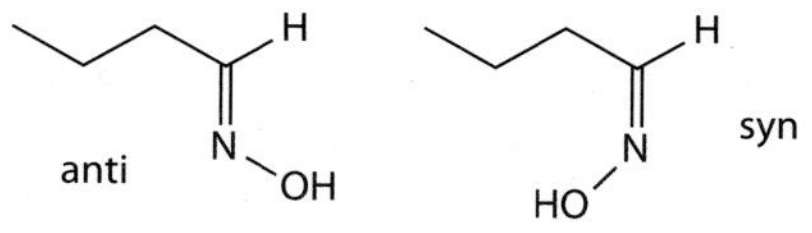

18.49

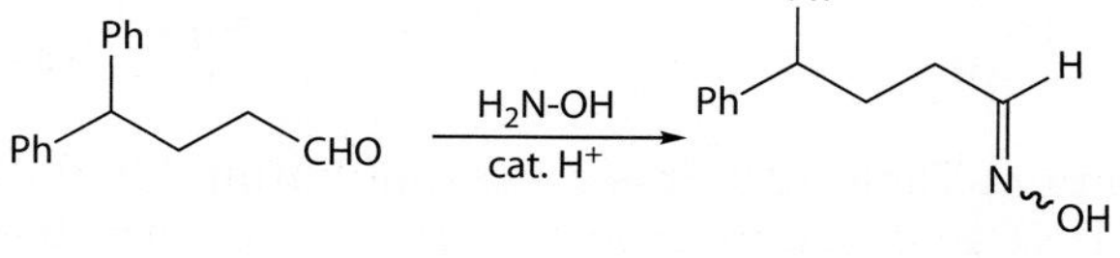

18.50

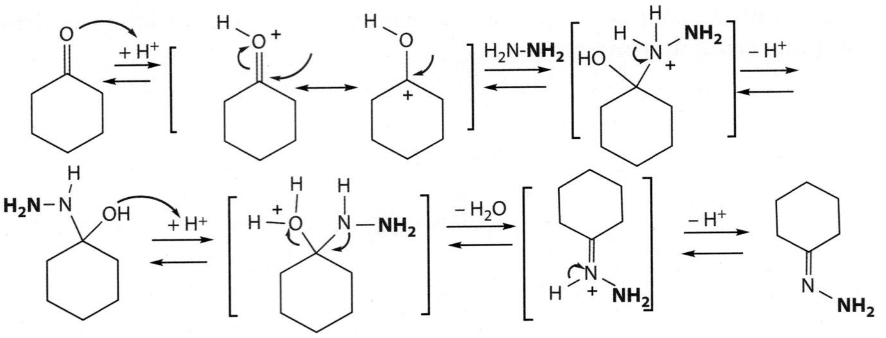

18.51 There are two products: the *E*- and *Z*-hydrazones shown.

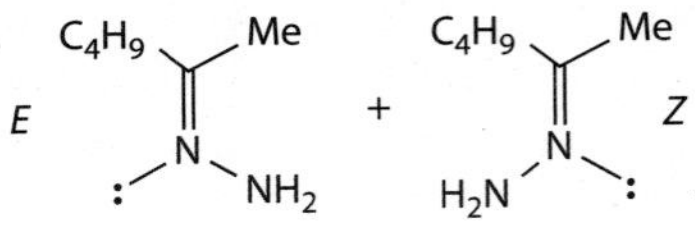

There are two products, the *E*- and *Z*-hydrazones shown.

18.52

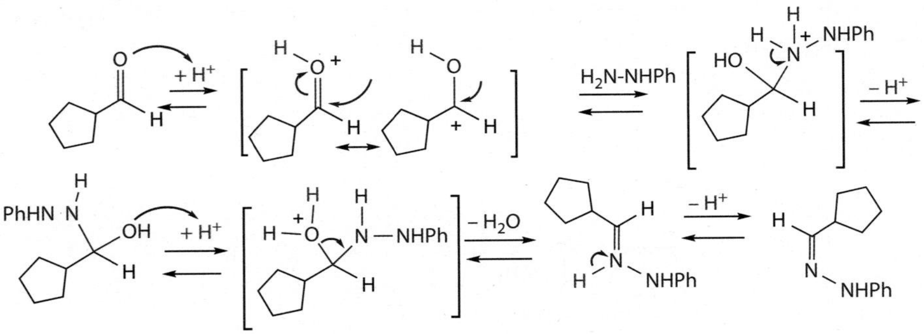

18.53

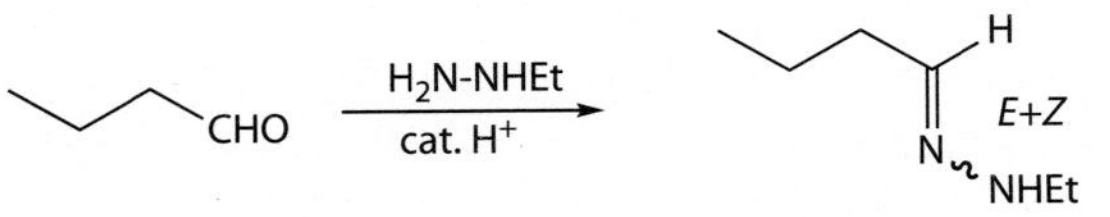

18.54

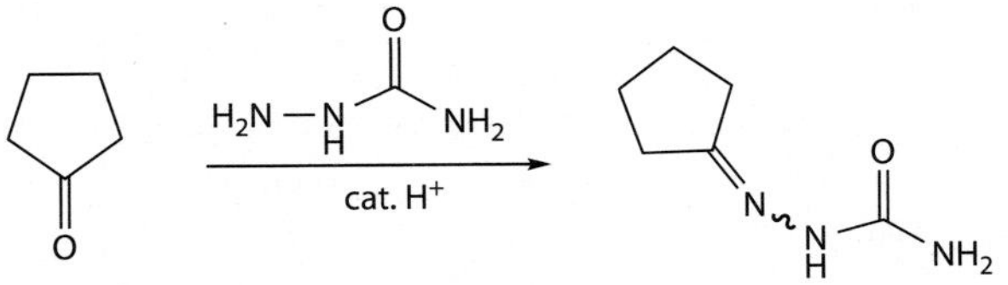

18.55 Using the data in Table 18.1, it is clear that butanal is the only molecule that forms a 2,4-DNP and a semicarbazone derivative with these melting points.

Correlation of Concepts with Homework

- Ketones and aldehydes react with nucleophiles to give substituted alkoxides by acyl addition to the carbonyl to give alcohols in a two-step process: (1) acyl addition and (2) hydrolysis. Nucleophilic acyl addition involves forming a new bond between the nucleophile and the acyl carbon, breaking the π-bond of the carbonyl with transfer of those electrons to the oxygen to give an alkoxide. Addition of an acid catalyst leads to an oxocarbenium ion that facilitates acyl addition: 1, 2, 3, 4, 8, 9, 23, 24, 25, 26, 27, 29, 63, 64, 68, 69, 74, 86.

- Grignard reagents and organolithium reagents react as carbanions with aldehydes and ketones to give alcohols in a two-step process: (1) acyl addition and (2) hydrolysis. The organometallic reagent reacts as a nucleophile with epoxides at the less substituted carbon atom: 10, 11, 12, 14, 15, 16, 56, 57, 58, 59, 66, 67, 72, 91, 93, 98, 109.

- Water adds reversibly to aldehydes and ketones to give an unstable hydrate: 18, 19, 20, 65, 69, 74.

- Alcohols add reversibly to aldehydes and ketones to give a transient hemiacetal or hemiketal, which then reacts with more alcohol to give an acetal or a ketal. Acetal and ketal formation is reversible. An excess of alcohol drives the reaction toward the acetal or ketal, but an excess of water will convert the acetal or ketal back to the aldehyde or ketone. 1,2-Diols react with aldehydes and ketones to give 1,3-dioxolane derivatives and 1,3- diols react to give 1,3-dioxane derivatives: 21, 22, 23, 24, 25, 26, 28, 29, 30, 31, 32, 33, 60, 70, 71, 73, 75, 76, 77, 78, 79, 82, 83, 84, 88, 90, 92, 93, 101.

- Thiols react with aldehydes and ketones to give dithioacetals or dithioketals. 1,2-Dithiols react with aldehydes and ketones to give 1,3-dithiolane derivatives and 1,3-dithiol reactions to give 1,3-dithiane derivatives: 34, 35, 36, 37, 38, 77, 78, 79, 80, 81.

- Primary amines react with aldehydes and ketones to give imines. Secondary amines react with ketones and sometimes with aldehydes to give enamines: 39, 40, 41, 42, 43, 44, 45, 46, 47, 61, 73, 84, 89, 93.

- Hydrazine reacts with aldehydes and ketones to give hydrazones. Hydrazone derivatives react with aldehydes and ketone to give N-substituted hydrazones. Hydroxylamine reacts with aldehydes and ketones to give oximes. Semicarbazone reacts with aldehydes and ketones to give semicarbazides: 48, 49, 50, 51, 52, 53, 54, 62, 64, 73, 84, 85, 93, 98.

- A molecule with a particular functional group can be prepared from molecules containing different functional groups by a

> series of chemical steps (reactions). This process is called synthesis: The new molecule is synthesized from the old one (see Chapter 25): 11, 13, 16, 94, 95, 96.
> - Spectroscopy can be used to determine the structure of a particular molecule and can distinguish the structure and functionality of one molecule when compared with another (see Chapter 14): 97, 98, 99, 100, 101, 102, 103, 104, 105, 106, 107, 108, 109.

Homework

18.56   Which of the following ketones reacts with CH_3MgBr to give 2-methyl-2-pentanol? Explain your answer.

18.57   Briefly explain why one cannot use ethanol as a solvent to form a Grignard reagent.

18.58   Draw and discuss the product or products expected from a reaction of 2S-bromohexane and 1. Mg/ether; 2. 2-pentanone; 3. hydrolysis.

18.59   Briefly explain why THF is used as a solvent to form the Grignard reagent of a vinyl halide when ether is used to form the Grignard reagent of a simple alkyl halide. (Hint: see Chapter 15, Section 15.2.)

18.60   The reaction of cyclopentanone and 1,8-octanediol, with an acid catalyst, does not give a ketal for the product. Suggest a reason for the failure of this reaction. (Hint: see Chapter 8, Section 8.7.)

18.61   Suggest a reason why the enamine formed by the reaction of 3,3-dimethyl-2-butanone and diisopropylamine is more likely to be **A** than **B**.

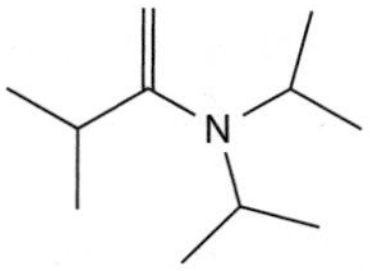

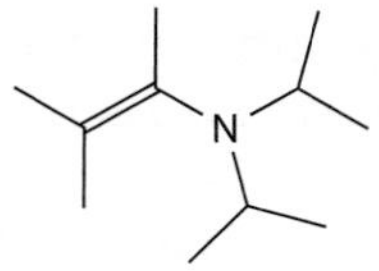

18.62   You have two vials and you know that one contains 5,5-dimethylcyclohexenone and the other contains 3,3-dimethylcyclohexanone. All of your NMR and IR equipment is broken (budget cuts). You have an inspiration and remember some chemistry that your parents used and find some 2,4-dinitrophenylhydrazine hidden away in the archives. Describe the experiment you must do and what to look for in order to distinguish these two compounds clearly.

18.63   The reaction of sodium cyanide and cyclohexanone gives a poor yield of the cyanohydrin. You attend a lecture and learn that silicon compounds such as Me_3SiCl react with methoxide to form Me_3SiOMe. In another flash of inspiration, you mix cyclohexanone, KCN, and Me_3SiCl. What product do you anticipate will be formed from this reaction? Draw a reaction sequence that leads to your product.

18.64   The reaction of 3-heptanone and butylamine leads to the corresponding imine. Draw it. In order to convert this imine back to 2-heptanone, you heat the imine with aqueous acid. Write out the complete mechanism for conversion of this imine back to 3-heptanone under these conditions.

18.65   Hydrates lose water to form an enol, which tautomerizes back to the carbonyl compound. It is known that most ketones exist in >99% as the keto form. It is also known that 2,4-pentanedione has a large percentage of the enol form in equilibrium with the keto form, depending on the solvent. Suggest a reason why this dione has such a high percentage of enol.

18.66   It is known that the reaction of ethylmagnesium bromide with 2,2,4,4-tetramethyl-3-pentanone does not give the expected alcohol product. In fact, the ketone is reduced to an alcohol and ethylmagnesium bromide is converted to ethene. Reductions are discussed in Chapter 19, but offer a reason for the fact that this ketone does not undergo normal acyl addition.

18.67   Give the major product formed when each of these starting materials reacts with 2-propylmagnesium chloride followed by hydrolysis. Draw the product formed when they react with ethyllithium followed by hydrolysis.

18.68   In a few words, describe the fundamental differences and similarities between the C=C unit in an alkene and the C=O unit of a ketone or aldehyde.

18.69   Offer a brief explanation why 1,1,1-trichloroethanal may be less reactive to nucleophilic acyl addition in aqueous solution.

18.70   Although sulfuric acid is a perfectly good catalyst for the acid-catalyzed formation of ketals from ketones, other acids such as *p*-toluenesulfonic acid are used more often. Suggest reasons for this.

18.71   Show the structure of the ketal or acetal and the hemiketal or hemiacetal for each of the following:
(a) 2,2-dimethyl-4-nonanone + propanol
(b) cyclobutanone + ethanol
(c) 1,5-dicyclohexyl-3-pentanone + methanol
(d) cyclooctanone + cyclohexanol

18.72 Give the product when each of the following nucleophiles reacts with cyclopentanone (there is no hydrolysis step):
(a) NaCN
(b) NaOEt
(c) sodium anion of 1-butyne
(d) MeMgBr
(e) MeLi
(f) ethylamine

18.73 Give the complete mechanism for each of the following transformations:

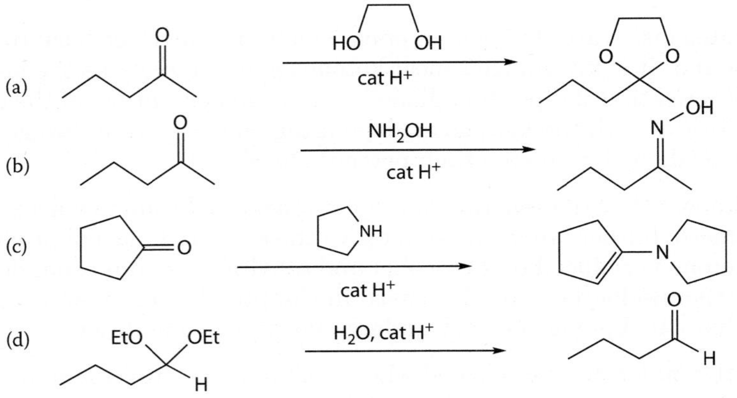

18.74 Benzoylformaldehyde (phenylglyoxal, see **A**) forms a relatively stable hydrate, particularly in aqueous solution. The hydrate formed from phenylacetaldehyde (**B**) is not very stable under the same conditions. Briefly explain.

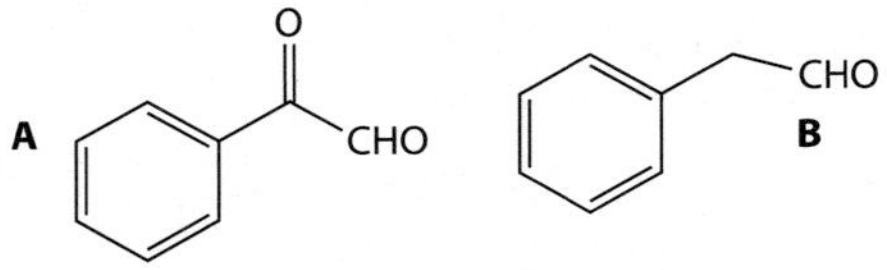

18.75 Molecule **A** (called talaromycin B) is a natural product that contains a ketal unit. Molecule **B** is a derivative of talaromycin **B**. Suggest an acyclic precursor that can be transformed to **B** by a simple chemical reaction. Show that reaction and briefly discuss why **B** could be formed from this precursor.

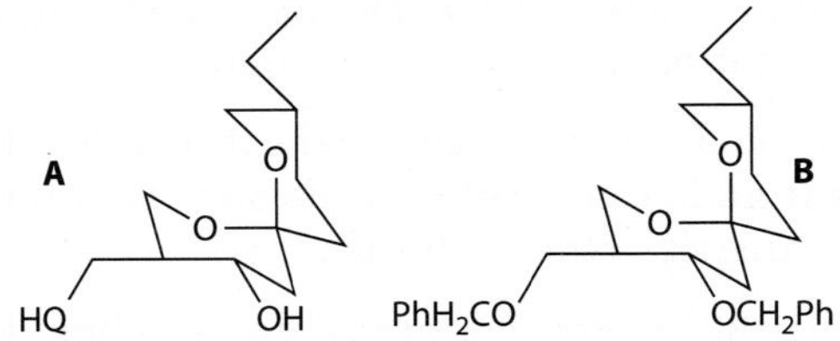

18.76 When cyclohexanone is treated with an acid catalyst in the presence of both 1,3-propanediol and 1,8-octanediol, one product predominates via reaction of only one of the diols. Draw this product and briefly discuss your choice (specifically, why it works and the other does not).

18.77   Show a chemical reaction, including reagents and reaction conditions, that
will transform each of the following back to the parent ketone or aldehyde:
(a) 1,1-diethoxycyclopentane
(b) 3,3-dimethoxyhexane
(c) 2,2-diethyl-1,3-dioxolane
(d) 5,5-dimethyl-1,1-dipropoxyhexane
(e) 2-phenyl-1,3-dithiane
(f) 2,2-diphenyl-1,3-dioxane
(g) 2,2-dithiomethylpentane

18.78   Give the complete mechanism for each of the following transformations:
(a) the reaction of cycloheptanone and methanol to give the ketal (acid
catalyzed)
(b) the reaction of 2-phenylethanal and water to give the hydrate (acid
catalyzed)
(c) the reaction of 2,2-diethoxyhexane to give 2-hexanone and ethanol (acid
catalyzed)
(d) the reaction of pentanal and 1,2-dimethyl-1,2-ethanediol to give the ketal
(acid catalyzed)
(e) the reaction of cyclobutanone and ethanethiol to give the dithioketal
(catalyzed with BF_3)

18.79   Draw the structure of each of the following molecules:
(a) 4-phenyl-1,1-dimethoxycyclohexane
(b) 2,2-dimethyl-4,4-diphenyl-1,3-dioxane
(c) 1,1-diethylthiopentane
(d) 2,2-(diisopropoxy)-3,4-dimethylhexane
(e) 2-butyl-1,3-dioxolane

18.80   A thioketone has a C=S unit rather than a C=O unit. Draw the structure of
thiocyclopentanone. Show the mechanism and final product if this thioketone
is treated ethanol and an acid catalyst.

18.81   It is known that *gem*-dithiols can be formed by the reaction of a ketone and
H_2S under pressure or under mild conditions using HCl as a catalyst. Give
a mechanism for the conversion of 2-butanone to the *gem*-dithiol using H_2S
with a HCl catalyst. If this dithiol is then mixed with cyclopentanone and an
acid catalyst, comment on the likelihood of forming a dithioketal.

18.82   Molecule **A** is derived from a diketone, but one carbonyl has been converted
to a ketal and the other to a dithioketal. Do you think it is possible to convert
the ketal or the dithioketal selectively to the corresponding carbonyl without
disturbing the other? Suggest reagents that might do this.

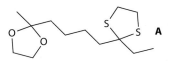

18.83   When we convert 3-pentanone to the ketal derived from ethanol, it is possible
to use ether as a solvent in this reaction. Why do we not see products derived
from ether attacking the carbonyl of the ketone?

18.84   Give the product formed from each of the following reactions:

(a) [structure] CHO $\xrightarrow{\text{NH}_2\text{OH}}_{\text{cat. H}^+}$

(f) [structure with Ph and O] [structure with OH] $\xrightarrow{\text{cat. H}^+}$

(b) [cyclopentanone structure] =O $\xrightarrow{\text{pyrrolidine, cat. H}^+}$

(c) [structure with O] [cyclopentyl–NH$_2$] $\xrightarrow{\text{cat. H}^+}$

(g) [structure] =O [morpholine structure with O and NH] $\xrightarrow{\text{cat. H}^+}$

(h) [dioxolane structure with O, O] $\xrightarrow{\text{H}_2\text{O}}_{\text{cat. H}^+}$

(d) [cyclohexenone structure] =O $\xrightarrow{\text{PhNHNH}_2, \text{cat. H}^+}$

(i) [cyclohexanone with O] $\xrightarrow{\text{2,4-dinitrophenylhydrazine}}_{\text{cat. H}^+}$

(e) [structure with CHO, Ph, Ph] [HO–OH structure] $\xrightarrow{\text{cat. H}^+}$

(j) [cyclopentene structure] $\xrightarrow{\begin{array}{l}\text{1. MeCO}_3\text{H}\\\text{2. PhMgBr}\\\text{3. H}_3\text{O}^+\end{array}}$

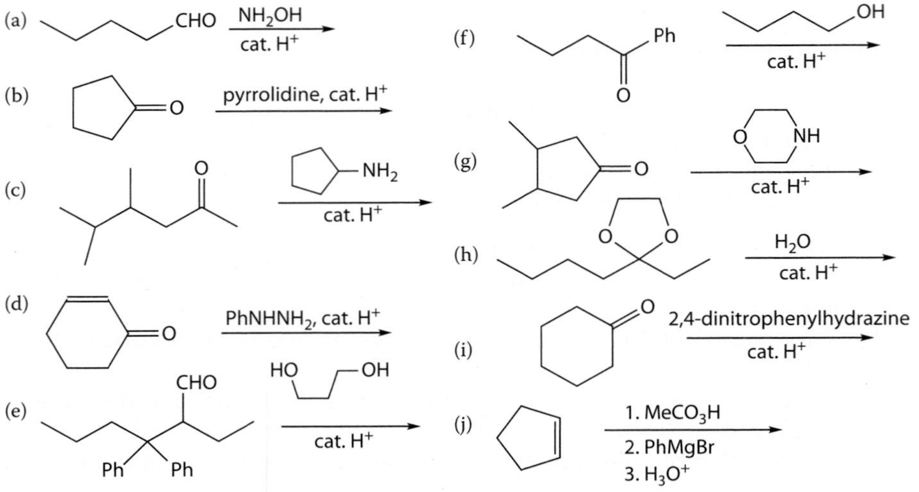

18.85   An unknown molecule is known to be a ketone and the 2,4-dinitrophenyl-hydrazone derivative of this ketone is bright yellow. What does this tell you about the structure? If the 2,4-DNP derivative had been a dark orange or red, what structural information would that convey?

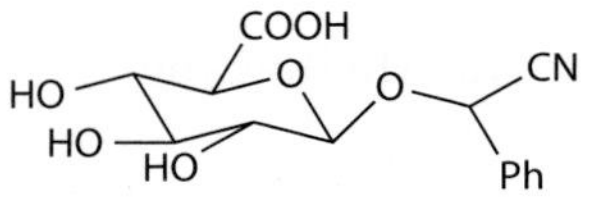

[structure of laetrile A with COOH, HO, HO, HO, O, O, CN, Ph labels]   **A**

18.86   Laetrile (**A**) is a compound that was believed to be effective against cancer. What do you think will happen to **A** when it is ingested and hits the 6N HCl in your stomach? Draw a mechanism and the products. Given this reaction, how do you think laetrile kills cells? Do you think it will be selective against cancer cells?

18.87   Draw the products formed for each of the following reactions:
(a)   cyclopentanone + HCN
(b)   3,3-diphenylpentanal + HCN
(c)   1,1-dimethoxybutane + HCN

18.88   Compare 2-ethyl-1,3-dioxane (**A**) and 2-ethyl-1,4-dioxane (**B**). Do you think **B** is as susceptible to reaction with aqueous acid as **A**? Why or why not?

A [1,3-dioxane structure with O, O and ethyl]   [1,4-dioxane structure with O, O and ethyl] **B**

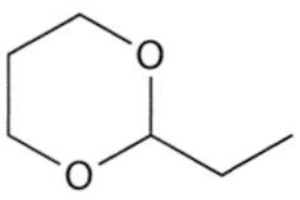

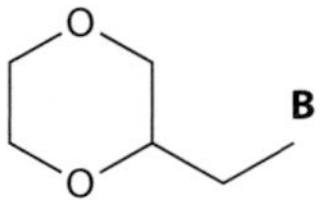

18.89   Conhydrine (**A**) has the structure shown. If it is reacted with cyclohexanone in the presence of an acid catalyst, draw the product expected and a mechanism for formation of that product.

A

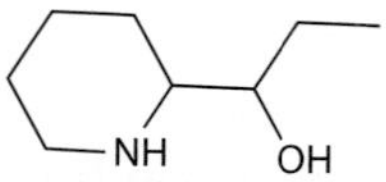

18.90 When 1,3,5-trioxane (**A**) is hydrolyzed, we obtain a common carbonyl derivative. Draw that product, name it, and show a mechanism for its formation.

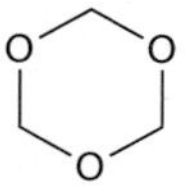

A

18.91 Draw the final product, if any, for each of the following reactions:
 (a) 2-iodopentane + 1. Mg/ether; 2. 2-butanone; 3. dilute aqueous acid
 (b) phenylmagnesium bromide + 1. cyclohexanone; 2. dilute aqueous acid
 (c) iodomethane + 1. Mg/ether; 2. 1-butyne
 (d) bromocyclopentane + 1. Mg/THF; 2. cyclopentanecarboxaldehyde; 3. dilute aqueous acid
 (e) butylmagnesium chloride + water
 (f) 2-methylhexylmagnesium bromide + 1. 2-hexanone; 2. dilute aqueous acid
 (g) 1-iodopentane + 1. Li; 2. acetone; 3. dilute aqueous acid
 (h) phenyllithium + 1. 4-ethyl-1-phenyl-1-heptanone; 2. dilute aqueous acid
 (i) n-butyllithium + 1. 1-propyne; 2. dilute aqueous acid
 (j) n-butyllithium + 2-phenyloxirane followed by an aqueous acid workup
 (k) methyllithium + 1. 3-phenylpropanal; 2. dilute aqueous acid

18.92 Briefly explain why removing water by azeotropic distillation with benzene drives the reaction of 2-butanone and excess ethanol to product when done in benzene as a solvent with an acid catalyst.

18.93 Give the major product for each reaction:

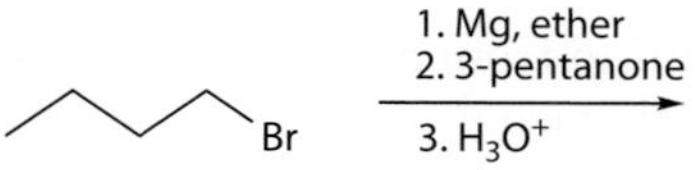

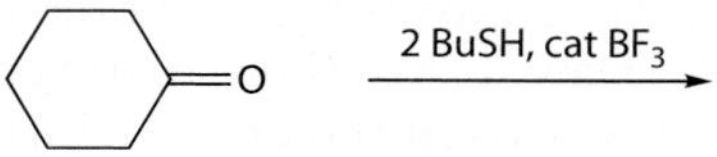

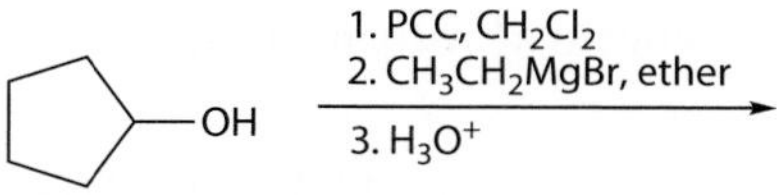

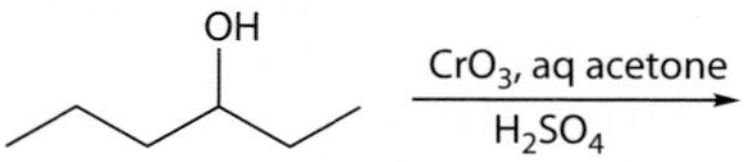

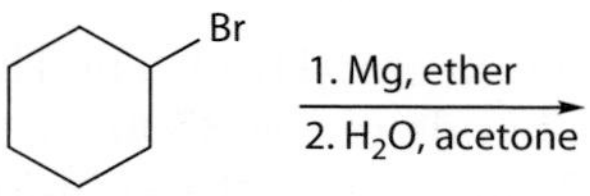

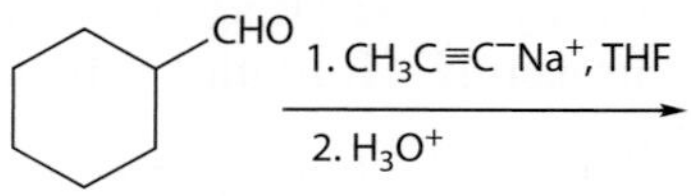

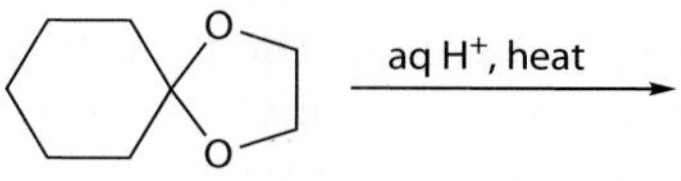

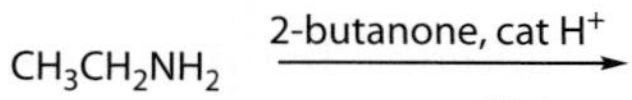

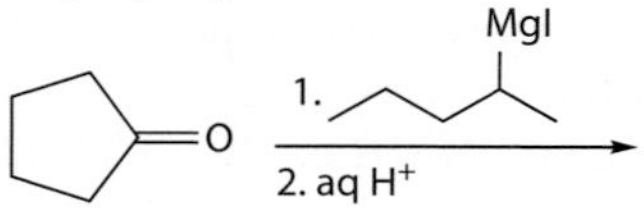

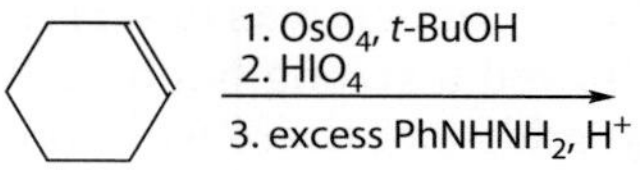

Synthesis Problems

Do not attempt these problems until you have read and understood Chapter 25.

18.94   Draw all reaction steps necessary, including all reagents and products, for
        each of the following transformations:
        (a) 3-phenyl-3-heptanol from 3-heptanone.
        (b) 1-(1-propenyl)cyclohexanol from 1-bromo-1-propene.
        (c) 5-methyldec-6-yn-5-ol from 1-pentyne.
        (d) 1-(1-methylethyl)cyclopentanol from 2-iodopropane.

18.95   Give a complete synthesis for each of the following:

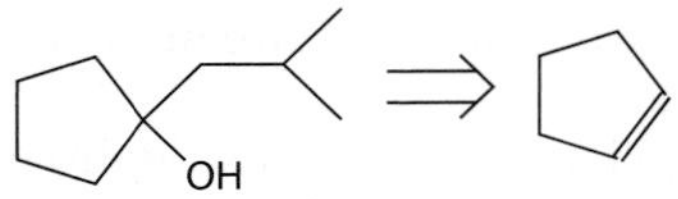

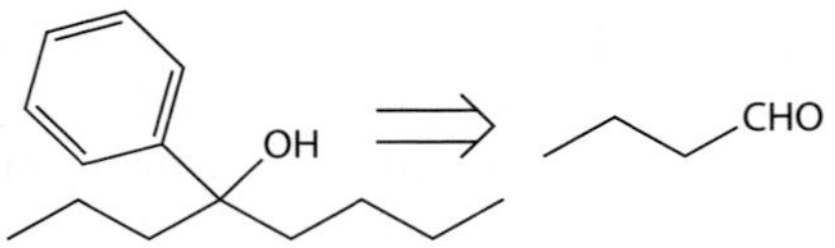

18.96   Suggest a synthesis of 1-(dimethylamino)cyclohexene from cyclohexene.

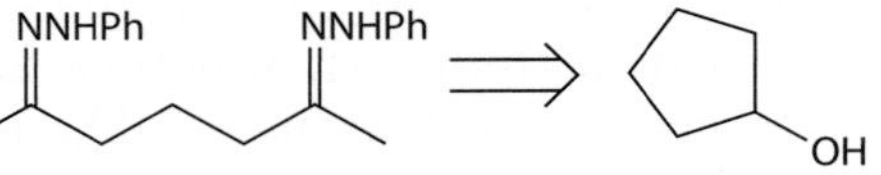

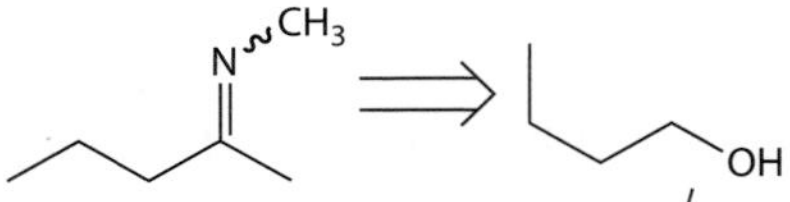

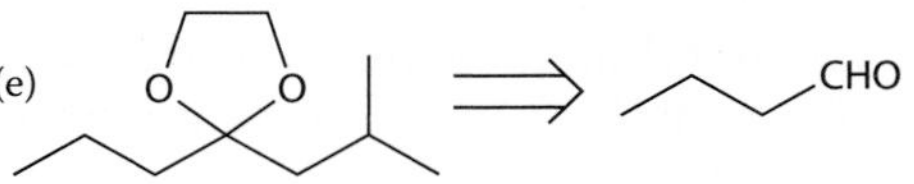

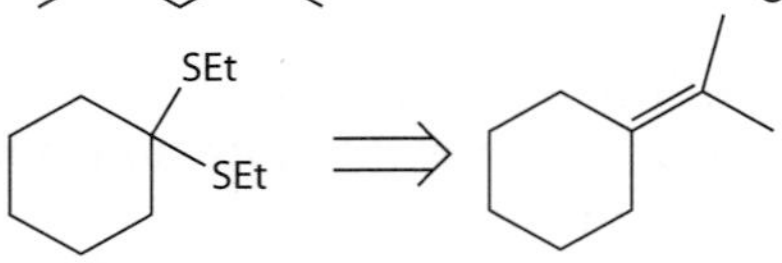

Spectroscopy Problems

Do not attempt these problems until you have read and understood Chapter 14.

18.97   Can you use IR alone to distinguish an imine from an enamine? Can you use
        proton NMR alone? Why or why not in both cases?

18.98   The product of the reaction between the Grignard reagent of cyclobutyl
        bromide and 3-pentanone is an alcohol. Draw it. Are there distinguish-
        ing features in the proton NMR? What features in the proton NMR might
        suggest a four-membered ring? Can structural information be obtained
        from the mass spectrum?

18.99   An unknown has an elemental analysis: C, 76.14; H, 11.18; O, 12.68. The infra-
        red shows strong peaks at 2930, 2858, and 1709 cm^{-1}. The proton NMR is: 1.43-
        1.53 (m, 4H), 1.47-1.49 (m, 2H), 1.55-1.80 (m, 4H), and 1.98 (s, 3H) ppm. The
        carbon NMR is: 212.0, 51.5, 28.5, 27.9, 26.0, and 25.7 ppm.

18.100 An unknown was determined to have a formula of $C_7H_{12}O$. The IR spectrum
        showed a strong band at 1665 cm^{-1} with no bands in the 3500 cm^{-1} region.
        In addition, there were no bands near 2800 cm^{-1} and nothing in the region
        around 2270 cm^{-1}. The 2,4-DNP derivative was an orange-red solid. Discuss
        possible structures for this molecule.

18.101 You have attempted to convert cyclopentanone to the corresponding ethylene glycol-ketal derivative. Your NMR is broken down and you cannot use the mass spectrum. Using only IR and any chemical test you like, describe how you can tell whether the reaction worked or not. (Note: you cannot use boiling point, melting point, etc.)

18.102 You have two bottles containing a liquid. You know that one is 3-pentenal and one is 4-pentenal. Describe how you can identify these using IR, NMR, and chemical transformations.

18.103 Given the following theoretical spectral data, give the structure: M = 148, 100%; M+1 = 149, 11.1%; M+2 = 0.82%.

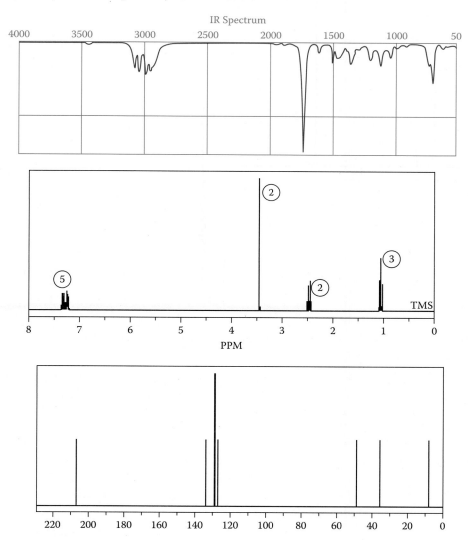

18.104 Given the following theoretical spectral data, give the structure: M = 98,
100%; M+1 = 99, 6.66%; M+2 = 100, 0.42%. The infrared spectrum shows a
strong peak at 1690 cm⁻¹ and a moderate peak at 1604 cm⁻¹. There are two
prominent peaks at 845 and 705 cm⁻¹.

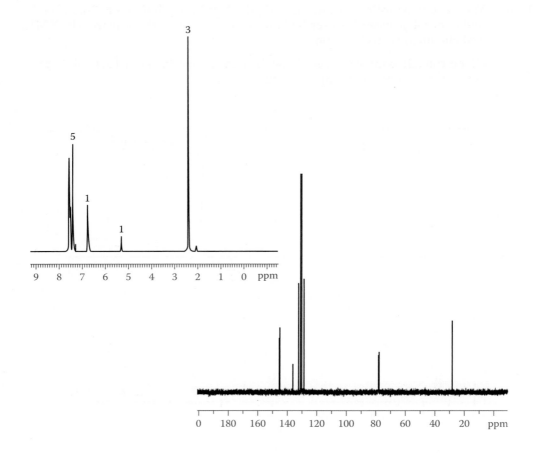

18.105 For the formula $C_{11}H_{16}O$, given the following spectral data, give the structure:

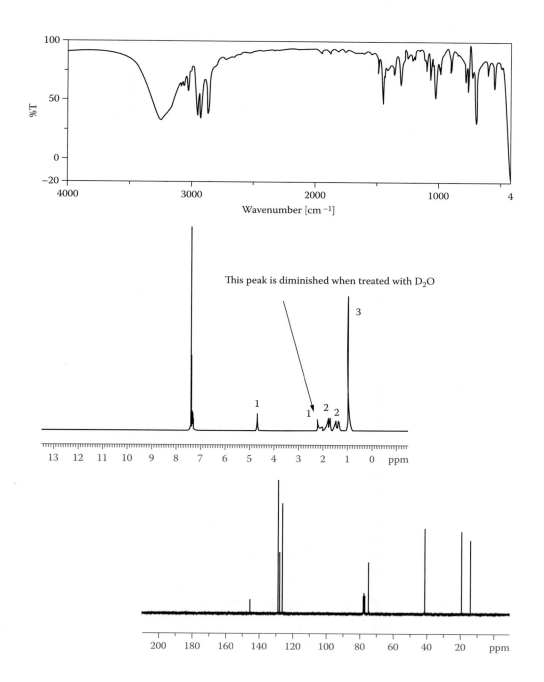

This peak is diminished when treated with D_2O

18.106 For the formula $C_{11}H_{16}O_2$, given the following theoretical spectral data, give the structure. The infrared spectrum shows no OH or C=O peaks, but there is strong absorption about 1050 cm^{-1}.

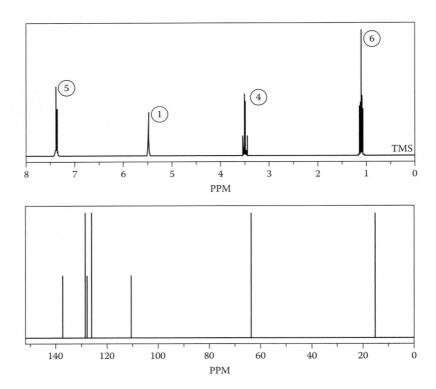

18.107   For the formula C_8H_7NO, given the following spectral data, give the structure:
IR: 3412–3036, 2261, 1691, 1658, 1698, 1584, 1485–1466, 1086–1003 cm^{-1}
^{1}H NMR: 7.70–7.22 (m, 5H), 5.46 (s, 1H), 3.75 (broad s, 1H; this peak is diminished when treated with D_2O) ppm

18.108   For the formula $C_6H_{10}O$, given the following spectral data, give the structure:
IR: 3310 (broad), 2962–2938, 2876, 2115 (weak), 1468, 1067–1010 cm^{-1}
^{1}H NMR: 4.38 (t, 1H), 2.7 (broad s, 1H; this peak is diminished when treated with D_2O), 2.46 (s, 1H), 1.67 (m, 2H), 1.51 (m, 2H), 0.96 (t, 3H) ppm

18.109   A molecule **A** that has the formula C_6H_{12} has the following spectra:
IR: a peak at 3178, 2960–2839, and a peak at 1643 cm^{-1}
^{1}H NMR: 5.77 (m, 1H), 4.97 (m, 1H), 4.96 (m, 1H), 1.94 (m, 2H), 1.63 (m, 1H), 0.90 (d, 6H) ppm
Reaction of **A** with 9-BBN followed by NaOH/H_2O_2 gives **B**, which has the following spectra:
IR: broad peak at 3325, 2957–2846, and a peak at 1071–1022 cm^{-1}
^{1}H NMR: 3.61 (broad t, 2H), 1.94 (broad s, 1H; this peak is diminished when treated with D_2O), 1.84–1.26 (broad m, 5H), 0.80 (d, 6H) ppm
When **B** is treated with PBr$_3$, **C** is formed. Subsequent reaction of **C** with magnesium metal in ether generates **D**, which reacts with 2-butanone to give **E**.
Unknown **E** has the formula $C_{10}H_{22}O$, and the following spectra:
IR: a broad peak at 3380, 2956–2871, and peaks at 1062–1018 cm^{-1}
^{1}H NMR: 1.55 (m, 1H), 1.48 (q, 2H), 1.42 (broad s, 1H; this peak is diminished when treated with D_2O), 1.41 (broad t, 2H), 1.32–1.17 (m, 4H), 1.14 (s, 3H), 0.89 (t, 3H), 0.878 (d, 6H) ppm
Show the structures of **A–E**.

Reduction

19

Oxidation reactions were discussed in Chapter 17; they involve the loss of two electrons, although oxidations are also associated with the loss of hydrogen atoms or the gain of oxygen atoms. The oxidation of one molecule is accompanied by the reduction of another. If an oxidation involves the loss of two electrons, a reduction involves the gain of two electrons. The structural changes in the reduction product are often measured by whether hydrogen is gained and/or if an oxygen atom is lost. Chapter 17 introduced several key oxidation reactions, including those that convert an alcohol to an aldehyde or a ketone. The conversion of an aldehyde or ketone back to an alcohol is formally a reduction, and several reagents used for this purpose are introduced in this chapter. In addition, hydrogenation of alkenes and alkynes to alkanes or alkenes is formally a reduction. This chapter will review known reactions that are classified as reductions, as well as the typical transformations associated with these reactions.

To begin this chapter, you should know the following:

- **basic rules of nomenclature for alcohols, aldehydes, ketones, carboxylic acids, and carboxylic acid derivatives (Chapter 5, Sections 5.6 and 5.9)**
- **bond polarization (Chapter 3, Section 3.7)**
- **CIP rules for prioritizing substituents, groups, and atoms (Chapter 9, Section 9.3)**
- **Lewis acids and Lewis bases (Chapter 6, Section 6.5)**
- **energetics and transition states (Chapter 7, Sections 7.5 and 7.6)**

- **rotamers and conformations (Chapter 8, Sections 8.1 and 8.4)**
- **chirality, enantiomers, and diastereomers (Chapter 9, Sections 9.1 and 9.5)**
- **alkene stereoisomers (Chapter 9, Section 9.4)**
- **fundamental reactivity of alkenes (Chapter 10, Sections 10.2–10.4)**
- **fundamental reactivity of carbonyl compounds (Chapter 16, Sections 16.3 and 16.7; Chapter 18, Section 18.1)**
- **how to identify acyl addition reactions with strong and weak nucleophiles (Chapter 18, Sections 18.3–18.7)**
- **fundamentals of electron transfer for an oxidation (Chapter 17, Section 17.1)**
- **reagents used to oxidize an alcohol to an aldehyde or ketone (Chapter 17, Section 17.2)**
- **oxidation reactions of alkenes (Chapter 17, Section 17.3)**

This chapter will discuss methods for the reduction of aldehydes, ketones, and carboxylic acid derivatives to the corresponding alcohol. Both hydride-reducing agents and catalytic hydrogenation will be discussed. Hydrogenation of alkenes to alkanes and hydrogenation of alkynes to alkenes will be discussed. A few specialized reduction techniques will also be introduced, including reduction of an aldehyde or ketone to the corresponding hydrocarbon, as well as reduction of an alkyl halide to an alkane.

When you have completed this chapter, you should understand the following points:

- **A reduction is defined as the gain of electrons or the gain of hydrogen atoms.**
- **Both sodium borohydride and lithium aluminum hydride reduce ketones or aldehydes to alkoxides, and hydrolysis gives the alcohol. Lithium aluminum hydride is a more powerful reducing agent than sodium borohydride.**
- **Lithium aluminum hydride will reduce acid derivatives to the corresponding alcohol and an amide to an amine. Nitriles are also reduced to an amine.**
- **In the presence of a transition metal catalyst, hydrogen gas converts an alkene to an alkane and an alkyne to an alkene. Hydrogenation of an alkyne with palladium in the presence of quinoline gives the Z-alkene in what is known as Lindlar reduction.**
- **In the presence of a transition metal catalyst, hydrogen gas converts a ketone or aldehyde to an alcohol.**

- Acid chlorides can be reduced by catalytic hydrogenation to give an aldehyde in what is called the Rosenmund reduction. Nitriles are reduced by catalytic hydrogenation to a primary amine.
- A ketone or aldehyde is reduced to an alcohol by reaction with sodium or lithium metal in liquid ammonia, in the presence of ethanol. This is called a dissolving metal reduction and it proceeds by an alkoxy-radical known as a ketyl.
- Alkynes are reduced to *E*-alkenes under dissolving metal conditions. Benzene is reduced to a cyclohexadiene under the same conditions.
- Zinc in HCl or acetic acid will reduce an alkyl halide to the corresponding hydrocarbon. Tin or iron in acid will reduce an aldehyde or ketone to an alcohol.
- A ketone or aldehyde is reduced to the corresponding hydrocarbon by treatment with zinc amalgam in HCl, in what is known as Clemmensen reduction. A ketone or aldehyde is reduced to the corresponding hydrocarbon by treatment with hydrazine and KOH, in what is known as a Wolff–Kishner reduction.
- A nitrile is reduced to an aldehyde by treatment with tin chloride and HCl, in what is known as a Stephen reduction.
- A molecule with a particular functional group can be prepared from molecules containing different functional groups by a series of chemical steps (reactions). This process is called synthesis: The new molecule is synthesized from the old one (see Chapter 25).
- Spectroscopy can be used to determine the structure of a particular molecule, and can distinguish the structure and functionality of one molecule when compared with another (see Chapter 14).
- Reduction is important in biological systems.

19.1 Defining a Reduction

Just as an oxidation is a reaction in which two electrons are lost, a reduction is defined as a reaction in which two electrons are gained. A practical definition of an oxidation in Chapter 17 based on structural changes is a reaction in which hydrogen atoms are lost or an oxygen atom is gained. In a similar manner, a practical definition of reduction is a reaction in which hydrogen atoms are gained or an oxygen atom or any heteroatom is lost. Several different transformations are categorized as reductions, including the conversion of a ketone to an alcohol, an alkyne to an alkene, or an alkene to an alkane.

Oxidation number was defined in Chapter 17 and it may be used to monitor electron loss or gain. When acetone (**1**) is converted to 2-propanol (**2**), the change in oxidation number at the acyl carbon is from +2 to 0. This means that the product has two more electrons than the starting material (gaining negative electrons makes it less positive), and the gain of two electrons is a reduction. The conversion of 1-butene (**3**) to butane (**4**) is also a reduction. An oxidation number of −1 is assigned to C1 of the C=C unit, and C2 has an oxidation number of −2 (0 for each carbon and −1 for each hydrogen). In butane, C1 and C2 have oxidation numbers of −2 and −3, respectively (there are now two hydrogen atoms on each carbon rather than one). The conversion of butene to butane therefore requires that one electron be gained on each carbon, for a net gain of two electrons: a reduction.

19.1   **Identify the conversion of 2-butyne to 2-butene as an oxidation or a reduction.**

19.2 Hydrides as Reducing Agents

The discussion of reduction begins with a reaction of aldehydes and ketones. Specifically, aldehydes are reduced to a primary alcohol and ketones are reduced to a secondary alcohol. The conversion of a C=O unit to a HC–OH unit requires transferring a hydrogen atom to the δ+ carbon of the carbonyl, as well as the transfer of a second hydrogen to the oxygen, as represented by the conversion of **5** to **6**. These transfers can occur in a synchronous manner or stepwise. The process that transfers the hydrogen atoms in a stepwise manner is used most often, and it requires a reagent that first transfers a hydrogen atom to the δ+ carbon atom of the C=O unit. This reaction is an acyl addition (Chapter 18) where a negatively polarized hydrogen atom, a "hydride" represented as $H^{\delta-}$, is transferred to the acyl carbon atom.

There are two common hydride reducing agents: sodium borohydride (**7**; $NaBH_4$) and lithium aluminum hydride (**8**; $LiAlH_4$). Borohydride is an ate complex (see Chapter 6, Section 6.5.4, for definition of an ate complex) of BH_3 (**7** is a tetrahydridoborate) and aluminum hydride is an ate complex of AlH_3 (**8** is a tetrahydridoaluminate; their respective IUPAC names are sodium tetrahydridoborate and lithium tetrahydridoaluminate. The borate anion is the Lewis acid–Lewis base complex of BH_3 (borane) and a hydride, whereas the aluminate anion is the Lewis acid–Lewis base complex of AlH_3 (aluminum hydride) and a hydride. Examination of the periodic table reveals that hydrogen is more electronegative than boron (see Chapter 3, Section 3.7.1), so the B–H unit in **7** is polarized such that boron is electropositive ($\delta+$) and hydrogen is electronegative ($\delta-$). Similarly, the hydrogen atom of the Al–H unit in **8** is electronegative. The key to any reaction of **7** or **8** with an aldehyde or ketone is to recognize that the $\delta-$ H of either hydride will be transferred to the $\delta+$ carbonyl carbon by an acyl addition-like process.

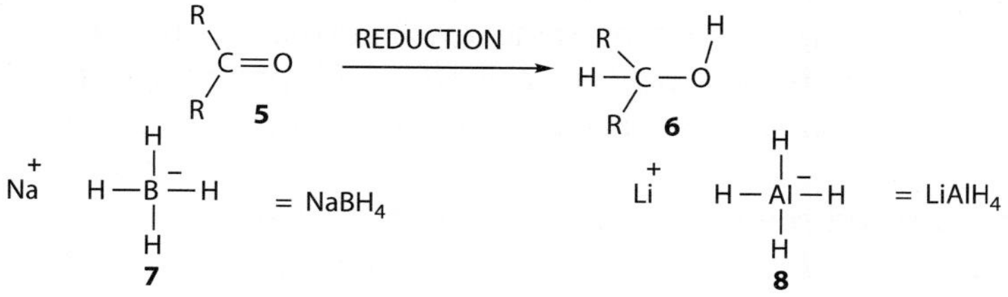

19.2.1 Reduction of Aldehydes and Ketones

Both sodium borohydride and lithium aluminum hydride contain the $\delta-$ hydrogen atom commonly known as a hydride. When **7** or **8** reacts with an aldehyde or ketone, the hydride is attracted to the $\delta+$ carbonyl carbon, suggesting an acyl addition reaction. In the experiment where 2-butanone (**9**) is treated with $NaBH_4$ in aqueous methanol, the product is alkoxide **11**—the acyl addition product expected if hydride attacks the acyl carbon. In **11**, the boron is attached to oxygen, and one hydrogen atom from the borohydride is attached to the acyl carbon. This means that a new σ-covalent C–H bond is formed. When **11** is treated with an aqueous solution of saturated ammonium chloride in a second step, 2-butanol (**12**) is isolated in 87% yield.[1]

In this and in other experiments using $NaBH_4$ with aldehydes and ketones, there are two steps: (1) acyl addition of the hydride and (2) aqueous acid workup. This result is entirely consistent with the acyl addition reactions discussed in Chapter 18. There has never been evidence for an intermediate, so the alkoxide product is formed directly. With knowledge that no intermediate is formed when **9** is transformed to **11**, this experiment must be explained by examining a transition state. The requisite transition state brings a B–H unit of sodium borohydride into close proximity to the polarized C=O unit of 2-butanone due to electrostatic attraction between the $\delta-$ H and the $\delta+$ C of the carbonyl. The hydride of the borohydride is transferred to the $\delta+$ carbon of the carbonyl, breaking the π-bond with transfer of those electrons to the boron. Structure **10** represents the requisite *four-centered transition state*.

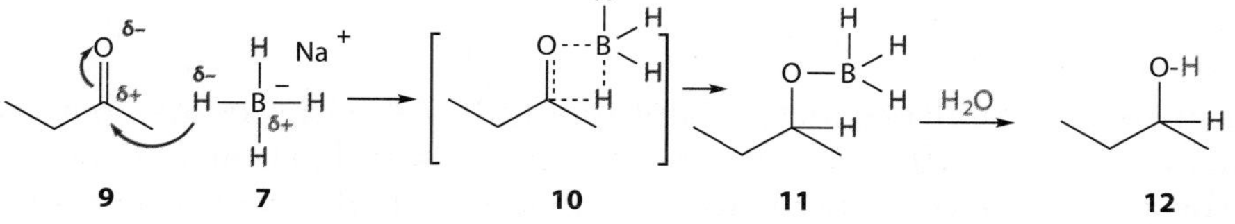

The product of the aldehyde-sodium borohydride reaction is **1**, which must be converted to a neutral species. The alkoxide is clearly a base, suggesting the need for an acid–base reaction. Because **11** is the conjugate base of an alcohol (see Chapter 6, Section 6.4.4), the reaction of **11** with an acid gives the conjugate acid 2-butanol (**12**). This step is analogous to the aqueous acid workup step required for the acyl addition products of nucleophiles such as a Grignard reagent (Chapter 18, Section 18.4). If product **12** is compared with **6**, the acyl addition to the carbonyl of the hydride incorporates only one hydrogen atom on the carbon. *One hydrogen atom in the HC–OH unit of 2-butanol comes from sodium borohydride, but the second one comes from the acid–base reaction with aqueous acid.*

Using a condensed reaction form, treatment of 2-butanone with $NaBH_4$ leads to 2-butanol, and the first step delivers the hydrogen marked in **blue** in the illustration, whereas aqueous ammonium chloride (a weakly acidic solution) is the second step that delivers the hydrogen marked in **red**. Sodium borohydride reduces both ketones and aldehydes. Indeed, the reduction of an aldehyde with $NaBH_4$ is somewhat easier than the similar reduction of a ketone. *Aldehydes are easier to reduce than ketones because the carbonyl unit is less sterically hindered.*

19.2 Draw the transition state and final product when cyclobutane-carboxaldehyde is treated with 1. $NaBH_4$ in ethanol; 2. aqueous ammonium chloride.

Lithium aluminum hydride (**8**) reacts with ketones and aldehydes in the same way as sodium borohydride, except that *$LiAlH_4$ is a more powerful reducing agent.* In one experiment, reaction of heptanal (**13**) with $LiAlH_4$ in diethyl ether, followed by aqueous acid workup, gave 1-heptanol (**16**) in 86% yield.[2] The mechanism is identical to that of borohydride in that heptanal reacts with the negatively polarized hydrogen of the Al–H unit in **8** via the four-centered transition state **14**. This leads to an alkoxyaluminate product, **15**, and subsequent treatment with dilute acid

(H_3O^+) gives the final product, 1-heptanol (**16**). As with sodium borohydride, lithium aluminum hydride reduces both aldehydes and ketones. As noted previously, the major difference between sodium borohydride and lithium aluminum hydride is the relative strength of each reducing agent. The Al–H unit is more polarized than the B–H unit, and lithium aluminum hydride is a much more powerful reducing agent. This is most easily seen in the choice of solvent for each reaction. In the sodium borohydride reduction of 2-butanone (**9** in illustration), the solvent is aqueous methanol; however, in the lithium aluminum hydride reduction of heptanal (**13**), the solvent is diethyl ether. Methanol is a polar protic solvent, with an acidic hydrogen (pK_a of about 15.3; see Chapter 6, Section 6.2) and water has a pK_a of 15.8, but $NaBH_4$ can reduce aldehydes and ketones in either methanol or water as the solvent.

Lithium aluminum hydride reacts *violently* with water or alcohols to release hydrogen gas. In effect, there is an acid–base reaction between the δ– hydrogen atoms of $LiAlH_4$ and the acidic protons in water. Therefore, an aprotic solvent such as ether or THF is used. Although $LiAlH_4$ is only partly soluble in ether solvents (typically, <5 g/100 g of solvent), the slurry (solid suspended in the solvent) has sufficient dissolved reagent to accomplish the reduction. The relative reactivity of each reagent with protic solvents is a good way to remember which is more reactive.

The hydride reactions that convert aldehydes or ketones to alcohols are as straightforward as this short section suggests. Aldehydes are reduced to primary alcohols and ketones to secondary alcohols. The key feature is to remember that *hydride reduction is a two-step procedure* in which the first step is treatment with $NaBH_4$ or $LiAlH_4$ and the second is hydrolysis of the alkoxide product.

19.3 **Draw the transition state and final product when phenylacetaldehyde is treated with 1. $LiAlH_4$ in ether; 2. aq. acid.**

19.4 **Draw the product when (a) 4,4-diphenylpentanal reacts with sodium borohydride and (b) 3-cyclopentylhexanal reacts with lithium aluminum hydride.**

19.5 **Draw the alcohol products formed by $LiAlH_4$ reduction of (a) 1-phenyl-3-octanone, (b) 4,4-diethylcycloheptanone, (c) benzophenone (1,1-diphenylmethanone), (d) cyclohexane carboxaldehyde, and (e) 3,4,6-trimethyloctanal.**

19.2.2 Hydride Reduction of Other Functional Groups

When compared to lithium aluminum hydride, sodium borohydride is a weaker reducing agent, but it easily reduces ketones and aldehydes to alcohols. However, sodium borohydride does not reduce other common functional groups, although esters may be reduced very slowly. Other acid derivatives, as well as amides or nitriles, are usually not reduced by sodium borohydride. See Chapter 20 for the nomenclature and structure of acid derivatives. *Lithium aluminum hydride, by comparison, will reduce virtually any heteroatom-containing functional group,* including the carbonyl of aldehydes and ketones. When pentanoic acid (**17**) is treated with $LiAlH_4$, hydrolysis of the initially formed product gives

1-pentanol. This contrasts with sodium borohydride, which reacts with **17** to give the sodium salt of the acid (the carboxylate anion) and does not give reduction to the alcohol. The reaction between sodium borohydride and **17** is formally an acid–base reaction.

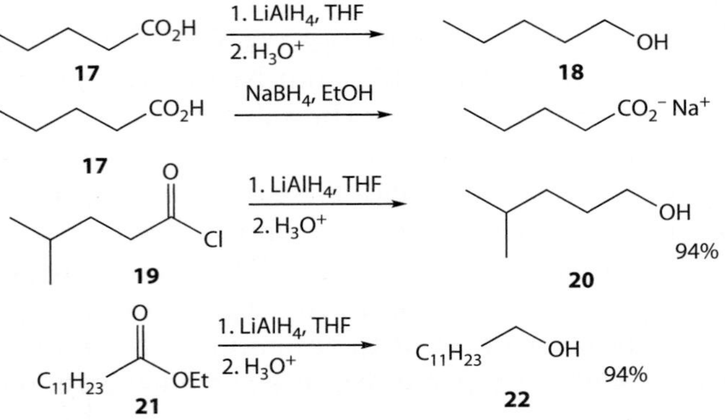

Carboxylic acid derivatives are reduced by $LiAlH_4$ to give the corresponding alcohol. 4-Methylpentanoyl chloride (**19**), for example, is reduced to 4-methyl-1-pentanol (**20**) in 94% yield by $LiAlH_4$.[3] Ethyl dodecanoate (**21**) is reduced in 94% yield to 1-dodecanol (**22**) using $LiAlH_4$.[3] Reduction of ethyl ester **21** also gives ethanol as a product (the "alcohol" part of the ester). Sodium borohydride can sometimes reduce esters, but in general this is often a slow reaction unless there are structural features that "activate" the ester to reduction. *For convenience, assume that $NaBH_4$ does not reduce esters, with the knowledge that sometimes it does.*

Amides such as *N,N*-dimethylcyclohexanecarboxamide (**23**; see Chapter 20, Section 20.7) are also carboxylic acid derivatives. Sodium borohydride does not reduce an amide. Lithium aluminum hydride reacts with **23**, but the product is an amine rather than an alcohol—specifically, 1-(*N,N*-dimethylaminomethyl)cyclohexane, **24**.[3] Amine **24** is isolated in 88% yield. Although the mechanism will not be discussed here in a formal manner, delivery of hydride to the acyl carbon of the C=O is followed by formation of an imine (C=N) that is further reduced to the amine. Nitriles such as octanenitrile (**25**) also react with $LiAlH_4$; in one experiment, reduction of **25** gave amine **26** (1-aminopentane) in 92% isolated yield.[4] This reduction also proceeds by delivery of hydride to the carbon of the nitrile, generating an imine that is further reduced to the amine. In general, $NaBH_4$ does not reduce amides or nitriles.

19.6 Draw the products formed when benzyl cyclopentanecarboxylate (see Chapter 20, Section 20.5, for the naming) is treated with 1. $LiAlH_4$; 2. aq. acid.

19.7 Draw the product of the reaction sequence 1. phenylacetic acid + $SOCl_2$; 2. $NaBH_4$, EtOH; 3. aq. NH_4Cl.

19.8 Draw the product formed when *N,N*-dimethylpentanamide is treated with 1. LiAlH₄ in THF; 2. aq. acid and then neutralization with base.

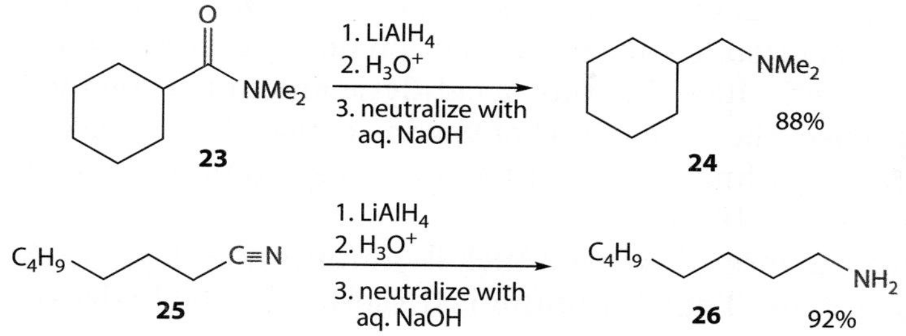

19.3 Catalytic Hydrogenation

Some functional groups are not reduced by lithium aluminum hydride or sodium hydride. Hydrogen gas is another reducing agent that can be used to reduce such functional groups, but only in the presence of a suitable catalyst. Catalytic hydrogenation of an alkene generates an alkane product, which is a formal reduction. Catalytic hydrogenation is also used to reduce a ketone or aldehyde to an alcohol. This section examines methods that use hydrogen gas as a reducing agent, the catalysts that are required to facilitate such reactions, and the transformations that are possible.

19.3.1 Mixing Hydrogen Gas and an Alkene Does Not Give Reduction

If a practical definition of reduction is addition of hydrogen atoms to a molecule, then hydrogen gas is a logical choice for a reducing agent. In fact, reduction of compounds that contain a π-bond is an important transformation in organic chemistry. If an alkene such as 1-heptene (**27**) is mixed with hydrogen gas in the solvent methanol, however, there is no reaction. It may seem surprising that H–H does not react with an alkene because Chapter 10 (Section 10.4.1) described the reaction of Br–Br as a Lewis base in the presence of an alkene. Halogens are polarizable, and that property is essential for the Lewis base–Lewis acid reaction. A comparison of diatomic bromine and diatomic hydrogen shows that diatomic hydrogen (H–H) is not polarizable. Therefore, it does not react directly with the alkene. In order to see a chemical reaction, something must be added first to break the H–H bond before there is a reaction with the alkene π-bond.

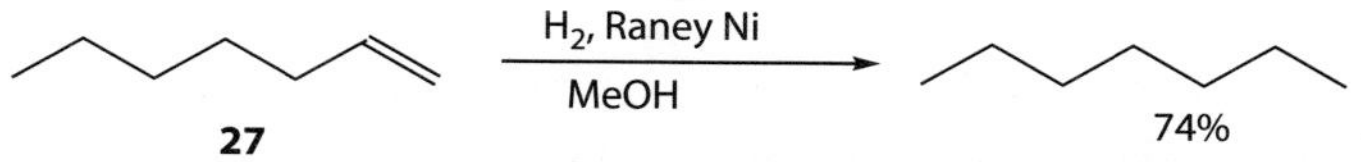

What can be added to break apart the H–H bond? In other words, what must be added to react with hydrogen gas to allow further reaction with an alkene? Based on many experiments, the answer is a transition metal such as nickel. Indeed, addition of a small amount of nickel to a mixture of hydrogen gas and 1-heptene (**27**) gives an excellent yield of heptane. Obviously, the metal plays a key role in this reaction because the reaction does not occur without the nickel. Further experiments show that only a catalytic amount of the nickel is required. In this specific case, the type of nickel used is called Raney nickel—abbreviated as Ni(R)—which is a finely divided form of nickel named after Murray Raney (United States; 1885–1966).

This catalyst is formed when a nickel–aluminum alloy (Ni–Al) is dissolved in aqueous hydroxide. The aluminum is dissolved by the hydroxide, and hydrogen gas is a by-product. The aluminum is converted to aluminum hydroxide, which is soluble in the aqueous hydroxide medium, but the nickel is insoluble and precipitates as a finely divided powder. Hydrogen gas is a by-product of this reaction, and nickel reacts with hydrogen gas to form a nickel species coated with hydrogen atoms. Therefore, the finely divided nickel powder is effectively a large surface coated with hydrogen atoms. Neutralization of the solution and washing provides the Raney nickel catalyst.

As pointed out earlier, hydrogen gas is a diatomic molecule (H–H) that is not polarized and not very polarizable. There is no reaction with an alkene. Addition of nickel causes the reaction to occur, however. The C=C unit does not react with hydrogen directly, so the nickel must react with H_2 first and that product reacts with the C=C unit. Indeed, transition metals react with diatomic hydrogen to produce hydrogen atoms (essentially H•) that are bound to the surface of the metal. These hydrogen atoms are very reactive and easily react with an alkene. The actual mechanism of this reaction is somewhat complex because the surface of the metal contains many reactive sites.

A cartoon version of the mechanism shows a metal as a surface (see **28** in Figure 19.1). The metal reacts with H–H to give a "hydrogen-loaded" metal, **29**. This means that hydrogen atoms are *adsorbed* on the surface of the metal. The transition metal also reacts with the π-bond of an alkene to form a coordination complex known as an η^2-complex. This nomenclature

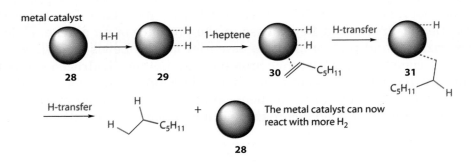

Figure 19.1   Heterogeneous catalytic hydrogenation.

is noted but will not be explained at this time. In this case, **29** reacts with 1-heptene (**27**) to form complex **30**, in which the metal is coordinated to the π-bond as well as the hydrogen atoms. At the surface of the metal, a hydrogen atom is transferred to the alkene π-bond to form a σ-covalent C–H bond in **31**. In **31**, hydrogen atoms remain on the surface of the metal, and one carbon of the partly reduced alkene remains coordinated to the metal. Transfer of a second atom of hydrogen from the metal to the coordinated carbon "releases" heptane and regenerates a "clean" surface of the catalyst (**28**) that can react with additional hydrogen gas. The net result of this process adds two hydrogen atoms to the C=C unit of an alkene. This process is called **catalytic hydrogenation**.

In this process, the metal is catalytic in that it serves as a template for reaction with hydrogen gas and the formation of two new σ-covalent C–H bonds. The catalyst facilitates hydrogen transfer to the alkene, but it is regenerated when the cycle is complete. Therefore, only a catalytic amount of the metal is required. The mechanism outlined in Figure 19.1 is used to explain the reduction of 1-heptene to heptane with hydrogen gas and nickel or certain other metals.

Other transition metals catalyze the reaction of hydrogen gas with an alkene, including palladium (Pd) and platinum (Pt) as well as nickel (Ni). Rhodium (Rh), ruthenium (Ru), or iridium (Ir) and compounds derived from these metals are also commonly used. These transition metals react with hydrogen gas to give a species whose surface is covered with hydrogen atoms and also coordinated to the alkene. *Because this reaction occurs on the surface of the metal, the larger the surface area is, the faster the process will be.* Finely divided nickel, palladium or platinum is used, not "chunks" of metal. These metals are rather expensive and once the hydrogen covers their surface they are very reactive. Indeed, they are often *pyrophoric* (spontaneously ignite when exposed to air).

Because of this, a solid material with a large surface area is usually added to "dilute" the metal. However, this solid must not be reactive to hydrogen gas or react with the organic substrates. The diluting solid is known as a ***solid support.*** Mixing a metal powder with a solid support makes the effective surface area of the metal larger, and it allows one to use less of the metal (common mixtures are 5–10% of the metal). Carbon black and Kieselgühr (diatomaceous earth and related inert solids), as well as inorganic chemicals such as calcium carbonate ($CaCO_3$) or barium carbonate ($BaCO_3$), are used. This means that a typical catalyst for this reaction with hydrogen is 5% palladium-on-carbon (Pd/C), or 10% platinum-on-calcium carbonate, etc. When these catalysts are mixed with hydrogen gas and a molecule that contains a π-bond, two hydrogen atoms "add" across the C=C unit, leading to reduction.

19.3.2 Hydrogenation of Alkenes

Figure 19.1 shows that the π-bond of an alkene is reduced by hydrogen gas to give an alkane in the presence of the proper catalyst. The most common metal used for reduction of alkenes is palladium because it binds alkenes better than many other catalysts. This is a straightforward and very general reaction and many alkenes are reduced to the alkane (C=C → HC–CH) via catalytic hydrogenation. When methylcyclopentene (**32**) reacts with hydrogen gas and Pd/C, the product is methylcyclopentane, **33**. Hydrogenation of 3,4-dimethyl-3*E*-heptene (**34**) gives **35**.

Interestingly, **34** is somewhat difficult to hydrogenate, and **32** reacts much faster than **34**. This result suggests a steric effect in the hydrogenation of alkenes, which is reasonable because the C=C unit must coordinate to the surface of the metal. Indeed, a tetrasubstituted alkene such as **34** is often difficult to hydrogenate at ambient temperatures, although heating using a large excess of hydrogen gas or using hydrogen gas under higher pressures will allow the reaction to proceed without a problem in most cases. Hydrogenation of **34** with Ni(R) in refluxing ethanol leads to **35**, for example.

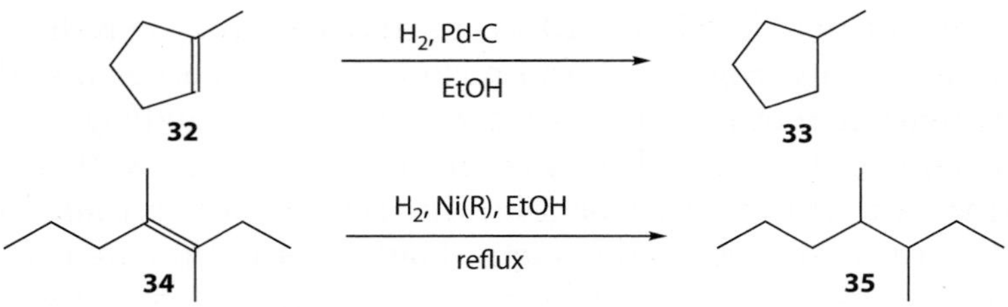

In the hydrogenation of 1,2-dimethylcyclohexene (**36**), a problem is encountered. Reduction may generate a mixture of diastereomers: *cis*-1,2-dimethylcyclohexane (**37**) and *trans*-1,2-dimethylcyclohexane (**38**). In most cases, catalytic hydrogenation of this type using nickel, palladium, or platinum catalysts gives a mixture of diastereomers favoring the *cis* derivative: in this case, **37**. This preference is due to the fact that transfer of the hydrogen atoms to the alkene occurs on the surface of the catalyst. The diastereoselectivity of the reaction depends on the catalyst, the solvent, the pressure of the hydrogen gas, and the alkene. Because predicting the actual amount of each product is difficult without doing the experiment, *assume that hydrogen leads to a mixture of the diastereomers with the cis product favored over the trans product.*

19.9 Draw the product when 3-methyl-2-phenyl-1-pentene is hydrogenated in the presence of a Raney nickel catalyst (abbreviated as Ni[R]).

19.10 Draw the product when methylenecyclohexane is treated with hydrogen gas and palladium metal in methanol.

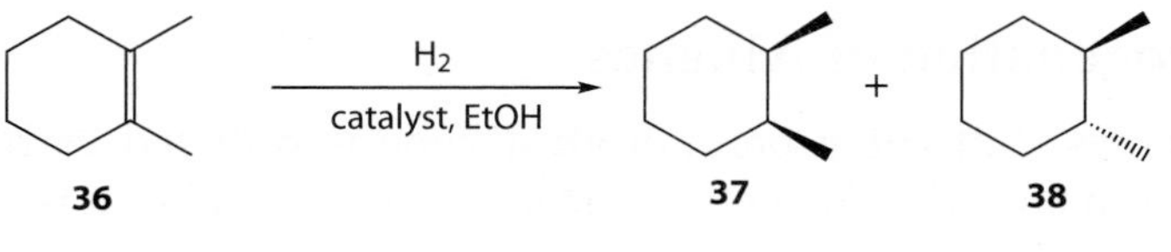

19.3.3 Hydrogenation of Alkynes

Chapter 10 pointed out the similarity in reactions of alkene and alkynes, as well as differences in acid–base reactions. Likewise, there are similarities and differences in the reduction of an alkyne versus an alkene. Alkynes have two π-bonds that may be reduced, and catalytic hydrogenation leads to an alkene or an alkane, depending of the stoichiometry of the hydrogen gas. There are two π-bonds and, in principle, both can react with hydrogen. If 1-heptyne (**39**) reacts with only one molar equivalent of hydrogen gas and a palladium-on-carbon catalyst, the product is 1-heptene (**40**). It is possible to stop the reaction at this point, but remember that an alkene can also react with hydrogen to give an alkane. Therefore, if an excess of hydrogen gas is used (two or more molar equivalents), then butene reacts further and the final product is heptane.

These two possibilities are illustrated by the reactions using one and two molar equivalents of hydrogen in methanol with a Raney nickel catalyst. The reaction of **39** with one equivalent of hydrogen gives **40** in 74% yield, whereas reaction with two equivalents of hydrogen gives heptane in 85% yield.[5] Hydrogenation of **40** with a palladium catalyst leads to butane, as expected. Many alkynes are reduced to alkenes or to alkanes depending on the amount of hydrogen used, as well as the catalyst. Palladium catalysts, especially Pd/C, are commonly used for the reduction of alkenes and alkynes and they are probably the catalyst of choice.

19.11 **Draw the product of the reaction between 1,2-diphenylethyne and an excess of hydrogen in the presence of 5% palladium-on-carbon in ethanol.**

The product of hydrogenation of an alkyne is an alkene, which is subject to further hydrogenation, so isolating the alkene uncontaminated by other products may be difficult. This means that catalytic hydrogenation of an alkyne often gives a mixture of the alkene and the alkane. There is a further complication. Hydrogenation of an internal alkyne leads to an alkene, but there is more than one possible product. Reduction of 5-decyne (**41**) with one equivalent of hydrogen, for example, gives two possible 5-decenes: 5*E*-decene (**42**) and 5*Z*-decene (**43**). In general, catalytic hydrogenation of an internal alkyne such as **41** will give both alkenes, but the catalyst often determines the major product.

Due to transfer of hydrogenation to the π-bond on the surface of the metal catalysts, the *Z*-compound usually predominates, although equilibration to a mixture of *E*- and *Z*-isomers is possible. In the example shown, Raney nickel leads to the *Z*-decene (**43**) in close to 90% yield, with only about 2–3% of **42**.[6] It can be difficult to predict the *E*:*Z* ratio without doing the experiment, so most of the time assume that the product will be a **mixture**. If Pd/C is used, the *Z*-isomer is favored, but with a platinum catalyst, increasing amounts of the *E*-isomer are formed. The variation in stereochemistry with the transition metal catalyst is due to the fact that different metals react differently with molecules containing a π-bond. *For the purposes of this book, assume that hydrogenation of an alkyne with one molar equivalent of hydrogen gas will give a mixture that favors the Z-isomer.*

Catalytic hydrogenation is slowed or even stopped altogether when certain contaminants are present. Such contaminants are usually specific compounds that are called **poisons** because they poison the hydrogenation reaction. *Poisons include amines, phosphines, sulfur compounds, carbon monoxide, etc.* These molecules coordinate to the metal catalyst and compete with the reaction of the metal with the hydrogen gas, the alkene, or the alkyne. In most cases, compounds known as poisons bind to the metal more or less irreversibly, and the metal is less reactive toward hydrogen and to the alkene or alkyne. In effect, the poison blocks the surface of the metal and prevents binding of hydrogen gas or the alkene or the alkyne.

To avoid this problem, hydrogenation of molecules containing sulfur, amine, or phosphine units should be avoided when Ni, Pt, or Pd catalysts are used. On the other hand, the fact that a "poisoned catalyst" is less reactive may be exploited if it is essential that the alkene product of alkyne hydrogenation not be hydrogenated further. Remember that catalytic hydrogenation of an alkyne usually gives a mixture of *E*- and *Z*-isomers, favoring the *Z*-isomer, although it is certainly useful to generate one or the other cleanly. During the course of many years, research has developed "*cis*-specific" (*Z*-specific) catalysts for alkynes (i.e., specific poisoned catalysts). It was discovered that a mixture of palladium on lead carbonate in the presence of a lead oxide (PbO) poison gave a catalyst that reproducibly reduced alkynes to *cis*-alkenes.

Hydrogenation of alkynes with this new catalyst gives the *Z*-alkene as the major product, with only trace amounts of the *E*-alkene. This selective catalyst is called the **Lindlar catalyst** and hydrogenation of alkynes using it is called **Lindlar hydrogenation** or **Lindlar reduction**, named for its discoverer, H. Lindlar (Switzerland).[7] This catalyst has been supplanted by another version that is easier to prepare and less prone to the oxidation of the lead carbonate and lead oxide constituents. It is now the standard and is formally called the **Rosenmund catalyst**, but it is commonly referred to as a "Lindlar catalyst." The new catalyst is composed of palladium-on-barium sulfate (BaSO$_4$) and it is poisoned with quinoline (**45**; see Chapter 26, Section 26.5, for a discussion of quinoline). Lindlar reduction of alkynes is the standard method for preparing Z-alkenes without significant amounts of the E-alkene. Lindlar hydrogenation of 2-butynoic acid (**44**), for example, gave Z-but-2-enoic acid (**46**) in 95% yield.[8]

19.12 **Draw the major product of a reaction between 3-methyl-1-phenyl-1-hexyne and hydrogen gas in the presence of palladium-on-barium sulfate and quinoline.**

19.13 **Draw the final product of the reaction sequence 1-propyne reacts with 1. NaNH$_2$; 2. benzyl bromide; 3. H$_2$, Lindlar catalyst.**

19.3.4 Hydrogenation of Aldehydes and Ketones

The carbonyl group of an aldehyde or a ketone has a π-bond, and it is reduced to an alcohol via catalytic hydrogenation, leading to the **5** → **6** transformation noted in Section 19.2. Recall that in the hydride reduction of a carbonyl, the hydrogen that attacks the carbon is from the hydride, but the second hydrogen (on oxygen) comes from the aqueous acid workup. In catalytic hydrogenation of a ketone or aldehyde, both hydrogen atoms in the alcohol product (**6**) arise from hydrogen gas (H$_2$). As with alkenes, a transition metal catalyst is required to react first with H$_2$ and also to coordinate the carbonyl in order to facilitate transfer of hydrogen (see Figure 19.1).

Although Pd, Ni, Ru, and Pd catalyze hydrogenation of ketones and aldehydes, a platinum catalyst is usually the best choice because the yields are better and there are fewer side reactions. Platinum oxide (PtO$_2$)—sometimes called **Adam's catalyst** after Roger Adams (United States; 1889–1971)—is commonly used. The choice of catalyst depends on the reaction conditions, the nature of the carbonyl compound, etc. The reaction of heptanal with hydrogen gas in the presence of a ruthenium catalyst (Ru/C), in 80% aqueous ethanol, gave 1-heptanol in near quantitative yield.[9]

Quantitative means that the yield is essentially 100%, less the amount lost during the workup and isolation procedure. This is a relatively straightforward reaction that works with most ketones and aldehydes. An example that uses PtO$_2$ reacts hydrogen gas with 4-phenyl-2-hexanone (**47**) to give alcohol **48**. Nowadays it is usually easier to reduce ketones and aldehydes with NaBH$_4$ or LiAlH$_4$, so catalytic hydrogenation is not used as much as it was many years ago. However, this reaction remains the method of choice for selected reductions.

19.14 **Using Figure 19.1 as a model, suggest a mechanism for the catalytic hydrogenation of acetone.**

19.15 **Write out this reaction with all starting materials, reagents, and final product.**

19.16 What is the IUPAC name of 48?

**19.17 Draw the final product of the following sequence: 1. cyclopen-
tanone + NaBH$_4$; 2. aq. H$_3$O$^+$; 3. PCC; 4. H$_2$, PtO$_2$.**

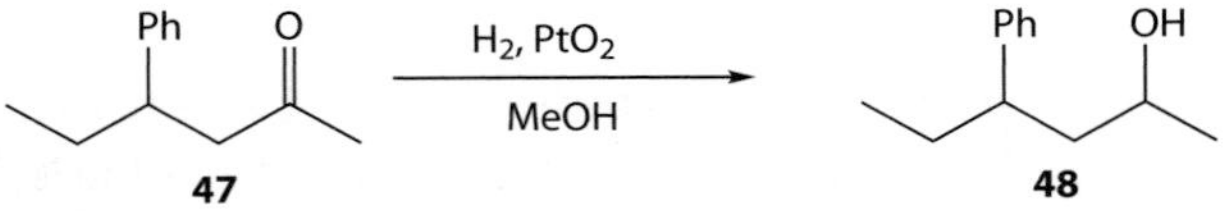

19.3.5 Hydrogenation of Other Functional Groups

It is possible to reduce a few other carbonyl compounds using catalytic hydrogenation, but not all. Partial reduction is also possible. Carboxylic acid derivatives
are an obvious choice, but carboxylic acids themselves, as well as esters, are difficult to reduce by catalytic hydrogenation. Likewise, amides are very difficult
to reduce, but acid chlorides are an exception. Acid chlorides are converted to
alcohols via catalytic hydrogenation with an excess of hydrogen gas. However, it
is also possible to reduce an acid chloride to an aldehyde with the proper catalyst and control of the number of molar equivalents of hydrogen gas.

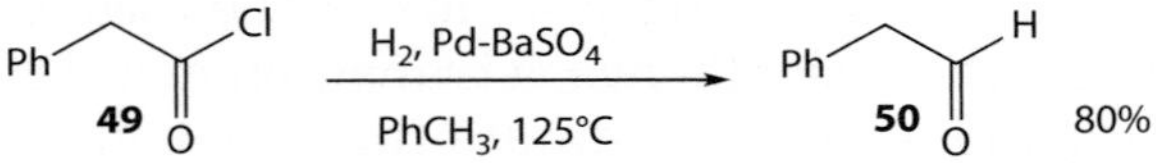

A classical reaction is the **Rosenmund reduction** (Karl Wilhelm Rosenmund,
Germany, 1884–1965), in which an acid chloride such as phenylacetyl chloride (**49**)
is reduced with one molar equivalent of hydrogen gas, a Pd–BaSO$_4$ catalyst (in toluene at 125°C in this example). In this experiment, **49** is reduced to the aldehyde
(**50**) in 80% yield.[10] As noted, using an excess hydrogen gas will usually reduce
most acid chlorides to the corresponding alcohol. Other acid derivatives, such as
esters, carboxylic acids, and amides, are not reduced by catalytic hydrogenation
or are reduced in low yield with great difficulty. This means that hydrogenation
of a molecule with an ester and a ketone group, such as ethyl 3-oxobutanoate **51**
(acetoacetic ester), should reduce the ketone unit but not the ester unit. In the
experiment shown in the illustration, hydrogenation of **51** with a platinum catalyst gives the hydroxy-ester (**52**; ethyl 3-hydroxybutanoate) in 88% yield.[11]

19.18 What is the IUPAC name of 50?

**19.19 Draw the product formed when pentanoyl chloride is treated
with hydrogen gas and a Pd/C catalyst.**

The general order of reactivity of carbonyl functional groups to hydrogenation is

acid chlorides > aldehydes ≈ ketones > anhydrides > esters
> carboxylic acids > amides

As mentioned, for all practical purposes, esters, acids, and amides are not reduced by this method.

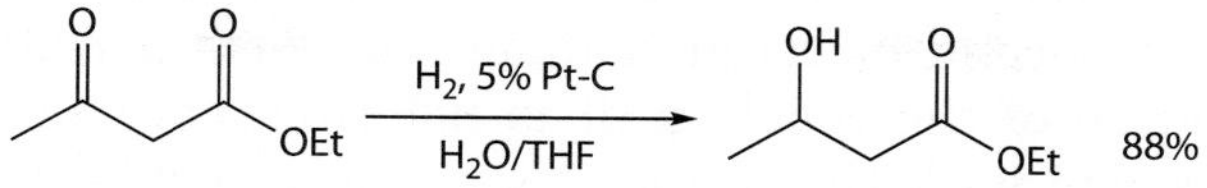

Catalytic hydrogenation of nitriles gives the corresponding amine, which is another exception in the reduction of acid derivatives. Nitriles are acid derivatives because they can be hydrolyzed to the parent carboxylic acid (see Chapter 20, Section 20.13). Conversion of the C≡N unit to a CH_2NH_2 unit requires two molar equivalents of hydrogen gas. In principle, it is possible to add one molar equivalent of hydrogen to convert the C≡N unit to an imine (–CH=NH), but imines are very difficult to isolate under these conditions. (For further information about imines, see Chapter 18, Section 18.7.1.) For that reason, an excess of hydrogen is typically used to ensure conversion of the nitrile to the amine. Indeed, this is an important preparation of amines.

Hydrogenation of the C≡N unit is somewhat difficult, however, and more vigorous reaction conditions (heat or longer reaction times) are usually required. This can be seen in the experimental conditions used for 2-phenylethanenitrile (**53**), which is hydrogenated at 130 atm of H_2 with a Raney nickel catalyst in ammonia at 120–130°C. Under these conditions, 2-phenyl-1-aminoethane (**54**) is isolated in 87% yield.[12]

19.20 Draw the product formed when 3,5-dimethylbenzonitrile is hydrogenated with two equivalents of hydrogen and a Raney nickel catalyst.

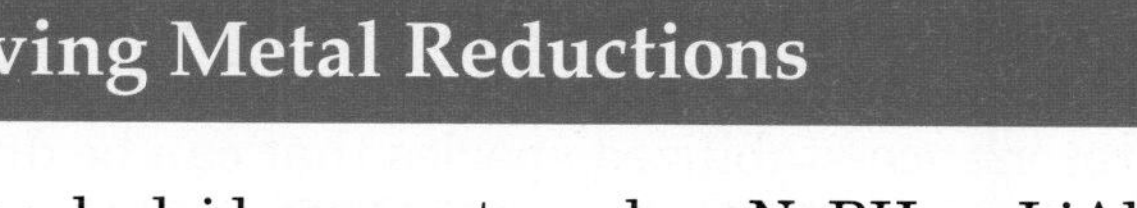

19.4 Dissolving Metal Reductions

In previous sections, hydride reagents such as $NaBH_4$ or $LiAlH_4$ reduce ketones or aldehydes and most acid derivatives to the corresponding alcohol. $LiAlH_4$ reduces amides to amines and also reduces nitriles to amines. Catalytic hydrogenation reduces alkenes and alkynes, as well as ketones, aldehydes, acid chlorides, and nitriles. Acid chlorides may be reduced to aldehydes, and nitriles are also reduced to amines by catalytic hydrogenation.

There is another way to reduce some functional groups. The technique involves the use of an alkali metal (group 1 or group 2), such as sodium, lithium, or potassium, in a protic solvent such as liquid ammonia with or without ethanol. These reductions occur by an initial electron transfer from the metal. Therefore, the mechanism involves formation of a radical species (see Chapter 11, Section 11.9). A second electron transfer converts a radical to a carbanion during the course of the reaction (it is not isolated; this is referred to as "*in situ*"), which reacts with the protic solvent in an acid–base reaction. Both hydrogen atoms incorporated in the reduced molecule therefore come from the solvent. Such reactions are termed dissolving metal reductions.

19.4.1 Reduction of Ketones and Aldehydes

In Chapter 3 (Section 3.2), lithium was identified as an element that can transfer an electron to a suitable receptor to form Li⁺ because it is in group 1. Many metals in group 1 or group 2 of the periodic table are capable of donating electrons to an appropriate functional group. This electron-donating ability is exploited for the reduction of some functional groups that can accept an electron from the metal.

When sodium metal is mixed with a ketone in liquid ammonia and an alcohol, a reduction occurs to give an alcohol. In one typical experiment, 4-*tert*-butylcyclohexanone was treated with Na metal in liquid ammonia and *tert*-butanol to give 4-*tert*-butylcyclohexanol (**59**) in 98% yield.[13] The only way an alcohol may be formed is via transfer of a hydrogen atom to both the carbon and the oxygen of the carbonyl unit. The only source of hydrogen, however, is either the ammonia (NH_3) or the alcohol (*t*-BuOH). These observations are explained by a mechanism that involves transfer of a single electron from sodium metal to the carbonyl unit of the ketone to give a *new type of intermediate:* a **radical anion**. In this case, the radical anion is **56A** and it is generically known as a *ketyl*.

This ketyl is a resonance-stabilized species that can be drawn with the negative charge on either oxygen (**56A**) or carbon (**56B**), and the single electron (the radical) can be drawn on either oxygen or carbon. The resonance form that has the negative charge on carbon (carbanion **56B**) functions as a carbanion base and removes a proton from either ammonia or the alcohol. In other words, ammonia is an acid in the presence of the powerful base **56B** to give **57**. It is more likely that **56B** will react with the more acidic alcohol to give alkoxy-radical **57**.

The alcohol is added to the reaction precisely because it is a stronger acid and facilitates formation of **57**. The reduction works in liquid ammonia without alcohol, but it is slower. Radical **57** reacts with another sodium atom and this electron transfer generates an alkoxide, **58**. The final step of this mechanism must give the alcohol product, so a protonation of the alkoxide occurs to give **59**. The overall reaction, shown in condensed reaction form, is reduction of the ketone (**55**) to an alcohol (**59**).

A variety of ketones or aldehydes may be reduced to an alcohol using dissolving metal reduction. Alcohols other than ethanol are used as solvents, and low boiling amines such as methylamine or dimethylamine can be used in place of ammonia. Other alkali metals such as lithium or calcium also work. Lithium in a mixture of methylamine and ethanol, and calcium metal in methylamine may also be used in these reduction reactions, primarily on large-scale reactions such as those found in industrial laboratories or factories.

19.21 **Draw the ketyl derived from treatment of acetone with sodium metal.**

19.22 **Draw the ketyl and final product when phenylacetaldehyde is treated with sodium metal in liquid ammonia containing ethanol.**

19.4.2 Reduction of Alkynes

Dissolving metal reductions works very well with aldehydes and ketones, but alkenes are not readily reduced under the same conditions. For example, 1-hexene is reduced to hexane in only 41% yield with Na/MeOH/liquid NH$_3$.[14] Alkynes, on the other hand, are reduced to alkenes in good yield using dissolving metal conditions, and the experimental evidence shows that the *E*-alkene is the major product. In a typical example, 4-octyne (**60**) is treated with sodium in liquid ammonia, and oct-4*E*-ene (**64**) is isolated in 90% yield. None of the *Z*-alkene is observed in this reaction.[12] The reaction with sodium in liquid ammonia is an electron transfer process similar to that observed with ketones and aldehydes, but how is the *E* geometry of the alkene product explained?

Electron transfer from the sodium to one π-bond of **60** leads to a radical anion, **61**. Transfer of the electron will generate an anion (the orbital containing the two electrons of the negative charge) and a radical on the same side of the remaining double bond (as in **61A**) or on opposite sides (as in **61B**). Formation of the observed *E* product is most easily explained by minimizing the electronic repulsion between the two orbitals, making **61B** the lowest energy species. If the radical and the anion orbitals are *"trans"* in **61B**, the reaction of **61B** with ethanol leads to a vinyl radical, **62**. Note that the stereochemistry of the double bond is set at this point. Another electron transfer to **62** from a second sodium atom leads to carbanion **63** and protonation of **63** gives the final product, alkene **64**. As a general statement, *dissolving metal reductions of alkynes give the E-alkene due to minimized electronic repulsion in the ketyl intermediate.* Dissolving metal reduction of a terminal alkyne such as 1-butyne gives the monosubstituted alkene (1-butene) and *E*- and *Z*-isomers are not possible.

19.23 **Draw the radical anion formed when 1,3-diphenyl-1-pentyne is treated with sodium metal in liquid ammonia and ethanol and then draw the final product of this reaction.**

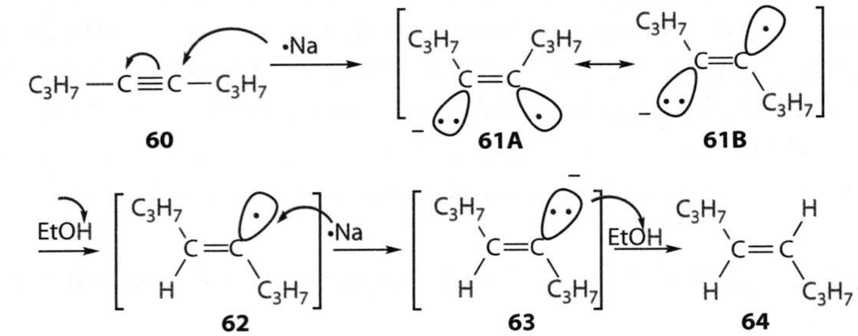

This method is a very interesting counterpoint to the Lindlar reduction discussed in Section 19.3.3. Whereas Lindlar reduction of an alkyne gives a *Z*-alkene, dissolving metal reduction of an alkyne gives an *E*-alkene. The reduction process can therefore be controlled, and either a *Z*- or an *E*-alkene can be prepared. Assume that alkenes are not reduced to the alkane under the conditions described using Na or Li in ammonia or ethanolic ammonia.

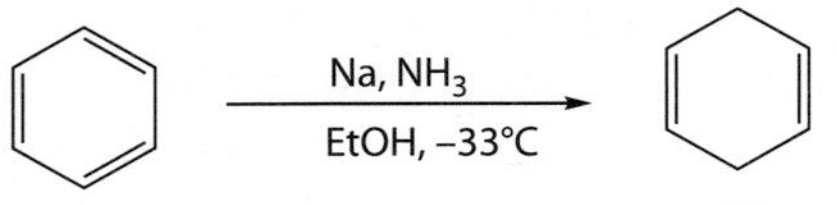

Benzene was introduced in Chapter 5 (Section 5.10). Chapter 21 will discuss many benzene derivatives, along with the chemical reactions that are characteristic of these compounds. In the context of dissolving metal reductions of aldehydes, ketones, and alkynes, however, one reaction of benzene must be introduced. When benzene (**65**) is treated with sodium metal in a mixture of liquid ammonia and ethanol, the product is 1,4-cyclohexadiene **66**. Note that the nonconjugated diene is formed. The reaction follows a similar mechanism to that presented for alkynes. Initial electron transfer from sodium metal to benzene leads to radical anion **67**. Resonance delocalization as shown should favor the resonance contributor **67B** due to charge separation.

In other words, the two electrons of the negative charge and the unpaired electron should be as far apart as possible in the delocalized electron cloud. If **67B** reacts with the acid (ethanol), reduction gives **68**. A second electron transfer from additional sodium metal converts radical **68** to carbanion **69**, which quickly reacts with the acidic proton of ethanol to give the final product **66**. The nonconjugated diene unit is determined by charge separation in **67** at the time of the initial electron transfer. This reduction of benzene and its derivatives is commonly known as **Birch reduction**, after Arthur John Birch (Australia; 1915–1995).

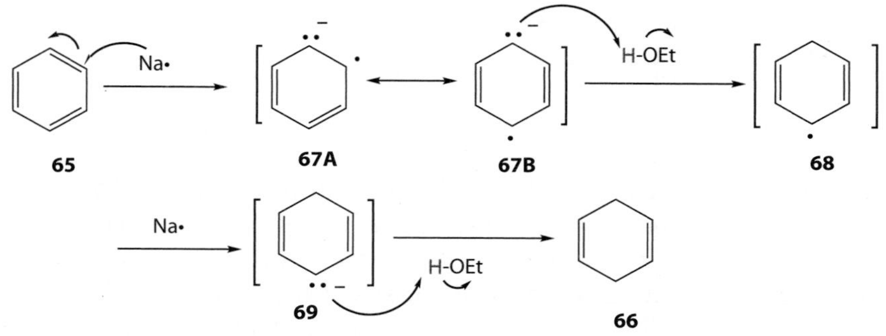

19.4.3 Reductions Using a Metal and an Acid

The dissolving metal reductions discussed in preceding sections used protic solvents and alkali metals. If a stronger proton transfer agent (i.e., a stronger acid) is used, other metals may be used for reductions, including zinc and tin. Zinc in acetic acid (ethanoic acid, CH_3CO_2H) is commonly used to reduce halides in a reaction that replaces the halogen atom with a hydrogen atom. The reaction of alkyl halides is slow in acetic acid, unless the halogen is conjugated to a carbonyl, but Zn in HCl is more effective. This is illustrated by the experiment in which 1-iodohexacosane (cetyl iodide, **70**; isolated from Chinese wax) reacts with Zn/HCl to give hexacosane (cerane, **71**) in 68% yield.[15]

19.24 **Draw the final product of a reaction between 1-bromo-1-phenylpropane and zinc powder in acetic acid.**

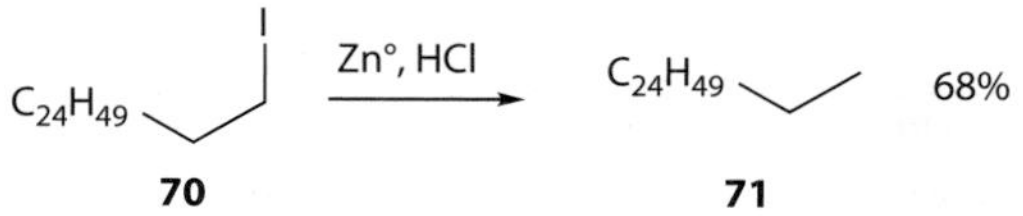

An alternative method for the reduction of an alkyl halide also uses a metal. In Chapter 15 (Sections 15.1 and 15.2), alkyl and vinyl halides were converted to the corresponding Grignard reagent by reaction with magnesium metal. In Section 15.3, it was pointed out that Grignard reagents are powerful bases. These facts may be used to reduce an alkyl halide selectively. The reaction of 3-chloro-3-methylhexane (**72**) and magnesium in diethyl ether generates the expected Grignard reagent **73**. When **73** is treated with water, the acid–base reaction generates the alkane, 3-methylhexane. This two-step sequence constitutes a formal reduction of **72** to 3-methylhexane. Similarly, vinyl bromide **74** is converted to **75** by reaction with magnesium in THF. Treatment with water gives alkene **76**. Note that this is a very mild reduction method in that only water is used as the acid.

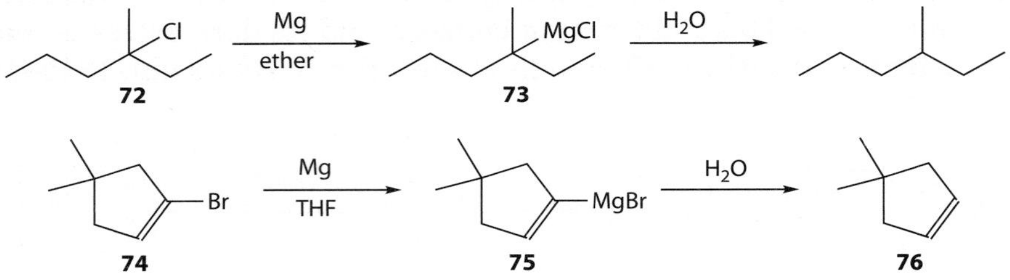

It is possible to reduce a carbonyl compound and remove the oxygen from the molecule. In one experiment, heptanal (**77**) was treated with zinc amalgam (Zn/Hg) in HCl to give heptane (**78**) as the final product in 72% yield.[16] This reduction is known as **Clemmensen reduction**, after E. Ch. Clemmensen (Denmark; 1876–1941). The starting material and the product must withstand the concentrated HCl used in the reaction, but it is a very effective method for the complete reduction of ketones to hydrocarbons. An alternative reductive method that treats ketones or aldehydes with hydrazine in aqueous KOH is called the **Wolff–Kishner reduction**, after Ludwig Wolff (Germany; 1857–1919) and N. M. Kishner (Russia; 1867–1935). For the purpose of comparison with the Clemmensen reduction, Wolff–Kishner reduction of heptanal (**77**) with hydrazine/KOH gives heptane (**78**) in 54% yield.[17] Clearly, the Clemmensen reduction works in acidic media, whereas the Wolff–Kishner reduction uses basic media. Ketones are also reduced to the corresponding hydrocarbon by these reagents. Wolff–Kishner reduction of ketone **79**, for example, leads to **80**.

19.25 Draw the product formed when 4,4-diphenyl-2-hexanone is treated with zinc amalgam in HCl.

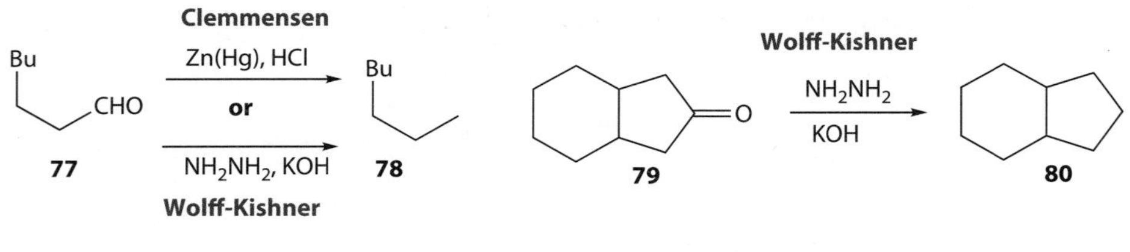

In the presence of an acid, tin is very effective for the reduction of some functional groups. A mixture of tin and concentrated HCl will reduce a ketone to an alcohol, after aqueous acid workup of the reaction. Cyclohexanone, for example, is reduced to cyclohexanol upon treatment with Sn/HCl. Stannous chloride (SnCl$_2$) in acid is another interesting reducing agent. When hexadecanenitrile (**81**) is treated with SnCl$_2$ and HCl, the HCl adds to the C≡N unit to give a chloro-imine, **82**. The chlorine is subsequently reduced and, in the acidic medium, replaced with hydrogen to give imine **83**. In the aqueous acid solution, the imine is converted to an aldehyde, **84**. In one reported experiment, **81** was converted to hexadecanal (**84**) in quantitative yield.[18] The conversion of an imine to a carbonyl group was discussed in Chapter 18 (Section 18.7.1) in connection with the reaction of a Grignard reagent RMgX and a nitrile, which gives an imine product. Subsequent reaction with aqueous acid converts the imine to an aldehyde or ketone. This particular reduction of a nitrile to an aldehyde is called the **Stephen reduction**, after Henry Stephen (England; 1889–1965).

19.26 **Write out the reaction with cyclohexanone with both starting materials and final product.**

19.27 **Draw the mechanism for conversion of 78 to the aldehyde 79 in aqueous acid (see Chapter 18, Section 18.7.1).**

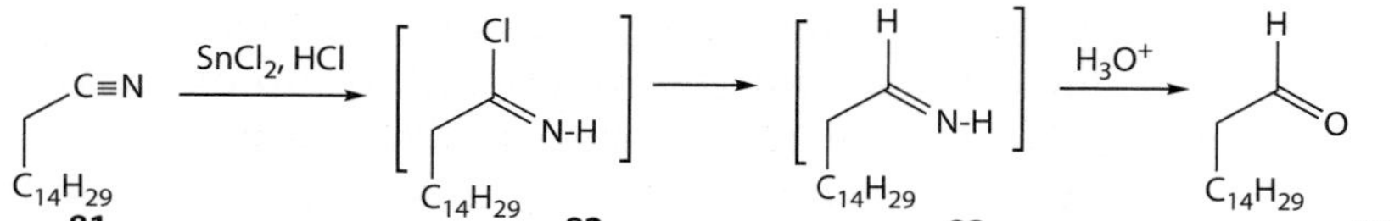

19.5 Summary of Functional Group Exchanges

Several new and interesting functional group transformations are possible via reduction reactions. To summarize these reactions, the functional group exchange will be shown for each. The functional group exchange for the reduction of a ketone or an aldehyde is exactly the opposite of that shown before for the oxidation of an alcohol. The alcohol is shown with the carbonyl precursor in the functional group exchange. Clearly, there are several different methods for the reduction of a ketone or aldehyde, but the functional group exchange reaction is the same for all of the different methods.

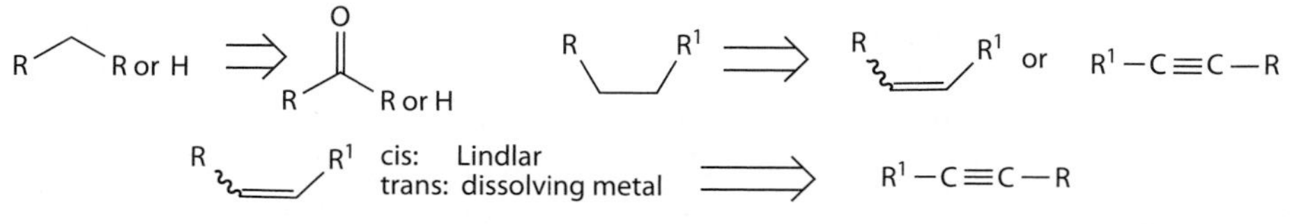

Complete removal of the oxygen from a carbonyl compound is possible via Clemmensen reduction or by Wolff–Kishner reduction. The functional group exchange for this process shows the hydrocarbon with a ketone or aldehyde precursor.

The reduction of an alkene leads to an alkane unit. Similarly, reduction of an alkyne leads to an alkene or an alkane. More than one method will accomplish this transformation, but the alkane unit is shown with its alkene precursor, and the alkene is shown with an alkyne precursor. In addition to the generic functional group exchange reactions shown, the exchanges for a *cis*-alkene with an alkyne precursor (via Lindlar reduction) as well as a *trans*-alkene with an alkyne precursor (via dissolving metal reduction) are shown.

The reduction of several functional groups in addition to ketones, aldehydes, alkenes, or alkynes is presented in this chapter. Alcohols are produced by reduction of acid derivatives, and the functional group exchange is shown for the alcohol with an acid derivative precursor. The X group can be Cl, OH, or OR. The reduction of an amide or a nitrile gives an amine. The functional group exchanges involve amines with an amide precursor in one case and a nitrile precursor in the other. A nitrile is formed by an S_N2 reaction of cyanide with a halide (see Chapter 11, Section 11.3.6) and the functional group exchange is expanded to include the halide precursor to the nitrile. Reduction of a halide leads to a hydrocarbon unit. The functional group exchange is simply the hydrocarbon shown with a halide precursor, where X = Cl, Br, I.

There is one remaining functional group exchange based on the Stephen reduction. The product is an aldehyde and the starting material is a nitrile, so the functional group exchange shows an aldehyde with a nitrile precursor. As with the other reactions, the nitrile is shown with a halide precursor, extending the functional group exchange process.

The new functional group exchange reactions presented in this chapter can be combined with reactions from previous chapters to expand the ability to synthesize molecules. Alkene **85** is synthesized from aldehyde **86**, for example. The first task is to identify the four carbons of **86** in **85**. It appears that the carbons marked in **blue** are the best candidates. Rather than disconnect the C–C=C unit marked in **blue**, first disconnect the ethyl group of **85** to give **87** and **88**. This choice is made because no reaction has been presented that will allow direct incorporation of EtCHCH to X–C–CMe₂. Disconnection of the ethyl group takes advantage of the fact that an alkyne anion reacts with an alkyl halide. However, before this reaction can be used, the alkene unit in **87** needs to be changed to an alkyne unit in **89**.

Disconnection of **89** generates the alkyne unit **90**, along with **91**, which is designated as an acceptor carbon because the alkyne anion is surely a donor carbon. The synthetic equivalent for this acceptor carbon is halide **91**, given as a bromide. A logical functional group exchange of bromide **91** is to alcohol **92**, which is simply the reduction product of the aldehyde starting material **86**. With this analysis, the synthesis is the reverse of the retrosynthesis, where appropriate reagents are added.

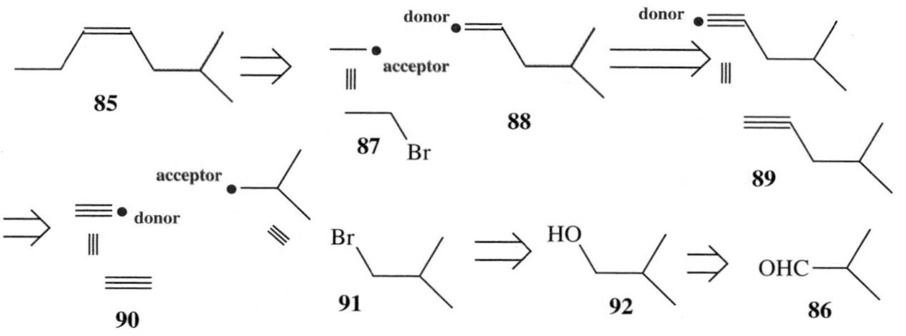

Reduction of aldehyde **86** with sodium borohydride leads to alcohol **92**. Another reduction method is acceptable for this reaction. Phosphorus tribromide or another halogenating agent converts **92** to primary bromide **91**, and an S_N2 reaction with sodium acetylide derived from **90** gives **89**. If this terminal alkyne reacts with a base such as sodium amide, the resulting alkyne anion reacts with bromoethane (**87**) to give alkyne **93**. This alkyne is the precursor to the product, but a specific reduction is required to give the Z-alkene **85**. Lindlar hydrogenation will accomplish this transformation to give the target. Note that reaction of **93** with sodium metal in ethanolic ammonia will generate the corresponding E-alkene if that is the target.

19.28 Draw the dissolving metal reaction for the conversion of 93 to the E-alkene.

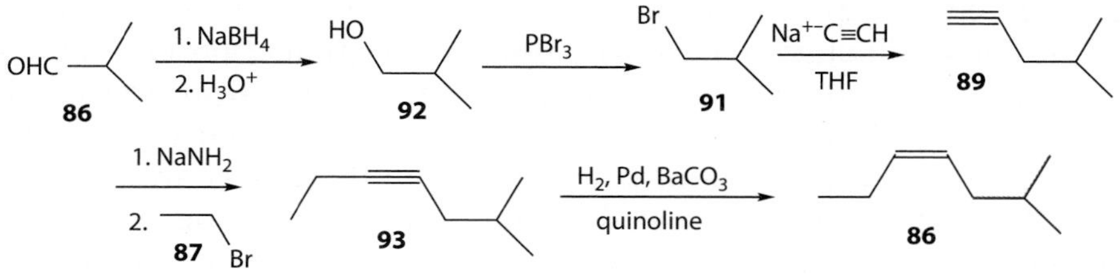

19.6 Biological Relevance

Fatty acid synthesis in bacteria and plants is a multistep process. One transformation in the process involves the reduction of the ketone unit in acetoacetyl ACP (**94**) with the enzyme β-ketoacyl-ACP reductase and NADPH (**91**) to give β-hydroxybutaryl-ACP (**96**) and NADP+ (**92**).[19] ACP is the acyl carrier protein, and it is bound to the acetoxy unit via a phosphopantetheine group (marked in **cyan** in **94**).

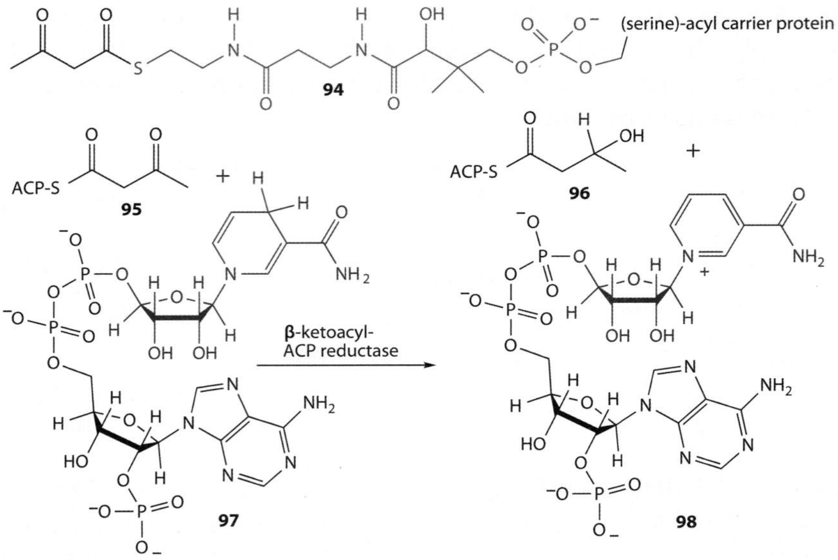

(Reprinted in part with permission from Lu, H., and Tonge, P. J. 2008. *Account of Chemical Research* 41:11. Copyright 2008 American Chemical Society.)

Examine NADP$^+$ (**97**) and NADPH (**98**). When **95** is reduced to **96**, **97** is oxidized to **98**. Just as the conversion of **97** to **98** is an oxidation, the conversion of **98** to **97** is a reduction. Look in Chapter 17 (Section 17.6) for a reaction in which NAD$^+$ is converted to NADH. Note that conversion of NAD$^+$ to NADH or **98** to **97** (both are reductions) involves conversion of the aromatic pyridine ring to a dihydropyridine. This transformation is formally analogous to the Birch reduction discussed in Section 19.4.

7-Dehydrocholesterol reductase (DHCR7) is associated with a human gene responsible for producing cholesterol (**100**) by the reduction of 7-dehydrocholesterol (**99**).[20] 7-Dehydrocholesterol is found in human skin and is essential for the production of vitamin D$_3$ when humans are exposed to the ultraviolet rays in sunlight. Cholesterol is produced in the human body and used to make important hormones and is involved in the production of bile acids important for digestion. Cholesterol is found in many foods consumed by humans, including egg yolks, meat, poultry, fish, and dairy products.

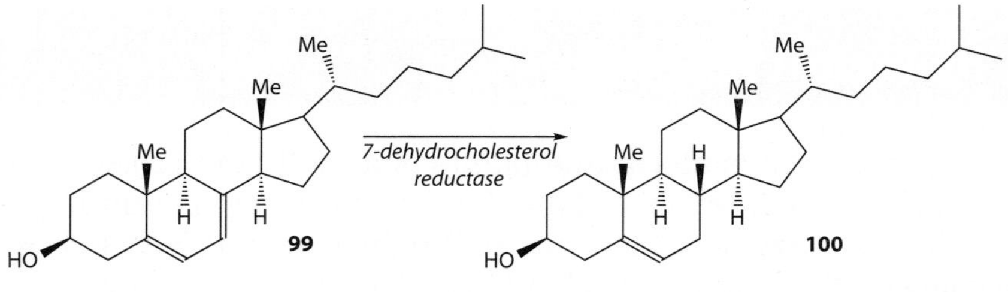

(Reproduced with permission from Waterham H. R., and Wanders R. J. A. 2000. *Biochimica et Biophysica Acta* 1529:340. Copyright 2000 Elsevier.)

References

1. Hudlicky, M. 1984. *Reductions in organic chemistry,* 108. Washington, D.C.: American Chemical Society.
2. Nystsrom, R. F., and Brown, W. G. 1947. *Journal of the American Chemical Society* 69:1197.
3. Micovic, V. M., and Mihailovic, M. L. J. 1953. *Journal of Organic Chemistry* 18, 1190.
4. Amundsen, L. H., and Nelson, L. S. 1951. *Journal of the American Chemical Society* 73: 242.
5. Campbell, K. N., and O'Connor, M. J. 1939. *Journal of the American Chemical Society* 61:2897.
6. Campbell, K. N., and Eby, L. T. 1941. *Journal of the American Chemical Society* 63:217.
7. Lindlar, H., and Dubuis, R. 1973. *Organic Syntheses* 5:880.
8. Furniss, B. S., Hannaford, A. J., Smith, P. W. G., and Tatchell, A. R., eds. 1994. *Vogel's textbook of practical organic chemistry,* 5th ed., 494, Exp. 5.16. New York: Longman.
9. Freifelder, M. 1978. *Catalytic hydrogenation in organic synthesis: Procedures and commentary,* 80. New York: John Wiley & Sons.
10. Mosettig, E., and Mozingo, R. 1948. *Organic Reactions* 4:362.
11. Rylander, P. 1979. *Catalytic hydrogenation in organic synthesis,* 86. New York: Academic Press.
12. House, H. O. 1972. *Modern synthetic reactions,* 2nd ed., 18. Menlo Park, CA: W. A. Benjamin (taken from Robinson J. C., Jr., and Snyder, H. R. *Organic Synthesis Collective* 3:720).
13. House, H. O. 1972. *Modern synthetic reactions,* 2nd ed., 18. Menlo Park, CA: W. A. Benjamin (taken from Huffman, J. W., and Charles, J. T. 1968. *Journal of the American Chemical Society* 90:6486).
14. Greenfield, H., Friedel, R. A., and Orchin, M. 1954. *Journal of the American Chemical Society* 76:1258.
15. Fieser, L. F., and Fieser, M. 1961. *Advanced organic chemistry,* 114. New York: Reinhold.
16. Martin, E. L. 1942. *Organic Reactions* 1:155.
17. Todd, D. 1948. *Organic Reactions* 4:478.
18. Stephen, H. 1925. *Journal of the Chemical Society* 127:1874.
19. Lu, H., and Tonge, P. J. 2008. *Account of Chemical Research* 41:11.
20. Waterham, H. R., and Wanders R. J. A. 2000. *Biochimica et Biophysica Acta* 1529:340.

Answers to Problems

19.1

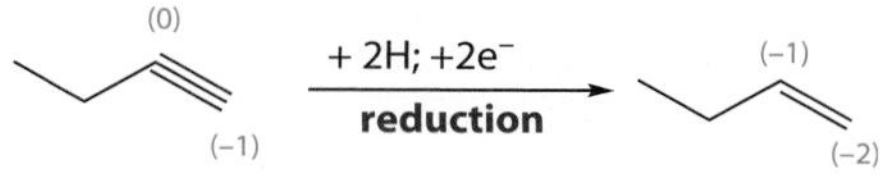

19.2

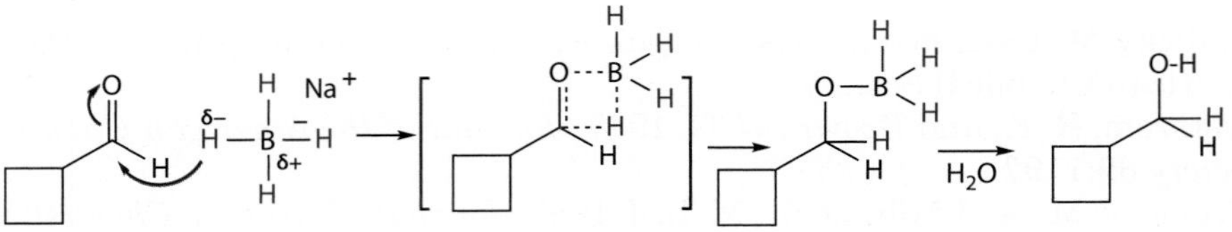

19.3

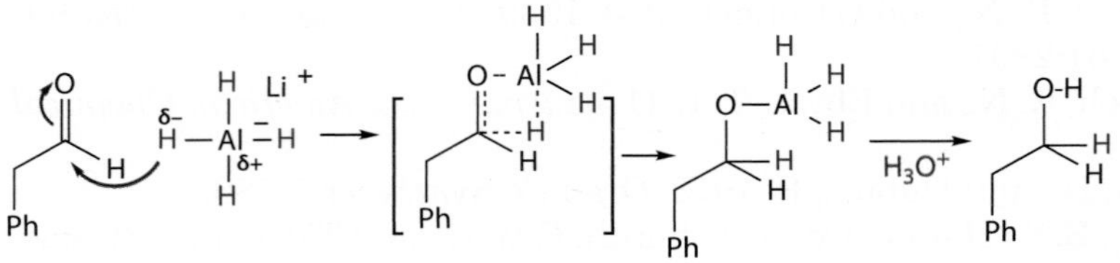

19.4

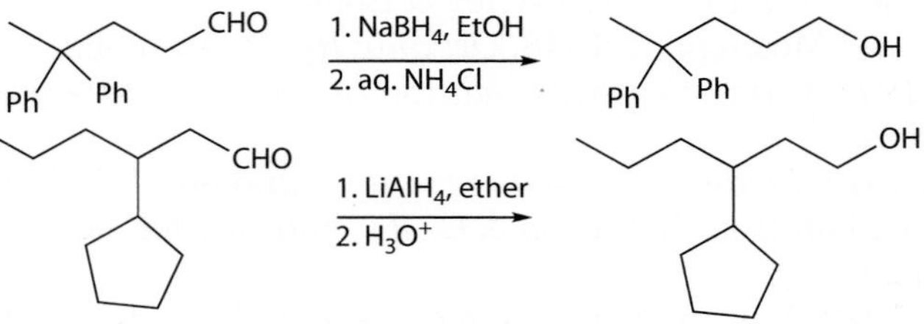

19.5

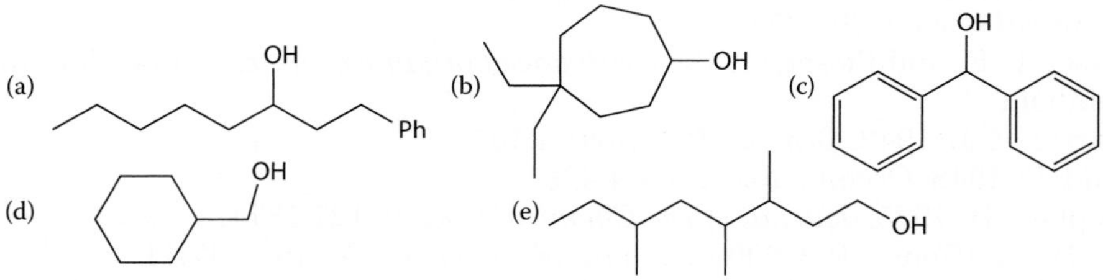

(a) (b) (c)

(d) (e)

19.6

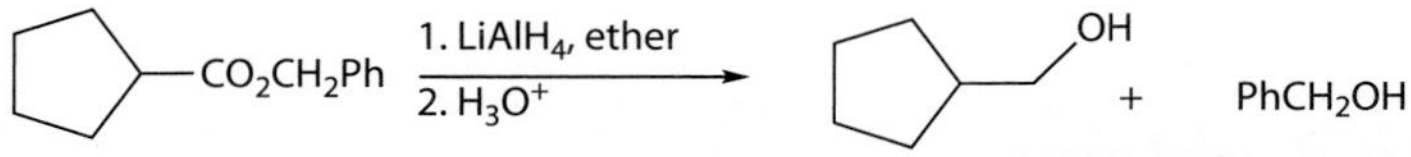

19.7

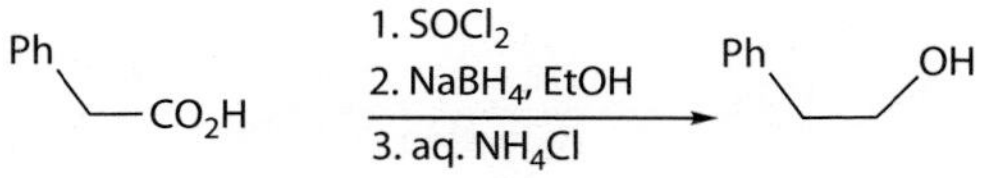

19.8

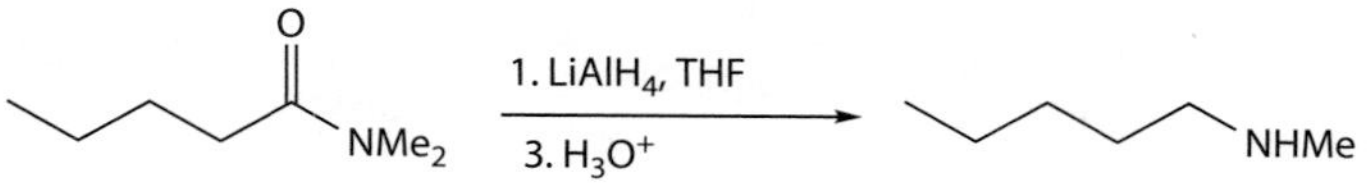

O=C(NMe₂) chain → (chain) NHMe

1. LiAlH₄, THF

3. H₃O⁺

19.9

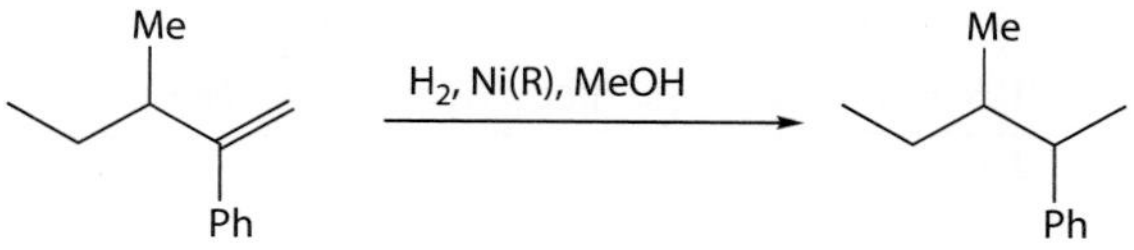

Me ... H₂, Ni(R), MeOH ... Me / Ph → Me ... Ph

19.10

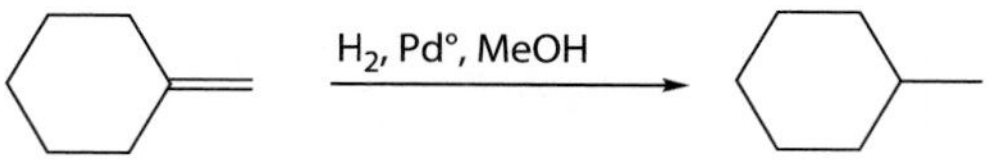

(methylenecyclohexane) → H₂, Pd°, MeOH → (methylcyclohexane)

19.11

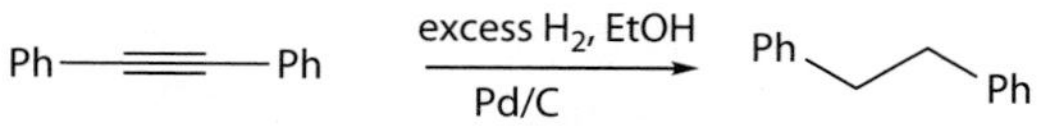

Ph—≡—Ph → excess H₂, EtOH / Pd/C → Ph⌒Ph

19.12

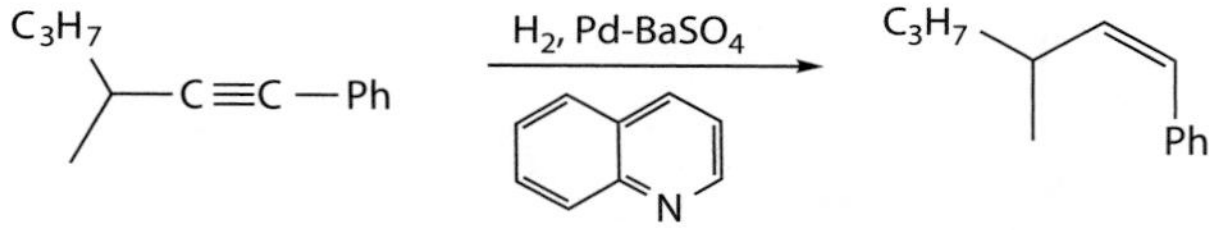

C₃H₇ —C≡C—Ph → H₂, Pd-BaSO₄ (quinoline) → C₃H₇ ⌐ Ph

19.13

Me—C≡C—H → 1. NaNH₂ 2. PhCH₂Br 3. H₂, Pd-BaSO₄ (quinoline) → Ph⌐Me

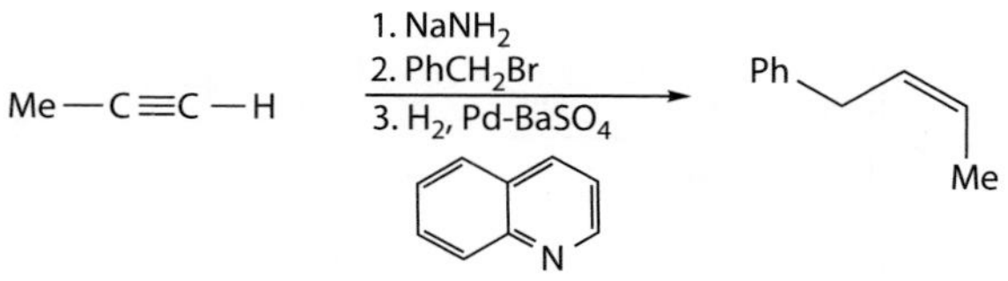

19.14

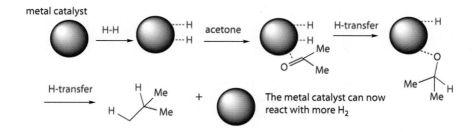

19.15

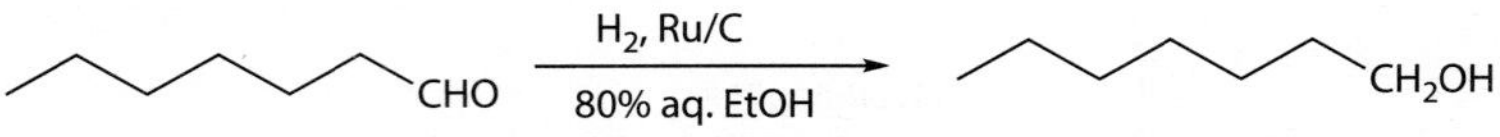

$$\underset{\text{CHO}}{\text{(hexanal)}} \xrightarrow[\text{80\% aq. EtOH}]{\text{H}_2, \text{Ru/C}} \underset{\text{CH}_2\text{OH}}{\text{(1-hexanol)}}$$

19.16 1-phenyl-2-hexanol.

19.17

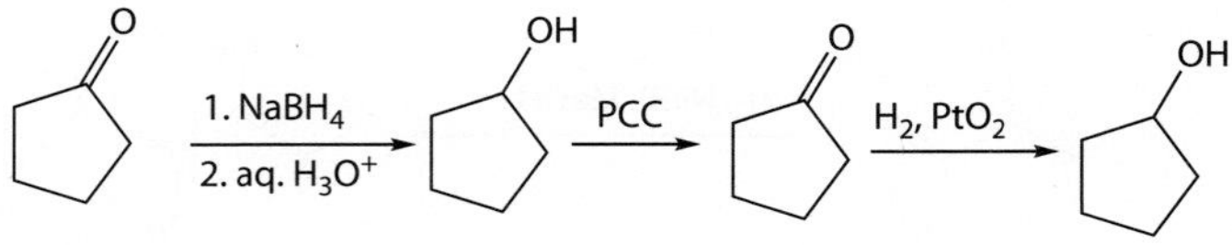

19.18 2-phenylethanal.

19.19

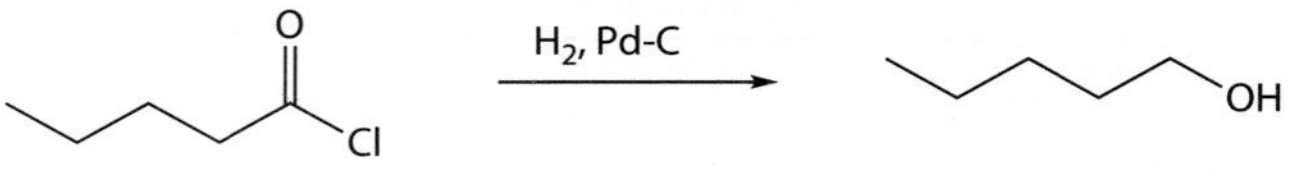

19.20

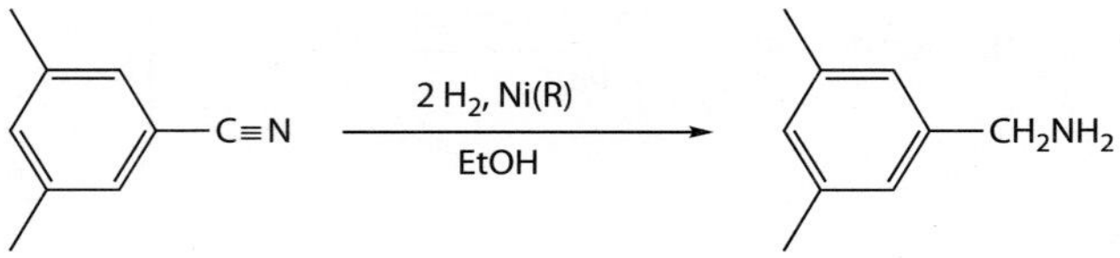

19.21

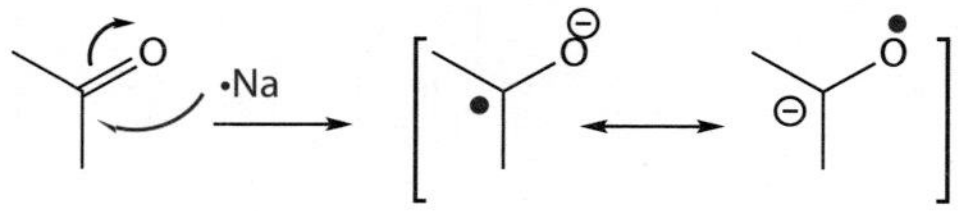

19.22

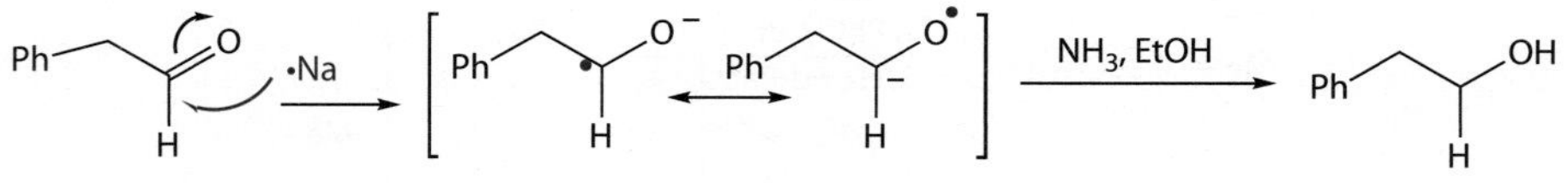

19.23

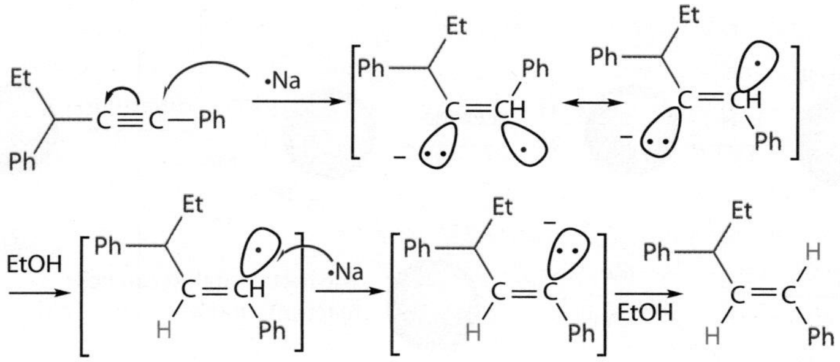

19.24

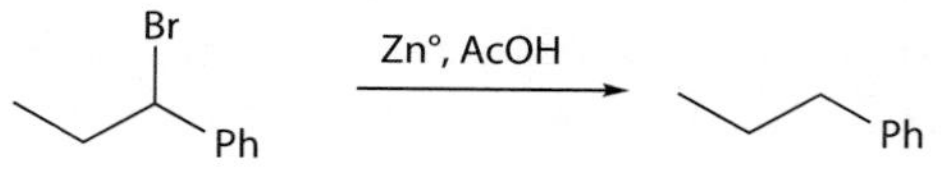

19.25

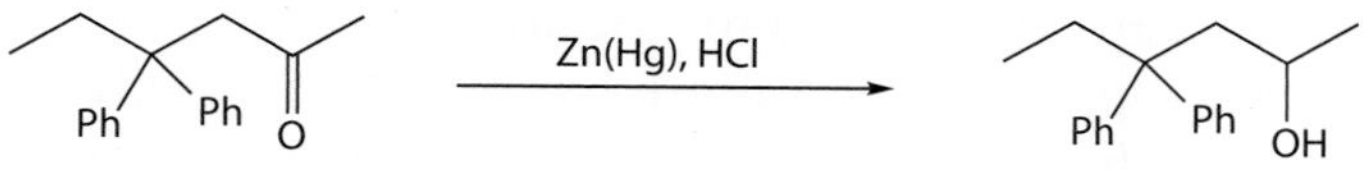

19.26

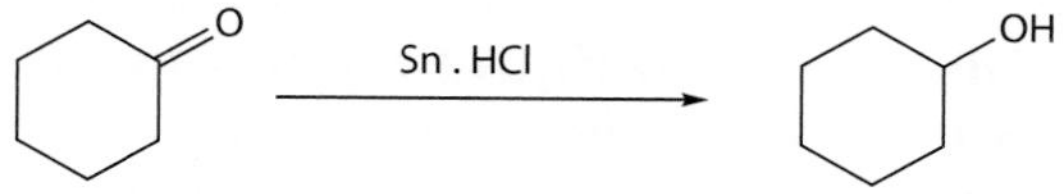

19.27

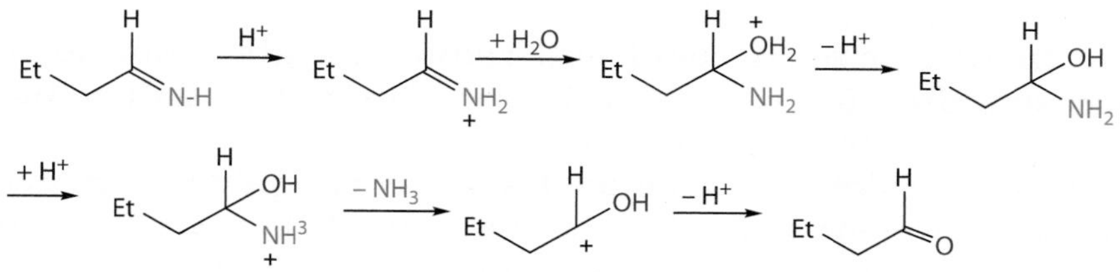

19.28

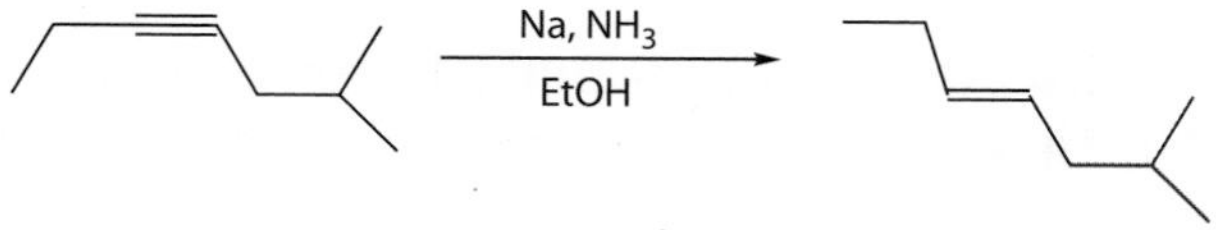

Correlation of Concepts with Homework

- A reduction is defined as the gain of electrons or the gain of hydrogen atoms: 1, 29.
- Both sodium borohydride and lithium aluminum hydride reduce ketones or aldehydes to alkoxides, and hydrolysis gives the alcohol. Lithium aluminum hydride is a more powerful reducing agent than sodium borohydride: 2, 4, 6, 30, 31, 35, 36, 37, 39, 40, 44, 45, 46, 50, 59, 60.
- Lithium aluminum hydride will reduce acid derivatives to the corresponding alcohol and an amide to an amine. Nitriles are also reduced to an amine: 3, 4, 5, 6, 7, 8, 32, 45, 46, 47, 50, 54, 55, 60.

- In the presence of a transition metal catalyst, hydrogen gas converts an alkene to an alkane and an alkyne to an alkene. Hydrogenation of an alkyne with palladium in the presence of quinoline gives the Z-alkene in what is known as Lindlar reduction: 9, 10, 11, 12, 13, 38, 47, 50, 58.
- In the presence of a transition metal catalyst, hydrogen gas converts a ketone or aldehyde to an alcohol: 14, 15, 16, 17, 37, 43, 44 46.
- Acid chlorides can be reduced by catalytic hydrogenation to give an aldehyde in what is called the Rosenmund reduction. Nitriles are reduced by catalytic hydrogenation to a primary amine: 18, 19, 20, 37.
- A ketone or aldehyde is reduced to an alcohol by reaction with sodium or lithium metal in liquid ammonia, in the presence of ethanol. This is called a dissolving metal reduction and it proceeds by an alkoxy-radical known as a ketyl: 21, 22, 46, 49.
- Alkynes are reduced to E-alkenes under dissolving metal conditions. Benzene is reduced to a cyclohexadiene under the same conditions: 23, 28, 33, 37, 41, 48.
- Zinc in HCl or acetic acid will reduce an alkyl halide to the corresponding hydrocarbon. Tin or iron in acid will reduce an aldehyde or ketone to an alcohol: 24, 26, 37, 42.
- A ketone or aldehyde is reduced to the corresponding hydrocarbon by treatment with zinc amalgam in HCl, in what is known as Clemmensen reduction. A ketone or aldehyde is reduced to the corresponding hydrocarbon by treatment with hydrazine and KOH, in what is known as a Wolff–Kishner reduction: 25, 34, 37.
- A nitrile is reduced to an aldehyde by treatment with tin chloride and HCl, in what is known as a Stephen reduction: 27, 50, 51, 60.
- A molecule with a particular functional group can be prepared from molecules containing different functional groups by a series of chemical steps (reactions). This process is called synthesis: The new molecule is synthesized from the old one (see Chapter 25): 51, 52.
- Spectroscopy can be used to determine the structure of a particular molecule, and can distinguish the structure and functionality of one molecule when compared with another (see Chapter 14): 53, 54, 55, 56, 57, 58, 59, 60.

Homework

19.29 Categorize each of the following transformations as an oxidation or a reduction using oxidation numbers:

(a) 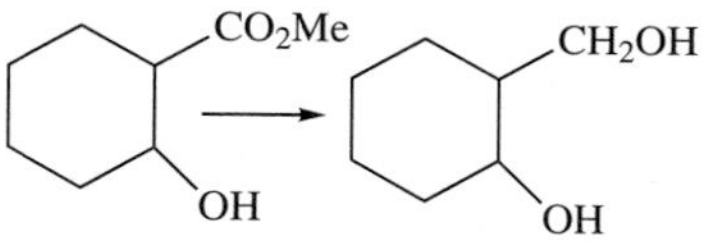 (b) 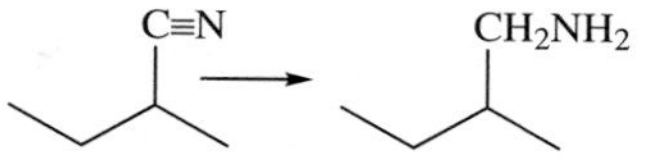 (c) 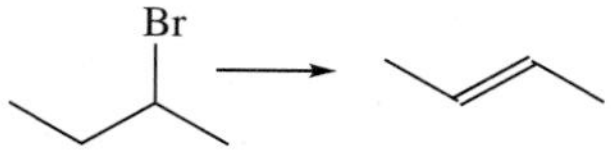 (d) 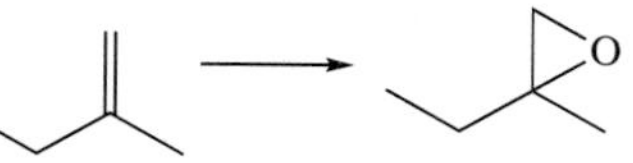 (e) 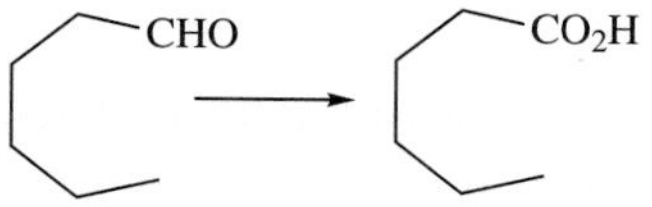 (f)

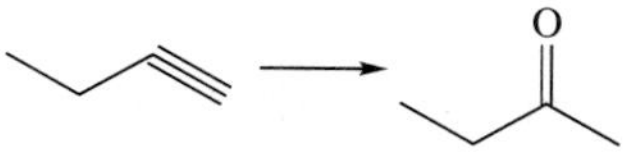

19.30 Briefly discuss whether you think LiAlH(OMe)$_3$ is stronger or weaker than LiAlH$_4$. Similarly, discuss the relative strength of NaBH$_4$ when compared to NaBHEt$_3$.

19.31 Why is water sufficient to hydrolyze a lithium aluminum hydride reduction, but aqueous ammonium chloride is used with the sodium borohydride reduction?

19.32 What is the reaction product when sodium borohydride reacts with butanoic acid?

19.33 Draw the ketyl derived from the reaction of benzophenone and sodium metal in ammonia. Draw all resonance forms for this intermediate.

19.34 What is an amalgam?

19.35 When 5-oxooctanal is treated with NaBH$_4$ in ethanol and then that product is subjected to aqueous acid workup, the product has only one hydroxyl. Draw the structure of this product and explain why only one carbonyl group is reduced in preference to the other one.

19.36 There is a reagent named lithium triethylborohydride and another named lithium trimethoxyaluminum hydride. Draw both reagents. Speculate on the strength of the first reagent relative to sodium borohydride and the second relative to lithium aluminum hydride.

19.37 Give the major product for each of the following:

(a) 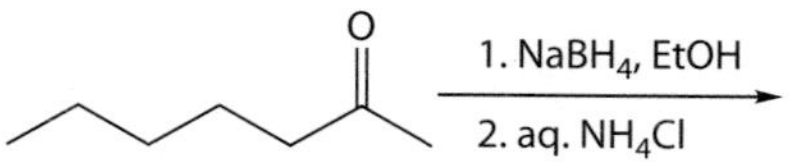 1. NaBH$_4$, EtOH 2. aq. NH$_4$Cl

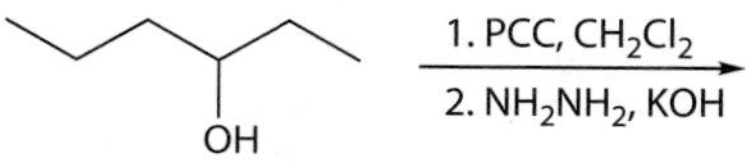

(b) 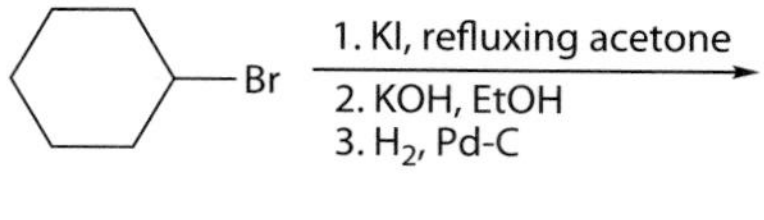 Br 1. KI, refluxing acetone 2. KOH, EtOH 3. H$_2$, Pd-C

(c) OH 1. PCC, CH$_2$Cl$_2$ 2. NH$_2$NH$_2$, KOH

(d) 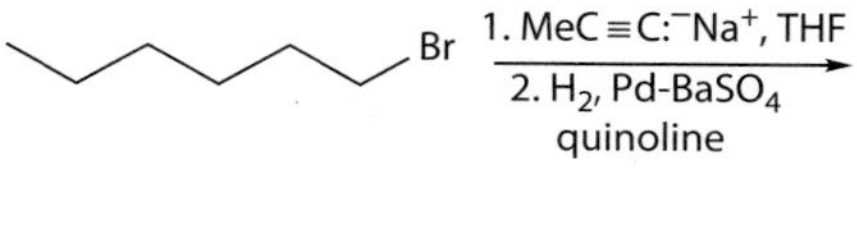 Br 1. MeC≡C:⁻Na⁺, THF 2. H$_2$, Pd-BaSO$_4$ quinoline

(e) 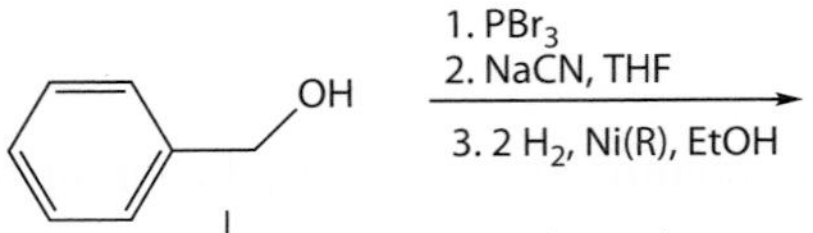 [benzyl alcohol structure, CH₂OH on benzene ring]   1. PBr₃
                                                         2. NaCN, THF
                                                         ——————————————→
                                                         3. 2 H₂, Ni(R), EtOH

(i) 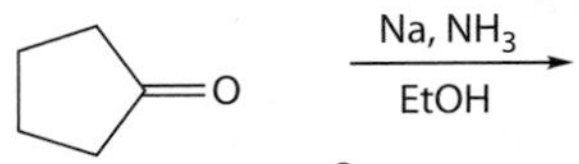 [cyclopentanone structure] ══O   Na, NH₃
                                      ——————————→
                                      EtOH

(f) 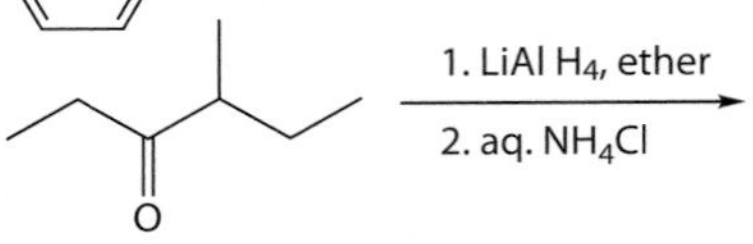 [ketone structure with methyl branch]   1. LiAl H₄, ether
                                             ——————————————→
                                             2. aq. NH₄Cl

(j) 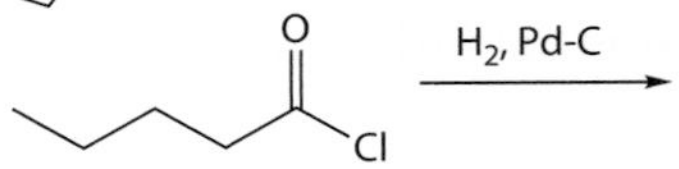 [acyl chloride structure]   H₂, Pd-C
                                 ——————————→

(g) 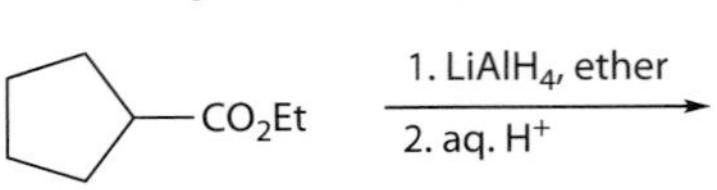 [cyclopentane with CO₂Et]—CO₂Et   1. LiAlH₄, ether
                                       ——————————————→
                                       2. aq. H⁺

(k) 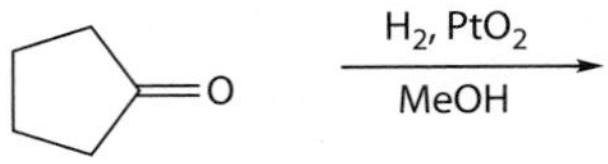 [cyclopentanone structure] ══O   H₂, PtO₂
                                      ——————————→
                                      MeOH

(h) [alkyne structure]   Na, NH₃
                          ——————————→
                          EtOH

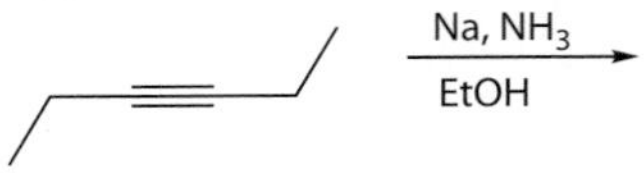

(l) 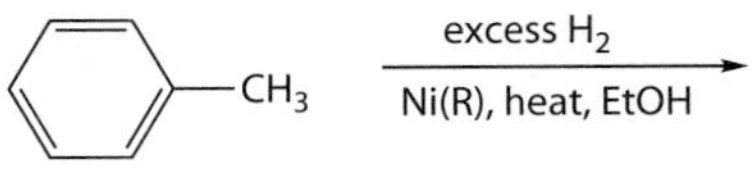 [ethylbenzene-like structure]—CH₃   excess H₂
                                         ——————————————→
                                         Ni(R), heat, EtOH

19.38 Suggest a reason why cyclopentene is hydrogenated with hydrogen gas and Pd-on-carbon at a rate that is much faster than 1,2-di-*tert*-butylcyclopentene.

19.39 What is the major product when cycloheptanone is reacted with LiAlH₄ and then water?

19.40 Draw the transition state involved in the reduction of 2-methyl-3-hexanone with LiAlH₄.

19.41 Draw the radical anion intermediate formed when methoxybenzene (anisole; see Chapter 21, Section 21.2) is treated with sodium in ethanol and ammonia (Birch reduction), and draw all resonance structures. Based on stability of the radical anion product, suggest the final product of this reduction.

19.42 What is the product formed when bromotriphenylmethane is treated with iron in ethanoic acid (acetic acid)?

19.43 In one experiment, cyclohexene was treated with hydrogen gas in ethanol in the presence of a lump of palladium. In a second experiment, everything was kept the same except that palladium powder was used. Briefly explain what differences in reactivity are expected between these two experiments and why there are differences.

19.44 You have a sample of hex-5-enal. Describe an experimental procedure that will allow you to prepare hex-5-en-1-ol without reduction of the C=C unit. Next, describe an experimental procedure that will allow you to prepare hexanal with minimum or no reduction of the aldehyde unit. Finally, describe an experimental procedure that will allow you to prepare 1-hexanol.

19.45 Describe an experimental procedure that will allow you to convert A to B:

A [ketone ester structure] $\xrightarrow{?}$ [hydroxy ester structure with OH]   **B**
     CO₂Et   CO₂Et

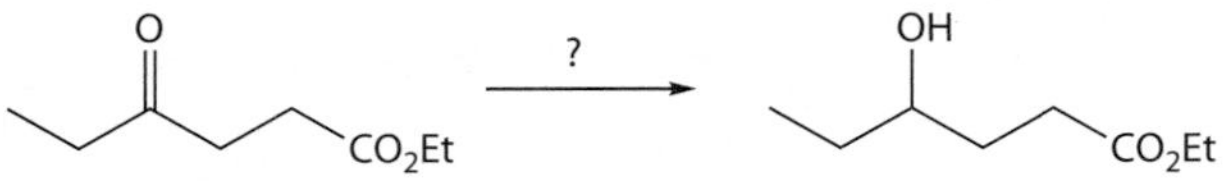

19.46 Give the major product for each of the following:

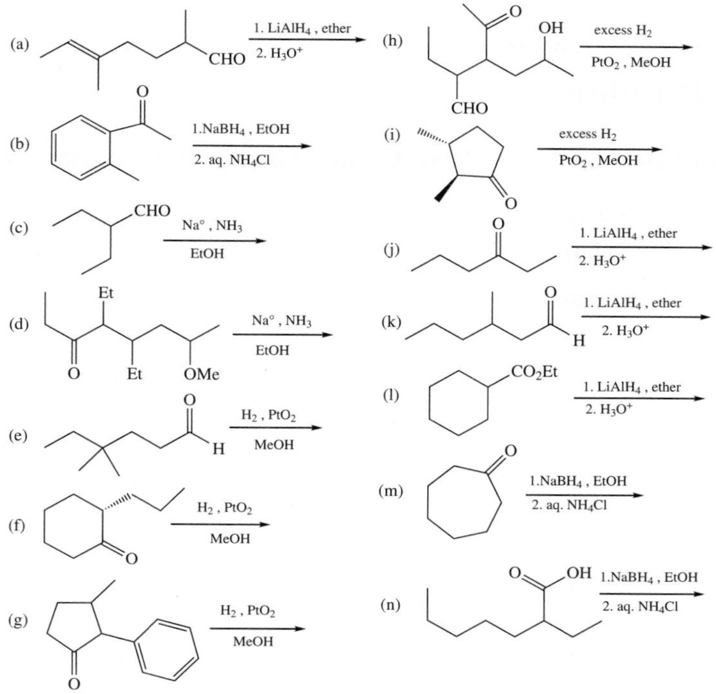

19.47 Give the product of each individual step where appropriate, and the final product for each of the following.
(a) 2-hexyne + Na, NH$_3$, EtOH
(b) 2-pentanol + LiAlH$_4$
(c) 3-heptyne + H$_2$, Pd/BaCO$_3$/quinoline
(d) ethyl butanoate + 1. LiAlH$_4$; 2. hydrolysis; 3. PCC
(e) 4-phenyl-1-pentane + 1. Mg, ether; 2. hot water
(f) cyclopentanone + 1. EtMgBr, ether; 2. H$_3$O$^+$; 3. Br$_3$; 4. Mg, ether; 5. hot water

19.48 Ethanol is added to a dissolving metal reduction of 2-pentyne with sodium in liquid ammonia. This reaction works if ethanol is not added, but it is slower. Explain.

19.49 The dissolving metal reduction of 3-pentanone gives 3-pentanol as the major product. Depending on how the reaction is done, one can sometimes see 3,4-diethyl-3,4-hexanediol formed as a secondary product after hydrolysis. Offer a mechanistic rationale for how this diol can be formed.

19.50 Give the major product for each of the following reactions:
(a) methyl 4-phenylbutanoate + 1. LiAlH$_4$; 2. hydrolysis
(b) 3-methyl-2-hexanol + 1. PCC; 2. NaBH$_4$; 3. aq. NH$_4$Cl
(c) 2R-bromopentane + 1. NaCN, THF; 2. SnCl$_2$/HCl
(d) cyclopentanecarboxaldehyde + 1. NaBH$_4$/EtOH; 2. hydrolysis
(e) 2-bromo-2-methylpentane + 1. KOH, EtOH; 2. EtOH, H$_2$, Pd/C

   (f) 3-phenylpentanal + 1. NaBH$_4$; 2. hydrolysis; 3. PCl$_5$; 4. KOH, EtOH, H$_2$,
       Pd/C

Synthesis Problems

Do not attempt these problems until Chapter 25 has been read and understood.

19.51 Discuss several different synthetic sequences in which an alkene may be converted to an aldehyde.

19.52 Provide a synthesis for each of the following:

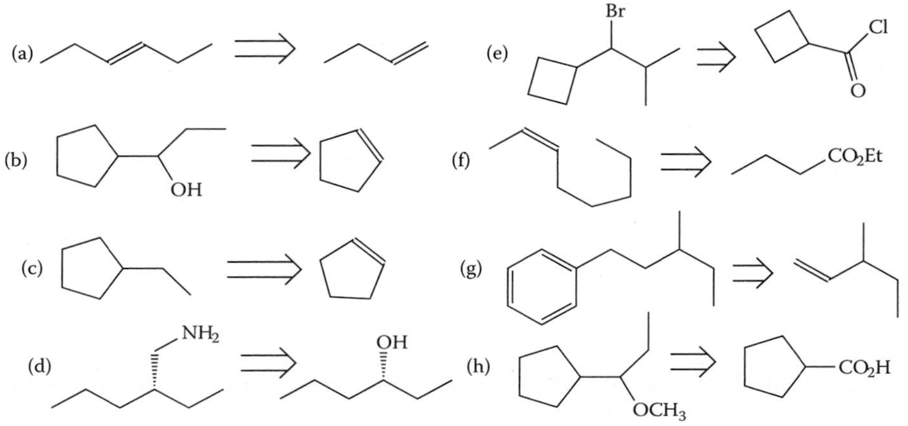

Spectroscopy Problems

Do not attempt these problems until Chapter 14 has been read and understood.

19.53 Describe differences in the IR and proton NMR that will allow you to distinguish 3-phenyl-z-pentenol and 3-phenyl-4-pentenol.

19.54 When ethyl 4-methyl-1-pentanoate is reduced with certain specialized reducing agents, it is possible to form an aldehyde and/or an alcohol. Describe differences in the IR and proton NMR that will allow you to distinguish these two products.

19.55 It is known that NaBH$_4$ reduces some, but not all, esters. If you try to reduce ethyl butanoate with NaBH$_4$ in EtOH, describe differences in the IR and proton NMR that will allow you to decide whether or not reduction to the alcohol occurred.

19.56 Briefly explain why the aldehyde H (of the O=C–H unit) appears at 9–10 ppm in the proton NMR but the hydrogen of an alkene (the C=C–H unit) appears at 4.5–5.5 ppm.

19.57 Describe spectroscopic differences between 2-hexanol and 3-hexanol that will allow you to distinguish them.

19.58 A molecule **A** with the formula C_5H_8 has the following spectra:
IR: broad peak at 3307, 2968–2843, 2120, 630 cm^{-1}
^{1}H NMR: 2.15 (t, 3H), 1.94 (s, 1H), 1.55 (m, 2H), 1.0 (t, 3H) ppm
When **A** is treated with NaNH$_2$ in THF, followed by 1-bromopropane, the product is **B**. Compound **B** has the formula C_8H_{14} and has the following spectra:
IR: 2963–2842, 1464–1435 cm^{-1}
^{1}H NMR: 2.12 (broad t, 4H), 1.50 (m, 4H), 0.98 (broad t, 6H) ppm
When **B** reacts with one molar equivalent of hydrogen gas in the presence of Pd-on-BaSO$_4$ and quinoline, the product is **C**, which has the following spectra:
IR: 3007, 2959–2874, 1656, 1455–1467, 1404 cm^{-1}
^{1}H NMR: 5.37 (m, 2H), 2.00 (m, 4H), 1.36 (m, 4H), 0.91 (broad t, 6H) ppm
Identify **A, B**, and **C**.

19.59 A molecule **A** has the formula C_7H_{14} and the following spectra:
IR: 2961–2869, 1675, 1467–1449, 1384–1362 cm^{-1}
^{1}H NMR: 4.94 (d, 1H), 2.48 (m, 1H), 1.66 (s, 3H), 1.61 (s, 3H), 0.92 (d, 6H) ppm
When **A** is treated with ozone at −78°C and then dimethyl sulfide, the two products are acetone and **B**. Compound **B** has the formula C_4H_8O and the following spectra:
IR: 2972–2877, 2714, 1737, 1468 cm^{-1}
^{1}H NMR: 2.39 (m, 1H), 1.06 (d, 6H), 9.6 (s, 1H) ppm
When **B** reacts with sodium borohydride in ethanol, followed by treatment with aqueous ammonium chloride, the final product is **C**. Compound **C** has the following spectra:
IR: broad peak at 3347–3326, 2968–2874, 1471–1463, 1042 cm^{-1}
^{1}H NMR: 3.39 (d, 2H), 2.07 (broad s, 1H; the peak is diminished when treated with D$_2$O), 1.75 (m, 1H), 0.92 (d, 6H) ppm
Identify **A, B**, and **C**.

19.60 Compound **A** has the formula C_5H_{10} and the following spectra:
IR: 2979–2884, 1716, 1461 cm^{-1}
^{1}H NMR: 2.44 (q, 4H), 1.06 (t, 6H) ppm
When **A** reacts with lithium aluminum hydride in ether, followed by reaction with water, the product **B** is formed, which has a strong peak in the IR at 3632 cm^{-1}. Reaction of **B** with PBr$_3$ leads to product **C**, which has the formula $C_5H_{11}Br$ and the following spectra: there are no significant peaks in the IR other than the normal C–H absorption peaks:
^{1}H NMR: 3.94 (m, 1H), 1.84 (m, 4H), 1.04 (broad t, 6H) ppm
Reaction of **C** with sodium cyanide in DMF leads to a product **D**, for which there are no spectral data. However, when **D** reacts with SnCl$_2$/aq. HCL, the product is **E**, with the formula $C_6H_{12}O$ and the following spectra:
IR: 2967–2879, 2693, 1708, 1462 cm^{-1}
^{1}H NMR: 9.6 (s, 1H), 2.12 (m, 1H), 1.67–1.53 (m, 4H), 0.92 (t, 6H) ppm
In addition, when **D** reacts with LiAlH$_4$ in THF, followed by treatment with dilute aqueous acid and then neutralization in aqueous NaOH, the product

is compound **F**—a slightly foul-smelling oil with the formula $C_6H_{15}N$ and the following spectra:

IR: 3376, 3297, 2962–2864, 1610, 1462, 966–742 cm^{-1}

^{1}H NMR: 2.61 (d, 2H), 1.35–1.30 (m, 6H; after treatment with D_2O; this peak is diminished and integrates to only 4H), 1.20 (m, 1H), 0.88 (broad t, 6H) ppm

Identify **A, B, C, D, E**, and **F**.

Carboxylic Acid Derivatives and Acyl Substitution

20

Chapter 18 discussed the acyl addition reactions of several nucleophiles with the carbonyl unit of aldehydes and ketones. As pointed out in Chapter 16, carboxylic acids and their derivatives also contain a carbonyl unit. Although these acid derivatives react with nucleophiles via attack at the acyl carbon, the presence of a leaving group attached to the acyl carbon leads to a subsequent reaction that is not possible with aldehydes and ketones. Acid derivatives react with nucleophiles via acyl substitution. This reaction pathway proceeds by an intermediate called a tetrahedral intermediate. The introduction to this reaction, presented in Chapter 16 (Section 16.8), will be explained and expanded here.

To begin this chapter, you should know the following:

- **basic rules of nomenclature for alcohols, amines, aldehydes and ketones, carboxylic acids, and carboxylic acid derivatives (Chapter 5, Sections 5.6, 5.9.2, and 5.9.3; Chapter 16, Section 16.4)**
- **basic rules of nomenclature for carboxylic acid derivatives (Chapter 16, Section 16.7)**
- **nomenclature of dicarboxylic acids (Chapter 16, Section 16.5)**
- **nomenclature for imines and nitriles (Chapter 5, Section 5.9.4)**
- **CIP rules for prioritizing substituents, groups, and atoms (Chapter 9, Section 9.3)**

- identification and nomenclature for *E*- and *Z*-isomers (Chapter 9, Section 9.4)
- bond polarization (Chapter 3, Section 3.7)
- fundamentals of acyl addition (Chapter 16, Section 16.3; also Chapter 18, Section 18.1)
- fundamentals of acyl substitution and the structure of a tetrahedral intermediate (Chapter 16, Section 16.8)
- acid–base reactions of the carbonyl group and formation of oxocarbenium ions (Chapter 18, Sections 18.1 and 18.5)
- how to identify a good leaving group (Chapter 11, Sections 11.1 and 11.2)
- how to identify relative nucleophilic strength (Chapter 11, Section 11.2; Chapter 18, Sections 18.3–18.7)
- characteristics and reactivity of Grignard reagents, and organolithium reagents (Chapter 15, Sections 15.1 and 15.5)
- characteristics and reactivity of organocuprates (Chapter 15, Section 15.6)
- nucleophilic acyl addition of organometallic reagents to carbonyl compounds (Chapter 18, Section 18.4)
- acid-catalyzed nucleophilic acyl addition of weak nucleophiles to carbonyl compounds (Chapter 18, Sections 18.5 and 18.6)
- the mechanism of acetal and ketal formation (Chapter 18, Section 18.6)
- oxidation reactions of peroxyacids (Chapter 10, Section 10.5; Chapter 17, Section 17.3)
- reagents for the conversion of alcohols to alkyl halides (Chapter 11, Section 11.7)

This chapter will discuss methods for the preparation of esters, acid chlorides, anhydrides, and amides from carboxylic acids, based on acyl substitution reactions. Acyl substitution reactions of carboxylic acid derivatives will include hydrolysis, interconversion of one acid derivative into another, and reactions with strong nucleophiles such as organometallic reagents. In addition, the chemistry of dicarboxylic acid derivatives will be discussed, as well as cyclic esters, amides, and anhydrides. Sulfonic acid derivatives will be introduced as well as sulfate esters and phosphate esters. Finally, nitriles will be shown to be acid derivatives by virtue of their reactivity.

When you have completed this chapter, you should understand the following points:

- **Many acid derivatives can be generated by "replacing" the OH unit of a carboxylic acid (COOH) with another atom or group: $-Cl$, $-Br$, $-O_2CR$, $-OR$, $-NR_2$. Common acid derivatives are acid chloride, acid anhydrides, esters, and amides.**

- Acid or base hydrolysis of acid chlorides, acid anhydrides, esters, or amides regenerates the parent carboxylic acid.
- Acid chlorides are prepared by reaction of a carboxylic acid with a halogenating agent such as thionyl chloride, phosphorus trichloride, phosphorus pentachloride, phosgene, or oxalyl chloride.
- Acid anhydrides can be prepared by reaction of an acid chloride with a carboxylic acid or by the dehydration of two carboxylic acids that leads to coupling.
- Esters can be prepared by the reaction of an acid chloride or acid anhydride with an alcohol as well as by direct reaction of an acid and an alcohol, with an acid catalyst, or by using a dehydrating agent such as DCC. Lactones are cyclic esters.
- Amides can be prepared by the reaction of an acid chloride, acid anhydride, or an ester with ammonia or an amine. Acids and amines can be coupled in the presence of a dehydrating agent such as DCC. Amines react with carboxylic acids to give ammonium salts. Heating these salts to around 200°C will usually give the amide. Lactams are cyclic amides.
- Imides are formed by the reaction of amides with other acid derivatives. Cyclic anhydrides and cyclic imides can be prepared from dicarboxylic acids or acid dichlorides.
- Acid derivatives react with Grignard reagents and organolithium reagents to give ketones, but this initial product reacts with a second equivalent of Grignard reagent or organolithium reagent to give a tertiary alcohol.
- Grignard reagents and organolithium reagents react with cadmium chloride to form dialkyl cadmium reagents, which react with acid chloride to give ketones. Organolithium reagents react with copper(I) salts to give lithium dialkyl cuprates, which react with acid chlorides to give ketones.
- Grignard reagents and organolithium reagents react with carbon dioxide to give carboxylic acids.
- Organolithium reagents react with carboxylic acids to give ketones.
- Grignard reagents and organolithium reagents react with nitriles to give imine anions, but aqueous acid hydrolysis usually converts the imine anion to a ketone as the final product.
- Peroxyacids react with ketones to give esters in the Baeyer–Villiger reaction.
- Acid dichlorides are prepared by reaction of the diacid with a halogenating agent. Diesters are prepared from dicarboxylic acids or diacid chlorides and alcohols. Diamides are prepared from diesters or diacid chlorides and amines.

- **Sulfonic acids can be converted to sulfonyl acid chlorides, sulfonate esters, and sulfonamides, similar to reactions of carboxylic acids.**
- **The ester derivatives of nitric acid, sulfuric acid, and phosphoric acid are important compounds, with some use in organic chemistry but greater use in biology.**
- **Nitriles are acid derivatives. Primary amides are dehydrated to nitriles, and nitriles are hydrolyzed to either an amide or a carboxylic acid.**
- **A molecule with a particular functional group can be prepared from molecules containing different functional groups by a series of chemical steps (reactions). This process is called synthesis: The new molecule is synthesized from the old one (see Chapter 25).**
- **Spectroscopy can be used to determine the structure of a particular molecule and can distinguish the structure and functionality of one molecule when compared with another (see Chapter 14).**
- **The biological relevance of key acid derivatives is significant.**

20.1 Chemical Reactivity of Carboxylic Acid Derivatives

Carboxylic acids were introduced in Chapter 16 (Section 16.4) and shown to have the general structure **1**, with an OH group directly attached to a carbonyl. The COOH group taken as a unit is considered to be a unique functional group: the carboxyl group. The salient feature of carboxylic acids is the acidity of the proton in the O–H unit and carboxylic acids are the quintessential "organic acid." The acidity of **1** varies with the nature of R, the solvent, and the base with which it reacts. The OH unit effectively dictates the chemistry.

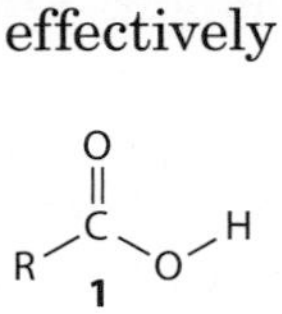

Carboxylic acid derivatives, introduced in Chapter 16 (Section 16.7), are categorized by the nature of the group that replaces OH in **1** to form **2**: OR for esters, Cl for an acid chloride, O_2CR for anhydrides, and NR_2 for amides. There are two fundamental reactions of carboxylic acid derivatives. Acid derivatives (**2**; X = OR, Cl, O_2CR, NR_2) react with an acid such as sulfuric acid, HCl, or *p*-toluenesulfonic acid to form an oxocarbenium ion, **3**. The electrophilic carbon subsequently reacts with a nucleophile to generate a tetrahedral intermediate that has a leaving group attached (see **4**). Loss of the leaving group regenerates the carbonyl in a new carboxylic acid derivative or a carboxylic acid, **5**. This reaction of an acid derivative with a nucleophile is known as ***nucleophilic acyl substitution***.

When acid derivative **2** reacts with sulfuric acid, the oxygen atom is the base and the conjugate acid product of this acid–base reaction is oxocarbenium ion **3**, which is resonance stabilized. When **2** is an acid chloride, anhydride, ester, or amide, a heteroatom is attached to the positive carbon in **3**. As in Chapter 18 (Section 18.1), the acid–base reaction of the carbonyl unit in **2** to give **3** facilitates reactions with nucleophiles. The reaction of intermediate **3** with a nucleophile (:Y) gives tetrahedral intermediate **4**; contrary to acyl addition, reaction **4** contains an X group that can function as a leaving group. Loss of X leads to the final product of this reaction: **5**. If the nucleophile (:Y) is hydroxide, compound **5** is the carboxylic acid (X = OH). If the nucleophile :Y is an alcohol, the product **5** is an ester, and if :Y is an amine, the product **5** is an amide. This first reaction is therefore the acid-catalyzed acyl substitution reaction of acid derivatives.

20.1 Show the steps required for the conversion of 4 to 5 that were just described.

20.2 Draw the product formed when X in 2 = Cl and :Y = ethoxide.

The other reaction of a carbonyl is the direct acyl addition of the nucleophile :Y to the acyl carbon of **2** to give tetrahedral intermediate **6** directly. Note the similarity of **6** with **4**; however, **6** is an alkoxide anion, which can displace the leaving group (X) to form a new carbonyl unit. If X is a better leaving group than the nucleophile Y, loss of X gives the acyl substitution product **5**. If the nucleophile (:Y) is hydroxide, the product is a carboxylic acid derivative **2** that reacts to give **5** (Y = OH) directly. Other nucleophiles lead to other acyl products. *The fundamental difference in chemical reactivity between acid derivatives and ketones or aldehydes is that acid derivatives undergo acyl substitution whereas aldehydes and ketones undergo acyl addition.* Acid chloride, anhydrides, esters, and amides undergo this reaction.

The acyl addition reactions described in this chapter will occur with carboxylic acid derivatives, but rarely with carboxylic acids. Remember that carboxylic acids **1** are good Brønsted–Lowry acids, with a pK_a between 1 and 5 for most derivatives. Therefore, if **1** reacts with a "nucleophile" that is also basic (sodium hydroxide, sodium methoxide, or sodium amide), the acid–base reaction dominates and the product is the anionic salt of a carboxylic acid. Butanoic acid (**7**) reacts with sodium hydroxide, for example, to give sodium butanoate (**8**) as the conjugate base and water as the conjugate acid. This means that if a carboxylic acid is formed as a product in the presence of a base such as sodium hydroxide, the initially formed carboxylic acid is converted to the carboxylate salt.

20.2 Acyl Substitution. Acid Derivatives React with Water: Hydrolysis

All acid derivatives react with water in the presence of an acid catalyst to generate the parent carboxylic acid used to prepare that derivative. In other words, the product is the carboxylic acid used to generate the acid derivative. This reaction is known as ***acid hydrolysis***. The reaction of an acid derivative with aqueous hydroxide also generates the parent carboxylic acid, but the basic conditions convert the acid to its carboxylate salt, as described before. Therefore, a second reaction with aqueous acid is required to convert the carboxylate salt to the carboxylic acid. This reaction is known as ***base hydrolysis***. Both reactions are discussed in this section.

If an acid chloride such as butanoyl chloride (**8**) is mixed with water in a neutral reaction medium, hydrolysis is very slow because water is a relatively weak nucleophile. To compensate, an acid catalyst is added, as was done with ketones and aldehydes in Chapter 18 in order to facilitate reactions with weak nucleophiles (see Sections 17.5 and 17.6). With this in mind, reaction of **8** with water and an acid catalyst gives butanoic acid (**7**) as the product. The OH unit in **7** must rise from water, which has displaced the chlorine leaving group. The acid chloride must react with the acid catalyst to facilitate this transformation and it is known that the intermediate is a tetrahedral intermediate. *Using this walk-through, the mechanism of the reaction is predictable based on an analogy with similar reactions with aldehydes and ketones (see Chapter 18, Section 18.6).* The carbonyl oxygen of **8** reacts with the acid catalyst to form a resonance-stabilized oxocarbenium ion **9**, which is attacked by the nucleophile (water) to form oxonium ion **10**. Either water or unreacted **8** reacts as a base with the acidic proton of oxonium ion **10** to give the conjugate base, **11**, and H_3O^+ is the conjugate acid.

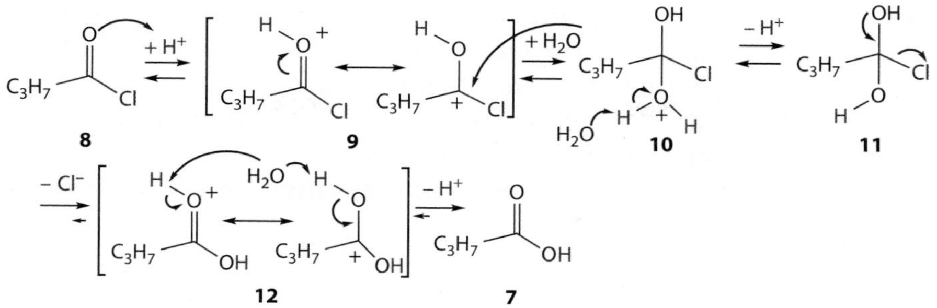

Compound **11** is the key tetrahedral intermediate and it has *two* leaving groups: **OH (from the water nucleophile)** and **Cl (from the acid chloride)**. Intermediate **11** is the tetrahedral intermediate analogous to **4** in Chapter 16, Section 16.7, with Cl and OH as potential leaving groups. Chlorine is a superior leaving group when compared to OH. Loss of chloride ion gives a protonated carbonyl intermediate that is one resonance contributor of oxocarbenium ion **12**. Either water or unreacted **8** reacts as a base with the acidic proton of intermediate **12** to give butanoic acid, **7**.

20.3 **Draw the structure of 3-cyclobutylhexanoyl chloride.**

20.4 **Why is the attacking nucleophile *not* hydroxide ion so that 9 can be converted directly to 11?**

20.5 **Write the mechanism for the acid-catalyzed hydrolysis of 2-methylpropanoyl chloride.**

An alternative method for the conversion of an acid chloride to an acid does not rely on the weak nucleophile water or on acid catalysis. Hydroxide is a strong base (it is the conjugate base of water), but it is also a nucleophile in reactions with carboxylic acid derivatives. If the basic $^-$OH ion reacts with acid chloride **8**, attack at the carbonyl carbon gives tetrahedral intermediate **13** directly. The alkoxide unit can donate electrons back to the carbon and regenerate the C=O unit, but there are two leaving groups. Either the C–O bond (OH is the leaving group) *or* the C–Cl bond (Cl is the leaving group) in **13** can be broken. Breaking the C–Cl bond is more facile than breaking the C–O bond, which is just another way to say that the chloride ion is a much better leaving group than hydroxide ion.

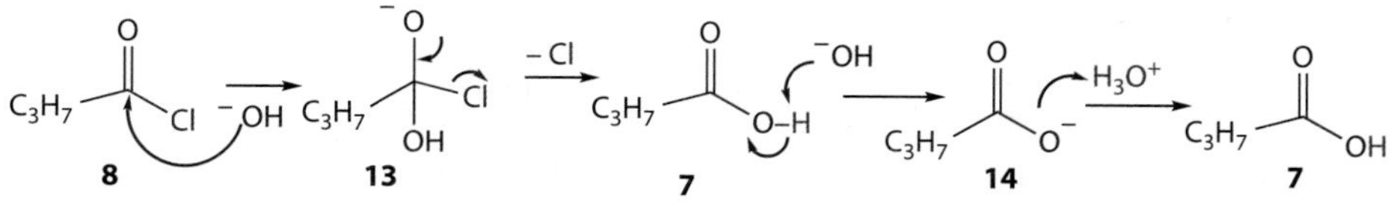

Loss of chloride ion from **13** leads directly to the acid, **7**. It is very important to realize, however, that once the acid is formed, *it is in a solution of hydroxide,* which is a strong base (as mentioned previously). The acid and the base react immediately to give the anionic salt of the carboxylic acid, **14**. *Carboxylate anion* *14 is the final product of this reaction.* In order to isolate **7**, an acid that is significantly stronger than **7** must be added to **14** (e.g., a mineral acid such as HCl) *in a second chemical step.* Therefore, treatment of **14** with aqueous HCl (shown as H_3O^+) gives **7**. The conversion of the acid chloride (**8**) to the free acid (**7**) under basic conditions is a **two-step process: (a) treatment with hydroxide, and (b) acid hydrolysis**.

The reaction of anhydrides with aqueous acid or aqueous base is essentially identical to that of acid chlorides (replace Cl with O_2CR in the mechanism for **1–6** or **8–7**), and it also gives the parent acid as the product. Anhydrides are converted to the anion of the carboxylic acid precursors by base hydrolysis. Base hydrolysis of dibutanoic anhydride with aqueous sodium hydroxide leads to two molar equivalents of sodium butanoate; similar reaction with butanoic ethanoic anhydride gives one molar equivalent of sodium ethanoate and one molar equivalent of sodium butanoate.

The base-induced reaction of the symmetrical dipropanoic anhydride (**15**) illustrates the mechanism of base hydrolysis with anhydrides, in a reaction that gives two equivalents of propanoic acid. Acyl addition of the nucleophilic hydroxide gives tetrahedral intermediate **16**. In this case, the best leaving group is the carboxylate anion **18**, which gives propanoic acid, **17**. Because **17** is formed in a basic solution, reaction with hydroxide gives **18**. A second step is required to convert **17** to **18**, using aqueous acid (aqueous sulfuric acid or aqueous HCl).

20.6 Write out the mechanism for the acid-catalyzed hydrolysis of propanoic butanoic anhydride, assuming that the propanoic acid part of the anhydride is protonated first.

20.7 Write out all reactants and products formed in the two base hydrolysis reactions just mentioned.

20.8 Write out the mechanism for base-catalyzed hydrolysis of the unsymmetrical anhydride butanoic propanoic anhydride.

Esters react with water under either acid or base conditions to give the acid, but the leaving group is now RO^- rather than Cl^- or ^-O_2CR, as in the previous reactions with acid chlorides or acid anhydrides. For basic hydrolysis conditions, the experiment requires two steps to convert the ester to the acid. Isopropyl acetate (**19**), for example, is heated to reflux in aqueous NaOH and then neutralized with aqueous sulfuric acid to give a quantitative yield of two products: acetic acid (**15**) and 2-propanol (isopropanol).[1] This reaction is explained by addition of hydroxide (in **blue** in the illustration) to the acyl carbon of **19** to give tetrahedral intermediate **20**.

The alkoxide unit (C–O–) can transfer electrons back to the electrophilic carbon, but there are two leaving groups, as in previous cases. In **20**, the leaving groups are ⁻OH and ⁻**OiPr**. Although it is not obvious, in water and with an excess of hydroxide ion, loss of ⁻**OiPr** (isopropoxide, in **green**) is more facile, and this generates the acid (acetic acid, **21**). Because it is formed in the presence of hydroxide (isopropoxide is also present), the acid–base reaction leads to the final product of the first chemical step, carboxylate ion **22**. To isolate acid **21**, **22** is treated with aqueous acid as shown in the illustration, in a second chemical step.

This two-step process, shown in reaction form for **19** → **21** + 2-propanol, is known as **saponification**. Saponification means "to make soap" and the term comes from the ancient practice of using wood ashes (rich in potassium hydroxide) to convert animal fat to soap. Animal fat as well as vegetable oils is usually a mixture of triglycerides (**23**, the triester derivative of fatty acids, and glycerol, **24**). Under these conditions, basic hydrolysis of all three ester units leads to formation of glycerol and the salt of the fatty acids. The salts of these fatty acids are solids and they are the fundamental constituent of what is known as "soap."

Two typical fats are the triesters of **23**, where the acid is octadecanoic acid (**25**; common name stearic acid) or hexadecanoic acid (**26**; common name palmitic acid). Interestingly, carboxylic acid **26** is a constituent of the white desert truffle, *Tirmania nivea*.[2] Oils, typically obtained from plant sources, are alkenoic acids rather than alkanoic acids. In other words, they have one or more C=C units in the carbon skeleton. An example is α-linolenic acid (**27**; all-*cis*-9,12,15-octadecatrienoic acid), which is found in many common vegetable oils and is a polyunsaturated omega-3 fatty acid. *cis*-9-Octadecenoic acid (**28**; known as oleic acid) is a monounsaturated omega-9 fatty acid. It is a major constituent of olive oil, and it is the most abundant fatty acid in human adipose tissue (body fat). The omega-3 and omega-9 nomenclature was discussed in Chapter 16, Section 16.10.

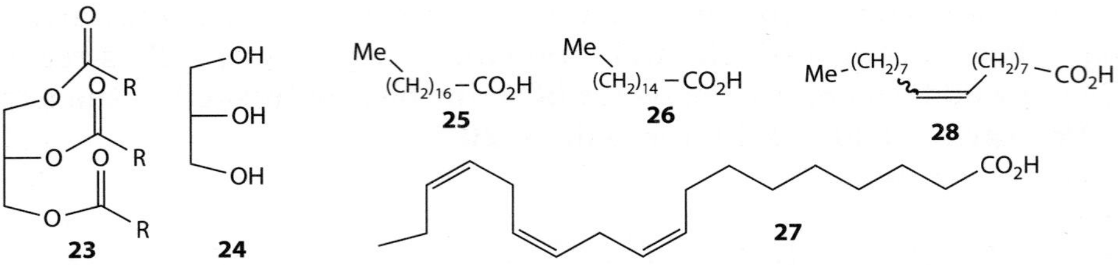

20.9 What is the IUPAC name of 19?

20.10 What is the IUPAC name of 24?

An ester is hydrolyzed with aqueous acid as well as with aqueous hydroxide. Under acidic conditions, the mechanism of ester hydrolysis is similar to that of acid chlorides and acid anhydrides, as shown previously. When ethyl butanoate (**29**) is treated with an acid catalyst in water, the products are carboxylic acid (butanoic acid, **6**) and ethanol. The OH unit in this acid is clearly derived from the water, and the OEt unit in ethanol is derived from the OEt unit of the ester. It is known that water does not react directly with **29** to give **7**, so the acid catalyst must facilitate the reaction. This, of course, indicates that **29** reacts as a base with the acid catalyst to give the resonance-stabilized oxocarbenium ion **30**.

Subsequent reaction with water (the nucleophile) gives oxonium ion **31**. Loss of the acidic proton leads to the "hydrate-like" intermediate **32**, which is a tetrahedral intermediate. As with **4** previously, in the presence of a large excess of water, **32** reacts with the acid catalyst to form oxonium ion **33**. Remember that the reaction converts **29** to **7** plus ethanol, and protonation of the OEt unit is necessary if ethanol is lost as a leaving group.

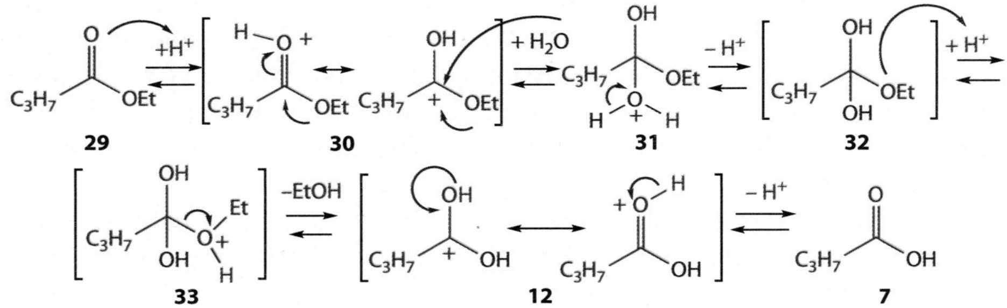

Loss of ethanol from **33** gives oxocarbenium ion **34**, and loss of the acidic proton from that species gives butanoic acid (**7**). The two products of the acid hydrolysis are the carboxylic acid (butanoic acid) and the alcohol (ethanol). *As with the hydrolysis of acid chlorides and acid anhydrides before, this mechanism is a series of acid–base reactions.*

Amides react under both acidic and basic conditions to give the parent carboxylic acid and an amine for secondary and tertiary amides, or ammonia from a primary amide. The mechanisms are essentially the same as those shown for the other acid derivatives, except that the tetrahedral intermediate will have NH_2 and OH as leaving groups. The $^-NH_2$ group (amide group) is a much worse leaving group when compared with the other groups in this section and, experimentally, the hydrolysis of an amide requires much more vigorous conditions than the other acid derivatives. This simply means that *it is more difficult to hydrolyze an amide compared to other acid derivatives.*

The acid-catalyzed hydrolysis of butanamide proceeds as expected to give butanoic acid, **7**. An example is the acid-catalyzed hydrolysis of the tertiary amide *N,N*-diethyl-3-phenylpentanamide (**34**) to give 3-phenylpentanoic acid (**35**). The initial product is *N,N*-diethylamine, but under the acid-catalyzed conditions, the basic amine reacts to form the observed product, ammonium salt, **36**. Secondary and primary amides react in an identical manner; the leaving group is an amine for the secondary amide, but it is ammonia for the primary amide. Acid hydrolysis of pentanamide, for example, gives pentanoic acid and ammonia (NH_3) and, under acid conditions, the ammonia is converted to the ammonium ion (NH_4^+). When amides are heated with aqueous hydroxide, the products are the salt of the carboxylic acid and the amine. Acid hydrolysis of the carboxylate salt generates the acid, as with saponification.

20.11 Write out the mechanism for the acid-catalyzed hydrolysis of methyl butanoate.

20.12 Based on the mechanism shown for converting 23 to 6, write out the mechanism for the acid-catalyzed hydrolysis of pentanamide.

20.13 Write out the reaction that converts *N,N*-diethylamine to 36.

20.14 Draw the structure of the amide that gives 4,4-diphenylcyclohexanecarboxylic acid and 1-aminobutane as the reaction products after acid hydrolysis.

This section discussed the acid and base hydrolysis of the derivatives of carboxylic acids. The mechanisms for all of these reactions are essentially identical, with the only real mechanistic difference being the leaving group. There is an order of reactivity for acid derivatives:

acid chlorides ≥ acid anhydrides >> esters >>> amides

This order or reactivity with nucleophiles such as water can be correlated to the ability of the X group (see **32**) to "leave." Because chlorine is the best leaving group and NR_2 is the poorest, the fact that an acid chloride is the most reactive and an amide is the least reactive is not surprising. This order of reactivity correlates with the relative stability of each acid derivative to acid or base hydrolysis. This simply means that it is difficult to hydrolyze an amide and rather easy to hydrolyze an acid chloride, which is apparent when specific reaction conditions are examined. Acid chlorides (**37**; X = Cl) can occasionally be hydrolyzed to the acid in water, whereas amides (X = NR_2) often require the use of strong acid (6N HCl) and prolonged heating for hydrolysis to occur. Acid anhydrides (X = O_2CR) are close to acid chlorides in reactivity and esters are "in the middle." Esters (X = OR) are hydrolyzed in dilute acid, sometimes with mild heating. In all cases, saponification is generally the more efficient method for hydrolysis because acid-catalyzed hydrolysis is always reversible.

20.15 Briefly discuss whether an acid chloride or an acid bromide is expected to be more reactive in acyl substitution reactions.

20.3 Preparation of Acid Chlorides

Carboxylic acids have an OH connected to the carbonyl, but acid chlorides have a chlorine (**37**; X = Cl). To prepare an acid chloride from an acid, a reagent must be used that exchanges OH for Cl. A number of reagents do this, and all are characterized by having one or more chlorine atoms in the reagent. Thionyl chloride ($SOCl_2$; see **38**) was utilized in Chapter 11 (Section 11.7.2) for the conversion of alcohols to alkyl chlorides; it also reacts with carboxylic acids to form acid chlorides. Other chlorinating reagents include phosphorus trichloride (PCl_3), phosphorus pentachloride (PCl_5), phosphorus oxychloride ($POCl_3$), and phosgene ($COCl_2$). The discussion will begin with thionyl chloride and then proceed to the other chlorinating agents. **The key point to remember is that a reagent rich in chlorine is used to transfer chlorine to another molecule.**

As for chemical reactivity, acid chlorides are highly reactive and susceptible to attack at the carbonyl by a variety of nucleophiles via acyl substitution. The more volatile acid chlorides tend to be lachrymators (they irritate the eyes and nasal membranes) and are sometimes corrosive. When they react, hydrochloric acid (HCl) is usually one of the by-products, and this may cause problems in many reactions that are sensitive to acids. Relative to the parent carboxylic acid, an acid chloride generally has a lower boiling point—in large part because carboxylic acids have the capability of extensive hydrogen bonding, whereas acid chlorides do not. The melting point of the acid chloride is also lower than the analogous carboxylic acid, but its density tends to be greater than that of the acid.

20.3.1 Reaction of Carboxylic Acids with Thionyl Chloride

Carboxylic acids react with thionyl chloride to give an acid chloride. In a typical experiment, heating butanoic acid (**7**) on a water bath with $SOCl_2$ (**38**) and then distilling the product gives butanoyl chloride (**8**) in 86% yield.[3] Careful analysis shows that sulfur dioxide (SO_2) and HCl are also produced as products during the reaction. Clearly, chlorine has been transferred from S to the acyl carbon, and the second chlorine atom on **38** has been converted to HCl. The proton in HCl logically comes from the acidic proton in **7**, suggesting an acid–base reaction. Moreover, the S=O unit in **38** is converted to SO_2, and the second oxygen atom must be the oxygen of the OH unit in **7**.

An explanation for these observations requires a mechanism that is related to the S_Ni mechanism introduced for the reaction of an alcohol and thionyl chloride in Chapter 11 (Section 11.7.2). The oxygen of the OH group in the acid (**7**) behaves as a Lewis base and donates electrons to the electrophilic sulfur of thionyl chloride to give **39**, which is an oxonium ion with a highly acidic proton. It is likely that the hydrogen atom and chloride are not lost as H^+ and Cl^-, however. As the new S=O bond is formed in **40**, chloride ion pulls off the acidic proton in **39** to form HCl as a product. Using an S_Ni mechanism, **40** undergoes a reaction in which the chlorine atom donates electrons to the acyl carbon in an intramolecular process that breaks one π-bond and forms a new one to the other oxygen, leading to loss of the O=S=O unit. Indeed, sulfur dioxide is a good leaving group (it is a gas and a neutral molecule) and the four curved arrows show this sequence in **40**.

Loss of SO_2 drives the reaction to the right to give butanoyl chloride (**8**) as the major organic product, along with SO_2 and HCl. Many carboxylic acids are converted to the corresponding acid chloride with thionyl chloride. In some cases, triethylamine (NEt_3) is added as a base to the reaction to "trap" the HCl as it is released in an acid–base reaction, forming triethylammonium hydrochloride (Et_3NH^+ Cl^-) and driving the reaction to the right toward the acid chloride. This reaction is also shown in reaction form to focus attention on reactants and products rather than the mechanism.

20.16 Draw the structures for all products generated by the reaction of cyclohexanecarboxylic acid and thionyl chloride in the presence of triethylamine.

20.3.2 Other Halogenating Agents

Thionyl chloride is among the most common halogenating agents for carboxylic acids because it is efficient and the by-products are gases that often escape from the reaction. However, several other reagents are available. At least three phosphorus reagents can be used: phosphorus trichloride (**41**), phosphorus pentachloride (**42**), and phosphorus oxychloride (**43**). All three reagents work by a mechanism that is related but not identical to the mechanism of thionyl chloride. The oxygen atom of the alcohol donates electrons to phosphorus to form a P–O bond, with transfer of chlorine from phosphorus to the carbonyl carbon and loss of HCl. As with thionyl chloride, an amine such as triethylamine (NEt_3) may be added as a base to react with the HCl that is formed in the reaction.

20.17 **Draw the final organic product or products of a reaction (a) between 3,3-dimethylpentanoic acid and PCl_3 and (b) between 2,4-diphenylcyclopentanecarboxylic acid and $POCl_3$ and triethylamine.**

Acid bromides are prepared by the reaction of carboxylic acids with brominating agents such as thionyl bromide (**44**; $SOBr_2$), phosphorus tribromide (PBr_3), or phosphorus pentabromide (PBr_5). An example is the conversion of pentanoic acid (**45**) to pentanoyl bromide (**46**) using **44**. Because acid chlorides are highly reactive, there is little need to prepare the more reactive acid bromide in most cases, and it will not be seen very often. If there is a need for the acid bromide, however, thionyl bromide is the most common reagent used for the preparation.

One last reagent is sometimes useful for preparing acid chlorides. Carbonic acid (**47**) is a very unstable compound (note that it is shown in brackets in the illustration) that decomposes to CO_2 and water, but it is formally a carboxylic acid. The monoacid chloride of **47** *would* be **48**, but this molecule is also highly unstable (it is also shown in brackets). The diacid chloride (**49**) is relatively stable, however, and can be isolated and used in chemical reactions. It is known as **phosgene** (usually abbreviated as $COCl_2$), but it is a highly toxic compound that was used as a "poison gas" in World War I with deadly effects. It can cause severe pulmonary edema (which can be fatal) or pneumonia, as well as choking, a constricted feeling in the chest, coughing, bloody sputum, and severe eye irritation. Clearly, any ***use of phosgene must be done with great care and special precautions***. For most of us, the use of phosgene is avoided unless no other method that will give the same results is available.

Phosgene reacts with carboxylic acids, usually in the presence of an amine base, to give the acid chloride. All by-products are gases (the HCl is trapped by an amine as the ammonium salt) and the reaction is very efficient. Most carboxylic acids react with phosgene, but, once again, the toxicity of phosgene makes it an unattractive reagent in most cases.

Oxalyl chloride is another reagent that reacts with carboxylic acids to give an acid chloride. This reagent and the pertinent reaction will be discussed in Section 20.9.1.

20.4 Preparation of Acid Anhydrides

Anhydrides of carboxylic acids are, as the name suggests, the product of two carboxylic acid units that couple together with loss of water. It is possible to convert carboxylic acids directly to anhydrides, but more commonly they are prepared from acid chlorides. Once an acid chloride has been prepared, it is highly reactive in the presence of a variety of nucleophiles, even relatively weak nucleophiles. In Section 20.2, for example, the reaction of an acid chloride with water gave the parent carboxylic acid. Acid chlorides react with other carboxylic acids to form anhydrides where the acid behaves as a nucleophile.

Anhydrides are highly reactive compounds, only slightly less reactive than acid chlorides. Acid anhydrides undergo hydrolysis under either acid or base conditions to give two equivalents of the carboxylic acid from a symmetrical anhydride or one equivalent of each acid from an unsymmetrical anhydride. They are also susceptible to nucleophilic attack to form new acyl derivatives. Acid anhydrides are relatively polar, but they cannot hydrogen bond as extensively as carboxylic acids. The melting point of an anhydride is lower, but the boiling point is significantly increased relative to the parent acids. This latter property is expected because the mass of the anhydride is roughly twice that of a simple acid.

When acetyl chloride is mixed with an equivalent of acetic acid, the product is acetic anhydride. This reaction proceeds by an acyl substitution mechanism similar to the transformation of $2 \rightarrow 5$ in Section 20.1. Because HCl is lost during the reaction, the medium will become acidic. To prevent reaction of HCl with the anhydride (see Section 20.5), amine bases such as triethylamine or pyridine (**52**) are added to react with HCl to form $Et_3NH^+Cl^-$ or pyridinium hydrochloride. Trapping the HCl product in this manner drives the reaction to the anhydride and inhibits the reverse reaction.

20.18 **Write out the reaction of acetyl chloride and acetic acid, showing both starting materials and the product.**

20.19 **Draw the structure of pyridinium hydrochloride.**

Acid chlorides are useful precursors to anhydrides, including both symmetrical and mixed (unsymmetrical) anhydrides. In one experiment, heptanoyl chloride (**50**) reacted with heptanoic acid (**51**), with benzene as a solvent and in the presence of pyridine (**53**), to give the symmetrical heptanoic anhydride (**53**) in 83% isolated yield.[4] Unsymmetrical anhydrides are prepared by choosing which half of the anhydride comes from the acid chloride and which comes from the carboxylic acid. This choice allows one to control formation of the mixed anhydride. If heptanoyl chloride (**50**) and butanoic acid (**7**) react in the presence of triethylamine as the base, anhydride **54** is formed. Anhydride **54** can also be prepared by the reaction of butanoyl chloride with hexanoic acid, but the easiest to obtain and cheapest combination is usually used.

20.20 **Draw the two starting materials that will produce 2-methylpropanoic pentanoic anhydride when they react.**

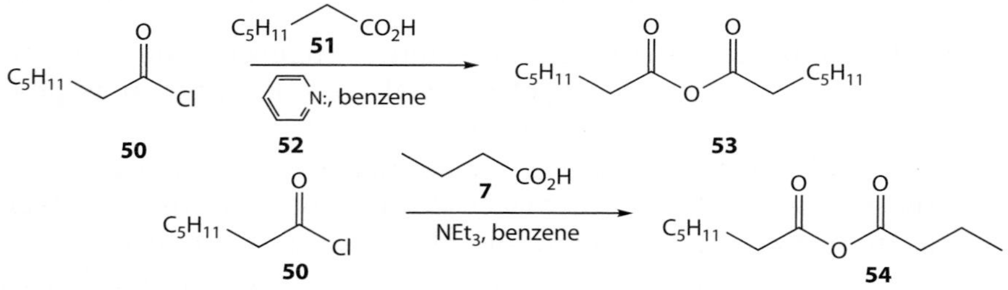

Anhydrides may be prepared by coupling two carboxylic acids under acidic conditions. If ethanoic acid (acetic acid, **21**) is heated with HCl, protonation to give an oxocarbenium ion is followed by reaction with a second equivalent of acetic acid to give a tetrahedral intermediate. This reaction is the usual acid-catalyzed acyl addition mechanism. Protonation of the OH unit leads to loss of water and formation of the anhydride. Each step in this process is reversible and steps must be taken to "drive" the equilibrium (see Chapter 7, Section 7.10, for a discussion of equilibria) toward the anhydride product by removing the water by-product (see Chapter 18, Section 18.6.3). Remember that such techniques are an application of Le Chatelier's principle (discussed in Section 18.3). Even when this is done, isolation of pure anhydrides by this method can be difficult. Unreacted acid may contaminate the product and attempts to remove the acid with aqueous base may induce hydrolysis of the anhydride.

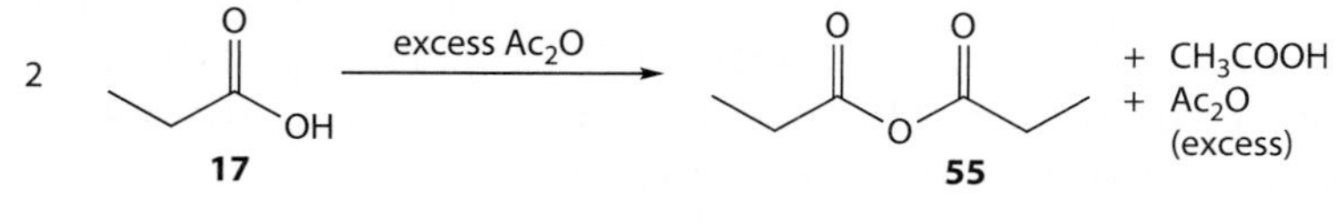

Because of these problems, an alternative method is used to prepare symmetrical anhydrides; it treats two equivalents of a carboxylic acid such as propanoic acid (**17**) with an excess of acetic anhydride. Acetic anhydride reacts with the **17** to give an intermediate that reacts with a second equivalent of **17** to give propanoic anhydride (**55**), along with acetic acid. Upon completion of the reaction, unreacted acetic anhydride will remain because it is in excess. Although it is not at all obvious, the anhydride acts as a chemical drying agent because it reacts with water to remove it from the reaction and drive the reaction to completion.

20.21 **Write out the mechanism and final product formed if acetic acid is the only reactant, mixed with a catalytic amount of HCl.**

20.22 **Draw the structure of acetic anhydride.**

20.23 **Based on knowledge of the hydrolysis of ester 29 in Section 20.2, write out the mechanism for the conversion of 17 to 55 using acetic anhydride.**

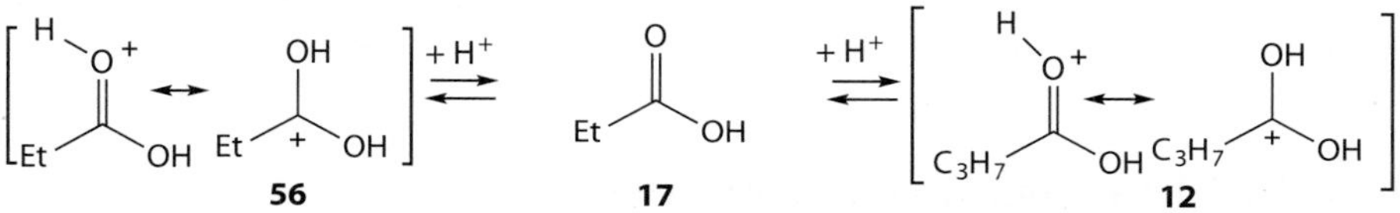

Formation of mixed anhydrides poses an even greater problem using this technique. If propanoic acid (**17**) and butanoic acid (**7**) react in the presence of an acid catalyst, propanoic acid may react with HCl to form **56**, but butanoic acid can also react to form **12**. In the absence of some unknown factor, there is an equal chance of forming either cation. If **56** reacts with another molecule of propanoic acid, the symmetrical **55** is formed, but if **56** reacts with a molecule of butanoic acid, **57** is formed. Similarly, cation **12** may react with propanoic acid to form **57**, but it can also react with another molecule of butanoic acid to form **58**. Forming either **55** or **58** has "one chance" and forming **57** has "two chances"; in fact, however, all three anhydrides are generated by this reaction in a ratio of 1:2:1 (**55:57:58**). **This is said to be a "statistical mixture" of products.** *The use of acid chlorides to form mixed anhydrides is a superior method to this statistical coupling of two acids.*

20.24 **Draw all three anhydrides formed when 2-methylpropanoic acid and 2,2-diphenylethanoic acid react under acidic conditions.**

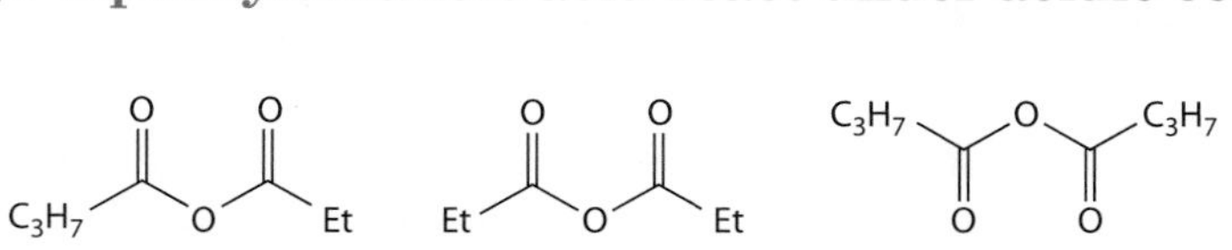

20.5 Preparation of Esters

Esters have two structural components: a carboxylic acid and an alcohol. An alcohol can react directly with carboxylic acids, in the presence of an acid cata-lyst, to give an ester. This is not always the most efficient preparative method, however. Both acid chlorides and acid anhydrides react with alcohols to give esters. In these reactions, the oxygen of the alcohol is a nucleophile, donating electrons to the acyl carbon to generate a new C–O bond.

Esters are derivatives of carboxylic acids and they are often oils (liquids) with a slightly sweet smell. They are less polar than acids and tend to have lower boil-ing points and lower melting points. Because an ester contains more carbons than the acid, it is generally less soluble in water and more soluble in organic solvents. The boiling point of an ester increases as the mass of the "alcohol part" of the ester increases (see methyl acetate, ethyl acetate, and propyl acetate); however, esters have lower boiling points than the corresponding carboxylic acids because carboxylic acids undergo extensive hydrogen bonding. The melting point gener-ally increases as the mass of the ester increases, but as seen with ethyl acetate and propyl acetate, this is not always predictable and more subtle structural fac-tors often influence the melting point.

20.5.1 Acid Chlorides React with Alcohols

One of the most efficient methods to prepare esters is the reaction of an alco-hol with an acid chloride. In a typical experiment, 2-methyl-2-propanol (**59**; *tert*-butyl alcohol), dissolved in diethyl ether in the presence of dimethylani-line (**61**) as an amine base, is heated to reflux and then treated with acetyl chloride (**60**) to give a 62% yield of 1,1-dimethylethyl ethanoate (*tert*-butyl acetate, **62**).[5] This reaction is explained by an acyl addition of the alcohol oxygen of **59** to the carbonyl carbon of 60 to give oxonium ion **63**. The added base (**61**) reacts with the acidic proton in **63** to give **64**, which is a tetrahe-dral intermediate. It is likely that, as **63** forms, the amine base (**61**) removes the proton to give **64** directly. Regeneration of the carbonyl in ester **62** with expulsion of the chloride in leaving group completes the acyl substitution to give the ester **62**.

The acid–base reaction of dimethylaniline with the HCl by-product gives $PhNMe_2H^+$ Cl^-, which helps drive the reaction toward **62**. This is an efficient and general method for the preparation of esters.

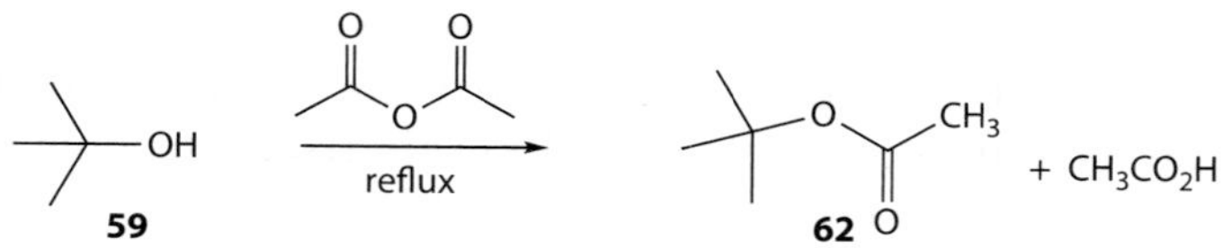

Acid anhydrides also react with alcohols, in the presence of amine bases, to give esters. The reaction is essentially the same as that for acid chlorides, and triethylamine is commonly added as a base. This is most useful when symmetrical anhydrides such as acetic anhydride are used. Heating acetic anhydride and isopropanol to reflux gives an 80% yield of isopropyl acetate (**19** in Section 20.2).[6] When *tert*-butyl alcohol **59** is heated with acetic anhydride, ester **62** is formed, along with acetic acid. Mixed anhydrides are expected to give a *mixture* of all possible ester products, so *this reaction is done most often with symmetrical anhydrides such as acetic (ethanoic) anhydride or propanoic anhydride.* When acetic anhydride reacts with methanol, for example, the products are methyl acetate and acetic acid. Formation of esters by reaction of an alcohol and an anhydride will be used only when a symmetrical anhydride can be used as a starting material.

20.25 Draw the two starting materials that can react to produce 2-methylethyl 3-phenylpentanoate as the final product.

20.26 Write the mechanism for the reaction of acetic anhydride and methanol.

20.5.2 The Acid-Catalyzed Reaction of Alcohols with Carboxylic Acids

Although esters are commonly prepared by the reaction of alcohols with acid chlorides, other methods are available. The two most important involve an acid-catalyzed reaction of an alcohol with an acid or the use of a "coupling reagent" that essentially removes water from the two components to form the ester.

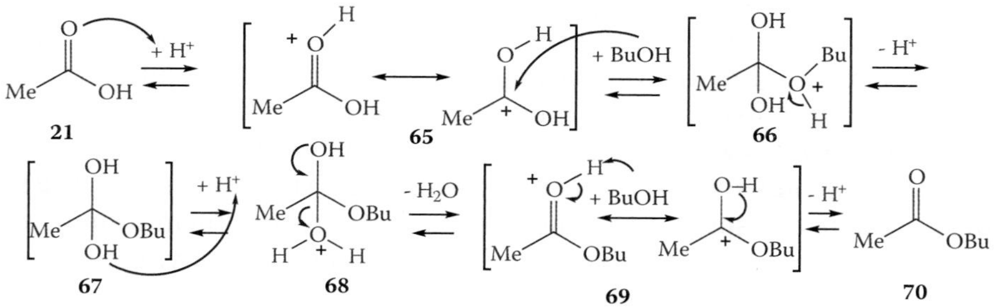

An old method for preparing esters simply reacts a carboxylic acid and an alcohol in the presence of an acid catalyst. Experimentally, the reaction of acetic acid (**21**) and butanol in the presence of a sulfuric acid catalyst gives butyl

20.30 Draw the product formed when 3,3-diphenylpentanoic acid reacts with 3,3-dimethylbutanol in the presence of DCC.

This type of "DCC coupling" of an acid and an alcohol is very attractive because it occurs *under neutral conditions* (no strong acid or base is present). It is a mild and very useful method for preparing esters, especially from "expensive" acids and "expensive" alcohols that might not give good yields under acid-catalyzed conditions. For relatively simple acids and alcohols, the use of DCC may be more expensive than other methods, however. Sometimes, an acid catalyst is added to the reaction of DCC, alcohols, and carboxylic acids to facilitate formation of the ester. Usually, the carboxylic acid is acidic enough without using another acid as catalyst.

20.5.5 Cyclic Esters (Lactones)

All esters discussed so far are acyclic molecules (this means they do not contain rings), but there are also cyclic esters, where both the acyl carbon and the "alcohol" oxygen are constituent members of a ring, as in **88–90**. These cyclic esters are called **lactones**. The common names for structurally simple lactones arise by using the suffix for the open chain carboxylic acid precursor and the word "lactone." Compound **88** is γ-butyrolactone and **89** is δ-valerolactone, where the name is taken from the common names for the four-carbon acid (butyric acid) and the five-carbon acid (valeric acid). The IUPAC names are based on the cyclic ether, so **88** is 2-oxoolane and **89** is 2-oxooxane.

The 14-membered ring lactone (**90**) does not have a simple name derived from the parent acid and is named as a "hydroxy acid": 14-hydroxytetradecanoic acid lactone. Using this hydroxy acid system, **88** is 5-hydroxypentanoic acid lactone, and **89** is 6-hydroxyhexanoic acid lactone. Note that lactone rings are often found in important biological compounds and that many antibiotics, such as methymycin (**91**), have a large (macrocyclic) lactone ring (highlighted in **red** in the illustration).

Just as esters react with aqueous acid or aqueous hydroxide to regenerate the acid and the alcohol precursors (see Section 20.2), lactones react to give the hydroxy acid. Hydrolysis of **89**, for example, gives 5-hydroxypentanoic acid. In the case of five-membered ring lactones, the lactone tends to be more stable than the hydroxy acid, so it is common for this equilibrium to favor the lactone.

In other words, it may be difficult to hydrolyze a five-membered ring lactone and isolate the hydroxy acid. For six-membered ring lactones and higher, the equilibrium usually favors the hydroxy acid. With most other lactones, however, the hydroxy acid can be isolated.

20.31 Draw the structure of 4-methyl-7-hydroxyheptanoic acid lactone.

20.32 Draw the structure of the lactone that would give 3,5,5-triethyl-7-hydroxyheptanoic acid after hydrolysis.

20.6 Amides

Amides are very interesting and important molecules that have a nitrogen atom attached to an acyl carbon (see **68, 69**, and **8**; also see **92–94**). There is a carboxylic acid component and an amine component. Therefore, it should be possible to combine a carboxylic acid and an amine in a chemical reaction to form an amide. This is partly correct, but because carboxylic acids are acids and amines are bases, complicating factors must be considered. Note that the important class of compounds called peptides in mammalian biology is linked together by "peptide bonds." These peptide bonds are actually amide bonds from a chemical viewpoint, and understanding the chemistry of amides allows one to understand the chemistry of peptides better.

Amides are polar compounds that are capable of hydrogen bonding. Primary amides (**92**) have two hydrogen atoms that can hydrogen bond, and they tend to be more polar than a tertiary amide (**94**) with no N–H units. Obviously, a secondary amide (**93**) with only one N–H unit will be intermediate in its polarity. Many amides are poorly soluble in organic solvents if the carbon chain of the parent acid is rather small. With two N–H units, the primary amides have the higher melting points and higher boiling points relative to the analogous secondary or tertiary amides. Note that the hydrogen bonding in carboxylic acids is far superior to the hydrogen bonding observed with amides. Adding alkyl groups to nitrogen dramatically decreases the melting point in most cases, as well as the boiling point.

Primary amides such as **92** are prepared by the reaction of an acid derivative with ammonia. Secondary amides (**93**) and tertiary amides (**94**) are prepared by the reaction of an acid derivative with a primary amine or a secondary amine, respectively. Several variations in these basic methods allow the preparation of a wide variety of amides.

20.6.1 Carboxylic Acids React with Ammonia and Amines: Pyrolysis of Amine Salts

Ammonia is a base, as are amines (see Chapter 6, Section 6.4.1), and carboxylic acids are obviously acids. When mixed together, the expected acid–base reaction occurs to form an ammonium carboxylate. When propanoic acid (**17**) is mixed with ammonia, ammonium propanoate (**95**) is formed. Similarly, dimethylamine (Me_2NH) reacts with cyclohexanecarboxylic acid (**97**) to give dimethylammonium cyclohexanecarboxylate (**98**). Because of this acid–base reaction, *mixing an amine or ammonia with an acid (and no other added reagents) at normal temperatures does not give an amide directly.*

20.33 Write out the complete acid–base reaction between butanoic acid and ammonia, labeling the acid, base, conjugate acid, and conjugate base.

20.34 Predict the products of a reaction between hexanoic acid and trimethylamine done at room temperature.

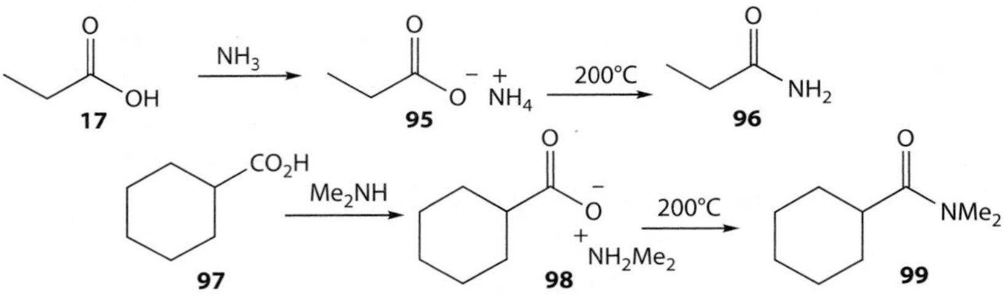

In order to prepare an amide after mixing an amine or ammonia with a carboxylic acid, the initially formed ammonium salt must be heated to a high temperature so that water is driven off (this is known as dehydration). Heating **95** (typically to 180–220°C) leads to propanamide (**96**), and *N,N*-diethylcyclohexanecarboxamide (**99**) is obtained similarly when **98** is heated. Note that **99** has the specialized name "carboxamide" because the amide unit is appended on a six-membered ring. This nomenclature is used when the carboxyl group is attached to a ring and this is formally analogous to nomenclature used when a –CHO or –COOH unit is attached to a ring (see cyclohexanecarboxylic acid in Chapter 16, Section 16.4). Clearly, any amide formed by this two-step method (mixing amine + acid; heating) must be reasonably stable to high temperatures.

Tertiary amines also react with carboxylic acids to form an ammonium salt. Triethylamine (NEt_3), for example, reacts with hexanoic acid to give an ammonium salt. Heating does ***not*** give an amide because there is no possibility for loss of water. Therefore, amides are formed only by heating the reaction products generated by mixing primary and secondary amines, or ammonia, with carboxylic acids.

20.35 Write out the starting materials and give the reaction conditions necessary to prepare *N*-propylpentanamide.

20.36 Write out the starting materials and the final product for the reaction of triethylamine and hexanoic acid.

20.6.2 Reaction of Amines with Acid Derivatives

Anhydrides are formed from the reaction of acid chlorides with acids, and an ester is formed from the reaction of acid chlorides or anhydrides with alcohols. Amides may be formed by the reaction of ammonia or an amine with acid chlorides, acid anhydrides, or esters. The nitrogen atom of ammonia or an amine is a good nucleophile, and it attacks the acyl carbon of an acid derivative in a now familiar sequence (see **60 → 62**) to give the amide. A typical example reacts hexanoyl chloride (**100**) with concentrated ammonia solution at 0°C, to give a 63% yield of hexanamide **101**.[8] Ammonium hydroxide is sometimes used as a source of aqueous ammonia.

20.37 What is the purpose of the second equivalent of ammonia in the conversion of 100 to 101?

20.38 Write out the reaction in which butanoyl chloride reacts with aqueous ammonia.

If the ammonia starting material is replaced with an amine—NHR_2 or NH_2R—in this mechanism, the product will be the tertiary or secondary amide, respectively. Amides are also formed by the reaction of an amine with an anhydride or an ester. The reaction of acetic anhydride and ethylamine, for example, gives *N*-ethyl ethanamide (**102**; also called *N*-acetylethanamine). Ethyl phenylacetate (**103**) reacts with ammonia in ethanol, at 175°C, to give phenylacetamide (**104**) in 75% yield.[9] If this reaction is done at temperatures much lower than this, however, the yield of **104** is rather poor. Heating to such a high temperature is not always necessary when converting an ester to an amide. When amines are used rather than ammonia, the reaction usually gives good yields at lower temperatures. There is a clear trend in reactivity, and acid chlorides are slightly more reactive than anhydrides, which are more reactive than the esters.

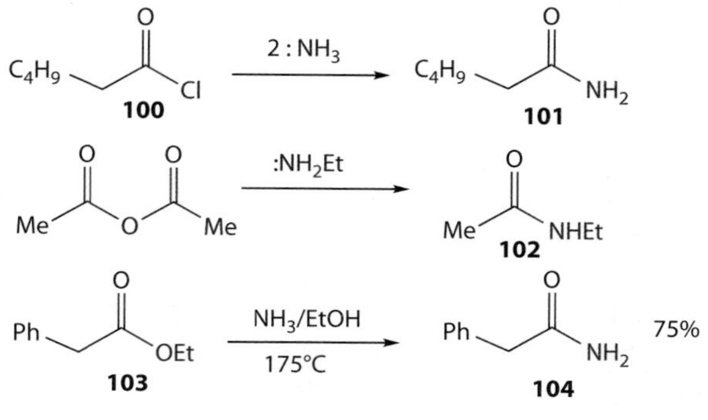

The reactions shown in the illustration are explained by noting that the NR_2 group is the poorest leaving group in this series of acid derivatives, so the reactivity order is $Cl \geq OOCR \gg OR \gg NR_2$. Note that an amine replaces Cl, O_2CMe, and OEt. *Experimentally, it is not possible to convert an amide to an ester, an anhydride, or an acid chloride by acyl substitution.* This means that amides are the **least** reactive species and acid chlorides are the **most** reactive species. This trend has been noted previously. The trend means that amides are at the bottom of the reactivity chain, so they are readily prepared from more reactive acid derivatives such as acid chlorides, acid anhydrides, or esters, as shown.

20.39 **Give the structure of one amine and *three different* starting materials as well as the reaction conditions that will give *N*-butyl-*N*-ethyl-3-phenylpentanamide.**

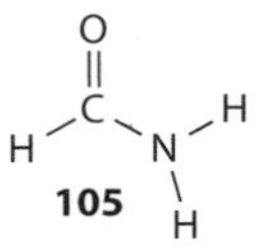

105 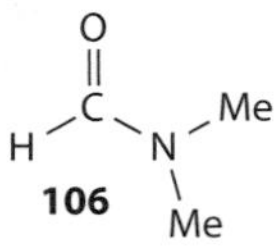   **106**

It is important to reiterate that tertiary amines do not give amides upon reaction with acid derivatives. It is also important to single out two simple amides that are particularly useful: formamide (**105**) and *N,N*-dimethylformamide (**106**; commonly known as **DMF**). Both are amides derived from formic acid and both are very useful as polar solvents in a variety of reactions. The amide DMF is used as a solvent in many reactions, particularly for the S_N2 reactions in Chapter 11.

20.6.3 DCC Coupling of Amines and Acids

Just as esters are formed from carboxylic acids and alcohols in the presence of dicyclohexylcarbodiimide (DCC; see Section 20.5.4), so amides are formed from amines (or ammonia) and carboxylic acids. The mechanism is identical to that presented for **77** → **80** using DCC, but the nucleophile is an amine rather than an alcohol. The initial sequence is identical to that described previously for alcohols in that the acid reacts with DCC (**78**) to form an intermediate that can lose dicyclohexyl urea as a leaving group.

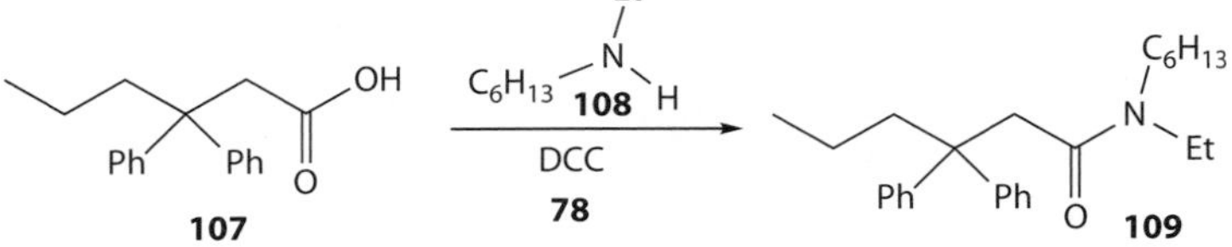

This reaction is useful for many acids, using both primary and secondary amines. An example is the reaction of *N*-ethyl-1-aminohexane (**108**) with 3,3-diphenylbutanoic acid (**107**) to give **109**.

20.40 Draw the mechanism for the DCC coupling of butanoic acid and *N,N*-dimethylamine.

20.41 Give the starting materials necessary to form *N*-cyclohexyl-3,3,4-trimethylhexanamide in the presence of DCC.

This DCC coupling procedure is especially useful when the amide product is sensitive to acid or base, when the acid chloride (or other acid derivative) is slow to react, or when traditional methods give poor yields. This DCC procedure is very useful for coupling two amino acids (see Chapter 27, Section 27.3.3) to form the corresponding amide (a peptide). An example is the reaction of the ethyl ester of phenylalanine (**110**) and the so-called *N-tert*-butylcarbamate derivative (the carbamate unit is N–CO–O; see the inset box in the following illustration) of alanine (**111**) to give **112**. The new amide bond is shown, but the product is called a dipeptide and this amide bond is often called a peptide bond. (As mentioned, this chemistry is discussed in Chapter 27.) Note that the *tert*-butyl carbamate unit is referred to as a *tert*-butoxycarbonyl, or **Boc** group, and **111** is said to have a N–Boc group.

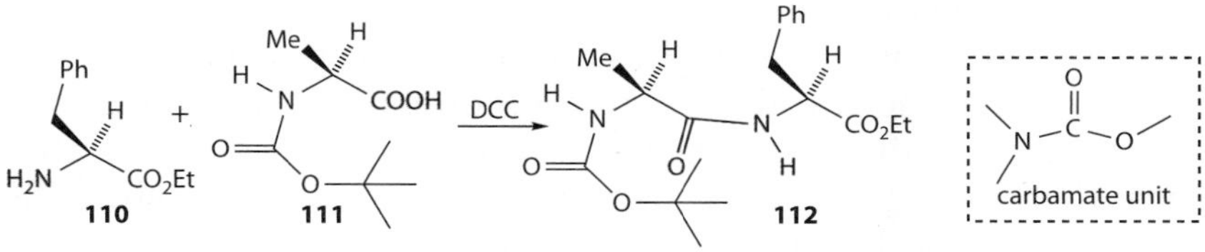

20.6.4 Cyclic Amides (Lactams)

Just as there are acyclic and cyclic esters, there are also acyclic and cyclic amides. *Cyclic amides* (see **113–115**) are called *lactams*, and they are characterized by an acyl carbon and a nitrogen atom as constituent atoms of a ring. The limited set of common names is based on the common name of the parent acid. Lactam **113** is γ-butyrolactam and **114** is δ-valerolactam (from butyric acid and valeric acid, respectively). Lactams containing a four-membered ring (such as **115**) are commonly called β-lactams. The IUPAC name for a four-membered ring containing nitrogen is azetidine and **115** is formally named azetidin-2-one. The IUPAC names for **113** and **114** are derived from the heterocycles pyrrole (five-membered ring) and pyridine (six-membered ring). They are 2-oxotetrahydropyrrole and 2-oxohexahydropyridine, respectively, and will be discussed in Chapter 26. For the time being, the common names are used.

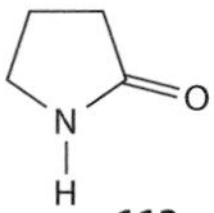

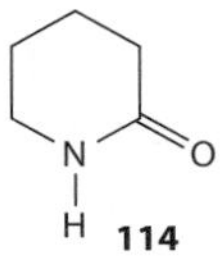

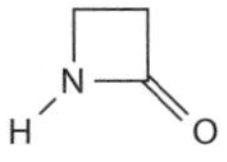

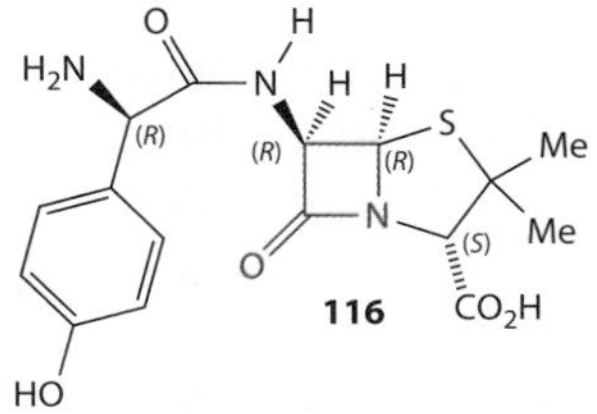

Lactams are relatively stable molecules, as are amides; however, heating a lactam with aqueous acid or aqueous base may lead to opening of the ring and formation of the parent amino acid. Lactams are important molecules for a variety of reactions, and β-lactams are constituents of an important class of antibiotics known as the penicillins. The important antibiotic amoxicillin (a penicillin, **116**) has the four-membered ring lactam unit (shown in **red** in the illustration).

20.42 **Give the structures of the lactams that have six carbons, seven carbons, and eight carbons in the ring.**

20.43 **Give the structure of the product formed when 114 is hydrolyzed with aqueous acid.**

20.6.5 Imides

Imides are a specialized class of compounds related to amides in that two acyl groups are attached to nitrogen, as in **117** and **118**. The name for such molecules is based on the longest chain carboxylic acid parent. In the case of **117**, there are two acyl portions: butanoic acid and ethanoic acid (acetic acid). The name is *N*-ethanoylbutanoic acid amide (or, more commonly, *N*-ethanoylbutanamide). It is also commonly called *N*-acetylbutanamide. In cases where the imide is symmetrical (both acyl groups are identical), the name can be simplified. Imide **118** is diacetamide (diethanamide). Imides will be discussed only occasionally in this book.

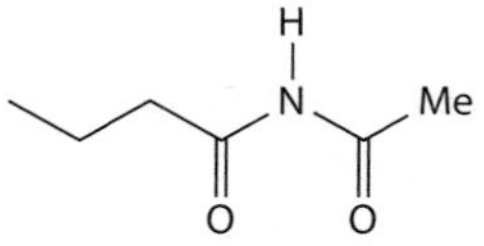

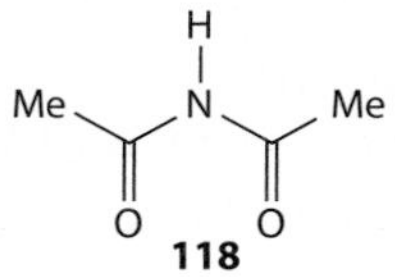

20.44 **Draw the structure of 3-methylbutanoyl 4,4-diphenylpentanamide.**

20.7 The Reaction of Carboxylic Acid Derivatives with Carbon Nucleophiles

Acid derivatives react with many nucleophiles by acyl substitution, including carbon nucleophiles. There are complications in this latter reaction because, in some cases, the compounds produced in the reaction are more reactive than the starting materials, and they compete for reaction with the nucleophiles. *Experimentally, cyanide and alkyne anions are not the best partners in this reaction, so the focus will be on Grignard reagents and organolithium reagents.*

20.7.1 Grignard Reagents and Organolithium Reagents

As noted in Chapter 18, Section 18.4, Grignard reagents are excellent nucleophiles in acyl addition reactions with aldehydes or ketones. Grignard reagents are also good nucleophiles for acyl substitution. The acyl substitution reaction of an acid derivative and a Grignard reagent should lead to a tetrahedral intermediate (see **6**, where Y = alkyl in Section 20.1), and loss of the X leaving group should give a ketone product (see **5**, where Y = alkyl). When this experiment is done, isolation of the ketone product is difficult, and **5** is not the major product.

Indeed, experiments over many years have discovered several things about this reaction. If a Grignard reaction with an acid chloride is done at ambient or higher temperatures, or with an excess of the Grignard reagent, an alcohol is the major, if not the exclusive, product. The reaction of octanoyl chloride (**119**) and *n*-butylmagnesium bromide does not give ketone **121** but rather tertiary alcohol **123** as the major product. The only way to form alcohol **123** is by reaction of the ketone product (**121**) with unreacted butylmagnesium bromide. (See Chapter 7, Section 7.11, for a review of rate of reaction.) After one half-life, for example, product, reagents, and unreacted starting material will be present. It is therefore reasonable to assume that the product may compete with the starting material for the reagent. In the reaction at hand, the reaction produces a product (the ketone) that may react with the Grignard reagent and will compete with the acid chloride for the Grignard reagent.

In the initial reaction, butylmagnesium bromide (BuMgBr) must first add to the carbonyl of octanoyl chloride (**119**) to give tetrahedral intermediate **120**. Chloride ion is a good leaving group, and the initially formed product is the expected **121** (5-undecanone) in an overall acyl substitution reaction. However, as noted in Chapter 18 (Section 18.4), a ketone reacts with butylmagnesium bromide to give an alcohol. *Although it is not obvious, a ketone reacts faster than an ester in most cases.* In this particular reaction, **121** is formed during the course of the reaction, and it competes with **119** for the available nucleophile (the Grignard reagent). In other words, the Grignard reagent reacts with *both* **119** *and* **121**. The reaction of butylmagnesium bromide and **121** gives **122**, and subsequent acid hydrolysis of the reaction mixture leads to the tertiary alcohol, **123** (5-butyl-5-undecanol).

20.45 Draw the ketone product expected when butanoyl chloride reacts with methylmagnesium bromide.

It is known that, at low temperatures and with a Lewis acid catalyst such as $FeCl_3$, good yields of ketone product are formed in this reaction. At lower temperatures, the acid chloride reacts much faster than the ketone, and addition of a Lewis acid essentially makes the reaction with the acid chloride even faster. For example, octanoyl chloride (**119**) reacts with butylmagnesium bromide in the presence of 2% $FeCl_3$, at −60°C, to give a 76% yield of 5-undecanone (**121**), along with 3% of a tertiary alcohol, **123**.[10]

Because of the possibility of this secondary reaction, acid chlorides are not always the best partner for generating ketones from reaction with Grignard reagents. If the product desired is a tertiary alcohol, the acid chloride should simply be reacted with an excess of the Grignard reagent. Reacting **119** with an excess of ethylmagnesium bromide gives alcohol **123** as the major product. Note that anhydrides react similarly to acid chlorides. An alternative reaction of acid chlorides that cleanly produces a ketone product is discussed in the next section.

Esters also react with Grignard reagents to give a ketone and an alcohol via acyl substitution. As with acid chlorides, the initially formed ketone can react with more Grignard reagent to form a tertiary alcohol. *It was noted in previous sections that an ester reacts more slowly than an acid chloride in acyl substitution reactions, and it is known from experiments that the rate of reaction of ketones with Grignard reagents is generally faster than that with esters.* In other words, it is very difficult to isolate good yields of ketone by the reaction of a Grignard reagent with an ester. For this reason, esters are often treated with two equivalents of Grignard reagent to ensure that the alcohol is the major product, as shown by the reaction of ethyl butanoate (**29**) and two equivalents of methylmagnesium iodide, which gave an 88% yield of 2-methyl-2-pentanol, **125**.[11] Clearly, the initially formed 2-pentanone (**124**) reacted with the second equivalent of MeMgI.

20.46 **Predict the product when the acid chloride of cyclopentanecar-boxylic acid reacts with an excess of 1-methylethylmagnesium iodide, followed by acid hydrolysis.**

20.47 **Draw the product expected when ethyl 3-phenylbutanoate reacts with ethylmagnesium bromide, followed by hydrolysis.**

Organolithium reagents are more reactive than Grignard reagents in reactions with aldehydes and ketones (see Chapter 18, Section 18.4), as well as with acid derivatives. Therefore, it is even more difficult to isolate a ketone product from reaction with an acid chloride, acid anhydride, or an ester. In general, this is correct. Due to the difficulty in controlling this reaction, it is not used here to prepare ketones, *but it is sometimes used to prepare alcohols when an excess of the organolithium reagent is added.* This is a convenient reaction when the organolithium is commercially available (methyllithium, butyllithium, *tert*-butyllithium, phenyllithium, etc.).

20.48 Draw the product of a reaction between methyl propanoate and 2.5 equivalents of phenyllithium, followed by hydrolysis.

20.7.2 Dialkyl Cuprate Reagents

The previous section described the reaction of an acid chloride and a Grignard reagent in the presence of a transition metal compound such as ferric chloride. Copper derivatives are more useful in this reaction—specifically, the organocuprates introduced in Chapter 15 (Section 15.6). Lithium dialkyl cuprates such as **126** are prepared from organolithium reagents and Cu(I) compounds such as cuprous iodide. The reaction of two molar equivalents of *n*-butyllithium and CuI gives lithium dibutylcuprate (**126**), which is usually abbreviated as Bu$_2$CuLi. In Chapter 15 (Section 15.6), cuprates reacted with alkyl halides to give the corresponding coupling product. Therefore, the reaction of **126** and benzyl bromide will give **127**. This overall transformation is shown again, with bromobutane as the starting material. Reaction of the alkyl halide with lithium metal gives butyllithium (see Chapter 15, Section 15.5.1); formation of the organocuprate (**126**; not shown in this sequence), followed by reaction with benzyl bromide, gives **127**.

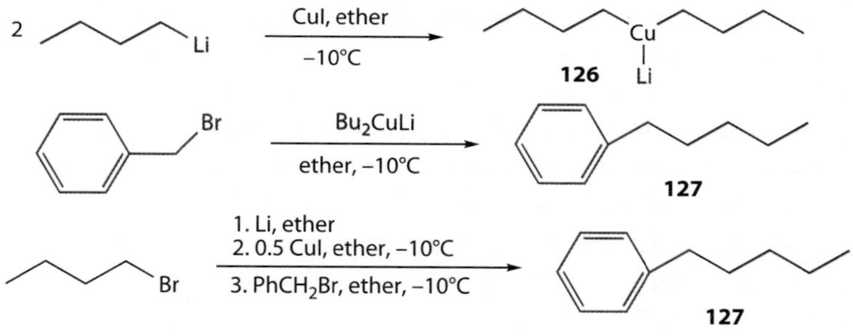

Organocuprates also react with acid chlorides to form a ketone. When **126** reacts with benzoyl chloride, for example, the product is **128**. *This reaction is a preferred method for the conversion of carboxylic acids to ketones.* A reaction sequence that begins with 3-methylpentanoic acid first forms the acid chloride via reaction with thionyl chloride (Section 20.3.1), and subsequent reaction with lithium diethyl cuprate leads to the ketone **129**.

20.49 What is the major product formed when lithium diethylcuprate reacts with 2-iodopentane in ether at –10°C?

20.50 What is the product formed by first heating 1-hexanol with Jones's reagent (see Chapter 17, Section 17.2.1) and the reaction with thionyl chloride, followed by reaction with lithium dimethyl cuprate?

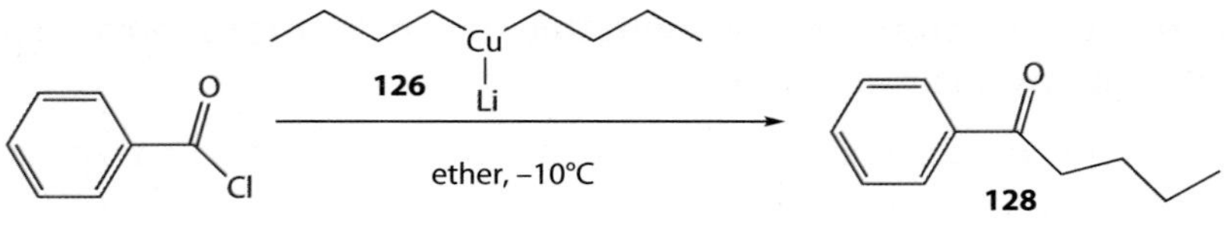

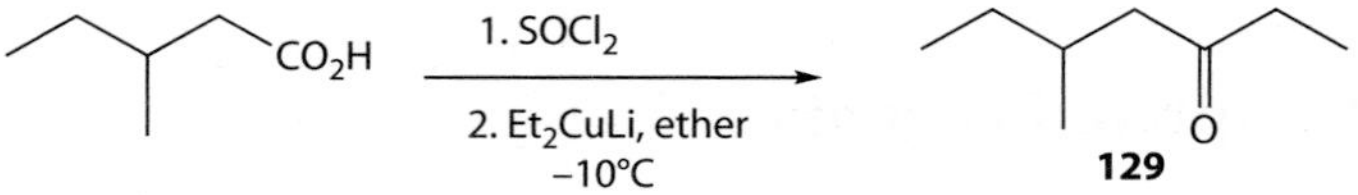

20.7.3 Dialkyl Cadmium Reagents

Another method that will produce a ketone from the Grignard reaction with an acid chloride *changes the organometallic reagent from RMgX or RLi to one that is less reactive.* In such a case, the organometallic reagent should react with the acid chloride but not with the ketone product.

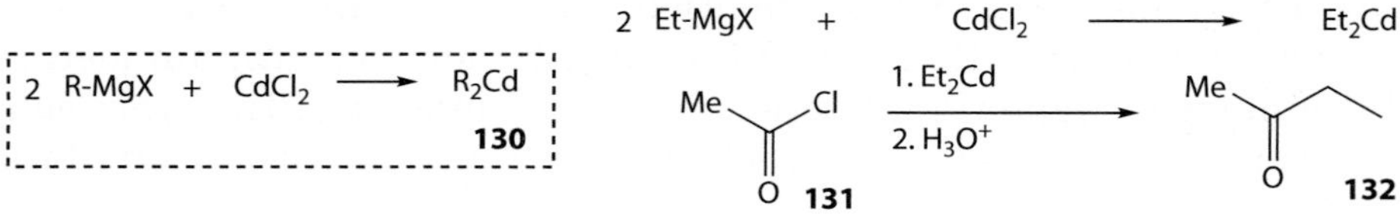

Many years ago, it was discovered that cadmium chloride ($CdCl_2$) reacts with Grignard reagents to form a new organometallic reagent, a dialkyl-cadmium (**130**). This reaction works only when the cadmium chloride is very dry (anhydrous) and not all Grignard reagents give good yields of **130**. Simple Grignard reagents generally work well and the reaction of $CdCl_2$ with two equivalents of ethylmagnesium iodide leads to diethylcadmium, Et_2Cd. The usefulness of dialkylcadmium reagents such as Et_2Cd is apparent from the experiment in which acetyl chloride (**131**) reacts with Et_2Cd to give a 46% yield of 2-butanone (**132**), with none of the tertiary alcohol (2-methyl-2-butanol).[12] Clearly, Et_2Cd reacted with the acid chloride and not with the ketone product, although the yield is not spectacular. *For this to work, the dialkylcadmium must be **less reactive** than a Grignard reagent.* In general, if **130** can be formed, reaction with an acid chloride should give the ketone.

Using this "trick," a variety of ketones can be prepared directly from the appropriate acid chloride. The most important thing to remember about this section is that dialkylcadmium reagents are *specialized reagents*. They are designed to convert acid chlorides to ketones and are used preferentially when the acid chloride or another acid derivative fails to give a reasonable yield of the ketone product, or if the ketone product cannot be isolated in a high state of purity. For the most part, the use of dialkylcadmium reagents has been superseded by the use of organocuprates, which is the preferred method for the preparation of ketones from acid chlorides.

20.51 Show the reagents necessary to make 5-methyl-6-phenyl-4-octanone using the dialkyl cadmium approach and then show how this ketone can be converted to 4,5-dimethyl-6-phenyl-4-octanol.

20.8 Reaction of Organometallics with Other Electrophilic "Carbonyl" Molecules

To this point, the only reactions discussed have been those of aldehydes, ketones, and acid derivatives (not carboxylic acids). Other molecules can have an electrophilic carbonyl or a reactive electrophilic atom. These include carbon dioxide and nitriles, which also react with carbon nucleophiles.

20.8.1 Reaction with Carbon Dioxide

Grignard reagents react with a variety of carbonyl compounds, including aldehydes, ketones, or acid derivatives. The structure of carbon dioxide is O=C=O, and it is apparent that there is a carbonyl unit. It is reasonable to ask if this carbonyl will react with a nucleophile such as a Grignard reagent. Based on the following experiment, the answer is yes. 2-Chlorobutane reacts with Mg metal in ether to give the Grignard reagent, **133**. The ether solution containing the Grignard reagent is poured carefully onto crushed dry ice (CO_2), and this slurry is then poured onto crushed ice containing concentrated HCl.

This reaction sequence gives 2-methylbutanoic acid (**135**) in 79% yield.[13] This reaction is explained by nucleophilic acyl addition of the Grignard reagent to a C=O unit in carbon dioxide to generate **134**, which is a carboxylate anion. This is simply the salt of a carboxylic acid, and acid hydrolysis (**134** is the base and hydronium ion is the acid) gives the carboxylic acid **135** (the conjugate acid of **134**). This reaction is reasonably general, although the yields of the reaction vary considerably (from <25% to >80%). *This reaction makes a carbon–carbon bond and extends the carbon chain by one carbon.*

Carbon dioxide is a gas at ambient temperatures, and its use in this reaction deserves some comment because using the gas poses several problems. Dry ice is solid CO_2 (chilled to below its freezing point, which is –78°C) and it is readily available. For convenience, a Grignard reagent is generated in a solvent (usually ether or THF) and then this reagent is added to crushed dry ice, as described in the experiment that gives **135**. After waiting a suitable period of time for the reaction to occur, the reaction is "quenched" by adding aqueous acid to hydrolyze the carboxylate anion to the carboxylic acid.

Organolithium reagents also react with carbon dioxide, analogous to Grignard reagents, and the product is also the carboxylic acid. The yields also vary from quite good to very poor. For convenience, *assume* that the major product of carbon dioxide and a Grignard reagent or an organolithium reagent is the carboxylic acid.

20.52 **Write out the reagents necessary to make 3,3-diphenylhexanoic acid and then convert this acid to its ethyl ester.**

20.53 **Write out the reaction and all reagents necessary to convert benzyl bromide to benzyllithium.**

20.54 **Describe all reactions necessary to make 2-(1-methylethyl)-3-phenylpent-2-enoic acid from an appropriate vinyl bromide.**

20.8.2 Reaction with Carboxylic Acids

Carboxylic acids are strong organic acids that react with sodium hydroxide or even with the relatively weak base, sodium bicarbonate, so they certainly react with the powerful bases methyllithium or methylmagnesium bromide (see Chapter 15, Sections 15.3 and 15.5.2, for a discussion of the basicity of these reagents). The acid–base reaction of a Grignard reagent and a carboxylic acid is much faster than nucleophilic attack at the acyl carbon. The product is a carboxylate salt (the conjugate base). When methylmagnesium bromide reacts with cyclohexanecarboxylic acid (**97**), for example, methane is the conjugate acid and magnesium carboxylate **136** is the conjugate base. When methyllithium reacts with **97**, however, ketone **140** is isolated. This is a different result from the reaction with Grignard reagents. Indeed, reaction of **97** with two equivalents of methyllithium gave an 83% yield of **140**.[14] How is this difference in reactivity explained?

20.55 **Draw the reaction between butanoic acid and methylmagnesium bromide.**

20.56 **What is the conjugate acid of butylmagnesium chloride?**

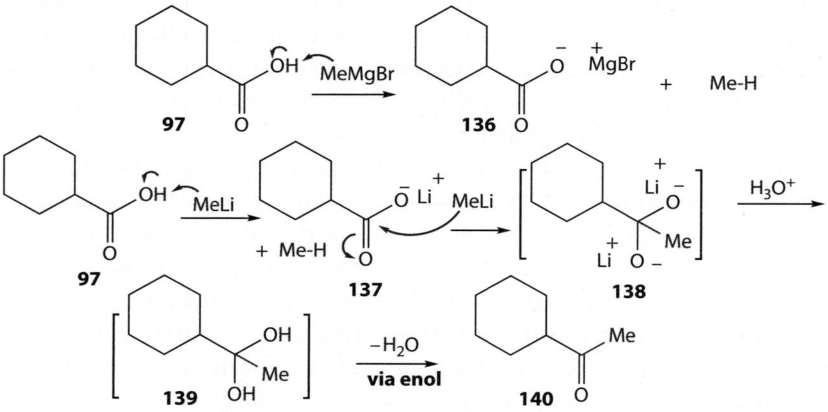

The initial acid–base reaction (the carbon of the methyl group is the base) leads to methane as the conjugate acid and the expected conjugate base, lithium butanoate (**137**). In previous comparisons of organolithium reagents and Grignard reagents (Chapter 15), an organolithium reagent is identified as a more powerful base and also a more powerful nucleophile. In this particular case, organolithium reagents *react with carboxylate anions,* whereas Grignard reagents cannot. Therefore, a second equivalent of methyllithium reacts with **137** to produce **138**. Because **138** is a base, reaction with aqueous acid leads to hydrolysis in which both alkoxide units are protonated to produce a hydrate (**139**). Hydrate **139** is the conjugate acid of **138**, but hydrates are inherently unstable (as discussed in Chapter 18, Section 18.5), losing water to form an enol, which tautomerizes to ketone **140** (1-cyclohexyl-1-ethanone, commonly called methyl cyclohexyl ketone) as the observed final product. In general, *organolithium reagents react with carboxylic acids to give ketones.*

20.57 Draw the enol formed by dehydration of 139.

20.58 Draw the reactions necessary to prepare 2,5-dimethyl-3-hexanone from the appropriate carboxylic acid.

20.8.3 Reaction with Nitriles

A nitrile has the cyanide functional group $C\equiv N$ (see Chapter 5, Section 5.9.4). Obviously, this is not a carbonyl. It is a polarized group, however, because nitrogen is more electronegative than carbon. Cyanide has a $\delta-$ nitrogen, a $\delta+$ carbon, and two π-bonds connecting the carbon to the nitrogen. Is it possible for a nucleophile to attack the electrophilic carbon of the nitrile to form a new bond to carbon, as one of the π-bonds is broken?

A simple experiment answers this question. When ethylmagnesium bromide is mixed with cyclopropylcyanide (**141**), the product is identified as an iminium salt, **142**. The iminium salt product is easily explained by acyl addition of the Grignard reagent to the nitrile carbon; as the π-bond breaks, these two electrons are transferred to the more electronegative nitrogen atom. The nitrogen atom of **142** is basic, and reaction with a suitable acid gives the conjugate acid, imine **143**. Hydrolysis of this iminium salt gives ketone **144** in 68% yield,[15] rather than the imine (**143**) expected by simple protonation of the nitrogen.

In aqueous solution, protonation of nitrogen in **142** first gives the expected imine, **143**. Examination of the ketone product (**144**) shows that nitrogen has been lost, and it is reasonable to assume that the nitrogen is lost as ammonia or ammonium. The oxygen of the carbonyl group in **144** must come from the only source of oxygen, water. Therefore, acyl addition of water to **143** is followed by proton transfer to give ammonia, which is a good leaving group. Indeed, in the presence of aqueous acid, protonation of the nitrogen atom in **143** is followed by acyl addition of water, proton transfer, loss of ammonia, and formation of the ketone, **144**. A Grignard reagent therefore reacts with a nitrile to give an

20.62 Draw the structure of diethyl malonate.

20.63 Draw the product expected when 151 reacts with an excess of ammonia.

20.64 Draw the product from the reaction of the acid dichloride of adipic acid and an excess of 1-aminopropane.

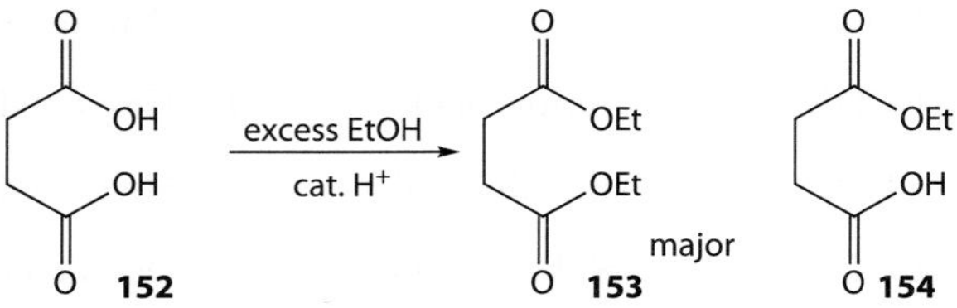

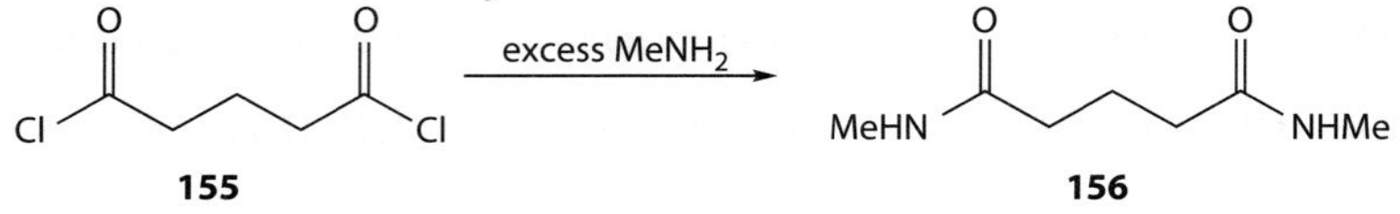

There is one acid dichloride that is extremely useful: oxalyl chloride (**157**). When **157** reacts with a carboxylic acid such as pentanoic acid (**45**) in the presence of an amine base (such as triethylamine), the acid is converted to an acid chloride (in this case, pentanoyl chloride, **158**). *Oxalyl chloride is a very efficient and very mild method for converting acids to acid chlorides.* This method should be added to the reactions in Section 20.3 for the preparation of acid chlorides. It is used primarily with acids that are sensitive to acids or bases or when the acid chloride product is too reactive in the presence of acids or bases.

20.65 Draw the product of the reaction between 3,3-diphenylbutanoic acid and oxalyl chloride.

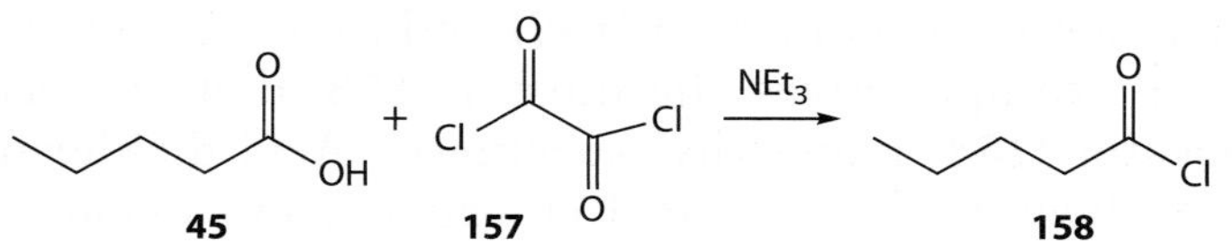

20.9.2 Cyclic Anhydrides and Imides

An anhydride is formed by coupling two COOH units, derived from two carboxylic acids (see Section 20.4). Because a dicarboxylic acid has two COOH units, it is possible to couple these units intramolecularly to form a *cyclic* anhydride. The smallest commonly observed cyclic anhydride formed contains five atoms formed by heating succinic acid (**152**). Loss of a molecule of water leads to formation of **159**, which is called succinic anhydride. The IUPAC name is 1,4-butanedioic anhydride. Anhydrides such as this find a variety of uses, primarily in generating other acid derivatives. Acid hydrolysis of **159**, for example, regenerates the dicarboxylic acid, **152**. Reaction of **159** with ethanol and an acid catalyst leads to diethyl succinate (**153**).

20.66 Draw all of the reactions necessary to convert 2,3-dimethylsuccinic acid first to the anhydride and then to the dimethyl ester of the diacid.

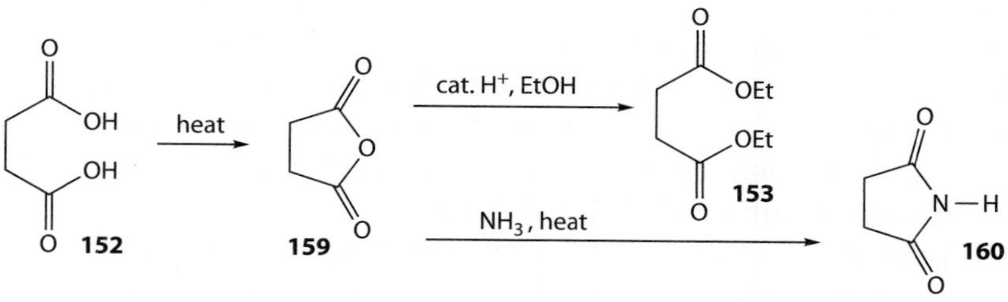

Imides (described in Section 20.7.5) are formed by the coupling of an acid unit with an amide unit. Cyclic imides are formed from dicarboxylic acid derivatives. If a cyclic anhydride such as succinic anhydride (**159**) is heated with ammonia, the result is a cyclic imide (**160**), which is named succinimide (1,4-butanedioic acid imide). Cyclic imides of this type appear occasionally in this book. Notable exceptions are the cyclic imides **161** and **162**. Both are five-membered ring imides with a halogen attached to nitrogen. Imide **161** is known as *N*-chlorosuccinimide (NCS) and **162** is *N*-bromosuccinimide (NBS). Imide **161** is used as a chlorinating reagent (converts a molecule to a chloride) and is particularly useful in radical substitution reactions. Similarly, **162** is used as a brominating reagent (converts a molecule to a bromide). The radical halogenation reactions of alkanes using NCS and NBS are discussed in Chapter 11 (Section 11.9.3).

20.67 Draw the structure of *N*-phenylglutarimide.

20.68 Write all of the reactions necessary to convert pentanedioic acid first to the acid dichloride and then to the *N*-ethyl imide.

20.69 Write out the reaction and product formed when diphenylmethane is reacted with 162.

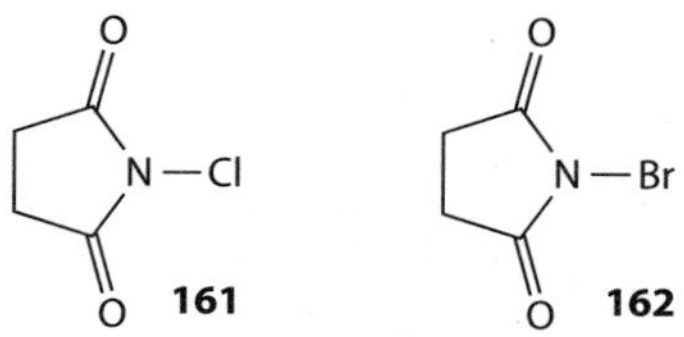

20.10 Baeyer–Villiger Oxidation

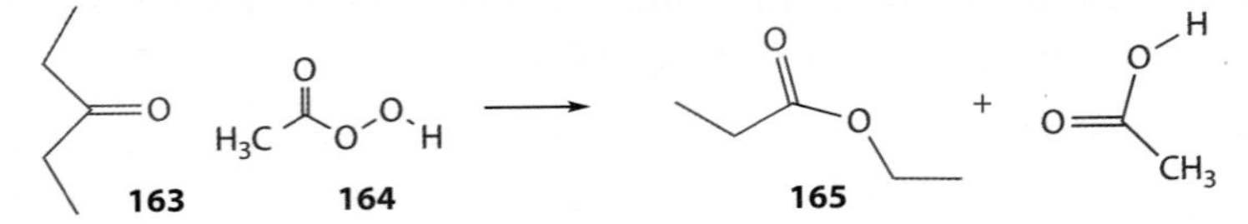

Although Chapter 17 discussed several oxidation reactions, an important reaction was not discussed in that chapter. Section 17.3 discussed the oxidation of an alkene to an epoxide using a peroxyacid such as **163** (also see Chapter 10, Section 10.5). Peroxyacids also react with ketones to form esters in what is known as the Baeyer–Villiger reaction. This reaction is named in honor of its discoverers, Johann Friedrich Wilhelm Adolf von Baeyer (Germany; 1835–1917) and Victor Villiger (Switzerland; 1868–1934).

When 3-pentanone (**163**) reacts with peroxyacetic acid (**164**), the products are ethyl propanoate (**165**) and acetic acid. The second oxygen atom in ester **165** (relative to ketone **163**) must come from the peroxyacid, which means that **164** must react with **163**, probably by an acyl addition. Moreover, there is a skeletal rearrangement in which one of the groups flanking the acyl carbon has migrated (rearranged) to an oxygen. This analysis suggests that ketones react with peroxyacids to give esters via a C → O rearrangement.

This acid-catalyzed mechanism proceeds via protonation of **163** to form the conjugate acid, oxonium ion **166**. The peroxyacid (**164**) reacts as a nucleophile via acyl addition to give **167**, which loses a proton to give the hemiketal type intermediate **168**. (Hemiketals and ketals were discussed in Chapter 18, Section 18.6.) At this point, protonation of the oxygen atom closest to the carbonyl generates an oxonium ion that allows one of the alkyl groups to migrate to give oxocarbenium ion **169** and acetic acid. Loss of a proton from **169** gives the ester product, ethyl propanoate.

20.70 **Draw the reaction and final products formed when cyclopentene reacts with peroxyformic acid.**

20.71 **What is the product formed when 2,4-imethyl-3-pentanoate reacts with peroxyformic acid?**

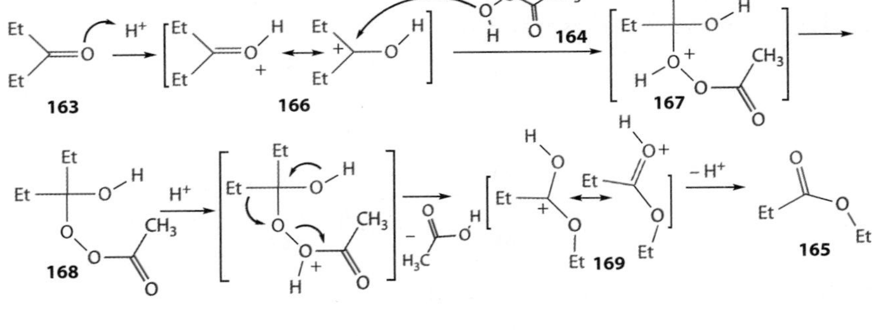

3-Pentanone is a symmetrical ketone, but rearrangement also occurs in unsymmetrical ketones, although there is a potential problem. When 2-methyl-3-pentanone (**170**) reacts with peroxyacetic acid, the key intermediate is **171**. Which alkyl group migrates? If the ethyl group migrates, the product is **172**; if the isopropyl group migrates, the product is **173**. The major product is **173**, so the secondary carbon group migrates in preference to the primary alkyl group. The order of migration for groups in unsymmetrical ketones is

tert-alkyl > cyclohexyl ≈ 2° alkyl ≈ benzyl ≈ phenyl > vinyl > 1° alkyl > methyl

The preference for group migration is consistent with the ability of each carbon to support a positive charge. In other words, *the group that can best form a carbocation migrates.* Therefore, a tertiary carbon migrates faster than a secondary, which migrates faster than a primary. This guide helps to predict that the product of a Baeyer–Villiger reaction of **170** to be **173** rather than **174**. Cyclic ketones undergo Baeyer–Villiger reaction to give the corresponding lactone. The reaction of cyclopentanone with peroxyacetic acid (**164**), for example, gives lactone **174**.

20.72 What is the product formed when 2,2-dimethyl-3-hexanone reacts with peroxyacetic acid?

20.73 What is the product formed when 2-ethylcyclohexanone reacts with peroxyformic acid?

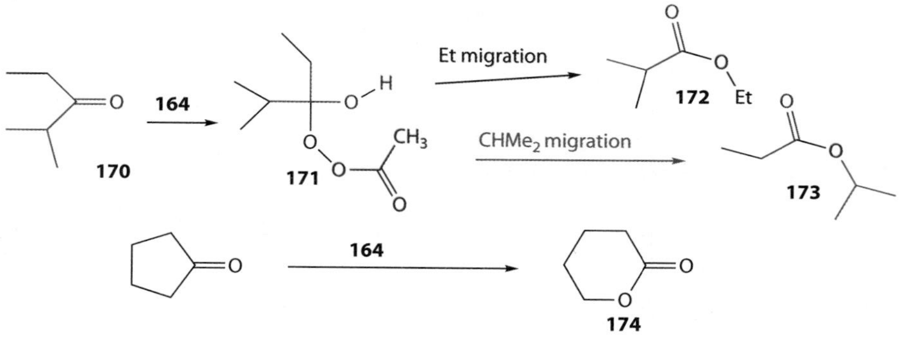

20.11 Sulfonic Acid Derivatives

Thus far, the chapter has been devoted to the chemistry of carboxylic acid derivatives. Another important class of organic acids is the sulfonic acids (see **175**; X = OH). In principle, sulfonic acids should give the same types of derivatives as the carboxylic acids: acid chlorides, esters, and amides. This section will explore some of these common sulfonic acid derivatives: sulfonyl chlorides (**175**;

X = Cl), sulfonate esters (X = OR¹), and sulfonamides (X = NR¹₂). Anhydride derivatives of sulfonic acids are known (X = OSO₂R), but they will be ignored because they are not used in this book.

20.11.1 Sulfonyl Chlorides

As seen in Section 20.3, the reaction of a chlorinating agent such as thionyl chloride (**38**) with a carboxylic acid gives an acid chloride in the presence of a base such as triethylamine to react with the HCl by-product of the reaction. Similarly, the reaction of **38** with a sulfonic acid such as butanesulfonic acid (**176**) gives the analogous acid chloride, butanesulfonyl chloride (**177**). This reaction is rather general, and benzenesulfonic acid (PhSO₃H) is converted to benzenesulfonyl chloride (PhSO₂Cl) just as methanesulfonic acid (MeSO₃H) is converted to methanesulfonyl chloride (MeSO₂Cl, **178**).

   Sulfonyl chloride **178** also has the common name of ***mesyl chloride***. Another very common derivative is 4-methylbenzenesulfonyl chloride (**179**), also called ***para*-toluenesulfonyl chloride**, which is also called ***tosyl chloride***) (nomenclature for benzene-containing compounds is discussed in Chapter 21). This particular compound is introduced here because it so common and clearly fits into this section. Thionyl bromide (SOBr₂) converts sulfonic acids to the corresponding sulfonyl bromide. This reaction is exactly analogous to the reactions with carboxylic acids.

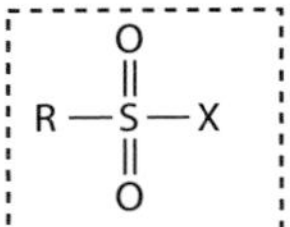

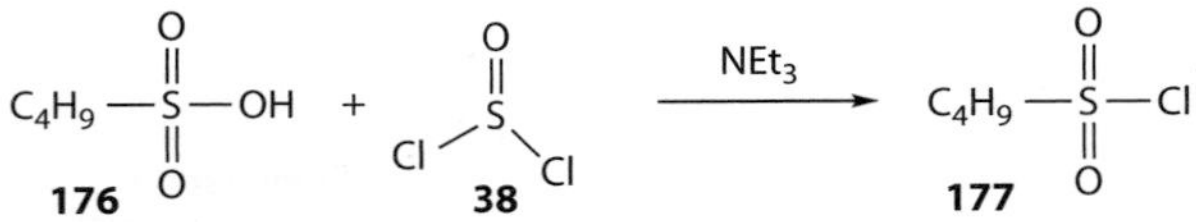

20.74 Draw the structure of benzenesulfonyl chloride.

20.75 Write out the reaction that converts methanesulfonic acid to 177.

20.76 Write out the reaction and final product formed when propanesulfonic acid reacts with thionyl bromide.

20.77 Write out the product of a reaction between pentanesulfonic acid and PCl₃.

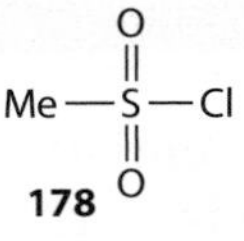

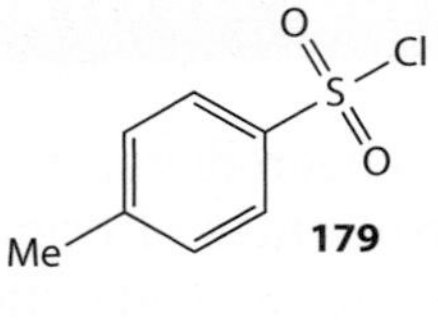

Sulfonyl chlorides are rather reactive and they are used to form other sulfonic acid derivatives (just as acid chlorides are used to form other carboxylic acid derivatives). If butanesulfonyl chloride (**177**) reacts with aqueous acid or aqueous base, the hydrolysis reaction occurs to give the sulfonic acid parent (**176**) as the product.

20.11.2 Sulfonate Esters

Esters are prepared from carboxylic acids by the reaction between an acid chloride and an alcohol or between a carboxylic acid and an alcohol under acidic conditions. Both sulfonic acids and sulfonyl chlorides react with alcohols to form sulfonate esters. When butanesulfonyl chloride (**177**) reacts with propanol, usually in the presence of a base such as triethylamine, propyl butanesulfonate (**180**) is formed. A wide range of sulfonyl esters can be formed this way from an alcohol and a sulfonic acid.

20.78 Draw the structure and give the name of the sulfonic acid precursors to ethyl benzenesulfonate, methyl methanesulfonate, and 1-methylethyl trifluoromethanesulfonate.

A sulfonate ester can also be prepared by the reaction of an alcohol and a sulfonic acid under acidic conditions, exactly analogous to the same reaction with a carboxylic acid. In the presence of an acid catalyst, 2-methylpropanesulfonic acid (**181**) reacts with ethanol to give ethyl 2-methylpropanesulfonate (**187**). The mechanism of this reaction is remarkably similar to that for the reaction of a carboxylic acid and an alcohol in the presence of an acid catalyst (see Section 20.5.2). The reaction proceeds by initial reaction of **181** with H^+ to give **182**. The oxygen of ethanol is the electron-donating species in this reaction, and the most electrophilic atom is *sulfur,* so reaction of **182** and ethanol generates a new S–O bond in **183**.

Loss of the acidic proton from the oxygen leads to **184**. An OH unit reacts with additional H^+ to give **185**, and water is lost (the leaving group) to give the protonated form (**186**) of the sulfonate ester product. Loss of a proton yields the final product, **187**. As mentioned, this mechanism is formally analogous to that shown for ester formation from carboxylic acids (Section 20.4), except that sulfur is now the electrophilic atom. As with carboxylic esters, this is a mechanism in which every step is reversible, and water must be removed or some other measure taken to drive the reaction to the right. Normally, using Le Chatelier's principle by providing a large excess of the alcohol (as the solvent) will suffice.

20.79 Draw the product of the reaction between benzenesulfonic acid and (1) $SOCl_2$ (2) cyclopentanol.

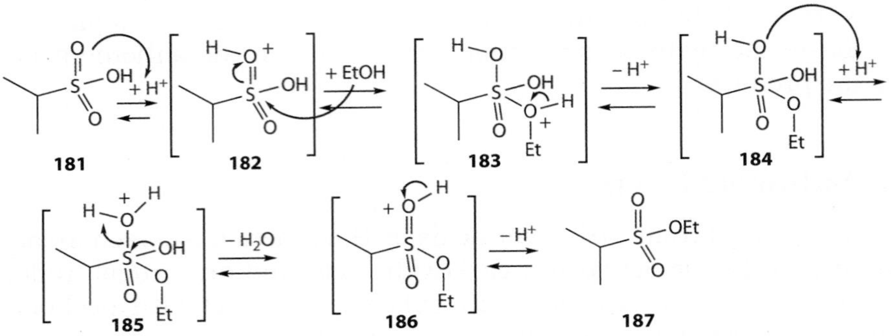

When sulfonate esters such as **187** react with aqueous acid or aqueous hydroxide, the parent sulfonic acid is regenerated. This is formally analogous to the saponification reaction discussed in Section 20.2. Acid or base hydrolysis of **187** gives 2-methylpropanesulfonic acid (**181**) as the product. The mechanism of this process is the exact reverse of that shown for converting the acid to the ester, except that an excess of water is used rather than an excess of alcohol. Sulfonate esters are useful compounds in a variety of applications.

20.11.3 Sulfonamides

Sulfonamides are the sulfonic acid analogs of carboxylic acid amides. They are formed in much the same way: by reaction of sulfonyl chlorides or sulfonate esters with amines or ammonia. 2-Methylbutanesulfonyl chloride (**188**) reacts with ammonia, for example, to give 2-methylbutanesulfonamide (**189**). Amines react to give *N*-substituted sulfonamides. Note the "shorthand" method for drawing the sulfonyl chloride unit in **188** as $-SO_2Cl$ and the sulfonamide unit as $-SO_2NH_2$ in **189**. A base is usually added to this reaction to drive it toward the sulfonamide product.

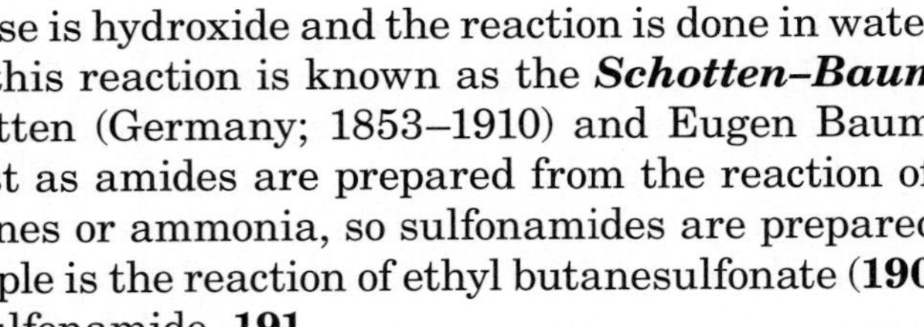

When the base is hydroxide and the reaction is done in water, as in the preparation of **189**, this reaction is known as the ***Schotten–Baumann reaction***, after Carl Schotten (Germany; 1853–1910) and Eugen Baumann (Germany; 1846–1896). Just as amides are prepared from the reaction of carboxylic acid esters with amines or ammonia, so sulfonamides are prepared from sulfonate esters. An example is the reaction of ethyl butanesulfonate (**190**) with ammonia to give butanesulfonamide, **191**.

20.80 Write out the mechanism for this transformation. Hint: the reactions are virtually identical to those shown for carboxylic acid amides from esters in Section 20.6.

192

193

When a sulfonamide reacts with aqueous acid or aqueous hydroxide, the amine (or ammonia) is "released" along with the parent sulfonic acid. Butanesulfonamide (**191**), for example, reacts with aqueous hydroxide (followed by an acid neutralization step) to give butanesulfonic acid (**190**) and ammonia. Sulfonamides are quite stable molecules and they are used in a variety of applications, particularly those sulfonamides derived from benzene derivatives (see Chapter 21). Sulfanilamide (**192**), for example, is a potent antibacterial agent and isobuzole (**193**), with the formal name of N-(5-isobutyl-1,3,4-thiadiazol-2-yl)-p-methoxy-benzenesulfonamide, has hypoglycemic (antidiabetic) properties.

20.12 Sulfate Esters and Phosphate Esters

All discussions of acid derivatives in preceding sections and chapters focused on "organic acids" such as carboxylic acids and sulfonic acids. Notably missing are derivatives of the mineral acids: nitric acid, sulfuric acid, and phosphoric acid. For the most part, such acid derivatives are used sparingly in organic chemical transformations. Phosphorus derivatives are widely observed in biological systems, however. This section offers a brief introduction to some of the organic chemistry of mineral acids. The structure of nitric acid is **194**, sulfuric acid is **195**, and phosphoric acid is **196**.

194

195

196

The structures of the mineral acids suggest that replacing one or more OH units will lead to acid derivatives. Esters are possible for all three mineral acids. Nitric acid gives a nitrate ester, **197**. Nitrate esters are often unstable, and the trinitrate ester of glycerol (**198**) is the well-known explosive nitroglycerin. However, highly specialized nitrate esters are known to relax vascular smooth muscle, which leads to vasodilation (relaxation of the muscle wall of blood vessels that leads to widening of those vessels), and they are important compounds used in the treatment of heart disease.[16] Nonetheless, nitrate esters will not be discussed further or used in this book. Amide derivatives O_3N-NR_2 are known as nitramides, but they will not be discussed.

Acid derivatives of sulfuric acids are more complicated because there are two acidic protons and two OH units. If one OH unit is replaced with Cl, this is the acid chloride equivalent **199**. However, the name of this compound is chlorosulfonic acid, and the chemistry is related more to the sulfonic acids found in Section 20.11. The amide derivative **200** is known as sulfamic acid. Sulfamic acid is used in the manufacture of some sweetening agents, and it finds use in some pharmaceutical preparations. Sulfamic acid derivatives will not be discussed further. The ester derivatives of sulfuric acid include the monoester **201** and the diester **202**. When R = methyl, **201** is methyl sulfate and **202** is dimethyl sulfate. Dimethyl sulfate is an alkylating agent, reacting with alcohols to give ethers. Organic sulfates are used in chemical transformations and are found in pharmaceutical preparations, but they will not be discussed further.

20.81 Give the product formed when dimethyl sulfate reacts with 2-butanol.

197	**198**	**199**	**200**	**201**	**202**

The main purpose of this section is to introduce derivatives of phosphoric acid. The acid chloride derivative is $ClPO(OH)_2$, chlorophosphoric acid, and the amide derivative is phosphoramic acid, $H_2N–PO(OH)_3$. These acid derivatives will not be discussed because they will not be used in this book. There are three ester derivatives of phosphoric acid: a monoester **203**, a diester **204**, and a triester **205**. There are also derivatives of pyrophosphoric acid (diphosphoric acid, **206**), which include various esters.

20.82 Give the structure of the dibenzyl ester of phosphoric acid.

203	**204**	**205**	**206**

In principle, the preparation of these compounds should be similar to the chemistry of carboxylic acids and sulfonic acids, but it is more complex and there are synthetic routes that are specific for the phosphorus compounds. For example, the reaction of phosphoric acid and an alcohol may give the ester, but the reaction is slow and this is not a good preparative method. An alternative is the reaction of phosphorous pentoxide with an excess of an alcohol to give a mixture of phosphate esters. The reaction of ethanol and P_4O_{10} (the dimeric form of phosphorus pentoxide), for example, gives a mixture of **207** and **208**. They are separable based on differences in solubility in alkaline earth salts.[17] The hydrolysis of pyrophosphate esters such as **209** gives the phosphate ester **208**.[18]

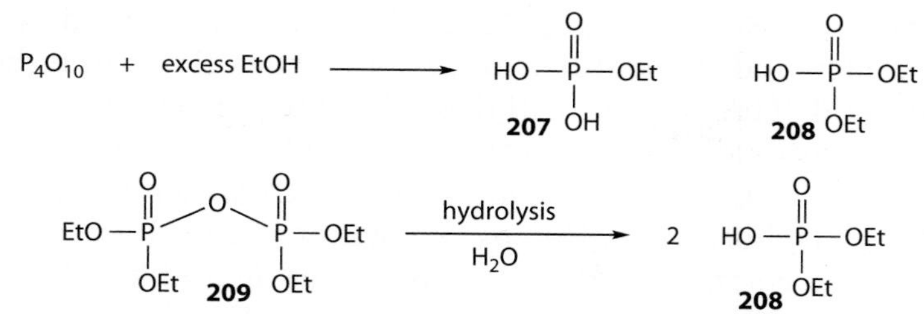

The chemistry of phosphate esters is rich and varied. Phosphate esters are important in biological systems. The phosphate ester of a nucleoside (a nucleobase attached to a ribose derivative; see Chapter 28, Section 28.5) is called a **nucleotide**. These are structural components used in DNA and RNA. Using adenosine (**210**) as an example, there are three possible monophosphate esters: **211, 212**, and **212**. The pyrophosphate (diphosphate) derivative is adenosine 5′-diphosphate (**214**). The symbol "A" is used to designate an adenosine derivative in biology, so **211** is abbreviated 5′-AMP (adenosine 5′-monophosphate) and **214** is 5′-ADP (adenosine 5′-diphosphate). The numbering is explained in Chapter 28, Section 28.5.

20.83 Give the structure of adenosine-5′-triphosphate.

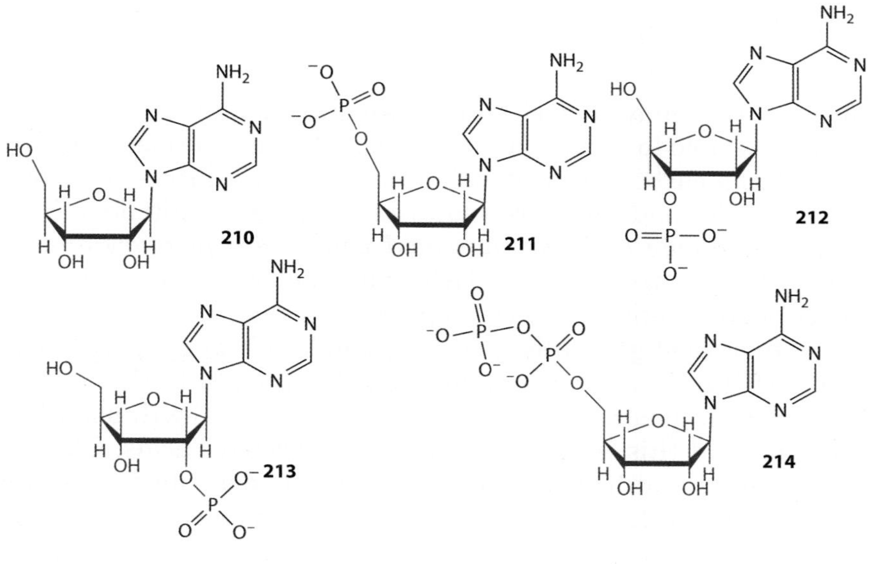

20.13 Nitriles Are Carboxylic Acid Derivatives

Nitriles are formed by an S_N2 reaction of an alkyl halide with NaCN or KCN (see Chapter 11, Section 11.3.6). Nitriles are not obviously acid derivatives when compared to the acid derivatives in previous sections of this chapter. They are indeed acid derivatives, however, because (a) nitriles are formed by dehydration reactions of amides and (b) acid hydrolysis of nitriles leads to either an amide or a carboxylic acid. This section will describe these reactions.

Phosphorus pentoxide has the structure P_2O_5, although it also exists in a dimeric form as P_4O_{10}. This reagent reacts with water to form phosphates and eventually phosphoric acid. Removal of water is usually necessary to drive certain reactions to completion. Formation of acetals and ketals in Chapter 18 (Section 18.6) required the using of an azeotropic distillation or addition of a drying agent such as macular sieves. Other reactions produce water as a by-product that must be removed, but the reagents described in Chapter 18 cannot be used. Heating a primary amide such as **215**, for example, leads to a nitrile by loss of water (O from the carbonyl and hydrogen atoms from the nitrogen), but the reaction is very inefficient. Addition of the drying agent P_2O_5, however, facilitates the dehydration of a primary amide to the corresponding nitrile. Heating **215** with P_2O_5, for example gives pentanenitrile (**216**). This reaction is specific for primary amides (see Chapter 16, Section 16.7).

20.84 Give the product formed when 3-pentanol is treated with (a) PBr₃; (b) NaCN in DMF.

Because a nitrile may be derived from an amide, a nitrile is an acid deriva-tive. It is also possible to convert a nitrile back to an amide (see **218**), which can be hydrolyzed to a carboxylic acid (see Section 20.2). A nitrile is rather stable, and strongly acidic conditions are required to convert a nitrile to an acid. In general, *heating nitrile with aqueous acid will give a primary amide*. When benzylnitrile (**217**) is treated with HCl and then with water, the product is the amide of phenylacetic acid, **218**. The reaction of a primary amide and con-centrated sulfuric acid generally gives the amide. If more vigorous conditions are used (6N HCl in hot water), the nitrile is hydrolyzed to the corresponding acid, **219**. Prolonged heating of an amide in aqueous acid will usually give the carboxylic acid, and this is an excellent way to prepare acids. Clearly, amide **218** can be isolated and hydrolyzed to acid **219**, as described in Section 20.2. Combined with the S_N2 reaction, formation of a nitrile followed by hydrolysis allows the conversion of an alkyl halide to a carboxylic acid.

20.85 Give the product formed when 1-pentene is treated with (a) HBr; (b) NaCN in DMF; (c) 6N HCl.

20.14 Carbon–Carbon Bond-Forming Reactions and Functional Group Exchanges of Acid Derivatives

Carboxylic acids are converted to acid chlorides, anhydrides, esters, and amides. In addition, acid chlorides are converted to esters and amides, and esters are converted to amides. All of these acid derivatives are hydrolyzed back to the parent carboxylic acid.

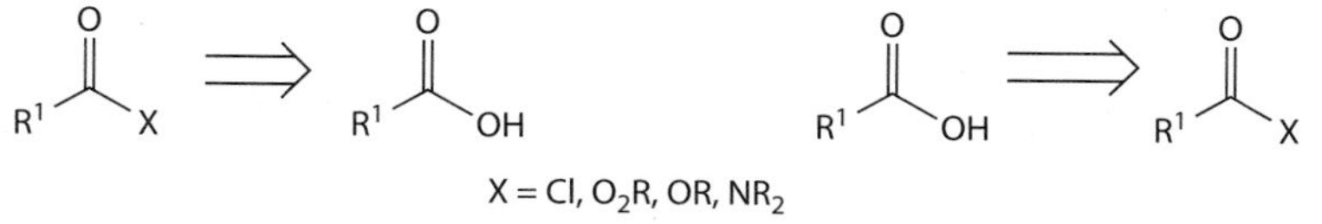

$$X = Cl, O_2R, OR, NR_2$$

Several disconnections are possible based on acyl substitution reactions with carbon nucleophiles. Tertiary alcohols are prepared from acid derivatives in reactions with Grignard reagents and organolithium reagents, and ketones are prepared from acid chlorides and acids. Carboxylic acids are prepared by the reaction of Grignard and organolithium reagents with CO_2. Nitriles are a source of ketones via reaction with Grignard or organolithium reagents.

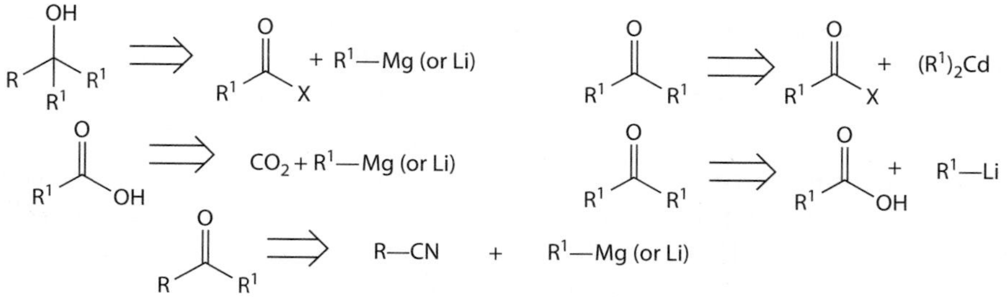

Transformations of carboxylic acid derivatives open several new possibilities for synthesis. One is the preparation of dimethyl glutarate (**220**) from cyclohexanone (**195**). Analysis of cyclohexanone indicates that its conversion to **220** requires loss of one carbonyl oxygen and oxidative cleavage of a carbon–carbon bond within the ring. The diester **220** is clearly a derivative of the dicarboxylic acid, glutaric acid. An oxidative cleavage is consistent with ozonolysis of a cyclic alkene (Chapter 10, Section 10.7.2; Chapter 17, Section 17.4.1) or, alternatively, by cleavage of a 1,2-diol (Chapter 17, Section 17.4.2). If cyclohexene is chosen as the precursor to glutaric acid, then cyclohexanol is the precursor to cyclohexene. Cyclohexanol is obtained by reduction of the carbonyl in **195**, and treatment of cyclohexanol with concentrated sulfuric acid gives cyclohexene. Ozonolysis gives glutaric acid, the immediate precursor to **220**.

With this analysis, the synthesis begins with the reduction of **195** with $NaBH_4$ to give cyclohexanol (see Chapter 19, Section 19.2.1) and reaction of that alcohol with sulfuric acid gives cyclohexene (see Chapter 12, Section 12.4).

Reaction of cyclohexene with ozone followed by hydrogen peroxide gives glutaric acid. Finally, the reaction of glutaric acid with an excess of methanol and an acid catalyst completes the synthesis to give **220**.

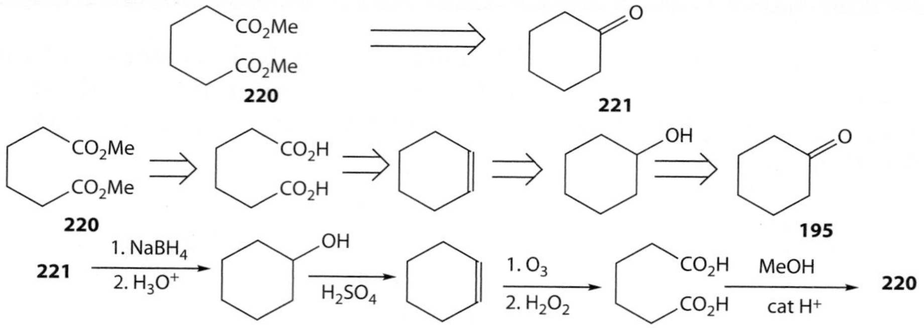

A second example prepares amide **222** from terminal alkyne **223**. It is clear by comparing the two structures that **222** is derived from a carboxylic acid and that the new acid has one more carbon than **223**. Therefore, a carbon–carbon bond-forming reaction is required. One suggestion prepares the carboxylic acid via a reaction of an appropriate Grignard reagent and carbon dioxide, which leads to 2-bromopentane as a precursor. The disconnection of **222** to give a donor atom on the pentane chain is consistent with this analysis, with the carbon of CO_2 as the acceptor atom. 2-Bromopentane is readily obtained from 1-pentene, which has an obvious relationship to the 1-pentyne starting material (**223**).

The retrosynthetic analysis shown for **220** allows the total synthesis. Catalytic hydrogenation of **223** gives 1-pentene (see Chapter 19, Section 19.3.2), which reacts with HBr to give 2-bromopentane (Chapter 10, Section 10.2). The reaction of 2-bromopentane and magnesium metal, followed by reaction with carbon dioxide and aqueous hydrolysis, leads to the requisite carboxylic acid. Reaction of the carboxylic acid with thionyl chloride leads to the acid chloride, which reacts with dimethylamine to give **222**.

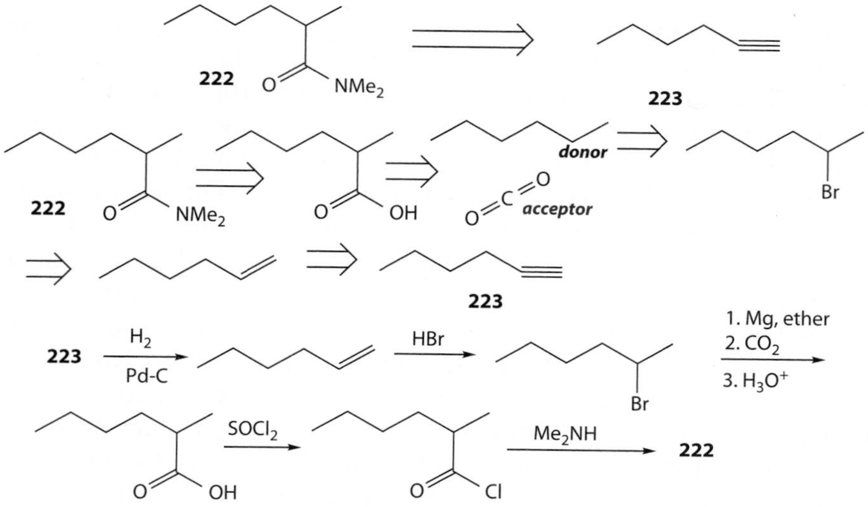

20.15 Biological Relevance

Amides and esters are particularly important in biological systems. The phosphate and pyrophosphate derivatives found in DNA and RNA were illustrated in Section 20.12. The structure and chemistry of these compounds will be elaborated upon in Chapter 28. Peptides (discussed in Chapter 27, Section 27.4) are polymers composed of discrete amino acid units. They are joined together by so-called peptide bonds, which are actually amide bonds. Many natural peptides are produced by bacteria via nonribosomal peptide synthesis. *Bacillus subtilis* produces the heptapeptide surfactin (**224**), which has antibiotic and antifungal activity.[18]

Nonribosomal peptide synthesis means that the peptide is not produced by the tRNA–mRNA mechanism described in Chapter 28, Section 28.6. Each amino acid found in **224** is directly selected for incorporation into the growing peptide chain by one of the domains of *surfactin synthetase,* shown with the pendant SH groups. Substrate activation occurs after binding the amino acid, and the enzyme catalyzes the formation of an aminoacyl adenylate intermediate using Mg^{2+}–ATP and release of a cofactor. Subsequently, the amino acid–O–AMP oxoester is converted into a thioester by a nucleophilic attack of the free thiol-bound cofactor of an adjacent PCP domain. (Note that ATP is adenosine triphosphate and AMP is adenosine monophosphate; see Chapter 28, Section 28.5.)

Formally, the binding domains attach two amino acids in **225** by an acyl substitution reaction in which the thiol unit attacks the acyl carbon, and each coupling proceeds by the appropriate tetrahedral intermediate, **226**. Loss of the O–AMP leaving group leads to **227A**, redrawn as **227B** to shown that the nucleophilic amine unit of one bound amino acid residue attacks the acyl unit of the other in an acyl substitution reaction to give another tetrahedral intermediate, **228**. The leaving group is the thiol-bound unit and formation of the bound dipeptide in **229**. The organism will produce the heptapeptide using similar coupling reactions and then complete the biosynthesis of **224**. This sequence shows that organisms can produce peptides by alternative mechanisms to ribosomal peptide synthesis.

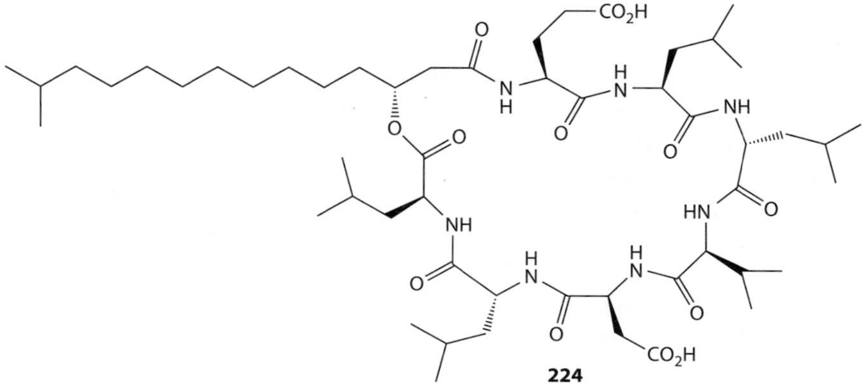

224

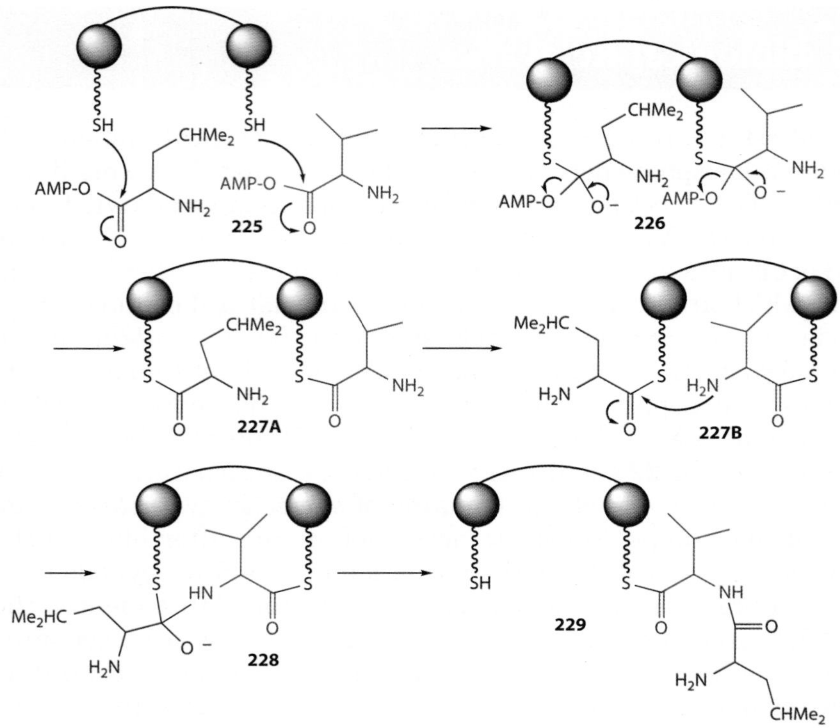

Reprinted in part with permission from Sieber, S.A., Marahiel, M.A. *Chemical Reviews* 2005, 105, 715. Copyright © 2005 American Chemiacl Society.

Note that enzymes are also available to cleave the amide bond (peptide bond) of a peptide linkage, as in an enzyme, to give the amino acid. Such enzymes are known as *peptidases.* An enzyme that cleaves an amino acid from the terminal position of a peptide is called an *exopeptidase,* whereas an enzyme that cleaves an amide bond at an internal position (within the peptide chain rather than at the ends) is called an *endopeptidase.*

Esters are prevalent in the body, often in the form of triglycerides (see Section 20.2). The cleavage of an ester into its alcohol and carboxylic acid components requires an enzyme known as an esterase. A *lipase* is a type of esterase, and there are pancreatic lipases as well as mitochondrial lipases. A lipase is a water-soluble enzyme that hydrolyzes the ester bonds in water-insoluble lipids, including glycerides. A triacylglycerol lipase will cleave a triglyceride, for example. Triglycerides can pass through the stomach into the duodenum, where alkaline pancreatic juice raises the pH and allows hydrolysis of the ester linkages that is mediated by an enzyme called pancreatic lipase as well as other esterases. One of the principal pancreatic lipases is known as steapsin. Cleavage occurs at the C1 and C3 positions:

The hydrolysis depends on bile salts (bile salts are carboxylic acid salts linked to a steroid), which act as detergents to emulsify the triglycerides and facilitate the ester cleavage reaction. The long-chain fatty acid products are converted to micelles via interaction with the bile salts, and carried to the surface of epithelial cells, where they react with glycerol to form new triglycerides, which aggregate with lipoproteins to form particles called chylomicrons.[19]

Chylomicrons are lipoprotein particles that transport dietary lipids from the intestines to other parts of the body. In the basic environment of the duodenum, the fatty acid is converted to the salt, which is emulsified by the bile salts.[19] Bile salts are commonly amide derivatives of steroids. Two bile acids are taurocholic acid (**230**) and glycocholic acid (**231**).

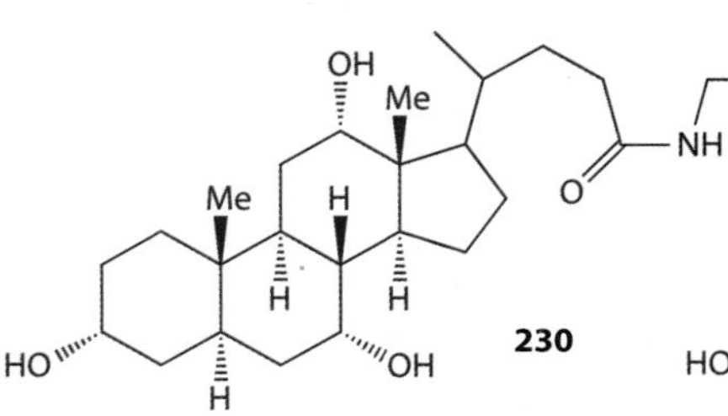

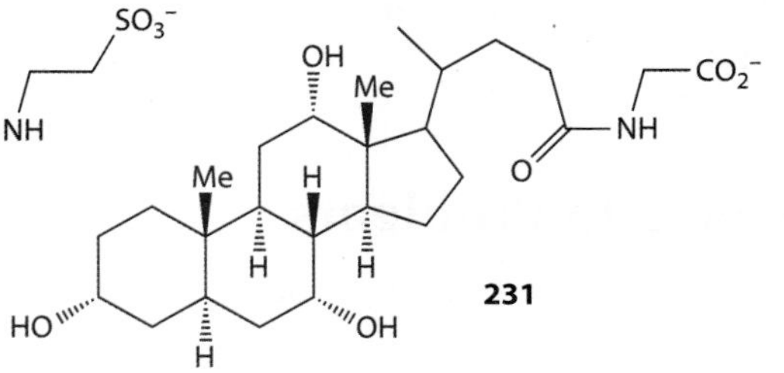

References

1. Hurd, C. D., and Blunck, F. H. 1938. *Journal of the American Chemical Society* 60:2419.
2. Omer, E. A., Smith, D. L., Wood, K. V., and el-Menshawi, B. S. *Plant Foods and Human Nutrition* 45:247–249.
3. Furniss, B. S., Hannaford, A. J., Smith, P. W. G., and Tatchell, A. R., eds. 1994. *Vogel's textbook of practical organic chemistry,* 5th ed., 692–693, Exp. 5.138. Essex, England: Longman.
4. Furniss, B. S., Hannaford, A. J., Smith, P. W. G., and Tatchell, A. R., eds. 1994. *Vogel's textbook of practical organic chemistry,* 5th ed., 694, Exp. 5.139. Essex, England: Longman.
5. Furniss, B. S., Hannaford, A. J., Smith, P. W. G., and Tatchell, A. R., eds. 1994. *Vogel's textbook of practical organic chemistry,* 5th ed., 704, Exp. 5.148. Essex, England: Longman.
6. Hassner, A., and Alexanian, V. 1978. *Tetrahedron Letters* 4475.
7. Furniss, B. S., Hannaford, A. J., Smith, P. W. G., and Tatchell, A. R., eds. 1994. *Vogel's textbook of practical organic chemistry,* 5th ed., 699, Exp. 5.142. Essex, England: Longman.
8. Furniss, B. S., Hannaford, A. J., Smith, P. W. G., and Tatchell, A. R., eds. 1994. *Vogel's textbook of practical organic chemistry,* 5th ed., 709, Exp. 5.154. Essex, England: Longman.
9. Fischer, E., and Dilthey, A. 1902. *Chemische Berichte* 35:844 (see p. 856).
10. Cason, J., and Kraus, K. W. 1961. *Journal of Organic Chemistry* 26:1768, 1772.
11. Furniss, B. S., Hannaford, A. J., Smith, P. W. G., and Tatchell, A. R., eds. 1994. *Vogel's textbook of practical organic chemistry,* 5th ed., 540, Exp. 5.42. Essex, England: Longman.
12. Shirley, D. A. 1954. *Organic Reactions* 8:28.

13. Furniss, B. S., Hannaford, A. J., Smith, P. W. G., and Tatchell, A. R., eds. 1994. *Vogel's textbook of practical organic chemistry,* 5th ed., 674, Exp. 5.129. Essex, England: Longman.
14. Jorgenson, M. J. 1970. *Organic Reactions* 18:1 (see p. 47).
15. Hrubiec, R. T., and Smith, M. B. 1984. *Journal of Organic Chemistry* 49:431.
16. Katzung B.G., and Chatterjee K. 1989. In *Basic and clinical pharmacology,* ed. Katzung B. G., 1017. Norwalk, CT: Appleton and Lange.
17. Corbridge, D. E. C. 2000. *Phosphorus 2000: Chemistry, biochemistry & technology,* 248–249. Amsterdam: Elsevier, Amsterdam.
18. Sieber, S. A., and Marahiel, M. A. 2005. *Chemical Reviews* 105:715.
19. Garrett, R. H., and Grisham, C. M. 1995. *Biochemistry,* 734–735. Ft. Worth, TX: Saunders.

Answers to Problems

20.1

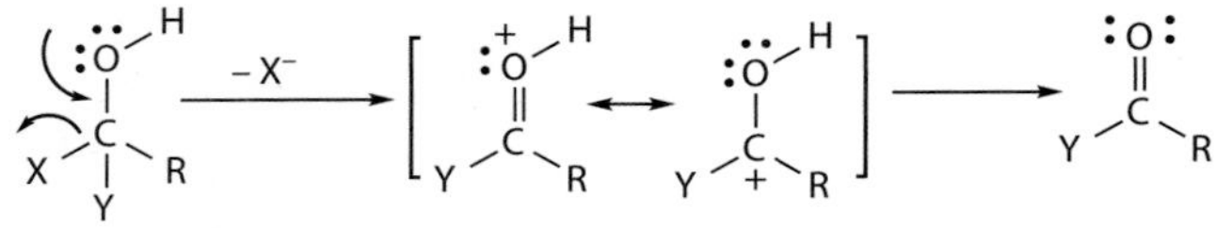

20.2

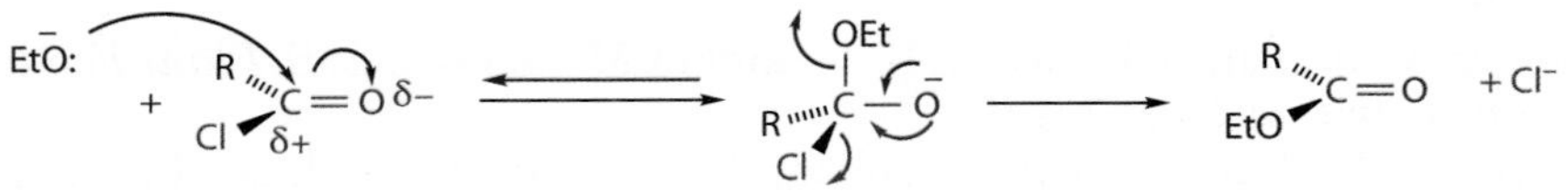

20.3

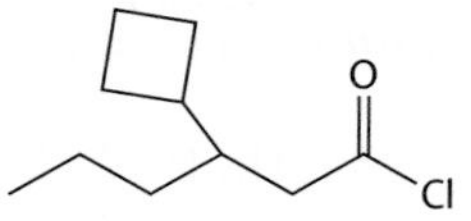

20.4   This is an aqueous acid solution. There is no hydroxide ion in the acidic medium, so the only oxygen nucleophile is water.

20.5

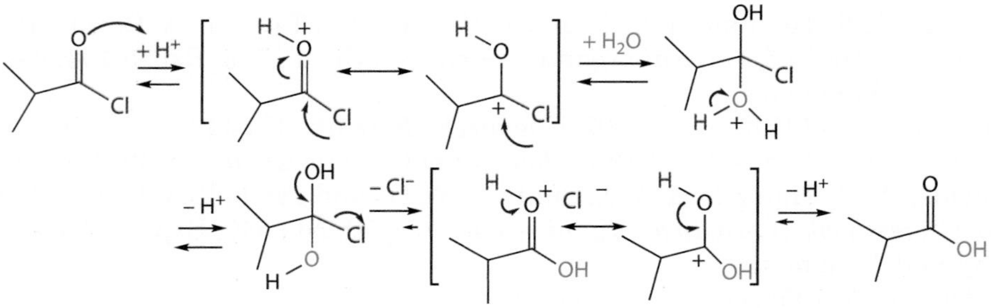

20.6

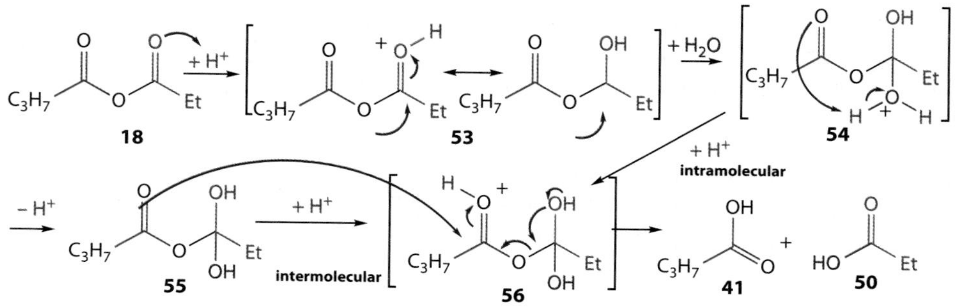

20.7

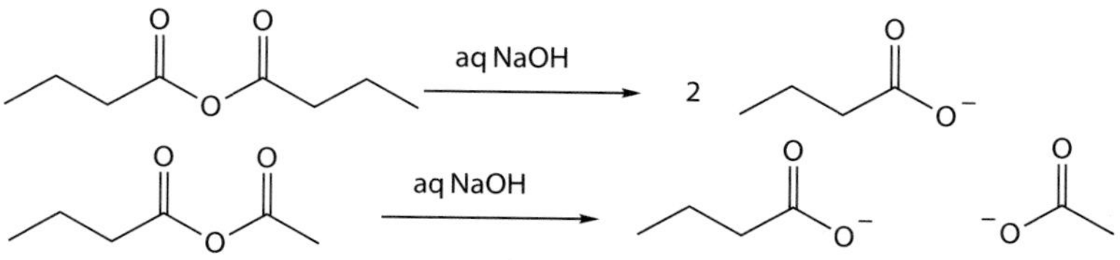

20.8

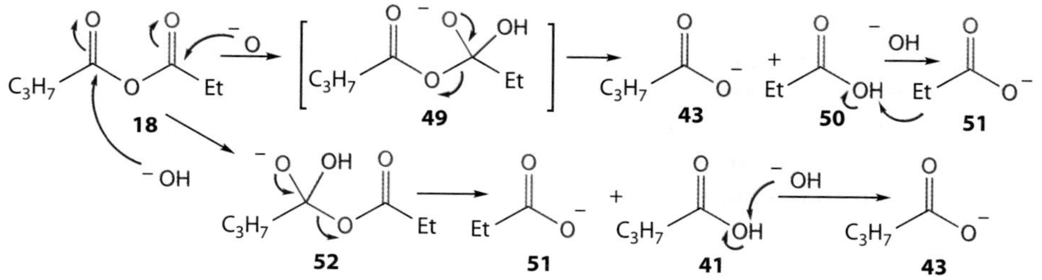

20.9   Isopropyl acetate is 1-methylethyl ethanoate.

20.10   Glycerol is 1,2,3-propanediol.

20.11

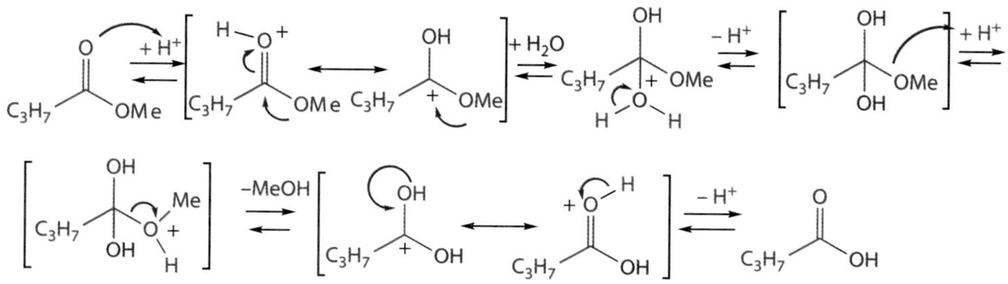

20.36

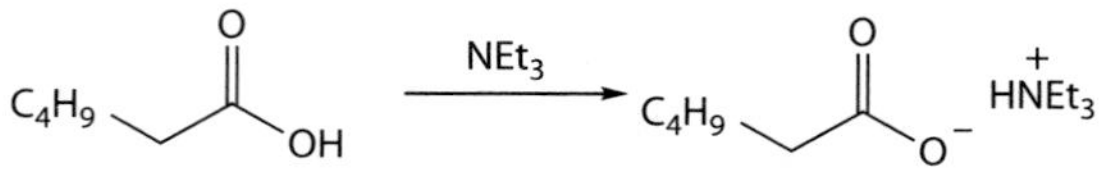

20.37 The reaction of ammonia with the acid chloride gives the amide, but also HCl as a by-product. The additional ammonia reacts as a base with the HCl to convert it to ammonium chloride and remove the HCl from the reaction.

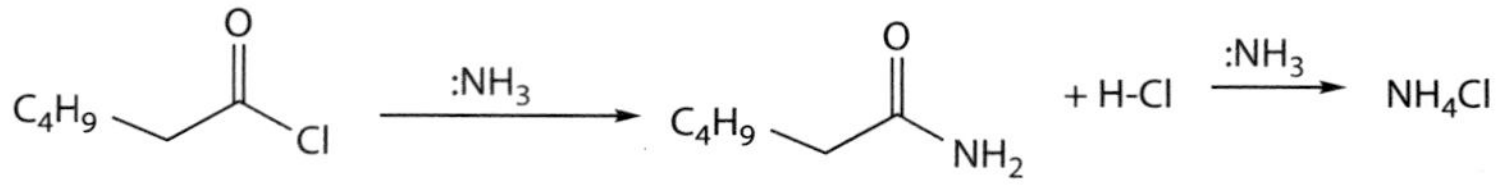

20.38

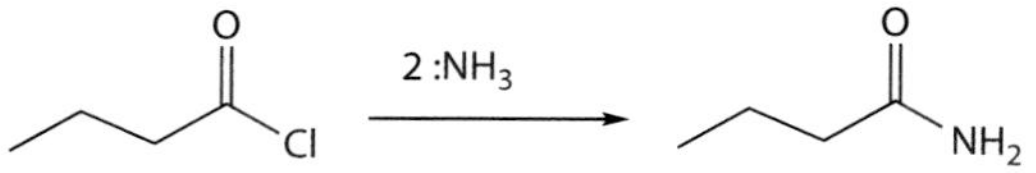

20.39

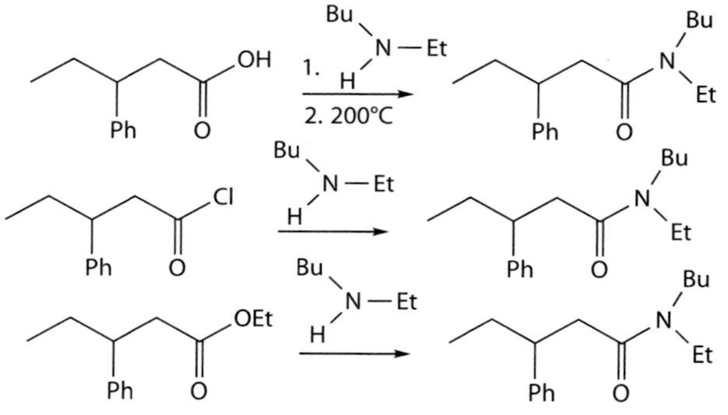

20.40

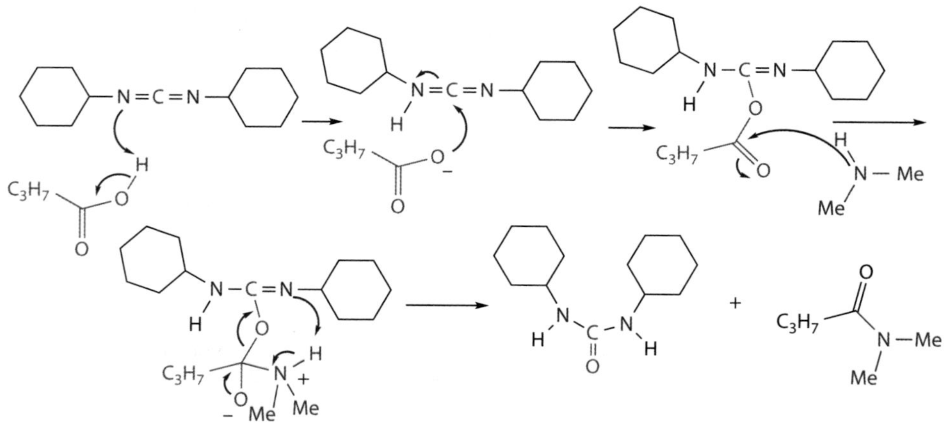

20.41

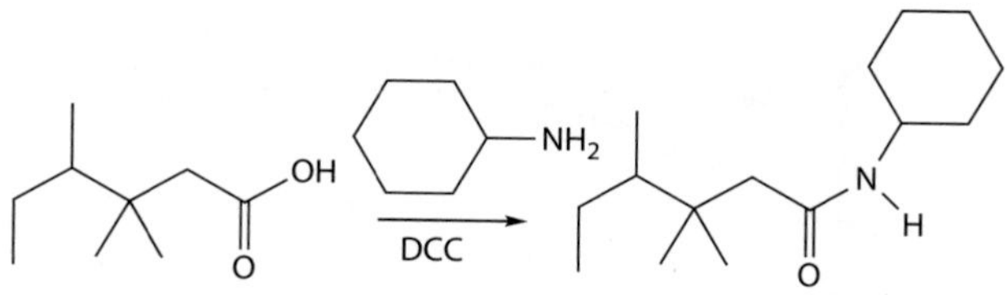
OH
NH2
DCC
O
N
H

20.42

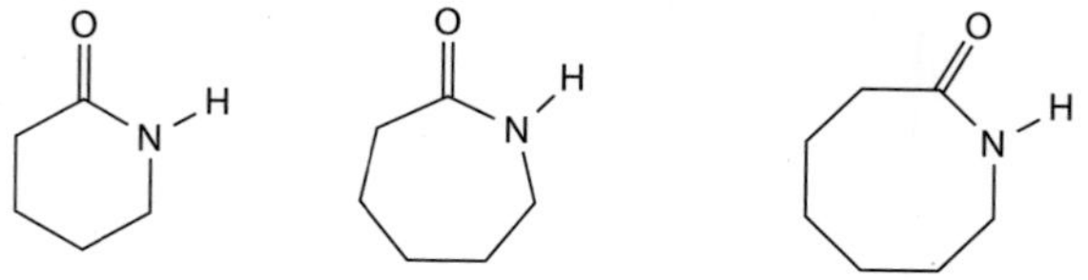
O
N
H

20.43

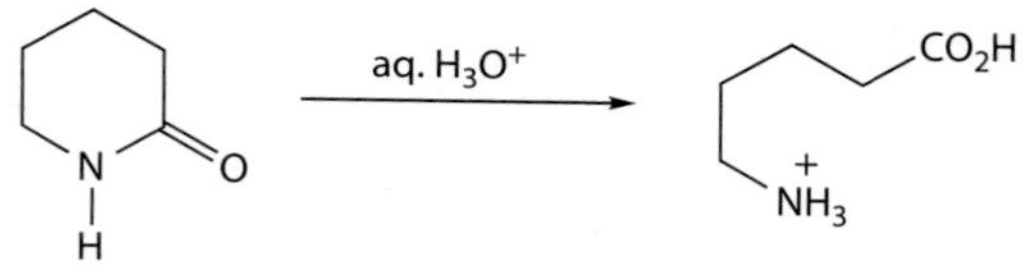
aq. H3O+
CO2H
NH3
N
H
O

20.44

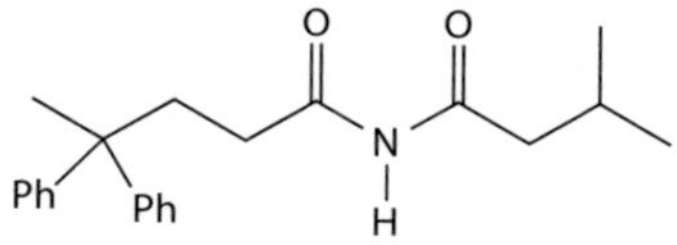
Ph
Ph
O
N
H

3-methylbutanoyl 4, 4-diphenylpentanamide

20.45

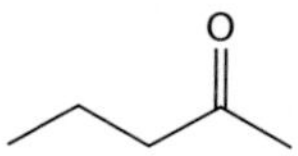

2-pentanone

20.46

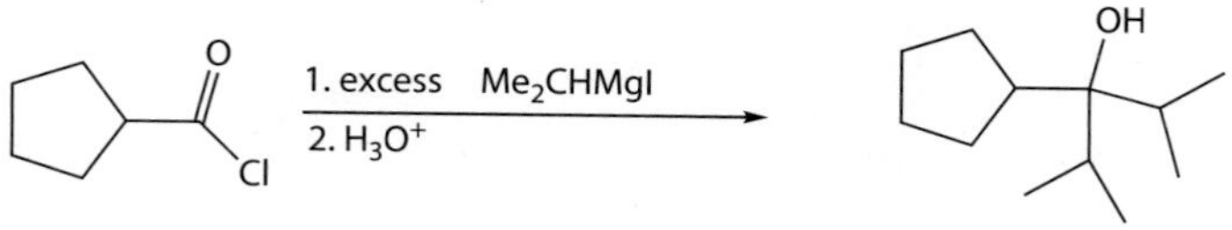
O
Cl
1. excess Me2CHMgI
2. H3O+
OH

20.47

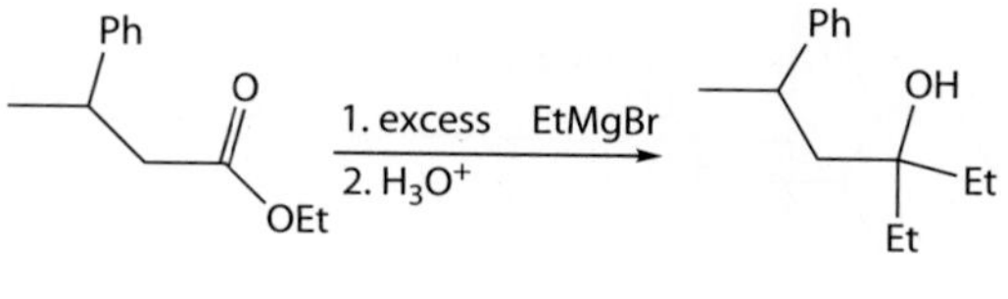
Ph
O
OEt
1. excess EtMgBr
2. H3O+
OH
Et
Et

20.48

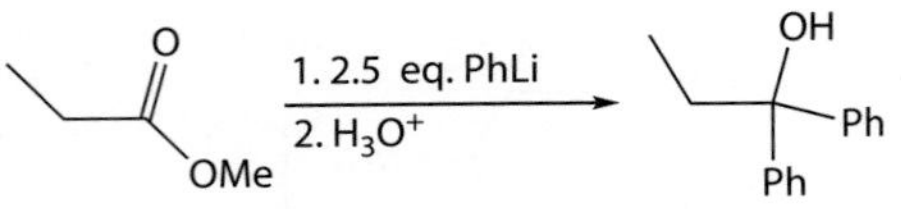

20.49

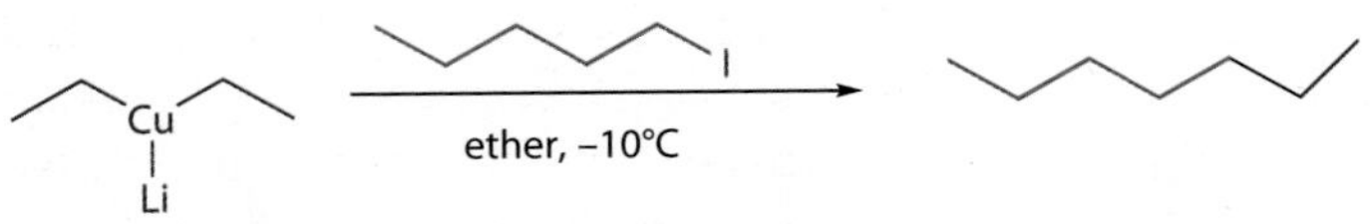

20.50

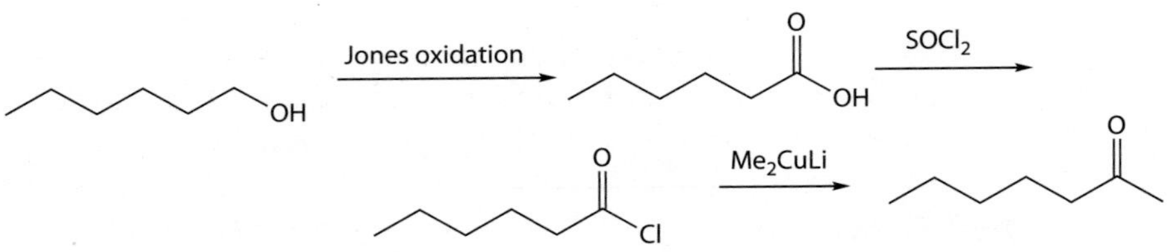

20.51

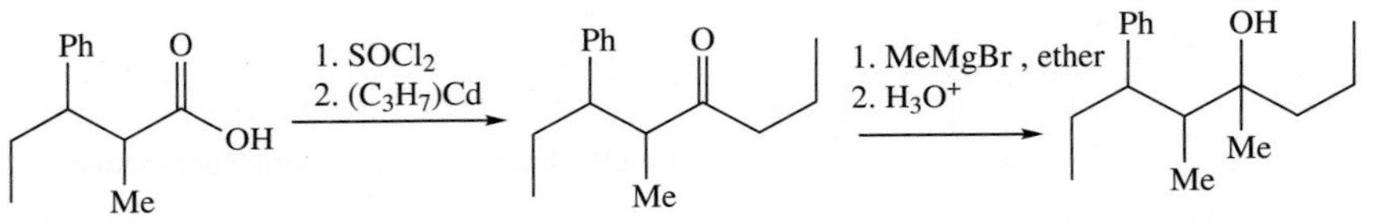

20.52

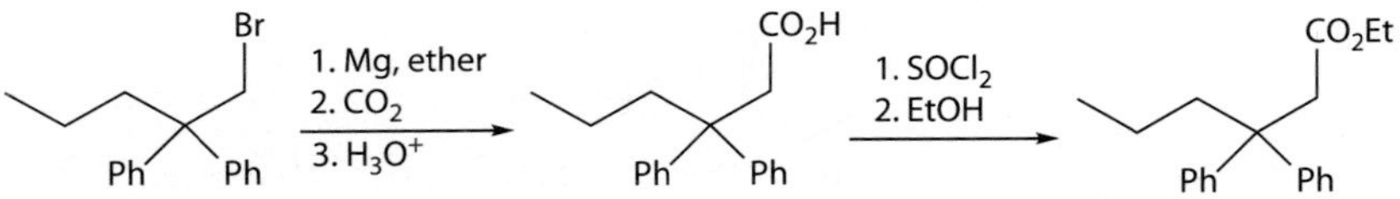

20.53

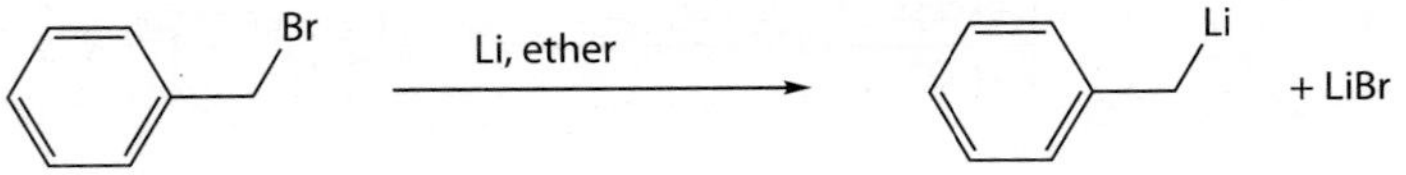

20.54

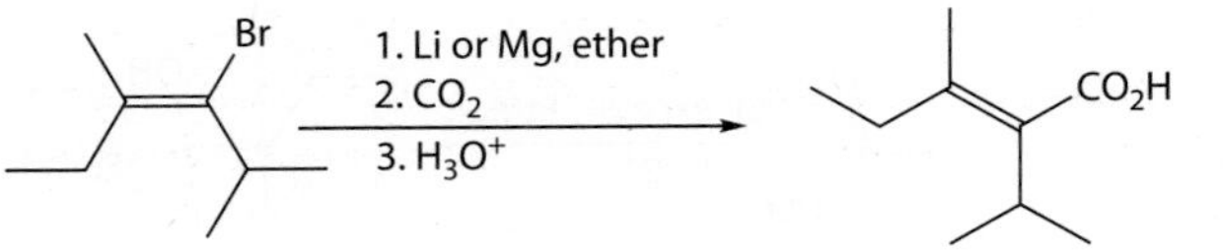

20.55

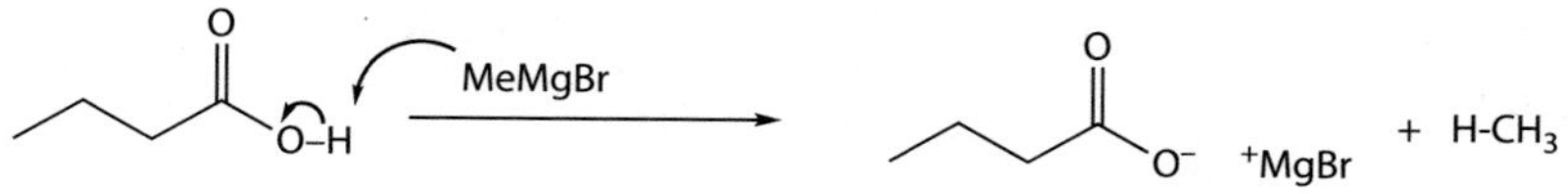

20.56   Butane.

20.57

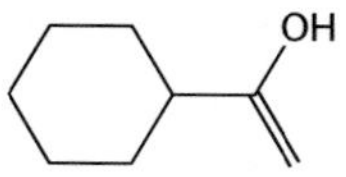

20.58

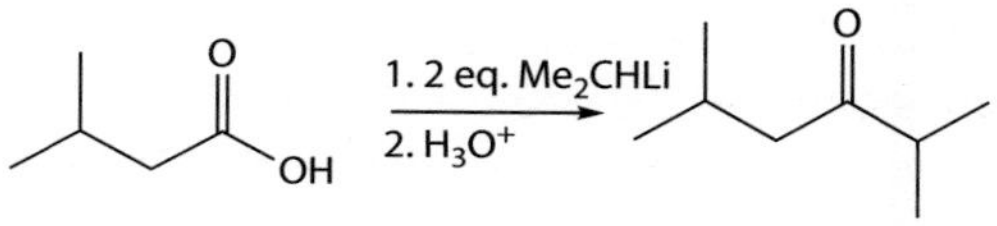

20.59

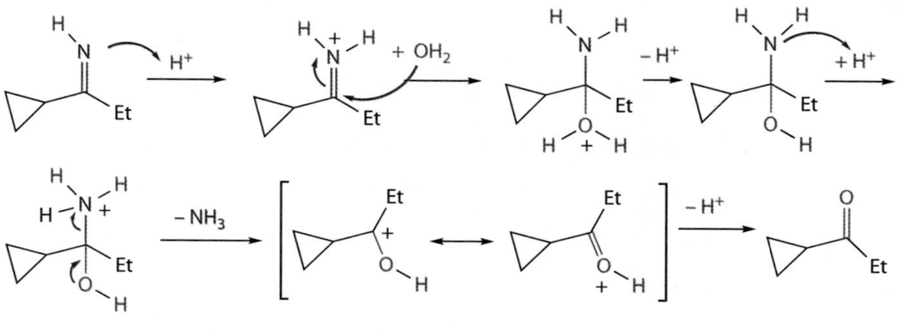

20.60

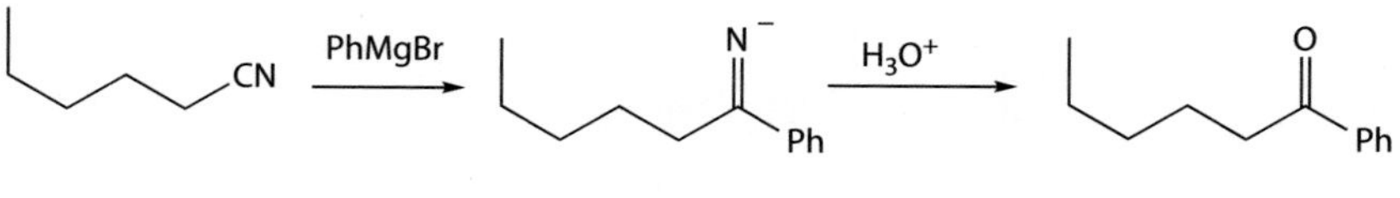

20.61

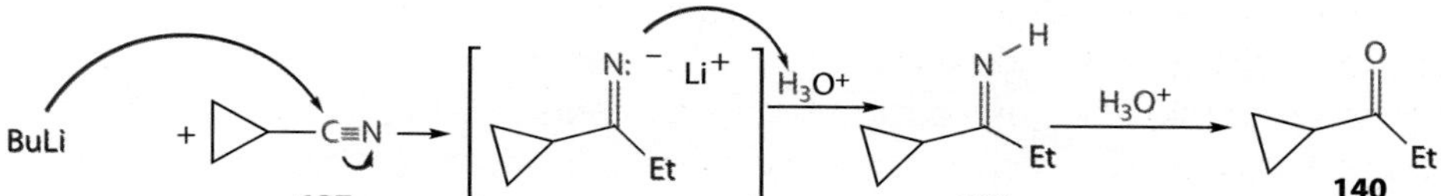

20.62

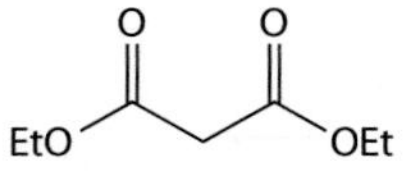

20.63

20.64

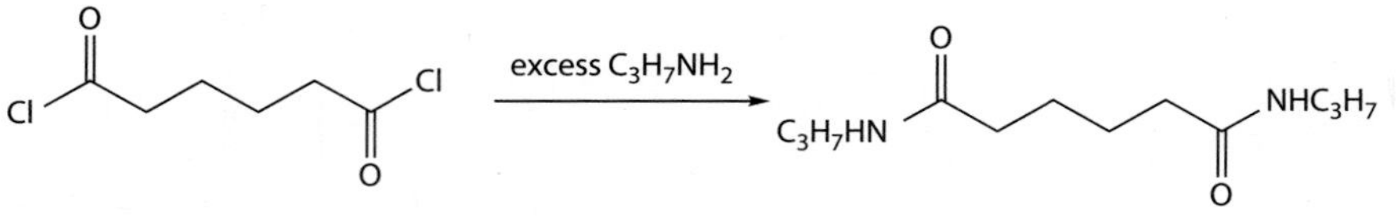

20.65

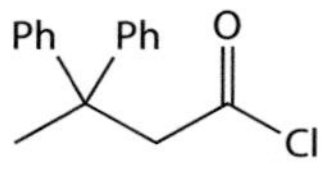

20.66

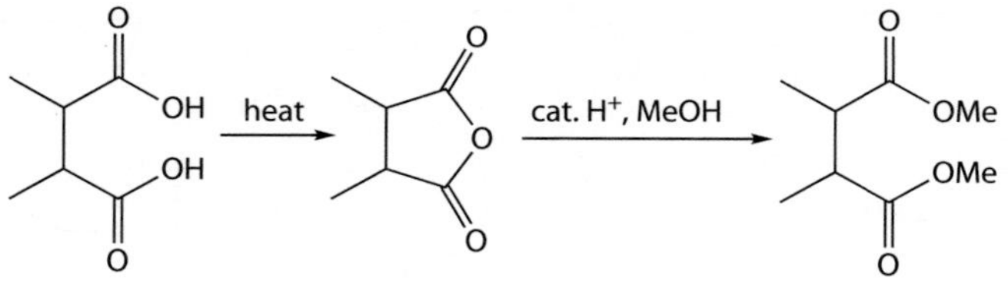

20.67

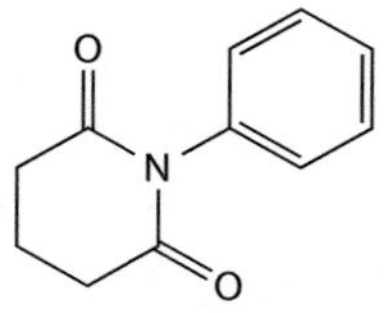

20.68

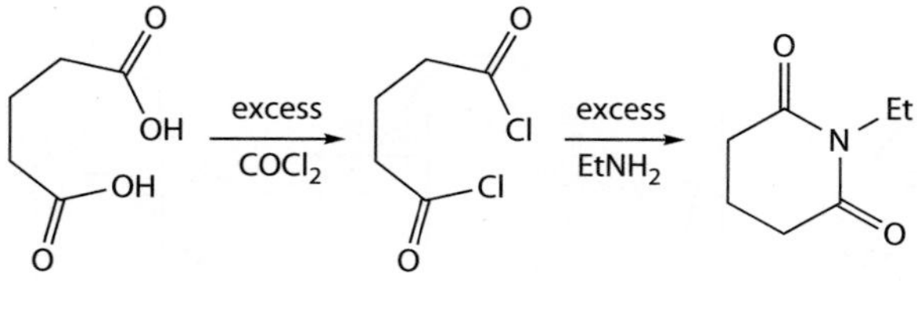

20.69

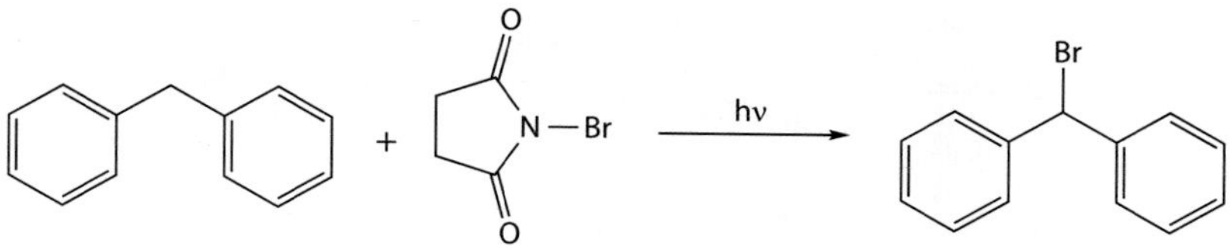

20.70

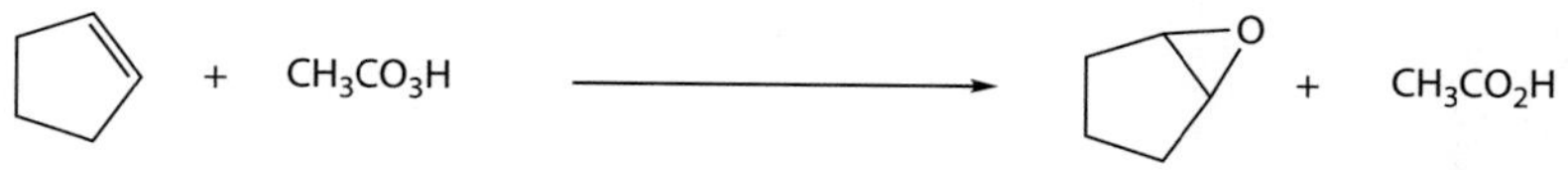

20.71

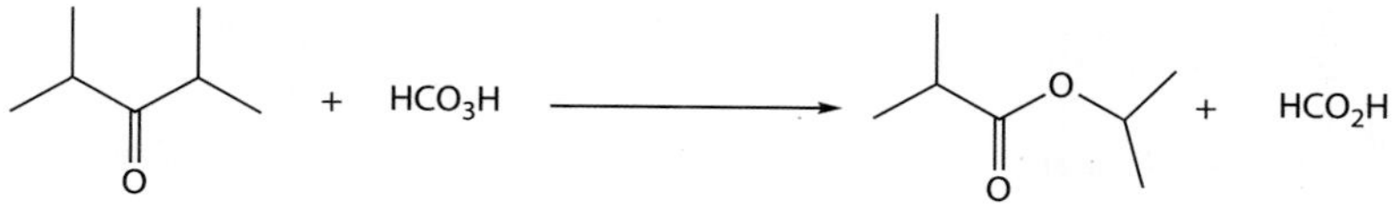

20.72

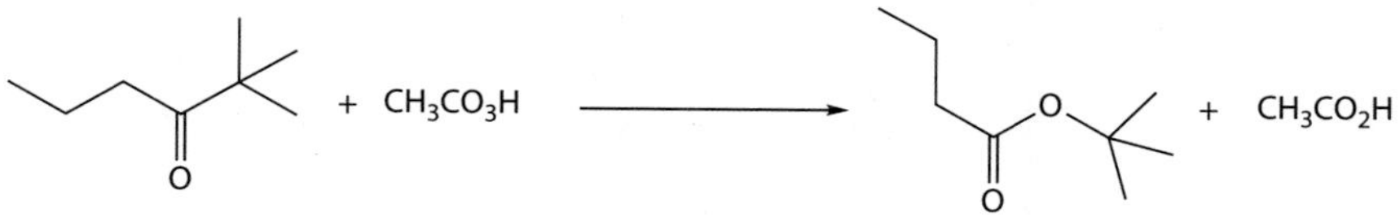

20.73

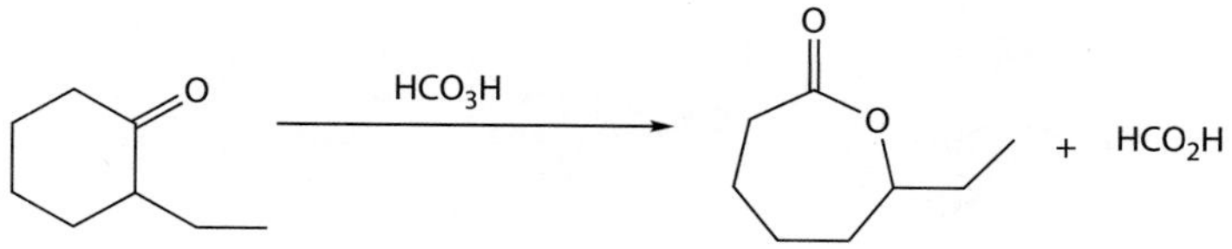

20.74

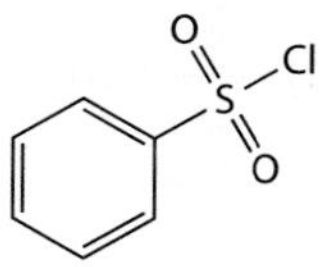

20.75

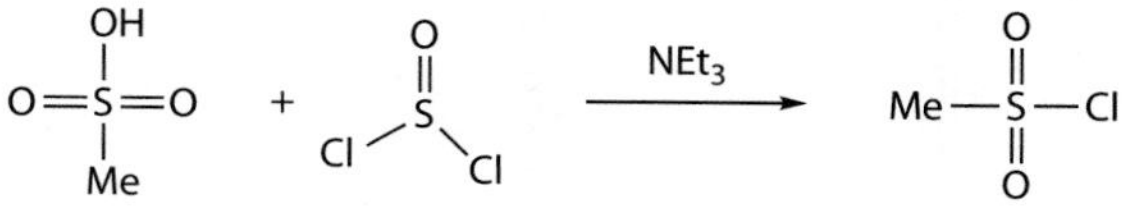

20.76

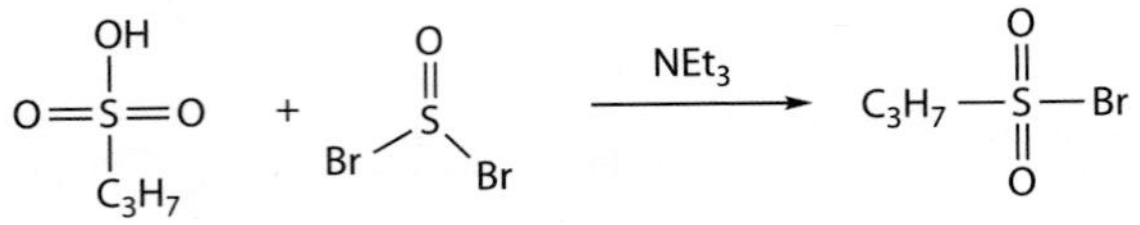

20.77

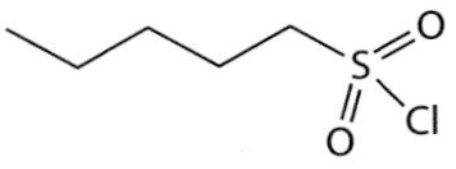

20.78

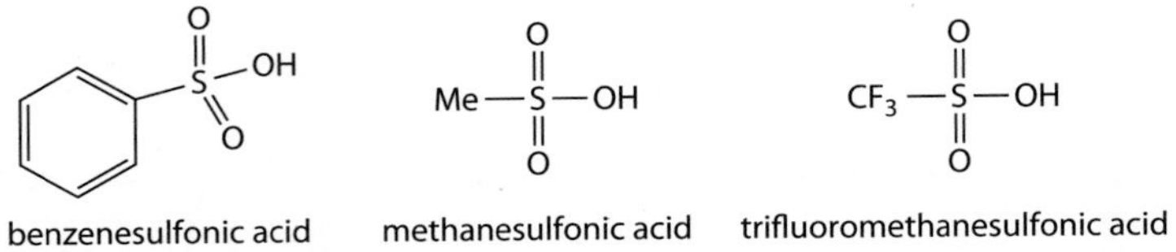

benzenesulfonic acid　　　methanesulfonic acid　　trifluoromethanesulfonic acid

20.79

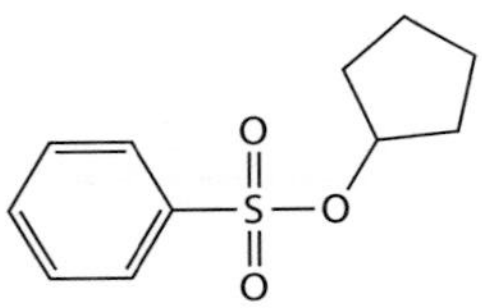

20.80

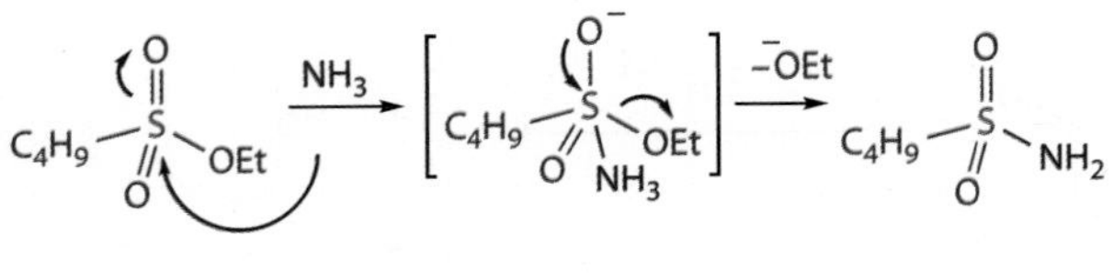

20.81

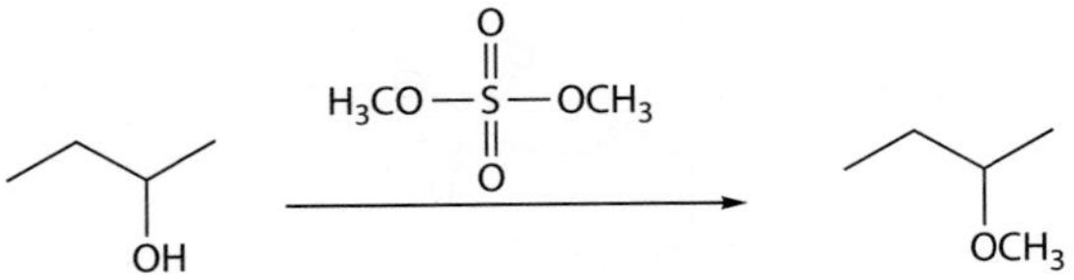

20.82

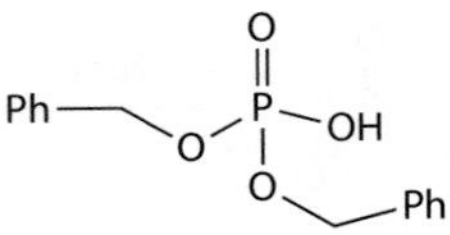

20.83

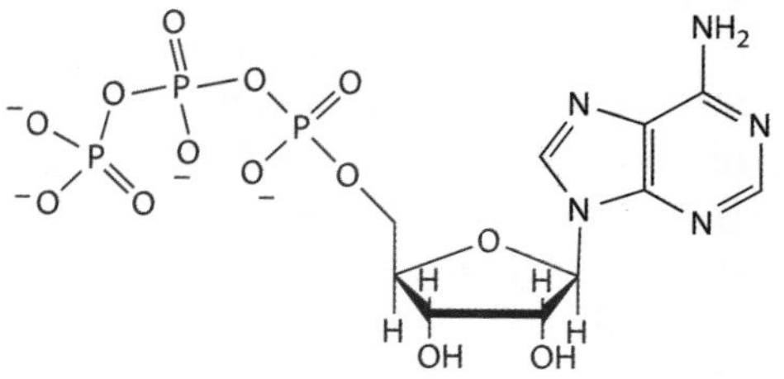

20.84

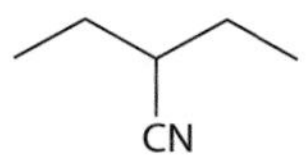

20.85

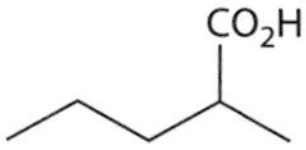

Correlation of Concepts with Homework

- Many acid derivatives can be generated by "replacing" the OH unit of a carboxylic acid (COOH) with another atom or group: $-Cl$, $-Br$, $-O_2CR$, $-OR$, $-NR_2$. Common acid derivatives are acid chloride, acid anhydrides, esters, and amides: 1, 2, 3, 9, 10, 15, 86, 87, 104, 116, 117, 136.
- Acid or base hydrolysis of acid chlorides, acid anhydrides, esters, or amides regenerates the parent carboxylic acid: 3, 4, 5, 6, 7, 8, 11, 12, 14, 43, 80, 96, 97, 99, 101, 102, 105, 109, 115, 129, 130.
- Acid chlorides are prepared by reaction of a carboxylic acid with a halogenating agent such as thionyl chloride, phosphorus trichloride, phosphorus pentachloride, phosgene, or oxalyl chloride: 16, 17, 65, 96, 106, 113, 114, 120.
- Acid anhydrides can be prepared by reaction of an acid chloride with a carboxylic acid or by the dehydration of two carboxylic acids that leads to coupling: 18, 19, 20, 21, 22, 23, 24, 66, 89, 91, 108, 109, 114, 123, 129.
- Esters can be prepared by the reaction of an acid chloride or acid anhydride with an alcohol as well as by direct reaction of an acid and an alcohol, with an acid catalyst, or by using a dehydrating agent such as DCC. Lactones are cyclic esters: 25, 26, 27, 28, 29, 30, 32, 52, 96, 98, 99, 100, 107, 108, 111, 112, 113, 114, 115, 118, 127, 129.

- Amides can be prepared by the reaction of an acid chloride, acid anhydride, or an ester with ammonia or an amine. Acids and amines can be coupled in the presence of a dehydrating agent such as DCC. Amines react with carboxylic acids to give ammonium salts. Heating these salts to around 200°C will usually give the amide. Lactams are cyclic amides: 33, 34, 35, 36, 37, 38, 39, 40, 41, 42, 63, 64, 88, 92, 96, 102, 108, 114, 118, 119, 120, 121, 129, 134.
- Imides are formed by the reaction of amides with other acid derivatives. Cyclic anhydrides and cyclic imides can be prepared from dicarboxylic acids or acid dichlorides: 44, 67, 68, 114, 123.
- Acid derivatives react with Grignard reagents and organolithium reagents to give ketones, but this initial product reacts with a second equivalent of Grignard reagent or organolithium reagent to give a tertiary alcohol: 45, 46, 47, 48, 55, 56, 93, 94, 95, 124, 125, 129, 140.
- Grignard reagents and organolithium reagents react with cadmium chloride to form dialkyl cadmium reagents, which react with acid chloride to give ketones. Organolithium reagents react with copper(I) salts to give lithium dialkyl cuprates, which react with acid chlorides to give ketones: 49, 50, 51, 53, 125, 126, 129.
- Grignard reagents and organolithium reagents react with carbon dioxide to give carboxylic acids: 52, 53, 54, 125, 126, 129.
- Organolithium reagents react with carboxylic acids to give ketones: 55, 56, 57, 58, 126, 141.
- Grignard reagents and organolithium reagents react with nitriles to give imine anions, but aqueous acid hydrolysis usually converts the imine anion to a ketone as the final product: 59, 60, 61, 126, 142.
- Peroxyacids react with ketones to give esters in the Baeyer–Villiger reaction: 70, 71, 72, 73, 89, 128, 129.
- Acid dichlorides are prepared by reaction of the diacid with a halogenating agent. Diesters are prepared from dicarboxylic acids or diacid chlorides and alcohols. Diamides are prepared from diesters or diacid chlorides and amines: 62, 63, 64, 65, 66, 67, 68, 69, 119, 131, 132, 133, 134.
- Sulfonic acids can be converted to sulfonic acid chlorides, sulfonate esters, and sulfonamides, similar to reactions of carboxylic acids: 74, 75, 76, 77, 78, 79, 80, 135, 136, 137.
- The ester derivatives of nitric acid, sulfuric acid, and phosphoric acid are important compounds, with some use in organic chemistry but greater use in biology: 81, 82, 83, 137, 138, 139, 140.
- Nitriles are acid derivatives. Primary amides are dehydrated to nitriles, and nitriles are hydrolyzed to either an amide or a carboxylic acid: 84, 85, 96, 108, 126, 129.

- **A molecule with a particular functional group can be prepared from molecules containing different functional groups by a series of chemical steps (reactions). This process is called synthesis: The new molecule is synthesized from the old one (see Chapter 25): 54, 58, 141, 142, 143.**
- **Spectroscopy can be used to determine the structure of a particular molecule and can distinguish the structure and functionality of one molecule when compared with another (see Chapter 14): 144, 145, 146, 147, 148, 149, 150, 151, 152, 153, 154, 155, 156, 157, 158, 159, 160.**

Homework

20.86   Draw the structure for each of the following:
   (a)   2,2-diethylcyclobutane-1-carboxylic acid
   (b)   16-phenylhexadecanoic acid
   (c)   1-butyl-1,4-butanedioic acid
   (d)   4-bromo-3-cyclopropyl-2-hydroxyhexanoic acid
   (e)   3,3-diethyloctanenitrile

20.87   Give the IUPAC name for each of the following molecules:

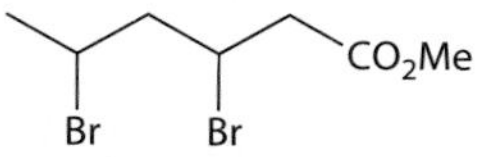

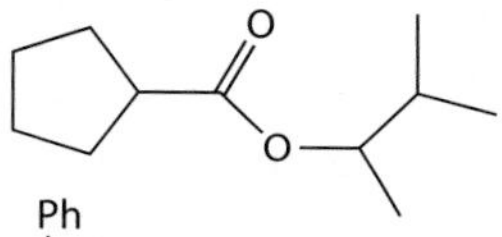

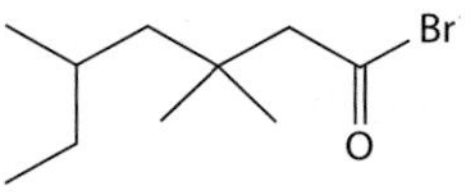

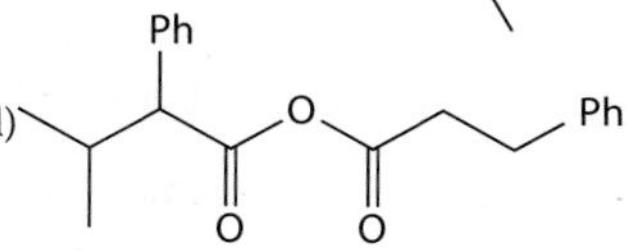

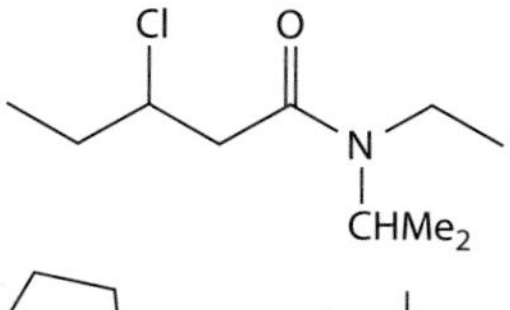

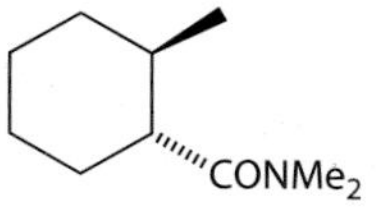

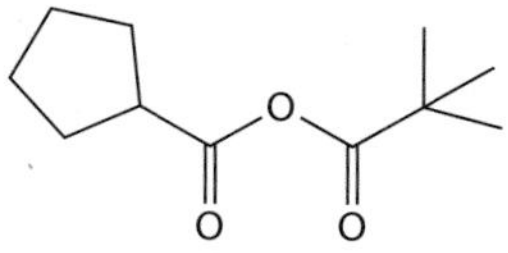

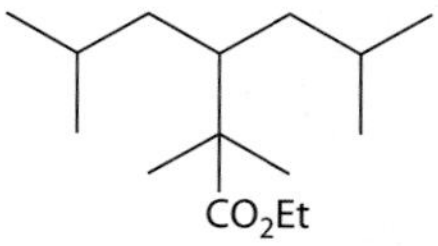

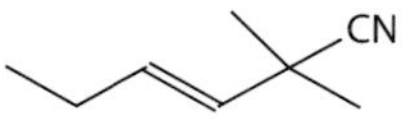

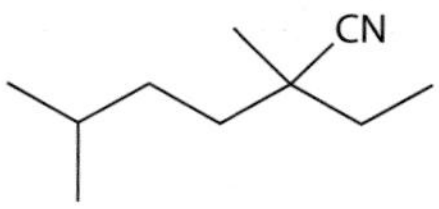

20.88   Indicate which of the following structures are lactams:

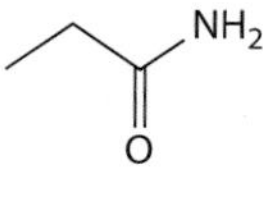

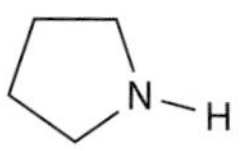

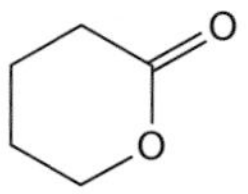

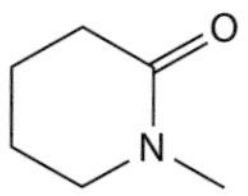

20.89   Indicate which of the following reagents will convert butanoic acid to
        butanoyl chloride:
        (a) PBr_3
        (b) $SOCl_2$
        (c) CH_3Cl
        (d) PCl_5
        (e) NaCl

20.90   In the Baeyer–Villiger reaction of 4-nonanone to butyl pentanoate with trif-
        luoroperoxyacetic acid, a large excess of sodium acetate is sometimes added
        as a buffer. Explain why this is necessary.

20.91   Mixing hexanol with HCl will convert the alcohol to 1-chlorohexane. Explain
        why mixing hexanoic acid with HCl does not give hexanoyl chloride.

20.92   Indicate which of the following sets of reagents will convert butanoic acid
        to butanamide:
        (a) NH_3, 25°
        (b) Me_2NH, 25°
        (c) 1. $SOCl_2$; 2. NH_3
        (d) 1. MeOH, H^+; 2. NH_3, heat
        (e) Me_2NH, 250°

20.93   Draw the product formed when 2-bromo-1-pentene is treated with (a) Mg,
        THF (b) 3-phenylbutanoyl chloride at –78°C.

20.94   Draw the product expected when ethyl pentanoate is heated with a large
        excess of phenylmagnesium bromide and then that product is treated with
        aqueous acid.

20.95   When the product from the reaction described in the preceding question
        is heated with acid, the product is commonly an alkene. Draw this alkene
        product and suggest a mechanism for its formation.

20.96   Give the major product for each of the following:

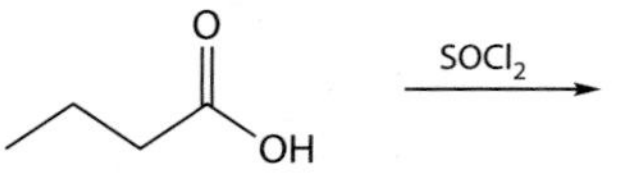

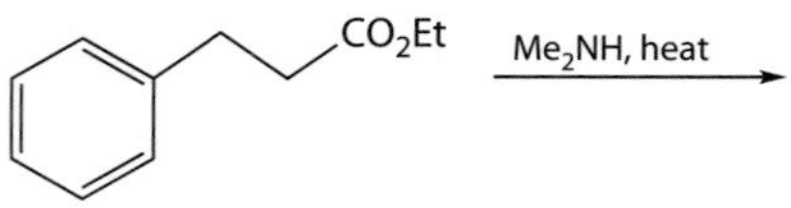

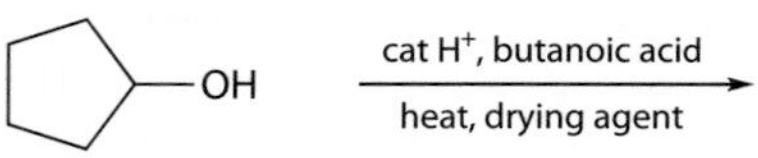

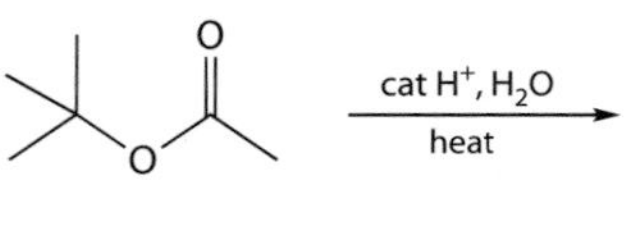

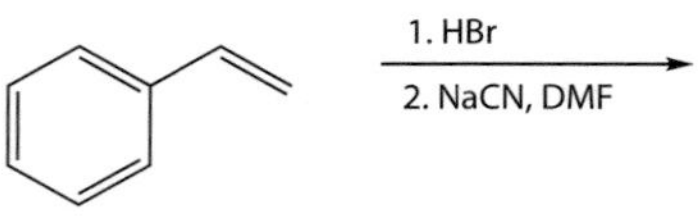

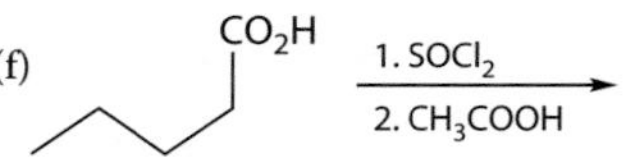

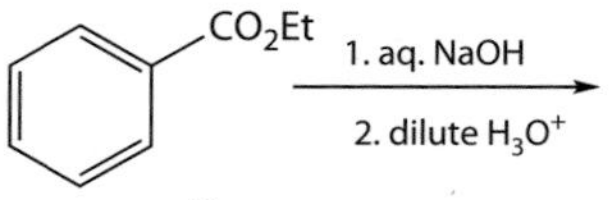

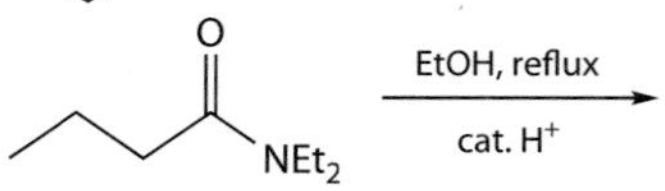

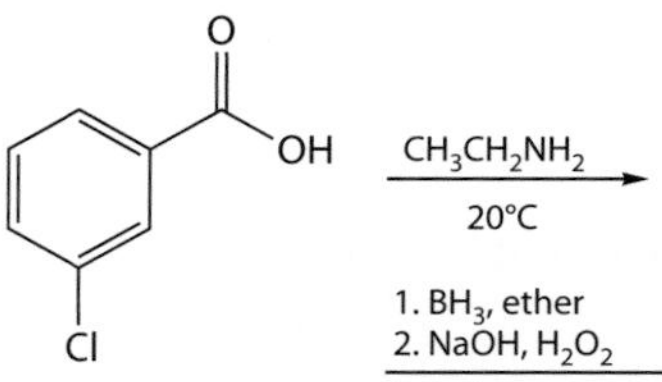

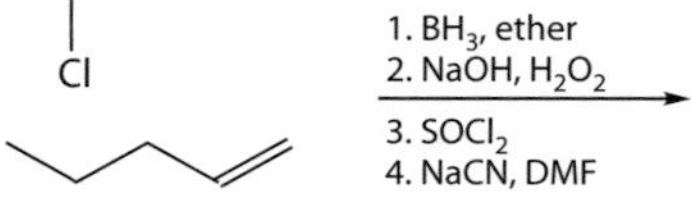

20.97 Briefly explain why saponification of 1,1-dimethylethyl butanoate is much slower than the saponification of methyl butanoate.

20.98 An attempt to convert ethyl butanoate to isopropyl butanoate was done as follows. Ethyl butanoate was dissolved in methanol and treated with five equivalents of isopropanol (2-propanol) and an acid catalyst. When the reaction was analyzed, there was essentially no ethyl butanoate, <10% of isopropyl butanoate, and about 90% of another product. What is that product? Describe how it might be formed. How can you modify this procedure to obtain isopropyl butanoate?

20.99 After hexanoic acid was successfully converted to ethyl ethanoate in ethanol solvent, the ethanol was removed and the product washed with aqueous sodium bicarbonate. What is the purpose of this last step?

20.100 Unless the ester has a rather simple structure, we commonly convert a carboxylic acid to its acid chloride and then add an alcohol to make an ester. We can make esters directly by adding the acid to an alcohol, in the presence of an acid catalyst. Why do we use the two-step procedure preferentially in most cases?

20.101 Can you think of any reason why the fundamental biological building blocks of the human body (amino acids and proteins) have amide links rather than ester links?

20.102 If we treat an ester with an amine, we obtain an amide. If we boil an amide in ethanol, we do not obtain an ester. Why not?

20.103 When we open a new bottle of acetyl chloride in an old lab during the summer, with the windows open, it is common to see it "fume." Assuming that acetyl chloride is a stable molecule, what is happening?

20.104 The weight of acetyl chloride is 78.50 and that of acetic acid is 60.65, so why is the boiling point of acetyl chloride lower than that of acetic acid?

20.105 When we prepare an amide by heating the ammonium salt of a carboxylic acid, it is common to wash the product with aqueous bicarbonate and then with dilute aqueous acid. Why? When we prepare an acid chloride, in most cases we take great pains not to wash the product with these solutions. Why?

20.106 Suggest five *different* reagents that will convert pentanoic acid to pentanoyl chloride.

20.107 When we react butanoyl chloride with ethanol in benzene solvent, it is common to add triethylamine to the reaction. What purpose does this serve and how does it help?

20.108 Give the major product for each of the following reactions:

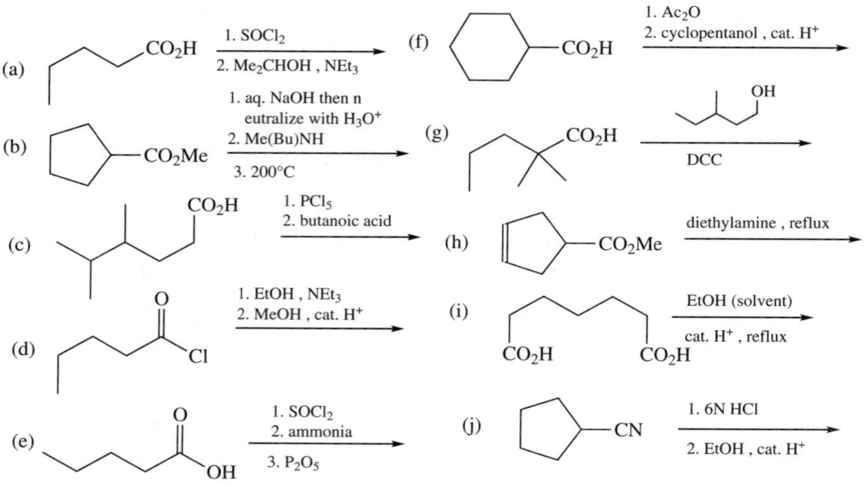

(a) [structure] CO₂H — 1. SOCl₂ / 2. Me₂CHOH , NEt₃

(b) [structure] CO₂Me — 1. aq. NaOH then neutralize with H₃O⁺ / 2. Me(Bu)NH / 3. 200°C

(c) [structure] CO₂H — 1. PCl₅ / 2. butanoic acid

(d) [structure with O and Cl] — 1. EtOH , NEt₃ / 2. MeOH , cat. H⁺

(e) [structure with O and OH] — 1. SOCl₂ / 2. ammonia / 3. P₂O₅

(f) [cyclohexyl] CO₂H — 1. Ac₂O / 2. cyclopentanol , cat. H⁺

(g) [structure] CO₂H — [structure with OH] / DCC

(h) [cyclopentene] CO₂Me — diethylamine , reflux

(i) [structure] CO₂H CO₂H — EtOH (solvent) / cat. H⁺ , reflux

(j) [cyclopentyl] CN — 1. 6N HCl / 2. EtOH , cat. H⁺

20.109 When phosgene (Cl₂C=O; dichloromethanone) was used as a weapon to gas troops in World War I, it was delivered in shells that exploded on contact and released the chemical. Phosgene is a colorless gas, but when the shell exploded a cloud of whitish gas soon appeared. Comment on what happened to the phosgene to cause this.

20.110 Several years ago, a tank car of thionyl chloride ruptured in a Boston rail yard, and a pall of whitish gas filled the area. The fire department was called in to clean up the mess. A few weeks later, an article in the paper described a problem that paint on the fire trucks was peeling off. Suggest a reason for this paint problem.

20.111 Given the mechanism for DCC coupling of carboxylic acids and alcohols to give the ester in Section 20.5.4, propose a mechanism for the DCC coupling of two equivalents of propanoic acid to give the anhydride.

20.112 A new molecule is produced when **A** is heated in the presence of an acid catalyst. Give the structure of that product and propose a mechanism for its formation.

A HO [structure] CO₂Me — heat / cat. H⁺

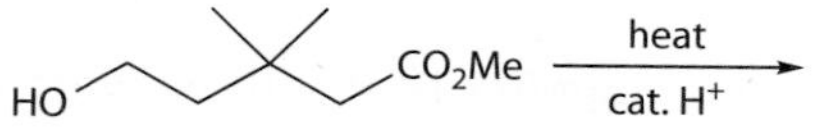

20.113 It is known that attempts to hydrolyze a five-membered ring lactone lead to products that have significant amounts of the starting lactone rather than the hydroxy acid. This sharply contrasts with the hydrolysis of seven-membered ring lactones, which give hydroxy acid products readily. Suggest a reason for this difference in reactivity.

20.114 Give the major product for each of the following reactions:

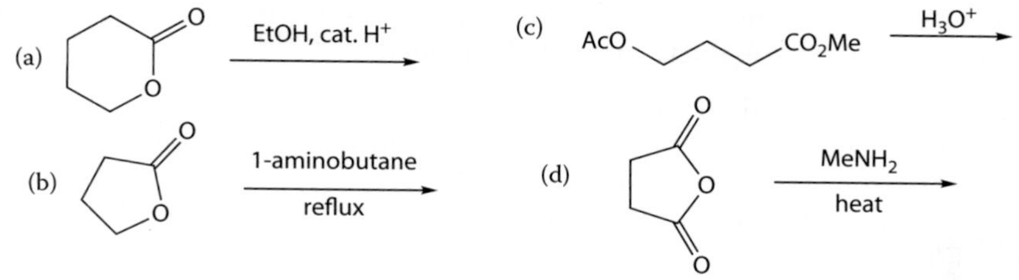

(a) [structure: δ-valerolactone] EtOH, cat. H⁺ →

(b) [structure: γ-butyrolactone] 1-aminobutane / reflux →

(c) AcO~~~CO₂Me H₃O⁺ →

(d) [structure: succinic anhydride] MeNH₂ / heat →

20.115 Heating 16-hydroxyhexadecanoic acid in ethanol with an acid catalyst does not give the lactone. What is the product? Briefly defend your choice.

20.116 The C–N bond length for a typical amine is 1.47 Å, but for a typical amide, it is 1.32 Å. Suggest a reason for this difference.

20.117 4-Aminobutanoic acid (also known as γ-aminobutyric acid or GABA) is an important mammalian neurotransmitter that exists as a neutral molecule. Analysis at neutral pH shows that it does not have a COOH unit or an NH₂ unit. Explain.

20.118 Draw the reaction products expected, if any, when piperidine reacts with
(a) butanoic acid at room temperature
(b) butanoyl chloride
(c) ethyl butanoate
(d) HCl
(e) acetic anhydride
(f) 1-aminobutane

20.119 One form of nylon is formed by the reaction of the acid dichloride of 1,6-hexanedioic acid (adipic acid) and 1,6-hexanediamine. Draw a reasonable structure for this form of nylon (called Nylon 66).

20.120 When glutamic acid (**A**) is heated with thionyl chloride and then with ethanol, a molecule called ethyl pyroglutamate is formed. Draw this molecule and give a mechanistic rationale for its formation.

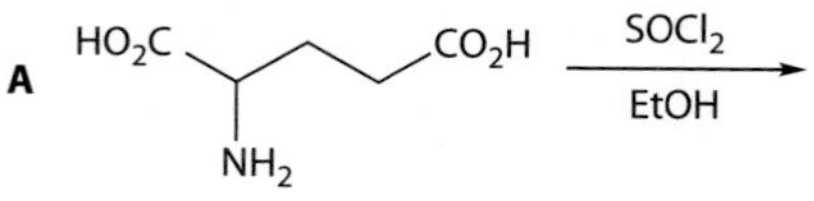

A HO₂C~~~CO₂H SOCl₂ / EtOH →
  |
  NH₂

20.121 Show three different ways to prepare propanoyl propanamide from three different starting materials.

20.122 Acetic anhydride is a common reagent used to convert amines to amides, but the imide known as acetyl acetamide is not used for this purpose. Why not?

20.123 Heating maleic acid leads to an anhydride, but heating fumaric acid does not. Both of these acids have the structure $HO_2CCH=CHCO_2H$, but one is *cis* and the other *trans*. Identify each acid based on its ability to form an anhydride. Draw the structure of the anhydride formed.

20.124 Draw the final product, if any, of the following reactions:
   (a) 1-bromobutane + 1. Mg/ether; 2. ethyl butanoate (0.5 equivalents)/ether; 3. hydrolysis
   (b) pentanoic acid + 1. oxalyl chloride; 2. Li enolate of acetophenone/THF; 3. hydrolysis
   (c) 3,3-diphenylpentanoic acid + 1. $SOCl_2$; 2. EtOH; 3. excess MeLi/ether; 4. hydrolysis
   (d) 1-butyne + 1. MeLi/THF; 2. propylpentanoate; 3. hydrolysis
   (e) *n*-butyllithium + 1. 0.5 $CdCl_2$; 2. 0.5 butanoyl chloride
   (f) butanoic anhydride + 1. excess phenylmagnesium bromide; 2. hydrolysis

20.125 We know that dimethylcadmium reacts with an acid chloride to form a ketone but does not react further to give an alcohol. This contrasts with methylmagnesium chloride, which reacts first to form the ketone but then reacts with the ketone to give an alcohol. What does this tell you about the relative reactivity of an acid chloride versus a ketone?

20.126 Draw the final product, if any, of each of the following reactions:
   (a) 2-bromopentane + 1. Mg/ether; 2. CO_2; 3. hydrolysis
   (b) 2-bromopentane + 1. Mg/ether; 2. butanenitrile; 3. H_3O^+
   (c) 2-bromopentane + 1. Mg/ether; 2. 1-butyne; 3. mild hydrolysis
   (d) *n*-butyllithium + 1. cyclopentanecarboxylic acid; 2. hydrolysis
   (e) butylmagnesium chloride + 1. cyclopentanecarboxylic acid; 2. hydrolysis
   (f) phenyllithium + 1. CO_2; 2. hydrolysis; 3. PCl_5; 4. diethylamine
   (g) pentanenitrile + 1. ethylmagnesium bromide; 2. H_3O^+; 3. MeMgBr; 4. hydrolysis
   (h) pentanenitrile + 1. 6N HCl/reflux; 2. $SOCl_2$; 3. Me_2Cd; 4. *n*-butyllithium; 5. hydrolysis
   (i) ethyl butanoate + 1. acid hydrolysis; 2. two equivalents of propyllithium; 3. hydrolysis
   (j) cyclopentylmagnesium bromide + 1. CO_2; 2. hydrolysis; 3. DCC/EtOH; 4. Li enolate of acetone
   (k) 1-hexanol + 1. Jones regent, heat; 2. $SOCl_2$; 3. NH_3; 4. P_2O_5.

20.127 Give the complete mechanism for the following transformation:

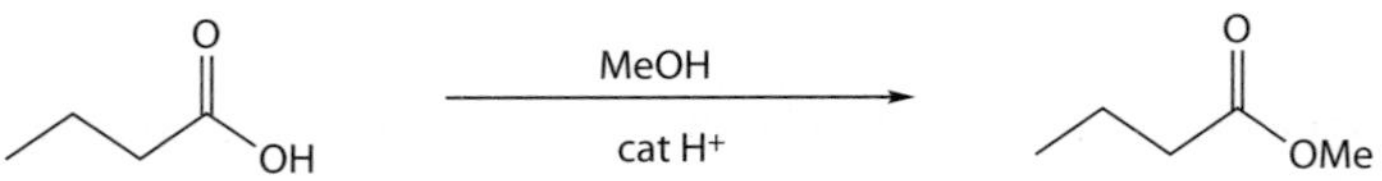

20.128 The reaction of 2,2-dimethyl-2-butanone and a peroxyacid leads to *tert*-butyl acetate. Draw both reactant and product. With this reaction in mind, suggest a reaction sequence that will convert 1-pentene to *tert*-butyl pentanoate.

20.129 Give the major product for each of the following:

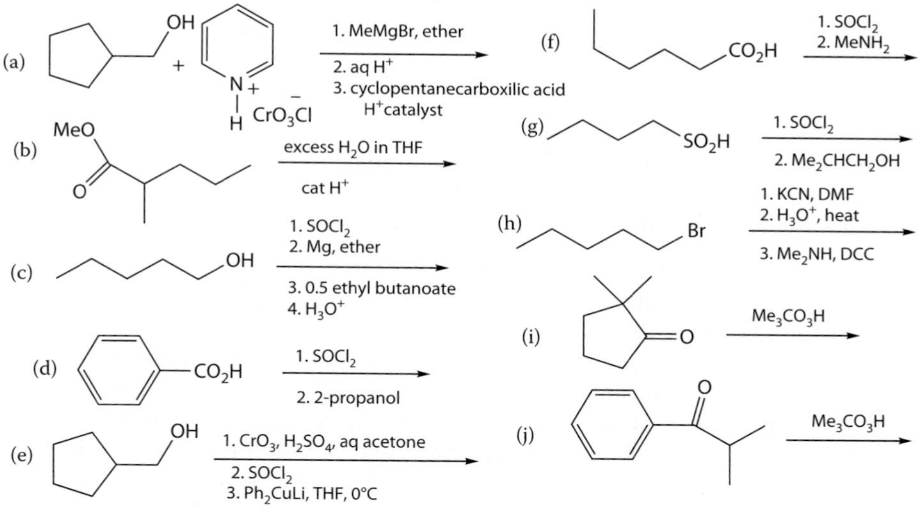
(a)
OH
CrO3Cl
1. MeMgBr, ether
2. aq H+
3. cyclopentanecarboxilic acid
H+catalyst
(b)
MeO
excess H2O in THF
cat H+
(c)
OH
1. SOCl2
2. Mg, ether
3. 0.5 ethyl butanoate
4. H3O+
(d)
CO2H
1. SOCl2
2. 2-propanol
(e)
OH
1. CrO3, H2SO4, aq acetone
2. SOCl2
3. Ph2CuLi, THF, 0°C
(f)
CO2H
1. SOCl2
2. MeNH2
(g)
SO2H
1. SOCl2
2. Me2CHCH2OH
(h)
Br
1. KCN, DMF
2. H3O+, heat
3. Me2NH, DCC
(i)
Me3CO3H
(j)
Me3CO3H

20.130 Give the complete mechanism for the following transformation:

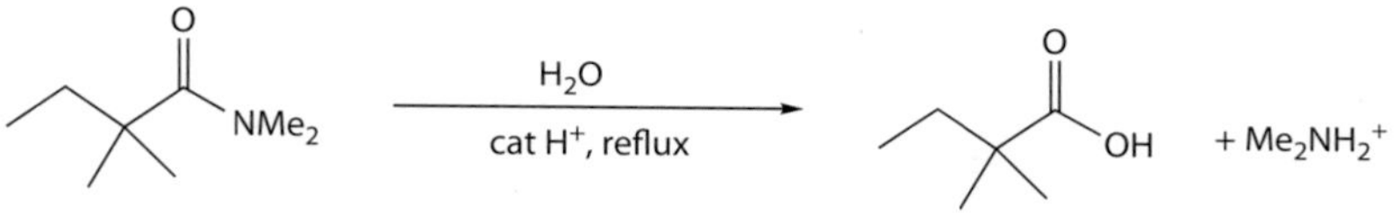
NMe2
H2O
cat H+, reflux
OH
+ Me2NH2+

20.131 Explain why it is relatively easy to convert succinic acid to the corresponding anhydride, but it is difficult to convert 1,8-ocatnedioic acid to the corresponding anhydride. Draw both anhydrides as part of your answer.

20.132 Give the major product formed when cyclooctene is treated with
(a) ozone at –78°C
(b) hydrogen peroxide
(c) an excess of ethanol with an acid catalyst

20.133 Write out the mechanism for the following transformation:

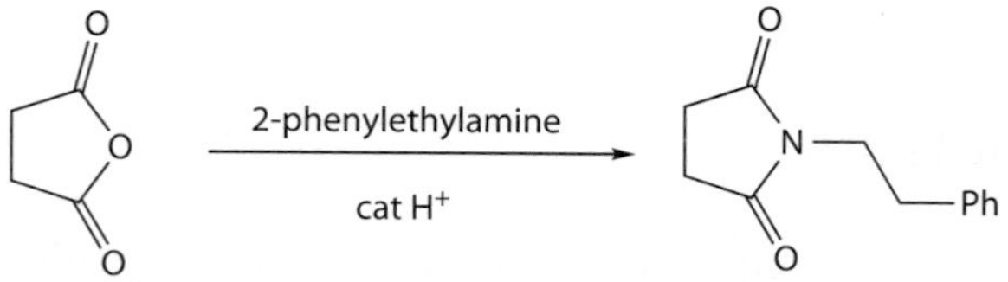
2-phenylethylamine
cat H+
N
Ph

20.134 When succinic anhydride is heated with ethanol in the presence of an acid catalyst, a new product is formed. Suggest that product based on a mechanism that is consistent with the anhydride, the alcohol, and the acid catalyst.

20.135 Draw the products that result from the reaction of methanesulfonyl chloride with each of the following:
    (a) ethanol
    (b) diethylamine
    (c) cyclohexanol
    (d) aqueous acid

20.136 The sulfonate salts of long-chain fatty acids can be used as detergents. Draw the structure of octadecanesulfonic acid. Now draw the product that is formed when this sulfonic acid reacts with sodium hydroxide.

20.137 We have seen the formation of sulfonate esters from sulfinic acids. Sulfuric acid (A) can form esters as well—either the monoesters (B) or the diester (C), which is known as a dialkyl sulfate. Esters of phosphoric acid (D) play a prominent role in biology. The phosphate linkages in nucleotides (see Chapter 28, Section 28.5) are important. Using D as a template, draw the mono- and diesters of phosphoric acid using cyclopentanol as the alcohol component. Take each of the phosphate esters you have drawn and draw the product you will obtain when they are treated with a base such as sodium hydride (NaH).

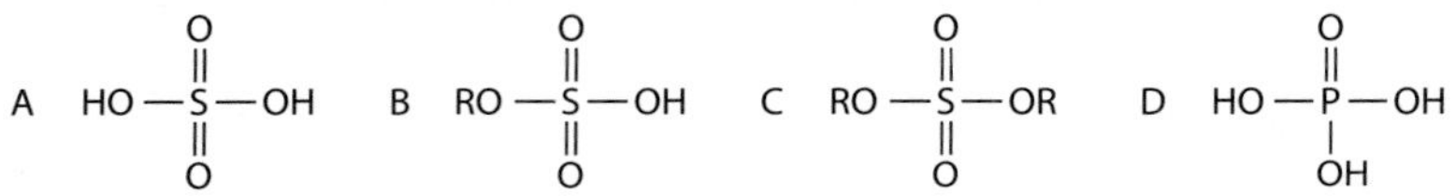

20.138 While alcohols are converted to alkyl sulfates in the presence of sulfuric acid or a derivative, diols such as *cis*-1,2-cyclohexanediol form a dialkyl sulfate with a different structure. Speculate on the structure of the dialkyl sulfate derived from this diol.

20.139 What is the product formed when the phosphorus compound with the structure $(EtO)_2P(=O)Cl$ reacts with ethanol? If this product reacts with dimethylamine, speculate on the structure.

20.140 Draw the structure of the 5′-triphosphate derivative of adenine, known as 5′-ATP.

Synthesis Problems

Do not attempt these problems until you have read and understood Chapter 25.
20.141 Show a method for the preparation of *N*-methylsuccinimide from succinic acid and provide all chemical steps and reagents.

20.142 Beginning with butanesulfonic acid, show how you would prepare
    (a) butanesulfonyl chloride
    (b) ethyl butanesulfonate
    (c) butanesulfonamide
    (d) sodium butanesulfonate

20.143 Show a complete synthesis for each of the following from the indicated starting material:

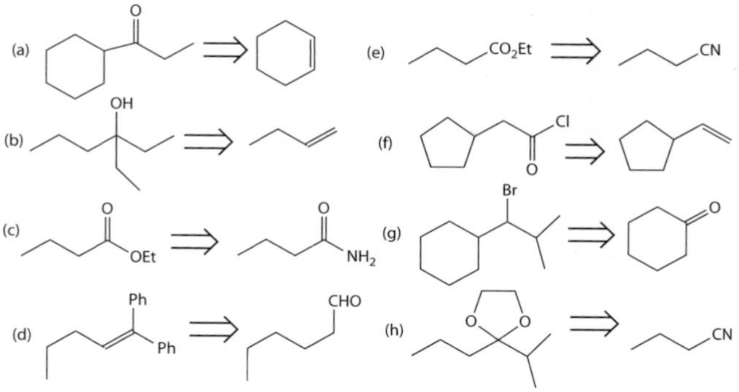

Spectroscopy Problems

Do not attempt these problems until you have read and understood Chapter 14.

20.144 An attempt to prepare pentanamide from ethyl pentanoate led to a mixture of products. Two molecules were separated: the ester and the amide. Describe the spectroscopic differences in the IR spectrum and proton NMR spectrum that will allow you to distinguish them.

20.145 Why does the hydrogen of the COOH unit appear so far downfield in the proton NMR spectrum relative to the hydrogen of an OH unit? Why is the position of this signal listed as 12–20 ppm rather than a single position?

20.146 The IR spectrum of an amide shows two "carbonyl" signals called the amide I and amide II bands. List typical values for each signal and briefly explain why the two signals appear.

20.147 Describe how to use the signals in the 1500–2000 cm^{-1} region of the IR spectrum to distinguish among a carboxylic acid, an ester, an acid chloride, and an anhydride (if this is possible). Can you use proton NMR spectroscopy to distinguish these molecules? Why or why not?

20.148 Describe how to use IR spectroscopy to distinguish among butanamide, *N*-methylbutanamide, and *N,N*-dimethylbutanamide.

20.149 You have two bottles containing liquids; you know one is hexanoic acid and the other is ethyl butanoate, but the labels have fallen off. The power is out, so you have no access to IR or NMR instruments. Devise a simple and unambiguous chemical test that will allow you to identify each unknown.

20.150 When you react ethylmagnesium bromide with ethyl butanoate, a ketone and a tertiary alcohol are possible products. Draw the structure of both products. Describe what you would see in the IR spectrum *and* the proton NMR spectrum *and* the mass spectrum that will allow you to distinguish these two products.

20.151 The reaction of *n*-butyllithium and carbon dioxide leads to pentanoic acid after hydrolysis. When you examine your product for confirmation that the reaction worked, what distinguishing features would you look for in the IR spectrum and in the proton NMR spectrum?

20.152 When butanenitrile reacts with methyllithium, a ketone is formed after hydrolysis of the iminium salt product. What would you look for in the IR spectrum to confirm that the nitrile has reacted completely? What would you look for in the IR to confirm that a ketone has been formed?

20.153 For a molecule with the formula $C_{12}H_{17}NO$, determine the structure, given the following theoretical spectral data:

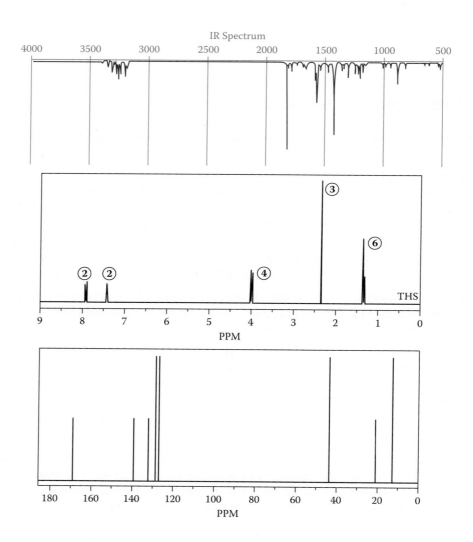

20.154 For a molecule with the formula $C_6H_{13}NO$, determine the structure, given the following theoretical spectral data:

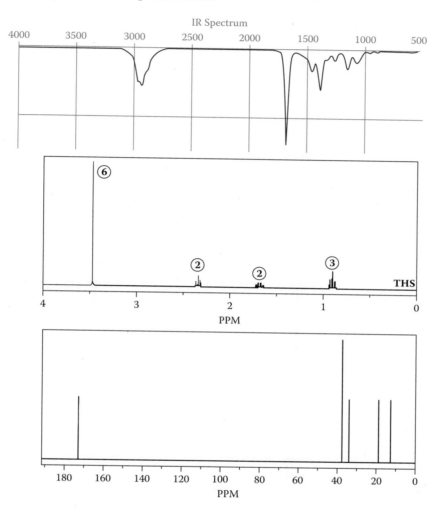

20.155 For a molecule with the formula $C_5H_{10}O$, determine the structure, given the following theoretical spectral data:

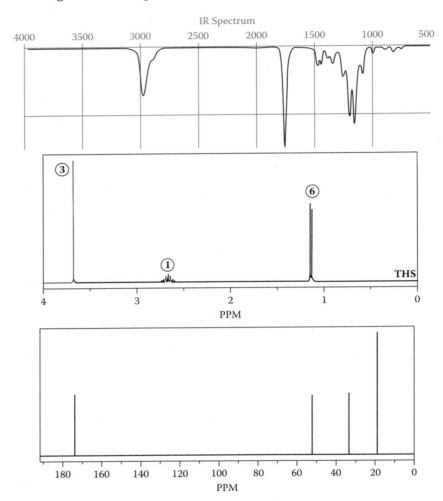

20.156 For a molecule with the formula $C_{13}H_{16}O_4$, determine the structure, given the following spectral data:

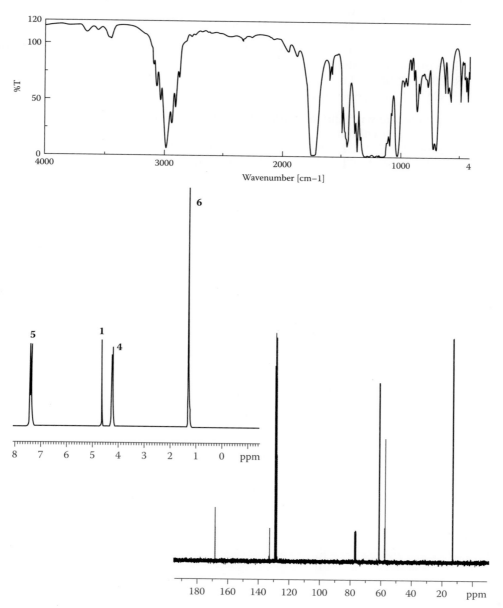

20.157 For a molecule with the formula $C_6H_8O_4$, determine the structure, given the following spectral data:
IR: 3057–2852, 1752, 1648, 1440, 1391, 1222–1164, 1008–994 cm⁻¹
¹H NMR: 3.73 (s, 2H), 6.20 (s, 6H) ppm

20.158 For a molecule with the formula $C_8H_{14}O_3$, determine the structure given the following spectral data:
IR: 2980–2881, 1813, 1747, 1471, 1389, 1020 cm^{-1}
^{1}H NMR: 2.66 (m, 2H), 1.24 (d, 12H) ppm

20.159 For a molecule with the formula $C_6H_{15}O_4P$, determine the structure given the following spectral data:
IR: 2986–2935, 2911–2874, 1273–1266, 1032, 976 cm^{-1}
^{1}H NMR: 4.11 (q, 6H), 1.35 (t, 9H) ppm

20.160 For a molecule with the formula C_5H_9ClO, determine the structure given the following spectral data:
IR: 2966–2877, 1802, 1469, 1403, 1134, 1010, 986 cm^{-1}
^{1}H NMR: 2.76 (d, 2H), 2.22 (m, 1H), 1.0 (d, 6H) ppm

Aromatic Compounds and Benzene Derivatives

21

In Chapter 5 (Section 5.10), a molecule called benzene (**1**) was identified as a special type of hydrocarbon. This chapter will discuss why benzene is a special hydrocarbon and also discuss derivatives formed by replacing the hydrogen atoms of **1** with substituents and/or functional groups. Hydrocarbons related to benzene have two, three, or more rings fused together (polycyclic compounds). The unifying concept of all these molecules is that they are aromatic, which means that they are especially stable with respect to their bonding and their structure. The discussion will begin with benzene.

To begin this chapter, you should know the following:

- **the concept of resonance and resonance stability (Chapter 5, Section 5.9.3)**
- **the structure of benzene (Chapter 5, Section 5.10)**
- **nomenclature and structure of alkenes, alkyl halides, alcohols, amines, aldehydes, and ketones (Chapter 5, Sections 5.1, 5.6, 5.9)**
- **nomenclature and structure of carboxylic acids (Chapter 5, Section 5.9.3; Chapter 16, Section 16.7)**
- **nomenclature and structure of carboxylic acid derivatives (Chapter 16, Section 16.7; Chapter 20, Sections 20.1, 20.3–20.6)**
- **the structure and nature of a π-bond (Chapter 5, Section 5.1.1)**

- the fundamental reactivity of alkenes (Chapter 6, Section 6.4.5; Chapter 10, Section 10.1)
- the formation, stability, and reactivity of carbocations (Chapter 7, Section 7.4.1; Chapter 10, Section 10.2)
- E2 elimination reactions of alkyl halides (Chapter 12, Sections 12.1 and 12.3)
- E1 elimination reactions (Chapter 12, Section 12.4)
- how to draw resonance contributors (Chapter 5, Section 5.9.3; Chapter 16, Section 16.3)
- CIP rules for prioritizing substituents, groups, and atoms (Chapter 9, Section 9.3)
- how to recognize electron-releasing and withdrawing substituents (Chapter 5, Section 5.4; Chapter 6, Section 6.2; Chapter 11, Section 11.3)
- Brønsted–Lowry acid–base reactions of alkenes (Chapter 10, Sections 10.1 and 10.2)
- factors that regulate carbocation stability (Chapter 10, Section 10.2; Chapter 11, Section 11.4)
- how to identify a good leaving group (Chapter 11, Section 11.2.4; Chapter 16, Section 16.8)
- how to identify relative nucleophilic strength (Chapter 11, Section 11.3)
- characteristics and reactivity of good Lewis acids (Chapter 6, Section 6.5)
- rate of reaction (Chapter 7, Section 7.11)
- oxidation of alcohols (Chapter 17, Section 17.2)
- reduction of carbonyl derivatives (Chapter 19, Sections 19.2 and 19.3)

This chapter will discuss benzene and related aromatic compounds, including aromaticity and their special stability. Reactions of aromatic compounds will be discussed, including electrophilic aromatic substitution and nucleophilic aromatic substitution.

When you have completed this chapter, you should understand the following points:

- **Benzene is a simple hydrocarbon with a unique structure of carbon–carbon bonds of equal length and strength. Overlap and delocalization of the π-electrons in benzene, coupled with its cyclic nature and continuous array of sp² atoms, leads to special stability called aromaticity, and benzene is an aromatic molecule. A molecule is aromatic if it is cyclic, has a continuous and** contiguous array of sp^2 hybridized atoms, and has $4n + 2$ π-electrons (the Hückel rule). Monocyclic aromatic compounds are called annulenes. Both anions and cations derived from cyclic hydrocarbons can be aromatic if they fit the usual criteria.

- Benzene derivatives with an alkyl substituent are called arenes. There are special names for benzene derivatives bearing heteroatom substituents. These include phenol (OH), aniline (NH_2), anisole (OMe), benzoic acid (COOH), benzonitrile (CN), benzaldehyde (CHO), acetophenone (COMe), and benzophenone (COPh).

- Because it is aromatic, benzene does not react directly with reagents such as HBr or HCl, or even with diatomic bromine or chlorine. Benzene reacts with cationic species to give a resonance-stabilized carbocation intermediate, which loses a hydrogen to give a substitution product. This reaction is called electrophilic aromatic substitution. The most common method for generating reactive cations in the presence of benzene is to treat certain reagents with strong Lewis acids. Lewis acids or mixtures of strong acids can be used to convert benzene to chlorobenzene, bromobenzene, nitrobenzene, or benzenesulfonic acid.

- The reaction of benzene with a carbocation leads to an arene in what is known as Friedel–Crafts alkylation. The reaction of an alkyl halide with a strong Lewis acid gives a carbocation, which is subject to rearrangement. Friedel–Crafts alkylation reactions are subject to polyalkylation because the arene is more reactive than benzene.

- The reaction of benzene with an acylium ion leads to a phenyl ketone in what is known as Friedel–Crafts acylation. The reaction of an acid chloride with a strong Lewis acid gives an acylium ion, which is not subject to rearrangement.

- Substituents on benzene lead either to a mixture of *ortho/para* products or to a *meta* product. Substituents that give *ortho/para* products react faster than benzene and are called activators; substituents that give *meta* products react more slowly than benzene and are called deactivators.

- Activating substituents are electron releasing and have excess electron density attached to the ring. These include alkyl groups and heteroatoms with a negative charge or a δ− atom. Activating substituents lead to a resonance-stabilized intermediate with a positive charge on a carbon atom that is adjacent to an electron-releasing substituent, which stabilizes the intermediate and makes the reaction proceed faster.

- Deactivating substituents are electron withdrawing and have diminished electron density attached to the ring. These include groups with a positive charge or a δ+ atom. Deactivating substituents lead to a resonance-stabilized intermediate with a positive charge on a carbon atom that is adjacent to an electron-withdrawing substituent, which destabilizes the intermediate and makes the reaction proceed more slowly.

- Benzene reacts with hydrogen gas and a suitable catalyst to give cyclohexane derivatives and, in some cases, cyclohexene or cyclohexadiene derivatives.

- Benzene reacts with sodium metal in liquid ammonia and ethanol to give nonconjugated cyclohexadiene derivatives in what is known as the Birch reduction.

- Substituents such as alkene units, alkyne units, and carbonyls can be reduced by catalytic hydrogenation. Lithium aluminum hydride reduces many heteroatom substituents, including nitrile and acid derivatives.

- Polycyclic aromatic compounds such as naphthalene, anthracene, and phenanthrene give electrophilic aromatic substitution reactions. The major product is determined by the number of resonance-stabilized intermediates for attack at a given carbon and the number of fully aromatic rings (intact rings) in the resonance structures.

- Aniline reacts with nitrous acid to give benzenediazonium salts, which react with a variety of reagents via a substitution reaction. These reagents include cuprous salts, aqueous acid, iodide, hypophosphorous acid, and activated benzene derivatives.

- Nucleophilic substitution at the sp^2 carbon of a halo-benzene derivative does not occur unless high heat and pressure are used. Electron-withdrawing substituents on the benzene ring significantly lower the temperature required for the reaction. Nucleophiles for this nucleophilic aromatic substitution reaction include water, hydroxide, alkoxide, and amines.

- When halobenzene derivatives are heated with powerful bases such as sodium amide, deprotonation is followed by elimination to give a benzyne, which reacts rapidly with the nucleophile.

- A molecule with a particular functional group can be prepared from molecules containing different functional groups by a series of chemical steps (reactions). This process is called synthesis: The new molecule is synthesized from the old one (see Chapter 25).

- Spectroscopy can be used to determine the structure of a particular molecule and can distinguish the structure and functionality of one molecule when compared with another (see Chapter 14).

- Aromatic compounds are important in biological systems.

21.1 Benzene and Aromaticity

Benzene (**1**) is a hydrocarbon with the formula C_6H_6. It is the parent of a large class of compounds known as aromatic hydrocarbons. The structure and chemical reactivity of aromatic hydrocarbons are so unique that benzene derivatives are given their own nomenclature system. The discussion will begin with the unique structure of benzene.

1A   **1B**

21.1.1 Structure of Benzene: Aromaticity

The usual representation of benzene is **1A** or **1B**, but neither adequately describes its structure. If benzene looked like either **1A** or **1B**, it should have three longer carbon–carbon single bonds and three shorter C=C units. It would then be called cyclohexatriene. *It is not called cyclohextatriene, but rather* ***benzene***. The C–C bond length in ethane is 1.53 Å (153 pm), and it is 1.536 Å (153.6 pm) in cyclohexane. The bond distance for the C=C bond in ethene is 1.34 Å (134 pm). These data indicate that a C=C unit has a shorter bond distance than a C–C unit. In benzene, all six carbon–carbon bonds have a measured bond distance of 1.397 Å (139.7 pm), a value that lies between those for the C–C bond in an alkane and the C=C unit of an alkene. This means that the carbon–carbon bonds in benzene are *not* single bonds and are not alkene-type bonds where the π-electrons are localized between two carbons in a π-bond. Each carbon in benzene is sp² hybridized, which means that each has a trigonal planar geometry (see **1C**). The planar geometry is seen more clearly in molecular model **1F**. Each carbon atom has a p-orbital perpendicular to the plane of the carbon and hydrogen atoms, and there is one π-electron associated with each p-orbital.

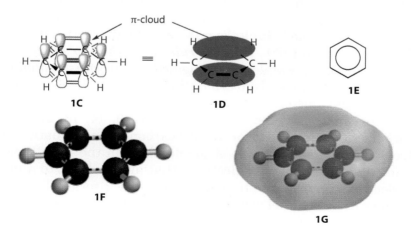

In the C=C unit of an alkene, two parallel p-orbitals on adjacent atoms share electron density to generate a π-bond (see Chapter 5, Section 5.1). If two adjacent and parallel orbitals can share electron density, then why cannot six adjacent and parallel p-orbitals? In **1C**, the p-orbitals are contiguous (one p-orbital on every carbon of the ring) and all are parallel for a total of six p-orbitals and six π-electrons. These six-electrons are delocalized by resonance in the six p-orbitals. Representation **1C** is drawn again as **1D**, showing the large π-cloud of electrons above and below the plane of the atoms in benzene. Because each p-orbital has an upper and a lower lobe, the π-cloud of benzene is both above and below the plane of the atoms.

In other words, electron density is concentrated above and below the plane of the carbon and hydrogen atoms. In the electron density map **1G**, the red area in the center of the ring indicates an area of high electron density, consistent with the π-cloud shown in cartoon form in **1C** and **1D**.

The point of this discussion is that the bonding in benzene is not properly described by either **1A** *or* **1B**, but rather by *both* structures. Each of these two resonance contributors is known as a Kekulé structure, named after August Kekulé (Germany; 1829–1896) and each is a six-membered ring with a C=C unit on alternate carbon atoms. When two Kekulé structures (**1**) are drawn, each structure is taken to be a **_resonance contributor_**, and *together* they represent the resonance delocalization of the π-electrons shown for **1D** and in **1G**. Benzene is sometimes drawn as a single six-membered ring with a circle in it (see **1E**), but it is more common that one of the Kekulé structures is used to represent benzene. *When one Kekulé structure is drawn, it is meant to represent the true resonance delocalized structure and the single structure is drawn for the sake of convenience.*

Benzene is resonance stabilized, but there is no charge. In benzene, six π-electrons are confined to a ring, and every carbon atom in that ring is sp^2 hybridized with a p-orbital attached. Further, the p-orbitals are contiguous and continuous. In other words, every carbon in the ring has a p-orbital and there are no intervening sp^3 atoms. These features lead to special stability for benzene, and benzene is said to be **_aromatic_**. Benzene is an **_aromatic hydrocarbon_**.

Aromatic hydrocarbons are well known, and to explain the special stability some rather simple rules have been developed to predict aromaticity. Perhaps the most important is called **Hückel's rule**, named after Erich Armand Arthur Hückel (Germany; 1896–1980). *Hückel's rule states that for planar, monocyclic hydrocarbons containing completely conjugated sp^2 hybridized atoms, the presence of (4n + 2) π-electrons leads to aromaticity ("n" is an integer in the series 0, 1, 2, 3, etc.).* As the value of "n" changes in the series 0, 1, 2, 3, 4, etc., a new series of numbers in generated by 4n + 2: 2, 6, 10, 14, 18, 22, etc. In other words, Hückel's rule states that for a hydrocarbon to be aromatic, the number of π-electrons must be equal to one of the numbers in the 4n + 2 series. Using Hückel's rule, benzene has six π-electrons that are confined to a planar ring, and six π-electrons corresponds to n = 1 in the 4n + 2 series. This means that six is a number in the Hückel series, 2, 6, 10, 14, etc. When a hydrocarbon fits these criteria and has a number of π-electrons that is equal to a number in the

4n + 2 series, it is aromatic and particularly stable. Aromatic compounds other than benzene will be discussed in later sections.

In previous chapters, resonance has been associated with stability (see Chapter 5, Section 5.9.3). An aromatic hydrocarbon is also particularly stable. This means that if electrons are not localized on a single atom, but rather delocalized on several atoms, that structure tends to be more stable. **The extent of this stability can be evaluated for benzene based on chemical reactivity.** From Chapter 10 (Section 10.2), it is known that the π-bond of an alkene reacts as a Brønsted–Lowry base with an acid HX (HCl, HBr, HI, etc.). This statement can be modified to say that the more easily a π-bond donates electrons to HX, the less tightly those electrons are retained by the alkene (it is a stronger base). The more tightly a π-bond holds its electrons, the less likely it is to donate them (it is a weaker base). If a molecule is more stable, it is less likely to donate electrons and it is a weaker base in reactions with HBr.

A crude comparison of the relative stability of cyclohexene (**2**), cyclohexadiene (**3**), and benzene (**1**; using one Kekulé structure implies both resonance forms **1A** *and* **1B**) is available by examining the reaction of each with HBr. The reaction of **2** and HBr is rapid and the product is carbocation **4**, which traps bromide to give bromocyclohexane **5** (see Chapter 10, Section 10.2). Cyclohexadiene (**3**) also reacts rapidly with HBr to give an allylic cation, **6**. This is a resonance-stabilized carbocation, and it reacts with bromide to give allylic bromide **7**. This reaction is more complicated due to the presence of the second C=C unit, which leads to resonance stability for **7**. Carbocations of this type will be discussed in detail in Chapter 23.

Returning to the problem at hand, both **2** and **3** react quickly and easily with HBr, so the π-bonds in each molecule are considered to be good Brønsted–Lowry bases in their reactions with mineral acids. *Benzene does not react with HBr, even with heating.* Benzene has six π-electrons; **2** has only two, so benzene is more electron rich relative to **2**. *The fact that benzene does not react suggests that it is too weak a base, which is a good indication that the six p-electrons are held tightly by the molecule and are not available for donation.* The poor basicity of benzene in a reaction with HBr is presented as evidence of the special stability of benzene, and the explanation for that stability is the resonance delocalization shown for **1C**.

21.1 Briefly discuss why **2** is much more reactive than **1** in catalytic hydrogenation.

21.1.2 Properties and Reactivity

Benzene is a liquid that was isolated from an oily condensate deposited from compressed illuminating gas in 1825 by Michael Faraday (England; 1791–1867). In 1834, Eilhard Mitscherlich (Germany; 1794–1863) prepared the same material by heating benzoic acid (**27**; see Section 21.2.2) with lime, and established the formula to be C_6H_6. He named the material "benzin," but Justus Liebig (Germany; 1803–1873) changed the name to benzol. In 1837, August Laurent (France; 1807–1853) proposed the name pheno (Greek: "I bear light") because it was isolated from illuminating gas. The name did not stick, but gave rise to the term phenyl when a benzene ring is used as a substituent (C_6H_5). (The phenyl substituent was introduced in Chapter 5, Section 5.10.2.)

Once the systematic nomenclature system established "ol" as a term for alcohols, benzol was changed to benzene so that benzol could be used for benzene rings that have an OH substituent (see Section 21.2.2). Benzene is a clear, nonpolar liquid with a boiling point of 80.1°C and a melting point of 5.5°C. It is readily soluble in most organic solvents, but not in water. Indeed, benzene has been used in the past as a solvent due to its stability to many reagents. *Benzene is now classified as a cancer suspect agent and should be handled only with special precautions and with great care.* Indeed, the shipment and use of benzene is largely avoided today due to this danger.

21.2 Functionalized Benzene Derivatives and a New Nomenclature System

Benzene is the parent of a large class of compounds known as aromatic hydrocarbons. Because of this, benzene derivatives are given a unique set of names in the IUPAC system, using a different suffix for each different benzene derivative. This section will briefly introduce the various types of functional groups attached to benzene and then discuss the nomenclature system that is used.

21.2.1 Alkyl Substituents on the Benzene Ring

Benzene derivatives may be viewed as molecules in which one or more hydrogen atoms have been replaced with heteroatoms, substituents, or functional groups. When a hydrogen atom on the benzene ring is replaced with an alkyl group such as methyl, ethyl, etc., the resulting molecule is called an **arene**. A simple method used to name an arene is to use the alkyl name of the substituent with benzene, so an arene becomes an alkyl benzene. Compound **8** is named methylbenzene, but its *common* name is toluene, a common constituent of paint solvents. Similarly, compound **9** is ethylbenzene. The alkyl substituent of an arene may have longer alkyl chains, as in **10**. However, compounds such as **10** pose a problem with respect to nomenclature. Is the parent compound benzene or is benzene a substituent? In general, if a benzene ring is attached to an alkane carbon chain of five or more carbon atoms, the benzene ring is treated as a substituent. Therefore, **10** is named 2-phenylpentane.

21.2 Draw the structures of (a) *n*-butylbenzene, (b) (3-phenyl-1-propyl) benzene, (c) cyclopentyl benzene, and (d) 2,5,-diphenyldecane.

Many arenes have two or more substituents on the ring: disubstituted, trisubstituted, etc. benzene derivatives. Three dimethylbenzene derivatives are known: 1,2-dimethylbenzene (**11**), 1,3-dimethylbenzene (**12**), and 1,4-dimethylbenzene (**13**). These are the IUPAC names, but the common name for the dimethylbenzenes is xylene. Xylenes are commonly found in paint solvents. For disubstituted benzene derivatives, a specialized *common* naming system— other than numbers—has been used for many years to designate on which carbon atoms the methyl groups appear.

In this common nomenclature, two substituents with a 1,2 relationship are said to be ortho; those with a 1,3 relationship are said to be meta, and those with a 1,4 relationship are said to be para. Using this system, **11** is called *ortho*-xylene (or *o*-xylene), **12** is called *meta*-xylene (or *m*-xylene), and **13** is called *para*-xylene (or *p*-xylene). The IUPAC system will be used most often, but *ortho*, *meta*, and *para* are so common that one must be familiar with this system. Indeed, *ortho*, *meta*, and *para* are used with many disubstituted benzene derivatives. *It is important to remember not to mix the IUPAC and common nomenclature systems.*

With three or more substituents on a benzene ring, the terms *ortho*, *meta*, and *para* are confusing and cannot be used, but the IUPAC system easily accommodates these names. Using the IUPAC system, arene **14** is 1,2,4-trimethylbenzene. Note that one carbon bearing a methyl is assigned C1, and the ring is numbered to give the lowest combination of numbers (1,2,4 rather than 1,2,5 or 1,3,6).

21.3 Draw the structures of *o*-diethylbenzene and of *m*-diisopropylbenzene.

21.4 Draw the structures of (a) 2,3-dimethylethylbenzene, (b) 3,4,5-triethylisopropylbenzene, and (c) 3-ethyl-4-methyl-(1,1-dimethylethyl)benzene.

There are hydrocarbon substituents other than the simple alkyl fragments just discussed. Alkene and alkyne groups may be attached to a benzene ring. In **15**, an ethenyl group is attached that is known as ethenylbenzne, but the common name is styrene. Compound **16** is *E*-1,2-diphenylethene, but it is commonly called *trans*-stilbene. Compound **17** is simply known as 1,2-diphenylethyne (or diphenyl acetylene, although it has the common name of tolane).

21.5 Draw the structure of *cis*-stilbene.

21.6 Draw the structure of (3,4-diethenyl)phenylcycloheptane.

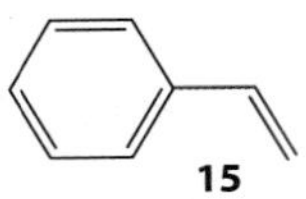

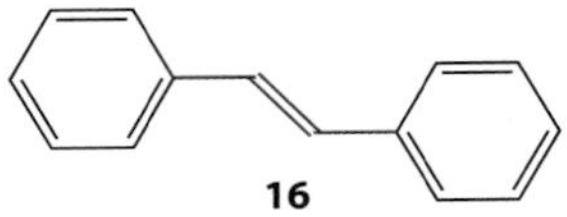

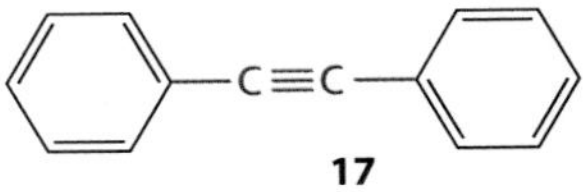

Many different heteroatoms or functional groups may be attached to a benzene ring, and the presence of a certain functional group often imparts a unique name to that family of compounds. Perhaps the simplest heteroatom derivatives are halobenzenes such bromobenzene (**18**). The halogens do not constitute a formal functional group with respect to nomenclature, but rather are treated as substituents. Therefore, the halobenzenes are named fluorobenzene, chlorobenzene, bromobenzene, and iodobenzene.

21.2.2 Functional Groups on the Benzene Ring

21.7 Draw the structures of *m*-dibromobenzene, 3-fluoroiodobenzene, and hexachlorobenzene.

Naming a molecule with an OH unit on a benzene ring requires a variation in the usual alcohol nomenclature used previously in Chapter 5 (Section 5.6.3). Rather than name compound **19** hydroxybenzene, where hydroxy is used when the OH group is a substituent, the OH is the dominant group and the "ol" ending for an alcohol from Chapter 5 is used. Therefore, the prefix is modified to accommodate benzene by dropping the "e" and replacing it with "ol"; thus, **19** *is named benzenol using the IUPAC system.* The common name of **19** is phenol, and its use is so widespread that this name is widely accepted. Phenol was called carbolic acid before the current system of nomenclature took effect. It was the first material used (in the late nineteenth century) to prevent infections in the operating room during surgery. It is still used as a disinfectant.

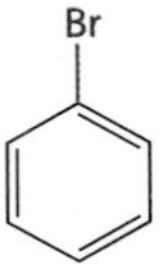

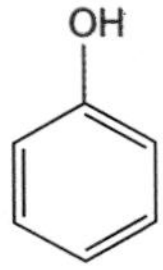

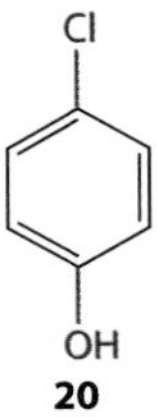

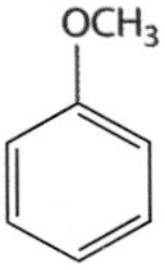

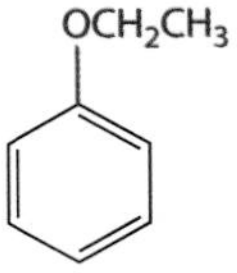

When substituents are on the ring, the carbon bearing the OH group is C1 and that derivative is named as a benzenol. Therefore, **20** is 4-chlorobenzenol. In common nomenclature, phenol is used as the parent name for this class of compounds, so **20** is 4-chlorophenol.

The ether unit is another oxygenated functional group that can be attached directly to the benzene ring and, in most cases, the OR is treated as an alkoxy substituent (methoxy, ethoxy, butoxy, etc.). Therefore, **21** is called methoxybenzene and **22** is ethoxybenzene. There are also common names for these compounds, so **21** is known as anisole and **22** is phenetole.

21.8 Give the structures for 3,5-dimethylbenzenol, 2-nitrobenzenol, and 3,4,5-trichlorobenzenol.

21.9 Draw the structures of 2,6-diiodophenol, 1,2-dimethoxybenzene, and *p*-iodoanisole.

When the group on the benzene ring is an amine unit such as $-NH_2$, the nomenclature is similar to that used for amines in Chapter 5 (Section 5.6.2). Compound **23** is named benzeneamine, but the common name is aniline, which is also an acceptable name. *The name aniline or benzeneamine defines the class of compounds bearing an $-NR_2$ unit on a benzene ring.* Substituents may appear on the benzene ring, as in 4-methylaniline (**24**; 4-methylbenzeneamine), or on the nitrogen, as in *N*-ethylaniline, **25** (*N*-ethylbenzeneamine). Note the distinction between a substituent on the benzene ring (indicated by a number) and one on the nitrogen (indicated by *N*-). It is also possible to convert an amine to an amide (see Chapter 20, Section 20.6) and **26** is known as *N*-acetylaniline (*N*-acetylbenzeneamine; note that the common name is acetanilide).

21.10 Draw the structures of 3,5-dichloroaniline, *N*-ethyl-*N*-methylaniline, and 4-methyl-*N*-propanoylaniline.

23   24   25   26   27   28

A molecule bearing the COOH (carboxyl) unit (see Chapter 16, Section 16.4) is a carboxylic acid, and the name of **27** must reflect this fact. Compound **27** is benzoic acid, the parent structure for this class. The usual acid derivatives, such as esters (benzoates), acid chlorides (benzoyl chlorides), and amides (benzamides), are all known, prepared by methods discussed in Chapter 20. There are also nitrile derivatives, and **28** is named benzonitrile.

21.11 Draw the structures of (a) 3-bromobenzoic acid, (b) benzyl 4-ethylbenzoate, (c) *N*,3,5-trimethylbenzamide, and (d) 2,6-dichlorobenzonitrile.

An aldehyde or a ketone unit may be attached to a benzene ring. In **29**, the CHO unit is attached directly to the benzene ring. Using the aldehyde nomenclature from Chapter 16 (Section 16.2.2), the "e" ending is dropped from benzene and replaced with "al," so **29** is named benzenal. The common name is benzaldehyde, and this name is widely used. Indeed, benzaldehyde has been used previously in several reactions in this book. Benzaldehyde is often abbreviated as PhCHO (Ph is the phenyl ring C_6H_5 and CHO is shorthand for the aldehyde unit).

Acetophenone is the common name for **30**, but the IUPAC system developed in Chapter 16 (Section 16.2.1) for ketones identifies the longest chain as two carbons and the C=O unit is at C1. Therefore, **30** is 1-phenylethanone. The normal IUPAC system accommodates ketones that have a longer alkyl fragment attached to the carbonyl. The phenyl unit is treated as a substituent in the IUPAC system. In the case of **31**, the carbonyl will receive the lowest possible number (in this case, C1), so the name is 1-phenyl-1-butanone. When the alkyl group is methyl, acetophenone is used more often than 1-phenylethanone for **30**, but the 1-phenyl-1 system is used for all other alkyl ketones of this type. An older system of nomenclature is available for such compounds, and **30** has been named methyl phenyl ketone and **31** would be phenyl propyl ketone. The IUPAC system will be used exclusively, except for acetophenone and one other derivative (see **32**). When the benzene ring of a ketone is functionalized, the nomenclature rules are straightforward, numbering the substituent relative to the carbonyl-bearing carbon of the benzene ring.

21.12 Give the names and draw the structures for ketones where the methyl group in 30 is replaced with (a) ethyl and (b) pentyl.

21.13 Draw the structure of benzyl propyl ketone and then give the IUPAC name using normal rules for naming an aliphatic ketone (Chapter 16, Section 16.2.1).

21.14 Draw the structure of 1-(2-chloro-4-methylphenyl)-4-bromo-1-pentanone.

A special case arises when two phenyl rings are attached to a carbonyl, as in **32**. This compound is named benzophenone (if named as an aliphatic ketone, the name would be diphenylmethanone). When both benzene rings of benzophenone are substituted, the name becomes a bit more complicated. The substituent in one ring is numbered 1,2,3, etc., but the substituents in the other ring are numbered with a prime, such as 1′,2′,3′, etc. Compound **33** is 4′-bromo-2,2′,

3-trimethylbenzophenone. Only a few simple examples of such compounds will be used—for simplicity and because they are not often encountered in relatively simple transformations.

21.15 Draw the structure of 3,3-dimethyl-1-(4′-ethylphenyl)pentanone.

It is obvious that more than one functional group may be attached to a benzene ring. In some cases, the presence of two or more functional groups leads to a special name for that compound. One such case occurs when an benzene ring contains two hydroxyl groups. Each of these compounds has an IUPAC name based on the rules just discussed, but many have a unique common name. The *ortho* compound (**34**) is named 1,2-benzenediol, but its common name is catechol. The *meta* compound is named 1,3-benzenediol (resorcinol) and the *para* compound is named 1,4-benzenediol (hydroquinone).

Three compounds also have two carboxyl groups: the dicarboxylic acids. 1,2-Benzenedioc acid (phthalic acid, **35**) has two carboxyl groups with an *ortho* relationship. In 1,3-benzenedioc acid (isophthalic acid), the carboxyl groups are *meta,* and in 1,4-benzenedioc acid (terephthalic acid) the carboxyl groups have a *para* relationship The dialdehyde compounds have related names: 1,2-benzenedial (phthalaldehyde **36**; also phthalic dicarboxaldehyde), 1,3-benzenedial (isophthalaldehyde; also isophthalic dicarboxaldehyde), and 1,4-benzenedial (terephthalaldehyde; also terephthalic dicarboxaldehyde).

21.16 Draw the structures of (a) 5-ethyl-1,3-benzenediol, (b) 4,5-dichlorophthalic acid, and (c) 2-ethyl-3-methylterephthaldehyde.

21.3 Electrophilic Aromatic Substitution

Benzene is particularly stable because of resonance, which means that it is not as reactive as other molecules containing π-bonds (i.e., alkenes and alkynes). This is clearly demonstrated by mixing benzene with HBr, where there is no reaction at normal temperatures. In other words, benzene is a very weak base due to the extra stability imparted by resonance (it is aromatic). Recall from Chapter 10 (Section 10.4.1) that cyclohexene reacts rapidly with Br_2 to give a *trans*-dibromide. It is known that benzene does not react directly with bromine, so it is easy to conclude that it is a weak Lewis base as well as a weak Brønsted–Lowry base. How can benzene be made to react? Can the bromine reagent be modified so that it can react with a weak Lewis base such as benzene? It is known that diatomic bromine reacts with a Lewis acid such as ferric bromide to generate an ate complex Br^+ $FeBr_4^-$. If the reaction of bromine and ferric bromide generates Br^+, the question is raised: Can benzene react with Br^+? If benzene is mixed with Br_2 in the presence of ferric bromide ($FeBr_3$), there is a reaction and bromobenzene (**18**) is formed in good yield, along with HBr. The answer is clearly yes!

21.17 Draw the product formed when cyclohexene reacts with diatomic bromine in carbon tetrachloride as a solvent.

21.18 Draw the ate complex formed when Br–Br reacts with $AlCl_3$, where the Al is the Lewis acid and one bromine is the Lewis base. Draw the ate complex formed when diatomic bromine reacts with ferric bromide ($FeBr_4$).

That **18** is formed as a product is testament to the fact that benzene reacts with Br^+, but formation of Br^+ requires that diatomic bromine first react with a Lewis acid. Once ferric bromide and bromine react to form Br^+ $FeBr_4^-$, the Br^+ is very electrophilic and may react with even a weak electron-donating species. Formation of **18** is explained by a reaction in which benzene donates electrons to Br^+ to form a new σ-covalent C–Br bond. This means that benzene reacts as a Lewis base ***after*** the cation is formed because benzene is not a strong enough Lewis base to react with Br_2 directly.

Analysis of the overall reaction shows that the bromine atom in **18** has replaced one H on the benzene ring. Therefore, this reaction is a substitution. Experiments have shown that the reaction proceeds by a ***cation intermediate*** (an electrophilic reaction), and it clearly involves an aromatic species such as benzene. If benzene reacts with bromine via Br^+ and there is a cation intermediate, it is reasonable to conclude that the hydrogen atom is lost in a subsequent step. This type of reaction is labeled ***electrophilic aromatic substitution*** and it accounts for most of the chemistry of benzene and its derivatives.

21.3.1 Generating Electrophilic Species and Aromatic Substitution

Bromine reacts with ferric bromide to form Br^+ $FeBr_4^-$, and benzene must react as a Lewis base with Br^+ in order to form **18**. It is known that an identical reaction occurs to give **18** when benzene reacts with Br_2 and the powerful Lewis acid aluminum chloride ($AlCl_3$). Reaction of Br_2 and $AlCl_3$ gives Br^+ $AlCl_3Br^-$, which then reacts with benzene. This example is provided to show that many Lewis acids may be used. This example also shows that Br^+ is formed despite the presence of chlorine atoms in $AlCl_3$.

Recall from Chapter 10 (Section 10.4.1) that cyclohexene reacts as a Lewis base with Br⁺ of Br–Br to form a bromonium ion, **37**, which then reacts with the nucleophilic bromide ion to form dibromocyclohexane, **38**. In Chapter 10, it was shown that a three-membered ring bromonium ion is favored rather than a carbocation due to back donation of electrons on the bromine atom to the developing positive charge. It is known that bromination of benzene proceeds by a cation intermediate. Remembering the cyclohexene transformation to **38**, will the reaction of benzene as a Lewis base with Br⁺ generate a carbocation or a bromonium ion?

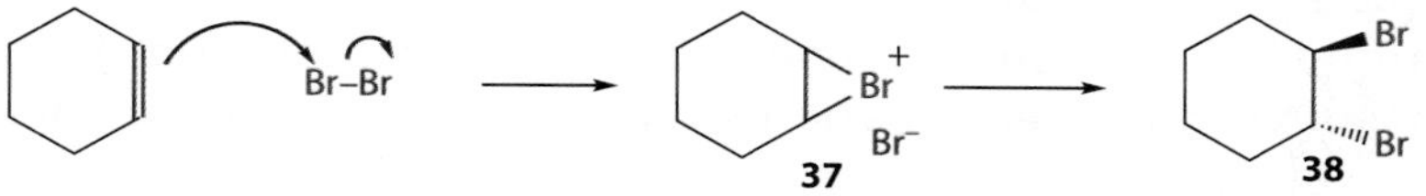

To probe this question, diatomic bromine is mixed with aluminum chloride (AlCl₃), a Lewis acid–Lewis base reaction forms the complex Br⁺AlCl₃Br⁻ as the product *before there is any reaction with benzene*. This cationic bromine atom is so reactive, however, that even a weak base like benzene will react. Benzene donates two electrons to Br⁺ to form a new C–Br bond, which disrupts the aromatic system of benzene and forms a carbocation, **40**.

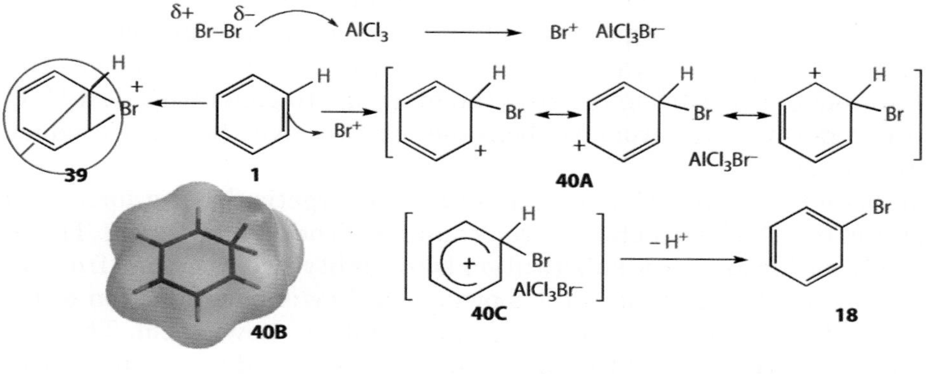

This high-energy intermediate is called a **Wheland intermediate**, named after George Willard Wheland (United States; 1907–1976). Such intermediates have also been called **Meisenheimer adducts**, after Jakob Meisenheimer (Germany; 1876–1934), σ-adducts, or **arenium ions**. The terms "Wheland intermediate" or "arenium ion" will be used most of the time in this book.

In the reaction with benzene, experiments show that the product is a carbocation and not a bromonium ion. A bromonium ion such as **39** is not the intermediate. The aromatic stability of benzene is disrupted to form carbocation **40**; it is an unstable intermediate, but **40** *is somewhat stabilized, as shown in **40A**; this leads to sufficient stability so that the reaction generates this intermediate*. The intact benzene ring is much more stable than the cyclohexadienyl carbocation, which is *resonance stabilized but not aromatic*. The electron potential model of arenium ion **40B** indicates **yellow-orange** areas over carbons 3 and 5 and light **blue** over carbons 2, 4, and 6. The **blue** color indicates less electron density (positive charge) relative to the **red** and **yellow-orange** regions. This electron potential map indicates that positive charge is distributed around the ring consistent with delocalization of the charge by resonance. The highest concentration of positive charge is at C2, C4, and C6 relative to the sp^3 carbon that bears the halogen.

Formation of arenium ion **40** requires disruption of the aromatic benzene ring. The special stability of the aromatic ring due to its aromatic nature is used to explain why benzene does not react directly with Br$_2$. The statement that benzene is too weak a Lewis base really states that the activation energy required to generate a cationic intermediate such as **40** directly from Br$_2$ is so high that the reaction does not occur. Formation of Br$^+$ and subsequent reaction with benzene really means that the activation energy for this new reaction is much lower and that the energy required for formation of the arenium ion intermediate is sufficiently low for reaction to occur at a reasonable rate.

Arenium ion **40** is a resonance-stabilized intermediate with the three resonance contributors, shown in **40A**, and it is sometimes written in an abbreviated form (**40C**) to represent the resonance. Note that **40** is a highly reactive intermediate, but what reactions are available in this reaction? It is known that the final product is **18**, which requires loss of hydrogen atom from **40**. There is no suitable nucleophile and, as pointed out in Chapter 12 (Section 12.4), carbocations undergo elimination reactions in the presence of a weak base. In this case, the aluminate anion (AlCl$_3$Br$^-$) probably removes the indicated proton (on the carbon bearing the bromine) to generate a new C=C unit, which regenerates the aromatic benzene ring to give the final product, bromobenzene (**18**).

This is formally an E1 reaction, and the energetic driving force for losing this proton (for the E1 reaction) is reformation of the aromatic ring. This means that electrophilic aromatic substitution is in reality two reactions from a mechanistic viewpoint. The first is the Lewis acid–Lewis base reaction of benzene with Br$^+$, which is generated from bromine and the Lewis acid. The second is an E1-type reaction to give the substitution product and it is also an acid–base reaction.

The reaction of benzene and bromine in the presence of aluminum chloride to give bromobenzene is an ***electrophilic aromatic substitution reaction***. The mechanism of this reaction is formally a two-step process: (1) Lewis base–Lewis acid reaction between benzene and Br$^+$, followed by (2) an E1 reaction that leads to loss of a proton to regenerate the aromatic benzene ring. Many other Lewis acids can be used, including aluminum bromide (AlBr$_3$), ferric bromide (FeBr$_3$), or ferric oxide (Fe$_2$O$_3$). Indeed, old lab experiments for this reaction suggest that adding a rusty nail (rust is ferric oxide) to benzene and then slowly adding bromine will give bromobenzene. As long as a suitable Lewis acid is present to generate Br$^+$ from Br$_2$, the subsequent reaction with benzene is straightforward. This reaction is not limited to bromine; if it is possible to form a cationic species X$^+$, electrophilic aromatic substitution should occur. The nature of X$^+$ may be varied, including extending the reaction observed with diatomic bromine to diatomic chlorine. The reaction of chlorine and benzene in the presence of $AlCl_3$ generates Cl^+ and subsequent reaction with benzene gives chlorobenzene.

21.19 Draw the product formed when Cl_2 reacts with $AlCl_3$. Draw the complete reaction of benzene, diatomic chlorine, and aluminum chloride, showing the arenium ion intermediate and the structure of the product.

Other X^+ cations can be generated, including nitrogen, sulfur, and carbon electrophiles. The nitronium ion NO_2^+ is formed when nitric acid is mixed with the stronger sulfuric acid; nitric acid functions as a base. The subsequent acid–base reaction generates the nitronium ion (NO_2^+), the hydrogen sulfate counterion and the hydronium ion as shown in the illustration. Note that this is an acid–base reaction in which *nitric acid is a base and sulfuric acid is the acid,* as shown in the reaction. Loss of water from the protonated nitric acid (the conjugate acid from the initial reaction) gives the nitronium ion. Once generated, benzene reacts as a Lewis base with the nitronium ion. The nitronium ion is classified as an electrophile in this reaction because it reacts with carbon. The reaction is the same as the reaction between benzene and Br^+ before and gives resonance-stabilized arenium ion intermediate **41**. The second step is the E1 reaction and loss of a proton to give nitrobenzene **42**.

The sulfonation of benzene by concentrated sulfuric acid is another reaction that involves the reaction of benzene with an electrophilic species. When benzene reacts with sulfuric acid, it attacks the electrophilic sulfur atom, as shown in the illustration, to give arenium ion intermediate **43**. This intermediate loses a proton via the E1 reaction, but it also loses a molecule of water to give benzenesulfonic acid (**44**). (See Chapter 16, Section 16.9, for an introduction to sulfonic acids.) Concentrated sulfuric acid reacts with the water byproduct, which effectively removes this product from the reaction, and drive the reaction toward **44**. In other words, concentrated sulfuric acid acts as a drying agent. If "fuming sulfuric acid" (which is simply concentrated sulfuric acid saturated with sulfur trioxide, SO_3) is used rather than sulfuric acid, the reaction proceeds faster to give **44** and with fewer problems. This latter reaction depends on the fact that the sulfur atom in SO_3 is more electrophilic than the sulfur atom in sulfuric acid.

21.20 Draw all resonance forms for 43.

21.21 Suggest a mechanism for the conversion of benzene to benzene-sulfonic acid with SO₃.

21.22 Draw 3-chloronitrobenzene and 3,4-diethylbenzenesulfonic acid.

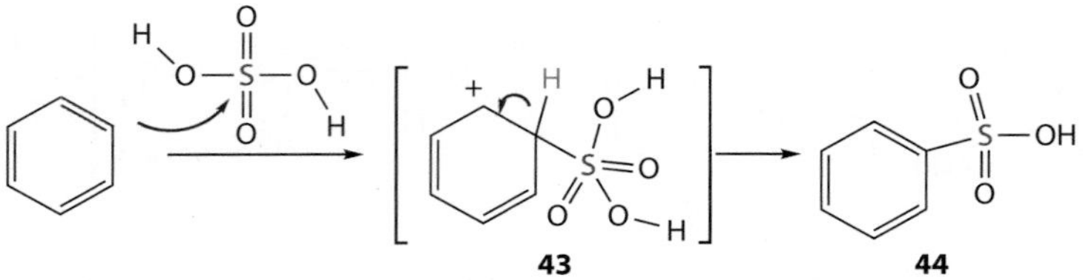

21.3.2 Friedel–Crafts Alkylation

In the preceding section, benzene reacted with cations to form substituted benzene derivatives. The cations of interest include Br⁺, Cl⁺, the nitronium ion, and sulfur trioxide or sulfuric acid, which react as electrophiles. In principle, benzene may react with any cation, including carbocations, once that cation is formed. Carbocations are generated by several different methods; they react with nucleophiles, as described for reactions of alkenes with acids such as HX (Chapter 10, Section 10.2) and for S_N1 reactions (Chapter 11, Section 11.4). If benzene reacts with a carbocation, a new carbon–carbon bond is formed, and electrophilic aromatic substitution will give an arene. The reaction of benzene and its derivatives with carbocations is generically called the **Friedel–Crafts reaction**, after the work of French chemist Charles Friedel (France; 1832–1899) and his American protege, James M. Crafts (1839–1917). The reaction takes two fundamental forms: Friedel–Crafts alkylation and Friedel–Crafts acylation. Both variations will be discussed, beginning with the alkylation reaction.

Friedel–Crafts alkylation involves formation of an alkyl carbocation in the presence of benzene or a benzene derivative. In previous chapters, these reactive intermediates are formed by reaction of an alkene with an acid (Chapter 10, Section 10.2), by reaction of an alcohol with an acid (Chapter 11, Section 11.7.1), or by the ionization of an alkyl halide in aqueous solution (Chapter 11, Section 11.4). For reactions with benzene and its derivatives, an alternative method that can generate primary carbocations that are particularly difficult to form by the other methods named is used to form carbocations.

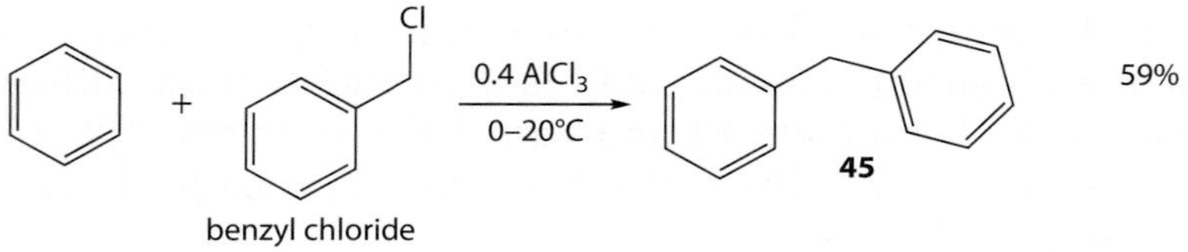

A simple example illustrates the reaction. When benzene reacts with benzyl chloride in the presence of 0.4 equivalents of $AlCl_3$, diphenylmethane (**45**) is isolated in 59% yield.[1] If this reaction proceeds by electrophilic aromatic substitution, then the sp^3 carbon of benzyl chloride is a precursor to a carbocation. To form a carbocation from benzyl chloride, the chlorine atom must react as a Lewis base with $AlCl_3$ to form $PhCH_2^+$ $AlCl^-$. Benzene reacts with this carbocation via electrophilic aromatic substitution in the same manner as the reaction with Br^+ in the previous section to form an arenium ion intermediate (see **40**) to give **45**.

21.23 **Write out the benzyl carbocation $PhCH_2^+$ and all of its resonance contributors.**

21.24 **Suggest a mechanism for reaction of $PhCH_2^+$ from benzene and benzyl chloride and formation of 45.**

If the preparation of **45** is typical, Friedel–Crafts alkylation appears to be just another example of electrophilic aromatic substitution where the reactive species is a carbocation. However, problems with a carbocation intermediate do not arise with other cations such as Br^+, Cl^+, or NO_2^+. If a stable carbocation such as the benzylic cation or a tertiary cation is the intermediate, the reaction is relatively straightforward. *If a primary or secondary carbocation is formed, however, rearrangement to a more stable cation may occur before the reaction with benzene.*

When 1-bromobutane reacts with aluminum chloride ($AlCl_3$) in the presence of benzene, the isolated product is **48**, *not* the expected *n*-butylbenzene (1-phenylbutane). If 1-bromobutane reacts with aluminum chloride, the expected product is the primary carbocation shown; however, to explain formation of **48**, benzene must react with a secondary carbocation. Therefore, the initially formed primary cation must rearrange to the more stable secondary cation, **46**, via a 1,2-hydride shift *before it reacts with benzene*. This type of 1,2-hydride shift was introduced in Chapter 10 (Section 10.2.2). Carbocation **46** reacts with benzene by the expected mechanism to give arenium ion **47**; once formed, loss of a proton by the E1 reaction gives **48**.

The first problem with Friedel–Crafts alkylation is therefore that alkyl cations are subject to rearrangement. *Assume that primary cations obtained from primary halides will always rearrange before they react with benzene.* Secondary halides lead to secondary cations, which may react directly or rearrange, depending of the substituent pattern. The reaction of 2-bromopropane and $AlCl_3$, for example, leads to isopropylbenzene. A second problem associated with Friedel–Crafts alkylation will be introduced in Section 21.4.

21.25 **Draw the structure of *n*-butylbenzene.**

21.26 **Draw the product of a reaction between benzene and 3-chloro-3-ethylpentane and $AlCl_3$ and between benzene and 1-bromo-2-methylpropane and $AlBr_3$.**

21.3.3 Friedel–Crafts Acylation

Another type of Friedel–Crafts reaction also involves a carbocation, but not the simple alkyl carbocations discussed in the previous section. When benzene reacts with butanoyl chloride (**49**) in the presence of $AlCl_3$, the isolated product is a ketone: butyrophenone (1-phenyl-1-butanone, **50**) in 51% yield.[2] It is known that benzene does not react with **49** without the presence of aluminum chloride. Clearly, the acyl unit has reacted with benzene to form the new carbon–carbon bond, with loss of the chlorine atom of **49**, but there must be a prior reaction between $AlCl_3$ and the acid chloride. This is an aromatic substitution and, based on knowledge of benzene, there must be an arenium ion intermediate. To form an arenium ion, benzene must react with a cationic species, which must arise from **49**. What is the nature of this cationic intermediate?

First, assume a Lewis base–Lewis acid reaction between **49** and aluminum chloride, as with the other reagents in this chapter. Specifically, the chlorine atom reacts as a Lewis base with the aluminum atom to form $AlCl_4^-$. When the O=C–Cl bond is broken, a new type of carbocation is formed: an acyl cation that is known as an ***acylium ion***. Specifically, when **49** reacts with $AlCl_3$, acylium ion **51** is formed, which is resonance stabilized. *An acylium ion is sufficiently stable so that it does not rearrange, but it is reactive enough to react with benzene.* In this case, acylium ion **51** reacts with benzene via electrophilic aromatic substitution as it would any other cation to give arenium ion **52**. Loss of a proton from **52** leads to the final product, 1-phenyl-1-butanone (**50**). The net result of this reaction is electrophilic aromatic substitution of a hydrogen atom on benzene with an acyl group, and it is called **Friedel–Crafts acylation**.

21.27 **Draw the acylium ion formed by the reaction of aluminum chloride with (a) acetyl bromide, (b) benzoyl chloride, (c) pentanoyl chloride.**

21.28 **Draw the product formed by the reaction of benzene and benzoyl chloride in the presence of aluminum chloride.**

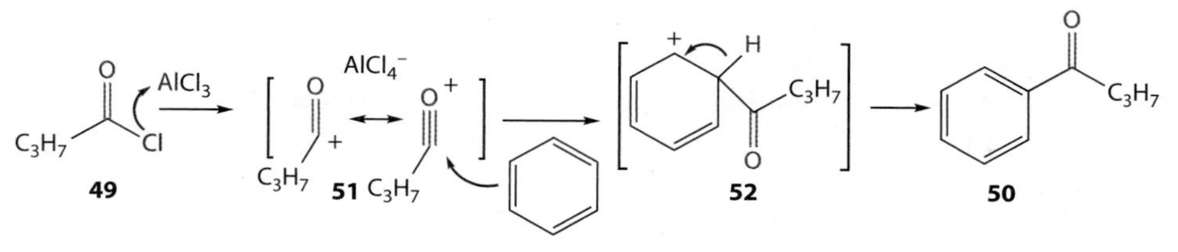

As noted before, an acylium ion is sufficiently stable that there is no rearrangement, and the carbonyl is attached directly to the benzene ring in **50**. This means that the alkyl chain on the other side of the carbonyl arises from the acid chloride and it may be a straight chain because there is no rearrangement. Remember that a primary alkyl cation will rearrange, so it is *not* possible to make straight-chain arenes via Friedel–Crafts alkylation. The fact that acylium ions do not rearrange, however, makes it possible to make straight-chain ketones. In Section 21.6.2 and in Chapter 19 (Section 19.4), methods to remove the carbonyl to prepare a straight-chain arene such as butylbenzene are described.

21.29 **Draw the product of the reaction of 3,3-dimethylpentanoyl chloride with AlCl$_3$ in the presence of benzene.**

21.4 Disubstituted Benzene Derivatives

Friedel–Crafts alkylation has the problem of cation rearrangement, but there is another problem with this reaction. When benzene reacts with 2-bromopropane and AlCl$_3$, it gives a mixture of **53** (1-methylethylbenzene or isopropylbenzene, also known as cumene), which is the expected product; however, it also gives the disubstituted product **54**. The latter may be the major product if an excess of 2-bromopropane is used. Therefore, the reaction with alkyl halides may lead to *polyalkylation* of the benzene ring, which is the second of the two problems noted for Friedel–Crafts alkylation. The only way to explain formation of **54** is via a Friedel–Crafts alkylation of the initially formed product **53** with 2-bromopropane. This result suggests that **53** must react more quickly with the carbocation derived from 2-bromopropane than does benzene. This point will be discussed in more detail later.

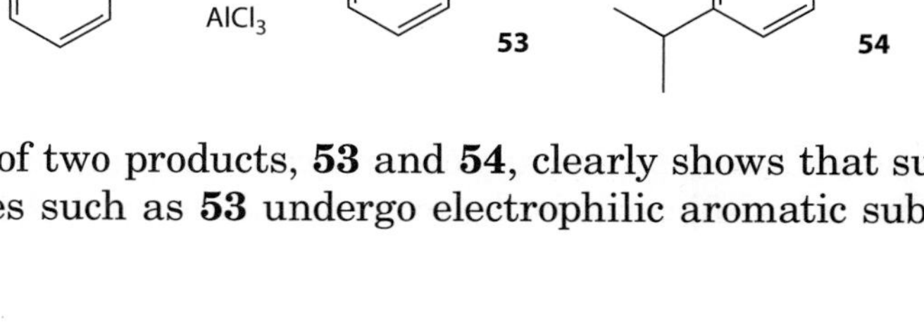

Formation of two products, **53** and **54**, clearly shows that substituted benzene derivatives such as **53** undergo electrophilic aromatic substitution. This raises another interesting question: Why does the second group add in the *para* position relative to the first one? In principle, substitution can occur at three sites: *ortho*, *meta*, and *para*. When a substituted benzene derivative reacts, it is possible to generate different isomers, so the issue of regioselectivity for this reaction must be addressed. In other words, **when one group is attached to the benzene ring, where will the second group be placed via electrophilic aromatic substitution?** If only one or two of these isomers are preferred in the reaction, as with the formation of **54**, then electrophilic aromatic substitution is *regioselective*. Another interesting experimental fact concerns Friedel–Crafts acylation. The reaction that gives ketone **50** does *not* give polyacylation. In other words, **50** does not react with acylium ion **51**. Why does **53** react with a carbocation to give **54**, but **50** does not react with **51**?

Before explaining the previous data, it is important to understand why **54** is formed in the Friedel–Crafts alkylation reaction. This means that the reactivity of benzene derivatives must be addressed. If **54** is formed by a reaction of **53**, then **53** must react with the intermediate carbocation *more quickly* than benzene. Why does **53** react more quickly than benzene? In addition, this discussion must address the question of why polyalkylation is a problem but polyacylation is not. The answers to these questions will also explain the regioselectivity of the reaction.

21.4.1 Activating Substituents

This discussion begins with an analysis of the reactivity of a simple benzene derivative such as toluene (**8**). In one experiment, toluene (**8**) is nitrated using HNO_3/H_2SO_4. *Three* products are isolated from this reaction: 43% of 2-nitrotoluene (**55**), 4% of 3-nitrotoluene (**56**), and 53% of 4-nitrotoluene (**57**).[3] There are three products, but clearly **55** and **57** are formed in nearly equal amounts, to the near exclusion of **56**. The reaction is regioselective. The 1,2 (*ortho*) and 1,4 (*para*) products constitute the major products with the 1,3 product (*meta*) being minor. There is a preference for the *ortho* and *para* products. **Why?** The *ortho* and *para* products must be formed more quickly than the meta product! Additional experiments show that the rate of reaction of toluene with the nitrating reagents is significantly *faster* than the identical reaction with benzene (see Chapter 7, Section 7.11, for an introduction to rate of reaction). Therefore, toluene reacts more quickly than benzene, and formation of **55** and **57** is faster than formation of **56**.

To explain these data, examine any differences in structure between toluene and benzene. The only difference is the presence of the methyl group. A methyl group is an alkyl substituent. How is the rate of reaction influenced by this substituent? To answer that, first compare the methyl group in toluene with the nitro group in nitrobenzene. In the experiment that reacts nitrobenzene (**40**) with bromine and ferric bromide, the major product of the reaction is the *meta* product 3-bromonitrobenzene (**58**), with only trace amounts of 2-bromonitrobenzene (**59**) and 4-bromonitrobenzene (**60**). Experimental results also show that this reaction is much *slower* than the nitration or bromination of benzene and almost 10^6 *slower* than nitration of toluene. This means that nitrobenzene reacts more slowly than benzene and much more slowly than toluene, and that formation of **60** is faster than formation of **58** and **59**.

These experiments show that toluene reacts faster than benzene to give primarily the *ortho* and *para* products, but that nitrobenzene reacts more slowly than benzene to give primarily the *meta* product. Toluene has a methyl group attached to the benzene ring and nitrobenzene has a nitro group. Is there a fundamental difference in these groups that will lead to such a fundamental difference in reactivity and regioselectivity? The question is addressed by examining the relative stability of the arenium ion intermediate formed in each reaction.

With respect to a positive carbon in an arenium ion, the methyl group is electron releasing and it will stabilize the positive charge (opposite charges on adjacent atoms attract). The nitro group has a positively charged nitrogen atom and is electron withdrawing, so it will destabilize a positive charge in an arenium ion (like charges on adjacent atoms). *Therefore, the results of these two experiments are explained by noting that when the electron-releasing group is attached to the benzene ring, the products are ortho/para and the reaction is faster than when compared to nitration of benzene. This contrasts with the presence of an electron-withdrawing group on the benzene ring, which gives a meta product and a reaction slower than the comparable bromination reaction of benzene.* These observations give the key to answering the questions that were posed previously for differences in rate and reactivity.

To explain electrophilic aromatic substitution of substituted benzene derivatives, a generic benzene derivative, **61**, is used in Figure 21.1, with a substituent X. This is used rather than a specific example in order to show the similarities and differences in reactivity for electron-releasing versus electron-withdrawing groups. The carbon of the benzene ring attached to the substituent is defined as the ***ipso carbon*** (•) in Figure 21.1. There are only three possible arenium ion intermediates for the reaction of any monosubstituted benzene derivative **61** with an electrophile such as Br^+: **62**, **63**, and **64**.

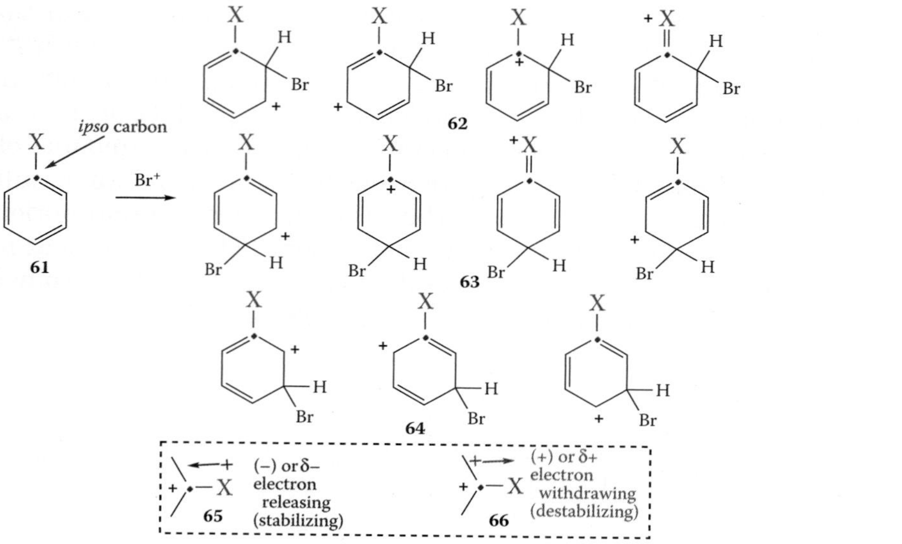

Figure 21.1 Arenium ion intermediates for electrophilic aromatic substitutions for *ortho*, *meta*, and *para* substitution.

Reaction at the *ortho* position generates arenium ion **62** with three resonance contributors within the six-membered ring, and if X has unshared electrons, there is one extra contributor with the positive charge on the *ipso* carbon (C=X⁺), as shown. Similarly, reaction at the *para* position generates arenium ion **63**, which also has three resonance contributors and one extra with a positive charge on the *ipso* carbon (when X has unshared electrons). These two intermediates contrast with reaction at the *meta* position, which gives arenium ion **64**, which has three resonance contributors, none of which has a resonance contributor with a positive charge on the *ipso* carbon.

Why is this difference important? If X is an electron-releasing group that contains a heteroatom such as O or N (unshared electrons, a negative charge, or polarized δ–), then the X group releases electrons to an adjacent positive center, as in **65**. Releasing electrons to a positive center will stabilize the charge and lead to a lower activation energy for the formation of that intermediate. In other words, that intermediate is easier to form and the reaction is faster. Remember that an alkyl group is also electron releasing with respect to the positive carbon. In arenium ions **62** and **63**, the positive charge resides on the carbon attached to the X group (the *ipso* carbon), so the electron-releasing effect will stabilize the charge, making those intermediates easier to form. The arenium ion intermediates are more stable (lower in energy); therefore, the activation energy for formation of these two intermediates is lower than formation of the intermediate for *meta* attack (**64**) because there is no resonance contributor for *meta* attack with a charge on the *ipso* carbon. If that intermediate is less stable and has a higher activation energy, it will form more slowly.

In other words, *the rate of reaction for generating the ortho and para products is faster than the rate for generating the meta product, so when X is electron releasing, the major products should be ortho and para.* Based on this simple analysis, **assume that the ortho and para arenium ions are about equal in energy; the ortho and para products should be formed in roughly equal amounts in the absence of other mitigating factors.** As noted in Figure 21.1, when the substituent X has unshared electrons (O, N, halogen), it is possible to draw a fourth resonance contributor if there is a resonance contributor with the positive charge on the *ipso* carbon (*ortho/para* attack). *The fourth resonance contributor is an indication of greater charge delocalization and greater stability for those arenium ions.* Therefore, when X is electron releasing, reaction at the *ortho* and *para* positions is favored over reaction at the *meta* position, and the reaction is faster than benzene because the arenium ions are more stable than the arenium ion derived from benzene (see **40** in Section 21.3.1).

If the X group in **61** (Figure 21.1) is electron withdrawing (a group with an atom that has positive charge or one that is polarized δ+), then reaction at the *ortho* or *para* position places a positive charge on the *ipso* carbon of **62** or **63**. However, the *ipso* carbon in those resonance contributors is proximal to a positive charge or a positively polarized atom. *When a positive charge resides on an atom that is adjacent to a positive charge (see 66), the repulsive interaction of two like charges will destabilize the arenium ion.* The net result is an electron-withdrawing effect that makes the carbocation less stable and more difficult to form (a higher activation barrier).

In other words, if arenium ions **62** and **63** have an electron-withdrawing group on the *ipso* carbon, the energy of those intermediates is increased, so they have a higher activation energy and their formation is much slower. For attack at the *meta* position, the proximity of the positive charge in the ring and that on the substituent X in **64** will increase the activation energy of the intermediate, but *that activation energy is lower than the energy of the arenium ions from ortho or para attack.* Therefore, when the X group is electron withdrawing, electrophilic aromatic substitution at the *meta* position will be faster than reaction at the *ortho* and *para* positions. Moreover, the reaction is slower than benzene because the arenium ion intermediates are higher in energy and more difficult to form.

Note that regardless of the nature of X, reaction at either the ortho or para position leads to a positive charge on the ipso carbon, but reaction at the meta position does not. These intermediates are common to all electrophilic substitution reactions, but the reaction is faster when X is electron releasing and slower when X is electron withdrawing. In addition, when X is electron releasing, formation of the ortho and para products is faster than formation of the meta product, so the major products of the reaction are ortho and para. When X is electron withdrawing, formation of the meta product is faster than formation of the ortho and para products, and the meta product is the major product.

Categorizing substituents (X in **61**) as electron releasing or electron withdrawing is the key to understanding whether a reaction is faster or slower than

Table 21.1
Electronic Characteristics of Groups in Electrophilic Aromatic Substitution

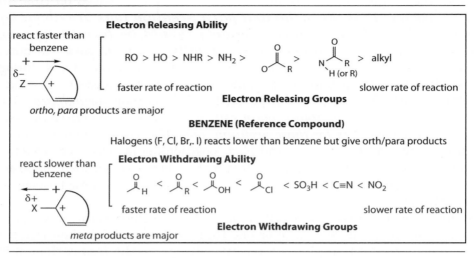

benzene and whether *ortho/para* are major products or *meta* is the major product. Several groups are categorized as electron releasing or electron withdrawing in Table 21.1. Note that all of the electron-withdrawing groups are electron rich and that the oxygen, nitrogen, or the alkyl group is δ– relative to C+. Those groups with a heteroatom also have an excess of electrons on the atom directly attached to the benzene ring. As seen in Chapter 10 (Section 10.2), an alkyl group is electron releasing relative to C+, although it does not have unshared electrons. When there is a δ– atom, a negative charge, or an electron-releasing group on the carbon atom of the benzene ring known as the *ipso* carbon, the reaction is faster and the *ortho* and *para* arenium ions form more quickly than the meta arenium ion. These substituents are labeled as **activators (faster) and *ortho/para* directors**.

All electron-withdrawing groups have a positive charge (as in nitro) or a δ+ atom (carbonyl carbon, the carbon of a cyano group, or the sulfur of sulfonic acid derivatives) on the atom directly attached to the benzene ring. All of the electron-withdrawing substituents lead to a *slower* reaction, but formation of the *meta* product is not as slow as formation of the *ortho* and *para* products. Therefore, the major product is *meta* and these substituents are labeled **deactivators (slower) and *meta* directors**.

The classification of a group as "deactivating" simply means that the rate of electrophilic aromatic substitution for a compound bearing that group is slower than the reaction with benzene itself. The category *meta* director means that the major product of electrophilic aromatic substitution is the *meta* product. *For activating substituents, the presence of an activating group (the electron-releasing group) on a benzene ring makes the rate of electrophilic aromatic substitution faster than that of benzene.* Classification as an *ortho/para* director means that electrophilic aromatic substitution gives a mixture of the *ortho* and the *para* products as the major products.

21.30 Categorize each of the following as activating or deactivating in electrophilic aromatic substitution: acetylbenzene (acetophenone), *tert*-butylbenzene, *N*-acetylbenzene, anisole, and nitrobenzene.

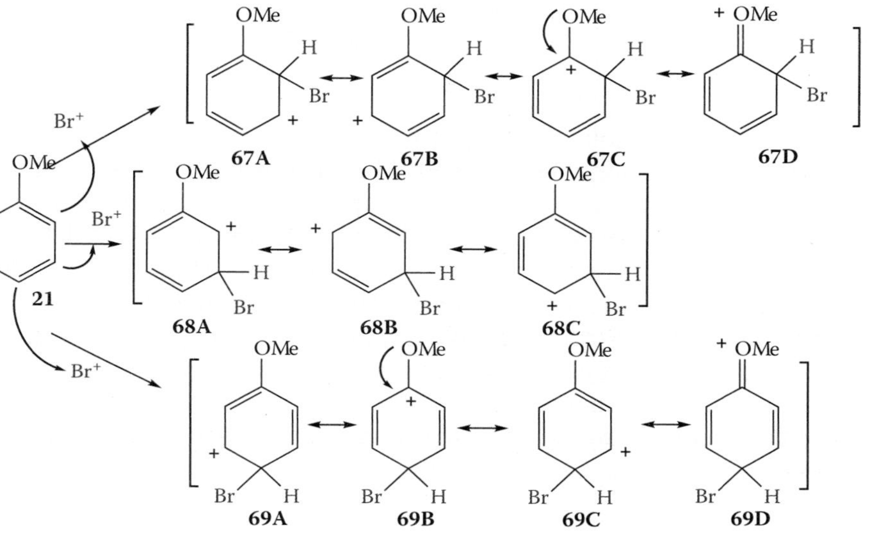

Figure 21.1 illustrates general arguments for electrophilic substitution using **61**. To make the point about differences between electron-releasing and electron-withdrawing groups, however, it is essential to examine specific molecules, using Figure 21.1 as a guide. First look at the reaction of anisole (**21**) with bromine/ferric bromide, which generates Br⁺. If **21** reacts with Br⁺ at the *ortho* position, arenium ion intermediate **67** is formed. Similarly, reaction at the *meta* position generates **68** and reaction at the *para* position generates **69**. In resonance contributor **67C**, the positive charge is on the *ipso* carbon that bears the electron-releasing methyl group. The same can be said for the cation generated by *para* attack (**69**) because **69B** also has the positive charge on the *ipso* carbon. The product of *meta* attack is **68** and the positive charge is *never on the* ipso *carbon*.

Note that in **67C** and **69B**, the oxygen atom attached to the *ipso* carbon has lone electron pairs, so it is δ– and electron releasing. As shown by the resonance contributor, those electrons may be donated back to the positive carbon to form a C=O unit with the positive charge on the oxygen (see **67D** and **69D**). *The presence of the electron-releasing group allows greater delocalization of the charge onto the oxygen atom; thus, the intermediate is more stable, leading to greater stability for the arenium ion formed by ortho or para attack.* Therefore, **67** and **69** will form at a faster rate than **68** because **68** has fewer resonance contributors and is less stable than **67** or **69**.

21.31 Draw the two major products formed when anisole reacts with Br₂/AlCl₃.

The analysis for anisole also explains why toluene gives **55** and **57** as the major products of nitration (see **67**– **69**). The arenium ions that lead to **55** and **57** have the electron-releasing methyl group on the *ipso* carbon, whereas the arenium ion that leads to **56** does not. Therefore, **55** and **57** will form faster. In the absence of any other factor, the arenium ions that lead to **55** and **57** should be roughly equal in energy and both products are equally favored. As stated previously, **55** is formed in 43% and **57** is formed in 53% yields, which is consistent with this simplistic prediction. Other factors influence the actual percentages of products, including the reaction time, temperature, etc. For the purposes of this book, ***assume that these reactions give a 1:1 mixture***, with the knowledge that the actual percentages will be close to that but will vary from reaction to reaction.

21.32 Draw all resonance forms for the reaction of phenol with Br⁺ (from bromine and AlBr₃) at the *ortho* position, the *meta* position, and the *para* position.

It is now possible to return to the reaction of benzene with 2-bromopropane to give **53** in the opening paragraph of Section 21.4. The isopropyl group is an alkyl substituent, and electron releasing (activating). Therefore, the initial product cumene (**53**) is more activated than benzene and it will react faster than benzene. Moreover, aromatic substitution of **53** leads to *ortho* and *para* substitution products (i.e., disubstituted product **54**). Because **54** is the *para* substituted product, it is assumed that placing two isopropyl groups on adjacent carbon atoms (*ortho*) is sterically hindered, which accounts for why no 1,2-diisopropyl product is reported.

21.33 Draw the structure of 1,2-diisopropybenzene.

21.4.2 Deactivating Substituents

Friedel–Crafts acylation does not suffer from polyacylation reactions. In the light of the discussion in Section 21.4.1, the initially formed ketone product must be less reactive than the benzene starting material. The product **50** from the reaction of butanoyl chloride and benzene is a ketone, and the carbonyl unit attached to benzene is an electron-withdrawing group. Electron-withdrawing groups deactivate the aromatic ring, so they react more slowly than benzene. In other words, the presence of an electron-withdrawing group makes the benzene ring a weaker Lewis base, so it reacts more slowly. If **50** is deactivated relative to benzene, it will react more slowly, and there is little chance that it can compete for reaction with the carbocation intermediate. Hence, there is no polyacylation.

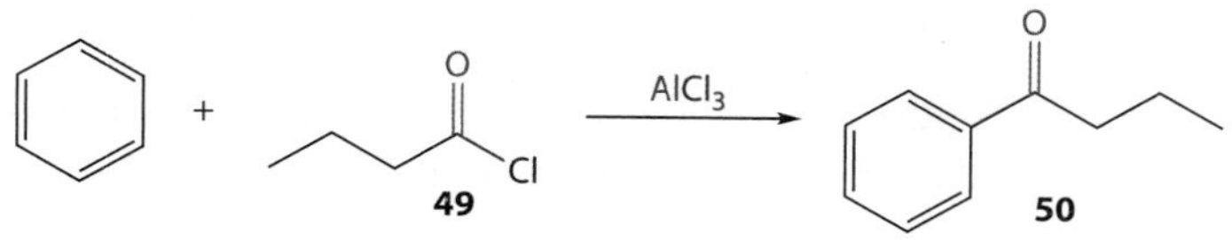

As with electron-releasing groups such as OMe in anisole, the influence of an electron-withdrawing group such as nitro can be related to **61** in Figure 21.1. In the reaction of Br$^+$ with nitrobenzene (**40**), the nitrogen atom of the nitro group bears a positive charge, making it electron withdrawing relative to the *ipso* carbon. Reaction of **40** with Br$^+$ gives three arenium ion intermediates via *ortho* attack (**70**), *meta* attack (**71**), and *para* attack (**72**). Note that there are only three resonance contributors for each intermediate because the charge cannot be delocalized on the nitro group.

For *ortho* attack (**70**) and for *para* attack (**72**), there is one resonance form where the positive charge resides on the *ipso* carbon that bears the nitro group (see **70C** and **72C**). Like charges strongly repel each other, so these resonance forms are particularly unstable and the activation energy for their formation is quite high. This energy barrier makes the rate of reaction to form the *ortho* and *para* products quite slow relative to benzene (nitro is deactivating). In **71**, the positive charge is never adjacent to the positive nitrogen, although repulsion of like charges makes this intermediate higher in energy; it will form more slowly than the arenium ion generated by reaction with benzene.

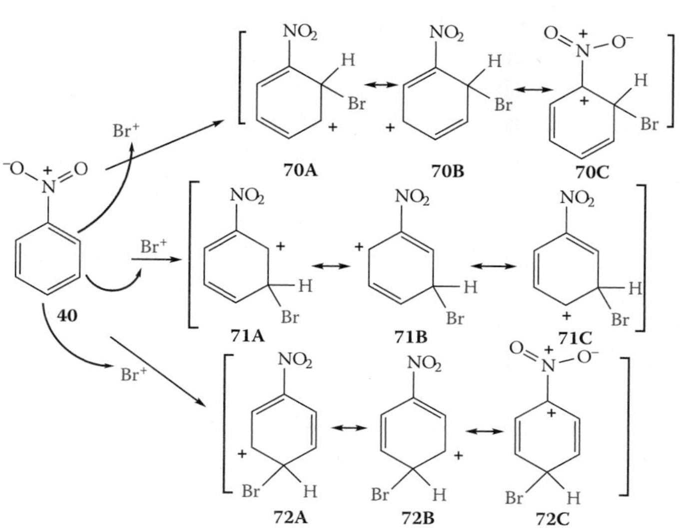

Because **70** and **72** are generated much more slowly than **67**, the major product is 3-bromonitrobenzene (**56**). Such an intermediate is very unstable, is difficult to form with a high activation energy barrier, and will react more slowly than benzene and much more slowly than derivatives bearing an electron-releasing group. Further, the arenium ion arising from attack at the *ortho* and *para* carbons is more stabilized and will form much more slowly than the less destabilized *meta* arenium ion, so the major product is *meta*.

21.34 **Draw all resonance forms for the reaction of acetophenone with Cl⁺ for attack at C3 and then draw the structure of the final product.**

One additional type of substituent must be considered for disubstituted benzene derivatives. When bromobenzene reacts with bromine and AlCl₃, the reaction proceeds *more slowly* than the bromination of benzene, but the major products are *ortho-* and *para*-dibromobenzene. The fact that bromobenzene reacts more slowly shows that the halogen is a deactivating substituent (see Table 21.1). However, the products are *ortho* and *para*. An explanation is that halogen substituents such as bromine are deactivating groups because they are slightly electron withdrawing. Remember from Chapter 10 that the halogens are polarizable. The benzene ring is electron rich, leading to an induced positive dipole of the adjacent halogen atom, as shown in **18**.

If the bromine (or any other halogen) is polarized δ+, electron density is donated from the benzene ring to the bromine atom. This means that benzene is less able to donate electrons, which is consistent with a slower rate of reaction for electrophilic substitution. Halogens have an excess of electrons, however, and, ***in the arenium ion intermediate*** for a reaction of bromobenzene with Br⁺ (from bromine and ferric bromide) at the *para* position (**73**), there is an interesting phenomenon. In resonance contributor **73B**, the positive charge resides on the carbon bearing the bromine atom. Bromine is polarizable and the *ipso* carbon bears a positive charge, so the *bromine will be polarized δ– in the arenium ion intermediate*. Polarization of the bromine changes from δ+ in the neutral bromobenzene **18** to δ– in arenium ion **73**. Just as importantly, the bromine has unshared electron pairs that may be *donated* to the positive carbon in **73B**, leading to a fourth resonance contributor where the positive charge is delocalized on the halogen atom (see **73C**).

The increased stabilization occurs for both *ortho* and *para* attack, but not for *meta* attack. Therefore, bromine deactivates the benzene ring to electrophilic substitution but it leads to *ortho* and *para* products. *In other words, the halogen is an ortho/para director* and the reaction of **18** with bromine and ferric bromide leads to a mixture of 1,2-dibromobenzene and 1,4-dibromobenzene as the major products. Fluorobenzene, chlorobenzene, bromobenzene, and iodobenzene react more slowly than benzene in electrophilic aromatic substitution reactions, but the major products of electrophilic aromatic substitution are the *ortho* and *para* disubstituted derivatives.

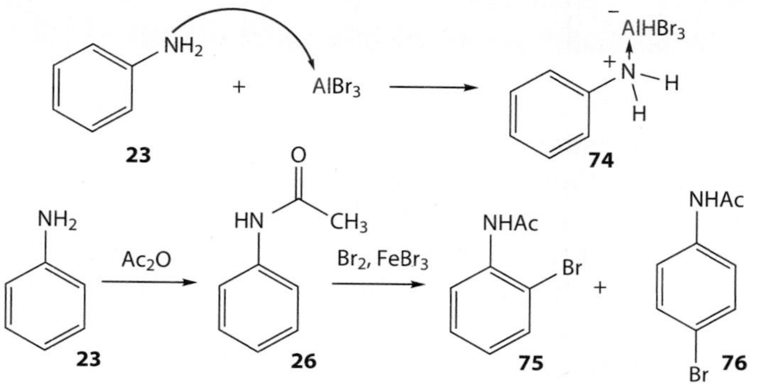

21.35 Draw the resonance forms for attack of Br⁺ at the *ortho* and *meta* positions of bromobenzene.

21.36 Draw 1,2-dibromobenzene and 1,4-dibromobenzene.

Another problem must be addressed before moving on. Aniline (**23**) has an amine unit with an electron pair that can be donated to a Lewis acid. In other words, aniline is a good Lewis base. The amine unit is a much stronger Lewis base than the benzene ring and, when aniline is mixed with $Br_2/AlBr_3$, the amine unit reacts preferentially with the Br⁺ cation to generate **74**, a Lewis acid/Lewis base ate complex. If the aluminum bromide is tied up as the ate complex, no Br⁺ is formed to react with the benzene ring. The positive charge on nitrogen in **74** makes that group electron withdrawing, so the benzene ring in **74** is deactivated. Therefore, electrophilic aromatic bromination does *not* occur. *A limitation of electrophilic aromatic substitution is that the substituent on benzene cannot be more basic than the benzene ring so that it reacts preferentially with the Lewis acid.* For the compounds used in this chapter, this problem is limited to aniline and its derivatives.

The problem relating to the basicity of aniline is largely counteracted by first reacting aniline (**23**) with acetic anhydride (see Chapter 20, Section 20.6) to give the amide *N*-acetyl aniline (**26**; known as acetamide). The acetyl group of the amide withdraws electron density from the nitrogen, making the nitrogen much less basic. Indeed, the Lewis basicity of the nitrogen is diminished sufficiently that normal electrophilic aromatic substitution reaction occurs, and the NAc unit is electron releasing, activating, and *ortho/para* directing (see Table 21.1). In effect, the acetyl group *protects* the nitrogen from reaction with the Lewis acid. When **26** is treated with bromine and ferric bromide, the products are 2-bromo *N*-acetylaniline (**75**) and 4-bromo *N*-acetylaniline (**76**).

21.37 Draw the arenium ion and all resonance contributors for reaction of 26 and Br⁺ at the *para* position.

21.5 Polysubstituted Benzene Derivatives

The presence of one substituent on a benzene ring leads to disubstituted products for electrophilic aromatic substitution reactions, and electron-releasing groups on a benzene ring lead to *ortho/para* disubstituted products. Electron-withdrawing groups on a benzene ring lead to *meta* disubstituted products. Moreover, electron-releasing groups lead to a faster reaction and electron-withdrawing groups lead to a slower reaction. If the starting material in an electrophilic aromatic substitution reaction has two substituents on the benzene ring, then the reaction places a third group on the ring. There are several possibilities for the starting material: (a) The benzene ring has two electron-releasing substituents, (b) it has two electron-withdrawing substituents, or (c) there is one electron-releasing and one electron-withdrawing group. All three cases will be addressed.

When a molecule has two activating substituents (electron-releasing groups), as with 4-methylanisole (**77**), both substituents are activating and *ortho* and *para* directing (see Table 21.1). In this case, the *para* position is blocked and the question becomes which *ortho* position will react with Cl⁺ from the Cl₂/FeCl₃, *ortho* to OMe or *ortho* to methyl. Because OMe is a very strong activator and methyl is a weak activator, the rate of reaction at the position *ortho* to OMe is much greater. In other words, the rate of electrophilic substitution directed by OMe is much faster than the rate of electrophilic substitution directed by Me. Therefore, the substitution reaction of **77** with chlorine and aluminum chloride gives 2-chloro-4-methylanisole (**78**) as the major product. This reaction is contrasted with that of 2-methylanisole, in which two chlorination products are formed, with a chlorine *ortho* and *para* to the methoxy group.

This analysis compares the activating or deactivating ability of each substituent and assumes that the stronger activating group will react at a faster rate, which leads to the major product. Determining the most activating substituent is particularly easy when the benzene ring contains one activator (electron-releasing group) and one deactivator (electron-withdrawing group), such as 3-nitroanisole, **79**. The OMe unit is a powerful activating group (faster), whereas nitro is a powerful deactivating group (slower). The difference in rate of reaction for these two groups may be greater than 10^{10}, so reaction of **79** with fuming sulfuric acid gives substitution *ortho* and *para* to the OMe group (compounds **80** and **81**) as the products. Note that the SO$_3$H unit in **80** is *ortho* to the nitro group, due to the activating influence of OMe for reaction at that carbon.

There is a third product (**82**), but it is formed in a much lower percentage than either **80** or **81**. The SO$_3$H unit in **82** is *ortho* to the OMe, but *sulfonation at that carbon is inhibited by significant steric hindrance because the group must be inserted between the adjacent OMe and NO$_2$ units.* In reactions where the product must be incorporated between two groups, the yield is usually much lower due to the increase in steric hindrance, which slows down the reaction that generates the requisite arenium ion.

21.38 Draw both products of the chlorination reaction of 2-methylanisole and draw all the resonance forms for the intermediate that leads to the product with the Cl *para* to OMe.

21.39 Draw the arenium ion intermediate that will lead to 82.

21.40 Draw the major product when 4-nitro-1,2,3-trimethylbenzene reacts with a mixture of nitric acid and sulfuric acid.

The reaction of a starting material in which the benzene ring contains two electron-withdrawing (deactivating) substituents will have two problems. The first is that the ring may be so deactivated that the reaction is too slow to take place at all. If it is *assumed* that a slow reaction does indeed occur, the idea of a most "activated" site is not appropriate, but the less deactivated site must be determined. *Substitution meta to the less deactivated group will be faster than substitution meta to the more deactivated group.* In **84** (4-nitrobenzoic acid), the nitro group is a more powerful deactivating group than the COOH unit, and substitution will occur *meta* to the least deactivating group (COOH). Therefore, reaction of **84** with bromine and AlCl$_3$ gives a very slow reaction, but the major product of that reaction is 3-bromo-4-nitrobenzoic acid, **85**.

21.41 Draw the product of a reaction between 3-chloronitrobenzene and bromine in the presence of aluminum bromide.

21.6 Reduction of Aromatic Compounds

There are several reactions of benzene derivatives other than electrophilic aromatic substitution. Some of them involve the benzene ring itself; others occur at the substituent attached to the benzene ring. This section will introduce several of these reactions and the functional group transformations that are available by them.

21.6.1 Reduction of the Benzene Ring

21.6.1.1 Catalytic Hydrogenation

The π-system in a benzene ring is particularly stable due to the aromatic nature of the ring. Because reduction of the benzene ring will disrupt the aromatic character, some difficulty is anticipated in such a reaction. This is correct, but at least two classical methods can reduce the π-system of the benzene ring. One is catalytic hydrogenation and the other is treatment with an alkali metal (Li, Na, K) in the presence of a weak acid such as ammonia and/or an alcohol. Catalytic hydrogenation of alkenes (in Chapter 19, Section 19.3.2) adds two hydrogen atoms to a C=C unit to give an alkane. The reaction with benzene is similar with two exceptions: Addition of one molar equivalent of hydrogen (H–H) will generate a diene and addition of two molar equivalents of H–H will give an alkene. Complete reduction of benzene to cyclohexane (addition of three molar equivalents of H–H) typically requires the use of an excess of hydrogen.

21.42 Draw the structures of 1,3-cyclohexadiene and cyclohexene.

There is a second issue for the reduction of benzene rings. The aromatic character of the benzene ring makes the reduction more difficult, so a hydrogenation reaction requires more vigorous conditions when compared to hydrogenation of an alkene. In other words, higher temperatures and/or higher pressures of hydrogen gas are necessary. When benzene reacts with three molar equivalents of hydrogen gas in the presence of a Raney nickel catalyst—abbreviated Ni(R)—reduction to cyclohexane occurs if the reaction is heated. Hydrogenation of benzene is also possible using a palladium catalyst or a rhodium (Rh) catalyst.

Another issue relates to reactivity. When benzene is treated with one molar equivalent of hydrogen gas, the product is expected to be cyclohexadiene (**3**), but **3** is much more reactive than benzene in the hydrogenation reaction. Cyclohexadiene will react with a molar equivalent of hydrogen gas to give cyclohexene or with two equivalents of hydrogen to give cyclohexane. Indeed, this secondary reaction may be competitive and, although it is possible to isolate **3**, it is often difficult and a mixture of products is common (cyclohexadiene, cyclohexene, and cyclohexane).

Careful control of the amount of hydrogen gas added, the reaction temperature, and care in the selection of the catalyst makes possible the isolation of **3**, however. With two molar equivalents of hydrogen gas, benzene is reduced to cyclohexene, which is also more reactive than benzene; once again, it may be difficult to stop the reaction at this stage. Control of the amount of hydrogen gas, the catalyst, and the reaction temperature also allows isolation of cyclohexene, however. Note that these conditions are not specified, so, for convenience, *assume that one equivalent of hydrogen will give cyclohexadiene (3) and that two equivalents of hydrogen will give cyclohexene*. With an excess of hydrogen gas (three or more molar equivalents), benzene is cleanly converted to cyclohexane.

21.43 Write out the reaction and draw the final product when 4-methylanisole is treated with three equivalents of hydrogen gas in the presence of a Raney nickel catalyst.

21.44 Write out a reaction sequence that generates 4-methyl-1-cyclohexanesulfonic acid from toluene.

21.6.1.2 *Birch Reduction*

An alternative method for the reduction of benzene rings uses alkali metals (group 1 or group 2) such as sodium or lithium in liquid ammonia, often in the presence of ethanol. This method was used in Chapter 19 (Section 19.4.2) for the reduction of alkynes to *E*-alkenes. When benzene is treated with sodium and ethanol in liquid ammonia, the product is 1,4-cyclohexadiene, **89**. This reaction is known as the **Birch reduction**, after Arthur John Birch (Australia; 1915–1995). In this reaction, two hydrogen atoms are incorporated into the benzene ring with a net reduction of one C=C unit. Note that the two hydrogen atoms are incorporated from the solvent and that the remaining C=C units are not conjugated (they are not directly attached one to the other; see this concept in Chapter 23).

To explain the formation of **89** and account for the nonconjugated nature of the product as well as the incorporation of hydrogen atoms from the solvent, a mechanism is proposed that involves single electron transfer from the alkali metal to the benzene ring. Remember that group 1 and group 2 metals can easily transfer a single electron from their outermost shell (see Chapter 3, Section 3.2). Initial reaction of benzene with sodium proceeds by transfer of one electron from sodium to a π-bond to give a radical anion, **86**. Although there are three resonance contributors to **86**, the lowest energy resonance contributor will arise by separation of the two electrons of the carbanion and the single electron of the radical in order to diminish electronic repulsion. This means that **86B** represents the lowest energy electronic distribution in the intermediate, and it will lead to the final product.

In the presence of an acidic solvent such as ethanol, the carbanion portion of **86B** reacts to form radical **87**. Note that the carbanion (C⁻) unit in
86B is a very powerful base (the conjugate acid is a hydrocarbon, which is a
very weak acid), so the proton of the N–H unit in liquid ammonia is a strong
acid in this system. Therefore, **86B** reacts with ammonia to give **87** and the
conjugate base, which is the amide anion, ⁻NH₂. Note that ethanol (pK$_a$ of
17) is a much stronger acid than ammonia (pK$_a$ of 25). Ethanol is added to
the reaction because it is a stronger acid, which makes the protonation reaction of **86B** much faster. If **86B** reacts with ethanol, the product is **87** and the
ethoxide ion (EtO⁻). Intermediate **87** is a radical, and reaction with a second
molar equivalent of sodium metal transfers one electron to **87** to generate a
new carbanion (**88**), which reacts with ethanol via another acid–base reaction
to give the product **89**.

21.45 Write out the acid–base reaction of 86 and ammonia. Write out
       the reaction of 86 with ethanol. Label the acid/base and conju-
       gate acid/conjugate base in both reactions.

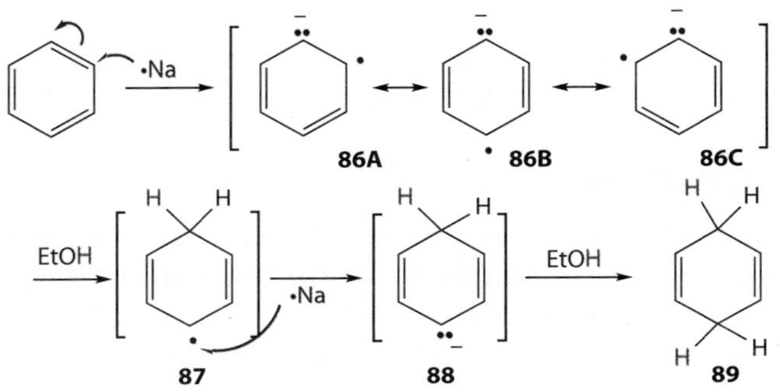

The electron transfer mechanism just described also applies to substituted
benzene derivatives, but different products are formed when there is an electron-withdrawing substituent versus an electron-releasing substituent. If anisole (**21**) is treated with sodium in ethanol and liquid ammonia, initial electron
transfer can give intermediates **90** or **91**. Intermediate **91** has a resonance contributor with the negative charge on the *ipso* carbon, but there is no such resonance contributor in **90**. Both are resonance stabilized, but in **91** the negative
charge on the *ipso* carbon leads to electron repulsion that greatly destabilizes
that resonance contributor.

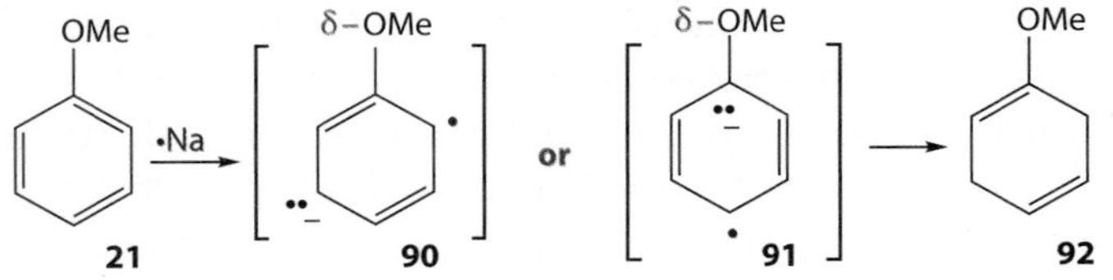

Electronic repulsion is minimized and the intermediate is more stable (it
will form faster) when the negative charge does not reside on the *ipso* carbon,

as in **90**. This intermediate will lead to the major product. Subsequent reaction with ethanol, electron transfer from a second equivalent of sodium metal, and protonation with ethanol will give **92**. Note that the oxygen atom of the methoxy group in **92** is attached to an sp² hybridized carbon atom.

21.46 Draw all resonance forms of 90 and 91 and briefly explain why 90 is preferred.

21.47 Draw a reaction sequence that converts phenol to anisole (21).

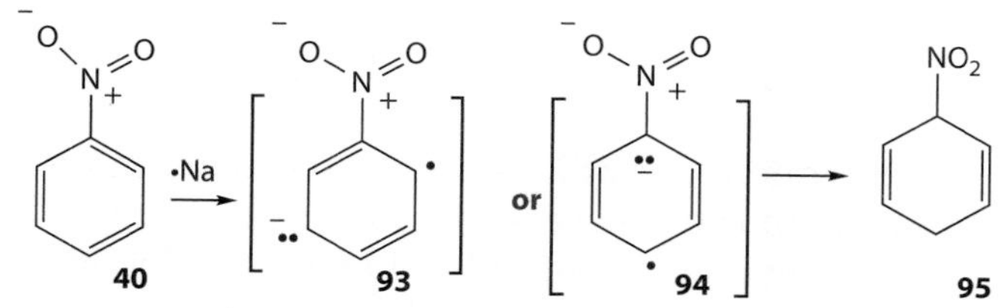

Birch reduction of nitrobenzene (**40**) differs in that the substituent is the electron-withdrawing nitro group, with a positively charged nitrogen atom. Initial electron transfer from sodium can generate radical anion **93** or **94** (both are resonance stabilized). In **94**, the negative charge is on the *ipso* carbon adjacent to the positively charged nitrogen atom of the nitro group. Clearly, the attraction of positive and negative charges is stabilizing and this intermediate is more stable and formed more quickly than the alternative, **93**, where the negative and positive charges are never adjacent to each other.

Therefore, **94** will be the major intermediate that leads to the major final product, **95**, via reaction with ethanol, electron transfer, and a final reaction with ethanol. In contrast to **92**, the electron-withdrawing substituent in **95** is attached to an sp³ hybridized carbon atom. In the Birch reduction, the position of the double bonds in the ring and the substituent attached to the ring are determined by electronic repulsion or attraction in the radical anion intermediate. *In general, an electron-releasing group (including alkyl groups such as methyl or ethyl) will be attached to an sp² carbon in the final product, whereas an electron-withdrawing group will be attached to an sp³ carbon in the final product.*

21.48 Draw all resonance forms for 94.

21.49 Draw the major product formed when 3-nitroanisole is treated with sodium metal in ethanol and liquid ammonia.

21.6.2 Reduction of Substituents Attached to the Benzene Ring

Many groups attached to a benzene ring are themselves subject to reduction (see Chapter 19). The C=C unit of an alkene and the C=O unit of an aldehyde or ketone are examples. If styrene (**15**) is treated with hydrogen gas and a palladium catalyst, for example, the C=C unit of the alkene is much more susceptible to hydrogenation than the benzene ring, and the product is ethylbenzene, **6**. If a large excess of hydrogen is used, ethylcyclohexane (**96**) can be formed.

In the case of benzaldehyde (**29**), catalytic hydrogenation leads to benzyl alcohol (**97**), particularly in the presence of a platinum catalyst (see Chapter 19, Section 19.3.4). Another reaction is possible when benzaldehyde is hydrogenated with an excess of hydrogen gas. The C–O bond in **97** is susceptible to cleavage and replacement of oxygen with hydrogen. This is called *hydrogenolysis*, and hydrogenolysis of **97** will give toluene. It is usually possible to minimize or prevent hydrogenolysis in the hydrogenation of benzaldehyde derivatives or 1-phenyl ketones by controlling the amount of hydrogen, the temperature, the catalyst, and the time the reaction is allowed to proceed. Therefore, *assume* that hydrogenation of these aldehydes or ketones will give the alcohol as the major product.

21.50 Write out a synthesis of 1-(3-bromophenyl)-1-bromoethane from benzene.

Nitrobenzene (**40**) is a particularly interesting compound for catalytic hydrogenation because the reduction product is aniline, **23**. Indeed, nitration of benzene followed by catalytic hydrogenation is an important method for the synthesis of aniline and many of its derivatives. Other reagents initiate reduction of nitrobenzene to aniline, including the metals tin or iron in acids such as acetic acid or HCl (see Chapter 19, Section 19.4). Note that although catalytic hydrogenation of nitrobenzene gives aniline (**23**), reaction with lithium aluminum hydride (LiAlH$_4$; see Chapter 19, Section 19.2.1) does *not* give **23**, but rather a diazo compound, **98**. Sodium borohydride is too weak to reduce a nitro group. The nitrile group can be reduced to an amine and catalytic hydrogenation of benzonitrile (**28**; cyanobenzene) gives aminomethyl benzene, **99** (called benzylamine). In this case, treatment of **28** with LiAlH$_4$ also gives **99** after hydrolysis.

21.51 Write out the reaction sequence that converts toluene to 4-nitrotoluene and then to 4-methylaniline, supplying all reagents that are required.

21.52 Write out a synthesis of *N*-benzoyl-4-(1,1-dimethylethyl)aniline from benzene.

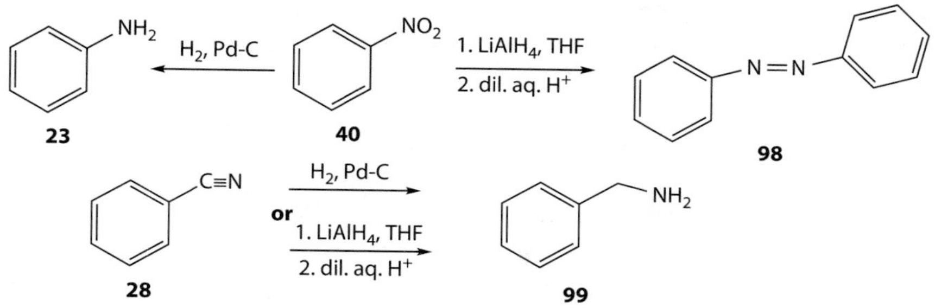

Aniline is a starting product for the manufacture of drugs such as acetamino-phen/paracetamol (Tylenol), which is *N*-acetyl-4-hydroxyaniline (**100**). Another commercial product is 4,4′-MDI (methylene diphenyl diisocyanate, **101**), which reacts with a polyol (**102**) to give polyurethanes such as **103**. Polyurethanes are used in applications ranging from the foam in upholstery to thermal insulation material in refrigerators.

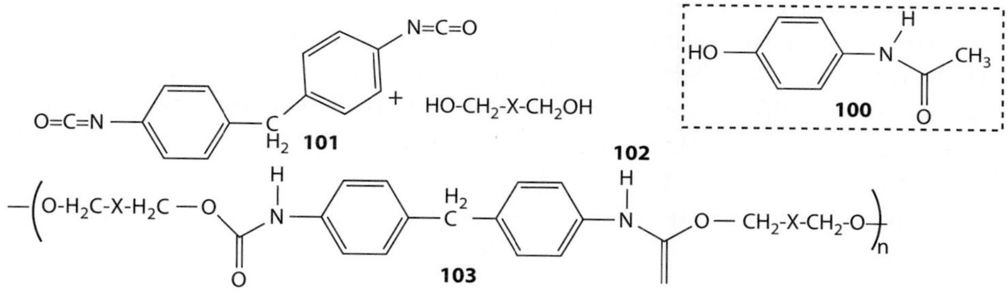

Benzoic acid derivatives have carboxylate units attached to a benzene ring, which can be reduced independently of the ring. When benzoic acid (**27**) is treated with LiAlH$_4$, the final product after hydrolysis is benzyl alcohol (**97**). Catalytic hydrogenation of **27** does *not* give benzyl alcohol because carboxylic acids are particularly resistant to hydrogenation. Similarly, catalytic hydrogenation of esters and amides is quite slow so ethyl benzoate (PhCO$_2$Et) and benzamide (**104**) are not reduced via catalytic hydrogenation. However, an ester such as ethyl benzoate is reduced to benzyl alcohol upon treatment with LiAlH$_4$ (see Chapter 19, Section 19.2.1). Similar reduction of the amide group in **104** gives aminomethyl benzene, **99**. Note that when amide **104** is heated with phosphorus pentoxide (P$_2$O$_5$), dehydration occurs to give benzonitrile, **28**. *Remember that dehydration of primary amides is an excellent route to nitriles, both aliphatic derivatives and benzo-nitrile derivatives.*

21.53 Write out the reaction for the LiAlH$_4$ reduction of ethyl benzoate to benzyl alcohol.

21.54 Write out a synthesis of 3-bromobenzonitrile from benzoic acid.

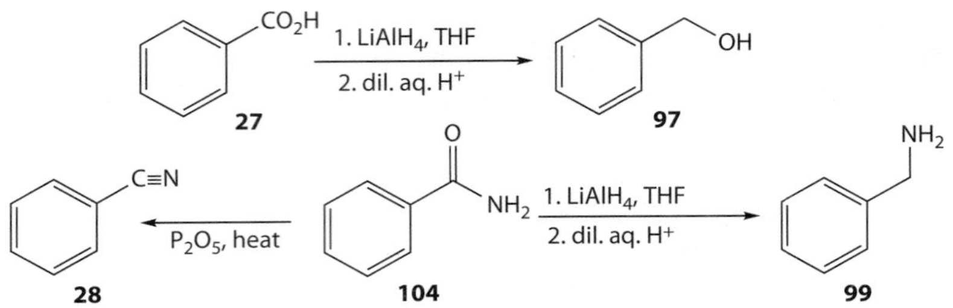

Another category of reductions involves aryl ketones. The Friedel–Crafts acylation reaction reacts benzene with an acid chloride such as butanoyl chloride (**49**) to give an aryl ketone, **50**. Complete removal of the oxygen from this ketone constitutes a method to make straight-chain arenes, which cannot be prepared via Friedel–Crafts alkylation (see Section 21.3.2). At least two classical methods are used to accomplish this reaction, which is formally a reduction. If **50** is treated with zinc metal in HCl, the product is 1-phenylbutane, **105**. This acidic reduction involves a mineral acid such as HCl and an active metal, and it is called the **Clemmensen reduction**.

Another method for reduction proceeds under basic conditions and treats **50** with hydrazine (NH₂NH₂) and potassium hydroxide (KOH). An aqueous workup produces **105**. This is called the **Wolff–Kishner reduction** (mentioned in Chapter 19, Section 19.4.1). By using either the Clemmensen or the Wolff–Kishner procedure, straight-chain arenes such as **105** can be produced in a two-step procedure: Friedel–Crafts acylation followed by reduction. Indeed, this two-step procedure is usually the preferred route to synthesize straight-chain arenes.

21.55 Write out a synthesis of 2-phenyl-3-pentanol from benzene.

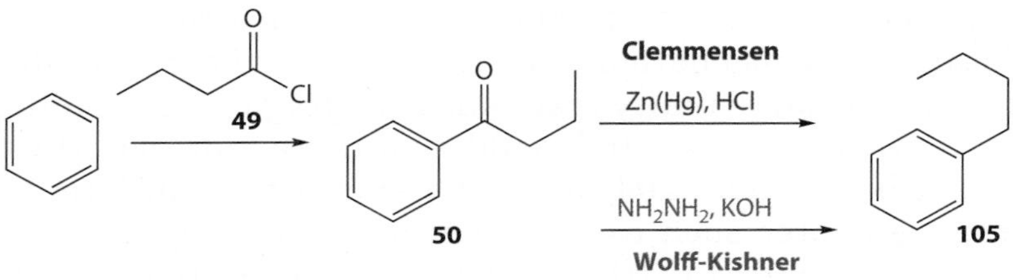

The mechanism of the Wolff–Kishner reduction was not presented in Chapter 19, but it is interesting because it involves carbanion intermediates and formation of imines (introduced in Chapter 18, Section 18.7.1). Initial reaction with hydrazine and the carbonyl of **50** gives a hydrazone (**106**; see Chapter 18, Section 18.7.3). Hydroxide is a strong enough base to remove a proton from the hydrazone, which is a weak acid here, to give a resonance-stabilized anion (**107A** and **107B**), where there is an N=N unit in **107B** along with the charge on the carbon.

This carbanion is a strong base and it reacts with water (an acid–base reaction) to generate the neutral molecule **108**, which has an N–H unit that is weakly acidic and reacts with more hydroxide to give anion **109**. Transfer of electrons from nitrogen leads to formation of diatomic nitrogen (N_2), which is a good leaving group; loss of nitrogen gives the strongly basic carbanion **110**, which quickly reacts with water (an acid–base reaction) to give the final product, **105**. This sequence results in reduction of the carbonyl group in **50** to the hydrocarbon **105**.

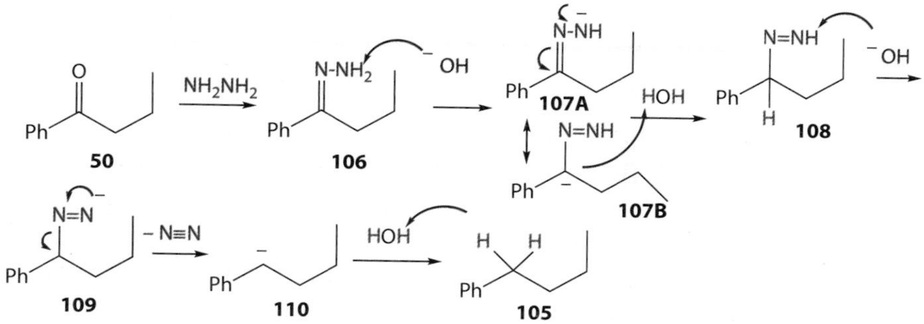

21.7 Aromaticity in Monocyclic Molecules Other Than Benzene

In Section 21.1, benzene was identified as an aromatic compound with the resonance-stabilized delocalization shown in **1**. Benzene is also identified as an aromatic compound because it meets certain unique criteria. There are six π-electrons confined to a ring, and every carbon atom in that ring is sp^2 hybridized with a p-orbital attached. Further, the p-orbitals are contiguous and continuous. In other words, every carbon in the ring has a p-orbital and there are no intervening sp^3 atoms. In principle, if a molecule other than benzene meets the criteria, it is aromatic. There are several such molecules, both neutral molecules and charged intermediates.

Hückel's rule (presented in Section 21.1) *states that for planar, monocyclic hydrocarbons containing completely conjugated sp^2 hybridized atoms, the presence of (4n + 2) π-electrons leads to aromaticity ("n" is an integer in the series 0, 1, 2, 3, etc.).* This rule is associated with particularly stable molecules. *Acyclic* molecules with a continuous array of sp^2 hybridized carbon atoms are known as *polyenes*. For aromaticity, there must be a ring, so *cyclic polyenes* are required. If the number of π-electrons does not equal 4n + 2, then aromaticity does not exist and the system is rather unstable (this is sometimes called an antiaromatic system). *If Hückel's rule (or the "4n + 2 rule," as it is called) is used, the presence of 2, 6, 10, 14, 18, etc. π-electrons in a ring where every atom is sp^2 hybridized is characteristic of an aromatic compound.* Benzene clearly fits this rule and so does compound **111**, which has 14 π-electrons.

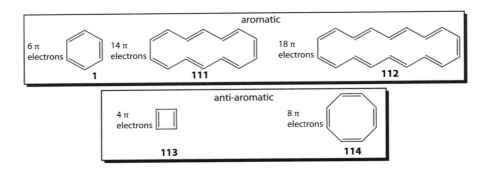

Compound **111** is a cyclic compound with 14 π-electrons, and every carbon in the ring is sp² hybridized. Because it fits the Hückel rule, it is aromatic. Larger cyclic polyenes may also be aromatic, and by definition they must be planar. Cyclic polyene **112** is a cyclic compound with 18 π-electrons; every carbon is sp² hybridized, so it also satisfies the Hückel rule and is aromatic. Compounds such as these are termed **annulenes** (completely conjugated aromatic hydrocarbons) and this becomes part of the name. Benzene could be named [6]-annulene, **111** is named [14]-annulene, and **112** is [18]-annulene.

Members of another set of cyclic polyenes have (4n) π-electrons, such as cyclobutadiene (**113**; 4 π-electrons) and cyclooctatetraene (**114**; 8 π-electrons), so they do not adhere to the Hückel rule. Compound **113** is a cyclic compound and every carbon is sp² hybridized, but it has only 4 π-electrons and does not satisfy the 4n + 2 rule (4 is not part of this series, so **113** does not have 2, 6, 10, 14, etc. π-electrons). Because **113** does not satisfy the Hückel rule, it is not aromatic. Likewise, **114** is cyclic and has a continuous array of sp² carbons, but 8 π-electrons do not fit the 4n + 2 series and thus do not satisfy the Hückel rule, and **114** is not aromatic. These compounds are *not* aromatic and are also very unstable and difficult to prepare. Because they are so difficult to prepare and the ring system is so unstable, cyclic compounds such as this with 4n π-electrons are called **antiaromatic compounds**. *For practical purposes, assume that such compounds cannot be prepared* (although they can be if extremely low temperatures and specialized conditions are used).

In addition to neutral molecules, certain cation and anion intermediates meet the criteria for aromaticity. If the cyclopropenyl cation (**115**) and the cycloheptatrienyl cation (**116**) are examined, both have a continuous array of p-orbitals confined to a ring and a number of π-electrons that fit the 4n + 2 series (two for **115** and six for **117**). Both of these carbocations are aromatic, which means that they are very stable, easy to form, and relatively long-lived intermediates. Compare these carbocations with the cyclopentadienyl cation (**117**), which meets the criterion of having a continuous array of p-orbitals confined to a ring, but has 4n π-electrons (not a number in the 4n + 2 series) and is *not* aromatic. Indeed, it is considered to be antiaromatic, is very unstable, and is very difficult to form.

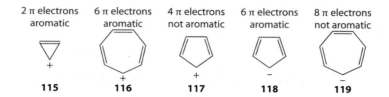

2 π electrons aromatic **115** / 6 π electrons aromatic **116** / 4 π electrons not aromatic **117** / 6 π electrons aromatic **118** / 8 π electrons not aromatic **119**

Anions are similarly examined, and the cyclopentadienyl anion (**118**) has six π-electrons and meets all criteria for aromaticity. It is aromatic, very easy to form, and quite stable. Formation of **118** from cyclopentadiene (**120**) is an acid–base reaction. It is known that **120** has a relatively low pK$_a$ that reflects the special aromatic stability of the aromatic conjugate base. The pK$_a$ of cyclopentadiene is 14–15 (compare that with a pK$_a$ of 15.8 for water).[4] This contrasts sharply to the cycloheptatrienyl anion (**119**), which has 4n π-electrons, is *not* aromatic, and is particularly unstable and difficult to form. As with **118**, formation of **119** is an acid–base reaction from cycloheptatriene, **121**. The pK$_a$ of **121** is about 36,[1] however, which reflects the great difficulty in forming the antiaromatic conjugate base.

21.56 Draw the structures of [26]-annulene and [12]-annulene and determine whether they are aromatic.

21.57 Suggest a reaction scheme to make 1-benzyl-2,4-cyclopentadiene from 120.

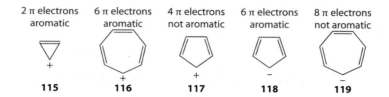

From these discussions, it is apparent that there are many aromatic compounds. Several of these involve useful individual compounds or intermediates. All of the cases discussed so far have involved monocyclic (one-ring) compounds. Many other aromatic compounds have two, three, or more aromatic rings fused together. These compounds are discussed in the next section.

21.8 Polynuclear Aromatic Hydrocarbons

Many aromatic compounds that meet the 4n + 2 rule have structures where the π-electrons are not confined to one ring but rather contained in several rings that are fused together. They are called ***polynuclear aromatic molecules***.

They are interesting structures and a variety of new compounds is available via electrophilic aromatic substitution.

21.8.1 Naphthalene, Anthracene, and Phenanthrene

Naphthalene is a bicyclic aromatic compound with the formula $C_{10}H_8$ and structure **122**. It was the main constituent of mothballs for many years, but toxicity led to its replacement with 1,4-dichlorobenzene (characterized as an insecticidal fumigant). Naphthalene is planar, with 10 π-electrons in a π-cloud above and below the plane of the 10 carbon atoms; like benzene, it is aromatic and particularly stable. Another polycyclic aromatic compound has three rings fused together, as in **123**; this aromatic molecule (14 π-electrons) is called anthracene (formula: $C_{14}H_{10}$). It is an important starting material in the synthesis of dyes such as the red dye alizarin (**125**) and is used in wood preservatives and insecticides.

There is an isomer of anthracene, called phenanthrene, in which the point of attachment of the third ring on the middle ring is different. Phenanthrene has the same empirical formula, but its structure is **124**. Like anthracene, phenanthrene is derived from coal tar and is used in the synthesis of dyes, explosives, and drugs. Both anthracene and phenanthrene are aromatic. The numbering of these ring systems is interesting. Note that in naphthalene only eight carbons on the periphery of the rings are numbered. The "bridgehead" carbons are not numbered because they cannot undergo substitution reactions (see the following section). Similarly, anthracene and phenanthrene have only 10 numbered carbons, each with four bridgehead positions.

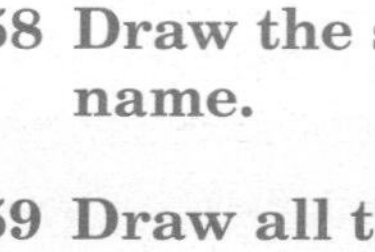

122 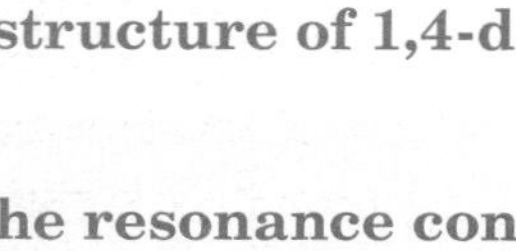   **123** 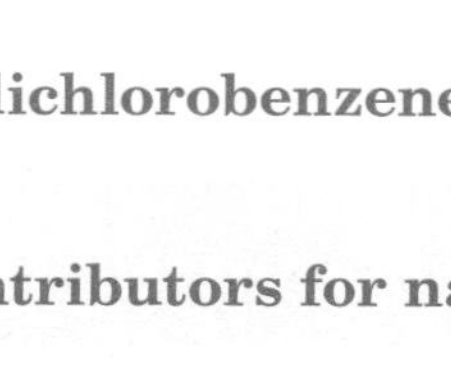   **124** 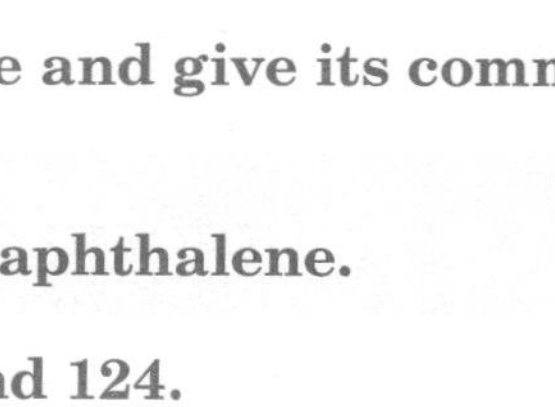   **125, alizarin**

21.58 Draw the structure of 1,4-dichlorobenzene and give its common name.

21.59 Draw all the resonance contributors for naphthalene.

21.60 Draw all resonance contributors to 123 and 124.

21.61 Draw the structures of (a) 2-nitronaphthalene, (b) 2-fluoro-anthracene, (c) 4-methylphenanthrene, (d) 5,9-dibromophenanthrene, (e) 1,4-dimethyl-6-ethylanthracene, and (f) 2,7-dihydroxynaphthalene.

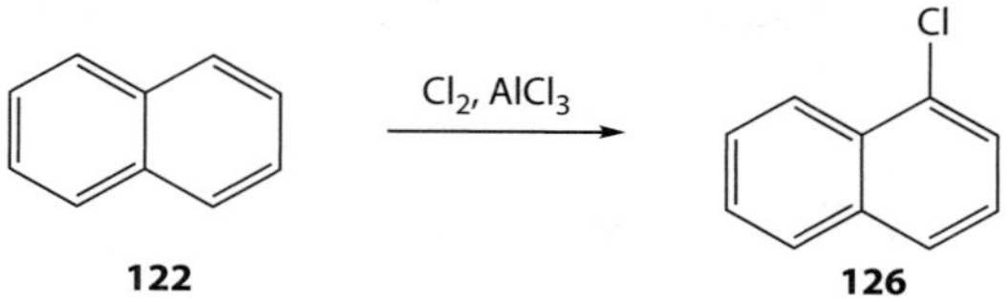

122　　　　　　　　　**126**

21.8.2 Substituted Derivatives

Polynuclear aromatic hydrocarbons such as naphthalene, anthracene, and phenanthrene undergo electrophilic aromatic substitution reactions in the same manner as benzene. A significant difference is that there are more carbon atoms, more potential sites for substitution, and more resonance structures to consider. In naphthalene, it is important to recognize that there are only two *different* positions: C1 and C2 (see **122**). This means that C1, C4, C5, and C8 are chemically identical and that C2, C3, C6, and C7 are chemically identical. In other words, if substitution occurs at C1, C4, C5, and C8 as labeled in **122**, only one product is formed: 1-chloronaphthalene (**121**), which is the actual product isolated from the chlorination reaction. Chlorination of naphthalene at C1 leads to the five resonance structures shown for arenium ion intermediate **127**.

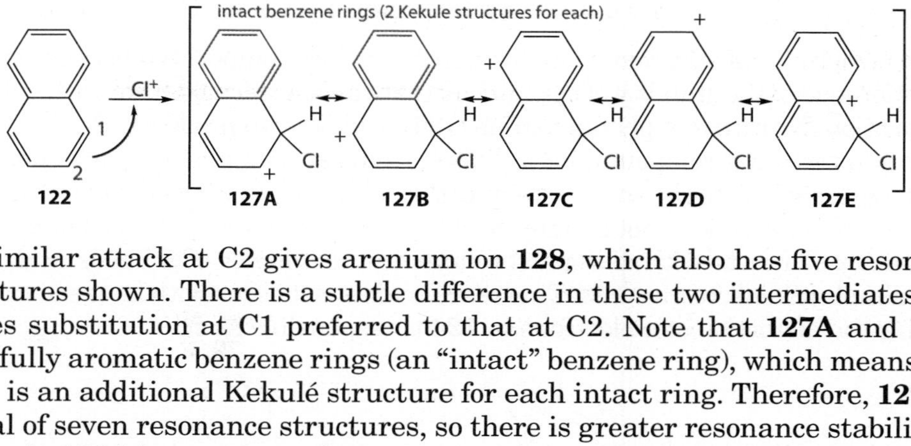

122　　　**127A**　　　**127B**　　　**127C**　　　**127D**　　　**127E**

Similar attack at C2 gives arenium ion **128**, which also has five resonance structures shown. There is a subtle difference in these two intermediates that makes substitution at C1 preferred to that at C2. Note that **127A** and **127B** have fully aromatic benzene rings (an "intact" benzene ring), which means that there is an additional Kekulé structure for each intact ring. Therefore, **127** has a total of seven resonance structures, so there is greater resonance stability. In **128**, only **128A** has an intact benzene ring, for a total of two Kekulé structures and a total of six resonance contributors. Therefore, **122** is more stable than **123**, and it will form faster and lead to the major product: 1-chloronaphthalene, **121**. Other methods are available to calculate or estimate resonance stability, but they are beyond the scope of this book.

21.62 Draw the two additional resonance contributors for 127 and the one extra contributor for 128.

21.63 Draw the product of a reaction of 122 with bromine/AlCl$_3$, of nitration of 122 with nitric acid/sulfuric acid, and of the reaction between 122 and benzoyl chloride, in the presence of aluminum chloride.

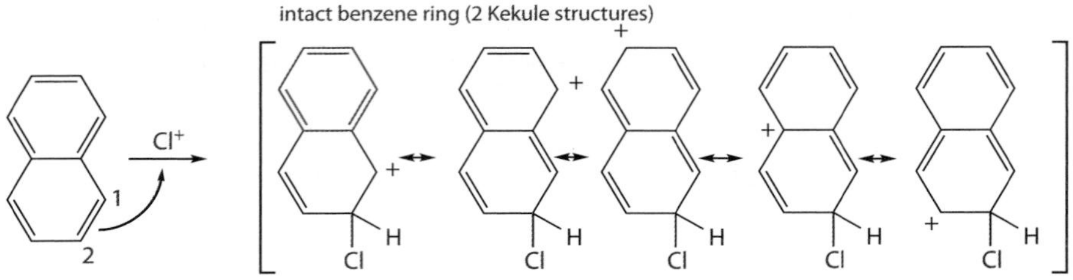

intact benzene ring (2 Kekule structures)

122   128A   128B   128C   128D   128E

In anthracene (**123**), there are three different positions (C1, C2, and C9) and there are five different positions (C1, C2, C4, C5, and C9) in phenanthrene (**124**). Electrophilic aromatic substitution of anthracene leads to substitution primarily at C9 because that gives an intermediate with the most resonance forms and the most intact benzene rings. A comparison of attack at C1 and at C2 in anthracene will show that there are more resonance forms for attack at C1 and more fully aromatic rings. Attack at C9 leads to an intermediate with even more resonance, and electrophilic substitution of anthracene leads to C9 and C1 products, with little reaction at C2.

21.64 Draw all resonance forms for attack of Cl$^+$ at C1 of anthracene.

21.65 Draw all resonance forms for attack by Cl$^+$ at C2 of anthracene.

Electrophilic substitution of phenanthrene is more complicated because there is less difference in the stability of the intermediate cations formed from each substitution position. Nitration of phenanthrene (**124**) leads to five products. Phenanthrene is a very interesting compound for a different reason. The middle ring is not stabilized by aromaticity to the same extent as the others (as mentioned earlier), making it susceptible to reactions not usually observed with aromatic systems. For example, phenanthrene reacts with diatomic bromine in the *absence* of a Lewis acid, much like a simple alkene, to give **129**. This reaction does not occur with benzene or naphthalene; the distortion of π-electrons in **124** is responsible for it.

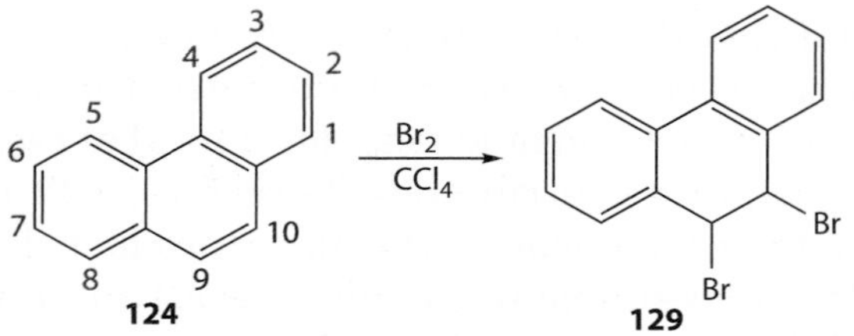

21.9 Aromatic Amines and Diazonium Salts

A class of highly reactive compounds derived from aromatic amines such as aniline is called diazonium salts. They are characterized by the presence of a –N$_2^+$ unit attached to the aromatic ring. Because nitrogen gas (N$_2$) is an excellent leaving group, diazonium salts undergo rapid substitution reactions. This makes possible the conversion of aromatic amines to several important aromatic derivatives.

21.9.1 Formation of Diazonium Compounds

When hydrochloric acid is mixed with sodium nitrite ($NaNO_2$), the resulting reaction generates a transient and highly reactive acid known as nitrous acid (HONO). Nitrous acid reacts rapidly with amines to generate a highly reactive product called a diazonium salt, which is also known as a diazo compound.

130 **131**

An aliphatic primary amine such as 1-aminobutane (**130**) reacts to give **131**, but this diazonium salt is extremely unstable and it loses nitrogen ($N_2 = N\equiv N$) to give a carbocation. This carbocation is primary and rapidly undergoes other reactions, including rearrangement (see Chapter 10, Section 10.2). In general, aliphatic amines react to give intermediates that are so unstable it is difficult to control their chemical reactions. Aromatic amines, however, generate diazonium intermediates that are more stable (in many cases) and several interesting chemical transformations are possible. When aniline (**23**) reacts with $NaNO_2$/HCl, the HONO generated *in situ* gives benzenediazonium chloride, **145** ("*in situ*" means that HONO is formed in the reaction mixture where reaction with amine occurs and is not isolated).

23 **132** **133**

Diazonium salt **145** (benzenediazonium chloride) is relatively stable *in aqueous solution*, and it will stay in solution long enough to give predictable chemical reactions. Note that many if not most *diazonium salts tend to be **dangerously explosive** if allowed to dry*. The cautionary note is added because most are unstable when dry and taking precautions with these reagents is usually a good idea. It is important to note that this reaction is restricted to primary amines ($-NH_2$) such as aniline. Reactions of secondary amines with HONO give *N*-nitrosoamines such as the reaction of *N*-methylbutanamine to give **146**, which also rapidly decomposes. Nitrosoamines derived from aliphatic secondary amines have been labeled as cancer suspect agents and as carcinogens. Nitrosamines may be present in many foodstuffs, such as processed fish and meat and perhaps even beer.

21.66 **Draw the product of a reaction between 1-aminonaphthalene and sodium nitrite in HCl.**

21.9.2 Reactions of Diazonium Salts

Once an aryl diazonium salt is formed in aqueous solution, reaction with several different reagents is possible. The most common is water itself to replace N_2 with OH, but the reaction must be heated in most cases and aqueous acid is used most of the time. An example is the treatment of aniline with HONO to give benzenediazonium chloride (**132**), as noted before. Heating this salt in aqueous sulfuric acid (at temperatures around 160°C) leads to substitution of N_2 by water to give phenol, **19**. *Note that N_2 is an extraordinary leaving group and that this type of substitution does not occur with other leaving groups on the benzene ring except under special circumstances.* The reaction works quite well with substituted aniline derivatives to produce substituted phenol derivatives.

Another common transformation of diazonium salts is their conversion to aryl halides by reaction with cuprous salts (CuX), in what is known as the **Sandmeyer reaction**, named after Traugott Sandmeyer (Switzerland; 1854–1922). This means that the Ar–NH_2 → ArX conversion is possible, where Ar = an aryl group. When **132** is treated with cuprous bromide (CuBr), the product is bromobenzene (**18**). The reaction works with many other cuprous salts as well, including cuprous chloride (CuCl). A variation of this reaction treats the diazonium salt with cuprous cyanide (CuCN) to give a nitrile. In this manner, 4-methylaniline (**24**) is treated with HCl and $NaNO_2$ and then with CuCN to give 4-methyl-1-cyanobenzene (**134**).

21.67 Draw the product of a reaction between 3,4-dimethylaniline and $NaNO_2$ in HCl, followed by treatment with aqueous sulfuric acid at 160°C.

21.68 Draw the product of the reaction between **132** and cuprous chloride.

21.69 Write out a synthesis of 3-ethylbenzonitrile from benzene.

In preceding sections, bromobenzene and chlorobenzene were generated by various methods, but not iodobenzene or fluorobenzene. Diazonium salts provide a route to both of these compounds. When **132** is treated with potassium iodide (KI), the product is iodobenzene, **135**. Fluorides can be prepared by changing the acid used to prepare the diazonium salt. When 3-nitroaniline (**136**) is treated with $NaNO_2$ and tetrafluoroboric acid (HBF_4), the product is the diazonium tetrafluoroborate, **137**. When this salt is isolated and heated to 100–150°C, the product is 3-fluoronitrobenzene, **138**. (Although these salts are much more stable and can often be isolated, care should be exercised.)

21.70 Write out a synthesis of 4-fluorobromobenzene from benzene.

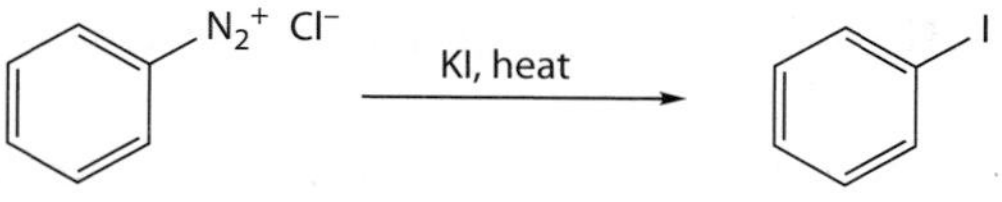

Finally, it is possible to remove nitrogen completely from the molecule and replace it with a hydrogen (a reduction). This reduction is commonly done by heating a diazonium salt with hypophosphorus acid (H_3PO_2).

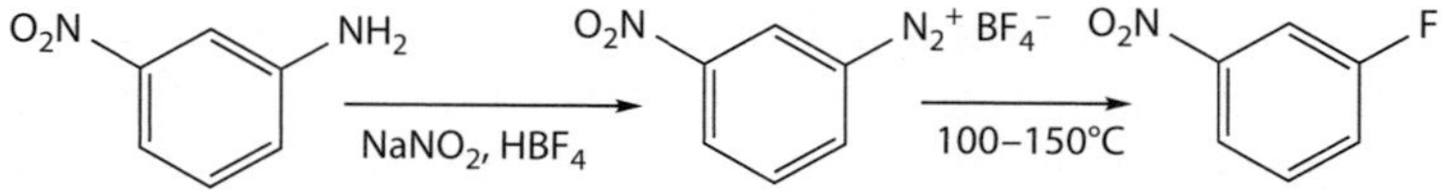

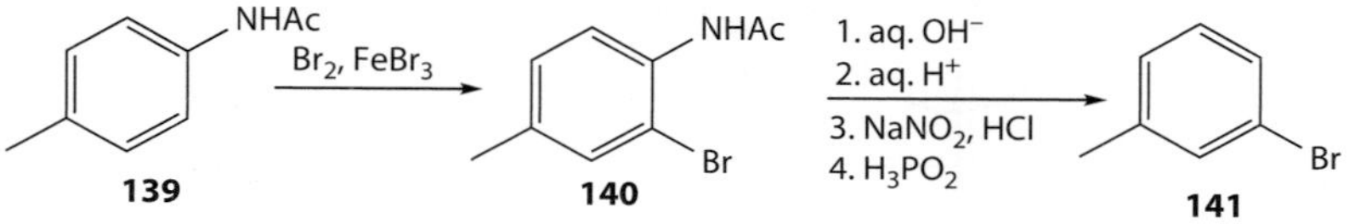

An example is the reaction of aniline with $HCl/NaNO_2$ to give **132**. Subsequent heating with H_3PO_2 gives benzene. This reaction is used to prepare compounds that are difficult to obtain by any other method. 4-Methyl *N*-acetylaniline (**139**) is brominated ($Br_2/AlBr_3$) to give **140**. Note that bromination occurs *ortho* to the more activating acetamide group. When treated with aqueous hydroxide and then aqueous acid, the amide is hydrolyzed to the amine (see Chapter 20, Section 20.2, for amide hydrolysis). Subsequent reaction with $NaNO_2/HCl$ and then H_3PO_2 gives 3-bromotoluene (**141**). Because both bromine and methyl are *ortho/para* directors, it would be impossible to obtain **141** from either bromobenzene or toluene by electrophilic aromatic substitution. The nitrogen group serves as a blocking group for that carbon in **141** that allows the bromine to be properly positioned. It is removed via the diazonium salt to give the desired product.

21.71 Write out this reaction sequence.

21.72 Draw the structure of this amine product.

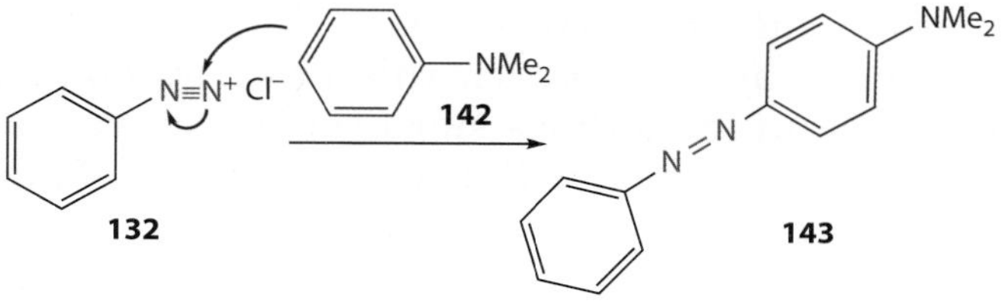

Using aryl diazonium salts, it is possible to make a series of highly colored and highly important compounds known as **azo dyes**. In Chapter 1, the dye mauveine A (see **148**) was described as one of the first commercial dyes. William Perkin (England; 1838–1907) prepared **148** in 1858 from aniline, and it was shown to be valuable for the preparation of many other dyes, including fuchsine, safranine, and the indulines. Mauveine as well as other diazo compounds is produced by the reaction of a diazonium salt with a benzene derivative that has an electron-releasing substituent.

One example is the reaction of **132** with *N,N*-dimethylaniline (**142**) to produce diazo compound **143**. The reaction works only when one benzene derivative is highly reactive (contains a powerful activating substituent); the mechanism involves attack of one benzene ring on the diazo unit of the second. Many such compounds are prepared in this manner. An example is the coupling reaction of the diazonium salt (**145**) derived from 4-nitroaniline (**144**) and the sodium salt of 2-hydroxynaphthalene (**146**) to give **147** (this has the common name of *para red* and is used as an acid–base indicator).

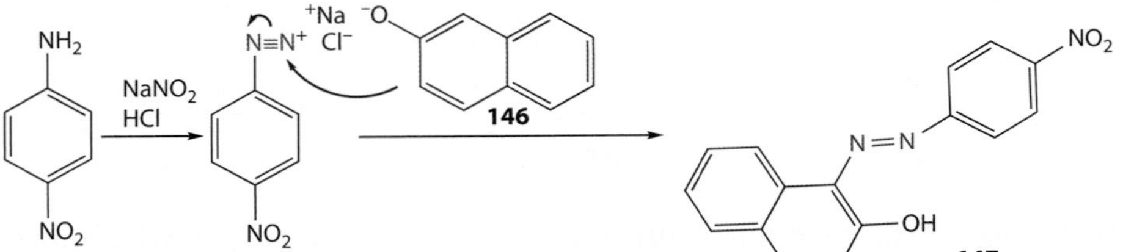

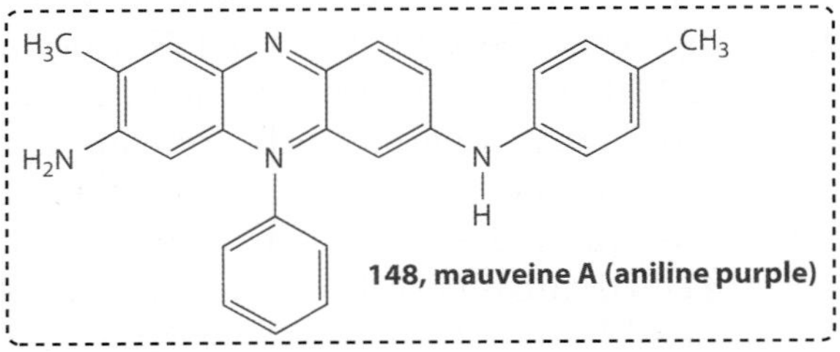

21.10 Nucleophilic Aromatic Substitution

Benzene undergoes nucleophilic substitution reactions *if* there is a good enough leaving group such as nitrogen gas (N_2), as introduced in Section 21.9. There is another way to approach this problem, however. If the leaving group cannot be changed, perhaps substituents can be modified to facilitate the substitution. This section will discuss reactions of nucleophiles with suitably substituted benzene derivatives in what is called nucleophilic aromatic substitution.

21.10.1 Nucleophilic Attack at an Aromatic Carbon

Chapter 11 (Section 11.2) discussed the S_N2 reaction of a nucleophile with an sp^3 hybridized carbon that contains a leaving group. At that time, it was assumed that the reaction could not occur at an sp^2 carbon. That is not entirely true, but vigorous conditions are required to facilitate this reaction. When the sp^2 carbon is part of a benzene ring, nucleophilic substitution is difficult, so when chlorobenzene (**149**) is heated to 50°C in aqueous hydroxide, there is no reaction. Hydroxide does not easily displace chloride by attack at the sp^2 carbon *unless* the reaction is placed in a sealed reaction vessel and heated to 350°C so that it is pressurized. Under these vigorous conditions, hydroxide attacks the sp^2 bearing the chlorine (called the ***ipso* carbon**) to generate a carbanionic intermediate, **159**.

Although this intermediate is resonance stabilized, it is quite high in energy and very reactive. Chloride ion is rapidly expelled to form phenol (**19**) initially; however, phenol is a moderately strong acid (pK_a of about 10) and it reacts with the aqueous hydroxide to form phenoxide **151** as the isolated product. After the substitution is complete, the reaction is cooled and the basic phenoxide ion is neutralized (an acid–base reaction) with dilute acid to regenerate phenol, **19**. This reaction is an example of an aromatic substitution process and is usually called **nucleophilic aromatic substitution**.

21.73 Write out a synthesis of 4-methylphenol from toluene.

This reaction requires high reaction temperatures because the benzene ring is electron rich and hydroxide is an electron-rich species. There is a natural tendency for these two electron-rich species to repel. To bring them together in a reaction requires energy, so the reaction has a rather high E_{act}. Therefore, the collision of these two species at normal temperatures is unproductive (i.e., there is no reaction). With sufficient heat, however, the so-called aromatic S_N2 reaction (termed S_NAr), in which Cl is replaced with OH, proceeds. In fact, this is the basis for the industrial preparation of phenol. Phenol (earlier called carbolic acid) was one of the first antiseptics used in operating rooms in the nineteenth century, and it is used today as an ingredient in many mouthwash formulas.

Another S_NAr reaction heats benzenesulfonic acid (**42**) with aqueous sodium hydroxide (to 300°C) in a reaction bomb. The fastest reaction is the deprotonation of the acid by hydroxide to give sodium benzenesulfonate (**152**), an acid–base reaction with the acidic phenol; **152** reacts with hydroxide to give the resonance-stabilized carbanionic intermediate, **153**. Loss of sodium sulfite (Na_2SO_3) gives phenoxide, **154**. Cooling the reaction and neutralization with acid give phenol, **19**.

This reaction is used to make aniline derivatives. When bromobenzene (**18**) is heated with ethylamine ($EtNH_2$) in aqueous solution (to about 300°C), the aromatic S_N2 reaction is slow, but the product is *N*-ethylaniline, **25**. In general, aromatic S_N2 reactions are slow in water or alcohol solvents and much faster when tertiary amide solvents such as dimethylformamide (DMF, **155**) or dimethylacetamide (**156**) are used. The reaction of **18** with ethylamine, for example, gives a good yield **25** at a reaction temperature of only 160–200°C when DMF is used as the solvent.

21.74 Write out a synthesis of *N,N*-dimethylaniline from benzene.

In some cases, aromatic substitution proceeds under very mild conditions. This occurs when the carbanion intermediate is particularly stable, but electron-withdrawing groups must be attached to the benzene ring. The reaction of 2,4,6-trinitrochlorobenzene (**157**) and aqueous hydroxide, for example, occurs at room temperature to give the trinitrophenol, **159**. The intermediate is **158**, and the negative charge is on a carbon atom that is adjacent to a nitro group in several resonance contributors. It is clear from the six resonance structures that the positive charge on the nitrogen of the nitro group stabilizes the negative charge in the ring by delocalization onto each of the three nitro groups. This is an extreme case where nitro groups are exactly positioned on the ring at those sites where the negative charge appears by resonance delocalization; this will maximize resonance delocalization and stability. Intermediate **158** is so stable that aromatic substitution is a very facile process (it can occur in warm water, without the need to add hydroxide). The presence of an electron-withdrawing group will make the rate of an aromatic substitution reaction faster.

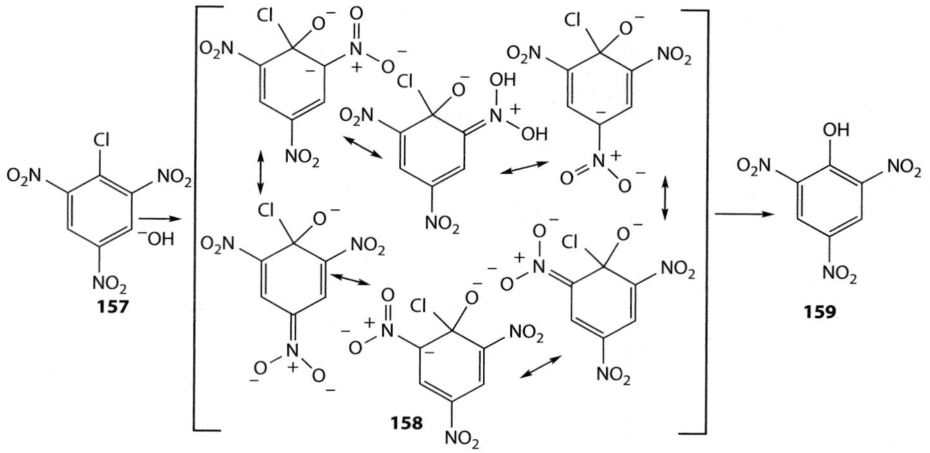

The more powerful the electron withdrawing group and the greater the number of those groups on the ring at the *ortho* and/or *para* positions, the faster the reaction will be (i.e., the reaction conditions are milder). Conversely, the presence of electron-releasing groups (OR, NR$_2$, alkyl, etc.) on the ring makes the rate of an aromatic substitution reaction slower and more difficult.

21.75 Determine why the aromatic substitution reaction of hydroxide with 3-nitro-1-bromobenzene is much slower than the identical reaction with 4-nitro-1-bromobenzene.

21.10.2 Aromatic Elimination Reactions. Benzyne

Chapter 12 described an elimination reaction that involves sp^3 atoms bearing a leaving group. Earlier in this chapter, the second step of electrophilic aromatic substitution involved an E1 reaction. *When an aryl halide reacts with a strong base such as methoxide or sodium amide, an acid–base reaction is possible in which a hydrogen atom on the benzene ring is removed.* When the aromatic halide contains an electron-withdrawing group, the rate of aromatic substitution reactions is relatively fast and dominates the reaction pathway. When the aromatic halide contains an electron-releasing group on the benzene ring, however, the aromatic substitution reaction is too slow. If such compounds are treated with a base, however, an elimination reaction can occur via an initial acid–base reaction.

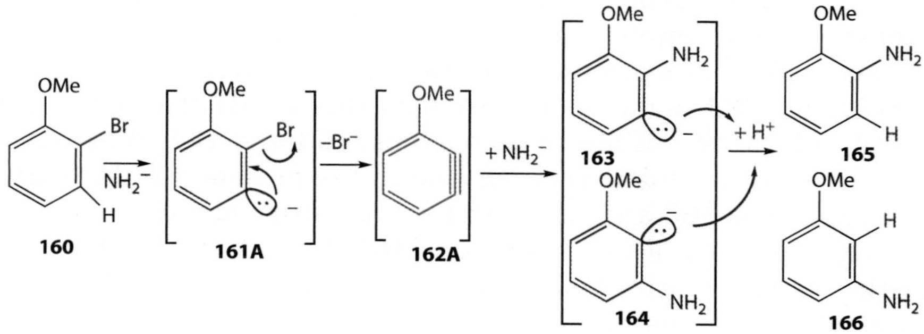

When 2-bromoanisole (**160**) reacts with sodium amide ($NaNH_2$) in liquid ammonia (at –33°C), a reaction occurs to give 2-aminoanisole (**165**) along with 3-aminoanisole (**166**). This appears to be a substitution reaction where a hydrogen is replaced by $-NH_2$. However, substitution occurs at two different carbon atoms, so *direct substitution is ruled out as a mechanism.* To rationalize these experimental observations, first recognize that sodium amide is a powerful base, and an acid–base reaction removes the hydrogen atom *ortho* to the bromine to give **161**.

To reach this conclusion without experimental evidence, it must be known that the *ortho* hydrogen is a weak acid. Although this fact is not necessarily obvious, this mechanism is supported by many experiments. Once **161A** is formed, it is highly reactive and the electrons of the carbanion unit expel the leaving group bromide, in an aromatic elimination reaction, to form a new π-bond in a molecule with the unusual structure shown for **162A**. Another view of the initially formed anion is **161B**, where it is clear that now two electrons of the carbanion are orthogonal to the π-electrons of the aromatic cloud. When these electrons are transferred to the carbon bearing the bromine (the leaving group), the new π-bond is formed that is orthogonal to the aromatic π-cloud (see **162B**).

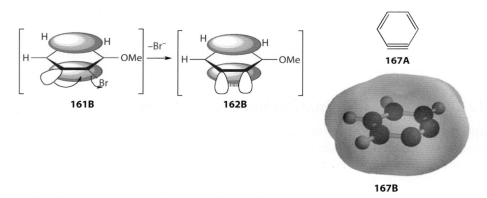

The triple bond shown in **162A** represents the normal Kekulé structure for the benzene ring plus a new π-bond that is orthogonal to the aromatic π-electrons. This benzene ring with a triple bond is known as a **benzyne** and it is highly susceptible to accepting electrons from a suitable nucleophile. Benzyne itself is **167A**, and the electron density map (see **167B**) clearly shows the red areas (higher electron density) in the middle of the ring, consistent with the usual aromatic cloud. However, a region of high electron density perpendicular to the aromatic cloud is consistent with the extra π-bond.

Returning to **162A**, a nucleophile such as amide (NH_2^-) attacks the orthogonal π-bond to form two different carbanions, **163** and **164**. When **163** reacts with ammonia (which functions as an acid in the presence of this carbanionic base), the product is **165** and when **164** reacts with ammonia, the product is **166**. The major product of this reaction is **165**. In **163**, the negative charge is closer to the electronegative OMe unit, which is destabilizing. This destabilization is much less in **164** because the negative charge on the ring is farther away from the OMe unit, and **164** leads to the major product. This is a somewhat specialized reaction, but it is quite useful for the preparation of aromatic derivatives that are not available by other methods.

21.76 Write out an acid–base reaction in which benzene is the acid and sodium amide is the base.

21.77 Write out a synthesis of 160 from benzene.

21.11 Aromatic Disconnections and Functional Group Exchange Reactions

This chapter provides two new carbon–carbon bond-forming reactions and several new functional group exchange reactions. The two carbon–carbon bond-forming reactions are the Friedel–Crafts alkylation and acylation reactions. The disconnection for both of these reactions is shown, giving a benzene ring and either an alkyl halide or an acid chloride, which is derived from the carboxylic acid.

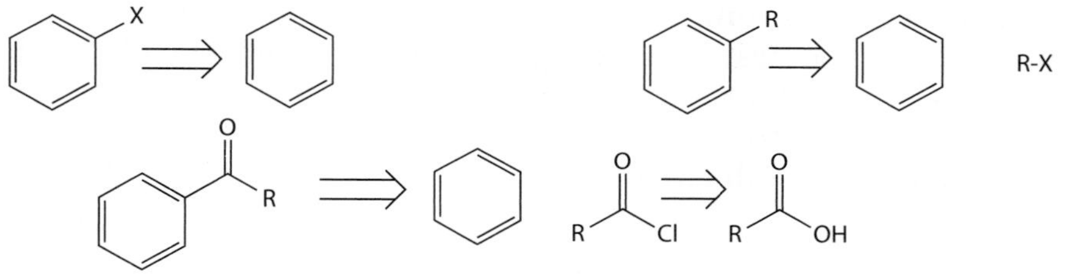

The functional group transformations are derived from either electrophilic aromatic substitution or nucleophilic aromatic substitution reactions. The electrophilic aromatic substitution functional group transform is shown with a simple "X" group, where X is chlorine, bromine, nitro, or sulfonyl. The reagents are different, but the basic principle for the formation of such compounds is the same.

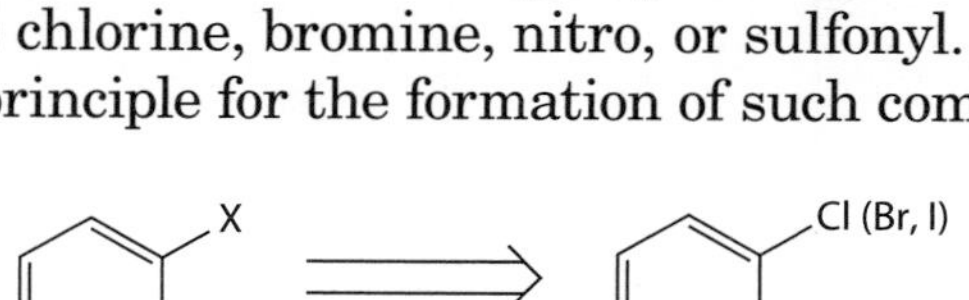

Nucleophilic aromatic substitution reactions have three distinct versions: displacement of diazonium salts, aromatic S_N2 reactions, and benzyne reactions. The transform involving diazonium salts involves X and leads back to an aniline derivative. In this transform, X is halogen, OH, CN, or –N=N–Ar. For aromatic S_N2 reactions, the X group is OH, OR, NR_2, and the "Y" groups on the benzene ring imply the presence of electron-withdrawing substituents. For the benzyne transform, the X group is OH, OR, NR_2; notice that the halide precursor is located *ortho* to the site of the X group.

21.12 Synthesis of Aromatic Compounds

A basic strategy for synthesizing aliphatic compounds is introduced in Chapter 25. This strategy can work with aromatic compounds, but it requires modification. In general, all disconnections occur between the benzene ring and the substituent. For electrophilic aromatic substitution, it is important to recognize whether the substituent is electron releasing or withdrawing, activating or deactivating, and an *ortho/para* director or a *meta* director. Because the diazonium group is a blocking/directing group (see Section 21.9), this must be factored into any strategy. This section will provide a few examples of the synthesis of substituted aromatic hydrocarbons to illustrate the general strategy used, which supplements the examples given in the problems following the references.

The first problem will prepare 4-bromoaniline (**168**) from benzene. Syntheses that transform one aromatic compound into another aromatic compound, such as this one, do not lend themselves to the retrosynthetic analysis approach presented in Chapter 25. Most of these syntheses involve functional group transformations. For the sake of continuity, a retrosynthesis is shown in which the amino group is removed to generate bromobenzene, which is obtained directly from benzene. The NH_2 unit probably comes from reduction of a nitro group (Section 21.6.2), so the first precursor is 4-bromonitrobenzene (**59**) and disconnection of the C–N bond leads to the preparation of **59** by reaction of bromobenzene (**35**) with nitric acid/sulfuric acid. Bromobenzene is prepared directly from benzene as shown in the illustration.

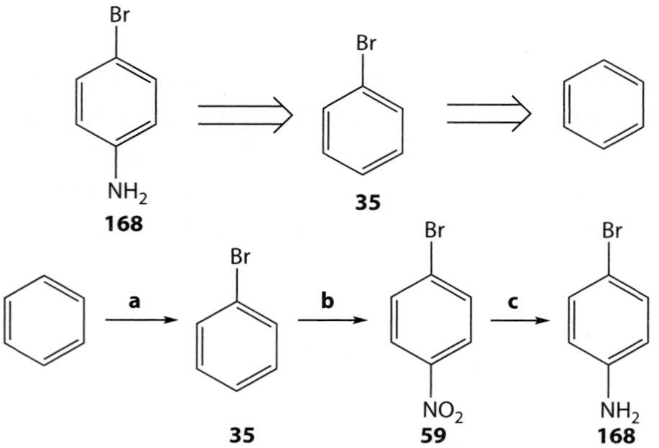

(a) Br_2, $FeBr_3$
(b) HNO_3/H_2SO_4
(separate *ortho*- compound)
(c) H_2, Pd-C

The second problem prepares ethoxybenzene (**21**; phenetole) from benzene, and the retrosynthetic analysis is **21** ⇒ **19** ⇒ **42** ⇒ benzene. Because **21** is an ether, the most likely precursor is phenol (**19**), which is prepared from aniline (**23**) via the diazonium salt. Aniline is derived from nitrobenzene (**42**), which is prepared directly from benzene. Therefore, the synthesis is that shown in the following illustration.

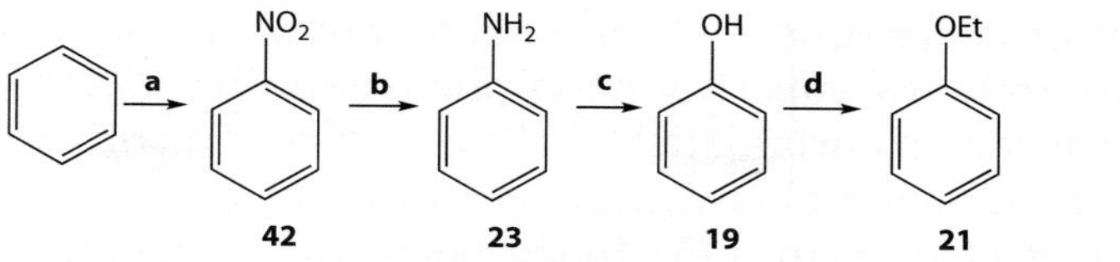

42 **23** **19** **21**

(a) HNO₃/H₂SO₄ (b) H₂, Pd-C (c) i. NaNO₂, HCl ii. aq. H₂SO₄,160°C (d) i. NaH ii. EtBr

In the third problem, 3-hydroxyaniline (**171**) is prepared from benzene, but the OH and NH₂ groups (both activating and *ortho/para* directors) are *meta* to each other. The retrosynthesis is **171** ⇒ **170** ⇒ **169** ⇒ **44** ⇒ benzene. This means that these groups must be derived from other groups that were *meta* directors and then chemically modified to the groups shown. The NH₂ group is derived from a nitro group, so one precursor is probably **170**. Because the OH group may also be derived from an NH₂ group (and therefore from a nitro group), one possible precursor is 1,3-dinitrobenzene. This is a problem because it is difficult to reduce both nitro groups to NH₂ and then selectively convert one of them to an OH. In Section 21.10, phenol was prepared from a sulfonic acid via an aromatic S_N2 reaction. Using this strategy, **170** is derived from sulfonic acid **169**, and this is prepared by nitration of benzenesulfonic acid, **44**. Because **44** is prepared directly from benzene, the synthesis is complete.

21.78 **Write out a synthesis of 3-chlorobromobenzene from benzene.**

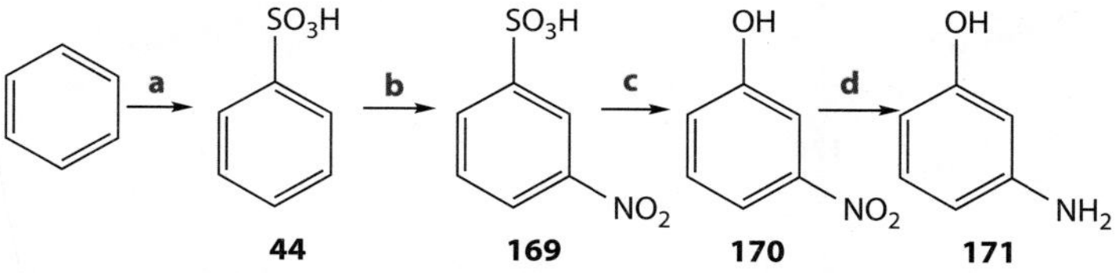

44 **169** **170** **171**

(a) SO₃/H₂SO₄ (b) HNO₃/H₂SO₄ (c) aq. NaOH, 200°C (d) H₂, Pd-C

21.13 Biological Relevance

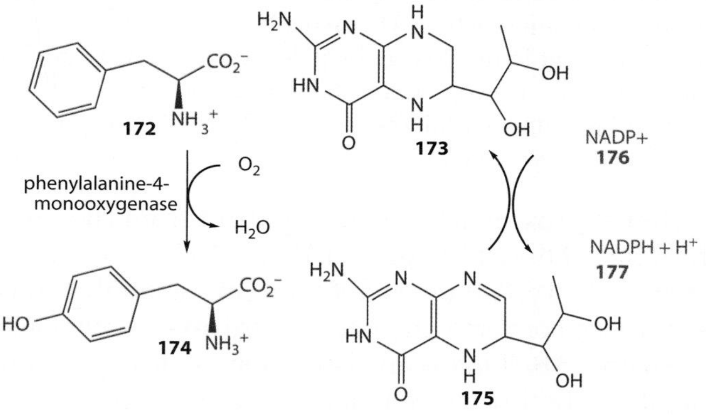

The biosynthesis of tyrosine (**174**) in some mammals involves an oxidation that is formally an aromatic substitution of phenylalanine (**172**).[5] Phenylalanine is obtained from food in a normal diet; in the presence of tetrahydrobiopterin (**173**), oxygen, and the enzyme phenylalanine-4-monooxygenase, tyrosine is formed along with dihydrobiopterin (**175**). In the body, dihydrobiopterin is converted by NADPH (**176**) back to **173** and NADP+ (**177**). Note that the conversion of **176** to **177** is for all practical purposes a Birch reduction (see Section 21.6.1).

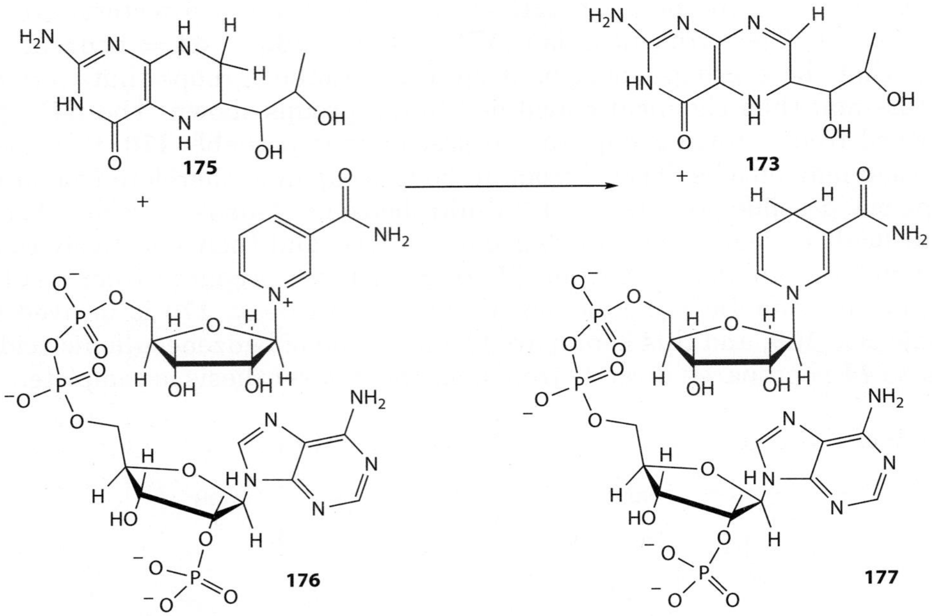

Thyroxine (3,5,3′,5′-tetraiodothyronine, **178**; abbreviated as **T4**) is the major hormone secreted by the follicular cells of the thyroid gland. In the hypothalamus, T4 is converted to triiodotyronine, T3 (**179**) via an enzyme called a deiodinase (tetraiodothyronine 5′ deiodinase). T3 is the main inhibitor of thyroid-stimulating hormone.

> Thyroxine (also known as Levothyroxine or T4) is the ultimate metabolism regulator. Its reactions and products influence carbohydrate metabolism, protein synthesis and breakdown, and cardiovascular, renal, and brain function. Thyroxine is essential to an animal's functions and it is essential for development in the young. Tadpoles won't develop into frogs, for example. Untreated human babies will develop cretinism, a condition marked by severe mental and physical retardation. Adult humans with low thyroxine levels (hypothyroidism) suffer mental slowness, weight gain, depression, and fatigue.[6]

It is believed that thyroxine is formed in nature from the amino acid tyrosine (**180**) through the stage of diiodotyrosine, **181**. When iodine is taken into the body, it is covalently bound to tyrosine residues in thyroglobulin molecules, and the enzyme thyroperoxidase converts the bound tyrosine to monoiodotyrosine (MIT, **181**) and diiodotyrosine, **18**. This transformation is clearly an enzyme-mediated electrophilic aromatic substitution reaction. Other enzymes convert **182** to **179**.

178 179

180 181 182

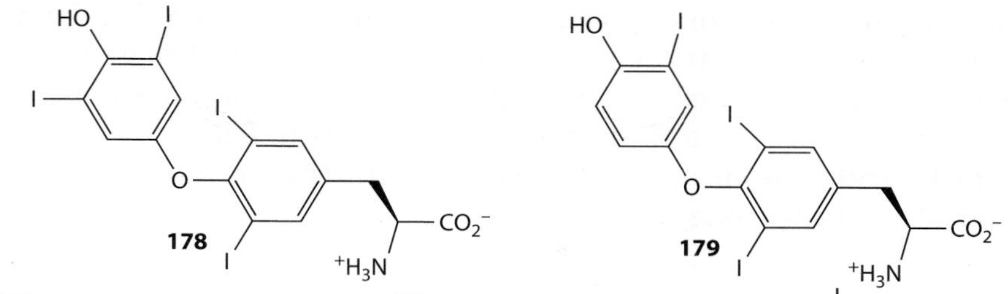

A number of important natural products (compounds that are produced in nature; see Chapter 1) contain chlorine atoms, including vancomycin (**183**) and cryptophycin A (**184**). Vancomycin is used to treat infections caused by Gram-positive bacteria and it has traditionally been reserved as a drug of "last resort," used only after treatment with other antibiotics has failed. A handful of other antibiotics have now been developed to fill this role. Their biosynthesis involves regioselective chlorination by flavin-dependent halogenases. One of these enzymes is tryptophan 7-halogenase (PrnA); it regioselectively chlorinates tryptophan (**184**),[7] using HOCl (see Chapter 10, Section 10.4.2), which is produced biosynthetically from chloride ion.

183 184

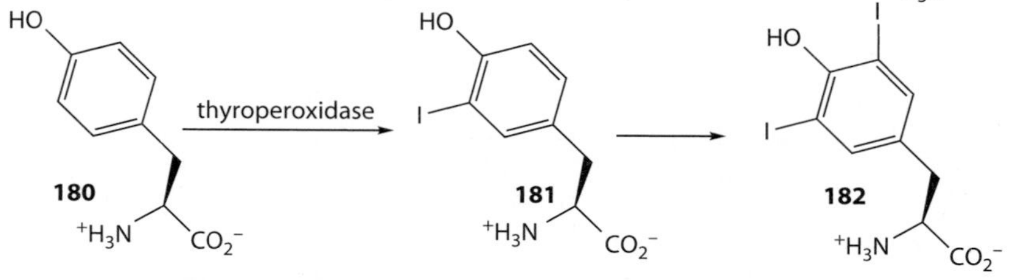

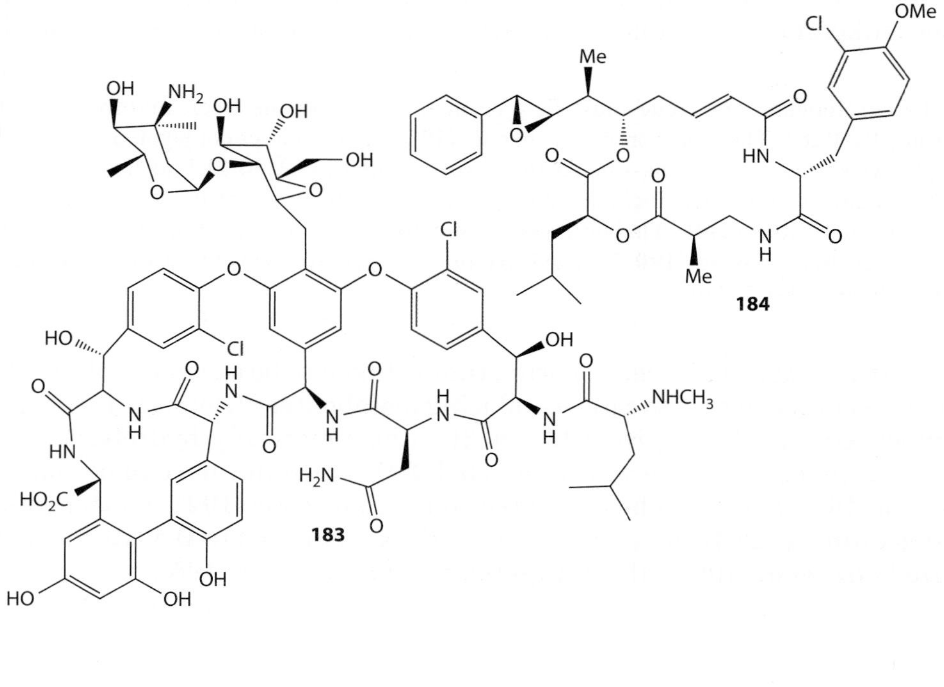

Using amino acid residue [79]K (a lysine residue, **187**), binding HOCl allows electrophilic substitution on the indole ring as shown in **184** to give an arenium ion, **185**. (Indole is discussed in Chapter 26, Section 26.5.) Loss of the proton to amino acid residue E[346] (a glutamic acid residue, **188**) leads to the chlorinated indole unit in **186**. The numbers associated with the amino acid residue refer to their positions in the enzyme.

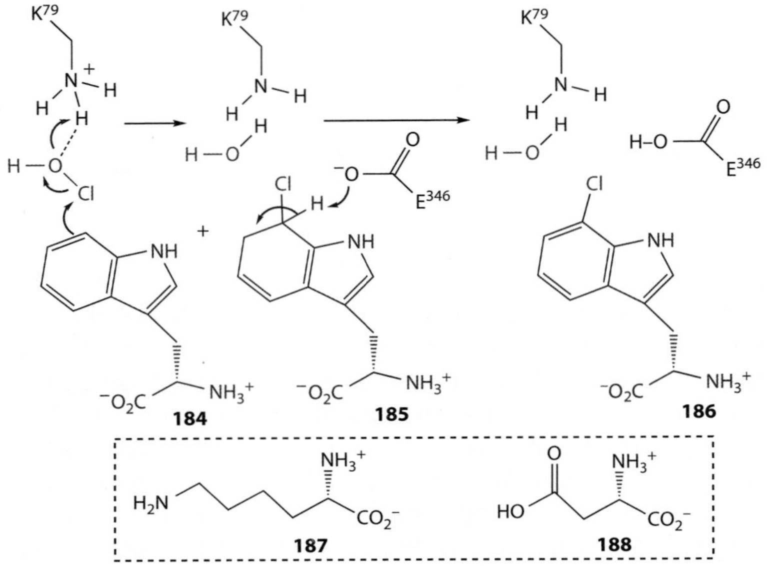

From Dong, C., Flecks, S., Unversucht, S., Haupt, C., ván Pée, K.-H., Naismith, J.H. *Science* 2005, 309, 2216–2219. Reprinted with permission from AAAS.

"Benzene is one of the best studied of the known human carcinogens. It causes leukemia in humans and a variety of solid tumors in rats and mice."[8]

To be carcinogenic, benzene must first be metabolized in the liver, mainly via cytochrome P4502E1. The major product is phenol (**19**), which is either conjugated—primarily to phenyl sulfate in humans—or further hydroxylated by P450 2E1 to hydroquinone. Other major metabolites include catechol (**34**) and *trans-trans*-muconic acid (**189**; 1,6-hexadienedioic acid). The latter is presumed to be formed from the ring opening of benzene epoxide (**190**; benzene oxepin), or perhaps benzene dihydrodiol (**191**; 3,5-cyclohexene-1,2-diol).[9]

A series of reactions has been reported that is taken to be the metabolic pathway for conversion of benzene to cancer-inducing metabolites.[10] Benzene is epoxidized by cytochrome P-450 to give **192** and then enzymatically hydrolyzed to **193**. This intermediate is converted to catechol (**34**), which in turn can be converted to phenol (**19**). Phenol can be converted to hydroquinone (**192**), which gives rise to benzoquinone (**201937**) and 1,2,4-trihydroxybenenze (**194**). Catechol can be oxidized enzymatically to the 1,2-quinone (*ortho*-quinone, **195**).

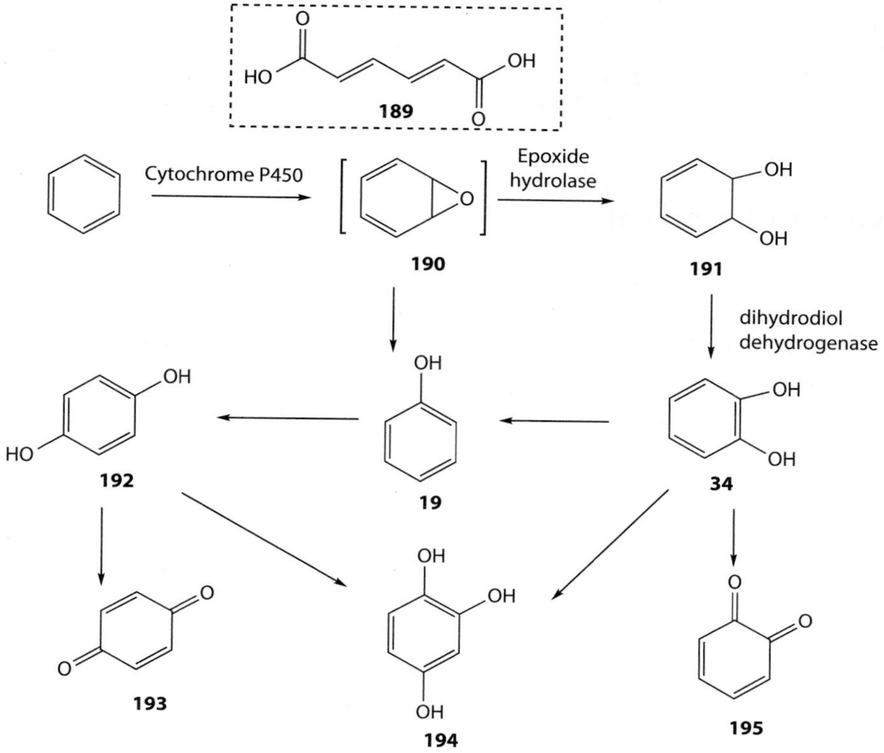

(Smith, M. T. 1996. *Environmental Health Perspectives* 104:1219–1225; Snyder, R., and Hedli, C. C. 1996. *Environmental Health Perspectives* 104, Supplement 6:1165–1171. Reproduced with permission from Environmental Health Perspectives.)

References

1. Fieser, L. F., and Fieser, M. 1961. *Advanced organic chemistry,* 650. New York: Reinhold Pub.
2. Furniss, B. S., Hannaford, A. J., Smith, P. W. G., and Tatchell, A. R., eds. 1994. *Vogel's textbook of practical organic chemistry,* 5th ed., 1008–1009, Exp. 6.121. Essex, England: Longman.
3. Fieser, L. F., and Fieser, M. 1961. *Advanced organic chemistry,* 635. New York: Reinhold Pub.
4. Cram, D. 1965. *Fundamentals of carbanion chemistry,* 4, 10, 13–14, 43, 48. New York: Academic Press.
5. Garrett, R. F., and Grisham, C. K. 1995. *Biochemistry,* 859–860. Ft. Worth, TX: Saunders.
6. The top pharmaceuticals that changed the world. 2005. *Chemical and Engineering News* 83(25). See http://pubs.acs.org/subscribe/journals/cen/83/i25/toc/toc_i25.html.
7. Dong, C., Flecks, S., Unversucht, S., Haupt, C., van Pée, K.-H., and Naismith, J. H. 2005. *Science* 309:2216–2219.

8. Cox, L. A., Jr. 1991. *Risk Analysis* 11:453–464.

9. Smith, M. T. 1996. *Environmental Health Perspectives* 104:1219–1225.

10. Snyder, R., and Hedli, C. C. 1996. *Environmental Health Perspectives* 104, Supplement 6:1165–1171.

Answers to Problems

21.1   The π-bond in **2** is localized between the two carbon atoms and the electrons are readily donated to a hydrogen atom on the hydrogenation catalyst. The π-electrons in **1** are delocalized in the aromatic π-cloud, making them less easily donated; hence, benzene is much less reactive.

21.2

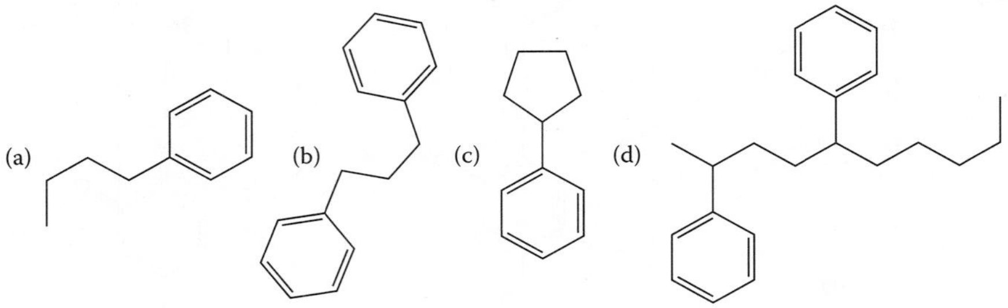

21.3

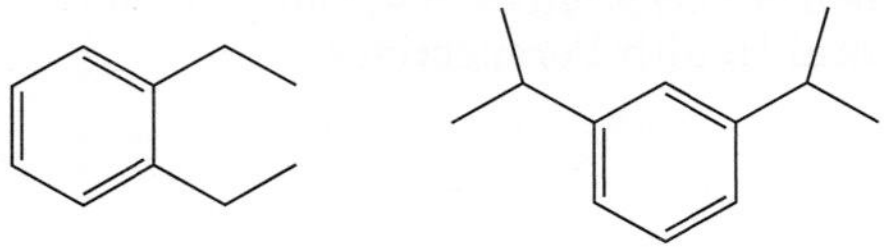

21.4

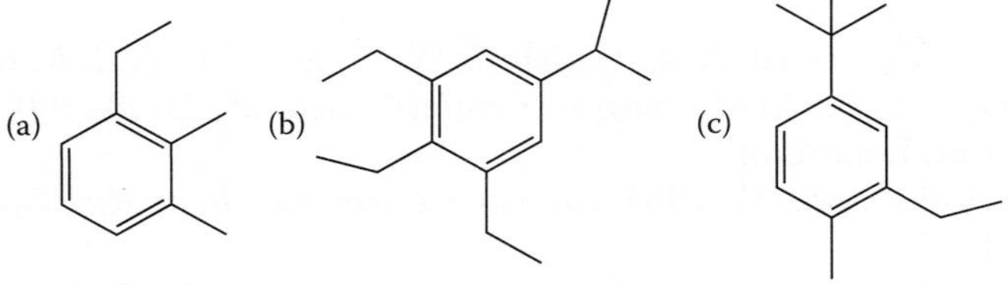

21.5

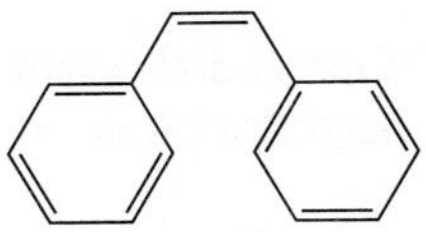

21.6

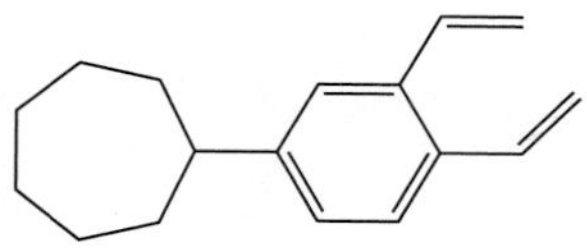

21.7

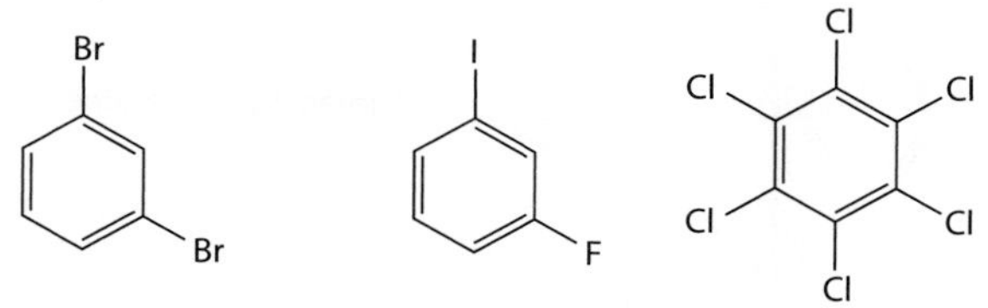

21.8

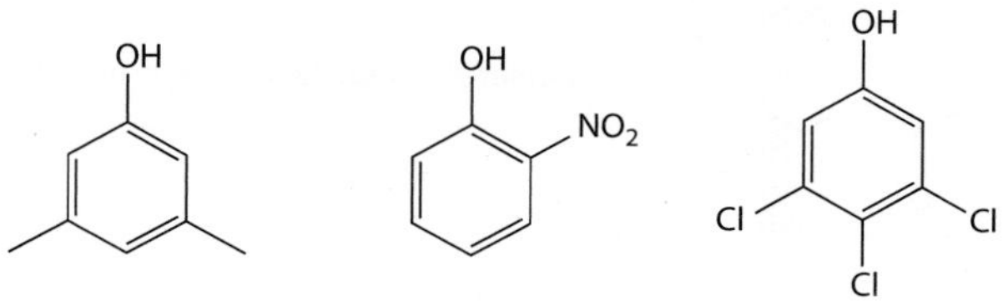

21.9

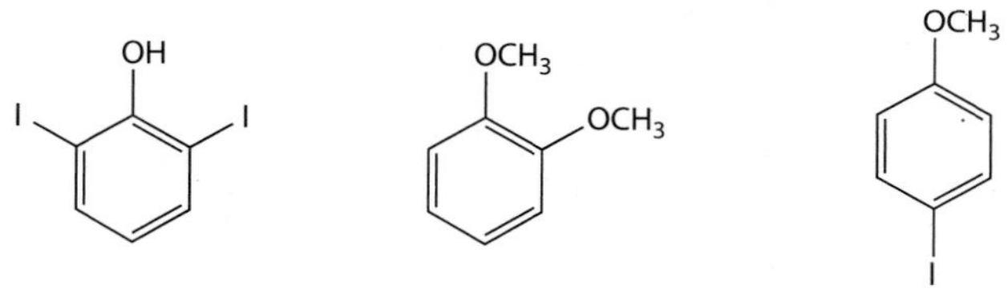

21.10

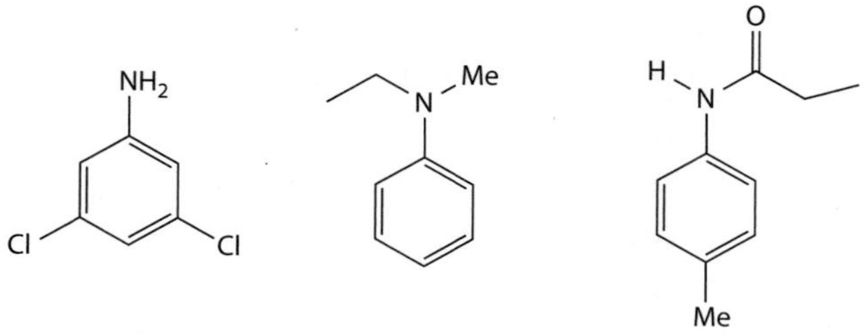

21.11

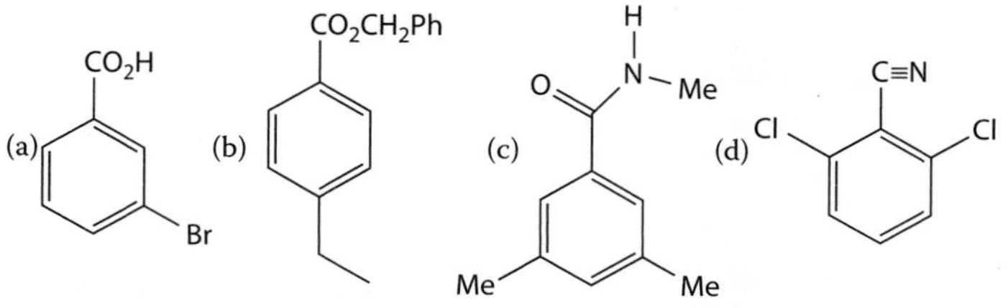

21.12

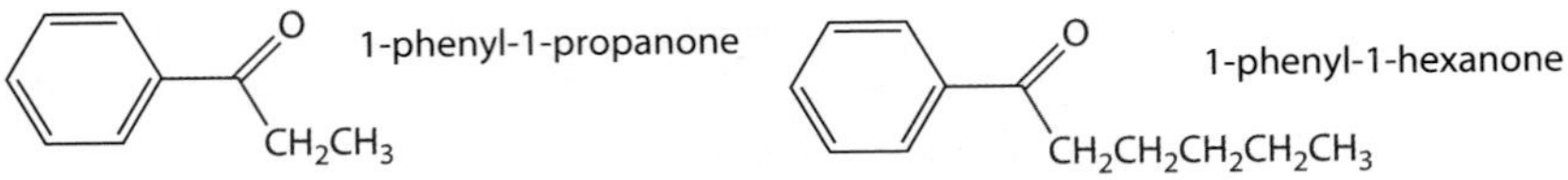
O
CH2CH3
1-phenyl-1-propanone
O
CH2CH2CH2CH2CH3
1-phenyl-1-hexanone

21.13

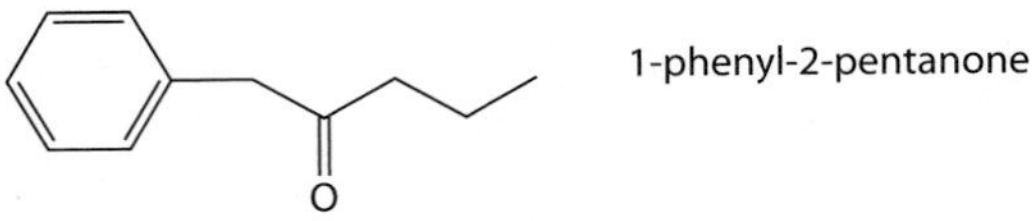
O
1-phenyl-2-pentanone

21.14

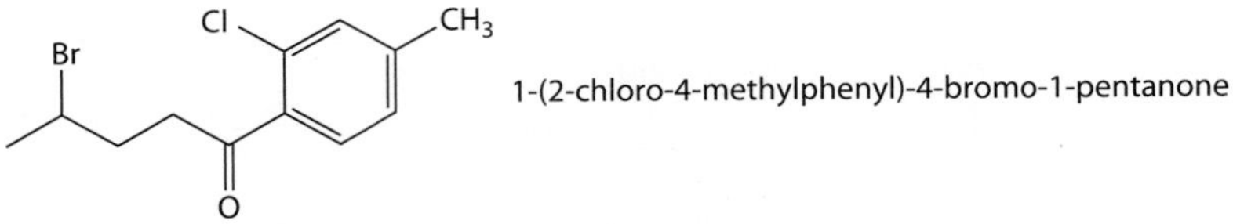
Br
Cl
CH3
O
1-(2-chloro-4-methylphenyl)-4-bromo-1-pentanone

21.15

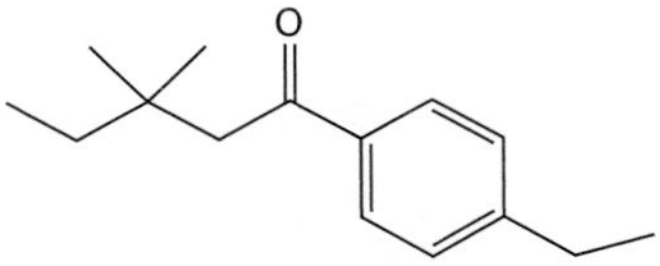
O

21.16

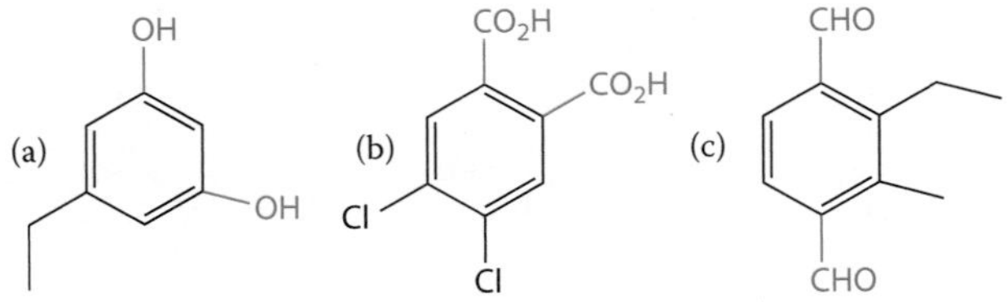
(a)
OH
OH
(b)
CO2H
CO2H
Cl
Cl
(c)
CHO
CHO

21.17

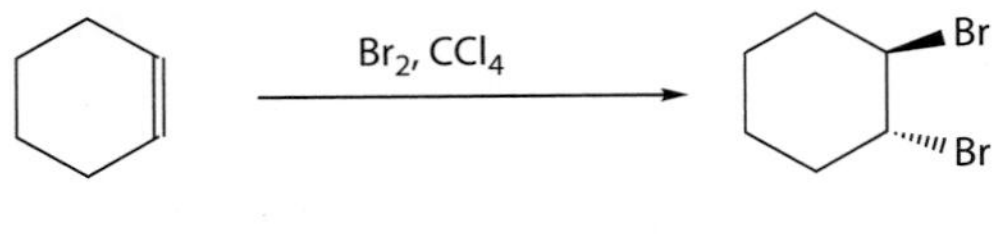
Br2, CCl4
Br
Br

21.18 $Br^+ \ AlCl_3Br^-$. $Br^+ \ FeBr_4^-$.

21.19

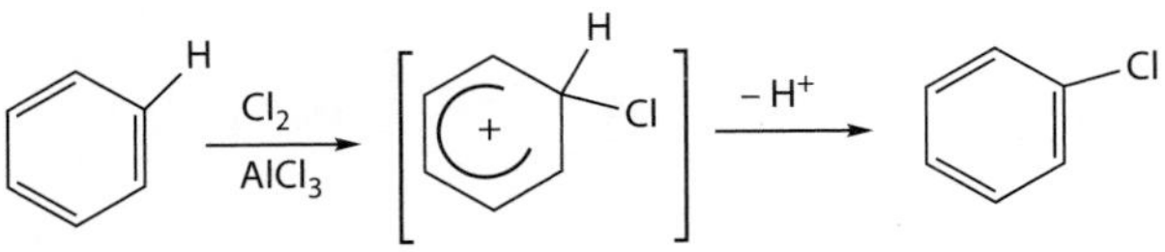
H
Cl2
AlCl3
H
+
Cl
– H+
Cl

21.20

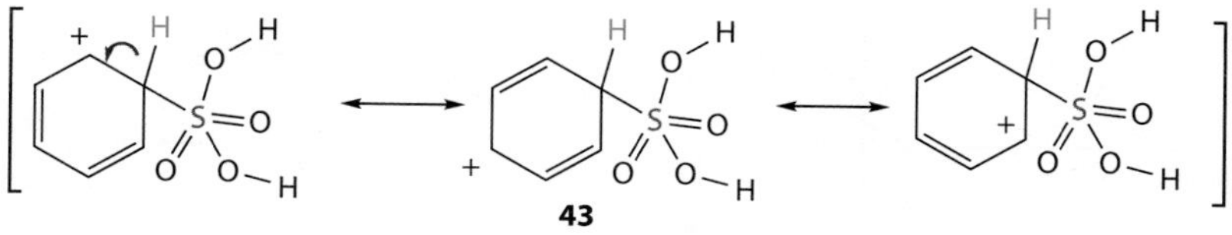

21.21

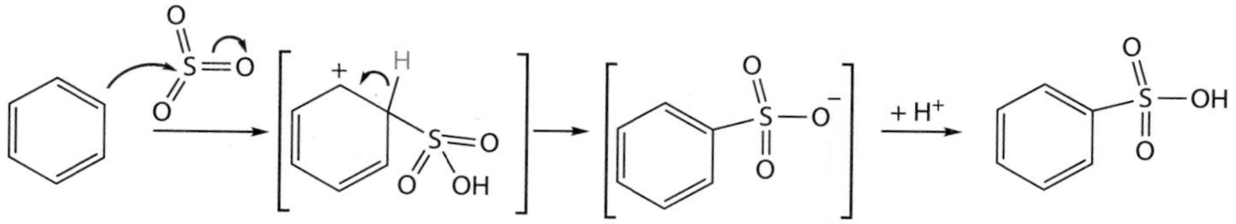

21.22

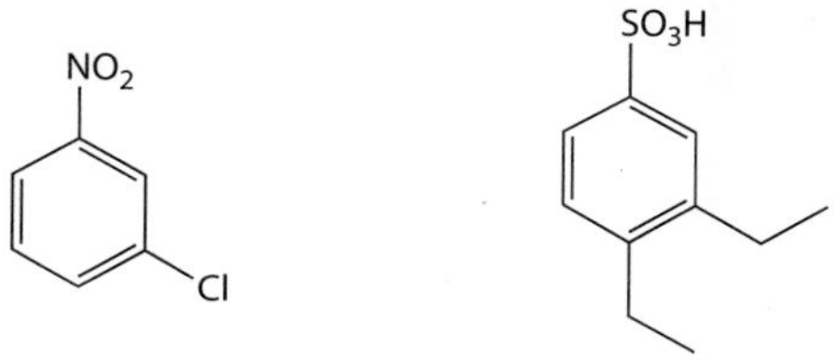

21.23

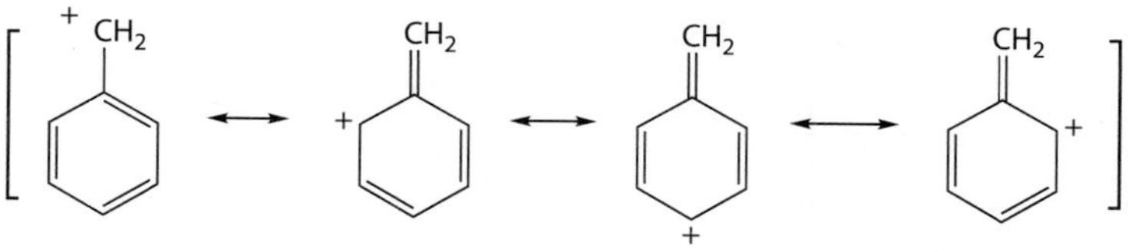

21.24

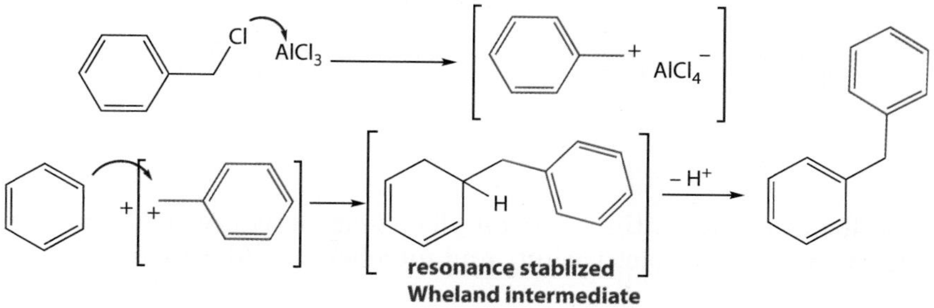

resonance stablized
Wheland intermediate

21.25

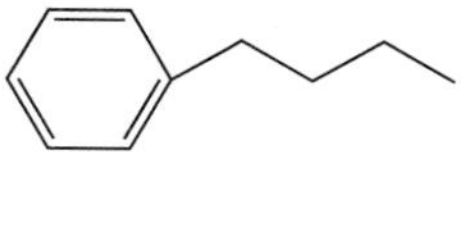

21.26

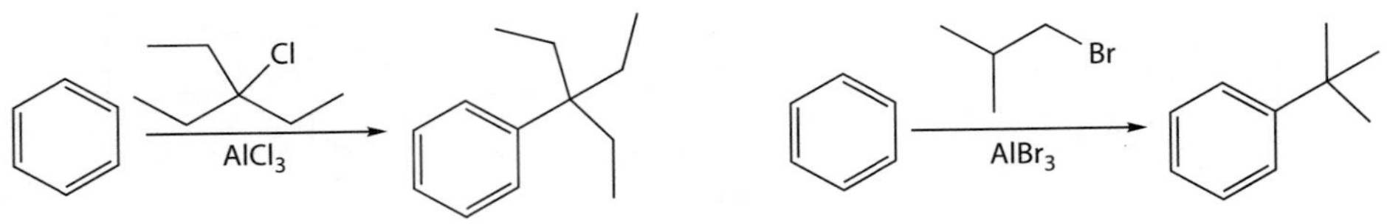

21.27

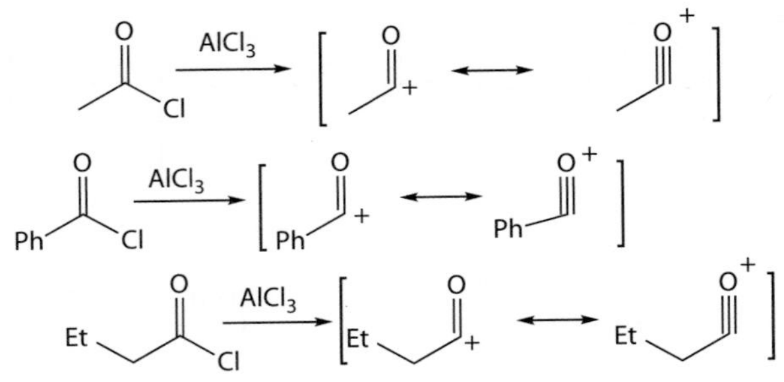

21.28

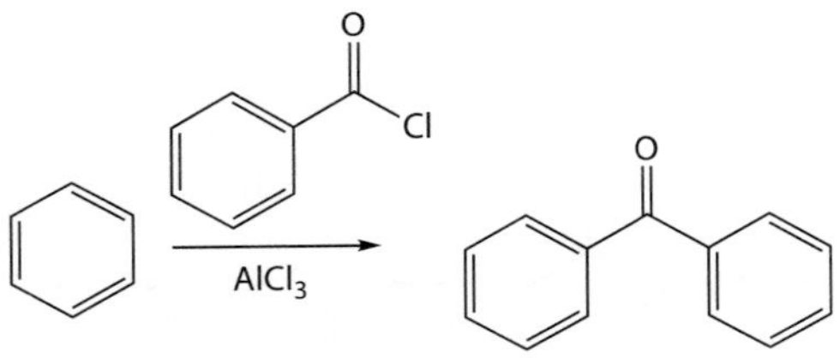

21.29

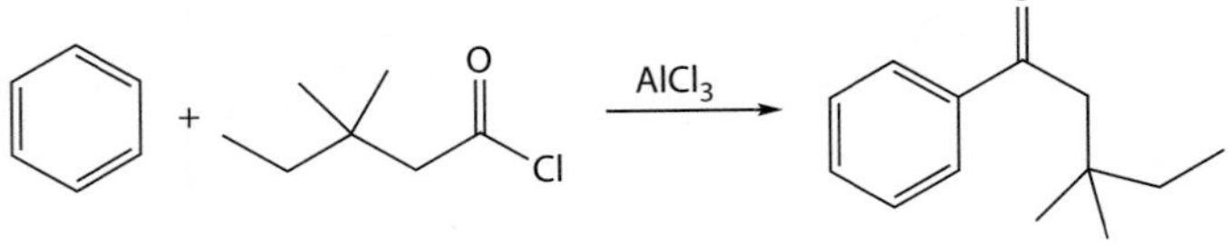

21.30 Acetylbenzene (activating), *tert*-butylbenzene (activating), *N*-acetylbenzene (activating), anisole (activating), and nitrobenzene (deactivating).

21.31

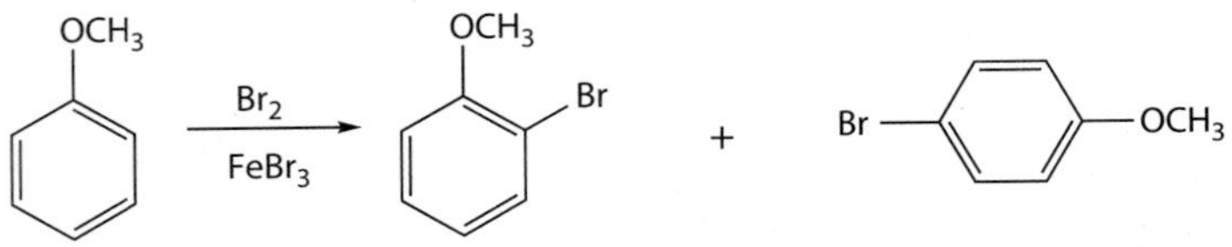

21.32

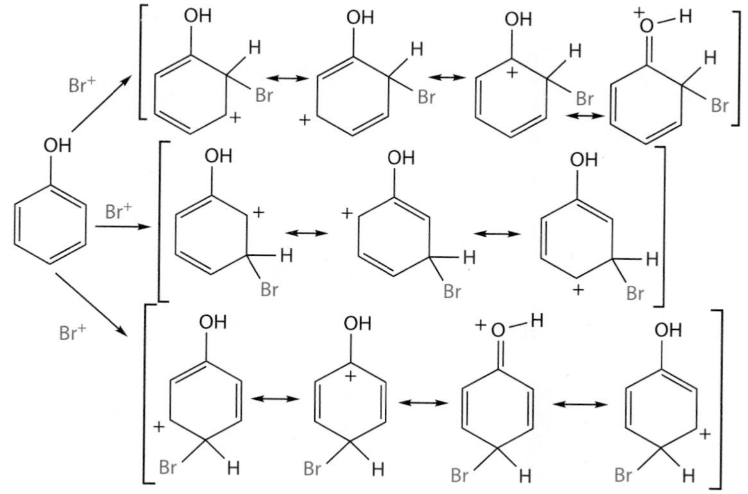

21.33

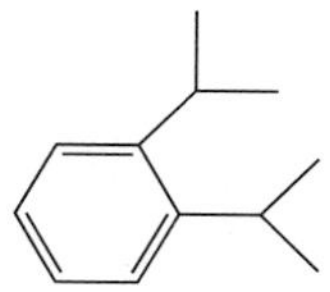

21.34

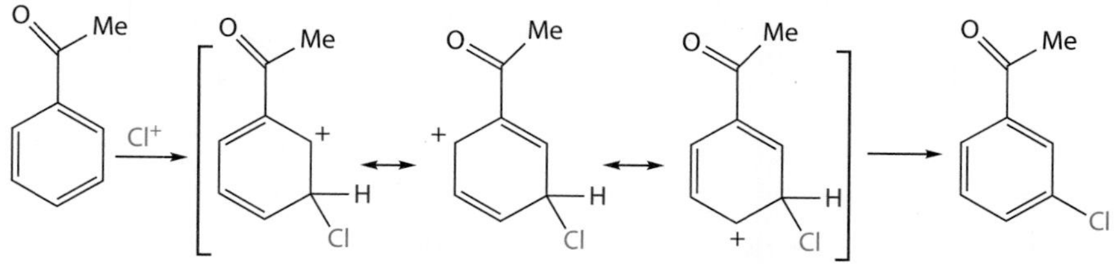

21.35

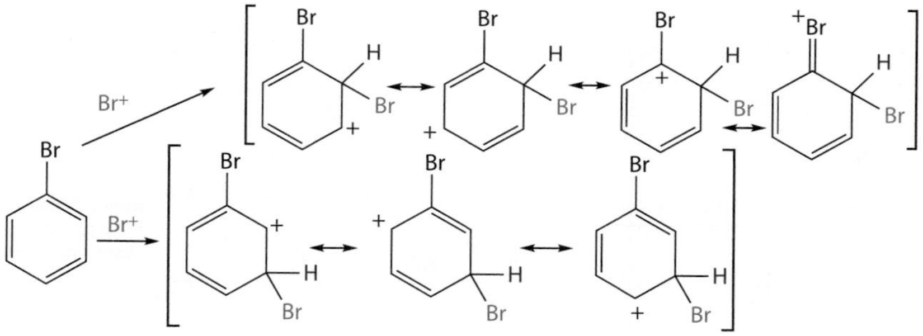

21.36

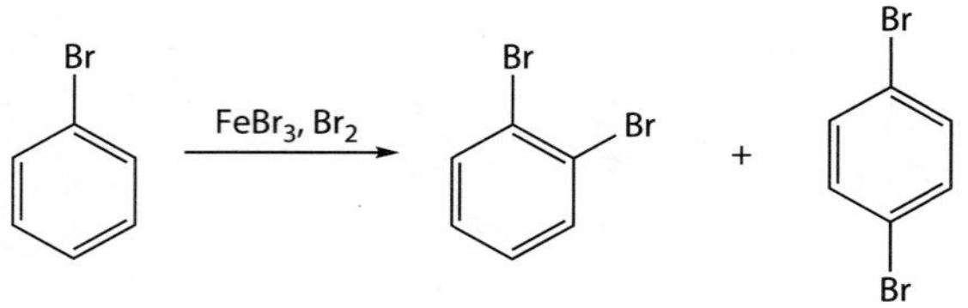

21.37

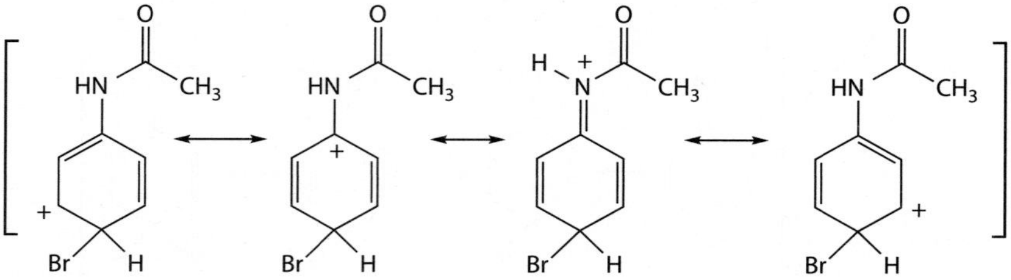

21.38

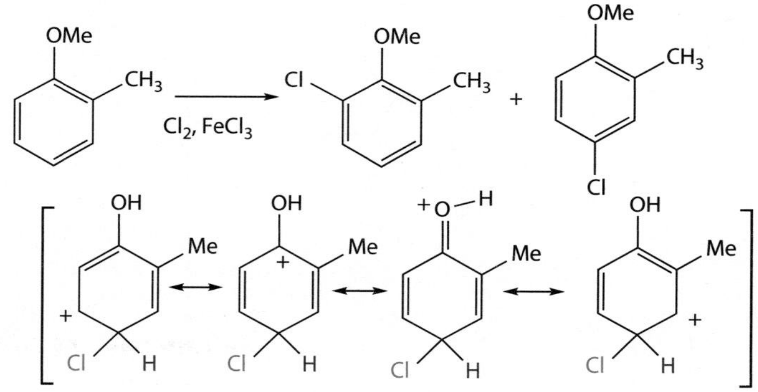

21.39

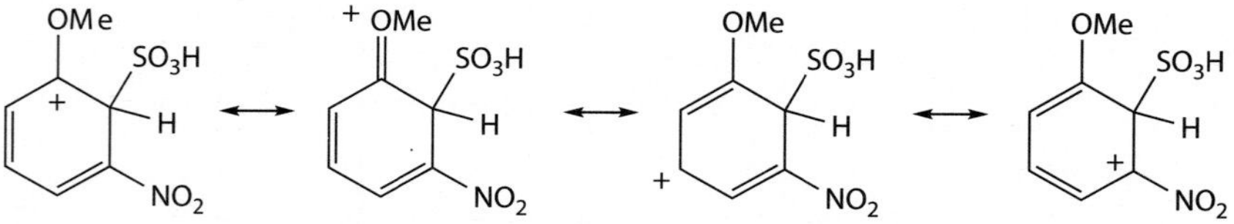

21.40

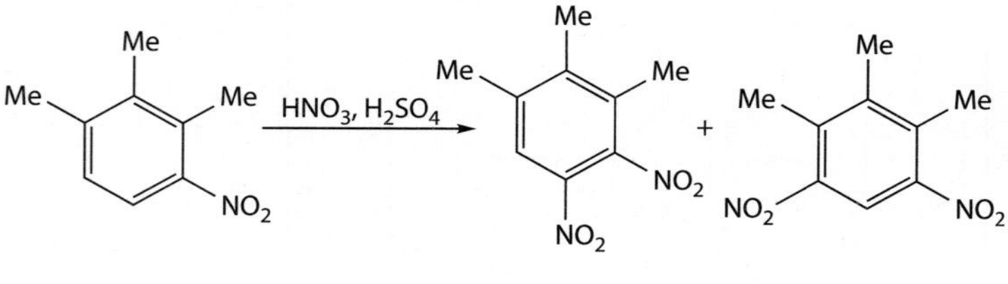

minor-because attack meta to NO_2 is slightly destabilized

21.41

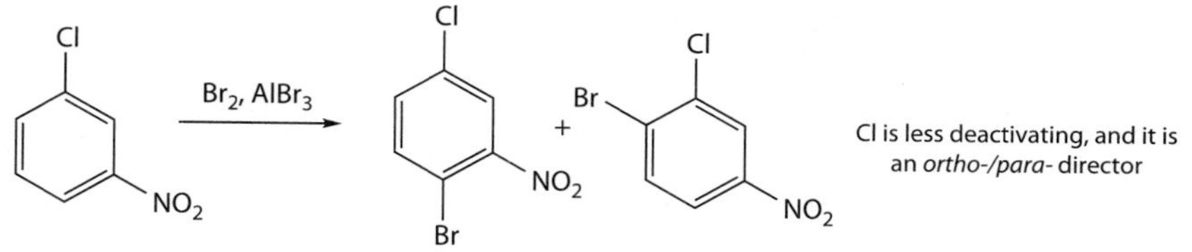

Cl is less deactivating, and it is an *ortho-/para-* director

21.42

21.43

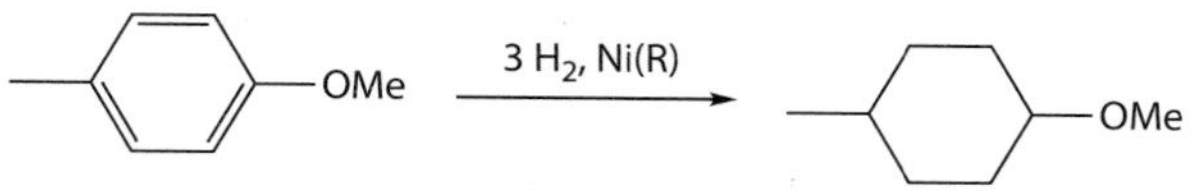

21.44

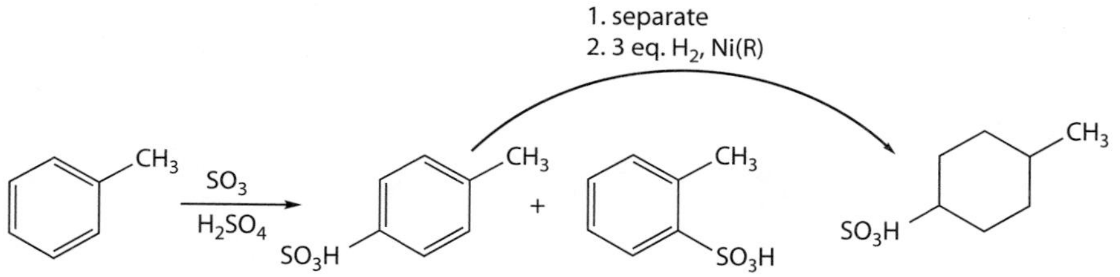

21.45

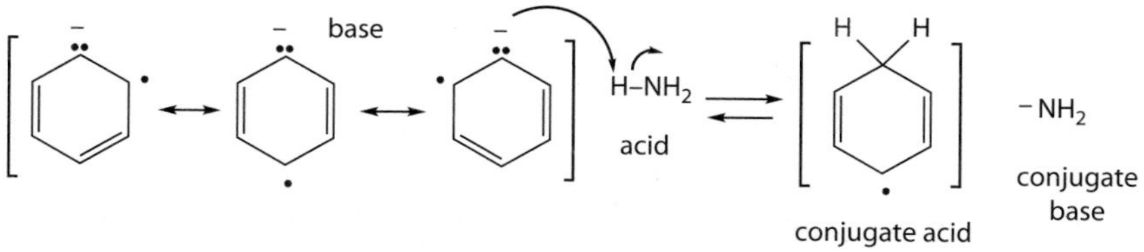

21.46 First, note that the position of the radical and the negative charge can be interchanged to give other resonance forms. These have been omitted because they are of the same energy. Notice that in B, D, and E the negative charge and the radical are on adjacent atoms; this is destabilizing and does not contribute much to the intermediate. Although D has the charge adjacent to the oxygen, it is one of the more destabilizing resonance forms and is not very important. Resonance forms G, I, and J also have radical and a negative charge on adjacent atoms and are very destabilizing. Resonance contributor J has the negative charge adjacent to the oxygen, but is not very important. In this case, however, resonance contributor F has the charge adjacent to oxygen and the electron repulsion makes it very destabilized. We cannot dismiss this contributor because the radical and negative charges are separated. The net result is that electron transfer to give F–J is unfavorable because these resonance forms are higher in energy than A–E.

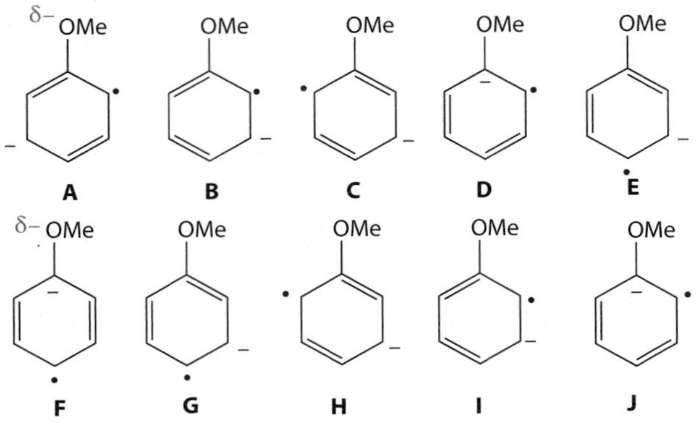

21.47

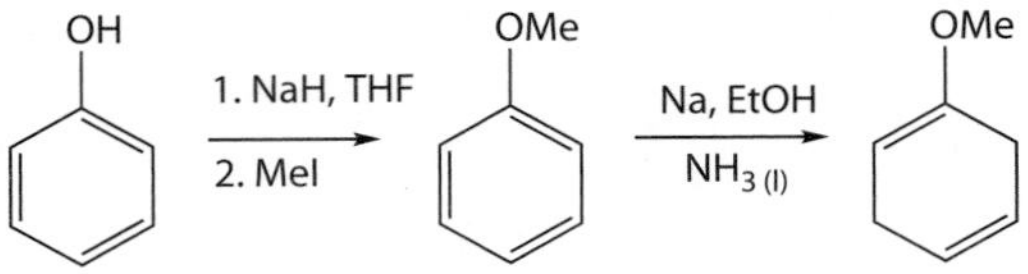

21.48

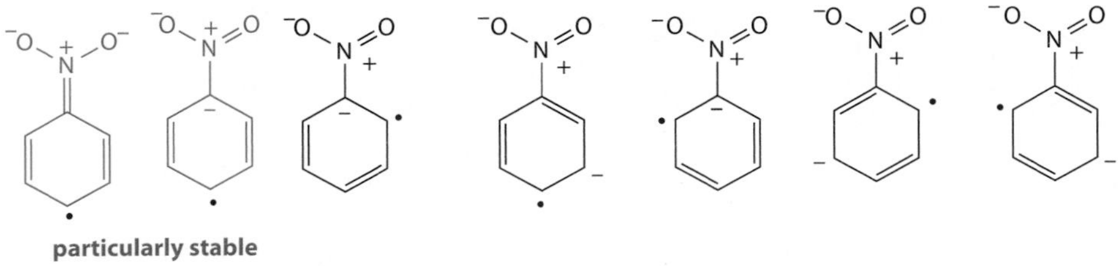

particularly stable

21.49

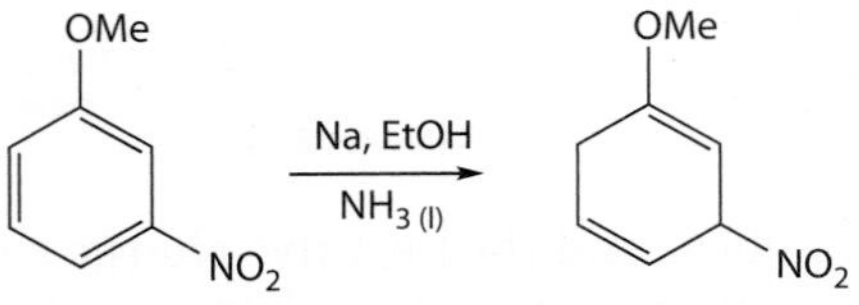

This product satisfies both substituents - OMe on sp² C and NO₂ on sp³ C

21.50

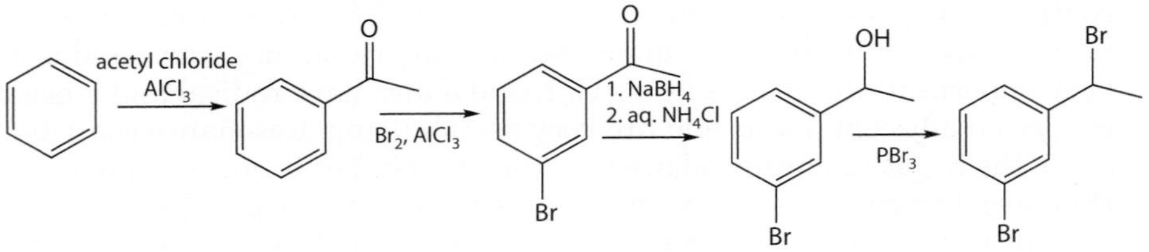

21.51

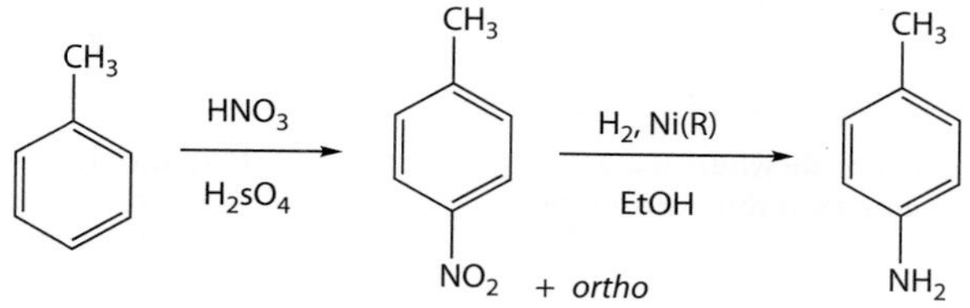

21.52

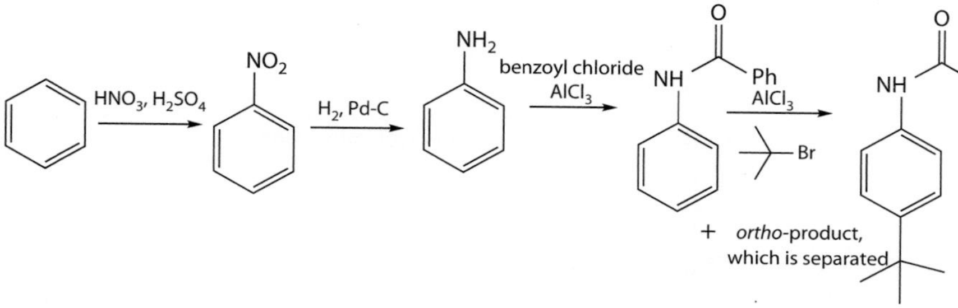

21.53

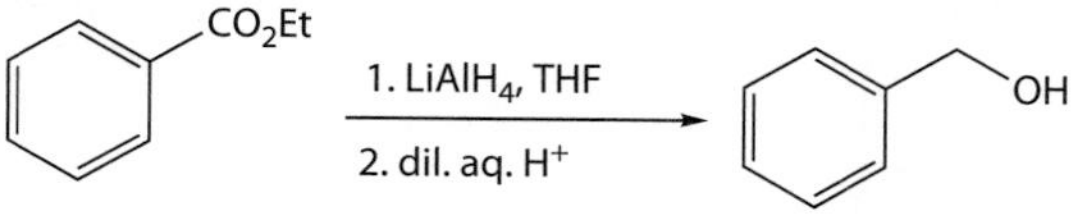

21.54

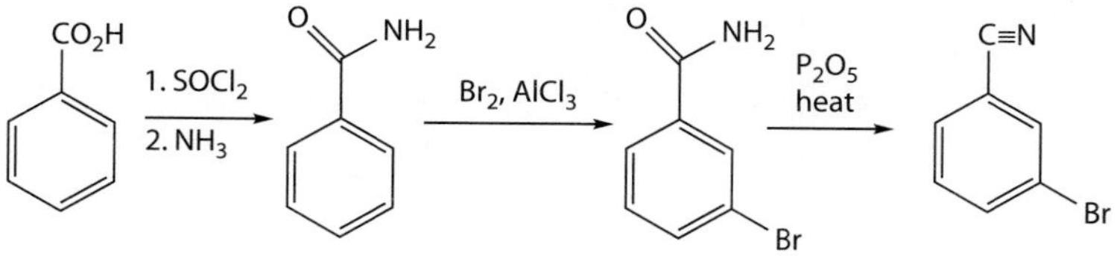

21.55

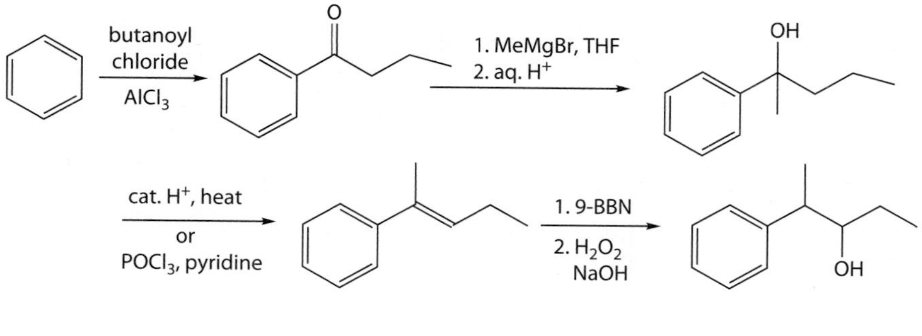

21.56

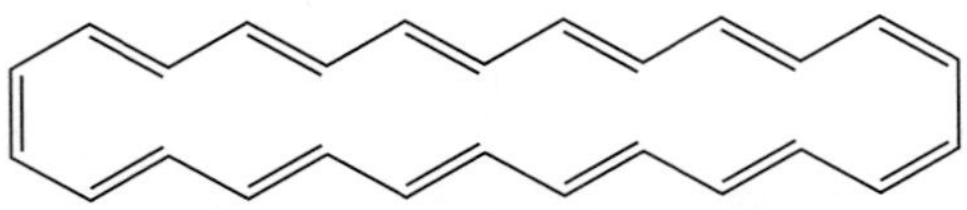

**[26]-annulene — fits 4n+2 rule where n = 6
Therefore, with 26 pi-electrons this is aromatic**

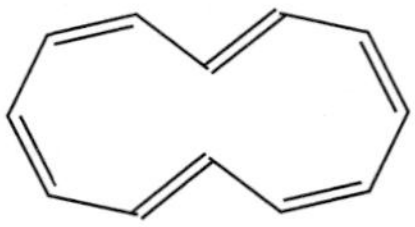

**[12]-annulene — does not
fit the 4n+2 rule so it is not aromatic.
It fits the 4n rule (n = 3) and is
actually anti-aromatic**

21.57

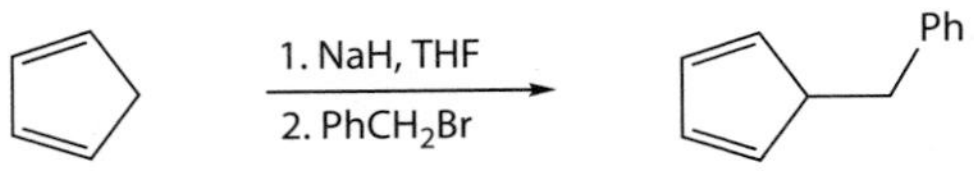

1. NaH, THF
2. PhCH$_2$Br

21.58

Cl — [   ] — Cl　　*para*-dichlorobenzene

21.59

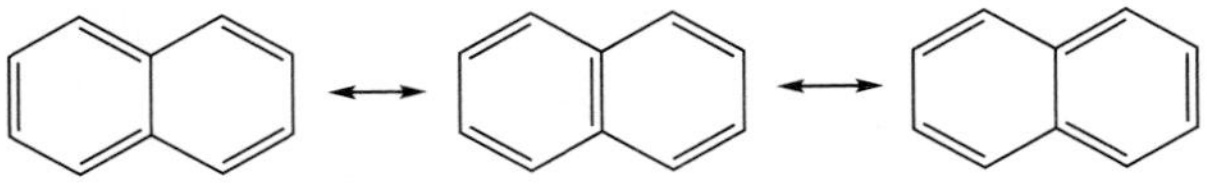

21.60

anthracene　　　　　　　　　　　　　**phenanthrene**

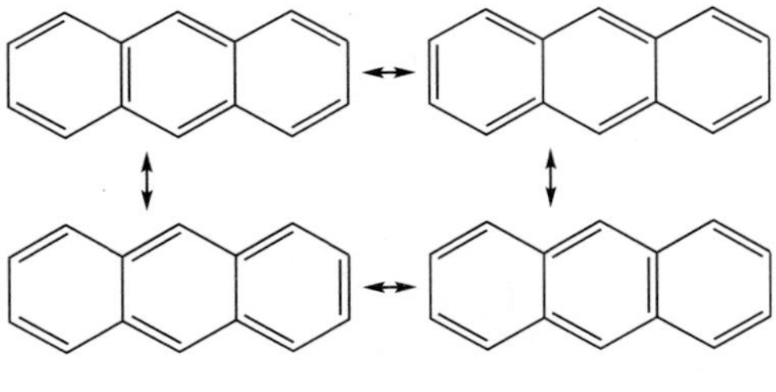

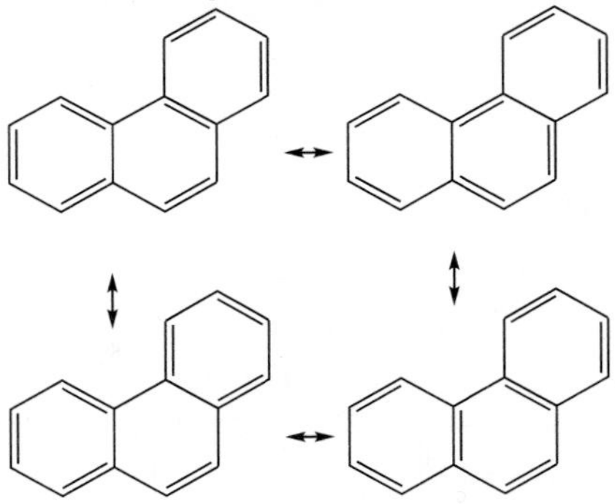

21.61

(a) NO$_2$

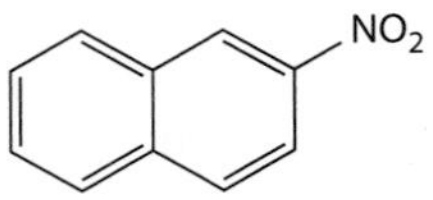

(b) F

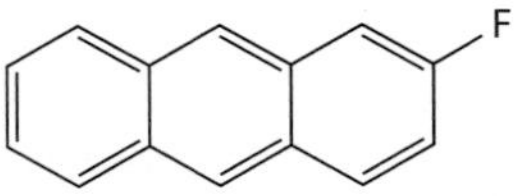

(c) Me

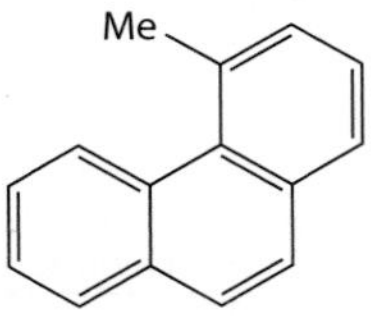

(d) Br ... Br

(e) Me ... Et ... Me

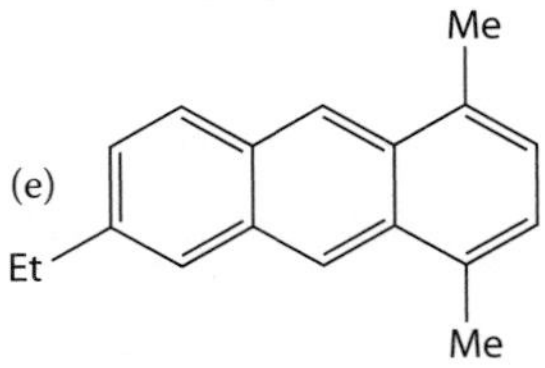

(f) HO ... OH

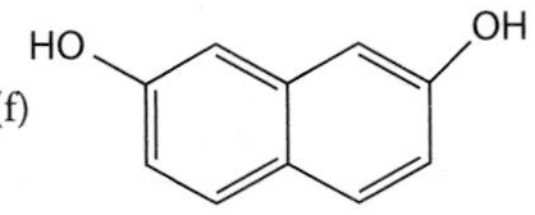

21.62

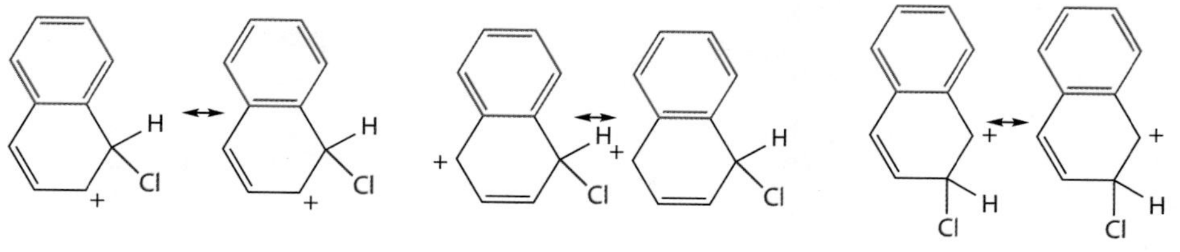

21.63

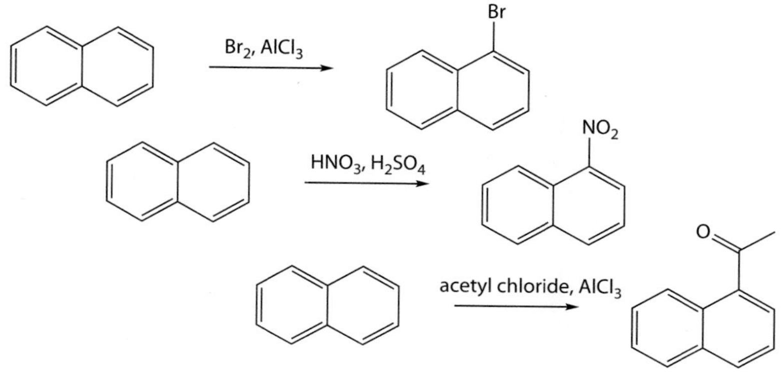
Br
Br_2, $AlCl_3$
NO_2
HNO_3, H_2SO_4
O
acetyl chloride, $AlCl_3$

21.64

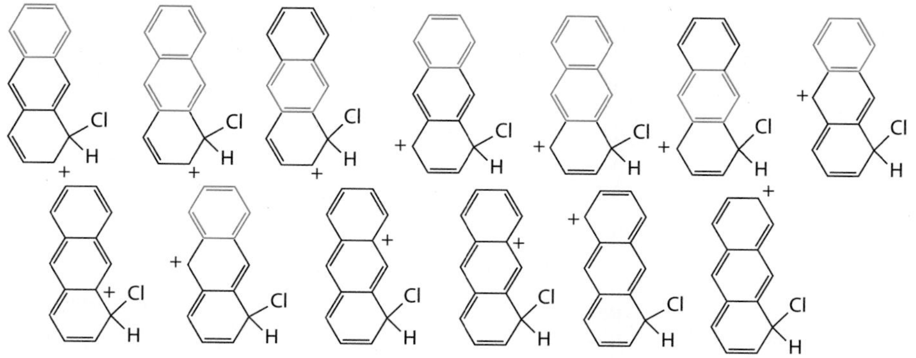

21.65

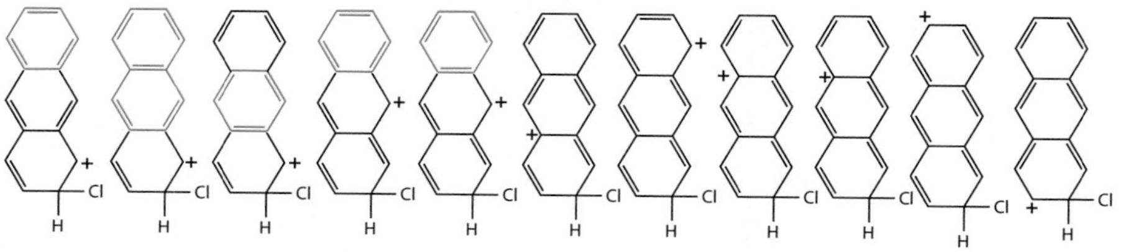

21.66

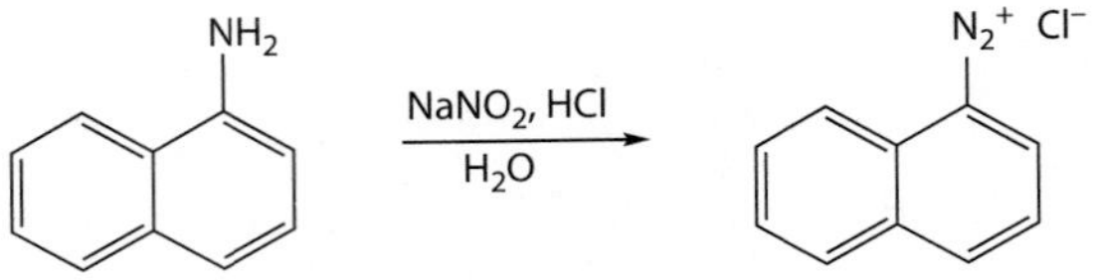

21.67

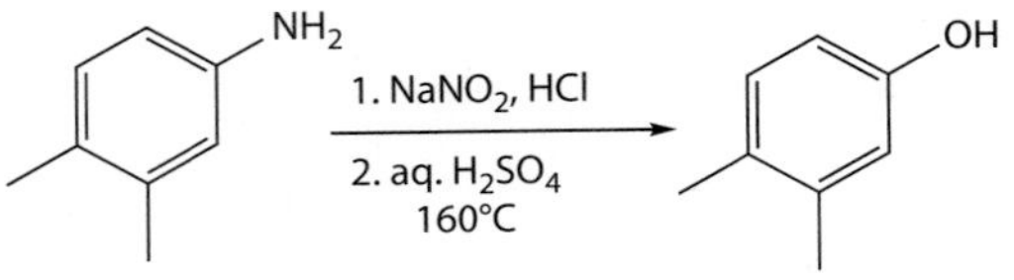

21.68

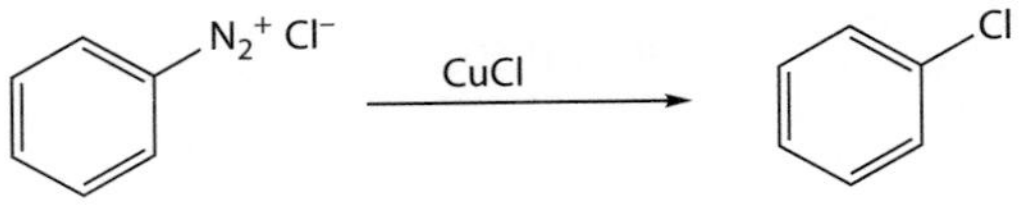

21.69

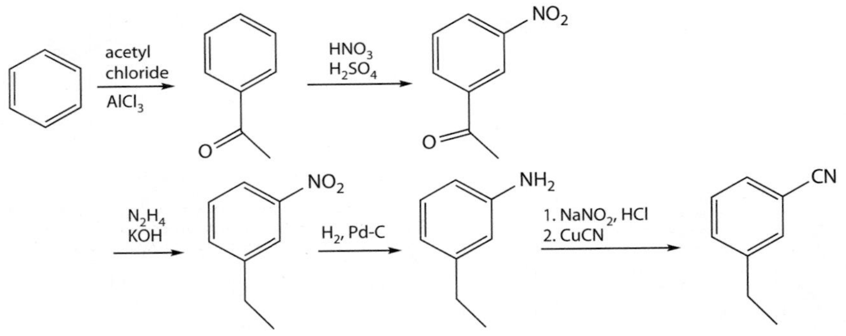

21.70

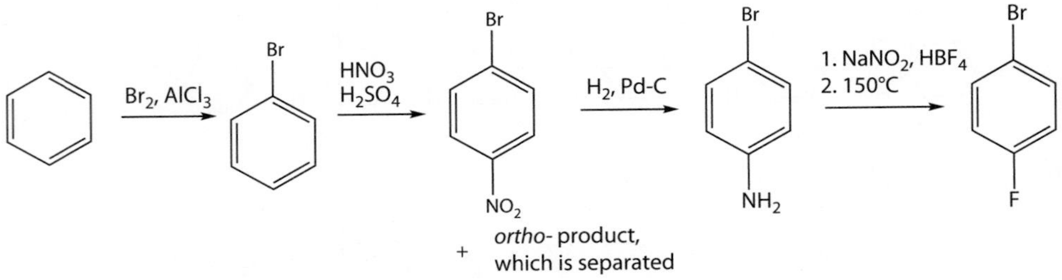

21.71

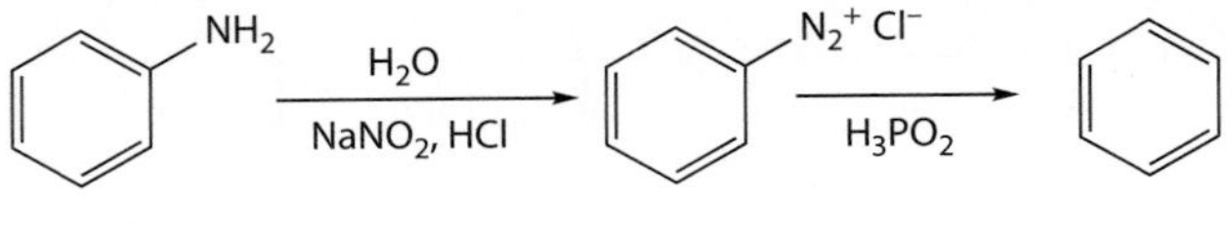

21.72

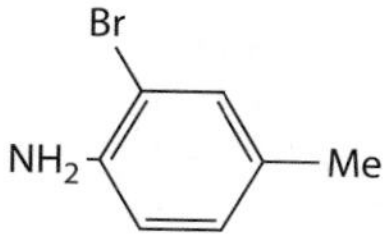

21.73

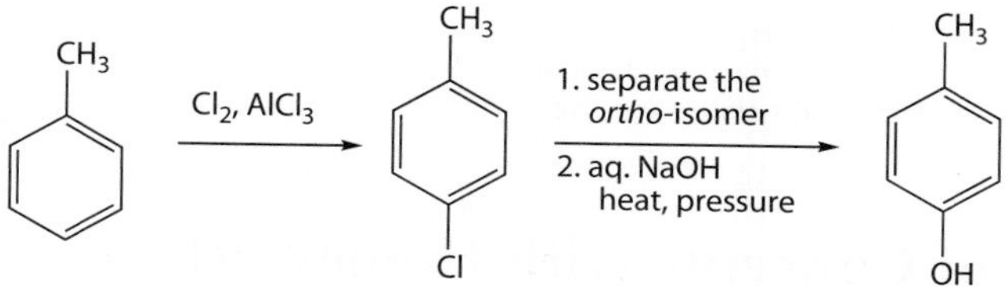

21.74

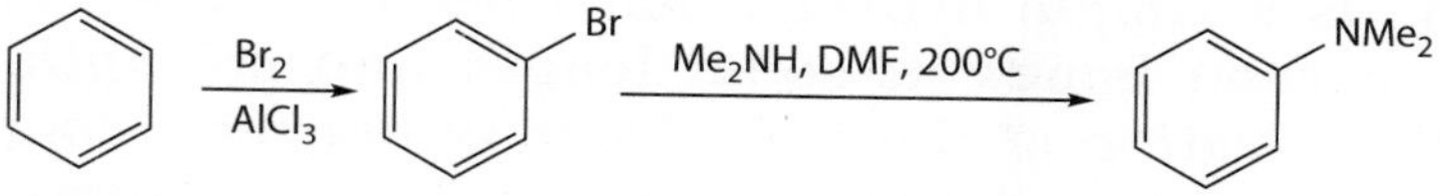

21.75

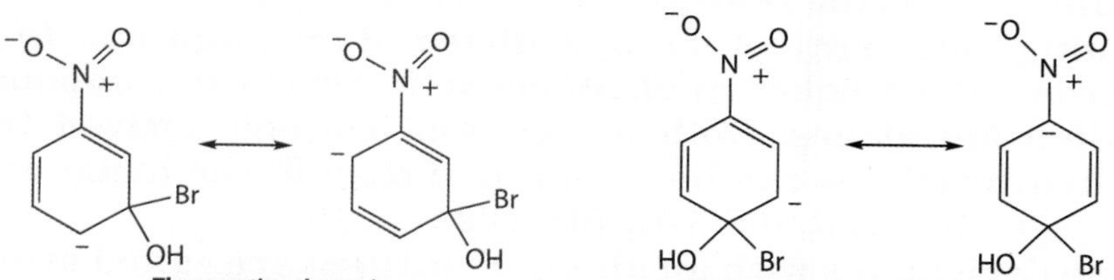

The negative charge is never on the carbon adjacent to the positive nitrogen atom.

The negative charge is on the carbon adjacent to the negative nitrogen atom in one resonance form. This makes this intermediate more stable, and it leads to the major product.

21.76

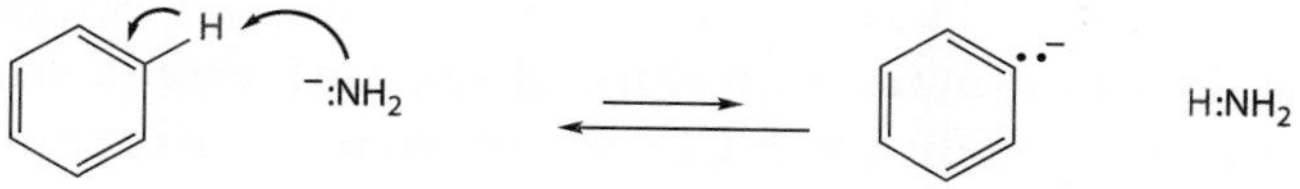

21.77

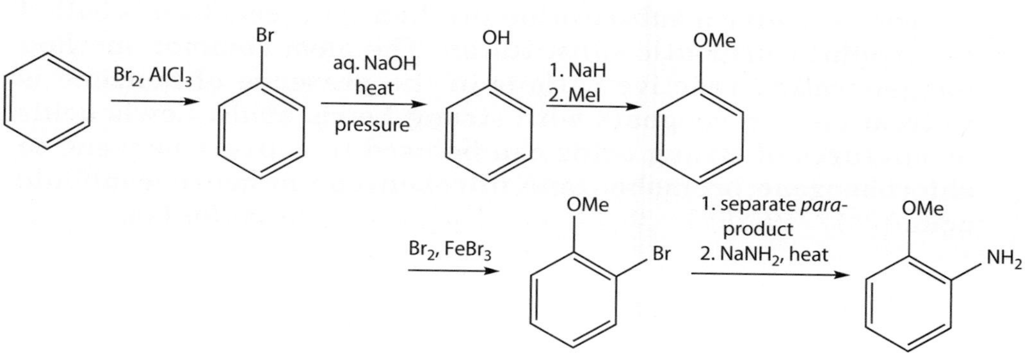

21.78

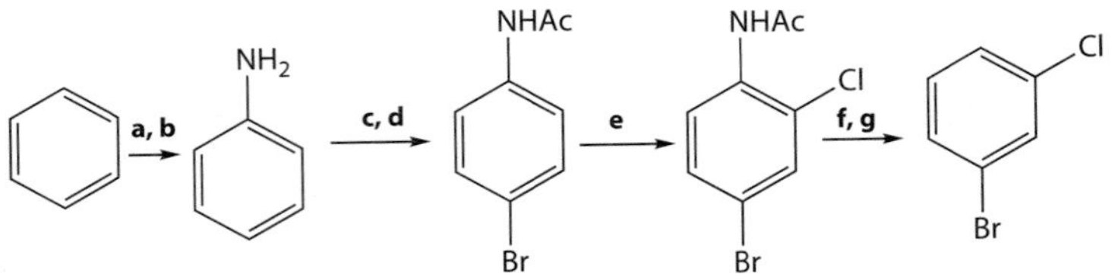

(a) HNO$_3$, H$_2$SO$_4$ (b)H$_2$, Pd-C (c) acetic anhydride
(d) Br$_2$, FeBr$_3$ (separate *ortho*-compound) (e) Cl$_2$, AlCl$_3$
(f) i. aq. NaOH ii. neutralize (g) i. NaNO$_2$, HCl ii. H$_3$PO$_2$

Correlation of Concepts with Homework

- Benzene is a simple hydrocarbon with a unique structure of carbon–carbon bonds of equal length and strength. Overlap and delocalization of the π-electrons in benzene, coupled with its cyclic nature and continuous array of sp^2 atoms, leads to special stability called aromaticity, and benzene is an aromatic molecule. A molecule is aromatic if it is cyclic, has a continuous and contiguous array of sp^2 hybridized atoms, and has 4n + 2 π-electrons (the Hückel rule). Monocyclic aromatic compounds are called annulenes. Both anions and cations derived from cyclic hydrocarbons can be aromatic if they fit the usual criteria: **1**, 56, 57, **104, 105, 106, 107, 108, 109**.

- Benzene derivatives with an alkyl substituent are called arenes. There are special names for benzene derivatives bearing heteroatom substituents. These include phenol (OH), aniline (NH$_2$), anisole (OMe), benzoic acid (COOH), benzonitrile (CN), benzaldehyde (CHO), acetophenone (COMe), and benzophenone (COPh): **2, 3, 4, 5, 6, 7, 8, 9, 10, 11, 12, 13, 14, 15, 16, 58, 79, 80, 88**.

- Because it is aromatic, benzene does not react directly with reagents such as HBr or HCl, or even with diatomic bromine or chlorine. Benzene reacts with cationic species to give a resonance-stabilized carbocation intermediate, which loses a hydrogen to give a substitution product. This reaction is called electrophilic aromatic substitution. The most common method for generating reactive cations in the presence of benzene is to treat certain reagents with strong Lewis acids. Lewis acids or mixtures of strong acids can be used to convert benzene to chlorobenzene, bromobenzene, nitrobenzene, or benzenesulfonic acid: **17, 18, 19, 20, 21, 22, 31, 32, 34, 35, 37, 81, 86, 93, 96 114**.

- The reaction of benzene with a carbocation leads to an arene in what is known as Friedel–Crafts alkylation. The reaction of an alkyl halide with a strong Lewis acid gives a carbocation, which is subject to rearrangement. Friedel–Crafts alkylation reactions are subject to polyalkylation because the arene is more reactive than benzene: 23, 24, 25, 26, 82, 86, 93, 94, 102, 113.
- The reaction of benzene with an acylium ion leads to a phenyl ketone in what is known as Friedel–Crafts acylation. The reaction of an acid chloride with a strong Lewis acid gives an acylium ion, which is not subject to rearrangement: 27, 28, 29, 82, 86, 102.
- Substituents on benzene lead to either a mixture of *ortho/para* products or to a *meta* product. Substituents that give *ortho/para* products react faster than benzene and are called activators; substituents that give *meta* products react more slowly than benzene and are called deactivators: 30, 31, 32, 33, 34, 35, 36, 37, 38, 83, 84, 88, 90, 91, 95, 96, 97, 98, 99, 100, 102, 105, 106, 107, 108, 109.
- Activating substituents are electron releasing and have excess electron density attached to the ring. These include alkyl groups and heteroatoms with a negative charge or a δ– atom. Activating substituents lead to a resonance-stabilized intermediate with a positive charge on a carbon atom that is adjacent to an electron-releasing substituent, which stabilizes the intermediate and makes the reaction proceed faster: 30, 31, 32, 37, 38, 39, 40, 41, 44, 81, 83, 84, 86, 89, 90, 95, 96, 98, 99, 100, 101, 102, 105, 107, 108, 109, 113, 114.
- Deactivating substituents are electron withdrawing and have diminished electron density attached to the ring. These include groups with a positive charge or a δ+ atom. Deactivating substituents lead to a resonance-stabilized intermediate with a positive charge on a carbon atom that is adjacent to an electron-withdrawing substituent, which destabilizes the intermediate and makes the reaction proceed more slowly: 30, 34, 35, 36, 39, 40, 41, 81, 86, 89, 91, 100, 102, 109, 113, 114.
- Benzene reacts with hydrogen gas and a suitable catalyst to give cyclohexane derivatives and, in some cases, cyclohexene or cyclohexadiene derivatives: 42, 43, 44, 113, 116, 117.
- Benzene reacts with sodium metal in liquid ammonia and ethanol to give nonconjugated cyclohexadiene derivatives in what is known as the Birch reduction: 45, 46, 48, 49, 103, 113, 118, 130.

- Substituents such as alkene units, alkyne units, and carbonyls can be reduced by catalytic hydrogenation. Lithium aluminum hydride reduces many heteroatom substituents, including nitrile and acid derivatives: 56, 57, 104, 105, 106, 107, 108, 109.
- Polycyclic aromatic compounds such as naphthalene, anthracene, and phenanthrene give electrophilic aromatic substitution reactions. The major product is determined by the number of resonance-stabilized intermediates for attack at a given carbon and the number of fully aromatic rings (intact rings) in the resonance structures: 59, 60, 61, 62, 63, 64, 65, 85, 104, 106, 107, 108, 109, 110, 113, 114, 118.
- Aniline reacts with nitrous acid to give benzenediazonium salts, which react with a variety of reagents via a substitution reaction. These reagents include cuprous salts, aqueous acid, iodide, hypophosphorous acid, and activated benzene derivatives: 66, 67, 68, 69, 70, 71, 72, 100, 113, 119, 120, 121, 126, 127.
- Nucleophilic substitution at the sp^2 carbon of a halo-benzene derivative does not occur unless high heat and pressure are used. Electron-withdrawing substituents on the benzene ring significantly lower the temperature required for the reaction. Nucleophiles for this nucleophilic aromatic substitution reaction include water, hydroxide, alkoxide, and amines: 73, 74, 75, 111, 112, 121, 126, 127.
- When halobenzene derivatives are heated with powerful bases such as sodium amide, deprotonation is followed by elimination to give a benzyne, which reacts rapidly with the nucleophile: 76, 77, 115, 125, 127, 129.
- A molecule with a particular functional group can be prepared from molecules containing different functional groups by a series of chemical steps (reactions). This process is called synthesis: The new molecule is synthesized from the old one (see Chapter 25): 47, 50, 51, 52, 54, 55, 69, 70, 73, 74, 77, 78, 131, 132, 133.
- Spectroscopy can be used to determine the structure of a particular molecule and can distinguish the structure and functionality of one molecule when compared with another (see Chapter 14): 134, 135, 136, 137, 138, 139, 140, 141, 142, 143, 144, 145.

Homework

21.79   Draw the structure of each of the following:
    (a)   1,3,5-trimethylbenzene
    (b)   *m*-chlorophenol
    (c)   3,5-dinitroanisole
    (d)   hexachlorobenzene

(e) 4-bromophthalic acid

(f) *p*-iodobenzenesulfonic acid

(g) 2-cyanobenzoic acid

(h) phenetole

(i) 4-bromo-3′-chlorobenzophenone

(j) 2,6-dinitrohydroquinone

(k) *o*-bromobenzonitrile

(l) *m*-xylene

(m) 2,2′-dimethylstilbene

(n) *N*-acetyl-3-methylaniline

(o) 2,2-dimethyl-4-phenylhexane

21.80 Give the correct IUPAC name to each of the following molecules:

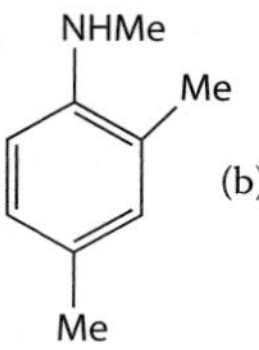

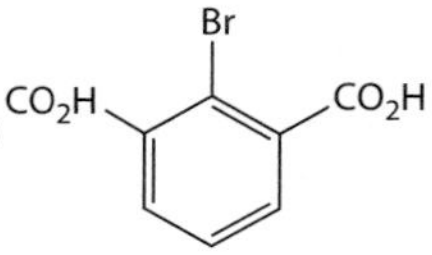

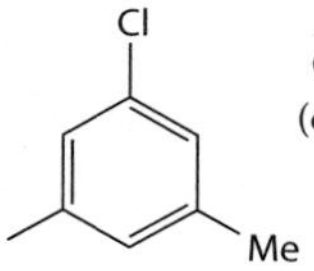

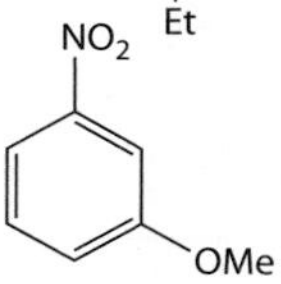

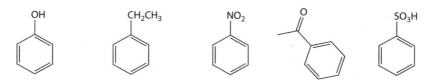

21.81 Draw the final product or products and all resonance forms for the arenium ion intermediate generated in each of the following reactions:

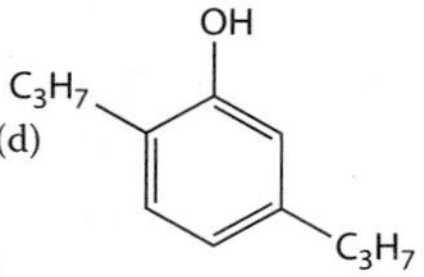

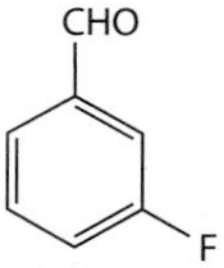

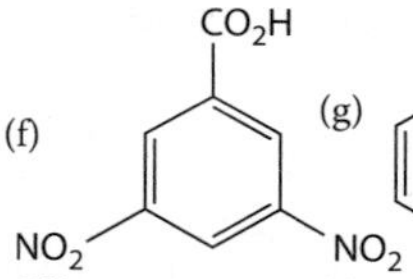

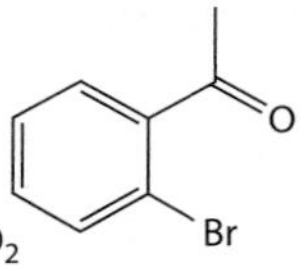

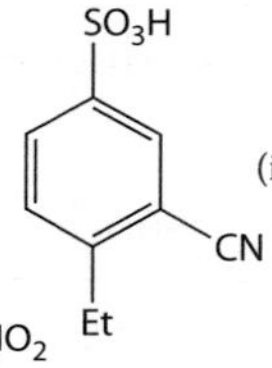

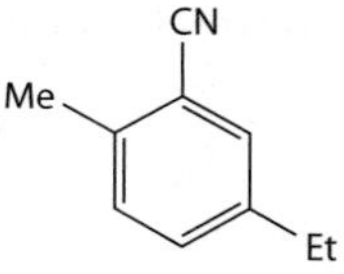

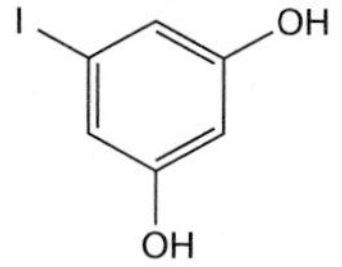

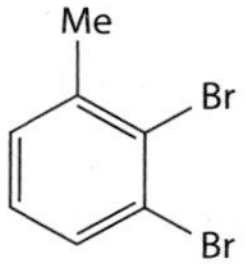

(a) anisole + bromine + ferric bromide

(b) nitrobenzene + chlorine + aluminum chloride

(c) *N*-acetylaniline + HNO$_3$/H$_2$SO$_4$

(d) bromobenzene + bromine + AlCl$_3$

21.82 Give the major product for each of the following:

(a) benzene + 2-bromo-3-methylbutane + AlCl$_3$

(b) toluene + phenylacetyl chloride + AlCl$_3$

(c) anisole + 2-bromo-2-methylpropane + AlCl$_3$

(d) nitrobenzene + propanoyl chloride + AlCl$_3$

(e) benzene + 1-bromopentane + AlCl$_3$

(f) *p*-xylene + acetyl chloride + AlCl$_3$

21.83 Which of the following react faster than benzene when treated with Br$_2$/AlCl$_3$? Why or why not do they react faster?

21.84 Which of the following arenium ions has the most resonance forms?

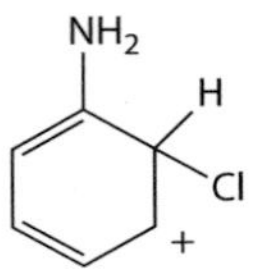

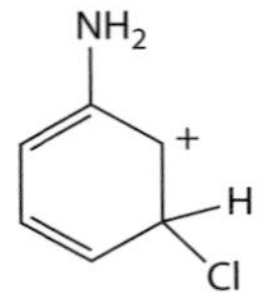

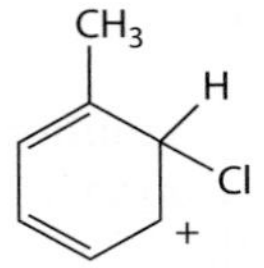

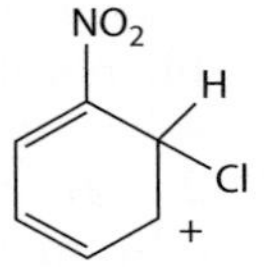

21.85 Identify the aromatic compounds in the following:

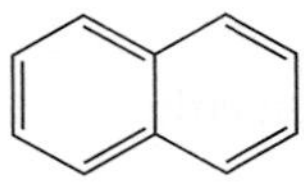
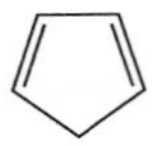
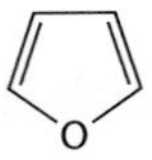

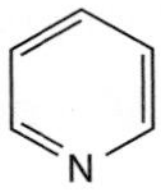

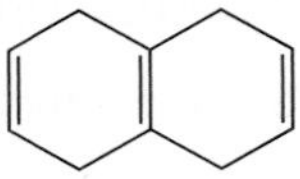

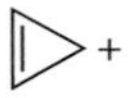

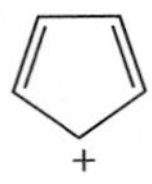

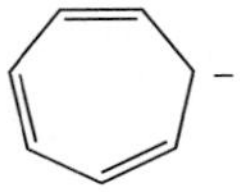

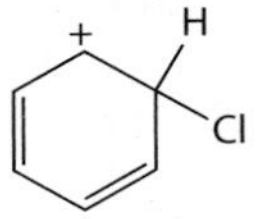

21.86 Give the major product for each of the following reactions:

(a) 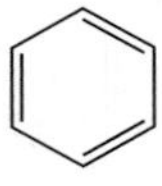 $\xrightarrow[H_2SO_4]{HNO_3}$

(b) 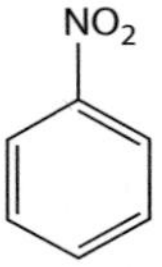

$\xrightarrow{Br_2,\ AlCl_3}$

(c) 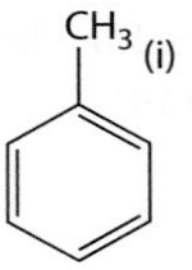

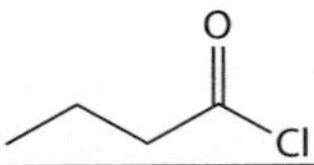

$\xrightarrow{AlCl_3}$

(d) 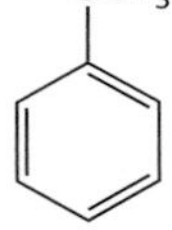

$\xrightarrow{Cl_2,\ FeCl_3}$

(e) 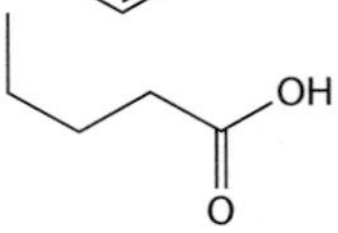

→ 1. oxalyl chloride 2. benzene, $AlCl_3$

(f) 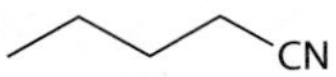

→ 1. H_3O^+, heat 2. $SOCl_2$ 3. 1-phenylhexane, $AlCl_3$

(g) 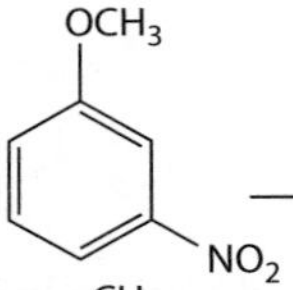

$\xrightarrow{Cl_2,\ AlCl_3}$

(h) 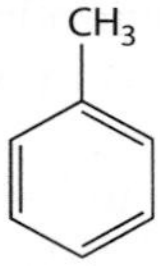

$\xrightarrow[AlCl_3]{Me_3CCl}$

(i) 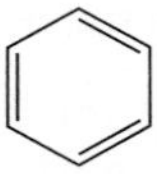 → 1. Br_2, $AlCl_3$ 2. Mg, THF 3. 3-methylcyclopentanone 4. aq H^+

(j) 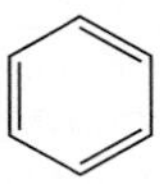 $\xrightarrow{SO_3,\ H_2SO_4}$

21.87 Which of the following molecules react faster than benzene with HNO_3/H_2SO_4? Justify your choice or choices.

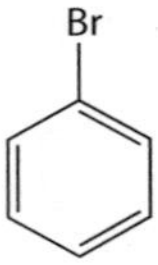

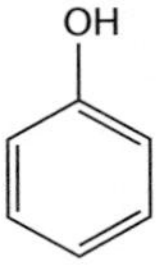

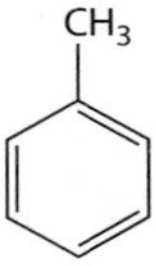

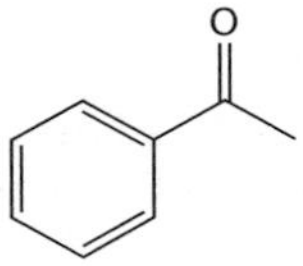

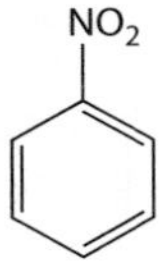

21.88 Which of the following molecules have a chlorine in the *meta* position relative to OMe?

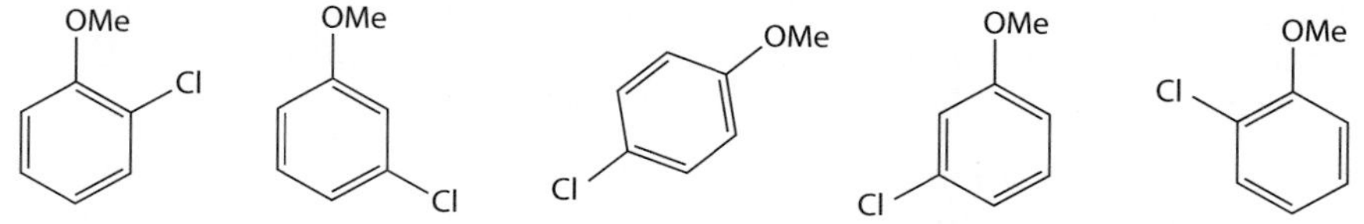

21.89 Draw all resonance forms for the arenium ion intermediate formed when chlorobenzene reacts at the *para* position with $Cl_2/AlCl_3$.

21.90 It is known that anisole reacts with bromine to give aromatic substitution products without the need for a Lewis acid, whereas benzene does not react without the Lewis acid. Offer an explanation for this fact.

21.91 Which of the following molecules react fastest with NaOH?

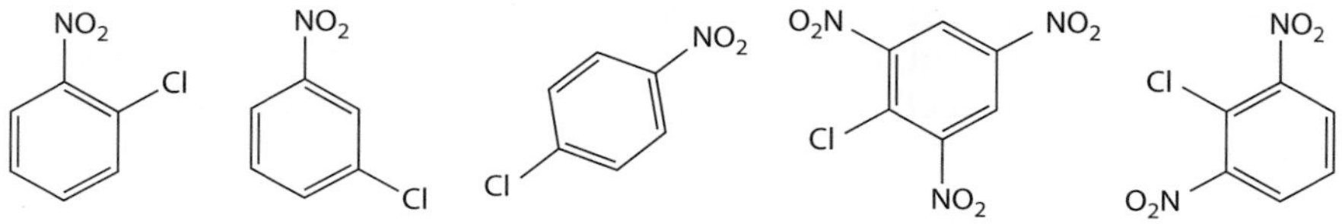

21.92 Friedel–Crafts type alkylations occur in systems other than alkyl halides. When 2-methyl-1-propene is treated with a catalytic amount of sulfuric acid in the presence of benzene, for example, an alkylbenzene is formed. Draw that product and draw a mechanism for its formation. Similarly, when *tert*-butanol is treated with a catalytic amount of sulfuric acid in the presence of toluene, a Friedel–Crafts type reaction occurs. Draw the product or products of this reaction and give a mechanism for its or their formation.

21.93 Benzene reacts slowly with sulfuric acid and more rapidly with fuming sulfuric acid to give benzenesulfonic acid. Benzene does not react similarly with acetic acid. Why not?

21.94 When 3-bromo-2-pentene reacts with benzene and $AlCl_3$, two products are formed. Draw them and account for their formation by a mechanism.

21.95 Draw all resonance forms for the intermediate formed in each of the following aromatic substitution reactions with bromine and $FeBr_3$. If there is a choice of *ortho* and *para* products, draw the intermediate for the *ortho* product.
(a) anisole
(b) 1,4-dimethoxybenzene
(c) *N*-acetylaniline
(d) ethylbenzene

21.96 1-Phenyl-1,3-butadiene reacts when treated with bromine and $AlCl_3$, but the product tends to be the polymer formed by cationic polymerization of the diene unit. Why?

21.97   Should triphenylborane (Ph_3B) undergo electrophilic aromatic substitution with benzene faster or more slowly than benzene? Are there any complicating factors? Explain your answer.

21.98   You have a sample of O-acetylphenol (phenyl acetate). Will this molecule undergo electrophilic aromatic substitution with bromine and $FeBr_3$ more quickly or more slowly than anisole? Explain your answer.

21.99   Explain why electrophilic aromatic substitution of styrene with bromine and ferric bromide is faster than the identical reaction with ethylbenzene.

21.100 Give the major product for each of the following reactions:
  (a) 3-nitroanisole + HNO_3/H_2SO_4
  (b) 1,3,5-trimethylbenzene + $Cl_2/AlCl_3$
  (c) 2-methylbenzoic acid + SO_3/H_2SO_4
  (d) 1,3-dinitrobenzene + $Br_2/AlCl_3$
  (e) 3-chloronitrobenzene + $Cl_2/AlCl_3$
  (f) 4-methylanisole + acetyl chloride + $AlCl_3$
  (g) N-acetyl-3,5-dimethylaniline + acetyl chloride/$AlCl_3$
  (h) 2-ethylanisole + t-butyl bromide/$AlCl_3$

21.101 In electrophilic substitution reactions of anisole, both the *ortho* and the *para* products are preferred. It is common to see that a higher percentage of the *ortho* product is formed. Using bromination as an example, offer an explanation for this "*ortho* effect."

21.102 Give the major product of the following reactions:
  (a) nitrobenzene + H_2 + Pd
  (b) 2-phenyl-2-hexene + H_2 + Pd/C
  (c) toluene + 3 H_2 + Rh, heat
  (d) p-xylene + $Na/NH_3/EtOH$
  (e) benzene + 1. acetyl chloride/$AlCl_3$; 2. $NaBH_4$; 3. hydrolysis
  (f) phenyllithium + 1. CO_2; 2. hydrolysis; 3. $Na/NH_3/EtOH$
  (g) butanoic acid + 1. thionyl chloride; 2. toluene/$AlCl_3$; 3. NH_2NH_2/KOH
  (h) nitrobenzene + 1. H_2, Pd/C; 2. acetic anhydride; 3. $LiAlH_4$; 4. hydrolysis
  (i) benzene + 1. $Br_2/FeBr_3$; 2. Mg, THF; 3. propanal; 4. PCC; 5. Zn(Hg), HCl
  (j) 1,4-diethylbenzene + 1. HNO_3/H_2SO_4; 2. $Na/NH_3/EtOH$
  (k) benzonitrile + 1. $LiAlH_4$; 2. hydrolysis; 3. ethyl butanoate, heat
  (l) p-xylene + 1. t-butyl bromide/$AlCl_3$; 2. 4 H_2, Pd, heat
  (m) benzoic acid + 1. $POCl_3$; 2. butanol/NEt_3; 3. $Na/NH_3/EtOH$
  (n) benzoic acid + 1. $POCl_3$; 2. butanol/NEt_3; 3. $LiAlH_4$; 4. hydrolysis;
       5. $SOCl_2$
  (o) bromobenzene + 1. Li, ether; 2. CO_2; 3. hydrolysis; 4. $SOCl_2$;
       5. NH_3 P_2O_5, heat

21.103 When phenol is treated with $Na/NH_3/EtOH$ and then hydrolyzed with 2N HCl, the product is cyclohex-2-en-1-one. Draw this product and also the one formed from the dissolving metal reduction. Provide a mechanism that converts the dissolving metal product to the final product in aqueous acid.

21.104 Categorize each of the following molecules. Give the number of π-electrons and indicate whether or not the molecule is aromatic.

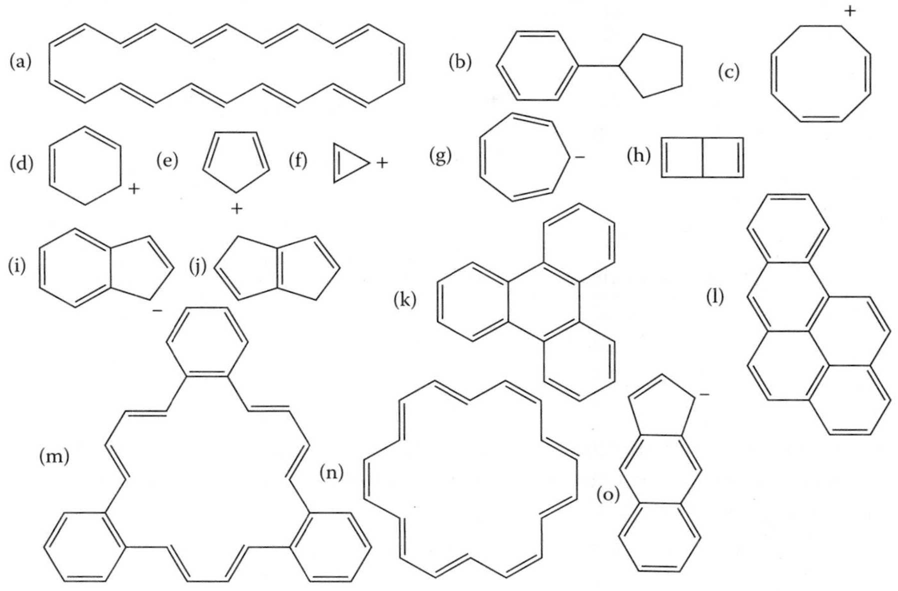

21.105 Determine the major product or products of the reaction between 1-methoxy-[14]-annulene and bromine/FeBr$_3$.

21.106 Explain why bromination of 1-methoxy-8-nitronaphthalene occurs preferentially in the ring bearing the OMe unit.

21.107 Draw all resonance forms for the intermediate generated when 2-methoxy-phenanthrene reacts with Br$_2$/FeBr$_3$, with attack at C7.

21.108 Compare the chlorination reaction at C$_a$ versus C$_b$ in the following molecule and determine whether chlorination will occur preferentially at C$_a$ or C$_b$:

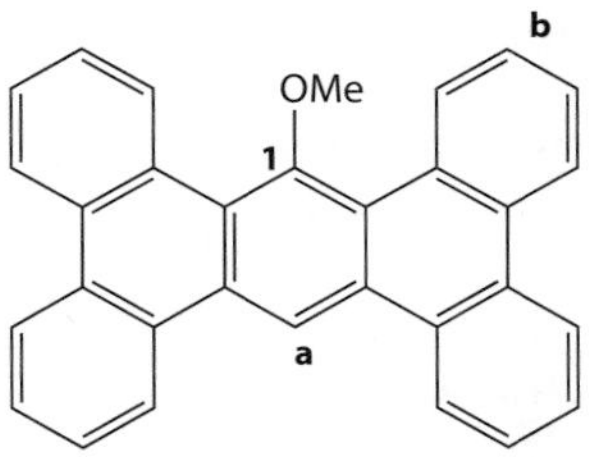

21.109 What is or are the major product or products from the following reactions?

(a) [structure with OMe on naphthalene] HNO₃, H₂SO₄ →

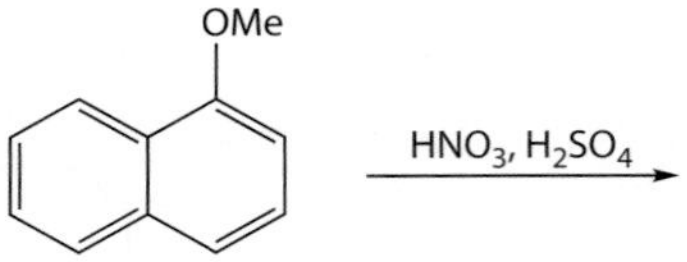

(d) [structure with OMe on anthracene, OMe below] Cl₂, AlCl₃ →

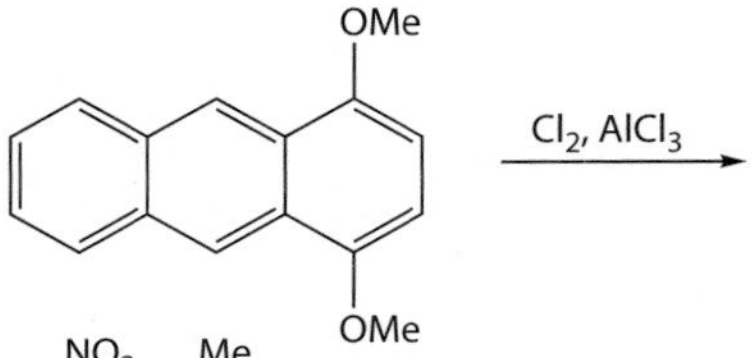

(b) [dimethylnaphthalene] Br₂, FeBr₃ →

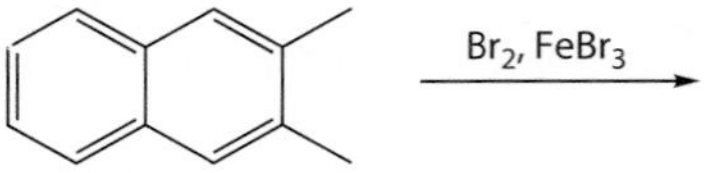

(e) [structure with NO₂, Me, Me, Me groups on naphthalene] Br₂, AlCl₃ →

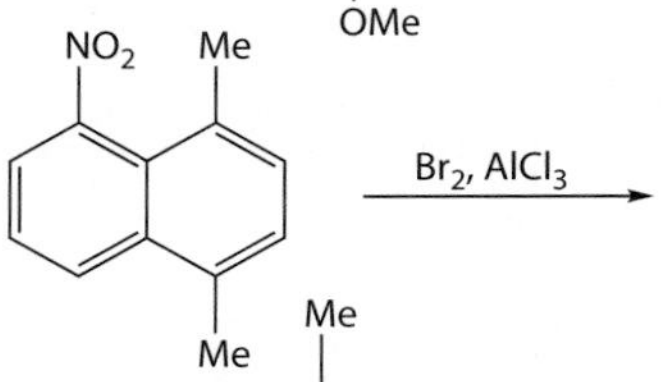

(c) [naphthalene with NO₂] acetyl chloride / AlCl₃ →

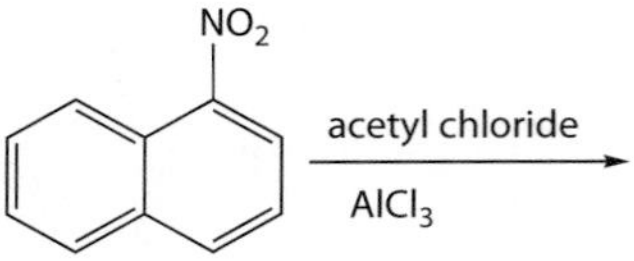

(f) [naphthalene with Me, Me, Me] Na, NH₃ / EtOH →

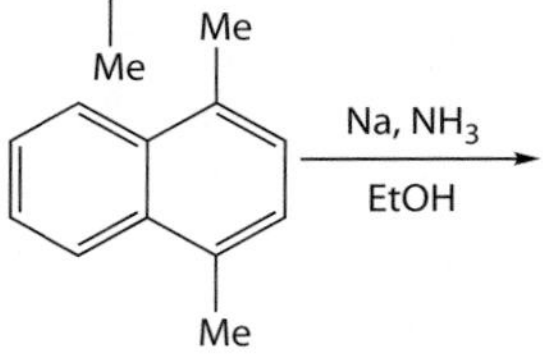

21.110 Give the major product formed when 1-aminonaphthalene is treated with HCl/NaNO₂ and then the following reagents:
  (a) 160°C, aq. H₂SO₄
  (b) CuCN
  (c) CuBr
  (d) CuCl
  (e) KI, heat
  (f) PhNH₂
  (g) H₃PO₂

21.111 Briefly explain which molecule reacts faster with aqueous NaOH at high temperature and pressure: 4-bromotoluene or 4-bromonitrobenzene.

21.112 Briefly explain which molecule reacts faster with aqueous NaOH at 300°C: 4-nitrobenzenesulfonic acid or 3-nitrobenzenesulfonic acid.

21.113 Give the major product of each of the following and show the intermediate product for each step:

(a) [benzene]
  1. HNO₃, H₂SO₄
  2. H₂, Pd-C
  3. NaNO₂, HBF₄
  4. 150°C

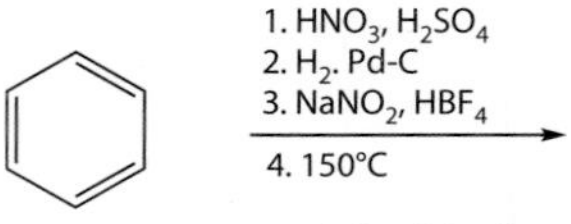

(b) [toluene, Me]
  1. Br₂, FeBr₃ (separate para product)
  2. HNO₃, HSO₄
  3. H₂, Pd-C
  4. NaNO₂, HCl
  5. CuBr

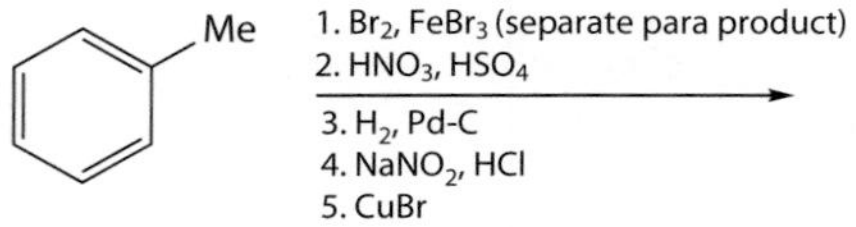

(c) [Br, NHAc substituted benzene]
  1. aq. NaOH; aq. HCl
  2. NaNO₂, HCl
  3. 160°C, aq. H₂SO₄

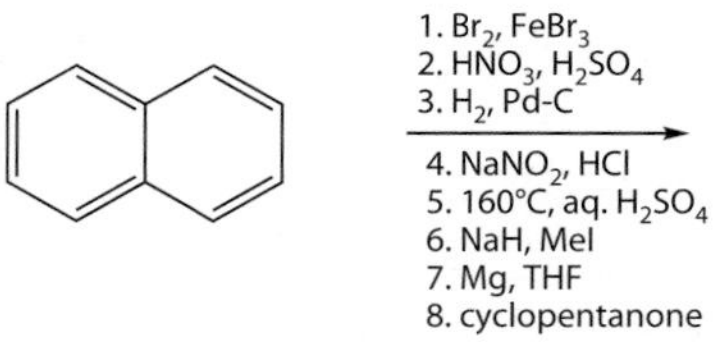

(d) [benzene with NH₂]
  1. NaNO₂, HCl
  2. 160°C, aq. H₂SO₄
  3. NaH, THF
  4. 2R-iodopentane

(e) [naphthalene]
  1. HNO₃, H₂SO₄
  2. H₂, Pd-C
  3. NaNO₂, HCl
  4. 1-aminonaphthalene

(f) [benzene with I]
  1. Mg, THF
  2. butanal
  3. PCC
  4. HNO₃, H₂SO₄
  5. NaNO₂, HCl
  6. CuBr

(g) [naphthalene]
  1. Br₂, FeBr₃
  2. HNO₃, H₂SO₄
  3. H₂, Pd-C
  4. NaNO₂, HCl
  5. 160°C, aq. H₂SO₄
  6. NaH, MeI
  7. Mg, THF
  8. cyclopentanone

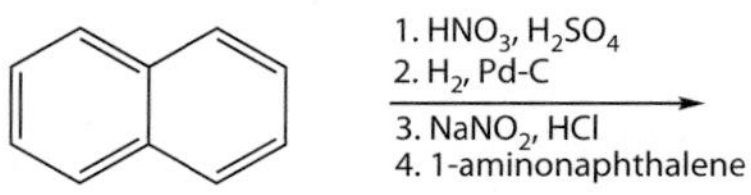

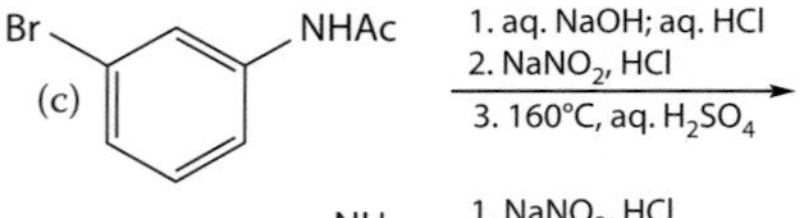

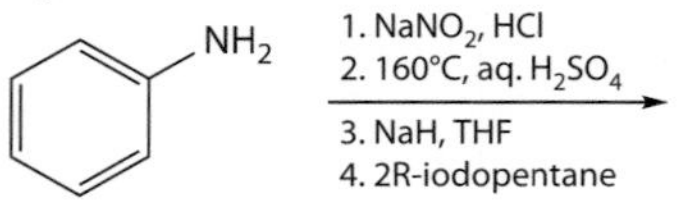

21.114 Give the major product formed from the following sequences:
  (a) naphthalene + 1. SO_3, H_2SO_4; 2. aq. NaOH, 300°C, reaction bomb
  (b) toluene + 1. SO_3, H_2SO_4; 2. separate *ortho* product; 3. aq. NaOH, 300°C, reaction bomb; 4. NaH, THF; 5. allyl bromide
  (c) benzene + 1. HNO_3/H_2SO_4; 2. H_2, Pd/C; 3. $NaNO_2/HCl$; 4. KI; 5. Mg, THF; 6. CO_2; 7. hydrolysis
  (d) 4-bromotoluene + methylamine, water, 300°C, reaction bomb

21.115 Draw the products expected when each of the following reacts with ammonia with heat and under pressure:
  (a) 2-bromotoluene
  (b) 3-bromotoluene
  (c) 4-iodoanisole
  (d) 2-bromophenetole
  (e) 2-bromonaphthalene

21.116 Draw the products formed when anisole is treated with three equivalents of hydrogen gas and a Pd/C catalyst.

21.117 Suggest a structure for the product formed when phenol is treated with exactly one equivalent of hydrogen gas with a suitable catalyst. Is there more than one possibility? If this product is then reacted with aqueous acid, suggest a mechanism for reaction and a reasonable final product.

21.118 Suggest the major product formed when 1-methoxynaphthalene is treated with sodium metal in a mixture of liquid ammonia and ethanol. What product is formed when 1-nitronaphthalene reacts under the same conditions?

21.119 What products result from each reaction and what is the final product of the following sequences?
  (a) nitrobenzene + $H_2/Ni(R)$
  (b) $HCl/NaNO_2$
  (c) KF, THF

21.120 Draw the diazonium salt formed when 2-aminonaphthalene reacts with acetic acid/$NaNO_2$. When this diazonium salt reacts with *N*-methyl diphenylamine, what is the structure of the resulting product? If the diazonium salt reacts with 1,3,5-trimethoxybenzene, what is the resulting product?

21.121 Which diazonium salt is expected to react faster with CuCN: 4-methoxybenzenediazonium chloride or 4-nitrobenzenediazonium chloride? Explain your choice.

21.122 What is the product formed when benzenediazonium chloride reacts with (a) CuCN, and (b) $LiAlH_4$ followed by aqueous acid workup?

21.123 What is the product formed when bromobenzene reacts with the following?
  (a) Mg, THF
  (b) carbon dioxide followed by an aqueous acid workup
  (c) ethanol with an acid catalyst
  (d) $LiAlH_4$ followed by an aqueous acid workup
  Draw out the structures for the product of each reaction.

21.142 Identify the molecule with the formula $C_8H_{10}O$ and the spectral data provided. The infrared spectrum shows a strong peak at about 3250 cm^{-1}.

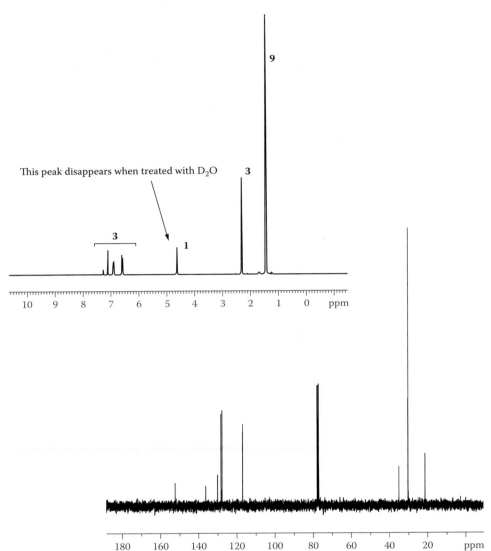

21.143 Identify the molecule with the formula $C_{11}H_{16}O$ and the theoretical spectral
data provided:

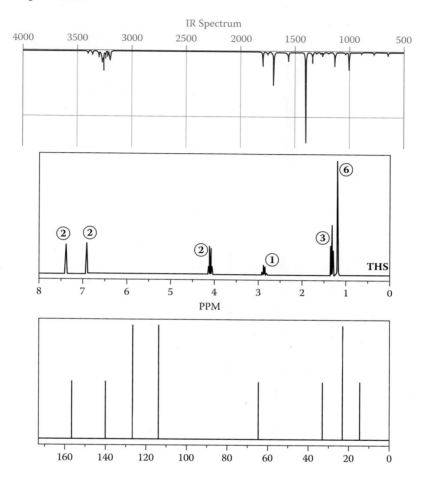

21.144 Identify the molecule with the formula $C_{13}H_{18}O_2$ and the spectral data provided:

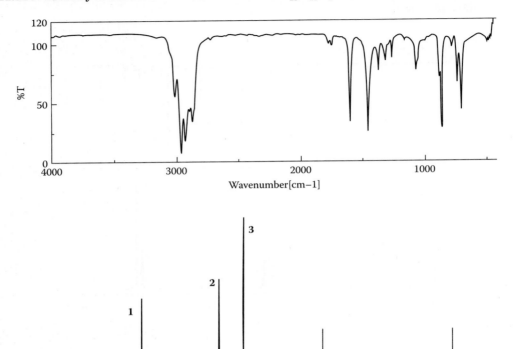

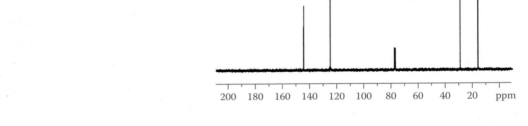

21.145 Identify the molecule with the formula C_7H_8O and the spectral data provided:
IR: broad peak at 3333, 3036, 2922, 1614, 1514, 1237, 816, 740 cm^{-1}
^{1}NMR: 7.00–6.72 (m, 4H), 5.20 (broad s, 1H; this peak is diminished when treated with D$_2$O), 2.25 (s, 3H) ppm

21.146 Identify the molecule with the formula $C_6H_4Cl_2$ and the spectral data provided:
IR: 1476, 1065, 1014, 817, 489 cm^{-1}
^{1}NMR: 7.26 (s, 4H) ppm

21.147 Identify the molecule with the formula $C_8H_{11}N$ and the spectral data provided:
IR: 3403, 3083–2872, 1604, 1509, 1321, 1256, 49, 693 cm^{-1}
^{1}NMR: 7.15–6.58 (m, 5H), 3.42 (broad s, 1H), 3.11 (q, 2H), 1.22 (t, 3H) ppm

Enolate Anions

Acyl Addition and Acyl Substitution

22

The reaction of carbon nucleophiles with ketones or aldehydes proceeds by acyl addition, as described in Chapter 18. The reaction of carbon nucleophiles with acid derivatives proceeds by acyl substitution, as described in Chapter 20. Carbon nucleophiles included cyanide, alkyne anions, Grignard reagents, organolithium reagents, and organocuprates. Alkyne anions are formed by an acid–base reaction with terminal alkynes ($RC\equiv C-H \rightarrow RC\equiv C{:}^-$). In this latter transformation, it is clear that formation of the alkyne anion relies on the fact that a terminal alkyne is a weak carbon acid. Other carbon acids specifically involve the proton on an α-carbon in aldehydes, ketones, or esters. With a suitable base, these carbonyl compounds generate a new type of carbon nucleophile called an enolate anion.

To begin this chapter, you should know the following:

- **nomenclature for aldehydes, ketones, and acid derivatives (Chapter 5, Section 5.9; Chapter 16, Sections 16.2, 16.4)**
- **the acid–base reaction and factors that contribute to the position of the acid–base equilibrium (Chapter 6, Sections 6.1 and 6.3)**
- **contributors to acid strength (Chapter 6, Section 6.3)**
- **contributors to base strength (Chapter 6, Section 6.4)**

1119

- the fundamentals of kinetic and thermodynamic controlled reactions (Chapter 7, Sections 7.6 and 7.7, 7.10)
- the fundamentals of equilibrium reactions and factors that influence the position of the equilibrium (Chapter 7, Section 7.10)
- the concept of resonance and resonance stability (Chapter 5, Section 5.9.3)
- how to draw resonance contributors and understand their significance (Chapter 5, Section 5.9.3; Chapter 21, Section 21.3)
- rate of reaction (Chapter 7, Section 7.11)
- rotamers and conformations (Chapter 8, Sections 8.1, 8.4)
- reactions that generate ketones and aldehydes (Chapter 17, Section 17.2; Chapter 21, Section 21.3.3)
- keto-enol tautomerism (Chapter 10, Sections 10.6.3 and 10.6.4; Chapter 18, Section 18.5)
- the E2 reaction (Chapter 12, Sections 12.1 and 12.2)
- the S_N2 reaction with alkyl halides (Chapter 11, Sections 11.1–11.3)
- differences in nucleophilic strength (Chapter 11, Section 11.3)
- formation of Grignard reagents and organolithium reagents (Chapter 15, Sections 15.1, 15.5)
- that Grignard reagents and organolithium reagents are strong bases (Chapter 15, Sections 15.3 and 15.5.2)
- that amines react as acids in the presence of a strong base (Chapter 15, Section 15.5.2; Chapter 6, Section 6.4.1)
- acyl addition reactions of nucleophiles (Chapter 16, Section 16.3; Chapter 18, Section 18.1)
- acyl addition of organometallic reagents to aldehydes and ketones (Chapter 18, Section 18.4)
- reaction of alcohols and amines with aldehydes and ketones (Chapter 18, Sections 18.6 and 18.7)
- reactions that generate carboxylic acids (Chapter 17, Section 17.4; Chapter 20, Section 20.2)
- reactions that generate acid derivatives (Chapter 20, Sections 20.3–20.6)
- acyl substitution reactions (Chapter 16, Section 16.8; Chapter 20, Section 20.1)
- electron-releasing and -withdrawing substituents (Chapter 3, Section 3.7; Chapter 21, Section 21.3)
- how to identify a good leaving group (Chapter 11, Section 11.2)
- how to convert alkoxide anions to the corresponding alcohol via an acid–base workup (Chapter 5, Section 5.7; Chapter 6, Section 6.2; Chapter 18, Section 18.1)
- *E* and *Z* nomenclature (Chapter 9, Section 9.4)

- **absolute configuration and stereogenic centers (Chapter 9, Sections 9.1 and 9.3)**
- **diastereomers (Chapter 9, Section 9.5)**
- **decarboxylation (Chapter 12, Section 12.6)**

This chapter will discuss carbanion-like reactions that utilize enolate anions. The acid–base reactions used to form enolate anions will be discussed. Formation of enolate anions from aldehyde, ketones, and esters will lead to substitution reactions, acyl addition reactions, and acyl substitution reactions. Several classical named reactions that arise from these three fundamental reactions of enolate anions are presented. In addition, phosphonium salts will be prepared from alkyl halides and converted to ylids, which react with aldehydes or ketones to form alkenes. These ylids are treated as phosphorus-stabilized carbanions in terms of their reactivity.

When you have completed this chapter, you should understand the following points:

- **The α-proton of a ketone or aldehyde is a weak acid and can be removed by a strong base to give a resonance-stabilized enolate anion. Deprotonation may occur via the enol tautomer. The α-proton of an aldehyde is slightly more acidic than that of a ketone.**
- **The α-proton of an aldehyde or ketone is less acidic as more carbon substituents are added. As more electron-withdrawing groups are added, the α-proton becomes more acidic, so a 1,3-diketone is more acidic than a ketone. The more acidic proton of an unsymmetrical ketone is the one attached to the less substituted carbon atom.**
- **Enolate anions react as nucleophiles. They give nucleophilic acyl addition reactions with aldehydes and ketones. The condensation reaction of an aldehyde or ketone enolate with another aldehyde or ketone is called an aldol condensation. Self-condensation of symmetrical aldehydes or ketones leads to a single product under thermodynamic conditions. Condensation between two different carbonyl compounds gives a mixture of products under thermodynamic conditions, but can give a single product under kinetic control conditions.**
- **Dialkyl amides are formed by the reaction of an amine with an organolithium reagent, and they are considered to be non-nucleophilic bases.**
- **Kinetic control is favored by a strong base that generates a conjugate acid weaker than the carbonyl compound, a polar aprotic solvent, short reaction times, and low temperatures.**

Thermodynamic control is favored by a strong base that generates a conjugate acid stronger than the carbonyl compound, a polar protic solvent, long reaction times, and high temperatures.

- Dehydration of enolate condensation products leads to a conjugated carbonyl compound.

- The intramolecular condensation of an α,ω-dialdehyde or diketone leads to a cyclic compound.

- An ester enolate is formed by reaction with a strong base, and the resulting enolate anion can condense with an aldehyde, a ketone, or another ester. Ester enolates react with aldehydes or ketones to form β-hydroxy esters. Aldehyde or ketone enolate anions react with esters to form β-hydroxy esters, 1,3-diketones, or β-keto aldehydes.

- Enolate anions react as nucleophiles. They give nucleophilic acyl substitution reactions with acid derivatives. The condensation reaction of one ester with another is called a Claisen condensation and it generates a β-keto ester. A mixed Claisen condensation under thermodynamic conditions leads to a mixture of products, but kinetic control conditions can give a single product.

- The intramolecular Claisen condensation is called a Dieckmann condensation, and it generates a cyclic compound.

- Malonic esters can be converted to the enolate anion and condensed with aldehydes, ketones, or acid derivatives. The reaction of malonic acid with an aldehyde using pyridine as a base is called the Knoevenagel condensation.

- A β-keto ester can be hydrolyzed to a β-keto acid, and heating leads to decarboxylation. Malonic acid derivatives, as well as β-ketone acids decarboxylate upon heating.

- Enolate anions react with alkyl halides by an S_N2 reaction to give alkylated carbonyl compounds.

- Triphenylphosphine reacts with alkyl halides to form phosphonium salts. Organolithium bases react with alkyltriphenylphosphonium salts to give phosphorus ylids, which react with aldehydes and ketones to give alkenes in what is known as the Wittig reaction.

- A molecule with a particular functional group can be prepared from molecules containing different functional groups by a series of chemical steps (reactions). This process is called synthesis: The new molecule is synthesized from the old one (see Chapter 25).

- Spectroscopy can be used to determine the structure of a particular molecule and can distinguish the structure and functionality of one molecule when compared with another (see Chapter 14).

- Enols and enolates are found in biology.

22.1 Aldehydes and Ketones Are Weak Acids

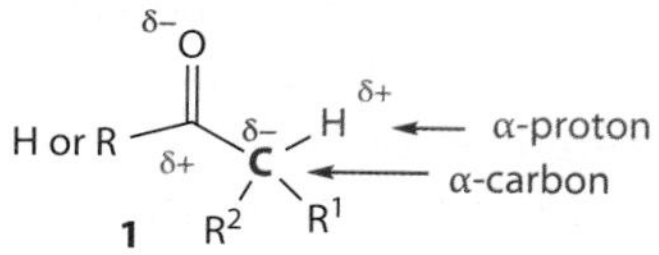

The carbonyl unit carbonyl (C=O) is the functional group for aldehydes and ketones as well as esters (see Chapter 5, Section 5.9). All discussions in previous chapters focused on the carbonyl group and acyl addition or acyl substitution reactions with nucleophiles. The carbonyl group is electron withdrawing with respect to the attached carbon atoms in an aldehyde or ketone. Therefore, the carbonyl group will induce a dipole in the adjacent carbon (the α-carbon is δ−), which in turn leads to a dipole between that carbon and its attached hydrogen (the α-proton), as shown in **1**. This so-called α-hydrogen or α-proton is δ+.

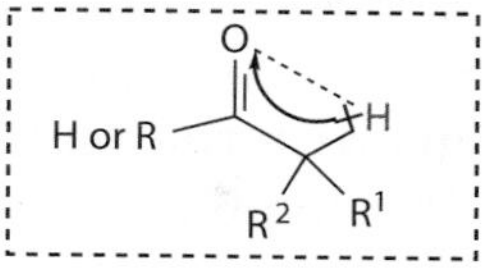

What is the significance of a δ+ hydrogen atom? It is a weak acid! In other words, that proton reacts with a base to give a carbanionic conjugate base. Remember that a "free" proton (H+) is often used as the generic representation of an acid. The larger the δ− is on an α-carbon atom, the greater the δ+ of the α-hydrogen atom is and the more it can be compared to H+ (i.e., it is a stronger acid). **Therefore, *the presence of a carbonyl group in an aldehyde or a ketone makes the α-proton slightly acidic*.** The α-proton in an aldehyde has an experimentally measured pK_a of about 15–16 (about the same as the OH unit in methanol or ethanol), and the pK_a of the α-proton of a ketone is about 19–21. *Aldehydes are slightly stronger acids than ketones*.

22.1 Indicate the acidic proton in propanal.

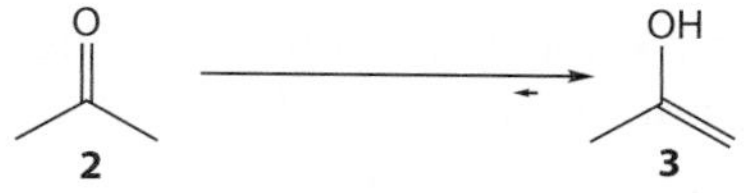

Keto-enol tautomerism was discussed in Chapter 10 (Section 10.6) in connection with the hydration reaction of alkynes, including oxymercuration or hydroboration. The carbonyl form is more stable, and for most aldehydes and ketones only a tiny amount of enol is present. Acetone (**2**), for example, exists primarily as the ketone, and experiments show only $1.5 \times 10^{-4}\%$ enol.[1] This experiment titrated acetone with diatomic bromine and measured the extent of reaction, which converted **2** to 1-bromo-2-propanone (α-bromoacetone, **4**). The second product in this reaction is HBr. By this experiment, the enol content of acetone—and presumably of other simple ketones—is remarkably small.

This experiment shows that **4** is generated from **2**, but how does this reaction occur? A telling experiment shows that direct bromination of acetone is relatively slow, but in the presence of an acid catalyst, the reaction is much faster. This reaction is shown to be independent of the concentration of the halogen, except at very high concentrations of acid. It is known that the acid-catalyzed reaction begins with the acid–base reaction of acetone to give oxocarbenium ion **5**, as seen in Chapter 18, Sections 18.5 and 18.6, but the presence of an acid catalyst also increases the percentage of enol **3**.

Enol **3** reacts with diatomic bromine as a nucleophile to give **6**, which loses a proton to give **4**. Experiments demonstrate that the key species in this reaction is the enol, and it is reasonable to assume that the reaction of **2** and bromine to give **4**, without the acid catalyst, also proceeds by the enol. Without the acid catalysts, only a tiny amount of enol **3** is available to react with bromine to give **4**, but once the enol reacts the acetone will tautomerize to give more **3**, which reacts with bromine. In this way, acetone is converted to **4**, but it is a slow process. The acid catalyst accelerates this process.

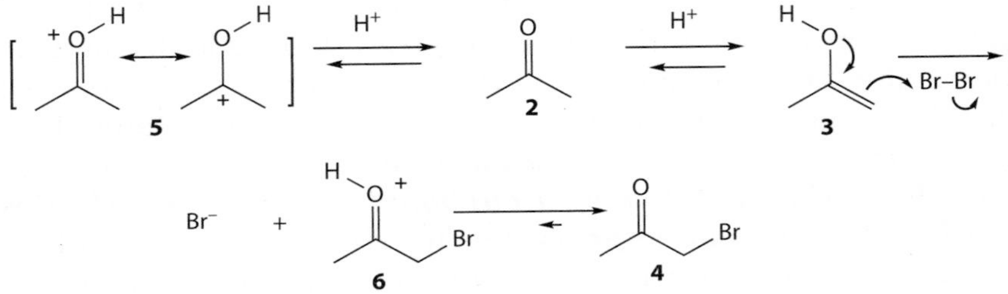

It is known that **4** reacts faster than **2** to form dibromoacetone, and this is usually associated with a higher percentage of enol for **4** when compared to **2**. In principle, the O–H bond of the enol is polarized such that H is δ+ and will be attracted to the δ– bromine atom by intramolecular hydrogen bonding. The net result is stabilization of the enol form. The presence of heteroatoms at the α-carbon should stabilize the enol form, and this is generally correct. Titration of 2,4-pentanedione (**7**; known as acetylacetone) with bromine, as with the reaction of **2**, gives 3-bromo-2,4-pentanedione. This experiment shows that there is 79.7% of the enol, **8**, which is consistent with the concept of enol stabilization by an adjacent heteroatom substituent.

22.2 Draw the structure of 1,1-dibromoacetone.

22.3 Draw the enol expected from 4 and indicate the intramolecular hydrogen bonding that is possible for this structure.

22.4 Draw the structure of 3-bromo-2,4-pentanedione.

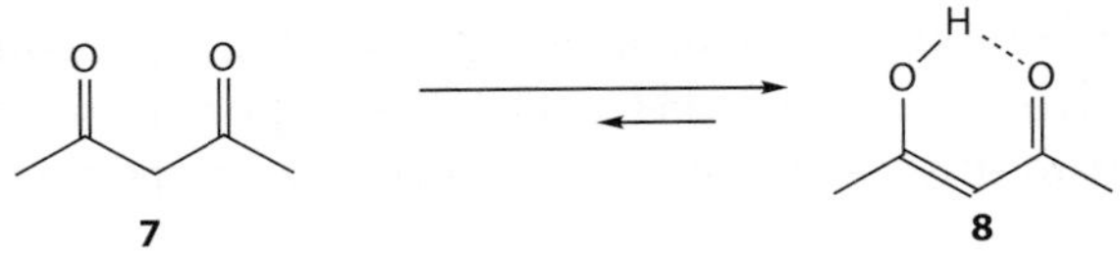

Why is the α-proton acidic? A carbonyl group is electron withdrawing, and there is a "through-bond" inductive effect in **1** (see Chapter 6, Section 6.2). There is also a "through-space" inductive effect where the negatively polarized carbonyl oxygen is attracted to the positively polarized α-hydrogen. These effects combine to gives a larger δ+ hydrogen. Therefore, inductive effects in the ketone or aldehydes are one contributing factor to the acidity.

Another major contributing factor is the charge delocalization (resonance) of the conjugate base. As in Chapter 6, a more stable conjugate base shifts the equilibrium of the acid–base reaction to the right (more product), leading to a large K_a (smaller pK_a), which is equated with greater acid strength. The conjugate base of the acid–base reaction with **1** is **9**, an ***enolate anion***. Enolate anion **9** is resonance stabilized and there are two resonance contributors. The negative charge is delocalized on oxygen in resonance contributor **9A** and on carbon in resonance contributor **9B**.

Using conventional line notation, *both* resonance contributors **9A** and **9B** are required to describe the enolate anion. The electron potential map **9C** shows the charge distribution, and more red (higher electron density) is concentrated on the carbon than on the oxygen, although this depends on the counter-ion. Note the similarity of **9** and the resonance-stabilized formate anion shown in Chapter 6, Section 6.2.4. This resonance delocalization makes the conjugate base **9** more stable, which shifts the acid–base equilibrium to the right. In other words, the α–proton in **2** is easier to remove (i.e., that hydrogen atom is more acidic).

The acid–base reaction of **2** clearly shows that the δ+ polarized proton on the α-carbon of acetone is the acid, and reaction with an undefined **base** gives a conjugate base (**9**) and a conjugate acid (H–base, as yet undefined). Deprotonation of carbon in **2** gives **9A**, but the resonance contributor is **9B**. There is an alternative way to form **9**. Without addition of any base, a small amount of enol **3** is present along with **2** due to tautomerization. Once added, the base may react with the acidic proton of the enol rather than with the α-proton to give **9B**, which is a resonance contributor with **9A**. The O–H bond of the enol is polarized such that the proton is δ+ (see Chapter 6, Section 6.2), so it is relatively acidic. Because the keto form (**2**) is the dominant species, deprotonation of **2** to give **9** is commonly used to describe this reaction, but deprotonation via the enol is a viable mechanistic pathway.

Except for the fact that the ketone is the acid, this reaction is just another acid–base reaction, analogous to those given in Chapters 2 and 6. As in any acid–base reaction, an equilibrium is established between the acid (ketone **2**), the base, the conjugate base (the enolate anion formed when the acidic proton is removed from the ketone), and the conjugate acid. When compared to carboxylic acids with a pK_a of 1–5 (Chapter 16), ketones and aldehydes (pK_a of about 15–21) are weak acids. Acetone (**2**) has a pK_a for the α-hydrogen of about 20. This means that a very strong base is required to remove the proton (in **red**) from the α-carbon. *The term "strong base" means that the conjugate acid generated from the base must be a much weaker acid than acetone, so the pK_a of that conjugate acid must be >22 on average.* The nature of the base is discussed in Section 22.3.

22.5 Draw the enol expected from 2-methylbutanal, as well as the enolate anion that is expected when this ketone reacts with a base.

22.6 Draw the enolate anion that results when the following compounds react with a suitable base: 3-pentanone, butanal, cyclopentanone.

As noted, the pK_a of the α-proton in acetone (**2**) is about 20.[2] In other ketones, different groups attached to the α-carbon lead to a different pK_a for the α-proton, depending on whether the group is electron releasing or electron withdrawing relative to the α-carbon. With a focus on the proton marked in **red** in the illustration, 3,3-dimethyl-2-butanone (**10**), for example, has a pK_a of about 20.8[1] and **11** has a pK_a of about 21.3.[3] These two compounds allow a direct comparison of the influence of an alkyl group on the α-carbon when compared with acetone.

It is known that *an alkyl group such as methyl is electron releasing relative to the carbonyl group,* and the presence of one methyl group on the α-carbon in **11** makes the α-proton *less* acidic by about 0.5 of a pK_a unit when compared with **10**. The methyl group releases electron density to the α-carbon, so the "**red**" α-proton is less polarized and less acidic than the proton attached to the α-carbon of **10**. If a phenyl group is attached to the α-carbon, as in 1-phenyl-1-ethanone (**12**; acetophenone, as described in Chapter 16, Section 16.2), the phenyl group enhances the electron-withdrawing capability of the carbonyl and makes the α-proton more polarized and more acidic. The pK_a of the proton in **12** is about 16.5.[4]

22.7 Estimate the pK_a of the hydrogen on C3 of 3-methyl-2-pentanone, based on this discussion, whether the proton of the methyl group in 2-pentanone has a pK_a of 20.

22.8 Predict whether acetone or 2,4,4-trimethyl-3-pentanone is more acidic.

22.9 Draw the reaction in which enolate anion 12 is deprotonated via the enol.

If an aldehyde and a ketone are compared, the aldehyde has a carbon group and a hydrogen atom attached to the carbonyl, whereas a ketone has two carbon groups. Because carbon groups are electron releasing and an aldehyde has only one attached carbon group, there is a difference in acidity of the α-proton. *The α-proton of an aldehyde is generally more acidic than that of a ketone, as noted above.* As an example, the measured pK_a of the proton in acetone (**2**, a ketone) is 20 and that of ethanal (acetaldehyde, **13**) is 16.5.[1]

22.10 Draw the structure of the enolate anion derived from the reaction of 7 with a strong base.

If an α-carbon has more than one electron-withdrawing group, the α-proton becomes increasingly acidic as the number of groups attached to the carbonyl increases. A simple illustration of this concept compares acetone (**2**) with chloroacetone (**14**) and dichloroacetone (**15**).[1] The pK_a of **2** is 20, whereas the pK_a of **14** is 16.5 and that of **15** is 15.[1] The electronegativity of the chlorine atom makes it electron withdrawing, so the α-proton in **14** is expected to be more polarized and more acidic relative to the α-proton in **2**, leading to an increase in acidity of about 4.5 pK_a units. Likewise, the second electron-withdrawing chlorine atom in **15** makes the α-proton even more acidic, but the effect on pK_a is not as large. *In general, the more powerful the electron withdrawing properties of a substituent at the α-carbon, the more acidic the α-proton, and the more powerful the electron releasing properties of the substituents, the weaker the acidity of the α-proton.* The enol content of chlorinated ketones **14** and **15** should be much higher, and it is reasonable to conclude that higher enol content will facilitate deprotonation to give the corresponding enol.

22.11 Write out the reaction for the enol of 9 deprotonated to give the corresponding enolate anion.

Functional groups other than the carbonyl group, such as the cyano group (–C≡N), will exert an electron-withdrawing effect on an adjacent carbon and on the α-proton. If the pK_a of methane is 58,[1] cyanomethane (**16**; this is commonly called acetonitrile) has a pK_a of 24, dicyanomethane (**17**) has a pK_a of 11.2, and tricyanomethane (**18**) has a pK_a of –5.5 Clearly, the presence of a cyano group leads to greater acidity, and the more cyano groups that are present, the greater the electron-withdrawing effect and the acidity (lower pK_a) will be. The first electron-withdrawing group has the most dramatic effect, but the second and third groups also have a large effect in this case. Note that the pK_a of **18** is quite low and that it is a stronger acid than formic acid (pK_a of 3.75) or even sulfuric acid (pK_a of −3).

As the proton in a molecule becomes increasingly acidic, a weaker base may be used to remove it. Reaction of **18** with $NaHCO_3$ (the weak base sodium bicarbonate) leads to the anion, but bicarbonate does *not* react with **16** under the same conditions to give the anion. It is important to point out that the electron-withdrawing influence of the substituents is not the only factor that influences acidity. The increase in acidity is also related to the stability of the anion formed, the conjugate base, as alluded to in the discussion of the resonance stability of enolate anions. In general, an electron-withdrawing group will stabilize an adjacent carbanion center in the enolate anion.

This point is illustrated with a molecule that does not contain a carbonyl, a cyano group, or any of the electron-withdrawing functional groups discussed. The benzene ring is capable of dispersing charge by resonance (see Chapter 21, Section 21.3). Triphenylmethane (**19**) is clearly a hydrocarbon, but the C–H bond is very slightly polarized because of the phenyl rings. When it is treated with sodium metal, the hydrogen of this alkane is removed to generate sodium triphenylmethide, **20**. The proximity of the phenyl rings allows the negative charge to be delocalized into all three rings, providing a great deal of resonance stabilization. This stabilization makes it relatively easy to form **20**, and the pK_a of **19** has been experimentally measured to be 31.48.[6] Alkane **19** is a very weak acid, but rather strong when compared to a simple alkane such as methane (CH_4), which has a pK_a of 58.1. What is the point of this example? Delocalization of charge in the conjugate base is a significant contributing factor to the acidity of a given molecule, including ketones and aldehydes.

22.12 Draw the anions formed from **16** and **18** and write out the acid–base reaction of **18** with sodium bicarbonate.

22.13 Draw all resonance forms for **14** to help understand how dispersal of the carbanion charge helps stabilize this anion.

The conjugate bases formed from **16–19** are carbanions, and the resonance contributor of enolate anion **9A** derived from acetone is also a carbanion. In terms of its reactivity, it is reasonable to classify **9** as a carbanion because the resonance-stabilized enolate anion has a larger concentration of electron density on the carbon relative to oxygen (see **9C**). This comparison makes a distinction between the resonance-stabilized enolate anion generated from aldehydes or ketones and a carbanion formed when another electron-withdrawing group is present, even if the carbanion is resonance stabilized. The formation and reactions of enolate anions are the main focus of this chapter.

22.14 Predict the ketone that has the more acidic proton: acetone or 2,5-pentanedione.

22.2 Enolate Anions Are Nucleophiles. The Aldol Condensation

The nineteenth century was a time when chemists were exploring chemical reactivity with no knowledge of mechanism and with a limited ability to identify products when compared to today's methodology (see Chapter 1, Section 1.1). Nonetheless, many reactions were discovered that are used today, and the structures of the products were accurately determined. L. Chiozza in 1856 (Italy) and Adolphe Wurtz (France; 1817–1884) in 1872 reported independent experiments in which an aldehyde was treated with an alkoxide base in an alcohol solvent heated to reflux (heated at the boiling point of the alcohol). When this reaction was cooled to room temperature and treated with dilute aqueous acid at low temperatures, a β-hydroxy aldehyde (known generically as an aldol) was isolated. Because of the type of product formed, this reaction has come to be called the **aldol condensation**.

A typical example treats butanal (**21**) with NaOEt in refluxing ethanol, and an aqueous acid second step (known as a reaction workup) gives **22** in 75% yield.[7] This new product contains eight carbons and it is clear that one molecule of the aldehyde must react with a second molecule of the same aldehyde. When the reaction is repeated, but treated in the second step with hot and concentrated acid, a conjugated aldehyde product (**23**) is formed in over 90% yield,[4] and it also contains eight carbons. If **22** is isolated, purified, and then heated with aqueous acid, **23** is the product.

How are these results explained? In **21**, the four carbons of one aldehyde unit are marked in **green** in the illustration and those of a second aldehyde molecule are marked in **violet**. Clearly, reaction occurred between the α-carbon of the "**green**" aldehyde and the carbonyl carbon of the "**violet**" aldehyde; the final product is an alcohol. Based on knowledge of nucleophilic acyl addition from Chapters 16 and 18, **21** must react as a nucleophile, attacking the acyl carbon of a second aldehyde molecule.

With respect to this sequence of observed reactions, experimental evidence shows that **21** reacts with the base (NaOEt) to form an enolate anion, and the nucleophilic carbon atom of that enolate anion attacks the carbonyl of a second aldehyde to give the alkoxide of **22**. This is a normal acyl addition reaction, and the nucleophile is the α-carbon of the enolate anion. Treatment of this initial alkoxide product with aqueous acid under mild conditions simply generates alcohol **22**, as with all other acyl addition reactions (see Chapter 18). Product **22** is called an ***aldol*** or an ***aldolate***. ***The reaction of an aldehyde or a ketone with a base generates an aldol product***. Vigorous acid hydrolysis led to protonation of the OH unit in **22** by the strong acid (to form an oxonium ion), which eliminated a molecule of water (dehydration) to give the alkene unit in **23**.

22.15 Draw the initial alkoxide product of the enolate of 21 reacting with a second molecule of 21 via acyl addition.

The process just described in the walk-through for the formation of **22** constitutes the mechanism of the reaction. Based on knowledge of the acidity of an α-hydrogen atom in aldehydes and ketones, an initial acid–base reaction of the ketone with ethoxide is the basis for that mechanism. When **21** reacts with NaOEt (a base in this reaction), the α-proton is removed to form enolate anion **23**. Enolate anion **23** is a resonance-stabilized anion with two resonance contributors: one with the negative charge on oxygen (**23A**) and the other with the charge on carbon (**23B**). Examination of the electron potential map of **9** in Section 22.1 shows that more electron density resides on the carbon (C2), so **23** will react as a carbon nucleophile. This means that the α-carbon in **23** (C2) is the nucleophile in an acyl addition reaction with the carbonyl of another molecule of **21** (C5); the product is **24**. Note that formation of **23** is an acid–base reaction and therefore an equilibrium reaction. This means that **23** (the conjugate base) is in equilibrium with **21** (the acid).

Experimentally, it has been shown that under these conditions, a small amount of **23** and a large amount of **21** are at equilibrium. As **23** reacts with **21**, the product is **24**. As **23** is removed by this reaction, the equilibrium shifts to give more **23**, which reacts with **21**. In other words, as **23** reacts with **21**, the equilibrium is reestablished and, eventually, all of **21** reacts to give **24**. A new carbon–carbon bond is formed in the product **24** (heavy bond marked in **blue**), analogous to acyl addition of carbon nucleophiles such as an alkyne anion or a Grignard reagent in Chapter 18 (Section 18.4).

22.16 What is the conjugate acid formed by the reaction of 21 and NaOEt?

22.17 Draw the product formed when butylmagnesium chloride reacts with cyclopentanone. What is the final product when the initial product of acyl addition is treated with aqueous acid?

The conversion of **21** to **24** is a "self-condensation" because butanal forms an enolate that reacts with another molecule of butanal. *This means that two identical molecules are joined together in the product.* Mild acid workup simply protonates the alkoxide unit in **24** to give the aldol, **22**. Self-condensation is also possible with symmetrical ketones such as 3-pentanone. In a self-condensation, one is tempted to ask where the second molecule of **21** comes from. Remember that structure **21** represents one molar equivalent (*1 mol* of molecules); because this is an equilibrium reaction, both **21** and enolate anion **23** will be present, so unreacted **21** is available to react with **23**. After one half-life (Chapter 7, Section 7.11.3), the solution will contain **21**, enolate anion **23**, and **24**. Therefore, as **24** is formed by the reaction, there is still plenty of **21** available for reaction.

It is clear from this last example that an enolate anion behaves as a carbon nucleophile and gives the same sort of acyl addition reaction as another previously discussed carbanion, an alkyne anion (see Chapter 18, Section 18.3.2). The products are different, of course, but *from the standpoint of comparing reactions, the only real difference between an enolate anion and an alkyne anion is the structure and complexity of the enolate anion as a carbon nucleophile and the functionality in the final acyl addition product*.

22.18 Draw the aldol sequence for self-condensation of 3-pentanone with sodium ethoxide (NaOEt) in ethanol.

22.19 Draw the reaction and product formed when the anion of 1-propyne reacts with 21.

The self-condensation of butanal involves a single compound, but it is also possible to convert one ketone or aldehyde to an enolate anion, and it will react with a different ketone or aldehyde. This is called a ***mixed aldol condensation***. If acetone (**2**) is treated with aqueous NaOH in the presence of another carbonyl molecule, such as benzaldehyde (**25**), enolate (**26**) is formed *in situ*. This enolate anion may react with itself (with another molecule of **2** in a self-condensation reaction), but it may also react with aldehyde **25** via acyl addition to give alkoxide **28**. Mild hydrolysis gives the ***mixed aldol*** product, **26**. There is a competition for the reaction of **27** with either **2** or **25**, so at least two products are possible in the reaction: **28** and the self-condensation product. Note that, in this reaction, the second carbonyl derivative (**25** in this case) does **not** have an acidic hydrogen atom, so **27** is the only possible enolate anion. Under these conditions, enolate **27** is formed and, if an excess of **25** is added, the major product will be **28** and then **26** after the aqueous acid workup. An excess of **25** may be added because it cannot form an enolate anion.

22.20 **Draw the aldol product formed by self-condensation of 27 and 2.**

22.21 **Draw the product or products of a reaction between cyclopentanone and benzaldehyde in refluxing ethanol and in the presence of sodium ethoxide (the initial product is treated with dilute aqueous acid).**

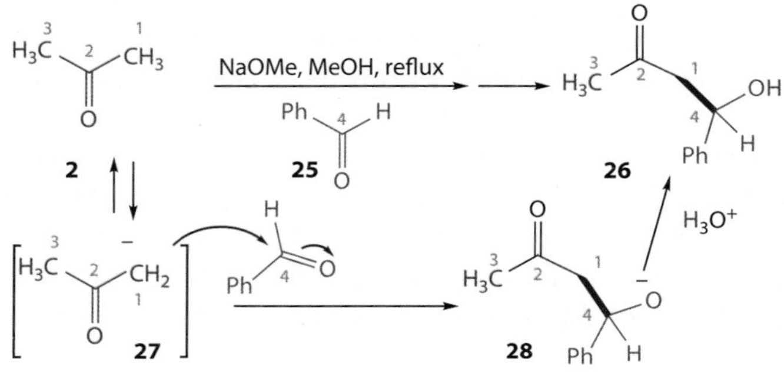

This simple example illustrates that mixed aldol condensation reactions may be problematic. If the enolate anion is formed in the presence of an aldehyde or ketone with no α-hydrogens, there is a short list of possible partners. This situation constituted the state of the art for nearly a century, until chemists eventually found experimental conditions that permit a mixed aldol to proceed under conditions that allow one to predict one product as the major product.

22.3 Non-Nucleophilic Bases

As just discussed, the condensation of an enolate derived from one aldehyde or ketone with a different aldehyde or ketone is a desirable reaction. In the reaction of acetone and benzaldehyde to give **26**, the reaction gave **26** because benzaldehyde has no α-proton. Therefore, the only enolate possible is the one from acetone (see **27**). Self-condensation of acetone may be a problem, but it is not possible from benzaldehyde because there is no enolate anion. The reaction of acetone and 3-pentanone is not this simple because an enolate anion may be formed from both ketones. Before this reaction is discussed (in Section 22.4.3), it is useful to describe a method that allows selective formation of the enolate anion of acetone or the enolate anion of 3-pentanone.

22.22 Draw the enolate anion generated from 3-pentanone.

22.23 Draw the enolate anions of acetone and 3-pentanone. Draw the self-condensation aldol products expected from acetone and from 3-pentanone.

When acetone reacts with NaOEt in ethanol to form enolate anion **27**, it is a reversible acid–base reaction. Therefore, unreacted ketone or aldehyde always remains in the reaction, and this fact allows self-condensation to occur. Is it possible to choose a base that will generate the enolate anion, but the equilibrium is pushed far to the right (toward the enolate anion product)? If such a base is available, self-condensation is much less of a problem, which is particularly important for mixed aldol condensation reactions. As chemists experimented to find such a base, it was discovered that amide bases (R_2N:⁻), derived from secondary amines (R_2NH) accomplished this goal.

A typical and highly useful amide base is lithium diisopropylamide (**30**; commonly abbreviated as LDA), which is formed by the acid–base reaction of a powerful base such as butyllithium (Chapter 15, Section 15.5.2) and the hydrogen atom attached to nitrogen in the parent amine, diisopropylamine (**29**). This means that the diisopropylamine is an acid in this reaction and diisopropylamide is its conjugate base. From experimental work, it is known that a secondary amine such as **29** is a weak acid (pK_a of about 25) and a powerful base is required to remove the proton from nitrogen.

In Chapter 15, organolithium reagents were shown to be powerful bases, and they are strong enough to react with an amine. Indeed, the reaction of n-butyllithium and diisopropylamine in THF (tetrahydrofuran) at low temperature gives a solution of **30**, and butane is the conjugate acid. The fact that butane is such a remarkably weak acid is an indication of the power of butyllithium as a base. In general, amide bases are *not isolated, but rather generated and used as they are needed.* In principle, a primary amine can be converted to an amide base, but in practice amide bases derived from secondary amines are used most of the time.

22.24 What is the conjugate acid formed when butyllithium reacts with 29? Is this conjugate acid a much stronger or much weaker acid than 29? How does this fact influence the reaction?

22.25 Draw the reaction and give the product formed when n-butyllithium reacts with diethylamine.

22.26 Draw and name the product of a reaction between 2,2,6,6-tetramethylpiperidine and butyllithium and then between hexamethyldisilazane [(Me₃Si)₂NH] and butyllithium.

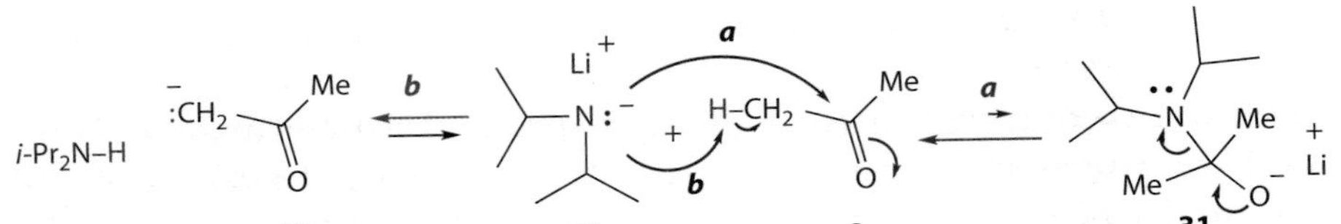

Amide bases have a formal charge of –1 on nitrogen, and they react as either a base or a nucleophile. If lithium diisopropylamide (**30**) reacts with the carbonyl carbon of acetone (**2**), the acyl addition product via *path "a" would be 31*. Based on the chemistry presented for other carbon nucleophiles, this should occur very easily. However, *the carbon groups around nitrogen (the basic atom) provide quite a bit of steric hindrance, which makes it difficult for the nitrogen to approach the sp² hybridized carbon atom via path a.* For this reason, *acyl addition to give 31 is very slow to the point of not being competitive with an acid-base reaction. This poor reactivity of the amide base in acyl addition is expressed by saying that **nitrogen is a poor nucleophile in this reaction**.*

By comparison, the acid–base reaction of **30** and **2** (*path "b"*) proceeds smoothly to give enolate anion **27** (the conjugate base) and diisopropylamine (the conjugate acid). There is little steric hindrance to the approach of the proton on the α-carbon, and **30** is a strong base. In other words, the acid–base reaction of **30** and **2** proceeds readily to form the enolate anion, but acyl addition to form **31** is so slow that virtually no **31** is formed. *Since **lithium diisopropylamide is a good base but a poor nucleophile, it is termed a non-nucleophilic base***. Non-nucleophilic amide bases such as **30** are used when the acid is very weak, as with simple ketones and aldehydes, and acyl addition is a competitive reaction to deprotonation. When the pK$_a$ of the aldehyde or ketone falls below 10, weaker bases can be used, but those reactions will be introduced when it is appropriate.

22.27 See Chapter 5, Section 5.5, and calculate the formal charge on N in 30, on N in 31, and on oxygen in 31.

22.28 Draw the products formed when 30 and 2 undergo an acid–base reaction.

22.29 Draw the acyl addition product or products of the reaction with lithium diethylamide and cyclohexanone.

22.30 Draw the product of the reaction between 3-cyano-2,4-pentanedione and sodium bicarbonate.

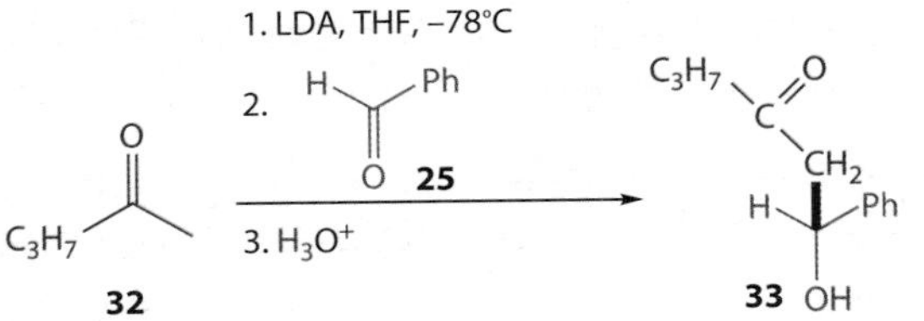

Because LDA is a non-nucleophilic base, it should react with a ketone to give an enolate anion. In an actual experiment, 2-pentanone (**32**) reacts first with LDA to form the enolate anion (not shown) and then with benzaldehyde (**25**) to form the aldol alkoxide product (also not shown). Subsequent mild acid hydrolysis gives **33** in 80% yield.[8] Virtually no self-condensation of **32** is observed in this experiment, which suggests that the reaction is largely irreversible. Assume that 2-pentanone reacts with LDA to give enolate anion **34** (the two resonance forms are **34A** and **34B**). As in all of these reactions, the "carbanion" form of the enolate **34A** is the nucleophile. To account for the observed lack of self-condensation, the equilibrium for this acid–base reaction must be pushed toward **34** (the reasons for this will be discussed in Section 22.4.2). If this statement is correct, it means that **32** has been converted almost entirely to **34**, so there is little or no **43** available to react. Therefore, a different carbonyl compound may be added in a second chemical step to give **35**. This is an overstatement of the facts, but it is a useful assumption that explains the results. Note that benzaldehyde is used, which has no α-protons and cannot form an enolate anion.

22.31 Give the IUPAC name for 27.

A third chemical step converts **35** to its conjugate acid (the alcohol **33**) using aqueous acid. There are three distinct chemical steps. Step 1 generates the enolate anion (**34**), which is then treated with benzaldehyde (**25**) in step 2 to give **32**. After this reaction, step 3 is an aqueous acid workup to give the final isolated product, **33**.

22.32 Draw the structure of S-33.

22.33 Draw the first product or products of the reaction between cyclopentanone and LDA, then the product that results when 3-pentanone is added, and then the final product when aqueous acid is added.

22.34 Draw the final product of a reaction of 4-phenyl-2-butanone with (a) LDA, (b) cyclopentanecarboxaldehyde, and (c) aqueous acid.

22.4 Enolate Anions from Unsymmetrical Ketones

For acetone (**2**), acetophenone (**12**), or acetaldehyde (**13**) from Section 22.1, only one α-carbon is available for reaction because each molecule is either symmetrical (as with acetone) or at least one group attached to the carbonyl does not have an α-hydrogen (as in aldehyde, **13**, or ketone, **12**). Because only one site is available for an acid–base reaction, only one enolate anion product is possible. 2-Pentanone (**32**) in the previous example, however, has two different α-carbon atoms, which means there are two different α-protons, H_a and H_b. If there are two different acidic protons, there are two sites for an acid–base reaction. In the reported experiment, the carbon–carbon bond in **33** clearly results from reaction of the enolate formed by removing H_a. Why does LDA remove H_a rather than H_b? Is this a problem worthy of concern? Both of these questions are addressed in this section.

22.4.1 Removing the More Acidic Proton
in Unsymmetrical Ketones

Examination of two acidic hydrogens in 2-pentanone (**36**) shows that the pK_a of H_a is about 20[1] and that of H_b is about 21. Proton H_b (in **violet**) **36** is attached to an α-carbon that also has an ethyl group attached to it. The α-carbon that bears H_a (in **red**) has only hydrogen atom attached to it. As noted earlier, an alkyl group is electron releasing relative to the α-carbon of a ketone or aldehyde, so H_b is less polarized than H_a (it has a smaller δ+). This observation is consistent with the observation that H_b is less acidic. 2-Pentanone (**32**) is called an *unsymmetrical ketone* because two different groups are attached to the carbonyl (the common name of **32** is ethyl methyl ketone, which reflects the fact that it is unsymmetrical). In an unsymmetrical ketone with different substituents on the different α-carbons, it is possible that those two α-protons will have different pK_a values and that one may be more acidic than the other.

Two different acid–base reactions may occur that will form two different conjugate bases (two different enolate anions). In the case of 2-pentanone, removal of H_a by a base generates enolate anion **34** (as in the reaction with benzaldehyde), but if H_b is removed, the product is enolate anion **36**. These are two **different products** and, if they react with an electrophile, they will give two different products. This is potentially a real problem. If **32** reacts with a base, **is the product 34 *or* 36, or is it a mixture of 34 *and* 36**? The answer depends on the reaction conditions and the base used in the reaction. In the reaction of **32** with LDA and then benzaldehyde, **34** is formed preferentially. This is an important clue that is used to explain this reaction.

22.35 Draw both enolate anion products formed when 2-methylcyclopentanone reacts with a base.

22.4.2 Kinetic versus Thermodynamic Control

Of the two different acidic protons in **32**, H_a is the more acidic, and that proton is attached to the less substituted carbon atom. When treated with a base, the more acidic proton (H_a) is always removed first—in this case, to give **34**. This statement is a staple of acid–base chemistry, where the most acidic proton will always be the first removed by a base. Because H_a is more acidic than H_b and always removed first, is it possible to remove H_b to generate **36**? If the conversion of **32** to **34** is *irreversible,* then **36** is never available, but *this is an acid–base reaction that by definition is reversible.* If the conversion of **32** to **34** is reversible, then at equilibrium some **34** will be present, as well as some unreacted **32** (in other words, **30** ⇄ **32**). It should be stated that H_b is acidic enough to be removed by the base, but H_a is removed faster.

If **34** is formed first, the rate of reaction for removal of H_a from **32** is faster than that for removing H_b. It is important to understand that the rate for removing H_b is slower, *but the rate is not zero,* so both H_a and H_b can be removed (to form **34** and **36**, respectively) at equilibrium. Note that **36** is drawn as the Z-isomer (the oxygen and the ethyl group have the higher priority, as in Section 9.4.3 of Chapter 9). The E-isomer is also possible, and in fact **36** exists as a mixture of E- and Z-isomers. *Assume that the enolate anions from unsymmetrical ketones exists as a mixture of E- and Z-isomers, and the stereochemical implications are ignored in this book.*

Removal of H_b via an acid–base reaction generates **36** and a new equilibrium will be established (**32** ⇄ **36**). The position of both equilibria is defined by a K_a (an equilibrium constant; see Chapter 6, Section 6.1) for each reaction. The magnitude of K_a for the equilibrium of **32** and **34** (K_a^1) when compared with K_a for the equilibrium of **32** and **36** (K_a^2) will determine whether the reaction gives **34**, primarily **36**, or a mixture of **34** and **36**. Because H_a is removed first, focus attention on K_a^1. If K_a^1 for this reaction is large, the equilibrium is shifted to the side of products and **34** is essentially the only product observed. If K_a^1 is unity (= 1), then some **34** is formed, but a significant amount of **32** remains and is converted to **36**. If K_a^1 is very small, there is only a small amount of **34** and a large amount of **32**, which reacts to give **36**.

22.36 Write the equation defining K_a for the conversion of 32 to 36.

For reaction conditions that make K_a^1 large, **34** is formed first and concentration of **32** is minimal. Under these conditions, **34** reacts as a carbon nucleophile in various reactions. For conditions that make K_a^1 small, **32** is always present in an equilibrium mixture and deprotonation can give **36**, so the overall equilibrium includes **32** along with *both* **34** and **36**. In other words, when K_a^1 is small, there is an equilibrium that includes **32**, **34**, and **36** in solution. ***An important point has been ignored***. The pK_a of H_a and H_b has been the only focus, but from Chapter 6 (Section 6.3), it is known that the stability of the conjugate base plays a major role in the magnitude of K_a. In other words, ***the structure of the enolate anion must be considered as a major influence on the position of the equilibrium***.

Differences in structure may make one enolate anion more stable than the other, which will shift the equilibrium toward the more stable enolate anion. ***Another way to state this fact is that an equilibrium reaction is under thermodynamic control, which will favor the more stable (lower energy) product***. An enolate anion has one resonance contributor that has a C=C unit, and anything that releases electrons to the π-bond will make that alkene unit have a stronger bond (see Chapter 5, Section 5.1). In general, the stronger the π-bond is, the more stable will be the enolate anion. Comparing **34** and **36**, the C=C unit in **34** has only two substituents: a carbon group and an oxygen. The C=C unit in **36** has three substituents: two carbon groups and an oxygen. Carbon groups are electron releasing, so more electron density is donated to the π-bond in **36** and it should be more stable. ***Assume that the more highly substituted enolate anion will be more stable and will be the major product in an equilibrium reaction***.

The previous discussion is summarized by saying that if the reaction is controlled to make K_a^1 large, the reaction will favor **34**. If the reaction is controlled to establish an equilibrium so that K_a^1 is small for **32–34**, then K_a^2 leads to **36** and the equilibrium will favor this more stable product. *Since **34** is formed fastest, it is referred to as the* **kinetic product**. *Since **36** is formed under equilibrium conditions and its thermodynamically more stable, it is referred to as the* ***thermodynamic product***.

What conditions favor a large K_a^1 and what conditions favor a small K_a^1? First, different bases may be used in separate reactions of **32**: one using LDA (**34**) and the second using sodium hydroxide. When LDA (the base) reacts with **26** (the acid), the acid–base reaction shown leads to the conjugate base (enolate anion **34**) and the conjugate acid (diisopropylamine, **29**). Similarly, reaction of **32** with NaOH (the base) leads to **3** and the conjugate acid water (HOH). In the reaction with LDA, compare the acid and the conjugate acid, 2-butanone with a pK_a for H_a of about 20,[1] and diisopropylamine, with a pK_a of about 25. 2-Pentanone is the stronger acid when compared to diisopropylamine, which favors the reaction of **32** as an acid rather than the reaction of **29** as an acid and is indicative of a larger K_a in this reaction (the equilibrium is to the right and K_a^1 is larger). Similarly, LDA is a stronger base than the enolate anion, which also favors a larger K_a^1. If 2-pentanone is compared with the conjugate acid of NaOH, water (pK$_a$ of 15.74)[9] in the second reaction, however, water is the stronger acid. This fact indicates that once **34** is formed, it will react with water to regenerate **32** and the hydroxide ion. Because water is the stronger acid, it is anticipated that the equilibrium is shifted to the left, which is consistent with a smaller K_a^1.

When LDA is the base, the equilibrium favors a large concentration for all practical purposes because K_a^1 is large, and there will be only a small amount of unreacted **32**. Under these conditions, formation of **36** is very slow relative to formation and subsequent reactions of **34**. If NaOH is the base, K_a^1 is small, so a significant amount of **32** is in solution and there is only a small amount of **34**. Therefore, formation of enolate **36** is competitive and, at equilibrium, there is a smaller concentration of **34** and a larger concentration of **36**. The major product of any subsequent reaction will arise from reaction of **36**. *If a strong base generates a conjugate acid that is a weaker acid than the ketone, K_a^1 will be large and the kinetic product is favored. If a strong base generates a conjugate acid that is a stronger acid than the ketone, K_a^1 will be small allowing K_a^2 to be established and the thermodynamic product will be favored.*

22.37 Draw the conjugate acid for a reaction of 2-butanone with each of the indicated bases and then indicate whether K$_a$ is large or small: NaOEt, NaNH$_2$, NaF.

The solvent plays a major role in establishing the position of both equilibria for K_a^1 and K_a^2. Assume that LDA reacts with **32** in a reaction with THF as the solvent. In the second reaction, ethanol is the solvent. In the THF reaction, the only acids present in the reaction are diisopropylamine and the starting ketone, 2-butanone. The solvent (THF) does not have an acidic proton and is classified as an **aprotic solvent** (no acidic hydrogens; see Chapter 13, Section 13.2). The equilibrium for the reaction of **32** and LDA favors **34** (K_a^1 is large), and *the solvent will have no influence on the position of this equilibrium*. **In an aprotic solvent the kinetic product is favored.**

If ethanol is the solvent, there is a problem. Ethanol has a pK$_a$ of about 15.9[4] and it is clearly much more acidic than 2-butanone. Once formed, the enolate anion (also a strong base) will react with ethanol to give 2-butanone as the conjugate acid. In other words, in the protic solvent, **34** will react with ethanol to regenerate **32**, and this reaction shifts the equilibrium back to the left (K_a^1 is small), which favors the thermodynamic process. *Therefore, an aprotic solvent will favor a large K_a^1 and kinetic control whereas a protic solvent will favor a small K_a^1 and thermodynamic control.*

22.38 Predict whether or not the use of 2-propanol as a solvent favors kinetic or thermodynamic control in a reaction of a base with 2-pentanone.

22.39 Predict whether the use of diethyl ether as a solvent favors kinetic or thermodynamic control in a reaction of a base with 2-pentanone.

Two more parameters of a reaction must be discussed: temperature and time. In some reactions, the product cannot survive temperatures above 0°C, whereas it can survive temperatures past 80°C in others. In general, *high temperatures favor equilibration and thermodynamic control whereas low temperatures favor kinetic control* because it is more difficult to attain an equilibrium at lower temperatures. Reaction time influences the enolate-forming reaction, but its effect is difficult to explain in general terms. A long reaction time is obviously a relative term. If the half-life of the reaction is very short (see Chapter 7, Section 7.11.3), 10 minutes may represent 30 half-lives and this is a long reaction time. If the half-life is long, then 2 days may represent only two half-lives and this is a short reaction time. The specifics of temperature and time are determined experimentally and will be ignored in this book. It is assumed that the reaction is allowed to proceed through five to eight half-lives.

What does all of this mean? **The reaction of 2-pentanone with LDA in THF at –78°C constitutes typical kinetic control conditions**. Therefore, formation of the kinetic enolate and subsequent reaction with benzaldehyde to give **34** is predictable based on the kinetic versus thermodynamic control arguments. In various experiments, the reaction with an unsymmetrical ketone under what are termed thermodynamic conditions leads to products derived from the more substituted (thermodynamic) enolate anion. **Thermodynamic control conditions typically use a base such as sodium methoxide or sodium amide in an alcohol solvent at reflux**. The yields of this reaction are not always good, as when 2-butanone (**37**) reacts with NaOEt in ethanol for 1 day. Self-condensation at the more substituted carbon occurs to give the dehydrated aldol product **38** in 14% yield.[10] Note that the second step uses aqueous acid and, under these conditions, elimination of water occurs.

22.40 Write the kinetic enolate of 2-ethylcycloheptanone and then write the product formed when it reacts with (a) 3-phenylpentanal, and (b) dilute aqueous acid.

22.41 Write out the reaction of 2-methylcyclohexanone under kinetic conditions (choose the base) that will lead to the kinetic aldol product in a subsequent reaction with 2-butanone.

22.42 Draw the aldol product that led to 32, before elimination of water.

Examination of **38** indicates that C3 of **37** attacks the acyl carbon in the self-condensation, so it is clear that the thermodynamic enolate anion is the nucleophilic species. Under the thermodynamic conditions shown (NaOEt, EtOH, reflux), the product is explained by deprotonation of **37** to give an equilibrium mixture of the thermodynamic enolate anion **39**. This enolate anion subsequently reacts with a second molecule of **35** to give the acyl addition product, **40**. Protonation of the alkoxide under mild acid conditions gives the aldol, **41**, which is clearly protonated by the acid and loses water to give the observed **38**.

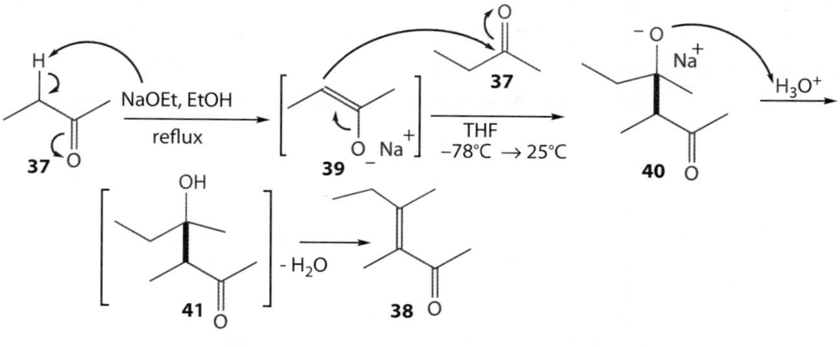

22.43 **Draw the product of a reaction when 3-pentanone is treated with NaOEt in refluxing EtOH.**

22.44 **Draw the product of the reaction with butanal and LDA followed by reaction with cyclohexanone.**

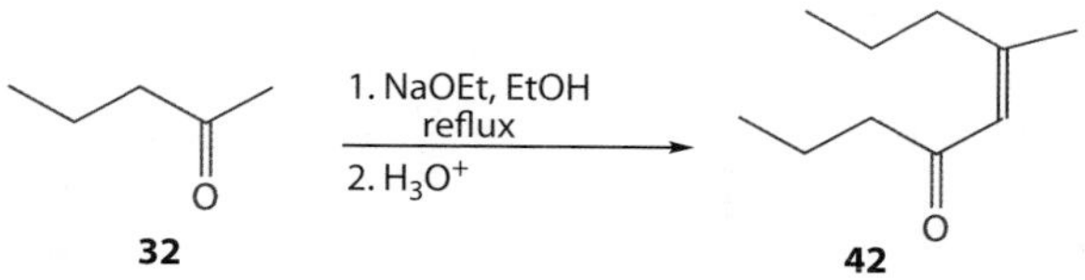

It is important to put the thermodynamic aldol condensation into perspective. The actual product as well as the yield depends on the solvent, the base used, and the nature of the carbonyl substrate. When both a kinetic and a thermodynamic enolate are present, the equilibrium may favor the thermodynamic enolate, but steric hindrance makes its reaction much slower. In such a case, the kinetic enolate product may be the isolated product. When 2-pentanone (**32**) is refluxed with NaOMe in methanol and then treated with 20% sulfuric acid, for example, conjugated ketone **42** is the only isolated product, but in only 8% yield.[11] *This clearly makes the reaction confusing to understand. It is presented only to show that this reaction is an equilibrium-driven process and that the equilibrium may be shifted between kinetic and thermodynamic enolates by several factors.* **For the purposes of doing homework, assume that thermodynamic conditions lead to the thermodynamic aldol as the major product and that kinetic control conditions lead to the kinetic aldol as the major product.**

22.4.3 Mixed Aldol Condensations under Thermodynamic Conditions

It is now apparent that the reaction of acetone and benzaldehyde in Section 22.2 used thermodynamic conditions, and that only one aldol is formed because only one aldehyde has a site for deprotonation (benzaldehyde has no α-hydrogen atoms). When two different ketones or aldehydes that have α-hydrogens are condensed, different enolate anions are possible and each can react to give different products, as suggested in Section 22.2. This potential problem is solved by using the kinetic control conditions presented in Section 22.4.2. However, the extent of the problem must be defined. It is presented after Section 22.4.2 because knowledge of thermodynamic conditions makes the explanation much easier to understand.

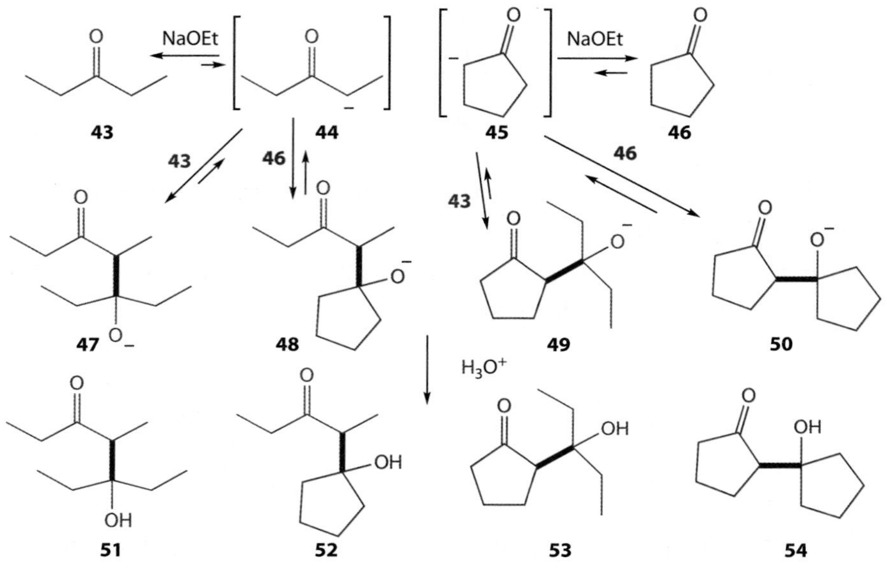

The reaction of two symmetrical ketones, 3-pentanone (**43**) and cyclopentanone (**46**), with sodium ethoxide in refluxing ethanol illustrates the problem. These are thermodynamic conditions, so both enolate anions are formed and they are in equilibrium with both starting materials. Only one enolate anion is possible from each ketone, but sodium ethoxide reacts with **both** ketones. The reaction of **43** leads to enolate **44** and **46** reacts to give enolate **45**. Under thermodynamic conditions, **all** of these species (**43**, **44**, **45**, and **46**) are in equilibrium and all will be in solution at the same time. The equilibrium will favor only small amounts of **44** and **45** with larger amounts of the more substituted enolate anions **43** and **46**. Enolate **44** reacts with *either* **43** or **46** to give **47** or **48**, respectively.

Similarly, enolate **45** reacts with *either* **43** (to give **49**) or **46** (to give **50**), so **under thermodynamic conditions there are *four alkoxide products***. Aqueous acid workup converts the alkoxide products to the aldol products **51**, **52**, **53**, and **54**. Clearly, trying to synthesize an unsymmetrical aldol (a **mixed aldol**) under these conditions is not very attractive. For the most part, kinetic control conditions are used and no attempt is made to synthesize mixed aldol products under thermodynamic conditions.

22.45 Write out the reaction, with both carbonyl reactants, that leads to **49**.

22.46 Draw all of the products when a mixture of 2-pentanone and 1-phenyl-3-pentanone is treated with sodium methoxide in refluxing methanol and then treated with dilute aqueous acid.

To conclude this section, it must be pointed out that a single product can often be isolated from reactions like the one just described despite the problems detailed. This usually has to do with the fact that one enolate is more stable than another and one ketone or aldehyde partner is more reactive. Even subtle differences can lead to significant differences in the product distribution. The yields of a single product may be low, however. A simple example is the condensation of acetone and cyclopentanone (**46**) with NaOMe in refluxing methanol. Vigorous hydrolysis leads to the isolation of **55** in 39% yield.[4]

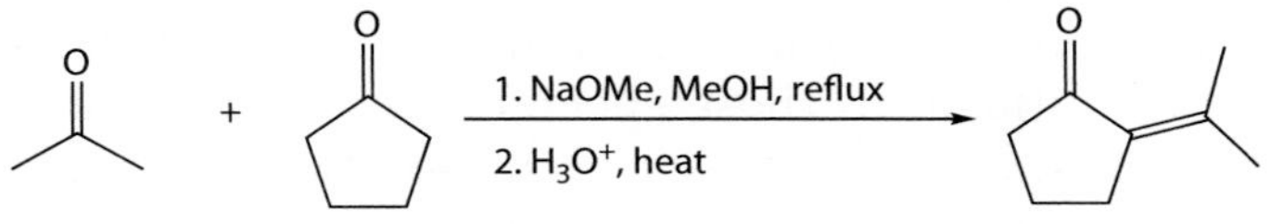

22.5 Dehydration of Aldol Products

In most of the aldol reactions thus far, the product is an aldol (a β-hydroxy ketone or aldehyde), although the dehydration reaction to give the alkene product has also been shown. In the first reaction in this chapter, vigorous hydrolysis of the aldol product **21** led to dehydration and formation of **23**. In most cases, it is relatively easy to isolate the aldol product if the aqueous acid workup step uses dilute aqueous acid or sometimes just water, and the temperature is kept relatively low. Heating with concentrated acid—particularly a concentrated acid solution—often leads to dehydration.

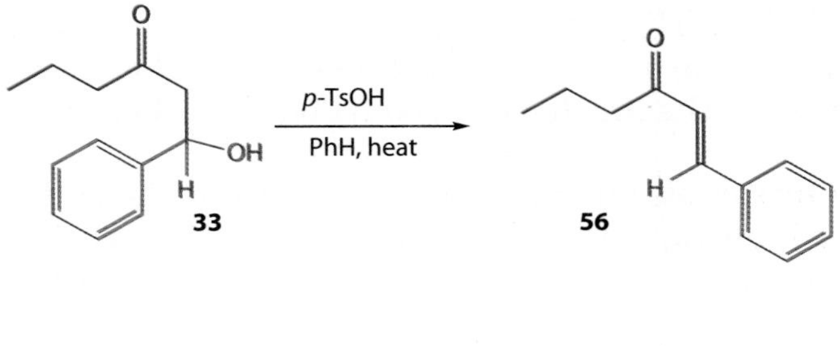

Recall aldol **33** from Section 22.3 and envision a reaction in which it is heated in benzene with a catalytic amount of *p*-toluenesulfonic acid (*p*-TsOH). The product under these conditions is **56**, where loss of water generates a carbon–carbon double bond (an alkene).[5] Aldehyde **5** is called a **conjugated ketone** (also called an α,β-unsaturated ketone; see Chapter 23, Section 23.2) because the carbonyl is attached directly to one of the sp² carbon atoms of the alkene unit. The major product is drawn as the *E*-isomer, which is usually more stable than the *Z*-isomer. Note that the phenyl group is also attached to an sp² carbon of the alkene, making **56** even more conjugated. When such **extended conjugation** occurs, dehydration is very easy and it is sometimes difficult to isolate the aldol. Many examples of conjugated compounds are known and some of their unique chemistry is discussed in Chapter 23.

The lesson of this section is straightforward. Mild acid hydrolysis will usually give the aldol product unless conjugating substituents are present. Heating and the use of strong acid can give dehydration and form a conjugated aldehyde or ketone. In some cases, extensive conjugation (as when phenyl groups are attached) leads to facile dehydration and formation of the conjugated system even under mild conditions. In those special cases, it may be very difficult to isolate the aldol product. *In general, however,* **assume** *that mild acid hydrolysis will give the aldol product rather than the alkene.*

22.47 Draw the structure of the *Z*-isomer of 56.

22.48 Draw the product that results from a reaction of 3-hexanone and LDA (–78°C, THF) that is treated first with 3-phenylpropanal and second with 2 *M* HCl at 100°C.

22.6 The Intramolecular Aldol Condensation

All of the examples shown for the aldol condensation in Sections 22.2 and 22.4 involved an intermolecular reaction (the reaction of two different molecules). If a molecule has two carbonyl units as well as at least one acidic α-hydrogen and is treated with base, an *intramolecular* aldol condensation can take place. A simple example is 1,6-hexanedial (**57**), which, when treated with lithium diisopropylamide in THF, gives enolate anion **58** (the carbanion is drawn at C2). This enolate anion reacts with the most available electrophilic center, which in this case is the aldehyde carbonyl on the other end of the molecule (labeled C6).

Why is the intramolecular reaction faster than the self-condensation of **57**? The answer is entropy. In Chapter 7 (Section 7.5), a very small value was assigned for the energy of the entropy term, and normally this is correct.

An intramolecular reaction is more ordered than the reaction between two molecules, so the entropy term is *smaller*. This energy difference is usually sufficient to favor the intramolecular process over the intermolecular process. *For the purposes of this book, assume that in systems such as 58, the intramolecular reaction is faster*.

This aldol condensation reaction forms a ring and gives the usual alkoxide product (**59**). Hydrolysis leads to the expected aldol product **60**. Because the aldol reaction occurs within the same molecule, it is an **intramolecular aldol** condensation. For convenience, the carbon chain in **57** is numbered, so it is clear where the reaction occur s and where the new bond is formed. *Note that 59 is drawn first with a peculiar looking "extended" bond in order to keep the relative shape the same, but it is then redrawn in its proper five-membered ring form*. This is done so that one can follow the bond-forming process and ring formation. Also note that the numbers used are arbitrary because the IUPAC name of **60** is 2-hydroxycyclopentanecarboxaldehyde (see Chapter 16, Section 16.2).

22.49 Draw the product when 2,5-hexanedione is treated with 1. LDA, THF, –78°C; 2. aq. H_3O^+.

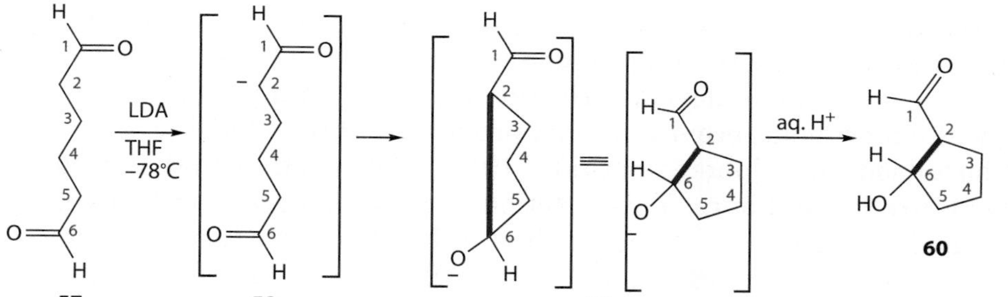

There is nothing unusual about the intramolecular aldol reaction other than formation of the ring. The reaction is not restricted to kinetic control conditions, and thermodynamic control conditions are used quite often. It is very easy to make five-, six-, and even seven-membered rings by an intramolecular aldol reaction, and three- or four-membered rings may be formed. Making rings of eight atoms or more is very difficult using this cyclization method due to the transannular strain involved in bringing the two ends of the molecule together (see Chapter 8, Section 8.7).

22.50 Draw the product of a reaction between 2,7-octanedione with 1. LDA, THF, –78°C; 2. aq. acid.

22.51 Draw the product of a reaction between 2,7-octanedione with 1. NaOEt in refluxing EtOH; 2. dilute aq. acid.

22.7 Ester Enolates

Until this point, the discussion has been focused on enolate anions generated from ketones and aldehydes. Esters and other acid derivatives also have a carbonyl group (see Chapter 16, Section 16.7) and the hydrogen atoms on the α-carbon are acidic. For practical reasons, only esters will be used in this discussion. Experiments have determined that the pK$_a$ of the α-hydrogen of an ester is about **24–25**, so esters are *weaker* acids than ketones. Therefore, the order of acidity is that aldehydes are more acidic than ketones, which are more acidic than esters. Nonetheless, treatment with the powerful bases discussed previously, particularly LDA and related amide bases, removes the acidic proton to give an ester enolate anion. However, even alkoxide bases will generate the enolate anion from an ester under equilibrium (thermodynamic) conditions, thus allowing reactions to occur. Once formed, ester enolate anions react as carbanion nucleophiles.

22.7.1 The Claisen Condensation

In the late nineteenth century, Ludwig Claisen (Germany; 1851–1930) treated an ester with a base; the isolated product was a β-keto ester. This product results from the enolate anion of one molecule of the ester condensing with a second molecule via an acyl substitution reaction (Chapter 16, Section 16.8). In a typical experiment, ethyl 2-methyl propionate (**60**) is treated with a specialized base (sodium triphenylmethide, **19**) in diethyl ether and stirred at room temperature for 60 hours. This reaction mixture is then acidified with glacial acetic acid (i.e., 100% acetic acid); the final isolated product is ethyl 2,2,4-trimethyl-3-oxopentanone (**61**) in 74% yield.[12] It is clear that *when an ester enolate reacts with another ester, the product is a β-keto ester and the reaction is now called the **Claisen condensation**.*

Triphenylmethide (**19**) is formed by the reaction of triphenylmethane (Ph$_3$CH) with sodium metal, as seen in Section 22.1. It is an unusual but effective base in this reaction because it is a relatively non-nucleophilic base (see Section 22.3). To explain the reaction with **60** and formation of product **61**, a mechanism requires that the base first remove the acidic α-proton on C2 from the ester to form enolate anion **62**. As with enolate anions derived from ketones and aldehydes, there are two resonance forms, and the carbanion form (**62A**) is the more nucleophilic. Therefore, resonance contribution **62A** will lead to the product. This is a self-condensation, and attack of C2 in **62** at the acyl carbon (C3) of a second molecule of **60** leads to **63**, the usual acyl substitution tetrahedral intermediate (Chapter 16, Section 16.8). Loss of ethoxide leads to formation of the ketone unit in keto-ester **61**.

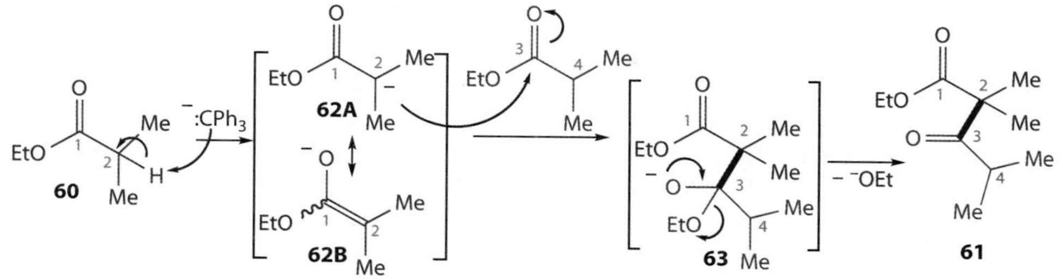

22.52 **Draw the product formed when ethyl pentanoate is heated with sodium ethoxide in ethanol and then hydrolyzed with aqueous acid.**

Ester enolates usually exist as mixtures of *E*- and *Z*-isomers (see Chapter 9, Section 9.4.3). An example is the enolate anion generated from ethyl propanoate, where the two priority groups (see Chapter 9, Section 9.3.1) of the C=C unit are the methyl group on one sp² carbon and the OEt group on the other sp² carbon atom. When the priority groups are on opposite sides, as in **64**, this is an *E*-enolate anion. The other enolate anion is **65**, the *Z*-isomer. Methods are available for the reaction that will lead to selectivity in the formation of *E*- or *Z*-isomers, but those methods are not discussed. **Assume that both 64 and 65 are present in the reaction. Assume also that formation of this mixture has no influence on the subsequent reaction.** This is not entirely correct because there may be stereochemical issues in the product, but it simplifies the fundamental concepts. As mentioned in Section 22.4.2, enolate anions from ketones may also exist as *E*- and *Z*-isomers. Enolate anion **39** in that section is drawn as the *Z*-isomer, but both are formed.

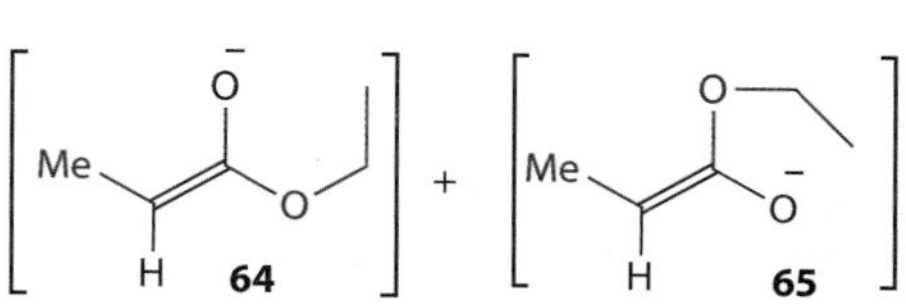

Keto-ester **61** results from self-condensation of the ester **60**, but the reaction of two different esters under these conditions gives a different result. Condensation between two different esters is called a **mixed Claisen condensation** and it occurs when the enolate anion of one ester condenses with the second ester. Under thermodynamic conditions, however, the ester is always present along with the ester enolate because it is an equilibrium process. At equilibrium, there are two esters in the medium as well as two different ester enolates, and each enolate anion will react with both esters to give a different product.

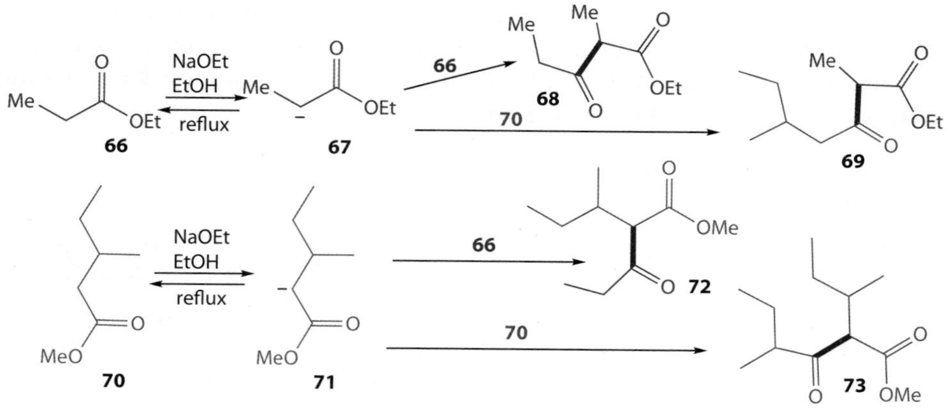

This means there are four possible Claisen condensation products, analogous to the problems encountered with the mixed aldol condensation in Section 22.4.3. Imagine that a mixture of esters **66** and **70** is refluxed in ethanol in the presence of sodium ethoxide. Ester **66** gives enolate **67** and ester **70** gives enolate **71**; at equilibrium, a mixture of **66**, **70**, **67**, and **71** is present. Enolate **67** reacts with **66** to give **68** and with ester **70** to give **69**. However, enolate **71** also reacts with **66** to give **72**. If **71** reacts with itself (ester **70**), the product is **72**. A mixed Claisen condensation under these conditions leads to four different products. Clearly, this is not very attractive. For this reaction and also because ester enolates are highly reactive, *kinetic control conditions to form ester enolates for mixed Claisen condensation reactions are preferred*. Note that there is only one α-carbon in an ester so kinetic or thermodynamic conditions give the same ester enolate anion.

22.53 **Draw the product of the reaction of benzyl pentanoate with LDA in THF at –78°C, when it is subsequently treated with methyl cyclohexanecarboxylate and then hydrolyzed with aqueous acid.**

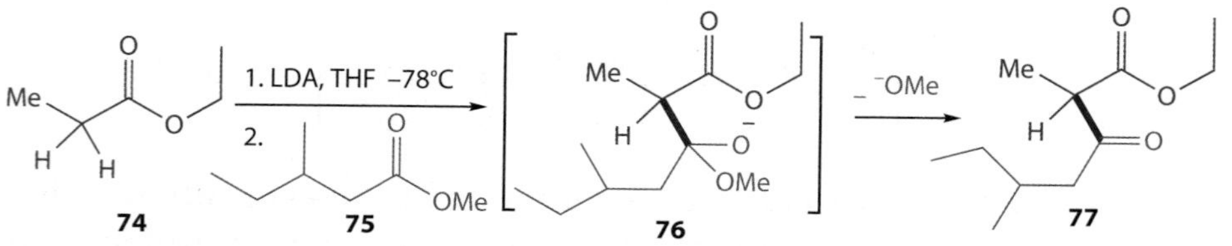

For the condensation reaction of **60**, **66**, and **71**, the thermodynamic reaction conditions constitute the traditional method of doing a Claisen condensation. This reaction may be modified to use kinetic control conditions using LDA as a base and THF as the solvent. An example is the reaction of **74** with LDA to form the ester enolate. Under these kinetic control conditions, assume that K_a is large and that the reaction will give primarily the enolate anion such that very little unreacted **74** remains. Therefore, self-condensation should not be a major problem. In a second chemical step, a different ester, methyl 2-methylpentanoate (**61**), is added as a reactant. The enolate attacks the carbonyl of ester **75** to give tetrahedral intermediate **76**; loss of the methoxy group (OMe) completes the acyl substitution reaction. The final product is β-keto ester **77**. The kinetic control conditions allowed a mixed Claisen condensation to prepare **76**.

There is one additional complicating factor. When an alcohol solvent is used in reactions with thermodynamic conditions, *the alcohol solvent should be the same as the alcohol part of the ester*. In other words, ethanol is used for an ethyl ester and methanol is used for a methyl ester. In Chapter 20 (Section 20.6.3), transesterification is the displacement of one alcohol unit (the OR group) of an ester by another alcohol unit in a nucleophilic acyl substitution reaction. If a methyl ester reacts with NaOEt in ethanol, the OEt unit displaces OMe to give an ethyl ester in the product. The product may be a mixture of ethyl and methyl esters. In general, make sure the OR portion of the ester matches the OR portion of the alcohol solvent (ROH).

22.54 Draw the structure of the enolate anion derived from 74.

22.55 Draw the final product if benzyl pentanoate is treated with NaOMe in refluxing methanol.

22.7.2 Reactions of Ester Enolates with Ketones and Aldehydes, and Aldehyde/Ketone Enolates with Esters

Once an ester enolate is generated, it can react with another ester in a Claisen condensation; however, it may also react with the carbonyl of an aldehyde or ketone. The ester enolate anion is a nucleophile and it reacts with an aldehyde or ketone via acyl addition. Kinetic control conditions are the most suitable for this reaction in order to minimize Claisen condensation of the ester with itself (self-condensation). If ester **74** (ethyl propanoate, in **green** in the illustration) is treated first with LDA and then with butanal (**21**, in **violet**), for example, the initial acyl addition product is **78**. The new carbon–carbon bond is marked in **blue** and treatment with dilute aqueous acid converts the alkoxide to an alcohol in the final product of this sequence, **79**. Compound **79** is a β-hydroxy ester, which is the usual product when an ester enolate reacts with an aldehyde or a ketone. Ester enolate anions react with ketones in the same way that they react with aldehydes.

Just as an ester enolate anion reacts with an aldehyde or ketone via acyl addition, it is also reasonable that the enolate anion of an aldehyde or a ketone may react with an ester via acyl substitution. In the former reaction, the ester enolate is the nucleophile; in the latter reaction, a ketone or aldehyde enolate is the nucleophile. When cyclohexanone (**80**) is treated with LDA (THF, –78°C) and then with methyl propanoate, the initial product is **81**. Loss of ⁻OMe completes the acyl substitution sequence to give diketone **82**. There is nothing special or unusual about these two variations. Virtually any ketone or aldehyde enolate reacts with an ester to form 1,3-diketones such as **82**.

22.56 Draw the final product of a reaction between methyl cyclopentanecarboxylate and LDA and then with acetone, followed by hydrolysis.

22.57 Give the reaction and final product for treatment of 2-pentanone with LDA followed by treatment with ethyl cyclopentanecarboxylate.

22.7.3 Intramolecular Condensation. The Dieckmann Condensation

Just as there is an intramolecular version of the aldol condensation (see Section 22.2), there is an intramolecular version of the Claisen condensation. A simple illustration is treatment of an α,ω-diester such as diethyl 1,6-hexanedioate (**77**, the diethyl ester of adipic acid called diethyl adipate; see Chapter 20, Section 20.9) with sodium ethoxide in refluxing ethanol. When this experiment is done, triphenylmethide (**19**; introduced in Section 22.1) is used as the base, and reaction in the solvent benzene gives the isolated final product (**86**) in 70% yield.[13] *Note that benzene is rarely used now-a-days because it is a suspected carcinogen and its use is being phased out.* This reaction sequence uses the same mechanism as the Claisen condensation.

Reaction of **83** with the base (*arbitrarily* labeled C2 because C2 and C5 are identical in this symmetrical molecule) gives the ester enolate **84**. Assume that the intramolecular reaction is faster than the intermolecular reaction (self-condensation), as with the intramolecular aldol condensation (Section 22.6); the only electrophile available for this enolate to react with is the carbonyl of the ester unit (C6) at the other end of the molecule. When C2 donates electrons to C6, this intramolecular Claisen condensation leads to tetrahedral intermediate **85**. In the usual manner, **85** loses ethoxide (⁻OEt) and the product is a cyclic β-ketone ester (ethyl 2-oxocyclopentanecarboxylate, **86**).

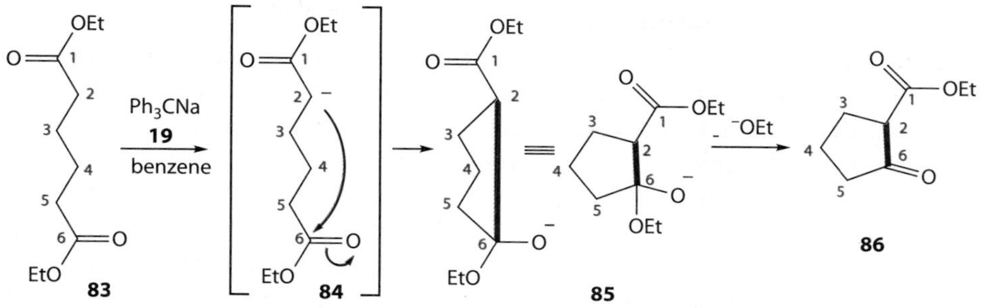

The carbon chain in **83** is numbered to follow which carbons are involved in forming the new bond (in **blue** in the illustration). Product **85** is first shown without changing the position of the atoms as they are drawn in **83**. Clearly, this is a distorted conformation and not real, but it shows exactly where the new bond is formed, and the structure is drawn a second time as the five-membered ring with the correct structure. This intramolecular Claisen condensation is formally known as the **Dieckmann condensation**, after Walter Dieckmann (Germany; 1869–1921). Diesters derived from four- to eight-carbon dicarboxylic acids are treated with base to form three- to seven-membered ring cyclic ketones, with a carboxylate group at the 2-position. Chapter 17 (Section 17.4) described the oxidative cleavage of cyclic alkenes, which is an excellent source of the dicarboxylic acid precursors to diesters such as **83**.

22.58 Show a reaction sequence that converts 1,8-octanedioic acid to 2-benzylcycloheptanone.

22.7.4 Malonic Ester Enolate Anions

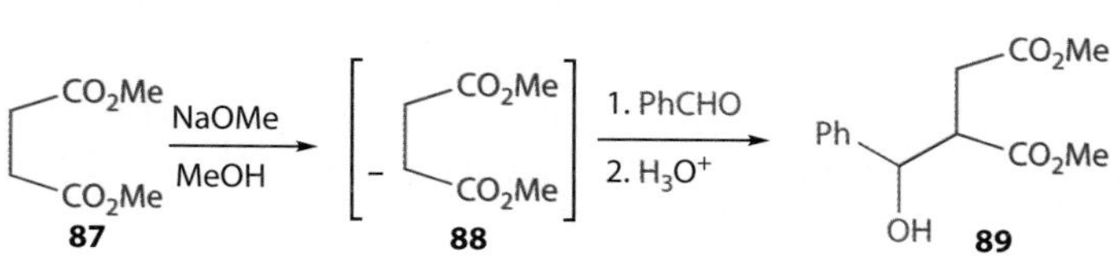

The esters derived from dicarboxylic acids in Chapter 20 (Section 20.9) behave more or less like all other esters in enolate anion reactions. Dimethyl succinimide (**87**; dimethyl 1,4-butanedioate), for example, has a pK_a relatively close to that of ethyl butanoate (about 25) and it reacts similarly. Treatment of **87** with NaOMe will give **88** and this enolate anion reacts with aldehyde, ketones, or another ester. If **88** is treated with benzaldehyde in a second chemical step, the final product is **89**, analogous to the conversion of **74** and **21** to **79** in Section 22.7.2.

22.59 Why was NaOMe chosen as a base rather than NaOEt?

In a different type of diester—malonic acid derivatives ($HOOCCH_2COOH$)—the two carboxyl units are attached to the same carbon. Esters of malonic acid (malonates) also form an enolate anion, but it has been experimentally determined that the α-proton is much more acidic than in a normal ester. Because two electron-withdrawing carbonyls are attached to a single carbon, the α-proton is significantly more acidic than the α-proton found in a simple ester such as ethyl acetate. As discussed in Chapter 6 (Section 6.3), inductive effects in the acid and the stability of the conjugate base lead to a pK_a for the α-hydrogen in diethyl malonate (**90**; diethyl 1,3-propanedioate) of about 15,[2] as compared to a pK_a of about 19–20 for acetone or about 25 for ethyl acetate. Interestingly, the enol content of **90** is reported to be very small, contrary to the acetylacetone (**7**) discussed in Section 22.1. The α-proton in **90** is sufficiently acidic that *a weaker base can be used to remove the α-hydrogen relative to that used for a simple ester.* These bases include sodium hydride (NaH), sodium carbonate, and alkoxide bases (NaOR).

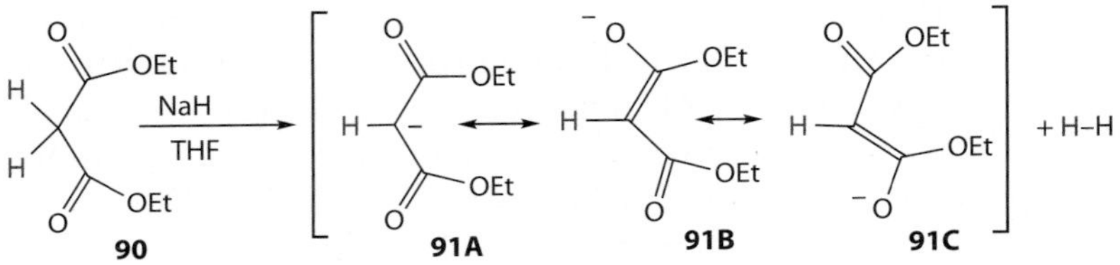

The reaction of diethyl malonate (**90**) with sodium hydride generates enolate anion **91** as the conjugate base, and hydrogen gas is the conjugate acid. It has the three resonance contributors shown in the illustration, although **91A** has the highest concentration of electron density, and **91** will react as a carbanion nucleophile. There is one extra resonance form in the malonate enolate anion relative to a simple ester due to the second carbonyl unit, and it means that **91** is more stable than the enolate derived from a monoester. In part, this accounts for the enhanced acidity and easier formation of the enolate anion using a weaker base. Once formed, **91** is a carbon nucleophile and it will react with both aldehydes and ketones, as well as with other esters.

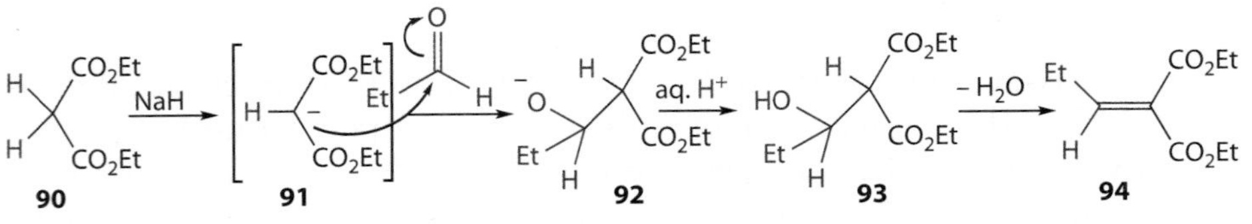

If propanal is added to **91**, for example, acyl addition to the aldehyde carbonyl leads to **92**, so **91** is just another carbon nucleophile with the aldehyde.

Aqueous acid workup of **92** gives the alcohol, **93**. *With malonic ester deriva-tives, loss of water to form **94** occurs very easily, with dilute acid or with gentle heating* because the C=C unit is conjugated to two carbonyl groups, facili-tating dehydration. Although it is possible to isolate **83**, it is more usually difficult. The enolate anion of malonate esters also reacts with ketones and may be condensed with other esters in acyl substitution reactions. When **90** is treated with NaOEt in ethanol and then with ethyl butanoate, the final product after mild hydrolysis is a keto-diester, **95**.

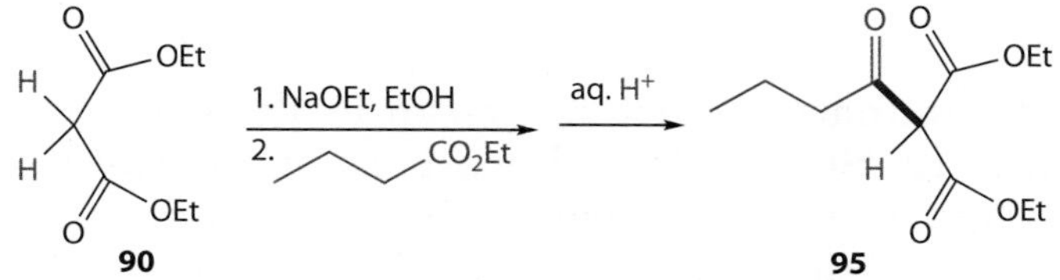

An interesting variation of this reaction treats malonic acid (**96**) with an aldehyde such as acetaldehyde, using pyridine as a base. The enolate anion is formed by reaction with the basic amine pyridine, and it condenses with the carbonyl group of the aldehyde. After treatment with aqueous acid, the final product isolated form this experiment was 2-butenoic acid, **98** (in 60% yield).[14] This specialized version of the malonic acid condensation is called the **Knoevenagel reaction**, after Emil Knoevenagel (Germany; 1865–1921). Acid **98** is drawn as the *E*-stereoisomer. The Knoevenagel reaction may give a mixture of *E*- and *Z*-isomers, but the more stable *E*-isomer is usually the major product. Note that dehydration occurs during the hydrolysis step. What hap-pened to the COOH unit? It is lost as CO_2. This new type of reaction, called decarboxylation, will be discussed in the next section.

22.60 Write out the reactants and product of the reaction of ethyl 2-methylmalonate with LDA, followed by treatment with cyclo-hexanone and hydrolysis.

22.61 Draw the reactants and product of the reaction of diethyl malonate and benzaldehyde under thermodynamic conditions.

22.62 Draw the structure of this enolate anion, including the struc-ture of the conjugate acid derived from pyridine.

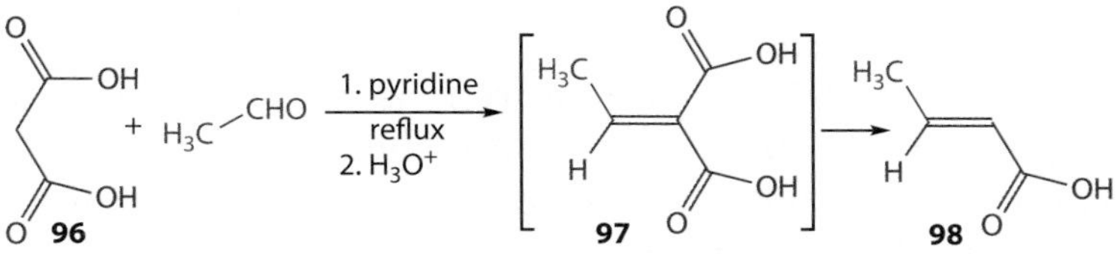

22.8 Decarboxylation

The specialized elimination reaction known as decarboxylation was introduced in Chapter 12, Section 12.6. Decarboxylation requires the presence of a carboxyl group (COOH) attached to a carbon that has another carbonyl substituent. With this structural motif, an internal acid–base reaction is possible that leads to cleavage of a C–C bond and loss of the neutral molecule carbon dioxide (CO_2 = O=C=O). Malonic acid has the requisite structural features (see **96**), and when malonic acid and related compounds are heated to 200°C, they lose carbon dioxide (CO_2); in other words, they **undergo decarboxylation**.

Another example is 2-methyl-1,3-propanedioic acid (**99**), which gives propanoic acid upon heating. The oxygen atom of one carboxyl unit acts as a base, donating electrons to the proton of the other carboxyl unit via the conformation shown that mimics a six-membered ring, as described in Chapter 12 (Section 12.6). The product is an enol, **100**. This particular compound is the enol of a carboxylic acid. In Chapter 10 (Section 10.4.), tautomerization of an enol led to a carbonyl. In this case, tautomerization of **100** leads to the carboxyl unit in propionic acid. This decarboxylation event occurs only when **99** is heated to a relatively high temperature. In other words, the activation energy for decarboxylation is relatively high.

The key feature in the decarboxylation of malonic acid derivatives is that they are 1,3-dicarbonyl compounds with the acidic proton of the acid in close proximity *to a carbonyl oxygen that is two carbon atoms away from the carboxyl unit*. The proximity of these units is essential, which is why decarboxylation occurs with 1,3-dicarbonyl compounds and *not* with 1,4-dicarbonyl compounds. This is a form of an elimination reaction (see Chapter 12). Going back to the expected product, **97** from **96**, it is clear that this is a 1,3-dicarbonyl compound capable of decarboxylation upon heating. That is precisely what occurs, so the product is **98** rather than **97**. Decarboxylation of malonic acid derivatives gives functionalized carboxylic acids.

Malonic acid derivatives are not the only compounds capable of decarboxylation. At least one COOH unit is required, as well as a suitable basic atom such as the oxygen of a carbonyl that is attached to the α-carbon of the acid. Reexamining the Claisen condensation shows that the β-keto ester products (**61** from Section 22.7.1) fit the criterion for decarboxylation. The requisite COOH unit is obtained by hydrolysis of the ester unit in **61** to give β-keto acid **101**.

This molecule contains a carbonyl unit that is two carbons away from the carboxyl unit, and it undergoes decarboxylation.

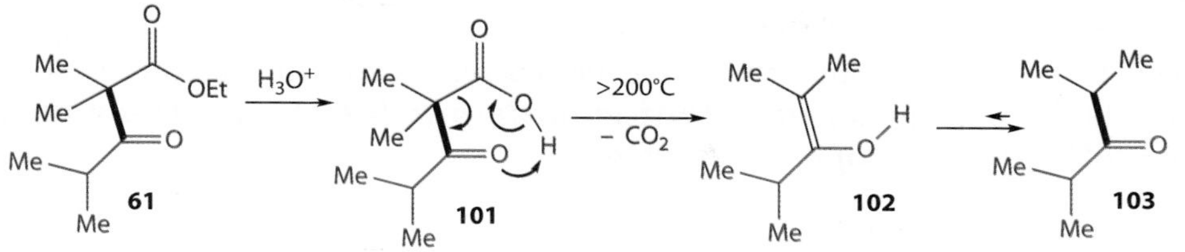

Indeed, heating **101** to >200°C leads to decarboxylation with formation
of enol **102**, which then tautomerizes to give 2,5-dimethyl-3-pentanone (**103**).
(Keto-enol tautomerism was discussed in Chapter 10, Section 10.4.5.) Note that
loss of CO_2 results in forming a C=C unit between the C=O carbon and the
α-carbon in **101**. Decarboxylation is slightly more difficult (requires a somewhat higher reaction temperature) for β-keto acids than for 1,3-dicarboxylic
acids, but it is still a very facile reaction at 200°C. Claisen condensation products can therefore potentially be converted to a ketone via decarboxylation.

22.63 **Write out a sequence of reactions that will produce 2-methyl-3-
heptanone from pentanoic acid.**

22.9 Enolate Alkylation

The acyl addition and acyl substitution reactions of enolate anions presented in
this chapter clearly show that enolate anions are nucleophiles. In Chapter 11
(Section 11.3), various nucleophiles reacted with primary and secondary alkyl
halides via S_N2 reactions. Enolate anions also react with alkyl halides via S_N2
reactions in what is known as enolate alkylation.

In one experiment using LDA (**30**), 2-benzylcyclohexanone (**104**) was converted to the enolate anion under kinetic conditions (the solvent is dimethoxyethane, which is abbreviated DME), and subsequent reaction with iodomethane
gave a 73% yield of **105** (via alkylation of the kinetic enolate anion) and 6% of
106 (via the thermodynamic enolate anion).[15] This example illustrates a typical
alkylation reaction as well as a typical result of enolate anion chemistry under
kinetic control conditions (Section 22.4.2). The main point is that an enolate
anion is a carbon nucleophile and reacts with primary and secondary alkyl
halides by what is a normal S_N2 pathway.

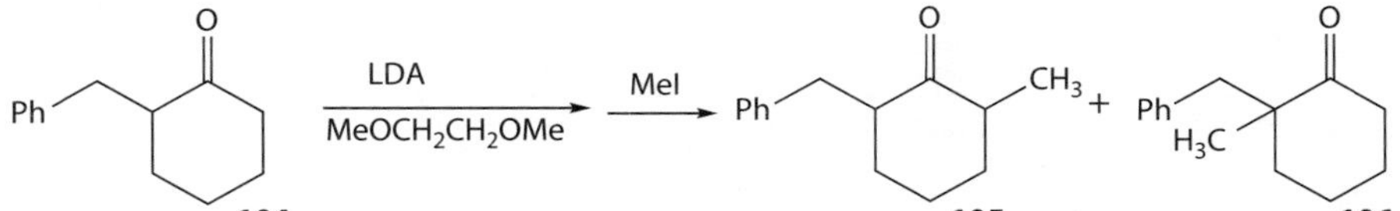

When 5,5-dimethyl-2-hexanone (**107**) is treated with LDA in THF at –78°C, the product is the kinetic enolate anion **108**. In a subsequent alkylation reaction, **108** reacts with 2*S*-bromobutane to give **109**. Because this is an S$_N$2 reaction, there is inversion of configuration at the C–Br bond (see Chapter 11, Section 11.2), as shown in **109**. Other factors play a role in the stereoselectivity of the alkylation, but *assume that enolate alkylation reactions proceed with 100% inversion of configuration*.

22.64 Draw the kinetic and thermodynamic enolate anions formed from 98.

22.65 Draw the product formed when the enolate anion of cyclopentanone reacts with 4*R*-benzyl-2*R*-iodoheptane.

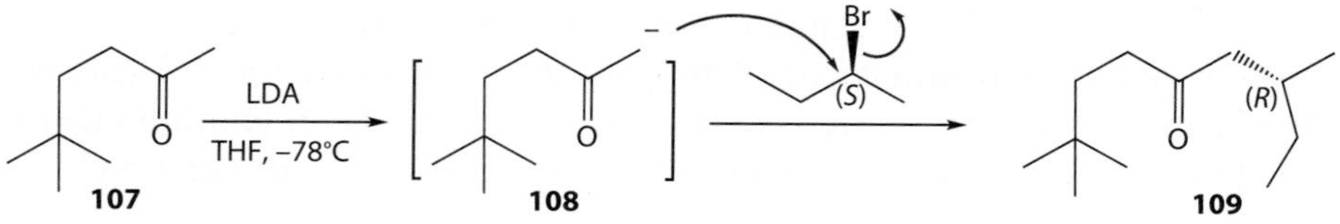

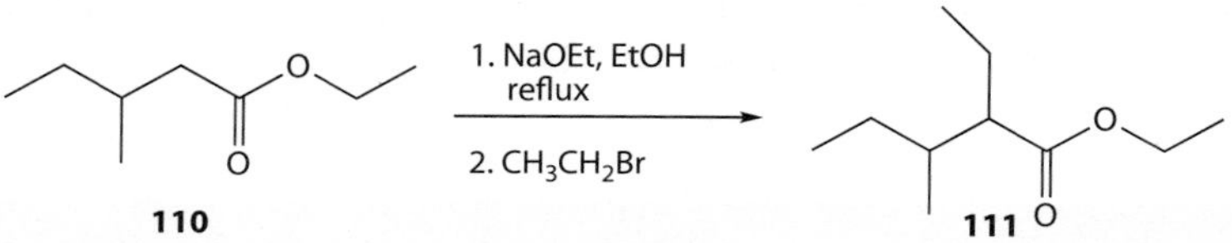

Ester enolates undergo alkylation reactions. When ethyl 3-methylpentanoate (**110**) reacts with sodium ethoxide in ethanol and then with bromoethane, the product is **111**. Alkylation of malonate derivatives leads to an interesting sequence of reactions that are useful in synthesis. The reaction of diethyl malonate (**90**) and NaOEt in ethanol, followed by reaction with benzyl bromide, gives **112**. In a second reaction, **112** reacts with NaOEt in ethanol and then with iodomethane to give **113**. Saponification of **113** (see Chapter 20, Section 20.2) gives the dicarboxylic acid, **114**, and heating leads to decarboxylation (Section 22.8) and formation of acid **115**. This overall sequence converted malonic acid via the diester to a substituted carboxylic acid, and it is known as the **malonic ester synthesis**.

22.66 Draw the enolate anion formed from 110 under these conditions.

22.67 Show a synthesis of 2-ethylhexanoic acid from 1,3-propanedioic acid.

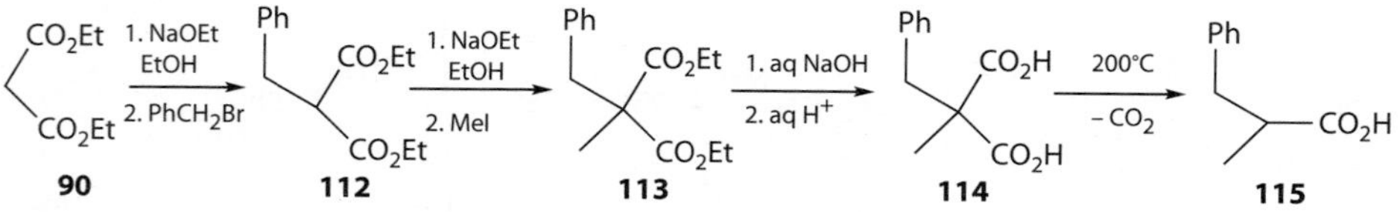

A variation of the malonic ester synthetic uses a β-keto ester such as **116**. In Section 22.7.1, the Claisen condensation generated β-keto esters via acyl substitution that employed ester enolate anions. When **116** is converted to the enolate anion with NaOEt in ethanol, reaction with benzyl bromide gives the alkylation product **117**. When **117** is saponified, the product is β-keto acid **118**, and decarboxylation via heating leads to 4-phenyl-2-butanone, **119**. This reaction sequence converts a β-keto ester, available from the ester precursors, to a substituted ketone in what is known as the **acetoacetic acid synthesis**. Both the malonic ester synthesis and the acetoacetic acid synthesis employ enolate alkylation reactions to build larger molecules from smaller ones, and they are quite useful in synthesis.

22.68 Show a synthesis of ester of 116 from a simple ester precursor.

22.69 Explain why the enolate anion is formed at the carbon between the two carbonyl groups in 116 rather than at the methyl carbon. If LDA were used in a reaction with 116 in THF, which enolate anion would be formed?

22.70 Show a synthesis of 1-phenyl-2,2,4-trimethyl-3-pentanone from 2-methylpropanoic acid.

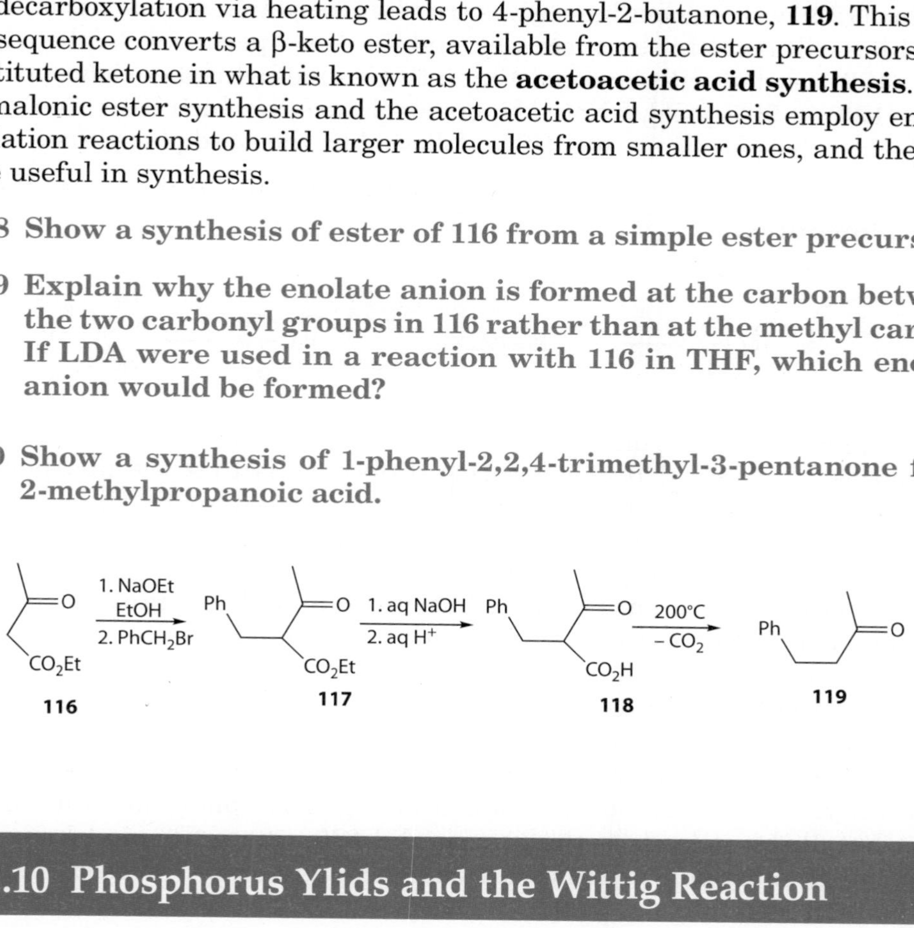

22.10 Phosphorus Ylids and the Wittig Reaction

The fundamental structural feature of an enolate anion is that the carbanion unit is stabilized by an electron-withdrawing group, the carbonyl. Phenyl groups stabilize an adjacent carbanion (see **19**) and the cyano group also stabilizes a carbanion (see **16–18**). A positively charged phosphorus atom that is connected to a carbanionic carbon will also stabilize the charge, constituting a new class of carbanions known as ylids. *An ylid is a compound with positive and negative charges on adjacent atoms.* (Note that these compounds may be spelled "ylide.") In the context of this chapter, phosphorus ylids will be discussed as if they are phosphorus-stabilized carbanions. A typical structure is $Ph_3P^+-CH_2^-$.

Phosphorus is in group 15, immediately under nitrogen in the periodic table, and it should have similar chemical properties. An amine is a trisubstituted nitrogen compound such as NR_3, and a phosphine is a trisubstituted phosphorous compound like PR_3. Likewise, a phosphonium salt (R_4P^+) is formally analogous to an ammonium salt (R_4N^+), formed by adding a fourth group to phosphorus. In Chapter 11 (Section 11.3.3), amines reacted with an alkyl halide to give an ammonium salt via an S_N2 reaction. The nitrogen is a nucleophile and the carbon of CH_3I has a δ+ carbon, so the reaction of triethylamine (**120**) with iodomethane gives triethylmethylammonium iodide, **121**. A phosphine will react in the same way and, in one experiment, triphenylphosphine (**122**) reacted with bromomethane to give methyltriphenylphosphonium iodide (**123**) in 99% yield.[16] Note the nomenclature in which the three groups attached to phosphorus are identified, followed by the word phosphine.

22.71 **Write the structures of trimethylphosphine and methylphenylphosphine.**

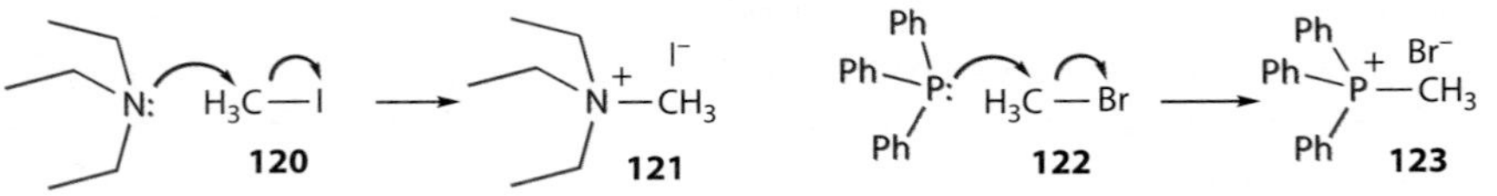

Phosphorus has d-orbitals and is larger than nitrogen; there are certain differences in chemical reactivity. One difference is that the phosphonium unit makes the α-hydrogen of the methyl unit more acidic than a proton attached to nitrogen, and **123** has a pK_a of about 22.[17] The α-proton in **123** is *not* as acidic as the α-proton in a ketone such as acetone, but it can be removed with a strong base such as butyllithium and, in many cases, by an alkoxide base. While experimenting with phosphonium salts, Georg Wittig (Germany; 1897–1987) found that if they are treated with a strong base, removal of the α-proton led to an ylid. A molecule that has a positive and a negative charge in the same molecule, as in **125**, is called a *zwitterion*. When these two charges are on adjacent atoms, as mentioned earlier, it is called an *ylid*. The acid–base reaction of a generic phosphonium salt **124** with *n*-butyllithium leads to the generic ylid **125**, where "R" can be any alkyl group, but R^1 must be a group *that does not contain an α-hydrogen*. This will be explained next.

22.72 **Draw the ylid that results from each of the following reactions: (a) PPh_3 + 2-bromobutane and then BuLi, and (b) PPh_3 + benzyl bromide and then BuLi.**

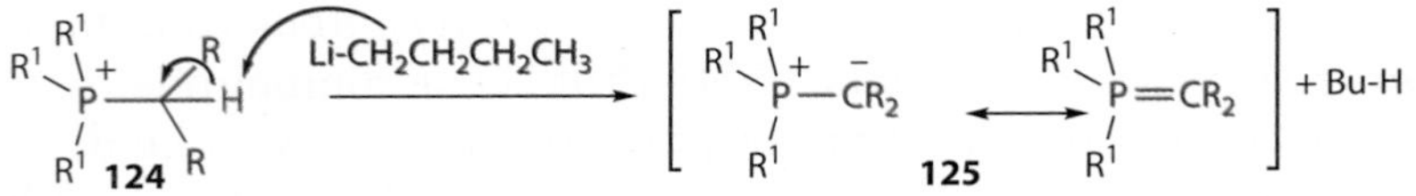

Phosphonium salt **123** is formed by the reaction of triphenylphosphine with an alkyl halide. Conversion of **123** to the corresponding ylid by reaction with *n*-butyllithium gives a product experimentally identified as **126**. The negatively charged carbon atom is adjacent to the positively charged phosphorus (the name is triphenylphosphonium methylid), and **126** is resonance stabilized with two contributing structures. In one, the P and C have adjacent and opposite charges, but in the other there is a P=C unit. In effect, having a (+) and a (−) on adjacent atoms is equivalent to a covalent bond.

   In terms of its chemical reactivity, an ylid such as **126** may be viewed as a *phosphorus-stabilized carbanion that will undergo acyl addition with an aldehyde or a ketone*. When this ylid is mixed with cyclohexanone (**80**), there are two isolated products. The one that is more interesting to an organic chemist is methylenecyclohexane (**129**), formed in 52% yield; the other is triphenylphosphine oxide, **130**.[18] It is obvious that **129** is not the expected acyl addition product. Formation of **130** indicates that the carbon atom of the ylid has been transferred to the ketone, but the oxygen atom of the ketone has been transferred to the phosphorous atom. Analysis of the reaction shows that the oxygen atom is lost from the ketone, and the CH₂ unit of ylid **126** is transferred to form a new C=C bond (in **green** in the illustration). What is the mechanism?

   If **126** is viewed as a phosphorus-stabilized carbanion, nucleophilic acyl addition to the carbonyl carbon of cyclohexanone (**80**) should give **127** as the product. This is known as a **betaine**, in which the phosphorus has a positive charge and the oxygen a negative charge. Because phosphorus has d-orbitals, it can expand its valence, which allows the alkoxide unit to attack phosphorus to form a new P–O bond in **128** (known as an **oxaphosphatane**). Both **127** and **128** are rather unstable intermediates and are highly reactive in this chemical reaction. Why? It is known that the P=O bond is rather strong, and if the P–C bond and the O–C bond in **128** are broken, a strong P=O bond is generated in the product triphenylphosphine oxide (**130**) in an exothermic process. The entire sequence is phosphonium iodide **123** converted to the ylid, which then reacts with cyclohexanone to give alkene **129**. The overall process converts a ketone or aldehyde to an alkene; this particular transformation, as developed by Wittig, is called the **Wittig reaction** or **Wittig olefination**. Phosphonium ylids such as **123** are often called **Wittig reagents**.

Many alkyl halides may be converted to the corresponding phosphine and then to ylids, where they react with a variety of aldehydes and ketones. The Wittig reaction is therefore quite versatile. Another example is shown in which phosphonium salt **131** is converted to the ylid, which reacts with 2-butanone to give **132** along with the phosphine oxide **130**. The reason for showing this second example is to illustrate that the alkene product is formed as a *mixture of E- and Z-isomers* and *this is typical of the Wittig reaction*. Although *the alkene formed in the greatest yield usually has the two largest groups attached to the C=C unit trans to each other, a mixture is almost always formed*. There are techniques for controlling the stereochemistry, but they are not addressed here. *Assume that the Wittig reaction always gives a mixture of alkene stereoisomers. This is not always correct, but it is a good working assumption*.

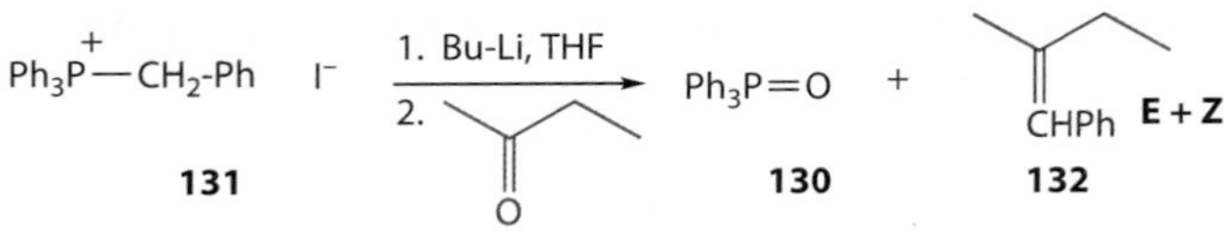

22.73 Draw the structure of the ylid (both resonance forms) generated from 131.

22.74 Write the reactions that will form each of the following alkenes from an appropriate halide, using the Wittig reaction: (a) 2-hexene, (b) 2-methyl-2-pentene, and (c) 1-phenyl-2-ethyl-2-pentene.

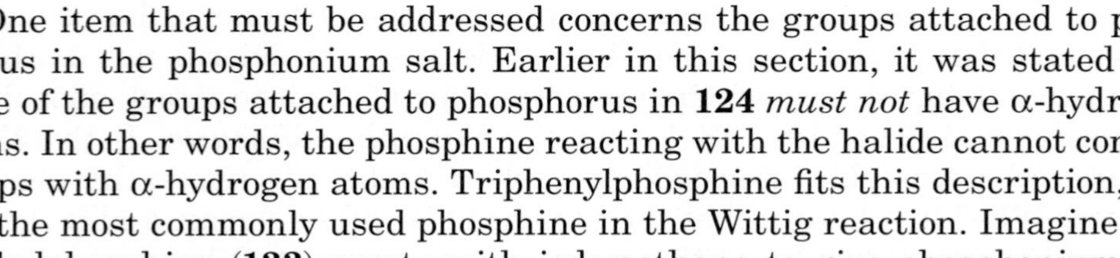

One item that must be addressed concerns the groups attached to phosphorus in the phosphonium salt. Earlier in this section, it was stated that three of the groups attached to phosphorus in **124** *must not* have α-hydrogen atoms. In other words, the phosphine reacting with the halide cannot contain groups with α-hydrogen atoms. Triphenylphosphine fits this description, and it is the most commonly used phosphine in the Wittig reaction. Imagine that triethylphosphine (**133**) reacts with iodomethane to give phosphonium salt **134**.

This is a perfectly reasonable reaction. Remember that in phosphonium salts, the α-hydrogens are now acidic and may be removed with a base. In **134**, there are *two different types of α-hydrogens:* H$_a$ on an ethyl and H$_b$ on a methyl.

Clearly, reaction with *n*-butyllithium removes either H$_a$ or H$_b$ to give two ylids: **135** and **136**. In this case, control of the reaction is lost and if this mixture is reacted with acetone, *two different alkenes are formed.* To form an alkene using only the carbon atoms derived from the halide, the phosphine starting material cannot have α-hydrogens that will interfere with formation of the targeted Wittig reagent. Again, triphenylphosphine has no α-hydrogens, is commercially available, reacts with most halides (see Chapter 11, Section 11.3.4, for the S$_N$2 reaction), and will give triphenylphosphine oxide (**130**), which is a solid that can be removed from the alkene product.

22.75 Draw the alkenes formed from ylids 135 and 136.

It is possible to use bases other than butyllithium in the Wittig reaction, as long as the base is sufficiently strong to remove the weakly acidic α-hydrogen for the phosphonium salt. In typical Wittig reactions, KO*t*-Bu, NaH, LiNH$_2$, LiNEt$_2$, and LiN(iPr)$_2$, as well as organolithium reagents such as butyllithium, methyllithium, and phenyllithium, have been used to generate the ylid.

22.11 Many New Synthetic Possibilities

This chapter explores several different carbon–carbon bond-forming reactions that lead to a variety of functionalized molecules: diketones, hydroxy-ketones and hydroxy-aldehydes, keto-esters, keto-aldehydes, keto-nitriles, cyano-esters, hydroxy-nitriles, and cyano-aldehydes and cyano-ketones. The reactions that form these products may be categorized by defining the relevant disconnection.

The product of an aldol condensation is either a hydroxy-aldehyde or a hydroxy-ketone. The disconnection for this process where the product is derived from a ketone or an aldehyde is shown. Another disconnection is shown for the aldol-dehydration product (loss of water from the aldol), where the conjugated aldehyde or ketone is formed from an aldehyde or ketone.

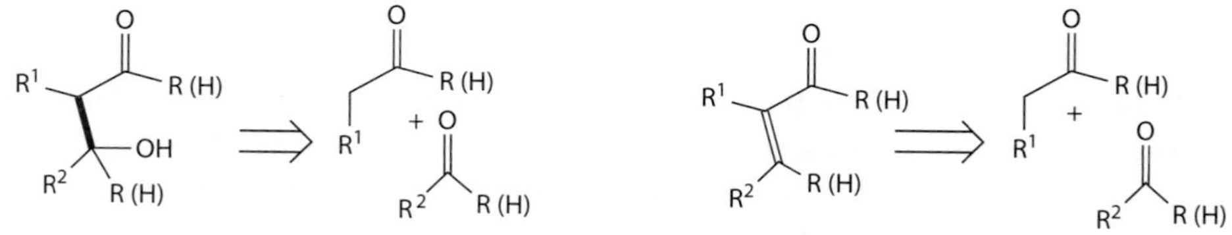

Several ester disconnections are covered in this chapter. The Claisen condensation gives a keto-ester and the "Claisen disconnection" is shown. The cyclic version of this reaction is the Dieckmann condensation, and the "Dieckmann disconnection" is also shown.

When an ester enolate reacts with an aldehyde or a ketone, the product is a hydroxy-ester. This disconnection is shown for both partners. If the reaction is "turned around," the reaction of an enolate derived from an aldehyde or a ketone and then with an ester gives a keto-aldehyde or a diketone. Both disconnections are shown. The enolate alkylation reaction involves disconnection of an alkyl halide fragment from an aldehyde, ketone, or ester. In addition, the malonic acid and acetoacetic acid syntheses have unique disconnections.

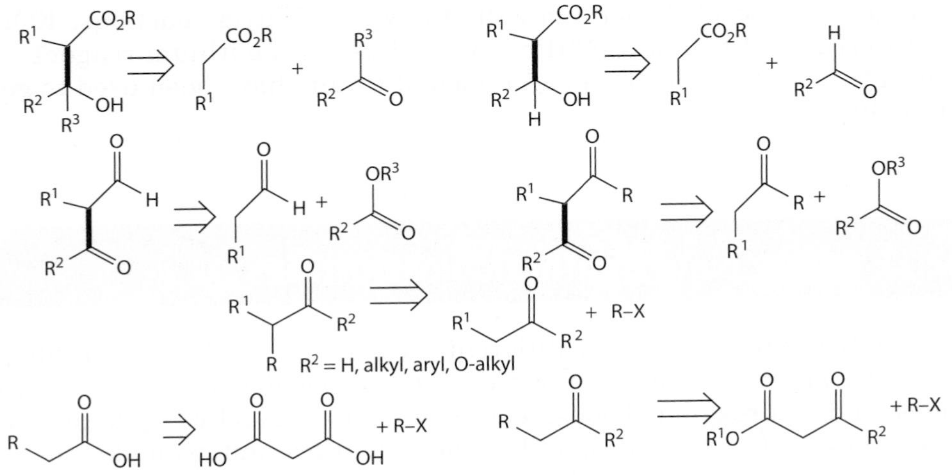

The Wittig disconnection is simply disconnection of the C=C unit of an alkene into a carbonyl component (ketone or aldehyde) and an ylid component, which arises from an alkyl halide.

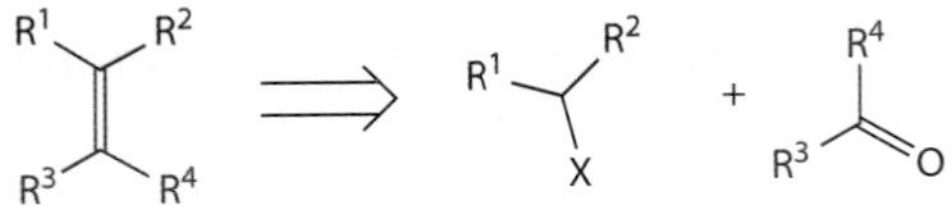

The new carbon–carbon bond-forming reactions presented in this chapter offer many new opportunities to synthesize molecules. Three examples are presented to illustrate some of these transformations. In the first example, 138 is synthesized from alkene **139**. Disconnection of **138** takes advantage of the five-carbon unit of **139** between the carbonyl unit and the hydroxyl unit. This disconnection is chosen because it leads to two fragments, and the ketone fragment is chosen as the donor because it is equivalent to an enolate anion, derived from 3-methyl-2-butanone.

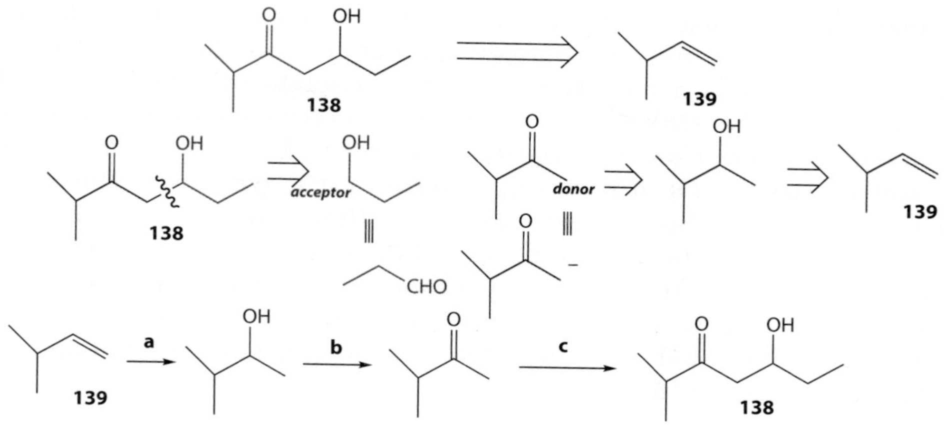

(a) i. Hg(OAc)$_2$/H$_2$O ii. NaBH$_4$ (b) PCC (c) i. LDA, THF, –78°C ii. propanal iii. H$_3$O$^+$

The second fragment has an acceptor site adjacent to the OH, which is the equivalent of an aldehyde—in this case, propanal. Therefore, plan on making **138** by an aldol reaction; the disconnection is chosen to exploit the aldol reaction. The five-carbon ketone is obtained from alkene **139**, which leads to 2-propanol as a precursor because that alcohol is directly available from **139**.

The synthesis is simply the reverse of the disconnection process, so alkene **139** is converted to alcohol **140** via oxymercuration–demercuration (Chapter 10, Section 10.4.5), and oxidation with pyridinium chlorochromate (PCC; Chapter 17, Section 17.2.3) leads to the requisite ketone. Treatment of the ketone with LDA to form the kinetic enolate (note the kinetic control conditions) is followed by addition of propanal and hydrolysis to give the final target, **138**.

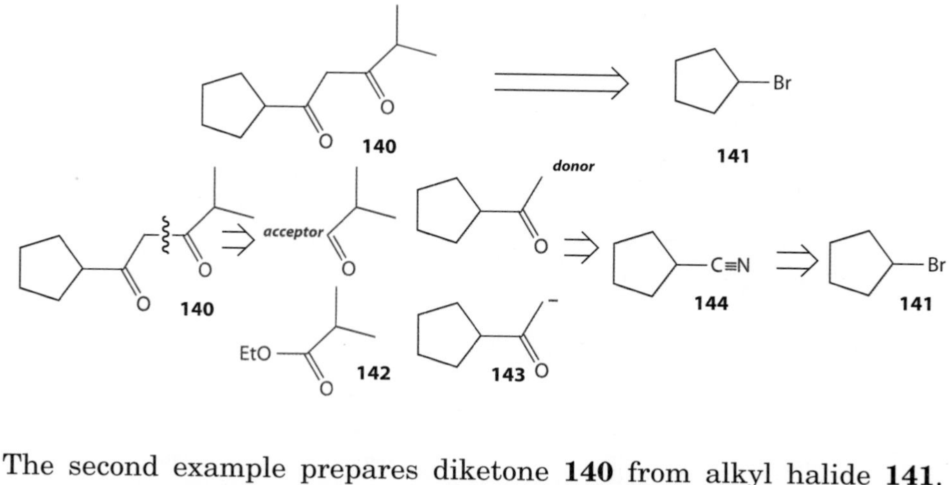

The second example prepares diketone **140** from alkyl halide **141**. The starting material has a cyclopentane unit, which is apparent in **140**, but a logical disconnection is between the two carbonyl units, suggesting the donor and acceptor sites shown. The donor site leads to a ketone derived from

1-cyclopentyl-1-ethanone and this fragment is derived from **141**. The acceptor fragment is more difficult to ascertain, but because **140** is a diketone, it probably comes from condensation of a ketone enolate **143** and an ester. *It is important to understand that the disconnection was chosen because this particular reaction is recognized, and the choice is made to exploit this carbon–carbon bond-forming reaction for a synthesis.* The requisite ester is ethyl 2-methyl-propanoate, **142**. Note that a methyl ester or another alcohol fragment may be chosen. Enolate **143** is derived from 1-cyclopentyl-1 -ethanone and differs from the cyclopentane fragment in **141** by the methyl ketone unit. Recognition that a ketone is prepared from a nitrile provides a link between **143** and **141**. Disconnection of the ketone to nitrile **144** allows the final disconnection to **146**.

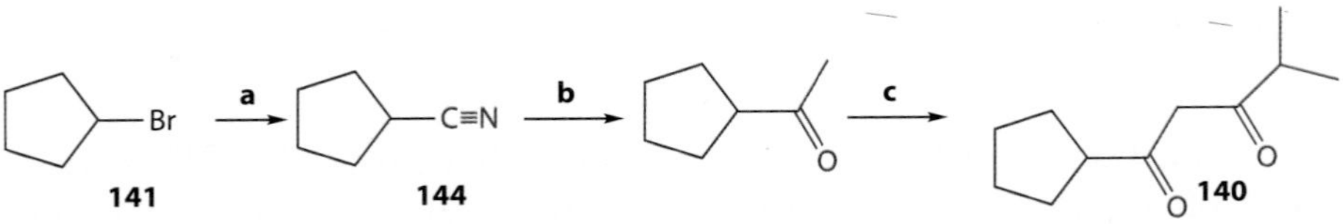

(a) KCN, DMF (b) i. CH₃MgBr ii. H₃O⁺, heat (c) i. LDA, THF, –78°C ii. ethyl 2-methylpropanoate

The synthesis begins with an S_N2 reaction (Chapter 10, Section 10.2) of bromide **141** with potassium cyanide to give **144**. Note the use of the aprotic solvent DMF to facilitate the S_N2 reaction. A Grignard reaction of the nitrile with methylmagnesium bromide followed by hydrolysis leads to the requisite ketone (see Chapter 20, Section 20.9.3). The final step simply reacts the methyl ketone with LDA under kinetic control conditions to give the enolate anion (**143**), which is condensed with the ester (**142**) to give the diketone target, **140** (Section 22.7.2).

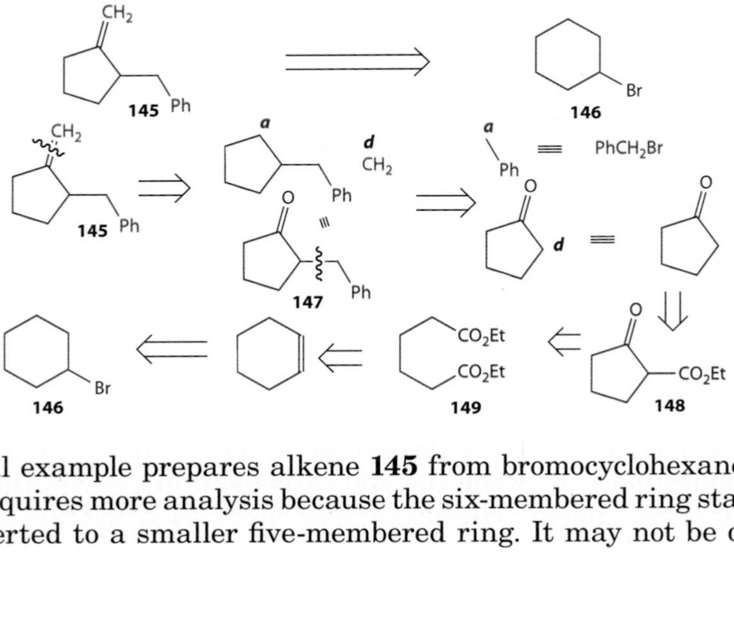

The final example prepares alkene **145** from bromocyclohexane, **146**. This synthesis requires more analysis because the six-membered ring starting material is converted to a smaller five-membered ring. It may not be obvious, but

such a transformation usually requires cleavage of the six-membered ring to an acyclic compound that is later cyclized to form the smaller ring. This suggests an oxidative cleavage reaction and an internal aldol or Dieckmann cyclization. The exocyclic alkene unit in **145** suggests a Wittig reaction from a cyclopentanone precursor, and the cyclic ketone is consistent with the idea of an aldol or Dieckmann ring closure.

With this analysis in mind, **145** is first disconnected at the alkene unit to give ketone **147** and "methylene," which suggests a Wittig reaction (Section 22.10). If a Wittig reaction is used to prepare **145**, the precursor must be **147**, which contains a five-membered ring. The benzyl group is not derived from **146**. The next step is not obvious unless one has extensive practice in many syntheses so that it is recognized that the five-membered ring comes from **146** and involves *the loss of one carbon*. If this structural relationship is recognized, a Dieckmann cyclization followed by decarboxylation is required. Therefore, the next disconnection is at the benzyl group of **147** to give cyclopentanone as the donor (the enolate anion) and the benzyl acceptor fragment, which corresponds to benzyl bromide (another halide or leaving group can be chosen).

(a) KOH, EtOH (b) i. O_3 ii. H_2O_2 (c) EtOH, H^+ (d) i. NaOEt, ETOH, reflux ii. hydrolysis (e) H_3O^+ (f) 200°C ($-CO_2$) (g) i. LDA, THF, $-78°C$ ii. $PhCH_2Br$ (h) $Ph_3P=CH_2$

This disconnection suggests that **147** is prepared by enolate alkylation of cyclopentanone (Section 22.9). If a Dieckmann condensation is planned, the precursor to cyclopentanone is **148**, which in turn is derived from diester **149**. Another ester may be chosen at this point (methyl, etc.). Diester **149** is derived from the dicarboxylic acid, which is prepared by oxidative cleavage (ozonolysis) of cyclohexene. Bromocyclohexane **146** is now the clear precursor to cyclohexene by an E2 reaction (Chapter 12, Section 12.1).

The synthesis based on this disconnection strategy is a long one. Initial E2 reaction of **146** gives cyclohexene (Chapter 12, Section 12.1). Ozonolysis with a hydrogen peroxide workup of cyclohexene gives the dicarboxylic acid (Chapter 17, Section 17.4.1). Esterification is straightforward (Chapter 20, Section 20.6.2) to give **148** and Dieckmann cyclization leads to **149** (Section 22.7.3). Acid hydrolysis of the ester group in **149** gives the keto acid and heating leads to decarboxylation (Section 22.8) to give cyclopentanone. Reaction with LDA under kinetic control conditions gives the enolate anion, and alkylation with benzyl bromide provides **147** (Section 22.9). The final step is the Wittig reaction (Section 22.10) to give **145**.

22.12 Biological Relevance

Perhaps the key reaction in this chapter was the aldol condensation in Section 22.2. Although it is not mentioned in the discussion, the aldol reaction is reversible under certain circumstances. Enzymes known as aldolases can catalyze both the forward and reverse aldol reactions. An example is the retro-aldol/ aldol mechanism employed by the enzyme *l-ribulose-5-phosphate-4-epimerase*, found in both prokaryotes and eukaryotes, to invert the stereochemistry of the hydroxyl-bearing carbon in **150** to that in **153**. In other words, l-ribulose-5-phosphate is epimerized to give d-xylulose-5-phosphate (see Chapter 28, Section 28.1, for a discussion of these carbohydrates).

The first step is a retro-aldol of the Zn^{+2}-coordinated l-ribulose-5-phosophate (**150**) induced by the enzyme to give the aldehyde **151A** (glycoaldehyde phosphate) and the zinc-coordinated enolate anion **152** (dihydroxyacetone enolate).[19] Before the aldol reaction occurs, there is a bond rotation to generate a different rotamer of the aldehyde, **151B** (see Chapter 8, Section 8.1, for a discussion of rotamers). The aldehyde unit is not positioned differently, such that an aldol reaction will give the aldolate **153**, but rather with a different absolute stereochemistry (Chapter 9, Section 9.3). The overall enzyme process leads to epimerization of the hydroxyl-bearing carbon.

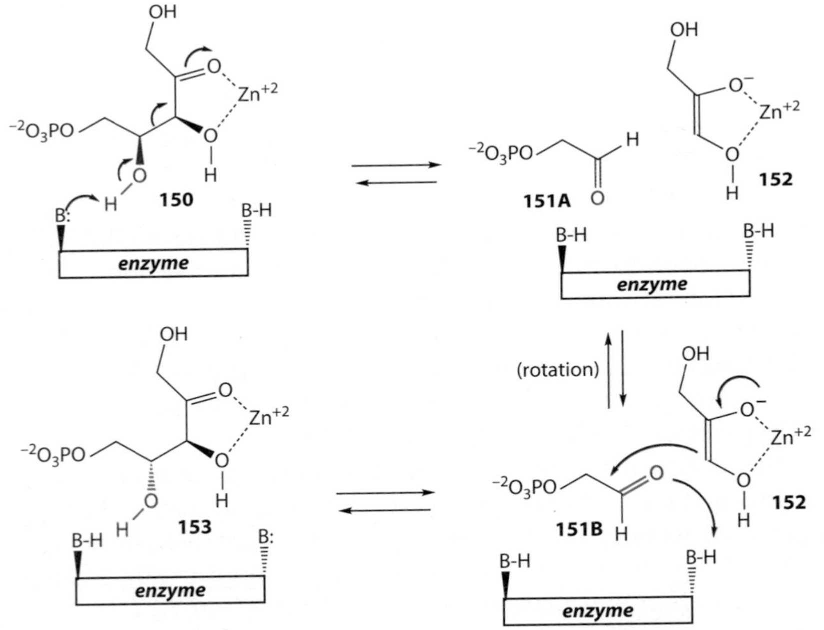

Reprinted in part with permission from Tanner, M.E. *Accounts of Chemical Research*, 2002, 35, 237. Copyright © 2002. American Chemical Society.

Epimerization or changing the absolute stereochemistry of a given stereogenic center is an important part of many enzymatic reactions. Enzymes known as enolases are categorized into what is known as the "enolase superfamily" of enzymes—structurally related proteins sharing the common ability to catalyze abstraction of the R-protons of carboxylic acids. These enzymes have the ability to catalyze the thermodynamically difficult step of proton abstraction, which often leads to epimerization. These proteins include enolase as well as more specialized enzymes such as mandelate racemase, galactonate dehydratase, glucarate dehydratase, muconate-lactonizing enzymes, N-acylamino acid racemase, $â$-methylaspartate ammonia-lyase, and o-succinylbenzoate synthase.[20]

Two examples are shown involving *N-acylamino acid racemase* (NAAAR) and β-*methylaspartate ammonia-lyase* (MAL). NAAAR is a *racemase* and it catalyzes a 1,1-proton transfer reaction that is stereorandom with respect to abstraction of the (R)-proton of the amino acid residue. NAAAR appears to contain two bases in its active site that are positioned for proton abstraction from either the (R)- or (S)-enantiomers of the amino acid. A simple example is epimerization of acyl valine derivative **154** to **155**, which proceeds by removal of the α-proton to form an enol or an enolate intermediate. Valine and other amino acids are discussed in Chapter 27, Section 27.3.2.

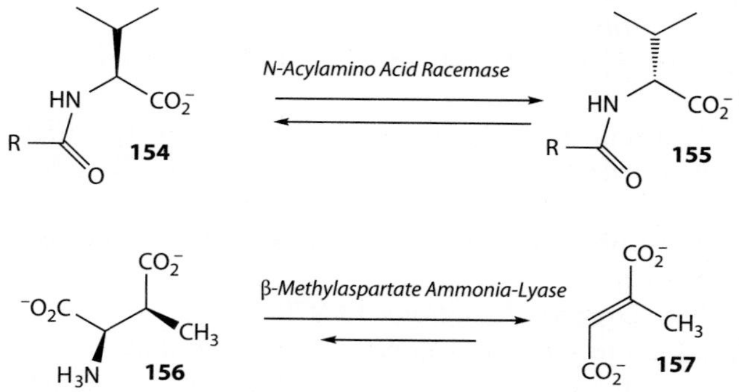

Reprinted in part with permission from Babbitt, P.C., Hasson, M.S., Wedekind, J.E., Palmer, D.R., Barrett, W.C., Reed, G.H., Rayment, J., Ringe, D., Kenyön, G.L., Gerlt, J.A. *Biochemistry* 1996, 35, 16489. Copyright © 1996 American Chemical Society.

22.76 Draw the enol derived from 154.

MAL initiates epimerization, but it involves the presence of oxygen leaving groups. Either diastereomer or 3-methyl-(2S)-aspartate is a substrate for this enzyme. An example is the conversion of **156** to **157** via loss of ammonia. Abstraction of the (R)-proton by the enzyme leads to loss of ammonia and formation of the C=C unit. Epimerization is completed by delivery of ammonia from another nitrogen source to the conjugate acid unit to regenerate the epimeric methyl asparate.

References

1. Gero, A. 1954. *Journal of Organic Chemistry* 19:469–471.
2. Stewart, R. 1985. *The proton: Applications to organic chemistry,* 73–74. New York: Academic Press.

3. Zook, H. D., Kelly, W. L., and Posey, I. Y. 1968. *Journal of Organic Chemistry* 33:3477.

4. Pearson, R. G., and Dillon, R. I. 1953. *Journal of the American Chemical Society* 75:2439–2443.

5. Stewart, R. 1985. *The proton: Applications to organic chemistry,* 54. New York: Academic Press.

6. Streitwieser, A., Jr., Ciuffarin, E., and Hammons, J. H. 1967. *Journal of the American Chemical Society* 89:63.

7. Nielsen, A. T., and Houlihan, W. J. 1968. *Organic Reactions* 16:1.

8. Stork, G., Kraus, G. A., and Garcia, G. A. 1974. *Journal of Organic Chemistry* 39:3459.

9. Stewart, R. 1985. *The proton: Applications to organic chemistry,* 43. New York: Academic Press.

10. Abbott, A. E., Kon, G. A. R., and Satchell, R. D. 1928. *Journal of the Chemical Society* 2514 (see p. 2521).

11. Faulk, D. D., and Fry, A. 1970. *Journal of Organic Chemistry* 35:364.

12. Furniss, B. S., Hannaford, A. J., Smith, P. W. G., and Tatchell, A. R., eds. 1994. *Vogel's textbook of practical organic chemistry,* 5th ed., Exp. 5.176, 741–742. Essex, England: Longman.

13. Schaefer, J. P., and Bloomfield, J. J. 1967. *Organic Reactions* 15:1.

14. Jones, G. 1967. *Organic Reactions* 15:204.

15. House, H. O., Gall, M., and Olmstead, H. D. 1971. *Journal of Organic Chemistry* 36:2361–237.

16. Furniss, B. S., Hannaford, A. J., Smith, P. W. G., and Tatchell, A. R., eds. 1994. *Vogel's textbook of practical organic chemistry,* 5th ed., Exp. 5.17, 498–499. Essex, England: Longman.

17. Zhang, X.-M., and Bordwell, F. G. 1994. *Journal of the American Chemical Society* 116:968–972.

18. Maercker, A. 1965. *Organic Reactions* 14:270.

19. Tanner, M. E. 2002. *Accounts of Chemical Research* 35:237.

20. Babbitt, P. C., Hasson, M. S., Wedekind, J. E., Palmer, D. R., Barrett, W. C., Reed, G. H., Rayment, I., Ringe, D., Kenyon, G. L., and Gerlt, J. A. 1996. *Biochemistry* 35:16489.

Answers to Problems

22.1

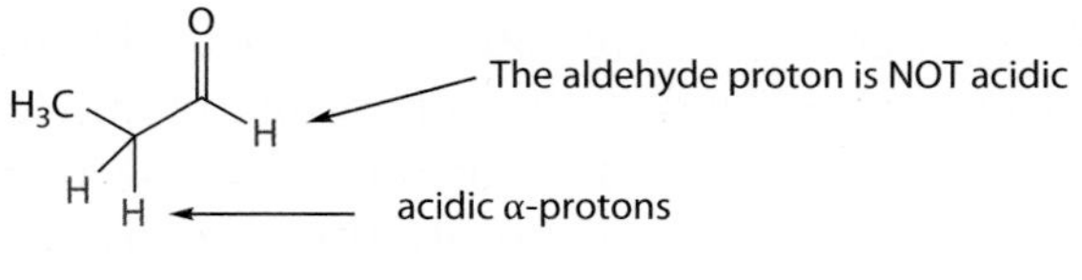

The aldehyde proton is NOT acidic

acidic α-protons

22.2

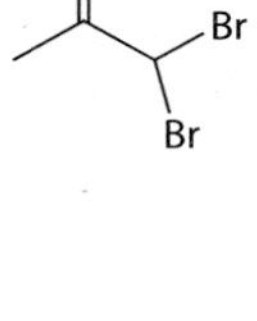

22.3

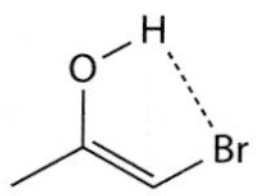

22.4

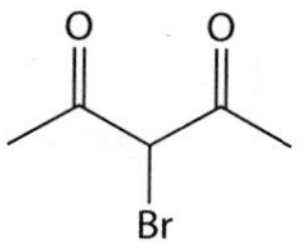

22.5

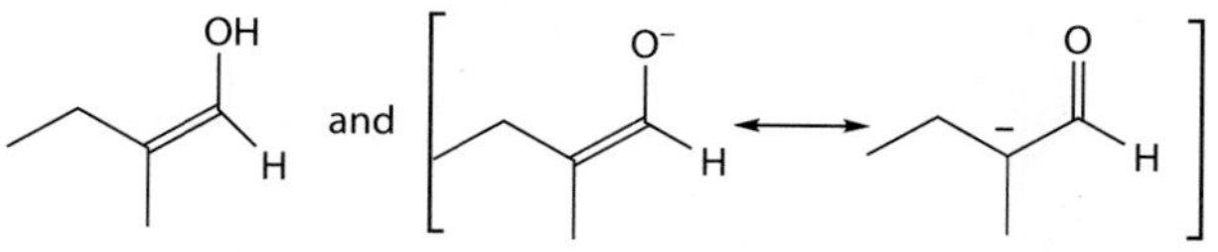

22.6

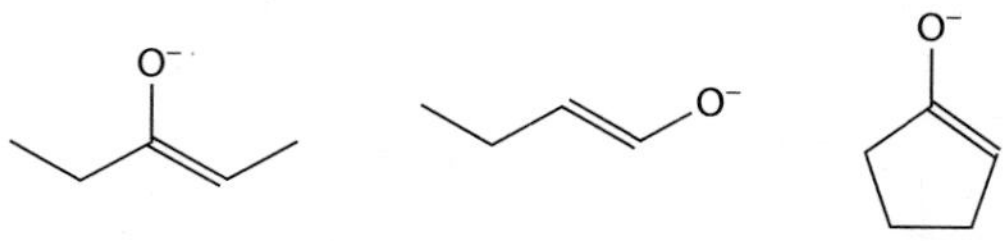

22.7   If the pK$_a$ of C$_2$ in 3-pentanone is about 20, another methyl group at C$_3$ in 3-methyl-2-pentanone should make the pK$_a$ of that hydrogen about 0.5–1 pK unit higher, or 21.

22.8   There is only one acidic hydrogen in 2,4,4-trimethyl-3-pentanone and the two electron-releasing methyl groups make it less acidic than the proton in acetone. An estimate of pK$_a$ is about 21.

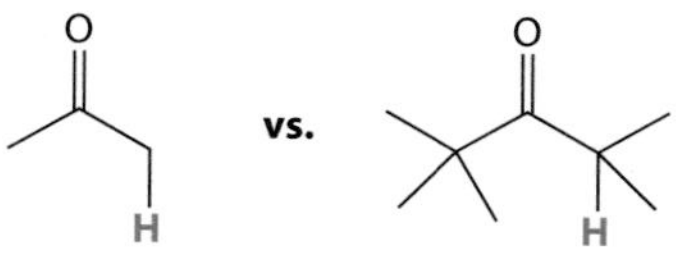

22.9

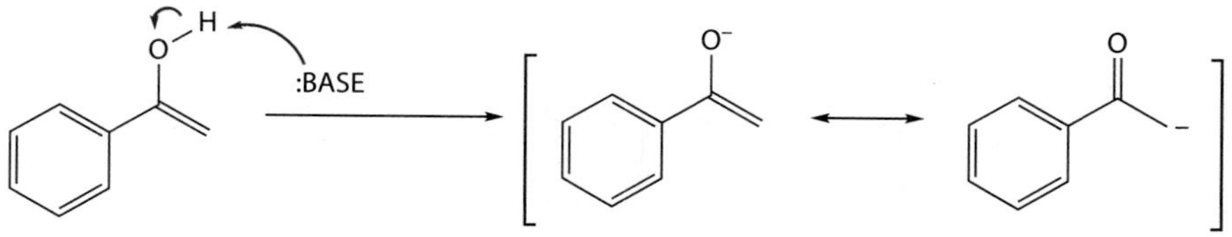

22.10

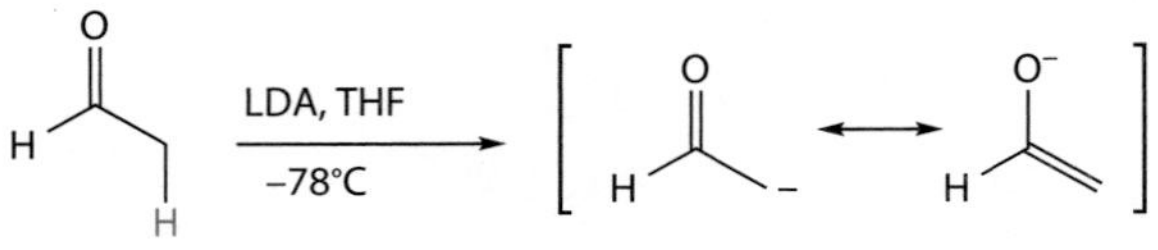

22.11

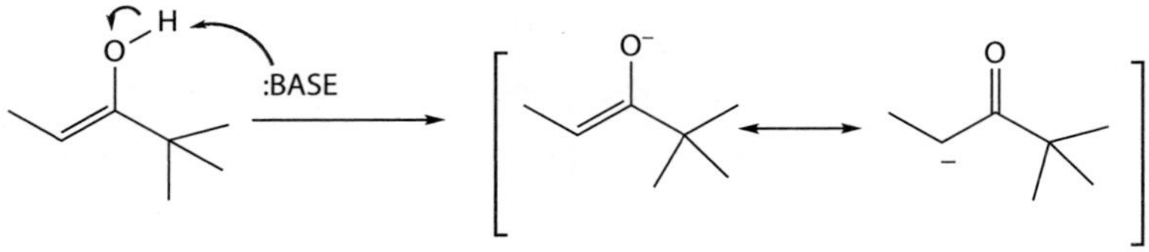

22.12

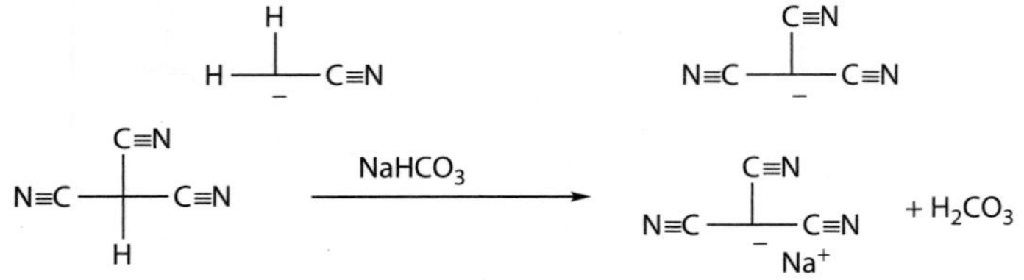

22.13

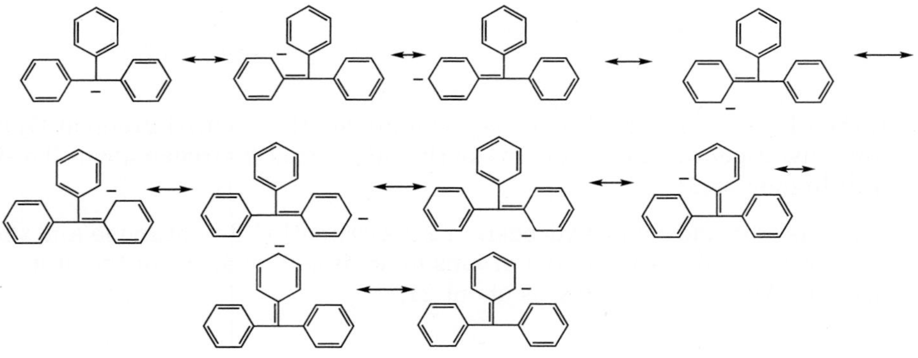

22.14

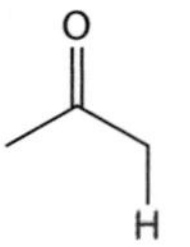

pKₐ is about 20

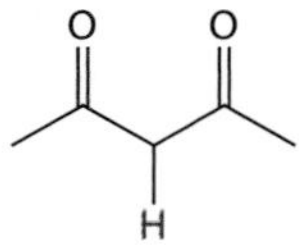

pKₐ is about 9–10

3 resonance contibutors for the enolate anion lead to enhancement of acidity

22.15

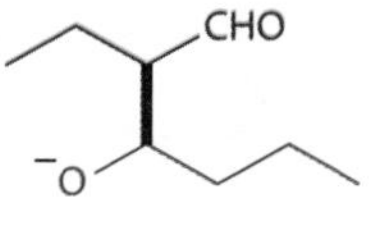

22.16 The base is NaOEt, so the conjugate acid is ethanol (EtOH).

22.17

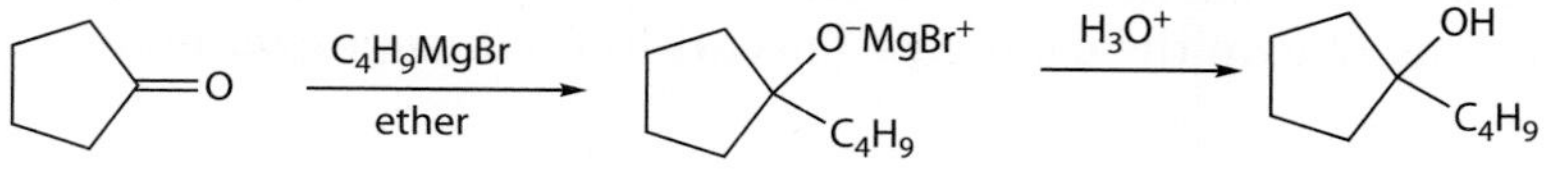

22.18

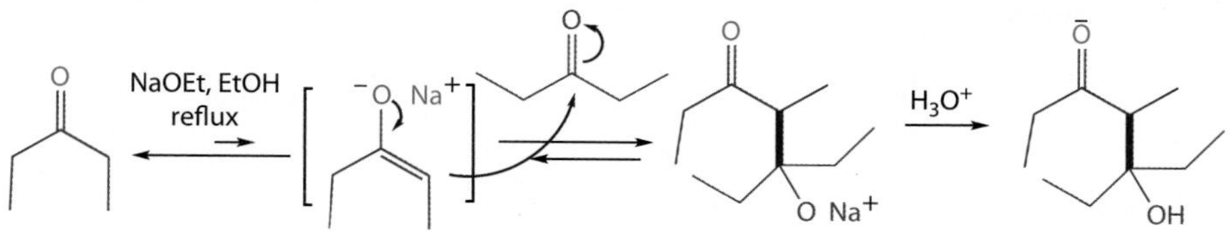

22.19

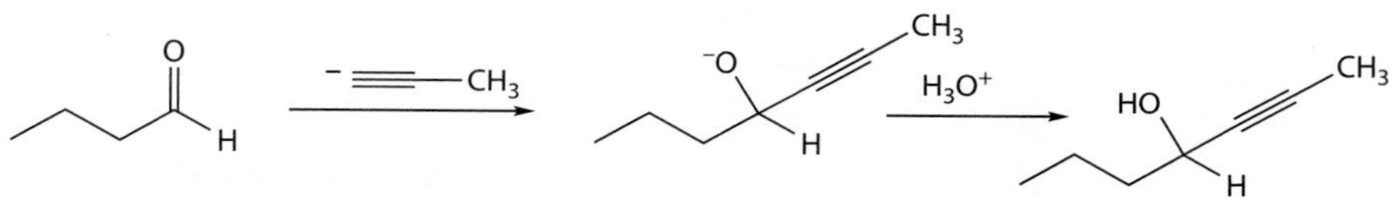

22.20

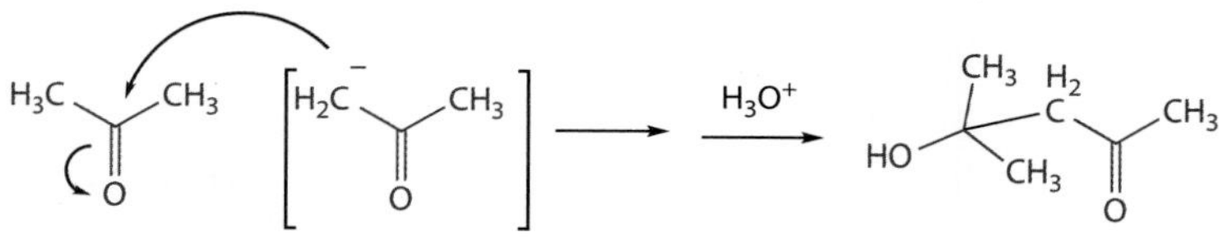

22.21

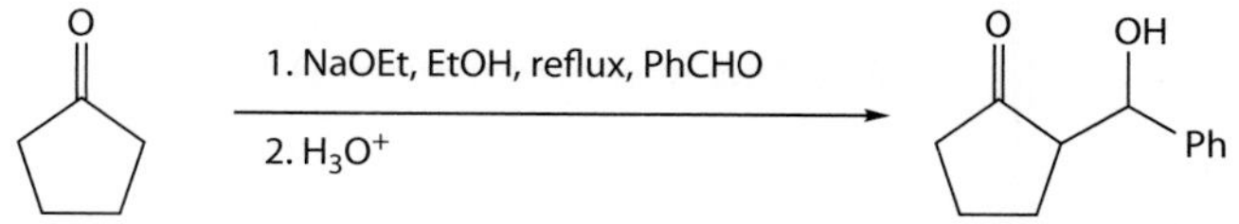

22.22

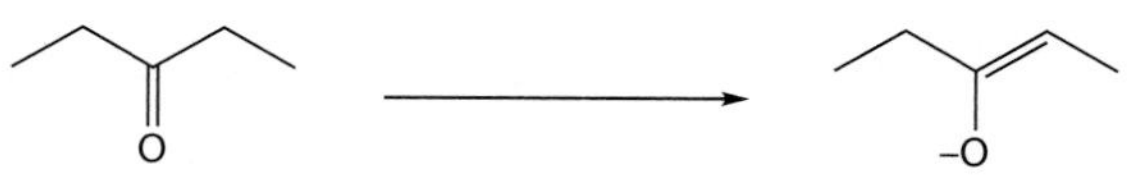

22.23

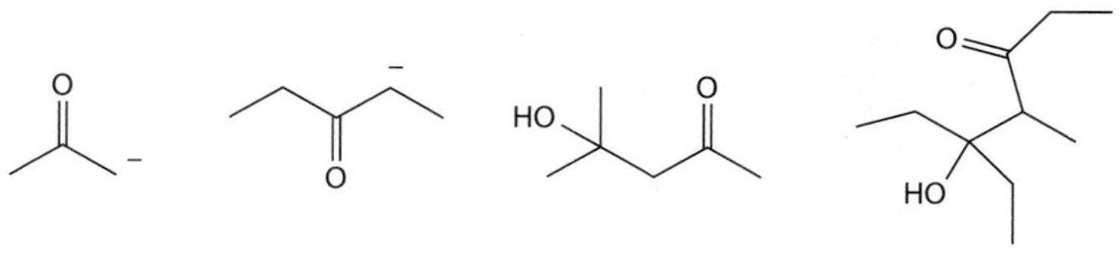

22.24   The conjugate acid of butyllithium is butane. Butane has a pK$_a$ > 40, so it is a much weaker acid than diisopropyl amine with a pK$_a$ of about 25. If butane is a much weaker acid than diisopropyl amine, K$_a$ for the reaction will be large (equilibrium is pushed to the right), facilitating the deprotonation reaction of the amine to give the conjugate base, lithium diisopropylamide.

22.25

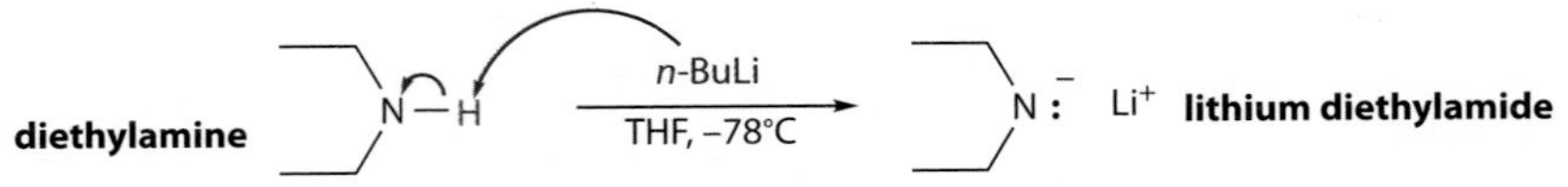

22.26

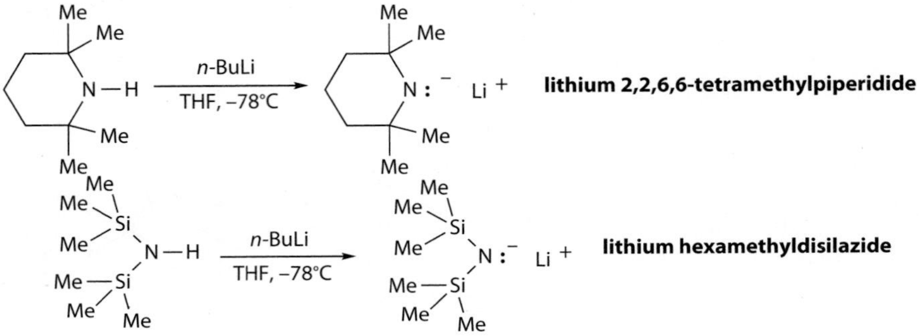

22.27

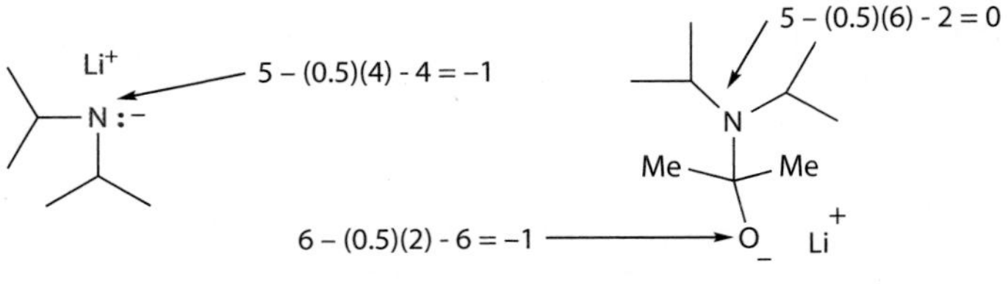

22.28

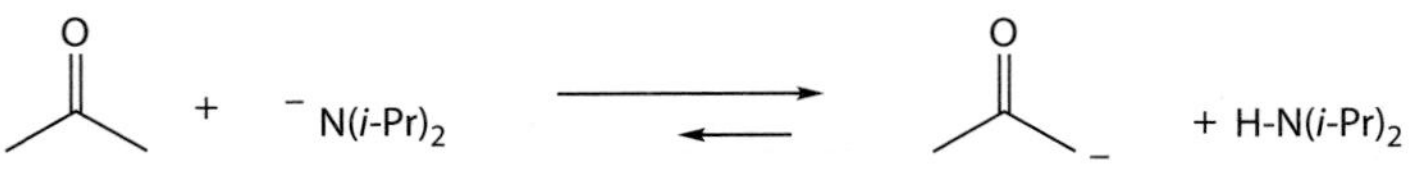

22.29

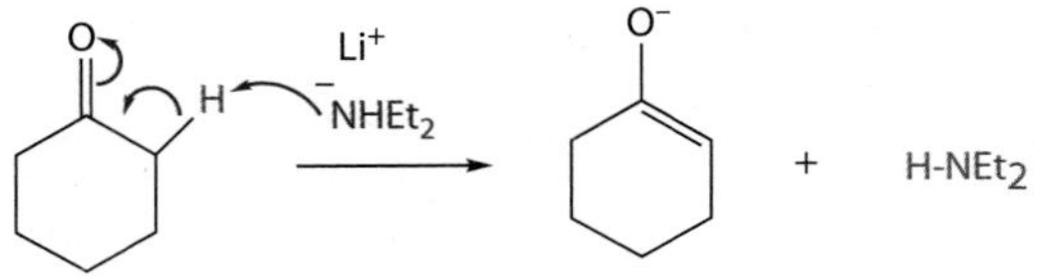

22.30

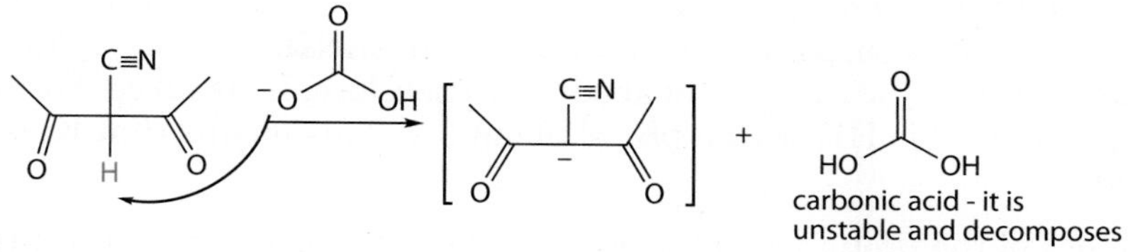

carbonic acid - it is
unstable and decomposes

22.31

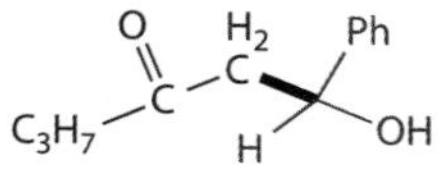

1-hydroxy-1-phenyl-3-hexanone

22.32

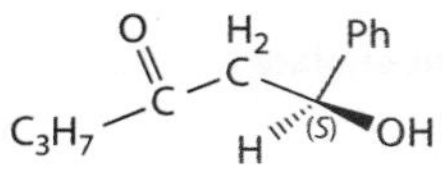

22.33

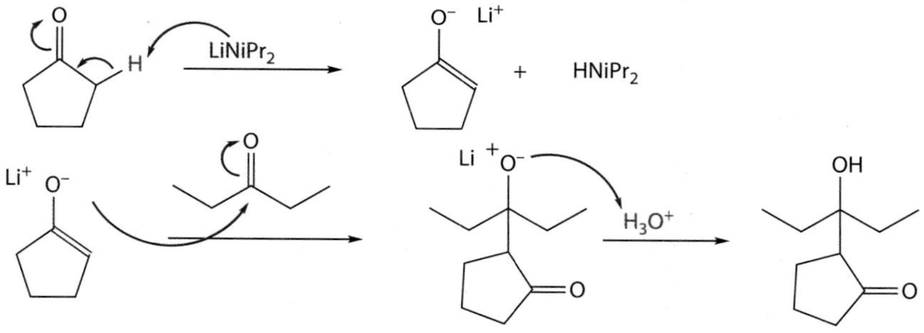

22.34

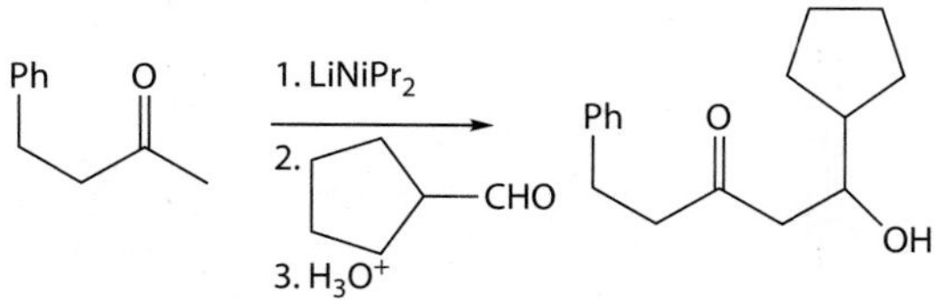

22.35

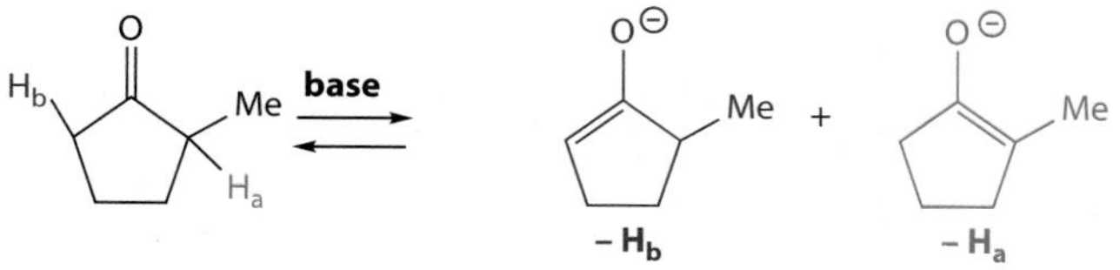

22.36   For the conversion of **32** to **36**, $K_a = [32/36]$.

22.37   For NaOEt, the conjugate acid is ethanol (EtOH), with a pK_a of about 17. Therefore, K_a is small. For the reaction with $NaNH_2$, the conjugate acid is ammonia (NH_3) with a pK_a of about 25. Therefore, K_a is large. For NaF, the conjugate acid is HF, with a pK_a of about 3.2. This means that K_a is very small.

22.38   The use of the protic solvent 2-propanol favors thermodynamic control.

22.39   The use of the aprotic solvent diethyl ether favors kinetic control.

22.40

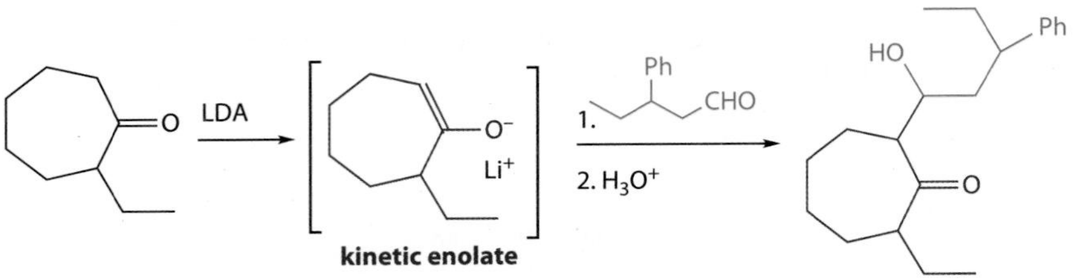

22.41

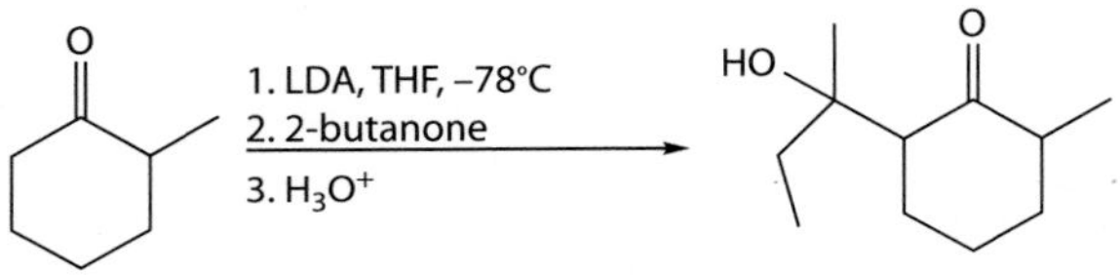

22.42

22.43

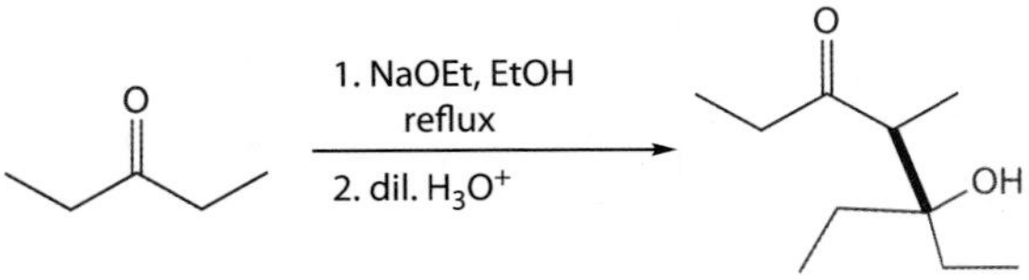

22.44

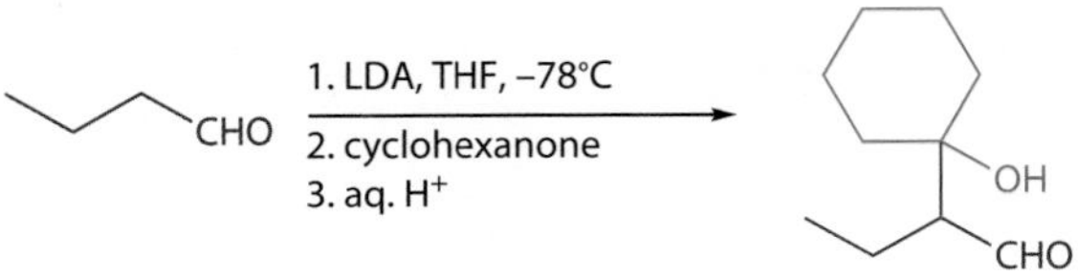

22.45

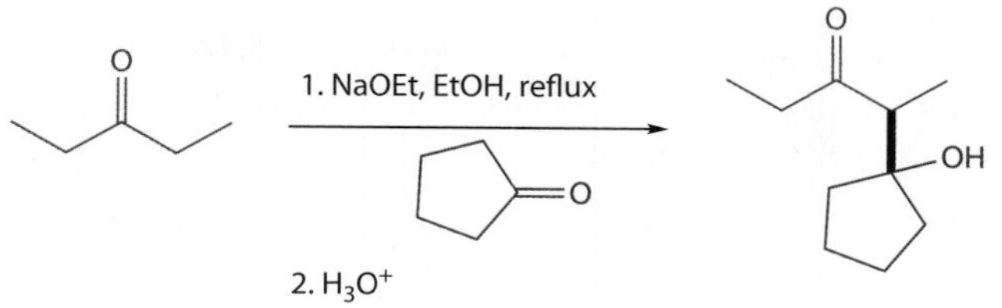

22.46 Both ketones are unsymmetrical, so there are four sites where an enolate anion can form a, b, a′, b′. An enolate at a can react with both ketones; an enolate at b can react with both ketones. Similarly, an enolate at a′ reacts with both ketones, as does the enolate at b′. These combinations lead to the eight aldol products shown.

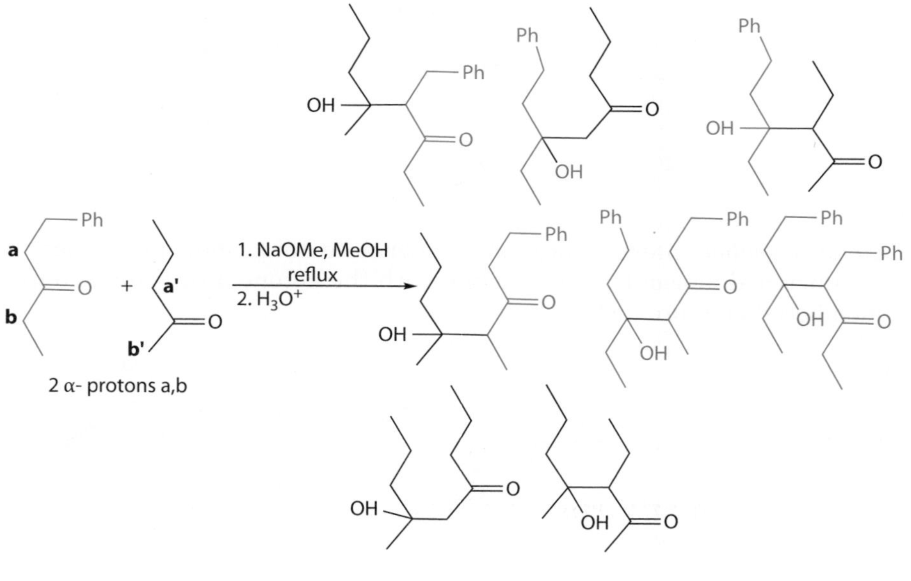

22.47

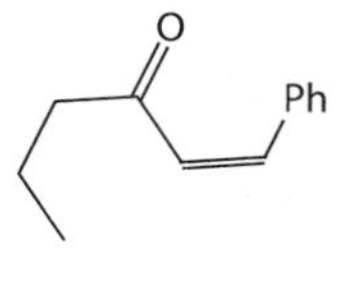

22.48

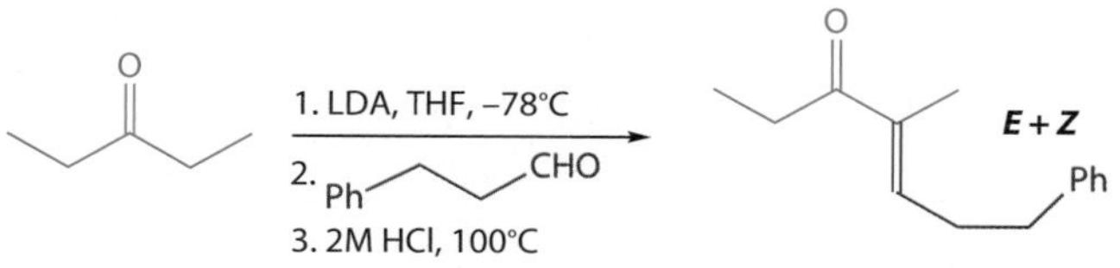

22.49

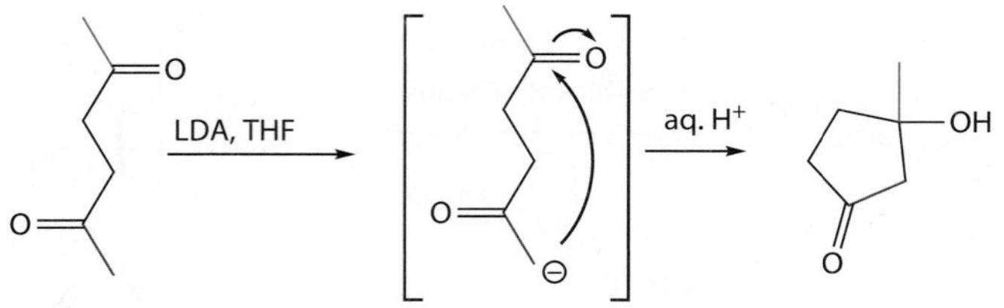

22.50   In the diketone, the more acidic proton is one of the methyl protons. The enolate produced attacks the other ketone–carbonyl unit to give the seven-membered ring product.

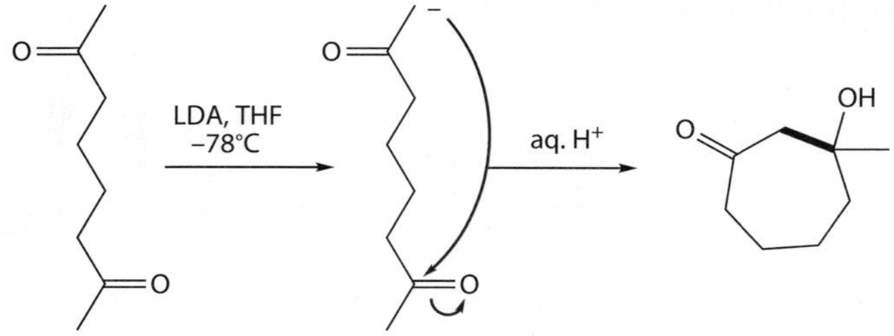

22.51   Under thermodynamic conditions, the enolate anion shown is formed in this symmetrical molecule, and reaction with the ketone unit leads to a five-membered ring product.

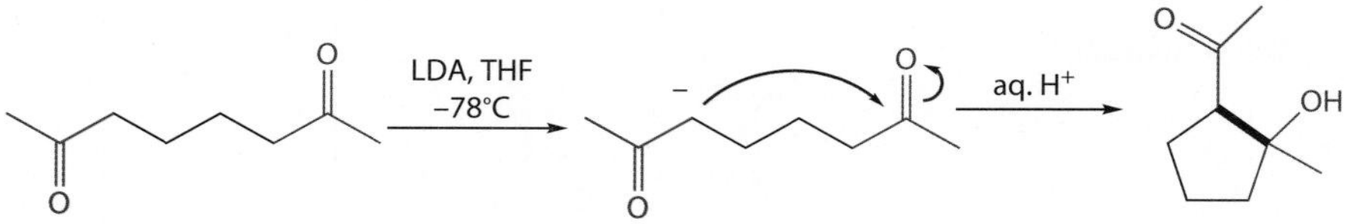

22.52

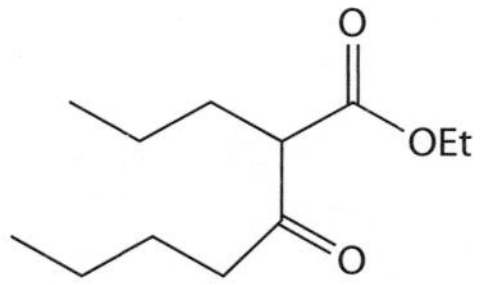

22.53

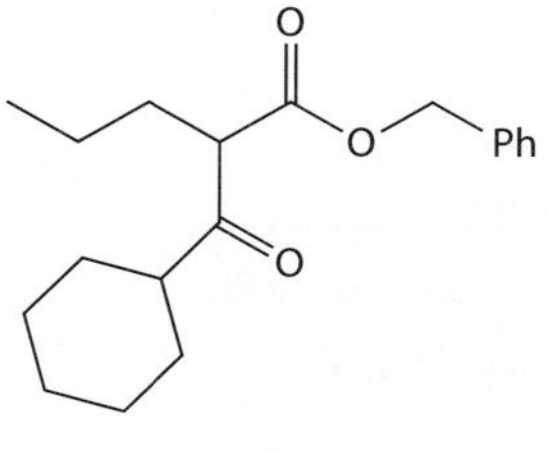

22.54

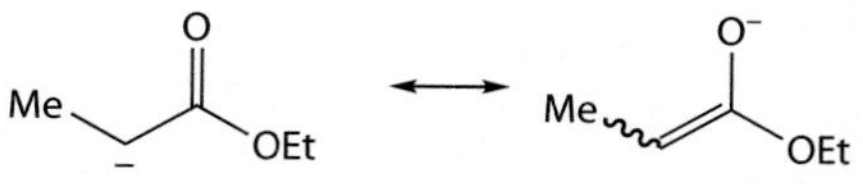

22.55

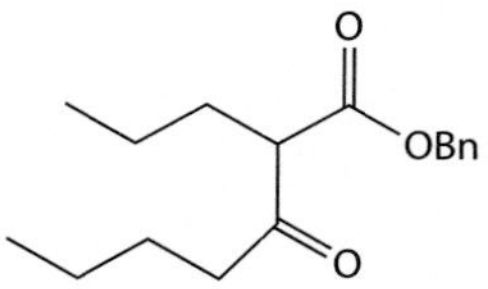

22.56

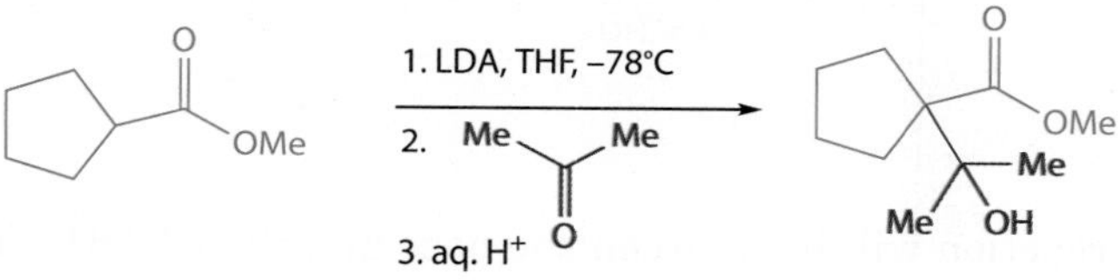

22.57

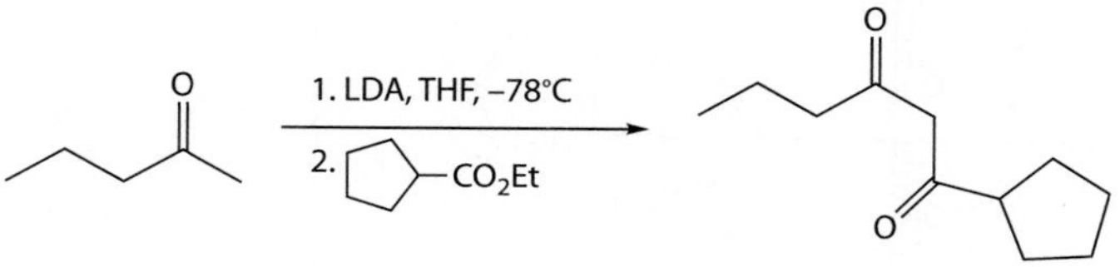

22.58

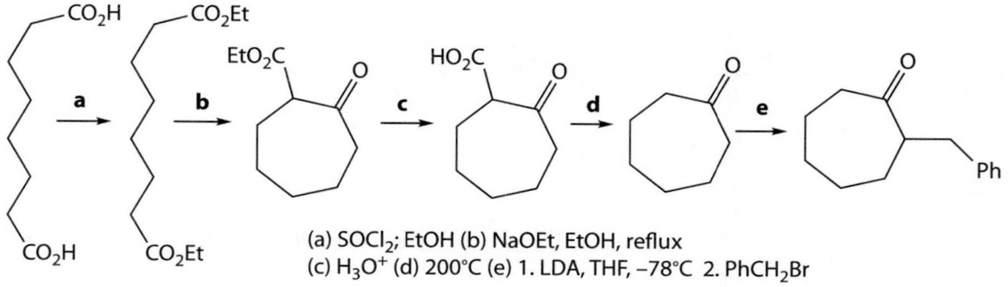

(a) SOCl$_2$; EtOH (b) NaOEt, EtOH, reflux
(c) H$_3$O$^+$ (d) 200°C (e) 1. LDA, THF, −78°C 2. PhCH$_2$Br

22.59 Diester **87** is a methyl ester. If we used sodium ethoxide, transesterification could occur to give **87** as a mixed methyl/ethyl ester, which could lead to extra products. To avoid the transesterification, the conjugate base of the ester is used. Therefore, a methyl ester uses sodium methoxide, an ethyl ester uses sodium ethoxide, etc.

22.60

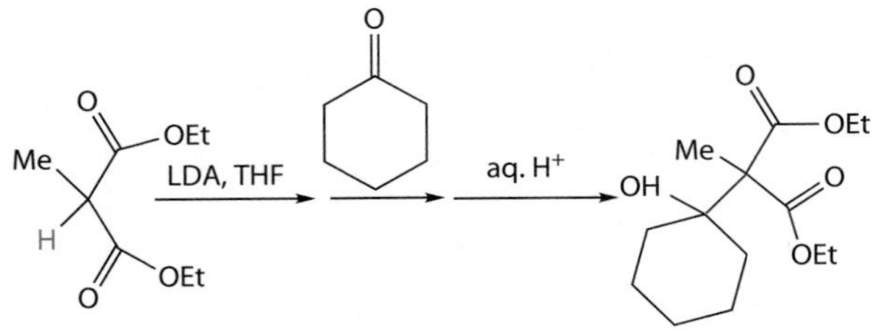

Me
H
O
OEt
OEt
O
LDA, THF
O
aq. H+
OH
Me
O
OEt
O
OEt

22.61

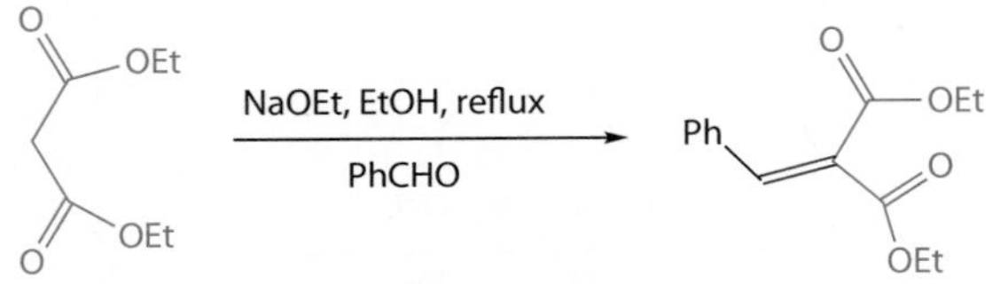

O
OEt
OEt
O
NaOEt, EtOH, reflux
PhCHO
Ph
O
OEt
O
OEt

22.62 The first reaction will be between the base and the COOH unit (the more acidic proton) to give the carboxylate salt (COO−). The α-proton is then removed to give the dianion shown.

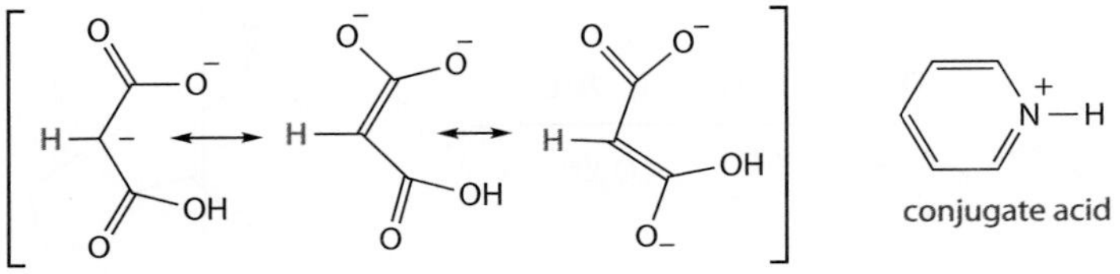

H
OH
O
OH
N—H
conjugate acid

22.63

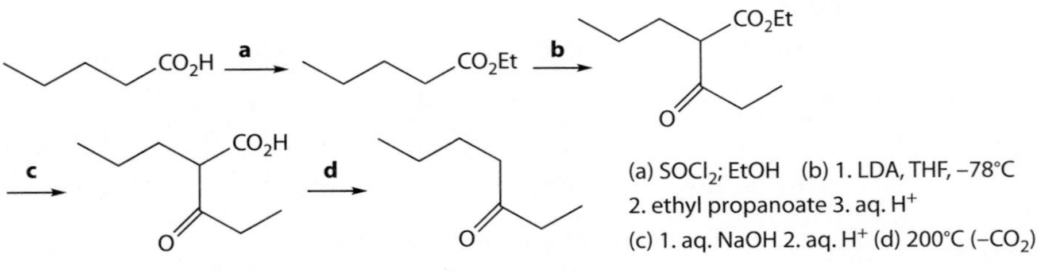

CO2H
a
CO2Et
b
CO2Et
c
CO2H
d
(a) SOCl2; EtOH (b) 1. LDA, THF, −78°C
2. ethyl propanoate 3. aq. H+
(c) 1. aq. NaOH 2. aq. H+ (d) 200°C (−CO2)

22.64

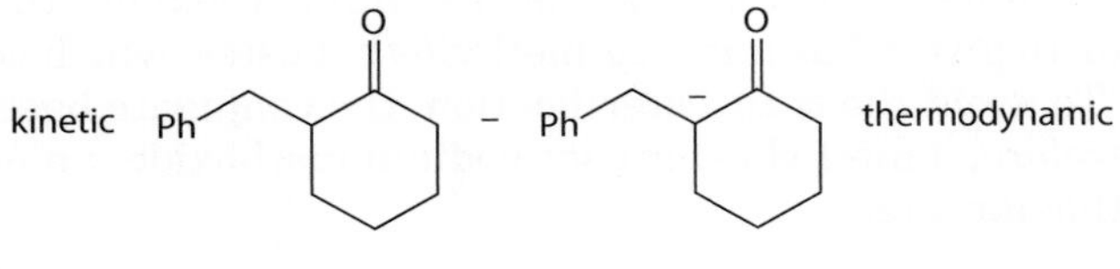

kinetic
Ph
Ph
thermodynamic

22.65

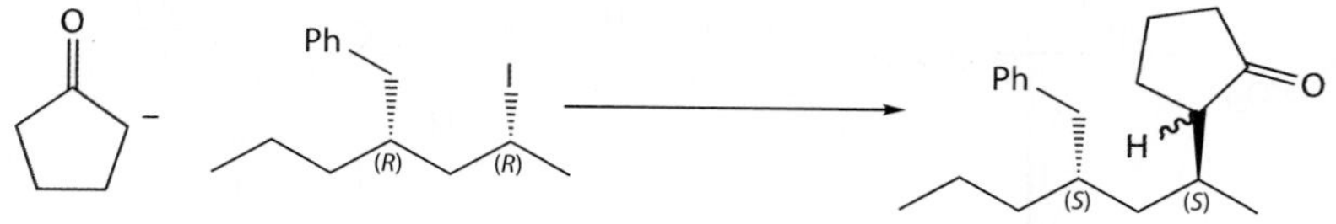

22.66

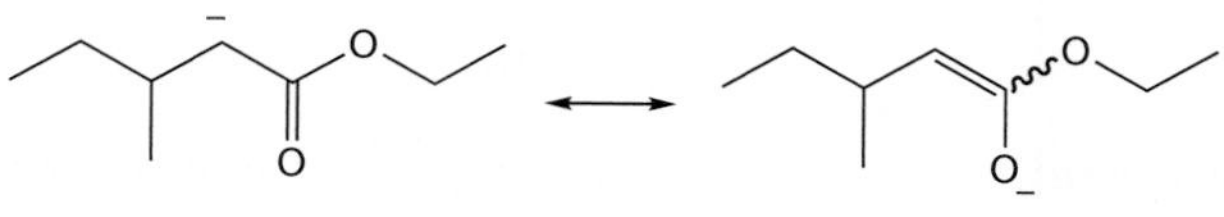

22.67

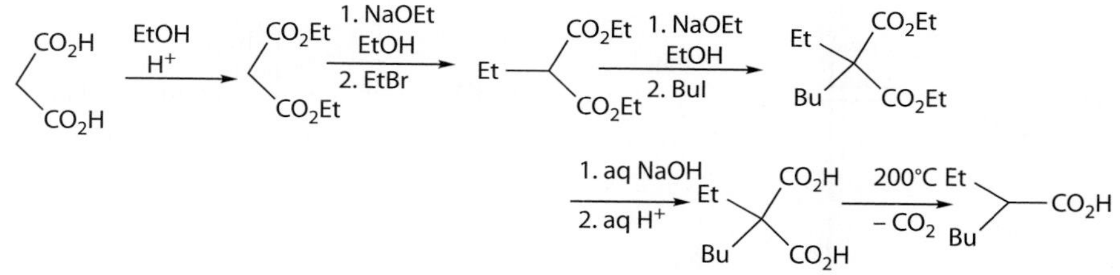

22.68

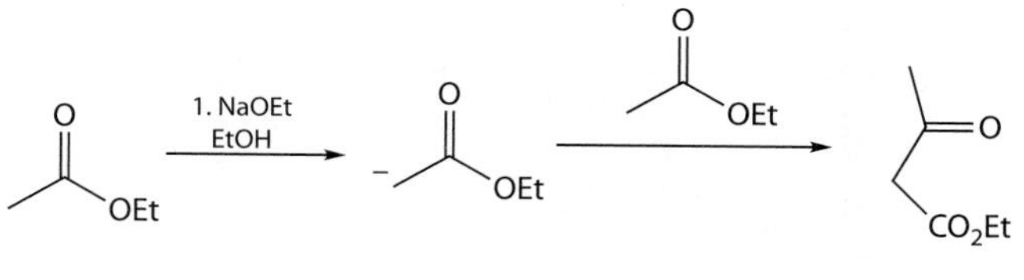

22.69 Removal of the "middle" proton leads to a resonance-stabilized enolate anion with three resonance structures, as shown. Removal of the methyl proton leads to an enolate anion with only two resonance contributors. The middle proton is significantly more acidic and will deprotonate to give the enolate shown. If LDA is used, the kinetic enolate is the one derived from removal of the more acidic proton, which is the same enolate anion.

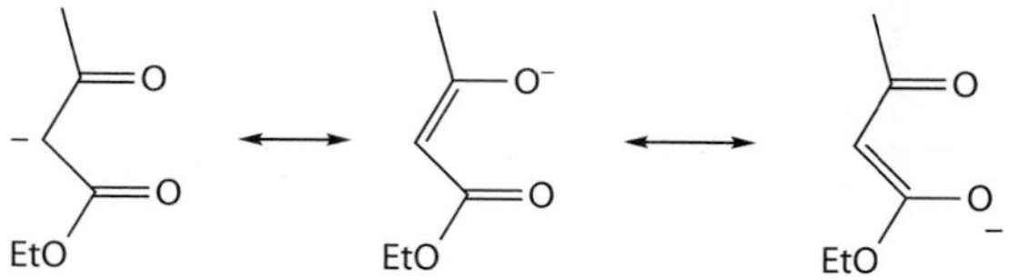

22.70

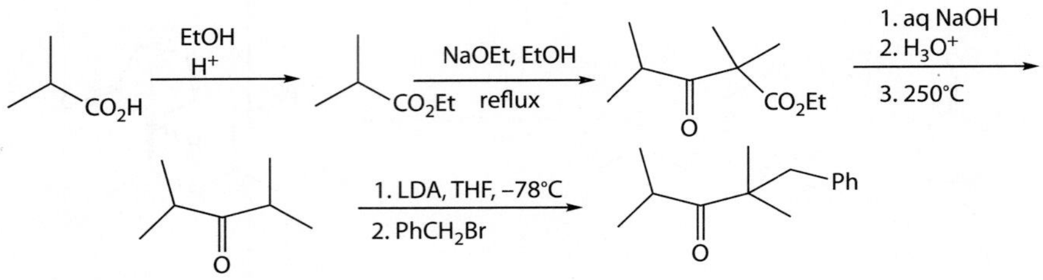

22.71

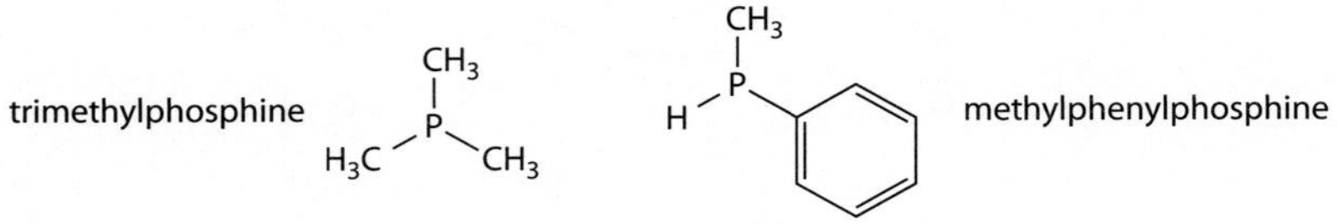

trimethylphosphine   methylphenylphosphine

22.72

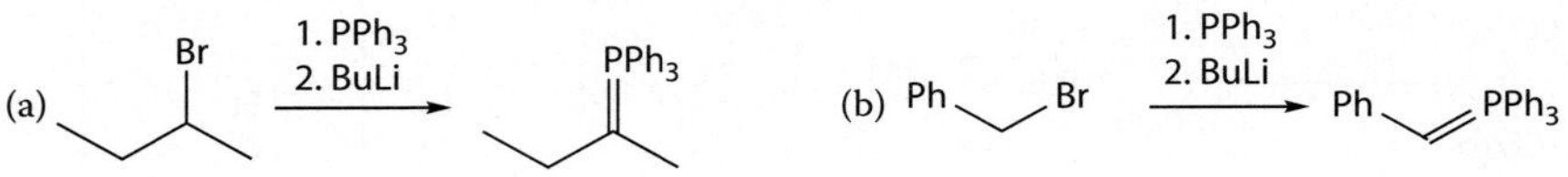

(a)   1. PPh₃ / 2. BuLi

(b) Ph — Br   1. PPh₃ / 2. BuLi   → Ph — PPh₃

22.73

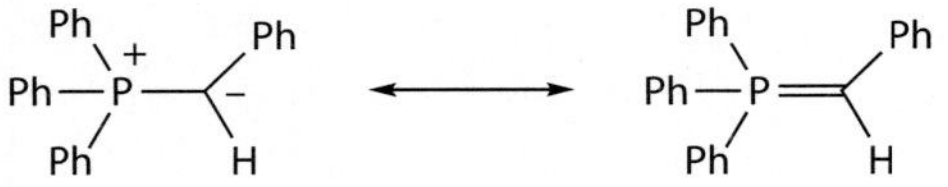

22.74

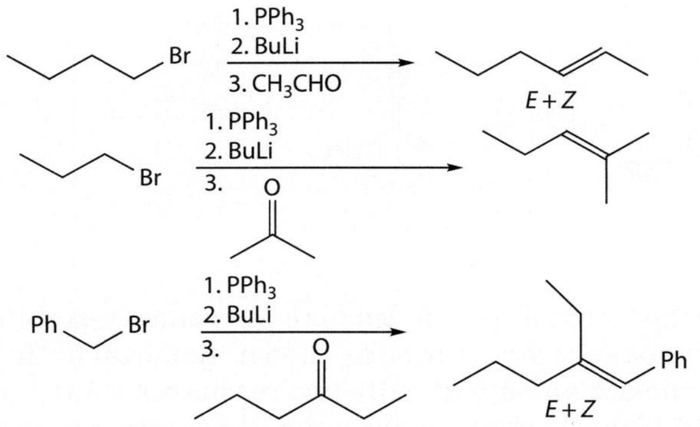

22.75

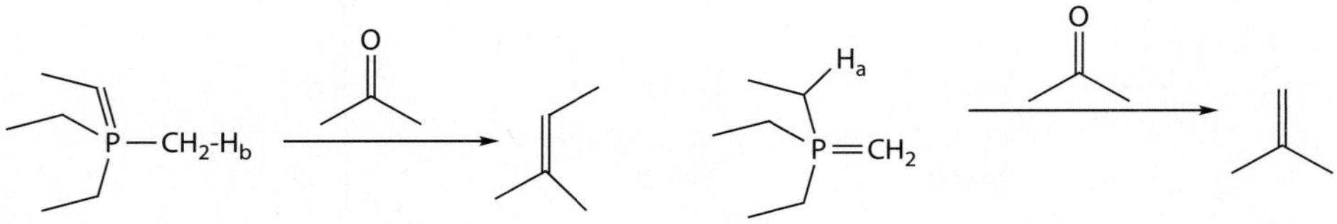

22.76

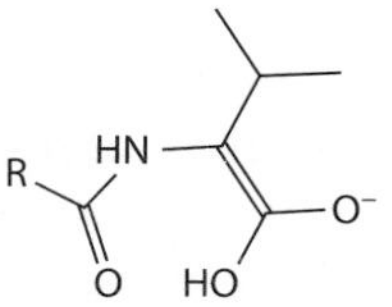

Correlation of Concepts with Homework

- The α-proton of a ketone or aldehyde is a weak acid and can be removed by a strong base to give a resonance-stabilized enolate anion. Deprotonation may occur via the enol tautomer. The α-proton of an aldehyde is slightly more acidic than that of a ketone: 1, 2, 3, 4, 5, 6, 7, 8, 9, 10, 11, 12, 22, 23, 28, 76, 77, 78, 81, 86, 89, 90.
- The α-proton of an aldehyde or ketone is less acidic as more carbon substituents are added. As more electron-withdrawing groups are added, the α-proton becomes more acidic, so a 1,3-diketone is more acidic than a ketone. The more acidic proton of an unsymmetrical ketone is the one attached to the less substituted carbon atom: 8, 12, 13, 14, 22, 23, 28, 30, 77, 81, 86, 89, 93.
- Enolate anions react as nucleophiles. They give nucleophilic acyl addition reactions with aldehydes and ketones. The condensation reaction of an aldehyde or ketone enolate with another aldehyde or ketone is called an aldol condensation. Self-condensation of symmetrical aldehydes or ketones leads to a single product under thermodynamic conditions. Condensation between two different carbonyl compounds gives a mixture of products under thermodynamic conditions, but can give a single product under kinetic control conditions: 5, 9, 11, 15, 16, 17, 18, 19, 20, 21, 23, 29, 30, 31, 32, 33, 34, 40, 41, 42, 43, 44, 45, 46, 49, 91, 92, 94, 102, 114, 115, 123, 134.
- Dialkyl amides are formed by the reaction of an amine with an organolithium reagent, and they are considered to be non-nucleophilic bases: 24, 25, 26, 27, 28, 33, 34, 85, 88, 93.
- Kinetic control is favored by a strong base that generates a conjugate acid weaker than the carbonyl compound, a polar aprotic solvent, short reaction times, and low temperatures. Thermodynamic control is favored by a strong base that generates a conjugate acid stronger than the carbonyl compound, a polar protic solvent, long reaction times, and high temperatures: 35, 36, 37, 38, 39, 40, 41, 42, 43, 44, 45, 46, 48, 64, 66, 79, 80, 82, 83, 87, 88, 89, 91, 92, 94.

- Dehydration of enolate condensation products leads to a conjugated carbonyl compound: 44, 46, 47, 48, 102, 114, 115.
- The intramolecular condensation of an α,ω-dialdehyde or diketone leads to a cyclic compound: 49, 50, 51, 95, 96, 97, 102, 118, 129.
- An ester enolate is formed by reaction with a strong base, and the resulting enolate anion can condense with an aldehyde, a ketone, or another ester. Ester enolates react with aldehydes or ketones to form β-hydroxy esters. Aldehyde or ketone enolate anions react with esters to form β-hydroxy esters, 1,3-diketones, or β-keto aldehydes: 56, 57, 84, 99, 100, 102, 108, 110, 114, 115.
- Enolate anions react as nucleophiles. They give nucleophilic acyl substitution reactions with acid derivatives. The condensation reaction of one ester with another is called a Claisen condensation and it generates a β-keto ester. A mixed Claisen condensation under thermodynamic conditions leads to a mixture of products, but kinetic control conditions can give a single product: 52, 53, 54, 55, 59, 68, 69, 98, 99, 101, 125.
- The intra-molecular Claisen condensation is called a Dieckmann condensation, and it generates a cyclic compound: 58, 99, 101, 118.
- Malonic esters can be converted to the enolate anion and condensed with aldehydes, ketones, or acid derivatives. The reaction of malonic acid with an aldehyde using pyridine as a base is called the Knoevenagel condensation: 59, 60, 61, 62, 69, 99, 108, 110, 112, 113, 119, 124.
- A β-keto ester can be hydrolyzed to a β-keto acid, and heating leads to decarboxylation. Malonic acid derivatives, as well as β-ketone acids decarboxylate upon heating: 63, 109, 111, 135.
- Enolate anions react with alkyl halides by an S_N2 reaction to give alkylated carbonyl compounds: 65, 67, 70, 84, 108, 116, 127, 128, 135.
- Triphenylphosphine reacts with alkyl halides to form phosphonium salts. Organolithium bases react with alkyltriphenylphosphonium salts to give phosphorus ylids, which react with aldehydes and ketones to give alkenes in what is known as the Wittig reaction: 71, 72, 73, 74, 75, 103, 104, 105, 106, 107, 117.
- A molecule with a particular functional group can be prepared from molecules containing different functional groups by a series of chemical steps (reactions). This process is called synthesis: the new molecule is synthesized from the old one (see Chapter 25): 63, 67, 70, 117, 118, 119.
- Spectroscopy can be used to determine the structure of a particular molecule and can distinguish the structure and functionality of one molecule when compared with another (see Chapter 14): 120, 121, 122, 123, 124, 125, 126, 127, 128, 129, 130, 131, 132, 133, 134, 135.

Homework

22.77 Contrast and compare the relative acidities of 1-phenyl-1-ethanone, 1-phenyl-1-propanone, 2-methyl-1-phenyl-1-propanone, and 2,2-dimethyl-1-phenyl-1-propanone.

22.78 Briefly discuss the relative acidity of the following: acetone, acetonitrile, and nitromethane.

22.79 If the enolate anion of 2-butanone is formed using sodium amide in ammonia, will the kinetic or the thermodynamic enolate anion be favored? Explain.

22.80 Rank order the following enolate anions in terms of their stability. Explain.

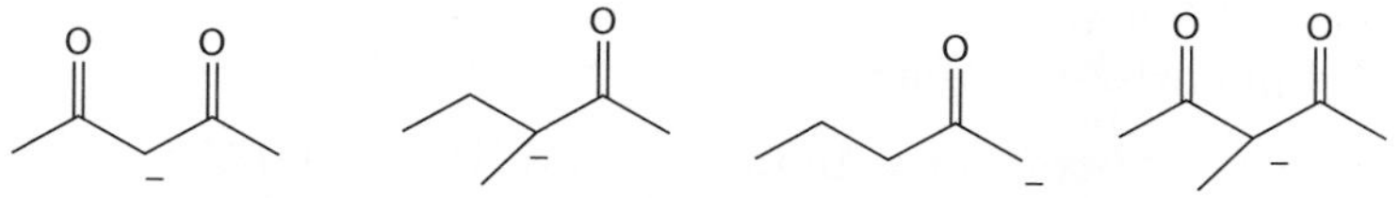

22.81 Draw the carbanion formed from dicyanomethane, $CH_2(CN)_2$. Draw the resonance contributors.

22.82 Which of the following solvents are compatible with kinetic control conditions? Which are compatible with thermodynamic control conditions?

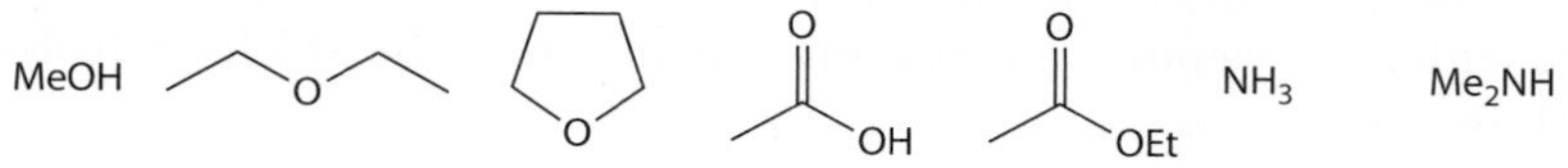

22.83 Draw the product formed when A is treated with LDA in THF at −78°C, followed by an aqueous acid workup.

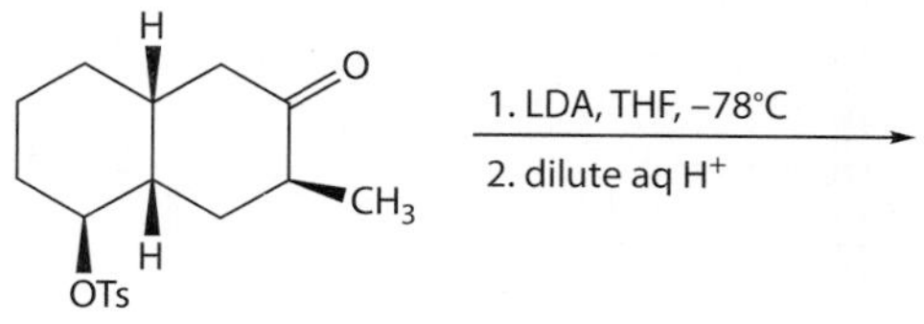

22.84 Draw the product formed when the kinetic enolate anon of 2-methyl-3-hexanone reacts with benzaldehyde; with ethyl butanoate; with 2S-iodohexane.

22.85 Draw the complete acid–base reaction, with all products, between 3-pentanone and:
   (a) lithium diisopropyl amide
   (b) lithium 2,2,6,6-tetramethylpiperidide
   (c) lithium amide
   (d) sodium hydride (assume that it is a strong enough base)

22.86   Indicate the more acidic proton in each of the following:
  (a) 2-butanone
  (b) butanal
  (c) 2,5-dimethyl-3-hexanone
  (d) 3-phenyl-2-butanone
  (e) 4-methyl-3-heptanone
  (f) 3,3-diphenyl-2-pentanone

22.87   Draw both the kinetic and thermodynamic enolate anion for each of the following; assume a Li cation:
  (a) 2-butanone
  (b) butanal
  (c) 2,5-dimethyl-3-hexanone
  (d) 3-phenyl-2-butanone
  (e) 4-methyl-3-heptanone
  (f) 3,3-diphenyl-2-pentanone

22.88   The reaction of 2-butanone with LiNMe$_2$ in THF at –100°C generates the enolate anion in 10 minutes. Identify the reaction conditions as kinetic or thermodynamic control and briefly discuss how each reactant and reaction condition influences this.

22.89   Pentanal has only one type of α-hydrogen. Does it matter whether we use kinetic or thermodynamic control conditions to form the enolate anion? Briefly justify your answer.

22.90   The aldehyde hydrogen of an aldehyde is not acidic, and reaction of an aldehyde with LDA removes the α-proton rather than the aldehyde hydrogen. Explain.

22.91   Draw the structure of all aldol condensation products for each of the following reactions:

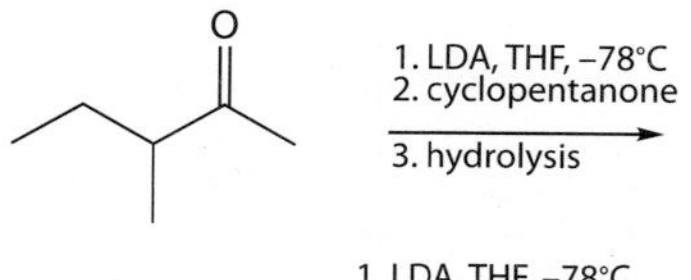

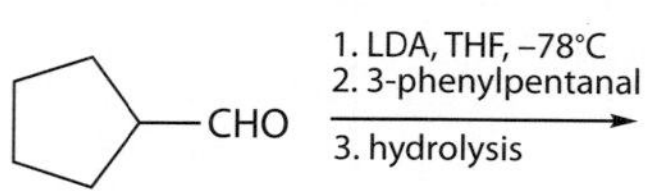

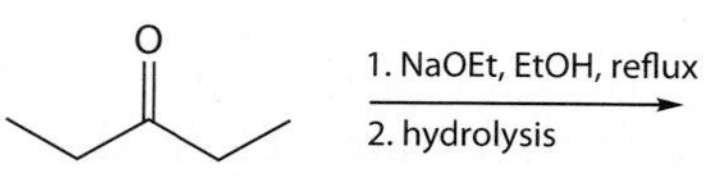

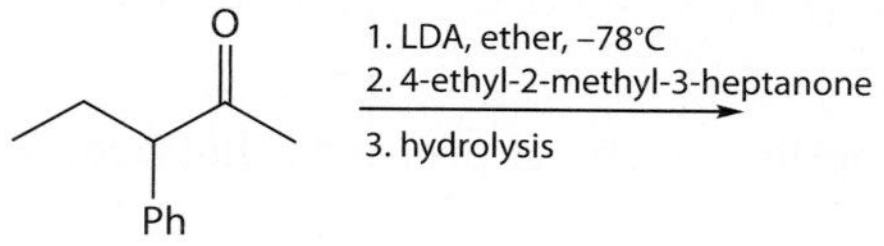

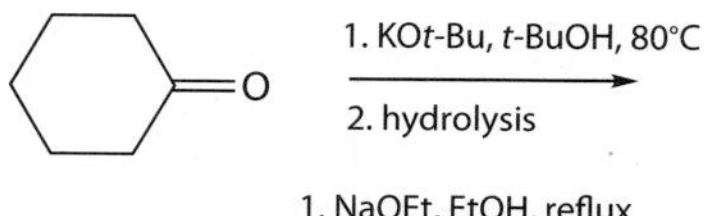

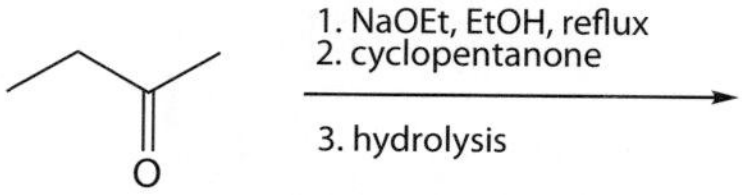

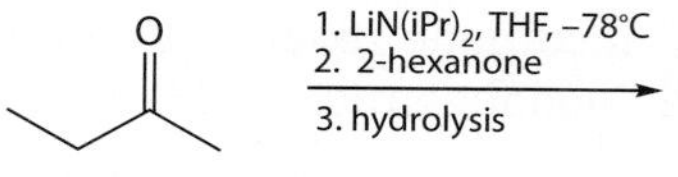

22.92   Briefly explain why the aldol condensation of acetophenone and benzaldehyde works when both reactants are mixed *together* in a flask with NaOEt in refluxing ethanol prior to hydrolysis. Specifically, why do we not use LDA in

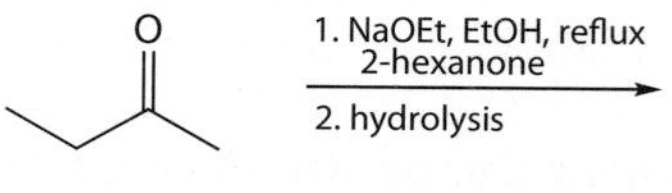

THF at low temperature to form the enolate independently before adding the other carbonyl compound?

22.93 Briefly discuss why the reaction of 3,5-diethyl-4-heptanone and lithium diisopropylamide generates very little enolate even after a few hours at −78°C.

22.94 You attempt a mixed aldol condensation reaction with pentanal as the enolate anion precursor and hexanal as the carbonyl reactant in the following way. You add pentanol to ethanol, add NaOEt, and bring the reaction to reflux. After 1 hour, you add hexanal and then reflux for an hour before cooling and hydrolyzing the solution. Your analysis shows that you have <5% of the mixed aldol. Why did you not form this product? Briefly discuss what product or products you do have. Suggest an alternative procedure that will improve your chances of obtaining the mixed aldol and draw the product.

22.95 Draw all aldol products formed from the following reactions:
(a) 2,5-hexanedione + 1. LDA (THF/−78°C); 2. hydrolysis
(b) 1,7-heptanedial + 1. NaOEt, EtOH, reflux; 2. hydrolysis
(c)

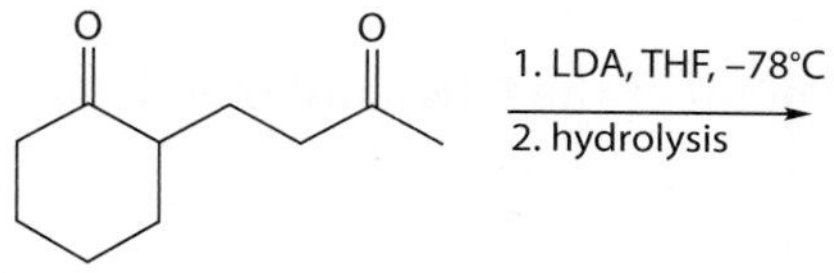

1. LDA, THF, −78°C
2. hydrolysis

(c)

(d) 1-phenyl-1-heptanone-7-carboxaldehyde + 1. LDA, THF, −78°C; 2. hydrolysis

22.96 Give the major product formed when 2,4,8-nonanetrione is treated with 1. LDA, THF, −78°C; 2. hydrolysis. Briefly explain your answer.

22.97 Give the major product of the following reaction and briefly explain your answer:

1. NaOEt, EtOH, reflux
2. hydrolysis

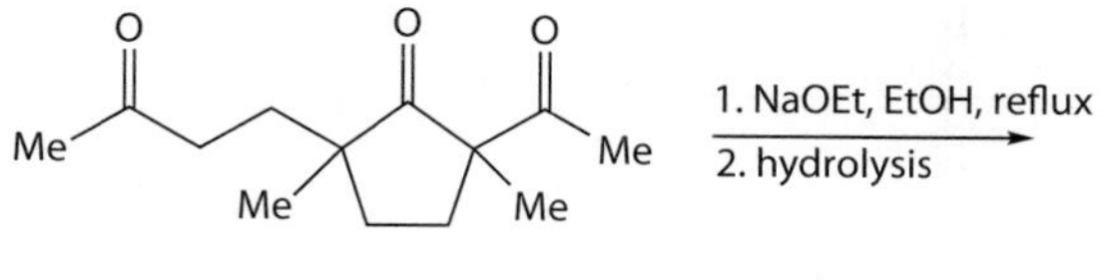

22.98 You attempt a Claisen condensation with ethyl butanoate by adding the ester to methanol containing NaOMe and you reflux the reaction before hydrolysis. Draw the normal Claisen product when ethyl butanoate is condensed with itself. Is this the product you obtain? Explain.

22.99 Give the major product for each of the following:
(a) diethyl malonate + 1. NaOEt, EtOH, reflux, benzaldehyde; 2. mild hydrolysis
(b) diethyl succinate + 1. LDA, THF, −78°C; 2. ethyl butanoate; 3. hydrolysis
(c) malonic acid + 1. oxalyl chloride; 2. MeOH, NEt$_3$; 3. LDA, THF, −78°C; 4. cyclopentanone; 5. hydrolysis
(d) glutaric acid + 1. excess EtOH, cat. H$^+$; 2. LiNEt$_2$, THF, −78°C; 3. methyl 4-phenylbutanoate; 4. hydrolysis
(e) adipic acid + 1. excess EtOH, cat. H$^+$; 2. NaOEt, EtOH, reflux; 3. hydrolysis

    (f)   cyclohexanone + 1. LDA, THF, –78°C; 2. 1 eq. diethyl succinate; 3. mild hydrolysis

    (g)   diethyl malonate + benzaldehyde + 1. NaOEt, EtOH, reflux; 2. hydrolysis and heat

22.100 Give the major product for the following reactions and explain your answer if there is more than one possibility:

(a)   1. LDA, THF, –78°C   2. mild hydrolysis

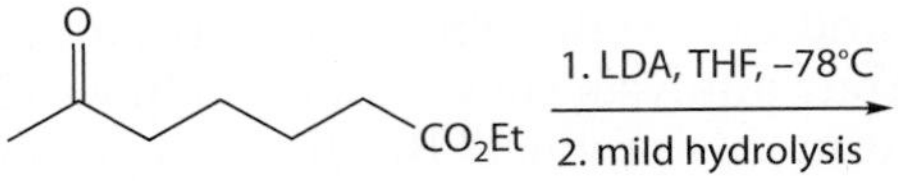

(b)   1. LDA, THF, –78°C   2. mild hydrolysis

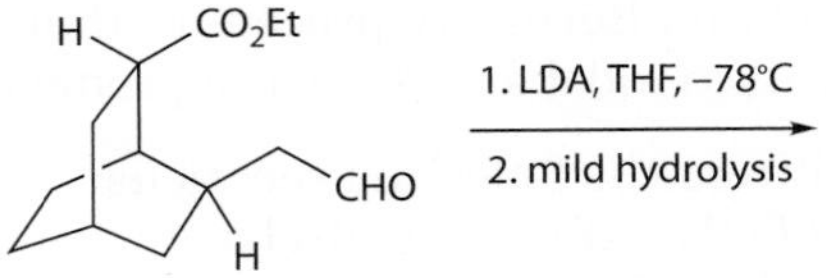

(c)   1. LDA, THF, –78°C   2. mild hydrolysis

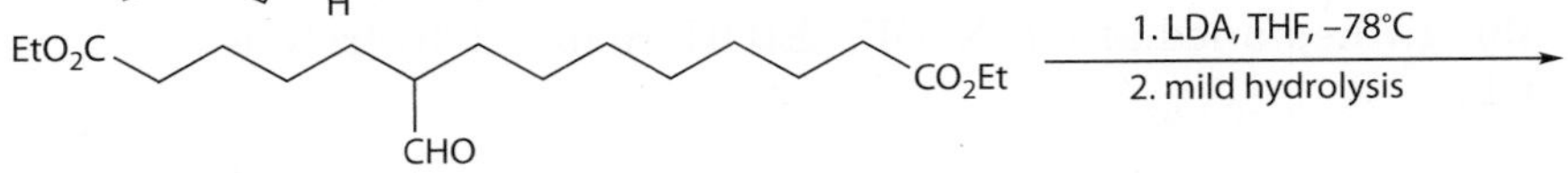

22.101 Give the products of all of the following reactions:

(a)   1. NaOEt, EtOH, reflux   2. hydrolysis

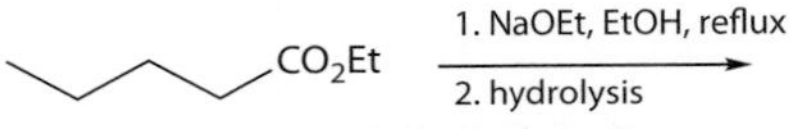

(b)   1. LDA, THF, –78°C   2. methyl pentanoate   3. hydrolysis

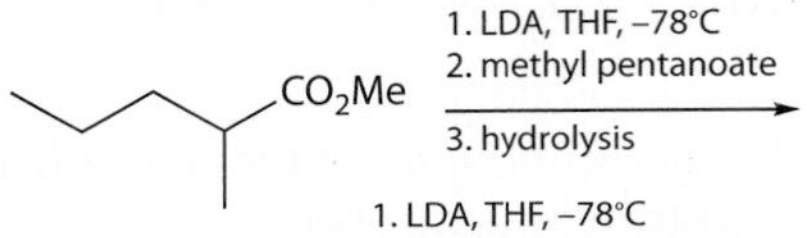

(c)   1. LDA, THF, –78°C   2. 2-methyl-4-octanone   3. hydrolysis

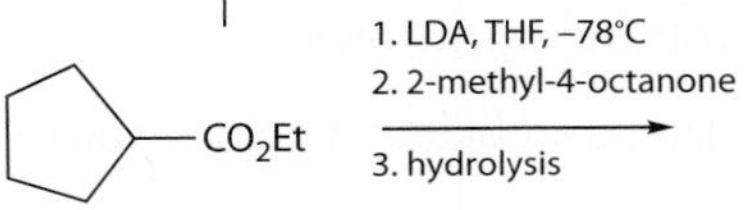

(d)   1. NaOEt, EtOH, reflux   Me₃C-CO₂Et   2. hydrolysis

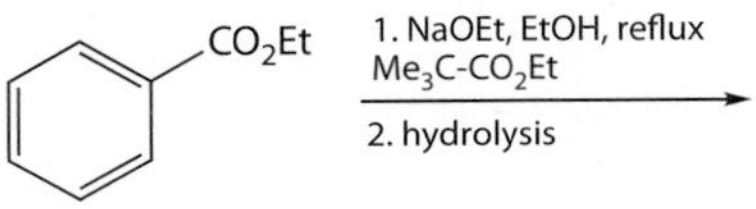

(e)   1. NaOEt, EtOH, reflux   ethyl 3-phenylpropanoate   2. hydrolysis

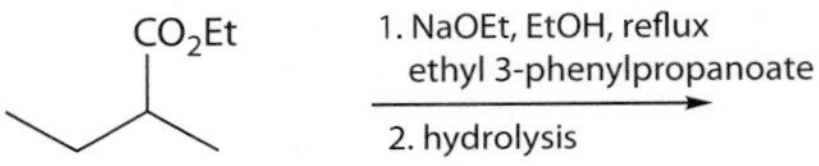

(f)   1. 2 eq. SOCl₂   2. excess MeOH, NEt₃   3. NaOMe, MeOH, reflux   4. hydrolysis

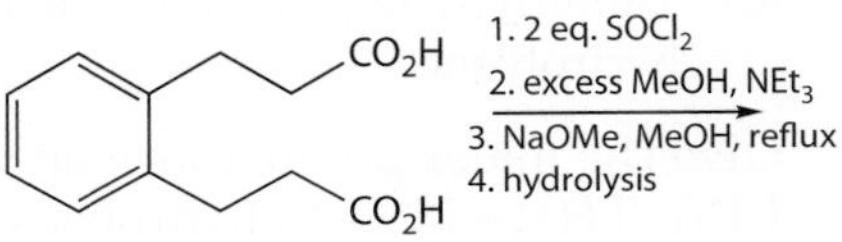

(g)   1. DCC, EtOH   2. LDA, THF, –78°C   3. ethyl cyclopentanecarboxylate   4. hydrolysis

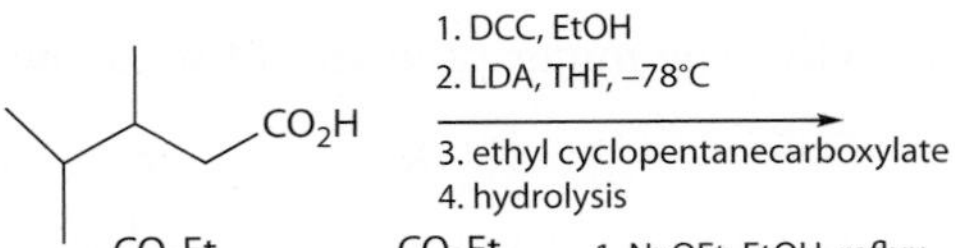

(h)   1. NaOEt, EtOH, reflux   2. hydrolysis

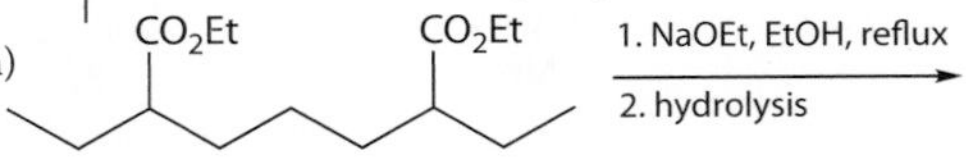

22.102 Give the major product of the following reactions:

(a)   1. LDA, THF, –78°C   2. ethyl 2-methylbutanoate   3. hydrolysis

(b)   1. NaOEt, EtOH, reflux, PhCHO   2. vigorous hydrolysis

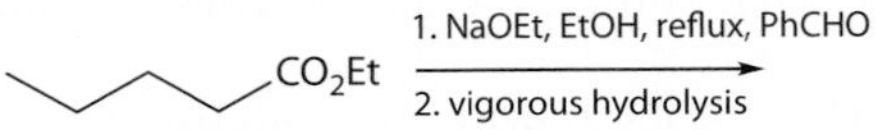

(c)   1. LDA, THF, –78°C   2. 2-pentanone   3. vigorous hydrolysis

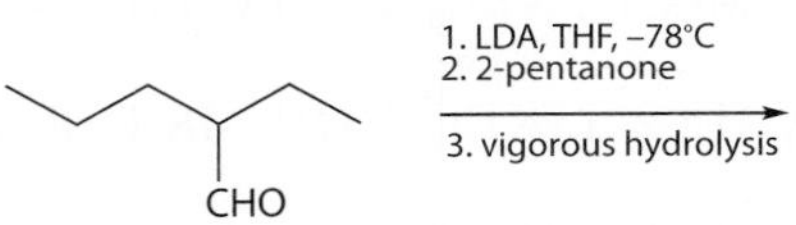

(d)   1. NaOEt, EtOH, reflux   2. vigorous hydrolysis

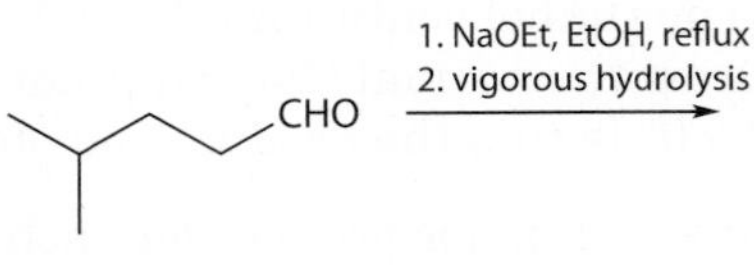

(e)   1. LDA, THF, –78°C   2. vigorous hydrolysis

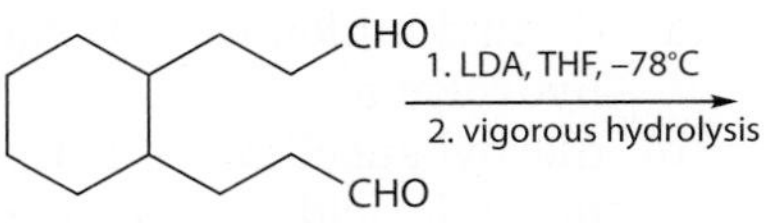

(f)   1. LDA, THF, –78°C   2. ethyl butanoate   3. 6N HCl

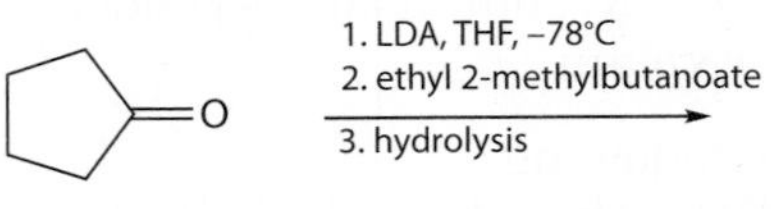

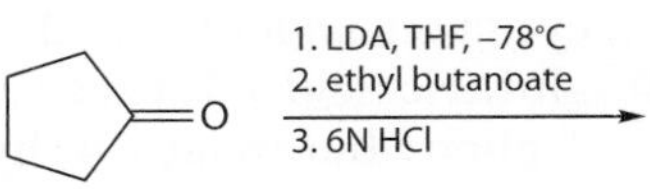

22.103 Draw the final product expected from each of the following reactions:
    (a) triphenylphosphine + 2-bromohexane
    (b) Me_3P + bromocyclobutane
    (c) Ph_3P + 1. bromoethane; 2. BuLi
    (d) Ph_3P + 1. 1-iodopentane; 2. BuLi; 3. acetone
    (e) tributylphosphine + iodomethane
    (f) Ph_3P + 3-iodopentane; 2. PhLi

22.104 In each case, give the final product of the reaction:
    (a) 4-phenyl-1-iodocyclohexane + 1. PPh_3; 2. BuLi; 3. cyclopentanone
    (b) cyclopentanecarbonitrile + 1. MeLi; 2. H_3O^+; 3. $Ph_3P=CMe_2$
    (c) 5-bromopentanoic acid + 1. $(COCl)_2$; 2. Bu_2Cd; 3. $C_5H_{11}CH=PPh_3$
    (d) 5-bromopentanoic acid + 1. $(COCl)_2$; 2. $EtOH/NEt_3$; 3. PPh_3
    (e) 1-iodobutane + 1. PPh_3; 2. BuLi; 3. aqueous acetone
    (f) butanoic acid + 1. excess MeLi; 2. hydrolysis; 3. $Ph_3P=CHCH_2CH_2CHPh$
    (g) $Ph_3P^+CH_2(CH_2)_5CH_2Br$ I^- + 1. BuLi; 2. 2-butanone; 3. PPh_3; 4. BuLi; 5. cyclobutanone

22.105 Draw the betaine and oxaphosphatane intermediates from the reaction of $Ph_3P=CMe_2$ and the following:
    (a) acetone
    (b) cyclopentanone
    (c) hexanal
    (d) cyclobutanecarboxaldehyde

22.106 It is known that dimethyl sulfide (MeSMe) reacts with 1-bromobutane to form a sulfonium salt. Draw it. It is also known that treatment of this salt with butyllithium leads to an ylid (draw it); however, when this ylid reacts with acetone, the product is an epoxide (1-butyl-2,2-dimethyloxirane) rather than an alkene. The coproduct of this reaction is dimethyl sulfide. Suggest a mechanism that accounts for formation of this epoxide and Me_2S from the ylid.

22.107 Explain why triphenylphosphine is a suitable reagent for a Wittig reaction, but diphenylphosphine is not.

22.108 Suggest a method that will generate 2,3-hexanedione from a suitable starting material. If this diketone is treated with LDA in THF and then with allyl bromide, what is the product?

22.109 Explain why heating 1,4-butanedioic acid derivatives does not lead to decarboxylation.

22.110 The so-called Doebner reaction condenses malonic acid and pyridine with an aldehyde such as benzaldehyde. Draw the product and suggest a mechanism for this reaction.

22.111 It is possible to decarboxylate benzoic acid, but the reaction requires much higher temperatures than are required for dicarboxylate malonic acid derivatives. Explain.

22.112 When dicyanomethane is heated with benzaldehyde in the presence of sodium ethoxide in refluxing ethanol, a new product is formed. Draw this product and suggest a mechanism for its formation.

22.113 Draw the product that is formed when 6,6-dicyano-2-hexanone is treated with LDA in THF at 0°C, and then subjected to an aqueous acid workup.

22.114 When the thermodynamic enolate of 2-methylcyclopentanone reacts with benzophenone, the reaction is sluggish, but assume that the aldol product is formed. Draw this product. Will heating this product lead to the alkene product? Now draw the product expected from the reaction of the kinetic enolate anion and benzophenone. Will heating this product lead to the alkene?

22.115 Dinitromethane reacts with sodium carbonate and then with butanal to give an alcohol after careful hydrolysis. Draw this product. Comment on why a base as weak as sodium carbonate can be used for this reaction. As part of your answer, draw all resonance contributors of the initially formed intermediate. Speculate as to whether or not the alcohol product is subject to facile dehydration.

22.116 When the thermodynamic enolate of 2-methyl-3-pentanone, formed with sodium ethoxide in ethanol, was treated with 2R-bromopentane, one product was the expected alkylated ketone. Draw this product. Two other products were isolated and identified as 2E-pentene and 2Z-pentene. Speculate on a mechanism that accounts for the formation of these two alkenes.

Synthesis Problems

Do not attempt these problems until Chapter 25 has been read and understood.

22.117 In each case, give the structure of the carbonyl derivative *and* a halide that would give the following alkenes via a Wittig reaction:
(a) 1-pentene
(b) 3-ethyl-3-octene
(c) 1-cyclopentyl-1-butene
(d) methylenecyclohexane
(e) 5-methyl-1-phenyl-2-heptene
(f) 2,3-diethyl-4-octene

22.118 In each case, give the final product:
(a) 4-methyl-1-pentene + 1. Hg(OAc)$_2$/H$_2$O, then NaBH$_4$; 2. Na$_2$Cr$_2$O7; 3. LDA, THF, –78°C; 4. MeI
(b) Cycloheptene + 1. OsO$_4$/NMO; 2. HIO$_4$; 3. LDA, THF, –78°C; 4. aq. acid workup
(c) Cycloheptene + 1. O$_3$; 2. H$_2$O$_2$; 3. H$^+$, EtOH; 4. NaOEt, EtOH, reflux; 5. aq. acid workup
(d) 2-methylcyclohexanone + 1. NaOEt, EtOH, reflux; 2. benzyl bromide

22.119 Propose a reasonable synthesis for each of the following from the designated starting material:

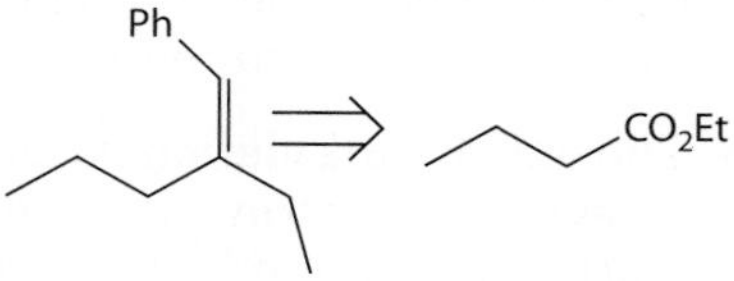

Spectroscopy Problems

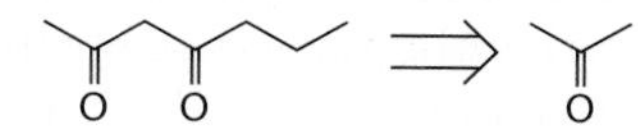

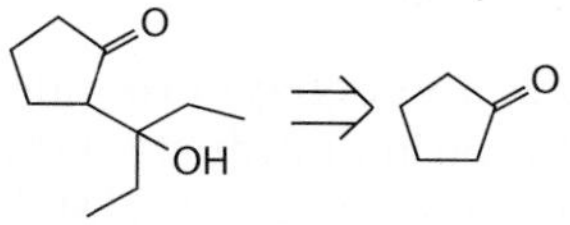

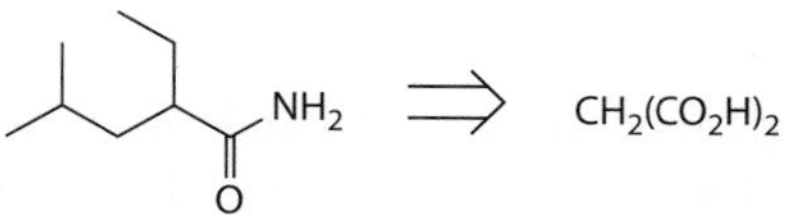

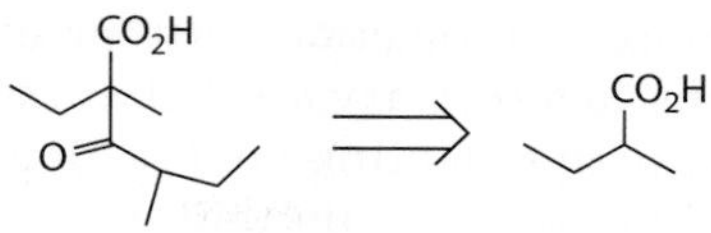

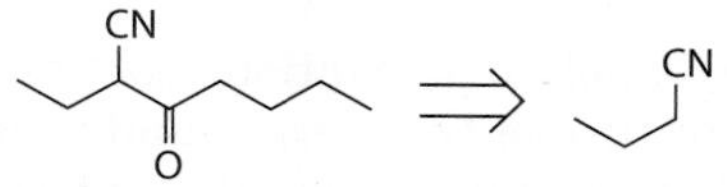

Do not attempt these problems until Chapter 14 has been read and understood.

22.120 Briefly discuss any differences in the IR spectrum or proton NMR spectrum that would allow you to distinguish between a methyl ester and an ethyl ester.

22.121 Briefly discuss any differences in the IR spectrum or proton NMR spectrum that would allow you to distinguish between a keto-ester and a diketone (assume a methyl ester).

22.122 After performing the reaction of ethylmagnesium bromide with butanenitrile, followed by vigorous acid hydrolysis, you analyze your product. Describe differences in the IR spectrum or proton NMR spectrum that would allow you to decide whether your product was an imine, a nitrile, or a ketone.

22.123 After you generate the enolate of 2-butanone with LDA in THF and react that enolate anion with cyclopentanone, you perform a mild hydrolysis step. You obtain the product mixture and you see a relatively sharp, medium-sized peak in the IR spectrum at about 3340 cm^{-1}. What molecule from this reaction might give rise to this peak?

22.124 When you react diethyl malonate with LDA and then with benzaldehyde, you obtain a product after treatment with dilute aqueous HCl. There are two possible products from this reaction. The IR spectrum shows no signal at about 2600 cm^{-1} and a strong signal at about 1670 cm^{-1}. What does this information tell you about the structure of the product?

22.125 You have done a Claisen condensation but are concerned that the subsequent acid hydrolysis was too vigorous and decarboxylation may have occurred. What characteristics in the IR spectrum and proton NMR spectrum will allow you to decide whether your product is a keto-ester or a keto-acid? Discuss differences in the carbon NMR spectra.

22.126 A Grignard reaction between 3-pentanone and phenylmagnesium bromide leads to a tertiary alcohol after hydrolysis. Draw the structure of this product and predict features of the IR spectrum, proton and carbon NMR spectra, and mass spectrum that will allow you to identify it.

22.127 An enolate alkylation reaction with 2-butanone and benzyl bromide gave a mixture of the kinetic and thermodynamic alkylation products. Draw both of them and describe differences in the IR spectrum and proton NMR spectrum that will allow you to distinguish them.

22.128 The enolate alkylation reaction of cyclopentanone and 2-iodopropane leads to a new ketone. Draw it and discuss differences in the IR spectrum and proton NMR spectrum of both starting material and product that will allow you to identify each.

22.129 When you reflux 2,5-octanedione, one possible product is an intramolecular aldol condensation. It is also possible to form a different product via an intermolecular aldol condensation. Discuss differences in the IR spectra and proton NMR spectra that would allow you to decide which of these two products you have formed.

22.130 Give the structure for a molecule with the formula $C_6H_{10}O_2$ and the spectra data given:
IR: 3002–2912, 1715, 1409–1401, 1382, 1160 cm^{-1}
^{1}H NMR: 2.71 (s, 4H), 2.19 (s, 6H) ppm
^{13}C NMR: 207.7, 35.7, 29.5 ppm

22.131 Give the structure for a molecule with the formula C_6H_{10} and the spectra data given:
IR: 3307, 2958–2840, 1657, 1433, 876 cm^{-1}
^{1}H NMR: 4.92 (d, 2H), 1.96 (broad t, 4H), 1.33 (broad t, 4H) ppm.
^{13}C NMR: 152.8, 107.9, 33.3, 26.3 ppm

22.132 Give the structure for a molecule with the formula $C_7H_{12}O_3$ and the spectra data given:
IR: 2984, 2909, 1745, 1716, 1647, 1629, 1313, 1250, 1160, 1028 cm^{-1}
^{1}H NMR: 4.2 (q, 2H), 3.46 (s, 2H), 2.58 (q, 2H), 1.28 (t, 3H), 1.08 (t, 3H) ppm
^{13}C NMR: 206.4, 168.1, 61.0, 48.6, 31.3, 14.1, 7.6 ppm

22.133 Give the structure for a molecule with the formula $C_6H_{12}O_2$ and the spectra data given:
IR: broad peak at 2876, 2936, 1717, 1471, 1361–1310, 1262–1183, 955, 916 cm^{-1}
^{1}H NMR: 3.81 (broad s, 1H; this peak is diminished when treated with D_2O), 2.64 (s, 2H), 2.18 (s, 3H), 1.26 (s, 6H) ppm
^{13}C NMR: 211.0, 71.1, 60.6, 30.4, 29.3 ppm

22.134 Molecule **A** has the formula C_5H_{10}. It exhibits a sharp peak in the IR at 1640 cm^{-1}, and the proton NMR shows peaks at 5.70 (m, 1H), 5.07 (m, 1H), 5.02 (m, 1H), 2.52 (m, 1H), and 1.06 (d, 6H) ppm. The reaction of **A** with (1) mercuric acetate in water followed by (2) sodium borohydride in ethanol and then (3) hydrolysis gives **B**. The reaction of **B** with pyridinium dichromate in dichloromethane gives **C**, which exhibits a strong peak in the IR at 1716 cm^{-1}. The ^{1}NMR is 1.11 (d, 6H), 2.141 (s, 3H), and 2.58 (m, 1H) ppm. The reaction of **C** with lithium diisiopropylamide in THF at −78°C gives an intermediate that reacts with propanal to give a product **D**. When **D** is treated with dilute aqueous acid at ambient temperatures, **E** is formed. Compound **E** has the formula $C_8H_{16}O_2$ and shows strong peaks at 3300 and 1725 cm^{-1} in the IR. It also shows peaks in the proton NMR: 0.90 (t, 3H), 1.06 (d, 6H), 1.48 (m, 2H), 2.48–2.73 (m, 4H), 3.44 (m, 1H), and 3.58 (broad s, 1H; this peak is diminished when treated with D_2O) ppm. ^{13}C NMR is 216.6, 68.8, 47.3, 41.3, 30.1, 17.6, and 9.5 ppm. Identify **A**, **B**, **C**, **D**, and **E**.

22.135 Compound **A** is refluxed in ethanol in the presence of an acid catalyst to give **B**. Compound **B** has the formula $C_7H_{12}O_4$ and shows peaks at about 1750 and at 1037 cm^{-1} in the IR; the proton NMR shows peaks at 4.21 (q, 4H), 1.29 (t, 6H), and 3.36 (s, 2H). When **B** is treated with sodium ethoxide in hot ethanol and then with 2-bromopropane, compound **C** is prepared. When **C** is heated with aqueous acid, compound **D** is isolated. When **D** is heated (neat) to 250°C, an oily compound with a sharp smell, **E**, is obtained as the final product. Compound **E** has the formula $C_5H_{10}O_2$ and shows a broad signal in the IR at 2964–2676, a strong peak at 1709, and a peak at 953 cm^{-1}. The proton NMR shows peaks at 11.9 (broad s, 1H; this peak is diminished when treated with D_2O), 2.36–2.15 (d, 2H), 1.90 (m 1H), 1.40–1.26 (m, 2H), 0.97 (d, 3H), and 0.91 (t, 3H) ppm. ^{13}C NMR is 178.9, 41.1, 32.3, 30.2, 20.1, and 11.3 ppm. Identify the structures of **A**, **B**, **C**, **D**, and **E**.

Difunctional Molecules

Dienes and Conjugated Carbonyl Compounds

23

Most of the molecules discussed in previous chapters have only one functional group, for the most part, although nomenclature rules include a handful of molecules with two or more functional groups. Most of the reactions discussed thus far show the transformation of one functional group into another or the formation of a carbon–carbon bond using molecules each of which contains a single functional group. The products of such reactions, however, often have two or more functional groups (diketone, ketone-ester, ketone-alcohol, ketone-alkene, etc.). The conjugated ketone and aldehyde products that result from dehydration in the aldol reaction (Chapter 22, Section 22.2) are one example.

Many molecules have two functional groups; sometimes there are two identical groups (diols, dienes, diketones, etc.) and sometimes the groups are different. The presence of two functional groups in one molecule leads to significant differences in common chemical reactions when compared to monofunctional compounds. Sometimes, one group influences the course of the reaction at the other functional group. At other times, one functional group prevents a reaction at the other functional group. This chapter will introduce the structures, properties, and reactions of some difunctional molecules (two functional groups). Specifically, it will discuss the chemistry of dienes (two C=C units in one molecule) and conjugated carbonyl compounds (a C=C unit connected to the carbon of a C=O unit).

To begin this chapter, you should know the following:

- **fundamental structure and nomenclature of alkenes and dienes (Chapter 5, Sections 5.1 and 5.3)**
- **fundamental structure and nomenclature of alkynes and diynes (Chapter 5, Sections 5.2 and 5.3)**
- **fundamental structure and nomenclature of ketones or aldehydes that also have an alkene unit in the molecule (Chapter 5, Section 5.9; Chapter 16, Section 16.2)**
- **concept of a molecular orbital (Chapter 3, Section 3.4)**
- **that the π-bond of an alkene can react as a Brønsted–Lowry or a Lewis base (Chapter 10, Sections 10.1 and 10.4)**
- **that an alkyne can react as a Brønsted–Lowry or a Lewis base (Chapter 10, Section 10.6)**
- **acyl addition reactions of aldehydes and ketones (Chapter 16, Section 16.3; Chapter 18, Sections 18.1, 18.3 and 18.4, 18.6 and 18.7)**
- **formation of Grignard reagents and organolithium reagents (Chapter 15, Sections 15.1 and 15.2, 15.5)**
- **structure and reactivity of organocuprates (Chapter 15, Section 15.6)**
- **structure and characteristics of carbocations (Chapter 7, Section 7.4.1; Chapter 10, Section 10.2)**
- **factors that contribute to acidity (Chapter 6, Section 6.3)**
- **factors that contribute to basicity (Chapter 6, Section 6.4)**
- **reversible reactions and reaction energetics (Chapter 7, Sections 7.5 and 7.6, 7.10)**
- **fundamentals of kinetic and thermodynamic control (Chapter 22, Section 22.4.2)**
- **concept of resonance and resonance stability, and how to draw resonance contributors (Chapter 5, Section 5.9.3; Chapter 18, Section 18.6; Chapter 21, Section 21.3; Chapter 22, Section 22.2)**
- **rate of reaction (Chapter 7, Section 7.11)**
- ***E*- and *Z*-stereoisomers (Chapter 9, Section 9.4)**
- **how to identify polymers (Chapter 10, Section 10.8.3)**

This chapter discusses dienes and alkene-ketones, alkene-aldehydes, or alkene-esters with a particular emphasis on those molecules that are conjugated. In conjugated molecules, the π-bonds are directly connected with no intervening sp^3 atoms. Conjugated dienes react similarly to other alkenes; however, due to conjugation, the intermediates formed from conjugated dienes are resonance stabilized. Such resonance-stabilized intermediates lead to differences in product distribution when compared to simple alkenes. An alkene unit may also be conjugated to ketones or aldehydes, and conjugated carbonyl compounds exhibit reactions different from those of nonconjugated ketones or aldehydes. Conjugated systems interact with ultraviolet (UV) light in ways that are different from interactions of nonconjugated compounds, and the UV spectra are used to identify conjugated compounds.

When you have completed this chapter, you should understand the following points:

- **Conjugated dienes are molecules that contain two C=C units connected together such that there is a continuous array of sp^2 carbons (C=C–C=C). Conjugate carbonyl compounds have a C=C unit directly attached to the carbon of a carbonyl (C=C–C=O). There are conjugated ketones, aldehydes, acids, and acid derivatives. When a C=C or a C=O unit is attached to a benzene ring, these are also conjugated alkenes or carbonyl derivatives.**
- **Conjugated dienes exhibit rotation about the C–C single bond of the C=C–C=C unit, generating two major rotamers—the cisoid and the transoid rotamers.**
- **When a molecule absorbs UV light, the most intense absorption peak (wavelength in nanometers) is labeled λ_{max} and the intensity of the absorption is given by the extinction coefficient, ε. Conjugated compounds absorb UV light at lower energy than nonconjugated compounds do because of the smaller energy gap between filled and unfilled molecular orbitals. The differences in UV spectra between conjugated and nonconjugated compounds allow them to be easily distinguished via UV spectroscopy.**
- **A set of empirical rules based on butadiene and acrolein allows prediction of λ_{max} for a compound.**
- **Dienes react with HX (HCl, HBr, HI) to give 1,2-addition and 1,4-addition products via formation of an allylic cation. The major product is determined by the temperature of the reaction, where low temperature favors 1,2-addition and high temperature favors 1,4-addition. Halogens such as Cl_2, Br_2, and I_2 also add to dienes to give a mixture of 1,2-and 1,4-addition products. Both HX and X_2 react with alkenes with the C=C unit conjugated to a benzene ring in essentially the same manner as with alkenes that do not contain a benzene ring.**
- **Dienes can be polymerized under radical and cationic conditions to give useful polymers such as rubber (polyisoprene).**
- **Conjugated carbonyl compounds react with HX to give 3-substituted derivatives (the X adds to the end of the C=C unit) via the more stable cation. Michael addition involves addition of a nucleophile to the end of the C=C unit of a conjugated carbonyl, forming an enoate anion. Grignard reagents and carbanionic nucleophiles tend to add to the C=O unit of conjugated aldehydes**

and of unhindered conjugated ketones. Organocuprates usually give conjugate addition to both conjugated aldehydes and ketones.

- The reaction of a ketone enolate with a conjugated ketone under thermodynamic conditions gives a cyclic product, in what is called the Robinson annulation.
- Conjugated carbonyl compounds can give primarily 1,2 reduction or 1,4 reduction, depending on the reagent.
- A molecule with a particular functional group can be prepared from molecules containing different functional groups by a series of chemical steps (reactions). This process is called synthesis: The new molecule is synthesized from the old one (see Chapter 25).
- Spectroscopy can be used to determine the structure of a particular molecule and can distinguish the structure and functionality of one molecule compared with another (see Chapter 14).
- Conjugation appears in biological systems.

23.1 Conjugated Dienes

A molecule containing one π-bond as part of a C=C unit is called an alkene (see Chapter 5, Section 5.1). When a molecule contains two π-bonds in two C=C units, it is called a diene. There are three fundamental structural variations for a diene: (*a*) those where the C=C units are separated by sp³ hybridized atoms, (*b*) those where the C=C units are connected together to form a C=C–C=C unit, and (*c*) those that contain two π-bonds that share an sp hybridized atom. Compounds (*a*) are called nonconjugated dienes; using the standard nomenclature rules for alkenes introduced in Chapter 5, an example is 1,5-hexadiene (**1**). Compounds (*b*) are called *conjugated* dienes, illustrated by 1,3-hexadiene (**2**). Note that **2** contains an *E* double bond (see Chapter 9, Section 9.4.3) and the name must reflect this structural feature. Therefore, **2** is 1,3*E*-hexadiene.

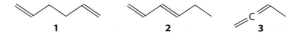

Compounds (*c*) are called allenes; an example is **3**, named 1,2-butadiene. Allene **3** is an example of a ***cumulative*** π-system. Benzene (Chapter 5, Section 5.10; Chapter 21) is a conjugated system, but the conjugated π-bonds in a six-membered ring lead to the aromatic stability of benzene (see Chapter 21, Section 21.1). For consistency with the chemistry presented throughout the book, this chapter will ignore allenes and focus attention on conjugated dienes.

23.1 Draw 2*E*,4*Z*-heptadiene.

23.2 Draw allene itself, which is 1,2-propadiene.

The nomenclature and identification of conjugated and nonconjugated dienes is straightforward based on concepts learned previously. The chemical reactions of such compounds are more complicated. Chapter 10 discussed several chemical reactions of simple alkenes that are primarily acid–base reactions and generate carbocations or other cation intermediates. In a nonconjugated diene, each π-bond behaves more or less independently, so **1** is expected to react with one molar equivalent of HBr to form an alkyl bromide that has a C=C unit elsewhere in the molecule. A reagent will react with one C=C unit without affecting the other or without the other C=C unit influencing the reaction.

The situation is quite different for a conjugated diene. Because the π-bonds are linked together, a reaction at one C=C unit is influenced by the presence of the other C=C unit. For this reason, these compounds have been separated into a different chapter.

23.3 Write out the reaction between 1 and one molar equivalent of HBr, including the intermediate carbocation and the product. What is the product if two molar equivalents of HBr are used?

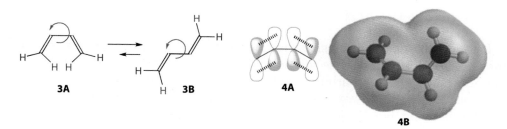

Before discussing the formal reactions of conjugated dienes, it is useful to analyze the diene unit (C=C–C=C) in a typical conjugated diene such as 1,3-butadiene (**3**). The first observation is that a carbon–carbon single bond connects the two C=C units. The bond length of the C–C unit is 1.48 Å (148 pm).[1,2] In a typical alkane, the C–C bond distance is about 1.54 Å (154 pm),[1] so the C–C unit in butadiene is shorter. The C=C bond distance in butadiene is 1.34 Å (134 pm),[2] which compares favorably with the 1.32 Å (132 pm) bond distance of the C=C unit in ethene[1] (a typical C=C unit has a bond distance of about 1.337 Å, 133.7 pm).[1] The shorter bond distance of the C–C unit in **3** is explained by structure **4A**, which is a different representation of butadiene that shows the p-orbitals of the π-bonds.

The p-orbitals on the "middle" carbons (the C–C unit) are relatively close, and it is reasonable that some electron density is shared. For the most part, the electrons are localized in the π-bonds of the C=C units, but there is a small amount of overlap that makes the bond shorter than expected. The electron-potential map for the molecular model of butadiene is shown in **4B**, where the **red** areas indicate high concentrations of electron density. The highest electron density is localized on the π-bonds, between C1–C2 and C3–C4 rather than between the single bond C3–C4, consistent with this analysis. Note the presence of some electron density between C3 and C4 in **4B**, however. This does not indicate resonance but rather an interaction of the π-bonds, which is consistent with a slightly shorter bond distance. Make no mistake, however: C3–C4 is a single covalent bond.

Although there is some overlap of the p-orbitals between C2 and C3, the C–C unit is a single covalent bond and there is rotation about that bond (see Chapter 8, Section 8.1). Indeed, two rotamers are in equilibrium that result from the high-energy and the low-energy interactions of the C=C units. When the C=C units have an eclipsed relationship (as in **3A**), butadiene is said to be in a ***cisoid conformation*** or ***s-cis* conformation**. When the two C=C units have a staggered relationship (as in **3B**), butadiene is said to be in a ***transoid conformation*** or ***s-trans* conformation**. The steric interaction of the hydrogen atoms (marked in **green** in **3A**) make rotamer **3A** higher in energy, which means there is a higher percentage of **3B** at equilibrium. Therefore, 1,3-butadiene (**3**) spends most of its time in the *s-trans* conformation.

23.4 Draw the cisoid (*s-cis*) and transoid (*s-trans*) conformations of 2*E*,4*E*-hexadiene and of 2*E*,4*E*-hexadiene; comment on which diene has the higher percentage of the transoid conformation.

It is important to emphasize that 1,3-butadiene is **not** resonance stabilized; that is, *there is **no** resonance delocalization in 1,3-butadiene*. Although there is some interaction of the p-orbitals on C2 and C3 (as discussed for **4B**), there is **no** resonance. A colloquial way to express this states that one C=C unit knows what the other is doing, but the π-bonds remain localized on each individual C=C unit.

Just as it is possible to have two C=C units conjugated to form a diene, two C≡C units may be conjugated to form a diyne. Conjugated diynes are not as common as conjugated dienes. It is also possible to have an ene-yne, with a C=C unit as well as a C≡C unit in the same molecule. For the most part, diynes and ene-ynes will be ignored in this book, and the focus will be on the chemistry of dienes.

5   6   7

In addition to the acyclic compounds just discussed, it is common to see cyclic molecules with conjugated π-bonds. An important cyclic diene is cyclopentadiene (**5**). The numbers are omitted from the name because there is only one possible structure. This is not the case for cyclohexadiene derivatives because there is the conjugated 1,3-cyclohexadiene (**6**) and the nonconjugated 1,4-cyclohexadiene, **7**. With all dienes that have larger rings, there is one conjugated isomer but there may be several possible nonconjugated isomers.

23.5 Draw **3,3,5-trimethylhex-5-en-1-yne** and **1-chloro-6-methyl-hept-1Z-en-3-yne.**

23.6 Draw all cyclooctadienes and label the conjugated compound.

23.2 Conjugated Carbonyl Compounds

Dienes are a well-known class of compounds, but as pointed out in the last section, diynes and ene-ynes are less common. By contrast, many molecules have a C=C unit as well as a C=O unit (a carbonyl). Nonconjugated molecules of this type include the ketone hex-5-ene-2-one (**8**) and the ester ethyl pent-4-enoate (**9**). For all practical purposes, the C=C unit and the C=O units in nonconjugated compounds behave independently of each other. This is not always true, but it is a good working assumption in most cases. Hydrogenation of **8** using a palladium catalyst, for example, generally gives the major product by reduction of the C=C unit, whereas switching to a PtO_2 catalyst gives reduction of the C=O unit as the major process.

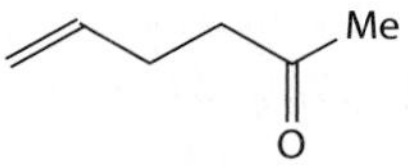

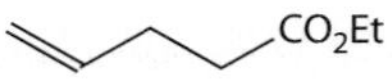

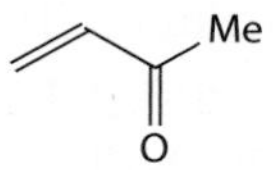

Although it is possible to reduce one or the other π-bond selectively, reduction of *both* groups is common unless care is taken. It is useful to *assume* first that the C=C and C=O units in a nonconjugated molecule behave independently in a chemical reaction, but always remember that, many times, this is not true, so care should be exercised.

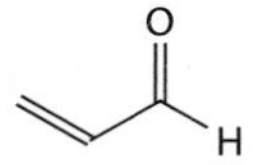

23.7 Draw (a) **5-phenylhex-3E-en-2-one,** (b) **methyl 2,2-diethylhept-4E-enoate,** and (c) **4-methylpent-2-ynal.**

23.8 Draw the two different hydrogenation products from 8.

Molecules that have a C=C unit and a C=O unit in a molecule but are not conjugated are common, but they are not the focus of this chapter. The C=C and C=O units are connected directly in conjugated carbonyl compounds. Such carbonyl compounds are often known as α,β-unsaturated carbonyl compounds (α,β-unsaturated ketones, α,β-unsaturated aldehydes, α,β-unsaturated esters, etc.). Typical examples include 3-buten-2-one (the common name is methyl vinyl ketone, **10**) and 2-propenal (the common name is acrolein, **11**). In these molecules, the presence of the carbonyl will influence reactions at the C=C unit and the presence of the alkene will influence reactions at the C=O unit. In other words, it is difficult to do a chemical reaction at one functional group without the other reacting or influencing the course of the reaction. In **10**, the bond distance for the C–C unit is about 1.44 Å (144 pm),[1] which is shorter than a typical C–C bond; the bond length of the C=O unit is about 1.36 Å (136 pm),[1] which is longer than the C=O bond distance of 1.145 Å (114.5 pm) found in a typical ketone.[1] This information suggests that *there is some overlap of the p-orbitals of C1 (the carbonyl carbon) and C2, but as with butadiene, there is no resonance.* Cyclic molecules may have a conjugated carbonyl unit within the ring.

23.9 Draw (a) 1-phenylhex-2Z-en-1-one, (b) 3-ethylpent-2-enal, and (c) 2,3,4-trimethylhex-4-en-3-one.

23.10 Draw the structures of cyclopent-2-en-1-one, cyclohex-2-en-1-one, cyclohex-3-en-1-one, 1-cyclohexene-1-carboxaldehyde, and 3-cyclohexene-1-carboxaldehyde.

23.11 Draw (a) 1-acetyl-1-cyclobutene, (b) 2,4-dimethylcyclohept-4-enone, and (c) cyclopent-3-en-1-carboxaldehyde.

Several difunctional compounds involve a C=C unit and the carbonyl unit of a carboxylic acid (or an acid chloride), ester, or amide. Typical conjugated carbonyl compounds are 2-propenoic acid (**12**; acrylic acid), but-2E-enoic acid (**13**; also known as crotonic acid), and 2-methylpropenoic acid (**14**; also known as methacrylic acid). There are cyclic derivatives of the carboxylic acids in which the carboxyl unit is attached to the ring. A C=C unit or a C=O unit is considered to be conjugated if it is connected to an aromatic ring. Ethenylbenzne (styrene, **15**) is a conjugated alkene and both benzaldehyde (**16**) and acetophenone (**17**) are conjugated.

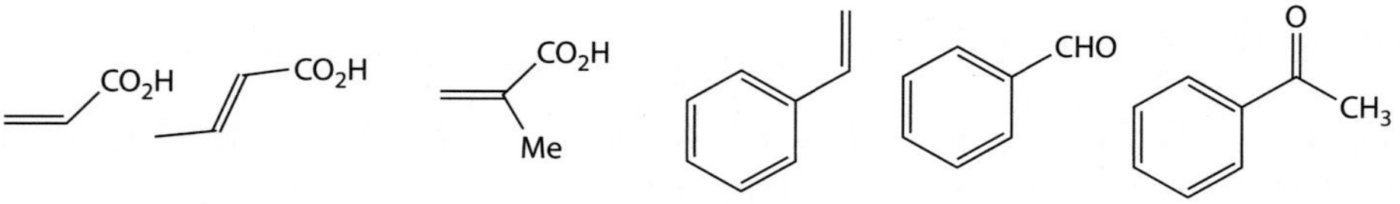

| 12 | 13 | 14 | 15 | 16 | 17 |

23.12 Draw (a) 2,3-diphenylhex-2Z-enoic acid, (b) 3-methylbut-2-enoyl chloride, and (c) methyl 4-phenylbut-2E-enoate.

23.13 Draw (a) ethyl 7,7-dimethylcyclopent-1-en-1-carboxylate, (b) cyclobut-2-en-1-carboxamide, and (c) 2-phenylcyclohex-1-en-1-oyl chloride.

23.14 Draw (a) anthracene-9-carboxaldehyde, (b) 3-bromo-4-ethylbenzaldehyde, and (c) 5-phenyl-3-hexanone. Label all nonconjugated molecules.

23.3 Detecting Conjugation: Ultraviolet Spectroscopy

Sections 23.1 and 23.2 showed many difunctional molecules containing C=C, C≡C, or C=O units; some are conjugated and others are not. Experimentally, a molecule may be identified as conjugated by shining ultraviolet light on it and observing how it interacts with that light. This section will introduce ultraviolet spectroscopy and how it is used to detect the presence of conjugation. This is another example of using a spectroscopic method to determine the identity of a molecule that will complement the methods described in Chapter 14.

23.3.1 UV Light and Ultraviolet Spectrophotometers

Humans see light that is part of the so-called visible spectrum, but it makes up only a tiny portion of the complete electromagnetic spectrum, which ranges from very low energy long waves called radio waves to extraordinarily high energy and short waves called cosmic rays. Figure 23.1 shows an abbreviated version of the electromagnetic spectrum that ranges from the ultraviolet (UV) to the infrared (IR); the low- and high-energy waves are labeled. The visible spectrum is more or less in the middle with respect to energy, and both the lower energy red and the higher energy violet lights are marked. Energy is expressed using several different units, including angstroms (Å), reciprocal centimeters (cm^{-1}), nanometers (nm), kilocalories (kcal), and kilojoules (kJ).

Infrared light is lower in energy than the visible spectrum, and the ultraviolet is higher in energy than the visible. Because it is higher in energy, ultraviolet light is capable of inducing chemical reactions, whereas infrared light is so low in energy that it generally does not. Absorption of infrared light usually causes molecules to vibrate, which is reflected by an increase in heat (see Chapter 14, Section 14.3). The effect of ultraviolet light will be examined when it interacts with molecules containing a C=C and a C=O unit.

The instrument used to measure the effect of ultraviolet light on a molecule is called an **ultraviolet spectrophotometer**. A stylized diagram for a typical instrument is shown in Figure 23.2.[3] There is an ultraviolet light source and a series of prisms and mirrors to split the light into two equal components. One beam is directed through a chamber that holds a clear vessel containing the molecule of interest (the sample). That sample is dissolved in a solvent that

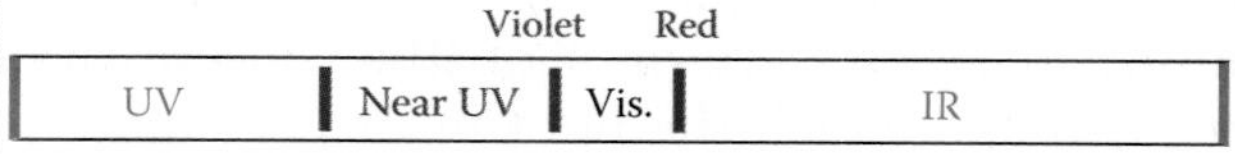

		Violet	Red			
UV		Near UV	Vis.		IR	
10	200	400	800	2860	28600	ν (mμ = nm)
2860	143	71.5	35.75	10	1	E (kcal mol^{-1})
11972	598.6	299.3	149.7	41.86	4.186	E (KJ mol^{-1})
100	2000	4000	8000	2.86×10^4	2.86×10^5	λ (Å)
1000×10^3	50×10^3	25×10^3	12.5×10^3	3530	353	ν (cm^{-1})

Figure 23.1 A portion of the electromagnetic spectrum.

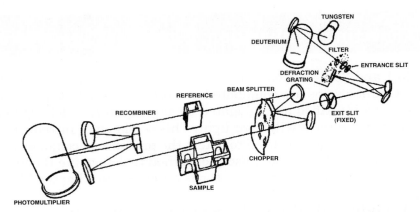

Figure 23.2 Diagram of UV spectrophotometer. Altemose, I. R. 1986. *Journal of Chemical Education,* (From 63(1), A216–222. Division of Chemical Education, Inc. With permission.)

does not absorb UV light in the region where the molecule of interest absorbs UV light. Typical solvents include water, ethanol, methanol, etc., which do not have conjugated π-bonds. The other beam is directed through a chamber that does not contain the sample and it is used as a reference.

After passage through the sample, the beams are recombined and analyzed to see whether light has been absorbed by the sample. The wavelength of UV light can be varied in the instrument during an analysis so that the detection system displays a spectrum that plots wavelength versus absorption that shows how much energy was absorbed for a given wavelength of UV light. This is the **ultraviolet spectrum** for that particular molecule. When light is absorbed, it registers as a "peak" at that wavelength, and the intensity of that peak is a measure of how much light is absorbed. *The intensity of the absorption is expressed as the extinction coefficient (ε) and the larger the extinction coefficient is, the stronger the signal will be.* The wavelength with the largest extinction coefficient (the largest peak) is referred to as λ_{max}. Both the wavelength of light absorbed and the extinction coefficient are measured by the UV spectrophotometer. A typical ultraviolet spectrum is illustrated in Figure 23.3.[4]

For a given wavelength of light, the amount of light that passes through the molecule is measured by the intensity (I) of that light. This is the amount of light transmitted through the solution (the transmittance). The amount of light used as a reference is labeled I_o, and the amount of light transmitted through the solution is I_o/I. The amount of light *absorbed* by the molecule is the inverse of this equation, which is I/I_o. The amount of light absorbed or transmitted is proportional to how many molecules are present (the concentration, c), and the length of the chamber through which the light must pass (path length, d). **Beer's law** shows this relationship as

$$\log I_o/I \propto c \bullet d.$$

However, a proportionality constant is required to set these terms equal. This constant is ε, **the extinction coefficient**. Therefore, Beer's law is

$$\log I_o/I = \varepsilon \bullet c \bullet d.$$

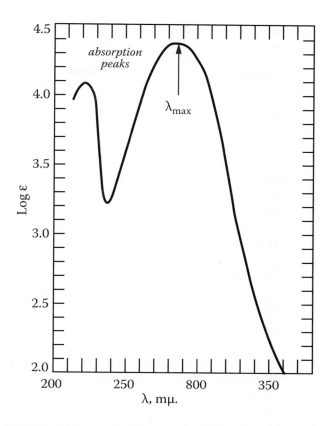

Figure 23.3 Typical ultraviolet spectrum. (From Wilds, A. L., Beck, L. W., Close, W. J., Djerassi, C., Johnson, Jr., J. A., Johnson, T. L., and Shunk, C. H. 1947. *Journal of the American Chemical Society*, 69, 1985. Copyright © 1947 American Chemical Society. Reprinted in part with permission.)

This equation means that if I_o and I are measured using a UV spectrophotometer, knowledge of the concentration and path length allows calculation of ε. If ε for a given wavelength is very large, it means that the molecule absorbs UV light readily; if ε is small for a wavelength, the molecule does *not* absorb UV light very well. The peak in the UV spectrum is simply the wavelength (or range of wavelengths) where a given molecule absorbs UV light. If there are several peaks, the values of ε may be compared to determine whether the absorption is strong or weak.

23.15 If $\varepsilon = 20$ for $c = 0.1$ g/mL (10 dm path length), calculate log I_o/I.

It is important to emphasize that a molecule absorbs UV light *only* if there is a functional group that interacts with the light. *Therefore, the UV spectrum is a measure of the presence or absence of certain functional groups in a molecule.*

23.3.2 Frontier Molecular Orbitals and Electron Promotion

Before discussing how molecules interact with ultraviolet light, it is important to differentiate high- and low-energy UV light. The normal range of ultraviolet radiation is 400–200 nm. This corresponds to 71.5 kcal mol⁻¹ for a frequency of 400 nm and 143 kcal mol⁻¹ for a frequency of 200 nm. Therefore, *a higher frequency in nanometers corresponds to lower energy radiation and a lower frequency in nanometers corresponds to a higher energy radiation.*

23.16 **Which is the higher energy absorption: an absorption at 310 nm or one at 210 nm?**

A discussion of how a molecule interacts with UV light requires a return to the concept of molecular orbitals first introduced in Chapter 3 (Section 3.4). The molecular orbital generated for a C–C bond of sp³ hybridized carbon atoms gives bonding and antibonding orbitals. The electrons used to form covalent bonds are found in the bonding molecular orbital (MO), whereas the antibonding orbital is "empty." **If an electron is made to leave the bonding orbital, where does it go?** It goes to the orbital lowest in energy that is able to accept an electron—in this case, the antibonding orbital. Promotion of an electron describes a molecule that absorbs UV light. An alkane such as butane has only sp³ hybridized (σ) bonds, and all electrons are in sp³ hybridized orbitals. Such electrons are bound very tightly and only applying a large amount of energy displaces them.

The bonding orbital in sp³ hybridized bonds is called a σ-orbital and the antibonding orbital a σ*-orbital. For this system, the wavelength of UV light required to make an electron leave the bonding MO and "jump" to the antibonding MO is out of the range of the spectrum in Figure 23.1. In other words, it is said that an alkane absorbs no UV light because the region where that σ→σ* absorption occurs is not in the UV-visible–IR region. A σ→σ* transition is illustrated in Figure 23.4.

If a substituent is added that has unshared electrons, such as a halogen, the situation changes because the unshared electrons on the halogen are much easier to displace. This means that less energy is required for the unshared electron to go from a bonding to an antibonding orbital. The orbital containing unshared electrons is known as an n-orbital (a nonbonding orbital); an n*-orbital *or* a σ*-orbital may be the antibonding orbital. If the halogen-containing compound absorbs UV light, an absorption corresponds to an n→σ* transition and another corresponds to an n→n* transition (see Figure 23.4). In fact, 1-iodobutane shows an absorption peak at 224 nm (160.2 kcal mol⁻¹) that corresponds to the n→σ* transition.

23.17 **Draw 1-iodobutane, with the unshared electrons.**

For the C=C unit of an alkene, the focus is on the π-electrons because they are the most reactive (most easily displaced) electrons. The π-electrons are found in two π-molecular orbitals: the bonding and antibonding orbitals shown in Figure 23.4. These molecular orbitals are associated with π-electrons, so they are labeled the π- and π*-orbitals, respectively. Absorption of UV light

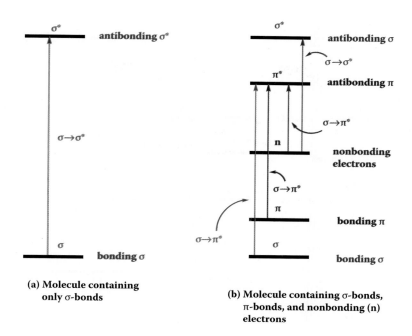

Figure 23.4 Transitions within electronic energy levels.

by the π-orbital will lead to a transfer of a π-electron from the π-orbital to a π*-orbital (known as a π→π* transition). Ethene shows a π→π* absorption at 165 nm.

Rather than call these orbitals bonding and antibonding, the π-orbital is labeled the **highest occupied molecular orbital (HOMO)** and the π*-orbital the **lowest unoccupied molecular orbital (LUMO)**. Therefore, absorption of UV light by an alkene causes an electron transfer from the HOMO to the LUMO. This is illustrated in Figure 23.5. For ethene, the HOMO is found at an energy of –10.52 eV and the LUMO is at +1.5 eV (where eV is an electron volt = 23.06 kcal mol^{-1}). Absorption of UV light leads to a transfer of an electron from the HOMO to the LUMO, so the difference in energy (ΔE, marked in Figure 23.5) is important. For ethane, ΔE is 12.02 eV (277.18 kcal mol^{-1}). This means that *the C=C unit of the alkenes must absorb UV light of the proper wavelength (equal to ΔE) before an electron is promoted from the HOMO to the LUMO.*

The carbonyl group also has a π-bond, so there is a π→π* transition for the C=O unit (see Figure 23.5). The oxygen of a carbonyl also has unshared electrons, whereas the C=C of an alkene does not, so there is an n→n* transition or an n→π* transition for the carbonyl. In addition to the n electrons and π-electrons, another transition is possible: the n→σ* transition. All of these possibilities are shown in Figure 23.5. In the case of *acetone, the π→π* transition occurs at 150 nm, the n→σ* transition at 188 nm, and the n→π* transition*

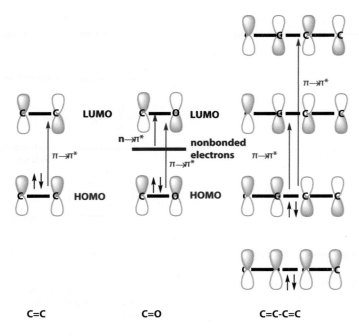

Figure 23.5 HOMO–LUMO transitions for C=C, C=O, and C=C–C=C.

at 279 nm. These numbers indicate that the n→π* transition is lower in energy. A typical molecular orbital diagram for a ketone is shown in Figure 23.5; the unshared electrons reside in a nonbonding orbital, which is midway in energy between the bonding and antibonding orbitals. The ΔE for the n→π* is smaller than that for any other transition, and this is generally true for functional groups containing both nonbonded electrons and π-electrons. Note also that the energy of UV light required for the π→π* transition of the carbonyl of acetone is higher than that for ethene (150 versus 165 nm).

23.3.3 Conjugation Shifts the Wavelength of Absorption

The premise of this section is that UV light is used to distinguish the structure of a conjugated compound because C=C and C=O units absorb UV light. A comparison of the π-molecular orbitals of ethene with those of 1,3-butadiene (3; shown in Figure 23.4) shows that it is possible to distinguish these molecules using UV. Absorption of UV light into the HOMO leads to promotion of an electron to the LUMO. The energy of the HOMO for 3 is −9.07 eV and the energy of LUMO is +1.0 eV. The other two molecular orbitals for the diene are shown, but the transition energy from the HOMO to the LUMO is the important one. *The ΔE for this transition is 10.07 eV (232.3 kcal mol⁻¹), which corresponds to the π→π* transition. The energy of the π→π* transition for ethene was 12.02 eV (277.18 kcal mol⁻¹)* in Section 23.3.2; this is higher in energy than that for the diene.

*Conjugation leads to more molecular orbitals with the net result that the HOMO is closer in energy to the LUMO than if the C=C unit were not conjugated. The actual UV absorption band for **3** is 217 nm for the π→π* transition, compared with 165 nm for ethene. Conjugation leads to a shift to lower energy for absorption of UV light (it is easier to remove an electron from a conjugated π-bond than from a nonconjugated π-bond).* This shift to a lower energy is associated with a shift to a higher value in nanometers (217 versus 165 nm). Therefore, *a conjugated diene is distinguished from a nonconjugated diene by the presence of an absorption band at around 217 nm for the conjugated system versus one at around 165 nm for the nonconjugated C=C units. The absorption of the conjugated molecule is at lower energy.*

23.18 **Which absorbs at shorter wavelength: 1,5-hexadiene or 1,3-hexadiene?**

A similar argument is made for conjugated carbonyl compounds. The π→π* transition for a nonconjugated C=C unit appears at 165 nm; in Section 23.3.2, the π→π* transition for a C=O unit (the carbonyl) was reported as 150 nm for acetone. The conjugated aldehyde acrolein (see **11**) combines both of these units and the π→π* absorption occurs at 210 nm.[5] The HOMO and LUMO values are −10.89 eV and +0.60 eV, respectively,[6] so the ΔE is 11.49 eV (264.96 kcal mol⁻¹) for the π→π* transition, compared with 10.07 eV (232.3 kcal mol⁻¹) for 1,3-butadiene (**3**). The energy required for acrolein to absorb UV light should be higher than for 1,3-butadiene, and this is correct—210 versus 217 nm. The unshared electrons on the oxygen atom of acrolein reside in a nonbonded orbital, and an n→π* transition for acrolein (a second absorption band) does *not* appear for 1,3-butadiene. This n→π* absorption appears at 315 nm,[4] which is much lower in energy than the π→π* absorption.

Sufficient information is now available to use UV spectroscopy to obtain structural information about molecules containing two C=C units or a C=C unit and a C=O unit. This is important for many reactions. When a product is formed in a chemical reaction, UV spectroscopy is used to determine what functional group might be present and whether or not there is conjugation.

23.3.4 Identifying Conjugated Molecules Using UV Spectroscopy

In Chapter 21 (Section 21.6.1), Birch reduction of benzene gave 1,4-cyclohexadiene (**7**). Experimental verification that the reaction gives a nonconjugated diene **7** and not the conjugated diene 1,3-cyclohexadiene (**6**) is possible

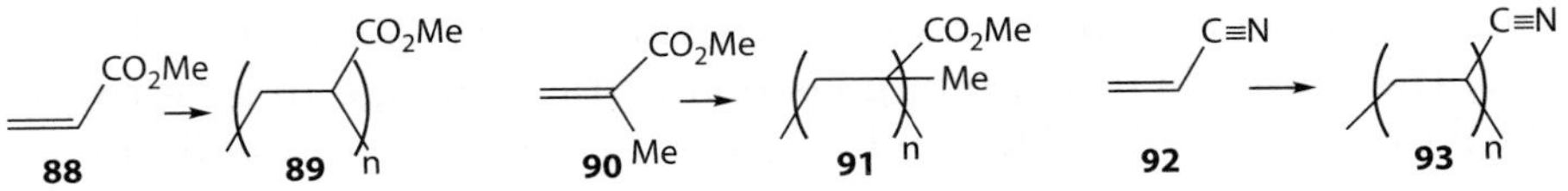

Some polymerization reactions of conjugated carbonyl compounds that give common commercial products are shown in the illustration. Conjugated carbonyl compounds such as methyl acrylate (**88**) or methyl methacrylate (**90**) are polymerized via the C=C unit, but in the final product the carbonyl unit is a substituent on the polymer and not part of the polymer chain as noted with dienes. The polymer derived from **88** is poly(methacrylate), **89**, whereas the polymer derived from **90** is poly(methyl methacrylate), **91**, which is known as PMMA but has the trade names Lucite® or Plexiglas®. This material is commonly used to fabricate dentures, among other things. Another useful polymer is derived from acrylonitrile (**92**) where the C=C unit is conjugated to the C≡N unit. The polymer is poly(acrylonitrile) (**93**) and it has the trade name of Orlon® or Creslan®. This is an interesting polymer because, when it is put into solution, it can be spun and the resulting material is woven into fabric.

23.6 Synthetic Possibilities

Several interesting functional groups exchange reactions in this chapter. Two transformations from dienes involve 1,2 and 1.4 addition of nucleophiles.

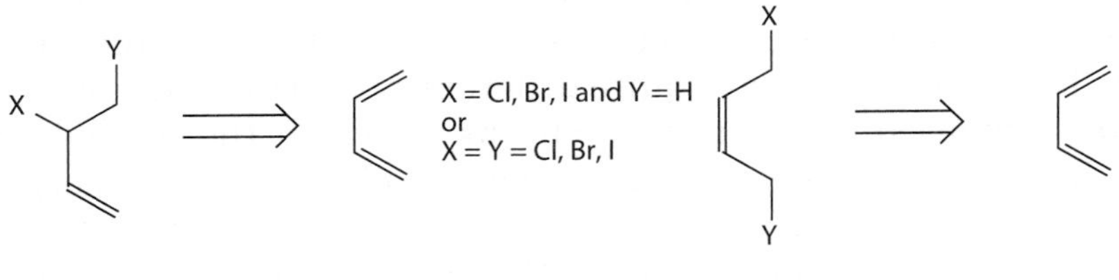

Halo-ketones have conjugated ketones as a synthetic precursor, and halo-acid derivatives have conjugated acid derivatives as a precursor. In both cases, the reaction relies on a reaction with HX (X = Cl, Br). Ketone-dihalides and acid derivative-dihalides also have conjugated ketone or conjugated acid derivatives as precursors, relying on a reaction with X_2 (X = Cl, Br).

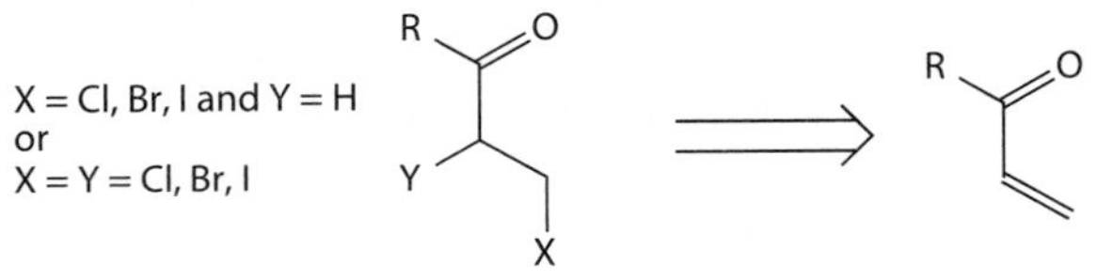

The conjugate addition disconnection is shown for incorporation of an R group, with a conjugated ketone, conjugated aldehyde, or conjugated acid derivative as a precursor. The R group is alkyl or aryl. The conjugate addition target can be more complex, leading to a disconnection to the conjugated carbonyl compound and a dicarbonyl compound. This disconnection is extended further to include formation of a conjugated cyclic ketone from a conjugated carbonyl compound and a second carbonyl compound. This is the Robinson annulation disconnection.

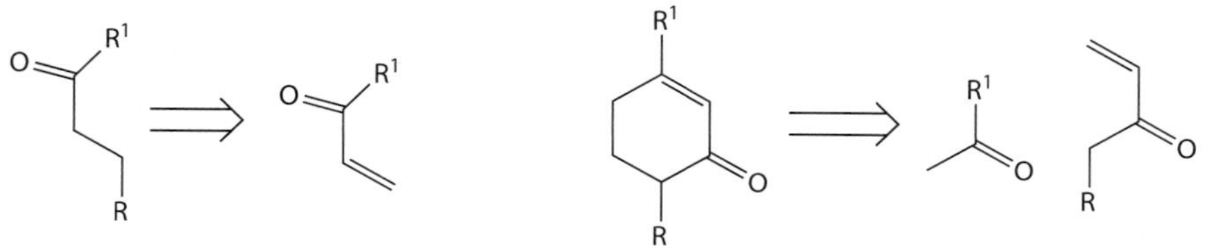

The more useful synthetic applications based on chemistry in this chapter are formation of allylic halides from conjugated dienes, Michael addition of organocuprates to conjugated compounds, and the Robinson annulation. The homework has synthetic problems that address these transformations.

23.7 Biological Relevance

Conjugate addition reactions occur in biological systems. An enzyme known as enoyl-CoA hydratase (also known as crotonase) facilitates the conjugate addition of water to the C=C unit of an acyl-CoA molecule such as **94** to give the 3-hydroxy thioester **95**. The fragment CoA is coenzyme A, **97**, which forms a thioester unit as seen in **94** and **95**. This process is essential for the metabolism of fatty acids and the production of energy[10] in which enoyl-CoA hydratase catalyzes the second step in the breakdown of fatty acids or the second step of β-oxidation in fatty acid metabolism.

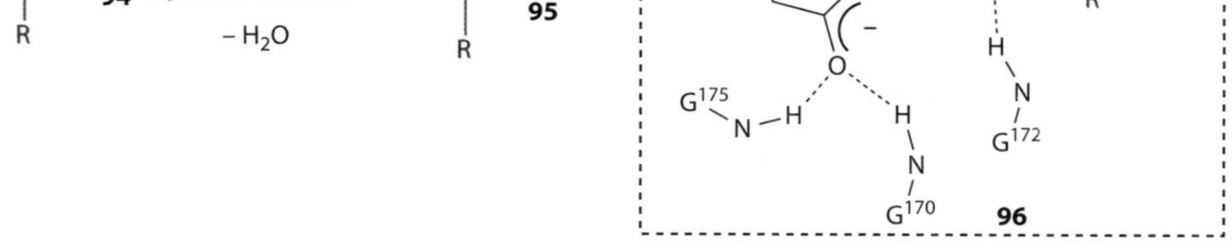

(Reprinted in part with permission from Bahnson, B. J., Anderson, V. E., and Petsko, G. A. 2002. *Biochemistry* 41:2621–2629. Copyright 2002 American Chemical Society.)

1228

Organic Chemistry: An Acid–Base Approach

23.14

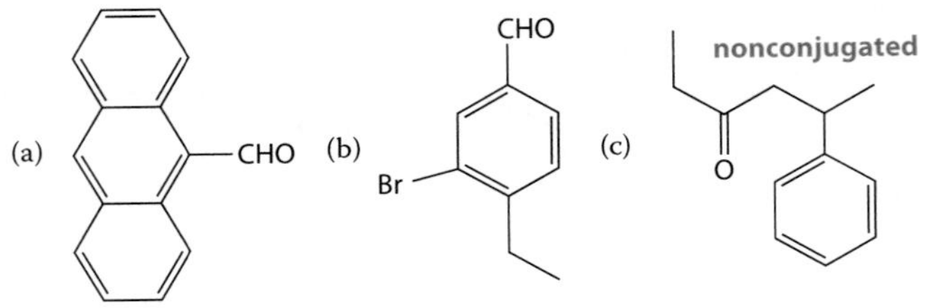

23.15 log $I_o/I = 20 \times 0.1 \times 10 = 20$. If log $I_o/I = 20$, then $I_o/I = 1 \times 10^{-20}$.

23.16 The lower value for frequency corresponds to higher energy, so an absorption at 210 nm is higher in energy than one at 310 nm.

23.17

23.18 Longer wavelength is lower energy and shorter wavelength is higher energy. The conjugated diene, 1,3-hexadiene, should absorb at lower energy (longer wavelength), so 1,5-hexadiene will absorb at shorter wavelength (higher energy).

23.19 (a) Nonconjugated: the rules do not apply; (b) $215 + 12 = 227$ nm; (c) $215 + 30 = 245$ nm; (d) $215 + 10 + 12 = 237$ nm.

23.20 (a) Use the base value for a conjugated ketone $= 215 + 12 = 227$ nm versus $215 + 10 + 12 = 237$ nm; (b) $215 + 12 = 227$ nm versus $214 + 5 = 219$ nm.

23.21

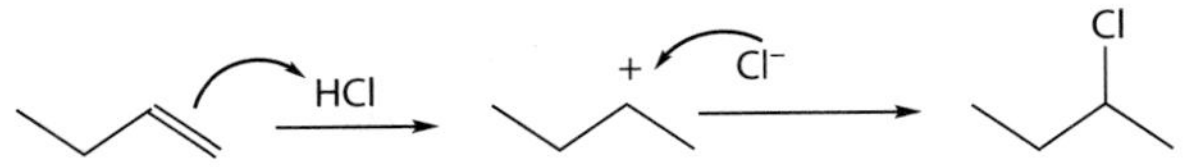

23.22

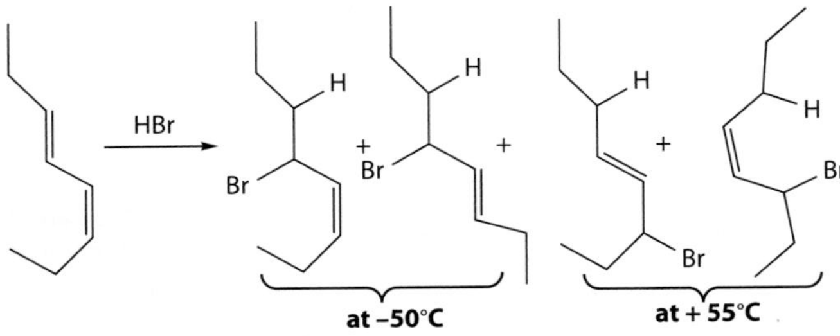

23.23

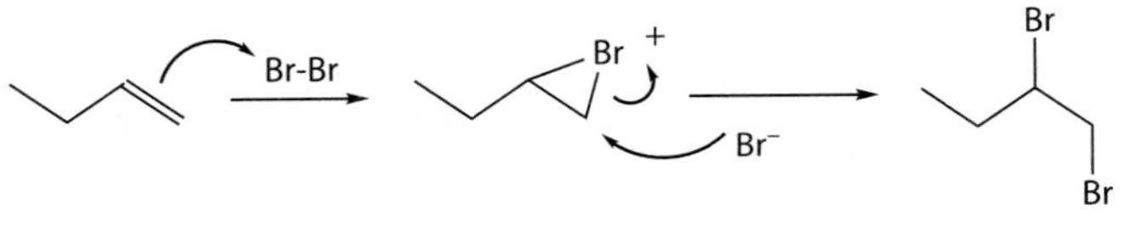

23.24

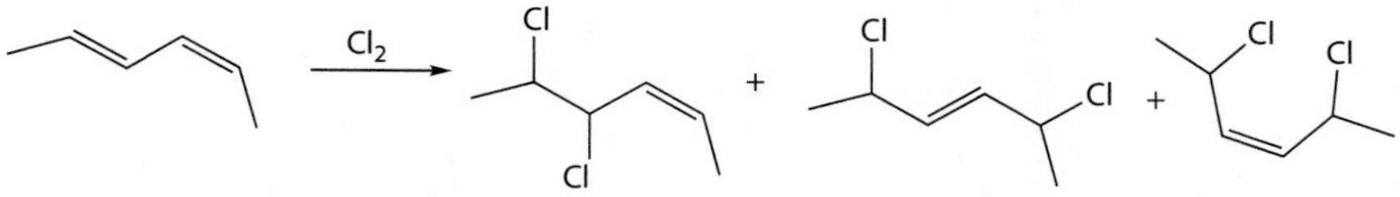
Cl2
Cl
Cl
Cl
Cl
Cl
Cl
+
+

23.25

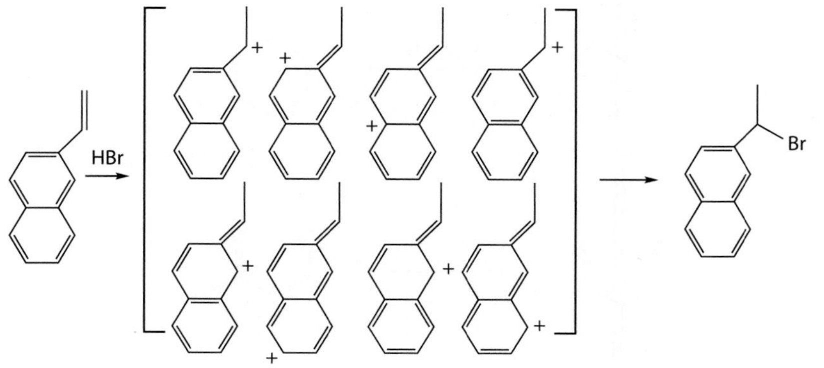
HBr
Br

23.26

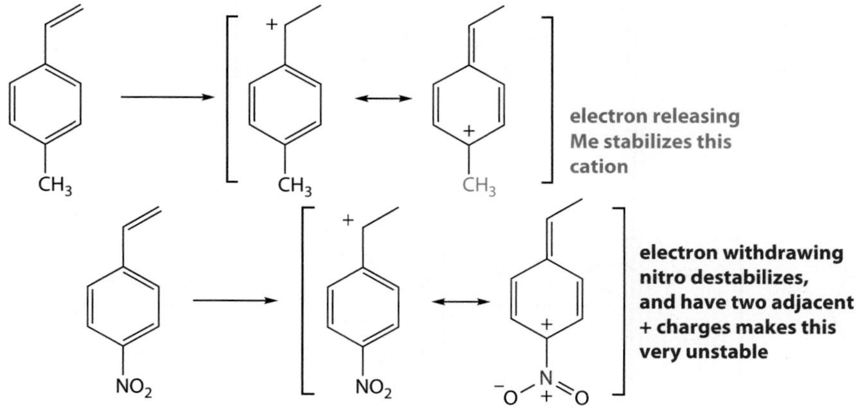
CH3
CH3
CH3
electron releasing
Me stabilizes this
cation
NO2
NO2
electron withdrawing
nitro destabilizes,
and have two adjacent
+ charges makes this
very unstable

23.27

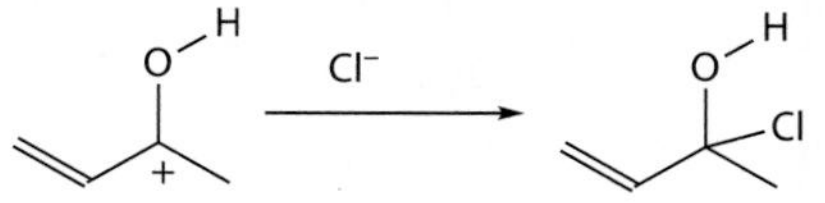
Cl⁻
Cl

23.28

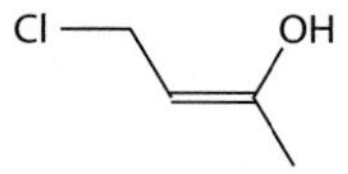
Cl
OH

23.29

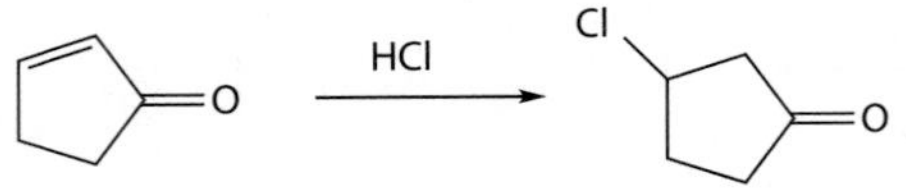

23.30

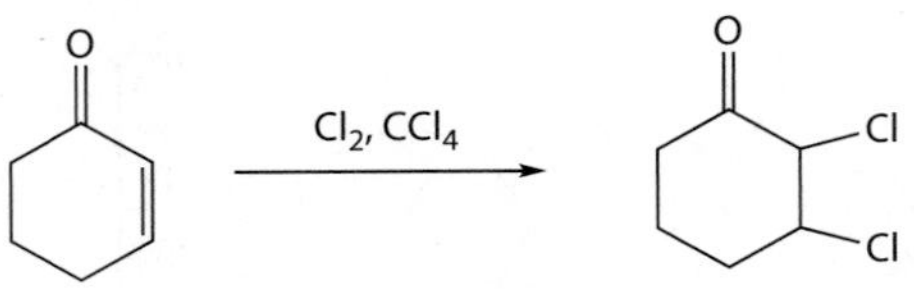

23.31

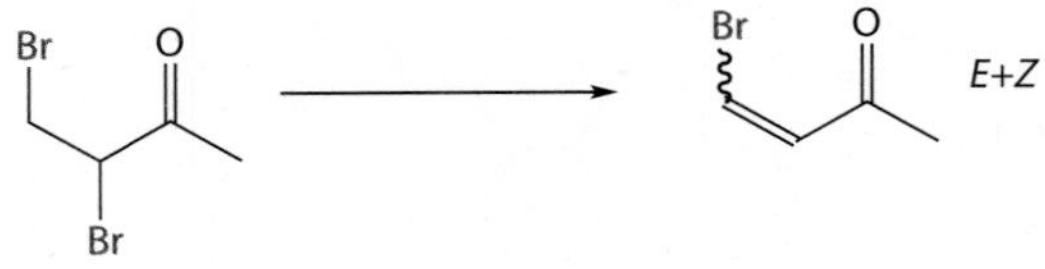

23.32

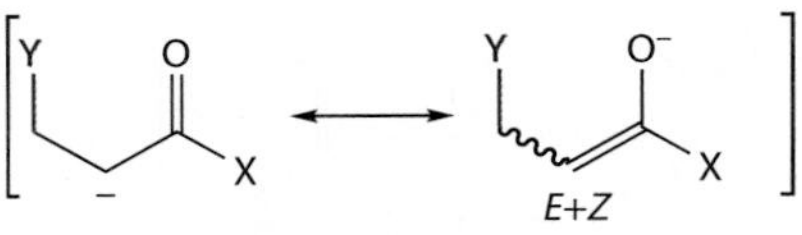

23.33

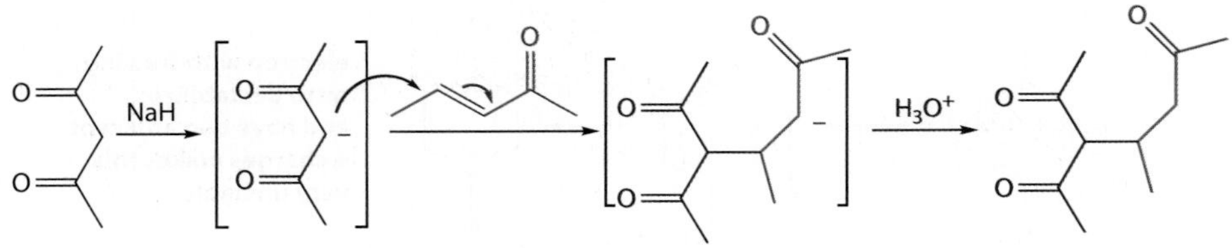

23.34

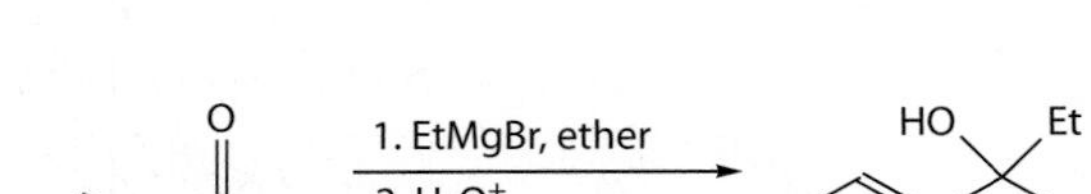

23.35

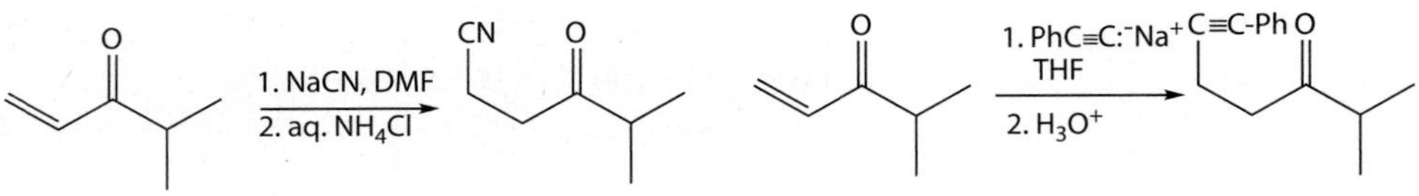

23.36

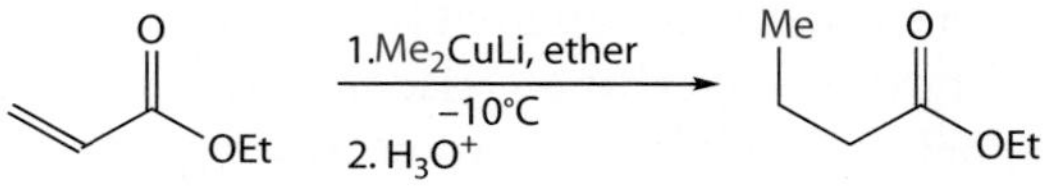

23.37

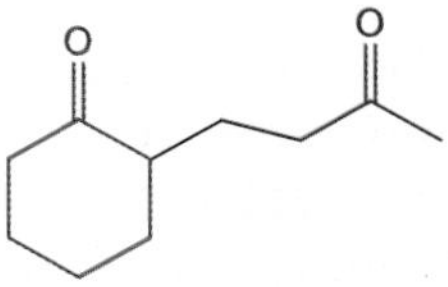

23.38

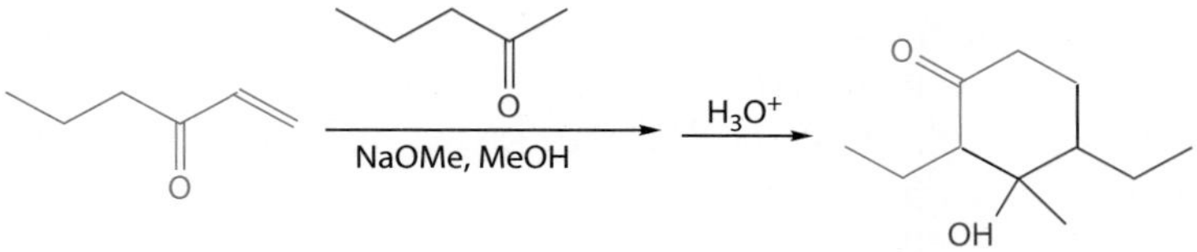

23.39

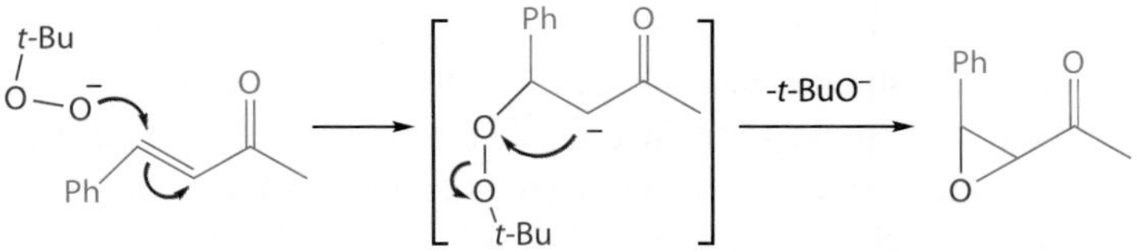

Correlation of Concepts with Homework

- **Conjugated dienes are molecules that contain two C=C units connected together such that there is a continuous array of sp² carbons (C=C–C=C). Conjugate carbonyl compounds have a C=C unit directly attached to the carbon of a carbonyl (C=C–C=O). There are conjugated ketones, aldehydes, acids, and acid derivatives. When a C=C or a C=O unit is attached to a benzene ring, these are also conjugated alkenes or carbonyl derivatives:** 1, 2, 5, 6, 7, 8, 9, 10, 11, 12, 13, 14, 40, 43.
- **Conjugated dienes exhibit rotation about the C–C single bond of the C=C–C=C unit, generating two major rotamers—the cisoid and the transoid rotamers:** 4, 41, 42, 61.

- When a molecule absorbs UV light, the most intense absorption peak (wavelength in nanometers) is labeled λ_{max} and the intensity of the absorption is given by the extinction coefficient, ε. Conjugated compounds absorb UV light at lower energy than nonconjugated compounds do because of the smaller energy gap between filled and unfilled molecular orbitals. The differences in UV spectra between conjugated and nonconjugated compounds allow them to be easily distinguished via UV spectroscopy: 15, 16, 17, 18, 47, 48, 49.
- A set of empirical rules based on butadiene and acrolein allows prediction of λ_{max} for a compound: 19, 20, 50, 51, 60.
- Dienes react with HX (HCl, HBr, HI) to give 1,2 addition and 1,4 addition products via formation of an allylic cation. The major product is determined by the temperature of the reaction, where low temperature favors 1,2 addition and high temperature favors 1,4 addition. Halogens such as Cl_2, Br_2, and I_2 also add to dienes to give a mixture of 1,2 and 1,4 addition products. Both HX and X_2 react with alkenes with the C=C unit conjugated to a benzene ring in essentially the same manner as with alkenes that do not contain a benzene ring: 3, 21, 22, 23, 24, 25, 26, 27, 28, 29, 30, 31, 39, 52, 53, 54.
- Dienes can be polymerized under radical and cationic conditions to give useful polymers such as rubber (polyisoprene): 55, 62.
- Conjugated carbonyl compounds react with HX to give 3-substituted derivatives (the X adds to the end of the C=C unit) via the more stable cation. Michael addition involves addition of a nucleophile to the end of the C=C unit of a conjugated carbonyl, forming an enoate anion. Grignard reagents and carbanionic nucleophiles tend to add to the C=O unit of conjugated aldehydes and of unhindered conjugated ketones. Organocuprates usually give conjugate addition to both conjugated aldehydes and ketones: 32, 33, 34, 35, 36, 45, 46, 56, 58.
- The reaction of a ketone enolate with a conjugated ketone under thermodynamic conditions gives a cyclic product, in what is called the Robinson annulation: 37, 38, 57, 59.
- Conjugated carbonyl compounds can give primarily 1,2 reduction or 1,4 reduction, depending on the reagent: 8, 44, 58.
- A molecule with a particular functional group can be prepared from molecules containing different functional groups by a series of chemical steps (reactions). This process is called synthesis: The new molecule is synthesized from the old one (see Chapter 25): 63, 64.
- Spectroscopy can be used to determine the structure of a particular molecule and can distinguish the structure and functionality of one molecule compared with another (see Chapter 14): 65, 66, 67, 68, 69, 70, 71, 72, 73.

Homework

23.40   Draw the structure of each of the following molecules:
    (a) 2E,5Z-nonadiene
    (b) 1,2-diethylcyclohexadiene
    (c) hex-1-en-3-yne
    (d) 2,3-dimethyl-1,3-butadiene
    (e) 2,4-pentadienoic acid
    (f) 3E,5E-dodecadienal
    (g) 1,5-cyclooctadiene
    (h) 1,4-diphenyl-1E,3E-butadiene
    (i) 2,3,4,5-tetramethyl-2,4-hexadiene

23.41   Explain why 2E,4E-hexadiene has a higher percentage of the cisoid conformation than 2Z,4Z-hexadiene.

23.42   Briefly comment on whether or not we have to consider *s-cis* versus *s-trans* conformations for 1,3-pentadiyne.

23.43   Draw the structure for each of the following molecules:
    (a) ethyl benzoate
    (b) hex-3E-en-2-one
    (c) pent-2-ynenitrile
    (d) 2,7-diethylcyclohept-2-en-1-one
    (e) acrolein
    (f) methyl vinyl ketone
    (g) dimethyl fumarate
    (h) hexa-3E-en-2,5-dione
    (i) 1,5-diphenyl-1-pentene
    (j) 2-methylhex-1-en-3-one
    (k) penta-1,4-dien-3-one
    (l) cyclopent-3-en-1-one
    (m) acrylic acid
    (n) cyclohexene-1-carboxylic acid
    (o) oct-4Z-enal

23.44   Heating cyclopentene with $LiAlH_4$ gives no reduction but heating it with cyclopent-2-en-1-one leads to reduction of both the C=C unit and the C=O unit and formation of cyclopentanol as a major product. Explain this experimental fact.

23.45   Briefly explain why an acid catalyst reacts with the carbonyl oxygen of methyl vinyl ketone. Draw the resultant intermediate.

23.46   Birch reduction of anisole (Chapter 21, Section 21.6.1) leads to 1-methoxycyclohexa-1,4-diene. Heating this with aqueous acid leads to cyclohex-2-en-1-one. Draw all of these products and provide a mechanism for the formation of cyclohexenone. Why should this product form?

23.47   Interconvert the following:
    (a) 345 nm to kcal; to cm^{-1}
    (b) 16×10^2 cm^{-1} to kJ; to nm
    (c) 1765 cm^{-1} to nm; to kcal; to Å

    (d) 325 kJ to kcal; to nm
    (e) 8000 Å to kcal; to cm^{-1}
    (f) 185 kcal to nm; to Å; to cm^{-1}

23.48 If $\varepsilon = 38{,}000$ for $c = 0.5$ g/mL (10 dm path length), calculate log I_o/I.

23.49 If $I_o/I = 3 \times 10^{-8}$, $c = 1.2$ g/mL, and the path length is 5 dm, calculate ε.

23.50 Indicate which of the following are expected to show strong UV absorption. Justify your choices.
    (a) methyl vinyl ketone
    (b) 3-hexanone
    (c) benzaldehyde
    (d) cyclohex-2-en-1-one
    (e) 1,5-heptadiene
    (f) 1,3-cyclohexadiene
    (g) ethyl hex-2-enoate
    (h) styrene
    (i) hex-3-en-2-ol

23.51 Predict the maximum UV absorption peak for each of the following:

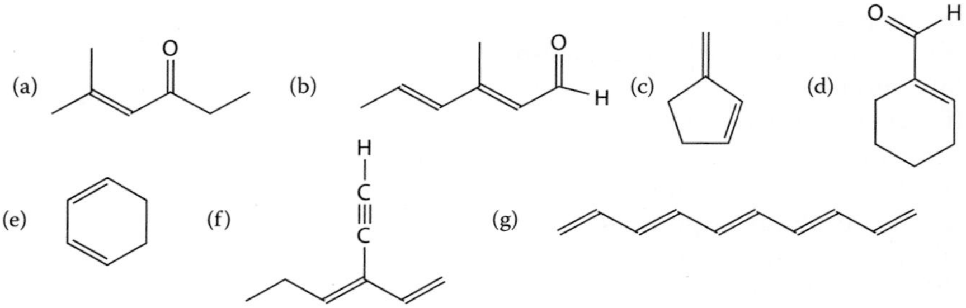

23.52 Give the major product for each of the following:

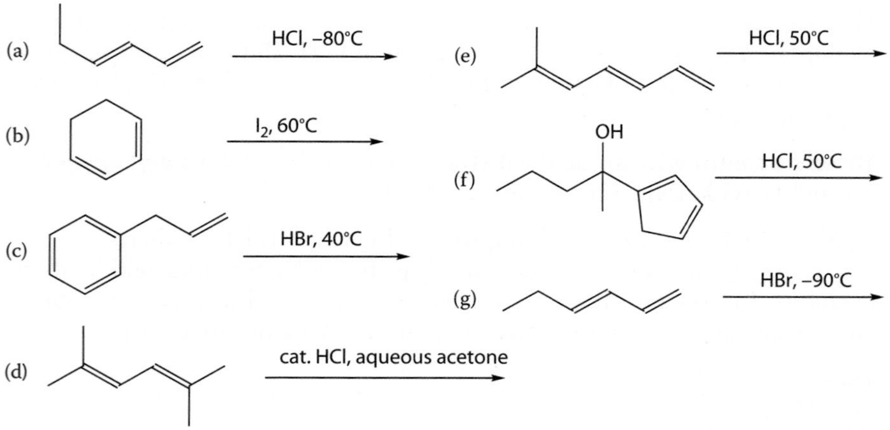

23.53   Draw the intermediate, all resonance contributors for that intermediate, and the final major product when 1-phenyl-1,3-butadiene reacts with HBr at −80°C.

23.54   Predict the major product of a reaction between methyl vinyl ketone and HCl at −80°C. Discuss your answer in terms of a mechanism for its formation. Do the same analysis for the reaction with Br_2 at −30°C.

23.55   Predict the general structure of the polymer formed when each of the following alkenes is treated with *t*-butyl hydroperoxide and heated:
(a)   2-methyl-1,3-butadiene
(b)   ethyl 3-methylbut-2-enoate
(c)   4-chloro-1-ethenylbenzene
(d)   methyl cyclopentene-1-carboxylate
(e)   acrolein
(f)   2-methylbut-2-enenitrile

23.56   Give the structure for the major product formed in each of the following reactions:

(a)   [structure] CHO   1. PhMgBr, THF / 2. hydrolysis →

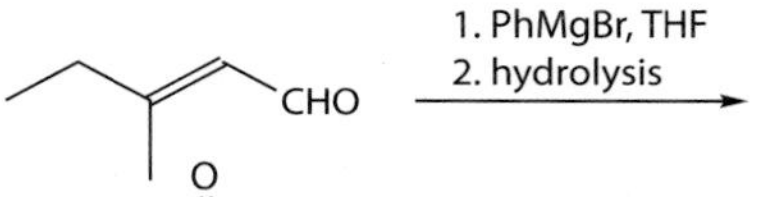

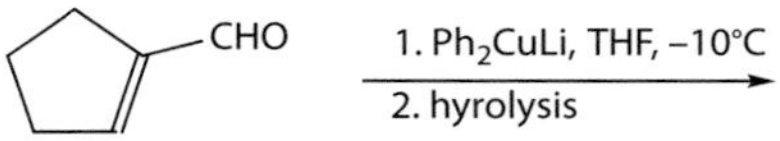

(d)   [structure] O   1. [structure] Li / ether / 2. hydrolysis →

(b)   [structure] O Ph   1. MeMgBr, ether / 2. hydrolysis →

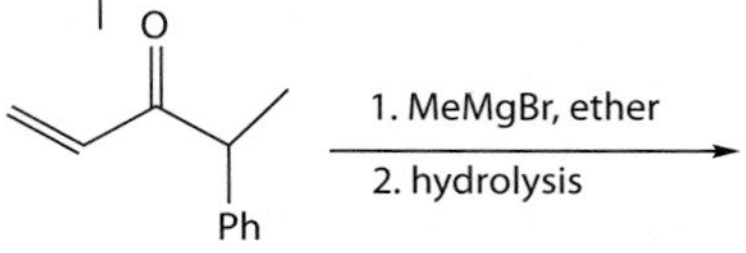

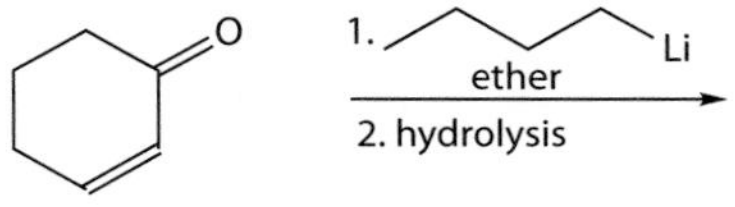

(e)   [structure] O   1. MeMgBr, ether / 2. hydrolysis →

(c)   [structure] CHO   1. Ph₂CuLi, THF, −10°C / 2. hyrolysis →

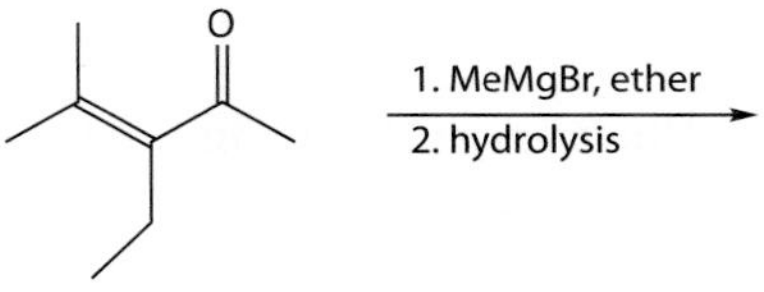

23.57   Give the major product for each of the following reactions:
(a)   cyclopent-2-en-1-one + 3-pentanone + NaOEt/ethanol/reflux followed by hydrolysis
(b)   hex-4-en-3-one + 2-butanone + NaOMe/methanol/reflux followed by hydrolysis
(c)   cyclohexanone + cyclopent-2-en-1-one + NaOEt/ethanol/reflux followed by hydrolysis

23.58   Give the major product for each of the following reactions:
(a)   2*E*,4*E*-hexadiene + lithium diphenylcuprate, THF, −10°C
(b)   hex-4-en-3-one + Br_2, −20°C
(c)   cyclopentene-1-carboxaldehyde + 1. NaBH₄/EtOH; 2. aq. NH₄Cl
(d)   ethylpent-2-enoate + H₂/Pd–C
(e)   hex-4-en-2-one + 1. LiAlH₄/THF; 2. hydrolysis
(f)   1,3-cyclopentadiene + 1. LiAlH₄/ether; 2. water
(g)   styrene + H₂/Pd-C
(h)   4-ethylhex-3-en-2-one + 1. NaBH₄/EtOH; 2. aq. NH₄Cl
(i)   acetophenone + 1. Ph₃P=CH₂; 2. H₂, Pd/C

23.59   What product is formed when cyclopentanone reacts with pent-1-en-3-one in refluxing ethanol with sodium ethoxide, followed by an aqueous acid workup? What is the product if cyclohexenone is used in the same reaction with cyclopentanone?

23.60   Suggest a method based on UV spectroscopic analysis of products that will distinguish whether 1,2 or 1,4 addition occurred when methylmagnesium bromide reacts with (E)-4-cyclopentyl-1-phenylbut-3-en-2-one.

23.61   Discuss the major rotamer of 1-phenyl-1E,3-butadiene and then 1-phenyl-1Z,3-butadiene.

23.62   Draw the polymer that is expected to form from (a) diethyl maleate, (b) cyclopentadiene, and (c) 1,1-dicyanoethene.

Synthesis Problems

Do not attempt these problems until Chapter 25 has been read and understood.

23.63   Using retrosynthetic analysis, devise a retrosynthetic scheme for the following molecules from a starting material with no more than six carbons:

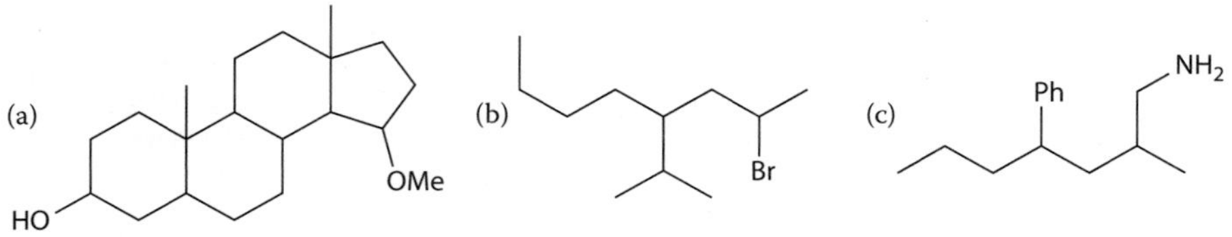

23.64   Show a synthesis for each of the following molecules from the designated starting material:

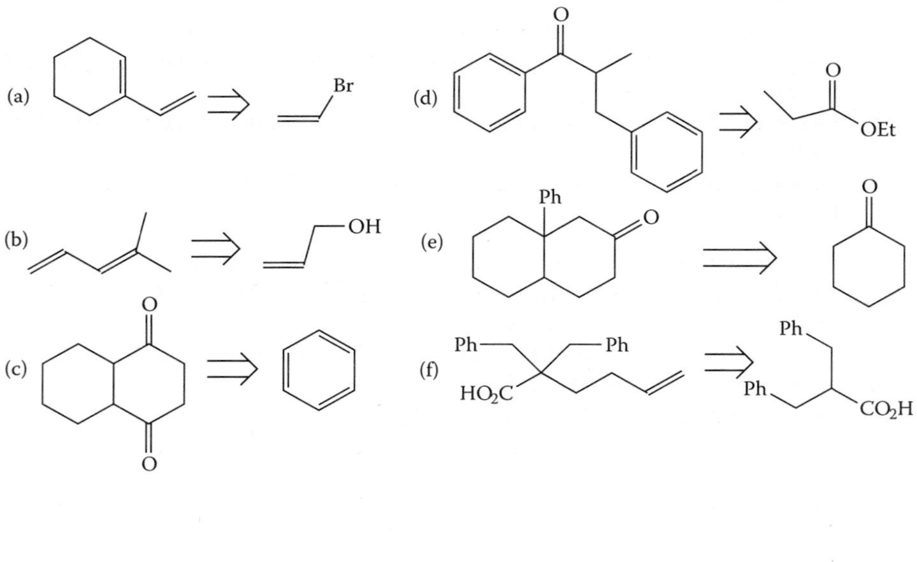

Spectroscopy Problems

Do not attempt these problems until Chapter 14 has been read and understood

23.65 Briefly discuss how you can distinguish pent-3-en-2-one from pent-4-en-2-one using UV spectroscopy. Are there differences in the proton and carbon NMR spectra?

23.66 Discuss the UV spectroscopic differences between 1,3-hexadiene and 1,5-hexadiene. Briefly discuss any differences in chemical reactivity.

23.67 Discuss the UV and IR spectroscopic differences between 2-pentanone and pent-3-en-2-one. Also discuss differences in the proton and carbon NMR spectra. Devise a chemical test that will distinguish these molecules based on your knowledge of the reactions of ketones.

23.68 When cyclohexadiene reacts with bromine, two products are possible. Draw both of them and describe UV and proton NMR spectroscopic characteristics that will allow you to distinguish them so that you can identify the major product.

23.69 Identify the molecule with the formula C_5H_8 with the following spectral data. There is a significant peak in the UV spectrum at 219 nm:
IR: 3088, 3055–2853, 1797, 1666, 163, 1002, 949, 897 cm^{-1}
^{1}NMR: 6.29 (m, 1H), 6.06 (m, 1H), 5.70 (m, 1H), 5.06 (m, 1H), 4.93 (m, 1H), 1.74 (d, 3H) ppm
J(B,C) = 15.1 Hz
J(A,B) = 10.4 Hz
J(A,D) = 17.0 Hz
J(A,E) = 10.1 Hz

23.70 Identify the molecule with the formula $C_{16}H_{14}$ with the following spectral data. There is a significant peak in the UV spectrum at 334 nm:

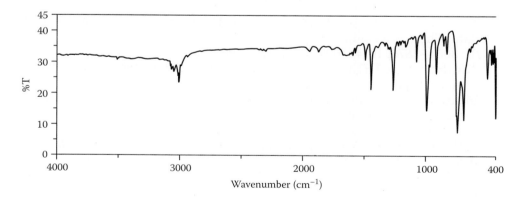

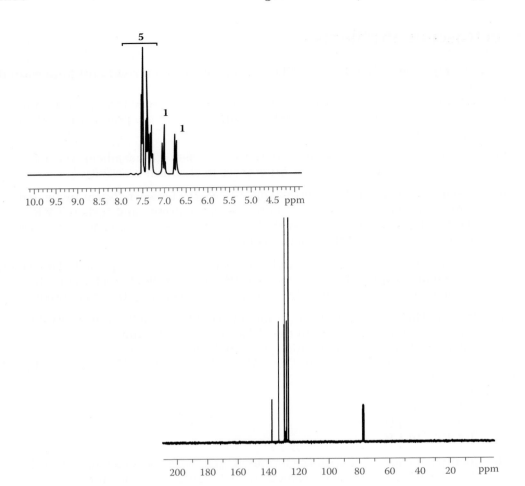

23.71   Identify the molecule with the formula $C_7H_{12}O$ with the following spectral
         data. There is a significant peak in the UV spectrum at 227 nm:
         IR: 3054–2876, 1674, 1629, 1362, 1254, 981 cm^{-1}
         ^{1}NMR: 6.82 (m, 1H), 6.08 (m, 1H), 2.25 (s, 3H), 2.22 (m, 2H), 1.51 (m, 2H), 0.95
            (t, 3H) ppm
23.72   Identify the molecule with the formula $C_6H_{10}O_2$ with the following spectral
         data. There is a significant peak in the UV spectrum at 225 nm:
         IR: 2984–2908, 1722, 1640, 1323, 1299, 1178, 1169, 1034, 942, 875 cm^{-1}
         ^{1}NMR: 6.1 (m, 1H), 5.54 (m, 1H), 4.21 (q, 2H), 1.94 (m, 3H), 1.30 (t, 3H) ppm
         J(A,B)= 1.7 Hz
         J(A,D) = –1.0 Hz
         J(B,D) = –1.6 Hz

23.73 Identify the molecule with the formula $C_5H_8O_2$ with the following spectral data. There are no significant peaks in the UV spectrum past 205 nm:

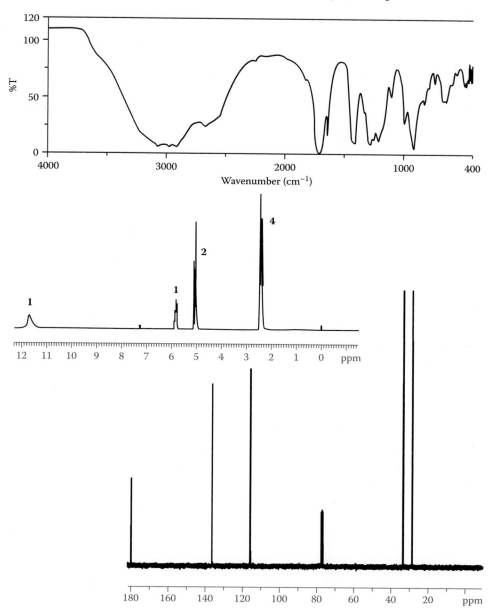

Difunctional Molecules

Pericyclic Reactions

24

Conjugated dienes and conjugated carbonyl compounds were discussed as more or less separate entities in the previous chapter. This chapter will show that one may react with the other. 1,3-Dienes react with alkenes, particularly with the C=C unit of a conjugated carbonyl compound, to form cyclohexene derivatives. This is a [4+2] cycloaddition, commonly known as the Diels–Alder reaction. The mechanism for this reaction requires an examination of the molecular orbitals of the reactants. This is one of the most important reactions in organic chemistry. It is also important to note that this reaction is not an acid–base reaction; it is a new type of reaction and a new type of mechanism. The product is often a difunctional molecule.

To begin this chapter, you should know the following:

- **nomenclature for alkenes and alkynes (Chapter 5, Sections 5.1 and 5.3)**
- ***E*- and *Z*-isomers of alkenes (Chapter 9, Section 9.4)**
- **fundamental structure and nomenclature of dienes (Chapter 5, Section 5.3; Chapter 23, Section 23.1)**
- **fundamental structure and nomenclature of ketones or aldehydes (Chapter 5, Section 5.9; Chapter 15, Section 15.2)**
- **nomenclature for conjugated carbonyl compounds (Chapter 23, Section 23.2)**
- **structure and nomenclature for bicyclic compounds (Chapter 9, Section 9.7.1)**

- how to assign molecular orbitals to π-bonds (Chapter 3, Section 3.4)
- HOMO and LUMO molecular orbitals (Chapter 3, Section 3.4; Chapter 23, Section 23.3.2)
- fundamentals of kinetic and thermodynamic control (Chapter 22, Section 22.4.2)
- rate of reaction (Chapter 7, Section 7.11)
- activation energy and transition states (Chapter 7, Sections 7.3 and 7.6)
- Lewis bases and reactions with carbonyl compounds (Chapter 6, Section 6.5)
- rotamers and conformations (Chapter 8, Section 8.1)
- that conjugated dienes are molecules that contain two C=C units connected together such that there is a continuous array of sp² carbons (C=C–C=C) (Chapter 23, Section 23.2)
- that conjugated dienes exhibit rotation about the C–C single bond of the C=C–C=C unit, generating two major rotamers: the s-cis and the s-trans rotamers (Chapter 23, Section 23.1; Chapter 8, Section 8.1)
- that conjugate carbonyl compounds have a C=C unit directly attached to the carbon of a carbonyl (C=C–C=O) and that there are conjugated ketones, aldehydes, acids, and acid derivatives (Chapter 23, Section 23.2)
- that when a C=C or a C=O unit is attached to a benzene ring, these are also conjugated alkenes or carbonyl derivatives (Chapter 23, Section 23.1)
- stereogenic centers and enantiomers (Chapter 9, Section 9.1)
- absolute configuration of stereogenic centers (Chapter 9, Section 9.3)
- diastereomers (Chapter 9, Section 9.5)

This chapter will discuss 1,3-dienes in a reaction with alkenes to give cyclohexene derivatives. This is a thermal reaction driven by interactions of molecular orbitals rather than ionic or polarized intermediates. In addition to the reaction of 1,3-dienes, 1,5-dienes undergo a rearrangement to a different 1,5-diene in what is known as a sigmatropic rearrangement. Similarly, allyl vinyl ethers rearrange to form alkenyl aldehydes or ketones. Both of these reactions tend to give difunctional molecules as products.

When you have completed this chapter, you should understand the following points:

- The thermal reaction between a conjugated diene and an alkene to give a cyclohexene derivative is a [4+2] cycloaddition, otherwise known as a Diels–Alder reaction.
- The Diels–Alder reaction is driven by the energy differences between the highest occupied molecular orbital (HOMO) of the

diene and the lowest unoccupied molecular orbital (LUMO) of the alkene. Alkenes bearing an electron-withdrawing group react faster because the LUMO is lowered, making the energy difference smaller. Conversely, alkenes bearing an electron-releasing group react more slowly because the LUMO is raised, making the energy difference large.

- Secondary orbital interactions of alkenes bearing substituents having a π-bond lead to a preference for the endo product in the Diels–Alder reaction. This reaction is regioselective due to interactions of orbital coefficients on the orbitals of the reacting atoms. Substituents at C1 and C4 of the diene move in a disrotatory manner and opposite, relative to the incoming alkene.

- A sigmatropic rearrangement is a reaction where a σ-bond moves across a conjugated π-system to a new site. The most common sigmatropic rearrangements are [3,3]-sigmatropic rearrangements, typified by the Cope rearrangement of 1,5-dienes and the Claisen rearrangement of allyl vinyl ethers. The Cope rearrangement generates a new 1,5-diene. The Claisen rearrangement generates a nonconjugated alkenyl aldehyde or ketone.

- A molecule with a particular functional group can be prepared from molecules containing different functional groups by a series of chemical steps (reactions). This process is called synthesis: The new molecule is synthesized from the old one (see Chapter 25).

- Spectroscopy can be used to determine the structure of a particular molecule and can distinguish the structure and functionality of one molecule when compared with another (see Chapter 14).

- Pericyclic reactions are rare in biological systems.

24.1 Frontier Molecular Orbitals: HOMOs and LUMOs

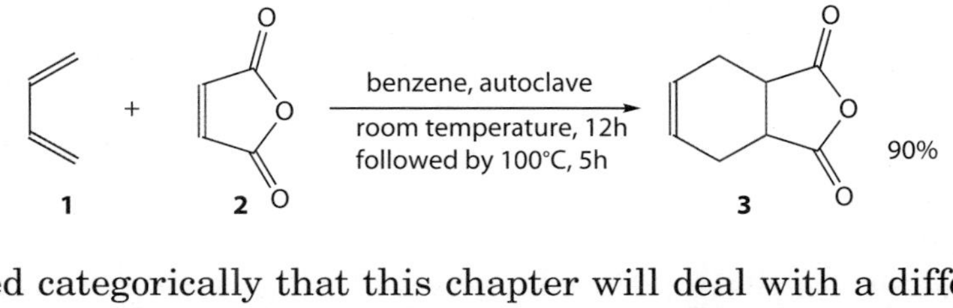

It was stated categorically that this chapter will deal with a different type of reaction. This is graphically illustrated by the reaction of 1,3-butadiene (**1**) and maleic anhydride (**2**; see Chapter 20, Section 20.4). When these two compounds are heated in an autoclave (a metal cylinder that can be charged with chemicals and sealed for heating, thus allowing reactions to be done at low to moderate pressures), the isolated product is **3**, in 90% yield.[1] No intermediates have been discovered for this type of cycloaddition reaction, and the reaction is believed to proceed by a *synchronous or concerted process* with only a transition state.

In this transition state, the π-bond of **2** is broken and both π-bonds between C1 and C2 and C3 and C4 in **1** are broken as *two* new carbon–carbon bonds are formed; a new π-bond between C2 and C3 of 1 is generated in a six-membered ring product. The failure to detect the presence of a cation, an anion, or a radical intermediate formed during this reaction leads to the conclusion that it is a synchronous or concerted mechanism. In other words, it is a "non-ionic" and a "nonradical" mechanism.

24.1 Give the major products formed when cyclohexadiene reacts with ethyl acrylate and when 2,3-dimethyl-1,3-butadiene reacts with methyl ethyl ketone.

The names associated with this reaction are taken from the two scientists who developed it, Otto Diels (Germany; 1876–1954) and Kurt Alder (Germany; 1902–1958). They were awarded the Nobel Prize in 1950, and the reaction is called the **Diels–Alder reaction** to honor their work. Many years of work led to a new mechanism that explains reactions of this type, based on what is known as *frontier molecular orbital theory*. This mechanism is based on the interaction of π-electrons in the diene with those of the alkene. Kenichi Fukui (Japan; 1918–1998) was awarded the Nobel Prize in 1981 for his contributions in this area.

A focus on π-electrons identifies diene **1** with four π-electrons in two C=C units, and alkene **2** has two π-electrons in the C=C unit. The overall reaction of the diene and the alkene makes a ring, so it is classified as a *cycloaddition*. The generic name of this reaction is a **[4+2] cycloaddition**, where the numbers refer to the four π-electrons of the diene reacting with the two π-electrons of the alkene; cycloaddition indicates the type of product that is formed. This is one of the most powerful methods known for making rings from acyclic compounds.

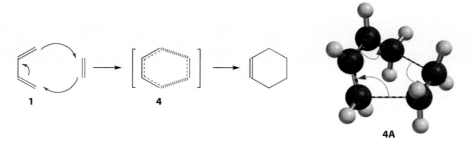

Perhaps the simplest example of a Diels–Alder reaction is between the conjugated diene 1,3-butadiene (**1**) and the simple alkene ethene. When they are heated to >200°C in a reaction bomb, the product is cyclohexene. Therefore, *the Diels–Alder reaction between a diene and an alkene leads to a*

cyclohexene product. As mentioned, no ionic or radical intermediates have been detected, and the reaction is said to proceed by a ***concerted mechanism***. This means that the bond-making/bond-breaking processes proceed simultaneously, so it is a synchronous reaction. To represent this concept, the transition state is represented by **4**, which is essentially a six-centered transition state involving reorganization of six π-electrons and the formation of two new σ-bonds. A molecular model of the transition state is shown as **4A**, indicating the electron flow, the elongated C–C single bonds being formed, and the transition of the C=C units. Representation **4** does not really explain why the reaction occurs, however. To do that, frontier molecular orbital theory suggests that the molecular orbitals for the two reactants must be examined.

24.2 What is the Diels–Alder product formed when 2,3-dimethyl-1,3-butadiene is heated with cyclopentene?

Figure 24.1 shows the molecular orbitals (MOs) for both ethene and 1,3-butadiene and identifies the highest occupied molecular orbitals (the HOMOs) and the lowest unoccupied molecular orbitals (the LUMOs) for both compounds. The MOs shown include the p-orbitals on each carbon atom. How are these molecular orbitals constructed? First, only the p-orbitals of the diene and the alkene are used because these orbitals are involved in the Diels–Alder reaction. An alkene has two π-electrons, so it will have two π-molecular orbitals. The two orbitals cannot be of the same energy, and the two electrons will reside, spin paired (see Chapter 3, Section 3.1.2), in the lowest energy orbital.

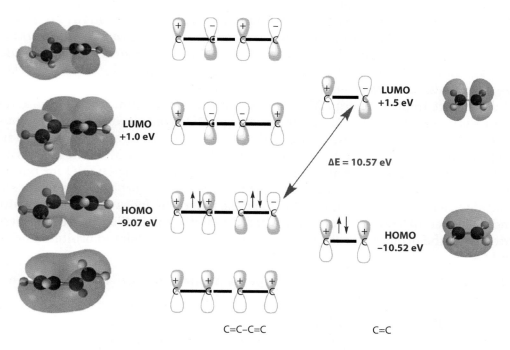

Figure 24.1 Frontier molecular orbitals for 1,3-butadiene and ethene.

For butadiene and for ethene, the orbitals are labeled "+" or "−" to represent the directional orientation of each orbital. These values arise from wave theory and the solution to the Schrödinger equation for this set of orbitals (see Chapter 3, Section 3.1), focusing *only* on the π-electrons. In the lowest energy orbital, all the "+" lobes are together (the same side); in the highest energy orbital, the "+" and "−" alternate. Another representation shows the lobes of the molecular orbitals with different colors and the molecular orbital of ethene that has "+ +" for the two colored lobes parallel one to the other. In the molecular orbital that has a "+ −" array, the colored lobes are on opposite sides. For reasons that will not be discussed here, overlap of lobes of the same symmetry is lower in energy, so the "+ +" MO is lower in energy than the "+ −" MO, and the lower energy orbital has both π-electrons of ethene.

At the point where a "+" lobe is adjacent to a "−" lobe, there is a **node** (the different color lobes are on opposite sides), which represents a point of zero electron density. The lowest energy orbital has zero nodes, whereas the highest energy orbital has one node. The higher the number of nodes is, the higher the energy of that molecular orbital will be. The highest energy molecular orbital that contains electrons for ethene is the highest occupied molecular orbital, the HOMO. The lowest energy molecular orbital that does not contain electrons is the lowest unoccupied molecular orbital, the LUMO. For ethane, the choices are obvious because there are only two MOs.

The same analysis is done for the molecular orbitals of butadiene, with the lowest energy orbital having zero nodes. Butadiene has four π-electrons, so there will be four π-molecular orbitals, each one capable of containing two electrons. Therefore, the two lower energy MOs will contain the four electrons—two each and spin paired. The lowest energy MO will have a "+ + + +" array so that all four colored lobes overlap and there are zero nodes. *It is important to understand that the four lobes shown for the "+ + + +" array constitute one molecular orbital, not four. There are four carbon atoms and each is sp² hybridized, so there are four lobes, but it is one of four molecular orbitals.*

The next highest energy MO should have one node and a "+ + − −" array, as in Figure 24.1. Both of these lower energy MOs contain electrons. The next highest MO should have two nodes with a "+ − − +" array, as shown, and will not contain electrons (it is an empty MO). The highest energy MO will have three nodes, with a "+ − + −" array, and it will also be empty. The HOMO for butadiene is the "+ + − −" orbital and the LUMO is the "+ − − +" orbital. The molecular orbitals are also shown as molecular models, where "+" lobes are shown in **blue** and "−" lobes in **red**. The nodes are readily distinguishable and easily correlated with the orbital cartoons drawn with the "yellow" colored orbitals and "+" and "−."

24.3 **Draw the lowest energy molecular orbital for 1,3,5-hexatriene that has zero nodes.**

The HOMO and the LUMO for both butadiene and ethene are labeled in Figure 24.1. *Note that the HOMOs for both ethene and butadiene are electron rich and the LUMOs for both are electron poor.* Experimentally, the first ionization potential (IP) for butadiene and for ethene is the energy of each HOMO, because IP represents the energy required to lose an electron (see Chapter 3, Section 3.3). Similarly, electron affinity (EA; see Chapter 3, Section 3.3) is the energy associated with gaining one electron, and it is taken to be the energy of the LUMO. A measurement of the first ionization potential for a diene or an alkene gives the energy of the HOMO, and measuring the first electron affinity gives the energy of the LUMO. Therefore, the HOMO is associated with electron donation and the LUMO with electron capture.

To explain the Diels–Alder reaction, make the assumption that in a reaction between a diene with four π-electrons and an alkene with two π-electrons, the diene donates electrons to the alkene because it has more electrons. For this to occur, the $HOMO_{diene}$ must donate electrons to the $LUMO_{alkene}$, which means that the energy between these two molecular orbitals is very important. In Figure 24.1, the energy of each HOMO and LUMO is provided (these are experimentally determined values) and, if the HOMO of the diene donates electrons to the LUMO of the alkene, we can use the energy values to determine the energy difference: the ΔE.

The ΔE for the $HOMO_{diene}$–$LUMO_{alkene}$ for butadiene/ethene is 10.57 eV (1 eV = 23.06 kcal mol^{-1}) because the $HOMO_{diene}$ energy is –9.07 eV and the $LUMO_{alkene}$ energy is +1.5 eV. *The ability of the diene to donate electrons to the alkene and the relatively low energy between the $HOMO_{diene}$ and $LUMO_{alkene}$ drives the Diels–Alder reaction.* Chapter 7 (Section 7.5) discussed the activation energy for a reaction, E_{act}, which is the energy required to begin the bond-making/bond-breaking process. Activation energy is formally the difference in energy between the starting materials and the transition state. *The value of ΔE just described for the Diels–Alder reaction is associated with the activation energy (E_{act}) for the reaction. A larger value of E_{act} (larger ΔE) indicates a slower reaction, and a lower E_{act} (smaller ΔE) indicates a faster reaction.*

24.4 Suggest a reason why the Diels–Alder reaction is driven by the $HOMO_{diene}$–$LUMO_{alkene}$ interaction rather than the $HOMO_{alkene}$–$LUMO_{diene}$ interaction, using Figure 24.1.

The C1 and C4 lobes of the HOMO of butadiene have opposite symmetry (one lobe is "+" and one is "–"). As shown in transition state **4**, the C1 and C4 carbons of butadiene are those where new carbon–carbon bonds are formed, so the symmetry of these orbitals is important. The orbitals of the diene must react with the molecular orbital of ethene in the LUMO, and the reactive LUMO has one "+" lobe and one "–" lobe. The symmetry of the reacting orbitals matches; this is important to the reaction. If the symmetry of the orbitals does not match, the reaction does not occur by this concerted mechanism.

The bottom line of this discussion is that the Diels–Alder reaction is driven by the interaction of the molecular orbitals of the diene and the alkene. The difference in energy between the $HOMO_{diene}$ and the $LUMO_{alkene}$ is effectively the activation energy for the reaction. If the energy difference is large, the activation energy is large; the reaction will be very slow and it will likely require high heat and long reaction times. If the energy difference is small, then the activation energy is small and the reaction will be much faster, perhaps occurring at very low reaction temperatures and/or short reaction times.

24.2 Reactivity of Dienes and Alkenes

There are many examples of the Diels–Alder reaction (a [4+2] cycloaddition). In the reaction of 1,3-butadiene (**1**) with ethene, high temperatures are required for the reaction to generate cyclohexene. If 1-methoxyethene (**5**) is heated with 1,3-butadiene, the reaction requires heating to >200°C in a reaction bomb for several hours; however, the yield of product (**6**) is lower than the yield of cyclohexene observed in the reaction with ethene. This result suggests that **5** is even *less* reactive than ethene. As shown in Section 24.1, the reaction of **1** and maleic anhydride (**2**) occurs in benzene at 100°C to give **3** (see the experiment from Section 24.1).

Clearly, the reaction of **2** and butadiene is much faster than the reaction of ethene because it does not require high temperatures and high pressure; the reaction with **5** is slower. According to frontier molecular orbital theory, a mechanistic explanation of these experimental observations must include a discussion of the molecular orbitals of 1,3-butadiene and ethene (see Figure 24.2) compared with the molecular orbitals of **2** and **5**. Figure 24.2 shows the energy level of the molecular orbitals but does not show the actual orbitals. Indeed, only the energy value is required once it is known that the HOMO of the diene reacts with the LUMO of each alkene.

The activation energy for the Diels–Alder reaction is taken to be the energy difference (ΔE) between the $HOMO_{diene}$ and the $LUMO_{alkene}$. The energies of the HOMO and LUMO for butadiene, ethene, **2**, and **5** are shown in Figure 24.2, using ΔE for butadiene-ethene of 10.57 eV as a standard for comparison with the other alkenes. Note that when compared to ethene, **2** has two carbonyl groups attached to the C=C unit, and the carbonyl groups are electron withdrawing. The OEt group attached to the C=C unit in **5** is electron releasing. Therefore, the π-bond in **2** is weaker (less electron density) than the π-bond in ethene, and the π-bond in **5** is stronger (more electron density).

Figure 24.2 allows one to estimate the influence of electron-releasing and electron-withdrawing groups when they are attached to a C=C unit. The ΔE for butadiene-maleic anhydride is smaller than that of ethene (only 8.5 eV), whereas the ΔE for butadiene-methyl vinyl ether is 11.07 eV, which is greater than that for ethene. By comparing the energy differences for all three alkenes with the reaction conditions, it is clear that the alkene that reacts fastest

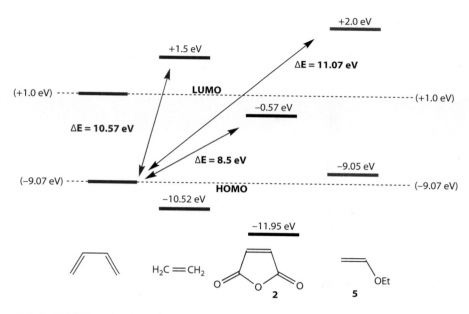

Figure 24.2 HOMO and LUMO energies for 1,3-butadiene, ethene, maleic anhydride, and methyl vinyl ether.

(maleic anhydride) has the smallest value for ΔE and the alkene that reacts slowest (methyl vinyl ether) has the largest ΔE. Ethene is in the middle. Maleic anhydride has electron-withdrawing carbonyl groups attached to the C=C unit and methyl vinyl ether has an electron-releasing OMe attached to the C=C unit; this difference correlates with the reactivity of each alkene.

If an alkene has an electron-withdrawing group attached, such as a carbonyl, the $LUMO_{alkene}$ is lower in energy than ethene because there is less electron density in the π-bond (it is weaker); this makes the ΔE value smaller. The smaller ΔE value leads to a faster reaction (milder reaction conditions). When the alkene has an electron-releasing group attached to it, such as OR, the $LUMO_{alkene}$ is higher in energy than ethene because the electron-releasing methoxy group increases the electron density in the π-bond (it is stronger); this makes the ΔE value larger. A larger ΔE value leads to a slower reaction (harsher reaction conditions). Butadiene and related conjugated dienes react best with alkenes bearing an electron-withdrawing group, which correlates with one theme of this chapter: the reaction of conjugated dienes and conjugated carbonyl compounds.

24.5 Determine which diene, (a) or (b), will react fastest with maleic anhydride (LUMO = −0.57 eV): (a) 1-phenyl-1,3-butadiene (HOMO = −8.16 eV) or (b) penta-2,4-dienoic acid (HOMO = −9.41 eV).

24.6 Draw the product formed by the Diels–Alder reaction of 2,3-diphenyl-1,3-butadiene with methyl acrylate.

24.7 Draw the product formed by the reaction of 1,3-butadiene (a) with methyl vinyl ketone, and (b) with acrolein (CH₂=CHCHO).

24.3 Selectivity

The Diels–Alder reaction generates new stereogenic centers from an alkene and a diene, which have no stereogenic centers (see Chapter 9, Section 9.1). The reaction will produce different diastereomers as well as different enantiomers. In addition, the orientation of substituents on the alkene or diene may lead to different positional isomers (regioisomers). It is therefore appropriate to discuss briefly the factors that influence the stereochemistry of the Diels–Alder product.

24.3.1 The Alder Endo Rule

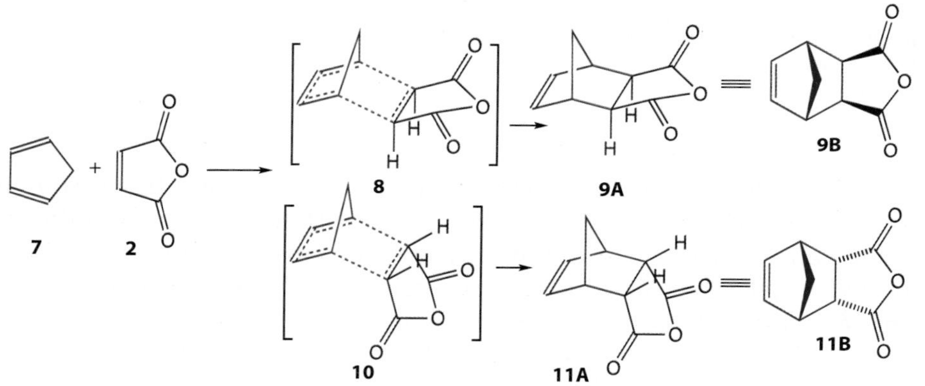

In addition to the acyclic dienes mentioned in the preceding section, cyclic dienes such as cyclopentadiene (**7**) and cyclohexadiene react with conjugated carbonyl derivatives to give a Diels–Alder product. A complicating factor, however, is quickly seen in the reaction of **7** with maleic anhydride (**2**). The –CH₂– of cyclopentadiene is incorporated in the cyclohexene products **9** and **11** as a –CH₂– "bridge." These compounds are drawn differently in **9B** and **11B** to call attention to the spatial relationships of the atoms in which the methylene is a bridge in the cyclohexene product. The anhydride unit is "up" relative to the –CH₂– bridge in **9B** (they have a *syn* relationship) and "down" in **11B** (they have an *anti* relationship). The carbonyl groups in **9A** are said to be *exo* to the bridge and the carbonyl groups in **11A** are said to be *endo* to the bridge; thus, **9** is the exo product and **11** is the endo product.

In this particular reaction, **11** is the major product by a factor of 3:1. To explain this experimental observation, the transition state for the formation of **9** from cyclopentadiene (shown as **8**) is compared with the transition state for the formation of **11** (shown as **10**). In **10**, the π-bonds of the carbonyl groups are underneath (endo) the π-bonds of the dienes. In this orientation, the p-orbitals of the carbonyls interact with the p-orbitals of the diene unit, and a small measure of stabilization is possible when compared to the exo orientation of the carbonyls in **8**. *This stabilization is called a secondary orbital interaction and it stabilizes transition state 10 relative to 8.* When the carbonyl units are projected away from the diene units (exo), there is no possibility for secondary orbital interactions, and the energy of that transition state is higher. The more stable transition state (**10**) leads to the major product, **11**.

24.8 Draw the structure of the conjugated isomer of cyclohexadiene.

24.9 Draw the major product of a reaction between cyclopentadiene and diethylmaleate (diethyl but-2Z-endioate).

Kurt Alder initially observed this preference for the endo product and it has come to be known as the **Alder endo rule**. Close scrutiny using modern mechanistic tools gives an understanding of this phenomenon as being driven by secondary orbital interactions. This preference for the endo product is also seen in reactions of acyclic dienes, but it is difficult to label the product unless there is a stereochemical marker (see Section 24.3.3).

24.3.2 Regioselectivity

All dienes used in the Diels–Alder reaction have been symmetrical, which means that they have either two identical substituents (on C2 and C3 as in 2,3-dimethyl-1,3-butadiene) or they have no substituents at all (1,3-butadiene, **1**; cyclopentadiene, **7**). Clearly, some dienes are unsymmetrical, with one substituent or different substituents on C2 or C3 or even on C1 or C4. This section discusses the Diels–Alder reactions of unsymmetrical dienes. The reaction of these dienes with alkenes leads to formation of regioisomers (isomers with a different structural attachment of atoms).

24.10 Draw the structure of the unsymmetrical conjugated diene 2-bromo-3-ethyl-2E,4Z-heptadiene.

In the Diels–Alder reaction of 2-methyl-1,3-butadiene (**12**; known as isoprene) and ethyl acrylate (**13**), there are two possible cycloadducts: **15** and **17**. These two compounds are regioisomers because the relative position of the methyl group and the CO_2Et group is 1,3 in **15** and 1,4 in **17**. The transition

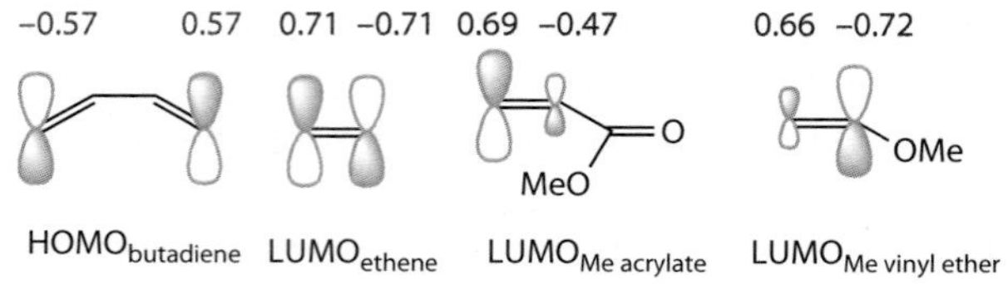

Figure 24.3 Orbital coefficients for butadiene, ethene, ethyl acrylate, and methyl vinyl ether. (Fleming, I. 1976. *Frontier orbitals and organic chemical reactions,* 132–138, Table 4-2. London: Wiley. Copyright Wiley-VCH Verlag GmbH & Co. KGaA. Reproduced with permission.)

state (**14**) for generating compound **15** has the C=C unit approach the diene with the CO$_2$Et group closest to C1 (the methyl is on C2). Similarly, if ethyl acrylate approaches the diene such that the CO$_2$Et group is farther away from C1, the product is **17** via transition state **16**. There is no obvious reason to believe that **14** is more or less sterically hindered than **16**, and one expects a mixture of both products.

When this reaction is analyzed, there is indeed a mixture, but **17** is the major product. Cycloadduct **17** is the major product because the transition state leading to that product is lower in energy than the transition state leading to **16**. Without molecular modeling tools, this is difficult to predict. An alternative approach recognizes that the major product is formed when the largest orbital (the orbital with the largest orbital coefficient) of the diene overlaps with that of the alkene in the transition state of the reaction. Each orbital lobe will be of a different size (increased or decreased electron density) in an unsymmetrical diene or alkene. In a symmetrical diene or alkene, the orbitals are typically of the same size. The orbital coefficient is a measure of the "size" of the orbital and thereby its electron density. Typical coefficients are illustrated in Figure 24.3.

With a completely symmetrical diene such as 1,3-butadiene (**1**), the orbitals at C1 and C4 are of equal magnitude. The same can be said for ethene. When a substituent is attached to the diene, however, the orbital coefficients closest to the substituent are often different from those farthest away. An electron-withdrawing group leads to a smaller coefficient on the closest orbital of the C=C unit. Conversely, an electron-releasing group leads to a larger coefficient on the closest orbital of the C=C unit. The actual regiochemistry depends on the magnitude of the coefficients, which is determined largely by computer calculations, although a few generalizations are possible.[2]

1. *A diene with a substituent at C1 (electron withdrawing or electron releasing) reacts with an electron-rich alkene to give the ortho product and it reacts with an electron-deficient alkene to give the 1,2 product.*
2. *A diene with a substituent at C2 (electron withdrawing or electron releasing) reacts with an electron-rich alkene to give the para product and it reacts with an electron-deficient alkene to give the 1,4 product, with one exception.*
3. *A diene with an electron-releasing substituent at C2 reacts with an alkene with an electron-releasing group to give a 1,3 product.*

In the absence of orbital coefficient data, experimental data are the only source that gives the actual regioselectivity, although the three rules presented can assist in making predictions. *Note that these rules are unreliable with some functional groups—particularly, weakly electron-releasing or -withdrawing groups. The rules are also unreliable when multiple substituents are present. However, for simple systems, predictions are reasonable.* When isoprene (**12**) reacts with ethyl acrylate (**13**), rule 2 predicts the 1,4 product (**17**) to be the major product. When 1-methoxy-1,3-butadiene (**18**) reacts with acrylonitrile (**19**), rule 1 predicts the 1,2 cycloadduct to be the major product. *In the absence of given coefficients or experimental data, any prediction of regioselectivity must rely on the three rules just given.*

There is another complicating factor with regard to stereochemistry. The two 1,2 cycloadducts in this reaction are **20** and **21** (the *syn-* and *anti*-diastereomers, respectively). This reaction generates stereoisomers in addition to regioisomers.

24.11 **Predict the structure of the major product resulting from the reaction of 2-methoxy-1,3-butadiene and ethyl vinyl ether.**

24.3.3 Diastereoselectivity

When a diene has a substituent at C1 and/or C2 and reacts with an alkene bearing a substituent, the product will have one or more stereogenic centers. If there are two or more stereogenic centers in the product, it may be formed as a mixture of diastereomers. In the reaction of **18** and **19**, for example, the cycloadduct has two stereogenic carbons; and the two diastereomers are **20**, where the two groups have a *syn* relationship, and **21**, where they have an *anti* relationship. Section 24.3.2 discussed the model of overlap of orbital coefficients to predict the "1,2" product from the reaction. Because acrylonitrile has π-bonds in the nitrile unit, secondary orbital interactions lead to a preferred "endo" transition state represented in Figure 24.4, although one "exo" transition state is also shown.

When the bond-making or -breaking process occurs, the groups on the diene marked in **red** and **blue** will be pushed in the direction *opposite* to the incoming alkene. The alkene is positioned on the "bottom," but, clearly, it may approach from either the top or the bottom. Attack from the bottom and the top will lead to two enantiomers—that is, to a racemic mixture. If the **red** and **blue** groups on the diene move away from the incoming alkene, they may move in the same direction (away from each other). If both groups move in the same direction to the right, **20** is predicted to be the product. This is the enantiomer of structure **20** shown previously. It is shown to reinforce the fact that the reaction gives a racemic product.

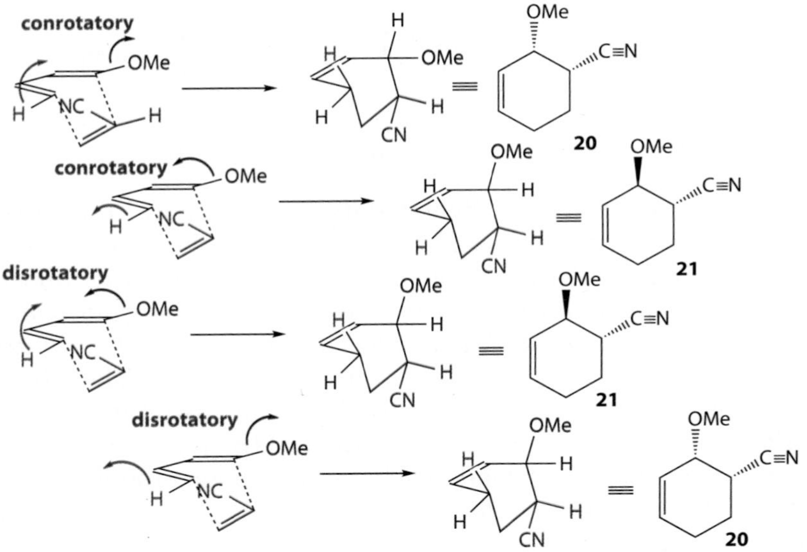

Figure 24.4 Conrotatory versus disrotatory cycloaddition.

If both groups move in the same direction to the left, we predict that **21** will be the product. Alternatively, the groups may move in opposite directions and, if the groups move toward each other, the product will also be **21**. If the groups move in opposite directions, but away from each other, the product will be **20**. If the groups move in the same direction, this is a ***conrotatory motion;*** if they move in the opposite direction, this is a ***disrotatory motion****. Diels–Alder reactions usually give the major product resulting from a disrotatory motion of the atoms marked in **red** and **blue**. For an endo approach, the groups usually move in a disrotatory motion toward each other; therefore, the major product of a reaction between **18** and **29** is predicted to be diastereomer **21**.*

24.12 Draw the two disrotatory products in which acrylonitrile approaches 1-methoxy-1,3-butadiene from the top, using Figure 24.4 as a guide.

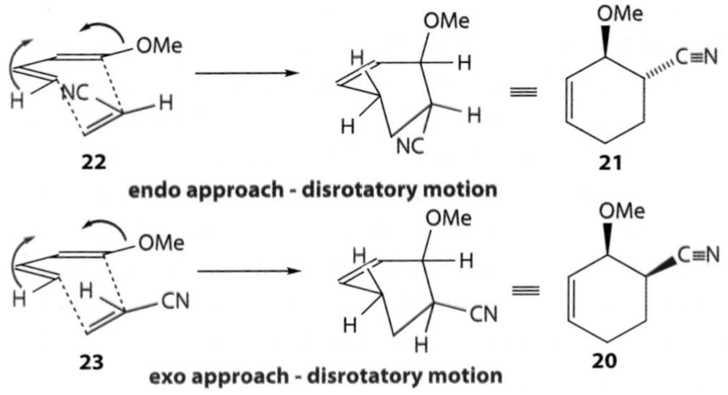

Section 24.3.1 noted that the Alder endo rule is applied to acyclic dienes as well as cyclic dienes, but it is more difficult to "see" the product in the former case. In order to form **21**, *an endo approach (**22**) of acrylonitrile to the diene is required*. An *exo* approach used transition state **23**, which leads to the other diastereomer, **20**. There is disrotatory motion for both *endo* and *exo* approaches. When the alkene has a substituent containing a π-bond—particularly, a C=O unit—endo approach is usually favored. An endo approach of the alkene combined with a disrotatory motion of the groups on the diene leads to formation of one diastereomer as the major product. *The Diels–Alder reaction is diastereoselective. Because the alkene may approach from either the bottom* (see Figure 24.4) or *the top, the product will be* **racemic**.

24.13 Assign *R/S* configuration to the stereogenic centers in both **20** and **21**.

24.14 Draw the proper transition state for this product.

24.15 Draw the structure of the major product when 1,3*E*-hexadiene reacts with methyl vinyl ketone.

24.16 What is the major product of the reaction of 2-cyano-1,3-butadiene and methyl vinyl ketone using these rules?

Three simple rules are available to predict the stereochemistry of the Diels–Alder reaction:

1. *The Alder endo rule predicts an endo cycloadduct as the major product.*
2. *The interaction of the largest orbital's coefficient of diene and alkene leads to the major regioisomer, as mentioned in Section 24.3.2.*
3. *Dienes react with alkenes via a disrotatory motion of the substituents at C1 and C4.*

24.4 Sigmatropic Rearrangements

Several important reactions involve transfer of electrons within a π-system, but they do not form rings. These reactions are categorized as *sigmatropic rearrangements. Sigmatropic rearrangements are defined as reactions in which a σ-bond (which means a substituent) moves across a conjugated system to a new site.*[3] Numbers attached to the length of each fragment associated with the move are placed in brackets. A [1,3]-shift involves migration of two fragments: One fragment is a single atom and the other fragment is three atoms in length. A [1,3]-shift of a proton is shown in **24**, where the hydrogen atom in **24A** migrates from C3 to C1 to give **24B**.

A [1,5]-shift involves migration of two fragments: One fragment is a single atom and the other fragment is five atoms in length. An example is the migration of the hydrogen atom from C5 in **25A** to C1 in **25B**. A [3,3]-shift involves migration of two fragments, each three atoms in length. In **26**, a three-carbon allylic unit (C1–C3) is connected to another three-carbon allylic unit (C4–C6). The rearrangement involves breaking the bond between C3 and C4 in **26** and forming a new bond between C1 and C6 in **27**, where the two three-carbon units have migrated. It is important to understand that all three of these sigmatropic rearrangements are reversible, as shown in the illustration.

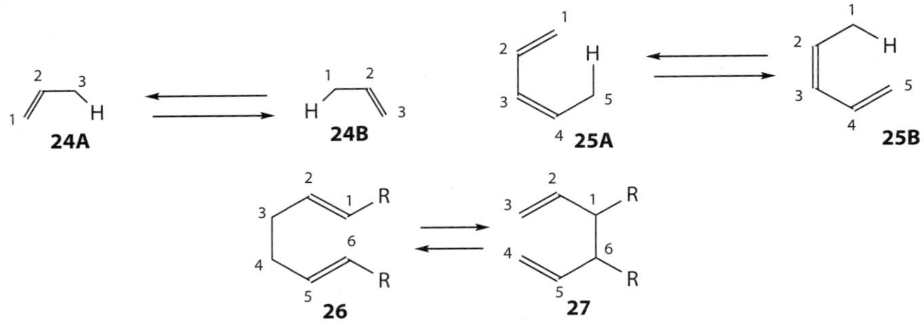

A recognizable molecule provides an example of a [1,5]-sigmatropic shift of a hydrogen atom. When 1-methylcyclopentadiene (**28**) is heated, the hydrogen marked in **violet** in **28** moves as shown to give **29** via a [1,5]-sigmatropic shift. The hydrogen atom is a one-atom fragment and it moves across a five-carbon fragment (C1 to C5). Similarly, the hydrogen in **29** moves to give **30** (from C5 to C4 as marked), and all of these isomers are in equilibrium with each other. The hydrogen atoms in **blue** and **violet** are marked to show the position movement of the hydrogen atom and the π-bonds during the [1,5]-sigmatropic rearrangement.

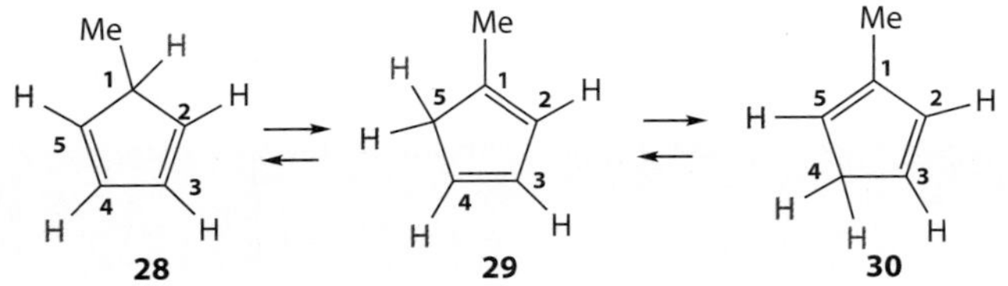

This rearrangement is called a ***suprafacial shift*** because ***the σ-bond to the hydrogen atom is made and broken on the same side of the conjugated system***. [1,7]-Sigmatropic shifts are observed in some systems. Although [1,5]-sigmatropic rearrangements are common, [1,3]-sigmatropic shifts are not because moving a hydrogen atom in a [1,3]-shift requires that the σ-bond be broken on one side of the conjugated system but made on the opposite side of that system. This is known as an *antarafacial shift*. Antarafacial shifts are generally not observed in this system and they will not be discussed in this book.

24.17 Draw the [1,7]-sigmatropic rearrangement product obtained when 1,3,5-dodecatriene is heated.

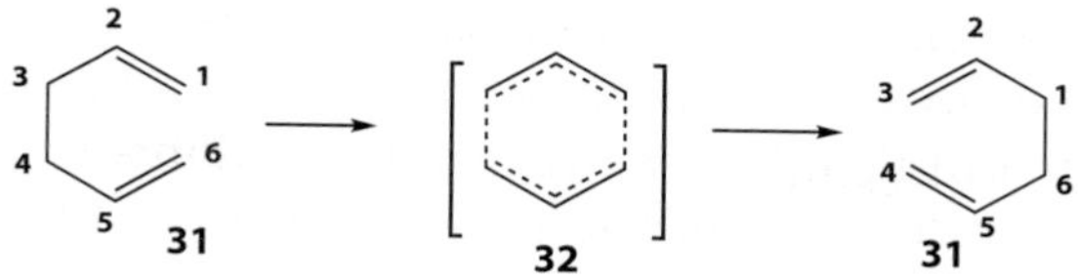

[3,3]-Sigmatropic rearrangements are extremely useful in the synthesis of organic molecules. The prototype [3,3]-sigmatropic rearrangement involves heating a 1,5-diene such as 1,5-hexadiene (**31**). The sigmatropic rearrangement proceeds by a concerted six-centered transition state represented by **32** and gives 1,5-hexadiene. The product is identical to the starting material. The C=C units of the diene are "labeled" with a substituent, as with 3,4-dimethyl-1,5-hexadiene (**33**); the [3,3]-sigmatropic rearrangement gives a different 1,5-diene: 2,6-octadiene, **34**.

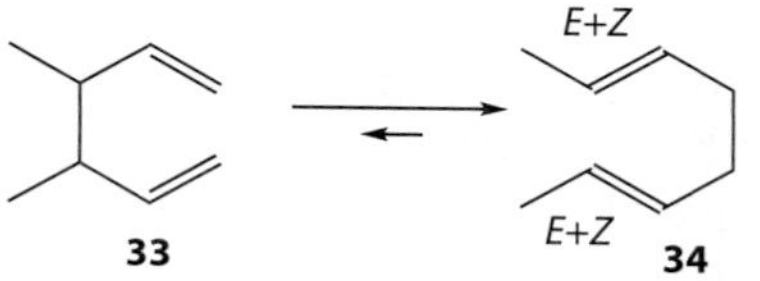

These two dienes (33 and 34) are in equilibrium, and the equilibrium favors the more thermodynamically stable compound. In this case, the more stable diene is **34**, where each C=C unit is disubstituted (see Chapter 12, Section 12.1). This rearrangement is called the **Cope rearrangement**, after Arthur C. Cope (United States; 1909–1966). The temperatures required for a Cope rearrangement are sometimes quite high, and the fact that an equilibrium is established between the two dienes is a problem. If the substitution pattern of the dienes is identical, the mixture of products may be difficult to separate. An example is the Cope rearrangement of 1,5-diene **35**, which, upon heating, gives **36**. There is one disubstituted C=C unit in **35** and one disubstituted C=C unit in **36**, so the equilibrium will not favor one over the other and close to a 1:1 mixture of these two dienes is expected. In other words, this reaction will produce a mixture and it is probably difficult to separate one from the other.

24.18 Give the IUPAC names for both 35 and 36.

24.19 Draw the Cope product formed when 37 is heated.

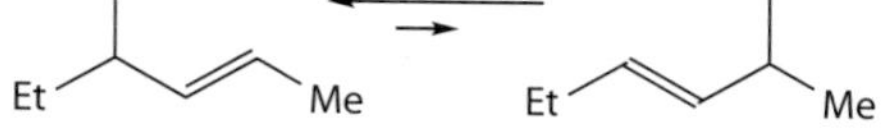

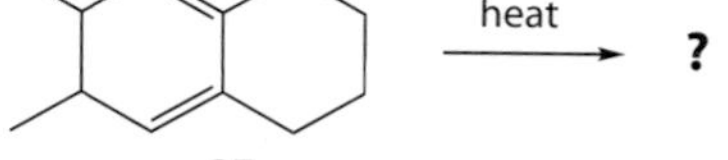

Another variation of a [3,3]-sigmatropic rearrangement is particularly useful. If one –CH$_2$– unit of a 1,5-diene is replaced with an oxygen atom, the structure is an allylic vinyl ether such as **38**. This compound has two C=C units in the 1,5-diene position, and heating leads to a [3,3]-sigmatropic rearrangement reaction that proceeds via transition state **39** to give aldehyde **40**. The equilibrium in this reaction favors **40** because the C=O bond is favored over a C=C bond, so the equilibrium favors formation of the aldehyde. The presence of the oxygen leads to an acceleration of the rate of reaction; therefore, the reaction temperature is relatively low when compared to the Cope rearrangement. This particular [3,3]-sigmatropic rearrangement is called the **Claisen rearrangement**, after Ludwig Claisen (Germany; 1851–1930).

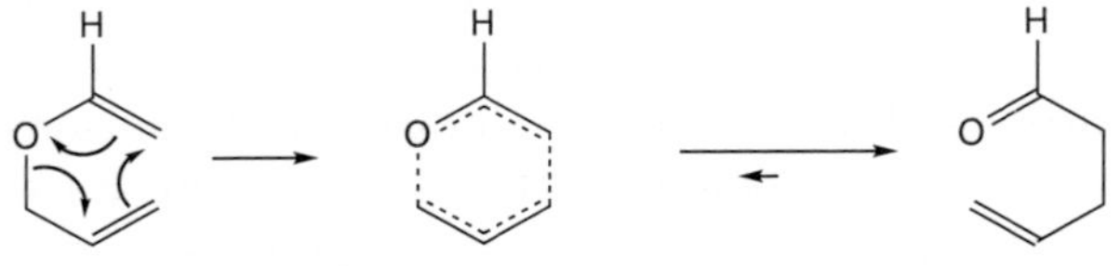

A modern version of this reaction is the so-called Ireland variant of the Claisen rearrangement (or the **Ireland–Claisen rearrangement**), after Robert E. Ireland (United States; 1929–). If an ester such as 2-propenyl acetate (**41**; the acetate of 2-propen-1-ol) is treated with lithium diisopropylamide (LDA) under kinetic control conditions, enolate anion **24** is formed (see Chapter 22, Section 22.4.2). The enolate anion is "trapped" by a reaction with chlorotrimethylsilane (Me₃SiCl) to give **43** because the silane reacts with the alkoxide of **42** rather than the carbanion. *Although not discussed previously and contrary to the assumptions made in Chapter 22, the enolate oxygen has a particular affinity for the silicon of chlorotrimethylsilane, which is where the reaction occurs.* The enolate oxygen displaces the chlorine from the S–Cl unit to give **43**. Enolate anion **43** has an allyl vinyl ether unit and heating leads to a [3,3]-sigmatropic rearrangement that gives **44**. Aqueous acid hydrolysis gives the carboxylic acid **45**.

24.20 Draw the structure of 7-methyl-4-oxadeca-2E, 7E-diene.

24.21 Give the IUPAC name of 40.

24.22 Draw both resonance contributors of 42 and highlight the "alkoxide" contributor.

24.23 Write out a synthesis of ethyl 2-phenylpent-4-enoate from acrolein.

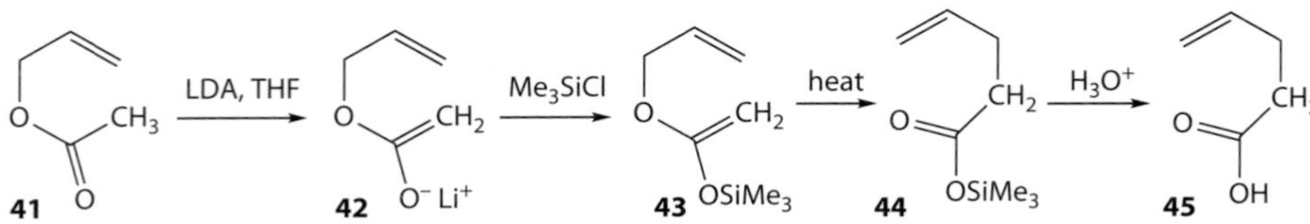

The Claisen rearrangement and the Cope rearrangement are very useful reactions. Note that the products are difunctional compounds. Dienes, alkene-aldehydes or alkene-ketones, and alkene-acids are all generated by these reactions.

24.5 Review of Synthetic Transformations

The Diels–Alder disconnection is actually a two-carbon disconnection. If a molecule has cyclohexene unit, the Diels–Alder disconnection leads to a diene and alkene for precursors as shown. A Cope disconnection has a 1,5-diene target with a 1,5-diene precursor. The oxy-Cope disconnection has an alkene aldehyde or an alkene-ketone target with a diene-alcohol precursor. The Claisen disconnection also has an alkene-aldehyde target, but the precursor is an allyl vinyl ether. The Ireland–Claisen disconnection is included, with an alkene-acid target and an alkene-ester precursor.

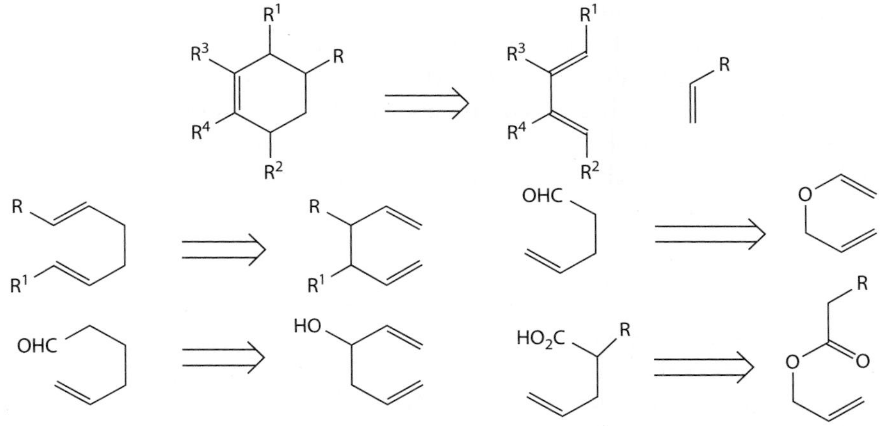

There are several interesting synthetic opportunities. The homework provides several examples that will illustrate synthesis using pericyclic reactions. One shown here is the synthesis of **46** from 1-methoxy-1,3-butadiene, **18**. The cyclohexene ring in the product suggests an Diels–Alder disconnection. It is reasonable that the carbon atoms of the starting material (**18**) are found in the cyclohexene ring. Disconnection of the alcohol side chain in **47** leads to disconnect product **47**, which most logically is an acceptor site leading to ketone **48**. The propyl fragment is designated as a donor, which correlates with a Grignard reagent. The Diels–Alder disconnection of **48** leads to the starting material **18**, along with methyl vinyl ketone.

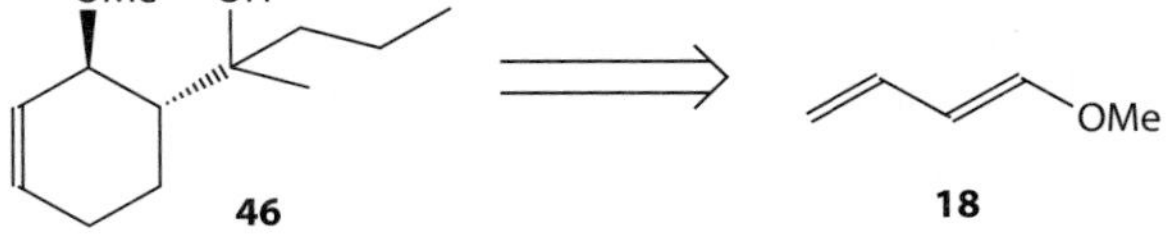

The synthesis of **46** therefore begins with a Diels–Alder reaction of **18** and methyl vinyl ketone to give **48** (see Section 24.3.2 for regiochemistry and Section 24.3.3 for stereochemistry). Only one step remains: **48** reacts with propylmagnesium bromide, obtained from 1-bromopropane (see Chapter 15, Section 15.1), and an aqueous acid workup to give the target, **46**.

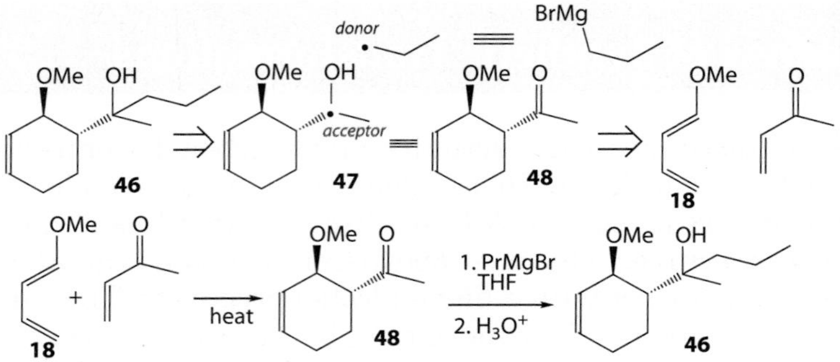

Another example is the synthesis of **49** from allyl vinyl ether **38**. Using **38** as a starting material clearly suggests a Claisen rearrangement, and the first disconnection recognizes that the aldehyde unit in **49** arises from a primary alcohol (see **50**). The primary alcohol can be prepared from a terminal alkene, as in **51**, and the phenyl ring in **51** arises by disconnection at the acyl carbon, which correlates with a Grignard addition to an aldehyde. These disconnections lead to **40**, which is the Claisen rearrangement product of **38**.

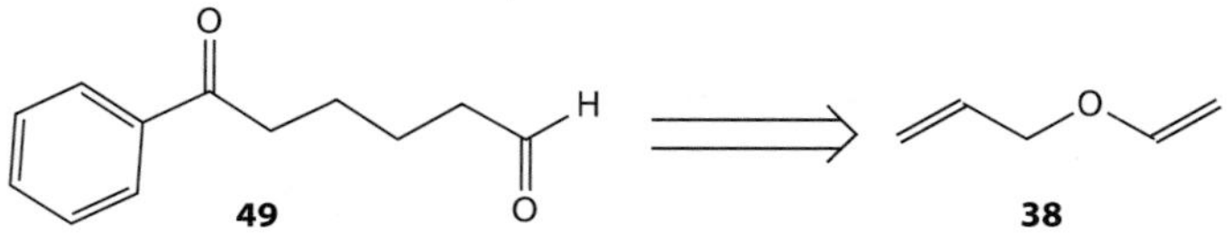

Therefore, the synthesis involves heating **38**, and Claisen rearrangement gives **40**. When **40** reacts with phenylmagnesium bromide, aqueous acid workup leads to alcohol **52** (see Chapter 18, Section 18.4, for the Grignard reaction). There are one or two options at this point, but hydroboration of the alkene and workup with NaOH and H_2O_2 (see Chapter 10, Section 10.4.4) give **53**. Treatment of this diol with an excess of PDC (Chapter 17, Section 17.2.3) gives the final target, **49**.

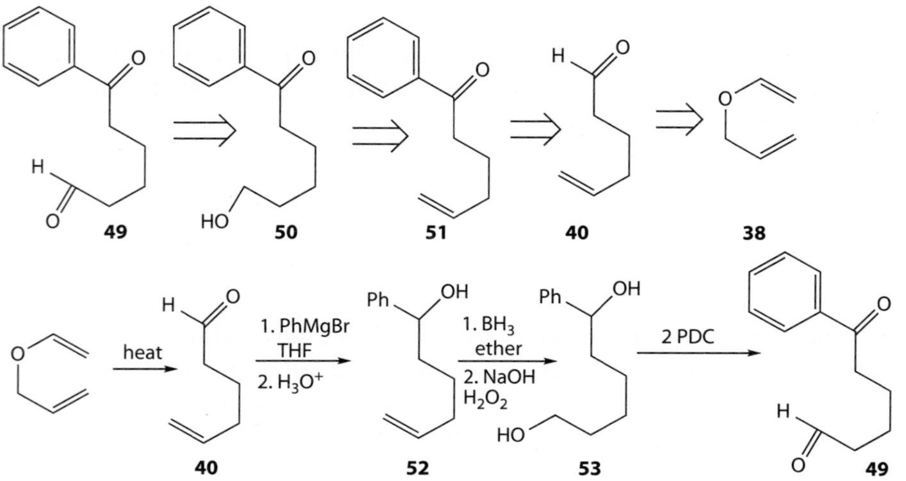

24.6 Biological Relevance

Pericyclic reactions are not common in biochemical transformations. One report shows that the last step in the aerobic biosynthesis of the corrin macrocycle of vitamin B12 in *Pseudomonas denitrificansis* is an enzyme-catalyzed reaction (precorrin-8x methyl mutase, abbreviated as CobH) in which the methyl group attached to C-11 of the substrate, precorrin-8x (**54**), which is bound to a protein, migrates from C-11 to C-12 to give the product hydrogenobyrinic acid (**55**).[4] This transformation is a 1,5 sigmatropic methyl shift.

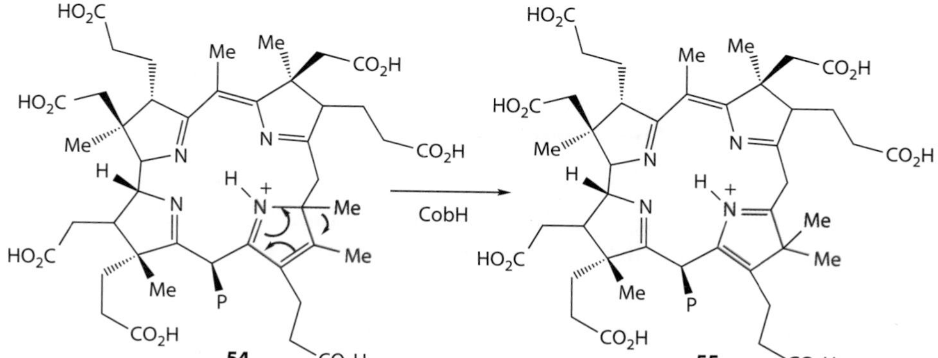

There are reports of Diels–Alderase enzymes in the biosynthesis of certain natural products, but their existence is somewhat controversial. For that reason, this discussion has been limited to the sigmatropic rearrangement example cited.

References

1. Fieser, L. F., and Novello, F. C. 1942. *Journal of the American Chemical Society* 64:802.
2. Fleming, I. 1976. *Frontier orbitals and organic chemical reactions,* 132–138. London: Wiley.
3. Fleming, I. 1976. *Frontier orbitals and organic chemical reactions,* 98. London: Wiley.
4. Shipman, L. W., Li, D., Roessner, C. A., Scott, A. I., and Sacchettini, J. C. 2001. *Structure* 9:587–596.

Answers to Problems

24.1

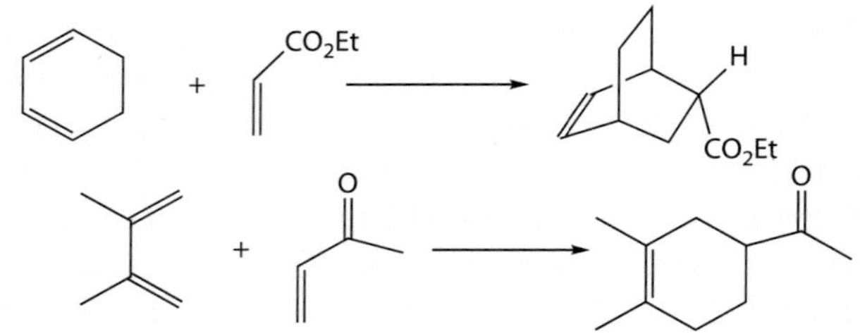

24.2

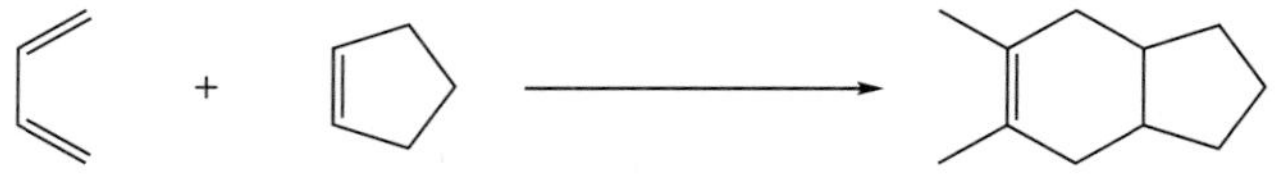

24.3

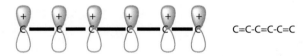

C=C-C=C-C=C

24.4   The ΔE for $HOMO_{alkene}$–$LUMO_{diene}$ is smaller than ΔE for $HOMO_{diene}$–$LUMO_{alkene}$, and the diene HOMO is more electron rich than the alkene HOMO.

24.5   The ΔE for (b) is larger, so the dienoic acid reacts faster than 1-phenylbutadiene.

24.6

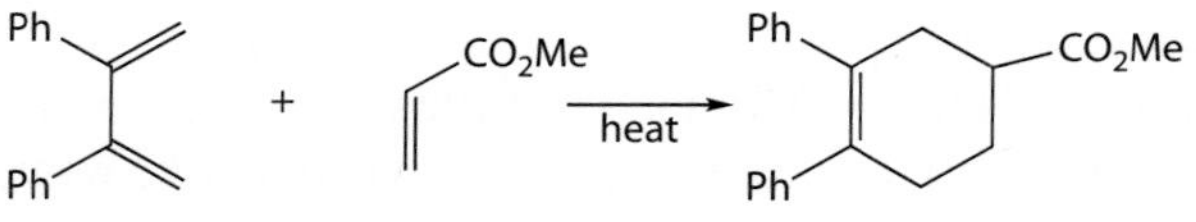

24.7

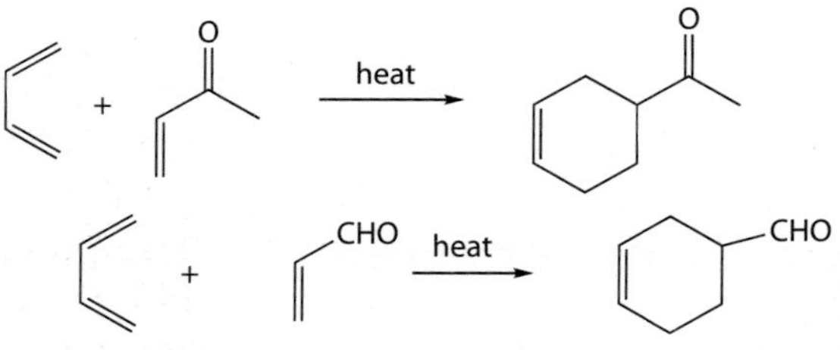

24.8

24.9

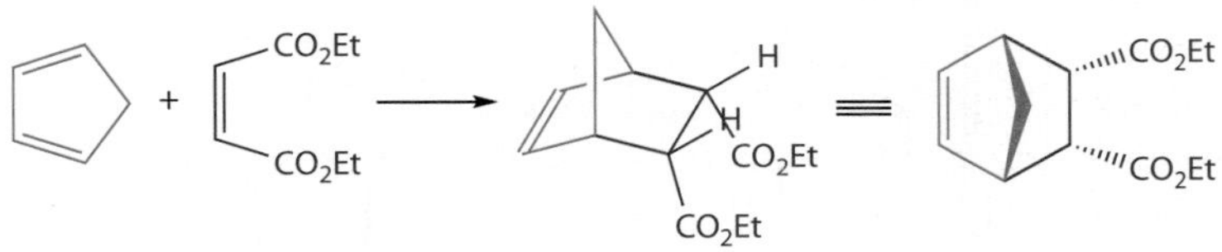

24.10

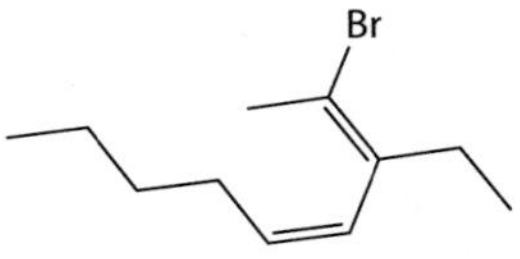

24.11

24.12

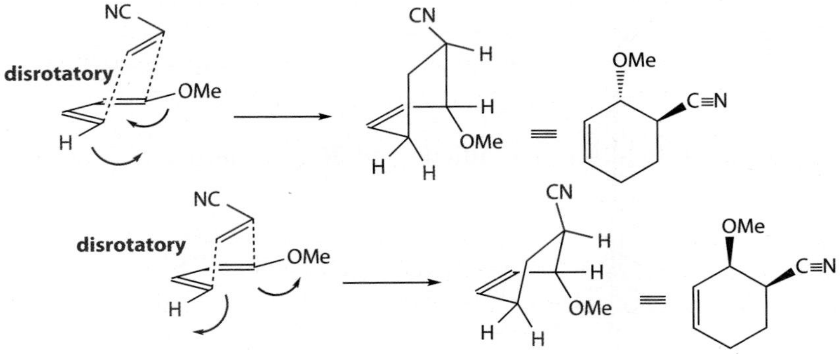

24.13

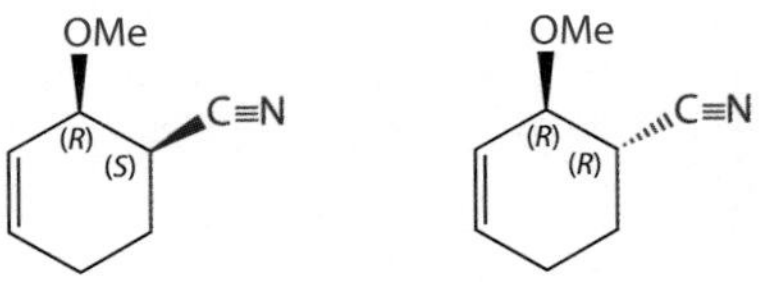

24.14

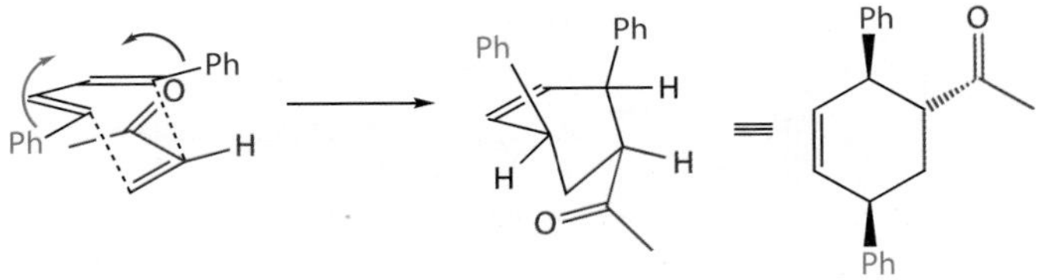

24.15

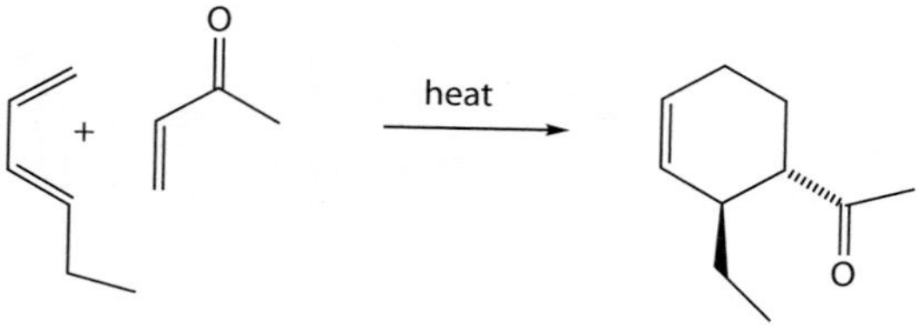

24.16

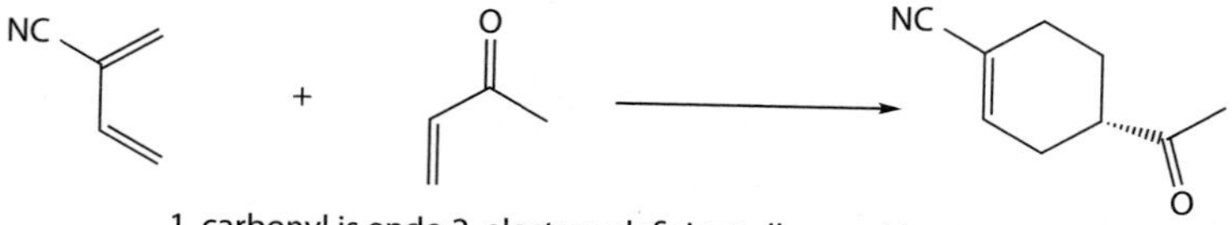

1. carbonyl is endo 2. electron deficient alkene with a group at C2
of diene, so there is a 1,4-relationship 3. not applicable

24.17

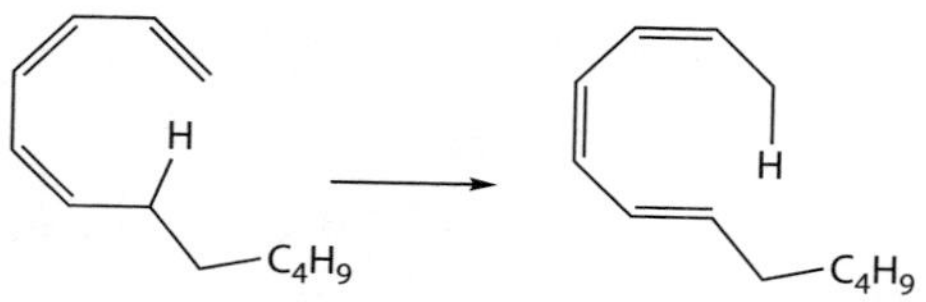

24.18   Diene **35** is 4-ethyl-2*E*,6-heptadiene and **36** is 5-methyl-3*E*,7-octadiene.

24.19

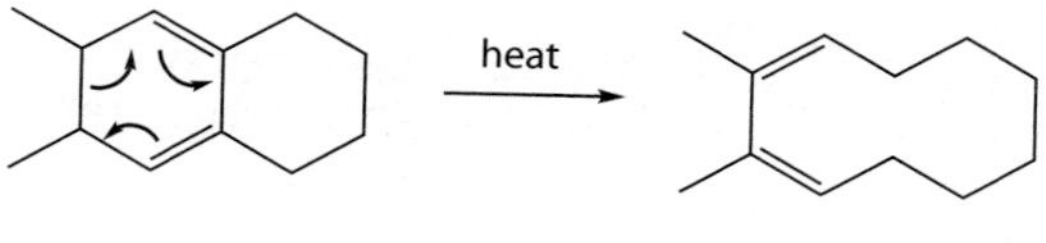

24.20

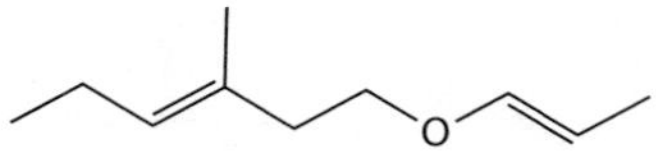

24.21   Pent-4-enal.

24.22

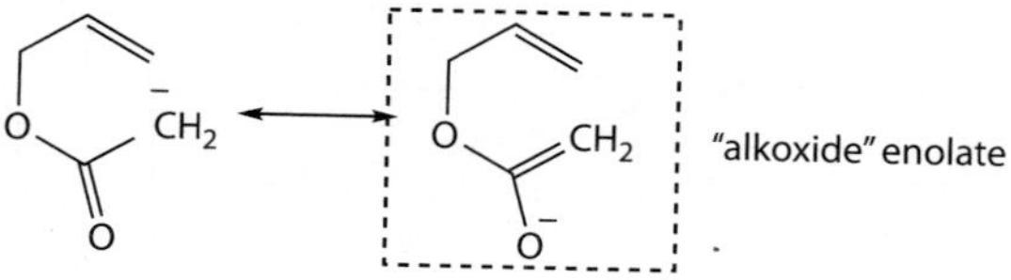

"alkoxide" enolate

24.23

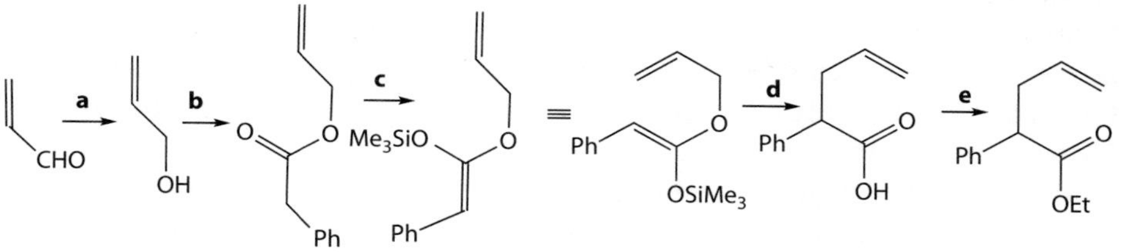

(a) NaBH$_4$, EtOH (b) PhCH$_2$COCl (c) 1. LDA, THF, –78°C 2. Me$_3$SiCl (d) heat (e) 1. SOCl$_2$ 2. EtOH

Correlation of Concepts with Homework

- **The thermal reaction between a conjugated diene and an alkene to give a cyclohexene derivative is a [4+2] cycloaddition, otherwise known as a Diels–Alder reaction:** 1, 2, 6, 7, 8, 24, 28, 31, 34, 36, 37, 39, 45.
- **The Diels–Alder reaction is driven by the energy differences between the highest occupied molecular orbital (HOMO) of the diene and the lowest unoccupied molecular orbital (LUMO) of the alkene. Alkenes bearing an electron-withdrawing group react faster because the LUMO is lowered, making the energy difference smaller. Conversely, alkenes bearing an electron-releasing group react more slowly because the LUMO is raised, making the energy difference large:** 3, 4, 5, 6, 7, 25, 26, 29, 31, 32, 33, 35, 37.
- **Secondary orbital interactions of alkenes bearing substituents having a π-bond lead to a preference for the endo product in the Diels–Alder reaction. This reaction is regioselective due to interactions of orbital coefficients on the orbitals of the reacting atoms. Substituents at C1 and C4 of the diene move in a disrotatory manner and opposite, relative to the incoming alkene:** 9, 10, 11, 12, 13, 14, 15, 16, 31, 37, 38, 39, 40.
- **A sigmatropic rearrangement is a reaction where a σ-bond moves across a conjugated π-system to a new site. The most common sigmatropic rearrangements are [3,3]-sigmatropic rearrangements, typified by the Cope rearrangement of 1,5 dienes and the Claisen rearrangement of allyl vinyl ethers. The Cope rearrangement generates a new 1,5 diene. The Claisen rearrangement generates a nonconjugated alkenyl aldehyde or ketone:** 17, 18, 19, 20, 21, 23, 24, 27, 30, 41, 42, 43, 44, 47, 48, 52.

- **A molecule with a particular functional group can be prepared from molecules containing different functional groups by a series of chemical steps (reactions). This process is called synthesis: The new molecule is synthesized from the old one (see Chapter 25): 46.**
- **Spectroscopy can be used to determine the structure of a particular molecule and can distinguish the structure and functionality of one molecule when compared with another (see Chapter 14): 47, 48, 49, 50, 51, 52, 53.**

Homework

24.24 Which of the following molecules cannot undergo a Diels–Alder reaction? Explain.

24.25 Which of the following molecules will require the mildest reaction conditions to react with 1,3-butadiene? Explain.

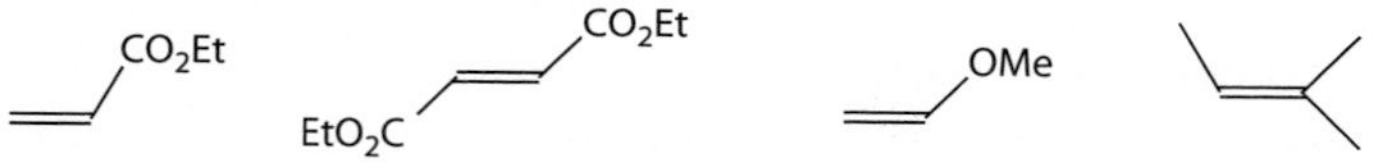

24.26 Which of the following molecules will require the harshest reaction conditions to react with 1,3-butadiene? Explain.

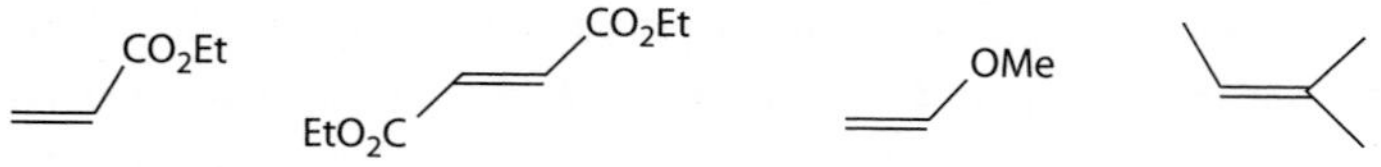

24.27 Molecule A is known to require much higher reaction temperatures to undergo a Claisen rearrangement when compared with B. Explain.

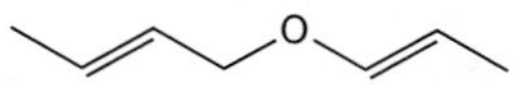

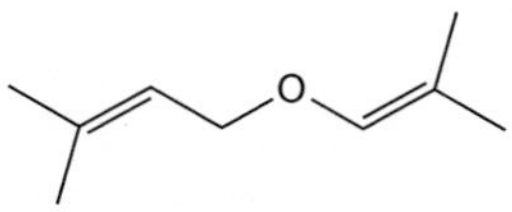

24.28 When cyclopentadiene reacts with $MeO_2C-C\equiv C-CO_2Me$, a Diels–Alder product is formed. Draw it. When this product reacts with one molar equivalent of hydrogen gas and a Pd catalyst, the nonconjugated double bond is reduced. Draw this product. When this new product is heated, ethylene gas is observed as a product and a new product is formed. Draw that product and offer a mechanism to account for its formation.

24.29 Cyclopentadiene reacts much faster in a Diels–Alder reaction than 1,3-butadiene. Offer a brief explanation for this observation.

24.30 Molecule A undergoes a Cope rearrangement significantly faster than B, and the reaction is largely irreversible. Suggest a reason for this difference and draw the product of both reactions.

24.31 Give the major product for each of the following reactions:

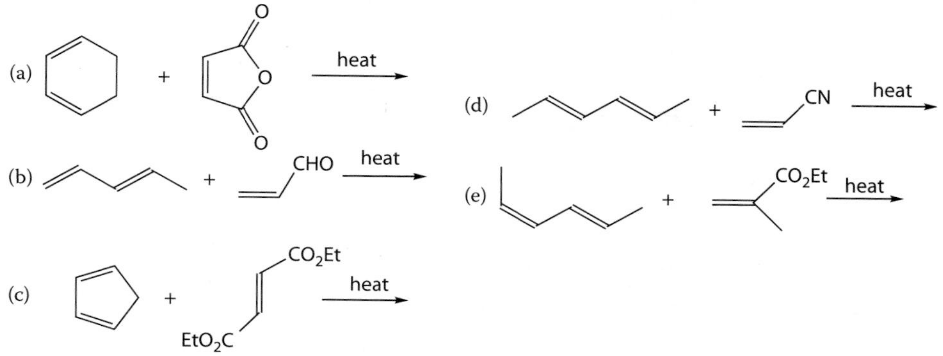

(a) ⬡ + [maleic anhydride] →heat→

(b) [pentadiene] + [CHO] →heat→

(c) ⬠ + [EtO₂C, CO₂Et] →heat→

(d) [diene] + [CN] →heat→

(e) [diene] + [CO₂Et] →heat→

24.32 Briefly explain why the reaction of acrolein and 1,3-butadiene requires much milder reaction conditions than the reaction with ethene.

24.33 Using your knowledge of Lewis acid–Lewis base reactions, suggest a reason why the addition of BF_3 to the Diels–Alder reaction of acrolein and 1,3-butadiene leads to a dramatic increase in reaction rate.

24.34 Briefly explain why the following molecules do not undergo a Diels–Alder reaction:

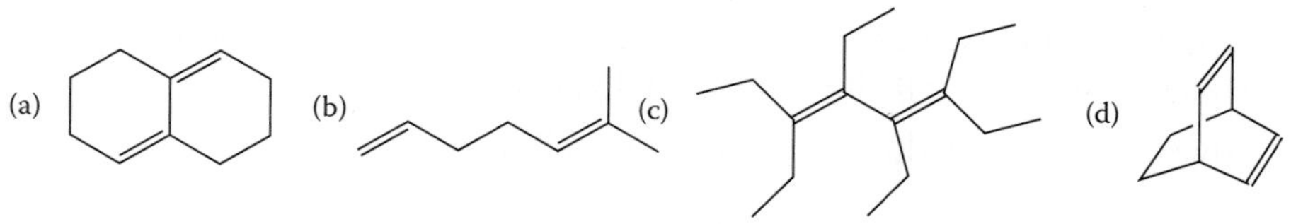

(a) [fused bicyclic] (b) [long chain alkene] (c) [branched diene] (d) [bicyclic]

24.35 The following molecule undergoes a Diels–Alder reaction with butadiene so slowly at temperatures below 100°C as to be essentially unreactive. Explain.

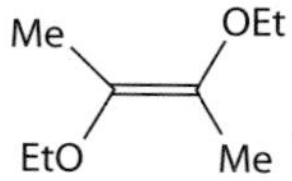

Me   OEt
   ⟍ ⟋
EtO   Me

24.36 Draw the major product for each of the following:

(a) 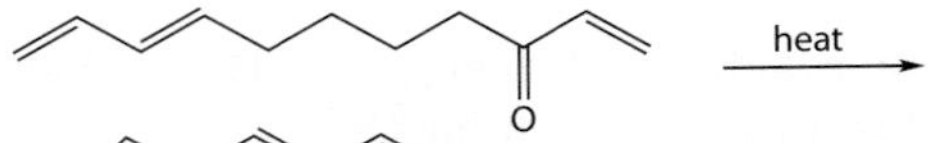 heat

(b) 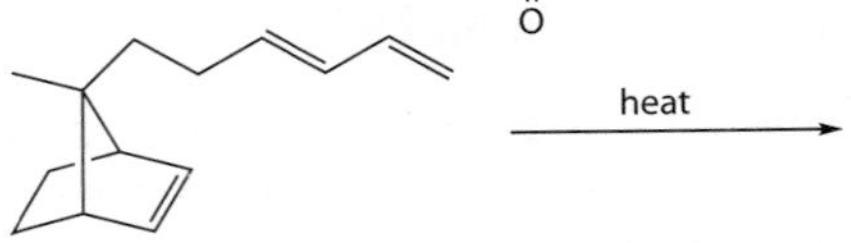 heat

(c) 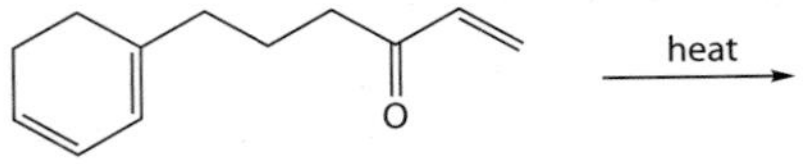 heat

24.37 Give the major product for each of the following reactions:

(a) 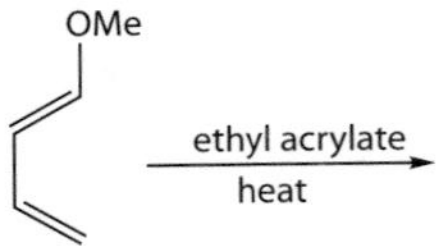   ethyl acrylate / heat

(d) 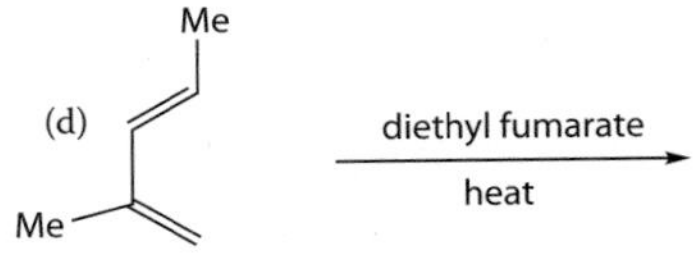   diethyl fumarate / heat

(b) 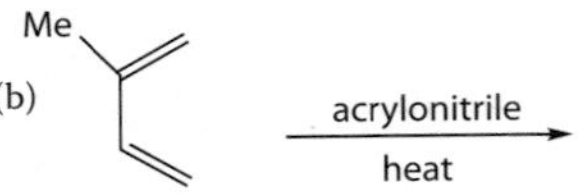   acrylonitrile / heat

(e) 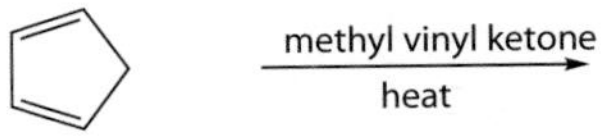   methyl vinyl ketone / heat

(c) 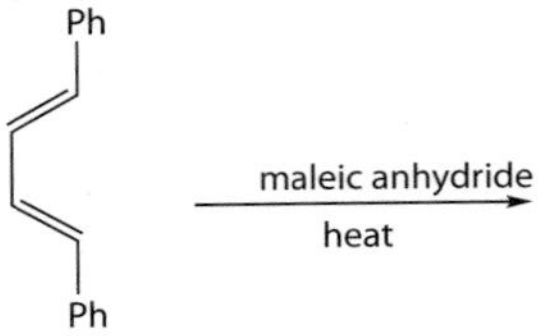   maleic anhydride / heat

(f) 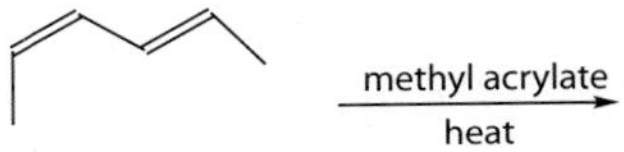   methyl acrylate / heat

24.38 Draw the products of both conrotatory and disrotatory addition for each of the following:
(a) 2E,4E-hexadiene + ethyl acrylate
(b) 2E,4Z-hexadiene + ethyl acrylate
(c) 1E-phenyl-1,3-butadiene + diethyl maleate
(d) 5-phenyl-2E-4E-pentadiene + diethyl fumarate

24.39 The Diels–Alder reaction of 2E,4E-hexadiene and diethyl maleate generates four new stereogenic carbon centers. How many possible stereoisomers are there? Draw all of them. Which one or ones can be expected to be the major product or products? Briefly explain.

24.40 Briefly explain why maleic anhydride shows greater "endo selectivity" in the Diels–Alder reaction than does ethyl acrylate.

24.41 Give the major product for each of the following:

(a) 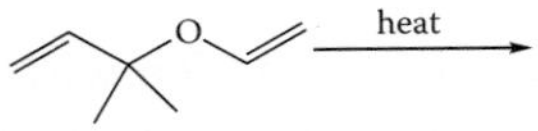   heat

(c)    heat

(b) 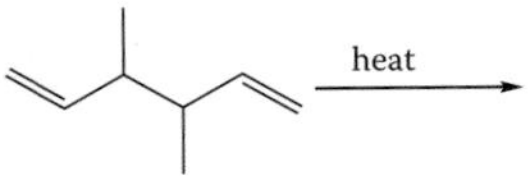   heat

(d) 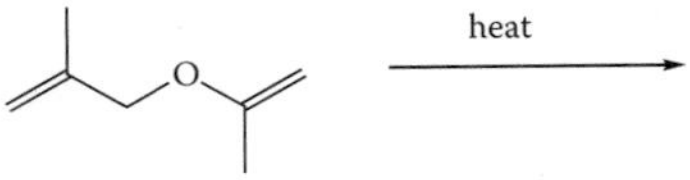   heat

24.42   The Cope rearrangement is a reversible process. When 3,4-diphenyl-1,5-hexadi-
         ene is heated, however, the equilibrium favors the rearrangement product. Why?

24.43   It is known that when an ester RCH_2CO_2R' is treated with 1. LDA; 2.
         Me_3SiCl, the product is $RCH=C(OR)-OSiMe_3$. Using this knowledge, predict
         the product of the following reaction:

1. LDA, THF, –78°C
2. Me₃SiCl
3. heat

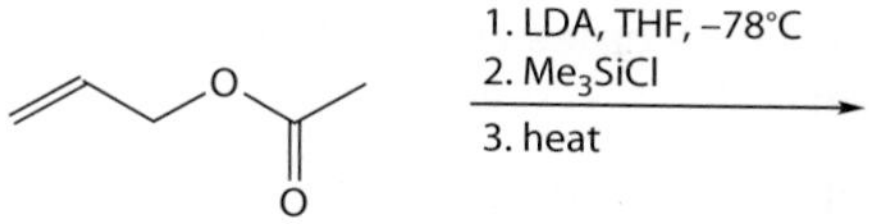

24.44   Draw the final product of the following reaction:

heat

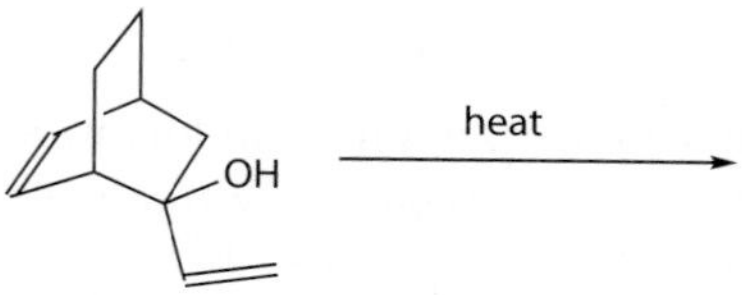

24.45   When you purchase a bottle of "cyclopentadiene," it is not really cyclopentadi-
         ene but rather another product. Heating this other product generates cyclo-
         pentadiene, which is distilled off and trapped. Suggest a structure for this
         other product based on the known chemistry of cyclopentadiene and suggest a
         reason why heating it generates cyclopentadiene.

Synthesis Problems

Do not attempt this problem until Chapter 25 has been read and understood.

24.46   Show a synthesis for each of the following molecules from the designated
         starting material:

(a)

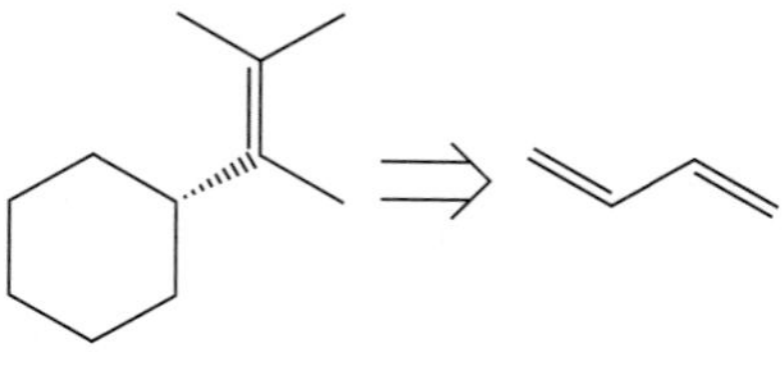

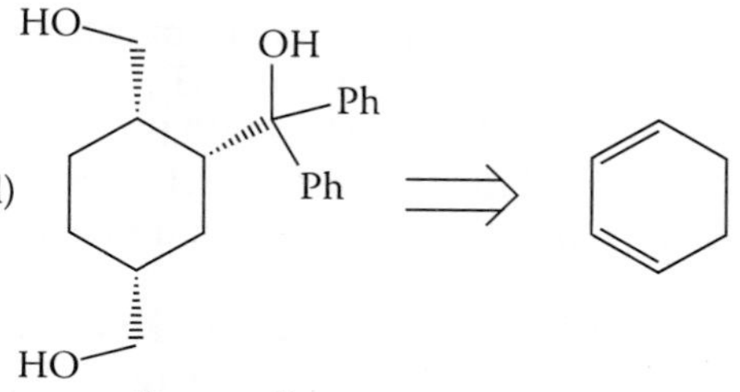

(b)

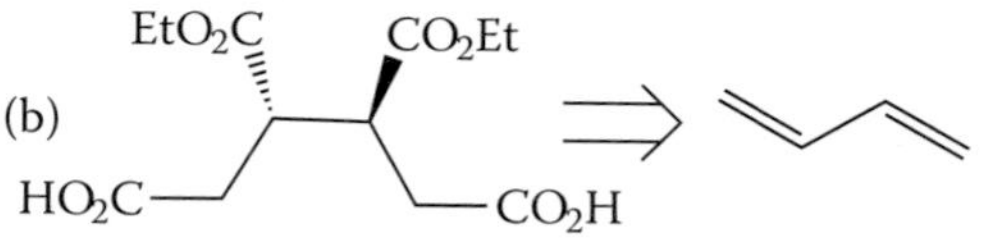

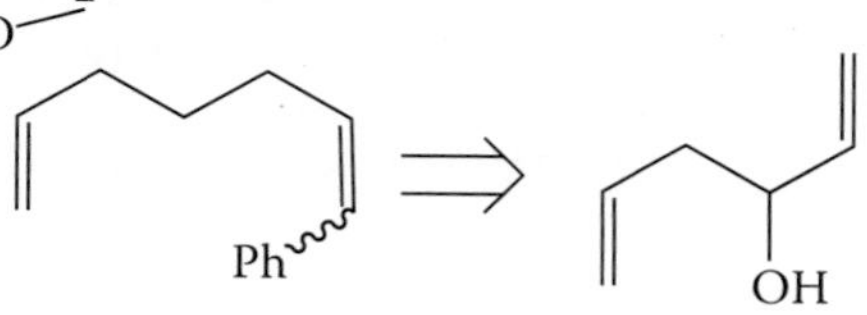

(c)

(d)

(e)

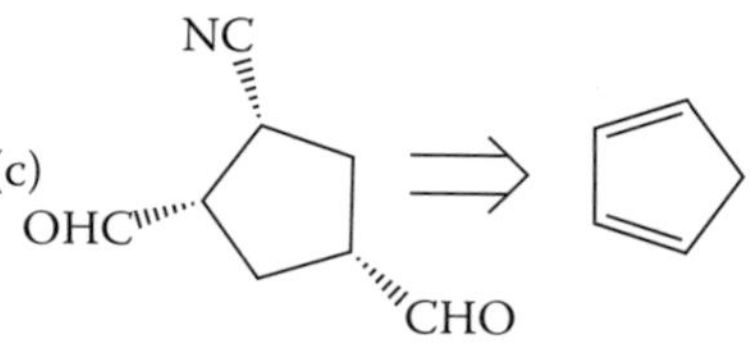

(f)

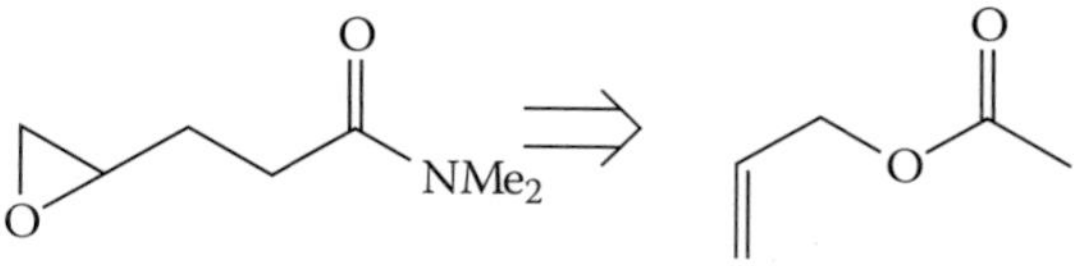

Spectroscopy Problems

Do not attempt these problems until Chapter 14 has been read and understood.

24.47   The Claisen rearrangement of allyl vinyl ether leads to an alkenyl aldehyde. Draw both the starting material and the product. Discuss spectroscopic differences in these two molecules that will allow you to determine whether or not the reaction worked.

24.48   When 3,4-dimethyl-1,5-hexadiene undergoes a Cope rearrangement, the product is 2,6-octadiene. This is an equilibrium reaction, so presumably both products are present, but one will be in excess. Describe any spectroscopic differences between these two molecules (draw both of them) that will help in your identification of the major product (i.e., which way the equilibrium lies).

24.49   Give the structure of a molecule with the formula C_7H_8 and the following spectral data. There is no significant UV absorption above 205 nm:
IR: 3067, 2986–2936, 2286, 1544, 1450, 1311, 728, 658 cm^{-1}
^1H NMR: 5.60 (m, 4H), 2.96 (m, 2H), 2.02 (d, 2H) ppm
^{13}C NMR: 143.2, 75.2, 54.4 ppm

24.50   Give the structure of a molecule with the formula C_6H_8O and the following spectral data:
IR: a sharp peak of medium intensity at about 1640 cm^{-1}
^1H NMR: 5.78 (m, 2H), 4.42 (m, 2H), 1.98–1.73 (m, 4H) ppm
^{13}C NMR: 132.8, 79.7, 23.4 ppm

24.51   Give the structure of a molecule with the formula $C_9H_8O_3$ and the following spectral data:
IR: very weak peak at 2982, 1842, 1777, 1229, 1090, 914, 843 cm^{-1}
^1H NMR: 6.31 (m, 2H), 3.59–3.52 (m, 4H), 1.77 (d, 1H), 1.59 (d, 1H) ppm
^{13}C NMR: 172.2, 135.9, 47.7, 472, 46.9 ppm

24.52   Give the structure of a molecule with the formula C_5H_8O and the following spectral data. When heated, this molecule gives a new compound that shows a strong peak in the IR at 1725 cm^{-1}. Identify both compounds:
IR: two sharp, medium-intensity peaks in the region of 1640 cm^{-1}
^1H NMR: 6.46 (m, 1H), 5.95 (m, 1H), 5.22 (m, 1H), 5.32 (m, 1H), 4.22 (d, 2H), 4.20 (m, 1H), 4.00 (m, 1H) ppm
^{13}C NMR: 152.4, 132.1, 118.2, 87.1, 70.6 ppm

24.53   Give the structure of a molecule with the formula C_7H_9N and the following spectral data:
IR: 3033, 2934, 2846, 2241, 1652, 656 cm^{-1}
^1H NMR: 5.70–5.59 (m, 2H), 2.77 (m, 1H), 2.34–2.05 (m, 4H), 1.93–1.84 (m, 2H) ppm
^{13}C NMR: 126.0, 123.9, 122.5, 29.2, 25.6, 25.3, 24.1 ppm

Disconnections and Synthesis

25

Is it possible to use all of the reactions shown in this book? Yes! Making carbon–carbon bonds and modifying functional groups, using a variety of reactions, is used to prepare many different individual compounds. One important area of organic chemistry strings together different reactions to make new and/or larger molecules from smaller ones. Many medicines or other important molecules possess a large number of carbon atoms and often several functional groups. If such a molecule is not readily available, it must be made. Normally this means choosing a molecule of fewer carbons as a starting material and building the molecule of interest by making the necessary carbon–carbon bonds and incorporating the functional groups. This process is called *synthesis*. This chapter will focus on rudimentary techniques for assembling molecules and provide methodology to analyze a molecule and determine what smaller molecule or molecules must be used for its synthesis.

In point of fact, the fastest way to study reactions is probably to attempt a synthesis that requires those reactions. Planning a synthesis via the disconnection approach instantly brings to light those reactions that are known and those that are not.

To begin this chapter, you should know the following:

- **the concept of bond polarization (Chapter 3, Section 3.7; Chapter 5, Section 5.4)**
- **the concept of strong and weak covalent bonds (Chapter 7, Sections 7.2 and 7.5)**
- **fundamental structure and nomenclature of all functional groups in this book (focus on Chapters 4, 5, 14, 15, and 20)**

- **all of the functional group transformations presented in this book (focus on Chapters 10–12, 16–21, 23)**
- **all of the carbon–carbon bond-forming reactions presented in this book (focus on Chapters 11, 14, 17, 19–21, 23)**
- **mechanisms (Chapter 7, Section 7.8; Chapters 10–12, 16–21, 23)**
- **all of the reagents used in this book (Chapters 10–12, 15, 17–21, 23)**
- **structure and bonding of organic molecules (focus on Chapters 3 and 5)**
- **the concept of isomers (Chapter 4, Section 4.5; Chapter 5)**
- **the concept of stereoisomers, including regioisomers, enantiomers, and diastereomers (Chapter 9, Sections 9.1, 9.3–9.6)**

This chapter will present a truncated version of the so-called disconnection approach to synthesis. The idea used here is to examine the molecule to be synthesized (the target) for bonds that are close to a functional group. Mentally breaking those bonds (a disconnection) will lead to fragments that are correlated with real molecules. This information allows one to correlate knowledge of chemical reactions associated with various functional groups with the disconnection fragments in order to make the bond that was disconnected. Working backward from the target to recognizable starting materials outlines the synthesis. In principle, following the reverse process and supplying reagents will give a synthesis. To accomplish a retrosynthesis–synthesis, all reactions must be considered from both directions. An example is an alcohol. An alcohol is oxidized to a ketone or aldehyde, but an aldehyde or ketone is reduced to an alcohol. Both reactions must be known.

When you have completed this chapter, you should understand the following points:

- **The target is the molecule to be synthesized. The starting material is the molecule used to begin the synthesis. Disconnection is the process of mentally breaking bonds in a target to generate simpler fragments as new targets to be used in the synthesis. The disconnection approach to synthesis is sometimes called retrosynthetic analysis.**
- **If a starting material is designated, try to identify the carbon atoms of the starting material in the target. The disconnections will occur at bonds connecting that fragment to the rest of the molecule.**
- **Assume that ionic reactions are used and convert each disconnect fragment into a donor (nucleophilic) or acceptor (electrophilic) site, if possible, based on the natural bond polarity of any heteroatoms that are present. A synthetic equivalent is the molecular fragment that correlates with the disconnect fragment in**

terms of the desired reactivity. In most retrosynthetic analyses, the bond α to the functional group and that β to the functional group are the most important for disconnection.

- **If no starting material is designated, use retrosynthetic analysis to find a commercially available or readily prepared starting material.**
- **Identify the relationship of functional groups and manipulate the functional group as required to complete the synthesis.**
- **The most efficient synthesis is usually a convergent strategy rather than a consecutive strategy.**
- **Disconnection of multifunctional targets requires a complete understanding of all reactions related to those functional groups.**
- **When one group interferes with another, it may be protected.**

25.1 What Is Synthesis?

Synthesis takes an available molecule (called the starting material) and transforms it by a series of reactions into a molecule that is required for some purpose (the target). The reactions employed in the synthesis include both carbon–carbon bond-forming reactions and functional group transformations. The disconnections and functional group transformations presented at the ends of several preceding chapters illustrated individual disconnections based on those reactions. The real point of those disconnections is to recognize the relationship between a target and disconnect products. This chapter will discuss a strategy for analyzing a target molecule and use the disconnect products in a manner that will determine how the target can be synthesized.

The place to start a synthesis is with the object at the end of a synthesis, the target molecule. If **1** is a target that must be prepared, several questions should be asked. **What is the starting material? What is the first chemical step? What reagents are used? How many chemical steps are required?** The answers to these questions may not be obvious if more than one or two steps are required. A protocol for analyzing the target will help answer these questions. In this chapter, the target will be examined and then simplified by a series of *mental bond-breaking steps* called **disconnections**. In effect, a disconnection is a thought experiment that breaks the target apart into smaller and simpler fragments, but always with the thought that a reaction is known to rejoin those fragments to generate the target.

Early pioneers who developed strategies for the synthesis of complex molecules include Robert B. Woodward (United States; 1917–1979) and Sir Robert Robinson (England; 1886–1975), but many others have contributed. Although the synthesis of complex molecules has been developed over many years,

establishing a set of cogent guidelines occurred in the latter part of the twentieth century. An important architect of such guidelines is Elias J. Corey (United
States; 1928–), although others also contributed to developing this approach.
The term **disconnection** implies a thought experiment that breaks a bond
within a molecule to generate simpler fragments. A disconnection is a mental
exercise, so no bonds are actually broken. If a bond is disconnected, *a chemical
process must be available to make that bond*. In other words, *choosing a specific
disconnection points toward a bond that must be made by a known chemical
reaction.*

There are 10 bonds in target **1**, labeled *a–j*, **not counting the bonds within
the phenyl group or the C≡C bond**. The reasons for discounting these bonds
are discussed later in this chapter, but the short answer is simple. Based on the
reactions presented in this book, it is easier to use them as intact units rather
than making them. Assume that **1** cannot be purchased, which means that it
must be synthesized. The task is to determine *how* to make **1** using the disconnection protocol. The key functional groups are the alkyne and the alcohol
units. Assume that the best bonds for disconnection are close to the functional
groups. The reason for this assumption lies in the fact that most reactions in
this book rely on reactions at or near a given functional group. Bond *f* is adjacent to the alkyne, but bond *d* is proximal to both the alkyne and the carbon
bearing the alcohol. These two bonds appear to be likely candidates for disconnection, but due to the proximity to both functional groups, bond *d* appears to
be more suitable than bond *f.*

Disconnection of bond *d* generates two smaller fragments: **2** and **3** ("smaller"
is defined as having fewer carbon atoms). In other words, disconnection has
simplified the target. If **1** is disconnected into fragments **2** and **3**, combining
those fragments in a known chemical reaction should generate **1** or a structurally related analog. This means that disconnection of bond *d* will point to a
chemical reaction by which **1** may be prepared from simpler fragments. This
is the fundamental principle behind the disconnection process. Note that the
reverse arrow symbol (⇒) is used with the disconnection.

Fragments **2** and **3** are not real because each carbon (marked in **red**) has
only three bonds. As noted before, there must be a protocol that converts these
fragments into real molecules. Disconnection of the indicated bond in **4** can be
correlated with two real molecules: **5** and **6**. A reaction in Chapter 18 prepared
the indicated bond of an alkyne–alcohol (**4**) from fragments **5** and **6** (Section
18.3.2). Note the similarity of **5** to **2** and of **6** to **3**. In terms of a disconnection,
this simply means that the bond in **4** is disconnected because that bond can be
formed in a synthesis by an acyl addition of the anion of **5** to **6**. Acyl addition

reactions of alkyne anions were described in Chapter 18, Section 18.3.2. With a real reaction in hand, the disconnection of **1** to give **2** and **3** is considered to be reasonable.

Alkyne **5** is a real molecule, whereas **2** is an imaginary fragment. Likewise, ketone **6** is real and **3** is an imaginary fragment. The **red** carbon in fragment **3** is missing a bond, so it is not real. Likewise, the alkyne fragment **2** is not real because it is missing the terminal hydrogen atom. What is the protocol that correlates **2** with alkyne **5**, or **3** with ketone **6**? The carbon of interest in **3** is connected to an oxygen atom, and oxygen is more electronegative than carbon. The C–O bond of **3** is polarized $C^{\delta+}–O^{\delta-}$ (see Chapter 3, Section 3.7). Any real molecule that correlates with **3** will also have a polarized $C^{\delta+}–O^{\delta-}$ bond, which means that it will have an electrophilic carbon. Remember that **3** must react with an analog of **2**.

If the carbon atom in 3 is electrophilic, then the carbon in 2 must be nucleophilic. This assumption is based on simple bond polarization and it makes it possible to correlate the imaginary **3** with a carbonyl compound that has an electrophilic carbon with a polarized C–O bond. Nucleophilic acyl addition to a ketone is a known reaction, so **3** correlates with a real molecule—acetone. If **3** correlates with an electrophilic center, then **2** must be a nucleophilic center and an alkyne anion is a reasonable choice. It is known that an alkyne anion will react with a ketone via acyl addition (see Chapter 18, Section 18.3.2). The correlation of **2** with an alkyne simply requires adding a hydrogen atom to the **red** carbon to give terminal alkyne, **7**. Conversion of alkyne **7** to the anion, followed by acyl addition to acetone, should lead to **1**. Disconnection of **1** generates acetone and **7**, and the reaction of **7** and acetone leads to **1**. Recognizing the forward and reverse relationships is essential for correlating the disconnection (retrosynthesis) with the reactions that make the bond (synthesis).

There is a simple test for this protocol. Is it possible to make molecule **1** by the reaction of the anion derived from **7** and acetone? The answer is yes! The lesson is to correlate disconnect fragments with real molecules. This often means recognizing the chemical relationships of the functional groups involved and adding appropriate functional group transformations.

25.1 Show a reasonable disconnection for (a) 3-pentanol, (b) 3-methylbutanoic acid, and (c) 2-ethyl-3-methylbutanenitrile.

The **first disconnection** of **1** leads to a chemical reaction that is the *last step of the synthesis*. The last step in a synthesis is always the one that generates the final target. An important lesson of the disconnection process is that *the first disconnection generates the last chemical step in the synthesis*. For a complex target, this disconnection approach is repeated until a synthesis of that molecule is constructed based on known reactions.

When compound **7** is converted to **1**, it is said that **1** is synthesized from **7**. The sequence proceeded *from* starting material **7** *to* target **1**. Therefore, working backward from **1** toward **7** (as was done before) is logically called a *retrosynthesis* (the reverse of the synthesis). However, *retrosynthesis* usually refers to the complete set of disconnections. The disconnection method used in Section 25.1.1 generated a retrosynthesis of **1**. The retrosynthesis generates key intermediate products and those structures suggest possible chemical reactions. The disconnection approach to synthesis is therefore sometimes called a *retrosynthetic analysis*. A retrosynthetic analysis will lead to a synthetic scheme. The following sections use a retrosynthetic analysis to generate a synthesis for various targets.

25.2 Specifying a Starting Material for a Given Target

Many times, a target molecule must be made from a specific starting material. This restriction arises when there is a large and cheap source of the starting material or when the starting material contains a stereogenic center that is essential for the preparation of a particular target. There are other reasons—some practical and some aesthetic. When forced to use a particular starting material, the retrosynthetic analysis approach to disconnect the target can be used, but the retrosynthesis must be biased toward the given starting material. In essence, this means that those carbons of the starting material must be identified in the target and then the bonds around the "hidden" starting material are disconnected in a retrosynthetic manner.

25.2.1 Retrosynthesis Assuming Ionic or Polarized Intermediates

A synthetic problem requires that starting material **8** be converted to target **9**. The first step is to identify those carbons in structure **8** that are part of the structure of target **9**. Close inspection shows that the carbon atoms highlighted in **red** and the OH unit in **9** correspond exactly to the carbon atoms and OH of **8**. Therefore, the bonds connected to the highlighted carbons in **9** (the green bond and the **blue** bond) are disconnected because those bonds must be formed in the synthesis This restriction leads specifically to bonds *a* and *b*. In other words, the retrosynthesis of **9** must be biased toward the designated starting material, **8**.

25.2 "Find" 2-propanol in (a) 2-ethyl-2-methyl-2-hexanol, (b) 2-methyl-3-phenyl-2-propanol, and (c) 4-phenyl-2-butanone.

What is the first bond in 9 to be disconnected? Either bond a or bond b may be chosen, based on the previous analysis. Before making a choice, a protocol will be developed to assist in the choice. *All or nearly all reactions presented in this book involve formal ionic intermediates (cations or anions) or they involve the reactions of molecules with positive or negative polarized atoms.* Exceptions are those reactions that involve a pericyclic mechanism (see Chapter 24). Reactions used to construct new carbon–carbon bonds usually involve carbocation intermediates, carbanion intermediates, compounds containing a nucleophilic (δ–) carbon, or those with an electrophilic (δ+) carbon atom. It can therefore be said that *most of the organic chemistry presented in this book involves ionic chemistry of one sort or another.*

25.3 Draw the structure of three molecules that react with a molecule that has a δ– C or is a formal carbanion. Identify three molecules that react with a molecule that has a δ + C or is a formal carbocation.

If most reactions involve ionic chemistry, *as a first step, disconnect those bonds that lend themselves to formation of ionic or polarized intermediates.* A polarized bond results from the presence of a heteroatom, as discussed in Chapter 3. Therefore, it is reasonable to assume that many reactive intermediates arise from a functional group that contains one or more heteroatoms. Therefore, *when a disconnection is made, that disconnection is probably adjacent or close to a functional group.* Disconnection of a polarized bond in a retrosynthesis should lead to a polarized or ionic fragment, and such fragments will form that bond in the synthesis via ionic or highly polarized chemical reactions.

In the case of **9**, **8** is "discovered" as part of the structure (highlighted in **red**). The **red** atoms include the heteroatom O, so the bonds connecting O to carbon are polarized. These include bonds a (in **blue**) and b (in **green**) in **9** and these polarized bonds are disconnected in preference to any other. Knowledge that those bonds are polarized makes it easier to make the correlation with real molecules.

Disconnection of bond a gives disconnect fragments **10** and **11**. Disconnection of bond b leads to fragments **12** and **13**. Fragments **10–13** are not real molecules because they are missing at least one bond (the atoms marked with **dots** have only three bonds connected to them). These fragments are pieces of a molecule (call them **disconnect products**; sometimes they are called **retrons**), but the fragments must be correlated with real molecules before determining whether the disconnection corresponds to a reaction that will regenerate the disconnected bond in a synthesis. *In order to determine the reaction required for the* *synthesis, convert the disconnect fragments into real molecules.* However, two possible retrosynthetic sequences are possible.

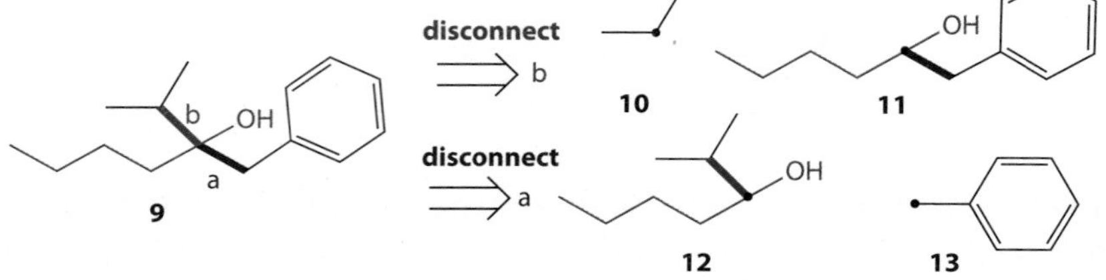

The first is the fragmentation of **9** to **10** and **11** and the second is the fragmentation of **9** to **12** and **13**. To determine whether one disconnection is preferred to the other, focus on translating the fragments into real molecules. This process will allow an evaluation of both chemical reactions in the context of this specific synthesis. This translation of disconnect fragments to real molecules involves the use of a ***synthetic equivalent***.

25.2.2　Making Real Molecules and Evaluating Reactions

The translation of disconnect fragments to real molecules requires an initial evaluation of each fragment in terms of its functionality. Disconnection of a polarized bond leads to fragments **12** and **13**; based on the assumption that key reactions are ionic in origin, it is logical to identify the polarization of each disconnect fragment as positive or negative. Rather than making cations and anions, use the idea that a negatively polarized carbon will donate electrons (a nucleophile) and a positively polarized carbon will accept electrons (an electrophile). *A nucleophilic carbon is a donor carbon and an electrophilic carbon is an acceptor carbon.*

　　Application of this donor/acceptor protocol to **13** leads to a problem, however. The carbon that was part of the disconnected bond, marked with a "•," may be either a donor or an acceptor. If that carbon is designated as a donor in **13**[donor], then the analogous carbon in **12**[acceptor] must be an acceptor. Conversely, if that carbon in **13**[acceptor] is designated as an acceptor, then the analogous carbon in **12**[donor] must be a donor. Examine both possibilities, but *use the natural bond polarization as a guide.* This donor–acceptor protocol is taken from work reported by Dieter Seebach (Germany; 1937–).[1]

　　In Figure 25.1, **12** and **13** are drawn as both donors and acceptors. To distinguish these possibilities, take advantage of the natural bond polarization

Figure 25.1　Donor/acceptor possibilities for fragments **12** and **13**.

for each key bond. Because O is more electronegative than C, the natural bond polarity is $C^{\delta+}-O^{\delta-}$. Using this as a guide, the fragment with an acceptor carbon (**12acceptor**) is more logical than the alternative. *The natural bond polarization of O–C is $^{\delta-}O-C^{\delta+}$and that carbon is naturally an electron acceptor. Attempts to make that carbon negative (a donor) are contrary to the normal bond polarization.* The synthetic equivalent of **12acceptor** must have an acceptor carbon, so the synthetic equivalent for **13donor** must have a donor carbon. This means that the key carbon is a nucleophilic carbon (an electron donor), which is essentially a carbanion.

What structural feature is required to make the donor carbon of **13donor** a real molecule? Fragments **12acceptor** and **13donor** must be converted into a real molecule using the concept of a *synthetic equivalent: a fragment that represents a real molecule.* The analysis using bond polarization makes it clear that the identity of a synthetic equivalent is based on its potential reactivity, taking into account that any donor carbon must react with an acceptor carbon in order to form a C–C bond.

25.4 Draw the disconnect fragments for disconnection of 2-(*N,N*-dimethylamino)-3-methylbutane at C2–C3 and label each as its most logical donor or acceptor.

25.5 Make a list of four common carbanions, all from different functional groups.

How are synthetic equivalents identified based on bond polarization? Many functional groups and molecules have donor carbons. These include Grignard reagents or organolithium reagents (Chapter 15), enolate anions (Chapter 22), cyanide (Chapter 11), and alkyne anions (Chapters 11 and 17). Of these, only Grignard reagents and organolithium reagents do not have another functional group or a heteroatom. Therefore, a simple **donor** carbon marked as c^d has the synthetic equivalent C–MgX or C–Li. Table 25.1 presents a correlation of

Table 25.1
Disconnect Fragments and Their Synthetic Equivalents

Disconnect Fragment		Synthetic Equivalent
	C^{donor}	C-MgX, X-Li
$\overset{O}{\underset{\parallel}{C}}-C^{donor}$	$\underset{\diagdown}{\overset{RO}{\diagup}}{}_{-C^{donor}}$	$\overset{O}{\underset{\parallel}{C}}-CH_2$
$N{\equiv}C^{donor}$	$N-C^{donor}$	$N{\equiv}C^-$
$R-C{\equiv}C^{donor}$ $R-C{=}C^{donor}$ $R-C-C^{donor}$		$R-C{\equiv}C^-$
	$C^{acceptor}$	C-X
$O{=}C^{acceptor}$	$O-C^{acceptor}$	$O{=}C$
$O-C-C^{acceptor}$		$\underset{O}{\overset{C-C}{\diagdown\diagup}}$

various real carbanions with an appropriate synthetic equivalent. In addition to the simple cases represented by Grignard reagents and organolithium reagents (Chapter 15, Sections 15.1, 15.5), $O=C-C^{donor}$ correlates with an enolate anion (Chapter 22), $N\equiv C^{donor}$ correlates with cyanide (Chapter 11, Section 11.3.6), and $C\equiv C^{donor}$ correlates with an alkyne anion (Chapter 11, Section 11.3.6).

This analysis is taken a step further by recognizing how functional groups are related. A carbonyl is reduced to an alcohol (Chapter 19) and an alcohol is oxidized to a carbonyl (Chapter 17). Both the forward and the backward relationships must be recognized. Using this analogy, an enolate anion equivalent is not limited to a C=O unit but may also include $RO-C-C^{donor}$. Because nitriles are reduced to amines (Chapter 19), $N-C^{donor}$ is also an equivalent for a nitrile. Finally, an alkyne can be reduced to an alkene (Chapter 19) or even to an alkane fragment, so $C=C^{donor}$ and $C-C^{donor}$ are included as synthetic equivalents. Drawing on a knowledge of reactions discussed in previous chapters, many possibilities can correlate the donor carbon in **13**^{donor} (Figure 25.1) to a real molecule. In this particular case, there are no oxygen, nitrogen, or double or triple bonds, so a "simple" donor carbon is required. Either the Grignard or the organolithium reagent is reasonable, so **13**^{donor} is converted to the Grignard reagent (PhCH$_2$MgBr, **14**; see Figure 25.2). **If the synthetic equivalent for 13**^{donor} **is a Grignard reagent, what is the synthetic equivalent for 12**^{acceptor}**?**

Because $C^{acceptor}$ is an electrophilic carbon, a reasonable synthetic equivalent should have a δ+ carbon. Several functional groups or substituents have this type of polarization. Two common classes of molecules polarized in this manner are alkyl halides and sulfonate esters, with a $^{δ-}X-C^{δ+}$ unit. These molecules undergo S$_N$2 reactions with loss of X$^-$ (Chapter 11, Section 11.2), so it is clear that the C–X carbon can accept electrons from a donor atom. The polarized carbonyl group ($^{δ-}O=C^{δ+}$) also has the necessary bond polarization, and acyl addition reactions (Chapter 18, Sections 18.3, 18.4) are consistent with a carbon accepting electrons from a donor.

Another molecule with this type of polarization is an epoxide with a $^{δ+}C-^{δ}O-C^{δ+}$ unit. (Epoxide reactions were discussed in Chapter 11, Section 11.8.3.) Armed with this knowledge, Table 25.1 includes these synthetic equivalents for $C^{acceptor}$. A "simple" $C^{acceptor}$ with no heteroatoms is best represented by an alkyl halide or a sulfonate ester. If an oxygen is connected directly to the acceptor carbon (O–$C^{acceptor}$ or O=$C^{acceptor}$), the synthetic equivalent is the carbonyl. If the oxygen is on the

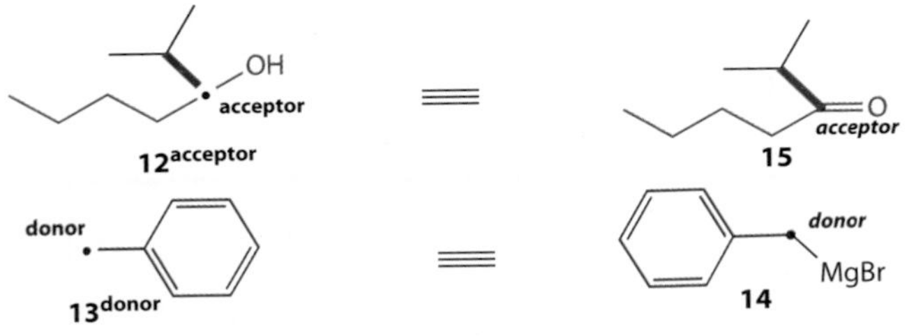

Figure 25.2 Donor/acceptor possibilities for fragments **12** and **13**.

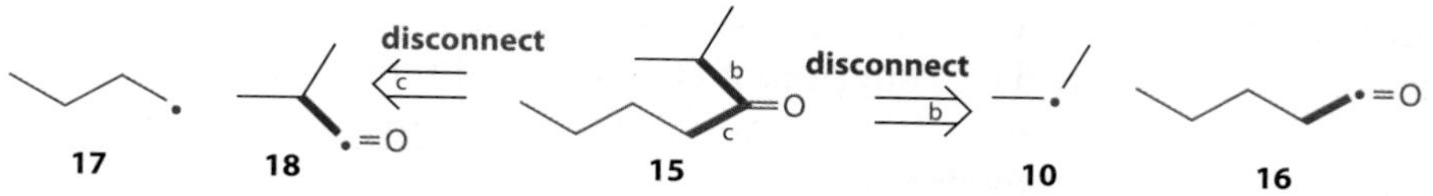

Figure 25.3 Disconnection of ketone **15**.

adjacent carbon (O–C–C^{acceptor}), then the equivalent is the epoxide. *Returning to* 12^{acceptor}, $C^{acceptor}$ *has an oxygen directly attached, so the synthetic equivalent is a carbonyl.* This means that the real molecule for **12**^{acceptor} is a ketone, **15** (2-methyl-3-heptanone; see Figure 25.2). This lengthy analysis of the first disconnection of **9** leads to the conclusion that the requisite reactants are Grignard **14** and ketone **15**. This is only the first disconnection.

25.6 Draw at least one actual molecule for each synthetic equivalent in Table 25.1.

Before evaluating disconnect fragments **10** and **11**, the retrosynthesis based on the disconnection to **15** will be completed. The same analysis will be done afterward with the other fragments. A comparison of both synthetic sequences will determine which, if either, is preferred. The first disconnection leads to **14** and **15**, but not the starting material, **8**. It is reasonable to assume that the five carbon atoms of the starting material (**8**) are contained in **15**. Therefore, disconnection of **15** is necessary to establish **8** as a starting material (see Figure 25.3). Analysis of **15** shows that the requisite carbon atoms of **8** are still there (in **red**).

Once again, disconnect the polarized bonds connected to the functional group (the carbonyl group) to give two possible fragmentations. Cleavage of bond *b* leads to **10** and **16**, whereas cleavage of bond *c* leads to **17** and **18**. Because the goal is to use the five carbon atoms in **8**, **17** and **18** are ruled out immediately because **17** contains four of the carbons found in **8** but the fifth carbon is part of **18**. Fragment **16** contains all five carbons of **8**, so that is the important disconnection. Figure 25.4 shows the oxygen atom attached to carbon, leading to **16**^{acceptor}; the synthetic equivalent is a carbonyl (Table 25.1). This analysis leads to pentanal (**20**) as the real molecule corresponding to **16**. Because **10** is a simple carbon donor, the Grignard equivalent is used, and **19** is the synthetic equivalent of **10**.

25.7 Disconnect 1-phenyl-1-propanone at C1–C2, show the disconnect fragments, indicate *donor* or *acceptor*, and supply a logical synthetic equivalent.

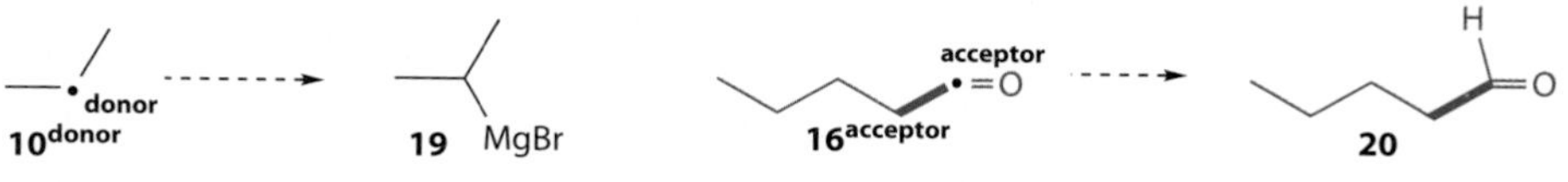

Figure 25.4 Disconnect fragments **10** and **16** and their synthetic equivalents.

Because the five carbons of **8** are accounted for in **20**, in principle, the retrosynthesis is complete. The reaction of the two real molecules (**19** and **20**) does not give **15** directly, however. Remember that functional group transformations associated with a disconnection must be accounted for in the real synthesis.

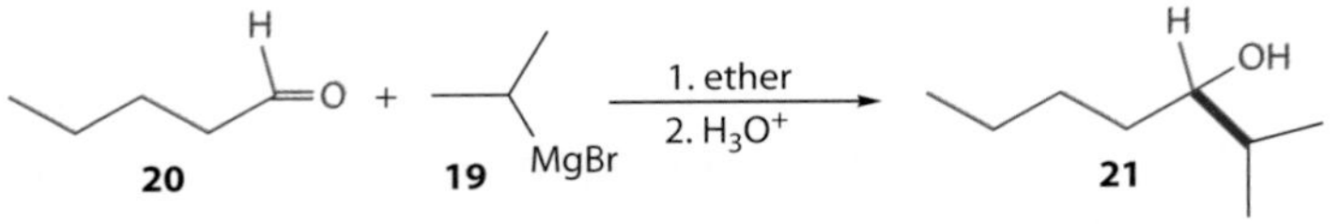

The reaction of **19** and **20** gives alcohol **21**, and oxidation (see Chapter 17, Section 17.2) is required to give ketone **15**. Understanding the relationship between aldehyde **20** and alcohol **8** leads to **20** as a precursor to **8** in the retrosynthesis, so the synthesis requires oxidation of **8** to give **20**. This final item allows the synthesis that begins with the oxidation of **8** with PCC to give **20**; subsequent treatment with **19** gives **21**. Oxidation of **21** with PCC gives **15**, which then reacts with Grignard reagent **14** to give the final target, **9**. Figure 25.5 shows the synthesis based on the retrosynthesis inspired by disconnection of bond *a* in **9**.

25.8 Show a retrosynthesis and a real synthesis for preparing 2-methyl-3-hexanone from 1-propanol.

The oxidation state of alcohol **21** and ketone **15** raises another point. The oxidation-reduction relationship of alcohols with ketones or aldehydes, or with alkynes, alkenes, and alkanes, indicates the need for another chemical reaction to convert one to the other. However, the key carbon atoms in halides, alcohols, ethers, alkenes, etc. are all in the same oxidation state. Although chemical transformations are required to convert one to the other, there is no need for an oxidation or a reduction. Recognition of these relationships can minimize the number of oxidations or reductions in a single synthesis.

Before proceeding to the next section, the alterative disconnection of **9**, which gives disconnect products **10** and **11**, must be examined (see Figure 25.1).

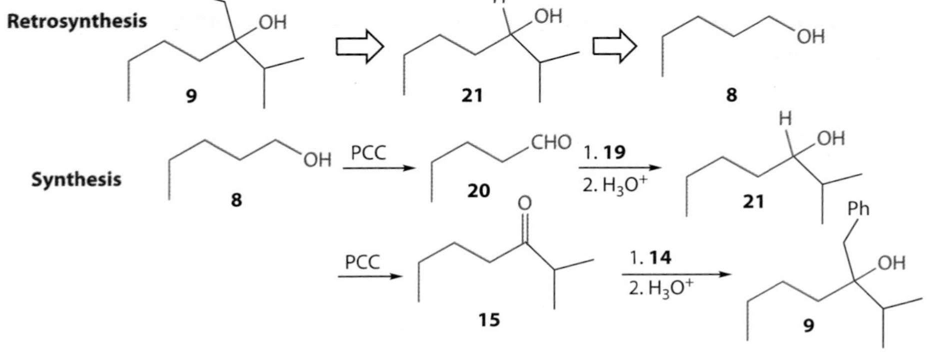

Figure 25.5   Synthesis of **9** and **8** based on the disconnection of bond *a*.

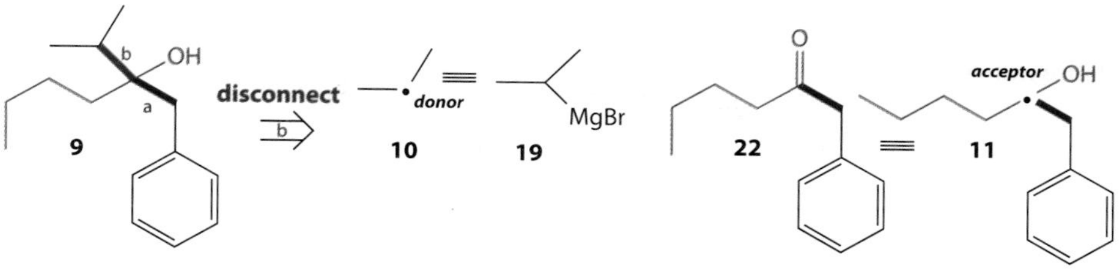

Figure 25.6 Disconnection of bond *b* in **9**.

An identical analysis of these fragments is applied (shown in Figure 25.6). The donor/acceptor analysis indicates that **11** has C^{acceptor} connected to an oxygen, so the equivalent is the carbonyl compound based on Table 25.1. This leads to ketone **22**. Fragment **10** must be a donor, and it correlates with Grignard reagent **19** (the same one used in Figure 25.4). With the first disconnection established, ketone **22** is disconnected because all five carbons found in **8** are present in **22**. The only reasonable disconnection is at bond *b* to give fragment **16** (which has all five carbons found in **8**) and fragment **13** (see Figure 25.7). Fragment **16** is identical to that seen previously and it has the O–C^{acceptor} unit.

The synthetic equivalent is aldehyde **20**, as seen before. The means that **13** is a donor fragment and it will correlate with the Grignard reagent, **14** (this is identical to the synthetic equivalent used previously). Factoring in the functional group transformation, this leads to an alternative synthesis based on a retrosynthesis beginning with disconnecting bond *b* (see Figure 25.8). The oxidation of **8** with PCC leads to **20** (as in the first synthesis; see Figure 25.5) and **20** reacts with **19** to give alcohol **23**. Oxidation of this alcohol product with PCC leads to ketone **22**, which reacts with **14** to give the target, **8**.

It is possible that one of the two routes (Figures 25.5 and 25.8) is more efficient than the other. In fact, *both* are reasonable, reliable, and straightforward. There is no "best" disconnection. The only difference is the order in which the bonds are formed and which intermediate ketone is formed. The reagents are the same. If it turns out that the yield is better in one sequence, then that is better. If it is easier to isolate and purify products from one sequence, that is better. If one of the intermediate ketones (**15** or **22**) is more reactive than the other, then that may be better. "Best" is determined by which route gives the best yield of product in the shortest and most facile manner. This is a very

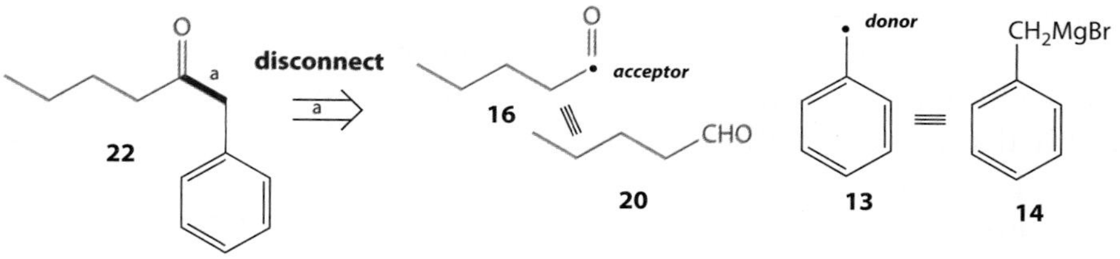

Figure 25.7 Disconnect fragments for bond *a* in **22**.

organocuprates. The equivalent for C≡C^{donor} or C=C^{donor} is an alkyne anion and that for NC^{donor} is cyanide. The equivalent for O–C–C^{donor} is an enolate anion. For simple C^{acceptor}, the synthetic equivalent is an alkyl halide or a sulfonate ester. The equivalent for O–C^{acceptor} is an aldehyde or a ketone and that for O–C–C^{acceptor} is an epoxide. Given these synthetic equivalents and the need to disconnect carbon–carbon bonds near the functional group, a real example will illustrate the protocol.

25.12 Give one real example for each category in Table 25.2.

Compound **33** is the target, and the bonds are labeled *a–h*. The π-bond of the carbonyl (C=O) is not labeled because it is a functional group, nor are the π-bonds of the benzene ring, which is also considered to be a functional group for the purposes of the analysis. *Assume that carbon–carbon π-bonds, including those in benzene rings, are not disconnected.* The benzene ring is used as an intact entity (a unique functional group).

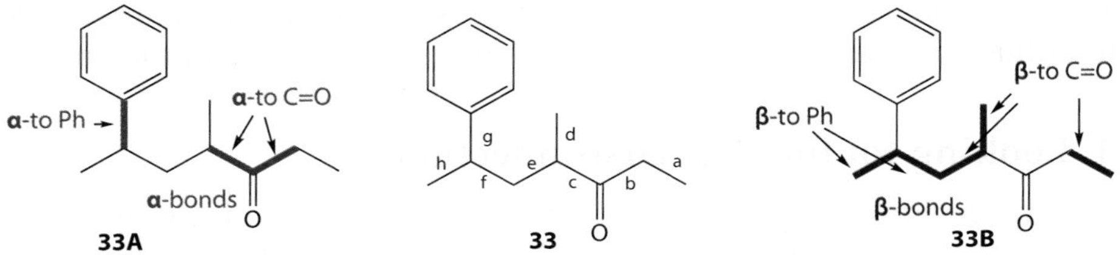

Disconnection of π-bonds in an alkene is possible, but there are only a few ways to directly form a C=C bond. One is the Wittig reaction, discussed in Chapter 22 (Section 22.10), but for now this disconnection is ignored. The labeled bonds are subdivided into two categories: *the α-bond that is connected directly to the functional groups and/or heteroatom and the β-bond that is one removed from the functional group or heteroatom.* In **33A**, the α-bonds to the carbonyl (C=O is one functional group and the phenyl ring is a second functional group) are labeled in **red**. In **33B**, the β-bonds to those two functional groups are labeled in **blue**. These bonds are identified in order to choose one for the first disconnection. In principle, any bond is suitable, but clues in the structure of the molecule suggest disconnection of some bonds is more productive.

25.13 Label all α- and β-bonds for each functional group in (a) 4-hydroxy-5-methyl-2-hexanone, (b) 2-cyano-1-phenyl-1-propanol, and (c) 2-methyl-6-phenyl-3-hexanone.

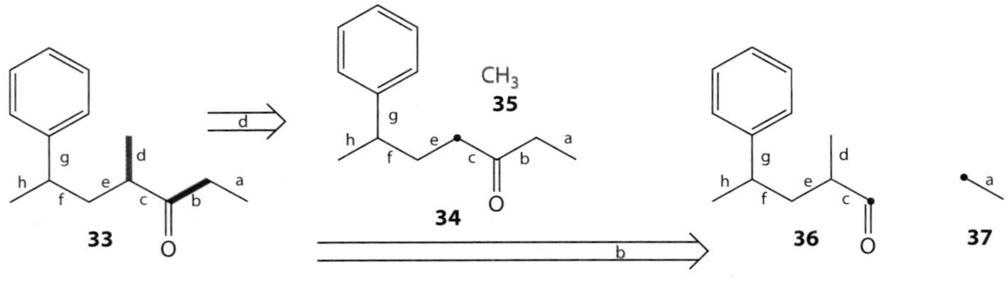

An important criterion for a disconnection is simplification of the target to the greatest extent possible. If either bond *b* or bond *d* in **33** is disconnected, the result is one very large fragment and one very small fragment. Disconnection of bond *d,* for example, gives **34** and a one-carbon fragment (**35**); disconnection of bond *b* leads to **36** and a two-carbon fragment (**37**). Compare these fragments with disconnection of bond *c,* also proximal to the functional group, which gives **38** and **39**. Disconnection of bond *e* gives fragments **40** and **41**. In the latter two cases, the fragments are close to the same size and these disconnections provide significant simplification.

A second important characteristic in a target is the presence of one or more stereogenic centers. Target **33** has two stereogenic centers, although *the racemic molecule is shown as the target.* Begin by focusing on bonds that are proximal to the functional groups, the carbonyl functional group, or the benzene ring. This choice is clear by simply recognizing that many reactions that make a carbon–carbon bond involve a carbonyl, but few if any reactions involve benzene.

Note that one bond α to C=O (bond c) and two bonds β to C=O (bonds e and d) are connected to the stereogenic center adjacent to the carbonyl. **When possible, disconnect a bond that is connected to a stereogenic center.** There is a good rationale for this statement. Formation of the stereogenic center during a reaction offers the potential to control the stereochemistry of that center. Disconnection of a bond that is not attached to a stereogenic center simply means that the stereogenic center must be made in another reaction. The stereogenic center in the fragment must be dealt with sooner or later. Sooner is better than later in a synthesis (see Section 25.4.5). Note that *the synthesis will generate racemic products and the focus is not on controlling absolute stereochemistry. Therefore, diastereoselectivity is an important issue but enantioselectivity is not.* Bonds *f, g,* and *h* are also connected to a stereogenic center. Less simplification occurs by disconnecting these bonds, so they are ignored until later in the synthesis.

25.14 Give one reasonable disconnection for (a) 3-methyl-1-phenyl-1-pentanone and (b) 3-cyclohexyl-2-methyl-3-hexanol.

25.3.2 Retrosynthetic Analysis

Based on the preceding section, it appears that bonds *c* and *e* in target **33** are the best candidates for a disconnection. The disconnection of bond *e* gives

40 and **41**, and disconnection of bond *c* gives fragments **38** and **39**. As noted in Section 25.2, **38–41** are disconnect fragments and not real molecules. Real molecules must be obtained to ascertain whether these are reasonable disconnections.

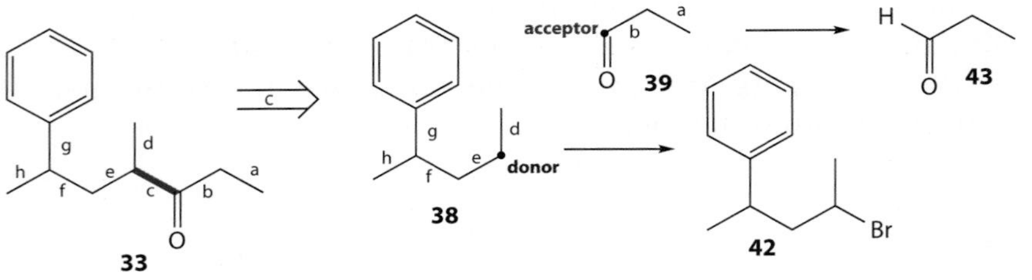

Using the method described in Section 25.2, each fragment is categorized as a donor or an acceptor. Fragment **39** is an O=C–C fragment, so it is logically an acceptor (see Table 25.1), O=C^a–C. This correlates well with a carbonyl—specifically, an aldehyde (propanal, **43**). This assignment means that **38** must be the donor fragment, leading to the Grignard reagent derived from halide **42**. Bromide **42** does not fit the criterion for a starting material because it has too many carbon atoms, and another disconnection is required. The halide is not obviously amenable to a disconnection because the halogen is treated as a substituent. However, changing the functional group to an alcohol (see Figure 25.9) generates a new target, **43**.

Several possible disconnections are now possible based on the C–O bond polarization, and one is cleavage of bond *f*, which leads to **44** and **45**. This was chosen because Table 25.1 contains the ^{acceptor}C–C–O fragment that correlates with an epoxide. This correlates with acceptor fragment **45**, which has epoxide **47** as a synthetic equivalent. If **45** is the acceptor fragment, then **44** is the donor, which correlates with the Grignard reagent derived from halide **46**. Halide **46** does not fit the criterion for a starting material because it has too many carbon atoms.

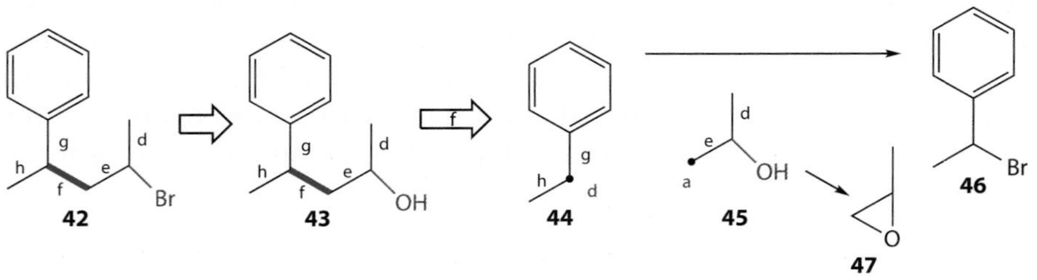

As in the case before, change the halide to an alcohol (Table 25.1), making alcohol **48** the new target. If bond *g* is disconnected, the result is phenyl fragment **49** and the C–C–OH fragment, **50**. If this latter fragment is identified as the acceptor, it correlates with C–C^{acceptor}–OH with a carbonyl as an equivalent—specifically, acetaldehyde, **52**. This means that **49** is the donor

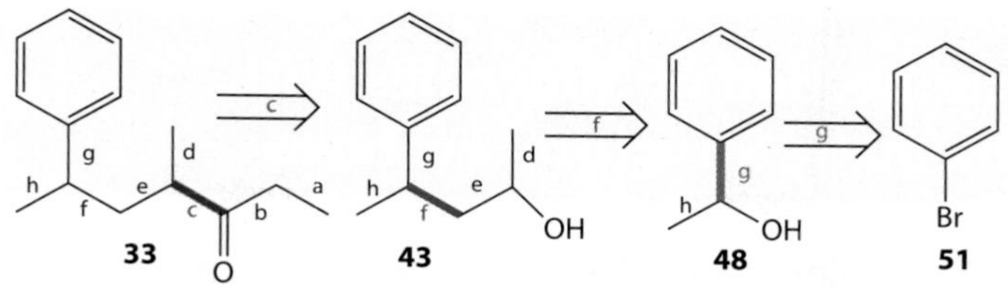

Figure 25.12   Retrosynthesis of **33** based on disconnection of bond *c*.

and its equivalent is the Grignard reagent derived from bromobenzene (**51**). Because **51** has six carbons, it fits the criterion for the starting material, and bromobenzene (**51**) is the starting material for the entire synthesis based on the retrosynthetic scheme shown in Figure 25.12.

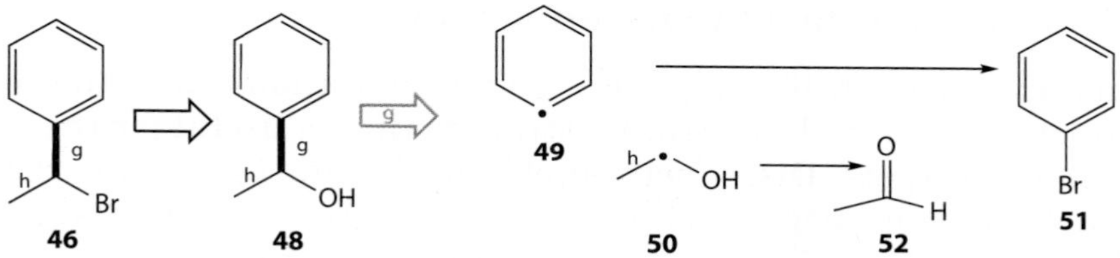

25.15 Write a reaction that prepares 47 from propanol.

25.16 Write a retrosynthesis for 1-cyclohexyl-2-methyl-1-butanone.

It is important to emphasize that *there is no "correct" disconnection for this molecule.* Choices are made by evaluation of the actual reactions used in each synthesis, and each choice is examined to see which might be easier and better suited to available resources. The synthesis based on the retrosynthesis in Figure 25.12 is shown in Figure 25.13. Note that there are a total of nine steps (*not* counting the hydrolysis steps), but all reactions are straightforward and there should be no major problems.

25.17 Devise an alternative to the reaction sequence in Figure 25.13 that converts 43 to 33.

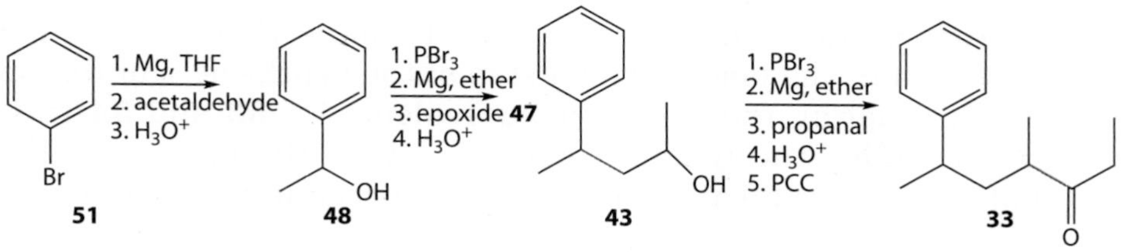

Figure 25.13   Synthesis of **33** based on disconnection of bond *c*.

25.4 Disconnection of Molecules with Problematic Structural Features

The synthesis of complex molecules requires an in-depth knowledge of chemical reactions, as well as an understanding of synthetic strategy for various types of structural motifs. An entire course is usually required to introduce the fundamentals of synthesis theory, and that is well beyond the scope of this book. However, a discussion of how to approach several common problems in the synthesis of complex molecules is appropriate. This will give a glimpse into the power and flexibility of the disconnection approach to develop a synthetic route to the most complex targets.

25.4.1 A Guide to Efficient Disconnections

Compound **52** is not significantly more complicated than some other problems in this chapter, but it will illustrate a strategy that is useful for the synthesis of very complex targets. Disconnection of bond b leads to disconnect products **53** and **54**. There is no attempt to translate these products into real molecules, as was done in Sections 25.2 and 25.3. The intent is to illustrate a strategy, not specific reactions. Compare this disconnection with the disconnection of bond a, which gives **10** and **55**. Disconnection of bond b cleaves **52** into two fragments that are close in mass (C_6O and C_9), but disconnection of bond a gives a small fragment (**10**) and a large fragment (**55**). In effect, disconnection of bond b leads to two equivalent fragments.

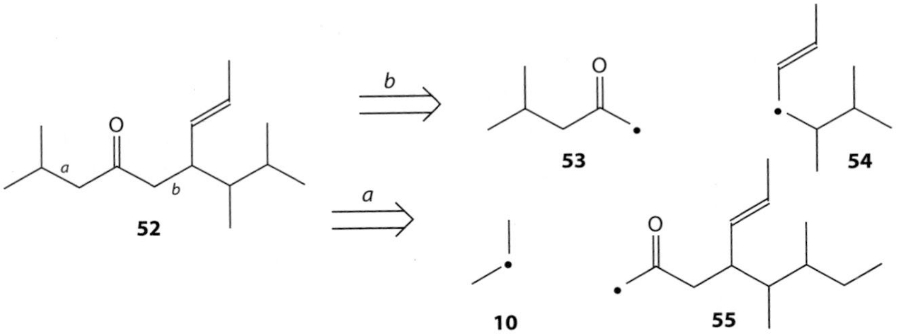

Why is the relative size of the disconnection fragments important? Remember from Section 25.1 that the first disconnection gives fragments that are combined in the last chemical step of the synthesis. For disconnection of bond b, coupling the synthetic equivalents for **53** and **54** will give **52**. Likewise, coupling the synthetic equivalents for **10** and **55** will give **52**. Assume that **53**, **54**, and **55** all require further disconnection. Disconnection and synthesis of **53** and **54** will require fewer steps than disconnection and synthesis of **55**. Therefore, a synthesis of **52** via fragments **53** and **54** should be shorter and more efficient. Another way to say this is that the goal is to correlate a disconnection with simplification of the target. Greater simplification of the target in a disconnection should lead to a more efficient synthesis. Based on this idea, the disconnection of bond *b* provides more simplification of **52** than the disconnection of bond *a,* which only trims a small piece from **52**.

Disconnection of target **52** to **53** and **54** gives more simplification, and it is stated that this leads to a more efficient synthesis. Why is this more efficient? An answer to this question requires two definitions: convergent synthesis and consecutive synthesis. One synthetic strategy is to work backward linearly from the target to a starting material. A synthesis based on a linear pathway is called a ***consecutive synthesis***. The consecutive approach was used in Sections 25.2 and 25.3 for all disconnections. Alternatively, the synthetic strategy may disconnect to several fragments, each with its own starting material. The resulting disconnect fragments are combined in a nonlinear fashion to prepare the target; a synthesis based on this branching pathway is called a ***convergent synthesis***. In a convergent synthesis, the several pieces of a molecule are synthesized individually, and the final target is sequentially assembled from the pieces. For **52**, fragments **53** and **54** may be synthesized individually and then combined to give **52**. This is a simple example of a convergent synthesis.

25.18 **Assume that a ketone is the synthetic equivalent for 53 and a bromide is the synthetic equivalent for 54. Show a synthesis of 52 from these equivalents. Now show a disconnection and synthesis of the ketone from 2-methyloxirane and a synthesis of the bromide from propyne.**

Figure 25.14 illustrates the difference between a consecutive and a convergent synthesis. In the consecutive synthesis, a single starting material is

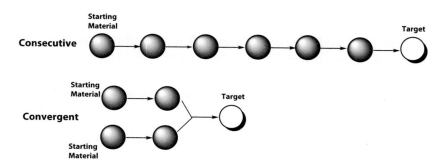

Figure 25.14 A comparison of a consecutive versus a convergent synthesis.

converted stepwise to the target in six steps. In the convergent synthesis, there are two starting materials and each is converted to a key fragment in one step. The two key fragments are combined in one step to give the target. This constitutes a shorter and more efficient synthesis and should yield a greater amount of product. It is suggested that, when it is possible, one use a convergent strategy, which requires disconnections that will provide the most simplification. In the case of **52**, disconnection of bond *b* is preferred.

25.19 If there is 1.0 mol of starting material and the yield of each step in the consecutive synthesis is 90%, how many moles of the target will be obtained?

25.20 If there is 1.0 mol of the two starting materials and the yield of each step in the convergent synthesis is 90%, how many moles of the target will be obtained?

25.4.2 Disconnection of Multifunctional Targets

The disconnection approach requires that a choice be made concerning which bonds will be disconnected. When a target contains more than one functional group, many different bonds may be choices for disconnection. As the number of choices increases, there is an increased requirement that one understand all possible chemical transformations for the functional groups that evolve from the disconnection. When many disconnections are possible, those that are most efficient and more likely to be successful are determined only by an extensive knowledge of different reactions.

Three relatively simple compounds—**56**, **57**, and **58**—illustrate the problems associated with multifunctional targets. Ketone **56** will have disconnections that focus on the carbonyl. Disconnection to **59** and **60** is shown because it provides the most simplification. This is a straightforward disconnection based on the principles discussed in this chapter. Keto-aldehyde **57** is a bit more challenging. There are two functional groups, but which is more important for the disconnection? There are several possibilities. Disconnection adjacent to the carbonyl leads to Grignard acyl addition reactions (Chapter 18, Section 18.4). Disconnection adjacent to the α-carbon leads to an aldol condensation reaction (Chapter 22, Section 22.2). Based on the principle of maximum simplification, **57** is disconnected to **61** and **60** based on an aldol condensation.

25.21 Show a disconnection for 57 based on an acyl addition reaction that involves the ketone unit and then show the appropriate synthetic equivalents and reactions.

25.22 Write out the synthetic equivalents and then the chemical reactions for formation of 57 from 1-propylcyclohexene.

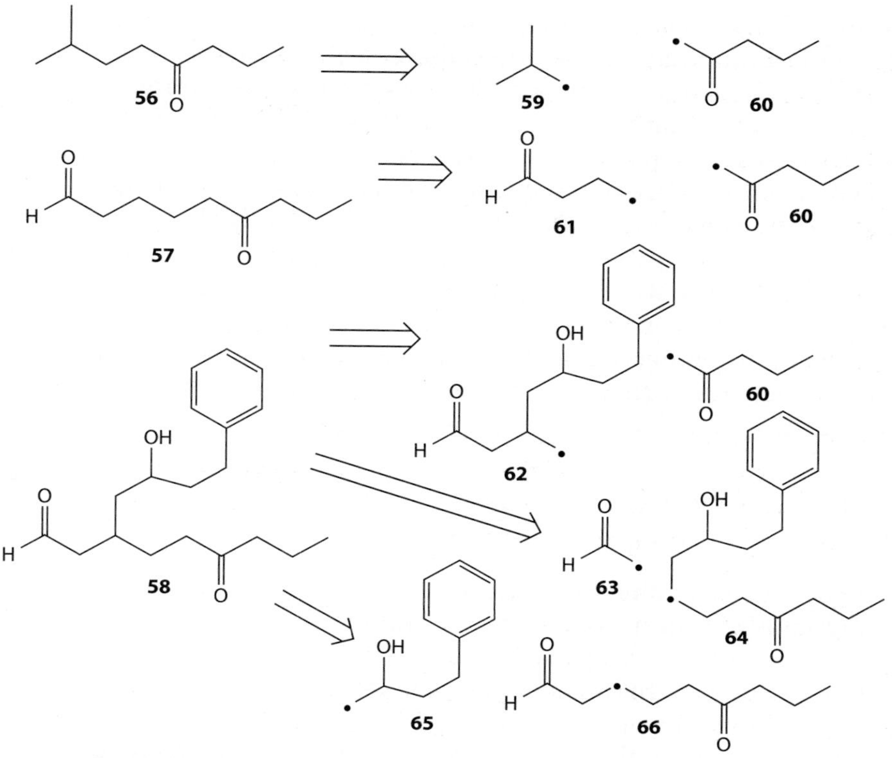

Compound **58** clearly offers more possibilities for disconnection. Disconnections are available at or near the carbon atom bearing the OH group, but also at or near both carbonyl carbons. The larger number of functional groups leads to more choices. Does the chemistry of the alcohol, the aldehyde, or the ketone offer the best choice for a disconnection? The chemistry of alcohols is associated with oxidation and reduction (Chapter 17, Section 17.2; Chapter 19, Sections 19.2, 19.3.4, 19.4.1), formation and reactions of alkoxides as nucleophiles (Chapter 11, Section 11.3.2) and as bases (Chapter 12, Section 12.1), and formation of esters (Chapter 20, Section 20.5). Alcohols are converted to alkyl halides (Chapter 11, Section 11.7). Aldehydes and ketones are formed by the oxidation of alcohols (Chapter 17, Section 17.2), are reduced to alcohols (Chapter 19, Sections 19.2, 19.3.4, 19.4.1), undergo acyl addition (Chapter 18, Sections 18.1–18.7), and participate in enolate anion reactions (Chapter 22, Sections 22.2, 22.4, 22.6). Based on these reactions, several disconnections are shown, but several more are possible.

Disconnection to **62** and **60** is reasonable and suggests an enolate alkylation condensation (Chapter 22, Section 22.9). Clearly, fragment **62** is complex and will require further disconnection to obtain reasonable starting materials. This is certainly a candidate for a convergent synthesis. Disconnection to **63** and **64** is not attractive because there is little simplification. Disconnection to **65** and **66** provides simplification. Fragment **65** may be correlated with a ketone, which will react via enolate alkylation (see Chapter 22, Section 22.9). This reaction requires that **66** must have a leaving group β to the carbonyl, which likely requires further synthetic manipulation and other disconnections to determine the appropriate precursor. The same can be said for fragment **62**, assuming that an enolate alkylation reaction (Chapter 22, Section 22.9) is used via fragment **60**.

In both cases, the placement of the required functional group is several bonds away from the functional groups, which makes the synthesis more difficult. Comparing the three disconnections suggests only that the disconnection to **63** and **64** is unlikely to be viable. Further disconnection of **62**, **65**, and **66** is required to determine which route is more viable. Indeed, this is an important consequence of this example. The only way to deal with such a target is to analyze each bond disconnection and the resultant disconnect products. Based on knowledge of chemical reactions for those functional groups, a choice may be made concerning the disconnection that will be used.

25.23 Draw the structure of the bromide suggested as a synthetic equivalent for **62**. Draw the structure of the bromide suggested as a synthetic equivalent for **66**.

To choose a reasonable disconnection, all possible chemical reactions for a given functional group must be known and considered. In the case of **58**, one must know all of the reactions that are possible for alcohols, for aldehydes, and for ketones, in addition to considering a consecutive versus convergent strategy. Perhaps more important is the statement that disconnect products **62**, **65**, and **66** require further disconnection before a route is chosen. When there are multiple functional groups, it is often difficult to choose a route based on the first disconnection because it usually gives a fragment that remains complicated. Further disconnection is required to simplify the products to a point where the most viable route may be chosen.

In other words, if there are three initial disconnections, several additional disconnections are required for each choice. After this is done, the different retrosynthetic routes are compared for viability. Therefore, knowledge of chemical reactions must extend far beyond the functional groups used for the first disconnection to those used for every step in the synthesis. A lack of understanding of all the reactions will stop a retrosynthetic analysis in its tracks.

25.4.3 The Use of Protecting Groups

Another feature of target **58** will complicate any chemical reaction chosen in a disconnection. If the disconnection requires an acyl addition reaction of a Grignard reagent (Chapter 18, Section 18.4) or formation of an enolate anion (Chapter 22, Sections 22.2, 22.4), there is a problem. Both Grignard reagents and enolate anions will react with an alcohol in an acid–base reaction that will interfere with the desired carbon–carbon bond-forming reaction (see Chapter 22, Section 22.2). Further, there are two sites for reaction. An aldehyde will undergo acyl addition or reaction with an enolate anion, but so will the ketone unit. Both the α-carbon of the aldehyde and the α-carbon of the ketone are candidates for enolate formation, which further complicates any reaction.

A disconnection strategy may plan reactions in such a way that these functional groups will not interfere with the synthetic sequence. In many cases, however, this is not possible. An alternative strategy converts the functional group that interferes with a planned reaction into an alternative group that will not interfere. Once the planned reaction is complete, the alternative group is converted back to the original functional group. The alternative group is called a **protecting group** and the process is called a **protecting group strategy**.

The use of protecting groups is shown by the disconcertion that converts **58** to **63**. Note that the secondary alcohol unit in **58** is converted to a benzyl ether in **63**, and the aldehyde unit in **58** is converted to a 1,3-dioxolane (an acetal) in **63**. With these two reactive groups protected, the use of an enolate alkylation reaction for the synthetic equivalents derived from **64** and **60** is more reasonable. Note that this disconnection is somewhat unusual in that protecting groups are part of the disconnection plan and not necessarily a unique disconnection. However, using **63**, an enolate anion generated from **60** can be employed without interference of the acidic proton of the OH unit or the more acidic α-proton of an aldehyde (see Chapter 22, Section 22.1).

This does not resolve the issue of further disconnection of **64** to determine whether the entire route is viable, but it does make the disconnection possible. This strategy is useless unless the benzyl ether can be converted back to an alcohol and the acetal converted back to an aldehyde at the appropriate time. In other words, if a functional group is transformed into a protecting group, chemical reactions must be known that will transform the protecting group back to the original functional group. To illustrate this point and also provide a short list of protecting groups for syntheses, a few selected protecting groups will be examined. The reaction that transforms alcohols, aldehydes or ketones, and also amines to an appropriate protecting group will be shown, as well as the reactions that transform the protecting group back to the original group.

Several reactions are consistent with the idea of the reaction with an alcohol to give a relatively unreactive product, followed by a second reaction at a later time to regenerate the original alcohol. One of the best choices for a protecting group converts a reactive alcohol to a relatively unreactive ether. Remember that a method must be known to convert the ether back to the alcohol. Two protecting groups are shown in **63** that satisfy these conditions. The first converts 2-butanol to benzyl ether **65**. The reaction of 2-butanol with the base, sodium hydride, generates the corresponding alkoxide. An S_N2 reaction of the alkoxide nucleophile with benzyl bromide gives **65** (see Chapter 11, Sections 11.2, 11.3.2).

Ethers are relatively unreactive and will not react in an acyl addition reaction or with the reactions of enolate anions. However, when a benzyl ether is treated with hydrogen gas and a palladium catalyst, hydrogenolysis leads to 2-butanol and toluene (see Chapter 19, Sections 19.3.4, 19.3.5, and Chapter 21, Section 21.6, for hydrogenolysis). The benzyl ether is a good projecting group for an alcohol and a few other ethers may also be used.

The alkoxide formed by the reaction of 2-butanol and sodium hydride also reacts with chlorotriethylsilane (Et₃SiCl) to give **66**, a silyl ether. Although this book does not discuss the reactions of silicon reagents in any detail, many nucleophiles react with trialkylsilyl halides (molecules with a Si–halogen bond such as R₃Si–Cl or R₃Si–Br) at the silicon atom. Silyl ether **66** is relatively stable to many reactions, although it is sensitive to acid–base reactions. Indeed, the reaction for **66** with aqueous acid will regenerate the alcohol. It is also true that fluoride ion has an affinity for silicon and, when **66** reacts with the ionic compound tetrabutylammonium fluoride, the product is the alcohol after treatment with water. Silyl ethers are also good protecting groups for alcohols.

25.24 It is known that the alkoxide anion derived from 2-butanol reacts with CH_3OCH_2Cl to form the corresponding ether. This ether is actually an acetal. It is useful as an alcohol-protecting group and it is removed by reaction with aqueous acid. Review the chemistry in Chapter 18, Section 18.6.5, and show a mechanism that converts this ether back to 2-butanol.

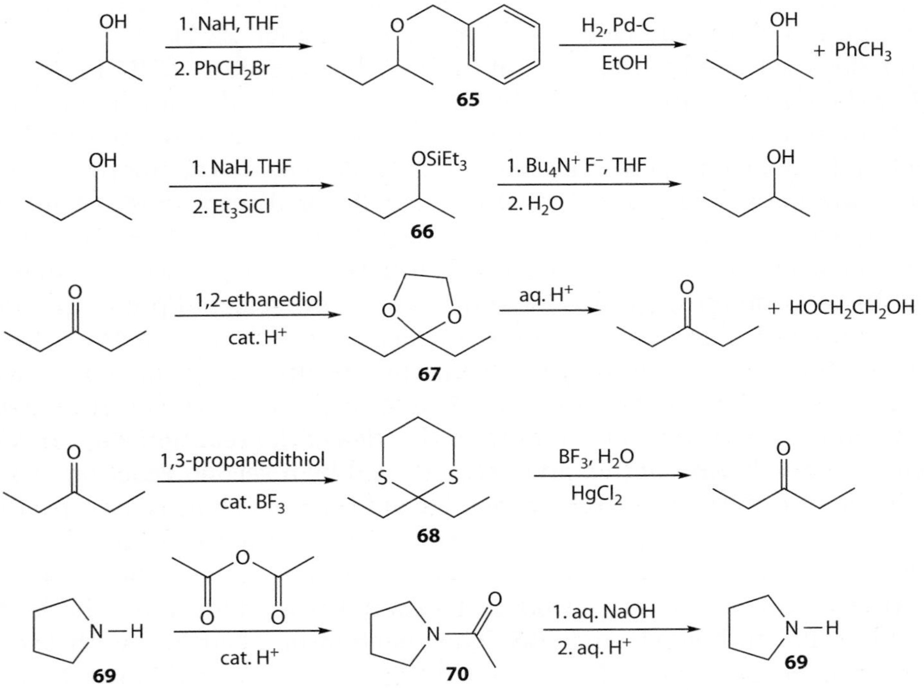

Both aldehydes and ketones are converted to the corresponding acetal or ketal by reaction with an alcohol or a diol (Chapter 18, Section 18.6). Treatment of an acetal or a ketal with aqueous acid regenerates the aldehyde or ketone (Chapter 18, Section 18.6.5). Therefore, acetals and ketals are good protecting groups for these carbonyl compounds. Two examples are shown using 3-pentanone (**67** and **68**). The reaction of 3-pentanone and 1,2-ethanediol generates the 1,3-dioxolane derivative **67**, a ketal. Ketals are unreactive to acyl addition reactions and enolate anion reactions, as well as many oxidation or reduction reactions; however, they are sensitive to aqueous acid or base. Therefore, the reaction of **67**, first with aqueous hydroxide and then with aqueous acid, will regenerate 3-pentaone and the diol.

It is also known that the reaction of a ketone or an aldehyde with thiols or dithiols will generate dithioketals or dithioacetals (Chapter 18, Section 18.6.7). A second protecting group is therefore possible by the reaction of 3-pentaone and 1,3-propanedithiol in the presence of a Lewis acid to give 1,3-dithiane derivative **67**, a dithioketal. Such protecting groups are stable to many reactions, but they are sensitive to Lewis acids in an aqueous medium. Therefore, the reaction of **67** and a mixture of BF_3 and $HgCl_2$ in water regenerates the ketone.

25.25 Write out the reaction that converts butanal to the corresponding diethyl acetal protecting group. Write out the reaction that converts the diethyl acetal back to butanal.

Amines have not been used in chemical transformations as frequently as other functional groups thus far in this book, but an important exception is the chemistry of aniline derivatives in Chapter 21. As shown in Section 21.3, aniline was converted to the corresponding amide in order to mediate electrophilic aromatic substitution reactions. The amide group serves as a good protecting group for amines. In the example shown, pyrrolidine (**69**) reacts with acetic anhydride to form the corresponding amide, **70**. Methods for the formation of amides from amines were described in Chapter 20, Section 20.6. Amides are not as basic as their amine precursors and this is an important issue for determining when an amine should be protected. Aqueous acid hydrolysis of **70** regenerates the amine, **69**.

Protecting groups are used sparingly in this book. However, the concept of using a protecting group is important. From time to time, protecting groups are important in our discussion, particularly in Chapters 27 and 28.

25.26 For 4-amino-2-heptanol, a reaction requires conversion of the alcohol to the corresponding bromide. The amine may interfere with this reaction. Briefly discuss why the amine might interfere and why bromination of the acetamide derivative of 4-amino-2-heptnaol may proceed without a problem.

25.4.4 Multiple Identical Functional Groups

When a target contains two or more identical functional groups, such as multiple OH units or multiple ketone and aldehyde units, there are many possible

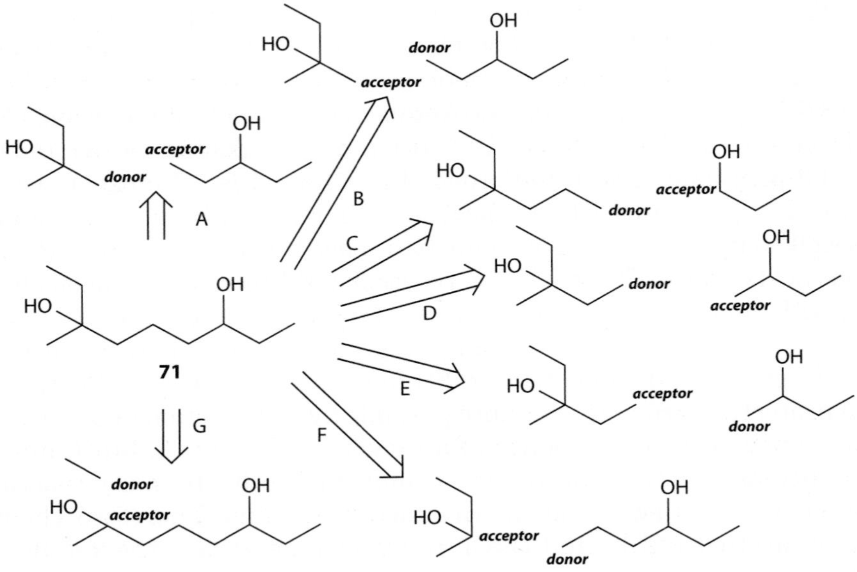

Figure 25.15   First disconnections of diol **71**.

disconnections. As with the analysis of **58** previously, the choice of disconnection demands an intimate knowledge of all possible reactions of that functional group. A simple example is diol **71**. Even with this simple diol, several possible disconnections are possible, as shown in Figure 25.15, **A–G**, based on breaking bonds that are α or β to a C–OH unit (see Section 25.2.2).

Why are these bonds chosen? It is known that secondary and primary alcohols are oxidized to ketones or aldehydes (Chapter 17, Section 17.2), and aldehydes or ketones are reduced to primary or secondary alcohols (Chapter 19, Sections 19.2, 19.3.4, 19.4.1). Once a ketone or aldehyde is generated, the chemistry of those functional groups becomes available. This includes enolate anion chemistry (Chapter 22, Sections 22.2, 22.4, 22.6). An alcohol can be formed by acyl addition reactions of Grignard reagents to aldehydes or ketones (Chapter 18, Section 18.4). Alcohols undergo acid–base reactions to form the corresponding alkoxides (Chapter 6, Section 6.2), which react as nucleophiles or as bases with alkyl halides to form ethers (Chapter 11, Section 11.3.2) or alkenes (Chapter 12, Section 12.1), respectively. Alcohols react with acid to give alkenes via dehydration reactions (Chapter 12, Section 12.3). Alcohols are converted to alkyl halides by reaction with mineral acids or specialized reagents such as thionyl chloride or phosphorus tribromide (Chapter 11, Section 11.7). Alcohols are also converted to esters (Chapter 20, Section 20.5). To understand the disconnections, all of these reactions must be reviewed and understood. There is no alternative.

Based on the concept of simplification and alcohol chemistry, disconnections C, D, and E are reasonable. Interestingly, disconnections A and B are quite reasonable if the tertiary alcohol is further disconnected to give a secondary alcohol and an ethyl fragment. The purpose of this section is not to analyze **71** in depth, but rather to point out that understanding the various reactions of the functional groups is essential to planning a disconnection.

25.27 Construct reactions based on disconnections C, D, and E.

25.4.5 Chiral versus Achiral Targets

Many interesting targets for synthesis are chiral molecules with stereogenic centers. Introduction of a stereogenic center makes the disconnection more interesting and more demanding. A disconnection may lead to chiral molecules that undergo bond formation, or the reaction suggested by the disconnection may generate the stereogenic center directly. A very simple example is **72**. This appears to be a trivial disconnection based on an aldol condensation (Chapter 22, Section 22.2). Disconnection of racemic **72** leads to disconnect products that have aldehyde **73** and ketone **74** as synthetic equivalents. An aldol condensation of the enolate anion of **74** with aldehyde **73** will give racemic **73**.

25.28 Show the disconnect products that lead to 73 and 74.

A similar analysis of chiral molecule **72**, with a 3*S*-stereogenic center, also gives **73** and **74** as precursors. The aldol reaction will generate the stereogenic center, but to form chiral **73**, the aldol condensation must give one enantiomer rather than the racemic mixture. In other words, the aldol condensation must be highly enantioselective and proceed with excellent stereoselectivity (see Chapter 9, Section 9.5). *This book does not discuss methodology that will allow such a transformation, but such methodology is available in modern organic chemistry.*

Another disconnection is shown for chiral **73** to give **75** and **76**. Note that **75** contains the identical stereogenic center, but it has the *R* configuration because the groups have changed. Although this disconnection is perfectly reasonable, it demands the use of a chiral compound. If **75** is not available in chiral form, then it must be prepared by a reaction that gives an enantiopure product. This may be difficult. Forming the stereogenic directly in a reaction (as from **73** and **74**) is usually preferable because it may be possible to control the stereoselectivity of that reaction. In other words, *when there is a choice, choose a disconnection that generates the stereogenic center directly rather than a disconnection that leaves a stereogenic center on the disconnect product.*

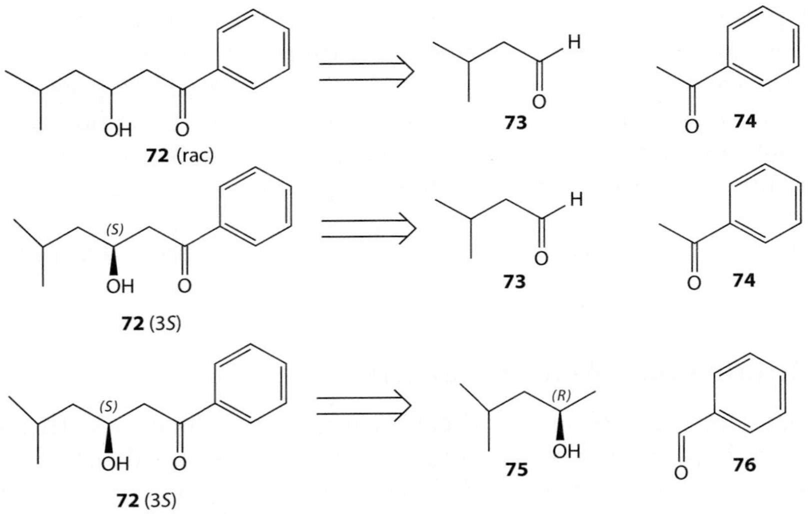

72 (rac)   **73**   **74**

72 (3S)   **73**   **74**

72 (3S)   **75**   **76**

The point of this discussion is not to introduce the details of enantioselective reactions. The point is simply to show that the synthesis presented in this chapter and in this book will give racemic compounds. ***The synthesis of chiral targets requires a significant escalation in the understanding in reactions and methodology.*** This statement is not necessarily obvious. Compound **72** is meant to show that changing the target from a racemic product to a chiral product poses a problem. Imagine the challenge if the target contains two or more stereogenic centers. If this point should be questioned, consider a retrosynthetic analysis of reserpine (**77**) that will lead to a synthesis of the single enantiomer shown. With seven stereogenic centers—including the nitrogen atom at the bridgehead position (Chapter 9, Section 9.7.3)—the 2^n rule (see Chapter 9, Section 9.5) indicates the possibility of $2^7 (= 128)$ stereoisomers for **77**.

The goal of a retrosynthesis is to synthesize one and only one of those stereoisomers. This is not an easy task. If you wish to give it a try, please discuss your retrosynthesis with your instructor or it will be a very frustrating exercise. A total synthesis by R. B. Woodward (United States; 1917–1979) can be found in the journal *Tetrahedron*.[2] Reserpine is an indole alkaloid isolated from the roots of *Rauwolfia serpentina* (also known as Indian snakeroot). This plant has been used in India for the treatment of insanity as well as for the treatment of fevers or snakebite. In pure form, it has been used in the treatment of high blood pressure and it has antipsychotic properties.

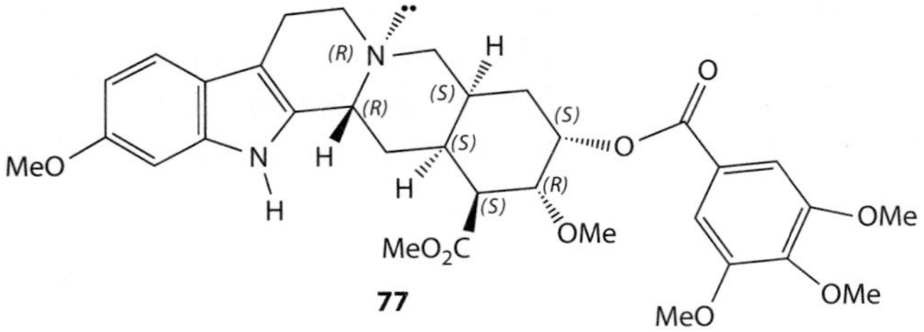

77

References

1. Seebach, D. 1979. *Angewandte Chemie International Edition* 18:239.
2. Woodward, R. B., Bader, F. E., Bickel, H., Frey, A. J., and Kierstead, R. W. 1958. *Tetrahedron* 2:1.

Answers to Problems

25.1

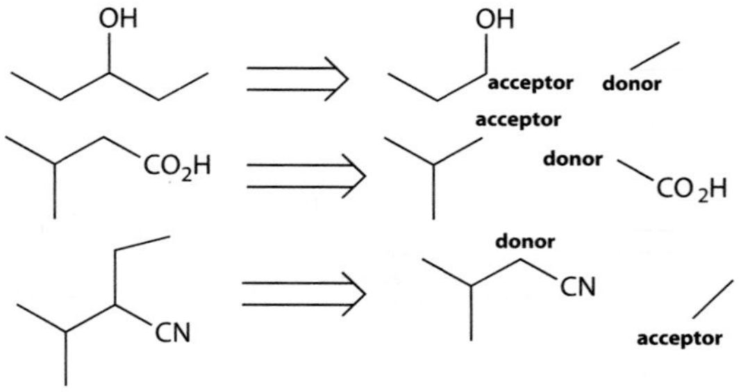

25.2

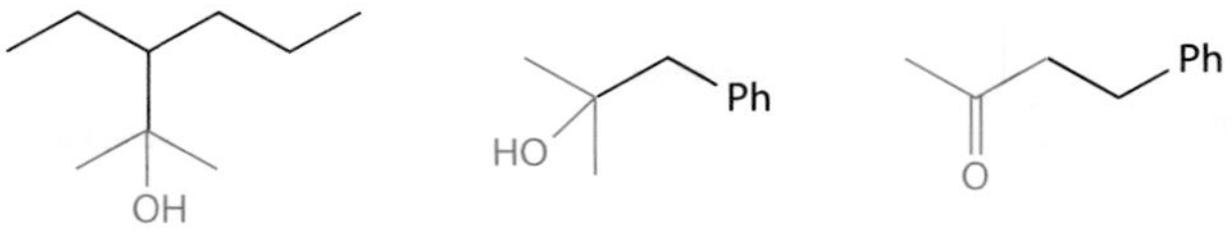

25.3

CH_3I H_2O

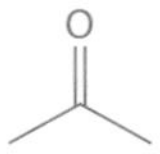

3 molecules that react with a δ– C or with a carbanion

3 molecules that react with a δ+ C or with a carbocation

CH_3MgBr

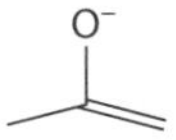

CN^-

25.4

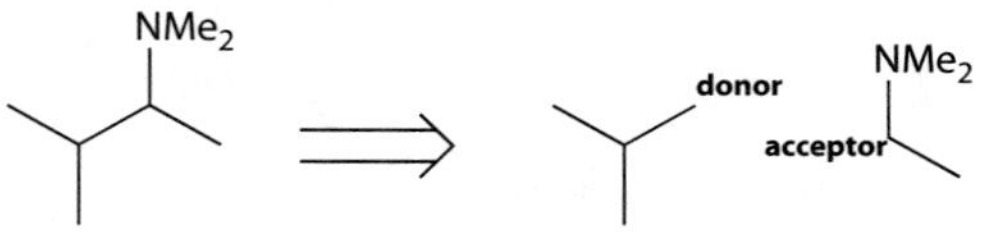

N has a δ– dipole, so the adjacent C should be an acceptor

25.5 Four common carbanions are Grignard reagents, enolate anions, alkyne anions, and cyanide.

25.44   Conversion of an aliphatic alcohol to a methyl ether constitutes a very stable protecting group. It is not used very often, however. Offer an explanation.

25.45   Show a retrosynthesis and then provide all reactions and reagents for the synthesis for the following. However, the synthesis must include the use of protecting groups:

(a)

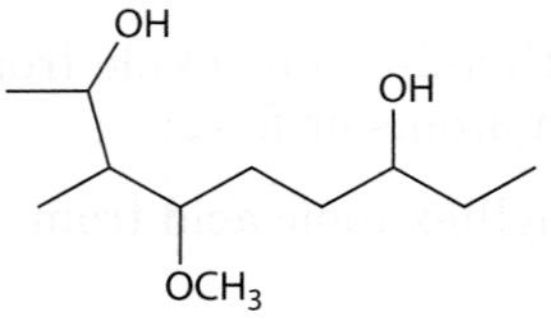

(b)

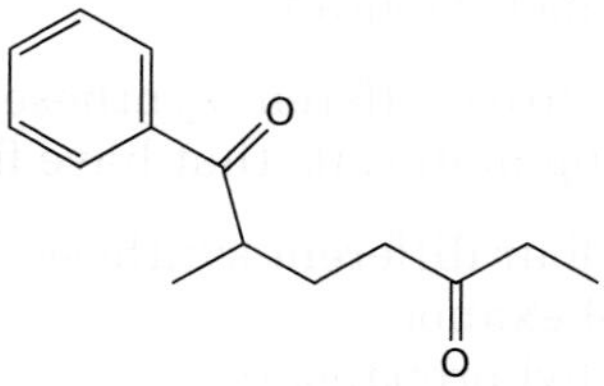

25.46   Provide a convergent synthesis for each of the following:

(a)

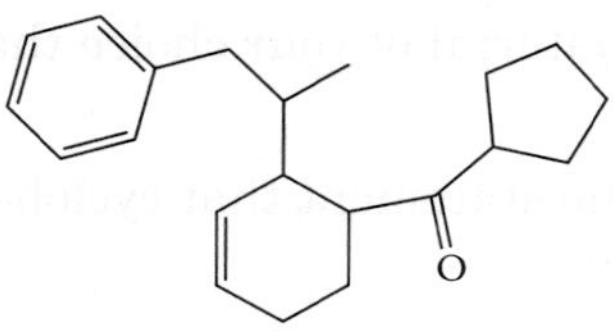

(b)

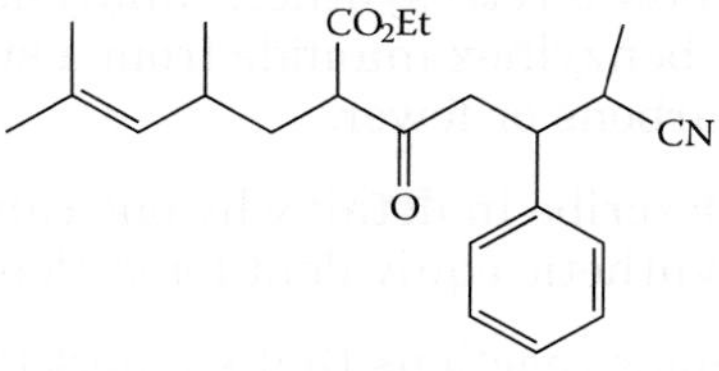

Heteroaromatic Compounds

26

Chapter 21 introduces aromatic hydrocarbons and their unique chemistry. In another class of aromatic compounds, heteroatoms replace one or more of the ring carbons. These compounds are collectively known as **heterocycles** or **heterocyclic aromatic compounds**, and they comprise a class of compounds so large that an entire course is easily built around their chemistry. Heterocycles are seen in several places in this book, including a brief introduction to their nomenclature in Chapter 5 (Section 5.6) and in Chapter 8 (Section 8.9). The most common heterocycles include five- and six-membered monocyclic derivatives that contain nitrogen, oxygen, or sulfur. Several important bicyclic derivatives contain nitrogen. The use of heterocycles in medicine and industry is extensive. This chapter will expand the aromatic chemistry from Chapter 21 and introduce the world of heterocyclic chemistry.

To begin this chapter, you should know the following:

- **heteroatom functional groups (Chapter 5, Section 5.6)**
- **nomenclature of ethers (Chapter 5, Section 5.6.4)**
- **nomenclature of amines (Chapter 5, Section 5.6.2)**
- **nomenclature of aldehydes and ketones (Chapter 5, Section 5.9.2)**
- **fundamental nomenclature of nonaromatic three- to six-membered rings that contain heteroatoms (Chapter 8, Section 8.9)**

- the concept of aromaticity (Chapter 21, Sections 21.1, 21.7)
- electrophilic aromatic substitution (Chapter 21, Section 21.3)
- nucleophilic aromatic substitution (Chapter 21, Section 21.10)
- the chemistry of alkenes (see Chapter 10)
- the chemistry of ketones and aldehydes (see Chapter 18)
- the chemistry of carboxylic acid derivatives (see Chapter 20)
- mechanism and chemistry of acyl addition (Chapter 18, Sections 18.1–18.4)
- acetal and ketal formation (Chapter 18, Section 18.6)
- imine and enamine formation (Chapter 18, Section 18.7)
- mechanism and chemistry of acyl substitution (Chapter 20, Sections 20.1–20.3)
- the concept of isomers (Chapter 4, Section 4.5)
- alcohols, ethers, and amines are bases (Chapter 6, Sections 6.4 and 6.5)
- the E1 reaction (Chapter 12, Section 12.4)
- reversible chemical reactions (Chapter 7, Section 7.10)

This chapter will present the most fundamental concepts for structure, nomenclature, and chemical reactions of simple heterocyclic compounds. The primary focus is on five- and six-membered ring compounds that contain one or two oxygen, nitrogen, of sulfur atoms. Simple bicyclic heterocyclic compounds will also be introduced. The chemical reactions for generating the heterocycles as well as the fundamental chemical reactions will be seen as relatively simple extensions of chemistry presented elsewhere in this book.

When you have completed this chapter, you should understand the following points:

- Heterocycles are cyclic aromatic molecules that contain one or more atoms other than carbon or hydrogen.
- Pyrrole is the five-membered ring heterocycle containing one nitrogen. There are two five-membered ring molecules with two nitrogen atoms—imidazole (1,3 N) and pyrazole (1,2 N).
- Pyridine is the six-membered ring heterocycle containing one nitrogen. Three six-membered ring molecules contain two nitrogen atoms—pyrazine (1,4 N), pyrimidine (1,3 N), and pyridazine (1,2 N).
- Furan is the five-membered ring heterocycle containing one oxygen. Thiophene is the five-membered ring heterocycle containing one sulfur. Pyran and thiopyran are the six-membered ring derivatives.

- **Five-membered ring heterocycles containing one heteroatom tend to undergo electrophilic aromatic substitution primarily at C2. Five-membered ring heterocycles containing one heteroatom tend to be very reactive in electrophilic aromatic substitution reactions.**
- **Six-membered ring heterocycles containing one heteroatom tend to be rather unreactive in electrophilic aromatic substitution reactions. Six-membered ring heterocycles containing one heteroatom tend to undergo electrophilic aromatic substitution at C3.**
- **Halogenated heterocycles undergo nucleophilic aromatic substitution reactions as expected.**
- **Many molecules that are not aromatic contain one or more heteroatoms. These are very common and include ring sizes of three to six containing N, O, or S. There are also derivatives containing two or more heteroatoms.**
- **Two bicyclic aromatic compounds contain one nitrogen: quinoline (N at position 1) and isoquinoline (N at position 2). Indole has a pyrrole unit fused to a benzene ring to give a bicyclic aromatic system with N at position 1.**
- **Quinoline and isoquinoline show that electrophilic aromatic substitution reactions are the more reactive benzene ring because the "pyridine" ring is less reactive. Indole undergoes electrophilic aromatic substitution primarily in the pyrrole ring because it is much more reactive than the benzene ring.**
- **Pyrroles, pyrazoles, imidazoles, furans, and thiophenes can be made from 1,3-dicarbonyl compounds by reaction with an appropriate amine, oxidizing agent, or P_4S_{10}. This process is called synthesis: The new molecule is synthesized from the old one (see Chapter 25).**
- **Spectroscopy can be used to determine the structure of a particular molecule and can distinguish the structure and functionality of one molecule when compared with another (see Chapter 14).**
- **Heterocyclic compounds are important in biological systems.**

26.1 Nitrogen in an Aromatic Ring

26.1.1 Five-Membered Rings

In Chapter 21 (Section 21.7) the cyclopentadienyl anion (1) was identified as an aromatic species with six π-electrons. It is possible to do a thought experiment to "replace" the C:⁻ unit of 1 with a N–H unit; the result is the aromatic compound **2**, which is named pyrrole. Pyrrole was introduced in Chapter 21

(Section 21.7) in connection with a discussion of the relative basicity of pyrrole versus pyridine. Pyrrole is a constituent of coal tar and it is found in bone oil. It is obtained by heating albumin or by the pyrolysis (heating to elevated temperatures) of gelatin.

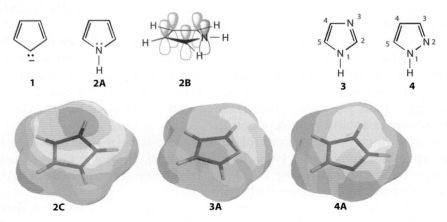

Nitrogen has an unshared electron pair in an orbital that is parallel with the four π-electrons in the C=C units for a six π-electron system in a π-framework, so pyrrole is aromatic. This requirement makes certain demands on the molecule that are important to its structure. In **2B**, the six aromatic π-electrons are above and below the plane of the carbon atoms and the nitrogen, with a hydrogen atom attached to the nitrogen that is perpendicular to the π-orbitals, but they in the same plane as the hydrogen atoms on the ring. The fact that the electron pair must be parallel with the other π-electrons forces the hydrogen atom to be coplanar with the carbon atoms and nitrogen, as shown.

Pyrrole is a planar molecule, consistent with its being an aromatic compound. Pyrrole is a secondary amine, but it is not very basic. The nitrogen of **2** must donate its electron pair to a proton to react as a Brønsted–Lowry base; however, that electron pair is "tied up" in the aromatic π-cloud, so those electrons are not available for donation. Another way to think about this is to understand that if pyrrole reacts as a base and donates two electrons, the aromaticity of the ring is disrupted. The energy cost for the reaction is simply too great and, consequently, pyrrole is a very weak base. This fact is supported by the electron density map for pyrrole (**2C**), which shows the concentration of electron density (**red**) above and below the ring—*not* on the nitrogen atom. The hydrogen atom on the nitrogen of pyrrole is actually somewhat acidic, with a pK_a of 17.5 in water. Note the **blue** color in **2C**, which indicates low electron density. The IUPAC numbering scheme for pyrrole begins with nitrogen (atom 1) and extends around the ring. This is typical for the names of most heterocycles.

26.1 Draw 1-ethylpyrrole, 4-nitropyrrole, and *N*-methyl-3-phenylpyrrole.

26.2 If the hydrogen on the nitrogen atom in 2 is removed, draw the anion and then comment on the positions of the two electron pairs on the nitrogen.

Some aromatic compounds have two nitrogen atoms in a five-membered ring. Compound **3** has the nitrogen atoms in a 1,3 relationship and it is known as imidazole. Imidazoles are also called azoles. Compound **4** has the nitrogen atoms in a 1,2 relationship and it is known as pyrazole. In both imidazole and pyrazole, one of the nitrogen atoms is sp^2 hybridized with a lone electron pair perpendicular to the π-cloud; the other nitrogen uses the electron pair as part of the π-cloud. The hydrogen atom on nitrogen is perpendicular to the π-cloud, as in pyrrole. Electron density maps for imidazole (**3A**) and for pyrazole (**4A**). It is clear that there is a high concentration of electron density on the ring nitrogen in both cases, suggesting that those molecules will react as bases far better than pyrrole.

The convention for naming these compounds makes the nitrogen bearing the hydrogen atom 1, and the numbering around the ring is such that the second nitrogen receives the lowest possible number (2 or 3). Both imidazole and pyrazole are important units in many pharmaceutical preparations as well as in naturally occurring compounds.

Histidine (**5**) is an important amino acid (see Chapter 27) and histamine (**6**) is a neurotransmitter that is important in cells during antigen–antibody reactions. Important physiological reactions attributed to histamine are anaphylaxes (a drop in blood pressure that can result in shock) and allergy. Pilocarpine (**7**) is used to treat glaucoma. Note the presence of a lactone ring in **7** (see Chapter 20, Section 20.5.5). The imidazole derivatives clotrimazole (Lotramin®, **8**) and miconazole (Monistat®, Micatin®, **9**) are antifungal agents that are applied topically. Both have been used in preparations to treat athlete's foot. In **5–9**, the imidazole ring is highlighted in **blue**.

26.3 Draw 1,3-dimethylpyrazole and 4-bromo-2-ethylimidazole.

26.1.2 Six-Membered Rings

The aromatic six-membered ring compound that contains one nitrogen atom is known as pyridine (**10**) and was first isolated in 1846 from coal tar. Pyridine has been used as a base from time to time in preceding chapters. Because it is aromatic (see Chapter 21, Section 21.7), pyridine is a planar compound with an aromatic π-cloud above and below the plane of the ring. The nitrogen atom is sp^2 hybridized, but the lone electron pair on nitrogen is *not* part of the aromatic π-cloud. That electron pair is perpendicular to that π-cloud (see **10B**), so it is available for donation in acid–base reactions. This means that, unlike pyrrole, pyridine is a good Brønsted–Lowry and Lewis base because the electron pair is readily available for donation.

The electron density map of pyridine (see **10C**) clearly shows a high concentration of electron density (more **red**) on the nitrogen. Note that pyridine has been implicated in male sterility, although some claim this is a myth. Nonetheless, it is reasonable to exercise caution when pyridine is used. Several important pyridine derivatives have substituents on the aromatic ring, including 2,6-lutidine (**11**) and picolinic acid (**12**). Many derivatives of pyridine are found in pharmaceutically active compounds. One is nicotinic acid (niacin, **13**), which is vitamin B$_3$ and is found in liver, yeast, and meat. A deficiency in this vitamin can lead to pellagra (a wasting disease). Nicotinamide (**14**, niacinamide) is one of the two principal forms of the B-complex vitamin niacin. Nicotinamide may be useful for individuals with type 1 (insulin-dependent) diabetes.

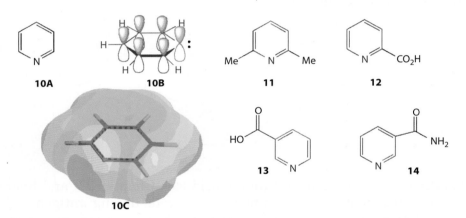

Three different aromatic six-membered ring compounds contain two nitrogen atoms in the ring: pyrazine, pyrimidine, and pyridazine. In pyrazine (**15**), the two nitrogen atoms have a 1,4 relationship; in pyrimidine (**16**), they have a 1,3 relationship, and, in pyridazine (**17**), they have a 1,2 relationship. In each case, both nitrogen atoms are sp^2 hybridized, and a lone electron pair on *both* nitrogen atoms is available for donation. In other words, both nitrogen atoms in **15–17** are basic. As with all aromatic compounds, these are planar compounds with an aromatic π-cloud above and below the plane of the ring.

26.4 Draw the product formed when pyridine is treated with sulfuric acid and name it.

26.5 Draw the structures of (a) 2-ethyl-5-nitropyrazine, (b) 2,6-dibromopyrazine, (c) 2,4,6-triethylpyrimidine, and (d) 4,5-dinitropyridazine.

The ring systems **15–17** are found in pharmaceutically important compounds. The pyrazine derivative **18** is called Acipimox®, and it is used to lower levels of cholesterol and triglycerides. (Note: it is an oxidized form of the amine known as an *N*-oxide.) Pyrazinamide (**19**) is an antibacterial agent. The pyrimidine ring system is found in thiamin (**20**, vitamin B$_1$). A deficiency of thiamin is associated with beriberi. Symptoms of beriberi include loss of appetite and fatigue, digestive irregularities, and a feeling of numbness and weakness in the limbs and extremities that may lead to nerve degeneration in extreme cases. Another pyrimidine, sulfamerazine (**21**), is a broad-spectrum antibacterial agent. Minoxidil® (Rogaine®, **22**) is classified as a vasodilator used to treat high blood pressure, but it is also used in hair-restoring preparations. A vasodilator relaxes the smooth muscle in blood vessels, causing the vessels to dilate.

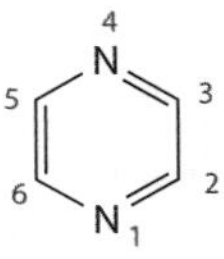

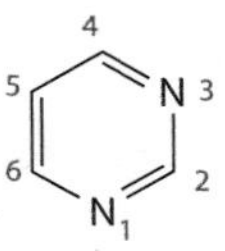

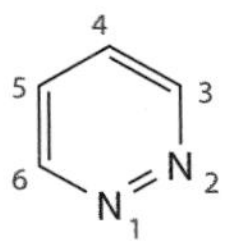

26.2 Oxygen and Sulfur in an Aromatic Ring

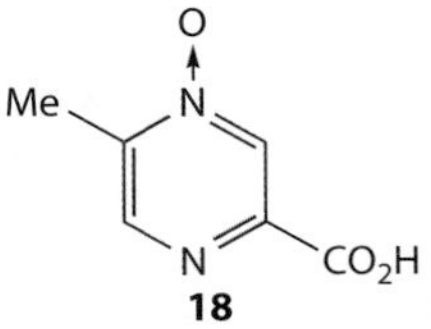

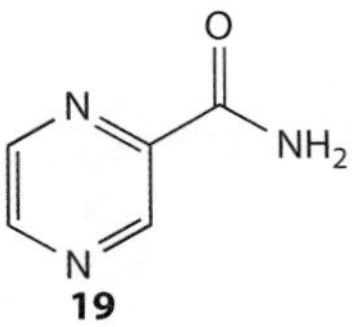

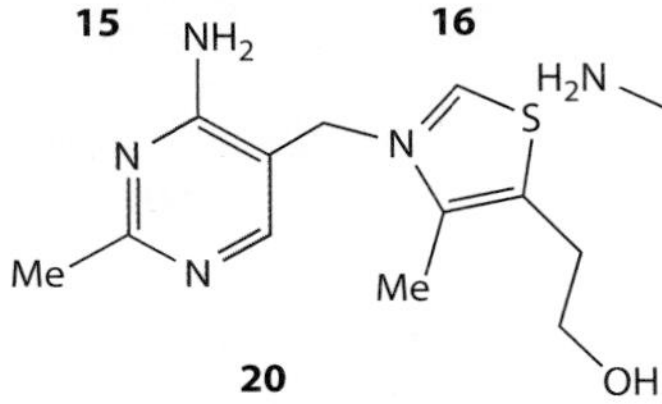

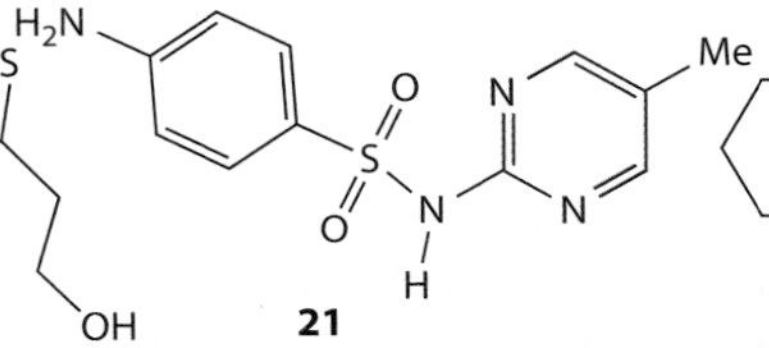

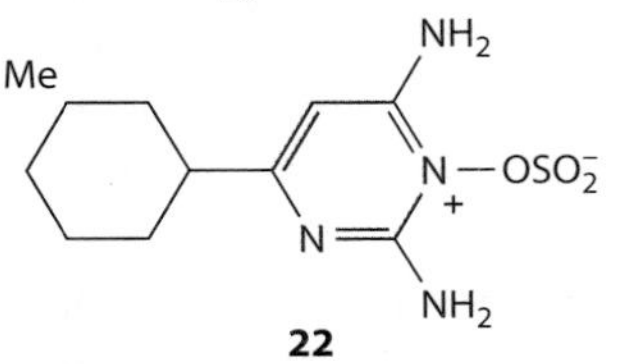

Some five-membered ring aromatic compounds contain an oxygen or a sulfur, but the analogous six-membered ring compounds are *not* aromatic. Furan (**23**) is an aromatic compound that is distilled from pine wood rosin; its vapors are narcotic. One of the two electron pairs on oxygen is involved in the aromatic π-cloud (those two electrons are needed to make a total of six), but the other lone electron pair is perpendicular to the π-cloud. This is illustrated in **23B**, where the second lone electron pair can function as a base. Furan should be a *stronger* base than pyrrole because of the availability of those electrons. The electron density map of furan (**23C**) shows some concentration of electron density on oxygen (more **red**)—slightly more than in pyrrole (see **2C**). Indeed, furan is a somewhat stronger base than pyrrole, although furan is categorized as a weak base. Furan is an aromatic ether and the furan ring plays a prominent role in many naturally occurring and synthetic compounds.

If the oxygen atom is replaced with sulfur, the resulting compound is the aromatic compound thiophene, **24A**, which behaves more or less as an aromatic thioether. As with furan, one electron pair is involved in the six π-electron aromatic cloud, and the other is perpendicular to that π-cloud. The electron density map of thiophene (see **24B**) suggests more aromatic character than furan, and the lessened electron density of sulfur suggests that thiophene is a weaker base than furan. Thiophene can be obtained from coal tar and coal gas, and it is used in the manufacture of dyes and pharmaceuticals. Both furan and thiophene are numbered in the IUPAC system with the heteroatom as 1 and numbering around the ring to give substituents (if any) the lowest numbers.

26.6 Draw the structures of (a) 3,4-dimethylfuran, (b) furan 3-carboxylic acid, (c) thiophene 2-carboxaldehyde, and (d) 2,5-dibromothiophene.

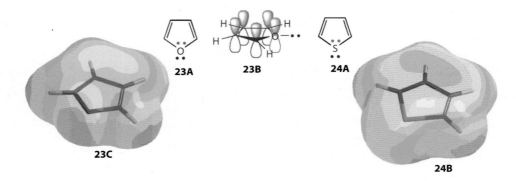

As noted, many important compounds contain furan or thiophene rings. Furfural (**25**), for example, contains the furan ring and is a common natural product found in cereal straws and brans. It is used in the manufacture of plastics and varnishes and is also used as an insecticide and fumigant. Furosemide® (Lasix®, **26**) has diuretic and antihypertensive properties. Lasix inhibits sodium ion reabsorption and is used to treat edema, hypertension, and cardiac insufficiency. The thiophene ring is important in many compounds, including the sodium salt of 2-thiophene carboxylic acid (**27**), which is used as a lubricant grease thickener. Thenium closylate (**28**) has anthelmintic properties (active against parasitic worms). In the compounds shown in the illustration, the furan ring is highlighted in **blue** and the thiophene ring in **purple**.

Because oxygen is divalent, incorporation into a six-membered ring does not allow an aromatic system to be generated without adding a charge to the oxygen. Pyran (**29**) is the six-membered ring analog, but for oxygen to remain neutral, one carbon must be sp^3 hybridized, as shown in the illustration. Making a double bond to oxygen would generate a pyrylium ion (**30**), which is a transient intermediate in some reactions. In order to focus on neutral compounds that are aromatic, the following discussions of aromatic oxygen and sulfur compounds are restricted to furan and thiophene.

26.7 Draw the sulfur analog of 29, known as thiopyran.

29   **30**

26.3 Substitution Reactions in Heterocyclic Aromatic Compounds

Heterocycles are aromatic compounds, and they undergo aromatic substitution reactions similar to reactions of aromatic hydrocarbons (see Chapter 21, Section 21.3). In some cases, electrophilic aromatic substitution reactions are faster than benzene due to the presence of the heteroatom, but in other cases the reaction is slower. In other words, the nature of the heteroatom and the size of the ring have a profound influence on the rate of reaction as well as the site of reaction. The basic principles of reactivity and regioselectivity in these cases are governed by the same fundamental principles discussed for benzene derivatives in Chapter 21. For electrophilic aromatic substitution reactions of heterocycles, a cationic intermediate is formed; however, the presence of the electron-rich heteroatom must be taken into account. The major site of substitution in this reaction is the one that gives the more stable intermediate.

26.3.1 Five-Membered Rings

The first reaction is an electrophilic substitution reaction of pyrrole and nitric acid in the presence of acetic anhydride, which generates NO_2^+ (the nitronium ion; see Chapter 21, Section 21.3.1) *in situ*. The nitronium ion reacts with the aromatic compound. Although pyrrole is a weak base, the use of a strong Lewis acid such as $AlCl_3$ or $FeBr_3$ or a Brønsted–Lowry acid such as sulfuric acid may react with the basic nitrogen atom. For this reason and because pyrrole is more reactive than benzene, a mixture of nitric acid and acetic anhydride is used to generate the NO_2^+ species.

Other than this modification, the principles of electronic aromatic substitution are the same as those introduced in Chapter 21 (Section 21.3). Pyrrole is a very electron-rich aromatic system. *It is much more reactive than benzene in electrophilic aromatic substitution reactions.* This means that it is an activated aromatic ring. To reiterate, the sulfuric acid/nitric acid combination used with benzene in Chapter 21 is not used here due to the presence of the nitrogen atom as well as the increased reactivity of the pyrrole ring.

There are two sites for substitution when an electrophile reacts with pyrrole (**2**): C2 and C3. If the pyrrole ring attacks the nitronium ion, NO_2^+ attaches to C2 to give intermediate **31**, whereas intermediate **32** is formed by attachment at C3. It is apparent that attack at C3 can generate only two resonance forms, but attack at C2 generates three. In both cases, a positive charge resides at the carbon adjacent to an sp^2 nitrogen, leading to a stable iminium ion structure. Because attack at C2 leads to a more stable intermediate, the activation energy for that reaction is lower, and it is the favored intermediate to the major product. Therefore, nitration of pyrrole gives 2-nitropyrrole (**33**) as the major product.

In a similar manner, the reaction of pyrrole with sulfuryl chloride (SO_2Cl_2), without the need for a Lewis acid, gives 2-chloropyrrole (**34**). 2,5-Dichloropyrrole (**35**) is also formed in this reaction. Note that diatomic chlorine is not used as a reagent, but once again the reaction conditions are modified to accommodate the increased reactivity of pyrrole as well as the weakly basic nitrogen atom. Bromination of pyrrole is done using *N*-bromosuccinimide (NBS; see Chapter 11, Section 11.9.3) as a brominating agent to give 2-bromopyrrole.

26.8 Draw the acid–base product of the reaction between pyrrole and boron trifluoride.

26.9 Draw the product of the reaction between pyrrole with sulfur trioxide (SO_3) using pyridine as a solvent at 100°C.

The presence of the nitrogen heteroatom in the five-membered aromatic ring leads to a preference for C2 substitution. The same should be true if the heteroatom is oxygen or sulfur. Indeed, both furan and thiophene give predominantly C2 substitution in electrophilic aromatic substitution, although there are differences. *Furan has less aromatic character than pyrrole* (see **23C**; it is less activated and reacts more slowly in electrophilic aromatic substitution), and it often reacts with electrophilic reagents such as bromine to give addition reactions to the C=C unit rather than substitution. This discussion will ignore such reactions and focus on electrophilic aromatic substitution reactions, in which furan gives primarily the 2-substituted product. More of the 3-substituted product is formed than in the similar reaction with pyrrole. *Although less reactive than pyrrole, furan is about 10^5 times more reactive than benzene in electrophilic aromatic substitution reactions.*

Catalysts such as concentrated acid or aluminum chloride are not used with furan because these reagents induce polymerization (see Chapter 10, Section 10.8.3); however, with proper reagents, substitution can occur. Treatment of furan (**23**) with nitrosonium tetrafluoroborate ($NO_2^+ BF_4^-$) gives a low yield of 2-nitrofuran, **36**. Furan reacts with SO_3 and pyridine to give primarily 2-furansulfonic acid and it is possible to prepare 2-bromofuran by reaction with a bromine•dioxane complex at −5°C using carbon disulfide (CS_2) as a solvent. Friedel–Crafts type reactions are possible, as illustrated by the reaction of furan with acetyl chloride and tin tetrachloride ($SnCl_4$), which gives 2-acetyl-furan, **37**. To reiterate, somewhat specialized reagents are used in these reactions because furan can react with strong Lewis acids.

Thiophene is less reactive than furan in electrophilic aromatic substitution, but it is 10^3–10^5 times more reactive than benzene. Thiophene has more aromatic character than furan (see **23C** and **24B**), and it tends to undergo substitution rather than addition. Thiophene is more stable to strong Lewis acids such as aluminum chloride than is furan (thiophene is a weaker base). The reaction conditions used for thiophene (**24**) are very similar to those used for furan, however. Treatment with nitric acid and acetic anhydride, for example, gives 2-nitrothiophene.

26.10 Briefly explain why furan is more reactive than benzene.

26.11 Determine the mechanism of bromination of furan at C_2, assuming that it proceeds via Br^+.

26.12 Draw the product formed in the reaction of thiophene with (a) sulfuryl chloride, (b) NBS, and (c) acetyl chloride and tin tetrachloride ($SnCl_4$).

26.3.2 Six-Membered Rings

Pyridine is a tertiary amine and a good base, as noted in Section 26.1.2. Because of this property, many of the electrophilic reagents used for aromatic substitution coordinate with the electron pair on nitrogen (an acid–base reaction). Specifically, the Lewis acids used in Chapter 21 for electrophilic aromatic substitution will coordinate with the electron pair on nitrogen, so they cannot be used. If electrophilic aromatic substitution does occur, the reaction is slow, and such reactions are difficult. Carbons **3** and **5**, relative to nitrogen, have the greatest π-electron density (see **10C**) and they are the major sites for reaction. The intermediates generated from pyridine in electrophilic aromatic substitution reactions are relatively high in energy, so the reaction is slow. Indeed, the pyridine ring is rather deactivated for electrophilic aromatic substitution.

26.13 Draw and name the acid–base product formed when pyridine is treated with HBr.

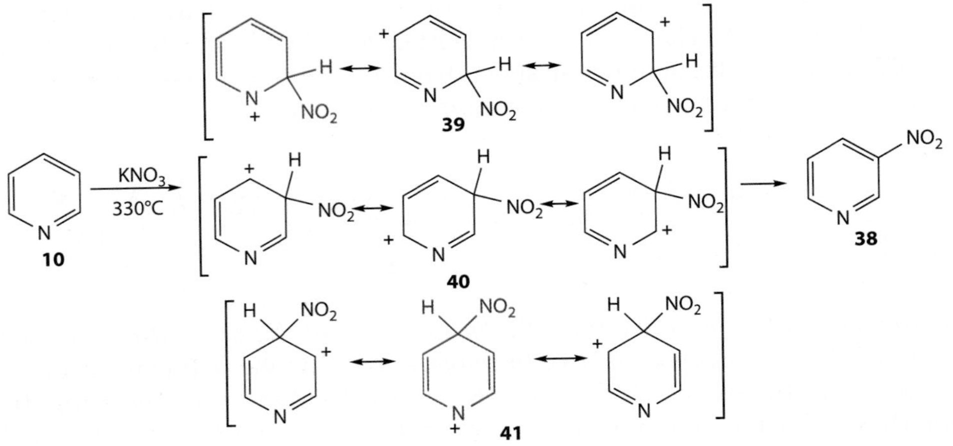

Aromatic substitution can occur if specialized reagents and vigorous reaction conditions are used. An example is the reaction of pyridine (**10**) with potassium nitrate (KNO_3) to give 3-nitropyridine (**38**), but this reaction requires heating to 330°C and the yield is poor. This example simply reinforces the statement that pyridine is resistant to electrophilic aromatic substitution and reacts much more slowly than benzene in contrast to the five-membered ring aromatic compounds. It also shows that substitution occurs at C3. If the reagent NO_2^+ is used in the reaction, attack at C2 gives intermediate **39**, attack at C3 gives **40**, and attack at C4 gives **41**. Attack at C2 leads to a resonance contributor of **39** that places a positive charge directly on the nitrogen (in **red**). *This is very unstable.* Similar attack at C4 puts a positive charge on nitrogen (in **red** in the illustration). *Attack at C3 gives three resonance forms that are not particularly destabilized, and the rate of forming 40 is faster than the rate for generating 39 or 41. Therefore, pyridine reacts with electrophilic reagents to give the 3-substituted derivative 38.*

In other words, attack at C3 leads to a less destabilized intermediate, which gives the major product. In addition to the nitration reaction, pyridine reacts with bromine (at a temperature near 300°C) to give mixture of 3-bromopyridine (**42**) and 3,5-dibromopyridine (**43**). It is clear that pyridine is not a particularly good partner in electrophilic aromatic substitution. For the same reasons given for pyridine, electrophilic aromatic substitution of simple pyrimidines is even more difficult. Although functionalized derivatives can give interesting reactions due to the presence of activating substituents (see Chapter 21, Section 21.4, for activating groups in reactions of benzene), they will not be discussed.

26.14 Draw the product of the reaction of pyridine with sulfur trioxide in sulfuric acid (in the presence of mercuric sulfate ($HgSO_4$)).

26.15 Draw the intermediates for attack at pyridine on Br^+ at C_2, C_3, and C_4 and determine why the preferred product is 3-bromopyridine.

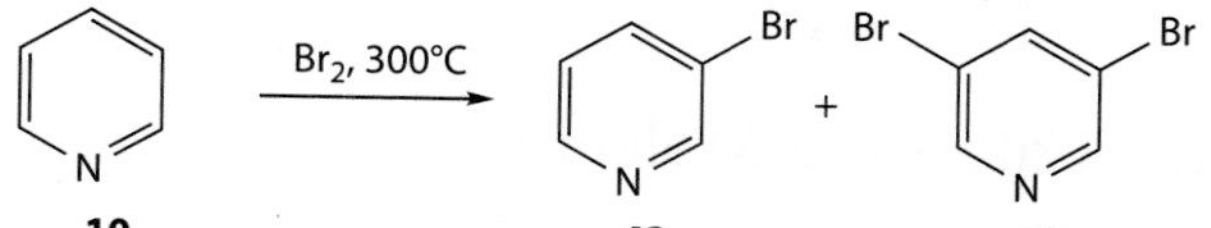

26.3.3 Nucleophilic Aromatic Substitution

Although pyridine is resistant to electrophilic aromatic substitution, it is susceptible to nucleophilic aromatic substitution. The reaction of 2-bromopyridine (**44**) and ammonia, for example, leads to 2-aminopyridine, **46**. This is a nucleophilic aromatic substitution reaction (see Chapter 21, Section 21.10), where NH_3 attacks the *ipso* carbon to generate a carbanionic intermediate (**45**), which loses bromine to give **46**. Pyridine (**10**) also reacts directly with sodium amide ($NaNH_2$) at 100°C to give **46** in what is known as the **Chichibabin reaction**, after Aleksei E. Chichibabin (Russia; 1871–1945).

Other reactive bases can be used in this reaction. Pyridine reacts with phenyllithium at 100°C, for example, to give 2-phenylpyridine. This reaction is limited in scope because powerful nucleophiles are required and the reaction conditions can be harsh. Nonetheless, several interesting transformations are observed. Five-membered ring heterocycles are less prone to nucleophilic aromatic substitution, in part because the reparation of the requisite halogen-substituted derivatives can be difficult.

26.16 Draw the structure of 2-phenylpyridine.

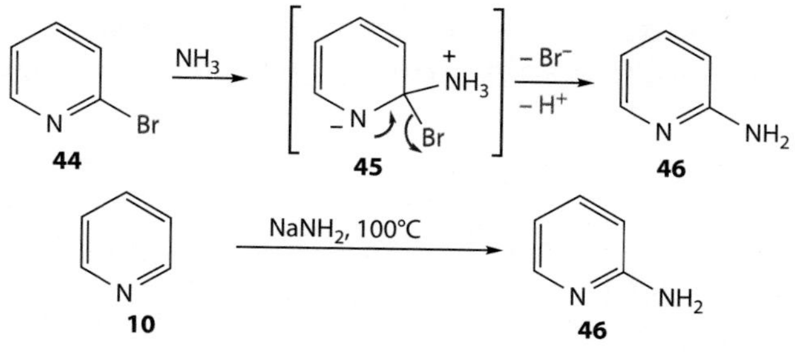

26.4 Reduced Forms of Heterocycles

This chapter formally discusses heterocycles, which are aromatic compounds that contain heteroatoms. Several important compounds, often identified as heterocyclic compounds, result from reduction of heterocycles. Most of these ring systems were introduced in previous chapters, particularly Chapter 5 (Section 5.6). These reduced heterocycles are commonly used in medicine and in industry. Several common monocyclic compounds that contain nitrogen, oxygen, or sulfur will be presented in this section. In part, these compounds are discussed because the IUPAC nomenclature system derives from the name of the parent heterocycle (introduced in preceding sections and in Chapter 5, Section 5.6).

26.4.1 Nitrogen-Containing Molecules

Nitrogen-containing ring systems are classified by the nature and number of heteroatoms, as well as by the size of the ring. For a monocyclic system with one nitrogen, the three-membered ring derivative (**47**) is called aziridine, the four-membered ring (**48**) is azetidine, the five-membered ring (**49**) is pyrrolidine, and the six-membered ring (**50**) is piperidine. Pyrrolidine is prepared by catalytic hydrogenation of pyrrole, and its IUPAC name is tetrahydropyrrole. Similarly, catalytic hydrogenation of pyridine leads to piperidine, which can be named hexahydropyridine.

These compounds were first introduced in Chapter 5 (Section 5.6) and many are key units in important pharmaceuticals. Compounds **47–50** are secondary amines and they react as bases in the presence of a suitable acid. One important compound that contains reduced heterocyclic rings is mitomycin C (**51**), an important drug used to treat cancer. In contains both an aziridine ring and dihydropyrrole units. 2-Azetidinecarboxylic acid (**52**) shows growth inhibitory activity on cultures of *Escherichia coli*. Several important compounds contain a pyrrolidine ring, including proline (**53**), which is a common amino acid (see Chapter 27). Nicotine (**54**) is found in tobacco, and the carboxylate salt of kainic acid (**55**) is a neurotoxin.

26.17 Draw *meso*-3,4-dimethylpyrrolidine and *trans*-3,5-diphenylpiperidine.

26.18 Draw the products formed when (a) 48 reacts with HCl, (b) 49 reacts with sulfuric acid, and (c) 50 reacts with BF_3.

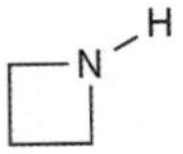

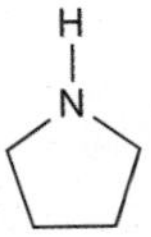

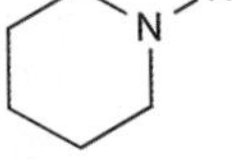

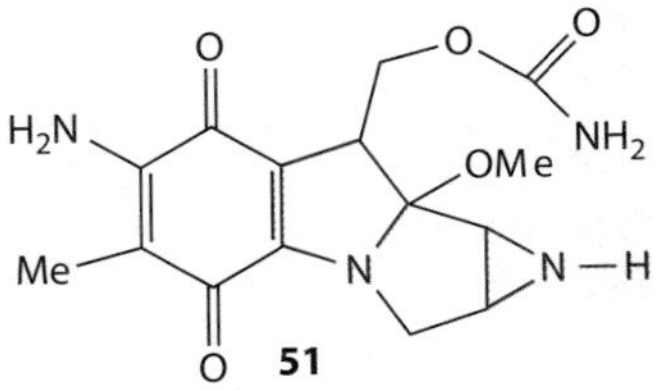

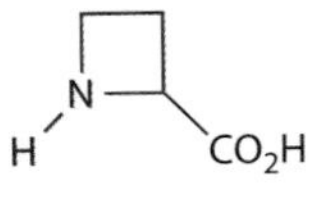

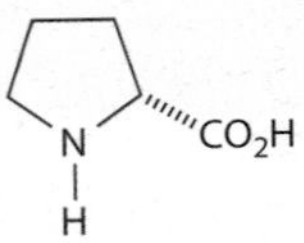

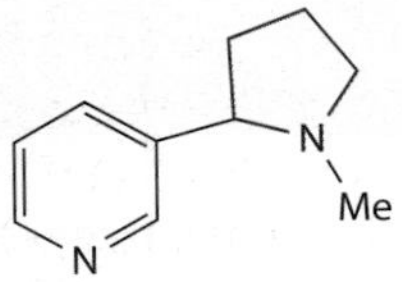

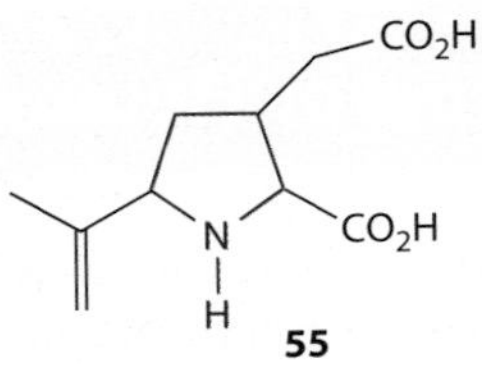

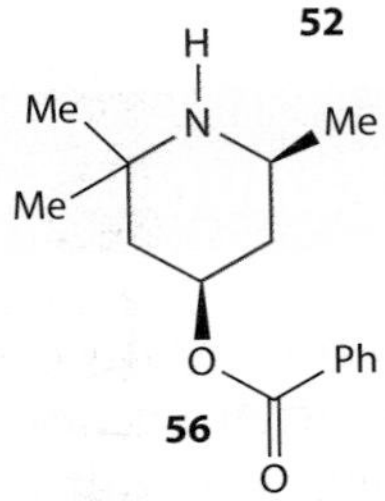

A piperidine derivative, β-eucaine (**56**), is used as a local anesthetic. The illegal designer drug MPTP (1-methyl-4-phenyl-1,2,3,6-tetrahydropyridine, **57**) contains the six-membered ring heterocycle and is particularly dangerous because it is converted to a metabolite, pyridinium salt **58** (MPP), in the body. This metabolite is associated with irreversible Parkinsonism. MPP is classified as a dopaminergic neurotoxin.

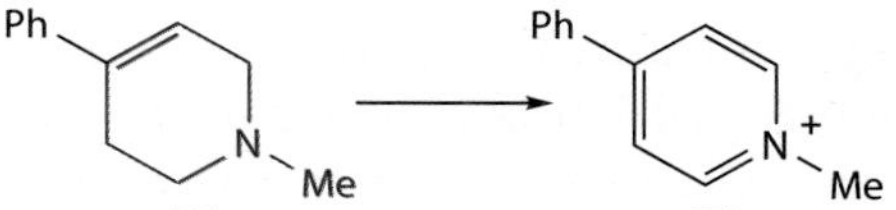

26.4.2 Oxygen-Containing Molecules

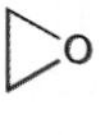

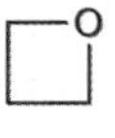

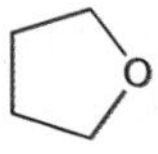

Oxygen-containing ring systems are important, and several were discussed in previous chapters. All are cyclic ethers. The three-membered ring ether (**59**) is oxirane and is the "parent" of the epoxides discussed in Chapter 10 (Section 10.5). The four-membered ring ether (**60**) is oxetane, and the five-membered ring ether (**61**) is oxolane, or tetrahydrofuran (THF). As first pointed out in Chapter 11 (Section 11.2.3), this important cyclic ether is prepared by catalytic hydrogenation of furan. Tetrahydrofuran is used many times in this book as a polar, aprotic solvent. The six-membered ring ether (**62**) is called oxane or tetrahydropyran and is obtained by catalytic hydrogenation of pyran (**29**).

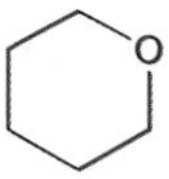

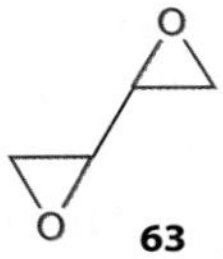

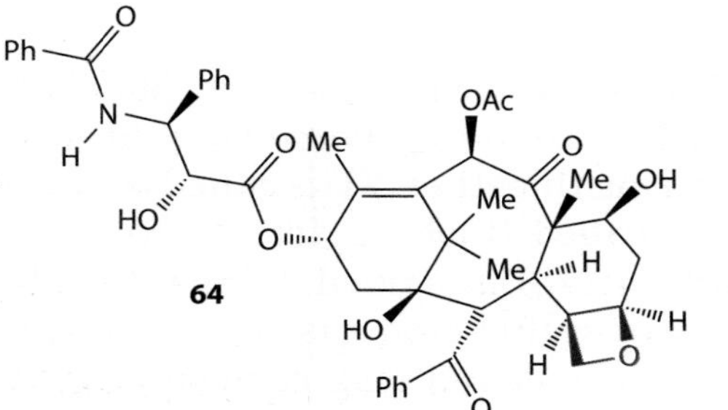

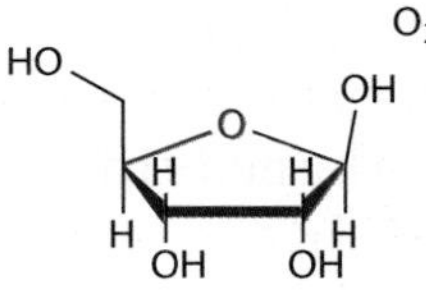

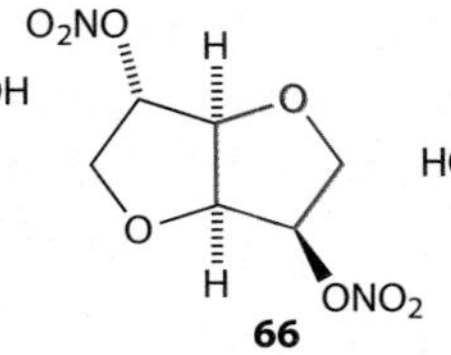

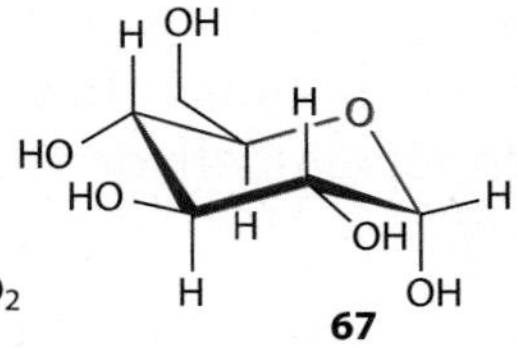

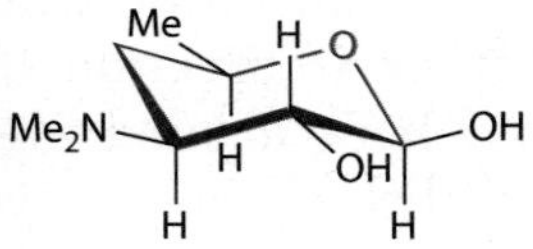

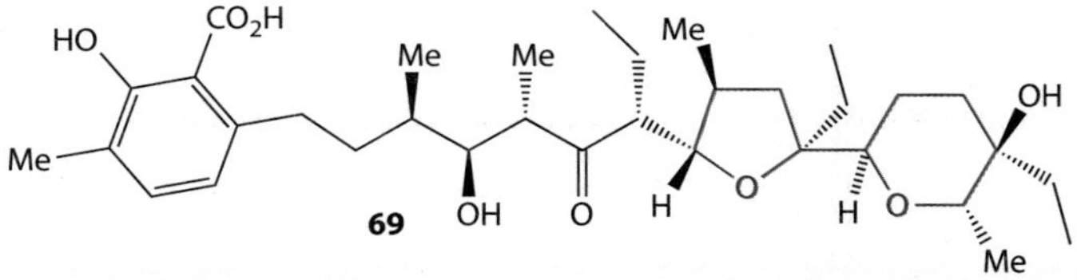

26.19 Draw 3-bromotetrahydrofuran and *cis*-2,5-diethyltetrahydropyran.

Cyclic ethers are components of many natural products as well as synthetic compounds. A simple one is erythritol anhydride (**63**), which contains the epoxide unit and is used to prevent microbial spoilage. The important anticancer drug taxol (**64**) contains an oxetane unit. The tetrahydrofuran unit is a component of the important carbohydrate ribose (**65**), which is the "sugar" portion of RNA (see Chapter 28). A tetrahydrofuran unit is found in isosorbide dinitrate (Isordil®, **66**), which is a vasodilator used in the prevention and treatment of angina. The tetrahydropyran unit is found in common sugars such as glucose (**67**) and in desosamine (**68**), which is the sugar component of many macrolide antibiotics. Both the tetrahydrofuran and tetrahydropyran units are found in the antibiotic Lasalocid® (**69**).

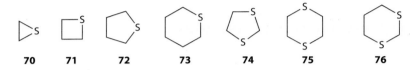

70   **71**   **72**   **73**   **74**   **75**   **76**

There are cyclic thioethers, but they are not encountered very often in this book. The three-membered ring thioether (**70**) is called thiirane, the four-membered ring thioether (**71**) is called thietane, and the five-membered ring thioether (**72**) is called thiolane (tetrahydrothiophene). Finally, the six-membered ring thioether (**73**) is named thiane (tetrahydrothiopyran). A five-membered ring with two sulfur atoms is called a dithiolane, and **74** is named 1,3-dithiolane. The six-membered ring compound with two sulfur atoms is a dithiane; **75** is named 1,3-dithiane, whereas **76** is 1,4-dithiane. Both **74** and **76** were noted in Chapter 18 (Section 18.6.7) because they are the dithioacetal derivatives of formaldehyde formed when formaldehyde reacts with 1,2-ethanedithiol or 1,3-propanedithiol, respectively.

26.20 Draw 3-ethylthietane, *cis*-2,6-dimethylthiane, and 2,3-dihydrothiophene.

26.21 Draw a synthesis of 2-benzyl-1,3-dithiane from benzene.

26.5 Heteroaromatic Compounds with More Than One Ring

Polycyclic aromatic hydrocarbons were discussed in Chapter 21 (Section 21.8), and two important examples are naphthalene and anthracene. There are also important polycyclic systems that contain heteroatoms. These particular heterocycles are extremely important because they are found in many biological systems and in medicines. The discussion will focus only on polycyclic heterocycles that contain one or more nitrogen atoms because they are arguably the more important ones.

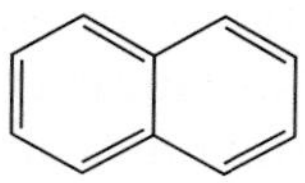

The aromatic compound with two fused benzene rings in Chapter 21 (Section 21.8) was naphthalene (**77**). If a thought experiment replaces carbon with a nitrogen, that nitrogen may be incorporated at two possible sites. If the nitrogen is at position "1," as in **78**, this is called *quinoline*. When the nitrogen is in position "2," the compound is **79** and is named *isoquinoline*. Isoquinoline is isolated from coal tar and used to prepare dyes, insecticides, and antimalarial compounds, among other things. Quinoline was mentioned briefly in Chapter 19 (Section 19.3.3) as a poison in the **Lindlar hydrogenation** of alkynes to give *cis*-alkenes. A large number of *alkaloids* (natural products containing nitrogen), with significant biological activity, contain either the quinoline unit, the isoquinoline ring system, or hydrogenated forms of these rings.

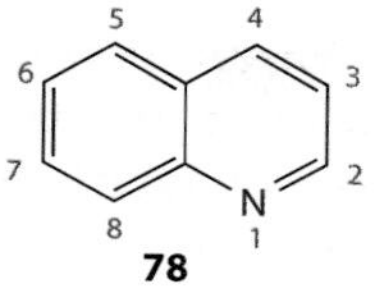

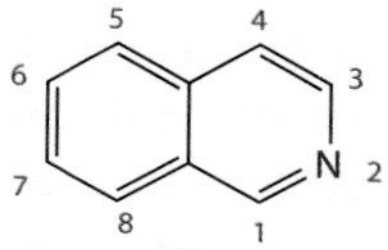

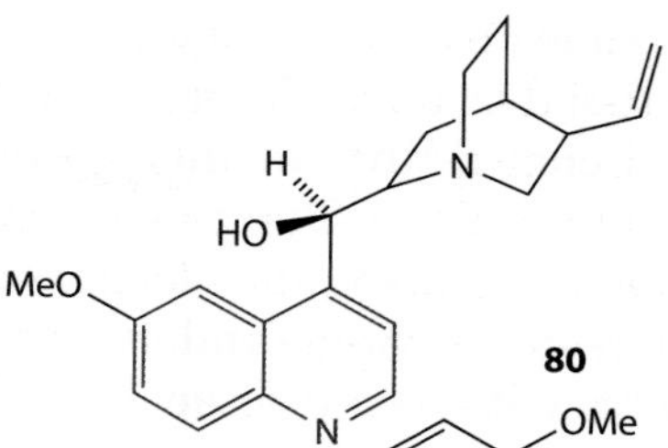

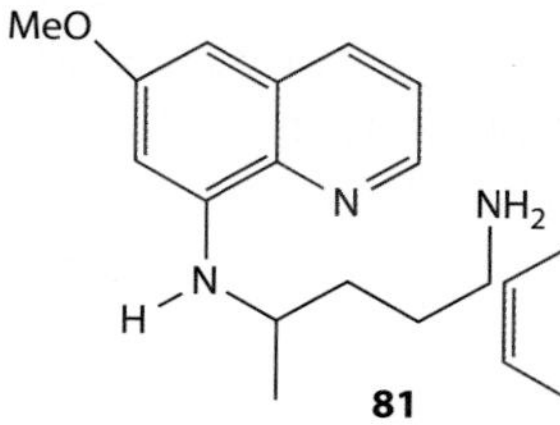

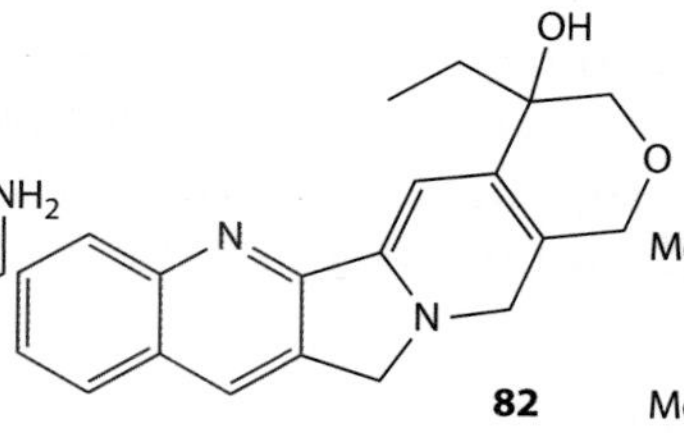

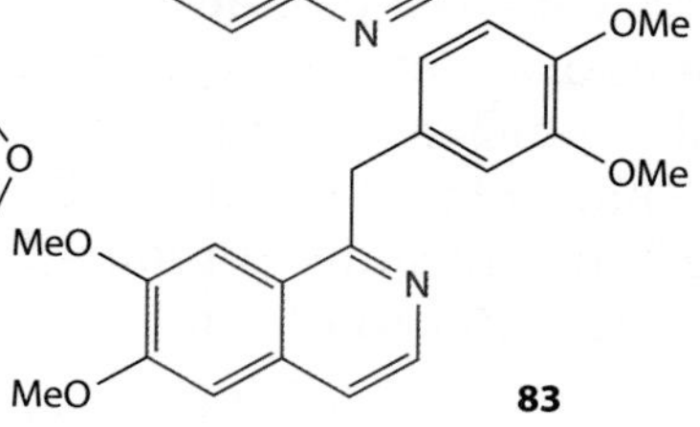

26.22 Draw 3-bromo-8-methylquinoline and 4,8-diethylisoquinoline.

Quinine (**80**) is a common quinoline derivative and an important antimalarial isolated from *Cinchona* bark (see Chapter 1 and compound number **1** in Section 1.1). Another antimalarial compound is Primaquine® (**81**). Camptothecin® (**82**) is an important anticancer drug that contains the quinoline unit. In all cases, the quinoline unit is highlighted in **blue** in the illustration. A typical isoquinoline derivative is papaverine (**83**), which is related to morphine and also isolated from the opium poppy. Papaverine® relaxes smooth muscle tissue in blood vessels. Once again, the isoquinoline unit is highlighted.

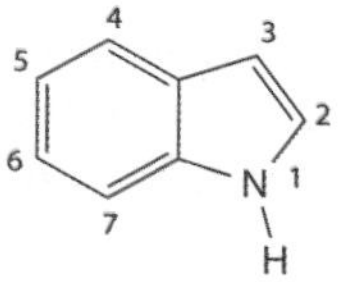

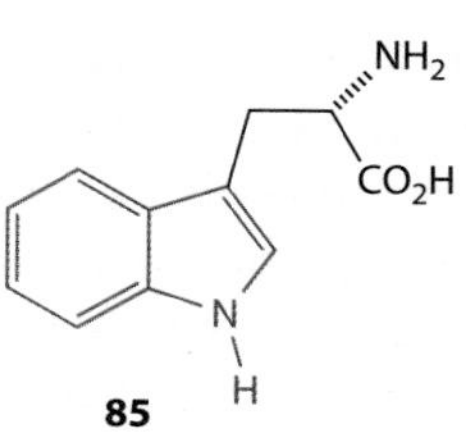

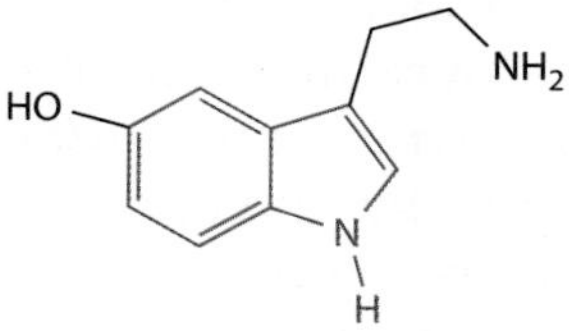

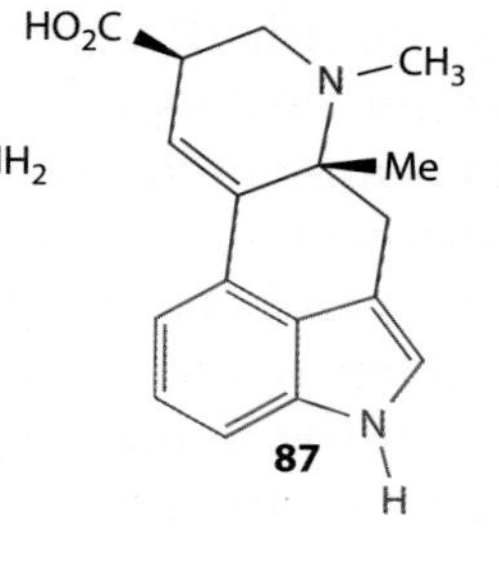

Indole (**84**) is a very important bicyclic heterocycle with a six-membered ring fused to a five-membered ring. This molecule is aromatic. A common compound containing an indole ring is the important amino acid tryptophan, **85** (see Chapter 27). The closely related compound serotonin (**86**) is a neurotransmitter that causes smooth muscle effects in the cardiovascular and gastrointestinal systems. Some indole-containing compounds have hallucinogenic effects, as seen with lysergic acid (**87**), which is found in a fungal parasite common to rye and wheat. Lysergic acid in combination with many other related compounds found in this fungus is responsible for "ergot poisoning," which killed more than 40,000 people in Europe in the tenth and twelfth centuries. Note that **87** also includes a reduced quinoline unit. The indole units in **85–87** are highlighted in **purple** in the illustration.

26.23 Draw 1-methyl-5-bromoindole.

Some derivatives have one or two nitrogen atoms at different positions in bicyclic six–six, six–five, or five–five fused aromatic rings. It is also possible to incorporate three, four, or even more nitrogen atoms into these rings. Most will not be seen in this book, but there is an important six–five heterocyclic ring system that contains four nitrogen atoms. This compound is called purine (**88**), and derivatives of this fundamental heterocycle include adenine (**89**) and guanine (**90**), which are components of DNA and RNA (see Chapter 28, Section 28.6). In addition, both uric acid (**91**; a component of urine) and caffeine (**92**; found in coffee and tea) have a purine skeleton.

26.6 Aromatic Substitution Reactions of Polycyclic Heterocycles

This discussion will be limited to a few of the polycyclic rings discussed previously: specifically, quinoline, isoquinoline, and indole. Many derivatives are prepared by electrophilic aromatic substitution reactions. Both the quinoline and isoquinoline contain a pyridine unit as well as a benzene ring unit. One ring (a benzene ring) has a nitrogen atom *attached* and it is expected to be activated, whereas in Section 26.5, the pyridine ring was shown to be deactivated relative to benzene. Therefore, electrophilic aromatic substitution will occur exclusively in the benzene ring unit. The indole system contains a pyrrole unit and a benzene ring unit. In Section 26.3.1, pyrrole was shown to be very activated to electrophilic aromatic substitution reactions. Therefore, indole reacts at the pyrrole unit rather than the benzene ring unit.

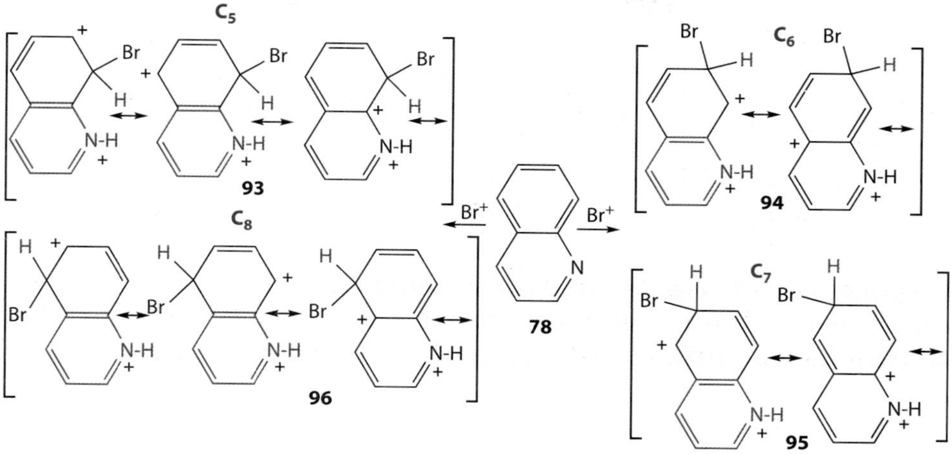

The reaction of quinoline (**78**) with bromine and sulfuric acid gives a brominated quinoline derivative via reaction with Br^+, but **where?** Note that quinoline is a base, and it will react with sulfuric acid to form an ammonium salt. Remember that pyridine is much less reactive than benzene in electrophilic aromatic substitution reactions. Therefore, assume that the ring containing nitrogen is much less reactive. This leaves C5–C8 as potential sites for electrophilic substitution. Indeed, **78** reacts with bromine and sulfuric acid to give a mixture of 5-bromoquinoline and 8-bromoquinoline, with 5-bromoquinoline being the major product.[1]

Because this reaction generates Br^+, reaction with quinoline at C5 gives intermediate **93**. Similar reaction at C6 gives **94**, reaction at C7 gives **95**, and reaction at C8 gives **96**. Comparing several key resonance structures leads to the conclusion that attack at C5 and C8 is preferred because there are more intact pyridine rings (different Kekulé structures) and therefore more resonance structures. Electrophilic aromatic substitution reactions of quinoline will give 5- and 8-substituted quinoline derivatives. In the bromination reaction mentioned for **78**, the major product is 5-bromoquinoline (**97**), but 8-bromoquinoline (**98**) is also formed. The sulfonation reaction is more selective. When quinoline is heated to 90°C with SO_3/H_2SO_4, the major product is 8-quinolinesulfonic acid, **99**. However, when the reaction is heated to 220°C, 5-quinolinesulfonic acid (**100**) is also formed.

26.24 Draw the structures of 5-bromoquinoline and 8-bromoquinoline.

26.25 Draw the intermediate for attack of Br^+ at C2 of quinoline.

26.26 Draw all the resonance structures for 94 and 97.

26.27 Draw the two major products formed when quinoline reacts with HNO_3/H_2SO_4.

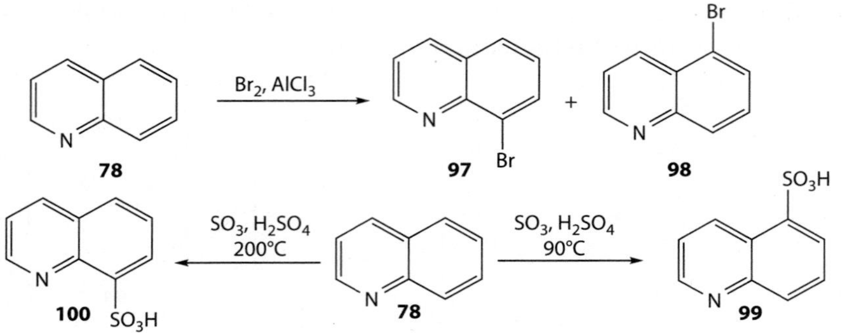

Isoquinoline (**79**) reacts similarly to quinoline. Reaction of isoquinoline with nitric acid/sulfuric acid gives a mixture of 5-nitroisoquinoline (**101**) and 8-nitroisoquinoline (**102**), but **101** is produced in about 90% yield.

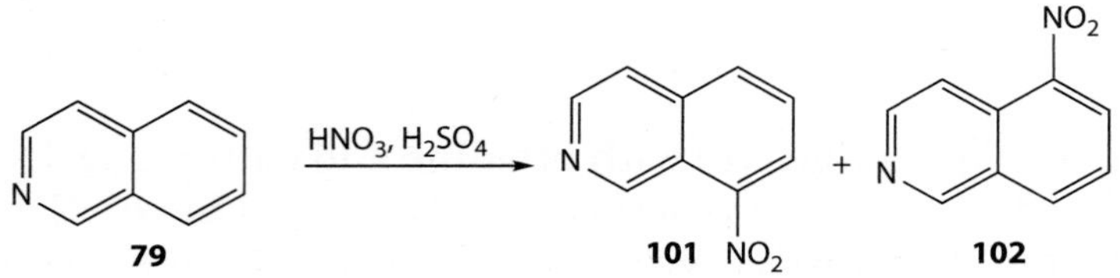

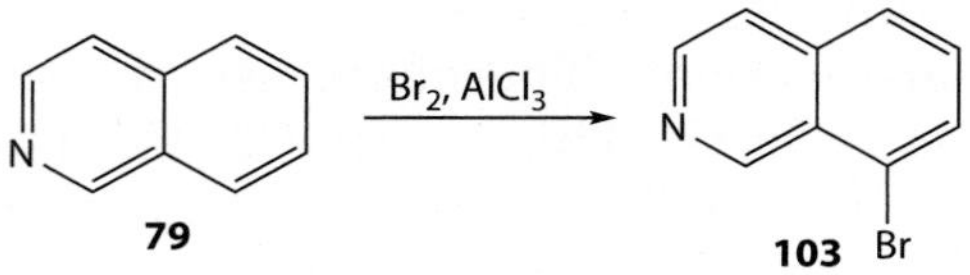

Similar reaction of isoquinoline with bromine and aluminum chloride gives a high yield of 5-bromoisoquinoline (**103**). The rationale for this selectivity is the same as that for quinoline. The nitrogen-containing ring is much less reactive, and substitution at C5 and C8 gives more stable intermediates.

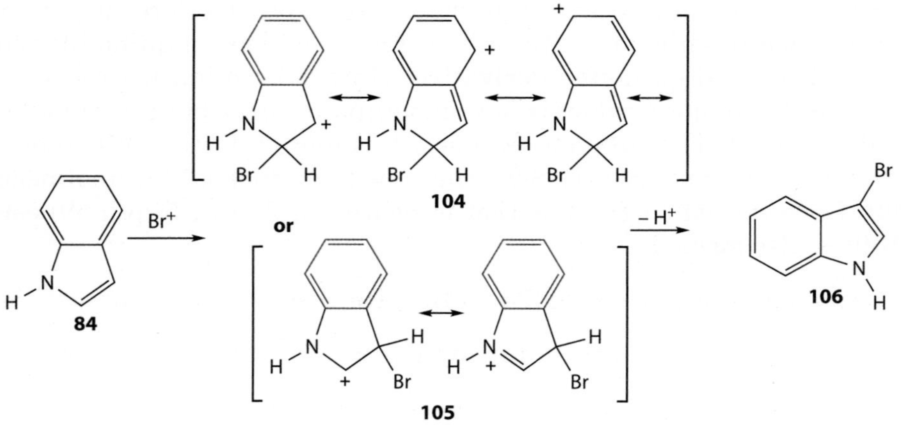

Indoles give good yields of electrophilic aromatic substitution. When indole (**84**) is treated with *N*-bromosuccinimide (NBS), the major reaction is 3-bromoindole, **106**. Because pyrrole is more reactive than benzene in electrophilic aromatic substitution (see Section 26.5), it is reasonable to assume that the pyrrole portion of indole is more reactive than the benzene portion. Attack of indole on Br⁺ at C2 generates intermediate **104** and attack at C3 generates **105**. Attack at C3 leads to a resonance structure that does not involve the benzene ring, whereas attack at C2 will delocalize the charge such that the aromatic character of the benzene ring is disrupted. This means that attack at C3 is preferred, and the major product of electrophilic aromatic substitution with indole is the 3-substituted indole **106**. This is a general observation, as when indole reacts with acetic anhydride to give 3-acetylindole.

26.28 Draw all resonance contributors for the intermediate formed when isoquinoline reacts with Cl⁺ at C8.

26.29 Draw the structure of 3-acetylindole.

26.30 Draw the product formed by the reaction of indole with (a) SO₃/pyridine, (b) *N*-chlorosuccinimide (NCS; see Chapter 11, Section 11.9.3), and (c) benzenediazonium chloride (PhN₂⁺ Cl⁻; see Chapter 21, Section 21.9).

26.7 Synthesis of Heterocycles

This section is in an unusual place to discuss the synthesis of heterocycles: next to last! Structures and some common reactions of several heterocycles were described in previous sections. The synthesis of these compounds is saved until the end of the chapter because much of the chemistry seen in previous chapters may be applied to heterocycles, with modifications that accommodate the heteroatoms. However, a heterocycle must be recognized before it can be synthesized. The idea is to link heterocycles—which may seem a little unusual when compared to other organic molecules in this book—to fundamental chemical reactions that have been used many times.

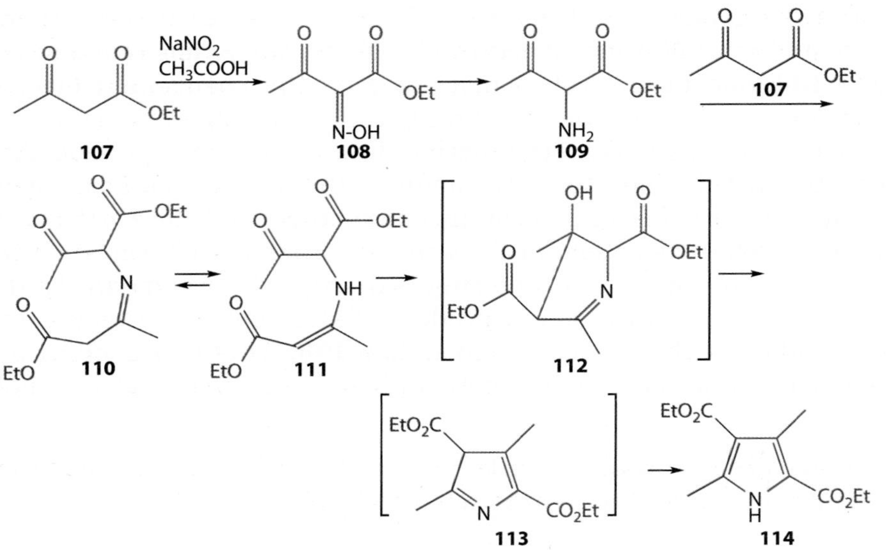

One of the most common methods for the preparation of pyrrole is called the **Knorr pyrrole synthesis**, after Ludwig Knorr (Germany; 1859–1921). It is a series of acyl addition–elimination reactions involving β-keto-esters and amines. When ethyl 3-oxobutanoate (**107**; better known as ethyl acetoacetate) reacts with NaNO$_2$ and acetic acid (to generate HONO), oxime **108** (see Chapter 18, Section 18.7.3) is formed. When the oxime is reduced with zinc metal and acetic acid, amino-ester **109** is formed and reacts with a second molecule of **107** to give imine **110**. Imine **110** is in equilibrium with its enamine form (**111**) and, as seen in Chapter 18 (Section 18.7.2), enamines react with ketones and aldehydes to form a new carbon–carbon bond. When the enamine unit (in **blue** in the illustration) of **111** attacks the ketone unit (in **red**; acyl addition), the product is a five-membered ring, **112**. Loss of water (elimination) leads to an imine unit in **113**, and this simply undergoes a sigmatropic rearrangement to **114**, which is the pyrrole. Remember the sigmatropic rearrangement reactions introduced in Chapter 24, Section 24.4. Formation of the aromatic pyrrole unit "drives" the reaction sequence.

26.31 Draw a synthesis of *N*-benzyl-2,5-diethylpyrrole from 1,2-diethylcyclobutene.

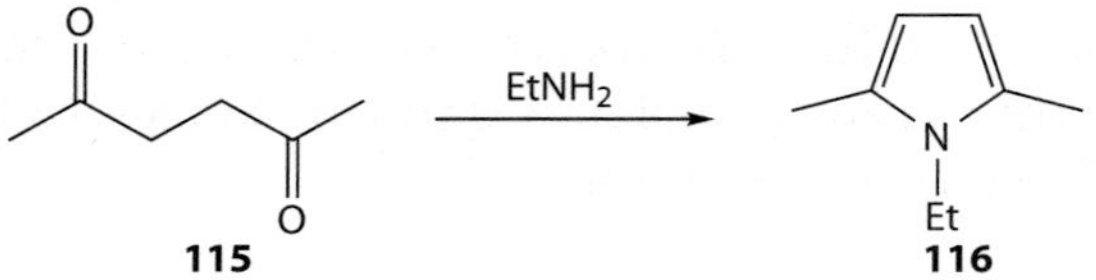

A variation of this reaction that involves a 1,4-diketone rather than a keto-ester is very useful. If a simple amine such as ethylamine (EtNH$_2$) reacts with 2,5-hexanedione (**115**), the product is pyrrole **116** in what is called the **Paal–Knorr synthesis**, after Carl Paal (Austria; 1860–1935) and Ludwig Knorr.

The reaction proceeds by acyl addition of the amine to one carbonyl, elimination of water, and attack of the nitrogen on the second carbonyl. Elimination of a second molecule of water gives **116**.

26.32 Write out a mechanism for this reaction.

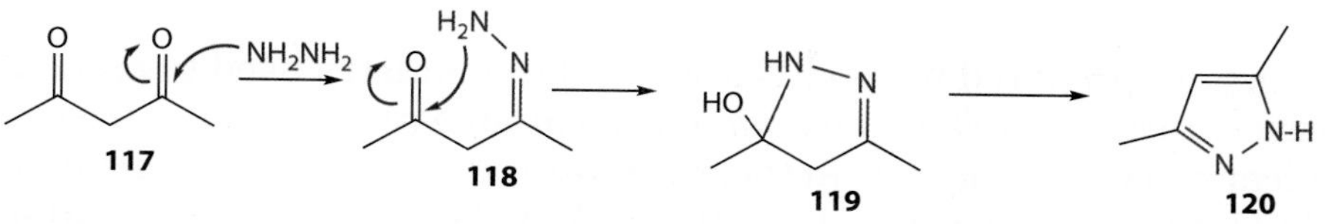

117 → 118 → 119 → 120

The synthesis of pyrazoles is somewhat related to the synthesis of pyrroles. When 2,4-pentanedione (**117**) reacts with hydrazine (NH_2NH_2) in the presence of a base such as hydroxide, the product is 3,5-dimethylpyrazole, **120**. In Chapter 18 (Section 18.7.3), hydrazine reacted with ketones and aldehydes in a manner similar to amines to form imine products called hydrazones. When hydrazine attacks one carbonyl group in **117**, the product is **118**, and the second $-NH_2$ unit attacks the second carbonyl to give **119**. Loss of water and a sigmatropic rearrangement gives the aromatic product, **120**. This modification of the pyrrole synthesis uses hydrazine (with two $-NH_2$ units) in the presence of a dicarbonyl compound to generate the pyrazole.

26.33 Draw a synthesis of 4-ethyl-1,3,5-trimethylpyrazole from 2,4-pentanedione.

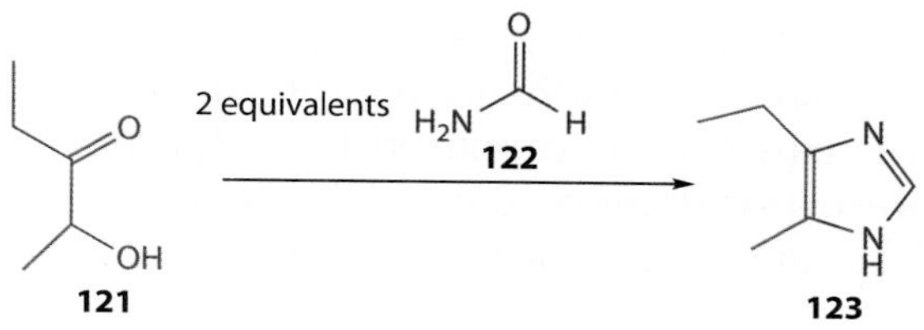

121 → 123

There is no simple, commonly accepted method for the preparation of imidazoles, but rather many different approaches. One approach, somewhat related to chemistry seen in previous chapters, involves the reaction of an α-hydroxy-ketone such as **121** with formamide, **122**. The $-NH_2$ unit of formamide attacks the carbonyl (acyl addition), and loss of water (elimination) gives an enol that tautomerizes to the ketone. (Keto-enol tautomerism was first discussed in Chapter 10, Section 10.4.5.) A *second molecule* of formamide reacts with this ketone via acyl addition to give a product, which loses water. An intramolecular attack of the nitrogen atom from this product to one $-CHO$ unit on the carbonyl of the other CHO unit, followed by loss of water under the reaction conditions, gives imidazole, **123**.

26.34 Provide a mechanism for the reaction of 121 with 122, based on the given mechanistic narrative.

26.35 Provide a name for 123.

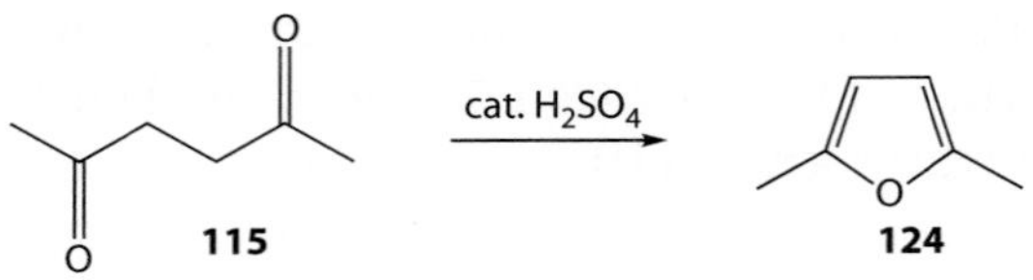

115 → cat. H_2SO_4 → **124**

   Furans are prepared by modification of the chemistry used to make pyrrole derivatives. Because furans have an oxygen atom rather than a nitrogen atom, a modification of the reactive partners is required. When 2,5-hexanedione (**115**) is treated with acid, the product is a furan, **124**. This is called the **Paal–Knorr furan synthesis**, and it begins with protonation of one carbonyl and attack of the oxygen atom of the second carbonyl to close the ring. Elimination of water leads to the furan because it generates an aromatic system. The reaction of protonated ketones and aldehydes with oxygen nucleophiles was discussed in Chapter 18 (Section 18.6). The mechanism for formation of **124** is therefore related to the chemistry presented in Chapter 18.

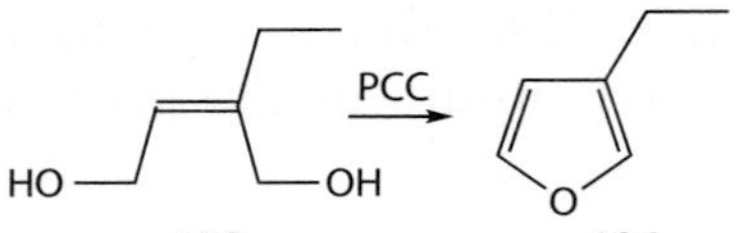

125 → PCC → **126**   **115** → P_4S_{10} → **127**

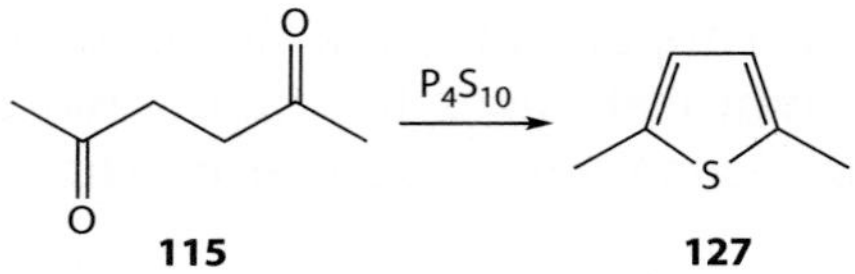

   Other interesting methods can generate furans. One method treats an allylic diol such as **125** with pyridinium chlorochromate (PCC; see Chapter 17, Section 17.2.3), in an oxidative-cyclization process, to give 3-ethylfuran, **126**. Thiophene derivatives can also be prepared from 1,4-dicarbonyl compounds. A **Paal–Knorr thiophene synthesis** reacts 2,4-hexanedione (**115**) with phosphorus pentasulfide (P_4S_{10}) to give 2,5-dimethylthiophene, **127**.

26.36 Write out the complete mechanism for formation of 124.

26.37 Write out a synthesis of 2,5-diphenylfuran from 1,2-diphenyl-1-cyclobutanol.

26.38 Write out a synthesis of 3-bromo-2,5-dimethyl-thiophene from 115.

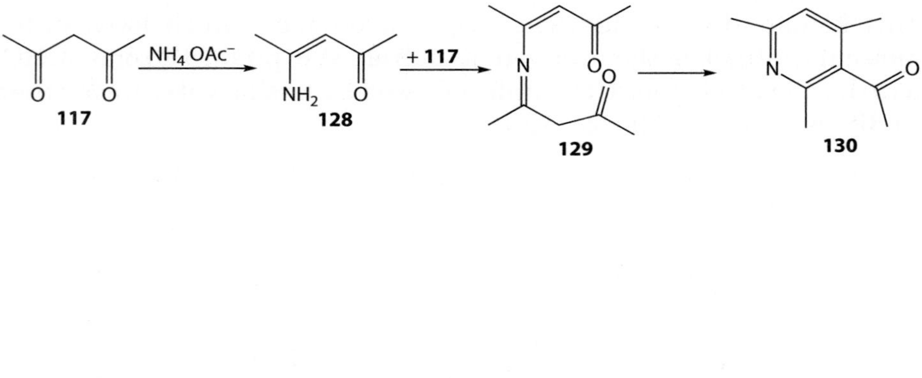

117 → $NH_4\,OAc^-$ → **128** → + **117** → **129** → **130**

There are methods for the preparation of pyridine derivatives, but they usually produce highly functionalized pyridines. One method is related to methods shown for the preparation of other heterocycles. When 2,4-pentanedione (**117**) is treated with ammonium acetate (NH_4OAc), the product is an enamine-ketone, **128**. This enamine subsequently reacts with a second equivalent of **117** to give the acyl addition product (the enamine attacks the carbonyl), which is imine **129**. The amine (NH_2) group attacks the second carbonyl and elimination gives the aromatic pyridine product (**130**). There are several variations of this fundamental approach, as well as other completely different approaches. The synthesis of pyrimidines will be delayed until the discussion of DNA and RNA in Chapter 28.

26.8 Biological Relevance

Throughout this chapter, the presence of heterocyclic rings in medicines and other biologically relevant molecules has been highlighted. Many of these biologically relevant heterocyclic compounds are discussed in Chapters 27 and 28. One interesting feature is the discussion of the biosynthesis of heterocyclic rings, which includes the biosynthesis of purine or pyrimidine units in DNA and RNA.

Other heterocyclic rings appear in biological systems. Thiamin (**131**) contains a thiazole ring as well as a pyrimidine ring, and it is an essential component of the human diet. Thiamin is a water-soluble vitamin of the B complex (vitamin B_1). Although humans cannot synthesize thiamin, many bacteria synthesize it and use it in the form of thiamin pyrophosphate. *Bacillus subtilis* produces thiamin by a biosynthetic route, but it also synthesizes the thiazole unit (**142**) found in **131** from the amino acid glycine (**133**; see Chapter 27, Section 27.3.2) and 1-deoxy-d-xylulose-5-phosophate, **132**.[2] (See Chapter 28, Section 28.1, for an introduction to sugars such as xylulose.)

The enzyme *DXP synthase* couples **131** and **132** to give iminium salt **134**. Note that this reaction is effectively an amine-ketone reaction to give an iminium salt, as described in Chapter 18, Section 18.7. Enzymes in the bacterium shift the double bond to generate the enamine form (**135**). (See Chapter 18, Section 18.7.2, for a discussion of enamines.) The enzyme also removes a hydrogen atom α to the carboxyl carbonyl (see Chapter 22, Sections 21.1 and 21.2, for a discussion of enols and enolate anions) to generate **136**.

26.5

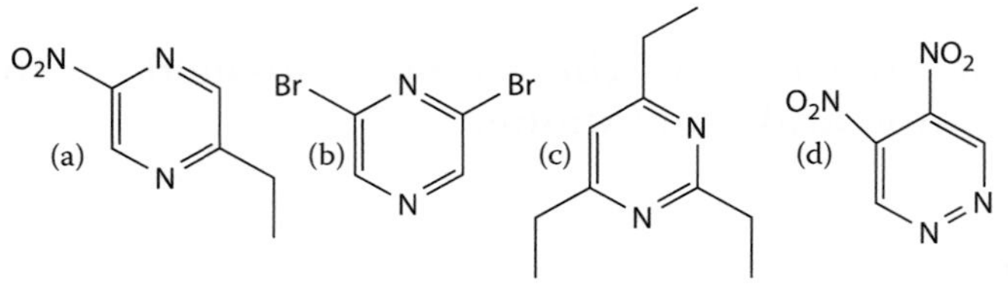

(a)   (b)   (c)   (d)

26.6

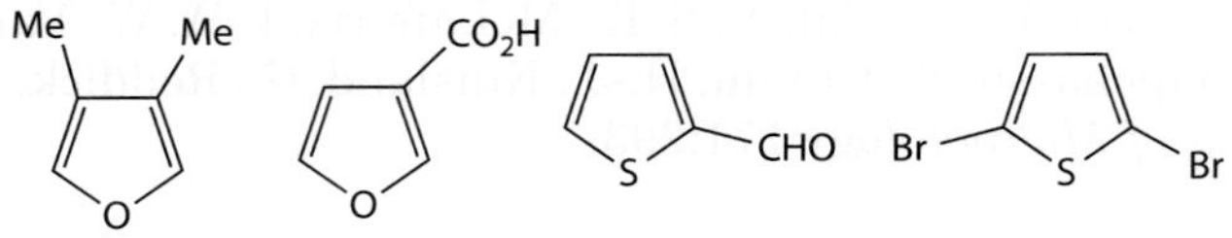

26.7

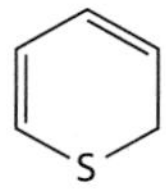

26.8

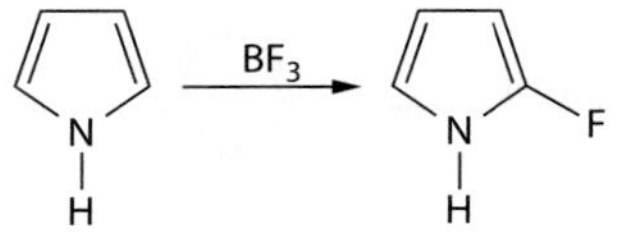

BF₃

26.9

SO₃, pyridine
100°C

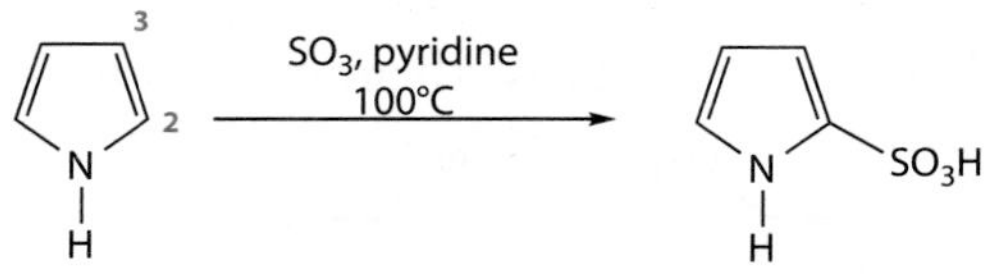

26.10   The electron pairs on the oxygen atom of furan make the molecule more electron rich. It is also possible to put the positive charge for the intermediate on
the oxygen atom. This means that furan reacts faster with the electrophile
and it generates a more stable intermediate.

26.11

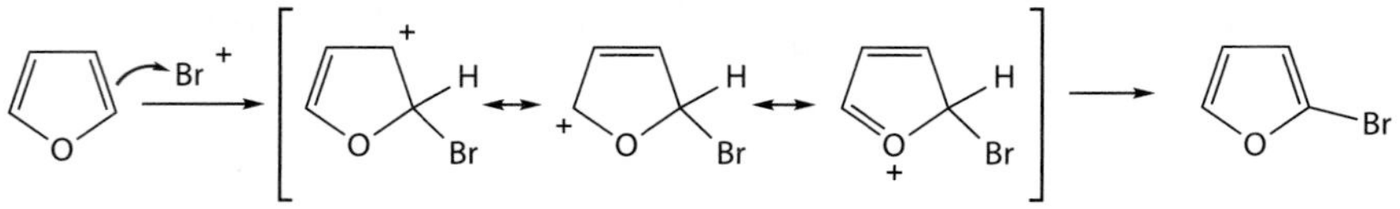

26.12

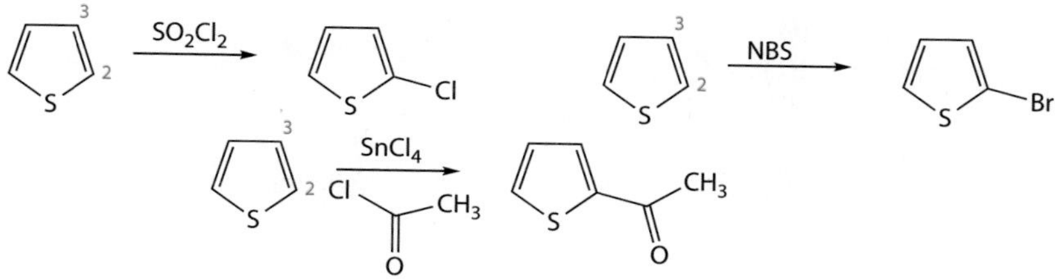

26.13

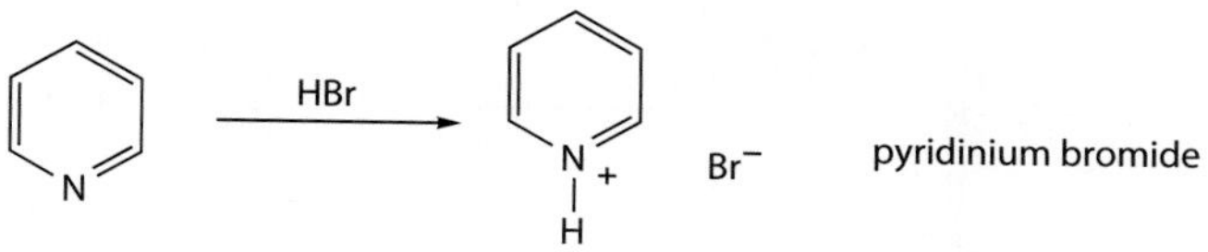

pyridinium bromide

26.14

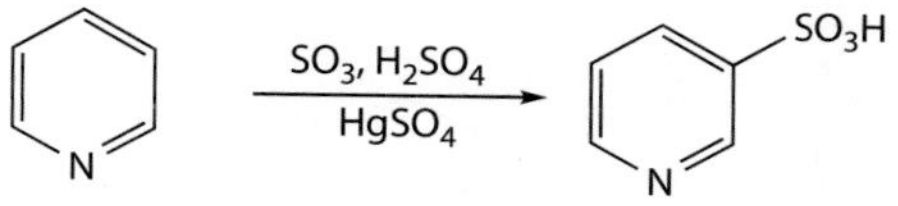

26.15   Attack at C_2 and at C_4 leads to a resonance contributor where the positive charge is deposited on nitrogen, which is very unstable. Because the charge is not deposited on nitrogen via attack at C_3, that is the more stable intermediate and it leads to the major product.

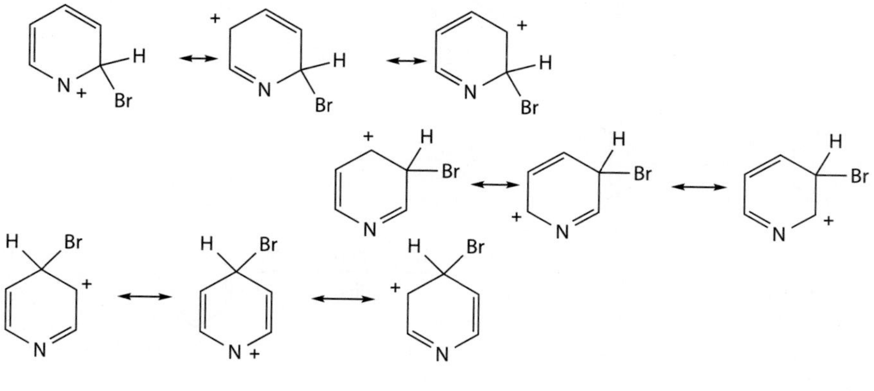

26.16

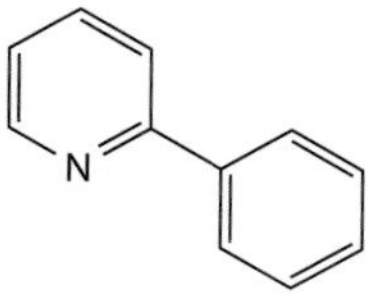

26.17

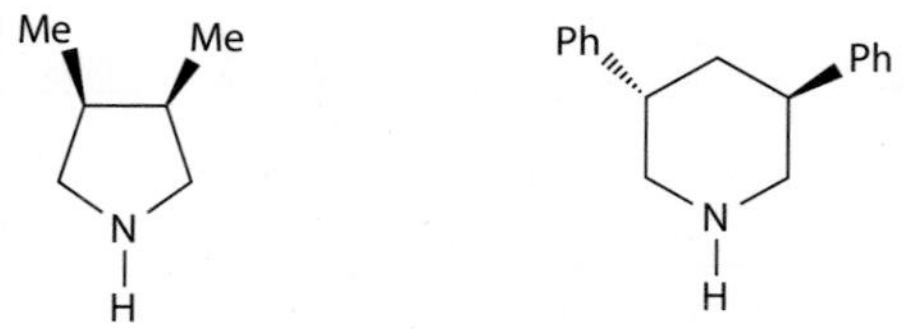

26.18

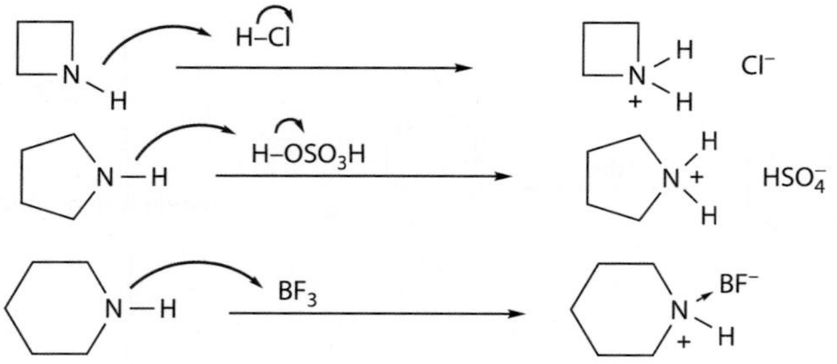

26.19

26.20

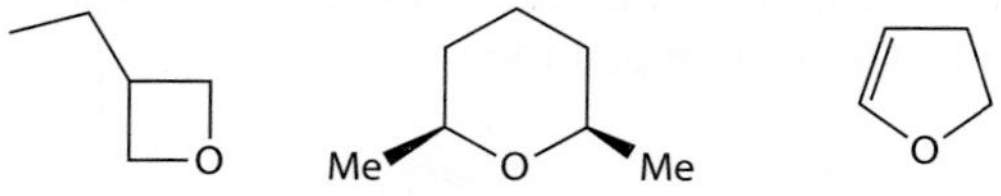

26.21

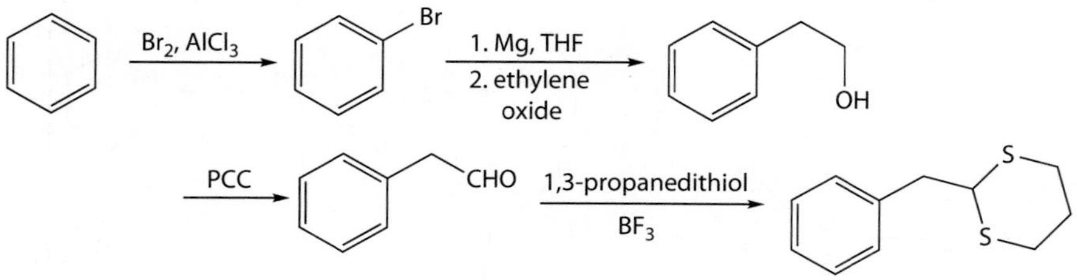

26.22

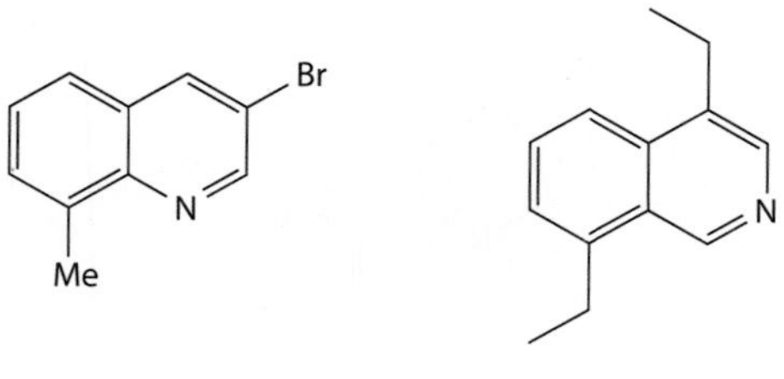

26.23

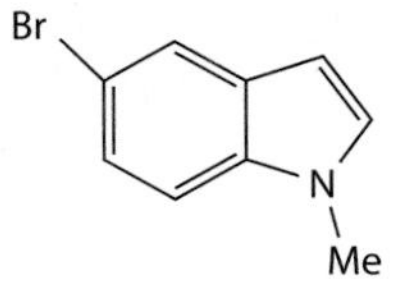

26.24

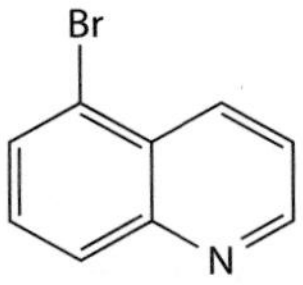

5-bromo

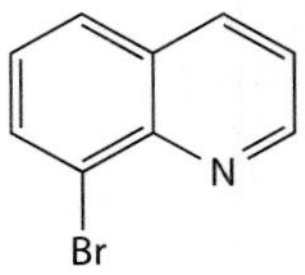

8-bromo

26.25

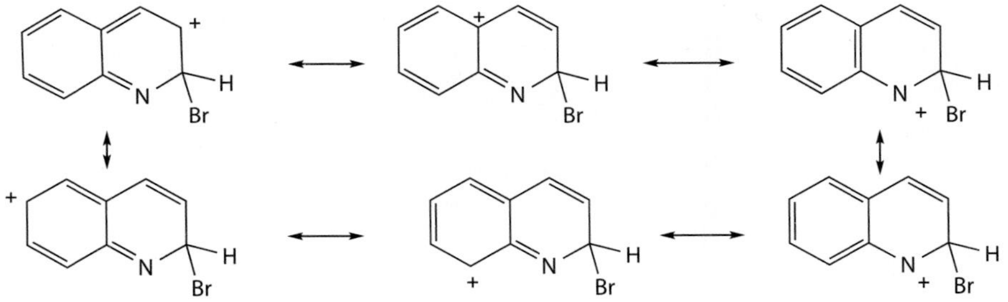

26.26

C_5

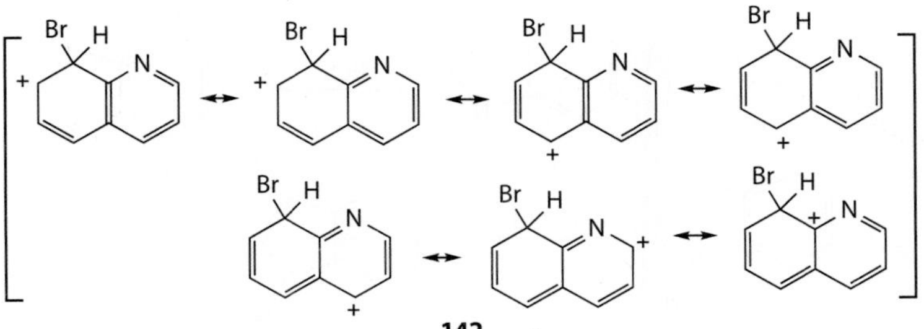

142

C_8

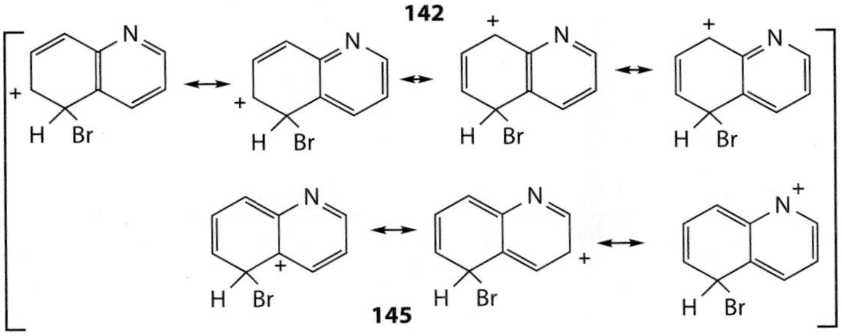

145

26.27

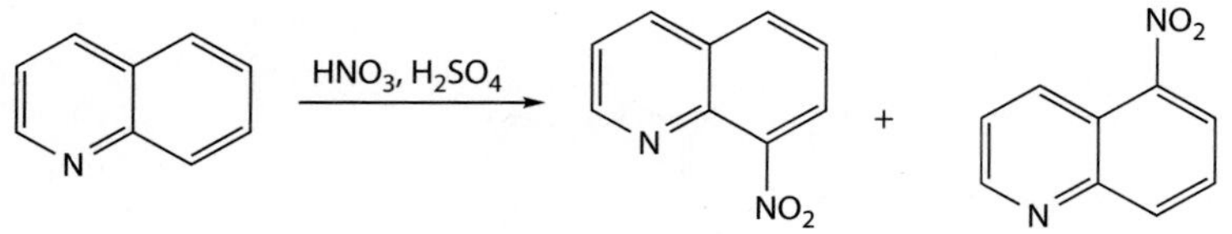

26.28

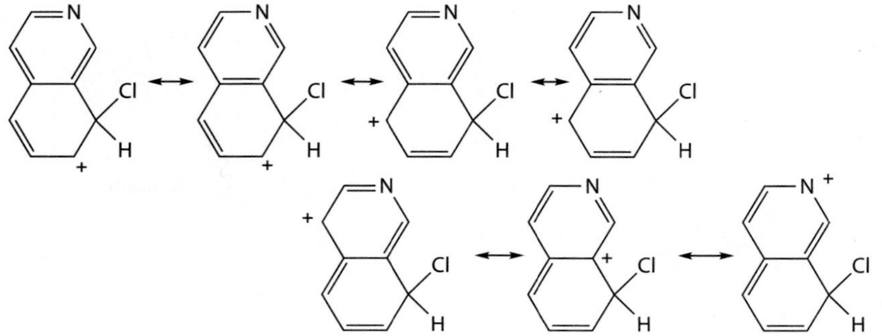

26.29

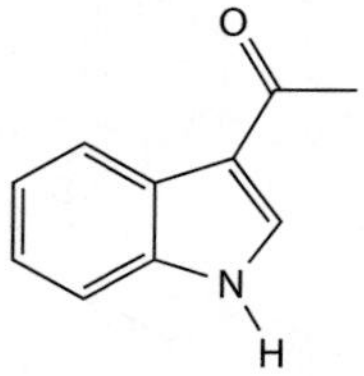

26.30

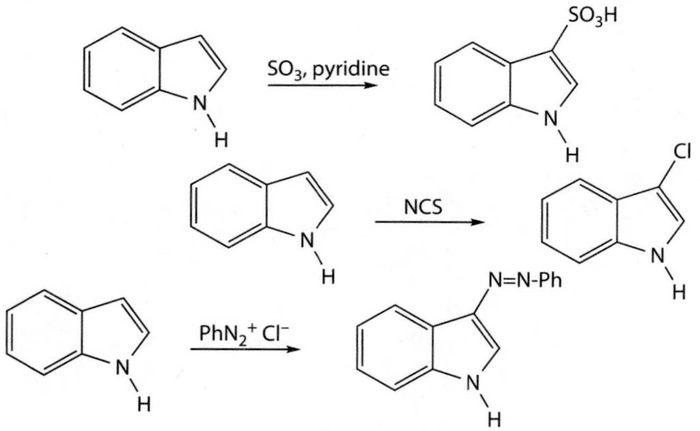

26.31

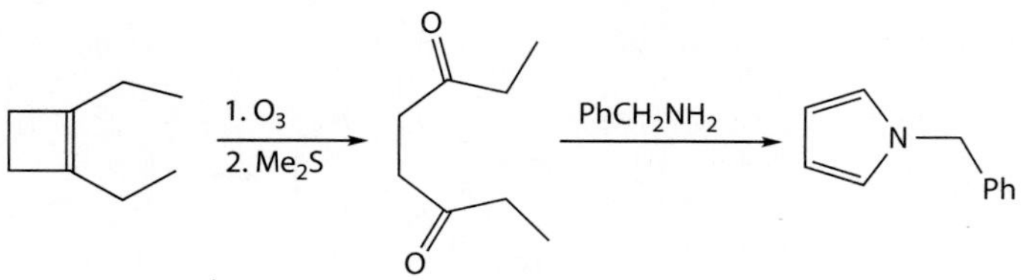

26.32

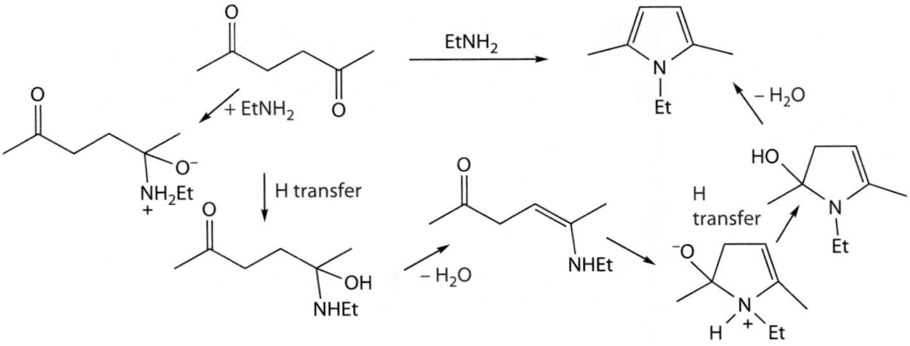

26.33

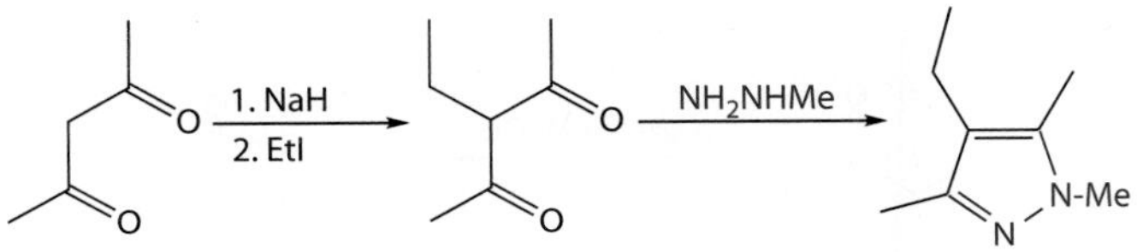

26.34

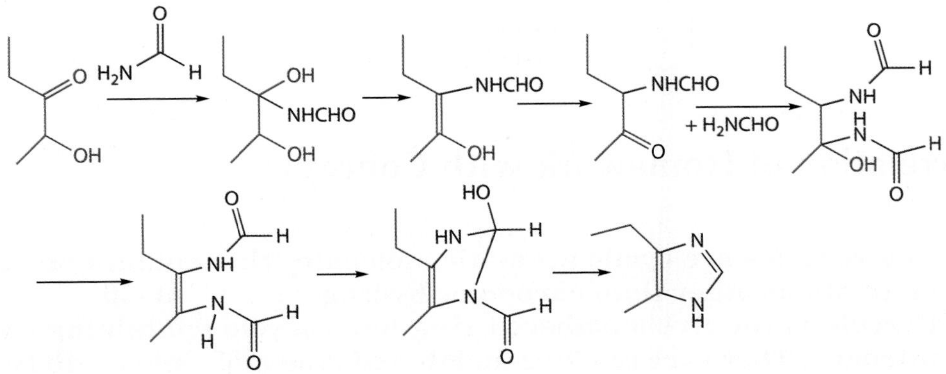

26.35 This compound is named 4-ethyl-5-methylpyrazole.

26.36

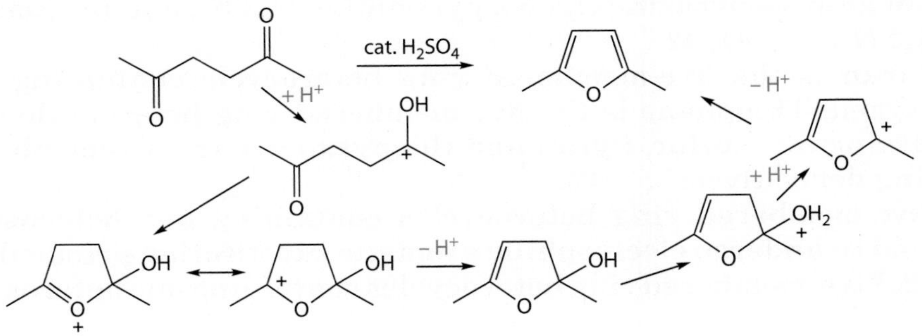

26.37

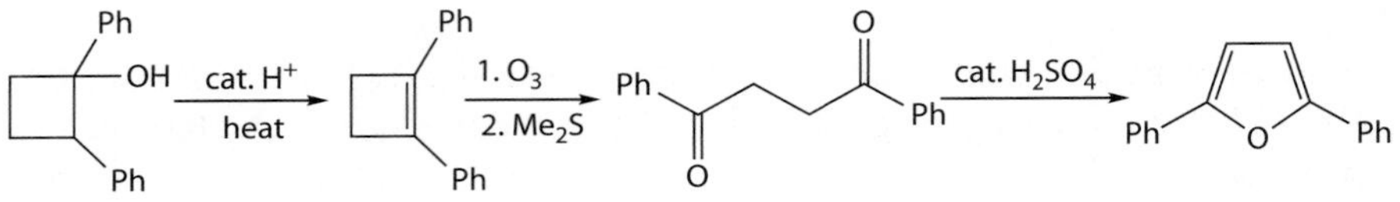

26.38

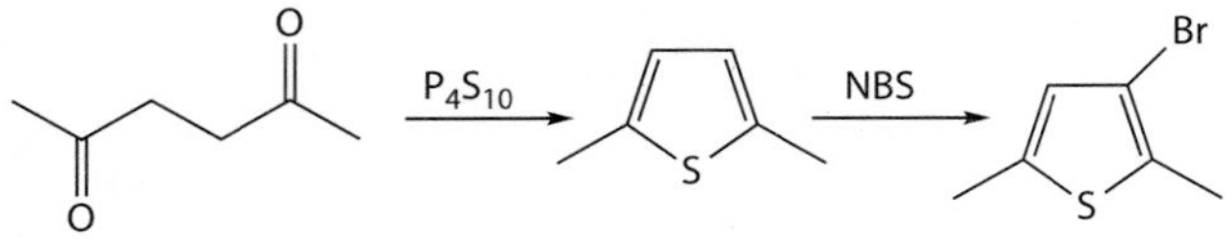

26.39

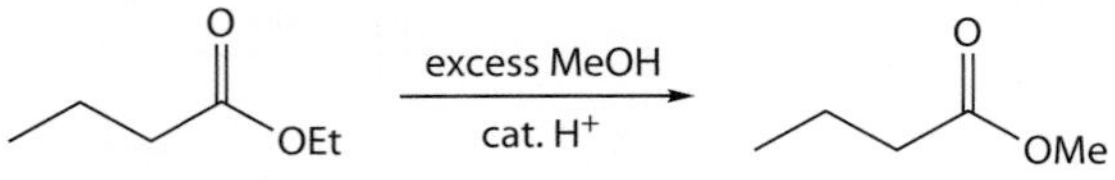

26.40

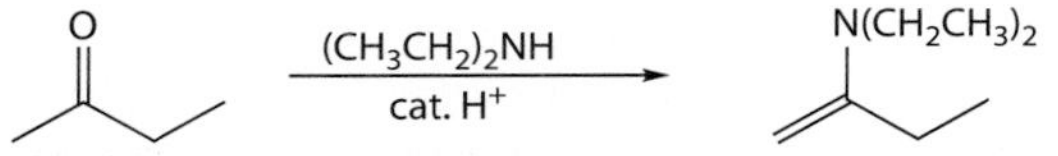

Correlation of Homework with Concepts

- **Heterocycles are cyclic aromatic molecules that contain one or more atoms other than carbon or hydrogen: 1–40, 41–59.**
- **Pyrrole is the five-membered ring heterocycle containing one nitrogen. There are two five-membered ring molecules with two nitrogen atoms—imidazole (1,3 N) and pyrazole (1,2 N): 1, 2, 3, 41, 42.**
- **Pyridine is the six-membered ring heterocycle containing one nitrogen. Three six-membered ring molecules contain two nitrogen atoms—pyrazine (1,4 N), pyrimidine (1,3 N), and pyridazine (1,2 N): 4, 5, 41, 42.**
- **Furan is the five-membered ring heterocycle containing one oxygen. Thiophene is the five-membered ring heterocycle containing one sulfur. Pyran and thiopyran are the six-membered ring derivatives: 6, 7, 42.**
- **Five-membered ring heterocycles containing one heteroatom tend to undergo electrophilic aromatic substitution primarily at C2. Five-membered ring heterocycles containing one heteroatom**

tend to be very reactive in electrophilic aromatic substitution reactions: 8, 9, 10, 11, 12, 43, 54.

- Six-membered ring heterocycles containing one heteroatom tend to be rather unreactive in electrophilic aromatic substitution reactions. Six-membered ring heterocycles containing one heteroatom tend to undergo electrophilic aromatic substitution at C3: 4, 13, 14, 15, 43, 44, 45.
- Halogenated heterocycles undergo nucleophilic aromatic substitution reactions as expected: 16, 46, 58, 59.
- Many molecules that are not aromatic contain one or more heteroatoms. These are very common and include ring sizes of three to six containing N, O, or S. There are also derivatives containing two or more heteroatoms: 17, 18, 19, 20, 21, 47, 52, 55, 56, 59.
- Two bicyclic aromatic compounds contain one nitrogen: quinoline (N at position 1) and isoquinoline (N at position 2). Indole has a pyrrole unit fused to a benzene ring to give a bicyclic aromatic system with N at position 1: 22, 23, 48, 49, 50.
- Quinoline and isoquinoline show that electrophilic aromatic substitution reactions are the more reactive benzene ring because the "pyridine" ring is less reactive. Indole undergoes electrophilic aromatic substitution primarily in the pyrrole ring because it is much more reactive than the benzene ring: 24, 25, 26, 27, 28, 29, 30, 51, 53.
- Pyrroles, pyrazoles, imidazoles, furans, and thiophenes can be made from 1,3-dicarbonyl compounds by reaction with an appropriate amine, oxidizing agent, or P_4S_{10}. This process is called synthesis: The new molecule is synthesized from the old one (see Chapter 25): 31, 32, 33, 34, 36, 37, 38, 52, 57.
- Spectroscopy can be used to determine the structure of a particular molecule and can distinguish the structure and functionality of one molecule when compared with another (see Chapter 14): 60, 61, 62, 63, 64, 65, 66, 67.

Homework

26.41 Draw the structures of each of the following molecules:
  (a) N,3-dimethylpyrrole
  (b) 3,4-diacetylpyrrole
  (c) 2,4-dichloropyrazole
  (d) 1-methyl-4-chloroimidazole
  (e) 2,4,6-trimethylpyridine
  (f) 5-aminopyrimidine

    (g) 3-nitropyrazine
    (h) 3,5-dibromopyridazine
    (i)   2-amino-5-methylpyrimidine

26.42   Give the IUPAC name for each of the following molecules:

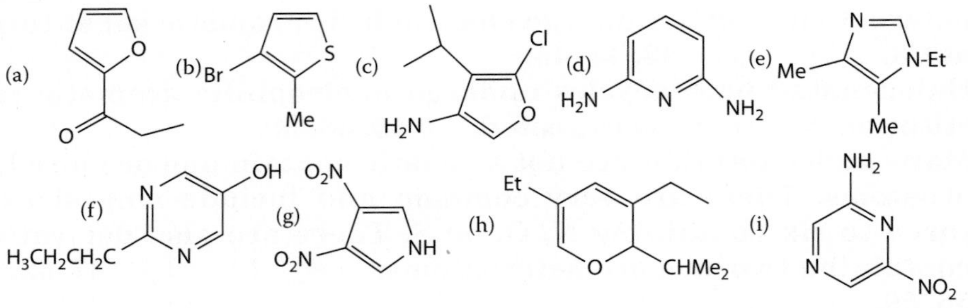

26.43   Give the major product expected from each of the following reactions:
    (a) *N*-methylpyrrole + HNO$_3$/acetic anhydride
    (b) 2-methylimidazole + SOCl$_2$
    (c) *N*,5-dimethylpyrazole + HNO$_3$/Ac$_2$O
    (d) furan + 1. butanoyl chloride/SnCl$_4$; 2. NaBH$_4$; 3. hydrolysis
    (e) 3-nitrofuran + SO$_3$/pyridine
    (f) furan + 1. NO$_2$$^+BF_4$$^-$; 2. H$_2$/Pd-C
    (g) furan + 1. dioxane-Br$_2$; 2. Mg, THF; 3. cyclohexanone; 4. hydrolysis
    (h) 3-ethylthiophene + SO$_3$/pyridine
    (i)   imidazole + 1. BuLi; 2. benzyl bromide, THF; 3. HNO$_3$/acetic anhydride

26.44   Give the major product for each of the following:
    (a) pyrimidine + 1. Br$_2$/300°C; 2. PhLi, heat
    (b) pyridine + 1. KNO$_3$/330°C; 2. H$_2$, Pd/C; 3. propanoyl chloride, NEt$_3$
    (c) pyridine + 1. Br$_2$/300°C; 2. MeNH$_2$, heat

26.45   Which of the following is expected to undergo reaction with bromine at elevated temperatures faster: **A** or **B?** Explain.

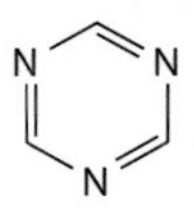

                                A 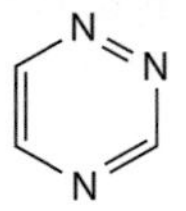   **B**

26.46   Which of the following is expected to react faster with an amide anion: 3-bromopyrrole or 3-bromofuran? Explain.

26.47   Give the correct IUPAC name for each of the following molecules:

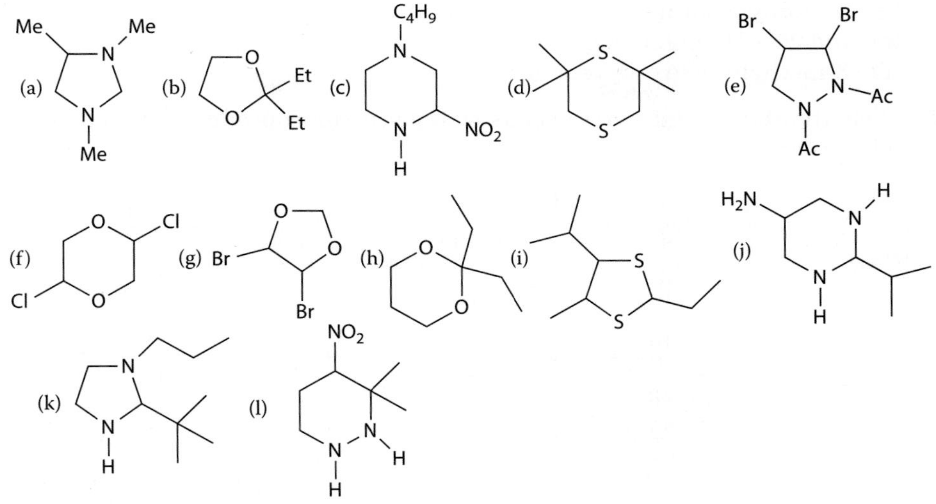

(a)   (b)   (c)   (d)   (e)

(f)   (g)   (h)   (i)   (j)

(k)   (l)

26.48 Draw the structure for each of the following:
(a) *N*-acetyl-3-ethylimidazoline
(b) 1,2,4,5-tetramethylpiperazine
(c) 3,5-dibromotetrahydrofuran
(d) *trans*-2,3-dimethyl-1,4-dioxane
(e) 1-ethyl-4-nitropyrazolidine
(f) *cis*-3,5-dinitropiperidine
(g) *N*-propylaziridine
(h) 2,3-diethyloxirane
(i) *N*,2-diethylazetidine
(j) 3-chlorothiane
(k) 2-phenylthiirane
(l) *cis*-2,3-diphenyloxetane

26.49 There are partly reduced forms of heterocycles that have not been discussed. Try to predict the structure of 1,2,3,4-tetrahydropyridine; 2,3-dihydrofuran; 1(H), 2(H)-dihydropyridine; and 1,2(H)-dihydropyrimidine.

26.50 Draw the structure for each of the following:
(a) 4,6-dibromoquinoline
(b) 6-methyl-7-nitroisoquinoline
(c) 6-ethylindole
(d) 3-butyl-5-nitroisoquinoline
(e) 8-chloroquinoline
(f) 1-methyl-7-cyanoindole

26.51 Give the major product for each of the following:
(a) quinoline + Br_2/$AlCl_3$
(b) 8-methoxyquinoline + Br_2/$AlCl_3$
(c) isoquinoline + acetyl chloride/$AlCl_3$

    (d) 5-methylisoquinoline + SO_3/H_2SO_4

    (e) indole + HNO_3/H_2SO_4

    (f) 3-methylindole + $Br_2/AlCl_3$

26.52 Give all intermediate products as well as the final product for each of the following:

(a) reaction of Ph diketone with NH₄OAc, heat

(b) reaction of Ph diketone with NH₂NH₂, heat

(c) Ph—CN with 1. H₂, Pd-C; 2. phenyl acetyl chloride; 3. POCl₃; 4. Pd°, heat

(d) cyclobutene with Et groups, 1. O₃; Me₂S; 2. 1-aminobutane, heat

(e) 2 eq. ketoester with CO₂Et, NaNO₂, AcOH, heat

(f) Ph—Br with 1. NaCN, THF; 2. H₂, Ni(R); 3. 4,4-dimethylheptanal; 4. polyphosphoric acid

(g) Ph diketone Ph with NH₂-aryl, H₂SO₄

(h) C₃H₇ diketone C₃H₇ with EtNH₂, heat

(i) Ph—C(OH)—Ph compound with H₂NCHO, heat

(j) aryl-NHNH₂ with Ph ketone Ph, heat

(k) cyclobutene with isopropyl groups, 1. O₃; Me₂S; 2. cat. H₂SO₄

(l) cyclobutene with CH₂Ph groups, 1. O₃; Me₂S; 2. P₄S₁₀, heat

(m) ketone with 1. LDA, THF, −78°C; 2. 4-phenylbutanal; 3. PDC, CH₂Cl₂; 4. NH₂NH₂, heat

26.53 Predict the major product or products formed when the molecule shown reacts with Br^+:

26.54 Suggest a reason why it is difficult to form 3-substituted furan derivatives.

26.55 Draw a structure for tetrahydrothiazole.

26.56 Suggest a reason why the molecule shown is unknown:

26.57   Explain why a Friedel–Crafts reaction based on the molecule shown is probably not a good way to make an isoquinoline derivative.

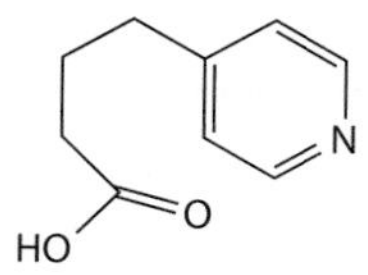

26.58   When 3-bromopyridine reacts with sodium amide, the products are 3-amino pyridine, along with 2-amino pyridine and 4-amino pyridine. Explain.

26.59   When 2-bromopyridine is heated with aqueous hydroxide, the product appears to be a mixture of a molecule that is aromatic with an OH group at the 2 position, but a lactam appears to be present. Explain this observation.

Spectroscopy Problems

Do not attempt these problems until Chapter 14 has been read and understood. All of them involve heterocycles.

26.60   Using only proton NMR, describe how you can distinguish between 2-methylpyridine and toluene.

26.61   Describe how you would distinguish 2-methylquinoline from 4-methylquinoline.

26.62   Describe how you could use proton NMR to distinguish between furan and pyrrole.

26.63   Describe differences in the proton NMR spectrum between pyrazine and pyrimidine.

26.64   Give the structure of this molecule, given the following spectral data:
MS: M (107) 100%, M+1 (108) 8.14%, M+2 (109) 0.3%
IR: 3096–287, 1603, 1571, 1492, 1453, 1380, 1032, 817, 727, 647, 641, 485 cm^{-1}
^{1}H NMR: 8.31 (d, 1H), 7.35 (d, 1H), 7.02 (d, 1H), 2.5 (s, 3H), 2.26 (s, 3H) ppm

26.65   Give the structure of this molecule given the following spectral data:
MS: exact mass of the molecular ion = 96.0575
IR: 3116, 2975–2850, 1697, 1509, 1454, 1089, 1006, 729 cm^{-1}
^{1}H NMR: 7.28 (d, 1H), 6.26 (m, 1H), 5.96 (d, 1H), 2.63 (q, 2H), 1.22 (t, 3H) ppm

26.66   A molecule B has the formula $C_4H_4N_2O_2$ and the following spectral data. This molecule is produced by the reaction of molecule A with a mixture of nitric acid and acetic anhydride. Identify both A and B:
IR: broad peak at 3260, 3139, 1543; strong peaks at 1469 and 1361, 756, 611, 589 cm^{-1}
^{1}H NMR: 12.0 (broad s, 1H), 7.9 (s, 1H), 6.9 (d, 1H), 6.7 (d, 1H) ppm

26.67   Give the structure of the molecule with the formula $C_9H_{13}N$ and the following
theoretical spectra:

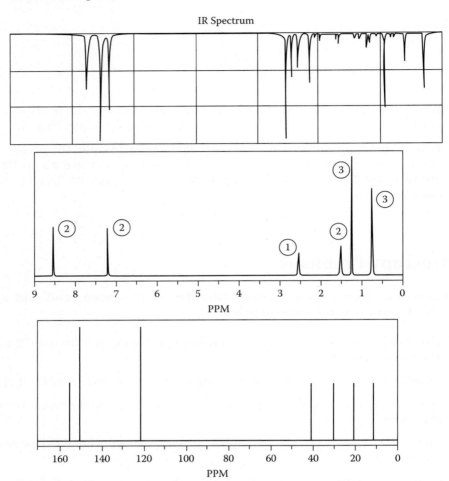

Multifunctional Compounds

Amino Acids and Peptides

27

Amines are used as bases and also as nucleophiles throughout portions of this book. The structure and nomenclature of simple amines were introduced in Chapter 5 (Section 5.6.2), but only a few reactions were discussed that form an amine. Exceptions are the formation and reactions of the aromatic amine, aniline (see Chapter 21, Sections 21.2 and 21.3), and the hydride reduction of nitriles and amides in Chapter 19 (Section 19.2.2). This chapter will first review some fundamentals of amine chemistry and then proceed into new areas.

Chapters 23, 24, and 26 may be viewed as discussions of multifunctional molecules, and this chapter will eventually focus attention on molecules that contain both an amine group and a carboxyl group. This is the important class of compounds called amino acids. It is also possible to prepare poly(amino acids) that are linked by amide bonds, known as peptide bonds. The poly(amides) are known as peptides, and some fundamental chemistry of peptides and the biologically important polypeptides called proteins will be presented.

To begin this chapter, you should know the following:

- **nomenclature of amines (Chapter 5, Section 5.6.2)**
- **nomenclature for aldehydes and ketones (Chapter 5, Section 5.9.2 and Chapter 16, Section 16.2)**

- **nomenclature of carboxylic acids (Chapter 5, Section 5.9.3; Chapter 25, Sections 15.4 and 15.5)**
- **nomenclature of carboxylic acid derivatives (Chapter 16, Section 16.7)**
- **nomenclature for heterocyclic compounds (Chapter 26, Sections 26.1–26.3)**
- **nomenclature for reduced forms of heterocycles (Chapter 26, Section 26.4)**
- **rotamers and conformations (Chapter 8, Sections 8.1–8.4, 8.9)**
- **transition states and reaction energetics (Chapter 7, Sections 7.3, 7.5, and 7.6)**
- **reversible chemical reactions (Chapter 7, Section 7.10)**
- **chemistry and mechanism of acyl addition reactions (Chapter 16, Section 16.3; Chapter 18, Sections 18.2 and 18.3)**
- **chemistry and mechanism of acyl substitution reactions (Chapter 16, Section 16.8; Chapter 20, Section 20.1)**
- **chemistry of amides (Chapter 20, Section 20.6)**
- **substitution reactions (Chapter 11, Sections 11.1–11.3)**
- **amines are bases in acid–base reactions (Chapter 6, Sections 6.4.1, 6.5.3)**
- **fundamentals of acid–base equilibria (Chapter 6, Section 6.1)**
- **the concept of isomers (Chapter 4, Section 4.5)**
- **absolute configuration (Chapter 9, Sections 9.1 and 9.3)**
- ***E*- and *Z*-isomers (Chapter 9, Section 9.4)**
- **stereochemistry, chirality, and stereoselectivity (Chapter 9, Sections 9.1 and 9.5)**
- **factors that influence the strength of acids (Chapter 6, Sections 6.3 and 6.5)**
- **oxidation reactions (Chapter 17, Sections 17.2–17.4)**
- **reduction reactions (Chapter 19, Sections 19.2–19.4)**

This chapter will present the most fundamental concepts for structure, nomenclature, and chemical reactions of amines. Biological applications will focus on the characteristics, formation, and reactions of amino acids. The use of amino acids in proteins and relatively simple reactions that form peptides will be discussed. In addition, several chemical reactions that lead to controlled degradation of peptides and proteins will be introduced.

When you have completed this chapter, you should understand the following points:

- Amines are prepared by the reaction of an alkyl halide with ammonia or another amine. Polyalkylation of an amine during reaction with alkyl halides is a serious problem.
- Amines are prepared by reductive amination of aldehydes and ketones, using sodium borohydride, catalytic hydrogenation, or dissolving metal conditions. Reaction of an amine with formaldehyde and formic acid leads to formation of an *N*-methyl amine.
- Amines are prepared by reduction of nitro compounds, nitriles, amides, azides, enamines, or imines.
- Amino acids are difunctional molecules that have a carboxylic acid unit and an amine unit in the same molecule. Amino acids exist as zwitterions under neutral conditions. The d- and l-stereochemical assignments for amino acids are based on the structure of d(+)-glyceraldehyde. Most of the important α-amino acids are l-amino acids and differences in structure focus on changing the substituent at the α-carbon.
- A zwitterionic amino acid can gain a proton at the carboxyl unit under acidic conditions or lose a proton from the ammonium unit under basic conditions to give two acid–base equilibria, K_1 and K_2. The isoelectric point is the pH at which the amino acid is neutral, which corresponds to formation of the zwitterion. When the amino acid contains an acidic side chain such as a carboxyl or a phenol, there is another acid–base equilibrium represented by K_3.
- For most reactions of amino acids, either the carboxyl or the amine unit must be protected to allow chemistry at the other unit. Amides and carbamates are the most common protecting groups for the amine unit of amino acids. Esters are the most common protecting group for the carboxyl group of amino acids.
- Amino acids are prepared from aldehydes and ketones via reaction with ammonia and HCN or via reactions of halo-acids with ammonia or amines. Phthalimide is a useful amine–nucleophile surrogate for the preparation of amino acids.
- Amino acids react with potassium thiocyanate to form a thiazolone derivative (a thiohydantoin). The azlactone of glycine can be used as a synthetic intermediate for the preparation of substituted amino acids.
- Peptides are composed of two or more amino acid units joined together by peptide (amide) bonds. The N-terminus of a peptide has an amino acid with a free amine group at that terminus. The C-terminus of a peptide has an amino acid with a free carboxyl group at that terminus. The amide bonds in a peptide assume a

relatively narrow range of rotational angles, which leads to relatively specific shapes for the peptide. These include the α-helix and the β-pleated sheet, which can exist in both parallel and antiparallel structures.

- Peptides are synthesized via coupling one protected amino acid with a second one to form a peptide bond. One terminus is deprotected and then coupled with another amino acid to elongate the peptide chain.

- Chemical reagents that induce degradation of a peptide allow various constituent amino acid residues to be identified. Ninhydrin is a common reagent used to indicate the presence of amino acids. Disulfide linkages can be cleaved with peroxyformic acid or with mercaptoethanol. Sanger's reagent forms a compound that allows the N-terminal amino acid to be identified. Dansyl chloride also reacts with the N-terminal amino acid to form a readily identifiable compound. Phenylisothiocyanate reacts with the N-terminal amino acid to form a phenylthiohydantoin derivative, which is readily identified.

- Carboxypeptidase enzymes can be used to identify the C-terminal amino acid. The enzymes trypsin and chymotrypsin cleave specific amino acid residues and this can be used to help identify the sequence of a peptide.

- A molecule with a particular functional group can be prepared from molecules containing different functional groups by a series of chemical steps (reactions). This process is called synthesis: The new molecule is synthesized from the old one (see Chapter 25).

- Spectroscopy can be used to determine the structure of a particular molecule, and can distinguish the structure and functionality of one molecule when compared with another (see Chapter 14).

27.1 A Review of Reactions That Form Amines

The nitrogen atom of an alkyl amine may have one, two, or three carbon groups attached, and these compounds are called primary amines (**1**), secondary amines (**2**), and tertiary amines (**3**), respectively. There are three important methods for making these compounds. Amines are prepared by the reaction of ammonia or an amine with an alkyl halide via an S_N2 process, by the reaction of an amine with an aldehyde or ketone under reducing conditions via an acyl addition–reduction process, and by the formal reduction of nitrogen-containing functional groups as seen in the conversion of nitriles to amines. Most of the amino acids in this chapter are considered to be derivatives of alkyl amines rather than aryl amines, so the focus will remain on the alkyl amines. The preparation and chemistry of aromatic amines were presented in Chapter 21 (Sections 21.3 and 21.6.2) and are not repeated here.

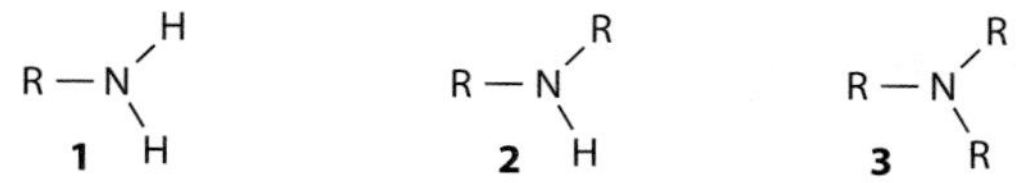

27.1.1 The Reaction of Ammonia or Amines with Alkyl Halides

Ammonia reacts as a nucleophile in the presence of compounds bearing an electrophilic carbon. When 1-bromobutane is heated with ammonia (a reaction bomb or autoclave is required because the boiling point of ammonia is $-33°C$), an initial S_N2 reaction (see Chapter 11, Section 11.2) gives butylammonium bromide, **4**. In the presence of ammonia, which is also a base, the weakly acidic ammonium salt is converted to its conjugate base, butanamine (also known as 1-aminobutane, **5**).

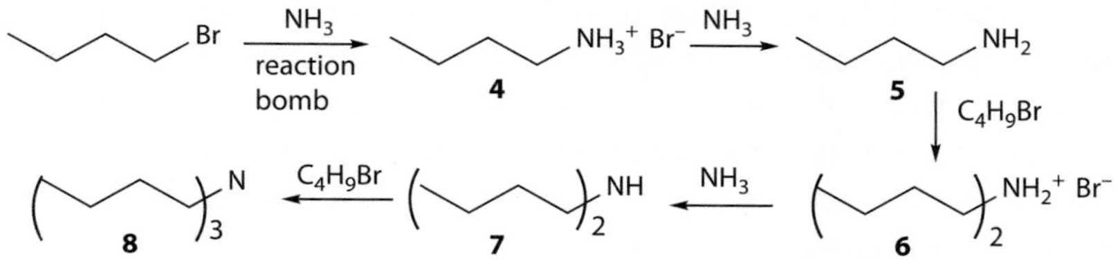

Unfortunately, there is a problem. The electron-releasing butyl group (Chapter 6, Section 6.4.1) makes **6** a stronger nucleophile than ammonia, which means that **5** will react with bromobutane faster than does ammonia. Therefore, the product is ammonium salt **6**. However, this is converted to secondary amine **7** by reaction with ammonia or **5** (ammonia reacts as a base). This secondary amine is more reactive than the primary amine, and **6** can react with bromobutane to give a tertiary amine (**8**) after reaction of the intermediate ammonia salt with ammonia. This reactivity of ammonia and amines in the S_N2 reaction was discussed in Section 11.3.3 of Chapter 11.

Formation of a primary amine such as **5** is difficult, but good yields are sometimes obtained by using a large excess of ammonia. In most cases, secondary and tertiary amine products are observed as minor constituents, and mixtures are common. This *overalkylation* process is particularly difficult to stop when a reactive halide such as iodomethane is used. The reaction of *N*-methyl-1-aminobutane (**9**) and iodomethane leads to a mixture of the tertiary amine *N,N*-dimethyl-1-aminobutane (**10**) and *N,N,N*-trimethylbutaneammonium iodide, **11**. The tendency of an amine to give a methylated ammonium salt upon reaction with methyl halides has given rise to the term **exhaustive methylation** to describe this reaction. The problems associated with the "over-reaction" of ammonia and amines with alkyl halides led to a search for other methods to make alkyl amines.

27.1 Write the structure and name of the ammonium salt precursor to 8.

27.2 Draw and name the product formed when (a) pyridine is treated with an excess of iodomethane, and (b) 1-aminobutane is similarly treated.

27.1.2 Reductive Amination

The reaction of an amine with an aldehyde or ketone will form an imine (see Chapter 18, Section 18.7.1), which may be reduced to a new amine in a second chemical step. In some cases, a reducing agent is mixed directly into the reaction of an amine and a carbonyl compound. If 1-aminobutane (**5**) reacts with 3-pentanone and Raney nickel (abbreviated Ni(R); see Chapter 19, Section 19.3) in the presence of hydrogen gas, the isolated product is amine **14**. Initial reaction of the amine and the ketone gives an amino-alcohol (**12**), which loses water to generate an imine, **13**. (See Chapter 19, Section 19.7.1, for a discussion of reactions between ketones or aldehydes and amines.)

    Because imine **13** is formed in the presence of hydrogen gas and the Raney nickel catalyst, catalytic hydrogenation converts it to amine **14**. This overall process is called reductive amination, and many different reducing agents have been used. Only one or two methods are presented here to focus on the method. For example, it is possible to mix a ketone and an amine in the presence of zinc metal (Zn°) and HCl to give the amine.

    Another synthesis of substituted amines mixes formic acid with an amine and formaldehyde. If 3-aminohexane (**15**) is treated with formaldehyde in the presence of formic acid, the initial product is iminium salt, **16**. In the presence of formic acid, the π-bond of the imine donates electrons to the acidic proton to give an *N*-methylamine product, 3-(*N*-methylamino)hexane, **17**. This particular reaction is called the **Eschweiler–Clarke reaction**, after Wilhelm Eschweiler (Germany; 1860–1936) and Hans Thacher Clarke (England; 1887–1972). It is also possible to mix formaldehyde and an amine in the presence of hydrogen gas and a catalyst, similarly to the reductive amination reaction described before, and generate an *N*-methyl amine.

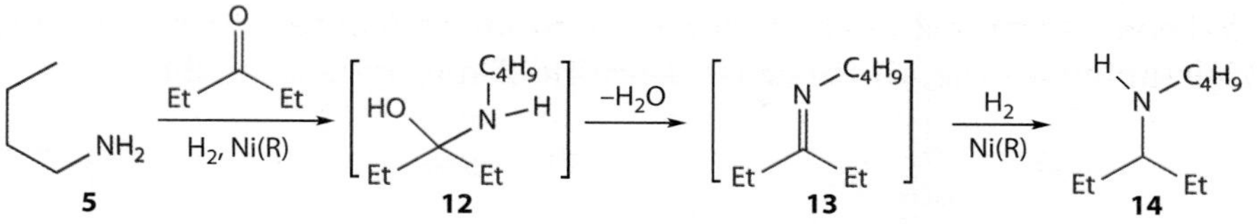

27.3 Draw the product of a reaction between 2-aminobutane and phenylacetaldehyde in the presence of hydrogen gas and Raney nickel.

27.4 Draw the product of a reaction between 2-amino-3-phenylbutane and a mixture of formaldehyde/formic acid.

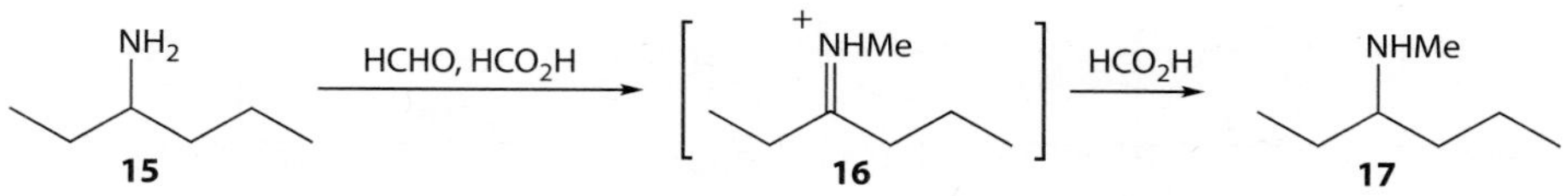

27.1.3 Direct Reduction of Functional Groups

Several functional groups contain nitrogen, and many may be reduced to give an amine, as seen in previous chapters. A nitro group on an alkyl chain was reduced to an amine by reaction with lithium aluminum hydride (LiAlH$_4$) in Chapter 19 (Section 19.2.1). LiAlH$_4$ reduction does not work well with aromatic nitro compounds, which are converted to diazo-compounds by this reaction. Catalytic hydrogenation of either an aliphatic nitro compound or an aromatic nitro compound, however, gives the amine. If nitrocyclopentane (**18**) is treated with LiAlH$_4$ or with H$_2$/Pd-C, the product is aminocyclopentane (**19**). Reduction of 3-nitrotoluene (**20**) with hydrogen and palladium gives 3-methylaniline (**21**). The synthesis of aliphatic nitro compounds is more difficult than for many other nitrogen-containing functional groups.

27.5 Write out a synthesis of *N*,4-dimethylaniline from toluene.

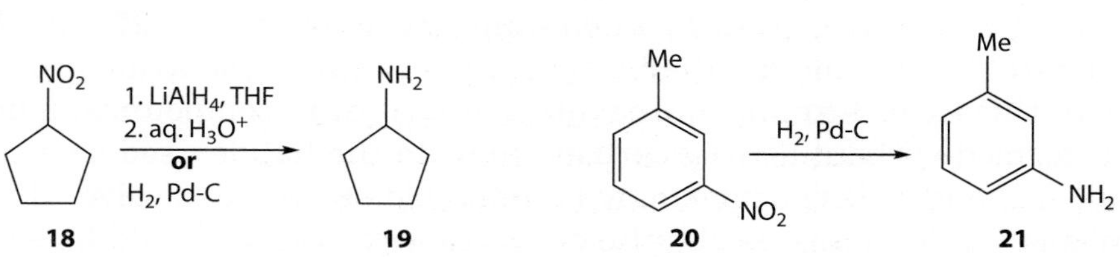

Nitriles are common nitrogen-containing compounds. In Chapter 11 (Section 11.3.6), cyanide ion reacted as a nucleophile in an S_N2 reaction with alkyl halides to give a nitrile. When 2-bromohexane (**22**) is treated with NaCN in DMF, the product is 2-cyanohexane, **23**. Reduction of this nitrile with LiAlH$_4$ gives a primary amine, in this case 1-amino-2-methylhexane **24**.

Because formation of nitriles by this method is an S_N2 process, it works best with primary and secondary halides, and reduction of the cyano unit will give a primary amine. There is one important difference, however. Note that conversion of **18** to **19** and conversion of **20** to **21** do not increase the number of carbon atoms. When bromide **22** is treated with cyanide to give **23** in a carbon–carbon bond-forming reaction, reduction leads to an amine (**24**) with one more carbon atom.

Carboxylic acids are converted to an amide by reaction of the corresponding acid chloride or ester with ammonia or an amine (see Chapter 20, Section 20.6). Reduction of the amide with LiAlH$_4$ gives a primary amine. Phenylacetyl chloride (**25**), for example, reacts with *N*-methyl-1-aminobutane (**9**) to give amide **26**. Subsequent treatment with LiAlH$_4$ gives amine **27**. This approach is somewhat more flexible than those given earlier because many carboxylic acids are available. Conversion to the corresponding amide via the acid chloride or ester, followed by reduction, gives the corresponding amine.

27.6 Draw a synthesis of *N*-methyl-3-benzyl-1-aminohexane from 1-pentanal.

27.7 Write out a synthesis of 1-aminopent-4-ene from malonic acid.

Sodium azide (NaN$_3$; **28**) is a resonance-stabilized nucleophilic species, and reaction with an alkyl halide gives an alkyl azide. Reaction of **28** with 1-iodobutane, for example, gives 1-azidobutane, **29**. Reduction of **29** with LiAlH$_4$ or with hydrogen gas and a palladium catalyst converts the azide group to an amine—in this case, 1-aminobutane (**30**). As with nitro compounds, a primary amine is formed without adding carbon atoms to the halide used as a starting material. Although alkyl azides may be unstable or even explosive, this route to amines is viable and is used quite often. However, care should be exercised whenever an azide is involved.

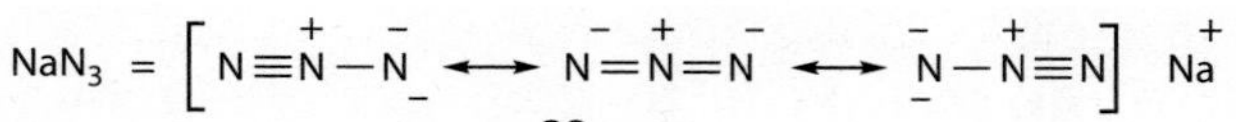

$$NaN_3 = \left[N\equiv \overset{+}{N} - \overset{-}{\underset{.}{N}} \longleftrightarrow \overset{-}{N} = \overset{+}{N} = \overset{-}{N} \longleftrightarrow \overset{-}{\underset{.}{N}} - \overset{+}{N}\equiv N \right] \overset{+}{Na}$$

28

27.8 Write out a synthesis of *N*-benzoyl-2*R*-aminopentane from 2*S*-bromopentane.

27.9 Write out a synthesis of 1-phenyl-1-aminobutane from benzyl alcohol.

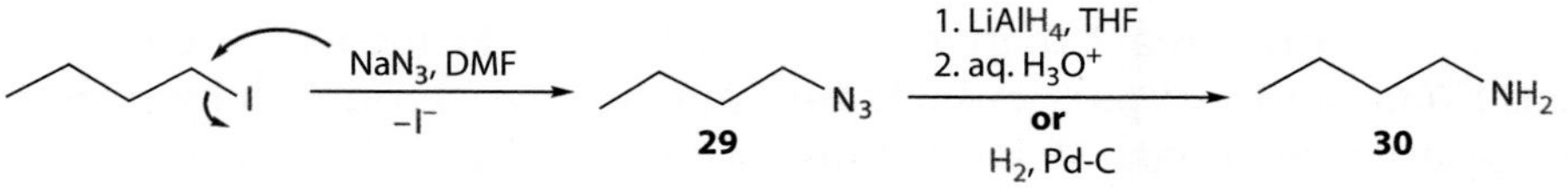

Chapter 11 (Section 11.3.3) described another method for the preparation of amines using an S_N2 reaction and a different amine surrogate. Phthalimide (**31**) is an imide (see Chapter 20, Section 20.7.5) derived from phthalic acid (**35**). If **31** is treated with a strong base such as butyllithium or sodium amide, the phthalimide anion (**32**) is formed as the conjugate base (ammonia is the conjugate acid of this reaction). Phthalimide is a good nucleophile in reactions with alkyl halides. Reaction of **32** with benzyl bromide gives *N*-benzylphthalimide (**33**) via a straightforward S_N2 displacement. Hydrolysis of the imide by acid–base reactions (1. aqueous base; 2. aqueous acid) gives amine **34** and phthalic acid (**35**). The reaction of **33** with aqueous hydroxide is an example of an acyl substitution reaction discussed in Chapter 20 (Section 20.2). A ***better*** procedure treats **33** with hydrazine (NH_2NH_2), which generates amine **34** and **36**. Compound **36** is called a ***hydrazide*** and is easily separated from the amine. Hydrazides will not be discussed further.

27.10 Draw a reaction sequence that prepares (2*S*)-aminopentane from (2*R*)-bromopentane.

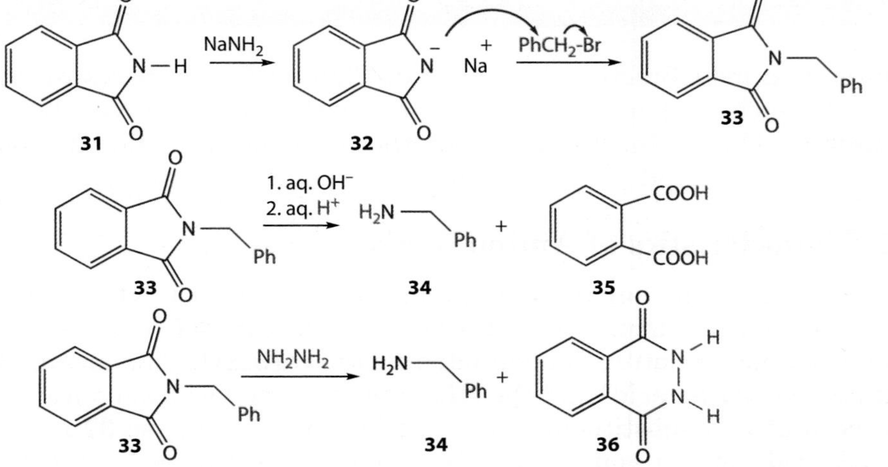

27.2 Reactions of Amines

Several reactions of amines have been discussed while discussing reactions that prepared amines. Aliphatic amines are nucleophiles and they react with alkyl halides, as shown in Section 27.1.1. In Chapter 19 (Section 19.7.2), aliphatic amines react as nucleophiles with ketones to give enamines, via acyl addition, and with aldehydes to give imines. In many chapters of this book, amines are used as Brønsted–Lowry bases or Lewis bases. In the presence of a strong acid such as HCl, amines react to give an ammonium chloride as the conjugate acid product. Some amines, particularly tertiary amines, induce dehydrohalogenation by reaction as a base (E2 reactions of alkyl halides, for example). Primary and secondary amines react as acids in the presence of a very powerful base such as organolithium reagent. In Chapter 22 (Section 22.3), the reaction of diisopropylamine and *n*-butyllithium gave lithium diisopropylamide, which is a much more powerful base than the amine (it is the conjugate base of the amine); it deprotonates even weak acids such as ketones, aldehydes, or esters.

27.11 Draw the product formed when (a) butanal reacts with 2-aminobutane, and (b) 3-pentanone reacts with pyrrolidine.

27.12 Draw the product of a reaction between isoquinoline and HBr.

27.13 Write out a synthesis of 1-ethylidene cyclopentane (*c*-C$_5$H$_8$=CHMe) from 1-cyclopentylethene.

27.14 Draw the product formed when diethylamine reacts with butyllithium in THF at –78°C.

27.3 Difunctional Molecules: Amino Acids

An amine is a monofunctional compound. Chapter 24 discusses difunctional compounds, such as dienes, and conjugated carbonyl derivatives. In this section, amino acids are introduced as another class of difunctional compounds with an amine unit and a carboxyl unit in the same molecule.

27.3.1 Characteristics of Amino Acids

An amino acid, as the name implies, has one amine unit (–NR$_2$) and one carboxylic acid unit (a carboxyl group, COOH). Structure **37** is a generic amino acid, and the nomenclature is dominated by the carboxyl. This means that the parent name of **37** is "acid" and the NR$_2$ unit is treated as a substituent. When an amine unit is a substituent, the name "amino" is used, so **37** is an amino carboxylic acid—or just amino acid. The amine unit is attached to C2 of the acid chain, the α-carbon of the carboxylic acid chain. Therefore, **37** is known as an α-amino acid. If the amine group is on C3, the β-carbon (as in **38**), the compound is known as a β-amino acid. Similarly, **39** is a γ-amino acid, **40** is a δ-amino acid, etc. Due to their biological importance, α-amino acids will be discussed most of the time. To distinguish α-amino acids from other amino acids, the term "non-α-amino acids" is used. Indeed, **38–40** are categorized as non-α-amino acids.

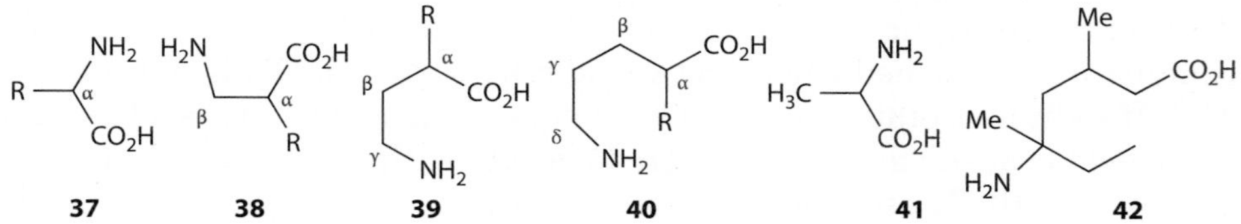

There are many possible amino acids and the important α-amino acids have common names. Amino acids are easily named using IUPAC nomenclature, however, where the carboxylic acid is the parent for each new compound. Using the IUPAC nomenclature, **41** is 2-aminopropanoic acid and **42** is 3,5-dimethyl-5-aminoheptanoic acid. Note than **42** is a non-α-amino acid. Amino acid **35** has the common name alanine, which is used most of the time. The common names of α-amino acids are presented in Section 27.3.2.

27.15 Draw the structure of (a) 2-aminomethyl-2-phenylpentanoic acid, (b) 1-amino-3,3-diethylcyclobutane-1-carboxylic acid, (c) 1-aminocyclohexane-1-carboxylic acid, and (d) *cis*-3-aminocyclohexane-1-carboxylic acid.

Perhaps the most important chemical feature of **37** is the presence of an amine (which is a base) *and* a carboxyl group (an acid) in the same molecule. The basic amine reacts with the acidic proton of acetic acid to form a conjugate base and a conjugate acid. If methylamine and acetic acid (R = CH$_3$) are mixed together, for example, the product is methylammonium acetate, **43**. Amino acid **44** reacts with the carboxylic acid unit in exactly the same way, but the acid–base reaction is an *internal reaction that gives **45***, with a positively charged ammonium unit (–NH$_3^+$) and a negatively charged carboxylate unit (–CO$_2^-$) in the same molecule. Such molecules are known as dipolar ionic molecules or **zwitterions**. The zwitterion form (**45**) is used for all amino acids discussed in this book, unless otherwise noted.

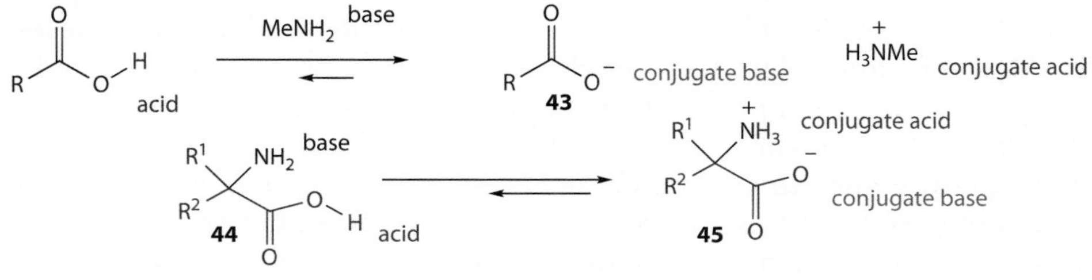

The zwitterionic amino acid structure **45** reacts as both an acid and a base under the right conditions. A simple example of this behavior is an ammonium salt such as ammonium chloride (NH_4Cl), which is the conjugate acid of ammonia. Ammonium salts are weak acids and reaction with a suitable base removes a proton from nitrogen to regenerate ammonia. Amino acid **45** has an ammonium unit and reaction with a base leads to loss of a proton and formation of **47**, the conjugate base of **45**. *The acid portion of a zwitterionic amino acid is the ammonium unit.*

The focus is changed to the carboxyl unit of an amino acid, and the conjugate base of a carboxylic acid, the carboxylate anion (Chapter 6, Section 6.2.4). The conjugate base formed when acetic acid reacts with NaOH is sodium acetate (NaOAc). If sodium acetate is treated with aqueous acid, acetate is a base and its conjugate acid is acetic acid. Zwitterion **45** reacts with aqueous acid in a similar manner to give **46**. These are acid–base reactions, so an equilibrium is shown between **46**, **45**, and **47**. Each equilibrium will have an equilibrium constant (K), labeled K_1 for conversion of **39** to **40** and K_2 for conversion of **39** to **41**.

The value of K_1 is defined in Equation 27.1 and the value of K_2 is defined in Equation 27.2. As with other acids (see Chapter 6, Section 6.2), the pK values are used to discuss acidity in amino acids. The value of pK_1 and pK_2 depends on the nature of the substituents (R^1 and R^2), as will be seen in Section 27.3.3.

$$K_1 = [45]/[46] \qquad\qquad (27.1)$$

$$K_2 = [47]/[45] \qquad\qquad (27.2)$$

27.16 Draw the equation that relates K_1 to pK_1.

If one equivalent of base is added to **46**, removal of the most acidic proton (the carboxyl proton) gives zwitterion **45**. This means that when exactly one equivalent of base has been added, the zwitterion form (**45**) is the species in solution. The pH of the solution at exactly this point is called the **isoelectric point**. As the R^1 and R^2 groups are changed, each new compound may have a *different* pK_1 value. As a practical matter, this means that *the isoelectric point for each different amino acid will change in relation to its structure.*

While adding base to **46**, the pH of the solution can be measured after one equivalent of base has been added (the isoelectric point), but the pH can also be measured when half of the molar equivalents of base have been added. This is the point where half of **46** is converted to **45**, and the pH of the solution at this point is equal to pK_1. Adding a second molar equivalent of base to **45** removes the ammonium proton to give **47**. Once again, the point at which half of the

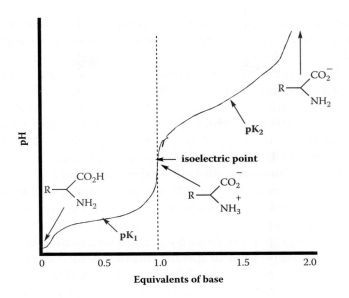

Figure 27.1 The position of pK and isoelectric points for a generic amino acid.

base has been added to **45** will have a pH equal to pK_2. As with pK_1, the value of pK_2 varies with the substituents attached to the amino acid. These values for a generic amino acid are shown in Figure 27.1.

27.3.2 Structure of α-Amino Acids

The chemistry of DNA is presented in Chapter 28 (Section 28.6). This includes the genetic code that provides a set of instructions for the formation of proteins (discussed in this chapter), which are made up of amino acids linked together into a polymer. The genetic code utilizes primarily about 20 α-amino acids, so this discussion is limited to those compounds. In principle, an infinite number of α-amino acids vary as R^1 and R^2 change in **45**. Examination of the generic structure (**45**) shows that the α-carbon is a stereogenic center, so these amino acids are chiral molecules. It is known that the 20 amino acids most commonly found in proteins have the identical absolute configuration. Amino acid **45** and all of the 20 amino acids contain a hydrogen atom and one alkyl group at the α-carbon rather than two alkyl groups. A compound called glyceraldehyde [$HOCH_2CH(OH)CHO$] is drawn in Fischer projection (see Chapter 9, Section 9.1.2, for the definition of a Fischer projection) and also as the usual line notation. Glyceraldehdye has one stereogenic center, so there are two enantiomers: one with a (+) specific rotation and one with a (−) specific rotation.

In an effort to standardize the way in which stereogenic centers are presented for amino acids (and especially carbohydrates; see Chapter 28), Emil Fischer (Germany; 1862–1919) invented a system to differentiate these enantiomers. The system draws structures in what we now call a Fischer projection. Using this system, Fischer assigned the structure of (+)-glyceraldehyde as **48**, with the CHO unit on "top" and the OH unit on the "right," as shown in the illustration. The distinguishing OH unit is on the right (*dexter* in Latin), so Fischer called this a d-configuration; **48** is d-(+)-glyceraldehyde. The other enantiomer is drawn as **49**, with the OH unit on the left (*lever* in Latin) and this is an l-configuration. Therefore, **49** is l-(–)-glyceraldehyde.

Other representations of **48** and **49** are also shown, including the normal line notation used with other molecules in this book. Fischer assigned (d) to the (+) enantiomer (which happens to be dextrorotatory), but this was an arbitrary choice (a guess that is properly called an assumption). Remember from Chapter 9 (Section 9.2) that (+) and (–) refer to specific rotation, which is a physical property, whereas d and l are names. In 1851, Johannes Martin Bijvoet (the Netherlands; 1892–1980) showed by x-ray analysis that Fischer's assignments were correct.

There are two enantiomers for each amino acid, based on the zwitterionic forms of amino acid **45** (with one alkyl group R and one hydrogen atom). Stereoisomers **50** and **51** are drawn in Fischer projection and also as the analogous line drawing. In the Fischer projections, the COOH unit is on the "top" just as the CHO unit is on top in glyceraldehyde. The distinguishing feature is the NH_3 unit; therefore, if the nitrogen group is on the "right," it is a d-amino acid (**50**), whereas an l-amino acid has the NH_3 unit on the left (**51**). In the line drawings, the (*R*) configuration has the amino group projected forward and the (*S*) configuration has the amino group projected to the rear. For the most part, the amino acids will be drawn using line notation.

27.17 Draw both d- and l-2-aminohexanoic acid in Fischer projection.

The 20 protein amino acids used most often in protein biosynthesis (see genetic code in Chapter 28, Section 28.6) are drawn by changing the R group in **50**, and all are l-amino acids. Imagine three categories of R: R is neutral, R has an acidic group as part of its structure, or R has a basic group as part of its structure. The subcategories of α-amino acids are therefore neutral, acidic, or basic. Table 27.1 shows the structure of the 20 protein amino acids, the R group in **50**, the name of the amino acid, and **a three-letter code used to abbreviate each compound**. The neutral amino acids include R groups that are simple alkyl substituents such as methyl or ethyl, but there are also groups that have polarized substituents such as OH, SH, or SR. If the OH or SH unit has a sufficiently low pK value, its acidity must be acknowledged.

Table 27.1
Structures and Names of the 20 Essential Amino Acids, Based on Structure 50

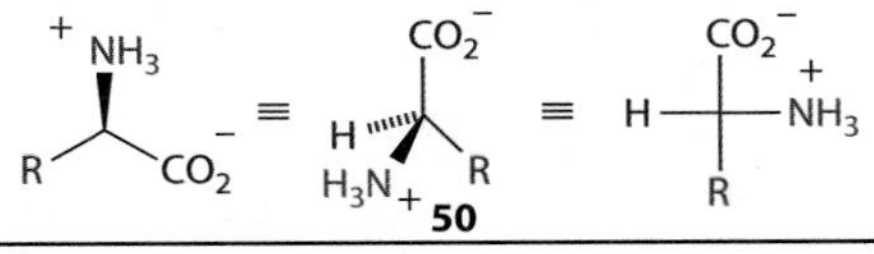

R	Name	Three-Letter Code	One-Letter Code
H	glycine (**52**)	gly	G
Me	alanine (**53**)	ala	A
$CHMe_2$	valine (**54**)	val	V
$CHMe_2$	leucine (**55**)	leu	L
CH(Me)Et	isoleucine (**56**)	ile	I
CH_2Ph	phenylalanine (**57**)	phe	F
CH_2OH	serine (**58**)	ser	S
CH(OH)Me	threonine (**59**)	thr	T
$CH_2(4\text{-hydroxy-}C_6H_4)$	tyrosine (**60**)	tyr	Y
CH_2SH	cysteine (**61**)	cys	C
CH_2CH_2SMe	methionine (**62**)	met	M
CH_2CONH_2	asparagine (**63**)	asn	N
$CH_2CH_2CONH_2$	glutamine (**64**)	gln	Q
CH_2COOH	aspartic acid (**65**)	asp	D
CH_2CH_2COOH	glutamic acid (**66**)	glu	E
$CH_2CH_2CH_2CH_2NH_2$	lysine (**67**)	lys	K
$CH_2(2\text{-indolyl})$	tryptophan (**68**)	trp	W
$CH_2(4\text{-imidazolyl})$	histidine (**69**)	his	H
$CH_2NHC(=NH)NH_2$	arginine (**70**)	arg	R
2-pyrrolidinyl	proline (**73**)	pro	P

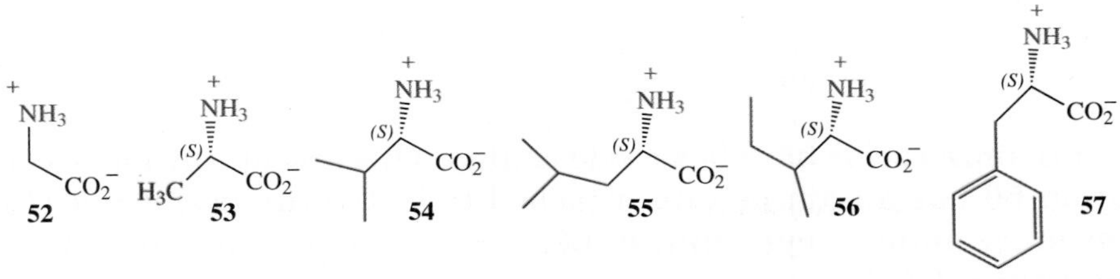

The first group of amino acids includes those where R in **50** is a simple alkyl fragment. This group includes **glycine** (**52**; R = H), **alanine** (**53**; R = Me), **valine** (**54**; R = isopropyl; $CHMe_2$), **leucine** (**55**; R = isobutyl), **isoleucine** (**56**; R = *sec*-butyl), and **phenylalanine** (**57**; R = benzyl; CH_2Ph). The second group of amino acids has a substituent that is considered to be neutral, but possesses a polarized group. These include **serine** (**58**; R = CH_2OH), **threonine** (**59**; R = CH[OH]Me), and **tyrosine** (**60**; R = CH_2–[4-OH)phenyl) with hydroxyl substituents. **Cysteine** (**61**; R = CH_2SH) has a thiol unit and **methionine** (**62**; R = CH_2CH_2SMe) has a thioether unit. Because the thiol unit in **61** and the phenolic unit in **60** are relatively acidic, these two amino acids are categorized as acidic amino acids. Also, two amino acids have an amide unit as part of the side chain, but they are considered to be neutral amino acids: **asparagine** (**63**; R=CH_2CONH_2) and **glutamine** (**64**; R = $CH_2CH_2CONH_2$).

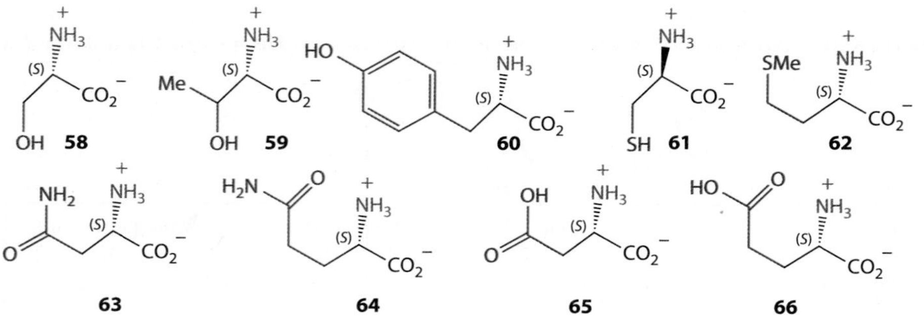

There are acidic amino acids that have a carboxyl group (COOH) as part of the side chain. These include **aspartic acid** (**65**; $R = CH_2COOH$) and **glutamic acid** (**66**; $R = CH_2CH_2COOH$). Note that asparagine is the amide derivative of aspartic acid and that glutamine is the amide derivative of glutamic acid. As mentioned before, the thiol unit in **cysteine** (**61**) and the phenol unit in **tyrosine** (**60**) have pK_a values of around 10, so these amino acids are considered to have acidic side chains.

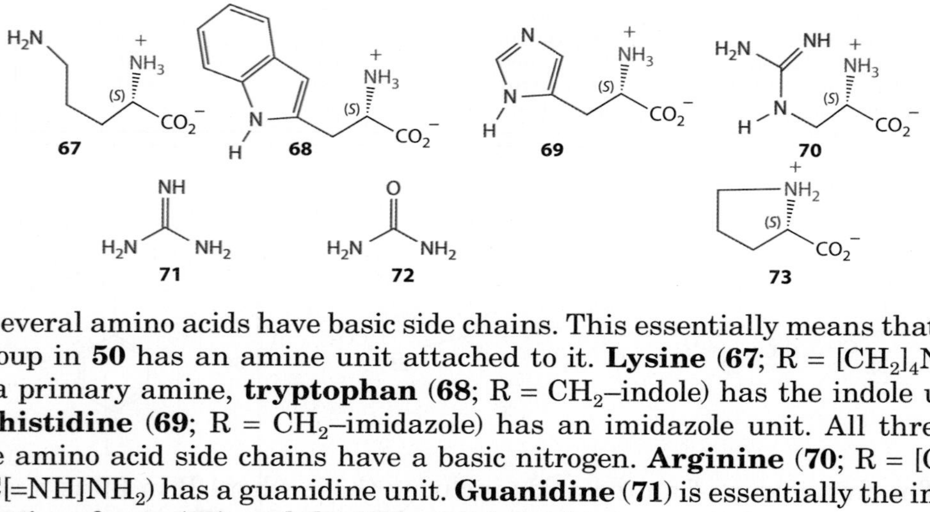

Several amino acids have basic side chains. This essentially means that the R group in **50** has an amine unit attached to it. **Lysine** (**67**; $R = [CH_2]_4NH_2$) has a primary amine, **tryptophan** (**68**; $R = CH_2$–indole) has the indole unit, and **histidine** (**69**; $R = CH_2$–imidazole) has an imidazole unit. All three of these amino acid side chains have a basic nitrogen. **Arginine** (**70**; $R = [CH_2]$ $NHC[=NH]NH_2$) has a guanidine unit. **Guanidine** (**71**) is essentially the imine derivative of urea (**72**), and the NH_2 unit is basic.

The remaining amino acid is **proline** (**73**) and it is unique for a couple of reasons. First, a pyrrolidine ring effectively connects the amine unit to the α-carbon. Second, proline is the only α-amino acid that has a secondary amine unit. This completes the structural evaluation of the 20 protein α-amino acids.

As will be seen in Section 27.6, these 20 amino acids are used to make proteins. The discovery of these amino acids is interesting, however, because it illustrates the variety of materials from which they have been isolated. All of the examples cited here are found in the excellent 1961 book *Advanced Organic Chemistry*.[1] The authors are Harvard professor Louis F. Fieser (United States; 1899–1978) and his wife and colleague, Mary Fieser (United States; 1909–1997), a research associate at Harvard. Many of the common names for amino acids are taken from the Greek word used to describe one of their properties.

An 1820 investigation of the hydrolysis of gelatin yielded glycine, which was so named because it had a sweet taste (*glykys*, sweet). Leucine was isolated in 1820 from muscle tissue as a white compound (*leukos*, white), and cysteine was isolated in 1810 from urinary calculi (*kystis*, bladder). Tyrosine was first isolated from cheese (*tyros*, cheese) and lysine was named as a product of hydrolysis (*lysis*, loosening). There are also Latin derivations for some amino acids. Serine is the most abundant constituent of silk (*serieus*, silken) and arginine was first isolated as a silver salt (*argentum*, silver). This paragraph is intended to show not only from where the names of the amino acids are derived, but also the types of materials in which they are found.

Before leaving this section, it is important to point out that each of the amino acids just discussed has a pK value and an isoelectric point. Table 27.2 shows the name of the amino acid, pK_1, pK_2, the isoelectric point, and pK_3, which is only observed when there is an acidic side chain or an amine group on the side chain that can form an ammonium salt.[2]

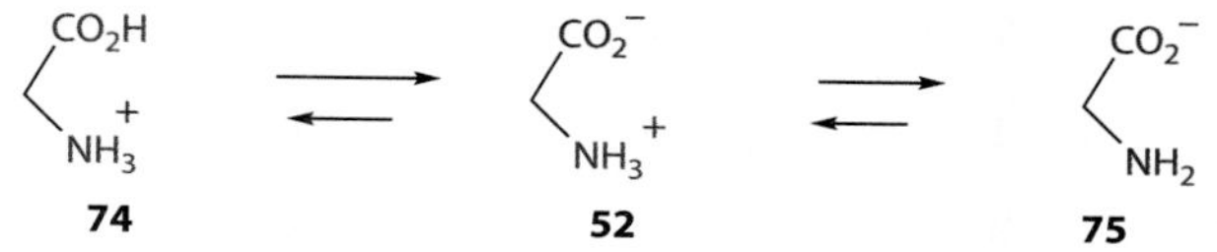

The acidic amino acids have a third pK value listed: pK_3 (shown in Table 27.2). This pK value refers to the acidity of the acidic unit on the side chain. The usual equilibrium for an amino acid is shown for glycine **52**, which is in equilibrium with **74** and **75**. The equilibrium for an amino acid with an acidic side chain is more complex. In the case of glutamic acid, the zwitterion form is **66**. The value for pK_1 results from loss of the carboxyl proton from the α-amino acid unit in **76** to form zwitterion **66**. The next most acidic proton is the carboxyl proton on the side chain, and loss of this proton generates **77** from **66**. The equilibrium between **66** and **77** is represented by K_3 and leads to the value of pK_3 in Table 27.2. Loss of the ammonium proton from **77** gives **78**, represented by pK_2. When the side chain has an acidic proton that is less acidic than the ammonium proton or the carboxyl proton—such as the phenolic proton in tyrosine (**60**)—the equilibrium is slightly different.

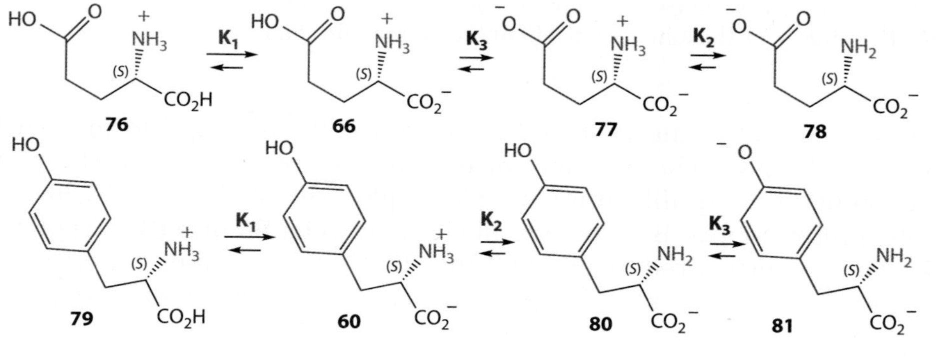

Table 27.2
pK Values of Amino Acid at the Isoelectric Point in Water at 25°C

Amino Acid		pK_1	pK_2	Isoelectric Point	pK_3	Solubility[a]
Glycine	52	2.34	9.60	5.97		251
Alanine	53	2.34	9.69	6.00		167
Valine	54	2.32	9.62	5.96		58
Leucine	55	2.36	9.60	5.98		23
Isoleucine	56	2.36	9.60	6.02		34
Phenylalanine	57	1.83	9.13	5.48		29
Serine	58	2.21	9.15	5.68		422
Threonine	59	2.09	9.10	5.60		97
Tyrosine	60	2.20	9.11	5.66	10.07	0.5
Systeine	61	1.96	10.28	5.07	8.18	—
Methionine	62	2.28	9.21	5.74		56
Asparagine	63	2.02	8.80	5.41		25
Glutamine	64	2.17	9.13	5.65		42
Aspartic acid	65	1.88	9.60	2.77	3.65	5
Glutamic acid	66	2.19	9.67	3.22	4.25	—
Lysine	67	2.18	8.95	9.74	10.53[b]	6
Tryptophan	68	2.83	9.39	5.89		12
Histidine	69	1.82	9.17	7.59	6.00[b]	43
Arginine	70	2.17	9.04	10.76	12.48[b]	181
Proline	73	1.99	10.60	6.30		1622

Source: Lide, D. R., ed. 1995. *Handbook of chemistry and physics,* 76th ed., 7-1. Boca Raton, FL: CRC Press.

a Solubility is measured in grams of amino acid per kilogram of water at 25°C.

b This pK_3 value is for the ammonium salt of the amine side chain.

Amino acid **60** is the zwitterionic form and it is in equilibrium with **79**, represented by pK_1. The ammonium proton is more acidic than the phenolic proton, so **60** is in equilibrium with **80** for pK_2; this is followed by loss of the phenolic proton to give **81**. The equilibrium between **80** and **81** is represented by K_3, which is measured as pK_3 in Table 27.2. *The real point is that pK_3 measures the acidity of the side chain acid, but does not necessarily indicate the* *position of that step in the equilibrium for a given amino acid.* Finally, lysine (**67**) has a third pK value for the ammonium salt of the amine unit on the side chain. The first equilibrium for lysine involves loss of the carboxyl proton from **82** to give **67**, which is pK_1. Because the ammonium salt on the amine closest to the carboxyl is more acidic, the next equilibrium involves loss of the ammonium proton from **67** to give **83**, defined by pK_2. The third equilibrium involves loss of the ammonium proton in **83** to give **84**.

27.18 Draw the zwitterion, acid, and base form of valine.

27.19 Write out the resonance forms associated with 81 for the phenoxide and for the carboxylate group.

27.20 Draw the product when 3-(4-hydroxyphenyl)propanoic acid is treated with (a) one equivalent of NaOH and (b) two equivalents of NaOH.

27.21 Write out the complete pK equilibrium for histidine, as was done for lysine.

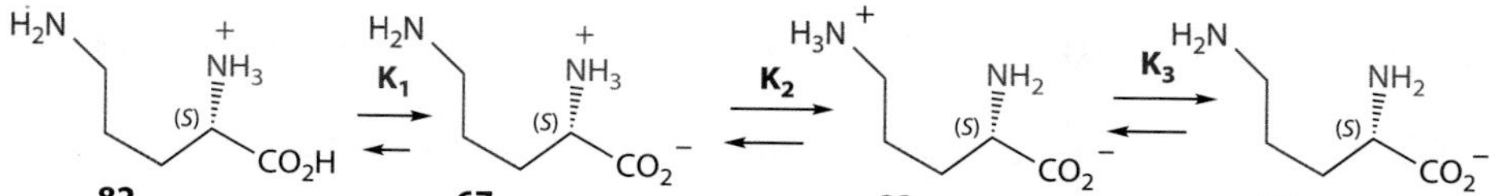

27.3.3 Reactions of α-Amino Acids

Amino acids undergo several important reactions dominated by the amine group and/or the carboxyl group. The fact that the amine and the carboxyl react with each other complicates things because one functional group will influence the reactivity of the other. The conversion of a simple carboxylic acid to an ester is quite easy, as in Chapter 20 (Section 20.5), either by reaction of the carboxylic acid with an alcohol under acidic conditions or by first converting the acid to an acid chloride and subsequent reaction with an alcohol. However, amines also react with acid chlorides or with esters to give the corresponding amide. In other words, trying to form an ester in the presence of an amine may be a problem. To target one functional group in an amino acid, the other must be taken into account. Because amino acids exist as zwitterions, however, the issues of dual reactivity are not always a major problem if the proper form of the molecule is used.

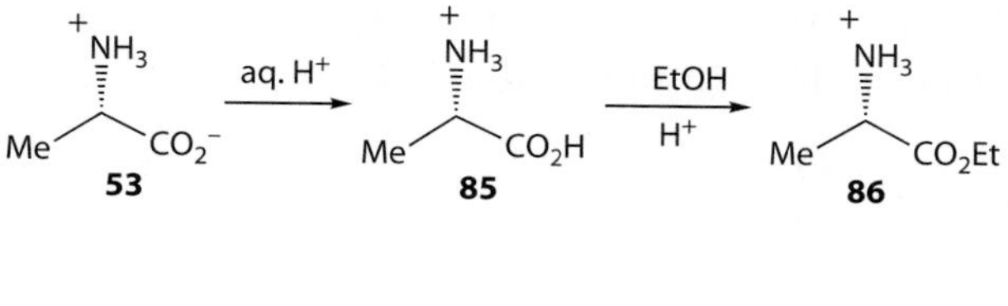

Using the zwitterion form of alanine (**53**) as an example, treatment with acid generates **85**. The amine unit already exists as its ammonium salt, and the carboxylate anion reacts with aqueous acid to give the COOH unit. There is no unshared electron pair on nitrogen in the ammonium salt, so there is no interference with reactions at the carboxyl group. Therefore, if **85** is treated with ethanol in this acidic solution, the product is the ammonium ester **86**.

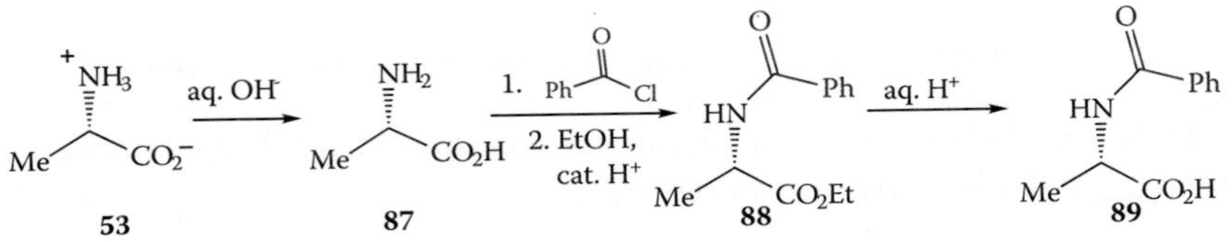

For reaction at the amine unit, the carboxyl unit must be blocked. A solution of **53** in aqueous hydroxide will give **87**, the carboxylate anion that has a free amine unit. If **87** reacts with an acid chloride such as benzoyl chloride, reaction with the nucleophilic amine group will generate an amide, **88** after esterification. Aqueous acid hydrolysis of the ester group in **88** gives the amide-acid **89**.

27.22 Write out a sequence that converts valine to its benzyl ester.

27.23 Write out a sequence that converts isoleucine to *N*-acetyl isoleucine.

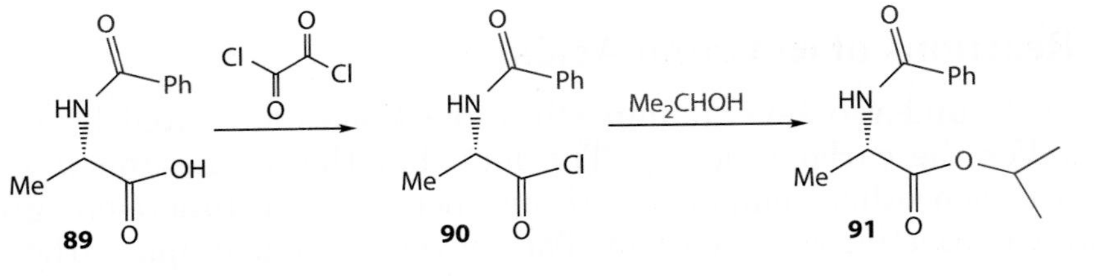

With **89** in hand, reaction with oxalyl chloride (Chapter 20, Section 20.9.1) will give acid chloride **90**, which reacts with 2-propanol to give amide-ester **91**. In effect, the amine group is ***protected*** as an amide, allowing a chemical reaction at the carboxyl unit. (See Chapter 25, Section 25.4.3, for an introduction to protecting groups.) By manipulating the acid–base equilibrium of the amino acid, either the amine unit or the carboxyl unit may be selectively targeted for reaction. Once the amine unit is protected as an amide, many of the normal reactions described for carboxylic acids in Chapter 20 are possible. For example, the reaction of **89** with ethanol and dicyclohexylcarbodiimide (DCC; see Chapter 20, Section 20.6.3) forms ester **92**, but if **89** reacts with ethylamine and DCC, the product is the amide-amide **93**.

27.24 Write out a sequence that converts phenylalanine to *N*-propanoyl phenylalanine ethyl ester.

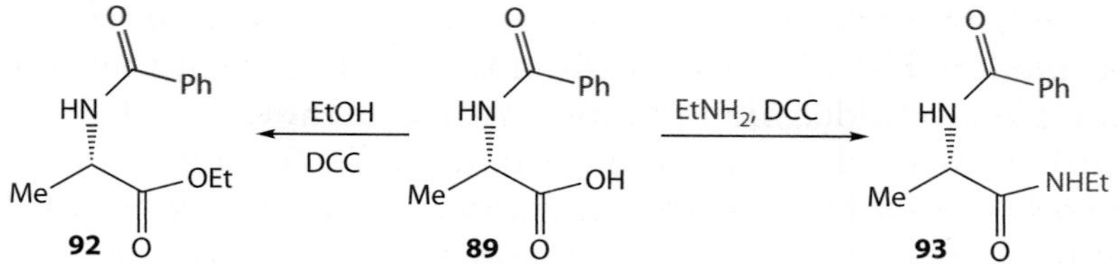

Several reactions of amino acids result in heterocyclic derivatives. In one example, the carbonyl unit of leucine (**55**) is converted to its *N*-acetyl derivative (**94**). Subsequent treatment with acetic anhydride and sodium acetate leads to a heterocycle, 5-oxazolone, **95**. *An oxazolone derived from amino acids has the common name of azlactone, so 95 is the azlactone of leucine.*

27.25 **Draw the structure for the azlactone derivatives of valine, serine, and tryptophan.**

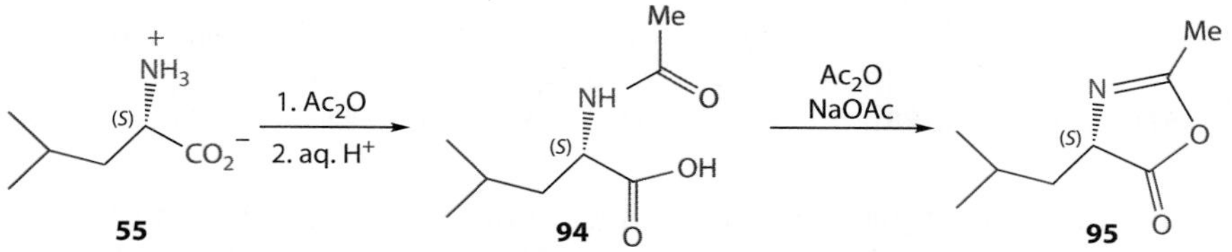

This section showed that reactions of amino acids are relatively normal with respect to the amine unit and the carboxyl unit, if the presence of both functional groups in the same molecule is taken into account. In addition to these "normal" reactions, several specialized transformations will be presented in Section 27.5.2 in connection with identification of protein structure. These are important reactions, but they will make more sense if presented in their proper context.

27.3.4 Synthesis of α-Amino Acids

Several methods are used for the synthesis of amino acids. A major problem with any synthesis is preparing enantiopure products. Most of the syntheses shown here give **racemic** amino acids, but methods are known that produce amino acids highly enriched in one enantiomer (see Chapter 9). Enantioselective synthetic methods will not be discussed. A method used quite often to obtain an enantiopure amino acid prepares the racemic compound, followed by isolation of the l-amino acid by resolution, as described in Chapter 9, Section 9.8.

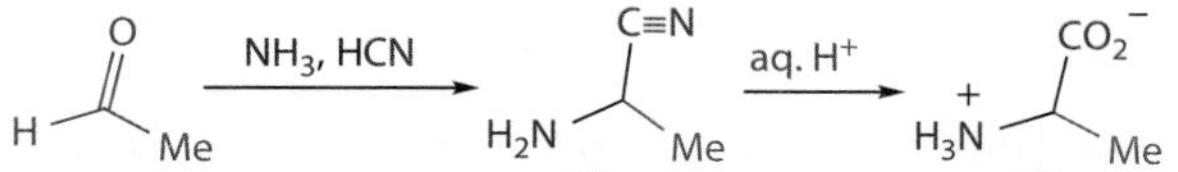

Adolph Strecker (Germany; 1822–1871) reported one of the first preparations of an amino acid, and the method, the **Strecker synthesis**, bears his name. If an aldehyde reacts with ammonia in the presence of HCN, an amino nitrile is formed. Acid hydrolysis of the nitrile unit gives an amino acid. In a specific example, acetaldehyde is heated with ammonia and HCN to give amino-nitrile, **96**, and acid hydrolysis leads to alanine, **53**. *The alanine produced by this method is racemic, as is any other amino acid made this way.* It is a general procedure, however, and many different amino acids can be prepared. The enantiopure amino acid is obtained by resolution.

27.26 Write out the mechanism for the formation of 96 based on the idea that ammonia and an aldehyde react to form an imine (see Chapter 19, Section 19.7.1).

27.27 Write out a synthesis of 2-amino-2-methyl-3-phenylpropanoic acid from 1-phenyl-2-propanol.

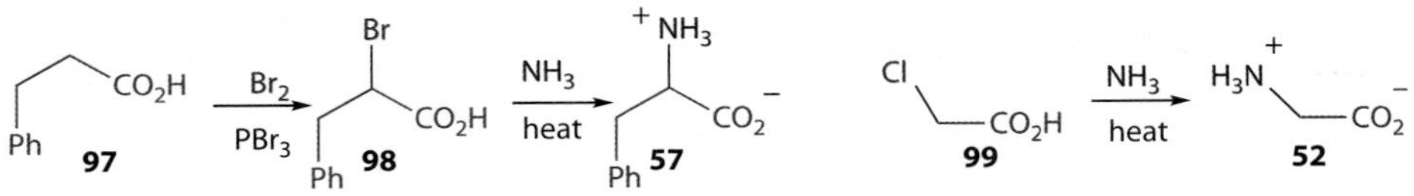

An α-bromo acid is prepared by reaction of the acid with bromine and phosphorus tribromide (this is known as the **Hell–Volhard–Zelinsky reaction**). This reaction was named for the work of C. Hell, Nikolai D. Zelinsky (Russia; 1861–1953), and J. Volhard. If 3-phenylpropanoic acid (**97**) is treated with these reagents, **91** is formed. If **98** is subsequently heated with ammonia (see Section 27.1.1) and the product is neutralized, amino acid **57** is formed (phenylalanine). This route was used as early as 1858, when glycine (**52**) was prepared by the reaction of chloroacetic acid (**99**) and ammonia. Amino acids formed this way are racemic.

27.28 Write out a synthesis of 2-amino-2-cyclopentylethanoic acid from bromomethylcyclopentane.

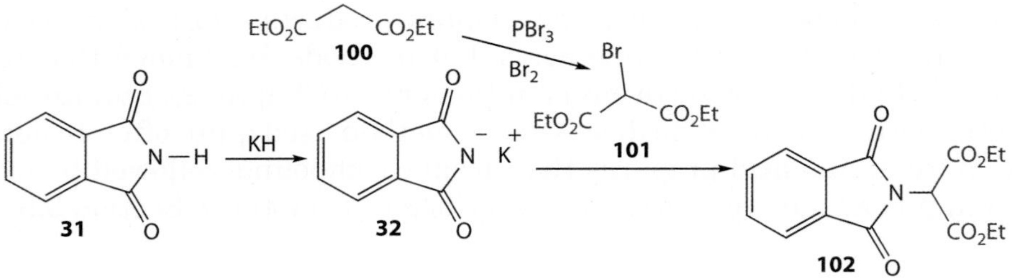

As discussed earlier in this chapter, ammonia is not always the best reagent to generate amines from halides. Amine surrogates such as azides or phthalimide usually give better yields in S_N2-type reactions, although their use requires an extra step to *liberate* the amine unit. In Chapter 22 (Section 22.7.4), esters of malonic acid were shown to be very reactive at C3. If diethyl malonate (**100**) is treated with bromine/PBr_3, the 3-bromo derivative (**101**) is readily formed. Reaction of this α-bromo ester with ammonia should lead to an amino acid precursor. Because ammonia also reacts with esters to form amides (see Chapter 20, Section 20.6), side reactions may occur. To ensure a good yield of the substitution product, a nitrogen surrogate is used. In Section 27.1, phthalimide (**31**) was treated with a base such as potassium hydride (KH) to give potassium phthalimide (**32**). If **32** is heated with bromomalonate, **101**, the product is **102**.

In Chapter 22 (Section 22.7.4), malonate derivatives were easily converted to the corresponding enolate anion, and reaction with alkyl halides or other electrophilic species gave the C3-alkylated product. Indeed, if **102** is treated with sodium metal (or NaH, LDA, etc.), enolate anion **103** is formed; it reacts with an alkyl halide such as benzyl bromide ($PhCH_2Br$) to give **104**. If **104** is heated with aqueous sodium hydroxide and then treated with aqueous HCl, phthalic acid (**35**) and the amino acid phenylalanine (**57**) are formed as the final products.

27.29 Draw the amino acid that will be produced if 103 reacts with (a) allyl bromide and is carried through the sequence, (b) 1-iodobutane, and (c) bromomethylcyclopentane.

27.30 Write out a complete synthesis of tyrosine using this approach with the appropriate alkyl halide and also show a synthesis of the requisite halide from phenol.

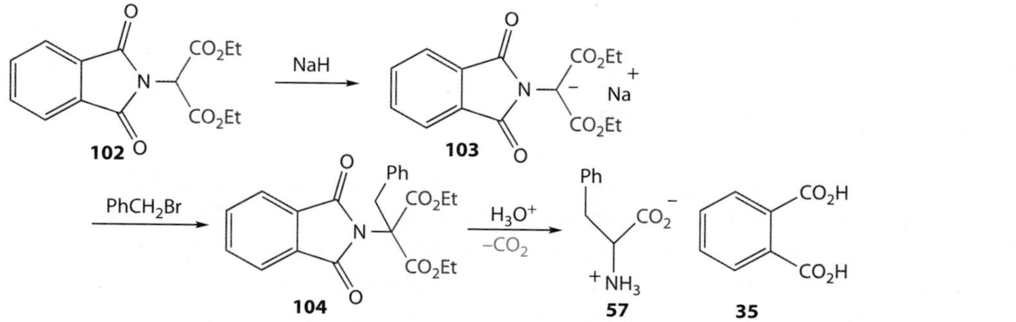

Amino acids that have an aromatic group as part of the side chain are prepared from aromatic aldehydes, but a special reagent called a ***thiohydantoin*** is often required. If glycine (**52**) is treated with acetic anhydride, the product is *N*-acetyl glycine (**105**). When reacted with potassium thiocyanate (KS–C≡N), the nitrogen atom in **105** attacks the carbon of thiocyanate to give **106** after hydrolysis. This compound, an acyl thiourea derivative, cyclizes to generate a heterocyclic species, **107** (a thiazolone). Thiazolones usually rearrange under the reaction conditions to give a new heterocycle, a thiohydantoin (**107**, 3-acetyl-2-thiohydantoin). The –CH$_2$– unit in **108**, derived from the original amino acid (in **blue**), is susceptible to enolate condensation reactions with aromatic aldehydes.

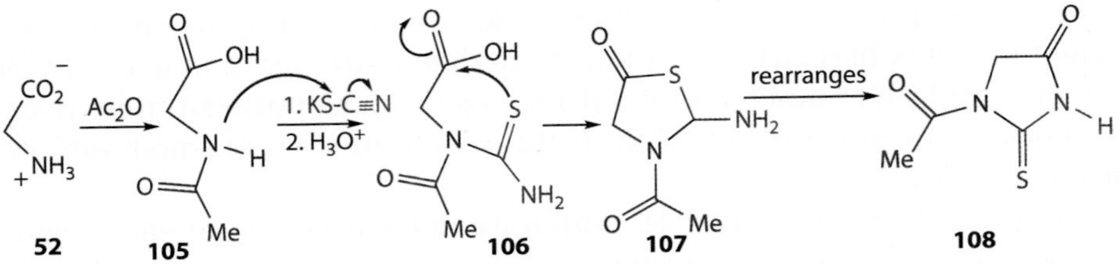

In one example, reaction of 4-hydroxybenzaldehyde (**109**) with the enolate anion of **108**, generated by reaction with the base pyridine, leads to a new thiohydantoin, **110**. When **110** is heated with thiourea (**111**) at 100°C for several hours, S=C=NH is lost and the other product is the amino acid tyrosine (**60**).

27.31 **Write out the mechanism for the reaction that converts 109 to 110.**

27.32 **Write out the synthesis of tryptophan from the appropriate indole aldehyde using this method.**

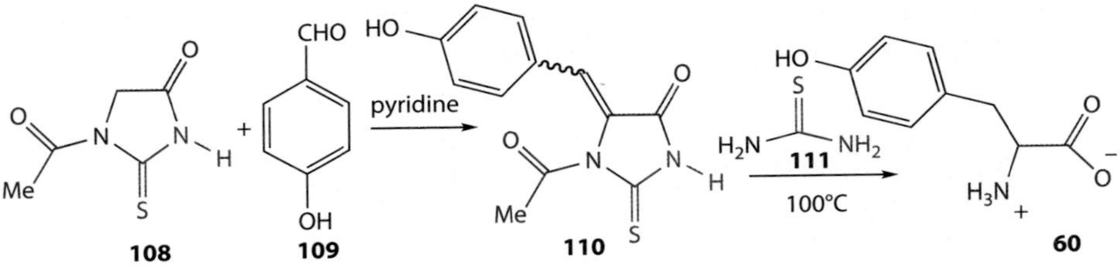

Another amino acid synthesis is called the **azlactone synthesis**. Remember from before that an azlactone is an oxazolone (see **95**). When glycine (**52**) is converted to its N-benzoyl derivative (**112**; known as hippuric acid) by reaction with benzoyl chloride, treatment with acetic anhydride (Ac$_2$O) gives the azlactone **113**. This is the reaction presented in the preceding section (see compound **95**). Compound **110** has the common name of hippuric acid azlactone. As with the thiohydantoin, the –CH$_2$– unit in **113** is susceptible to an enolate anion condensation reaction with aldehydes (Chapter 22, Section 22.7.2), and reaction with 2-methylpropanal in the presence of pyridine gives azlactone **114**. Catalytic hydrogenation of the alkene unit (Chapter 19, Section 19.3.2) and acid hydrolysis lead to the amino acid leucine (**55**).

27.33 **Write out an azlactone synthesis of phenylalanine using the appropriate aldehyde.**

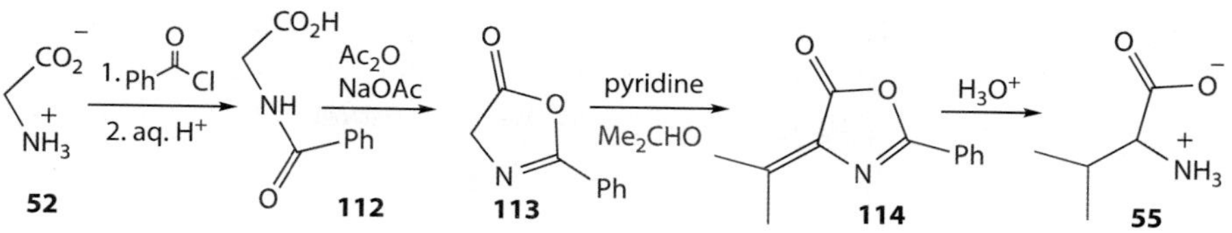

As pointed out previously, all of the amino acids prepared in this section are **racemic**. To obtain an enantiopure amino acid requires separation of the enantiomers via resolution. As discussed in Chapter 9 (Section 9.2), the physical properties of enantiomers are identical except for specific rotation. Because separation methods rely on differences in physical properties, this is a problem. It is overcome if the racemic amino acid mixture reacts with a reagent that has a stereogenic center. The resulting product will be a mixture of diastereomers, which have different physical properties and may be separated.

A second chemical reaction is required to remove the first reagent and "release" the enantiopure amino acid. This process is called **optical resolution** (see Chapter 9, Section 9.8). Chemicals obtained from nature as a single enantiomer are used most often as the reactive agent, but they must have a functional group that is able to react with one of the functional groups in a racemic molecule.

Most of these chiral compounds have an amine, a carboxylic acid, or an alcohol unit that can interact or react with the racemic mixture. Such compounds are called chiral **resolving agents**. Three different resolving agents are shown that may be used for optical resolution. The first two reagents are naturally occurring alkaloids brucine (**115**) and strychnine (**116**). The third is an optically pure amine, (*R*)-phenylethylamine (**117**). The method will be illustrated using the simper regent **117**.

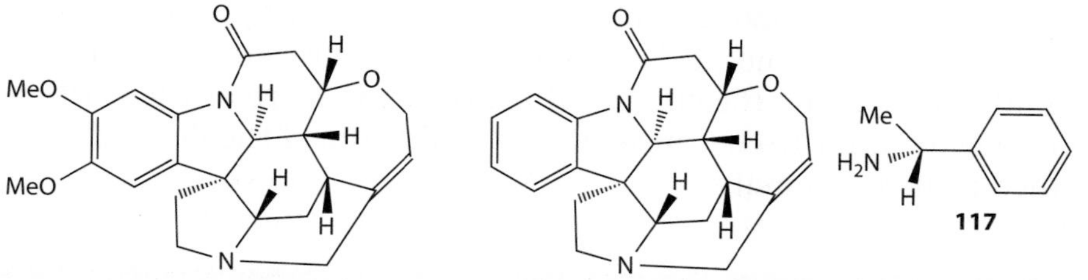

Assume that the racemic alanine (**47**) must be resolved into pure l-alanine. If racemic **53** (**53rac**) reacts with enantiopure **117**, the salt of each enantiomer is formed: **118** and **119**. These diastereomeric salts are separated by fractional crystallization, assuming that one salt is more or less soluble in a given solvent than the other. If **118** can be crystallized from a solution, a highly purified **119** remains in solution. The salt **119** is isolated and recrystallized from a different solvent to high enantiopurity. When this process is complete, **118** is treated with dilute base to regenerate **117** and l-alanine, **53**. The other enantiomer, d-alanine, is similarly obtained from **119**.

27.34 Draw the structure of d-alanine in Fischer projection.

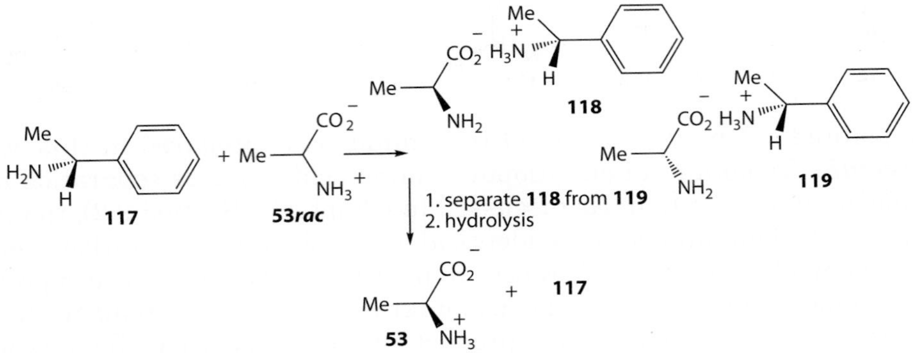

Brucine is used in a similar manner and the carboxyl unit of the amino acid is coordinated to the tertiary amine unit in brucine rather than the poorly basic amide nitrogen. Selective crystallization of these salts leads to their separation, and basic hydrolysis leads to an enantiopure amino acid. It is now possible to separate many racemic mixtures into their enantiomeric components by using **high-pressure liquid chromatography (HPLC)** fitted with a column that contains a chiral compound bound to an adsorbent (known as chiral HPLC columns). Such columns have been developed by William H. Pirkle (United States; 1934–). The chiral HPLC column is prepared by coating a chiral chemical compound on an inert material; when a solution of the racemic mixture passes through this column, one enantiomer is adsorbed to the column material better than the other. These are sometimes called **Pirkle columns**.

This column works because one enantiomer will travel more slowly through the column than the other, allowing them to be separated. As methods for the synthesis of organic compounds have grown more sophisticated, it has become possible to prepare amino acids directly or purify racemic amino acids with high enantiopurity. An efficient chiral synthesis is more efficient than resolution of a racemic mixture because, at best, only 50% of the desired enantiomer is obtained from a racemate.

27.4 Biological Relevance. Peptides Are Polyamides of Amino Acid Residues

Chapter 20 (Section 20.6) discussed the reaction in which a carboxylic acid or an acid derivative reacts with an amine to form an amide. Amino acids have an amine unit and an acid unit, so it is conceivable that two amino acids may be coupled together to form an amide; however, there are two possible products. Using **120** as an example, coupling to another amino acid (**121**) may generate either **122** or **123**. In **123**, amino acid **121** (in **blue**) is coupled via its amine unit to the carboxyl unit of **120**. In **122**, amino acid **121** is coupled via its carboxyl group to the amine group of **120**. Clearly, a third, a fourth, or a larger number of new amino acids may be added to the carboxyl end or the amino end. Dimers, trimers, tetramers, etc. that contain several amide bonds are known as **poly(amides)** or just **polyamides**. Extension of this process will make a lengthy polyamide, but each new reaction must be controlled to occur at the carboxyl or the amine unit.

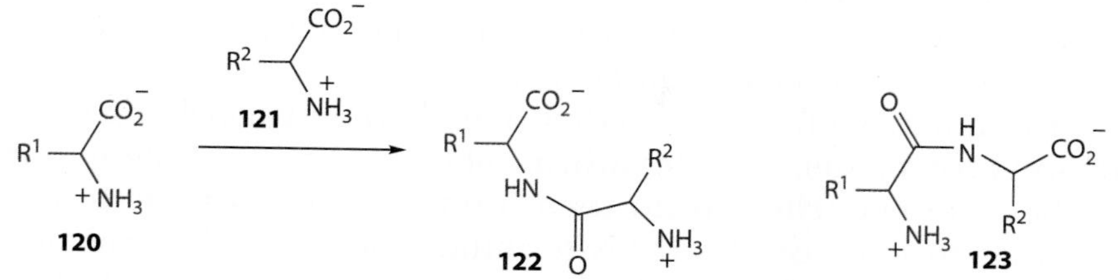

Polyamides derived from amino acids are known as **peptides**, and the amide bonds within the peptide are called **peptide bonds**. Both **122** and **123** are composed of two amino acids, so they are called dipeptides. A peptide with three amino acid units is a tripeptide and a peptide with 15 amino acid units is called a pentadecapeptide. The amino acid components of a peptide are known as **amino acid residues**, so **122** or **123** has two amino acid residues, and a decapeptide has 10 amino acid residues. Large peptides (hundreds or thousand of amino acid residues) that use primarily the α-amino acids in Table 27.2 constitute the important biological molecules called proteins. Proteins control many life processes and they will be briefly discussed in the following section. At first, the structures of peptides are discussed, followed by a discussion of synthetic methods used to prepare peptides from their amino acid components.

27.4.1 Structure and Properties of Peptides

A peptide is a polyamide composed of amino acids coupled together by amide bonds. All peptides in this section are derived from the amino acids listed in Table 27.2. As more amino acids are coupled together, drawing the structure of the resulting polypeptide becomes increasingly difficult. An example is the nonapeptide (nine amino acid residues) **124**. This nonapeptide is composed (reading from left to right) of **alanine-valine-serine-leucine-alanine-phenylalanine-glutamic acid-methionine-histidine**, using only (S)-amino acids. The name of **124** is rather long, and to facilitate communicating the structure of peptides, a shorthand notation gives a three-letter code or a one-letter code to each amino acid residue.

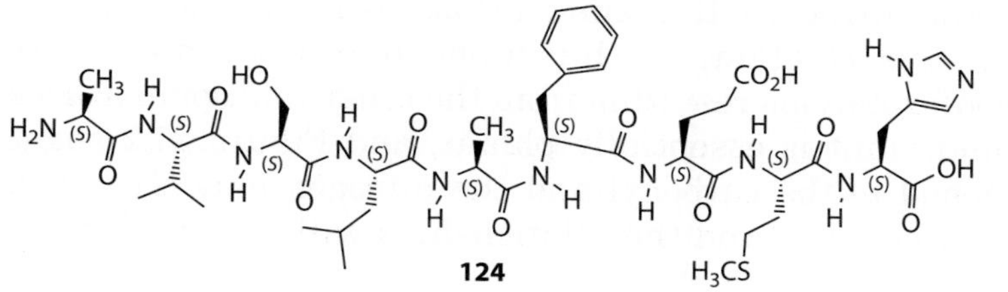

124

The three-letter abbreviations for each of the amino acids are listed in Table 27.1. In most cases, these codes take the first three letters of the name of the amino acid; however, isoleucine is abbreviated ile, glutamine is abbreviated gln, and tryptophan is abbreviated trp. A one-letter code is used for particularly large peptides, and those codes are also listed in Table 27.1. Using the three-letter codes from Table 27.1, **124** is **ala-val-ser-leu-ala-phe-glu-met-his**; it is **A-V-S-L-A-F-E-M-H**, using the one-letter codes. This nomenclature system makes it easier to write the structure of a peptide, but it demands memorization of the code to go with each amino acid.

Closer examination of **124** shows that there is one carboxyl group on one end (**the carboxyl terminus, or C-terminus**) of the peptide and there is an amino group at the other end (**the amino terminus, or N-terminus**). In the case of **124**, an alanine residue occupies the N-terminus and a histidine residue occupies the C-terminus. *By convention, the N-terminus is always drawn on the left and the C-terminus is drawn on the right.* This is an important convention because when **ala-val-ser-leu-ala-phe-glu-met-his** is read for **124**, it indicates that alanine is the N-terminus and histidine is the C-terminus. The order in which the amino acids are connected together is called the **primary structure of a peptide**.

27.35 Write the primary structure of a peptide in three-letter code if the one-letter code is S-S-L-N-C-D-G-A-F-W-H.

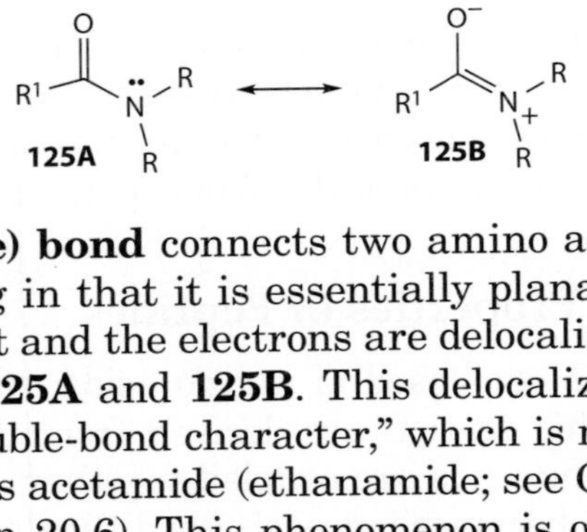

An amide (peptide) bond connects two amino acid residues. The amide unit is quite interesting in that it is essentially planar. Structure **125** shows an amide bond fragment and the electrons are delocalized as shown by the two resonance structures, **125A** and **125B**. This delocalization leads to the C–N unit having "partial double-bond character," which is normal for the C–N unit in simple amides such as acetamide (ethanamide; see Chapter 16, Section 16.7, and Chapter 20, Section 20.6). This phenomenon is observed in the infrared spectrum of primary and secondary amides, which exhibit two absorptions (1640 cm^{-1} [C–O stretch] and 1650–1515 cm^{-1} [imine N–H bend of NH$_2$ or NH] for the amide I and amide II bands; see Chapter 14, Section 14.3.4).

As a practical matter, there is little rotation around the C–N bond and the significant amount of sp^2 hybridization leads to a planar geometry for the amide unit. Figure 27.2[3] shows rotation about the C–N bond (between the carbonyl carbon and the nitrogen). It is known that the groups on carbon and nitrogen can be *trans-* to each other, which is lower in energy than the arrangement of groups in conformations resulting from the other rotational angles. It is known that the amide unit is essentially planar, and Figure 27.2 suggests that the groups attached to the carbonyl and the nitrogen may have different stereochemical relationships, and the relationships will vary with the nature of the R groups.

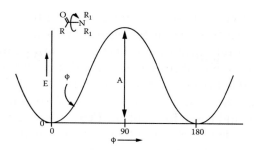

Figure 27.2 Approximation of the potential energy surface for amide band rotation as a function of the twist angle ϕ.[3] See Figure 6 in Pluth, N. D.; Bergman, R.G.; Raymond, K. N., *Journal of Organic Chemistry,* 2008, *73,* 7132. (Reprinted in part with permission from Pluth, N. D., Bergman, R. G., and Raymond, K. N. *Journal of Organic Chemistry, 2008, 73,* 7132. Copyright © 2008 American Chemical Society.)

    If all amide bonds in a peptide have a planar geometry, the attached groups have a *cis* or *trans* relationship. This means that the nature of the substituent at the α-carbon has an important influence on the conformation of the entire molecule. A tripeptide example uses the first three amino acids of **124**, which is tripeptide (**126**, ala-val-ser). If the C–N unit of the amide bond is planar, rotation occurs only around the C–N bond marked in **green** and the C–C bond marked in **blue** (see **126A**). Each $C_{carbonyl}$–N–C_α unit defines a plane and, because of the restricted rotation, a rotational angle is defined for the atoms connected to the **green** bonds and to the **blue** bonds. In **126B**, the tripeptide is drawn in three-dimensional form using a wire-frame molecular model. The rotational angle for the C^α–C=O bond is labeled by ψ, which is the angle defined by rotation about that bond. The C^α is the carbon of the amino acid that bears the substituent (methyl, isopropyl, hydroxymethyl, etc).

The angle ϕ is defined as the angle for rotation about the C^α–N bond. Each peptide bond in a peptide tends to be planar, which is a consequence of the planar nature of the amide unit. The alkyl groups in each amino acid residue have a great influence on the magnitude of angles ψ and ϕ, and these angles of rotation define the conformation for that portion of the peptide. Structure **126B** shows that the amide unit of one amino acid residue is *anti* to the amide unit of the adjacent amino acid residue. Peptide **126B** also shows the carbonyl of the valine residue is *anti* to the carbonyl of the valine residue, which is *anti* to the carbonyl of the serine residue. The consequence of this observation is that a peptide chain assumes this alternating or *anti* pattern.

127   **128**

If the two dipeptide units are brought into close proximity, as shown in **127**, the oxygen of one carbonyl can form a hydrogen bond to the proton on the amide nitrogen of the second dipeptide. For this to occur, however, one dipeptide must have an *anti* orientation. This type of hydrogen bonding occurs in long-chain peptides and, when combined with the planar nature of the amide units, the *anti* orientation of adjacent amide carbonyls, and the magnitude of the angles ψ and ϕ, leads to a rather unique structure for the peptide. It forms an α-helical structure as shown in the poly(alanine) peptide **128**.

The α-helix structure was proposed by Linus Pauling (United States; 1901–1995) and Robert Brainard Corey (United States; 1897–1971) in 1951. The hydrogen atoms have been omitted from **128** so that the helical structure is easier to see. In a long-chain peptide composed of l-amino acid residues, a right-handed helix (called an α-helix) is formed, where the hydrogen atom on the amide nitrogen is hydrogen bonded to the oxygen of the carbonyl on the fourth amino acid residue. The hydrogen bonds stabilize the α-helix structure, which is an example of the **secondary structure** of a peptide. Formally, *the secondary structure is the amount of structural regularity in a peptide that results from hydrogen bonding between the peptide bonds.*

Analysis of the α-helix reveals some interesting data (Figure 27.3). First, the hydrogen bonds occur between a carbonyl of one amino acid residue and

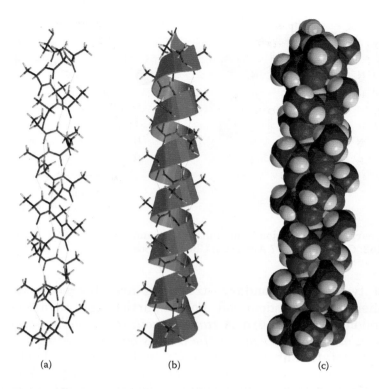

(a) (b) (c)

Figure 27.3 α-Helix of a peptide, poly(alanine): (a) tube model; (b) the ribbon drawing (see Figure 27.4); (c) space-filling model.

an amide proton, which is usually on the fourth amino acid residue. In formal terms, *one turn of the helix represents 3.6 amino acid residues and there are 13 atoms involved in the turn, starting with the carbonyl oxygen and ending with the amide proton.* The number of turns in the α-helix of the peptide depends on the number of amino acid residues in that peptide. *The pitch of the α-helix is defined as the "rise" in the amino acid residue in the helix for each turn. Because there are 3.6 residues per turn and the rise per amino acid is measured to be about 1.5 Å (150 pm), the pitch of each turn is about 5.4 Å (540 pm).* The presence of the side chains on the amino acid residues may introduce steric hindrance, which will influence the pitch of each turn.

Some peptides have a high percentage of their secondary structure as an α-helix, whereas others have a low percentage. This percentage is largely determined by the nature of the side chain groups of the amino acid residues. Some amino acid residues actually destabilize the α-helix, including glutamate, aspartate, lysine, arginine, glycine serine, isoleucine, and threonine. One amino acid residue (proline) actually creates a bend in the α-helix. A peptide composed entirely of leucine (polyleucine) will exist almost entirely in an α-helix secondary structure, whereas a peptide composed only of aspartic acid residues (polyaspartic acid) will form another secondary structure. Note that a left-handed helix can be formed when d-amino acids are used to make a peptide.

129   **130**

(Reprinted in part with permission from Smith, C. K., and Regan, L. 1997. *Accounts of Chemical Research* 30:153. Copyright 1997 American Chemical Society.)

At least two other secondary structures are observed with peptides: a β-pleated sheet and a random coil. Poly(aspartic) acid, mentioned previously, forms a random coil structure. *A random coil, as its name implies, does not assume a regular structure such as the α-helix because hydrogen bonds are not easily formed.* Rotation about the ψ and φ angles (see **126**) leads to a random orientation of the various amino acid residues. *The β-pleated sheet, on the other hand, does involve intramolecular hydrogen bonding. In other words, there are hydrogen bonds between two different peptide chains rather than within a single peptide chain.*

An α-helix and a β-pleated sheet are contrasted by saying that the α-helix is held together by ***intramolecular*** hydrogen bonds, whereas the β-pleated sheet is held together by ***intermolecular*** hydrogen bonds. There are parallel β-pleated sheets (**129**)[4] or antiparallel β-pleated sheets (**130**), and these variations are defined by having the two C-termini aligned (in **129**) or the C-terminus of one chain aligned with the N-terminus of the other chain (in **130**).[4]

In a shorthand method for drawing complex peptide structures, β-strands are shown as "thick" arrows, an α-helix is a spiral ribbon, and nonrepetitive structures are shown as ropes. These shorthand structures are called "ribbon drawings" or ***Richardson diagrams***, after Jane S. Richardson (United States; 1941–).[5–7] To illustrate this method, ribonuclease A **131** is shown in Figure 27.4,[8] along with the helical ribbon and "tube" used to represent a helix. The "wide arrow" used to represent a β-strand in a ribbon structure is also shown.

Another feature of peptide structure must be considered. The overall structure of a long polypeptide is determined by its amino acid sequence (primary structure) and whether it forms an α-helix or a β-pleated sheet (secondary structure). It is also possible to fold or coil the peptide chain into a complex, globular structure that is known as its tertiary structure. This is illustrated by the ribbon diagram for ribonuclease A, **131**.[8] The peptide chain folds and coils into a very complex structure in **131**. This is the **tertiary structure** of the peptide.

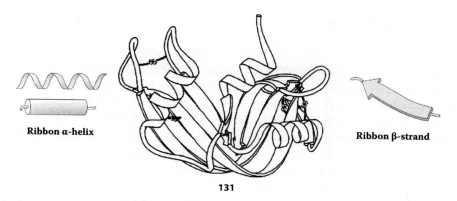

131

Figure 27.4 Richardson drawings. Ribonuclease A, illustrating a variety of junctions between loops and helices and between loops and β-strand arrows. (Reprinted from Richardson, J. S. 1985. *Methods in Enzymology: Macromolecular Crystallography Part B* 115:359–380. Copyright 1985, with permission from Elsevier.)

27.4.2 Synthesis of Peptides

The retrosynthetic analysis (see Chapter 25) of a peptide is easy because polypeptides are chains of amino acids coupled together. Each of the peptide (amide) bonds is disconnected to give a collection of individual amino acids. The synthesis is not trivial, however, because the amino acids must be coupled together in the correct sequence: N- to C-terminus, the primary structure. To do this requires a strategy, which is outlined in Figure 27.5.

Beginning with two amino acids (**53** and **57**; alanine and phenylalanine), the target is a dipeptide. Because the amine unit of one amino acid may react

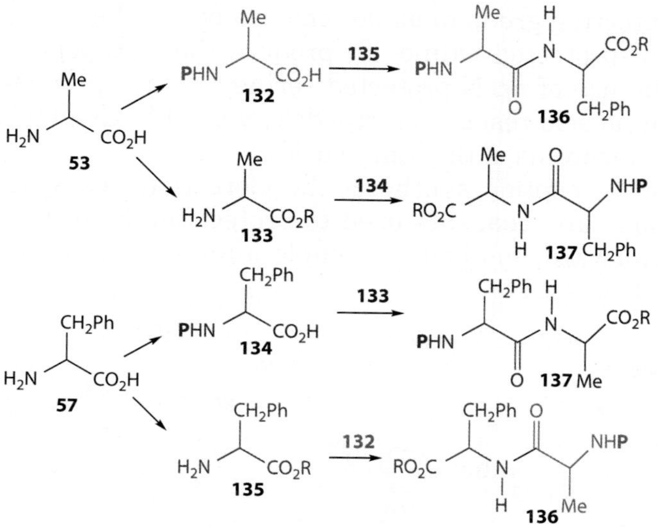

Figure 27.5 Strategy for peptide synthesis.

with the carboxyl group of the other amino acid, the amine group of one amino acid and the acid group of the other must be blocked with an unspecified ***protecting group*** (**P**), in order to form the designated amide bond. Protecting group strategy was introduced in Chapter 25, Section 25.4.3. For each amino acid, there are two possible protected compounds. Protection of the nitrogen in **53** (the protecting group is marked P) leads to **132** and protection of the carboxyl group leads to **133**. Similarly, **57** is converted to **134** or **135**.

If the protected derivatives are coupled together, there are four possible combinations but only two different products: **136** and **137**. Therefore, mixing two amino acids will lead to two distinct dipeptides via reaction of an N-protected amino acid with a C-protected amino acid. Dipeptide **136** is formed by coupling **132** with **135**; **137** is prepared by coupling **133** with **134**; **137** is prepared by coupling **134** with **133**; and **136** is prepared by coupling **135** with **132**. Removal of the protecting groups in each case gives the dipeptide. This approach is straightforward for a dipeptide because there are only two possible structures. Specifying the N-terminus and C-terminus of the dipeptide gives the two synthetic options.

$$\text{PNH-ser-CO}_2\text{H} \xrightarrow{\text{NH}_2\text{-leu-CO}_2\text{P}} \text{PNH-ser-leu-CO}_2\text{P} \longrightarrow \text{PNH-ser-leu-CO}_2\text{H}$$

$$\xrightarrow{\text{NH}_2\text{-val-CO}_2\text{P}} \text{PNH-ser-leu-val-CO}_2\text{P} \longrightarrow \text{PNH-ser-leu-val-CO}_2\text{H}$$

$$\xrightarrow{\text{NH}_2\text{-ala-CO}_2\text{P}} \text{PNH-ser-leu-val-ala-CO}_2\text{P} \longrightarrow \underset{\textbf{138}}{\text{H}_2\text{N-ser-leu-val-ala-CO}_2\text{H}}$$

A tetrapeptide target such as ser-leu-val-ala (**137**) poses a more complex problem. The N-terminus is serine, which requires the use of an N-protected serine residue. The C-terminus is alanine, which requires the use a C-protected alanine residue. To couple leucine and valine, there is a problem. To couple valine to N-protected serine requires a C-protected leucine, but after coupling, the carboxyl protecting group must be removed before the resulting peptide can react with the C-protected valine. To produce the tetrapeptide, further reactions require the use of an N-protected serine and a C-protected alanine. For the "middle" amino acid residues, methodology must be available to protect and deprotect either terminus of the amino acid.

A discussion of peptide synthesis therefore requires a discussion of the chemical reactions and reagents used to protect the N- or C-terminus of any amino acid. Afterward, methods to couple amino acids together to form the peptide will be discussed.

An ester is the most common protecting group for the COOH unit of an amino acid (see reactions that form amides in Chapter 20, Section 20.6). Methyl or ethyl esters are prepared most of the time, but benzyl, *p*-nitrobenzyl, and *tert*-butyl esters are also used. In Section 27.3.3, an ester was prepared by adjusting the pH of an amino acid followed by reaction with HCl and an alcohol. In acidic solution, alanine (**85**) reacts with ethanol (HCl is the acid) to give ethyl alanate, **86**. Methyl, ethyl, and benzyl esters are prepared this way.

The *p*-nitrobenzyl ester of valine is prepared by reacting the silver salt of valine (**139**) with 4-bromomethylnitrobenzene (**140**), and the product is **141**. The *tert*-butyl ester is usually prepared by reacting the carboxylic acid (glycine) with 2-methyl-2-propene (isobutylene) in the presence of an acid catalyst. The alkene reacts with the acid catalyst to give a carbocation (see Chapter 10, Section 10.2), which traps the carboxyl oxygen to form the *tert*-butyl ester, **142**. An alternative method reacts the acid unit of an amino acid (glycine) with *tert*-butylacetate in the presence of an acid catalyst. Transfer of the ester unit via a transesterification reaction (see Chapter 20, Section 20.6.3) gives **142**.

27.36 What type of reaction is the reaction of 139 and 140?

27.37 Write out the mechanism of the transformation for the formation of 142.

27.38 Write out the mechanism of this transformation.

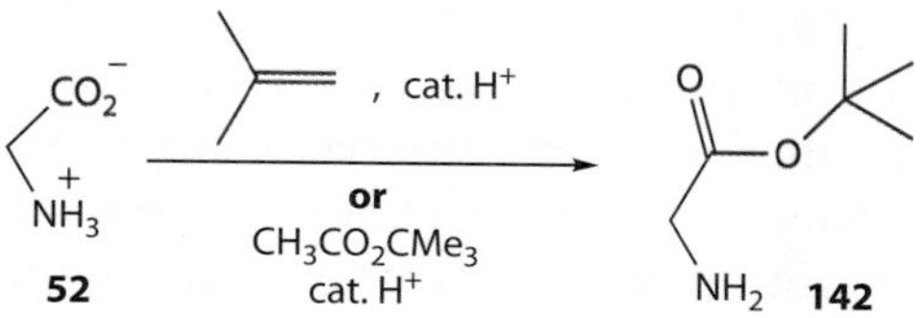

Many groups are available for the protection of the nitrogen unit of an amino acid. Sulfonyl chlorides are commonly used; an example is the reaction of isoleucine with *p*-toluenesulfonyl chloride under basic conditions to give the *N*-sulfonyl derivative. Another *N* protecting group is a tertiary halide called triphenylmethyl chloride (Ph$_3$CCl; called trityl chloride). When it reacts with valine (**55**) under basic conditions, the product is the *N*-trityl amino acid **143**. The mechanism of this reaction is probably a modified S$_N$1, but it will not be discussed here.[9]

27.39 Write out the reaction that gives the *N*-sulfonyl derivative of isoleucine, including starting materials and final product.

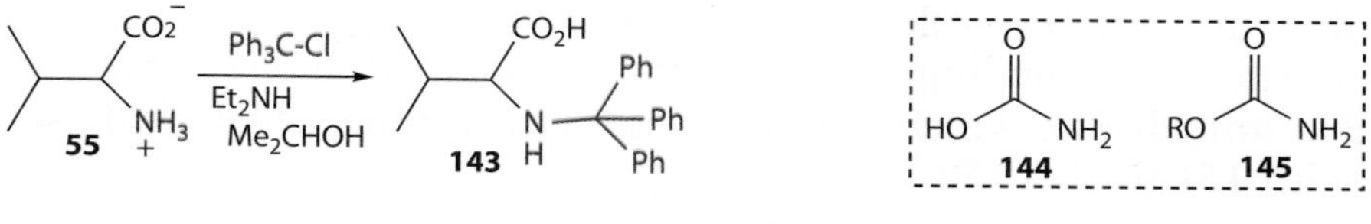

A peptide is composed of amide bonds, so protecting the $-NH_2$ group as an amide (NHCOR) group is not a good choice unless it is a different sort of amide. Therefore, converting an amino unit to its *N*-acetyl or *N*-benzyl derivative is of limited value in the synthesis of peptides. A specialized amide-type protecting group called a **carbamate** has become very popular. Carbamic acid (**144**) is characterized by an OH unit on one side of a carbonyl and a NR_2 group on the other side. When the OH unit is converted to an ester (OR), that compound is a urethane, **145**. The amine unit of an amino acid is converted to a urethane by using several specialized reagents. The two most common are the benzyloxy-carbonyl group (Cbz; benzyl carbamate) and the *tert*-butoxycarbonyl group (Boc or *t*-Boc; *tert*-butyl carbamate). When benzyl alcohol reacts with phosgene (**146**), the product is benzyloxycarbonyl chloride, **147**.

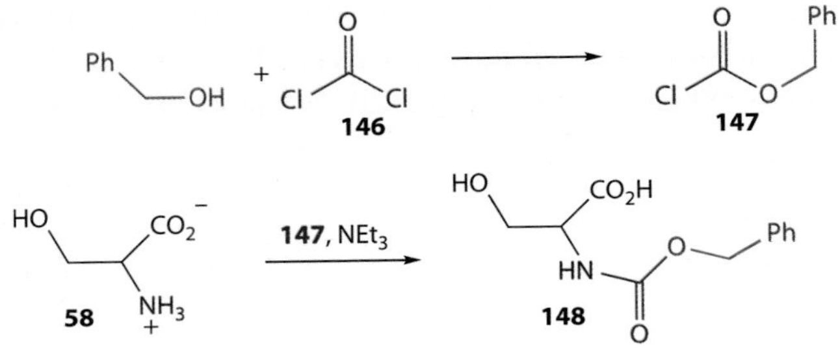

When **147** reacts with serine (**58**), in the presence of an amine base, the product is **148**, which is called Cbz-valine. The Cbz–NH unit is actually a benzyl carbamate, and this is an excellent way to protect the N-terminus of an amino acid. The corresponding *tert*-butoxycarbonyl chloride derivative (**149**) **cannot** be isolated because it is too unstable. Attempts to prepare it from *tert*-butyl alcohol lead to formation of isobutylene (2-methylpropene).

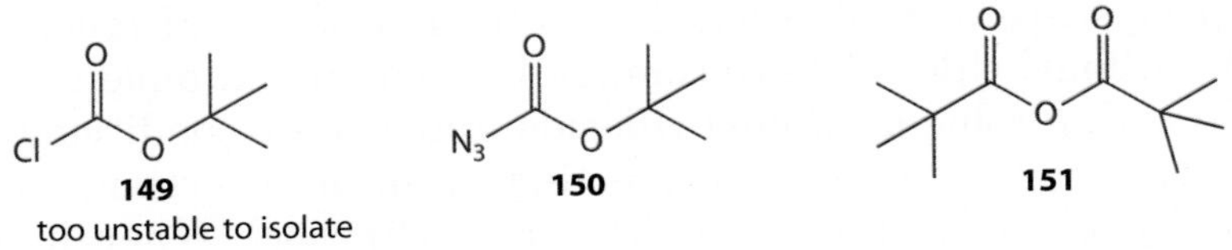

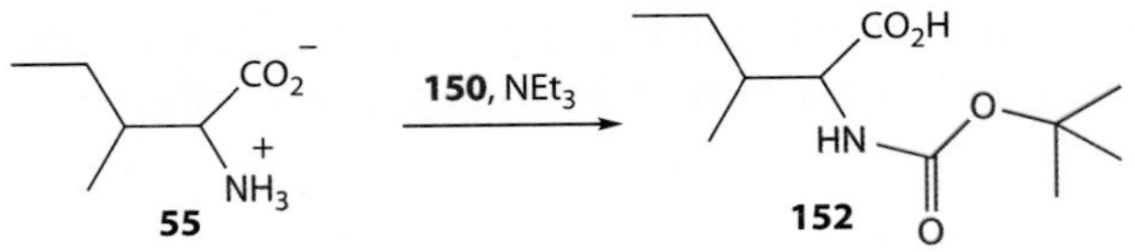

It is possible to prepare *tert*-butoxycarbonyl azide (**150**), however, and this reacts with the amine unit of an amino acid such as isoleucine (**55**) to give the *N*-Boc derivative, *N*-Boc-isoleucine (**152**). A common reagent used to convert an amine to the corresponding *N*-Boc derivative is called di-*tert*-butyl anhydride (**151**; Boc anhydride), which easily reacts with amino acids, and isoleucine reacts with **151** to give **152**.

27.40 Write out a synthesis of *N*-Cbz isoleucine from isoleucine.

If either the nitrogen or the carboxyl unit is protected, there must be a reaction that will **remove** the protecting group (**deprotection**). The ester groups are usually removed by saponification (Chapter 20, Section 20.2), but *p*-nitrobenzyl esters are removed by reaction with HBr in acetic acid, and *tert*-butyl esters are removed by treatment with trifluoroacetic acid. Removal of the nitrogen protecting groups can be more problematic. Cleavage of a trityl group requires reaction with dilute aqueous acetic acid. The Cbz group is removed by catalytic hydrogenation with a palladium catalyst, whereas the Boc group is removed by reaction with trifluoroacetic acid.

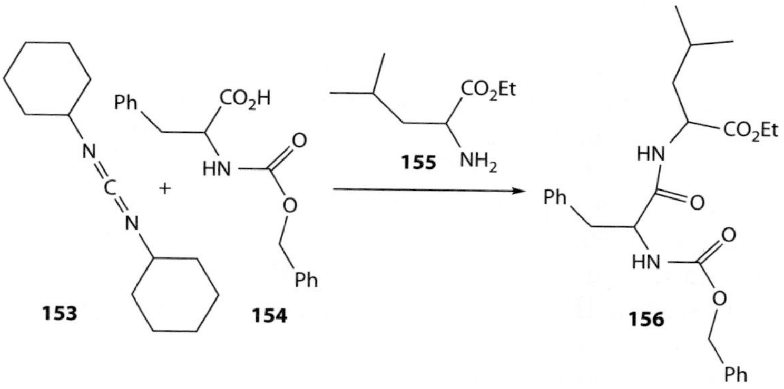

Because selective protection and deprotection of the amine and carboxyl groups of amino acids are possible, the coupling reaction of two amino acids is straightforward. The first peptide coupling reaction, reported in 1902 by Theodor Curtius (Germany; 1857–1928), involved the reaction of an acyl azide with an amino acid. Probably the most common method for making a peptide bond involves coupling an amine unit with a carboxyl unit in the presence of dicyclohexylcarbodiimide (**153**; called DCC). This method was used to prepare amides in Chapter 20 (Section 20.6.4). When *N*-Cbz-phenylalanine (**154**) reacts with ethyl leucinate (**155**) in the presence of DCC and an acid catalyst, the product is the protected dipeptide **155**. Catalytic hydrogenation and saponification yield the dipeptide phe-leu.

27.41 Write out a synthesis of gly-leu using DCC.

An important discovery in the field of peptide synthesis was made by R. Bruce Merrifield (United States; 1921–2006) in 1962. He was able to bind an amino acid to a polymer bead via the C-terminus and subsequently synthesized the peptide bradykinin (arg-pro-pro-gly-phe-ser-pro-phe-arg; a potent hypotensive agent) by doing the chemical reactions, including the protection and deprotection steps, *on a polymer bead*. When completed, the nonapeptide was "released" from the polymer by a simple chemical reaction. This basic approach is now called the **Merrifield synthesis** and is one of the most important advancements in peptide synthesis.

Chapter 10 (Section 10.8.3) introduced the reaction in which styrene is converted to polystyrene. It is also possible to prepare a mixed polymer by polymerizing styrene (**157**) and 4-chloromethylstyrene (**158**) to produce a copolymer (represented as **159**) that has chloromethyl units on many of the benzene rings of the polymer chain.

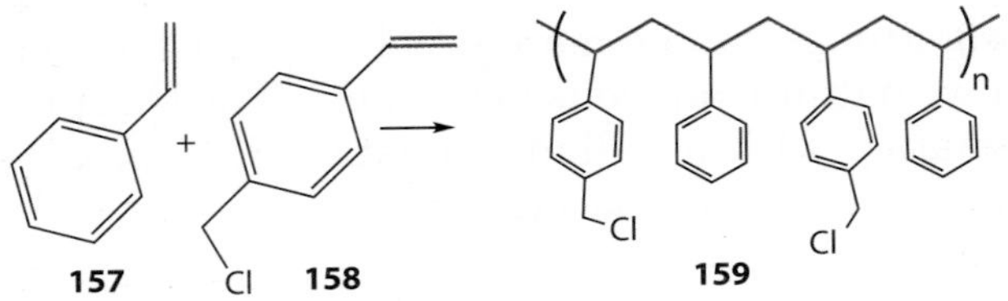

When *N*-Boc-alanine (**160**) reacts with this polymer, **161** is formed (an alanine-loaded polymer). The remainder of the synthesis is essentially a DCC-coupling procedure. When **161** is treated with HCl, the Boc group is removed, and treatment with DCC and *N*-Boc-valine leads to **162**. This sequence of reactions is repeated, using other protected amino acids, until the desired peptide is prepared. In the example at hand, the final product is gly-pro-ser-ala-ala-his-val-leu-val-ala-**P**, **163** (**P** represents the polymer bead). When **163** is treated with HBr and trifluoroacetic acid, the peptide gly-pro-ser-ala-ala-val-leu-val-ala (**164**) is released from the bead and isolated. The polymer is left behind with bromomethyl fragments (Br–CH$_2$–**P**), which can be recycled.

One of the more interesting features of the Merrifield synthesis is that it can be automated. Once the original amino acid is attached to the bead, it can be washed with reagents and then treated with the next amino acid. This is subsequently hydrolyzed and treated with the next amino acid, etc. Indeed, there are peptide synthesis machines based on the Merrifield synthesis.

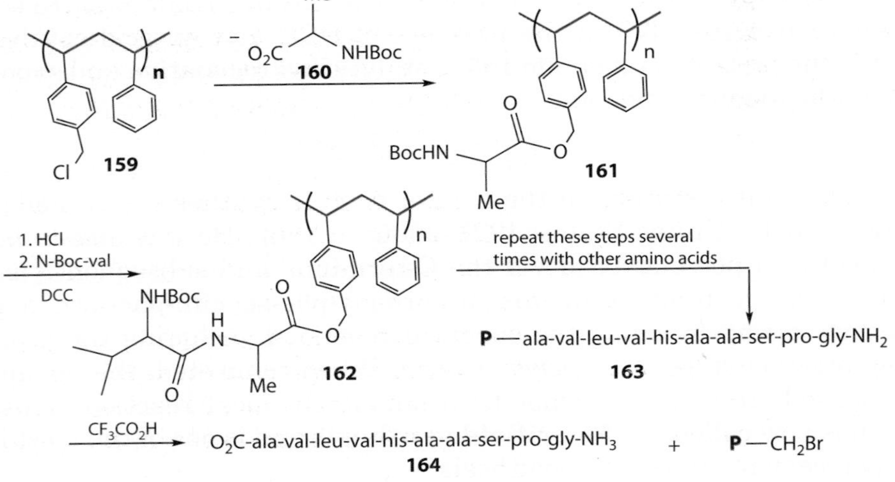

27.5 Biological Relevance. Proteins and Enzymes Are Polypeptides

Proteins constitute one of the most important classes of biomolecules, and they are polymeric organic molecules. Proteins are important for virtually everything connected with cell structure and cell function. Chemically, proteins are polymers of amino acid residues, linked from the amino group to the carboxyl group. In other words, they are large polypeptides. The l-amino acids are the most important contributors to biologically important proteins, although d-amino acids are sometimes incorporated. This section will introduce some important proteins.

A key category of polypeptides is enzymes, which are biological catalysts that accelerate the rate of biological reactions. A few key enzymes will be discussed. This introduction to proteins and enzymes will focus primarily on simple examples and the primary functions of these complex molecules. The other part of this section will concentrate on chemical methods that allow one to identify the structure of proteins.

27.5.1 Important Polypeptides

A proper discussion of proteins and their functions would occupy at least one complete course and is beyond the scope of this chapter. Proteins are critical to life functions. Regulatory proteins control the ability of other proteins to carry out their functions: Insulin regulates glucose metabolism in animals, and somatotropin and thyrotropin stimulate the thyroid gland. Hemoglobin is an example of a transport protein, which transports specific substances from one place to another. Specifically, hemoglobin transports oxygen from the lungs to tissues. Some nutrients are stored in special proteins called storage proteins. Examples are casein, the most abundant protein found in milk, and ferritin, which stores iron in animal tissues. The very structure of cells and tissues is due to structural proteins such as the α-keratins, collagen, elastin, and fibroin. The α-keratins make up hair, horn, and fingernails. Collagen is found in bone, tendons, and cartilage; elastin is a component of ligaments. Fibroin is a major component of the silk used to make cocoons and of spider webs.

Some proteins are important for cell protection. The immunoglobulins (antibodies) are produced by lymphocytes and defend the body against bacteria, viruses, etc. Thrombin and fibrinogen are important blood-clotting proteins. Glycoproteins contain carbohydrates (see Chapter 28) and lipoproteins contain lipids, and nucleoproteins that are important for storage and transmission of genetic information. Proteins are critical to the life process in plants, insects, and animals; as a group, they may well be the most important organic chemicals known.

A description of all of these proteins (and the many others that are important) is beyond the scope of this book and is best left to a course in biochemistry. All of these proteins have a primary structure that is a polypeptide composed of l-amino acids, primarily those appearing in Table 27.1 (Section 27.3.2). When the peptides in Section 27.4 are prepared, the structure of those peptides is pre-dictable because the process used to form the peptides is under control. When a protein is isolated from an animal or a plant, how is it possible to determine which amino acids are present, and in what order those amino acids are con-nected in the polypeptide? This problem is addressed in the following section.

27.5.2 Peptide Identification

The process of identifying the chemical structure of a protein (or any other pep-tide) usually begins by heating it in 6N HCl at 105°C for 24 hours. This reaction should completely hydrolyze the peptide into its constituent amino acids, which are then separated. This approach has some problems. Tryptophan (**68**) has an indole unit, which is acid sensitive, and it is partially destroyed by harsh acidic conditions. Glutamine (**64**), asparagine (**63**), glutamic acid (**66**), and aspartic acid (**65**) decompose with loss of ammonia when heated in 6N HCl. When the amino acid residue has a sterically hindered side chain, as in valine (**54**) or isoleucine (**56**), hydrolysis may be incomplete and heating for a longer period of time may be necessary. These problems mean that after the 24-hour hydrolysis, 100% of the amino acid residues in the original peptide or protein may not be available.

Once all of the amino acids have been obtained, however, several techniques can be used to give structural information. The first method must identify each individual amino acid in order to calculate a percentage of each residue in the protein. For example, if the protein consists of 21% alanine (**53**) and 5% methi-onine (**62**), this knowledge is important for determining the primary structure. Chromatography columns will separate individual amino acids; this makes the identification of each constituent amino acid possible. As these separated amino acids are isolated, one method for their identification relies on heating them to 100°C in the presence of ninhydrin. Ninhydrin can exist as the hydrate (**165**) or the tri-ketone (**166**) as the amount of water in the medium is changed. (See Chapter 18, Section 18.6, for a discussion of hydrates.) When an amino acid such as serine (**58**) reacts with ninhydrin, the initial reaction between a ketone unit and the amine unit produces imine **167** along with ammonia. Under the reaction conditions, decarboxylation occurs to give **168**, which reacts with a second molecule of ninhydrin to give **169**.

This compound is produced as a product from the reaction of *ninhydrin, which reacts with all amino acids except proline.* Imine **169** has a characteristic **purple** color (λ_{max} = 570 nm; also see Chapter 23, Section 23.3) that is easily detected. This reaction is used to detect the presence of an amino acid as it is isolated from the hydrolysis mixture, and the intensity of the color due to **169** is a measure of the relative amount of that acid in the total mixture. As mentioned, proline (**73**) reacts with ninhydrin to give a different compound (**170**) that has a distinctive **yellow** color (λ_{max} = 440 nm).

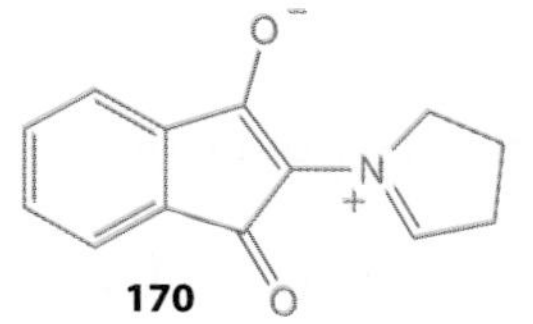

170

The information gathered from this process allows one to identify the relative number of amino acids in a protein. An example subjects a protein to acid hydrolysis. One experiment shows a total of 19 amino acid residues for a nonadecapeptide structure after acid hydrolysis. There are 3 lys, 4 gly, 2 val, 3 ile, and 2 phe, along with one each of arg, his, trp, ser, and met. The order in which these amino acid residues are connected must now be determined.

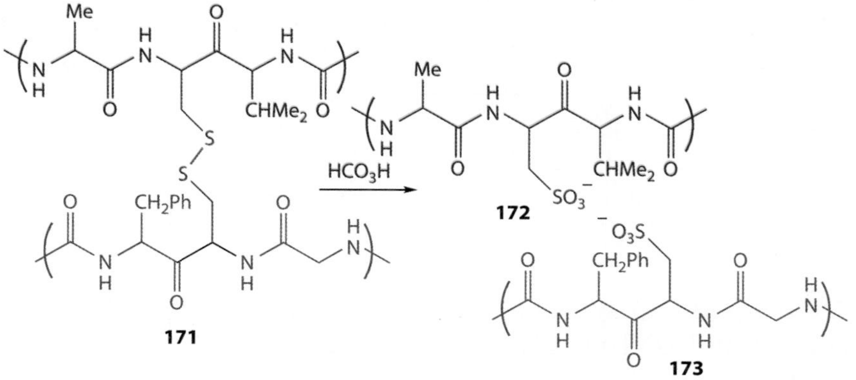

171 **172** **173**

A general strategy for identifying the sequence of amino acids is first to identify the N-terminal and C-terminal amino acid residues. The protein is then treated with specialized reagents that cleave the peptide chain at known amino acid residues, requiring a test that will identify the terminus of each fragment. With thought and care, the various fragments constitute a puzzle whose solution is the amino acid sequence for the peptide. To begin the process, focus on a specific amino acid residue. When two cysteine residues (**61**) are in close proximity, they can exist as two different strands or in one strand that has coiled around in a manner that brings them together. In either case, they react to form a disulfide bond (see **171**). These bonds may be formed intramolecularly to create coils of the tertiary protein structure, or they may occur intermolecularly to help bind two peptides together in a β-sheet type structure.

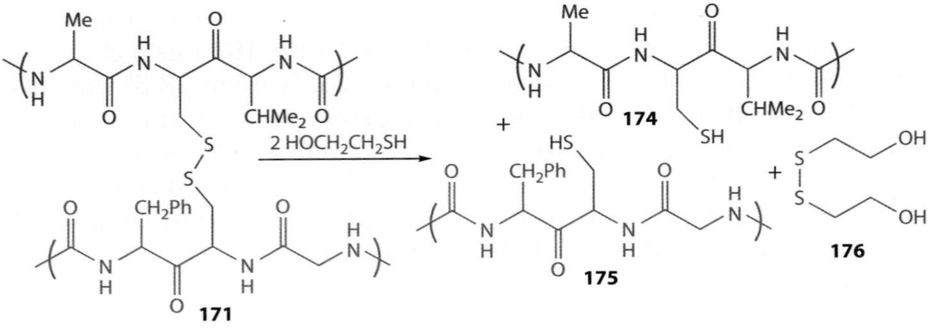

Disulfide bonds are cleaved if the peptide is treated with peroxyformic acid (see Chapter 10, Section 10.5); in the case of **171**, the products are cysteic acid residues in **172** and **173**. Alternatively, **171** reacts with 2-mercaptoethanol to give two free cysteine units in **174** and **175**, as well as the disulfide, **176**. To prevent the cysteine residues from recombining to form a new disulfide linkage, **174** and **175** are quickly treated with **iodoacetic acid** to give **177** and **178**. Once the disulfide bridges (if any) are removed, the process of identifying the termini of the peptide can begin; this is called *end group analysis*.

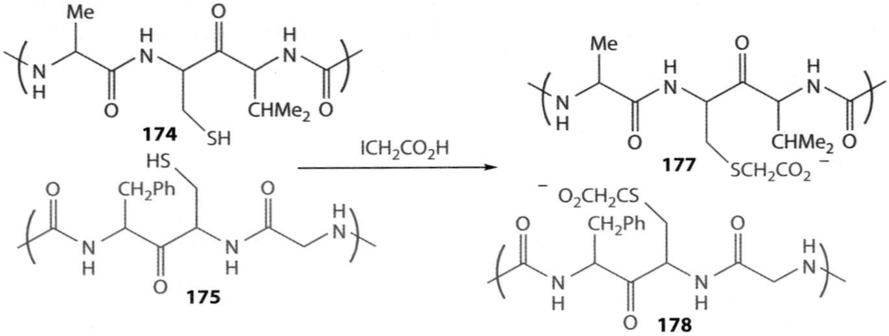

The N-terminal amino acid of a peptide will react with **1-fluoro-2,4-dinitrobenzene** (**180**; FDNB) to give an *N*-aryl derivative. Reagent **180** is known as *Sanger's reagent,* after Frederick Sanger (England; 1918–). If a peptide terminates in an alanine residue (see **179**), it reacts with **180** in aqueous ethanol that is buffered with sodium bicarbonate to give **181**. When this *N*-aryl peptide is hydrolyzed with 6N HCl, the individual amino acids are released, and one (and only one) has the FDNB group attached—the N-terminal amino acid. In this case, complete hydrolysis of peptide **181** releases free amino acids (including the valine residue shown) and a single molecule of *N*-(2,4-dinitrophenyl) alanine (**182**). This compound is **yellow** and is easily separated and identified. There are some problems, however, because FDNB reacts with the amine unit on the side chain of lysine (**67**), the imidazole nitrogen on the side chain of histidine (**69**), and the sulfur unit on the side chain of cysteine (**61**).

27.42 Draw the product formed when valine ethyl ester is treated with Sanger's reagent.

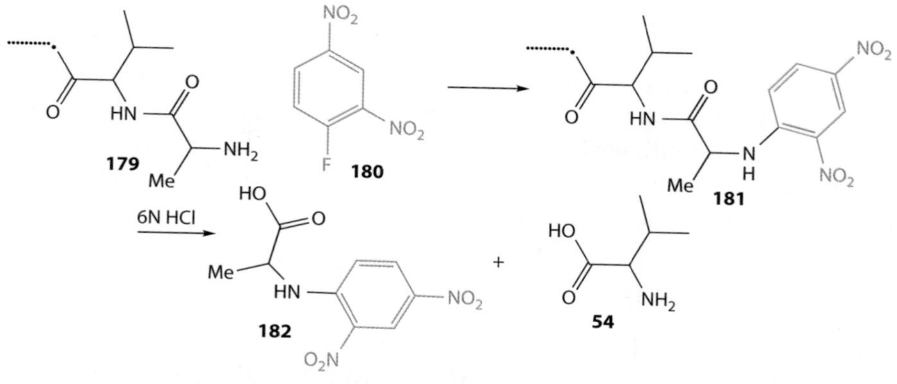

Another method for determining the identity of the N-terminal amino acid is to react the peptide with **dimethylaminonaphthalenesulfonyl chloride (183**, known as **dansyl chloride**). When **183** reacts with a peptide such as **179**, the amine unit reacts with the sulfonyl chloride to give the sulfonamide **184** (see Chapter 20, Section 20.11.3, for an introduction to sulfonamides). Acid hydrolysis of this *N*-dansyl peptide leads to release of the amino acids, with the N-terminal amino acid tagged with the dansyl group, in this case *N*-dansyl alanine, **185**. It is easily isolated and the presence of the dansyl group makes **185** highly *fluorescent*. This means that lower concentrations can be detected than can be detected by using Sanger's reagent.

27.43 Draw the product formed when dansyl chloride reacts with serine *tert*-butyl ester.

Another N-terminus identification technique is more versatile than Sanger's reagent or dansyl chloride. The problem with both **180** and **183** is that the peptide must be destroyed by hydrolysis after tagging the N-terminal amino acid residue, so only the amino acid at the N-terminus position can be located. Another reagent identifies the N-terminus, but it also allows sequencing of the remainder of the peptide. This reagent is **phenyl isothiocyanate (186)**, which is known as **Edman's reagent**, after Pehr Victor Edman (Sweden; 1916–1977). This method of identification is known as the **Edman degradation**.[10]

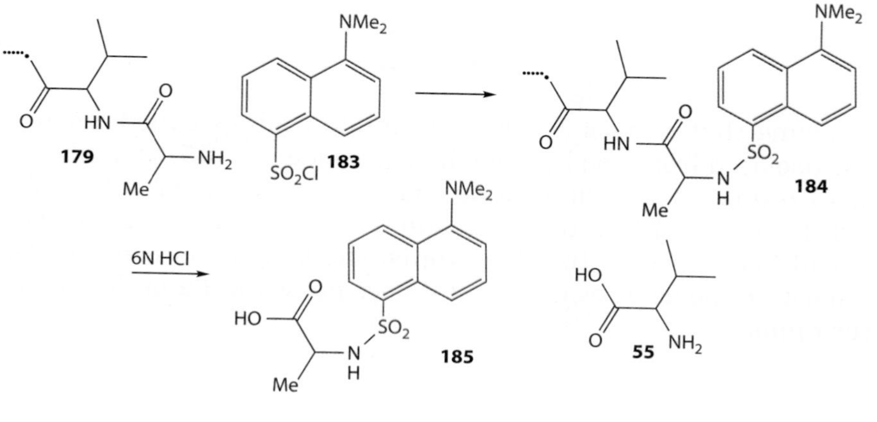

When peptide **179** reacts with **186**, the isothiocyanate reacts with the amide unit (at pH 8–9) of the alanine residue to give **187**. This is not isolated, however, because the sulfur atom attracts the amide carbonyl to release the phenylthiohydantoin derivative of alanine (**188**) and the original peptide minus the alanine residue (see **189**), which is a new N-terminus. The phenylthiohydantoin is soluble in organic solvents, so it is easily removed from the peptide and can be identified using various techniques.

27.44 Draw the product formed when phenylalanine ethyl ester reacts with 185 and is then treated with trifluoroacetic acid.

The remaining peptide has another amino unit, and it is treated with more phenyl isothiocyanate and then cleaved to examine and identify the next amino acid. If a pentapeptide (ala-val-ser-leu-ile) is subjected to the sequential Edman degradations by treatment with Edman's reagent and loss of the phenylthiohydantoin, each amino acid can be identified, in order. Because the process begins at the N-terminus and progresses toward the C-terminus, both the identity of the amino acids and their exact sequence in the peptide are known. This method can be used to identify from 30 to 60 amino acid residues in a long peptide under the right conditions.

27.45 Draw the phenylthiohydantoin formed when 189 is treated with 185 and then trifluoroacetic acid.

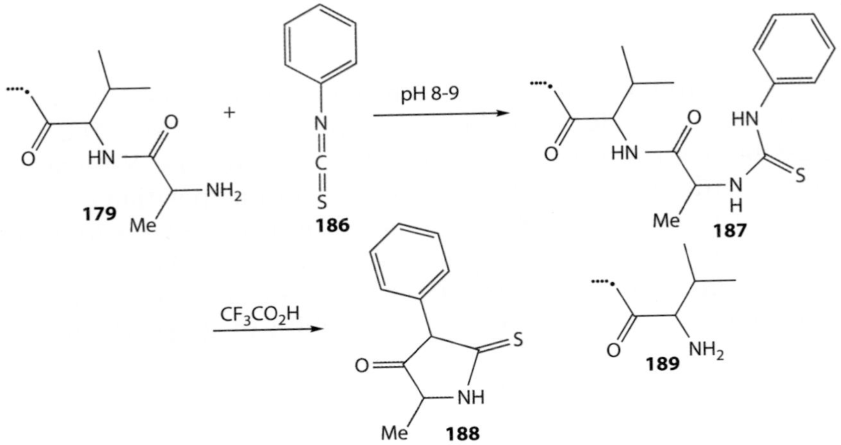

An enzyme that cleaves N-terminal amino acids is called leucine aminopeptidase (isolated from hog kidney). It cleaves leucine (**55**) and other *nonpolar amino acid residues* from the peptide chain, allowing them to be isolated and identified. If proline (**73**) is the N-terminal amino acid residue, however, this enzyme will *not* cleave it. In other words, the enzyme is used to cleave and identify most amino acid residues, but it does *not* work if a proline residue is at the N-terminus.

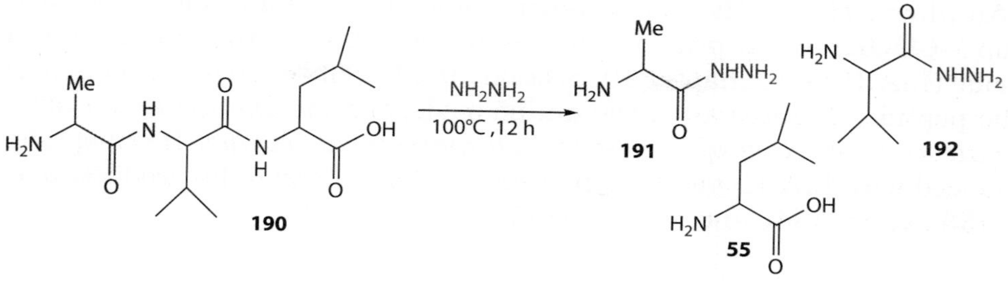

Fewer methods are available to identify the C-terminal amino acid residue of a peptide. When the peptide is heated to 100°C (for about 12 hours) with hydrazine (NH$_2$NH$_2$), the amide bond of each residue is attacked and cleaved. The products are *amino hydrazides*. If tripeptide ala-val-leu (**190**) is heated with hydrazine, there are two amide bonds and the products are alanine hydrazide (**191**), valine hydrazide (**192**), and leucine (**55**). The C-terminal amino acid is *not* converted to the hydrazide because hydrazine reacts with the amide rather than with the free carboxyl group.

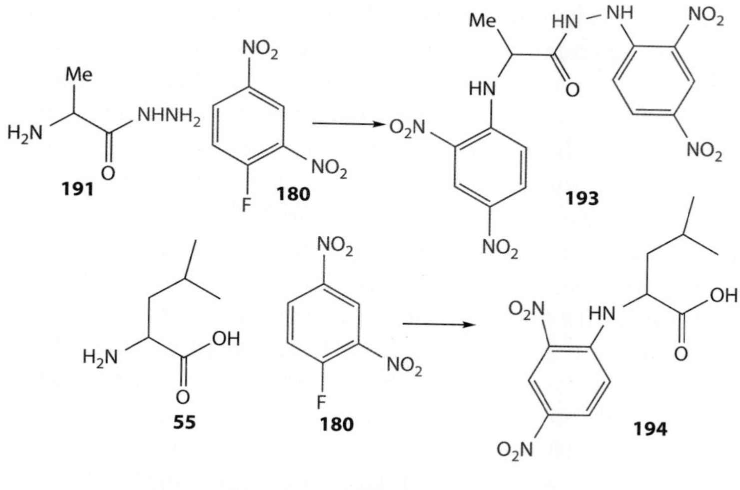

By this analysis, leucine was identified as the C-terminal amino acid residue. This approach can be applied to large peptides, but the peptide is destroyed, and it is difficult in some cases to separate one amino acid from many amino acid hydrazides. To overcome this problem, the mixture of hydrazides and amino acid is treated with Sanger's reagent (**180**), which converts the hydrazide compounds to a *bis*(dinitrophenyl) derivative. In the case of **191**, treatment with Sanger's reagent gives **193**. The free amino acid is also converted to its FDNB derivative (**194** from leucine). The *bis*(2,4-dinitrophenyl) derivatives such as **193** are soluble in organic solvents but insoluble in aqueous bicarbonate; **194** is soluble in aqueous bicarbonate because it has the carboxyl group. Amino acid **194** is easily removed from the mixture, making identification relatively easy.

An alternative method for C-terminus identification reduces the carboxyl group (–COOH) of a peptide to an alcohol (–CH$_2$OH) with lithium aluminum hydride (LiAlH$_4$; see Chapter 19, Section 19.2.1). Subsequently, acid hydrolysis of the peptide liberates all of the amino acids, *but one amino alcohol was also generated by reduction of the C-terminal amino acid residue.* If tripeptide **190** is reduced with LiAlH$_4$ and then hydrolyzed, for example, the products are alanine (**53**), valine (**54**), and leucinol (**195**).

An enzymatic procedure can be used to identify the C-terminal amino acid residue. Four common enzymes cleave the C-terminal amino acid residue from a peptide. They are carboxypeptidase A (from bovine pancreas), carboxypeptidase B (from hog pancreas), carboxypeptidase C (from citrus leaves), and carboxypeptidase Y (from yeast). *Carboxypeptidase A cleaves all C-terminal amino acid residues except proline, arginine, and lysine. Carboxypeptidase B cleaves only an arginine or a lysine. Carboxypeptidase C cleaves all amino acid residues from the C-terminus, as does carboxypeptidase Y.* In principle, a combination of these enzymes can be used to gain information about the C-terminus.

The next step in the analysis is to cleave the polypeptide selectively to produce small fragments that can be analyzed. This is important because the Edman degradation, for example, cannot be used to give the linear sequence of a polypeptide containing several hundred amino acid residues. It can be used for 30–60 residues, however. Fragmentation of a large peptide into pieces no larger than 30–60 amino acid residues allows each fragment to be sequenced by the Edman degradation. For the most part, these selective cleavages are done with enzymes, although a handful of chemicals induce cleavage as well.

The enzyme trypsin (which is a digestive enzyme) cleaves peptide bonds, but only when the amino acid residue has a carbonyl unit that is part of arginine (70) or lysine (67). In other words, it cleaves specifically on the C-side of arginine or lysine of a peptide to give fragment peptides that have an arginine or lysine at the C-terminus. *Chymotrypsin cleaves amino acid residues that have a carbonyl that is part of an aromatic amino acid such as phenylalanine (57), tyrosine (60), or tryptophan (68).* This enzyme produces fragment peptides that have one of these three amino acids at the C-terminus. Because chymotrypsin can cleave many other amino acids if given enough time to react, care must be exercised when using this particular analysis. *Staphylococcal protease cleaves acidic amino acid residues such as aspartic acid (65) and glutamic acid (66).*

This enzyme produces fragment peptides that have aspartic acid or glutamic acid at the C-terminus.

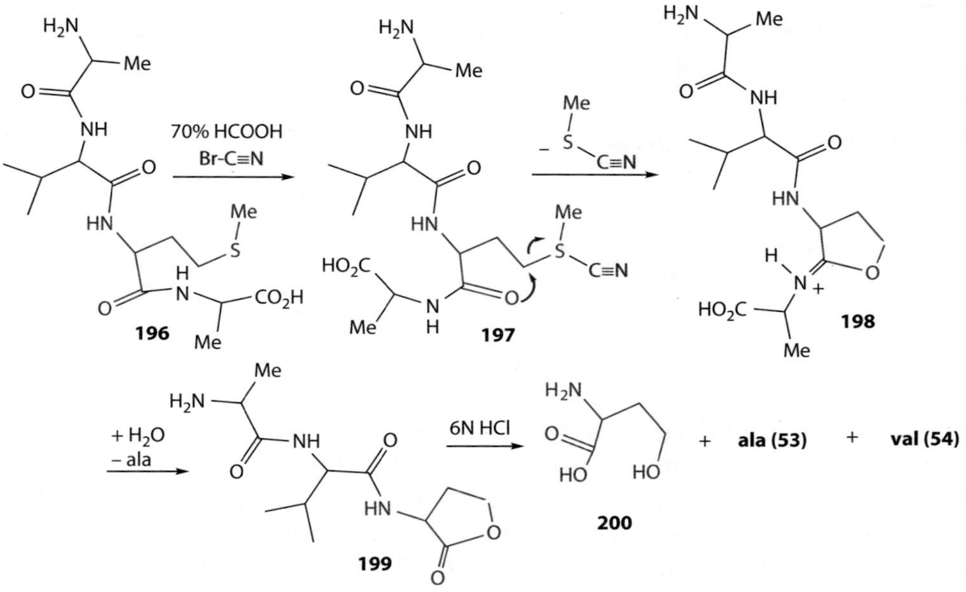

A chemical method for fragmenting peptides uses **cyanogen bromide** (BrC≡N), and it reacts specifically with methionine residues. In the tetrapeptide ala-val-met-ala (**196**), cyanogen bromide reacts with the sulfur atom to produce an *S*-cyano complex, **197**. Loss of methyl thiocyanate (CH₃S–C≡N) leads to formation of **198**, and cleavage of the iminium unit with water leads to loss of an alanine residue. The result is formation of a peptide fragment (**199**) that terminates with a residue containing a **lactone**. Hydrolysis leads to release of the alanine and valine residues along with a homoserine unit (**200**), which is easily identified. *The term "homoserine" refers to the fact that **200** has one more carbon atom in its side chain than does serine (**58**); serine has a –CH₂OH unit and homoserine has a –CH₂CH₂OH unit.*

These fragmentation techniques are used to identify the primary structure of a peptide. As an example, ala-ala-val-met-phe-ile-glu-gly-ser-val-ile-leu-asp-lys-ala-ala-trp-met-gly-val-lys (**201**) is subjected to end group analysis to reveal the presence of ala at the N-terminus and lys at the C-terminus. When **201** is treated with trypsin, cleavage at the carbonyl of arg or lys leads to two fragments ending in lys (no arg residues) at the C-terminus. End group analysis shows that one fragment ends in ala and the other terminates in ala, so the fragments are **202** and **203**. Treatment of **201** with chymotrypsin leads to carbonyl cleavage of phe, trp, or tyr. This gives three fragments, one with a C-terminus of phe, one with a trp C-terminus, and one with a lys C-terminus. End group analysis shows these three fragments are **204**, **205**, and **206**.

Application of the Edman degradation to **204–206** indicates that **202** is ala-ala-val-met-phe-ile-glu-gly-ser-val-ile-leu-asp-lys, and **203** is ala-ala-trp-met-gly-val-lys. Because **201** has an ala N-terminus and a lys C-terminus, as do **202** and **203**, the C-terminal lys of **202** together with the N-terminal ala of **203** yields **201**. After chymotrypsin fragmentation, Edman degradation shows that **204** is ala-ala-val-met-phe, **205** is ile-glu-gly-ser-val-ile-leu-asp-lys-ala-ala-trp, and **206** is met-gly-val-lys. Because **201** has ala for the N-terminus, **205** must contain the N-terminus. Because **201** has a lys C-terminus, **206** must contain the C-terminus. This analysis puts **204** in the middle to give the complete sequence.

This rather simplistic analysis is straightforward, and just one analysis coupled with an Edman degradation gave the sequence. In fact, **201** is sufficiently small that an Edman degradation may be done directly. A more challenging problem arises when the answer is unknown before the analysis. For an unknown peptide, hydrolysis with 6N HCl and analysis of the amino acid residues reveals a total of 15 amino acids with the following distribution: 1 ala, 1 asp, 1 gly, 2 ile, 1 leu, 1 met, 2 phe, 1 pro, 2 ser, 1 trp, 2 val. End group analysis shows that pro is the N-terminus and that leu is the C-terminus. Digestion of the unknown with trypsin is now useless because it is known that there are no arg or lys residues in the peptide. Digestion with chymotrypsin leads to four fragments. Identification of the first reveals it to be pro-ile-ile-ser-val-met-trp (A), and the second is ala-asp-phe (B). The third fragment is a single amino acid residue, phe. The fourth fragment is ser-gly-val-leu (C).

Because the unknown has a pro N-terminus, the first fragment must be A. Likewise, the C-terminus is leu, which means that this fragment is C. This leaves two possibilities: A-B-phe-C or A-phe-B-C. Chymotrypsin is known to cleave aromatic amino acids; therefore, the only way to obtain phe as a single amino acid is if phe is attached to another phe, a tyr, or a trp. Because A has a trp terminus and B has a phe terminus, a decision is not possible. When the unknown is treated with cyanogen bromide, two fragments are obtained, and Edman degradation identifies them as pro-ile-ile-ser-val-homo-ser (D) and trp-ala-asp-phe-phe-ser-gly-val-leu (E). The homoserine residue must be derived from methionine (met). Because pro is the N-terminus of the unknown and leu is the C-terminus, the unknown must be D or E.

Is there a trp-phe sequence or a phe-phe sequence in either of the cyanogen bromide fragments? There is a phe-phe sequence in E and that fragment also contains trp as its N-terminus. In addition, A is known to contain trp as its C-terminus and B contains ala as its N-terminus. From fragment E, the sequence from chymotrypsin must be A-B, which makes it A-B-phe-C. *The final sequence is pro-ile-ile-ser-val-met-trp-ala-asp-phe-phe-ser-gly-val-leu.*

　Using such procedures to sequence a long and complex protein is obviously much more difficult. These basic techniques, especially when all of the cleavage procedures are used and correlated, lead to identification of the primary structure (the amino acid sequence) of the protein. Several hundred proteins have been sequenced, and other proteins are constantly being analyzed. *Note that modern technology includes automated protein sequencers to identify the amino acid residues in the protein sequences by tagging and removing one amino acid at a time. Each is then analyzed and identified. Modern NMR techniques also provide a powerful tool for poly(peptide) and protein structure determination.* Modern NMR is probably used more often than many of the chemical techniques described here. However, the chemical techniques remain a powerful tool in protein research. Indeed, several of the reagents noted in this section are used in automated peptide sequencers.

27.6 New Synthetic Methodology

There are several functional group transformations in this chapter. Amines have been prepared from halides, nitriles, amides, or ketones/aldehydes, as shown in the diagram.

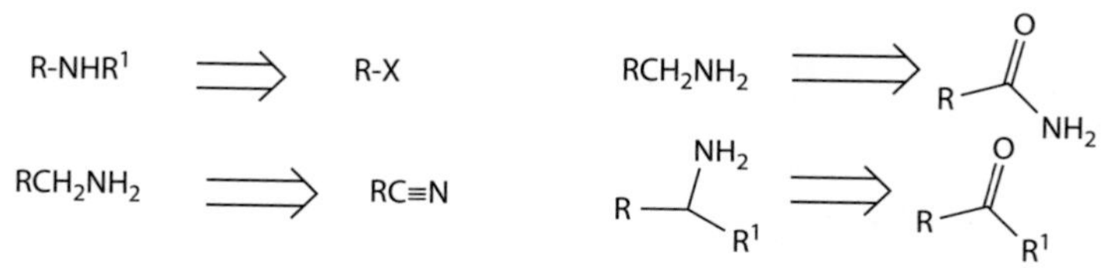

　Amines are also prepared from amino acids via decarboxylation. Halo-acids can be prepared from amino acids, as shown.

　Several routes will prepare an amino acid. Carboxylic acids are precursors to amino acids, as are aldehydes. Malonic acid derivatives may also be used.

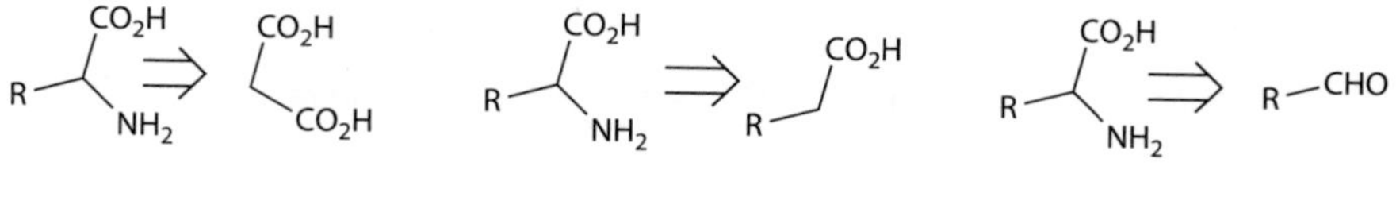

References

1. Fieser, L. F., and Fieser, M. 1961. *Advanced organic chemistry,* 1014–1019. New York: Reinhold Publishers.
2. Lide, D. R., ed. 1995. *Handbook of chemistry and physics,* 76th ed., 7-1. Boca Raton, FL: CRC Press.
3. See Figure 6 in Pluth, N. D., Bergman, R. G., and Raymond, K. N. 2008. *Journal of Organic Chemistry* 73:7132.
4. Smith, C. K., and Regan, L. 1997. *Accounts of Chemical Research* 30:153.
5. Richardson, J. S. 1981. *Advances in Protein Chemistry* 34:167–339.
6. Richardson, J. S. 1985. *Methods in Enzymology: Macromolecular Crystallography Part B* 115:359–380.
7. Richardson, J. S. 2000. *Nature Structural and Molecular Biology* 7:624.
8. Richardson, J. S. 1985. *Methods in Enzymology: Macromolecular Crystallography Part B* 115:359–380, Figure 13, p. 374.
9. Swain, C. G. 1948. *Journal of the American Chemical Society* 70:1119–1128.
10. Edman, P. 1950. *Acta Chemica Scandinavica* 4:283.

Answers to Problems

27.1

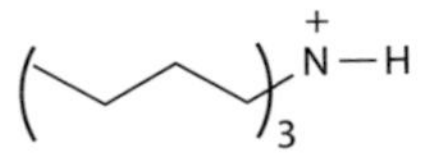

tributylammonium bromide

27.2

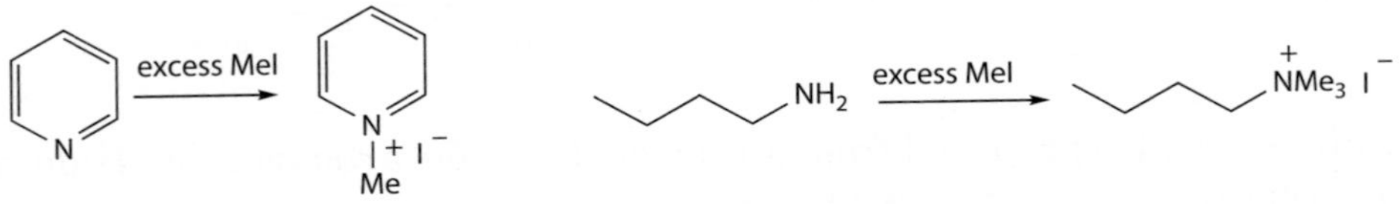

27.3

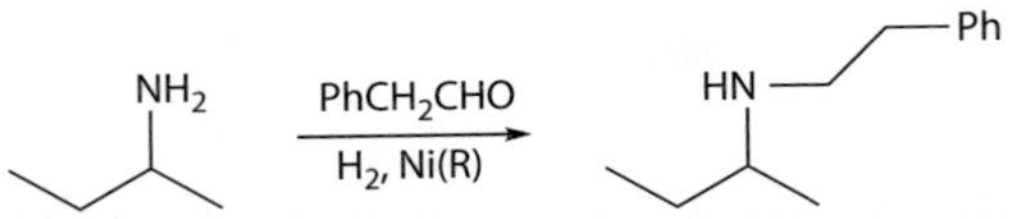

27.4

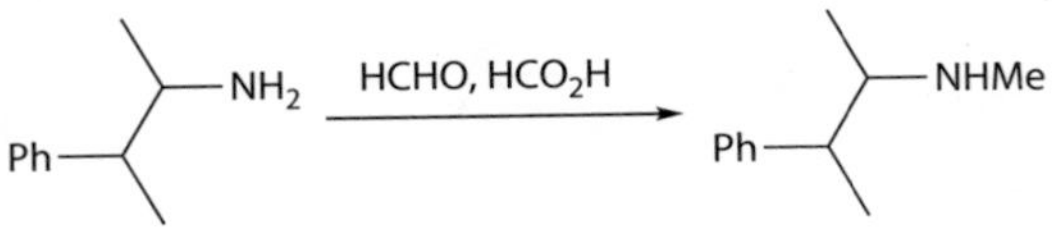

27.5

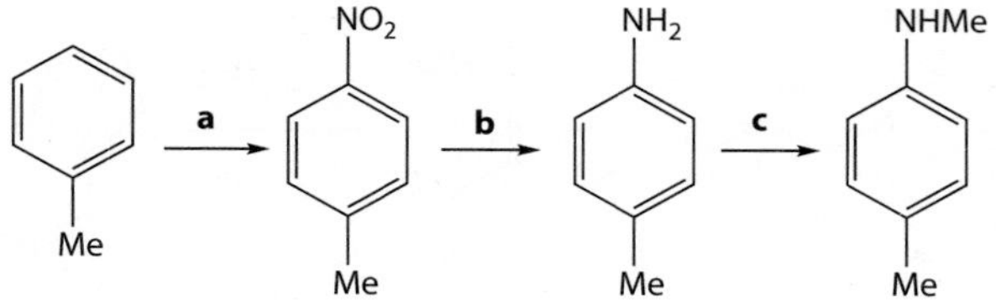

(a) NHO$_3$/H$_2$SO$_4$; separate out ortho-isomer (b) H$_2$, Pd-C
(c) HCHO, HCO$_2$H

27.6

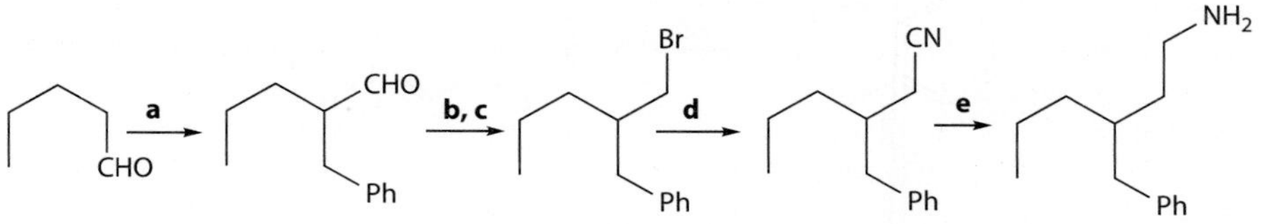

(a) 1. LDA 2. PhCH$_2$Br (b) NaBH$_4$ (c) PBr$_3$ (d) NaCN, DMF (e) LiAlH$_4$

27.7

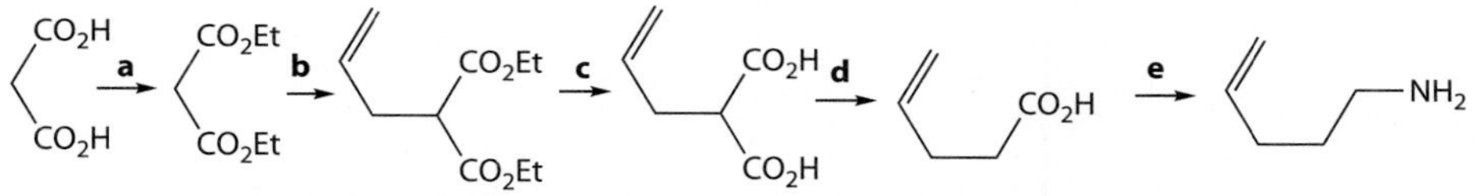

(a) SOCl$_2$; EtOH   (b) NaH; allyl bromide (c) saponify (d)200°C (–CO$_2$) (e) 1. SOCl$_2$ 2. NH$_3$ 3. LiAlH$_4$

27.8

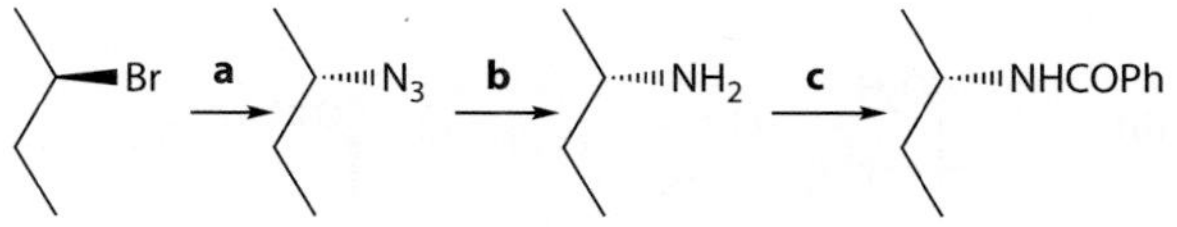

(a) NaN$_3$, DMF (b) LiAlH$_4$
(c) benzoyl chloride

27.9

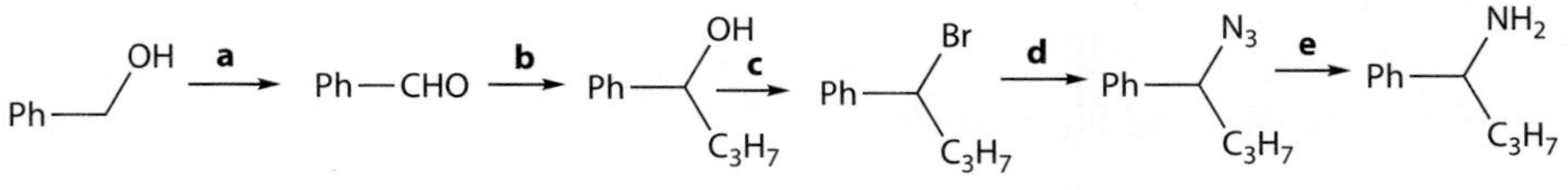

(a) PCC (b) C$_3$H$_7$MgBr; H$_3$O$^+$ (c) PBr$_3$ (d) NaN$_3$, DMF (e) LiAlH$_4$

27.10

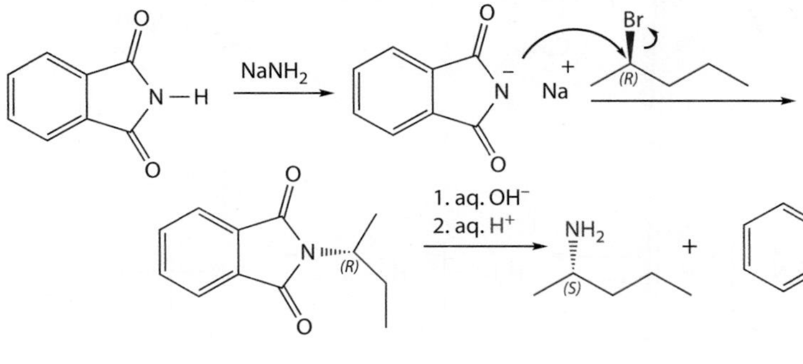

27.11

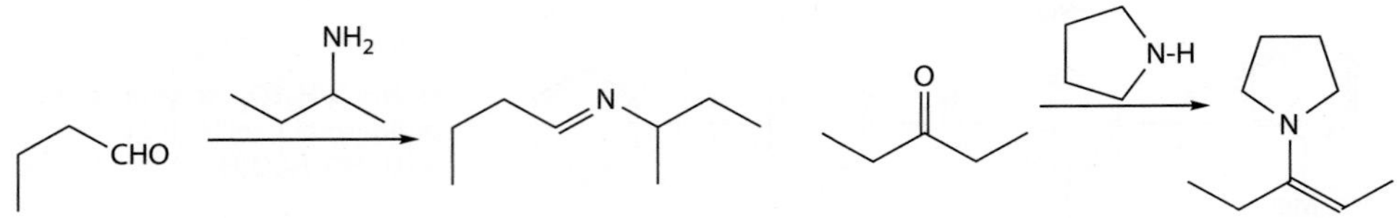

27.12

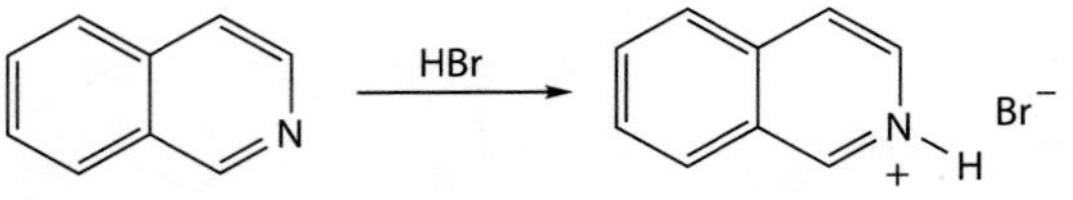

27.13

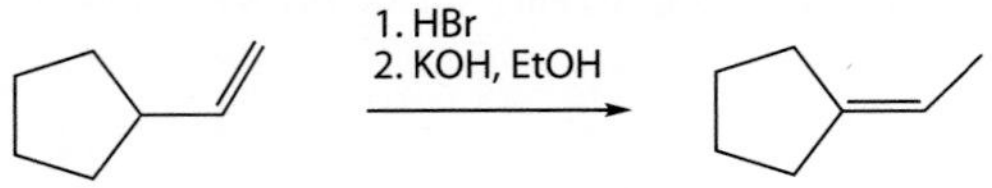

27.14

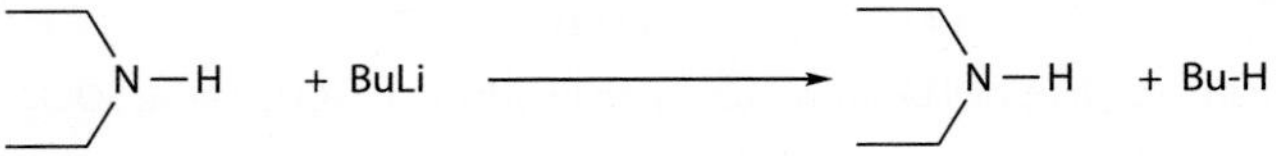

27.15

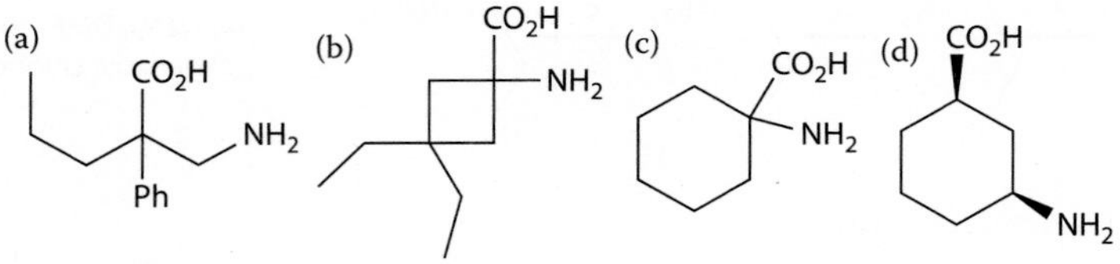

27.16 $pK_1 = -\log K_1$, and $K_1 = 10^{-pK_1}$

27.17

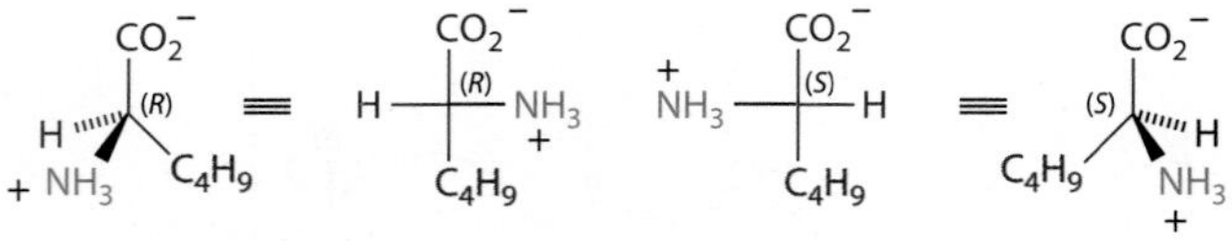

27.18

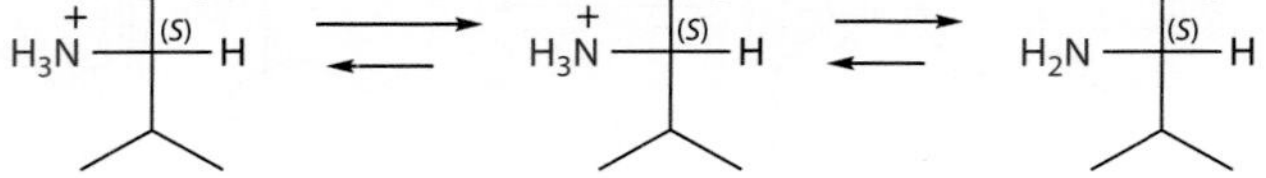

27.19

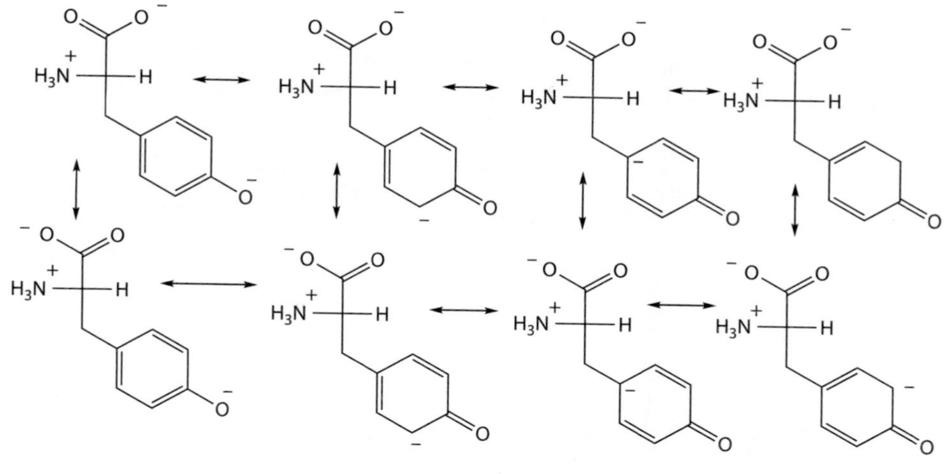

27.20

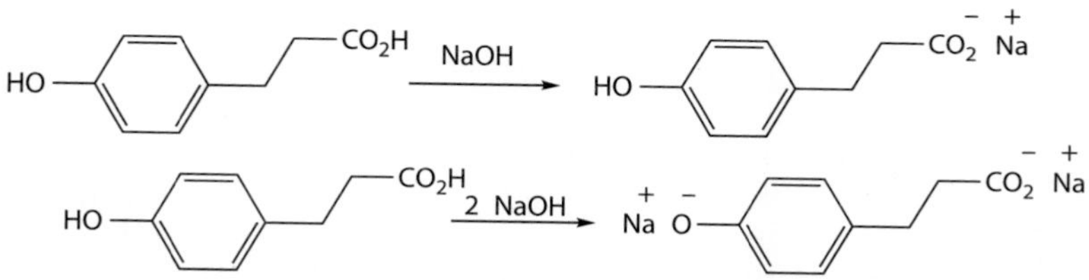

27.21

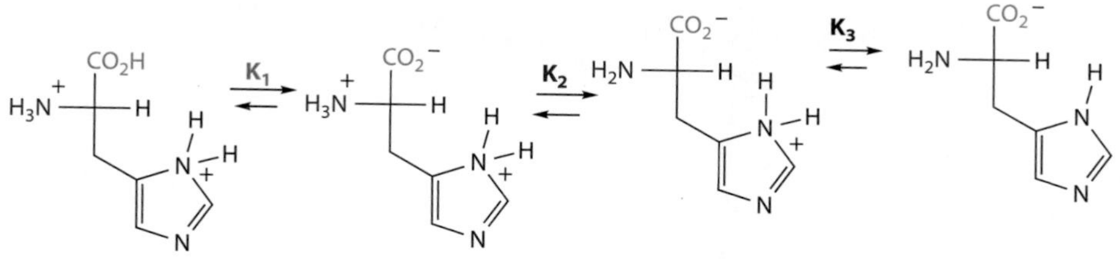

27.22

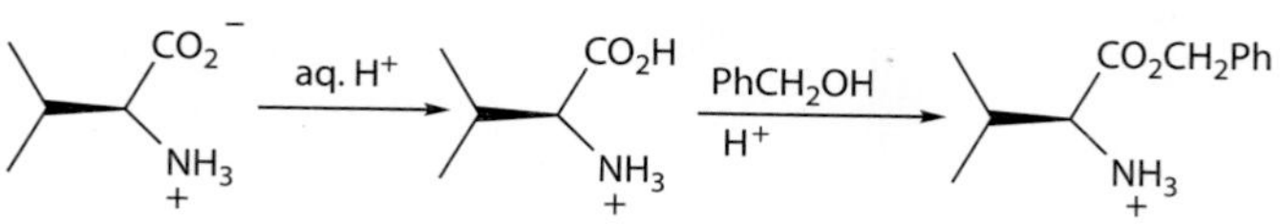

27.23

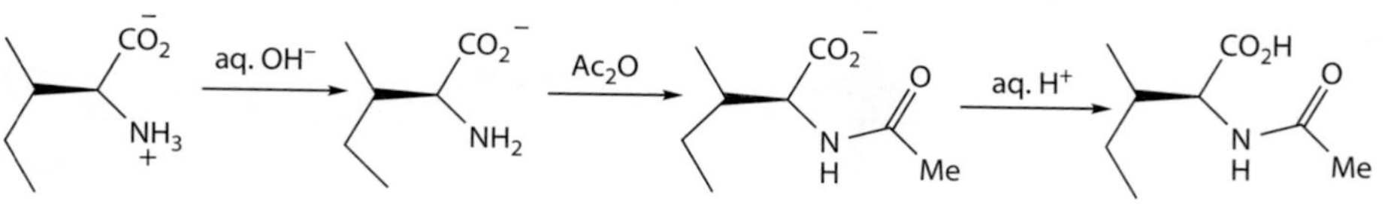

27.24

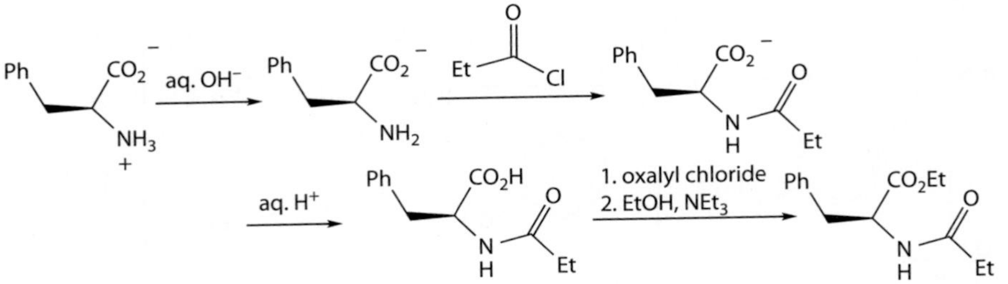

27.25

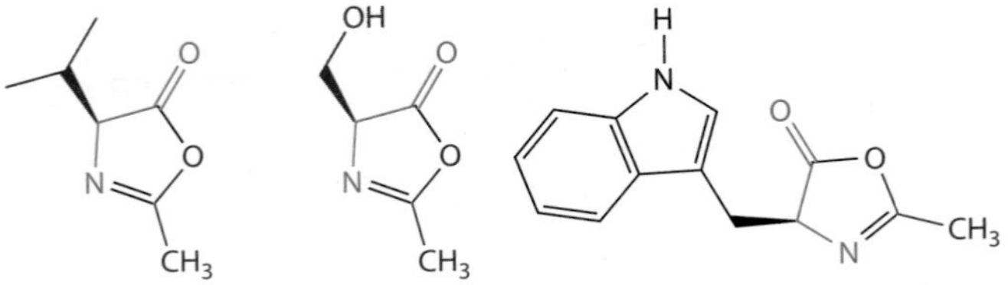

27.26

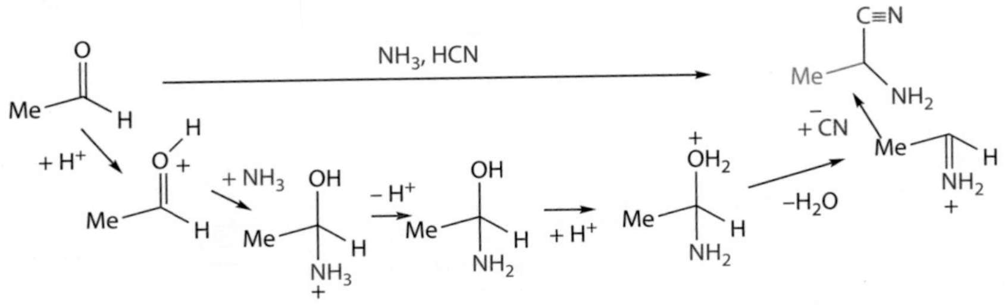

27.27

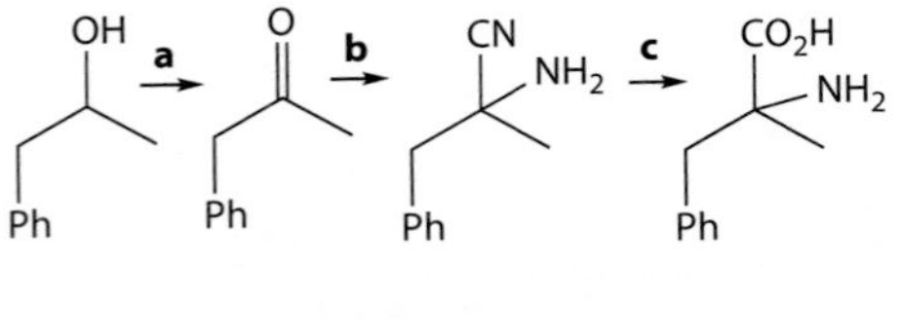

(a) PCC (b) HCN, NH$_3$ (c) aq. H$^+$

27.28

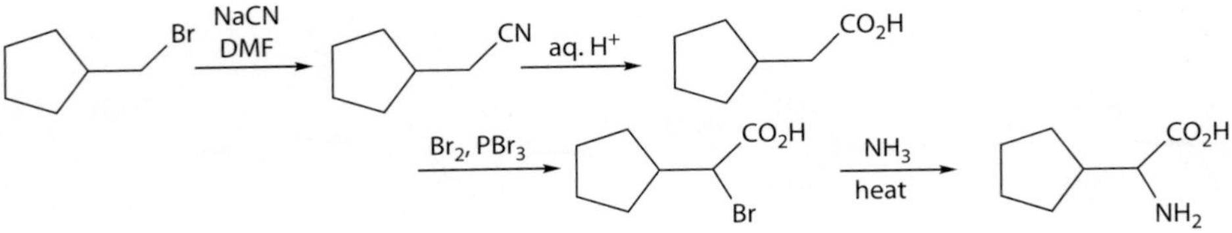

27.29

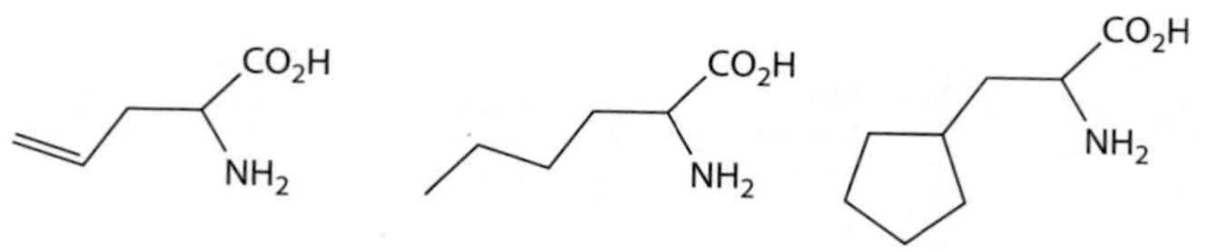

27.30

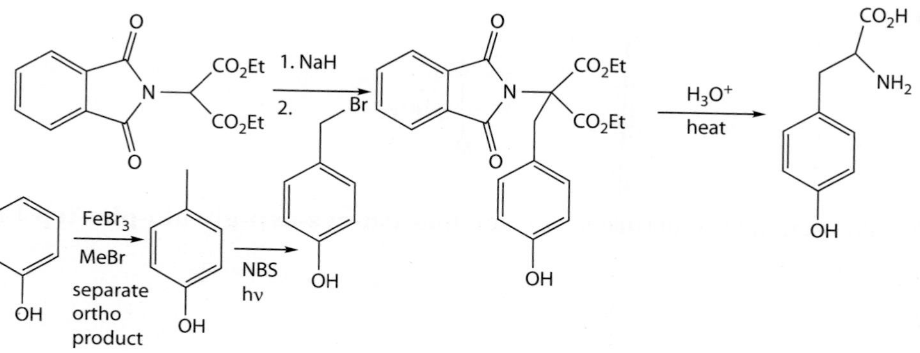

27.31

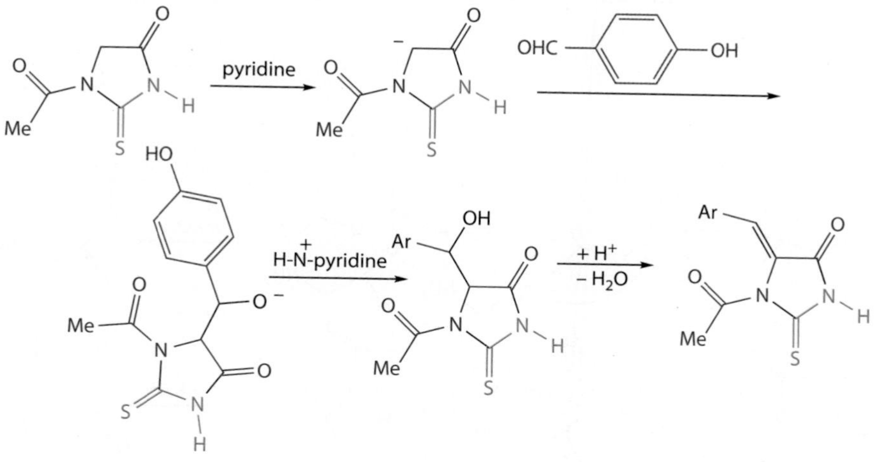

27.32

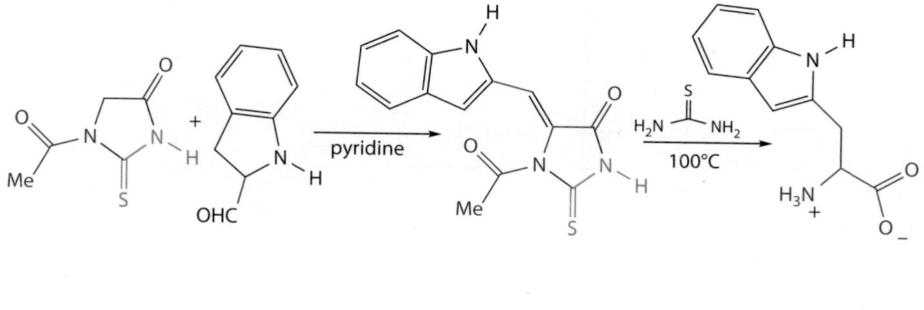

27.33

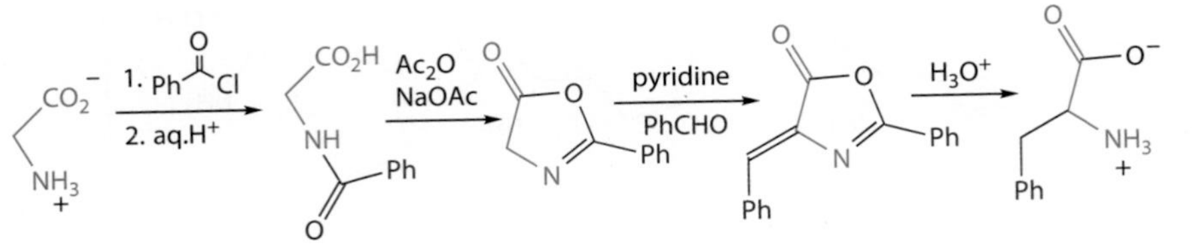

27.34

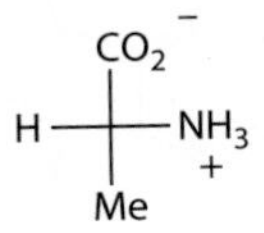

27.35 The primary structure is **ser-ser-leu-asn-cys-asp-gly-ala-phe-trp-his**.

27.36 S_N2.

27.37

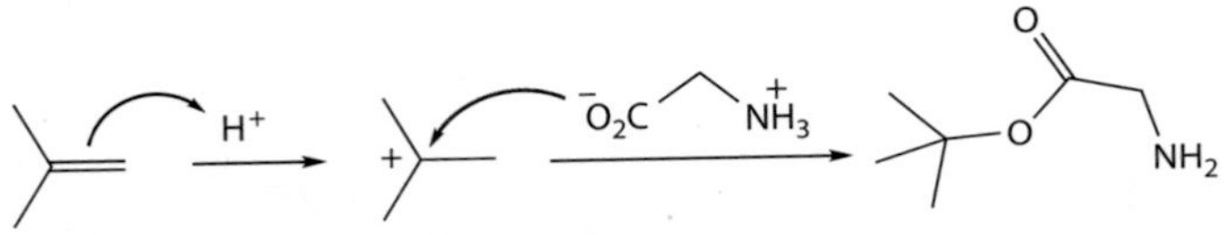

27.38

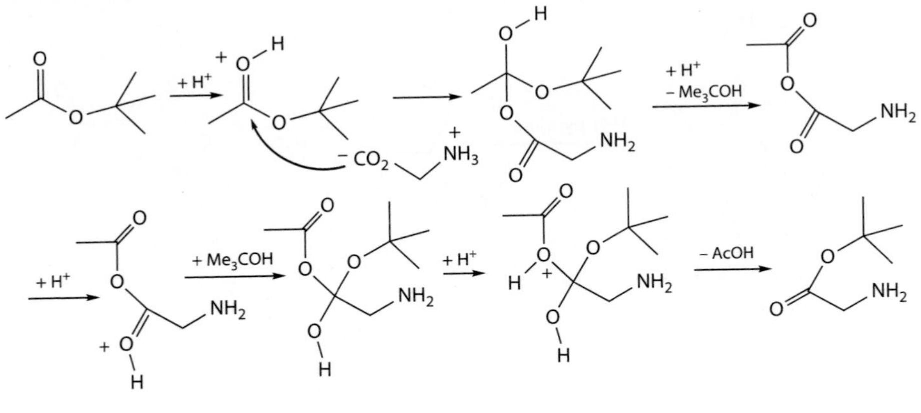

27.39

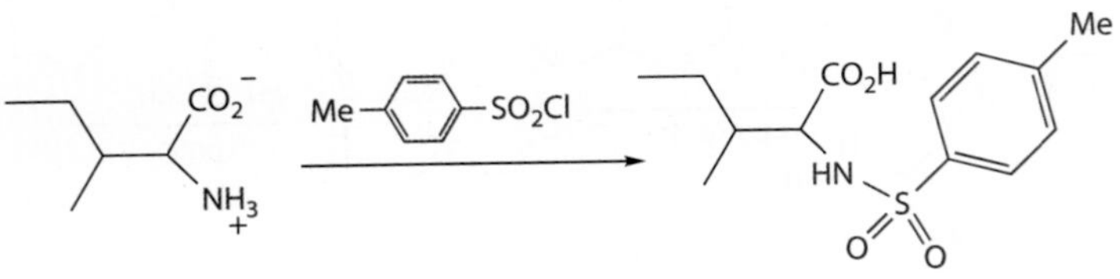

27.40

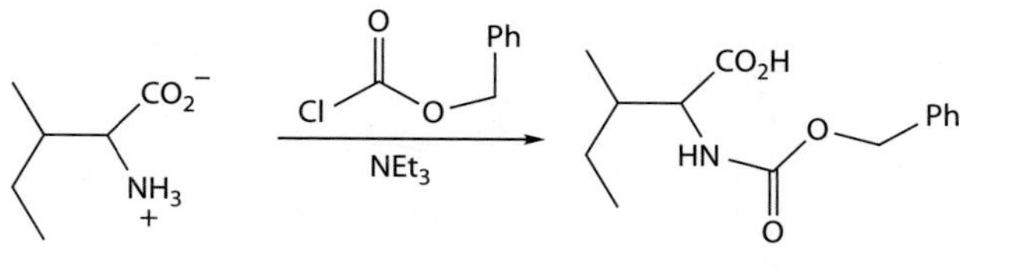

27.41

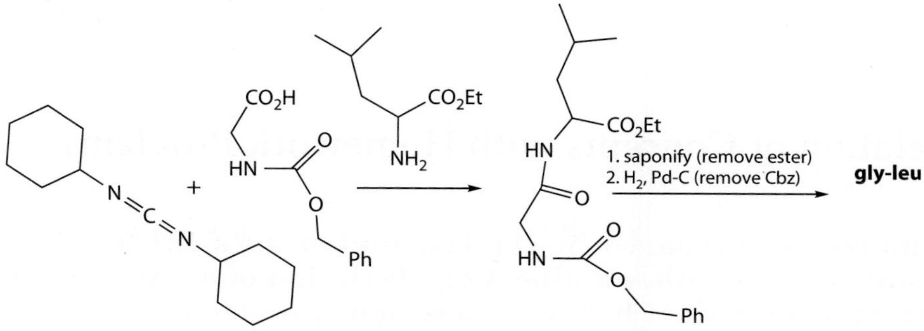

27.42

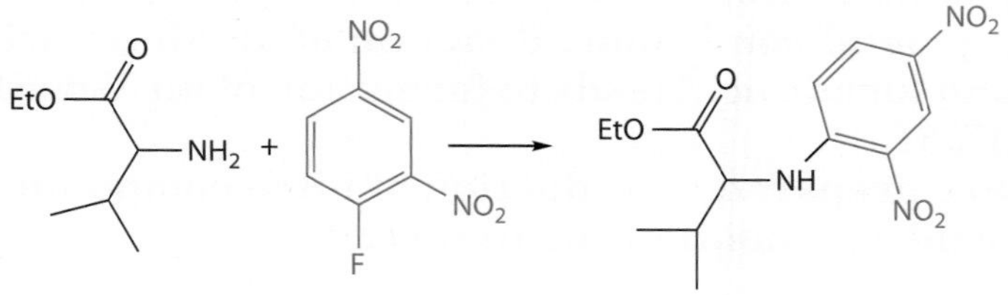

27.43

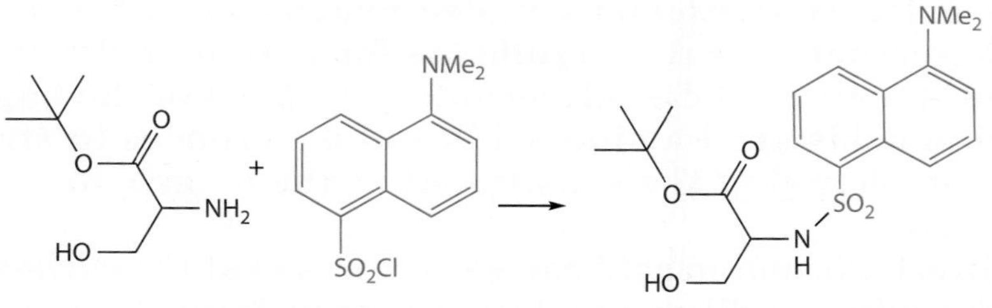

27.44

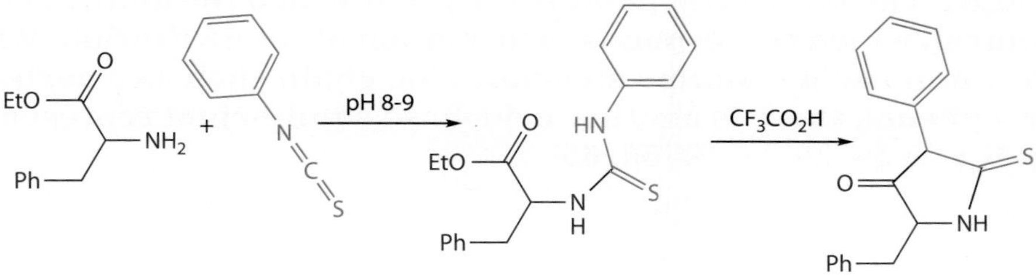

27.45

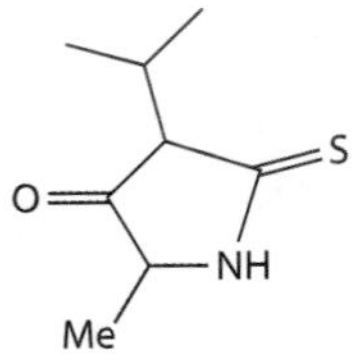

Correlation of Concepts with Homework Problems

- **Amines are prepared by the reaction of an alkyl halide with ammonia or another amine. Polyalkylation of an amine during reaction with alkyl halides is a serious problem:** 1, 2, 12, 14, 34, 46, 48, 49, 50, 51, 52, 64.
- **Amines are prepared by reductive amination of aldehydes and ketones, using sodium borohydride, catalytic hydrogenation, or dissolving metal conditions. Reaction of an amine with formaldehyde and formic acid leads to formation of an *N*-methyl amine:** 3, 4, 26, 47, 51.
- **Amines are prepared by reduction of nitro compounds, nitriles, amides, azides, enamines, or imines:** 5, 6, 8, 9, 10, 11, 13, 47, 51, 72, 84.
- **Amino acids are difunctional molecules that have a carboxylic acid unit and an amine unit in the same molecule. Amino acids exist as zwitterions under neutral conditions. The d- and l-stereochemical assignments for amino acids are based on the structure of d(+)-glyceraldehyde. Most of the important α-amino acids are l-amino acids and differences in structure focus on changing the substituent at the α-carbon:** 15, 17, 18, 19, 53, 55.
- **A zwitterionic amino acid can gain a proton at the carboxyl unit under acidic conditions or lose a proton from the ammonium unit under basic conditions to give two acid–base equilibria, K_1 and K_2. The isoelectric point is the pH at which the amino acid is neutral, which corresponds to formation of the zwitterion. When the amino acid contains an acidic side chain such as a carboxyl or a phenol, there is another acid–base equilibrium represented by K_3:** 16, 18, 20, 21, 54, 56, 65.

- For most reactions of amino acids, either the carboxyl or the amine unit must be protected to allow chemistry at the other unit. Amides and carbamates are the most common protecting groups for the amine unit of amino acids. Esters are the most common protecting group for the carboxyl group of amino acids: 22, 23, 24, 36, 37, 38, 39, 57, 59, 78, 79, 86.

- Amino acids are prepared from aldehydes and ketones via reaction with ammonia and HCN or via reactions of halo-acids with ammonia or amines. Phthalimide is a useful amine-nucleophile surrogate for the preparation of amino acids: 26, 27, 28, 29, 30, 61, 81, 82.

- Amino acids react with potassium thiocyanate to form a thiazolone derivative (a thiohydantoin). The azlactone of glycine can be used as a synthetic intermediate for the preparation of substituted amino acids: 25, 31, 32, 33, 61, 79, 80, 85.

- Peptides are composed of two or more amino acid units joined together by peptide (amide) bonds. The N-terminus of a peptide has an amino acid with a free amine group at that terminus. The C-terminus of a peptide has an amino acid with a free carboxyl group at that terminus. The amide bonds in a peptide assume a relatively narrow range of rotational angles, which leads to relatively specific shapes for the peptide. These include the α-helix and the β-pleated sheet, which can exist in both parallel and antiparallel structures: 35, 66, 74, 75, 76.

- Peptides are synthesized via coupling one protected amino acid with a second one to form a peptide bond. One terminus is deprotected and then coupled with another amino acid to elongate the peptide chain: 36, 37, 38, 39, 40, 41, 62, 63, 72.

- Chemical reagents that induce degradation of a peptide allow various constituent amino acid residues to be identified. Ninhydrin is a common reagent used to indicate the presence of amino acids. Disulfide linkages can be cleaved with peroxyformic acid or with mercaptoethanol. Sanger's reagent forms a compound that allows the N-terminal amino acid to be identified. Dansyl chloride also reacts with the N-terminal amino acid to form a readily identifiable compound. Phenylisothiocyanate reacts with the N-terminal amino acid to form a phenylthiohydantoin derivative, which is readily identified: 42, 43, 44, 45, 67, 68, 69, 70, 71.

- Carboxypeptidase enzymes can be used to identify the C-terminal amino acid. The enzymes trypsin and chymotrypsin cleave specific amino acid residues and this can be used to help identify the sequence of a peptide: 73, 77.

- **A molecule with a particular functional group can be prepared from molecules containing different functional groups by a series of chemical steps (reactions). This process is called synthesis: The new molecule is synthesized from the old one (see Chapter 25): 5, 6, 7, 8, 9, 13, 27, 28, 30, 32, 33, 40, 41, 57, 58, 59, 60, 61, 62, 63, 82, 85, 86.**
- **Spectroscopy can be used to determine the structure of a particular molecule, and can distinguish the structure and functionality of one molecule when compared with another (see Chapter 14): 83, 84, 85, 86, 87, 88, 89, 90, 91, 92.**

Homework

27.46 Give the IUPAC name for each of the following amines:

(a) 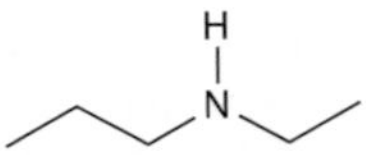

(b) 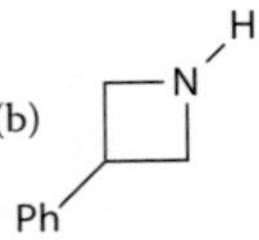

(c) 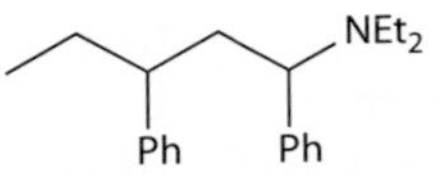

(d) 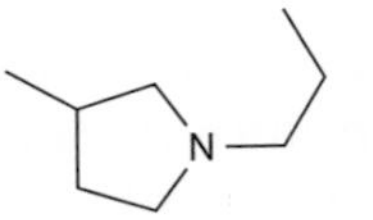

(e) 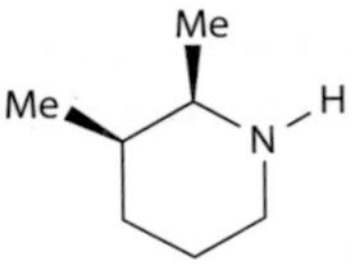

(f) 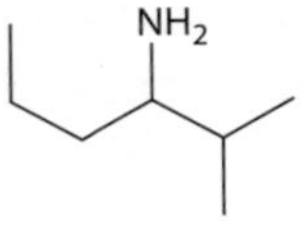

(g) 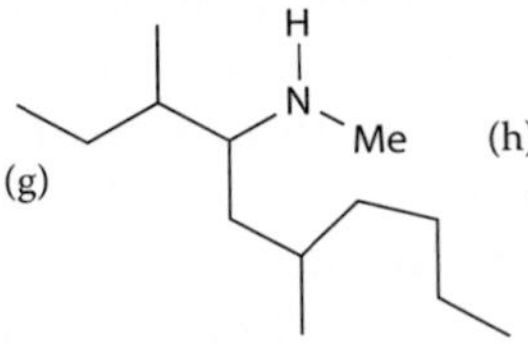

(h) 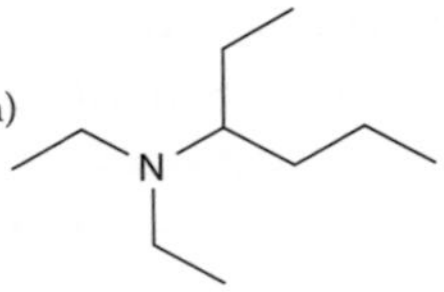

(i) 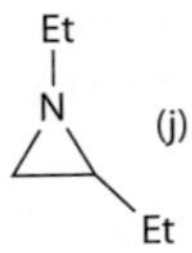

(j) 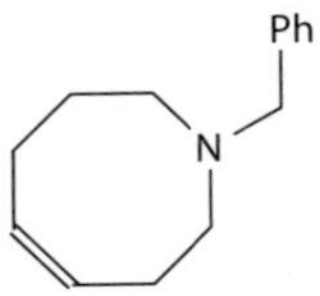

(k) 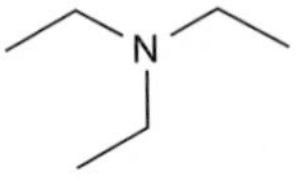

27.47 Give the major product of each of the following reactions:

(a) 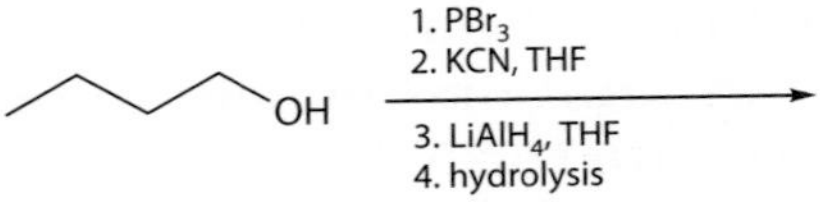

(b) 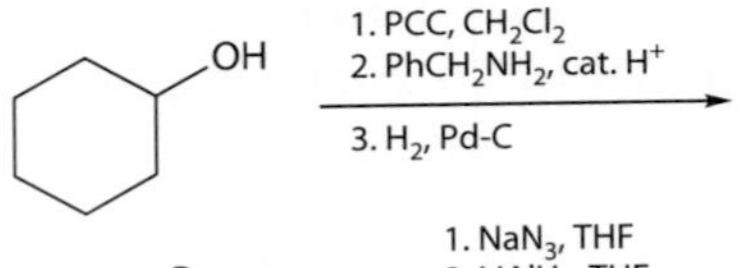

(c) 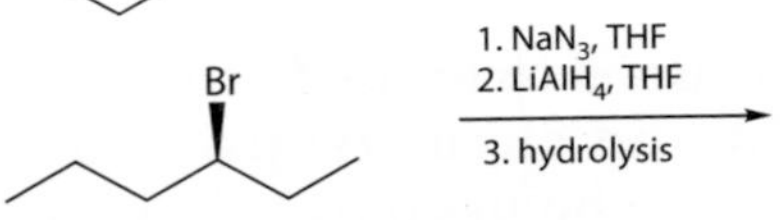

(d) 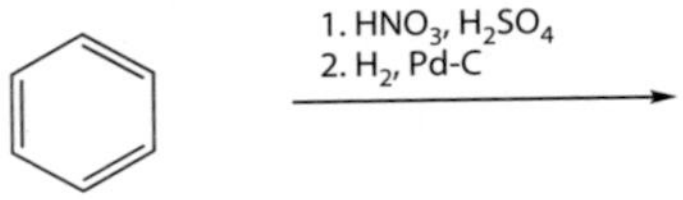

(e) 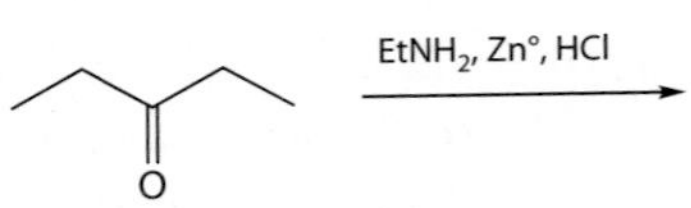

(f) 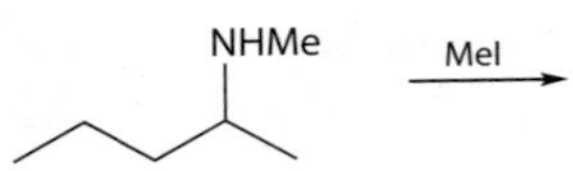

(g) 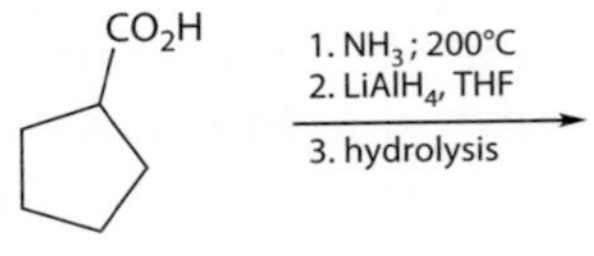

27.48   Apart from the problem of polyalkylation when an amine reacts with an alkyl halide, there is another potential problem. Speculate on potential side reactions that might occur when diethylamine reacts with 2-bromopentane in ethanol solvent.

27.49   In reactions with *n*-butyllithium, which is the stronger acid: diisopropylamine or diphenylamine? Justify your answer.

27.50   Which is the stronger acid: the conjugate acid of triethylamine or the conjugate acid of ethylamine? Justify your answer.

27.51   Give the major product for each of the following reactions:

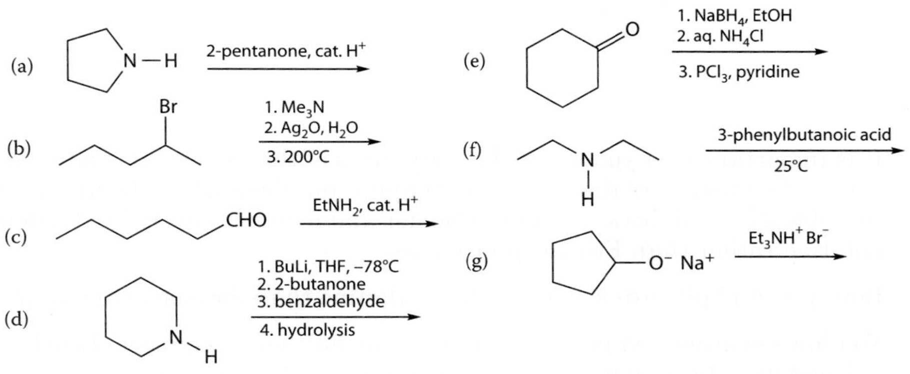

27.52   In the Hofmann elimination reaction, we used a trimethylammonium derivative rather than a triethylammonium derivative. Why?

27.53   Draw the structure for each of the following, using line notation:
   (a) 3,4-diphenyl-5-aminohexanoic acid
   (b) 4-aminohex-5-enoic acid
   (c) aziridine-2-carboxylic acid
   (d) *N*-methylpiperidine-4-carboxylic acid
   (e) 3-aminobenzoic acid
   (f) 2*R*-amino-3*R*,4*S*-dihydroxyhexanoic acid
   (g) *N*-ethyl-3-amino-1, 5-pentanedioic acid
   (h) 2*S*-amino-3-phenylpropanoic acid
   (i) pyrrolidine-2*S*-carboxylic acid
   (j) *N*,3-dimethyl-2*S*-aminobutanoic acid

27.72   Show the product formed when each of the following reacts with 1. LiAlH$_4$; 2. hydrolysis:
       (a)   ser-ile
       (b)   ala-val
       (c)   cys-leu

27.73   Show the product formed when each of the following reacts with trypsin and with chymotrypsin:
       (a)   ala-ala-thr-cys-asn-val-phe-leu-thr-his-arg-pro-phe
       (b)   tyr-ile-ile-ile-arg-gln-asp-val-his-his-phe-ile-tyr

27.74   Draw both the *syn* and *anti* conformations for the dipeptide phe-ala.

27.75   Draw out the structure of the peptide ala-gln-phe-ser in the extended conformation using line notation, with all l-amino acids. Repeat the structure, but use d-phenylalanine with the other three amino acids having the l configuration.

27.76   Aplidine, also known as dehydrodidemnin B, was isolated from a species of tunicate called *Aplidium albicans* and interferes with DNA and protein synthesis. Identify each amino acid residue and label it as an l- or a d-amino acid.

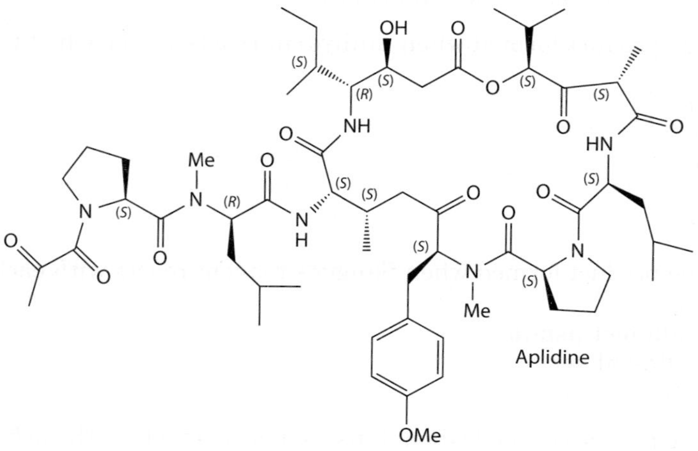

Aplidine

27.77   For the peptide gly-lys-ser-phe-phe-ala-ile-ile-trp-leu-asp-met-pro-arg-glu-tyr-ile-lys-arg, draw the following:
       (a)   all fragments that result from treatment with trypsin
       (b)   all fragments that result from treatment with chymotrypsin
       (c)   all fragments that result from treatment with carboxypeptidase B
       (d)   all fragments that result from treatment with staphylococcal protease

27.78   Draw the product formed when EtO-ile-ser-phe-NH$_2$ reacts with MeO$_2$CCl.

27.79   Explain why an amino acid cannot be prepared from phthalimide and the ethyl ester of 2-bromo-2-methylbutanoic acid.

27.80   Draw the amino acid formed by reductive amination of 3-phenylhexanal and the ethyl ester of glycine.

27.81   Write out the structures of the thiohydantoin derivatives of phenylalanine, serine, and histidine.

27.82   Use the azlactone of glycine to prepare the allyl, benzyl, and 4,4-diphenylbutyl derivatives. Convert each to the appropriate amino acid.

Spectroscopy Problems

Do not attempt these problems until Chapter 14 has been read and understood. These problems involve amines, amino acids, and derivatives of amino acids.

27.83   Describe spectroscopic differences that would allow you to distinguish *N*-methyl-1-amino butane from 2-aminopentane.

27.84   Selective reduction of 3-cyanobutanoic acid can give either 4-aminobutanoic acid or 4-cyano-1-butanol. Describe spectroscopic differences in these molecules that will allow you to distinguish them and determine the identity of the product.

27.85   The synthesis of phenylalanine from malonic acid, using phthalimide as the nitrogen source, relies on decarboxylation in a key step. Draw phenylalanine as well as the product that has *not* decarboxylated. Describe spectroscopic differences that will allow you to determine whether or not decarboxylation has occurred.

27.86   Treatment of pent-4-enoic acid with 1. HBr; 2. NaN$_3$, H$_2$/Pd-C leads to an amino acid. Two regioisomers are possible based on the initial reaction with HBr. Describe spectroscopic differences between the two regioisomeric bromides that will allow you to determine which is formed. Also describe spectroscopic differences between the two regioisomeric amino acids that will allow you to distinguish them.

27.87   Discuss spectroscopic differences between the *N*-acetyl ethyl ester of ala-gly and the *N*-ethyl-dimethyl amide of ala-gly.

27.88 Give the structure for the molecule with a formula of $C_{11}H_{13}NO_3$ and the following spectral data:

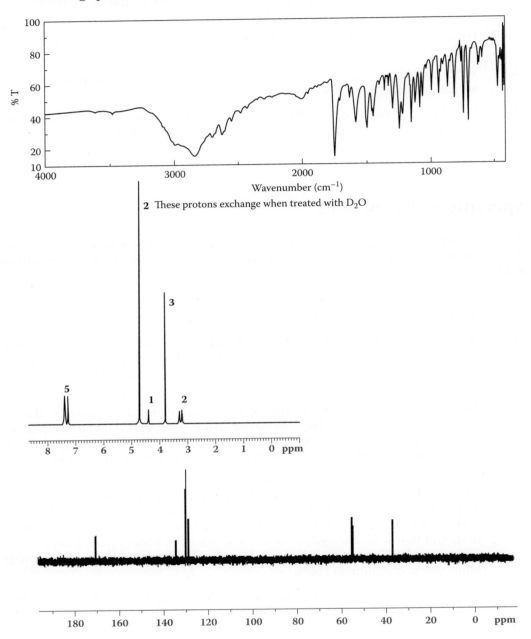

2 These protons exchange when treated with D_2O

27.89   Give the structure for the molecule with a formula of $C_6H_{11}NO_2$ and the following theoretical spectral data:

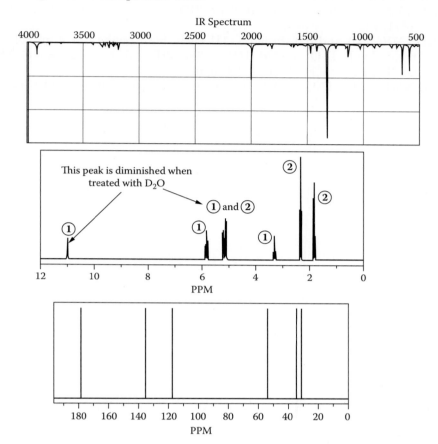

27.90   Give the structure for the molecule with a formula of $C_8H_{17}NO_2$ and the following spectral data:

IR: 3198, 3092 and 3076, 2950–2860, 1681, 1632, 1463, 805 cm^{-1}

^{1}H NMR: 5.11 (broad s, 2H; this peak is diminished when treated with D_2O), 4.25 (d, 1H), 4.21 (q, 2H), 2.14 (m, 1H), 1.55 (m, 2H), 1.29 (t, 3H), 1.11 (d, 3H), 0.90 (t, 3H) ppm

^{13}C NMR: 171.5, 61.3, 56.9, 36.3, 25.1, 15.0, 14.1, 11.3 ppm

27.91   Give the structure for the molecule with a formula of $C_3H_7BrO_4$ and the following spectral data:

IR: weak peaks at 3353, 3309, 3250, and 1974; 1941, 1459, 1394, 1324, 1132 cm^{-1}

^{1}H NMR: 5.10 (s, 1H), 3.68 (s, 6H) ppm

^{13}C NMR: 163.5, 50.9, 41.2 ppm

27.92   Give the structure for the molecule with a formula of $C_{12}H_{20}N_2O_5$ and the following spectral data:

IR: 3588, 3282, weak peaks at 3279–3176 and 1932, 1903, 1853, 1694, 1427, 1341, 776 cm^{-1}

^{1}H NMR: 10.0 and 8.0 (broad s, 1H each; both peaks are diminished when treated with D_2O), 4.52 (d, 1H), 4.13 (q, 2H), 3.53 (s, 2H), 2.65 (m, 1H), 2.05 (s, 3H), 1.29 (t, 3H), 0.91 (d, 6H) ppm

^{13}C NMR: 173.6, 170.7, 169.2, 61.6, 60.6, 37.1, 30.8, 22.9, 18.5, 14.1 ppm

Multifunctional Compounds

Carbohydrates

28

The number of functional groups contained in a single molecule will increase in this chapter. Multifunctional compounds containing several hydroxyl (OH) units, which may also contain a ketone, an aldehyde, or a carboxyl unit, are called **carbohydrates** (hydrates of carbon). Carbohydrates are commonly known as **sugars** and they have the generic formula (CH_2O_n) where the value of "n" is typically three to six, although many carbohydrates have values of "n" that are multiples of six. These compounds are extremely important in mammalian biology, as well as in the biology of plants and of insects.

To illustrate how important carbohydrates are to life itself, sunlight is essential to all life, and plants convert carbon dioxide (CO_2) to carbohydrates in the presence of water and sunlight (photosynthesis). Carbohydrates store energy within living systems, and this process is required for life. For all animals, including humans, their very essence is carried in the DNA found in genes, and DNA is simply a polymeric sugar with heteroaromatic substituents. This chapter will introduce carbohydrate structure, nomenclature, preparation, and reaction.

To begin this chapter, you should know the following:

- **nomenclature of alcohols (Chapter 5, Section 5.6.3)**
- **nomenclature of aldehydes and ketones (Chapter 5, Section 5.9.2; Chapter 16, Section 16.2)**

- nomenclature of carboxylic acids (Chapter 5, Section 5.9.3; Chapter 16, Section 16.4)
- nomenclature of carboxylic derivatives (Chapter 16, Section 16.7)
- nomenclature of amines (Chapter 5, Section 5.6.2)
- reduced forms of heterocycles (Chapter 26, Section 26.4)
- formation and reactions of acetals and ketals (Chapter 18, Section 18.6)
- relative strength of acids (Chapter 6, Section 6.3)
- transition states and reaction energetics (Chapter 7, Sections 7.3 and 7.6)
- reversible reactions (Chapter 7, Section 7.10)
- fundamentals of acid–base equilibria (Chapter 6, Section 6.1)
- conformations of cyclic compounds (Chapter 8, Sections 8.5, 8.6, and 8.9)
- stereogenic centers and R/S configuration (Chapter 9, Sections 9.1 and 9.3)
- diastereomers (Chapter 9, Section 9.5)
- stereoisomers in cyclic compounds (Chapter 9, Sections 9.6 and 9.7)
- carbocations (Chapter 10, Section 10.2)
- oxocarbenium ion intermediates (Chapter 16, Section 16.3; Chapter 18, Section 18.1; Chapter 20, Sections 20.1 and 20.2)
- S_N2 reactions (Chapter 11, Sections 11.2 and 11.3)
- S_N1 reactions (Chapter 11, Sections 11.4–11.6)
- that amines react as nucleophiles (Chapter 11, Section 11.3.3; Chapter 18, Section 18.7; Chapter 20, Section 20.6)
- reactions of amines (Chapter 27, Sections 27.1 and 27.2)
- acyl substitution reactions (Chapter 16, Section 16.3; Chapter 18, Sections 18.1 and 18.2)
- acyl addition reactions (Chapter 16, Section 16.8; Chapter 20, Section 20.1)
- oxidation reactions (Chapter 17, Sections 17.2–17.4)
- reduction reactions (Chapter 19, Sections 19.2–19.4)
- formation and reactions of esters (Chapter 20, Section 20.5)
- formation and reactions of amides (Chapter 20, Section 20.6)
- protecting groups (Chapter 25, Section 15.4.3)
- reactions and synthesis of amino acids (Chapter 27, Sections 27.3.3 and 27.3.4)
- peptides (Chapter 28, Sections 27.4 and 27.5)

The most fundamental concepts for structure, nomenclature, and chemical reactions of carbohydrates will be presented. The chapter will focus primarily on the characteristics, formation, reactions, and synthesis of simple carbohydrates. The use of carbohydrates in DNA and RNA will also be introduced. In addition, several chemical reactions that lead to the synthesis of DNA or RNA will be discussed.

When you have completed this chapter, you should understand the following points:

- Carbohydrates are polyhydroxy aldehydes and ketones that are commonly known as sugars. A monosaccharide is a carbohydrate composed of one sugar unit and a disaccharide is a carbohydrate composed of two monosaccharide units. An oligosaccharide is composed of 2–10 monosaccharide units; a polysaccharide is composed of 10 or more monosaccharide units. An aldose is a carbohydrate that contains an aldehyde unit and a ketose is a carbohydrate that contains a ketone unit. Monosaccharides are categorized by the total number of carbons in the structure: triose, tetrose, pentose, hexose, etc. The d and l configurations of a monosaccharide are based on the Fischer projection of d-glyceraldehyde. A Fischer projection is an older representation of sugars.

- The OH unit derived from the carbonyl group, formed by cyclization to a furanose or a pyranose, is attached to the so-called anomeric carbon. An aldose will cyclize to form a stable hemiacetal structure: a furanose or a pyranose. A ketose will cyclize to form a stable hemiketal structure: a furanose or a pyranose. The Haworth projection is an older representation for carbohydrates that is based on a planar pyranose or furanose ring.

- Cyclization of an aldose or a ketose to a furanose or pyranose is accompanied by mutarotation at the anomeric carbon. Mutarotation is the change in optical rotation of a pure furanose or pyranose derivative to that resulting from a mixture of isomers at the anomeric carbon.

- The anomeric effect occurs in 2-alkoxy pyranose derivatives, and it describes the stabilization of the electronegative substituent in an axial position rather than an equatorial position.

- An aldose is oxidized to a carboxylic acid derivative via treatment with aqueous copper sulfate, silver nitrate in ammonia, bromine, aqueous potassium permanganate, or nitric acid.

- The carbonyl unit of an aldose or a ketose is reduced to the corresponding alcohol with all common reducing agents; these include sodium borohydride, lithium aluminum hydride, and catalytic hydrogenation.

- Disaccharides are formed by coupling monosaccharides to generate different regioisomers. Trisaccharides and higher are prepared by similar coupling reactions. To react one hydroxyl group of a carbohydrate selectively, all or some of the other OH units in the molecule must be protected. Common hydroxyl protecting groups include acetate and benzoate esters, methyl ethers, acetals, and ketals.

- The Ruff degradation and the Wohl degradation both cleave the carbonyl-bearing carbon from a monosaccharide to form a new monosaccharide of one fewer carbon. The Nef reaction adds one carbon to a monosaccharide to form a new monosaccharide, as does the Fischer–Kiliani synthesis.

- The Koenigs–Knorr synthesis couples two monosaccharides together to form a disaccharide.

- Several heterocyclic compounds, including purines and pyrimidines, can be coupled to monosaccharides at the anomeric carbon to form a ribonucleoside (or simply a nucleoside). The most common nucleosides are ribose and 2-deoxyribose derivatives containing adenine, guanine, cytosine, thymine, or uracil. The phosphate ester of a nucleoside is called a nucleotide.

- Polynucleotides include ribonucleic acid (RNA) and deoxyribonucleic acid (DNA). Under physiological conditions, DNA normally exists in the β-form, which consists of two antiparallel strands: the Watson–Crick double helix. Interstrand hydrogen bonding connects the two strands. An adenosine in one strand can hydrogen bond with a thymine in the second, and a guanosine can hydrogen bond with a cytosine in the other strand. The strands must be antiparallel.

- DNA stores genetic information; RNA transcribes and translates this information for the cell. There are three major types of RNA: transfer RNA, ribosomal RNA, and messenger RNA. Transfer RNA contains a loop of nucleotides with three base pairs that bind to a messenger RNA, allowing a specific amino acid to be transferred to a protein that is being made. These three base pair sequences that correlate with specific amino acids are known as the genetic code.

- Polynucleotides are prepared by coupling individual nucleotide units containing activated and protected phosphate linkages. The hydroxyl and amine units of the heterocycle structure must be protected during this process.

- A molecule with a particular functional group can be prepared from molecules containing different functional groups by a series of chemical steps (reactions). This process is called synthesis: The new molecule is synthesized from the old one (see Chapter 25).

- Spectroscopy can be used to determine the structure of a particular molecule and can distinguish the structure and functionality of one molecule when compared with another (see Chapter 14).

28.1 Polyhydroxy Carbonyl Compounds

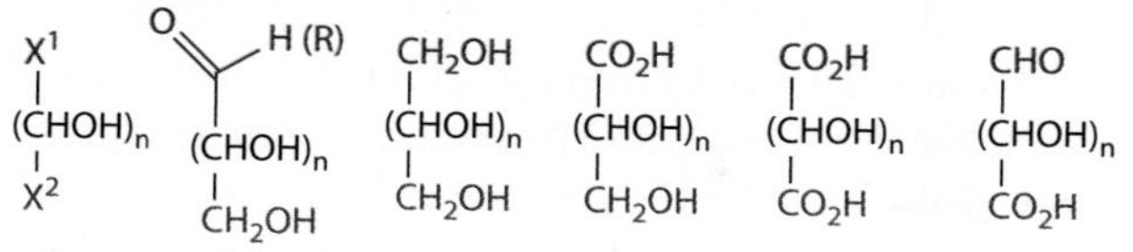

Most of the carbohydrates discussed in this chapter are polyhydroxylated aldehydes or ketones, although there are other polyhydroxylated derivatives. Carbohydrates can be classified in two fundamental ways: The first is based on the type of functional group that accompanies the hydroxyl units and the second on the molecular size of the carbohydrate. If the fundamental unit of a carbohydrate is **1**, carbohydrates are classified by the nature of the functional groups X^1 and X^2. In structure **1**, several repeating CHOH units are defined by the integer n, where $n = 3$, $n = 5$, $n = 6$, etc. When $n = 3$, for example, **1** is X^1–CHOH–CHOH–CHOH–X^2. Most carbohydrates are defined by making X^1 and/or $X^2 = CH_2OH$, CHO, COR (a ketone), or COOH.

Compound **2** with a CH_2OH and an aldehyde or ketone unit is a **glycose** and **3**, which has two CH_2OH units, is called a **glycitol** (sometimes called an **alditol**). When one group is a carboxylic acid and the other is CH_2OH, as in **4**, this is a **glyconic acid** (sometimes called an **aldonic acid**). In **5**, there are two carboxyl units and it is called a **glycaric acid** (or an **aldaric acid**); in a carboxyl group and an aldehyde group, as in **6**, the compound is classified as a **uronic acid**.

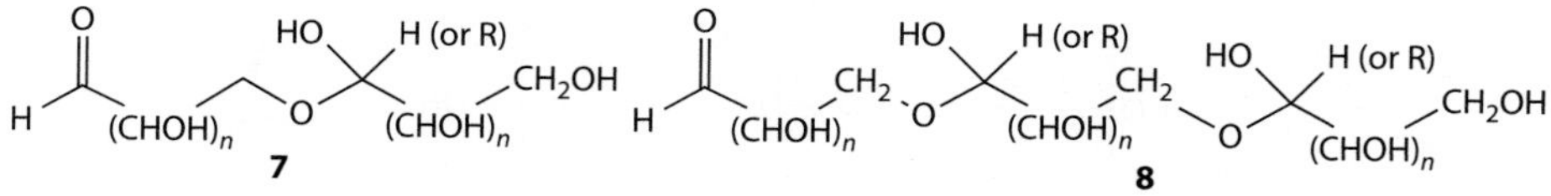

The other way to categorize carbohydrates is by the number of sugar units the carbohydrate contains. Examination of **2**, for example, shows that it contains one carbohydrate unit and thus is categorized as a monosaccharide. If two monosaccharides are coupled together, as in **7**, the resulting molecule is a disaccharide; **8**, with three monosaccharide units, is a trisaccharide. Linking 5–15 monosaccharides gives an oligosaccharide. A polysaccharide has more than 15 monosaccharides linked together. Note that in **7** and **8** the monosaccharides are linked together by an acetal linkage (see Chapter 18, Section 18.6, for a discussion of acetals and ketals).

28.1.1 Monosaccharides

Monosaccharide glycose derivatives are the main focus here, but the other categories of carbohydrates will be introduced as reaction products. Glycoses (**2**) are further subdivided into **aldoses**, where the carbonyl unit is an aldehyde, and

ketoses, where the carbonyl unit is a ketone. This discussion begins with the characteristics and properties of glycoses, their structures, and how to name them.

The basic unit of a glycose is **2**, and there are several monosaccharide glycoses. The carbon of each hydroxymethyl (CHOH) unit in **2** is a stereogenic center (see Chapter 9, Section 9.1), so there are several diastereomers for most carbohydrates. Each time a CH_2OH unit is added, another stereogenic center is added, which increases the number of diastereomers. Each diastereomer will have a (+) and a (–) form (enantiomers). For **2** ($n = 1$; three carbons), there is one stereogenic center and there are two stereoisomers (the two enantiomers). When $n = 2$ (four carbons), there are four stereoisomers (two diastereomers, each with an enantiomer); when $n = 3$ (five carbons), there are eight stereoisomers (four diastereomers, each with an enantiomer). When $n = 4$ (six carbons), there are 16 stereoisomers (eight diastereomers, each with an enantiomer). *A discussion of carbohydrates must include the chirality inherent to these molecules, which means discussing absolute configuration, diastereomers, and enantiomers.*

Glycoses are classified by the total number of carbon atoms. The glycose with three carbon atoms (**2**, $n = 1$) is called a **triose**, and when there are four carbon atoms (**2**, $n = 2$), it is a **tetrose**. A **pentose** has five carbon atoms (**2**, $n = 3$) and a **hexose** has six carbon atoms (**2**, n = 4). If the carbonyl unit in **2** is an aldehyde, it is an aldose (see preceding discussion); thus, a three-carbon aldose is an *aldotriose* and a six-carbon aldose is an *aldohexose*. If the carbonyl unit in **2** is a ketone unit, the molecule is a ketotetrose, a ketohexose, etc.

28.1 Draw the structure of a ketopentose and an aldotetrose.

A simple carbohydrate that fits the generic formula for **2** is the aldotriose known as glyceraldehyde, **9**. This compound has one stereogenic center and two enantiomers. Glyceraldehyde was introduced in Chapter 27 in connection with the nomenclature scheme for amino acids. Emil Fischer used it as the basis for his d,l system (Section 27.3) and this model is also applied to the structures of carbohydrates. The absolute configuration of d-(+)-glyceraldehyde is shown in **9** and that of l-(–)-glyceraldehyde is shown in **10**. Both structures are shown in Fischer projection and using the normal line notation. *Note that 9 is 2(S)-glyceraldehyde and 10 is 2(R)-glyceraldehyde.*

When Fischer examined the structures of many carbohydrates that are important in biological systems, he found a correlation with the –CHOH–CH_2OH unit of d-glyceraldehyde (C2 and C3 of that compound). This means that if C2 in **10** had an (R) absolute configuration (review Chapter 9, Section 9.3), all of the carbohydrates examined also had an (*R*) configuration. Remember that the *R*/*S* nomenclature was devised many years after this work by Fischer. Once Fischer saw the correlation, however, he labeled the carbohydrates as d if they had the same –CHOH–CH_2OH unit as d-glyceraldehyde and l if they correlated with l-glyceraldehyde.

11A 12A 13A 14A

15A 16A 17A

11B 12B 13B 14B 15B 16B 17B

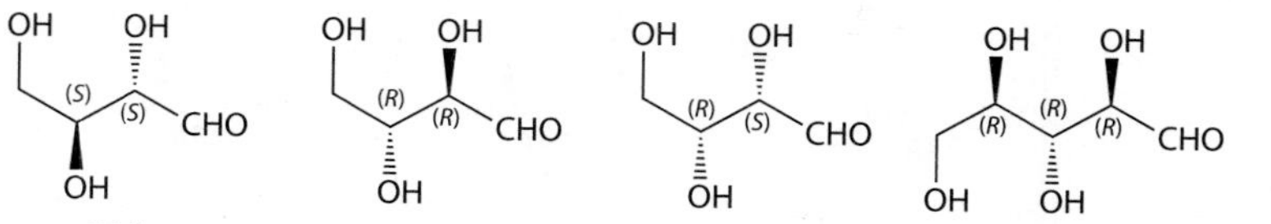

There are two aldotetroses—erythrose (**12A**) and threose (**13A**)—and the enantiomeric form of **12A** is **11A**. In the aldopentoses, there are four diastereomers; each has an enantiomer. With a focus on the d-diastereomers, the four diastereomers are d-ribose (**14**), d-arabinose (**15**), d-xylose (**16**), and d-lyxose (**17**). All of these compounds are also drawn as their Fischer projection, primarily for comparative purposes.

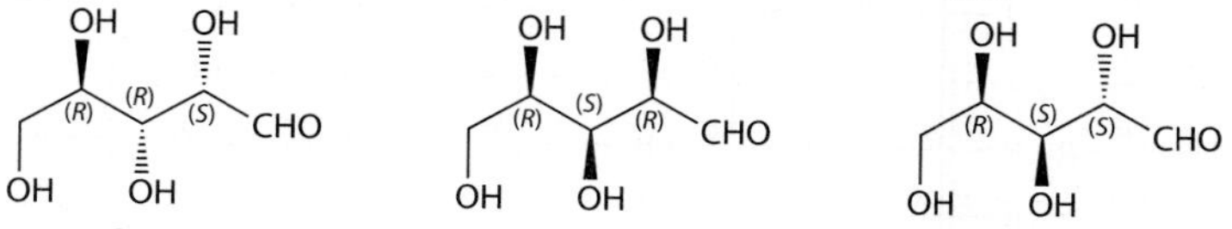

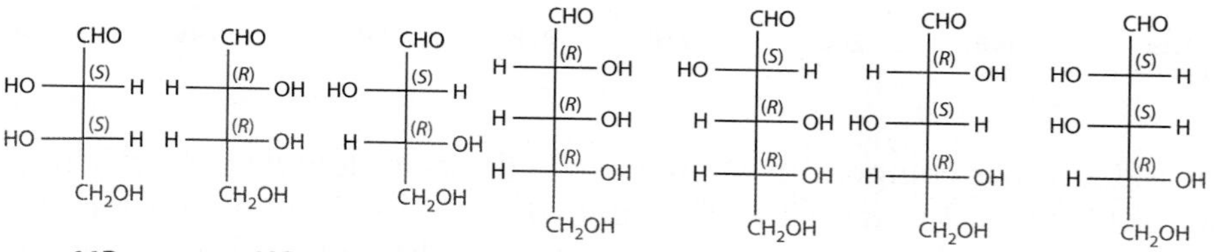

To continue, there are eight diastereomeric aldohexoses, each with an enantiomer, and the d compounds are d-allose (**18A**), d-altrose (**19A**), d-glucose (**20A**), d-mannose (**21A**), d-gulose (**22A**), d-idose (**23A**), d-galactose (**24A**), and d-talose (**25A**). For all of the d-sugars shown, the last –CHOH–CH_2OH unit (in **violet**) has the same (*R*) absolute configuration. The Fischer projections for each sugar are also shown (**18B–25B**). All of the l-sugars will have the (*S*) configuration at that carbon. *Today, Fischer projections are not used as much, and the remainder of this text will use the line notation that has been used throughout the book* (i.e., **11A–17A; 18A–25A**).

28.2 Draw the structure of l-erythrose and l-threose in line notation.

28.3 Determine the number of diastereomers and the total number of stereoisomers for (a) an aldoheptose, and (b) an aldooctose.

28.4 Draw the structure of two d-aldoheptoses in line notation.

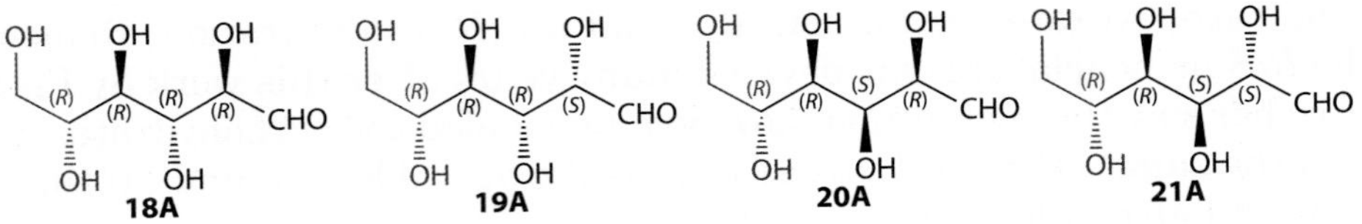

18A 19A 20A 21A

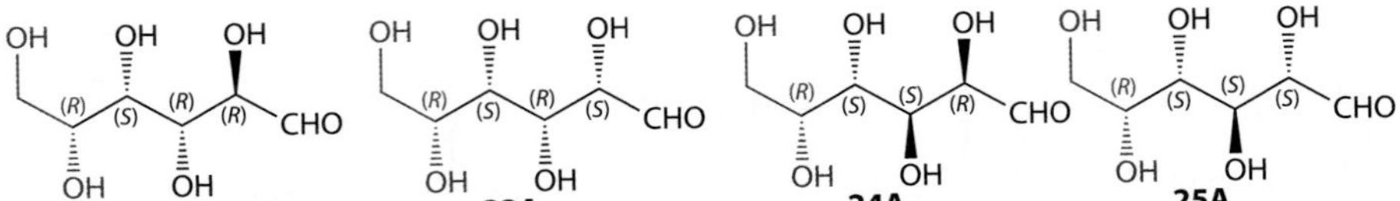

22A 23A 24A 25A

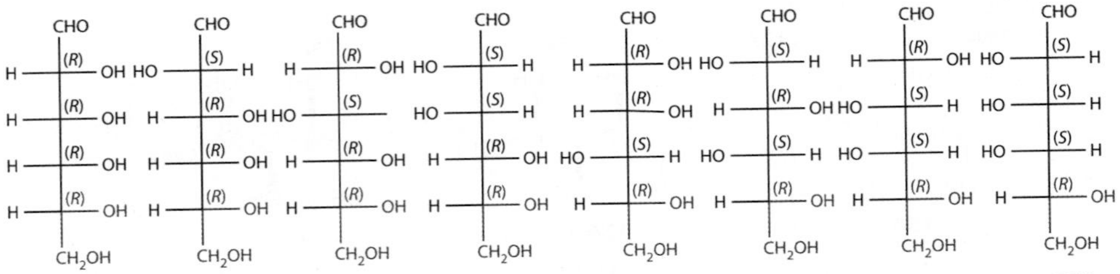

18B 19B 20B 21B 22B 23B 24B 25B

The aldoglycose compounds shown are hydroxy-aldehydes. From Chapter 18 (Section 18.6), it is known that an aldehyde and an alcohol react to form an acetal. An example is the reaction of pentanal (**26**) with ethanol, in the presence of an acid catalyst, to give the acetal product 1,1-diethoxypentane (**28**). Formation of **28** requires formation of a hemiacetal intermediate, **27**. Based on this reaction, it is reasonable to assume that hydroxy-aldehydes **9–26** might form acetals. *In fact, aldoglycoses do not form acetals directly but rather form hemiacetals.* In general terms, an aldoglycose such as mannose (**21**) will cyclize by reaction of a secondary alcohol unit (*not the primary alcohol unit of the CH$_2$OH group*) to give hemiacetal **29A**.

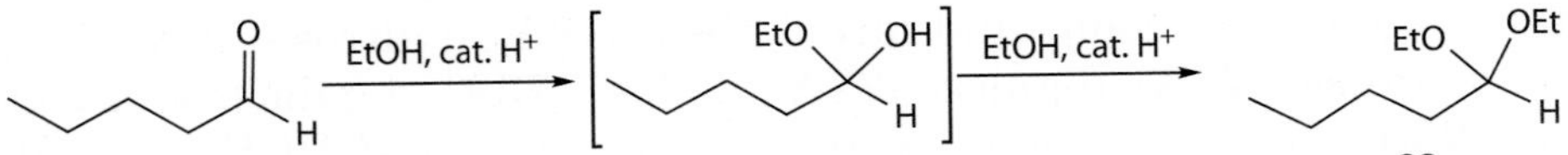

26   EtOH, cat. H$^+$ 27   EtOH, cat. H$^+$ 28

This is actually a six-membered ring and the C1 carbon and oxygen on the C5 carbon are joined together (see the exaggerated **green** bond in **29A**, which represents a single bond between C and O). This hemiacetal is drawn again as **29B** to show its cyclic structure (the single **green** bond is between C and O). The hemiacetal OH unit is drawn as a mixture of diastereomers at the acyl carbon (squiggle lines). This point will be addressed in Section 28.1.2. In order to form such a hemiacetal, a ring must be formed (a cyclic hemiacetal) as shown for **29**. Any attempt to form a cyclic hemiacetal from d-glyceraldehyde (**10**) will generate **30**. This is a three-membered ring, however, and the equilibrium usually favors **10**.

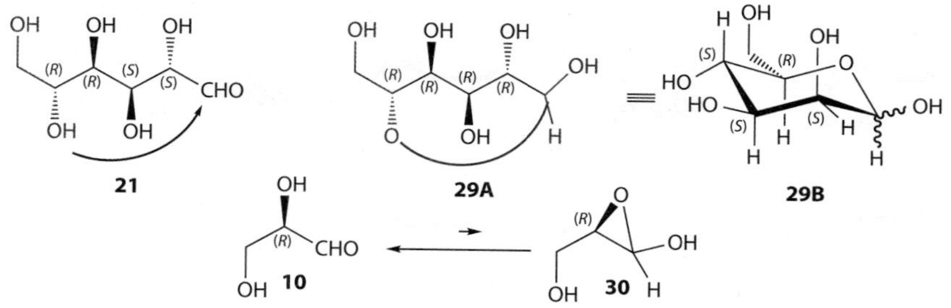

In other words, **10** does **not** form a cyclic hemiacetal. Both d-erythrose (**12B**) and d-threose (**13B**) form cyclic hemiacetals, but the terminal CH_2OH group is used to form a five-membered ring rather than a four-membered ring. As with three-membered rings, the equilibrium for a four-membered ring acetal usually favors the aldehyde, although three- and four-membered rings are also energetically unfavorable relative to the five- and six-membered rings. d-Erythrose cyclizes to give **31A** and d-threose cyclizes to give **32A**. Both are drawn in a hyperextended conformation to emphasize which bond is formed to which atoms, *but such a conformation cannot exist.* Therefore, the structures are drawn a second time to show the five-membered tetrahydrofuran ring (**31B** and **32B**, respectively). The term "ose" is used for **31** and **32** because they are carbohydrates, but they are also derived from furan. This leads to the name **furanose** for a five-membered ring hemiacetal derived from an aldose. Therefore, **31** is named d-erythrofuranose and **32** is d-threofuranose.

28.5 Write out the mechanism of formation and the hemiacetal product when benzaldehyde reacts with methanol in the presence of an acid catalyst.

Hemiacetal formation is found in the aldopentoses, with formation of a cyclic hemiacetal using the C4–OH unit to generate a five-membered ring. Examination of d-ribose (**14**) reveals that the hydroxyl group on the "next to last" carbon attacks the acyl carbon of the aldehyde to form two possible furanose structures: **33** or **34**. In both cases, the cyclization product is shown in the same fundamental conformation as ribose, with hyperextended bonds to emphasize the bond that is formed and which atoms are involved.

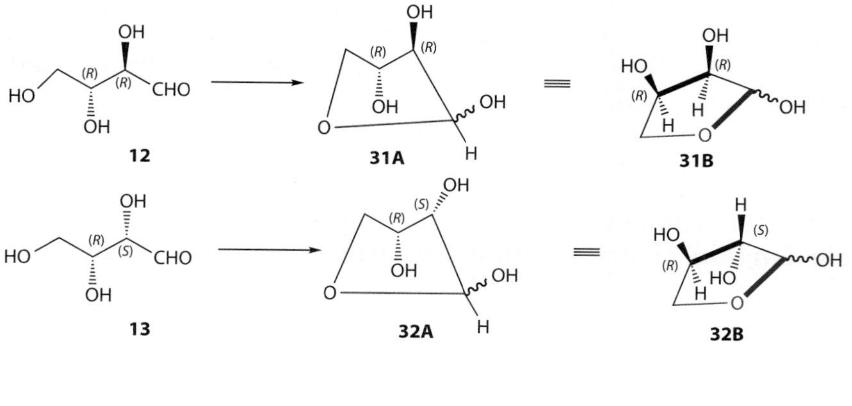

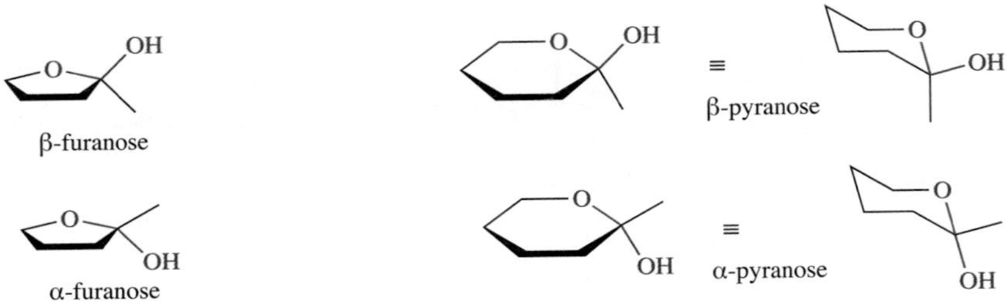

Figure 28.1   Convention for assigning α and β configurations for d sugars.

Such as structure is absurd, of course; it is drawn again with a normal tet-rahydrofuran ring, and two products are shown. In **33**, the OH group at C2 is on the same side of the ring as the hydroxymethyl group, whereas in **34** that OH group is on the opposite side of the ring. These compounds are isomers and different compounds. In the previous reactions, the product was shown as a mixture because two possible cyclized hemiacetals can be formed. Based on the nomenclature scheme for erythrose and threose, both **33** and **34** are riboses, so **33** and **34** are named as ribofuranoses.

The hydroxyl group of interest may also be connected to the other hydroxyl group rather than the hydroxymethyl group. In **33**, the OH of the hemiacetal unit is *syn* to the CH$_2$OH unit, but in **34** that OH is *anti* to the CH$_2$OH unit. Formally, the five-membered ring is drawn such that the ring oxygen is away from the viewer (to the rear) and the ring carbons are forward (to the front), with the anomeric carbon drawn on the right side. In this drawing, the upper bond on the anomeric carbon is the β-carbon and the lower bond is α. The six-membered pyranose ring is similarly drawn with the ring oxygen to the rear and the anomeric carbon on the right. *Therefore, α means down and β means up, but only when the ring is properly drawn for the d series.* This convention is outlined in Figure 28.1. *Note: these α and β designations are for the d series. In the l series, the definitions of α and β are reversed in that the anomeric carbon is positioned to the left of the ring oxygen.*

28.6 Draw the structure of d-xylofuranose.

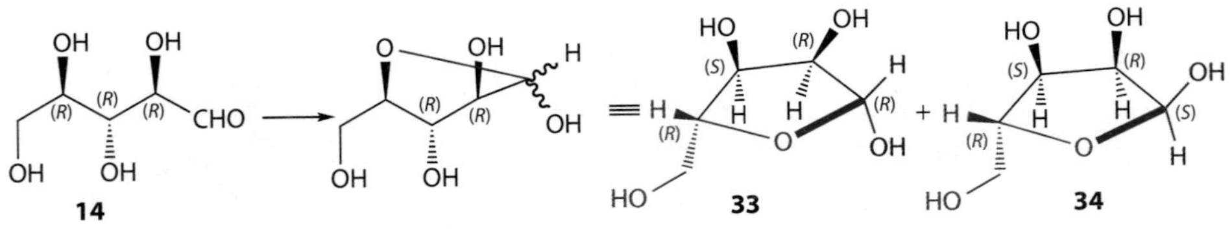

The carbon bearing this acetal–OH unit is the hemiacetal carbon atom derived from the acyl carbon of the aldehyde unit in **14**, and it is called the **anomeric carbon** or an **anomeric center**. Isomer **33** is identified as the α-anomer and **34** is the β-anomer for the d series. The formal name of **33** is α-d-ribofuranose and the name of **34** is β-d-ribofuranose. These designations are simply an alternative to stating that formation of the hemiacetal generates a mixture of the *R*-stereogenic and *S*-stereogenic centers (see Chapter 9, Section 9.1) at the former acyl carbon.

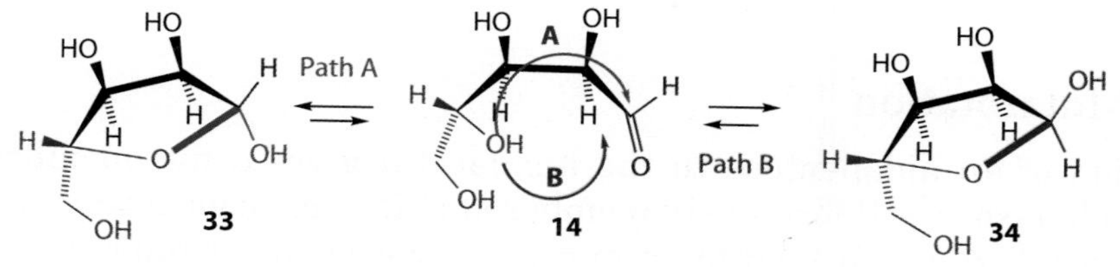

In other words, because there are other stereogenic centers, cyclization produces two diastereomers. The anomeric carbon gives a mixture because there is an equilibrium between the hemiacetal and the aldehyde, as shown for **33↔14↔34**. In an equilibrium process, the OH may attack the acyl carbon from one face (path **A** in **cyan**) or from the other face (see path **B** in **green**). Attack from faces **A** and **B** occurs because there is free rotation around the C1–C2 bond in **14** and both rotamers are present. Cyclization via **path A** leads to **33** and cyclization via **path B** leads to **34**. *Conversion of an acyclic aldoglycose to the cyclic hemiacetal may lead to formation of an anomeric center.*

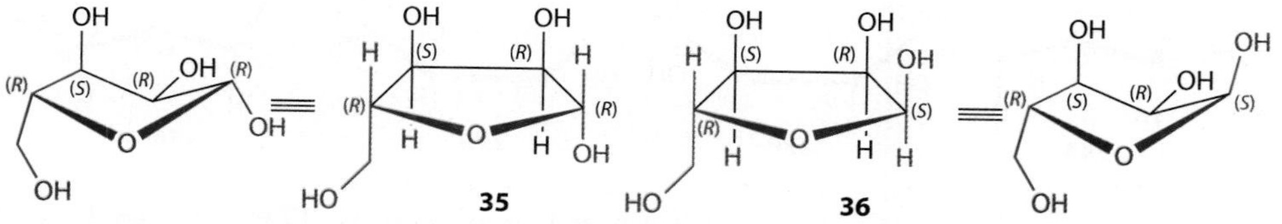

α-d-Ribofuranose **33** is in equilibrium with the acyclic aldopentose, **14**, which is in equilibrium with the β-d-ribofuranose, **34**. Another way to represent these rings is called the **Haworth formulas**, after Norman Haworth (England; 1883–1950). In effect, the ring is drawn as a flat structure with the bonds "up" or "down" at each carbon. Using this model, **34** is drawn as **35** in Haworth projection and **33** as **36**. Because furanoses are five-membered ring compounds, assume they will take on a conformation that approximates the envelope conformation observed for cyclopentane (see Chapter 8, Section 8.5.3), which is essentially correct. The envelope conformation is also provided as a more modern representation of furanose derivatives.

28.7 **Draw the furanose structures for β-d-arabinose and α-d-xylose in Haworth projection.**

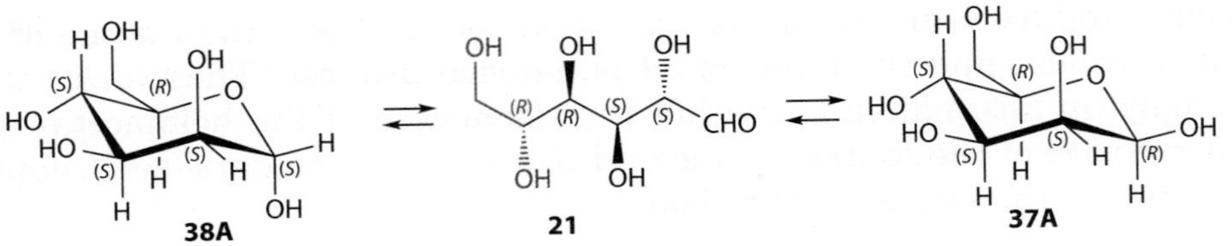

28.1.2 Mutarotation

Cyclization of an aldopentose to the hemiacetal gives a five-membered ring and an aldohexose will form a six-membered ring, each containing one oxygen (a tetrahydrofuran or a tetrahydropyran derivative; see Chapter 26, Section 26.4.2). The six-membered ring derivative is called a **pyranose**. Analogous to the conformation of cyclohexane, a six-membered ring compound should assume a chair conformation as the lowest energy form (see Chapter 8, see Section 8.5.4). Using d-mannose (**21**) as an example, hemiacetal formation will lead to the α-isomer **37** and the β-isomer **38**. Pyranose **37** is named α-d-mannopyranose and **38** is β-d-mannopyranose. The equilibrium among these three forms is similar to the equilibrium we observed for ribopyranose. In **37** and **38**, the hydroxyl groups are axial or equatorial, which is the proper representation.

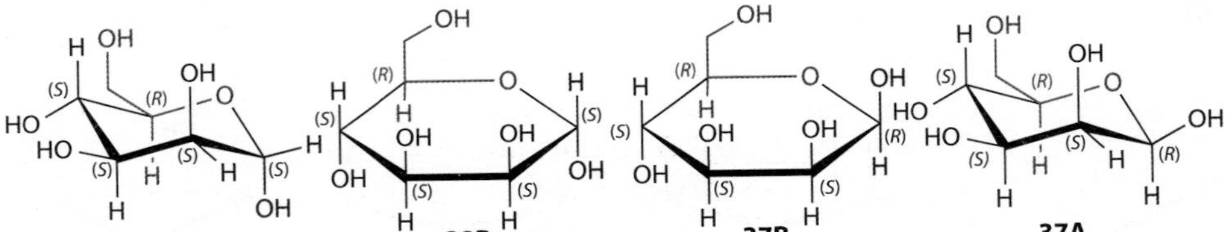

In some cases, it is useful to use Haworth projections **37B** and **38B** to illustrate the "sidedness" of the OH groups. The anomeric OH is axial in **38** and equatorial in **37** because the OH units are on opposite sides of the ring.

28.8 **Using chair representations, write the structure of α-d-allopyranose, β-d-altropyranose, α-d-gulopyranose, β-d-idopyranose, and α-d-talopyranose.**

Mannose has three species in equilibrium: α-mannopyranose, the aldehyde form, and β-mannopyranose. The position of this equilibrium can be measured, so it is possible to determine the relative percentage of these three species. This discussion must begin with an *assumption*, however. *Assume that the open-chain aldehyde form of mannose (21) accounts for less than 1% of the equilibrium (for most sugars in this chapter, it is about 0.1–0.5% or less); therefore, the pyranose forms are the major conformations.* Because these are chiral compounds, one way to measure the equilibrium is to measure the specific rotation (see Chapter 9, Section 9.2). The experimentally measured specific rotation for a pure sample of α-d-mannopyranose (**37**) is +29.3°. The experimentally measured specific rotation of a pure sample of β-d-mannopyranose is –16.3°. The specific rotation of the mixture can also be measured, and once the specific rotations of each pure enantiomer are known, the relative percentage of **37** and **38** may be determined (see Chapter 9, Section 9.2.4).

There is a curious phenomenon, however, called mutarotation. When pure α-d-mannopyranose is allowed to strand in aqueous solution, the specific rotation changes and eventually stabilizes at +14.5°. Similarly, when a sample of pure β-d-mannopyranose is allowed to strand in solution, the specific rotation of the final mixture is also measured to be +14.5°. This observation indicates two things. First, the sign of specific rotation is positive, which means that there is more of the α-d-mannopyranose than the β-d-mannopyranose, indicating that there is more of the axial anomer than the equatorial anomer. Remember from Chapter 8 (Section 8.6.1) that a substituent in the equatorial position has less $A^{1,3}$-strain than when that substituent is in the axial position, so it is expected to be lower in energy. *Because α-d-mannopyranose is the major conformation, a different effect must dominate the equilibrium.*

Second, a reaction must occur at the anomeric center of both isomers in aqueous solution. *This change in optical rotation of a carbohydrate when it is dissolved in a solution is called mutarotation.* Mutarotation occurs at the anomeric carbon as a direct result of the equilibrium between the open-chain aldehyde form and the two pyranose forms of mannose. When pure α-d-mannopyranose is in solution, an equilibrium is established with the aldehyde **21**. Once **21** is formed, it cyclizes to **37** or to **38**, establishing a new equilibrium.

If pure β-d-mannopyranose (**38**) is used, it also establishes an equilibrium with **21**, which means that **37** is also in solution. Therefore, one pure pyranose generates the aldehyde form and, once the aldehyde is in solution, both α- and β-d-mannopyranose are formed. Because there is an equilibrium among all three species, the final value of the specific rotation will reflect the relative percentages of **37** and **38** at equilibrium (*if the percentage of aldehyde is very small*). If this equilibrium value is +14.5° for d-mannopyranose, an equation is available (see Chapter 9, Section 9.2.4) to calculate the percentage:

%**37** (+29.3) + %**38** (–16.3) = +14.5, and because %**37** + %**38** = 1, %**37** = 1 – %**38**

Therefore, (1 – %**38**)(+29.3) + %**38**(–16.3) = +14.5 = 29.3 –29.3 %**38** – 16.3 %**38** = +14.5

%**38** (–29.3–16.3) = 14.5–29.3

%**38** (–45.6) = –14.8, so %**38** = –14.8/–45.6 = 0.3246

Therefore, %**38** = 32.46% and %**37** = 67.54%

This calculation, based on the fact that both **37** and **38** undergo mutarotation and on the assumption that very little **21** is present in the equilibrium, leads to the conclusion that at equilibrium about 67.5% of α-d-mannopyranose (**37**; the axial anomer) and about 32.5% of β-d-mannopyranose (**38**; the equatorial anomer) are present. The equilibrium should favor the more stable isomer, which in this case has the OH in the axial position.

28.9 **Calculate the equilibrium mixture for mannopyranose *if* the specific rotation of the final solution is –3.2°.**

Why is there more of the axial anomer? In Chapter 8, Section 8.6.1, it was clear that an axial substituent should have more $A^{1,3}$-strain, which makes it less stable when compared to the chair conformation with that substituent in an equatorial position. Before answering the question, another example will be examined to ascertain whether this is a general phenomenon.

Glucose (**20**) has two cyclic hemiacetal forms: α-d-glucopyranose and β-d-glucopyranose. If α-d-glucopyranose has a specific rotation of +112.2° and β-d-glucopyranose has a specific rotation of +18.7°, mutarotation occurs to give an equilibrium value of +52.5°. This corresponds to about 36% of the α-d-glucopyranose and 64% of the β-d-glucopyranose, in which the OH is in an equatorial position. In fact, mutarotation is a general phenomenon. The question relating to the axial anomer essentially asks whether the anomeric OH preferred an equatorial or an axial position. Clearly, in mannopyranose, the OH unit prefers an axial orientation, but in glucopyranose, the OH unit prefers an equatorial orientation.

To resolve this conflict, all of the substituents on the six-membered ring must be considered. For α-d-mannopyranose, there are two equatorial OH groups and two axial OH groups, whereas β-d-mannopyranose has three equatorial OH groups and one axial OH. For α-d-glucopyranose, there are four equatorial OH groups and zero axial groups, whereas β-d-glucopyranose has three equatorial OH groups and one axial OH. The preponderance of equatorial groups in β-d-glucopyranose is consistent with the view of cyclohexane derivatives that equatorial substituents have less $A^{1,3}$-strain (see Chapter 8, Section 8.6), which leads to a higher percentage of that conformation. ***This does not explain the observations with α-d-mannopyranose, however****. Simply looking at axial versus equatorial substituents does lead to a solution to the problem. Remember, however, that these are hydroxyl substituents and this is a pyran ring—not cyclohexane.*

Nonbonded interactions between the groups play an important role in pyranose derivatives. A phenomenon called the *anomeric effect is the tendency for an electronegative substituent in a pyran ring to prefer an axial orientation when attached to an anomeric carbon.* The anomeric effect arises from the designation of the C1 carbon of a pyranose as the anomeric carbon. *Isomers that differ only in the configuration at the anomeric carbon are called anomers.* Therefore, **40A** and **40B** are anomers. Further, **37A** and **38A** are anomers.

One explanation for the anomeric effect calls attention to the dipole moments of both heteroatoms. In an equatorial configuration, the dipoles of both heteroatoms are partially aligned and they repel. In an axial configuration, these dipoles are opposed and represent a lower energy state. Another explanation invokes a stabilizing interaction known as ***hyperconjugation*** between the unshared electron pair on the oxygen atom in the ring and the σ*-orbital for the axial C–O bond. (Note that σ*-orbitals were discussed in Chapter 23, Section 23.3.2, in connection with ultraviolet spectroscopy.) The oxygen of the C–O bond also has a lone pair of electrons and interaction occurs between that unshared electron pair and the σ*-orbital of the oxygen atom of the ring.

In 1-methoxycyclohexane (**39**), the equatorial conformation (**39A**) is preferred to the axial conformation, **39B**. In 1-methoxypyranose (**40**), however, the anomeric effect makes **40B** (with an axial OMe) in preference to **40A**, which has the equatorial OMe group. This means that a pyranose with an axial OH group at the anomeric carbon is preferred in the pyranose–aldehyde equilibrium. This certainly explains why α-d-mannopyranose is preferred, but it does not explain why β-d-glucopyranose is preferred. For glucose, the magnitude of the anomeric effect is small because of the equatorial OH at C2, and the lower conformational energy arises by having all-equatorial substituents. This leads to a preference for β-d-glucopyranose. *In most cases, however, mutarotation in a pyranose derivative leads to an equilibrium preference for the α-pyranose.*

28.10 Draw the structures of these two glucopyranoses.

28.11 If α-d-galactopyranose has a specific rotation of +150.7, α-d-galactopyranose has a specific rotation of +52.8, and the mutarotation value is +80.2°, draw all three equilibrium structures and calculate the relative percentages of the two galactopyranose structures.

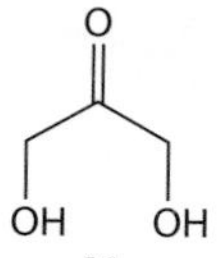

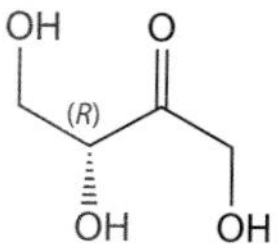

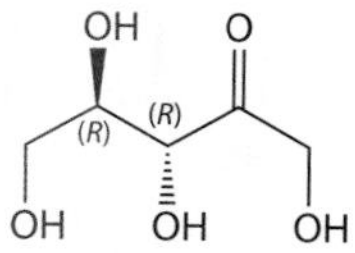

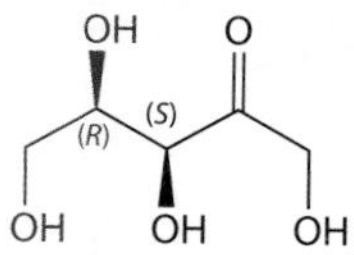

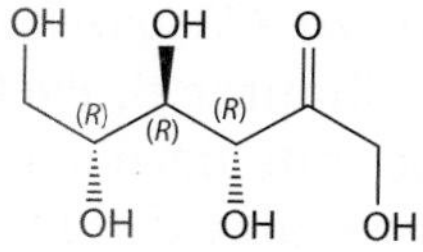

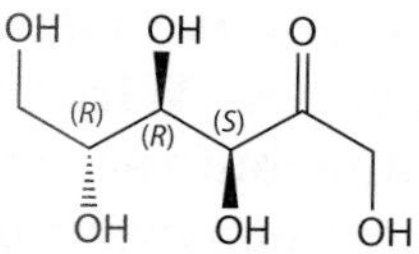

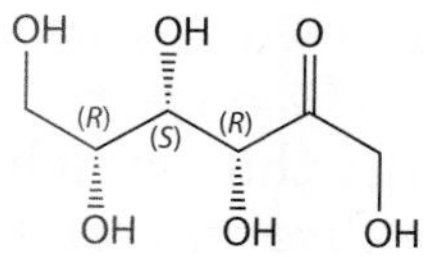

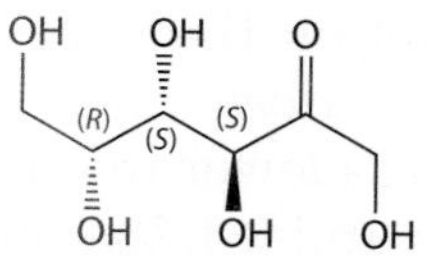

28.1.3 Ketose Monosaccharides

There are obviously many aldose monosaccharides and many ketose monosaccharides, but the discussion will focus only on the d diastereomers, as with the aldoses. The triose is 1,3-dihydroxy-2-propanone (**41**; also called glycerone) and the tetrose is d-glycero-tetrulose (**42**). Two pentoses are named d-ribulose (**43**) and d-xylulose (**44**) and four hexoses are named d-psicose (**45**), d-fructose (**46**), d-sorbose (**47**), and d-tagatose (**48**). All of these compounds are ketoses and are further classified according to the number of carbon atoms, as noted in Section 28.1. For example, **42** is a ketotetrose, **43** is a ketopentose, and **45** is a ketohexose. For **43–48**, cyclization is possible to form a hemiketal. Just as aldehydes and alcohols react to form a hemiacetal, ketones and alcohols react to form a hemiketal (see Chapter 18, Section, 18.6). For **43**, a five-membered ring (a furanose) is formed if the terminal CH₂OH unit (in **violet**) reacts with the ketone carbonyl. The two anomers formed are **49** (α-d-ribulofuranose) and **50** (β-d-ribulofuranose).

28.12 Draw l-glycerotetrulose and l-xylulose.

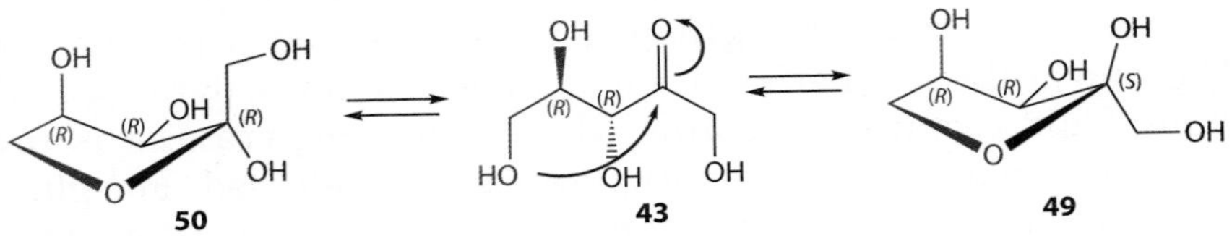

These furanose derivatives undergo mutarotation in a manner similar to the aldofuranose compounds. The ketohexoses form pyranose derivatives via the OH unit at C6 (in **violet**). d-Sorbose (**47**), for example, will form α-d-sorbopyranose (**51**) and β-d-sorbopyranose (**52**), and d-fructose (**46**) will form α-d-fructopyranose (**53**) and β-d-fructopyranose (**54**).

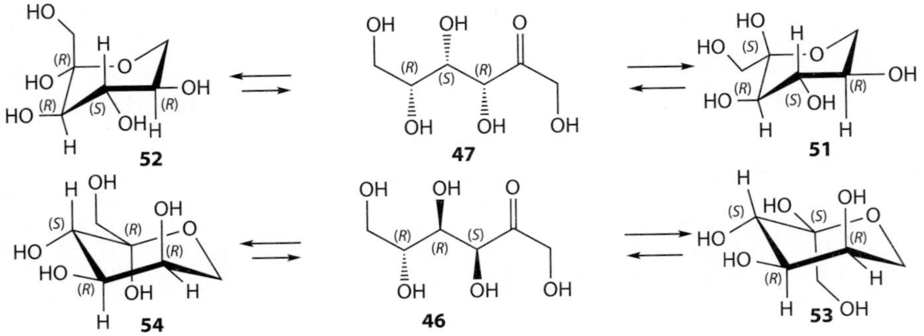

For keto-hexoses, however, the pyranose derivatives are *not* the major hemiketals that are formed. *Cyclization via the OH unit at C5 will lead to furanose derivatives as the preferred structure.* **Why?** In Chapter 8, cyclopentane was lower in energy than cyclohexane, but the compounds here are cyclic ethers. Indeed, the tetrahydropyran ring is somewhat more stable than the tetrahydrofuran ring. Stabilization of the hydroxyl groups and the hydroxymethyl groups must account for the difference in stability in this case. Comparing **46** and **47** and using the OH at C5 of d-sorbose, α-d-sorbofuranose (**55**) and β-d-sorbofuranose (**56**) are formed, as are α-d-fructofuranose (**57**) and β-d-fructofuranose (**58**) from d-fructose.

As mentioned, these compounds undergo mutarotation, but the equilibrium is much more complex than that observed with most of the aldopyranoses. Examine fructose (**46**): two pyranose compounds, **53** and **54**, and two furanose compounds, **57** and **58**, are all in equilibrium. The relative percentages of each are 2.5% of **53**, 65% of **54**, 6.5% of **57**, and 25% of **58**.[1] When β-d-fructopyranose (**51**) with a specific rotation of −133.5° is dissolved in water, it undergoes mutarotation to give an equilibrium specific rotation of −92°, which is the value obtained when the listed percentages of **53**, **54**, **55**, and **56** are present at equilibrium. Not all keto-derivatives give significant amounts of both furanose and pyranose derivatives. For example, d-xylose forms 36.5% of α-d-xylopyranose and 63% of β-d-xylopyranose, but <1% of xylofuranose derivatives. Specifics for an individual compound should be obtained from the literature rather than attempting to use a generalized rule of thumb.

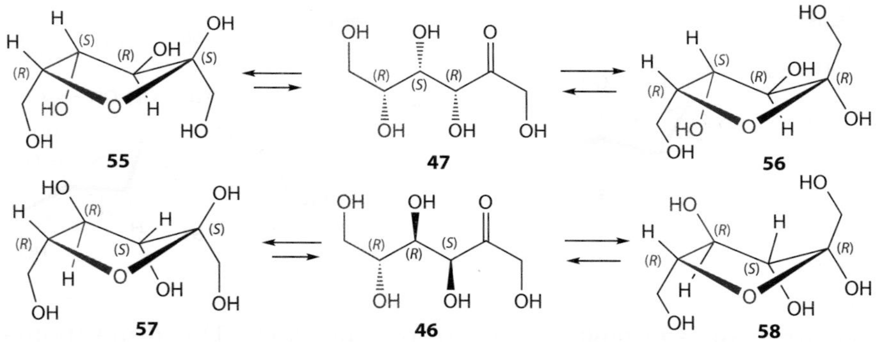

28.13 Draw the structure of α-d-tagatofuranose, β-d-psicopyranose, and β-d-psicofuranose.

28.2 Biological Relevance. Oligosaccharides and Polysaccharides

If two monosaccharides are coupled together, the result is a disaccharide (a molecule containing two sugar units). Two monosaccharides are coupled together by reaction of a hydroxyl unit of one saccharide with a carbonyl of the second saccharide to give an acetal linkage or a ketal linkage. The anomeric partner is usually the "glycosyl donor" and the nonanomeric partner is the "acceptor."

The reaction of α-d-altropyranose (**59**) and α-d-mannopyranose (**37**) can, in principle, couple via any of the OH units. There are usually two coupling

Using this system, O-α-d-ribofuranosyl-(1→4)-α-d-glucopyranosyl-(1→1)-α-d-altropyranose becomes O-α-d-Rib$f$-(1→4)-α-d-Glc$p$-(1→1)-α-d-Alt$p$. With this system, the structure for a hexasaccharide used as an example is O-α-d-Fru$f$-(1→4)-α-d-Gcl$p$-(1→1)-β-d-Gal$p$-β-d-Tal$p$-(1→1)-α-d-Ido$p$-(1→2)-α-d-All$p$.

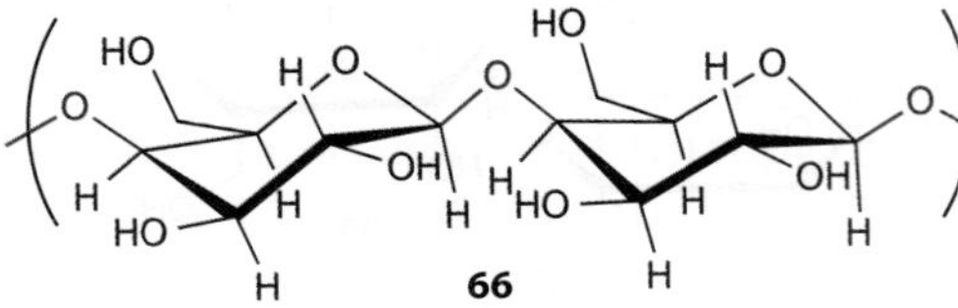

cellulose   amylose

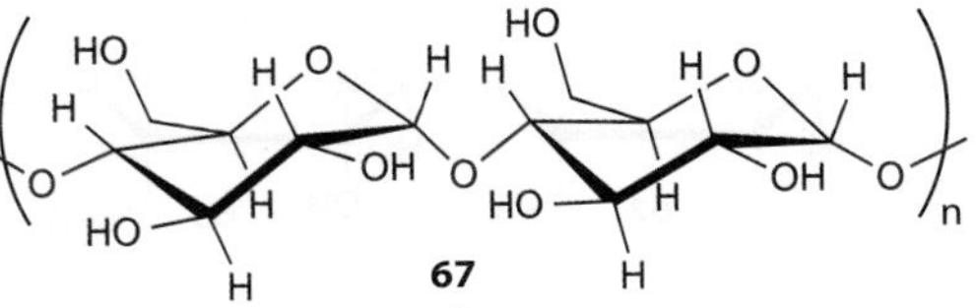

For polysaccharides, even the shorthand notation is unwieldy. Most of these compounds are given common names and, because they are usually polymeric, the repeating monosaccharide or disaccharide units are often shown. Many of these are **homopolymers**; that is, the poly(saccharide) is formed by coupling only one monosaccharide unit. Cellulose, for example, is a linear poly(glucopyranose); coupled in a (1→4)-β-d manner as shown in **66**, it is a major constituent of plant cell walls. Amylose (a constituent of starch) is a linear poly(glucopyranose) coupled (1→4)-α-d (see **67**), and inulin (found in dandelions) is a linear fructofuranose coupled (2→1)-β-d (see **68**).

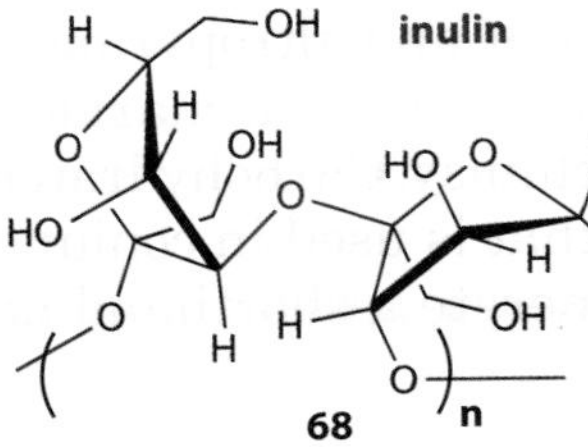

inulin   amylopectin

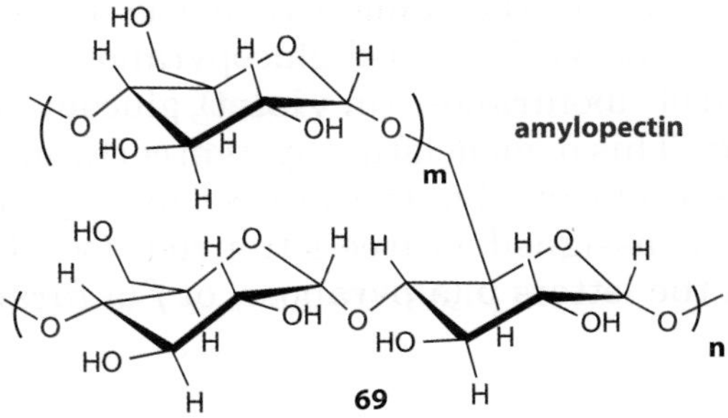

There are branched polysaccharides as well as linear polysaccharides. This simply means that there is a linear chain with other polymeric chains branching from the main one. An example is amylopectin (another constituent of starch), which has a linear (1→4)-α-d-glucopyranosyl chain with (1→6)-α-d-glucopyranosyl branches (see **69**). Similarly, fucoidan (found in brown seaweed) has a (1→2)-α-l-fucosofuranosyl chain with (1→4)-α-l-fucosofuranosyl branches (see **70**). Recall that α and β are reversed for the l series relative to the d series. The sugar fucose is 6-deoxygalactose. Therefore, fucofuranose has a methyl group at C6, as shown in **blue** in structure **70**.

In addition to homopolymeric polysaccharides, there are polysaccharides composed of more than one monosaccharide unit. For example, coniferous woods contain a polysaccharide that is a linear chain of d-glucopyranosyl and d-mannopyranosyl units. Plant cell walls also contain a polysaccharide that is a branched structure containing l-arabinofuranosyl and d-xylofuranosyl units.

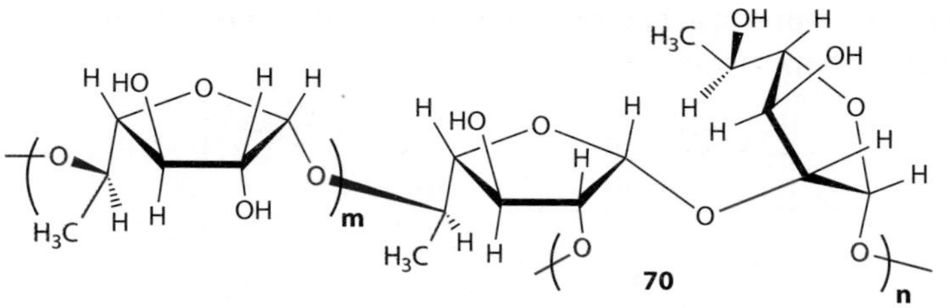

fucoidan

28.3 Reactions of Carbohydrates

Reactions of carbohydrates are closely related to the normal reactions of alcohols, aldehydes, and ketones seen in previous chapters. The presence of multiple functional groups is problematic, however, because one group may interact with another. The formation of furanose and pyranose rings in the previous sections via a hemiacetal or a hemiketal reaction and of disaccharides via an acetal or a ketal reaction is one example where the functional groups interact with each other. In many other reactions, the hydroxyl units behave independently or the aldehyde or ketone unit behaves independently. This section will explore several fundamental carbohydrate reactions.

28.3.1 Oxidation and Reduction

A discussion of reactions will begin by introducing another classification scheme for carbohydrates. *There are reducing sugars and nonreducing sugars.* These are generally categorized by whether or not they react with Fehling's solution (aqueous copper (II) sulfate and sodium tartrate, **72**) or with Tollens' solution (silver nitrate in ammonia). These reagents are named after Hermann von Fehling (Germany; 1812–1885) and Bernhard Tollens (Germany; 1841–1918). *A carbohydrate must have an aldehyde unit in its structure in order to react with these reagents because the reaction oxidizes the aldehyde to a carbonyl compound* (see Chapter 17, Section 17.2). Therefore, aldoses, but not ketoses, may be categorized as reducing sugars.

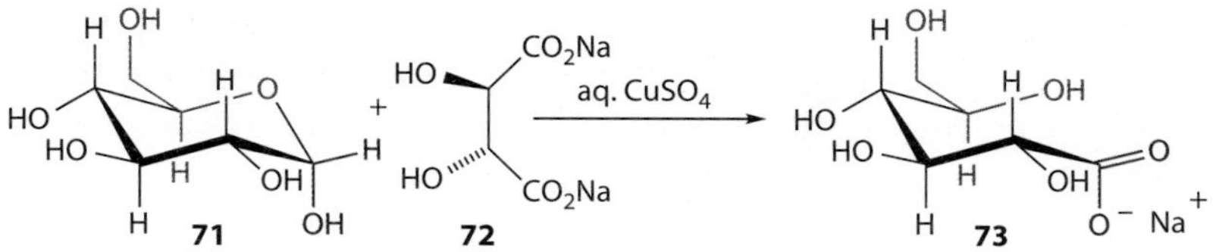

A disaccharide that is coupled 1→4 may have an aldehyde unit and be a reducing sugar, but a disaccharide coupled 1→1 cannot have an aldehyde unit available and is not a reducing sugar. An example is the reaction of α-d-glucopyranose (**71**) with Fehling's solution. The products are the carboxylate

salt, **73**, and a brick-red precipitate of copper oxide (Cu_2O), which is the characteristic sign that the reaction worked.

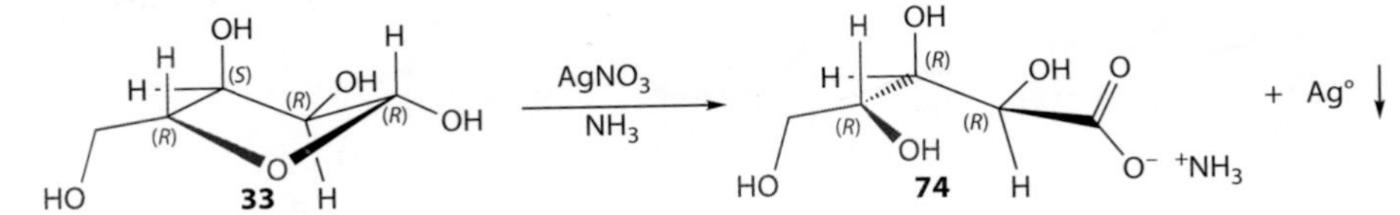

Monosaccharide **71** is said to be a reducing sugar because it is oxidized by
Fehling's solution. The reaction of α-d-ribofuranose (**33**) with Tollens' reagent
also gives a carboxylate salt (**74**), along with metallic silver (which coats the reaction vessel as a silver mirror and is diagnostic of a positive test). Monosaccharide
33 is a reducing sugar. If α-d-fructofuranose (**57**) is mixed with either Fehling's
solution or Tollens' reagent, no reaction occurs and it is not a reducing sugar.

**28.17 Draw the products of a reaction between erythrose and Tollens'
reagent.**

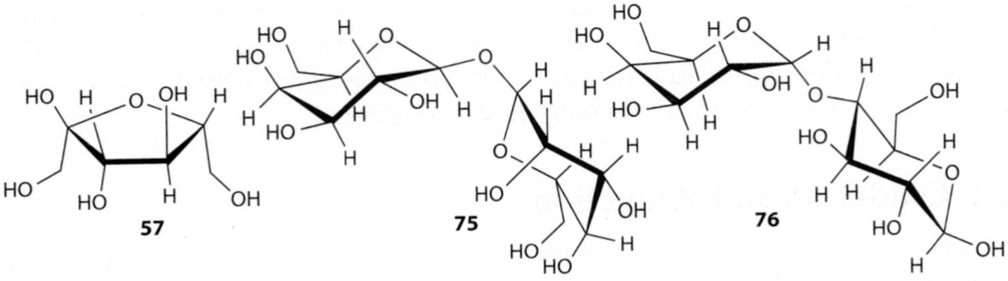

Disaccharides are potentially more difficult to classify. When *O*-α-d-
glucopyranosyl-(1→4)-α-d-glucopyranose (**75**) is mixed with Fehling's solution,
the glucopyranose unit can undergo mutarotation, which means that an aldehyde unit is available and it is a reducing sugar. When *O*-α-d-glucopyranosyl-
(1→1)-α-d-glucopyranose (**76**) is mixed with Fehling's solution, however, both
anomeric carbons are tied up in the ketal linkage and mutarotation cannot
occur. If no aldehyde unit is available to react, **76** is not a reducing sugar.

**28.18 Categorize each of the following as a reducing sugar or a nonre-
ducing sugar: *O*-α-d-Ara*p*-(1→4)-α-d-Glc*p*; *O*-α-d-Gul*p*-(1→1)-α-
d-Man*p*; *O*-α-d-Fru*f*-(1→3)-α-d-ido*p*; *O*-α-d-Ara*f*-(1→1)-α-d-Glc*p*;
and *O*-α-d-Gul*p*-(1→1)-α-d-Glc*p*.**

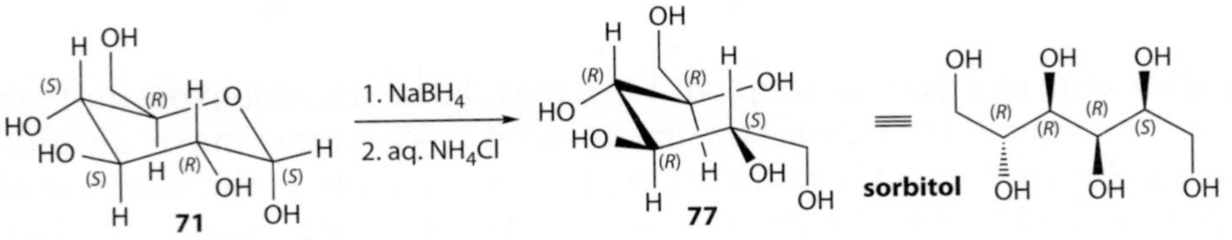

Carbohydrates are easily reduced under the proper conditions. The aldehyde unit is converted to a primary alcohol. When glucopyranose (**71**) is reduced with sodium borohydride (Chapter 19, Section 19.2.1), the product is d-glucitol (**77**), which is an example of a glycitol (see **3**). This compound is also known as sorbitol, and it is a naturally occurring sweetening agent found in many berries, plums, and apples and in seaweed and algae. It is sold commercially and added to many foods, particularly candy, as a natural sweetening agent. It has also been used as a sugar substitute for those who have diabetes.

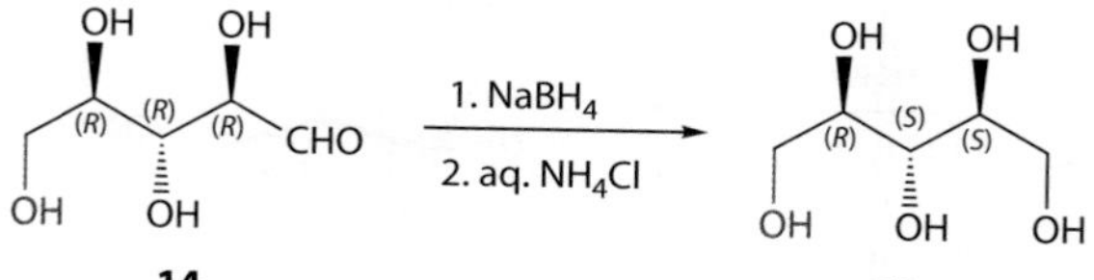

Most of the aldoses are similarly reduced, and *the name of the product is generated by dropping "ose" and adding "itol."* Therefore, reduction of d-ribose (**14**) gives d-ribitol (**78**) and d-mannose (**21**) is reduced to give d-mannitol (**79**). Mannitol is found in many plants, including seaweed, and is used to make artificial resins and plasticizers. The aldehyde unit of an aldose is reduced with sodium amalgam (Na/Hg), with Raney nickel (see Chapter 19, Section 19.3.4) in refluxing ethanol, or by catalytic hydrogenation. In all cases, the product is a glycitol. Reduction of a ketose leads to some stereochemical problems. Hydrogenation of d-fructose (**46**), for example, gives both the R- and the S-alcohols when the carbonyl is reduced, so the products are a mixture of d-glucitol (**80**) and d-mannitol (**79**).

28.19 **Draw the product formed when l-aribinose is reduced.**

28.20 **Draw and name the product formed by the catalytic hydrogenation of d-threose.**

28.21 **Briefly explain why d-erythritol has a specific rotation of zero.**

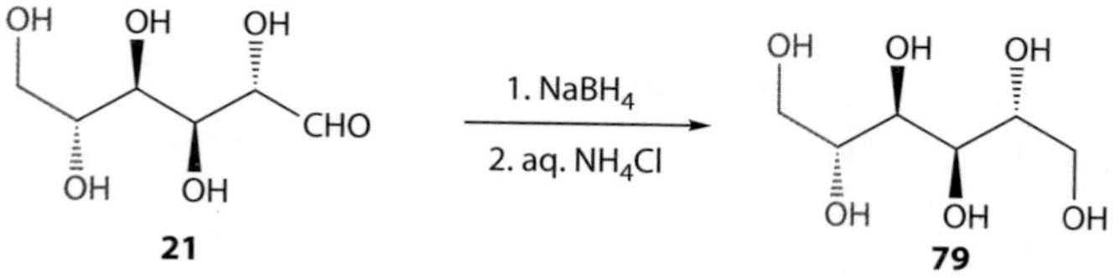

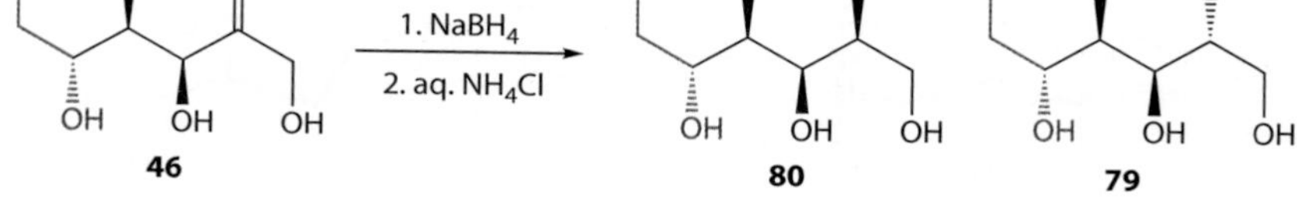

28.3.2 Blocking Selected OH Units. Protecting Groups

There are many OH units in a carbohydrate; using the proper techniques, one may be selected for reaction, leaving the others untouched. Most of the time, however, reagents react with all available hydroxyl units. Many reactions of carbohydrates demand control of the reactivity for individual hydroxyl groups. When β-d-allopyranose (**81**) reacts with acetic anhydride and sodium acetate, the product is the pentaacetate (penta-*O*-acetyl-β-d-allopyranose, **82**).

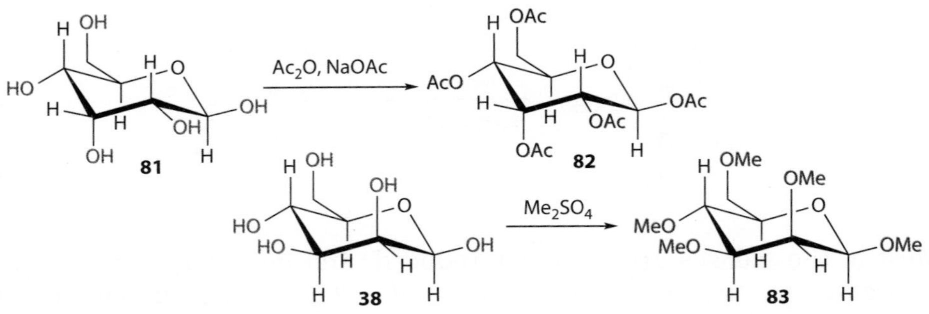

Similar results are obtained when benzoyl chloride and pyridine are used. When β-d-mannopyranose (**38**) reacts with dimethyl sulfate (Me₂SO₄), the product is the pentaether, **84** (methyl tetra-*O*-methyl-β-d-mannopyranoside). Dimethyl sulfate is the dimethyl ester of sulfuric acid (see Chapter 20, Section 20.12), and it is a common reagent for converting an OH unit to an OMe unit. *The suffix "ide" is used rather than "ose" for the ether.* If **38** is treated with methanol and an acidic catalyst, the product is the monoether, methyl β-d-glucopyranoside (**84**).

28.22 Draw the structure of the pentabenzoyl derivative of α-d-glucopyranose.

28.23 Draw the product of the reaction of β-d-glucopyranose with benzyl alcohol and an acid catalyst.

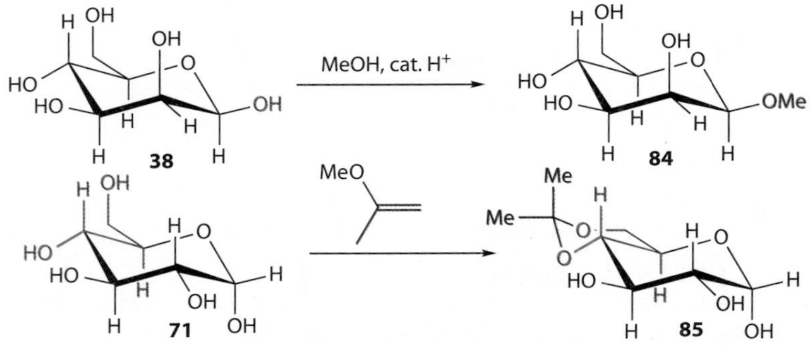

When a carbohydrate reacts with a vinyl ether such as 2-methoxy-1-propene, the product is a ketal. The reaction of α-d-glucopyranose (**71**) with this vinyl ether in the presence of an acid catalyst leads to **85**. This reaction essentially "protects" the C4- and C6-OH units as a 1,3-dioxane (see Chapter 18, Section 18.6.6). (Protecting groups were introduced in Chapter 25, Section 25.4.3.) The name of **102** reflects the presence of this ketal; it is 4,6-*O*-isopropylidene-α-d-glucopyranose. The isopropylidene group is the CMe_2 unit, and the *O*-isopropylidene name refers to the presence of the acetone ketal unit. When **71** reacts with acetaldehyde in the presence of an acid catalyst, the product is 4,6-*O*-ethylidene-α-d-glucopyranose, **86**. The ethylidene term refers to the CHEt unit and benzylidene refers to the CHPh unit.

28.24 Draw the product of a reaction between benzaldehyde and 4,6-*O*-benzylidene-α-d-glucopyranose.

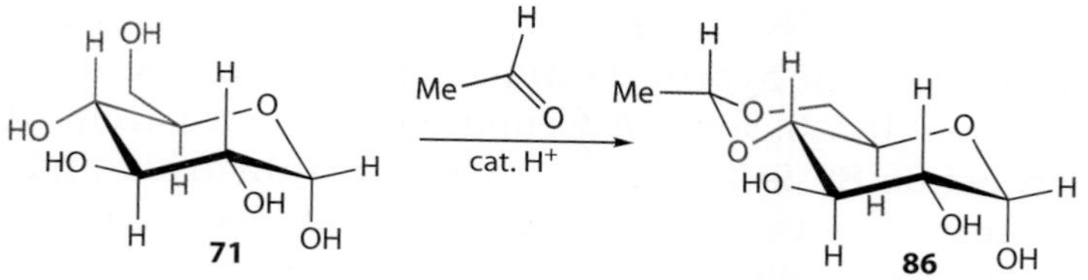

28.4 Synthesis of Carbohydrates

This section will discuss several of the more common methods for the synthesis of carbohydrates, particularly those that use reactions seen in previous chapters. Because carbohydrates are multifunctional molecules, the chemistry in this section will illustrate the power and the problems that arise with such compounds. A common starting material for the synthesis of a carbohydrate is another carbohydrate. Many carbohydrates are inexpensive and available in large quantities, which makes them ideal.

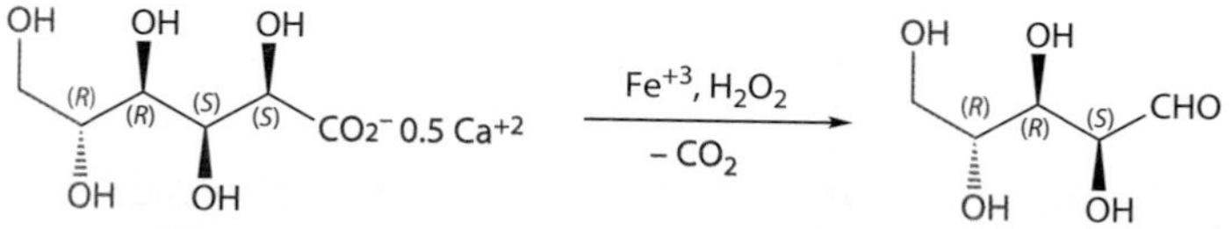

In Chapter 19 (Section 19.2.4), a carboxylic acid was reduced to an aldehyde. When the salt of an aldonic acid such as the calcium salt of d-gluconic acid (**87**) is treated with hydrogen peroxide and iron(III), the product is an aldose *of one fewer carbon atom*—in this case, **14** (d-ribose). This reaction, which oxidatively cleaves one carbon (lost as CO_2), is called the **Ruff degradation**, after Otto Ruff (Germany;

1871–1939). This oxidative cleavage reaction is a good synthesis of "smaller" carbon aldoses from "larger" carbohydrates (a chain-shortening reaction).

28.25 Draw the product of a Ruff degradation of d-altronic acid.

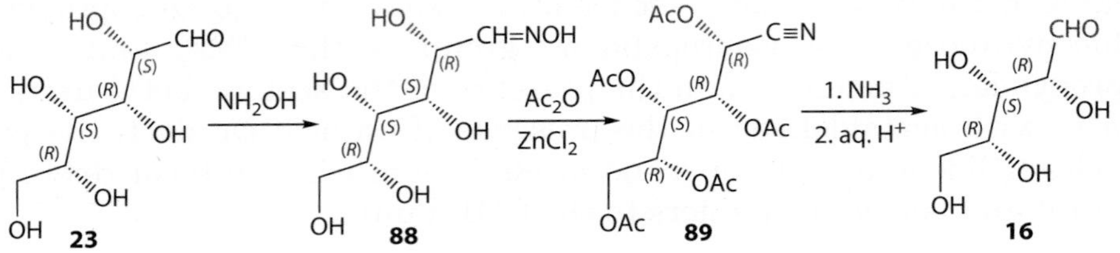

Another method for decreasing the number of carbon atoms in a carbohydrate while producing another sugar is called the **Wohl degradation**, after Alfred Wohl (Germany; 1863–1939). In this sequence, a monosaccharide such as d-idose (**23**) reacts with hydroxylamine (NH₂OH) to give an oxime (**88**; see Chapter 18, Section 18.7.3, for the formation of oximes from aldehydes and ketones). When oxime **88** is treated with acetic anhydride and zinc chloride (ZnCl₂), all OH units are converted to their acetate ester and the oxime unit is converted to a nitrile (**89**). This is a dehydration reaction. In Chapter 19 (Section 19.4.3), nitriles were reduced to an aldehyde; when **89** is treated with ammonia and then with aqueous acid, the net result is reduction of the nitrile to an aldehyde unit—in this case, d-xylose (**16**).

28.26 Draw the product of a Wohl degradation of d-talose.

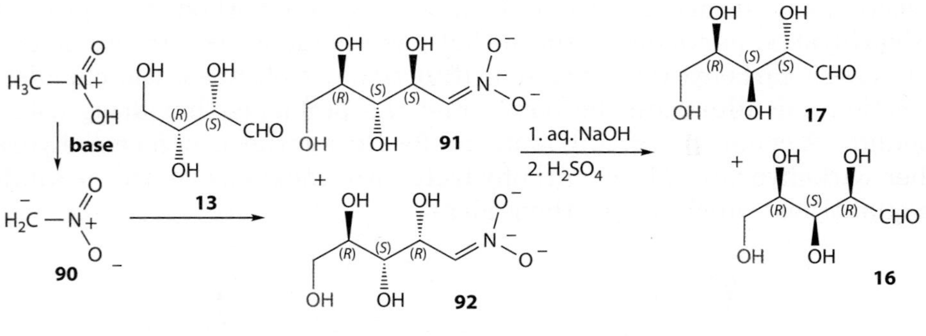

Chain-lengthening reactions can be applied to carbohydrates. As the term implies, *the carbohydrate chain is extended by one carbon.* One important method uses a transformation called the **Nef reaction**, after John U. Nef (Switzerland/ United States; 1862–1915). When d-threose (**13**) reacts with nitromethane in the presence of sodium methoxide as a base, the enolate anion of nitromethane (**90**) reacts with the aldehyde unit of **13** (an enolate-aldehyde condensation reaction; see Chapter 22, Section 22.2, for related reactions) to give two products: **91** and **92**.

The two products arise from the fact that the enolate-aldehyde reaction generates both the R- and the S-alcohol units in the product. Compound **91** is 1-deoxy-1-nitro-d-arabinitol and **92** is 1-deoxy-1-nitro-d-xylitol. Deoxy indicates that an oxygen atom has been removed from C1 relative to **13**. When **91** is treated with NaOH and then with sulfuric acid, the product is d-lyxose (**17**). Under the same conditions, **92** is converted to d-xylose, **16**. In principle, this process can be repeated indefinitely to produce even longer chain aldose monosaccharides.

28.27 Draw the product of a Nef reaction with l-lyxose.

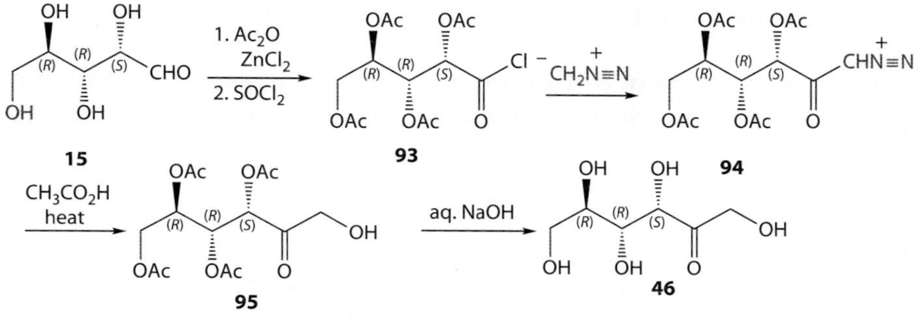

Another chain-extension method is called the **Fischer–Kiliani synthesis**. This reaction is named for Emil Fischer, whose work was described earlier, and Heinrich Kiliani (Germany; 1855–1945). The synthesis generates a ketose from an aldose. If d-arabinose (**15**) is reacted first with acetic anhydride and zinc chloride ($ZnCl_2$) and then with thionyl chloride, the hydroxyl groups are acetylated and the aldehyde unit is oxidized to an acid, which is converted to acid chloride (**93**). Oxidation of an aldehyde to a carboxylic acid was described in Chapter 17 (Section 17.2.2), and conversion of a carboxylic acid to an acid chloride was seen in Chapter 20 (Section 20.3). When the acid chloride reacts with diazomethane (CH_2N_2), a carbon is added to the molecule to give **94**, which is called a diazoketone. When the diazoketone is heated with acetic acid, the diazo unit is converted to an alcohol (**95**) and saponification (Chapter 20, Section 20.2) of the acetate unit gives the ketose, d-fructose (**46**). Diazomethane, a relatively specialized reagent, has not been mentioned previously in this book.

28.28 Draw the final product of a Fischer–Kiliani synthesis that begins with l-xylose.

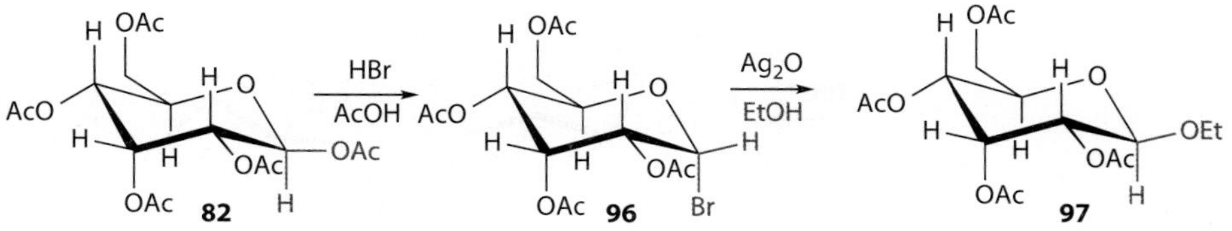

Disaccharides are prepared by coupling two monosaccharide units together. This may be done in several ways. An effective method is the **Koenigs–Knorr synthesis**,[3] after Wilhelm Koenigs (Germany; 1851–1906) and Edward Knorr (Germany; d. 1901). This method couples a carbohydrate and an alcohol. β-d-Allopyranose pentaacetate (**82**) is prepared previously from d-allose (**18**) and, if reacted with HBr in acetic acid, the α-bromide (**96**) is formed (2,3,4,6-tetra-*O*-acetyl-α-d-allopyranosyl bromide). When this bromide reacts with ethanol in the presence of silver oxide (Ag$_2$O), the product is the β-anomer, ethyl 2,3,4,6-tetra-*O*-acetyl-α-d-allopyranoside (**97**). It is possible to modify the procedure to make disaccharides. If bromide **97** is reacted with tetraacetate **98** (derived from α-d-idose), reaction with silver oxide and then sodium methoxide leads to the disaccharide **99** (*O*-α-d-allopyranosyl-[1→6]-α-d-idosopyranose).

28.29 Draw a synthesis of α-d-glucose-(1→1)-α-d-idosopyranose from α-d-glucopyranose pentaacetate using the Koenigs–Knorr method.

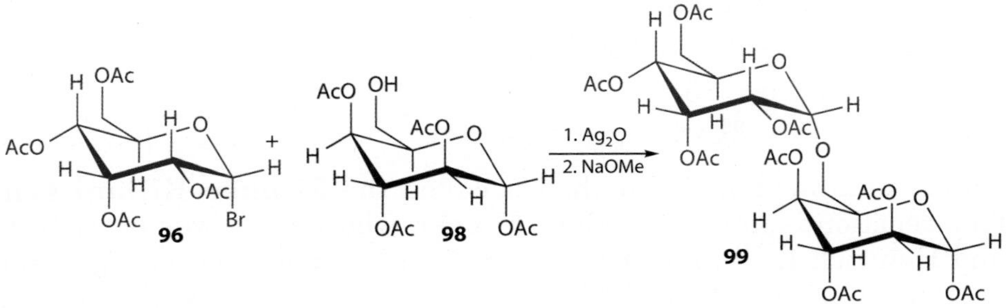

A newer but related method for the preparation of disaccharides or even poly(saccharides) uses a carbohydrate that has a trichloroacetimidate group (Cl$_3$C–C=NH) at the anomeric carbon rather than a halide. If 2,3,4,6-tetra-*O*-acetyl-α-d-glucopyranosyl bromide (**100**) reacts with trichloroacetonitrile, the product is *O*-α-d-glucopyranosyl trichloroacetamidate **101**. Chloroimidates such as **101** react with a variety of nucleophiles, including alcohols. If **101** reacts with **100** and if that product is treated with barium hydroxide, the product is the disaccharide *O*-β-d-glucopyranosyl-[1→1]-β-d-glucopyranose (**102**).

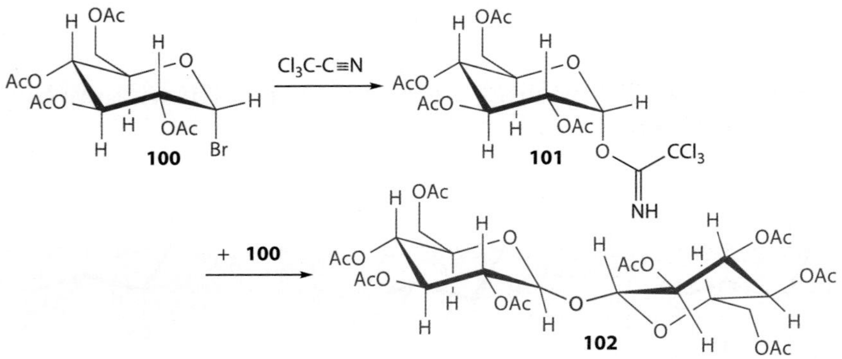

28.5 Biological Relevance. Nucleosides and Nucleotides (Heterocycles Combined with Sugars)

It is clear that carbohydrates are multifunctional compounds because they contain hydroxyl units, aldehyde or ketone units, and sometimes carboxyl units. Carbohydrates may also contain an amine unit to give amino sugars. Examples are 2-amino-2-deoxy-β-d-mannopyranose (d-mannosamine, **103**), 3-amino-3,6-dideoxy-β-d-glucopyranose (**104**), and 3-deoxy-β-d-idopyranose (**105**). These compounds are known as deoxy sugars because they have an OH unit replaced with NH$_2$ or H, or another group.

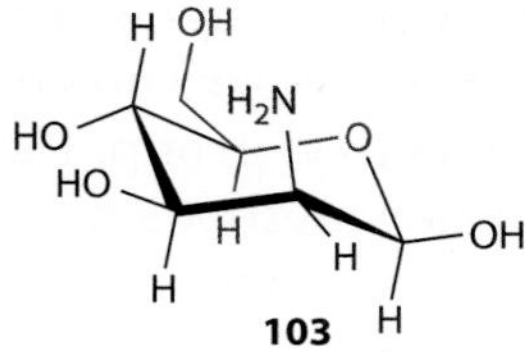

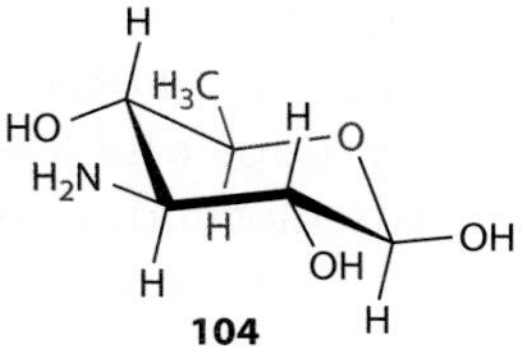

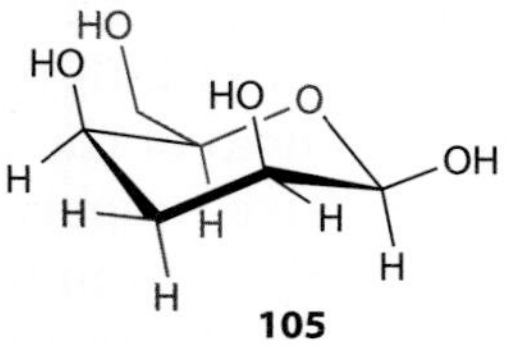

It is also possible to attach more complex amines to sugar molecules, especially pyrimidine bases such as thymine (**106**) and purine bases such as adenine (**107**). These heterocyclic bases were introduced in Chapter 26 (Sections 26.1.2 and 26.4.1). Of particular interest are the compounds that arise when a purine or pyrimidine base is attached to a ribofuranose unit (illustrated by structure **108**, which contains a thymine unit).

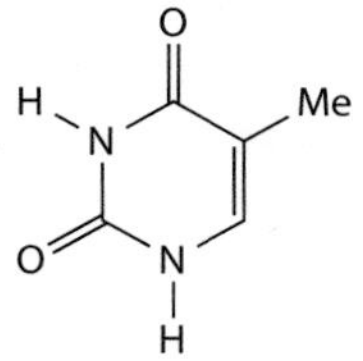

Other sugars may be used, however. The first reported synthesis of this type of compound involved the reaction of 2,3,4,6-tetra-O-acetyl-α-d-glucopyranosyl bromide (**100**) with the silver salt of a purine called theophylline (**109**) in refluxing xylene. The reaction coupled these two fragments, and subsequent removal of the acetate groups by saponification gave **110**. This product is called 7-β-d-glucopyranosyltheophylline, and it was first synthesized by Emil Fischer and B. Helferich (Germany) in 1914.[4] This early work did not give good results in the reaction if pyrimidines were used because O-glycosides were formed rather than the desired N-glycosides.

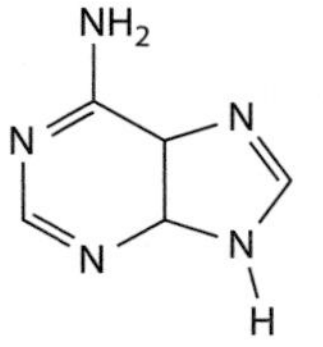

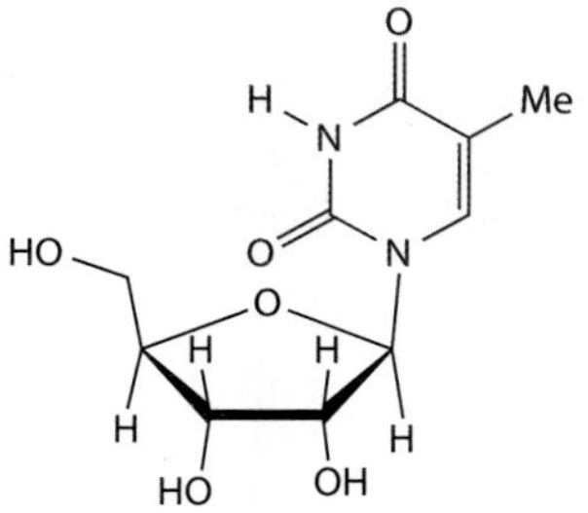

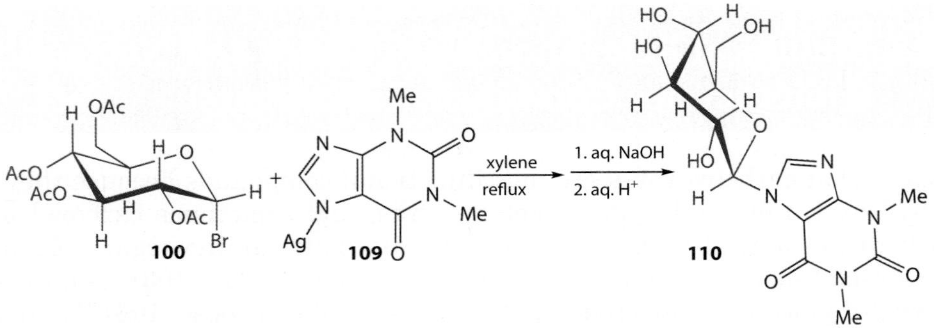

A later improvement on this approach led to the reaction of a chloromercury salt of the heterocyclic amine with a halo-sugar. The reaction of chloromercuric-6-benzamidoadenine (**112**) and 2,3,5-tri-*O*-acetyl-d-ribofuranoyl chloride (**111**), for example, gives **113**, 6-(β-d-ribofuranoyl)adenine (otherwise known as aden-osine), after removal of the acetate groups and the *N*-benzyl group. Pyrimidine derivatives are also prepared by this method, and other methods are available that may be used to prepare these compounds. In one, a purine (cytosine, **114**) reacts with **111** in the presence of mercuric cyanide and nitromethane to give 1-(β-d-ribofuranoyl)cytosine (otherwise known as cytidine, **115**), after removal of the protecting groups.

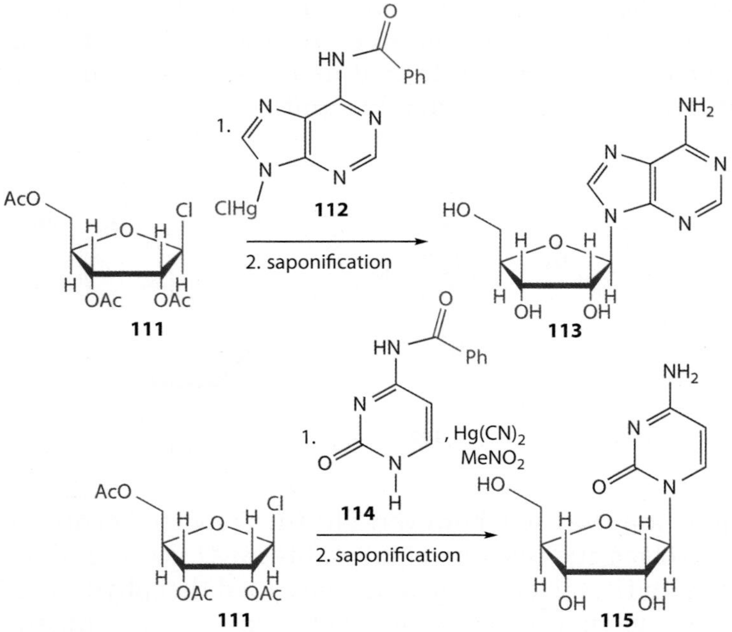

An important reaction is used by many today to prepare many nucleosides. The Vorbrüggen glycosylation reaction[5,6] (H. Vorbrüggen, Germany) couples a highly protected heterocycle, such as **116**, protected as the trimethysilyloxy derivative. Reaction of **116** and tetraacetyl ribofuranose **117** in the presence of a Lewis acid such as SnCl4 or TiCl$_4$ leads to nucleoside derivative **118**. Note that the O(SiMe$_3$)$_3$ groups are removed during the course of the reaction to reveal the base in **118** as uridine. This means that **116** is a protected form of uracil. Other heterocyclic bases may be used in the reaction; each base requires a unique set of protecting groups that will not be discussed here.

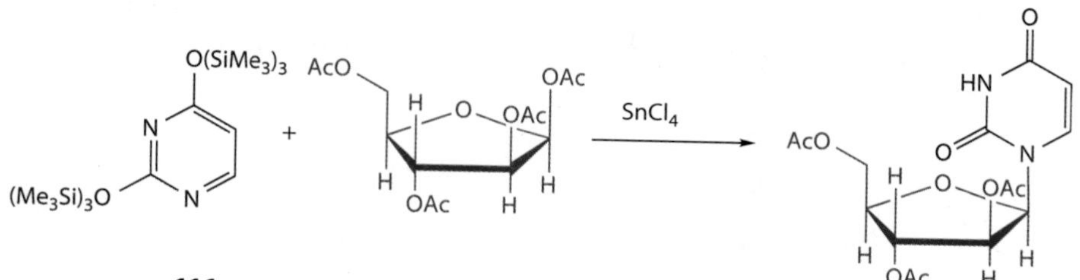

Ribose-purines and ribose-pyrimidines are particularly important because these structures are found in DNA and RNA. DNA and RNA are known as nucleic acids, and they are biopolymers made up of repeating deoxyribofura-nose or ribofuranose units. *Nucleic acids are oligo (sugar phosphates) because the backbone of these compounds is phosphorous.* This monomeric unit is called a *nucleoside, which is a ribose derivative coupled with a purine or a pyrimidine.* A ribofuranose with a purine or pyrimidine at C1 is called a **ribonucleoside**, or simply a **nucleoside**.

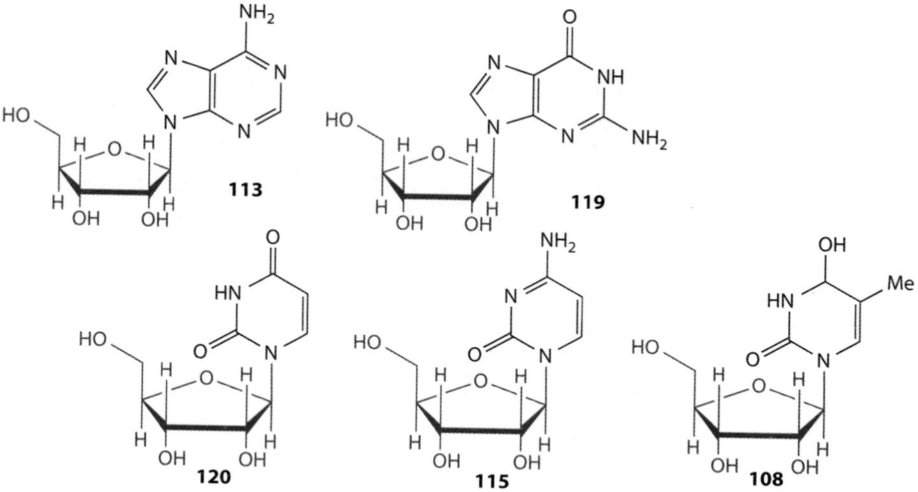

Five particularly important nucleosides are derived from d-ribofuranose: two purine derivatives—adenosine (**113**) with an adenine unit on the ribose and guanosine (**119**) with a guanine unit on the ribose—and three pyrimidine derivatives—uridine (**120**) with a uracil unit, cytidine (**115**) with a cytosine unit, and thymidine (**108**) with a thymine unit. Two important deoxyribofura-nose derivatives are formed by using a ribose unit without an OH at C2. Purine derivative **121** is named 2′-deoxyadenosine, **122** is 2′-deoxyguanosine, **123** is 2′-deoxyuridine, **124** is 2′-deoxycytidine, and **125** is 2′-deoxythymidine. The ribofuranose derivatives are the monomeric units found in RNA (ribonucleic acids) and the 2′-deoxyribofuranose derivatives are the monomeric units found in DNA (deoxyribonucleic acids).

28.30 Draw (a) the arabinofuranose formed when it is coupled to cytosine and (b) the xylofuranose formed when it is coupled with guanine.

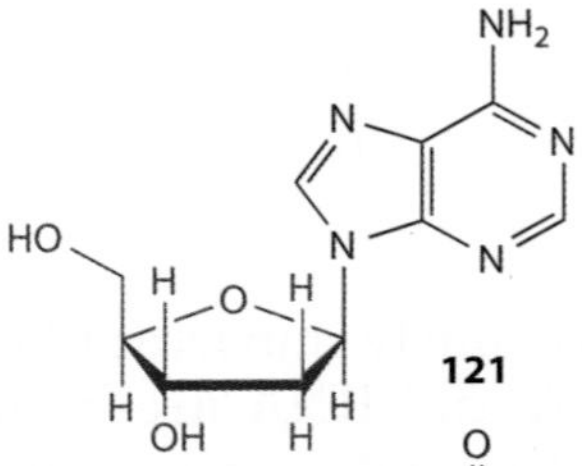

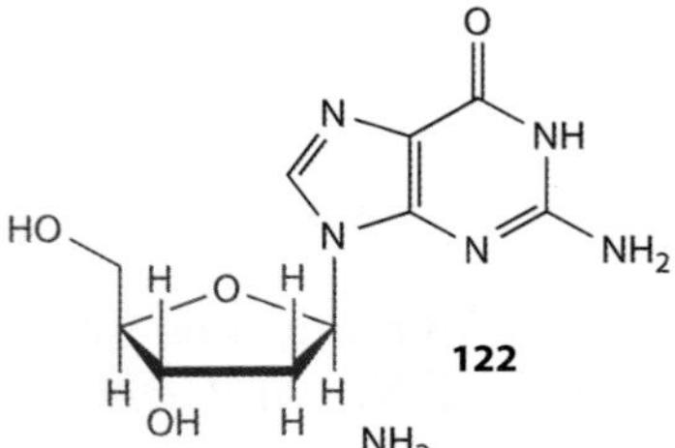

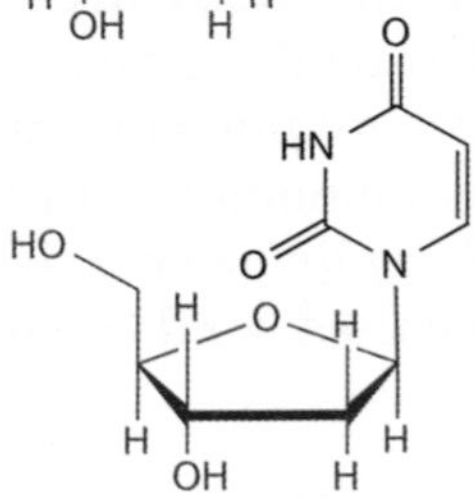

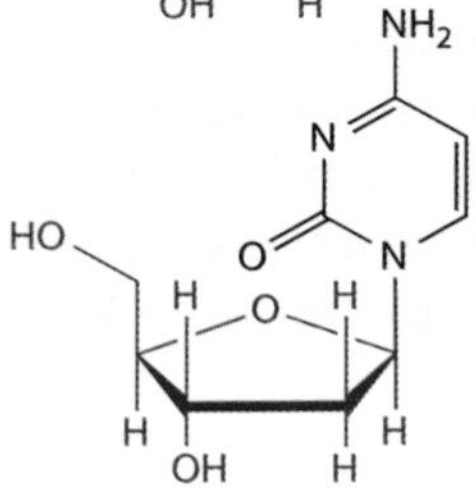

Each nucleoside can react with phosphoric acid or a phosphoric acid derivative to form a phosphate ester (see Chapter 20, Section 20.12). *The phosphate ester of a nucleoside is called a nucleotide.* Using adenosine (**113**) as an example, there are three possible monophosphate esters: **126**, **127**, and **128**. These compounds are named adenosine 5′-monophosphate, adenosine 3′-monophosphate, and adenosine 2′-monophosphate. It is also possible to attach a second phosphate unit to the first to make a diphosphate such as adenosine 5′-diphosphate (**129**), and three phosphate units may be attached as in adenosine 5′-triphosphate (**130**).

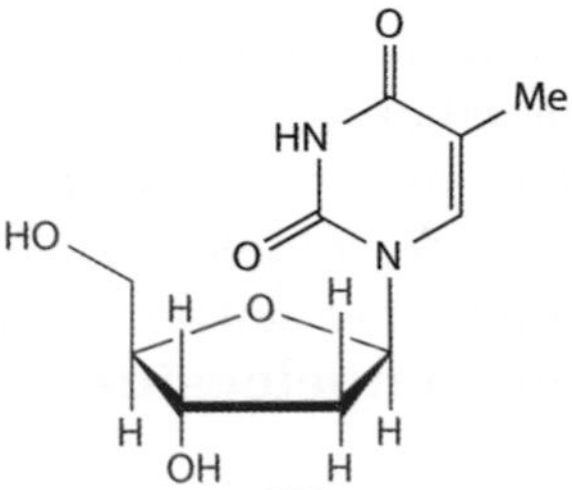

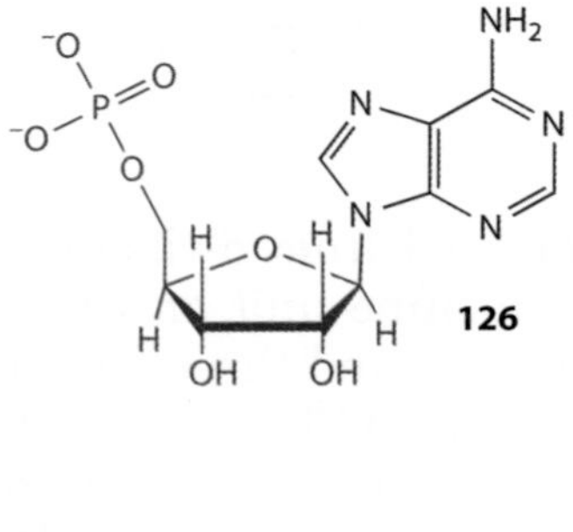

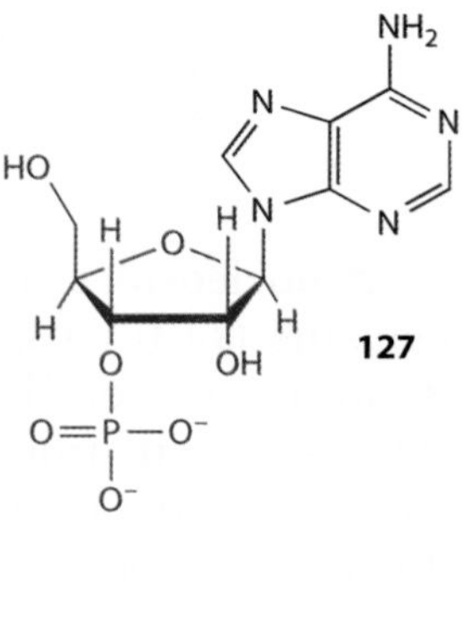

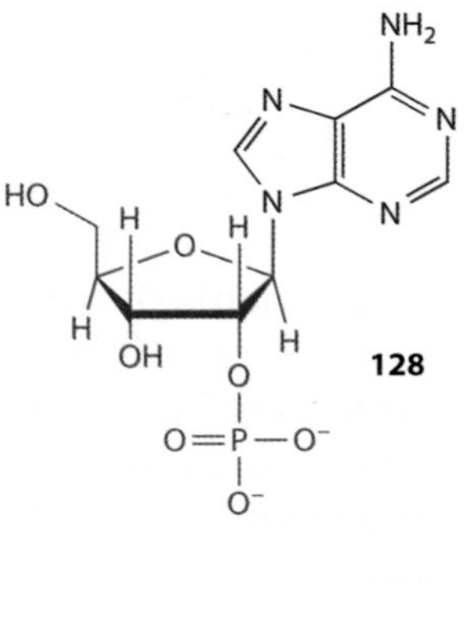

Table 28.2
One-Letter Codes for Heterocyclic Amines

Heterocycles (Bases, B)		One-Letter Code
adenine		A
guanine		G
uracil		U
cytosine		C
thymine		T

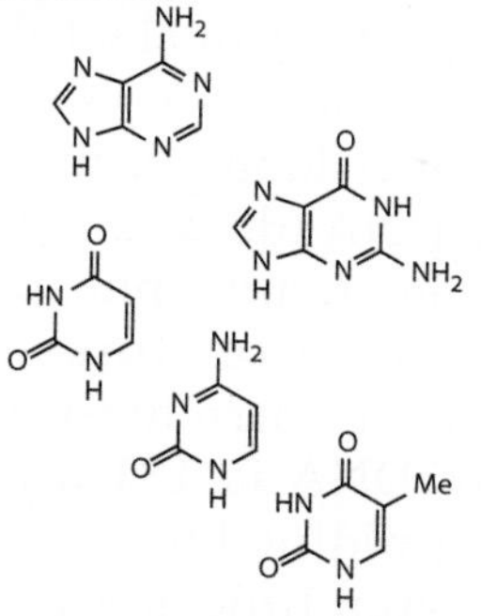

These three compounds have abbreviations that are commonly used: **126** is abbreviated AMP, **129** is ADP, and **130** is ATP. These abbreviations stem from a one-letter code for each heterocycle used to form the nucleotide—in this case, A for adenine. These codes are shown in Table 28.2, but they are trivial; the code simply takes the first letter from each heterocycle.

28.31 Draw the structure of (a) GMP, (b) UTP, and (c) TDP.

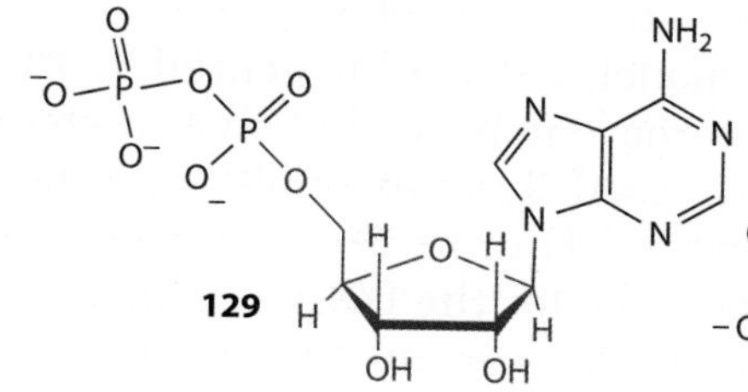

129

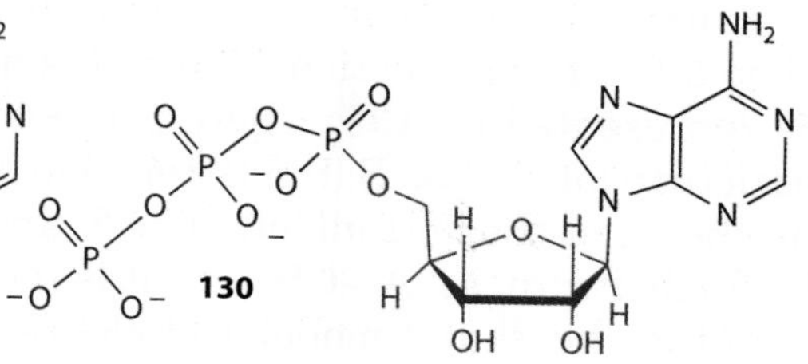

130

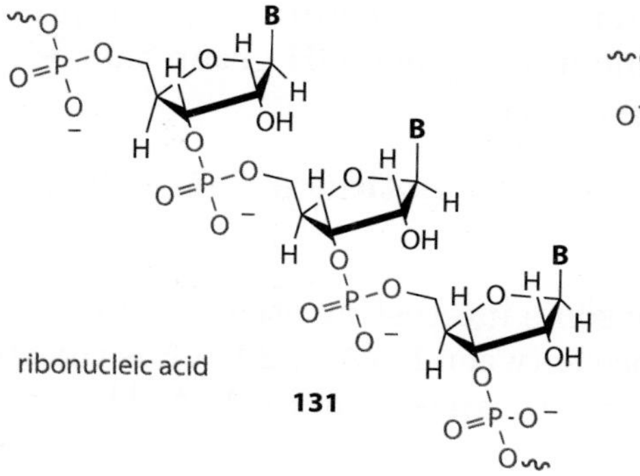

ribonucleic acid

131

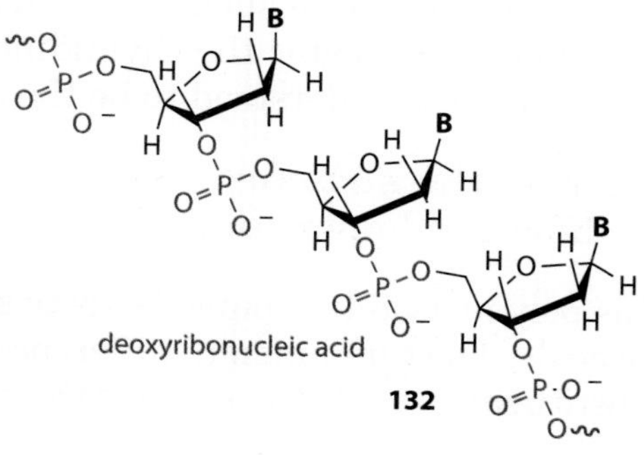

deoxyribonucleic acid

132

28.6 Biological Relevance. Polynucleotides

Without question, the most important nucleotides are the purine and pyrimidine derivatives shown in Section 28.5. Nucleotides are linked together by phosphate linkages to form a long-chain nucleotide (a polynucleotide), resulting in the structures of DNA (deoxyribonucleic acids) and RNA (ribonucleic acids). Both of these oligo (sugar phosphates) are formed by linking nucleotides together by phosphodiester links at the 3′–5′ positions of the nucleotides. Phosphodiesters were introduced in Chapter 20, Section 20.12. The fundamental structure of DNA is **131** and that of RNA is **132**, where "B" represents a purine or pyrimidine heterocycle. Both are linear polymers that begin with a phosphodiester link at the 3′ position of one nucleotide that is linked to the 5′ position of the next nucleotide. This 3′→5′ linkage continues as the polynucleotide chain grows. This structure leads to a polymer with phosphate backbone and a free 5′-terminus at the "beginning" of the polymer and a free 3′-terminus at the "end" of the polymer, as illustrated in **131** and **132**.

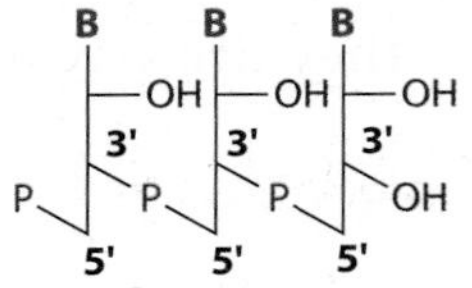

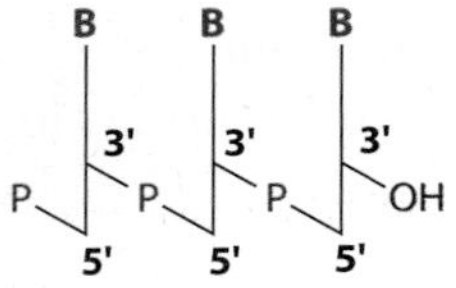

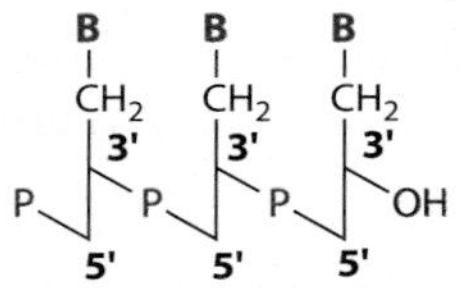

A shorthand notation represents the structures of nucleotides, roughly based on a Fischer projection. Using this model, **131** is represented as **133**, and **134** is represented by **135**, where the heterocyclic base is shown at the top of the structure labeled as B (the letter code is used: T, for example), and the 5′ OH unit is shown at the bottom left. The 2′ and 3′ OH units are shown on the right for the d-ribofuranose. A –CH$_2$– unit is included for the DNA strand (as in **135**) to make the shorthand model **134** clearer if there is any chance of confusion. An individual nucleotide such as thymidine-deoxyribofuranose is represented as **136**, where T indicates the thymidine and the one OH group that is shown in the structure is understood to be the 3′ OH.

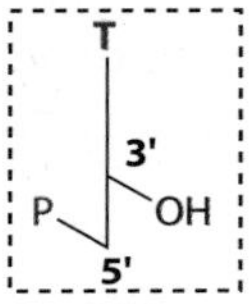

28.32 Draw the structure of the tetranucleotide A-G-G-U using the shorthand notation.

This polynucleotide linkage leads to some interesting structural features for nucleic acids. The fundamental difference between DNA and RNA is the absence of a substituent at the C2′ position of the ribofuranose unit in DNA. The structure of DNA usually consists of nucleotides containing the purines adenine (**A**) and guanine (**G**), as well as the pyrimidines cytosine (**C**) and thymine (**T**). The same heterocycles are incorporated in RNA, but uracil (**U**), rather than thymine (**T**), is usually found. These differences typically lead to a different conformation for different nucleic acids. Both DNA and RNA have a highly charged and polar sugar-phosphate "backbone" that is relatively hydrophilic (soluble in water); however, the heterocyclic bases attached to the ribofuranose or deoxyribofuranose units are relatively hydrophobic (relatively insoluble in water). These interactions lead to specialized secondary and tertiary structures of DNA and RNA.

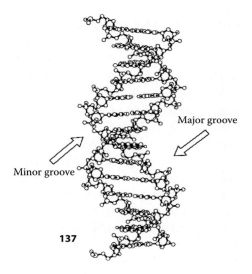

137

(Hecht, S. M., ed. 1996. *Bioorganic chemistry: Nucleic acids*, 9, Figure 1-8. New York: Oxford University Press. By permission of Oxford University Press.)

DNA is the repository of genetic information in cells. The major form of DNA under physiological conditions (those found in a normal cell) is the β-form, which consists of two antiparallel strands. This is the **Watson–Crick double helix structure,** after James D. Watson (United States; 1928–) and Francis H. C. Crick (England; 1916–2004), who first reported this property of DNA. This β-form is shown in **137**,[7] where the two strands are connected by hydrogen bonds between particular nucleotides on each strand (intermolecular hydrogen bonding). An adenosine (A) nucleotide (called a base in DNA and RNA) can hydrogen bond with a T of the other strand (an A–T base pair) and a G base in one strand can hydrogen bond with a C base in the other strand (a G–C base pair). The strands must be antiparallel (one strand is 3′→5′ and the other strand is 5′→3′) to maximize hydrogen bonding because the stereogenic nature of the ribofuranose and deoxyribofuranose units leads to a "twist" in the polynucleotide backbone.

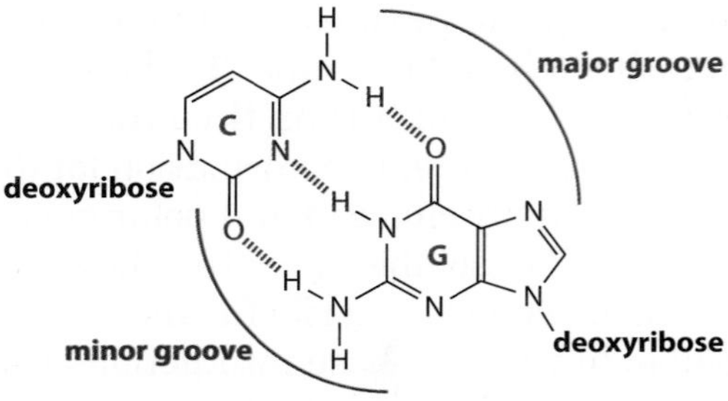

138 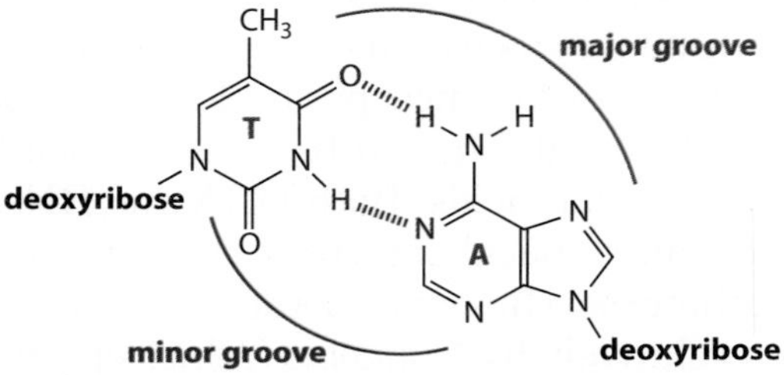   **139**

(Hecht, S. M., ed. 1996. *Bioorganic chemistry: Nucleic acids*, 6–7, Figure 1-4. New York: Oxford University Press. By permission of Oxford University Press.)

These hydrogen-bonding base pairs are called ***Watson–Crick base pairs;*** the C–G pair is shown in **138** and the A–T base pair in **139**.[8] The inherent chirality of the d-ribofuranose and the d-deoxyribofuranose leads the β-form of DNA to adopt a right-handed helix (see **137**) and each base-pair plane is rotated about 36° relative to the one preceding it. This leads to a complete right-handed turn for every 10 contiguous base pairs and a helical pitch of about 34 Å (34 pm).[8] The β-form leads to the creation of two helical grooves: the **major groove** and the **minor grove** (marked in **137, 138**, and **139**). The minor groove is narrow and the major grove is wide, but both are of about the same depth. When DNA interacts with small molecules or proteins, it is usually in the major or minor groove, which provides a microenvironment for bonding and recognition of these compounds (called ligands).[8] As the specific nature of the nucleotides in the strands of DNA changes, the "floor" of each groove will change, creating sequence-selective sites that can be specific for certain ligands.

Two other forms of DNA are the A-form and the Z-form (see Figure 28.2).[9] If the relative humidity of the β-form of DNA decreases to 75% and the sodium chloride concentration drops to below 10%, the β-form is transformed into the A-form helix.[8] The A-form is a right-handed helix with 11 base pairs per complete turn and a helical pitch of about 28°; however, a 20° tilt to the base-pair planes leads to a displacement from the central axis (see Figure 28.2). The Z-form of DNA adopts a left-handed helix, *but it is not simply the mirror image of the β-form or the A form helices.* The Z-form has 12 base pairs per complete turn and a helical pitch of about 45 Å; it has a wide, shallow major groove and a narrow, deep minor groove.[8]

DNA stores genetic information, and RNA transcribes and translates this information for the cell. There are several unique structures for RNA. Although DNA tends to form linear repeating structures, RNA generates diverse structures have a unique function. Three main classes of RNA are found in a cell: transfer RNA (tRNA), ribosomal RNA (rRNA), and messenger RNA (mRNA). Transfer RNA tends to be single strands of nucleic acid with 60–96 nucleotides.

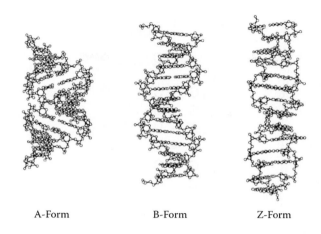

A-Form   B-Form   Z-Form

Figure 28.2 A-, B-, and Z-forms of DNA. (Hecht, S. M., ed. 1996. *Bioorganic chemistry: Nucleic acids,* 10–11, Figure 1-9. New York: Oxford University Press. By permission of Oxford University Press.)

The function is to transport an amino acid, covalently attached to a specific proton of the polynucleotide chain, to a ribosome where it can be incorporated into its proper sequence in a protein. This will be described later.

The primary and secondary structure of alanine tRNA (the tRNA used to transport the amino acid alanine) is shown as a "cloverleaf" structure (see Figure 28.3)[10] that contains double-stranded stems connected to single-stranded loops. The structure in Figure 28.3 consists of one nucleotide strand, but base pairing in self-complementary regions of this strand leads to double-stranded loops. This strand is effectively held together by intramolecular hydrogen bonding. The 5′-termini are phosphorylated and the 3′-terminus of tRNA is where the amino acid is attached; this molecule always terminates in the sequence CCA-3′ with a free OH group that can bind the amino acid. Almost all tRNAs have a seven-base-pair structure that is called the **acceptor stem** (the seven base pairs prior to the CCA-3′-terminus). In **169**, the acceptor stem is A-C-C-U-G-C-U. *The loop of nucleotides at the "bottom" of the cloverleaf has the three base pairs that will bind this tRNA to a messenger RNA, which allows the amino acid to be transferred to the protein.*

Ribosomal RNA (rRNA) usually has a complex structure, and rRNA is the main component of the ribosome. The ribosome for *Escherichia coli* (70S) has two rRNA units: the 50S and the 30S subunits. The 30S subunit has one large rRNA (16S RNA) and 21 individual proteins; the 50S subunit has two RNAs (5S and 23S) and 32 different proteins.[11] The structure of the 16S rRNA is shown in Figure 28.4.[11] *The ribosome is the body that an organism uses to manufacture proteins.* The ribosome binds messenger RNA (mRNA), which then interacts with tRNA.

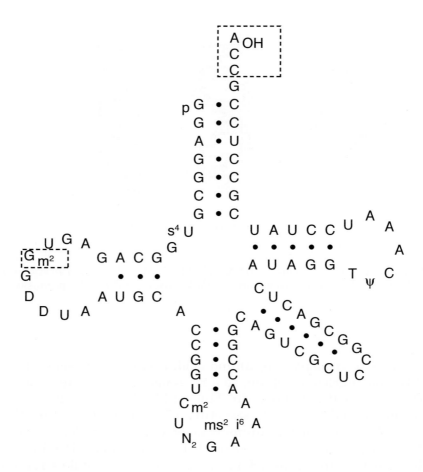

Figure 28.3 Cloverleaf structure of the alanine transfer RNA. (Reprinted in part with permission from McClain, W. H. 1977. *Accounts of Chemical Research* 10:418. Taken from Figure 1. Copyright 1977 American Chemical Society.)

Messenger RNA (mRNA) is a single-stranded nucleic acid with unique secondary structures. It carries the genetic code from DNA to the ribosome, and this code is read via forming a hydrogen bond to the three base pairs of a tRNA mentioned before (the triplet anticodon). To ensure that this genetic code is easily "read," mRNA is generally less structured than tRNA or rRNA. A strand of mRNA may be categorized as having groups of three base pairs, where each group is the genetic code for a specific amino acid. These three base pairs (called codons) will match a complementary set of three base pairs on a tRNA (an anticodon). The anticodon is at the bottom loop of tRNA in Figure 28.3.

The codon and anticodon are complementary, which means that a G-C-C codon on mRNA will match a C-G-G anticodon on tRNA. This is important because the C-G, G-C, G-C base pairs are capable of hydrogen bonding (see **138** and **139**). This hydrogen bonding allows the tRNA to attach itself to

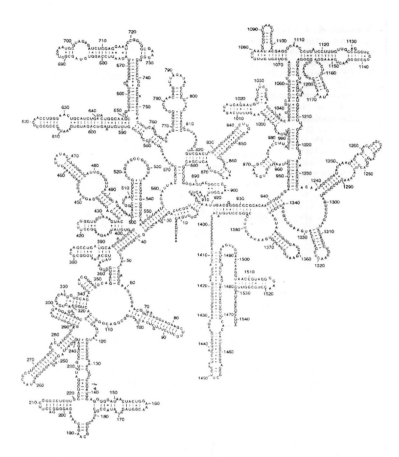

Figure 28.4 The 16S RNA unit of *E. coli* for the 30S subunit of the 70S subunit. (From Hecht, S.M. (Editor), *Bioorganic chemistry: Nucleic acids, Oxford University Press,* New York, 1996, see Figures 1–12, pp. 16. With permission.)

the mRNA at the proper place; the amino acid carried by the tRNA is then released to a growing peptide chain being synthesized on the ribosome. This process is illustrated in Figure 28.5[12] and the tRNA anticodon is shown as C-G-C.

The process illustrated in this figure is the mechanism by which an organism synthesizes proteins from amino acids. Each amino acid will have a tRNA with a unique anticodon, and that amino acid will attach itself only to that tRNA. The ribosome will bind a strand of mRNA in order to build a specific protein, which has a specific sequence of amino acids. The genetic code on the mRNA (the set of codons) matches the amino acid sequence of the protein. Therefore, the mRNA will bind tRNA molecules in a particular sequence, allowing that amino acid to be released to the growing protein sequence associated with the ribosome. The genetic code for the amino acids used to make proteins is shown in Figure 28.6.[13] Although only five heterocyclic bases have been cited thus far, several others are also known.

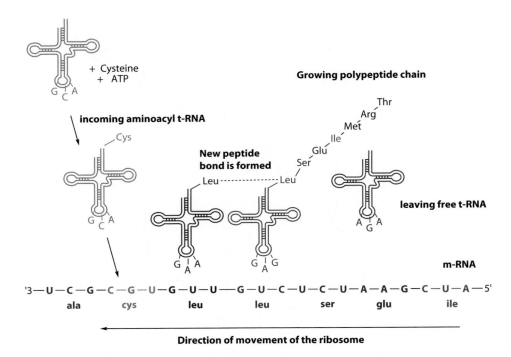

Figure 28.5 Protein synthesis. (From Lehninger, A. L. 1970. *Biochemistry,* 692, Figure 30.1. New York: Worth Publishers, Inc. With permission from W.H. Freeman and Company.)

	U	C	A	G	
U	UUU Phe UUC UUA Leu UUG	UCU UCC Ser UCA UCG	UAU Tyr AUC UAA stop UAG	UGU Cys UGC UGA stop UGG Trp	U C A G
C	CUU CUC Leu CUA CUG	CCU CCC Pro CCA CCG	CAU His CAC CAA Gln CAG	CGU CGC Arg CGA CGG	U C A G
A	AUU AUC Ile AUA AUG Met	ACU ACC Thr ACA ACG	AAU Asn AAC AAA Lys AAG	AGU Ser AGC AGA Arg AGG	U C A G
G	GUU GUC Val GUA GUG	GCU GCC Ala GCA GCG	GAU Asp GAC GAA Glu GAG	GGU GGC Gly GGA GGG	U C A G

5' Base (left) · 3' Base (right)

The genetic code comprises 64 codons, which are permutations of four bases taken in threes. Note the importance of sequence: three bases, each used once per triplet codon, give six permutations: ACG, AGC, GAC, GCA, CAG, and CGA, for threonine, serine, aspartate, alanine, glutamime, and arginine, respectively.

Figure 28.6 The genetic code composed of 64 codons. (Devlin, T. M. 1986. *Textbook of Biochemistry with clinical correlations,* 2nd ed., 738, Table 19.2. New York: John Wiley & Sons. Copyright Wiley-VCH Verlag GmbH & Co. KGaA. Reproduced with permission.)

28.7 Synthesis of Polynucleotides

An important theme of this chapter is the recognition that nucleotides, nucleosides, and polynucleosides are simply organic chemicals. The end of this chapter will continue that theme by showing how these compounds are synthesized using variations of standard organic reactions found throughout this book.

Both DNA and RNA are poly(nucleosides) where any two nucleotide fragments are connected in a 5′–3′ manner by a phosphate linkage. The disconnections are obvious in that the 5′-O-phosphate unit or the 3′-O-phosphate unit can be disconnected. If the 5′ position has a free OH, then the 3′ unit must contain the phosphate unit. Conversely, if the 3′ position has a free OH, then the 5′ unit must contain the phosphate (P indicates the phosphate—not a protecting group). In practical terms, **140** may be coupled to **141** or **142** may be coupled to **143**. This strategy raises several problems.

With either approach, a phosphate unit must be coupled to an alcohol. Because the ribofuranose or deoxyribofuranose unit of RNA or DNA will have several hydroxyl units, all OH groups except the one at 3′ or at 5′ may require protection. Several of the purine and pyrimidine heterocyclic bases have an –NH$_2$ unit, which must be protected most of the time before coupling can occur. Methodology that will convert an alcohol into a phosphate ester must also be introduced.

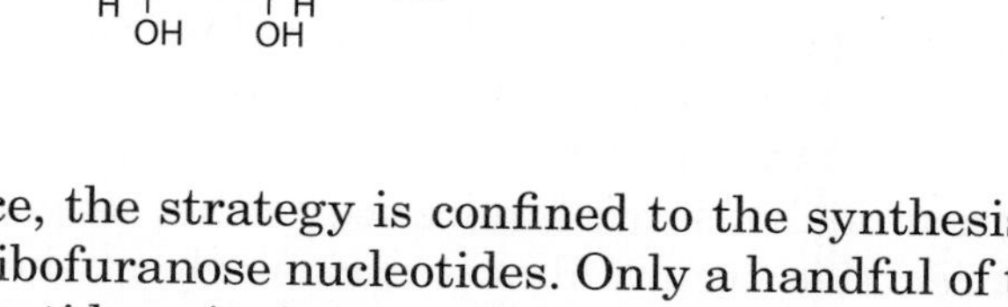

For convenience, the strategy is confined to the synthesis of a polynucleoside composed of ribofuranose nucleotides. Only a handful of the methods used to protect the nucleotide units is presented, and only a few methods for coupling nucleotide fragments together are discussed. This section is meant to be an introduction showing that poly(nucleosides) are synthesized using fundamental

chemical reactions. The methods used to prepare poly(ribofuranside) derivatives can also be used to prepare poly(deoxyribofuronoside) derivatives. The
ribofuranose units have an OH at 2′, whereas the deoxyribofuranose units
do not.

The ribofuranose units have one extra OH group that requires protection.
If *N*-benzoyl A-ribofuranose (**144**) is treated with a catalytic amount of *p*-tolu-
enesulfonic acid and trimethyl orthoacetate, (MeO)$_3$C–Me, the C2′–C3′ diol unit
is converted to an orthoester (in **violet**; an aldehyde is converted to an acetal
and an ester is converted to an orthoester), as in **145**. An orthoester has the
structure RC(OR)$_3$, but this class of compounds will not be discussed further.
This compound has the 5′ OH free.

The "free OH" can be modified by further chemical manipulation, and reaction with benzoyl chloride (BzCl) and pyridine (Py) gives **146** (the *O*-benzyl
group is in **blue** in the illustration). Treatment of **146** with 10% aqueous acetic
acid cleaves the orthoester and gives a 1:1 mixture of hydroxy acetates **147** and
148, where the acetate units are marked in **cyan**. If **147** is isolated and treated
with dihydropyran (see Chapter 26, Section 26.4.2) and an acid catalyst, the
2′-hydroxyl is converted to a dihydropyranyl ether (**149**) where the pyranyl
ether unit is marked in **green**. It is therefore possible to protect and deprotect
the 5′-, 3′-, and 2′-hydroxyl groups selectively, although the ability to differentiate the 2′- and 3′-hydroxyls is more difficult to achieve.

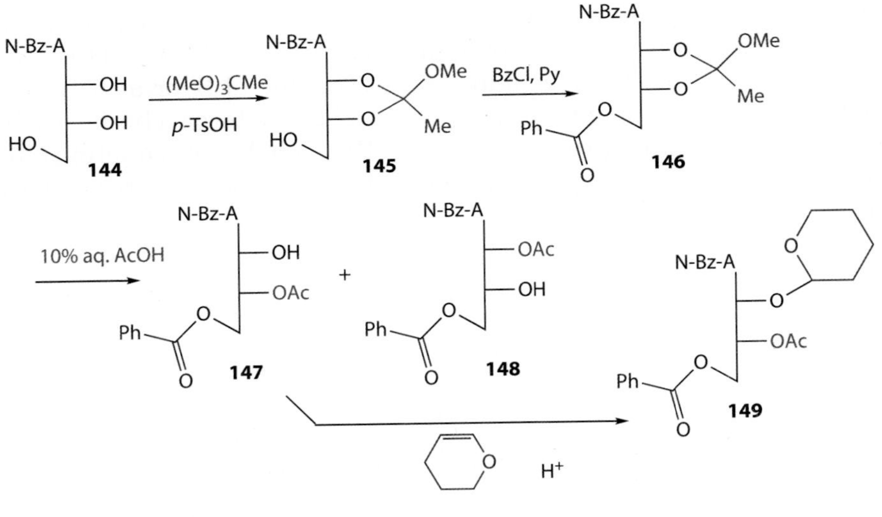

A common protecting group for the 5′-hydroxyl is triphenylmethyl (–CPh$_3$),
which has the common name of trityl. When chlorotriphenylmethane (**150**; trityl chloride) reacts with **144**, the product is **151**. Because the CPh$_3$ unit is very
bulky, there is a great deal of steric hindrance as this reagent approaches **144**.
Consequently, the reaction occurs almost exclusively at the less hindered C5′
primary alcohol, although there may be minor amounts of product based on
reaction at the C2′- or C3′-hydroxyls. Treatment with strong acid is usually
required to remove the trityl group and restore the free OH unit.

28.33 **Draw a synthesis of 2-(2-hydroxyethyl)-cyclohexanol from 2-(2-hydroxyethyl)cyclohexanone using a trityl protecting group.**

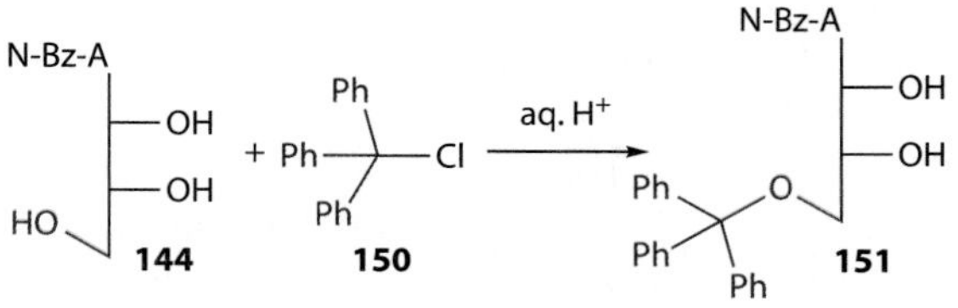

Conversion of a nucleotide such as **152** into a phosphate derivative requires reaction with a derivative of phosphoric acid, but phosphoric acid itself cannot be used because there are three reactive oxygen sites. The phosphoric acid unit must be protected. An early approach is shown for the reaction of **152** with **153** (2-cyanoethyl phosphate). In the presence of methanesulfonyl chloride, **153** is coupled to the protected phosphate to give **154**. The alkoxide unit in **154** can now be coupled to a second nucleotide (**155**), in the presence of an aromatic sulfonyl chloride, to give a dinucleotide, **156**. Removing protecting groups and adding a new phosphate activating group allows preparation of the polynucleotide. The phosphate protecting groups are removed at the end of the synthesis.

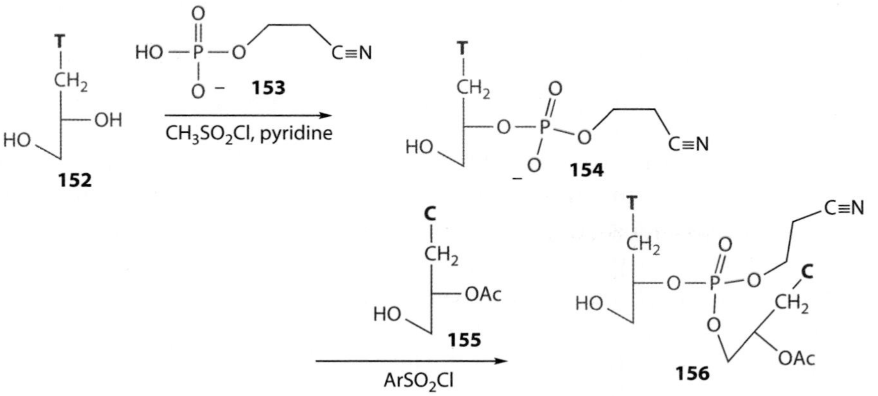

Much of the approach outlined here was pioneered by Har G. Khorana (India/United States; 1922–); more recent developments have showed that an aromatic phosporamidate such as **158** gives better results. The reaction of **158** and nucleotide **157** (the 5′-hydroxyl has a protecting group, P) in the presence of 1-methylimidazole (see imidazoles in Chapter 26, Section 26.1.1) gives **159**. When deprotected and coupled to a second nucleotide, the resulting aryloxy–phosphate linkage is much more stable to hydrolysis and other chemical manipulations. 2-Cyanoethyl phosphate derivatives such as **159** are too reactive under the reaction conditions required to prepare a large polynucleotide. Other protecting groups are known, but these two will suffice.

28.34 Draw the structure of 1-methylimidazole.

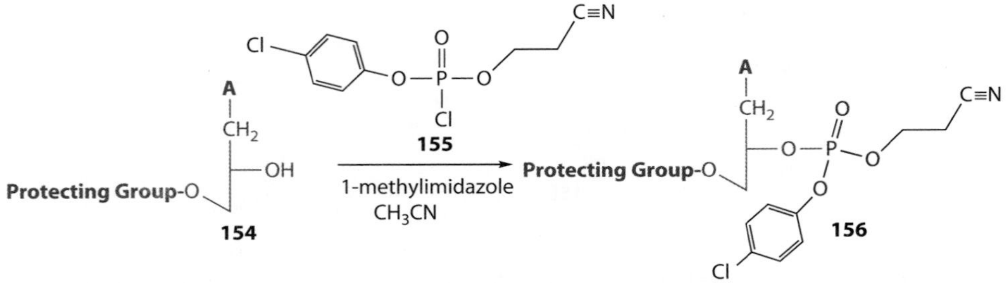

When coupling nucleotides such as adenosine (**132**; contains adenine), guanosine (**136**; contains guanine), or cytidine (**135**; contains cytosine), the free amine group (–NH$_2$) of the heterocycle is usually protected. Although in several instances the unprotected amine unit does *not* interfere with formation of a polynucleoside, *assume* that protecting groups are required. Conversion to an amide is a common method to protect the amine unit (see Chapter 20, Section 20.6). Although it is possible to prepare many different amides, one structural type of amide protecting group is usually the best choice for each heterocyclic base. *The benzoyl group is preferred for adenine, the anisoyl group for cytosine, and the isobutryl group for guanine.* In all cases, the C2′ hydroxyl of these compounds is protected as the tetrahydropyranyl ether.

28.35 Draw the structures of these three protected nucleotides.

Once the sugar unit, the phosphoric acid, and the heterocyclic base are protected in the proper manner, the two nucleotides must be linked together to begin the synthesis of a polynucleoside. Beginning with an adenine derivative (**145**), where the 5′-hydroxyl is free, the 2′–3′-hydroxyls are blocked as an orthoester, and the adenine is protected as benzoyl, reaction must occur with a 3′-phosphate derivative. That partner is thymidine derivative **157**, which has the 3′-phosphate unit protected with a 2-chlorophenoxy unit.

How are they coupled together? Several **condensing agents** make a new phosphate ester bond. One of the most common is the dicyclohexylcarbodiimide used in Chapter 20 (Section 20.6.4; DCC; c-C$_6$H$_{11}$–N=C=N-c-C$_6$H$_{11}$) to couple alcohols and acids to make an ester or an amine and an acid to make an amide (see Chapter 20, Section 20.6). In addition, DCC was used in Chapter 27 (Section 27.4.2) to prepare peptides. If **142** and **157** are mixed with DCC, the product is a dinucleoside, **158**. Once the coupling reaction is complete, the 5′ position of the T-nucleoside in **158** is free, and it can react with **159** (in the presence of DCC) to give the trinucleoside, **160**. If this is the final target, all that remains is to remove the protecting groups.

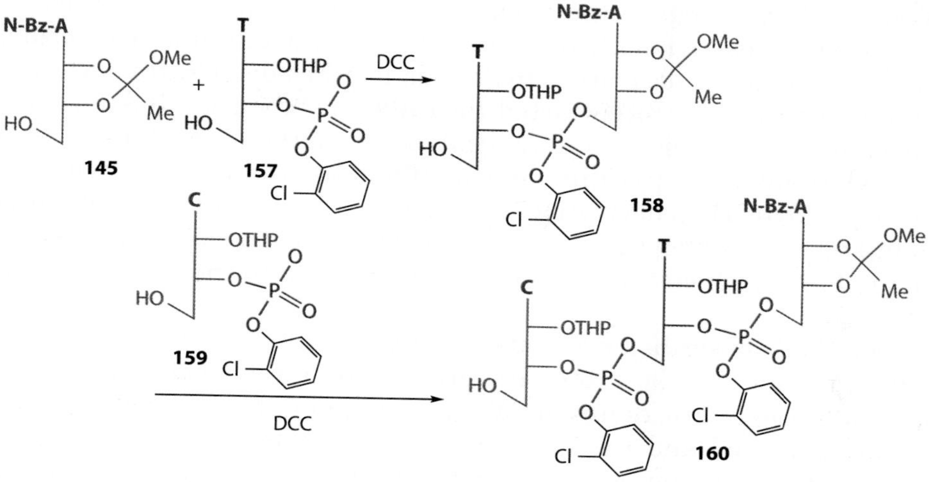

Alternative reagents used to couple nucleosides include mesitylenesulfonyl chloride (**161**) and 2,4,6-triisopropylbenzenesulfonyl chloride (**162**). Of the three reagents mentioned so far, DCC is the *least* reactive. In some cases, the chloride unit of the sulfonyl chloride is not reactive enough and that unit must be replaced with something more reactive. *p*-Toluenesulfonyl imidazolide (**165**) and mesitylenesulfonyl imidazolide (**165**) are used in such cases. These reagents are prepared by reacting either *p*-toluenesulfonyl chloride (**163**) or mesitylenesulfonyl chloride (**161**) with imidazole (**166**).

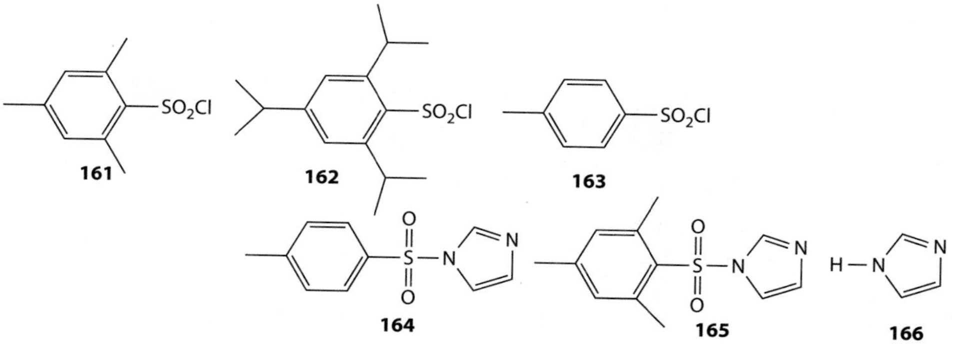

Just as the Merrifield synthesis in Chapter 27 (Section 27.4.2) is used to prepare polyamides by using a polymer bead to anchor the growing polyamide chain, there is a method for making nucleic acids using a polymer support. Of the several polymer supports used, styrene or copolymers of styrene are used most of the time. If polystyrene (see Chapter 10, Section 10.8.3) is used as a support, a chemical reaction is required to attach the first nucleotide. This is accomplished by preparing a polystyrene derivative in which the benzene rings are functionalized.

When a phenyl ring on the backbone of polystyrene (represented by the solid sphere) is reacted with benzoyl chloride and aluminum chloride, the product is the benzoyl derivative **167**[14] (this is a Friedel–Crafts acylation; see Chapter 21, Section 21.3.3). If reacted with **168**, the product is **169** (an acyl addition reaction of an organolithium reagent; see Chapter 18, Section 18.4). When the methyl group of the pyridine unit in **169** is treated with phenyllithium to form the (pyridyl)CH$_2$Li derivative, reaction with formaldehyde leads to **170**. If **170** is linked to *N*-benzoyl 2′-*O*-isobutyladenosine-3′-monophosphate (**171**), the product is **172**, in which the first nucleotide is bound to the polymer via the 3′ position.

The methods described previously can be used to build a polynucleoside. In actual practice, special protecting groups and a relatively standard set of reaction conditions are required. When completed, however, the protecting groups are removed and the nucleic acid unit is cleaved from the polymer. This sequence can be automated and machines can prepare DNA and RNA strands if the reagents for coupling, as well as protection and deprotection reagents, are supplied and, of course, there is a supply of each individual nucleotide.

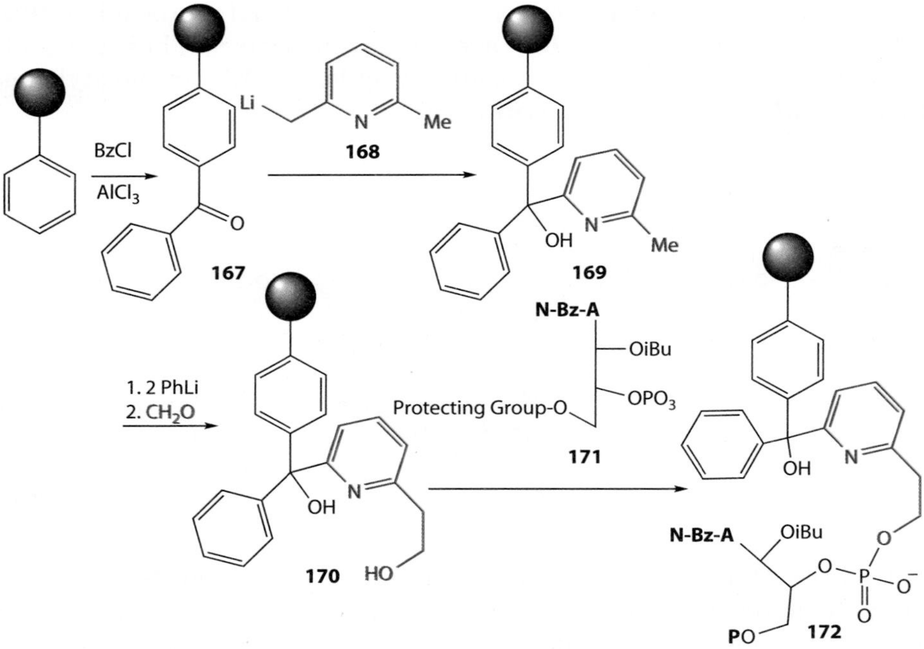

Other methods for attaching the nucleotide to the polymer exist. An amine unit may be used to bind the growing nucleoside to the polymer, as in **173**, rather than the alcohol unit found in **172**. Using these fundamental techniques, large nucleic acids can be synthesized.

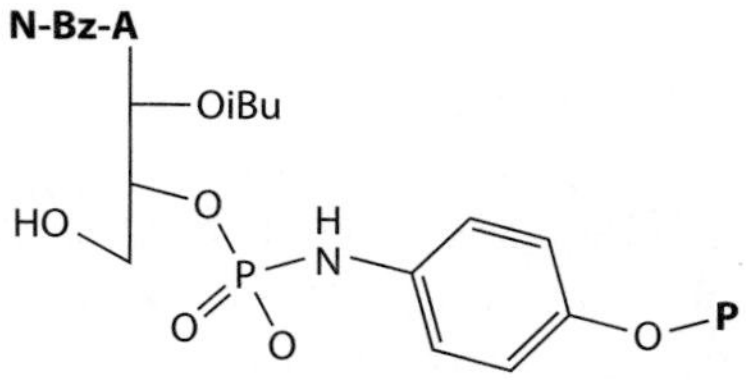

N-Bz-A

173

References

1. Binkley, R. W. 1988. *Modern carbohydrate chemistry,* 80. New York: Marcel Dekker.
2. Kennedy, J. F., and White, C. A. 1983. *Bioactive carbohydrates,* 41. Chichester, England: Ellis Horwood Ltd.
3. Koenigs, W., and Knorr, E. 1901. *Berichte* 34:957.
4. Fischer, E., and Helferich, B. 1914. *Berichte* 47:210, 1377.
5. Niedballa, U., and Vorbrüggen, H. 1970. *Angewandte Chemie International Edition* 9:461.
6. Vorbrüggen, H., Krolikiewicz, K., and Bennua, B. 1981. *Chemische Berichte* 114:1234.
7. Hecht, S. M., ed. 1996. *Bioorganic chemistry: Nucleic acids,* 9, Figure 1-8. New York: Oxford University Press.
8. Hecht, S. M., ed. 1996. *Bioorganic chemistry: Nucleic acids,* 6. New York: Oxford University Press.
9. Hecht, S. M., ed. 1996. *Bioorganic chemistry: Nucleic acids,* 10–11, Figure 1-9. New York: Oxford University Press. See Figures 1–12.
10. McClain, W. H. 1977. *Accounts of Chemical Research* 10:418. Taken from Figure 1.
11. Hecht, S. M., ed. 1996. *Bioorganic chemistry: Nucleic acids,* 15–17. New York: Oxford University Press. See Figures 1–12.
12. Lehninger, A. L. 1970. *Biochemistry,* 692, Figure 30.1. New York: Worth Publishers, Inc.
13. Devlin, T. M. 1986. *Textbook of biochemistry with clinical correlations,* 2nd ed., 738, Table 19.2. New York: John Wiley & Sons.
14. Amarnath, V., and Broom, A. D. 1977. *Chemical Reviews* 77:183 (see p. 210).

Answers to Problems

28.1

a ketopentose

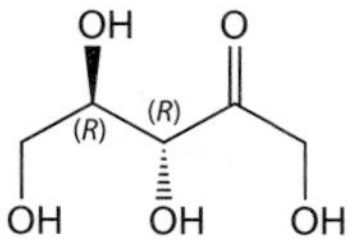

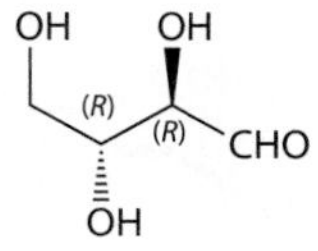

an aldotetrose

28.2

L-erythrose

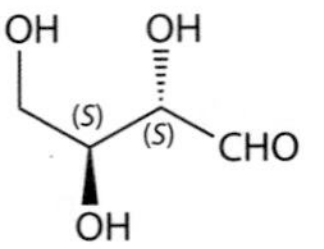

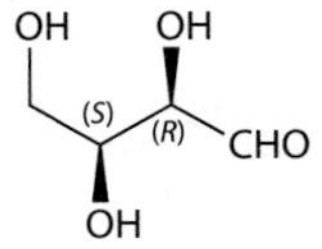

L-threose

28.3 There are five stereogenic centers in a typical aldoheptose. Discounting the possibility of *meso* compounds, there should be $2^5 = 32$ total stereoisomers, which means there would be 16 diastereomers.

28.4

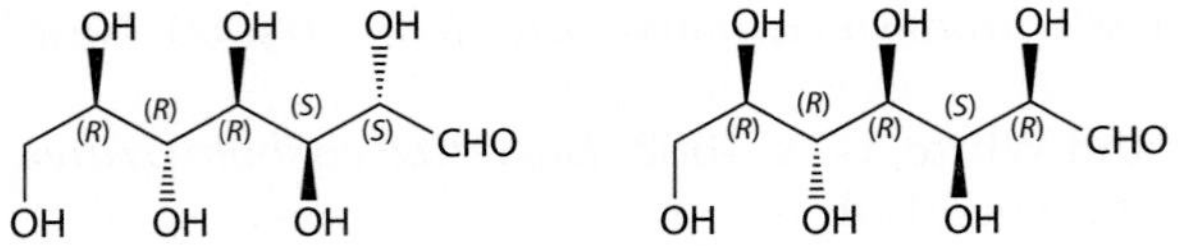

28.5

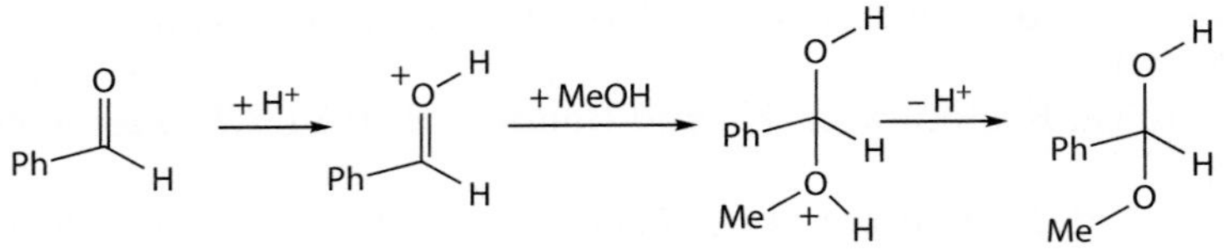

28.6

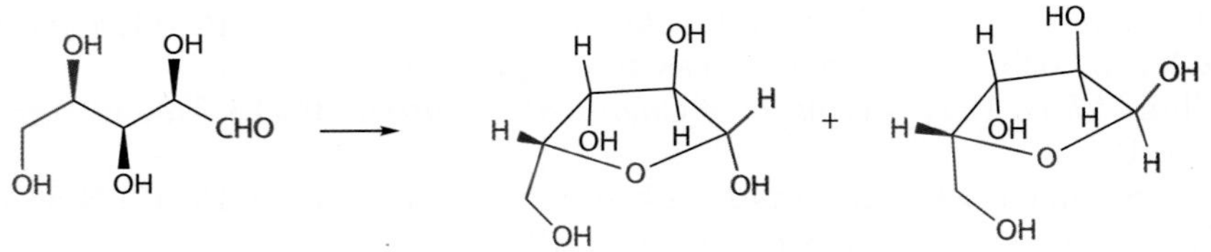

28.7

α-D-arabinose

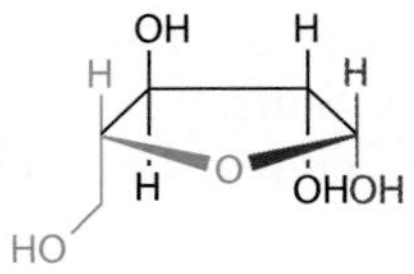

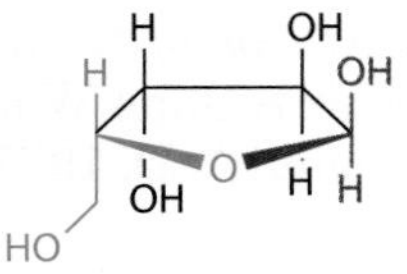

β-D-xylose

28.8

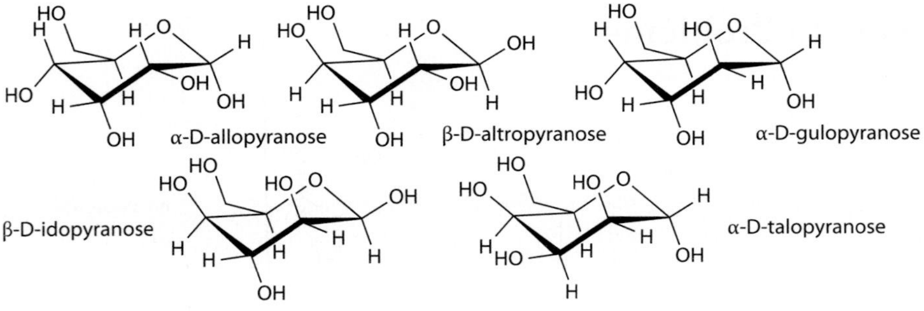

α-D-allopyranose

β-D-altropyranose

α-D-gulopyranose

β-D-idopyranose

α-D-talopyranose

28.9   x(+14.5) + y(−14.5) = −3.2, where x = relative % of d and y = relative % of l.
x + y = 1, so x = 1 − y, so (1 − y)(+14.5) − 14.5y = −3.2
14.5 − 14.5y − 14.5y = −3.2
−29y = −17.7, so y = −17.7/−29 = 0.61, making x = −0.39
Therefore, 61% of l and 39% of d

28.10

α-D-glucopyranose 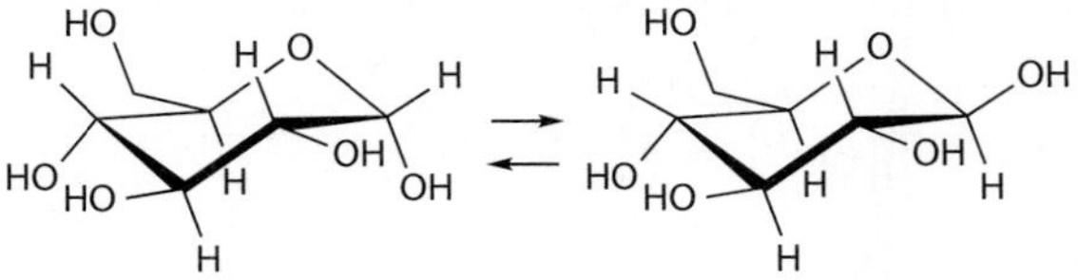   β-D-glucopyranose

28.11

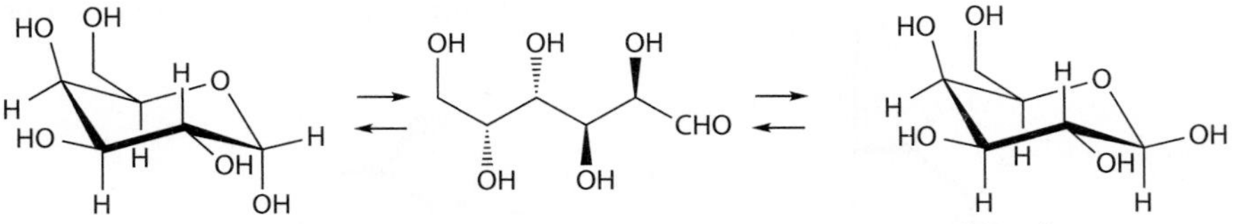

α-D-galactopyranose "x"   "y" β-D-galactopyranose

Assume that the aldehyde form does not contribute for this calculation
x(+150.7) + y(+52.8) = +80.2   and   x=y =1
(1−y)(150.7) + y(52.8) = 80.2
150.7 − 150.7y + 52.8y = 80.2
−97.9y = −70.5
y = −70.5/−97.9 = 0.72   Therefore 72% of β-D and 28% of α-D

28.12

L-glycerotetrulose 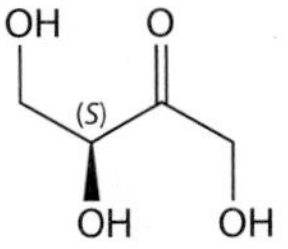 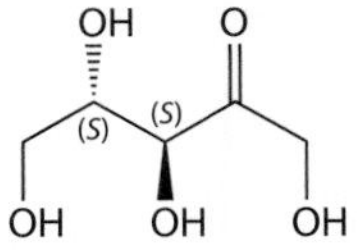   L-xylolose

28.13

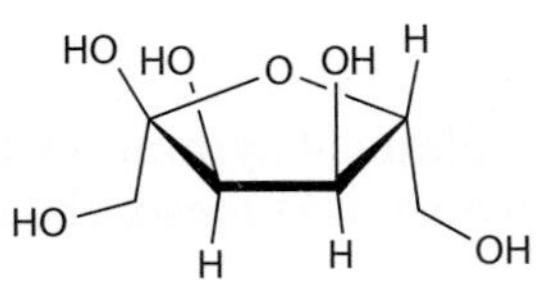

α-D-tagatofuranose 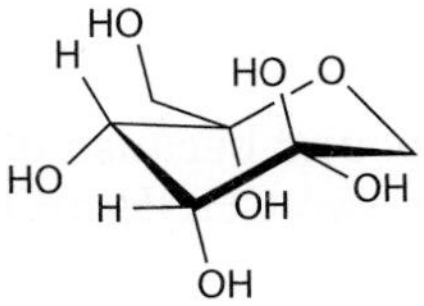   β-D-psicopyranose 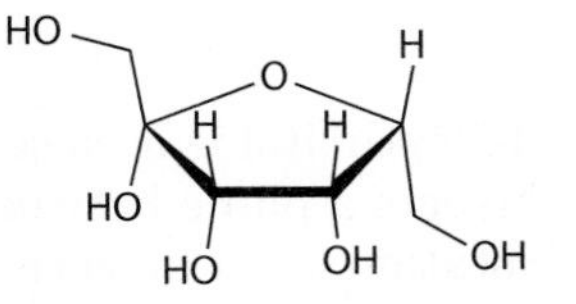   β-D-psicofuranose

28.14

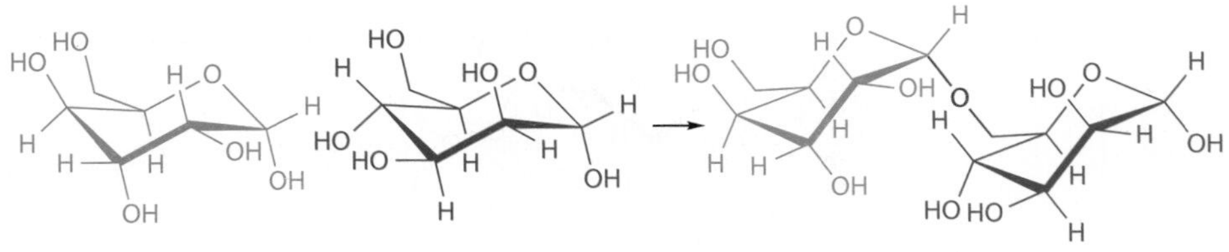

28.15

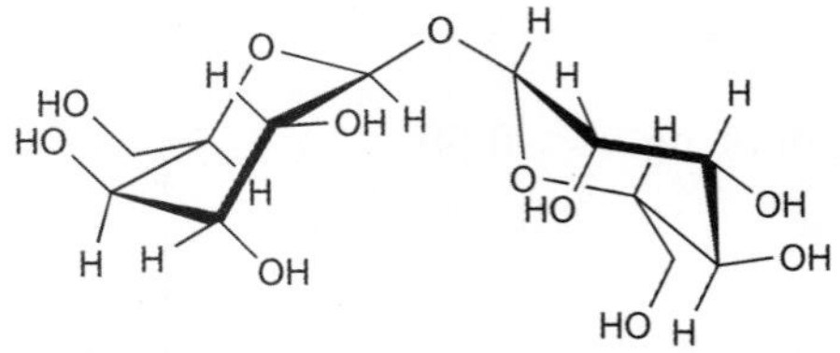

O-β-D-altropyranosyl-(1→1)-α-D-mannopyranose

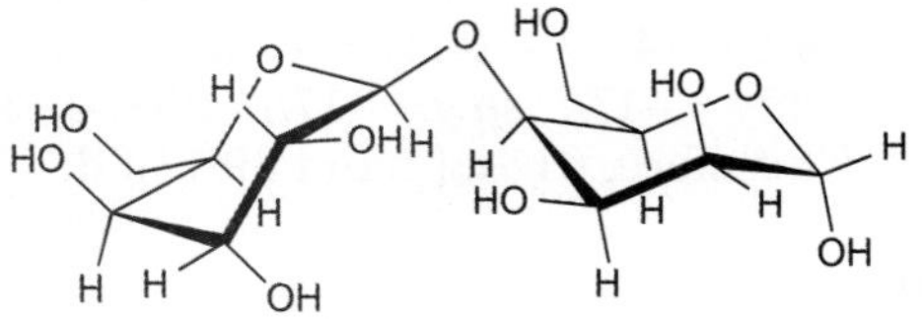

O-β-D-altropyranosyl-(1→4)-α-D-mannopyranose

28.16

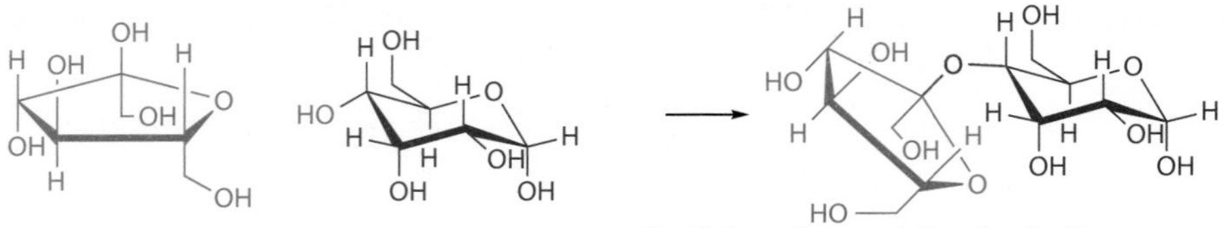

O-α-D-fructofuranosyl-(1→4)-α-D-allopyranose

28.17

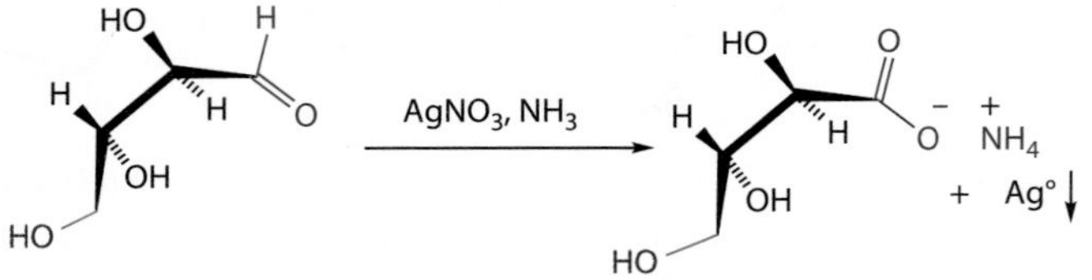

28.18 (a) Reducing; (b) not reducing; (c) reducing; (d) not reducing; and (e) reducing.

28.19

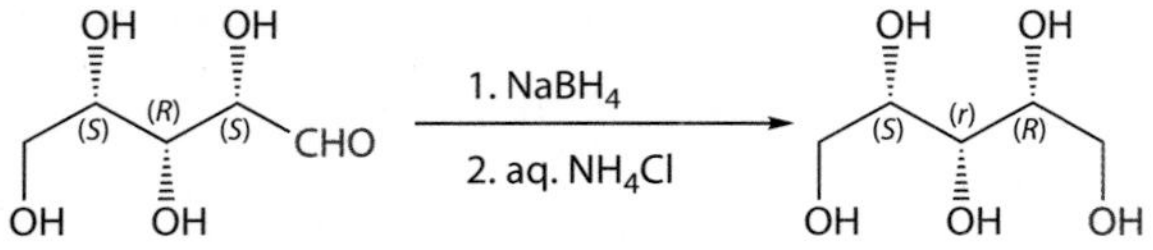

28.20

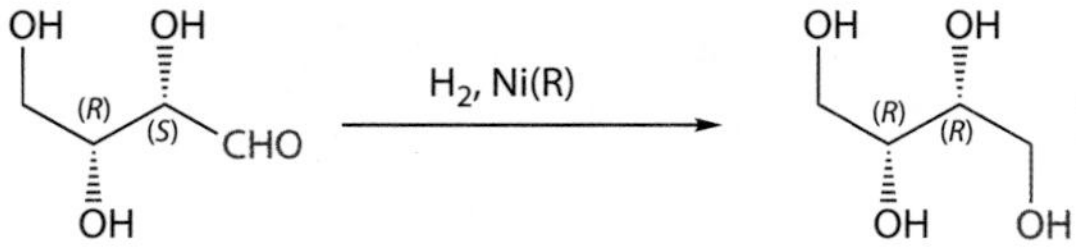

D-threitol

28.21 d-Erythritol is a *meso* compound because it has a plane of symmetry that bisects a plane between the two CHOH units. Because it is *meso,* the specific rotation must be zero.

28.22

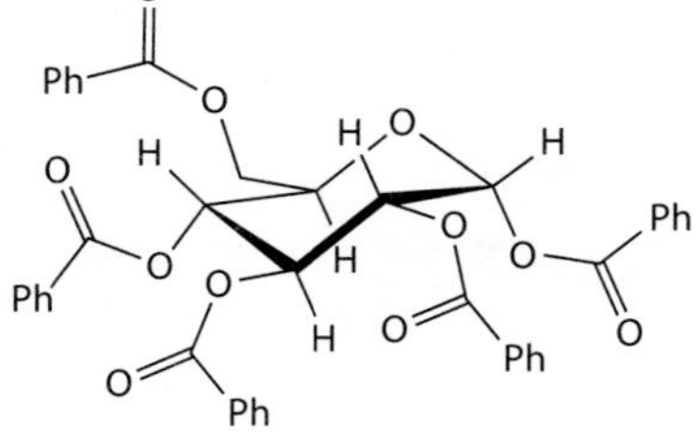

28.23

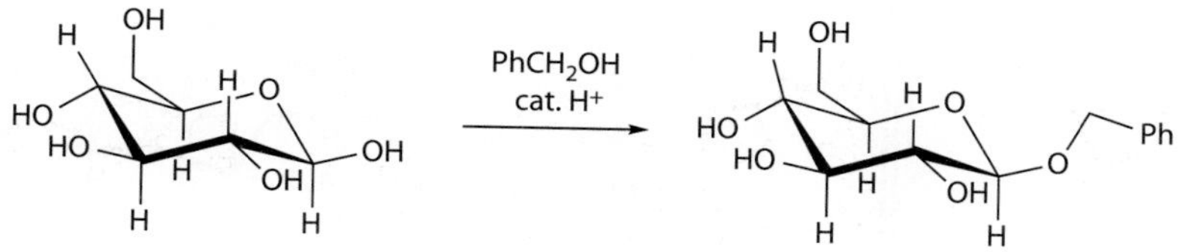

28.24

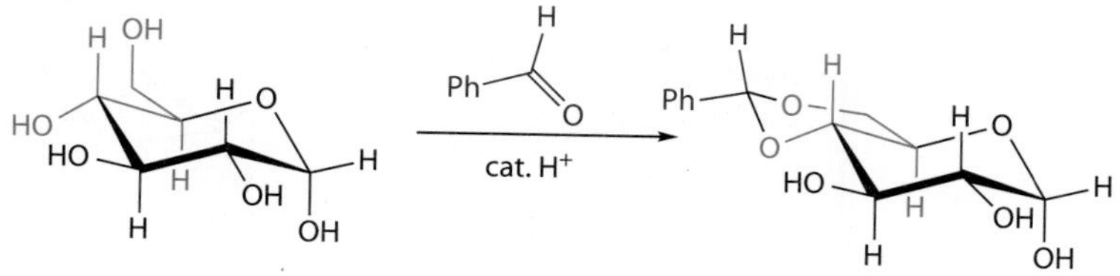

28.25

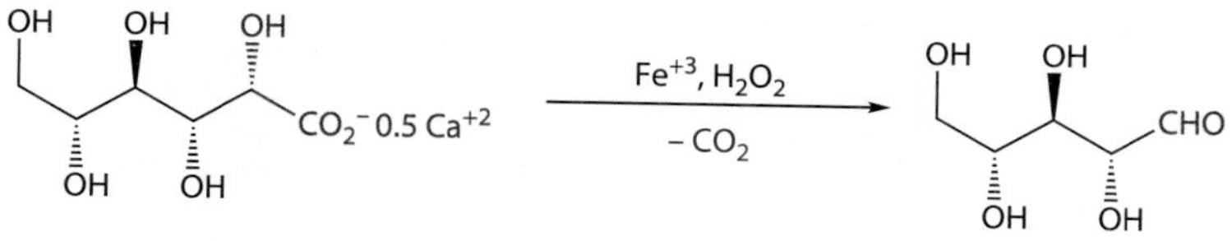

28.26

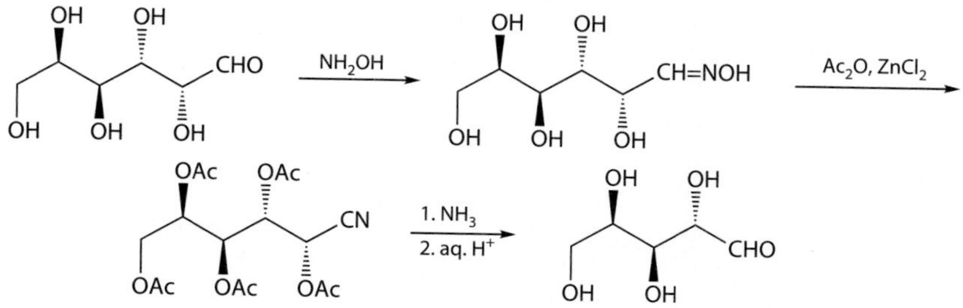

28.27

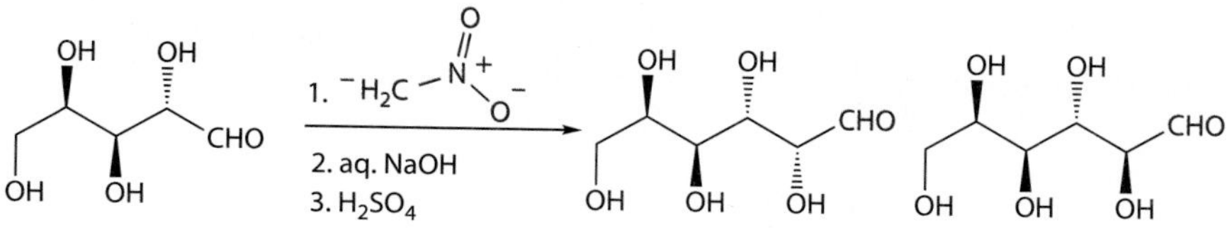

28.28

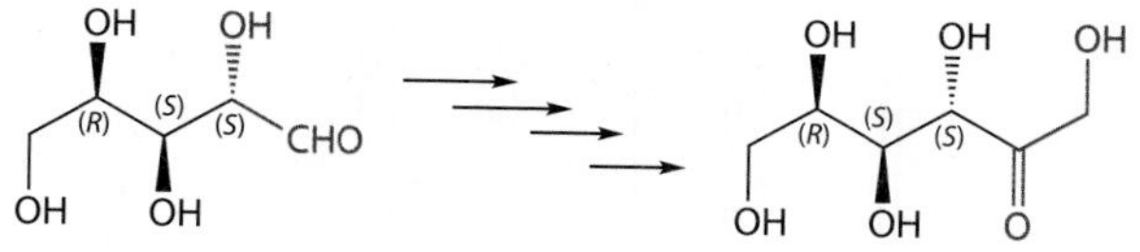

28.29

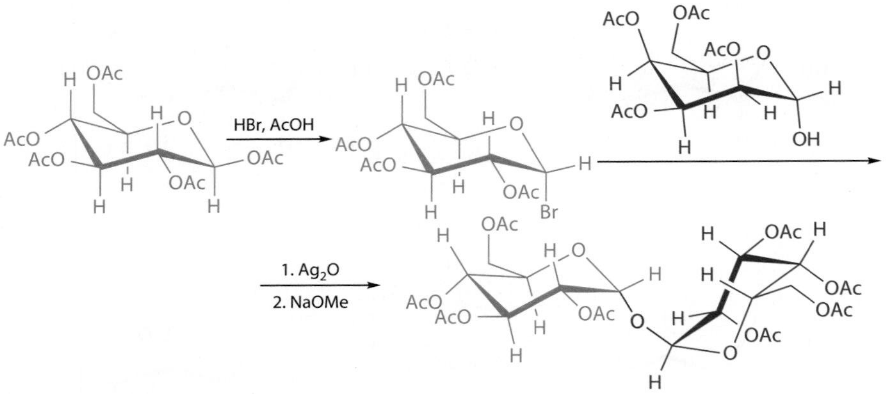

28.30

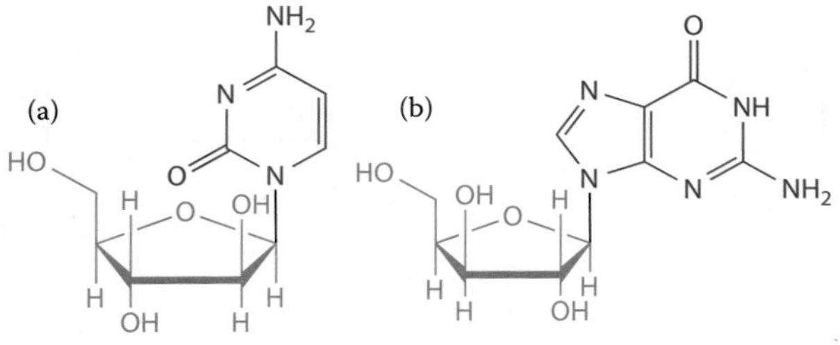

28.31

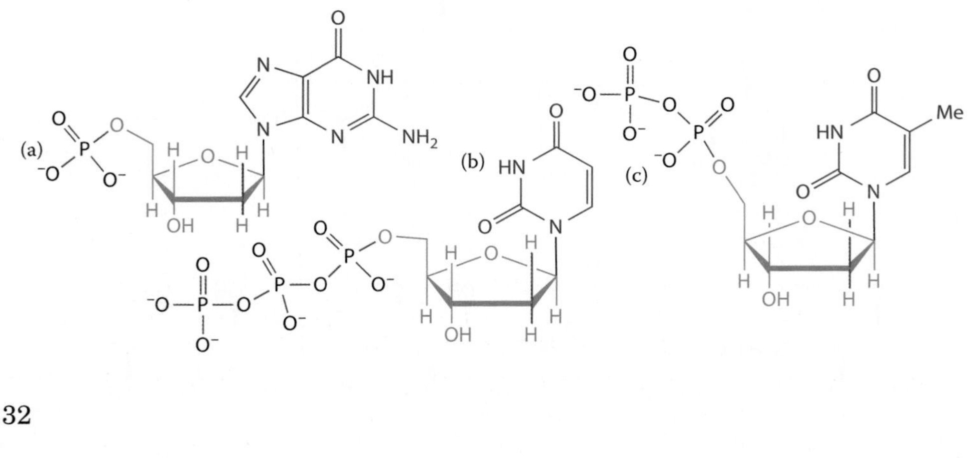

28.32

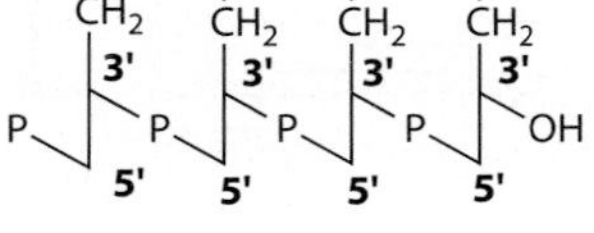

28.33

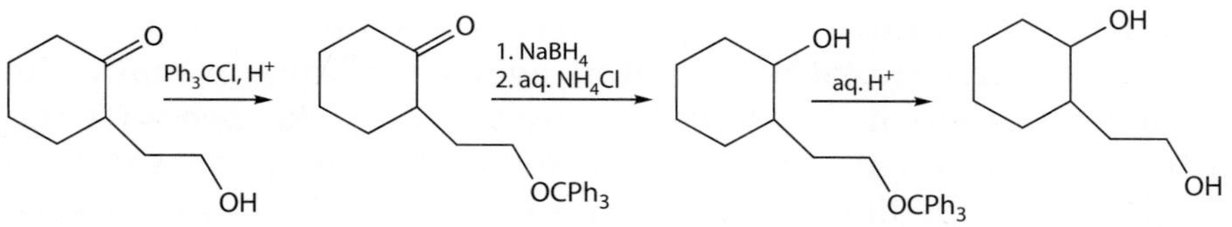
O
Ph$_3$CCl, H$^+$
OH
O
1. NaBH$_4$
2. aq. NH$_4$Cl
OCPh$_3$
OH
aq. H$^+$
OCPh$_3$
OH
OH

28.34

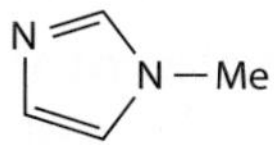
N
N—Me

28.35

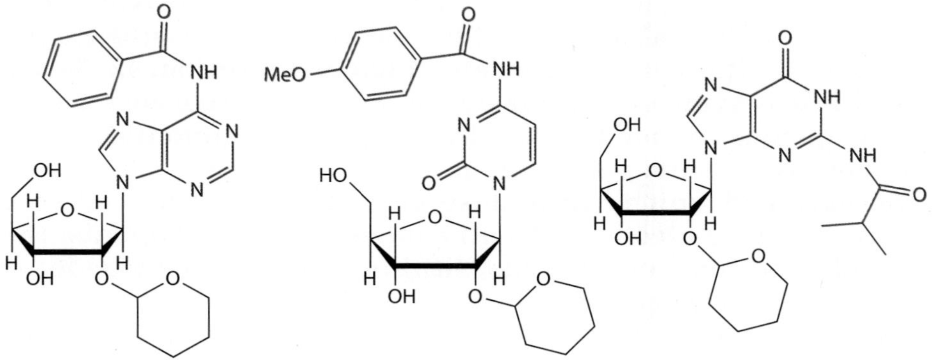
O
NH
N
N
N
N
OH
H
O
H
H
H
OH
O
O
MeO
O
NH
N
O
N
HO
H
O
H
H
H
OH
O
O
O
N
NH
OH
N
N
NH
O
H
O
H
H
H
OH
O
O

Correlation of Homework with Concepts

- **Carbohydrates are polyhydroxy aldehydes and ketones that are commonly known as sugars. A monosaccharide is a carbohydrate composed of one sugar unit and a disaccharide is a carbohydrate composed of two monosaccharide units. An oligosaccharide is composed of 2–10 monosaccharide units; a polysaccharide is composed of 10 or more monosaccharide units. An aldose is a carbohydrate that contains an aldehyde unit and a ketose is a carbohydrate that contains a ketone unit. Monosaccharides are categorized by the total number of carbons in the structure: triose, tetrose, pentose, hexose, etc. The d and l configurations of a monosaccharide are based on the Fischer projection of d-glyceraldehyde. A Fischer projection is an older representation of sugars: 1, 2, 3, 4, 12, 36, 37, 38.**

- The OH unit derived from the carbonyl group, formed by cyclization to a furanose or a pyranose, is attached to the so-called anomeric carbon. An aldose will cyclize to form a stable hemiacetal structure: a furanose or a pyranose. A ketose will cyclize to form a stable hemiketal structure: a furanose or a pyranose. The Haworth projection is an older representation for carbohydrates that is based on a planar pyranose or furanose ring: 5, 6, 7, 10, 11, 13, 39, 40, 43, 67.

- Cyclization of an aldose or a ketose to a furanose or pyranose is accompanied by mutarotation at the anomeric carbon. Mutarotation is the change in optical rotation of a pure furanose or pyranose derivative to that resulting from a mixture of isomers at the anomeric carbon: 8, 9, 10, 11, 13, 21, 41, 60, 61, 62.

- The anomeric effect occurs in 2-alkoxy pyranose derivatives, and it describes the stabilization of the electronegative substituent in an axial position rather than an equatorial position: 42, 54, 55.

- An aldose is oxidized to a carboxylic acid derivative via treatment with aqueous copper sulfate, silver nitrate in ammonia, bromine, aqueous potassium permanganate, or nitric acid: 17, 18, 47, 58.

- The carbonyl unit of an aldose or a ketose is reduced to the corresponding alcohol with all common reducing agents; these include sodium borohydride, lithium aluminum hydride, and catalytic hydrogenation: 19, 20, 21, 47, 57.

- Disaccharides are formed by coupling monosaccharides to generate different regioisomers. Trisaccharides and higher are prepared by similar coupling reactions. To react one hydroxyl group of a carbohydrate selectively, all or some of the other OH units in the molecule must be protected. Common hydroxyl protecting groups include acetate and benzoate esters, methyl ethers, acetals, and ketals: 14, 15, 16, 22, 23, 24, 44, 45, 46, 47, 50, 56, 67.

- The Ruff degradation and the Wohl degradation both cleave the carbonyl-bearing carbon from a monosaccharide to form a new monosaccharide of one fewer carbon. The Nef reaction adds one carbon to a monosaccharide to form a new monosaccharide, as does the Fischer–Kiliani synthesis: 25, 26, 27, 28, 48.

- The Koenigs–Knorr synthesis couples two monosaccharides together to form a disaccharide: 29, 49.

- Several heterocyclic compounds, including purines and pyrimidines, can be coupled to monosaccharides at the anomeric carbon to form a ribonucleoside (or simply a nucleoside). The most common nucleosides are ribose and 2-deoxyribose derivatives containing adenine, guanine, cytosine, thymine, or uracil. The phosphate ester of a nucleoside is called a nucleotide: 30, 31, 34, 51, 59, 64, 66.

- **Polynucleotides include ribonucleic acid (RNA) and deoxyribonucleic acid (DNA). Under physiological conditions, DNA normally exists in the β-form, which consists of two antiparallel strands: the Watson–Crick double helix. Interstrand hydrogen bonding connects the two strands. An adenosine in one strand can hydrogen bond with a thymine in the second, and a guanosine can hydrogen bond with a cytosine in the other strand. The strands must be antiparallel: 32, 52, 63, 64.**
- **DNA stores genetic information; RNA transcribes and translates this information for the cell. There are three major types of RNA: transfer RNA, ribosomal RNA, and messenger RNA. Transfer RNA contains a loop of nucleotides with three base pairs that bind to a messenger RNA, allowing a specific amino acid to be transferred to a protein that is being made. These three base pair sequences that correlate with specific amino acids are known as the genetic code.**

No homework for the preceding item!

- **Polynucleotides are prepared by coupling individual nucleotide units containing activated and protected phosphate linkages. The hydroxyl and amine units of the heterocycle structure must be protected during this process: 33, 34, 35, 53.**
- **A molecule with a particular functional group can be prepared from molecules containing different functional groups by a series of chemical steps (reactions). This process is called synthesis: The new molecule is synthesized from the old one (see Chapter 25): 29, 48, 49, 50, 53, 56, 57.**
- **Spectroscopy can be used to determine the structure of a particular molecule and can distinguish the structure and functionality of one molecule when compared with another (see Chapter 14): 68, 69, 70, 71, 72, 73, 74.**

Homework

28.36   Draw the structure for each of the following:
     (a)   a four-carbon glycose
     (b)   a three-carbon glycitol
     (c)   a six-carbon glyconic acid
     (d)   a four-carbon glycaric acid
     (e)   a seven-carbon uronic acid

28.37   Draw the structure for each of the following:
   (a) an aldotetrose
   (b) a ketopentose
   (c) an aldopentose
   (d) an aldoheptose
   (e) a ketooctose
   (f) an aldotriose

28.38   Draw the structure for each of the following using line notation (the extended or zigzag representation):
   (a) l-threose
   (b) l-ribose
   (c) d-xylose
   (d) l-allose
   (e) l-mannose
   (f) d-idose
   (g) l-talose

28.39   Draw the complete mechanism for conversion of pentanal to the hemiacetal under acid-catalyzed conditions. Complete the mechanism by converting the hemiacetal to the acetal.

28.40   Draw the structure for
   (a) α-d-arabinopyranose
   (b) α-l-xylofuranose
   (c) β-d-allofuranose
   (d) β-l-mannopyranose (**21**)
   (e) α-d-idopyranose
   (f) β-l-talofuranose

28.41   The α-form of a carbohydrate has a specific rotation of $-38°$ and the β-form has a specific rotation of $+90°$. Mutarotation occurs to give a mixture of α and β with a specific rotation of $+18°$. What is the percentage composition of this mixture, assuming <0.1% of the open-chain form?

28.42   Draw both chair conformations for 3-methoxytetrahydropyran. Indicate which should be favored as the primary conformation and briefly explain your answer.

28.43   Draw the structure for each of the following:
   (a) the furanose form of α-l-ribulose
   (b) the pyranose form of α-d-psicose
   (c) the furanose form of β-l-fructose
   (d) the pyranose form of β-d-tagatose

28.44   Draw the structure for each of the following:
   (a) O-α-d-glucopyranosyl-(1→1)-β-d-allopyranose
   (b) O-β-d-altropyranosyl-(1→4)-β-d-galactopyranose
   (c) O-α-d-talopyranosyl-(1→1)-α-d-glucopyranose
   (d) O-β-d-idopyranosyl-(1→4)-α-d-mannopyranose

28.45   Draw the structure of each of the following:
   (a) O-α-d-Fuc$f$-(1→4)-α-d-Lyx$p$-(1→1)-α-d-Alt$p$
   (b) O-α-d-Rib$f$-(1→4)-α-d-Xyl$p$-(1→1)-α-d-Fru$p$
   (c) O-α-d-Alt$f$-(1→4)-α-d-All$p$-(1→1)-α-d-Ido$p$

28.46 Give the major product of the following reactions:
 (a) O-α-d-Ara$f$-(1→4)-α-d-Gul$p$ + AgNO$_3$/NH$_3$
 (b) O-α-d-Fuc$f$-(1→4)-α-d-Man$p$ + aq. CuSO$_4$/sodium tartrate
 (c) O-α--d-Gal$f$-(1→1)-α-d-Gal$p$ + AgNO$_3$/NH$_3$
 (d) α-d-Ara$f$ + aq. CuSO$_4$/sodium tartrate
 (e) α-d-Gal$p$ + AgNO$_3$/NH$_3$
 (f) α-d-Lyx$p$ + aq. KMnO$_4$
 (g) α-d-Rha$p$ + 1. NaBH$_4$; 2. aq. NH$_4$Cl
 (h) α-d-Tal$f$ + Na/Hg
 (i) α-d-All$p$ + Br$_2$/pH 5
 (j) α-d-Glc$p$ + H$_2$, Ni(R)
 (k) α-d-Ido$p$ + HNO$_3$

28.47 Draw the major product for each of the following reactions:
 (a) α-d-Glc$p$ + acetic anhydride/NaOAc
 (b) α-d-Alt$p$ + Et$_2$SO$_4$
 (c) α-d-Tal$p$ + EtOH, cat. H$^+$
 (d) α-d-Ido$f$ + acetone, cat. H$^+$
 (e) α-d-Ara$p$ + benzaldehyde, cat. H$^+$

28.48 Show a complete synthesis of each carbohydrate from the designated starting material:

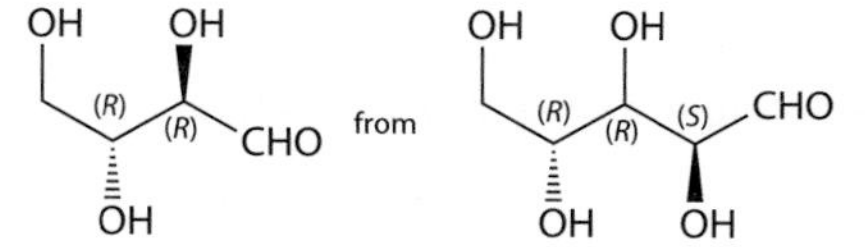

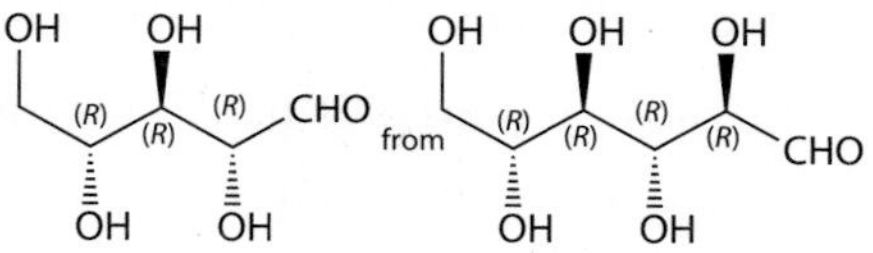

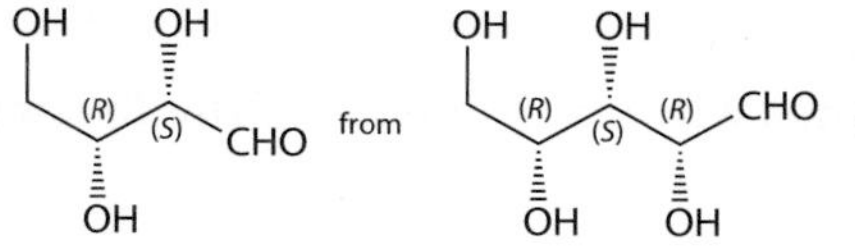

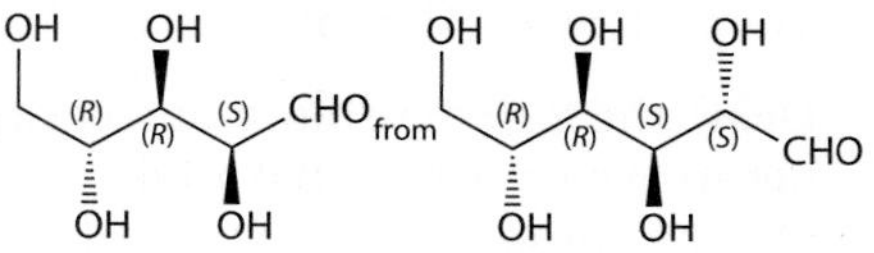

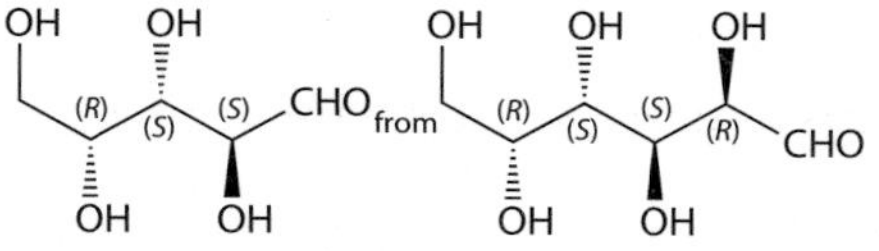

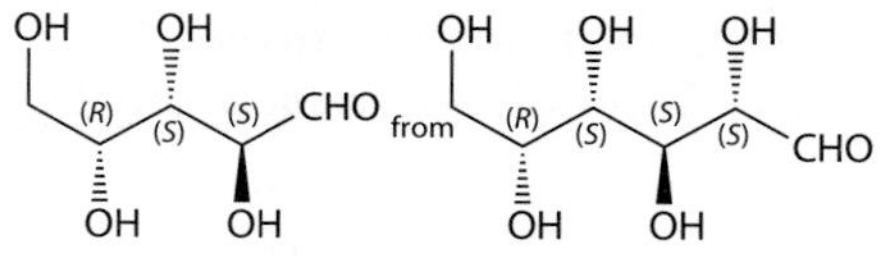

28.49 Show a complete synthesis of each target from the designated starting material:

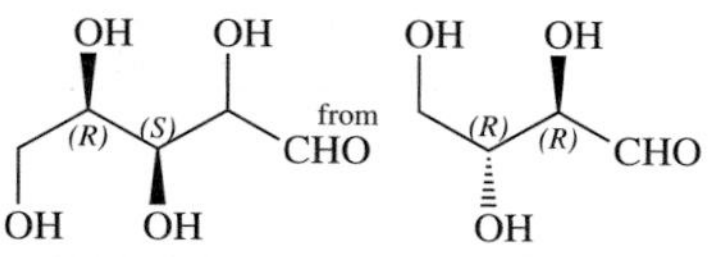

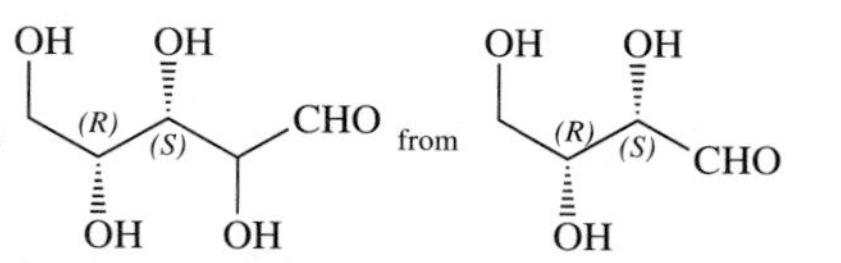

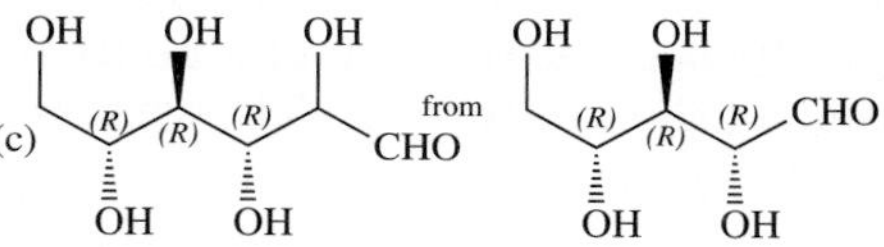

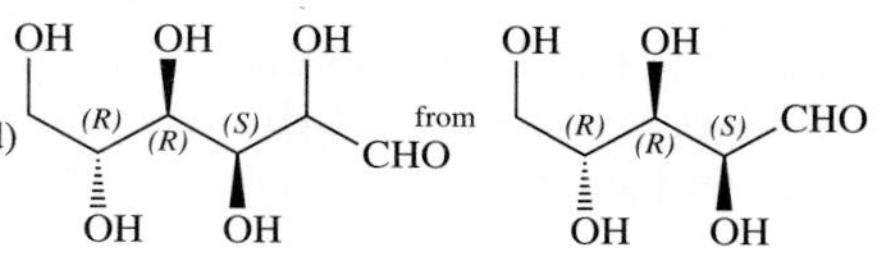

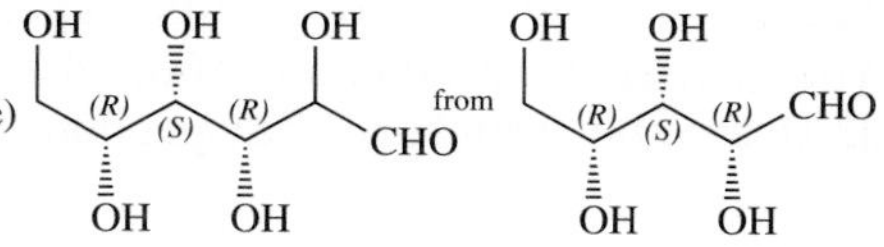

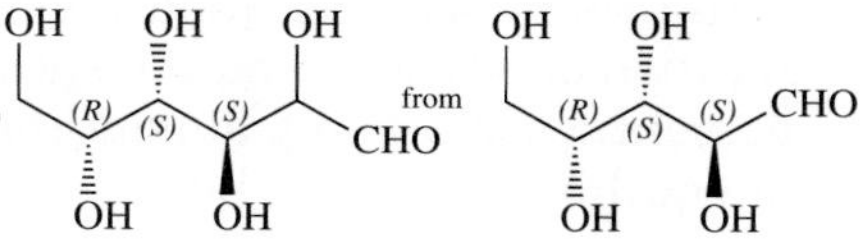

28.50   Show a synthesis
        O-α-d-Ara*f*-(1→4)-α-d-Ido*p*

28.51   Draw the structure of each of the following:
        (a) 9-(β-d-ribofuranoyl)guanine
        (b) 1-(β-d-ribofuranoyl)thymine
        (c) 9-(β-d-ribofuranoyl)adenine
        (d) 1-(β-d-ribofuranoyl)uracil

28.52   Draw each of the following using shorthand notation:
        (a) *d*A-*d*A-*d*G-*d*T
        (b) G-U-C-T-T-A
        (c) G-C-A-C-C

28.53   Write out the structures of the following DNA sequences:
        (a) A-A-C
        (b) U-G-T
        (c) T-T-T

28.54   When an-l-glucopyranose is converted to the 2-methoxy derivative, is the
        methoxy group likely to be axial or equatorial in the major product? Explain.

28.55   Draw the compound formed if the OH unit at the anomeric carbon of
        α-d-mannose is replaced with an amine group (NH_2). Is the amine group
        expected to be primarily axial or equatorial?

28.56   Suggest a synthesis of 3-β-amino-α-d-xylose from α-d-xylose.

28.57   Suggest two different reactions that will convert α-d-xylose to xylitol.

28.58   Suggest two different reactions that will convert sorbitol to glucoaldaric acid.

28.59   Draw the 2-deoxyribose nucleotide that will result if each of the following
        heterocycles is incorporated:
        (a) 5-fluorouracil
        (b) 5-cyclopropylcytosine
        (c) 6-amino-2-chloropurine

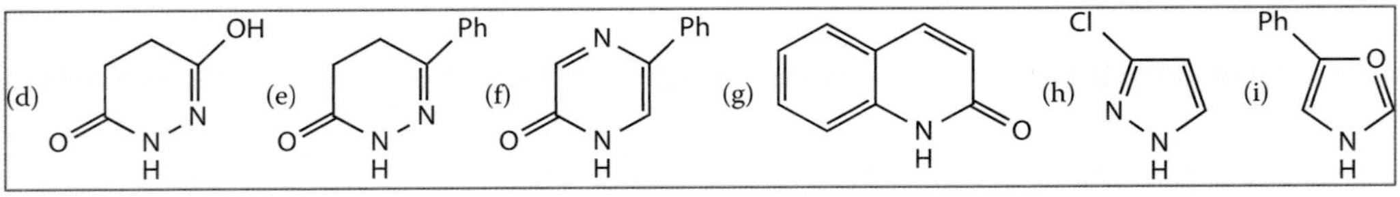

28.60   If α-d-glucopyranose has a specific rotation of +112° and β-d-glucopyranose
        has a specific rotation of +19°C, what is the specific rotation of a mixture of
        80% α and 20% β? If the specific rotation of the mixture is +53°, what is the
        percentage of α and β at equilibrium?

28.61   If α-d-rhamnose has a specific rotation of –17°, β-d-rhamnose has a specific
        rotation of +31.5°C, and the specific rotation of the mixture is +53, what is the
        percentage of α and β at equilibrium? Draw the structure that predominates
        at equilibrium.

28.62 The specific rotation of α-xylose is +79° and, after mutarotation, it falls to +19°. Draw the pyranose form that predominates at equilibrium. Similarly, the specific rotation of α-arabinose is +175° and it falls to +104° after mutarotation. Draw the pyranose form that predominates at equilibrium.

28.63 Draw the two base-paired antiparallel strands for A-G-G-T-A.

28.64 ATP is utilized in an enzymatic reaction that produces AMP. Draw both of these compounds. Draw the structures of uracil diphosphate and uracil monophosphate.

28.65 Inosine monophosphate (IMP) is a key component in the biosynthesis of nucleosides, and it contains 3,7-dihydropurin-6-one (also known as hypoxanthine) as the base. Draw the structure of this compound.

28.66 Caffeine is 1,3,7-trimethyl-1H-purine-2,6(3H,7H)-dione. Draw the structure of this compound and then draw the structure of a compound in which the 1-methyl group of caffeine is replaced with a deoxyribose.

28.67 When α-d-glucose is heated with p-toluenesulfonic acid, the final product is **A**, which is believed to arise via conversion of glucose to fructose followed by cyclization and dehydration. Draw a mechanism that accounts for these observations.

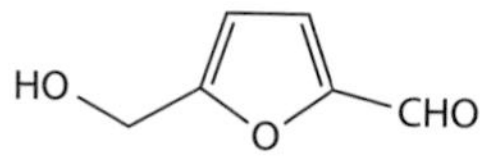

Spectroscopy Problems

Read and understand Chapter 14 before attempting these problems, which involve carbohydrates or nucleotide derivatives.

28.68 Describe spectroscopic features that will allow one to distinguish between the open-chain form of a sugar and a pyranose ring form.

28.69 Describe spectroscopic features in the proton NMR spectroscopy that will allow one to distinguish between adenine ribonucleoside and adenine deoxyribonucleoside.

28.70 Give the structure of the molecule with the formula $C_6H_{14}O_6$ and the spectral data shown:
   IR: broad peak at 3550–3070, 2980–2930, 1419, 1309, 1265, 1098, 1064, 1048, 1000, 646 cm^{-1}
   ^{1}H NMR: 4.56–4.12 (very broad s, 6H; this peak is diminished when treated with D$_2$O), 3.67 (m, 1H), 3.58–3.37 (d, 2H), 3.55 (m, 1H), 3.48 (m, 1H), 3.44–3.35 (d, 2H), 3.39 (m, 1H) ppm (Note: most of these peaks overlap.)
   ^{13}C NMR: 74.0, 72.3, 72.2, 70.8, 63.9, 63.6 ppm

28.71   Give the structure of the molecule with the formula $C_5H_{10}O_5$ and the spectral data shown:
IR: broad peak 3400–3230, 2970–2900, 1041, 934 cm^{-1}
^1H NMR: 5.2 (d, 1H), 4.6 (broad s, 4H; this peak is diminished when treated with D_2O); 3.69–3.63 (m, 4H), 3.52 (m, 1H) ppm

28.72   Give the structure of the molecule with the formula $C_{10}H_{14}N_2O_5$ and the spectral data shown:
IR: broad peak at 3312, 3158, 3026–2975, 1709, 1660, 1478, 1273, 1057 cm^{-1}
^1H NMR: 11.3 (broad s, 1H; diminished when treated with D_2O), 7.71 (s, 1H), 6.18 (overlapping doublets, 1H), 5.25 and 5.04 (broad s, 2H; diminished when treated with D_2O), 4.26 (m, 1H), 3.77 (m, 1H), 3.60–3.55 (d, 2H), 2.08 (m, 2H), 1.78 (s, 3H) ppm
^{13}C NMR: 167.2, 152.4, 138.3, 112.2, 87.3, 85.9, 71.3, 62.0, 39.4, 12.3 ppm

28.73   Give the structure of the molecule with the formula $C_7H_{14}O_6$ and the spectral data shown:
IR: 3648, broad peak at 3290–3001, 250–289, 1074–1036, 994, 900 cm^{-1}
^1H NMR: several peaks at 4.86 (s, 1H), 4.75 (s, 1H), 4.71(s, 1H), and 4.47 (s, 1H); peaks diminished when treated with D_2O; in addition, 4.52 (d, 1H), 3.62–3.44 (d, 2H), 3.38 (m, 1H), 3.3 (m, 1H), 3.26 (s, 3H), 3.18 (m, 1H), 3.05 (m, 1H) ppm
^{13}C NMR: 100.3, 74.2, 72.6, 72.3, 70.7, 61.7, 56.1 ppm

Index

A